Microelectronic Circuits

SIXTH EDITION

Microelectronic Circuits

Adel S. Sedra
University of Waterloo

Kenneth C. Smith
University of Toronto

New York Oxford
OXFORD UNIVERSITY PRESS
2010

Oxford University Press, Inc., publishes works that further Oxford University's
objective of excellence in research, scholarship, and education.

Oxford New York
Auckland Cape Town Dar es Salaam Hong Kong Karachi
Kuala Lumpur Madrid Melbourne Mexico City Nairobi
New Delhi Shanghai Taipei Toronto

With offices in
Argentina Austria Brazil Chile Czech Republic France Greece
Guatemala Hungary Italy Japan Poland Portugal Singapore
South Korea Switzerland Thailand Turkey Ukraine Vietnam

Published by Oxford University Press, Inc.
198 Madison Avenue, New York, New York 10016
http://www.oup.com

Library of Congress Cataloging-in-Publication Data

Sedra, Adel S.
Microelectronic circuits / Adel S. Sedra, Kenneth C. Smith.—6th ed.
p. cm.—(The Oxford series in electrical and computer engineering)
ISBN 978-0-19-532303-0
1. Electronic circuits. 2. Integrated circuits. I. Smith, Kenneth Carless. II. Title.
TK7867.S39 2010
621.3815—dc22

2009042633

Cover Photo: The device is a fully integrated triple-band, dual-arm WiMAX RFIC targeted at broadband wireless access applications, including fixed and mobile terminals, as well as pico and femto base stations. The multiple frequency bands enable equipment to be readily adapted to different regional requirements, while the dual-arm (dual-channel) arrangement allows the use of Multiple-Input/Multiple-Output (MIMO) technology. It illustrates the high degree of integration required by the latest wireless standards, incorporating high-quality Phase-Locked Loops, Radio Frequency Low-Noise Amplifiers, Mixers, and Power Amplifier stages, as well as programmable baseband filters and digital circuitry for control and calibration. (Photo credit: PMC-Sierra, the premier Internet infrastructure solutions provider. NASDAQ:PMCS) **Circuits:** Analog (Active-loaded Differential Amplifier). Digital (CMOS Inverter).

Printing number: 9 8 7 6 5 4 3 2

Printed in the United States of America
on acid-free paper

BRIEF TABLE OF CONTENTS

TABLE OF CONTENTS

6 Bipolar Junction Transistors (BJTs) 350

9 Frequency Response 686

10 Feedback 802

PART IV FILTERS AND OSCILLATORS 1252

16 Filters and Tuned Amplifiers 1254

TABLES

FOR REFERENCE AND STUDY

PREFACE

Microelectronic Circuits, sixth edition, is intended as a text for the core courses in electronic circuits taught to majors in electrical and computer engineering. It should also prove useful to engineers and other professionals wishing to update their knowledge through self-study.

As was the case with the first five editions, the objective of this book is to develop in the reader the ability to analyze and design electronic circuits, both analog and digital, discrete and integrated. While the application of integrated circuits is covered, emphasis is placed on transistor circuit design. This is done because of our belief that even if the majority of those studying this book were not to pursue a career in IC design, knowledge of what is inside the IC package would enable intelligent and innovative application of such chips. Furthermore, with the advances in VLSI technology and design methodology, IC design itself is becoming accessible to an increasing number of engineers.

Prerequisites

The prerequisite for studying the material in this book is a first course in circuit analysis. As a review, some linear circuits material is included here in the appendices: specifically, two-port network parameters in Appendix C; some useful network theorems in Appendix D; single-time-constant circuits in Appendix E; and *s*-domain analysis in Appendix F. No prior knowledge of physical electronics is assumed. All required semiconductor device physics is included, and Appendix A provides a brief description of IC fabrication. All these appendices can be found on the DVD that accompanies this book.

Emphasis on Design

It has been our philosophy that circuit design is best taught by pointing out the various trade-offs available in selecting a circuit configuration and in selecting component values for a given configuration. The emphasis on design has been increased in this edition by including more design examples, simulation examples, exercise problems, and end-of-chapter problems. Those exercises and end-of-chapter problems that are considered "design-oriented" are indicated with a D. Also, considerable material is provided on the most valuable design aid, SPICE, including Appendix B, which is available on the DVD so that it can be offered in searchable format, and in the full detail it deserves while not crowding other topics out of the text.

New to This Edition

Although the philosophy and pedagogical approach of the first five editions have been retained, several changes have been made to both organization and coverage.

1. **Four-Part Organization.** The book has been reorganized into four Parts. Part I: *Devices and Basic Circuits* (Chapters 1-6) provides a coherent and comprehensive single-semester introductory course in electronics. Similarly, Part II: *Integrated-Circuit Amplifiers* (Chapters 7-12) presents a rich package of material suitable for a second course. Part III: *Digital Integrated Circuits* (Chapters 13-15) represents a nearly self-contained coverage of digital electronics that can be studied after Chapters 5 (MOSFETs) and 6 (BJTs), or even only 5 if the emphasis is on MOS digital circuits—extremely helpful for teaching Computer Engineering students. Finally, Part IV: *Filters and Oscillators* (Chapters 16-17), deals with more specific application-oriented material that can be used to supplement a second course on analog circuits, be part of a third course, or used as reading and reference material to support student design projects. More on course design is given below.
2. **Flexible organization.** The most important feature of this edition is its flexible organization. Some manifestations of this flexibility are:
    - **MOSFETs and BJTs**. Chapter 5 (MOSFETs) and Chapter 6 (BJTs) are written to be completely independent of each other and thus can be taught in whatever order the instructor desires. Because the two chapters have identical structures, the chapter taught second can be covered much faster.
    - **Robust Digital Coverage**. The digital material has been grouped together in the new Part III, updated, and expanded. It can be covered at various points in the first or second course. All that is needed by way of background is the material on the two transistor types (Chapters 5 and 6) or even just Chapter 5 since most digital electronics today is MOS-based.
    - **Semiconductors as Needed**. The required material on semiconductor physics has been grouped together in a short chapter (Chapter 3) that can be taught, skipped, or assigned as reading material, depending on the background of the students and the instructor's teaching philosophy. This chapter serves as a primer on the basics, or as a refresher, depending on whether students have had a prior course in semiconductors.
    - **Op-amps Anywhere**. The op-amp chapter (Chapter 2) can be taught at any point in the first or second course, or skipped altogether if this material is taught in other courses.
    - **Frequency Response**. The material on amplifier frequency response has been grouped together into a single chapter (Chapter 9). The chapter is organized in a way that allows coverage of as few sections as the instructor deems necessary. Also, some of the basic material (Sections 9.1 to 9.3) can be covered earlier (after Chapters 5 or 6) as part of the first course.
    - **"Must-Cover" Topics First.** Each chapter is organized so that the essential "must-cover" topics are placed first, and the more specialized material appears last. More specialized material that can be skipped on a first reading, while the student is first learning the basics, is marked with a ⊕. Once the students understand the core concepts, they can return to these important but specialized topics.

3. **Streamlined MOSFETs and BJTs**. Chapters 5 (MOSFETs) and 6 (BJTs) have been rewritten to increase the clarity of presentation and emphasize essential topics. Also, these chapters are now shorter and can be covered faster.
4. **Cascode Configuration**. A novel and intuitively appealing approach is used to introduce the cascode configuration in Chapter 7.
5. **Comparison of MOSFETs and BJTs**. The insightful comparison of the MOSFET and the BJT has been moved to an appendix attached to Chapter 7. The appendix also includes an update of the device parameter values corresponding to various generations of fabrication process technologies. This appendix provides a good review and a reference that can be consulted at various points in a second course.
6. **Feedback**. The feedback chapter (Chapter 10) has been rewritten to increase clarity. Also, a large number of new examples, mostly MOS-based, are included.
7. **Class AB Amplifiers**. New material on MOSFET class AB amplifiers is included in Chapter 11.
8. **Low-Voltage Bipolar Design**. While the classical 741 op-amp circuit is retained, a new section on modern techniques for the design of low-voltage bipolar op amps has been added to Chapter 12.
9. **Deep-Submicron Design**. In addition to augmenting and consolidating the material on digital electronics in Part III, a new section on technology scaling (Moore's Law) and deep-submicron design issues has been added (Chapter 13).
10. **MOS Emphasis**. Throughout the book, greater emphasis is placed on MOS circuits to reflect the current dominance of the MOSFET in electronics.
11. **Bonus Reading on DVD**. Supplementary material on a wide variety of topics that were included in previous editions is made available on the DVD accompanying the book (see a listing below).
12. **Examples, Exercises, and Problems**. The number of Examples has been increased. Also, the in-chapter Exercises and end-of-chapter Problems have been updated with parameter values of current technologies so students work with a real-world perspective on technology. More Exercises and Problems, of a greater variety, have been added.
13. **Summary Tables**. As a study aid and for easy reference, many summary tables are included. See the complete List of Summary Tables after the Table of Contents.
14. **Learning Objectives**. A new section (In This Chapter You Will Learn…) has been added at the beginning of each chapter to focus attention on the major learning objectives of the chapter.
15. **SPICE**. A significant number of new simulation examples using National Instruments™ Multisim™ are added to the Cadence PSpice® simulation examples. Together with a section describing the SPICE device models, these design and simulation examples are grouped together in Appendix B. They can also be found together with other simulation files in the Lab-on-a-Disc on the DVD.
16. **Simulation**. A number of end-of-chapter Problems in each chapter are marked with the SIM icon **SIM** as simulation problems. Students attempting these problems will find considerable additional guidance on the DVD.
17. **Key Equations**. All equations that will be cross-referenced and used again are numbered. Particularly important equations are marked with a special icon.

As well as the structural differences described above, new coverage is included on all of the following technical topics.

- Entirely rewritten coverage of semiconductors (Chapter 3)
- MOSFET and BJT chapters extensively rewritten and restructured, with new figures and examples (Chapter 5 and 6)
- The basic gain cell (Chapter 7)
- The cascode amplifier (Chapter 7)
- CC-CE, CD-CS, and CD-CE transistor configurations (Chapter 7)
- CMRR (Chapter 8)
- The differential amplifier with active load (Chapter 8)
- Determining the output resistance R_o (Chapter 8)
- All new sections on frequency response (Chapter 9)
- Many, many new MOS examples of feedback (Chapter 10)
- CMOS class AB output stages (Chapter 11)
- Rejection ratios (CMRR and PSRR) (Chapter 12)
- Modern techniques for the design of BJT op amps (Section 12.7)
- Digital logic inverters (Chapter 13)
- The CMOS inverter (Chapter 13)
- Deep submicron design and technology scaling (Moore's Law) (Section 13.5)

The DVD and the Website

A DVD accompanies this book. It contains much useful supplementary information and material intended to enrich the student's learning experience. These include

1. Student versions of both Cadence PSpice® and National Instruments™ Multisim™.
2. The input files for all the PSpice® and Multisim™ examples in this book.
3. Step-by-step guidance to help with the simulation Examples and end-of-chapter Problems identified with a SIM icon.
4. A link to the book's website, offering PowerPoint slides of every figure in this book that students can print and carry to class to facilitate taking notes.
5. Bonus text material of specialized topics not covered in the current edition of the textbook. These include:
   - Junction Field-Effect Transistors (JFETs)
   - Gallium Arsenide (GaAs) devices and circuits
   - Transistor-Transistor Logic (TTL) circuits
   - Analog-to-Digital and Digital-to-Analog converter circuits
6. Appendices for the book:
   - Appendix A: VLSI Fabrication Technology
   - Appendix B: SPICE Device Models and Design and Simulation Examples Using PSpice® and Multisim™
   - Appendix C: Two-Port Network Parameters
   - Appendix D: Some Useful Network Theorems
   - Appendix E: Single-Time-Constant Circuits
   - Appendix F: *s*-domain Analysis: Poles, Zeroes, and Bode Plots
   - Appendix G: Bibliography

A website for the book has been set up (www.oup.com/us/sedrasmith, or www.sedrasmith.org). Its content will change frequently to reflect new developments in the field. On the site, PowerPoint-based slides of all the figures in the text are available for easy note-taking. The website also features datasheets for hundreds of useful devices to help in laboratory experiments, links to industrial and academic websites of interest, and a message center to communicate with the authors and with Oxford University Press.

Exercises and End-of-Chapter Problems

Over 475 Exercises are integrated throughout the text. The answer to each exercise is given below the exercise so students can check their understanding of the material as they read. Solving these exercises should enable the reader to gauge his or her grasp of the preceding material. In addition, more than 1450 end-of-chapter Problems, 55% of which are new or revised in this edition, are provided. The problems are keyed to the individual chapter sections and their degree of difficulty is indicated by a rating system: difficult problems are marked with an asterisk (*); more difficult problems with two asterisks (**); and very difficult (and/or time consuming) problems with three asterisks (***). We must admit, however, that this classification is by no means exact. Our rating no doubt depended to some degree on our thinking (and mood!) at the time a particular problem was created. Answers to sample problems are given in Appendix I, so students have a checkpoint to tell if they are working out the problems correctly. Complete solutions for all exercises and problems are included in the *Instructor's Solutions Manual*, which is available from the publisher to those instructors who adopt the book.

As in the previous five editions, many examples are included. The examples, and indeed most of the problems and exercises, are based on real circuits and anticipate the applications encountered in designing real-life circuits. This edition continues the use of numbered solution steps in the figures for many examples, as an attempt to recreate the dynamics of the classroom.

Course Organization

The book contains sufficient material for a sequence of two single-semester courses (each of 40-50 lecture hours). The organization of the book provides considerable flexibility for course design. In the following, we suggest various possibilities for the two courses. This is also laid out in an easy-to-follow visual form at the beginning of the Instructor's Edition of the book.

The First Course

At the core of the first course are Chapters 4 (Diodes), 5 (MOSFETs), and 6 (BJTs). Of these three, the MOSFET chapter is the one that has to be covered most thoroughly. If it is covered before the BJT, and we recommend that it should be, then the BJT chapter can be covered much faster. If time does not permit, some of the later sections in Chapter 4 can be skipped. Chapter 1 (Signals and Amplifiers) deserves some treatment in class. Although the signal concepts can be assigned as out-of-class reading, the amplifier material should be discussed. However, if frequency response is not emphasized in the first course, Section 1.6 can be skipped.

Around this core, one can build three possible curricula for the first course:

1. *Standard*: Chapters 1–6. Here, some or all of Chapter 2 (Op Amps) can be delayed. Also, the decision as to how much to cover of Chapter 3 (Semiconductors) will

depend on the students' background and the instructor's philosophy. If desired, this course can be supplemented by the material on amplifier frequency response in Sections 9.1–9.3.

2. *Digital Orientation*: Chapters 1 (without Section 1.6), 4 (without the later applications sections), all of 5, 6 (perhaps focusing only on the early sections), Section 9.2, and Chapters 13, 14, and 15. If time constraints are a concern, coverage of 6 can be shortened; Section 13.5 on Moore's Law and deep-submicron design can be skipped, and Sections 14.4 and 14.5 that depend on BJTs can be omitted. This course is ideal for Computer Engineering students.
3. *Analog Orientation*. Chapters 1, 4 (perhaps without all of the later, more application-oriented sections), 5, 6, 7 (without the advanced material in 7.6), 8, 9 (including at least 9.1–9.3, and the instructor's selection of other topics), and 10 (a selection of topics). This is a heavy course, and assumes that the students have previously covered op amps and maybe diodes, as well as device physics. This course is ideal where the first electrical engineering course is a hybrid of circuits and basic electronics, and where students have taken a semiconductor device physics course.

The Second Course

There are three possibilities for the second course:

1. *Standard*: Chapters 7–12. If time does not permit, some of the later sections in Chapter 9 can be skipped. Also, some of the more advanced topics in Chapters 11 and 12 can be skipped. If desired, some material from Chapter 16 (Filters) and Chapters 17 (Oscillators) can be included. This course ideally follows the "Standard First Course" outlined above.
2. *Analog and Digital Combination*: Chapters 7, 8, 9 (selection of topics); 10 (selection of topics), 13 (perhaps without Section 13.5 on technology scaling), 14 (omitting 14.4 and 14.5 if time is short), and 15 (selection of topics).
3. *Electrical Follow-up*: Chapters 6, 7, 8, 9, 10, and a choice of topics as time allows, selected from Chapters 11 and 12. This course is ideal for Electrical Engineering students who took a first semester with a "Digital Orientation" outlined above to accommodate Computer Engineering students.

Supplementary Material/Third Course

Chapters 16 (Filters) and 17 (Oscillators) contain material that can be used to supplement a third course on analog circuits. As well, this material is highly design-oriented and can be used to aid students who are pursuing design projects.

Chapters 13, 14, and 15 can be used as about half (15 hours of lecture) of a senior level course on digital IC design.

An Outline for the Reader

Part I, *Devices and Basic Circuits*, includes the most fundamental and essential topics for the study of electronic circuits. At the same time, it constitutes a complete package for a first course on the subject.

Chapter 1. The book starts with an introduction to the basic concepts of electronics in Chapter 1. Signals, their frequency spectra, and their analog and digital forms are presented.

Amplifiers are introduced as circuit building blocks and their various types and models are studied. This chapter also establishes some of the terminology and conventions used throughout the text.

Chapter 2. Chapter 2 deals with operational amplifiers, their terminal characteristics, simple applications, and practical limitations. We chose to discuss the op amp as a circuit building block at this early stage simply because it is easy to deal with and because the student can experiment with op-amp circuits that perform nontrivial tasks with relative ease and with a sense of accomplishment. We have found this approach to be highly motivating to the student. We should point out, however, that part or all of this chapter can be skipped and studied at a later stage (for instance, in conjunction with Chapter 8, Chapter 10, and/or Chapter 12) with no loss of continuity.

Chapter 3. Chapter 3 provides an overview of semiconductor concepts at a level sufficient for understanding the operation of diodes and transistors in later chapters. Coverage of this material is useful in particular for students who have had no prior exposure to device physics. Even those with such a background would find a review of Chapter 3 beneficial as a refresher. The instructor can choose to cover this material in class or assign it for outside reading.

Chapter 4. The first electronic device, the diode, is studied in Chapter 4. The diode terminal characteristics, the circuit models that are used to represent it, and its circuit applications are presented. Depending on the time available in the course, some of the diode applications (e.g., Section 4.6) can be skipped. Also, the brief description of special diode types (Section 4.7) can be left for the student to read.

Chapters 5 and 6. The foundation of electronic circuits is established by the study of the two transistor types in use today: the MOS transistor in Chapter 5 and the bipolar transistor in Chapter 6. These are the two most important chapters of the book. ***These two chapters have been written to be completely independent of one another and thus can be studied in either order, as desired.*** Furthermore, the two chapters have the same structure, making it easier and faster to study the second device, as well as to draw comparisons between the two device types.

Each of Chapters 5 and 6 begins with a study of the device structure and its physical operation, leading to a description of its terminal characteristics. Then, to allow the student to become very familiar with the operation of the transistor as a circuit element, a large number of examples are presented of dc circuits utilizing the device. We then ask: How can the transistor be used as an amplifier? To answer the question we consider the large-signal operation of the basic common-source (common-emitter) circuit and use it to delineate the regions over which the device can be used as a linear amplifier, from those regions where it can be used as a switch. We then pursue the small-signal operation of the transistor and develop circuit models for its representation. The various configurations in which the transistor can be used as an amplifier are then studied and contrasted. This is followed by a study of methods to bias the transistor to operate as an amplifier in discrete-circuit applications. We then put everything together by presenting complete practical discrete-circuit transistor amplifiers. The last section of each of Chapters 5 and 6 deals with second-order effects that are included for completeness, but that can be skipped if time does not permit detailed coverage.

After the study of Part I, the reader will be fully prepared to study either integrated-circuit amplifiers in Part II, or digital integrated circuits in Part III.

Part II, *Integrated-Circuit Amplifiers*, is devoted to the study of practical amplifier circuits that can be fabricated in the integrated-circuit (IC) form. Its six chapters constitute a coherent treatment of IC amplifier design and can thus serve as a second course in electronic circuits.

Chapter 7. Beginning with a brief introduction to the philosophy of IC design, Chapter 7 presents the basic circuit building blocks that are used in the design of IC amplifiers. We start with the basic gain cell comprising a common-source (common-emitter) transistor loaded with a current source, and ask: How can we increase its voltage gain? This leads naturally to the concept of cascoding and its use in the cascode amplifier and the cascode current source. We then consider the methods used for biasing IC amplifiers. The chapter concludes, as do most chapters in the book, with advanced topics (Sections 7.5 and 7.6) that can be skipped if the instructor is pressed for time.

Chapter Appendix 7.A. Chapter 7 includes an appendix that provides a comprehensive compilation and comparison of the properties of the MOSFET and the BJT. The comparison is aided by the inclusion of typical parameter values of devices fabricated with modern process technologies. This appendix can be consulted at any point from Chapter 7 on, and should serve as a concise review of the important characteristics of both transistor types.

MOS and Bipolar. Throughout Part II, both MOS and bipolar circuits are presented side-by-side. Because the MOSFET is by far the dominant device, its circuits are presented first. Bipolar circuits are discussed to the same depth but occasionally more briefly.

Chapter 8. The most important IC building block, the differential pair, is the main topic of Chapter 8. The last section of Chapter 8 is devoted to the study of multistage amplifiers.

Chapter 9. Chapter 9 presents a comprehensive treatment of the important subject of amplifier frequency response. Here, Sections 9.1, 9.2, and 9.3 contain essential material; Sections 9.4 and 9.5 provide an in-depth treatment of very useful new tools; and Sections 9.6 to 9.10 present the frequency response analysis of a variety of amplifier configurations that can be studied as and when needed. A selection of the latter sections can be made depending on the time available and the instructor's preference.

Chapter 10. The fourth of the essential topics of Part II, feedback, is the subject of Chapter 10. Both the theory of negative feedback and its application in the design of practical feedback amplifiers are presented. We also discuss the stability problem in feedback amplifiers and treat frequency compensation in some detail.

Chapter 11. In Chapter 11 we switch gears from dealing with small-signal amplifiers to those that are required to handle large signals and large amounts of power. Here we study the different amplifier classes—A, B, and AB—and their realization in bipolar and CMOS technologies. We also consider power BJTs and power MOSFETs, and study representative IC power amplifiers. Depending on the availability of time, some of the later sections (e.g., 11.8–11.10 on special applications) can be skipped in a first reading.

Chapter 12. Finally, Chapter 12 brings together all the topics of Part II in an important application; namely, the design of operational amplifier circuits. We study both CMOS and bipolar op amps. In the latter category, besides the classical and still timely 741 circuit, we present modern techniques for the design of low-voltage op amps (Section 12.7).

Part III, *Digital Integrated Circuits*, provides a brief but nonetheless comprehensive and sufficiently detailed study of digital IC design. Our treatment is almost self-contained, requiring for the most part only a thorough understanding of the MOSFET material presented in Chapter 5. Thus, Part III can be studied right after Chapter 5. The only exceptions to this are the last two sections in Chapter 14 which require knowledge of the BJT (Chapter 6). Also, knowledge of the MOSFET internal capacitances (Section 9.2.2) will be needed.

Chapter 13. Chapter 13 is the foundation of Part III. It begins with digital logic inverters (Section 13.1), and then concentrates on the bread-and-butter topics of digital IC design: the CMOS inverter (Sections 13.2 and 13.3) and CMOS logic gates (Section 13.4). The last section (13.5) deals with the implications of technology scaling (Moore's law) and discusses important issues in deep-submicron technologies. With the possible exception of Section 13.5, the material in Chapter 13 is the minimum needed to learn something meaningful about digital circuits.

Chapter 14. Chapter 14 builds on the foundation established in Chapter 13 and presents three important types of MOS logic circuits. As well, a significant family of bipolar logic circuits, emitter-coupled logic, is studied. The chapter concludes with an interesting digital circuit technology that attempts to combine the best attributes of bipolar and CMOS: BiCMOS.

Chapter 15. Digital circuits can be broadly divided into logic and memory circuits. The latter is the subject of Chapter 15.

Part IV, *Filters and Oscillators*, is intentionally oriented toward applications and systems. The two topics illustrate powerfully and dramatically the application of both negative and positive feedback.

Chapter 16. Chapter 16 deals with the design of filters, which are important building blocks of communication and instrumentation systems. A comprehensive, design-oriented treatment of the subject is presented. The material provided should allow the reader to perform a complete filter design, starting from specification and ending with a complete circuit realization. A wealth of design tables is included.

Chapter 17. Chapter 17 deals with circuits for the generation of signals with a variety of waveforms: sinusoidal, square, and triangular. We also present circuits for the nonlinear shaping of waveforms.

Appendices. The eight appendices contain much useful background and supplementary material. We wish to draw the reader's attention in particular to the first two: Appendix A provides a concise introduction to the important topic of IC fabrication technology including IC layout. Appendix B provides SPICE device models as well as a large number of design and simulation examples in PSpice® and Multisim™. The examples are keyed to the book chapters. These Appendices and a great deal more material on these simulation examples can be found on the DVD accompanying the book.

Ancillaries

A complete set of ancillary materials is available with this text to support your course.

For the Instructor

The *Instructor's Solutions Manual* provides complete worked solutions to all the exercises in each chapter and all the end-of-chapter problems in the text.

The *Instructor's Resource CD* is bound into the *Instructor's Solutions Manual* so instructors can find all their support materials in one place. The *Resource CD* contains PowerPoint-based slides of every figure in the book and each corresponding caption. The slides can be projected in class, added to a course management system, printed as overhead transparencies, or used as handouts. The CD also contains complete solutions and instructor's support for the Lab-on-a-Disc simulation problems. (ISBN 9780195340303)

For the Student and Instructor

The *DVD* included with every new copy of the textbook contains Lab-on-a-Disc simulation activities in Multisim™ and PSpice® for many of the simulation Examples and Problems in the text. It also contains a Student Edition of Cadence PSpice® v. 16.2 Demo software, and a Student Edition of National Instruments™ Multisim™ version 10.1.1, both of which can be run by students on their own computers so they can practice their coursework wherever they happen to study. Bonus text topics, the Appendices, and a link to the book's website featuring manufacturer datasheets and PowerPoint-based slides of all of the book's illustrations, complete the DVD.

Acknowledgments

Many of the changes in this sixth edition were made in response to feedback received from instructors who adopted the fifth edition. We are grateful to all those who took the time to write to us. In addition, dozens of reviewers provided detailed commentary on the fifth edition and suggested many of the changes that we have incorporated in this revision. They are listed later; to all of them, we extend our sincere thanks.

A number of individuals made significant contributions to this edition. Sam Emaminejad and Muhammad Faisal prepared the Multisim™ and new PSpice® simulations and helped with many aspects of the manuscript preparation. Olivier Trescases of the University of Toronto and his students helped immensely, independently testing all the simulations in the Lab-on-a-Disc. Wai-Tung Ng of the University of Toronto rewrote Appendix A. Gordon Roberts of McGill University gave us permission to use some of the examples from the book *SPICE* 2nd edition, by Roberts and Sedra. Sima Dimitrijev of Griffith University undertook a detailed review of Chapter 3 on semiconductor devices, and David Pulfrey of the University of British Columbia offered suggestions as well. As in the previous edition, Anas Hamoui of McGill University was the source of many good ideas. Jim Somers of Sonora Designworks prepared discs for the student and instructor support materials. Jennifer Rodrigues typed all the revisions with skill and good humor and assisted with many of the logistics. Linda Lyman assisted with more details than we can possibly list here, and has been invaluable. Laura Fujino assisted in proofreading, and perhaps most importantly, in keeping one of us (KCS) focused. To all of these friends and colleagues we say thank you.

We are also grateful to the following colleagues and friends who have provided many helpful suggestions: Anthony Chan-Carusone, University of Toronto; Roman Genov, University of Toronto; David Johns, University of Toronto; Ken Martin, University of Toronto; David Nairn, University of Waterloo; Wai-Tung Ng, University of Toronto; Khoman Phang, University of Toronto; Gordon Roberts, McGill University; and Ali Sheikholeslami, University of Toronto.

The authors would like to thank Cadence and National Instruments for allowing Oxford University Press to distribute the PSpice® and Multisim™ software with this book. Mark Walters of National Instruments in particular has been very supportive. We are grateful to PMC Sierra for the excellent cover photo (which is fully described on the copyright page of this book, for readers who are interested in the intriguing technology shown here).

A large number of people at Oxford University Press contributed to the development of this edition and its various ancillaries. We would like to specifically mention Art Director Paula Schlosser and designers Dan Niver, Binbin Li, and Annika Sarin, Senior Copywriter Jill Crosson, as well as Susanne Arrington, Andy Battle, Brian Black, Sonya Borders, Gigi Brienza, Jim Brooks, Chris Critelli, Michael Distler, Diane Erickson, Ned Escobar, Adam Glazer, Chris Hellstrom, Andrea Hill, Adriana Hurtado, Holly Lewis, Jenny Lupica, Johanna Marcelino, Bill Marting, Laura Mahoney, Joella Molway, Preeti Parasharami, Emily Pillars, Terry Retchless, Kim Rimmer, Linda Roths, Sarah Smith, Patrick Thompson, Adam Tyrell, Euan White, and David Wright.

We wish to extend special thanks to our Publisher at Oxford University Press, John Challice, and to the hardworking editorial team of Engineering Associate Editor Rachael Zimmermann and Editorial Director Patrick Lynch, who have meticulously prepared all the ancillary support for this book. Steve Cestaro, Director of Editorial, Design, and Production, pulled out all the stops on this edition. Barbara Mathieu, Senior Production Editor, worked quietly, cheerfully, and tirelessly to bring this book to completion under significant pressure, making a difficult job look easy with grace and creativity. And last but certainly not least, a special note of thanks and gratitude to our Development Editor, Danielle Christensen, who was our main point of contact with OUP throughout the entire project and who managed the project with creativity, thoughtfulness, and dedication.

Finally, we wish to thank our families for their support and understanding, and to thank all the students and instructors who have valued this book throughout its history.

Adel S. Sedra
Kenneth C. (KC) Smith

Problem Solvers and Accuracy Checkers, Solutions Manual

Mandana Amiri, University of British Columbia, BC
Alok Berry, George Mason University, VA
Marc Cahay, University of Cincinnati, OH
Yun Chiu, University of Illinois–Urbana-Champaign, IL
Norman Cox, Missouri University of Science and Technology, MO
John Davis, University of Texas–Austin, TX
Michael Green, University of California–Irvine, CA
Roger King, University of Toledo, OH
Clark Kinnaird, Southern Methodist University, TX
Robert Krueger, University of Wisconsin–Milwaukee, WI
Shahriar Mirabbasi, University of British Columbia, BC
Daniel Moore, Rose-Hulman Institute of Technology, IN
Kathleen Muhonen, The Pennsylvania State University, PA
Angela Rasmussen, University of Utah, UT
Roberto Rosales, University of British Columbia, BC
John Wilson, Royal Military College, ON

Reviewers of the Sixth Edition

Elizabeth Brauer, Northern Arizona University, AZ
Martin Brooke, Duke University, NC
Yun Chiu, University of Illinois–Urbana-Champaign, IL
Norman Cox, Missouri University of Science and Technology, MO
Robert Bruce Darling, University of Washington, WA
John Davis, University of Texas–Austin, TX
Christopher DeMarco, University of Wisconsin–Madison, WI
Robert Engelken, Arkansas State University, AR
Ethan Farquhar, University of Tennessee, TN
Patrick Fay, University of Notre Dame, IN
George Giakos, University of Akron, OH
John Gilmer, Wilkes University, PA
Tayeb Giuma, University of North Florida, FL
Michael Green, University of California–Irvine, CA
Steven de Haas, California State University–Sacramento, CA
Anas Hamoui, McGill University, QC
William Harrell, Clemson University, SC
Reid Harrison, University of Utah, UT
Timothy Horiuchi, University of Maryland–College Park, MD
Mohammed Ismail, The Ohio State University, OH
Paul Israelson, Utah State University, UT
Zhenhua Jiang, University of Miami, FL
Seongsin M. Kim, University of Alabama, AL
Roger King, University of Toledo, OH
Clark Kinnaird, Southern Methodist University, TX
Tsu-Jae King Liu, University of California–Berkeley, CA
Yicheng Lu, Rutgers University, NJ
David Nairn, University of Waterloo, ON
Thomas Matthews, California State University–Sacramento, CA
Ken Noren, University of Idaho, ID
Brita Olson, California Polytechnic University–Pomona, CA
Martin Peckerar, University of Maryland–College Park, MD
Khoman Phang, University of Toronto, ON
Mahmudur Rahman, Santa Clara University, CA
John Ringo, Washington State University, WA
Norman Scheinberg, City College, NY

Kuang Sheng, Rutgers University, NJ
Andrew Szeto, San Diego State University, CA
Joel Therrien, University of Massachusetts–Lowell, MA
Len Trombetta, University of Houston, TX
Mustapha C.E. Yagoub, University of Ottawa, ON
Donna Yu, North Carolina State University, NC
Jiann-Shiun Yuan, University of Central Florida, FL
Sandra Yost, University of Detroit–Mercy, MI
Jianhua (David) Zhang, University of Illinois–Urbana-Champaign, IL

Reviewers of Prior Editions

Maurice Aburdene, Bucknell University, PA
Michael Bartz, University of Memphis, TN
Patrick L. Chapman, University of Illinois—Urbana-Champaign, IL
Roy H. Cornely, New Jersey Institute of Technology, NJ
Dale L. Critchlow, University of Vermont, VT
Artice Davis, San Jose State University, CA
Eby G. Friedman, University of Rochester, NY
Paul M. Furth, New Mexico State University, NM
Rhett T. George, Jr., Duke University, NC
Roobik Gharabagi, St. Louis University, MO
Steven de Haas, California State University—Sacramento, CA
Reza Hashemian, Northern Illinois University, IL
Ward J. Helms, University of Washington, WA
Richard Hornsey, York University, ON
Hsiung Hsu, The Ohio State University, OH
Robert Irvine, California State Polytechnic University—Pomona, CA
Steve Jantzi, Broadcom
Marian Kazimierczuk, Wright State University, OH
John Khoury, Columbia University, NY
Jacob B. Khurgin, The Johns Hopkins University, MD
Roger King, University of Toledo, OH
Robert J. Krueger, University of Wisconsin—Milwaukee, WI
Joy Laskar, Georgia Institute of Technology, GA
David Luke, University of New Brunswick, NB
Un-Ku Moon, Oregon State University, OR
Bahram Nabet, Drexel University, PA
Dipankar Nagchoudhuri, Indian Institute of Technology—Delhi, India
David Nairn, Analog Devices
Joseph H. Nevin, University of Cincinnati, OH
Rabin Raut, Concordia University, QC
John A. Ringo, Washington State University, WA
Zvi S. Roth, Florida Atlantic University, FL
Mulukutla Sarma, Northeastern University, MA
John Scalzo, Louisiana State University, LA
Pierre Schmidt, Florida International University, FL
Richard Schreier, Analog Devices
Dipankar Sengupta, Royal Melbourne Institute of Technology, Australia
Ali Sheikholeslami, University of Toronto, ON
Michael L. Simpson, University of Tennessee, TN
Karl A. Spuhl, Washington University in St. Louis, MO
Charles Sullivan, Dartmouth College, NH
Daniel van der Weide, University of Delaware, DE
Gregory M. Wierzba, Michigan State University, MI
Alex Zaslavsky, Brown University, RI

Microelectronic Circuits

PART I

Devices and Basic Circuits

Part I, *Devices and Basic Circuits*, includes the most fundamental and essential topics for the study of electronic circuits. At the same time, it constitutes a complete package for a first course on the subject.

The heart of Part I is the study of the three basic semiconductor devices: the diode (Chapter 4); the MOS transistor (Chapter 5); and the bipolar transistor (Chapter 6). In each case, we study the device operation, its characterization, and its basic circuit applications. For those who have not had a prior course on device physics, Chapter 3 provides an overview of semiconductor concepts at a level sufficient for the study of electronic circuits. A review of Chapter 3 should prove useful even for those with prior knowledge of semiconductors.

Since the purpose of electronic circuits is the processing of signals, an understanding is essential of signals, their characterization in the time and frequency domains, and their analog and digital representations. This is provided in Chapter 1, which also introduces the most common signal-processing function, *amplification*, and the characterization and types of *amplifiers*.

Besides diodes and transistors, the basic electronic devices, the op amp is studied in Part I. Although not an electronic device in the most fundamental sense, the op amp is commercially available as an integrated circuit (IC) package and has well-defined terminal characteristics. Thus, despite the fact that the op amp's internal circuit is complex, typically incorporating 20 or more transistors, its almost-ideal terminal behavior makes it possible to treat the op amp as a circuit element and to use it in the design of powerful circuits, as we do in Chapter 2, without any knowledge of its internal construction. We should mention, however, that the study of op amps can be delayed to a later point, and Chapter 2 can be skipped with no loss of continuity.

The foundation of this book, and of any electronics course, is the study of the two transistor types in use today: the MOS transistor in Chapter 5 and the bipolar transistor in Chapter 6. These two chapters have been written to be completely independent of one another and thus can be studied in either order as desired. Furthermore, the two chapters have the same structure, making it easier and faster to study the second device, as well as to draw comparisons between the two device types.

After the study of Part I, the reader will be fully prepared to undertake the study of either integrated-circuit amplifiers in Part II or digital integrated circuits in Part III.

CHAPTER 1

Signals and Amplifiers

IN THIS CHAPTER YOU WILL LEARN

1. That electronic circuits process signals, and thus understanding electrical signals is essential to appreciating the material in this book.
2. The Thévenin and Norton representations of signal sources.
3. The representation of a signal as the sum of sine waves.
4. The analog and digital representations of a signal.
5. The most basic and pervasive signal-processing function: signal amplification, and correspondingly, the signal amplifier.
6. How amplifiers are characterized (modeled) as circuit building blocks independent of their internal circuitry.
7. How the frequency response of an amplifier is measured, and how it is calculated, especially in the simple but common case of a single-time-constant (STC) type response.

Introduction

The subject of this book is modern electronics, a field that has come to be known as **microelectronics**. **Microelectronics** refers to the integrated-circuit (IC) technology that at the time of this writing is capable of producing circuits that contain hundreds of millions of components in a small piece of silicon (known as a **silicon chip**) whose area is on the order of 100 mm^2. One such microelectronic circuit, for example, is a complete digital computer, which accordingly is known as a **microcomputer** or, more generally, a **microprocessor**.

In this book we shall study electronic devices that can be used singly (in the design of **discrete circuits**) or as components of an **integrated-circuit (IC)** chip. We shall study the design and analysis of interconnections of these devices, which form discrete and integrated circuits of varying complexity and perform a wide variety of functions. We shall also learn about available IC chips and their application in the design of electronic systems.

The purpose of this first chapter is to introduce some basic concepts and terminology. In particular, we shall learn about signals and about one of the most important signal-processing functions electronic circuits are designed to perform, namely, signal amplification. We shall then look at circuit representations or models for linear amplifiers. These models will be employed in subsequent chapters in the design and analysis of actual amplifier circuits.

In addition to motivating the study of electronics, this chapter serves as a bridge between the study of linear circuits and that of the subject of this book: the design and analysis of electronic circuits.

1.1 Signals

Signals contain information about a variety of things and activities in our physical world. Examples abound: Information about the weather is contained in signals that represent the air temperature, pressure, wind speed, etc. The voice of a radio announcer reading the news into a microphone provides an acoustic signal that contains information about world affairs. To monitor the status of a nuclear reactor, instruments are used to measure a multitude of relevant parameters, each instrument producing a signal.

To extract required information from a set of signals, the observer (be it a human or a machine) invariably needs to **process** the signals in some predetermined manner. This **signal processing** is usually most conveniently performed by electronic systems. For this to be possible, however, the signal must first be converted into an electrical signal, that is, a voltage or a current. This process is accomplished by devices known as **transducers**. A variety of transducers exist, each suitable for one of the various forms of physical signals. For instance, the sound waves generated by a human can be converted into electrical signals by using a microphone, which is in effect a pressure transducer. It is not our purpose here to study transducers; rather, we shall assume that the signals of interest already exist in the electrical domain and represent them by one of the two equivalent forms shown in Fig. 1.1. In Fig. 1.1(a) the signal is represented by a voltage source $v_s(t)$ having a source resistance R_s. In the alternate representation of Fig. 1.1(b) the signal is represented by a current source $i_s(t)$ having a source resistance R_s. Although the two representations are equivalent, that in Fig. 1.1(a) (known as the Thévenin form) is preferred when R_s is low. The representation of Fig. 1.1(b) (known as the Norton form) is preferred when R_s is high. The reader will come to appreciate this point later in this chapter when we study the different types of amplifiers. For the time being, it is important to be familiar with Thévenin's and Norton's theorems (for a brief review, see Appendix D) and to note that for the two representations in Fig. 1.1 to be equivalent, their parameters are related by

$$v_s(t) = R_s i_s(t)$$

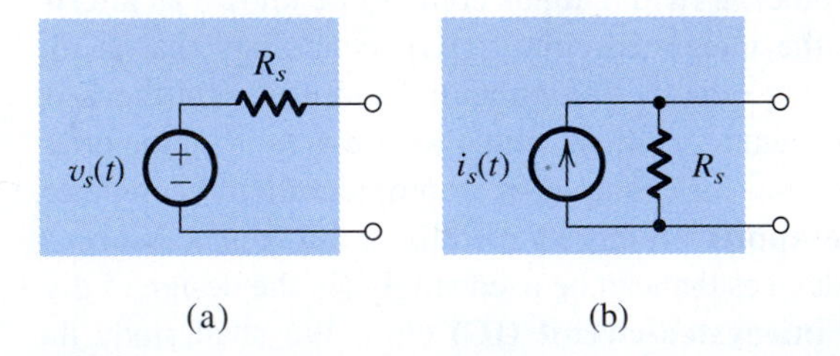

Figure 1.1 Two alternative representations of a signal source: **(a)** the Thévenin form; **(b)** the Norton form.

Example 1.1

The output resistance of a signal source, although inevitable, is an imperfection that limits the ability of the source to deliver its full signal strength to a **load**. To see this point more clearly, consider the signal source when connected to a load resistance R_L as shown in Fig. 1.2. For the case in which the source is represented by its Thévenin equivalent form, find the voltage v_o that appears across R_L, and hence the condition that R_s must satisfy for v_o to be close to the value of v_s. Repeat for the Norton-represented source; in this case finding the current i_o that flows through R_L and hence the condition that R_s must satisfy for i_o to be close to the value of i_s.

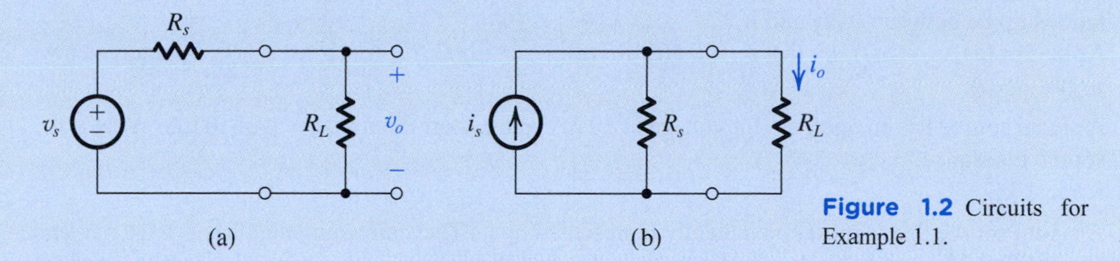

Figure 1.2 Circuits for Example 1.1.

Solution

For the Thévenin-represented signal source shown in Fig. 1.2(a), the output voltage v_o that appears across the load resistance R_L can be found from the ratio of the voltage divider formed by R_s and R_L,

$$v_o = v_s \frac{R_L}{R_L + R_s}$$

From this equation we see that for

$$v_o \simeq v_s$$

the source resistance R_s must be much lower than the load resistance R_L,

$$R_s \ll R_L$$

Thus, for a source represented by its Thévenin equivalent, ideally $R_s = 0$, and as R_s is increased, relative to the load resistance R_L with which this source is intended to operate, the voltage v_o that appears across the load becomes smaller, not a desirable outcome.

Next, we consider the Norton-represented signal source in Fig. 1.2(b). To obtain the current i_o that flows through the load resistance R_L, we utilize the ratio of the current divider formed by R_s and R_L,

$$i_o = i_s \frac{R_s}{R_s + R_L}$$

From this relationship we see that for

$$i_o \simeq i_s$$

the source resistance R_s must be much larger that R_L,

$$R_s \gg R_L$$

Thus for a signal source represented by its Norton equivalent, ideally $R_s = \infty$, and as R_s is reduced, relative to the load resistance R_L with which this source is intended to operate, the current i_o that flows through the load becomes smaller, not a desirable outcome.

Finally, we note that although circuit designers cannot usually do much about the value of R_s; they may have to devise a circuit solution that minimizes or eliminates the loss of signal strength that results when the source is connected to the load.

EXERCISES

1.1 For the signal-source representations shown in Figs. 1.1(a) and 1.1(b), what are the open-circuit output voltages that would be observed? If, for each, the output terminals are short-circuited (i.e., wired together), what current would flow? For the representations to be equivalent, what must the relationship be between v_s, i_s, and R_s?
Ans. For (a), $v_{oc} = v_s(t)$; for (b), $v_{oc} = R_s i_s(t)$; for (a), $i_{sc} = v_s(t)/R_s$; for (b), $i_{sc} = i_s(t)$; for equivalency, $v_s(t) = R_s i_s(t)$

1.2 A signal source has an open-circuit voltage of 10 mV and a short-circuit current of 10 μA. What is the source resistance?
Ans. 1 kΩ

1.3 A signal source that is most conveniently represented by its Thévenin equivalent has $v_s = 10$ mV and $R_s = 1$ kΩ. If the source feeds a load resistance R_L, find the voltage v_o that appears across the load for $R_L = 100$ kΩ, 10 kΩ, 1 kΩ, and 100 Ω. Also, find the lowest permissible value of R_L for which the output voltage is at least 80% of the source voltage.
Ans. 9.9 mV; 9.1 mV; 5 mV; 0.9 mV; 4 kΩ

1.4 A signal source that is most conveniently represented by its Norton equivalent form has $i_s = 10$ μA and $R_s = 100$ kΩ. If the source feeds a load resistance R_L, find the current i_o that flows through the load for $R_L = 1$ kΩ, 10 kΩ, 100 kΩ, and 1 MΩ. Also, find the largest permissible value of R_L for which the load current is at least 80% of the source current.
Ans. 9.9 μA; 9.1 μA; 5 μA; 0.9 μA; 25 kΩ

From the discussion above, it should be apparent that a signal is a time-varying quantity that can be represented by a graph such as that shown in Fig. 1.3. In fact, the information content of the signal is represented by the changes in its magnitude as time progresses; that is, the information is contained in the "wiggles" in the signal waveform. In general, such waveforms are difficult to characterize mathematically. In other words, it is not easy to describe succinctly an arbitrary-looking waveform such as that of Fig. 1.3. Of course, such a description is of great importance for the purpose of designing appropriate signal-processing circuits that perform desired functions on the given signal. An effective approach to signal characterization is studied in the next section.

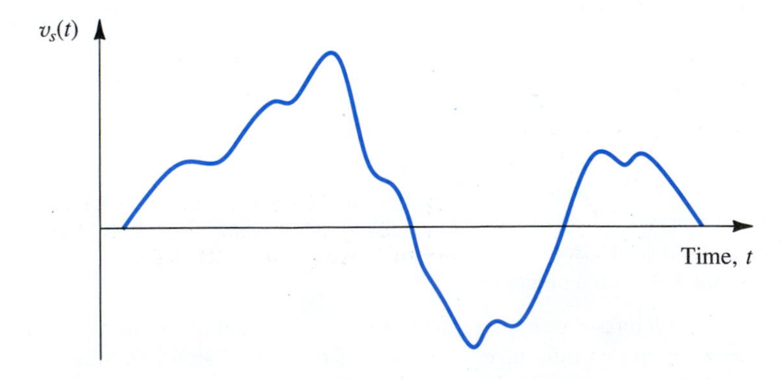

Figure 1.3 An arbitrary voltage signal $v_s(t)$.

1.2 Frequency Spectrum of Signals

An extremely useful characterization of a signal, and for that matter of any arbitrary function of time, is in terms of its **frequency spectrum**. Such a description of signals is obtained through the mathematical tools of **Fourier series** and **Fourier transform**.[1] We are not interested here in the details of these transformations; suffice it to say that they provide the means for representing a voltage signal $v_s(t)$ or a current signal $i_s(t)$ as the sum of sine-wave signals of different frequencies and amplitudes. This makes the sine wave a very important signal in the analysis, design, and testing of electronic circuits. Therefore, we shall briefly review the properties of the sinusoid.

Figure 1.4 shows a sine-wave voltage signal $v_a(t)$,

$$v_a(t) = V_a \sin \omega t \tag{1.1}$$

where V_a denotes the peak value or amplitude in volts and ω denotes the angular frequency in radians per second; that is, $\omega = 2\pi f$ rad/s, where f is the frequency in hertz, $f = 1/T$ Hz, and T is the period in seconds.

The sine-wave signal is completely characterized by its peak value V_a, its frequency ω, and its phase with respect to an arbitrary reference time. In the case depicted in Fig. 1.4, the time origin has been chosen so that the phase angle is 0. It should be mentioned that it is common to express the amplitude of a sine-wave signal in terms of its root-mean-square (rms) value, which is equal to the peak value divided by $\sqrt{2}$. Thus the rms value of the sinusoid $v_a(t)$ of Fig. 1.4 is $V_a/\sqrt{2}$. For instance, when we speak of the wall power supply in our homes as being 120 V, we mean that it has a sine waveform of $120\sqrt{2}$ volts peak value.

Returning now to the representation of signals as the sum of sinusoids, we note that the Fourier series is utilized to accomplish this task for the special case of a signal that is a periodic function of time. On the other hand, the Fourier transform is more general and can be used to obtain the frequency spectrum of a signal whose waveform is an arbitrary function of time.

The Fourier series allows us to express a given periodic function of time as the sum of an infinite number of sinusoids whose frequencies are harmonically related. For instance, the symmetrical square-wave signal in Fig. 1.5 can be expressed as

$$v(t) = \frac{4V}{\pi}(\sin \omega_0 t + \tfrac{1}{3}\sin 3\omega_0 t + \tfrac{1}{5}\sin 5\omega_0 t + \cdots) \tag{1.2}$$

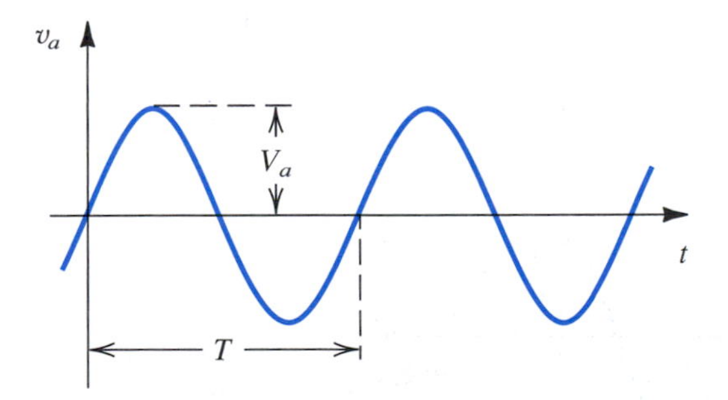

Figure 1.4 Sine-wave voltage signal of amplitude V_a and frequency $f = 1/T$ Hz. The angular frequency $\omega = 2\pi f$ rad/s.

[1]The reader who has not yet studied these topics should not be alarmed. No detailed application of this material will be made until Chapter 9. Nevertheless, a general understanding of Section 1.2 should be very helpful in studying early parts of this book.

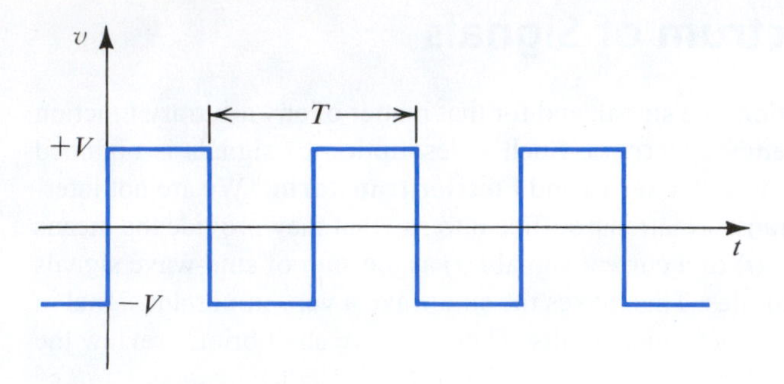

Figure 1.5 A symmetrical square-wave signal of amplitude V.

where V is the amplitude of the square wave and $\omega_0 = 2\pi/T$ (T is the period of the square wave) is called the **fundamental frequency**. Note that because the amplitudes of the harmonics progressively decrease, the infinite series can be truncated, with the truncated series providing an approximation to the square waveform.

The sinusoidal components in the series of Eq. (1.2) constitute the frequency spectrum of the square-wave signal. Such a spectrum can be graphically represented as in Fig. 1.6, where the horizontal axis represents the angular frequency ω in radians per second.

The Fourier transform can be applied to a nonperiodic function of time, such as that depicted in Fig. 1.3, and provides its frequency spectrum as a continuous function of frequency, as indicated in Fig. 1.7. Unlike the case of periodic signals, where the spectrum consists of discrete frequencies (at ω_0 and its harmonics), the spectrum of a nonperiodic signal contains in general all possible frequencies. Nevertheless, the essential parts of the spectra of practical signals are usually confined to relatively short segments of the frequency (ω) axis—an observation that is very useful in the processing of such signals. For instance, the spectrum of audible sounds such as speech and music extends from about 20 Hz to about 20 kHz—a frequency range known as the **audio band**. Here we should note that although some musical tones have frequencies above 20 kHz, the human ear is incapable of hearing frequencies that are much above 20 kHz. As another example, analog video signals have their spectra in the range of 0 MHz to 4.5 MHz.

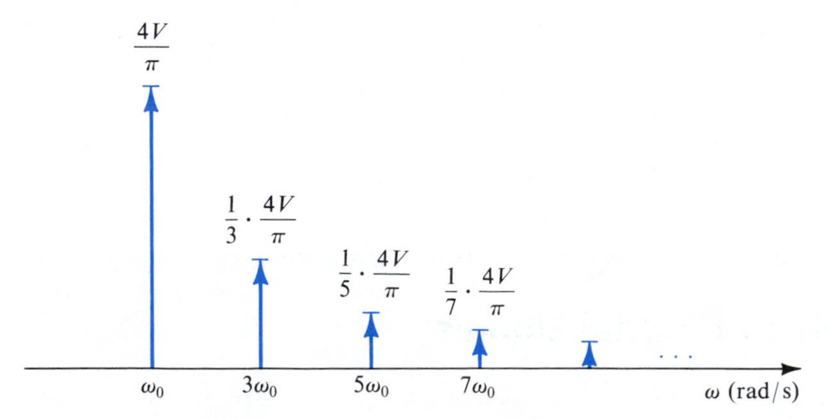

Figure 1.6 The frequency spectrum (also known as the **line spectrum**) of the periodic square wave of Fig. 1.5.

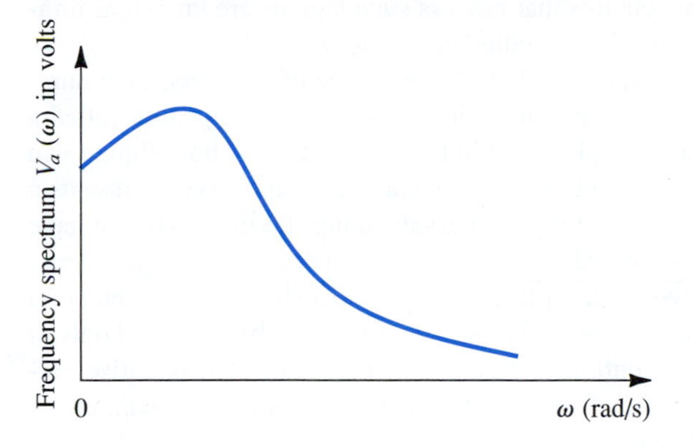

Figure 1.7 The frequency spectrum of an arbitrary waveform such as that in Fig. 1.3.

We conclude this section by noting that a signal can be represented either by the manner in which its waveform varies with time, as for the voltage signal $v_a(t)$ shown in Fig. 1.3, or in terms of its frequency spectrum, as in Fig. 1.7. The two alternative representations are known as the time-domain representation and the frequency-domain representation, respectively. The frequency-domain representation of $v_a(t)$ will be denoted by the symbol $V_a(\omega)$.

EXERCISES

1.5 Find the frequencies f and ω of a sine-wave signal with a period of 1 ms.
Ans. $f = 1000$ Hz; $\omega = 2\pi \times 10^3$ rad/s

1.6 What is the period T of sine waveforms characterized by frequencies of (a) $f = 60$ Hz? (b) $f = 10^{-3}$ Hz? (c) $f = 1$ MHz?
Ans. 16.7 ms; 1000 s; 1 µs

1.7 The UHF (ultra high frequency) television broadcast band begins with channel 14 and extends from 470 MHz to 806 MHz. If 6 MHz is allocated for each channel, how many channels can this band accommodate?
Ans. 56; channels 14 to 69

1.8 When the square-wave signal of Fig. 1.5, whose Fourier series is given in Eq. (1.2), is applied to a resistor, the total power dissipated may be calculated directly using the relationship $P = 1/T \int_0^T (v^2/R)\, dt$ or indirectly by summing the contribution of each of the harmonic components, that is, $P = P_1 + P_3 + P_5 + \ldots$, which may be found directly from rms values. Verify that the two approaches are equivalent. What fraction of the energy of a square wave is in its fundamental? In its first five harmonics? In its first seven? First nine? In what number of harmonics is 90% of the energy? (Note that in counting harmonics, the fundamental at ω_0 is the first, the one at $2\omega_0$ is the second, etc.)
Ans. 0.81; 0.93; 0.95; 0.96; 3

1.3 Analog and Digital Signals

The voltage signal depicted in Fig. 1.3 is called an **analog signal**. The name derives from the fact that such a signal is analogous to the physical signal that it represents. The magnitude of an analog signal can take on any value; that is, the amplitude of an analog signal exhibits a continuous variation over its range of activity. The vast majority of signals in the world around us are analog. Electronic circuits that process such signals are known as **analog circuits**. A variety of analog circuits will be studied in this book.

An alternative form of signal representation is that of a sequence of numbers, each number representing the signal magnitude at an instant of time. The resulting signal is called a **digital signal**. To see how a signal can be represented in this form—that is, how signals can be converted from analog to digital form—consider Fig. 1.8(a). Here the curve represents a voltage signal, identical to that in Fig. 1.3. At equal intervals along the time axis, we have marked the time instants t_0, t_1, t_2, and so on. At each of these time instants, the magnitude of the signal is measured, a process known as **sampling**. Figure 1.8(b) shows a representation of the signal of Fig. 1.8(a) in terms of its samples. The signal of Fig. 1.8(b) is defined only at the sampling instants; it no longer is a continuous function of time; rather, it is a **discrete-time signal**. However, since the magnitude of each sample can take any value in a continuous range, the signal in Fig. 1.8(b) is still an analog signal.

Now if we represent the magnitude of each of the signal samples in Fig. 1.8(b) by a number having a finite number of digits, then the signal amplitude will no longer be continuous; rather, it is said to be **quantized**, **discretized**, or **digitized**. The resulting digital signal then is simply a sequence of numbers that represent the magnitudes of the successive signal samples.

The choice of number system to represent the signal samples affects the type of digital signal produced, and has a profound effect on the complexity of the digital circuits required to process the signals. It turns out that the **binary** number system results in the simplest possible digital signals and circuits. In a binary system, each digit in the number takes on one of

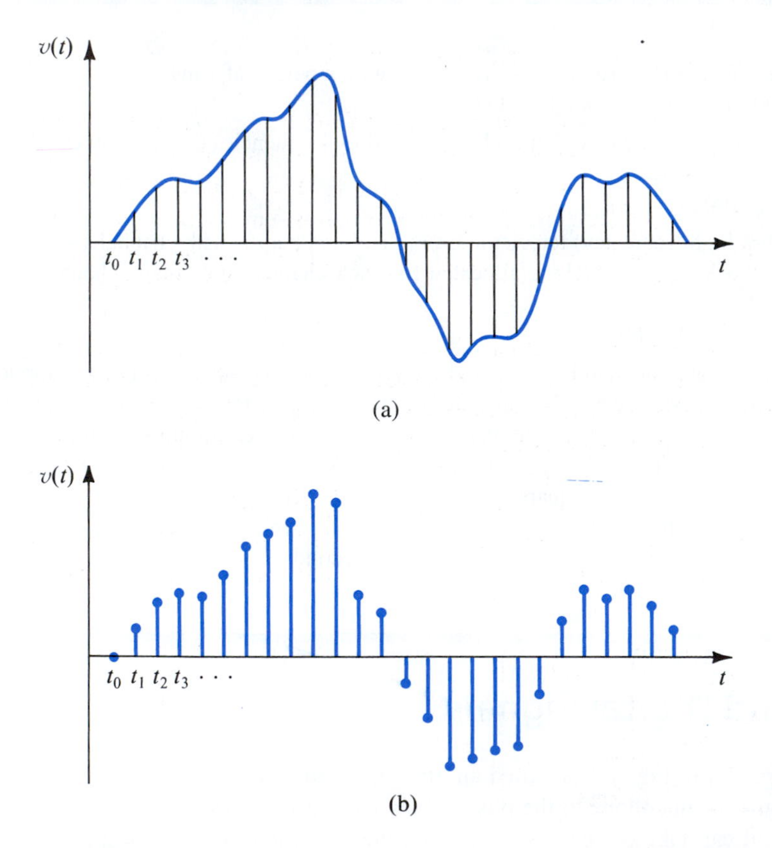

Figure 1.8 Sampling the continuous-time analog signal in **(a)** results in the discrete-time signal in **(b).**

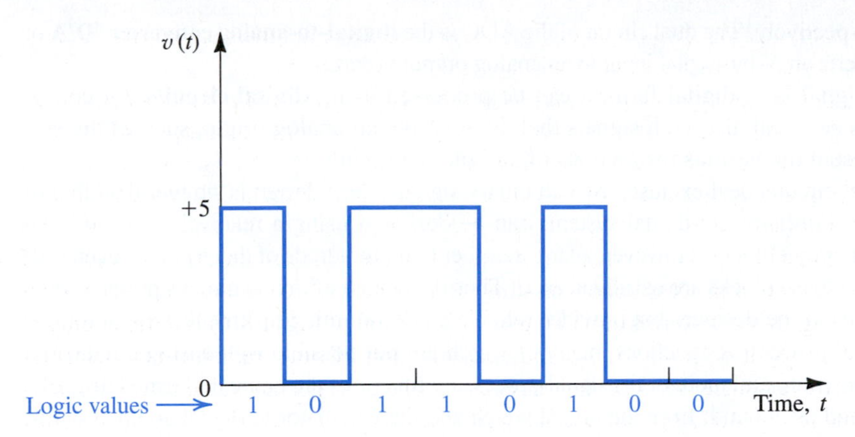

Figure 1.9 Variation of a particular binary digital signal with time.

only two possible values, denoted 0 and 1. Correspondingly, the digital signals in binary systems need have only two voltage levels, which can be labeled low and high. As an example, in some of the digital circuits studied in this book, the levels are 0 V and +5 V. Figure 1.9 shows the time variation of such a digital signal. Observe that the waveform is a pulse train with 0 V representing a 0 signal, or logic 0, and +5 V representing logic 1.

If we use N *b*inary dig*its* (bits) to represent each sample of the analog signal, then the digitized sample value can be expressed as

$$D = b_0 2^0 + b_1 2^1 + b_2 2^2 + \cdots + b_{N-1} 2^{N-1} \quad (1.3)$$

where b_0, b_1, …, b_{N-1}, denote the N bits and have values of 0 or 1. Here bit b_0 is the **least significant bit (LSB)**, and bit b_{N-1} is the **most significant bit (MSB)**. Conventionally, this binary number is written as $b_{N-1}\, b_{N-2} \ldots b_0$. We observe that such a representation quantizes the analog sample into one of 2^N levels. Obviously the greater the number of bits (i.e., the larger the N), the closer the digital word D approximates the magnitude of the analog sample. That is, increasing the number of bits reduces the *quantization error* and increases the resolution of the analog-to-digital conversion. This improvement is, however, usually obtained at the expense of more complex and hence more costly circuit implementations. It is not our purpose here to delve into this topic any deeper; we merely want the reader to appreciate the nature of analog and digital signals. Nevertheless, it is an opportune time to introduce a very important circuit building block of modern electronic systems: the **analog-to-digital converter (A/D** or **ADC**) shown in block form in Fig. 1.10 The ADC accepts at its input the samples of an analog signal and provides for each input sample the corresponding N-bit digital representation (according to Eq. 1.3) at its N output terminals. Thus although the voltage at the input might be, say, 6.51 V, at each of the output terminals (say, at the ith terminal), the voltage will be either low (0 V) or high (5 V) if b_i is supposed

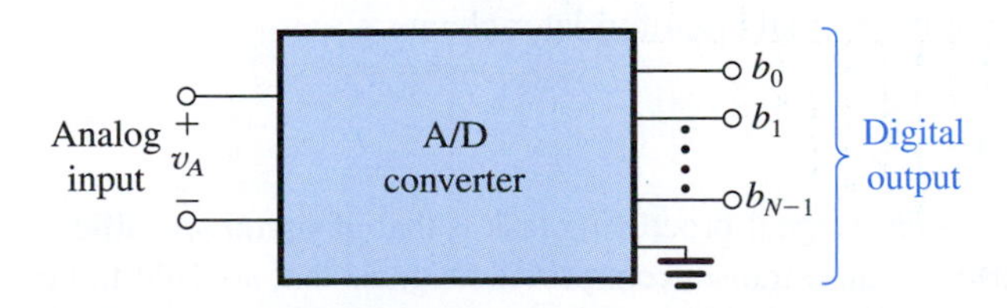

Figure 1.10 Block-diagram representation of the analog-to-digital converter (ADC).

to be 0 or 1, respectively. The dual circuit of the ADC is the **digital-to-analog converter** (**D/A** or **DAC**). It converts an N-bit digital input to an analog output voltage.

Once the signal is in digital form, it can be processed using **digital circuits.** Of course digital circuits can deal also with signals that do not have an analog origin, such as the signals that represent the various instructions of a digital computer.

Since digital circuits deal exclusively with binary signals, their design is simpler than that of analog circuits. Furthermore, digital systems can be designed using a relatively few different kinds of digital circuit blocks. However, a large number (e.g., hundreds of thousands or even millions) of each of these blocks are usually needed. Thus the design of digital circuits poses its own set of challenges to the designer but provides reliable and economic implementations of a great variety of signal-processing functions, many of which are not possible with analog circuits. At the present time, more and more of the signal-processing functions are being performed digitally. Examples around us abound: from the digital watch and the calculator to digital audio systems, digital cameras and, more recently, digital television. Moreover, some longstanding analog systems such as the telephone communication system are now almost entirely digital. And we should not forget the most important of all digital systems, the digital computer.

The basic building blocks of digital systems are logic circuits and memory circuits. We shall study both in this book, beginning in Chapter 13.

One final remark: Although the digital processing of signals is at present all-pervasive, there remain many signal-processing functions that are best performed by analog circuits. Indeed, many electronic systems include both analog and digital parts. It follows that a good electronics engineer must be proficient in the design of both analog and digital circuits, or **mixed-signal** or **mixed-mode** design as it is currently known. Such is the aim of this book.

EXERCISE

1.9 Consider a 4-bit digital word $D = b_3b_2b_1b_0$ (see Eq. 1.3) used to represent an analog signal v_A that varies between 0 V and +15 V.
(a) Give D corresponding to v_A = 0 V, 1 V, 2 V, and 15 V.
(b) What change in v_A causes a change from 0 to 1 in (i) b_0, (ii) b_1, (iii) b_2, and (iv) b_3?
(c) If v_A = 5.2 V, what do you expect D to be? What is the resulting error in representation?
Ans. (a) 0000, 0001, 0010, 1111; (b) +1 V, +2 V, +4 V, +8 V; (c) 0101, −4%

1.4 Amplifiers

In this section, we shall introduce the most fundamental signal-processing function, one that is employed in some form in almost every electronic system, namely, signal amplification. We shall study the amplifier as a circuit building-block; that is, we shall consider its external characteristics and leave the design of its internal circuit to later chapters.

1.4.1 Signal Amplification

From a conceptual point of view the simplest signal-processing task is that of **signal amplification**. The need for amplification arises because transducers provide signals that are said to be "weak," that is, in the microvolt (μV) or millivolt (mV) range and possessing little energy. Such signals are too small for reliable processing, and processing is much easier if the signal magnitude is made larger. The functional block that accomplishes this task is the **signal amplifier**.

It is appropriate at this point to discuss the need for **linearity** in amplifiers. Care must be exercised in the amplification of a signal, so that the information contained in the signal is not changed and no new information is introduced. Thus when we feed the signal shown in Fig. 1.3 to an amplifier, we want the output signal of the amplifier to be an exact replica of that at the input, except of course for having larger magnitude. In other words, the "wiggles" in the output waveform must be identical to those in the input waveform. Any change in waveform is considered to be **distortion** and is obviously undesirable.

An amplifier that preserves the details of the signal waveform is characterized by the relationship

$$v_o(t) = Av_i(t) \tag{1.4}$$

where v_i and v_o are the input and output signals, respectively, and A is a constant representing the magnitude of amplification, known as **amplifier gain**. Equation (1.4) is a linear relationship; hence the amplifier it describes is a **linear amplifier**. It should be easy to see that if the relationship between v_o and v_i contains higher powers of v_i, then the waveform of v_o will no longer be identical to that of v_i. The amplifier is then said to exhibit **nonlinear distortion**.

The amplifiers discussed so far are primarily intended to operate on very small input signals. Their purpose is to make the signal magnitude larger and therefore are thought of as **voltage amplifiers**. The **preamplifier** in the home stereo system is an example of a voltage amplifier.

At this time we wish to mention another type of amplifier, namely, the **power amplifier**. Such an amplifier may provide only a modest amount of voltage gain but substantial current gain. Thus while absorbing little power from the input signal source to which it is connected, often a preamplifier, it delivers large amounts of power to its load. An example is found in the power amplifier of the home stereo system, whose purpose is to provide sufficient power to drive the loudspeaker, which is the amplifier load. Here we should note that the loudspeaker is the output transducer of the stereo system; it converts the electric output signal of the system into an acoustic signal. A further appreciation of the need for linearity can be acquired by reflecting on the power amplifier. A linear power amplifier causes both soft and loud music passages to be reproduced without distortion.

1.4.2 Amplifier Circuit Symbol

The signal amplifier is obviously a two-port network. Its function is conveniently represented by the circuit symbol of Fig. 1.11(a). This symbol clearly distinguishes the input and output ports and indicates the direction of signal flow. Thus, in subsequent diagrams it will not be necessary to label the two ports "input" and "output." For generality we have shown the amplifier to have two input terminals that are distinct from the two output terminals. A more common situation is illustrated in Fig. 1.11(b), where a common terminal exists between the input and output ports of the amplifier. This common terminal is used as a reference point and is called the **circuit ground**.

1.4.3 Voltage Gain

A linear amplifier accepts an input signal $v_I(t)$ and provides at the output, across a load resistance R_L (see Fig. 1.12(a)), an output signal $v_O(t)$ that is a magnified replica of $v_I(t)$. The **voltage gain** of the amplifier is defined by

$$\text{Voltage gain } (A_v) \equiv \frac{v_O}{v_I} \tag{1.5}$$

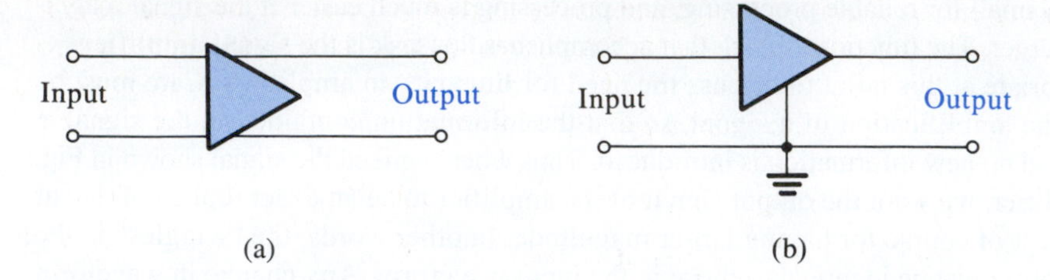

Figure 1.11 **(a)** Circuit symbol for amplifier. **(b)** An amplifier with a common terminal (ground) between the input and output ports.

Fig. 1.12(b) shows the **transfer characteristic** of a linear amplifier. If we apply to the input of this amplifier a sinusoidal voltage of amplitude $\hat{V}$, we obtain at the output a sinusoid of amplitude $A_v\hat{V}$.

1.4.4 Power Gain and Current Gain

An amplifier increases the signal power, an important feature that distinguishes an amplifier from a transformer. In the case of a transformer, although the voltage delivered to the load could be greater than the voltage feeding the input side (the primary), the power delivered to the load (from the secondary side of the transformer) is less than or at most equal to the power supplied by the signal source. On the other hand, an amplifier provides the load with power greater than that obtained from the signal source. That is, amplifiers have power gain. The **power gain** of the amplifier in Fig. 1.12(a) is defined as

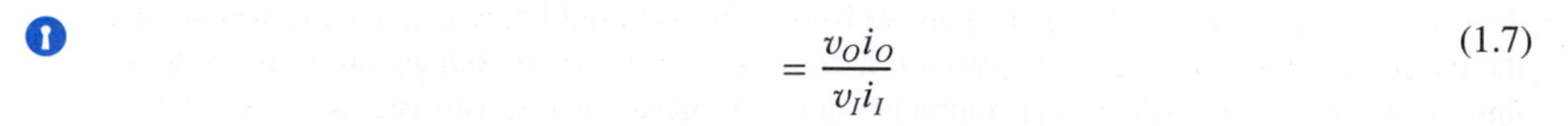

$$\text{Power gain } (A_p) \equiv \frac{\text{load power } (P_L)}{\text{input power } (P_I)} \tag{1.6}$$

$$= \frac{v_O i_O}{v_I i_I} \tag{1.7}$$

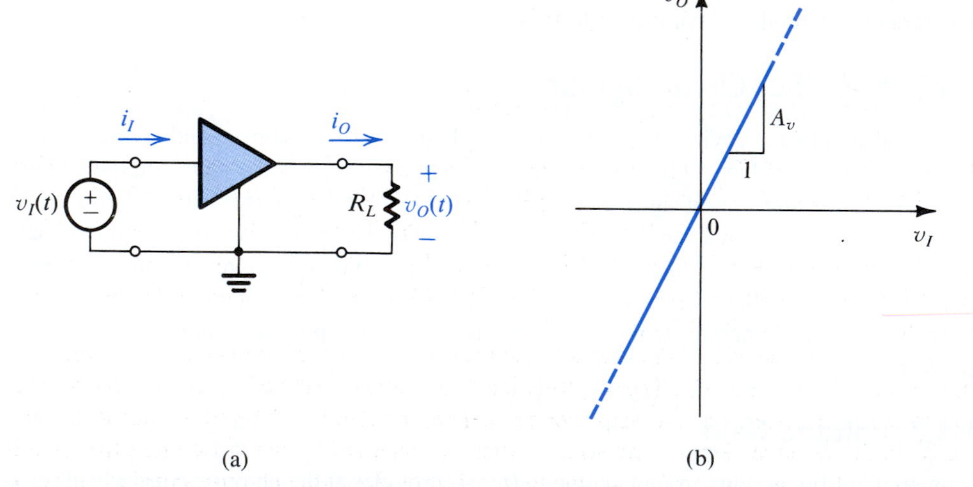

Figure 1.12 **(a)** A voltage amplifier fed with a signal $v_I(t)$ and connected to a load resistance R_L. **(b)** Transfer characteristic of a linear voltage amplifier with voltage gain A_v.

where i_O is the current that the amplifier delivers to the load (R_L), $i_O = v_O/R_L$, and i_I is the current the amplifier draws from the signal source. The **current gain** of the amplifier is defined as

$$\text{Current gain } (A_i) \equiv \frac{i_O}{i_I} \tag{1.8}$$

From Eqs. (1.5) to (1.8) we note that

$$A_p = A_v A_i \tag{1.9}$$

1.4.5 Expressing Gain in Decibels

The amplifier gains defined above are ratios of similarly dimensioned quantities. Thus they will be expressed either as dimensionless numbers or, for emphasis, as V/V for the voltage gain, A/A for the current gain, and W/W for the power gain. Alternatively, for a number of reasons, some of them historic, electronics engineers express amplifier gain with a logarithmic measure. Specifically the voltage gain A_v can be expressed as

$$\text{Voltage gain in decibels} = 20 \log|A_v| \quad \text{dB}$$

and the current gain A_i can be expressed as

$$\text{Current gain in decibels} = 20 \log|A_i| \quad \text{dB}$$

Since power is related to voltage (or current) squared, the power gain A_p can be expressed in decibels as

$$\text{Power gain in decibels} = 10 \log A_p \quad \text{dB}$$

The absolute values of the voltage and current gains are used because in some cases A_v or A_i will be a negative number. A negative gain A_v simply means that there is a 180° phase difference between input and output signals; it does not imply that the amplifier is **attenuating** the signal. On the other hand, an amplifier whose voltage gain is, say, –20 dB is in fact attenuating the input signal by a factor of 10 (i.e., $A_v = 0.1$ V/V).

1.4.6 The Amplifier Power Supplies

Since the power delivered to the load is greater than the power drawn from the signal source, the question arises as to the source of this additional power. The answer is found by observing that amplifiers need dc power supplies for their operation. These dc sources supply the extra power delivered to the load as well as any power that might be dissipated in the internal circuit of the amplifier (such power is converted to heat). In Fig. 1.12(a) we have not explicitly shown these dc sources.

Figure 1.13(a) shows an amplifier that requires two dc sources: one positive of value V_{CC} and one negative of value V_{EE}. The amplifier has two terminals, labeled V^+ and V^-, for connection to the dc supplies. For the amplifier to operate, the terminal labeled V^+ has to be connected to the positive side of a dc source whose voltage is V_{CC} and whose negative side is connected to the circuit ground. Also, the terminal labeled V^- has to be connected to the negative side of a dc source whose voltage is V_{EE} and whose positive side is connected to the circuit ground. Now, if the current drawn from the positive supply is denoted I_{CC} and that from the negative supply is I_{EE} (see Fig. 1.13a), then the dc power delivered to the amplifier is

$$P_{dc} = V_{CC}I_{CC} + V_{EE}I_{EE}$$

If the power dissipated in the amplifier circuit is denoted $P_{\text{dissipated}}$, the power-balance equation for the amplifier can be written as

$$P_{\text{dc}} + P_I = P_L + P_{\text{dissipated}}$$

where P_I is the power drawn from the signal source and P_L is the power delivered to the load. Since the power drawn from the signal source is usually small, the amplifier power **efficiency** is defined as

$$\eta \equiv \frac{P_L}{P_{\text{dc}}} \times 100 \tag{1.10}$$

The power efficiency is an important performance parameter for amplifiers that handle large amounts of power. Such amplifiers, called power amplifiers, are used, for example, as output amplifiers of stereo systems.

In order to simplify circuit diagrams, we shall adopt the convention illustrated in Fig. 1.13(b). Here the V^+ terminal is shown connected to an arrowhead pointing upward and the V^- terminal to an arrowhead pointing downward. The corresponding voltage is indicated next to each arrowhead. Note that in many cases we will not explicitly show the connections of the amplifier to the dc power sources. Finally, we note that some amplifiers require only one power supply.

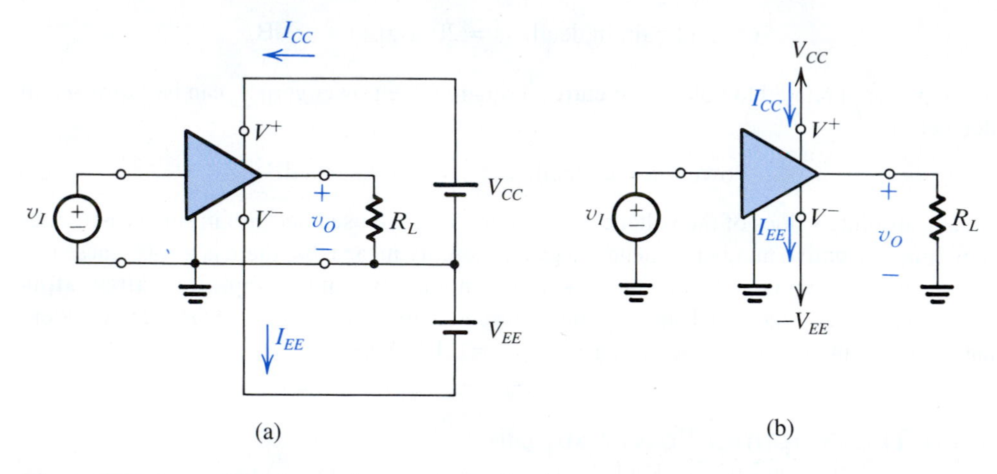

Figure 1.13 An amplifier that requires two dc supplies (shown as batteries) for operation.

Example 1.2

Consider an amplifier operating from ±10-V power supplies. It is fed with a sinusoidal voltage having 1 V peak and delivers a sinusoidal voltage output of 9 V peak to a 1-kΩ load. The amplifier draws a current of 9.5 mA from each of its two power supplies. The input current of the amplifier is found to be sinusoidal with 0.1 mA peak. Find the voltage gain, the current gain, the power gain, the power drawn from the dc supplies, the power dissipated in the amplifier, and the amplifier efficiency.

Solution

$$A_v = \frac{9}{1} = 9 \text{ V/V}$$

or

$$A_v = 20 \log 9 = 19.1 \text{ dB}$$

$$\hat{I}_o = \frac{9 \text{ V}}{1 \text{ k}\Omega} = 9 \text{ mA}$$

$$A_i = \frac{\hat{I}_o}{\hat{I}_i} = \frac{9}{0.1} = 90 \text{ A/A}$$

or

$$A_i = 20 \log 90 = 39.1 \text{ dB}$$

$$P_L = V_{o_{\text{rms}}} I_{o_{\text{rms}}} = \frac{9}{\sqrt{2}} \frac{9}{\sqrt{2}} = 40.5 \text{ mW}$$

$$P_I = V_{i_{\text{rms}}} I_{i_{\text{rms}}} = \frac{1}{\sqrt{2}} \frac{0.1}{\sqrt{2}} = 0.05 \text{ mW}$$

$$A_p = \frac{P_L}{P_I} = \frac{40.5}{0.05} = 810 \text{ W/W}$$

or

$$A_p = 10 \log 810 = 29.1 \text{ dB}$$

$$P_{\text{dc}} = 10 \times 9.5 + 10 \times 9.5 = 190 \text{ mW}$$

$$P_{\text{dissipated}} = P_{\text{dc}} + P_I - P_L$$

$$= 190 + 0.05 - 40.5 = 149.6 \text{ mW}$$

$$\eta = \frac{P_L}{P_{\text{dc}}} \times 100 = 21.3\%$$

From the above example we observe that the amplifier converts some of the dc power it draws from the power supplies to signal power that it delivers to the load.

1.4.7 Amplifier Saturation

Practically speaking, the amplifier transfer characteristic remains linear over only a limited range of input and output voltages. For an amplifier operated from two power supplies the output voltage cannot exceed a specified positive limit and cannot decrease below a specified negative limit. The resulting transfer characteristic is shown in Fig. 1.14, with the positive and negative saturation levels denoted L_+ and L_-, respectively. Each of the two saturation levels is usually within a fraction of a volt of the voltage of the corresponding power supply.

Obviously, in order to avoid distorting the output signal waveform, the input signal swing must be kept within the linear range of operation,

$$\frac{L_-}{A_v} \le v_I \le \frac{L_+}{A_v}$$

In Fig. 1.14, which shows two input waveforms and the corresponding output waveforms, the peaks of the larger waveform have been clipped off because of amplifier saturation.

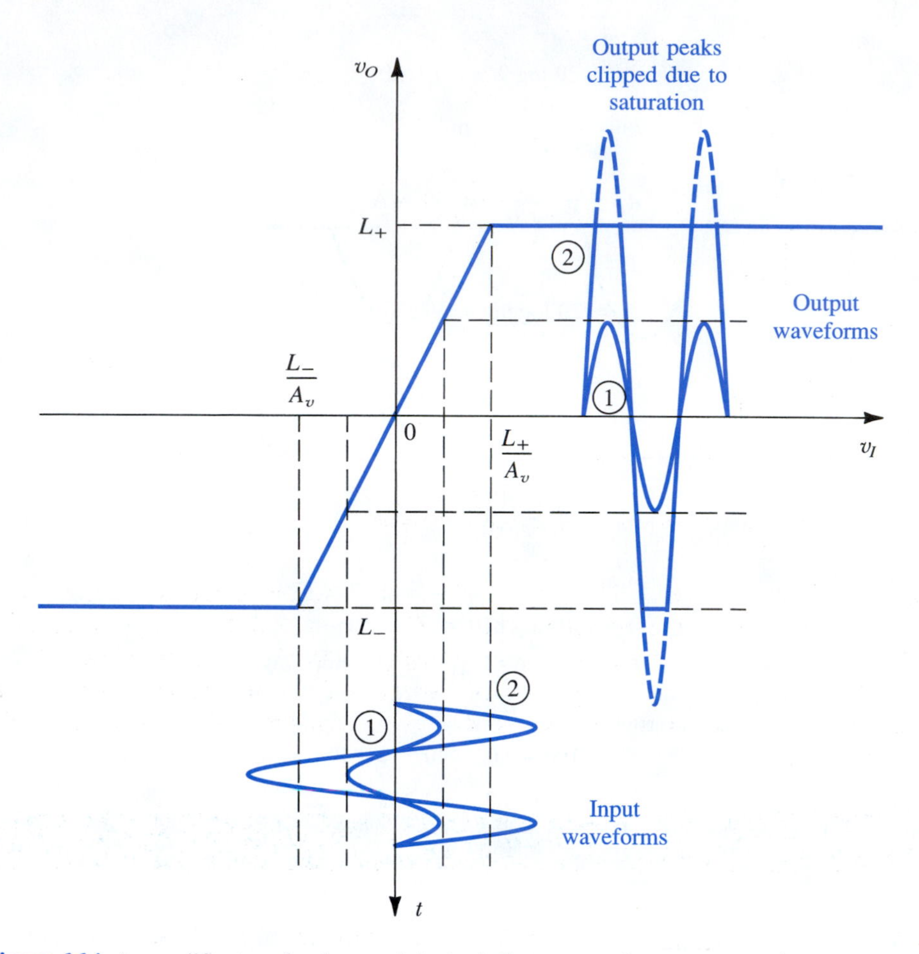

Figure 1.14 An amplifier transfer characteristic that is linear except for output saturation.

1.4.8 Symbol Convention

At this point, we draw the reader's attention to the terminology we shall employ throughout the book. To illustrate the terminology, Fig. 1.15 shows the waveform of a current $i_C(t)$ that is flowing through a branch in a particular circuit. The current $i_C(t)$ consists of a dc component I_C on which is superimposed a sinusoidal component $i_c(t)$ whose peak amplitude is I_c. Observe that at a time t, the **total instantaneous** current $i_C(t)$ is the sum of the dc current I_C and the signal current $i_c(t)$,

$$i_C(t) = I_C + i_c(t) \tag{1.11}$$

where the signal current is given by

$$i_c(t) = I_c \sin \omega t$$

Thus, we state some conventions: Total instantaneous quantities are denoted by a lowercase symbol with uppercase subscript(s), for example, $i_C(t)$, $v_{DS}(t)$. Direct-current (dc) quantities are denoted by an uppercase symbol with uppercase subscript(s), for example I_C, V_{DS}. Incremental

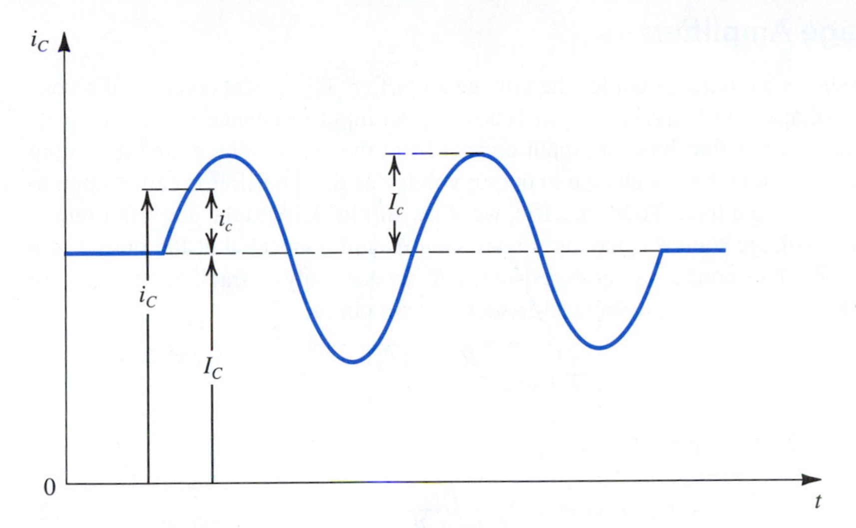

Figure 1.15 Symbol convention employed throughout the book.

signal quantities are denoted by a lowercase symbol with lowercase subscript(s), for example, $i_c(t)$, $v_{gs}(t)$. If the signal is a sine wave, then its amplitude is denoted by an uppercase symbol with lowercase subscript(s), for example I_c, V_{gs}. Finally, although not shown in Fig. 1.15, dc power supplies are denoted by an uppercase letter with a double-letter uppercase subscript, for example, V_{CC}, V_{DD}. A similar notation is used for the dc current drawn from the power supply, for example, I_{CC}, I_{DD}.

EXERCISES

1.10 An amplifier has a voltage gain of 100 V/V and a current gain of 1000 A/A. Express the voltage and current gains in decibels and find the power gain.
Ans. 40 dB; 60 dB; 50 dB

1.11 An amplifier operating from a single 15-V supply provides a 12-V peak-to-peak sine-wave signal to a 1-kΩ load and draws negligible input current from the signal source. The dc current drawn from the 15-V supply is 8 mA. What is the power dissipated in the amplifier, and what is the amplifier efficiency?
Ans. 102 mW; 15%

1.5 Circuit Models for Amplifiers

A substantial part of this book is concerned with the design of amplifier circuits that use transistors of various types. Such circuits will vary in complexity from those using a single transistor to those with 20 or more devices. In order to be able to apply the resulting amplifier circuit as a building block in a system, one must be able to characterize, or **model**, its terminal behavior. In this section, we study simple but effective amplifier models. These models apply irrespective of the complexity of the internal circuit of the amplifier. The values of the model parameters can be found either by analyzing the amplifier circuit or by performing measurements at the amplifier terminals.

1.5.1 Voltage Amplifiers

Figure 1.16(a) shows a circuit model for the voltage amplifier. The model consists of a voltage-controlled voltage source having a gain factor A_{vo}, an input resistance R_i that accounts for the fact that the amplifier draws an input current from the signal source, and an output resistance R_o that accounts for the change in output voltage as the amplifier is called upon to supply output current to a load. To be specific, we show in Fig. 1.16(b) the amplifier model fed with a signal voltage source v_s having a resistance R_s and connected at the output to a load resistance R_L. The nonzero output resistance R_o causes only a fraction of $A_{vo}v_i$ to appear across the output. Using the voltage-divider rule we obtain

$$v_o = A_{vo} v_i \frac{R_L}{R_L + R_o}$$

Thus the voltage gain is given by

$$A_v \equiv \frac{v_o}{v_i} = A_{vo} \frac{R_L}{R_L + R_o} \tag{1.12}$$

It follows that in order not to lose gain in coupling the amplifier output to a load, the output resistance R_o should be much smaller than the load resistance R_L. In other words, for a given R_L one must design the amplifier so that its R_o is much smaller than R_L. Furthermore, there are applications in which R_L is known to vary over a certain range. In order to keep the output voltage v_o as constant as possible, the amplifier is designed with R_o much smaller than the lowest value of R_L. An ideal voltage amplifier is one with $R_o = 0$. Equation (1.12) indicates also that for $R_L = \infty$, $A_v = A_{vo}$. Thus A_{vo} is the voltage gain of the unloaded amplifier, or the **open-circuit voltage gain**. It should also be clear that in specifying the voltage gain of an amplifier, one must also specify the value of load resistance

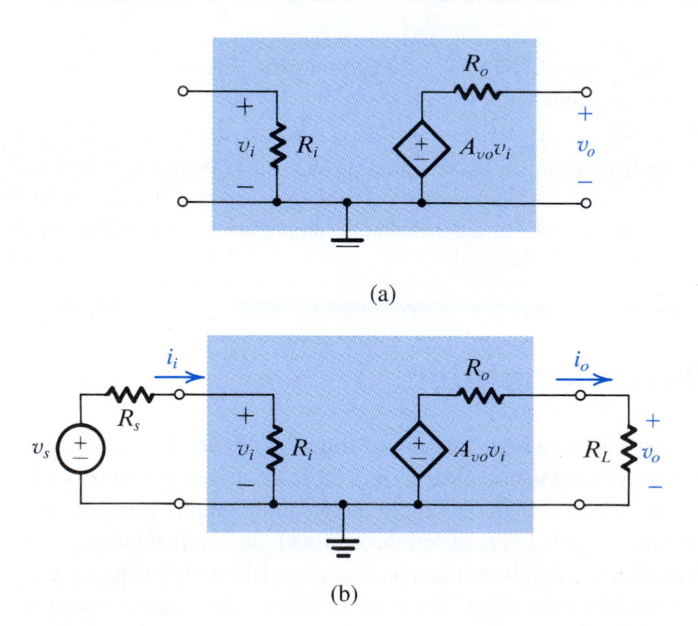

Figure 1.16 **(a)** Circuit model for the voltage amplifier. **(b)** The voltage amplifier with input signal source and load.

at which this gain is measured or calculated. If a load resistance is not specified, it is normally assumed that the given voltage gain is the open-circuit gain A_{vo}.

The finite input resistance R_i introduces another voltage-divider action at the input, with the result that only a fraction of the source signal v_s actually reaches the input terminals of the amplifier; that is,

$$v_i = v_s \frac{R_i}{R_i + R_s} \tag{1.13}$$

It follows that in order not to lose a significant portion of the input signal in coupling the signal source to the amplifier input, the amplifier must be designed to have an input resistance R_i much greater than the resistance of the signal source, $R_i \gg R_s$. Furthermore, there are applications in which the source resistance is known to vary over a certain range. To minimize the effect of this variation on the value of the signal that appears at the input of the amplifier, the design ensures that R_i is much greater than the largest value of R_s. An ideal voltage amplifier is one with $R_i = \infty$. In this ideal case both the current gain and power gain become infinite.

The overall voltage gain (v_o/v_s) can be found by combining Eqs. (1.12) and (1.13),

$$\frac{v_o}{v_s} = A_{vo} \frac{R_i}{R_i + R_s} \frac{R_L}{R_L + R_o}$$

There are situations in which one is interested not in voltage gain but only in a significant power gain. For instance, the source signal can have a respectable voltage but a source resistance that is much greater than the load resistance. Connecting the source directly to the load would result in significant signal attenuation. In such a case, one requires an amplifier with a high input resistance (much greater than the source resistance) and a low output resistance (much smaller than the load resistance) but with a modest voltage gain (or even unity gain). Such an amplifier is referred to as a **buffer amplifier**. We shall encounter buffer amplifiers often throughout this book.

EXERCISES

1.12 A transducer characterized by a voltage of 1 V rms and a resistance of 1 MΩ is available to drive a 10-Ω load. If connected directly, what voltage and power levels result at the load? If a unity-gain (i.e., $A_{vo} = 1$) buffer amplifier with 1-MΩ input resistance and 10-Ω output resistance is interposed between source and load, what do the output voltage and power levels become? For the new arrangement, find the voltage gain from source to load, and the power gain (both expressed in decibels).
Ans. 10 μV rms; 10^{-11} W; 0.25 V; 6.25 mW; −12 dB; 44 dB

1.13 The output voltage of a voltage amplifier has been found to decrease by 20% when a load resistance of 1 kΩ is connected. What is the value of the amplifier output resistance?
Ans. 250 Ω

1.14 An amplifier with a voltage gain of +40 dB, an input resistance of 10 kΩ, and an output resistance of 1 kΩ is used to drive a 1-kΩ load. What is the value of A_{vo}? Find the value of the power gain in decibels.
Ans. 100 V/V; 44 dB

1.5.2 Cascaded Amplifiers

To meet given amplifier specifications, we often need to design the amplifier as a cascade of two or more stages. The stages are usually not identical; rather, each is designed to serve a specific purpose. For instance, in order to provide the overall amplifier with a large input resistance, the first stage is usually required to have a large input resistance. Also, in order to equip the overall amplifier with a low output resistance, the final stage in the cascade is usually designed to have a low output resistance. To illustrate the analysis and design of cascaded amplifiers, we consider a practical example.

Example 1.3

Figure 1.17 depicts an amplifier composed of a cascade of three stages. The amplifier is fed by a signal source with a source resistance of 100 kΩ and delivers its output into a load resistance of 100 Ω. The first stage has a relatively high input resistance and a modest gain factor of 10. The second stage has a higher gain factor but lower input resistance. Finally, the last, or output, stage has unity gain but a low output resistance. We wish to evaluate the overall voltage gain, that is, v_L/v_s, the current gain, and the power gain.

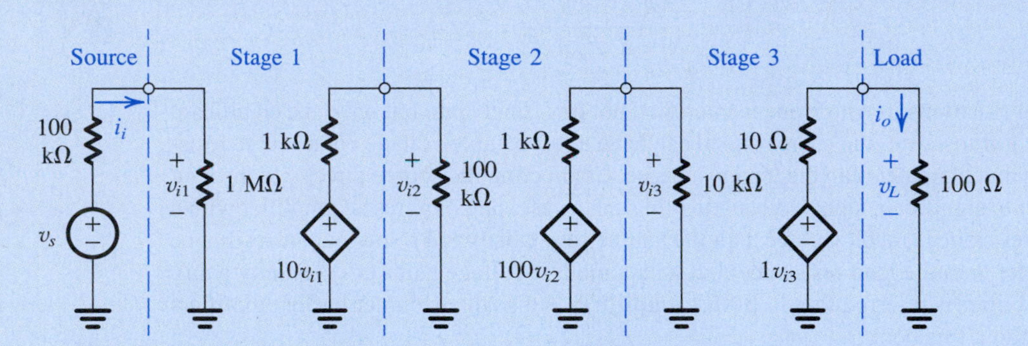

Figure 1.17 Three-stage amplifier for Example 1.3.

Solution

The fraction of source signal appearing at the input terminals of the amplifier is obtained using the voltage-divider rule at the input, as follows:

$$\frac{v_{i1}}{v_s} = \frac{1\text{ M}\Omega}{1\text{ M}\Omega + 100\text{ k}\Omega} = 0.909\text{ V/V}$$

The voltage gain of the first stage is obtained by considering the input resistance of the second stage to be the load of the first stage; that is,

$$A_{v1} \equiv \frac{v_{i2}}{v_{i1}} = 10\frac{100\text{ k}\Omega}{100\text{ k}\Omega + 1\text{ k}\Omega} = 9.9\text{ V/V}$$

Similarly, the voltage gain of the second stage is obtained by considering the input resistance of the third stage to be the load of the second stage,

$$A_{v2} \equiv \frac{v_{i3}}{v_{i2}} = 100\frac{10\text{ k}\Omega}{10\text{ k}\Omega + 1\text{ k}\Omega} = 90.9\text{ V/V}$$

Finally, the voltage gain of the output stage is as follows:

$$A_{v3} \equiv \frac{v_L}{v_{i3}} = 1\frac{100\ \Omega}{100\ \Omega + 10\ \Omega} = 0.909\ \text{V/V}$$

The total gain of the three stages in cascade can be now found from

$$A_v \equiv \frac{v_L}{v_{i1}} = A_{v1}A_{v2}A_{v3} = 818\ \text{V/V}$$

or 58.3 dB.

To find the voltage gain from source to load, we multiply A_v by the factor representing the loss of gain at the input; that is,

$$\frac{v_L}{v_s} = \frac{v_L}{v_{i1}}\frac{v_{i1}}{v_s} = A_v\frac{v_{i1}}{v_s}$$
$$= 818 \times 0.909 = 743.6\ \text{V/V}$$

or 57.4 dB.

The current gain is found as follows:

$$A_i \equiv \frac{i_o}{i_i} = \frac{v_L/100\ \Omega}{v_{i1}/1\ \text{M}\Omega}$$
$$= 10^4 \times A_v = 8.18 \times 10^6\ \text{A/A}$$

or 138.3 dB.

The power gain is found from

$$A_p \equiv \frac{P_L}{P_I} = \frac{v_L i_o}{v_{i1} i_i}$$
$$= A_v A_i = 818 \times 8.18 \times 10^6 = 66.9 \times 10^8\ \text{W/W}$$

or 98.3 dB. Note that

$$A_p(\text{dB}) = \tfrac{1}{2}[A_v(\text{dB}) + A_i(\text{dB})]$$

A few comments on the cascade amplifier in the above example are in order. To avoid losing signal strength at the amplifier input where the signal is usually very small, the first stage is designed to have a relatively large input resistance (1 MΩ), which is much larger than the source resistance. The trade-off appears to be a moderate voltage gain (10 V/V). The second stage does not need to have such a high input resistance; rather, here we need to realize the bulk of the required voltage gain. The third and final, or output, stage is not asked to provide any voltage gain; rather, it functions as a buffer amplifier, providing a relatively large input resistance and a low output resistance, much lower than R_L. It is this stage that enables connecting the amplifier to the 10-Ω load. These points can be made more concrete by solving the following exercises. In so doing, observe that in finding the gain of an amplifier stage in a cascade amplifier, the loading effect of the succeeding amplifier stage must be taken into account as we have done in the above example.

EXERCISES

1.15 What would the overall voltage gain of the cascade amplifier in Example 1.3 be without stage 3?
Ans. 81.8 V/V

1.16 For the cascade amplifier of Example 1.3, let v_s be 1 mV. Find v_{i1}, v_{i2}, v_{i3}, and v_L.
Ans. 0.91 mV; 9 mV; 818 mV; 744 mV

1.17 (a) Model the three-stage amplifier of Example 1.3 (without the source and load), using the voltage amplifier model. What are the values of R_i, A_{vo}, and R_o?
(b) If R_L varies in the range 10 Ω to 1000 Ω, find the corresponding range of the overall voltage gain, v_o/v_s.
Ans. 1 MΩ, 900 V/V, 10 Ω; 409 V/V to 810 V/V

1.5.3 Other Amplifier Types

In the design of an electronic system, the signal of interest—whether at the system input, at an intermediate stage, or at the output—can be either a voltage or a current. For instance, some transducers have very high output resistances and can be more appropriately modeled as current sources. Similarly, there are applications in which the output current rather than the voltage is of

Table 1.1 The Four Amplifier Types

Type	Circuit Model	Gain Parameter	Ideal Characteristics
Voltage Amplifier	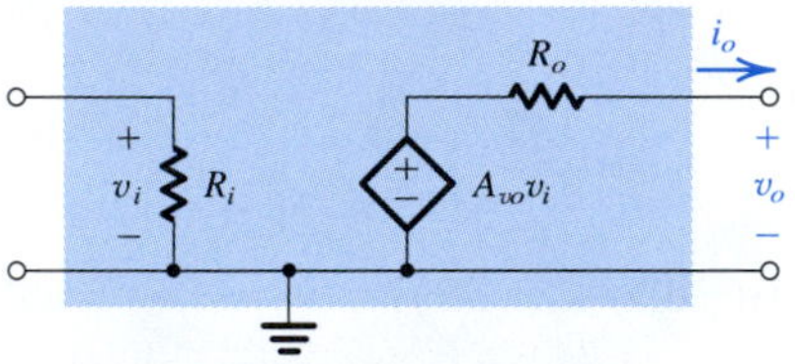	Open-Circuit Voltage Gain $A_{vo} \equiv \left.\frac{v_o}{v_i}\right\|_{i_o=0}$ (V/V)	$R_i = \infty$ $R_o = 0$
Current Amplifier	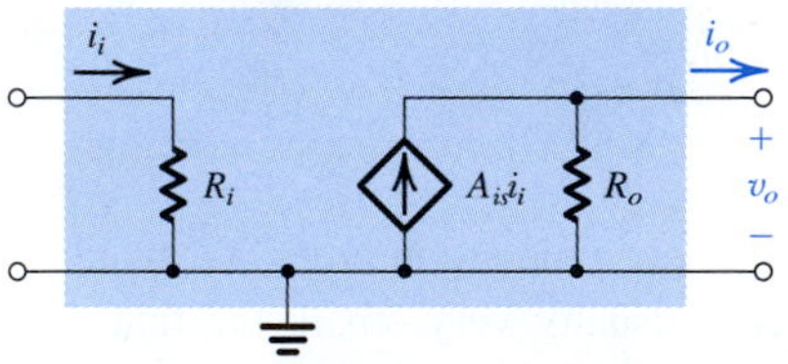	Short-Circuit Current Gain $A_{is} \equiv \left.\frac{i_o}{i_i}\right\|_{v_o=0}$ (A/A)	$R_i = 0$ $R_o = \infty$
Transconductance Amplifier	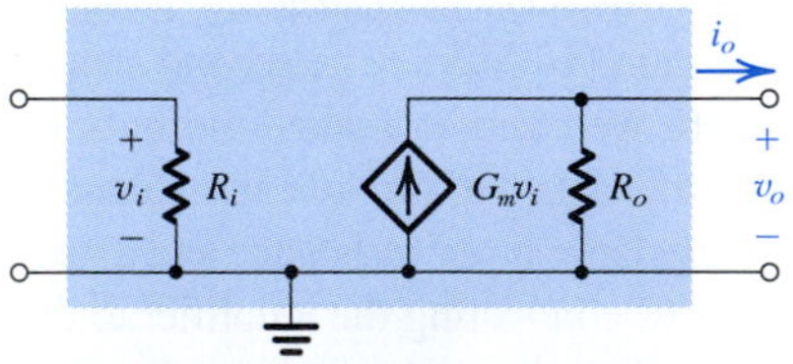	Short-Circuit Transconductance $G_m \equiv \left.\frac{i_o}{v_i}\right\|_{v_o=0}$ (A/V)	$R_i = \infty$ $R_o = \infty$
Transresistance Amplifier	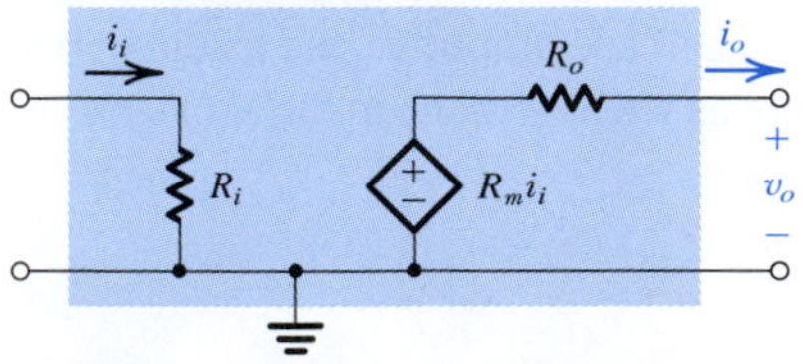	Open-Circuit Transresistance $R_m \equiv \left.\frac{v_o}{i_i}\right\|_{i_o=0}$ (V/A)	$R_i = 0$ $R_o = 0$

interest. Thus, although it is the most popular, the voltage amplifier considered above is just one of four possible amplifier types. The other three are the current amplifier, the transconductance amplifier, and the transresistance amplifier. Table 1.1 shows the four amplifier types, their circuit models, the definition of their gain parameters, and the ideal values of their input and output resistances.

1.5.4 Relationships between the Four Amplifier Models

Although for a given amplifier a particular one of the four models in Table 1.1 is most preferable, *any of the four can be used to model any amplifier.* In fact, simple relationships can be derived to relate the parameters of the various models. For instance, the open-circuit voltage gain A_{vo} can be related to the short-circuit current gain A_{is} as follows: The open-circuit output voltage given by the voltage amplifier model of Table 1.1 is $A_{vo}v_i$. The current amplifier model in the same table gives an open-circuit output voltage of $A_{is}i_iR_o$. Equating these two values and noting that $i_i = v_i/R_i$ gives

$$A_{vo} = A_{is}\left(\frac{R_o}{R_i}\right) \tag{1.14}$$

Similarly, we can show that

$$A_{vo} = G_mR_o \tag{1.15}$$

and

$$A_{vo} = \frac{R_m}{R_i} \tag{1.16}$$

The expressions in Eqs. (1.14) to (1.16) can be used to relate any two of the gain parameters A_{vo}, A_{is}, G_m, and R_m.

1.5.5 Determining R_i and R_o

From the amplifier circuit models given in Table 1.1, we observe that the input resistance R_i of the amplifier can be determined by applying an input voltage v_i and measuring (or calculating) the input current i_i; that is, $R_i = v_i/i_i$. The output resistance is found as the ratio of the open-circuit output voltage to the short-circuit output current. Alternatively, the output resistance can be found by eliminating the input signal source (then i_i and v_i will both be zero) and applying a voltage signal v_x to the output of the amplifier, as shown in Fig. 1.18. If we denote the current drawn from v_x *into* the output terminals as i_x (note that i_x is opposite in direction to i_o), then $R_o = v_x/i_x$. Although these techniques are conceptually correct, in actual practice more refined methods are employed in measuring R_i and R_o.

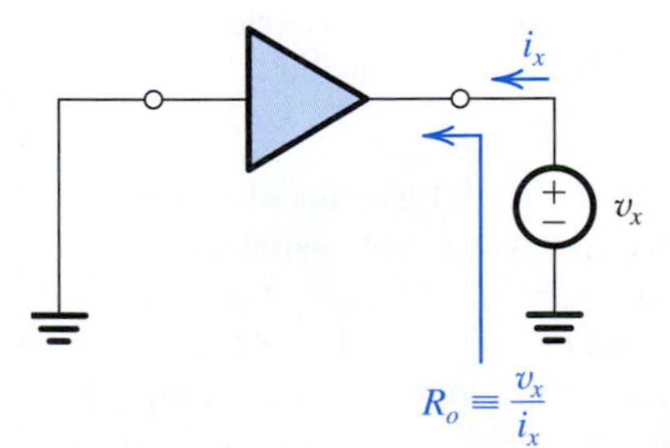

Figure 1.18 Determining the output resistance.

1.5.6 Unilateral Models

The amplifier models considered above are **unilateral**; that is, signal flow is unidirectional, from input to output. Most real amplifiers show some reverse transmission, which is usually undesirable but must nonetheless be modeled. We shall not pursue this point further at this time except to mention that more complete models for linear two-port networks are given in Appendix C. Also, in later chapters, we will find it necessary in certain cases to augment the models of Table 1.1 to take into account the nonunilateral nature of some transistor amplifiers.

Example 1.4

The **bipolar junction transistor (BJT)**, which will be studied in Chapter 6, is a three-terminal device that when powered-up by a dc source (battery) and operated with small signals can be modeled by the linear circuit shown in Fig. 1.19(a). The three terminals are the **base (B)**, the **emitter (E)**, and the **collector (C)**. The heart of the model is a transconductance amplifier represented by an input resistance between B and E (denoted r_π), a short-circuit transconductance g_m, and an output resistance r_o.

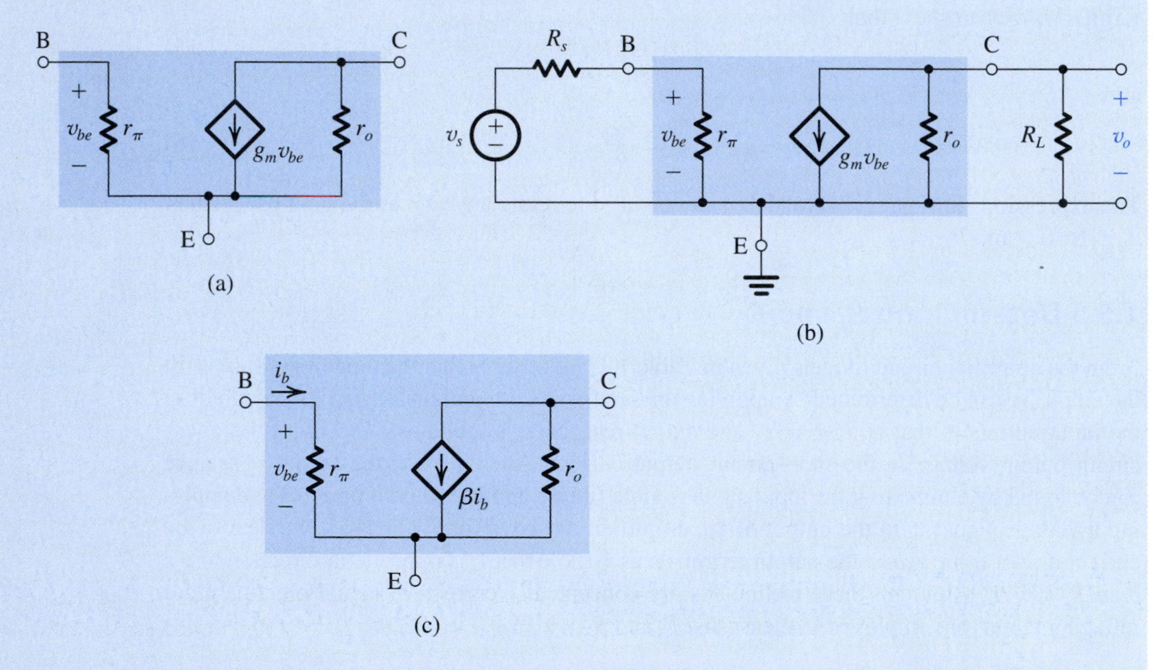

Figure 1.19 **(a)** Small-signal circuit model for a bipolar junction transistor (BJT). **(b)** The BJT connected as an amplifier with the emitter as a common terminal between input and output (called a common-emitter amplifier). **(c)** An alternative small-signal circuit model for the BJT.

(a) With the emitter used as a common terminal between input and output, Fig. 1.19(b) shows a transistor amplifier known as a **common-emitter** or **grounded-emitter** circuit. Derive an expression for the voltage gain v_o/v_s, and evaluate its magnitude for the case $R_s = 5\ \text{k}\Omega$, $r_\pi = 2.5\ \text{k}\Omega$, $g_m = 40$ mA/V, $r_o = 100\ \text{k}\Omega$, and $R_L = 5\ \text{k}\Omega$. What would the gain value be if the effect of r_o were neglected?

(b) An alternative model for the transistor in which a current amplifier rather than a transconductance amplifier is utilized is shown in Fig. 1.19(c). What must the short-circuit current gain β be? Give both an expression and a value.

Solution

(a) Refer to Fig. 1.19(b). We use the voltage-divider rule to determine the fraction of input signal that appears at the amplifier input as

$$v_{be} = v_s \frac{r_\pi}{r_\pi + R_s} \tag{1.17}$$

Next we determine the output voltage v_o by multiplying the current $(g_m v_{be})$ by the resistance $(R_L \parallel r_o)$,

$$v_o = -g_m v_{be}(R_L \parallel r_o) \tag{1.18}$$

Substituting for v_{be} from Eq. (1.17) yields the voltage-gain expression

$$\frac{v_o}{v_s} = -\frac{r_\pi}{r_\pi + R_s} g_m (R_L \parallel r_o) \tag{1.19}$$

Observe that the gain is negative, indicating that this amplifier is inverting. For the given component values,

$$\frac{v_o}{v_s} = -\frac{2.5}{2.5+5} \times 40 \times (5 \parallel 100)$$

$$= -63.5 \text{ V/V}$$

Neglecting the effect of r_o, we obtain

$$\frac{v_o}{v_s} \simeq -\frac{2.5}{2.5+5} \times 40 \times 5$$

$$= -66.7 \text{ V/V}$$

which is quite close to the value obtained including r_o. This is not surprising, since $r_o \gg R_L$.

(b) For the model in Fig. 1.19(c) to be equivalent to that in Fig. 1.19(a),

$$\beta i_b = g_m v_{be}$$

But $i_b = v_{be}/r_\pi$; thus,

$$\beta = g_m r_\pi$$

For the values given,

$$\beta = 40 \text{ mA/V} \times 2.5 \text{ k}\Omega$$

$$= 100 \text{ A/A}$$

EXERCISES

1.18 Consider a current amplifier having the model shown in the second row of Table 1.1. Let the amplifier be fed with a signal current-source i_s having a resistance R_s, and let the output be connected to a load resistance R_L. Show that the overall current gain is given by

$$\frac{i_o}{i_s} = A_{is} \frac{R_s}{R_s + R_i} \frac{R_o}{R_o + R_L}$$

1.19 Consider the transconductance amplifier whose model is shown in the third row of Table 1.1. Let a voltage signal source v_s with a source resistance R_s be connected to the input and a load resistance R_L be connected to the output. Show that the overall voltage-gain is given by

$$\frac{v_o}{v_s} = G_m \frac{R_i}{R_i + R_s}(R_o \parallel R_L)$$

1.20 Consider a transresistance amplifier having the model shown in the fourth row of Table 1.1. Let the amplifier be fed with a signal current-source i_s having a resistance R_s, and let the output be connected to a load resistance R_L. Show that the overall gain is given by

$$\frac{v_o}{i_s} = R_m \frac{R_s}{R_s + R_i} \frac{R_L}{R_L + R_o}$$

1.21 Find the input resistance between terminals B and G in the circuit shown in Fig. E1.21. The voltage v_x is a test voltage with the input resistance R_{in} defined as $R_{in} \equiv v_x/i_x$.

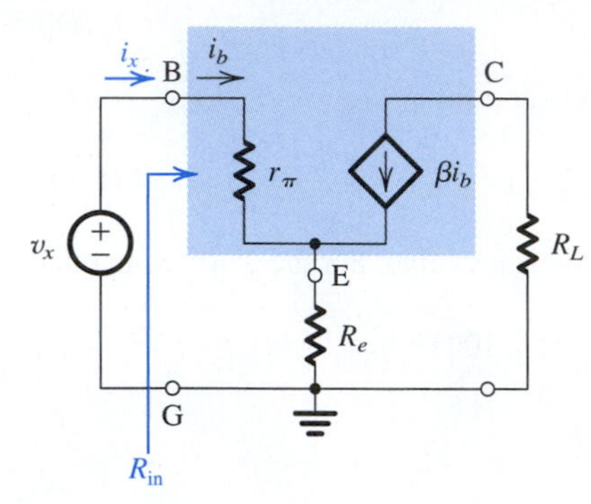

Figure E1.21

Ans. $R_{in} = r_\pi + (\beta + 1)R_e$

1.6 Frequency Response of Amplifiers[2]

From Section 1.2 we know that the input signal to an amplifier can always be expressed as the sum of sinusoidal signals. It follows that an important characterization of an amplifier is in terms of its response to input sinusoids of different frequencies. Such a characterization of amplifier performance is known as the amplifier frequency response.

1.6.1 Measuring the Amplifier Frequency Response

We shall introduce the subject of amplifier frequency response by showing how it can be measured. Figure 1.20 depicts a linear voltage amplifier fed at its input with a sine-wave signal of amplitude V_i and frequency ω. As the figure indicates, the signal measured at the amplifier output also is sinusoidal with exactly the same frequency ω. This is an important point to note: *Whenever a sine-wave signal is applied to a linear circuit, the resulting output is sinusoidal with the same frequency as the input.* In fact, the sine wave is the only signal that does not change shape as it passes through a linear circuit. Observe, however, that the output sinusoid will in general have a different amplitude and will be shifted in phase relative to the input. The ratio of the amplitude of the output sinusoid (V_o) to the amplitude of the input sinusoid (V_i) is the magnitude of the amplifier gain (or transmission) at the test frequency ω. Also, the angle ϕ is the phase of the amplifier transmission at the test frequency ω. If we denote the **amplifier transmission**, or **transfer function** as it is more commonly

[2]Except for its use in the study of the frequency response of op-amp circuits in Sections 2.5 and 2.7, the material in this section will not be needed in a substantial manner until Chapter 9.

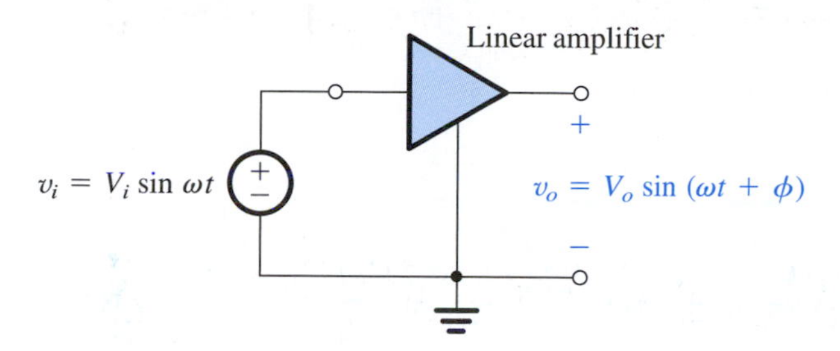

Figure 1.20 Measuring the frequency response of a linear amplifier: At the test frequency ω, the amplifier gain is characterized by its magnitude (V_o/V_i) and phase ϕ.

known, by $T(\omega)$, then

$$|T(\omega)| = \frac{V_o}{V_i}$$

$$\angle T(\omega) = \phi$$

The response of the amplifier to a sinusoid of frequency ω is completely described by $|T(\omega)|$ and $\angle T(\omega)$. Now, to obtain the complete frequency response of the amplifier we simply change the frequency of the input sinusoid and measure the new value for $|T|$ and $\angle T$. The end result will be a table and/or graph of gain magnitude [$|T(\omega)|$] versus frequency and a table and/or graph of phase angle [$\angle T(\omega)$] versus frequency. These two plots together constitute the frequency response of the amplifier; the first is known as the **magnitude** or **amplitude response**, and the second is the **phase response**. Finally, we should mention that it is a common practice to express the magnitude of transmission in decibels and thus plot $20 \log |T(\omega)|$ versus frequency.

1.6.2 Amplifier Bandwidth

Figure 1.21 shows the magnitude response of an amplifier. It indicates that the gain is almost constant over a wide frequency range, roughly between ω_1 and ω_2. Signals whose frequencies are below ω_1 or above ω_2 will experience lower gain, with the gain decreasing as we move farther away from ω_1 and ω_2. The band of frequencies over which the gain of the amplifier is almost constant, to within a certain number of decibels (usually 3 dB), is called the **amplifier bandwidth**. Normally the amplifier is designed so that its bandwidth coincides with the spectrum of the signals it is required to amplify. If this were not the case, the amplifier would *distort* the frequency spectrum of the input signal, with different components of the input signal being amplified by different amounts.

1.6.3 Evaluating the Frequency Response of Amplifiers

Above, we described the method used to measure the frequency response of an amplifier. We now briefly discuss the method for analytically obtaining an expression for the frequency response. What we are about to say is just a preview of this important subject, whose detailed study is in Chapter 9.

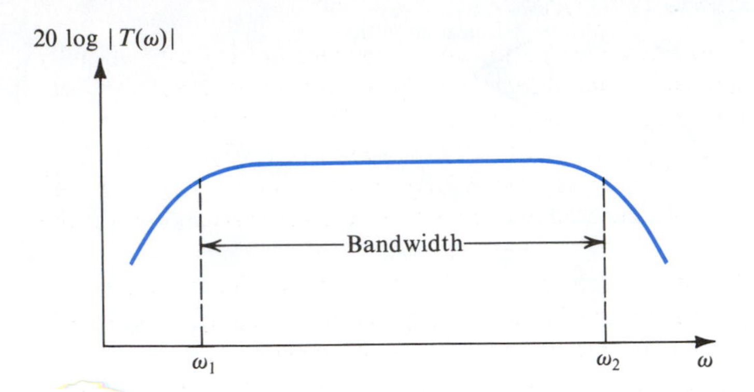

Figure 1.21 Typical magnitude response of an amplifier: $|T(\omega)|$ is the magnitude of the amplifier transfer function—that is, the ratio of the output $V_o(\omega)$ to the input $V_i(\omega)$.

To evaluate the frequency response of an amplifier, one has to analyze the amplifier equivalent circuit model, taking into account all reactive components.[3] Circuit analysis proceeds in the usual fashion but with inductances and capacitances represented by their reactances. An inductance L has a reactance or impedance $j\omega L$, and a capacitance C has a reactance or impedance $1/j\omega C$ or, equivalently, a susceptance or admittance $j\omega C$. Thus in a *frequency-domain* analysis we deal with impedances and/or admittances. The result of the analysis is the amplifier transfer function $T(\omega)$

$$T(\omega) = \frac{V_o(\omega)}{V_i(\omega)}$$

where $V_i(\omega)$ and $V_o(\omega)$ denote the input and output signals, respectively. $T(\omega)$ is generally a complex function whose magnitude $|T(\omega)|$ gives the magnitude of transmission or the magnitude response of the amplifier. The phase of $T(\omega)$ gives the phase response of the amplifier.

In the analysis of a circuit to determine its frequency response, the algebraic manipulations can be considerably simplified by using the **complex frequency variable** s. In terms of s, the impedance of an inductance L is sL and that of a capacitance C is $1/sC$. Replacing the reactive elements with their impedances and performing standard circuit analysis, we obtain the transfer function $T(s)$ as

$$T(s) \equiv \frac{V_o(s)}{V_i(s)}$$

Subsequently, we replace s by $j\omega$ to determine the transfer function for **physical frequencies**, $T(j\omega)$. Note that $T(j\omega)$ is the same function we called $T(\omega)$ above[4]; the additional j is included in order to emphasize that $T(j\omega)$ is obtained from $T(s)$ by replacing s with $j\omega$.

[3]Note that in the models considered in previous sections no reactive components were included. These were simplified models and cannot be used alone to predict the amplifier frequency response.
[4]At this stage, we are using s simply as a shorthand for $j\omega$. We shall not require detailed knowledge of s-plane concepts until Chapter 9. A brief review of s-plane analysis is presented in Appendix F.

1.6.4 Single-Time-Constant Networks

In analyzing amplifier circuits to determine their frequency response, one is greatly aided by knowledge of the frequency-response characteristics of single-time-constant (STC) networks. An STC network is one that is composed of, or can be reduced to, one reactive component (inductance or capacitance) and one resistance. Examples are shown in Fig. 1.22. An STC network formed of an inductance L and a resistance R has a time constant $\tau = L/R$. The time constant τ of an STC network composed of a capacitance C and a resistance R is given by $\tau = CR$.

Appendix E presents a study of STC networks and their responses to sinusoidal, step, and pulse inputs. Knowledge of this material will be needed at various points throughout this book, and the reader will be encouraged to refer to the appendix. At this point we need in particular the frequency response results; we will, in fact, briefly discuss this important topic, now.

Most STC networks can be classified into two categories,[5] **low pass (LP)** and **high pass (HP)**, with each of the two categories displaying distinctly different signal responses. As an example, the STC network shown in Fig. 1.22(a) is of the *low-pass* type and that in Fig. 1.22(b) is of the *high-pass* type. To see the reasoning behind this classification, observe that the transfer function of each of these two circuits can be expressed as a voltage-divider ratio, with the divider composed of a resistor and a capacitor. Now, recalling how the impedance of a capacitor varies with frequency ($Z = 1/j\omega C$), it is easy to see that the transmission of the circuit in Fig. 1.22(a) will decrease with frequency and approach zero as ω approaches ∞. Thus the circuit of Fig. 1.22(a) acts as a **low-pass filter**[6]; it passes low-frequency, sine-wave inputs with little or no attenuation (at $\omega = 0$, the transmission is unity) and attenuates high-frequency input sinusoids. The circuit of Fig. 1.22(b) does the opposite; its transmission is unity at $\omega = \infty$ and decreases as ω is reduced, reaching 0 for $\omega = 0$. The latter circuit, therefore, performs as a **high-pass filter**.

Table 1.2 provides a summary of the frequency-response results for STC networks of both types.[7] Also, sketches of the magnitude and phase responses are given in Figs. 1.23 and 1.24. These frequency-response diagrams are known as **Bode plots** and the **3-dB frequency**

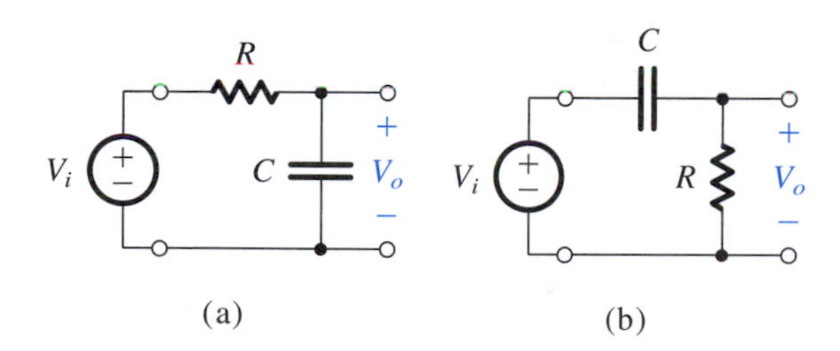

Figure 1.22 Two examples of STC networks: **(a)** a low-pass network and **(b)** a high-pass network.

[5]An important exception is the **all-pass** STC network studied in Chapter 16.

[6]A filter is a circuit that passes signals in a specified frequency band (the filter passband) and stops or severely attenuates (filters out) signals in another frequency band (the filter stopband). Filters will be studied in Chapter 16.

[7]The transfer functions in Table 1.2 are given in general form. For the circuits of Fig. 1.22, $K = 1$ and $\omega_0 = 1/CR$.

Table 1.2 Frequency Response of STC Networks

	Low-Pass (LP)	High-Pass (HP)
Transfer Function $T(s)$	$\frac{K}{1+(s/\omega_0)}$	$\frac{Ks}{s+\omega_0}$
Transfer Function (for physical frequencies) $T(j\omega)$	$\frac{K}{1+j(\omega/\omega_0)}$	$\frac{K}{1-j(\omega_0/\omega)}$
Magnitude Response $\lvert T(j\omega)\rvert$	$\frac{\lvert K\rvert}{\sqrt{1+(\omega/\omega_0)^2}}$	$\frac{\lvert K\rvert}{\sqrt{1+(\omega_0/\omega)^2}}$
Phase Response $\angle T(j\omega)$	$-\tan^{-1}(\omega/\omega_0)$	$\tan^{-1}(\omega_0/\omega)$
Transmission at $\omega = 0$ (dc)	K	0
Transmission at $\omega = \infty$	0	K
3-dB Frequency	$\omega_0 = 1/\tau$; $\tau \equiv$ time constant $\tau = CR$ or L/R	
Bode Plots	in Fig. 1.23	in Fig. 1.24

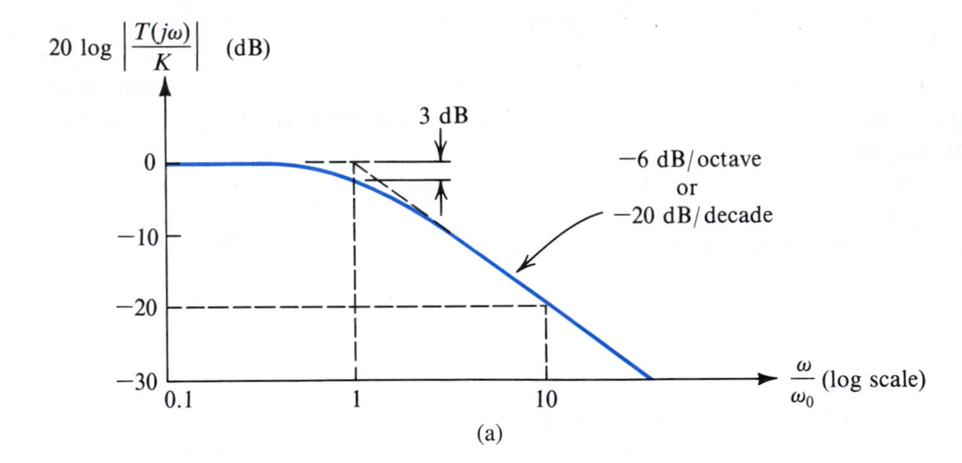

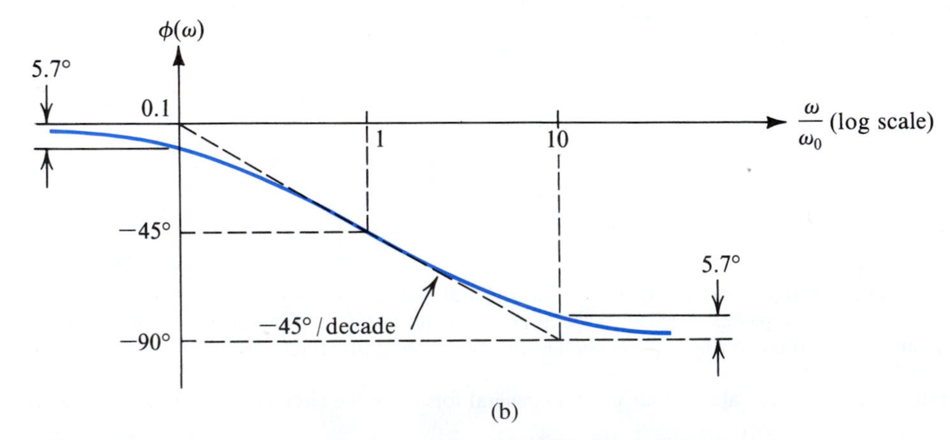

Figure 1.23 **(a)** Magnitude and **(b)** phase response of STC networks of the low-pass type.

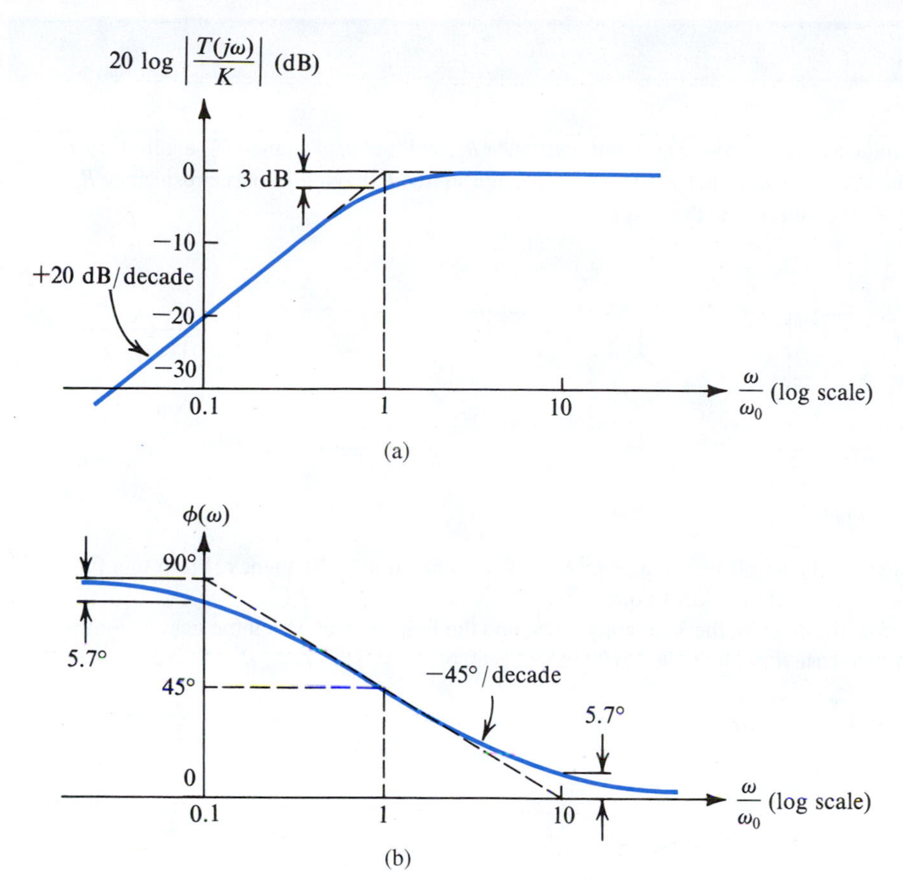

Figure 1.24 **(a)** Magnitude and **(b)** phase response of STC networks of the high-pass type.

(ω_0) is also known as the **corner frequency**, **break frequency**, or **pole frequency**. The reader is urged to become familiar with this information and to consult Appendix E if further clarifications are needed. In particular, it is important to develop a facility for the rapid determination of the time constant τ of an STC circuit. The process is very simple: Set the independent voltge or current source to zero; "grab hold" of the two terminals of the reactive element (capacitor C or inductor L); and determine the equivalent resistance R that appears between these two terminals. The time-constant is then CR or L/R.

Example 1.5

Figure 1.25 shows a voltage amplifier having an input resistance R_i, an input capacitance C_i, a gain factor μ, and an output resistance R_o. The amplifier is fed with a voltage source V_s having a source resistance R_s, and a load of resistance R_L is connected to the output.

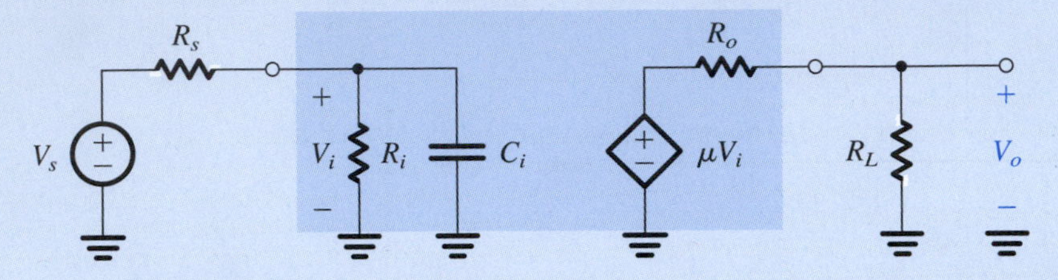

Figure 1.25 Circuit for Example 1.5.

(a) Derive an expression for the amplifier voltage gain V_o/V_s as a function of frequency. From this find expressions for the dc gain and the 3-dB frequency.

(b) Calculate the values of the dc gain, the 3-dB frequency, and the frequency at which the gain becomes 0 dB (i.e., unity) for the case $R_s = 20$ kΩ, $R_i = 100$ kΩ, $C_i = 60$ pF, $\mu = 144$ V/V, $R_o = 200$ Ω, and $R_L = 1$ kΩ.

(c) Find $v_o(t)$ for each of the following inputs:
(i) $v_i = 0.1 \sin 10^2 t$, V
(ii) $v_i = 0.1 \sin 10^5 t$, V
(iii) $v_i = 0.1 \sin 10^6 t$, V
(iv) $v_i = 0.1 \sin 10^8 t$, V

Solution

(a) Utilizing the voltage-divider rule, we can express V_i in terms of V_s as follows

$$V_i = V_s \frac{Z_i}{Z_i + R_s}$$

where Z_i is the amplifier input impedance. Since Z_i is composed of two parallel elements, it is obviously easier to work in terms of $Y_i = 1/Z_i$. Toward that end we divide the numerator and denominator by Z_i, thus obtaining

$$V_i = V_s \frac{1}{1 + R_s Y_i}$$

$$= V_s \frac{1}{1 + R_s[(1/R_i) + sC_i]}$$

Thus,

$$\frac{V_i}{V_s} = \frac{1}{1 + (R_s/R_i) + sC_iR_s}$$

This expression can be put in the standard form for a low-pass STC network (see the top line of Table 1.2) by extracting $[1 + (R_s/R_i)]$ from the denominator; thus we have

$$\frac{V_i}{V_s} = \frac{1}{1 + (R_s/R_i)} \frac{1}{1 + sC_i[(R_sR_i)/(R_s + R_i)]} \tag{1.20}$$

At the output side of the amplifier we can use the voltage-divider rule to write

$$V_o = \mu V_i \frac{R_L}{R_L + R_o}$$

This equation can be combined with Eq. (1.20) to obtain the amplifier transfer function as

$$\frac{V_o}{V_s} = \mu \frac{1}{1 + (R_s/R_i)} \frac{1}{1 + (R_o/R_L)} \frac{1}{1 + sC_i[(R_sR_i)/(R_s + R_i)]} \tag{1.21}$$

We note that only the last factor in this expression is new (compared with the expression derived in the last section). This factor is a result of the input capacitance C_i, with the time constant being

$$\begin{aligned} \tau &= C_i \frac{R_sR_i}{R_s + R_i} \\ &= C_i(R_s \| R_i) \end{aligned} \tag{1.22}$$

We could have obtained this result by inspection: From Fig. 1.25 we see that the input circuit is an STC network and that its time constant can be found by reducing V_s to zero, with the result that the resistance seen by C_i is R_i in parallel with R_s. The transfer function in Eq. (1.21) is of the form $K/(1 + (s/\omega_0))$, which corresponds to a low-pass STC network. The dc gain is found as

$$K \equiv \frac{V_o}{V_s}(s = 0) = \mu \frac{1}{1 + (R_s/R_i)} \frac{1}{1 + (R_o/R_L)} \tag{1.23}$$

The 3-dB frequency ω_0 can be found from

$$\omega_0 = \frac{1}{\tau} = \frac{1}{C_i(R_s \| R_i)} \tag{1.24}$$

Since the frequency response of this amplifier is of the low-pass STC type, the Bode plots for the gain magnitude and phase will take the form shown in Fig. 1.23, where K is given by Eq. (1.23) and ω_0 is given by Eq. (1.24).

(b) Substituting the numerical values given into Eq. (1.23) results in

$$K = 144 \frac{1}{1 + (20/100)} \frac{1}{1 + (200/1000)} = 100 \text{ V/V}$$

Thus the amplifier has a dc gain of 40 dB. Substituting the numerical values into Eq. (1.24) gives the 3-dB frequency

$$\begin{aligned} \omega_0 &= \frac{1}{60 \text{ pF} \times (20 \text{ k}\Omega // 100 \text{ k}\Omega)} \\ &= \frac{1}{60 \times 10^{-12} \times (20 \times 100/(20 + 100)) \times 10^3} = 10^6 \text{ rad/s} \end{aligned}$$

Example 1.5 *continued*

Thus,

$$f_0 = \frac{10^6}{2\pi} = 159.2 \text{ kHz}$$

Since the gain falls off at the rate of –20 dB/decade, starting at ω_0 (see Fig. 1.23a) the gain will reach 0 dB in two decades (a factor of 100); thus we have

$$\text{Unity-gain frequency} = 100 \times \omega_0 = 10^8 \text{ rad/s or } 15.92 \text{ MHz}$$

(c) To find $v_o(t)$ we need to determine the gain magnitude and phase at 10^2, 10^5, 10^6, and 10^8 rad/s. This can be done either approximately utilizing the Bode plots of Fig. 1.23 or exactly utilizing the expression for the amplifier transfer function,

$$T(j\omega) \equiv \frac{V_o}{V_s}(j\omega) = \frac{100}{1 + j(\omega/10^6)}$$

We shall do both:

(i) For $\omega = 10^2$ rad/s, which is $(\omega_0/10^4)$, the Bode plots of Fig. 1.23 suggest that $|T| = K = 100$ and $\phi = 0°$. The transfer function expression gives $|T| \simeq 100$ and $\phi = -\tan^{-1} 10^{-4} \simeq 0°$. Thus,

$$v_o(t) = 10 \sin 10^2 t, \text{ V}$$

(ii) For $\omega = 10^5$ rad/s, which is $(\omega_0/10)$, the Bode plots of Fig. 1.23 suggest that $|T| \simeq K = 100$ and $\phi = -5.7°$. The transfer function expression gives $|T| = 99.5$ and $\phi = -\tan^{-1} 0.1 = -5.7°$. Thus,

$$v_o(t) = 9.95 \sin(10^5 t - 5.7°), \text{ V}$$

(iii) For $\omega = 10^6$ rad/s $= \omega_0$, $|T| = 100/\sqrt{2} = 70.7$ V/V or 37 dB and $\phi = -45°$. Thus,

$$v_o(t) = 7.07 \sin(10^6 t - 45°), \text{ V}$$

(iv) For $\omega = 10^8$ rad/s, which is $(100\omega_0)$, the Bode plots suggest that $|T| = 1$ and $\phi = -90°$. The transfer function expression gives

$$|T| \simeq 1 \quad \text{and} \quad \phi = -\tan^{-1} 100 = -89.4°$$

Thus,

$$v_o(t) = 0.1 \sin(10^8 t - 89.4°), \text{ V}$$

1.6.5 Classification of Amplifiers Based on Frequency Response

Amplifiers can be classified based on the shape of their magnitude-response curve. Figure 1.26 shows typical frequency-response curves for various amplifier types. In Fig. 1.26(a) the gain remains constant over a wide frequency range, but falls off at low and high frequencies. This type of frequency response is common in audio amplifiers.

As will be shown in later chapters, **internal capacitances** in the device (a transistor) cause the falloff of gain at high frequencies, just as C_i did in the circuit of Example 1.5. On the other hand, the falloff of gain at low frequencies is usually caused by **coupling capacitors** used to connect one amplifier stage to another, as indicated in Fig. 1.27. This practice is usually adopted to simplify the design process of the different stages. The coupling capacitors

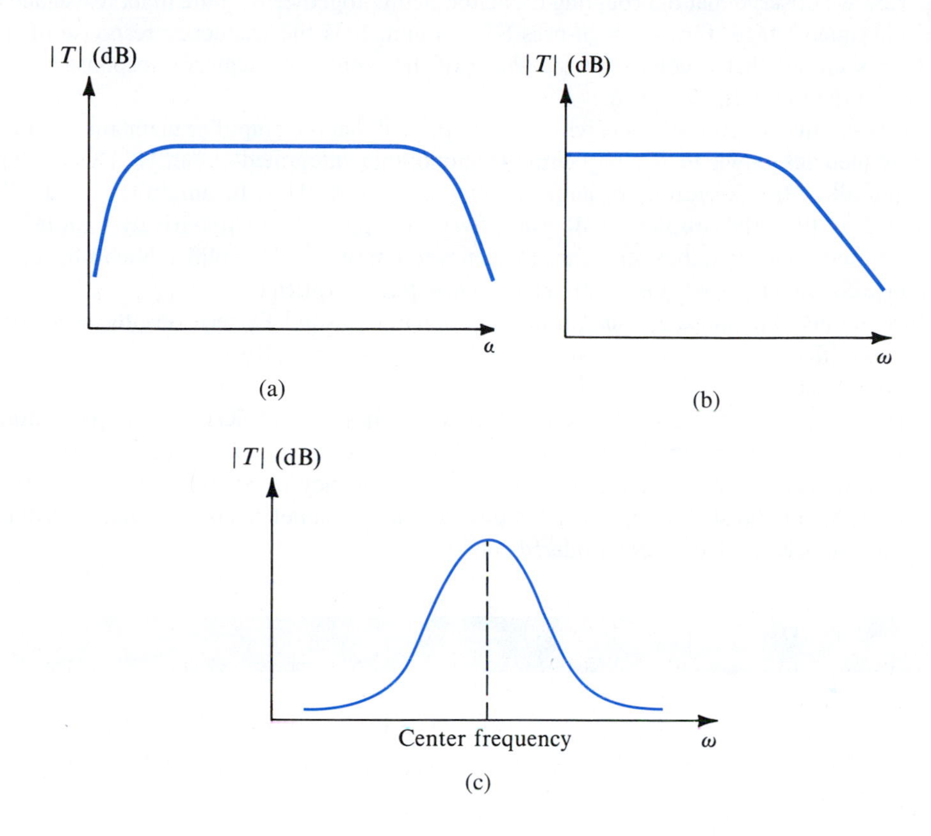

Figure 1.26 Frequency response for **(a)** a capacitively coupled amplifier, **(b)** a direct-coupled amplifier, and **(c)** a tuned or bandpass amplifier.

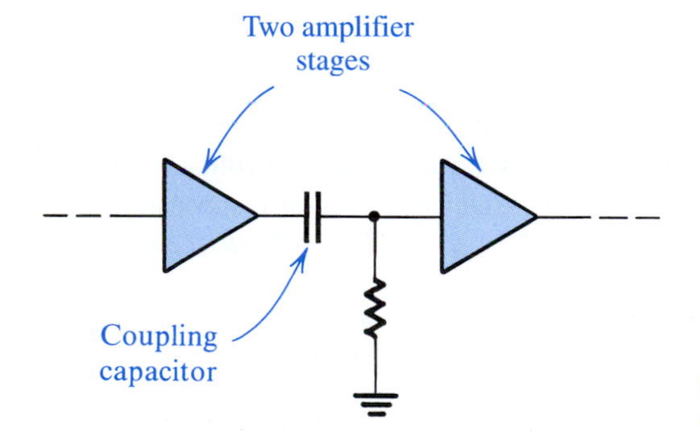

Figure 1.27 Use of a capacitor to couple amplifier stages.

are usually chosen quite large (a fraction of a microfarad to a few tens of microfarads) so that their reactance (impedance) is small at the frequencies of interest. Nevertheless, at sufficiently low frequencies the reactance of a coupling capacitor will become large enough to cause part of the signal being coupled to appear as a voltage drop across the coupling capacitor, thus not reaching the subsequent stage. Coupling capacitors will thus cause loss of gain at low frequencies and cause the gain to be zero at dc. This is not at all surprising, since from

Fig. 1.27 we observe that the coupling capacitor, acting together with the input resistance of the subsequent stage, forms a high-pass STC circuit. It is the frequency response of this high-pass circuit that accounts for the shape of the amplifier frequency response in Fig. 1.26(a) at the low-frequency end.

There are many applications in which it is important that the amplifier maintain its gain at low frequencies down to dc. Furthermore, monolithic integrated-circuit (IC) technology does not allow the fabrication of large coupling capacitors. Thus IC amplifiers are usually designed as **directly coupled** or **dc amplifiers** (as opposed to **capacitively coupled**, or **ac amplifiers**). Figure 1.26(b) shows the frequency response of a dc amplifier. Such a frequency response characterizes what is referred to as a **low-pass amplifier**.

In a number of applications, such as in the design of radio and TV receivers, the need arises for an amplifier whose frequency response peaks around a certain frequency (called the **center frequency**) and falls off on both sides of this frequency, as shown in Fig. 1.26(c). Amplifiers with such a response are called **tuned amplifiers**, **bandpass amplifiers**, or **bandpass filters**. A tuned amplifier forms the heart of the front-end or tuner of a communication receiver; by adjusting its center frequency to coincide with the frequency of a desired communications channel (e.g., a radio station), the signal of this particular channel can be received while those of other channels are attenuated or filtered out.

EXERCISES

1.22 Consider a voltage amplifier having a frequency response of the low-pass STC type with a dc gain of 60 dB and a 3-dB frequency of 1000 Hz. Find the gain in dB at $f = 10$ Hz, 10 kHz, 100 kHz, and 1 MHz.
Ans. 60 dB; 40 dB; 20 dB; 0 dB

D1.23 Consider a transconductance amplifier having the model shown in Table 1.1 with $R_i = 5$ kΩ, $R_o = 50$ kΩ, and $G_m = 10$ mA/V. If the amplifier load consists of a resistance R_L in parallel with a capacitance C_L, convince yourself that the voltage transfer function realized, V_o/V_i, is of the low-pass STC type. What is the lowest value that R_L can have while a dc gain of at least 40 dB is obtained? With this value of R_L connected, find the highest value that C_L can have while a 3-dB bandwidth of at least 100 kHz is obtained.
Ans. 12.5 kΩ; 159.2 pF

D1.24 Consider the situation illustrated in Fig. 1.27. Let the output resistance of the first voltage amplifier be 1 kΩ and the input resistance of the second voltage amplifier (including the resistor shown) be 9 kΩ. The resulting equivalent circuit is shown in Fig. E1.24 where V_s and R_s are the output voltage and output resistance of the first amplifier, C is a coupling capacitor, and R_i is the input resistance of the second amplifier. Convince yourself that V_2/V_s is a high-pass STC function. What is the smallest value for C that will ensure that the 3-dB frequency is not higher than 100 Hz?
Ans. 0.16 μF

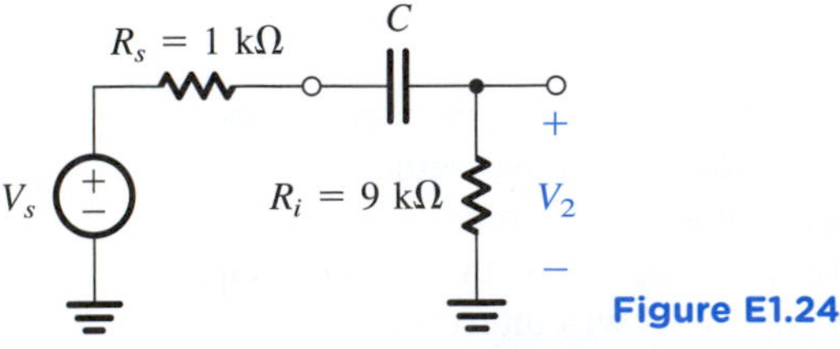

Figure E1.24

Summary

- An electrical signal source can be represented in either the Thévenin form (a voltage source v_s in series with a source resistance R_s) or the Norton form (a current source i_s in parallel with a source resistance R_s). The Thévenin voltage v_s is the open-circuit voltage between the source terminals; the Norton current i_s is equal to the short-circuit current between the source terminals. For the two representations to be equivalent, v_s and $R_s i_s$ must be equal.
- A signal can be represented either by its waveform versus time or as the sum of sinusoids. The latter representation is known as the frequency spectrum of the signal.
- The sine-wave signal is completely characterized by its peak value (or rms value which is the peak $/\sqrt{2}$), its frequency (ω in rad/s or f in Hz; $\omega = 2\pi f$ and $f = 1/T$, where T is the period in seconds), and its phase with respect to an arbitrary reference time.
- Analog signals have magnitudes that can assume any value. Electronic circuits that process analog signals are called analog circuits. Sampling the magnitude of an analog signal at discrete instants of time and representing each signal sample by a number results in a digital signal. Digital signals are processed by digital circuits.
- The simplest digital signals are obtained when the binary system is used. An individual digital signal then assumes one of only two possible values: low and high (say, 0 V and +5 V), corresponding to logic 0 and logic 1, respectively.
- An analog-to-digital converter (ADC) provides at its output the digits of the binary number representing the analog signal sample applied to its input. The output digital signal can then be processed using digital circuits. Refer to Fig. 1.10 and Eq. (1.3).
- The transfer characteristic, v_O versus v_I, of a linear amplifier is a straight line with a slope equal to the voltage gain. Refer to Fig. 1.12.
- Amplifiers increase the signal power and thus require dc power supplies for their operation.
- The amplifier voltage gain can be expressed as a ratio A_v in V/V or in decibels, $20 \log|A_v|$, dB. Similarly, for current gain: A_i A/A or $20 \log|A_i|$, dB. For power gain: A_p W/W or $10 \log A_p$, dB.
- Depending on the signal to be amplified (voltage or current) and on the desired form of output signal (voltage or current), there are four basic amplifier types: voltage, current, transconductance, and transresistance amplifiers. For the circuit models and ideal characteristics of these four amplifier types, refer to Table 1.1. A given amplifier can be modeled by any one of the four models, in which case their parameters are related by the formulas in Eqs. (1.14) to (1.16).
- A sinusoid is the only signal whose waveform is unchanged through a linear circuit. Sinusoidal signals are used to measure the frequency response of amplifiers.
- The transfer function $T(s) \equiv V_o(s)/V_i(s)$ of a voltage amplifier can be determined from circuit analysis. Substituting $s = j\omega$ gives $T(j\omega)$, whose magnitude $|T(j\omega)|$ is the magnitude response, and whose phase $\phi(\omega)$ is the phase response, of the amplifier.
- Amplifiers are classified according to the shape of their frequency response, $|T(j\omega)|$. Refer to Fig. 1.26.
- Single-time-constant (STC) networks are those networks that are composed of, or can be reduced to, one reactive component (L or C) and one resistance (R). The time constant τ is either L/R or CR.
- STC networks can be classified into two categories: low-pass (LP) and high-pass (HP). LP networks pass dc and low frequencies and attenuate high frequencies. The opposite is true for HP networks.
- The gain of an LP (HP) STC circuit drops by 3 dB below the zero-frequency (infinite-frequency) value at a frequency $\omega_0 = 1/\tau$. At high frequencies (low frequencies) the gain falls off at the rate of 6 dB/octave or 20 dB/decade. Refer to Table 1.2 on page 34 and Figs. 1.23 and 1.24. Further details are given in Appendix E.

Problems involving design are marked with D throughout the text. As well, problems are marked with asterisks to describe their degree of difficulty. Difficult problems are marked with an asterisk (*); more difficult problems with two asterisks (**); and very challenging and/or time-consuming problems with three asterisks (***).

Circuit Basics

As a review of the basics of circuit analysis and in order for the readers to gauge their preparedness for the study of electronic circuits, this section presents a number of relevant circuit analysis problems. For a summary of Thévenin's and Norton's theorems, refer to Appendix D. The problems are grouped in appropriate categories.

Resistors and Ohm's Law

1.1 Ohm's law relates *V*, *I*, and *R* for a resistor. For each of the situations following, find the missing item:

(a) $R = 1\ \text{k}\Omega$, $V = 10$ V
(b) $V = 10$ V, $I = 1$ mA
(c) $R = 10\ \text{k}\Omega$, $I = 10$ mA
(d) $R = 100\ \Omega$, $V = 10$ V

1.2 Measurements taken on various resistors are shown below. For each, calculate the power dissipated in the resistor and the power rating necessary for safe operation using standard components with power ratings of 1/8 W, 1/4 W, 1/2 W, 1 W, or 2 W:

(a) 1 kΩ conducting 30 mA
(b) 1 kΩ conducting 40 mA
(c) 10 kΩ conducting 3 mA
(d) 10 kΩ conducting 4 mA
(e) 1 kΩ dropping 20 V
(f) 1 kΩ dropping 11 V

1.3 Ohm's law and the power law for a resistor relate *V*, *I*, *R*, and *P*, making only two variables independent. For each pair identified below, find the other two:

(a) $R = 1\ \text{k}\Omega$, $I = 10$ mA
(b) $V = 10$ V, $I = 1$ mA
(c) $V = 10$ V, $P = 1$ W
(d) $I = 10$ mA, $P = 0.1$ W
(e) $R = 1\ \text{k}\Omega$, $P = 1$ W

Combining Resistors

1.4 You are given three resistors whose values are 10 kΩ, 20 kΩ, and 40 kΩ. How many different resistances can you create using series and parallel combinations of these three? List them in value order, lowest first. Be thorough and organized. (*Hint:* In your search, first consider all parallel combinations, then consider series combinations, and then consider series-parallel combinations, of which there are two kinds).

1.5 In the analysis and test of electronic circuits, it is often useful to connect one resistor in parallel with another to obtain a nonstandard value, one which is smaller than the smaller of the two resistors. Often, particularly during circuit testing, one resistor is already installed, in which case the second, when connected in parallel, is said to "shunt" the first. If the original resistor is 10 kΩ, what is the value of the shunting resistor needed to reduce the combined value by 1%, 5%, 10%, and 50%? What is the result of shunting a 10-kΩ resistor by 1 MΩ? By 100 kΩ? By 10 kΩ?

Voltage Dividers

1.6 Figure P1.6(a) shows a two-resistor voltage divider. Its function is to generate a voltage V_O (smaller than the power-supply voltage V_{DD}) at its output node X. The circuit looking back at node X is equivalent to that shown in Fig. P1.6(b). Observe that this is the Thévenin equivalent of the voltage divider circuit. Find expressions for V_O and R_O.

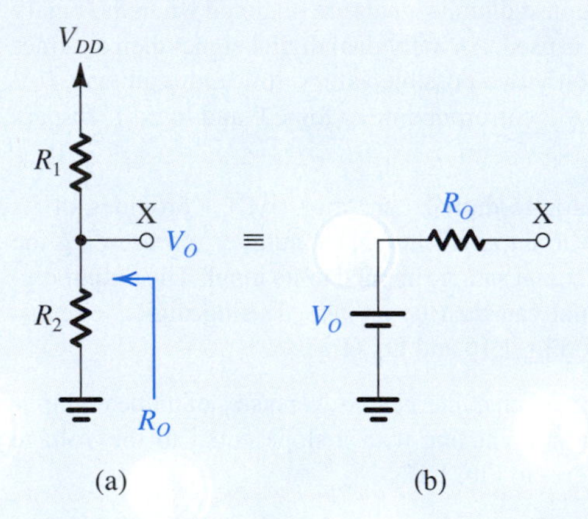

Figure P1.6

1.7 A two-resistor voltage divider employing a 3.3-kΩ and a 6.8-kΩ resistor is connected to a 9-V ground-referenced power supply to provide a relatively low voltage (close to 3V). Sketch the circuit. Assuming exact-valued resistors, what output voltage (measured to ground) and equivalent output resistance result? If the resistors used are not ideal but have a ±5% manufacturing tolerance, what are the extreme output voltages and resistances that can result?

1.8 You are given three resistors, each of 10 kΩ, and a 9-V battery whose negative terminal is connected to ground. With a voltage divider using some or all of your resistors, how many positive-voltage sources of magnitude less than 9 V can you design? List them in order, smallest first. What is the output resistance (i.e., the Thévenin resistance) of each?

D *1.9 Two resistors, with nominal values of 4.7 kΩ and 10 kΩ, are used in a voltage divider with a +15-V supply to create a nominal +10-V output. Assuming the resistor values to be exact, what is the actual output voltage produced? Which resistor must be shunted (paralleled) by what third resistor to create a voltage-divider output of 10.00 V? If an output resistance of exactly 3.33 kΩ is also required, what do you suggest? What should be done if the original 4.7-kΩ and 10-kΩ resistors are used but the requirement is 10.00 V and 3.00 kΩ?

Current Dividers

1.10 Current dividers play an important role in circuit design. Therefore it is important to develop a facility for dealing with current dividers in circuit analysis. Figure P1.10 shows a two-resistor current divider fed with an ideal current source I. Show that

$$I_1 = \frac{R_2}{R_1 + R_2} I$$

$$I_2 = \frac{R_1}{R_1 + R_2} I$$

and find the voltage V that develops across the current divider.

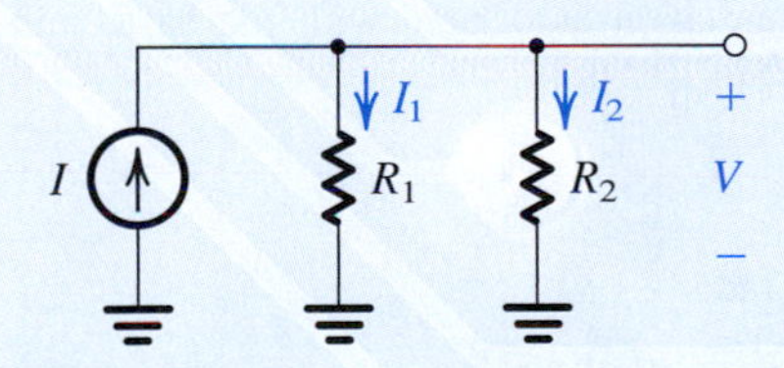

Figure P1.10

D 1.11 Design a simple current divider that will reduce the current provided to a 1-kΩ load to 20% of that available from the source.

D 1.12 A designer searches for a simple circuit to provide one-third of a signal current I to a load resistance R. Suggest a solution using one resistor. What must its value be? What is the input resistance of the resulting current divider? For a particular value R, the designer discovers that the otherwise-best-available resistor is 10% too high. Suggest two circuit topologies using one additional resistor that will solve this problem. What is the value of the resistor required? What is the input resistance of the current divider in each case?

D 1.13 A particular electronic signal source generates currents in the range 0 mA to 1 mA under the condition that its load voltage not exceed 1 V. For loads causing more than 1 V to appear across the generator, the output current is no longer assured but will be reduced by some unknown amount. This circuit limitation, occurring, for example, at the peak of a sine-wave signal, will lead to undesirable signal distortion that must be avoided. If a 10-kΩ load is to be connected, what must be done? What is the name of the circuit you must use? How many resistors are needed? What is (are) the(ir) value(s)?

Thévenin Equivalent Circuits

1.14 For the circuit in Fig. P1.14, find the Thévenin equivalent circuit between terminals (a) 1 and 2, (b) 2 and 3, and (c) 1 and 3.

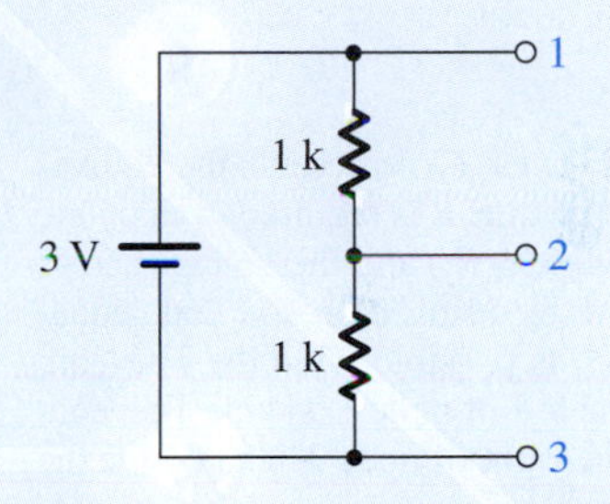

Figure P1.14

1.15 Through repeated application of Thévenin's theorem, find the Thévenin equivalent of the circuit in Fig. P1.15 between node 4 and ground, and hence find the current that flows through a load resistance of 1.5 kΩ connected between node 4 and ground.

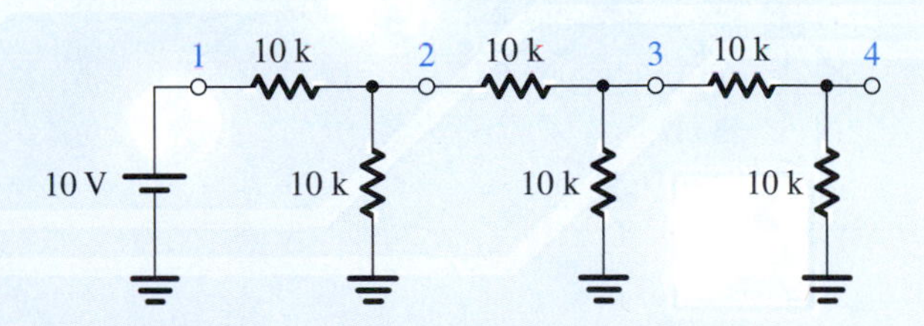

Figure P1.15

Circuit Analysis

1.16 For the circuit shown in Fig. P1.16, find the current in all resistors and the voltage (with respect to ground) at their common node using two methods:

(a) Current: Define branch currents I_1 and I_2 in R_1 and R_2, respectively; identify two equations; and solve them.
(b) Voltage: Define the node voltage V at the common node; identify a single equation; and solve it.

Which method do you prefer? Why?

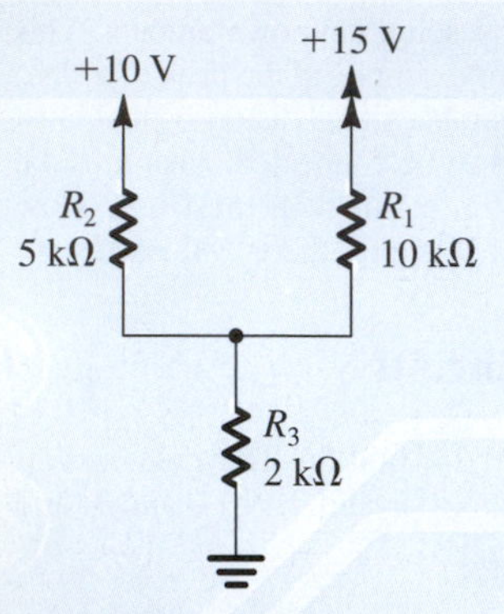

Figure P1.16

1.17 The circuit shown in Fig. P1.17 represents the equivalent circuit of an unbalanced bridge. It is required to calculate the current in the detector branch (R_5) and the voltage across it. Although this can be done by using loop and node equations, a much easier approach is possible: Find the Thévenin equivalent of the circuit to the left of node 1 and the Thévenin equivalent of the circuit to the right of node 2. Then solve the resulting simplified circuit.

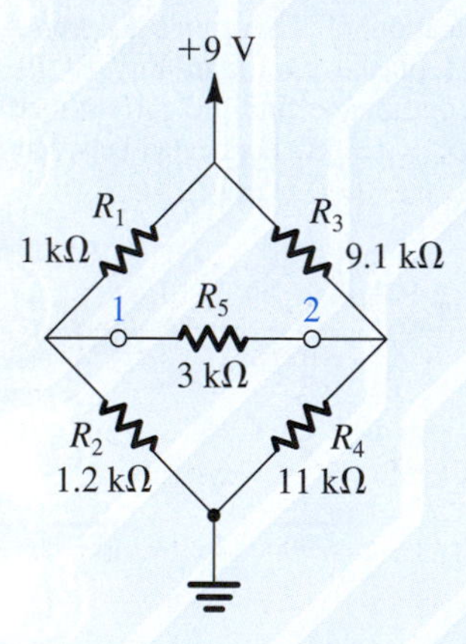

Figure P1.17

1.18 For the circuit in Fig. P1.18, find the equivalent resistance to ground, R_{eq}. To do this, apply a voltage V_x between terminal X and ground and find the current drawn from V_x. Note that you can use particular special properties of the circuit to get the result directly! Now, if R_4 is raised to 1.2 kΩ, what does R_{eq} become?

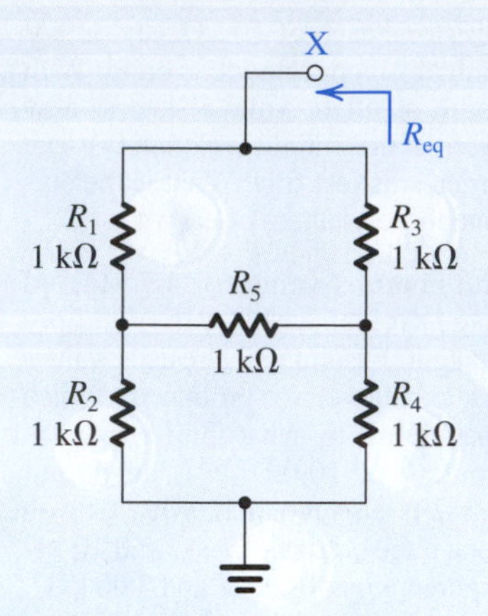

Figure P1.18

AC Circuits

1.19 The periodicity of recurrent waveforms, such as sine waves or square waves, can be completely specified using only one of three possible parameters: radian frequency, ω, in radians per second (rad/s); (conventional) frequency, f, in hertz (Hz); or period T, in seconds (s). As well, each of the parameters can be specified numerically in one of several ways: using letter prefixes associated with the basic units, using scientific notation, or using some combination of both. Thus, for example, a particular period may be specified as 100 ns, 0.1 μs, 10^{-1} μs, 10^5 ps, or 1×10^{-7} s. (For the definition of the various prefixes used in electronics, see Appendix H) For each of the measures listed below, express the trio of terms in scientific notation associated with the basic unit (e.g., 10^{-7} s rather than 10^{-1} μs).

(a) $T = 10^{-4}$ ms
(b) $f = 1$ GHz
(c) $\omega = 6.28 \times 10^2$ rad/s
(d) $T = 10$ s
(e) $f = 60$ Hz
(f) $\omega = 1$ krad/s
(g) $f = 1900$ MHz

1.20 Find the complex impedance, Z, of each of the following basic circuit elements at 60 Hz, 100 kHz, and 1 GHz:

(a) $R = 1$ kΩ
(b) $C = 10$ nF
(c) $C = 2$ pF
(d) $L = 10$ mH
(e) $L = 1$ nH

1.21 Find the complex impedance at 10 kHz of the following networks:

(a) 1 kΩ in series with 10 nF
(b) 1 kΩ in parallel with 0.01 μF

(c) 100 kΩ in parallel with 100 pF
(d) 100 Ω in series with 10 mH

Section 1.1: Signals

1.22 Any given signal source provides an open-circuit voltage, v_{oc}, and a short-circuit current i_{sc}. For the following sources, calculate the internal resistance, R_s; the Norton current, i_s; and the Thévenin voltage, v_s:

(a) $v_{oc} = 10$ V, $i_{sc} = 100$ μA
(b) $v_{oc} = 0.1$ V, $i_{sc} = 10$ μA

1.23 A particular signal source produces an output of 30 mV when loaded by a 100-kΩ resistor and 10 mV when loaded by a 10-kΩ resistor. Calculate the Thévenin voltage, Norton current, and source resistance.

1.24 A temperature sensor is specified to provide 2 mV/°C. When connected to a load resistance of 10 kΩ, the output voltage was measured to change by 10 mV, corresponding to a change in temperature of 10°C. What is the source resistance of the sensor?

1.25 Refer to the Thévenin and Norton representations of the signal source (Fig. 1.1). If the current supplied by the source is denoted i_o and the voltage appearing between the source output terminals is denoted v_o, sketch and clearly label v_o versus i_o for $0 \le i_o \le i_s$.

1.26 The connection of a signal source to an associated signal processor or amplifier generally involves some degree of signal loss as measured at the processor or amplifier input. Considering the two signal-source representations shown in Fig. 1.1, provide two sketches showing each signal-source representation connected to the input terminals (and corresponding input resistance) of a signal processor. What signal-processor input resistance will result in 90% of the open-circuit voltage being delivered to the processor? What input resistance will result in 90% of the short-circuit signal current entering the processor?

Section 1.2: Frequency Spectrum of Signals

1.27 To familiarize yourself with typical values of angular frequency ω, conventional frequency f, and period T, complete the entries in the following table:

Case	ω (rad/s)	f (Hz)	T (s)
a		1×10^9	
b	1×10^9		
c			1×10^{-10}
d		60	
e	6.28×10^3		
f			1×10^{-6}

1.28 For the following peak or rms values of some important sine waves, calculate the corresponding other value:

(a) 117 V rms, a household-power voltage in North America
(b) 33.9 V peak, a somewhat common peak voltage in rectifier circuits
(c) 220 V rms, a household-power voltage in parts of Europe
(d) 220 kV rms, a high-voltage transmission-line voltage in North America

1.29 Give expressions for the sine-wave voltage signals having:

(a) 10-V peak amplitude and 10-kHz frequency
(b) 120-V rms and 60-Hz frequency
(c) 0.2-V peak-to-peak and 1000-rad/s frequency
(d) 100-mV peak and 1-ms period

1.30 Using the information provided by Eq. (1.2) in association with Fig. 1.5, characterize the signal represented by $v(t) = 1/2 + 2/\pi\,(\sin 2000\pi t + \frac{1}{3}\sin 6000\pi t + \frac{1}{5}\sin 10{,}000\pi t + \cdots)$. Sketch the waveform. What is its average value? Its peak-to-peak value? Its lowest value? Its highest value? Its frequency? Its period?

1.31 Measurements taken of a square-wave signal using a frequency-selective voltmeter (called a spectrum analyzer) show its spectrum to contain adjacent components (spectral lines) at 98 kHz and 126 kHz of amplitudes 63 mV and 49 mV, respectively. For this signal, what would direct measurement of the fundamental show its frequency and amplitude to be? What is the rms value of the fundamental? What are the peak-to-peak amplitude and period of the originating square wave?

1.32 What is the fundamental frequency of the highest-frequency square wave for which the fifth harmonic is barely audible by a relatively young listener? What is the fundamental frequency of the lowest-frequency square wave for which the fifth and some of the higher harmonics are directly heard? (Note that the psychoacoustic properties of human hearing allow a listener to sense the lower harmonics as well.)

1.33 Find the amplitude of a symmetrical square wave of period T that provides the same power as a sine wave of peak amplitude $\hat{V}$ and the same frequency. Does this result depend on equality of the frequencies of the two waveforms?

Section 1.3: Analog and Digital Signals

1.34 Give the binary representation of the following decimal numbers: 0, 5, 8, 25, and 57.

1.35 Consider a 4-bit digital word $b_3b_2b_1b_0$ in a format called signed-magnitude, in which the most significant bit, b_3, is interpreted as a sign bit—0 for positive and 1 for negative values. List the values that can be represented by this scheme. What is peculiar about the representation of zero? For a particular analog-to-digital converter (ADC), each change in b_0 corresponds to a 0.5-V change in the analog input. What is the full range of the analog signal that can be represented? What signed-magnitude digital code results for an input of +2.5 V? For −3.0 V? For +2.7 V? For −2.8 V?

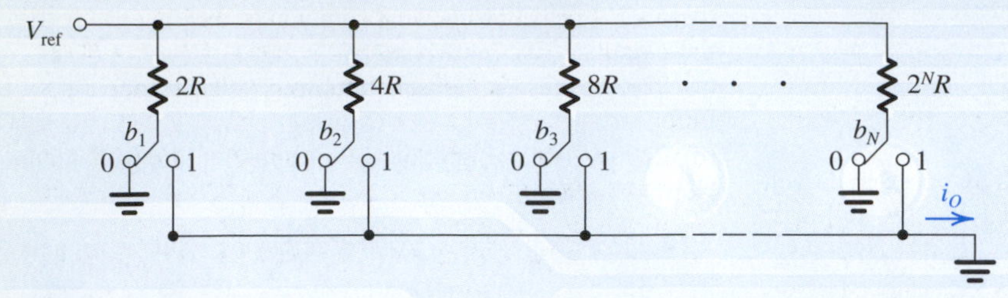

Figure P1.37

1.36 Consider an N-bit ADC whose analog input varies between 0 and V_{FS} (where the subscript FS denotes "full scale").

(a) Show that the least significant bit (LSB) corresponds to a change in the analog signal of $V_{FS}/(2^N - 1)$. This is the resolution of the converter.
(b) Convince yourself that the maximum error in the conversion (called the quantization error) is half the resolution; that is, the quantization error = $V_{FS}/2(2^N - 1)$.
(c) For $V_{FS} = 10$ V, how many bits are required to obtain a resolution of 5 mV or better? What is the actual resolution obtained? What is the resulting quantization error?

1.37 Figure P1.37 shows the circuit of an N-bit digital-to-analog converter (DAC). Each of the N bits of the digital word to be converted controls one of the switches. When the bit is 0, the switch is in the position labeled 0; when the bit is 1, the switch is in the position labeled 1. The analog output is the current i_O. V_{ref} is a constant reference voltage.

(a) Show that

$$i_O = \frac{V_{ref}}{R}\left(\frac{b_1}{2^1} + \frac{b_2}{2^2} + \cdots + \frac{b_N}{2^N}\right)$$

(b) Which bit is the LSB? Which is the MSB?
(c) For $V_{ref} = 10$ V, $R = 5$ kΩ, and $N = 6$, find the maximum value of i_O obtained. What is the change in i_O resulting from the LSB changing from 0 to 1?

1.38 In compact-disc (CD) audio technology, the audio signal is sampled at 44.1 kHz. Each sample is represented by 16 bits. What is the speed of this system in bits per second?

Section 1.4: Amplifiers

1.39 Various amplifier and load combinations are measured as listed below using rms values. For each, find the voltage, current, and power gains (A_v, A_i, and A_p, respectively) both as ratios and in dB:

(a) $v_I = 100$ mV, $i_I = 100$ μA, $v_O = 10$ V, $R_L = 100$ Ω
(b) $v = 10$ μV, $i_I = 100$ nA, $v_O = 2$ V, $R_L = 10$ kΩ
(c) $v_I = 1$ V, $i_I = 1$ mA, $v_O = 10$ V, $R_L = 10$ Ω

1.40 An amplifier operating from ±3-V supplies provides a 2.2-V peak sine wave across a 100-Ω load when provided with a 0.2-V peak input from which 1.0 mA peak is drawn. The average current in each supply is measured to be 20 mA. Find the voltage gain, current gain, and power gain expressed as ratios and in decibels as well as the supply power, amplifier dissipation, and amplifier efficiency.

1.41 An amplifier using balanced power supplies is known to saturate for signals extending within 1.2 V of either supply. For linear operation, its gain is 500 V/V. What is the rms value of the largest undistorted sine-wave output available, and input needed, with ±5-V supplies? With ±10-V supplies? With ±15-V supplies?

1.42 Symmetrically saturating amplifiers, operating in the so-called clipping mode, can be used to convert sine waves to pseudo-square waves. For an amplifier with a small-signal gain of 1000 and clipping levels of ±9 V, what peak value of input sinusoid is needed to produce an output whose extremes are just at the edge of clipping? Clipped 90% of the time? Clipped 99% of the time?

Section 1.5: Circuit Models for Amplifiers

1.43 Consider the voltage-amplifier circuit model shown in Fig. 1.16(b), in which $A_{vo} = 10$ V/V under the following conditions:

(a) $R_i = 10R_s$, $R_L = 10R_o$
(b) $R_i = R_s$, $R_L = R_o$
(c) $R_i = R_s/10$, $R_L = R_o/10$

Calculate the overall voltage gain v_o/v_s in each case, expressed both directly and in decibels.

1.44 An amplifier with 40 dB of small-signal, open-circuit voltage gain, an input resistance of 1 MΩ, and an output resistance of 10 Ω, drives a load of 100 Ω. What voltage and power gains (expressed in dB) would you expect with the load connected? If the amplifier has a peak output-current limitation of 100 mA, what is the rms value of the largest sine-wave input for which an undistorted output is possible? What is the corresponding output power available?

1.45 A 10-mV signal source having an internal resistance of 100 kΩ is connected to an amplifier for which the input resistance is 10 kΩ, the open-circuit voltage gain is 1000 V/V, and the output resistance is 1 kΩ. The amplifier is connected in turn to a 100-Ω load. What overall voltage gain results as measured from the source internal voltage to the load? Where did all the gain go? What would the gain be if the source was connected directly to the load? What is the ratio of these two gains? This ratio is a useful measure of the benefit the amplifier brings.

1.46 A buffer amplifier with a gain of 1 V/V has an input resistance of 1 MΩ and an output resistance of 10 Ω. It is connected between a 1-V, 100-kΩ source and a 100-Ω load. What load voltage results? What are the corresponding voltage, current, and power gains (in dB)?

1.47 Consider the cascade amplifier of Example 1.3. Find the overall voltage gain v_o/v_s obtained when the first and second stages are interchanged. Compare this value with the result in Example 1.3, and comment.

1.48 You are given two amplifiers, A and B, to connect in cascade between a 10-mV, 100-kΩ source and a 100-Ω load. The amplifiers have voltage gain, input resistance, and output resistance as follows: for A, 100 V/V, 10 kΩ, 10 kΩ, respectively; for B, 1 V/V, 100 kΩ, 100 Ω, respectively. Your problem is to decide how the amplifiers should be connected. To proceed, evaluate the two possible connections between source S and load L, namely, SABL and SBAL. Find the voltage gain for each both as a ratio and in decibels. Which amplifier arrangement is best?

D *1.49 A designer has available voltage amplifiers with an input resistance of 10 kΩ, an output resistance of 1 kΩ, and an open-circuit voltage gain of 10. The signal source has a 10-kΩ resistance and provides a 10-mV rms signal, and it is required to provide a signal of at least 2 V rms to a 1-kΩ load. How many amplifier stages are required? What is the output voltage actually obtained.

D *1.50 Design an amplifier that provides 0.5 W of signal power to a 100-Ω load resistance. The signal source provides a 30-mV rms signal and has a resistance of 0.5 MΩ. Three types of voltage-amplifier stages are available:

(a) A high-input-resistance type with $R_i = 1$ MΩ, $A_{vo} = 10$, and $R_o = 10$ kΩ

(b) A high-gain type with $R_i = 10$ kΩ, $A_{vo} = 100$, and $R_o = 1$ kΩ

(c) A low-output-resistance type with $R_i = 10$ kΩ, $A_{vo} = 1$, and $R_o = 20$ Ω

Design a suitable amplifier using a combination of these stages. Your design should utilize the minimum number of stages and should ensure that the signal level is not reduced below 10 mV at any point in the amplifier chain. Find the load voltage and power output realized.

D *1.51 It is required to design a voltage amplifier to be driven from a signal source having a 10-mV peak amplitude and a source resistance of 10 kΩ to supply a peak output of 3 V across a 1-kΩ load.

(a) What is the required voltage gain from the source to the load?

(b) If the peak current available from the source is 0.1 μA, what is the smallest input resistance allowed? For the design with this value of R_i, find the overall current gain and power gain.

(c) If the amplifier power supply limits the peak value of the output open-circuit voltage to 5 V, what is the largest output resistance allowed?

(d) For the design with R_i as in (b) and R_o as in (c), what is the required value of open-circuit voltage gain $\left(\text{i.e., } \left.\frac{v_o}{v_i}\right|_{R_L=\infty}\right)$ of the amplifier?

(e) If, as a possible design option, you are able to increase R_i to the nearest value of the form 1×10^n Ω and to decrease R_o to the nearest value of the form 1×10^m Ω, find (i) the input resistance achievable; (ii) the output resistance achievable; and (iii) the open-circuit voltage gain now required to meet the specifications.

D 1.52 A voltage amplifier with an input resistance of 10 kΩ, an output resistance of 200 Ω, and a gain of 1000 V/V is connected between a 100-kΩ source with an open-circuit voltage of 10 mV and a 100-Ω load. For this situation:

(a) What output voltage results?

(b) What is the voltage gain from source to load?

(c) What is the voltage gain from the amplifier input to the load?

(d) If the output voltage across the load is twice that needed and there are signs of internal amplifier overload, suggest the location and value of a single resistor that would produce the desired output. Choose an arrangement that would cause minimum disruption to an operating circuit. (*Hint:* Use parallel rather than series connections.)

1.53 A current amplifier for which $R_i = 1$ kΩ, $R_o = 10$ kΩ, and $A_{is} = 100$ A/A is to be connected between a 100-mV source with a resistance of 100 kΩ and a load of 1 kΩ. What are the values of current gain i_o/i_i, of voltage gain v_o/v_s, and of power gain expressed directly and in decibels?

1.54 A transconductance amplifier with $R_i = 2$ kΩ, $G_m = 40$ mA/V, and $R_o = 20$ kΩ is fed with a voltage source having a source resistance of 2 kΩ and is loaded with a 1-kΩ resistance. Find the voltage gain realized.

D **1.55 A designer is required to provide, across a 10-kΩ load, the weighted sum, $v_O = 10v_1 + 20v_2$, of input signals v_1 and v_2, each having a source resistance of 10 kΩ. She has a number of transconductance amplifiers for which the input and output resistances are both 10 kΩ and $G_m = 20$ mA/V, together with a selection of suitable resistors. Sketch an appropriate amplifier topology with additional resistors selected to provide the desired result. (*Hint*: In your design, arrange to add currents.)

1.56 Figure P1.56 shows a transconductance amplifier whose output is *fed back* to its input. Find the input resistance

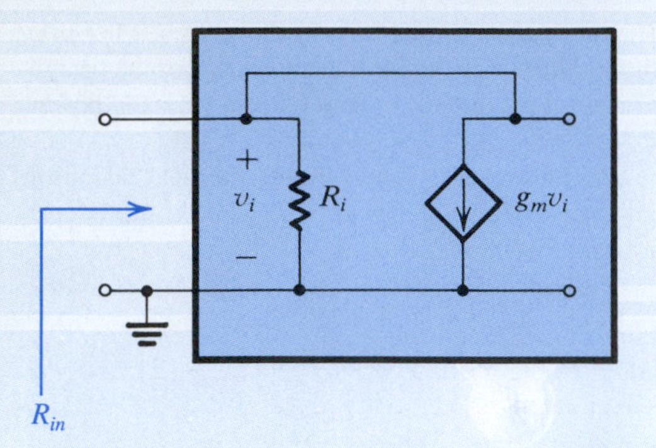

Figure P1.56

R_{in} of the resulting one-port network. (Hint: Apply a test voltage v_x between the two input terminals, and find the current i_x drawn from the source. Then, $R_{in} \equiv v_x / i_x$.)

D 1.57 It is required to design an amplifier to sense the open-circuit output voltage of a transducer and to provide a proportional voltage across a load resistor. The equivalent source resistance of the transducer is specified to vary in the range of 1 kΩ to 10 kΩ. Also, the load resistance varies in the range of 1 kΩ to 10 kΩ. The change in load voltage corresponding to the specified change in R_s should be 10% at most. Similarly, the change in load voltage corresponding to the specified change in R_L should be limited to 10%. Also, corresponding to a 10-mV transducer open-circuit output voltage, the amplifier should provide a minimum of 1 V across the load. What type of amplifier is required? Sketch its circuit model, and specify the values of its parameters. Specify appropriate values for R_i and R_o of the form 1×10^m Ω.

D 1.58 It is required to design an amplifier to sense the short-circuit output current of a transducer and to provide a proportional current through a load resistor. The equivalent source resistance of the transducer is specified to vary in the range of 1 kΩ to 10 kΩ. Similarly, the load resistance is known to vary over the range of 1 kΩ to 10 kΩ. The change in load current corresponding to the specified change in R_s is required to be limited to 10%. Similarly, the change in load current corresponding to the specified change in R_L should be 10% at most. Also, for a nominal short-circuit output current of the transducer of 10 μA, the amplifier is required to provide a minimum of 1 mA through the load. What type of amplifier is required? Sketch the circuit model of the amplifier, and specify values for its parameters. Select appropriate values for R_i and R_o in the form 1×10^m Ω.

D 1.59 It is required to design an amplifier to sense the open-circuit output voltage of a transducer and to provide a proportional current through a load resistor. The equivalent source resistance of the transducer is specified to vary in the range of 1 kΩ to 10 kΩ. Also, the load resistance is known to vary in the range of 1 kΩ to 10 kΩ. The change in the current supplied to the load corresponding to the specified change in R_s is to be 10% at most. Similarly, the change in load current corresponding to the specified change in R_L is to be 10% at most. Also, for a nominal transducer open-circuit output voltage of 10 mV, the amplifier is required to provide a minimum of 1 mA current through the load. What type of amplifier is required? Sketch the amplifier circuit model, and specify values for its parameters. For R_i and R_o, specify values in the form 1×10^m Ω.

D 1.60 It is required to design an amplifier to sense the short-circuit output current of a transducer and to provide a proportional voltage across a load resistor. The equivalent source resistance of the transducer is specified to vary in the range of 1 kΩ to 10 kΩ. Similarly, the load resistance is known to vary in the range of 1 kΩ to 10 kΩ. The change in load voltage corresponding to the specified change in R_s should be 10% at most. Similarly, the change in load voltage corresponding to the specified change in R_L is to be limited to 10%. Also, for a nominal transducer short-circuit output current of 10 μA, the amplifier is required to provide a minimum voltage across the load of 1 V. What type of amplifier is required? Sketch its circuit model, and specify the values of the model parameters. For R_i and R_o, specify appropriate values in the form 1×10^m Ω.

1.61 For the circuit in Fig. P1.61, show that

$$\frac{v_c}{v_b} = \frac{-\beta R_L}{r_\pi + (\beta + 1)R_E}$$

and

$$\frac{v_e}{v_b} = \frac{R_E}{R_E + [r_\pi / (\beta + 1)]}$$

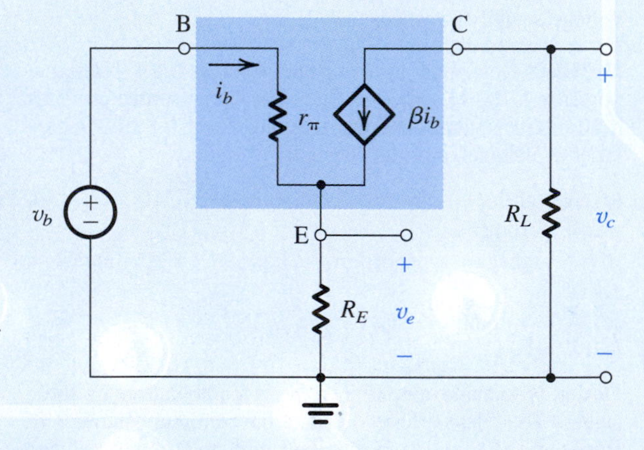

Figure P1.61

1.62 An amplifier with an input resistance of 10 kΩ, when driven by a current source of 1 μA and a source resistance of 100 kΩ, has a short-circuit output current of 10 mA and an open-circuit output voltage of 10 V. The device is driving a 4-kΩ load. Give the values of the voltage gain, current gain, and power gain expressed as ratios and in decibels?

1.63 Figure P1.63(a) shows two transconductance amplifiers connected in a special configuration. Find v_o in terms of v_1 and v_2. Let $g_m = 100$ mA/V and $R = 5$ kΩ. If $v_1 = v_2 = 1$ V, find the value of v_o. Also, find v_o for the case $v_1 = 1.01$ V and $v_2 = 0.99$ V. (Note: This circuit is called a **differential amplifier** and is given the symbol shown in Fig. P1.63(b). A particular type of differential amplifier known as an **operational amplifier** will be studied in Chapter 2.)

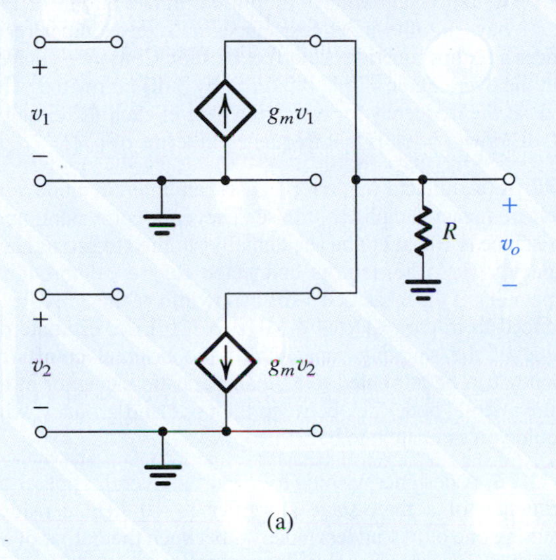

(a)

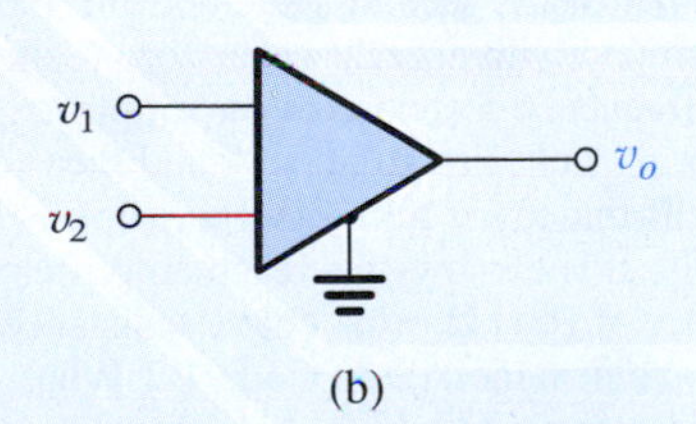

(b)

Figure P1.63

1.64 Any linear two-port network including linear amplifiers can be represented by one of four possible parameter sets, given in Appendix C. For the voltage amplifier, the most convenient representation is in terms of the g parameters. If the amplifier input port is labeled as port 1 and the output port as port 2, its g-parameter representation is described by the two equations:

$$I_1 = g_{11}V_1 + g_{12}I_2$$

$$V_2 = g_{21}V_1 + g_{22}I_2$$

Figure P1.64 shows an equivalent circuit representation of these two equations. By comparing this equivalent circuit to that of the voltage amplifier in Fig. 1.16(a), identify corresponding currents and voltages as well as the correspondence between the parameters of the amplifier equivalent circuit and the g parameters. Hence give the g parameter that corresponds to each of R_i, A_{vo} and R_o. Notice that there is an additional g parameter with no correspondence in the amplifier equivalent circuit. Which one? What does it signify? What assumption did we make about the amplifier that resulted in the absence of this particular g parameter from the equivalent circuit in Fig. 1.16(a)?

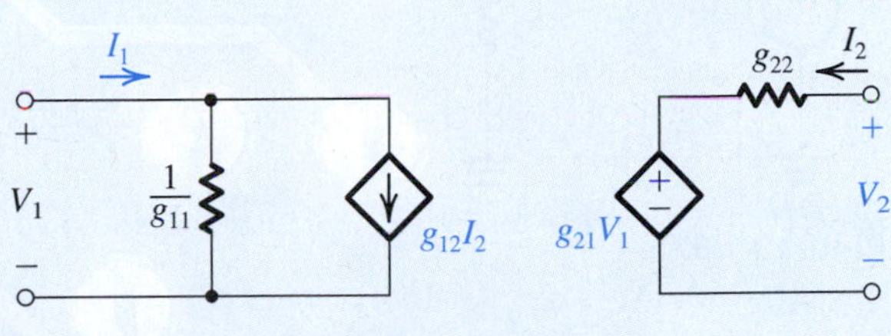

Figure P1.64

Section 1.6: Frequency Response of Amplifiers

1.65 Use the voltage-divider rule to derive the transfer functions $T(s) \equiv V_o(s)/V_i(s)$ of the circuits shown in Fig. 1.22, and show that the transfer functions are of the form given at the top of Table 1.2.

1.66 Figure P1.66 shows a signal source connected to the input of an amplifier. Here R_s is the source resistance, and R_i and C_i are the input resistance and input capacitance, respectively, of the amplifier. Derive an expression for $V_i(s)/V_s(s)$, and show that it is of the low-pass STC type. Find the 3-dB frequency for the case $R_s = 20$ kΩ, $R_i = 80$ kΩ, and $C_i = 5$ pF.

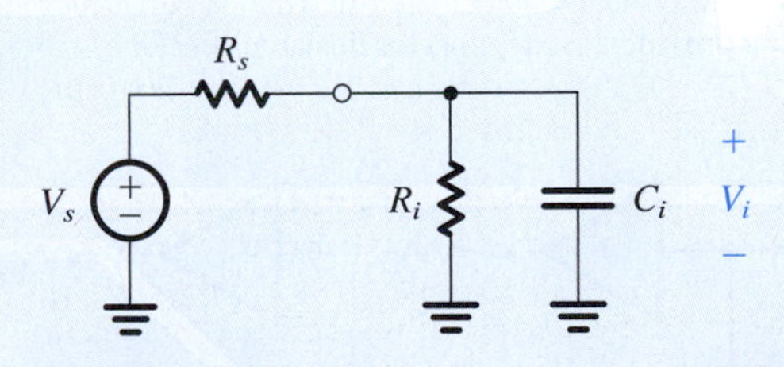

Figure P1.66

1.67 For the circuit shown in Fig. P1.67, find the transfer function $T(s) = V_o(s)/V_i(s)$, and arrange it in the appropriate standard form from Table 1.2. Is this a high-pass or a low-pass network? What is its transmission at very high frequencies? [Estimate this directly, as well as by letting $s \rightarrow \infty$ in your expression for $T(s)$.] What is the corner frequency ω_0? For $R_1 = 10\ \text{k}\Omega$, $R_2 = 40\ \text{k}\Omega$, and $C = 0.1\ \mu\text{F}$, find f_0. What is the value of $|T(j\omega_0)|$?

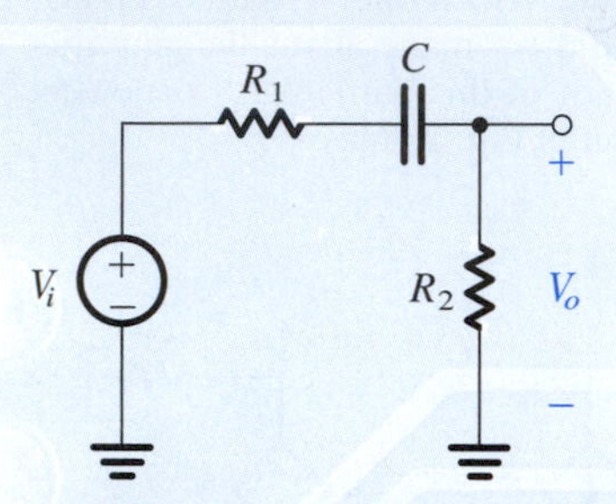

Figure P1.67

D 1.68 It is required to couple a voltage source V_s with a resistance R_s to a load R_L via a capacitor C. Derive an expression for the transfer function from source to load (i.e., V_L/V_s), and show that it is of the high-pass STC type. For $R_s = 5\ \text{k}\Omega$ and $R_L = 20\ \text{k}\Omega$, find the smallest coupling capacitor that will result in a 3-dB frequency no greater than 10 Hz.

1.69 Measurement of the frequency response of an amplifier yields the data in the following table:

f (Hz)	$\|T\|$ (dB)	$\angle T$ (°)
0	40	0
100	40	0
1000		
10^4	37	−45
10^5	20	
	0	

Provide plausible approximate values for the missing entries. Also, sketch and clearly label the magnitude frequency response (i.e., provide a Bode plot) for this amplifier.

1.70 Measurement of the frequency response of an amplifier yields the data in the following table:

f (Hz)		10	10^2	10^3	10^4	10^5	10^6	10^7	
$\|T\|$ (dB)	0	20	37	40			37	20	0

Provide approximate plausible values for the missing table entries. Also, sketch and clearly label the magnitude frequency response (Bode plot) of this amplifier.

1.71 The unity-gain voltage amplifiers in the circuit of Fig. P1.71 have infinite input resistances and zero output resistances and thus function as perfect buffers. Convince yourself that the overall gain V_o/V_i will drop by 3 dB below the value at dc at the frequency for which the gain of each RC circuit is 1.0 dB down. What is that frequency in terms of CR?

1.72 A manufacturing error causes an internal node of a high-frequency amplifier whose Thévenin-equivalent node resistance is 100 kΩ to be accidentally shunted to ground by a capacitor (i.e., the node is connected to ground through a capacitor). If the measured 3-dB bandwidth of the amplifier is reduced from the expected 6 MHz to 120 kHz, estimate the value of the shunting capacitor. If the original cutoff frequency can be attributed to a small parasitic capacitor at the same internal node (i.e., between the node and ground), what would you estimate it to be?

D *1.73 A designer wishing to lower the overall upper 3-dB frequency of a three-stage amplifier to 10 kHz considers shunting one of two nodes: Node A, between the output of the first stage and the input of the second stage, and Node B, between the output of the second stage and the input of the third stage, to ground with a small capacitor. While measuring the overall frequency response of the amplifier, she connects a capacitor of 1 nF, first to node A and then to node B, lowering the 3-dB frequency from 2 MHz to 150 kHz and 15 kHz, respectively. If she knows that each amplifier stage has an input resistance of 100 kΩ, what output resistance must the driving stage have at node A? At node B? What capacitor value should she connect to which node to solve her design problem most economically?

D 1.74 An amplifier with an input resistance of 100 kΩ and an output resistance of 1 kΩ is to be capacitor-coupled to a 10-kΩ source and a 1-kΩ load. Available capacitors have values only of the form 1×10^{-n} F. What are the values of the

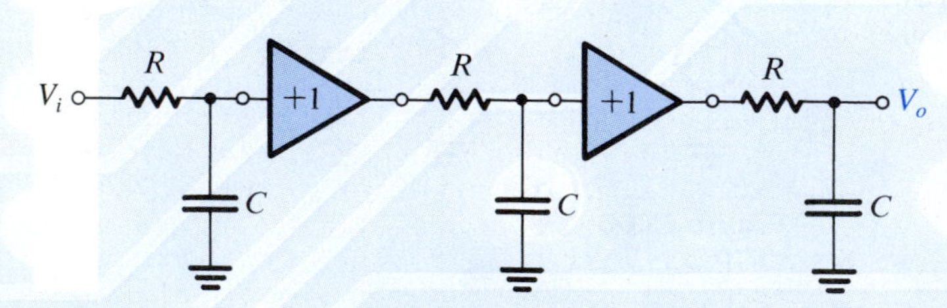

Figure P1.71

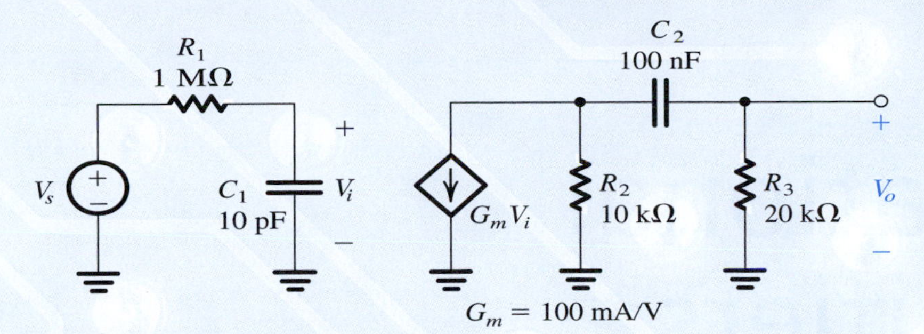

Figure P1.76

smallest capacitors needed to ensure that the corner frequency associated with each is less than 100 Hz? What actual corner frequencies result? For the situation in which the basic amplifier has an open-circuit voltage gain (A_{vo}) of 100 V/V, find an expression for $T(s) = V_o(s)/V_s(s)$.

***1.75** A voltage amplifier has the transfer function

$$A_v = \frac{100}{\left(1 + j\frac{f}{10^4}\right)\left(1 + \frac{10^2}{jf}\right)}$$

Using the Bode plots for low-pass and high-pass STC networks (Figs. 1.23 and 1.24), sketch a Bode plot for $|A_v|$. Give approximate values for the gain magnitude at $f = 10$ Hz, 10^2 Hz, 10^3 Hz, 10^4 Hz, 10^5 Hz, 10^6 Hz, and 10^7 Hz. Find the bandwidth of the amplifier (defined as the frequency range over which the gain remains within 3 dB of the maximum value).

***1.76** For the circuit shown in Fig. P1.76 first, evaluate $T_i(s) = V_i(s)/V_s(s)$ and the corresponding cutoff (corner) frequency. Second, evaluate $T_o(s) = V_o(s)/V_i(s)$ and the corresponding cutoff frequency. Put each of the transfer functions in the standard form (see Table 1.2), and combine them to form the overall transfer function, $T(s) = T_i(s) \times T_o(s)$. Provide a Bode magnitude plot for $|T(j\omega)|$. What is the bandwidth between 3-dB cutoff points?

D ****1.77** A transconductance amplifier having the equivalent circuit shown in Table 1.1 is fed with a voltage source V_s having a source resistance R_s, and its output is connected to a load consisting of a resistance R_L in parallel with a capacitance C_L. For given values of R_s, R_L, and C_L, it is required to specify the values of the amplifier parameters R_i, G_m, and R_o to meet the following design constraints:

(a) At most, x% of the input signal is lost in coupling the signal source to the amplifier (i.e., $V_i \geq [1 - (x/100)]V_s$).
(b) The 3-dB frequency of the amplifier is equal to or greater than a specified value $f_{3\text{dB}}$.
(c) The dc gain V_o/V_s is equal to or greater than a specified value A_0.

Show that these constraints can be met by selecting

$$R_i \geq \left(\frac{100}{x} - 1\right)R_s$$

$$R_o \leq \frac{1}{2\pi f_{3\text{dB}} C_L - (1/R_L)}$$

$$G_m \geq \frac{A_0/[1 - (x/100)]}{(R_L \parallel R_o)}$$

Find R_i, R_o, and G_m for $R_s = 10$ kΩ, $x = 20\%$, $A_o = 80$, $R_L = 10$ kΩ, $C_L = 10$ pF, and $f_{3\text{dB}} = 3$ MHz.

***1.78** Use the voltage-divider rule to find the transfer function $V_o(s)/V_i(s)$ of the circuit in Fig. P1.78. Show that the transfer function can be made independent of frequency if the condition $C_1R_1 = C_2R_2$ applies. Under this condition the circuit is called a **compensated attenuator** and is frequently employed in the design of oscilloscope probes. Find the transmission of the compensated attenuator in terms of R_1 and R_2.

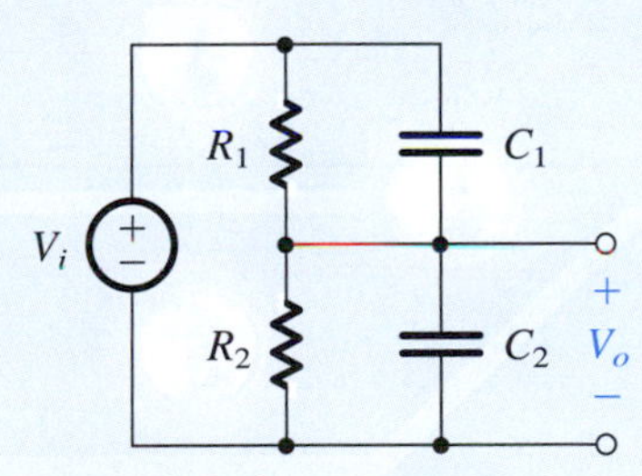

Figure P1.78

***1.79** An amplifier with a frequency response of the type shown in Fig. 1.21 is specified to have a phase shift of magnitude no greater than 11.4° over the amplifier bandwidth, which extends from 100 Hz to 1 kHz. It has been found that the gain falloff at the low-frequency end is determined by the response of a high-pass STC circuit and that at the high-frequency end it is determined by a low-pass STC circuit. What do you expect the corner frequencies of these two circuits to be? What is the drop in gain in decibels (relative to the maximum gain) at the two frequencies that define the amplifier bandwidth? What are the frequencies at which the drop in gain is 3 dB?

CHAPTER 2

Operational Amplifiers

IN THIS CHAPTER YOU WILL LEARN

1. The terminal characteristics of the ideal op amp.
2. How to analyze circuits containing op amps, resistors, and capacitors.
3. How to use op amps to design amplifiers having precise characteristics.
4. How to design more sophisticated op-amp circuits, including summing amplifiers, instrumentation amplifiers, integrators, and differentiators.
5. Important nonideal characteristics of op amps and how these limit the performance of basic op-amp circuits.

Introduction

Having learned basic amplifier concepts and terminology, we are now ready to undertake the study of a circuit building block of universal importance: The operational amplifier (op amp). Op amps have been in use for a long time, their initial applications being primarily in the areas of analog computation and sophisticated instrumentation. Early op amps were constructed from discrete components (vacuum tubes and then transistors, and resistors), and their cost was prohibitively high (tens of dollars). In the mid-1960s the first integrated-circuit (IC) op amp was produced. This unit (the μA 709) was made up of a relatively large number of transistors and resistors all on the same silicon chip. Although its characteristics were poor (by today's standards) and its price was still quite high, its appearance signaled a new era in electronic circuit design. Electronics engineers started using op amps in large quantities, which caused their price to drop dramatically. They also demanded better-quality op amps. Semiconductor manufacturers responded quickly, and within the span of a few years, high-quality op amps became available at extremely low prices (tens of cents) from a large number of suppliers.

One of the reasons for the popularity of the op amp is its versatility. As we will shortly see, one can do almost anything with op amps! Equally important is the fact that the IC op amp has characteristics that closely approach the assumed ideal. This implies that it is quite easy to design circuits using the IC op amp. Also, op-amp circuits work at performance levels that are quite close to those predicted theoretically. It is for this reason that we are studying op amps at this early stage. It is expected that by the end of this chapter the reader should be able to design nontrivial circuits successfully using op amps.

As already implied, an IC op amp is made up of a large number (tens or more) of transistors, resistors, and (usually) one capacitor connected in a rather complex circuit. Since we have not yet studied transistor circuits, the circuit inside the op amp will not be discussed in this chapter. Rather, we will treat the op amp as a circuit building block and study its terminal characteristics and its applications. This approach is quite satisfactory in many op-amp applications. Nevertheless, for the more difficult and demanding applications it is quite useful to know what is inside the op-amp package. This topic will be studied in Chapter 12. More advanced applications of op amps will appear in later chapters.

2.1 The Ideal Op Amp

2.1.1 The Op-Amp Terminals

From a signal point of view the op amp has three terminals: two input terminals and one output terminal. Figure 2.1 shows the symbol we shall use to represent the op amp. Terminals 1 and 2 are input terminals, and terminal 3 is the output terminal. As explained in Section 1.4, amplifiers require dc power to operate. Most IC op amps require two dc power supplies, as shown in Fig. 2.2. Two terminals, 4 and 5, are brought out of the op-amp package and connected to a positive voltage V_{CC} and a negative voltage $-V_{EE}$, respectively. In Fig. 2.2(b) we explicitly show the two dc power supplies as batteries with a common ground. It is interesting to note that the reference grounding point in op-amp circuits is just the common terminal of the two power supplies; that is, no terminal of the op-amp package is physically connected to ground. In what follows we will not, for simplicity, explicitly show the op-amp power supplies.

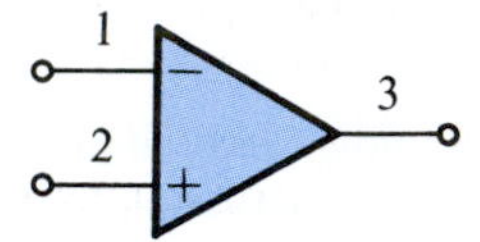

Figure 2.1 Circuit symbol for the op amp.

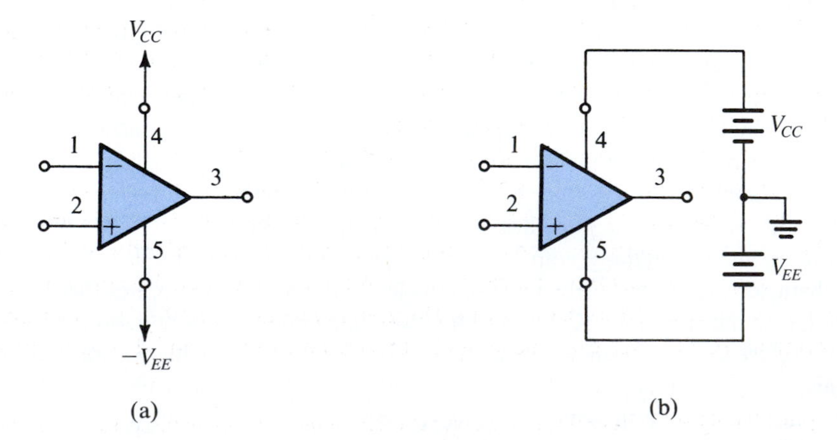

Figure 2.2 The op amp shown connected to dc power supplies.

In addition to the three signal terminals and the two power-supply terminals, an op amp may have other terminals for specific purposes. These other terminals can include terminals for frequency compensation and terminals for offset nulling; both functions will be explained in later sections.

EXERCISE

2.1 What is the minimum number of terminals required by a single op amp? What is the minimum number of terminals required on an integrated-circuit package containing four op amps (called a quad op amp)?
Ans. 5; 14

2.1.2 Function and Characteristics of the Ideal Op Amp

We now consider the circuit function of the op amp. The op amp is designed to sense the difference between the voltage signals applied at its two input terminals (i.e., the quantity $v_2 - v_1$), multiply this by a number A, and cause the resulting voltage $A(v_2 - v_1)$ to appear at output terminal 3. Thus $v_3 = A(v_2 - v_1)$. Here it should be emphasized that when we talk about the voltage at a terminal we mean the voltage between that terminal and ground; thus v_1 means the voltage applied between terminal 1 and ground.

The ideal op amp is not supposed to draw any input current; that is, the signal current into terminal 1 and the signal current into terminal 2 are both zero. In other words, *the input impedance of an ideal op amp is supposed to be infinite.*

How about the output terminal 3? This terminal is supposed to act as the output terminal of an ideal voltage source. That is, the voltage between terminal 3 and ground will always be equal to $A(v_2 - v_1)$, independent of the current that may be drawn from terminal 3 into a load impedance. In other words, *the output impedance of an ideal op amp is supposed to be zero.*

Putting together all of the above, we arrive at the equivalent circuit model shown in Fig. 2.3. Note that the output is in phase with (has the same sign as) v_2 and is out of phase with (has the opposite sign of) v_1. For this reason, input terminal 1 is called the **inverting input terminal** and is distinguished by a "–" sign, while input terminal 2 is called the **noninverting input terminal** and is distinguished by a "+" sign.

As can be seen from the above description, the op amp responds only to the *difference* signal $v_2 - v_1$ and hence ignores any signal *common* to both inputs. That is, if $v_1 = v_2 = 1$ V, then the output will (ideally) be zero. We call this property **common-mode rejection**, and we conclude that an ideal op amp has zero common-mode gain or, equivalently, infinite common-mode rejection. We will have more to say about this point later. For the time being note that the op amp is a **differential-input**, **single-ended-output** amplifier, with the latter term referring to the fact that the output appears between terminal 3 and ground.[1]

[1]Some op amps are designed to have differential outputs. This topic will not be discussed in this book. Rather, we confine ourselves here to single-ended-output op amps, which constitute the vast majority of commercially available op amps.

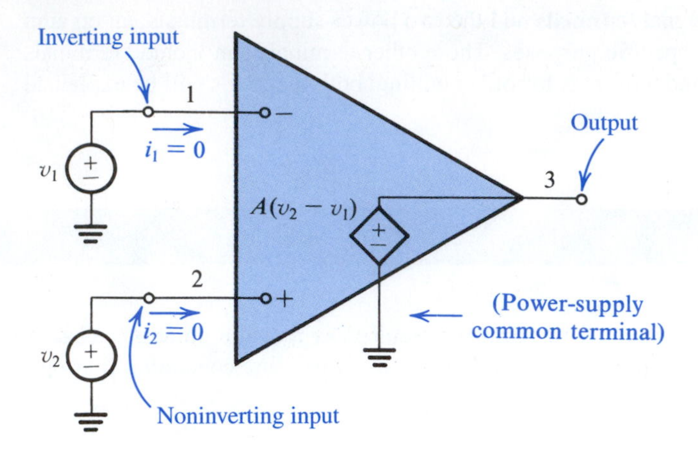

Figure 2.3 Equivalent circuit of the ideal op amp.

Furthermore, gain A is called the **differential gain**, for obvious reasons. Perhaps not so obvious is another name that we will attach to A: the **open-loop gain**. The reason for this name will become obvious later on when we "close the loop" around the op amp and define another gain, the closed-loop gain.

An important characteristic of op amps is that they are **direct-coupled** or **dc amplifiers**, where dc stands for direct-coupled (it could equally well stand for direct current, since a direct-coupled amplifier is one that amplifies signals whose frequency is as low as zero). The fact that op amps are direct-coupled devices will allow us to use them in many important applications. Unfortunately, though, the direct-coupling property can cause some serious practical problems, as will be discussed in a later section.

How about bandwidth? The ideal op amp has a gain A that remains constant down to zero frequency and up to infinite frequency. That is, ideal op amps will amplify signals of any frequency with equal gain, and are thus said to have *infinite bandwidth.*

We have discussed all of the properties of the ideal op amp except for one, which in fact is the most important. This has to do with the value of A. *The ideal op amp should have a gain* A *whose value is very large and ideally infinite*. One may justifiably ask: If the gain A is infinite, how are we going to use the op amp? The answer is very simple: In almost all applications the op amp will *not* be used alone in a so-called open-loop configuration. Rather, we will use other components to apply feedback to close the loop around the op amp, as will be illustrated in detail in Section 2.2.

For future reference, Table 2.1 lists the characteristics of the ideal op amp.

Table 2.1 Characteristics of the Ideal Op Amp

1. Infinite input impedance
2. Zero output impedance
3. Zero common-mode gain or, equivalently, infinite common-mode rejection
4. Infinite open-loop gain A
5. Infinite bandwidth

2.1.3 Differential and Common-Mode Signals

The differential input signal v_{Id} is simply the difference between the two input signals v_1 and v_2; that is,

$$v_{Id} = v_2 - v_1 \tag{2.1}$$

The common-mode input signal v_{Icm} is the average of the two input signals v_1 and v_2; namely,

$$v_{Icm} = \tfrac{1}{2}(v_1 + v_2) \tag{2.2}$$

Equations (2.1) and (2.2) can be used to express the input signals v_1 and v_2 in terms of their differential and common-mode components as follows:

$$v_1 = v_{Icm} - v_{Id}/2 \tag{2.3}$$

and

$$v_2 = v_{Icm} + v_{Id}/2 \tag{2.4}$$

These equations can in turn lead to the pictorial representation in Fig. 2.4.

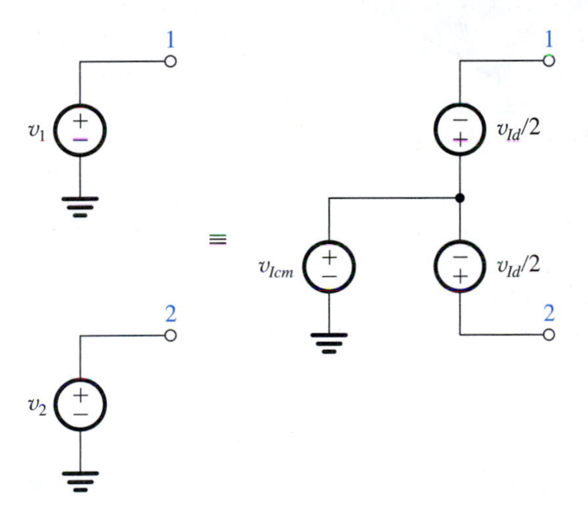

Figure 2.4 Representation of the signal sources v_1 and v_2 in terms of their differential and common-mode components.

EXERCISES

2.2 Consider an op amp that is ideal except that its open-loop gain $A = 10^3$. The op amp is used in a feedback circuit, and the voltages appearing at two of its three signal terminals are measured. In each of the following cases, use the measured values to find the expected value of the voltage at the third terminal. Also give the differential and common-mode input signals in each case. (a) $v_2 = 0$ V and $v_3 = 2$ V; (b) $v_2 = +5$ V and $v_3 = -10$ V; (c) $v_1 = 1.002$ V and $v_2 = 0.998$ V; (d) $v_1 = -3.6$ V and $v_3 = -3.6$ V.
Ans. (a) $v_1 = -0.002$ V, $v_{Id} = 2$ mV, $v_{Icm} = -1$ mV; (b) $v_1 = +5.01$ V, $v_{Id} = -10$ mV, $v_{Icm} = 5.005 \simeq 5$ V; (c) $v_3 = -4$ V, $v_{Id} = -4$ mV, $v_{Icm} = 1$ V; (d) $v_2 = -3.6036$ V, $v_{Id} = -3.6$ mV, $v_{Icm} \simeq -3.6$ V

2.3 The internal circuit of a particular op amp can be modeled by the circuit shown in Fig. E2.3. Express v_3 as a function of v_1 and v_2. For the case $G_m = 10$ mA/V, $R = 10$ kΩ, and $\mu = 100$, find the value of the open-loop gain A.

Ans. $v_3 = \mu G_m R(v_2 - v_1)$; $A = 10{,}000$ V/V or 80 dB

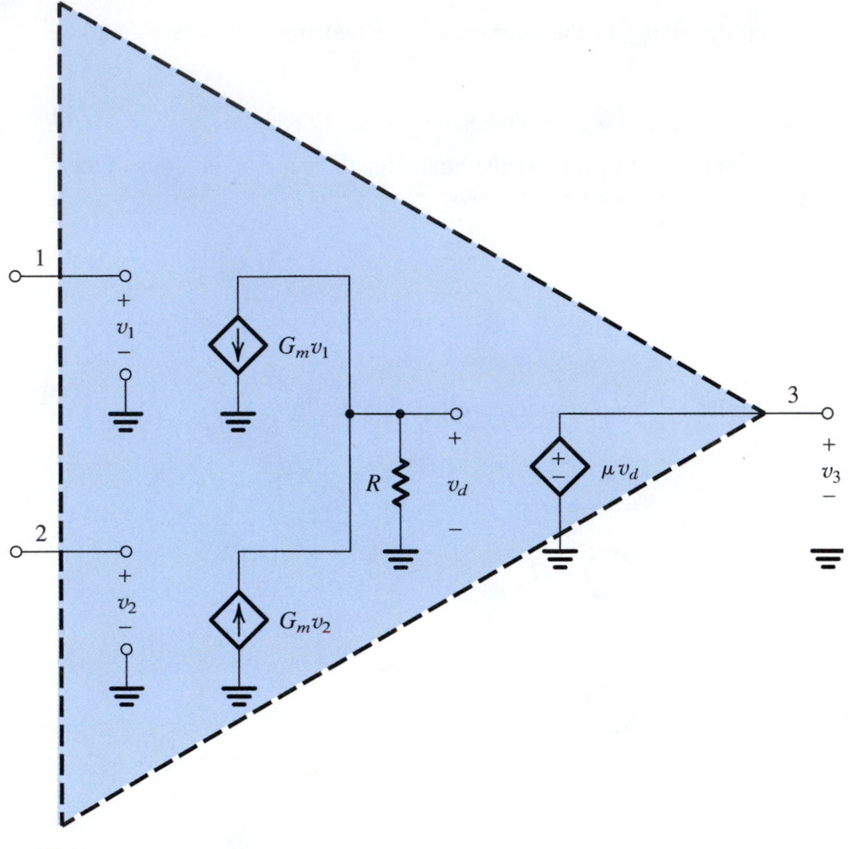

Figure E2.3

2.2 The Inverting Configuration

As mentioned above, op amps are not used alone; rather, the op amp is connected to passive components in a feedback circuit. There are two such basic circuit configurations employing an op amp and two resistors: the inverting configuration, which is studied in this section, and the noninverting configuration, which we shall study in the next section.

Figure 2.5 shows the inverting configuration. It consists of one op amp and two resistors R_1 and R_2. Resistor R_2 is connected from the output terminal of the op amp, terminal 3, *back* to the *inverting* or *negative* input terminal, terminal 1. We speak of R_2 as applying **negative feedback**; if R_2 were connected between terminals 3 and 2 we would have called this **positive feedback**. Note also that R_2 *closes the loop* around the op amp. In addition to adding R_2, we have grounded terminal 2 and connected a resistor R_1 between terminal 1 and an input signal source with a voltage v_I. The output of the overall circuit is taken at terminal 3 (i.e., between terminal 3 and

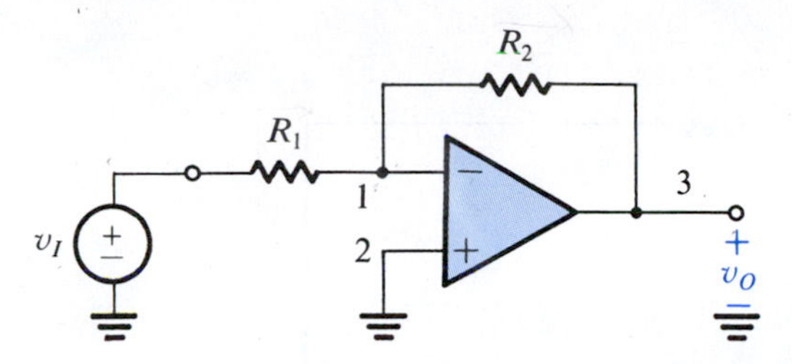

Figure 2.5 The inverting closed-loop configuration.

ground). Terminal 3 is, of course, a convenient point from which to take the output, since the impedance level there is ideally zero. Thus the voltage v_O will not depend on the value of the current that might be supplied to a load impedance connected between terminal 3 and ground.

2.2.1 The Closed-Loop Gain

We now wish to analyze the circuit in Fig. 2.5 to determine the **closed-loop gain** G, defined as

$$G \equiv \frac{v_O}{v_I}$$

We will do so assuming the op amp to be ideal. Figure 2.6(a) shows the equivalent circuit, and the analysis proceeds as follows: The gain A is very large (ideally infinite). If we assume that the circuit is "working" and producing a finite output voltage at terminal 3, then the voltage between the op-amp input terminals should be negligibly small and ideally zero. Specifically, if we call the output voltage v_O, then, by definition,

$$v_2 - v_1 = \frac{v_O}{A} = 0$$

It follows that the voltage at the inverting input terminal (v_1) is given by $v_1 = v_2$. That is, because the gain A approaches infinity, the voltage v_1 approaches and ideally equals v_2. We speak of this as the two input terminals "tracking each other in potential." We also speak of a "virtual short circuit" that exists between the two input terminals. Here the word *virtual* should be emphasized, and one should *not* make the mistake of physically shorting terminals 1 and 2 together while analyzing a circuit. A **virtual short circuit** means that whatever voltage is at 2 will automatically appear at 1 because of the infinite gain A. But terminal 2 happens to be connected to ground; thus $v_2 = 0$ and $v_1 = 0$. We speak of terminal 1 as being a **virtual ground**—that is, having zero voltage but not physically connected to ground.

Now that we have determined v_1 we are in a position to apply Ohm's law and find the current i_1 through R_1 (see Fig. 2.6) as follows:

$$i_1 = \frac{v_I - v_1}{R_1} = \frac{v_I - 0}{R_1} = \frac{v_I}{R_1}$$

Where will this current go? It cannot go into the op amp, since the ideal op amp has an infinite input impedance and hence draws zero current. It follows that i_1 will have to flow through R_2 to the low-impedance terminal 3. We can then apply Ohm's law to R_2 and determine v_O; that is,

$$v_O = v_1 - i_1 R_2$$
$$= 0 - \frac{v_I}{R_1} R_2$$

Thus,

$$\frac{v_O}{v_I} = -\frac{R_2}{R_1}$$

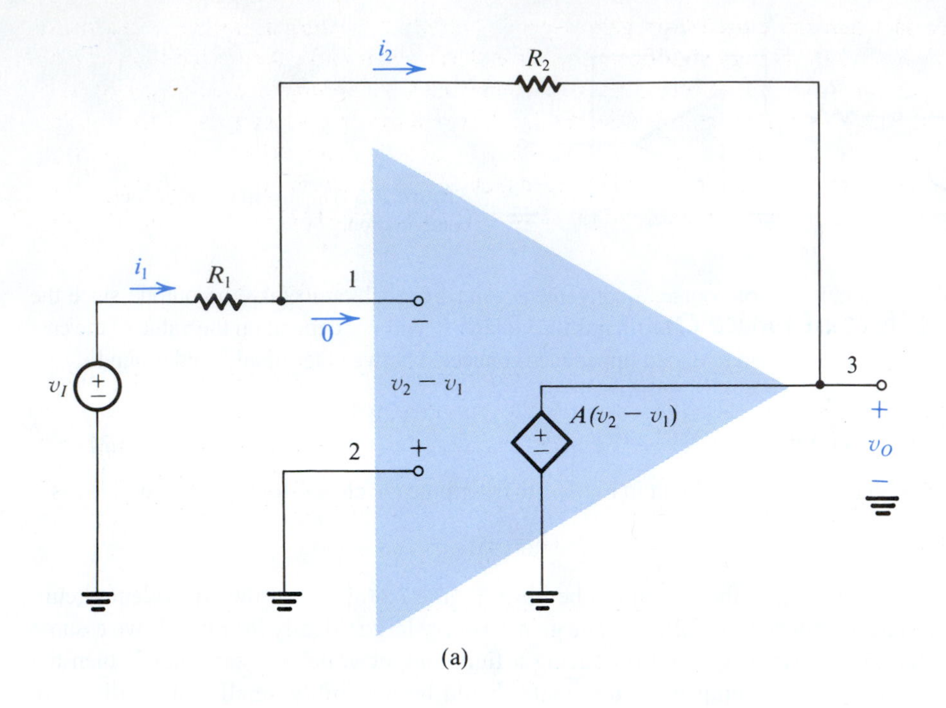

(a)

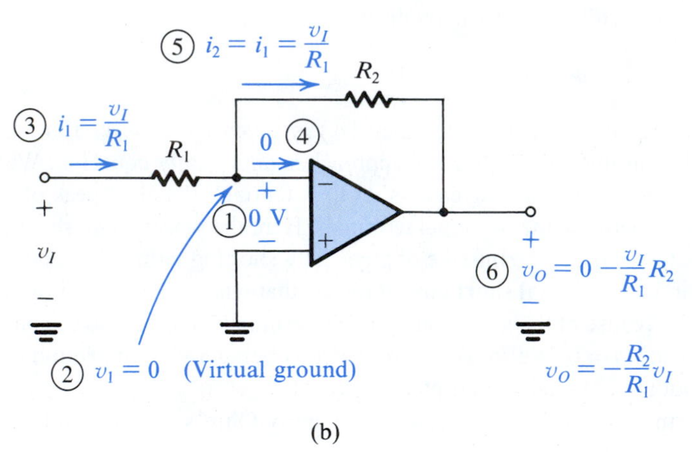

(b)

Figure 2.6 Analysis of the inverting configuration. The circled numbers indicate the order of the analysis steps.

which is the required closed-loop gain. Figure 2.6(b) illustrates these steps and indicates by the circled numbers the order in which the analysis is performed.

We thus see that the closed-loop gain is simply the ratio of the two resistances R_2 and R_1. The minus sign means that the closed-loop amplifier provides signal inversion. Thus if $R_2/R_1 = 10$ and we apply at the input (v_I) a sine-wave signal of 1 V peak-to-peak, then the output v_O will be a sine wave of 10 V peak-to-peak and phase-shifted 180° with respect to the input sine wave. Because of the minus sign associated with the closed-loop gain, this configuration is called the **inverting configuration**.

The fact that the closed-loop gain depends entirely on external passive components (resistors R_1 and R_2) is very significant. It means that we can make the closed-loop gain as accurate as we want by selecting passive components of appropriate accuracy. It also means that the closed-loop gain is (ideally) independent of the op-amp gain. This is a dramatic illustration of negative feedback: We started out with an amplifier having very large gain A, and through applying negative feedback we have obtained a closed-loop gain R_2/R_1 that is much smaller than A but is stable and predictable. That is, we are trading gain for accuracy.

2.2.2 Effect of Finite Open-Loop Gain

The points just made are more clearly illustrated by deriving an expression for the closed-loop gain under the assumption that the op-amp open-loop gain A is finite. Figure 2.7 shows the analysis. If we denote the output voltage v_O, then the voltage between the two input terminals of the op amp will be v_O/A. Since the positive input terminal is grounded, the voltage at the negative input terminal must be $-v_O/A$. The current i_1 through R_1 can now be found from

$$i_1 = \frac{v_I - (-v_O/A)}{R_1} = \frac{v_I + v_O/A}{R_1}$$

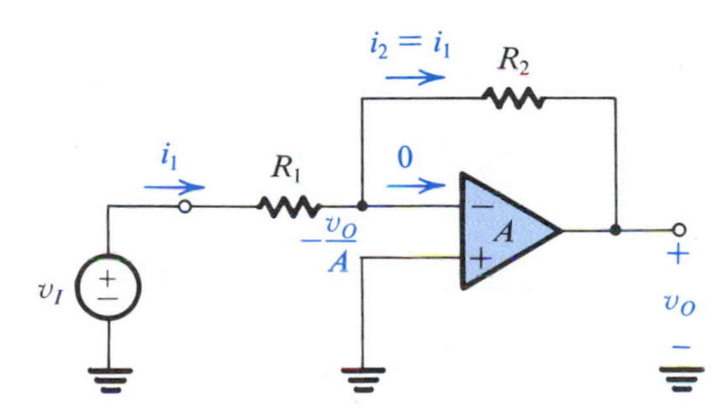

Figure 2.7 Analysis of the inverting configuration taking into account the finite open-loop gain of the op amp.

The infinite input impedance of the op amp forces the current i_1 to flow entirely through R_2. The output voltage v_O can thus be determined from

$$v_O = -\frac{v_O}{A} - i_1 R_2$$

$$= -\frac{v_O}{A} - \left(\frac{v_I + v_O/A}{R_1}\right) R_2$$

Collecting terms, the closed-loop gain G is found as

$$G \equiv \frac{v_O}{v_I} = \frac{-R_2/R_1}{1 + (1 + R_2/R_1)/A} \tag{2.5}$$

We note that as A approaches ∞, G approaches the ideal value of $-R_2/R_1$. Also, from Fig. 2.7 we see that as A approaches ∞, the voltage at the inverting input terminal approaches zero. This is the virtual-ground assumption we used in our earlier analysis when the op amp was assumed to be ideal. Finally, note that Eq. (2.5) in fact indicates that to minimize the dependence of the closed-loop gain G on the value of the open-loop gain A, we should make

$$1 + \frac{R_2}{R_1} \ll A$$

Example 2.1

Consider the inverting configuration with $R_1 = 1\ \text{k}\Omega$ and $R_2 = 100\ \text{k}\Omega$.

(a) Find the closed-loop gain for the cases $A = 10^3$, 10^4, and 10^5. In each case determine the percentage error in the magnitude of G relative to the ideal value of R_2/R_1 (obtained with $A = \infty$). Also determine the voltage v_1 that appears at the inverting input terminal when $v_I = 0.1$ V.

(b) If the open-loop gain A changes from 100,000 to 50,000 (i.e., drops by 50%), what is the corresponding percentage change in the magnitude of the closed-loop gain G?

Solution

(a) Substituting the given values in Eq. (2.5), we obtain the values given in the following table, where the percentage error ε is defined as

$$\varepsilon \equiv \frac{|G| - (R_2/R_1)}{(R_2/R_1)} \times 100$$

The values of v_1 are obtained from $v_1 = -v_O/A = Gv_I/A$ with $v_I = 0.1$ V.

A	$\lvert G \rvert$	ε	v_1
10^3	90.83	−9.17%	−9.08 mV
10^4	99.00	−1.00%	−0.99 mV
10^5	99.90	−0.10%	−0.10 mV

(b) Using Eq. (2.5), we find that for $A = 50{,}000$, $|G| = 99.80$. Thus a −50% change in the open-loop gain results in a change of only −0.1% in the closed-loop gain!

2.2.3 Input and Output Resistances

Assuming an ideal op amp with infinite open-loop gain, the input resistance of the closed-loop inverting amplifier of Fig. 2.5 is simply equal to R_1. This can be seen from Fig. 2.6(b), where

$$R_i \equiv \frac{v_I}{i_1} = \frac{v_I}{v_I/R_1} = R_1$$

Now recall that in Section 1.5 we learned that the amplifier input resistance forms a voltage divider with the resistance of the source that feeds the amplifier. Thus, to avoid the loss of signal strength, voltage amplifiers are required to have high input resistance. In the case of the inverting op-amp configuration we are studying, to make R_i high we should select a high value for R_1. However, if the required gain R_2/R_1 is also high, then R_2 could become impractically large (e.g., greater than a few megohms). We may conclude that the inverting configuration suffers from a low input resistance. A solution to this problem is discussed in Example 2.2 below.

Since the output of the inverting configuration is taken at the terminals of the ideal voltage source $A(v_2 - v_1)$ (see Fig. 2.6a), it follows that the output resistance of the closed-loop amplifier is zero.

Example 2.2

Assuming the op amp to be ideal, derive an expression for the closed-loop gain v_O/v_I of the circuit shown in Fig. 2.8. Use this circuit to design an inverting amplifier with a gain of 100 and an input resistance of 1 MΩ. Assume that for practical reasons it is required not to use resistors greater than 1 MΩ. Compare your design with that based on the inverting configuration of Fig. 2.5.

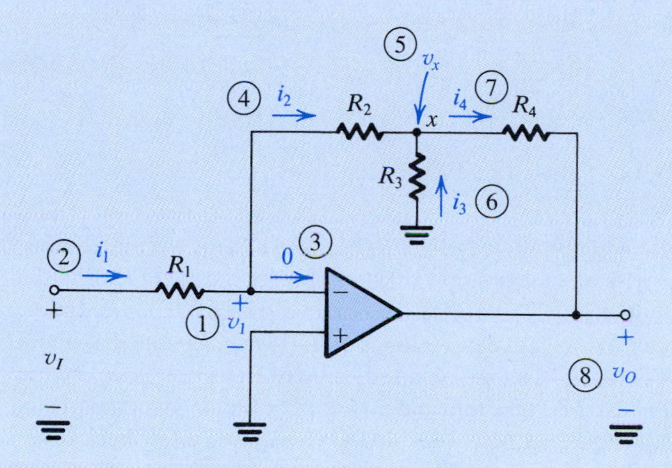

Figure 2.8 Circuit for Example 2.2. The circled numbers indicate the sequence of the steps in the analysis.

Solution

The analysis begins at the inverting input terminal of the op amp, where the voltage is

$$v_1 = \frac{-v_O}{A} = \frac{-v_O}{\infty} = 0$$

Here we have assumed that the circuit is "working" and producing a finite output voltage v_O. Knowing v_1, we can determine the current i_1 as follows:

$$i_1 = \frac{v_I - v_1}{R_1} = \frac{v_I - 0}{R_1} = \frac{v_I}{R_1}$$

Since zero current flows into the inverting input terminal, all of i_1 will flow through R_2, and thus

$$i_2 = i_1 = \frac{v_I}{R_1}$$

Now we can determine the voltage at node x:

$$v_x = v_1 - i_2 R_2 = 0 - \frac{v_I}{R_1} R_2 = -\frac{R_2}{R_1} v_I$$

Example 2.2 *continued*

This in turn enables us to find the current i_3:

$$i_3 = \frac{0 - v_x}{R_3} = \frac{R_2}{R_1 R_3} v_I$$

Next, a node equation at x yields i_4:

$$i_4 = i_2 + i_3 = \frac{v_I}{R_1} + \frac{R_2}{R_1 R_3} v_I$$

Finally, we can determine v_O from

$$v_O = v_x - i_4 R_4$$

$$= -\frac{R_2}{R_1} v_I - \left(\frac{v_I}{R_1} + \frac{R_2}{R_1 R_3} v_I\right) R_4$$

Thus the voltage gain is given by

$$\frac{v_O}{v_I} = -\left[\frac{R_2}{R_1} + \frac{R_4}{R_1}\left(1 + \frac{R_2}{R_3}\right)\right]$$

which can be written in the form

$$\frac{v_O}{v_I} = -\frac{R_2}{R_1}\left(1 + \frac{R_4}{R_2} + \frac{R_4}{R_3}\right)$$

Now, since an input resistance of 1 MΩ is required, we select $R_1 = 1$ MΩ. Then, with the limitation of using resistors no greater than 1 MΩ, the maximum value possible for the first factor in the gain expression is 1 and is obtained by selecting $R_2 = 1$ MΩ. To obtain a gain of –100, R_3 and R_4 must be selected so that the second factor in the gain expression is 100. If we select the maximum allowed (in this example) value of 1 MΩ for R_4, then the required value of R_3 can be calculated to be 10.2 kΩ. Thus this circuit utilizes three 1-MΩ resistors and a 10.2-kΩ resistor. In comparison, if the inverting configuration were used with $R_1 = 1$ MΩ we would have required a feedback resistor of 100 MΩ, an impractically large value!

Before leaving this example it is insightful to inquire into the mechanism by which the circuit is able to realize a large voltage gain without using large resistances in the feedback path. Toward that end, observe that because of the virtual ground at the inverting input terminal of the op amp, R_2 and R_3 are in effect in parallel. Thus, by making R_3 lower than R_2 by, say, a factor k (i.e., where $k > 1$), R_3 is forced to carry a current k-times that in R_2. Thus, while $i_2 = i_1$, $i_3 = ki_1$ and $i_4 = (k+1)i_1$. It is the current multiplication by a factor of $(k+1)$ that enables a large voltage drop to develop across R_4 and hence a large v_O without using a large value for R_4. Notice also that the current through R_4 is independent of the value of R_4. It follows that the circuit can be used as a current amplifier as shown in Fig. 2.9.

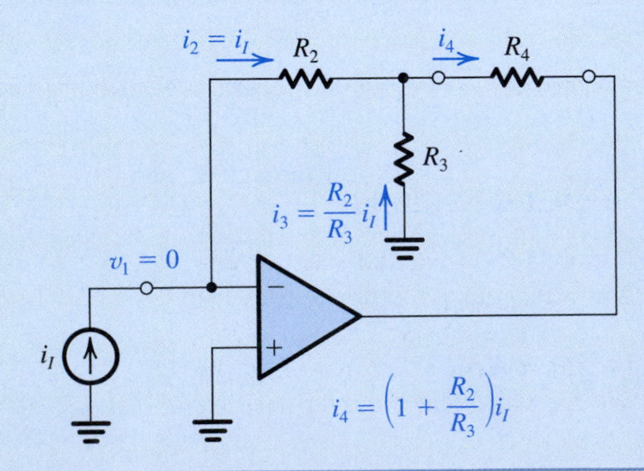

Figure 2.9 A current amplifier based on the circuit of Fig. 2.8. The amplifier delivers its output current to R_4. It has a current gain of $(1 + R_2/R_3)$, a zero input resistance, and an infinite output resistance. The load (R_4), however, must be floating (i.e., neither of its two terminals can be connected to ground).

EXERCISES

D2.4 Use the circuit of Fig. 2.5 to design an inverting amplifier having a gain of -10 and an input resistance of 100 kΩ. Give the values of R_1 and R_2.
Ans. $R_1 = 100$ kΩ; $R_2 = 1$ MΩ

2.5 The circuit shown in Fig. E2.5(a) can be used to implement a transresistance amplifier (see Table 1.1 in Section 1.5). Find the value of the input resistance R_i, the transresistance R_m, and the output resistance R_o of the transresistance amplifier. If the signal source shown in Fig. E2.5(b) is connected to the input of the transresistance amplifier, find its output voltage.
Ans. $R_i = 0$; $R_m = -10$ kΩ; $R_o = 0$; $v_O = -5$ V

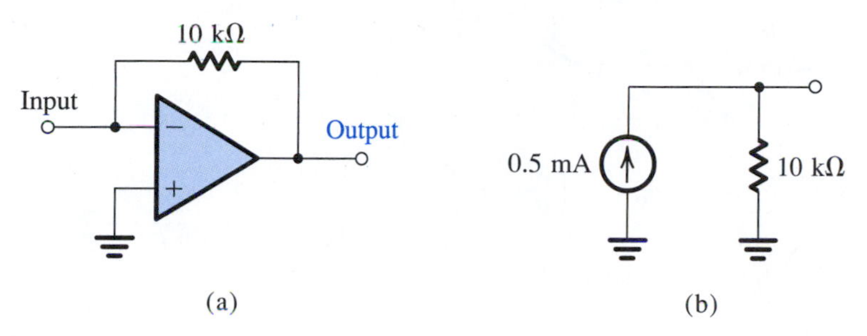

Figure E2.5

2.6 For the circuit in Fig. E2.6 determine the values of v_1, i_1, i_2, v_O, i_L, and i_O. Also determine the voltage gain v_O/v_I, current gain i_L/i_I, and power gain P_O/P_I.
Ans. 0 V; 1 mA; 1 mA; -10 V; -10 mA; -11 mA; -10 V/V (20 dB), -10 A/A (20 dB); 100 W/W (20 dB)

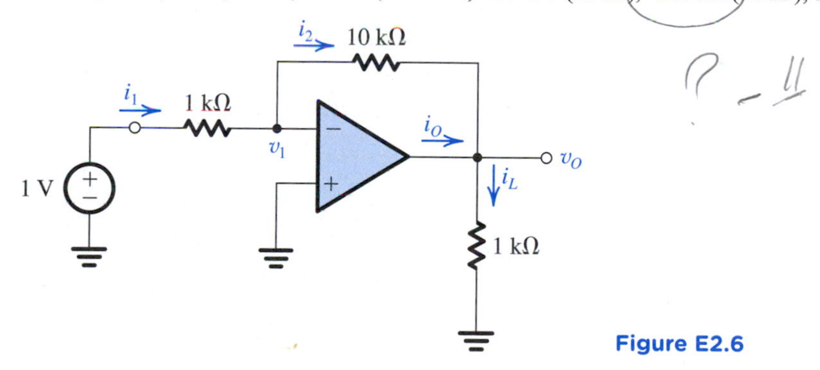

Figure E2.6

2.2.4 An Important Application—The Weighted Summer

A very important application of the inverting configuration is the weighted-summer circuit shown in Fig. 2.10. Here we have a resistance R_f in the negative-feedback path (as before); but we have a number of input signals $v_1, v_2, \ldots, v_n$ each applied to a corresponding resistor $R_1, R_2, \ldots, R_n$, which are connected to the inverting terminal of the op amp. From our previous discussion, the ideal op amp will have a virtual ground appearing at its negative input terminal. Ohm's law then tells us that the currents $i_1, i_2, \ldots, i_n$ are given by

$$i_1 = \frac{v_1}{R_1}, \qquad i_2 = \frac{v_2}{R_2}, \qquad \ldots, \qquad i_n = \frac{v_n}{R_n}$$

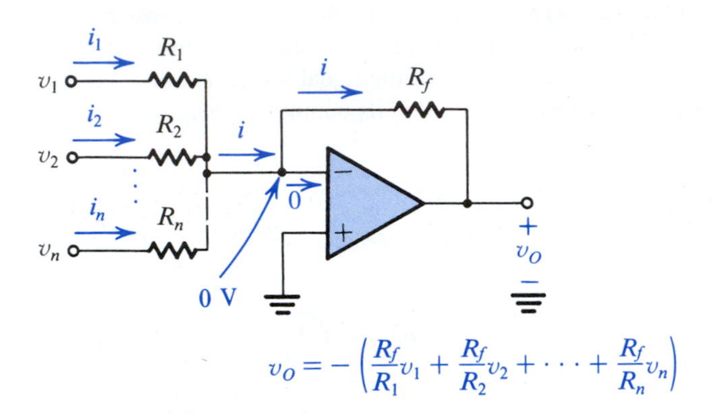

$$v_O = -\left(\frac{R_f}{R_1}v_1 + \frac{R_f}{R_2}v_2 + \cdots + \frac{R_f}{R_n}v_n\right)$$

Figure 2.10 A weighted summer.

All these currents sum together to produce the current i; that is,

$$i = i_1 + i_2 + \cdots + i_n \tag{2.6}$$

will be forced to flow through R_f (since no current flows into the input terminals of an ideal op amp). The output voltage v_O may now be determined by another application of Ohm's law,

$$v_O = 0 - iR_f = -iR_f$$

Thus,

$$v_O = -\left(\frac{R_f}{R_1}v_1 + \frac{R_f}{R_2}v_2 + \cdots + \frac{R_f}{R_n}v_n\right) \tag{2.7}$$

That is, the output voltage is a weighted sum of the input signals $v_1, v_2, \ldots, v_n$. This circuit is therefore called a **weighted summer**. Note that each summing coefficient may be independently adjusted by adjusting the corresponding "feed-in" resistor (R_1 to R_n). This nice property, which greatly simplifies circuit adjustment, is a direct consequence of the virtual ground that exists at the inverting op-amp terminal. As the reader will soon come to appreciate, virtual grounds are extremely "handy." In the weighted summer of Fig. 2.10 all the summing coefficients must be of the same sign. The need occasionally arises for summing signals with opposite signs. Such a function can be implemented, however, using two op amps as shown in Fig. 2.11. Assuming ideal op amps, it can be easily shown that the output voltage is given by

$$v_O = v_1\left(\frac{R_a}{R_1}\right)\left(\frac{R_c}{R_b}\right) + v_2\left(\frac{R_a}{R_2}\right)\left(\frac{R_c}{R_b}\right) - v_3\left(\frac{R_c}{R_3}\right) - v_4\left(\frac{R_c}{R_4}\right) \tag{2.8}$$

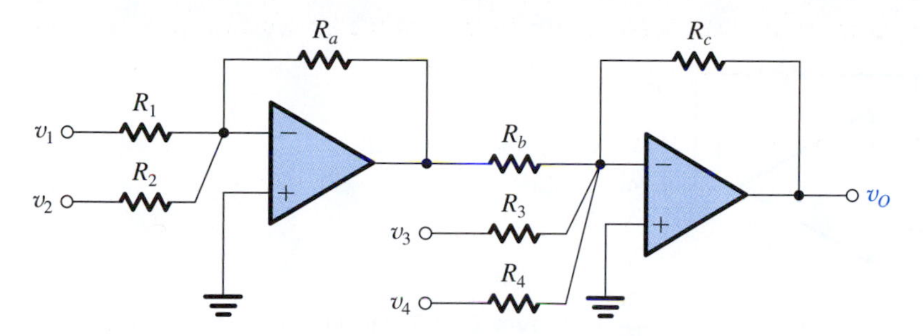

Figure 2.11 A weighted summer capable of implementing summing coefficients of both signs.

EXERCISES

D2.7 Design an inverting op-amp circuit to form the weighted sum v_O of two inputs v_1 and v_2. It is required that $v_O = -(v_1 + 5v_2)$. Choose values for R_1, R_2, and R_f so that for a maximum output voltage of 10 V the current in the feedback resistor will not exceed 1 mA.
Ans. A possible choice: $R_1 = 10\ \text{k}\Omega$, $R_2 = 2\ \text{k}\Omega$, and $R_f = 10\ \text{k}\Omega$

D2.8 Use the idea presented in Fig. 2.11 to design a weighted summer that provides

$$v_O = 2v_1 + v_2 - 4v_3$$

Ans. A possible choice: $R_1 = 5\ \text{k}\Omega$, $R_2 = 10\ \text{k}\Omega$, $R_a = 10\ \text{k}\Omega$, $R_b = 10\ \text{k}\Omega$, $R_3 = 2.5\ \text{k}\Omega$, $R_c = 10\ \text{k}\Omega$.

2.3 The Noninverting Configuration

The second closed-loop configuration we shall study is shown in Fig. 2.12. Here the input signal v_I is applied directly to the positive input terminal of the op amp while one terminal of R_1 is connected to ground.

2.3.1 The Closed-Loop Gain

Analysis of the noninverting circuit to determine its closed-loop gain (v_O/v_I) is illustrated in Fig. 2.13. Again the order of the steps in the analysis is indicated by circled numbers. Assuming that the op amp is ideal with infinite gain, a virtual short circuit exists between its two input terminals. Hence the difference input signal is

$$v_{Id} = \frac{v_O}{A} = 0 \qquad \text{for } A = \infty$$

Thus the voltage at the inverting input terminal will be equal to that at the noninverting input terminal, which is the applied voltage v_I. The current through R_1 can then be determined as v_I/R_1. Because of the infinite input impedance of the op amp, this current will flow through R_2, as shown in Fig. 2.13. Now the output voltage can be determined from

$$v_O = v_I + \left(\frac{v_I}{R_1}\right)R_2$$

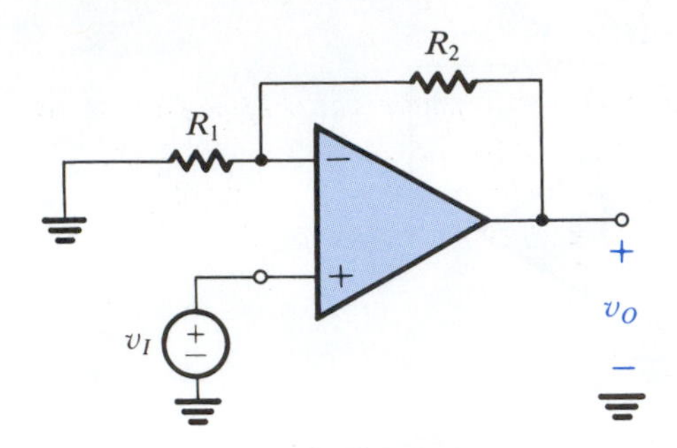

Figure 2.12 The noninverting configuration.

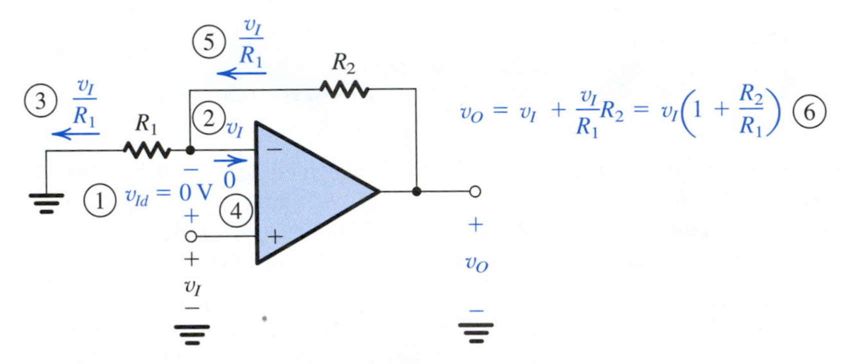

Figure 2.13 Analysis of the noninverting circuit. The sequence of the steps in the analysis is indicated by the circled numbers.

which yields

$$\frac{v_O}{v_I} = 1 + \frac{R_2}{R_1} \tag{2.9}$$

Further insight into the operation of the noninverting configuration can be obtained by considering the following: Since the current into the op-amp inverting input is zero, the circuit composed of R_1 and R_2 acts in effect as a voltage divider feeding a fraction of the output voltage back to the inverting input terminal of the op amp; that is,

$$v_1 = v_O\left(\frac{R_1}{R_1 + R_2}\right) \tag{2.10}$$

Then the infinite op-amp gain and the resulting virtual short circuit between the two input terminals of the op amp forces this voltage to be equal to that applied at the positive input terminal; thus,

$$v_O\left(\frac{R_1}{R_1 + R_2}\right) = v_I$$

which yields the gain expression given in Eq. (2.9).

This is an appropriate point to reflect further on the action of the negative feedback present in the noninverting circuit of Fig. 2.12. Let v_I increase. Such a change in v_I will cause v_{Id} to increase, and v_O will correspondingly increase as a result of the high (ideally infinite) gain of the op amp. However, a fraction of the increase in v_O will be fed back to the inverting input terminal of the op amp through the (R_1, R_2) voltage divider. The result of this feedback will be to counteract the increase in v_{Id}, driving v_{Id} back to zero, albeit at a higher value of v_O that corresponds to the increased value of v_I. This *degenerative* action of negative feedback gives it the alternative name **degenerative feedback**. Finally, note that the argument above applies equally well if v_I decreases. A formal and detailed study of feedback is presented in Chapter 10.

2.3.2 Effect of Finite Open-Loop Gain

As we have done for the inverting configuration, we now consider the effect of the finite op-amp open-loop gain A on the gain of the noninverting configuration. Assuming the op amp to be ideal except for having a finite open-loop gain A, it can be shown that the closed-loop gain of the noninverting amplifier circuit of Fig. 2.12 is given by

$$G \equiv \frac{v_O}{v_I} = \frac{1 + (R_2/R_1)}{1 + \dfrac{1 + (R_2/R_1)}{A}} \tag{2.11}$$

Observe that the denominator is identical to that for the case of the inverting configuration (Eq. 2.5). This is no coincidence; it is a result of the fact that both the inverting and the noninverting configurations have the same feedback loop, which can be readily seen if the input signal source is eliminated (i.e., short-circuited). The numerators, however, are different, for the numerator gives the ideal or nominal closed-loop gain ($-R_2/R_1$ for the inverting configuration, and $1 + R_2/R_1$ for the noninverting configuration). Finally, we note (with reassurance) that the gain expression in Eq. (2.11) reduces to the ideal value for $A = \infty$. In fact, it approximates the ideal value for

$$A \gg 1 + \frac{R_2}{R_1}$$

This is the same condition as in the inverting configuration, except that here the quantity on the right-hand side is the nominal closed-loop gain.The expressions for the actual and ideal values of the closed-loop gain G in Eqs. (2.11) and (2.9), respectively, can be used to determine the percentage error in G resulting from the finite op-amp gain A as

$$\text{Percent gain error} = -\frac{1 + (R_2/R_1)}{A + 1 + (R_2/R_1)} \times 100 \tag{2.12}$$

Thus, as an example, if an op amp with an open-loop gain of 1000 is used to design a noninverting amplifier with a nominal closed-loop gain of 10, we would expect the closed-loop gain to be about 1% below the nominal value.

2.3.3 Input and Output Resistance

The gain of the noninverting configuration is positive—hence the name *noninverting*. The input impedance of this closed-loop amplifier is ideally infinite, since no current flows into the positive input terminal of the op amp. The output of the noninverting amplifier is taken at the terminals of the ideal voltage source $A(v_2 - v_1)$ (see the op-amp equivalent circuit in Fig. 2.3), thus the output resistance of the noninverting configuration is zero.

2.3.4 The Voltage Follower

The property of high input impedance is a very desirable feature of the noninverting configuration. It enables using this circuit as a buffer amplifier to connect a source with a high impedance to a low-impedance load. We have discussed the need for buffer amplifiers in Section 1.5. In many applications the buffer amplifier is not required to provide any voltage gain; rather, it is used mainly as an impedance transformer or a power amplifier. In such cases we may make $R_2 = 0$ and $R_1 = \infty$ to obtain the **unity-gain amplifier** shown in Fig. 2.14(a). This circuit is commonly referred to as a **voltage follower**, since the output "follows" the input. In the ideal case, $v_O = v_I$, $R_{in} = \infty$, $R_{out} = 0$, and the follower has the equivalent circuit shown in Fig. 2.14(b).

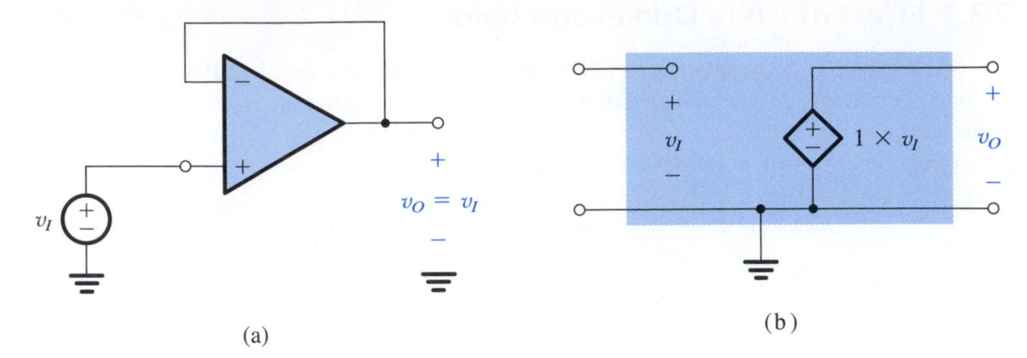

Figure 2.14 **(a)** The unity-gain buffer or follower amplifier. **(b)** Its equivalent circuit model.

Since in the voltage-follower circuit the entire output is fed back to the inverting input, the circuit is said to have 100% negative feedback. The infinite gain of the op amp then acts to make $v_{Id} = 0$ and hence $v_O = v_I$. Observe that the circuit is elegant in its simplicity!

Since the noninverting configuration has a gain greater than or equal to unity, depending on the choice of R_2/R_1, some prefer to call it "a follower with gain."

EXERCISES

2.9 Use the superposition principle to find the output voltage of the circuit shown in Fig. E2.9.
Ans. $v_O = 6v_1 + 4v_2$

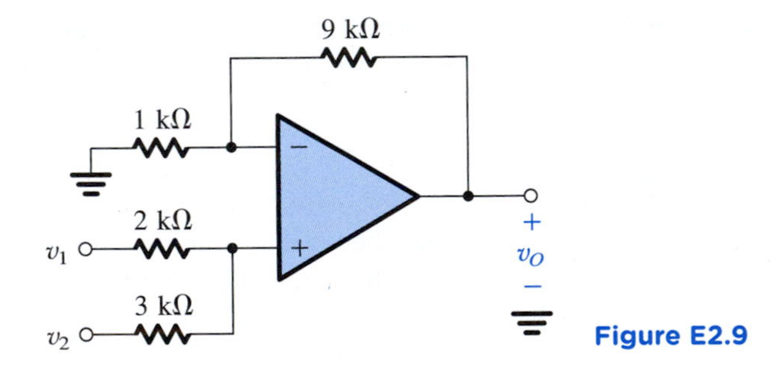

Figure E2.9

2.10 If in the circuit of Fig. E2.9 the 1-kΩ resistor is disconnected from ground and connected to a third signal source v_3, use superposition to determine v_O in terms of v_1, v_2, and v_3.
Ans. $v_O = 6v_1 + 4v_2 - 9v_3$

D2.11 Design a noninverting amplifier with a gain of 2. At the maximum output voltage of 10 V the current in the voltage divider is to be 10 μA.
Ans. $R_1 = R_2 = 0.5$ MΩ

2.12 (a) Show that if the op amp in the circuit of Fig. 2.12 has a finite open-loop gain A, then the closed-loop gain is given by Eq. (2.11). (b) For $R_1 = 1$ kΩ and $R_2 = 9$ kΩ find the percentage deviation ε of the closed-loop gain from the ideal value of $(1 + R_2/R_1)$ for the cases $A = 10^3$, 10^4, and 10^5. For $v_I = 1$ V, find in each case the voltage between the two input terminals of the op amp.
Ans. $\varepsilon = -1\%, -0.1\%, -0.01\%$; $v_2 - v_1 = 9.9$ mV, 1 mV, 0.1 mV

2.13 For the circuit in Fig. E2.13 find the values of i_I, v_1, i_1, i_2, v_O, i_L, and i_O. Also find the voltage gain v_O/v_I, the current gain i_L/i_I, and the power gain P_L/P_I.
Ans. 0; 1 V; 1 mA; 1 mA; 10 V; 10 mA; 11 mA; 10 V/V (20 dB); ∞; ∞

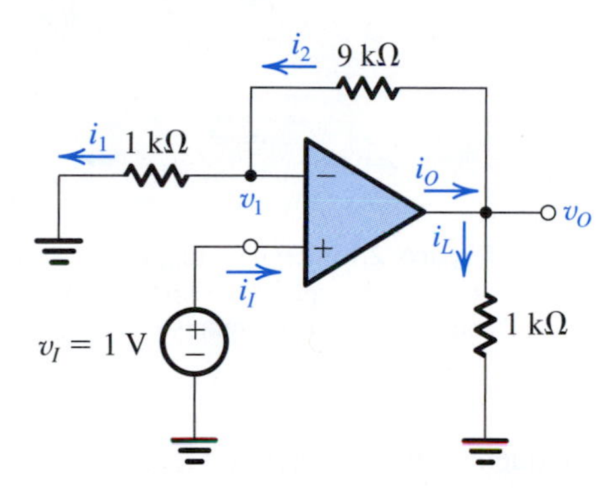

Figure E2.13

2.14 It is required to connect a transducer having an open-circuit voltage of 1 V and a source resistance of 1 MΩ to a load of 1-kΩ resistance. Find the load voltage if the connection is done (a) directly and (b) through a unity-gain voltage follower.
Ans. (a) 1 mV; (b) 1 V

2.4 Difference Amplifiers

Having studied the two basic configurations of op-amp circuits together with some of their direct applications, we are now ready to consider a somewhat more involved but very important application. Specifically, we shall study the use of op amps to design difference or differential amplifiers.[2] A difference amplifier is one that responds to the difference between the two signals applied at its input and ideally rejects signals that are common to the two inputs. The representation of signals in terms of their differential and common-mode components was given in Fig. 2.4. It is repeated here in Fig. 2.15 with slightly different symbols to serve as the input signals for the difference amplifiers we are about to design. Although ideally the difference amplifier will amplify only the differential input signal v_{Id} and reject completely the common-mode input signal v_{Icm}, practical circuits will have an output voltage v_O given by

$$v_O = A_d v_{Id} + A_{cm} v_{Icm} \tag{2.13}$$

where A_d denotes the amplifier differential gain and A_{cm} denotes its common-mode gain (ideally zero). The efficacy of a differential amplifier is measured by the degree of its rejection of common-mode signals in preference to differential signals. This is usually quantified by a measure known as the **common-mode rejection ratio (CMRR)**, defined as

$$\text{CMRR} = 20 \log \frac{|A_d|}{|A_{cm}|} \tag{2.14}$$

[2]The terms *difference* and *differential* are usually used to describe somewhat different amplifier types. For our purposes at this point, the distinction is not sufficiently significant. We will be more precise near the end of this section.

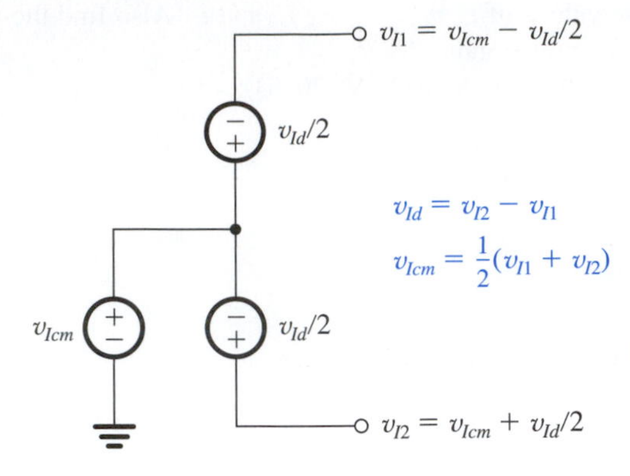

Figure 2.15 Representing the input signals to a differential amplifier in terms of their differential and common-mode components.

The need for difference amplifiers arises frequently in the design of electronic systems, especially those employed in instrumentation. As a common example, consider a transducer providing a small (e.g., 1 mV) signal between its two output terminals while each of the two wires leading from the transducer terminals to the measuring instrument may have a large interference signal (e.g., 1 V) relative to the circuit ground. The instrument front end obviously needs a difference amplifier.

Before we proceed any further we should address a question that the reader might have: The op amp is itself a difference amplifier; why not just use an op amp? The answer is that the very high (ideally infinite) gain of the op amp makes it impossible to use by itself. Rather, as we did before, we have to devise an appropriate feedback network to connect to the op amp to create a circuit whose closed-loop gain is finite, predictable, and stable.

2.4.1 A Single-Op-Amp Difference Amplifier

Our first attempt at designing a difference amplifier is motivated by the observation that the gain of the noninverting amplifier configuration is positive, $(1 + R_2/R_1)$, while that of the inverting configuration is negative, $(-R_2/R_1)$. Combining the two configurations together is then a step in the right direction—namely, getting the difference between two input signals. Of course, we have to make the two gain magnitudes equal in order to reject common-mode signals. This, however, can be easily achieved by attenuating the positive input signal to reduce the gain of the positive path from $(1 + R_2/R_1)$ to (R_2/R_1). The resulting circuit would then look like that shown in Fig. 2.16, where the attenuation in the positive input path is achieved by the voltage divider (R_3, R_4). The proper ratio of this voltage divider can be determined from

$$\frac{R_4}{R_4 + R_3}\left(1 + \frac{R_2}{R_1}\right) = \frac{R_2}{R_1}$$

which can be put in the form

$$\frac{R_4}{R_4 + R_3} = \frac{R_2}{R_2 + R_1}$$

This condition is satisfied by selecting

$$\frac{R_4}{R_3} = \frac{R_2}{R_1} \tag{2.15}$$

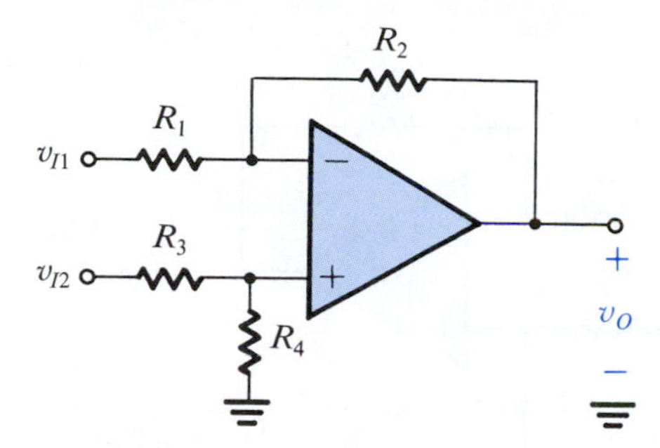

Figure 2.16 A difference amplifier.

This completes our work. However, we have perhaps proceeded a little too fast! Let's step back and verify that the circuit in Fig. 2.16 with R_3 and R_4 selected according to Eq. (2.15) does in fact function as a difference amplifier. Specifically, we wish to determine the output voltage v_O in terms of v_{I1} and v_{I2}. Toward that end, we observe that the circuit is linear, and thus we can use superposition.

To apply superposition, we first reduce v_{I2} to zero—that is, ground the terminal to which v_{I2} is applied—and then find the corresponding output voltage, which will be due entirely to v_{I1}. We denote this output voltage v_{O1}. Its value may be found from the circuit in Fig. 2.17(a), which we recognize as that of the inverting configuration. The existence of R_3 and R_4 does not affect the gain expression, since no current flows through either of them. Thus,

$$v_{O1} = -\frac{R_2}{R_1}v_{I1}$$

Next, we reduce v_{I1} to zero and evaluate the corresponding output voltage v_{O2}. The circuit will now take the form shown in Fig. 2.17(b), which we recognize as the noninverting configuration with an additional voltage divider, made up of R_3 and R_4, connected to the input v_{I2}. The output voltage v_{O2} is therefore given by

$$v_{O2} = v_{I2}\frac{R_4}{R_3 + R_4}\left(1 + \frac{R_2}{R_1}\right) = \frac{R_2}{R_1}v_{I2}$$

where we have utilized Eq. (2.15).

The superposition principle tells us that the output voltage v_O is equal to the sum of v_{O1} and v_{O2}. Thus we have

$$v_O = \frac{R_2}{R_1}(v_{I2} - v_{I1}) = \frac{R_2}{R_1}v_{Id} \tag{2.16}$$

Thus, as expected, the circuit acts as a difference amplifier with a differential gain A_d of

$$A_d = \frac{R_2}{R_1} \tag{2.17}$$

Of course this is predicated on the op amp being ideal and furthermore on the selection of R_3 and R_4 so that their ratio matches that of R_1 and R_2 (Eq. 2.15). To make this matching requirement a little easier to satisfy, we usually select

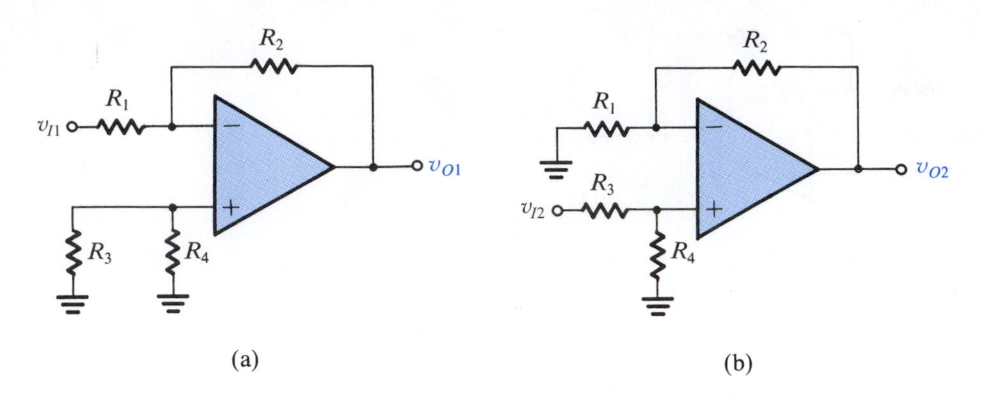

Figure 2.17 Application of superposition to the analysis of the circuit of Fig. 2.16.

$$R_3 = R_1 \quad \text{and} \quad R_4 = R_2$$

Let's next consider the circuit with only a common-mode signal applied at the input, as shown in Fig. 2.18. The figure also shows some of the analysis steps. Thus,

$$i_1 = \frac{1}{R_1}\left[v_{Icm} - \frac{R_4}{R_4+R_3}v_{Icm}\right]$$

$$= v_{Icm}\frac{R_3}{R_4+R_3}\frac{1}{R_1} \tag{2.18}$$

The output voltage can now be found from

$$v_O = \frac{R_4}{R_4+R_3}v_{Icm} - i_2R_2$$

Substituting $i_2 = i_1$ and for i_1 from Eq. (2.18),

$$v_O = \frac{R_4}{R_4+R_3}v_{Icm} - \frac{R_2}{R_1}\frac{R_3}{R_4+R_3}v_{Icm}$$

$$= \frac{R_4}{R_4+R_3}\left(1 - \frac{R_2}{R_1}\frac{R_3}{R_4}\right)v_{Icm}$$

Thus,

$$A_{cm} \equiv \frac{v_O}{v_{Icm}} = \left(\frac{R_4}{R_4+R_3}\right)\left(1 - \frac{R_2}{R_1}\frac{R_3}{R_4}\right) \tag{2.19}$$

For the design with the resistor ratios selected according to Eq. (2.15), we obtain

$$A_{cm} = 0$$

as expected. Note, however, that any mismatch in the resistance ratios can make A_{cm} nonzero, and hence CMRR finite.

In addition to rejecting common-mode signals, a difference amplifier is usually required to have a high input resistance. To find the input resistance between the two input terminals

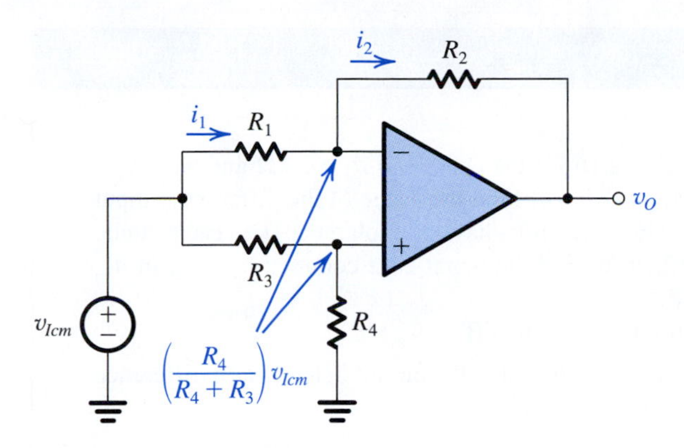

Figure 2.18 Analysis of the difference amplifier to determine its common-mode gain $A_{cm} \equiv v_O/v_{Icm}$.

(i.e., the resistance seen by v_{Id}), called the **differential input resistance** R_{id}, consider Fig. 2.19. Here we have assumed that the resistors are selected so that

$$R_3 = R_1 \quad \text{and} \quad R_4 = R_2$$

Now

$$R_{id} \equiv \frac{v_{Id}}{i_I}$$

Since the two input terminals of the op amp track each other in potential, we may write a loop equation and obtain

$$v_{Id} = R_1 i_I + 0 + R_1 i_I$$

Thus,

$$R_{id} = 2R_1 \tag{2.20}$$

Note that if the amplifier is required to have a large differential gain (R_2/R_1), then R_1 of necessity will be relatively small and the input resistance will be correspondingly low, a drawback of this circuit. Another drawback of the circuit is that it is not easy to vary the differential gain of the amplifier. Both of these drawbacks are overcome in the instrumentation amplifier discussed next.

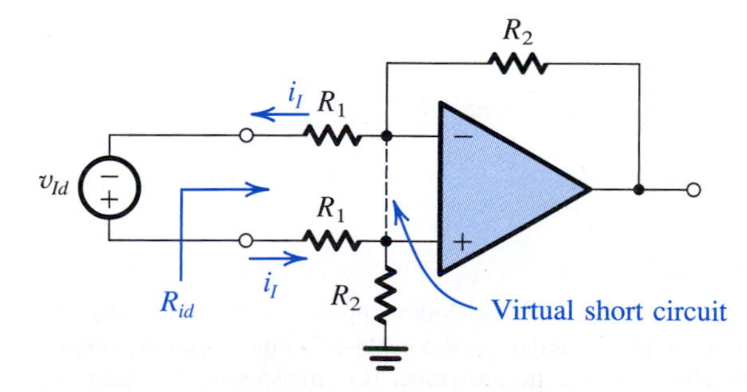

Figure 2.19 Finding the input resistance of the difference amplifier for the case $R_3 = R_1$ and $R_4 = R_2$.

EXERCISES

2.15 Consider the difference-amplifier circuit of Fig. 2.16 for the case $R_1 = R_3 = 2$ kΩ and $R_2 = R_4 = 200$ kΩ. (a) Find the value of the differential gain A_d. (b) Find the value of the differential input resistance R_{id} and the output resistance R_o. (c) If the resistors have 1% tolerance (i.e., each can be within ±1% of its nominal value), use Eq. (2.19) to find the worst-case common-mode gain A_{cm} and hence the corresponding value of CMRR.
Ans. (a) 100 V/V (40 dB); (b) 4 kΩ, 0 Ω; (c) 0.04 V/V, 68 dB

D2.16 Find values for the resistances in the circuit of Fig. 2.16 so that the circuit behaves as a difference amplifier with an input resistance of 20 kΩ and a gain of 10.
Ans. $R_1 = R_3 = 10$ kΩ; $R_2 = R_4 = 100$ kΩ

2.4.2 A Superior Circuit—The Instrumentation Amplifier

The low-input-resistance problem of the difference amplifier of Fig. 2.16 can be solved by using voltage followers to buffer the two input terminals; that is, a voltage follower of the type in Fig. 2.14 is connected between each input terminal and the corresponding input terminal of the difference amplifier. However, if we are going to use two additional op amps, we should ask the question: Can we get more from them than just impedance buffering? An obvious answer would be that we should try to get some voltage gain. It is especially interesting that we can achieve this without compromising the high input resistance simply by using followers with gain rather than unity-gain followers. Achieving some or indeed the bulk of the required gain in this new first stage of the

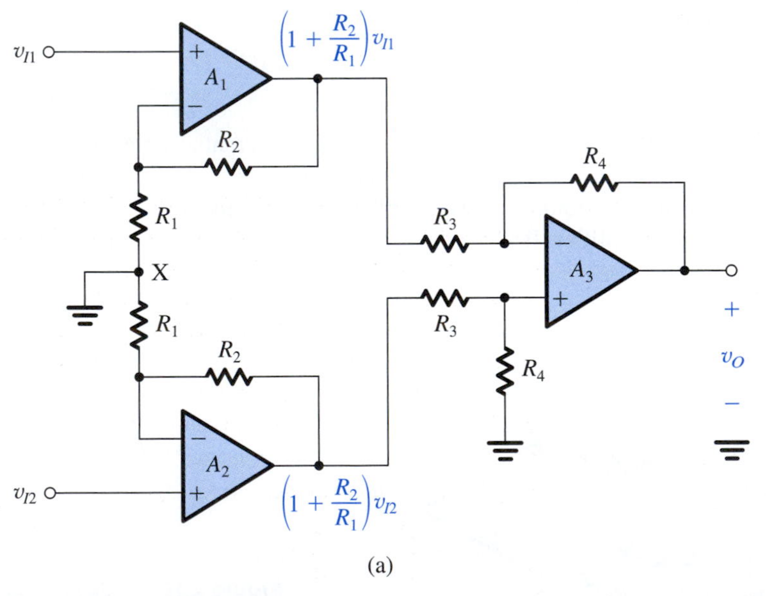

(a)

Figure 2.20 A popular circuit for an instrumentation amplifier. **(a)** Initial approach to the circuit **(b)** The circuit in **(a)** with the connection between node X and ground removed and the two resistors R_1 and R_1 lumped together. This simple wiring change dramatically improves performance. **(c)** Analysis of the circuit in **(b)** assuming ideal op amps.

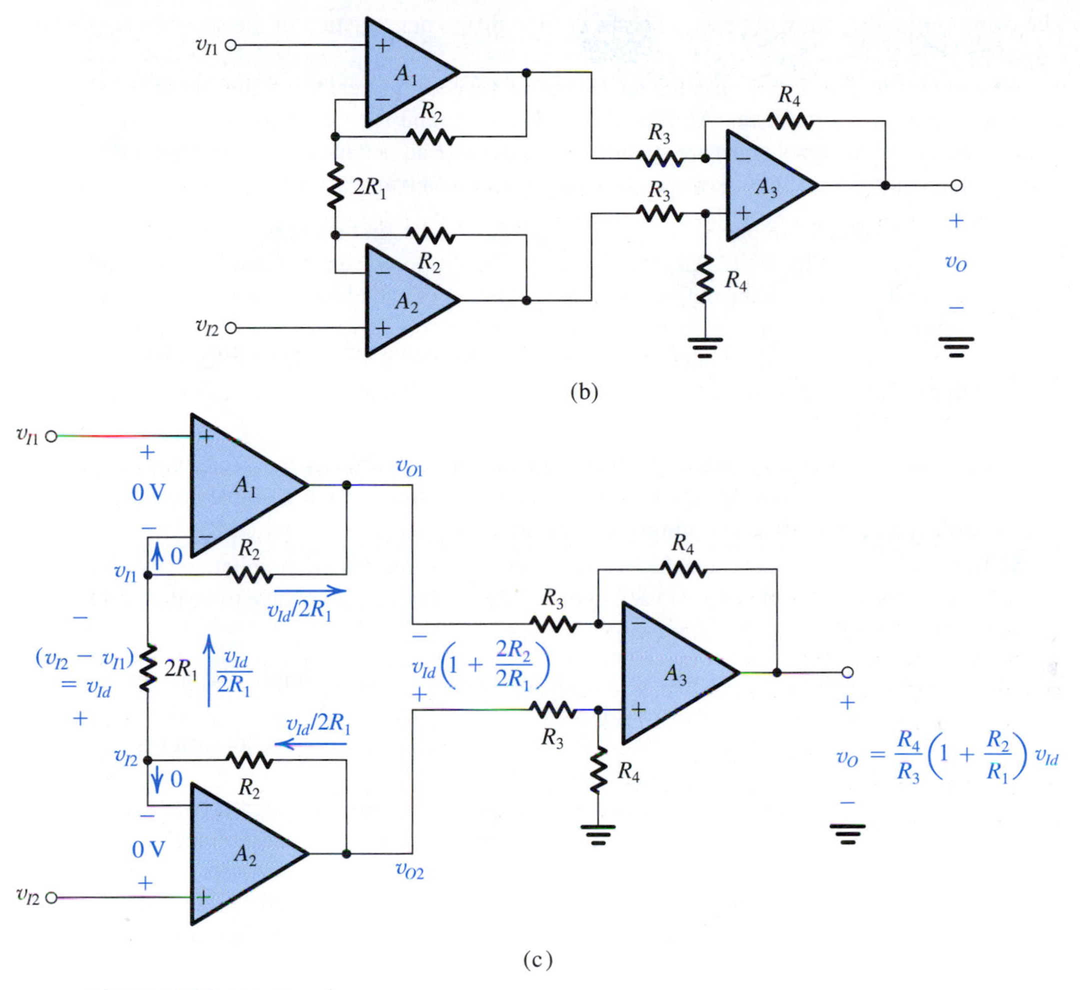

Figure 2.20 *(Continued)*

differential amplifier eases the burden on the difference amplifier in the second stage, leaving it to its main task of implementing the differencing function and thus rejecting common-mode signals.

The resulting circuit is shown in Fig. 2.20(a). It consists of two stages in cascade. The first stage is formed by op amps A_1 and A_2 and their associated resistors, and the second stage is the by-now-familiar difference amplifier formed by op amp A_3 and its four associated resistors. Observe that as we set out to do, each of A_1 and A_2 is connected in the noninverting configuration and thus realizes a gain of $(1 + R_2/R_1)$. It follows that each of v_{I1} and v_{I2} is amplified by this factor, and the resulting amplified signals appear at the outputs of A_1 and A_2, respectively.

The difference amplifier in the second stage operates on the difference signal $(1 + R_2/R_1)(v_{I2} - v_{I1}) = (1 + R_2/R_1)v_{Id}$ and provides at its output

$$v_O = \frac{R_4}{R_3}\left(1 + \frac{R_2}{R_1}\right) v_{Id}$$

Thus the differential gain realized is

$$A_d = \left(\frac{R_4}{R_3}\right)\left(1 + \frac{R_2}{R_1}\right) \tag{2.21}$$

The common-mode gain will be zero because of the differencing action of the second-stage amplifier.

The circuit in Fig. 2.20(a) has the advantage of very high (ideally infinite) input resistance and high differential gain. Also, provided A_1 and A_2 and their corresponding resistors are matched, the two signal paths are symmetric—a definite advantage in the design of a differential amplifier. The circuit, however, has three major disadvantages:

1. The input common-mode signal v_{Icm} is amplified in the first stage by a gain equal to that experienced by the differential signal v_{Id}. This is a very serious issue, for it could result in the signals at the outputs of A_1 and A_3 being of such large magnitudes that the op amps saturate (more on op-amp saturation in Section 2.8). But even if the op amps do not saturate, the difference amplifier of the second stage will now have to deal with much larger common-mode signals, with the result that the CMRR of the overall amplifier will inevitably be reduced.
2. The two amplifier channels in the first stage have to be perfectly matched, otherwise a spurious signal may appear between their two outputs. Such a signal would get amplified by the difference amplifier in the second stage.
3. To vary the differential gain A_d, two resistors have to be varied simultaneously, say the two resistors labeled R_1. At each gain setting the two resistors have to be perfectly matched: a difficult task.

All three problems can be solved with a very simple wiring change: Simply disconnect the node between the two resistors labeled R_1, node X, from ground. The circuit with this small but functionally profound change is redrawn in Fig. 2.20(b), where we have lumped the two resistors (R_1 and R_1) together into a single resistor ($2R_1$).

Analysis of the circuit in Fig. 2.20(b), assuming ideal op amps, is straightforward, as is illustrated in Fig. 2.20(c). The key point is that the virtual short circuits at the inputs of op amps A_1 and A_2 cause the input voltages v_{I1} and v_{I2} to appear at the two terminals of resistor ($2R_1$). Thus the differential input voltage $v_{I2} - v_{I1} \equiv v_{Id}$ appears across $2R_1$ and causes a current $i = v_{Id}/2R_1$ to flow through $2R_1$ and the two resistors labeled R_2. This current in turn produces a voltage difference between the output terminals of A_1 and A_2 given by

$$v_{O2} - v_{O1} = \left(1 + \frac{2R_2}{2R_1}\right) v_{Id}$$

The difference amplifier formed by op amp A_3 and its associated resistors senses the voltage difference ($v_{O2} - v_{O1}$) and provides a proportional output voltage v_O:

$$\begin{aligned} v_O &= \frac{R_4}{R_3}(v_{O2} - v_{O1}) \\ &= \frac{R_4}{R_3}\left(1 + \frac{R_2}{R_1}\right) v_{Id} \end{aligned}$$

Thus the overall differential voltage-gain is given by

$$A_d \equiv \frac{v_O}{v_{Id}} = \frac{R_4}{R_3}\left(1 + \frac{R_2}{R_1}\right) \tag{2.22}$$

Observe that proper differential operation does *not* depend on the matching of the two resistors labeled R_2. Indeed, if one of the two is of different value, say R_2', the expression for A_d becomes

$$A_d = \frac{R_4}{R_3}\left(1 + \frac{R_2 + R_2'}{2R_1}\right) \tag{2.23}$$

Consider next what happens when the two input terminals are connected together to a common-mode input voltage v_{Icm}. It is easy to see that an equal voltage appears at the negative input terminals of A_1 and A_2, causing the current through $2R_1$ to be zero. Thus there will be no current flowing in the R_2 resistors, and the voltages at the output terminals of A_1 and A_2 will be equal to the input (i.e., v_{Icm}). Thus the first stage no longer amplifies v_{Icm}; it simply propagates v_{Icm} to its two output terminals, where they are subtracted to produce a zero common-mode output by A_3. The difference amplifier in the second stage, however, now has a much improved situation at its input: The difference signal has been amplified by $(1 + R_2/R_1)$ while the common-mode voltage remained unchanged.

Finally, we observe from the expression in Eq. (2.22) that the gain can be varied by changing only one resistor, $2R_1$. We conclude that this is an excellent differential amplifier circuit and is widely employed as an instrumentation amplifier; that is, as the input amplifier used in a variety of electronic instruments.

Example 2.3

Design the instrumentation amplifier circuit in Fig. 2.20(b) to provide a gain that can be varied over the range of 2 to 1000 utilizing a 100-kΩ variable resistance (a potentiometer, or "pot" for short).

Solution

It is usually preferable to obtain all the required gain in the first stage, leaving the second stage to perform the task of taking the difference between the outputs of the first stage and thereby rejecting the common-mode signal. In other words, the second stage is usually designed for a gain of 1. Adopting this approach, we select all the second-stage resistors to be equal to a practically convenient value, say 10 kΩ. The problem then reduces to designing the first stage to realize a gain adjustable over the range of 2 to 1000. Implementing $2R_1$ as the series combination of a fixed resistor R_{1f} and the variable resistor R_{1v} obtained using the 100-kΩ pot (Fig. 2.21), we can write

$$1 + \frac{2R_2}{R_{1f} + R_{1v}} = 2 \text{ to } 1000$$

Thus,

$$1 + \frac{2R_2}{R_{1f}} = 1000$$

and

$$1 + \frac{2R_2}{R_{1f} + 100\text{ k}\Omega} = 2$$

These two equations yield $R_{1f} = 100.2\ \Omega$ and $R_2 = 50.050$ kΩ. Other practical values may be selected; for instance, $R_{1f} = 100\ \Omega$ and $R_2 = 49.9$ kΩ (both values are available as standard 1%-tolerance metal-film resistors; see Appendix H) results in a gain covering approximately the required range.

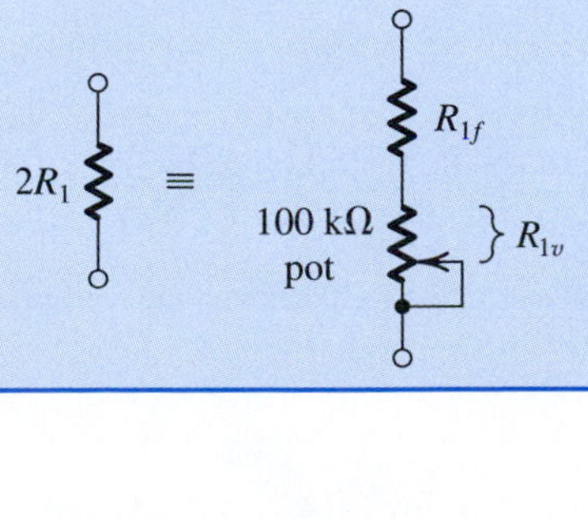

Figure 2.21 To make the gain of the circuit in Fig. 2.20(b) variable, $2R_1$ is implemented as the series combination of a fixed resistor R_{1f} and a variable resistor R_{1v}. Resistor R_{1f} ensures that the maximum available gain is limited.

EXERCISE

2.17 Consider the instrumentation amplifier of Fig. 2.20(b) with a common-mode input voltage of +5 V (dc) and a differential input signal of 10-mV-peak sine wave. Let $(2R_1) = 1\text{ k}\Omega$, $R_2 = 0.5\text{ M}\Omega$, and $R_3 = R_4 = 10\text{ k}\Omega$. Find the voltage at every node in the circuit.
Ans. $v_{I1} = 5 - 0.005 \sin \omega t$; $v_{I2} = 5 + 0.005 \sin \omega t$; v_- (op amp A_1) $= 5 - 0.005 \sin \omega t$; v_- (op amp A_2) $= 5 + 0.005 \sin \omega t$; $v_{O1} = 5 - 5.005 \sin \omega t$; $v_{O2} = 5 + 5.005 \sin \omega t$; $v_-(A_3) = v_+(A_3) = 2.5 + 2.5025 \sin \omega t$; $v_O = 10.01 \sin \omega t$ (all in volts)

2.5 Integrators and Differentiators

The op-amp circuit applications we have studied thus far utilized resistors in the op-amp feedback path and in connecting the signal source to the circuit, that is, in the feed-in path. As a result, circuit operation has been (ideally) independent of frequency. By allowing the use of capacitors together with resistors in the feedback and feed-in paths of op-amp circuits, we open the door to a very wide range of useful and exciting applications of the op amp. We begin our study of op-amp–*RC* circuits by considering two basic applications, namely, signal integrators and differentiators.

2.5.1 The Inverting Configuration with General Impedances

To begin with, consider the inverting closed-loop configuration with impedances $Z_1(s)$ and $Z_2(s)$ replacing resistors R_1 and R_2, respectively. The resulting circuit is shown in Fig. 2.22 and, for an ideal op amp, has the closed-loop gain or, more appropriately, the closed-loop transfer function

$$\frac{V_o(s)}{V_i(s)} = -\frac{Z_2(s)}{Z_1(s)} \tag{2.24}$$

As explained in Section 1.6, replacing s by $j\omega$ provides the transfer function for physical frequencies ω, that is, the transmission magnitude and phase for a sinusoidal input signal of frequency ω.

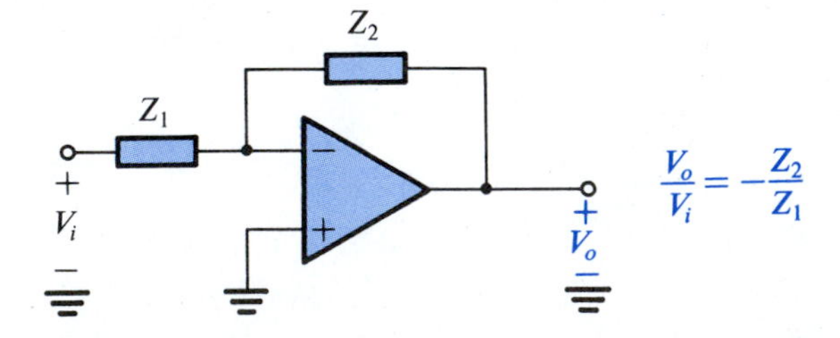

Figure 2.22 The inverting configuration with general impedances in the feedback and the feed-in paths.

Example 2.4

For the circuit in Fig. 2.23, derive an expression for the transfer function $V_o(s)/V_i(s)$. Show that the transfer function is that of a low-pass STC circuit. By expressing the transfer function in the standard form shown in Table 1.2 on page 34, find the dc gain and the 3-dB frequency. Design the circuit to obtain a dc gain of 40 dB, a 3-dB frequency of 1 kHz, and an input resistance of 1 kΩ. At what frequency does the magnitude of transmission become unity? What is the phase angle at this frequency?

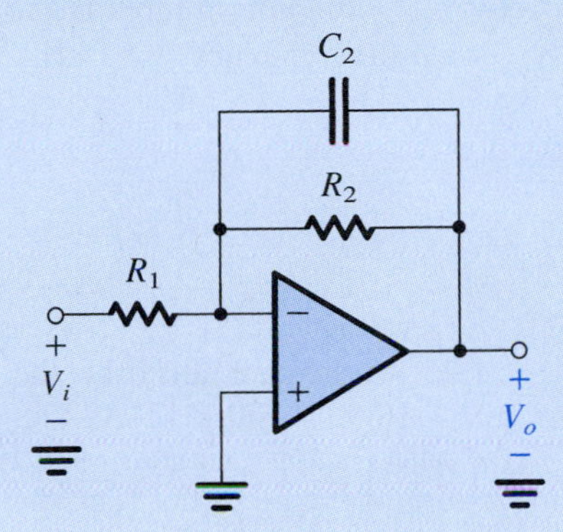

Figure 2.23 Circuit for Example 2.4.

Solution

To obtain the transfer function of the circuit in Fig. 2.23, we substitute in Eq. (2.24), $Z_1 = R_1$ and $Z_2 = R_2 \| (1/sC_2)$. Since Z_2 is the parallel connection of two components, it is more convenient to work in terms of Y_2; that is, we use the following alternative form of the transfer function:

$$\frac{V_o(s)}{V_i(s)} = -\frac{1}{Z_1(s)Y_2(s)}$$

and substitute $Z_1 = R_1$ and $Y_2(s) = (1/R_2) + sC_2$ to obtain

$$\frac{V_o(s)}{V_i(s)} = -\frac{1}{\dfrac{R_1}{R_2} + sC_2R_1}$$

This transfer function is of first order, has a finite dc gain (at $s = 0$, $V_o/V_i = -R_2/R_1$), and has zero gain at infinite frequency. Thus it is the transfer function of a low-pass STC network and can be expressed in the standard form of Table 1.2 as follows:

$$\frac{V_o(s)}{V_i(s)} = \frac{-R_2/R_1}{1 + sC_2R_2}$$

from which we find the dc gain K to be

$$K = -\frac{R_2}{R_1}$$

Example 2.4 *continued*

and the 3-dB frequency ω_0 as

$$\omega_0 = \frac{1}{C_2 R_2}$$

We could have found all this from the circuit in Fig. 2.23 by inspection. Specifically, note that the capacitor behaves as an open circuit at dc; thus at dc the gain is simply $(-R_2/R_1)$. Furthermore, because there is a virtual ground at the inverting input terminal, the resistance seen by the capacitor is R_2, and thus the time constant of the STC network is C_2R_2.

Now to obtain a dc gain of 40 dB, that is, 100 V/V, we select $R_2/R_1 = 100$. For an input resistance of 1 kΩ, we select $R_1 = 1$ kΩ, and thus $R_2 = 100$ kΩ. Finally, for a 3-dB frequency $f_0 = 1$ kHz, we select C_2 from

$$2\pi \times 1 \times 10^3 = \frac{1}{C_2 \times 100 \times 10^3}$$

which yields $C_2 = 1.59$ nF.

The circuit has gain and phase Bode plots of the standard form in Fig. 1.23. As the gain falls off at the rate of –20 dB/decade, it will reach 0 dB in two decades, that is, at $f = 100f_0 = 100$ kHz. As Fig. 1.23(b) indicates, at such a frequency which is much greater than f_0, the phase is approximately –90°. To this, however, we must add the 180° arising from the inverting nature of the amplifier (i.e., the negative sign in the transfer function expression). Thus at 100 kHz, the total phase shift will be –270° or, equivalently, +90°.

2.5.2 The Inverting Integrator

By placing a capacitor in the feedback path (i.e., in place of Z_2 in Fig. 2.22) and a resistor at the input (in place of Z_1), we obtain the circuit of Fig. 2.24(a). We shall now show that this circuit realizes the mathematical operation of integration. Let the input be a time-varying function $v_I(t)$. The virtual ground at the inverting op-amp input causes $v_I(t)$ to appear in effect across R, and thus the current $i_1(t)$ will be $v_I(t)/R$. This current flows through the capacitor C, causing charge to accumulate on C. If we assume that the circuit begins operation at time $t = 0$, then at an arbitrary time t the current $i_1(t)$ will have deposited on C a charge equal to $\int_0^t i_1(t)\,dt$. Thus the capacitor voltage $v_C(t)$ will change by $\frac{1}{C}\int_0^t i_1(t)\,dt$. If the initial voltage on C (at $t = 0$) is denoted V_C, then

$$v_C(t) = V_C + \frac{1}{C}\int_0^t i_1(t)\,dt$$

Now the output voltage $v_O(t) = -v_C(t)$; thus,

$$v_O(t) = -\frac{1}{CR}\int_0^t v_I(t)\,dt - V_C \tag{2.25}$$

Thus the circuit provides an output voltage that is proportional to the time integral of the input, with V_C being the initial condition of integration and CR the **integrator time constant**. Note that, as expected, there is a negative sign attached to the output voltage, and thus this

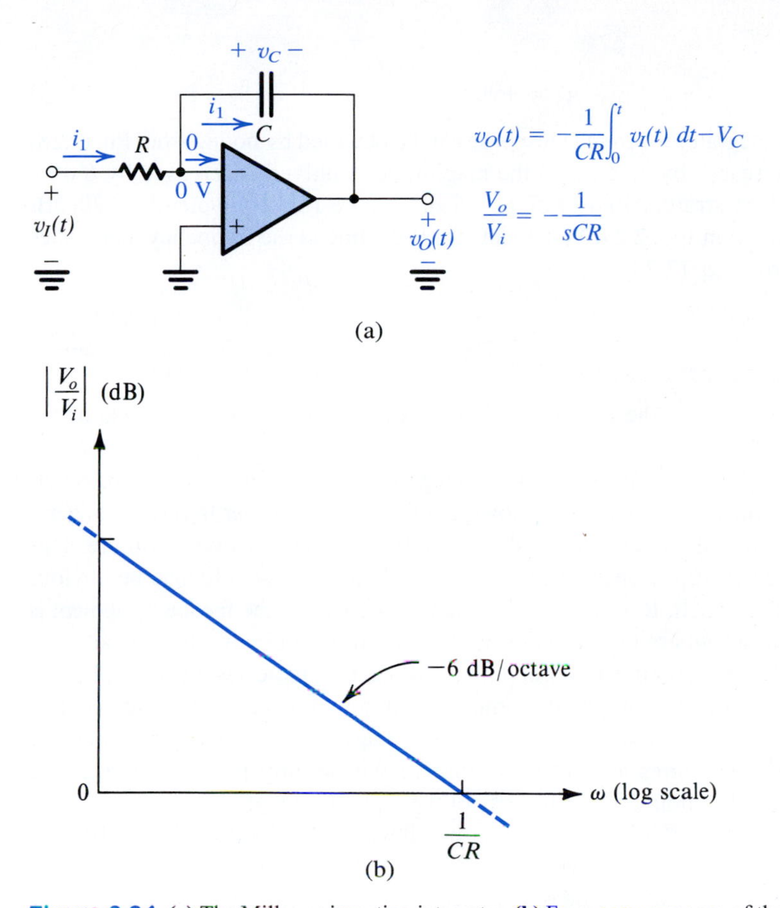

Figure 2.24 **(a)** The Miller or inverting integrator. **(b)** Frequency response of the integrator.

integrator circuit is said to be an **inverting integrator**. It is also known as a **Miller integrator** after an early worker in this field.

The operation of the integrator circuit can be described alternatively in the frequency domain by substituting $Z_1(s) = R$ and $Z_2(s) = 1/sC$ in Eq. (2.24) to obtain the transfer function

$$\frac{V_o(s)}{V_i(s)} = -\frac{1}{sCR} \tag{2.26}$$

For physical frequencies, $s = j\omega$ and

$$\frac{V_o(j\omega)}{V_i(j\omega)} = -\frac{1}{j\omega CR} \tag{2.27}$$

Thus the integrator transfer function has magnitude

$$\left|\frac{V_o}{V_i}\right| = \frac{1}{\omega CR} \tag{2.28}$$

and phase

$$\phi = +90° \tag{2.29}$$

The Bode plot for the integrator magnitude response can be obtained by noting from Eq. (2.28) that as ω doubles (increases by an octave) the magnitude is halved (decreased by 6 dB). Thus the Bode plot is a straight line of slope –6 dB/octave (or, equivalently, –20 dB/decade). This line (shown in Fig. 2.24b) intercepts the 0-dB line at the frequency that makes $|V_o/V_i| = 1$, which from Eq. (2.28) is

$$\omega_{int} = \frac{1}{CR} \tag{2.30}$$

The frequency ω_{int} is known as the **integrator frequency** and is simply the inverse of the integrator time constant.

Comparison of the frequency response of the integrator to that of an STC low-pass network indicates that the integrator behaves as a low-pass filter with a corner frequency of zero. Observe also that at $\omega = 0$, the magnitude of the integrator transfer function is infinite. This indicates that at dc the op amp is operating with an open loop. This should also be obvious from the integrator circuit itself. Reference to Fig. 2.24(a) shows that the feedback element is a capacitor, and thus at dc, where the capacitor behaves as an open circuit, there is no negative feedback! This is a very significant observation and one that indicates a source of problems with the integrator circuit: Any tiny dc component in the input signal will theoretically produce an infinite output. Of course, no infinite output voltage results in practice; rather, the output of the amplifier saturates at a voltage close to the op-amp positive or negative power supply (L_+ or L_-), depending on the polarity of the input dc signal.

The dc problem of the integrator circuit can be alleviated by connecting a resistor R_F across the integrator capacitor C, as shown in Fig. 2.25 and thus the gain at dc will be $-R_F/R$ rather than infinite. Such a resistor provides a dc feedback path. Unfortunately, however, the integration is no longer ideal, and the lower the value of R_F, the less ideal the integrator circuit becomes. This is because R_F causes the frequency of the integrator pole to move from its ideal location at $\omega = 0$ to one determined by the corner frequency of the STC network (R_F, C). Specifically, the integrator transfer function becomes

$$\frac{V_o(s)}{V_i(s)} = -\frac{R_F/R}{1 + sCR_F}$$

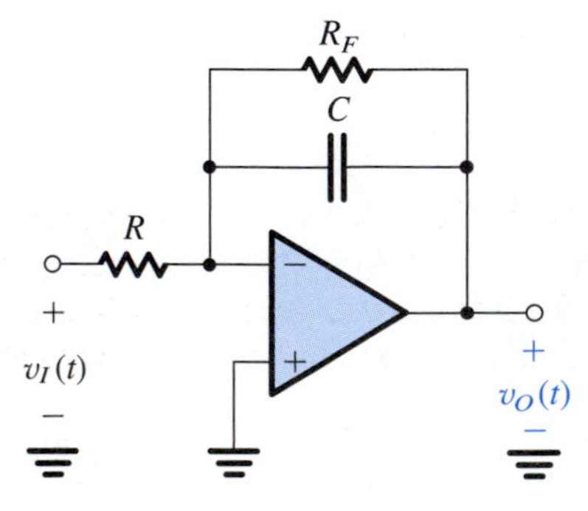

Figure 2.25 The Miller integrator with a large resistance R_F connected in parallel with C in order to provide negative feedback and hence finite gain at dc.

as opposed to the ideal function of $-1/sCR$. The lower the value we select for R_F, the higher the corner frequency $(1/CR_F)$ will be and the more nonideal the integrator becomes. Thus selecting a value for R_F presents the designer with a trade-off between dc performance and signal performance. The effect of R_F on integrator performance is investigated further in the Example 2.5.

Example 2.5

Find the output produced by a Miller integrator in response to an input pulse of 1-V height and 1-ms width [Fig. 2.26(a)]. Let $R = 10\ \text{k}\Omega$ and $C = 10$ nF. If the integrator capacitor is shunted by a 1-MΩ resistor, how will the response be modified? The op amp is specified to saturate at ±13 V.

Solution

In response to a 1-V, 1-ms input pulse, the integrator output will be

$$v_O(t) = -\frac{1}{CR}\int_0^t 1\,dt, \qquad 0 \le t \le 1 \text{ ms}$$

where we have assumed that the initial voltage on the integrator capacitor is 0. For $C = 10$ nF and $R = 10\ \text{k}\Omega$, $CR = 0.1$ ms, and

$$v_O(t) = -10t, \qquad 0 \le t \le 1 \text{ ms}$$

which is the linear ramp shown in Fig. 2.26(b). It reaches a magnitude of −10 V at $t = 1$ ms and remains constant thereafter.

That the output is a linear ramp should also be obvious from the fact that the 1-V input pulse produces a constant current through the capacitor of 1 V/10 kΩ = 0.1 mA. This constant current $I = 0.1$ mA supplies the capacitor with a charge It, and thus the capacitor voltage changes linearly as (It/C), resulting in $v_O = -(I/C)t$. It is worth remembering that charging a capacitor with a constant current produces a linear voltage across it.

Next consider the situation with resistor $R_F = 1\ \text{M}\Omega$ connected across C. As before, the 1-V pulse will provide a constant current $I = 0.1$ mA. Now, however, this current is supplied to an STC network composed of R_F in parallel with C. Thus, the output will be an exponential heading toward −100 V with a time constant of $CR_F = 10 \times 10^{-9} \times 1 \times 10^6 = 10$ ms,

$$v_O(t) = -100(1 - e^{-t/10}), \qquad 0 \le t \le 1 \text{ ms}$$

Of course, the exponential will be interrupted at the end of the pulse, that is, at $t = 1$ ms, and the output will reach the value

$$v_O(1 \text{ ms}) = -100(1 - e^{-1/10}) = -9.5 \text{ V}$$

The output waveform is shown in Fig. 2.26(c), from which we see that including R_F causes the ramp to be slightly rounded such that the output reaches only −9.5 V, 0.5 V short of the ideal value of −10 V. Furthermore, for $t > 1$ ms, the capacitor discharges through R_F with the relatively long time-constant of 10 ms. Finally, we note that op amp saturation, specified to occur at ±13 V, has no effect on the operation of this circuit.

Example 2.5 *continued*

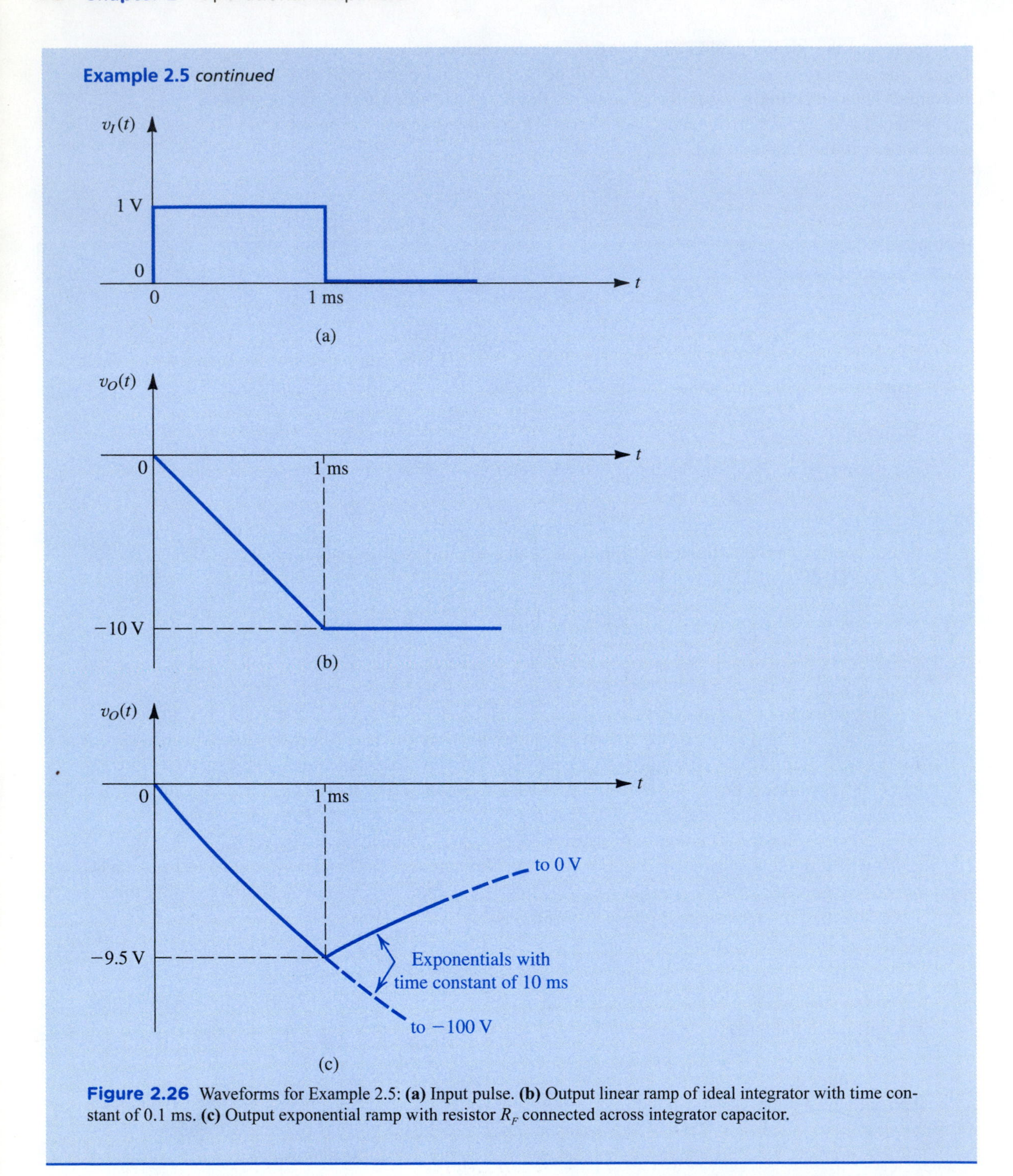

Figure 2.26 Waveforms for Example 2.5: **(a)** Input pulse. **(b)** Output linear ramp of ideal integrator with time constant of 0.1 ms. **(c)** Output exponential ramp with resistor R_F connected across integrator capacitor.

The preceding example hints at an important application of integrators, namely, their use in providing triangular waveforms in response to square-wave inputs. This application is explored in Exercise 2.18. Integrators have many other applications, including their use in the design of filters (Chapter 16).

2.5.3 The Op-Amp Differentiator

Interchanging the location of the capacitor and the resistor of the integrator circuit results in the circuit in Fig. 2.27(a), which performs the mathematical function of differentiation. To see how this comes about, let the input be the time-varying function $v_I(t)$, and note that the virtual ground at the inverting input terminal of the op amp causes $v_I(t)$ to appear in effect across the capacitor C. Thus the current through C will be $C(dv_I/dt)$, and this current flows through the feedback resistor R providing at the op-amp output a voltage $v_O(t)$,

$$v_O(t) = -CR\frac{dv_I(t)}{dt} \tag{2.31}$$

The frequency-domain transfer function of the differentiator circuit can be found by substituting in Eq. (2.24), $Z_1(s) = 1/sC$ and $Z_2(s) = R$ to obtain

$$\frac{V_o(s)}{V_i(s)} = -sCR \tag{2.32}$$

which for physical frequencies $s = j\omega$ yields

$$\frac{V_o(j\omega)}{V_i(j\omega)} = -j\omega CR \tag{2.33}$$

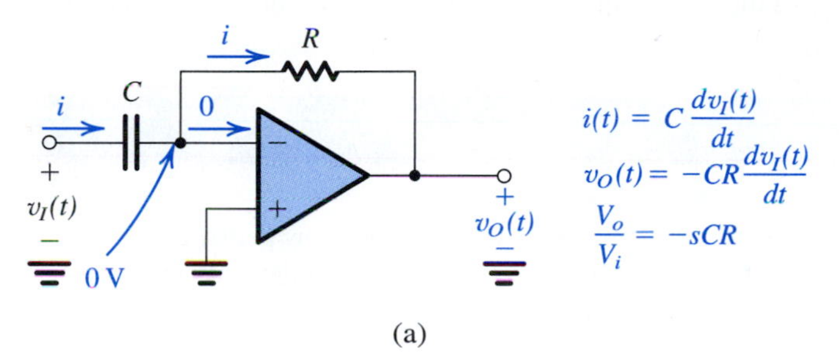

(a)

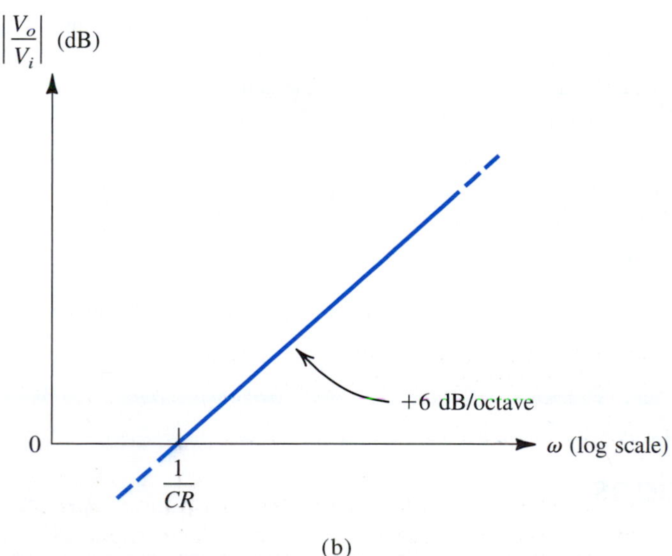

(b)

Figure 2.27 **(a)** A differentiator. **(b)** Frequency response of a differentiator with a time-constant CR.

Thus the transfer function has magnitude

$$\left|\frac{V_o}{V_i}\right| = \omega CR \tag{2.34}$$

and phase

$$\phi = -90° \tag{2.35}$$

The Bode plot of the magnitude response can be found from Eq. (2.34) by noting that for an octave increase in ω, the magnitude doubles (increases by 6 dB). Thus the plot is simply a straight line of slope +6 dB/octave (or, equivalently, +20 dB/decade) intersecting the 0-dB line (where $|V_o/V_i| = 1$) at $\omega = 1/CR$, where CR is the **differentiator time-constant** [see Fig. 2.27(b)].

The frequency response of the differentiator can be thought of as that of an STC highpass filter with a corner frequency at infinity (refer to Fig. 1.24). Finally, we should note that the very nature of a differentiator circuit causes it to be a "noise magnifier." This is due to the spike introduced at the output every time there is a sharp change in $v_I(t)$; such a change could be interference coupled electromagnetically ("picked up") from adjacent signal sources. For this reason and because they suffer from stability problems (Chapter 10), differentiator circuits are generally avoided in practice. When the circuit of Fig. 2.27(a) is used, it is usually necessary to connect a small-valued resistor in series with the capacitor. This modification, unfortunately, turns the circuit into a nonideal differentiator.

EXERCISES

2.18 Consider a symmetrical square wave of 20-V peak-to-peak, 0 average, and 2-ms period applied to a Miller integrator. Find the value of the time constant CR such that the triangular waveform at the output has a 20-V peak-to-peak amplitude.
Ans. 0.5 ms

D2.19 Use an ideal op amp to design an inverting integrator with an input resistance of 10 kΩ and an integration time constant of 10^{-3} s. What is the gain magnitude and phase angle of this circuit at 10 rad/s and at 1 rad/s? What is the frequency at which the gain magnitude is unity?
Ans. $R = 10$ kΩ, $C = 0.1$ μF; at $\omega = 10$ rad/s: $|V_o/V_i| = 100$ V/V and $\phi = +90°$; at $\omega = 1$ rad/s: $|V_o/V_i| = 1{,}000$ V/V and $\phi = +90°$; 1000 rad/s

D2.20 Design a differentiator to have a time constant of 10^{-2} s and an input capacitance of 0.01 μF. What is the gain magnitude and phase of this circuit at 10 rad/s, and at 10^3 rad/s? In order to limit the high-frequency gain of the differentiator circuit to 100, a resistor is added in series with the capacitor. Find the required resistor value.
Ans. $C = 0.01$ μF; $R = 1$ MΩ; at $\omega = 10$ rad/s: $|V_o/V_i| = 0.1$ V/V and $\phi = -90°$; at $\omega = 1000$ rad/s: $|V_o/V_i| = 10$ V/V and $\phi = -90°$; 10 kΩ

2.6 DC Imperfections

Thus far we have considered the op amp to be ideal. The only exception has been a brief discussion of the effect of the op-amp finite gain A on the closed-loop gain of the inverting and noninverting configurations. Although in many applications the assumption of an ideal op amp is not a bad one, a circuit designer has to be thoroughly familiar with the characteristics of practical op amps and the effects of such characteristics on the performance of op-amp circuits. Only then will the designer be able to use the op amp intelligently, especially if the application at hand is not a straightforward one. The nonideal properties of op amps will, of course, limit the range of operation of the circuits analyzed in the previous examples.

In this and the two sections that follow, we consider some of the important nonideal properties of the op amp.[3] We do this by treating one nonideality at a time, beginning in this section with the dc problems to which op amps are susceptible.

2.6.1 Offset Voltage

Because op amps are direct-coupled devices with large gains at dc, they are prone to dc problems. The first such problem is the dc offset voltage. To understand this problem consider the following *conceptual* experiment: If the two input terminals of the op amp are tied together and connected to ground, it will be found that despite the fact that $v_{Id} = 0$, a finite dc voltage exists at the output. In fact, if the op amp has a high dc gain, the output will be at either the positive or negative saturation level. The op-amp output can be brought back to its ideal value of 0 V by connecting a dc voltage source of appropriate polarity and magnitude between the two input terminals of the op amp. This external source balances out the input offset voltage of the op amp. It follows that the **input offset voltage** (V_{OS}) must be of equal magnitude and of opposite polarity to the voltage we applied externally.

The input offset voltage arises as a result of the unavoidable mismatches Present in the input differential stage inside the op amp. In later chapters (in particular Chapters 8 and 12) we shall study this topic in detail. Here, however, our concern is to investigate the effect of V_{OS} on the operation of closed-loop op-amp circuits. Toward that end, we note that general-purpose op amps exhibit V_{OS} in the range of 1 mV to 5 mV. Also, the value of V_{OS} depends on temperature. The op-amp data sheets usually specify typical and maximum values for V_{OS} at room temperature as well as the temperature coefficient of V_{OS} (usually in μV/°C). They do not, however, specify the polarity of V_{OS} because the component mismatches that give rise to V_{OS} are obviously not known a priori; different units of the same op-amp type may exhibit either a positive or a negative V_{OS}.

To analyze the effect of V_{OS} on the operation of op-amp circuits, we need a circuit model for the op amp with input offset voltage. Such a model is shown in Fig. 2.28. It consists of a

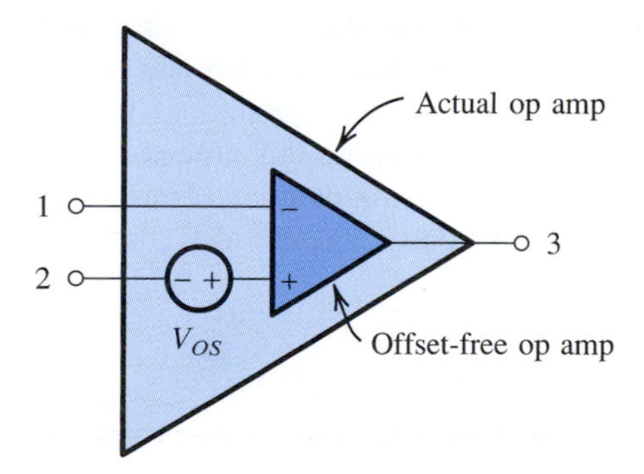

Figure 2.28 Circuit model for an op amp with input offset voltage V_{OS}.

[3]We should note that real op amps have nonideal effects additional to those discussed in this chapter. These include finite (nonzero) common-mode gain or, equivalently, noninfinite CMRR, noninfinite input resistance, and nonzero output resistance. The effect of these, however, on the performance of most of the closed-loop circuits studied here is not very significant, and their study will be postponed to later chapters (in particular Chapters 8, 9, and 12).

dc source of value V_{OS} placed in series with the positive input lead of an offset-free op amp. The justification for this model follows from the description above.

EXERCISE

2.21 Use the model of Fig. 2.28 to sketch the transfer characteristic v_O versus v_{Id} ($v_O \equiv v_3$ and $v_{Id} \equiv v_2 - v_1$) of an op amp having an open-loop dc gain $A_0 = 10^4$ V/V, output saturation levels of ±10 V, and V_{OS} of +5 mV.

Ans. See Fig. E2.21. Observe that true to its name, the input offset voltage causes an offset in the voltage-transfer characteristic; rather than passing through the origin it is now shifted to the left by V_{OS}.

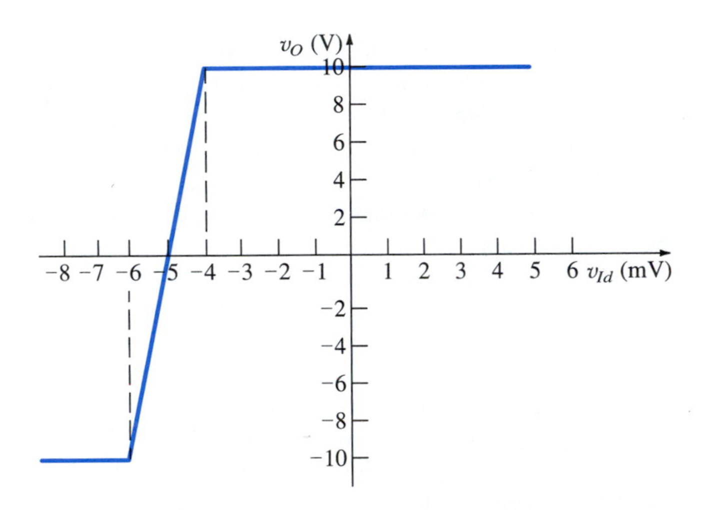

Figure E2.21 Transfer characteristic of an op amp with $V_{OS} = 5$ mV.

Analysis of op-amp circuits to determine the effect of the op-amp V_{OS} on their performance is straightforward: The input voltage signal source is short-circuited and the op amp is replaced with the model of Fig. 2.28. (Eliminating the input signal, done to simplify matters, is based on the principle of superposition.) Following this procedure, we find that both the inverting and the noninverting amplifier configurations result in the same circuit, that shown in Fig. 2.29, from which the output dc voltage due to V_{OS} is found to be

$$V_O = V_{OS}\left[1 + \frac{R_2}{R_1}\right] \tag{2.36}$$

This output dc voltage can have a large magnitude. For instance, a noninverting amplifier with a closed-loop gain of 1000, when constructed from an op amp with a 5-mV input offset voltage, will have a dc output voltage of +5 V or −5 V (depending on the polarity of V_{OS}) rather than the ideal value of 0 V. Now, when an input signal is applied to the

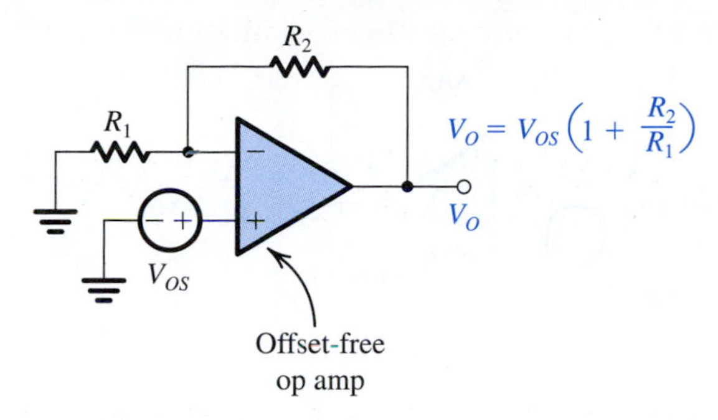

Figure 2.29 Evaluating the output dc offset voltage due to V_{OS} in a closed-loop amplifier.

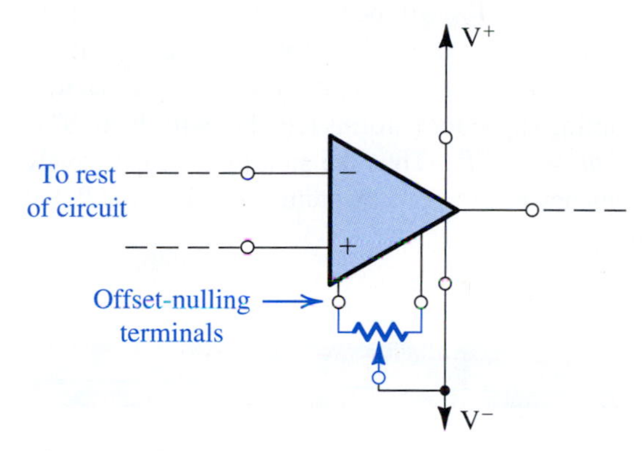

Figure 2.30 The output dc offset voltage of an op amp can be trimmed to zero by connecting a potentiometer to the two offset-nulling terminals. The wiper of the potentiometer is connected to the negative supply of the op amp.

amplifier, the corresponding signal output will be superimposed on the 5-V dc. Obviously then, the allowable signal swing at the output will be reduced. Even worse, if the signal to be amplified is dc, we would not know whether the output is due to V_{OS} or to the signal!

Some op amps are provided with two additional terminals to which a specified circuit can be connected to trim to zero the output dc voltage due to V_{OS}. Figure 2.30 shows such an arrangement that is typically used with general-purpose op amps. A potentiometer is connected between the offset-nulling terminals with the wiper of the potentiometer connected to the op-amp negative supply. Moving the potentiometer wiper introduces an imbalance that counteracts the asymmetry present in the internal op-amp circuitry and that gives rise to V_{OS}. We shall return to this point in the context of our study of the internal circuitry of op amps in Chapter 12. It should be noted, however, that even though the dc output offset can be trimmed to zero, the problem remains of the variation (or drift) of V_{OS} with temperature.

One way to overcome the dc offset problem is by capacitively coupling the amplifier. This, however, will be possible only in applications where the closed-loop amplifier is not required to amplify dc or very-low-frequency signals. Figure 2.31(a) shows a capacitively coupled amplifier. Because of its infinite impedance at dc, the coupling capacitor will cause the gain to be zero at dc. As a result the equivalent circuit for determining the dc output

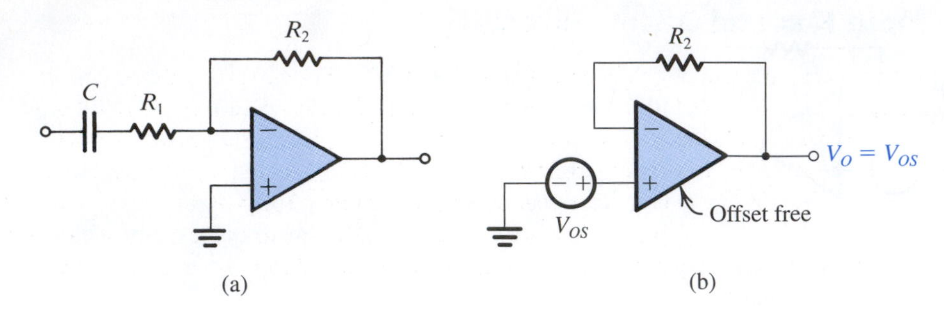

Figure 2.31 **(a)** A capacitively coupled inverting amplifier. **(b)** The equivalent circuit for determining its dc output offset voltage V_O.

voltage resulting from the op-amp input offset voltage V_{OS} will be that shown in Fig. 2.31(b). Thus V_{OS} sees in effect a unity-gain voltage follower, and the dc output voltage V_O will be equal to V_{OS} rather than $V_{OS}(1 + R_2/R_1)$, which is the case without the coupling capacitor. As far as input signals are concerned, the coupling capacitor C forms together with R_1 an STC high-pass circuit with a corner frequency of $\omega_0 = 1/CR_1$. Thus the gain of the capacitively coupled amplifier will fall off at the low-frequency end [from a magnitude of $(1 + R_2/R_1)$ at high frequencies] and will be 3 dB down at ω_0.

EXERCISES

2.22 Consider an inverting amplifier with a nominal gain of 1000 constructed from an op amp with an input offset voltage of 3 mV and with output saturation levels of ±10 V. (a) What is (approximately) the peak sine-wave input signal that can be applied without output clipping? (b) If the effect of V_{OS} is nulled at room temperature (25°C), how large an input can one now apply if: (i) the circuit is to operate at a constant temperature? (ii) the circuit is to operate at a temperature in the range 0°C to 75°C and the temperature coefficient of V_{OS} is 10 μV/°C?

Ans. (a) 7 mV; (b) 10 mV, 9.5 mV

2.23 Consider the same amplifier as in Exercise 2.22—that is, an inverting amplifier with a nominal gain of 1000 constructed from an op amp with an input offset voltage of 3 mV and with output saturation levels of ±10 V—except here let the amplifier be capacitively coupled as in Fig. 2.31(a). (a) What is the dc offset voltage at the output, and what (approximately) is the peak sine-wave signal that can be applied at the input without output clipping? Is there a need for offset trimming? (b) If $R_1 = 1$ kΩ and $R_2 = 1$ MΩ, find the value of the coupling capacitor C_1 that will ensure that the gain will be greater than 57 dB down to 100 Hz.

Ans. (a) 3 mV, 10 mV, no need for offset trimming; (b) 1.6 μF

2.6.2 Input Bias and Offset Currents

The second dc problem encountered in op amps is illustrated in Fig. 2.32. In order for the op amp to operate, its two input terminals have to be supplied with dc currents, termed the **input bias currents.**[4] In Fig. 2.32 these two currents are represented by two current sources, I_{B1} and I_{B2}, connected to the two input terminals. It should be emphasized that the input bias currents are independent of the fact that a real op amp has finite (though large) input resistance (not shown in Fig. 2.32). The op-amp manufacturer usually specifies the average value of I_{B1} and I_{B2} as well as their expected difference. The average value I_B is called the **input bias current**,

$$I_B = \frac{I_{B1} + I_{B2}}{2}$$

and the difference is called the **input offset current** and is given by

$$I_{OS} = |I_{B1} - I_{B2}|$$

Typical values for general-purpose op amps that use bipolar transistors are $I_B = 100$ nA and $I_{OS} = 10$ nA.

We now wish to find the dc output voltage of the closed-loop amplifier due to the input bias currents. To do this we ground the signal source and obtain the circuit shown in Fig. 2.33 for both the inverting and noninverting configurations. As shown in Fig. 2.33, the output dc voltage is given by

$$V_O = I_{B1}R_2 \simeq I_B R_2 \tag{2.37}$$

This obviously places an upper limit on the value of R_2. Fortunately, however, a technique exists for reducing the value of the output dc voltage due to the input bias currents. The method consists of introducing a resistance R_3 in series with the noninverting input lead, as

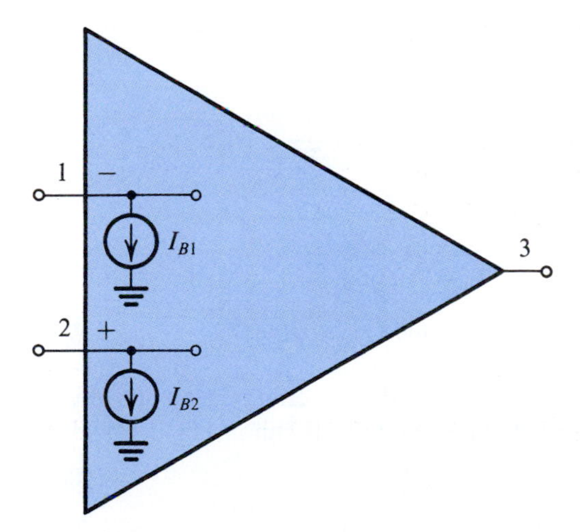

Figure 2.32 The op-amp input bias currents represented by two current sources I_{B1} and I_{B2}.

[4]This is the case for op amps constructed using bipolar junction transistors (BJTs). Those using MOSFETs in the first (input) stage do not draw an appreciable input bias current; nevertheless, the input terminals should have continuous dc paths to ground. More on this in later chapters.

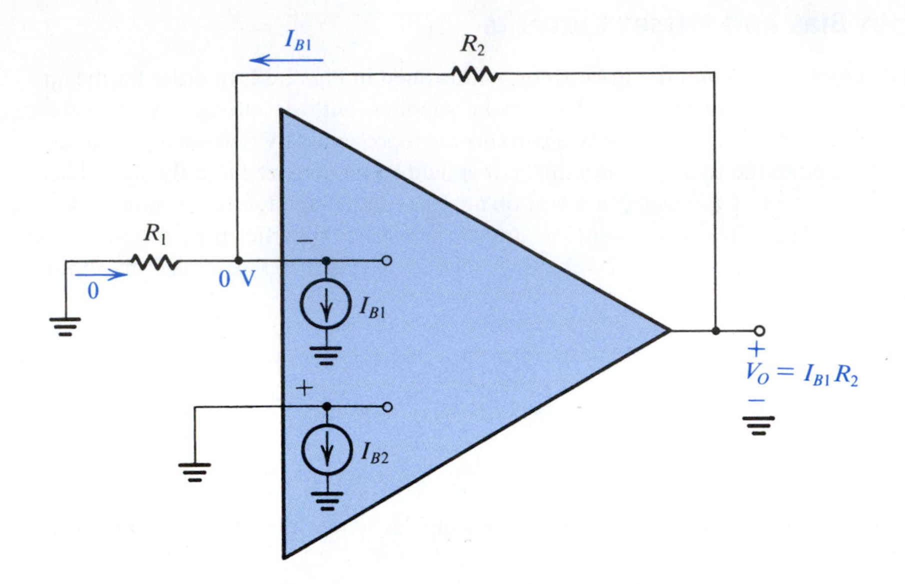

Figure 2.33 Analysis of the closed-loop amplifier, taking into account the input bias currents.

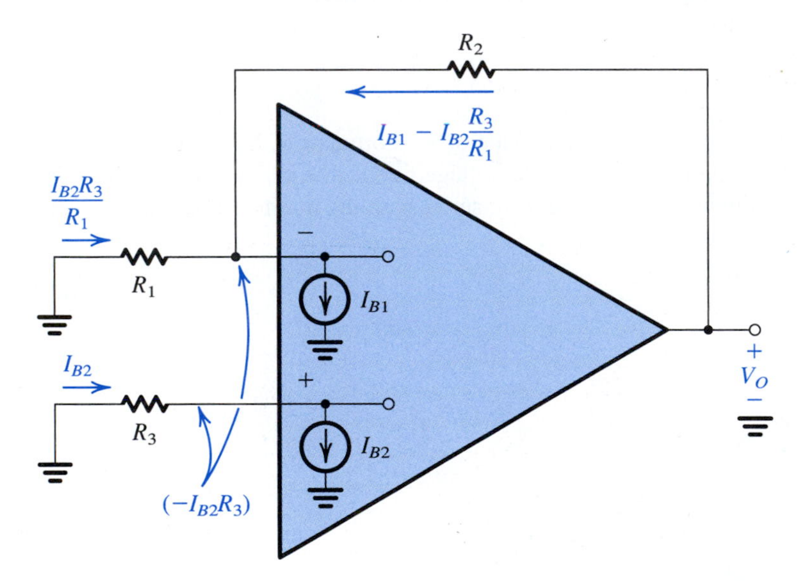

Figure 2.34 Reducing the effect of the input bias currents by introducing a resistor R_3.

shown in Fig. 2.34. From a signal point of view, R_3 has a negligible effect (ideally no effect). The appropriate value for R_3 can be determined by analyzing the circuit in Fig. 2.34, where analysis details are shown, and the output voltage is given by

$$V_O = -I_{B2}R_3 + R_2(I_{B1} - I_{B2}R_3/R_1) \tag{2.38}$$

Consider first the case $I_{B1} = I_{B2} = I_B$, which results in

$$V_O = I_B[R_2 - R_3(1 + R_2/R_1)]$$

Thus we can reduce V_O to zero by selecting R_3 such that

$$R_3 = \frac{R_2}{1 + R_2/R_1} = \frac{R_1 R_2}{R_1 + R_2} \tag{2.39}$$

That is, R_3 should be made equal to the parallel equivalent of R_1 and R_2.

Having selected R_3 as above, let us evaluate the effect of a finite offset current I_{OS}. Let $I_{B1} = I_B + I_{OS}/2$ and $I_{B2} = I_B - I_{OS}/2$, and substitute in Eq. (2.38). The result is

$$V_O = I_{OS} R_2 \tag{2.40}$$

which is usually about an order of magnitude smaller than the value obtained without R_3 (Eq. 2.37). We conclude that to minimize the effect of the input bias currents, one should *place in the positive lead a resistance equal to the equivalant* dc *resistance seen by the inverting terminal.* We emphasize the word *dc* in the last statement; note that if the amplifier is ac-coupled, we should select $R_3 = R_2$, as shown in Fig. 2.35.

While we are on the subject of ac-coupled amplifiers, we should note that one must always provide a continuous dc path between each of the input terminals of the op amp and ground. This is the case no matter how small I_B is. For this reason the ac-coupled noninverting amplifier of Fig. 2.36 will *not* work without the resistance R_3 to ground. Unfortunately, including R_3 lowers considerably the input resistance of the closed-loop amplifier.

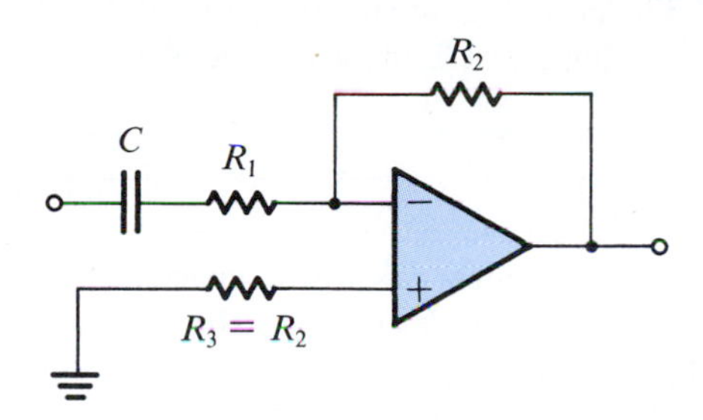

Figure 2.35 In an ac-coupled amplifier the dc resistance seen by the inverting terminal is R_2; hence R_3 is chosen equal to R_2.

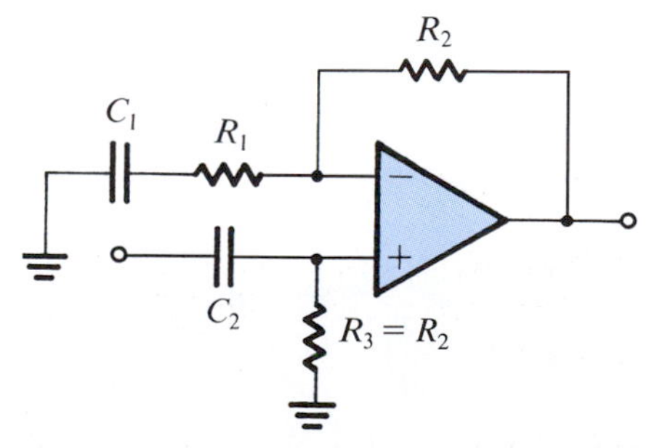

Figure 2.36 Illustrating the need for a continuous dc path for each of the op-amp input terminals. Specifically, note that the amplifier will *not* work without resistor R_3.

EXERCISE

2.24 Consider an inverting amplifier circuit designed using an op amp and two resistors, $R_1 = 10\ \text{k}\Omega$ and $R_2 = 1\ \text{M}\Omega$. If the op amp is specified to have an input bias current of 100 nA and an input offset current of 10 nA, find the output dc offset voltage resulting and the value of a resistor R_3 to be placed in series with the positive input lead in order to minimize the output offset voltage. What is the new value of V_O?
Ans. 0.1 V; 9.9 kΩ (≃ 10 kΩ); 0.01 V

2.6.3 Effect of V_{OS} and I_{OS} on the Operation of the Inverting Integrator

Our discussion of the inverting integrator circuit in Section 2.5.2 mentioned the susceptibility of this circuit to saturation in the presence of small dc voltages or currents. It behooves us therefore to consider the effect of the op-amp dc offsets on its operation. As will be seen, these effects can be quite dramatic.

To see the effect of the input dc offset voltage V_{OS}, consider the integrator circuit in Fig. 2.38, where for simplicity we have short-circuited the input signal source. Analysis of the circuit is straightforward and is shown in Fig. 2.37. Assuming for simplicity that at time $t = 0$ the voltage across the capacitor is zero, the output voltage as a function of time is given by

$$v_O = V_{OS} + \frac{V_{OS}}{CR}t \tag{2.41}$$

Thus v_O increases linearly with time until the op amp saturates—clearly an unacceptable situation! As should be expected, the dc input offset current I_{OS} produces a similar problem. Figure 2.38 illustrates the situation. Observe that we have added a resistance R in the op-amp positive-input lead in order to keep the input bias current I_B from flowing through C. Nevertheless, the offset current I_{OS} will flow through C and cause v_O to ramp linearly with time until the op amp saturates.

As mentioned in Section 2.5.2 the dc problem of the integrator circuit can be alleviated by connecting a resistor R_F across the integrator capacitor C, as shown in Fig. 2.25.

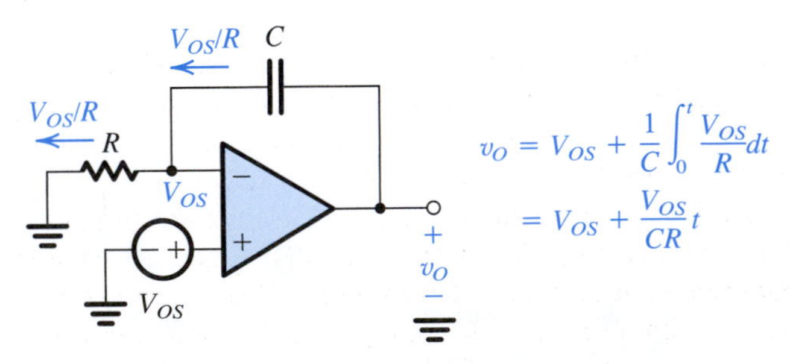

$$v_O = V_{OS} + \frac{1}{C}\int_0^t \frac{V_{OS}}{R}dt$$
$$= V_{OS} + \frac{V_{OS}}{CR}t$$

Figure 2.37 Determining the effect of the op-amp input offset volage V_{OS} on the Miller integrator circuit. Note that since the output rises with time, the op amp eventually saturates.

Such a resistor provides a dc path through which the dc currents (V_{OS}/R) and I_{OS} can flow, with the result that v_O will now have a dc component $[V_{OS}(1+R_F/R)+I_{OS}R_F]$ instead of rising linearly. To keep the dc offset at the output small, one would select a low value for R_F. Unfortunately, however, the lower the value of R_F, the less ideal the integrator circuit becomes.

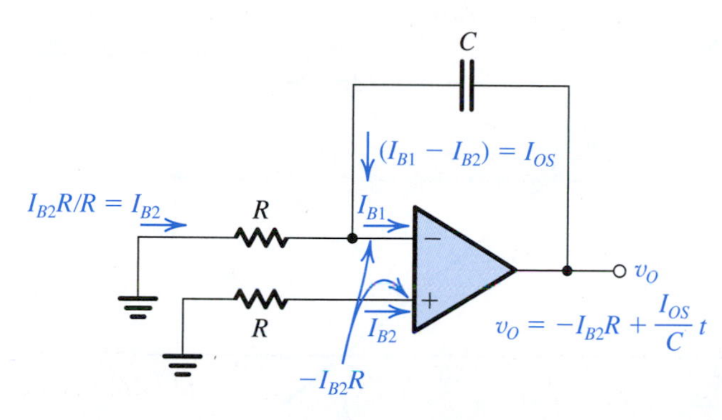

Figure 2.38 Effect of the op-amp input bias and offset currents on the performance of the Miller integrator circuit.

EXERCISE

2.25 Consider a Miller integrator with a time constant of 1 ms and an input resistance of 10 kΩ. Let the op amp have V_{OS} = 2 mV and output saturation voltages of ±12 V. (a) Assuming that when the power supply is turned on the capacitor voltage is zero, how long does it take for the amplifier to saturate? (b) Select the largest possible value for a feedback resistor R_F so that at least ±10 V of output signal swing remains available. What is the corner frequency of the resulting STC network?

Ans. (a) 6 s; (b) 10 MΩ, 0.16 Hz

2.7 Effect of Finite Open-Loop Gain and Bandwidth on Circuit Performance

2.7.1 Frequency Dependence of the Open-Loop Gain

The differential open-loop gain A of an op amp is not infinite; rather, it is finite and decreases with frequency. Figure 2.39 shows a plot for $|A|$, with the numbers typical of some commercially available general-purpose op amps (such as the popular 741-type op amp, available from many semiconductor manufacturers; its internal circuit is studied in Chapter 12).

Note that although the gain is quite high at dc and low frequencies, it starts to fall off at a rather low frequency (10 Hz in our example). The uniform –20-dB/decade gain rolloff shown is typical of **internally compensated** op amps. These are units that have a network (usually a single capacitor) included within the same IC chip whose function is to cause the op-amp gain to have the single-time-constant (STC) low-pass response shown. This

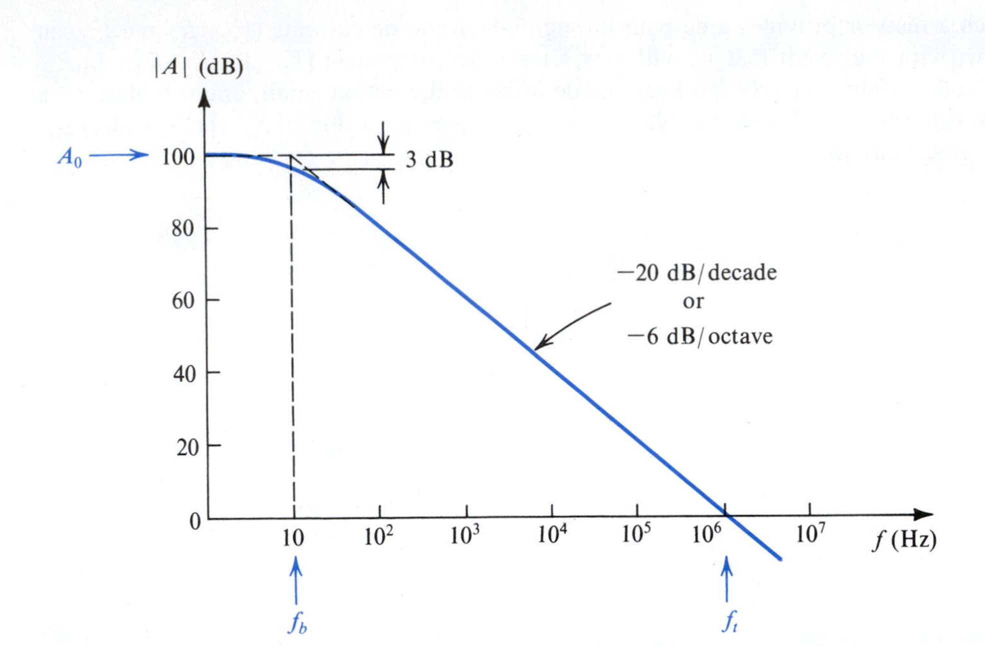

Figure 2.39 Open-loop gain of a typical general-purpose internally compensated op amp.

process of modifying the open-loop gain is termed **frequency compensation**, and its purpose is to ensure that op-amp circuits will be stable (as opposed to oscillatory). The subject of stability of op-amp circuits—or, more generally, of feedback amplifiers—will be studied in Chapter 10.

By analogy to the response of low-pass STC circuits (see Section 1.6 and, for more detail, Appendix E), the gain $A(s)$ of an internally compensated op amp may be expressed as

$$A(s) = \frac{A_0}{1 + s/\omega_b} \tag{2.42}$$

which for physical frequencies, $s = j\omega$, becomes

$$A(j\omega) = \frac{A_0}{1 + j\omega/\omega_b} \tag{2.43}$$

where A_0 denotes the dc gain and ω_b is the 3-dB frequency (corner frequency or "break" frequency). For the example shown in Fig. 2.39, $A_0 = 10^5$ and $\omega_b = 2\pi \times 10$ rad/s. For frequencies $\omega \gg \omega_b$ (about 10 times and higher) Eq. (2.43) may be approximated by

$$A(j\omega) \simeq \frac{A_0\omega_b}{j\omega} \tag{2.44}$$

Thus,

$$|A(j\omega)| = \frac{A_0\omega_b}{\omega} \tag{2.45}$$

from which it can be seen that the gain $|A|$ reaches unity (0 dB) at a frequency denoted by ω_t and given by

$$\omega_t = A_0\omega_b \tag{2.46}$$

Substituting in Eq. (2.44) gives

$$A(j\omega) \simeq \frac{\omega_t}{j\omega} \qquad (2.47)$$

The frequency $f_t = \omega_t/2\pi$ is usually specified on the data sheets of commercially available op amps and is known as the **unity-gain bandwidth**.[5] Also note that for $\omega \gg \omega_b$ the open-loop gain in Eq. (2.42) becomes

$$A(s) \simeq \frac{\omega_t}{s} \qquad (2.48)$$

The gain magnitude can be obtained from Eq. (2.47) as

$$|A(j\omega)| \simeq \frac{\omega_t}{\omega} = \frac{f_t}{f} \qquad (2.49)$$

Thus if f_t is known (10^6 Hz in our example), one can easily determine the magnitude of the op-amp gain at a given frequency f. Furthermore, observe that this relationship correlates with the Bode plot in Fig. 2.39. Specifically, for $f \gg f_b$, doubling f (an octave increase) results in halving the gain (a 6-dB reduction). Similarly, increasing f by a factor of 10 (a decade increase) results in reducing $|A|$ by a factor of 10 (20 dB).

As a matter of practical importance, we note that the production spread in the value of f_t between op-amp units of the same type is usually much smaller than that observed for A_0 and f_b. For this reason f_t is preferred as a specification parameter. Finally, it should be mentioned that an op amp having this uniform –6-dB/octave (or equivalently –20-dB/decade) gain rolloff is said to have a **single-pole model**. Also, since this single pole *dominates* the amplifier frequency response, it is called a *dominant pole*. For more on poles (and zeros), the reader may wish to consult Appendix F.

EXERCISE

2.26 An internally compensated op amp is specified to have an open-loop dc gain of 106 dB and a unity-gain bandwidth of 3 MHz. Find f_b and the open-loop gain (in dB) at f_b, 300 Hz, 3 kHz, 12 kHz, and 60 kHz.

Ans. 15 Hz; 103 dB; 80 dB; 60 dB; 48 dB; 34 dB

2.7.2 Frequency Response of Closed-Loop Amplifiers

We next consider the effect of limited op-amp gain and bandwidth on the closed-loop transfer functions of the two basic configurations: the inverting circuit of Fig. 2.5 and the noninverting circuit of Fig. 2.12. The closed-loop gain of the inverting amplifier, assuming a finite op-amp open-loop gain A, was derived in Section 2.2 and given in Eq. (2.5), which we repeat here as

$$\frac{V_o}{V_i} = \frac{-R_2/R_1}{1 + (1 + R_2/R_1)/A} \qquad (2.50)$$

[5]Since f_t is the product of the dc gain A_0 and the 3-dB bandwidth f_b (where $f_b = \omega_b/2\pi$), it is also known as the **gain–bandwidth product** (GB). The reader is cautioned, however, that in some amplifiers, the unity-gain frequency and the gain-bandwidth product are *not* equal.

Substituting for A from Eq. (2.42) and using Eq. (2.46) gives

$$\frac{V_o(s)}{V_i(s)} = \frac{-R_2/R_1}{1+\frac{1}{A_0}\left(1+\frac{R_2}{R_1}\right)+\frac{s}{\omega_t/(1+R_2/R_1)}} \tag{2.51}$$

For $A_0 \gg 1+R_2/R_1$, which is usually the case,

$$\frac{V_o(s)}{V_i(s)} \approx \frac{-R_2/R_1}{1+\frac{s}{\omega_t/(1+R_2/R_1)}} \tag{2.52}$$

which is of the same form as that for a low-pass STC network (see Table 1.2, page 34). Thus the inverting amplifier has an STC low-pass response with a dc gain of magnitude equal to R_2/R_1. The closed-loop gain rolls off at a uniform –20-dB/decade slope with a corner frequency (3-dB frequency) given by

$$\omega_{3\text{dB}} = \frac{\omega_t}{1+R_2/R_1} \tag{2.53}$$

Similarly, analysis of the noninverting amplifier of Fig. 2.12, assuming a finite open-loop gain A, yields the closed-loop transfer function

$$\frac{V_o}{V_i} = \frac{1+R_2/R_1}{1+(1+R_2/R_1)/A} \tag{2.54}$$

Substituting for A from Eq. (2.42) and making the approximation $A_0 \gg 1+R_2/R_1$ results in

$$\frac{V_o(s)}{V_i(s)} \simeq \frac{1+R_2/R_1}{1+\frac{s}{\omega_t/(1+R_2/R_1)}} \tag{2.55}$$

Thus the noninverting amplifier has an STC low-pass response with a dc gain of $(1+R_2/R_1)$ and a 3-dB frequency given also by Eq. (2.53).

Example 2.6

Consider an op amp with $f_t = 1$ MHz. Find the 3-dB frequency of closed-loop amplifiers with nominal gains of +1000, +100, +10, +1, −1, −10, −100, and −1000. Sketch the magnitude frequency response for the amplifiers with closed-loop gains of +10 and −10.

Solution

We use Eq. (2.53) to obtain the results given in the following table.

Closed-Loop Gain	R_2/R_1	$f_{3\text{dB}} = f_t/(1+R_2/R_1)$
+1000	999	1 kHz
+100	99	10 kHz
+10	9	100 kHz
+1	0	1 MHz
−1	1	0.5 MHz
−10	10	90.9 kHz
−100	100	9.9 kHz
−1000	1000	≃1 kHz

Figure 2.40 shows the frequency response for the amplifier whose nominal dc gain is +10 (20 dB), and Fig. 2.41 shows the frequency response for the –10 (also 20 dB) case. An interesting observation follows from the table above: The unity-gain inverting amplifier has a 3-dB frequency of $f_t/2$ as compared to f_t for the unity-gain noninverting amplifier (the unity-gain voltage follower).

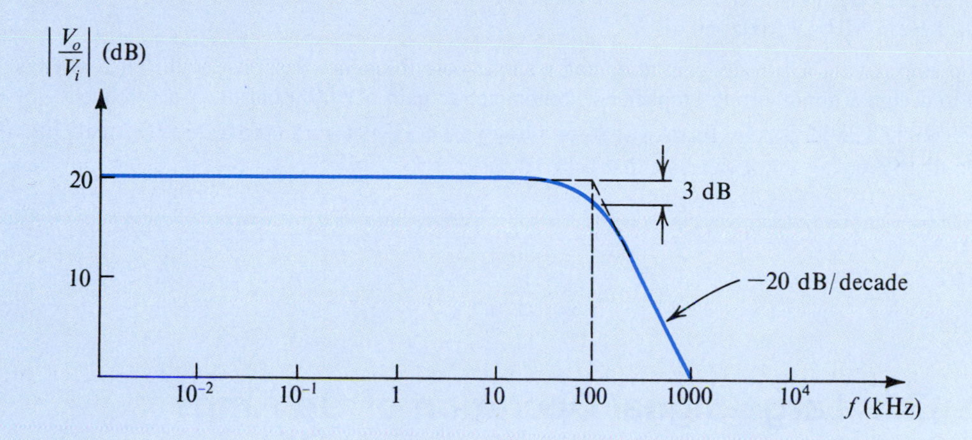

Figure 2.40 Frequency response of an amplifier with a nominal gain of +10 V/V.

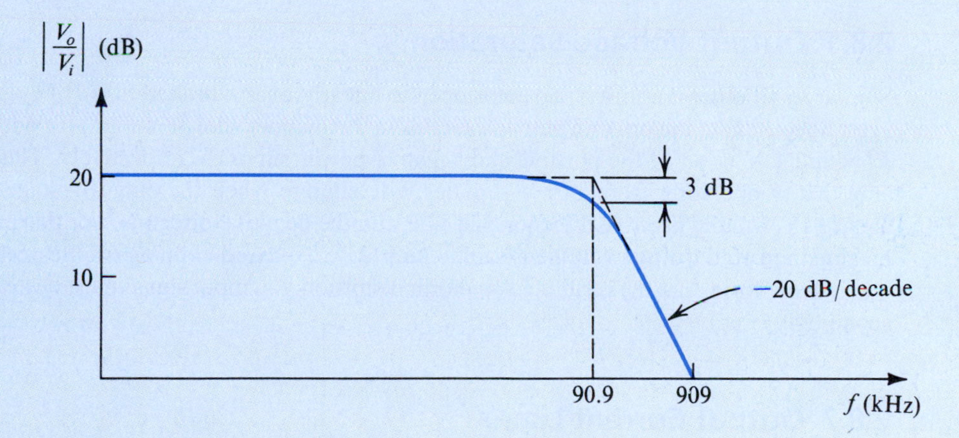

Figure 2.41 Frequency response of an amplifier with a nominal gain of –10 V/V.

The table in Example 2.6 above clearly illustrates the trade-off between gain and bandwidth: For a given op amp, the lower the closed-loop gain required, the wider the bandwidth achieved. Indeed, the noninverting configuration exhibits a constant **gain–bandwidth product** equal to f_t of the op amp. An interpretation of these results in terms of feedback theory will be given in Chapter 10.

EXERCISES

2.27 An internally compensated op amp has a dc open-loop gain of 10^6 V/V and an ac open-loop gain of 40 dB at 10 kHz. Estimate its 3-dB frequency, its unity-gain frequency, its gain–bandwidth product, and its expected gain at 1 kHz.
Ans. 1 Hz; 1 MHz; 1 MHz; 60 dB

2.28 An op amp having a 106-dB gain at dc and a single-pole frequency response with $f_t = 2$ MHz is used to design a noninverting amplifier with nominal dc gain of 100. Find the 3-dB frequency of the closed-loop gain.
Ans. 20 kHz

2.8 Large-Signal Operation of Op Amps

In this section, we study the limitations on the performance of op-amp circuits when large output signals are present.

2.8.1 Output Voltage Saturation

Similar to all other amplifiers, op amps operate linearly over a limited range of output voltages. Specifically, the op-amp output saturates in the manner shown in Fig. 1.14 with L_+ and L_- within 1 V or so of the positive and negative power supplies, respectively. Thus, an op amp that is operating from ±15-V supplies will saturate when the output voltage reaches about +13 V in the positive direction and –13 V in the negative direction. For this particular op amp the **rated output voltage** is said to be ±13 V. To avoid clipping off the peaks of the output waveform, and the resulting waveform distortion, the input signal must be kept correspondingly small.

2.8.2 Output Current Limits

Another limitation on the operation of op amps is that their output current is limited to a specified maximum. For instance, the popular 741 op amp is specified to have a maximum output current of ±20 mA. Thus, in designing closed-loop circuits utilizing the 741, the designer has to ensure that under no condition will the op amp be required to supply an output current, in either direction, exceeding 20 mA. This, of course, has to include both the current in the feedback circuit as well as the current supplied to a load resistor. If the circuit requires a larger current, the op-amp output voltage will saturate at the level corresponding to the maximum allowed output current.

Example 2.7

Consider the noninverting amplifier circuit shown in Fig. 2.42. As shown, the circuit is designed for a nominal gain $(1 + R_2/R_1) = 10$ V/V. It is fed with a low-frequency sine-wave signal of peak voltage V_p and is connected to a load resistor R_L. The op amp is specified to have output saturation voltages of ±13 V and output current limits of ±20 mA.

(a) For $V_p = 1$ V and $R_L = 1$ kΩ, specify the signal resulting at the output of the amplifier.
(b) For $V_p = 1.5$ V and $R_L = 1$ kΩ, specify the signal resulting at the output of the amplifier.
(c) For $R_L = 1$ kΩ, what is the maximum value of V_p for which an undistorted sine-wave output is obtained?
(d) For $V_p = 1$ V, what is the lowest value of R_L for which an undistorted sine-wave output is obtained?

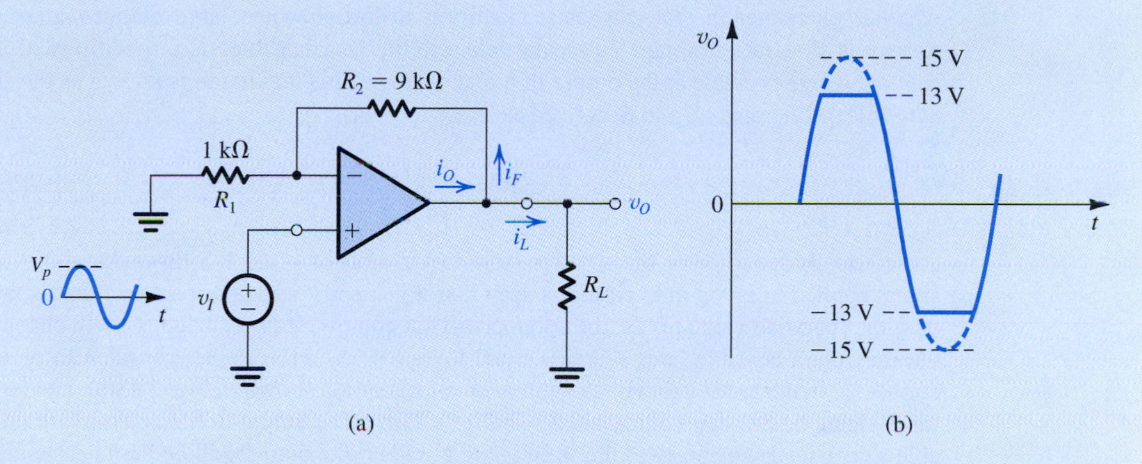

Figure 2.42 **(a)** A noninverting amplifier with a nominal gain of 10 V/V designed using an op amp that saturates at ±13-V output voltage and has ±20-mA output current limits. **(b)** When the input sine wave has a peak of 1.5 V, the output is clipped off at ±13 V.

Solution

(a) For $V_p = 1$ V and $R_L = 1$ kΩ, the output will be a sine wave with peak value of 10 V. This is lower than output saturation levels of ±13 V, and thus the amplifier is not limited that way. Also, when the output is at its peak (10 V), the current in the load will be 10 V/1 kΩ = 10 mA, and the current in the feedback network will be 10 V/(9 + 1) kΩ = 1 mA, for a total op-amp output current of 11 mA, well under its limit of 20 mA.

(b) Now if V_p is increased to 1.5 V, ideally the output would be a sine wave of 15-V peak. The op amp, however, will saturate at ±13 V, thus clipping the sine-wave output at these levels. Let's next check on the op-amp output current: At 13-V output and $R_L = 1$ kΩ, $i_L = 13$ mA and $i_F = 1.3$ mA; thus $i_O = 14.3$ mA, again under the 20-mA limit. Thus the output will be a sine wave with its peaks clipped off at ±13 V, as shown in Fig. 2.42(b).

Example 2.7 *continued*

(c) For $R_L = 1\ \text{k}\Omega$, the maximum value of V_p for undistorted sine-wave output is 1.3 V. The output will be a 13-V peak sine wave, and the op-amp output current at the peaks will be 14.3 mA.

(d) For $V_p = 1$ V and R_L reduced, the lowest value possible for R_L while the output is remaining an undistorted sine wave of 10-V peak can be found from

$$i_{O\max} = 20\ \text{mA} = \frac{10\ \text{V}}{R_{L\min}} + \frac{10\ \text{V}}{9\ \text{k}\Omega + 1\ \text{k}\Omega}$$

which results in

$$R_{L\min} = 526\ \Omega$$

2.8.3 Slew Rate

Another phenomenon that can cause nonlinear distortion when large output signals are present is slew-rate limiting. The name refers to the fact that there is a specific *maximum rate of change* possible at the output of a real op amp. This maximum is known as the **slew rate** (SR) of the op amp and is defined as

$$\text{SR} = \left.\frac{dv_O}{dt}\right|_{\max} \tag{2.56}$$

and is usually specified on the op-amp data sheet in units of V/μs. It follows that if the input signal applied to an op-amp circuit is such that it demands an output response that is faster than the specified value of SR, the op amp will not comply. Rather, its output will change at the maximum possible rate, which is equal to its SR. As an example, consider an op amp connected in the unity-gain voltage-follower configuration shown in Fig. 2.43(a), and let the input signal be the step voltage shown in Fig. 2.43(b). The output of the op amp will not be able to rise instantaneously to the ideal value V; rather, the output will be the linear ramp of slope equal to SR, shown in Fig. 2.43(c). The amplifier is then said to be **slewing**, and its output is **slew-rate limited**.

In order to understand the origin of the slew-rate phenomenon, we need to know about the internal circuit of the op amp, and we will study it in Chapter 12. For the time being, however, it is sufficient to know about the phenomenon and to note that it is distinct from the finite op-amp bandwidth that limits the frequency response of the closed-loop amplifiers, studied in the previous section. The limited bandwidth is a linear phenomenon and does not result in a change in the shape of an input sinusoid; that is, it does not lead to nonlinear distortion. The slew-rate limitation, on the other hand, can cause nonlinear distortion to an input sinusoidal signal when its frequency and amplitude are such that the corresponding ideal output would require v_O to change at a rate greater than SR. This is the origin of another related op-amp specification, its full-power bandwidth, to be explained later.

Before leaving the example in Fig. 2.43, however, we should point out that if the step input voltage V is sufficiently small, the output can be the exponentially rising ramp shown in Fig. 2.43(d). Such an output would be expected from the follower if the only limitation on its dynamic performance were the finite op-amp bandwidth. Specifically, the transfer function of the follower can be found by substituting $R_1 = \infty$ and $R_2 = 0$ in Eq. (2.55) to obtain

$$\frac{V_o}{V_i} = \frac{1}{1 + s/\omega_t} \tag{2.57}$$

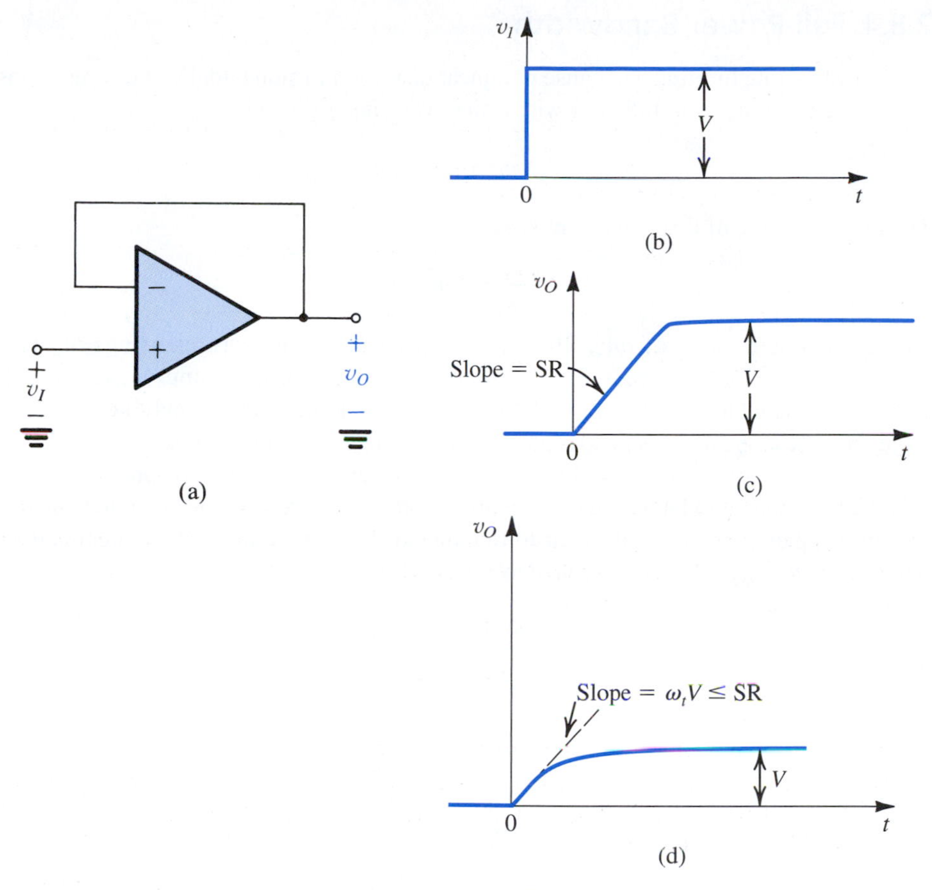

Figure 2.43 **(a)** Unity-gain follower. **(b)** Input step waveform. **(c)** Linearly rising output waveform obtained when the amplifier is slew-rate limited. **(d)** Exponentially rising output waveform obtained when V is sufficiently small so that the initial slope ($\omega_t V$) is smaller than or equal to SR.

which is a low-pass STC response with a time constant $1/\omega_t$. Its step response would therefore be (see Appendix E)

$$v_O(t) = V(1 - e^{-\omega_t t}) \tag{2.58}$$

The initial slope of this exponentially rising function is ($\omega_t V$). Thus, as long as V is sufficiently small so that $\omega_t V \leq$ SR, the output will be as in Fig. 2.43(d).

EXERCISE

2.29 An op amp that has a slew rate of 1 V/μs and a unity-gain bandwidth f_t of 1 MHz is connected in the unity-gain follower configuration. Find the largest possible input voltage step for which the output waveform will still be given by the exponential ramp of Eq. (2.58). For this input voltage, what is the 10% to 90% rise time of the output waveform? If an input step 10 times as large is applied, find the 10% to 90% rise time of the output waveform.
Ans. 0.16 V; 0.35 μs; 1.28 μs

2.8.4 Full-Power Bandwidth

Op-amp slew-rate limiting can cause nonlinear distortion in sinusoidal waveforms. Consider once more the unity-gain follower with a sine-wave input given by

$$v_I = \hat{V}_i \sin \omega t$$

The rate of change of this waveform is given by

$$\frac{dv_I}{dt} = \omega \hat{V}_i \cos \omega t$$

with a maximum value of $\omega\hat{V}_i$. This maximum occurs at the zero crossings of the input sinusoid. Now if $\omega\hat{V}_i$ exceeds the slew rate of the op amp, the output waveform will be distorted in the manner shown in Fig. 2.44. Observe that the output cannot keep up with the large rate of change of the sinusoid at its zero crossings, and the op amp slews.

The op-amp data sheets usually specify a frequency f_M called the **full-power bandwidth**. It is the frequency at which an output sinusoid with amplitude equal to the rated output voltage of the op amp begins to show distortion due to slew-rate limiting. If we denote the rated output voltage $V_{o\max}$, then f_M is related to SR as follows:

$$\omega_M V_{o\max} = \text{SR}$$

Thus,

$$f_M = \frac{\text{SR}}{2\pi V_{o\max}} \tag{2.59}$$

It should be obvious that output sinusoids of amplitudes smaller than $V_{o\max}$ will show slew-rate distortion at frequencies higher than ω_M. In fact, at a frequency ω higher than ω_M, the maximum amplitude of the undistorted output sinusoid is given by

$$V_o = V_{o\max}\left(\frac{\omega_M}{\omega}\right) \tag{2.60}$$

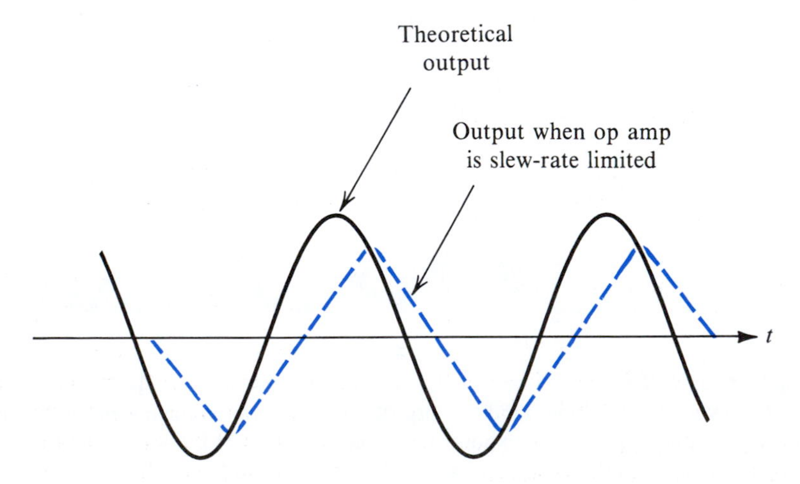

Figure 2.44 Effect of slew-rate limiting on output sinusoidal waveforms.

EXERCISE

2.30 An op amp has a rated output voltage of ±10 V and a slew rate of 1 V/μs. What is its full-power bandwidth? If an input sinusoid with frequency $f = 5f_M$ is applied to a unity-gain follower constructed using this op amp, what is the maximum possible amplitude that can be accommodated at the output without incurring SR distortion?
Ans. 15.9 kHz; 2 V (peak)

Summary

- The IC op amp is a versatile circuit building block. It is easy to apply, and the performance of op-amp circuits closely matches theoretical predictions.
- The op-amp terminals are the inverting input terminal (1), the noninverting input terminal (2), the output terminal (3), the positive-supply terminal (4) to be connected to the positive power supply (V_{CC}), and the negative-supply terminal (5) to be connected to the negative supply ($-V_{EE}$). The common terminal of the two supplies is the circuit ground.
- The ideal op amp responds only to the difference input signal, that is, $(v_2 - v_1)$; providing at the output, between terminal 3 and ground, a signal $A(v_2 - v_1)$, where A, the open-loop gain, is very large (10^4 to 10^6) and ideally infinite; and has an infinite input resistance and a zero output resistance. (See Table 3.1.)
- Negative feedback is applied to an op amp by connecting a passive component between its output terminal and its inverting (negative) input terminal. Negative feedback causes the voltage between the two input terminals to become very small and ideally zero. Correspondingly, a virtual short circuit is said to exist between the two input terminals. If the positive input terminal is connected to ground, a virtual ground appears on the negative input terminal.
- The two most important assumptions in the analysis of op-amp circuits, presuming negative feedback exists and the op amps are ideal, are as follows: the two input terminals of the op amp are at the same voltage, and zero current flows into the op-amp input terminals.
- With negative feedback applied and the loop closed, the closed-loop gain is almost entirely determined by external components: For the inverting configuration, $V_o/V_i = -R_2/R_1$; and for the noninverting configuration, $V_o/V_i = 1 + R_2/R_1$.
- The noninverting closed-loop configuration features a very high input resistance. A special case is the unity-gain follower, frequently employed as a buffer amplifier to connect a high-resistance source to a low-resistance load.
- The difference amplifier of Fig. 2.16 is designed with $R_4/R_3 = R_2/R_1$, resulting in $v_O = (R_2/R_1)(v_{I2} - v_{I1})$.
- The instrumentation amplifier of Fig. 2.20(b) is a very popular circuit. It provides $v_O = (1 + R_2/R_1)(R_4/R_3)(v_{I2} - v_{I1})$. It is usually designed with $R_3 = R_4$, and R_1 and R_2 selected to provide the required gain. If an adjustable gain is needed, part of R_1 can be made variable.
- The inverting Miller integrator of Fig. 2.24 is a popular circuit, frequently employed in analog signal-processing functions such as filters (Chapter 16) and oscillators (Chapter 17).
- The input offset voltage, V_{OS}, is the magnitude of dc voltage that when applied between the op amp input terminals, with appropriate polarity, reduces the dc offset voltage at the output to zero.
- The effect of V_{OS} on performance can be evaluated by including in the analysis a dc source V_{OS} in series with the op-amp positive input lead. For both the inverting and the noninverting configurations, V_{OS} results in a dc offset voltage at the output of $V_{OS}(1 + R_2/R_1)$.
- Capacitively coupling an op amp reduces the dc offset voltage at the output considerably.
- The average of the two dc currents, I_{B1} and I_{B2}, that flow in the input terminals of the op amp, is called the input bias current, I_B. In a closed-loop amplifier, I_B gives rise to a dc offset voltage at the output of magnitude $I_B R_2$. This voltage can be reduced to $I_{OS} R_2$ by connecting a resistance in series with the positive input terminal equal to the total dc resistance seen by the negative input terminal. I_{OS} is the input offset current; that is, $I_{OS} = |I_{B1} - I_{B2}|$.

- Connecting a large resistance in parallel with the capacitor of an op-amp inverting integrator prevents op-amp saturation (due to the effect of V_{OS} and I_B).
- For most internally compensated op amps, the open-loop gain falls off with frequency at a rate of −20 dB/decade, reaching unity at a frequency f_t (the unity-gain bandwidth). Frequency f_t is also known as the gain–bandwidth product of the op amp: $f_t = A_0 f_b$, where A_0 is the dc gain, and f_b is the 3-dB frequency of the open-loop gain. At any frequency f ($f \gg f_b$), the op-amp gain $|A| \simeq f_t/f$.
- For both the inverting and the noninverting closed-loop configurations, the 3-dB frequency is equal to $f_t/(1 + R_2/R_1)$.
- The maximum rate at which the op-amp output voltage can change is called the slew rate. The slew rate, SR, is usually specified in V/μs. Op-amp slewing can result in nonlinear distortion of output signal waveforms.
- The full-power bandwidth, f_M, is the maximum frequency at which an output sinusoid with an amplitude equal to the op-amp rated output voltage (V_{omax}) can be produced without distortion: $f_M = SR/2\pi V_{omax}$.

PROBLEMS

Computer Simulation Problems

SIM Problems identified by this icon are intended to demonstrate the value of using SPICE simulation to verify hand analysis and design, and to investigate important issues such as allowable signal swing and amplifier nonlinear distortion. Instructions to assist in setting up PSpice and Multism simulations for all the indicated problems can be found in the corresponding files on the disc. Note that if a particular parameter value is not specified in the problem statement, you are to make a reasonable assumption. * difficult problem; ** more difficult; *** very challenging and/or time-consuming; D: design problem.

Section 2.1: The Ideal Op Amp

2.1 What is the minimum number of pins required for a so-called dual-op-amp IC package, one containing two op amps? What is the number of pins required for a so-called quad-op-amp package, one containing four op-amps?

2.2 The circuit of Fig. P2.2 uses an op amp that is ideal except for having a finite gain A. Measurements indicate $v_O = 4.0$ V when $v_I = 2.0$ V. What is the op-amp gain A?

2.3 Measurement of a circuit incorporating what is thought to be an ideal op amp shows the voltage at the op-amp output to be −2.000 V and that at the negative input to be −1.000 V. For the amplifier to be ideal, what would you expect the voltage at the positive input to be? If the measured voltage at the positive input is −1.010 V, what is likely to be the actual gain of the amplifier?

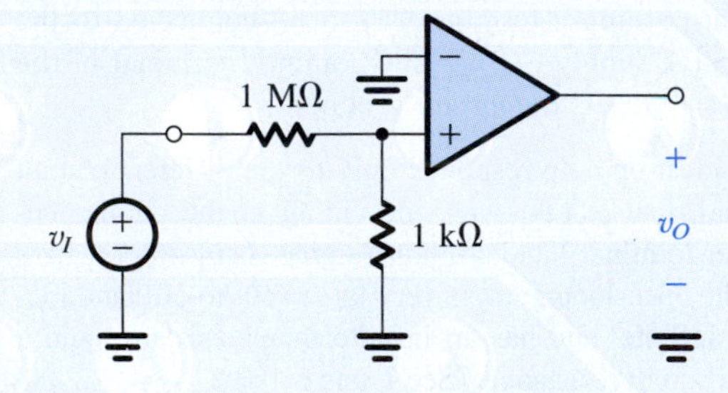

Figure P2.2

2.4 A set of experiments is run on an op amp that is ideal except for having a finite gain A. The results are tabulated below. Are the results consistent? If not, are they reasonable, in view of the possibility of experimental error? What do they show the gain to be? Using this value, predict values of the measurements that were accidentally omitted (the blank entries).

Experiment #	v_1	v_2	v_O
1	0.00	0.00	0.00
2	1.00	1.00	0.00
3		1.00	1.00
4	1.00	1.10	10.1
5	2.01	2.00	−0.99
6	1.99	2.00	1.00
7	5.10		−5.10

2.5 Refer to Exercise 2.3. This problem explores an alternative internal structure for the op amp. In particular, we wish to model the internal structure of a particular op amp using two transconductance amplifiers and one transresistance amplifier. Suggest an appropriate topology. For equal transconductances G_m and a transresistance R_m, find an expression for the open-loop gain A. For $G_m = 10$ mA/V and $R_m = 2 \times 10^6\ \Omega$, what value of A results?

2.6 The two wires leading from the output terminals of a transducer pick up an interference signal that is a 60-Hz, 1-V sinusoid. The output signal of the transducer is sinusoidal of 10-mV amplitude and 1000-Hz frequency. Give expressions for v_{cm}, v_d, and the total signal between each wire and the system ground.

2.7 Nonideal (i.e., real) operational amplifiers respond to both the differential and common-mode components of their input signals (refer to Fig. 2.4 for signal representation). Thus the output voltage of the op amp can be expressed as

$$v_O = A_d v_{Id} + A_{cm} v_{Icm}$$

where A_d is the differential gain (referred to simply as A in the text) and A_{cm} is the common-mode gain (assumed to be zero in the text). The op amp's effectiveness in rejecting common-mode signals is measured by its CMRR, defined as

$$\text{CMRR} = 20 \log \left|\frac{A_d}{A_{cm}}\right|$$

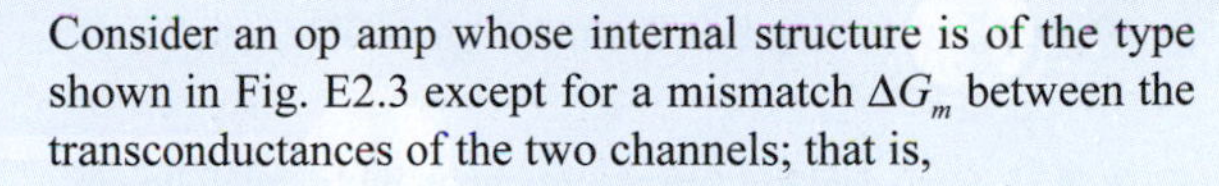

Consider an op amp whose internal structure is of the type shown in Fig. E2.3 except for a mismatch ΔG_m between the transconductances of the two channels; that is,

$$G_{m1} = G_m - \tfrac{1}{2}\Delta G_m$$

$$G_{m2} = G_m + \tfrac{1}{2}\Delta G_m$$

Find expressions for A_d, A_{cm}, and CMRR. If A_d is 80 dB and the two transconductances are matched to within 0.1% of each other, calculate A_{cm} and CMRR.

Section 2.2: The Inverting Configuration

2.8 Assuming ideal op amps, find the voltage gain v_o/v_i and input resistance R_{in} of each of the circuits in Fig. P2.8.

2.9 A particular inverting circuit uses an ideal op amp and two 10-kΩ resistors. What closed-loop gain would you expect? If a dc voltage of +1.00 V is applied at the input, what output result? If the 10-kΩ resistors are said to be "1% resistors," having values somewhere in the range (1 ± 0.01) times the nominal value, what range of outputs would you expect to actually measure for an input of precisely 1.00 V?

2.10 You are provided with an ideal op amp and three 10-kΩ resistors. Using series and parallel resistor combinations, how many different inverting-amplifier circuit topologies are possible? What is the largest (noninfinite) available voltage

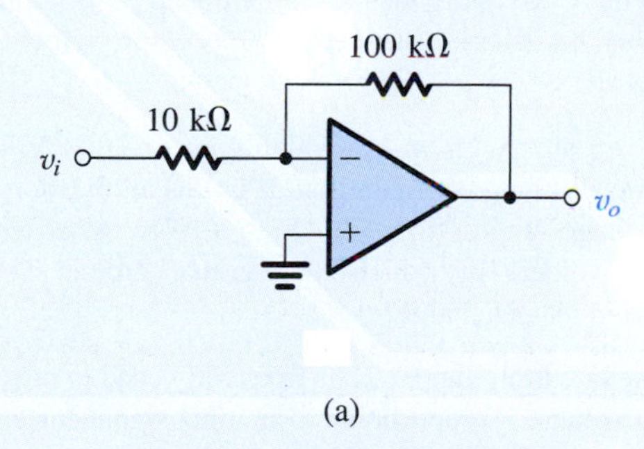

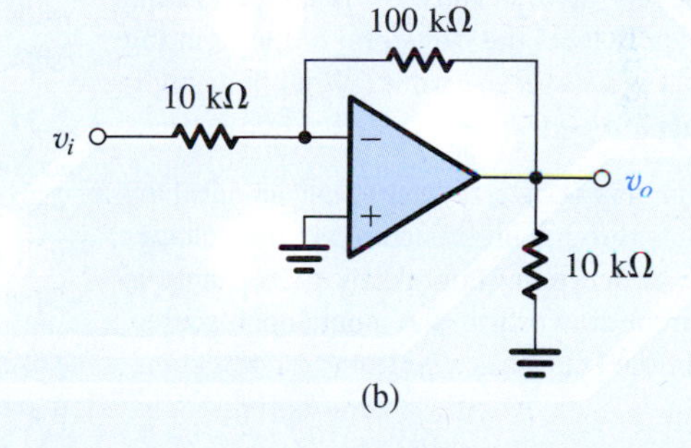

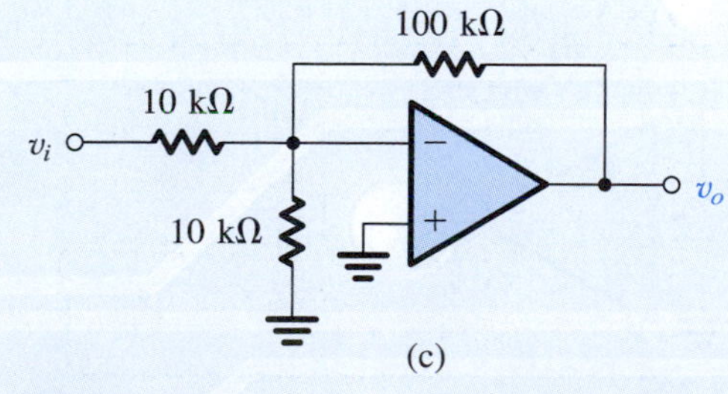

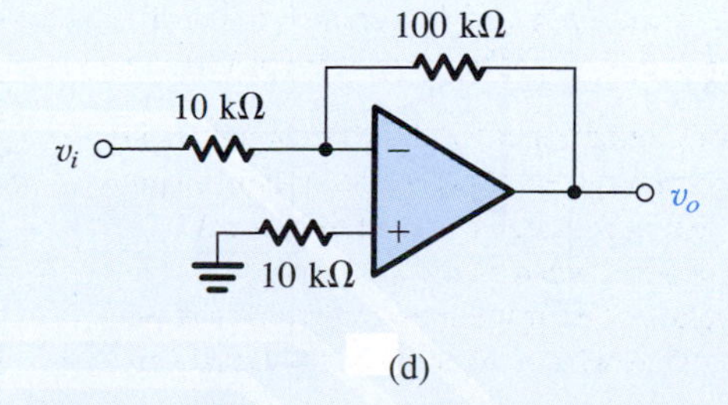

Figure P2.8

gain? What is the smallest (nonzero) available gain? What are the input resistances in these two cases?

SIM **2.11** For ideal op amps operating with the following feedback networks in the inverting configuration, what closed-loop gain results?

(a) $R_1 = 10\ \text{k}\Omega$, $R_2 = 10\ \text{k}\Omega$
(b) $R_1 = 10\ \text{k}\Omega$, $R_2 = 100\ \text{k}\Omega$
(c) $R_1 = 10\ \text{k}\Omega$, $R_2 = 1\ \text{k}\Omega$
(d) $R_1 = 100\ \text{k}\Omega$, $R_2 = 10\ \text{M}\Omega$
(e) $R_1 = 100\ \text{k}\Omega$, $R_2 = 1\ \text{M}\Omega$

D 2.12 Given an ideal op amp, what are the values of the resistors R_1 and R_2 to be used to design amplifiers with the closed-loop gains listed below? In your designs, use at least one 10-kΩ resistor and another equal or larger resistor.

(a) −1 V/V
(b) −2 V/V
(c) −0.5 V/V
(d) −100 V/V

D 2.13 Design an inverting op-amp circuit for which the gain is −4 V/V and the total resistance used is 100 kΩ.

D 2.14 Using the circuit of Fig. 2.5 and assuming an ideal op amp, design an inverting amplifier with a gain of 26 dB having the largest possible input resistance under the constraint of having to use resistors no larger than 1 MΩ. What is the input resistance of your design?

2.15 An ideal op amp is connected as shown in Fig. 2.5 with $R_1 = 10\ \text{k}\Omega$ and $R_2 = 100\ \text{k}\Omega$. A symmetrical square-wave signal with levels of 0 V and 1 V is applied at the input. Sketch and clearly label the waveform of the resulting output voltage. What is its average value? What is its highest value? What is its lowest value?

2.16 For the circuit in Fig. P2.16, assuming an ideal op amp, find the currents through all branches and the voltages at all nodes. Since the current supplied by the op amp is greater than the current drawn from the input signal source, where does the additional current come from?

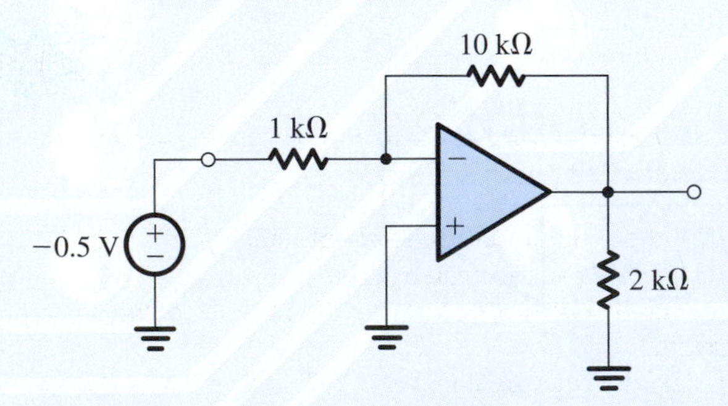

Figure P2.16

2.17 An inverting op-amp circuit is fabricated with the resistors R_1 and R_2 having x% tolerance (i.e., the value of each resistance can deviate from the nominal value by as much as $\pm x$%). What is the tolerance on the realized closed-loop gain? Assume the op amp to be ideal. If the nominal closed-loop gain is −100 V/V and $x = 1$, what is the range of gain values expected from such a circuit?

2.18 An ideal op amp with 5-kΩ and 15-kΩ resistors is used to create a +5-V supply from a −15-V reference. Sketch the circuit. What are the voltages at the ends of the 5-kΩ resistor? If these resistors are so-called 1% resistors, whose actual values are the range bounded by the nominal value ±1%, what are the limits of the output voltage produced? If the −15-V supply can also vary by ±1%, what is the range of the output voltages that might be found?

2.19 An inverting op-amp circuit for which the required gain is −50 V/V uses an op amp whose open-loop gain is only 300 V/V. If the larger resistor used is 100 kΩ, to what must the smaller be adjusted? With what resistor must a 2-kΩ resistor connected to the input be shunted to achieve this goal? (Note that a resistor R_a is said to be shunted by resistor R_b when R_b is placed in parallel with R_a.)

D 2.20 (a) Design an inverting amplifier with a closed-loop gain of −100 V/V and an input resistance of 1 kΩ.
(b) If the op amp is known to have an open-loop gain of 2000 V/V, what do you expect the closed-loop gain of your circuit to be (assuming the resistors have precise values)?
(c) Give the value of a resistor you can place in parallel (shunt) with R_1 to restore the closed-loop gain to its nominal value. Use the closest standard 1% resistor value (see Appendix H).

2.21 An op amp with an open-loop gain of 2000 V/V is used in the inverting configuration. If in this application the output voltage ranges from −10 V to +10 V, what is the maximum voltage by which the "virtual ground node" departs from its ideal value?

2.22 The circuit in Fig. P2.22 is frequently used to provide an output voltage v_o proportional to an input signal current i_i.

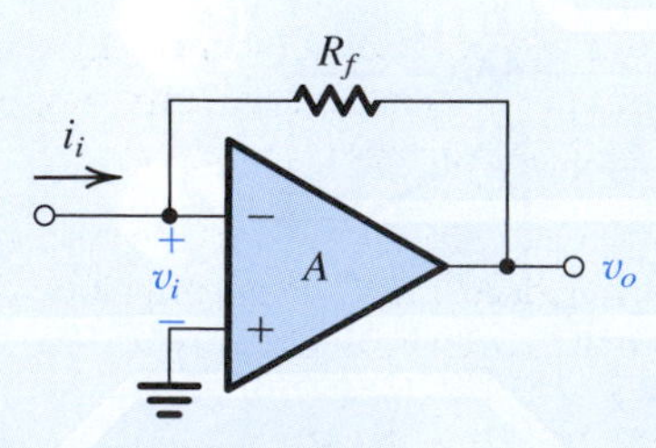

Figure P2.22

Derive expressions for the transresistance $R_m \equiv v_o/i_i$ and the input resistance $R_i \equiv v_i/i_i$ for the following cases:

(a) A is infinite.
(b) A is finite.

2.23 Show that for the inverting amplifier if the op-amp gain is A, the input resistance is given by

$$R_{\text{in}} = R_1 + \frac{R_2}{A+1}$$

***2.24** For an inverting amplifier with nominal closed-loop gain R_2/R_1, find the minimum value that the op-amp open-loop gain A must have (in terms of R_2/R_1) so that the gain error is limited to 0.1%, 1%, and 10%. In each case find the value of a resistor R_{Ia} such that when it is placed in shunt with R_i, the gain is restored to its nominal value.

***2.25** Figure P2.25 shows an op amp that is ideal except for having a finite open-loop gain and is used to realize an inverting amplifier whose gain has a nominal magnitude $G = R_2/R_1$. To compensate for the gain reduction due to the finite A, a resistor R_c is shunted across R_1. Show that perfect compensation is achieved when R_c is selected according to

$$\frac{R_c}{R_1} = \frac{A-G}{1+G}$$

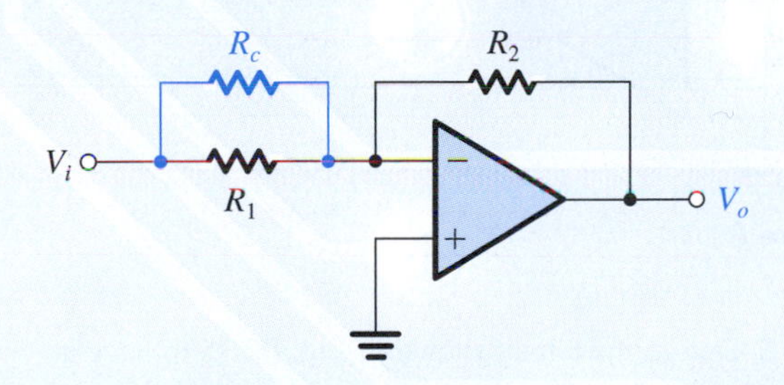

Figure P2.25

***D 2.26** (a) Use Eq. (2.5) to obtain the amplifier open-loop gain A required to realize a specified closed-loop gain ($G_{\text{nominal}} = -R_2/R_1$) within a specified gain error ε,

$$\varepsilon \equiv \left| \frac{G - G_{\text{nominal}}}{G_{\text{nominal}}} \right|$$

(b) Design an inverting amplifer for a nominal closed-loop gain of −100, an input resistance of 2 kΩ, and a gain error of ≤10%. Specify R_1, R_2, and the minimum A required.

***2.27** (a) Use Eq. (2.5) to show that a reduction ΔA in the op-amp gain A gives rise to a reduction $\Delta|G|$ in the magnitude of the closed-loop gain G with $\Delta|G|$ and ΔA related by

$$\frac{\Delta|G|/|G|}{\Delta A/A} = \frac{1 + R_2/R_1}{A}$$

(b) If in a closed-loop amplifier with a nominal gain (i.e, R_2/R_1) of 100, A decreases by 50%, what is the minimum nominal A required to limit the percentage change in $|G|$ to 0.5%?

2.28 Consider the circuit in Fig. 2.8 with $R_1 = R_2 = R_4 =$ 1 MΩ, and assume the op amp to be ideal. Find values for R_3 to obtain the following gains:

(a) −200 V/V
(b) −20 V/V
(c) −2 V/V

D 2.29 An inverting op-amp circuit using an ideal op amp must be designed to have a gain of −1000 V/V using resistors no larger than 100 kΩ.

(a) For the simple two-resistor circuit, what input resistance would result?
(b) If the circuit in Fig. 2.8 is used with three resistors of maximum value, what input resistance results? What is the value of the smallest resistor needed?

2.30 The inverting circuit with the T network in the feedback is redrawn in Fig. P2.30 in a way that emphasizes the observation that R_2 and R_3 in effect are in parallel (because the ideal op amp forces a virtual ground at the inverting input terminal). Use this observation to derive an expression for the gain (v_O/v_I) by first finding (v_X/v_I) and (v_O/v_X). For the latter use the voltage-divider rule applied to R_4 and ($R_2 \parallel R_3$).

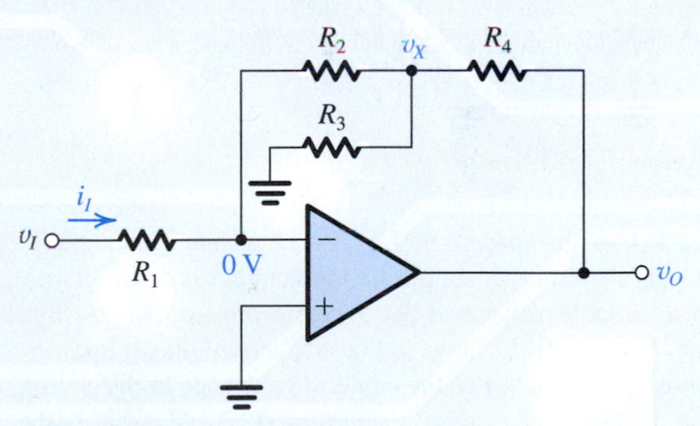

Figure P2.30

***2.31** The circuit in Fig. P2.31 can be considered to be an extension of the circuit in Fig. 2.8.

(a) Find the resistances looking into node 1, R_1; node 2, R_2; node 3, R_3; and node 4, R_4.
(b) Find the currents I_1, I_2, I_3, and I_4, in terms of the input current I.
(c) Find the voltages at nodes 1, 2, 3, and 4, that is, V_1, V_2, V_3, and V_4 in terms of (IR).

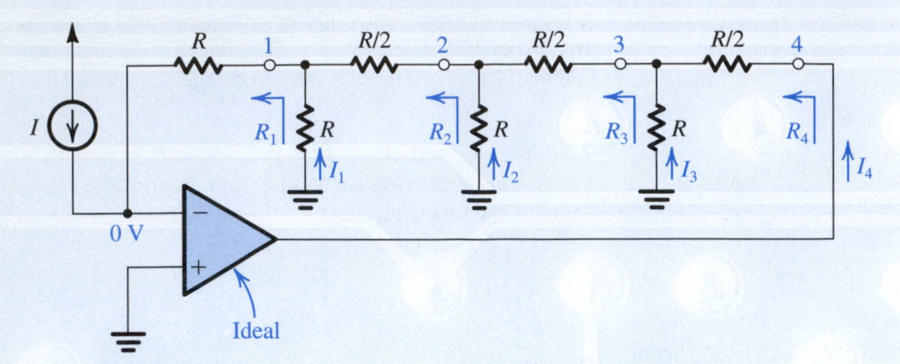

Figure P2.31

2.32 The circuit in Fig. P2.32 utilizes an ideal op amp.

(a) Find I_1, I_2, I_3, I_L, and V_x.
(b) If V_O is not to be lower than −13 V, find the maximum allowed value for R_L.
(c) If R_L is varied in the range 100 Ω to 1 kΩ, what is the corresponding change in I_L and in V_O?

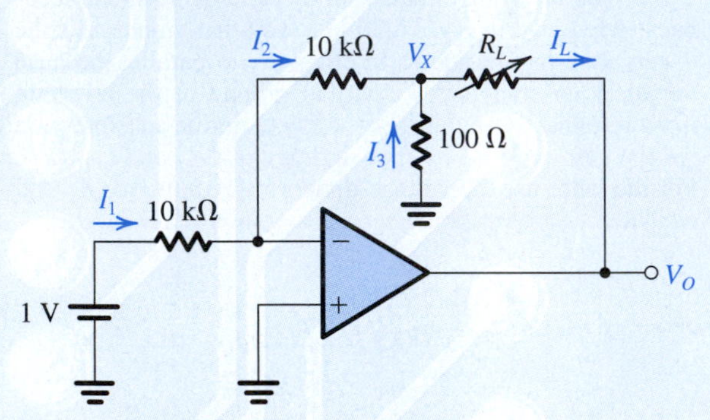

Figure P2.32

2.33 Use the circuit in Fig. P2.32 as an inspiration to design a circuit that supplies a constant current I of 3.1 mA to a variable resistance R_L. Assume the availability of a 1.5 V battery and design so that the current drawn from the battery is 0.1 mA. For the smallest resistance in the circuit, use 500 Ω. If the op amp saturates at ±12 V, what is the maximum value that R_L can have while the current-source supplying it operates properly?

D 2.34 Assuming the op amp to be ideal, it is required to design the circuit shown in Fig. P2.34 to implement a current amplifier with gain $i_L/i_I = 10$ A/A.

(a) Find the required value for R.
(b) What are the input and the output resistance of this current amplifier?
(c) If $R_L = 1$ kΩ and the op amp operates in an ideal manner as long as v_O is in the range ±12 V, what range of i_I is possible?
(d) If the amplifier is fed with a current source having a current of 0.2 mA and a source resistance of 10 kΩ, find i_L.

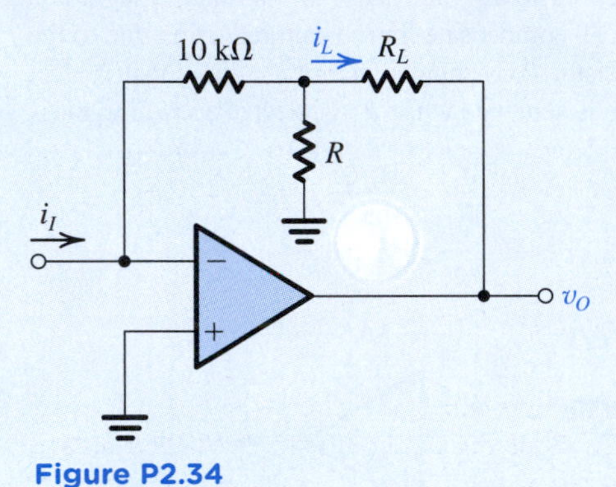

Figure P2.34

D 2.35 Design the circuit shown in Fig. P2.35 to have an input resistance of 100 kΩ and a gain that can be varied

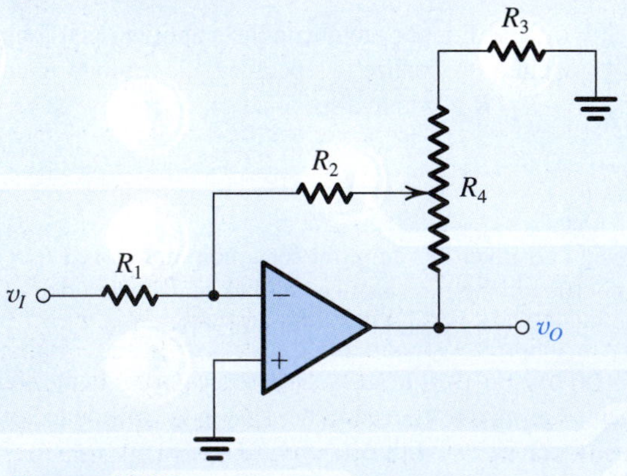

Figure P2.35

from −1 V/V to −10 V/V using the 10-kΩ potentiometer R_4. What voltage gain results when the potentiometer is set exactly at its middle value?

2.36 A weighted summer circuit using an ideal op amp has three inputs using 100-kΩ resistors and a feedback resistor of 50 kΩ. A signal v_1 is connected to two of the inputs while a signal v_2 is connected to the third. Express v_O in terms of v_1 and v_2. If $v_1 = 2$ V and $v_2 = -2$ V, what is v_O?

D 2.37 Design an op amp circuit to provide an output $v_O = -[2v_1 + (v_2/2)]$. Choose relatively low values of resistors but ones for which the input current (from each input signal source) does not exceed 0.1 mA for 1-V input signals.

D 2.38 Use the scheme illustrated in Fig. 2.10 to design an op-amp circuit with inputs v_1, v_2, and v_3, whose output is $v_O = -(2v_1 + 4v_2 + 8v_3)$ using small resistors but no smaller than 10 kΩ.

D 2.39 An ideal op amp is connected in the weighted summer configuration of Fig. 2.10. The feedback resistor R_f = 10 kΩ, and six 10-kΩ resistors are connected to the inverting input terminal of the op amp. Show, by sketching the various circuit configurations, how this basic circuit can be used to implement the following functions:

(a) $v_O = -(v_1 + 2v_2 + 3v_3)$
(b) $v_O = -(v_1 + v_2 + 2v_3 + 2v_4)$
(c) $v_O = -(v_1 + 5v_2)$
(d) $v_O = -6v_1$

In each case find the input resistance seen by each of the signal sources supplying v_1, v_2, v_3, and v_4. Suggest at least two additional summing functions that you can realize with this circuit. How would you realize a summing coefficient that is 0.5?

D 2.40 Give a circuit, complete with component values, for a weighted summer that shifts the dc level of a sine-wave signal of 3 sin(ωt) V from zero to −3 V. Assume that in addition to the sine-wave signal you have a dc reference voltage of 1.5 V available. Sketch the output signal waveform.

D 2.41 Use two ideal op amps and resistors to implement the summing function

$$v_O = v_1 + 2v_2 - 3v_3 - 4v_4$$

D *2.42 In an instrumentation system, there is a need to take the difference between two signals, one of $v_1 = 2\sin(2\pi \times 60t) + 0.01\sin(2\pi \times 1000t)$ volts and another of $v_2 = 2\sin(2\pi \times 60t) - 0.01\sin(2\pi \times 1000t)$ volts. Draw a circuit that finds the required difference using two op amps and mainly 100-kΩ resistors. Since it is desirable to amplify the 1000-Hz component in the process, arrange to provide an overall gain of 100 as well. The op amps available are ideal except that their output voltage swing is limited to ±10 V.

***2.43** Figure P2.43 shows a circuit for a digital-to-analog converter (DAC). The circuit accepts a 4-bit input binary word $a_3a_2a_1a_0$, where a_0, a_1, a_2, and a_3 take the values of 0 or 1, and it provides an analog output voltage v_O proportional to the value of the digital input. Each of the bits of the input word controls the correspondingly numbered switch. For instance, if a_2 is 0 then switch S_2 connects the 20-kΩ resistor to ground, while if a_2 is 1 then S_2 connects the 20-kΩ resistor to the +5-V power supply. Show that v_O is given by

$$v_O = -\frac{R_f}{16}[2^0a_0 + 2^1a_1 + 2^2a_2 + 2^3a_3]$$

where R_f is in kilohms. Find the value of R_f so that v_O ranges from 0 to −12 volts.

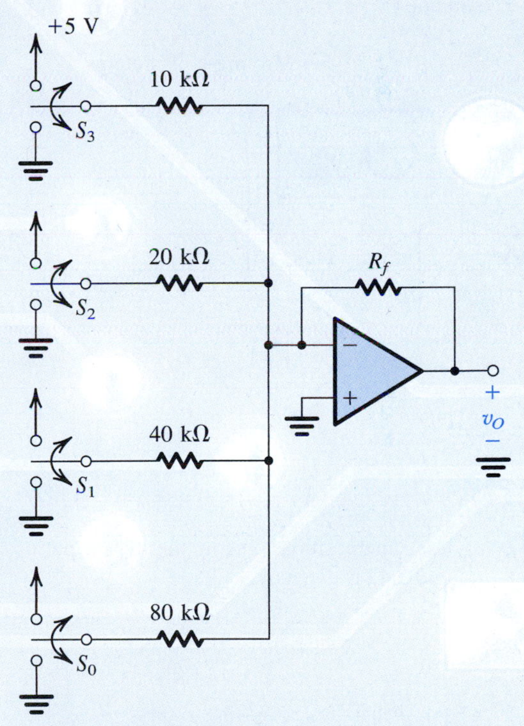

Figure P2.43

Section 2.3: The Noninverting Configuration

D 2.44 Given an ideal op amp to implement designs for the following closed-loop gains, what values of resistors (R_1, R_2) should be used? Where possible, use at least one 10-kΩ resistor as the smallest resistor in your design.

(a) +1 V/V
(b) +2 V/V
(c) +11 V/V
(d) +100 V/V

D 2.45 Design a circuit based on the topology of the noninverting amplifier to obtain a gain of +1.5 V/V, using only 10-kΩ resistors. Note that there are two possibilities. Which of these can be easily converted to have a gain of either +1.0 V/V or +2.0 V/V simply by short-circuiting a single resistor in each case?

D 2.46 Figure P2.46 shows a circuit for an analog voltmeter of very high input resistance that uses an inexpensive moving-coil meter. The voltmeter measures the voltage V applied between the op amp's positive-input terminal and ground. Assuming that the moving coil produces full-scale deflection when the current passing through it is 100 μA, find the value of R such that full-scale reading is obtained when V is +10 V. Does the meter resistance shown affect the voltmeter calibration?

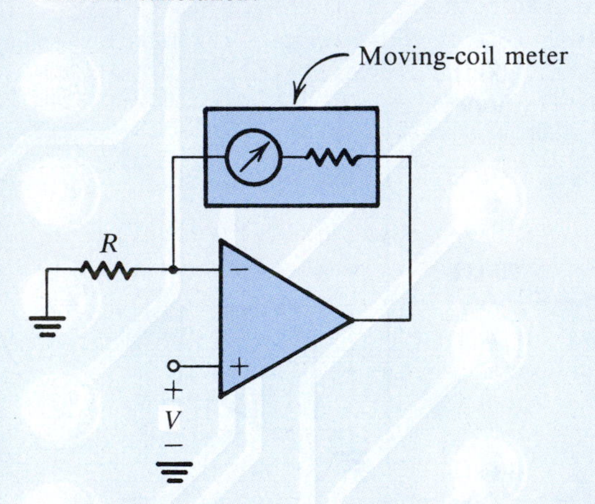

Figure P2.46

D *2.47 (a) Use superposition to show that the output of the circuit in Fig. P2.47 is given by

$$v_O = -\left[\frac{R_f}{R_{N1}}v_{N1} + \frac{R_f}{R_{N2}}v_{N2} + \cdots + \frac{R_f}{R_{Nn}}v_{Nn}\right]$$
$$+\left[1 + \frac{R_f}{R_N}\right]\left[\frac{R_P}{R_{P1}}v_{P1} + \frac{R_P}{R_{P2}}v_{P2} + \cdots + \frac{R_P}{R_{Pn}}v_{Pn}\right]$$

where $R_N = R_{N1} \| R_{N2} \| \cdots \| R_{Nn}$ and

$$R_P = R_{P1} \| R_{P2} \| \cdots \| R_{Pn} \| R_{P0}$$

(b) Design a circuit to obtain

$$v_O = -3v_{N1} + v_{P1} + 2v_{P2}$$

The smallest resistor used should be 10 kΩ.

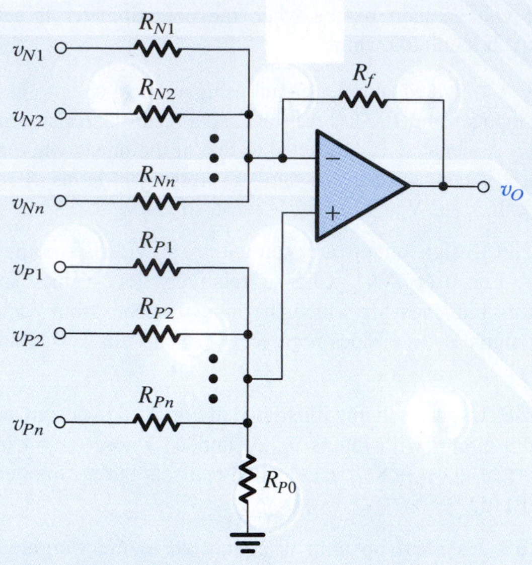

Figure P2.47

D 2.48 Design a circuit, using one ideal op amp, whose output is $v_O = v_{I1} + 3v_{I2} - 2(v_{I3} + 3v_{I4})$. (*Hint:* Use a structure similar to that shown in general form in Fig. P2.47.)

2.49 Derive an expression for the voltage gain, v_O/v_I, of the circuit in Fig. P2.49.

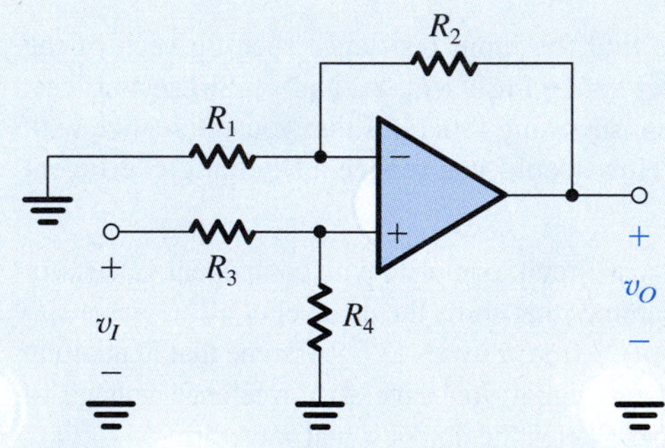

Figure P2.49

2.50 For the circuit in Fig. P2.50, use superposition to find v_O in terms of the input voltages v_1 and v_2. Assume an ideal op amp. For

$$v_1 = 10\sin(2\pi \times 60t) - 0.1\sin(2\pi \times 1000t), \text{ volts}$$
$$v_2 = 10\sin(2\pi \times 60t) + 0.1\sin(2\pi \times 1000t), \text{ volts}$$

find v_O.

D 2.51 The circuit shown in Fig. P2.51 utilizes a 10-kΩ potentiometer to realize an adjustable-gain amplifier. Derive an expression for the gain as a function of the potentiometer

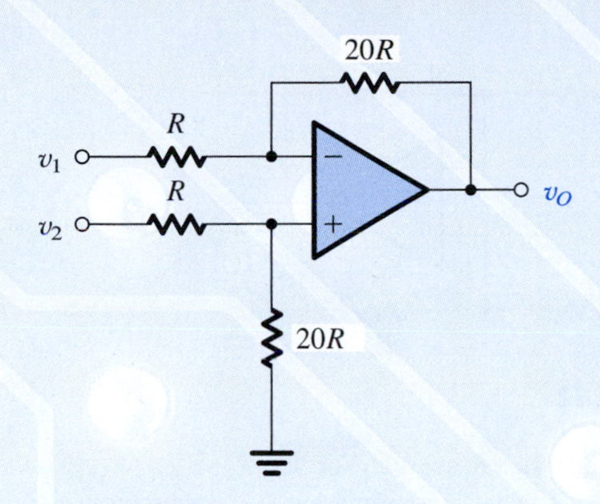

Figure P2.50

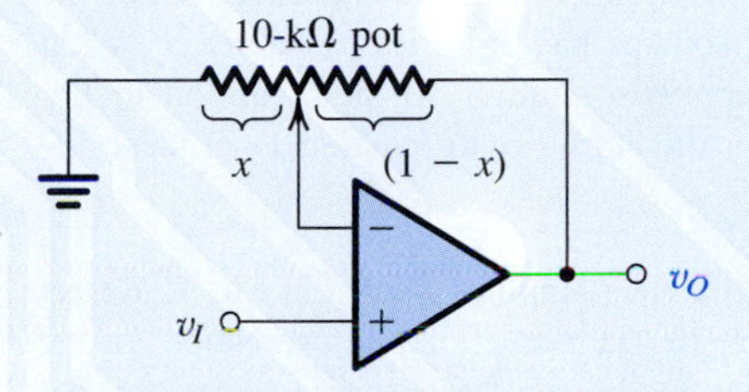

Figure P2.51

setting x. Assume the op amp to be ideal. What is the range of gains obtained? Show how to add a fixed resistor so that the gain range can be 1 to 11 V/V. What should the resistor value be?

D 2.52 Given the availability of resistors of value 1 kΩ and 10 kΩ only, design a circuit based on the noninverting configuration to realize a gain of +10 V/V.

2.53 It is required to connect a 10-V source with a source resistance of 100 kΩ to a 1-kΩ load. Find the voltage that will appear across the load if:

(a) The source is connected directly to the load.
(b) A unity-gain op-amp buffer is inserted between the source and the load.

In each case find the load current and the current supplied by the source. Where does the load current come from in case (b)?

2.54 Derive an expression for the gain of the voltage follower of Fig. 2.14, assuming the op amp to be ideal except for having a finite gain A. Calculate the value of the closed-loop gain for A = 1000, 100, and 10. In each case find the percentage error in gain magnitude from the nominal value of unity.

2.55 Complete the following table for feedback amplifiers created using one ideal op amp. Note that R_{in} signifies input resistance and R_1 and R_2 are feedback-network resistors as labelled in the inverting and noninverting configurations.

Case	Gain	R_{in}	R_1	R_2
a	−10 V/V	10 kΩ		
b	−1 V/V		100 kΩ	
c	−2 V/V			100 kΩ
d	+1 V/V	∞		
e	+2 V/V		10 kΩ	
f	+11 V/V			100 kΩ
g	−0.5 V/V	10 kΩ		

D 2.56 A noninverting op-amp circuit with nominal gain of 10 V/V uses an op amp with open-loop gain of 50 V/V and a lowest-value resistor of 10 kΩ. What closed-loop gain actually results? With what value resistor can which resistor be shunted to achieve the nominal gain? If in the manufacturing process, an op amp of gain 100 V/V were used, what closed-loop gain would result in each case (the uncompensated one, and the compensated one)?

2.57 Use Eq. (2.11) to show that if the reduction in the closed-loop gain G from the nominal value $G_0 = 1 + R_2/R_1$ is to be kept less than x% of G_0, then the open-loop gain of the op amp must exceed G_0 by at least a factor $F = (100/x) - 1 \simeq 100/x$. Find the required F for x = 0.01, 0.1, 1, and 10. Utilize these results to find for each value of x the minimum required open-loop gain to obtain closed-loop gains of 1, 10, 10^2, 10^3, and 10^4 V/V.

2.58 For each of the following combinations of op-amp open-loop gain A and nominal closed-loop gain G_0, calculate the actual closed-loop gain G that is achieved. Also, calculate the percentage by which $|G|$ falls short of the nominal gain magnitude $|G_0|$.

Case	G_0 (V/V)	A (V/V)
a	−1	10
b	+1	10
c	−1	100
d	+10	10
e	−10	100
f	−10	1000
g	+1	2

2.59 Figure P2.59 shows a circuit that provides an output voltage v_O whose value can be varied by turning the wiper of the 100-kΩ potentiometer. Find the range over which v_O can be varied. If the potentiometer is a "20-turn" device, find the change in v_O corresponding to each turn of the pot.

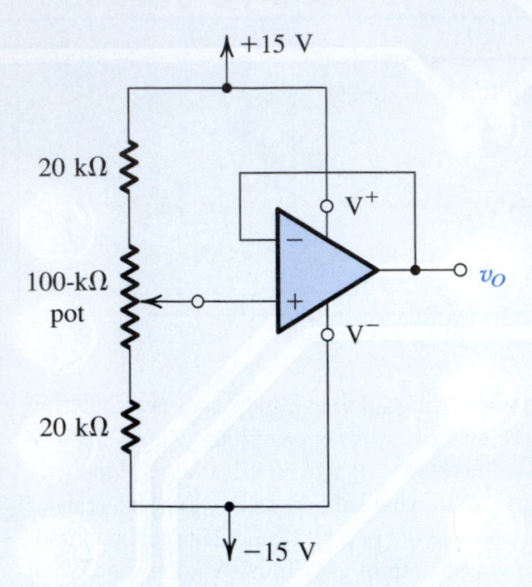

Figure P2.59

Section 2.4: Difference Amplifiers

2.60 Find the voltage gain v_O/v_{Id} for the difference amplifier of Fig. 2.16 for the case $R_1 = R_3 = 10$ kΩ and $R_2 = R_4 = 100$ kΩ. What is the differential input resistance R_{id}? If the two key resistance ratios (R_2/R_1) and (R_4/R_3) are different from each other by 1%, what do you expect the common-mode gain A_{cm} to be? Also, find the CMRR in this case. Neglect the effect of the ratio mismatch on the value of A_d.

D 2.61 Using the difference amplifier configuration of Fig. 2.16 and assuming an ideal op amp, design the circuit to provide the following differential gains. In each case, the differential input resistance should be 20 kΩ.

(a) 1 V/V
(b) 2 V/V
(c) 100 V/V
(d) 0.5 V/V

2.62 For the circuit shown in Fig. P2.62, express v_O as a function of v_1 and v_2. What is the input resistance seen by v_1 alone? By v_2 alone? By a source connected between the two input terminals? By a source connected to both input terminals simultaneously?

2.63 Consider the difference amplifier of Fig. 2.16 with the two input terminals connected together to an input common-mode signal source. For $R_2/R_1 = R_4/R_3$, show that the input common-mode resistance is $(R_3 + R_4) \| (R_1 + R_2)$.

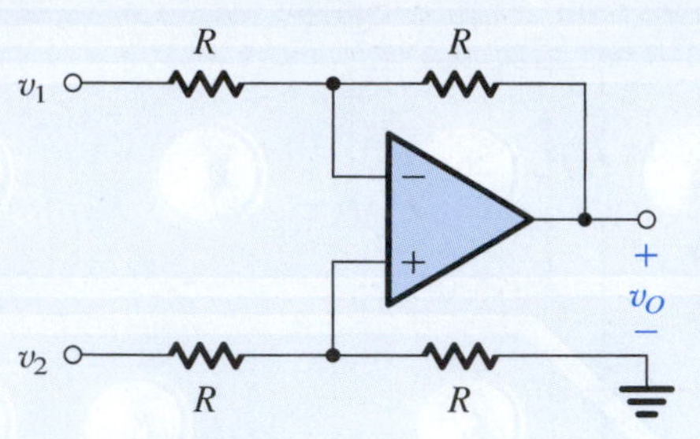

Figure P2.62

2.64 Consider the circuit of Fig. 2.16, and let each of the v_{I1} and v_{I2} signal sources have a series resistance R_s. What condition must apply in addition to the condition in Eq. (2.15) in order for the amplifier to function as an ideal difference amplifier?

***2.65** For the difference amplifier shown in Fig. P2.62, let all the resistors be 10 kΩ ± x%. Find an expression for the worst-case common-mode gain that results. Evaluate this for $x = 0.1$, 1, and 5. Also, evaluate the resulting CMRR in each case. Neglect the effect of resistor tolerances on A_d.

2.66 For the difference amplifier of Fig. 2.16, show that if each resistor has a tolerance of ±100 ε% (i.e., for, say, a 5% resistor, $\varepsilon = 0.05$) then the worst-case CMRR is given approximately by

$$\text{CMRR} \simeq 20 \log\left[\frac{K+1}{4\varepsilon}\right]$$

where K is the nominal (ideal) value of the ratios (R_2/R_1) and (R_4/R_3). Calculate the value of worst-case CMRR for an amplifier designed to have a differential gain of ideally 100 V/V, assuming that the op amp is ideal and that 1% resistors are used.

D *2.67 Design the difference amplifier circuit of Fig. 2.16 to realize a differential gain of 100, a differential input resistance of 20 kΩ, and a minimum CMRR of 80 dB. Assume the op amp to be ideal. Specify both the resistor values and their required tolerance (e.g., better than x%).

***2.68** (a) Find A_d and A_{cm} for the difference amplifier circuit shown in Fig. P2.68.
(b) If the op amp is specified to operate properly as long as the common-mode voltage at its positive and negative inputs falls in the range ±2.5 V, what is the corresponding limitation on the range of the input common-mode signal v_{Icm}? (This is known as the **common-mode range** of the differential amplifier.)

(c) The circuit is modified by connecting a 10-kΩ resistor between node A and ground, and another 10-kΩ resistor between node B and ground. What will now be the values of A_d, A_{cm}, and the input common-mode range?

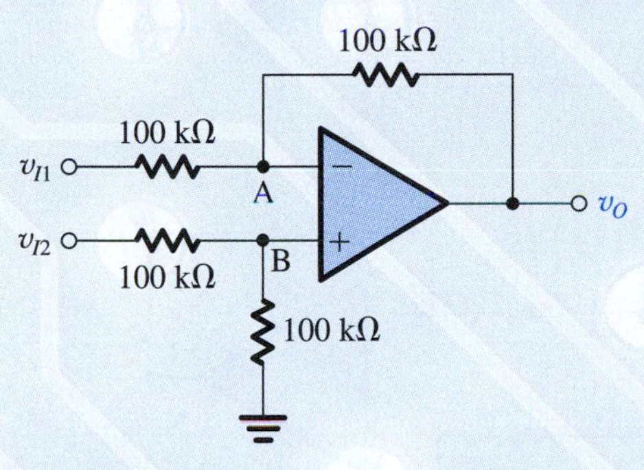

Figure P2.68

****2.69** To obtain a high-gain, high-input-resistance difference amplifier, the circuit in Fig. P2.69 employs positive feedback, in addition to the negative feedback provided by the resistor R connected from the output to the negative input of the op amp. Specifically, a voltage divider (R_5, R_6) connected across the output feeds a fraction β of the output, that is, a voltage βv_O, back to the positive-input terminal of the op amp through a resistor R. Assume that R_5 and R_6 are much smaller than R so that the current through R is much lower than the current in the voltage divider, with the result that $\beta \simeq R_6 | (R_5 + R_6)$. Show that the differential gain is given by

$$A_d \equiv \frac{v_O}{v_{Id}} = \frac{1}{1-\beta}$$

(*Hint:* Use superposition.)

Design the circuit to obtain a differential gain of 10 V/V and differential input resistance of 2 MΩ. Select values for R, R_5, and R_6, such that $(R_5 + R_6) \le R/100$.

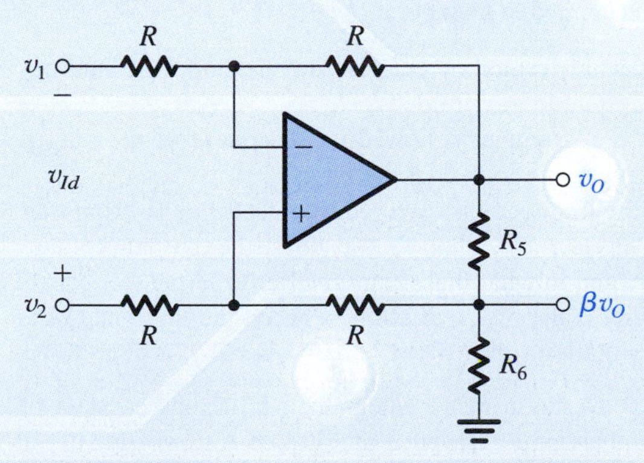

Figure P2.69

***2.70** Figure P2.70 shows a modified version of the difference amplifier. The modified circuit includes a resistor R_G, which can be used to vary the gain. Show that the differential voltage gain is given by

$$\frac{v_O}{v_{Id}} = -2\frac{R_2}{R_1}\left[1 + \frac{R_2}{R_G}\right]$$

(*Hint:* The virtual short circuit at the op-amp input causes the current through the R_1 resistors to be $v_{Id}/2R_1$.)

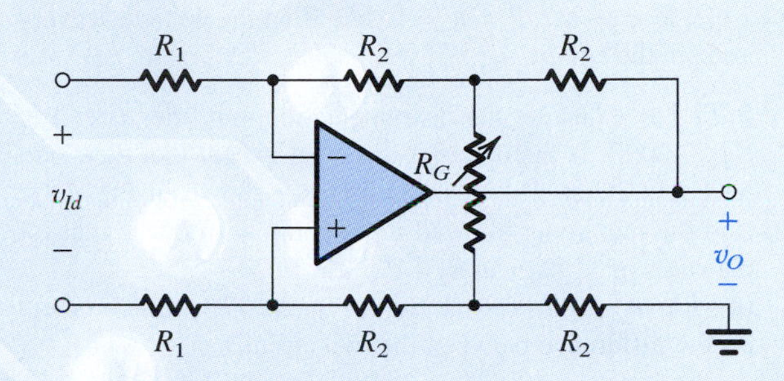

Figure P2.70

D *2.71 The circuit shown in Fig. P2.71 is a representation of a versatile, commercially available IC, the INA105, manufactured by Burr-Brown and known as a differential amplifier module. It consists of an op amp and precision, laser-trimmed, metal-film resistors. The circuit can be configured for a variety of applications by the appropriate connection of terminals A, B, C, D, and O.

(a) Show how the circuit can be used to implement a difference amplifier of unity gain.

(b) Show how the circuit can be used to implement single-ended amplifiers with gains:

(i) −1 V/V
(ii) +1 V/V
(iii) +2 V/V
(iv) +1/2 V/V

Avoid leaving a terminal open-circuited, for such a terminal may act as an "antenna," picking up interference and noise

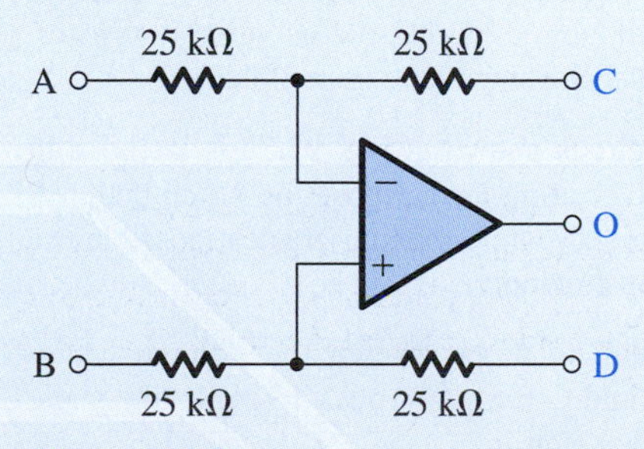

Figure P2.71

through capacitive coupling. Rather, find a convenient node to connect such a terminal in a redundant way. When more than one circuit implementation is possible, comment on the relative merits of each, taking into account such considerations as dependence on component matching and input resistance.

2.72 Consider the instrumentation amplifier of Fig. 2.20(b) with a common-mode input voltage of +2 V (dc) and a differential input signal of 80-mV peak sine wave. Let $2R_1 = 2\ \text{k}\Omega$, $R_2 = 50\ \text{k}\Omega$, $R_3 = R_4 = 10\ \text{k}\Omega$. Find the voltage at every node in the circuit.

2.73 (a) Consider the instrumentation amplifier circuit of Fig. 2.20(a). If the op amps are ideal except that their outputs saturate at ±14 V, in the manner shown in Fig. 1.14, find the maximum allowed input common-mode signal for the case $R_1 = 1\ \text{k}\Omega$ and $R_2 = 100\ \text{k}\Omega$.
(b) Repeat (a) for the circuit in Fig. 2.20(b), and comment on the difference between the two circuits.

2.74 (a) Expressing v_{I1} and v_{I2} in terms of differential and common-mode components, find v_{O1} and v_{O2} in the circuit in Fig. 2.20(a) and hence find their differential component $v_{O2} - v_{O1}$ and their common-mode component $\frac{1}{2}(v_{O1} + v_{O2})$. Now find the differential gain and the common-mode gain of the first stage of this instrumentation amplifier and hence the CMRR.
(b) Repeat for the circuit in Fig. 2.20(b), and comment on the difference between the two circuits.

***2.75** For an instrumentation amplifier of the type shown in Fig. 2.20(b), a designer proposes to make $R_2 = R_3 = R_4 = 100\ \text{k}\Omega$, and $2R_1 = 10\ \text{k}\Omega$. For ideal components, what difference-mode gain, common-mode gain, and CMRR result? Reevaluate the worst-case values for these for the situation in which all resistors are specified as ±1% units. Repeat the latter analysis for the case in which $2R_1$ is reduced to 1 kΩ. What do you conclude about the effect of the gain of the first stage on CMRR? (*Hint:* Eq. (2.19) can be used to evaluate A_{cm} of the second stage.)

D 2.76 Design the instrumentation-amplifier circuit of Fig. 2.20(b) to realize a differential gain, variable in the range 1 to 100, utilizing a 100-kΩ pot as variable resistor. (*Hint:* Design the second stage for a gain of 0.5.)

SIM ***2.77** The circuit shown in Fig. P2.77 is intended to supply a voltage to floating loads (those for which both terminals are ungrounded) while making greatest possible use of the available power supply.

(a) Assuming ideal op amps, sketch the voltage waveforms at nodes B and C for a 1-V peak-to-peak sine wave applied at A. Also sketch v_O.
(b) What is the voltage gain v_O/v_I?

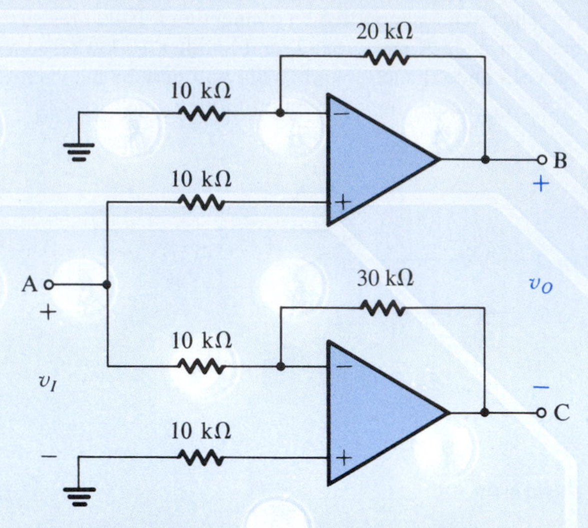

Figure P2.77

(c) Assuming that the op amps operate from ±15-V power supplies and that their output saturates at ±14 V (in the manner shown in Fig. 1.14), what is the largest sine-wave output that can be accommodated? Specify both its peak-to-peak and rms values.

***2.78** The two circuits in Fig. P2.78 are intended to function as voltage-to-current converters; that is, they supply the load impedance Z_L with a current proportional to v_I and independent of the value of Z_L. Show that this is indeed the case, and find for each circuit i_O as a function of v_I. Comment on the differences between the two circuits.

Section 2.5: Integrators and Differentiators

2.79 A Miller integrator incorporates an ideal op amp, a resistor R of 100 kΩ, and a capacitor C of 1 nF. A sine-wave signal is applied to its input.

(a) At what frequency (in Hz) are the input and output signals equal in amplitude?
(b) At that frequency, how does the phase of the output sine wave relate to that of the input?
(c) If the frequency is lowered by a factor of 10 from that found in (a), by what factor does the output voltage change, and in what direction (smaller or larger)?
(d) What is the phase relation between the input and output in situation (c)?

D 2.80 Design a Miller integrator with a time constant of 0.1 s and an input resistance of 100 kΩ. A dc voltage of −1 volt is applied at the input at time 0, at which moment $v_O = -10$ V. How long does it take the output to reach 0 V? +10 V?

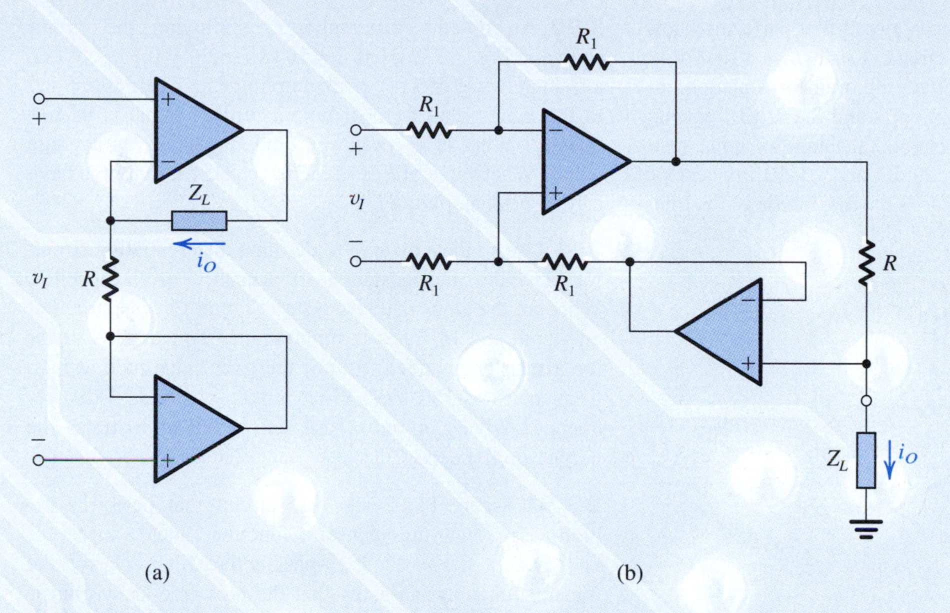

Figure P2.78

2.81 An op-amp-based inverting integrator is measured at 1 kHz to have a voltage gain of −100 V/V. At what frequency is its gain reduced to −1 V/V? What is the integrator time constant?

D 2.82 Design a Miller integrator that has a unity-gain frequency of 1 krad/s and an input resistance of 100 kΩ. Sketch the output you would expect for the situation in which, with output initially at 0 V, a 2-V, 2-ms pulse is applied to the input. Characterize the output that results when a sine wave 2 sin 1000t is applied to the input.

D 2.83 Design a Miller integrator whose input resistance is 20 kΩ and unity-gain frequency is 10 kHz. What components are needed? For long-term stability, a feedback resistor is introduced across the capacitor, limits the dc gain to 40 dB. What is its value? What is the associated lower 3-dB frequency? Sketch and label the output that results with a 0.1-ms, 1-V positive-input pulse (initially at 0 V) with (a) no dc stabilization (but with the output initially at 0 V) and (b) the feedback resistor connected.

***2.84** A Miller integrator whose input and output voltages are initially zero and whose time constant is 1 ms is driven by the signal shown in Fig. P2.84. Sketch and label the output waveform that results. Indicate what happens if the input levels are ±2 V, with the time constant the same (1 ms) and with the time constant raised to 2 ms.

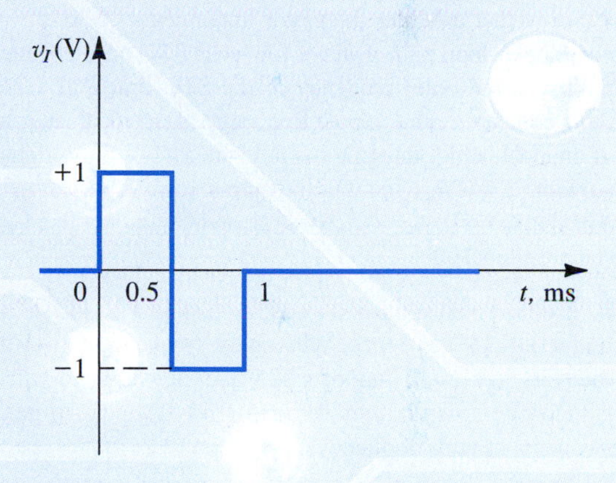

Figure P2.84

2.85 Consider a Miller integrator having a time constant of 1 ms and an output that is initially zero, when fed with a string of pulses of 10-µs duration and 1-V amplitude rising from 0 V (see Fig. P2.85). Sketch and label the output wave form resulting. How many pulses are required for an output voltage change of 1 V?

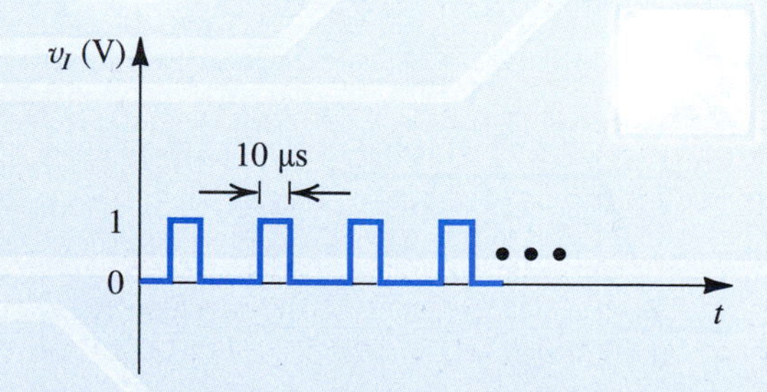

Figure P2.85

D 2.86 Figure P2.86 shows a circuit that performs a low-pass STC function. Such a circuit is known as a first-order, low-pass active filter. Derive the transfer function and show that the dc gain is $(-R_2/R_1)$ and the 3-dB frequency $\omega_0 = 1/CR_2$. Design the circuit to obtain an input resistance of 10 kΩ, a dc gain of 20 dB, and a 3-dB frequency of 10 kHz. At what frequency does the magnitude of the transfer function reduce to unity?

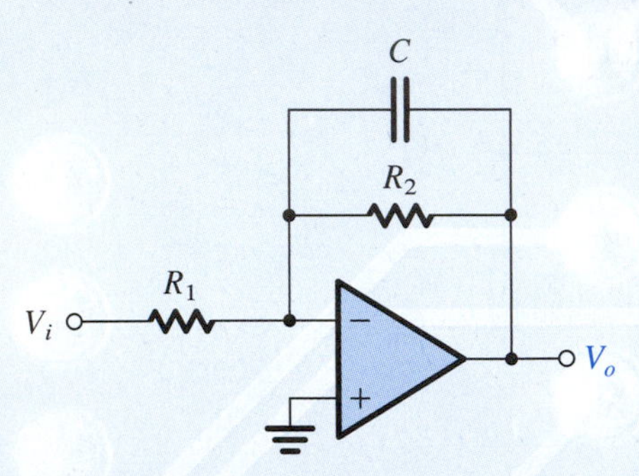

Figure P2.86

2.87 Show that a Miller integrator implemented with an op amp with open-loop gain A_0 has a low-pass STC transfer function. What is the pole frequency of the STC function? How does this compare with the pole frequency of the ideal integrator? If an ideal Miller integrator is fed with a –1-V pulse signal with a width $T = CR$, what will the output voltage be at $t = T$? Assume that at $t = 0$, $v_O = 0$. Repeat for an integrator with an op amp having $A_0 = 1000$.

2.88 A differentiator utilizes an ideal op amp, a 10-kΩ resistor, and a 0.01-μF capacitor. What is the frequency f_0 (in Hz) at which its input and output sine-wave signals have equal magnitude? What is the output signal for a 1-V peak-to-peak sine-wave input with frequency equal to $10f_0$?

2.89 An op-amp differentiator with 1-ms time constant is driven by the rate-controlled step shown in Fig. P2.89. Assuming v_O to be zero initially, sketch and label its waveform.

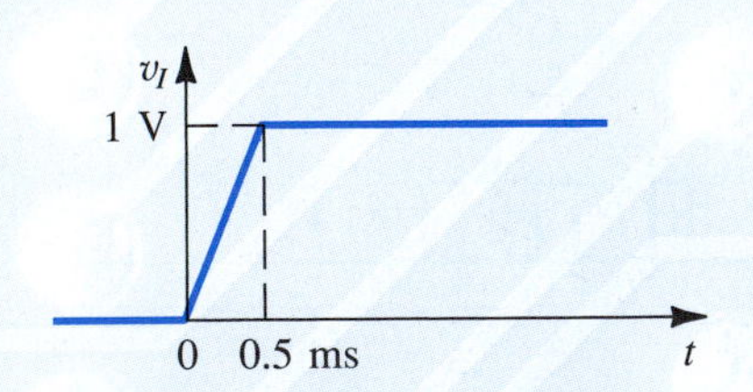

Figure P2.89

2.90 An op-amp differentiator, employing the circuit shown in Fig. 2.27(a), has $R = 10$ kΩ and $C = 0.1$ μF. When a triangle wave of ±1-V peak amplitude at 1 kHz is applied to the input, what form of output results? What is its frequency? What is its peak amplitude? What is its average value? What value of R is needed to cause the output to have a 10-V peak amplitude?

2.91 Use an ideal op amp to design a differentiation circuit for which the time constant is 10^{-3} s using a 10-nF capacitor. What are the gains and phase shifts found for this circuit at one-tenth and 10 times the unity-gain frequency? A series input resistor is added to limit the gain magnitude at high frequencies to 100 V/V. What is the associated 3-dB frequency? What gain and phase shift result at 10 times the unity-gain frequency?

D 2.92 Figure P2.92 shows a circuit that performs the high-pass, single-time-constant function. Such a circuit is known as a first-order high-pass active filter. Derive the transfer function and show that the high-frequency gain is $(-R_2/R_1)$ and the 3-dB frequency $\omega_0 = 1/CR_1$. Design the circuit to obtain a high-frequency input resistance of 10 kΩ, a high-frequency gain of 40 dB, and a 3-dB frequency of 500 Hz. At what frequency does the magnitude of the transfer function reduce to unity?

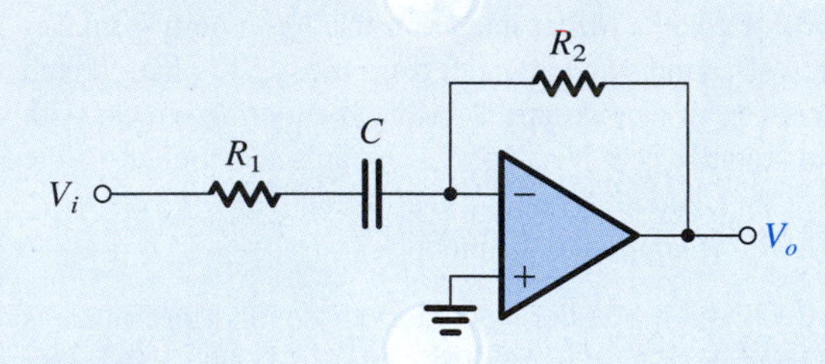

Figure P2.92

D **2.93 Derive the transfer function of the circuit in Fig. P2.93 (for an ideal op amp) and show that it can be written in the form

$$\frac{V_o}{V_i} = \frac{-R_2/R_1}{[1 + (\omega_1/j\omega)][1 + j(\omega/\omega_2)]}$$

where $\omega_1 = 1/C_1R_1$ and $\omega_2 = 1/C_2R_2$. Assuming that the circuit is designed such that $\omega_2 \gg \omega_1$, find approximate expressions for the transfer function in the following frequency regions:

(a) $\omega \ll \omega_1$
(b) $\omega_1 \ll \omega \ll \omega_2$
(c) $\omega \gg \omega_2$

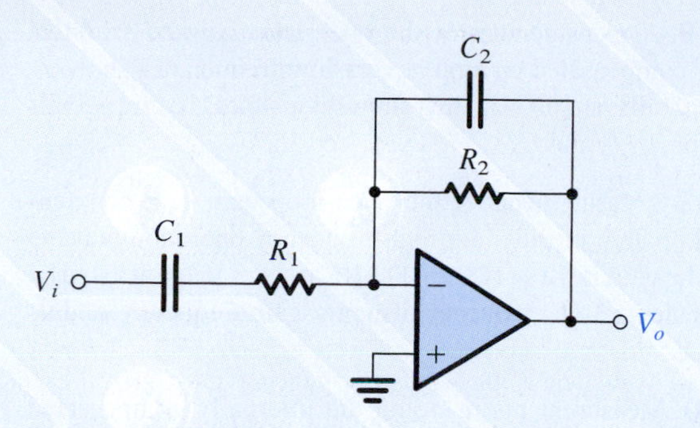

Figure P2.93

Use these approximations to sketch a Bode plot for the magnitude response. Observe that the circuit performs as an amplifier whose gain rolls off at the low-frequency end in the manner of a high-pass STC network, and at the high-frequency end in the manner of a low-pass STC network. Design the circuit to provide a gain of 40 dB in the "middle frequency range," a low-frequency 3-dB point at 100 Hz, a high-frequency 3-dB point at 100 kHz, and an input resistance (at $\omega \gg \omega_1$) of 1 kΩ.

Section 2.6: DC Imperfections

2.94 An op amp wired in the inverting configuration with the input grounded, having $R_2 = 100$ kΩ and $R_1 = 1$ kΩ, has an output dc voltage of –0.4 V. If the input bias current is known to be very small, find the input offset voltage.

2.95 A noninverting amplifier with a gain of 200 uses an op amp having an input offset voltage of ±2 mV. Find the output when the input is 0.01 sin ωt, volts.

2.96 A noninverting amplifier with a closed-loop gain of 1000 is designed using an op amp having an input offset voltage of 5 mV and output saturation levels of ±13 V. What is the maximum amplitude of the sine wave that can be applied at the input without the output clipping? If the amplifier is capacitively coupled in the manner indicated in Fig. 2.36, what would the maximum possible amplitude be?

2.97 An op amp connected in a closed-loop inverting configuration having a gain of 1000 V/V and using relatively small-valued resistors is measured with input grounded to have a dc output voltage of –1.4 V. What is its input offset voltage? Prepare an offset-voltage-source sketch resembling that in Fig. 2.28. Be careful of polarities.

2.98 A particular inverting amplifier with nominal gain of –100 V/V uses an imperfect op amp in conjunction with 100-kΩ and 10-MΩ resistors. The output voltage is found to be +9.31 V when measured with the input open and +9.09 V with the input grounded.

(a) What is the bias current of this amplifier? In what direction does it flow?
(b) Estimate the value of the input offset voltage.
(c) A 10-MΩ resistor is connected between the positive-input terminal and ground. With the input left floating (disconnected), the output dc voltage is measured to be –0.8 V. Estimate the input offset current.

D *2.99 A noninverting amplifier with a gain of +10 V/V using 100 kΩ as the feedback resistor operates from a 5-kΩ source. For an amplifier offset voltage of 0 mV, but with a bias current of 1 μA and an offset current of 0.1 μA, what range of outputs would you expect? Indicate where you would add an additional resistor to compensate for the bias currents. What does the range of possible outputs then become? A designer wishes to use this amplifier with a 15-kΩ source. In order to compensate for the bias current in this case, what resistor would you use? And where?

D 2.100 The circuit of Fig. 2.36 is used to create an ac-coupled noninverting amplifier with a gain of 200 V/V using resistors no larger than 100 kΩ. What values of R_1, R_2, and R_3 should be used? For a break frequency due to C_1 at 100 Hz, and that due to C_2 at 10 Hz, what values of C_1 and C_2 are needed?

***2.101** Consider the difference amplifier circuit in Fig. 2.16. Let $R_1 = R_3 = 10$ kΩ and $R_2 = R_4 = 1$ MΩ. If the op amp has $V_{OS} = 4$ mV, $I_B = 0.5$ μA, and $I_{OS} = 0.1$ μA, find the worst-case (largest) dc offset voltage at the output.

***2.102** The circuit shown in Fig. P2.102 uses an op amp having a ±4-mV offset. What is its output offset voltage? What does the output offset become with the input ac coupled through a capacitor C? If, instead, a large capacitor is placed in series with 1-kΩ resistor, what does the output offset become?

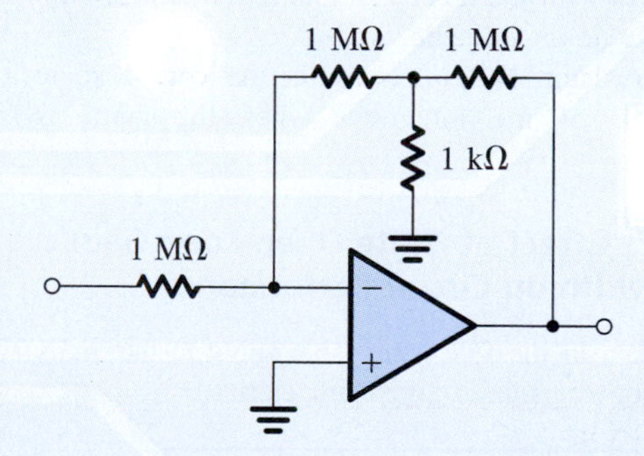

Figure P2.102

2.103 Using offset-nulling facilities provided for the op amp, a closed-loop amplifier with gain of +1000 is adjusted at 25°C to produce zero output with the input grounded. If the input offset-voltage drift of the op amp is specified to be 10 μV/°C, what output would you expect at 0°C and at 75°C? While nothing can be said separately about the polarity of the output offset at either 0 or 75°C, what would you expect their relative polarities to be?

2.104 An op amp is connected in a closed loop with gain of +100 utilizing a feedback resistor of 1 MΩ.

(a) If the input bias current is 100 nA, what output voltage results with the input grounded?
(b) If the input offset voltage is ±1 mV and the input bias current as in (a), what is the largest possible output that can be observed with the input grounded?
(c) If bias-current compensation is used, what is the value of the required resistor? If the offset current is no more than one-tenth the bias current, what is the resulting output offset voltage (due to offset current alone)?
(d) With bias-current compensation as in (c) in place what is the largest dc voltage at the output due to the combined effect of offset voltage and offset current?

***2.105** An op amp intended for operation with a closed-loop gain of –100 V/V uses resistors of 10 kΩ and 1 MΩ with a bias-current-compensation resistor R_3. What should the value of R_3 be? With input grounded, the output offset voltage is found to be +0.21 V. Estimate the input offset current assuming zero input offset voltage. If the input offset voltage can be as large as 1 mV of unknown polarity, what range of offset current is possible?

2.106 A Miller integrator with $R = 10$ kΩ and $C = 10$ nF is implemented by using an op amp with $V_{OS} = 3$ mV, $I_B = 0.1$ μA, and $I_{OS} = 10$ nA. To provide a finite dc gain, a 1-MΩ resistor is connected across the capacitor.

(a) To compensate for the effect of I_B, a resistor is connected in series with the positive-input terminal of the op amp. What should its value be?
(b) With the resistor of (a) in place, find the worst-case dc output voltage of the integrator when the input is grounded.

Section 2.7: Effect of Finite Open-Loop Gain and Bandwidth on Circuit Performance

2.107 The data in the following table apply to internally compensated op amps. Fill in the blank entries.

A_0	f_b (Hz)	f_t (Hz)
10^5	10^2	
10^6		10^6
	10^3	10^8
	10^{-1}	10^6
2×10^5	10	

2.108 A measurement of the open-loop gain of an internally compensated op amp at very low frequencies shows it to be 92 dB; at 100 kHz, this shows it is 40 dB. Estimate values for A_0, f_b, and f_t.

2.109 Measurements of the open-loop gain of a compensated op amp intended for high-frequency operation indicate that the gain is 5.1×10^3 at 100 kHz and 8.3×10^3 at 10 kHz. Estimate its 3-dB frequency, its unity-gain frequency, and its dc gain.

2.110 Measurements made on the internally compensated amplifiers listed below provide the dc gain and the frequency at which the gain has dropped by 20 dB. For each, what are the 3 dB and unity-gain frequencies?

(a) 3×10^5 V/V and 6×10^2 Hz
(b) 50×10^5 V/V and 10 Hz
(c) 1500 V/V and 0.1 MHz
(d) 100 V/V and 0.1 GHz
(e) 25 V/mV and 25 kHz

2.111 An inverting amplifier with nominal gain of –20 V/V employs an op amp having a dc gain of 10^4 and a unity-gain frequency of 10^6 Hz. What is the 3-dB frequency f_{3dB} of the closed-loop amplifier? What is its gain at 0.1 f_{3dB} and at 10 f_{3dB}?

2.112 A particular op amp, characterized by a gain–bandwidth product of 10 MHz, is operated with a closed-loop gain of +100 V/V. What 3-dB bandwidth results? At what frequency does the closed-loop amplifier exhibit a –6° phase shift? A –84° phase shift?

2.113 Find the f_t required for internally compensated op amps to be used in the implementation of closed-loop amplifiers with the following nominal dc gains and 3-dB bandwidths:

(a) –100 V/V; 100 kHz
(b) +100 V/V; 100 kHz
(c) +2 V/V; 10 MHz
(d) –2 V/V; 10 MHz
(e) –1000 V/V; 20 kHz
(f) +1 V/V; 1 MHz
(g) –1 V/V; 1 MHz

2.114 A noninverting op-amp circuit with a gain of 96 V/V is found to have a 3-dB frequency of 8 kHz. For a particular system application, a bandwidth of 24 kHz is required. What is the highest gain available under these conditions?

2.115 Consider a unity-gain follower utilizing an internally compensated op amp with $f_t = 1$ MHz. What is the 3-dB frequency of the follower? At what frequency is the gain of the follower 1% below its low-frequency magnitude? If the input to the follower is a 1-V step, find the 10% to 90% rise time of the output voltage. (*Note:* The step response of STC low-pass networks is discussed in Appendix E.)

D *2.116 It is required to design a noninverting amplifier with a dc gain of 10. When a step voltage of 100 mV is applied at the input, it is required that the output be within 1% of its final value of 1 V in at most 100 ns. What must the f_t of the op amp be? (*Note:* The step response of STC low-pass networks is discussed in Appendix E.)

D *2.117 This problem illustrates the use of cascaded closed-loop amplifiers to obtain an overall bandwidth greater than can be achieved using a single-stage amplifier with the same overall gain.

(a) Show that cascading two identical amplifier stages, each having a low-pass STC frequency response with a 3-dB frequency f_1, results in an overall amplifier with a 3-dB frequency given by

$$f_{3\text{dB}} = \sqrt{\sqrt{2} - 1}\, f_1$$

(b) It is required to design a noninverting amplifier with a dc gain of 40 dB utilizing a single internally compensated op amp with $f_t = 1$ MHz. What is the 3-dB frequency obtained?
(c) Redesign the amplifier of (b) by cascading two identical noninverting amplifiers each with a dc gain of 20 dB. What is the 3-dB frequency of the overall amplifier? Compare this to the value obtained in (b) above.

D **2.118 A designer, wanting to achieve a stable gain of 100 V/V at 5 MHz, considers her choice of amplifier topologies. What unity-gain frequency would a single operational amplifier require to satisfy her need? Unfortunately, the best available amplifier has an f_t of 40 MHz. How many such amplifiers connected in a cascade of identical noninverting stages would she need to achieve her goal? What is the 3-dB frequency of each stage she can use? What is the overall 3-dB frequency?

2.119 Consider the use of an op amp with a unity-gain frequency f_t in the realization of:

(a) An inverting amplifier with dc gain of magnitude *K*.
(b) A noninverting amplifier with a dc gain of *K*.

In each case find the 3-dB frequency and the gain-bandwidth product (GBP $\equiv$ |Gain| $\times f_{3\text{dB}}$). Comment on the results.

***2.120** Consider an inverting summer with two inputs V_1 and V_2 and with $V_o = -(V_1 + 2V_2)$. Find the 3-dB frequency of each of the gain functions V_o/V_1 and V_o/V_2 in terms of the op amp f_t. (*Hint:* In each case, the other input to the summer can be set to zero—an application of superposition.)

Section 2.8: Large-Signal Operation of Op Amps

2.121 A particular op amp using ±15-V supplies operates linearly for outputs in the range −12 V to +12 V. If used in an inverting amplifier configuration of gain −100, what is the rms value of the largest possible sine wave that can be applied at the input without output clipping?

2.122 Consider an op amp connected in the inverting configuration to realize a closed-loop gain of −100 V/V utilizing resistors of 1 kΩ and 100 kΩ. A load resistance R_L is connected from the output to ground, and a low-frequency sine-wave signal of peak amplitude V_p is applied to the input. Let the op amp be ideal except that its output voltage saturates at ±10 V and its output current is limited to the range ±20 mA.

(a) For $R_L = 1$ kΩ, what is the maximum possible value of V_p while an undistorted output sinusoid is obtained?
(b) Repeat (a) for $R_L = 100$ Ω.
(c) If it is desired to obtain an output sinusoid of 10-V peak amplitude, what minimum value of R_L is allowed?

2.123 An op amp having a slew rate of 10 V/μs is to be used in the unity-gain follower configuration, with input pulses that rise from 0 to 5 V. What is the shortest pulse that can be used while ensuring full-amplitude output? For such a pulse, describe the output resulting.

2.124 For operation with 10-V output pulses with the requirement that the sum of the rise and fall times represent only 20% of the pulse width (at half amplitude), what is the slew-rate requirement for an op amp to handle pulses 2 μs wide? (*Note:* The rise and fall times of a pulse signal are usually measured between the 10%- and 90%-height points.)

2.125 What is the highest frequency of a triangle wave of 20-V peak-to-peak amplitude that can be reproduced by an op amp whose slew rate is 10 V/μs? For a sine wave of the same frequency, what is the maximum amplitude of output signal that remains undistorted?

2.126 For an amplifier having a slew rate of 60 V/μs, what is the highest frequency at which a 20-V peak-to-peak sine wave can be produced at the output?

D *2.127 In designing with op amps one has to check the limitations on the voltage and frequency ranges of operation of the closed-loop amplifier, imposed by the op-amp finite bandwidth (f_t), slew rate (SR), and output saturation (V_{omax}). This problem illustrates the point by considering the use of an op amp with $f_t = 2$ MHz, SR = 1 V/μs, and $V_{omax} = 10$ V in the design of a noninverting amplifier with a nominal gain of 10. Assume a sine-wave input with peak amplitude V_i.

(a) If $V_i = 0.5$ V, what is the maximum frequency before the output distorts?
(b) If $f = 20$ kHz, what is the maximum value of V_i before the output distorts?
(c) If $V_i = 50$ mV, what is the useful frequency range of operation?
(d) If $f = 5$ kHz, what is the useful input voltage range?

CHAPTER 3

Semiconductors

IN THIS CHAPTER YOU WILL LEARN

1. The basic properties of semiconductors and in particular silicon, which is the material used to make most of today's electronic circuits.
2. How doping a pure silicon crystal dramatically changes its electrical conductivity, which is the fundamental idea underlying the use of semiconductors in the implementation of electronic devices.
3. The two mechanisms by which current flows in semiconductors: drift and diffusion of charge carriers.
4. The structure and operation of the *pn* junction; a basic semiconductor structure that implements the diode and plays a dominant role in transistors.

Introduction

Thus far we have dealt with electronic circuits, and notably amplifiers, as system building blocks. For instance, in Chapter 2 we learned how to use op amps to design interesting and useful circuits, taking advantage of the terminal characteristics of the op amp and without any knowledge of what is inside the op amp package. Though interesting and motivating, this approach has its limitations. Indeed, to achieve our goal of preparing the reader to become a proficient circuit designer, we have to go beyond this black-box or system-level abstraction and learn about the basic devices from which electronic circuits are assembled, namely, diodes (Chapter 4) and transistors (Chapters 5 and 6). These solid-state devices are made using semiconductor materials, predominantly silicon.

In this chapter, we briefly introduce the properties and physics of semiconductors. The objective is to provide a basis for understanding the physical operation of diodes and transistors in order to enable their effective use in the design of circuits. Although many of the concepts studied in this chapter apply to semiconductor materials in general, our treatment is heavily biased toward silicon, simply because it is the material used in the vast majority of microelectronic circuits. To complement the material presented here, Appendix A provides a description of the integrated-circuit fabrication process. As discussed in Appendix A, whether our circuit consists of a single transistor or is an **integrated circuit** containing more than 2 billion transistors, it is fabricated in a single silicon crystal, which gives rise to the name **monolithic circuit**. This chapter therefore begins with a study of the crystal structure of semiconductors and introduces the two types of charge carriers available for current conduction: electrons and holes. The most significant property of semiconductors is that their conductivity can be varied over a very wide range through the introduction of controlled amounts of impurity atoms into the semiconductor crystal in a process called **doping**. Doped semiconductors are discussed in Section 3.2. This is followed by the study in Section 3.3 of the two mechanisms for current flow in semiconductors, namely, carrier drift and carrier diffusion.

Armed with these basic semiconductor concepts, we spend the remainder of the chapter on the study of an important semiconductor structure; the *pn* junction. In addition to being essentially a diode, the *pn* junction is the basic element of the bipolar junction transistor (BJT, Chapter 6) and plays an important role in the operation of field-effect transistors (FETs, Chapter 5).

3.1 Intrinsic Semiconductors

As their name implies, semiconductors are materials whose conductivity lies between that of conductors, such as copper, and insulators, such as glass. There are two kinds of semiconductors: single-element semiconductors, such as germanium and silicon, which are in group IV in the periodic table; and compound semiconductors, such as gallium-arsenide, which are formed by combining elements from groups III and V or groups II and VI. Compound semiconductors are useful in special electronic circuit applications as well as in applications that involve light, such as light-emitting diodes (LEDs). Of the two elemental semiconductors, germanium was used in the fabrication of very early transistors (late 1940s, early 1950s). It was quickly supplanted, however, with silicon, on which today's integrated-circuit technology is almost entirely based. For this reason, we will deal mostly with silicon devices throughout this book.[1]

A silicon atom has four valence electrons, and thus it requires another four to complete its outermost shell. This is achieved by sharing one of its valence electrons with each of its four neighboring atoms. Each pair of shared electrons forms a **covalent bond**. The result is that a crystal of pure or intrinsic silicon has a regular lattice structure, where the atoms are held in their position by the covalent bonds. Figure 3.1 shows a two-dimensional representation of such a structure.

At sufficiently low temperatures, approaching absolute zero (0 K), all the covalent bonds are intact and no electrons are available to conduct electric current. Thus, at such low temperatures, the intrinsic silicon crystal behaves as an insulator.

At room temperature, sufficient thermal energy exists to break some of the covalent bonds, a process known as thermal generation. As shown in Fig. 3.2, when a covalent bond is broken, an electron is freed. The **free electron** can wander away from its parent atom, and it becomes available to conduct electric current if an electric field is applied to the crystal. As the electron leaves its parent atom, it leaves behind a net positive charge, equal to the magnitude of the electron charge. Thus, an electron from a neighboring atom may be attracted to this positive charge, and leaves its parent atom. This action fills up the "hole" that existed in the ionized atom but creates a new hole in the other atom. This process may repeat itself, with the result that we effectively have a positively charged carrier, or **hole**, moving through the silicon crystal structure and being available to conduct electric current. The charge of a hole is equal in magnitude to the charge of an electron. We can thus see that as temperature increases, more covalent bonds are broken and electron–hole pairs are generated. The increase in the numbers of free electrons and holes results in an increase in the conductivity of silicon.

[1]An exception is the subject of gallium arsenide (GaAs) circuits, which though not covered in this edition of the book, is studied in some detail in material provided on the text website and on the disc accompanying the text.

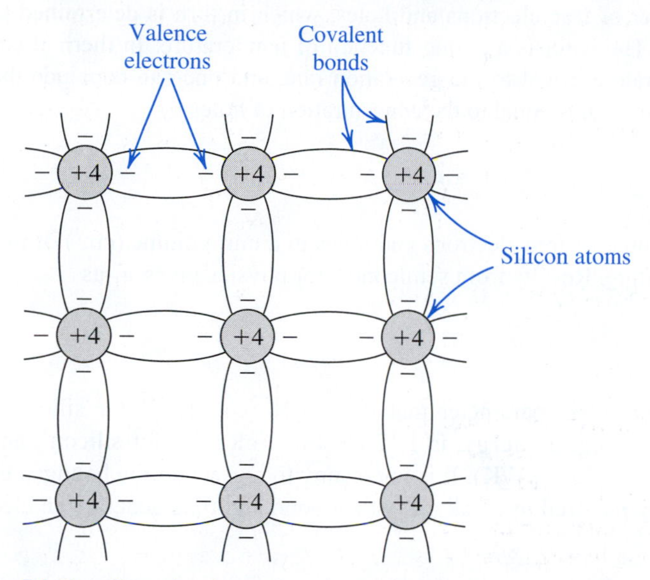

Figure 3.1 Two-dimensional representation of the silicon crystal. The circles represent the inner core of silicon atoms, with +4 indicating its positive charge of $+4q$, which is neutralized by the charge of the four valence electrons. Observe how the covalent bonds are formed by sharing of the valence electrons. At 0 K, all bonds are intact and no free electrons are available for current conduction.

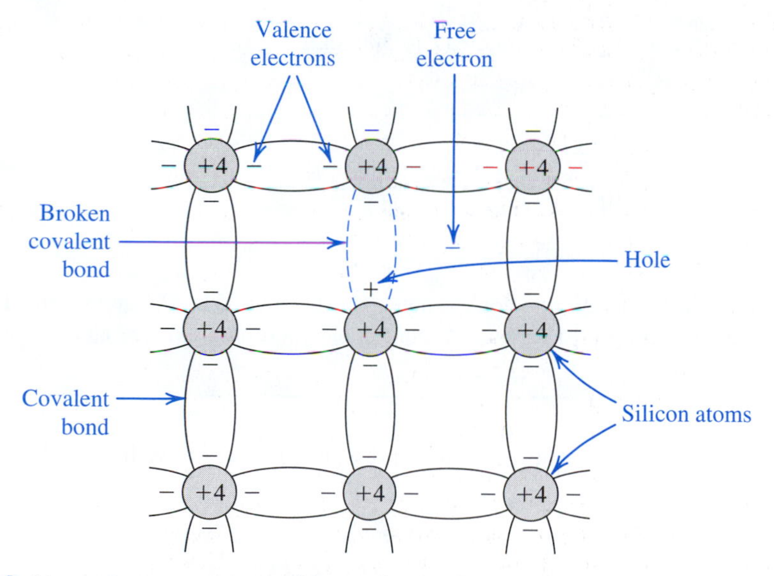

Figure 3.2 At room temperature, some of the covalent bonds are broken by thermal generation. Each broken bond gives rise to a free electron and a hole, both of which become available for current conduction.

Thermal generation results in free electrons and holes in equal numbers and hence equal concentrations, where concentration refers to the number of charge carriers per unit volume (cm^3). The free electrons and holes move randomly through the silicon crystal structure, and in the process some electrons may fill some of the holes. This process, called **recombination**, results in the disappearance of free electrons and holes. The recombination rate is proportional to the number of free electrons and holes, which in turn is determined by the thermal **generation** rate. The latter is a strong function of temperature. In thermal equilibrium, the recombination rate is equal to the generation rate, and one can conclude that the concentration of free electrons n is equal to the concentration of holes p,

$$n = p = n_i \tag{3.1}$$

where n_i denotes the number of free electrons and holes in a unit volume (cm^3) of intrinsic silicon at a given temperature. Results from semiconductor physics gives n_i as

$$n_i = BT^{3/2} e^{-E_g/2kT} \tag{3.2}$$

where B is a material-dependent parameter that is $7.3 \times 10^{15} cm^{-3} K^{-3/2}$ for silicon; E_g, a parameter known as the **bandgap energy**, is 1.12 electron volt (eV) for silicon[2]; and k is Boltzmann's constant (8.62×10^{-5} eV/K). It is interesting to know that the bandgap energy E_g is the minimum energy required to break a covalent bond and thus generate an electron-hole pair.

Example 3.1

Calculate the value of n_i for silicon at room temperature ($T \simeq 300$ K).

Solution

Substituting the values given above in Eq. (3.1) provides

$$n_i = 7.3 \times 10^{15} (300)^{3/2} e^{-1.12/(2 \times 8.62 \times 10^{-5} \times 300)}$$

$$= 1.5 \times 10^{10} \text{carriers/cm}^3$$

Although this number seems large, to place it into context note that silicon has 5×10^{22} atoms/cm^3. Thus at room temperature only one in about 5×10^{12} atoms is ionized and contributing a free electron and a hole!

Finally, it is useful for future purposes to express the product of the hole and free-electron concentration as

$$pn = n_i^2 \tag{3.3}$$

where for silicon at room temperature, $n_i \simeq 1.5 \times 10^{10}/\text{cm}^3$. As will be seen shortly, this relationship extends to extrinsic or doped silicon as well.

[2] Note that 1 eV = 1.6×10^{-19} J.

EXERCISE

3.1 Calculate the intrinsic carrier density n_i for silicon at $T = 50$ K and 350 K.
Ans. $9.6 \times 10^{-39}/\text{cm}^3$; $4.15 \times 10^{11}/\text{cm}^3$

3.2 Doped Semiconductors

The intrinsic silicon crystal described above has equal concentrations of free electrons and holes, generated by thermal generation. These concentrations are far too small for silicon to conduct appreciable current at room temperature. Also, the carrier concentrations and hence the conductivity are strong functions of temperature, not a desirable property in an electronic device. Fortunately, a method was developed to change the carrier concentration in a semiconductor crystal substantially and in a precisely controlled manner. This process is known as doping, and the resulting silicon is referred to as **doped silicon**.

Doping involves introducing impurity atoms into the silicon crystal in sufficient numbers to substantially increase the concentration of either free electrons or holes but with little or no change in the crystal properties of silicon. To increase the concentration of free electrons, n, silicon is doped with an element with a valence of 5, such as phosphorus. The resulting doped silicon is then said to be of ***n* type**. To increase the concentration of holes, p, silicon is doped with an element having a valence of 3, such as boron, and the resulting doped silicon is said to be of ***p* type**.

Figure 3.3 shows a silicon crystal doped with phosphorus impurity. The dopant (phosphorus) atoms replace some of the silicon atoms in the crystal structure. Since the phosphorus atom has five electrons in its outer shell, four of these electrons form covalent bonds with the

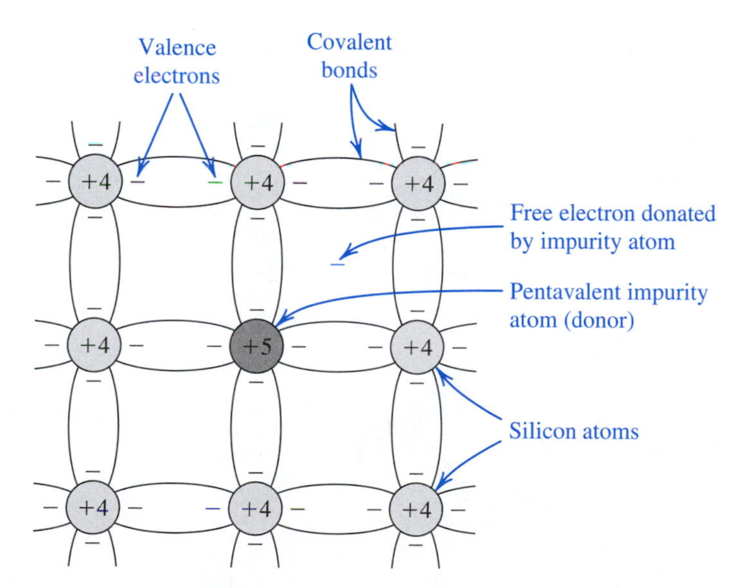

Figure 3.3 A silicon crystal doped by a pentavalent element. Each dopant atom donates a free electron and is thus called a donor. The doped semiconductor becomes n type.

neighboring atoms, and the fifth electron becomes a free electron. Thus each phosphorus atom *donates* a free electron to the silicon crystal, and the phosphorus impurity is called a **donor**. It should be clear, though, that no holes are generated by this process. The positive charge associated with the phosphorus atom is a **bound charge** that does not move through the crystal.

If the concentration of donor atoms is N_D, where N_D is usually much greater than n_i, the concentration of free electrons in the n-type silicon will be

$$n_n \simeq N_D \tag{3.4}$$

where the subscript n denotes n-type silicon. Thus n_n is determined by the doping concentration and not by temperature. This is not the case, however, for the hole concentration. All the holes in the n-type silicon are those generated by thermal ionization. Their concentration p_n can be found by noting that the relationship in Eq. (3.3) applies equally well for doped silicon, provided thermal equilibrium is achieved. Thus for n-type silicon

$$p_n n_n = n_i^2$$

Substituting for n_n from Eq. (3.4), we obtain for p_n

$$p_n \simeq \frac{n_i^2}{N_D} \tag{3.5}$$

Thus p_n will have the same dependence on temperature as that of n_i^2. Finally, we note that in n-type silicon the concentration of free electrons n_n will be much larger than that of holes. Hence electrons are said to be the **majority** charge carriers and holes the **minority** charge carriers in n-type silicon.

To obtain p-type silicon in which holes are the majority charge carriers, a trivalent impurity such as boron is used. Figure 3.4 shows a silicon crystal doped with boron. Note that the boron atoms replace some of the silicon atoms in the silicon crystal structure. Since each

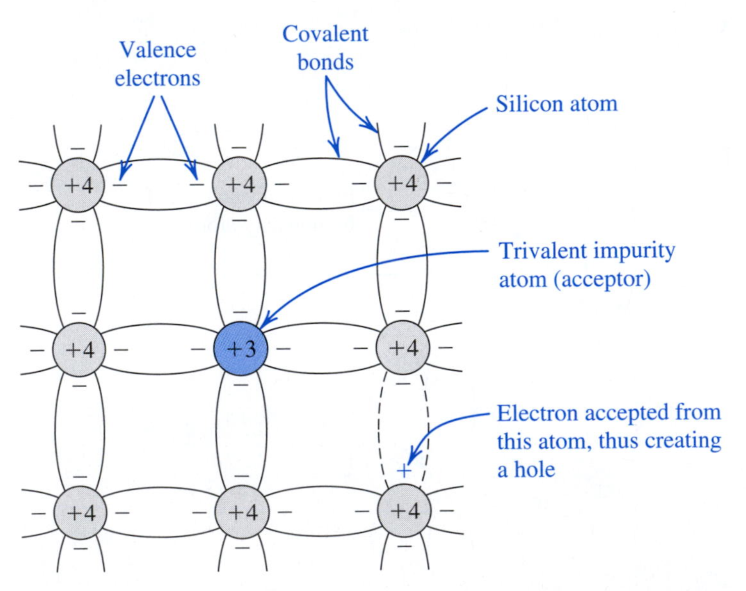

Figure 3.4 A silicon crystal doped with a trivalent impurity. Each dopant atom gives rise to a hole, and the semiconductor becomes p type.

boron atom has three electrons in its outer shell, it **accepts** an electron from a neighboring atom, thus forming covalent bonds. The result is a hole in the neighboring atom and a bound negative charge at the **acceptor** (boron) atom. It follows that each acceptor atom provides a hole. If the acceptor doping concentration is N_A, where $N_A \gg n_i$; the hole concentration becomes

$$p_p \simeq N_A \tag{3.6}$$

where the subscript p denotes p-type silicon. Thus, here the majority carriers are holes and their concentration is determined by N_A. The concentration of minority electrons can be found by using the relationship

$$p_p n_p = n_i^2$$

and substituting for p_p from Eq. (3.6),

$$n_p \simeq \frac{n_i^2}{N_A} \tag{3.7}$$

Thus, the concentration of the minority electrons will have the same temperature dependence as that of n_i^2.

It should be emphasized that a piece of n-type or p-type silicon is electrically neutral; the charge of the majority free carriers (electrons in the n-type and holes in the p-type silicon) are neutralized by the bound charges associated with the impurity atoms.

Example 3.2

Consider an n-type silicon for which the dopant concentration $N_D = 10^{17}/\text{cm}^3$. Find the electron and hole concentrations at $T = 300$ K.

Solution

The concentration of the majority electrons is

$$n_n \simeq N_D = 10^{17}/\text{cm}^3$$

The concentration of the minority holes is

$$p_n \simeq \frac{n_i^2}{N_D}$$

In Example 3.1 we found that at $T = 300$ K, $n_i = 1.5 \times 10^{10}/\text{cm}^3$. Thus,

$$p_n = \frac{(1.5 \times 10^{10})^2}{10^{17}}$$

$$= 2.25 \times 10^3/\text{cm}^3$$

Observe that $n_n \gg n_i$ and that n_n is vastly higher than p_n.

EXERCISES

3.2 For the situation in Example 3.2, find the electron and hole concentrations at 350 K. You may use the value of n_i at $T = 350$ K found in Exercise 3.1.
Ans. $n_n = 10^{17}/\text{cm}^3$, $p_n = 1.72 \times 10^6/\text{cm}^3$

3.3 For a silicon crystal doped with boron, what must N_A be if at $T = 300$ K the electron concentration drops below the intrinsic level by a factor of 10^6?
Ans. $N_A = 1.5 \times 10^{16}/\text{cm}^3$

3.3 Current Flow in Semiconductors

There are two distinctly different mechanisms for the movement of charge carriers and hence for current flow in semiconductors: drift and diffusion.

3.3.1 Drift Current

When an electrical field E is established in a semiconductor crystal, holes are accelerated in the direction of E, and free electrons are accelerated in the direction opposite to that of E. This situation is illustrated in Fig. 3.5. The holes acquire a velocity $v_{p\text{-drift}}$ given by

$$v_{p\text{-drift}} = \mu_p E \tag{3.8}$$

where μ_p is a constant called the **hole mobility**: It represents the degree of ease by which holes move through the silicon crystal in response to the electrical field E. Since velocity has the units of centimeters per second and E has the units of volts per centimeter, we see from Eq. (3.8) that the mobility μ_p must have the units of centimeters squared per volt-second ($\text{cm}^2/\text{V}\cdot\text{s}$). For intrinsic silicon $\mu_p = 480\ \text{cm}^2/\text{V}\cdot\text{s}$.

The free electrons acquire a drift velocity $v_{n\text{-drift}}$ given by

$$v_{n\text{-drift}} = -\mu_n E \tag{3.9}$$

where the result is negative because the electrons move in the direction opposite to E. Here μ_n is the **electron mobility**, which for intrinsic silicon is about 1350 $\text{cm}^2/\text{V}\cdot\text{s}$. Note that μ_n is about 2.5 times μ_p, signifying that electrons move with much greater ease through the silicon crystal than do holes.

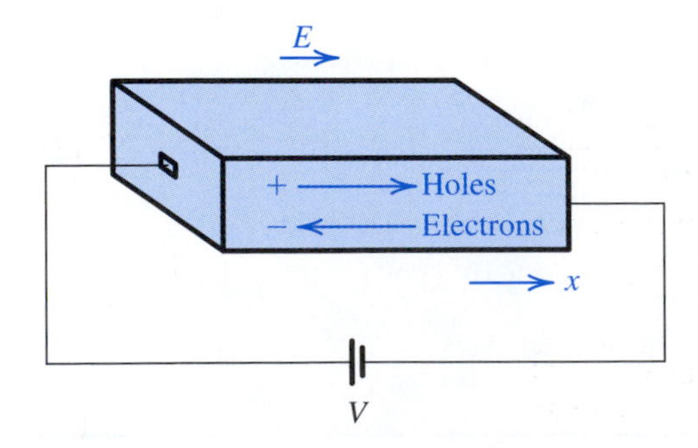

Figure 3.5 An electric field E established in a bar of silicon causes the holes to drift in the direction of E and the free electrons to drift in the opposite direction. Both the hole and electron drift currents are in the direction of E.

Let's now return to the single-crystal silicon bar shown in Fig. 3.5. Let the concentration of holes be p and that of free electrons n. We wish to calculate the current component due to the flow of holes. Consider a plane perpendicular to the x direction. In one second, the hole charge that crosses that plane will be $(Aqpv_{p\text{-drift}})$ coulombs, where A is the cross-sectional area of the silicon bar and q is the magnitude of electron charge. This then must be the hole component of the drift current flowing through the bar,

$$I_p = Aqpv_{p\text{-drift}} \tag{3.10}$$

Substituting for $v_{p\text{-drift}}$ from Eq. (3.9), we obtain

$$I_p = Aqp\mu_p E$$

We are usually interested in the current density J_p, which is the current per unit cross-sectional area,

$$J_p = \frac{I_p}{A} = qp\mu_p E \tag{3.11}$$

The current component due to the drift of free electrons can be found in a similar manner. Note, however, that electrons drifting from right to left result in a current component from left to right. This is because of the convention of taking the direction of current flow as the direction of flow of positive charge and opposite to the direction of flow of negative charge. Thus,

$$I_n = -Aqnv_{n\text{-drift}}$$

Substituting for $v_{n\text{-drift}}$ from Eq. (3.9), we obtain the current density $J_n = I_n/A$ as

$$J_n = qn\mu_n E \tag{3.12}$$

The total drift current density can now be found by summing J_p and J_n from Eqs. (3.11) and (3.12),

$$J = J_p + J_n = q(p\mu_p + n\mu_n)E \tag{3.13}$$

This relationship can be written as

$$J = \sigma E \tag{3.14}$$

or

$$J = E/\rho \tag{3.15}$$

where the **conductivity** σ is given by

$$\sigma = q(p\mu_p + n\mu_n) \tag{3.16}$$

and the **resistivity** ρ is given by

$$\rho \equiv \frac{1}{\sigma} = \frac{1}{q(p\mu_p + n\mu_n)} \tag{3.17}$$

Observe that Eq. (3.15) is a form of Ohm's law and can be written alternately as

$$\rho = \frac{E}{J} \tag{3.18}$$

Thus the units of ρ are ohm-centimeters $\left(\Omega \cdot \text{cm} = \dfrac{\text{V/cm}}{\text{A/cm}^2}\right)$.

Example 3.3

Find the resistivity of (a) intrinsic silicon and (b) p-type silicon with $N_A = 10^{16}/\text{cm}^3$. Use $n_i = 1.5 \times 10^{10}/\text{cm}^3$, and assume that for intrinsic silicon $\mu_n = 1350\ \text{cm}^2/\text{V}\cdot\text{s}$ and $\mu_p = 480\ \text{cm}^2/\text{V}\cdot\text{s}$, and for the doped silicon $\mu_n = 1110\ \text{cm}^2/\text{V}\cdot\text{s}$ and $\mu_p = 400\ \text{cm}^2/\text{V}\cdot\text{s}$. (Note that doping results in reduced carrier mobilities).

Solution

(a) For intrinsic silicon,

$$p = n = n_i = 1.5 \times 10^{10}/\text{cm}^3$$

Thus,

$$\rho = \frac{1}{q(p\mu_p + n\mu_n)}$$

$$\rho = \frac{1}{1.6 \times 10^{-19}(1.5 \times 10^{10} \times 480 + 1.5 \times 10^{10} \times 1350)}$$

$$= 2.28 \times 10^5\ \Omega\cdot\text{cm}$$

(b) For the p-type silicon

$$p_p \simeq N_A = 10^{16}/\text{cm}^3$$

$$n_p \simeq \frac{n_i^2}{N_A} = \frac{(1.5 \times 10^{10})^2}{10^{16}} = 2.25 \times 10^4/\text{cm}^3$$

Thus,

$$\rho = \frac{1}{q(p\mu_p + n\mu_n)}$$

$$= \frac{1}{1.6 \times 10^{-19}(10^{16} \times 400 + 2.25 \times 10^4 \times 1110)}$$

$$\simeq \frac{1}{1.6 \times 10^{-19} \times 10^{16} \times 400} = 1.56\ \Omega\cdot\text{cm}$$

Observe that the resistivity of the p-type silicon is determined almost entirely by the doping concentration. Also observe that doping the silicon reduces its resistivity by a factor of about 10^5, a truly remarkable change.

EXERCISE

3.4 A uniform bar of n-type silicon of 2 µm length has a voltage of 1 V applied across it. If $N_D = 10^{16}/\text{cm}^3$ and $\mu_n = 1350\ \text{cm}^2/\text{V}\cdot\text{s}$, find (a) the electron drift velocity, (b) the time it takes an electron to cross the 2-µm length, (c) the drift-current density, and (d) the drift current in the case the silicon bar has a cross sectional area of 0.25 µm^2.
Ans. 6.75×10^6 cm/s; 30 ps; $1.08 \times 10^4 \text{A/cm}^2$; 27 µA

3.3.2 Diffusion Current

Carrier diffusion occurs when the density of charge carriers in a piece of semiconductor is not uniform. For instance, if by some mechanism the concentration of, say, holes, is made higher in one part of a piece of silicon than in another, then holes will diffuse from the region of high concentration to the region of low concentration. Such a diffusion process is like that observed if one drops a few ink drops in a water-filled tank. The diffusion of charge carriers gives rise to a net flow of charge, or **diffusion current**.

As an example, consider the bar of silicon shown in Fig. 3.6(a): By some unspecified process, we have arranged to inject holes into its left side. This continuous hole injection gives rise to and maintains a hole **concentration profile** such as that shown in Fig. 3.6(b). This profile in turn causes holes to diffuse from left to right along the silicon bar, resulting in a hole current in the x direction. The magnitude of the current at any point is proportional to the slope of the concentration profile, or the **concentration gradient**, at that point,

$$J_p = -qD_p\frac{dp(x)}{dx} \tag{3.19}$$

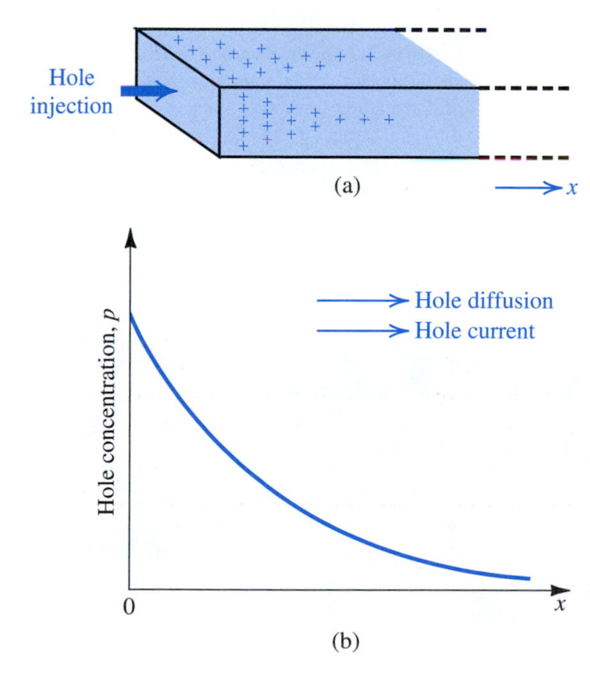

Figure 3.6 A bar of silicon **(a)** into which holes are injected, thus creating the hole concentration profile along the x axis, shown in **(b)**. The holes diffuse in the positive direction of x and give rise to a hole-diffusion current in the same direction. Note that we are not showing the circuit to which the silicon bar is connected.

where J_p is the hole-current density (A/cm^2), q is the magnitude of electron charge, D_p is a constant called the **diffusion constant** or **diffusivity** of holes; and $p(x)$ is the hole concentration at point x. Note that the gradient (dp/dx) is negative, resulting in a positive current in the x direction, as should be expected.

In the case of electron diffusion resulting from an electron concentration gradient (see Fig. 3.7), a similar relationship applies, giving the electron-current density,

$$J_n = qD_n \frac{dn(x)}{dx} \tag{3.20}$$

where D_n is the diffusion constant or diffusivity of electrons. Observe that a negative (dn/dx) gives rise to a negative current, a result of the convention that the positive direction of current is taken to be that of the flow of positive charge (and opposite to that of the flow of negative charge). For holes and electrons diffusing in intrinsic silicon, typical values for the diffusion constants are $D_p = 12\ \text{cm}^2/\text{s}$ and $D_n = 35\ \text{cm}^2/\text{s}$.

At this point the reader is probably wondering where the diffusion current in the silicon bar in Fig. 3.6(a) goes. A good question as we are not showing how the right-side end of the bar is connected to the rest of the circuit. We will address this and related questions in detail in our discussion of the *pn* junction in later sections.

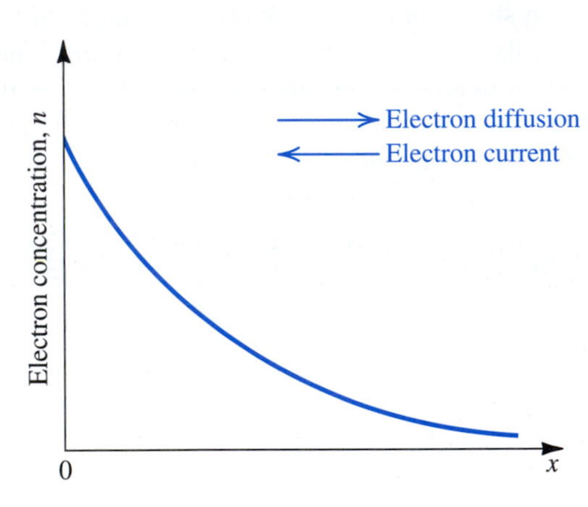

Figure 3.7 If the electron-concentration profile shown is established in a bar of silicon, electrons diffuse in the x direction, giving rise to an electron-diffusion current in the negative $-x$ direction.

Example 3.4

Consider a bar of silicon in which a hole concentration profile described by

$$p(x) = p_0\, e^{-x/L_p}$$

is established. Find the hole-current density at $x = 0$. Let $p_0 = 10^{16}/\text{cm}^3$ and $L_p = 1\ \mu\text{m}$. If the cross-sectional area of the bar is $100\ \mu\text{m}^2$, find the current I_p.

Solution

$$J_p = -qD_p \frac{dp(x)}{dx}$$

$$= -qD_p \frac{d}{dx}[p_0 e^{-x/L_p}]$$

Thus,

$$J_p(0) = q\frac{D_p}{L_p}p_0$$

$$= 1.6 \times 10^{-19} \times \frac{12}{1 \times 10^{-4}} \times 10^{16}$$

$$= 192 \ \text{A/cm}^2$$

The current I_p can be found from

$$I_p = J_p \times A$$

$$= 192 \times 100 \times 10^{-8}$$

$$= 192 \ \mu\text{A}$$

EXERCISE

3.5 The linear electron-concentration profile shown in Fig. E3.5 has been established in a piece of silicon. If $n_0 = 10^{17}/\text{cm}^3$ and $W = 1\ \mu\text{m}$, find the electron-current density in micro amperes per micron squared (μA/μm^2). If a diffusion current of 1 mA is required what must the cross-sectional area (in a direction perpendicular to the page) be?

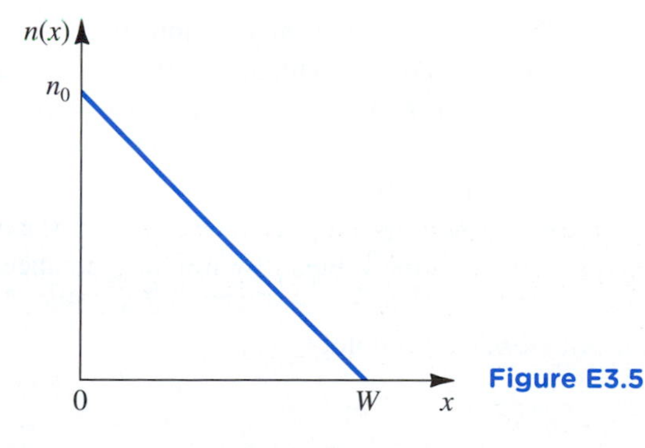

Figure E3.5

Ans. 56 μA/μm^2; 18 μm^2

3.3.3 Relationship between D and μ

A simple but powerful relationship ties the diffusion constant with the mobility,

$$\frac{D_n}{\mu_n} = \frac{D_p}{\mu_p} = V_T \quad (3.21)$$

where $V_T = kT/q$. The parameter V_T is known as the **thermal voltage**. At room temperature, $T \simeq 300$ K and $V_T = 25.9$ mV. We will encounter V_T repeatedly throughout this book. The relationship in Eq. (3.21) is known as the **Einstein relationship**.

EXERCISE

3.6 Use the Einstein relationship to find D_n and D_p for intrinsic silicon using $\mu_n = 1350\ \text{cm}^2/\text{V}\cdot\text{s}$ and $\mu_p = 480\ \text{cm}^2/\text{V}\cdot\text{s}$.
Ans. 35 cm^2/s; 12.4 cm^2/s

3.4 The *pn* Junction with Open-Circuit Terminals (Equilibrium)

Having learned important semiconductor concepts, we are now ready to consider our first practical semiconductor structure—the *pn* junction. As mentioned previously, the *pn* junction implements the diode (Chapter 4) and plays the dominant role in the structure and operation of the bipolar junction transistor (BJT). As well, understanding *pn* junctions is very important to the study of the MOSFET operation (Chapter 5).

3.4.1 Physical Structure

Figure 3.8 shows a simplified physical structure of the *pn* junction. It consists of *p*-type semiconductor (e.g., silicon) brought into close contact with an *n*-type semiconductor material (also silicon). In actual practice, both the *p* and *n* regions are part of the same silicon crystal; that is, the *pn* junction is formed within a single silicon crystal by creating regions of different dopings (*p* and *n* regions). Appendix A provides a description of the fabrication process of integrated circuits including *pn* junctions. As indicated in Fig. 3.8, external wire connections are made to the *p* and *n* regions through metal (aluminum) contacts. If the *pn* junction is used as a diode, these constitute the diode terminals and are therefore labeled "anode" and "cathode" in keeping with diode terminology.[3]

[3]This terminology in fact is a carryover from that used with vacuum-tube technology, which was the technology for making diodes and other electronic devices until the invention of the transistor in 1947. This event ushered in the era of solid-state electronics, which changed not only electronics, communications, and computers but indeed the world!

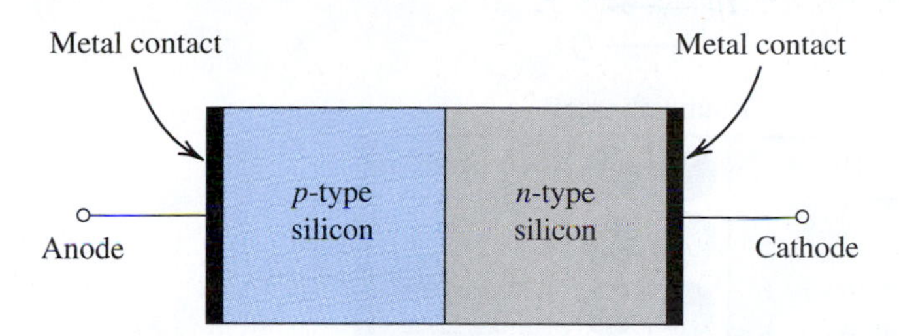

Figure 3.8 Simplified physical structure of the *pn* junction. (Actual geometries are given in Appendix A.) As the *pn* junction implements the junction diode, its terminals are labeled anode and cathode.

3.4.2 Operation with Open-Circuit Terminals

Figure 3.9 shows a *pn* junction under open-circuit conditions—that is, the external terminals are left open. The "+" signs in the *p*-type material denote the majority holes. The charge of these holes is neutralized by an equal amount of bound negative charge associated with the acceptor atoms. For simplicity, these bound charges are not shown in the diagram. Also not shown are the minority electrons generated in the *p*-type material by thermal ionization.

In the *n*-type material the majority electrons are indicated by "–" signs. Here also, the bound positive charge, which neutralizes the charge of the majority electrons, is not shown in order to keep the diagram simple. The *n*-type material also contains minority holes generated by thermal ionization but not shown in the diagram.

The Diffusion Current I_D Because the concentration of holes is high in the *p* region and low in the *n* region, holes diffuse across the junction from the *p* side to the *n* side; similarly, electrons diffuse across the junction from the *n* side to the *p* side. These two current components add together to form the diffusion current I_D, whose direction is from the *p* side to the *n* side, as indicated in Fig. 3.9.

The Depletion Region The holes that diffuse across the junction into the *n* region quickly recombine with some of the majority electrons present there and thus disappear from the scene. This recombination process results also in the disappearance of some free electrons from the *n*-type material. Thus some of the bound positive charge will no longer be neutralized by free electrons, and this charge is said to have been **uncovered**. Since recombination takes place close to the junction, there will be a region close to the junction that is *depleted of free electrons* and contains uncovered bound positive charge, as indicated in Fig. 3.9.

The electrons that diffuse across the junction into the *p* region quickly recombine with some of the majority holes there, and thus disappear from the scene. This results also in the disappearance of some majority holes, causing some of the bound negative charge to be uncovered (i.e., no longer neutralized by holes). Thus, in the *p* material close to the junction, there will be a region *depleted of holes* and containing uncovered bound negative charge, as indicated in Fig. 3.9.

From the above it follows that a **carrier-depletion region** will exist on both sides of the junction, with the *n* side of this region positively charged and the *p* side negatively charged. This carrier-depletion region—or, simply, **depletion region**—is also called the **space-charge region**. The charges on both sides of the depletion region cause an electric field E to be

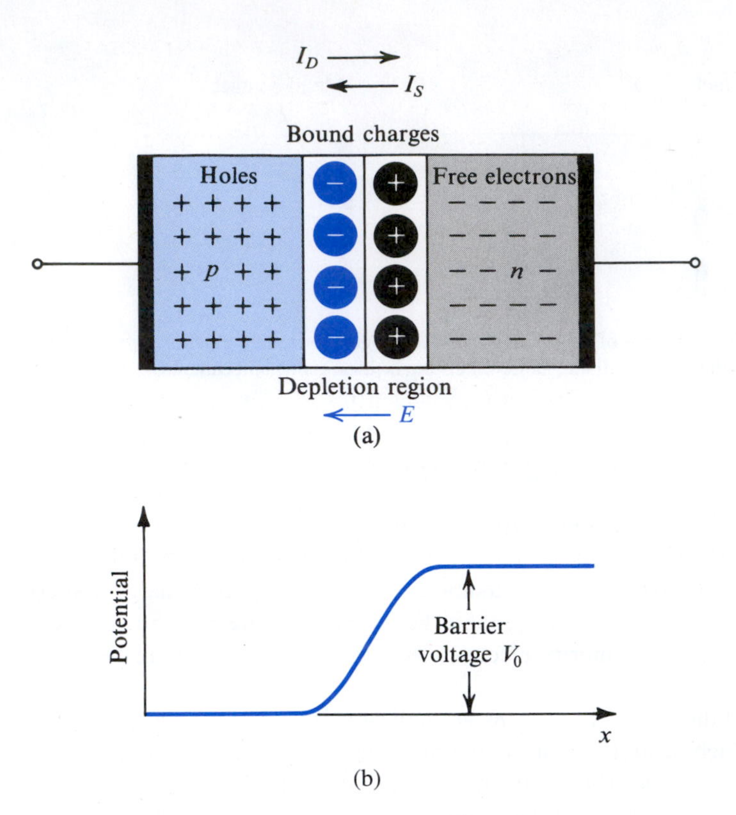

Figure 3.9 **(a)** The *pn* junction with no applied voltage (open-circuited terminals). **(b)** The potential distribution along an axis perpendicular to the junction.

established across the region in the direction indicated in Fig. 3.9. Hence a potential difference results across the depletion region, with the *n* side at a positive voltage relative to the *p* side, as shown in Fig. 3.9(b). Thus the resulting electric field opposes the diffusion of holes into the *n* region and electrons into the *p* region. In fact, the voltage drop across the depletion region acts as a **barrier** that has to be overcome for holes to diffuse into the *n* region and electrons to diffuse into the *p* region. The larger the barrier voltage, the smaller the number of carriers that will be able to overcome the barrier and hence the lower the magnitude of diffusion current. Thus it is the appearance of the barrier voltage V_0 that limits the carrier diffusion process. It follows that the diffusion current I_D depends strongly on the voltage drop V_0 across the depletion region.

The Drift Current I_S and Equilibrium In addition to the current component I_D due to majority-carrier diffusion, a component due to minority carrier drift exists across the junction. Specifically, some of the thermally generated holes in the *n* material move toward the junction and reach the edge of the depletion region. There, they experience the electric field in the depletion region, which sweeps them across that region into the *p* side. Similarly, some of the minority thermally generated electrons in the *p* material move to the edge of the depletion region and get swept by the electric field in the depletion region across that region into the *n* side. These two current components—electrons moved by drift from *p* to *n* and holes moved by drift from *n* to *p*—add together to form the drift current I_S, whose direction is from the *n* side to the *p* side of the junction, as indicated in Fig. 3.9. Since the current I_S is carried by thermally generated minority carriers, its value is strongly dependent on temperature; however, it is independent of the value of the depletion-layer voltage V_0. This is due to the fact that the drift current is determined by the number of minority carriers that make it to the edge of the depletion region; any minority carriers that manage to get to the edge of the depletion region will be swept across by E irrespective of the value of E or, correspondingly, of V_0.

Under open-circuit conditions (Fig. 3.9) no external current exists; thus the two opposite currents across the junction must be equal in magnitude:

$$I_D = I_S$$

This equilibrium condition[4] is maintained by the barrier voltage V_0. Thus, if for some reason I_D exceeds I_S, then more bound charge will be uncovered on both sides of the junction, the depletion layer will widen, and the voltage across it (V_0) will increase. This in turn causes I_D to decrease until equilibrium is achieved with $I_D = I_S$. On the other hand, if I_S exceeds I_D, then the amount of uncovered charge will decrease, the depletion layer will narrow, and the voltage across it (V_0) will decrease. This causes I_D to increase until equilibrium is achieved with $I_D = I_S$.

The Junction Built-In Voltage With no external voltage applied, the barrier voltage V_0 across the *pn* junction can be shown to be given by[5]

$$V_0 = V_T \ln\left(\frac{N_A N_D}{n_i^2}\right) \tag{3.22}$$

where N_A and N_D are the doping concentrations of the p side and n side of the junction, respectively. Thus V_0 depends both on doping concentrations and on temperature. It is known as the **junction built-in voltage**. Typically, for silicon at room temperature, V_0 is in the range of 0.6 V to 0.9 V.

When the *pn* junction terminals are left open-circuited, the voltage measured between them will be zero. That is, the voltage V_0 across the depletion region *does not* appear between the junction terminals. This is because of the contact voltages existing at the metal-semiconductor junctions at the terminals, which counter and exactly balance the barrier voltage. If this were not the case, we would have been able to draw energy from the isolated *pn* junction, which would clearly violate the principle of conservation of energy.

Width of and Charge Stored in the Depletion Region Figure 3.10 provides further illustration of the situation that obtains in the *pn* junction when the junction is in equilibrium. In Fig. 3.10(a) we show a junction in which $N_A > N_D$, a typical situation in practice. This is borne out by the carrier concentration on both sides of the junction, as shown in Fig 3.10(b). Note that we have denoted the minority carrier concentrations in both sides by n_{p0} and p_{n0}, with the additional subscript "0" signifying equilibrium (i.e., before external voltages are applied as will be seen in the next section). Observe that the depletion region extends in both the p and n materials and that equal amounts of charge exist on both sides (Q_+ and Q_- in Fig. 3.10c). However, since usually unequal dopings N_A and N_D are used, as in the case illustrated in Fig. 3.10, the width of the depletion layer will not be the same on the two sides. Rather, to uncover the same amount of charge, the depletion layer will extend deeper into the more lightly doped material. Specifically, if we denote the width of the

[4]In fact, in equilibrum the equality of drift and diffusion currents applies not just to the total currents but also to their individual components. That is, the hole drift current must equal the hole diffusion current and, similarly, the electron drift current must equal the electron diffusion current.

[5]The derivation of this formula and of a number of others in this chapter can be found in textbooks dealing with devices, such as that by Streetman and Bannerjee (see the reading list in Appendix G).

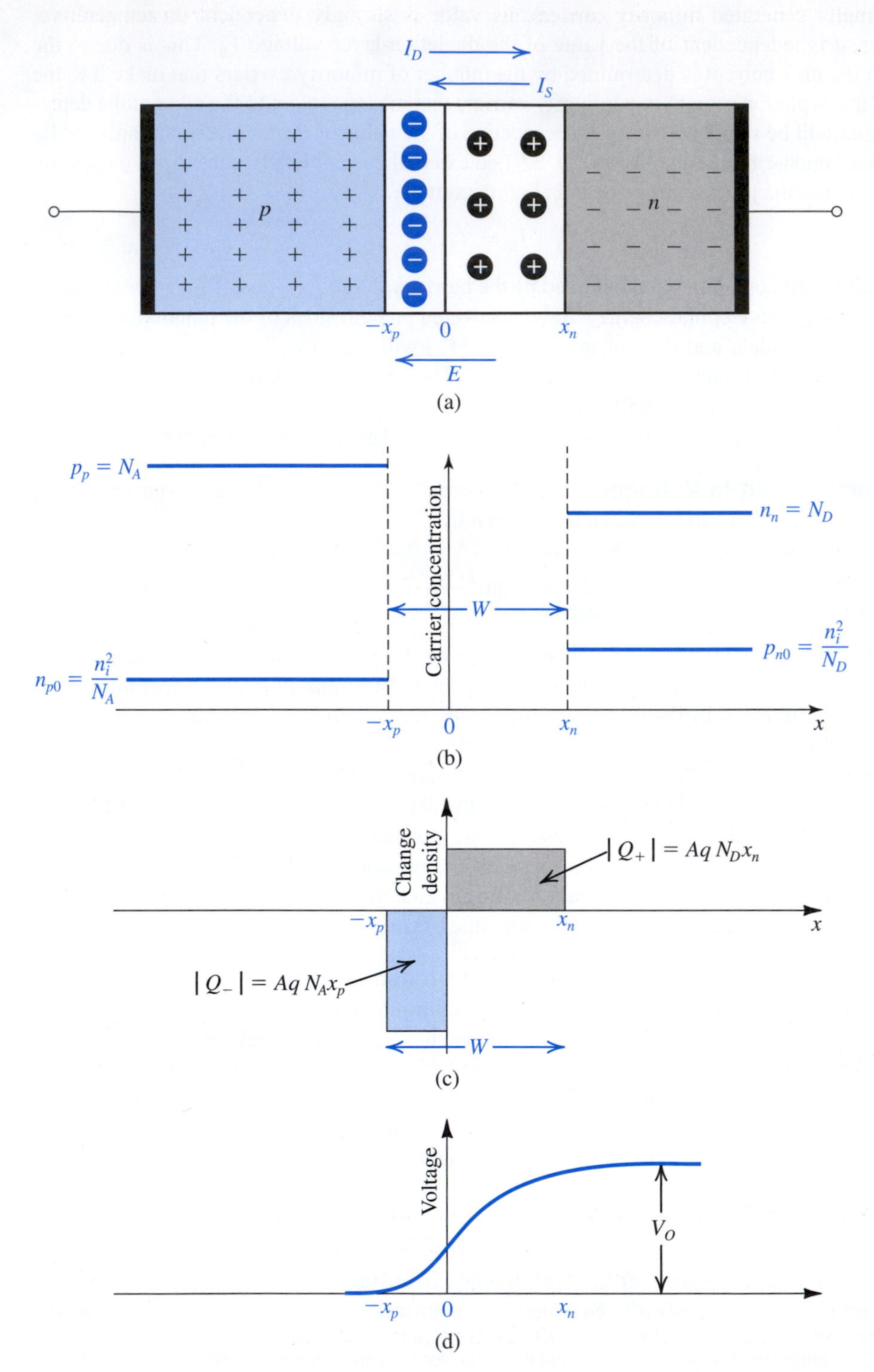

Figure 3.10 **(a)** A *pn* junction with the terminals open circuited. **(b)** Carrier concentrations; note that $N_A \gg N_D$. **(c)** The charge stored in both sides of the depletion region; $Q_J = |Q_+| = |Q_-|$. **(d)** The built-in voltage V_0.

depletion region in the p side by x_p and in the n side by x_n, we can express the magnitude of the charge on the n side of the junction as

$$|Q_+| = qAx_nN_D \tag{3.23}$$

and that on the p side of the junction as

$$|Q_-| = qAx_pN_A \tag{3.24}$$

where A is the cross-sectional area of the junction in the plane perpendicular to the page. The charge equality condition can now be written as

$$qAx_nN_D = qAx_pN_A$$

which can be rearranged to yield

$$\frac{x_n}{x_p} = \frac{N_A}{N_D} \tag{3.25}$$

In actual practice, it is usual for one side of the junction to be much more heavily doped than the other, with the result that the depletion region exists almost entirely on one side (the lightly doped side).

The width W of the depletion layer can be shown to be given by

$$W = x_n + x_p = \sqrt{\frac{2\varepsilon_s}{q}\left(\frac{1}{N_A} + \frac{1}{N_D}\right)V_0} \tag{3.26}$$

where ε_s is the electrical permittivity of silicon $= 11.7\varepsilon_0 = 11.7 \times 8.85 \times 10^{-14}$ F/cm $= 1.04 \times 10^{-12}$ F/cm. Typically W is in the range 0.1 μm to 1 μm. Eqs. (3.25) and (3.26) can be used to obtain x_n and x_p in terms of W as

$$x_n = W\frac{N_A}{N_A + N_D} \tag{3.27}$$

$$x_p = W\frac{N_D}{N_A + N_D} \tag{3.28}$$

The charge stored on either side of the depletion region can be expressed in terms of W by utilizing Eqs. (3.23) and (3.27) to obtain

$$Q_J = |Q_+| = |Q_-|$$

$$Q_J = Aq\left(\frac{N_AN_D}{N_A + N_D}\right)W \tag{3.29}$$

Finally, we can substitute for W from Eq. (3.26) to obtain

$$Q_J = A\sqrt{2\varepsilon_s q\left(\frac{N_AN_D}{N_A + N_D}\right)V_0} \tag{3.30}$$

These expressions for Q_J will prove useful in subsequent sections.

Example 3.5

Consider a *pn* junction in equilibrium at room temperature ($T = 300$ K) for which the doping concentrations are $N_A = 10^{18}/\text{cm}^3$ and $N_D = 10^{16}/\text{cm}^3$ and the cross-sectional area $A = 10^{-4}$ cm^2. Calculate p_p, n_{p0}, n_n, p_{n0}, V_0, W, x_n, x_p, and Q_J. Use $n_i = 1.5 \times 10^{10}/\text{cm}^3$.

Solution

$$p_p \simeq N_A = 10^{18}\ \text{cm}^{-3}$$

$$n_{p0} = \frac{n_i^2}{p_p} \simeq \frac{n_i^2}{N_A} = \frac{(1.5 \times 10^{10})^2}{10^{18}} = 2.25 \times 10^2\ \text{cm}^{-3}$$

$$n_n \simeq N_D = 10^{16}\ \text{cm}^{-3}$$

$$p_{n0} = \frac{n_i^2}{n_n} \simeq \frac{n_i^2}{N_D} = \frac{(1.5 \times 10^{10})^2}{10^{16}} = 2.25 \times 10^4\ \text{cm}^{-3}$$

To find V_0 we use Eq. (3.22),

$$V_O = V_T \ln\left(\frac{N_A N_D}{n_i^2}\right)$$

where

$$V_T = \frac{kT}{q} = \frac{8.62 \times 10^{-5} \times 300\ (\text{eV})}{q\qquad\qquad(\text{e})}$$

$$= 25.9 \times 10^{-3}\ \text{V}$$

Thus,

$$V_0 = 25.9 \times 10^{-3} \ln\left(\frac{10^{18} \times 10^{16}}{2.25 \times 10^{20}}\right)$$

$$= 0.814\ \text{V}$$

To determine W we use Eq. (3.26):

$$W = \sqrt{\frac{2 \times 1.04 \times 10^{-12}}{1.6 \times 10^{-19}}\left(\frac{1}{10^{18}} + \frac{1}{10^{16}}\right) \times 0.814}$$

$$= 3.27 \times 10^{-5}\ \text{cm} = 0.327\ \mu\text{m}$$

To determine x_n and x_p we use Eq. (3.27) and (3.28), respectively:

$$x_n = W\frac{N_A}{N_A + N_D}$$

$$= 0.327\frac{10^{18}}{10^{18} + 10^{16}} = 0.324\ \mu\text{m}$$

$$x_p = W\frac{N_D}{N_A + N_D}$$

$$= 0.327\frac{10^{16}}{10^{18} + 10^{16}} = 0.003\ \mu\text{m}$$

Finally, to determine the charge stored on either side of the depletion region, we use Eq. (3.29)

$$Q_J = 10^{-4} \times 1.6 \times 10^{-19}\left(\frac{10^{18} \times 10^{16}}{10^{18} + 10^{16}}\right) \times 0.327 \times 10^{-4}$$

$$= 5.18 \times 10^{-12}\ \text{C} = 5.18\ \text{pC}$$

EXERCISES

3.7 Show that

$$V_0 = \frac{1}{2}\left(\frac{q}{\varepsilon_s}\right)\left(\frac{N_A N_D}{N_A + N_D}\right)W^2$$

3.8 Show that for a *pn* junction in which the *p* side is much more heavily doped than the *n* side, (i.e. $N_A \gg N_D$), referred to as a p^+n diode, Eqs. (3.26), (3.27), (3.28), (3.29), and (3.30) can be simplified as follows:

$$W \simeq \sqrt{\frac{2\varepsilon_s}{qN_D}V_0} \qquad (3.26')$$

$$x_n \simeq W \qquad (3.27')$$

$$x_p \simeq (W/(N_A/N_D)) \qquad (3.28')$$

$$Q_J \simeq AqN_D W \qquad (3.29')$$

$$Q_J \simeq A\sqrt{2\varepsilon_s qN_D V_0} \qquad (3.30')$$

3.9 If in the fabrication of the *pn* junction in Example 3.5, it is required to increase the minority carrier concentration in the *n* region by a factor of 2, what must be done?
Ans. Lower N_D by a factor of 2.

3.5 The *pn* Junction with an Applied Voltage

Having studied the open-circuited *pn* junction in detail, we are now ready to apply a dc voltage between its two terminals to find its electrical conduction properties. If the voltage is applied so that the *p* side is made more positive than the *n* side, it is referred to as a forward-bias[6] voltage. Conversely, if our applied dc voltage is such that it makes the *n* side more positive than the *p* side, it is said to be a reverse-bias voltage. As will be seen, the *pn* junction exhibits vastly different conduction properties in its forward and reverse directions.

Our plan is as follows. We begin by a simple qualitative description in Section 3.5.1 and then consider an analytical description of the i–v characteristic of the junction in Section 3.5.2.

3.5.1 Qualitative Description of Junction Operation

Figure 3.11 shows the *pn* junction under three different conditions: (a) the open-circuit or equilibrium condition studied in the previous section; (b) the reverse-bias condition, where a dc voltage V_R is applied; and (c) the forward-bias condition where a dc voltage V_F is applied. Observe that in the open-circuit case, a barrier voltage V_0 develops, making *n* more positive than *p*, and limiting the diffusion current I_D to a value exactly equal to the drift current I_S,

[6]For the time being, we take the term *bias* to refer simply to the application of a dc voltage. We will see in later chapters that it has a deeper meaning in the design of electronic circuits.

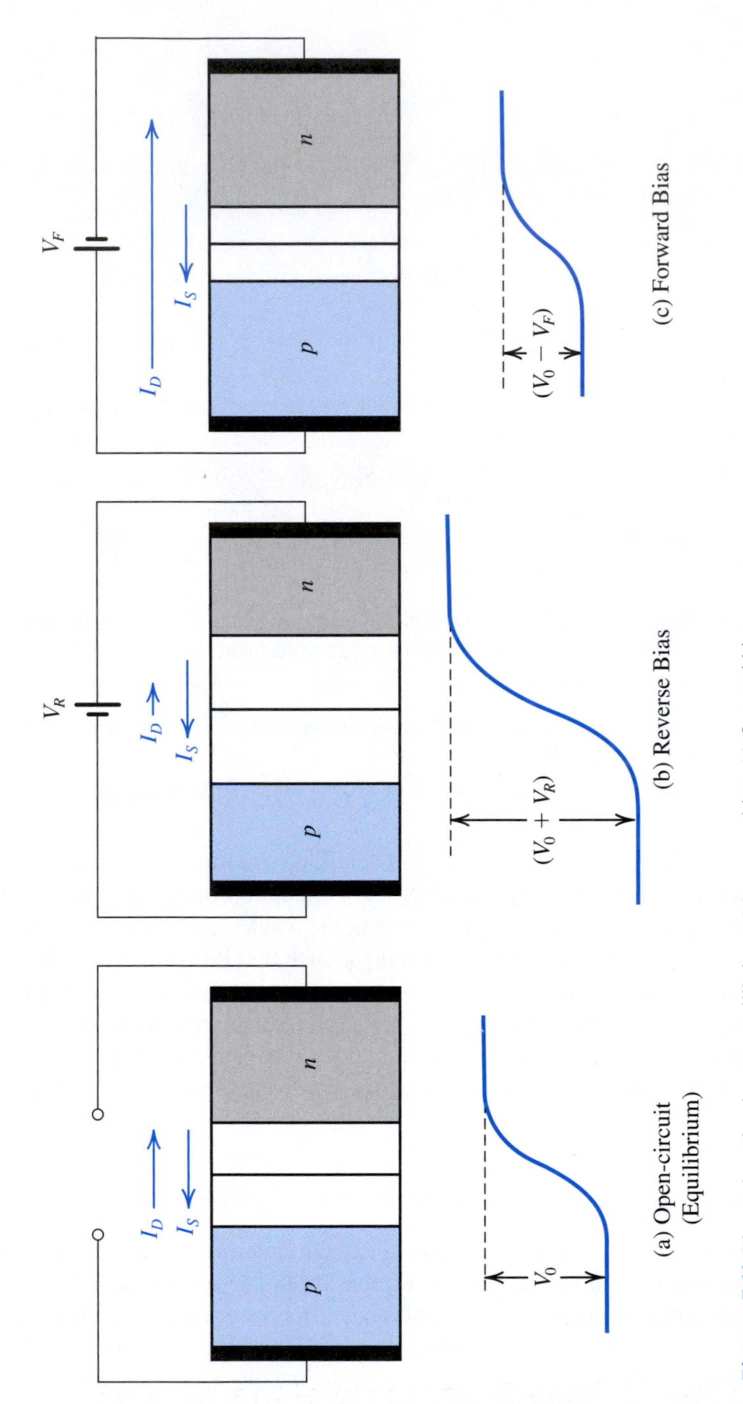

Figure 3.11 The *pn* junction in: **(a)** equilibrium; **(b)** reverse bias; **(c)** forward bias.

thus resulting in a zero current at the junction terminals, as should be the case since the terminals are open circuited. Also, as mentioned previously, the barrier voltage V_0, though it establishes the current equilibrium across the junction, does *not* in fact appear between the junction terminals.

Consider now the reverse-bias case in (b). The externally applied reverse-bias voltage V_R is in the direction to add to the barrier voltage, and it does, thus increasing the effective barrier voltage to $(V_0 + V_R)$ as shown. This reduces the number of holes that diffuse into the n region and the number of electrons that diffuse into the p region. The end result is that the diffusion current I_D is dramatically reduced. As will be seen shortly, a reverse-bias voltage of a volt or so is sufficient to cause $I_D \simeq 0$, and the current across the junction and through the external circuit will be equal to I_S. Recalling that I_S is the current due to the drift across the depletion region of the thermally generated minority carriers, we expect I_S to be very small and to be strongly dependent on temperature. We will show this to be the case very shortly. We thus conclude that in the reverse direction, the pn junction conducts a very small and almost-constant current equal to I_S.

Before leaving the reverse-bias case, observe that the increase in barrier voltage will be accompanied by a corresponding increase in the stored uncovered charge on both sides of the depletion region. This in turn means a wider depletion region, needed to uncover the additional charge required to support the larger barrier voltage $(V_0 + V_R)$. Analytically, these results can be obtained easily by a simple extension of the results of the equilibrium case. Thus the width of the depletion region can be obtained by replacing V_0 in Eq. (3.26) by $(V_0 + V_R)$,

$$W = x_n + x_p = \sqrt{\frac{2\varepsilon_s}{q}\left(\frac{1}{N_A} + \frac{1}{N_D}\right)(V_0 + V_R)} \tag{3.31}$$

and the magnitude of the charge stored on either side of the depletion region can be determined by replacing V_0 in Eq. (3.30) by $(V_0 + V_R)$,

$$Q_J = A\sqrt{2\varepsilon_s q\left(\frac{N_A N_D}{N_A + N_D}\right)(V_0 + V_R)} \tag{3.32}$$

We next consider the forward-bias case shown in Fig. 3.11(c). Here the applied voltage V_F is in the direction that subtracts from the built-in voltage V_0, resulting in a reduced barrier voltage $(V_0 - V_F)$ across the depletion region. This reduced barrier voltage will be accompanied by reduced depletion-region charge and correspondingly narrower depletion-region width W. Most importantly, the lowering of the barrier voltage will enable more holes to diffuse from p to n and more electrons to diffuse from n to p. Thus the diffusion current I_D increases substantially and, as will be seen shortly, can become many orders of magnitude larger than the drift current I_S. The current I in the external circuit is of course the difference between I_D and I_S,

$$I = I_D - I_S$$

and it flows in the forward direction of the junction, from p to n. We thus conclude that the pn junction can conduct a substantial current in the forward-bias region and that current is mostly a diffusion current whose value is determined by the forward-bias voltage V_F.

3.5.2 The Current–Voltage Relationship of the Junction

We are now ready to find an analytical expression that describes the current–voltage relationship of the pn junction. In the following we consider a junction operating with a forward applied voltage V and derive an expression for the current I that flows in the forward direction (from p to n). However, our derivation is general and will be seen to yield the reverse current when the applied voltage V is made negative.

From the qualitative description above we know that a forward-bias voltage V subtracts from the built-in voltage V_0, thus resulting in a lower barrier voltage $(V_0 - V)$. The lowered barrier in turn makes it possible for a greater number of holes to overcome the barrier and diffuse into the n region. A similar statement can be made about electrons from the n region diffusing into the p region.

Let us now consider the holes injected into the n region. The concentration of holes in the n region at the edge of the depletion region will increase considerably. In fact, an important result from device physics shows that the steady-state concentration at the edge of the depletion region will be

$$p_n(x_n) = p_{n0}e^{V/V_T} \tag{3.33}$$

That is, the concentration of the minority holes increases from the equilibrium value of p_{n0} (see Fig. 3.10) to the much larger value determined by the value of V, given by Eq. (3.33).

We describe this situation as follows: The forward-bias voltage V results in an **excess concentration** of minority holes at $x = x_n$, given by

$$\begin{aligned} \text{Excess concentration} &= p_{n0}e^{V/V_T} - p_{n0} \\ &= p_{n0}(e^{V/V_T} - 1) \end{aligned} \tag{3.34}$$

The increase in minority carrier concentration in Eqs. (3.33) and (3.34) occurs at the edge of the depletion region $(x = x_n)$. As the injected holes diffuse into the n material, some will recombine with the majority electrons and disappear. Thus, the excess hole concentration will decay exponentially with distance. As a result, in the total hole concentration in the n material will be given by

$$p_n(x) = p_{n0} + (\text{Excess concentration})e^{-(x-x_n)/L_p}$$

Substituting for the "Excess concentration" from Eq. (3.34) gives

$$p_n(x) = p_{n0} + p_{n0}(e^{V/V_T} - 1)e^{-(x-x_n)/L_p} \tag{3.35}$$

The exponential decay is characterized by the constant L_p, which is called the **diffusion length** of holes in the n material. The smaller the value of L_p, the faster the injected holes will recombine with the majority electrons, resulting in a steeper decay of minority carrier concentration.

Figure 3.12 shows the steady-state minority carrier concentration profiles on both sides of a pn junction in which $N_A \gg N_D$. Let's stay a little longer with the diffusion of holes into the n region. Note that the shaded region under the exponential represents the excess minority carriers (holes). From our study of diffusion in Section 3.3, we know that the establishment of a carrier concentration profile such as that in Fig. 3.12 is essential to support a steady-state diffusion current. In fact, we can now find the value of the hole–diffusion current density by applying Eq. (3.19),

$$J_p(x) = -qD_p\frac{dp_n(x)}{dx}$$

Substituting for $p_n(x)$ from Eq. (3.35) gives

$$J_p(x) = q\left(\frac{D_p}{L_p}\right)p_{n0}(e^{V/V_T} - 1)e^{-(x-x_n)/L_p} \tag{3.36}$$

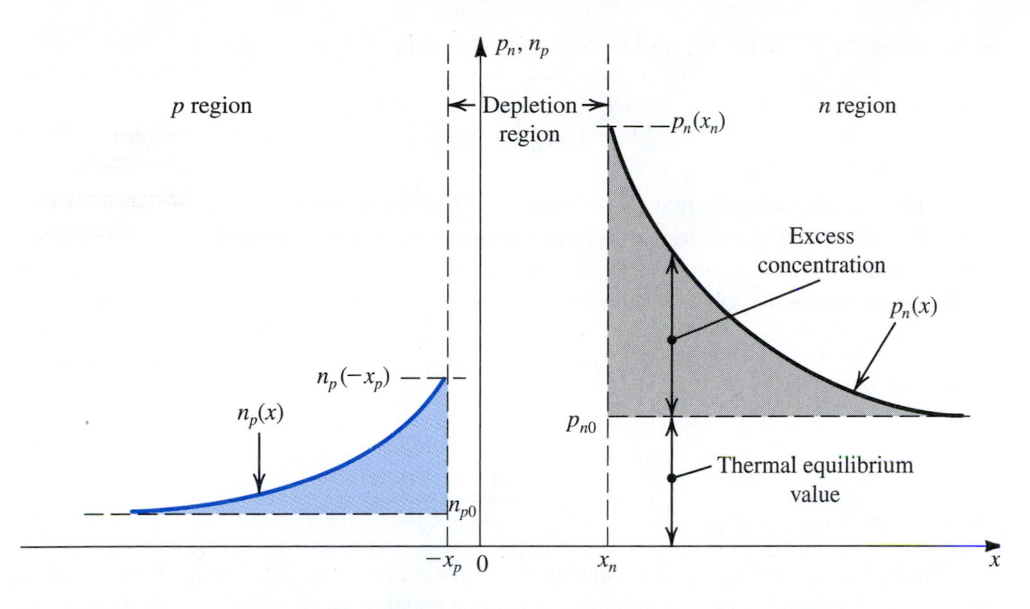

Figure 3.12 Minority-carrier distribution in a forward-biased *pn* junction. It is assumed that the *p* region is more heavily doped than the *n* region; $N_A \gg N_D$.

As expected, $J_p(x)$ is highest at $x = x_n$,

$$J_p(x_n) = q\left(\frac{D_p}{L_p}\right)p_{n0}(e^{V/V_T} - 1) \tag{3.37}$$

and decays exponentially for $x > x_n$, as the minority holes recombine with the majority electrons. This recombination, however, means that the majority electrons will have to be replenished by a current that injects electrons from the external circuit into the *n* region of the junction. This latter current component has the same direction as the hole current (because electrons moving from right to left give rise to current in the direction from left to right). It follows that as $J_p(x)$ decreases, the electron current component increases by exactly the same amount, making the total current in the *n* material constant at the value given by Eq. (3.37).

An exactly parallel development can be applied to the electrons that are injected from the *n* to the *p* region, resulting in an electron diffusion current given by a simple adaptation of Eq. (3.37),

$$J_n(-x_p) = q\left(\frac{D_n}{L_n}\right)n_{p0}(e^{V/V_T} - 1) \tag{3.38}$$

Now, although the currents in Eqs. (3.37) and (3.38) are found at the two edges of the depletion region, their values do not change in the depletion region. Thus we can drop the location descriptors (x_n), $(-x_p)$, add the two current densities, and multiply by the junction area A to obtain the total current I as

$$I = A(J_p + J_n)$$

$$I = Aq\left(\frac{D_p}{L_p}\,p_{n0} + \frac{D_n}{L_n}\,n_{p0}\right)(e^{V/V_T} - 1)$$

Substituting for $p_{n0} = n_i^2/N_D$ and for $n_{p0} = n_i^2/N_A$ gives

$$I = Aqn_i^2\left(\frac{D_p}{L_pN_D} + \frac{D_n}{L_nN_A}\right)(e^{V/V_T} - 1) \tag{3.39}$$

From this equation we note that for a negative V (reverse bias) with a magnitude of a few times V_T (25.9 mV), the exponential term becomes essentially zero, and the current across the junction becomes negative and constant. From our qualitative description in Section 3.5.1, we know that this current must be I_S. Thus,

$$I = I_S(e^{V/V_T} - 1) \tag{3.40}$$

where

$$I_S = Aqn_i^2\left(\frac{D_p}{L_pN_D} + \frac{D_n}{L_nN_A}\right) \tag{3.41}$$

Figure (3.13) shows the I–V characteristic of the *pn* junction (Eq. 3.40). Observe that in the reverse direction the current saturates at a value equal to $-I_S$. For this reason, I_S is given the name **saturation current**. From Eq. (3.41) we see that I_S is directly proportional to the cross-sectional area A of the junction. Thus, another name for I_S, one we prefer to use in this book, is the junction **scale current**. Typical values for I_S, for junctions of various areas, range from 10^{-18} to 10^{-12} A.

Besides being proportional to the junction area A, the expression for I_S in Eq. (3.41) indicates that I_S is proportional to n_i^2 which is a very strong function of temperature (see Eq. 3.2).

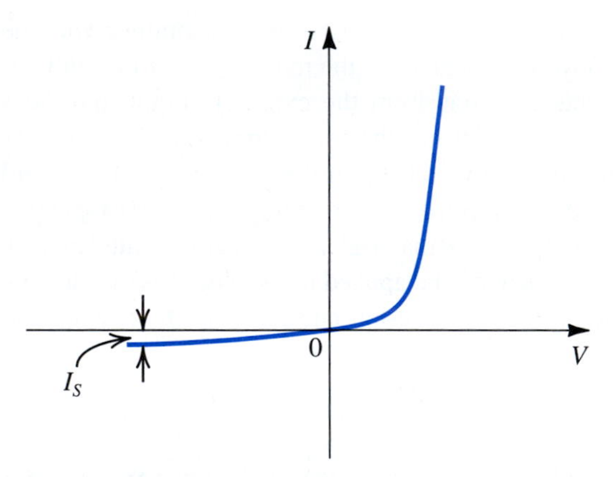

Figure 3.13 The *pn* junction I–V characteristic.

Example 3.6

For the *pn* junction considered in Example 3.5 for which $N_A = 10^{18}/\text{cm}^3$, $N_D = 10^{16}/\text{cm}^3$, $A = 10^{-4}\,\text{cm}^2$, $n_i = 1.5 \times 10^{10}/\text{cm}^3$, let $L_p = 5\ \mu\text{m}$, $L_n = 10\ \mu\text{m}$, D_p (in the *n* region) $= 10\ \text{cm}^2/\text{V}\cdot\text{s}$, and D_n (in the *p* region) $= 18\ \text{cm}^2/\text{V}\cdot\text{s}$. The *pn* junction is forward biased and conducting a current $I = 0.1$ mA. Calculate: (a) I_S; (b) the forward-bias voltage V; and (c) the component of the current I due to hole injection and that due to electron injection across the junction.

Solution

(a) Using Eq. (3.41), we find I_S as

$$\begin{aligned} I_S &= 10^{-4} \times 1.6 \times 10^{-19} \times (1.5 \times 10^{10})^2 \times \\ &\quad \left(\frac{10}{5 \times 10^{-4} \times 10^{16}} + \frac{18}{10 \times 10^{-4} \times 10^{18}}\right) \\ &= 7.3 \times 10^{-15}\ \text{A} \end{aligned}$$

(b) In the forward direction,

$$\begin{aligned} I &= I_S(e^{V/V_T} - 1) \\ &\simeq I_S e^{V/V_T} \end{aligned}$$

Thus,

$$V = V_T \ln\left(\frac{I}{I_S}\right)$$

For $I = 0.1$ mA,

$$\begin{aligned} V &= 25.9 \times 10^{-3} \ln\left(\frac{0.1 \times 10^{-3}}{7.3 \times 10^{-15}}\right) \\ &= 0.605\ \text{V} \end{aligned}$$

(c) The hole-injection component of I can be found using Eq. (3.37)

$$\begin{aligned} I_p &= Aq\frac{D_p}{L_p}\, p_{n0}(e^{V/V_T} - 1) \\ &= Aq\frac{D_p}{L_p}\,\frac{n_i^2}{N_D}(e^{V/V_T} - 1) \end{aligned}$$

Similarly I_n can be found using Eq. (3.39),

$$I_n = Aq\frac{D_n}{L_n}\,\frac{n_i^2}{N_A}(e^{V/V_T} - 1)$$

Thus,

$$\frac{I_p}{I_n} = \left(\frac{D_p}{D_n}\right)\left(\frac{L_n}{L_p}\right)\left(\frac{N_A}{N_D}\right)$$

For our case,

$$\frac{I_p}{I_n} = \frac{10}{18} \times \frac{10}{5} \times \frac{10^{18}}{10^{16}} = 1.11 \times 10^2 = 111$$

Example 3.6 *continued*

Thus most of the current is conducted by holes injected into the *n* region.

Specifically,

$$I_p = \frac{111}{112} \times 0.1 = 0.0991 \text{ mA}$$

$$I_n = \frac{1}{112} \times 0.1 = 0.0009 \text{ mA}$$

This stands to reason, since the *p* material has a doping concentration 100 times that of the *n* material.

EXERCISES

3.10 Show that if $N_A \gg N_D$,

$$I_S \simeq Aqn_i^2 \frac{D_p}{L_p N_D}$$

3.11 For the *pn* junction in Example 3.6, find the value of I_S and that of the current I at $V = 0.605$ V (same voltage found in Example 3.6 at a current $I = 0.1$ mA) if N_D is reduced by a factor of 2.
Ans. 1.46×10^{-14} A; 0.2 mA

3.12 For the *pn* junction considered in Examples 3.5 and 3.6, find the width of the depletion region W corresponding to the forward-bias voltage found in Example 3.6. (*Hint*: Use the formula in Eq. (3.31) with V_R replaced with $-V_F$.)
Ans. 0.166 μm

3.13 For the *pn* junction considered in Examples 3.5 and 3.6, find the width of the depletion region W and the charge stored in the depletion region Q_J when a 2-V reverse bias is applied. Also find the value of the reverse current I.
Ans. 0.608 μm; 9.63 pC; 7.3×10^{-15} A

3.5.3 Reverse Breakdown

The description of the operation of the *pn* junction in the reverse direction, and the $I-V$ relationship of the junction in Eq. (3.40), indicate that at a reverse-bias voltage $-V$, with $V \gg V_T$, the reverse current that flows across the junction is approximately equal to I_S and thus is very small. However, as the magnitude of the reverse-bias voltage V is increased, a value is reached at which a very large reverse current flows as shown in Fig. 3.14. Observe that as V reaches the value V_Z, the dramatic increase in reverse current is accompanied by a very small increase in the reverse voltage; that is, the reverse voltage across the junction remains very close to the value V_Z. The phenomenon that occurs at $V = V_Z$ is known as **junction breakdown**. It is not a destructive phenomenon. That is, the *pn* junction can be repeatedly operated in the breakdown region without a permanent effect on its characteristics. This, however, is predicated on the assumption that the magnitude of the reverse-breakdown current is

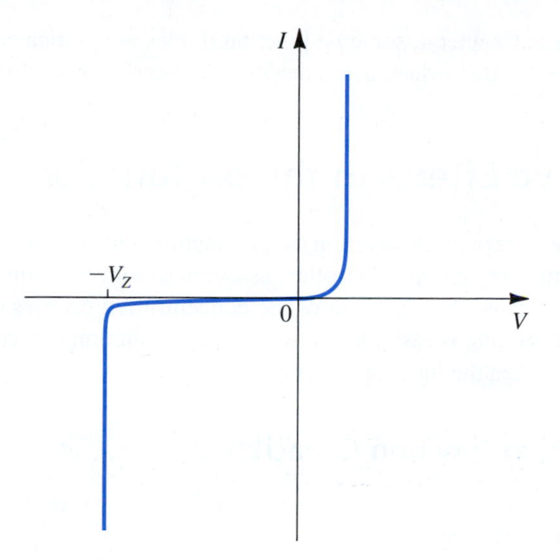

Figure 3.14 The *I-V* characteristic of the *pn* junction showing the rapid increase in reverse current in the breakdown region.

limited by the external circuit to a "safe" value. The "safe" value is one that results in the limitation of the power dissipated in the junction to a safe, allowable level.

There are two possible mechanisms for *pn* junction breakdown: the **zener effect**[7] and the **avalanche effect**. If a *pn* junction breaks down with a breakdown voltage $V_Z < 5$ V, the breakdown mechanism is usually the zener effect. Avalanche breakdown occurs when V_Z is greater than approximately 7 V. For junctions that break down between 5 V and 7 V, the breakdown mechanism can be either the zener or the avalanche effect or a combination of the two.

Zener breakdown occurs when the electric field in the depletion layer increases to the point of breaking covalent bonds and generating electron-hole pairs. The electrons generated in this way will be swept by the electric field into the *n* side and the holes into the *p* side. Thus these electrons and holes constitute a reverse current across the junction. Once the zener effect starts, a large number of carriers can be generated, with a negligible increase in the junction voltage. Thus the reverse current in the breakdown region will be large and its value must be determined by the external circuit, while the reverse voltage appearing between the diode terminals will remain close to the specified breakdown voltage V_Z.

The other breakdown mechanism, avalanche breakdown, which occurs when the minority carriers that cross the depletion region under the influence of the electric field gain sufficient kinetic energy to be able to break covalent bonds in atoms with which they collide. The carriers liberated by this process may have sufficiently high energy to be able to cause other carriers to be liberated in another ionizing collision. This process keeps repeating in the fashion of an avalanche, with the result that many carriers are created that are able to support any value of reverse current, as determined by the external circuit, with a negligible change in the voltage drop across the junction.

[7]Named after an early worker in the area. Note that the subscript *Z* in V_Z denotes *zener*. We will use V_Z to denote the breakdown voltage whether the breakdown mechanism is the zener effect or the avalanche effect.

As will be seen in Chapter 4, some *pn* junction diodes are fabricated to operate specifically in the breakdown region, where use is made of the nearly constant voltage V_Z.

3.6 Capacitive Effects in the *pn* Junction

There are two charge storage mechanisms in the *pn* junction. One is associated with the charge stored in the depletion region, and the other associated with the minority carrier charge stored in the *n* and *p* materials as a result of the concentration profiles established by carrier injection. While the first is easier to see when the *pn* junction is reverse biased, the second is in effect only when the junction is forward biased.

3.6.1 Depletion or Junction Capacitance

When a *pn* junction is reverse biased with a voltage V_R, the charge stored on either side of the depletion region is given by Eq. (3.32),

$$Q_J = A\sqrt{2\varepsilon_s q \frac{N_A N_D}{N_A + N_D}(V_0 + V_R)}$$

Thus, for a given *pn* junction,

$$Q_J = \alpha\sqrt{V_0 + V_R} \tag{3.42}$$

where α is given by

$$\alpha = A\sqrt{2\varepsilon_s q \frac{N_A N_D}{N_A + N_D}} \tag{3.43}$$

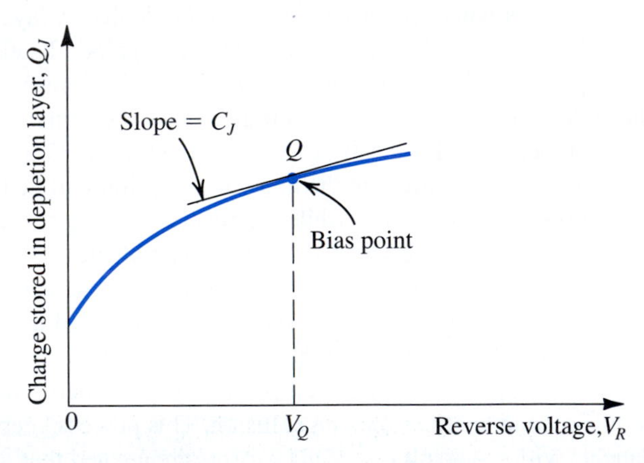

Figure 3.15 The charge stored on either side of the depletion layer as a function of the reverse voltage V_R.

Thus Q_J is nonlinearly related to V_R, as shown in Fig. (3.15). This nonlinear relationship makes it difficult to define a capacitance that accounts for the need to change Q_J whenever V_R is changed. We can, however, assume that the junction is operating at a point such as Q, as indicated in Fig. 3.15, and define a capacitance C_j that relates the change in the charge Q_J to a change in the voltage V_R,

$$C_j = \left.\frac{dQ_J}{dV_R}\right|_{V_R=V_Q} \tag{3.44}$$

This incremental-capacitance approach turns out to be quite useful in electronic circuit design, as we shall see throughout this book.

Using Eq. (3.44) together with Eq. (3.42) yields

$$C_j = \frac{\alpha}{2\sqrt{V_0 + V_R}} \tag{3.45}$$

The value of C_j at zero reverse-bias can be obtained from Eq. (3.45) as

$$C_{j0} = \frac{\alpha}{2\sqrt{V_0}} \tag{3.46}$$

which enables us to express C_j as

$$C_j = \frac{C_{j0}}{\sqrt{1 + \frac{V_R}{V_0}}} \tag{3.47}$$

where C_{j0} is given by Eq. (3.46) or alternatively if we substitute for α from Eq. (3.43) by

$$C_{j0} = A\sqrt{\left(\frac{\varepsilon_s q}{2}\right)\left(\frac{N_A N_D}{N_A + N_D}\right)\left(\frac{1}{V_0}\right)} \tag{3.48}$$

Before leaving the subject of depletion-region or junction capacitance we point out that in the *pn* junction we have been studying, the doping concentration is made to change abruptly at the junction boundary. Such a junction is known as an **abrupt junction**. There is another type of *pn* junction in which the carrier concentration is made to change gradually from one side of the junction to the other. To allow for such a **graded junction**, the formula for the junction capacitance (Eq. 3.47) can be written in the more general form

$$C_j = \frac{C_{j0}}{\left(1 + \frac{V_R}{V_0}\right)^m} \tag{3.49}$$

where m is a constant called the **grading coefficient**, whose value ranges from 1/3 to 1/2 depending on the manner in which the concentration changes from the p to the n side.

EXERCISE

3.12 For the *pn* junction considered in Examples 3.5 and 3.6, find C_{j0} and C_j at $V_R = 2$ V. Recall that $V_0 = 0.814$ V, $N_A = 10^{18}/\text{cm}^3$, $N_D = 10^{16}/\text{cm}^3$ and $A = 10^{-4}\text{cm}^2$.

Ans. 3.2 pF; 1.7 pF

3.6.2 Diffusion Capacitance

Consider a forward-biased *pn* junction. In steady-state, minority carrier distributions in the *p* and *n* materials are established, as shown in Fig. 3.12. Thus a certain amount of excess minority carrier charge is stored in each of the *p* and *n* bulk regions (outside the depletion region). If the terminal voltage V changes, this charge will have to change before a new steady state is achieved. This charge-storage phenomenon gives rise to another capacitive effect, distinctly different from that due to charge storage in the depletion region.

To calculate the excess minority carrier charge, refer to Fig. 3.12. The excess hole charge stored in the *n* region can be found from the shaded area under the exponential as follows:[8]

$$Q_p = Aq \times \text{shaded area under the } p_n(x) \text{ curve}$$

$$= Aq[p_n(x_n) - p_{n0}]L_p$$

substituting for $p_n(x_n)$ from Eq. (3.33) and using Eq. (3.37) enables us to express Q_p as

$$Q_p = \frac{L_p^2}{D_p} I_p \tag{3.50}$$

The factor (L_p^2/D_p) that relates Q_p to I_p is a useful device parameter that has the dimension of time (s) and is denoted τ_p

$$\tau_p = \frac{L_p^2}{D_p} \tag{3.51}$$

Thus,

$$Q_p = \tau_p I_p \tag{3.52}$$

The time constant τ_p is known as the excess **minority carrier (hole) lifetime**. *It is the average time it takes for a hole injected into the n region to recombine with a majority electron*. This definition of τ_p implies that the entire charge Q_p disappears and has to be replenished every τ_p seconds. The current that accomplishes the replenishing is $I_p = Q_p/\tau_p$. This is an alternate derivation for Eq. (3.52).

A relationship similar to that in Eq. (3.52) can be developed for the electron charge stored in the *p* region,

$$Q_n = \tau_n I_n \tag{3.53}$$

where τ_n is the electron lifetime in the *p* region. The total excess minority carrier charge can be obtained by adding together Q_p and Q_n,

$$Q = \tau_p I_p + \tau_n I_n \tag{3.54}$$

This charge can be expressed in terms of the diode current $I = I_p + I_n$ as

$$Q = \tau_T I \tag{3.55}$$

where τ_T is called the **mean transit time** of the junction. Obviously, τ_T is related to τ_p and τ_n. Furthermore, for most practical devices, one side of the junction is much more heavily doped than the other. For instance, if $N_A \gg N_D$, one can show that $I_p \gg I_n$, $I \simeq I_p$, $Q_p \gg Q_n$, $Q \simeq Q_p$, and thus $\tau_T \simeq \tau_p$.

[8]Recall that the area under an exponential curve $Ae^{-x/B}$ is equal to AB.

For small changes around a bias point, we can define an **incremental diffusion capacitance** C_d as

$$C_d = \frac{dQ}{dV} \tag{3.56}$$

and can show that

$$C_d = \left(\frac{\tau_T}{V_T}\right)I \tag{3.57}$$

where I is the forward-bias current. Note that C_d is directly proportional to the forward current I and thus is negligibly small when the diode is reverse biased. Also note that to keep C_d small, the transit time τ_T must be made small, an important requirement for a *pn* junction intended for high-speed or high-frequency operation.

EXERCISES

3.15 Use the definition of C_d in Eq. (3.56) to derive the expression in Eq. (3.57) by means of Eqs. (3.55) and (3.40).

3.16 For the *pn* junction considered in Examples 3.5 and 3.6 for which $D_p = 10\ \text{cm}^2/\text{V}\cdot\text{s}$, and $L_p = 5\ \mu\text{m}$, find τ_p and C_d at a forward-bias current of 0.1 mA. Recall that for this junction, $I_p \simeq I$.
Ans. 25 ns; 96.5 pF

Summary

- Today's microelectronics technology is almost entirely based on the semiconductor material silicon. If a circuit is to be fabricated as a monolithic integrated circuit (IC) it is made using a single silicon crystal, no matter how large the circuit is [a recent chip (2009) contains 2.3 billion transistors].
- In a crystal of intrinsic or pure silicon, the atoms are held in position by covalent bonds. At very low temperatures, all the bonds are intact, and no charge carriers are available to conduct electrical current. Thus, at such low temperatures, silicon behaves as an insulator.
- At room temperature, thermal energy causes some of the covalent bonds to break, thus generating free electrons and holes that become available for current conduction.
- Current in semiconductors is carried by free electrons and holes. Their numbers are equal and relatively small in intrinsic silicon.
- The conductivity of silicon can be increased dramatically by introducing small amounts of appropriate impurity materials into the silicon crystal in a process called doping.
- There are two kinds of doped semiconductor: *n*-type, in which electrons are abundant, and *p*-type, in which holes are abundant.
- There are two mechanisms for the transport of charge carriers in semiconductor: drift and diffusion.
- Carrier drift results when an electric field E is applied across a piece of silicon. The electric field accelerate the holes in the direction of E and the electrons in the direction opposite to E. These two current components add together to produce a drift current in the direction of E.
- Carrier diffusion occurs when the concentration of charge carriers is made higher in one part of the silicon crystal than in other parts. To establish a steady-state diffusion current, a carrier concentration gradient must be maintained in the silicon crystal.
- A basic semiconductor structure is the *pn* junction. It is fabricated in a silicon crystal by creating a p region in close proximity to an n region. The *pn* junction is a diode and plays a dominant role in the structure and operation of transistors.
- When the terminals of the *pn* junction are left open, no current flows externally. However, two equal and opposite currents, I_D and I_S, flow across the junction, and equilibrium is maintained by a built-in voltage V_0 that develops across the junction, with the n side positive relative to the p side. Note, however, that the voltage across an open junction is 0 V, since V_0 is cancelled by potentials appearing at the metal-to-semiconductor connection interfaces.
- The voltage V_0 appears across the depletion region, which extends on both sides of the junction.
- The diffusion current I_D is carried by holes diffusing from p to n and electrons diffusing from n to p. I_D flows from p to n, which is the forward direction of the junction. Its value depends on V_0.
- The drift current I_S is carried by thermally generated minority electrons in the p material that are swept across the depletion layer into the n side, and by thermally generated minority holes in the n side that are swept across the depletion region into the p side. I_S flows from n to p, in the reverse direction of the junction, and its value is a strong function of temperature but independent of V_0.
- Forward biasing the *pn* junction, that is, applying an external voltage V that makes p more positive than n, reduces the barrier voltage to $V_0 - V$ and results in an exponential increase in I_D while I_S remains unchanged. The net result is a substantial current $I = I_D - I_S$ that flows across the junction and through the external circuit.
- Applying a negative V reverse-biases the junction and increases the barrier voltage, with the result that I_D is reduced to almost zero and the net current across the junction becomes the very small reverse current I_S.
- If the reverse voltage is increased in magnitude to a value V_Z specific to the particular junction, the junction breaks down, and a large reverse current flows. The value of the reverse current must be limited by the external circuit.
- Whenever the voltage across a *pn* junction is changed, some time has to pass before steady state is reached. This is due to the charge-storage effects in the junction, which are modeled by two capacitances: the junction capacitance C_j and the diffusion capacitance C_d.
- For future reference, we present in Table 3.1 a summary of pertinent relationships and the values of physical constants.

Table 3.1 Summary of Important Equations

Quantity	Relationship	Values of Constants and Parameters (for Intrinsic Si at $T = 300$ K)
Carrier concentration in intrinsic silicon (cm^{-3})	$n_i = BT^{3/2}e^{-E_g/2kT}$	$B = 7.3 \times 10^{15}\text{cm}^{-3}\text{K}^{-3/2}$ $E_g = 1.12$ eV $k = 8.62 \times 10^{-5}$eV/K $n_i = 1.5 \times 10^{10}/\text{cm}^3$
Diffusion current density (A/cm^2)	$J_p = -qD_p\dfrac{dp}{dx}$ $J_n = qD_n\dfrac{dn}{dx}$	$q = 1.60 \times 10^{-19}$ coulomb $D_p = 12$ cm^2/s $D_n = 34$ cm^2/s
Drift current density (A/cm^2)	$J_{drift} = q(p\mu_p + n\mu_n)E$	$\mu_p = 480$ cm^2/V·s $\mu_n = 1350$ cm^2/V·s
Resistivity (Ω·cm)	$\rho = 1/[q(p\mu_p + n\mu_n)]$	μ_p and μ_n decrease with the increase in doping concentration
Relationship between mobility and diffusivity	$\dfrac{D_n}{\mu_n} = \dfrac{D_p}{\mu_p} = V_T$	$V_T = kT/q \simeq 25.8$ mV
Carrier concentration in *n*-type silicon (cm^{-3})	$n_{n0} \simeq N_D$ $p_{n0} = n_i^2/N_D$	
Carrier concentration in *p*-type silicon (cm^{-3})	$p_{p0} \simeq N_A$ $n_{p0} = n_i^2/N_A$	
Junction built-in voltage (V)	$V_0 = V_T \ln\left(\dfrac{N_A N_D}{n_i^2}\right)$	
Width of depletion region (cm)	$\dfrac{x_n}{x_p} = \dfrac{N_A}{N_D}$ $W = x_n + x_p$ $= \sqrt{\dfrac{2\varepsilon_s}{q}\left(\dfrac{1}{N_A} + \dfrac{1}{N_D}\right)(V_0 + V_R)}$	$\varepsilon_s = 11.7\varepsilon_0$ $\varepsilon_0 = 8.854 \times 10^{-14}$ F/cm
Charge stored in depletion layer (coulomb)	$Q_J = q\dfrac{N_A N_D}{N_A + N_D}AW$	
Forward current (A)	$I = I_p + I_n$ $I_p = Aqn_i^2\dfrac{D_p}{L_p N_D}(e^{V/V_T} - 1)$ $I_n = Aqn_i^2\dfrac{D_n}{L_n N_A}(e^{V/V_T} - 1)$	
Saturation current (A)	$I_S = Aqn_i^2\left(\dfrac{D_p}{L_p N_D} + \dfrac{D_n}{L_n N_A}\right)$	
I-V Relationship	$I = I_S(e^{V/V_T} - 1)$	

Table 3.1 *continued*

Quantity	Relationship	Values of Constants and Parameters (for Intrinsic Si at $T = 300$ K)
Minority-carrier lifetime (s)	$\tau_p = L_p^2/D_p \qquad \tau_n = L_n^2/D_n$	$L_p, L_n = 1\ \mu\text{m to } 100\ \mu\text{m}$ $\tau_p, \tau_n = 1 \text{ ns to } 10^4 \text{ ns}$
Minority-carrier charge storage (coulomb)	$Q_p = \tau_p I_p \qquad Q_n = \tau_n I_n$ $Q = Q_p + Q_n = \tau_T I$	
Depletion capacitance (F)	$C_{j0} = A\sqrt{\left(\frac{\varepsilon_s q}{2}\right)\left(\frac{N_A N_D}{N_A + N_D}\right)\frac{1}{V_0}}$ $C_j = C_{j0}\Big/\left(1 + \frac{V_R}{V_0}\right)^m$	 $m = \frac{1}{3} \text{ to } \frac{1}{2}$
Diffusion capacitance (F)	$C_d = \left(\frac{\tau_T}{V_T}\right)I$	

PROBLEMS

Problems are marked with asterisks to describe their degree of difficulty. Difficult problems are marked with an asterisk (*); more difficult problems with two asterisks (**); and very challenging and/or time-consuming problems with three asterisks (***). Also, if in the following problems the need arises for the values of particular parameters or physical constants that are not stated, please consult Table 3.1.

Section 3.1: Intrinsic Semiconductors

3.1 Find values of the intrinsic carrier concentration n_i for silicon at –70°C, 0°C, 20°C, 100°C, and 125°C. At each temperature, what fraction of the atoms is ionized? Recall that a silicon crystal has approximately 5×10^{22} atoms/cm^3.

3.2 Calculate the value of n_i for gallium arsenide (GaAs) at $T = 300$ K. The constant $B = 3.56 \times 10^{14}$ (cm^{-3}K$^{-3/2}$) and the bandgap voltage $E_g = 1.42$ eV.

Section 3.2: Doped Semiconductors

3.3 For a *p*-type silicon in which the dopant concentration $N_A = 10^{18}/\text{cm}^3$, find the hole and electron concentrations at $T = 300$ K.

3.4 For a silicon crystal doped with phosphorus, what must N_D be if at $T = 300$ K the hole concentration drops below the intrinsic level by a factor of 10^7?

3.5 In a phosphorus-doped silicon layer with impurity concentration of $10^{16}/\text{cm}^3$, find the hole and electron concentrations at 27°C and 125°C.

Section 3.3: Current Flow in Semiconductors

3.6 A young designer, aiming to develop intuition concerning conducting paths within an integrated circuit, examines the end-to-end resistance of a connecting bar 10 µm long, 3 µm wide, and 1 µm thick, made of various materials. The designer considers:

(c) n-doped silicon with $N_D = 10^{18}/\text{cm}^3$

(d) p-doped silicon with $N_A = 10^{16}/\text{cm}^3$

(e) aluminum with resistivity of $2.8\ \mu\Omega\cdot\text{cm}$

Find the resistance in each case. For intrinsic silicon, use the data in Table 3.1. For doped silicon, assume $\mu_n = 2.5\mu_p = 1200\ \text{cm}^2/\text{V}\cdot\text{s}$. (Recall that $R = \rho L/A$)

3.7 Contrast the electron and hole drift velocities through a 10-μm layer of intrinsic silicon across which a voltage of 5 V is imposed. Let $\mu_n = 1350\ \text{cm}^2/\text{V}\cdot\text{s}$ and $\mu_p = 480\ \text{cm}^2/\text{V}\cdot\text{s}$.

3.8 Find the current that flows in a silicon bar of 10-μm length having a 5-μm × 4-μm cross section and having free electron and hole densities of $10^5/\text{cm}^3$ and $10^{15}/\text{cm}^3$, respectively, when a 1 V is applied end-to-end. Use $\mu_n = 1200\ \text{cm}^2/\text{V}\cdot\text{s}$ and $\mu_p = 500\ \text{cm}^2/\text{V}\cdot\text{s}$.

3.9 In a 10-μm long bar of donor-doped silicon, what donor concentration is needed to realize a current density of $1\ \text{mA}/\mu\text{m}^2$ in response to an applied voltage of 1 V. (*Note*: Although the carrier mobilities change with doping concentration, as a first approximation you may assume μ_n to be constant and use the value for intrinsic silicon, $1350\ \text{cm}^2/\text{V}\cdot\text{s}$).

3.10 Holes are being steadily injected into a region of n-type silicon (connected to other devices, the details of which are not important for this question). In the steady state, the excess-hole concentration profile shown in Fig. P3.10 is established in the n-type silicon region. Here "excess" means over and above the thermal-equilibrium concentration (in the absence of hole injection), denoted p_{n0}. If $N_D = 10^{16}/\text{cm}^3$, $n_i = 1.5 \times 10^{10}/\text{cm}^3$, $D_n = 12\ \text{cm}^2/\text{s}$, and $W = 0.1\ \mu\text{m}$, find the density of the current that will flow in the x direction.

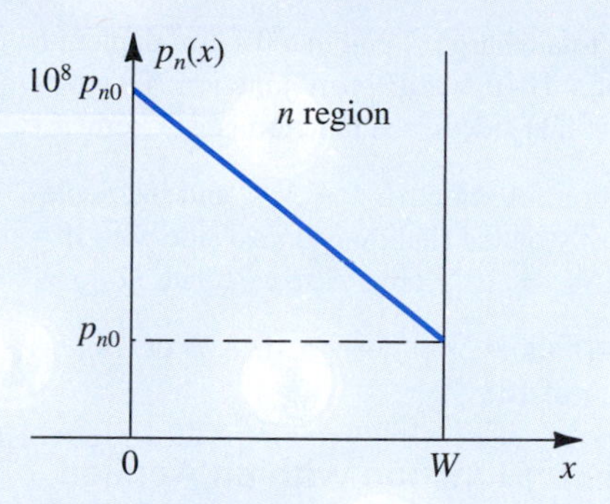

Figure P3.10

3.11 Both the carrier mobility and diffusivity decrease as the doping concentration of silicon is increased. The table below provides a few data points for μ_n and μ_p versus doping concentration. Use the Einstein relationship to obtain the corresponding values for D_n and D_p.

Section 3.4: The *pn* Junction with Open-Circuit Terminals (Equilibrium)

3.12 Calculate the built-in voltage of a junction in which the p and n regions are doped equally with 10^{16} atoms/cm³. Assume $n_i = 1.5 \times 10^{10}/\text{cm}^3$. With the terminals left open, what is the width of the depletion region, and how far does it extend into the p and n regions? If the cross-sectional area of the junction is 100 μm², find the magnitude of the charge stored on either side of the junction.

3.13 If, for a particular junction, the acceptor concentration is $10^{16}/\text{cm}^3$ and the donor concentration is $10^{15}/\text{cm}^3$, find the junction built-in voltage. Assume $n_i = 1.5 \times 10^{10}/\text{cm}^3$. Also, find the width of the depletion region (W) and its extent in each of the p and n regions when the junction terminals are left open. Calculate the magnitude of the charge stored on either side of the junction. Assume that the junction area is 400 μm².

Doping Concentration (carriers/cm³)	μ_n (cm²/V·s)	μ_p (cm²/V·s)	D_n (cm²/s)	D_p (cm²/s)
Intrinsic	1350	480		
10^{16}	1100	400		
10^{17}	700	260		
10^{18}	360	150		

Table P3.11

3.14 Estimate the total charge stored in a 0.1-μm depletion layer on one side of a 10-μm × 10-μm junction. The doping concentration on that side of the junction is $10^{16}/\text{cm}^3$.

3.15 In a *pn* junction for which $N_A \gg N_D$, and the depletion layer exists mostly on the shallowly doped side with W = 0.3 μm, find V_0 if $N_D = 10^{16}/\text{cm}^3$. Also calculate Q_J.

3.16 By how much does V_0 change if N_A or N_D is increased by a factor of 10?

Section 3.5: The *pn* Junction with an Applied Voltage

3.17 If a 5-V reverse-bias voltage is applied across the junction specified in Problem 3.13, find W and Q_J.

3.18 Show that for a *pn* junction reverse-biased with a voltage V_R, the depletion-layer width W and the charge stored on either side of the junction, Q_J, can be expressed as

$$W = W_0\sqrt{1+\frac{V_R}{V_0}}$$

$$Q_J = Q_{J0}\sqrt{1+\frac{V_R}{V_0}}$$

where W_0 and Q_{J0} are the values in equilibrium.

3.19 In a forward-biased *pn* junction show that the ratio of the current component due to hole injection across the junction to the component due to electron injection is given by

$$\frac{I_p}{I_n} = \frac{D_p}{D_n}\frac{L_n}{L_p}\frac{N_A}{N_D}$$

Evaluate this ratio for the case $N_A = 10^{18}/\text{cm}^3$, $N_D = 10^{16}/\text{cm}^3$, $L_p = 5$ μm, $L_n = 10$ μm, $D_p = 10\ \text{cm}^2/\text{s}$, and $D_n = 20\ \text{cm}^2/\text{s}$, and hence find I_p and I_n for the case in which the *pn* junction is conducting a forward current $I = 1$ mA.

3.20 Calculate I_S and the current I for $V = 700$ mV for a *pn* junction for which $N_A = 10^{17}/\text{cm}^3$, $N_D = 10^{16}/\text{cm}^3$, $A = 200\ \mu\text{m}^2$, $n_i = 1.5\times10^{10}/\text{cm}^3$, $L_p = 5$ μm, $L_n = 10$ μm, $D_p = 10\ \text{cm}^2/\text{s}$, and $D_n = 18\ \text{cm}^2/\text{s}$.

3.21 Assuming that the temperature dependence of I_S arises mostly because I_S is proportional to n_i^2, use the expression for n_i in Eq. (3.2) to determine the factor by which n_i^2 changes as T changes from 300 K to 305 K. This will be approximately the same factor by which I_S changes for a 5°C rise in temperature. What is the factor?

3.22 A p^+n junction is one in which the doping concentration in the *p* region is much greater than that in the *n* region. In such a junction, the forward current is mostly due to hole injection across the junction. Show that

$$I \simeq I_p = Aqn_i^2\frac{D_p}{L_pN_D}(e^{V/V_T}-1)$$

For the specific case in which $N_D = 10^{16}/\text{cm}^3$, $D_p = 10\ \text{cm}^2/\text{s}$, $L_p = 10$ μm, and $A = 10^4\ \mu\text{m}^2$, find I_S and the voltage V obtained when $I = 0.5$ mA. Assume operation at 300 K where $n_i = 1.5\times10^{10}/\text{cm}^3$.

3.23 A *pn* junction for which the breakdown voltage is 12 V has a rated (i.e., maximum allowable) power dissipation of 0.25 W. What continuous current in the breakdown region will raise the dissipation to half the rated value? If breakdown occurs for only 10 ms in every 20 ms, what average breakdown current is allowed?

Section 3.6: Capacitive Effects in the *pn* Junction

3.24 For the *pn* junction specified in Problem 3.13, find C_{j0} and C_j at $V_R = 5$ V.

3.25 For a particular junction for which $C_{j0} = 0.6$ pF, $V_0 = 0.75$ V, and $m = 1/3$, find C_j at reverse-bias voltages of 1 V and 10 V.

3.26 The junction capacitance C_j can be thought of as that of a parallel-plate capacitor and thus given by

$$C_j = \frac{\varepsilon A}{W}$$

Show that this approach leads to a formula identical to that obtained by combining Eqs. (3.43) and (3.45) [or equivalently, by combining Eqs. (3.47) and (3.48)].

3.27 A *pn* junction operating in the forward-bias region with a current I of 1 mA is found to have a diffusion capacitance of 10 pF. What diffusion capacitance do you expect this junction to have at $I = 0.1$ mA? What is the mean transit time for this junction?

3.28 For the p^+n junction specified in Problem 3.22, find τ_p and calculate the excess minority carrier charge and the value of the diffusion capacitance at $I = 0.2$ mA.

3.29 A **short-base diode** is one where the widths of the p and n regions are much smaller than L_n and L_p, respectively. As a result, the excess minority carrier distribution in each region is a straight line rather than the exponentials shown in Fig. 3.12.

(a) For the short-base diode, sketch a figure corresponding to Fig. 3.12 and assume as in Fig. 3.12 that $N_A \gg N_D$.

(b) Following a derivation similar to that given in Section 3.5.2, show that if the widths of the p and n regions are denoted W_p and W_n then

$$I = Aqn_i^2\left[\frac{D_p}{(W_n - x_n)N_D} + \frac{D_n}{(W_p - x_p)N_A}\right](e^{V/V_T} - 1)$$

and

$$Q_p = \frac{1}{2}\frac{(W_n - x_n)^2}{D_p}I_p$$

$$\simeq \frac{1}{2}\frac{W_n^2}{D_p}I_p, \text{ for } W_n \gg x_n$$

(c) Also, assuming $Q \simeq Q_p$, $I \simeq I_p$, show that

$$C_d = \frac{\tau_T}{V_T}I$$

where

$$\tau_T = \frac{1}{2}\frac{W_n^2}{D_p}$$

(d) If a designer wishes to limit C_d to 8 pF at $I = 1$ mA, what should W_n be? Assume $D_p = 10\ \text{cm}^2/\text{s}$.

CHAPTER 4

Diodes

IN THIS CHAPTER YOU WILL LEARN

1. The characteristics of the ideal diode and how to analyze and design circuits containing multiple ideal diodes together with resistors and dc sources to realize useful and interesting nonlinear functions.

2. The details of the i–v characteristic of the junction diode (which was derived in Chapter 3) and how to use it to analyze diode circuits operating in the various bias regions: forward, reverse, and breakdown.

3. A simple but effective model of the diode i–v characteristic in the forward direction; the constant-voltage-drop model.

4. A powerful technique for the application and modeling of the diode (and in later chapters, transistors): dc-biasing the diode and modeling its operation for small signals around the dc operating point by means of the small-signal model.

5. The use of a string of forward-biased diodes and of diodes operating in the breakdown region (zener diodes), to provide constant dc voltages (voltage regulators).

6. Application of the diode in the design of rectifier circuits, which convert ac voltages to dc as needed for powering electronic equipment.

7. A number of other practical and important applications of diodes.

Introduction

In Chapters 1 and 2 we dealt almost entirely with linear circuits; any nonlinearity, such as that introduced by amplifier output saturation, was treated as a problem to be solved by the circuit designer. However, there are many other signal-processing functions that can be implemented only by nonlinear circuits. Examples include the generation of dc voltages from the ac power supply, and the generation of signals of various waveforms (e.g., sinusoids, square waves, pulses). Also, digital logic and memory circuits constitute a special class of nonlinear circuits.

The simplest and most fundamental nonlinear circuit element is the diode. Just like a resistor, the diode has two terminals; but unlike the resistor, which has a linear (straight-line) relationship between the current flowing through it and the voltage appearing across it, the diode has a nonlinear i–v characteristic.

This chapter is concerned with the study of diodes. In order to understand the essence of the diode function, we begin with a fictitious element, the ideal diode. We then introduce the silicon junction diode, explain its terminal characteristics, and provide techniques for the analysis of diode circuits. The latter task involves the important subject of device modeling. Our study of modeling the diode characteristics will lay the foundation for our study of modeling transistor operation in the next two chapters.

Of the many applications of diodes, their use in the design of rectifiers (which convert ac to dc) is the most common. Therefore we shall study rectifier circuits in some detail and briefly look at a number of other diode applications. Further nonlinear circuits that utilize diodes and other devices will be found throughout the book, but particularly in Chapter 17.

The junction diode is nothing more than the *pn* junction we studied in Chapter 3, and most of this chapter is concerned with the study of silicon *pn*-junction diodes. In the last section, however, we briefly consider some specialized diode types, including the photodiode and the light-emitting diode.

4.1 The Ideal Diode

4.1.1 Current–Voltage Characteristic

The ideal diode may be considered to be the most fundamental nonlinear circuit element. It is a two-terminal device having the circuit symbol of Fig. 4.1(a) and the i–v characteristic shown in Fig. 4.1(b). The terminal characteristic of the ideal diode can be interpreted as follows: If a negative voltage (relative to the reference direction indicated in Fig. 4.1a) is applied to the diode, no current flows and the diode behaves as an open circuit (Fig. 4.1c). Diodes operated in this mode are said to be **reverse biased**, or operated in the reverse direction. An ideal diode has zero current when operated in the reverse direction and is said to be **cut off**, or simply **off**.

On the other hand, if a positive current (relative to the reference direction indicated in Fig. 4.1a) is applied to the ideal diode, zero voltage drop appears across the diode. In other words, the ideal diode behaves as a short circuit in the *forward* direction (Fig. 4.1d); it passes any current with zero voltage drop. A **forward-biased** diode is said to be **turned on**, or simply **on**.

From the above description it should be noted that the external circuit must be designed to limit the forward current through a conducting diode, and the reverse voltage across a cutoff diode, to predetermined values. Figure 4.2 shows two diode circuits that illustrate this point. In the circuit of Fig. 4.2(a) the diode is obviously conducting. Thus its voltage drop will be zero, and the current through it will be determined by the +10-V supply and the 1-kΩ resistor as 10 mA. The diode in the circuit of Fig. 4.2(b) is obviously cut off, and thus its current will be zero, which in turn means that the entire 10-V supply will appear as reverse bias across the diode.

The positive terminal of the diode is called the **anode** and the negative terminal the **cathode**, a carryover from the days of vacuum-tube diodes. The i–v characteristic of the ideal diode (conducting in one direction and not in the other) should explain the choice of its arrow like circuit symbol.

As should be evident from the preceding description, the i–v characteristic of the ideal diode is highly nonlinear; although it consists of two straight-line segments, they are at 90° to one another. A nonlinear curve that consists of straight-line segments is said to be **piecewise linear**. If a device having a piecewise-linear characteristic is used in a particular application in such a way that the signal across its terminals swings along only one of the linear

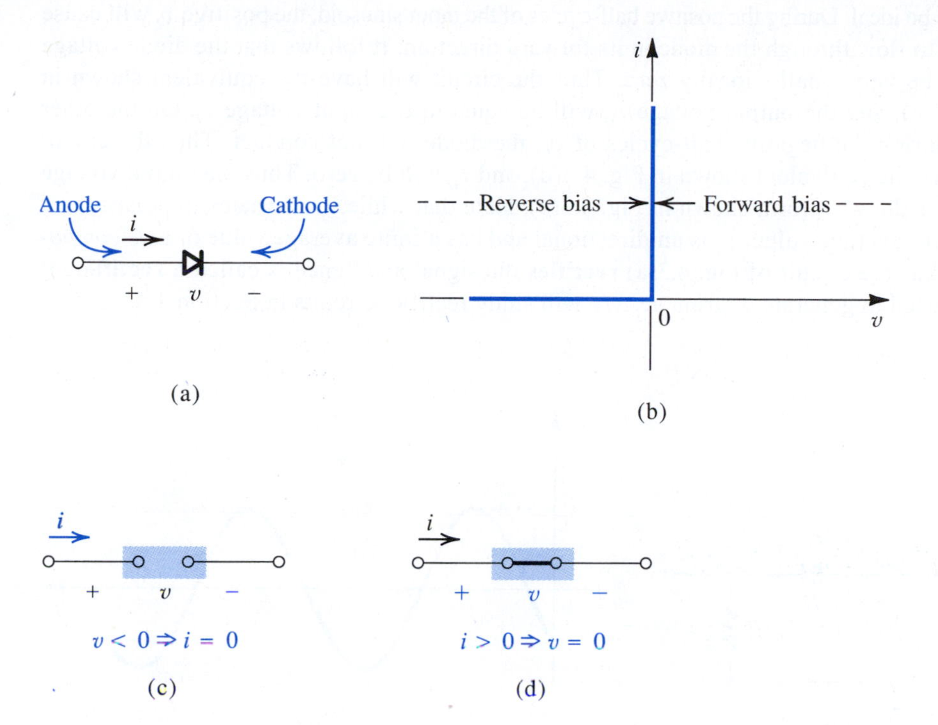

Figure 4.1 The ideal diode: **(a)** diode circuit symbol; **(b)** i–v characteristic; **(c)** equivalent circuit in the reverse direction; **(d)** equivalent circuit in the forward direction.

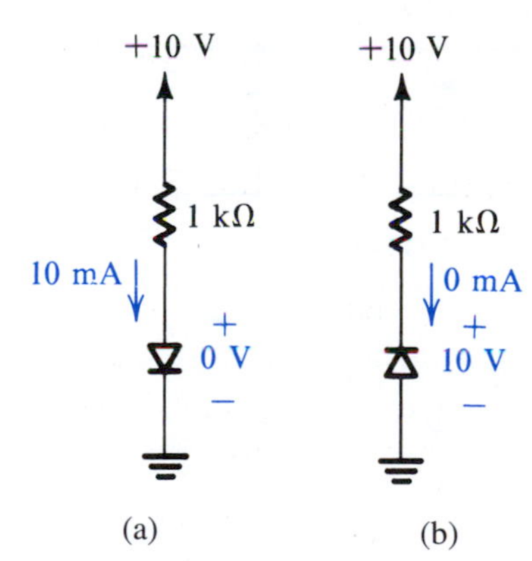

Figure 4.2 The two modes of operation of ideal diodes and the use of an external circuit to limit **(a)** the forward current and **(b)** the reverse voltage.

segments, then the device can be considered a linear circuit element as far as that particular circuit application is concerned. On the other hand, if signals swing past one or more of the break points in the characteristic, linear analysis is no longer possible.

4.1.2 A Simple Application: The Rectifier

A fundamental application of the diode, one that makes use of its severely nonlinear i–v curve, is the rectifier circuit shown in Fig. 4.3(a). The circuit consists of the series connection of a diode D and a resistor R. Let the input voltage v_I be the sinusoid shown in Fig. 4.3(b), and assume the diode to be ideal. During the positive half-cycles of the input sinusoid, the positive v_I will cause current to flow through the diode in its forward direction. It follows that the diode voltage v_D will be very small—ideally zero. Thus the circuit will have the equivalent shown in Fig. 4.3(c), and the output voltage v_O will be equal to the input voltage v_I. On the other hand, during the negative half-cycles of v_I, the diode will not conduct. Thus the circuit will have the equivalent shown in Fig. 4.3(d), and v_O will be zero. Thus the output voltage will have the waveform shown in Fig. 4.3(e). Note that while v_I alternates in polarity and has a zero average value, v_O is unidirectional and has a finite average value or a *dc component*. Thus the circuit of Fig. 4.3(a) **rectifies** the signal and hence is called a **rectifier**. It can be used to generate dc from ac. We will study rectifier circuits in Section 4.5.

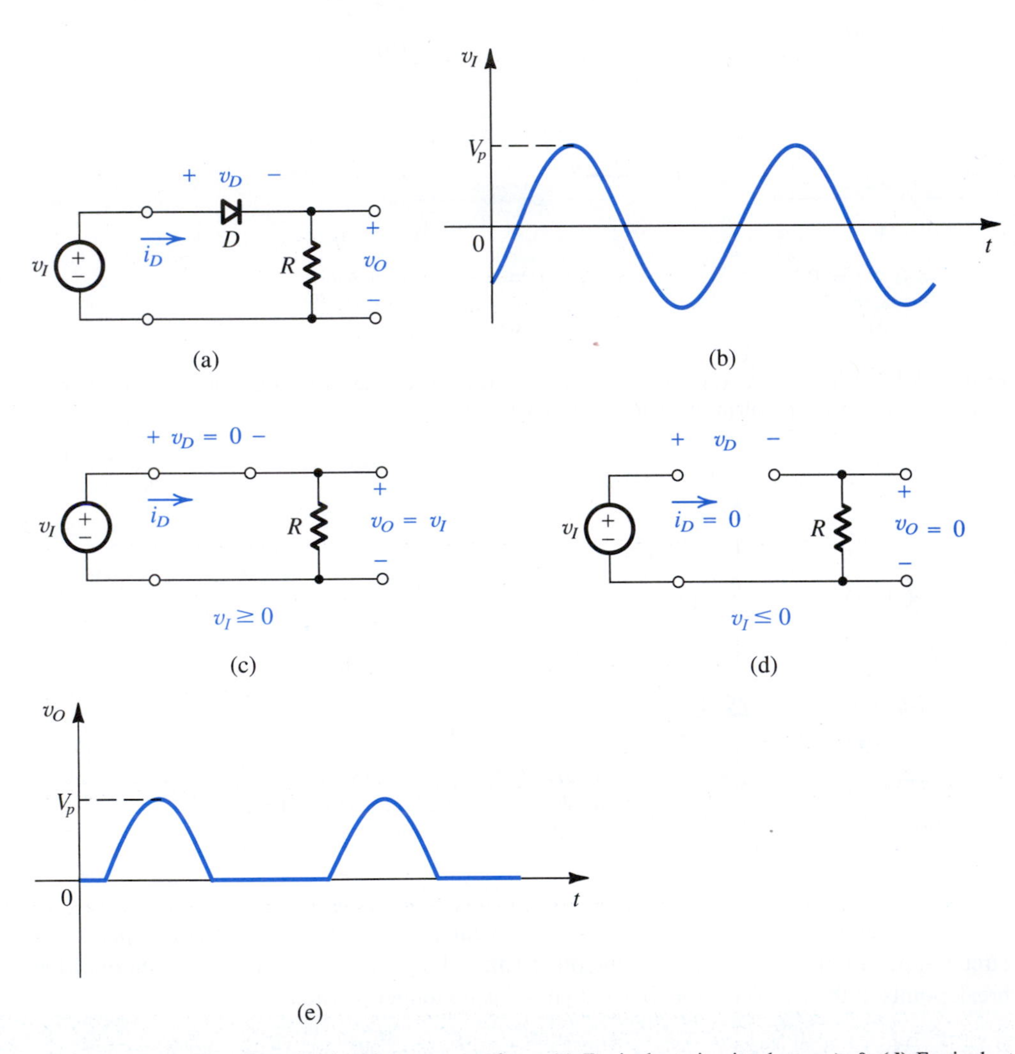

Figure 4.3 **(a)** Rectifier circuit. **(b)** Input waveform. **(c)** Equivalent circuit when $v_I \geq 0$. **(d)** Equivalent circuit when $v_I \leq 0$. **(e)** Output waveform.

EXERCISES

4.1 For the circuit in Fig. 4.3(a), sketch the transfer characteristic v_O versus v_I.
Ans. See Fig. E4.1.

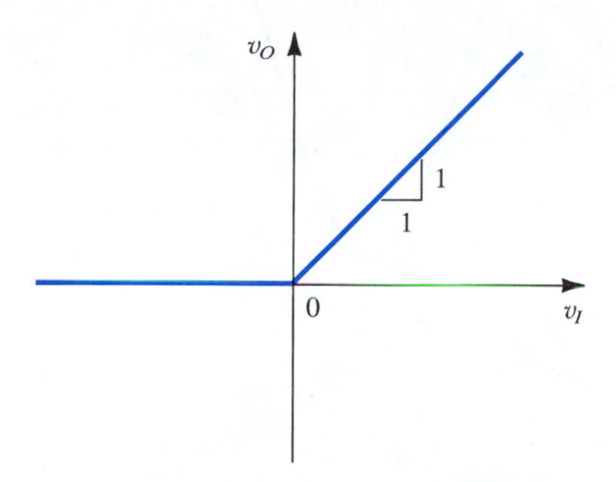

Figure E4.1

4.2 For the circuit in Fig. 4.3(a), sketch the waveform of v_D.
Ans. $v_D = v_I - v_O$, resulting in the waveform in Fig. E4.2

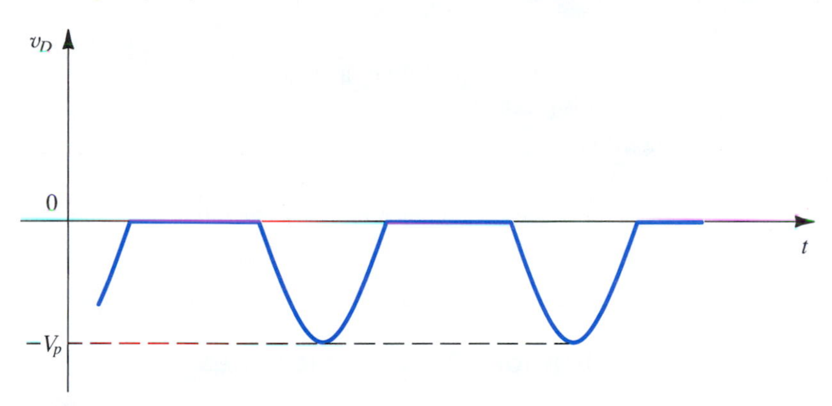

Figure E4.2

4.3 In the circuit of Fig. 4.3(a), let v_I have a peak value of 10 V and $R = 1\ \text{k}\Omega$. Find the peak value of i_D and the dc component of v_O.
Ans. 10 mA; 3.18 V

Example 4.1

Figure 4.4(a) shows a circuit for charging a 12-V battery. If v_S is a sinusoid with 24-V peak amplitude, find the fraction of each cycle during which the diode conducts. Also, find the peak value of the diode current and the maximum reverse-bias voltage that appears across the diode.

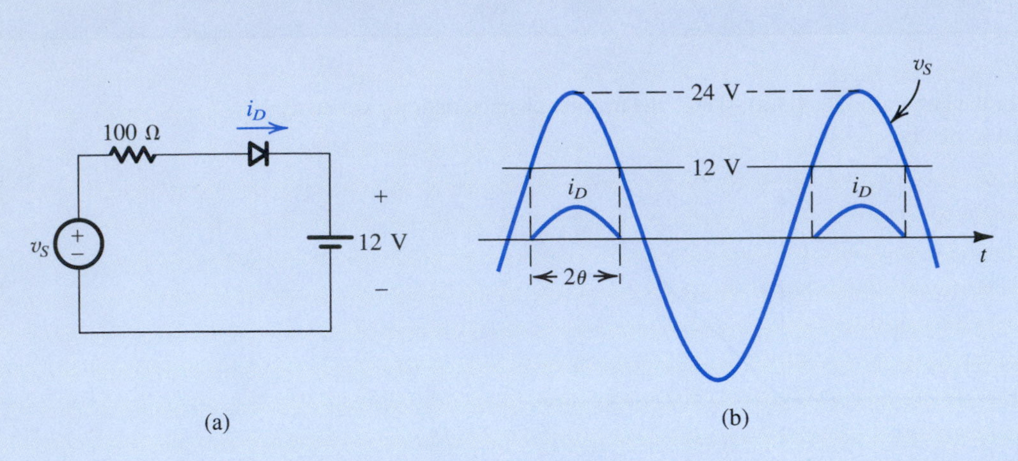

Figure 4.4 Circuit and waveforms for Example 4.1.

Solution

The diode conducts when v_S exceeds 12 V, as shown in Fig. 4.4(b). The conduction angle is 2θ, where θ is given by

$$24 \cos\theta = 12$$

Thus $\theta = 60°$ and the conduction angle is 120°, or one-third of a cycle.

The peak value of the diode current is given by

$$I_d = \frac{24 - 12}{100} = 0.12 \text{ A}$$

The maximum reverse voltage across the diode occurs when v_S is at its negative peak and is equal to $24 + 12 = 36$ V.

4.1.3 Another Application: Diode Logic Gates

Diodes together with resistors can be used to implement digital logic functions. Figure 4.5 shows two diode logic gates. To see how these circuits function, consider a positive-logic system in which voltage values close to 0 V correspond to logic 0 (or low) and voltage values close to +5 V correspond to logic 1 (or high). The circuit in Fig. 4.5(a) has three inputs, v_A, v_B, and v_C. It is easy to see that diodes connected to +5-V inputs will conduct, thus clamping the output v_Y to a value equal to +5 V. This positive voltage at the output will keep the diodes whose inputs are low (around 0 V) cut off. Thus *the output will be high if one or more of the inputs are high*. The circuit therefore implements the **logic OR function**, which in Boolean notation is expressed as

$$Y = A + B + C$$

Similarly, the reader is encouraged to show that using the same logic system mentioned above, the circuit of Fig. 4.5(b) implements the **logic AND function**,

$$Y = A \cdot B \cdot C$$

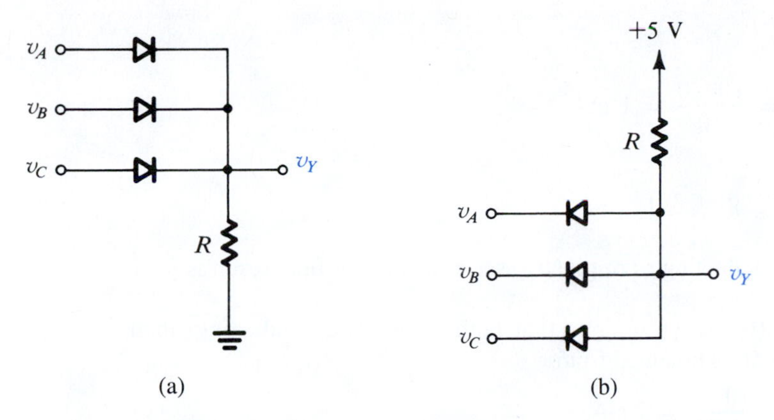

Figure 4.5 Diode logic gates: **(a)** OR gate; **(b)** AND gate (in a positive-logic system).

Example 4.2

Assuming the diodes to be ideal, find the values of I and V in the circuits of Fig. 4.6.

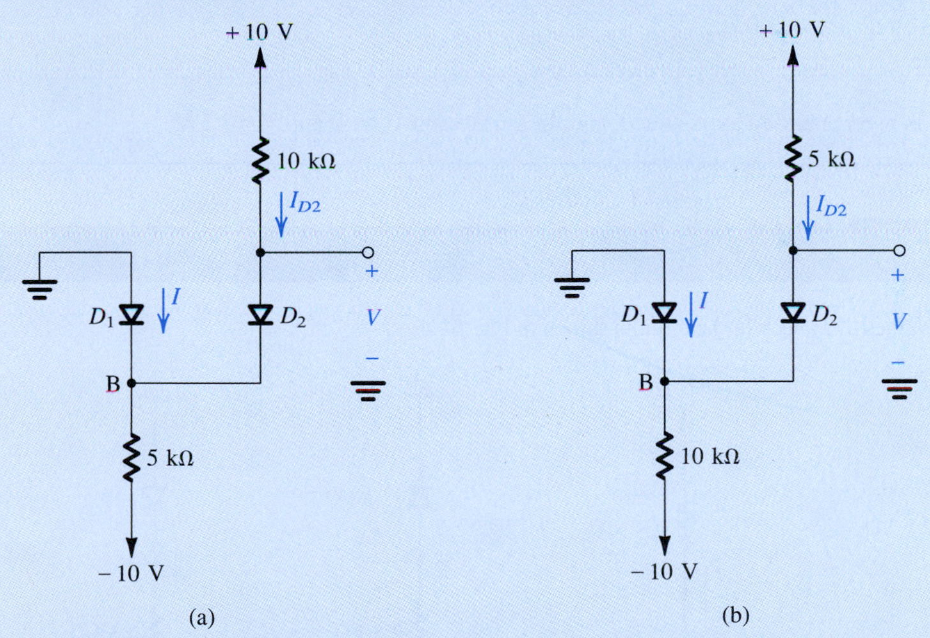

Figure 4.6 Circuits for Example 4.2.

Solution

In these circuits it might not be obvious at first sight whether none, one, or both diodes are conducting. In such a case, *we make a plausible assumption, proceed with the analysis, and then check whether we end up with a consistent solution*. For the circuit in Fig. 4.6(a), we shall assume that both diodes are conducting. It follows that $V_B = 0$ and $V = 0$. The current through D_2 can now be determined from

Example 4.2 *continued*

$$I_{D2} = \frac{10-0}{10} = 1 \text{ mA}$$

Writing a node equation at B,

$$I+1 = \frac{0-(-10)}{5}$$

results in $I = 1$ mA. Thus D_1 is conducting as originally assumed, and the final result is $I = 1$ mA and $V = 0$ V.

For the circuit in Fig. 4.6(b), if we assume that both diodes are conducting, then $V_B = 0$ and $V = 0$. The current in D_2 is obtained from

$$I_{D2} = \frac{10-0}{5} = 2 \text{ mA}$$

The node equation at B is

$$I+2 = \frac{0-(-10)}{10}$$

which yields $I = -1$ mA. Since this is not possible, our original assumption is *not* correct. We start again, assuming that D_1 is off and D_2 is on. The current I_{D2} is given by

$$I_{D2} = \frac{10-(-10)}{15} = 1.33 \text{ mA}$$

and the voltage at node B is

$$V_B = -10 + 10 \times 1.33 = +3.3 \text{ V}$$

Thus D_1 is reverse biased as assumed, and the final result is $I = 0$ and $V = 3.3$ V.

EXERCISES

4.4 Find the values of I and V in the circuits shown in Fig. E4.4.

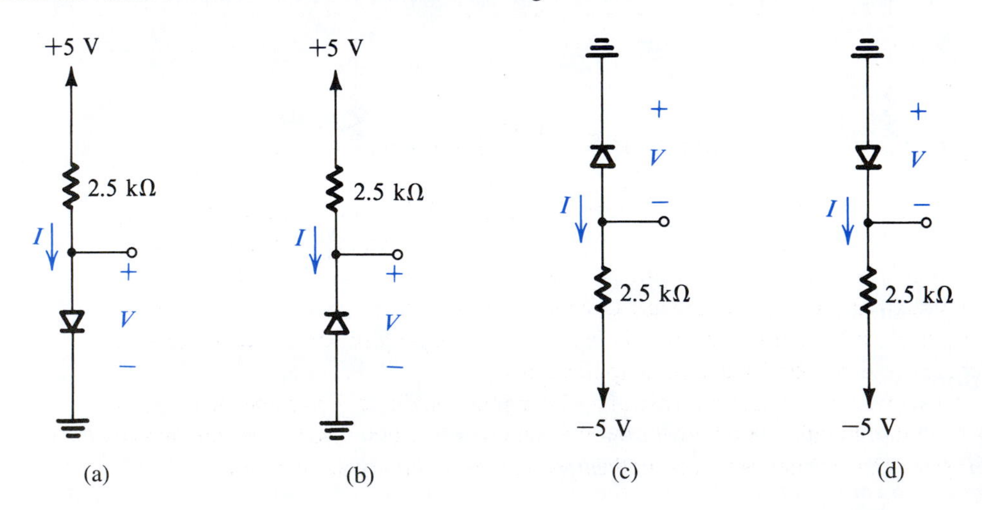

Figure E4.4

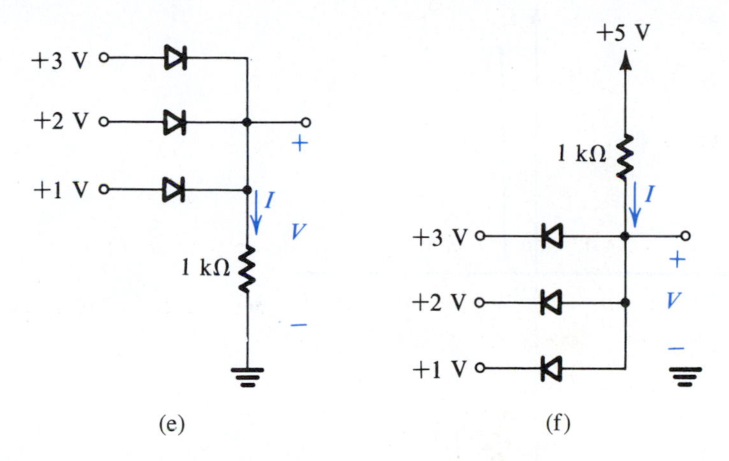

Figure E4.4 (Continued)

Ans. (a) 2 mA, 0 V; (b) 0 mA, 5 V; (c) 0 mA, 5 V; (d) 2 mA, 0 V; (e) 3 mA, +3 V; (f) 4 mA, +1 V

4.5 Figure E4.5 shows a circuit for an ac voltmeter. It utilizes a moving-coil meter that gives a full-scale reading when the *average* current flowing through it is 1 mA. The moving-coil meter has a 50-Ω resistance.

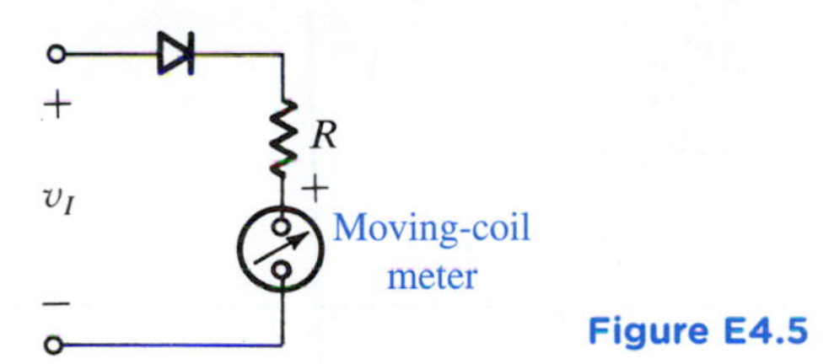

Figure E4.5

Find the value of R that results in the meter indicating a full-scale reading when the input sine-wave voltage v_I is 20 V peak-to-peak. (*Hint:* The average value of half-sine waves is V_p/π.)

Ans. 3.133 kΩ

4.2 Terminal Characteristics of Junction Diodes

The most common implementation of the diode utilizes a *pn* junction. We have studied the physics of the *pn* junction and derived its *i–v* characteristic in Chapter 3. That the *pn* junction is used to implement the diode function should come as no surprise: the *pn* junction can conduct substantial current in the forward direction and almost no current in the reverse direction. In this section we study the *i–v* characteristic of the *pn* junction diode in detail in order to prepare ourselves for diode circuit applications.

Figure 4.7 shows the *i–v* characteristic of a silicon junction diode. The same characteristic is shown in Fig. 4.8 with some scales expanded and others compressed to reveal details. Note that the scale changes have resulted in the apparent discontinuity at the origin.

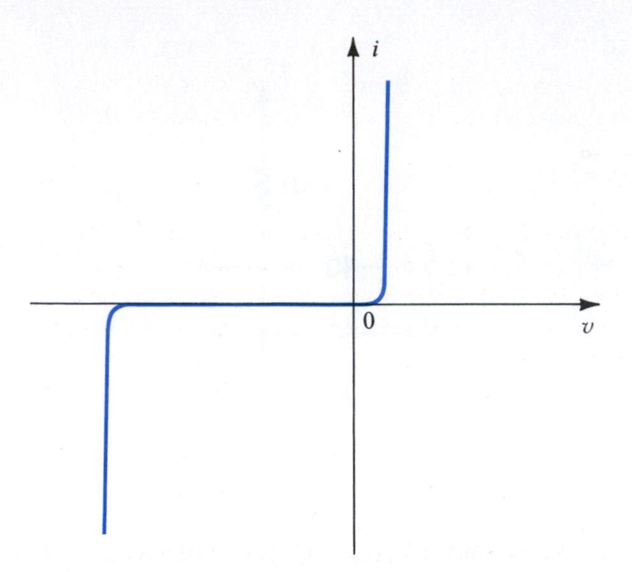

Figure 4.7 The i–v characteristic of a silicon junction diode.

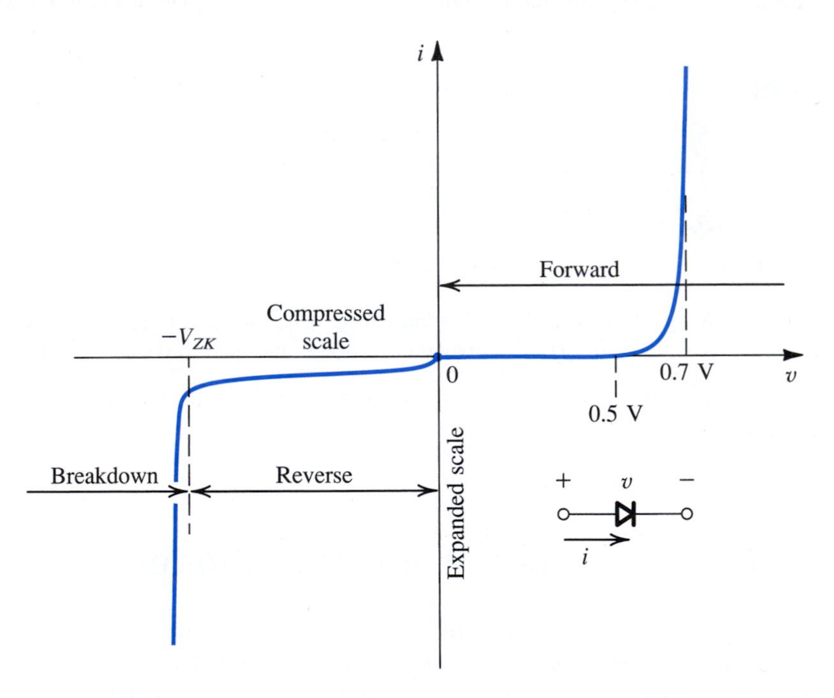

Figure 4.8 The diode i–v relationship with some scales expanded and others compressed in order to reveal details.

As indicated, the characteristic curve consists of three distinct regions:

1. The forward-bias region, determined by $v > 0$
2. The reverse-bias region, determined by $v < 0$
3. The breakdown region, determined by $v < -V_{ZK}$

These three regions of operation are described in the following sections.

4.2.1 The Forward-Bias Region

The forward-bias—or simply forward—region of operation is entered when the terminal voltage v is positive. In the forward region the i–v relationship is closely approximated by

$$i = I_S(e^{v/V_T} - 1) \tag{4.1}$$

In this equation[1] I_S is a constant for a given diode at a given temperature. A formula for I_S in terms of the diode's physical parameters and temperature was given in Eq.(3.41). The current I_S is usually called the **saturation current** (for reasons that will become apparent shortly). Another name for I_S, and one that we will occasionally use, is the **scale current**. This name arises from the fact that I_S is directly proportional to the cross-sectional area of the diode. Thus doubling of the junction area results in a diode with double the value of I_S and, as the diode equation indicates, double the value of current i for a given forward voltage v. For "small-signal" diodes, which are small-size diodes intended for low-power applications, I_S is on the order of 10^{-15} A. The value of I_S is, however, a very strong function of temperature. As a rule of thumb, I_S doubles in value for every 5°C rise in temperature.

The voltage V_T in Eq. (4.1) is a constant called the **thermal voltage** and is given by

$$V_T = \frac{kT}{q} \tag{4.2}$$

where

k = Boltzmann's constant = 8.62×10^{-5} eV/K = 1.38×10^{-23} joules/kelvin

T = the absolute temperature in kelvins = 273 + temperature in °C

q = the magnitude of electronic charge = 1.60×10^{-19} coulomb

Substituting $k = 8.62 \times 10^{-5}$ eV/K into Eq. (4.2) gives

$$V_T = 0.0862T, \text{mV} \tag{4.2a}$$

Thus, at room temperature (20°C) the value of V_T is 25.3 mV. In rapid approximate circuit analysis we shall use $V_T \simeq 25$ mV at room temperature.[2]

For appreciable current i in the forward direction, specifically for $i \gg I_S$, Eq. (4.1) can be approximated by the exponential relationship

$$i \simeq I_S e^{v/V_T} \tag{4.3}$$

This relationship can be expressed alternatively in the logarithmic form

$$v = V_T \ln \frac{i}{I_S} \tag{4.4}$$

where ln denotes the natural (base e) logarithm.

---

[1]Equation (4.1), the diode equation, is sometimes written to include a constant n in the exponential,

$$i = I_S(e^{v/nkt} - 1)$$

with n having a value between 1 and 2, depending on the material and the physical structure of the diode. Diodes using the standard integrated-circuit fabrication process exhibit $n = 1$ when operated under normal conditions. For simplicity, we shall use $n = 1$ throughout this book, unless otherwise specified.

[2]A slightly higher ambient temperature (25°C or so) is usually assumed for electronic equipment operating inside a cabinet. At this temperature, $V_T \simeq 25.8$ mV. Nevertheless, for the sake of simplicity and to promote rapid circuit analysis, we shall use the more arithmetically convenient value of $V_T \simeq 25$ mV throughout this book.

The exponential relationship of the current i to the voltage v holds over many decades of current (a span of as many as seven decades—i.e., a factor of 10^7—can be found). This is quite a remarkable property of junction diodes, one that is also found in bipolar junction transistors and that has been exploited in many interesting applications.

Let us consider the forward i–v relationship in Eq. (4.3) and evaluate the current I_1 corresponding to a diode voltage V_1:

$$I_1 = I_S e^{V_1/V_T}$$

Similarly, if the voltage is V_2, the diode current I_2 will be

$$I_2 = I_S e^{V_2/V_T}$$

These two equations can be combined to produce

$$\frac{I_2}{I_1} = e^{(V_2 - V_1)/V_T}$$

which can be rewritten as

$$V_2 - V_1 = V_T \ln \frac{I_2}{I_1}$$

or, in terms of base-10 logarithms,

$$V_2 - V_1 = 2.3 V_T \log \frac{I_2}{I_1} \tag{4.5}$$

This equation simply states that for a decade (factor of 10) change in current, the diode voltage drop changes by $2.3V_T$, which is approximately 60 mV. This also suggests that the diode i–v relationship is most conveniently plotted on semilog paper. Using the vertical, linear axis for v and the horizontal, log axis for i, one obtains a straight line with a slope of 60 mV per decade of current.

A glance at the i–v characteristic in the forward region (Fig. 4.8) reveals that the current is negligibly small for v smaller than about 0.5 V. This value is usually referred to as the **cut-in voltage**. It should be emphasized, however, that this apparent threshold in the characteristic is simply a consequence of the exponential relationship. Another consequence of this relationship is the rapid increase of i. Thus, for a "fully conducting" diode, the voltage drop lies in a narrow range, approximately 0.6 V to 0.8 V. This gives rise to a simple "model" for the diode where it is assumed that a conducting diode has approximately a 0.7-V drop across it. Diodes with different current ratings (i.e., different areas and correspondingly different I_S) will exhibit the 0.7-V drop at different currents. For instance, a small-signal diode may be considered to have a 0.7-V drop at $i = 1$ mA, while a higher-power diode may have a 0.7-V drop at $i = 1$ A. We will study the topics of diode-circuit analysis and diode models in the next section.

Example 4.3

A silicon diode said to be a 1-mA device displays a forward voltage of 0.7 V at a current of 1 mA. Evaluate the junction scaling constant I_S. What scaling constants would apply for a 1-A diode of the same manufacture that conducts 1 A at 0.7 V?

Solution

Since

$$i = I_S e^{v/V_T}$$

then

$$I_S = i e^{-v/V_T}$$

For the 1-mA diode:

$$I_S = 10^{-3} e^{-700/25} = 6.9 \times 10^{-16} \text{ A}$$

The diode conducting 1 A at 0.7 V corresponds to one-thousand 1-mA diodes in parallel with a total junction area 1000 times greater. Thus I_S is also 1000 times greater,

$$I_S = 6.9 \times 10^{-13} \text{ A}$$

Since both I_S and V_T are functions of temperature, the forward i–v characteristic varies with temperature, as illustrated in Fig. 4.9. At a given constant diode current, the voltage drop across the diode decreases by approximately 2 mV for every 1°C increase in temperature. The change in diode voltage with temperature has been exploited in the design of electronic thermometers.

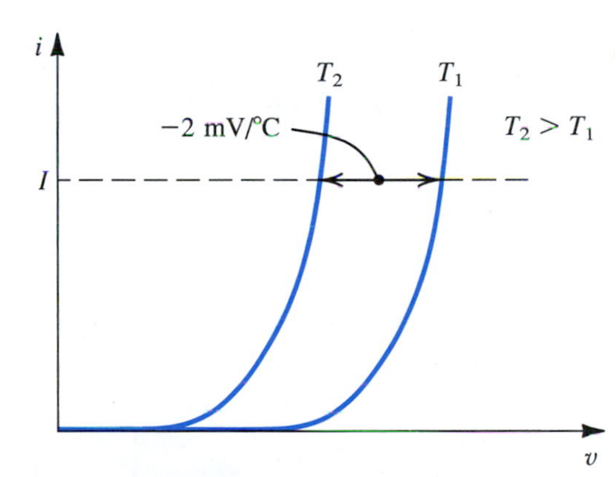

Figure 4.9 Temperature dependence of the diode forward characteristic. At a constant current, the voltage drop decreases by approximately 2 mV for every 1°C increase in temperature.

EXERCISES

4.6 Find the change in diode voltage if the current changes from 0.1 mA to 10 mA.
Ans. 120 mV

4.7 A silicon junction diode has $v = 0.7$ V at $i = 1$ mA. Find the voltage drop at $i = 0.1$ mA and $i = 10$ mA.
Ans. 0.64 V; 0.76 V

4.8 Using the fact that a silicon diode has $I_S = 10^{-14}$ A at 25°C and that I_S increases by 15% per °C rise in temperature, find the value of I_S at 125°C.
Ans. 1.17×10^{-8} A

4.2.2 The Reverse-Bias Region

The reverse-bias region of operation is entered when the diode voltage v is made negative. Equation (4.1) predicts that if v is negative and a few times larger than V_T (25 mV) in magnitude, the exponential term becomes negligibly small compared to unity, and the diode current becomes

$$i \simeq -I_S$$

That is, the current in the reverse direction is constant and equal to I_S. This constancy is the reason behind the term *saturation current*.

Real diodes exhibit reverse currents that, though quite small, are much larger than I_S. For instance, a small-signal diode whose I_S is on the order of 10^{-14} A to 10^{-15} A could show a reverse current on the order of 1 nA. The reverse current also increases somewhat with the increase in magnitude of the reverse voltage. Note that because of the very small magnitude of the current, these details are not clearly evident on the diode i–v characteristic of Fig. 4.8.

A large part of the reverse current is due to leakage effects. These leakage currents are proportional to the junction area, just as I_S is. Their dependence on temperature, however, is different from that of I_S. Thus, whereas I_S doubles for every 5°C rise in temperature, the corresponding rule of thumb for the temperature dependence of the reverse current is that it doubles for every 10°C rise in temperature.

EXERCISE

4.9 The diode in the circuit of Fig. E4.9 is a large high-current device whose reverse leakage is reasonably independent of voltage. If $V = 1$ V at 20°C, find the value of V at 40°C and at 0°C.

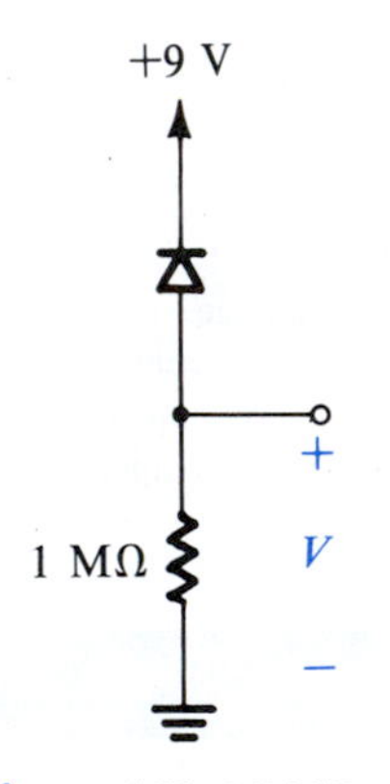

Figure E4.9

Ans. 4 V; 0.25 V

4.2.3 The Breakdown Region

The third distinct region of diode operation is the breakdown region, which can be easily identified on the diode i–v characteristic in Fig. 4.8. The breakdown region is entered when the magnitude of the reverse voltage exceeds a threshold value that is specific to the particular diode, called the **breakdown voltage**. This is the voltage at the "knee" of the i–v curve in Fig. 4.8 and is denoted V_{ZK}, where the subscript Z stands for zener (see Section 3.5.3) and K denotes knee.

As can be seen from Fig. 4.8, in the breakdown region the reverse current increases rapidly, with the associated increase in voltage drop being very small. Diode breakdown is normally not destructive, provided the power dissipated in the diode is limited by external circuitry to a "safe" level. This safe value is normally specified on the device data sheets. It therefore is necessary to limit the reverse current in the breakdown region to a value consistent with the permissible power dissipation.

The fact that the diode i–v characteristic in breakdown is almost a vertical line enables it to be used in voltage regulation. This subject will be studied in Section 4.5.

4.3 Modeling the Diode Forward Characteristic

Having studied the diode terminal characteristics we are now ready to consider the analysis of circuits employing forward-conducting diodes. Figure 4.10 shows such a circuit. It consists of a dc source V_{DD}, a resistor R, and a diode. We wish to analyze this circuit to determine the diode voltage V_D and current I_D. Toward that end we consider developing a variety of models for the operation of the diode. We already know of two such models: the ideal-diode model, and the exponential model. In the following discussion we shall assess the suitability of these two models in various analysis situations. Also, we shall develop and comment on other models. This material, besides being useful in the analysis and design of diode circuits, establishes a foundation for the modeling of transistor operation that we will study in the next two chapters.

4.3.1 The Exponential Model

The most accurate description of the diode operation in the forward region is provided by the exponential model. Unfortunately, however, its severely nonlinear nature makes this model the most difficult to use. To illustrate, let's analyze the circuit in Fig. 4.10 using the exponential diode model.

Assuming that V_{DD} is greater than 0.5 V or so, the diode current will be much greater than I_S, and we can represent the diode i–v characteristic by the exponential relationship, resulting in

$$I_D = I_S e^{V_D/V_T} \tag{4.6}$$

The other equation that governs circuit operation is obtained by writing a Kirchhoff loop equation, resulting in

$$I_D = \frac{V_{DD} - V_D}{R} \tag{4.7}$$

Assuming that the diode parameter I_S is known, Eqs. (4.6) and (4.7) are two equations in the two unknown quantities I_D and V_D. Two alternative ways for obtaining the solution are graphical analysis and iterative analysis.

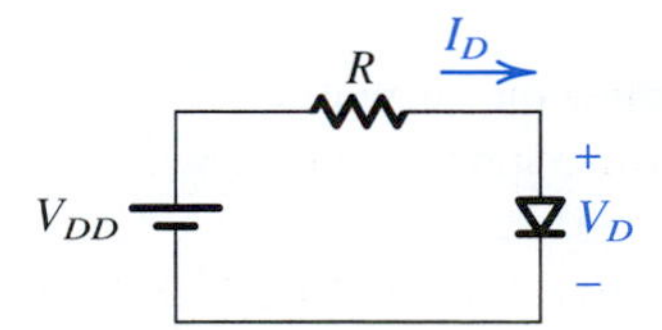

Figure 4.10 A simple circuit used to illustrate the analysis of circuits in which the diode is forward conducting.

4.3.2 Graphical Analysis Using the Exponential Model

Graphical analysis is performed by plotting the relationships of Eqs. (4.6) and (4.7) on the i–v plane. The solution can then be obtained as the coordinates of the point of intersection of the two graphs. A sketch of the graphical construction is shown in Fig. 4.11. The curve represents the exponential diode equation (Eq. 4.6), and the straight line represents Eq. (4.7). Such a straight line is known as the **load line**, a name that will become more meaningful in later chapters. The load line intersects the diode curve at point Q, which represents the **operating point** of the circuit. Its coordinates give the values of I_D and V_D.

Graphical analysis aids in the visualization of circuit operation. However, the effort involved in performing such an analysis, particularly for complex circuits, is too great to be justified in practice.

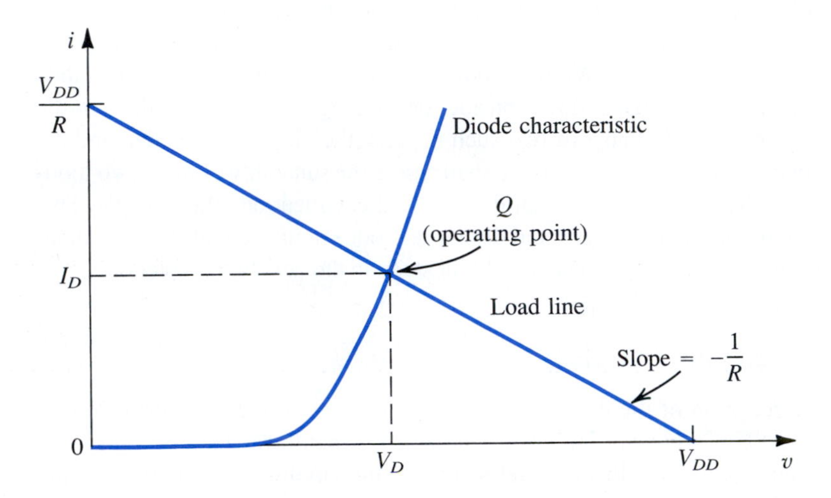

Figure 4.11 Graphical analysis of the circuit in Fig. 4.10 using the exponential diode model.

4.3.3 Iterative Analysis Using the Exponential Model

Equations (4.6) and (4.7) can be solved using a simple iterative procedure, as illustrated in the following example.

Example 4.4

Determine the current I_D and the diode voltage V_D for the circuit in Fig. 4.10 with $V_{DD} = 5$ V and $R = 1$ kΩ. Assume that the diode has a current of 1 mA at a voltage of 0.7 V.

Solution

To begin the iteration, we assume that $V_D = 0.7$ V and use Eq. (4.7) to determine the current,

$$I_D = \frac{V_{DD} - V_D}{R}$$

$$= \frac{5 - 0.7}{1} = 4.3 \text{ mA}$$

We then use the diode equation to obtain a better estimate for V_D. This can be done by employing Eq. (4.5), namely,

$$V_2 - V_1 = 2.3 V_T \log \frac{I_2}{I_1}$$

Subsituting $2.3\, V_T = 60$ mV, we have

$$V_2 = V_1 + 0.06 \log \frac{I_2}{I_1}$$

Substituting $V_1 = 0.7$ V, $I_1 = 1$ mA, and $I_2 = 4.3$ mA results in $V_2 = 0.738$ V. Thus the results of the first iteration are $I_D = 4.3$ mA and $V_D = 0.738$ V. The second iteration proceeds in a similar manner:

$$I_D = \frac{5 - 0.738}{1} = 4.262 \text{ mA}$$

$$V_2 = 0.738 + 0.06 \log\left[\frac{4.262}{4.3}\right]$$

$$= 0.738 \text{ V}$$

Thus the second iteration yields $I_D = 4.262$ mA and $V_D = 0.738$ V. Since these values are very close to the values obtained after the first iteration, no further iterations are necessary, and the solution is $I_D = 4.262$ mA and $V_D = 0.738$ V.

4.3.4 The Need for Rapid Analysis

The iterative analysis procedure utilized in the example above is simple and yields accurate results after two or three iterations. Nevertheless, there are situations in which the effort and time required are still greater than can be justified. Specifically, if one is doing a pencil-and-paper design of a relatively complex circuit, rapid circuit analysis is a necessity. Through quick analysis, the designer is able to evaluate various possibilities before deciding on a suitable circuit design. To speed up the analysis process one must be content with less precise results. This, however, is seldom a problem, because the more accurate analysis can be postponed until a final or almost-final design is obtained. Accurate analysis of the almost-final design can be performed with the aid of a computer circuit-analysis program such as SPICE (see Appendix B and the disc). The results of such an analysis can then be used to further refine or "fine-tune" the design.

To speed up the analysis process, we must find a simpler model for the diode forward characteristic.

4.3.5 The Constant-Voltage-Drop Model

The simplest and most widely used diode model is the constant-voltage-drop model. This model is based on the observation that a forward-conducting diode has a voltage drop that varies in a relatively narrow range, say 0.6 to 0.8 V. The model assumes this voltage to be constant at a value, say, 0.7 V. This development is illustrated in Fig. 4.12.

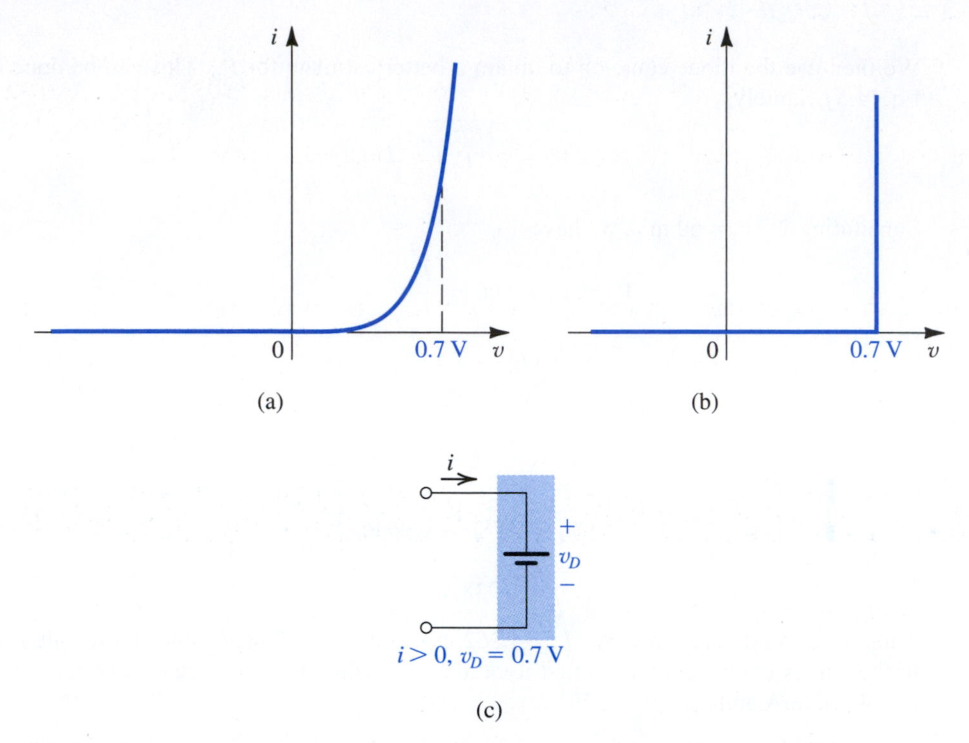

Figure 4.12 Development of the diode constant-voltage-drop model: **(a)** the exponential characteristic; **(b)** approximating the exponential characteristic by a constant voltage, usually about 0.7 V$_i$; **(c)** the resulting model of the forward–conducting diodes.

The constant-voltage-drop model is the one most frequently employed in the initial phases of analysis and design. This is especially true if at these stages one does not have detailed information about the diode characteristics, which is often the case.

Finally, note that if we employ the constant-voltage-drop model to solve the problem in Example 4.4, we obtain

$$V_D = 0.7 \text{ V}$$

and

$$I_D = \frac{V_{DD} - 0.7}{R}$$

$$= \frac{5 - 0.7}{1} = 4.3 \text{ mA}$$

which are not very different from the values obtained before with the more elaborate exponential model.

4.3.6 The Ideal-Diode Model

In applications that involve voltages much greater than the diode voltage drop (0.6 V–0.8 V), we may neglect the diode voltage drop altogether while calculating the diode current. The result is the ideal-diode model, which we studied in Section 4.1. For the circuit in Examples 4.4 (i.e., Fig. 4.10 with $V_{DD} = 5$ V and $R = 1$ kΩ), utilization of the ideal-diode model leads to

$$V_D = 0 \text{ V}$$

$$I_D = \frac{5-0}{1} = 5 \text{ mA}$$

which for a very quick analysis would not be bad as a gross estimate. However, with almost no additional work, the 0.7-V-drop model yields much more realistic results. We note, however, that the greatest utility of the ideal-diode model is in determining which diodes are on and which are off in a multidiode circuit, such as those considered in Section 4.1.

EXERCISES

4.10 For the circuit in Fig. 4.10, find I_D and V_D for the case $V_{DD} = 5$ V and $R = 10$ kΩ. Assume that the diode has a voltage of 0.7 V at 1-mA current. Use (a) iteration and (b) the constant-voltage-drop model with $V_D = 0.7$ V.
Ans. (a) 0.43 mA, 0.68 V; (b) 0.43 mA, 0.7 V

D4.11 Design the circuit in Fig. E4.11 to provide an output voltage of 2.4 V. Assume that the diodes available have 0.7-V drop at 1 mA.

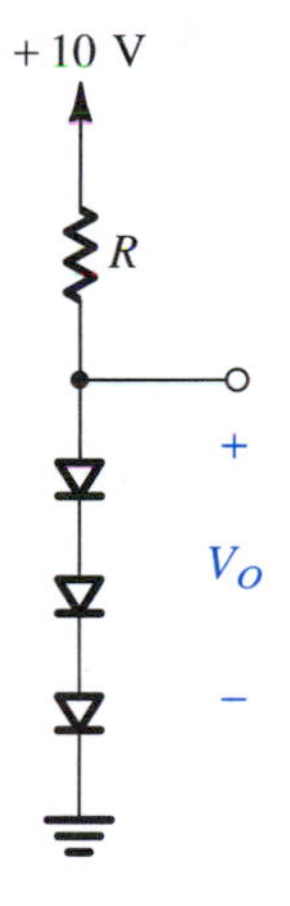

Figure E4.11

Ans. $R = 139$ Ω

4.12 Repeat Exercise 4.4 using the 0.7-V-drop model to obtain better estimates of I and V than those found in Exercise 4.4 (using the ideal-diode model).
Ans. (a) 1.72 mA, 0.7 V; (b) 0 mA, 5 V; (c) 0 mA, 5 V; (d) 1.72 mA, 0.7 V; (e) 2.3 mA, +2.3 V; (f) 3.3 mA, +1.7 V

4.3.7 The Small-Signal Model

There are applications in which a diode is biased to operate at a point on the forward i–v characteristic and a small ac signal is superimposed on the dc quantities. For this situation, we first have to determine the dc operating point (V_D and I_D) of the diode using one of the models discussed above. Most frequently, the 0.7-V-drop model is utilized. Then, for small-signal operation around the dc bias point, the diode is modeled by a resistance equal to the inverse of the slope of the tangent to the exponential i–v characteristic at the bias point. The technique of biasing a nonlinear device and restricting signal excursion to a short, almost-linear segment of its characteristic around the bias point is central to designing linear amplifiers using transistors, as will be seen in the next two chapters. In this section, we develop such a small-signal model for the junction diode and illustrate its application.

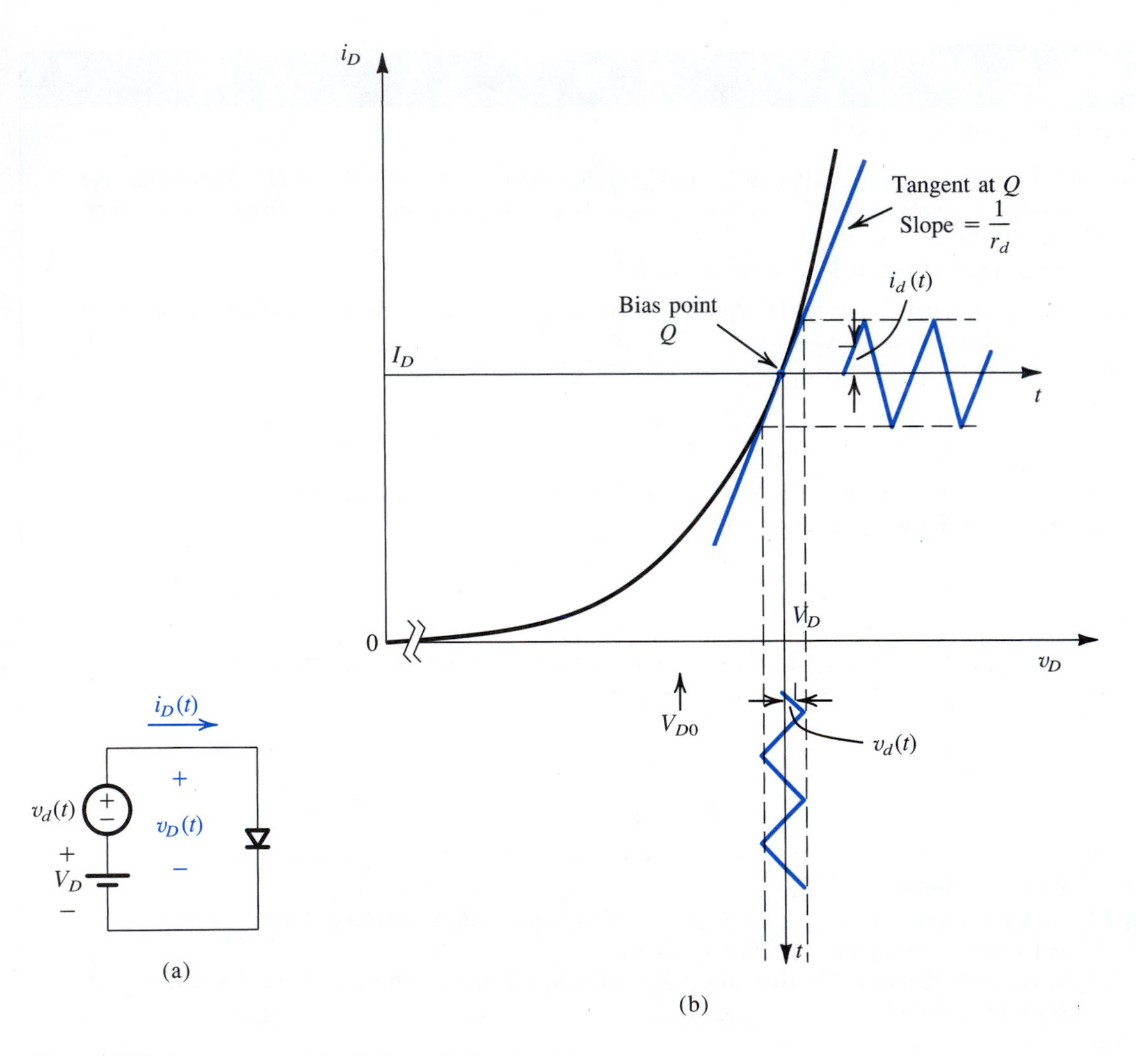

Figure 4.13 Development of the diode small-signal model.

Consider the conceptual circuit in Fig. 4.13(a) and the corresponding graphical representation in Fig. 4.13(b). A dc voltage V_D, represented by a battery, is applied to the diode, and a time-varying signal $v_d(t)$, assumed (arbitrarily) to have a triangular waveform, is superimposed on the dc voltage V_D. In the absence of the signal $v_d(t)$, the diode voltage is equal to V_D, and correspondingly, the diode will conduct a dc current I_D given by

$$I_D = I_S e^{V_D/V_T} \tag{4.8}$$

When the signal $v_d(t)$ is applied, the total instantaneous diode voltage $v_D(t)$ will be given by

$$v_D(t) = V_D + v_d(t) \tag{4.9}$$

Correspondingly, the total instantaneous diode current $i_D(t)$ will be

$$i_D(t) = I_S e^{v_D/V_T} \tag{4.10}$$

Substituting for v_D from Eq. (4.9) gives

$$i_D(t) = I_S e^{(V_D+v_d)/V_T} \tag{4.11}$$

which can be rewritten

$$i_D(t) = I_S e^{V_D/V_T} e^{v_d/V_T}$$

Using Eq. (4.8) we obtain

$$i_D(t) = I_D e^{v_d/V_T} \tag{4.12}$$

Now if the amplitude of the signal $v_d(t)$ is kept sufficiently small such that

$$\frac{v_d}{V_T} \ll 1 \tag{4.13}$$

then we may expand the exponential of Eq. (4.12) in a series and truncate the series after the first two terms to obtain the approximate expression

$$i_D(t) \simeq I_D\left(1 + \frac{v_d}{V_T}\right) \tag{4.14}$$

This is the **small-signal approximation**. It is valid for signals whose amplitudes are smaller than about 5 mV (see Eq. 4.13, and recall that $V_T = 25$ mV).[3]

From Eq. (4.14) we have

$$i_D(t) = I_D + \frac{I_D}{V_T} v_d \tag{4.15}$$

Thus, superimposed on the dc current I_D, we have a signal current component directly proportional to the signal voltage v_d. That is,

$$i_D = I_D + i_d \tag{4.16}$$

where

$$i_d = \frac{I_D}{V_T} v_d \tag{4.17}$$

[3]For $v_d = 5$ mV, $v_d/V_T = 0.2$. Thus the next term in the series expansion of the exponential will be $\frac{1}{2} \times 0.2^2 = 0.02$, a factor of 10 lower than the linear term we kept.

The quantity relating the signal current i_d to the signal voltage v_d has the dimensions of conductance, mhos (℧), and is called the **diode small-signal conductance**. The inverse of this parameter is the **diode small-signal resistance**, or **incremental resistance**, r_d,

$$r_d = \frac{V_T}{I_D} \tag{4.18}$$

Note that the value of r_d is inversely proportional to the bias current I_D.

Let us return to the graphical representation in Fig. 4.13(b). It is easy to see that using the small-signal approximation is equivalent to assuming that *the signal amplitude is sufficiently small such that the excursion along the i–v curve is limited to a short almost-linear segment.* The slope of this segment, which is equal to the slope of the tangent to the i–v curve at the operating point Q, is equal to the small-signal conductance. The reader is encouraged to prove that the slope of the i–v curve at $i = I_D$ is equal to I_D/V_T, which is $1/r_d$; that is,

$$r_d = 1\bigg/\left[\frac{\partial i_D}{\partial v_D}\right]_{i_D=I_D} \tag{4.19}$$

From the preceding we conclude that superimposed on the quantities V_D and I_D that define the dc bias point, or **quiescent point**, of the diode will be the small-signal quantities $v_d(t)$ and $i_d(t)$, which are related by the diode small-signal resistance r_d evaluated at the bias point (Eq. 4.18). Thus the small-signal analysis can be performed separately from the dc bias analysis, a great convenience that results from the linearization of the diode characteristics inherent in the small-signal approximation. Specifically, after the dc analysis is performed, the small-signal equivalent circuit is obtained by eliminating all dc sources (i.e., short-circuiting dc voltage sources and open-circuiting dc current sources) and replacing the diode by its small-signal resistance. The following example should illustrate the application of the small-signal model.

Example 4.5

Consider the circuit shown in Fig. 4.14(a) for the case in which $R = 10\ \text{k}\Omega$. The power supply V^+ has a dc value of 10 V on which is superimposed a 60-Hz sinusoid of 1-V peak amplitude. (This "signal" component of the power-supply voltage is an imperfection in the power-supply design. It is known as the **power-supply ripple**. More on this later.) Calculate both the dc voltage of the diode and the amplitude of the sine-wave signal appearing across it. Assume the diode to have a 0.7-V drop at 1-mA current.

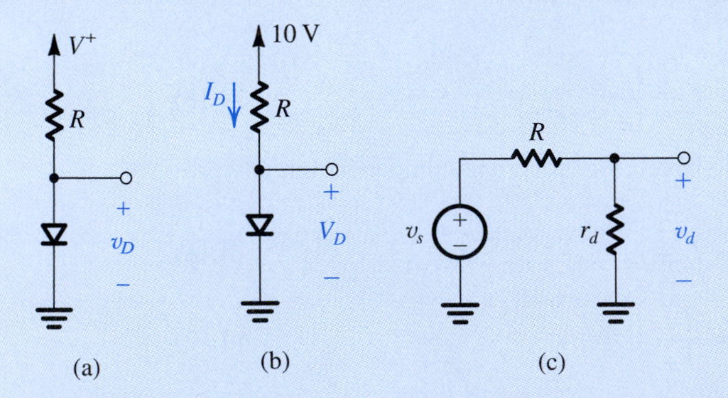

Figure 4.14 **(a)** Circuit for Example 4.5. **(b)** Circuit for calculating the dc operating point. **(c)** Small-signal equivalent circuit.

Solution

Considering dc quantities only, we assume $V_D \simeq 0.7$ V and calculate the diode dc current

$$I_D = \frac{10 - 0.7}{10} = 0.93 \text{ mA}$$

Since this value is very close to 1 mA, the diode voltage will be very close to the assumed value of 0.7 V. At this operating point, the diode incremental resistance r_d is

$$r_d = \frac{V_T}{I_D} = \frac{25}{0.93} = 26.9 \ \Omega$$

The signal voltage across the diode can be found from the small-signal equivalent circuit in Fig. 4.14(c). Here v_s denotes the 60-Hz 1-V peak sinusoidal component of V^+, and v_d is the corresponding signal across the diode. Using the voltage-divider rule provides the peak amplitude of v_d as follows:

$$v_d \text{ (peak)} = \hat{V}_s \frac{r_d}{R + r_d}$$

$$= 1 \frac{0.0269}{10 + 0.0269} = 2.68 \text{ mV}$$

Finally we note that since this value is quite small, our use of the small-signal model of the diode is justified.

Finally, we note that while r_d models the small-signal operation of the diode at low frequencies, its dynamic operation is modeled by the capacitances C_j and C_d, which we studied in Section 3.6 and which also are small-signal parameters. A complete model of the diode includes C_j and C_d in parallel with r_d.

4.3.8 Use of the Diode Forward Drop in Voltage Regulation

A further application of the diode small-signal model is found in a popular diode application, namely, the use of diodes to create a regulated voltage. A **voltage regulator** is a circuit whose purpose is to provide a constant dc voltage between its output terminals. The output voltage is required to remain as constant as possible in spite of (a) changes in the load current drawn from the regulator output terminal and (b) changes in the dc power-supply voltage that feeds the regulator circuit. Since the forward-voltage drop of the diode remains almost constant at approximately 0.7 V while the current through it varies by relatively large amounts, a forward-biased diode can make a simple voltage regulator. For instance, we have seen in Example 4.5 that while the 10-V dc supply voltage had a ripple of 2 V peak-to-peak (a ±10% variation), the corresponding ripple in the diode voltage was only about ±2.7 mV (a ±0.4% variation). Regulated voltages greater than 0.7 V can be obtained by connecting a number of diodes in series. For example, the use of three forward-biased diodes in series provides a voltage of about 2 V. One such circuit is investigated in the following example, which utilizes the diode small-signal model to quantify the efficacy of the voltage regulator that is realized.

Example 4.6

Consider the circuit shown in Fig. 4.15. A string of three diodes is used to provide a constant voltage of about 2.1 V. We want to calculate the percentage change in this regulated voltage caused by (a) a ±10% change in the power-supply voltage and (b) connection of a 1-kΩ load resistance.

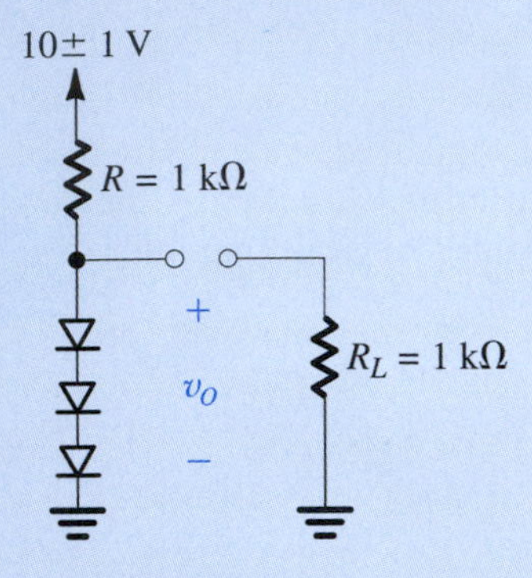

Figure 4.15 Circuit for Example 4.6.

Solution

With no load, the nominal value of the current in the diode string is given by

$$I = \frac{10 - 2.1}{1} = 7.9 \text{ mA}$$

Thus each diode will have an incremental resistance of

$$r_d = \frac{V_T}{I}$$

Thus,

$$r_d = \frac{25}{7.9} = 3.2 \ \Omega$$

The three diodes in series will have a total incremental resistance of

$$r = 3r_d = 9.6 \ \Omega$$

This resistance, along with the resistance R, forms a voltage divider whose ratio can be used to calculate the change in output voltage due to a ±10% (i.e., ±1-V) change in supply voltage. Thus the peak-to-peak change in output voltage will be

$$\Delta v_O = 2\frac{r}{r+R} = 2\frac{0.0096}{0.0096+1} = 19 \text{ mV peak-to-peak}$$

That is, corresponding to the ±1-V (±10%) change in supply voltage, the output voltage will change by ±9.5 mV or ±0.5%. Since this implies a change of about ±3.2 mV per diode, our use of the small-signal model is justified.

When a load resistance of 1 kΩ is connected across the diode string, it draws a current of approximately 2.1 mA. Thus the current in the diodes decreases by 2.1 mA, resulting in a decrease in voltage across the diode string given by

$$\Delta v_O = -2.1 \times r = -2.1 \times 9.6 = -20 \text{ mV}$$

Since this implies that the voltage across each diode decreases by about 6.7 mV, our use of the small-signal model is not entirely justified. Nevertheless, a detailed calculation of the voltage change using the exponential model results in $\Delta v_O = -23$ mV, which is not too different from the approximate value obtained using the incremental model.

EXERCISES

4.13 Find the value of the diode small-signal resistance r_d at bias currents of 0.1 mA, 1 mA, and 10 mA.
Ans. 250 Ω; 25 Ω; 2.5 Ω

4.14 Consider a diode biased at 1 mA. Find the change in current as a result of changing the voltage by (a) −10 mV, (b) −5 mV, (c) +5 mV, and (d) +10 mV. In each case, do the calculations (i) using the small-signal model and (ii) using the exponential model.
Ans. (a) −0.40, −0.33 mA; (b) −0.20, −0.18 mA; (c) +0.20, +0.22 mA; (d) +0.40, +0.49 mA

D4.15 Design the circuit of Fig. E4.15 so that $V_O = 3$ V when $I_L = 0$, and V_O changes by 20 mV per 1 mA of load current.

(a) Use the small-signal model of the diode to find the value of R.

(b) Specify the value of I_S of each of the diodes.

(c) For this design, use the diode exponential model to determine the actual change in V_O when a current $I_L = 1$ mA is drawn from the regulator.

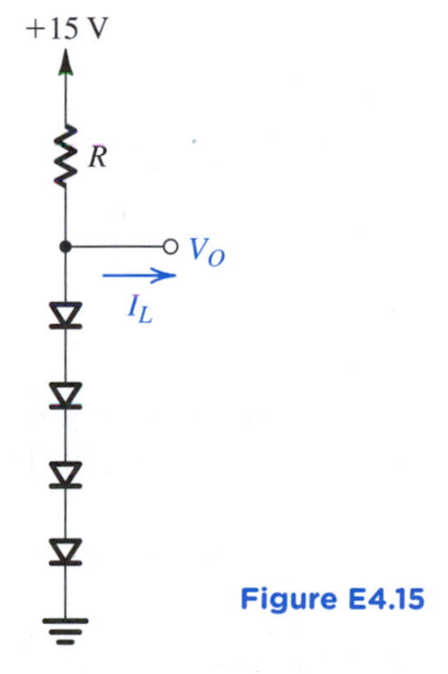

Figure E4.15

Ans. (a) $R = 2.4$ kΩ; (b) $I_S = 4.7 \times 10^{-16}$ A; (c) −22.3 mV

4.4 Operation in the Reverse Breakdown Region—Zener Diodes

The very steep i–v curve that the diode exhibits in the breakdown region (Fig. 4.8) and the almost-constant voltage drop that this indicates, suggest that diodes operating in the breakdown region can be used in the design of voltage regulators. From the previous section, the reader will recall that voltage regulators are circuits that provide a constant dc output voltage in the face of changes in their load current and in the system power-supply voltage. This in fact turns out to be an important application of diodes operating in the reverse-breakdown region, and special diodes are manufactured to operate specifically in the breakdown region. Such diodes are called **breakdown diodes** or, more commonly, as noted earlier, **zener diodes**.

Figure 4.16 shows the circuit symbol of the zener diode. In normal applications of zener diodes, current flows into the cathode, and the cathode is positive with respect to the anode. Thus I_Z and V_Z in Fig. 4.16 have positive values.

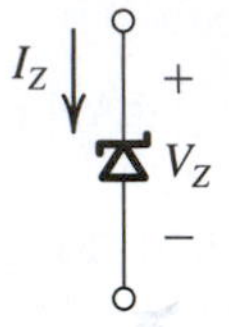

Figure 4.16 Circuit symbol for a zener diode.

4.4.1 Specifying and Modeling the Zener Diode

Figure 4.17 shows details of the diode i–v characteristic in the breakdown region. We observe that for currents greater than the **knee current** I_{ZK} (specified on the data sheet of the zener diode), the i–v characteristic is almost a straight line. The manufacturer usually specifies the voltage across the zener diode V_Z at a specified test current, I_{ZT}. We have indicated these parameters in Fig. 4.17 as the coordinates of the point labeled Q. Thus a 6.8-V

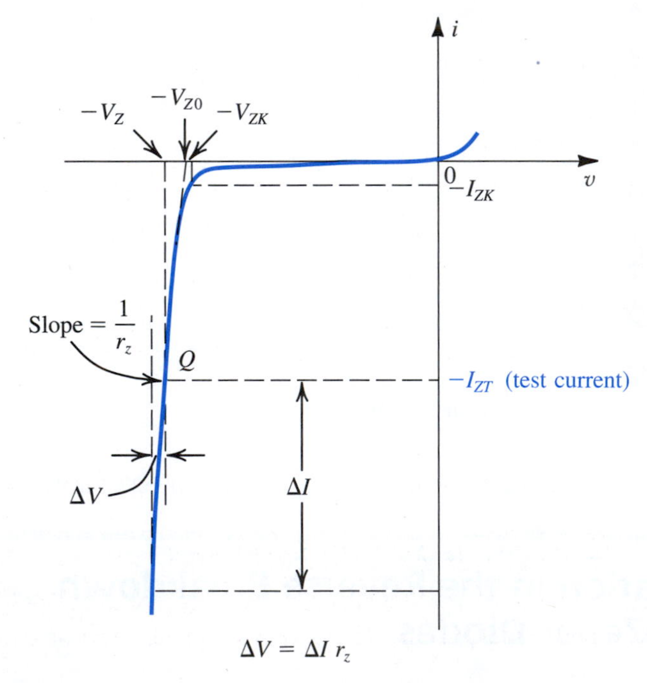

Figure 4.17 The diode i–v characteristic with the breakdown region shown in some detail.

zener diode will exhibit a 6.8-V drop at a specified test current of, say, 10 mA. As the current through the zener deviates from I_{ZT}, the voltage across it will change, though only slightly. Figure 4.17 shows that corresponding to current change ΔI the zener voltage changes by ΔV, which is related to ΔI by

$$\Delta V = r_z \Delta I$$

where r_z is the inverse of the slope of the almost-linear i–v curve at point Q. Resistance r_z is the **incremental resistance** of the zener diode at operating point Q. It is also known as the **dynamic resistance** of the zener, and its value is specified on the device data sheet. Typically, r_z is in the range of a few ohms to a few tens of ohms. Obviously, the lower the value of r_z is, the more constant the zener voltage remains as its current varies, and thus the more ideal its performance becomes in the design of voltage regulators. In this regard, we observe from Fig. 4.17 that while r_z remains low and almost constant over a wide range of current, its value increases considerably in the vicinity of the knee. Therefore, as a general design guideline, one should avoid operating the zener in this low-current region.

Zener diodes are fabricated with voltages V_Z in the range of a few volts to a few hundred volts. In addition to specifying V_Z (at a particular current I_{ZT}), r_z, and I_{ZK}, the manufacturer also specifies the maximum power that the device can safely dissipate. Thus a 0.5-W, 6.8-V zener diode can operate safely at currents up to a maximum of about 70 mA.

The almost-linear i–v characteristic of the zener diode suggests that the device can be modeled as indicated in Fig. 4.18. Here V_{Z0} denotes the point at which the straight line of slope $1/r_z$ intersects the voltage axis (refer to Fig. 4.17). Although V_{Z0} is shown in Fig. 4.17 to be slightly different from the knee voltage V_{ZK}, in practice their values are almost equal. The equivalent circuit model of Fig. 4.18 can be analytically described by

$$V_Z = V_{Z0} + r_z I_Z \qquad (4.20)$$

and it applies for $I_Z > I_{ZK}$ and, obviously, $V_Z > V_{Z0}$.

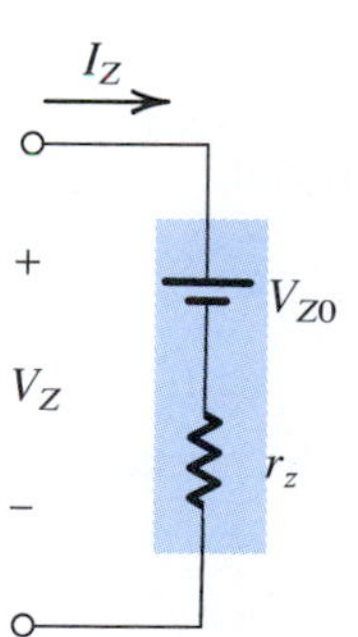

Figure 4.18 Model for the zener diode.

4.4.2 Use of the Zener as a Shunt Regulator

We now illustrate, by way of an example, the use of zener diodes in the design of shunt regulators, so named because the regulator circuit appears in parallel (shunt) with the load.

Example 4.7

The 6.8-V zener diode in the circuit of Fig. 4.19(a) is specified to have $V_Z = 6.8$ V at $I_Z = 5$ mA, $r_z = 20\ \Omega$, and $I_{ZK} = 0.2$ mA. The supply voltage V^+ is nominally 10 V but can vary by ±1 V.

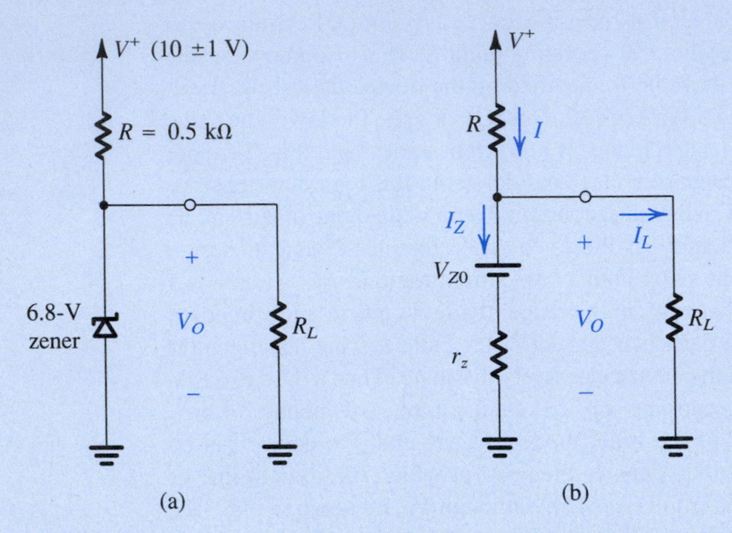

Figure 4.19 **(a)** Circuit for Example 4.7. **(b)** The circuit with the zener diode replaced with its equivalent circuit model.

(a) Find V_O with no load and with V^+ at its nominal value.
(b) Find the change in V_O resulting from the ±1-V change in V^+. Note that $(\Delta V_O/\Delta V^+)$, usually expressed in mV/V, is known as **line regulation**.
(c) Find the change in V_O resulting from connecting a load resistance R_L that draws a current $I_L =$ 1 mA, and hence find the **load regulation** $(\Delta V_O/\Delta I_L)$ in mV/mA.
(d) Find the change in V_O when $R_L = 2$ kΩ.
(e) Find the value of V_O when $R_L = 0.5$ kΩ.
(f) What is the minimum value of R_L for which the diode still operates in the breakdown region?

Solution

First we must determine the value of the parameter V_{Z0} of the zener diode model. Substituting $V_Z =$ 6.8 V, $I_Z = 5$ mA, and $r_z = 20\ \Omega$ in Eq. (4.20) yields $V_{Z0} = 6.7$ V. Figure 4.19(b) shows the circuit with the zener diode replaced with its model.

(a) With no load connected, the current through the zener is given by

$$I_Z = I = \frac{V^+ - V_{Z0}}{R + r_z}$$

$$= \frac{10 - 6.7}{0.5 + 0.02} = 6.35 \text{ mA}$$

Thus,

$$V_O = V_{Z0} + I_Z r_z$$

$$= 6.7 + 6.35 \times 0.02 = 6.83 \text{ V}$$

(b) For a ±1-V change in V^+, the change in output voltage can be found from

$$\Delta V_O = \Delta V^+ \frac{r_z}{R + r_z}$$

$$= \pm 1 \times \frac{20}{500 + 20} = \pm 38.5 \text{ mV}$$

Thus,

$$\text{Line regulation} = 38.5 \text{ mV/V}$$

(c) When a load resistance R_L that draws a load current $I_L = 1$ mA is connected, the zener current will decrease by 1 mA. The corresponding change in zener voltage can be found from

$$\Delta V_O = r_z \Delta I_Z$$

$$= 20 \times -1 = -20 \text{ mV}$$

Thus the load regulation is

$$\text{Load regulation} \equiv \frac{\Delta V_O}{\Delta I_L} = -20 \text{ mV/mA}$$

(d) When a load resistance of 2 kΩ is connected, the load current will be approximately 6.8 V/2 kΩ = 3.4 mA. Thus the change in zener current will be $\Delta I_Z = -3.4$ mA, and the corresponding change in zener voltage (output voltage) will thus be

$$\Delta V_O = r_z \Delta I_Z$$

$$= 20 \times -3.4 = -68 \text{ mV}$$

This calculation, however, is approximate, because it neglects the change in the current I. A more accurate estimate of ΔV_O can be obtained by analyzing the circuit in Fig. 4.19(b). The result of such an analysis is $\Delta V_O = -70$ mV.

(e) An R_L of 0.5 kΩ would draw a load current of 6.8/0.5 = 13.6 mA. This is not possible, because the current I supplied through R is only 6.4 mA (for $V^+ = 10$ V). Therefore, the zener must be cut off. If this is indeed the case, then V_O is determined by the voltage divider formed by R_L and R (Fig. 4.19a),

$$V_O = V^+ \frac{R_L}{R + R_L}$$

$$= 10 \frac{0.5}{0.5 + 0.5} = 5 \text{ V}$$

Since this voltage is lower than the breakdown voltage of the zener, the diode is indeed no longer operating in the breakdown region.

(f) For the zener to be at the edge of the breakdown region, $I_Z = I_{ZK} = 0.2$ mA and $V_Z \simeq V_{ZK} \simeq 6.7$ V. At this point the lowest (worst-case) current supplied through R is $(9 - 6.7)/0.5 = 4.6$ mA, and thus the load current is $4.6 - 0.2 = 4.4$ mA. The corresponding value of R_L is

$$R_L = \frac{6.7}{4.4} \simeq 1.5 \text{ k}\Omega$$

4.4.3 Temperature Effects

The dependence of the zener voltage V_Z on temperature is specified in terms of its **temperature coefficient TC**, or **temco** as it is commonly known, which is usually expressed in mV/°C. The value of TC depends on the zener voltage, and for a given diode the TC varies with the operating current. Zener diodes whose V_Z are lower than about 5 V exhibit a negative TC. On the other hand, zeners with higher voltages exhibit a positive TC. The TC of a zener diode with a V_Z of about 5 V can be made zero by operating the diode at a specified current. Another commonly used technique for obtaining a reference voltage with low temperature coefficient is to connect a zener diode with a positive temperature coefficient of about 2 mV/°C in series with a forward-conducting diode. Since the forward-conducting diode has a voltage drop of $\simeq$0.7 V and a TC of about −2 mV/°C, the series combination will provide a voltage of $(V_Z + 0.7)$ with a TC of about zero.

EXERCISES

4.16 A zener diode whose nominal voltage is 10 V at 10 mA has an incremental resistance of 50 Ω. What voltage do you expect if the diode current is halved? Doubled? What is the value of V_{Z0} in the zener model?
Ans. 9.75 V; 10.5 V; 9.5 V

4.17 A zener diode exhibits a constant voltage of 5.6 V for currents greater than five times the knee current. I_{ZK} is specified to be 1 mA. The zener is to be used in the design of a shunt regulator fed from a 15-V supply. The load current varies over the range of 0 mA to 15 mA. Find a suitable value for the resistor R. What is the maximum power dissipation of the zener diode?
Ans. 470 Ω; 112 mW

4.18 A shunt regulator utilizes a zener diode whose voltage is 5.1 V at a current of 50 mA and whose incremental resistance is 7 Ω. The diode is fed from a supply of 15-V nominal voltage through a 200-Ω resistor. What is the output voltage at no load? Find the line regulation and the load regulation.
Ans. 5.1 V; 33.8 mV/V; −7 mV/mA

4.4.4 A Final Remark

Though simple and useful, zener diodes have lost a great deal of their popularity in recent years. They have been virtually replaced in voltage-regulator design by specially designed integrated circuits (ICs) that perform the voltage regulation function much more effectively and with greater flexibility than zener diodes.

4.5 Rectifier Circuits

One of the most important applications of diodes is in the design of rectifier circuits. A diode rectifier forms an essential building block of the dc power supplies required to power electronic equipment. A block diagram of such a power supply is shown in Fig. 4.20. As indicated, the power supply is fed from the 120-V (rms) 60-Hz ac line, and it delivers a dc voltage V_O (usually in the range of 5 V to 20 V) to an electronic circuit represented by the

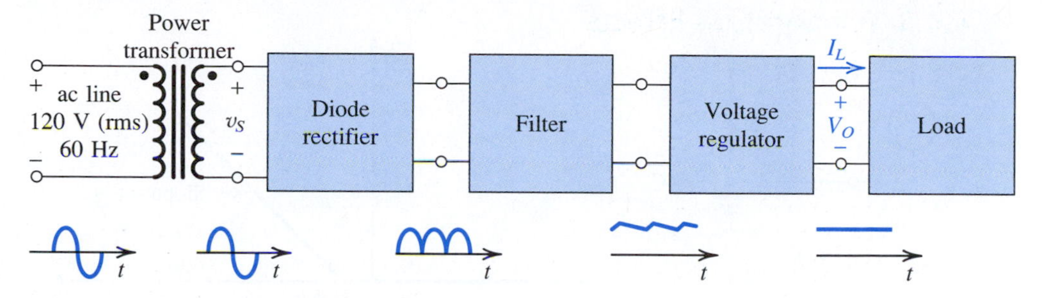

Figure 4.20 Block diagram of a dc power supply.

load block. The dc voltage V_O is required to be as constant as possible in spite of variations in the ac line voltage and in the current drawn by the load.

The first block in a dc power supply is the **power transformer**. It consists of two separate coils wound around an iron core that magnetically couples the two windings. The **primary winding**, having N_1 turns, is connected to the 120-V ac supply, and the **secondary winding**, having N_2 turns, is connected to the circuit of the dc power supply. Thus an ac voltage v_S of $120(N_2/N_1)$ V (rms) develops between the two terminals of the secondary winding. By selecting an appropriate turns ratio (N_1/N_2) for the transformer, the designer can step the line voltage down to the value required to yield the particular dc voltage output of the supply. For instance, a secondary voltage of 8-V rms may be appropriate for a dc output of 5 V. This can be achieved with a 15:1 turns ratio.

In addition to providing the appropriate sinusoidal amplitude for the dc power supply, the power transformer provides electrical isolation between the electronic equipment and the power-line circuit. This isolation minimizes the risk of electric shock to the equipment user.

The diode rectifier converts the input sinusoid v_S to a unipolar output, which can have the pulsating waveform indicated in Fig. 4.20. Although this waveform has a nonzero average or a dc component, its pulsating nature makes it unsuitable as a dc source for electronic circuits, hence the need for a filter. The variations in the magnitude of the rectifier output are considerably reduced by the filter block in Fig. 4.20. In the following sections we shall study a number of rectifier circuits and a simple implementation of the output filter.

The output of the rectifier filter, though much more constant than without the filter, still contains a time-dependent component, known as **ripple**. To reduce the ripple and to stabilize the magnitude of the dc output voltage of the supply against variations caused by changes in load current, a voltage regulator is employed. Such a regulator can be implemented using the zener shunt regulator configuration studied in Section 4.4. Alternatively, and much more commonly at present, an integrated-circuit regulator can be used.

4.5.1 The Half-Wave Rectifier

The half-wave rectifier utilizes alternate half-cycles of the input sinusoid. Figure 4.21(a) shows the circuit of a half-wave rectifier. This circuit was analyzed in Section 4.1 (see Fig. 4.3) assuming an ideal diode. Using the more realistic constant-voltage-drop diode model, we obtain

$$v_O = 0, \qquad v_S < V_D \tag{4.21a}$$

$$v_O = v_S - V_D, \qquad v_S \geq V_t \tag{4.21b}$$

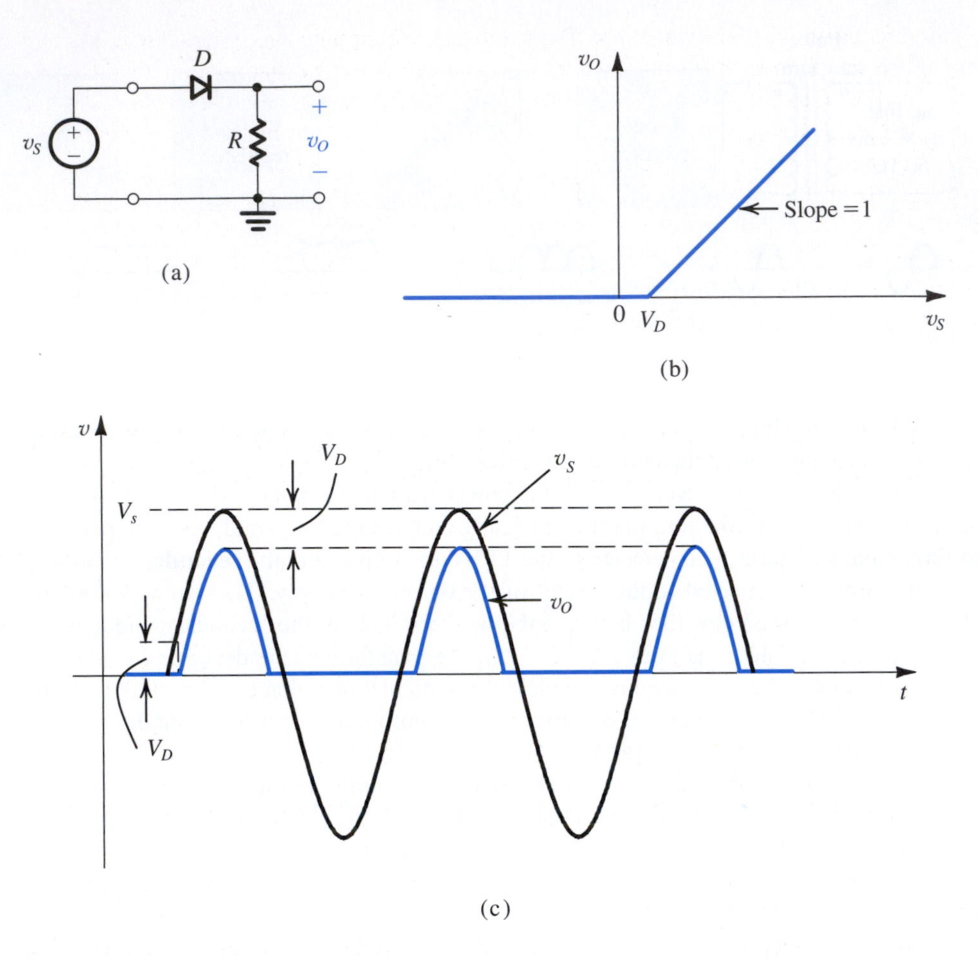

Figure 4.21 **(a)** Half-wave rectifier. **(b)** Transfer characteristic of the rectifier circuit. **(c)** Input and output waveforms.

The transfer characteristic represented by these equations is sketched in Fig. 4.21(b), where $V_D = 0.7$ V or 0.8 V. Figure 4.21(c) shows the output voltage obtained when the input v_S is a sinusoid.

In selecting diodes for rectifier design, two important parameters must be specified: the current-handling capability required of the diode, determined by the largest current the diode is expected to conduct, and the **peak inverse voltage** (PIV) that the diode must be able to withstand without breakdown, determined by the largest reverse voltage that is expected to appear across the diode. In the rectifier circuit of Fig. 4.21(a), we observe that when v_S is negative the diode will be cut off and v_O will be zero. It follows that the PIV is equal to the peak of v_S,

$$\text{PIV} = V_s \tag{4.22}$$

It is usually prudent, however, to select a diode that has a reverse breakdown voltage at least 50% greater than the expected PIV.

Before leaving the half-wave rectifier, the reader should note two points. First, it is possible to use the diode exponential characteristic to determine the exact transfer characteristic of the rectifier (see Problem 4.65). However, the amount of work involved is usually too great to be justified in practice. Of course, such an analysis can be easily done using a computer circuit-analysis program such as SPICE.

Second, whether we analyze the circuit accurately or not, it should be obvious that this circuit does not function properly when the input signal is small. For instance, this circuit cannot be used to rectify an input sinusoid of 100-mV amplitude. For such an application one resorts to a so-called precision rectifier, a circuit utilizing diodes in conjunction with op amps. One such circuit is presented in Section 4.5.5.

EXERCISE

4.19 For the half-wave rectifier circuit in Fig. 4.21(a), show the following: (a) For the half-cycles during which the diode conducts, conduction begins at an angle $\theta = \sin^{-1}(V_D/V_s)$ and terminates at $(\pi - \theta)$, for a total conduction angle of $(\pi - 2\theta)$. (b) The average value (dc component) of v_O is $V_O \simeq (1/\pi)V_s - V_D/2$. (c) The peak diode current is $(V_s - V_D)/R$).
Find numerical values for these quantities for the case of 12-V (rms) sinusoidal input, $V_D \simeq 0.7$ V, and $R = 100\ \Omega$. Also, give the value for PIV.
Ans. (a) $\theta = 2.4°$, conduction angle $= 175°$; (b) 5.05 V; (c) 163 mA; 17 V

4.5.2 The Full-Wave Rectifier

The full-wave rectifier utilizes both halves of the input sinusoid. To provide a unipolar output, it inverts the negative halves of the sine wave. One possible implementation is shown in Fig. 4.22(a). Here the transformer secondary winding is **center-tapped** to provide two equal voltages v_S across the two halves of the secondary winding with the polarities indicated. Note that when the input line voltage (feeding the primary) is positive, both of the signals labeled v_S will be positive. In this case D_1 will conduct and D_2 will be reverse biased. The current through D_1 will flow through R and back to the center tap of the secondary. The circuit then behaves like a half-wave rectifier, and the output during the positive half-cycles when D_1 conducts will be identical to that produced by the half-wave rectifier.

Now, during the negative half-cycle of the ac line voltage, both of the voltages labeled v_S will be negative. Thus D_1 will be cut off while D_2 will conduct. The current conducted by D_2 will flow through R and back to the center tap. It follows that during the negative half-cycles while D_2 conducts, the circuit behaves again as a half-wave rectifier. The important point, however, is that the current through R always flows in the same direction, and thus v_O will be unipolar, as indicated in Fig. 4.22(c). The output waveform shown is obtained by assuming that a conducting diode has a constant voltage drop V_D. Thus the transfer characteristic of the full-wave rectifier takes the shape shown in Fig. 4.22(b).

The full-wave rectifier obviously produces a more "energetic" waveform than that provided by the half-wave rectifier. In almost all rectifier applications, one opts for a full-wave type of some kind.

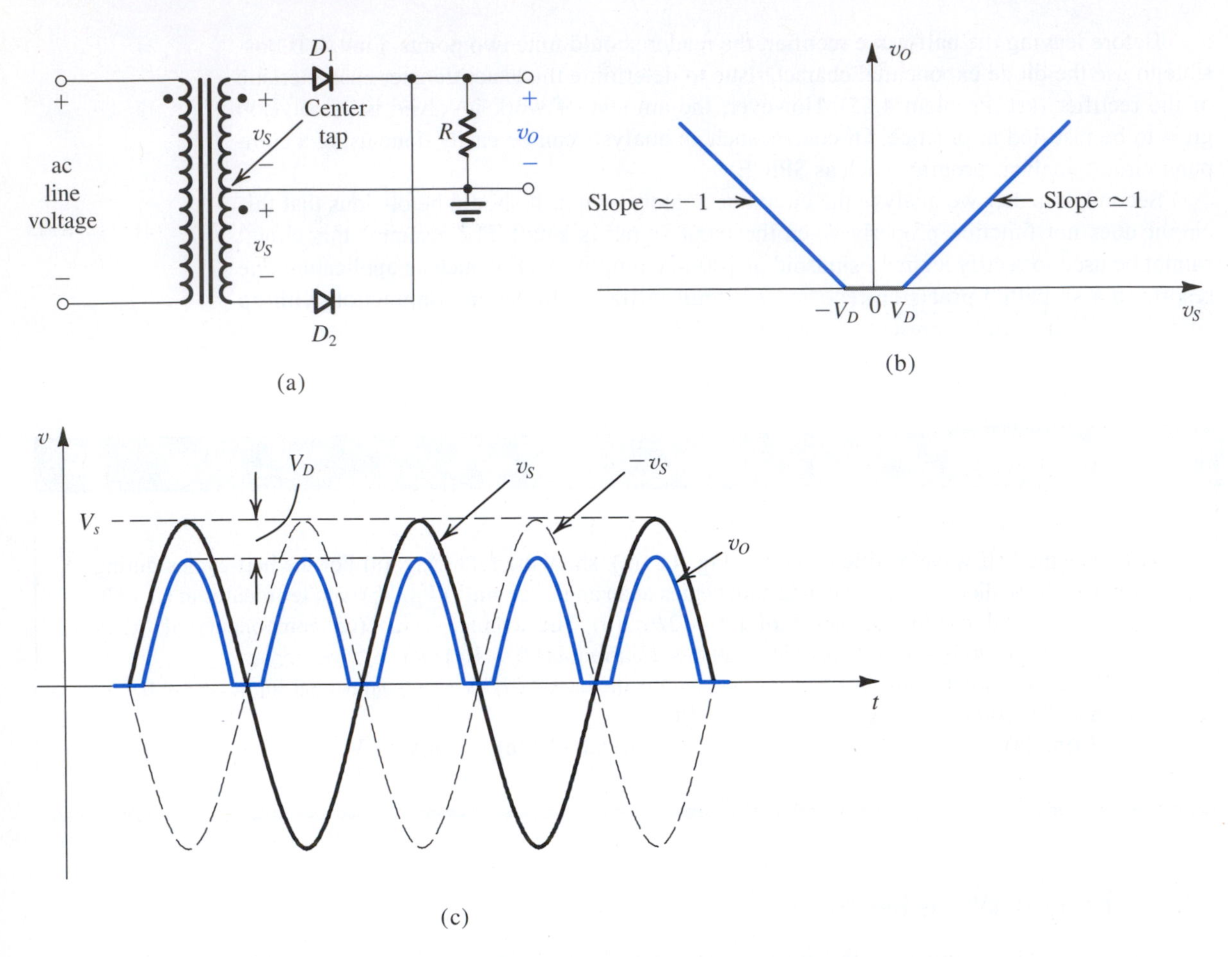

Figure 4.22 Full-wave rectifier utilizing a transformer with a center-tapped secondary winding: **(a)** circuit; **(b)** transfer characteristic assuming a constant-voltage-drop model for the diodes; **(c)** input and output waveforms.

To find the PIV of the diodes in the full-wave rectifier circuit, consider the situation during the positive half-cycles. Diode D_1 is conducting, and D_2 is cut off. The voltage at the cathode of D_2 is v_O, and that at its anode is $-v_S$. Thus the reverse voltage across D_2 will be $(v_O + v_S)$, which will reach its maximum when v_O is at its peak value of $(V_s - V_D)$, and v_S is at its peak value of V_s; thus,

$$\text{PIV} = 2V_s - V_D$$

which is approximately twice that for the case of the half-wave rectifier.

EXERCISE

4.20 For the full-wave rectifier circuit in Fig. 4.22(a), show the following: (a) The output is zero for an angle of $2\sin^{-1}(V_D/V_s)$ centered around the zero-crossing points of the sine-wave input. (b) The average value (dc component) of v_O is $V_O \simeq (2/\pi)V_s - V_D$. (c) The peak current through each diode is $(V_s - V_D)/R$. Find the fraction (percentage) of each cycle during which $v_O > 0$, the value of V_O, the peak diode current, and the value of PIV, all for the case in which v_S is a 12-V (rms) sinusoid, $V_D \simeq 0.7$ V, and $R = 100\ \Omega$.

Ans. 97.4%; 10.1 V; 163 mA; 33.2 V

4.5.3 The Bridge Rectifier

An alternative implementation of the full-wave rectifier is shown in Fig. 4.23(a). This circuit, known as the bridge rectifier because of the similarity of its configuration to that of the Wheatstone bridge, does not require a center-tapped transformer, a distinct advantage over the full-wave rectifier circuit of Fig. 4.22. The bridge rectifier, however, requires four diodes as compared to two in the previous circuit. This is not much of a disadvantage, because diodes are inexpensive and one can buy a diode bridge in one package.

The bridge rectifier circuit operates as follows: During the positive half-cycles of the input voltage, v_S is positive, and thus current is conducted through diode D_1, resistor R, and diode D_2. Meanwhile, diodes D_3 and D_4 will be reverse biased. Observe that there are two diodes in series in the conduction path, and thus v_O will be lower than v_S by two diode drops (compared to one drop in the circuit previously discussed). This is somewhat of a disadvantage of the bridge rectifier.

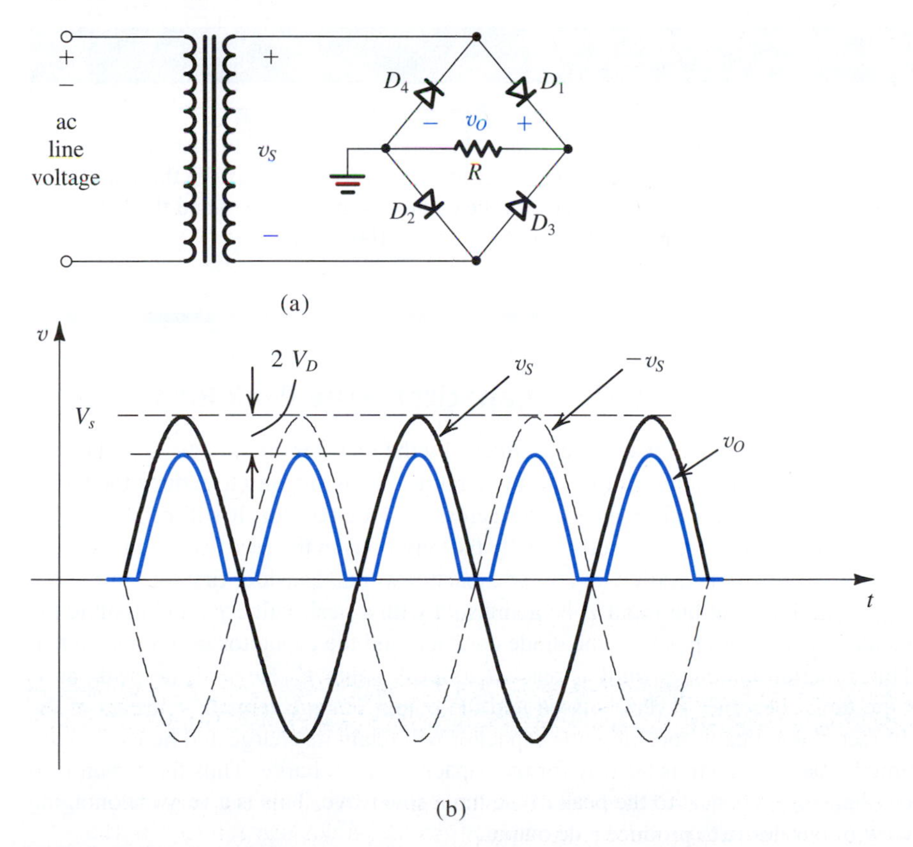

Figure 4.23 The bridge rectifier: (a) circuit; (b) input and output waveforms.

Next, consider the situation during the negative half-cycles of the input voltage. The secondary voltage v_S will be negative, and thus $-v_S$ will be positive, forcing current through D_3, R, and D_4. Meanwhile, diodes D_1 and D_2 will be reverse biased. The important point to note, though, is that during both half-cycles, current flows through R in the same direction (from right to left), and thus v_O will always be positive, as indicated in Fig. 4.23(b).

To determine the peak inverse voltage (PIV) of each diode, consider the circuit during the positive half-cycles. The reverse voltage across D_3 can be determined from the loop formed by D_3, R, and D_2 as

$$v_{D3}\text{ (reverse)} = v_O + v_{D2}\text{ (forward)}$$

Thus the maximum value of v_{D3} occurs at the peak of v_O and is given by

$$\text{PIV} = V_s - 2V_D + V_D = V_s - V_D$$

Observe that here the PIV is about half the value for the full-wave rectifier with a center-tapped transformer. This is another advantage of the bridge rectifier.

Yet one more advantage of the bridge rectifier circuit over that utilizing a center-tapped transformer is that only about half as many turns are required for the secondary winding of the transformer. Another way of looking at this point can be obtained by observing that each half of the secondary winding of the center-tapped transformer is utilized for only half the time. These advantages have made the bridge rectifier the most popular rectifier circuit configuration.

EXERCISE

4.21 For the bridge rectifier circuit of Fig. 4.23(a), use the constant-voltage-drop diode model to show that (a) the average (or dc component) of the output voltage is $V_O \simeq (2/\pi)V_s - 2V_D$ and (b) the peak diode current is $(V_s - 2V_D)/R$. Find numerical values for the quantities in (a) and (b) and the PIV for the case in which v_S is a 12-V (rms) sinusoid, $V_D \simeq 0.7$ V, and $R = 100\ \Omega$.
Ans. 9.4 V; 156 mA; 16.3 V

4.5.4 The Rectifier with a Filter Capacitor—The Peak Rectifier

The pulsating nature of the output voltage produced by the rectifier circuits discussed above makes it unsuitable as a dc supply for electronic circuits. A simple way to reduce the variation of the output voltage is to place a capacitor across the load resistor. It will be shown that this **filter capacitor** serves to reduce substantially the variations in the rectifier output voltage.

To see how the rectifier circuit with a filter capacitor works, consider first the simple circuit shown in Fig. 4.24. Let the input v_I be a sinusoid with a peak value V_p, and assume the diode to be ideal. As v_I goes positive, the diode conducts and the capacitor is charged so that $v_O = v_I$. This situation continues until v_I reaches its peak value V_p. Beyond the peak, as v_I decreases the diode becomes reverse biased and the output voltage remains constant at the value V_p. In fact, theoretically speaking, the capacitor will retain its charge and hence its voltage indefinitely, because there is no way for the capacitor to discharge. Thus the circuit provides a dc voltage output equal to the peak of the input sine wave. This is a very encouraging result in view of our desire to produce a dc output.

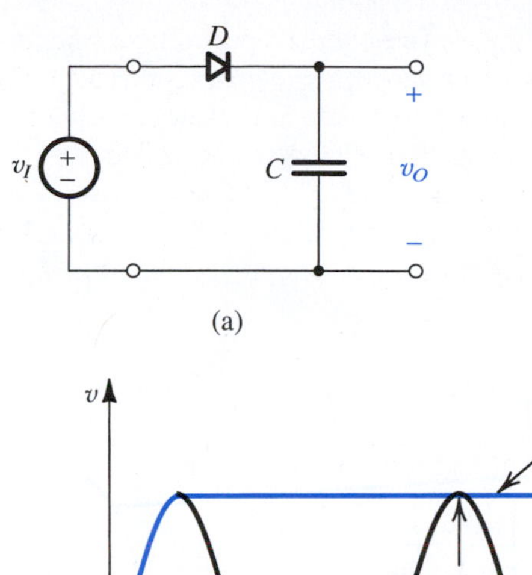

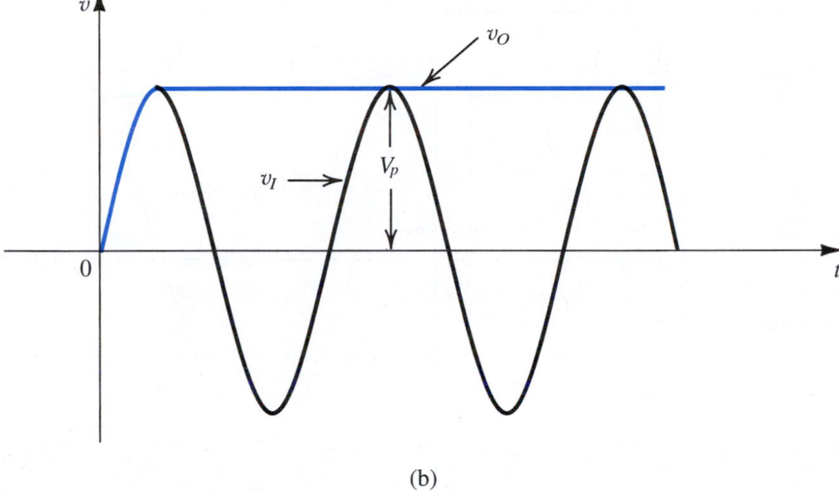

Figure 4.24 **(a)** A simple circuit used to illustrate the effect of a filter capacitor. **(b)** Input and output waveforms assuming an ideal diode. Note that the circuit provides a dc voltage equal to the peak of the input sine wave. The circuit is therefore known as a *peak rectifier* or a *peak detector.*

Next, we consider the more practical situation where a load resistance R is connected across the capacitor C, as depicted in Fig. 4.25(a). However, we will continue to assume the diode to be ideal. As before, for a sinusoidal input, the capacitor charges to the peak of the input V_p. Then the diode cuts off, and the capacitor discharges through the load resistance R. The capacitor discharge will continue for almost the entire cycle, until the time at which v_I exceeds the capacitor voltage. Then the diode turns on again and charges the capacitor up to the peak of v_I, and the process repeats itself. Observe that to keep the output voltage from decreasing too much during capacitor discharge, one selects a value for C so that the time constant CR is much greater than the discharge interval.

We are now ready to analyze the circuit in detail. Figure 4.25(b) shows the steady-state input and output voltage waveforms under the assumption that $CR \gg T$, where T is the period of the input sinusoid. The waveforms of the load current

$$i_L = v_O/R \tag{4.23}$$

and of the diode current (when it is conducting)

$$i_D = i_C + i_L \tag{4.24}$$

$$= C\frac{dv_I}{dt} + i_L \tag{4.25}$$

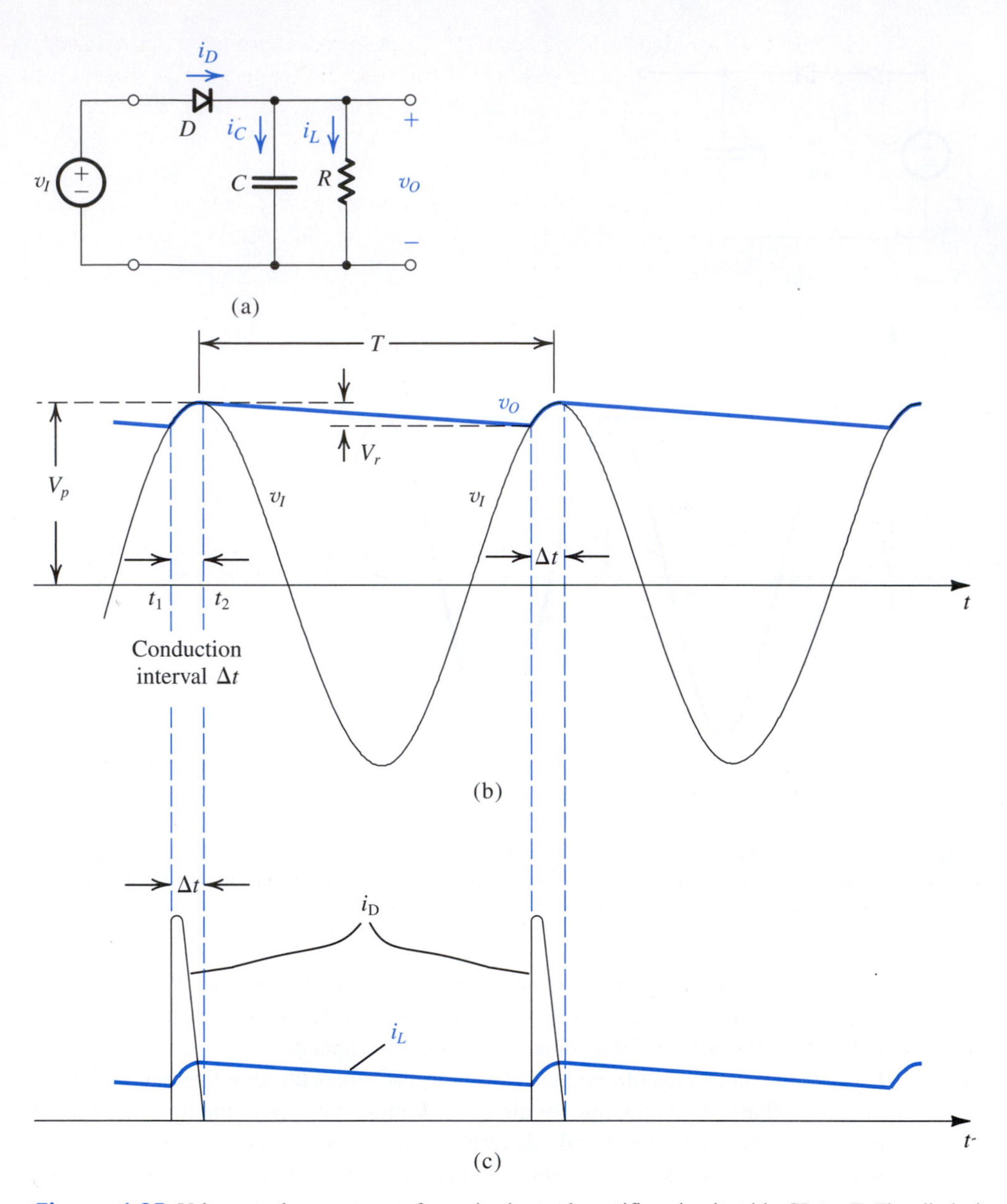

Figure 4.25 Voltage and current waveforms in the peak rectifier circuit with $CR \gg T$. The diode is assumed ideal.

are shown in Fig. 4.25(c). The following observations are in order:

1. The diode conducts for a brief interval, Δt, near the peak of the input sinusoid and supplies the capacitor with charge equal to that lost during the much longer discharge interval. The latter is approximately equal to the period T.
2. Assuming an ideal diode, the diode conduction begins at time t_1, at which the input v_I equals the exponentially decaying output v_O. Conduction stops at t_2 shortly after the peak of v_I; the exact value of t_2 can be determined by setting $i_D = 0$ in Eq. (4.25).

3. During the diode-off interval, the capacitor C discharges through R, and thus v_O decays exponentially with a time constant CR. The discharge interval begins just past the peak of v_I. At the end of the discharge interval, which lasts for almost the entire period T, $v_O = V_p - V_r$, where V_r is the peak-to-peak ripple voltage. When $CR \gg T$, the value of V_r is small.
4. When V_r is small, v_O is almost constant and equal to the peak value of v_I. Thus the dc output voltage is approximately equal to V_p. Similarly, the current i_L is almost constant, and its dc component I_L is given by

$$I_L = \frac{V_p}{R} \tag{4.26}$$

If desired, a more accurate expression for the output dc voltage can be obtained by taking the average of the extreme values of v_O,

$$V_O = V_p - \tfrac{1}{2}V_r \tag{4.27}$$

With these observations in hand, we now derive expressions for V_r and for the average and peak values of the diode current. During the diode-off interval, v_O can be expressed as

$$v_O = V_p e^{-t/CR}$$

At the end of the discharge interval we have

$$V_p - V_r \simeq V_p e^{-T/CR}$$

Now, since $CR \gg T$, we can use the approximation $e^{-T/CR} \simeq 1 - T/CR$ to obtain

$$V_r \simeq V_p \frac{T}{CR} \tag{4.28}$$

We observe that to keep V_r small we must select a capacitance C so that $CR \gg T$. The **ripple voltage** V_r in Eq. (4.28) can be expressed in terms of the frequency $f = 1/T$ as

$$V_r = \frac{V_p}{fCR} \tag{4.29a}$$

Using Eq. (4.26) we can express V_r by the alternate expression

$$V_r = \frac{I_L}{fC} \tag{4.29b}$$

Note that an alternative interpretation of the approximation made above is that the capacitor discharges by means of a constant current $I_L = V_p/R$. This approximation is valid as long as $V_r \ll V_p$.

Assuming that diode conduction ceases almost at the peak of v_I, we can determine the **conduction interval** Δt from

$$V_p \cos(\omega \Delta t) = V_p - V_r$$

where $\omega = 2\pi f = 2\pi/T$ is the angular frequency of v_I. Since $(\omega\Delta t)$ is a small angle, we can employ the approximation $\cos(\omega\Delta t) \simeq 1 - \frac{1}{2}(\omega\Delta t)^2$ to obtain

$$\omega \Delta t \simeq \sqrt{2V_r/V_p} \tag{4.30}$$

We note that when $V_r \ll V_p$, the conduction angle $\omega\Delta t$ will be small, as assumed.

To determine the average diode current during conduction, i_{Dav}, we equate the charge that the diode supplies to the capacitor,

$$Q_{\text{supplied}} = i_{Cav}\,\Delta t$$

where from Eq. (4.24),

$$i_{Cav} = i_{Dav} - I_L$$

to the charge that the capacitor loses during the discharge interval,

$$Q_{\text{lost}} = CV_r$$

to obtain, using Eqs. (4.30) and (4.29a),

$$i_{Dav} = I_L(1 + \pi\sqrt{2V_p/V_r}) \tag{4.31}$$

Observe that when $V_r \ll V_p$, the average diode current during conduction is much greater than the dc load current. This is not surprising, since the diode conducts for a very short interval and must replenish the charge lost by the capacitor during the much longer interval in which it is discharged by I_L.

The peak value of the diode current, i_{Dmax}, can be determined by evaluating the expression in Eq. (4.25) at the onset of diode conduction—that is, at $t = t_1 = -\Delta t$ (where $t = 0$ is at the peak). Assuming that i_L is almost constant at the value given by Eq. (4.26), we obtain

$$i_{Dmax} = I_L(1 + 2\pi\sqrt{2V_p/V_r}) \tag{4.32}$$

From Eqs. (4.31) and (4.32), we see that for $V_r \ll V_p$, $i_{Dmax} \simeq 2i_{Dav}$, which correlates with the fact that the waveform of i_D is almost a right-angle triangle (see Fig. 4.25c).

Example 4.8

Consider a peak rectifier fed by a 60-Hz sinusoid having a peak value $V_p = 100$ V. Let the load resistance $R = 10$ kΩ. Find the value of the capacitance C that will result in a peak-to-peak ripple of 2 V. Also, calculate the fraction of the cycle during which the diode is conducting and the average and peak values of the diode current.

Solution

From Eq. (4.29a) we obtain the value of C as

$$C = \frac{V_p}{V_r fR} = \frac{100}{2 \times 60 \times 10 \times 10^3} = 83.3\ \mu\text{F}$$

The conduction angle $\omega\Delta t$ is found from Eq. (4.30) as

$$\omega\Delta t = \sqrt{2 \times 2/100} = 0.2 \text{ rad}$$

Thus the diode conducts for $(0.2/2\pi) \times 100 = 3.18\%$ of the cycle. The average diode current is obtained from Eq. (4.31), where $I_L = 100/10 = 10$ mA, as

$$i_{Dav} = 10(1 + \pi\sqrt{2 \times 100/2}) = 324 \text{ mA}$$

The peak diode current is found using Eq. (4.32),

$$i_{Dmax} = 10(1 + 2\pi\sqrt{2 \times 100/2}) = 638 \text{ mA}$$

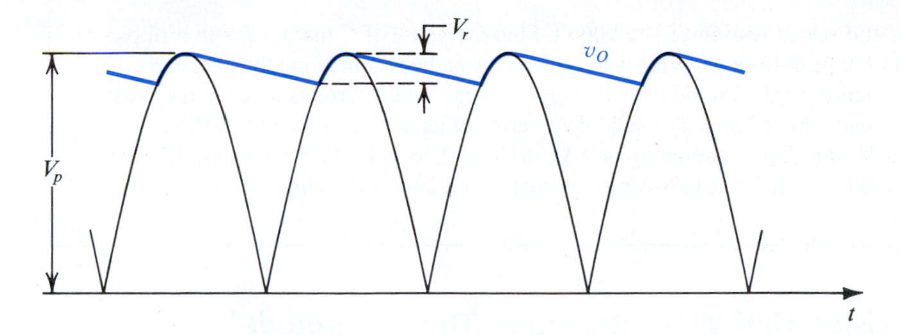

Figure 4.26 Waveforms in the full-wave peak rectifier.

The circuit of Fig. 4.25(a) is known as a half-wave **peak rectifier**. The full-wave rectifier circuits of Figs. 4.22(a) and 4.23(a) can be converted to peak rectifiers by including a capacitor across the load resistor. As in the half-wave case, the output dc voltage will be almost equal to the peak value of the input sine wave (Fig. 4.26). The ripple frequency, however, will be twice that of the input. The peak-to-peak ripple voltage, for this case, can be derived using a procedure identical to that above but with the discharge period T replaced by $T/2$, resulting in

$$V_r = \frac{V_p}{2fCR} \tag{4.33}$$

While the diode conduction interval, Δt, will still be given by Eq. (4.30), the average and peak currents in each of the diodes will be given by

$$i_{D\text{av}} = I_L(1 + \pi\sqrt{V_p/2V_r}) \tag{4.34}$$

$$i_{D\text{max}} = I_L(1 + 2\pi\sqrt{V_p/2V_r}) \tag{4.35}$$

Comparing these expressions with the corresponding ones for the half-wave case, we note that for the same values of V_p, f, R, and V_r (and thus the same I_L), we need a capacitor half the size of that required in the half-wave rectifier. Also, the current in each diode in the full-wave rectifier is approximately half that which flows in the diode of the half-wave circuit.

The analysis above assumed ideal diodes. The accuracy of the results can be improved by taking the diode voltage drop into account. This can be easily done by replacing the peak voltage V_p to which the capacitor charges with $(V_p - V_D)$ for the half-wave circuit and the full-wave circuit using a center-tapped transformer and with $(V_p - 2V_D)$ for the bridge-rectifier case.

We conclude this section by noting that peak-rectifier circuits find application in signal-processing systems where it is required to detect the peak of an input signal. In such a case, the circuit is referred to as a **peak detector**. A particularly popular application of the peak detector is in the design of a demodulator for amplitude-modulated (AM) signals. We shall not discuss this application further here.

EXERCISES

4.22 Derive the expressions in Eqs. (4.33), (4.34), and (4.35).

4.23 Consider a bridge-rectifier circuit with a filter capacitor C placed across the load resistor R for the case in which the transformer secondary delivers a sinusoid of 12 V (rms) having a 60-Hz frequency and

assuming $V_D = 0.8$ V and a load resistance $R = 100\ \Omega$. Find the value of C that results in a ripple voltage no larger than 1 V peak-to-peak. What is the dc voltage at the output? Find the load current. Find the diodes' conduction angle. Provide the average and peak diode currents What is the peak reverse voltage across each diode? Specify the diode in terms of its peak current and its PIV.
Ans. 1281 μF; 15.4 V or (a better estimate) 14.9 V; 0.15 A; 0.36 rad (20.7°); 1.45 A; 2.74 A; 16.2 V. Thus select a diode with 3.5-A to 4-A peak current and a 20-V PIV rating.

4.5.5 Precision Half-Wave Rectifier—The Superdiode[4]

The rectifier circuits studied thus far suffer from having one or two diode drops in the signal paths. Thus these circuits work well only when the signal to be rectified is much larger than the voltage drop of a conducting diode (0.7 V or so). In such a case, the details of the diode forward characteristics or the exact value of the diode voltage do not play a prominent role in determining circuit performance. This is indeed the case in the application of rectifier circuits in power-supply design. There are other applications, however, where the signal to be rectified is small (e.g., on the order of 100 mV or so) and thus clearly insufficient to turn on a diode. Also, in instrumentation applications, the need arises for rectifier circuits with very precise and predictable transfer characteristics. For these applications, a class of circuits has been developed utilizing op amps (Chapter 2) together with diodes to provide precision rectification. In the following discussion, we study one such circuit, leaving a more comprehensive study of op amp–diode circuits to Chapter 17.

Figure 4.27(a) shows a precision half-wave rectifier circuit consisting of a diode placed in the negative-feedback path of an op amp, with R being the rectifier load resistance. The op amp, of course, needs power supplies for its operation. For simplicity, these are not shown in the circuit diagram. The circuit works as follows: If v_I goes positive, the output voltage v_A of the op amp will go positive and the diode will conduct, thus establishing a closed feedback path between the op amp's output terminal and the negative input terminal. This negative-feedback path will cause a virtual short circuit to appear between the two input terminals of

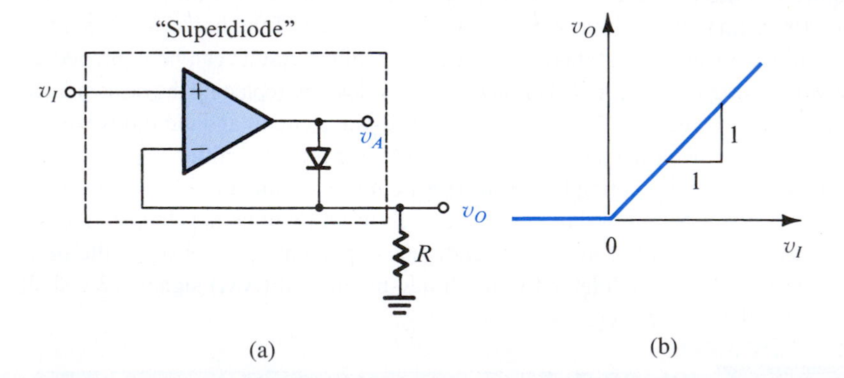

Figure 4.27 The "superdiode" precision half-wave rectifier and its almost-ideal transfer characteristic. Note that when $v_I > 0$ and the diode conducts, the op amp supplies the load current, and the source is conveniently buffered, an added advantage. Not shown are the op-amp power supplies.

[4]This section requires knowledge of operational amplifiers (Chapter 2).

the op amp. Thus the voltage at the negative input terminal, which is also the output voltage v_O, will equal (to within a few millivolts) that at the positive input terminal, which is the input voltage v_I,

$$v_O = v_I \qquad v_I \geq 0$$

Note that the offset voltage ($\simeq$0.7 V) exhibited in the simple half-wave rectifier circuit of Fig. 4.21 is no longer present. For the op-amp circuit to start operation, v_I has to exceed only a negligibly small voltage equal to the diode drop divided by the op amp's open-loop gain. In other words, the straight-line transfer characteristic v_O–v_I almost passes through the origin. This makes this circuit suitable for applications involving very small signals.

Consider now the case when v_I goes negative. The op amp's output voltage v_A will tend to follow and go negative. This will reverse-bias the diode, and no current will flow through resistance R, causing v_O to remain equal to 0 V. Thus, for $v_I < 0$, $v_O = 0$. Since in this case the diode is off, the op amp will be operating in an open-loop fashion, and its output will be at its negative saturation level.

The transfer characteristic of this circuit will be that shown in Fig. 4.27(b), which is almost identical to the ideal characteristic of a half-wave rectifier. The nonideal diode characteristics have been almost completely masked by placing the diode in the negative-feedback path of an op amp. This is another dramatic application of negative feedback, a subject we will study formally in Chapter 10. The combination of diode and op amp, shown in the dotted box in Fig. 4.27(a), is appropriately referred to as a "superdiode."

EXERCISES

4.24 Consider the **operational rectifier** or superdiode circuit of Fig. 4.27(a), with $R = 1\text{ k}\Omega$. For $v_I = 10$ mV, 1 V, and −1 V, what are the voltages that result at the rectifier output and at the output of the op amp? Assume that the op amp is ideal and that its output saturates at ±12 V. The diode has a 0.7-V drop at 1-mA current.
Ans. 10 mV, 0.59 V; 1 V, 1.7 V; 0 V, −12 V

4.25 If the diode in the circuit of Fig. 4.27(a) is reversed, find the transfer characteristic v_O as a function of v_I.
Ans. $v_O = 0$ for $v_I \geq 0$; $v_O = v_I$ for $v_I \leq 0$

4.6 Limiting and Clamping Circuits

In this section, we shall present additional nonlinear circuit applications of diodes.

4.6.1 Limiter Circuits

Figure 4.28 shows the general transfer characteristic of a limiter circuit. As indicated, for inputs in a certain range, $L_-/K \leq v_I \leq L_+/K$, the limiter acts as a linear circuit, providing an output proportional to the input, $v_O = Kv_I$. Although in general K can be greater than 1, the circuits discussed in this section have $K \leq 1$ and are known as **passive limiters**. (Examples of active limiters will be presented in Chapter 17.) If v_I exceeds the upper *threshold* (L_+/K), the output voltage is *limited* or clamped to the upper limiting level L_+. On the other hand, if v_I is

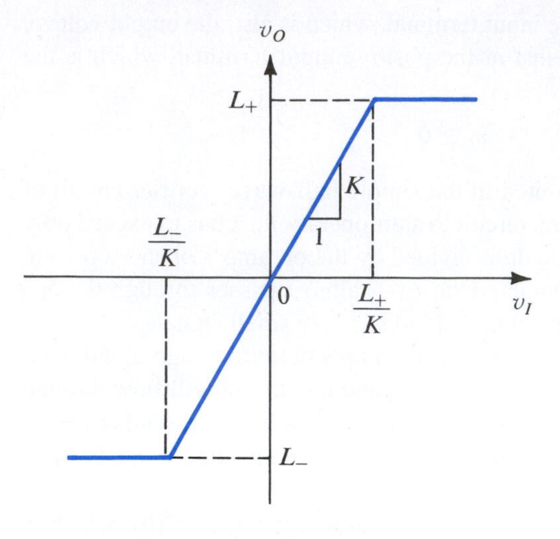

Figure 4.28 General transfer characteristic for a limiter circuit.

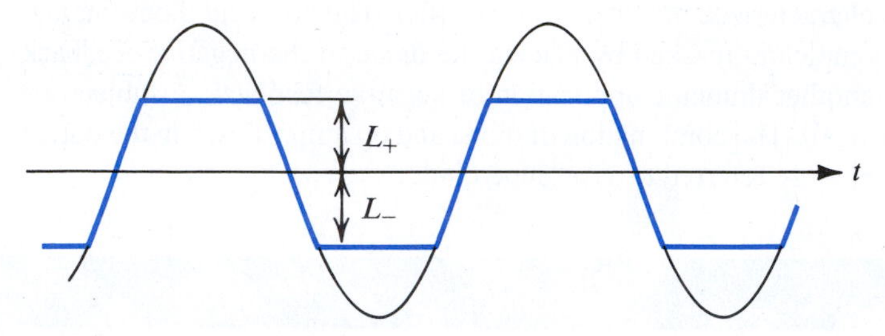

Figure 4.29 Applying a sine wave to a limiter can result in clipping off its two peaks.

reduced below the lower limiting threshold (L_-/K), the output voltage v_O is limited to the lower limiting level L_-.

The general transfer characteristic of Fig. 4.28 describes a **double limiter**—that is, a limiter that works on both the positive and negative peaks of an input waveform. **Single limiters**, of course, exist. Finally, note that if an input waveform such as that shown in Fig. 4.29 is fed to a double limiter, its two peaks will be *clipped off*. Limiters therefore are sometimes referred to as **clippers**.

The limiter whose characteristics are depicted in Fig. 4.28 is described as a **hard limiter**. **Soft limiting** is characterized by smoother transitions between the linear region and the saturation regions and a slope greater than zero in the saturation regions, as illustrated in Fig. 4.30. Depending on the application, either hard or soft limiting may be preferred.

Limiters find application in a variety of signal-processing systems. One of their simplest applications is in limiting the voltage between the two input terminals of an op amp to a value lower than the breakdown voltage of the transistors that make up the input stage of the op-amp circuit. We will have more to say on this and other limiter applications at later points in this book.

Diodes can be combined with resistors to provide simple realizations of the limiter function. A number of examples are depicted in Fig. 4.31. In each part of the figure both the circuit and its transfer characteristic are given. The transfer characteristics are obtained using the constant-voltage-drop (V_D = 0.7 V) diode model but assuming a smooth transition between the linear and saturation regions of the transfer characteristic.

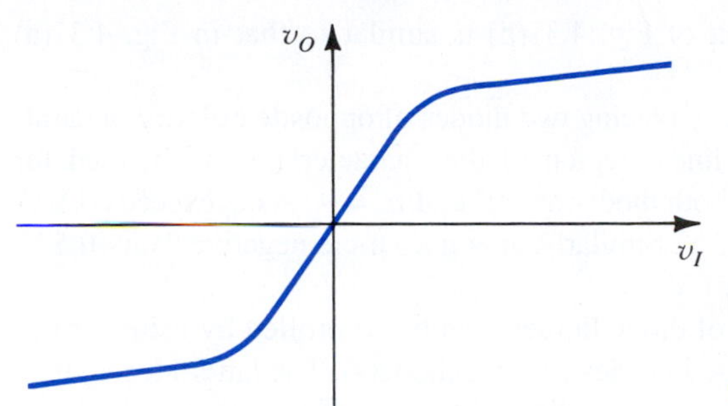

Figure 4.30 Soft limiting.

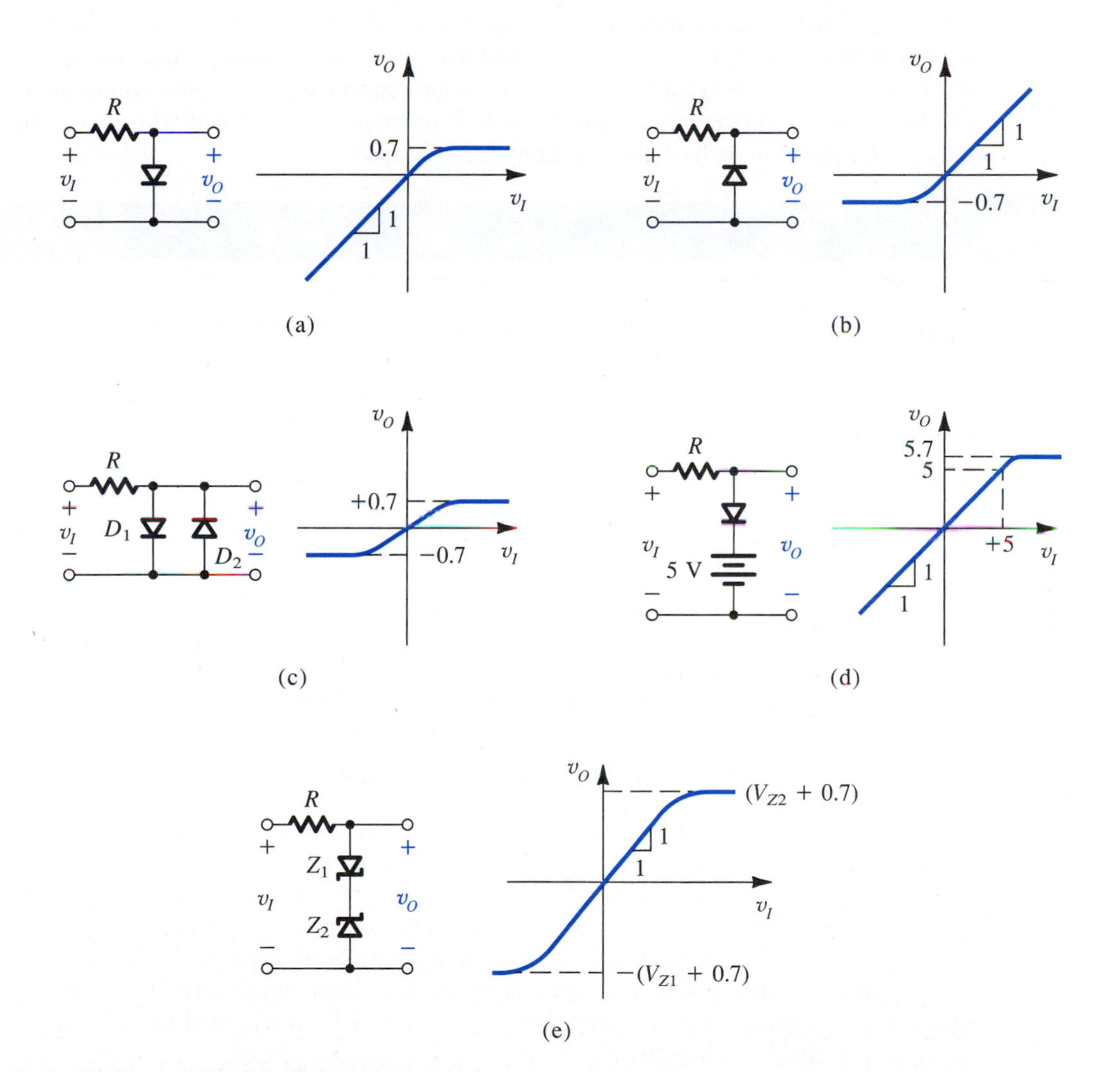

Figure 4.31 A variety of basic limiting circuits.

The circuit in Fig. 4.31(a) is that of the half-wave rectifier except that here the output is taken across the diode. For $v_I < 0.5$ V, the diode is cut off, no current flows, and the voltage drop across R is zero; thus $v_O = v_I$. As v_I exceeds 0.5 V, the diode turns on, eventually limiting

v_O to one diode drop (0.7 V). The circuit of Fig. 4.31(b) is similar to that in Fig. 4.31(a) except that the diode is reversed.

Double limiting can be implemented by placing two diodes of opposite polarity in parallel, as shown in Fig. 4.31(c). Here the linear region of the characteristic is obtained for $-0.5\text{ V} \le v_I \le 0.5\text{ V}$. For this range of v_I, both diodes are off and $v_O = v_I$. As v_I exceeds 0.5 V, D_1 turns on and eventually limits v_O to +0.7 V. Similarly, as v_I goes more negative than −0.5 V, D_2 turns on and eventually limits v_O to −0.7 V.

The thresholds and saturation levels of diode limiters can be controlled by using strings of diodes and/or by connecting a dc voltage in series with the diode(s). The latter idea is illustrated in Fig. 4.31(d). Finally, rather than strings of diodes, we may use two zener diodes in series, as shown in Fig. 4.31(e). In this circuit, limiting occurs in the positive direction at a voltage of $V_{Z2} + 0.7$, where 0.7 V represents the voltage drop across zener diode Z_1 when conducting in the *forward* direction. For negative inputs, Z_1 acts as a zener, while Z_2 conducts in the forward direction. It should be mentioned that pairs of zener diodes connected in series are available commercially for applications of this type under the name **double-anode zener**.

More flexible limiter circuits are possible if op amps are combined with diodes and resistors. Examples of such circuits are discussed in Chapter 17.

EXERCISE

4.26 Assuming the diodes to be ideal, describe the transfer characteristic of the circuit shown in Fig. E4.26.

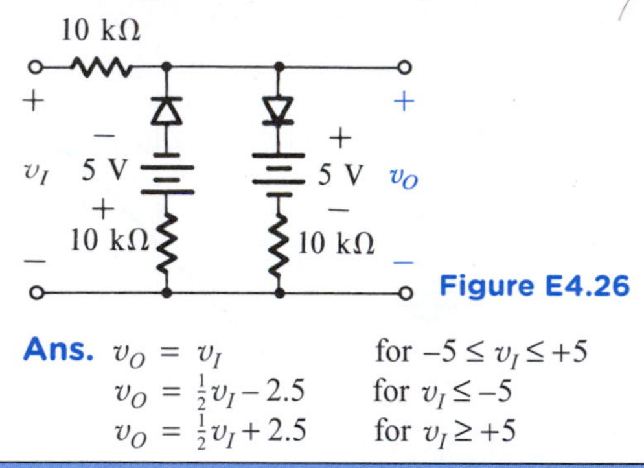

Figure E4.26

Ans. $v_O = v_I$ for $-5 \le v_I \le +5$
$v_O = \frac{1}{2}v_I - 2.5$ for $v_I \le -5$
$v_O = \frac{1}{2}v_I + 2.5$ for $v_I \ge +5$

4.6.2 The Clamped Capacitor or DC Restorer

If in the basic peak-rectifier circuit, the output is taken across the diode rather than across the capacitor, an interesting circuit with important applications results. The circuit, called a dc restorer, is shown in Fig. 4.32 fed with a square wave. Because of the polarity in which the diode is connected, the capacitor will charge to a voltage v_C with the polarity indicated in Fig. 4.32 and equal to the magnitude of the most negative peak of the input signal. Subsequently, the diode turns off and the capacitor retains its voltage indefinitely. If, for instance, the input square wave has the arbitrary levels −6 V and +4 V, then v_C will be equal to 6 V. Now, since the output voltage v_O is given by

$$v_O = v_I + v_C$$

it follows that the output waveform will be identical to that of the input, except that it is shifted upward by v_C volts. In our example the output will thus be a square wave with levels of 0 V and +10 V.

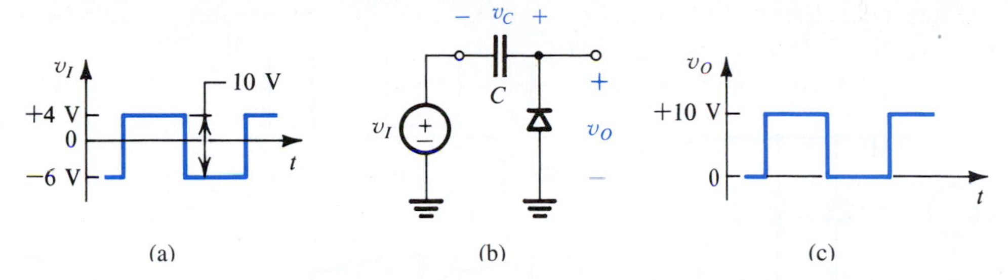

Figure 4.32 The clamped capacitor or dc restorer with a square-wave input and no load.

Another way of visualizing the operation of the circuit in Fig. 4.32 is to note that because the diode is connected across the output with the polarity shown, it prevents the output voltage from going below 0 V (by conducting and charging up the capacitor, thus causing the output to rise to 0 V), but this connection will not constrain the positive excursion of v_O. The output waveform will therefore have its lowest peak *clamped* to 0 V, which is why the circuit is called a **clamped capacitor**. It should be obvious that reversing the diode polarity will provide an output waveform whose highest peak is clamped to 0 V. In either case, the output waveform will have a finite average value or dc component. This dc component is entirely unrelated to the average value of the input waveform. As an application, consider a pulse signal being transmitted through a capacitively coupled or ac-coupled system. The capacitive coupling will cause the pulse train to lose whatever dc component it originally had. Feeding the resulting pulse waveform to a clamping circuit provides it with a well-determined dc component, a process known as **dc restoration**. This is why the circuit is also called a **dc restorer**.

Restoring dc is useful because the dc component or average value of a pulse waveform is an effective measure of its **duty cycle**.[5] The duty cycle of a pulse waveform can be modulated (in a process called **pulsewidth modulation**) and made to carry information. In such a system, detection or demodulation could be achieved simply by feeding the received pulse waveform to a dc restorer and then using a simple *RC* low-pass filter to separate the average of the output waveform from the superimposed pulses.

When a load resistance R is connected across the diode in a clamping circuit, as shown in Fig. 4.33, the situation changes significantly. While the output is above ground, a current must flow in R. Since at this time the diode is off, this current obviously comes from the capacitor, thus causing the capacitor to discharge and the output voltage to fall. This is shown in Fig. 4.33 for a square-wave input. During the interval t_0 to t_1, the output voltage falls exponentially with time constant CR. At t_1 the input decreases by V_a volts, and the output attempts to follow. This causes the diode to conduct heavily and to quickly charge the capacitor. At the end of the interval t_1 to t_2, the output voltage would normally be a few tenths of a volt negative (e.g., −0.5 V). Then, as the input rises by V_a volts (at t_2), the output follows, and the cycle repeats itself. In the steady state the charge lost by the capacitor during the interval t_0 to t_1 is recovered during the interval t_1 to t_2. This charge equilibrium enables us to calculate the average diode current as well as the details of the output waveform.

[5]The duty cycle of a pulse waveform is the proportion of each cycle occupied by the pulse. In other words, it is the pulse width expressed as a fraction of the pulse period.

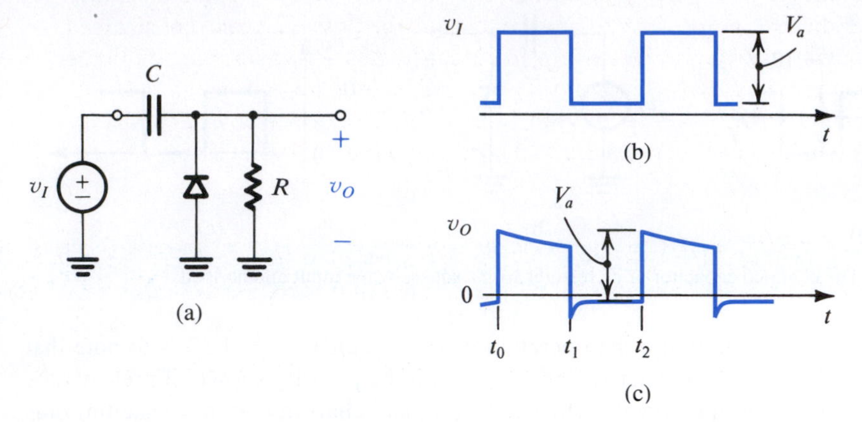

Figure 4.33 The clamped capacitor with a load resistance R.

4.6.3 The Voltage Doubler

Figure 4.34(a) shows a circuit composed of two sections in cascade: a clamped capacitor formed by C_1 and D_1, and a peak rectifier formed by D_2 and C_2. When excited by a

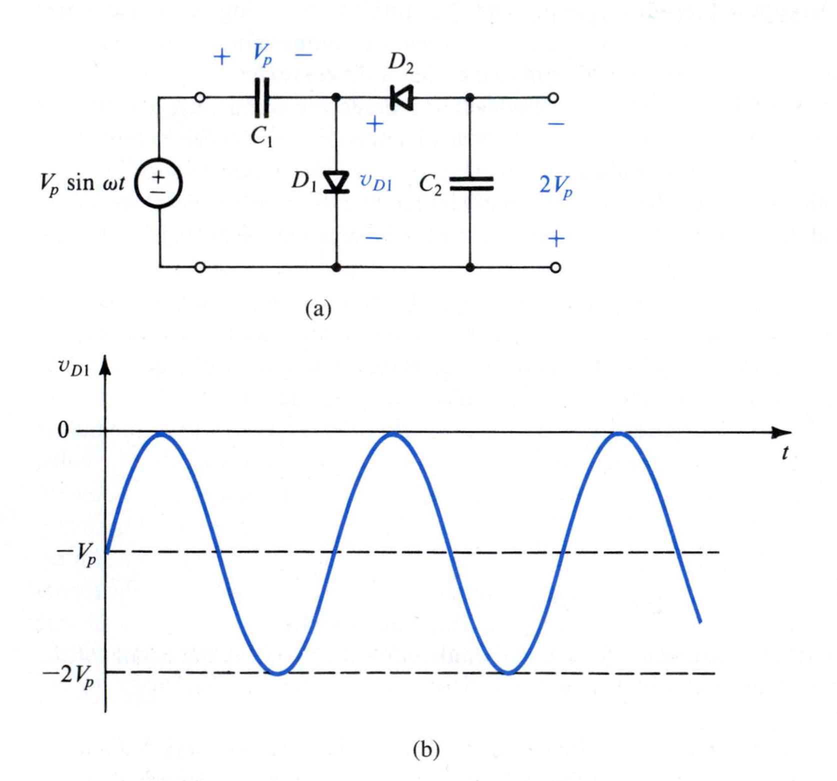

Figure 4.34 Voltage doubler: **(a)** circuit; **(b)** waveform of the voltage across D_1.

sinusoid of amplitude V_p the clamping section provides the voltage waveform shown, assuming ideal diodes, in Fig. 4.34(b). Note that while the positive peaks are clamped to 0 V, the negative peak reaches $-2V_p$. In response to this waveform, the peak-detector section provides across capacitor C_2 a negative dc voltage of magnitude $2V_p$. Because the output voltage is double the input peak, the circuit is known as a voltage doubler. The technique can be extended to provide output dc voltages that are higher multiples of V_p.

EXERCISE

4.27 If the diode in the circuit of Fig. 4.32 is reversed, what will the dc component of v_O become?
Ans. –5 V

4.7 Special Diode Types

In this section, we discuss briefly some important special types of diodes.

4.7.1 The Schottky-Barrier Diode (SBD)

The Schottky-barrier diode (SBD) is formed by bringing metal into contact with a moderately doped *n*-type semiconductor material. The resulting metal–semiconductor junction behaves like a diode, conducting current in one direction (from the metal anode to the semiconductor cathode) and acting as an open circuit in the other, and is known as the Schottky-barrier diode or simply the Schottky diode. In fact, the current–voltage characteristic of the SBD is remarkably similar to that of a *pn*-junction diode, with two important exceptions:

1. In the SBD, current is conducted by majority carriers (electrons). Thus the SBD does not exhibit the minority-carrier charge-storage effects found in forward-biased *pn* junctions. As a result, Schottky diodes can be switched from on to off, and vice versa, much faster than is possible with *pn*-junction diodes.
2. The forward voltage drop of a conducting SBD is lower than that of a *pn*-junction diode. For example, an SBD made of silicon exhibits a forward voltage drop of 0.3 V to 0.5 V, compared to the 0.6 V to 0.8 V found in silicon *pn*-junction diodes. SBDs can also be made of gallium arsenide (GaAs) and, in fact, play an important role in the design of GaAs circuits.[6] Gallium-arsenide SBDs exhibit forward voltage drops of about 0.7 V.

Apart from GaAs circuits, Schottky diodes find application in the design of a special form of bipolar-transistor logic circuits, known as Schottky-TTL, where TTL stands for transistor-transistor logic.

Before leaving the subject of Schottky-barrier diodes, it is important to note that not every metal–semiconductor contact is a diode. In fact, metal is commonly deposited on

⊕[6] The disc accompanying this text contains material on GaAs circuits.

the semiconductor surface in order to make terminals for the semiconductor devices and to connect different devices in an integrated-circuit chip. Such metal–semiconductor contacts are known as **ohmic contacts** to distinguish them from the rectifying contacts that result in SBDs. Ohmic contacts are usually made by depositing metal on very heavily doped (and thus low-resistivity) semiconductor regions. (Recall that SBDs use moderately doped material.)

4.7.2 Varactors

In Chapter 3 we learned that reverse-biased *pn* junctions exhibit a charge-storage effect that is modeled with the depletion-layer or junction capacitance C_j. As Eq. (3.44) indicates, C_j is a function of the reverse-bias voltage V_R. This dependence turns out to be useful in a number of applications, such as the automatic tuning of radio receivers. Special diodes are therefore fabricated to be used as voltage-variable capacitors known as **varactors**. These devices are optimized to make the capacitance a strong function of voltage by arranging that the grading coefficient m is 3 or 4.

4.7.3 Photodiodes

If a reverse-biased *pn* junction is illuminated—that is, exposed to incident light—the photons impacting the junction cause covalent bonds to break, and thus electron–hole pairs are generated in the depletion layer. The electric field in the depletion region then sweeps the liberated electrons to the n side and the holes to the p side, giving rise to a reverse current across the junction. This current, known as photocurrent, is proportional to the intensity of the incident light. Such a diode, called a photodiode, can be used to convert light signals into electrical signals.

Photodiodes are usually fabricated using a compound semiconductor[7] such as gallium arsenide. The photodiode is an important component of a growing family of circuits known as **optoelectronics** or **photonics**. As the name implies, such circuits utilize an optimum combination of electronics and optics for signal processing, storage, and transmission. Usually, electronics is the preferred means for signal processing, whereas optics is most suited for transmission and storage. Examples include fiber-optic transmission of telephone and television signals and the use of optical storage in CD-ROM computer disks. Optical transmission provides very wide bandwidths and low signal attenuation. Optical storage allows vast amounts of data to be stored reliably in a small space.

Finally, we should note that without reverse bias, the illuminated photodiode functions as a **solar cell**. Usually fabricated from low-cost silicon, a solar cell converts light to electrical energy.

4.7.4 Light-Emitting Diodes (LEDs)

The light-emitting diode (LED) performs the inverse of the function of the photodiode; it converts a forward current into light. The reader will recall from Chapter 3 that in a forward-biased *pn* junction, minority carriers are injected across the junction and diffuse into the p

[7]Whereas an elemental semiconductor, such as silicon, uses an element from column IV of the periodic table, a compound semiconductor uses a combination of elements from columns III and V or II and VI. For example, GaAs is formed of gallium (column III) and arsenic (column V) and is thus known as a III-V compound.

and n regions. The diffusing minority carriers then recombine with the majority carriers. Such recombination can be made to give rise to light emission. This can be done by fabricating the pn junction using a semiconductor of the type known as direct-bandgap materials. Gallium arsenide belongs to this group and can thus be used to fabricate light-emitting diodes.

The light emitted by an LED is proportional to the number of recombinations that take place, which in turn is proportional to the forward current in the diode.

LEDs are very popular devices. They find application in the design of numerous types of displays, including the displays of laboratory instruments such as digital voltmeters. They can be made to produce light in a variety of colors. Furthermore, LEDs can be designed so as to produce coherent light with a very narrow bandwidth. The resulting device is a **laser diode**. Laser diodes find application in optical communication systems and in CD players, among other things.

Combining an LED with a photodiode in the same package results in a device known as an **optoisolator**. The LED converts an electrical signal applied to the optoisolator into light, which the photodiode detects and converts back to an electrical signal at the output of the optoisolator. Use of the optoisolator provides complete electrical isolation between the electrical circuit that is connected to the isolator's input and the circuit that is connected to its output. Such isolation can be useful in reducing the effect of electrical interference on signal transmission within a system, and thus optoisolators are frequently employed in the design of digital systems. They can also be used in the design of medical instruments to reduce the risk of electrical shock to patients.

Note that the optical coupling between an LED and a photodiode need not be accomplished inside a small package. Indeed, it can be implemented over a long distance using an optical fiber, as is done in fiber-optic communication links.

Summary

- In the forward direction, the ideal diode conducts any current forced by the external circuit while displaying a zero voltage drop. The ideal diode does not conduct in the reverse direction; any applied voltage appears as reverse bias across the diode.

- The unidirectional-current-flow property makes the diode useful in the design of rectifier circuits.

- The forward conduction of practical silicon-junction diodes is accurately characterized by the relationship $i = I_S e^{v/V_T}$.

- A silicon diode conducts a negligible current until the forward voltage is at least 0.5 V. Then the current increases rapidly, with the voltage drop increasing by 60 mV for every decade of current change.

- In the reverse direction, a silicon diode conducts a current on the order of 10^{-9} A. This current is much greater than I_S and increases with the magnitude of reverse voltage.

- Beyond a certain value of reverse voltage (that depends on the diode), breakdown occurs, and current increases rapidly with a small corresponding increase in voltage.

- Diodes designed to operate in the breakdown region are called zener diodes. They are employed in the design of voltage regulators whose function is to provide a constant dc voltage that varies little with variations in power supply voltage and/or load current.

- In many applications, a conducting diode is modeled as having a constant voltage drop, usually approximately 0.7 V.

- A diode biased to operate at a dc current I_D has a small-signal resistance $r_d = V_T/I_D$.

- Rectifiers convert ac voltages into unipolar voltages. Half-wave rectifiers do this by passing the voltage in half of each cycle and blocking the opposite-polarity voltage in the other half of the cycle. Full-wave rectifiers accomplish the task by passing the voltage in half of each cycle and inverting the voltage in the other half-cycle.

- The bridge-rectifier circuit is the preferred full-wave rectifier configuration.
- The variation of the output waveform of the rectifier is reduced considerably by connecting a capacitor C across the output load resistance R. The resulting circuit is the peak rectifier. The output waveform then consists of a dc voltage almost equal to the peak of the input sine wave, V_p, on which is superimposed a ripple component of frequency $2f$ (in the full wave case) and of peak-to-peak amplitude $V_r = V_p/2fCR$. To reduce this ripple voltage further a voltage regulator is employed.
- Combination of diodes, resistors, and possibly reference voltages can be used to design voltage limiters that prevent one or both extremities of the output waveform from going beyond predetermined values, the limiting level(s).
- Applying a time-varying waveform to a circuit consisting of a capacitor in series with a diode and taking the output across the diode provides a clamping function. Specifically, depending on the polarity of the diode either the positive or negative peaks of the signal will be clamped to the voltage at the other terminal of the diode (usually ground). In this way the output waveform has a non zero average or dc component and the circuit is known as a dc restorer.
- By cascading a clamping circuit with a peak-rectifier circuit, a voltage doubler is realized.

PROBLEMS

Computer Simulation Problems

SIM Problems identified by this icon are intended to demonstrate the value of using SPICE simulation to verify hand analysis and design, and to investigate important issues such as allowable signal swing and nonlinear distortion. Instructions to assist in setting up PSpice and Multisim simulations for all the indicated problems can be found in the corresponding files on the disc. Note that if a particular parameter value is not specified in the problem statement, you are to make a reasonable assumption.
* difficult problem; ** more difficult; *** very challenging and/or time-consuming; D: design problem.

Section 4.1: The Ideal Diode

4.1 An AA flashlight cell, whose Thévenin equivalent is a voltage source of 1.5 V and a resistance of 1 Ω, is connected to the terminals of an ideal diode. Describe two possible situations that result. What are the diode current and terminal voltage when (a) the connection is between the diode cathode and the positive terminal of the battery and (b) the anode and the positive terminal are connected?

4.2 For the circuits shown in Fig. P4.2 using ideal diodes, find the values of the voltages and currents indicated.

4.3 For the circuits shown in Fig. P4.3 using ideal diodes, find the values of the labeled voltages and currents

4.4 In each of the ideal-diode circuits shown in Fig. P4.4, v_I is a 1-kHz, 10-V peak sine wave. Sketch the waveform resulting at v_O. What are its positive and negative peak values?

4.5 The circuit shown in Fig. P4.5 is a model for a battery charger. Here v_I is a 10-V peak sine wave, D_1 and D_2 are

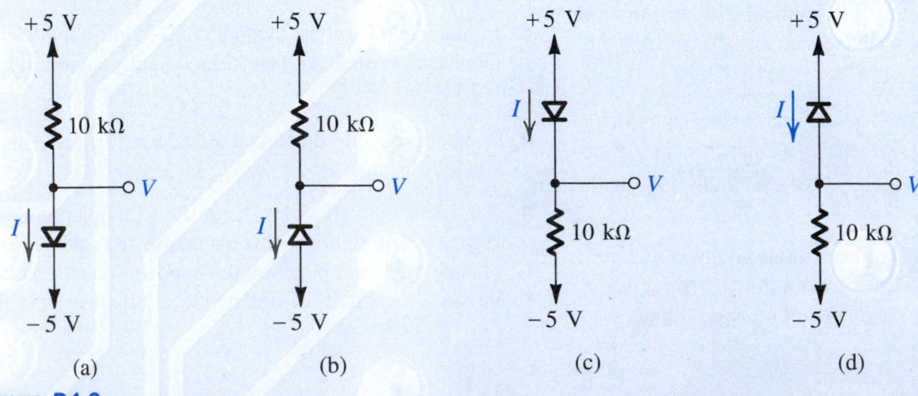

Figure P4.2

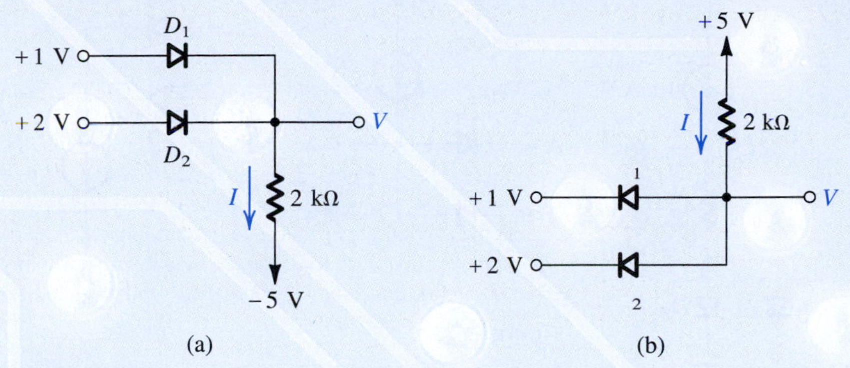

Figure P4.3

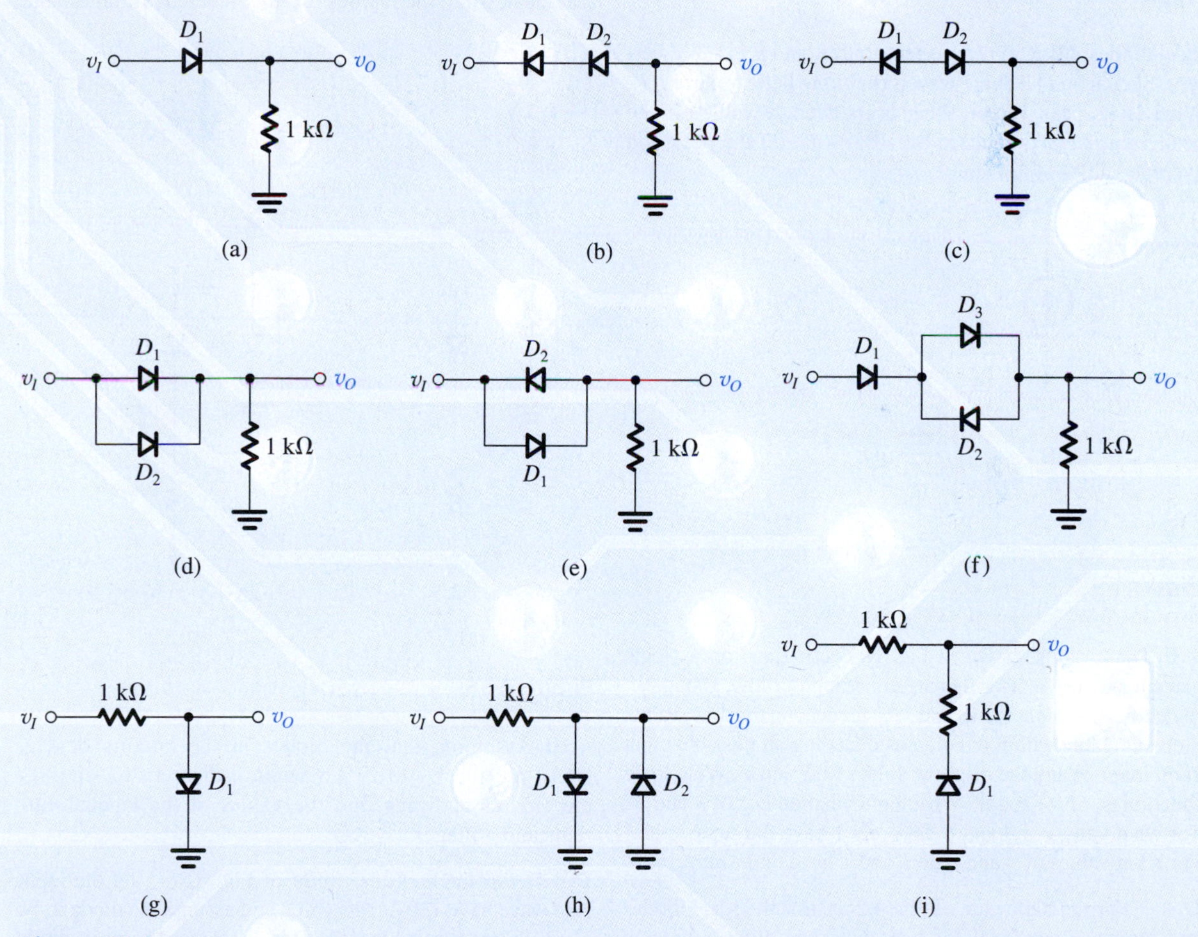

Figure P4.4

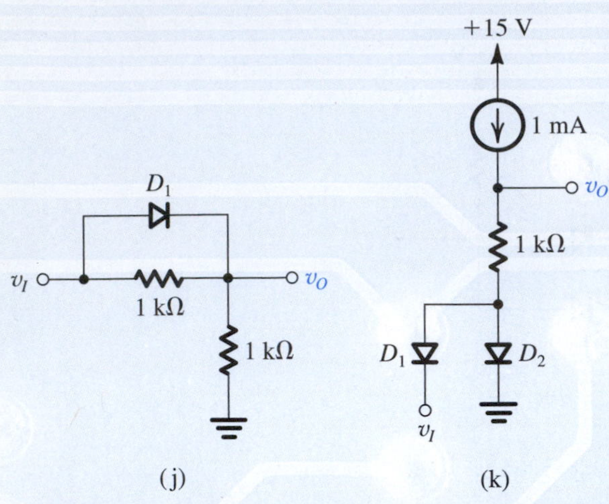

Figure P4.4 *(Contd.)*

ideal diodes, I is a 60-mA current source, and B is a 3-V battery. Sketch and label the waveform of the battery current i_B. What is its peak value? What is its average value? If the peak value of v_I is reduced by 10%, what do the peak and average values of i_B become?

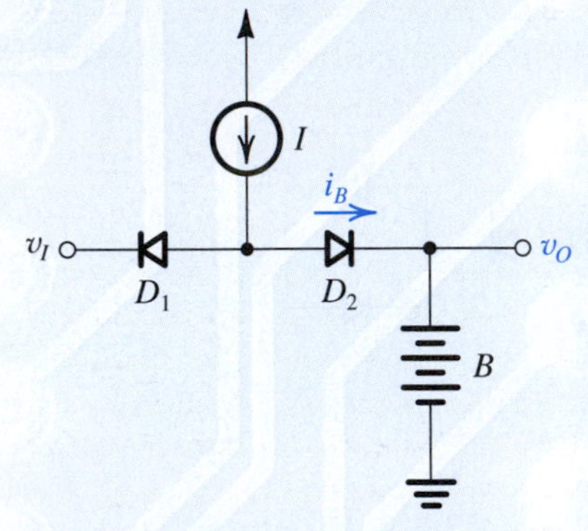

Figure P4.5

4.6 The circuits shown in Fig. P4.6 can function as logic gates for input voltages that are either high or low. Using "1" to denote the high value and "0" to denote the low value, prepare a table with four columns including all possible input combinations and the resulting values of X and Y. What logic function is X of A and B? What logic function is Y of A and B? For what values of A and B do X and Y have the same value? For what values of A and B do X and Y have opposite values?

D 4.7 For the logic gate of Fig. 4.5(a), assume ideal diodes and input voltage levels of 0 V and +5 V. Find a suitable value for R so that the current required from each of the input signal sources does not exceed 0.2 mA.

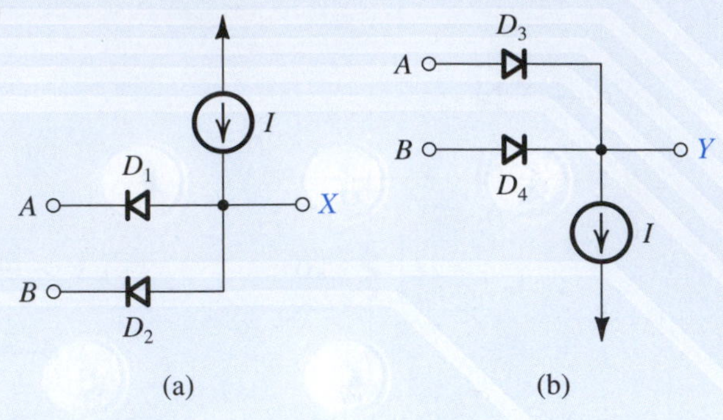

Figure P4.6

D 4.8 Repeat Problem 4.7 for the logic gate of Fig. 4.5(b).

4.9 Assuming that the diodes in the circuits of Fig. P4.9 are ideal, find the values of the labeled voltages and currents.

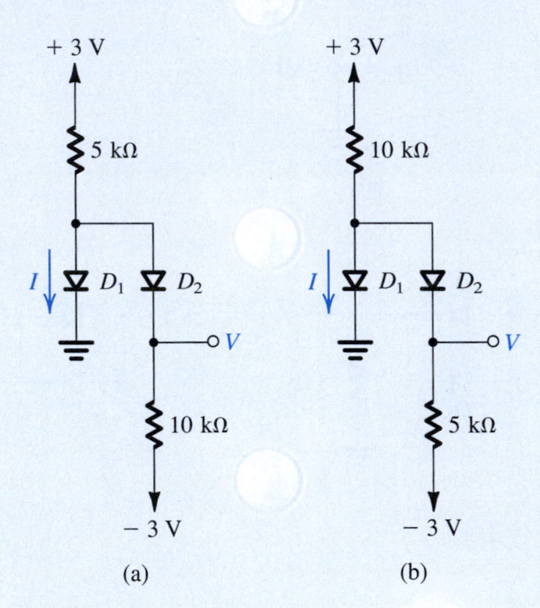

Figure P4.9

4.10 Assuming that the diodes in the circuits of Fig. P4.10 are ideal, utilize Thévenin's theorem to simplify the circuits and thus find the values of the labeled currents and voltages.

D 4.11 For the rectifier circuit of Fig. 4.3(a), let the input sine wave have 120-V rms value and assume the diode to be ideal. Select a suitable value for R so that the peak diode current does not exceed 50 mA. What is the greatest reverse voltage that will appear across the diode?

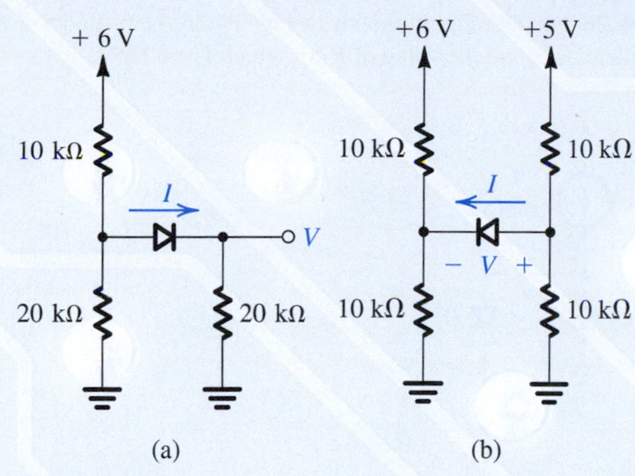

Figure P4.10

4.12 Consider the rectifier circuit of Fig. 4.3 in the event that the input source v_I has a source resistance R_s. For the case $R_s = R$ and assuming the diode to be ideal, sketch and clearly label the transfer characteristic v_O versus v_I.

4.13 A symmetrical square wave of 4-V peak-to-peak amplitude and zero average is applied to a circuit resembling that in Fig. 4.3(a) and employing a 100-Ω resistor. What is the peak output voltage that results? What is the average output voltage that results? What is the peak diode current? What is the average diode current? What is the maximum reverse voltage across the diode?

4.14 Repeat Problem 4.13 for the situation in which the average voltage of the square wave is 1 V, while its peak-to-peak value remains at 4 V.

D *4.15 Design a battery-charging circuit, resembling that in Fig. 4.4 and using an ideal diode, in which current flows to the 12-V battery 20% of the time with an average value of 100 mA. What peak-to-peak sine-wave voltage is required? What resistance is required? What peak diode current flows? What peak reverse voltage does the diode endure? If resistors can be specified to only one significant digit, and the peak-to-peak voltage only to the nearest volt, what design would you choose to guarantee the required charging current? What fraction of the cycle does diode current flow? What is the average diode current? What is the peak diode current? What peak reverse voltage does the diode endure?

4.16 The circuit of Fig. P4.16 can be used in a signalling system using one wire plus a common ground return. At any moment, the input has one of three values: +3 V, 0 V, –3 V. What is the status of the lamps for each input value? (Note that the lamps can be located apart from each other and that there may be several of each type of connection, all on one wire!)

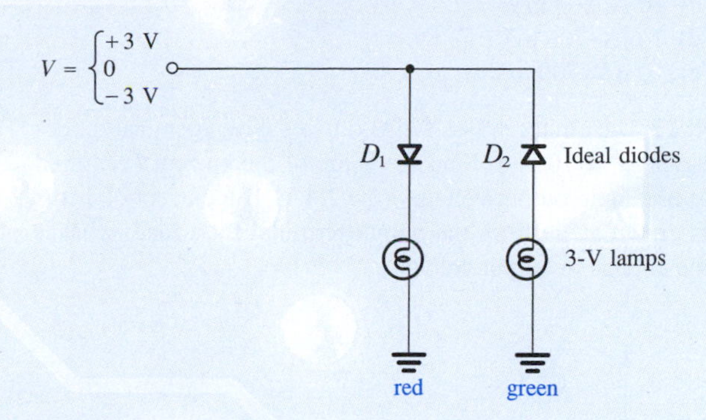

Figure P4.16

Section 4.2: Terminal Characteristics of Junction Diodes

4.17 Calculate the value of the thermal voltage, V_T, at –40°C, 0°C, +40°C, and +150°C. At what temperature is V_T exactly 25 mV?

4.18 At what forward voltage does a diode conduct a current equal to $1000 I_S$? In terms of I_S, what current flows in the same diode when its forward voltage is 0.7 V?

4.19 A diode for which the forward voltage drop is 0.7 V at 1.0 mA is operated at 0.5 V. What is the value of the current?

4.20 A particular diode is found to conduct 0.5 mA with a junction voltage of 0.7 V. What is its saturation current I_S? What current will flow in this diode if the junction voltage is raised to 0.71 V? To 0.8 V? If the junction voltage is lowered to 0.69 V? To 0.6 V? What change in junction voltage will increase the diode current by a factor of 10?

4.21 The following measurements are taken on particular junction diodes for which V is the terminal voltage and I is the diode current. For each diode, estimate values of I_S and the terminal voltage at 10% of the measured current.

(a) $V = 0.700$ V at $I = 1.00$ A
(b) $V = 0.650$ V at $I = 1.00$ mA
(c) $V = 0.650$ V at $I = 10$ μA
(d) $V = 0.700$ V at $I = 10$ mA

4.22 Listed below are the results of measurements taken on several different junction diodes. For each diode, the data provided are the diode current I and the corresponding diode voltage V. In each case, estimate I_S, and the diode voltage at $10I$ and $I/10$.

(a) 10.0 mA, 700 mV
(b) 1.0 mA, 700 mV
(c) 10 A, 800 mV
(d) 1 mA, 700 mV
(e) 10 μA, 700 mV

4.23 The circuit in Fig. P4.23 utilizes three identical diodes having $I_S = 10^{-16}$ A. Find the value of the current I required to obtain an output voltage $V_O = 2.4$ V. If a current of 1 mA is drawn away from the output terminal by a load, what is the change in output voltage?

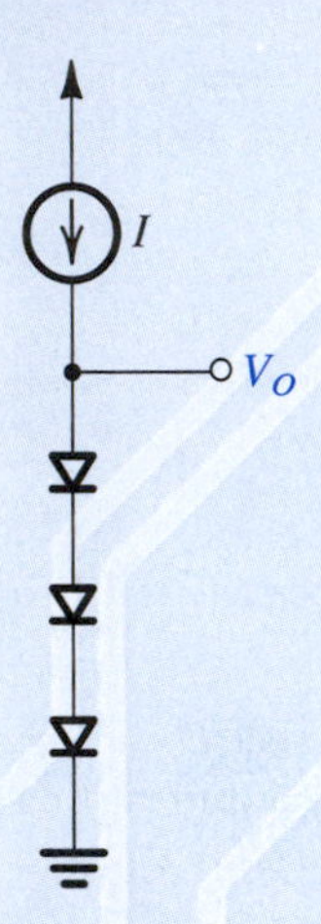

Figure P4.23

4.24 A junction diode is operated in a circuit in which it is supplied with a constant current I. What is the effect on the forward voltage of the diode if an identical diode is connected in parallel?

4.25 In the circuit shown in Fig. P4.25, D_1 has 10 times the junction area of D_2. What value of V results? To obtain a value for V of 50 mV, what current I_2 is needed?

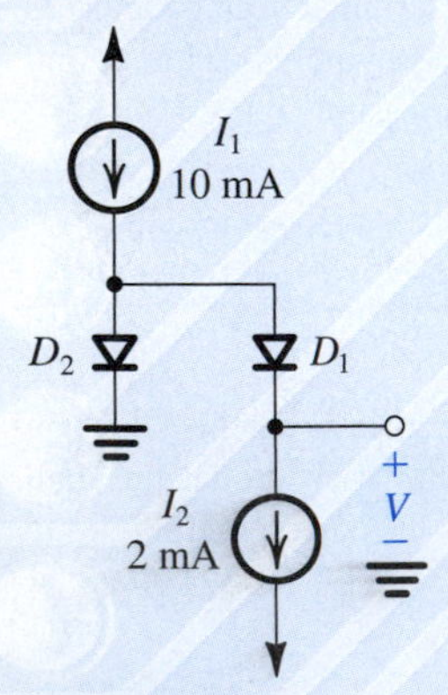

Figure P4.25

4.26 For the circuit shown in Fig. P4.26, both diodes are identical. Find the value of R for which $V = 80$ mV.

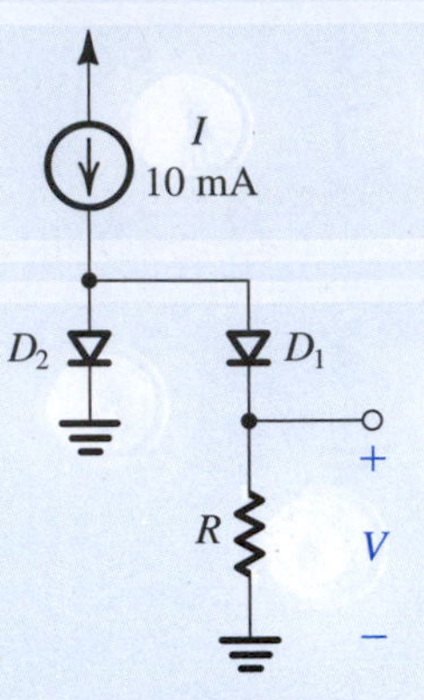

Figure P4.26

4.27 A diode fed with a constant current $I = 1$ mA has a voltage $V = 690$ mV at 20° C. Find the diode voltage at –20° C and at +70° C.

4.28 In the circuit shown in Fig. P4.28, D_1 is a large-area, high-current diode whose reverse leakage is high and independent of applied voltage, while D_2 is a much smaller, low-current diode. At an ambient temperature of 20°C, resistor R_1 is adjusted to make $V_{R1} = V_2 = 520$ mV. Subsequent measurement indicates that R_1 is 520 kΩ. What do you expect the voltages V_{R1} and V_2 to become at 0°C and at 40°C?

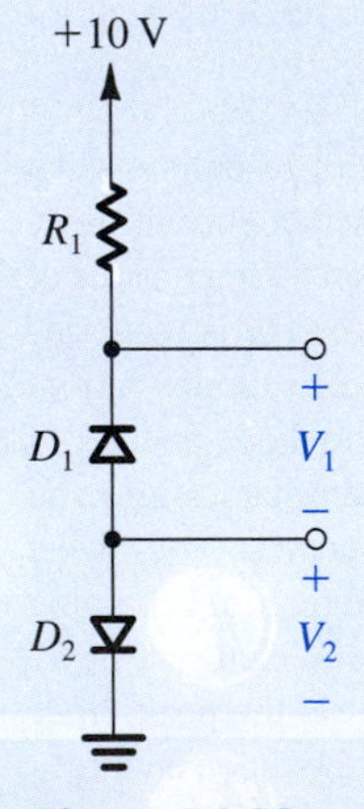

Figure P4.28

4.29 When a 15-A current is applied to a particular diode, it is found that the junction voltage immediately becomes 700 mV. However, as the power being dissipated in the diode raises its temperature, it is found that the voltage decreases and eventually reaches 600 mV. What is the apparent rise in junction temperature? What is the power dissipated in the diode in its final state? What is the temperature rise per watt of power dissipation? (This is called the thermal resistance.)

***4.30** A designer of an instrument that must operate over a wide supply-voltage range, noting that a diode's junction-voltage drop is relatively independent of junction current, considers the use of a large diode to establish a small relatively constant voltage. A power diode, for which the nominal current at 0.8 V is 10 A, is available. If the current source feeding the diode changes in rhe range 0.5 mA to 1.5 mA and if, in addition, the temperature changes by ±25°C, what is the expected range of diode voltage?

***4.31** As an alternative to the idea suggested in Problem 4.30, the designer considers a second approach to producing a relatively constant small voltage from a variable current supply: It relies on the ability to make quite accurate copies of any small current that is available (using a process called current mirroring). The designer proposes to use this idea to supply two diodes of different junction areas with the same current and to measure their junction-voltage difference. Two types of diodes are available; for a forward voltage of 700 mV, one conducts 0.1 mA, while the other conducts 1 A. Now, for identical currents in the range of 0.5 mA to 1.5 mA supplied to each, what range of difference voltages result? What is the effect of a temperature change of ±25°C on this arrangement?

Section 4.3: Modeling the Diode Forward Characteristic

***4.32** Consider the graphical analysis of the diode circuit of Fig. 4.10 with $V_{DD} = 1$ V, $R = 1$ kΩ, and a diode having $I_S = 10^{-15}$ A. Calculate a small number of points on the diode characteristic in the vicinity of where you expect the load line to intersect it, and use a graphical process to refine your estimate of diode current. What value of diode current and voltage do you find? Analytically, find the voltage corresponding to your estimate of current. By how much does it differ from the graphically estimated value?

4.33 Use the iterative-analysis procedure to determine the diode current and voltage in the circuit of Fig. 4.10 for $V_{DD} = 1$ V, $R = 1$ kΩ, and a diode having $I_S = 10^{-15}$ A.

4.34 A "1-mA diode" (i.e., one that has $v_D = 0.7$ V at $i_D = 1$ mA) is connected in series with a 200-Ω resistor to a 1.0-V supply.

(a) Provide a rough estimate of the diode current you would expect.

(b) Estimate the diode current more closely using iterative analysis.

D 4.35 Assuming the availability of diodes for which $v_D = 0.7$ V at $i_D = 1$ mA, design a circuit that utilizes four diodes connected in series, in series with a resistor R connected to a 10-V power supply. The voltage across the string of diodes is to be 3.0 V.

4.36 A diode operates in a series circuit with R and V. A designer, considering using a constant-voltage model, is uncertain whether to use 0.7 V or 0.6 V for V_D. For what value of V is the difference in the calculated values of current only 1%? For $V = 2$ V and $R = 1$ kΩ, what two currents would result from the use of the two values of V_D? What is their percentage difference?

4.37 A designer has a supply of diodes for which a current of 2 mA flows at 0.7 V. Using a 1-mA current source, the designer wishes to create a reference voltage of 1.25 V. Suggest a combination of series and parallel diodes that will do the job as well as possible. How many diodes are needed? What voltage is actually achieved?

4.38 Solve the problems in Example 4.2 using the constant-voltage-drop ($V_D = 0.7$ V) diode model.

4.39 For the circuits shown in Fig. P4.2, using the constant-voltage-drop ($V_D = 0.7$ V) diode model, find the voltages and currents indicated.

4.40 For the circuits shown in Fig. P4.3, using the constant-voltage-drop ($V_D = 0.7$ V) diode model, find the voltages and currents indicated.

4.41 For the circuits in Fig. P4.9, using the constant-voltage-drop ($V_D = 0.7$ V) diode model, find the values of the labeled currents and voltages.

4.42 For the circuits in Fig. P4.10, utilize Thévenin's theorem to simplify the circuits and find the values of the labeled currents and voltages. Assume that conducting diodes can be represented by the constant-voltage-drop model ($V_D = 0.7$ V).

D 4.43 Repeat Problem 4.11, representing the diode by the constant-voltage-drop ($V_D = 0.7$ V) model. How different is the resulting design?

4.44 The small-signal model is said to be valid for voltage variations of about 5 mV. To what percentage current change does this correspond? (Consider both positive and negative signals.) What is the maximum allowable voltage signal (positive or negative) if the current change is to be limited to 10%?

4.45 In a particular circuit application, ten "20-mA diodes" (a 20-mA diode is a diode that provides a 0.7-V drop when the current through it is 20 mA) connected in parallel operate at a total current of 0.1 A. For the diodes closely matched, what current flows in each? What is the corresponding small-signal resistance of each diode and of the combination? Compare this with the incremental resistance of a single diode conducting 0.1 A. If each of the 20-mA diodes has a series resistance of 0.2 Ω associated with the wire bonds to the junction, what is the equivalent resistance of the 10 parallel-connected diodes? What connection resistance would a single diode need in order to be totally equivalent? (*Note:* This is why the parallel connection of real diodes can often be used to advantage.)

4.46 In the circuit shown in Fig. P4.46, I is a dc current and v_s is a sinusoidal signal. Capacitors C_1 and C_2 are very large; their function is to couple the signal to and from the diode but block the dc current from flowing into the signal source or the load (not shown). Use the diode small-signal model to show that the signal component of the output voltage is

$$v_o = v_s \frac{V_T}{V_T + IR_s}$$

If $v_s = 10$ mV, find v_o for $I = 1$ mA, 0.1 mA, and 1 μA. Let $R_s = 1$ kΩ. At what value of I does v_o become one-half of v_s? Note that this circuit functions as a signal attenuator with the attenuation factor controlled by the value of the dc current I.

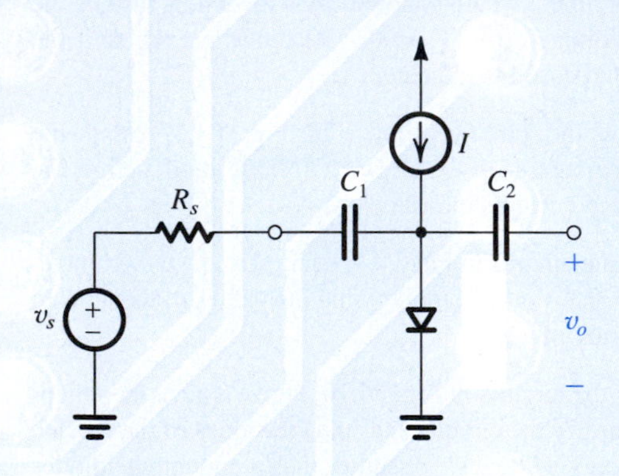

Figure P4.46

4.47 In the attenuator circuit of Fig. P4.46, let $R_s = 10$ kΩ. The diode is a 1-mA device; that is, it exhibits a voltage drop of 0.7 V at a dc current of 1 mA. For small input signals, what value of current I is needed for $v_o/v_s = 0.50$? 0.10? 0.01? 0.001? In each case, what is the largest input signal that can be used while ensuring that the signal component of the diode current is limited to ±10% of its dc current? What output signals correspond?

4.48 In the capacitor-coupled attenuator circuit shown in Fig. P4.48, I is a dc current that varies from 0 mA to 1 mA, and C_1 and C_2 are large coupling capacitors. For very small input signals, so that the diodes can be representedby their small-signal resistances r_{d1} and r_{d2}, show that $\frac{v_o}{v_i} = \frac{r_{d2}}{r_{d1} + r_{d2}}$ and hence that $\frac{v_o}{v_i} = I$, where I is in mA. Find v_o/v_i for $I = 0$ μA, 1 μA, 10 μA, 100 μA, 500 μA, 600 μA, 900 μA, 990 μA, and 1 mA.

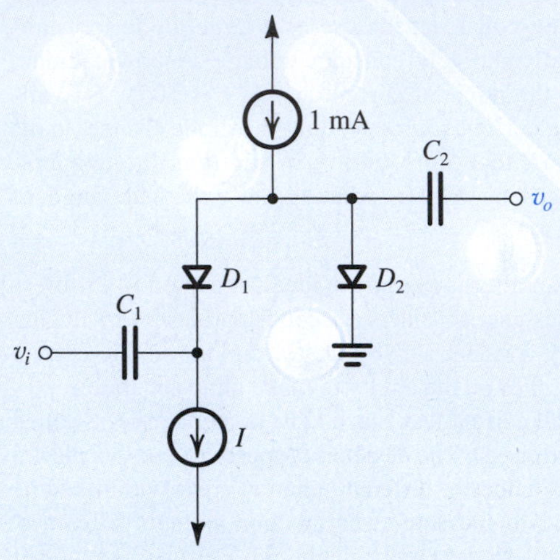

Figure P4.48

***4.49** In the circuit shown in Fig. P4.49, diodes D_1 through D_4 are identical and each exhibits a voltage drop of 0.7 V at a 1-mA current.

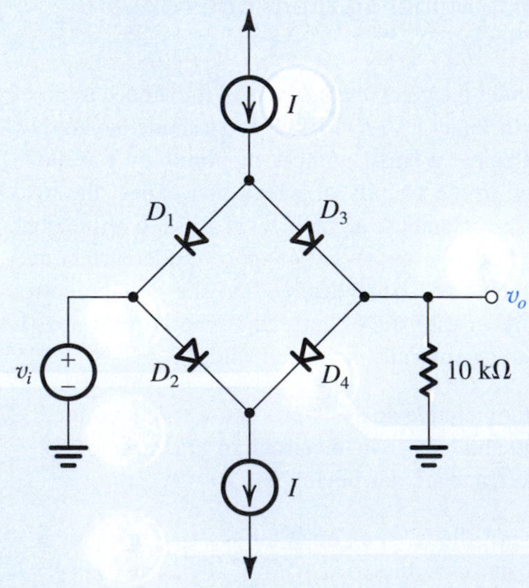

Figure P4.49

(a) For small input signals (e.g., 10 mV peak), find values of the small-signal transmission v_o/v_i for various values of I: 0 μA, 1 μA, 10 μA, 100 μA, 1 mA, and 10 mA.
(b) For a forward-conducting diode, what is the largest signal-voltage magnitude that it can support while the corresponding signal current is limited to 10% of the dc bias current. Now, for the circuit in Fig. P4.49, for 10-mV peak input, what is the smallest value of I for which the diode currents remain within ±10% of their dc value?
(c) For $I = 1$ mA, what is the largest possible output signal for which the diode currents deviate by at most 10% of their dc values? What is the corresponding peak input? What is the total current in each diode?

****4.50** In Problem 4.49 we investigated the operation of the circuit in Fig. P4.49 for small input signals. In this problem we wish to find the voltage transfer characteristic (VTC) v_O versus v_I for $-12\text{ V} \le v_I \le 12\text{ V}$ for the case $I = 1$ mA and each of the diodes exhibits a voltage drop of 0.7 V at a current of 1 mA. Toward this end, use the diode exponential characteristic to construct a table that gives the values of: the current i_O in the 10-kΩ resistor, the current in each of the four diodes, the voltage drop across each of the four diodes, and the input voltage v_I, for $v_O = 0$, +1 V, +2 V, +5 V, +9 V, +9.9 V, +9.99 V, +10.5 V, +11 V, and +12 V. Use these data, with extrapolation to negative values of v_I and v_O, to sketch the required VTC. Also sketch the VTC that results if I is reduced to 0.5 mA.

SIM ***4.51** In the circuit shown in Fig. P4.51, I is a dc current and v_i is a sinusoidal signal with small amplitude (less than 10 mV) and a frequency of 100 kHz. Representing the diode by its small-signal resistance r_d, which is a function of I, sketch the circuit for determining the sinusoidal output voltage V_o, and thus find the phase shift between V_i and V_o. Find the value of I that will provide a phase shift of −45°, and find the range of phase shift achieved as I is varied over the range of 0.1 times to 10 times this value.

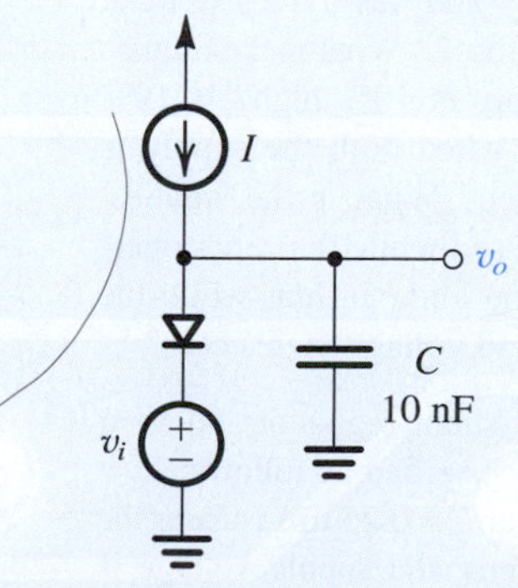

Figure P4.51

***4.52** Consider the voltage-regulator circuit shown in Fig. P4.52. The value of R is selected to obtain an output voltage V_O (across the diode) of 0.7 V.

(a) Use the diode small-signal model to show that the change in output voltage corresponding to a change of 1 V in V^+ is

$$\frac{\Delta V_O}{\Delta V^+} = \frac{V_T}{V^+ + V_T - 0.7}$$

This quantity is known as the line regulation and is usually expressed in mV/V.
(b) Generalize the expression above for the case of m diodes connected in series and the value of R adjusted so that the voltage across each diode is 0.7 V (and $V_O = 0.7m$ V).
(c) Calculate the value of line regulation for the case $V^+ = 10$ V (nominally) and (i) $m = 1$ and (ii) $m = 3$.

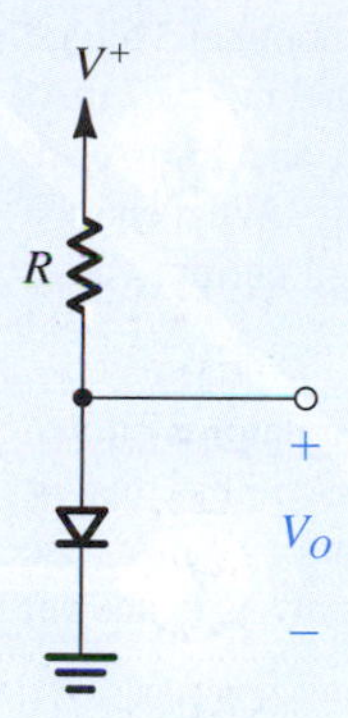

Figure P4.52

***4.53** Consider the voltage-regulator circuit shown in Fig P4.52 under the condition that a load current I_L is drawn from the output terminal.

(a) If the value of I_L is sufficiently small that the corresponding change in regulator output voltage ΔV_O is small enough to justify using the diode small-signal model, show that

$$\frac{\Delta V_O}{I_L} = -(r_d \parallel R)$$

This quantity is known as the load regulation and is usually expressed in mV/mA.
(b) If the value of R is selected such that at no load the voltage across the diode is 0.7 V and the diode current is I_D, show that the expression derived in (a) becomes

$$\frac{\Delta V_O}{I_L} = -\frac{V_T}{I_D}\,\frac{V^+ - 0.7}{V^+ - 0.7 + V_T}$$

Select the lowest possible value for I_D that results in a load regulation ≤ 5 mV/mA. If V^+ is nominally 10 V, what value

of R is required? Also, specify the diode required in terms of its I_S.
(c) Generalize the expression derived in (b) for the case of m diodes connected in series and R adjusted to obtain $V_O = 0.7m$ V at no load.

***4.54** Design a diode voltage regulator to supply 1.5 V to a 150-Ω load. Use two diodes specified to have a 0.7-V drop at a current of 10 mA. The diodes are to be connected to a +5-V supply through a resistor R. Specify the value for R. What is the diode current with the load connected? What is the increase resulting in the output voltage when the load is disconnected? What change results if the load resistance is reduced to 100 Ω? To 75 Ω? To 50 Ω? (*Hint*: Use the small-signal diode model to calculate all changes in ouput voltage.)

***4.55** A voltage regulator consisting of two diodes in series fed with a constant-current source is used as a replacement for a single carbon-zinc cell (battery) of nominal voltage 1.5 V. The regulator load current varies from 2 mA to 7 mA. Constant-current supplies of 5 mA, 10 mA, and 15 mA are available. Which would you choose, and why? What change in output voltage would result when the load current varies over its full range?

****4.56** A particular design of a voltage regulator is shown in Fig. P4.56. Diodes D_1 and D_2 are 10-mA units; that is, each has a voltage drop of 0.7 V at a current of 10 mA. Use the diode exponential model and iterative analysis to answer the following questions:

(a) What is the regulator output voltage V_O with the 150-Ω load connected?
(b) Find V_O with no load.
(c) With the load connected, to what value can the 5-V supply be lowered while maintaining the loaded output voltage within 0.1 V of its nominal value?
(d) What does the loaded output voltage become when the 5-V supply is raised by the same amount as the drop found in (c)?

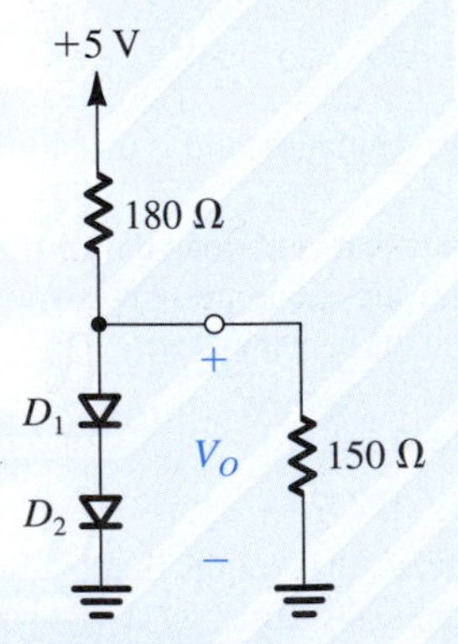

Figure P4.56

(e) For the range of changes explored in (c) and (d), by what percentage does the output voltage change for each percentage change of supply voltage in the worst case?

Section 4.4: Operation in the Reverse Breakdown Region—Zener Diodes

4.57 Partial specifications of a collection of zener diodes are provided below. For each, identify the missing parameter, and estimate its value. Note from Fig. 4.17 that $V_{ZK} \simeq V_{Z0}$.

(a) $V_Z = 10.0$ V, $V_{ZK} = 9.6$ V, and $I_{ZT} = 50$ mA
(b) $I_{ZT} = 10$ mA, $V_Z = 9.1$ V, and $r_z = 30$ Ω
(c) $r_z = 2$ Ω, $V_Z = 6.8$ V, and $V_{ZK} = 6.6$ V
(d) $V_Z = 18$ V, $I_{ZT} = 5$ mA, and $V_{ZK} = 17.6$ V
(e) $I_{ZT} = 200$ mA, $V_Z = 7.5$ V, and $r_z = 1.5$ Ω

Assuming that the power rating of a breakdown diode is established at about twice the specified zener current (I_{ZT}), what is the power rating of each of the diodes described above?

D 4.58 A designer requires a shunt regulator of approximately 20 V. Two kinds of zener diodes are available: 6.8-V devices with r_z of 10 Ω and 5.1-V devices with r_z of 30 Ω. For the two major choices possible, find the load regulation. In this calculation neglect the effect of the regulator resistance R.

4.59 A shunt regulator utilizing a zener diode with an incremental resistance of 5 Ω is fed through an 82-Ω resistor. If the raw supply changes by 1.0 V, what is the corresponding change in the regulated output voltage?

4.60 A 9.1-V zener diode exhibits its nominal voltage at a test current of 28 mA. At this current the incremental resistance is specified as 5 Ω. Find V_{Z0} of the zener model. Find the zener voltage at a current of 10 mA and at 100 mA.

D 4.61 Design a 7.5-V zener regulator circuit using a 7.5-V zener specified at 12 mA. The zener has an incremental resistance $r_z = 30$ Ω and a knee current of 0.5 mA. The regulator operates from a 10-V supply and has a 1.2-kΩ load. What is the value of R you have chosen? What is the regulator output voltage when the supply is 10% high? Is 10% low? What is the output voltage when both the supply is 10% high and the load is removed? What is the smallest possible load resistor that can be used while the zener operates at a current no lower than the knee current while the supply is 10% low? What is the load voltage in this case?

***D 4.62** Provide two designs of shunt regulators utilizing the 1N5235 zener diode, which is specified as follows: $V_Z = 6.8$ V and $r_z = 5$ Ω for $I_Z = 20$ mA; at $I_Z = 0.25$ mA (nearer the knee), $r_z = 750$ Ω. For both designs, the supply voltage is nominally 9 V and varies by ±1 V. For the first design, assume that the availability of supply current is not a problem, and thus operate the diode at 20 mA. For the second design, assume that the current from the raw supply is limited, and therefore you are forced to operate the diode at 0.25 mA. For the purpose of these initial designs, assume no load. For each design find the value of R and the line regulation.

D *4.63 A zener shunt regulator employs a 9.1-V zener diode for which $V_Z = 9.1$ V at $I_Z = 9$ mA, with $r_z = 30\ \Omega$ and $I_{ZK} = 0.3$ mA. The available supply voltage of 15 V can vary as much as ±10%. For this diode, what is the value of V_{Z0}? For a nominal load resistance R_L of 1 kΩ and a nominal zener current of 10 mA, what current must flow in the supply resistor R? For the nominal value of supply voltage, select a value for resistor R, specified to one significant digit, to provide at least that current. What nominal output voltage results? For a ±10% change in the supply voltage, what variation in output voltage results? If the load current is reduced by 50%, what increase in V_O results? What is the smallest value of load resistance that can be tolerated while maintaining regulation when the supply voltage is low? What is the lowest possible output voltage that results? Calculate values for the line regulation and for the load regulation for this circuit using the numerical results obtained in this problem.

D *4.64 It is required to design a zener shunt regulator to provide a regulated voltage of about 10 V. The available 10-V, 1-W zener of type 1N4740 is specified to have a 10-V drop at a test current of 25 mA. At this current, its r_z is 7 Ω. The raw supply, V_S, available has a nominal value of 20 V but can vary by as much as ±25%. The regulator is required to supply a load current of 0 mA to 20 mA. Design for a minimum zener current of 5 mA.

(a) Find V_{Z0}.
(b) Calculate the required value of R.
(c) Find the line regulation. What is the change in V_O expressed as a percentage, corresponding to the ±25% change in V_S?
(d) Find the load regulation. By what percentage does V_O change from the no-load to the full-load condition?
(e) What is the maximum current that the zener in your design is required to conduct? What is the zener power dissipation under this condition?

Section 4.5: Rectifier Circuits

4.65 Consider the half-wave rectifier circuit of Fig. 4.21(a) with the diode reversed. Let v_S be a sinusoid with 12-V peak amplitude, and let $R = 1.5$ kΩ. Use the constant-voltage-drop diode model with $V_D = 0.7$ V.

(a) Sketch the transfer characteristic.
(b) Sketch the waveform of v_O.
(c) Find the average value of v_O.
(d) Find the peak current in the diode.
(e) Find the PIV of the diode.

4.66 Using the exponential diode characteristic, show that for v_S and v_O both greater than zero, the circuit of Fig. 4.21(a) has the transfer characteristic

$$v_O = v_S - v_D\ (\text{at } i_D = 1\text{ mA}) - V_T \ln (v_O/R)$$

where v_S and v_O are in volts and R is in kilohms. Note that this relationship can be used to obtain the voltage transfer characteristic v_O vs v_S by finding v_S corresponding to various values of v_O.

SIM 4.67 Consider a half-wave rectifier circuit with a triangular-wave input of 5-V peak-to-peak amplitude and zero average, and with $R = 1$ kΩ. Assume that the diode can be represented by the constant-voltage-drop model with $V_D =$ 0.7 V. Find the average value of v_O.

4.68 A half-wave rectifier circuit with a 1-kΩ load operates from a 120-V (rms) 60-Hz household supply through a 10-to-1 step-down transformer. It uses a silicon diode that can be modeled to have a 0.7-V drop for any current. What is the peak voltage of the rectified output? For what fraction of the cycle does the diode conduct? What is the average output voltage? What is the average current in the load?

4.69 A full-wave rectifier circuit with a 1-kΩ load operates from a 120-V (rms) 60-Hz household supply through a 5-to-1 transformer having a center-tapped secondary winding. It uses two silicon diodes that can be modeled to have a 0.7-V drop for all currents. What is the peak voltage of the rectified output? For what fraction of a cycle does each diode conduct? What is the average output voltage? What is the average current in the load?

4.70 A full-wave bridge rectifier circuit with a 1-kΩ load operates from a 120-V (rms) 60-Hz household supply through a 10-to-1 step-down transformer having a single secondary winding. It uses four diodes, each of which can be modeled to have a 0.7-V drop for any current. What is the peak value of the rectified voltage across the load? For what fraction of a cycle does each diode conduct? What is the average voltage across the load? What is the average current through the load?

4.71 It is required to design a full-wave rectifier circuit using the circuit of Fig. 4.22 to provide an average output voltage of:

(a) 10 V
(b) 100 V

In each case find the required turns ratio of the transformer. Assume that a conducting diode has a voltage drop of 0.7 V. The ac line voltage is 120 V rms.

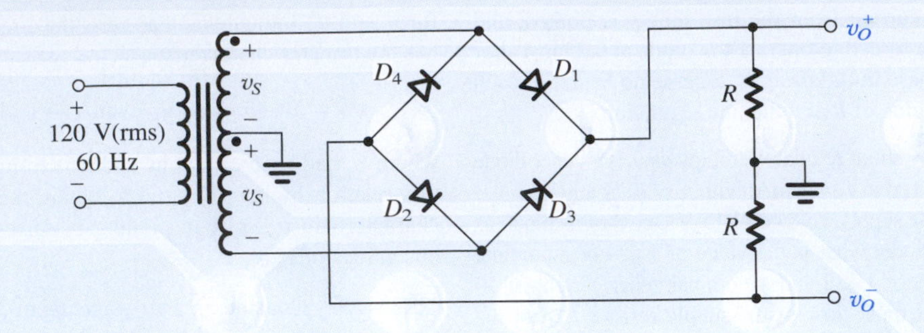

Figure P4.74

4.72 Repeat Problem 4.71 for the bridge rectifier circuit of Fig. 4.23.

D 4.73 Consider the full-wave rectifier in Fig. 4.22 when the transformer turns ratio is such that the voltage across the entire secondary winding is 24 V rms. If the input ac line voltage (120 V rms) fluctuates by as much as ±10%, find the required PIV of the diodes. (Remember to use a factor of safety in your design.)

4.74 The circuit in Fig. P4.74 implements a complementary-output rectifier. Sketch and clearly label the waveforms of v_O^+ and v_O^-. Assume a 0.7-V drop across each conducting diode. If the magnitude of the average of each output is to be 15 V, find the required amplitude of the sine wave across the entire secondary winding. What is the PIV of each diode?

4.75 Augment the rectifier circuit of Problem 4.68 with a capacitor chosen to provide a peak-to-peak ripple voltage of (i) 10% of the peak output and (ii) 1% of the peak output. In each case:

(a) What average output voltage results?
(b) What fraction of the cycle does the diode conduct?
(c) What is the average diode current?
(d) What is the peak diode current?

4.76 Repeat Problem 4.75 for the rectifier in Problem 4.69.

4.77 Repeat Problem 4.75 for the rectifier in Problem 4.70.

D *4.78 It is required to use a peak rectifier to design a dc power supply that provides an average dc output voltage of 15 V on which a maximum of ±1-V ripple is allowed. The rectifier feeds a load of 150 Ω. The rectifier is fed from the line voltage (120 V rms, 60 Hz) through a transformer. The diodes available have 0.7-V drop when conducting. If the designer opts for the half-wave circuit:

(a) Specify the rms voltage that must appear across the transformer secondary.
(b) Find the required value of the filter capacitor.
(c) Find the maximum reverse voltage that will appear across the diode, and specify the PIV rating of the diode.
(d) Calculate the average current through the diode during conduction.
(e) Calculate the peak diode current.

D *4.79 Repeat Problem 4.78 for the case in which the designer opts for a full-wave circuit utilizing a center-tapped transformer.

D *4.80 Repeat Problem 4.78 for the case in which the designer opts for a full-wave bridge rectifier circuit.

D *4.81 Consider a half-wave peak rectifier fed with a voltage v_S having a triangular waveform with 20-V peak-to-peak amplitude, zero average, and 1-kHz frequency. Assume that the diode has a 0.7-V drop when conducting. Let the load resistance $R = 100\ \Omega$ and the filter capacitor $C = 100\ \mu$F. Find the average dc output voltage, the time interval during which the diode conducts, the average diode current during conduction, and the maximum diode current.

D *4.82 Consider the circuit in Fig. P4.74 with two equal filter capacitors placed across the load resistors R. Assume that the diodes available exhibit a 0.7-V drop when conducting. Design the circuit to provide ±15-V dc output voltages with a peak-to-peak ripple no greater than 1 V. Each supply should be capable of providing 200 mA dc current to its load resistor R. Completely specify the capacitors, diodes and the transformer.

4.83 The op amp in the precision rectifier circuit of Fig. P4.83 is ideal with output saturation levels of ±12 V. Assume that when conducting the diode exhibits a constant voltage drop of 0.7 V. Find v_-, v_O, and v_A for:

(a) $v_I = +1$ V
(b) $v_I = +2$ V
(c) $v_I = -1$ V
(d) $v_I = -2$ V

Also, find the average output voltage obtained when v_I is a symmetrical square wave of 1-kHz frequency, 3-V amplitude, and zero average.

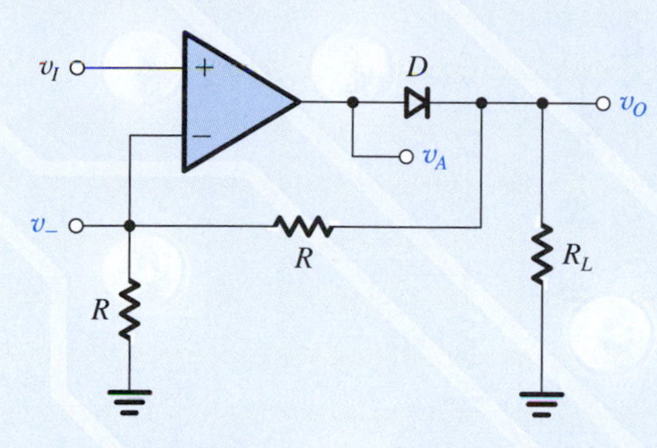

Figure P4.83

4.84 The op amp in the circuit of Fig. P4.84 is ideal with output saturation levels of ±12 V. The diodes exhibit a constant 0.7-V drop when conducting. Find v_-, v_A, and v_O for:

(a) $v_I = +1$ V
(b) $v_I = +2$ V
(c) $v_I = -1$ V
(d) $v_I = -2$ V

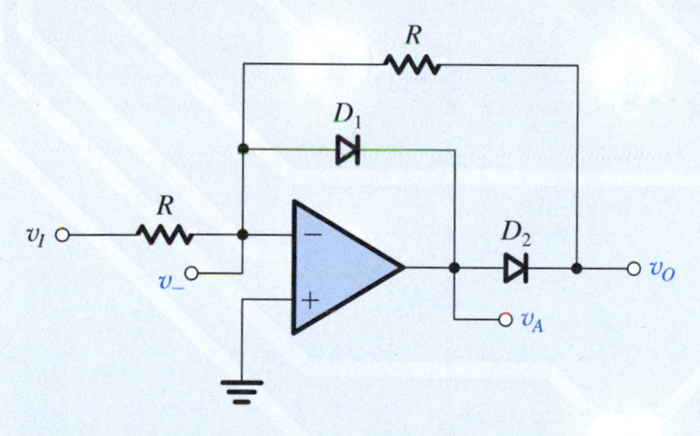

Figure P4.84

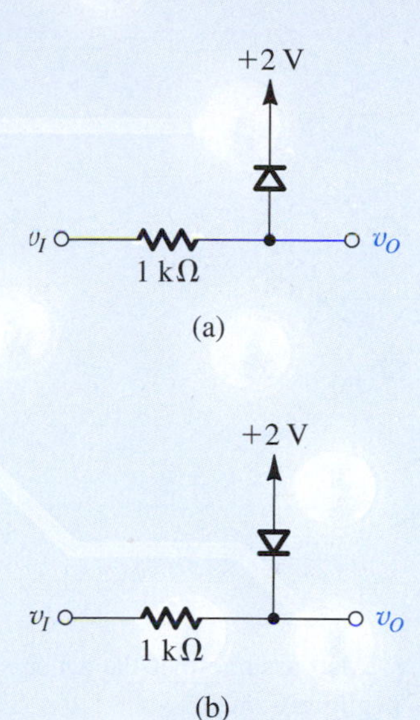

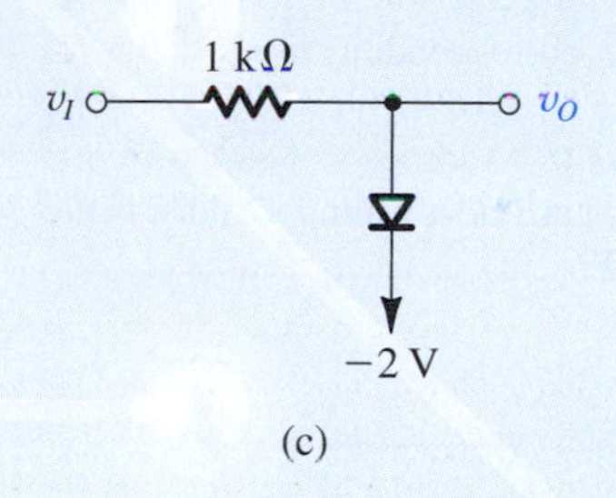

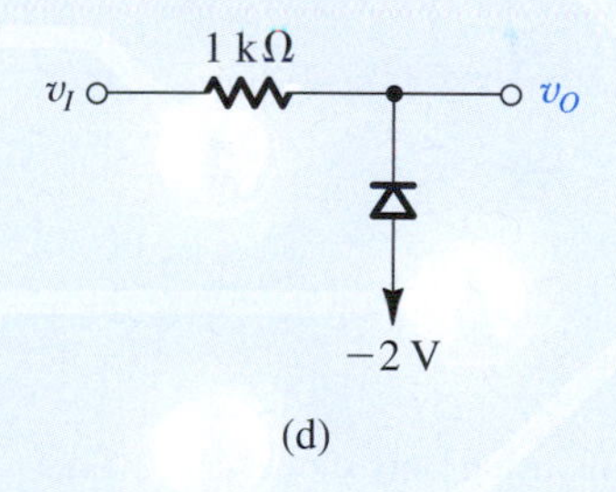

Figure P4.85

Section 4.6: Limiting and Clamping Circuits

4.85 Sketch the transfer characteristic v_O versus v_I for the limiter circuits shown in Fig. P4.85. All diodes begin conducting at a forward voltage drop of 0.5 V and have voltage drops of 0.7 V when conducting a current $i_D \geq 1$ mA.

4.86 The circuits in Fig. P4.85(a) and (d) are connected as follows: The two input terminals are tied together, and the two output terminals are tied together. Sketch the transfer characteristic of the circuit resulting, assuming that the cut-in voltage of the diodes is 0.5 V and their voltage drop when conducting a current $i_D \geq 1$ mA is 0.7 V.

4.87 Repeat Problem 4.86 for the two circuits in Fig. P4.85(a) and (b) connected together as follows: The two input terminals are tied together, and the two output terminals are tied together.

4.88 Sketch and clearly label the transfer characteristic of the circuit in Fig. P4.88 for $-20\text{ V} \leq v_I \leq +20$ V. Assume that the diodes can be represented by the constant-voltage-

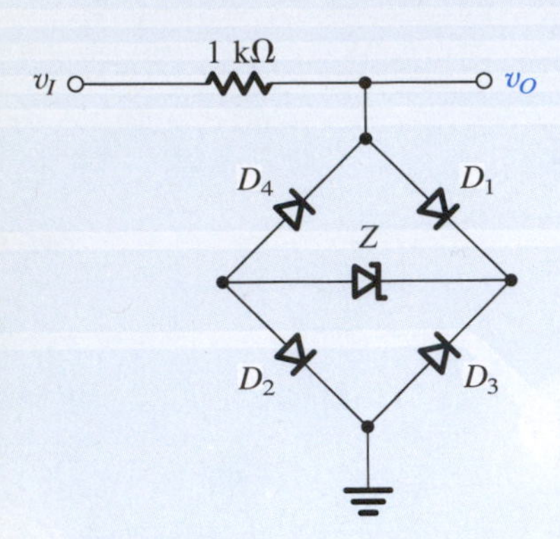

Figure P4.88

drop model with $V_D = 0.7$ V. Also assume that the zener voltage is 8.2 V and that r_z is negligibly small.

***4.89** Plot the transfer characteristic of the circuit in Fig. P4.89 by evaluating v_I corresponding to $v_O = 0.5$ V, 0.6 V, 0.7 V, 0.8 V, 0 V, −0.5 V, −0.6 V, −0.7 V, and −0.8 V. Assume that the diodes have 0.7-V drops at 1-mA currents. Characterize the circuit as a hard or soft limiter. What is the value of K? Estimate L_+ and L_-.

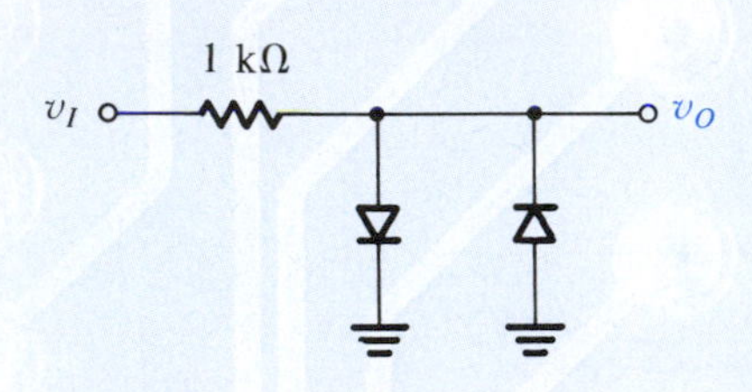

Figure P4.89

4.90 Design limiter circuits using only diodes and 10-kΩ resistors to provide an output signal limited to the range:

(a) −0.7 V and above
(b) −2.1 V and above
(c) ±1.4 V

Assume that each diode has a 0.7-V drop when conducting.

4.91 Design a two-sided limiting circuit using a resistor, two diodes, and two power supplies to feed a 1-kΩ load with nominal limiting levels of ±3 V. Use diodes modeled by a constant 0.7 V. In the nonlimiting region, the voltage gain should be at least 0.95 V/V.

***4.92** In the circuit shown in Fig. P4.92, the diodes exhibit a 0.7-V drop at 0.1 mA. For inputs over the range of ±5 V, provide a calibrated sketch of the voltages at outputs B and C versus v_A. For a 5-V peak, 100-Hz sinusoid applied at A, sketch the signals at nodes B and C.

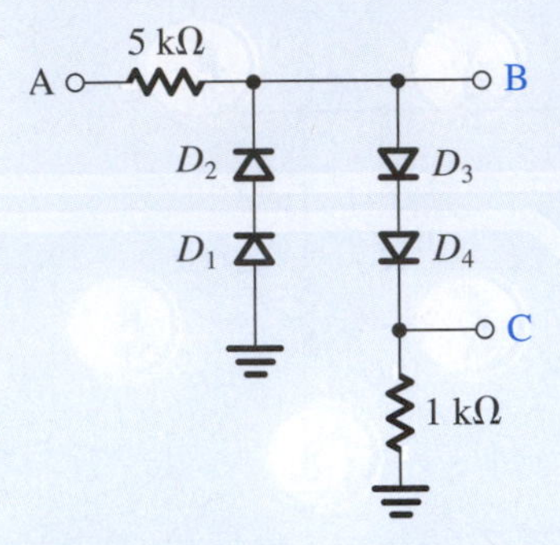

Figure P4.92

****4.93** Sketch and label the voltage transfer characteristic v_O versus v_I of the circuit shown in Fig. P4.93 over a ±10-V range of input signals. All diodes are 1-mA units (i.e., each exhibits a 0.7-V drop at a current of 1 mA). What are the slopes of the characteristic at the extreme ±10-V levels?

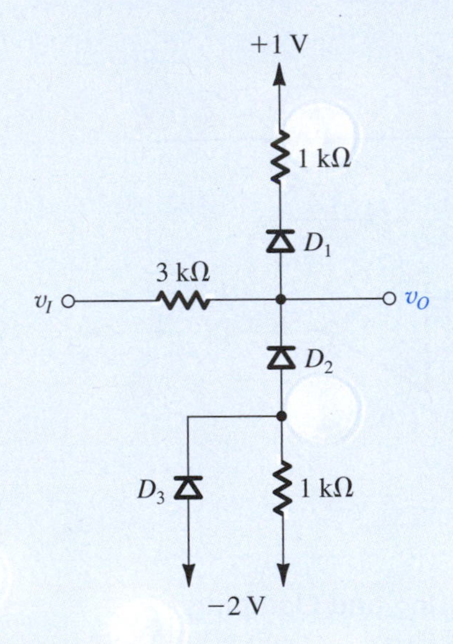

Figure P4.93

4.94 A clamped capacitor using an ideal diode with cathode grounded is supplied with a sine wave of 10-V rms. What is the average (dc) value of the resulting output?

***4.95** For the circuits in Fig. P4.95, each utilizing an ideal diode (or diodes), sketch the output for the input shown. Label the most positive and most negative output levels. Assume $CR \gg T$.

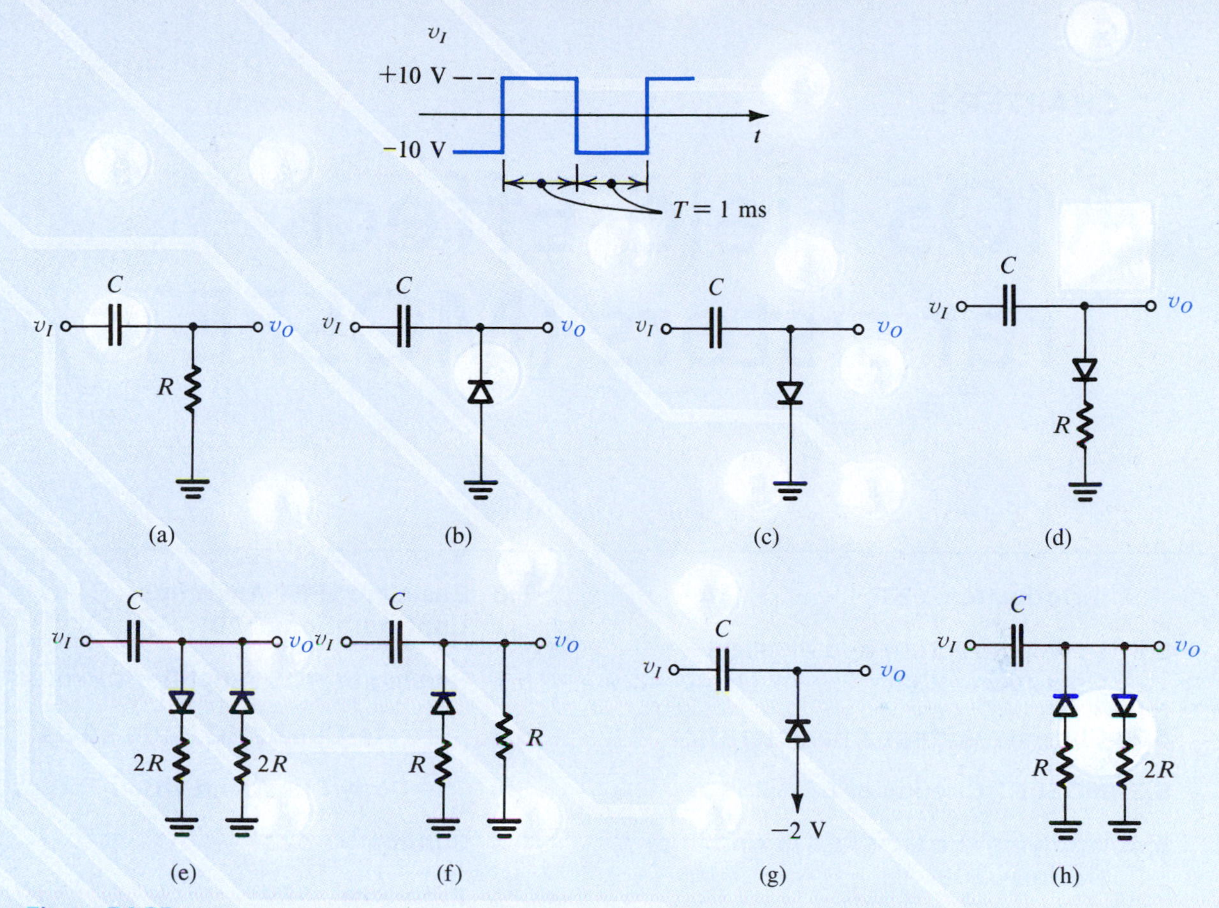

Figure P4.95

CHAPTER 5

MOS Field-Effect Transistors (MOSFETs)

IN THIS CHAPTER YOU WILL LEARN

1. The physical structure of the MOS transistor and how it works.
2. How the voltage between two terminals of the transistor controls the current that flows through the third terminal, and the equations that describe these current-voltage characteristics.
3. How to analyze and design circuits that contain MOS transistors, resistors, and dc sources.
4. How the transistor can be used to make an amplifier, and how it can be used as a switch in digital circuits.
5. How to obtain linear amplification from the fundamentally nonlinear MOS transistor.
6. The three basic ways for connecting a MOSFET to construct amplifiers with different properties.
7. Practical circuits for MOS-transistor amplifiers that can be constructed using discrete components.

Introduction

Having studied the junction diode, which is the most basic two-terminal semiconductor device, we now turn our attention to three-terminal semiconductor devices. Three-terminal devices are far more useful than two-terminal ones because they can be used in a multitude of applications, ranging from signal amplification to digital logic and memory. The basic principle involved is the use of the voltage between two terminals to control the current flowing in the third terminal. In this way a three-terminal device can be used to realize a controlled source, which as we have learned in Chapter 1 is the basis for amplifier design. Also, in the extreme, the control signal can be used to cause the current in the third terminal to change from zero to a large value, thus allowing the device to act as a switch. As we shall see in Chapter 13, the switch is the basis for the realization of the logic inverter, the basic element of digital circuits.

There are two major types of three-terminal semiconductor devices: the metal-oxide-semiconductor field-effect transistor (MOSFET), which is studied in this chapter, and the bipolar junction transistor (BJT), which we shall study in Chapter 6. Although each of the two transistor types offers unique features and areas of application, the MOSFET has become by far the most widely used electronic device, especially in the design of integrated circuits (ICs), which are entire circuits fabricated on a single silicon chip.

Compared to BJTs, MOSFETs can be made quite small (i.e., requiring a small area on the silicon IC chip), and their manufacturing process is relatively simple (see Appendix A). Also, their operation requires comparatively little power. Furthermore, circuit designers have found ingenious ways to implement digital and analog functions utilizing MOSFETs almost exclusively (i.e., with very few or no resistors). All of these properties have made it possible to pack large numbers of MOSFETs (as many as 2 billion!) on a single IC chip to implement very sophisticated, very-large-scale-integrated (VLSI) digital circuits such as those for memory and microprocessors. Analog circuits such as amplifiers and filters can also be implemented in MOS technology, albeit in smaller, less-dense chips. Also, both analog and digital functions are increasingly being implemented on the same IC chip, in what is known as mixed-signal design.

The objective of this chapter is to develop in the reader a high degree of familiarity with the MOSFET: its physical structure and operation, terminal characteristics, circuit models, and basic circuit applications. Although discrete MOS transistors exist, and the material studied in this chapter will enable the reader to design discrete MOS circuits, our study of the MOSFET is strongly influenced by the fact that most of its applications are in integrated-circuit design. The design of IC analog and digital MOS circuits occupies a large proportion of the remainder of this book.

5.1 Device Structure and Physical Operation

The enhancement-type MOSFET is the most widely used field-effect transistor. Except for the last section, this chapter is devoted to the study of the enhancement-type MOSFET. We begin in this section by learning about its structure and physical operation. This will lead to the current–voltage characteristics of the device, studied in the next section.

5.1.1 Device Structure

Figure 5.1, shows the physical structure of the *n*-channel enhancement-type MOSFET. The meaning of the names "enhancement" and "*n*-channel" will become apparent shortly. The transistor is fabricated on a *p*-type substrate, which is a single-crystal silicon wafer that provides physical support for the device (and for the entire circuit in the case of an integrated circuit). Two heavily doped *n*-type regions, indicated in the figure as the n^+ **source**[1] and the n^+ **drain** regions, are created in the substrate. A thin layer of silicon dioxide (SiO_2) of thickness t_{ox} (typically 1 to 10 nm),[2] which is an excellent electrical insulator, is grown on the surface of the substrate, covering the area between the source and drain regions. Metal is deposited on top of the oxide layer to form the **gate electrode** of the device. Metal contacts are also made to the source region, the drain region, and the substrate, also known as the

[1]The notation n^+ indicates heavily doped *n*-type silicon. Conversely, n^- is used to denote lightly doped *n*-type silicon. Similar notation applies for *p*-type silicon.

[2]A nanometer (nm) is 10^{-9} m or 0.001 μm. A micrometer (μm), or micron, is 10^{-6} m. Sometimes the oxide thickness is expressed in angstroms. An angstrom (Å) is 10^{-1} nm, or 10^{-10} m.

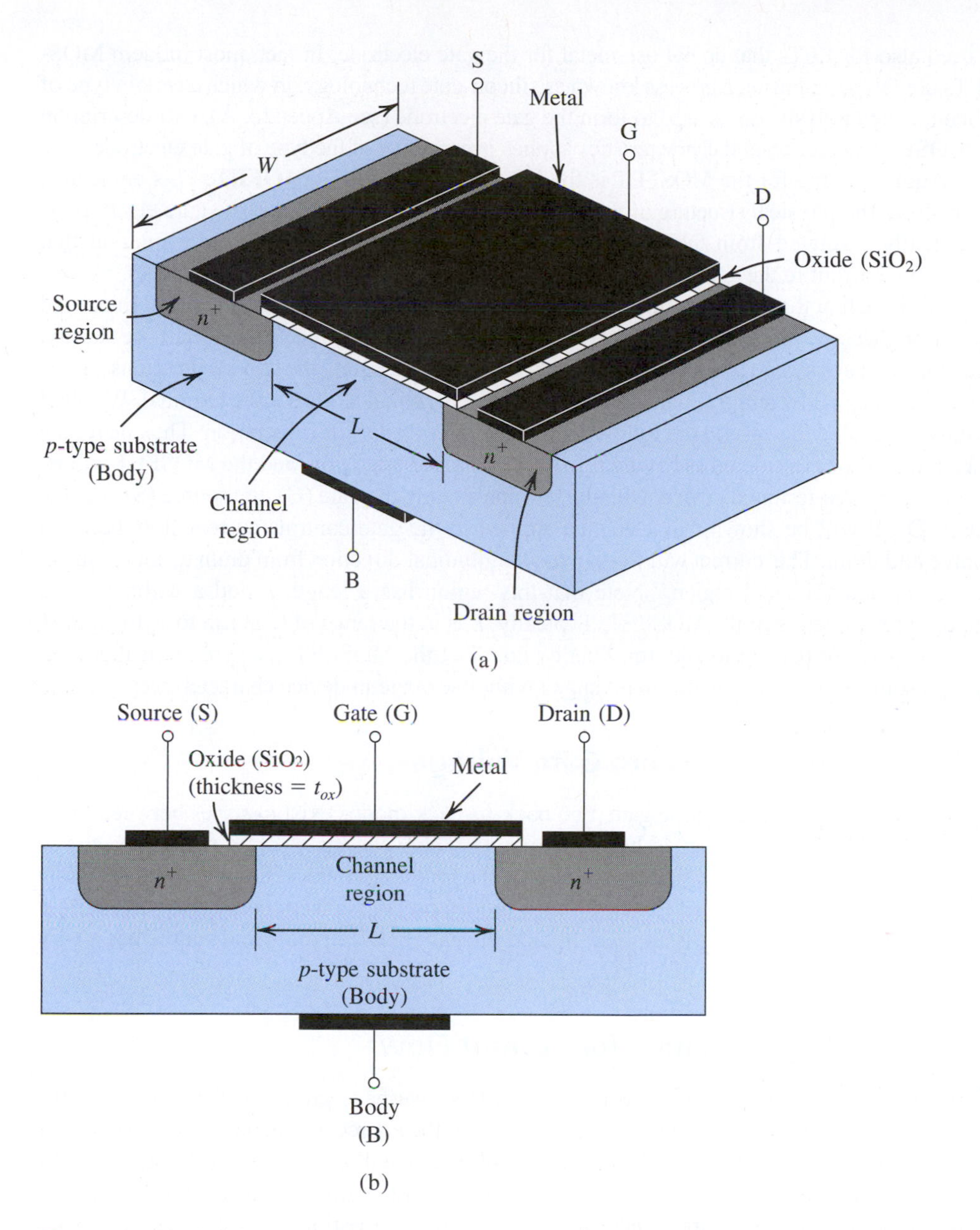

Figure 5.1 Physical structure of the enhancement-type NMOS transistor: **(a)** perspective view; **(b)** cross section. Typically $L = 0.03$ μm to 1 μm, $W = 0.1$ μm to 100 μm, and the thickness of the oxide layer (t_{ox}) is in the range of 1 to 10 nm.

body.[3] Thus four terminals are brought out: the gate terminal (G), the source terminal (S), the drain terminal (D), and the substrate or body terminal (B).

At this point it should be clear that the name of the device (metal-oxide-semiconductor FET) is derived from its physical structure. The name, however, has become a general one and

[3]In Fig. 5.1, the contact to the body is shown on the bottom of the device. This will prove helpful in Section 5.9 in explaining a phenomenon known as the "body effect." It is important to note, however, that in actual ICs, contact to the body is made at a location on the top of the device.

is used also for FETs that do not use metal for the gate electrode. In fact, most modern MOSFETs are fabricated using a process known as silicon-gate technology, in which a certain type of silicon, called polysilicon, is used to form the gate electrode (see Appendix A). Our description of MOSFET operation and characteristics applies irrespective of the type of gate electrode.

Another name for the MOSFET is the **insulated-gate FET** or **IGFET**. This name also arises from the physical structure of the device, emphasizing the fact that the gate electrode is electrically insulated from the device body (by the oxide layer). It is this insulation that causes the current in the gate terminal to be extremely small (of the order of 10^{-15} A).

Observe that the substrate forms *pn* junctions with the source and drain regions. In normal operation these *pn* junctions are kept reverse-biased at all times. Since, as we shall see shortly, the drain will always be at a positive voltage relative to the source, the two *pn* junctions can be effectively cut off by simply connecting the substrate terminal to the source terminal. We shall assume this to be the case in the following description of MOSFET operation. Thus, here, the substrate will be considered as having no effect on device operation, and the MOSFET will be treated as a three-terminal device, with the terminals being the gate (G), the source (S), and the drain (D). It will be shown that a voltage applied to the gate controls current flow between source and drain. This current will flow in the longitudinal direction from drain to source in the region labeled "channel region." Note that this region has a length L and a width W, two important parameters of the MOSFET. Typically, L is in the range of 0.03 μm to 1 μm, and W is in the range of 0.1 μm to 100 μm. Finally, note that the MOSFET is a symmetrical device; thus its source and drain can be interchanged with no change in device characteristics.

5.1.2 Operation with Zero Gate Voltage

With zero voltage applied to the gate, two back-to-back diodes exist in series between drain and source. One diode is formed by the *pn* junction between the n^+ drain region and the *p*-type substrate, and the other diode is formed by the *pn* junction between the *p*-type substrate and the n^+ source region. These back-to-back diodes prevent current conduction from drain to source when a voltage v_{DS} is applied. In fact, the path between drain and source has a very high resistance (of the order of 10^{12} Ω).

5.1.3 Creating a Channel for Current Flow

Consider next the situation depicted in Fig. 5.2. Here we have grounded the source and the drain and applied a positive voltage to the gate. Since the source is grounded, the gate voltage appears in effect between gate and source and thus is denoted v_{GS}. The positive voltage on the gate causes, in the first instance, the free holes (which are positively charged) to be repelled from the region of the substrate under the gate (the channel region). These holes are pushed downward into the substrate, leaving behind a carrier-depletion region. The depletion region is populated by the bound negative charge associated with the acceptor atoms. These charges are "uncovered" because the neutralizing holes have been pushed downward into the substrate.

As well, the positive gate voltage attracts electrons from the n^+ source and drain regions (where they are in abundance) into the channel region. When a sufficient number of electrons accumulate near the surface of the substrate under the gate, an *n* region is in effect created, connecting the source and drain regions, as indicated in Fig. 5.2. Now if a voltage is applied between drain and source, current flows through this induced *n* region, carried by the mobile electrons. The *induced n* region thus forms a **channel** for current flow from drain to source and *is aptly called so*. Correspondingly, the MOSFET of Fig. 5.2 is called an ***n*-channel MOSFET** or, alternatively, an **NMOS transistor**. Note that an *n*-channel MOSFET is

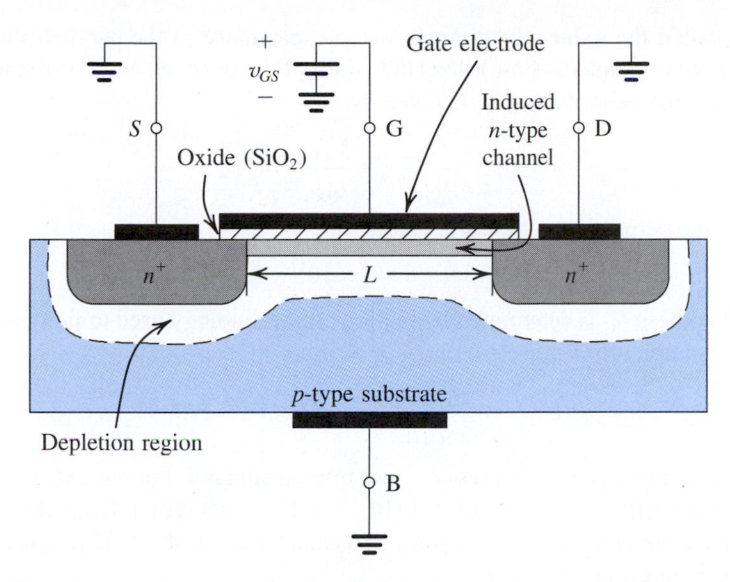

Figure 5.2 The enhancement-type NMOS transistor with a positive voltage applied to the gate. An *n* channel is induced at the top of the substrate beneath the gate.

formed in a *p*-type substrate: The channel is created by *inverting* the substrate surface from *p* type to *n* type. Hence the induced channel is also called an **inversion layer**.

The value of v_{GS} at which a sufficient number of mobile electrons accumulate in the channel region to form a conducting channel is called the **threshold voltage** and is denoted V_t.[4] Obviously, V_t for an *n*-channel FET is positive. The value of V_t is controlled during device fabrication and typically lies in the range of 0.3 V to 1.0 V.

The gate and the channel region of the MOSFET form a parallel-plate capacitor, with the oxide layer acting as the capacitor dielectric. The positive gate voltage causes positive charge to accumulate on the top plate of the capacitor (the gate electrode). The corresponding negative charge on the bottom plate is formed by the electrons in the induced channel. An electric field thus develops in the vertical direction. It is this field that controls the amount of charge in the channel, and thus it determines the channel conductivity and, in turn, the current that will flow through the channel when a voltage v_{DS} is applied. This is the origin of the name "field-effect transistor" (FET).

The voltage across this parallel-plate capacitor, that is, the voltage across the oxide, must exceed V_t for a channel to form. When $v_{DS} = 0$, as in Fig. 5.2, the voltage at every point along the channel is zero, and the voltage across the oxide (i.e., between the gate and the points along the channel) is uniform and equal to v_{GS}. The excess of v_{GS} over V_t is termed the **effective voltage** or the **overdrive voltage** and is the quantity that determines the charge in the channel. In this book, we shall denote $(v_{GS} - V_t)$ by v_{OV},

$$v_{GS} - V_t \equiv v_{OV} \tag{5.1}$$

We can express the magnitude of the electron charge in the channel by

$$|Q| = C_{ox}(WL)v_{OV} \tag{5.2}$$

[4]Some texts use V_T to denote the threshold voltage. We use V_t to avoid confusion with the thermal voltage V_T.

where C_{ox}, called the **oxide capacitance**, is the capacitance of the parallel-plate capacitor per unit gate area (in units of F/m²), W is the width of the channel, and L is the length of the channel. The oxide capacitance C_{ox} is given by

$$C_{ox} = \frac{\varepsilon_{ox}}{t_{ox}} \tag{5.3}$$

where ε_{ox} is the permittivity of the silicon dioxide,

$$\varepsilon_{ox} = 3.9\varepsilon_0 = 3.9 \times 8.854 \times 10^{-12} = 3.45 \times 10^{-11} \text{ F/m}$$

The oxide thickness t_{ox} is determined by the process technology used to fabricate the MOSFET. As an example, for a process with $t_{ox} = 4$ nm,

$$C_{ox} = \frac{3.45 \times 10^{-11}}{4 \times 10^{-9}} = 8.6 \times 10^{-3} \text{ F/m}^2$$

It is much more convenient to express C_{ox} per micron squared. For our example, this yields 8.6 fF/μm², where fF denotes femtofarad (10^{-15} F). For a MOSFET fabricated in this technology with a channel length $L = 0.18$ μm and a channel width $W = 0.72$ μm, the total capacitance between gate and channel is

$$C = C_{ox}WL = 8.6 \times 0.18 \times 0.72 = 1.1 \text{ fF}$$

Finally, note from Eq. (5.2) that as v_{OV} is increased, the magnitude of the channel charge increases proportionately. Sometimes this is depicted as an increase in the depth of the channel; that is, the larger the overdrive voltage, the deeper the channel.

5.1.4 Applying a Small v_{DS}

Having induced a channel, we now apply a positive voltage v_{DS} between drain and source, as shown in Fig. 5.3. We first consider the case where v_{DS} is small (i.e., 50 mV or so). The voltage v_{DS} causes a current i_D to flow through the induced n channel. Current is carried by free electrons traveling from source to drain (hence the names source and drain). By convention, the direction of current flow is opposite to that of the flow of negative charge. Thus the current in the channel, i_D, will be from drain to source, as indicated in Fig. 5.3.

We now wish to calculate the value of i_D. Toward that end, we first note that because v_{DS} is small, we can continue to assume that the voltage between the gate and various points along the channel remains approximately constant and equal to the value at the source end, v_{GS}. Thus, the effective voltage between the gate and the various points along the channel remains equal to v_{OV}, and the channel charge Q is still given by Eq. (5.2). Of particular interest in calculating the current i_D is the charge per unit channel length, which can be found from Eq. (5.2) as

$$\frac{|Q|}{\text{unit channel length}} = C_{ox}Wv_{OV} \tag{5.4}$$

The voltage v_{DS} establishes an electric field E across the length of the channel,

$$|E| = \frac{v_{DS}}{L} \tag{5.5}$$

This electric field in turn causes the channel electrons to drift toward the drain with a velocity given by

$$\text{Electron drift velocity} = \mu_n |E| = \mu_n \frac{v_{DS}}{L} \tag{5.6}$$

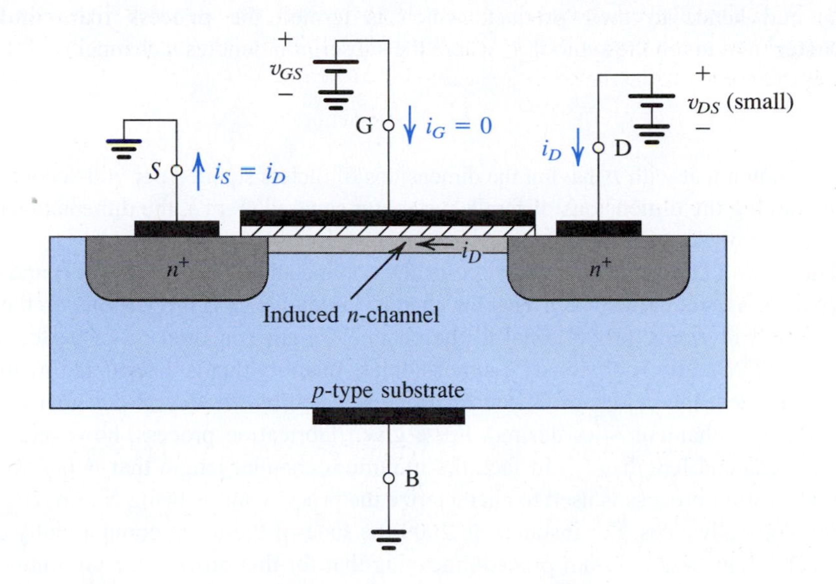

Figure 5.3 An NMOS transistor with $v_{GS} > V_t$ and with a small v_{DS} applied. The device acts as a resistance whose value is determined by v_{GS}. Specifically, the channel conductance is proportional to $v_{GS} - V_t$, and thus i_D is proportional to $(v_{GS} - V_t)v_{DS}$. Note that the depletion region is not shown (for simplicity).

where μ_n is the mobility of the electrons at the surface of the channel. It is a physical parameter whose value depends on the fabrication process technology. The value of i_D can now be found by multiplying the charge per unit channel length (Eq. 5.4) by the electron drift velocity (Eq. 5.6),

$$i_D = \left[(\mu_n C_{ox})\left(\frac{W}{L}\right)v_{OV}\right]v_{DS} \tag{5.7}$$

Thus, for small v_{DS}, the channel behaves as a linear resistance whose value is controlled by the overdrive voltage v_{OV}, which in turn is determined by v_{GS}:

$$i_D = \left[(\mu_n C_{ox})\left(\frac{W}{L}\right)(v_{GS} - V_t)\right]v_{DS} \tag{5.8}$$

The conductance g_{DS} of the channel can be found from Eq. (5.7) or (5.8) as

$$g_{DS} = (\mu_n C_{ox})\left(\frac{W}{L}\right)v_{OV} \tag{5.9}$$

or

$$g_{DS} = (\mu_n C_{ox})\left(\frac{W}{L}\right)(v_{GS} - V_t) \tag{5.10}$$

Observe that the conductance is determined by the product of three factors: $(\mu_n C_{ox})$, (W/L), and v_{OV} (or equivalently, $v_{GS} - V_t$). To gain insight into MOSFET operation, we consider each of the three factors in turn.

The first factor, $(\mu_n C_{ox})$, is determined by the process technology used to fabricate the MOSFET. It is the product of the electron mobility, μ_n, and the oxide capacitance, C_{ox}. It makes physical sense for the channel conductance to be proportional to each of μ_n and C_{ox}

(why?) and hence to their product, which is termed the **process transconductance parameter**[5] and given the symbol k_n' where the subscript n denotes n channel,

$$k_n' = \mu_n C_{ox} \tag{5.11}$$

It can be shown that with μ_n having the dimensions of meters squared per volt-second ($m^2/V{\cdot}s$) and C_{ox} having the dimensions of farads per meter squared (F/m^2), the dimensions of k_n' are amperes per volt squared (A/V^2).

The second factor in the expression for the conductance g_{DS} in Eqs. (5.9) and (5.10) is the transistor **aspect ratio** (W/L). That the channel conductance is proportional to the channel width W and inversely proportional to the channel length L should make perfect physical sense. The (W/L) ratio is obviously a dimensionless quantity that is determined by the device designer. Indeed, the values of W and L can be selected by the device designer to give the device the i–v characteristics desired. For a given fabrication process, however, there is a minimum channel length, L_{min}. In fact, the minimum channel length that is possible with a given fabrication process is used to characterize the process and is being continually reduced as technology advances. For instance, in 2009 the state-of-the-art in commercially available MOS technology was a 45-nm process, meaning that for this process the minimum channel length possible was 45 nm. Finally, we should note that the oxide thickness t_{ox} scales down with L_{min}. Thus, for a 0.13-μm technology, t_{ox} is 2.7 nm, but for the modern 45-nm technology t_{ox} is about 1.4 nm.

The product of the process transconductance parameter k_n' and the transistor aspect ratio (W/L) is the **MOSFET transconductance parameter** k_n,

$$k_n = k_n'(W/L) \tag{5.12a}$$

or

$$k_n = (\mu_n C_{ox})(W/L) \tag{5.12b}$$

The MOSFET parameter k_n has the dimensions of A/V^2.

The third term in the expression of the channel conductance g_{DS} is the overdrive voltage v_{OV}. This is hardly surprising since v_{OV} directly determines the magnitude of electron charge in the channel. As will be seen, v_{OV} is a very important circuit-design parameter. In this book, we will use v_{OV} and $v_{GS} - V_t$ interchangeably.

We conclude this subsection by noting that with v_{DS} kept small, the MOSFET behaves as a linear resistance r_{DS} whose value is controlled by the gate voltage v_{GS},

$$r_{DS} = \frac{1}{g_{DS}}$$

$$r_{DS} = \frac{1}{(\mu_n C_{ox})(W/L)v_{OV}} \tag{5.13a}$$

$$r_{DS} = \frac{1}{(\mu_n C_{ox})(W/L)(v_{GS} - V_t)} \tag{5.13b}$$

The operation of the MOSFET as a voltage-controlled resistance is further illustrated in Fig. 5.4, which is a sketch of i_D versus v_{DS} for various values of v_{GS}. Observe that the

[5]This name arises from the fact that $(\mu_n C_{ox})$ determines the transconductance of the MOSFET, as will be seen shortly.

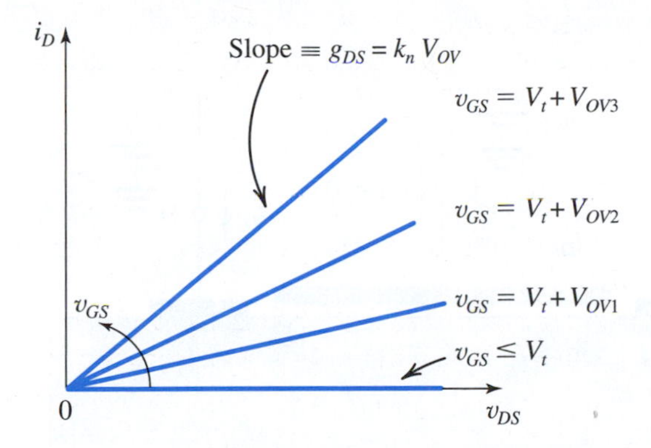

Figure 5.4 The i_D–v_{DS} characteristics of the MOSFET in Fig. 5.3 when the voltage applied between drain and source, v_{DS}, is kept small. The device operates as a linear resistance whose value is controlled by v_{GS}.

resistance is infinite for $v_{GS} \leq V_t$ and decreases as v_{GS} is increased above V_t. It is interesting to note that although v_{GS} is used as the parameter for the set of graphs in Fig. 5.4, the graphs in fact depend only on v_{OV} (and, of course, k_n).

The description above indicates that for the MOSFET to conduct, a channel has to be induced. Then, increasing v_{GS} above the threshold voltage V_t enhances the channel, hence the names **enhancement-mode operation** and **enhancement-type MOSFET**. Finally, we note that the current that leaves the source terminal (i_S) is equal to the current that enters the drain terminal (i_D), and the gate current $i_G = 0$.

EXERCISE

5.1 A 0.18-μm fabrication process is specified to have t_{ox} = 4 nm, μ_n = 450 cm²/V·s, and V_t = 0.5 V. Find the value of the process transconductance parameter k'_n. For a MOSFET with minimum length fabricated in this process, find the required value of W so that the device exhibits a channel resistance r_{DS} of 1 kΩ at v_{GS} = 1 V.
Ans. 388 μA/V²; 0.93 μm

5.1.5 Operation as v_{DS} Is Increased

We next consider the situation as v_{DS} is increased. For this purpose, let v_{GS} be held constant at a value greater than V_t; that is, let the MOSFET be operated at a constant overdrive voltage V_{OV}. Refer to Fig. 5.5, and note that v_{DS} appears as a voltage drop across the length of the channel. That is, as we travel along the channel from source to drain, the voltage (measured relative to the source) increases from zero to v_{DS}. Thus the voltage between the gate and points along the channel decreases from $v_{GS} = V_t + V_{OV}$ at the source end to $v_{GD} = v_{GS} - v_{DS} = V_t + V_{OV} - v_{DS}$ at the drain end. Since the channel depth depends on this voltage, and specifically on the amount by which this voltage exceeds V_t, we find that the channel is no longer of uniform depth; rather, the channel will take the tapered shape shown in Fig. 5.5, being deepest at the source end (where the depth is proportional to V_{OV}) and shallowest at the drain end[6] (where the depth is proportional to $V_{OV} - v_{DS}$). This point is further illustrated in Fig. 5.6.

[6]For simplicity, we do not show in Fig. 5.5 the depletion region. Physically speaking, it is the widening of the depletion region as a result of the increased v_{DS} that makes the channel shallower near the drain.

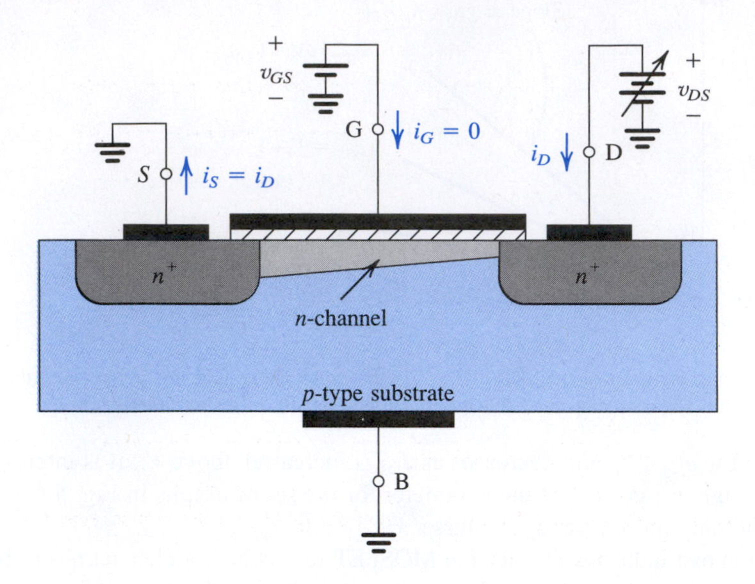

Figure 5.5 Operation of the enhancement NMOS transistor as v_{DS} is increased. The induced channel acquires a tapered shape, and its resistance increases as v_{DS} is increased. Here, v_{GS} is kept constant at a value $> V_t$; $v_{GS} = V_t + V_{OV}$.

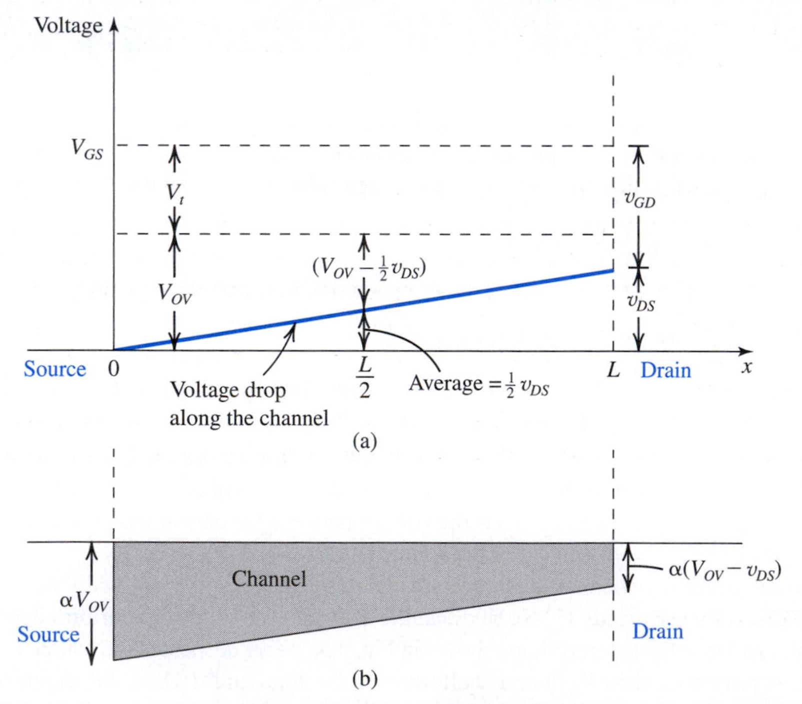

Figure 5.6 **(a)** For a MOSFET with $v_{GS} = V_t + V_{OV}$, application of v_{DS} causes the voltage drop along the channel to vary linearly, with an average value of $\frac{1}{2}v_{DS}$ at the midpoint. Since $v_{GD} > V_t$, the channel still exists at the drain end. **(b)** The channel shape corresponding to the situation in (a). While the depth of the channel at the source end is still proportional to V_{OV}, that at the drain end is proporational to $(V_{OV} - v_{DS})$.

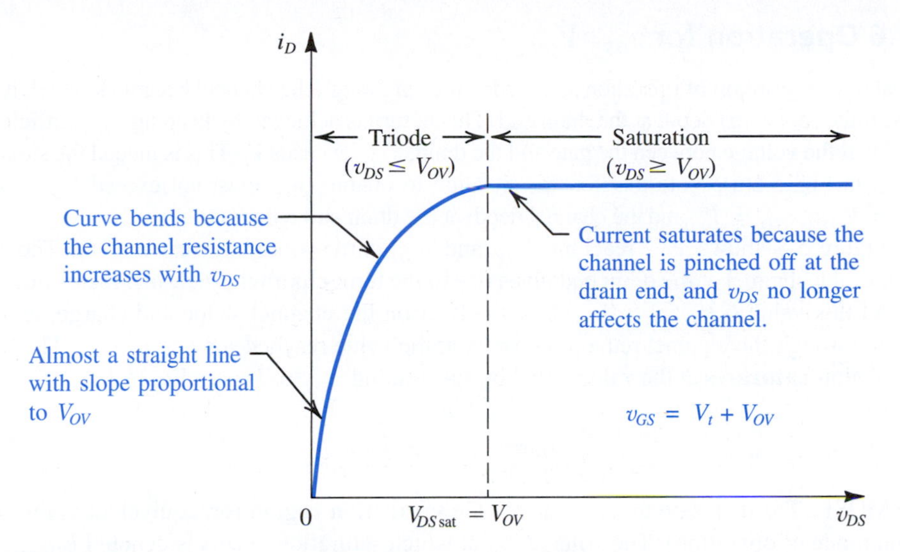

Figure 5.7 The drain current i_D versus the drain-to-source voltage v_{DS} for an enhancement-type NMOS transistor operated with $v_{GS} = V_t + V_{OV}$.

As v_{DS} is increased, the channel becomes more tapered and its resistance increases correspondingly. Thus, the i_D–v_{DS} curve does not continue as a straight line but bends as shown in Fig. 5.7. The equation describing this portion of the i_D-v_{DS} curve can be easily derived by utilizing the information in Fig. 5.6. Specifically, note that the charge in the tapered channel is proportional to the channel cross-sectional area shown in Fig. 5.6(b). This area in turn can be easily seen as proportional to $\frac{1}{2}[V_{OV} + (V_{OV} - v_{DS})]$ or $(V_{OV} - \frac{1}{2}v_{DS})$. Thus, the relationship between i_D and v_{DS} can be found by replacing V_{OV} in Eq. (5.7) by $(V_{OV} - \frac{1}{2}v_{DS})$,

$$i_D = k'_n\left(\frac{W}{L}\right)\left(V_{OV} - \frac{1}{2}v_{DS}\right)v_{DS} \tag{5.14}$$

This relationship describes the semiparabolic portion of the i_D–v_{DS} curve in Fig. 5.7. It applies to the entire segment down to $v_{DS} = 0$. Specifically, note that as v_{DS} is reduced, we can neglect $\frac{1}{2}v_{DS}$ relative to V_{OV} in the factor in parentheses, and the expression reduces to that in Eq. (5.7). The latter of course is an approximation and applies only for small v_{DS} (i.e., near the origin).

There is another useful interpretation of the expression in Eq. (5.14). From Fig. 5.6(a) we see that the average voltage along the channel is $\frac{1}{2}v_{DS}$. Thus, the average voltage that gives rise to channel charge and hence to i_D is no longer V_{OV} but $(V_{OV} - \frac{1}{2}v_{DS})$, which is indeed the factor that appears in Eq. (5.14). Finally, we note that Eq. (5.14) is frequently written in the alternate form

$$i_D = k'_n\left(\frac{W}{L}\right)\left(V_{OV}\, v_{DS} - \frac{1}{2}v_{DS}^2\right) \tag{5.15}$$

Furthermore, for an arbitrary value of V_{OV}, we can replace V_{OV} by $(v_{GS} - V_t)$ and rewrite Eq. (5.15) as

$$i_D = k'_n\left(\frac{W}{L}\right)\left[(v_{GS} - V_t)v_{DS} - \frac{1}{2}v_{DS}^2\right] \tag{5.16}$$

5.1.6 Operation for $v_{DS} \geq V_{OV}$

The above description of operation assumed that even though the channel became tapered, it still had a finite (nonzero) depth at the drain end. This in turn is achieved by keeping v_{DS} sufficiently small that the voltage between the gate and the drain, v_{GD}, exceeds V_t. This is indeed the situation shown in Fig. 5.6(a). Note that for this situation to obtain, v_{DS} must not exceed V_{OV}, for as $v_{DS} = V_{OV}$, $v_{GD} = V_t$, and the channel depth at the drain end reduces to zero.

Figure 5.8 shows v_{DS} reaching V_{OV} and v_{GD} correspondingly reaching V_t. The zero depth of the channel at the drain end gives rise to the term **channel pinch-off**. Increasing v_{DS} beyond this value (i.e., $v_{DS} > V_{OV}$) has no effect on the channel shape and charge, and the current through the channel remains constant at the value reached for $v_{DS} = V_{OV}$. The drain current thus **saturates** at the value found by substituting $v_{DS} = V_{OV}$ in Eq. (5.14),

$$i_D = \frac{1}{2}k_n'\left(\frac{W}{L}\right)V_{OV}^2 \tag{5.17}$$

The MOSFET is then said to have entered the **saturation region** (or, equivalently, the saturation mode of operation). The voltage v_{DS} at which saturation occurs is denoted $V_{DS\text{sat}}$,

$$V_{DS\text{sat}} = V_{OV} = V_{GS} - V_t \tag{5.18}$$

It should be noted that channel pinch-off does *not* mean channel blockage: Current continues to flow through the pinched-off channel, and the electrons that reach the drain end of the

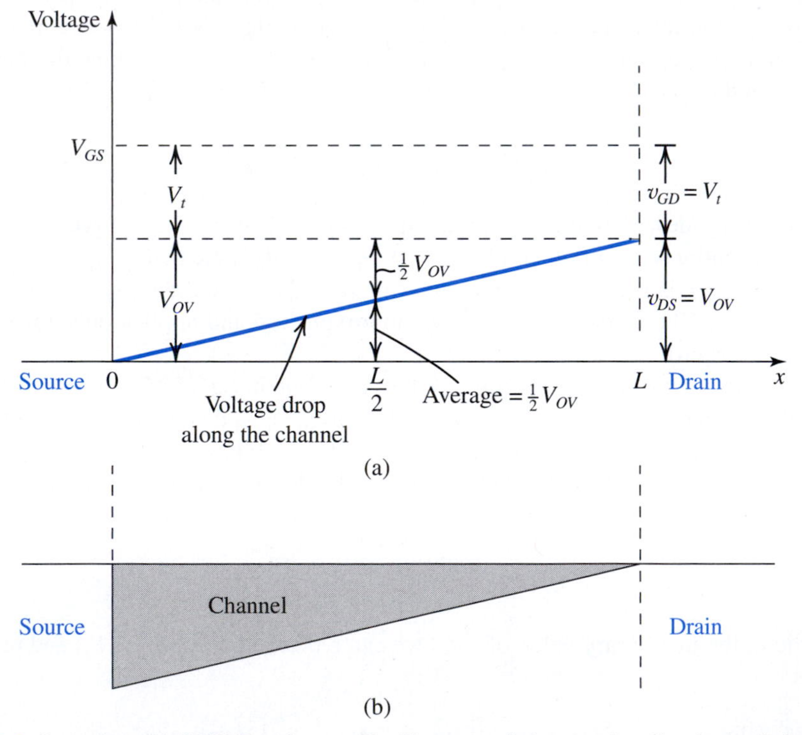

Figure 5.8 Operation of MOSFET with $v_{GS} = V_t + V_{OV}$, as v_{DS} is increased to V_{OV}. At the drain end, v_{GD} decreases to V_t and the channel depth at the drain end reduces to zero (pinch off). At this point, the MOSFET enters the saturation mode of operation. Further increasing v_{DS} (beyond $V_{D\text{sat}} = V_{OV}$) has no effect on the channel shape and i_D remains constant.

channel are accelerated through the depletion region that exists there (not shown in Fig. 5.5) and into the drain terminal. Any increase in v_{DS} above V_{DSsat} appears as a voltage drop across the depletion region. Thus, both the current through the channel and the voltage drop across it remain constant in saturation.

The saturation portion of the $i_D - v_{DS}$ curve is, as expected, a horizontal straight line, as indicated in Fig. 5.7. Also indicated in Fig. 5.7 is the name of the region of operation obtained with a continuous (non-pinched-off) channel, the **triode region**. This name is a carryover from the days of vacuum-tube devices, whose operation a FET resembles.

Finally, we note that the $i_D - v_{DS}$ relationship in saturation can be generalized by replacing the constant overdrive voltage V_{OV} by a variable one, v_{OV}:

$$i_D = \frac{1}{2}k'_n\left(\frac{W}{L}\right)v_{OV}^2 \tag{5.19}$$

Also, v_{OV} can be replaced by $(v_{GS} - V_t)$ to obtain the alternate expression for saturation-mode i_D,

$$i_D = \frac{1}{2}k'_n\left(\frac{W}{L}\right)(v_{GS} - V_t)^2 \tag{5.20}$$

Example 5.1

Consider a process technology for which $L_{min} = 0.4\ \mu m$, $t_{ox} = 8$ nm, $\mu_n = 450\ cm^2/V \cdot s$, and $V_t = 0.7$ V.

(a) Find C_{ox} and k'_n.

(b) For a MOSFET with $W/L = 8\ \mu m/0.8\ \mu m$, calculate the values of V_{OV}, V_{GS}, and V_{DSmin} needed to operate the transistor in the saturation region with a dc current $I_D = 100\ \mu A$.

(c) For the device in (b), find the values of V_{OV} and V_{GS} required to cause the device to operate as a 1000-Ω resistor for very small v_{DS}.

Solution

(a)

$$C_{ox} = \frac{\varepsilon_{ox}}{t_{ox}} = \frac{3.45 \times 10^{-11}}{8 \times 10^{-9}} = 4.32 \times 10^{-3}\ F/m^2$$

$$= 4.32\ fF/\mu m^2$$

$$k'_n = \mu_n C_{ox} = 450\ (cm^2/V \cdot s) \times 4.32\ (fF/\mu m^2)$$

$$= 450 \times 10^8\ (\mu m^2/V \cdot s) \times 4.32 \times 10^{-15} (F/\mu m^2)$$

$$= 194 \times 10^{-6}\ (F/V \cdot s)$$

$$= 194\ \mu A/V^2$$

(b) For operation in the saturation region,

$$i_D = \frac{1}{2}k'_n\frac{W}{L}v_{OV}^2$$

Thus,

$$100 = \frac{1}{2} \times 194 \times \frac{8}{0.8}V_{OV}^2$$

which results in

$$V_{OV} = 0.32\ V$$

Example 5.1 *continued*

Thus,

$$V_{GS} = V_t + V_{OV} = 1.02 \text{ V}$$

and

$$V_{DS\min} = V_{OV} = 0.32 \text{ V}$$

(c) For the MOSFET in the triode region with v_{DS} very small,

$$r_{DS} = \frac{1}{k'_n \frac{W}{L} V_{OV}}$$

Thus

$$1000 = \frac{1}{194 \times 10^{-6} \times 10 \times V_{OV}}$$

which yields

$$V_{OV} = 0.52 \text{ V}$$

Thus,

$$V_{GS} = 1.22 \text{ V}$$

EXERCISES

5.2 For a 0.8-μm process technology for which $t_{ox} = 15$ nm and $\mu_n = 550$ cm^2/V·s, find C_{ox}, k'_n, and the overdrive voltage V_{OV} required to operate a transistor having $W/L = 20$ in saturation with $I_D = 0.2$ mA. What is the minimum value of V_{DS} needed?

Ans. 2.3 fF/μm^2; 127 μA/V^2; 0.40 V; 0.40 V

D5.3 A circuit designer intending to operate a MOSFET in saturation is considering the effect of changing the device dimensions and operating voltages on the drain current I_D. Specifically, by what factor does I_D change in each of the following cases?

(a) The channel length is doubled.
(b) The channel width is doubled.
(c) The overdrive voltage is doubled.
(d) The drain-to-source voltage is doubled.
(e) Changes (a), (b), (c), and (d) are made simultaneously.

Which of these cases might cause the MOSFET to leave the saturation region?

Ans. 0.5; 2; 4; no change; 4; case (c) if v_{DS} is smaller than 2 V_{OV}

5.1.7 The *p*-Channel MOSFET

Figure 5.9(a) shows a cross-sectional view of a *p*-channel enhancement-type MOSFET. The structure is similar to that of the NMOS device except that here the substrate is *n* type and the source and the drain regions are p^+ type; that is, all semiconductor regions are reversed in polarity relative to their counterparts in the NMOS case. The PMOS and NMOS transistors are said to be *complementary* devices.

To induce a channel for current flow between source and drain, a negative voltage is applied to the gate, that is, between gate and source, as indicated in Fig. 5.9(b). By increasing the magni-

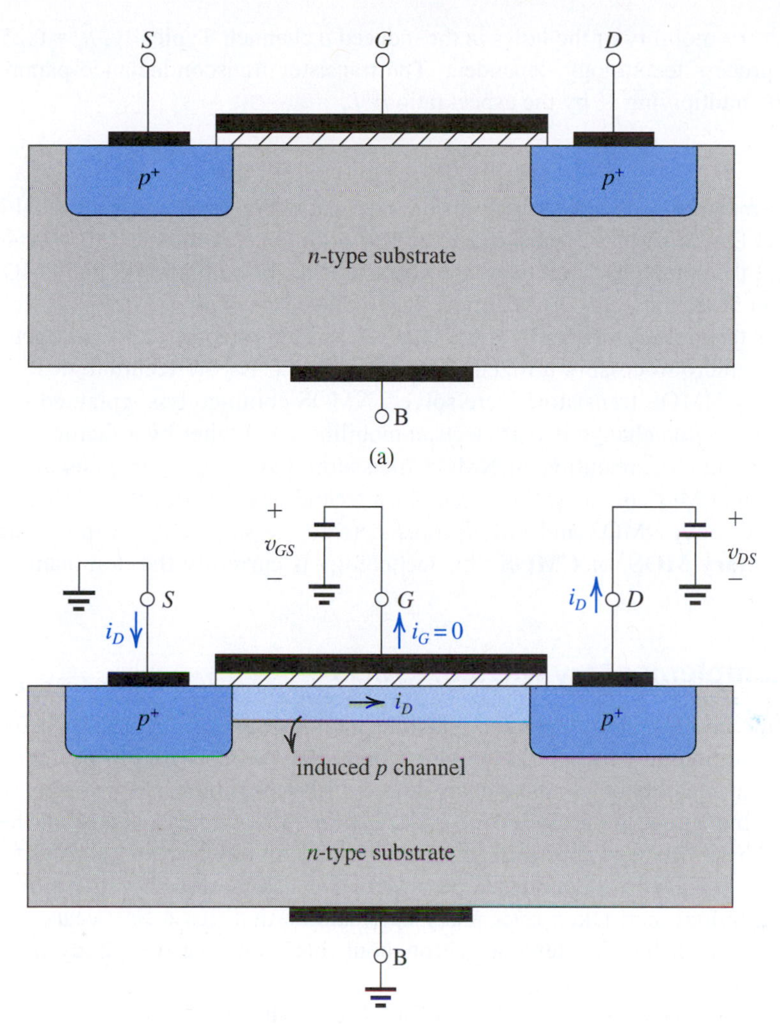

Figure 5.9 (**a**) Physical structure of the PMOS transistor. Note that it is similar to the NMOS transistor shown in Fig. 5.1(b) except that all semiconductor regions are reversed in polarity. (**b**) A negative voltage v_{GS} of magnitude greater than $|V_{tp}|$ induces a p channel, and a negative v_{DS} causes a current i_D to flow from source to drain.

tude of the negative v_{GS} beyond the magnitude of the threshold voltage V_{tp}, which by convention is negative, a p channel is established as shown in Fig. 5.9(b). This condition can be described as

$$v_{GS} \leq V_{tp}$$

or, to avoid dealing with negative signs,

$$|v_{GS}| \geq |V_{tp}|$$

Now, to cause a current i_D to flow in the p channel, a negative voltage v_{DS} is applied to the drain. The current i_D is carried by holes and flows through the channel from source to drain. As we have done for the NMOS transistor, we define the process transconductance parameter for the PMOS device as

$$k'_p = \mu_p C_{ox}$$

where μ_p is the mobility of the holes in the induced p channel. Typically, $\mu_p = 0.25\ \mu_n$ to 0.5 μ_n and is process technology dependent. The transistor transconductance parameter k_p is obtained by multiplying k_p' by the aspect ratio W/L,

$$k_p = k_p'(W/L)$$

The remainder of the description of the physical operation of the p-channel MOSFET follows that for the NMOS device, except of course for the sign reversals of all voltages. We will present the complete current–voltage characteristics of both NMOS and PMOS transistors in the next section.

PMOS technology originally dominated MOS integrated-circuit manufacturing, and the original microprocessors utilized PMOS transistors. As the technological difficulties of fabricating NMOS transistors were solved, NMOS completely supplanted PMOS. The main reason for this change is that electron mobility μ_n is higher by a factor of 2 to 4 than the hole mobility μ_p, resulting in NMOS transistors having greater gains and speeds of operation than PMOS devices. Subsequently, a technology was developed that permits the fabrication of both NMOS and PMOS transistors on the same chip. Appropriately called **complementary MOS**, or **CMOS**, this technology is currently the dominant electronics technology.

5.1.8 Complementary MOS or CMOS

As the name implies, complementary MOS technology employs MOS transistors of both polarities. Although CMOS circuits are somewhat more difficult to fabricate than NMOS, the availability of complementary devices makes possible many powerful circuit configurations. Indeed, at the present time CMOS is the most widely used of all the IC technologies. This statement applies to both analog and digital circuits. CMOS technology has virtually replaced designs based on NMOS transistors alone. Furthermore, by 2009 CMOS technology had taken over many applications that just a few years earlier were possible only with bipolar devices. Throughout this book, we will study many CMOS circuit techniques.

Figure 5.10 shows a cross section of a CMOS chip illustrating how the PMOS and NMOS transistors are fabricated. Observe that while the NMOS transistor is implemented directly in the p-type substrate, the PMOS transistor is fabricated in a specially created n region, known as an ***n* well**. The two devices are isolated from each other by a thick region of oxide that functions as an insulator. Not shown on the diagram are the connections made to the p-type body and to the n well. The latter connection serves as the body terminal for the PMOS transistor.

5.1.9 Operating the MOS Transistor in the Subthreshold Region

The above description of the n-channel MOSFET operation implies that for $v_{GS} < V_t$, no current flows and the device is cut off. This is not entirely true, for it has been found that for values of v_{GS} smaller than but close to V_t, a small drain current flows. In this **subthreshold region** of operation, the drain current is exponentially related to v_{GS}, much like the i_C–v_{BE} relationship of a BJT, as will be shown in the next chapter.

Although in most applications the MOS transistor is operated with $v_{GS} > V_t$, there are special, but a growing number of, applications that make use of subthreshold operation. In Chapter 13, we will briefly consider subthreshold operation.

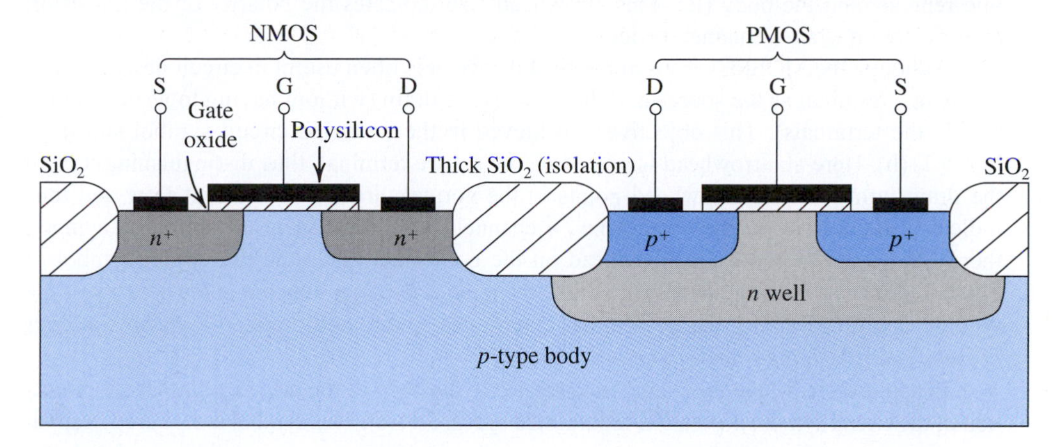

Figure 5.10 Cross-section of a CMOS integrated circuit. Note that the PMOS transistor is formed in a separate n-type region, known as an n well. Another arrangement is also possible in which an n-type body is used and the n device is formed in a p well. Not shown are the connections made to the p-type body and to the n well; the latter functions as the body terminal for the p-channel device.

5.2 Current–Voltage Characteristics

Building on the physical foundation established in the previous section for the operation of the enhancement MOS transistor, in this section we present its complete current–voltage characteristics. These characteristics can be measured at dc or at low frequencies and thus are called static characteristics. The dynamic effects that limit the operation of the MOSFET at high frequencies and high switching speeds will be discussed in Chapter 9.

5.2.1 Circuit Symbol

Figure 5.11(a) shows the circuit symbol for the n-channel enhancement-type MOSFET. Observe that the spacing between the two vertical lines that represent the gate and the channel indicates the fact that the gate electrode is insulated from the body of the device. The polarity of the p-type substrate (body) and the n channel is indicated by the arrowhead on the

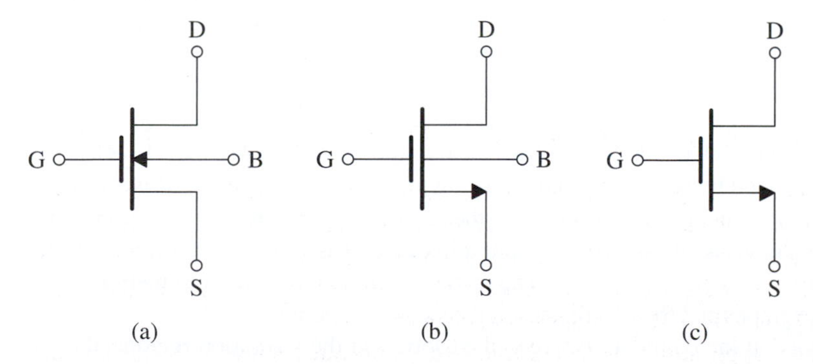

Figure 5.11 **(a)** Circuit symbol for the n-channel enhancement-type MOSFET. **(b)** Modified circuit symbol with an arrowhead on the source terminal to distinguish it from the drain and to indicate device polarity (i.e., n channel). **(c)** Simplified circuit symbol to be used when the source is connected to the body or when the effect of the body on device operation is unimportant.

line representing the body (B). This arrowhead also indicates the polarity of the transistor, namely, that it is an n-channel device.

Although the MOSFET is a symmetrical device, it is often useful in circuit design to designate one terminal as the source and the other as the drain (without having to write S and D beside the terminals). This objective is achieved in the modified circuit symbol shown in Fig. 5.11(b). Here an arrowhead is placed on the source terminal, thus distinguishing it from the drain terminal. The arrowhead points in the normal direction of current flow and thus indicates the polarity of the device (i.e., n channel). Observe that in the modified symbol, there is no need to show the arrowhead on the body line. Although the circuit symbol of Fig. 5.11(b) clearly distinguishes the source from the drain, in practice it is the polarity of the voltage impressed across the device that determines source and drain; *the drain is always positive relative to the source in an n-channel FET.*

In applications where the source is connected to the body of the device, a further simplification of the circuit symbol is possible, as indicated in Fig. 5.11(c). This symbol is also used in applications when the effect of the body on circuit operation is not important, as will be seen later.

5.2.2 The i_D–v_{DS} Characteristics

Table 5.1 provides a compilation of the conditions and the formulas for the operation of the NMOS transistor in each of the three possible regions: the cutoff region, the triode region, and the saturation region. The first two are useful if the MOSFET is to be utilized as a switch. On the other hand, if the MOSFET is to be used to design an amplifier, it must be operated in the saturation region. The rationale for these choices will be addressed in Section 5.4.

At the top of Table 5.1 we show a circuit consisting of an NMOS transistor and two dc supplies providing v_{GS} and v_{DS}. This conceptual circuit can be used to measure the i_D–v_{DS} characteristic curves of the NMOS transistor. Each curve is measured by setting v_{GS} to a desired constant value, varying v_{DS}, and measuring the corresponding i_D. Two of these characteristic curves are shown in the accompanying diagram: one for $v_{GS} < V_{tn}$ and the other for $v_{GS} = V_{tn} + v_{OV}$. (Note that we now use V_{tn} to denote the threshold voltage of the NMOS transistor, to distinguish it from that of the PMOS transistor, denoted V_{tp}.)

As Table 5.1 shows, the boundary between the triode region and the saturation region is determined by whether v_{DS} is less or greater than the overdrive voltage v_{OV} at which the transistor is operating. An equivalent way to check for the region of operation is to examine the relative values of the drain and gate voltages. To operate in the triode region, the gate voltage must exceed the drain voltage by at least V_{tn} volts, which ensures that the channel remains continuous (not pinched off). On the other hand, to operate in saturation, the channel must be pinched off at the drain end; pinch-off is achieved here by keeping v_D higher than $v_G - V_{tn}$, that is, not allowing v_D to fall below v_G by more than V_{tn} volts. The graphical construction of Fig. 5.12 should serve to remind the reader of these conditions.

A set of $i_D - v_{DS}$ characteristics for the NMOS transistor is shown in Fig. 5.13. Observe that each graph is obtained by setting v_{GS} above V_{tn} by a specific value of overdrive voltage, denoted V_{OV1}, V_{OV2}, V_{OV3}, and V_{OV4}. This in turn is the value of v_{DS} at which the corresponding graph saturates, and the value of the resulting saturation current is directly determined by the value of v_{OV}, namely, $\frac{1}{2}k_n' V_{OV1}^2$, $\frac{1}{2}k_n' V_{OV2}^2$, . . . The reader is advised to commit to memory both the structure of these graphs and the coordinates of the saturation points.

Finally, observe that the boundary between the triode and the saturation regions, that is, the locus of the saturation points, is a parabolic curve described by

$$i_D = \frac{1}{2}k_n'\left(\frac{W}{L}\right)v_{DS}^2$$

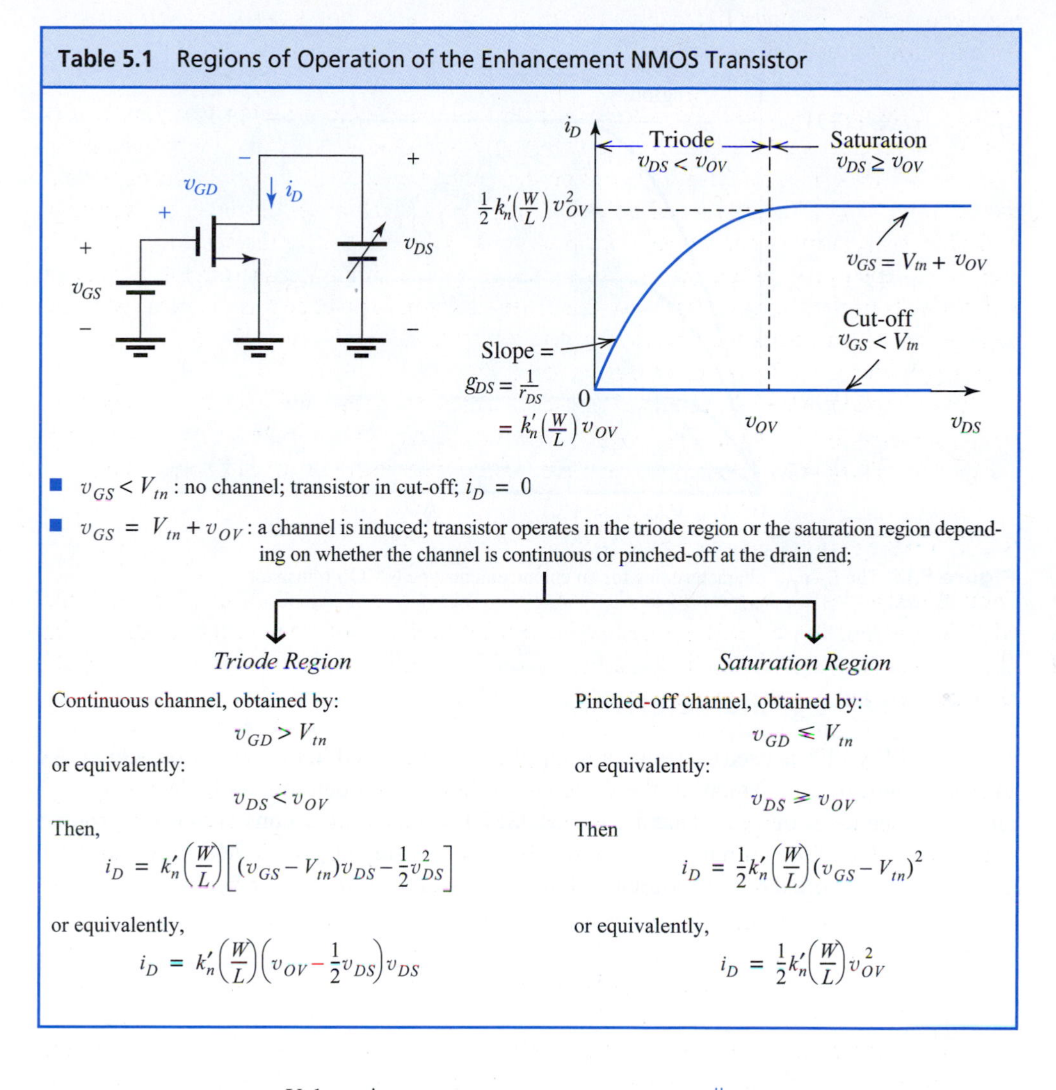

Table 5.1 Regions of Operation of the Enhancement NMOS Transistor

- $v_{GS} < V_{tn}$: no channel; transistor in cut-off; $i_D = 0$
- $v_{GS} = V_{tn} + v_{OV}$: a channel is induced; transistor operates in the triode region or the saturation region depending on whether the channel is continuous or pinched-off at the drain end;

Triode Region

Continuous channel, obtained by:

$$v_{GD} > V_{tn}$$

or equivalently:

$$v_{DS} < v_{OV}$$

Then,

$$i_D = k_n'\left(\frac{W}{L}\right)\left[(v_{GS} - V_{tn})v_{DS} - \frac{1}{2}v_{DS}^2\right]$$

or equivalently,

$$i_D = k_n'\left(\frac{W}{L}\right)\left(v_{OV} - \frac{1}{2}v_{DS}\right)v_{DS}$$

Saturation Region

Pinched-off channel, obtained by:

$$v_{GD} \leq V_{tn}$$

or equivalently:

$$v_{DS} \geq v_{OV}$$

Then

$$i_D = \frac{1}{2}k_n'\left(\frac{W}{L}\right)(v_{GS} - V_{tn})^2$$

or equivalently,

$$i_D = \frac{1}{2}k_n'\left(\frac{W}{L}\right)v_{OV}^2$$

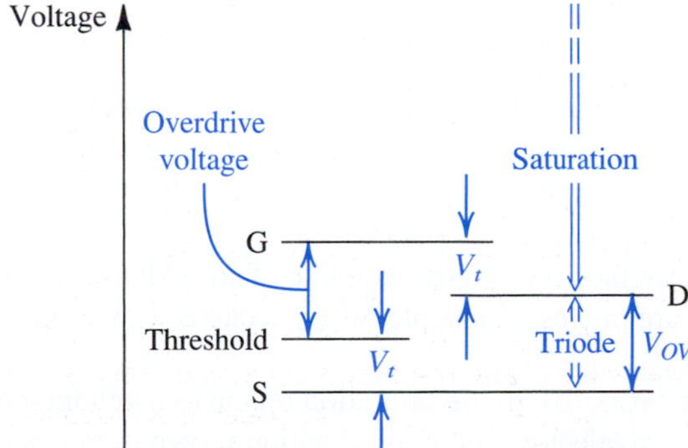

Figure 5.12 The relative levels of the terminal voltages of the enhancement NMOS transistor for operation in the triode region and in the saturation region.

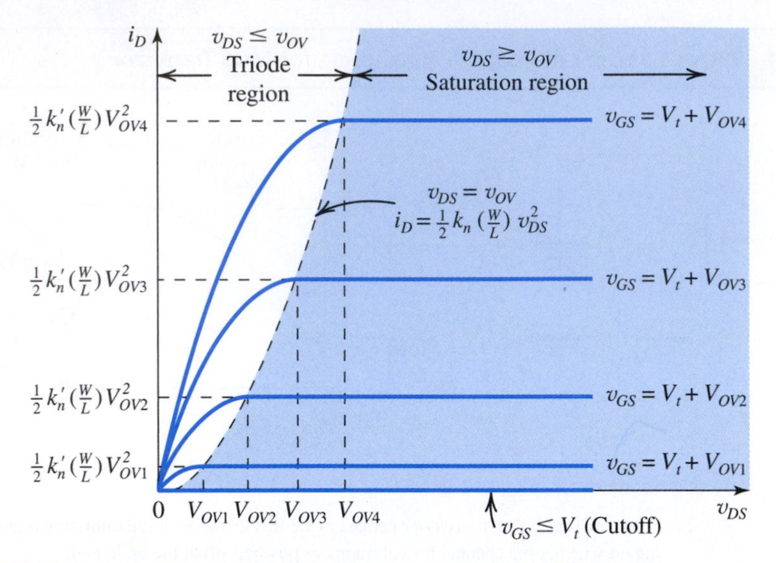

Figure 5.13 The i_D–v_{DS} characteristics for an enhancement-type NMOS transistor.

5.2.3 The i_D–v_{GS} Characteristic

When the MOSFET is used to design an amplifier, it is operated in the saturation region. As Fig. 5.13 indicates, in saturation the drain current is constant determined by v_{GS} (or v_{OV}) and is independent of v_{DS}. That is, the MOSFET operates as a constant-current source where the value of the current is determined by v_{GS}. In effect, then, the MOSFET operates as a voltage-controlled current source with the control relationship described by

$$i_D = \frac{1}{2}k_n'\left(\frac{W}{L}\right)(v_{GS} - V_{tn})^2 \tag{5.21}$$

or in terms of v_{OV},

$$i_D = \frac{1}{2}k_n'\left(\frac{W}{L}\right)v_{OV}^2 \tag{5.22}$$

This is the relationship that underlies the application of the MOSFET as an amplifier. That it is nonlinear should be of concern to those interested in designing linear amplifiers. Nevertheless, later in this chapter, we will see how one can obtain linear amplification from this nonlinear control or transfer characteristic.

Figure 5.14 shows the i_D–v_{GS} characteristic of an NMOS transistor operating in saturation. Note that if we are interested in a plot of i_D versus v_{OV}, we simply shift the origin to the point $v_{GS} = V_{tn}$.

The view of the MOSFET in the saturation region as a voltage-controlled current source is illustrated by the equivalent-circuit representation shown in Fig. 5.15. For reasons that will become apparent shortly, the circuit in Fig. 5.15 is known as a **large-signal equivalent circuit**. Note that the current source is ideal, with an infinite output resistance representing the independence, in saturation, of i_D from v_{DS}. This, of course, has been assumed in the idealized model of device operation utilized thus far. We are about to rectify an important shortcoming of this model. First, however, we present an example.

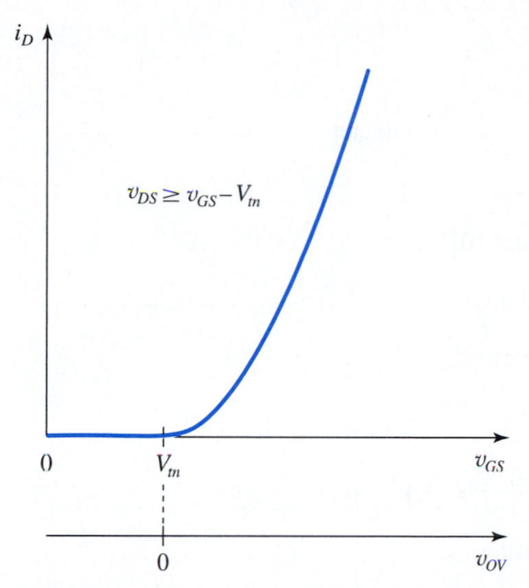

Figure 5.14 The i_D–v_{GS} characteristic of an NMOS transistor operating in the saturation region. The i_D–v_{OV} characteristic can be obtained by simply re-labelling the horizontal axis; that is, shifting the origin to the point $v_{GS} = V_{tn}$.

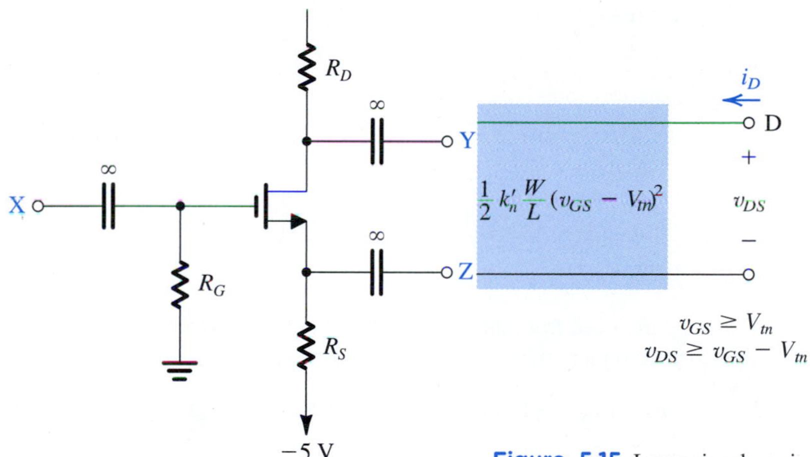

Figure 5.15 Large-signal equivalent-circuit model of an n-channel MOSFET operating in the saturation

Example 5.2

Consider an NMOS transistor fabricated in a 0.18-μm process with $L = 0.18$ μm and $W = 2$ μm. The process technology is specified to have $C_{ox} = 8.6$ fF/μm², $\mu_n = 450$ cm²/V·s, and $V_{tn} = 0.5$ V.

(a) Find V_{GS} and V_{DS} that result in the MOSFET operating at the edge of saturation with $I_D = 100$ μA.
(b) If V_{GS} is kept constant, find V_{DS} that results in $I_D = 50$ μA.
(c) To investigate the use of the MOSFET as a linear amplifier, let it be operating in saturation with $V_{DS} = 0.3$ V. Find the change in i_D resulting from v_{GS} changing from 0.7 V by +0.01 V and by −0.01 V.

Solution

First we determine the process transconductance parameter k_n'

$$k_n' = \mu_n C_{ox}$$
$$= 450 \times 10^{-4} \times 8.6 \times 10^{-15} \times 10^{12} \text{ A/V}^2$$
$$= 387 \text{ µA/V}^2$$

and the transistor transconductance parameter k_n,

$$k_n = k_n'\left(\frac{W}{L}\right)$$
$$= 387\left(\frac{2}{0.18}\right) = 4.3 \text{ mA/V}^2$$

(a) With the transistor operating in saturation,

$$I_D = \frac{1}{2}k_n V_{OV}^2$$

Thus,

$$100 = \frac{1}{2} \times 4.3 \times 10^3 \times V_{OV}^2$$

which results in

$$V_{OV} = 0.22 \text{ V}$$

Thus,

$$V_{GS} = V_{tn} + V_{OV} = 0.5 + 0.22 = 0.72 \text{ V}$$

and since operation is at the edge of saturation,

$$V_{DS} = V_{OV} = 0.22 \text{ V}$$

(b) With V_{GS} kept constant at 0.72 V and I_D reduced from the value obtained at the edge of saturation, the MOSFET will now be operating in the triode region, thus

$$I_D = k_n\left[V_{OV} V_{DS} - \frac{1}{2}V_{DS}^2\right]$$
$$50 = 4.3 \times 10^3\left[0.22 V_{DS} - \frac{1}{2}V_{DS}^2\right]$$

which can be rearranged to the form

$$V_{DS}^2 - 0.44 V_{DS} + 0.023 = 0$$

This quadratic equation has two solutions

$$V_{DS} = 0.06 \text{ V} \quad \text{and} \quad V_{DS} = 0.39 \text{ V}$$

The second answer is greater than V_{OV} and thus is physically meaningless, since we know that the transistor is operating in the triode region. Thus we have

$$V_{DS} = 0.06 \text{ V}$$

Example 5.2 *continued*

(c) For $v_{GS} = 0.7$ V, $V_{OV} = 0.2$ V, and since $V_{DS} = 0.3$ V, the transistor is operating in saturation and

$$I_D = \frac{1}{2}k_n V_{OV}^2$$
$$= \frac{1}{2} \times 4300 \times 0.04$$
$$= 86\ \mu\text{A}$$

Now for $v_{GS} = 0.710$ V, $v_{OV} = 0.21$ V and

$$i_D = \frac{1}{2} \times 4300 \times 0.21^2 = 94.8\ \mu\text{A}$$

and for $v_{GS} = 0.690$ V, $v_{OV} = 0.19$ V, and

$$i_D = \frac{1}{2} \times 4300 \times 0.19^2 = 77.6\ \mu\text{A}$$

Thus, with $\Delta V_{GS} = +0.01$ V, $\Delta i_D = 8.8\ \mu$A; and for $\Delta V_{GS} = -0.01$ V, $\Delta i_D = -8.4\ \mu$A.

We conclude that the two changes are almost equal, an indication of almost-linear operation when the changes in v_{GS} are kept small. This is just a preview of the "small-signal operation" of the MOSFET studied in Sections 5.4 and 5.5.

EXERCISES

5.4 An NMOS transistor is operating at the edge of saturation with an overdrive voltage V_{OV} and a drain current I_D. If V_{OV} is doubled, and we must maintain operation at the edge of saturation, what should V_{DS} be changed to? What value of drain current results?
Ans. 2 V_{OV}; 4 I_D

5.5 An n-channel MOSFET operating with $V_{OV} = 0.5$ V exhibits a linear resistance $r_{DS} = 1$ kΩ when v_{DS} is very small. What is the value of the device transconductance parameter k_n? What is the value of the current I_D obtained when v_{DS} is increased to 0.5 V? and to 1 V?
Ans. 2 mA/V^2; 0.25 mA; 0.25 mA

5.2.4 Finite Output Resistance in Saturation

Equation (5.21) and the corresponding large-signal equivalent circuit in Fig. 5.15, as well as the graphs in Fig. 5.13, indicate that in saturation, i_D is independent of v_{DS}. Thus, a change Δv_{DS} in the drain-to-source voltage causes a zero change in i_D, which implies that the incremental resistance looking into the drain of a saturated MOSFET is infinite. This, however, is an idealization based on the premise that once the channel is pinched off at the drain end, further increases in v_{DS} have no effect on the channel's shape. But, in practice, increasing v_{DS} beyond v_{OV} does affect the channel somewhat. Specifically, as v_{DS} is increased, the channel pinch-off point is moved slightly away from the drain, toward the source. This is illustrated in Fig. 5.16, from which we note that the voltage across the channel remains constant at v_{OV}, and the additional voltage applied to the drain appears as a voltage drop across the narrow depletion region between the end of the channel and the drain region. This voltage accelerates the electrons that reach the drain end of the channel and sweeps them across the depletion region into the drain. Note, however, that (with depletion-layer widening) the channel

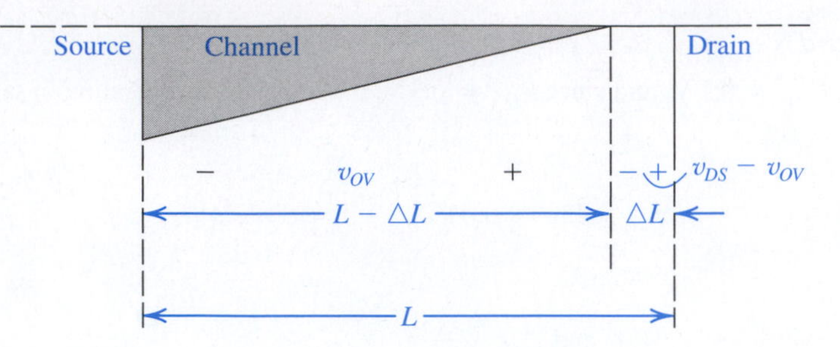

Figure 5.16 Increasing v_{DS} beyond v_{DSsat} causes the channel pinch-off point to move slightly away from the drain, thus reducing the effective channel length (by ΔL).

length is in effect reduced, from L to $L - \Delta L$, a phenomenon known as **channel-length modulation**. Now, since i_D is inversely proportional to the channel length (Eq. 5.21), i_D increases with v_{DS}.

This effect can be accounted for in the expression for i_D by including a factor $1 + \lambda(v_{DS} - v_{OV})$ or, for simplicity, $(1 + \lambda v_{DS})$,

$$i_D = \frac{1}{2}k'_n\left(\frac{W}{L}\right)(v_{GS} - V_{tn})^2\ (1 + \lambda v_{DS}) \tag{5.23}$$

Here λ is a device parameter having the units of reciprocal volts (V^{-1}). The value of λ depends both on the process technology used to fabricate the device and on the channel length L that the circuit designer selects. Specifically, the value of λ is much larger for newer submicron technologies than for older technologies. This makes intuitive sense: Newer technologies have very short channels, and are thus much greatly impacted by the channel-length modulation effect. Also, for a given process technology, λ is inversely proportional to L.

A typical set of i_D–v_{DS} characteristics showing the effect of channel-length modulation is displayed in Fig. 5.17. The observed linear dependence of i_D on v_{DS} in the saturation region is represented in Eq. (5.23) by the factor $(1 + \lambda v_{DS})$. From Fig. 5.17 we observe that when the straight-line i_D–v_{DS} characteristics are extrapolated, they intercept the v_{DS} axis at the point, $v_{DS} = -V_A$, where V_A is a positive voltage. Equation (5.23), however, indicates that $i_D = 0$ at $v_{DS} = -1/\lambda$. It follows that

$$V_A = \frac{1}{\lambda}$$

and thus V_A is a device parameter with the dimensions of V. For a given process, V_A is proportional to the channel length L that the designer selects for a MOSFET. We can isolate the dependence of V_A on L by expressing it as

$$V_A = V'_A L$$

where V'_A is entirely process-technology dependent with the dimensions of volts per micron. Typically, V'_A falls in the range of 5 V/μm to 50 V/μm. The voltage V_A is usually referred to as the Early voltage, after J. M. Early, who discovered a similar phenomenon for the BJT (Chapter 6).

Equation (5.23) indicates that when channel-length modulation is taken into account, the saturation values of i_D depend on v_{DS}. Thus, for a given v_{GS}, a change Δv_{DS} yields a corresponding change Δi_D in the drain current i_D. It follows that the output resistance of the current source representing i_D in saturation is no longer infinite. Defining the output resistance r_o as[7]

[7]In this book we use r_o to denote the output resistance in saturation, and r_{DS} to denote the drain-to-source resistance in the triode region, for small v_{DS}.

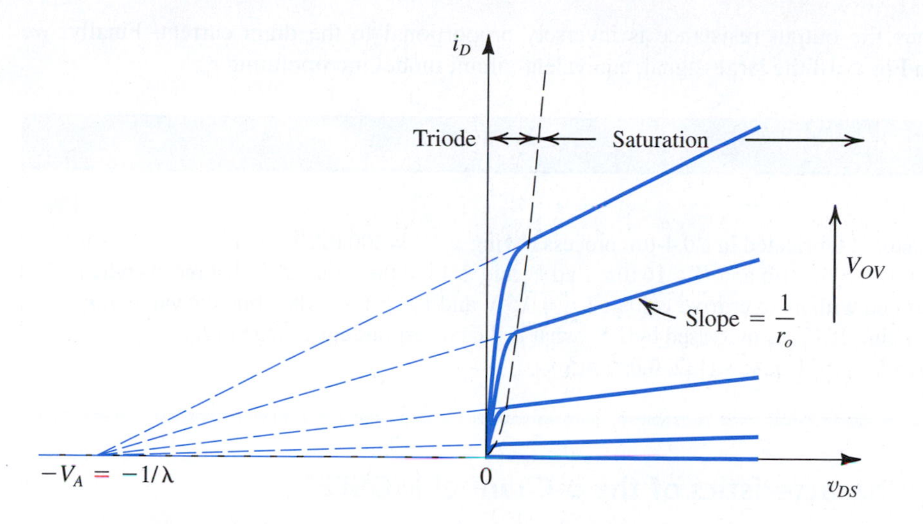

Figure 5.17 Effect of v_{DS} on i_D in the saturation region. The MOSFET parameter V_A depends on the process technology and, for a given process, is proportional to the channel length L.

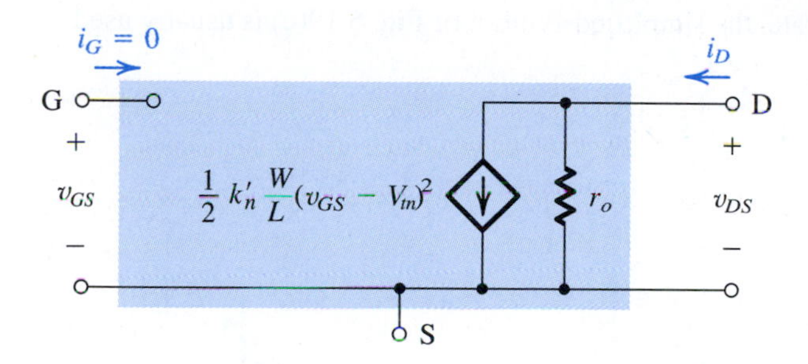

Figure 5.18 Large-signal equivalent circuit model of the n-channel MOSFET in saturation, incorporating the output resistance r_o. The output resistance models the linear dependence of i_D on v_{DS} and is given by Eq. (5.23).

$$r_o \equiv \left[\frac{\partial i_D}{\partial v_{DS}}\right]^{-1}_{v_{GS}\text{ constant}} \tag{5.24}$$

and using Eq. (5.23) results in

$$r_o = \left[\lambda \frac{k_n'}{2}\frac{W}{L}(V_{GS} - V_{tn})^2\right]^{-1} \tag{5.25}$$

which can be written as

$$r_o = \frac{1}{\lambda I_D} \tag{5.26}$$

or, equivalently,

$$r_o = \frac{V_A}{I_D} \tag{5.27}$$

where I_D is the drain current *without* channel-length modulation taken into account; that is,

$$I_D = \frac{1}{2}k_n' \frac{W}{L}(V_{GS} - V_{tn})^2 \tag{5.27'}$$

Thus the output resistance is inversely proportional to the drain current. Finally, we show in Fig. 5.18 the large-signal, equivalent-circuit model incorporating r_o.

EXERCISE

5.6 An NMOS transistor is fabricated in a 0.4-μm process having $\mu_n C_{ox} = 200\ \mu A/V^2$ and $V_A' = 50\ V/\mu m$ of channel length. If $L = 0.8\ \mu m$ and $W = 16\ \mu m$, find V_A and λ. Find the value of I_D that results when the device is operated with an overdrive voltage $V_{OV} = 0.5$ V and $V_{DS} = 1$ V. Also, find the value of r_o at this operating point. If V_{DS} is increased by 2 V, what is the corresponding change in I_D?
Ans. 40 V; 0.025 V^{-1}; 0.51 mA; 80 kΩ; 0.025 mA

5.2.5 Characteristics of the *p*-Channel MOSFET

The circuit symbol for the *p*-channel enhancement-type MOSFET is shown in Fig. 5.19(a). Figure 5.19(b) shows a modified circuit symbol in which an arrowhead pointing in the normal direction of current flow is included on the source terminal. For the case where the source is connected to the substrate, the simplified symbol of Fig. 5.19(c) is usually used.

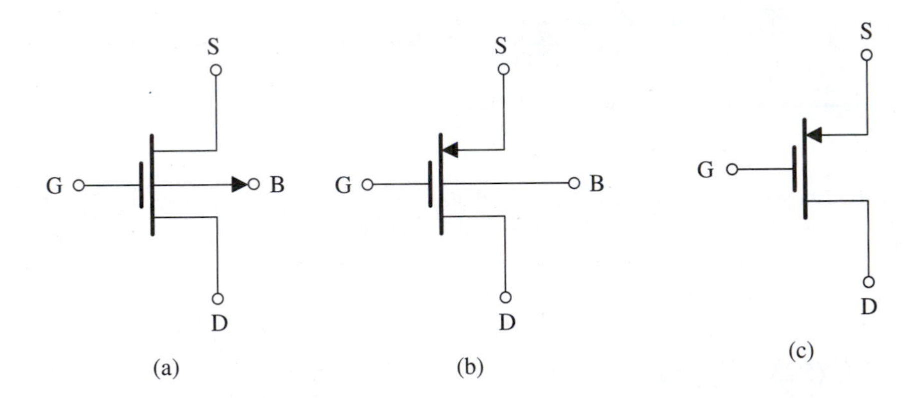

Figure 5.19 **(a)** Circuit symbol for the *p*-channel enhancement-type MOSFET. **(b)** Modified symbol with an arrowhead on the source lead. **(c)** Simplified circuit symbol for the case where the source is connected to the body.

The regions of operation of the PMOS transistor and the corresponding conditions and expression for i_D are shown in Table 5.2. Observe that the equations are written in a way that emphasizes physical intuition and avoids the confusion of negative signs. Thus while V_{tp} is by convention negative, we use $|V_{tp}|$, and the voltages v_{SG} and v_{SD} are positive. Also, in all of our circuit diagrams we will always draw *p*-channel devices with their sources on top so that current flows from top to bottom. Finally, we note that PMOS devices also suffer from the channel-length modulation effect. This can be taken into account by including a factor $(1 + |\lambda| v_{SD})$ in the saturation-region expression for i_D as follows

$$i_D = \frac{1}{2} k_p' \left(\frac{W}{L}\right) (v_{SG} - |V_{tp}|)^2 (1 + |\lambda| v_{SD}) \tag{5.28}$$

or equivalently

$$i_D = \frac{1}{2}k_p'\left(\frac{W}{L}\right)(v_{SG} - |V_{tp}|)^2\left(1 + \frac{v_{SD}}{|V_A|}\right) \tag{5.29}$$

where λ and V_A (the Early voltage for the PMOS transistor) are by convention negative quantities, hence we use $|\lambda|$ and $|V_A|$.

Finally, we should note that for a given CMOS fabrication process λ_n and $|\lambda_p|$ are generally not equal, and similarly for V_{An} and $|V_{Ap}|$.

To recap, to turn a PMOS transistor on, the gate voltage has to be made lower than that of the source by at least $|V_{tp}|$. To operate in the triode region, the drain voltage has to exceed that of the gate by at least $|V_{tp}|$; otherwise, the PMOS operates in saturation. Finally, Fig. 5.20 provides a pictorial representation of these operating conditions.

Table 5.2 Regions of Operation of the Enhancement PMOS Transistor

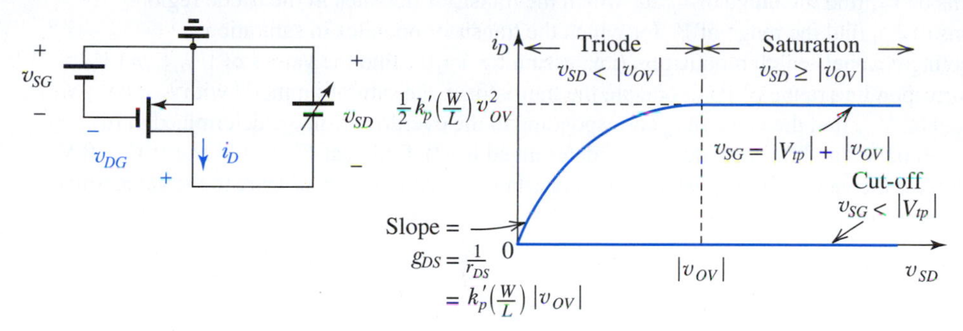

- $v_{SG} < |V_{tp}|$: no channel; transistor in cut-off; $i_D = 0$
- $v_{SG} = |V_{tp}| + |v_{OV}|$: a channel is induced; transistor operates in the triode region or in the saturation region depending on whether the channel is continuous or pinched-off at the drain end;

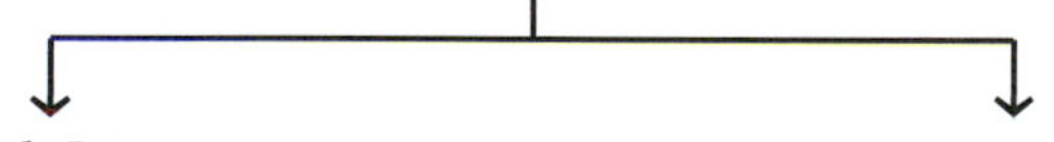

Triode Region

Continuous channel, obtained by:

$$v_{DG} > |V_{tp}|$$

or equivalently:

$$v_{SD} < |v_{OV}|$$

Then,

$$i_D = k_p'\left(\frac{W}{L}\right)\left[(v_{SG} - |V_{tp}|)v_{SD} - \frac{1}{2}v_{SD}^2\right]$$

or equivalently

$$i_D = k_p'\left(\frac{W}{L}\right)\left(|v_{OV}| - \frac{1}{2}v_{SD}\right)v_{SD}$$

Saturation Region

Pinched-off channel, obtained by:

$$v_{DG} \leqslant |V_{tp}|$$

or equivalently

$$v_{SD} \geqslant |v_{OV}|$$

Then

$$i_D = \frac{1}{2}k_p'\left(\frac{W}{L}\right)(v_{SG} - |V_{tp}|)^2$$

or equivalently

$$i_D = \frac{1}{2}k_p'\left(\frac{W}{L}\right)v_{OV}^2$$

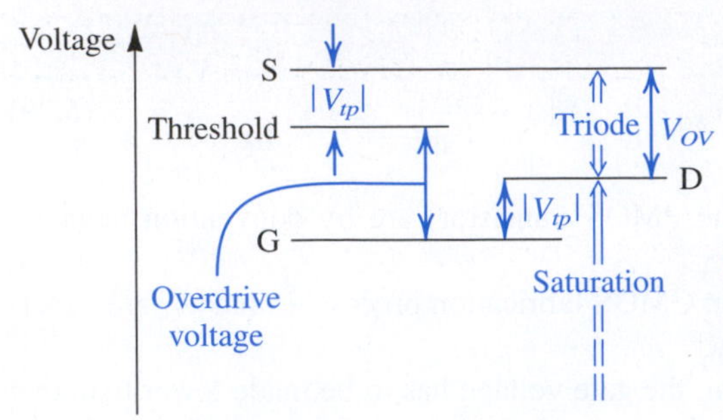

Figure 5.20 The relative levels of the terminal voltages of the enhancement-type PMOS transistor for operation in the triode region and in the saturation region.

EXERCISE

5.7 The PMOS transistor shown in Fig. E5.7 has $V_{tp} = -1$ V, $k'_p = 60\ \mu A/V^2$, and $W/L = 10$.
(a) Find the range of V_G for which the transistor conducts.
(b) In terms of V_G, find the range of V_D for which the transistor operates in the triode region.
(c) In terms of V_G, find the range of V_D for which the transistor operates in saturation.
(d) Neglecting channel-length modulation (i.e., assuming $\lambda = 0$), find the values of $|V_{OV}|$ and V_G and the corresponding range of V_D to operate the transistor in the saturation mode with $I_D = 75\ \mu A$.
(e) If $\lambda = -0.02\ V^{-1}$, find the value of r_o corresponding to the overdrive voltage determined in (d).
(f) For $\lambda = -0.02\ V^{-1}$ and for the value of V_{OV} determined in (d), find I_D at $V_D = +3$ V and at $V_D = 0$ V; hence, calculate the value of the apparent output resistance in saturation. Compare to the value found in (e).

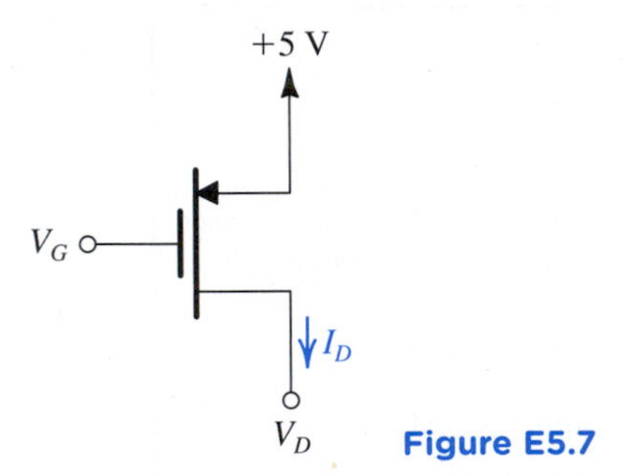

Figure E5.7

Ans. (a) $V_G \leq +4$ V; (b) $V_D \geq V_G + 1$; (c) $V_D \leq V_G + 1$; (d) 0.5 V, 3.5 V, ≤4.5 V; (e) 0.67 MΩ; (f) 78 μA, 82.5 μA, 0.67 MΩ (same)

5.3 MOSFET Circuits at DC

Having studied the current–voltage characteristics of MOSFETs, we now consider circuits in which only dc voltages and currents are of concern. Specifically, we shall present a series of design and analysis examples of MOSFET circuits at dc. The objective is to instill in the reader a familiarity with the device and the ability to perform MOSFET circuit analysis both rapidly and effectively.

In the following examples, to keep matters simple and thus focus attention on the essence of MOSFET circuit operation, we will generally neglect channel-length modulation; that is, we will assume $\lambda = 0$. We will find it convenient to work in terms of the overdrive voltage; $V_{OV} = V_{GS} - V_{tn}$ for NMOS and $|V_{OV}| = V_{SG} - |V_{tp}|$ for PMOS.

Example 5.3

Design the circuit of Fig. 5.21, that is, determine the values of R_D and R_S, so that the transistor operates at $I_D = 0.4$ mA and $V_D = +0.5$ V. The NMOS transistor has $V_t = 0.7$ V, $\mu_n C_{ox} = 100\ \mu\text{A/V}^2$, $L = 1\ \mu$m, and $W = 32\ \mu$m. Neglect the channel-length modulation effect (i.e., assume that $\lambda = 0$).

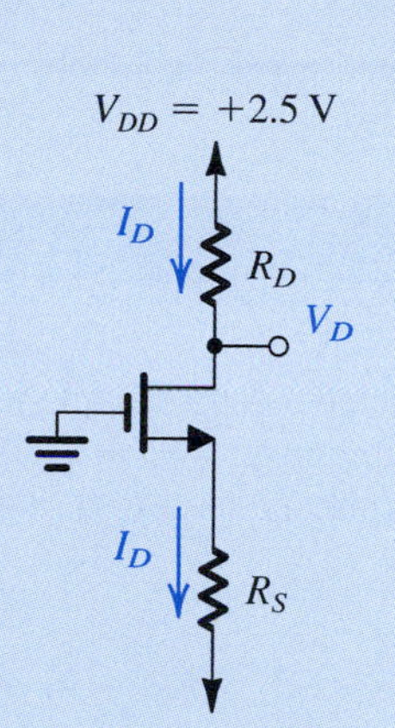

Figure 5.21 Circuit for Example 5.3.

Solution

To establish a dc voltage of +0.5 V at the drain, we must select R_D as follows:

$$R_D = \frac{V_{DD} - V_D}{I_D}$$

$$= \frac{2.5 - 0.5}{0.4} = 5\ \text{k}\Omega$$

To determine the value required for R_S, we need to know the voltage at the source, which can be easily found if we know V_{GS}. This in turn can be determined from V_{OV}. Toward that end, we note that since $V_D =$ 0.5 V is greater than V_G, the NMOS transistor is operating in the saturation region, and we can use the saturation-region expression of i_D to determine the required value of V_{OV},

$$I_D = \frac{1}{2}\mu_n C_{ox}\frac{W}{L}V_{OV}^2$$

Then substituting $I_D = 0.4\ \text{mA} = 400\ \mu\text{A}$, $\mu_n C_{ox} = 100\ \mu\text{A/V}^2$, and $W/L = 32/1$ gives

$$400 = \frac{1}{2} \times 100 \times \frac{32}{1}V_{OV}^2$$

which results in

$$V_{OV} = 0.5\ \text{V}$$

Thus,

$$V_{GS} = V_t + V_{OV} = 0.7 + 0.5 = 1.2\ \text{V}$$

Referring to Fig. 5.21, we note that the gate is at ground potential. Thus, the source must be at −1.2 V, and the required value of R_S can be determined from

$$R_S = \frac{V_S - V_{SS}}{I_D}$$

$$= \frac{-1.2 - (-2.5)}{0.4} = 3.25\ \text{k}\Omega$$

EXERCISE

D5.8 Redesign the circuit of Fig. 5.21 for the following case: $V_{DD} = -V_{SS} = 2.5$ V, $V_t = 1$ V, $\mu_n C_{ox} = 60$ μA/V², $W/L = 120$ μm/3 μm, $I_D = 0.3$ mA, and $V_D = +0.4$ V.
Ans. $R_D = 7$ kΩ; $R_S = 3.3$ kΩ

Example 5.4

Figure 5.22 shows an NMOS transistor with its drain and gate terminals connected together. Find the $i-v$ relationship of the resulting two-terminal device in terms of the MOSFET parameters $k_n = k_n'(W/L)$ and V_{tn}. Neglect channel-length modulation (i.e., $\lambda = 0$). Note that this two-terminal device is known as a **diode-connected transistor**.

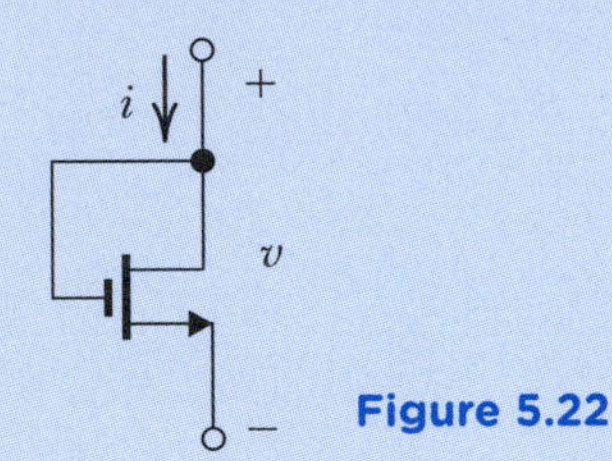

Figure 5.22

Solution

Since $v_D = v_G$ implies operation in the saturation mode,

$$i_D = \frac{1}{2}k_n'\left(\frac{W}{L}\right)(v_{GS} - V_{tn})^2$$

Now, $i = i_D$ and $v = v_{GS}$, thus

$$i = \frac{1}{2}k_n'\left(\frac{W}{L}\right)(v - V_{tn})^2$$

Replacing $k_n'\left(\frac{W}{L}\right)$ by k_n results in

$$i = \frac{1}{2}k_n(v - V_{tn})^2$$

EXERCISES

D5.9 For the circuit in Fig. E5.9, find the value of R that results in $V_D = 0.8$ V. The MOSFET has $V_{tn} = 0.5$ V, $\mu_n C_{ox} = 0.4$ mA/V^2, $W/L = \dfrac{0.72\ \mu\text{m}}{0.18\ \mu\text{m}}$, and $\lambda = 0$.

Ans. 13.9 kΩ

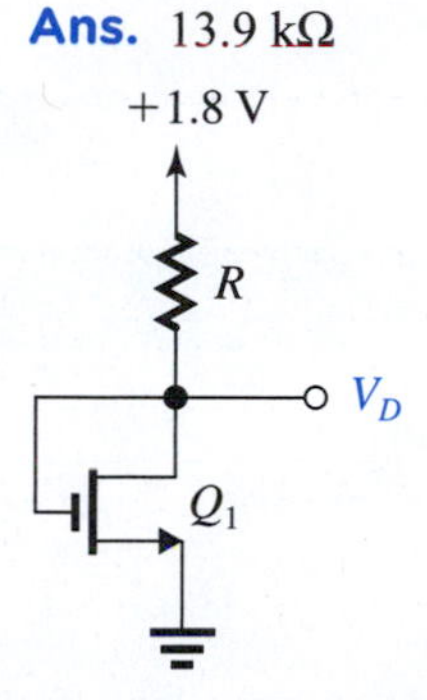

Figure E5.9

D5.10 Figure E5.10, shows a circuit obtained by augmenting the circuit of Fig. E5.9 considered in Exercise 5.9 with a transistor Q_2 identical to Q_1 and a resistance R_2. Find the value of R_2 that results in Q_2 operating at the edge of the saturation region. Use your solution to Exercise 5.9.

Ans. 20.8 kΩ

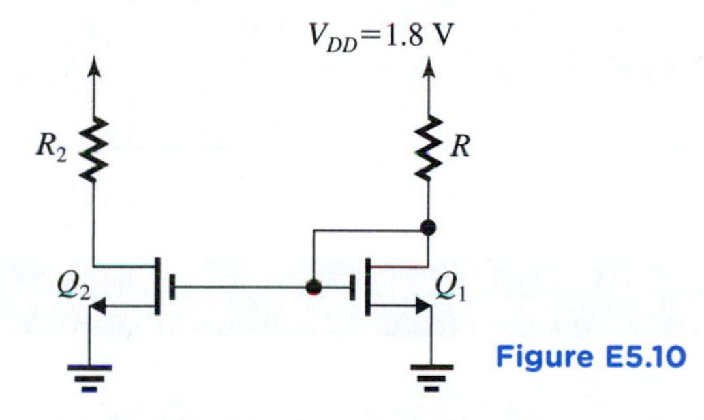

Figure E5.10

Example 5.5

Design the circuit in Fig. 5.23 to establish a drain voltage of 0.1 V. What is the effective resistance between drain and source at this operating point? Let $V_{tn} = 1$ V and $k_n'(W/L) = 1$ mA/V^2.

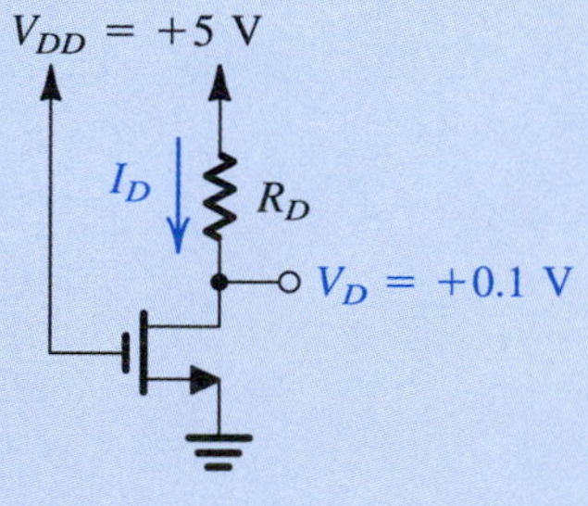

Figure 5.23 Circuit for Example 5.5.

Example 5.5 *continued*

Solution

Since the drain voltage is lower than the gate voltage by 4.9 V and $V_{tn} = 1$ V, the MOSFET is operating in the triode region. Thus the current I_D is given by

$$I_D = k_n' \frac{W}{L}\left[(V_{GS} - V_{tn})V_{DS} - \frac{1}{2}V_{DS}^2\right]$$

$$I_D = 1 \times \left[(5-1) \times 0.1 - \frac{1}{2} \times 0.01\right]$$

$$= 0.395 \text{ mA}$$

The required value for R_D can be found as follows:

$$R_D = \frac{V_{DD} - V_D}{I_D}$$

$$= \frac{5 - 0.1}{0.395} = 12.4 \text{ k}\Omega$$

In a practical discrete-circuit design problem, one selects the closest standard value available for, say, 5% resistors—in this case, 12 kΩ; see Appendix G. Since the transistor is operating in the triode region with a small V_{DS}, the effective drain-to-source resistance can be determined as follows:

$$r_{DS} = \frac{V_{DS}}{I_D}$$

$$= \frac{0.1}{0.395} = 253 \ \Omega$$

EXERCISE

5.11 If in the circuit of Example 5.5 the value of R_D is doubled, find approximate values for I_D and V_D.
Ans. 0.2 mA; 0.05 V

Example 5.6

Analyze the circuit shown in Fig. 5.24(a) to determine the voltages at all nodes and the currents through all branches. Let $V_{tn} = 1$ V and $k_n'(W/L) = 1$ mA/V^2. Neglect the channel-length modulation effect (i.e., assume $\lambda = 0$).

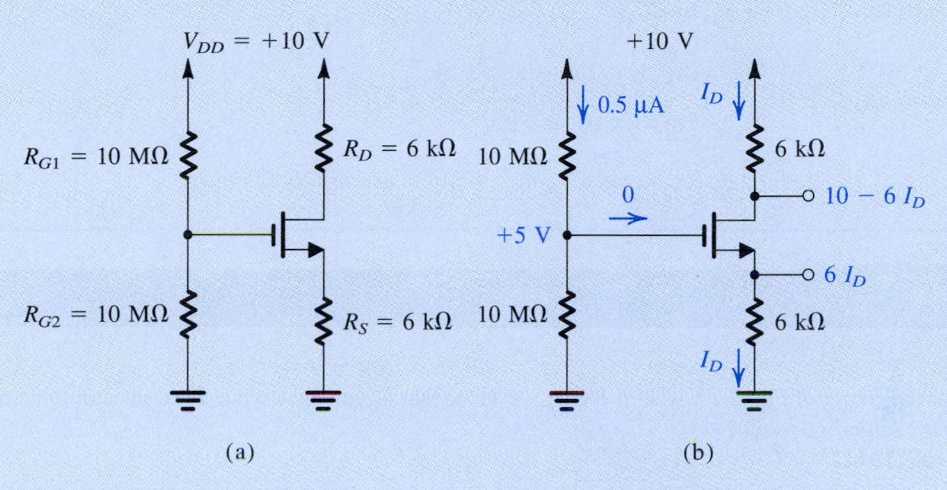

Figure 5.24 **(a)** Circuit for Example 5.6. **(b)** The circuit with some of the analysis details shown.

Solution

Since the gate current is zero, the voltage at the gate is simply determined by the voltage divider formed by the two 10-MΩ resistors,

$$V_G = V_{DD}\frac{R_{G2}}{R_{G2}+R_{G1}} = 10 \times \frac{10}{10+10} = +5 \text{ V}$$

With this positive voltage at the gate, the NMOS transistor will be turned on. We do not know, however, whether the transistor will be operating in the saturation region or in the triode region. We shall assume saturation-region operation, solve the problem, and then check the validity of our assumption. Obviously, if our assumption turns out not to be valid, we will have to solve the problem again for triode-region operation.

Refer to Fig. 5.24(b). Since the voltage at the gate is 5 V and the voltage at the source is $I_D\,(\text{mA}) \times 6\,(\text{k}\Omega) = 6I_D$, we have

$$V_{GS} = 5 - 6I_D$$

Thus, I_D is given by

$$I_D = \frac{1}{2}k_n'\frac{W}{L}(V_{GS} - V_{tn})^2$$

$$= \frac{1}{2} \times 1 \times (5 - 6I_D - 1)^2$$

which results in the following quadratic equation in I_D:

$$18I_D^2 - 25I_D + 8 = 0$$

Example 5.6 *continued*

This equation yields two values for I_D: 0.89 mA and 0.5 mA. The first value results in a source voltage of $6 \times 0.89 = 5.34$ V, which is greater than the gate voltage and does not make physical sense as it would imply that the NMOS transistor is cut off. Thus,

$$I_D = 0.5 \text{ mA}$$

$$V_S = 0.5 \times 6 = +3 \text{ V}$$

$$V_{GS} = 5 - 3 = 2 \text{ V}$$

$$V_D = 10 - 6 \times 0.5 = +7 \text{ V}$$

Since $V_D > V_G - V_{tn}$, the transistor is operating in saturation, as initially assumed.

EXERCISES

5.12 For the circuit of Fig. 5.24, what is the largest value that R_D can have while the transistor remains in the saturation mode?
Ans. 12 kΩ

D5.13 Redesign the circuit of Fig. 5.24 for the following requirements: $V_{DD} = +5$ V, $I_D = 0.32$ mA, $V_S = 1.6$ V, $V_D = 3.4$ V, with a 1-μA current through the voltage divider R_{G1}, R_{G2}. Assume the same MOSFET as in Example 5.6.
Ans. $R_{G1} = 1.6$ MΩ; $R_{G2} = 3.4$ MΩ, $R_S = R_D = 5$ kΩ

Example 5.7

Design the circuit of Fig. 5.25 so that the transistor operates in saturation with $I_D = 0.5$ mA and $V_D = +3$ V. Let the enhancement-type PMOS transistor have $V_{tp} = -1$ V and $k_p'(W/L) = 1$ mA/V^2. Assume $\lambda = 0$. What is the largest value that R_D can have while maintaining saturation-region operation?

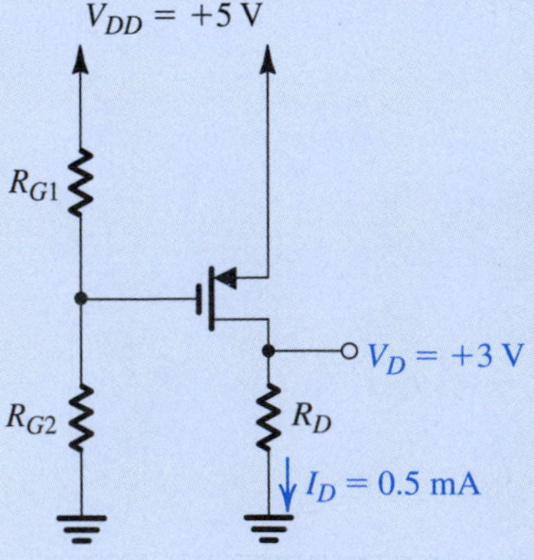

Figure 5.25 Circuit for Example 5.7.

Solution

Since the MOSFET is to be in saturation, we can write

$$I_D = \frac{1}{2}k_p'\frac{W}{L}|V_{OV}|^2$$

Substituting $I_D = 0.5$ mA and $k_p'W/L = 1\ \text{mA/V}^2$, we obtain

$$|V_{OV}| = 1\text{ V}$$

and

$$V_{SG} = |V_{tp}| + |V_{OV}| = 1 + 1 = 2\text{ V}$$

Since the source is at +5 V, the gate voltage must be set to +3 V. This can be achieved by the appropriate selection of the values of R_{G1} and R_{G2}. A possible selection is $R_{G1} = 2\ \text{M}\Omega$ and $R_{G2} = 3\ \text{M}\Omega$.

The value of R_D can be found from

$$R_D = \frac{V_D}{I_D} = \frac{3}{0.5} = 6\text{ k}\Omega$$

Saturation-mode operation will be maintained up to the point that V_D exceeds V_G by $|V_{tp}|$; that is, until

$$V_{D_{\max}} = 3 + 1 = 4\text{ V}$$

This value of drain voltage is obtained with R_D given by

$$R_D = \frac{4}{0.5} = 8\text{ k}\Omega$$

EXERCISE

D5.14 For the circuit in Fig. E5.14, find the value of R that results in the PMOS transistor operating with an overdrive voltage $|V_{OV}| = 0.6$ V. The threshold voltage is $V_{tp} = -0.4$ V, the process transconductance parameter $k_p' = 0.1\text{mA/V}^2$, and $W/L = 10\ \mu\text{m}/0.18\ \mu\text{m}$.
Ans. 800 Ω

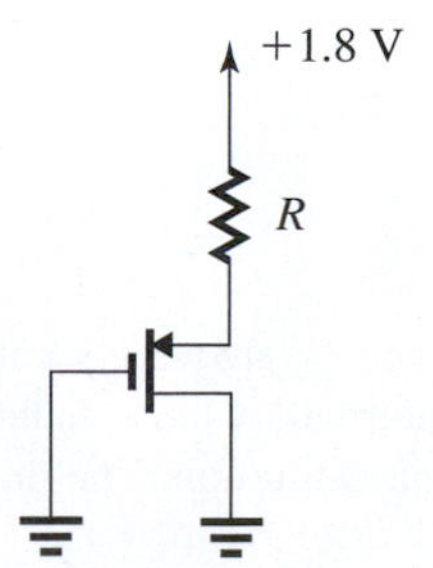

Figure E5.14

Example 5.8

The NMOS and PMOS transistors in the circuit of Fig. 5.26(a) are matched, with $k_n'(W_n/L_n) = k_p'(W_p/L_p) = 1\ \text{mA/V}^2$ and $V_{tn} = -V_{tp} = 1\ \text{V}$. Assuming $\lambda = 0$ for both devices, find the drain currents i_{DN} and i_{DP}, as well as the voltage v_O, for $v_I = 0$ V, +2.5 V, and −2.5 V.

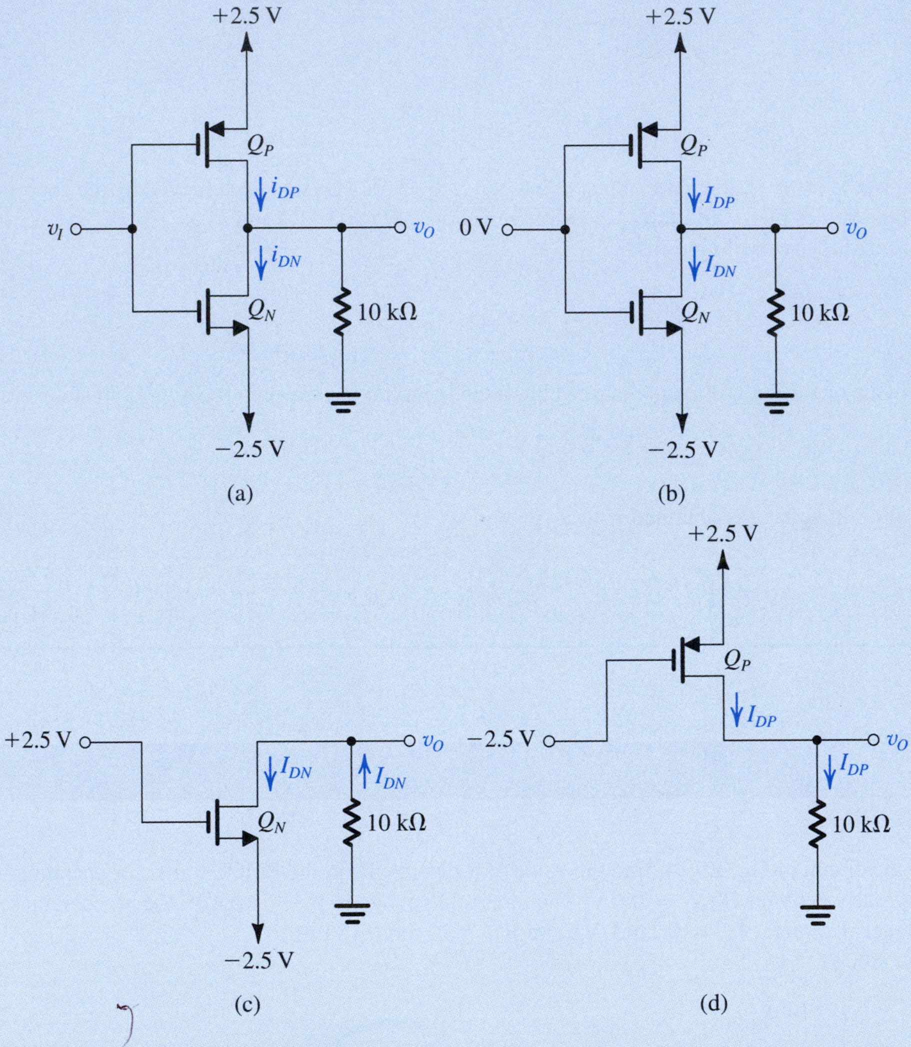

Figure 5.26 Circuits for Example 5.8.

Solution

Figure 5.26(b) shows the circuit for the case $v_I = 0$ V. We note that since Q_N and Q_P are perfectly matched and are operating at equal values of $|V_{GS}|$ (2.5 V), the circuit is symmetrical, which dictates that $v_O = 0$ V. Thus both Q_N and Q_P are operating with $|V_{DG}| = 0$ and, hence, in saturation. The drain currents can now be found from

$$I_{DP} = I_{DN} = \frac{1}{2} \times 1 \times (2.5 - 1)^2 = 1.125\ \text{mA}$$

Next, we consider the circuit with $v_I = +2.5$ V. Transistor Q_P will have a V_{SG} of zero and thus will be cut off, reducing the circuit to that shown in Fig. 5.26(c). We note that v_O will be negative, and thus v_{GD} will be greater than V_{tn}, causing Q_N to operate in the triode region. For simplicity we shall assume that v_{DS} is small and thus use

$$I_{DN} \simeq k_n'(W_n/L_n)(V_{GS} - V_{tn})V_{DS}$$
$$= 1[2.5 - (-2.5) - 1][v_O - (-2.5)]$$

From the circuit diagram shown in Fig. 5.26(c), we can also write

$$I_{DN}\,(\text{mA}) = \frac{0 - v_O}{10\,(\text{k}\Omega)}$$

These two equations can be solved simultaneously to yield

$$I_{DN} = 0.244 \text{ mA} \qquad v_O = -2.44 \text{ V}$$

Note that $V_{DS} = -2.44 - (-2.5) = 0.06$ V, which is small as assumed.

Finally, the situation for the case $v_I = -2.5$ V [Fig. 5.26(d)] will be the exact complement of the case $v_I = +2.5$ V: Transistor Q_N will be off. Thus $I_{DN} = 0$, Q_P will be operating in the triode region with $I_{DP} = 2.44$ mA and $v_O = +2.44$ V.

EXERCISE

5.15 The NMOS and PMOS transistors in the circuit of Fig. E5.15 are matched with $k_n'(W_n/L_n) = k_p'(W_p/L_p) = 1 \text{ mA/V}^2$ and $V_{tn} = -V_{tp} = 1$ V. Assuming $\lambda = 0$ for both devices, find the drain currents i_{DN} and i_{DP} and the voltage v_O for $v_I = 0$ V, +2.5 V, and −2.5 V.
Ans. $v_I = 0$ V: 0 mA, 0 mA, 0 V; $v_I = +2.5$ V: 0.104 mA, 0 mA, 1.04 V; $v_I = -2.5$ V: 0 mA, 0.104 mA, −1.04 V

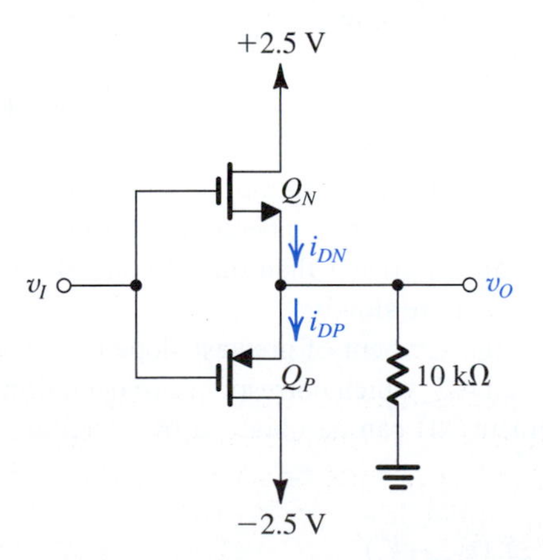

Figure E5.15

5.4 Applying the MOSFET in Amplifier Design

We now begin our study of the utilization of the MOSFET in the design of amplifiers. The basis for this important application is that when operated in saturation, the MOSFET functions as voltage-controlled current source: The gate-to-source voltage v_{GS} controls the drain current i_D. Although the control relationship is nonlinear (square law), we will shortly devise a method for obtaining almost-linear amplification from this fundamentally nonlinear device.

5.4.1 Obtaining a Voltage Amplifier

In the introduction to amplifier circuits in Section 1.5, we learned that a voltage-controlled current source can serve as a transconductance amplifier; that is, an amplifier whose input signal is a voltage and whose output signal is a current. More commonly, however, one is interested in voltage amplifiers. A simple way to convert a transconductance amplifier to a voltage amplifier is to pass the output current through a resistor and take the voltage across the resistor as the output. Doing this for a MOSFET results in the simple amplifier circuit shown in Fig. 5.27(a). Here v_{GS} is the input voltage, R_D (known as a **load resistance**) converts the drain current i_D to a voltage ($i_D R_D$), and V_{DD} is the supply voltage that powers up the amplifier and, together with R_D, establishes operation in the saturation region, as will be shown shortly.

In the amplifier circuit of Fig. 5.27(a) the output voltage is taken between the drain and ground, rather than simply across R_D. This is done because of the need to maintain a ground reference throughout the circuit. The output voltage v_{DS} is given by

$$v_{DS} = V_{DD} - i_D R_D \tag{5.30}$$

Thus it is an inverted version (note the minus sign) of $i_D R_D$ that is shifted by the constant value of the supply voltage V_{DD}.

5.4.2 The Voltage Transfer Characteristic (VTC)

A very useful tool that yields great insight into the operation of an amplifier circuit is its voltage transfer characteristic (VTC). This is simply a plot (or a clearly labeled sketch) of the output voltage versus the input voltage. For the MOS amplifier in Fig. 5.27(a), this is the plot of v_{DS} versus v_{GS} shown in Fig. 5.27(b).

Observe that for $v_{GS} < V_t$, the transistor is cut off, $i_D = 0$ and, from Eq. (5.30), $v_{DS} = V_{DD}$. As v_{GS} exceeds V_t, the transistor turns on and v_{DS} decreases. However, since initially v_{DS} is still high, the MOSFET will be operating in saturation. This continues as v_{GS} is increased until the value of v_{GS} is reached that results in v_{DS} becoming lower than v_{GS} by V_t volts (point B on the VTC in Fig. 5.27b). For v_{GS} greater than that at point B, the transistor operates in the triode region and v_{DS} decreases more slowly.

The VTC in Fig. 5.27(b) indicates that the segment of greatest slope (and hence potentially the largest amplifier gain) is that labeled AB, which corresponds to operation in the saturation region. An expression for the segment AB can be obtained by substituting for i_D in Eq. (5.30) by its saturation-region value

$$i_D = \frac{1}{2}k_n(v_{GS} - V_t)^2 \tag{5.31}$$

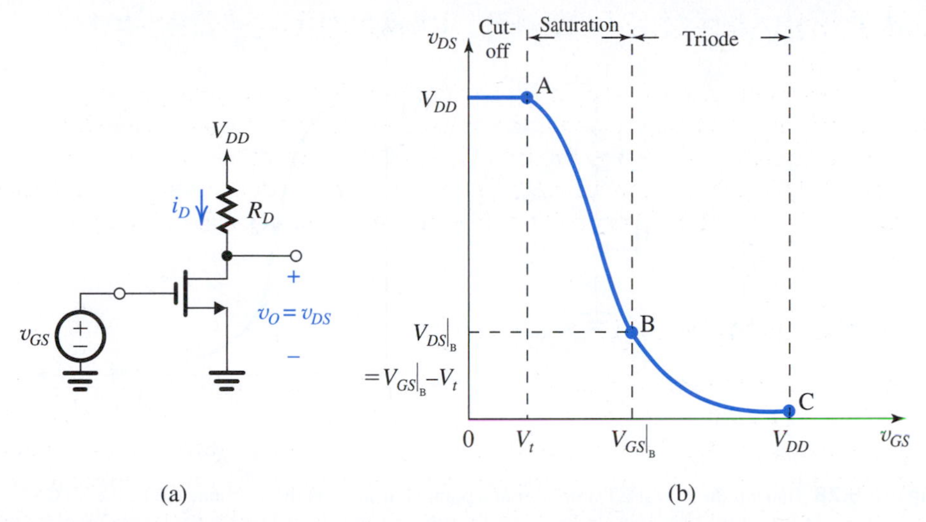

Figure 5.27 **(a)** Simple MOSFET amplifier with input v_{GS} and output v_{DS}. **(b)** The voltage transfer characteristic (VTC) of the amplifier in **(a)**. The three segments of the VTC correspond to the three regions of operation of the MOSFET.

where we have for simplicity neglected channel-length modulation. The result is

$$v_{DS} = V_{DD} - \frac{1}{2}k_n R_D (v_{GS} - V_t)^2 \tag{5.32}$$

This is obviously a nonlinear relationship. Nevertheless, linear (or almost-linear) amplification can be obtained by using the technique of biasing the MOSFET. Before considering biasing, however, it is useful to determine the coordinates of point B, which is at the boundary between the saturation and the triode regions of operation. These can be obtained by substituting in Eq. (5.32), $v_{GS} = V_{GS}|_B$ and $v_{DS} = V_{DS}|_B = V_{GS}|_B - V_t$. The result is

$$V_{GS}|_B = V_t + \frac{\sqrt{2k_n R_D V_{DD} + 1} - 1}{k_n R_D} \tag{5.33}$$

EXERCISE

5.16 Consider the amplifier of Fig. 5.27(a) with $V_{DD} = 1.8$ V, $R_D = 17.5$ kΩ, and with a MOSFET specified to have $V_t = 0.4$ V, $k_n = 4$ mA/V², and $\lambda = 0$. Determine the coordinates of the end points of the saturation-region segment of the VTC. Also, determine $V_{DS}|_C$ assuming $V_{GS}|_C = V_{DD}$.

Ans. A: 0.4 V, 1.8 V; B: 0.613 V, 0.213 V; $V_{DS}|_C = 18$ mV

5.4.3 Biasing the MOSFET to Obtain Linear Amplification

Biasing enables us to obtain almost-linear amplification from the MOSFET. The technique is illustrated in Fig. 5.28(a). A dc voltage V_{GS} is selected to obtain operation at a point Q on the segment AB of the VTC. How to select an appropriate location for the bias point Q will be discussed shortly. For the time being, observe that the coordinates of Q are the dc

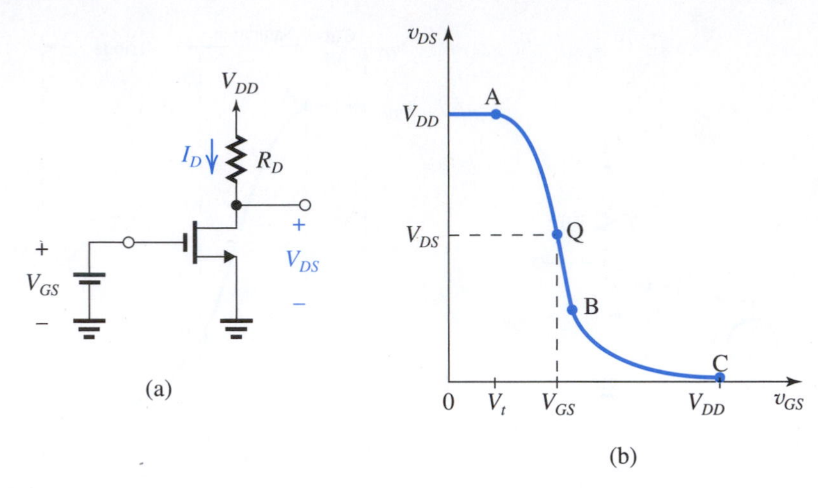

Figure 5.28 Biasing the MOSFET amplifier at a point Q located on the segment AB of the VTC.

voltages V_{GS} and V_{DS}, which are related by

$$V_{DS} = V_{DD} - \frac{1}{2}k_n R_D (V_{GS} - V_t)^2 \tag{5.34}$$

Point Q is known as the **bias point** or the **dc operating point**. Also, since at Q no signal component is present, it is also known as the **quiescent point** (which is the origin of the symbol Q).

Next, the signal to be amplified, v_{gs}, a function of time t, is superimposed on the bias voltage V_{GS}, as shown in Fig. 5.29(a). Thus the total instantaneous value of v_{GS} becomes

$$v_{GS}(t) = V_{GS} + v_{gs}(t)$$

The resulting $v_{DS}(t)$ can be obtained by substituting for $v_{GS}(t)$ into Eq. (5.32). Graphically, we can use the VTC to obtain $v_{DS}(t)$ point-by-point, as illustrated in Fig. 5.29(b). Here we show the case of v_{gs} being a triangular wave of "small" amplitude. Specifically, the amplitude of v_{gs} is small enough to restrict the excursion of the instantaneous operating point to a short, almost-linear segment of the VTC around the bias point Q. The shorter the segment, the greater the linearity achieved, and the closer to an ideal triangular wave the signal component at the output, v_{ds}, will be. This is the essence of obtaining linear amplification from the nonlinear MOSFET.

5.4.4 The Small-Signal Voltage Gain

If the input signal v_{gs} is kept small, the corresponding signal at the output v_{ds} will be nearly proportional to v_{gs} with the constant of proportionality being the slope of the almost-linear segment of the VTC around Q. This is the voltage gain of the amplifier, and its value can be determined by evaluating the slope of the tangent to the VTC at the bias point Q,

$$A_v \equiv \left.\frac{dv_{DS}}{dv_{GS}}\right|_{v_{GS}=V_{GS}} \tag{5.35}$$

Utilizing Eq. (5.32) we obtain

$$A_v = -k_n(V_{GS} - V_t)R_D \tag{5.36}$$

which can be expressed in terms of the overdrive voltage at the bias point V_{OV} as

$$A_v = -k_n V_{OV} R_D \tag{5.37}$$

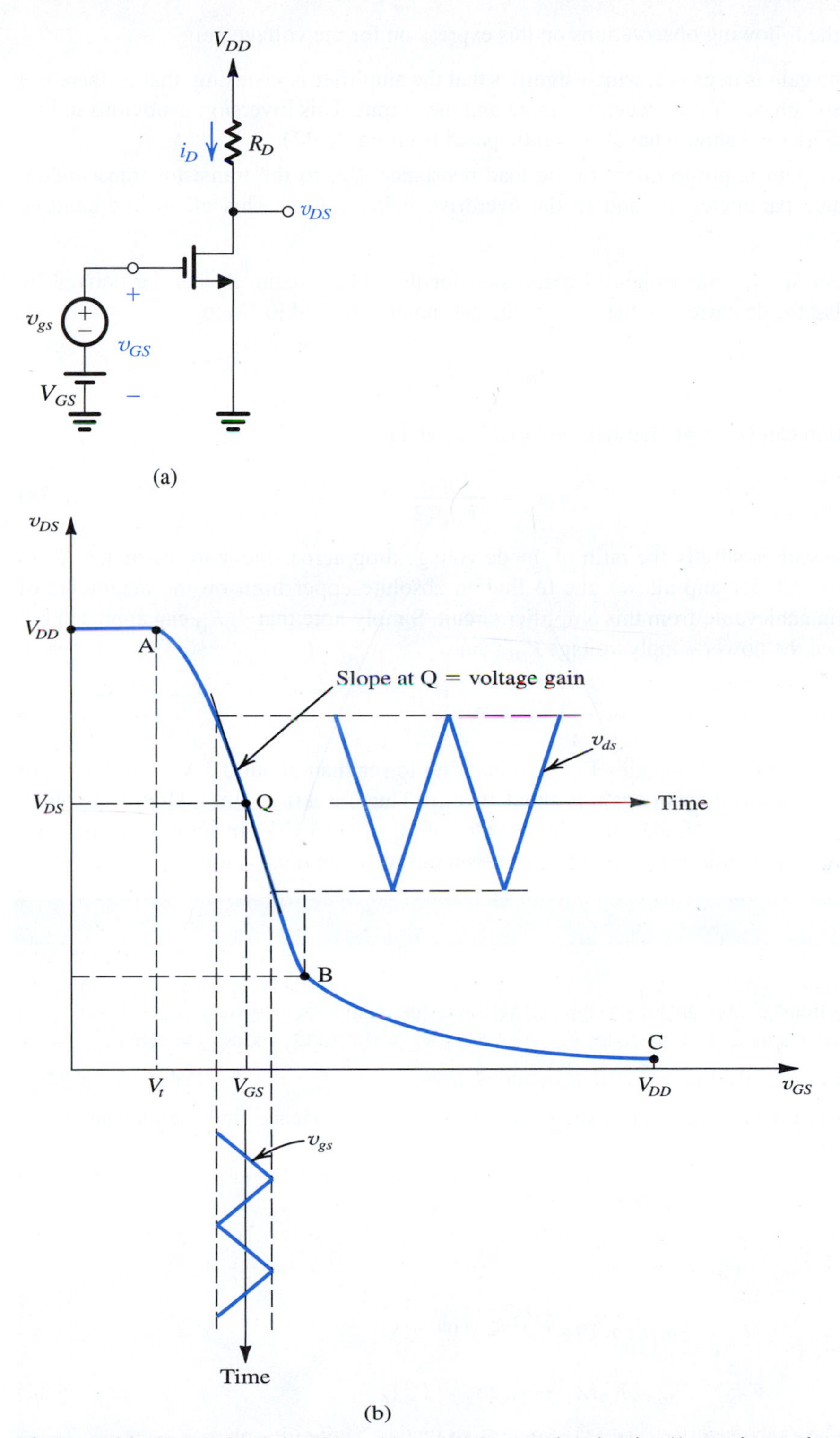

Figure 5.29 The MOSFET amplifier with a small time-varying signal $v_{gs}(t)$ superimposed on the dc bias voltage V_{GS}. The MOSFET operates on a short almost-linear segment of the VTC around the bias point Q and provides an output voltage $v_{ds} = A_v v_{gs}$.

We make the following observations on this expression for the voltage gain.

1. The gain is negative, which signifies that the amplifier is inverting; that is, there is a 180° phase shift between the input and the output. This inversion is obvious in Fig. 5.29(b) and should have been anticipated from Eq. (5.32).
2. The gain is proportional to the load resistance R_D, to the transistor transconductance parameter k_n, and to the overdrive voltage V_{OV}. This all makes intuitive sense.

Another simple and insightful expression for the voltage gain A_v can be derived by recalling that the dc current in the drain at the bias point is related to V_{OV} by

$$I_D = \frac{1}{2}k_n V_{OV}^2$$

This equation can be combined with Eq. (5.37) to yield

$$A_v = -\frac{I_D R_D}{V_{OV}/2} \tag{5.38}$$

That is, the gain is simply the ratio of the dc voltage drop across the load resistance R_D to $V_{OV}/2$. This relationship allows one to find an absolute upper limit on the magnitude of voltage gain achievable from this amplifier circuit. Simply note that $I_D R_D$ can approach but never exceed the power-supply voltage V_{DD}; thus,

$$|A_{v\max}| = \frac{V_{DD}}{V_{OV}/2}$$

For modern CMOS technologies V_{OV} is usually no lower than about 0.2 V, with the result that the maximum achievable gain is about 10 V_{DD}. Thus for a 0.13-μm CMOS technology that utilizes V_{DD} = 1.3 V, the approximate value of $|A_{\max}|$ is 13 V/V. In actual circuits, however, the maximum gain achievable is lower than this absolute maximum.

Example 5.9

Consider the amplifier circuit shown in Fig. 5.29(a). The transistor is specified to have V_t = 0.4 V, k_n' = 0.4 mA/V², $W/L = 10$, and $\lambda = 0$. Also, let V_{DD} = 1.8 V, R_D = 17.5 kΩ, and V_{GS} = 0.6 V.

(a) For $v_{gs} = 0$ (and hence $v_{ds} = 0$), find V_{OV}, I_D, V_{DS}, and A_v.

(b) What is the maximum symmetrical signal swing allowed at the drain? Hence find the maximum allowable amplitude of a sinusoidal v_{gs}.

Solution

(a) With V_{GS} = 0.6 V, V_{OV} = 0.6 – 0.4 = 0.2 V.

Thus,

$$I_D = \frac{1}{2} \times 0.4 \times 10 \times 0.2^2 = 0.08 \text{ mA}$$

$$V_{DS} = V_{DD} - R_D I_D$$

$$= 1.8 - 17.5 \times 0.08 = 0.4 \text{ V}$$

Since V_{DS} is greater than V_{OV}, the transistor is indeed operating in saturation. The voltage gain can be found from Eq. (5.37),

$$\begin{aligned} A_v &= -k_n V_{OV} R_D \\ &= -0.4 \times 10 \times 0.2 \times 17.5 \\ &= -14 \text{ V/V} \end{aligned}$$

An identical result can be found using Eq. (5.38).

(b) Since $V_{OV} = 0.2$ V and $V_{DS} = 0.4$ V, we see that the maximum allowable negative signal swing at the drain is 0.2 V. In the positive direction, a swing of +0.2 V would not cause the transistor to cut off and thus is allowed. Thus the maximum symmetrical signal swing allowable at the drain is ±0.2 V. The corresponding amplitude of v_{gs} can be found from

$$\hat{v}_{gs} = \frac{\hat{v}_{ds}}{|A_v|} = \frac{0.2 \text{ V}}{14} = 14.2 \text{ mV}$$

Since $\hat{v}_{gs} \ll V_{OV}$, the operation will be reasonably linear (more on this in later sections).

Greater insight into the issue of allowable signal swing can be obtained by examining the signal waveforms shown in Fig. 5.30. Note that for the MOSFET to remain in saturation at the negative peak of v_{ds}, we must ensure that

$$v_{DS\text{min}} \geq v_{GS\text{max}} - V_t$$

that is,

$$0.4 - |A_v|\hat{v}_{gs} \geq 0.6 + \hat{v}_{gs} - 0.4$$

which results in

$$\hat{v}_{gs} \leq \frac{0.2}{|A_v| + 1} = 13.3 \text{ mV}$$

This is a more precise result than the one obtained earlier.

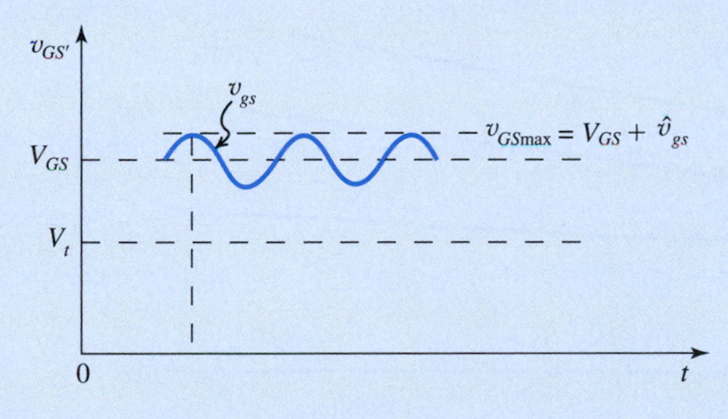

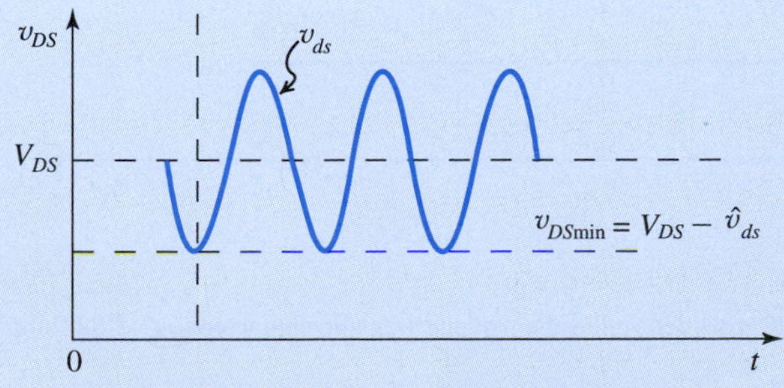

Figure 5.30 Signal waveforms at gate and drain for the amplifier in Example 5.9. Note that to ensure operation in the saturation region at all times, $v_{DS\text{min}} \geq v_{GS\text{max}} - V_t$.

EXERCISE

5.17 For the amplifier circuit studied in Example 5.9, provide two alternative designs, each providing a voltage gain of 10 by (a) changing R_D while keeping V_{OV} constant, and (b) changing V_{OV} while keeping R_D constant. For each design, specify V_{GS}, I_D, R_D, and V_{DS}.
Ans. (a) 0.6 V, 0.08 mA, 12.5 kΩ, 0.8 V; (b) 0.54 V, 0.04 mA, 17.5 kΩ, 1.1 V

5.4.5 Determining the VTC by Graphical Analysis

Figure 5.31 shows a graphical method for determining the VTC of the amplifier of Fig. 5.29(a). Although graphical analysis of transistor circuits is rarely employed in practice, it is useful for us at this stage for gaining greater insight into circuit operation, especially in answering the question of where to locate the bias point Q.

The graphical analysis is based on the observation that for each value of v_{GS}, the circuit will be operating at the point of intersection of the i_D–v_{DS} graph corresponding to the particular value of v_{GS} and the straight line representing Eq. (5.30), which can be rewritten in the form

$$i_D = \frac{V_{DD}}{R_D} - \frac{1}{R_D}v_{DS} \tag{5.39}$$

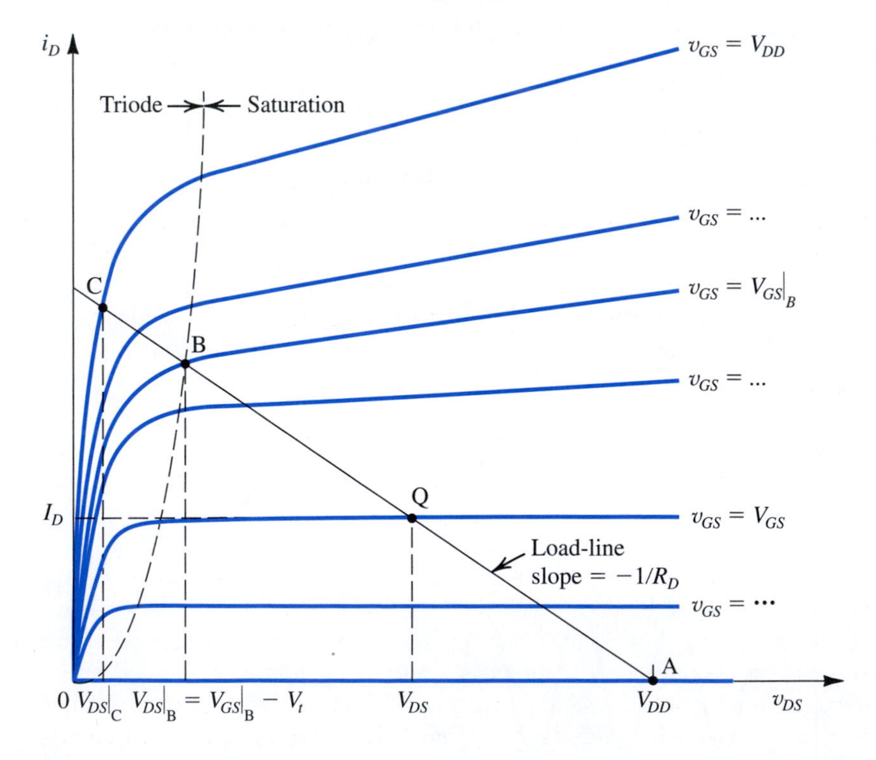

Figure 5.31 Graphical construction to determine the voltage transfer characteristic of the amplifier in Fig. 5.29(a).

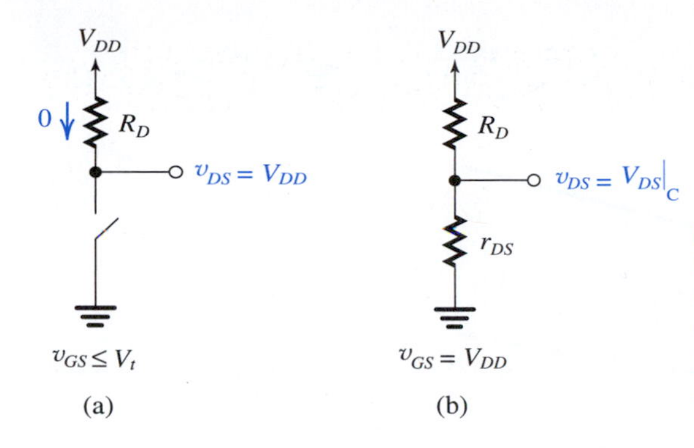

Figure 5.32 Operation of the MOSFET in Figure 5.29(a) as a switch: **(a)** Open, corresponding to point A in Figure 5.31; **(b)** Closed, corresponding to point C in Figure 5.31. The closure resistance is approximately equal to r_{DS} because V_{DS} is usually very small.

The straight line representing this relationship is superimposed on the i_D–v_{DS} characteristics in Fig. 5.31. It intersects the horizontal axis at $v_{DS} = V_{DD}$ and has a slope of $-1/R_D$. Since this straight line represents in effect the load resistance R_D, it is called the **load line**. The VTC is then determined point by point. Note that we have labeled four important points: point A at which $v_{GS} = V_t$, point Q at which the MOSFET can be biased for amplifier operation ($v_{GS} = V_{GS}$ and $v_{DS} = V_{DS}$), point B at which the MOSFET leaves saturation and enters the triode region, and point C, which is deep into the triode region and for which $v_{GS} = V_{DD}$. If the MOSFET is to be used as a switch, then operating points A and C are applicable: At A the transistor is off (open switch), and at C the transistor operates as a low-valued resistance r_{DS} and has a small voltage drop (closed switch). The incremental resistance at point C is also known as the **closure resistance**. The operation of the MOSFET as a switch is illustrated in Fig. 5.32. A detailed study of the application of the MOSFET as a switch is undertaken in Chapter 13 dealing with CMOS digital logic circuits.

5.4.6 Locating the Bias Point Q

The bias point Q is determined by the value of V_{GS} and that of the load resistance R_D. Two important considerations in deciding on the location of Q are the required gain and the allowable signal swing at the output. To illustrate, consider the VTC shown in Fig. 5.29(b). Here the value of R_D is fixed and the only variable remaining is the value of V_{GS}. Since the slope increases as we move closer to point B, we obtain higher gain by locating Q as close to B as possible. However, the closer Q is to the boundary point B, the smaller the allowable magnitude of negative signal swing. Thus, as often happens in engineering design, we encounter a situation requiring a trade-off.

In deciding on a value for R_D, it is useful to refer to the i_D–v_{DS} plane. Figure 5.33 shows two load lines resulting in two extreme bias points: Point Q_1 is too close to V_{DD}, resulting in a severe constraint on the positive signal swing of v_{ds}. Exceeding the allowable positive maximum results in the positive peaks of the signal being clipped off, since the MOSFET will turn off for the part of each cycle near the positive peak. We speak of this situation by saying that the circuit does not have sufficient "headroom." Similarly, point Q_2 is too close to the boundary of the triode region, thus severely limiting the allowable negative signal swing of v_{ds}. Exceeding this limit would result in the transistor entering the triode region for part of each cycle near the negative peaks, resulting in a distorted output signal. In this situation we say that the circuit does not have sufficient "legroom." We will have more to say on bias design in the Section 5.7.

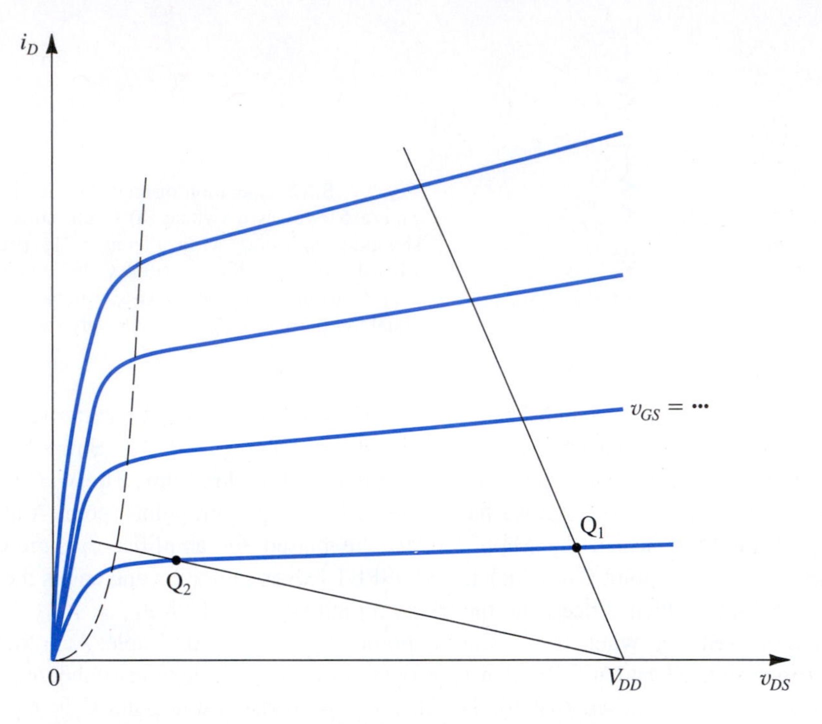

Figure 5.33 Two load lines and corresponding bias points. Bias point Q_1 does not leave sufficient room for positive signal swing at the drain (too close to V_{DD}). Bias point Q_2 is too close to the boundary of the triode region and might not allow for sufficient negative signal swing.

5.5 Small-Signal Operation and Models

In our study of the operation of the MOSFET amplifier in Section 5.4 we learned that linear amplification can be obtained by biasing the MOSFET to operate in the saturation region and by keeping the input signal small. In this section, we explore the small-signal operation in some detail. For this purpose we utilize the conceptual amplifier circuit shown in Fig. 5.34. Here the MOS transistor is biased by applying a dc voltage[8] V_{GS}, and the input signal to be amplified, v_{gs}, is superimposed on the dc bias voltage V_{GS}. The output voltage is taken at the drain.

5.5.1 The DC Bias Point

The dc bias current I_D can be found by setting the signal v_{gs} to zero; thus,

$$I_D = \frac{1}{2}k_n(V_{GS} - V_t)^2 = \frac{1}{2}k_n V_{OV}^2 \tag{5.40}$$

where we have neglected channel-length modulation (i.e., we have assumed $\lambda = 0$). Here $V_{OV} = V_{GS} - V_t$ is the overdrive voltage at which the MOSFET is biased to operate. The dc

[8]Practical biasing arrangments will be studied in Section 5.7.

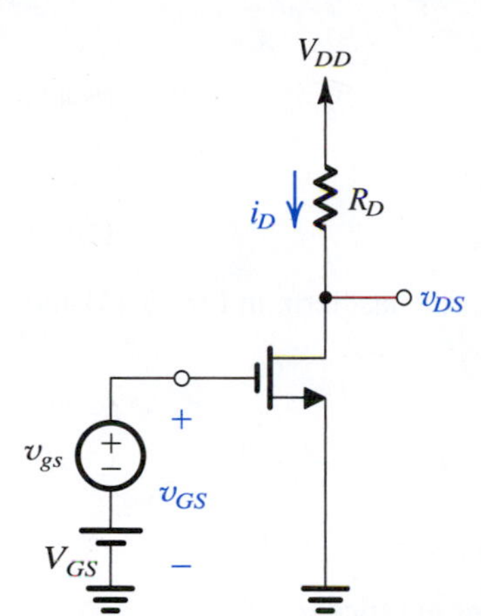

Figure 5.34 Conceptual circuit utilized to study the operation of the MOSFET as a small-signal amplifier.

voltage at the drain, V_{DS}, will be

$$V_{DS} = V_{DD} - R_D I_D \tag{5.41}$$

To ensure saturation-region operation, we must have

$$V_{DS} > V_{OV}$$

Furthermore, since the total voltage at the drain will have a signal component superimposed on V_{DS}, V_{DS} has to be sufficiently greater than (V_{OV}) to allow for the required signal swing.

5.5.2 The Signal Current in the Drain Terminal

Next, consider the situation with the input signal v_{gs} applied. The total instantaneous gate-to-source voltage will be

$$v_{GS} = V_{GS} + v_{gs} \tag{5.42}$$

resulting in a total instantaneous drain current i_D,

$$\begin{aligned} i_D &= \frac{1}{2}k_n(V_{GS} + v_{gs} - V_t)^2 \\ &= \frac{1}{2}k_n(V_{GS} - V_t)^2 + k_n(V_{GS} - V_t)v_{gs} + \frac{1}{2}k_n\, v_{gs}^2 \end{aligned} \tag{5.43}$$

The first term on the right-hand side of Eq. (5.43) can be recognized as the dc bias current I_D (Eq. 5.40). The second term represents a current component that is directly proportional to the input signal v_{gs}. The third term is a current component that is proportional to the square of the input signal. This last component is undesirable because it represents *nonlinear distortion*. To reduce the nonlinear distortion introduced by the MOSFET, the input signal should be kept small so that

$$\frac{1}{2}k_n v_{gs}^2 \ll k_n(V_{GS} - V_t)v_{gs}$$

resulting in

$$v_{gs} \ll 2(V_{GS} - V_t) \tag{5.44}$$

or, equivalently,

$$v_{gs} \ll 2V_{OV} \tag{5.45}$$

If this **small-signal condition** is satisfied, we may neglect the last term in Eq. (5.43) and express i_D as

$$i_D \simeq I_D + i_d \tag{5.46}$$

where

$$i_d = k_n(V_{GS} - V_t)v_{gs}$$

The parameter that relates i_d and v_{gs} is the MOSFET **transconductance** g_m,

$$g_m \equiv \frac{i_d}{v_{gs}} = k_n(V_{GS} - V_t) \tag{5.47}$$

or in terms of the overdrive voltage V_{OV},

$$g_m = k_n V_{OV} \tag{5.48}$$

Figure 5.35 presents a graphical interpretation of the small-signal operation of the MOSFET amplifier. Note that g_m is equal to the slope of the $i_D - v_{GS}$ characteristic at the bias point,

$$g_m \equiv \left.\frac{\partial i_D}{\partial v_{GS}}\right|_{v_{GS} = V_{GS}} \tag{5.49}$$

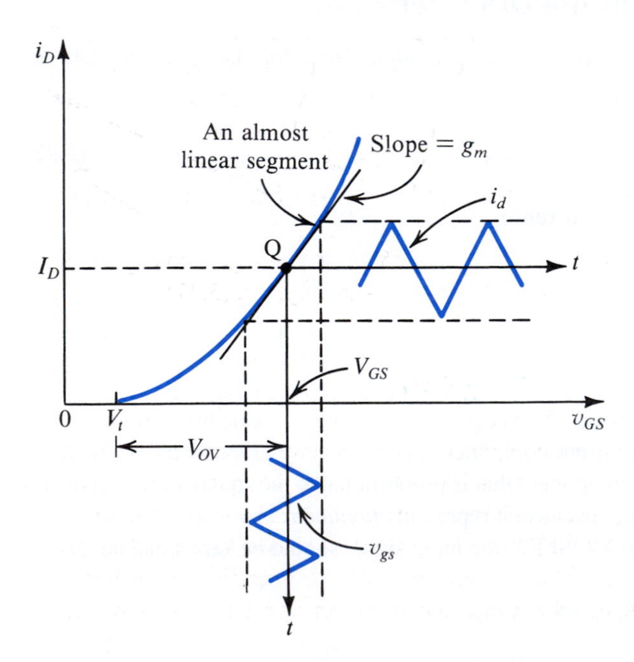

Figure 5.35 Small-signal operation of the MOSFET amplifier.

This is the formal definition of g_m, which can be shown to yield the expressions given in Eqs. (5.47) and (5.48).

5.5.3 The Voltage Gain

Returning to the circuit of Fig. 5.34, we can express the total instantaneous drain voltage v_{DS} as follows:

$$v_{DS} = V_{DD} - R_D i_D$$

Under the small-signal condition, we have

$$v_{DS} = V_{DD} - R_D(I_D + i_d)$$

which can be rewritten as

$$v_{DS} = V_{DS} - R_D i_d$$

Thus the signal component of the drain voltage is

$$v_{ds} = -i_d R_D = -g_m v_{gs} R_D \tag{5.50}$$

which indicates that the voltage gain is given by

$$A_v \equiv \frac{v_{ds}}{v_{gs}} = -g_m R_D \tag{5.51}$$

The minus sign in Eq. (5.51) indicates that the output signal v_{ds} is 180° out of phase with respect to the input signal v_{gs}. This is illustrated in Fig. 5.36, which shows v_{GS} and v_{DS}. The input signal is assumed to have a triangular waveform with an amplitude much smaller than $2(V_{GS} - V_t)$, the small-signal condition in Eq. (5.44), to ensure linear operation. For operation in the saturation region at all times, the minimum value of v_{DS} should not fall below the corresponding value of v_{GS} by more than V_t. Also, the maximum value of v_{DS} should be smaller than V_{DD}; otherwise the FET will enter the cutoff region and the peaks of the output signal waveform will be clipped off.

Finally, we note that by substituting for g_m from Eq. (5.48) the voltage gain expression in Eq. (5.51) becomes identical to that derived in Section 5.4—namely, Eq. (5.37).

5.5.4 Separating the DC Analysis and the Signal Analysis

From the preceding analysis, we see that under the small-signal approximation, signal quantities are superimposed on dc quantities. For instance, the total drain current i_D equals the dc current I_D plus the signal current i_d, the total drain voltage $v_{DS} = V_{DS} + v_{ds}$, and so on. It follows that the analysis and design can be greatly simplified by separating dc or bias calculations from small-signal calculations. That is, once a stable dc operating point has been established and all dc quantities calculated, we may then perform signal analysis ignoring dc quantities.

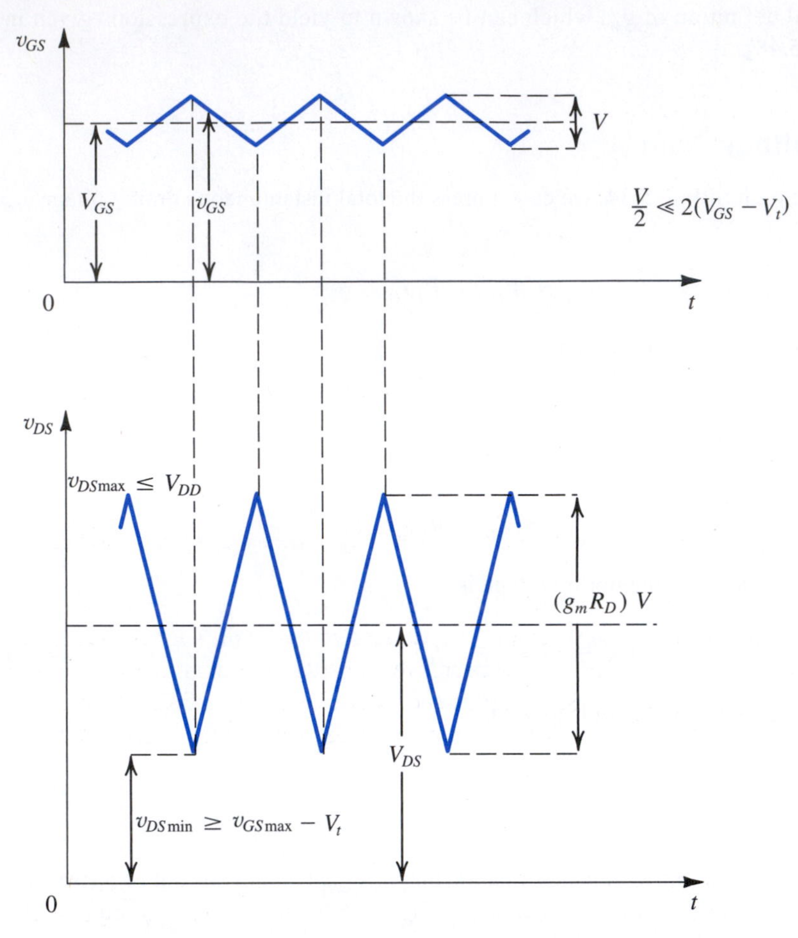

Figure 5.36 Total instantaneous voltages v_{GS} and v_{DS} for the circuit in Fig. 5.34.

5.5.5 Small-Signal Equivalent-Circuit Models

From a signal point of view, the FET behaves as a voltage-controlled current source. It accepts a signal v_{gs} between gate and source and provides a current $g_m v_{gs}$ at the drain terminal. The input resistance of this controlled source is very high—ideally, infinite. The output resistance—that is, the resistance looking into the drain—also is high, and we have assumed it to be infinite thus far. Putting all of this together, we arrive at the circuit in Fig. 5.37(a), which represents the small-signal operation of the MOSFET and is thus a **small-signal model** or a **small-signal equivalent circuit**.

In the analysis of a MOSFET amplifier circuit, the transistor can be replaced by the equivalent circuit model shown in Fig. 5.37(a). The rest of the circuit remains unchanged except that *ideal constant dc voltage sources are replaced by short circuits*. This is a result of the fact that the voltage across an ideal constant dc voltage source does not change, and thus there will always be a zero voltage signal across a constant dc voltage source. A dual statement applies for constant dc current sources; namely, the signal current of an ideal constant dc current source will always be zero, and thus *an ideal constant dc current source can be replaced by an open circuit* in the small-signal equivalent circuit of the amplifier. The

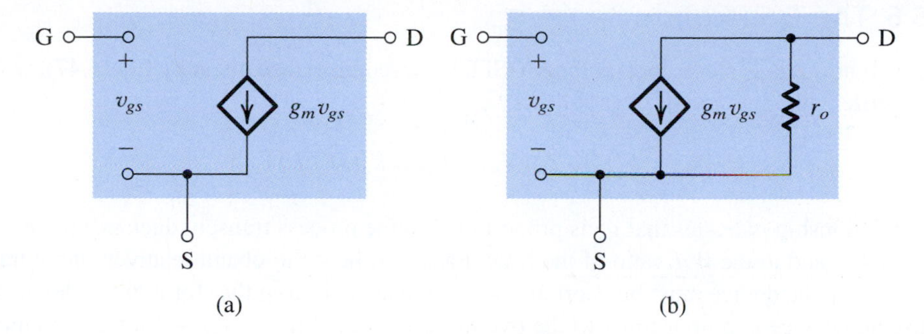

Figure 5.37 Small-signal models for the MOSFET: **(a)** neglecting the dependence of i_D on v_{DS} in saturation (the channel-length modulation effect); and **(b)** including the effect of channel-length modulation, modeled by output resistance $r_o = |V_A|/I_D$.

circuit resulting can then be used to perform any required signal analysis, such as calculating voltage gain.

The most serious shortcoming of the small-signal model of Fig. 5.37(a) is that it assumes the drain current in saturation to be independent of the drain voltage. From our study of the MOSFET characteristics in saturation, we know that the drain current does in fact depend on v_{DS} in a linear manner. Such dependence was modeled by a finite resistance r_o between drain and source, whose value was given by Eq. (5.27) in Section 5.2.4, which we repeat here as

$$r_o = \frac{|V_A|}{I_D} \tag{5.52}$$

where $V_A = 1/\lambda$ is a MOSFET parameter that either is specified or can be measured. It should be recalled that for a given process technology, V_A is proportional to the MOSFET channel length. The current I_D is the value of the dc drain current without the channel-length modulation taken into account; that is,

$$I_D = \frac{1}{2}k_n V_{OV}^2 \tag{5.53}$$

Typically, r_o is in the range of 10 kΩ to 1000 kΩ. It follows that the accuracy of the small-signal model can be improved by including r_o in parallel with the controlled source, as shown in Fig. 5.37(b).

It is important to note that the small-signal model parameters g_m and r_o depend on the dc bias point of the MOSFET.

Returning to the amplifier of Fig. 5.34, we find that replacing the MOSFET with the small-signal model of Fig. 5.37(b) results in the voltage-gain expression

$$A_v = \frac{v_{ds}}{v_{gs}} = -g_m(R_D \| r_o) \tag{5.54}$$

Thus, the finite output resistance r_o results in a reduction in the magnitude of the voltage gain.

Although the analysis above is performed on an NMOS transistor, the results, and the equivalent circuit models of Fig. 5.37, apply equally well to PMOS devices, except for using $|V_{GS}|$, $|V_t|$, $|V_{OV}|$, and $|V_A|$ and replacing k_n with k_p.

5.5.6 The Transconductance g_m

We shall now take a closer look at the MOSFET transconductance given by Eq. (5.47), which we rewrite with $k_n = k_n'(W/L)$ as follows:

$$g_m = k_n'(W/L)(V_{GS} - V_t) = k_n'(W/L)V_{OV} \tag{5.55}$$

This relationship indicates that g_m is proportional to the process transconductance parameter $k_n' = \mu_n C_{ox}$ and to the W/L ratio of the MOS transistor; hence to obtain relatively large transconductance the device must be short and wide. We also observe that for a given device the transconductance is proportional to the overdrive voltage, $V_{OV} = V_{GS} - V_t$, the amount by which the bias voltage V_{GS} exceeds the threshold voltage V_t. Note, however, that increasing g_m by biasing the device at a larger V_{GS} has the disadvantage of reducing the allowable voltage signal swing at the drain.

Another useful expression for g_m can be obtained by substituting for V_{OV} in Eq. (5.55) by $\sqrt{2I_D/(k_n'(W/L))}$ [from Eq. (5.40)]:

$$g_m = \sqrt{2k_n'}\sqrt{W/L}\sqrt{I_D} \tag{5.56}$$

This expression shows two things:

1. For a given MOSFET, g_m is proportional to the square root of the dc bias current.
2. At a given bias current, g_m is proportional to $\sqrt{W/L}$.

In contrast, the transconductance of the bipolar junction transistor (BJT) studied in Chapter 6 is proportional to the bias current and is independent of the physical size and geometry of the device.

To gain some insight into the values of g_m obtained in MOSFETs consider an integrated-circuit device operating at $I_D = 0.5$ mA and having $k_n' = 120\ \mu\text{A/V}^2$. Equation (5.56) shows that for $W/L = 1$, $g_m = 0.35$ mA/V, whereas a device for which $W/L = 100$ has $g_m = 3.5$ mA/V. In contrast, a BJT operating at a collector current of 0.5 mA has $g_m =$ 20 mA/V.

Yet another useful expression for g_m of the MOSFET can be obtained by substituting for $k_n'(W/L)$ in Eq. (5.55) by $2I_D/(V_{GS} - V_t)^2$:

$$g_m = \frac{2I_D}{V_{GS} - V_t} = \frac{2I_D}{V_{OV}} \tag{5.57}$$

A convenient graphical construction that clearly illustrates this relationship is shown in Fig. 5.38.

In summary, there are three different relationships for determining g_m—Eqs. (5.55), (5.56), and (5.57)—and there are three design parameters—(W/L), V_{OV}, and I_D, any two of which can be chosen independently. That is, the designer may choose to operate the MOSFET with a certain overdrive voltage V_{OV} and at a particular current I_D; the required W/L ratio can then be found and the resulting g_m determined.

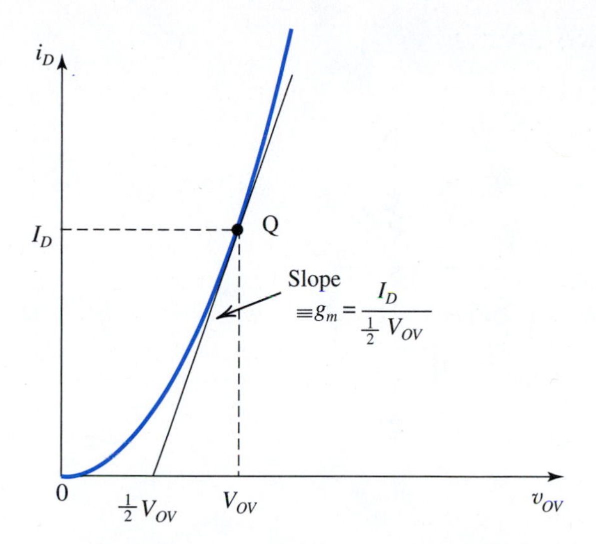

Figure 5.38 The slope of the tangent at the bias point Q intersects the v_{OV} axis at $\frac{1}{2} V_{OV}$. Thus, $g_m = I_D/(\frac{1}{2} V_{OV})$.

Example 5.10

Figure 5.39(a) shows a discrete common-source MOSFET amplifier utilizing a drain-to-gate resistance R_G for biasing purposes. Such a biasing arrangement will be studied in Section 5.7. The input signal v_i is coupled to the gate via a large capacitor, and the output signal at the drain is coupled to the load resistance R_L via another large capacitor. We wish to analyze this amplifier circuit to determine its small-signal voltage gain, its input resistance, and the largest allowable input signal. The transistor has $V_t = 1.5$ V, $k_n'(W/L) = 0.25$ mA/V^2, and $V_A = 50$ V. Assume the coupling capacitors to be sufficiently large so as to act as short circuits at the signal frequencies of interest.

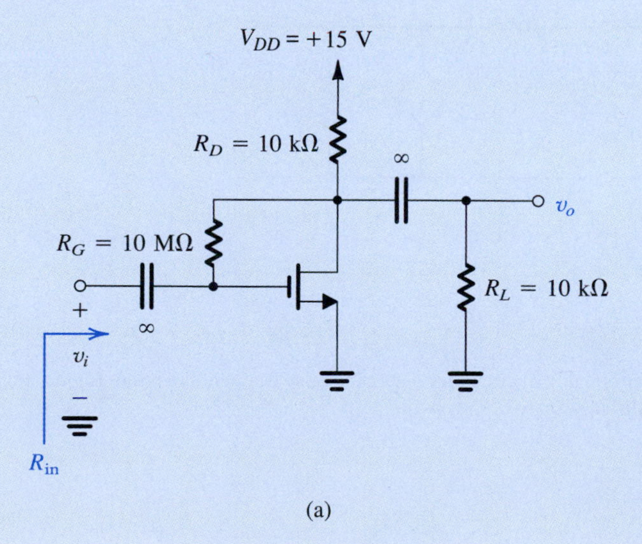

(a)

Example 5.10 *continued*

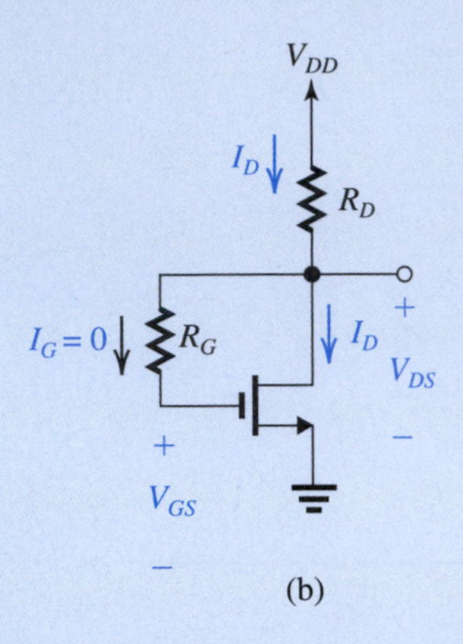

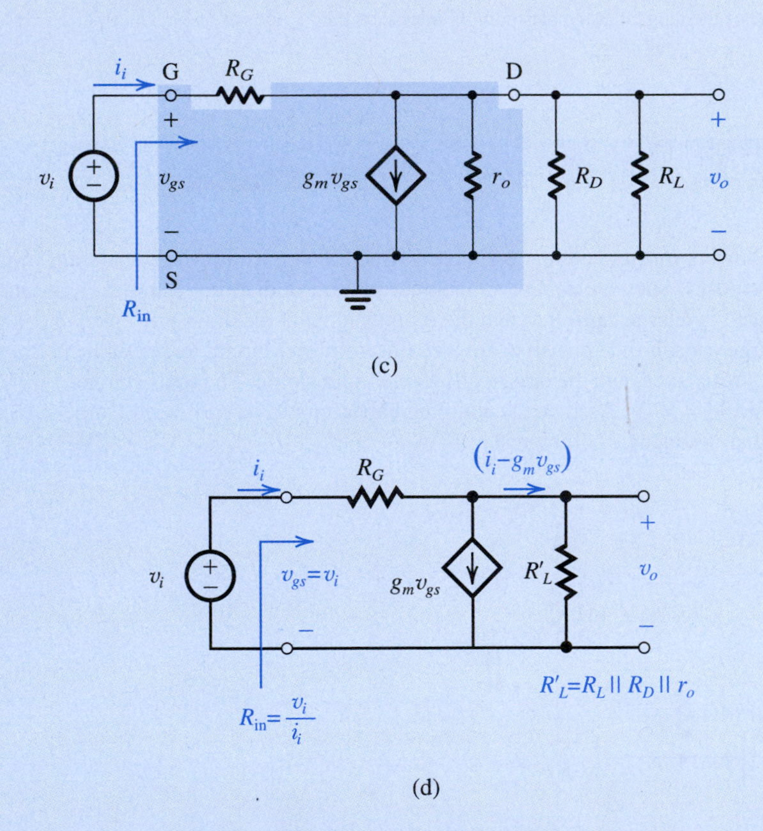

Figure 5.39 Example 5.10: **(a)** amplifier circuit; **(b)** circuit for determining the dc operating point; **(c)** the amplifier small-signal equivalent circuit; **(d)** a simplified version of the circuit in (**c**).

Solution

We first determine the dc operating point. For this purpose, we eliminate the input signal v_i, and open-circuit the two coupling capacitors (since they block dc currents). The result is the circuit shown in Fig. 5.39(b). We note that since $I_G = 0$, the dc voltage drop across R_G will be zero, and

$$V_{GS} = V_{DS} = V_{DD} - R_D I_D \tag{5.58}$$

With $V_{DS} = V_{GS}$, the NMOS transistor will be operating in saturation. Thus,

$$I_D = \frac{1}{2} k_n (V_{GS} - V_t)^2 \tag{5.59}$$

where, for simplicity, we have neglected the effect of channel-length modulation on the dc operating point. Substituting $V_{DD} = 15$ V, $R_D = 10$ kΩ, $k_n = 0.25$ mA/V^2, and $V_t = 1.5$ V in Eqs. (5.58) and (5.59), and substituting for V_{GS} from Eq. (5.58) into Eq. (5.59) results in a quadratic equation in I_D. Solving the latter and discarding the root that is not physically meaningful yields the solution

$$I_D = 1.06 \text{ mA}$$

which corresponds to

$$V_{GS} = V_{DS} = 4.4 \text{ V}$$

and

$$V_{OV} = 4.4 - 1.5 = 2.9 \text{ V}$$

Next we proceed with the small-signal analysis of the amplifier. Toward that end we replace the MOSFET with its small-signal model to obtain the small-signal equivalent circuit of the amplifier, shown in Fig. 5.39(c). Observe that we have replaced the coupling capacitors with short circuits. The dc voltage supply V_{DD} has also been replaced with a short circuit to ground.

The values of the transistor small-signal parameters g_m and r_o can be determined by using the dc bias quantities found above, as follows:

$$g_m = k_n V_{OV}$$

$$= 0.25 \times 2.9 = 0.725 \text{ mA/V}$$

$$r_o = \frac{V_A}{I_D} = \frac{50}{1.06} = 47 \text{ k}\Omega$$

Next we use the equivalent circuit of Fig. 5.39(c) to determine the input resistance $R_{in} \equiv v_i/i_i$ and the voltage gain $A_v = v_o/v_i$. Toward that end we simplify the circuit by combining the three parallel resistances r_o, R_D, and R_L in a single resistance R_L',

$$R_L' = R_L \| R_D \| r_o$$

$$= 10 \| 10 \| 47 = 4.52 \text{ k}\Omega$$

as shown in Fig. 5.39(d). For the latter circuit we can write the two equations

$$v_o = (i_i - g_m v_{gs}) R_L' \tag{5.60}$$

Example 5.10 *continued*

and

$$i_i = \frac{v_{gs} - v_o}{R_G} \tag{5.61}$$

Substituting for i_i from Eq. (5.61) into Eq. (5.60) results in the following expression for the voltage gain $A_v \equiv v_o/v_i = v_o/v_{gs}$:

$$A_v = -g_m R_L' \frac{1-(1/g_m R_G)}{1+(R_L'/R_G)}$$

Since R_G is very large, $g_m R_G \gg 1$ and $R_L'/R_G \ll 1$ (the reader can easily verify this), and the gain expression can be approximated as

$$A_v \simeq -g_m R_L' \tag{5.62}$$

Substituting, $g_m = 0.725$ mA/V and $R_L' = 4.52$ kΩ yields

$$A_v = -3.3 \text{ V/V}$$

To obtain the input resistance, we substitute in Eq. (5.61) for $v_o = A_v v_{gs} = -g_m R_L' v_{gs}$, then use $R_{\text{in}} \equiv v_i/i_i = v_{gs}/i_i$ to obtain

$$R_{\text{in}} = \frac{R_G}{1 + g_m R_L'} \tag{5.63}$$

This is an interesting relationship: The input resistance decreases as the gain $(g_m R_L')$ is increased. The value of R_{in} can now be determined; it is

$$R_{\text{in}} = \frac{10 \text{ M}\Omega}{1+3.3} = 2.33 \text{ M}\Omega$$

which is still very large.

The largest allowable input signal $\hat{v}_i$ is constrained by the need to keep the transistor in saturation at all times; that is,

$$v_{DS} \geq v_{GS} - V_t$$

Enforcing this condition with equality at the point v_{GS} is maximum and v_{DS} is minimum, we write

$$v_{DS\text{min}} = v_{GS\text{max}} - V_t$$

$$V_{DS} - |A_v|\hat{v}_i = V_{GS} + \hat{v}_i - V_t$$

Since $V_{DS} = V_{GS}$, we obtain

$$\hat{v}_i = \frac{V_t}{|A_v| + 1}$$

This is a general relationship that applies to this circuit irrespective of the component values. Observe that it simply states that the maximum signal swing is determined by the fact that the bias arrangement makes $V_D = V_G$ and thus, to keep the MOSFET out of the triode region, the signal between D and G is constrained to be equal to V_t. For our particular design,

$$\hat{v}_i = \frac{1.5}{3.3+1} = 0.35 \text{ V}$$

A modification of this circuit that increases the allowable signal swing is investigated in Problem 5.80.

EXERCISE

D5.18 Consider the amplifier circuit of Fig. 5.39(a) without the load resistance R_L and with channel length modulation neglected. Let $V_{DD} = 5$ V, $V_t = 0.7$ V, and $k_n = 1$ mA/V². Find V_{OV}, I_D, R_D, and R_G to obtain a voltage gain of 25 V/V and an input resistance of 0.5 MΩ. What is the maximum allowable input signal, $\hat{v}_i$?

Ans. 0.319 V; 50.7 μA; 78.5 kΩ; 13 MΩ; 27 mV

5.5.7 The T Equivalent-Circuit Model

Through a simple circuit transformation it is possible to develop an alternative equivalent-circuit model for the MOSFET. The development of such a model, known as the T model, is illustrated in Fig. 5.40. Figure 5.40(a) shows the equivalent circuit studied

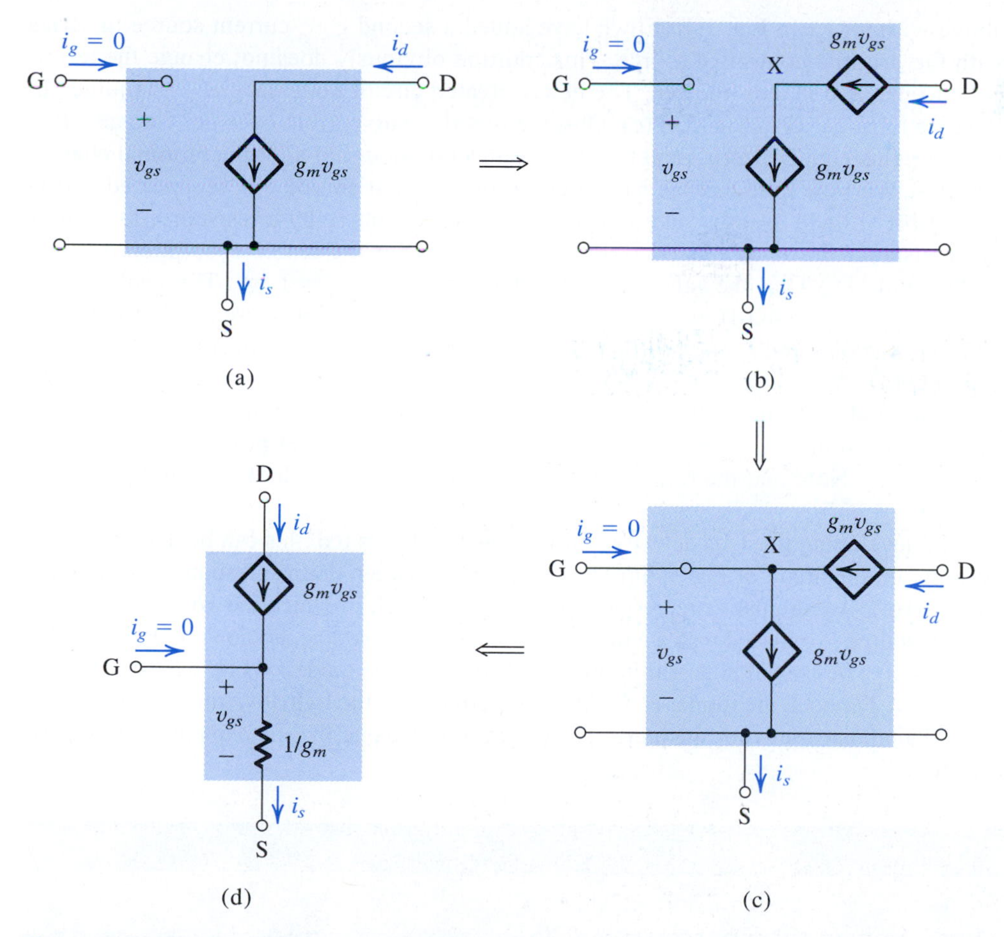

Figure 5.40 Development of the T equivalent-circuit model for the MOSFET. For simplicity, r_o has been omitted; however, it may be added between D and S in the T model of **(d)**.

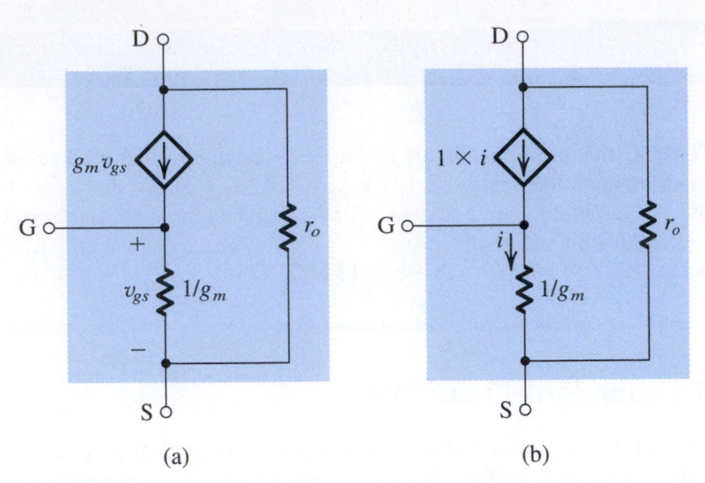

Figure 5.41 **(a)** The T model of the MOSFET augmented with the drain-to-source resistance r_o. **(b)** An alternative representation of the T model.

above without r_o. In Fig. 5.40(b) we have added a second $g_m v_{gs}$ current source in series with the original controlled source. This addition obviously does not change the terminal currents and is thus allowed. The newly created circuit node, labeled X, is joined to the gate terminal G in Fig. 5.40(c). Observe that the gate current does not change—that is, it remains equal to zero—and thus this connection does not alter the terminal characteristics. We now note that we have a controlled current source $g_m v_{gs}$ connected across its control voltage v_{gs}. We can replace this controlled source by a resistance as long as this resistance draws an equal current as the source. (See the source-absorption theorem in Appendix D.) Thus the value of the resistance is $v_{gs}/g_m v_{gs} = 1/g_m$. This replacement is shown in Fig. 5.40(d), which depicts the alternative model. Observe that i_g is still zero, $i_d = g_m v_{gs}$, and $i_s = v_{gs}/(1/g_m) = g_m v_{gs}$, all the same as in the original model in Fig. 5.40(a).

The model of Fig. 5.40(d) shows that the resistance between gate and source looking into the source is $1/g_m$. This observation and the T model prove useful in many applications. Note that the resistance between gate and source, looking into the gate, is infinite.

In developing the T model we did not include r_o. If desired, this can be done by incorporating in the circuit of Fig. 5.40(d) a resistance r_o between drain and source, as shown in Fig. 5.41(a). An alternative representation of the T model, in which the voltage-controlled current source is replaced with a current-controlled current source, is shown in Fig. 5.41(b).

Finally, we should note that in order to distinguish the model of Fig. 5.37(b) from the equivalent T model, the former is sometimes referred to as the **hybrid-π model**, a carryover from the bipolar transistor literature. The origin of this name will be explained in the next chapter.

Example 5.11

Figure 5.42(a) shows a MOSFET amplifier biased by a constant-current source I. Assume that the values of I and R_D are such that the MOSFET operates in the saturation region. The input signal v_i is coupled to the source terminal by utilizing a large capacitor C_{C1}. Similarly, the output signal at the drain is taken through a large coupling capacitor C_{C2}. Find the input resistance R_{in} and the voltage gain v_o/v_i. Neglect channel-length modulation.

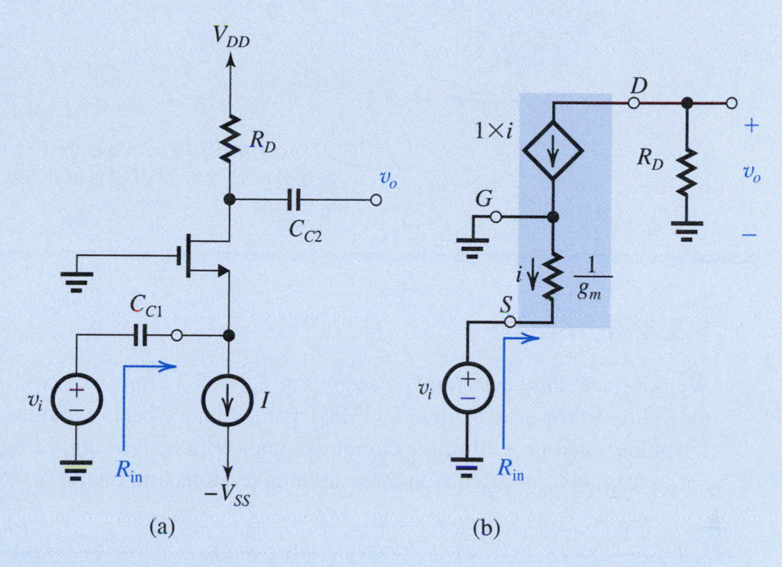

Figure 5.42 (**a**) Amplifier circuit for Example 5.11; (**b**) Small-signal equivalent circuit of the amplifier in (a).

Solution

Replacing the MOSFET with its T equivalent-circuit model results in the amplifier equivalent circuit shown in Fig. 5.42(b). Observe that the dc current source I is replaced with an open circuit and the dc voltage source V_{DD} is replaced by a short circuit. The large coupling capacitors have been replaced by short circuits. From the equivalent circuit-model we determine

$$R_{in} = \frac{v_i}{-i} = 1/g_m$$

and

$$v_o = -iR_D = \left(\frac{v_i}{1/g_m}\right)R_D = g_m R_D v_i$$

Thus,

$$A_v \equiv \frac{v_o}{v_i} = g_m R_D$$

We note that this amplifier, known as the common-gate amplifier because the gate at ground potential is common to both the input and output ports, has a low input resistance $(1/g_m)$ and a noninverting gain. We shall study this amplifier type in Section 5.6.5.

EXERCISE

5.19 Use the T model of Fig. 5.41(b) to show that a MOSFET whose drain is connected to its gate exhibits an incremental resistance equal to $[(1/g_m) \| r_o]$.
Ans. See Fig. E5.19.

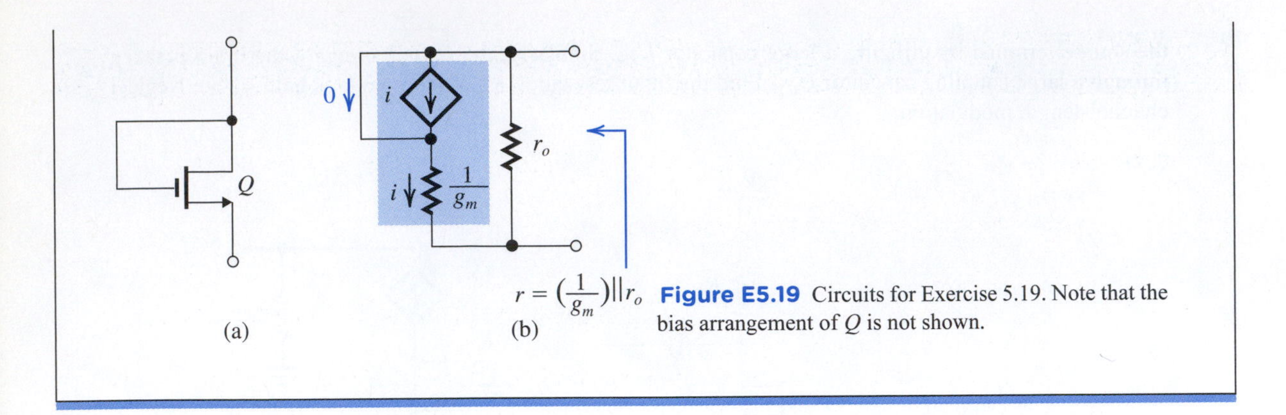

Figure E5.19 Circuits for Exercise 5.19. Note that the bias arrangement of Q is not shown.

5.5.8 Summary

We conclude this section by presenting in Table 5.3 a summary of the formulas for calculating the values of the small-signal MOSFET parameters. Observe that for g_m we have three different formulas, each providing the circuit designer with insight regarding design choices. We shall make frequent comments on these in later sections and chapters.

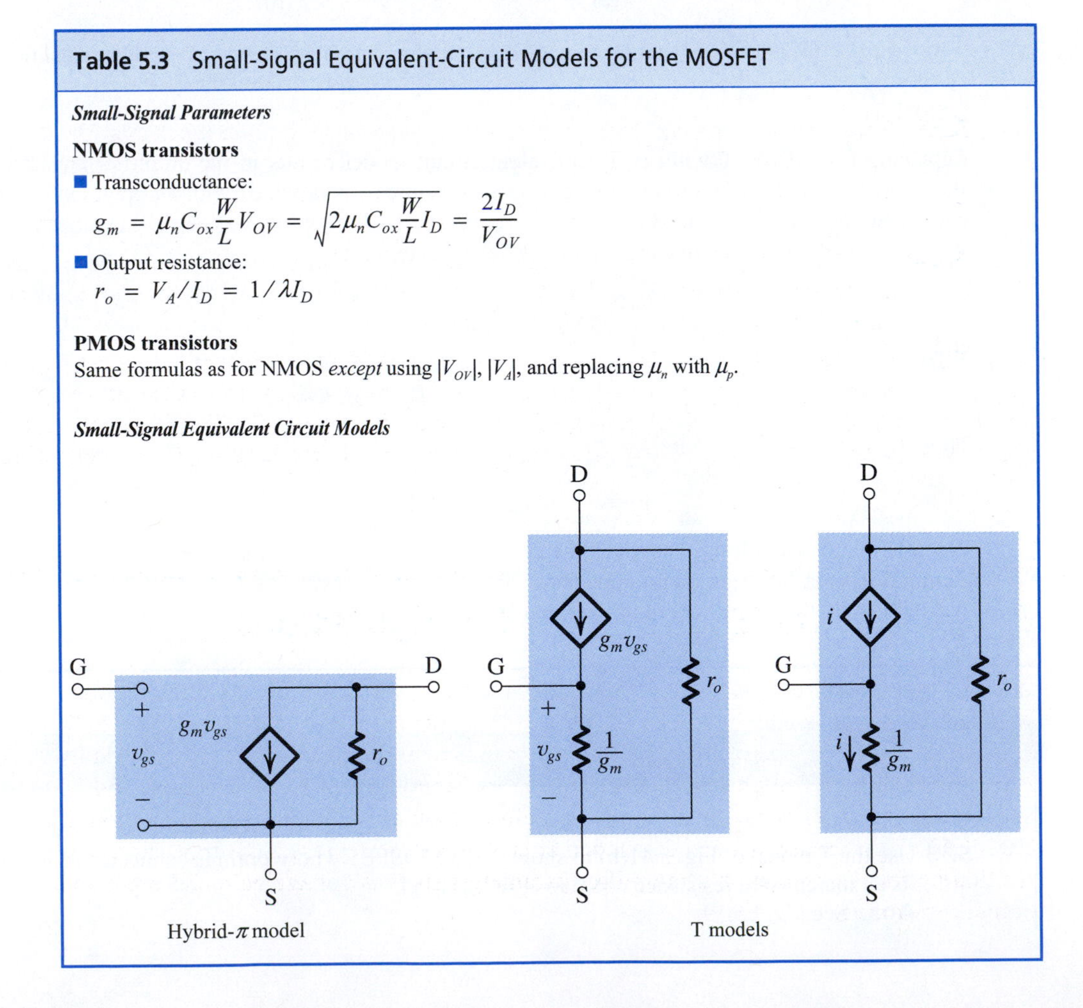

Table 5.3 Small-Signal Equivalent-Circuit Models for the MOSFET

Small-Signal Parameters

NMOS transistors

- Transconductance:

$$g_m = \mu_n C_{ox} \frac{W}{L} V_{OV} = \sqrt{2\mu_n C_{ox} \frac{W}{L} I_D} = \frac{2I_D}{V_{OV}}$$

- Output resistance:

$$r_o = V_A / I_D = 1/\lambda I_D$$

PMOS transistors

Same formulas as for NMOS *except* using $|V_{OV}|$, $|V_A|$, and replacing μ_n with μ_p.

Small-Signal Equivalent Circuit Models

EXERCISES

5.20 For the amplifier in Fig. 5.34, let $V_{DD} = 5$ V, $R_D = 10$ kΩ, $V_t = 1$ V, $k'_n = 20$ μA/V², $W/L = 20$, $V_{GS} = 2$ V, and $\lambda = 0$.
(a) Find the dc current I_D and the dc voltage V_{DS}.
(b) Find g_m.
(c) Find the voltage gain.
(d) If $v_{gs} = 0.2 \sin \omega t$ volts, find v_{ds} assuming that the small-signal approximation holds. What are the minimum and maximum values of v_{DS}?
(e) Use Eq. (5.43) to determine the various components of i_D. Using the identity ($\sin^2 \omega t = \frac{1}{2} - \frac{1}{2} \cos 2\omega t$), show that there is a slight shift in I_D (by how much?) and that there is a second-harmonic component (i.e., a component with frequency 2ω). Express the amplitude of the second-harmonic component as a percentage of the amplitude of the fundamental. (This value is known as the second-harmonic distortion.)
Ans. (a) 200 μA, 3 V; (b) 0.4 mA/V; (c) −4 V/V; (d) $v_{ds} = -0.8 \sin \omega t$ volts, 2.2 V, 3.8 V; (e) $i_D = (204 + 80 \sin \omega t - 4 \cos 2\omega t)$ μA, 5%

5.21 An NMOS transistor has $\mu_n C_{ox} = 60$ μA/V², $W/L = 40$, $V_t = 1$ V, and $V_A = 15$ V. Find g_m and r_o when (a) the bias voltage $V_{GS} = 1.5$ V, (b) the bias current $I_D = 0.5$ mA.
Ans. (a) 1.2 mA/V, 50 kΩ; (b) 1.55 mA/V, 30 kΩ

5.22 A MOSFET is to operate at $I_D = 0.1$ mA and is to have $g_m = 1$ mA/V. If $k'_n = 50$ μA/V², find the required W/L ratio and the overdrive voltage.
Ans. 100; 0.2 V

5.23 For a fabrication process for which $\mu_p \simeq 0.4\mu_n$, find the ratio of the width of a PMOS transistor to the width of an NMOS transistor so that the two devices have equal g_m for the same bias conditions. The two devices have equal channel lengths.
Ans. 2.5

5.24 A PMOS transistor has $V_t = -1$ V, $k'_p = 60$ μA/V², and $W/L = 16\ \mu\text{m}/0.8\ \mu\text{m}$. Find I_D and g_m when the device is biased at $V_{GS} = -1.6$ V. Also, find the value of r_o if λ (at $L = 1$ μm) $= -0.04$ V⁻¹.
Ans. 216 μA; 0.72 mA/V; 92.6 kΩ

5.25 Use the formulas in Table 5.3 to derive an expression for $(g_m r_o)$ in terms of V_A and V_{OV}. As we shall see in Chapter 7, this is an important transistor parameter and is known as the intrinsic gain. Evaluate the value of $g_m r_o$ for an NMOS transistor fabricated in a 0.8-μm CMOS process for which $V'_A = 12.5$ V/μm of channel length. Let the device have minimum channel length and be operated at an overdrive voltage of 0.2 V.
Ans. $g_m r_o = 2V_A/V_{OV}$; 100 V/V

5.6 Basic MOSFET Amplifier Configurations

It is useful at this point to take stock of where we are and where we are going in our study of MOSFET amplifiers. In Section 5.4 we examined the essence of the use of the MOSFET as an amplifier. There we found that almost-linear amplification can be obtained by biasing the MOSFET at an appropriate point in its saturation region of operation and by keeping the signal v_{gs} small. We then took a closer look at the small-signal operation of the MOSFET in Section 5.5 and developed circuit models to represent the transistor, thus facilitating the determination of amplifier parameters such as voltage gain and input and output resistances.

We are now ready to consider the various possible configurations of MOSFET amplifiers, and we will do that in the present section. To focus our attention on the salient features of the various configurations, we shall present them in their most simple, or "stripped down" version. Thus, we will not show the dc biasing arrangements, leaving the study of bias design to the next section. Finally, in Section 5.8 we will bring everything together and present practical circuits for discrete-circuit MOSFET amplifiers; namely, those amplifer circuits that can be constructed using discrete components. The study of integrated-circuit amplifiers begins in Chapter 7.

5.6.1 The Three Basic Configurations

There are three basic configurations for connecting the MOSFET as an amplifier. Each of these configurations is obtained by connecting one of the three MOSFET terminals to ground, thus creating a two-port network with the grounded terminal being *common* to the input and output ports. Figure 5.43 shows the resulting three configurations with the biasing arrangements omitted.

In the circuit of Fig. 5.43(a) the source terminal is connected to ground, the input voltage signal v_i is applied between the gate and ground, and the output voltage signal v_o is taken between the drain and ground, across the resistance R_D. This configuration, therefore, is called the grounded-source or **common-source (CS)** amplifier. It is by far the most popular MOS amplifier configuration and is the one we utilized in Sections 5.4 and 5.5 to study MOS amplifier operation.

The **common-gate (CG)** or grounded-gate amplifier is shown in Fig. 5.43(b). It is obtained by connecting the gate to ground, applying the input v_i between the source and

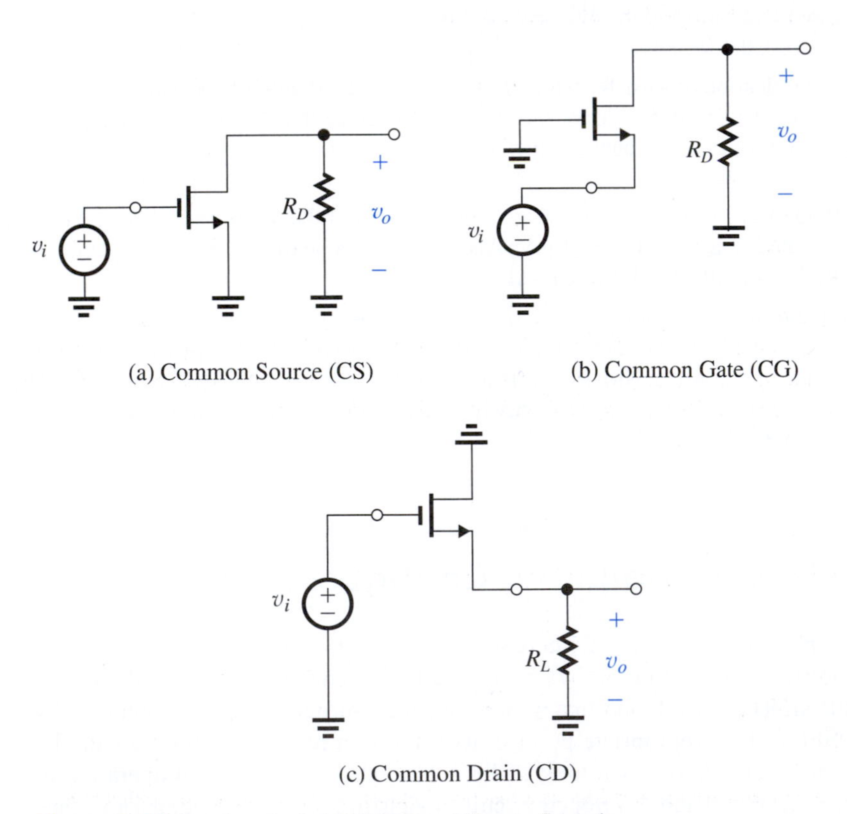

Figure 5.43 The three basic MOSFET amplifier configurations.

ground, and taking the output v_o across the resistance R_D connected between the drain and ground. We encountered a CG amplifier in Example 5.11.

Finally, Fig. 5.43(c) shows the **common-drain (CD)** or grounded-drain amplifier. It is obtained by connecting the drain terminal to ground, applying the input voltage signal v_i between gate and ground, and taking the output voltage signal between the source and ground, across a load resistance R_L. For reasons that will become apparent shortly, this configuration is more commonly called the **source follower**.

Our study of the three basic MOS amplifier configurations will reveal that each has distinctly different attributes and hence areas of application.

5.6.2 Characterizing Amplifiers

Before we begin our study of the different MOSFET amplifier configurations, we consider how to characterize the performance of an amplifier as a circuit building block. An introduction to this topic was presented in Section 1.5.

Figure 5.44(a) shows an amplifier fed with a signal source having an open-circuit voltage v_{sig} and an internal resistance R_{sig}. These can be the parameters of an actual signal source or, in a cascade amplifier, the Thévenin equivalent of the output circuit of another amplifier stage preceding the one under study. The amplifier is shown with a load resistance R_L

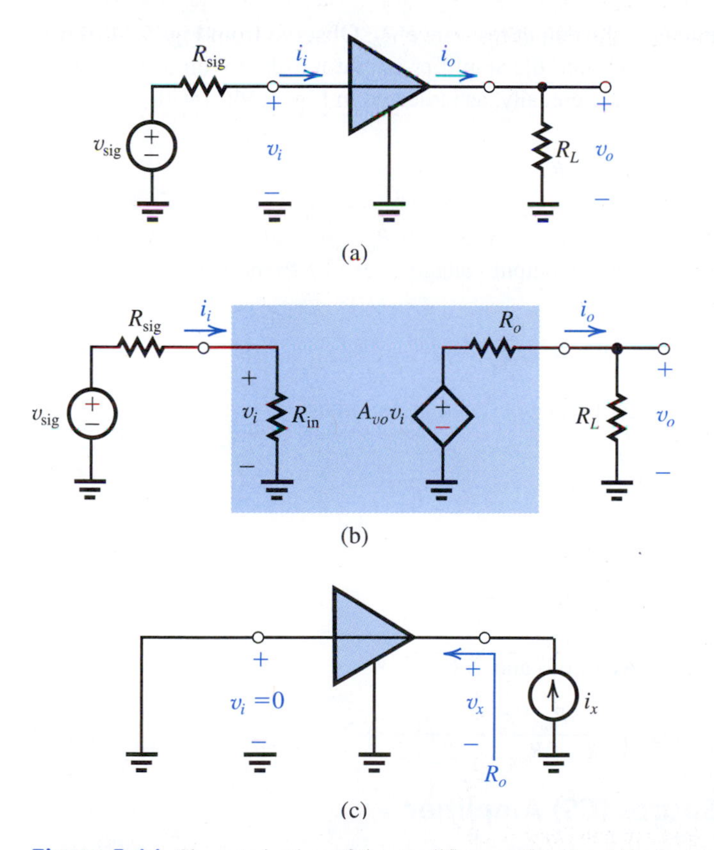

Figure 5.44 Characterization of the amplifier as a functional block: (**a**) An amplifier fed with a voltage signal v_{sig} having a source resistance R_{sig}, and feeding a load resistance R_L; (**b**) Equivalent-circuit representation of the circuit in (a); (**c**) Determining the amplifier output resistance R_o.

connected to the output terminal. Here, R_L can be an actual load resistance or the input resistance of a succeeding amplifier stage in a cascade amplifier.

Figure 5.44(b) shows the amplifier circuit with the amplifier block replaced by its equivalent-circuit model. The input resistance R_{in} represents the loading effect of the amplifier input on the signal source. It is found from

$$R_{in} \equiv \frac{v_i}{i_i}$$

and together with the resistance R_{sig} forms a voltage divider that reduces v_{sig} to the value v_i that appears at the amplifier input,

$$v_i = \frac{R_{in}}{R_{in} + R_{sig}} v_{sig} \tag{5.65}$$

All the amplifier circuits studied in this section are **unilateral**. That is, they do not contain internal feedback, and thus R_{in} will be independent of R_L. However, as will be seen in subsequent chapters, this is not always the case.

The second parameter in characterizing amplifier performance is the **open-circuit voltage gain** A_{vo}, defined as

$$A_{vo} \equiv \left.\frac{v_o}{v_i}\right|_{R_L=\infty}$$

The third and final parameter is the output resistance R_o. Observe from Fig. 5.44(b) that R_o is the resistance seen looking back into the amplifier output terminal with v_i set to zero. Thus R_o can be determined, at least conceptually, as indicated in Fig. 5.44(c) with

$$R_o = \frac{v_x}{i_x}$$

The controlled source $A_{vo}v_i$ and the output resistance R_o represent the Thévenin equivalent of the amplifier output circuit, and the output voltage v_o can be found from

$$v_o = \frac{R_L}{R_L + R_o} A_{vo} v_i \tag{5.66}$$

Thus the voltage gain of the amplifier proper, A_v, can be found as

$$A_v \equiv \frac{v_o}{v_i} = A_{vo} \frac{R_L}{R_L + R_o} \tag{5.67}$$

and the overall voltage gain G_v,

$$G_v \equiv \frac{v_o}{v_{sig}}$$

can be determined by combining Eqs. (5.65) and (5.67):

$$G_v = \frac{R_{in}}{R_{in} + R_{sig}} A_{vo} \frac{R_L}{R_L + R_o} \tag{5.68}$$

5.6.3 The Common-Source (CS) Amplifier

Of the three basic MOS amplifier configurations, the common source is the most widely used. Typically, in an amplifier formed by cascading a number of stages, the bulk of the voltage gain is obtained by using one or more common-source stages in the cascade.

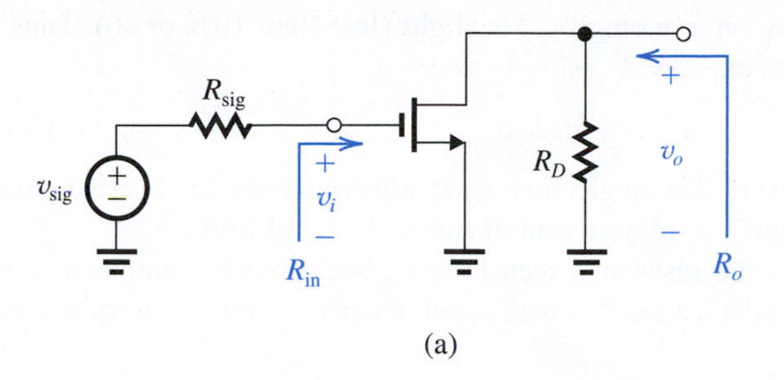

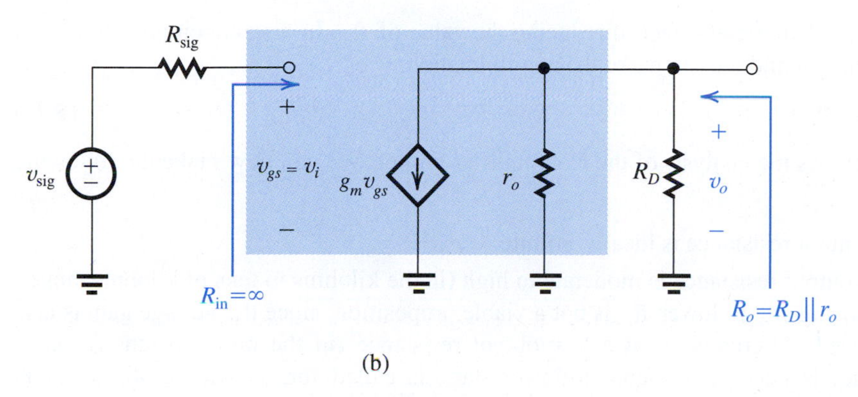

Figure 5.45 (**a**) Common-source amplifier fed with a signal v_{sig} from a generator with a resistance R_{sig}. The bias circuit is omitted. (**b**) The common-source amplifier with the MOSFET replaced with its hybrid-π model.

Figure 5.45(a) shows a common-source amplifier (with the biasing arrangement omitted) fed with a signal source v_{sig} having a source resistance R_{sig}. We wish to analyze this circuit to determine R_{in}, A_{vo}, R_o, and G_v. For this purpose we shall assume that R_D is part of the amplifier; thus if a load resistance R_L is connected to the amplifier output, it appears in parallel with R_D.

Characteristic Parameters of the CS Amplifier Replacing the MOSFET with its hybrid-π model, we obtain the CS amplifier equivalent circuit shown in Fig 5.45(b). We shall use this equivalent circuit to determine the characteristic parameters R_{in}, A_{vo}, and R_o as follows.

The input resistance R_{in} is obviously infinite,

$$R_{in} = \infty \tag{5.69}$$

The output voltage v_o is found by multiplying the current $(g_m v_{gs})$ by the total resistance between the output node and ground,

$$v_o = -(g_m v_{gs})(R_D \| r_o)$$

Since $v_{gs} = v_i$, the open-circuit voltage gain $A_{vo} \equiv v_o/v_i$ can be obtained as

$$A_{vo} = -g_m(R_D \| r_o) \tag{5.70}$$

Observe that the transistor output resistance r_o reduces the magnitude of the voltage gain. In discrete-circuit amplifiers, which are of interest to us in this chapter, R_D is usually much lower than r_o and the effect of r_o on reducing $|A_{vo}|$ is slight (less than 10% or so). Thus in many cases we can neglect r_o and express A_{vo} simply as

$$A_{vo} \simeq (-g_m R_D) \tag{5.71}$$

The reader is cautioned, however, that neglecting r_o is allowed only in discrete-circuit design. As will be seen in Chapter 7, r_o plays a central role in IC amplifiers.

The output resistance R_o is the resistance seen looking back into the output terminal with v_i set to zero. From Fig. 5.45(b) we see that with v_i set to zero, v_{gs} will be zero, and thus $g_m v_{gs}$ will be zero, resulting in

$$R_o = R_D \| r_o \tag{5.72}$$

Here, r_o has the beneficial effect of reducing the value of R_o. In discrete circuits, however, this effect is slight and we can make the approximation

$$R_o \simeq R_D \tag{5.73}$$

This concludes the analysis of the CS amplifier proper. We can now make the following observations.

1. The input resistance is ideally infinite.
2. The output resistance is moderate to high (in the kilohms to tens of kilohms range). Reducing R_D to lower R_o is not a viable proposition, since the voltage gain is also reduced. Alternatively, if a low output resistance (in the ohms to tens of ohms range) is needed, a source follower stage is called for, as will be discussed in Section 5.6.6.
3. The open-circuit voltage gain A_{vo} can be high, making the CS configuration the workhorse in MOS amplifier design. Unfortunately, however, the bandwidth of the CS amplifier is severely limited. We shall study amplifier frequency response in Chapter 9.

Overall Voltage Gain To determine the overall voltage gain G_v, we first note that the infinite input resistance will make the entire signal v_{sig} appear at the amplifier input,

$$v_i = v_{sig} \tag{5.74}$$

an obviously ideal situation. At this point we should remind the reader that to maintain a reasonably linear operation, v_i and hence v_{sig} should be kept much smaller than $2V_{OV}$.

If a load resistance R_L is connected to the output terminal of the amplifier, this resistance will appear in parallel with R_D. It follows that the voltage gain A_v can be obtained by simply replacing R_D in the expression for A_{vo} in Eq. (5.70) by $R_D \| R_L$,

$$A_v = -g_m(R_D \| R_L \| r_o) \tag{5.75}$$

This expression together with the fact that $v_i = v_{sig}$, provides the overall voltage gain,

$$G_v = A_v = -g_m(R_D \| R_L \| r_o) \tag{5.76}$$

EXERCISE

5.26 Use A_{vo} in Eq. (5.70) together with R_o in Eq. (5.72) to obtain A_v. Show that the result is identical to that in Eq. (5.75).

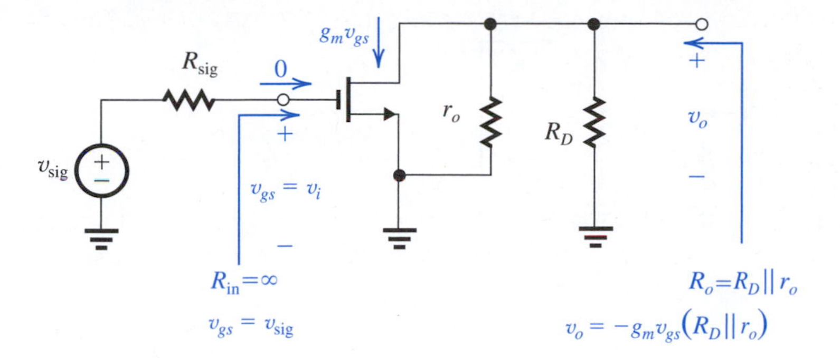

Figure 5.46 Performing the analysis directly on the circuit diagram with the MOSFET model used implicitly.

Performing the Analysis Directly on the Circuit Diagram Although small-signal, equivalent-circuit models provide a systematic process for the analysis of any amplifier circuit, the effort involved in drawing the equivalent circuit is sometimes not justified. That is, in simple situations and after a lot of practice, one can perform the small-signal analysis directly on the circuit schematic. Because in this way one remains closer to the actual circuit, the direct analysis can yield greater insight into circuit operation. Figure 5.46 shows the direct analysis of the CS amplifier. Observe that we have "pulled out" the resistance r_o from the transistor, thus making the transistor drain conduct $g_m v_{gs}$ while still accounting for the effect of r_o.

EXERCISE

5.27 A CS amplifier utilizes a MOSFET biased at $I_D = 0.25$ mA with $V_{OV} = 0.25$ V and $R_D = 20$ kΩ. The device has $V_A = 50$ V. The amplifier is fed with a source having $R_{sig} = 100$ kΩ, and a 20-kΩ load is connected to the output. Find R_{in}, A_{vo}, R_o, A_v, and G_v. If to maintain reasonable linearity, the peak of the input sine-wave signal is limited to 10% of $(2V_{OV})$ what is the peak of the sine-wave voltage at the output?
Ans. ∞; −36.4 V/V; 18.2 kΩ; −19 V/V; −19 V/V; 0.95 V

5.6.4 The Common-Source Amplifier with a Source Resistance

It is often beneficial to insert a resistance R_s in the source lead of the common-source amplifier as shown in Fig. 5.47(a). The corresponding small-signal equivalent circuit is shown in Fig. 5.47(b), where we note that the MOSFET has been replaced with its T equivalent-circuit model. The T model is used in preference to the π model because it makes the analysis in this case somewhat simpler. In general, *whenever a resistance is connected in the source lead, the T model is preferred.* The source resistance then simply appears in series with the resistance $1/g_m$ and can be added to it.

It should be noted that we have not included r_o in the equivalent-circuit model. Including r_o would complicate the analysis considerably; r_o would connect the output node of the amplifier to the input side and thus would make the amplifier *nonunilateral.*

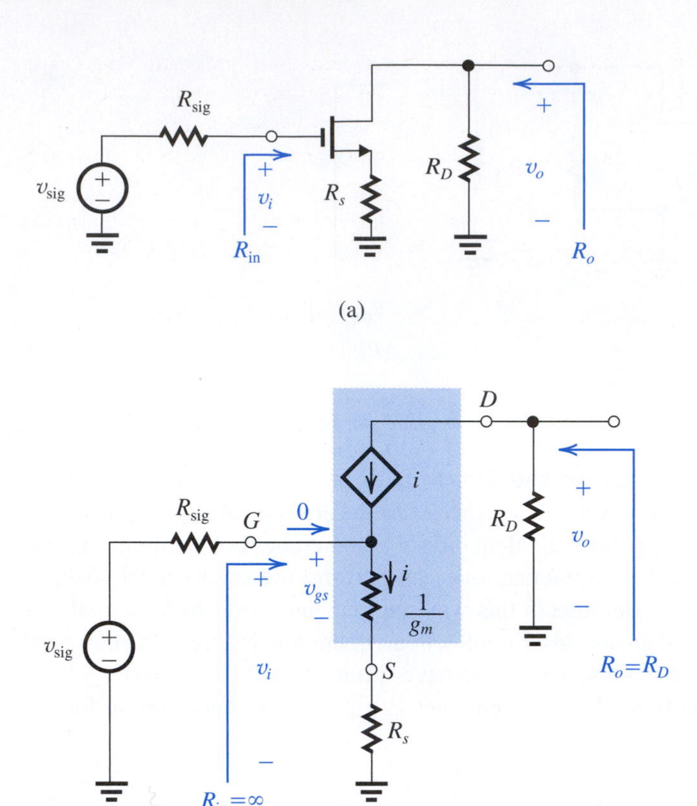

Figure 5.47 The CS amplifier with a source resistance R_s: (**a**) Circuit without bias details; (**b**) Equivalent circuit with the MOSFET represented by its T model.

Fortunately, it turns out that the effect of r_o on the operation of the discrete-circuit amplifier is not important. This can be verified by computer simulation, using for instance SPICE. This is not the case, however, for the integrated-circuit version of the circuit, where r_o plays a major role and must be taken into account, as we shall do in Chapter 7.

From Fig. 5.47(b) we see that the input resistance R_{in} is infinite and thus $v_i = v_{sig}$. Unlike the CS amplifier, however, here only a fraction of v_i appears between gate and source as v_{gs}. It can be determined from the voltage divider composed of $1/g_m$ and R_s that appears across the amplifier input, as follows:

$$v_{gs} = v_i \frac{1/g_m}{1/g_m + R_s} = \frac{v_i}{1 + g_m R_s} \tag{5.77}$$

Thus we can use the value of R_s to control the magnitude of the signal v_{gs} and thereby ensure that v_{gs} does not become too large and cause unacceptably high nonlinear distortion. This is the first benefit of including resistor R_s. Other benefits will be encountered in later sections and chapters. For instance, it will be shown in Chapter 9 that R_s causes the useful bandwidth of the amplifier to be extended. The mechanism by which R_s causes such improvements in amplifier performance is negative feedback. To see how R_s introduces negative feedback, refer to Fig. 5.47(a): If while keeping v_i constant, for some reason the drain current increases, the source current also will increase, resulting in an increased voltage drop across R_s. Thus the source voltage rises, and the gate-to-source voltage decreases. The latter effect causes the drain current to decrease, counteracting the initially assumed change, an indication of the presence of negative feedback. In Chapter 10 we shall study negative feedback formally. There we will learn that the improvements that negative feedback provides are obtained at the expense of a reduction in gain. We will now show this to be the case in the circuit of Fig. 5.47.

The output voltage v_o is obtained by multiplying the controlled-source current i by R_D,

$$v_o = -i\ R_D$$

The current i in the source lead can be found by dividing v_i by the total resistance in the source,

$$i = \frac{v_i}{1/g_m + R_s} = \left(\frac{g_m}{1 + g_m R_s}\right) v_i \tag{5.78}$$

Thus, the voltage gain A_{vo} can be found as

$$A_{vo} = \frac{v_o}{v_i} = -\frac{R_D}{1/g_m + R_s} \tag{5.79}$$

which can also be expressed as

$$A_{vo} = -\frac{g_m R_D}{1 + g_m R_s} \tag{5.80}$$

Equation (5.80) indicates that including the resistance R_s reduces the voltage gain by the factor $(1 + g_m R_s)$. This is the price paid for the improvements that accrue as a result of R_s. It is interesting to note that in Chapter 10, we will find that the factor $(1 + g_m R_s)$ is the "amount of negative feedback" introduced by R_s. It is also the same factor by which bandwidth and other performance parameters improve. Because of the negative-feedback action of R_s it is known as a **source-degeneration resistance**.

There is another useful interpretation of the expression for the drain current in Eq. (5.78): The quantity between brackets on the right-hand side can be thought of as the "effective transconductance with R_s included." Thus, including R_s reduces the transconductance by the factor $(1 + g_m R_s)$. This, of course, is simply the result of the fact that only a fraction $1/(1 + g_m R_s)$ of v_i appears as v_{gs} (see Eq. 5.77.).

The alternative gain expression in Eq. (5.79) has a powerful and insightful interpretation: The voltage gain between gate and drain is equal to the ratio of the total resistance in the drain (R_D) to the total resistance in the source $(1/g_m + R_s)$,

$$\text{Voltage gain from gate to drain} = -\frac{\text{Total resistance in drain}}{\text{Total resistance in source}} \tag{5.81}$$

This is a general expression. For instance, setting $R_s = 0$ in Eq. (5.79) yields A_{vo} of the CS amplifier.

Finally, we consider the situation of a load resistance R_L connected at the output. We can obtain the gain A_v using the open-circuit voltage gain A_{vo} together with the output resistance R_o, which can be found by inspection to be

$$R_o = R_D$$

Alternatively, A_v can be obtained by simply replacing R_D in Eq. (5.79) or (5.80) by $(R_D \| R_L)$; thus,

$$A_v = -\frac{R_D \| R_L}{1/g_m + R_s} \tag{5.82}$$

or

$$A_v = -\frac{g_m(R_D \| R_L)}{1 + g_m R_s} \tag{5.83}$$

Observe that Eq. (5.82) is a direct application of the ratio of total resistance rule of Eq. (5.81). Finally, note that because R_{in} is infinite, $v_i = v_{sig}$ and the overall voltage gain G_v is equal to A_v.

EXERCISE

5.28 In Exercise 5.27 we applied an input signal v_{sig} of 50 mV peak and obtained an output signal of approximately 1 V peak. Assume that for some reason we now have an input signal v_{sig} that is 0.2 V peak and that we wish to modify the circuit to keep v_{gs} unchanged, and thus keep the nonlinear distortion from increasing. What value should we use for R_s? What value of G_v will result? What will the peak signal at the output become? Assume $r_o = \infty$.
Ans. 1.5 kΩ; –5 V/V; 1 V

5.6.5 The Common-Gate (CG) Amplifier

Figure 5.48(a) shows a common-gate amplifier with the biasing circuit omitted. The amplifier is fed with a signal source characterized by v_{sig} and R_{sig}. Since R_{sig} appears in series with the source, it is more convenient to represent the transistor with the T model than with the π model. Doing this, we obtain the amplifier equivalent circuit shown in Fig. 5.48(b). Note that we have not included r_o: This would have complicated the analysis considerably, for r_o would have appeared between the output and the input side of the amplifier. Fortunately, it

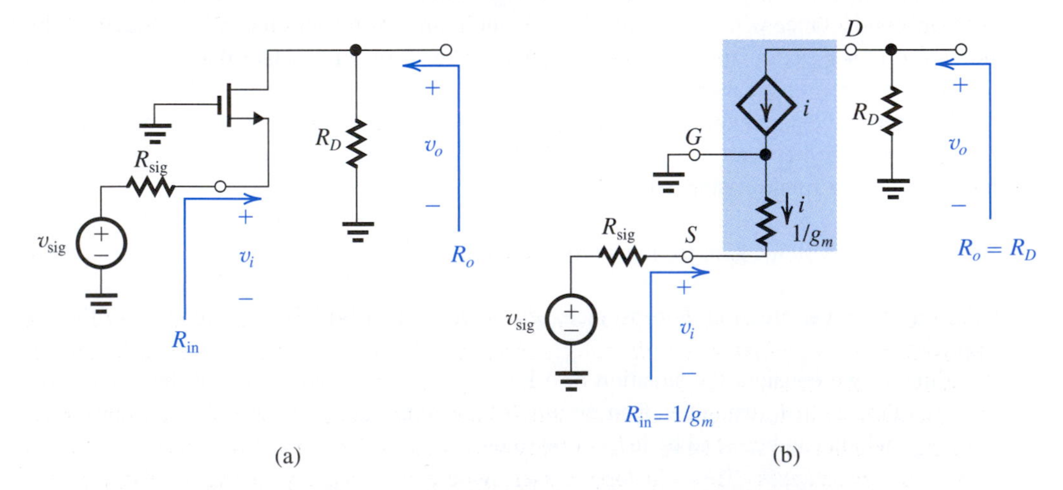

Figure 5.48 **(a)** Common-gate (CG) amplifier with bias arrangement omitted. **(b)** Equivalent circuit of the CG amplifier with the MOSFET replaced with its T model.

turns out that the effect of r_o on the performance of a discrete CG amplifier is very small. We will consider the effect of r_o when we study the IC form of the CG amplifier in Chapter 7.

From inspection of the equivalent circuit of Fig. 5.48(b), we see that the input resistance

$$R_{in} = \frac{1}{g_m} \tag{5.84}$$

This should have been expected, since we are looking into the source and the gate is grounded. Typically $1/g_m$ is a few hundred ohms; thus the CG amplifier has a low input resistance.

To determine the voltage gain A_{vo}, we write at the drain node

$$v_o = -iR_D$$

and substitute for the source current i from

$$i = -\frac{v_i}{1/g_m}$$

to obtain

$$A_{vo} \equiv \frac{v_o}{v_i} = g_m R_D \tag{5.85}$$

which except for the positive sign is identical to the expression for A_{vo} of the CS amplifier (when r_o is neglected).

The output resistance of the CG circuit can be found by inspection of the circuit in Fig. 5.48(b) as

$$R_o = R_D \tag{5.86}$$

which is the same as in the case of the CS amplifier (with r_o neglected).

Although the gain of the CG amplifier proper has the same magnitude as that of the CS amplifier, this is usually not the case as far as the overall voltage gain is concerned. The low input resistance of the CG amplifier can cause the input signal to be severely attenuated. Specifically,

$$\frac{v_i}{v_{sig}} = \frac{R_{in}}{R_{in} + R_{sig}} = \frac{1/g_m}{1/g_m + R_{sig}} \tag{5.87}$$

from which we see that except for situations in which R_{sig} is on the order of $1/g_m$, the signal transmission factor v_i/v_{sig} can be very small and the overall voltage gain G_v can be correspondingly small. Specifically, with a resistance R_L connected at the output

$$G_v = \frac{1/g_m}{R_{sig} + 1/g_m}[g_m(R_D \| R_L)]$$

Thus,

$$G_v = \frac{(R_D \| R_L)}{R_{sig} + 1/g_m} \tag{5.88}$$

Observe that *the overall voltage gain is simply the ratio of the total resistance in the drain circuit to the total resistance in the source circuit.* If R_{sig} is of the same order as R_D and R_L, G_v will be very small.

Because of its low input resistance, the CG amplifier alone has very limited application. One such application is to amplify high-frequency signals that come from sources with relatively low resistances. These include cables, where it is usually necessary for the input resistance of the amplifier to match the characteristic resistance of the cable. As will be shown in Chapter 9, the CG amplifier has excellent high-frequency response. Thus it can be combined with the CS amplifier in a very beneficial way that takes advantage of the best features of each of the two configurations. A very significant circuit of this kind will be studied in Chapter 7.

EXERCISE

5.29 A CG amplifier is required to match a signal source with $R_{sig} = 100\ \Omega$. At what current I_D should the MOSFET be biased if it is operated at an overdrive voltage of 0.20 V? If the total resistance in the drain circuit is 2 kΩ, what overall voltage gain is realized?
Ans. 1 mA; 10 V/V

5.6.6 The Common-Drain Amplifier or Source Follower

The last of the basic MOSFET amplifier configurations is the common-drain amplifier, an important circuit that finds application in the design of both small-signal amplifiers as well as amplifiers that are required to handle large signals and deliver substantial amounts of signal power to a load. This latter variety will be studied in Chapter 11. The common drain amplifier is more commonly known as the *source follower*. The reason behind this name will become apparent shortly.

The Need for Voltage Buffers Before embarking on the analysis of the source follower, it is useful to look at one of its more common applications. Consider the situation depicted in Fig. 5.49(a). A signal source delivering a signal of reasonable strength (1 V)

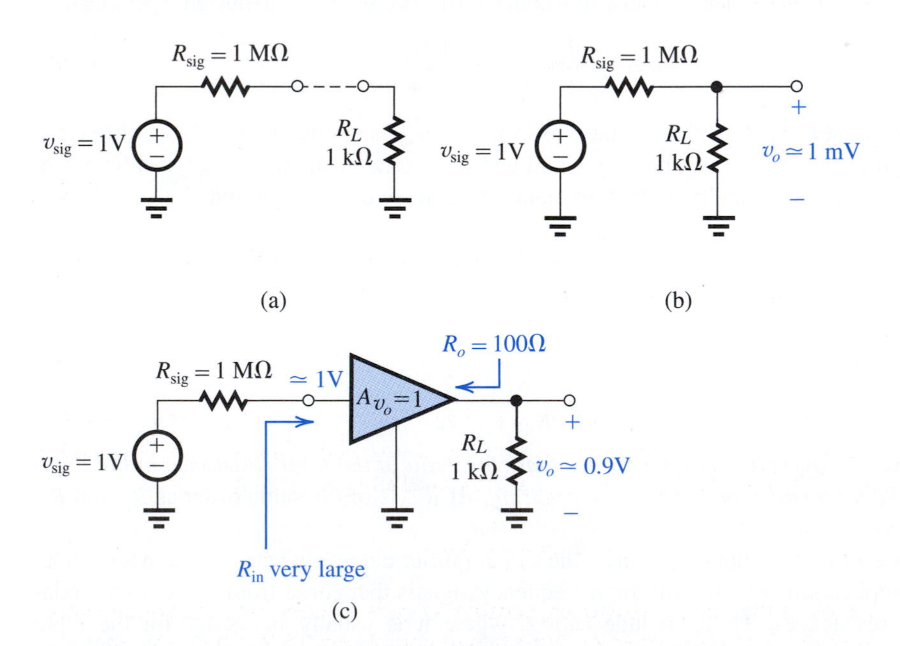

Figure 5.49 Illustrating the need for a unity-gain buffer amplifier.

with an internal resistance of 1 MΩ is to be connected to a 1-kΩ load resistance. Connecting the source to the load directly as in Fig. 5.49(b) would result in severe attenuation of the signal; the signal appearing across the load will be only $1/(1000+1)$ of the input signal or about 1 mV. An alternative course of action is suggested in Fig. 5.49(c). Here we have interposed an amplifier between the source and the load. Our amplifier, however, is unlike the amplifiers we have been studying in this chapter thus far; it has a voltage gain of only unity. This is because our signal is already of sufficient strength and we do not need to increase its amplitude. Note, however, that our amplifier has a very large input resistance, thus almost all of v_{sig} (i.e., 1 V) will appear at the input of the amplifier proper. Since the amplifier has a low output resistance (100 Ω), 90% of this signal (0.9 V) will appear at the output, obviously a very significant improvement over the situation without the amplifier. As will be seen shortly, the source follower can easily implement the unity-gain buffer amplifier shown in Fig. 5.49(c).

Characteristic Parameters of the Source Follower Figure 5.50(a) shows a source follower with the bias circuit omitted. The source follower is fed with a signal generator (v_{sig}, R_{sig}) and has a load resistance R_L connected between the source terminal and ground. We shall assume that R_L includes both the actual load and any other resistance that may be present between the source terminal and ground (e.g., for biasing purposes). Normally, the actual load resistance would be much lower in value than such other resistances and thus would dominate.

Since the MOSFET has a resistance R_L connected in its source terminal, it is most convenient to use the T model, as shown in Fig. 5.50(b). Note that we have included r_o, simply because it is very easy to do so. However, since r_o in effect appears in parallel with R_L, and since in discrete circuits $r_o \gg R_L$, we can neglect r_o and obtain the simplified equivalent circuit shown in Fig. 5.50(c). From the latter circuit we can write by inspection

$$R_{in} = \infty$$

and obtain A_v from the voltage divider formed by $1/g_m$ and R_L as

$$A_v \equiv \frac{v_o}{v_i} = \frac{R_L}{R_L + 1/g_m} \tag{5.89}$$

Setting $R_L = \infty$ we obtain

$$A_{vo} = 1 \tag{5.90}$$

The output resistance R_o is found by setting $v_i = 0$ (i.e., by grounding the gate). Now looking back into the output terminal, excluding R_L, we simply see $1/g_m$, thus

$$R_o = 1/g_m \tag{5.91}$$

The unity open-circuit voltage gain together with R_o in Eq. (5.91) can be used to find A_v when a load resistance R_L is connected. The result is simply the expression in Eq. (5.89). Finally, because of the infinite R_{in}, $v_i = v_{sig}$, and the overall voltage gain is

$$G_v = A_v = \frac{R_L}{R_L + 1/g_m} \tag{5.92}$$

Thus G_v will be lower than unity. However, because $1/g_m$ is usually low, the voltage gain can be close to unity. The unity open-circuit voltage gain in Eq. (5.90) indicates that the

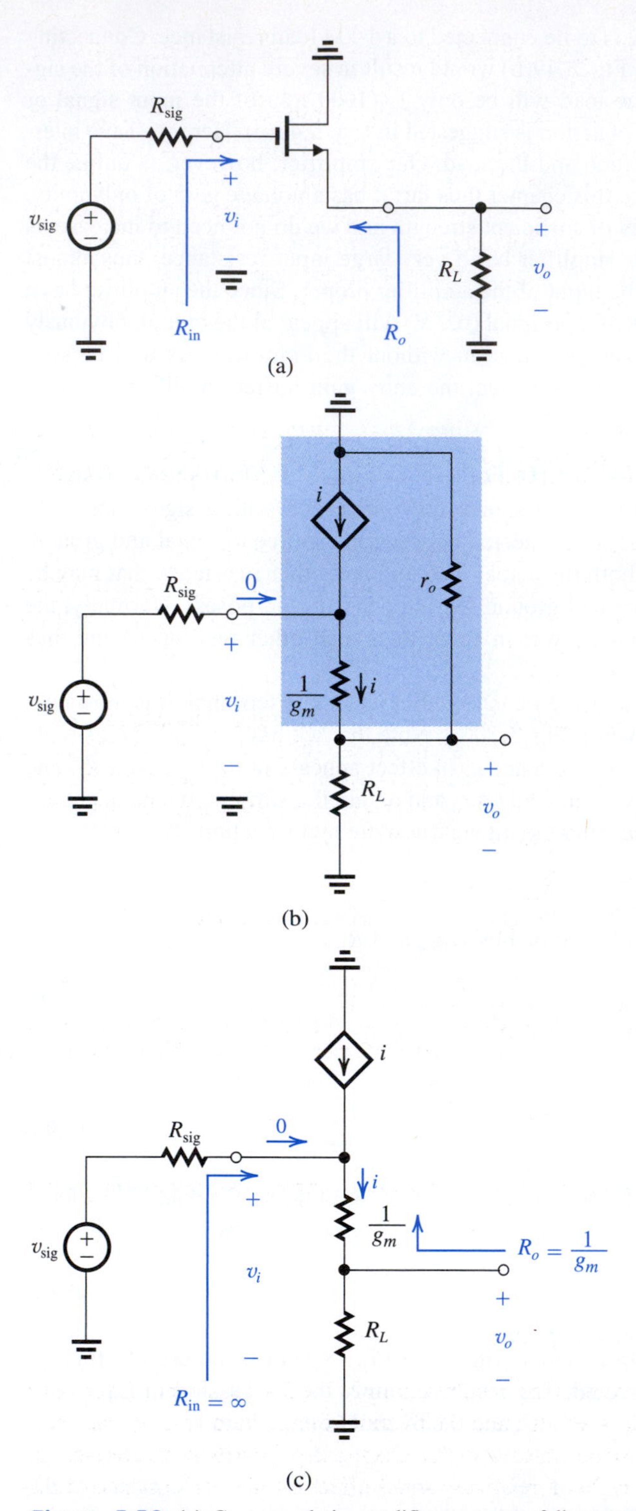

Figure 5.50 (**a**) Common-drain amplifier or source follower. (**b**) Equivalent circuit of the source follower obtained by replacing the MOSFET with its T model. Note that r_o appears in parallel with R_L and in discrete circuits, $r_o \gg R_L$. Neglecting r_o, we obtain the simplified equivalent circuit in (**c**).

voltage at the source terminal will follow that at the input, hence the name *source follower*.

In conclusion, the source follower features a very high input resistance (ideally, infinite), a relatively low output resistance, and an open-circuit voltage gain that is near unity (ideally, unity). Thus the source follower is ideally suited for implementing the unity-gain voltage buffer of Fig. 5.49(c). The source follower is also used as the output (i.e., last) stage in a multistage amplifier, where its function is to equip the overall amplifier with a low output resistance, thus enabling it to supply relatively large load currents without loss of gain (i.e., with little reduction of output signal level). The design of output stages is studied in Chapter 11.

EXERCISES

D5.30 It is required to design a source follower that implements the buffer amplifier shown in Fig. 5.49(c). If the MOSFET is operated with an overdrive voltage $V_{OV} = 0.25$ V, at what drain current should it be biased? Find the output signal amplitude and the signal amplitude between gate and source.

Ans. 1.25 mA; 0.91 V; 91 mV

D5.31 A MOSFET is connected in the source-follower configuration and employed as the output stage of a cascade amplifier. It is required to provide an output resistance of 200 Ω. If the MOSFET has $k'_n = 0.4$ mA/V^2 and is operated at $V_{OV} = 0.25$ V, find the required W/L ratio. Also specify the dc bias current I_D. If the amplifier load resistance varies over the range 1 kΩ to 10 kΩ, what is the range of G_v of the source follower?

Ans. 50; 0.625 mA; 0.83 V/V to 0.98 V/V

5.32 Refer to Fig. 5.50(b). Show that taking r_o into account results in

$$A_{vo} = \frac{r_o}{r_o + 1/g_m}$$

Now, recalling that $r_o = V_A/I_D$ and $g_m = 2I_D/V_{OV}$, find A_{vo} in terms of V_A and V_{OV}. For a technology for which $V_A = 20$ V, what is the maximum V_{OV} at which the transistor can be operated while obtaining $A_{vo} \geq 0.99$ V/V?

Ans. $A_{vo} = 1/[1 + V_{OV}/2V_A]$; 0.4 V

5.6.7 Summary and Comparisons

For easy reference and to enable comparisons, we present in Table 5.4 the formulas for determining the characteristic parameters of discrete MOS amplifiers. Note that r_o has been neglected throughout. This is because our interest in this chapter is primarily in discrete-circuit amplifiers. As already mentioned, r_o has a relatively small effect on the performance of discrete-circuit amplifiers and can usually be neglected. In some cases, however, it is very easy to take r_o into account, such as in the case of the CS and CD amplifiers, and one is encouraged to do so. For integrated-circuit amplifiers, r_o *must* always be taken into account.

Table 5.4 Characteristics of MOSFET Amplifiers

	Characteristics[a, b]				
Amplifier type	R_{in}	A_{vo}	R_o	A_v	G_v
Common source (Fig. 5.45)	∞	$-g_m R_D$	R_D	$-g_m(R_D \parallel R_L)$	$-g_m(R_D \parallel R_L)$
Common source with R_s (Fig. 5.47)	∞	$-\dfrac{g_m R_D}{1+g_m R_s}$	R_D	$\dfrac{-g_m(R_D \parallel R_L)}{1+g_m R_s}$ $-\dfrac{R_D \parallel R_L}{1/g_m + R_s}$	$-\dfrac{g_m(R_D \parallel R_L)}{1+g_m R_s}$ $-\dfrac{R_D \parallel R_L}{1/g_m + R_s}$
Common gate (Fig. 5.48)	$\dfrac{1}{g_m}$	$g_m R_D$	R_D	$g_m(R_D \parallel R_L)$	$\dfrac{R_D \parallel R_L}{R_{sig} + 1/g_m}$
Source follower (Fig. 5.50)	∞	1	$\dfrac{1}{g_m}$	$\dfrac{R_L}{R_L + 1/g_m}$	$\dfrac{R_L}{R_L + 1/g_m}$

[a] For the interpretation of R_{in}, A_{vo}, and R_o, refer to Fig. 5.44(b).

[b] The MOSFET output resistance r_o has been neglected, as is permitted in the discrete-circuit amplifiers studied in this chapter. For IC amplifiers, r_o must always be taken into account.

In addition to the remarks already made throughout this section about the characteristics and areas of applicability of the various configurations, we make the following concluding points:

1. The CS configuration is the best suited for realizing the bulk of the gain required in an amplifier. Depending on the magnitude of the gain required, either a single stage or a cascade of two or three stages can be used.
2. Including a resistor R_s in the source lead of the CS stage provides a number of performance improvements at the expense of gain reduction.
3. The low input resistance of the CG amplifier makes it useful only in specific applications. As we shall see in Chapter 9, it has a much better high-frequency response than the CS amplifier. This superiority makes it useful as a high-frequency amplifier, especially when combined with the CS circuit. We shall see one such combination in Chapter 7.
4. The source follower finds application as a voltage buffer for connecting a high-resistance source to a low-resistance load and as the output stage in a multistage amplifier where its purpose is to equip the amplifier with a low output resistance.

5.7 Biasing in MOS Amplifier Circuits

As discussed in Section 5.4, an essential step in the design of a MOSFET amplifier circuit is the establishment of an appropriate dc operating point for the transistor. This is the step known as biasing or bias design. An appropriate dc operating point or bias point is characterized by a stable and predictable dc drain current I_D and by a dc drain-to-source voltage V_{DS} that ensures operation in the saturation region for all expected input-signal levels.

5.7.1 Biasing by Fixing V_{GS}

The most straightforward approach to biasing a MOSFET is to fix its gate-to-source voltage V_{GS} to the value required[9] to provide the desired I_D. This voltage value can be derived from the power-supply voltage V_{DD} through the use of an appropriate voltage divider. Alternatively, it can be derived from another suitable reference voltage that might be available in the system. Independent of how the voltage V_{GS} may be generated, this is *not* a good approach to biasing a MOSFET. To understand the reason for this statement, recall that

$$I_D = \frac{1}{2}\mu_n C_{ox}\frac{W}{L}(V_{GS} - V_t)^2$$

and note that the values of the threshold voltage V_t, the oxide-capacitance C_{ox}, and (to a lesser extent) the transistor aspect ratio W/L vary widely among devices of supposedly the same size and type. This is certainly the case for discrete devices, in which large spreads in the values of these parameters occur among devices of the same manufacturer's part number. The spread is also large in integrated circuits, especially among devices fabricated on different wafers and certainly between different batches of wafers. Furthermore, both V_t and μ_n depend on temperature, with the result that if we fix the value of V_{GS}, the drain current I_D becomes very much temperature dependent.

To emphasize the point that biasing by fixing V_{GS} is not a good technique, we show in Fig. 5.51 two i_D–v_{GS} characteristic curves representing extreme values in a batch of MOSFETs of the same type. Observe that for the fixed value of V_{GS}, the resultant spread in the values of the drain current can be substantial.

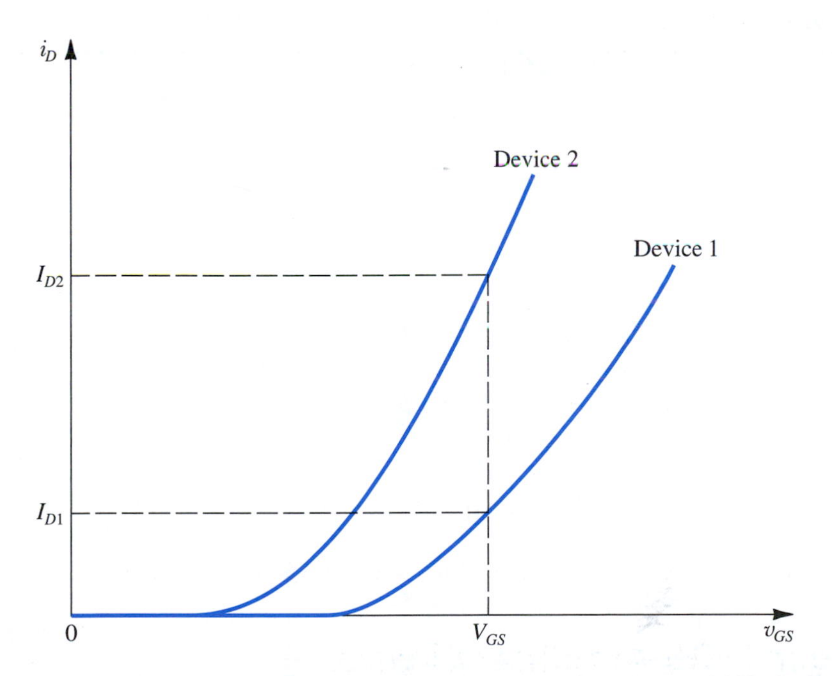

Figure 5.51 The use of fixed bias (constant V_{GS}) can result in a large variability in the value of I_D. Devices 1 and 2 represent extremes among units of the same type.

[9] That is indeed what we were doing in Section 5.4. However, the amplifier circuits studied there were conceptual ones, not actual practical circuits. Our purpose in this section is to study the latter.

5.7.2 Biasing by Fixing V_G and Connecting a Resistance in the Source

An excellent biasing technique for discrete MOSFET circuits consists of fixing the dc voltage at the gate, V_G, and connecting a resistance in the source lead, as shown in Fig. 5.52(a). For this circuit we can write

$$V_G = V_{GS} + R_S I_D \tag{5.93}$$

Now, if V_G is much greater than V_{GS}, I_D will be mostly determined by the values of V_G and R_S. However, even if V_G is not much larger than V_{GS}, resistor R_S provides *negative feedback,* which acts to stabilize the value of the bias current I_D. To see how this comes about, consider what happens when I_D increases for whatever reason. Equation (5.93) indicates that since V_G is constant, V_{GS} will have to decrease. This in turn results in a decrease in I_D, a change that is

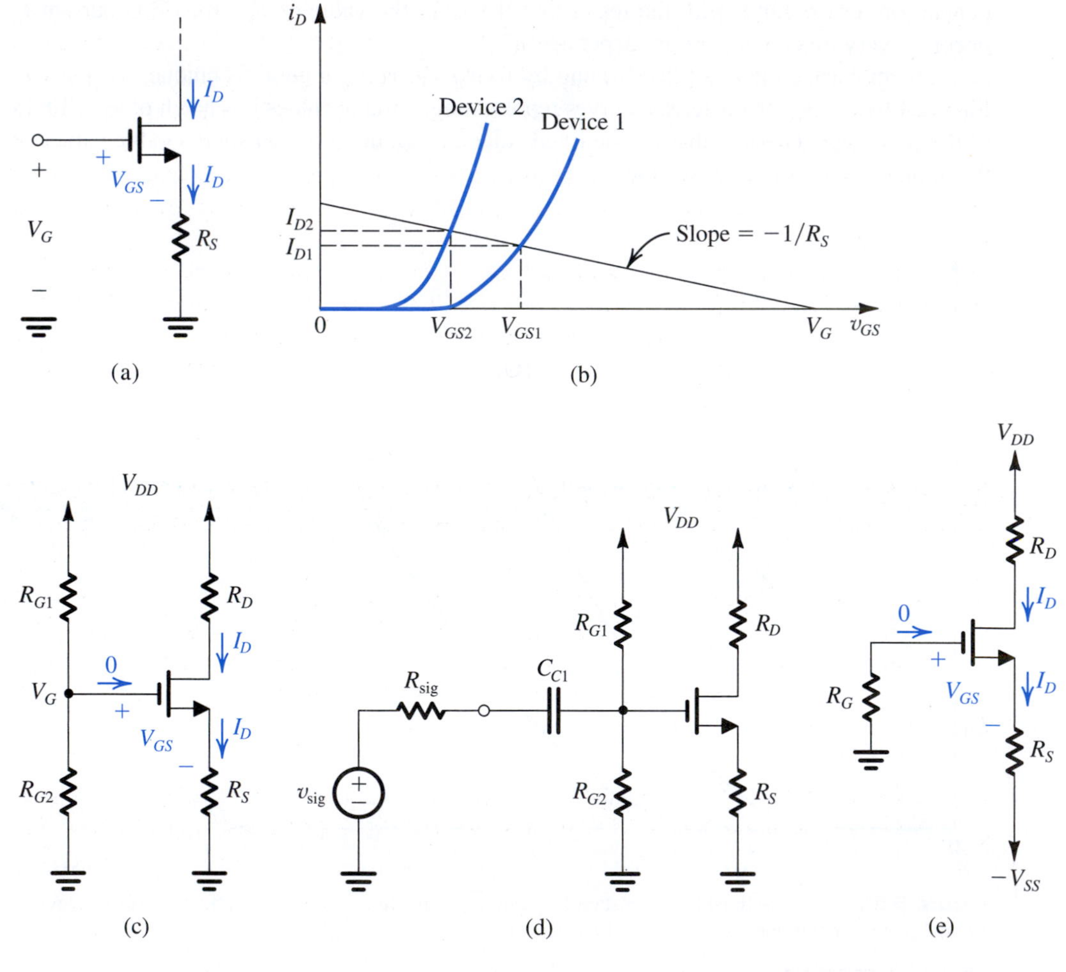

Figure 5.52 Biasing using a fixed voltage at the gate, V_G, and a resistance in the source lead, R_S: **(a)** basic arrangement; **(b)** reduced variability in I_D; **(c)** practical implementation using a single supply; **(d)** coupling of a signal source to the gate using a capacitor C_{C1}; **(e)** practical implementation using two supplies.

opposite to that initially assumed. Thus the action of R_S works to keep I_D as constant as possible. This negative feedback action of R_S gives it the name **degeneration resistance**, a name that we will appreciate much better at a later point in this text.[10]

Figure 5.52(b) provides a graphical illustration of the effectiveness of this biasing scheme. Here too we show the i_D–v_{GS} characteristics for two devices that represent the extremes of a batch of MOSFETs. Superimposed on the device characteristics is a straight line that represents the constraint imposed by the bias circuit—namely, Eq. (5.93). The intersection of this straight line with the i_D–v_{GS} characteristic curve provides the coordinates (I_D and V_{GS}) of the bias point. Observe that compared to the case of fixed V_{GS}, here the variability obtained in I_D is much smaller. Also, note that the variability decreases as V_G and R_S are made larger (thus providing a bias line that is less steep).

Two possible practical discrete implementations of this bias scheme are shown in Fig. 5.52(c) and (e). The circuit in Fig. 5.52(c) utilizes one power-supply V_{DD} and derives V_G through a voltage divider (R_{G1}, R_{G2}). Since $I_G = 0$, R_{G1} and R_{G2} can be selected to be very large (in the megohm range), allowing the MOSFET to present a large input resistance to a signal source that may be connected to the gate through a coupling capacitor, as shown in Fig. 5.52(d). Here capacitor C_{C1} blocks dc and thus allows us to couple the signal v_{sig} to the amplifier input without disturbing the MOSFET dc bias point. The value of C_{C1} should be selected large enough to approximate a short circuit at all signal frequencies of interest. We shall study capacitively coupled MOSFET amplifiers, which are suitable only in discrete circuit design, in Section 5.8. Finally, note that in the circuit of Fig. 5.52(c), resistor R_D is selected to be as large as possible to obtain high gain but small enough to allow for the desired signal swing at the drain while keeping the MOSFET in saturation at all times.

When two power supplies are available, as is often the case, the somewhat simpler bias arrangement of Fig. 5.52(e) can be utilized. This circuit is an implementation of Eq. (5.93), with V_G replaced by V_{SS}. Resistor R_G establishes a dc ground at the gate and presents a high input resistance to a signal source that may be connected to the gate through a coupling capacitor.

Example 5.12

It is required to design the circuit of Fig. 5.52(c) to establish a dc drain current $I_D = 0.5$ mA. The MOSFET is specified to have $V_t = 1$ V and $k_n'W/L = 1\ \text{mA/V}^2$. For simplicity, neglect the channel-length modulation effect (i.e., assume $\lambda = 0$). Use a power-supply $V_{DD} = 15$ V. Calculate the percentage change in the value of I_D obtained when the MOSFET is replaced with another unit having the same $k_n'W/L$ but $V_t = 1.5$ V.

Solution

As a rule of thumb for designing this classical biasing circuit, we choose R_D and R_S to provide one-third of the power-supply voltage V_{DD} as a drop across each of R_D, the transistor (i.e., V_{DS}) and R_S. For $V_{DD} = 15$ V,

[10]The action of R_S in stabilizing the value of the bias current I_D is not unlike that of the resistance R_s, which we included in the source lead of a CS amplifier in Section 5.6.4. In the latter case also, R_s works to reduce the change in i_D with the result that the amplifier gain is reduced.

Example 5.12 *continued*

this choice makes $V_D = +10$ V and $V_S = +5$ V. Now, since I_D is required to be 0.5 mA, we can find the values of R_D and R_S as follows:

$$R_D = \frac{V_{DD} - V_D}{I_D} = \frac{15 - 10}{0.5} = 10\ \text{k}\Omega$$

$$R_S = \frac{V_S}{R_S} = \frac{5}{0.5} = 10\ \text{k}\Omega$$

The required value of V_{GS} can be determined by first calculating the overdrive voltage V_{OV} from

$$I_D = \tfrac{1}{2}k_n'(W/L)V_{OV}^2$$

$$0.5 = \tfrac{1}{2} \times 1 \times V_{OV}^2$$

which yields $V_{OV} = 1$ V, and thus,

$$V_{GS} = V_t + V_{OV} = 1 + 1 = 2\ \text{V}$$

Now, since $V_S = +5$ V, V_G must be

$$V_G = V_S + V_{GS} = 5 + 2 = 7\ \text{V}$$

To establish this voltage at the gate we may select $R_{G1} = 8$ MΩ and $R_{G2} = 7$ MΩ. The final circuit is shown in Fig. 5.53. Observe that the dc voltage at the drain (+10 V) allows for a positive signal swing of +5 V (i.e., up to V_{DD}) and a negative signal swing of –4 V [i.e., down to $(V_G - V_t)$].

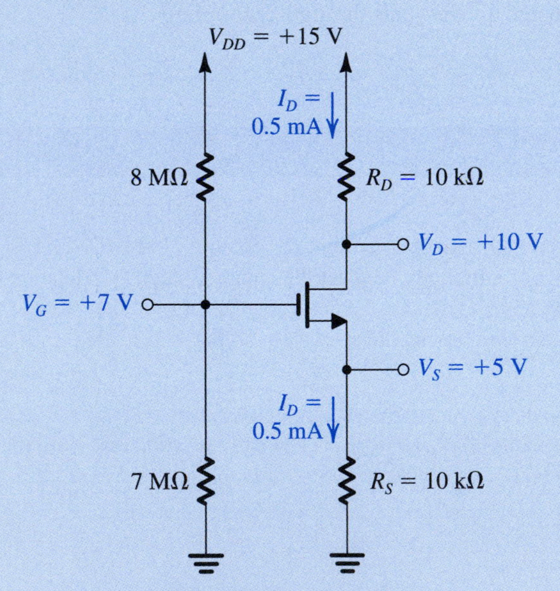

Figure 5.53 Circuit for Example 5.12.

If the NMOS transistor is replaced with another having $V_t = 1.5$ V, the new value of I_D can be found as follows:

$$I_D = \frac{1}{2} \times 1 \times (V_{GS} - 1.5)^2 \tag{5.94}$$

$$V_G = V_{GS} + I_D R_S$$

$$7 = V_{GS} + 10 I_D \tag{5.95}$$

Solving Eqs. (5.94) and (5.95) together yields

$$I_D = 0.455 \text{ mA}$$

Thus the change in I_D is

$$\Delta I_D = 0.455 - 0.5 = -0.045 \text{ mA}$$

which is $\dfrac{-0.045}{0.5} \times 100 = -9\%$ change.

EXERCISES

5.33 Consider the MOSFET in Example 5.12 when fixed-V_{GS} bias is used. Find the required value of V_{GS} to establish a dc bias current $I_D = 0.5$ mA. Recall that the device parameters are $V_t = 1$ V, $k_n' W/L = 1$ mA/V^2, and $\lambda = 0$. What is the percentage change in I_D obtained when the transistor is replaced with another having $V_t = 1.5$ V?
Ans. $V_{GS} = 2$ V; -75%

D5.34 Design the circuit of Fig. 5.52(e) to operate at a dc drain current of 0.5 mA and $V_D = +2$ V. Let $V_t = 1$ V, $k_n' W/L = 1$ mA/V^2, $\lambda = 0$, $V_{DD} = V_{SS} = 5$ V. Use standard 5% resistor values (see Appendix G), and give the resulting values of I_D, V_D, and V_S.
Ans. $R_D = R_S = 6.2$ kΩ; $I_D = 0.49$ mA, $V_S = -1.96$ V, and $V_D = +1.96$ V. R_G can be selected in the range of 1 MΩ to 10 MΩ.

5.7.3 Biasing Using a Drain-to-Gate Feedback Resistor

A simple and effective discrete-circuit biasing arrangement utilizing a feedback resistor connected between the drain and the gate is shown in Fig. 5.54. Here the large feedback resistance R_G (usually in the megohm range) forces the dc voltage at the gate to be equal to that at the drain (because $I_G = 0$). Thus we can write

$$V_{GS} = V_{DS} = V_{DD} - R_D I_D$$

which can be rewritten in the form

$$V_{DD} = V_{GS} + R_D I_D \tag{5.96}$$

which is identical in form to Eq. (5.93), which describes the operation of the bias scheme discussed above [that in Fig. 5.52(a)]. Thus, here too, if I_D for some reason changes, say increases, then Eq. (5.96) indicates that V_{GS} must decrease. The decrease in V_{GS} in turn causes a decrease in I_D, a change that is opposite in direction to the one originally assumed. Thus the negative feedback or degeneration provided by R_G works to keep the value of I_D as constant as possible.

The circuit of Fig. 5.54 can be utilized as an amplifier by applying the input voltage signal to the gate via a coupling capacitor so as not to disturb the dc bias conditions already established. The amplified output signal at the drain can be coupled to another part of the circuit, again via a capacitor. We have considered such an amplifier circuit in Section 5.5 (Example 5.10).

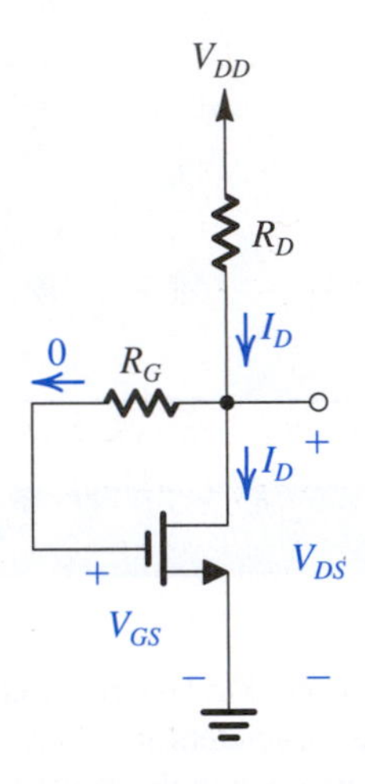

Figure 5.54 Biasing the MOSFET using a large drain-to-gate feedback resistance, R_G.

EXERCISE

D5.35 Design the circuit in Fig. 5.54 to operate at a dc drain current of 0.5 mA. Assume V_{DD} = +5 V, $k_n'W/L = 1$ mA/V^2, $V_t = 1$ V, and $\lambda = 0$. Use a standard 5% resistance value for R_D, and give the actual values obtained for I_D and V_D.
Ans. $R_D = 6.2$ kΩ; $I_D \simeq 0.49$ mA; $V_D \simeq 1.96$ V

5.7.4 Biasing Using a Constant-Current Source

The most effective scheme for biasing a MOSFET amplifier is that using a constant-current source. Figure 5.55(a) shows such an arrangement applied to a discrete MOSFET. Here R_G (usually in the megohm range) establishes a dc ground at the gate and presents a large resistance to an input signal source that can be capacitively coupled to the gate. Resistor R_D establishes an appropriate dc voltage at the drain to allow for the required output signal swing while ensuring that the transistor always remains in the saturation region.

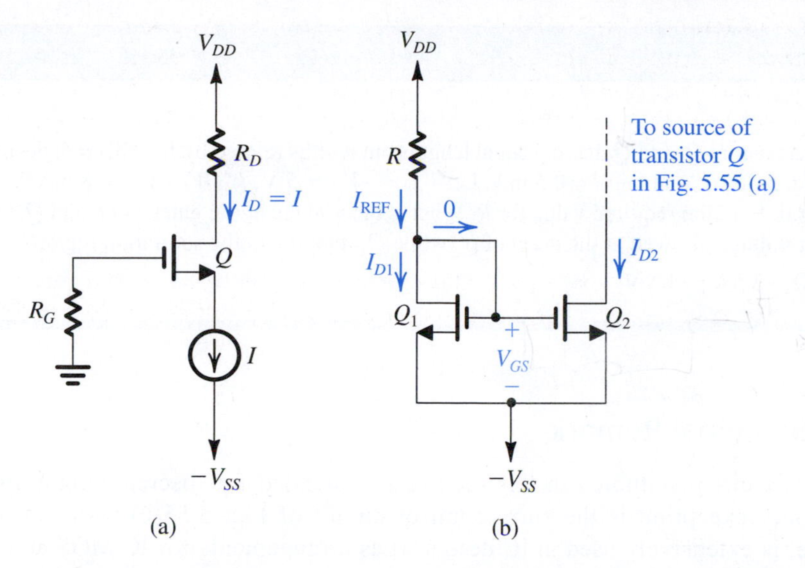

Figure 5.55 **(a)** Biasing the MOSFET using a constant-current source I. **(b)** Implementation of the constant-current source I using a current mirror.

A circuit for implementing the constant-current source I is shown in Fig. 5.55(b). The heart of the circuit is transistor Q_1, whose drain is shorted to its gate, and thus is operating in the saturation region, such that

$$I_{D1} = \frac{1}{2}k_n'\left(\frac{W}{L}\right)_1(V_{GS} - V_t)^2 \tag{5.97}$$

where we have neglected channel-length modulation (i.e., assumed $\lambda = 0$). The drain current of Q_1 is supplied by V_{DD} through resistor R. Since the gate currents are zero,

$$I_{D1} = I_{REF} = \frac{V_{DD} + V_{SS} - V_{GS}}{R} \tag{5.98}$$

where the current through R is considered to be the *reference current* of the current source and is denoted I_{REF}. Given the parameter values of Q_1 and a desired value for I_{REF}, Eqs. (5.97) and (5.98) can be used to determine the value of R. Now consider transistor Q_2: It has the same V_{GS} as Q_1; thus if we assume that it is operating in saturation, its drain current, which is the desired current I of the current source, will be

$$I = I_{D2} = \frac{1}{2}k_n'\left(\frac{W}{L}\right)_2(V_{GS} - V_t)^2 \tag{5.99}$$

where we have neglected channel-length modulation. Equations (5.98) and (5.99) enable us to relate the current I to the reference current I_{REF},

$$I = I_{REF}\frac{(W/L)_2}{(W/L)_1} \tag{5.100}$$

Thus I is related to I_{REF} by the ratio of the aspect ratios of Q_1 and Q_2. This circuit, known as a **current mirror**, is very popular in the design of IC MOS amplifiers and will be studied in great detail in Chapter 7.

EXERCISE

D5.36 Using two transistors Q_1 and Q_2 having equal lengths but widths related by $W_2/W_1 = 5$, design the circuit of Fig. 5.55(b) to obtain $I = 0.5$ mA. Let $V_{DD} = -V_{SS} = 5$ V, $k_n'(W/L)_1 = 0.8$ mA/V^2, $V_t = 1$ V, and $\lambda = 0$. Find the required value for R. What is the voltage at the gates of Q_1 and Q_2? What is the lowest voltage allowed at the drain of Q_2 while Q_2 remains in the saturation region?

Ans. 85 kΩ; −3.5 V; −4.5 V

5.7.5 A Final Remark

The bias circuits studied in this section are intended for discrete-circuit applications. The only exception is the current mirror circuit of Fig. 5.55(b) which, as mentioned above, is extensively used in IC design. Bias arrangements for IC MOS amplifiers will be studied in Chapter 7.

5.8 Discrete-Circuit MOS Amplifiers

With our study of MOS amplifier basics complete, we now put everything together by presenting practical circuits for discrete-circuit amplifiers. These circuits, which utilize the amplifier configurations studied in Section 5.6 and one of the biasing methods of Section 5.7, can be assembled using off-the-shelf discrete transistors, resistors, and capacitors. Though practical and carefully selected to illustrate some important points, the circuits presented in this section should be regarded only as examples of discrete-circuit MOS amplifiers. Indeed, there is a great variety of such circuits, a number of which are explored in the end-of-chapter problems. We should, however, caution the reader that MOS transistors are primarily used in integrated circuit design, as we shall see in Chapter 7 and beyond.

In this section we present a series of exercise problems, Exercises 5.37 to 5.41, that are carefully designed to illustrate important aspects of the amplifier circuits studied. These exercises are also intended to enable the reader to see more clearly the differences between the various circuit configurations. We strongly urge the reader to solve these exercises. As usual, the answers are provided.

5.8.1 The Basic Structure

Figure 5.56 shows the basic circuit we shall utilize to implement the various configurations of discrete-circuit MOS amplifiers. Among the various schemes for biasing MOS amplifiers (Section 5.7), we have selected, for both its effectiveness and its simplicity, the one employing constant-current biasing. Figure 5.56 indicates the dc current and the dc voltages resulting at various nodes.

EXERCISE

5.37 Consider the circuit of Fig. 5.56 for the case $V_{DD} = V_{SS} = 10$ V, $I = 0.5$ mA, $R_G = 4.7$ MΩ, $R_D = 15$ kΩ, $V_t = 1.5$ V, and $k_n'(W/L) = 1$ mA/V^2. Find V_{OV}, V_{GS}, V_G, V_S, and V_D. Also, calculate the values of g_m and r_o, assuming that $V_A = 75$ V. What is the maximum possible signal swing at the drain for which the MOSFET remains in saturation?

Ans. See Fig. E5.37; without taking into account the signal swing at the gate, the drain can swing to −1.5 V, a negative signal swing of 4 V

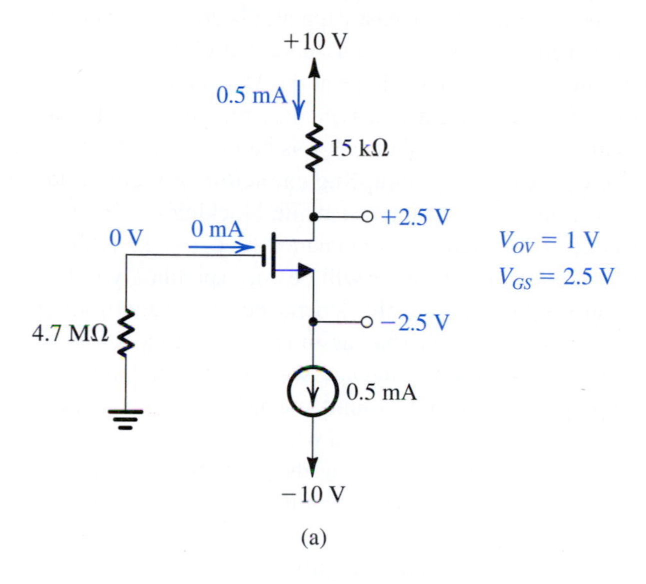

(a)

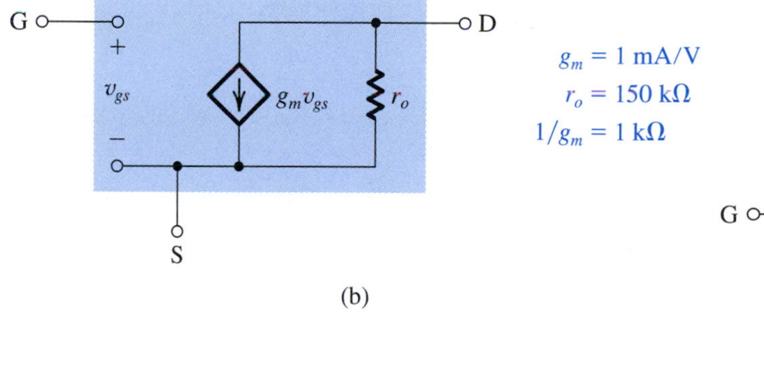

(b)

D
G
i
i
$\frac{1}{g_m}$
r_o
S

(c)

Figure E5.37

5.8.2 The Common-Source (CS) Amplifier

As mentioned in Section 5.6, the common-source (CS) configuration is the most widely used of all MOSFET amplifier circuits. A common-source amplifier realized using the circuit of Fig. 5.56 is shown in Fig. 5.57(a). Observe that to establish a **signal ground**, or an **ac ground** as it is sometimes called, at the source, we have connected a large capacitor, C_S, between the source and ground. This capacitor, usually in the microfarad range, is required to provide a very small impedance (ideally, zero impedance; i.e., in effect, a short circuit) at all signal frequencies of interest. In this way, the signal current passes through C_S to ground and thus *bypasses* the output resistance of current source I (and any other circuit component that might be connected to the MOSFET source); hence, C_S is called a **bypass capacitor**. Obviously, the lower the signal frequency, the less effective the bypass capacitor becomes. This issue will be studied in Section 9.1. For our purposes here we shall assume that C_S is acting as a perfect short circuit and thus is establishing a zero signal voltage at the MOSFET source.

In order not to disturb the dc bias current and voltages, the signal to be amplified, shown as voltage source v_{sig} with an internal resistance R_{sig}, is connected to the gate through a large capacitor C_{C1}. Capacitor C_{C1}, known as a **coupling capacitor**, is required to act as a perfect short circuit at all signal frequencies of interest while blocking dc. Here again, we note that as the signal frequency is lowered, the impedance of C_{C1} (i.e., $1/j\omega C_{C1}$) will increase and its effectiveness as a coupling capacitor will be correspondingly reduced. This problem too will be considered in Section 9.1 when the dependence of the amplifier operation on frequency is studied. For our purposes here we shall assume C_{C1} is acting as a perfect short circuit as far as the signal is concerned. Before leaving C_{C1}, we should point out that when the signal source can provide an appropriate dc path to ground, the gate can be connected directly to the signal source and both R_G and C_{C1} can be dispensed with.

The voltage signal resulting at the drain is coupled to the load resistance R_L via another coupling capacitor C_{C2}. We shall assume that C_{C2} acts as a perfect short circuit at all signal frequencies of interest and thus that the output voltage $v_o = v_d$. Note that R_L can be either an actual load resistor, to which the amplifier is required to provide its output voltage signal, or it can be the input resistance of another amplifier stage in cases where more than one stage of amplification is needed. (We will study multistage amplifiers in Chapter 8.)

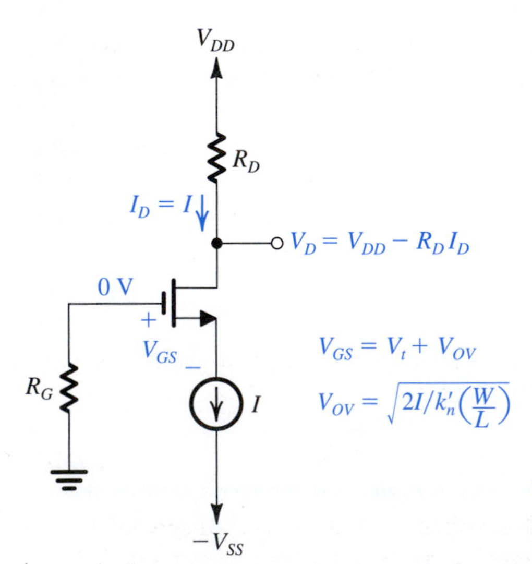

Figure 5.56 Basic structure of the circuit used to realize single-stage, discrete-circuit MOS amplifier configurations.

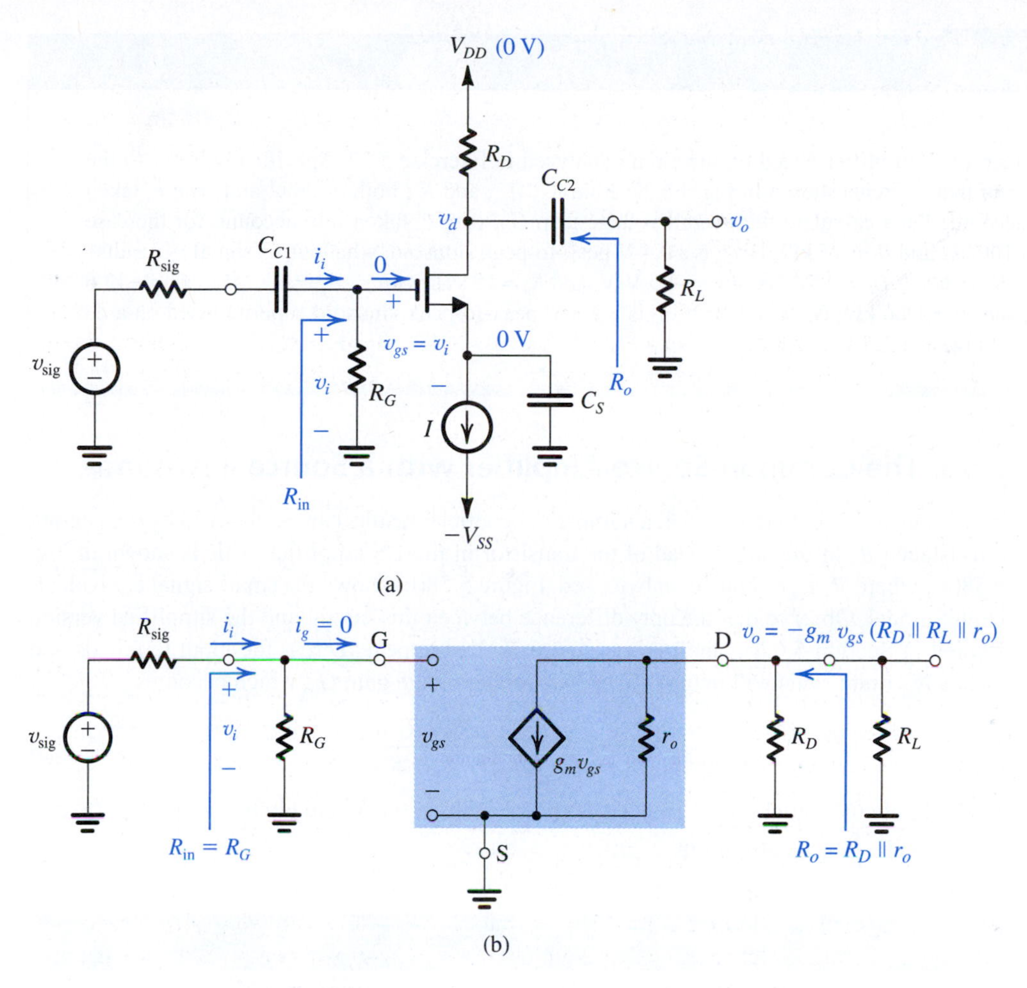

Figure 5.57 **(a)** Common-source amplifier based on the circuit of Fig. 5.56. **(b)** Equivalent circuit of the amplifier for small-signal analysis.

To determine the terminal characteristics of the CS amplifier—that is, its input resistance, voltage gain, and output resistance—we replace the MOSFET with its small-signal model. The resulting circuit is shown in Fig. 5.57(b).

We observe that the only difference between this circuit and the stripped-down version studied in Section 5.6.3 (Fig. 5.45) is that here we have the bias resistance R_G. Since R_G appears across the input terminals of the amplifier, the input resistance will no longer be infinite, rather

$$R_{in} = R_G$$

To keep R_{in} high, a large value of R_G (in the megohm range) is usually selected. The finite R_{in} will affect the overall voltage gain G_v which becomes

$$G_v = -\frac{R_G}{R_G + R_{sig}} g_m(R_D \| R_L \| r_o) \tag{5.101}$$

Finally, to encourage the reader to do the analysis directly on the circuit diagram, with the MOSFET model used implicitly, we show some of the analysis on the circuit in Fig. 5.57(a).

EXERCISE

5.38 Consider a CS amplifier based on the circuit analyzed in Exercise 5.37. Specifically, refer to the results of that exercise shown in Fig. E5.37. Find R_{in}, A_{vo}, and R_o, both without and with r_o taken into account. Then calculate the overall voltage gain G_v, with r_o taken into account, for the case $R_{sig} = 100$ kΩ and $R_L = 15$ kΩ. If v_{sig} is a 0.4-V peak-to-peak sinusoid, what output signal v_o results?
Ans. Without r_o: $R_{in} = 4.7$ MΩ, $A_{vo} = -15$ V/V, and $R_o = 15$ kΩ; with r_o: $R_{in} = 4.7$ MΩ, $A_{vo} = -13.6$ V/V, and $R_o = 13.6$ kΩ; $G_v = -7$ V/V; v_o is a 2.8-V peak-to-peak sinusoid superimposed on a dc drain voltage of +2.5 V.

5.8.3 The Common-Source Amplifier with a Source Resistance

As demonstrated in Section 5.6.4, a number of beneficial results can be obtained by connecting a resistance R_s in the source lead of the transistor in the CS amplifier. This is shown in Fig. 5.58(a), where R_s is, of course, unbypassed. Figure 5.58(b) shows the small-signal equivalent-circuit model. Observe that the only difference between this circuit and the simplified version studied in Section 5.6.4 is the bias resistance R_G that appears across the input terminals and makes R_{in} finite. This will in turn affect the overall voltage gain G_v, which becomes

$$G_v = -\frac{R_G}{R_G + R_{sig}} \frac{R_D \| R_L}{1/g_m + R_s} \tag{5.102}$$

Finally, note that much of the analysis is shown both on the actual circuit in Fig. 5.58(a) and on the equivalent circuit in Fig. 5.58(b).

EXERCISE

5.39 In Exercise 5.38 we applied an input signal of 0.4 V peak-to-peak, which resulted in an output signal of the CS amplifier of 2.8 V peak-to-peak. Assume that for some reason we now have an input signal three times as large as before (i.e., 1.2 V p-p) and that we wish to modify the circuit to keep the output signal level unchanged. What value should we use for R_s?
Ans. 2.15 kΩ

5.8.4 The Common-Gate (CG) Amplifier

Figure 5.59(a) shows a CG amplifier obtained from the circuit of Fig. 5.56. Observe that since both the dc and ac voltages at the gate are to be zero, we have connected the gate directly to ground, thus eliminating resistor R_G altogether. Coupling capacitors C_{C1} and C_{C2} perform similar functions to those in the CS circuit.

The small-signal, equivalent circuit model of the CG amplifier is shown in Fig. 5.59(b). We note that this circuit is identical to the equivalent circuit of the stripped-down version of the CG amplifier, in Fig. 5.48(b). Thus the analysis performed and the results obtained in Section 5.6.5 apply directly here. A substantial portion of the analysis is also shown in Fig. 5.59.

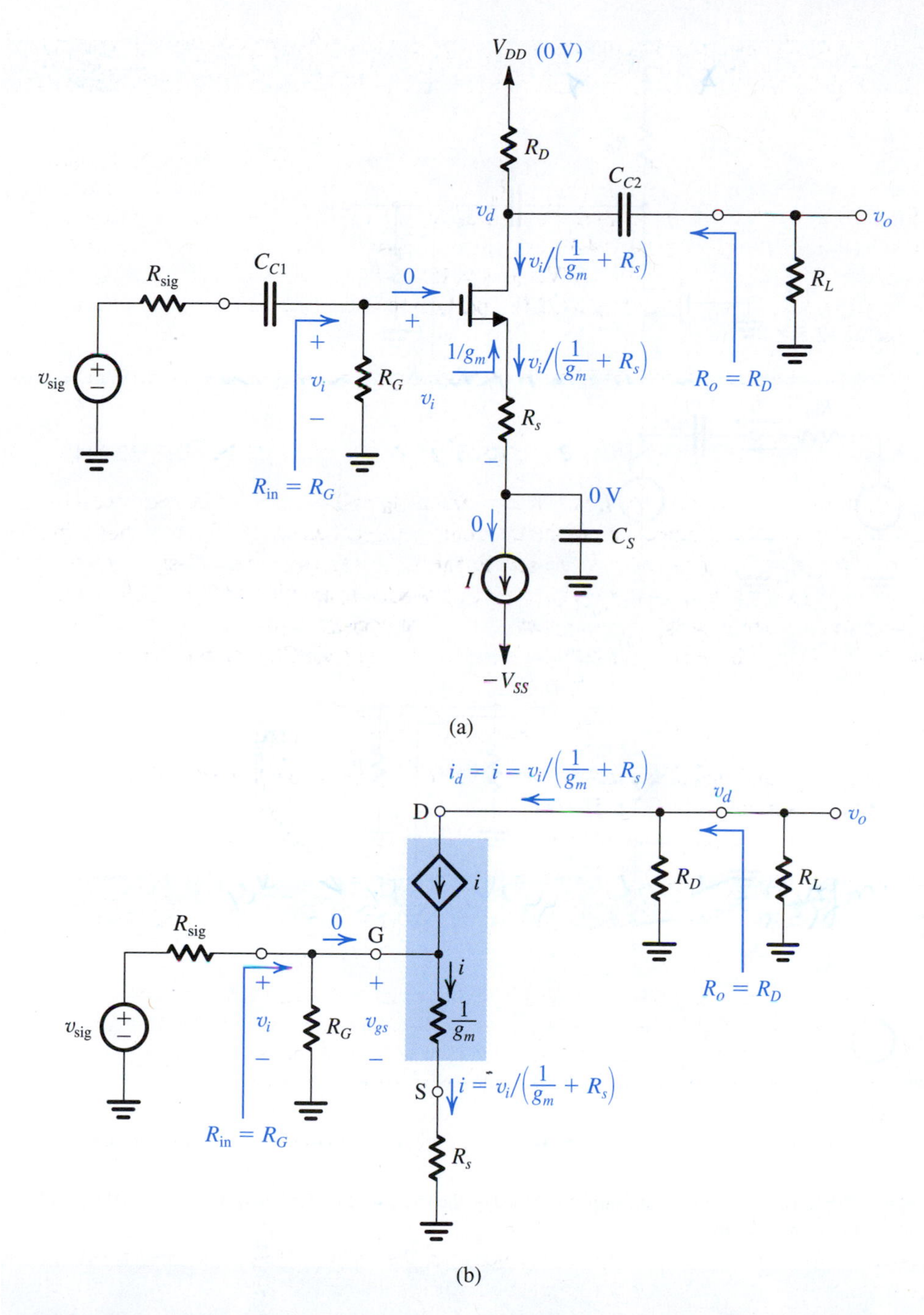

Figure 5.58 **(a)** Common-source amplifier with a resistance R_S in the source lead. **(b)** Small-signal equivalent circuit with r_o neglected.

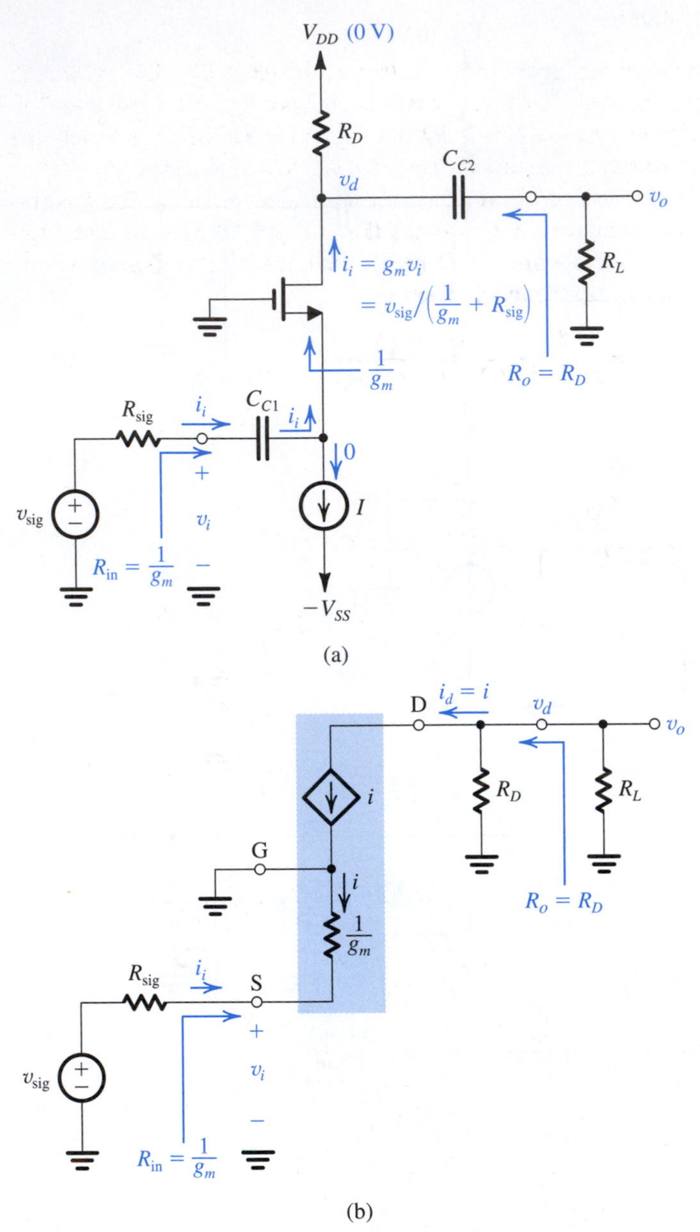

Figure 5.59 **(a)** A common-gate amplifier based on the circuit of Fig. 5.56. **(b)** A small-signal equivalent circuit of the amplifier in (a).

EXERCISE

5.40 Consider a CG amplifier designed using the circuit of Fig. 5.56, which is analyzed in Exercise 5.37 with the analysis results displayed in Fig. E5.37. Note that $g_m = 1$ mA/V and $R_D = 15$ kΩ. Find R_{in}, R_o, A_{vo}, A_v, and G_v for $R_L = 15$ kΩ and $R_{sig} = 50$ Ω. What will the overall voltage gain become for $R_{sig} = 1$ kΩ? 10 kΩ? 100 kΩ?

Ans. 1 kΩ, 15 kΩ, +15 V/V, +7.5 V/V, +7.1 V/V; +3.75 V/V; 0.68 V/V; 0.07 V/V

5.8.5 The Source Follower

Figure 5.60(a) shows a common-drain amplifier based on the circuit of Fig. 5.56. Since the drain is to function as a signal ground, there is no need for resistor R_D, and it has therefore been eliminated. The input signal is coupled via capacitor C_{C1} to the MOSFET gate, and the output signal at the MOSFET source is coupled via capacitor C_{C2} to a load resistor R_L.

Replacing the MOSFET with its T model results in the equivalent circuit in Fig. 5.60(b). We note that the only difference between this circuit and that in Fig. 5.50(b) is the bias resistance R_G that appears across the input terminals. Thus, here too, the input resistance will no longer be infinite and the overall voltage gain will become

$$G_v = \frac{R_G}{R_G + R_{sig}} \frac{(R_L \| r_o)}{(R_L \| r_o) + 1/g_m} \tag{5.103}$$

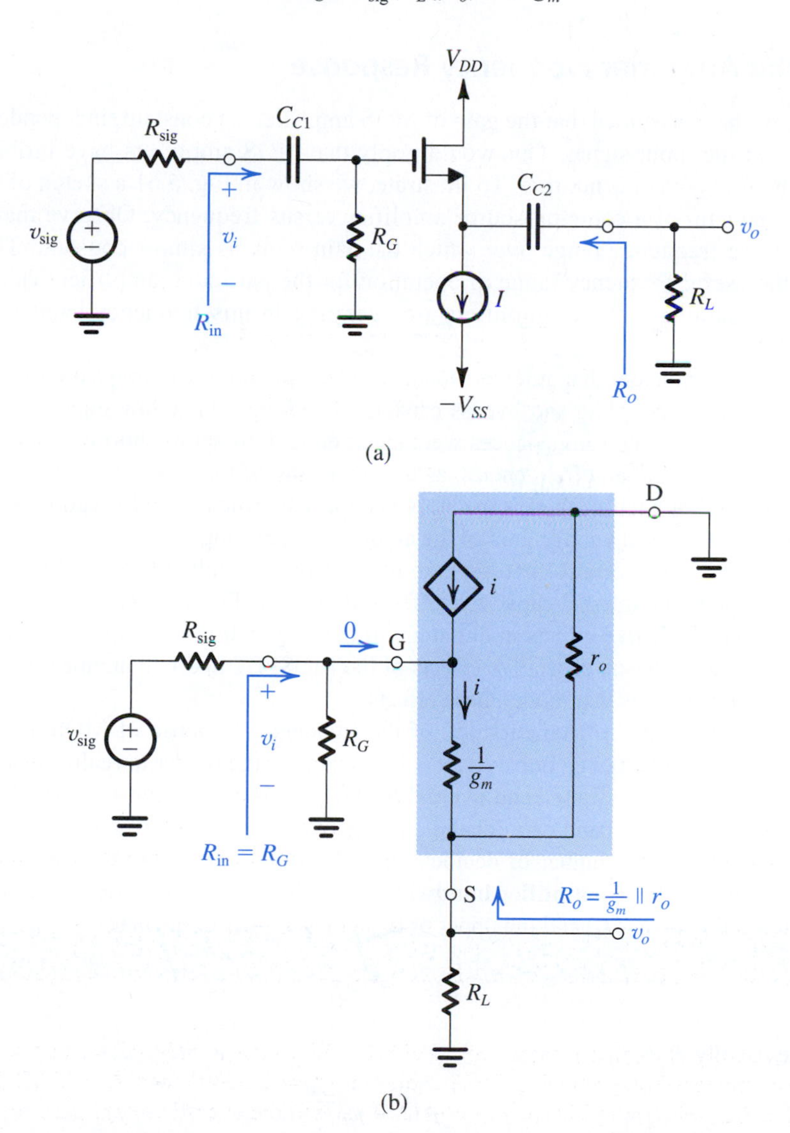

Figure 5.60 **(a)** A source-follower amplifier. **(b)** Small-signal, equivalent-circuit model.

EXERCISE

5.41 Consider a source follower such as that in Fig. 5.60(a) designed on the basis of the circuit of Fig. 5.56, the results of whose analysis are displayed in Fig. E5.37. Specifically, note that $g_m = 1$ mA/V and $r_o = 150$ kΩ. Let $R_{sig} = 1$ MΩ and $R_L = 15$ kΩ. (a) Find R_{in}, A_{vo}, A_v, and R_o without and with r_o taken into account. (b) Find the overall small-signal voltage gain G_v with r_o taken into account.

Ans. (a) $R_{in} = 4.7$ MΩ; $A_{vo} = 1$ V/V (without r_o), 0.993 V/V (with r_o); $A_v = 0.938$ (without r_o), 0.932 V (with r_o); $R_o = 1$ kΩ (without r_o), 0.993 kΩ (with r_o); (b) 0.768 V/V

5.8.6 The Amplifier Frequency Response

Thus far, we have assumed that the gain of MOS amplifiers is constant, independent of the frequency of the input signal. This would imply that MOS amplifiers have infinite bandwidth, which of course is not true. To illustrate, we show in Fig. 5.61 a sketch of the magnitude of the gain of a common-source amplifier versus frequency. Observe that there is indeed a wide frequency range over which the gain remains almost constant. This obviously is the useful frequency range of operation for the particular amplifier. Thus far, we have been assuming that our amplifiers are operating in this frequency band, called **the midband**.

Figure 5.61 indicates that at lower frequencies, the magnitude of amplifier gain falls off. This is because the coupling and bypass capacitors no longer have low impedances. Recall that we assumed that their impedances were small enough to act as short circuits. Although this can be true at midband frequencies, as the frequency of the input signal is lowered, the reactance $1/j\omega C$ of each of these capacitors becomes significant, and it can be shown that this results in the overall voltage gain of the amplifier decreasing.

Figure 5.61 indicates also that the gain of the amplifier falls off at the high-frequency end. This is due to the internal capacitive effects in the MOSFET. We have had a brief introduction to such capacitive effects in our study of the *pn* junction in Chapter 3. In Chapter 9, we shall study the internal capacitive effects of the MOSFET and will augment the hybrid-π model with capacitances that model these effects.

We will undertake a detailed study of the frequency response of MOS amplifiers in Chapter 9. For the time being, however, it is important for the reader to realize that for every MOS amplifier there is a finite band over which the gain is almost constant. The boundaries of this useful frequency band or midband, are the two frequencies f_L and f_H, at which the gain drops by a certain number of decibels (usually 3 dB) below its value at midband. As indicated in Fig. 5.61, the **amplifier bandwidth**, or 3-dB bandwidth, is defined as the difference between the lower (f_L) and the upper or higher (f_H) 3-dB frequencies:

$$BW = f_H - f_L \tag{5.104}$$

and since usually $f_L \ll f_H$,

$$BW \simeq f_H \tag{5.105}$$

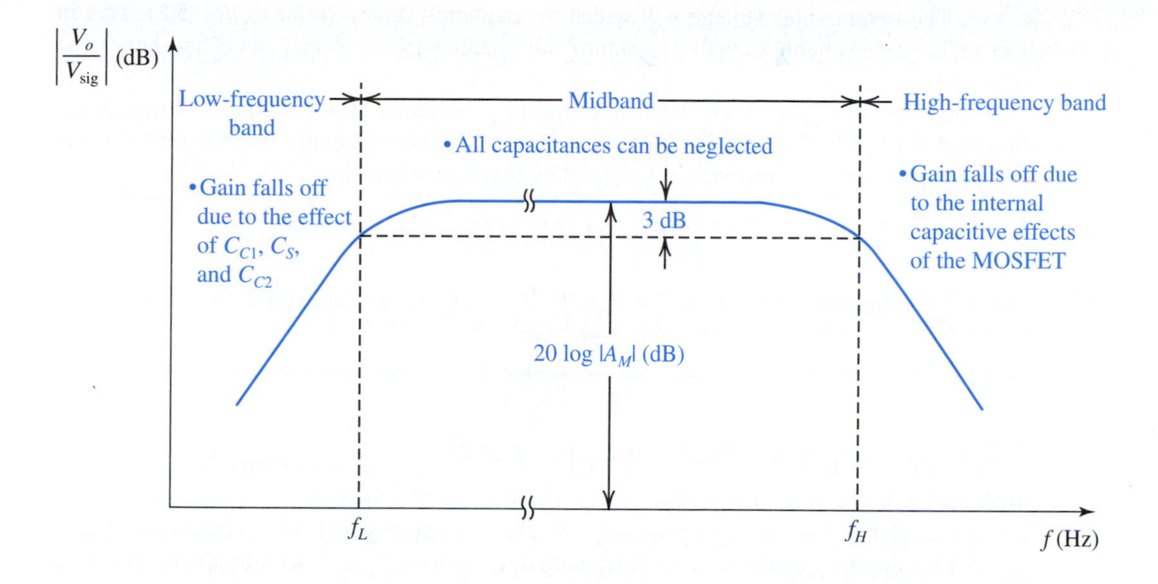

Figure 5.61 A sketch of the frequency response of a CS amplifier delineating the three frequency bands of interest.

A figure of merit for the amplifier is its gain–bandwidth product, defined as

$$GB = |A_M|\, BW \tag{5.106}$$

where $|A_M|$ is the magnitude of the amplifier gain in the midband. It will be seen in Chapter 9 that in amplifier design it is usually possible to trade off gain for bandwidth. One way to accomplish this, for instance, is by including resistance R_s in the source of the CS amplifier.

5.9 The Body Effect and Other Topics[11]

In this section we briefly consider a number of important though secondary issues.

5.9.1 The Role of the Substrate—The Body Effect

In many applications the source terminal is connected to the substrate (or body) terminal B, which results in the *pn* junction between the substrate and the induced channel (review Fig. 5.5) having a constant zero (cutoff) bias. In such a case the substrate does not play any role in circuit operation and its existence can be ignored altogether.

In integrated circuits, however, the substrate is usually common to many MOS transistors. In order to maintain the cutoff condition for all the substrate-to-channel junctions, the substrate is usually connected to the most negative power supply in an NMOS circuit (the most positive in a PMOS circuit). The resulting reverse-bias voltage between source and body (V_{SB} in an *n*-channel device) will have an effect on device operation. To appreciate this fact, consider an NMOS transistor and let its substrate be made negative relative to the

[11] This section can be omitted in a first reading with little or no loss of continuity. Some of this material, however, will be required for the study of digital circuits in Chapter 13.

source. The reverse-bias voltage will widen the depletion region (refer to Fig. 5.2). This in turn reduces the channel depth. To return the channel to its former state, v_{GS} has to be increased.

The effect of V_{SB} on the channel can be most conveniently represented as a change in the threshold voltage V_t. Specifically, it has been shown that increasing the reverse substrate bias voltage V_{SB} results in an increase in V_t according to the relationship

$$V_t = V_{t0} + \gamma[\sqrt{2\phi_f + V_{SB}} - \sqrt{2\phi_f}] \tag{5.107}$$

where V_{t0} is the threshold voltage for $V_{SB} = 0$; ϕ_f is a physical parameter with $(2\phi_f)$ typically 0.6 V; γ is a fabrication-process parameter given by

$$\gamma = \frac{\sqrt{2qN_A\varepsilon_s}}{C_{ox}} \tag{5.108}$$

where q is the electron charge (1.6×10^{-19} C), N_A is the doping concentration of the p-type substrate, and ε_s is the permittivity of silicon ($11.7\varepsilon_0 = 11.7 \times 8.854 \times 10^{-14} = 1.04 \times 10^{-12}$ F/cm). The parameter γ has the dimension of $\sqrt{\text{V}}$ and is typically 0.4 $\text{V}^{1/2}$. Finally, note that Eq. (5.107) applies equally well for p-channel devices with V_{SB} replaced by the reverse bias of the substrate, V_{BS} (or, alternatively, replace V_{SB} by $|V_{SB}|$) and note that γ is negative. Also, in evaluating γ, N_A must be replaced with N_D, the doping concentration of the n well in which the PMOS is formed. For p-channel devices, $2\phi_f$ is typically 0.75 V, and γ is typically $-0.5\ \text{V}^{1/2}$.

EXERCISE

5.42 An NMOS transistor has $V_{t0} = 0.8$ V, $2\phi_f = 0.7$ V, and $\gamma = 0.4\ \text{V}^{1/2}$. Find V_t when $V_{SB} = 3$ V.
Ans. 1.23 V

Equation (5.107) indicates that an incremental change in V_{SB} gives rise to an incremental change in V_t, which in turn results in an incremental change in i_D even though v_{GS} might have been kept constant. It follows that the body voltage controls i_D; thus the body acts as another gate for the MOSFET, a phenomenon known as the **body effect**. Here we note that the parameter γ is known as the **body-effect parameter**.

5.9.2 Modeling the Body Effect

As mentioned above the body effect occurs in a MOSFET when the source is not tied to the substrate (which is always connected to the most negative power supply in the integrated circuit for n-channel devices and to the most positive for p-channel devices). Thus the substrate (body) will be at signal ground, but since the source is not, a signal voltage v_{bs} develops between the body (B) and the source (S). The substrate then acts as a "second gate" or a **backgate** for the MOSFET. Thus the signal v_{bs} gives rise to a drain-current component, which we shall write as $g_{mb}v_{bs}$, where g_{mb} is the **body transconductance**, defined as

$$g_{mb} \equiv \left.\frac{\partial i_D}{\partial v_{BS}}\right|_{\substack{v_{GS} = \text{constant} \\ v_{DS} = \text{constant}}} \tag{5.109}$$

Recalling that i_D depends on v_{BS} through the dependence of V_t on V_{BS}, we can show that

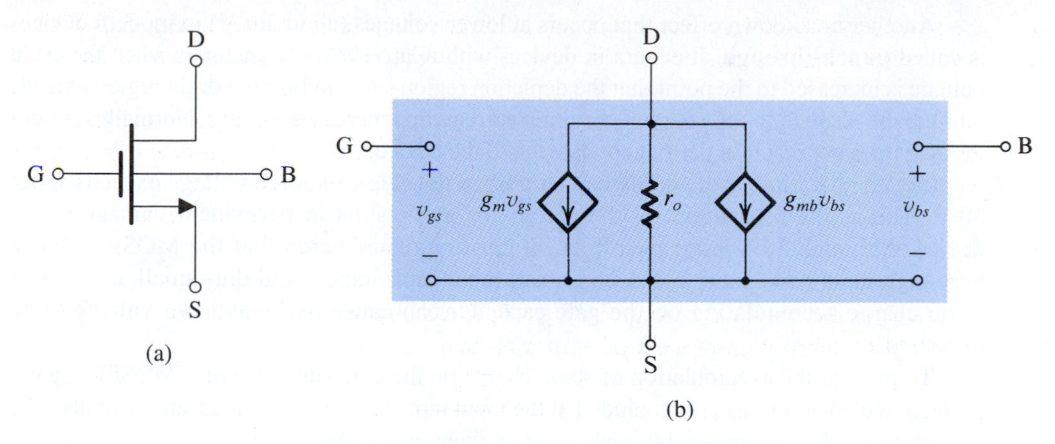

Figure 5.62 Small-signal, equivalent-circuit model of a MOSFET in which the source is not connected to the body.

$$g_{mb} = \chi g_m \tag{5.110}$$

where

$$\chi \equiv \frac{\partial V_t}{\partial V_{SB}} = \frac{\gamma}{2\sqrt{2\phi_f + V_{SB}}} \tag{5.111}$$

Typically the value of χ lies in the range 0.1 to 0.3.

Figure 5.62 shows the MOSFET model augmented to include the controlled source $g_{mb}v_{bs}$ that models the body effect. Ideally, this is the model to be used whenever the source is not connected to the substrate. It has been found, however, that except in some very particular situations, the body effect can generally be ignored in the initial, pencil-and-paper design of MOSFET amplifiers.

Finally, although the analysis above was performed on a NMOS transistor, the results and the equivalent circuit of Fig. 5.62 apply equally well to PMOS transistors, except for using $|V_{GS}|$, $|V_t|$, $|V_{OV}|$, $|V_A|$, $|V_{SB}|$, $|\gamma|$, and $|\lambda|$ and replacing k_n' with k_p' in the appropriate formula.

5.9.3 Temperature Effects

Both V_t and k' are temperature sensitive. The magnitude of V_t decreases by about 2 mV for every 1°C rise in temperature. This decrease in $|V_t|$ gives rise to a corresponding increase in drain current as temperature is increased. However, because k' decreases with temperature and its effect is a dominant one, the overall observed effect of a temperature increase is a *decrease* in drain current. This very interesting result is put to use in applying the MOSFET in power circuits (Chapter 11).

5.9.4 Breakdown and Input Protection

As the voltage on the drain is increased, a value is reached at which the *pn* junction between the drain region and substrate suffers avalanche breakdown (see Section 3.5.3). This breakdown usually occurs at voltages of 20 V to 150 V and results in a somewhat rapid increase in current (known as a **weak avalanche**).

Another breakdown effect that occurs at lower voltages (about 20 V) in modern devices is called **punch-through**. It occurs in devices with relatively short channels when the drain voltage is increased to the point that the depletion region surrounding the drain region extends through the channel to the source. The drain current then increases rapidly. Normally, punch-through does not result in permanent damage to the device.

Yet another kind of breakdown occurs when the gate-to-source voltage exceeds about 30 V. This is the breakdown of the gate oxide and results in permanent damage to the device. Although 30 V may seem high, it must be remembered that the MOSFET has a very high input resistance, and a very small input capacitance, and thus small amounts of static charge accumulating on the gate capacitor can cause its breakdown voltage to be exceeded.

To prevent the accumulation of static charge on the gate capacitor of a MOSFET, gate-protection devices are usually included at the input terminals of MOS integrated circuits. The protection mechanism invariably makes use of clamping diodes.

5.9.5 Velocity Saturation

At high longitudinal electric fields, the drift velocity of charge carriers in the channel reaches an upper limit (approximately 10^7 cm/s for electrons and holes in silicon). This effect, which in modern very-short-channel devices can occur for v_{DS} lower than 1 V, is called velocity saturation. It can be shown be that when velocity saturation occurs, the current i_D will no longer be related to v_{GS} by the square-law relationship. Rather, i_D becomes linearly dependent on v_{GS} and the transconductance g_m becomes constant and independent of v_{GS}. In Chapter 13, we shall consider velocity saturation in our study of deep submicron (i.e., $L < 0.25$ μm) CMOS digital circuits.

5.9.6 The Depletion-Type MOSFET

We conclude this section with a brief discussion of another type of MOSFET, the depletion-type MOSFET. Its structure is similar to that of the enhancement-type MOSFET with one important difference: The depletion MOSFET has a physically implanted channel. Thus an *n*-channel depletion-type MOSFET has an *n*-type silicon region connecting the n^+ source and the n^+ drain regions at the top of the *p*-type substrate. Thus if a voltage v_{DS} is applied between drain and source, a current i_D flows for $v_{GS} = 0$. In other words, there is no need to induce a channel, unlike the case of the enhancement MOSFET.

The channel depth and hence its conductivity can be controlled by v_{GS} in exactly the same manner as in the enhancement-type device. Applying a positive v_{GS} enhances the channel by attracting more electrons into it. Here, however, we also can apply a negative v_{GS}, which causes electrons to be repelled from the channel, and thus the channel becomes shallower and its conductivity decreases. The negative v_{GS} is said to **deplete** the channel of its charge carriers, and this mode of operation (negative v_{GS}) is called **depletion mode**. As the magnitude of v_{GS} is increased in the negative direction, a value is reached at which the channel is completely depleted of charge carriers and i_D is reduced to zero even though v_{DS} may be still applied. This negative value of v_{GS} is the threshold voltage of the *n*-channel depletion-type MOSFET.

The description above suggests (correctly) that a depletion-type MOSFET can be operated in the enhancement mode by applying a positive v_{GS} and in the depletion mode by applying a negative v_{GS}. This is illustrated in Fig. 5.63, which shows both the circuit symbol for the depletion NMOS transistor (Fig. 5.63a) and its i_D–v_{GS} characteristic. Observe that here the threshold voltage V_{tn} is negative. The i_D–v_{DS} characteristics (not shown) are similar to those for the enhancement-type MOSFET except for the negative V_{tn}. Finally, note that the device symbol denotes the existing channel via the shaded area next to the vertical line.

Depletion-type MOSFETs can be fabricated on the same IC chip as enhancement-type devices, resulting in circuits with improved characteristics, as will be shown in a later chapter.

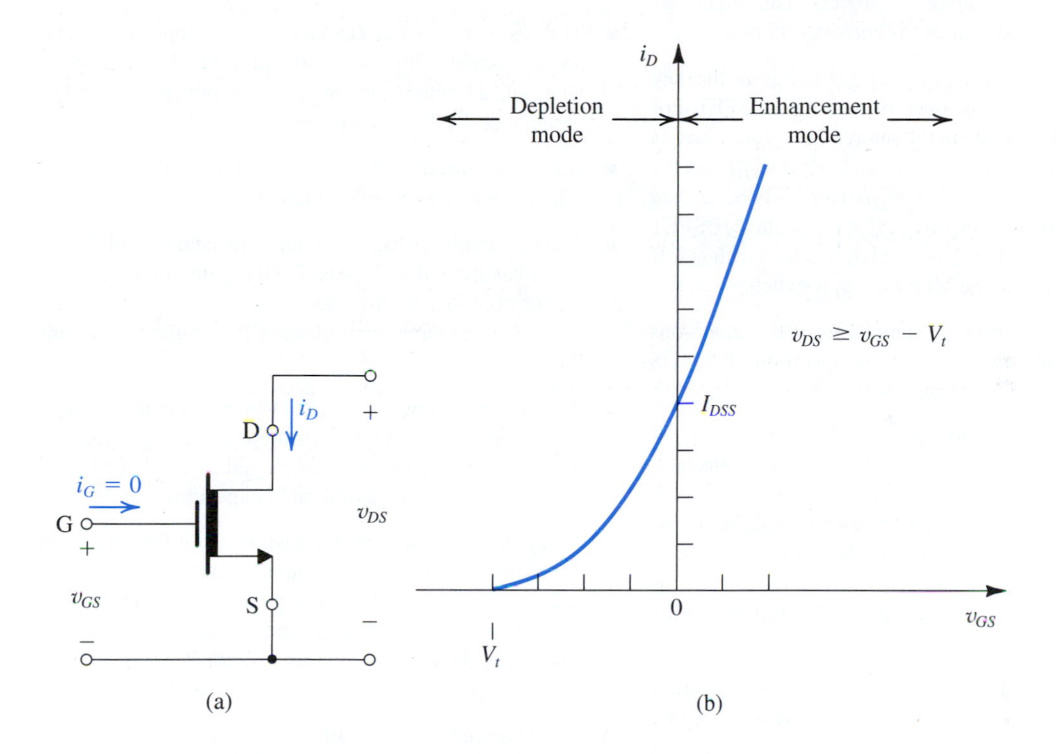

Figure 5.63 The circuit symbol (**a**) and the i_D–v_{GS} characteristic in saturation; (**b**) for an n-channel depletion-type MOSFET.

EXERCISE

5.43 For a depletion-type NMOS transistor with $V_t = -2$ V and $k_n'(W/L) = 2$ mA/V^2, find the minimum v_{DS} required to operate in the saturation region when $v_{GS} = +1$ V. What is the corresponding value of i_D?

Ans. 3 V; 9 mA

Summary

- The enhancement-type MOSFET is currently the most widely used semiconductor device. It is the basis of CMOS technology, which is the most popular IC fabrication technology at this time. CMOS provides both *n*-channel (NMOS) and *p*-channel (PMOS) transistors, which increases design flexibility. The minimum MOSFET channel length achievable with a given CMOS process is used to characterize the process. This figure has been continually reduced and is currently 45 nm.

- The overdrive voltage, $|v_{OV}| \equiv |v_{GS}| - |V_t|$, is the key quantity that governs the operation of the MOSFET. For the MOSFET to operate in the saturation region, which is the region for amplifier application, $|v_{DS}| \geq |v_{OV}|$, and the resulting $i_D = \frac{1}{2}\mu_n C_{ox}(W/L)v_{OV}^2$ (for NMOS; replace μ_n with μ_p for PMOS). If $|v_{DS}| < |v_{OV}|$, the MOSFET operates in the triode region, which together with cutoff is used for operating the MOSFET as a switch.

- Tables 5.1 and 5.2 provide summaries of the conditions and relationships that describe the operation of NMOS and PMOS transistors, respectively.

- In saturation, i_D shows some linear dependence on v_{DS} as a result of the change in channel length. This channel-length modulation phenomenon becomes more pronounced as L decreases. It is modeled by ascribing an output resistance $r_o = |V_A|/I_D$ to the MOSFET model. Although the effect of r_o on the operation of discrete-circuit MOS amplifiers is small, that is not the case in IC amplifiers (Chapter 7).

- The essence of the use of the MOSFET as an amplifier is that in saturation v_{GS} controls i_D in the manner of a voltage-controlled current source. When the device is dc biased in the saturation region and the signal v_{gs} is kept small, the operation of the MOSFET becomes almost linear.

- A systematic procedure to analyze a MOS amplifier circuit consists of replacing the MOSFET with one of its small-signal, equivalent-circuit models (Refer to Table 5.3). DC voltage sources are replaced by short circuits, and dc current sources by open circuits. The analysis is then performed on the resulting equivalent circuit.

- In cases where a resistance is connected in series with the source lead of the MOSFET, the T model is the most convenient to use.

- The three basic configurations of MOS amplifiers are shown in Fig. 5.43 (without the bias arrangements). Their characteristic parameter values are provided in Table 5.4.

- The CS amplifier has (ideally) infinite input resistance and a reasonably high gain but a rather high output resistance and a limited high-frequency response. It is used to obtain most of the gain in a cascade amplifier.

- Adding a resistance R_s in the source lead of the CS amplifier can lead to beneficial results.

- The CG amplifier has a low input resistance and thus it alone has limited and specialized applications. However, its excellent high-frequency response makes it attractive in combination with the CS amplifier (Chapters 7 and 9).

- The source follower has (ideally) infinite input resistance, a voltage gain lower than but close to unity, and a low output resistance. It is employed as a voltage buffer and as the output stage of a multistage amplifier.

- A key step in the design of transistor amplifiers is to bias the transistor to operate at an appropriate point in the saturation region. A good bias design ensures that the parameters of the bias point, I_D, V_{OV}, and V_{DS}, are predictable and stable, and do not vary by a large amount when the transistor is replaced by another of the same type.

- As evidenced by the example circuits given in Section 5.8, discrete-circuit MOS amplifiers utilize large coupling and bypass capacitors. As will be seen in Chapter 7, this is *not* the case in IC amplifiers.

- The depletion-type MOSFET has an implanted channel and thus can be operated in either the depletion or enhancement modes. It is characterized by the same equations used for the enhancement device except for having a negative V_t (positive V_t for depletion PMOS transistors).

Computer Simulation Problems

SIM Problems identified by this icon are intended to demonstrate the value of using SPICE simulation to verify hand analysis and design, and to investigate important issues such as allowable signal swing and amplifier nonlinear distortion. Instructions to assist in setting up PSpice and Multisim simulations for all the indicated problems can be found in the corresponding files on the disc. Note that if a particular parameter value is not specified in the problem statement, you are to make a reasonable assumption.
* difficult problem; ** more difficult; *** very challenging and/or time-consuming; D: design problem.

Section 5.1: Device Structure and Physical Operation

5.1 MOS technology is used to fabricate a capacitor, utilizing the gate metallization and the substrate as the capacitor electrodes. Find the area required per 1-pF capacitance for oxide thickness ranging from 2 nm to 10 nm. For a square plate capacitor of 10 pF, what dimensions are needed?

5.2 Calculate the total charge stored in the channel of an NMOS transistor having $C_{ox} = 6\ \text{fF}/\mu\text{m}^2$, $L = 0.25\ \mu\text{m}$, and $W = 2.5\ \mu\text{m}$, and operated at $V_{OV} = 0.5$ V and $V_{DS} = 0$ V.

5.3 Use dimensional analysis to show that the units of the process transconductance parameter k_n' are A/V^2. What are the dimensions of the MOSFET transconductance parameter k_n?

5.4 An NMOS transistor that is operated with a small v_{DS} is found to exhibit a resistance r_{DS}. By what factor will r_{DS} change in each of the following situations?

(a) V_{OV} is doubled.
(b) The device is replaced with another fabricated in the same technology but with double the width.
(c) The device is replaced with another fabricated in the same technology but with both the width and length doubled.
(d) The device is replaced with another fabricated in a more advanced technology for which the oxide thickness is halved and similarly for W and L (assume μ_n remains unchanged).

D 5.5 An NMOS transistor fabricated in a technology for which $k_n' = 400\ \mu\text{A/V}^2$ and $V_t = 0.4$ V is required to operate with a small v_{DS} as a variable resistor ranging in value from 200 Ω to 1 kΩ. Specify the range required for the control voltage V_{GS} and the required transistor width W. It is required to use the smallest possible device, as limited by the minimum channel length of this technology ($L_{\min} = 0.18\ \mu\text{m}$) and the maximum allowed voltage of 1.8 V.

5.6 Sketch a set of i_D–v_{DS} characteristic curves for an NMOS transistor operating with a small v_{DS} (in the manner shown in Fig. 5.4). Let the MOSFET have $k_n = 5\ \text{mA/V}^2$ and $V_t = 0.5$ V. Sketch and clearly label the graphs for $V_{GS} = 0.5, 1.0, 1.5, 2.0$, and 2.5 V. Let V_{DS} be in the range 0 to 50 mV. Give the value of r_{DS} obtained for each of the five values of V_{GS}. Although only a sketch, your diagram should be drawn to scale as much as possible.

D 5.7 An n-channel MOS device in a technology for which oxide thickness is 20 nm, minimum channel length is 1 μm, $k_n' = 100\ \mu\text{A/V}^2$, and $V_t = 0.8$ V operates in the triode region, with small v_{DS} and with the gate–source voltage in the range 0 V to +5 V. What device width is needed to ensure that the minimum available resistance is 1 kΩ?

5.8 Consider an NMOS transistor operating in the triode region with an overdrive voltage V_{OV}. Find an expression for the incremental resistance

$$r_{ds} \equiv 1\Big/\left.\frac{\partial i_D}{\partial v_{DS}}\right|_{v_{DS}=V_{DS}}$$

Give the values of r_{ds} in terms of k_n and V_{OV} for $V_{DS} = 0$, $0.5\,V_{OV}$, $0.8\,V_{OV}$, and V_{OV}.

5.9 An NMOS transistor with $k_n = 1\ \text{mA/V}^2$ and $V_t = 1$ V is operated with $V_{GS} = 2.5$ V. At what value of V_{DS} does the transistor enter the saturation region? What value of I_D is obtained in saturation?

5.10 Consider a CMOS process for which $L_{\min} = 0.25\ \mu\text{m}$, $t_{ox} = 6$ nm, $\mu_n = 460\ \text{cm}^2/\text{V}\cdot\text{s}$, and $V_t = 0.5$ V.

(a) Find C_{ox} and k_n'.
(b) For an NMOS transistor with $W/L = 15\ \mu\text{m}/0.25\ \mu\text{m}$, calculate the values of V_{OV}, V_{GS}, and $V_{DS\min}$ needed to operate the transistor in the saturation region with a dc current $I_D = 0.8$ mA.
(c) For the device in (b), find the value of V_{OV} and V_{GS} required to cause the device to operate as a 500-Ω resistor for very small v_{DS}.

5.11 A p-channel MOSFET with a threshold voltage $V_{tp} = -0.7$ V has its source connected to ground.

(a) What should the gate voltage be for the device to operate with an overdrive voltage of $|V_{OV}| = 0.5$ V?
(b) With the gate voltage as in (b), what is the highest voltage allowed at the drain while the device operates in the saturation region?
(c) If the drain current obtained in (b) is 1 mA, what would the current be for $V_D = -10$ mV and for $V_D = -2$V?

5.12 With the knowledge that $\mu_p \simeq 0.4\mu_n$, what must be the relative width of n-channel and p-channel devices if they are to have equal drain currents when operated in the saturation mode with overdrive voltages of the same magnitude?

5.13 An n-channel device has $k'_n = 50\ \mu A/V^2$, $V_t = 0.8$ V, and $W/L = 20$. The device is to operate as a switch for small v_{DS}, utilizing a control voltage v_{GS} in the range 0 V to 5 V. Find the switch closure resistance, r_{DS}, and closure voltage, V_{DS}, obtained when $v_{GS} = 5$ V and $i_D = 1$ mA. Recalling that $\mu_p \simeq 0.4\mu_n$, what must W/L be for a p-channel device that provides the same performance as the n-channel device in this application?

5.14 Consider an n-channel MOSFET with $t_{ox} = 9$ nm, $\mu_n = 500\ cm^2/V \cdot s$, $V_t = 0.7$ V, and $W/L = 10$. Find the drain current in the following cases:

(a) $v_{GS} = 5$ V and $v_{DS} = 1$ V
(b) $v_{GS} = 2$ V and $v_{DS} = 1.3$ V
(c) $v_{GS} = 5$ V and $v_{DS} = 0.2$ V
(d) $v_{GS} = v_{DS} = 5$ V

***5.15** This problem illustrates the central point in the electronics revolution that has been in effect for the past four decades: By continually reducing the MOSFET size, we are able to pack more devices on an IC chip. Gordon Moore, co-founder of Intel Corporation, predicted this exponential growth of chip-packing density very early in the history of the development of the integrated circuit in the formulation that has become known as **Moore's law**.

The table below shows four technology generations, each characterized by the minimum possible MOSFET channel length (row 1). In going from one generation to another, both L and t_{ox} are scaled by the same factor. The power supply utilized V_{DD} is also scaled by the same factor, to keep the magnitudes of all electrical fields within the device unchanged. Unfortunately, but for good reasons, V_t cannot be scaled similarly.

Complete the table entries, noting that row 5 asks for the transconductance parameter of an NMOS transistor with $W/L = 10$; row 9 asks for the value of I_D obtained with $V_{GS} = V_{DS} = V_{DD}$; row 10 asks for the power $P = V_{DD}I_D$ dissipated in the circuit. An important quantity is the power density, P/A, asked for in row 11. Finally, you are asked to find the number of transistors that can be placed on an IC chip fabricated in each of the technologies in terms of the number obtained with the 0.5-μm technology (n).

1	L (μm)	0.5	0.25	0.18	0.13
2	t_{ox} (nm)	10			
3	C_{ox} (fF/μm^2)				
4	k'_n (μA/V^2) ($\mu_n = 500\ cm^2/V \cdot s$)				
5	k_n (mA/V^2) for $W/L = 10$				
6	Device area, A (μm^2)				
7	V_{DD} (V)	5			
8	V_t (V)	0.7	0.5	0.4	0.4
9	I_D (mA) For $V_{GS} = V_{DS} = V_{DD}$				
10	P (mW)				
11	P/A (mW/ μm^2)				
12	Devices per chip	n			

Section 5.2: Current–Voltage Characteristics

In the following problems, when λ is not specified, assume it is zero.

5.16 Show that when channel-length modulation is neglected (i.e., $\lambda = 0$), plotting i_D/k_n versus v_{DS} for various values of v_{OV}, and plotting i_D/k_n versus v_{OV} for $v_{DS} \geq v_{OV}$, results in universal representation of the $i_D - v_{DS}$ and $i_D - v_{GS}$ characteristics of the NMOS transistor. That is, the resulting graphs are both technology and device independent. Furthermore, these graphs apply equally well to the PMOS transistor by a simple relabeling of variables. (How?) What is the slope at $v_{DS} = 0$ of each of the i_D/k_n versus v_{DS} graphs? For the i_D/k_n versus v_{GS} graph, find the slope at a point $v_{OV} = V_{OV}$.

5.17 An NMOS transistor having $V_t = 1$ V is operated in the triode region with v_{DS} small. With $V_{GS} = 1.5$ V, it is found to have a resistance r_{DS} of 1 kΩ. What value of V_{GS} is required to obtain $r_{DS} = 200$ Ω? Find the corresponding resistance values obtained with a device having twice the value of W.

5.18 A particular enhancement MOSFET for which $V_t = 0.5$ V and $k_n'(W/L) = 0.1$ mA/V^2 is to be operated in the saturation region. If i_D is to be 12.5 μA, find the required v_{GS} and the minimum required v_{DS}. Repeat for $i_D = 50$ μA.

5.19 A particular n-channel enhancement MOSFET is measured to have a drain current of 0.4 mA at $V_{GS} = V_{DS} = 2$ V and of 0.1 mA at $V_{GS} = V_{DS} = 1.5$ V. What are the values of k_n and V_t for this device?

D 5.20 For a particular IC-fabrication process, the transconductance parameter $k_n' = 400$ μA/V^2, and $V_t = 0.4$ V. In an application in which $v_{GS} = v_{DS} = V_{\text{supply}} = 1.8$ V, a drain current of 2 mA is required of a device of minimum length of 0.18 μm. What value of channel width must the design use?

5.21 An NMOS transistor, operating in the linear-resistance region with $v_{DS} = 0.1$ V, is found to conduct 60 μA for $v_{GS} = 2$ V and 160 μA for $v_{GS} = 4$ V. What is the apparent value of threshold voltage V_t? If $k_n' = 50$ μA/V^2, what is the device W/L ratio? What current would you expect to flow with $v_{GS} = 3$ V and $v_{DS} = 0.15$ V? If the device is operated at $v_{GS} = 3$ V, at what value of v_{DS} will the drain end of the MOSFET channel just reach pinch-off, and what is the corresponding drain current?

5.22 For an NMOS transistor, for which $V_t = 0.5$ V, operating with v_{GS} in the range of 0.8 V to 1.8 V, what is the largest value of v_{DS} for which the channel remains continuous?

5.23 An NMOS transistor, fabricated with $W = 100$ μm and $L = 5$ μm in a technology for which $k_n' = 50$ μA/V^2 and $V_t = 1$ V, is to be operated at very low values of v_{DS} as a linear resistor. For v_{GS} varying from 1.1 V to 11 V, what range of resistor values can be obtained? What is the available range if

(a) the device width is halved?
(b) the device length is halved?
(c) both the width and length are halved?

5.24 When the drain and gate of a MOSFET are connected together, a two-terminal device known as a "diode-connected transistor" results. Figure P5.24 shows such devices obtained from MOS transistors of both polarities. Show that

(a) the i–v relationship is given by

$$i = \frac{1}{2}k'\frac{W}{L}(v - |V_t|)^2$$

(b) the incremental resistance r for a device biased to operate at $v = |V_t| + V_{OV}$ is given by

$$r \equiv 1\Big/\left[\frac{\partial i}{\partial v}\right] = 1\Big/\left(k'\frac{W}{L}V_{OV}\right)$$

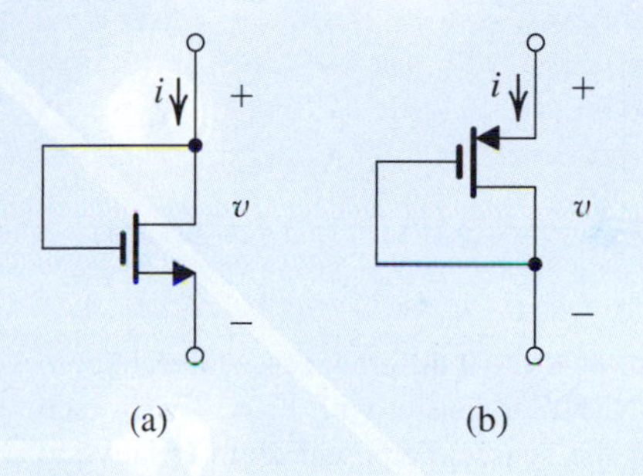

Figure P5.24

5.25 For the circuit in Fig. P5.25, sketch i_D versus v_S for v_S varying from 0 to V_{DD}. Clearly label your sketch.

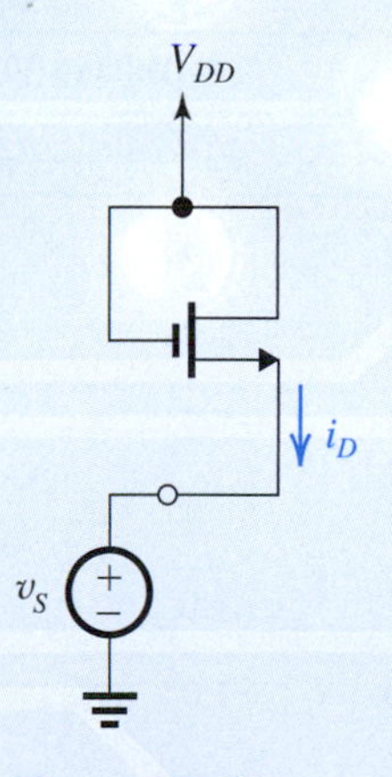

Figure P5.25

5.26 For the circuit in Fig. P5.26, find an expression for v_{DS} in terms of i_D. Sketch and clearly label a graph for v_{DS} versus i_D.

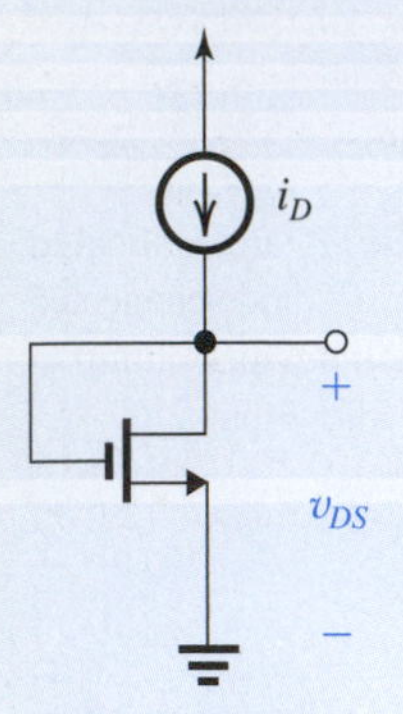

Figure P5.26

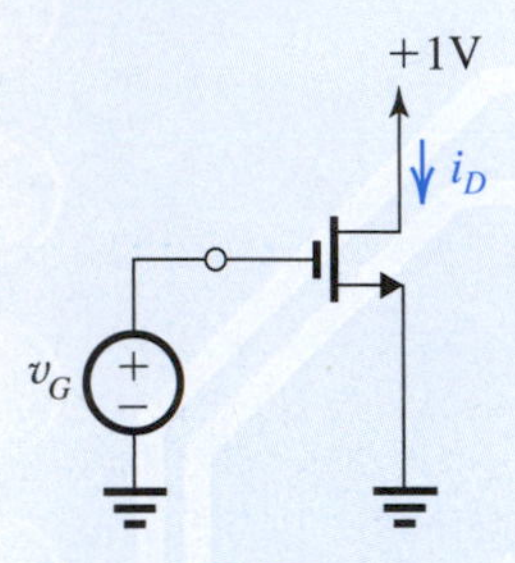

Figure P5.28

***5.27** The table below lists 10 different cases labeled (a) to (j) for operating an NMOS transistor with $V_t = 1$ V. In each case the voltages at the source, gate, and drain (relative to the circuit ground) are specified. You are required to complete the table entries. Note that if you encounter a case for which v_{DS} is negative, you should exchange the drain and source before solving the problem. You can do this because the MOSFET is a symmetric device.

5.28 The NMOS transistor in Fig. P5.28 has $V_t = 0.4$ V and $k_n'(W/L) = 1$ mA/V^2. Sketch and clearly label i_D versus v_G with v_G varying in the range 0 to +1.8 V. Give equations for the various portions of the resulting graph.

5.29 Fig. P5.29 shows two NMOS transistors operating in saturation at equal V_{GS} and V_{DS}.

(a) If the two devices are matched except for a maximum possible mismatch in their W/L ratios of 2%, what is the maximum resulting mismatch in the drain currents?

(b) If the two devices are matched except for a maximum possible mismatch in their V_t values of 10 mV, what is the maximum resulting mismatch in the drain currents? Assume that the nominal value of V_t is 1 V.

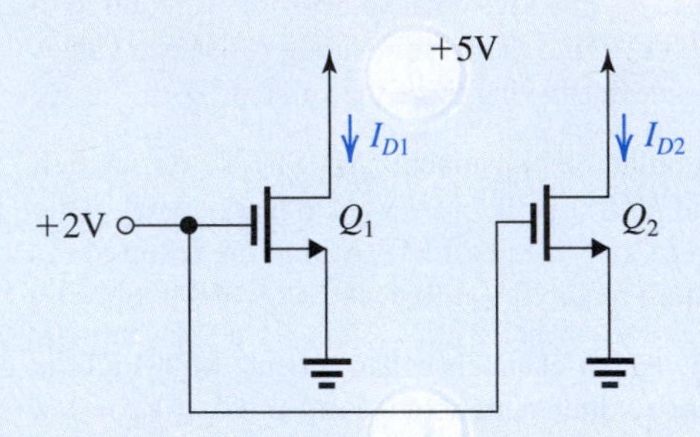

Figure P5.29

5.30 For a particular MOSFET operating in the saturation region at a constant v_{GS}, i_D is found to be 1 mA for

Case	Voltage (V) V_S	V_G	V_D	V_{GS}	V_{OV}	V_{DS}	Region of operation
a	+1.0	+1.0	+2.0				
b	+1.0	+2.5	+2.0				
c	+1.0	+2.5	+1.5				
d	+1.0	+1.5	0				
e	0	+2.5	1.0				
f	+1.0	+1.0	+1.0				
g	−1.0	0	0				
h	−1.5	0	0				
i	−1.0	0	+1.0				
j	+0.5	+2.0	+0.5				

$v_{DS} = 1$ V and 1.05 mA for $v_{DS} = 2$ V. What values of r_o, V_A, and λ correspond?

5.31 A particular MOSFET has $V_A = 50$ V. For operation at 0.1 mA and 1 mA, what are the expected output resistances? In each case, for a change in v_{DS} of 1 V, what percentage change in drain current would you expect?

D 5.32 In a particular IC design in which the standard channel length is 2 µm, an NMOS device with W/L of 5 operating at 100 µA is found to have an output resistance of 0.5 MΩ, about $\frac{1}{4}$ of that needed. What dimensional change can be made to solve the problem? What is the new device length? The new device width? The new W/L ratio? What is V_A for the standard device in this IC? The new device?

D 5.33 For a particular n-channel MOS technology, in which the minimum channel length is 1 µm, the associated value of λ is 0.02 V^{-1}. If a particular device for which L is 3 µm operates at $v_{DS} = 1$ V with a drain current of 80 µA, what does the drain current become if v_{DS} is raised to 5 V? What percentage change does this represent? What can be done to reduce the percentage by a factor of 2?

5.34 An NMOS transistor is fabricated in a 0.8-µm process having $k_n' = 130$ µA/V^2 and $V_A' = 20$ V/µm of channel length. If $L = 1.6$ µm and $W = 16$ µm, find V_A and λ. Find the value of I_D that results when the device is operated with an overdrive voltage of 0.5 V and $V_{DS} = 2$ V. Also, find the value of r_o at this operating point. If V_{DS} is increased by 1 V, what is the corresponding change in I_D?

5.35 If in an NMOS transistor, both W and L are quadrupled and V_{OV} is halved, by what factor does r_o change?

D 5.36 Consider the circuit in Fig. P5.29 with both transistors perfectly matched but with the dc voltage at the drain of Q_1 lowered to +2 V. If the two drain currents are to be matched within 1% (i.e., the maximum difference allowed between the two currents is 1%), what is the minimum required value of V_A? If the technology is specified to have $V_A' = 100$ V/µm, what is the minimum channel length the designer must use?

5.37 Complete the missing entries in the following table, which describes characteristics of suitably biased NMOS transistors:

MOS	1	2	3	4
λ(V^{-1})		0.01		
V_A (V)	10			200
I_D (mA)	1		0.1	
r_o (kΩ)		30	100	1000

5.38 An enhancement PMOS transistor has $k_p'(W/L) = 80$ µA/V^2, $V_t = -1.5$ V, and $\lambda = -0.02$ V^{-1}. The gate is connected to ground and the source to +5 V. Find the drain current for $v_D = +4$ V, +1.5 V, 0 V, and −5 V.

5.39 A p-channel transistor for which $|V_t| = 1$ V and $|V_A| = 50$ V operates in saturation with $|v_{GS}| = 3$ V, $|v_{DS}| = 4$ V, and $i_D = 3$ mA. Find corresponding signed values for v_{GS}, v_{SG}, v_{DS}, v_{SD}, V_t, V_A, λ, and $k_p'(W/L)$.

5.40 The table below lists the terminal voltages of a PMOS transistor in six cases, labeled a, b, c, d, e, and f. The transistor has $V_{tp} = -1$ V. Complete the table entries.

	V_S	V_G	V_D	V_{SG}	$\lvert V_{OV}\rvert$	V_{SD}	Region of operation
a	+2	+2	0				
b	+2	+1	0				
c	+2	0	0				
d	+2	0	+1				
e	+2	0	+1.5				
f	+2	0	+2				

5.41 The PMOS transistor in Fig. P5.41 has $V_{tp} = -0.5$ V. As the gate voltage v_G is varied from +2.5 V to 0 V, the transistor moves through all of its three possible modes of operation. Specify the value of v_G at which the device changes modes of operation.

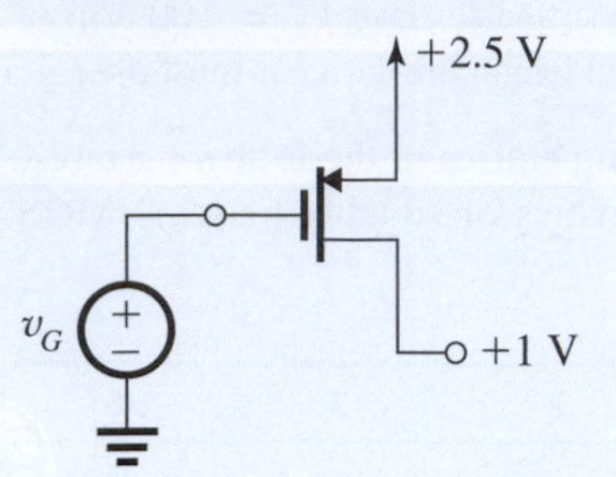

Figure P5.41

***5.42** (a) Using the expression for i_D in saturation and neglecting the channel-length modulation effect (i.e., let $\lambda = 0$), derive an expression for the per unit change in i_D per °C $[(\partial i_D/i_D)/\partial T]$ in terms of the per unit change in k_n' per °C $[(\partial k_n'/k_n')/\partial T]$, the temperature coefficient of V_t in V/°C $(\partial V_t/\partial T)$, and V_{GS} and V_t.
(b) If V_t decreases by 2 mV for every °C rise in temperature, find the temperature coefficient of k_n' that results in i_D decreasing by 0.2%/°C when the NMOS transistor with $V_t = 1$ V is operated at $V_{GS} = 5$ V.

***5.43** Various NMOS and PMOS transistors, numbered 1 to 4, are measured in operation, as shown in the table at the bottom of the page. For each transistor, find the value of $\mu C_{ox}W/L$ and V_t that apply and complete the table, with V in volts, I in μA, and $\mu C_{ox}W/L$ in μA/V^2.

***5.44** All the transistors in the circuits shown in Fig. P5.44 have the same values of $|V_t|$, k', W/L, and λ. Moreover, λ is negligibly small. All operate in saturation at $I_D = I$ and $|V_{GS}| = |V_{DS}| = 1$ V. Find the voltages V_1, V_2, V_3, and V_4. If $|V_t| = 0.5$ V and $I = 0.1$ mA, how large a resistor can be inserted in series with each drain connection while maintaining saturation? What is the largest resistor that can be placed in series with each gate? If the current source I requires at least 0.5 V between its terminals to operate properly, what is the largest resistor that can be placed in series with each MOSFET source while ensuring saturated-mode operation of each transistor at $I_D = I$? In the latter limiting situation, what do V_1, V_2, V_3, and V_4 become?

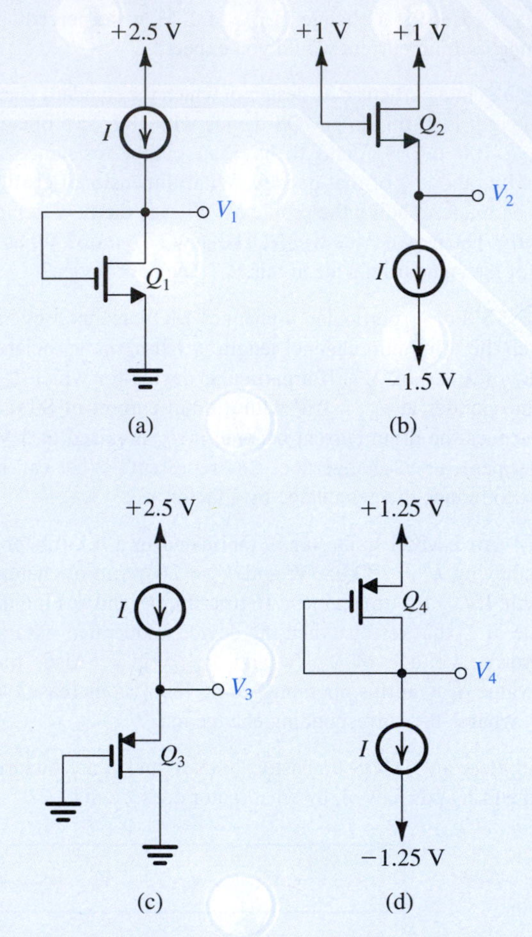

Figure P5.44

Case	Transistor	V_S	V_G	V_D	I_D	Type	Mode	$\mu C_{ox}W/L$	V_t
a	1	0	2	5	100				
	1	0	3	5	400				
b	2	5	3	−4.5	50				
	2	5	2	−0.5	450				
c	3	5	3	4	200				
	3	5	2	0	800				
d	4	−2	0	0	72				
	4	−4	0	−3	270				

Section 5.3: MOSFET Circuits at DC

Note: If λ is not specified, assume it is zero.

D 5.45 Design the circuit of Fig. 5.21 to establish a drain current of 0.25 mA and a drain voltage of 0 V. The MOSFET has $V_t = 1$ V, $\mu_n C_{ox} = 60\ \mu\text{A/V}^2$, $L = 3\ \mu\text{m}$, and $W = 100\ \mu\text{m}$.

D 5.46 For the circuit in Fig. E5.10, assume that Q_1 and Q_2 are matched except for having different widths, W_1 and W_2. Let $V_t = 0.5$ V, $k_n' = 0.4\ \text{mA/V}^2$, $L_1 = L_2 = 0.36\ \mu\text{m}$, $W_1 = 1.8\ \mu\text{m}$, and $\lambda = 0$.

(a) Find the value of R required to establish a current of 90 μA in Q_1.
(b) Find W_2 and R_2 so that Q_2 operates at the edge of saturation with a current of 0.9 mA.

5.47 The transistor in the circuit of Fig. P5.47 has $k_n' = 0.4\ \text{mA/V}^2$, $V_t = 0.5$ V, and $\lambda = 0$. Show that operation at the edge of saturation is obtained when the following condition is satisfied:

$$\left(\frac{W}{L}\right) R_D = 1.5\ \text{k}\Omega$$

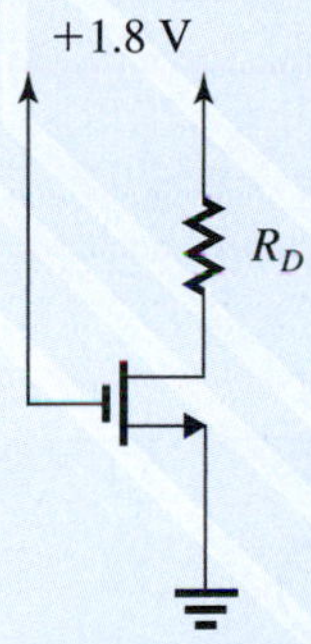

Figure P5.47

5.48 It is required to operate the transistor in the circuit of Fig. P5.47 at the edge of saturation with $I_D = 1$ mA. If $V_t = 0.5$ V, find the required value of R_D.

D 5.49 The PMOS transistor in the circuit of Fig. P5.49 has $V_t = -0.6$ V, $\mu_p C_{ox} = 100\ \mu\text{A/V}^2$, $L = 0.25\ \mu\text{m}$, and $\lambda = 0$. Find the values required for W and R in order to establish a drain current of 0.8 mA and a voltage V_D of 1.5 V.

D 5.50 The NMOS transistors in the circuit of Fig. P5.50 have $V_t = 0.5$ V, $\mu_n C_{ox} = 250\ \mu\text{A/V}^2$, $\lambda = 0$, and $L_1 = L_2 = 0.25\ \mu\text{m}$. Find the required values of gate width for each of Q_1 and Q_2, and the value of R, to obtain the voltage and current values indicated.

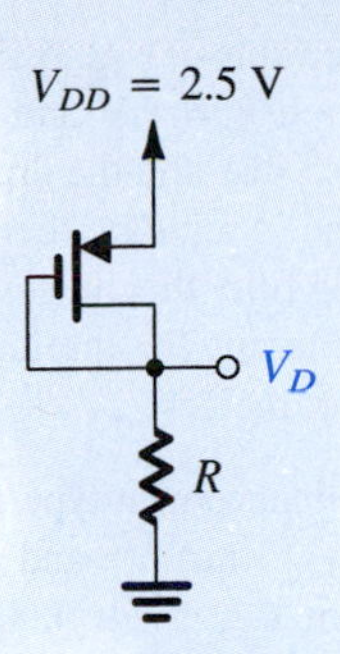

Figure P5.49

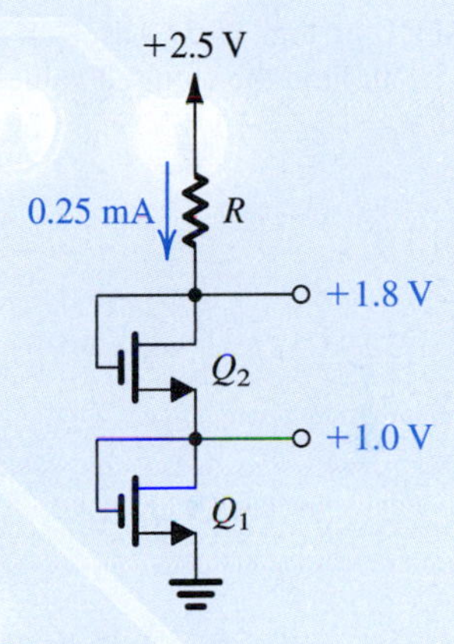

Figure P5.50

D 5.51 The NMOS transistors in the circuit of Fig. P5.51 have $V_t = 1$ V, $\mu_n C_{ox} = 120\ \mu\text{A/V}^2$, $\lambda = 0$, and $L_1 = L_2 = L_3 = 1\ \mu\text{m}$. Find the required values of gate width for each of Q_1, Q_2, and Q_3 to obtain the voltage and current values indicated.

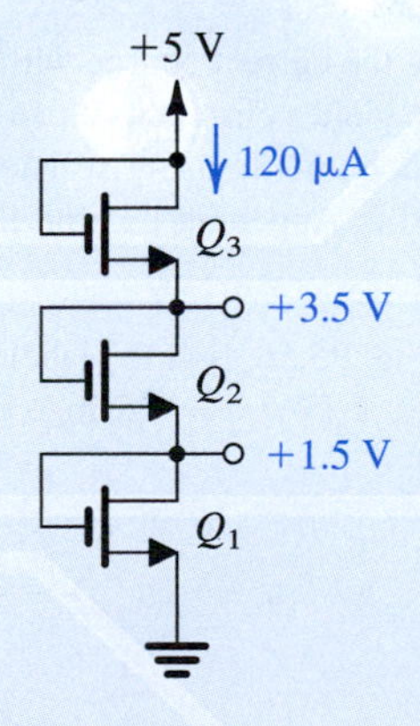

Figure P5.51

5.52 Consider the circuit of Fig. 5.24(a). In Example 5.5 it was found that when $V_t = 1$ V and $k_n'(W/L) = 1$ mA/V^2, the drain current is 0.5 mA and the drain voltage is +7 V. If the transistor is replaced with another having $V_t = 2$ V and $k_n'(W/L) = 2$ mA/V^2, find the new values of I_D and V_D. Comment on how tolerant (or intolerant) the circuit is to changes in device parameters.

D 5.53 Using an enhancement-type PMOS transistor with $V_t = -1.5$ V, $k_p'(W/L) = 1$ mA/V^2, and $\lambda = 0$, design a circuit that resembles that in Fig. 5.24(a). Using a 10-V supply, design for a gate voltage of +6 V, a drain current of 0.5 mA, and a drain voltage of +5 V. Find the values of R_S and R_D.

5.54 The MOSFET in Fig. P5.54 has $V_t = 0.5$ V, $k_n' = 400$ μA/V^2, and $\lambda = 0$. Find the required values of W/L and of R so that when $v_I = V_{DD} = +1.8$ V, $r_{DS} = 50$ Ω, and $v_O = 50$ mV.

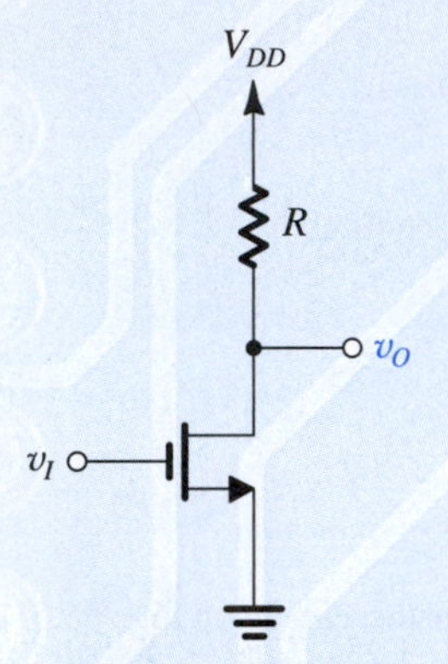

Figure P5.54

5.55 In the circuits shown in Fig. P5.55, transistors are characterized by $|V_t| = 2$ V, $k'W/L = 1$ mA/V^2, and $\lambda = 0$.

(a) Find the labeled voltages V_1 through V_7.
(b) In each of the circuits, replace the current source with a resistor. Select the resistor value to yield a current as close to that of the current source as possible, while using resistors specified in the 1% table provided in Appendix G. Find the new values of V_1 to V_7.

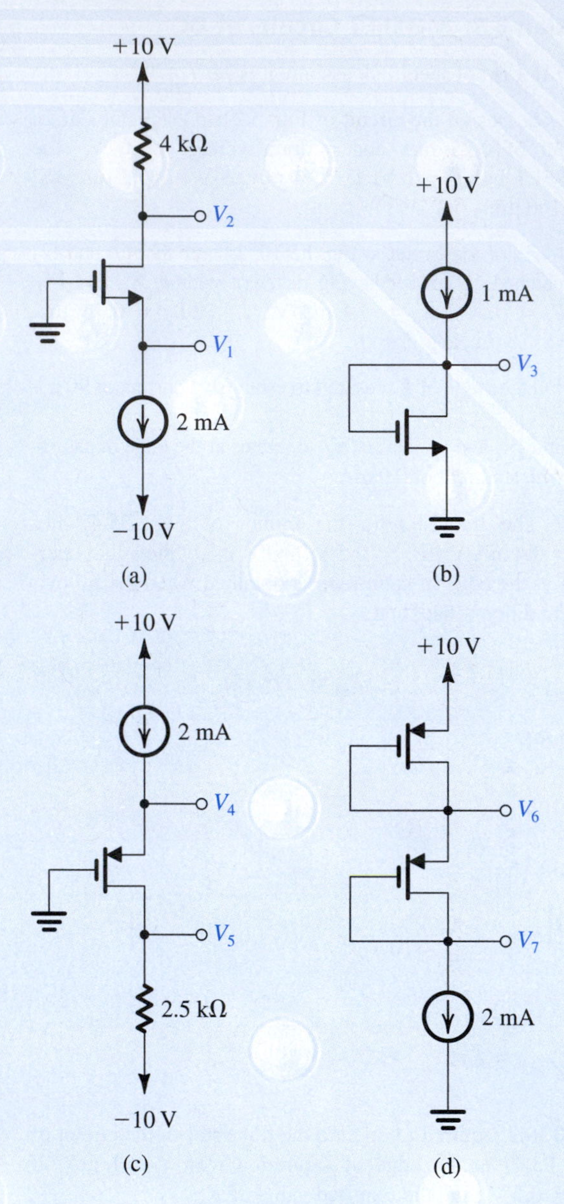

Figure P5.55

5.56 For each of the circuits in Fig. P5.56, find the labeled node voltages. For all transistors, $k_n'(W/L) = 0.5$ mA/V^2, $V_t = 0.8$ V, and $\lambda = 0$.

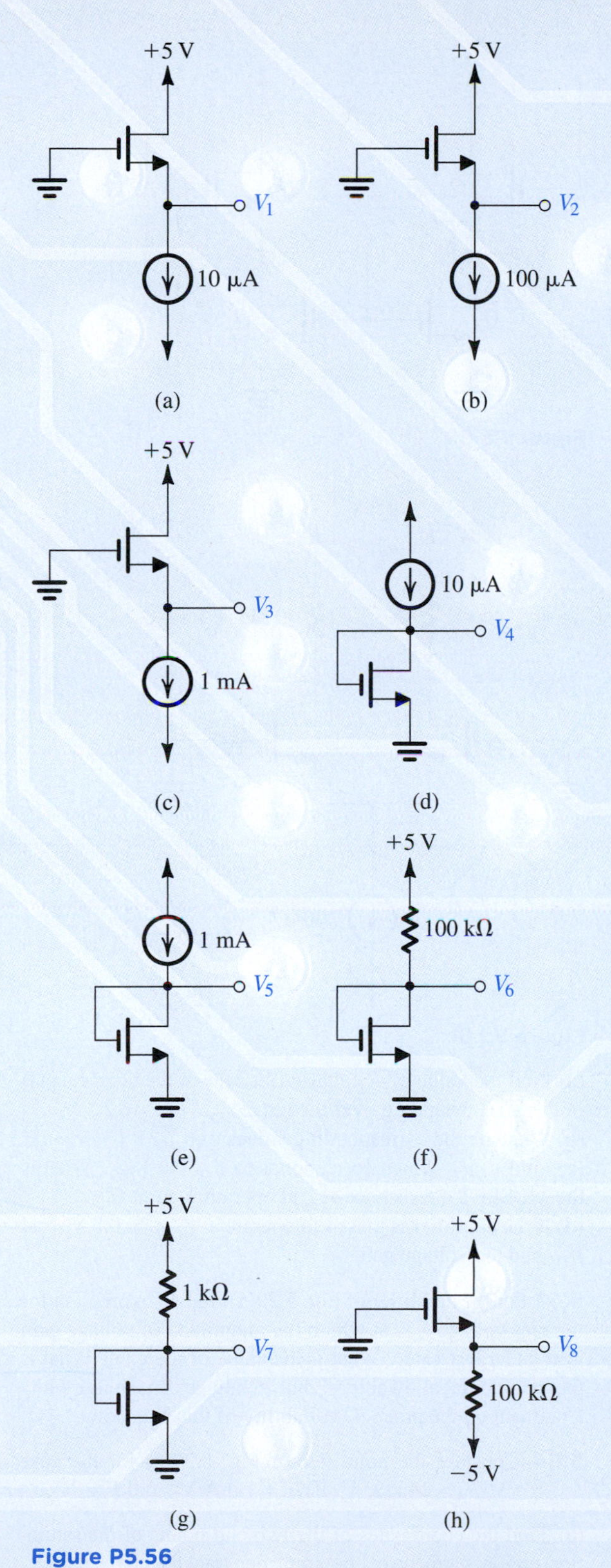

Figure P5.56

5.57 For each of the circuits shown in Fig. P5.57, find the labeled node voltages. The NMOS transistors have $V_t = 1$ V and $k_n' W/L = 5$ mA/V^2.

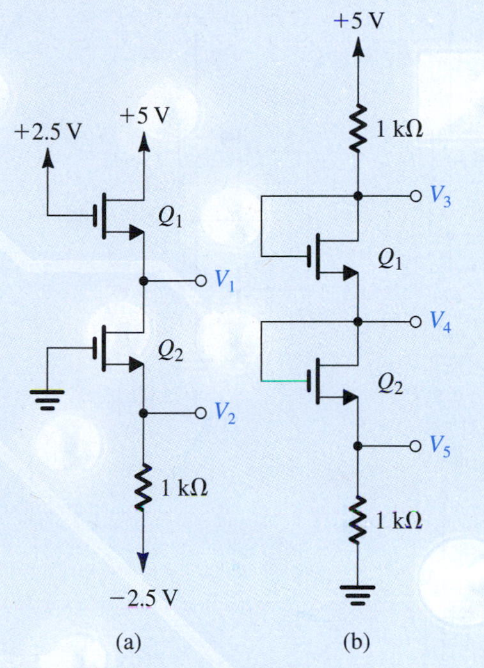

Figure P5.57

***5.58** For the PMOS transistor in the circuit shown in Fig. P5.58, $k_p' = 8$ μA/V^2, $W/L = 25$, and $|V_{tp}| = 1$ V. For $I = 100$ μA, find the voltages V_{SD} and V_{SG} for $R = 0$, 10 kΩ, 30 kΩ, and 100 kΩ. For what value of R is $V_{SD} = V_{SG}$? $V_{SD} = V_{SG}/2$? $V_{SD} = V_{SG}/10$?

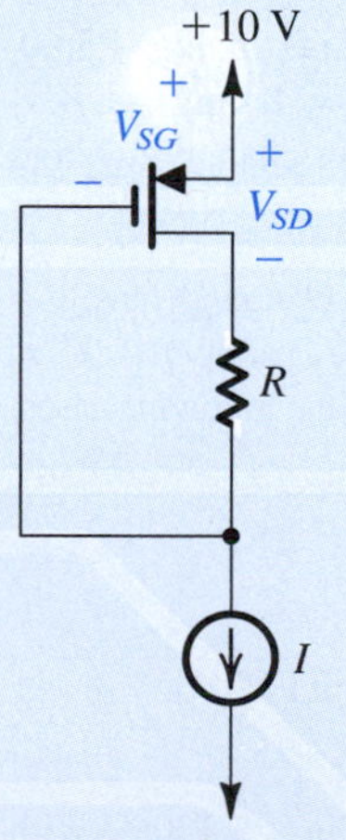

Figure P5.58

5.59 For the circuits in Fig. P5.59, $\mu_n C_{ox} = 2.5\ \mu_p C_{ox} = 20\ \mu\text{A/V}^2$, $|V_t| = 1$ V, $\lambda = 0$, $L = 10\ \mu$m, and $W = 30\ \mu$m, unless otherwise specified. Find the labeled currents and voltages.

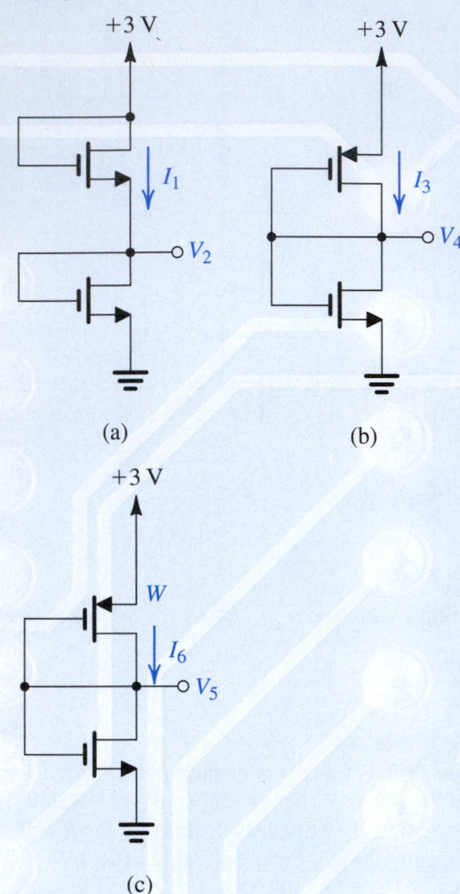

Figure P5.59

SIM ***5.60** For the devices in the circuits of Fig. P5.60, $|V_t| = 1$ V, $\lambda = 0$, $\mu_n C_{ox} = 50\ \mu\text{A/V}^2$, $L = 1\ \mu$m, and $W = 10\ \mu$m. Find V_2 and I_2. How do these values change if Q_3 and Q_4 are made to have $W = 100\ \mu$m?

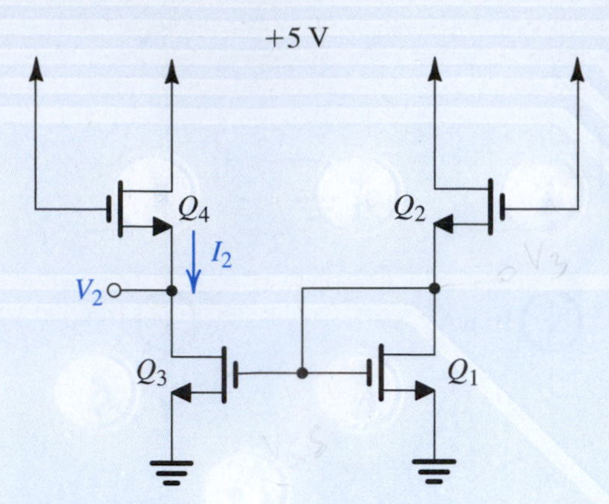

Figure P5.60

5.61 In the circuit of Fig. P5.61, transistors Q_1 and Q_2 have $V_t = 1$ V, and the process transconductance parameter $k_n' = 100\ \mu\text{A/V}^2$. Find V_1, V_2, and V_3 for each of the following cases:

(a) $(W/L)_1 = (W/L)_2 = 20$
(b) $(W/L)_1 = 1.5(W/L)_2 = 20$

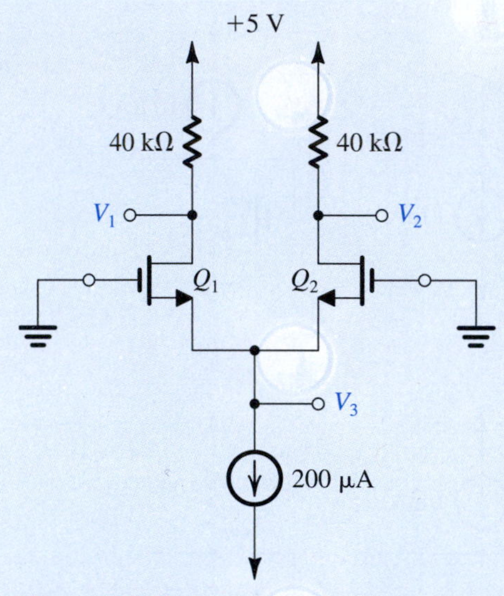

Figure P5.61

Section 5.4: Applying the MOSFET in Amplifier Design

5.62 Consider the amplifier of Fig. 5.27(a) with $V_{DD} = 2.5$ V and with the MOSFET having $V_t = 0.5$ V, $k_n' = 0.25$ mA/V² and $W/L = 40$.

(a) Find the value of R_D that will result in the segment AB of the VTC extending over the range $v_{DS} = 0.5$ to 2.5 V.
(b) What are the corresponding values of v_{GS}?
(c) Find $v_{DS}|_C$ which corresponds to $v_{GS} = V_{DD}$. What is the MOSFET's resistance r_{DS} at operating point C?
(d) If the amplifier is biased to operate at $V_{GS} = 0.8$ V, find V_{DS} and the voltage gain.

5.63 For the amplifier of Fig. 5.29(a) find an expression for the bias voltage V_{GS} at which the magnitude of voltage gain is at its largest value. What is the value of the gain? What is the maximum allowable signal swing at this bias point? Comment on the practical suitability of this bias point.

5.64 Consider the amplifier of Fig. 5.29(a) for the case $V_{DD} = 5$ V, $R_D = 24$ kΩ, $k_n'(W/L) = 1$ mA/V², and $V_t = 1$ V.

(a) Find the coordinates of the two end points of the saturation-region segment of the amplifier transfer characteristic, that is, points A and B on the sketch of Fig. 5.29(b).

(b) If the amplifier is biased to operate with an overdrive voltage V_{OV} of 0.5 V, find the coordinates of the bias point Q on the transfer characteristic. Also, find the value of I_D and of the incremental gain A_v at the bias point.

(c) For the situation in (b), and disregarding the distortion caused by the MOSFET's square-law characteristic, what is the largest amplitude of a sine-wave voltage signal that can be applied at the input while the transistor remains in saturation? What is the amplitude of the output voltage signal that results? What gain value does the combination of these amplitudes imply? By what percentage is this gain value different from the incremental gain value calculated above? Why is there a difference?

5.65 Various measurements are made on an NMOS amplifier for which the drain resistor R_D is 20 kΩ. First, dc measurements show the voltage across the drain resistor, V_{RD}, to be 1.5 V and the gate-to-source bias voltage to be 0.7 V. Then, ac measurements with small signals show the voltage gain to be –10 V/V. What is the value of V_t for this transistor? If the process transconductance parameter k_n' is 200 μA/V², what is the MOSFET's W/L?

***D 5.66** Refer to the expression for the incremental voltage gain in Eq. (5.38). Various design considerations place a lower limit on the value of the overdrive voltage V_{OV}. For our purposes here, let this lower limit be 0.2 V. Also, assume that $V_{DD} = 5$ V.

(a) Without allowing any room for output voltage swing, what is the maximum voltage gain achievable?

(b) If we are required to allow for an output voltage swing of ±0.5 V, what dc bias voltage should be established at the drain to obtain maximum gain? What gain value is achievable? What input signal results in a ±0.5-V output swing?

(c) For the situation in (b), find W/L of the transistor to establish a dc drain current of 100 μA. For the given process technology, $k_n' = 100$ μA/V².

(d) Find the required value of R_D.

5.67 The expression for the incremental voltage gain A_v given in Eq. (5.38) can be written in as

$$A_v = -\frac{2(V_{DD} - V_{DS})}{V_{OV}}$$

where V_{DS} is the bias voltage at the drain. This expression indicates that for given values of V_{DD} and V_{OV}, the gain magnitude can be increased by biasing the transistor at a lower V_{DS}. This, however, reduces the allowable output signal swing in the negative direction. Assuming linear operation around the bias point, show that the largest possible negative output signal peak $\hat{v}_o$ that is achievable while the transistor remains saturated is

$$\hat{v}_o = (V_{DS} - V_{OV})\Big/\left(1 + \frac{1}{|A_v|}\right)$$

For $V_{DD} = 5$ V and $V_{OV} = 0.5$ V, provide a table of values for A_v, $\hat{v}_o$, and the corresponding $\hat{v}_i$ for $V_{DS} = 1$ V, 1.5 V, 2 V, and 2.5 V. If $k_n'W/L = 1$ mA/V², find I_D and R_D for the design for which $V_{DS} = 1$ V.

***5.68** Figure P5.68 shows an amplifier in which the load resistor R_D has been replaced with another NMOS transistor Q_2 connected as a two-terminal device. Note that because v_{DG} of Q_2 is zero, it will be operating in saturation at all times, even when $v_I = 0$ and $i_{D2} = i_{D1} = 0$. Note also that the two transistors conduct equal drain currents. Using $i_{D1} = i_{D2}$, show that for the range of v_I over which Q_1 is operating in saturation, that is, for

$$V_{t1} \le v_I \le v_O + V_{t1}$$

the output voltage will be given by

$$v_O = V_{DD} - V_t + \sqrt{\frac{(W/L)_1}{(W/L)_2}}\,V_t - \sqrt{\frac{(W/L)_1}{(W/L)_2}}\,v_I$$

where we have assumed $V_{t1} = V_{t2} = V_t$. Thus the circuit functions as a linear amplifier, even for large input signals. For $(W/L)_1 = (50\ \mu\text{m}/0.5\ \mu\text{m})$ and $(W/L)_2 = (5\ \mu\text{m}/0.5\ \mu\text{m})$, find the voltage gain.

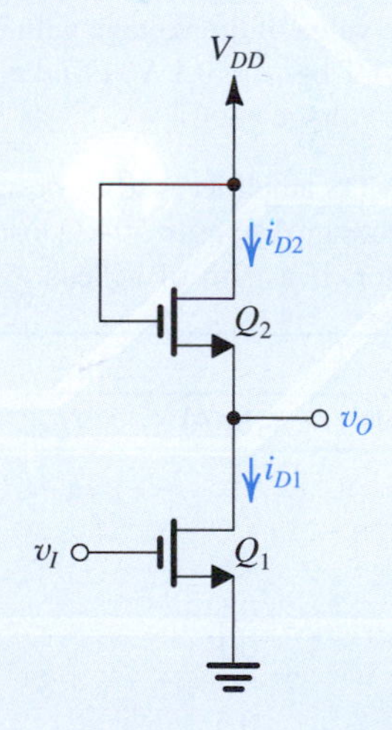

Figure P5.68

Section 5.5: Small-Signal Operation and Models

***5.69** This problem investigates the nonlinear distortion introduced by a MOSFET amplifier. Let the signal v_{gs} be a sine wave with amplitude V_{gs}, and substitute $v_{gs} = V_{gs} \sin \omega t$ in Eq. (5.43). Using the trigonometric identity $\sin^2 \theta = \frac{1}{2} - \frac{1}{2}\cos 2\theta$, show that the ratio of the signal at frequency 2ω to that at frequency ω, expressed as a percentage (known as the second-harmonic distortion) is

$$\text{Second-harmonic distortion} = \frac{1}{4}\frac{V_{gs}}{V_{OV}} \times 100$$

If in a particular application V_{gs} is 10 mV, find the minimum overdrive voltage at which the transistor should be operated so that the second-harmonic distortion is kept to less than 1%.

5.70 Consider an NMOS transistor having $k_n = 10\text{ mA/V}^2$. Let the transistor be biased at $V_{OV} = 0.5$ V. For operation in saturation, what dc bias current I_D results? If a 0.05-V signal is superimposed on V_{GS}, find the corresponding increment in collector current by evaluating the total collector current i_D and subtracting the dc bias current I_D. Repeat for a –0.05-V signal. Use these results to estimate g_m of the FET at this bias point. Compare with the value of g_m obtained using Eq. (5.48).

5.71 Consider the FET amplifier of Fig. 5.34 for the case $V_t = 0.4$ V, $k_n = 4\text{ mA/V}^2$, $V_{GS} = 0.65$ V, $V_{DD} = 1.8$ V, and $R_D = 8\text{ k}\Omega$.

(a) Find the dc quantities I_D and V_D.
(b) Calculate the value of g_m at the bias point.
(c) Calculate the value of the voltage gain.
(d) If the MOSFET has $\lambda = 0.1\text{ V}^{-1}$, find r_o at the bias point and calculate the voltage gain.

D *5.72 An NMOS amplifier is to be designed to provide a 0.50-V peak output signal across a 50-kΩ load that can be used as a drain resistor. If a gain of at least 5 V/V is needed, what g_m is required? Using a dc supply of 1.8 V, what values of I_D and V_{OV} would you choose? What W/L ratio is required if $\mu_n C_{ox} = 200\text{ µA/V}^2$? If $V_t = 0.4$ V, find V_{GS}.

D *5.73 In this problem we investigate an optimum design of the CS amplifier circuit of Fig. 5.34. First, use the voltage gain expression $A_v = -g_m R_D$ together with Eq. (5.57) for g_m to show that

$$A_v = -\frac{2I_D R_D}{V_{OV}} = -\frac{2(V_{DD} - V_D)}{V_{OV}}$$

Next, let the maximum positive input signal be $\hat{v}_i$. To keep the second-harmonic distortion to an acceptable level, we bias the MOSFET to operate at an overdrive voltage $V_{OV} \gg \hat{v}_i$. Let $V_{OV} = m\hat{v}_i$. Now, to maximize the voltage gain $|A_v|$, we design for the lowest possible V_D. Show that the minimum V_D that is consistent with allowing a negative signal voltage swing at the drain of $|A_v|\hat{v}_i$ while maintaining saturation-mode operation is given by

$$V_D = \frac{V_{OV} + \hat{v}_i + 2V_{DD}(\hat{v}_i / V_{OV})}{1 + 2(\hat{v}_i / V_{OV})}$$

Now, find V_{OV}, V_D, A_v, and $\hat{v}_o$ for the case $V_{DD} = 2.5$ V, $\hat{v}_i = 20$ mV, and $m = 15$. If it is desired to operate this transistor at $I_D = 100$ µA, find the values of R_D and W/L, assuming that for this process technology $k'_n = 100\text{ µA/V}^2$.

5.74 In the table below, for enhancement MOS transistors operating under a variety of conditions, complete as many entries as possible. Although some data is not available, it is always possible to calculate g_m using one of Eqs. (5.55), (5.56) or (5.57). Assume $\mu_n = 500\text{ cm}^2/\text{V·s}$, $\mu_p = 250\text{ cm}^2/\text{V·s}$, and $C_{ox} = 0.4\text{ fF/µm}^2$.

5.75 An NMOS technology has $\mu_n C_{ox} = 250\text{ µA/V}^2$ and $V_t = 0.5$ V. For a transistor with $L = 0.5$ µm, find the value of W that results in $g_m = 1$ mA/V at $I_D = 0.25$ mA. Also, find the required V_{GS}.

			Voltages (V)			Dimensions (µm)				
Case	Type	I_D (mA)	$\lvert V_{GS}\rvert$	$\lvert V_t\rvert$	V_{OV}	W	L	W/L	$k'(W/L)$	g_m (mA/V)
a	N	1	3	2			1			
b	N	1		0.7	0.5	50				
c	N	10			2		1			
d	N	0.5			0.5					
e	N	0.1				10	2			
f	N		1.8	0.8		40	4			
g	P	0.5			2			25		
h	P		3	1					0.5	
i	P	10				4000	2			
j	P	10			4					
k	P				1	30	3			
l	P	0.1			5				0.008	

5.76 For the NMOS amplifier in Fig. P5.76, replace the transistor with its T equivalent circuit, assuming $\lambda = 0$. Derive expressions for the voltage gains v_s/v_i and v_d/v_i.

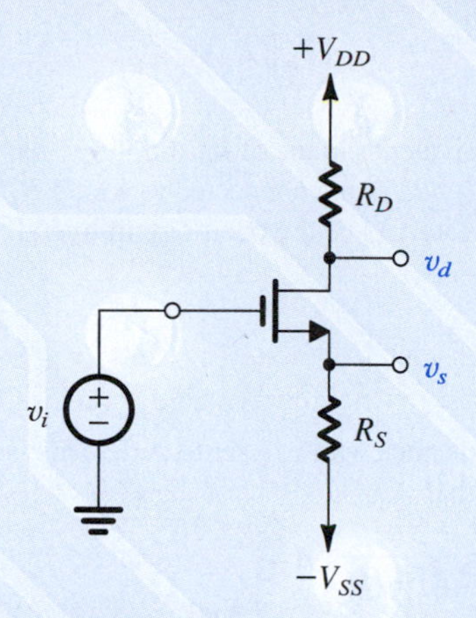

Figure P5.76

SIM 5.77 In the circuit of Fig. P5.77, the NMOS transistor has $|V_t| = 0.5$ V and $V_A = 50$ V and operates with $V_D = 1$ V. What is the voltage gain v_o/v_i? What do V_D and the gain become for I increased to 1 mA?

5.78 For a 0.8-μm CMOS fabrication process: $V_{tn} = 0.8$ V, $V_{tp} = -0.9$ V, $\mu_n C_{ox} = 90\ \mu\text{A/V}^2$, $\mu_p C_{ox} = 30\ \mu\text{A/V}^2$, $C_{ox} = 1.9\ \text{fF}/\mu\text{m}^2$, V_A (*n*-channel devices) = $8L$ (μm), and $|V_A|$ (*p*-channel devices) = $12L$ (μm). Find the small-signal model parameters (g_m and r_o) for both an NMOS and a PMOS transistor having W/L = 20 μm/2 μm and operating at $I_D = 100\ \mu$A. Also, find the overdrive voltage at which each device must be operating.

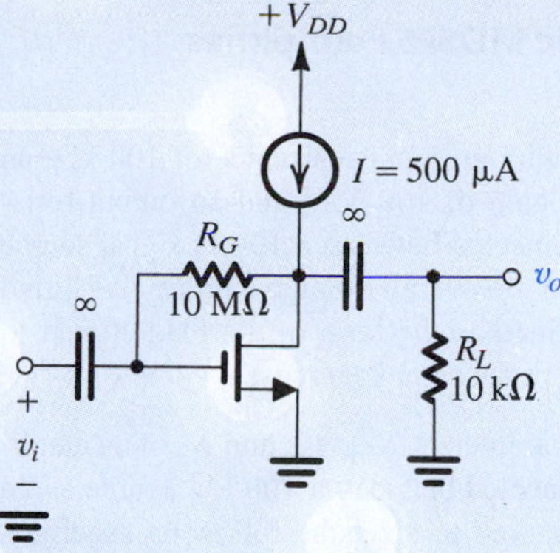

Figure P5.77

***5.79** Figure P5.79 shows a discrete-circuit amplifier. The input signal v_{sig} is coupled to the gate through a very large capacitor (shown as infinite). The transistor source is connected to ground at signal frequencies via a very large capacitor (shown as infinite). The output voltage signal that develops at the drain is coupled to a load resistance via a very large capacitor (shown as infinite).

(a) If the transistor has $V_t = 1$ V, and $k_n = 2\ \text{mA/V}^2$, verify that the bias circuit establishes $V_{GS} = 2$ V, $I_D = 1$ mA, and $V_D = +7.5$ V. That is, assume these values, and verify that they are consistent with the values of the circuit components and the device parameters.

(b) Find g_m and r_o if $V_A = 100$ V.

(c) Draw a complete small-signal equivalent circuit for the amplifier, assuming all capacitors behave as short circuits at signal frequencies.

(d) Find R_{in}, v_{gs}/v_{sig}, v_o/v_{gs}, and v_o/v_{sig}.

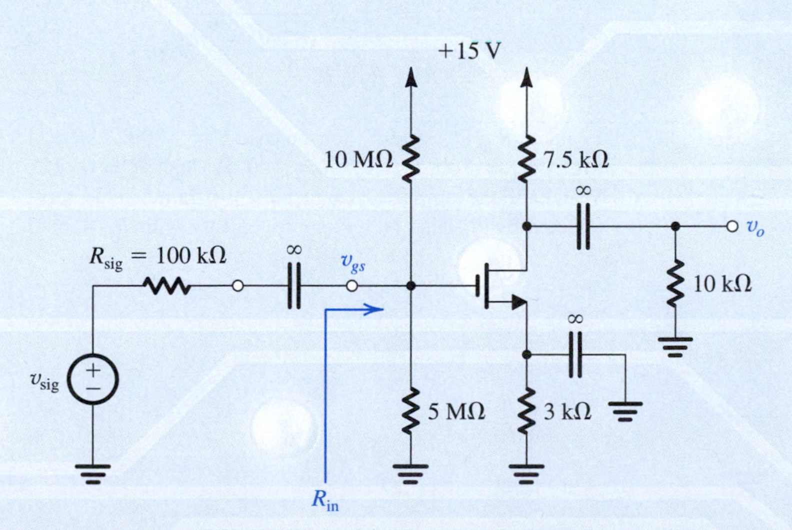

Figure P5.79

Section 5.6: Basic MOSFET Amplifier Configurations*

5.80 An amplifier with an input resistance of 100 kΩ, an open-circuit voltage gain of 100 V/V and an output resistance of 100 Ω is connected between a 10-kΩ signal source and a 1-kΩ load. Find the overall voltage gain G_v. Also find the current gain, defined as the ratio of the load current to the current drawn from the signal source.

D 5.81 Specify the parameters R_{in}, A_{vo} and R_o of an amplifier that is to be connected between a 100-kΩ source and a 2-kΩ load and is required to meet the following specifications:

(a) No more than 10% of the signal strength is lost in the connection to the amplifier input;
(b) If the load resistance changes from the nominal value of 2 kΩ to a low value of 1 kΩ, the change in output voltage is limited to 10% of nominal value; and
(c) The nominal overall voltage gain is 10 V/V.

5.82 Figure P5.82 shows an alternative equivalent circuit representation of an amplifier. If this circuit is to be equivalent to that in Fig. 5.44(b) show that $G_m = A_{vo}/R_o$. Also convince yourself that the transconductance G_m is defined as

$$G_m = \left.\frac{i_o}{v_i}\right|_{R_L=0}$$

and hence is known as the short-circuit transconductance. Now, if the amplifier is fed with a signal source (v_{sig}, R_{sig}) and is connected to a load resistance R_L show that the gain of the amplifier proper A_v is given by $A_v = G_m(R_o \| R_L)$ and the overall voltage gain G_v is given by

$$G_v = \frac{R_{in}}{R_{in} + R_{sig}} G_m(R_o \| R_L)$$

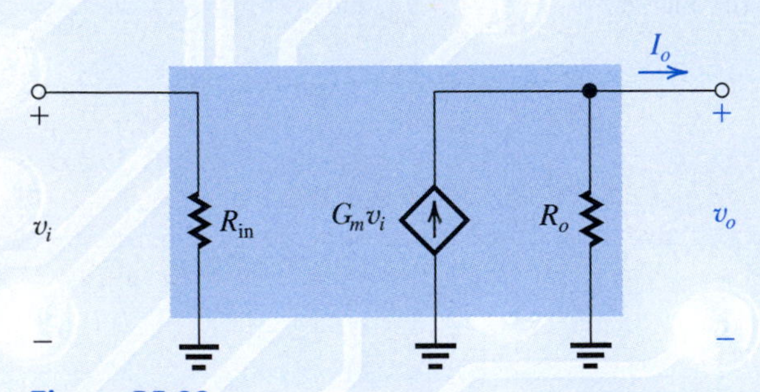

Figure P5.82

5.83 An alternative equivalent circuit of an amplifier fed with a signal source (v_{sig}, R_{sig}) and connected to a load R_L is shown in Fig. P5.83. Here G_{vo} is the open-circuit overall voltage gain,

$$G_{vo} = \left.\frac{v_o}{v_{sig}}\right|_{R_L=\infty}$$

and R_{out} is the output resistance with v_{sig} set to zero. This is different than R_o. Show that

$$G_{vo} = \frac{R_i}{R_i + R_{sig}} A_{vo}$$

where $R_i = R_{in}|_{R_L=\infty}$.

Also show that the overall voltage gain is

$$G_v = G_{vo}\frac{R_L}{R_L + R_{out}}$$

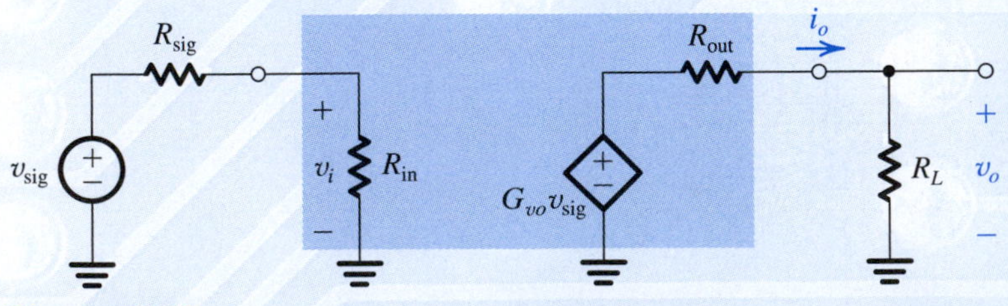

Figure P5.83

****5.84** Most practical amplifiers have internal feedback that make them non-unilateral. In such a case, R_{in} depends on R_L. To illustrate this point we show in Fig. P5.84 the equivalent circuit of an amplifier where a feedback resistance R_f models the internal feedback mechanism that is present in this amplifier. It is R_f that makes the amplifier non-unilateral. Show that

$$R_{in} = R_1 \| \left[\frac{R_f + (R_2 \| R_L)}{1 + g_m(R_2 \| R_L)}\right]$$

$$A_{vo} = -g_m R_2 \frac{1 - 1/(g_m R_f)}{1 + (R_2/R_f)}$$

$$R_o = R_2 \| R_f$$

Evaluate R_{in}, A_{vo} and R_o for the case $R_1 = 100$ kΩ, $R_f = 1$ MΩ, $g_m = 100$ mA/V, $R_2 = 100$ Ω and $R_L = 1$ kΩ. Which of the amplifier characteristic parameters is most affected by R_f (that is, relative to the case with $R_f = \infty$)?

* Problems 5.80 to 5.84 are identical to Problems 6.107 to 6.111.

For R_{sig}= 100 kΩ determine the overall voltage gain, G_v, with and without R_f present.

5.85 Calculate the overall voltage gain of a CS amplifier fed with a 1-MΩ source and connected to a 20-kΩ load. The MOSFET has g_m = 2 mA/V and r_o = 50 kΩ, and a drain resistance R_D = 10 kΩ is utilized.

5.86 A CS amplifier utilizes a MOSFET with $\mu_n C_{ox}$ = 400 μA/V², W/L = 10, and V_A = 10 V. It is biased at I_D = 0.2 mA and uses R_D = 6 kΩ. Find R_{in}, A_{vo}, and R_o. Also, if a load resistance of 10 kΩ is connected to the output, what overall voltage gain G_v is realized? Now, if a 0.2-V peak sine-wave signal is required at the output, what must the peak amplitude of v_{sig} be?

5.87 A common-source amplifier utilizes a MOSFET for which V_A = 12.5 V and is operated at V_{OV} = 0.25 V. What is the value of its $(g_m r_o)$? The amplifier feeds a load resistance R_L = 15 kΩ. The designer selects $R_D = 2R_L$. If it is required to realize an overall voltage gain G_v of –10 V/V what g_m is needed? Also specify the bias current I_D. If, to increase the output signal swing, R_D is reduced to $R_D = R_L$, what does G_v become?

5.88 Two identical CS amplifiers are connected is cascade. The first stage is fed with a source v_{sig} having a resistance R_{sig} = 100 kΩ. A load resistance R_L = 10 kΩ is connected to the drain of the second stage. Each MOSFET is biased at I_D = 0.25 mA and operates with V_{OV} = 0.25 V. Assume V_A is very large. Each stage utilizes a drain resistance R_D = 10 kΩ.

(a) Sketch the equivalent circuit of the two-stage amplifier.
(b) Calculate the overall voltage gain G_v.

5.89 In discrete-circuit amplifiers, $(R_D \| R_L)$ is usually much smaller than r_o, and thus r_o can be neglected in determining the voltage gain of the CS amplifier. Nevertheless, it is useful to note that r_o poses an absolute upper limit on the voltage gain of a CS amplifier. Find this upper limit by letting $R_D \| R_L = \infty$. Express the maximum achievable gain in terms of V_A and V_{OV}.

5.90 A MOSFET connected in the CS configuration has a transconductance g_m = 5 mA/V. When a resistance R_s is connected in the source lead, the effective transconductance is reduced to 1 mA/V. What do you estimate the value of R_s to be?

5.91 A CS amplifier using an NMOS transistor with g_m = 4 mA/V is found to have an overall voltage gain of –16 V/V. What value should a resistance R_s inserted in the source lead have to reduce the overall voltage gain to –8 V/V?

5.92 The overall voltage gain of a CS amplifier with a resistance R_s = 1 kΩ in the source lead was measured and found to be –15 V/V. When R_s is shorted, but the circuit operation remained linear, the gain doubled. What must g_m be? What value of R_s is needed to obtain an overall voltage gain of –10 V/V?

5.93 A CG amplifier using an NMOS transistor for which g_m = 4 mA/V has a 5-kΩ drain resistance R_D and a 5-kΩ load resistance R_L. The amplifier is driven by a voltage source having a 500 Ω resistance. What is the input resistance of the amplifier? What is the overall voltage gain G_v? By what factor must the bias current I_D of the MOSFET be changed so that R_{in} matches R_{sig}?

5.94 A CG amplifier when fed with a signal source having R_{sig} = 200 Ω is found to have an overall voltage gain of 10 V/V. When a 200-Ω resistance is added in series with the signal generator the overall voltage gain decreased to 8 V/V. What must g_m of the MOSFET be? If the MOSFET is biased at I_D = 0.2 mA, at what overdrive voltage it must be operating?

D 5.95 A source follower is required to connect a high-resistance source to a load whose resistance is nominally 2 kΩ but can be as low as 1 kΩ and as high as 3 kΩ. What is the maximum output resistance that the source follower must have if the output voltage is to remain within ±20% of nominal value? If the MOSFET has k_n = 16 mA/V², at

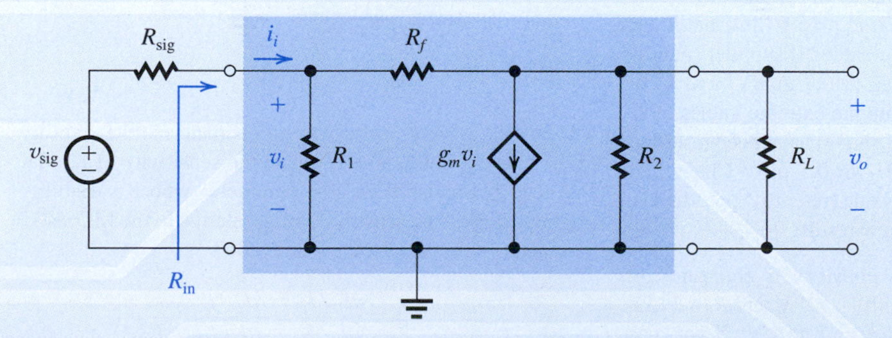

Figure P5.84

what current I_D must it be biased? At what overdrive voltage is the MOSFET operating?

***5.96** Refer to the source-follower equivalent circuit shown in Fig. 5.50(b). Show that

$$G_v \equiv \frac{v_o}{v_{sig}} = \frac{R_L \| r_o}{(R_L \| r_o) + \dfrac{1}{g_m}}$$

Now, with R_L removed, the voltage gain is carefully measured and found to be 0.98. Then, when R_L is connected and its value is varied, it is found that the gain is halved at R_L = 500 Ω. If the amplifier remained linear throughout this measurement, what must the values of g_m and r_o be?

D 5.97 A source follower is required to deliver a 0.5-V peak sinusoid to 2-kΩ load. If the peak amplitude of v_{gs} is to be limited to 50 mV, what is the lowest value of I_D at which the MOSFET can be biased? At this bias current, what are the maximum and minimum currents that the MOSFET will be conducting (at the positive and negative peaks of the output sine wave)? What must the peak amplitude of v_{sig} be?

Section 5.7: Biasing in MOS Amplifier Circuits

D 5.98 Consider the classical biasing scheme shown in Fig. 5.52(c), using a 9-V supply. For the MOSFET, V_t = 1 V, λ = 0, and k_n = 2 mA/V^2. Arrange that the drain current is 1 mA, with about one-third of the supply voltage across each of R_S and R_D. Use 22 MΩ for the larger of R_{G1} and R_{G2}. What are the values of R_{G1}, R_{G2}, R_S, and R_D that you have chosen? Specify them to two significant digits. For your design, how far is the drain voltage from the edge of saturation?

D 5.99 Using the circuit topology displayed in Fig. 5.52(e), arrange to bias the NMOS transistor at I_D = 1 mA with V_D midway between cutoff and the beginning of triode operation. The available supplies are ±5 V. For the NMOS transistor, V_t = 1.0 V, λ = 0, and k_n = 2 mA/V^2. Use a gate-bias resistor of 10 MΩ. Specify R_S and R_D to two significant digits.

D *5.100 In an electronic instrument using the biasing scheme shown in Fig. 5.52(c), a manufacturing error reduces R_S to zero. Let V_{DD} = 12 V, R_{G1} = 5.6 MΩ, and R_{G2} = 2.2 MΩ. What is the value of V_G created? If supplier specifications allow k_n to vary from 0.2 to 0.3 mA/V^2 and V_t to vary from 1.0 V to 1.5 V, what are the extreme values of I_D that may result? What value of R_S should have been installed to limit the maximum value of I_D to 0.5 mA? Choose an appropriate standard 5% resistor value (refer to Appendix G). What extreme values of current now result?

5.101 An enhancement NMOS transistor is connected in the bias circuit of Fig. 5.52(c), with V_G = 4 V and R_S = 2 kΩ. The transistor has V_t = 1 V and k_n = 2 mA/V^2. What bias current results? If a transistor for which k_n is 50% higher is used, what is the resulting percentage increase in I_D?

SIM **5.102** The bias circuit of Fig. 5.52(c) is used in a design with V_G = 5 V and R_S = 2 kΩ. For an enhancement MOSFET with k_n = 2 mA/V^2, the source voltage was measured and found to be 2 V. What must V_t be for this device? If a device for which V_t is 0.5 V less is used, what does V_S become? What bias current results?

D 5.103 Design the circuit of Fig. 5.52(e) for an enhancement MOSFET having V_t = 1 V and k_n = 2 mA/V^2. Let V_{DD} = V_{SS} = 5 V. Design for a dc bias current of 1 mA and for the largest possible voltage gain (and thus the largest possible R_D) consistent with allowing a 2-V peak-to-peak voltage swing at the drain. Assume that the signal voltage on the source terminal of the FET is zero.

SIM **D 5.104** Design the circuit in Fig. P5.104 so that the transistor operates in saturation with V_D biased 1 V from the edge of the triode region, with I_D = 1 mA and V_D = 3 V, for each of the following two devices (use a 10-μA current in the voltage divider):

(a) $|V_t|$ = 1 V and $k_p'W/L$ = 0.5 mA/V^2
(b) $|V_t|$ = 2 V and $k_p'W/L$ = 1.25 mA/V^2

For each case, specify the values of V_G, V_D, V_S, R_1, R_2, R_S, and R_D.

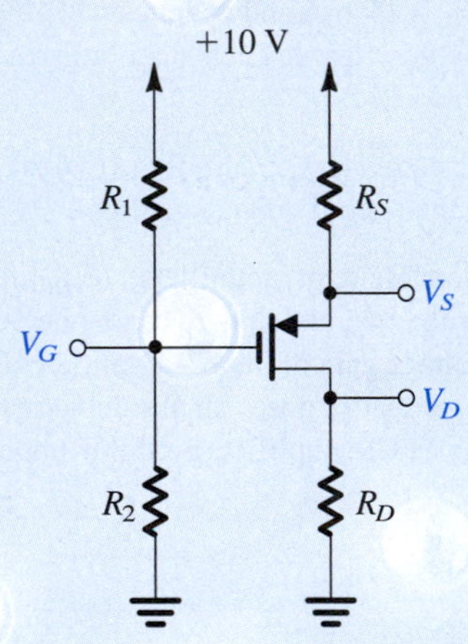

Figure P5.104

****D 5.105** A very useful way to characterize the stability of the bias current I_D is to evaluate the sensitivity of I_D relative to a particular transistor parameter whose variability might be large. The sensitivity of I_D relative to the MOSFET parameter $K \equiv \frac{1}{2}k'(W/L)$ is defined as

$$S_K^{I_D} \equiv \frac{\partial I_D / I_D}{\partial K / K} = \frac{\partial I_D}{\partial K}\frac{K}{I_D}$$

and its value, when multiplied by the variability (or tolerance) of K, provides the corresponding expected variability of I_D. The purpose of this problem is to investigate the use of the sensitivity function in the design of the bias circuit of Fig. 5.52(e).

(a) Show that for V_t constant,

$$S_K^{I_D} = 1/(1 + 2\sqrt{KI_D}R_S)$$

(b) For a MOSFET having $K = 100\ \mu A/V^2$ with a variability of ±10% and $V_t = 1$ V, find the value of R_S that would result in $I_D = 100\ \mu A$ with a variability of ±1%. Also, find V_{GS} and the required value of V_{SS}.
(c) If the available supply $V_{SS} = 5$ V, find the value of R_S for $I_D = 100\ \mu A$. Evaluate the sensitivity function, and give the expected variability of I_D in this case.

5.106 For the circuit in Fig. 5.55(a) with $I = 0.2$ mA, $R_G = 0$, $R_D = 10\ k\Omega$, and $V_{DD} = 2.5$ V, consider the behavior in each of the following two cases. In each case, find the voltages V_S, V_D, and V_{DS} that result.

(a) $V_t = 1$ V and $k_n = 1.6\ mA/V^2$
(b) $V_t = 0.8$ V and $k_n = 1.25\ mA/V^2$

SIM **5.107** In the circuit of Fig. 5.54, let $R_G = 10\ M\Omega$, $R_D = 10\ k\Omega$, and $V_{DD} = 10$ V. For each of the following two transistors, find the voltages V_D and V_G.

(a) $V_t = 1$ V and $k_n = 0.5\ mA/V^2$
(b) $V_t = 2$ V and $k_n = 1.25\ mA/V^2$

D 5.108 Using the feedback bias arrangement shown in Fig. 5.54 with a 5-V supply and an NMOS device for which $V_t = 1$ V and $k_n = 0.6\ mA/V^2$, find R_D to establish a drain current of 0.2 mA. If resistor values are limited to those on the 5% resistor scale (see Appendix G), what value would you choose? What values of current and V_D result?

D 5.109 Figure P5.109 shows a variation of the feedback-bias circuit of Fig. 5.54. Using a 5-V supply with an NMOS transistor for which $V_t = 1$ V, $k_n = 6.25\ mA/V^2$ and $\lambda = 0$, provide a design that biases the transistor at $I_D = 2$ mA, with V_{DS} large enough to allow saturation operation for a 2-V negative signal swing at the drain. Use 22 MΩ as the largest resistor in the feedback-bias network. What values of R_D, R_{G1}, and R_{G2} have you chosen? Specify all resistors to two significant digits.

Section 5.8: Discrete-Circuit MOS Amplifiers

5.110 Calculate the overall voltage gain G_v of a common-source amplifier for which $g_m = 2$ mA/V, $r_o = 50\ k\Omega$, $R_D = 10\ k\Omega$, and $R_G = 10\ M\Omega$. The amplifier is fed from a signal source with a Thévenin resistance of 0.5 MΩ, and the amplifier output is coupled to a load resistance of 20 kΩ.

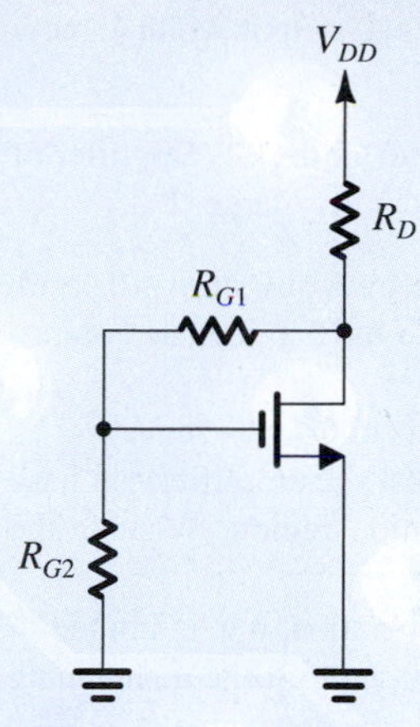

Figure P5.109

D 5.111 This problem investigates a redesign of the common-source amplifier of Exercise 5.38 whose bias design was done in Exercise 5.37 and shown in Fig. E5.37. Please refer to these two exercises.

(a) The open-circuit voltage gain of the CS amplifier can be written as

$$A_{vo} = -\frac{2(V_{DD} - V_D)}{V_{OV}}$$

Verify that this expression yields the results in Exercise 5.38 (i.e., $A_{vo} = -15$ V/V).
(b) A_{vo} can be doubled by reducing V_{OV} by a factor of 2, (i.e., from 1 V to 0.5 V) while V_D is kept unchanged. What corresponding values for I_D, R_D, g_m, and r_o apply?
(c) Find A_{vo} and R_o with r_o taken into account.
(d) For the same value of signal-generator resistance $R_{sig} = 100\ k\Omega$, the same value of gate-bias resistance $R_G = 4.8\ M\Omega$, and the same value of load resistance $R_L = 15\ k\Omega$, evaluate the new value of overall voltage gain G_v with r_o taken into account.
(e) Compare your results to those obtained in Exercises 5.37 and 5.38, and comment.

SIM **5.112** The NMOS transistor in the CS amplifier shown in Fig. P5.112 has $V_t = 0.7$ V and $V_A = 50$ V.

(a) Neglecting the Early effect, verify that the MOSFET is operating in saturation with $I_D = 0.5$ mA and $V_{OV} = 0.3$ V. What must the MOSFET's k_n be? What is the dc voltage at the drain?
(b) Find R_{in} and G_v.
(c) If v_{sig} is a sinusoid with a peak amplitude $\hat{v}_{sig}$, find the maximum allowable value of $\hat{v}_{sig}$ for which the transistor remains in saturation. What is the corresponding amplitude of the output voltage?
(d) What is the value of resistance R_s that needs to be inserted in series with capacitor C_S in order to allow us to

double the input signal $\hat{v}_{sig}$? What output voltage now results?

SIM **D *5.113** The PMOS transistor in the CS amplifier of Fig. P5.113 has $V_{tp} = -0.7$ V and a very large $|V_A|$.

(a) Select a value for R_S to bias the transistor at $I_D = 0.3$ mA and $|V_{OV}| = 0.3$ V. Assume v_{sig} to have a zero dc component.
(b) Select a value for R_D that results in $G_v = -10$ V/V.
(c) Find the largest sinusoid $\hat{v}_{sig}$ that the amplifier can handle while remaining in the saturation region. What is the corresponding signal at the output?
(d) If to obtain reasonably linear operation, $\hat{v}_{sig}$ is limited to 50 mV, what value can R_D be increased to while maintaining saturation-region operation? What is the new value of G_v?

5.114 Figure P5.114 shows a scheme for coupling and amplifying a high-frequency pulse signal. The circuit utilizes two MOSFETs whose bias details are not shown and a 50-Ω coaxial cable. Transistor Q_1 operates as a CS amplifier and Q_2 as a CG amplifier. For proper operation, transistor Q_2 is required to present a 50-Ω resistance to the cable. This situation is known as "proper termination" of the cable and ensures that there will be no signal reflection coming back on the cable. When the cable is properly terminated, its input resistance is 50 Ω. What must g_{m2} be? If Q_1 is biased at the same point as Q_2, what is the amplitude of the current pulses in the drain of Q_1? What is the amplitude of the voltage pulses at the drain of Q_1? What value of R_D is required to provide 1-V pulses at the drain of Q_2?

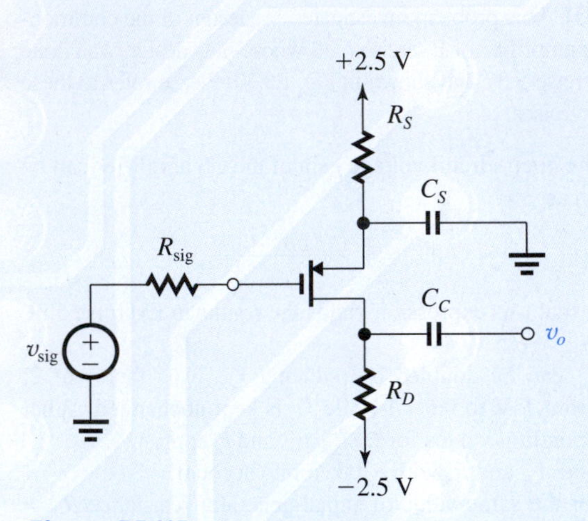

Figure P5.113

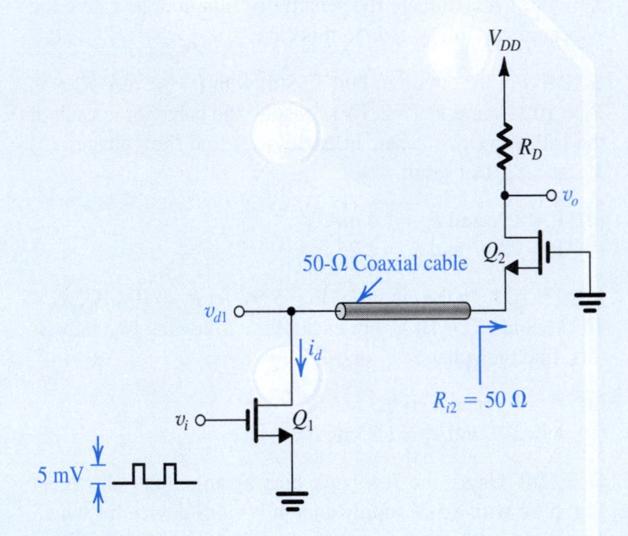

Figure P5.114

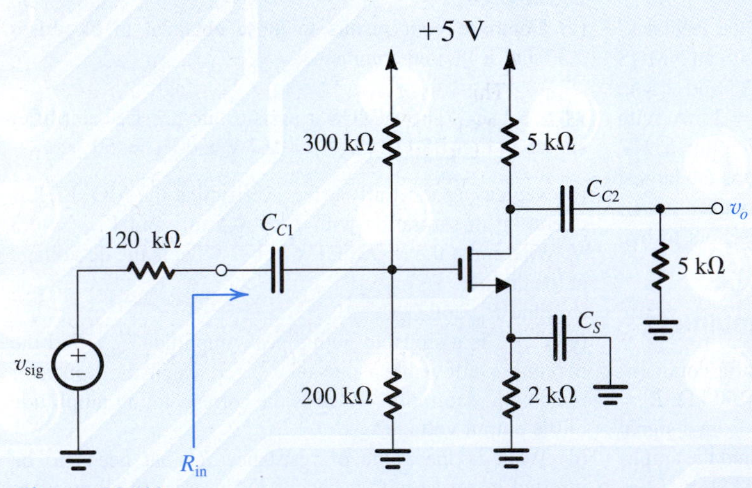

Figure P5.112

D *5.115 The MOSFET in the circuit of Fig. P5.115 has $V_t = 1$ V, $k_n' = 0.8$ mA/V^2, and $V_A = 40$ V.

(a) Find the values of R_S, R_D, and R_G so that $I_D = 0.1$ mA, the largest possible value for R_D is used while a maximum signal swing at the drain of ±1 V is possible, and the input resistance at the gate is 10 MΩ. Neglect the Early effect.
(b) Find the values of g_m and r_o at the bias point.
(c) If terminal Z is grounded, terminal X is connected to a signal source having a resistance of 1 MΩ, and terminal Y is connected to a load resistance of 40 kΩ, find the voltage gain from signal source to load.
(d) If terminal Y is grounded, find the voltage gain from X to Z with Z open-circuited. What is the output resistance of the source follower?
(e) If terminal X is grounded and terminal Z is connected to a current source delivering a signal current of 10 μA and having a resistance of 100 kΩ, find the voltage signal that can be measured at Y. For simplicity, neglect the effect of r_o.

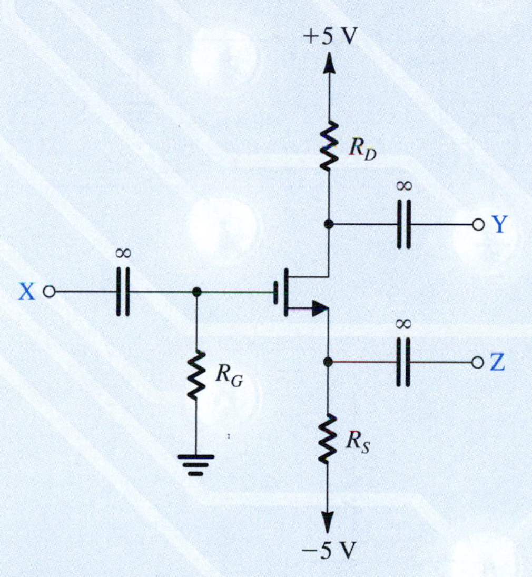

Figure P5.115

***5.116** (a) The NMOS transistor in the source-follower circuit of Fig. P5.116(a) has $g_m = 5$ mA/V and a large r_o. Find the open-circuit voltage gain and the output resistance.
(b) The NMOS transistor in the common-gate amplifier of Fig. P5.116(b) has $g_m = 5$ mA/V and a large r_o. Find the input resistance and the voltage gain.
(c) If the output of the source follower in (a) is connected to the input of the common-gate amplifier in (b), use the results of (a) and (b) to obtain the overall voltage gain v_o/v_i.

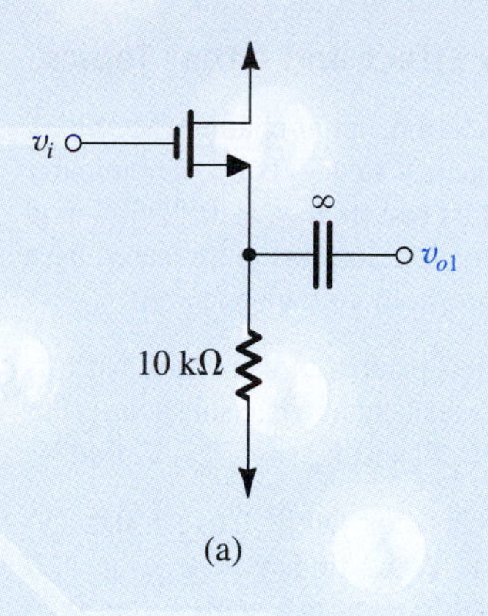

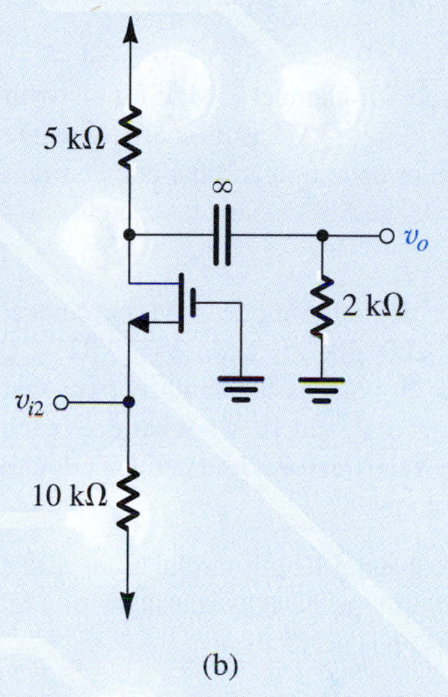

Figure P5.116

***5.117** In this problem we investigate the large-signal operation of the source follower of Fig. 5.60(a). Specifically, consider the situation when negative input signals are applied. Let the negative signal voltage at the output be $-V$. The current in R_L will flow away from ground and will have a value of V/R_L. This current will subtract from the bias current I, resulting in a transistor current of $(I - V/R_L)$. One can use this current value to determine v_{GS}. Now, the signal at the transistor source terminal will be $-V$, superimposed on the dc voltage, which is $-V_{GS}$ (corresponding to a drain current of I). We can thus find the signal voltage at the gate v_i. For the circuit analyzed in Exercise 5.41, find v_i for $v_o = -1$ V, −5 V, −6 V, and −7 V. At each point, find the voltage gain v_o/v_i and compare to the small-signal value found in Exercise 5.41. What is the largest possible negative-output signal?

Section 5.9: The Body Effect and Other Topics

5.118 In a particular application, an n-channel MOSFET operates with V_{SB} in the range 0 V to 4 V. If V_{t0} is nominally 1.0 V, find the range of V_t that results if $\gamma = 0.5\ \text{V}^{1/2}$ and $2\phi_f = 0.6$ V. If the gate oxide thickness is increased by a factor of 4, what does the threshold voltage become?

5.119 A p-channel transistor operates in saturation with its source voltage 3 V lower than its substrate. For $\gamma = 0.5\ \text{V}^{1/2}$, $2\phi_f = 0.75$ V, and $V_{t0} = -0.7$ V, find V_t.

5.120 For an NMOS transistor with $2\phi_f = 0.6$ V, $\gamma = 0.5\ \text{V}^{1/2}$,, and $V_{SB} = 4$ V, find $\chi = g_{mb}/g_m$. If the transistor is biased at $I_D = 0.5$ mA with $V_{OV} = 0.25$ V, find g_m and g_{mb}.

5.121 A depletion-type n-channel MOSFET with $k_n'W/L = 2\ \text{mA/V}^2$ and $V_t = -3$ V has its source and gate grounded. Find the region of operation and the drain current for $v_D = 0.1$ V, 1 V, 3 V, and 5 V. Neglect the channel-length-modulation effect.

5.122 For a particular depletion-mode NMOS device, $V_t = -2$ V, $k_n'W/L = 200\ \mu\text{A/V}^2$, and $\lambda = 0.02\ \text{V}^{-1}$. When operated at $v_{GS} = 0$, what is the drain current that flows for $v_{DS} = 1$ V, 2 V, 3 V, and 10 V? What does each of these currents become if the device width is doubled with L the same? With L also doubled?

5.123 Neglecting the channel-length-modulation effect show that for the depletion-type NMOS transistor of Fig. P5.123, the $i - v$ relationship is given by

$$i = \frac{1}{2}k_n'(W/L)(v^2 - 2V_t v), \quad \text{for } v \geq V_t$$
$$i = -\frac{1}{2}k_n'(W/L)V_t^2 \quad \text{for } v \leq V_t$$

(Recall that V_t is negative). Sketch the $i - v$ relationship for the case: $V_t = -2$ V and $k_n'(W/L) = 2\ \text{mA/V}^2$.

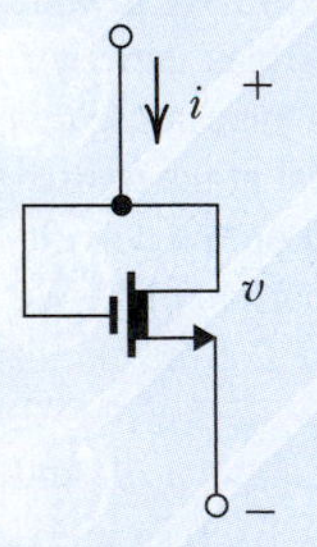

Figure P5.123

General Problems

****5.124** The circuits shown in Fig. P5.124 employ negative feedback, a subject we shall study in detail in Chapter 10. Assume that each transistor is sized and biased so that $g_m = 1$ mA/V and $r_o = 100$ kΩ. Otherwise, ignore all dc biasing detail and concentrate on small-signal operation resulting in response to the input signal v_{sig}. For $R_L = 10$ kΩ, $R_1 = 500$ kΩ, and $R_2 = 1$ MΩ, find the overall voltage gain v_o/v_{sig} and the input resistance R_{in} for each circuit. Neglect the body effect. Do these circuits remind you of op-amp circuits? Comment.

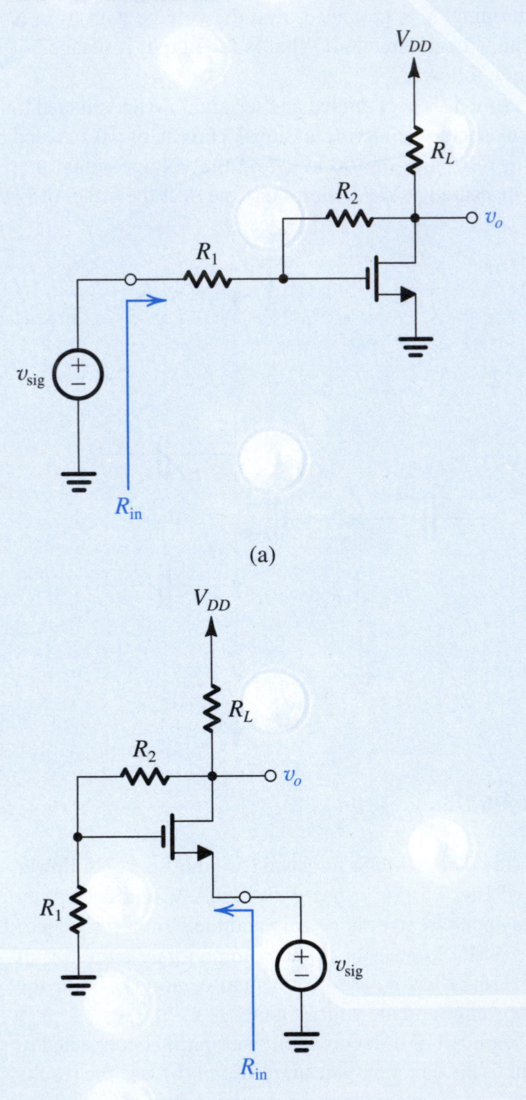

Figure P5.124

5.125 For the two circuits in Problem 5.124 (shown in Fig. P5.124), we wish to consider their dc bias design. Since v_{sig} has a zero dc component, we short-circuit its generator. For NMOS transistors with $V_t = 0.6$ V, find V_{OV}, $k_n'(W/L)$, and V_A to bias each device at $I_D = 0.1$ mA and to obtain the values of g_m and r_o specified in Problem 5.124: namely, $g_m = 1$ mA/V and $r_o = 100$ kΩ. For $R_1 = 0.5$ MΩ, $R_2 = 1$ MΩ, and $R_L = 10$ kΩ, find the required value of V_{DD}.

****5.126** In the amplifier shown in Fig. P5.126, transistors having $V_t = 0.6$ V and $V_A = 20$ V are operated at $V_{GS} = 0.8$ V using the appropriate choice of W/L ratio. In a particular application, Q_1 is to be sized to operate at 10 μA, while Q_2 is intended to operate at 1 mA. For $R_L = 2$ kΩ, the (R_1, R_2) network sized to consume only 1% of the current in R_L, v_{sig}, having zero dc component, and $I_1 = 10$ μA, find the values of R_1 and R_2 that satisfy all the requirements. (*Hint*: V_O must be +2 V.) What is the voltage gain v_o/v_i? Using a result from a theorem known as Miller's theorem (Chapter 9), find the input resistance R_{in} as $R_2/(1 - v_o/v_i)$. Now, calculate the value of the overall voltage gain v_o/v_{sig}. Does this result remind you of the inverting configuration of the op amp? Comment. How would you modify the circuit at the input by using an additional resistor and a very large capacitor to raise the gain v_o/v_{sig} to −5 V/V? Neglect the body effect.

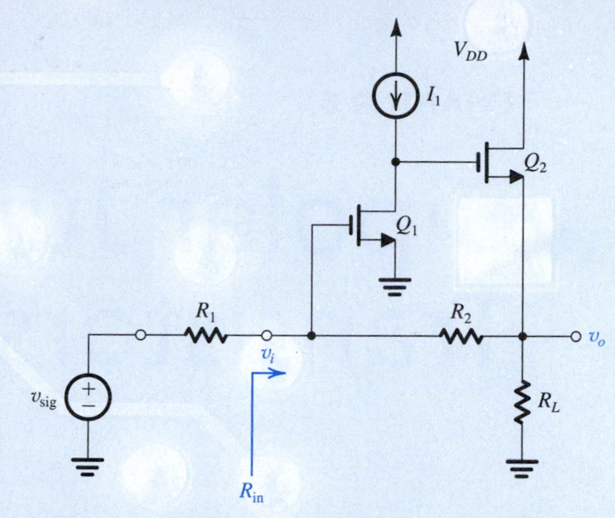

Figure P5.126

5.127 Consider the bias design of the circuit of Problem 5.126 (shown in Fig. P5.126). For $k_n' = 200$ μA/V² and $V_{DD} = 3.3$ V, find $(W/L)_1$ and $(W/L)_2$ to obtain the operating conditions specified in Problem 5.126.

CHAPTER 6

Bipolar Junction Transistors (BJTs)

IN THIS CHAPTER YOU WILL LEARN

1. The physical structure of the bipolar transistor and how it works.
2. How the voltage between two terminals of the transistor controls the current that flows through the third terminal, and the equations that describe these current-voltage characteristics.
3. How to analyze and design circuits that contain bipolar transistors, resistors, and dc sources.
4. How the transistor can be used to make an amplifier.
5. How to obtain linear amplification from the fundamentally nonlinear BJT.
6. The three basic ways for connecting a BJT to be able to construct amplifiers with different properties.
7. Practical circuits for bipolar-transistor amplifiers that can be constructed by using discrete components.

Introduction

In this chapter, we study the other major three-terminal device: the bipolar junction transistor (BJT). The presentation of the material in this chapter parallels but does not rely on that for the MOSFET in Chapter 5; thus, if desired, the BJT can be studied before the MOSFET.

Three-terminal devices are far more useful than two-terminal ones, such as the diodes studied in Chapter 4, because they can be used in a multitude of applications, ranging from signal amplification to the design of digital logic and memory circuits. The basic principle involved is the use of the voltage between two terminals to control the current flowing in the third terminal. In this way, a three-terminal device can be used to realize a controlled source, which as we learned in Chapter 1 is the basis for amplifier design. Also, in the extreme, the control signal can be used to cause the current in the third terminal to change from zero to a large value, thus allowing the device to act as a switch. The switch is the basis for the realization of the logic inverter, the basic element of digital circuits.

The invention of the BJT in 1948 at the Bell Telephone Laboratories ushered in the era of solid-state circuits, which led to electronics changing the way we work, play, and indeed, live. The invention of the BJT also eventually led to the dominance of information technology and the emergence of the knowledge-based economy.

The bipolar transistor enjoyed nearly three decades as the device of choice in the design of both discrete and integrated circuits. Although the MOSFET had been known very early on, it was not until the 1970s and 1980s that it became a serious competitor to the BJT. By 2009, the MOSFET was undoubtedly the most widely used electronic device, and CMOS technology the technology of choice in the design of integrated circuits. Nevertheless, the BJT remains a significant device that excels in certain applications. For instance, the reliability of BJT circuits under severe environmental conditions makes them the dominant device in certain automotive applications.

The BJT remains popular in discrete-circuit design, in which a very wide selection of BJT types are available to the designer. Here we should mention that the characteristics of the bipolar transistor are so well understood that one is able to design transistor circuits whose performance is remarkably predictable and quite insensitive to variations in device parameters.

The BJT is still the preferred device in very demanding analog circuit applications, both integrated and discrete. This is especially true in very-high-frequency applications, such as radio-frequency (RF) circuits for wireless systems. A very-high-speed digital logic-circuit family based on bipolar transistors, namely, emitter-coupled logic, is still in use. Finally, bipolar transistors can be combined with MOSFETs to create innovative circuits that take advantage of the high-input-impedance and low-power operation of MOSFETs and the very-high-frequency operation and high-current-driving capability of bipolar transistors. The resulting technology is known as BiCMOS, and it is finding increasingly larger areas of application (see Chapters 7, 8, 12, and 14).

In this chapter, we shall start with a description of the physical operation of the BJT. Though simple, this physical description provides considerable insight regarding the performance of the transistor as a circuit element. We then quickly move from describing current flow in terms of electrons and holes to a study of the transistor terminal characteristics. Circuit models for transistor operation in different modes will be developed and utilized in the analysis and design of transistor circuits. The main objective of this chapter is to develop in the reader a high degree of familiarity with the BJT. Thus, by the end of the chapter, the reader should be able to perform rapid first-order analysis of transistor circuits and to design single-stage transistor amplifiers.

6.1 Device Structure and Physical Operation

6.1.1 Simplified Structure and Modes of Operation

Figure 6.1 shows a simplified structure for the BJT. A practical transistor structure will be shown later (see also Appendix A, which deals with fabrication technology).

As shown in Fig. 6.1, the BJT consists of three semiconductor regions: the emitter region (*n* type), the base region (*p* type), and the collector region (*n* type). Such a transistor is called an *npn* transistor. Another transistor, a dual of the *npn* as shown in Fig. 6.2, has a *p*-type emitter, an *n*-type base, and a *p*-type collector, and is appropriately called a *pnp* transistor.

A terminal is connected to each of the three semiconductor regions of the transistor, with the terminals labeled **emitter** (E), **base** (B), and **collector** (C).

The transistor consists of two *pn* junctions, the **emitter–base junction** (EBJ) and the **collector–base junction** (CBJ). Depending on the bias condition (forward or reverse) of each of these junctions, different modes of operation of the BJT are obtained, as shown in Table 6.1. The **active mode** is the one used if the transistor is to operate as an amplifier. Switching applications (e.g., logic circuits) utilize both the **cutoff mode** and the **saturation mode**. As the name implies, in the cutoff mode no current flows because both junctions are reverse biased.

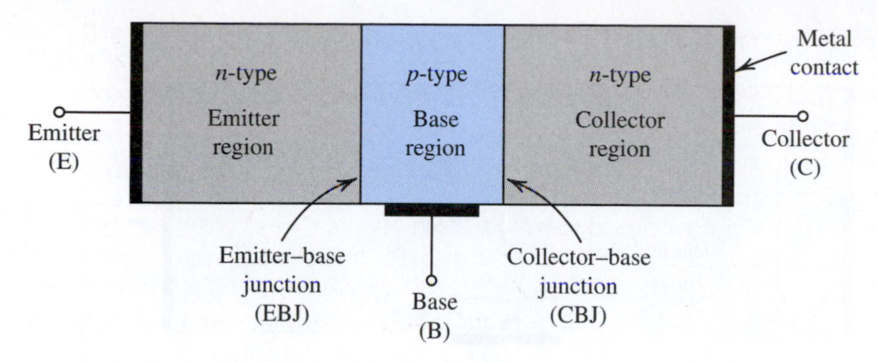

Figure 6.1 A simplified structure of the *npn* transistor.

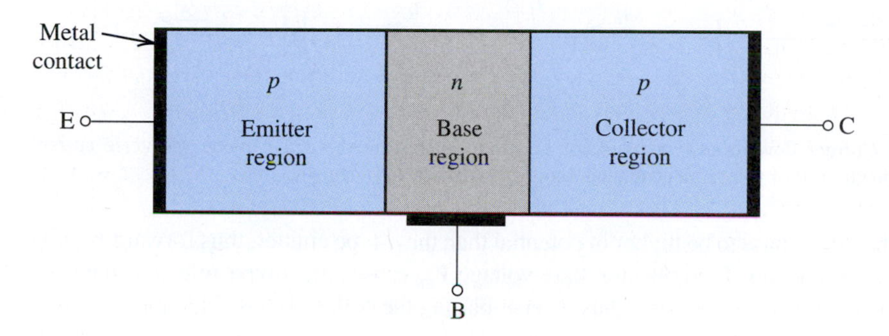

Figure 6.2 A simplified structure of the *pnp* transistor.

Table 6.1 BJT Modes of Operation

Mode	EBJ	CBJ
Cutoff	Reverse	Reverse
Active	Forward	Reverse
Saturation	Forward	Forward

As we will see shortly, charge carriers of both polarities—that is, electrons and holes—participate in the current-conduction process in a bipolar transistor, which is the reason for the name *bipolar*.[1]

6.1.2 Operation of the *npn* Transistor in the Active Mode

Of the three modes of operation of the BJT, the active mode is the most important. Therefore, we begin our study of the BJT by considering its physical operation in the active mode.[2] This situation is illustrated in Fig. 6.3 for the *npn* transistor. Two external voltage sources (shown as batteries) are used to establish the required bias conditions for active-mode operation. The voltage

[1]This should be contrasted with the situation in the MOSFET, where current is conducted by charge carriers of one type only; electrons in n-channel devices or holes in p-channel devices. In earlier days, some referred to FETs as unipolar devices.

[2]The material in this section assumes that the reader is familiar with the operation of the *pn* junction under forward-bias conditions (Section 3.5).

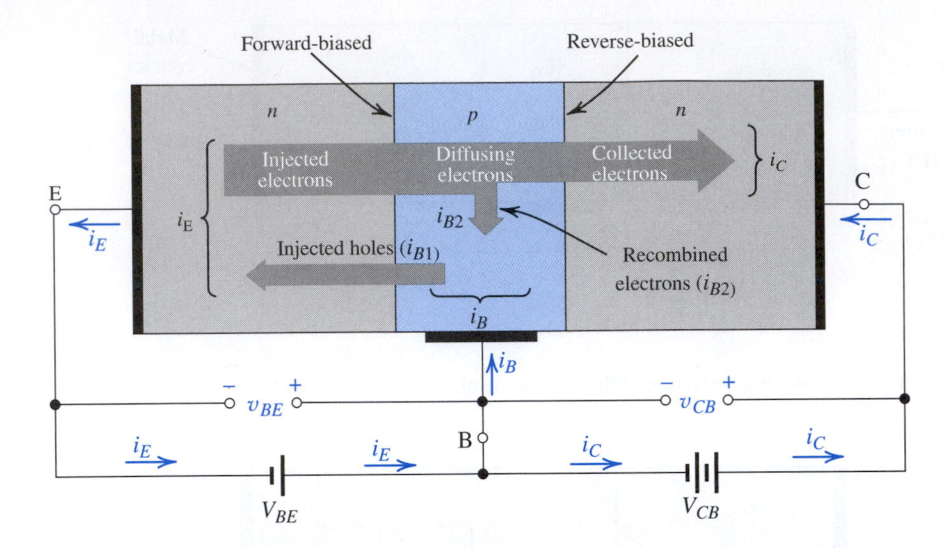

Figure 6.3 Current flow in an *npn* transistor biased to operate in the active mode. (Reverse current components due to drift of thermally generated minority carriers are not shown.)

V_{BE} causes the *p*-type base to be higher in potential than the *n*-type emitter, thus forward-biasing the emitter–base junction. The collector–base voltage V_{CB} causes the *n*-type collector to be at a higher potential than the *p*-type base, thus reverse-biasing the collector–base junction.

Current Flow The forward bias on the emitter–base junction will cause current to flow across this junction. Current will consist of two components: electrons injected from the emitter into the base, and holes injected from the base into the emitter. As will become apparent shortly, it is highly desirable to have the first component (electrons from emitter to base) at a much higher level than the second component (holes from base to emitter). This can be accomplished by fabricating the device with a heavily doped emitter and a lightly doped base; that is, the device is designed to have a high density of electrons in the emitter and a low density of holes in the base.

The current that flows across the emitter–base junction will constitute the emitter current i_E, as indicated in Fig. 6.3. The direction of i_E is "out of" the emitter lead, which, following the usual conventions, is in the direction of the positive-charge flow (hole current) and opposite to the direction of the negative-charge flow (electron current), with the emitter current i_E being equal to the sum of these two components. However, since the electron component is much larger than the hole component, the emitter current will be dominated by the electron component.

Let us now consider the electrons injected from the emitter into the base. These electrons will be **minority carriers** in the *p*-type base region. Because the base is usually very thin, in the steady state the excess minority carrier (electron) concentration in the base will have an almost-straight-line profile, as indicated by the solid straight line in Fig. 6.4. The electron concentration will be highest [denoted by $n_p(0)$] at the emitter side and lowest (zero) at the collector side.[3] As in the case of any forward-biased *pn* junction (Section 3.5), the concentration $n_p(0)$ will be proportional to e^{v_{BE}/V_T},

[3]This minority carrier distribution in the base results from the boundary conditions imposed by the two junctions. It is not an exponentially decaying distribution, which would result if the base region were infinitely thick. Rather, the thin base causes the distribution to decay linearly. Furthermore, the reverse bias on the collector–base junction causes the electron concentration at the collector side of the base to be zero.

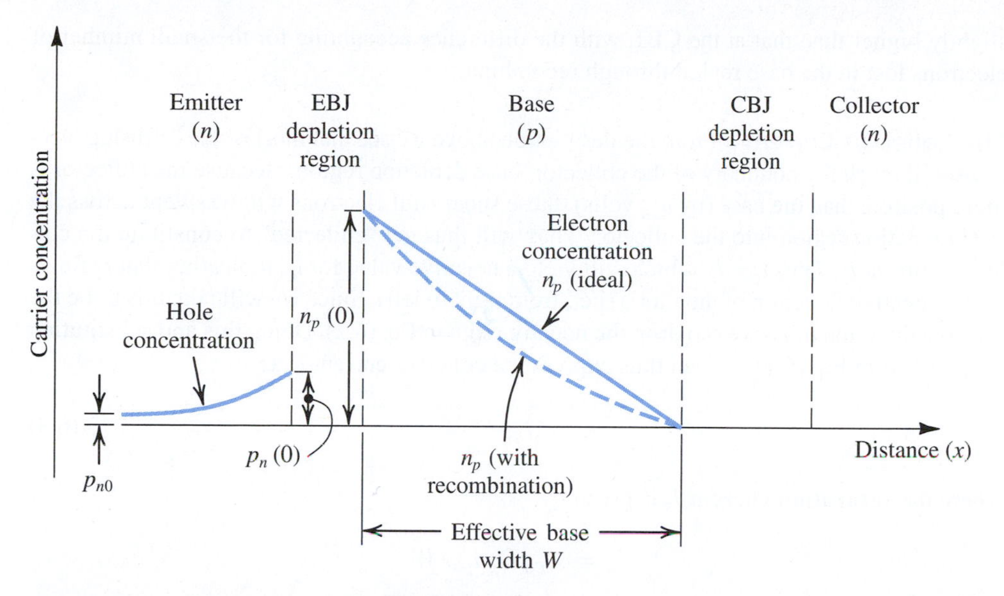

Figure 6.4 Profiles of minority-carrier concentrations in the base and in the emitter of an *npn* transistor operating in the active mode: $v_{BE} > 0$ and $v_{CB} \geq 0$.

$$n_p(0) = n_{p0}e^{v_{BE}/V_T} \tag{6.1}$$

where n_{p0} is the thermal-equilibrium value of the minority carrier (electron) concentration in the base region, v_{BE} is the forward base–emitter bias voltage, and V_T is the thermal voltage, which is equal to approximately 25 mV at room temperature. The reason for the zero concentration at the collector side of the base is that the positive collector voltage v_{CB} causes the electrons at that end to be swept across the CBJ depletion region.

The tapered minority-carrier concentration profile (Fig. 6.4) causes the electrons injected into the base to diffuse through the base region toward the collector. This electron diffusion current I_n is directly proportional to the slope of the straight-line concentration profile,

$$\begin{aligned} I_n &= A_E q D_n \frac{dn_p(x)}{dx} \\ &= A_E q D_n \left(-\frac{n_p(0)}{W}\right) \end{aligned} \tag{6.2}$$

where A_E is the cross-sectional area of the base–emitter junction (in the direction perpendicular to the page), q is the magnitude of the electron charge, D_n is the electron diffusivity in the base, and W is the effective width of the base. Observe that the negative slope of the minority carrier concentration results in a negative current I_n across the base; that is, I_n flows from right to left (in the negative direction of x), which corresponds to the usual convention, namely, opposite to the direction of electron flow.

Some of the electrons that are diffusing through the base region will combine with holes, which are the majority carriers in the base. However, since the base is usually very thin and lightly doped, the proportion of electrons "lost" through this **recombination process** will be quite small. Nevertheless, the recombination in the base region causes the excess minority carrier concentration profile to deviate from a straight line and take the slightly concave shape indicated by the broken line in Fig. 6.4. The slope of the concentration profile at the EBJ is slightly higher than that at the CBJ, with the difference accounting for the small number of electrons lost in the base region through recombination.

The Collector Current From the description above we see that most of the diffusing electrons will reach the boundary of the collector–base depletion region. Because the collector is more positive than the base (by v_{CB} volts), these successful electrons will be swept across the CBJ depletion region into the collector. They will thus get "collected" to constitute the collector current i_C. Thus $i_C = I_n$, which will yield a negative value for i_C, indicating that i_C flows in the negative direction of the x axis (i.e., from right to left). Since we will take this to be the positive direction of i_C, we can drop the negative sign in Eq. (6.2). Doing this and substituting for $n_p(0)$ from Eq. (6.1), we can thus express the collector current i_C as

$$i_C = I_S e^{v_{BE}/V_T} \tag{6.3}$$

where the **saturation current** I_S is given by

$$I_S = A_E q D_n n_{p0}/W$$

Substituting $n_{p0} = n_i^2/N_A$, where n_i is the intrinsic carrier density and N_A is the doping concentration in the base, we can express I_S as

$$I_S = \frac{A_E q D_n n_i^2}{N_A W} \tag{6.4}$$

An important observation to make here is that the magnitude of i_C is independent of v_{CB}. That is, as long as the collector is positive with respect to the base, the electrons that reach the collector side of the base region will be swept into the collector and register as collector current.

The saturation current I_S is inversely proportional to the base width W and is directly proportional to the area of the EBJ. Typically I_S is in the range of 10^{-12} A to 10^{-18} A (depending on the size of the device). Because I_S is proportional to n_i^2, it is a strong function of temperature, approximately doubling for every 5°C rise in temperature. (For the dependence of n_i^2 on temperature, refer to Eq. 3.37.)

Since I_S is directly proportional to the junction area (i.e., the device size), it will also be referred to as the **scale current**. Two transistors that are identical except that one has an EBJ area, say, twice that of the other will have saturation currents with that same ratio (i.e., 2). Thus for the same value of v_{BE} the larger device will have a collector current twice that in the smaller device. This concept is frequently employed in integrated-circuit design.

The Base Current The base current i_B is composed of two components. The first component i_{B1} is due to the holes injected from the base region into the emitter region. This current component is proportional to e^{v_{BE}/V_T}. The second component of base current, i_{B2}, is due to holes that have to be supplied by the external circuit in order to replace the holes lost from the base through the recombination process. Because i_{B2} is proportional to the number of electrons injected into the base, it also will be proportional to e^{v_{BE}/V_T}. Thus the total base current, $i_B = i_{B1} + i_{B2}$, will be proportional to e^{v_{BE}/V_T}, and can be expressed as a fraction of the collector current i_C as follows:

$$i_B = \frac{i_C}{\beta} \tag{6.5}$$

That is,

$$i_B = \left(\frac{I_S}{\beta}\right) e^{v_{BE}/V_T} \tag{6.6}$$

where β is a transistor parameter.

For modern *npn* transistors, β is in the range 50 to 200, but it can be as high as 1000 for special devices. For reasons that will become clear later, the parameter β is called the **common-emitter current gain.**

The above description indicates that the value of β is highly influenced by two factors: the width of the base region, W, and the relative dopings of the base region and the emitter region, N_A/N_D. To obtain a high β (which is highly desirable since β represents a gain parameter) the base should be thin (W small) and lightly doped and the emitter heavily doped (making N_A/N_D small). For modern integrated circuit fabrication technologies, W is in the nanometer range.

The Emitter Current Since the current that enters a transistor must leave it, it can be seen from Fig. 6.3 that the emitter current i_E is equal to the sum of the collector current i_C and the base current i_B; that is,

$$i_E = i_C + i_B \tag{6.7}$$

Use of Eqs. (6.5) and (6.7) gives

$$i_E = \frac{\beta+1}{\beta} i_C \tag{6.8}$$

That is,

$$i_E = \frac{\beta+1}{\beta} I_S e^{v_{BE}/V_T} \tag{6.9}$$

Alternatively, we can express Eq. (6.8) in the form

$$i_C = \alpha i_E \tag{6.10}$$

where the constant α is related to β by

$$\alpha = \frac{\beta}{\beta+1} \tag{6.11}$$

Thus the emitter current in Eq. (6.9) can be written

$$i_E = (I_S/\alpha) e^{v_{BE}/V_T} \tag{6.12}$$

Finally, we can use Eq. (6.11) to express β in terms of α, that is,

$$\beta = \frac{\alpha}{1-\alpha} \tag{6.13}$$

It can be seen from Eq. (6.11) that α is a constant (for a particular transistor) that is less than but very close to unity. For instance, if $\beta = 100$, then $\alpha \simeq 0.99$. Equation (6.13) reveals an important fact: Small changes in α correspond to very large changes in β. This mathematical observation manifests itself physically, with the result that transistors of the same type may have widely different values of β. For reasons that will become apparent later, α is called the **common-base current gain**.

Recapitulation and Equivalent-Circuit Models We have presented a first-order model for the operation of the *npn* transistor in the active mode. Basically, the forward-bias voltage v_{BE} causes an exponentially related current i_C to flow in the collector terminal. The collector current i_C is independent of the value of the collector voltage as long as the collector–base junction remains reverse biased; that is, $v_{CB} \geq 0$. Thus in the active mode the collector terminal behaves as an ideal constant-current source where the value of the current is determined by v_{BE}. The base current i_B is a factor $1/\beta$ of the collector current, and the emitter current is equal to the sum of the collector and base currents. Since i_B is much smaller than i_C (i.e., $\beta \gg 1$), $i_E \simeq i_C$. More precisely, the collector current is a fraction α of the emitter current, with α smaller than, but close to, unity.

This first-order model of transistor operation in the active mode can be represented by the equivalent circuit shown in Fig. 6.5(a). Here, diode D_E has a scale current I_{SE} equal to (I_S/α) and thus provides a current i_E related to v_{BE} according to Eq. (6.12). The current of the controlled source, which is equal to the collector current, is controlled by v_{BE} according to the exponential relationship indicated, a restatement of Eq. (6.3). This model is in essence a nonlinear voltage-controlled current source. It can be converted to the current-controlled current-source model shown in Fig. 6.5(b) by expressing the current of the controlled source as αi_E. Note that this model is also nonlinear because of the exponential

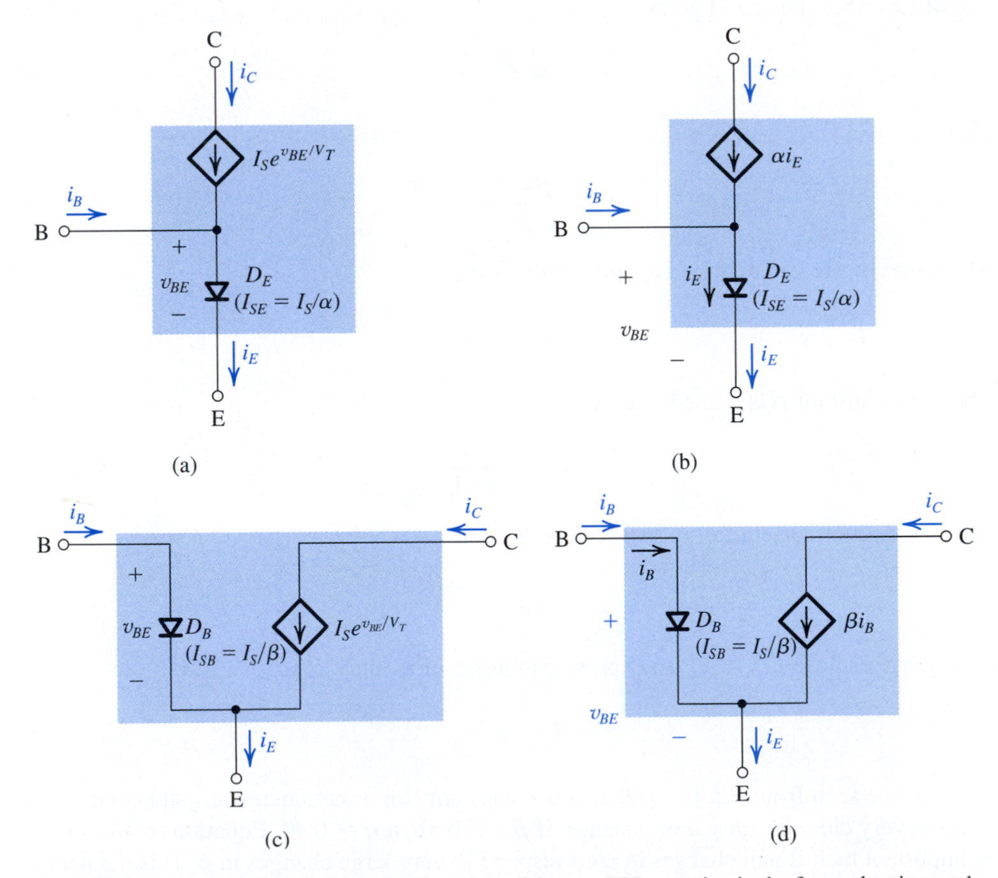

Figure 6.5 Large-signal equivalent-circuit models of the *npn* BJT operating in the forward active mode.

relationship of the current i_E through diode D_E and the voltage v_{BE}. From this model we observe that if the transistor is used as a two-port network with the input port between E and B and the output port between C and B (i.e., with B as a common terminal), then the current gain observed is equal to α. Thus α is called the common-base current gain.

Two other equivalent circuit models, shown in Fig. 6.5(c) and (d), may be used to represent the operation of the BJT. The model of Fig. 6.5(c) is essentially a voltage-controlled current source. However, here diode D_B conducts the base current and thus its current scale factor is I_S/β, resulting in the i_B–v_{BE} relationship given in Eq. (6.6). By simply expressing the collector current as βi_B we obtain the current-controlled current-source model shown in Fig. 6.5(d). From this latter model we observe that if the transistor is used as a two-port network with the input port between B and E and the output port between C and E (i.e., with E as the common terminal), then the current gain observed is equal to β. Thus β is called the common-emitter current gain.

Finally, we note that the models in Fig. 6.5 apply for any positive value of v_{BE}. That is, unlike the models we will be discussing in Section 6.5, here there is no limitation on the size of v_{BE}, and thus these models are referred to as **large-signal models**.

Example 6.1

An *npn* transistor having $I_S = 10^{-15}$A and $\beta = 100$ is connected as follows: The emitter is grounded, the base is fed with a constant-current source supplying a dc current of 10 μA, and the collector is connected to a 5-V dc supply via a resistance R_C of 3 kΩ. Assuming that the transistor is operating in the active mode, find V_{BE} and V_{CE}. Use these values to verify active-mode operation. Replace the current source with a resistance connected from the base to the 5-V dc supply. What resistance value is needed to result in the same operating conditions?

Solution

If the transistor is operating in the active mode, it can be represented by one of the four possible equivalent-circuit models shown in Fig. 6.5. Because the emitter is grounded, either the model in Fig. 6.5(c) or that in Fig. 6.5(d) would be suitable. Since we know the base current I_B, the model of Fig. 6.5(d) is the most suitable.

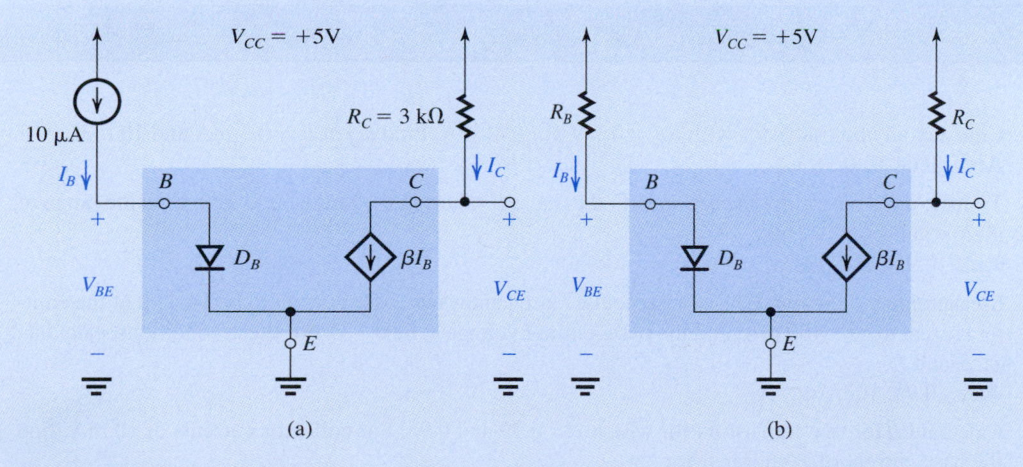

Figure 6.6 Circuits for Example 6.1.

Example 6.1 *continued*

Figure 6.6(a) shows the circuit as described with the transistor represented by the model of Fig. 6.5(d). We can determine V_{BE} from the exponential characteristic of D_B as follows:

$$V_{BE} = V_T \ln \frac{I_B}{I_S/\beta}$$

$$= 25 \ln \left(\frac{10 \times 10^{-6}}{10^{-17}} \right)$$

$$= 690 \text{ mV} = 0.69 \text{ V}$$

Next we determine the value of V_{CE} from

$$V_{CE} = V_{CC} - R_C I_C$$

where

$$I_C = \beta I_B = 100 \times 10 \times 10^{-6} = 10^{-3} \text{ A} = 1 \text{ mA}$$

Thus,

$$V_{CE} = 5 - 3 \times 1 = +2 \text{ V}$$

Since V_C at +2 V is higher than V_B at 0.69 V, the transistor is indeed operating in the active mode.

Now, replacing the 10-μA current source with a resistance R_B connected from the base to the 5-V dc supply V_{CC}, as in Fig. 6.6(b), the value of R_B must be

$$R_B = \frac{V_{CC} - V_{BE}}{I_B}$$

$$= \frac{5 - 0.69}{10 \ \mu\text{A}} = 431 \text{ k}\Omega$$

EXERCISES

6.1 Consider an *npn* transistor with $v_{BE} = 0.7$ V at $i_C = 1$ mA. Find v_{BE} at $i_C = 0.1$ mA and 10 mA.
Ans. 0.64 V; 0.76 V

6.2 Transistors of a certain type are specified to have β values in the range 50 to 150. Find the range of their α values.
Ans. 0.980 to 0.993

6.3 Measurement of an *npn* BJT in a particular circuit shows the base current to be 14.46 μA, the emitter current to be 1.460 mA, and the base–emitter voltage to be 0.7 V. For these conditions, calculate α, β, and I_S.
Ans. 0.99; 100; 10^{-15} A

6.4 Calculate β for two transistors for which $\alpha = 0.99$ and 0.98. For collector currents of 10 mA, find the base current of each transistor.
Ans. 99; 49; 0.1 mA; 0.2 mA

6.5 A transistor for which $I_S = 10^{-16}$ A and $\beta = 100$ is conducting a collector current of 1 mA. Find v_{BE}. Also, find I_{SE} and I_{SB} for this transistor.
Ans. 747.5 mV; 1.01×10^{-16}A ; 10^{-18}A

6.6 For the circuit in Fig. 6.6(a) analyzed in Example 6.1, find the maximum value of R_C that will still result in active-mode operation.
Ans. 4.31 kΩ

6.1.3 Structure of Actual Transistors

Figure 6.7 shows a more realistic (but still simplified) cross section of an *npn* BJT. Note that the collector virtually surrounds the emitter region, thus making it difficult for the electrons injected into the thin base to escape being collected. In this way, the resulting α is close to unity and β is large. Also, observe that the device is *not* symmetrical, and thus the emitter and collector cannot be interchanged.[4] For more detail on the physical structure of actual devices, the reader is referred to Appendix A.

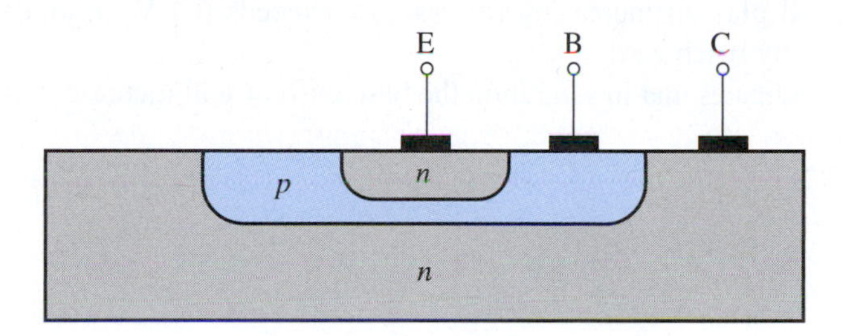

Figure 6.7 Cross-section of an *npn* BJT.

The structure in Fig. 6.7 indicates also that the CBJ has a much larger area than the EBJ. Thus the CB diode D_C has a saturation current I_{SC} that is much larger than the saturation current of the EB diode D_E. Typically, I_{SC} is 10 to 100 times larger than I_{SE} (recall that $I_{SE} = I_S/\alpha \simeq I_S$).

EXERCISE

6.7 A particular transistor has $I_S = 10^{-15}$A and $\alpha \simeq 1$. If the CBJ area is 100 times the area of the EBJ, find the collector scale current I_{SC}.
Ans. 10^{-13} A

[4]If the emitter and collector are reversed—that is, the CBJ is forward biased and the EBJ is reverse biased—the device operates in a mode called the "reverse-active mode." The resulting values of α and β, denoted α_R and β_R (with R denoting reverse), are much lower than the values of α and β, respectively, obtained in the "forward" active mode discussed above. Hence, the reverse-active mode has no practical application. The MOSFET, on the other hand, being a perfectly symmetrical device, can operate equally well with its drain and source terminals interchanged.

6.1.4 Operation in the Saturation Mode[5]

As mentioned above, for the BJT to operate in the active mode, the CBJ must be reverse biased. Thus far, we have stated this condition for the *npn* transistor as $v_{CB} \geq 0$. However, we know that a *pn* junction does not effectively become forward biased until the forward voltage across it exceeds approximately 0.4 V. It follows that one can maintain active-mode operation of an *npn* transistor for negative v_{CB} down to approximately –0.4 V. This is illustrated in Fig. 6.8, which is a sketch of i_C versus v_{CB} for an *npn* transistor operated with a constant emitter current I_E. As expected, i_C is independent of v_{CB} in the active mode, a situation that extends for v_{CB} going negative to approximately –0.4 V. Below this value of v_{CB}, the CBJ begins to conduct sufficiently that the transistor leaves the active mode and enters the saturation mode of operation, where i_C decreases.

To see why i_C decreases in saturation, we can construct a model for the saturated *npn* transistor as follows. We augment the model of Fig. 6.5(c) with the forward-conducting CBJ diode D_C, as shown in Fig. 6.9. Observe that the current i_{BC} will subtract from the controlled-source current, resulting in the reduced collector current i_C given by

$$i_C = I_S e^{v_{BE}/V_T} - I_{SC} e^{v_{BC}/V_T} \tag{6.14}$$

The second term will play an increasing role as v_{BC} exceeds 0.4 V or so, causing i_C to decrease and eventually reach zero.

Figure 6.9 also indicates that in saturation the base current will increase to the value

$$i_B = (I_S/\beta) e^{v_{BE}/V_T} + I_{SC} e^{v_{BC}/V_T} \tag{6.15}$$

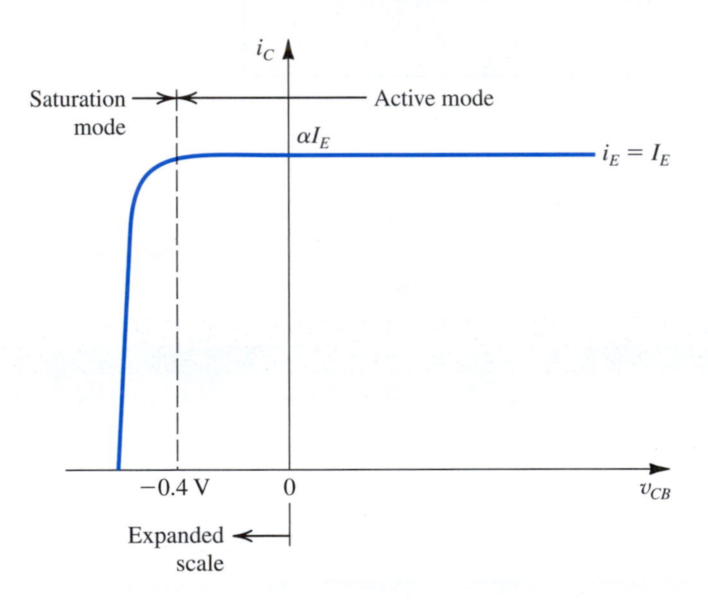

Figure 6.8 The i_C–v_{CB} characteristic of an *npn* transistor fed with a constant emitter current I_E. The transistor enters the saturation mode of operation for $v_{CB} < -0.4$ V, and the collector current diminishes.

[5]Saturation in a BJT means something completely different from that in a MOSFET. The saturation mode of operation of the BJT is analogous to the triode region of operation of the MOSFET. On the other hand, the saturation region of operation of the MOSFET corresponds to the active mode of BJT operation.

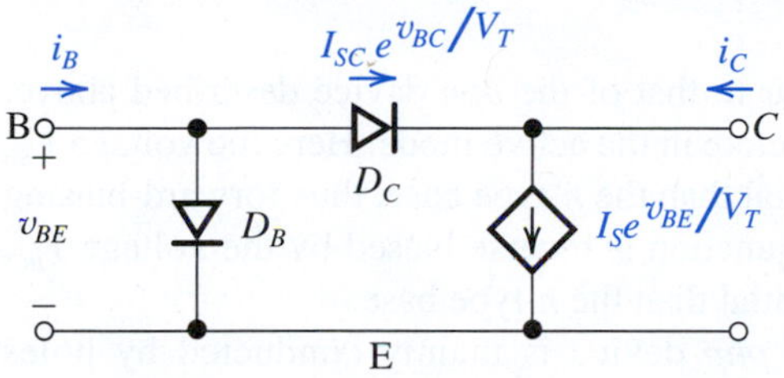

Figure 6.9 Modeling the operation of an *npn* transistor in saturation by augmenting the model of Fig. 6.5(c) with a forward conducting diode D_C. Note that the current through D_C increases i_B and reduces i_C.

Equations (6.14) and (6.15) can be combined to obtain the ratio i_C/i_B for a saturated transistor. We observe that this ratio will be *lower* than the value of β. Furthermore, the ratio will decrease as v_{BC} is increased and the transistor is driven deeper into saturation. Because i_C/i_B of a saturated transistor can be set to any desired value lower than β by adjusting v_{BC}, this ratio is known as **forced** β and denoted β_{forced},

$$\beta_{\text{forced}} = \left.\frac{i_C}{i_B}\right|_{\text{saturation}} \le \beta \tag{6.16}$$

As will be shown later, in analyzing a circuit we can determine whether the BJT is in the saturation mode by either of the following two tests:

1. Is the CBJ forward biased by more than 0.4 V?
2. Is the ratio i_C/i_B lower than β?

The collector-to-emitter voltage v_{CE} of a saturated transistor can be found from Fig. 6.9 as the difference between the forward-bias voltages of the EBJ and the CBJ,

$$V_{CE\text{sat}} = V_{BE} - V_{BC} \tag{6.17}$$

Recalling that the CBJ has a much larger area than the EBJ, V_{BC} will be smaller than V_{BE} by 0.1 to 0.3 V. Thus,

$$V_{CE\text{sat}} \simeq 0.1 \text{ to } 0.3 \text{ V}$$

Typically we will assume that a transistor at the edge of saturation has $V_{CE\text{sat}} = 0.3$ V, while a transistor deep in saturation has $V_{CE\text{sat}} = 0.2$ V.

EXERCISES

6.8 Use Eq. (6.14) to show that i_C reaches zero at

$$V_{CE} = V_T \ln(I_{SC}/I_S)$$

Calculate V_{CE} for a transistor whose CBJ has 100 times the area of the EBJ.
Ans. 115 mV

6.9 Use Eqs. (6.14), (6.15), and (6.16) to show that a BJT operating in saturation with $V_{CE} = V_{CE\text{sat}}$ has a forced β given by

$$\beta_{\text{forced}} = \beta \frac{e^{V_{CE\text{sat}}/V_T} - I_{SC}/I_S}{e^{V_{CE\text{sat}}/V_T} + \beta I_{SC}/I_S}$$

Find β_{forced} for $\beta = 100$, $I_{SC}/I_S = 100$, and $V_{CE\text{sat}} = 0.2$ V.
Ans. 22.2

6.1.5 The *pnp* Transistor

The *pnp* transistor operates in a manner similar to that of the *npn* device described above. Figure 6.10 shows a *pnp* transistor biased to operate in the active mode. Here the voltage V_{EB} causes the *p*-type emitter to be higher in potential than the *n*-type base, thus forward-biasing the emitter–base junction. The collector–base junction is reverse biased by the voltage V_{BC}, which keeps the *p*-type collector lower in potential than the *n*-type base.

Unlike the *npn* transistor, current in the *pnp* device is mainly conducted by holes injected from the emitter into the base as a result of the forward-bias voltage V_{EB}. Since the component of emitter current contributed by electrons injected from base to emitter is kept small by using a lightly doped base, most of the emitter current will be due to holes. The electrons injected from base to emitter give rise to the first component of base current, i_{B1}. Also, a number of the holes injected into the base will recombine with the majority carriers in the base (electrons) and will thus be lost. The disappearing base electrons will have to be replaced from the external circuit, giving rise to the second component of base current, i_{B2}. The holes that succeed in reaching the boundary of the depletion region of the collector–base junction will be attracted by the negative voltage on the collector. Thus these holes will be swept across the depletion region into the collector and appear as collector current.

It can easily be seen from the above description that the current–voltage relationship of the *pnp* transistor will be identical to that of the *npn* transistor except that v_{BE} has to be replaced by v_{EB}. Also, the large-signal, active-mode operation of the *pnp* transistor can be modeled by any of four equivalent circuits similar to those for the *npn* transistor in Fig. 6.5. Two of these four circuits are shown in Fig. 6.11. Finally, we note that the *pnp* transistor can operate in the saturation mode in a manner analogous to that described for the *npn* device.

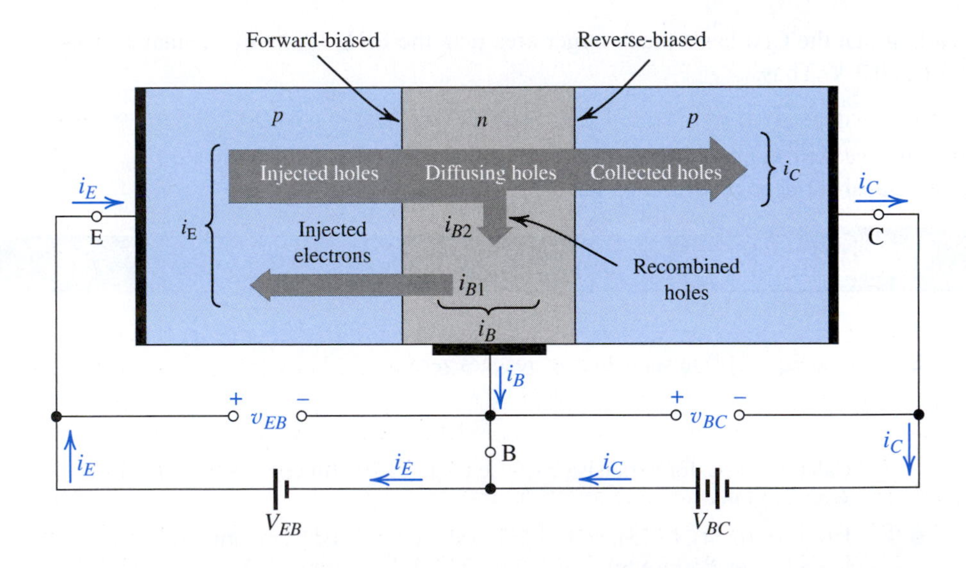

Figure 6.10 Current flow in a *pnp* transistor biased to operate in the active mode.

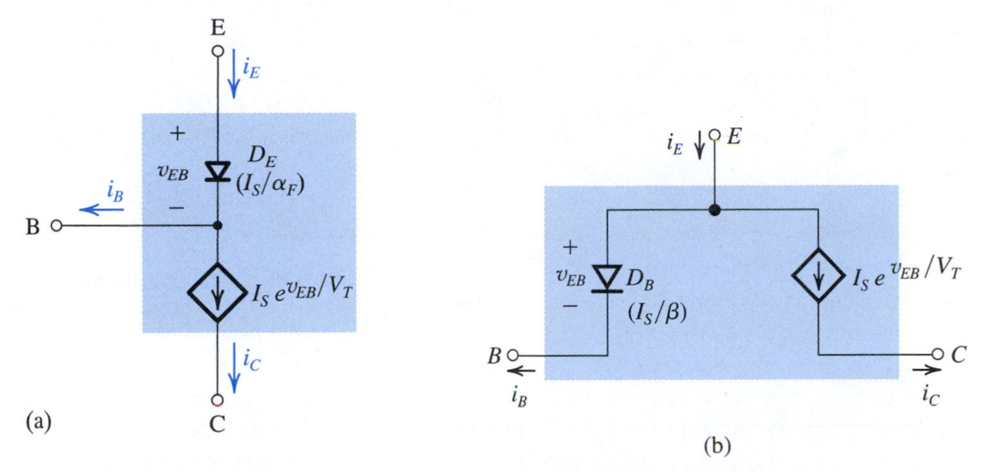

Figure 6.11 Two large-signal models for the *pnp* transistor operating in the active mode.

EXERCISES

6.10 Consider the model in Fig. 6.11(a) applied in the case of a *pnp* transistor whose base is grounded, the emitter is fed by a constant-current source that supplies a 2-mA current into the emitter terminal, and the collector is connected to a −10-V dc supply. Find the emitter voltage, the base current, and the collector current if for this transistor $\beta = 50$ and $I_S = 10^{-14}$ A.
Ans. 0.650 V; 39.2 μA; 1.96 mA

6.11 For a *pnp* transistor having $I_S = 10^{-11}$ A and $\beta = 100$, calculate v_{EB} for $i_C = 1.5$ A.
Ans. 0.643 V

6.2 Current–Voltage Characteristics

6.2.1 Circuit Symbols and Conventions

The physical structure used thus far to explain transistor operation is rather cumbersome to employ in drawing the schematic of a multitransistor circuit. Fortunately, a very descriptive and convenient circuit symbol exists for the BJT. Figure 6.12(a) shows the symbol for the *npn* transistor; the *pnp* symbol is given in Fig. 6.12(b). In both symbols the emitter is distinguished by an arrowhead. This distinction is important because, as we have seen in the last section, practical BJTs are not symmetric devices.

The polarity of the device—*npn* or *pnp*—is indicated by the direction of the arrowhead on the emitter. This arrowhead points in the direction of normal current flow in the emitter, which is also the forward direction of the base–emitter junction. Since we have adopted a drawing convention by which currents flow from top to bottom, we will always draw *pnp* transistors in the manner shown in Fig. 6.12(b) (i.e., with their emitters on top).

Figure 6.13 shows *npn* and *pnp* transistors biased to operate in the active mode. It should be mentioned in passing that the biasing arrangement shown, utilizing two dc voltage sources,

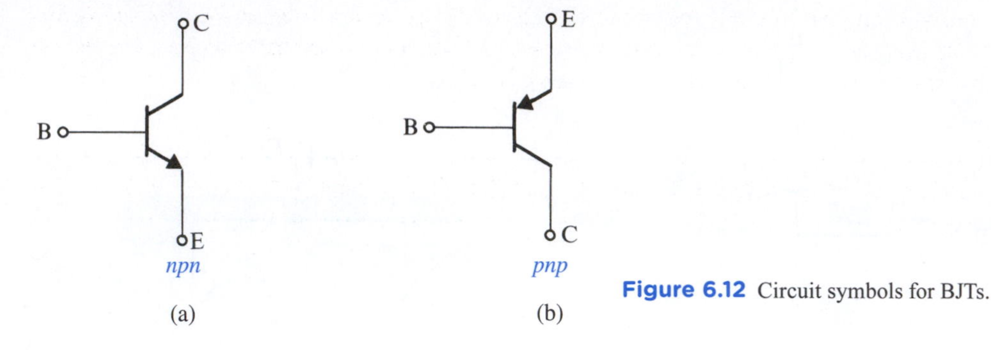

Figure 6.12 Circuit symbols for BJTs.

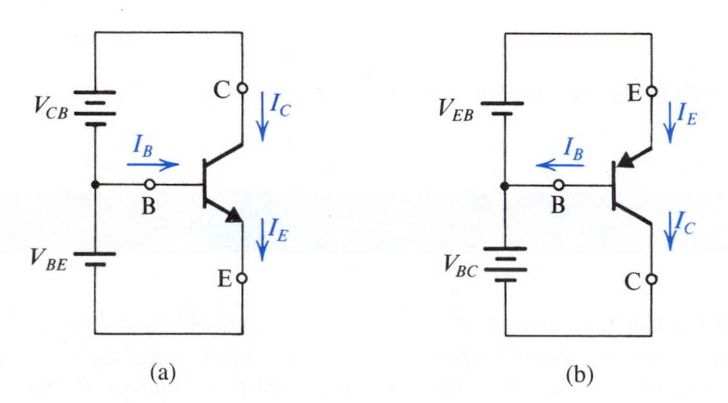

Figure 6.13 Voltage polarities and current flow in transistors biased in the active mode.

is not a usual one and is used here merely to illustrate operation. Practical biasing schemes will be presented in Section 6.7. Figure 6.13 also indicates the reference and actual directions of current flow throughout the transistor. Our convention will be to take the reference direction to coincide with the normal direction of current flow. Hence, normally, we should not encounter a negative value for i_E, i_B, or i_C.

The convenience of the circuit-drawing convention that we have adopted should be obvious from Fig. 6.13. Note that currents flow from top to bottom and that voltages are higher at the top and lower at the bottom. The arrowhead on the emitter also implies the polarity of the emitter–base voltage that should be applied in order to forward bias the emitter–base junction. Just a glance at the circuit symbol of the *pnp* transistor, for example, indicates that we should make the emitter higher in voltage than the base (by v_{EB}) in order to cause current to flow into the emitter (downward). Note that the symbol v_{EB} means the voltage by which the emitter (E) is higher than the base (B). Thus for a *pnp* transistor operating in the active mode v_{EB} is positive, while in an *npn* transistor v_{BE} is positive.

From the discussion of Section 6.1 it follows that an *npn* transistor whose EBJ is forward biased will operate in the active mode *as long as the collector voltage does not fall below that of the base by more than approximately 0.4 V*. Otherwise, the transistor leaves the active mode and enters the saturation region of operation.[6]

Table 6.2 Summary of the BJT Current–Voltage Relationships in the Active Mode

$$i_C = I_S e^{v_{BE}/V_T}$$

$$i_B = \frac{i_C}{\beta} = \left(\frac{I_S}{\beta}\right) e^{v_{BE}/V_T}$$

$$i_E = \frac{i_C}{\alpha} = \left(\frac{I_S}{\alpha}\right) e^{v_{BE}/V_T}$$

Note: For the *pnp* transistor, replace v_{BE} with v_{EB}.

$$i_C = \alpha i_E \qquad i_B = (1-\alpha)i_E = \frac{i_E}{\beta+1}$$

$$i_C = \beta i_B \qquad i_E = (\beta+1)i_B$$

$$\beta = \frac{\alpha}{1-\alpha} \qquad \alpha = \frac{\beta}{\beta+1}$$

V_T = thermal voltage = $\dfrac{kT}{q} \simeq 25$ mV at room temperature

In a parallel manner, the *pnp* transistor will operate in the active mode *if the EBJ is forward biased and the collector voltage is not allowed to rise above that of the base by more than 0.4 V or so*. Otherwise, the CBJ becomes forward biased, and the *pnp* transistor enters the saturation region of operation.

For easy reference, we present in Table 6.2 a summary of the BJT current–voltage relationships in the active mode of operation.

The Collector–Base Reverse Current (I_{CBO}) In our discussion of current flow in transistors we ignored the small reverse currents carried by thermally generated minority carriers. Although such currents can be safely neglected in modern transistors, the reverse current across the collector–base junction deserves some mention. This current, denoted I_{CBO}, is the reverse current flowing from collector to base with the emitter open-circuited (hence the subscript O). This current is usually in the nanoampere range, a value that is many times higher than its theoretically predicted value. As with the diode reverse current, I_{CBO} contains a substantial leakage component, and its value is dependent on v_{CB}. I_{CBO} depends strongly on temperature, approximately doubling for every 10°C rise.[7]

[6]It is interesting to contrast the active-mode operation of the BJT with the corresponding mode of operation of the MOSFET: The BJT needs a minimum v_{CE} of about 0.3 V, and the MOSFET needs a minimum v_{DS} equal to V_{OV}, which for modern technologies is in the range 0.2 V to 0.3 V. Thus we see a great deal of similarity! Also note that reverse biasing the CBJ of the BJT corresponds to pinching off the channel of the MOSFET. This condition results in the collector current (drain current in the MOSFET) being independent of the collector voltage (the drain voltage in the MOSFET).

[7]The temperature coefficient of I_{CBO} is different from that of I_S because I_{CBO} contains a substantial leakage component.

Example 6.2

The transistor in the circuit of Fig. 6.14(a) has $\beta = 100$ and exhibits a v_{BE} of 0.7 V at $i_C = 1$ mA. Design the circuit so that a current of 2 mA flows through the collector and a voltage of +5 V appears at the collector.

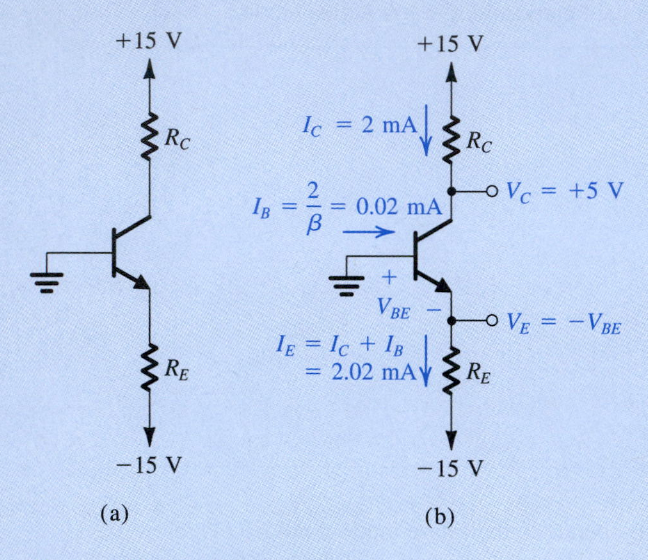

Figure 6.14 Circuit for Example 6.2.

Solution

Refer to Fig. 6.14(b). We note at the outset that since we are required to design for $V_C = +5$ V, the CBJ will be reverse biased and the BJT will be operating in the active mode. To obtain a voltage $V_C = +5$ V, the voltage drop across R_C must be $15 - 5 = 10$ V. Now, since $I_C = 2$ mA, the value of R_C should be selected according to

$$R_C = \frac{10 \text{ V}}{2 \text{ mA}} = 5 \text{ k}\Omega$$

Since $v_{BE} = 0.7$ V at $i_C = 1$ mA, the value of v_{BE} at $i_C = 2$ mA is

$$V_{BE} = 0.7 + V_T \ln\left(\frac{2}{1}\right) = 0.717 \text{ V}$$

Since the base is at 0 V, the emitter voltage should be

$$V_E = -0.717 \text{ V}$$

For $\beta = 100$, $\alpha = 100/101 = 0.99$. Thus the emitter current should be

$$I_E = \frac{I_C}{\alpha} = \frac{2}{0.99} = 2.02 \text{ mA}$$

Now the value required for R_E can be determined from

$$R_E = \frac{V_E - (-15)}{I_E}$$

$$= \frac{-0.717 + 15}{2.02} = 7.07 \text{ k}\Omega$$

This completes the design. We should note, however, that the calculations above were made with a degree of precision that is usually neither necessary nor justified in practice in view, for instance, of the expected tolerances of component values. Nevertheless, we chose to do the design precisely in order to illustrate the various steps involved.

EXERCISES

D6.12 Repeat Example 6.2 for a transistor fabricated in a modern integrated-circuit process. Such a process yields devices that exhibit larger v_{BE} at the same i_C because they have much smaller junction areas. The dc power supplies utilized in modern IC technologies fall in the range of 1 V to 3 V. Design a circuit similar to that shown in Fig. 6.14 except that now the power supplies are ±1.5 V and the BJT has $\beta = 100$ and exhibits v_{BE} of 0.8 V at $i_C = 1$ mA. Design the circuit so that a current of 2 mA flows through the collector and a voltage of +0.5 V appears at the collector.
Ans. $R_C = 500\ \Omega$; $R_E = 338\ \Omega$

6.13 In the circuit shown in Fig. E6.13, the voltage at the emitter was measured and found to be –0.7 V. If $\beta = 50$, find I_E, I_B, I_C, and V_C.

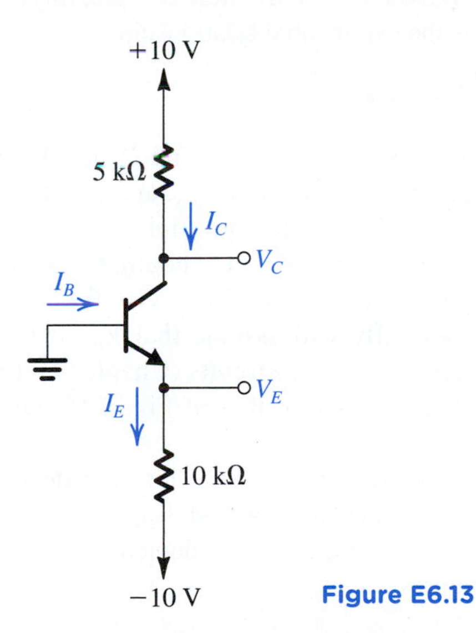

Figure E6.13

Ans. 0.93 mA; 18.2 μA; 0.91 mA; +5.45 V

6.14 In the circuit shown in Fig. E6.14, measurement indicates V_B to be +1.0 V and V_E to be +1.7 V. What are α and β for this transistor? What voltage V_C do you expect at the collector?

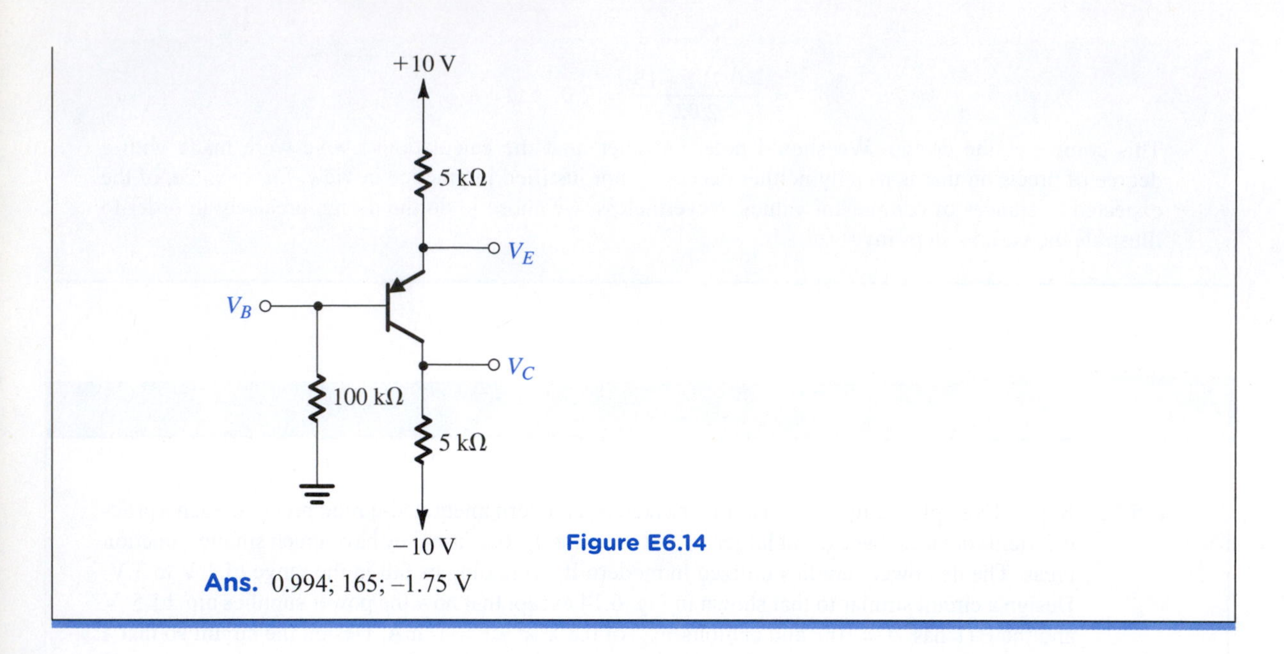

Figure E6.14

Ans. 0.994; 165; –1.75 V

6.2.2 Graphical Representation of Transistor Characteristics

It is sometimes useful to describe the transistor i–v characteristics graphically. Figure 6.15 shows the i_C–v_{BE} characteristic, which is the exponential relationship

$$i_C = I_S e^{v_{BE}/V_T}$$

which is identical to the diode i–v relationship. The i_E–v_{BE} and i_B–v_{BE} characteristics are also exponential but with different scale currents: I_S/α for i_E, and I_S/β for i_B. Since the constant of the exponential characteristic, $1/V_T$, is quite high ($\simeq$ 40), the curve rises very sharply. For v_{BE} smaller than about 0.5 V, the current is negligibly small.[8] Also, over most of the normal current range v_{BE} lies in the range of 0.6 V to 0.8 V. In performing rapid first-order dc calculations, we normally will assume that $V_{BE} \simeq 0.7$ V, which is similar to the approach used in the analysis of diode circuits (Chapter 4). For a *pnp* transistor, the i_C–v_{EB} characteristic will look identical to that of Fig. 6.15 with v_{BE} replaced with v_{EB}.

As in silicon diodes, the voltage across the emitter–base junction decreases by about 2 mV for each rise of 1°C in temperature, provided the junction is operating at a constant current. Figure 6.16 illustrates this temperature dependence by depicting i_C–v_{BE} curves for an *npn* transistor at three different temperatures.

[8]The i_C–v_{BE} characteristic is the BJT's counterpart of the i_D–v_{GS} characteristic of the MOSFET. They share an important attribute: In both cases the voltage has to exceed a "threshold" for the device to conduct appreciably. In the case of the MOSFET, there is a formal threshold voltage, V_t, which lies typically in the range of 0.4 V to 0.8 V. For the BJT, there is an "apparent threshold" of approximately 0.5 V. The i_D–v_{GS} characteristic of the MOSFET is parabolic, and thus is less steep than the i_C–v_{BE} characteristic of the BJT. This difference has a direct and significant implication for the value of transconductance g_m realized with each device.

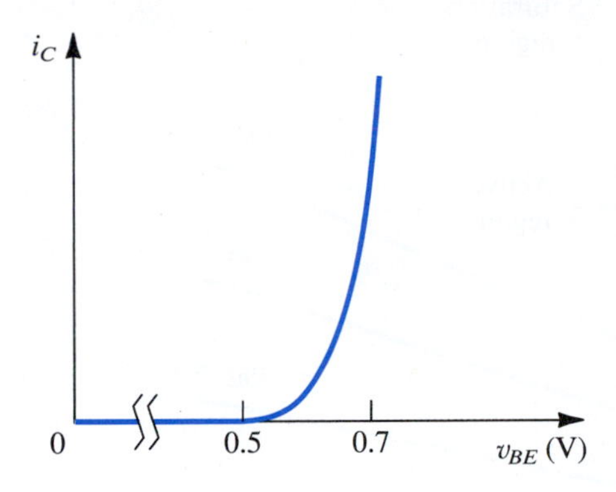

Figure 6.15 The i_C–v_{BE} characteristic for an *npn* transistor.

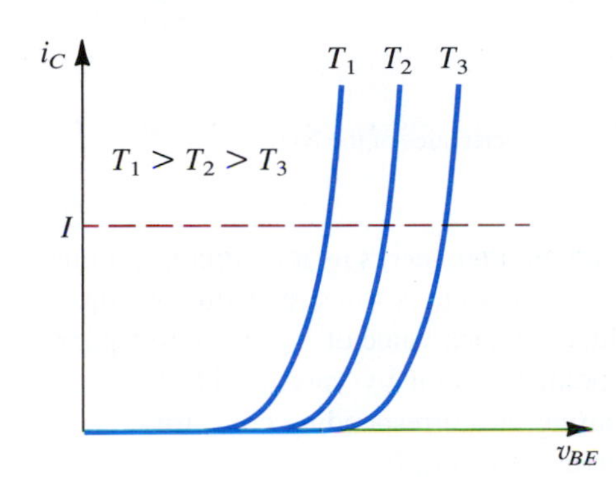

Figure 6.16 Effect of temperature on the i_C–v_{BE} characteristic. At a constant emitter current (broken line), v_{BE} changes by –2 mV/°C.

EXERCISE

6.15 Consider a *pnp* transistor with $v_{EB} = 0.7$ V at $i_E = 1$ mA. Let the base be grounded, the emitter be fed by a 2-mA constant-current source, and the collector be connected to a –5-V supply through a 1-kΩ resistance. If the temperature increases by 30°C, find the changes in emitter and collector voltages. Neglect the effect of I_{CBO}.
Ans. –60 mV; 0 V.

6.2.3 Dependence of i_C on the Collector Voltage—The Early Effect

When operated in the active region, practical BJTs show some dependence of the collector current on the collector voltage, with the result that, unlike the graph shown in Fig. 6.8, their i_C–v_{CB} characteristics are not perfectly horizontal straight lines. To see this dependence more clearly, consider the conceptual circuit shown in Fig. 6.17(a). The transistor is connected in

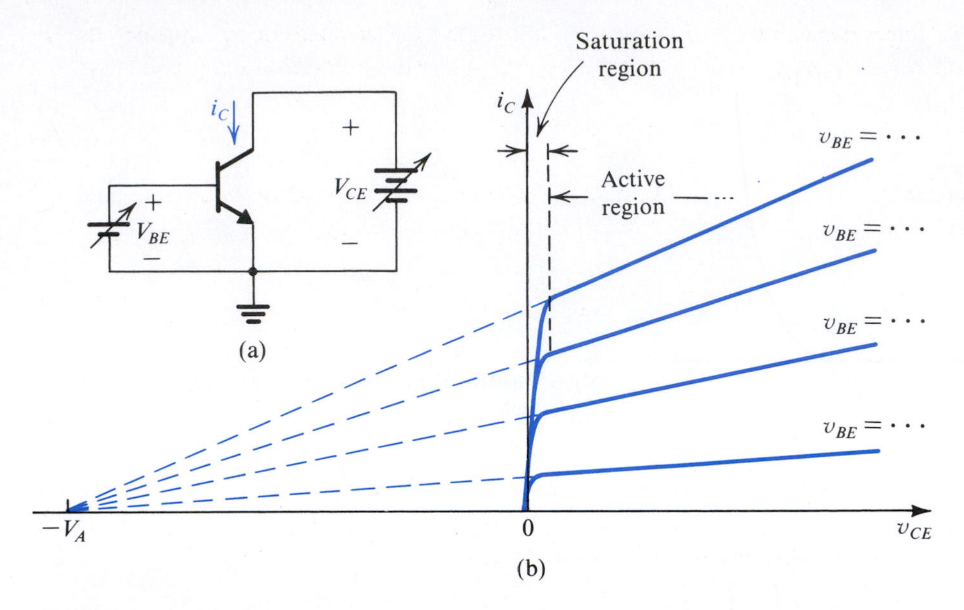

Figure 6.17 **(a)** Conceptual circuit for measuring the i_C–v_{CE} characteristics of the BJT. **(b)** The i_C–v_{CE} characteristics of a practical BJT.

the **common-emitter configuration**; that is, here the emitter serves as a common terminal between the input and output ports. The voltage V_{BE} can be set to any desired value by adjusting the dc source connected between base and emitter. At each value of V_{BE}, the corresponding i_C–v_{CE} characteristic curve can be measured point by point by varying the dc source connected between collector and emitter and measuring the corresponding collector current. The result is the family of i_C–v_{CE} characteristic curves shown in Fig. 6.17(b) and known as **common-emitter characteristics**.

At low values of v_{CE} (lower than about 0.3 V), as the collector voltage goes below that of the base by more than 0.4 V, the collector–base junction becomes forward biased and the transistor leaves the active mode and enters the saturation mode. Shortly, we shall look at the details of the i_C–v_{CE} curves in the saturation region. At this time, however, we wish to examine the characteristic curves in the active region in detail. We observe that the characteristic curves, though still straight lines, have finite slope. In fact, when extrapolated, the characteristic lines meet at a point on the negative v_{CE} axis, at $v_{CE} = -V_A$. The voltage V_A, a positive number, is a parameter for the particular BJT, with typical values in the range of 10 V to 100 V. It is called the **Early voltage**, after J. M. Early, the engineering scientist who first studied this phenomenon.

At a given value of v_{BE}, increasing v_{CE} increases the reverse-bias voltage on the collector–base junction, and thus increases the width of the depletion region of this junction (refer to Fig. 6.3). This in turn results in a decrease in the **effective base width** W. Recalling that I_S is inversely proportional to W (Eq. 6.4), we see that I_S will increase and that i_C increases proportionally. This is the Early effect. For obvious reasons, it is also known as the **base-width moduation effect**.[9]

[9] Recall that the MOSFET's counterpart is the channel-length modulation effect. These two effects are remarkably similar and have been assigned the same name, Early effect.

The linear dependence of i_C on v_{CE} can be explicitly accounted for by assuming that I_S remains constant and including the factor $(1 + v_{CE}/V_A)$ in the equation for i_C as follows:

$$i_C = I_S e^{v_{BE}/V_T}\left(1 + \frac{v_{CE}}{V_A}\right) \tag{6.18}$$

The nonzero slope of the i_C–v_{CE} straight lines indicates that the **output resistance** looking into the collector is not infinite. Rather, it is finite and defined by

$$r_o \equiv \left[\left.\frac{\partial i_C}{\partial v_{CE}}\right|_{v_{BE} = \text{constant}}\right]^{-1} \tag{6.19}$$

Using Eq. (6.18) we can show that

$$r_o = \frac{V_A + V_{CE}}{I_C} \tag{6.20}$$

where I_C and V_{CE} are the coordinates of the point at which the BJT is operating on the particular i_C–v_{CE} curve (i.e., the curve obtained for v_{BE} equal to constant value V_{BE} at which Eq. (6.19) is evaluated). Alternatively, we can write

$$r_o = \frac{V_A}{I'_C} \tag{6.21}$$

where I'_C is the value of the collector current with the Early effect neglected; that is,

$$I'_C = I_S e^{V_{BE}/V_T} \tag{6.22}$$

It is rarely necessary to include the dependence of i_C on v_{CE} in dc bias design and analysis that is performed by hand. Such an effect, however, can be easily included in the SPICE simulation of circuit operation, which is frequently used to “fine-tune” pencil-and-paper analysis or design.

The finite output resistance r_o can have a significant effect on the gain of transistor amplifiers. This is particularly the case in integrated–circuit amplifiers, as will be shown in chapter 7. Fortunately, there are many situations in which r_o can be included relatively easily in pencil-and-paper analysis.

The output resistance r_o can be included in the circuit model of the transistor. This is illustrated in Fig. 6.18, where we show the two large-signal circuit models of a common-emitter *npn* transistor operating in the active mode, those in Fig 6.5(c) and (d), with the resistance r_o connected between the collector and the emitter terminals.

EXERCISES

6.16 Use the circuit model in Fig. 6.18(a) to express I_C in terms of e^{v_{BE}/V_T} and v_{CE} and thus show that this circuit is a direct representation of Eq. (6.18).

6.17 Find the output resistance of a BJT for which $V_A = 100$ V at $I_C = 0.1$, 1, and 10 mA.
Ans. 1 MΩ; 100 kΩ; 10 kΩ

6.18 Consider the circuit in Fig. 6.17(a). At $V_{CE} = 1$ V, V_{BE} is adjusted to yield a collector current of 1 mA. Then, while V_{BE} is kept constant, V_{CE} is raised to 11 V. Find the new value of I_C. For this transistor, $V_A = 100$ V.
Ans. 1.1 mA

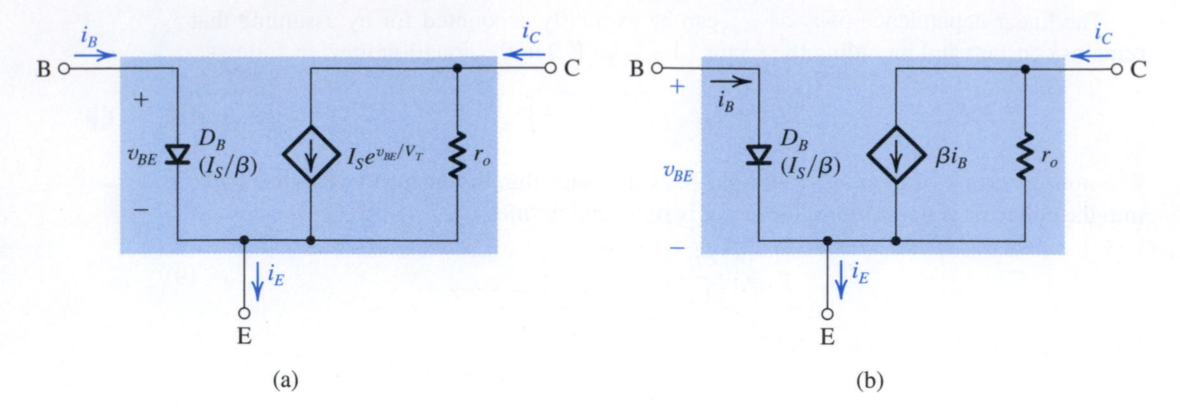

Figure 6.18 Large-signal equivalent-circuit models of an *npn* BJT operating in the active mode in the common-emitter configuration with the output resistance r_o included.

6.2.4 An Alternative Form of the Common-Emitter Characteristics

An alternative way of expressing the transistor common-emitter characteristics is illustrated in Fig. 6.19. Here the base current i_B rather than the base–emitter voltage v_{BE} is used as a parameter. That is, each i_C–v_{CE} curve is measured with the base fed with a constant current I_B. The resulting characteristics, shown in Fig. 6.19(b), look similar to those in Fig. 6.17. Figure 6.19(c) shows an expanded view of the characteristics in the saturation region.

The Common-Emitter Current Gain β In the active region of the characteristics shown in Fig. 6.19(b) we have identified a particular point Q. Note that this operating point for the transistor is characterized by a base current I_B, a collector current I_C, and a collector–emitter voltage V_{CE}. The ratio I_C/I_B is the transistor β. However, there is another way to measure β: change the base current by an increment Δi_B and measure the resulting increment Δi_C, while keeping V_{CE} constant. This is illustrated in Fig. 6.19(b). The ratio $\Delta i_C/\Delta i_B$ should, according to our study thus far, yield an identical value for β. It turns out, however, that the latter value of β (called *incremental*, or ac, β) is a little different from the dc β (i.e., I_C/I_B). Such a distinction, however, is too subtle for our needs in this book. We shall use β to denote both dc and incremental values.[10]

The Saturation Voltage V_{CEsat} and Saturation Resistance R_{CEsat} Refer next to the expanded view of the common-emitter characteristics in the saturation region shown in Fig. 6.19(c). The "bunching together" of the curves in the saturation region implies that the incremental β is lower there than in the active region. A possible operating point in the saturation region is that labeled X. It is characterized by a base current I_B, a collector current I_{Csat}, and a collector–emitter voltage V_{CEsat}. From our previous discussion of saturation, recall that $I_{Csat} = \beta_{forced} I_B$, where $\beta_{forced} < \beta$.

The i_C–v_{CE} curves in saturation are rather steep, indicating that the saturated transistor exhibits a low collector-to-emitter resistance R_{CEsat},

$$R_{CEsat} \equiv \left.\frac{\partial v_{CE}}{\partial i_C}\right|_{\substack{i_B = I_B \\ i_C = I_{Csat}}} \tag{6.23}$$

Typically, R_{CEsat} ranges from a few ohms to a few tens of ohms.

[10] Manufacturers of bipolar transistors use h_{FE} to denote the dc value of β and h_{fe} to denote the incremental β. These symbols come from the *h*-parameter description of two-port networks (see Appendix C), with the subscript *F*(*f*) denoting forward and *E*(*e*) denoting common emitter.

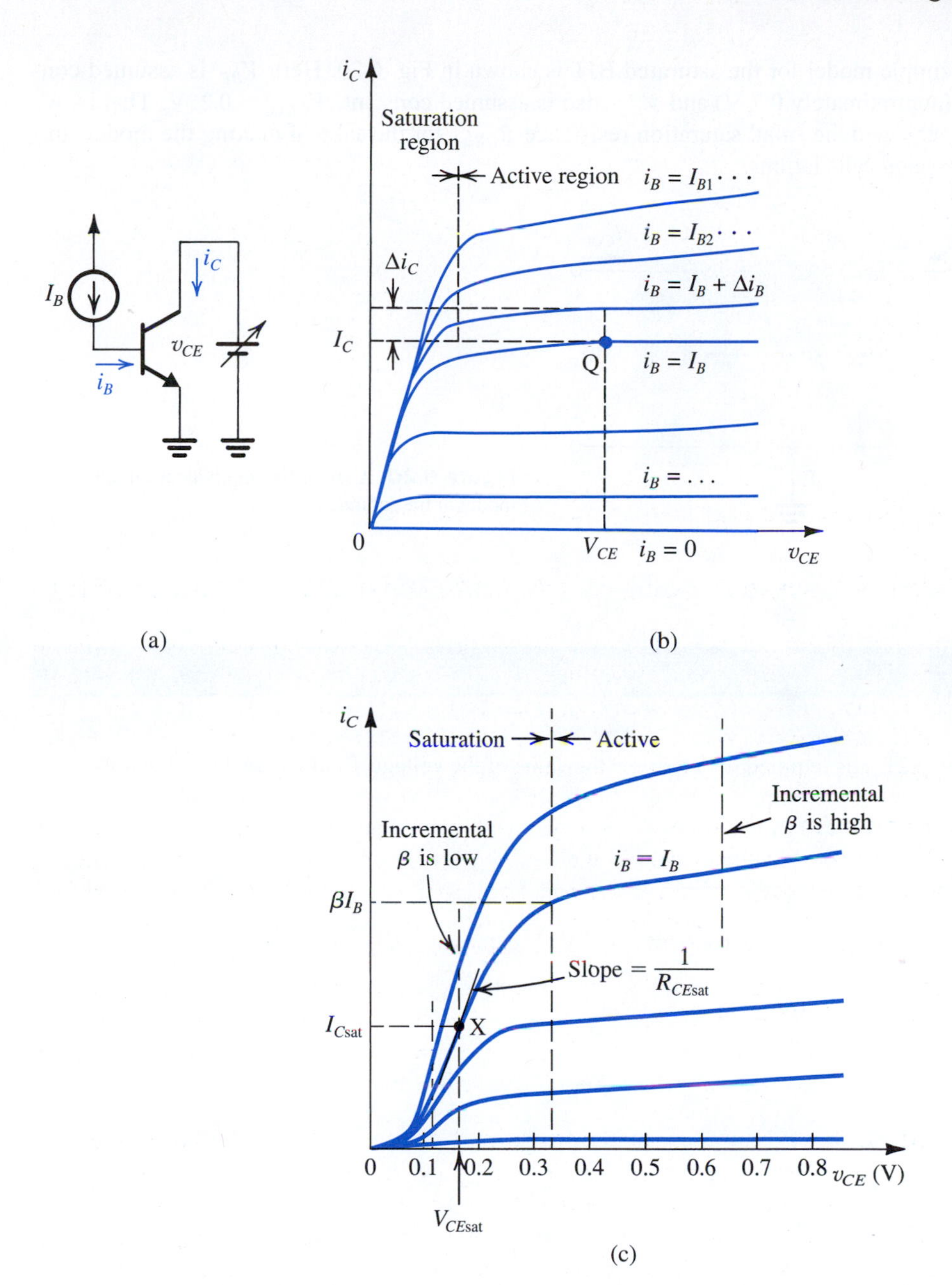

Figure 6.19 Common-emitter characteristics. (**a**) Basic CE circuit; note that in (**b**) the horizontal scale is expanded around the origin to show the saturation region in some detail. A much greater expansion of the saturation region is shown in (**c**).

That the collector-to-emitter resistance of a saturated BJT is small should have been anticipated from the fact that between C and E we now have two forward-conducting diodes in series[11] (see also Fig. 6.9).

[11]In the corresponding mode of operation for the MOSFET, the triode region, the resistance between drain and source is small because it is the resistance of the continuous (non-pinched-off) channel.

A simple model for the saturated BJT is shown in Fig. 6.20. Here V_{BE} is assumed constant (approximately 0.7 V) and V_{CE} also is assumed constant, $V_{CEsat} \simeq 0.2$ V. That is, we have neglected the small saturation resistance R_{CEsat} for the sake of making the model simple for hand calculations.

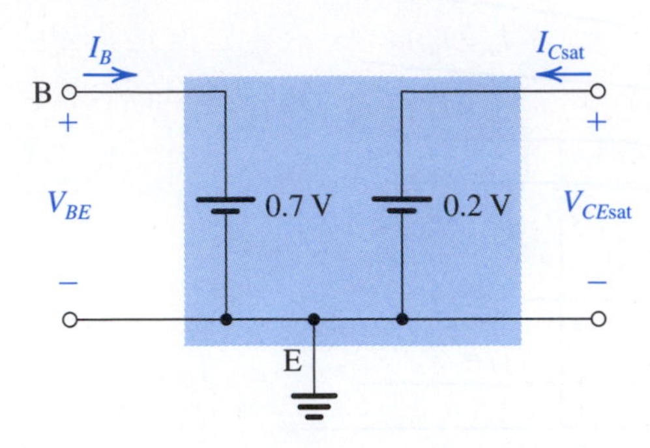

Figure 6.20 A simplified equivalent-circuit model of the saturated transistor.

Example 6.3

For the circuit in Fig. 6.21, it is required to determine the value of the voltage V_{BB} that results in the transistor operating

(a) in the active mode with $V_{CE} = 5$ V
(b) at the edge of saturation
(c) deep in saturation with $\beta_{forced} = 10$

For simplicity, assume that V_{BE} remains constant at 0.7 V. The transistor β is specified to be 50.

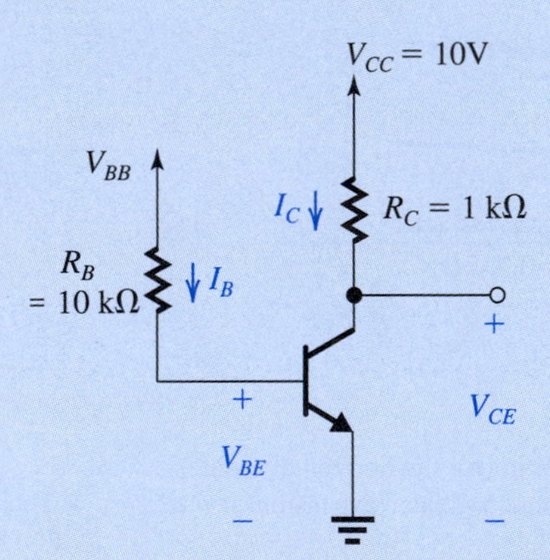

Figure 6.21 Circuit for Example 6.3.

Solution

(a) To operate in the active mode with $V_{CE} = 5$ V,

$$I_C = \frac{V_{CC} - V_{CE}}{R_C}$$

$$= \frac{10 - 5}{1\text{ k}\Omega} = 5\text{ mA}$$

$$I_B = \frac{I_C}{\beta} = \frac{5}{50} = 0.1 \text{ mA}$$

Now the required value of V_{BB} can be found as follows:

$$V_{BB} = I_B R_B + V_{BE}$$

$$= 0.1 \times 10 + 0.7 = 1.7 \text{ V}$$

(b) Operation at the edge of saturation is obtained with $V_{CE} = 0.3$ V. Thus

$$I_C = \frac{10 - 0.3}{1} = 9.7 \text{ mA}$$

Since, at the edge of saturation, I_C and I_B are still related by β,

$$I_B = \frac{9.7}{50} = 0.194 \text{ mA}$$

The required value of V_{BB} can be determined as

$$V_{BB} = 0.194 \times 10 + 0.7 = 2.64 \text{ V}$$

(c) To operate deep in saturation,

$$V_{CE} = V_{CEsat} \simeq 0.2 \text{ V}$$

Thus,

$$I_C = \frac{10 - 0.2}{1} = 9.8 \text{ mA}$$

We then use the value of forced β to determine the required value of I_B as

$$I_B = \frac{I_C}{\beta_{\text{forced}}} = \frac{9.8}{10} = 0.98 \text{ mA}$$

and the required V_{BB} can now be found as

$$V_{BB} = 0.98 \times 10 + 0.7 = 10.5 \text{ V}$$

Observe that once the transistor is in saturation, increasing V_{BB} and thus I_B results in negligible change in I_C since V_{CEsat} will change only slightly. Thus I_C is said to *saturate*, which is the origin of the name "saturation mode of operation."

EXERCISES

6.19 Repeat Example 6.3 for $R_C = 10 \text{ k}\Omega$.
Ans. 0.8 V; 0.894 V; 1.68 V

6.20 For the circuit in Fig. 6.21, find V_{CE} for $V_{BB} = 0$ V.
Ans. + 10 V

6.21 For the circuit in Fig. 6.21, let V_{BB} be set to the value obtained in Example 6.3, part (a), namely, $V_{BB} = 1.7$ V. Verify that the transistor is indeed operating in the active mode. Now, while keeping V_{BB} constant, find that value to which R_C should be increased in order to obtain (a) operation at the edge of saturation, and (b) operation deep in saturation with $\beta_{\text{forced}} = 10$.
Ans. (a) 1.94 kΩ; (b) 9.8 kΩ

6.3 BJT Circuits at DC

We are now ready to consider the analysis of BJT circuits to which only dc voltages are applied. In the following examples we will use the simple model in which $|V_{BE}|$ of a conducting transistor is 0.7 V and $|V_{CE}|$ of a saturated transistor is 0.2 V, and we will neglect the Early effect. Better models can, of course, be used to obtain more accurate results. This, however, is usually achieved at the expense of speed of analysis, and more importantly, it could impede the circuit designer's ability to gain insight regarding circuit behavior. Accurate results using elaborate models can be obtained using circuit simulation with SPICE. This is almost always done in the final stages of a design and certainly before circuit fabrication. Computer simulation, however, is not a substitute for quick pencil-and-paper circuit analysis, an essential ability that aspiring circuit designers must muster. The following series of examples is a step in that direction.

As will be seen, in analyzing a circuit the first question that one must answer is: *In which mode is the transistor operating*? In some cases, the answer will be obvious. For instance, a quick check of the terminal voltages will indicate whether the transistor is cut off or conducting. If it is conducting, we have to determine whether it is operating in the active mode or in saturation. In some cases, however, this may not be obvious. Needless to say, as the reader gains practice and experience in transistor circuit analysis and design, the answer will be apparent in a much larger proportion of problems. The answer, however, can always be determined by utilizing the following procedure:

Assume that the transistor is operating in the active mode, and proceed to determine the various voltages and currents that correspond. Then check for consistency of the results with the assumption of active-mode operation; that is, is v_{CB} of an *npn* transistor greater than -0.4 V (or v_{CB} of a *pnp* transistor lower than 0.4 V)? If the answer is yes, then our task is complete. If the answer is no, assume saturation-mode operation, and proceed to determine currents and voltages and then to check for consistency of the results with the assumption of saturation-mode operation. Here the test is usually to compute the ratio I_C/I_B and to verify that it is lower than the transistor β (i.e., $\beta_{\text{forced}} < \beta$). Since β for a given transistor type varies over a wide range,[12] one must use the lowest specified β for this test. Finally, note that the order of these two assumptions can be reversed. As a further aid to the reader, we provide in Table 6.3 a summary of the conditions and models for the operation of the BJT in its three possible modes.

[12]That is, if one buys BJTs of a certain part number, the manufacturer guarantees only that their values of β fall within a certain range, say 50 to 150.

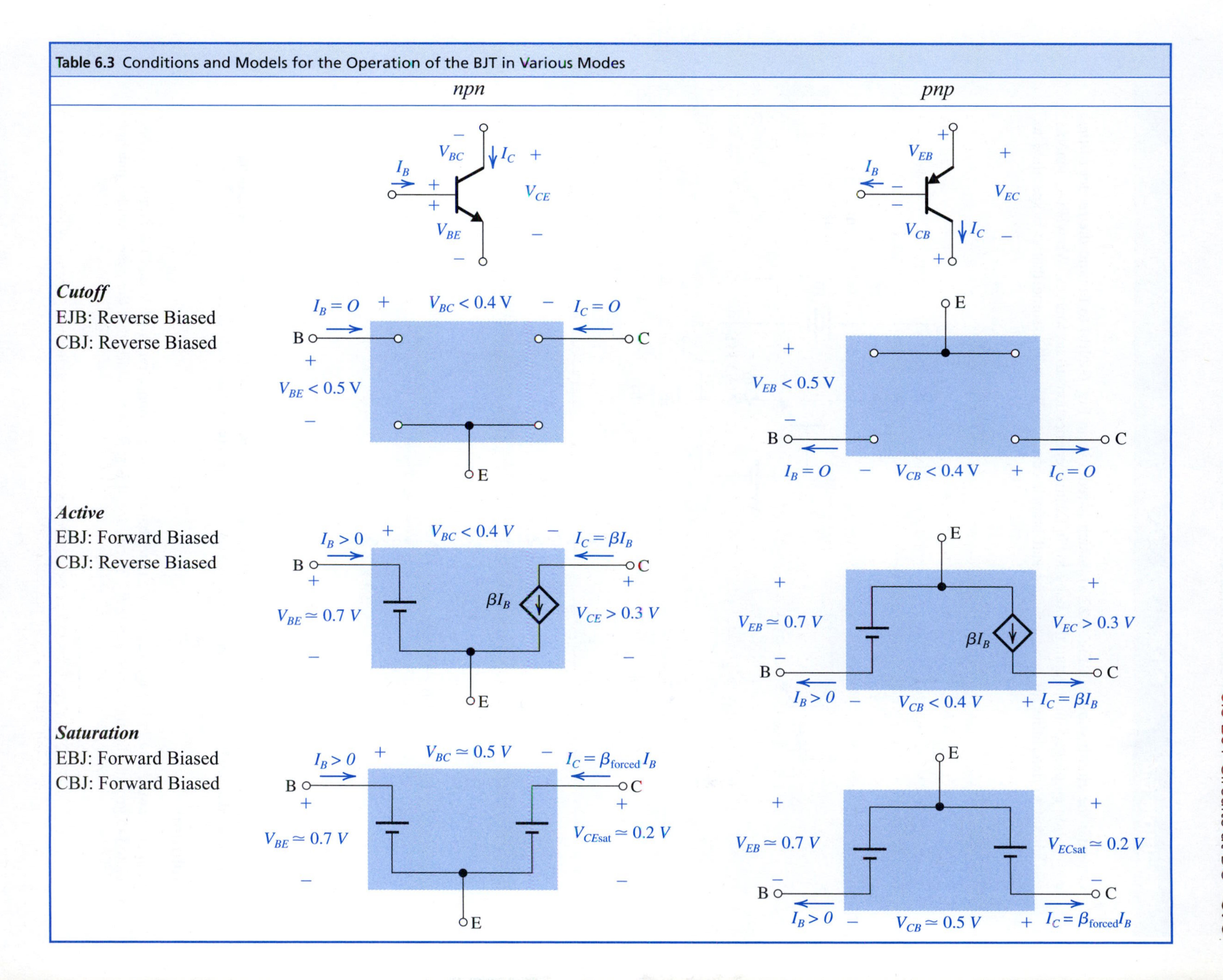
Table 6.3 Conditions and Models for the Operation of the BJT in Various Modes
npn
pnp
I_B V_{BC} I_C V_{CE} V_{BE}
I_B V_{EB} V_{EC} V_{CB} I_C
Cutoff
EJB: Reverse Biased
CBJ: Reverse Biased
$I_B = O$ $V_{BC} < 0.4$ V $I_C = O$
$V_{BE} < 0.5$ V
$V_{EB} < 0.5$ V
$I_B = O$ $V_{CB} < 0.4$ V $I_C = O$
Active
EBJ: Forward Biased
CBJ: Reverse Biased
$I_B > 0$ $V_{BC} < 0.4$ V $I_C = \beta I_B$
$V_{BE} \simeq 0.7$ V βI_B $V_{CE} > 0.3$ V
$V_{EB} \simeq 0.7$ V βI_B $V_{EC} > 0.3$ V
$I_B > 0$ $V_{CB} < 0.4$ V $I_C = \beta I_B$
Saturation
EBJ: Forward Biased
CBJ: Forward Biased
$I_B > 0$ $V_{BC} \simeq 0.5$ V $I_C = \beta_{\text{forced}} I_B$
$V_{BE} \simeq 0.7$ V $V_{CEsat} \simeq 0.2$ V
$V_{EB} \simeq 0.7$ V $V_{ECsat} \simeq 0.2$ V
$I_B > 0$ $V_{CB} \simeq 0.5$ V $I_C = \beta_{\text{forced}} I_B$
B
C
E

Example 6.4

Consider the circuit shown in Fig. 6.22(a), which is redrawn in Fig. 6.22(b) to remind the reader of the convention employed throughout this book for indicating connections to dc sources. We wish to analyze this circuit to determine all node voltages and branch currents. We will assume that β is specified to be 100.

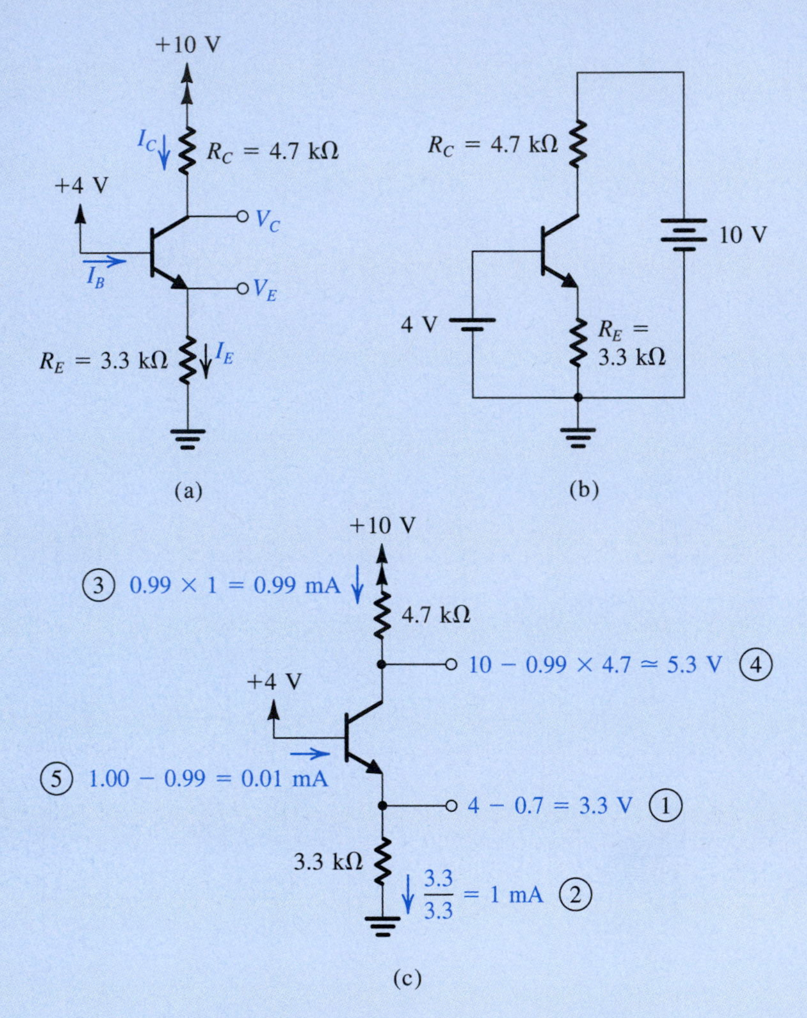

Figure 6.22 Analysis of the circuit for Example 6.4: **(a)** circuit; **(b)** circuit redrawn to remind the reader of the convention used in this book to show connections to the power supply; **(c)** analysis with the steps numbered.

Solution

Glancing at the circuit in Fig. 6.22(a), we note that the base is connected to +4 V and the emitter is connected to ground through a resistance R_E. Therefore, it is safe to conclude that the base–emitter junction will be forward biased. Assuming that this is the case and assuming that V_{BE} is approximately 0.7 V, it follows that the emitter voltage will be

$$V_E = 4 - V_{BE} \simeq 4 - 0.7 = 3.3 \text{ V}$$

We are now in an opportune position; we know the voltages at the two ends of R_E and thus can determine the current I_E through it,

$$I_E = \frac{V_E - 0}{R_E} = \frac{3.3}{3.3} = 1 \text{ mA}$$

Since the collector is connected through R_C to the +10-V power supply, it appears possible that the collector voltage will be higher than the base voltage, which implies active-mode operation. Assuming that this is the case, we can evaluate the collector current from

$$I_C = \alpha I_E$$

The value of α is obtained from

$$\alpha = \frac{\beta}{\beta + 1} = \frac{100}{101} \simeq 0.99$$

Thus I_C will be given by

$$I_C = 0.99 \times 1 = 0.99 \text{ mA}$$

We are now in a position to use Ohm's law to determine the collector voltage V_C,

$$V_C = 10 - I_C R_C = 10 - 0.99 \times 4.7 \simeq +5.3 \text{ V}$$

Since the base is at +4 V, the collector–base junction is reverse biased by 1.3 V, and the transistor is indeed in the active mode as assumed.

It remains only to determine the base current I_B, as follows:

$$I_B = \frac{I_E}{\beta + 1} = \frac{1}{101} \simeq 0.01 \text{ mA}$$

Before leaving this example we wish to emphasize strongly the value of carrying out the analysis directly on the circuit diagram. Only in this way will one be able to analyze complex circuits in a reasonable length of time. Figure 6.22(c) illustrates the above analysis on the circuit diagram, with the order of the analysis steps indicated by the circled numbers.

Example 6.5

We wish to analyze the circuit of Fig. 6.23(a) to determine the voltages at all nodes and the currents through all branches. Note that this circuit is identical to that of Fig. 6.22 except that the voltage at the base is now +6 V. Assume that the transistor β is specified to be *at least* 50.

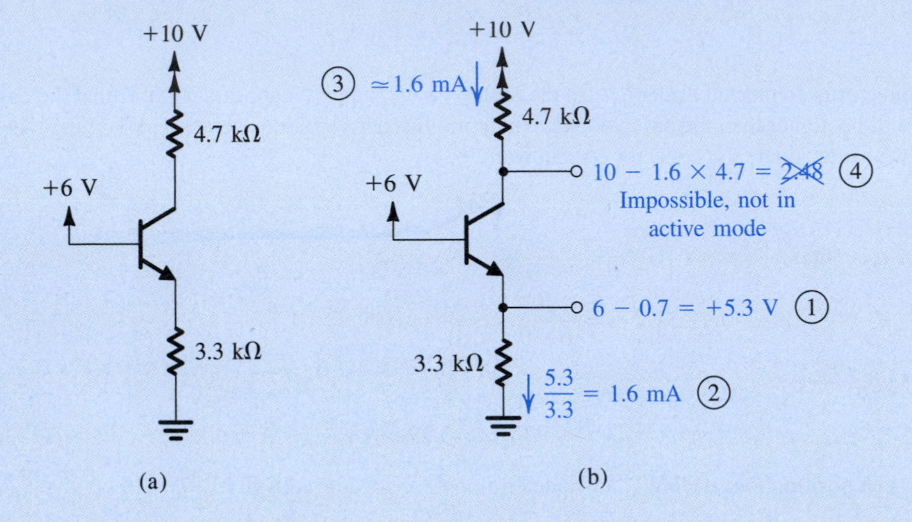

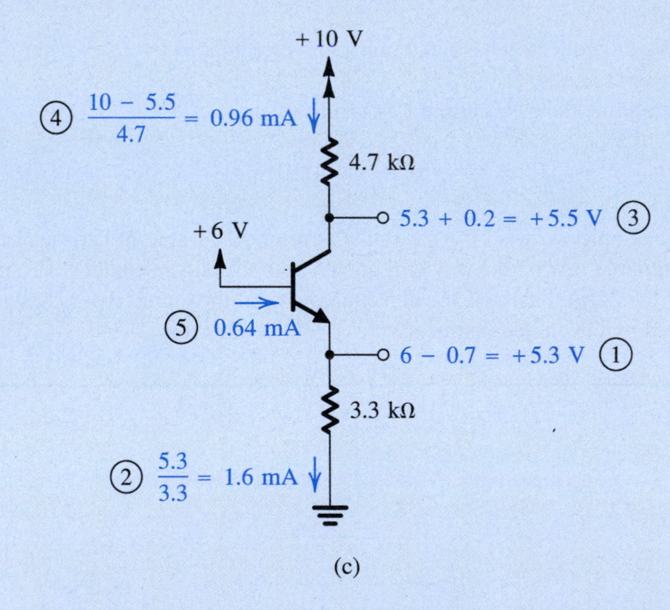

Figure 6.23 Analysis of the circuit for Example 6.5. Note that the circled numbers indicate the order of the analysis steps.

Solution

With +6 V at the base, the base–emitter junction will be forward biased; thus,

$$V_E = +6 - V_{BE} \simeq 6 - 0.7 = 5.3 \text{ V}$$

and

$$I_E = \frac{5.3}{3.3} = 1.6 \text{ mA}$$

Now, assuming active-mode operation, $I_C = \alpha I_E \simeq I_E$; thus,

$$V_C = +10 - 4.7 \times I_C \simeq 10 - 7.52 = 2.48 \text{ V}$$

The details of the analysis performed above are illustrated in Fig. 6.23(b).

Since the collector voltage calculated appears to be less than the base voltage by 3.52 V, it follows that our original assumption of active-mode operation is incorrect. In fact, the transistor has to be in the *saturation* mode. Assuming this to be the case, the values of V_E and I_E will remain unchanged. The collector voltage, however, becomes

$$V_C = V_E + V_{CE\text{sat}} \simeq +5.3 + 0.2 = +5.5 \text{ V}$$

from which we can determine I_C as

$$I_C = \frac{+10 - 5.5}{4.7} = 0.96 \text{ mA}$$

and I_B can now be found as

$$I_B = I_E - I_C = 1.6 - 0.96 = 0.64 \text{ mA}$$

Thus the transistor is operating at a forced β of

$$\beta_{\text{forced}} = \frac{I_C}{I_B} = \frac{0.96}{0.64} = 1.5$$

Since β_{forced} is less than the *minimum* specified value of β, the transistor is indeed saturated. We should emphasize here that in testing for saturation the minimum value of β should be used. By the same token, if we are designing a circuit in which a transistor is to be saturated, the design should be based on the minimum specified β. Obviously, if a transistor with this minimum β is saturated, then transistors with higher values of β will also be saturated. The details of the analysis are shown in Fig. 6.23(c), where the order of the steps used is indicated by the circled numbers.

Example 6.6

We wish to analyze the circuit in Fig. 6.24(a) to determine the voltages at all nodes and the currents through all branches. Note that this circuit is identical to that considered in Examples 6.4 and 6.5 except that now the base voltage is zero.

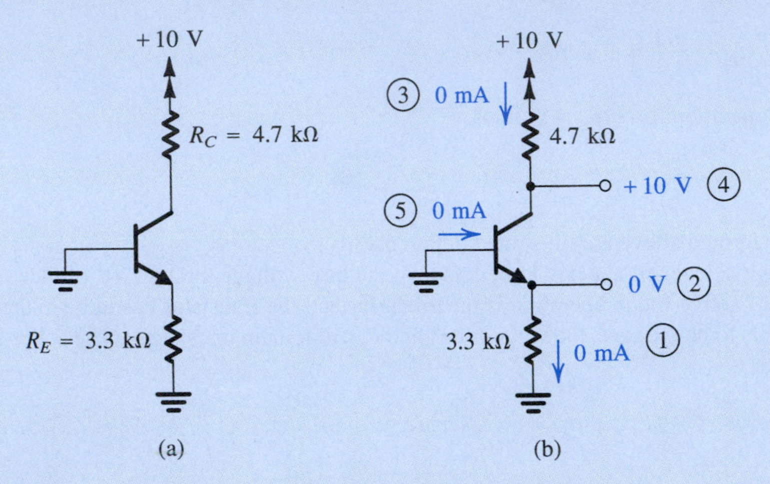

Figure 6.24 Example 6.6: **(a)** circuit; **(b)** analysis, with the order of the analysis steps indicated by circled numbers.

Solution

Since the base is at zero volts and the emitter is connected to ground through R_E, the base–emitter junction cannot conduct and the emitter current is zero. Note that this situation will obtain as long as the voltage at the base is less than 0.5 V or so. Also, the collector–base junction cannot conduct, since the *n*-type collector is connected through R_C to the positive power supply while the *p*-type base is at ground. It follows that the collector current will be zero. The base current will also have to be zero, and the transistor is in the *cutoff* mode of operation.

The emitter voltage will be zero, while the collector voltage will be equal to +10 V, since the voltage drops across R_E and R_C are zero. Figure 6.24(b) shows the analysis details.

EXERCISES

D6.22 For the circuit in Fig. 6.22(a), find the highest voltage to which the base can be raised while the transistor remains in the active mode. Assume $\alpha \simeq 1$.
Ans. +4.7 V

D6.23 Redesign the circuit of Fig. 6.22(a) (i.e., find new values for R_E and R_C) to establish a collector current of 0.5 mA and a reverse-bias voltage on the collector–base junction of 2 V. Assume $\alpha \simeq 1$.
Ans. $R_E = 6.6$ kΩ; $R_C = 8$ kΩ

6.24 For the circuit in Fig. 6.23(a), find the value to which the base voltage should be changed so that the transistor operates in saturation with a forced β of 5.
Ans. +5.18 V

Example 6.7

We want to analyze the circuit of Fig. 6.25(a) to determine the voltages at all nodes and the currents through all branches.

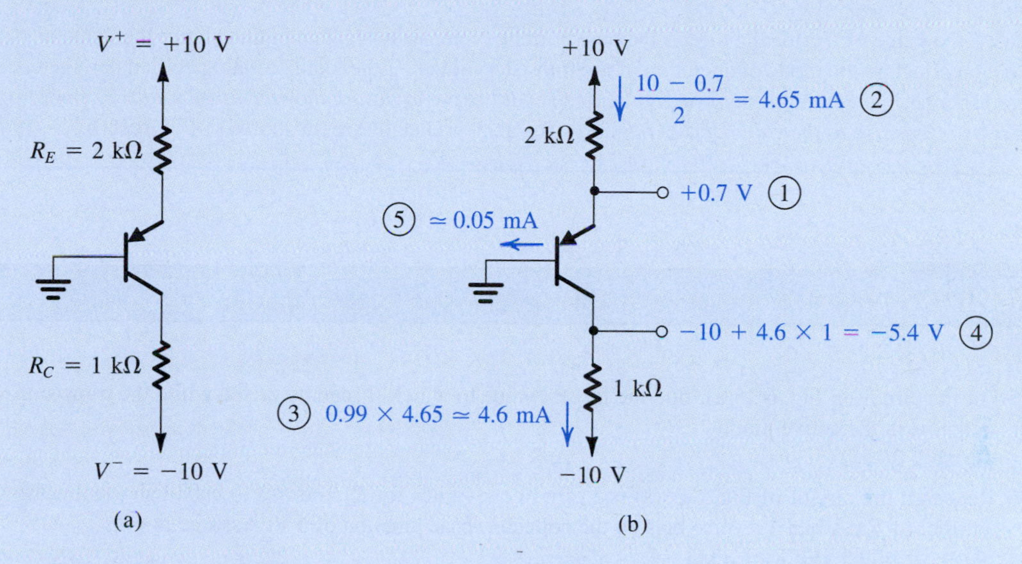

Figure 6.25 Example 6.7: **(a)** circuit; **(b)** analysis, with the steps indicated by circled numbers.

Solution

The base of this *pnp* transistor is grounded, while the emitter is connected to a positive supply ($V^+ = +10$ V) through R_E. It follows that the emitter–base junction will be forward biased with

$$V_E = V_{EB} \simeq 0.7 \text{ V}$$

Thus the emitter current will be given by

$$I_E = \frac{V^+ - V_E}{R_E} = \frac{10 - 0.7}{2} = 4.65 \text{ mA}$$

Since the collector is connected to a negative supply (more negative than the base voltage) through R_C, it is *possible* that this transistor is operating in the active mode. Assuming this to be the case, we obtain

$$I_C = \alpha I_E$$

Since no value for β has been given, we shall assume $\beta = 100$, which results in $\alpha = 0.99$. Since large variations in β result in small differences in α, this assumption will not be critical as far as determining the value of I_C is concerned. Thus,

$$I_C = 0.99 \times 4.65 = 4.6 \text{ mA}$$

The collector voltage will be

$$\begin{aligned} V_C &= V^- + I_C R_C \\ &= -10 + 4.6 \times 1 = -5.4 \text{ V} \end{aligned}$$

Thus the collector–base junction is reverse biased by 5.4 V, and the transistor is indeed in the active mode, which supports our original assumption.

It remains only to calculate the base current,

$$I_B = \frac{I_E}{\beta+1} = \frac{4.65}{101} \simeq 0.05 \text{ mA}$$

Obviously, the value of β critically affects the base current. Note, however, that in this circuit the value of β will have no effect on the mode of operation of the transistor. Since β is generally an ill-specified parameter, this circuit represents a good design. As a rule, one should strive to *design the circuit such that its performance is as insensitive to the value of β as possible.* The analysis details are illustrated in Fig. 6.25(b).

EXERCISES

D6.25 For the circuit in Fig. 6.25(a), find the largest value to which R_C can be raised while the transistor remains in the active mode.

Ans. 2.26 kΩ

D6.26 Redesign the circuit of Fig. 6.25(a) (i.e., find new values for R_E and R_C) to establish a collector current of 1 mA and a reverse bias on the collector–base junction of 4 V. Assume $\alpha \simeq 1$.

Ans. $R_E = 9.3$ kΩ; $R_C = 6$ kΩ

Example 6.8

We want to analyze the circuit in Fig. 6.26(a) to determine the voltages at all nodes and the currents in all branches. Assume $\beta = 100$.

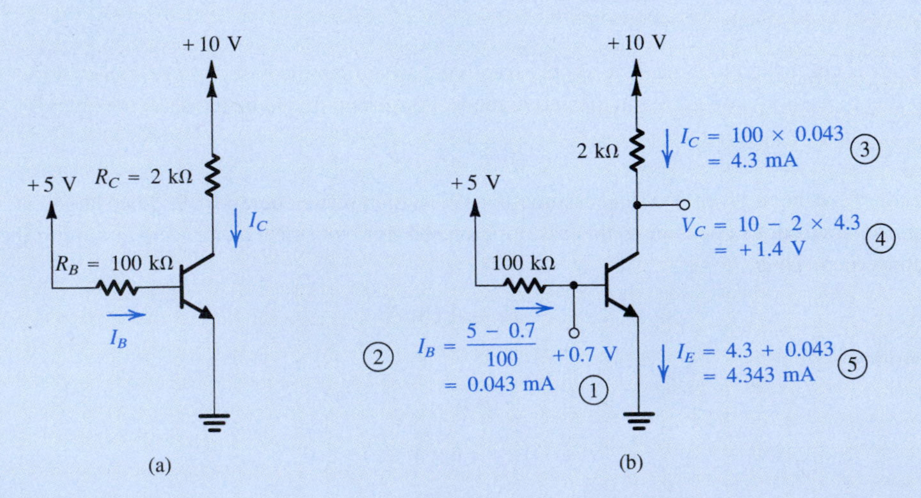

Figure 6.26 Example 6.8: **(a)** circuit; **(b)** analysis, with the steps indicated by the circled numbers.

Solution

The base–emitter junction is clearly forward biased. Thus,

$$I_B = \frac{+5 - V_{BE}}{R_B} \simeq \frac{5 - 0.7}{100} = 0.043 \text{ mA}$$

Assume that the transistor is operating in the active mode. We now can write

$$I_C = \beta I_B = 100 \times 0.043 = 4.3 \text{ mA}$$

The collector voltage can now be determined as

$$V_C = +10 - I_C R_C = 10 - 4.3 \times 2 = +1.4 \text{ V}$$

Since the base voltage V_B is

$$V_B = V_{BE} \simeq +0.7 \text{ V}$$

it follows that the collector–base junction is reverse-biased by 0.7 V and the transistor is indeed in the active mode. The emitter current will be given by

$$I_E = (\beta + 1) I_B = 101 \times 0.043 \simeq 4.3 \text{ mA}$$

We note from this example that the collector and emitter currents depend critically on the value of β. In fact, if β were 10% higher, the transistor would leave the active mode and enter saturation. Therefore this clearly is a *bad* design. The analysis details are illustrated in Fig. 6.26(b).

EXERCISE

D6.27 The circuit of Fig. 6.26(a) is to be fabricated using a transistor type whose β is specified to be in the range of 50 to 150. That is, individual units of this same transistor type can have β values anywhere in this range. Redesign the circuit by selecting a new value for R_C so that all fabricated circuits are guaranteed to be in the active mode. What is the range of collector voltages that the fabricated circuits may exhibit?

Ans. $R_C = 1.5 \text{ k}\Omega$; $V_C = 0.3$ V to 6.8 V

Example 6.9

We want to analyze the circuit of Fig. 6.27 to determine the voltages at all nodes and the currents through all branches. The minimum value of β is specified to be 30.

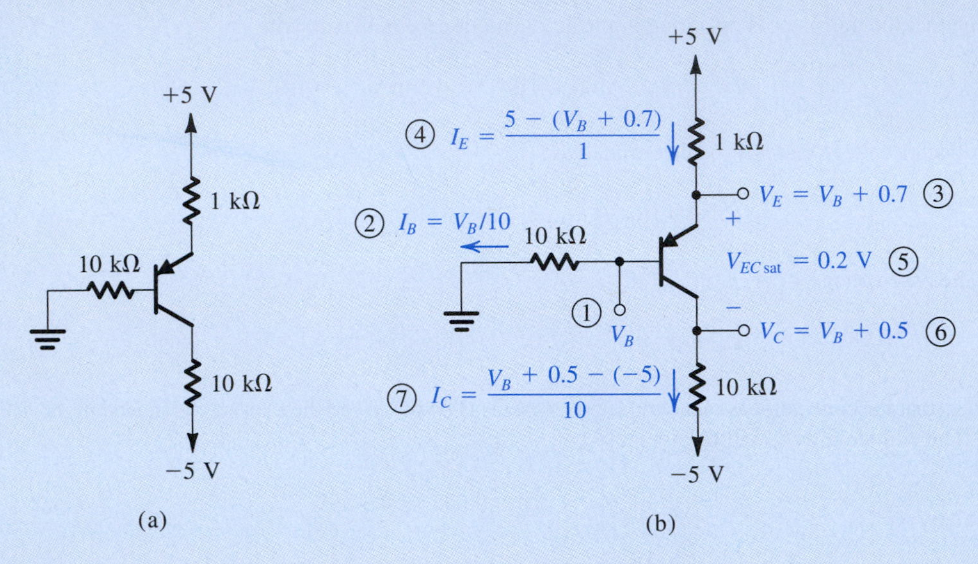

Figure 6.27 Example 6.9: **(a)** circuit; **(b)** analysis with steps numbered.

Solution

A quick glance at this circuit reveals that the transistor will be either active or saturated. Assuming active-mode operation and neglecting the base current, we see that the base voltage will be approximately zero volts, the emitter voltage will be approximately +0.7 V, and the emitter current will be approximately 4.3 mA. Since the maximum current that the collector can support while the transistor remains in the active mode is approximately 0.5 mA, it follows that the transistor is definitely saturated.

Assuming that the transistor is saturated and denoting the voltage at the base by V_B (refer to Fig. 6.27b), it follows that

$$V_E = V_B + V_{EB} \simeq V_B + 0.7$$

$$V_C = V_E - V_{EC\text{sat}} \simeq V_B + 0.7 - 0.2 = V_B + 0.5$$

$$I_E = \frac{+5 - V_E}{1} = \frac{5 - V_B - 0.7}{1} = 4.3 - V_B \quad \text{mA}$$

$$I_B = \frac{V_B}{10} = 0.1V_B \quad \text{mA}$$

$$I_C = \frac{V_C - (-5)}{10} = \frac{V_B + 0.5 + 5}{10} = 0.1V_B + 0.55 \quad \text{mA}$$

Using the relationship $I_E = I_B + I_C$, we obtain

$$4.3 - V_B = 0.1V_B + 0.1V_B + 0.55$$

which results in

$$V_B = \frac{3.75}{1.2} \simeq 3.13 \text{ V}$$

Substituting in the equations above, we obtain

$$V_E = 3.83 \text{ V}$$

$$V_C = 3.63 \text{ V}$$

$$I_E = 1.17 \text{ mA}$$

$$I_C = 0.86 \text{ mA}$$

$$I_B = 0.31 \text{ mA}$$

from which we see that the transistor is saturated, since the value of forced β is

$$\beta_{\text{forced}} = \frac{0.86}{0.31} \simeq 2.8$$

which is much smaller than the specified minimum β.

Example 6.10

We want to analyze the circuit of Fig. 6.28(a) to determine the voltages at all nodes and the currents through all branches. Assume $\beta = 100$.

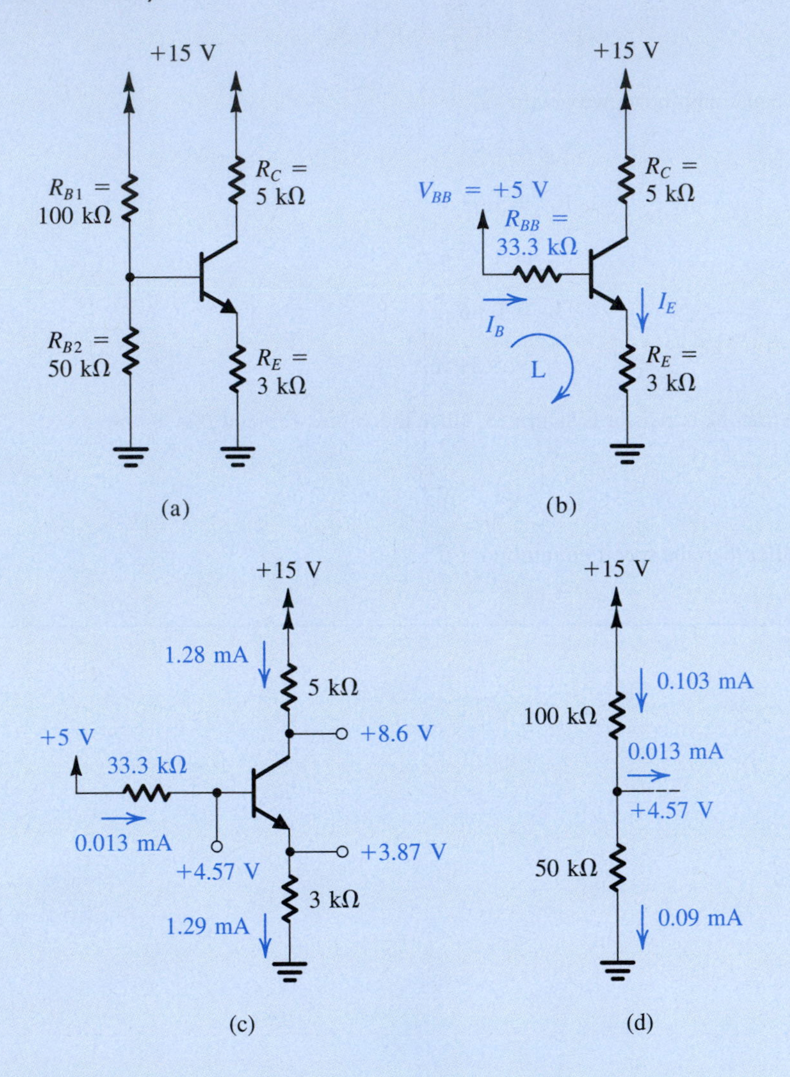

Figure 6.28 Circuits for Example 6.10.

Solution

The first step in the analysis consists of simplifying the base circuit using Thévenin's theorem. The result is shown in Fig. 6.28(b), where

$$V_{BB} = +15\frac{R_{B2}}{R_{B1} + R_{B2}} = 15\frac{50}{100 + 50} = +5 \text{ V}$$

$$R_{BB} = R_{B1} \| R_{B2} = 100 \| 50 = 33.3 \text{ k}\Omega$$

To evaluate the base or the emitter current, we have to write a loop equation around the loop labeled L in Fig. 6.28(b). Note, however, that the current through R_{BB} is different from the current through R_E. The loop equation will be

$$V_{BB} = I_B R_{BB} + V_{BE} + I_E R_E$$

Now, assuming active-mode operation, we replace I_B with

$$I_B = \frac{I_E}{\beta + 1}$$

and rearrange the equation to obtain

$$I_E = \frac{V_{BB} - V_{BE}}{R_E + [R_{BB}/(\beta + 1)]}$$

For the numerical values given we have

$$I_E = \frac{5 - 0.7}{3 + (33.3/101)} = 1.29 \text{ mA}$$

The base current will be

$$I_B = \frac{1.29}{101} = 0.0128 \text{ mA}$$

The base voltage is given by

$$V_B = V_{BE} + I_E R_E$$
$$= 0.7 + 1.29 \times 3 = 4.57 \text{ V}$$

We can evaluate the collector current as

$$I_C = \alpha I_E = 0.99 \times 1.29 = 1.28 \text{ mA}$$

The collector voltage can now be evaluated as

$$V_C = +15 - I_C R_C = 15 - 1.28 \times 5 = 8.6 \text{ V}$$

It follows that the collector is higher in potential than the base by 4.03 V, which means that the transistor is in the active mode, as had been assumed. The results of the analysis are given in Fig. 6.28(c, d).

EXERCISE

6.28 If the transistor in the circuit of Fig. 6.28(a) is replaced with another having half the value of β (i.e., $\beta = 50$), find the new value of I_C, and express the change in I_C as a percentage.
Ans. $I_C = 1.15$ mA; -10%

Example 6.11

We want to analyze the circuit in Fig. 6.29(a) to determine the voltages at all nodes and the currents through all branches.

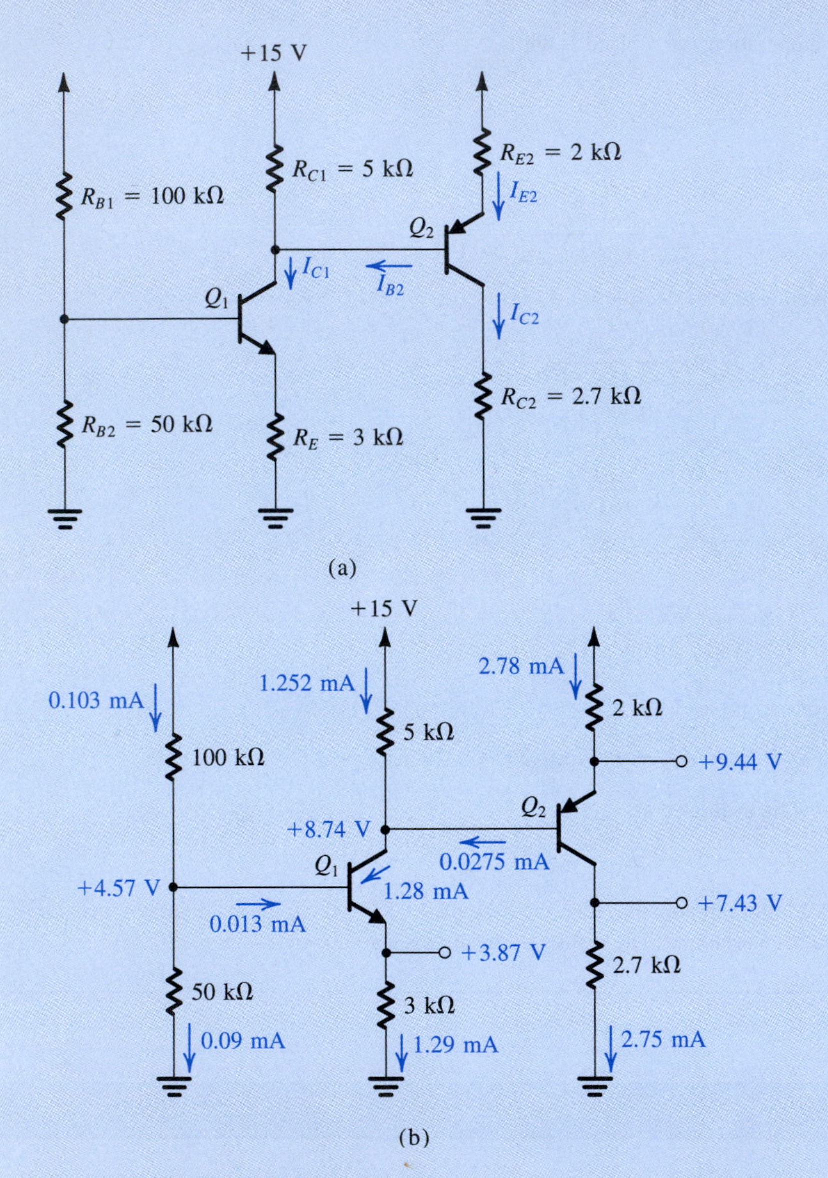

Figure 6.29 Circuits for Example 6.11.

Solution

We first recognize that part of this circuit is identical to the circuit we analyzed in Example 6.10—namely, the circuit of Fig. 6.28(a). The difference, of course, is that in the new circuit we have an additional transistor Q_2 together with its associated resistors R_{E2} and R_{C2}. Assume that Q_1 is still in the active mode. The following values will be identical to those obtained in the previous example:

$$V_{B1} = +4.57 \text{ V} \qquad I_{E1} = 1.29 \text{ mA}$$

$$I_{B1} = 0.0128 \text{ mA} \qquad I_{C1} = 1.28 \text{ mA}$$

However, the collector voltage will be different than previously calculated, since part of the collector current I_{C1} will flow in the base lead of Q_2 (I_{B2}). As a first approximation we may assume that I_{B2} is much smaller than I_{C1}; that is, we may assume that the current through R_{C1} is almost equal to I_{C1}. This will enable us to calculate V_{C1}:

$$\begin{aligned} V_{C1} &\simeq +15 - I_{C1}R_{C1} \\ &= 15 - 1.28 \times 5 = +8.6 \text{ V} \end{aligned}$$

Thus Q_1 is in the active mode, as had been assumed.

As far as Q_2 is concerned, we note that its emitter is connected to +15 V through R_{E2}. It is therefore safe to assume that the emitter–base junction of Q_2 will be forward biased. Thus the emitter of Q_2 will be at a voltage V_{E2} given by

$$V_{E2} = V_{C1} + V_{EB}|_{Q_2} \simeq 8.6 + 0.7 = +9.3 \text{ V}$$

The emitter current of Q_2 may now be calculated as

$$I_{E2} = \frac{+15 - V_{E2}}{R_{E2}} = \frac{15 - 9.3}{2} = 2.85 \text{ mA}$$

Since the collector of Q_2 is returned to ground via R_{C2}, it is possible that Q_2 is operating in the active mode. Assume this to be the case. We now find I_{C2} as

$$\begin{aligned} I_{C2} &= \alpha_2 I_{E2} \\ &= 0.99 \times 2.85 = 2.82 \text{ mA} \quad (\text{assuming } \beta_2 = 100) \end{aligned}$$

The collector voltage of Q_2 will be

$$V_{C2} = I_{C2}R_{C2} = 2.82 \times 2.7 = 7.62 \text{ V}$$

which is lower than V_{B2} by 0.98 V. Thus Q_2 is in the active mode, as assumed.

It is important at this stage to find the magnitude of the error incurred in our calculations by the assumption that I_{B2} is negligible. The value of I_{B2} is given by

$$I_{B2} = \frac{I_{E2}}{\beta_2 + 1} = \frac{2.85}{101} = 0.028 \text{ mA}$$

which is indeed much smaller than I_{C1} (1.28 mA). If desired, we can obtain more accurate results by iterating one more time, assuming I_{B2} to be 0.028 mA. The new values will be

$$\begin{aligned} \text{Current in } R_{C1} &= I_{C1} - I_{B2} = 1.28 - 0.028 = 1.252 \text{ mA} \\ V_{C1} &= 15 - 5 \times 1.252 = 8.74 \text{ V} \\ V_{E2} &= 8.74 + 0.7 = 9.44 \text{ V} \\ I_{E2} &= \frac{15 - 9.44}{2} = 2.78 \text{ mA} \end{aligned}$$

Example 6.11 *continued*

$$I_{C2} = 0.99 \times 2.78 = 2.75 \text{ mA}$$

$$V_{C2} = 2.75 \times 2.7 = 7.43 \text{ V}$$

$$I_{B2} = \frac{2.78}{101} = 0.0275 \text{ mA}$$

Note that the new value of I_{B2} is very close to the value used in our iteration, and no further iterations are warranted. The final results are indicated in Fig. 6.29(b).

The reader justifiably might be wondering about the necessity for using an iterative scheme in solving a linear (or linearized) problem. Indeed, we can obtain the exact solution (if we can call anything we are doing with a first-order model exact!) by writing appropriate equations. The reader is encouraged to find this solution and then compare the results with those obtained above. It is important to emphasize, however, that in most such problems it is quite sufficient to obtain an approximate solution, provided we can obtain it quickly and, of course, correctly.

In the above examples, we frequently used a precise value of α to calculate the collector current. Since $\alpha \simeq 1$, the error in such calculations will be very small if one assumes $\alpha = 1$ and $I_C = I_E$. Therefore, except in calculations that depend critically on the value of α (e.g., the calculation of base current), one usually assumes $\alpha \simeq 1$.

EXERCISES

6.29 For the circuit in Fig. 6.29, find the total current drawn from the power supply. Hence find the power dissipated in the circuit.
Ans. 4.135 mA; 62 mW

6.30 The circuit in Fig. E6.30 is to be connected to the circuit in Fig. 6.29(a) as indicated; specifically, the base of Q_3 is to be connected to the collector of Q_2. If Q_3 has $\beta = 100$, find the new value of V_{C2} and the values of V_{E3} and I_{C3}.

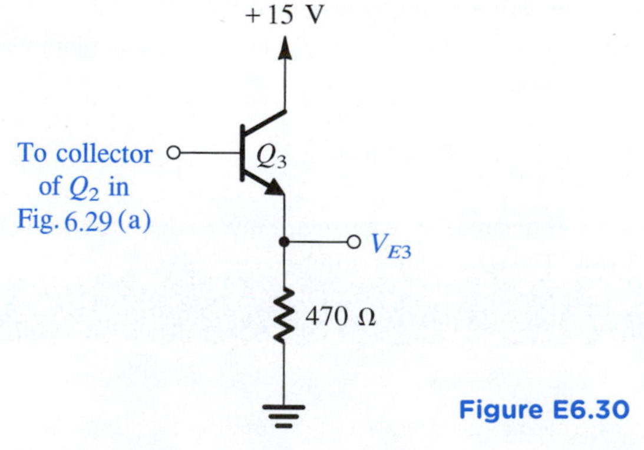

Figure E6.30

Ans. +7.06 V; +6.36 V; 13.4 mA

Example 6.12

We desire to evaluate the voltages at all nodes and the currents through all branches in the circuit of Fig. 6.30(a). Assume $\beta = 100$.

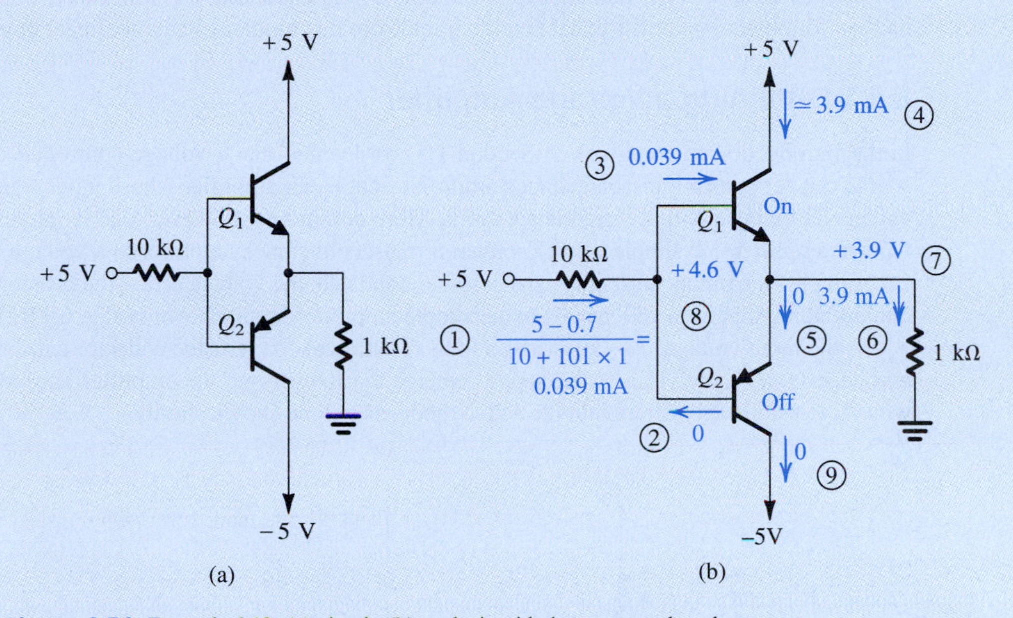

Figure 6.30 Example 6.12: **(a)** circuit; **(b)** analysis with the steps numbered.

Solution

By examining the circuit, we conclude that the two transistors Q_1 and Q_2 cannot be simultaneously conducting. Thus if Q_1 is on, Q_2 will be off, and vice versa. Assume that Q_2 is on. It follows that current will flow from ground through the 1-kΩ resistor into the emitter of Q_2. Thus the base of Q_2 will be at a negative voltage, and base current will be flowing out of the base through the 10-kΩ resistor and into the +5-V supply. This is impossible, since if the base is negative, current in the 10-kΩ resistor will have to flow into the base. Thus we conclude that our original assumption—that Q_2 is on—is incorrect. It follows that Q_2 will be off and Q_1 will be on.

The question now is whether Q_1 is active or saturated. The answer in this case is obvious: Since the base is fed with a +5-V supply and since base current flows into the base of Q_1, it follows that the base of Q_1 will be at a voltage lower than +5 V. Thus the collector–base junction of Q_1 is reverse biased and Q_1 is in the active mode. It remains only to determine the currents and voltages using techniques already described in detail. The results are given in Fig. 6.30(b).

EXERCISES

6.31 Solve the problem in Example 6.12 for the case of a voltage of –5 V feeding the bases. What voltage appears at the emitters?
Ans. –3.9 V

6.32 Solve the problem in Example 6.12 with the voltage feeding the bases changed to +10 V. Assume that $\beta_{min} = 30$, and find V_E, V_B, I_{C1}, and I_{C2}.
Ans. +4.8 V; +5.5 V; 4.35 mA; 0

6.4 Applying the BJT in Amplifier Design

We now begin our study of the utilization of the BJT in the design of amplifiers.[13] The basis for this important application is that when operated in the active mode, the BJT functions as a voltage-controlled current source: the base–emitter voltage v_{BE} controls the collector current i_C. Although the control relationship is nonlinear (exponential), we will shortly devise a method for obtaining almost-linear amplification from this fundamentally nonlinear device.

6.4.1 Obtaining a Voltage Amplifier

In the introduction to amplifiers in Section 1.5, we learned that a voltage-controlled current source can serve as a transconductance amplifier, that is, an amplifier whose input signal is a voltage and whose output signal is a current. More commonly, however, one is interested in voltage amplifiers. A simple way to convert a transconductance amplifier to a voltage amplifier is to pass the output current through a resistor and take the voltage across the resistor as the output. Doing this for a BJT results in the simple amplifier circuit shown in Fig. 6.31(a). Here v_{BE} is the input voltage, R_C (known as a **load resistance**) converts the collector current i_C to a voltage ($i_C R_C$), and V_{CC} is the supply voltage that powers up the amplifier and, together with R_C, establishes operation in the active mode, as will be shown shortly.

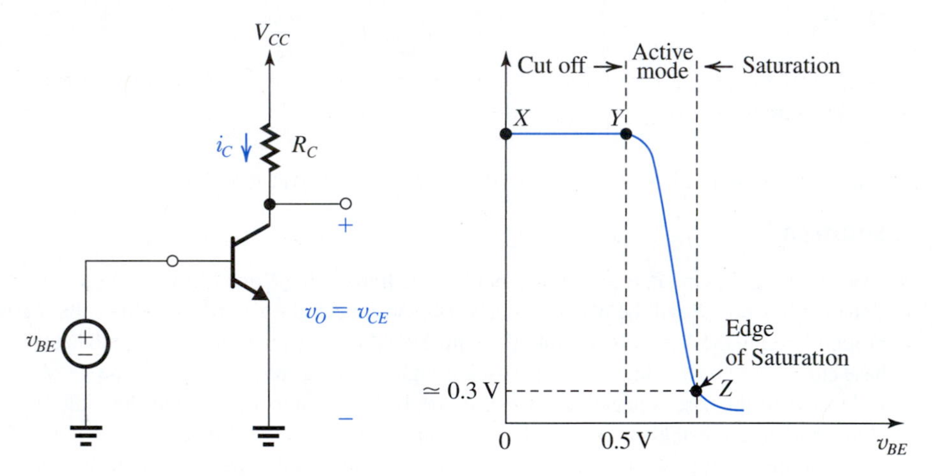

Figure 6.31 **(a)** Simple BJT amplifier with input v_{BE} and output v_{CE}. **(b)** The voltage transfer characteristic (VTC) of the amplifier in **(a)**. The three segments of the VTC correspond to the three modes of operation of the BJT.

In the amplifier circuit of Fig. 6.31(a), the output voltage is taken between the collector and ground, rather than simply across R_C. This is done because of the need to maintain a ground reference throughout the circuit. The output voltage v_{CE} is given by

$$v_{CE} = V_{CC} - i_C R_C \tag{6.24}$$

Thus it is an inverted version (note the minus sign) of $i_C R_C$ that is shifted by the constant value of the supply voltage V_{CC}.

[13]An introduction to amplifiers from an external terminals perspective is presented in Sections 1.4 and 1.5. It would be helpful for readers unfamiliar with basic amplifier concepts to review this material before proceeding with the study of BJT amplifiers.

6.4.2 The Voltage Transfer Characteristic (VTC)

A very useful tool that yields great insight into the operation of an amplifier circuit is its voltage transfer characteristic (VTC). This is simply a plot (or a clearly labeled sketch) of the output voltage versus the input voltage. For the BJT amplifier in Fig. 6.31(a), this is the plot of v_{CE} versus v_{BE} shown in Fig. 6.31(b).

Observe that for v_{BE} lower than about 0.5 V, the transistor is cut off, $i_C = 0$, and, from Eq. (6.24), $v_{CE} = V_{CC}$. As v_{BE} rises, the transistor turns on and v_{CE} decreases. However, since initially v_{CE} will still be high, the BJT will be operating in the active mode. This continues as v_{BE} is increased until it reaches a value that results in v_{CE} becoming lower than v_{BE} by 0.4 volt or so (point Z on the VTC in Fig. 6.31b). For v_{BE} greater than that at point Z, the transistor operates in the saturation region and v_{CE} decreases very slowly.

The VTC in Fig. 6.31(b) indicates that the segment of greatest slope (and hence potentially the largest amplifier gain) is that labeled YZ, which corresponds to operation in the active mode. An expression for the segment YZ can be obtained by substituting for i_C in Eq. (6.24) by its active-mode value

$$i_C = I_S e^{v_{BE}/V_T} \tag{6.25}$$

where we have for simplicity neglected base-width modulation (the Early effect). The result is

$$v_{CE} = V_{CC} - R_C I_S e^{v_{BE}/V_T} \tag{6.26}$$

This is obviously a nonlinear relationship. Nevertheless, linear (or almost-linear) amplification can be obtained by using the technique of biasing the BJT.

6.4.3 Biasing the BJT to Obtain Linear Amplification

Biasing enables us to obtain almost-linear amplification from the BJT. The technique is illustrated in Fig. 6.32(a). A dc voltage V_{BE} is selected to obtain operation at a point Q on the segment YZ of the VTC. How to select an appropriate location for the bias point Q will be discussed shortly. For the time being, observe that the coordinates of Q are the dc

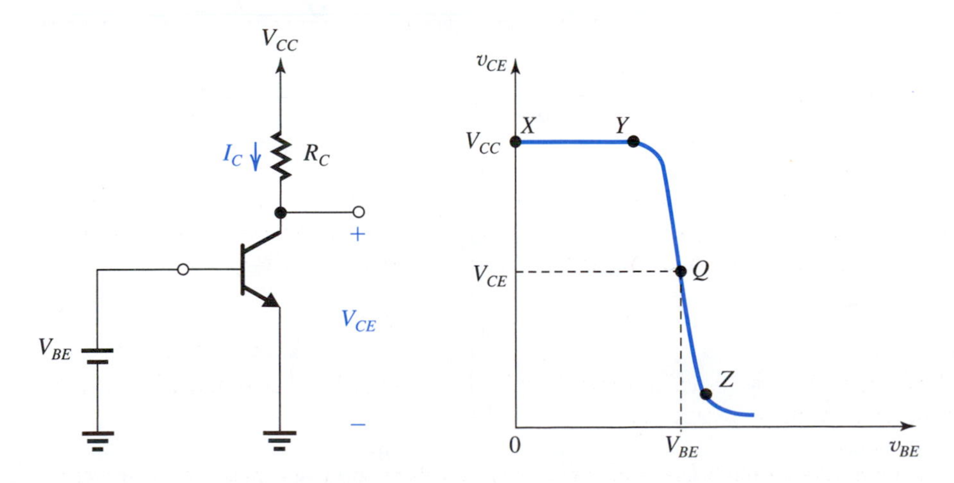

Figure 6.32 Biasing the BJT amplifier at a point Q located on the active-mode segment of the VTC.

voltages V_{BE} and V_{CE}, which are related by

$$V_{CE} = V_{CC} - R_C I_S e^{V_{BE}/V_T} \tag{6.27}$$

Point Q is known as the **bias point** or the **dc operating point**. Also, since at Q no signal component is present, it is also known as the **quiescent point** (which is the origin of the symbol Q). Note that a transistor operating at Q will have a collector current I_C given by

$$I_C = I_S e^{V_{BE}/V_T} \tag{6.28}$$

Next, the signal to be amplified v_{be}, a function of time t, is superimposed on the bias voltage V_{BE}, as shown in Fig. 6.33(a). Thus the total instantaneous value of v_{BE} becomes

$$v_{BE}(t) = V_{BE} + v_{be}(t)$$

The resulting $v_{CE}(t)$ can be obtained by substituting this expression for $v_{BE}(t)$ into Eq. (6.25). Graphically, we can use the VTC to obtain $v_{CE}(t)$, point by point, as illustrated in Fig. 6.33(b).

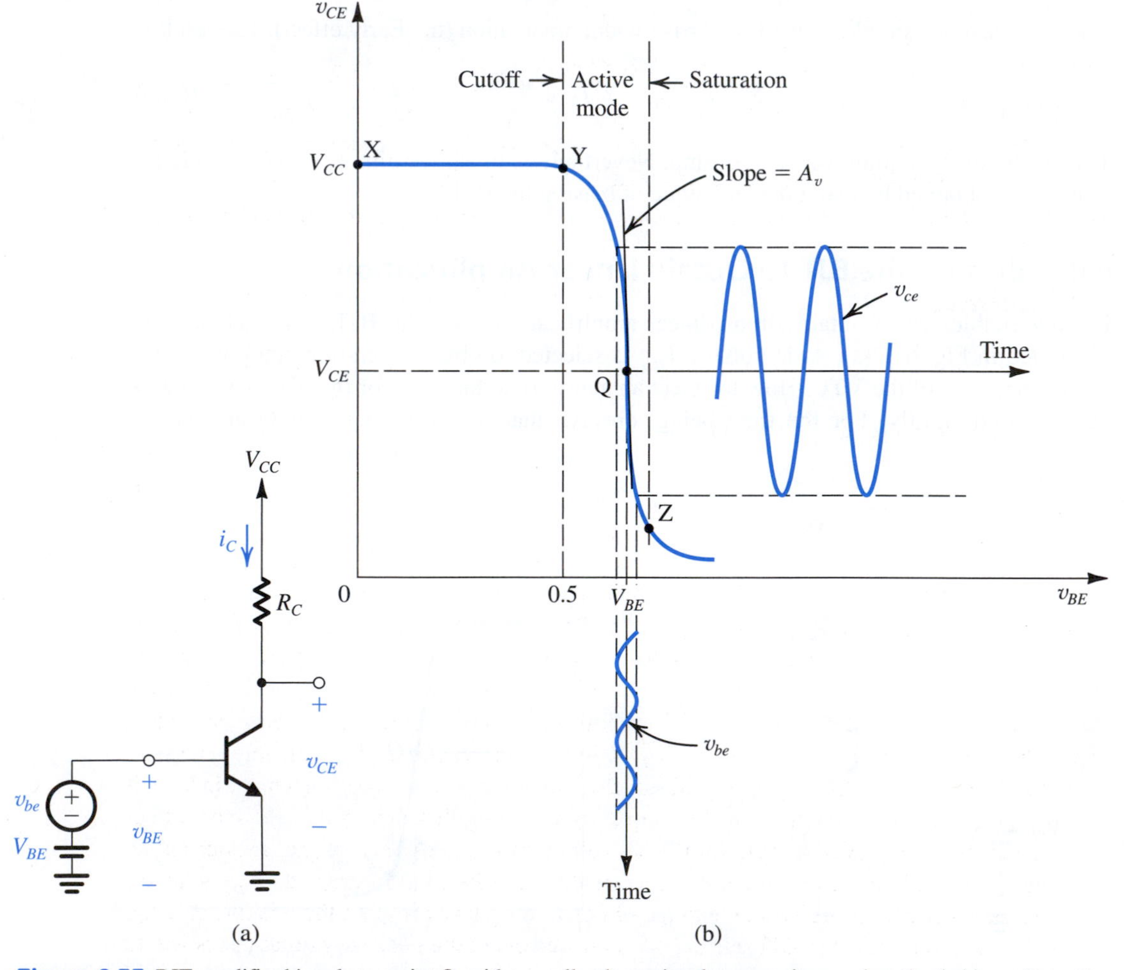

Figure 6.33 BJT amplifier biased at a point Q, with a small voltage signal v_{be} superimposed on the dc bias voltage V_{BE}. The resulting output signal v_{ce} appears superimposed on the dc collector voltage V_{CE}. The amplitude of v_{ce} is larger than that of v_{be} by the voltage gain A_v.

Here we show the case when v_{be} is a sine wave of "small" amplitude. Specifically, the amplitude of v_{be} is small enough to restrict the excursion of the instantaneous operating point to a short almost-linear segment of the VTC around the bias point Q. The shorter the segment, the greater the linearity achieved, and the closer to an ideal sine wave the signal component at the output, v_{ce}, will be. This is the essence of obtaining linear amplification from the nonlinear BJT.

6.4.4 The Small-Signal Voltage Gain

If the input signal v_{be} is kept small, the corresponding signal at the output v_{ce} will be nearly proportional to v_{be} with the constant of proportionality being the slope of the almost-linear segment of the VTC around Q. This is the voltage gain of the amplifier, and its value can be determined by evaluating the slope of the tangent to the VTC at the bias point Q,

$$A_v \equiv \left.\frac{dv_{CE}}{dv_{BE}}\right|_{v_{BE}=V_{BE}} \tag{6.29}$$

Utilizing Eq. (6.26) together with Eq. (6.28), we obtain

$$A_v = -\left(\frac{I_C}{V_T}\right)R_C \tag{6.30}$$

We make the following observations on this expression for the voltage gain:

1. The gain is negative, which signifies that the amplifier is inverting; that is, there is a 180° phase shift between the input and the output. This inversion is obvious in Fig. 6.33(b) and should have been anticipated from Eq. (6.26).
2. The gain is proportional to the collector bias current I_C and to the load resistance R_C.

Additional insight into the voltage gain A_v can be obtained by expressing Eq. (6.30) as

$$A_v = -\frac{I_C R_C}{V_T} = -\frac{V_{RC}}{V_T} \tag{6.31}$$

where V_{RC} is the dc voltage drop across R_C,

$$V_{RC} = V_{CC} - V_{CE} \tag{6.32}$$

The simple expression in Eq. (6.31) indicates that the voltage gain of the amplifier is the ratio of the dc voltage drop across R_C to the thermal voltage V_T ($\simeq$ 25 mV at room temperature). It follows that to maximize the voltage gain we should use as large a voltage drop across R_C as possible. For a given value of V_{CC}, Eq. (6.32) indicates that to increase V_{RC} we have to operate at a lower V_{CE}. However, reference to Fig. 6.33(b) shows that a lower V_{CE} means a bias point Q close to the end of the active-region segment, which might not leave sufficient room for the negative-output signal swing without the amplifier entering the saturation region. If this happens, the negative peaks of the waveform of v_{ce} will be flattened. Indeed, it is the need to allow sufficient room for output signal swing that determines the most effective placement of the bias point Q on the active-region segment, YZ, of the transfer curve. Placing Q too high on this segment not only results in reduced gain (because V_{RC} is lower) but could possibly limit the available range of positive signal swing. At the positive end, the limitation is imposed by the BJT cutting off, in which event the positive-output peaks would be clipped off at a level equal to V_{CC}. Finally, it is useful to note that the theoretical maximum gain A_v is obtained by biasing the BJT at the edge of saturation, which of course would not leave any room for negative signal swing. The resulting gain is given by

$$A_v = -\frac{V_{CC} - V_{CEsat}}{V_T} \tag{6.33}$$

Thus,

$$|A_{v\max}| \simeq \frac{V_{CC}}{V_T} \tag{6.34}$$

Although the gain can be increased by using a larger supply voltage, other considerations come into play when one is determining an appropriate value for V_{CC}. In fact, the trend has been toward using lower and lower supply voltages, currently approaching 1 V or so. At such low supply voltages, large gain values can be obtained by replacing the resistance R_C with a constant-current source, as will be seen in Chapter 7.

Example 6.13

Consider an amplifier circuit using a BJT having $I_S = 10^{-15}$ A, a collector resistance $R_C = 6.8$ kΩ, and a power supply $V_{CC} = 10$ V.

(a) Determine the value of the bias voltage V_{BE} required to operate the transistor at $V_{CE} = 3.2$ V. What is the corresponding value of I_C?
(b) Find the voltage gain A_v at this bias point. If an input sine-wave signal of 5-mV peak amplitude is superimposed on V_{BE}, find the amplitude of the output sine-wave signal (assume linear operation).
(c) Find the positive increment in v_{BE} (above V_{BE}) that drives the transistor to the edge of saturation, where $v_{CE} = 0.3$ V.
(d) Find the negative increment in v_{BE} that drives the transistor to within 1% of cutoff (i.e., to $v_{CE} = 0.99V_{CC}$).

Solution

(a)

$$I_C = \frac{V_{CC} - V_{CE}}{R_C}$$

$$= \frac{10 - 3.2}{6.8} = 1 \text{ mA}$$

The value of V_{BE} can be determined from

$$1 \times 10^{-3} = 10^{-15} e^{V_{BE}/V_T}$$

which results in

$$V_{BE} = 690.8 \text{ mV}$$

(b)

$$A_v = -\frac{V_{CC} - V_{CE}}{V_T}$$

$$= -\frac{10 - 3.2}{0.025} = -272 \text{ V/V}$$

$$\hat{V}_{ce} = 272 \times 0.005 = 1.36 \text{ V}$$

(c) For $v_{CE} = 0.3$ V,

$$i_C = \frac{10 - 0.3}{6.8} = 1.617 \text{ mA}$$

To increase i_C from 1 mA to 1.617 mA, v_{BE} must be increased by

$$\Delta v_{BE} = V_T \ln\left(\frac{1.617}{1}\right)$$

$$= 12 \text{ mV}$$

(d) For $v_{CE} = 0.99V_{CC} = 9.9$ V,

$$i_C = \frac{10 - 9.9}{6.8} = 0.0147 \text{ mA}$$

To decrease i_C from 1 mA to 0.0147 mA, v_{BE} must change by

$$\Delta v_{BE} = V_T \ln\left(\frac{0.0147}{1}\right)$$

$$= -105.5 \text{ mV}$$

EXERCISE

6.33 For the situation described in Example 6.13, while keeping I_C unchanged at 1 mA, find the value of R_C that will result in a voltage gain of –320 V/V. What is the largest negative signal swing allowed at the output (assume that v_{CE} is not to decrease below 0.3 V)? What (approximately) is the corresponding input signal amplitude? (Assume linear operation.)
Ans. 8 kΩ; 1.7 V; 5.3 mV

6.4.5 Determining the VTC by Graphical Analysis

Figure 6.34 shows a graphical method for determining the VTC of the amplifier of Figure 6.33(a). Although graphical analysis of transistor circuits is rarely employed in practice, it is useful for us at this stage in gaining greater insight into circuit operation, especially in answering the question of where to locate the bias point Q.

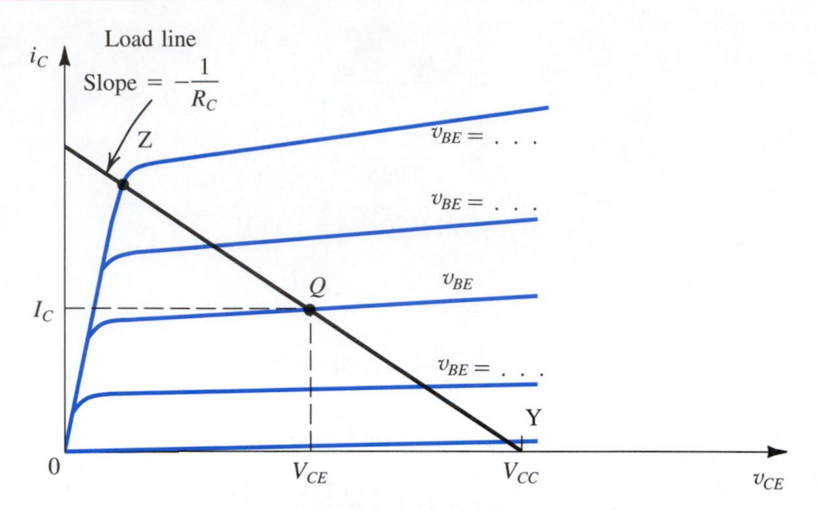

Figure 6.34 Graphical construction for determining the VTC of the amplifier circuit of Fig. 6.33(a).

The graphical analysis is based on the observation that for each value of v_{BE}, the circuit will be operating at the point of intersection of the corresponding $i_C - v_{CE}$ graph and the straight line representing Eq. (6.24), which can be rewritten in the form

$$i_C = \frac{V_{CC}}{R_C} - \frac{1}{R_C}v_{CE} \tag{6.35}$$

The straight line representing this relationship is superimposed on the $i_C - v_{CE}$ characteristics in Fig. 6.34. It intersects the horizontal axis at $v_{CE} = V_{CC}$ and has a slope of $-1/R_C$. Since this straight line represents in effect the load resistance R_C, it is called the **load line**. The VTC is then determined point by point. Note that we have labeled three important points: point Y at which $v_{BE} = 0.5$ V, point Q at which the BJT can be biased for amplifier operation ($v_{BE} = V_{BE}$ and $v_{CE} = V_{CE}$), and point Z at which the BJT leaves the active mode and enters the saturation region. If the BJT is to be used as a switch, then operating points Y and Z are applicable: At Y the transistor is off (open switch), and at Z the transistor operates as a low valued resistance R_{CEsat} and has a small voltage drop (closed switch). It should be noted, however, that because of the long delay time needed to turn off a saturated BJT, modern digital integrated circuits no longer utilize the saturated mode of operation. Nonsaturated BJT digital circuits will be studied in Chapter 14.

6.4.6 Locating the Bias Point Q

The bias point Q is determined by the value of V_{BE} and that of the load resistance R_C. Two important considerations in deciding on the location of Q are the gain and the allowable signal swing at the output. To illustrate, consider the VTC shown in Fig. 6.33(b). Here the value of R_C is fixed, and the only variable remaining is the value of V_{BE}. Since the slope increases as we move closer to point Z, we obtain higher gain by locating Q as close to Z as possible. However, the closer Q is to the boundary point Z, the smaller the allowable magnitude of negative signal swing. Thus, as usual in engineering design, we encounter a situation requiring a trade-off.

In deciding on a value for R_C it is useful to refer to the $i_C - v_{CE}$ plane. Figure 6.35 shows two load lines resulting in two extreme bias points: Point Q_A, is too close to V_{CC}, resulting in a severe constraint on the positive signal swing of v_{ce}. Exceeding the allowable positive

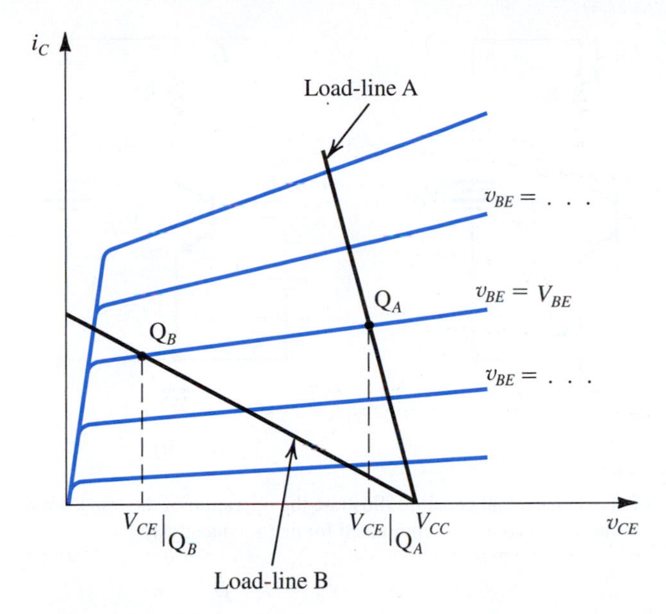

Figure 6.35 Effect of bias-point location on allowable signal swing: Load line A results in bias point Q_A with a corresponding V_{CE} that is too close to V_{CC} and thus limits the positive swing of v_{CE}. At the other extreme, load line B results in an operating point, Q_B, too close to the saturation region, thus limiting the negative swing of v_{CE}.

maximum results in the positive peaks of the signal being clipped off, since the BJT will turn off for the part of each cycle near the positive peak. We speak of this situation as the circuit not having sufficient "headroom." Similarly, point Q_B is too close to the boundary of the saturation region, thus severely limiting the allowable negative signal swing of v_{ce}. Exceeding this limit would result in the transistor entering the saturation region for part of each cycle near the negative peaks, resulting in a distorted output signal. We speak of this situation as the circuit not having sufficient "legroom." We will have more to say on bias design in Section 6.7.

6.5 Small-Signal Operation and Models

Having learned the basis for the operation of the BJT as an amplifier, we now take a closer look at the small-signal operation of the transistor. Toward that end, consider once more the *conceptual* amplifier circuit shown in Fig. 6.36(a). Here the base–emitter junction is forward biased by a dc voltage V_{BE} (battery). The reverse bias of the collector–base junction is established by connecting the collector to another power supply of voltage V_{CC} through a resistor R_C. The input signal to be amplified is represented by the voltage source v_{be} that is superimposed on V_{BE}.

We consider first the dc bias conditions by setting the signal v_{be} to zero. The circuit reduces to that in Fig. 6.36(b), and we can write the following relationships for the dc currents and voltages:

$$I_C = I_S e^{V_{BE}/V_T} \tag{6.36}$$

$$I_E = I_C/\alpha \tag{6.37}$$

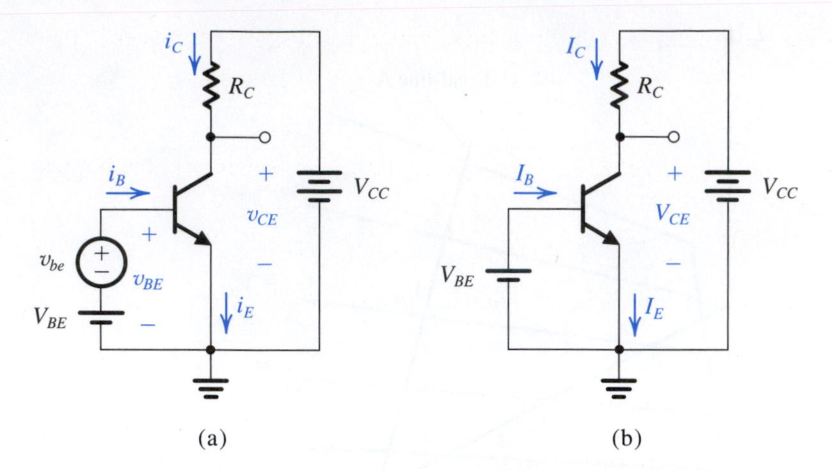

Figure 6.36 **(a)** Conceptual circuit to illustrate the operation of the transistor as an amplifier. **(b)** The circuit of **(a)** with the signal source v_{be} eliminated for dc (bias) analysis.

$$I_B = I_C/\beta \tag{6.38}$$

$$V_{CE} = V_{CC} - I_C R_C \tag{6.39}$$

Obviously, for active-mode operation, V_C should be greater than $(V_B - 0.4)$ by an amount that allows for the required signal swing at the collector.

6.5.1 The Collector Current and the Transconductance

If a signal v_{be} is applied as shown in Fig. 6.36(a), the total instantaneous base–emitter voltage v_{BE} becomes

$$v_{BE} = V_{BE} + v_{be}$$

Correspondingly, the collector current becomes

$$i_C = I_S e^{v_{BE}/V_T} = I_S e^{(V_{BE}+v_{be})/V_T}$$

$$= I_S e^{V_{BE}/V_T} e^{v_{be}/V_T}$$

Use of Eq. (6.36) yields

$$i_C = I_C e^{v_{be}/V_T} \tag{6.40}$$

Now, if $v_{be} \ll V_T$, we may approximate Eq. (6.40) as

$$i_C \simeq I_C\left(1 + \frac{v_{be}}{V_T}\right) \tag{6.41}$$

Here we have expanded the exponential in Eq. (6.40) in a series and retained only the first two terms. This approximation, which is valid only for v_{be} less than approximately 10 mV, is referred to as the **small-signal approximation**. Under this approximation, the total collector current is given by Eq. (6.41) and can be rewritten

$$i_C = I_C + \frac{I_C}{V_T} v_{be} \tag{6.42}$$

Thus the collector current is composed of the dc bias value I_C and a signal component i_c,

$$i_c = \frac{I_C}{V_T} v_{be} \tag{6.43}$$

This equation relates the signal current in the collector to the corresponding base–emitter signal voltage. It can be rewritten as

$$i_c = g_m v_{be} \tag{6.44}$$

where g_m is called the **transconductance**, and from Eq. (6.43), it is given by

$$g_m = \frac{I_C}{V_T} \tag{6.45}$$

We observe that the transconductance of the BJT is directly proportional to the collector bias current I_C. Thus to obtain a constant predictable value for g_m, we need a constant predictable I_C. Finally, we note that BJTs have relatively high transconductance (as compared to MOSFETs, which we studied in Chapter 5); for instance, at $I_C = 1$ mA, $g_m \simeq 40$ mA/V.

A graphical interpretation for g_m is given in Fig. 6.37, where it is shown that g_m is equal to the slope of the i_C–v_{BE} characteristic curve at $i_C = I_C$ (i.e., at the bias point Q). Thus,

$$g_m = \left.\frac{\partial i_C}{\partial v_{BE}}\right|_{i_C = I_C} \tag{6.46}$$

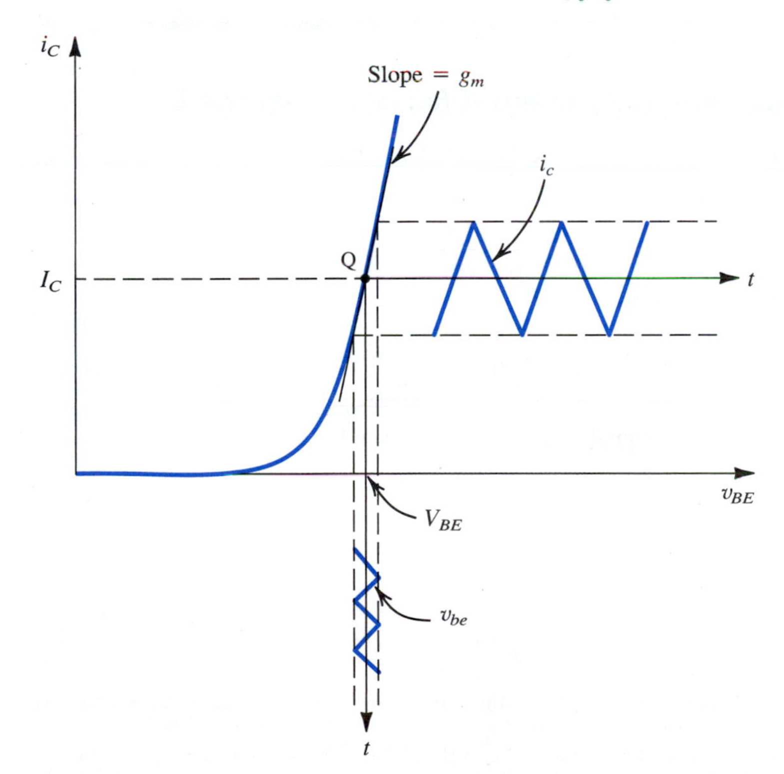

Figure 6.37 Linear operation of the transistor under the small-signal condition: A small signal v_{be} with a triangular waveform is superimposed on the dc voltage V_{BE}. It gives rise to a collector signal current i_c, also of triangular waveform, superimposed on the dc current I_C. Here, $i_c = g_m v_{be}$, where g_m is the slope of the i_C–v_{BE} curve at the bias point Q.

The small-signal approximation implies keeping the signal amplitude sufficiently small that *operation is restricted to an almost-linear segment of the i_C–v_{BE} exponential curve.* Increasing the signal amplitude will result in the collector current having components nonlinearly related to v_{be}. This, of course, is the same approximation that we discussed in the context of the amplifier transfer curve in Section 6.4.

The analysis above suggests that for small signals ($v_{be} \ll V_T$), the transistor behaves as a voltage-controlled current source. The input port of this controlled source is between base and emitter, and the output port is between collector and emitter. The transconductance of the controlled source is g_m, and the output resistance is infinite. The latter ideal property is a result of our first-order model of transistor operation in which the collector voltage has no effect on the collector current in the active mode. As we have seen in Section 6.2, practical BJTs have finite output resistance because of the Early effect. The effect of the output resistance on amplifier performance will be considered later.

EXERCISES

6.34 Use Eq. (6.46) to derive the expression for g_m in Eq. (6.45).

6.35 Calculate the value of g_m for a BJT biased at $I_C = 0.5$ mA.
Ans. 20 mA/V

6.5.2 The Base Current and the Input Resistance at the Base

To determine the resistance seen by v_{be}, we first evaluate the total base current i_B using Eq. (6.42), as follows:

$$i_B = \frac{i_C}{\beta} = \frac{I_C}{\beta} + \frac{1}{\beta}\frac{I_C}{V_T}v_{be}$$

Thus,

$$i_B = I_B + i_b \tag{6.47}$$

where I_B is equal to I_C/β and the signal component i_b is given by

$$i_b = \frac{1}{\beta}\frac{I_C}{V_T}v_{be} \tag{6.48}$$

Substituting for I_C/V_T by g_m gives

$$i_b = \frac{g_m}{\beta}v_{be} \tag{6.49}$$

The small-signal input resistance between base and emitter, *looking into the base,* is denoted by r_π and is defined as

$$r_\pi \equiv \frac{v_{be}}{i_b} \tag{6.50}$$

Using Eq. (6.49) gives

$$r_\pi = \frac{\beta}{g_m} \tag{6.51}$$

Thus r_π is directly dependent on β and is inversely proportional to the bias current I_C. Substituting for g_m in Eq. (6.51) from Eq. (6.45) and replacing I_C/β by I_B gives an alternative expression for r_π,

$$r_\pi = \frac{V_T}{I_B} \tag{6.52}$$

EXERCISE

6.36 A BJT amplifier is biased to operate at a constant collector current $I_C = 0.5$ mA irrespective of the value β. If the transistor manufacturer specifies β to range from 50 to 200, give the expected range of g_m, I_B, and r_π.
Ans. g_m is constant at 20 mA/V; $I_B = 10$ μA to 2.5 μA; $r_\pi = 2.5$ kΩ to 10 kΩ

6.5.3 The Emitter Current and the Input Resistance at the Emitter

The total emitter current i_E can be determined from

$$i_E = \frac{i_C}{\alpha} = \frac{I_C}{\alpha} + \frac{i_c}{\alpha}$$

Thus,

$$i_E = I_E + i_e \tag{6.53}$$

where I_E is equal to I_C/α and the signal current i_e is given by

$$i_e = \frac{i_c}{\alpha} = \frac{I_C}{\alpha V_T} v_{be} = \frac{I_E}{V_T} v_{be} \tag{6.54}$$

If we denote the small-signal resistance between base and emitter *looking into the emitter* by r_e, it can be defined as

$$r_e \equiv \frac{v_{be}}{i_e} \tag{6.55}$$

Using Eq. (6.54) we find that r_e, called the **emitter resistance**, is given by

$$r_e = \frac{V_T}{I_E} \tag{6.56}$$

Comparison with Eq. (6.45) reveals that

$$r_e = \frac{\alpha}{g_m} \simeq \frac{1}{g_m} \tag{6.57}$$

The relationship between r_π and r_e can be found by combining their respective definitions in Eqs. (6.50) and (6.55) as

$$v_{be} = i_b r_\pi = i_e r_e$$

Thus,

$$r_\pi = (i_e/i_b) r_e$$

which yields

$$r_\pi = (\beta + 1) r_e \tag{6.58}$$

Figure 6.38 illustrates the definition of r_π and r_e.

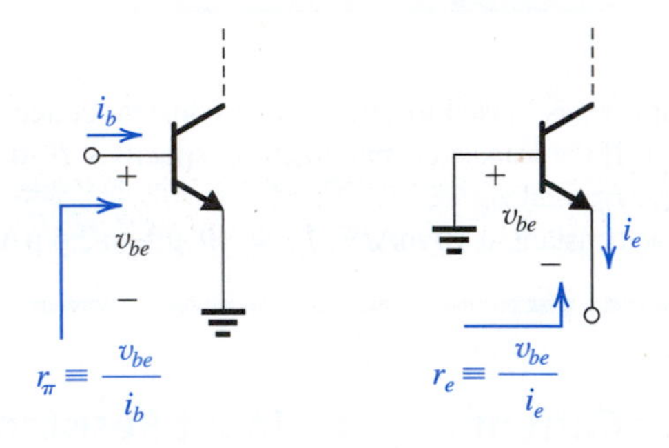

Figure 6.38 Illustrating the definition of r_π and r_e.

EXERCISE

6.37 A BJT having $\beta = 100$ is biased at a dc collector current of 1 mA. Find the value of g_m, r_e, and r_π at the bias point.
Ans. 40 mA/V; 25 Ω; 2.5 kΩ

6.5.4 Voltage Gain

We have established above that the transistor senses the base–emitter signal v_{be} and causes a proportional current $g_m v_{be}$ to flow in the collector lead at a high (ideally infinite) impedance level. In this way the transistor is acting as a voltage-controlled current source. To obtain an output voltage signal, we may force this current to flow through a resistor, as is done in Fig. 6.36(a). Then the total collector voltage v_{CE} will be

$$\begin{aligned} v_{CE} &= V_{CC} - i_C R_C \\ &= V_{CC} - (I_C + i_c) R_C \\ &= (V_{CC} - I_C R_C) - i_c R_C \\ &= V_{CE} - i_c R_C \end{aligned} \tag{6.59}$$

Here the quantity V_{CE} is the dc bias voltage at the collector, and the signal voltage is given by

$$v_{ce} = -i_c R_C = -g_m v_{be} R_C \qquad (6.60)$$

$$= (-g_m R_C) v_{be}$$

Thus the voltage gain of this amplifier A_v is

$$A_v \equiv \frac{v_{ce}}{v_{be}} = -g_m R_C \qquad (6.61)$$

Here again we note that because g_m is directly proportional to the collector bias current, the gain will be as stable as the collector bias current is made. Substituting for g_m from Eq. (6.45) enables us to express the gain in the form

$$A_v = -\frac{I_C R_C}{V_T} \qquad (6.62)$$

which is identical to the expression we derived in Section 6.4 (Eq. 6.31).

EXERCISE

6.38 In the circuit of Fig. 6.36(a), V_{BE} is adjusted to yield a dc collector current of 1 mA. Let $V_{CC} = 15$ V, $R_C = 10$ kΩ, and $\beta = 100$. Find the voltage gain v_{ce}/v_{be}. If $v_{be} = 0.005 \sin \omega t$ volt, find $v_C(t)$ and $i_B(t)$.
Ans. −400 V/V; $5 - 2 \sin \omega t$ volts; $10 + 2 \sin \omega t$ μA

6.5.5 Separating the Signal and the DC Quantities

The analysis above indicates that every current and voltage in the amplifier circuit of Fig. 6.36(a) is composed of two components: a dc component and a signal component. For instance, $v_{BE} = V_{BE} + v_{be}$, $I_C = I_C + i_c$, and so on. The dc components are determined from the dc circuit given in Fig. 6.36(b) and from the relationships imposed by the transistor (Eqs. 6.36 through 6.38). On the other hand, a representation of the signal operation of the BJT can be obtained by eliminating the dc sources, as shown in Fig. 6.39. Observe that since the voltage of an ideal dc supply does not change, the signal voltage across it will be zero. For this reason we have replaced V_{CC} and V_{BE} with short circuits. Had the circuit contained ideal dc

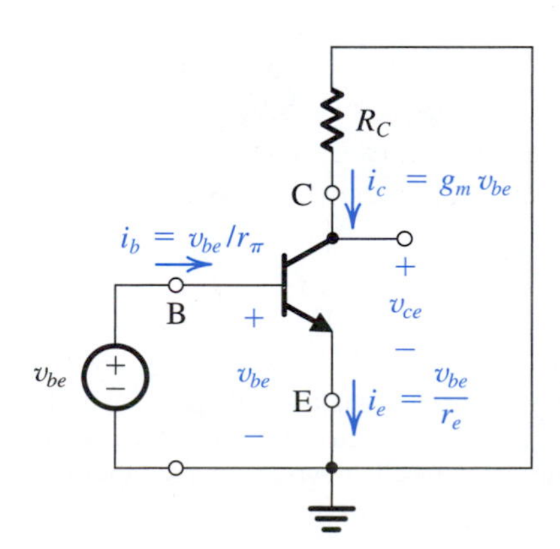

Figure 6.39 The amplifier circuit of Fig. 6.36(a) with the dc sources (V_{BE} and V_{CC}) eliminated (short-circuited). Thus only the signal components are present. Note that this is a representation of the signal operation of the BJT and not an actual amplifier circuit.

current sources, these would have been replaced by open circuits. Note, however, that the circuit of Fig. 6.39 is useful only in so far as it shows the various signal currents and voltages; it is *not* an actual amplifier circuit, since the dc bias circuit is not shown.

Figure 6.39 also shows the expressions for the current increments (i_c, i_b, and i_e) obtained when a small signal v_{be} is applied. These relationships can be represented by a circuit. Such a circuit should have three terminals—C, B, and E—and should yield the same terminal currents indicated in Fig. 6.39. The resulting circuit is then *equivalent to the transistor as far as small-signal operation is concerned,* and thus it can be considered an equivalent small-signal circuit model.

6.5.6 The Hybrid-π Model

An equivalent circuit model for the BJT is shown in Fig. 6.40(a). This model represents the BJT as a voltage-controlled current source and explicitly includes the input resistance looking into the base, r_π. The model obviously yields $i_c = g_m v_{be}$ and $i_b = v_{be}/r_\pi$. Not so obvious, however, is the fact that the model also yields the correct expression for i_e. This can be shown as follows: At the emitter node we have

$$
\begin{aligned}
i_e &= \frac{v_{be}}{r_\pi} + g_m v_{be} = \frac{v_{be}}{r_\pi}(1 + g_m r_\pi) \\
&= \frac{v_{be}}{r_\pi}(1 + \beta) = v_{be} \bigg/ \left(\frac{r_\pi}{1+\beta}\right) \\
&= v_{be}/r_e
\end{aligned}
$$

A slightly different equivalent-circuit model can be obtained by expressing the current of the controlled source ($g_m v_{be}$) in terms of the base current i_b as follows:

$$
\begin{aligned}
g_m v_{be} &= g_m(i_b r_\pi) \\
&= (g_m r_\pi) i_b = \beta i_b
\end{aligned}
$$

This results in the alternative equivalent-circuit model shown in Fig. 6.40(b). Here the transistor is represented as a current-controlled current source, with the control current being i_b.

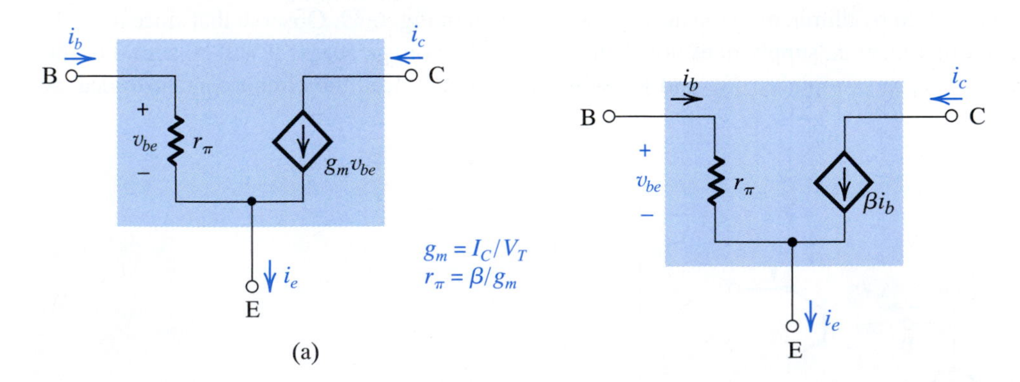

Figure 6.40 Two slightly different versions of the hybrid-π model for the small-signal operation of the BJT. The equivalent circuit in **(a)** represents the BJT as a voltage-controlled current source (a transconductance amplifier), and that in **(b)** represents the BJT as a current-controlled current source (a current amplifier).

The two models of Fig. 6.40 are simplified versions of what is known as the hybrid-π model. This is the most widely used model for the BJT.

It is important to note that the small-signal equivalent circuits of Fig. 6.40 model the operation of the BJT *at a given bias point.* This should be obvious from the fact that the model parameters g_m and r_π depend on the value of the dc bias current I_C, as indicated in Fig. 6.40. It is interesting and useful to note that the models of Fig. 6.40 (a) and (b) are the small-signal versions of the models of Fig. 6.5(c) and (d), respectively. Specifically, observe that r_π is the incremental resistance of D_B.

EXERCISE

6.39 For the model in Fig. 6.40(b) show that $i_c = g_m v_{be}$ and $i_e = v_{be}/r_e$.

6.5.7 The T Model

Although the hybrid-π model (in one of its two variants shown in Fig. 6.40) can be used to carry out small-signal analysis of any transistor circuit, there are situations in which an alternative model, shown in Fig. 6.41, is much more convenient. This model, called the **T model**, is shown in two versions in Fig. 6.41. The model of Fig. 6.41(a) represents the BJT as a voltage-controlled current source with the control voltage being v_{be}. Here, however, the resistance between base and emitter, looking into the emitter, is explicitly shown. From Fig. 6.41(a) we see clearly that the model yields the correct expressions for i_c and i_e. For i_b we note that at the base node we have

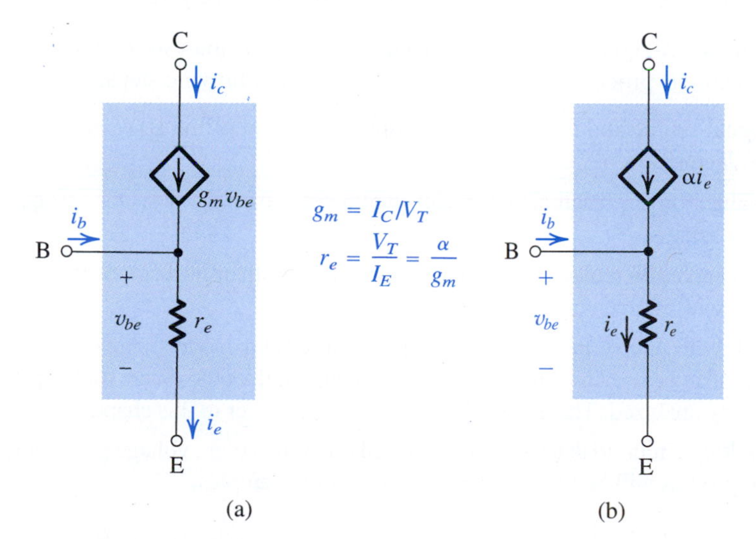

Figure 6.41 Two slightly different versions of what is known as the *T model* of the BJT. The circuit in **(a)** is a voltage-controlled current source representation and that in **(b)** is a current-controlled current source representation. These models explicitly show the emitter resistance r_e rather than the base resistance r_π featured in the hybrid-π model.

$$i_b = \frac{v_{be}}{r_e} - g_m v_{be} = \frac{v_{be}}{r_e}(1 - g_m r_e)$$

$$= \frac{v_{be}}{r_e}(1 - \alpha) = \frac{v_{be}}{r_e}\left(1 - \frac{\beta}{\beta + 1}\right)$$

$$= \frac{v_{be}}{(\beta + 1)r_e} = \frac{v_{be}}{r_\pi}$$

as should be the case.

If in the model of Fig. 6.41(a) the current of the controlled source is expressed in terms of the emitter current as

$$g_m v_{be} = g_m(i_e r_e)$$

$$= (g_m r_e) i_e = \alpha i_e$$

we obtain the alternative T model shown in Fig. 6.41(b). Here the BJT is represented as a current-controlled current source but with the control signal being i_e.

It is interesting and useful to note that the models of Fig. 6.41(a) and (b) are the small-signal versions of the models in Fig. 6.5(a) and (b), respectively. Specifically observe that r_e is the incremental resistance of D_E.

6.5.8 Small-Signal Models of the *pnp* Transistor

Although the small-signal models in Figs. 6.40 and 6.41 were developed for the case of the *npn* transistor, they apply equally well to the *pnp* transistor *with no change in polarities*.

6.5.9 Application of the Small-Signal Equivalent Circuits

The availability of the small-signal BJT circuit models makes the analysis of transistor amplifier circuits a systematic process. The process consists of the following steps:

1. Eliminate the signal source and determine the dc operating point of the BJT and in particular the dc collector current I_C.
2. Calculate the values of the small-signal model parameters: $g_m = I_C/V_T$, $r_\pi = \beta/g_m$, and $r_e = V_T/I_E = \alpha/g_m$.
3. Eliminate the dc sources by replacing each dc voltage source with a short circuit and each dc current source with an open circuit.
4. Replace the BJT with one of its small-signal equivalent circuit models. Although any one of the models can be used, one might be more convenient than the others for the particular circuit being analyzed. This point will be made clearer later in this chapter.
5. Analyze the resulting circuit to determine the required quantities (e.g., voltage gain, input resistance). The process will be illustrated by the following examples.

Example 6.14

We wish to analyze the transistor amplifier shown in Fig. 6.42(a) to determine its voltage gain v_o/v_i. Assume $\beta = 100$.

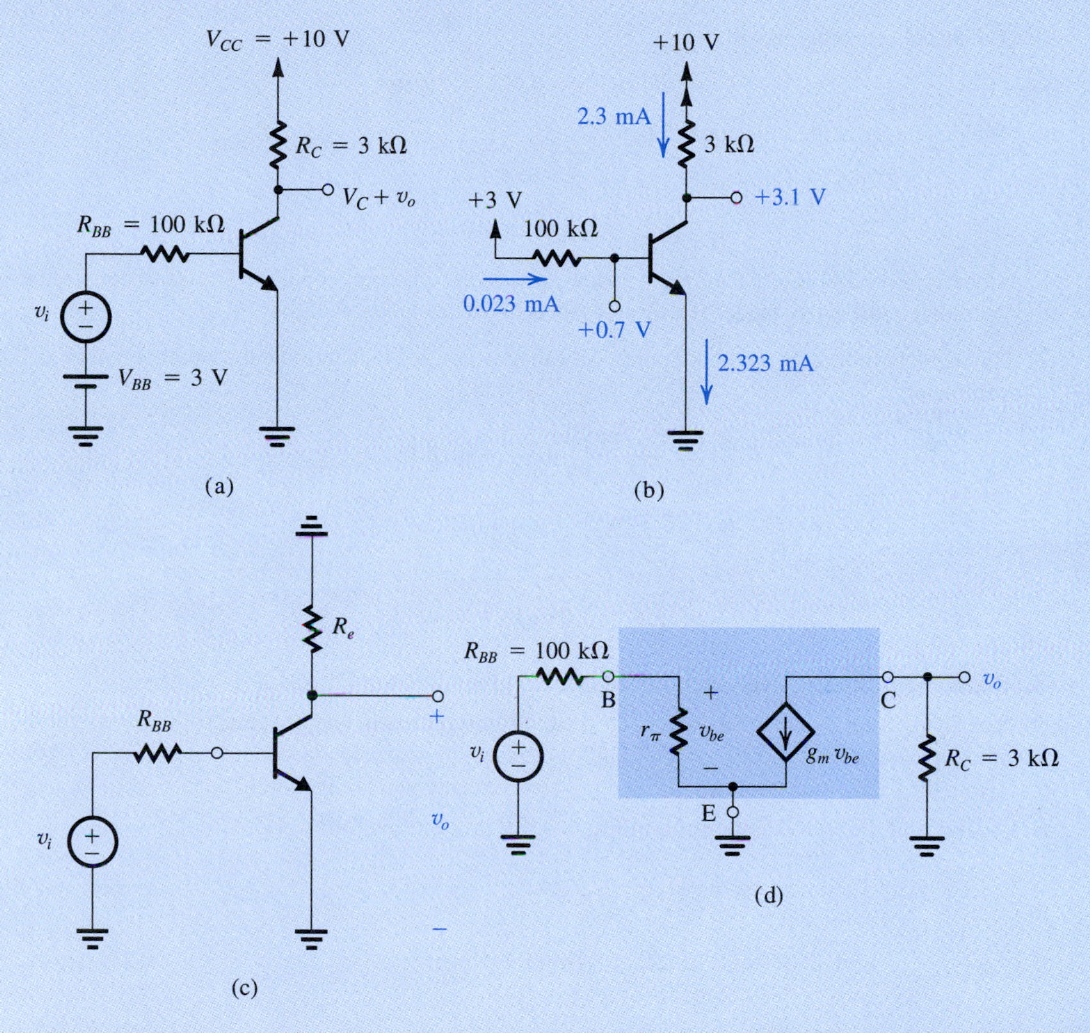

Figure 6.42 Example 6.14: **(a)** amplifier circuit; **(b)** circuit for dc analysis; **(c)** amplifier circuit with dc sources replaced by short circuits; **(d)** amplifier circuit with transistor replaced by its hybrid-π, small-signal models.

Solution

We shall follow the five-step process outlined above:

1. The first step in the analysis consists of determining the quiescent operating point. For this purpose we assume that $v_i = 0$ and thus obtain the dc circuit shown in Fig. 6.42(b). The dc base current will be

Example 6.14 *continued*

$$I_B = \frac{V_{BB} - V_{BE}}{R_{BB}}$$

$$\simeq \frac{3 - 0.7}{100} = 0.023 \text{ mA}$$

The dc collector current will be

$$I_C = \beta I_B = 100 \times 0.023 = 2.3 \text{ mA}$$

The dc voltage at the collector will be

$$V_C = V_{CC} - I_C R_C$$

$$= +10 - 2.3 \times 3 = +3.1 \text{ V}$$

Since V_B at +0.7 V is less than V_C, it follows that in the quiescent condition the transistor will be operating in the active mode. The dc analysis is illustrated in Fig. 6.42(b).

2. Having determined the operating point, we can now proceed to determine the small-signal model parameters:

$$r_e = \frac{V_T}{I_E} = \frac{25 \text{ mV}}{(2.3/0.99) \text{ mA}} = 10.8 \ \Omega$$

$$g_m = \frac{I_C}{V_T} = \frac{2.3 \text{ mA}}{25 \text{ mV}} = 92 \text{ mA/V}$$

$$r_\pi = \frac{\beta}{g_m} = \frac{100}{92} = 1.09 \text{ k}\Omega$$

3. Replacing V_{BB} and V_{CC} with short circuits results in the circuit in Fig. 6.42(c).

4. To carry out the small-signal analysis, it is equally convenient to employ either of the two hybrid-π, equivalent-circuit models of Fig. 6.40 to replace the transistor in the circuit of Fig. 6.42(c). Using the first results in the amplifier equivalent circuit given in Fig. 6.42(d).

5. Analysis of the equivalent circuit in Fig. 6.42(d) proceeds as follows:

$$v_{be} = v_i \frac{r_\pi}{r_\pi + R_{BB}}$$

$$= v_i \frac{1.09}{101.09} = 0.011 v_i \tag{6.63}$$

The output voltage v_o is given by

$$v_o = -g_m v_{be} R_C$$

$$= -92 \times 0.011 v_i \times 3 = -3.04 v_i$$

Thus the voltage gain will be

$$A_v = \frac{v_o}{v_i} = -3.04 \text{ V/V} \tag{6.64}$$

Example 6.15

To gain more insight into the operation of transistor amplifiers, we wish to consider the waveforms at various points in the circuit analyzed in the previous example. For this purpose assume that v_i has a triangular waveform. First determine the maximum amplitude that v_i is allowed to have. Then, with the amplitude of v_i set to this value, give the waveforms of the total quantities $i_B(t)$, $v_{BE}(t)$, $i_C(t)$, and $v_C(t)$.

Solution

One constraint on signal amplitude is the small-signal approximation, which stipulates that v_{be} should not exceed about 10 mV. If we take the triangular waveform v_{be} to be 20 mV peak-to-peak and work backward, Eq. (6.63) can be used to determine the maximum possible peak of v_i,

$$\hat{v}_i = \frac{\hat{v}_{be}}{0.011} = \frac{10}{0.011} = 0.91 \text{ V}$$

To check whether the transistor remains in the active mode with v_i having a peak value $\hat{v}_i = 0.91$ V, we have to evaluate the collector voltage. The voltage at the collector will consist of a triangular wave v_o superimposed on the dc value $V_C = 3.1$ V. The peak voltage of the triangular waveform will be

$$\hat{v}_o = \hat{v}_i \times \text{gain} = 0.91 \times 3.04 = 2.77 \text{ V}$$

It follows that when the output swings negative, the collector voltage reaches a minimum of 3.1 – 2.77 = 0.33 V, which is lower than the base voltage by less than 0.4 V. Thus the transistor will remain in the active mode with v_i having a peak value of 0.91 V. Nevertheless, to be on the safe side, we will use a somewhat lower value for $\hat{v}_i$ of approximately 0.8 V, as shown in Fig. 6.43(a), and complete the analysis of this problem utilizing the equivalent circuit in Fig. 6.42(d). The signal current in the base will be triangular, with a peak value $\hat{i}_b$ of

$$\hat{i}_b = \frac{\hat{v}_i}{R_{BB} + r_\pi} = \frac{0.8}{100 + 1.09} = 0.008 \text{ mA}$$

This triangular-wave current will be superimposed on the quiescent base current I_B, as shown in Fig. 6.43(b). The base–emitter voltage will consist of a triangular-wave component superimposed on the dc V_{BE} that is approximately 0.7 V. The peak value of the triangular waveform will be

$$\hat{v}_{be} = \hat{v}_i \frac{r_\pi}{r_\pi + R_{BB}} = 0.8 \frac{1.09}{100 + 1.09} = 8.6 \text{ mV}$$

The total v_{BE} is sketched in Fig. 6.43(c).

The signal current in the collector will be triangular in waveform, with a peak value $\hat{i}_c$ given by

$$\hat{i}_c = \beta \hat{i}_b = 100 \times 0.008 = 0.8 \text{ mA}$$

This current will be superimposed on the quiescent collector current I_C (=2.3 mA), as shown in Fig. 6.43(d).

The signal voltage at the collector can be obtained by multiplying v_i by the voltage gain; that is,

$$\hat{v}_o = 3.04 \times 0.8 = 2.43 \text{ V}$$

Figure 6.43(e) shows a sketch of the total collector voltage v_C versus time. Note the phase reversal between the input signal v_i and the output signal v_o.

Finally, we observe that each of the total quantities is the sum of a dc quantity (found from the dc circuit in Fig. 6.42b), and a signal quantity (found from the circuit in Fig. 6.42d).

Example 6.15 *continued*

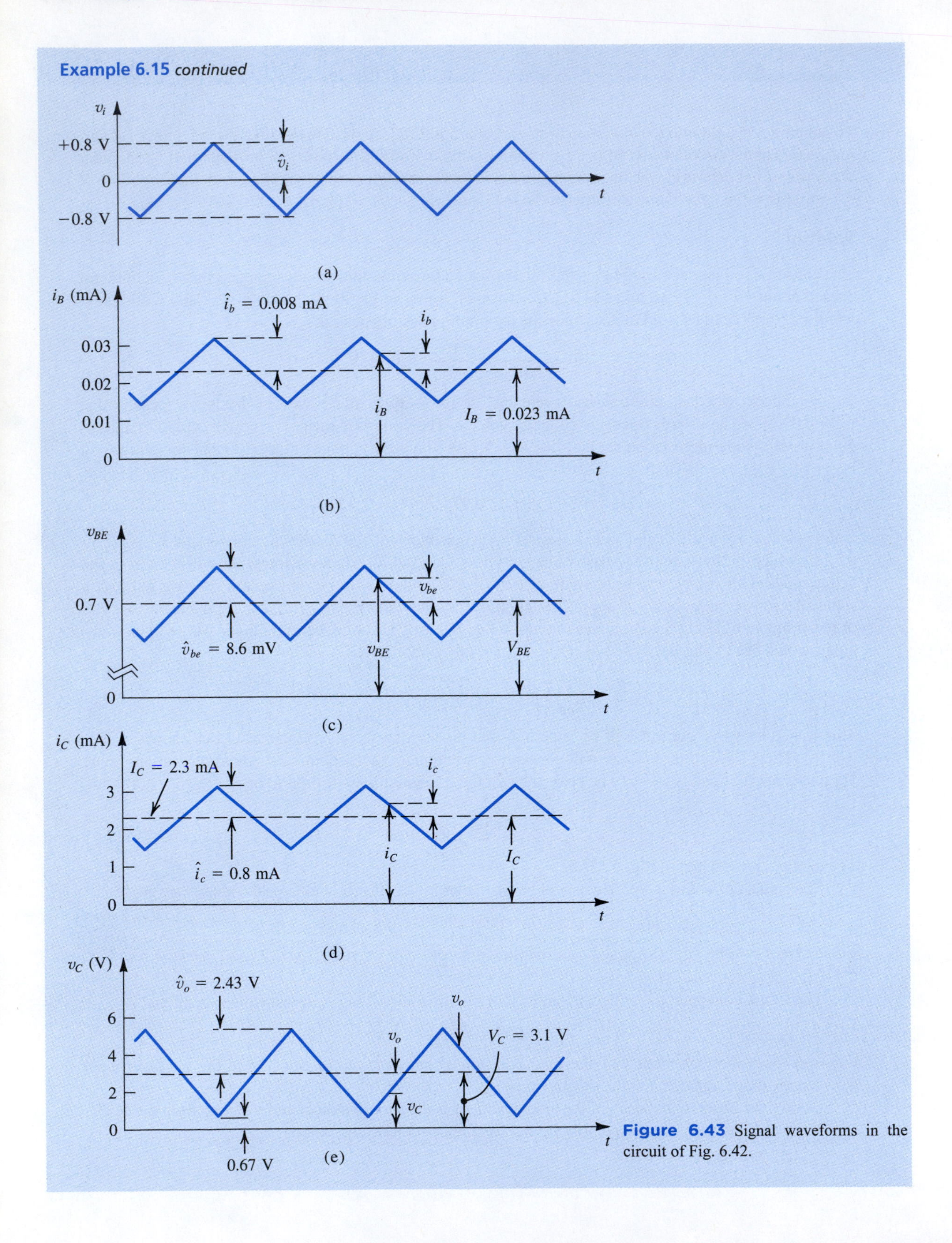

Figure 6.43 Signal waveforms in the circuit of Fig. 6.42.

Example 6.16

We need to analyze the circuit of Fig. 6.44(a) to determine the voltage gain and the signal waveforms at various points. The capacitor C_{C1} is a coupling capacitor whose purpose is to couple the signal v_i to the emitter while blocking dc. In this way the dc bias established by V^+ and V^- together with R_E and R_C will not be disturbed when the signal v_i is connected. For the purpose of this example, C_{C1} will be assumed to be very large so as to act as a perfect short circuit at signal frequencies of interest. Similarly, another very large capacitor C_{C2} is used to couple the output signal v_o to other parts of the system.

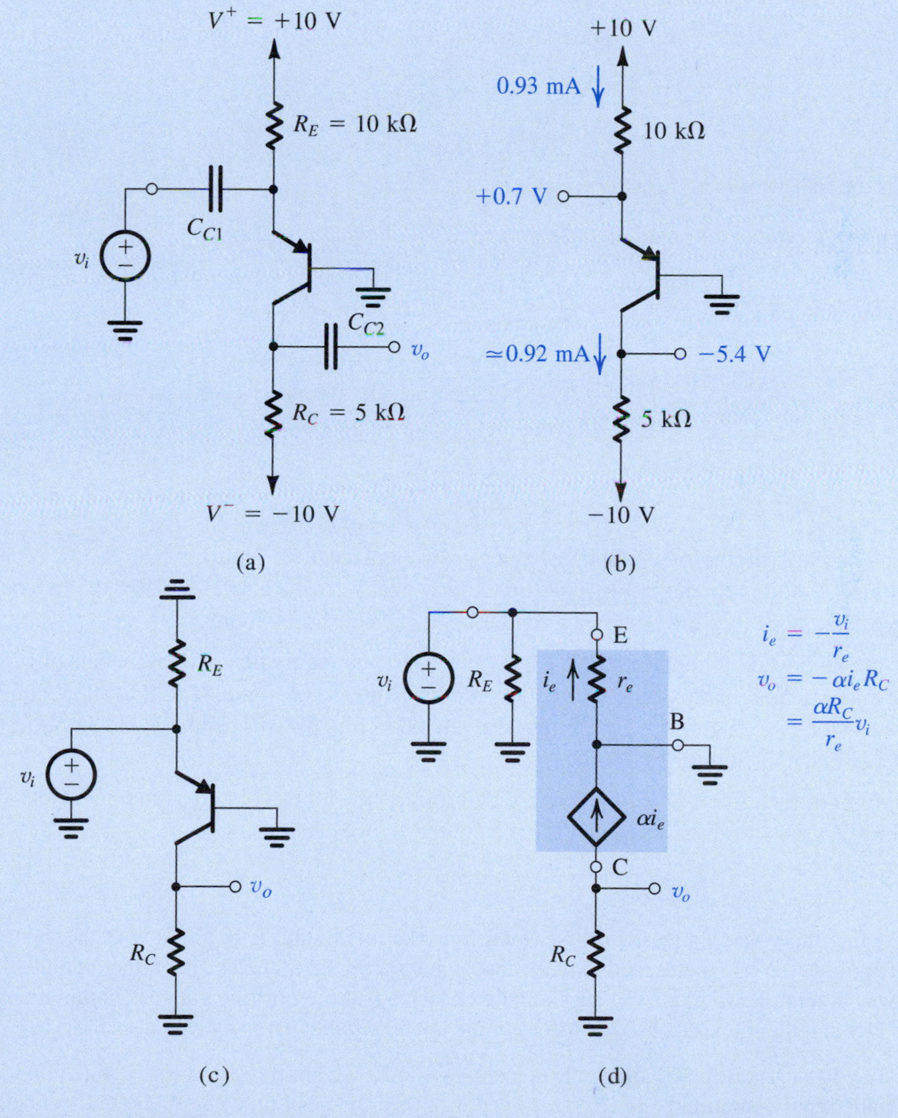

Figure 6.44 **(a)** circuit; **(b)** dc analysis; **(c)** circuit with the dc sources eliminated; **(d)** small-signal analysis using the T model for the BJT.

Example 6.16 *continued*

Solution

Here again we shall follow the five-step process outlined at the beginning of Section 6.5.9:

1. Figure 6.44(b) shows the circuit with the signal source and the coupling capacitors eliminated. The dc operating point can be determined as follows:

$$I_E = \frac{+10 - V_E}{R_E} \simeq \frac{+10 - 0.7}{10} = 0.93 \text{ mA}$$

Assuming $\beta = 100$, then $\alpha = 0.99$, and

$$I_C = 0.99 I_E = 0.92 \text{ mA}$$
$$V_C = -10 + I_C R_C$$
$$= -10 + 0.92 \times 5 = -5.4 \text{ V}$$

Thus the transistor is in the active mode.

2. We now determine the small-signal parameters as follows:

$$g_m = \frac{I_C}{V_T} = \frac{0.92}{0.025} = 36.8 \text{ mA/V}$$
$$r_e = \frac{V_T}{I_E} = \frac{0.025}{0.92} = 27.2 \ \Omega$$
$$\beta = 100 \qquad \alpha = 0.99$$
$$r_\pi = \frac{\beta}{g_m} = \frac{100}{36.8} = 2.72 \text{ k}\Omega$$

3. To prepare the circuit for small-signal analysis, we replace the dc sources with short circuits. The resulting circuit is shown in Fig. 6.44(c). Observe that we have also eliminated the two coupling capacitors, since they are assumed to be acting as perfect short circuits.
4. We are now ready to replace the BJT with one of the four equivalent circuit models of Figs. 6.40 and 6.41. Although any of the four will work, the T models of Fig. 6.41 will be more convenient because the base is grounded. Selecting the version in Fig. 6.41(b) results in the amplifier equivalent circuit shown in Fig. 6.44(d).
5. Analysis of the circuit in Fig. 6.44(d) to determine the output voltage v_o and hence the voltage gain v_o/v_i is straightforward and is given in the figure. The result is

$$A_v = \frac{v_o}{v_i} = 183.3 \text{ V/V}$$

Note that the voltage gain is positive, indicating that the output is in phase with the input signal. This property is due to the fact that the input signal is applied to the emitter rather than to the base, as was done in Example 6.14. We should emphasize that the positive gain has nothing to do with the fact that the transistor used in this example is of the *pnp* type.

Returning to the question of allowable signal magnitude, we observe from Fig. 6.44(d) that $v_{eb} = v_i$. Thus, if small-signal operation is desired (for linearity), then the peak of v_i should be limited to approximately 10 mV. With $\hat{V}_i$ set to this value, as shown for a sine-wave input in Fig. 6.45, the peak amplitude at the collector, $\hat{V}_o$, will be

$$\hat{V}_o = 183.3 \times 0.01 = 1.833 \text{ V}$$

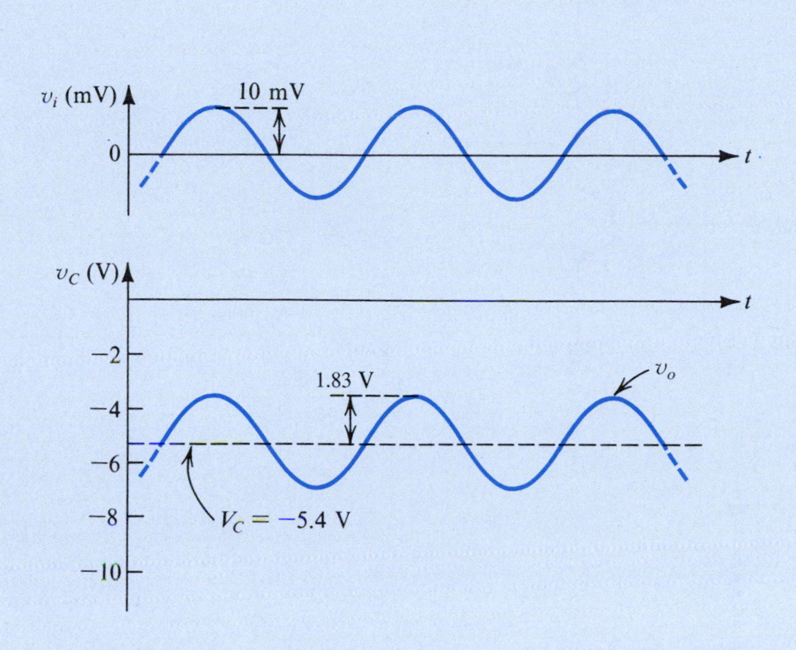

Figure 6.45 Input and output waveforms for the circuit of Fig. 6.44. Observe that this amplifier is noninverting, a property of the grounded base configuration.

EXERCISE

6.40 To increase the voltage gain of the amplifier analyzed in Example 6.16, the collector resistance R_C is increased to 7.5 kΩ. Find the new values of V_C, A_v, and the peak amplitude of the output sine wave corresponding to an input sine wave v_i of 10-mV peak.
Ans. –3.1 V; 275 V/V; 2.75 V

6.5.10 Performing Small-Signal Analysis Directly on the Circuit Diagram

In most cases one should explicitly replace each BJT with its small-signal model and analyze the resulting circuit, as we have done in the examples above. This systematic procedure is particularly recommended for beginning students. Experienced circuit designers, however, often perform a first-order analysis directly on the circuit. Figure 6.46 illustrates this process for the two circuits we analyzed in Examples 6.14 and 6.16. The reader is urged to follow this direct analysis procedure (the steps are numbered). Observe that the equivalent-circuit model is *implicitly* utilized; we are only saving the step of drawing the circuit with the BJT replaced by its model. Direct analysis, however, has an additional very important benefit: It

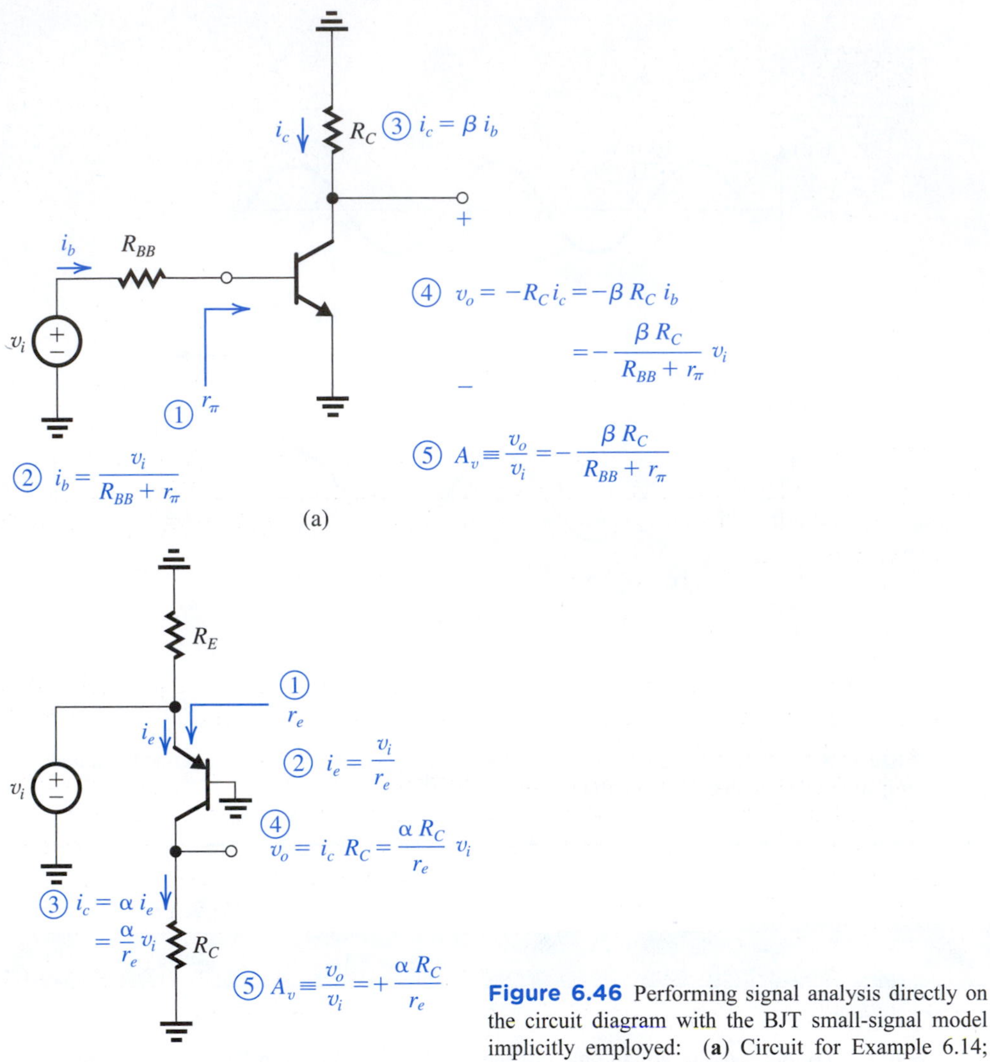

Figure 6.46 Performing signal analysis directly on the circuit diagram with the BJT small-signal model implicitly employed: (**a**) Circuit for Example 6.14; (**b**) Circuit for Example 6.16.

provides insight regarding the signal transmission through the circuit. Such insight can prove invaluable in design, particularly at the stage of selecting a circuit configuration appropriate for a given application.

6.5.11 Augmenting the Small-Signal Models to Account for the Early Effect

The Early effect, discussed in Section 6.2, causes the collector current to depend not only on v_{BE} but also on v_{CE}. The dependence on v_{CE} can be modeled by assigning a finite output resistance to the controlled current source in the hybrid-π model, as shown in Fig. 6.47. The output resistance r_o was defined in Eq. (6.19); its value is given by $r_o = V_A/I'_C$, where V_A is the Early voltage and I'_C is the dc bias current without taking the Early effect into account. We will normally drop the prime and just use $r_o = V_A/I_C$. Note that in the models of Fig. 6.47 we have renamed v_{be} as v_π, in order to conform with the literature.

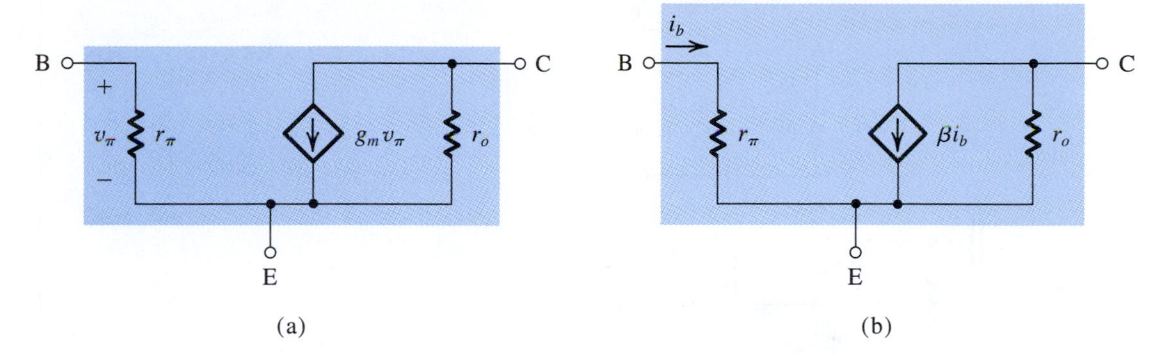

Figure 6.47 The hybrid-π small-signal model, in its two versions, with the resistance r_o included.

The question arises as to the effect of r_o on the operation of the transistor as an amplifier. In amplifier circuits in which the emitter is grounded (as in the circuit of Fig. 6.42), r_o simply appears in parallel with R_C. Thus, if we include r_o in the equivalent circuit of Fig. 6.42(d), for example, the output voltage v_o becomes

$$v_o = -g_m v_{be}(R_C \| r_o)$$

Thus the gain will be somewhat reduced. Obviously if $r_o \gg R_C$, the reduction in gain will be negligible, and one can ignore the effect of r_o. In general, in such a configuration r_o can be neglected if it is greater than $10R_C$.

When the emitter of the transistor is not grounded, including r_o in the model can complicate the analysis. We will make comments regarding r_o and its inclusion or exclusion on frequent occasions throughout the book. We should also note that in integrated-circuit BJT amplifiers, r_o plays a dominant role and *cannot* be neglected, as will be seen in Chapter 7. Of course, if one is performing an accurate analysis of an almost-final design using computer-aided analysis, then r_o can be easily included.

Finally, it should be noted that either of the T models in Fig. 6.41 can be augmented to account for the Early effect by including r_o between collector and emitter.

EXERCISE

6.41 The transistor in Fig. E6.41 is biased with a constant current source $I = 1$ mA and has $\beta = 100$ and $V_A = 100$ V.

(a) Find the dc voltages at the base, emitter, and collector.

(b) Find g_m, r_π, and r_o.

(c) If terminal Z is connected to ground, X to a signal source v_{sig} with a source resistance $R_{sig} = 2$ kΩ, and Y to an 8-kΩ load resistance, use the hybrid-π model of Fig. 6.47(a), to draw the small-signal equivalent circuit of the amplifier. (Note that the current source I should be replaced with an open circuit.) Calculate the overall voltage gain v_y/v_{sig}. If r_o is neglected, what is the error in estimating the gain magnitude? (*Note:* An infinite capacitance is used to indicate that the capacitance is sufficiently large that it acts as a short circuit at all signal frequencies of interest. However, the capacitor still blocks dc.)

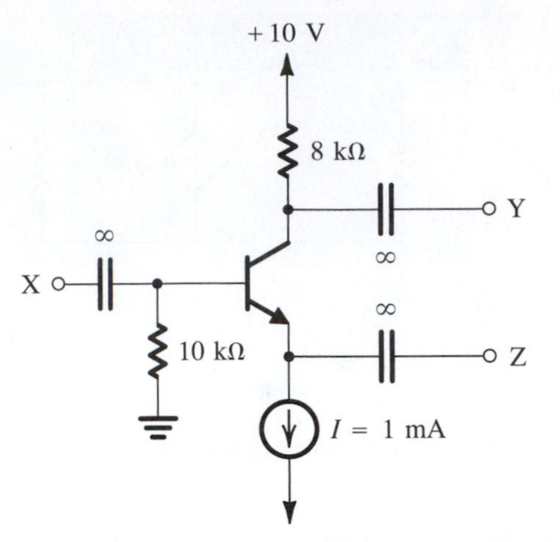

Figure E6.41

Ans. (a) –0.1 V, –0.8 V, +2 V; (b) 40 mA/V, 2.5 kΩ, 100 kΩ; (c) –77 V/V, +3.9%

6.5.12 Summary

The analysis and design of BJT amplifier circuits is greatly facilitated if the relationships between the various small-signal model parameters are at your fingertips. For easy reference, these are summarized in Table 6.4. Over time, however, we expect the reader to be able to recall these from memory. Finally, note that the material in Table 6.4 applies equally well to both the *npn* and the *pnp* transistors with no change in polarities.

6.6 Basic BJT Amplifier Configurations

It is useful at this point to take stock of where we are and where we are going in our study of BJT amplifiers. In Section 6.4 we examined the essence of the use of the BJT as an amplifier. There we found that almost-linear amplification can be obtained by biasing the BJT at an appropriate point in its active region of operation and by keeping the signal v_{be} (or v_π) small. Then in Section 6.5 we took a closer look at the small-signal operation of the BJT and developed circuit models to represent the transistor, thus facilitating the determination of amplifier parameters such as voltage gain and input and output resistances.

We are now ready to consider the various possible configurations of BJT amplifiers, and we will do that in the present section. To focus our attention on the salient features of the various configurations, we shall present them in their most simple, or "stripped-down," version. Thus, we will not show the dc biasing arrangements, leaving the study of bias design to the next section. Finally, in Section 6.8 we will bring everything together and present practical circuits for discrete-circuit BJT amplifiers, namely, those amplifier circuits that can be constructed using discrete components. The study of integrated-circuit amplifiers begins in Chapter 7.

Table 6.4 Small-Signal Models of the BJT

Hybrid-π Model

■ $(g_m v_\pi)$ Version

■ (βi_b) Version

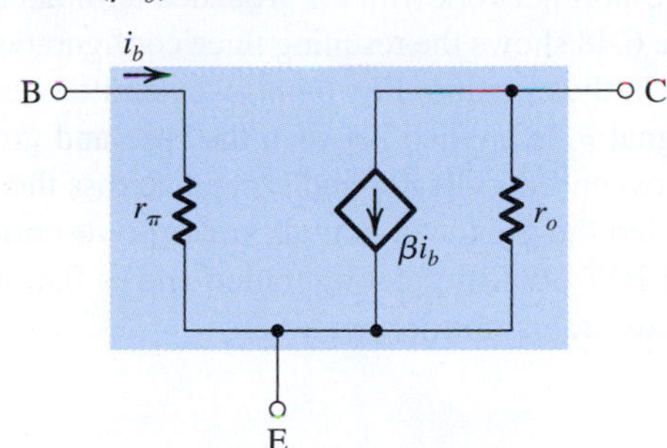

■ $(g_m v_\pi)$ Version

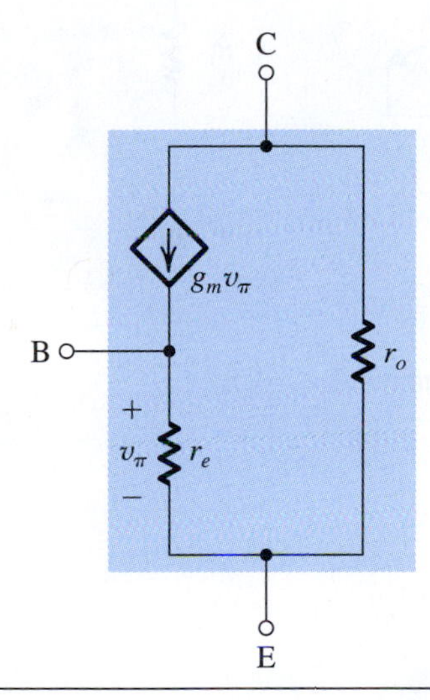

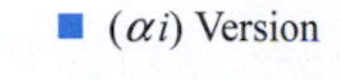

■ (αi) Version

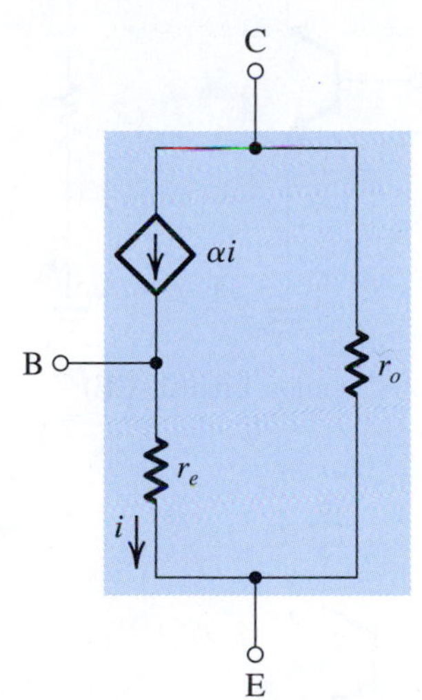

Model Parameters in Terms of DC Bias Currents

$$g_m = \frac{I_C}{V_T} \qquad r_e = \frac{V_T}{I_E} = \alpha\frac{V_T}{I_C} \qquad r_\pi = \frac{V_T}{I_B} = \beta\frac{V_T}{I_C} \qquad r_o = \frac{|V_A|}{I_C}$$

In Terms of g_m

$$r_e = \frac{\alpha}{g_m} \qquad r_\pi = \frac{\beta}{g_m}$$

In Terms of r_e

$$g_m = \frac{\alpha}{r_e} \qquad r_\pi = (\beta + 1)r_e \qquad g_m + \frac{1}{r_\pi} = \frac{1}{r_e}$$

Relationships between α and β

$$\beta = \frac{\alpha}{1-\alpha} \qquad \alpha = \frac{\beta}{\beta+1} \qquad \beta + 1 = \frac{1}{1-\alpha}$$

6.6.1 The Three Basic Configurations

There are three basic configurations for connecting the BJT as an amplifier. Each of these configurations is obtained by connecting one of the three BJT terminals to ground, thus creating a two-port network with the grounded terminal being *common* to the input and output ports. Figure 6.48 shows the resulting three configurations with the biasing arrangements omitted.

In the circuit of Fig. 6.48(a) the emitter terminal is connected to ground, the input voltage signal v_i is applied between the base and ground, and the output voltage signal v_o is taken between the collector and ground, across the resistance R_C. This configuration, therefore, is called the grounded-emitter or **common-emitter (CE)** amplifier. It is by far the most popular BJT amplifier configuration and is the one we have utilized in Sections 6.4 and 6.5 to study BJT amplifier operation.

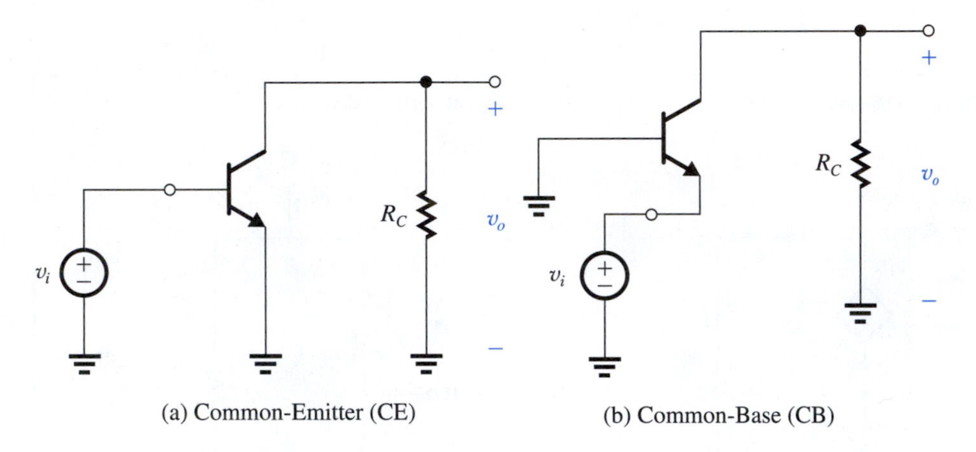

(a) Common-Emitter (CE)

(b) Common-Base (CB)

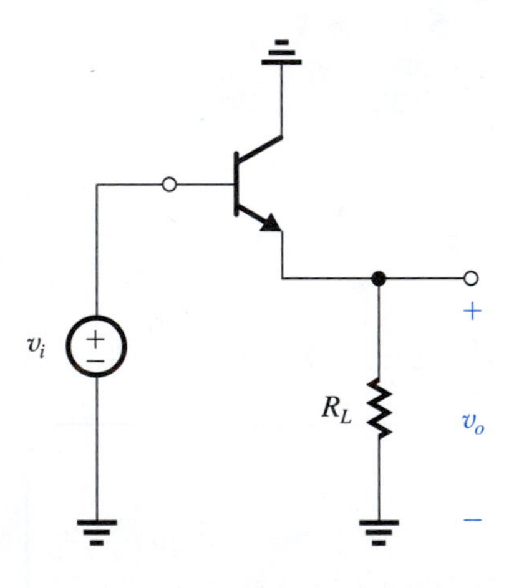

(c) Common-Collector (CC)
or Emitter Follower

Figure 6.48 The three basic configurations of BJT amplifier. The biasing arrangements are not shown.

The **common-base (CB)** or grounded-base amplifier is shown in Fig. 6.48(b). It is obtained by connecting the base to ground, applying the input v_i between the emitter and ground, and taking the output v_o across the resistance R_C connected between the collector and ground. We have encountered a CB amplifier in Example 6.14.

Finally, Fig. 6.48(c) shows the **common-collector (CC)** or grounded-collector amplifier. It is obtained by connecting the collector terminal to ground, applying the input voltage signal v_i between base and ground, and taking the output voltage signal v_o between the emitter and ground, across a load resistance R_L. For reasons that will become apparent shortly, this configuration is more commonly called the **emitter follower**.

Our study of the three basic BJT amplifier configurations will reveal that each has distinctly different attributes and hence areas of application.[14]

6.6.2 Characterizing Amplifiers[15]

Before we begin our study of the different BJT amplifier configurations, we consider how to characterize the performance of an amplifier as a circuit building block. An introduction to this topic was presented in Section 1.5.

Figure 6.49(a) shows an amplifier fed with a signal source having an open-circuit voltage v_{sig} and an internal resistance R_{sig}. These can be the parameters of an actual signal source or, in a cascade amplifier, the Thévenin equivalent of the output circuit of another stage preceding the one under study. The amplifier is shown with a load resistance R_L connected to the output terminal. Here, R_L can be an actual load resistance or the input resistance of a succeeding amplifier stage in a cascade amplifier.

Figure 6.49(b) shows the amplifier circuit with the amplifier block replaced by its equivalent-circuit model. The input resistance R_{in} represents the loading effect of the amplifier input on the signal source. It is found from

$$R_{in} \equiv \frac{v_i}{i_i}$$

and together with the resistance R_{sig} forms a voltage divider that reduces v_{sig} to the value v_i that appears at the input of the amplifier proper,

$$v_i = \frac{R_{in}}{R_{in} + R_{sig}} v_{sig} \tag{6.65}$$

It is important to note that in general R_{in} may depend on the load resistance R_L. One of the three configurations we are studying in this section, the emitter follower, exhibits such dependence.

The second parameter for characterizing amplifier performance is the **open-circuit voltage gain** A_{vo}, defined as

$$A_{vo} \equiv \left. \frac{v_o}{v_i} \right|_{R_L=\infty}$$

The third and final parameter is the output resistance R_o. Observe from Fig. 6.49(b) that R_o is the resistance seen looking back into the amplifier output terminal with v_i set to zero. Thus R_o can be determined, at least conceptually, as indicated in Fig. 6.49(c) with

$$R_o = \frac{v_x}{i_x}$$

[14]The CE, CB, and CC configurations are the BJT counterparts of the MOSFET CS, CG, and CD configurations, respectively.

[15]This section can be skipped if the reader has already studied Section 5.6.2; it presents substantially the same material.

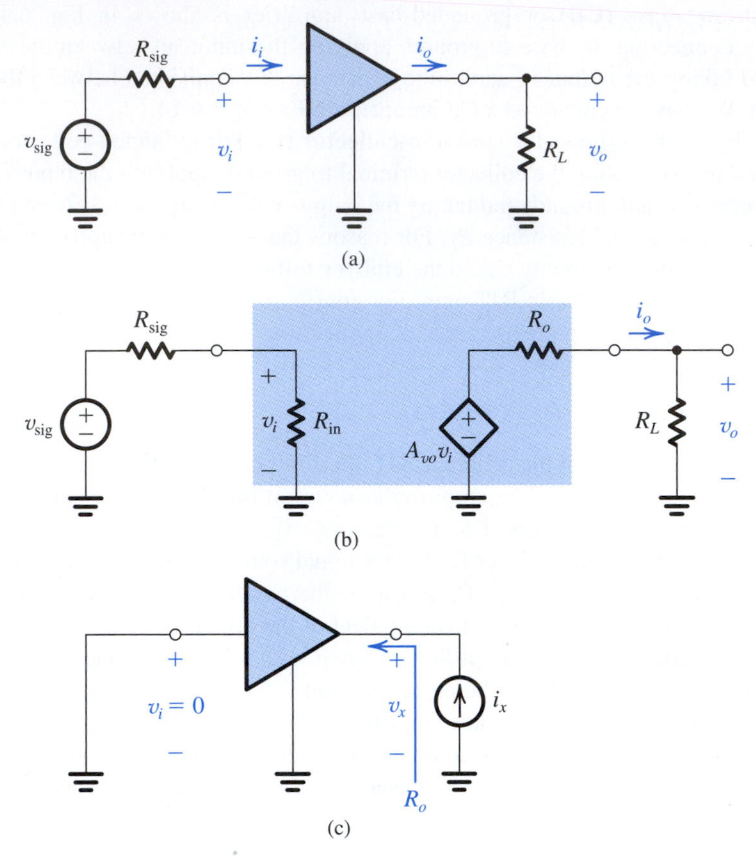

Figure 6.49 **(a)** An amplifier fed with a signal source (v_{sig}, R_{sig}) and providing its output across a load resistance R_L. **(b)** The circuit in **(a)** with the amplifier represented by its equivalent circuit model. **(c)** Determining the output resistance R_o of the amplifier.

Because R_o is determined with $v_i = 0_1$ its value does not depend on R_{sig}.

The controlled source $A_{vo}v_i$ and the output resistance R_o represent the Thévenin equivalent of the amplifier output circuit, and the output voltage v_o can be found from

$$v_o = \frac{R_L}{R_L + R_o} A_{vo} v_i \tag{6.66}$$

Thus the voltage gain of the amplifier proper, A_v, can be found as

$$A_v \equiv \frac{v_o}{v_i} = A_{vo} \frac{R_L}{R_L + R_o} \tag{6.67}$$

and the overall voltage gain G_v,

$$G_v \equiv \frac{v_o}{v_{sig}}$$

can be determined by combining Eqs. (6.65) and (6.66),

$$G_v = \frac{R_{in}}{R_{in} + R_{sig}} A_v \tag{6.68}$$

6.6.3 The Common-Emitter (CE) Amplifier

Of the three basic BJT amplifier configurations, the common emitter is the most widely used. Typically, in an amplifier formed by cascading a number of stages, the bulk of the voltage gain is obtained by using one or more common-emitter stages in the cascade.

Figure 6.50(a) shows a common-emitter amplifier (with the biasing arrangement omitted) fed with a signal source v_{sig} having a source resistance R_{sig}. We wish to analyze the circuit to determine R_{in}, A_{vo}, R_o, and G_v. For this purpose we shall assume that R_C is part of the amplifier; thus if a load resistance R_L is connected to the amplifier output, it appears in parallel with R_C.

Characteristic Parameters of the CE Amplifier Replacing the BJT with its hybrid-π model, we obtain the CE amplifier equivalent circuit shown in Fig. 6.50(b). We shall use this equivalent circuit to determine the characteristic parameters of the amplifier R_{in}, A_{vo}, and R_o as follows.

The input resistance R_{in} is found by inspection to be

$$R_{in} = r_\pi \tag{6.69}$$

Observe that R_{in} does not depend on the output side of the amplifier; hence, this amplifier is said to be **unilateral**.

The output voltage v_o can be found by multiplying the current $(g_m v_\pi)$ by the total resistance between the output node and ground,

$$v_o = -(g_m v_\pi)(R_C \parallel r_o)$$

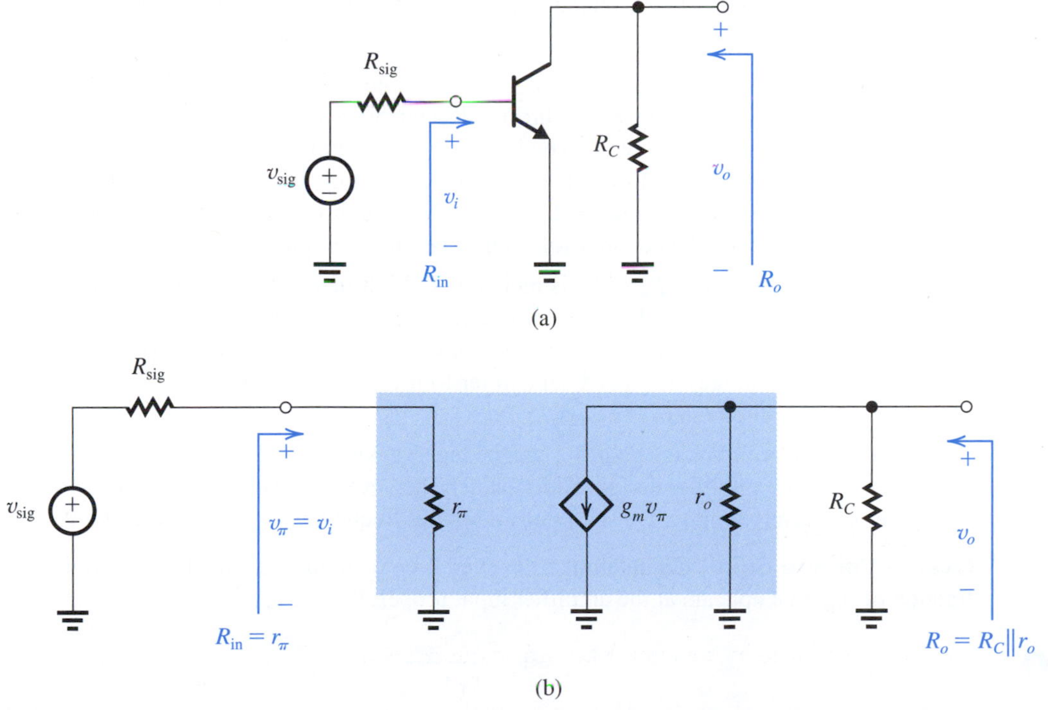

Figure 6.50 **(a)** Common-emitter amplifier fed with a signal v_{sig} from a generator with a resistance R_{sig}. **(b)** The common-emitter amplifier circuit with the BJT replaced with its hybrid-π model.

Since $v_\pi = v_i$, the open-circuit voltage gain $A_{vo} \equiv v_o/v_i$ can be obtained as

$$A_{vo} = -g_m(R_C \| r_o) \tag{6.70}$$

Observe that the transistor output resistance r_o reduces the magnitude of the voltage gain. In discrete-circuit amplifiers, which are of interest to us in this chapter, R_C is usually much lower than r_o and the effect of r_o on reducing $|A_{vo}|$ is slight (less than 10% or so). Thus in many cases we can neglect r_o and express A_{vo} simply as

$$A_{vo} \simeq (-g_m R_C) \tag{6.71}$$

The reader is cautioned, however, that neglecting r_o is allowed only in discrete-circuit design. As will be seen in Chapter 7, r_o plays a central role in IC amplifiers.

The output resistance R_o is the resistance seen looking back into the output terminal with v_i set to zero. From Fig. 6.50(b) we see that with v_i set to zero, v_π will be zero and $g_m v_\pi$ will be zero, resulting in

$$R_o = R_C \| r_o \tag{6.72}$$

Here r_o has the beneficial effect of reducing the value of R_o. In discrete circuits, however, this effect is slight and we can make the approximation

$$R_o \simeq R_C \tag{6.73}$$

This concludes the analysis of the amplifier proper. Now, we can make the following observations:

1. The input resistance $R_{\text{in}} = r_\pi = \beta/g_m$ is moderate to low in value (typically, in the kilohm range). Obviously R_{in} is directly dependent on β and is inversely proportional to the collector bias current I_C. To obtain a higher input resistance, the bias current can be lowered, but this also lowers the gain. This is a significant design trade-off. If a much higher input resistance is desired, then a modification of the CE configuration (to be discussed shortly) or an emitter-follower stage can be employed.
2. The output resistance $R_o \simeq R_C$ is moderate to high in value (typically, in the kilohm range). Reducing R_C to lower R_o is usually not a viable proposition because the voltage gain is also reduced. Alternatively, if a very low output resistance (in the ohms to tens of ohms range) is needed, an emitter-follower stage is called for, as will be discussed in Section 6.6.6.
3. The open-circuit voltage gain A_{vo} can be high, making the CE configuration the workhorse in BJT amplifier design. Unfortunately, however, the bandwidth of the CE amplifier is severely limited. We shall study amplifier frequency response in Chapter 9.

Overall Voltage Gain To determine the overall voltage gain G_v we first determine the fraction of v_{sig} that appears at the amplifier input proper, that is, v_i;

$$v_i = v_{\text{sig}} \frac{r_\pi}{r_\pi + R_{\text{sig}}} \tag{6.74}$$

Depending on the relative values of r_π and R_{sig}, significant loss of signal strength can occur at the input, which is obviously undesirable and can be avoided by raising the input resistance, as discussed above. At this point, we should remind the reader that to maintain a reasonably linear operation, v_i should not exceed about 5 mV to 10 mV, which poses a constraint on the value of v_{sig}.

If a load resistance R_L is connected to the output terminal of the amplifier, this resistance will appear in parallel with R_C. It follows that the voltage gain A_v can be obtained by simply replacing R_C in the expression of A_{vo} in Eq. (6.70) by $R_C \| R_L$,

$$A_v = -g_m(R_C \| R_L \| r_o) \tag{6.75}$$

We can now use this expression for A_v together with (v_i/v_{sig}) from Eq. (6.74) to obtain the overall voltage gain G_v as

$$G_v \equiv \frac{v_o}{v_{sig}} = -\frac{r_\pi}{r_\pi + R_{sig}} g_m(R_C \| R_L \| r_o) \tag{6.76}$$

EXERCISE

6.41 Use A_{vo} in Eq. (6.70) together with R_o in Eq. (6.72) to obtain A_v. Show that the result is identical to that in Eq. (6.75).

Alternative Gain Expressions There are alternative forms for A_v and G_v that can yield considerable insight besides being intuitive and easy to remember. The expression for A_v can be obtained by replacing g_m in Eq. (6.75) with α/r_e;

$$A_v = -\alpha \frac{(R_C \| R_L \| r_o)}{r_e} \tag{6.77}$$

Observing that $(R_C \| R_L \| r_o)$ is the total resistance in the collector and r_e is the total resistance in the emitter, this expression simply states that the voltage gain from base to collector is given by

$$A_v = -\alpha \frac{\text{Total resistance in collector}}{\text{Total resistance in emitter}} \tag{6.78}$$

The reason for the factor α is that the collector current is α times the emitter current. Of course $\alpha \simeq 1$ and can usually be neglected, and the expression in Eq. (6.78) is simply stated as a resistance ratio. This expression is a general one and applies to any BJT amplifier circuit for finding the voltage gain from base to collector.

A corresponding expression for G_v can be obtained by replacing $(g_m r_\pi)$ in the numerator of Eq. (6.76) with β,

$$G_v = -\beta \frac{(R_C \| R_L \| r_o)}{R_{sig} + r_\pi} \tag{6.79}$$

which can be expressed in words as

$$G_v = -\beta \frac{\text{Total resistance in collector}}{\text{Total resistance in base}} \tag{6.80}$$

Observe that here the multiplicative factor is β, which is the ratio of i_c to i_b; this makes sense because we are using the ratio of resistances in the collector and the base. The reader is urged to reflect on these expressions while referring to Fig. 6.50.

Performing the Analysis Directly on the Circuit As mentioned in Section 6.5, with practice one can dispense with the *explicit* use of the BJT equivalent circuit and perform the analysis directly on the circuit schematic. Because in this way one remains closer to the actual circuit, this direct analysis can yield greater insight into circuit operation. Although at this stage in learning electronic circuits it is perhaps a little early to follow this direct analysis route, we show in Fig. 6.51 the CE amplifier circuit prepared for direct analysis. Observe that we have "pulled out" the resistance r_o from the transistor, thus making the transistor collector conduct $g_m v_\pi$ while still accounting for the effect of r_o.

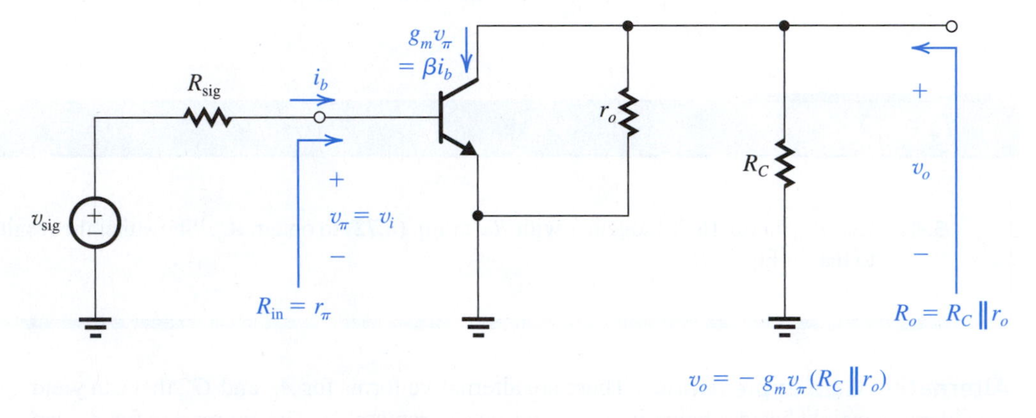

Figure 6.51 Performing the analysis directly on the circuit with the BJT model used implicitly.

Example 6.17

A CE amplifier utilizes a BJT with $\beta = 100$ and $V_A = 100$ V, is biased at $I_C = 1$ mA and has a collector resistance $R_C = 5$ kΩ. Find R_{in}, R_o, and A_{vo}. If the amplifier is fed with a signal source having a resistance of 5 kΩ, and a load resistance $R_L = 5$ kΩ is connected to the output terminal, find the resulting A_v and G_v. If $\hat{v}_\pi$ is to be limited to 5 mV, what are the corresponding $\hat{v}_{sig}$ and $\hat{v}_o$ with the load connected?

Solution

At $I_C = 1$ mA,

$$g_m = \frac{I_C}{V_T} = \frac{1\text{ mA}}{0.025\text{ V}} = 40\text{ mA/V}$$

$$r_\pi = \frac{\beta}{g_m} = \frac{100}{40\text{ mA/V}} = 2.5\text{ k}\Omega$$

$$r_o = \frac{V_A}{I_C} = \frac{100\text{ V}}{1\text{ mA}} = 100\text{ k}\Omega$$

The amplifier characteristic parameters can now be found as

$$R_{\text{in}} = r_\pi = 2.5 \text{ k}\Omega$$

$$\begin{aligned} A_{vo} &= -g_m(R_C \,\|\, r_o) \\ &= -40 \text{ mA/V} \quad (5 \text{ k}\Omega \,\|\, 100 \text{ k}\Omega) \\ &= -190.5 \text{ V/V} \end{aligned}$$

$$\begin{aligned} R_o &= R_C \,\|\, r_o \\ &= 5 \,\|\, 100 \;=\; 4.76 \text{ k}\Omega \end{aligned}$$

With a load resistance $R_L = 5 \text{ k}\Omega$ connected at the output, we can find A_v by either of the following two approaches:

$$\begin{aligned} A_v &= A_{vo} \frac{R_L}{R_L + R_o} \\ &= -190.5 \times \frac{5}{5 + 4.76} = -97.6 \text{ V/V} \end{aligned}$$

or

$$\begin{aligned} A_v &= -g_m(R_C \,\|\, R_L \,\|\, r_o) \\ &= -40(5 \,\|\, 5 \,\|\, 100) = -97.6 \text{ V/V} \end{aligned}$$

The overall voltage gain G_v can now be determined as

$$\begin{aligned} G_v &= \frac{R_{\text{in}}}{R_{\text{in}} + R_{\text{sig}}} A_v \\ &= \frac{2.5}{2.5 + 5} \times -97.6 = -32.5 \text{ V/V} \end{aligned}$$

If the maximum amplitude of v_π is to be 5 mV, the corresponding value of $\hat{v}_{\text{sig}}$ will be

$$\hat{v}_{\text{sig}} = \left(\frac{R_{\text{in}} + R_{\text{sig}}}{R_{\text{in}}}\right)\hat{v}_\pi = \frac{2.5 + 5}{2.5} \times 5 = 15 \text{ mV}$$

and the amplitude of the signal at the output will be

$$\hat{v}_o = G_v \hat{v}_{\text{sig}} = 32.5 \times 0.015 = 0.49 \text{ V}$$

EXERCISE

6.42. The designer of the amplifier in Example 6.17 decides to lower the bias current to half its original value in order to raise the input resistance and hence increase the fraction of v_{sig} that appears at the input of the amplifier proper. In an attempt to maintain the voltage gain, the designer decides to double the value of R_C. For the new design, determine R_{in}, A_{vo}, R_o, A_v, and G_v. If the peak amplitude of v_π is to be limited to 5 mV, what are the corresponding values of $\hat{v}_{\text{sig}}$ and $\hat{v}_o$ (with the load connected)?
Ans. 5 kΩ; −190.5 V/V; 9.5 kΩ; −65.6 V/V; −32.8 V/V; 10 mV; 0.33 V

6.6.4 The Common-Emitter Amplifier with an Emitter Resistance

Including a resistance in the emitter as shown in Fig. 6.52(a) can lead to significant changes in the amplifier characteristics. Thus, such a resistor can be an effective design tool for tailoring the amplifier characteristics to fit the design requirements.

Analysis of the circuit in Fig. 6.52(a) can be performed by replacing the BJT with one of its small-signal models. Although any one of the models of Figs. 6.40 and 6.41 can be used, the most convenient for this application is one of the two T models. This is because the resistance R_e in the emitter will appear in series with the emitter resistance r_e of the T model and can thus be added to it, simplifying the analysis considerably. In fact, whenever there is a resistance in the emitter lead, the T model should prove more convenient to use than the hybrid-π model.

Replacing the BJT with the T model of Fig. 6.41(b) results in the amplifier small-signal, equivalent-circuit model shown in Fig. 6.52(b). Note that we have not included the BJT output resistance r_o; because this would complicate the analysis considerably. Since for the discrete amplifier at hand it turns out that the effect of r_o on circuit performance is small, we shall not include it in the analysis here. This is not the case, however, for the IC version of this circuit, and we shall indeed take r_o into account in the analysis in Chapter 7.

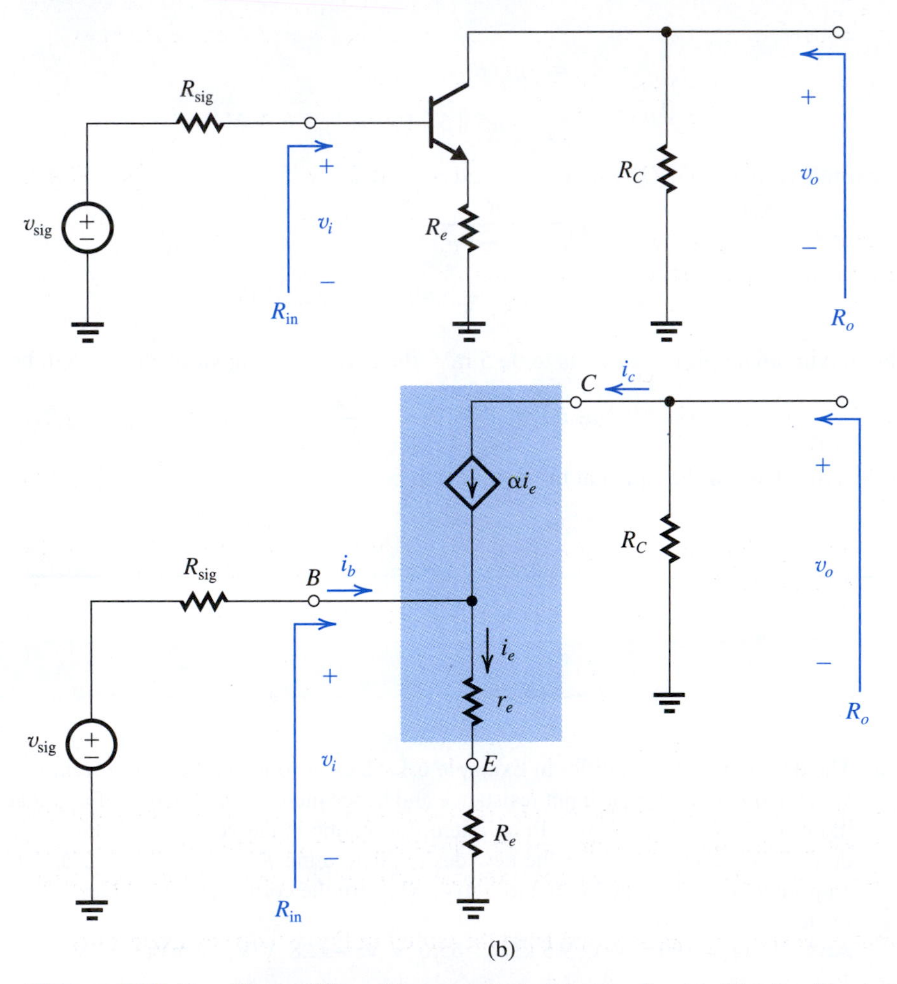

Figure 6.52 The CE amplifier with an emitter resistance R_e; **(a)** Circuit without bias details; **(b)** Equivalent circuit with the BJT replaced with its T model.

To determine the amplifier input resistance R_{in}, we note from Fig. 6.52(b) that

$$R_{in} \equiv \frac{v_i}{i_b}$$

where

$$i_b = (1-\alpha)i_e = \frac{i_e}{\beta+1} \tag{6.81}$$

and

$$i_e = \frac{v_i}{r_e + R_e} \tag{6.82}$$

Thus,

$$R_{in} = (\beta+1)(r_e + R_e) \tag{6.83}$$

This is a very important result. It states that *the input resistance looking into the base is* $(\beta+1)$ *times the total resistance in the emitter*, and is known as the **resistance-reflection rule**. The factor $(\beta+1)$ arises because the base current is $1/(\beta+1)$ times the emitter current. The expression for R_{in} in Eq. (6.83) shows clearly that including a resistance R_e in the emitter can substantially increase R_{in}. Indeed, the value of R_{in} is increased by the ratio

$$\frac{R_{in}(\text{with } R_e \text{ included})}{R_{in}(\text{without } R_e)} = \frac{(\beta+1)(r_e+R_e)}{(\beta+1)r_e}$$

$$= 1 + \frac{R_e}{r_e} \simeq 1 + g_m R_e \tag{6.84}$$

Thus the circuit designer can use the value of R_e to control the value of R_{in}.

To determine the voltage gain A_{vo}, we see from Fig. 6.52(b) that

$$v_o = -i_c R_C$$

$$= -\alpha i_e R_C$$

Substituting for i_e from Eq. (6.82) gives

$$A_{vo} = -\alpha \frac{R_C}{r_e + R_e} \tag{6.85}$$

which is a simple application of the general expression in Eq. (6.78). Here, of course, the total resistance in the emitter is $r_e + R_e$.

The open-circuit voltage gain in Eq. (6.85) can be expressed alternatively as

$$A_{vo} = -\frac{\alpha}{r_e} \frac{R_C}{1 + R_e/r_e}$$

$$A_{vo} = -\frac{g_m R_C}{1 + R_e/r_e} \simeq -\frac{g_m R_C}{1 + g_m R_e} \tag{6.86}$$

Thus, including R_e reduces the voltage gain by the factor $(1 + g_m R_e)$, which is the same factor by which R_{in} is increased. This points out an interesting trade-off between gain and input resistance, a trade-off that the designer can exercise through the choice of an appropriate value for R_e.

The output resistance R_o can be found from the circuit in Fig. 6.52(b) by inspection:

$$R_o = R_C$$

If a load resistance R_L is connected at the amplifier output, A_v can be found as

$$
\begin{aligned}
A_v &= A_{vo}\frac{R_L}{R_L+R_o} \\
&= -\alpha\frac{R_C}{r_e+R_e}\frac{R_L}{R_L+R_C} \\
&= -\alpha\frac{R_C \| R_L}{r_e+R_e} \qquad (6.87)
\end{aligned}
$$

which could have been written directly using Eq. (6.78). The overall voltage gain G_v can now be found:

$$G_v = \frac{R_{\text{in}}}{R_{\text{in}}+R_{\text{sig}}} \times -\alpha\frac{R_C \| R_L}{r_e+R_e}$$

Substituting for R_{in} from Eq. (6.83) and replacing α with $\beta/(\beta+1)$ results in

$$G_v = -\beta\frac{R_C \| R_L}{R_{\text{sig}}+(\beta+1)(r_e+R_e)} \qquad (6.88)$$

which is a direct application of the general expression presented in Eq. (6.80). We observe that the overall voltage gain G_v is lower than the value without R_e because of the additional term $(\beta+1)R_e$ in the denominator. The gain, however, is less sensitive to the value of β, a desirable result.

Another important consequence of including the resistance R_e in the emitter is that it enables the amplifier to handle larger input signals without incurring nonlinear distortion. This is because only a fraction of the input signal at the base, v_i, appears between the base and the emitter. Specifically, from the circuit in Fig. 6.52(b), we see that

$$\frac{v_\pi}{v_i} = \frac{r_e}{r_e+R_e} \simeq \frac{1}{1+g_m R_e} \qquad (6.89)$$

Thus, for the same v_π, the signal at the input terminal of the amplifier, v_i, can be greater than for the CE amplifier by the factor $(1+g_m R_e)$.

To summarize, including a resistance R_e in the emitter of the CE amplifier results in the following characteristics:

1. The input resistance R_{in} is increased by the factor $(1+g_m R_e)$.
2. The voltage gain from base to collector, A_v, is reduced by the factor $(1+g_m R_e)$.
3. For the same nonlinear distortion, the input signal v_i can be increased by the factor $(1+g_m R_e)$.
4. The overall voltage gain is less dependent on the value of β.
5. The high-frequency response is significantly improved (as we shall see in Chapter 9).

With the exception of gain reduction, these characteristics represent performance improvements. Indeed, the reduction in gain is the price paid for obtaining the other performance improvements. In many cases this is a good bargain; it is the underlying philosophy for the use of negative feedback. That the resistance R_e introduces negative feedback in the amplifier circuit can be seen by reference to Fig. 6.52(a): While keeping v_i constant, assume that for some reason the collector current increases; the emitter current also will increase, resulting in an increased voltage drop across R_e. Thus the emitter voltage rises, and the base–emitter voltage decreases. The latter effect causes the collector current to decrease, counteracting the initially assumed change, an indication of the presence of negative feedback. In Chapter 10, where we shall study negative feedback formally, we will find that the factor $(1 + g_m R_e)$, which appears repeatedly, is the "amount of negative-feedback" introduced by R_e. Finally, we note that the negative-feedback action of R_e gives it the name **emitter degeneration resistance.**

Example 6.18

For the CE amplifier specified in Example 6.17, what value of R_e is needed to raise R_{in} to a value four times that of R_{sig}? With R_e included, find A_{vo}, R_o, A_v, and G_v. Also, if $\hat{v}_\pi$ is limited to 5 mV, what are the corresponding values of $\hat{v}_{sig}$ and $\hat{v}_o$?

Solution

To obtain $R_{in} = 4R_{sig} = 4 \times 5 = 20$ kΩ, the required R_e is found from

$$20 = (\beta + 1)\ (r_e + R_e)$$

With $\beta = 100$,

$$r_e + R_e \simeq 200\ \Omega$$

Thus,

$$R_e = 200 - 25 = 175\ \Omega$$

$$A_{vo} = -\alpha\ \frac{R_C}{r_e + R_e}$$

$$\simeq \left(-\frac{5000}{25 + 175}\right) = -25\ \text{V/V}$$

$$R_o = R_C = 5\ \text{k}\Omega\ \text{(unchanged)}$$

$$A_v = A_{vo}\ \frac{R_L}{R_L + R_o} = -25 \times \frac{5}{5+5} = -12.5\ \text{V/V}$$

$$G_v = \frac{R_{in}}{R_{in} + R_{sig}}\ A_v = -\frac{20}{20+5} \times 12.5 = -10\ \text{V/V}$$

For $\hat{v}_\pi = 5$ mV,

$$\hat{v}_i = \hat{v}_\pi\left(\frac{r_e + R_e}{r_e}\right)$$

$$= 5\left(1 + \frac{175}{25}\right) = 40\ \text{mV}$$

$$\hat{v}_{sig} = \hat{v}_i\ \frac{R_{in} + R_{sig}}{R_{in}}$$

$$= 40\left(1 + \frac{5}{20}\right) = 50\ \text{mV}$$

$$\hat{v}_o = \hat{v}_{sig} \times |G_v|$$

$$= 50 \times 10 = 500\ \text{mV} = 0.5\ \text{V}$$

Thus, while $|G_v|$ has decreased to about a third of its original value, the amplifier is able to produce as large an output signal as before for the same nonlinear distortion.

EXERCISE

6.43 Show that with R_e included, and v_π limited to a maximum value $\hat{v}_\pi$, the maximum allowable input signal, $\hat{v}_{sig}$, is given by

$$\hat{v}_{sig} = \hat{v}_\pi\left(1 + \frac{R_e}{r_e} + \frac{R_{sig}}{r_\pi}\right)$$

If the transistor is biased at $I_C = 0.5$ mA and has a β of 100, what value of R_e is needed to permit an input signal $\hat{v}_{sig}$ of 100 mV from a source with a resistance $R_{sig} = 10$ kΩ while limiting $\hat{v}_\pi$ to 10 mV? What is R_{in} for this amplifier? If the total resistance in the collector is 10 kΩ, what G_v value results?

Ans. 350 Ω; 40.4 kΩ; –19.8 V/V

6.6.5 The Common-Base (CB) Amplifier

Figure 6.53(a) shows a common-base amplifier with the biasing circuit omitted. The amplifier is fed with a signal source characterized by v_{sig} and R_{sig}. Since R_{sig} appears in series with the emitter, it is more convenient to represent the transistor with the T model than with the hybrid-π model. Doing this, we obtain the amplifier equivalent circuit shown in Fig. 6.53(b). Note that we have not included r_o. This is because including r_o would complicate the analysis considerably, for it would appear between the output and input of the amplifier. Fortunately, it turns out that the effect of r_o on the performance of a discrete CB amplifier is very small. We will consider the effect of r_o when we study the IC form of the CB amplifier in Chapter 7.

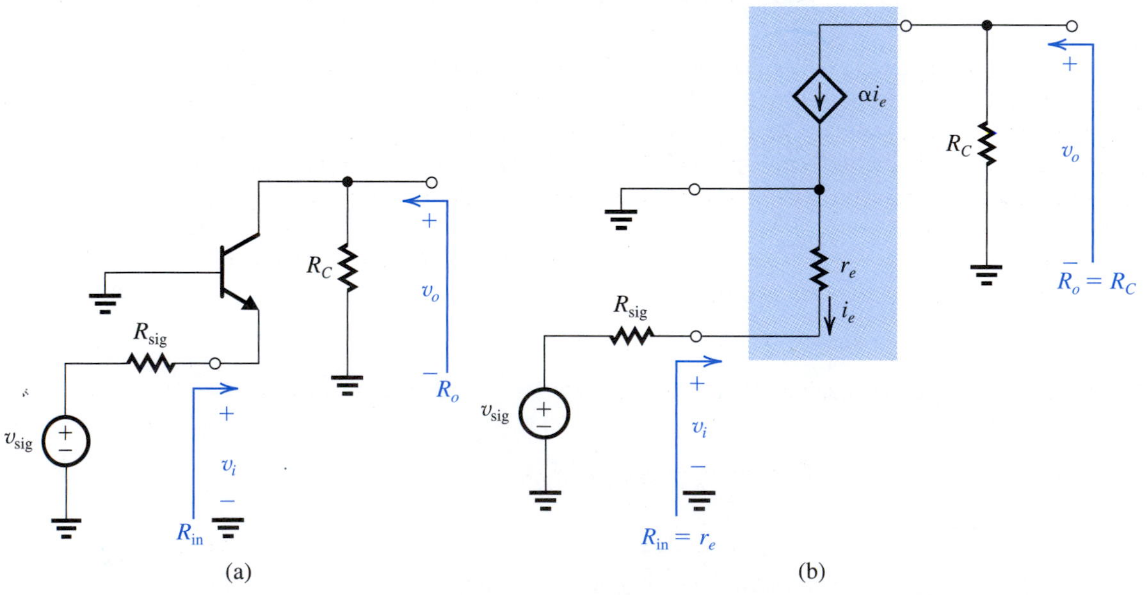

Figure 6.53 (a) CB amplifier with bias details omitted; (b) Amplifier equivalent circuit with the BJT represented by its T Model.

From inspection of the equivalent circuit in Fig. 6.53(b), we see that the input resistance is

$$R_{in} = r_e \tag{6.90}$$

This should have been expected, since we are looking into the emitter and the base is grounded. Typically r_e is a few ohms to a few tens of ohms; thus the CB amplifier has a low input resistance.

To determine the voltage gain, we write at the collector node

$$v_o = -\alpha i_e R_C$$

and substitute for the emitter current from

$$i_e = -\frac{v_i}{r_e}$$

to obtain

$$A_{vo} \equiv \frac{v_o}{v_i} = \frac{\alpha}{r_e} R_C = g_m R_C \tag{6.91}$$

which except for its positive sign is identical to the expression for A_{vo} for the CE amplifier.

The output resistance of the CB circuit can be found by inspection of the circuit in Fig. 6.53(b) as

$$R_o = R_C \tag{6.92}$$

which is the same as in the case of the CE amplifier (with r_o neglected).

Although the gain of the CB amplifier proper has the same magnitude as that of the CE amplifier, this is usually not the case for the overall voltage gain. The low input resistance of the CB amplifier can cause the input signal to be severely attenuated, specifically,

$$\frac{v_i}{v_{sig}} = \frac{R_{in}}{R_{sig} + R_{in}} = \frac{r_e}{R_{sig} + r_e} \tag{6.93}$$

from which we see that except for situations in which R_{sig} is on the order of r_e, the signal transmission factor v_i/v_{sig} can be very small. It is useful at this point to mention that one of the applications of the CB circuit is to amplify high-frequency signals that appear on a coaxial cable. To prevent signal reflection on the cable, the CB amplifier is required to have an input resistance equal to the characteristic resistance of the cable, which is usually in the range of 50 Ω to 75 Ω.

If a load resistance R_L is connected to the amplifier output terminal, it will appear in parallel with R_C and thus A_v can be determined as

$$A_v = g_m(R_C \| R_L)$$

The overall voltage gain G_v can now be obtained by multiplying A_v with the expression for v_i/v_{sig} in Eq. (6.93),

$$G_v = \frac{r_e}{R_{sig} + r_e} g_m(R_C \| R_L)$$

$$= \alpha \frac{R_C \| R_L}{R_{sig} + r_e} \tag{6.94}$$

Since $\alpha \simeq 1$, we see that the overall voltage gain is simply the ratio of the total resistance in the collector circuit to the total resistance in the emitter circuit. We also note that the overall voltage gain is almost independent of the value of β (except through the small dependence of α on β), a desirable property. Observe that for R_{sig} of the same order as R_C and R_L, the gain will be very small.

In summary, the CB amplifier exhibits a very low input resistance (r_e), an open-circuit voltage gain that is positive and equal in magnitude to that of the CE amplifier $(g_m R_C)$, and, like the CE amplifier, a relatively high output resistance (R_C). Because of its very low input resistance, the CB circuit *alone* is not attractive as a voltage amplifier except in specialized applications, such as the cable amplifier mentioned above. The CB amplifier has excellent high-frequency performance, which as we shall see in Chapters 7 and 9, makes it useful in combination with other circuits in the implementation of high-frequency amplifiers.

EXERCISES

6.44 Consider a CB amplifier utilizing a BJT biased at $I_C = 1$ mA and with $R_C = 5$ kΩ. Determine R_{in}, A_{vo}, and R_o, If the amplifier is loaded in $R_L = 5$ kΩ, what value of A_v results? What G_v is obtained if $R_{sig} = 5$ kΩ?
Ans. 25 Ω; 200 V/V; 5 kΩ; 100 V/V; 0.5 V/V

6.45 A CB amplifier is required to amplify a signal delivered by a coaxial cable having a characteristic resistance of 50 Ω. What bias current I_C should be utilized to obtain R_{in} that is matched to the cable resistance? To obtain an overall voltage gain of G_v of 40 V/V, what should the total resistance in the collector (i.e., $R_C \| R_L$) be?
Ans. 0.5 mA; 4 kΩ

6.6.6 The Common-Collector Amplifier or Emitter Follower

The last of the basic BJT amplifier configurations is the common-collector amplifier, a very important circuit that finds frequent application in the design of both small-signal amplifiers and amplifiers that are required to handle large signals and deliver substantial amounts of signal power to a load. This latter variety will be studied in Chapter 11. As well, the common-collector amplifier is utilized in a significant family of digital logic circuits (Chapter 14). The circuit is more commonly known by the alternative name *emitter follower*; the reason for this will become apparent shortly.

The Need for Voltage Buffers Before delving into the analysis of the emitter follower, it is useful to look at one of its most common applications. Consider the situation depicted in Fig. 6.54(a). A signal source delivering a signal of reasonable strength (200 mV) with an internal resistance of 100 kΩ, is to be connected to a 1-kΩ load resistance. Connecting the source to the load directly as in Fig. 6.54(b) would result in severe attenuation of the signal; the signal appearing across the load will be only $1/(100+1)$ of the input signal, or about 2 mV.

An alternative course of action is suggested in Fig. 6.54(c). Here we have interposed an amplifier between the source and the load. Our amplifier, however, is unlike the amplifiers we have been studying in this chapter thus far; it has a voltage gain of unity. This is because our signal is already of sufficient strength and we do not need to increase its amplitude. Note, however, that our amplifier has an input resistance of 100 kΩ; thus half the input signal (100 mV) will appear at the input of the amplifier proper. Since the amplifier has a low

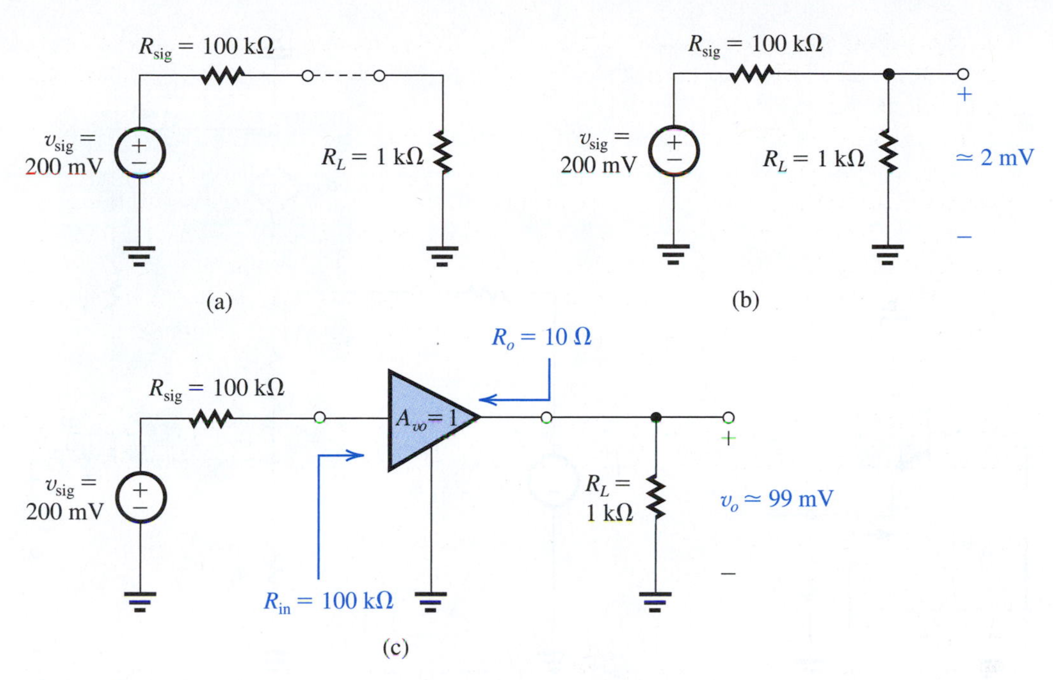

Figure 6.54 Illustrating the need for a unity-gain buffer amplifer.

output resistance (10 Ω), 99% of this signal (99 mV) will appear at the output. This is a significant improvement over the situation with the source connected directly to the load. As will be seen shortly, the emitter follower can easily implement the unity-gain **buffer amplifier** shown in Fig. 6.54(c).

Characteristic Parameters of the Emitter Follower Figure 6.55(a) shows a common-collector amplifier or emitter follower, as we will refer to it henceforth. Note that the biasing circuit is not shown. The emitter follower is fed with a signal source (v_{sig}, R_{sig}) and has a load resistance R_L connected between emitter and ground. To keep things simple, we are assuming that R_L includes both the actual load and any other resistance that may be present between emitter and ground. Normally the actual R_L would be much lower in value than such other resistances and thus would dominate.

Since the BJT has a resistance R_L connected in its emitter, it is most convenient to use the T model to represent the BJT. Doing this results in the emitter-follower equivalent circuit shown in Fig. 6.55(b). We have included r_o simply because it is very easy to do so. However, note that r_o appears in parallel with R_L, and in discrete circuits is much larger than R_L and can thus be neglected. The resulting simplified circuit shown in Fig. 6.55(c), can now be used to determine the characteristic parameters of the amplifier.

The input resistance R_{in} is found from

$$R_{in} = \frac{v_i}{i_b}$$

Substituting for $i_b = i_e/(\beta + 1)$ where i_e is given by

$$i_e = \frac{v_i}{r_e + R_L}$$

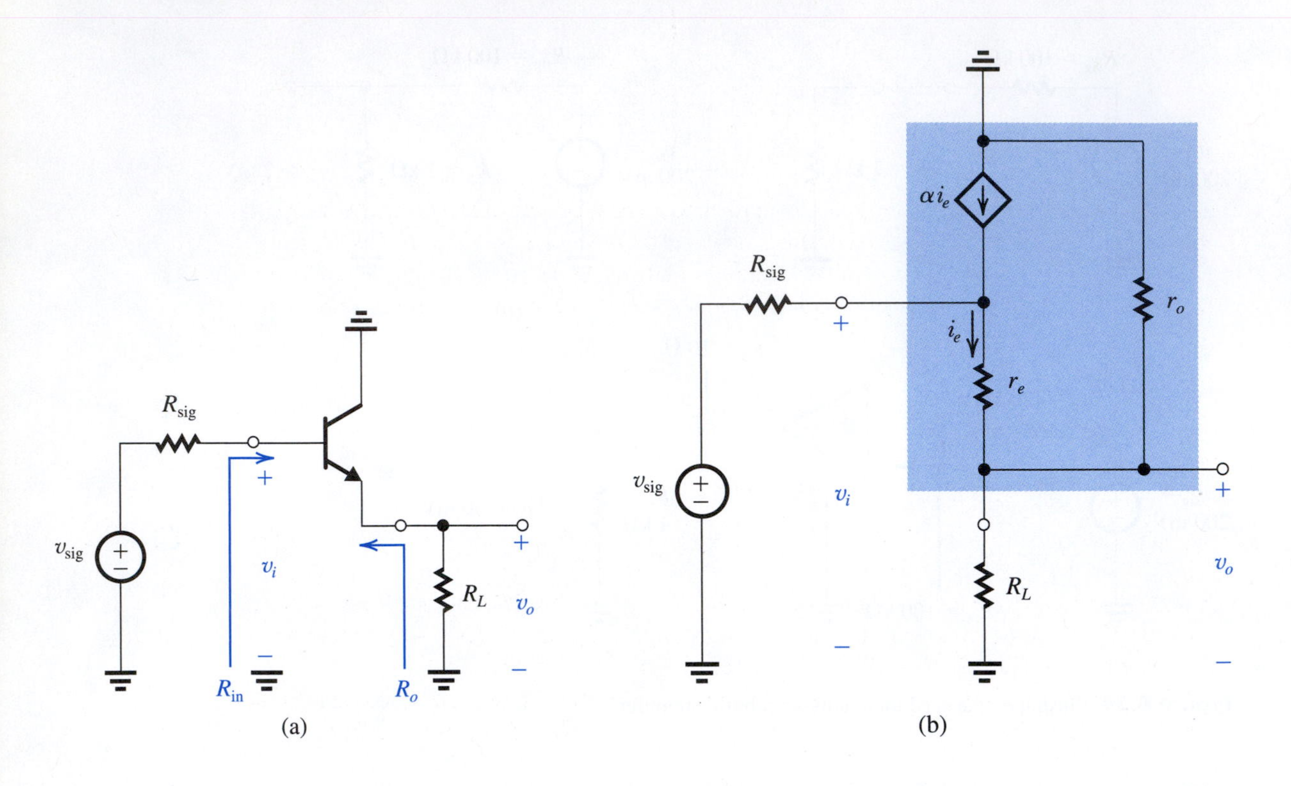

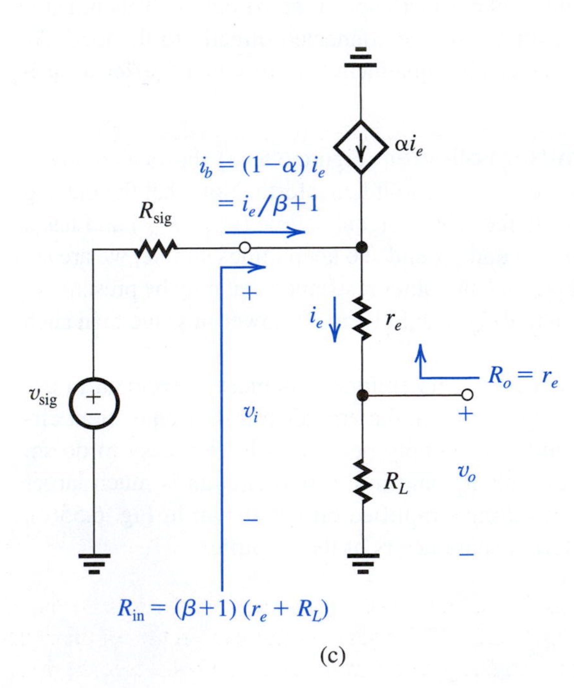

Figure 6.55 **(a)** Common-collector amplifier or emitter-follower. **(b)** Equivalent circuit obtained by replacing the BJT with its T model. Note that r_o appears in parallel with R_L. Since in discrete circuits $r_0 \gg R_L$, we shall neglect it, thus obtaining the simplified circuit in **(c)**.

we obtain

$$R_{in} = (\beta + 1)(r_e + R_L) \tag{6.95}$$

a result that we could have written directly, utilizing the resistance-reflection rule. Note that as expected the emitter follower takes the low load resistance and reflects it to the base side, where the signal source is, after increasing its value by a factor $(\beta + 1)$. It is this impedance transformation property of the emitter follower that makes it useful in connecting a low-resistance load to a high-resistance source, that is, to implement a buffer amplifier.

The voltage gain A_v is given by

$$A_v \equiv \frac{v_o}{v_i} = \frac{R_L}{R_L + r_e} \tag{6.96}$$

Setting $R_L = \infty$ yields A_{vo},

$$A_{vo} = 1 \tag{6.97}$$

Thus, as expected, the open-circuit voltage gain of the emitter follower proper is unity[16] which means that the signal voltage at the emitter *follows* that at the base; which is the origin of the name "emitter follower."

To determine R_o, refer to Fig. 6.55(c) and look back into the emitter (i.e., behind or excluding R_L) while setting $v_i = 0$ (i.e., grounding the base). You will see r_e of the BJT, thus

$$R_o = r_e \tag{6.98}$$

This result together with $A_{vo} = 1$ yields A_v in Eq. (6.96), thus confirming our earlier analysis.

Overall Voltage Gain We now proceed to determine the overall voltage gain G_v, as follows:

$$\frac{v_i}{v_{sig}} = \frac{R_{in}}{R_{in} + R_{sig}}$$

$$= \frac{(\beta + 1)(r_e + R_L)}{(\beta + 1)(r_e + R_L) + R_{sig}}$$

$$G_v \equiv \frac{v_o}{v_{sig}} = \frac{v_i}{v_{sig}} \times A_v$$

Substituting for A_v from Eq. (6.96), results in

$$G_v = \frac{(\beta + 1)R_L}{(\beta + 1)R_L + (\beta + 1)r_e + R_{sig}} \tag{6.99}$$

This equation indicates that the overall gain, though lower than one, can be close to one if $(\beta + 1)R_L$ is larger or comparable in value to R_{sig}. This again confirms the action of the emitter follower in delivering a large proportion of v_{sig} to a low-valued load resistance R_L even though R_{sig} can be much larger than R_L. The key point is that R_L is multiplied by $(\beta + 1)$

[16]In practice, the value of A_{vo} will be lower than, but close to unity. For one thing, r_o, which we have neglected, would make $A_{vo} = r_o/(r_o + r_e)$. Also, as already mentioned, there may be other resistances (e.g., for biasing purposes) attached to the emitter.

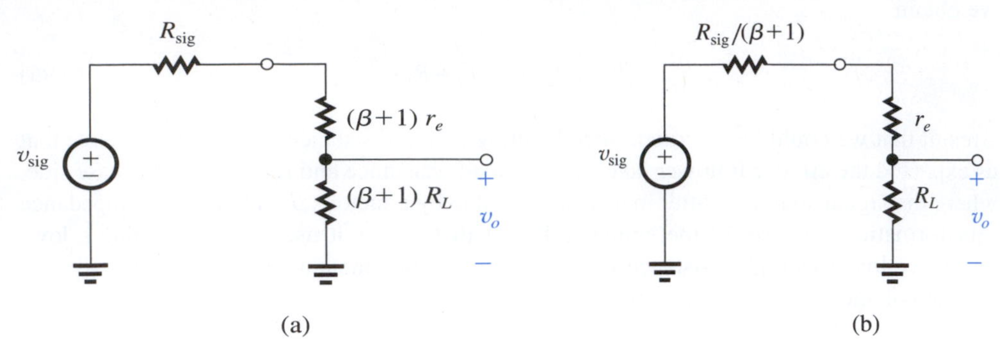

Figure 6.56 Simple equivalent circuits for the emitter follower obtained by **(a)** reflecting r_e and R_L to the base side, and **(b)** reflecting v_{sig} and R_{sig} to the emitter side. Note that the circuit in **(b)** can be obtained from that in **(a)** by simply dividing all resistances by $(\beta + 1)$.

before it is "presented to the source." Figure 6.56(a) shows an equivalent circuit of the emitter follower obtained by simply reflecting r_e and R_L to the base side. The overall voltage gain $G_v \equiv v_o/v_{sig}$ can be determined directly and very simply from this circuit by using the voltage divider rule. The result is the expression for G_v already given in Eq. (6.99).

Dividing all resistances in the circuit of Fig. 6.56(a) by $\beta + 1$ does not change the voltage ratio v_o/v_{sig}. Thus we obtain another equivalent circuit, shown in Fig. 6.56(b), that can be used to determine $G_v \equiv v_o/v_{sig}$ of the emitter follower. A glance at this circuit reveals that it is simply the equivalent circuit obtained by reflecting v_{sig} and R_{sig} from the base side to the emitter side. In this reflection, v_{sig} does not change, but R_{sig} is divided by $\beta + 1$. Thus, we either reflect to the base side and obtain the circuit in Fig. 6.56(a) or reflect to the emitter side and obtain the circuit in Fig. 6.56(b). From the latter, G_v can be found as

$$G_v \equiv \frac{v_o}{v_{sig}} = \frac{R_L}{R_L + r_e + R_{sig}/(\beta + 1)} \tag{6.100}$$

Observe that this expression is the same as that in Eq. (6.99) except for dividing both the numerator and denominator by $\beta + 1$.

The expression for G_v in Eq. (6.100) has an interesting interpretation: The emitter follower reduces R_{sig} by the factor $(\beta + 1)$ before "presenting it to the load resistance R_L": an impedance transformation that has the same buffering effect.

At this point it is important to note that although the emitter follower does not provide voltage gain it has a current gain of $\beta + 1$.

Thévenin Representation of the Emitter-Follower Output A more general representation of the emitter-follower output is shown in Fig. 6.57(a). Here G_{vo} is the overall open-circuit voltage gain that can be obtained by setting $R_L = \infty$ in the circuit of Fig. 6.56(b), as illustrated in Fig. 6.57(b). The result is $G_{vo} = 1$. The output resistance R_{out} is *different* from R_o. To determine R_{out} we set v_{sig} to zero (rather than setting v_i to zero). Again we can use the equivalent circuit in Fig. 6.56(b) to do this, as illustrated in Fig. 6.57(c). We see that

$$R_{out} = r_e + \frac{R_{sig}}{\beta + 1} \tag{6.101}$$

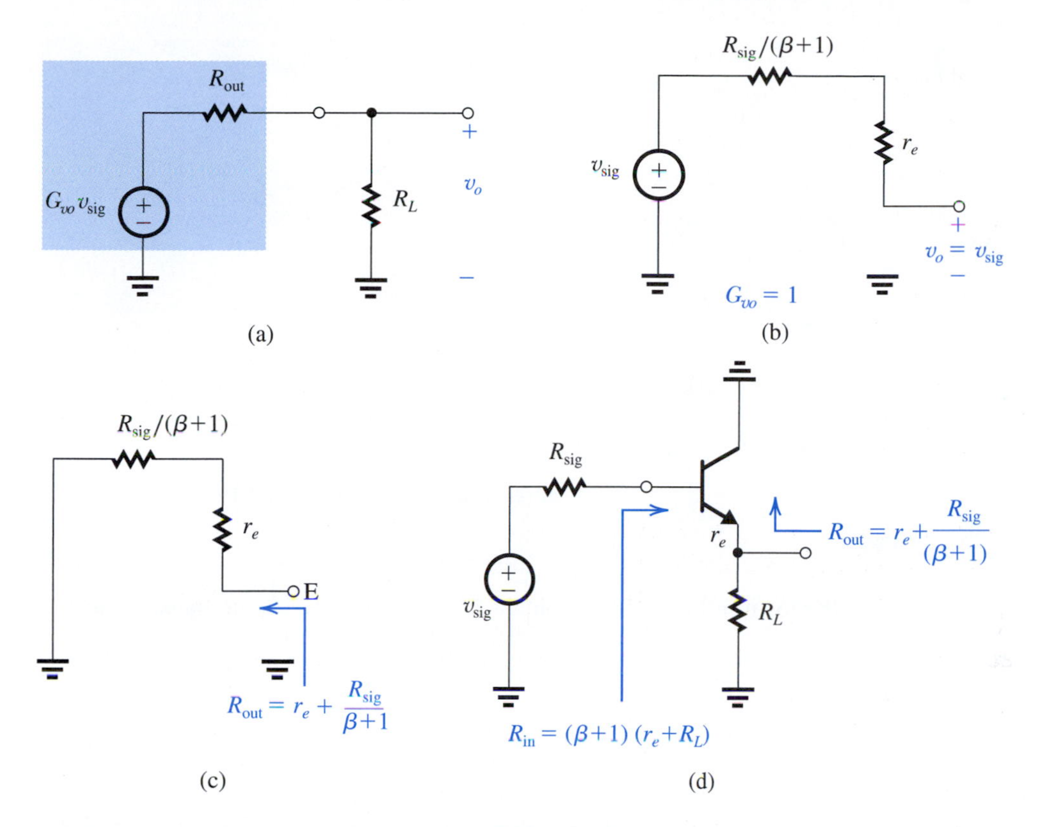

Figure 6.57 **(a)** Thévenin representation of the output of the emitter follower. **(b)** Obtaining G_{vo} from the equivalent circuit in Fig. 6.56(b). **(c)** Obtaining R_{out} from the equivalent circuit in Fig. 6.56(b) with v_{sig} set to zero. **(d)** The emitter follower with R_{in} and R_{out} determined simply by looking into the input and output terminals, respectively.

Finally, we show in Fig. 6.57(d) the emitter-follower circuit together with its R_{in} and R_{out}. Observe that R_{in} is determined by reflecting r_e and R_L to the base side (by multiplying their values by $\beta + 1$). To determine R_{out}, grab hold of the emitter and walk (or just look!) backward while $v_{sig} = 0$. You will see r_e in series with R_{sig}, which because it is in the base must be divided by $(\beta + 1)$.

We note that unlike the CE and CB amplifiers we studied earlier, the emitter follower is *not* unilateral. This is manifested by the fact that R_{in} depends on R_L and R_{out} depends on R_{sig}.

Example 6.19

It is required to design an emitter follower to implement the buffer amplifier of Fig. 6.54(c). Specify the required bias current I_E and the minimum value the transistor β must have. Determine the maximum allowed value of v_{sig} if v_π is to be limited to 5 mV in order to obtain reasonably linear operation. With $v_{sig} = 200$ mV, determine the signal voltage at the output if R_L is changed to 2 kΩ, and to 0.5 kΩ.

Example 6.19 *continued*

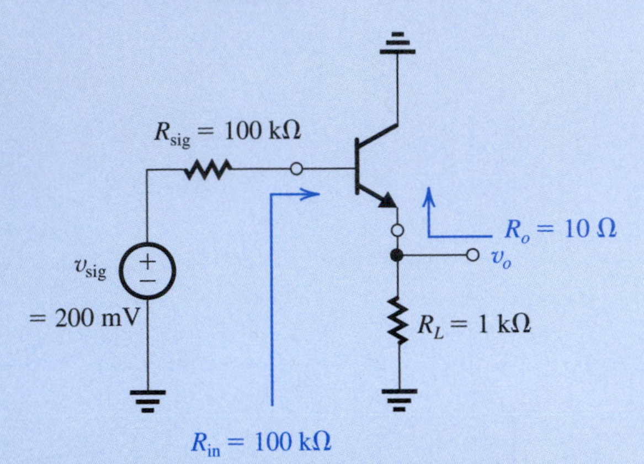

Figure 6.58 Circuit for Example 6.19.

Solution

The emitter-follower circuit is shown in Fig. 6.58. To obtain $R_o = 10\ \Omega$, we bias the transistor to obtain $r_e = 10\ \Omega$. Thus,

$$10\ \Omega = \frac{V_T}{I_E}$$

$$I_E = 2.5\ \text{mA}$$

The input resistance R_{in} will be

$$R_{in} = (\beta + 1)(r_e + R_L)$$

$$100 = (\beta + 1)(0.01 + 1)$$

Thus, the BJT should have a β with a minimum value of 98. A higher β would obviously be beneficial.

The overall voltage gain can be determined from

$$G_v \equiv \frac{v_o}{v_{sig}} = \frac{R_L}{R_L + r_e + \dfrac{R_{sig}}{(\beta + 1)}}$$

Assuming $\beta = 100$, the value of G_v obtained is

$$G_v = 0.5$$

Thus when $v_{sig} = 200$ mV, the signal at the output will be 100 mV. Since the 100 mV appears across the 1-kΩ load, the signal across the base–emitter junction can be found from

$$v_\pi = \frac{v_o}{R_L} \times r_e$$

$$= \frac{100}{1000} \times 10 = 1\ \text{mV}$$

If $\hat{v}_\pi = 5$ mV then v_{sig} can be increased by a factor of 5, resulting in $\hat{v}_{sig} = 1$ V.

To obtain v_o as the load is varied, we use the Thévenin equivalent of the emitter follower, shown in Fig. 6.57(a) with $G_{vo} = 1$ and

$$R_{out} = \frac{R_{sig}}{\beta + 1} + r_e = \frac{100}{101} + 0.01 = 1\ \text{k}\Omega$$

to obtain

$$v_o = v_{sig} \frac{R_L}{R_L + R_{out}}$$

For $R_L = 2\ \text{k}\Omega$,

$$v_o = 200\ \text{mV} \times \frac{2}{2+1} = 133.3\ \text{mV}$$

and for $R_L = 0.5\ \text{k}\Omega$,

$$v_o = 200\ \text{mV} \times \frac{0.5}{0.5+1} = 66.7\ \text{mV}$$

EXERCISE

6.46 An emitter follower utilizes a transistor with $\beta = 100$ and is biased at $I_C = 5$ mA. It operates between a source having a resistance of 10 kΩ and a load of 1 kΩ. Find R_{in}, G_{vo}, R_{out}, and G_v. What is the peak amplitude of v_{sig} that results in v_π having a peak amplitude of 5 mV? Find the resulting peak amplitude at the output.

Ans. 101.5 kΩ; 1 V/V; 104 Ω; 0.91 V/V; 1.1 V; 1 V

6.6.7 Summary and Comparisons

For easy reference and to enable comparisons, we present in Table 6.5 the formulas for determining the characteristic parameters of discrete BJT amplifiers. Note that r_o has been neglected throughout. As has already been mentioned, this is possible in discrete-circuit amplifiers. In addition to the remarks made throughout this section about the characteristics and applicability of the various configurations, we make the following concluding points.

1. The CE configuration is the one best suited for realizing the bulk of the gain required in an amplifier. Depending on the magnitude of the gain required, either a single stage or a cascade of two or three stages can be used.
2. Including a resistor R_e in the emitter lead of the CE stage provides a number of performance improvements at the expense of gain reduction.
3. The low input resistance of the CB amplifier makes it useful only in specific applications. As we shall see in Chapter 9, it has a much better high-frequency response than the CE amplifier. This superiority will make it useful as a high-frequency amplifier, especially when combined with the CE circuit. We shall see one such combination in Chapter 7.
4. The emitter follower finds application as a voltage buffer for connecting a high-resistance source to a low-resistance load and as the output stage in a multistage amplifier, where its purpose is to equip the amplifier with a low output-resistance.

TABLE 6.5 Characteristics of BJT Amplifiers[a, b, c]

	R_{in}	A_{vo}	R_o	A_v	G_v
Common emitter (Fig. 6.50)	$(\beta+1)r_e$	$-g_m R_C$	R_C	$-g_m(R_C \parallel R_L)$ $-\alpha\frac{R_C \parallel R_L}{r_e}$	$-\beta\frac{R_C \parallel R_L}{R_{sig}+(\beta+1)r_e}$
Common emitter with R_e (Fig. 6.52)	$(\beta+1)(r_e+R_e)$	$-\frac{g_m R_C}{1+g_m R_e}$	R_C	$\frac{-g_m(R_C \parallel R_L)}{1+g_m R_e}$ $-\alpha\frac{R_C \parallel R_L}{r_e+R_e}$	$-\beta\frac{R_C \parallel R_L}{R_{sig}+(\beta+1)(r_e+R_e)}$
Common base (Fig. 6.53)	r_e	$g_m R_C$	R_C	$g_m(R_C \parallel R_L)$ $\alpha\frac{R_C \parallel R_L}{r_e}$	$\alpha\frac{R_C \parallel R_L}{R_{sig}+r_e}$
Emitter follower (Fig. 6.55)	$(\beta+1)(r_e+R_L)$	1	r_e	$\frac{R_L}{R_L+r_e}$	$\frac{R_L}{R_L+r_e+R_{sig}/(\beta+1)}$ $G_{vo}=1$ $R_{out}=r_e+\frac{R_{sig}}{\beta+1}$

[a] For the interpretation of R_{in}, A_{vo}, and R_o refer to Fig. 6.49.
[b] The BJT output resistance r_o has been neglected, which is permitted in the discrete-circuit amplifiers studied in this chapter. For integrated-circuit amplifiers (Chapter 7), r_o must always be taken into account.
[c] Setting $\beta=\infty$ ($\alpha=1$) and replacing r_e with $1/g_m$, R_C with R_D, and R_e with R_s results in the corresponding formulas for MOSFET amplifiers (Table 5.4).

6.7 Biasing in BJT Amplifier Circuits

Having studied the various configurations of BJT amplifiers, we now address the important question of biasing and its relationship to small-signal behavior. The biasing problem is that of establishing a constant dc current in the collector of the BJT. This current has to be calculable, predictable, and insensitive to variations in temperature and to the large variations in the value of β encountered among transistors of the same type. Another important consideration in bias design is locating the dc bias point in the i_C–v_{CE} plane to allow for maximum output signal swing (see the discussion in Section 6.4.6). In this section, we shall deal with various approaches to solving the bias problem in transistor circuits designed with discrete devices. Bias methods for integrated-circuit design are presented in Chapter 7.

Before presenting the "good" biasing schemes, we should point out why two obvious arrangements are *not* good. First, attempting to bias the BJT by fixing the voltage V_{BE} by, for instance, using a voltage divider across the power supply V_{CC}, as shown in Fig. 6.59(a), is not a viable approach: The very sharp exponential relationship i_C–v_{BE} means that any small and inevitable differences in V_{BE} from the desired value will result in large differences in I_C and in V_{CE}. Second, biasing the BJT by establishing a constant current in the base, as shown in Fig. 6.59(b), where $I_B \simeq ((V_{CC}-0.7)/R_B)$, is also not a recommended approach. Here the typically large variations in the value of β among units of the same device type will result in correspondingly large variations in I_C and hence in V_{CE}.

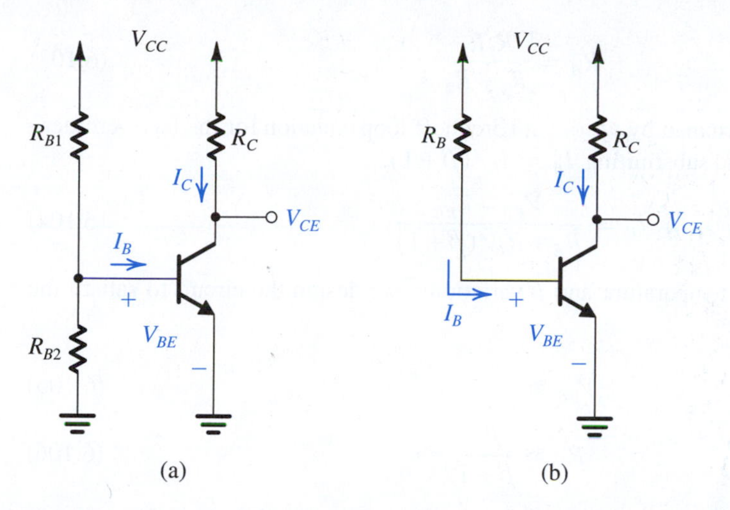

Figure 6.59 Two obvious schemes for biasing the BJT: **(a)** by fixing V_{BE}; **(b)** by fixing I_B. Both result in wide variations in I_C and hence in V_{CE} and therefore are considered to be "bad." Neither scheme is recommended.

6.7.1 The Classical Discrete-Circuit Bias Arrangement

Figure 6.60(a) shows the arrangement most commonly used for biasing a discrete-circuit transistor amplifier if only a single power supply is available. The technique consists of supplying the base of the transistor with a fraction of the supply voltage V_{CC} through the voltage divider R_1, R_2. In addition, a resistor R_E is connected to the emitter.

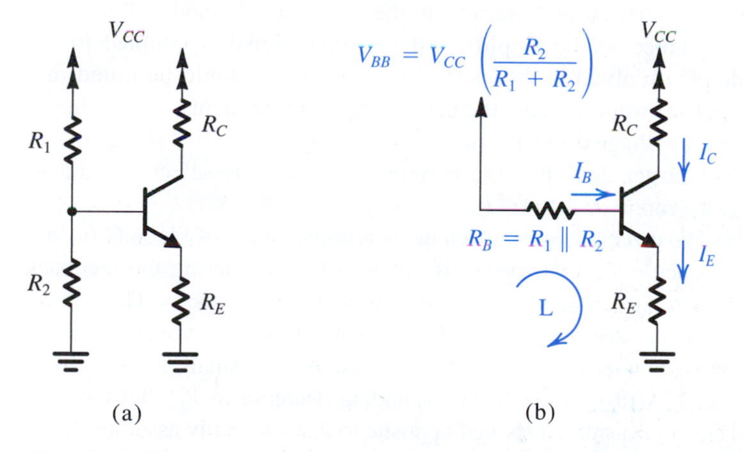

Figure 6.60 Classical biasing for BJTs using a single power supply: **(a)** circuit; **(b)** circuit with the voltage divider supplying the base replaced with its Thévenin equivalent.

Figure 6.60(b) shows the same circuit with the voltage divider network replaced by its Thévenin equivalent,

$$V_{BB} = \frac{R_2}{R_1 + R_2} V_{CC} \tag{6.102}$$

$$R_B = \frac{R_1 R_2}{R_1 + R_2} \tag{6.103}$$

The current I_E can be determined by writing a Kirchhoff loop equation for the base–emitter–ground loop, labeled L, and substituting $I_B = I_E/(\beta+1)$:

$$I_E = \frac{V_{BB} - V_{BE}}{R_E + R_B/(\beta+1)} \tag{6.104}$$

To make I_E insensitive to temperature and β variation,[17] we design the circuit to satisfy the following two constraints:

$$V_{BB} \gg V_{BE} \tag{6.105}$$

$$R_E \gg \frac{R_B}{\beta+1} \tag{6.106}$$

Condition (6.105) ensures that small variations in V_{BE} ($\simeq$ 0.7 V) will be swamped by the much larger V_{BB}. There is a limit, however, on how large V_{BB} can be: For a given value of the supply voltage V_{CC}, the higher the value we use for V_{BB}, the lower will be the sum of voltages across R_C and the collector–base junction (V_{CB}). On the other hand, we want the voltage across R_C to be large in order to obtain high voltage gain and large signal swing (before transistor cutoff). We also want V_{CB} (or V_{CE}) to be large to provide a large signal swing (before transistor saturation). Thus, as is the case in any design, we have a set of conflicting requirements, and the solution must be a trade-off. As a rule of thumb, one designs for V_{BB} about $\frac{1}{3}V_{CC}$, V_{CB} (or V_{CE}) about $\frac{1}{3}V_{CC}$, and $I_C R_C$ about $\frac{1}{3}V_{CC}$.

Condition (6.106) makes I_E insensitive to variations in β and could be satisfied by selecting R_B small. This in turn is achieved by using low values for R_1 and R_2. Lower values for R_1 and R_2, however, will mean a higher current drain from the power supply, and will result in a lowering of the input resistance of the amplifier (if the input signal is coupled to the base),[18] which is the trade-off involved in this part of the design. It should be noted that condition (6.106) means that we want to make the base voltage independent of the value of β and determined solely by the voltage divider. This will obviously be satisfied if the current in the divider is made much larger than the base current. Typically one selects R_1 and R_2 such that their current is in the range of I_E to $0.1I_E$.

Further insight regarding the mechanism by which the bias arrangement of Fig. 6.60(a) stabilizes the dc emitter (and hence collector) current is obtained by considering the feedback action provided by R_E. Consider that for some reason the emitter current increases. The voltage drop across R_E, and hence V_E will increase correspondingly. Now, if the base voltage is determined primarily by the voltage divider R_1, R_2, which is the case if R_B is small, it will remain constant, and the increase in V_E will result in a corresponding decrease in V_{BE}. This in turn reduces the collector (and emitter) current, a change opposite to that originally assumed. Thus R_E provides a *negative feedback* action that stabilizes the bias current. This should remind the reader of the resistance R_e that we included in the emitter lead of the CE amplifier in Section 6.6.4. We shall study negative feedback formally in Chapter 10.

[17]Bias design seeks to stabilize either I_E or I_C since $I_C = \alpha I_E$ and α varies very little. That is, a stable I_E will result in an equally stable I_C, and vice versa.

[18]If the input signal is coupled to the transistor base, the two bias resistances R_1 and R_2 effectively appear in parallel between the base and ground. Thus, low values for R_1 and R_2 will result in lowering R_{in}.

Example 6.20

We wish to design the bias network of the amplifier in Fig. 6.60 to establish a current $I_E = 1$ mA using a power supply $V_{CC} = +12$ V. The transistor is specified to have a nominal β value of 100.

Solution

We shall follow the rule of thumb mentioned above and allocate one-third of the supply voltage to the voltage drop across R_2 and another one-third to the voltage drop across R_C, leaving one-third for possible negative signal swing at the collector. Thus,

$$V_B = +4 \text{ V}$$

$$V_E = 4 - V_{BE} \simeq 3.3 \text{ V}$$

and R_E is determined from

$$R_E = \frac{V_E}{I_E} = \frac{3.3}{1} = 3.3 \text{ k}\Omega$$

From the discussion above we select a voltage divider current of $0.1I_E = 0.1 \times 1 = 0.1$ mA. Neglecting the base current, we find

$$R_1 + R_2 = \frac{12}{0.1} = 120 \text{ k}\Omega$$

and

$$\frac{R_2}{R_1 + R_2} V_{CC} = 4 \text{ V}$$

Thus $R_2 = 40$ kΩ and $R_1 = 80$ kΩ.

At this point, it is desirable to find a more accurate estimate for I_E, taking into account the nonzero base current. Using Eq. (6.104),

$$I_E = \frac{4 - 0.7}{3.3(\text{k}\Omega) + \dfrac{(80 \parallel 40)(\text{k}\Omega)}{101}} = 0.93 \text{ mA}$$

This is quite a bit lower than 1 mA, the value we are aiming for. It is easy to see from the above equation that a simple way to restore I_E to its nominal value would be to reduce R_E from 3.3 kΩ by the magnitude of the second term in the denominator (0.267 kΩ). Thus a more suitable value for R_E in this case would be $R_E = 3$ kΩ, which results in $I_E = 1.01 \text{ mA} \simeq 1$ mA.[19]

It should be noted that if we are willing to draw a higher current from the power supply and to accept a lower input resistance for the amplifier, then we may use a voltage-divider current equal, say, to I_E (i.e., 1 mA), resulting in $R_1 = 8$ kΩ and $R_2 = 4$ kΩ. We shall refer to the circuit using these latter values as design 2, for which the actual value of I_E using the initial value of R_E of 3.3 kΩ will be

$$I_E = \frac{4 - 0.7}{3.3 + 0.027} = 0.99 \simeq 1 \text{ mA}$$

[19]Although reducing R_E restores I_E to the design value of 1 mA, it does not solve the problem of the dependence of the value of I_E on β. See Exercise 6.47.

Example 6.20 *continued*

In this case, design 2, we need not change the value of R_E.

Finally, the value of R_C can be determined from

$$R_C = \frac{12 - V_C}{I_C}$$

Substituting $I_C = \alpha I_E = 0.99 \times 1 = 0.99 \text{ mA} \simeq 1 \text{ mA}$ results, for both designs, in

$$R_C = \frac{12 - 8}{1} = 4 \text{ k}\Omega$$

EXERCISE

6.47 For design 1 in Example 6.20, calculate the expected range of I_E if the transistor used has β in the range of 50 to 150. Express the range of I_E as a percentage of the nominal value ($I_E \simeq 1$ mA) obtained for $\beta = 100$. Repeat for design 2.

Ans. For design 1: 0.94 mA to 1.04 mA, a 10% range; for design 2: 0.984 mA to 0.995 mA, a 1.1% range.

6.7.2 A Two-Power-Supply Version of the Classical Bias Arrangement

A somewhat simpler bias arrangement is possible if two power supplies are available, as shown in Fig. 6.61. Writing a loop equation for the loop labeled L gives

$$I_E = \frac{V_{EE} - V_{BE}}{R_E + R_B/(\beta + 1)} \tag{6.107}$$

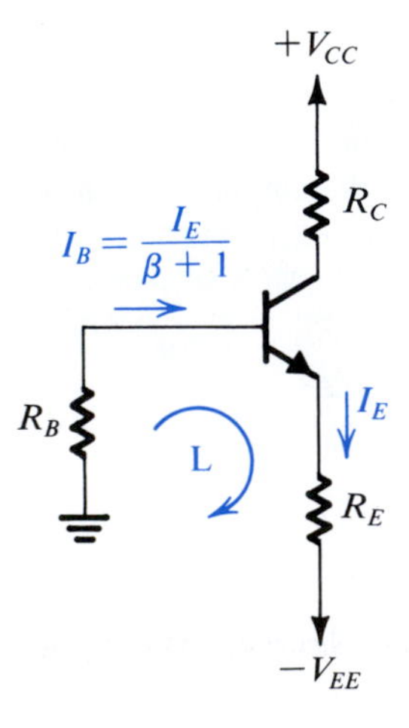

Figure 6.61 Biasing the BJT using two power supplies. Resistor R_B is needed only if the signal is to be capacitively coupled to the base. Otherwise, the base can be connected directly to ground, or to a grounded signal source, resulting in almost total β-independence of the bias current.

This equation is identical to Eq. (6.104) except for V_{EE} replacing V_{BB}. Thus the two constraints of Eqs. (6.105) and (6.106) apply here as well. Note that if the transistor is to be used with the base grounded (i.e., in the common-base configuration), then R_B can be eliminated altogether. On the other hand, if the input signal is to be coupled to the base, then R_B is needed. We shall study complete circuits of the various BJT amplifier configurations in Section 6.8.

EXERCISE

D6.48 The bias arrangement of Fig. 6.61 is to be used for a common-base amplifier. Design the circuit to establish a dc emitter current of 1 mA and provide the highest possible voltage gain while allowing for a maximum signal swing at the collector of ±2 V. Use +10-V and –5-V power supplies.
Ans. $R_B = 0$; $R_E = 4.3\ \text{k}\Omega$; $R_C = 8.4\ \text{k}\Omega$

6.7.3 Biasing Using a Collector-to-Base Feedback Resistor

Figure 6.62(a) shows a simple but effective alternative biasing arrangement suitable for common-emitter amplifiers. The circuit employs a resistor R_B connected between the collector and the base. Resistor R_B provides negative feedback, which helps to stabilize the bias point of the BJT. We shall study feedback formally in Chapter 10.

Analysis of the circuit is shown in Fig. 6.62(b), from which we can write

$$V_{CC} = I_E R_C + I_B R_B + V_{BE}$$
$$= I_E R_C + \frac{I_E}{\beta + 1} R_B + V_{BE}$$

Thus the emitter bias current is given by

$$I_E = \frac{V_{CC} - V_{BE}}{R_C + R_B/(\beta + 1)} \tag{6.108}$$

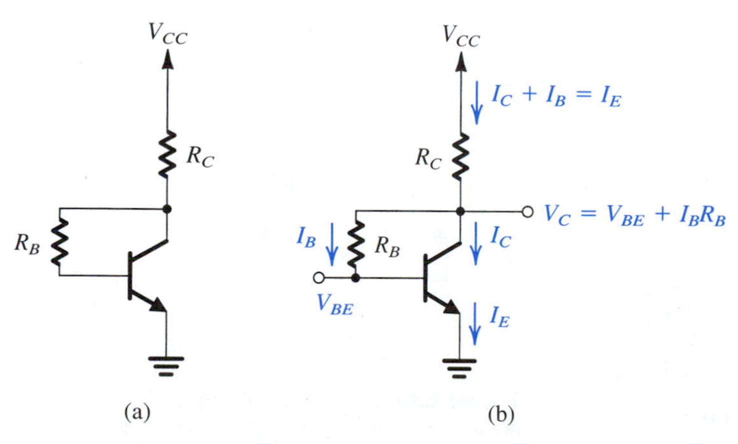

Figure 6.62 **(a)** A common-emitter transistor amplifier biased by a feedback resistor R_B. **(b)** Analysis of the circuit in **(a)**.

It is interesting to note that this equation is identical to Eq. (6.109), which governs the operation of the traditional bias circuit, except that V_{CC} replaces V_{BB} and R_C replaces R_E. It follows that to obtain a value of I_E that is insensitive to variation of β, we select $R_B/(\beta+1) \ll R_C$. Note, however, that the value of R_B determines the allowable negative signal swing at the collector since

$$V_{CB} = I_B R_B = I_E \frac{R_B}{\beta+1} \tag{6.109}$$

EXERCISE

D6.49 Design the circuit of Fig. 6.62 to obtain a dc emitter current of 1 mA, maximum gain, and a ±2-V signal swing at the collector; that is, design for $V_{CE} = +2.3$ V. Let $V_{CC} = 10$ V and $\beta = 100$.
Ans. $R_B = 162$ kΩ; $R_C = 7.7$ kΩ. Note that if standard 5% resistor values are used (Appendix G) we select $R_B = 160$ kΩ and $R_C = 7.5$ kΩ. This results in $I_E = 1.02$ mA and $V_C = +2.3$ V.

6.7.4 Biasing Using a Constant-Current Source

The BJT can be biased using a constant-current source I as indicated in the circuit of Fig. 6.63(a). This circuit has the advantage that the emitter current is independent of the values of β and R_B. Thus R_B can be made large, enabling an increase in the input resistance at the base without adversely affecting bias stability. Further, current-source biasing leads to significant design simplification, as will become obvious in later sections and chapters.

A simple implementation of the constant-current source I is shown in Fig. 6.63(b). The circuit utilizes a pair of matched transistors Q_1 and Q_2, with Q_1 connected as a diode by shorting its collector to its base. If we assume that Q_1 and Q_2 have high β values, we can neglect their base currents. Thus the current through Q_1 will be approximately equal to I_{REF},

$$I_{REF} = \frac{V_{CC} - (-V_{EE}) - V_{BE}}{R} \tag{6.110}$$

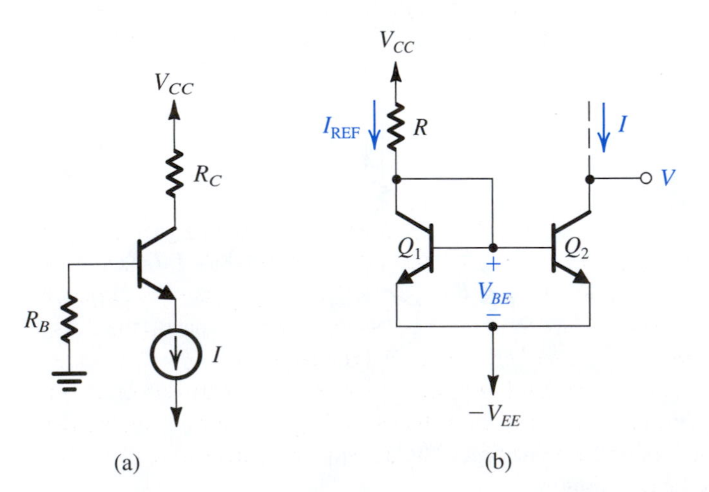

Figure 6.63 **(a)** A BJT biased using a constant-current source I. **(b)** Circuit for implementing the current source I.

Now, since Q_1 and Q_2 have the same V_{BE}, their collector currents will be equal, resulting in

$$I = I_{\text{REF}} = \frac{V_{CC} + V_{EE} - V_{BE}}{R} \tag{6.111}$$

Neglecting the Early effect in Q_2, the collector current will remain constant at the value given by this equation as long as Q_2 remains in the active region. This can be guaranteed by keeping the voltage at the collector, V, greater than that at the emitter ($-V_{EE}$) by at least 0.3V. The connection of Q_1 and Q_2 in Fig. 6.63(b) is known as a **current mirror**. We will study current mirrors in detail in Chapter 7.

EXERCISE

6.50 For the circuit in Fig. 6.63(a) with $V_{CC} = 10$ V, $I = 1$ mA, $\beta = 100$, $R_B = 100$ kΩ, and $R_C = 7.5$ kΩ, find the dc voltage at the base, the emitter, and the collector. For $V_{EE} = 10$ V, and neglecting base currents, find the required value of R in order for the circuit of Fig. 6.63(b) to implement the current source I.
Ans. −1 V; −1.7 V; +2.6 V; 19.3 kΩ

6.8 Discrete-Circuit BJT Amplifiers

With our study of BJT amplifier basics complete, we now put everything together by presenting practical circuits for discrete-circuit amplifiers. These circuits, which utilize the amplifier configurations studied in Section 6.6 and one of the biasing methods of Section 6.7, can be assembled using off-the-shelf discrete transistors, resistors, and capacitors. Though practical and carefully selected to illustrate some important points, the circuits presented in this section should be regarded as examples of discrete-circuit, bipolar-transistor amplifiers. Indeed, there is a great variety of such circuits, a number of which are explored in the end-of-chapter problems.

In this section we present a series of exercise problems, Exercises 6.51 to 6.55, which are carefully designed to illustrate important aspects of the amplifier circuits studied. These exercises are also intended to enable the reader to see more clearly the differences between the various circuit configurations. We strongly urge the reader to solve these exercises. As usual, the answers are provided.

6.8.1 The Basic Structure

Figure 6.64 shows the basic circuit that we shall utilize to implement the various configurations of discrete BJT amplifiers. Among the various biasing schemes possible for discrete BJT amplifiers (Section 6.7), we have selected, for simplicity and effectiveness, the one employing constant-current biasing. Figure 6.64 indicates the dc currents in all branches and the dc voltages at all nodes. We should note that one would want to select a large value for R_B in order to keep the input resistance at the base large. However, we also want to limit the dc voltage drop across R_B and even more importantly the variability of this dc voltage resulting from the variation in β values among transistors of the same type. The dc voltage V_B determines the allowable negative signal swing at the collector.

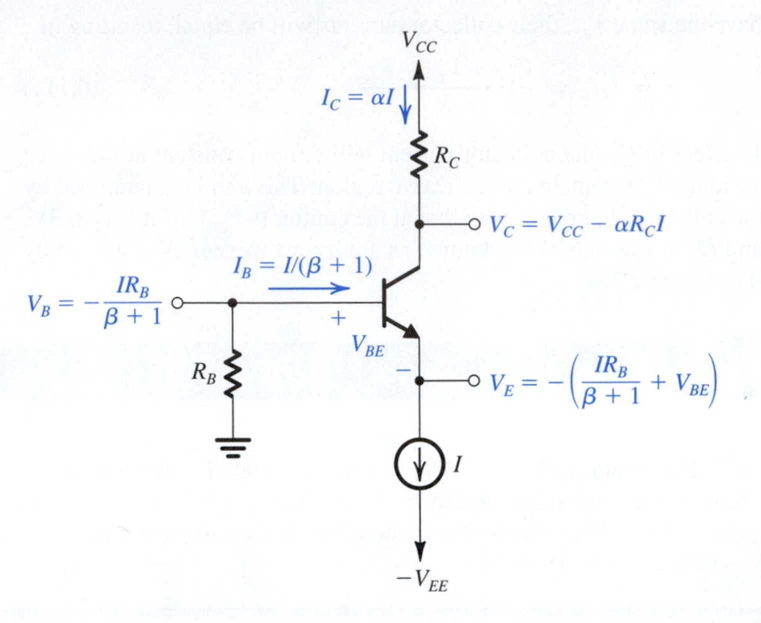

Figure 6.64 Basic structure of the circuit used to realize single-stage, discrete-circuit BJT amplifier configurations.

EXERCISE

6.51 Consider the circuit of Fig. 6.64 for the case $V_{CC} = V_{EE} = 10$ V, $I = 1$ mA, $R_B = 100$ kΩ, $R_C = 8$ kΩ, and $\beta = 100$. Find all dc currents and voltages. What are the allowable signal swings at the collector in both directions? How do these values change as β is changed to 50? To 200? Find the values of the BJT small-signal parameters at the bias point (with $\beta = 100$). The Early voltage $V_A = 100$ V.
Ans. See Fig. E6.51. Signal swing: for $\beta = 100$, +8 V, –3.4 V; for $\beta = 50$, +8 V, –4.4 V; for $\beta = 200$, +8 V, –2.9 V.

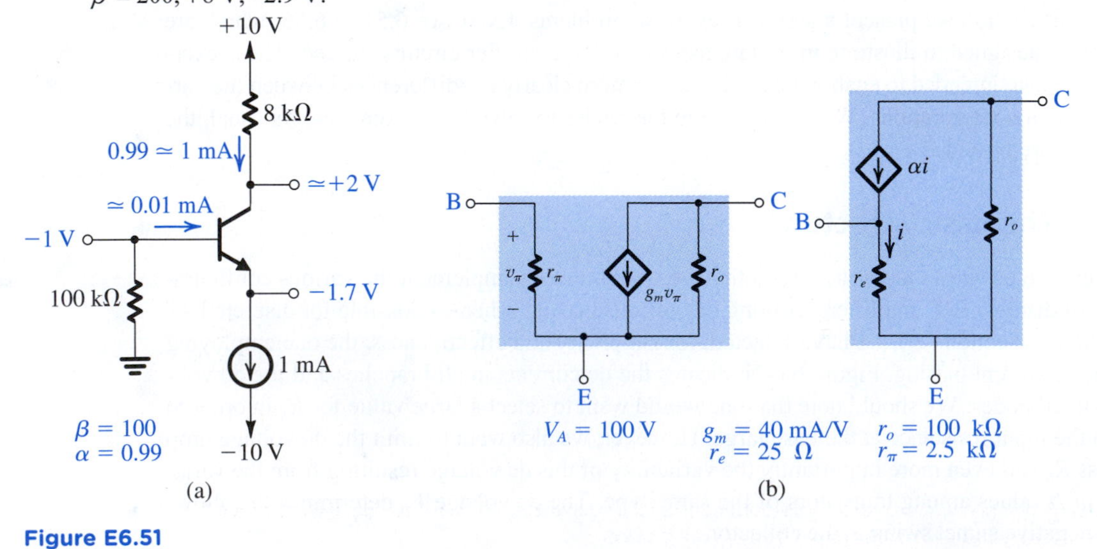

Figure E6.51

6.8.2 The Common-Emitter (CE) Amplifier

As mentioned in Section 6.6, the CE configuration is the most widely used of all BJT amplifier circuits. Figure 6.65(a) shows a CE amplifier implemented using the circuit of Fig. 6.64. To establish a **signal ground** (or an **ac ground**, as it is sometimes called) at the emitter, a large capacitor C_E, usually in the range of microfarads or tens of microfarads is connected between emitter and ground. This capacitor is required to provide a very low impedance to ground (ideally, zero impedance; i.e., in effect, a short circuit) at all signal frequencies of interest. In this way, the emitter signal current passes through C_E to ground and thus *bypasses* the output resistance of the current source I (and any other circuit component that might be connected to the emitter); hence C_E is called a **bypass capacitor.** Obviously, the lower the signal frequency, the less effective the bypass capacitor becomes. This issue will be studied in Section 9.1.2. For our purposes here we shall assume that C_E

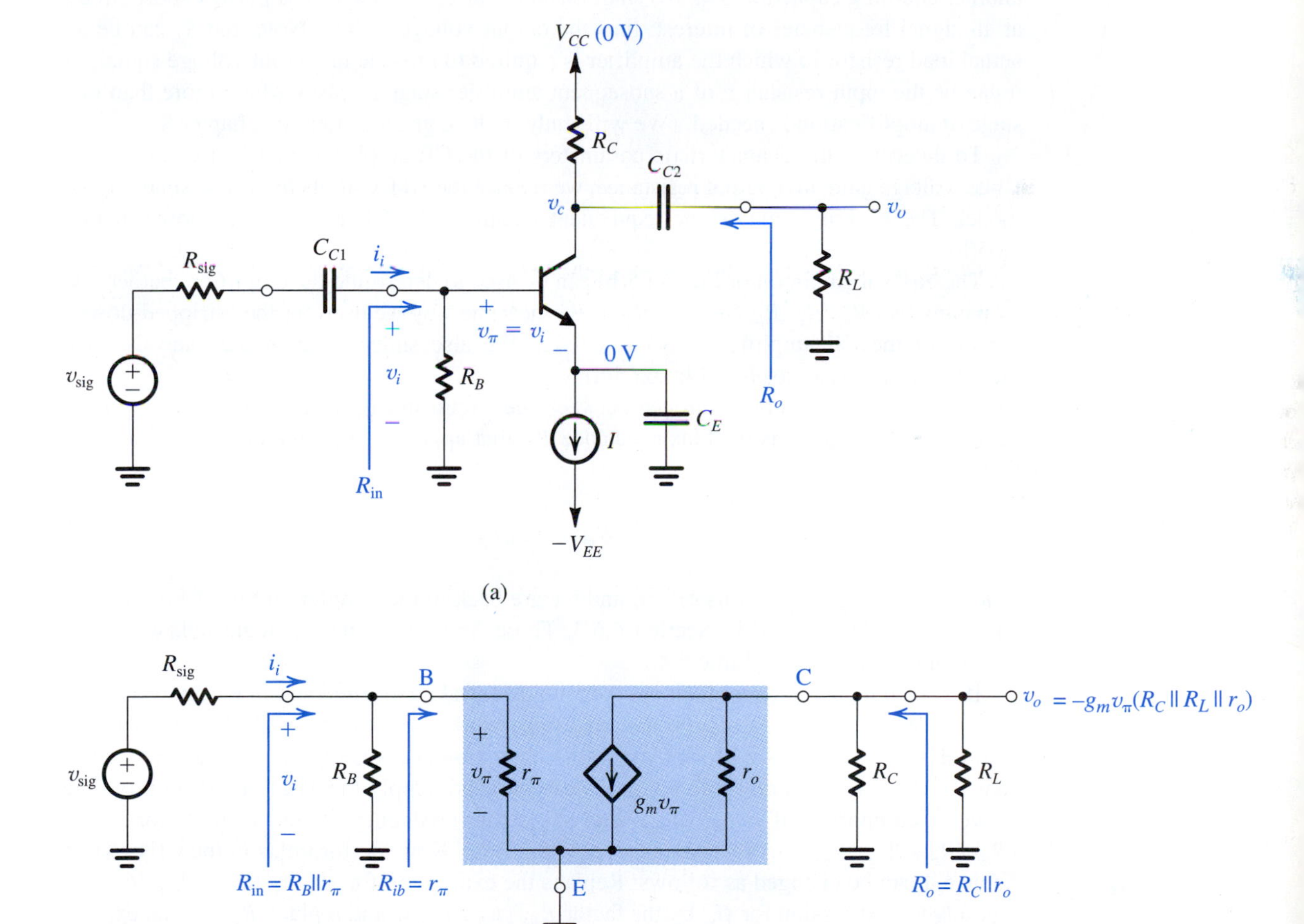

Figure 6.65 **(a)** A common-emitter amplifier using the structure of Fig. 6.64. **(b)** Equivalent circuit obtained by replacing the transistor with its hybrid-π model.

is acting as a perfect short circuit and thus is establishing a zero signal voltage at the emitter.

In order not to disturb the dc bias currents and voltages, the signal to be amplified, shown as a voltage source v_{sig} with an internal resistance R_{sig}, is connected to the base through a large capacitor C_{C1}. Capacitor C_{C1}, known as a **coupling capacitor**, is required to act as a perfect short circuit at all signal frequencies of interest while blocking dc. Here again we shall assume this to be the case and defer discussion of imperfect signal coupling, arising as a result of the rise of the impedance of C_{C1} at low frequencies, to Section 9.1.2. At this juncture, we should point out that in situations where the signal source can provide a dc path for the dc base current I_B without significantly changing the bias point, we may connect the source directly to the base, thus dispensing with C_{C1} as well as R_B. Eliminating R_B has the added beneficial effect of raising the input resistance of the amplifier.

The voltage signal resulting at the collector, v_c, is coupled to the load resistance R_L via another coupling capacitor C_{C2}. We shall assume that C_{C2} also acts as a perfect short circuit at all signal frequencies of interest; thus the output voltage $v_o = v_c$. Note that R_L can be an actual load resistor to which the amplifier is required to provide its output voltage signal, or it can be the input resistance of a subsequent amplifier stage in cases where more than one stage of amplification is needed. (We will study multistage amplifiers in Chapter 8.)

To determine the characteristic parameters of the CE amplifier, that is, its input resistance, voltage gain, and output resistance, we replace the BJT with its hybrid-π, small-signal model. The resulting small-signal equivalent circuit of the CE amplifier is shown in Fig. 6.65(b).

The equivalent circuit of Fig. 6.65(b) can be used to determine the amplifier characteristic parameters R_{in}, A_v, R_o, and G_v in exactly the same way we used for the "stripped-down" version of the CE amplifier in Section 6.6.3. We also show some of the analysis done directly on the circuit itself in Fig. 6.65(a).

Observe that the only difference between the circuit in Fig. 6.65(b) and the simplified version in Fig. 6.50(b) is the bias resistance R_B that appears across the amplifier input and thus changes R_{in} to

$$R_{in} = R_B \| r_\pi \tag{6.112}$$

If $R_B \gg r_\pi$ we can neglect its effect, and we are back to the simpler circuit of Fig. 6.50(b) and the formulas derived in Section 6.6.3. Those formulas, with r_o neglected, were presented in the CE entry in Table 6.4.

If R_B is not much greater than r_π, then it must be taken into account in the analysis. This is a simple task, and we urge the readers to just work their way through the circuit rather than relying on memorized formulas. As a check, however, there is a simple approach to adapt the CE formulas of Table 6.4 to the case at hand: Apply the Thévenin theorem to the network composed of v_{sig}, R_{sig}, and R_B, thus reducing it to a generator $v'_{sig} = (R_B/(R_B + R_{sig}))v_{sig}$ and a resistance $R'_{sig} = R_{sig} \| R_B$. Now the formulas in the CE entry in Table 6.4 can be changed as follows: Replace the expression for R_{in} by that in Eq. (6.112); multiply the expression for G_v by the factor $R_B/(R_B + R_{sig})$; and replace R_{sig} in that expression by $(R_{sig} \| R_B)$.

EXERCISE

6.52 Consider the CE amplifier of Fig. 6.65(a) when biased as in Exercise 6.51. In particular, refer to Fig. E6.51 for the bias currents and the values of the elements of the BJT model at the bias point. Evaluate R_{in} (without and with R_B taken into account), A_{vo} (without and with r_o taken into account), and R_o (without and with r_o taken into account). For $R_L = 5\ \text{k}\Omega$, find A_v. If $R_{sig} = 5\ \text{k}\Omega$, find the overall voltage gain G_v. If the sine-wave v_π is to be limited to 5 mV peak, what is the maximum allowed peak amplitude of v_{sig} and the corresponding peak amplitude of v_o?
Ans. 2.5 kΩ, 2.4 kΩ; –320 V/V, –296 V/V; 8 kΩ, 7.4 kΩ; –119 V/V; –39 V/V; 15 mV; 0.6 V

6.8.3 The Common-Emitter Amplifier with an Emitter Resistance

As demonstrated in Section 6.6.4, a number of beneficial results can be obtained by connecting a resistance R_e in the emitter of the transistor. This is shown in Fig. 6.66(a) where R_e is, of course, unbypassed. Figure 6.66(b) shows the small-signal, equivalent-circuit model. Observe that the only difference between this circuit and the simplified version studied in Section 6.6.4 is the inclusion of the bias resistance R_B, which unfortunately can limit the increase in R_{in} due to R_e, since

$$R_{in} = R_B \| [(\beta + 1)(r_e + R_e)] \tag{6.113}$$

The analysis of the circuits in Fig. 6.66 is straightforward and is illustrated in the figure. The formulas given in Table 6.4 can be adapted to apply to the circuit here by replacing the formula for R_{in} with that in Eq. (6.113), replacing R_{sig} by $R'_{sig} = R_{sig} \| R_B$, and multiplying the expression for G_v by the factor $R_B/(R_B + R_{sig})$. Once again, we do not recommend this approach of plugging into formulas; rather, since each circuit the reader will encounter will be different, it is much more useful to work one's way through the circuit using the analysis methods studied as a guide.

EXERCISE

6.53 Consider the emitter-degenerated CE circuit of Fig. 6.66 when biased as in Exercise 6.51. In particular, refer to Fig. E6.51 for the bias currents and for the values of the elements of the BJT model at the bias point. Let the amplifier be fed from a source having $R_{sig} = 5\ \text{k}\Omega$, and let $R_L = 5\ \text{k}\Omega$. Find the value of R_e that results in R_{in} equal to four times the source resistance R_{sig}. For this value of R_e, find A_{vo}, R_o, A_v, and G_v. If v_π is to be limited to 5 mV, what is the maximum value v_{sig} can have with and without R_e included? Find the corresponding v_o.
Ans. 225 Ω; –32 V/V; 8 kΩ; –12.3 V/V; –9.8 V/V; 62.5 mV; 15 mV; 0.6 V

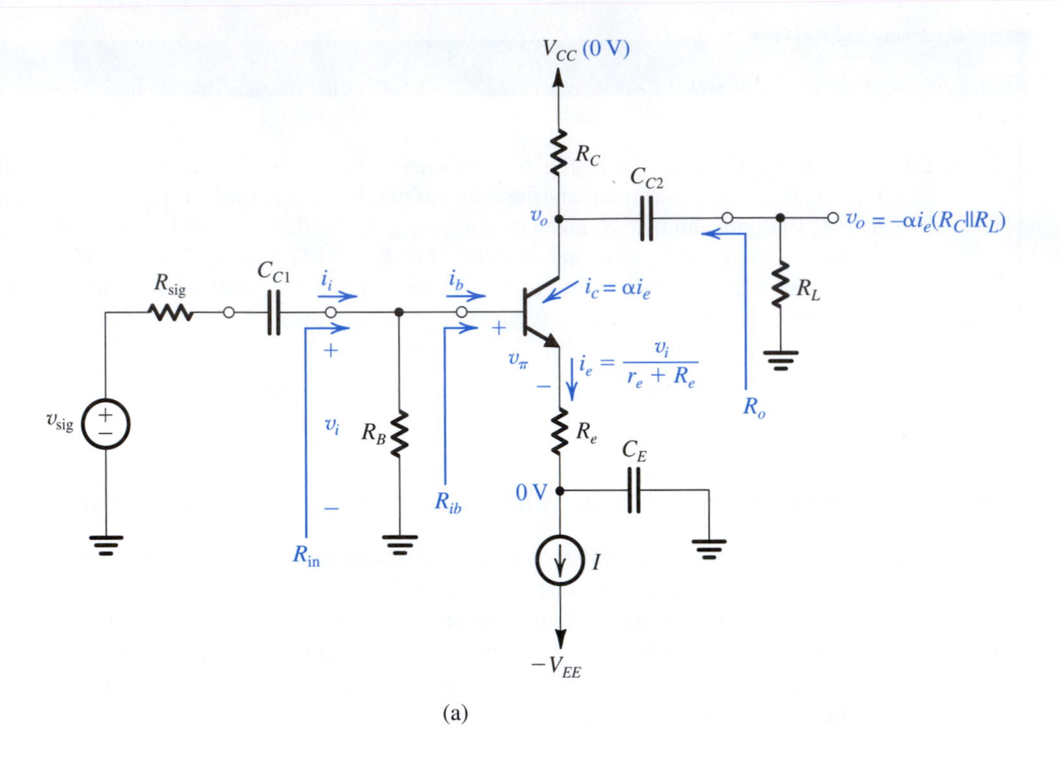

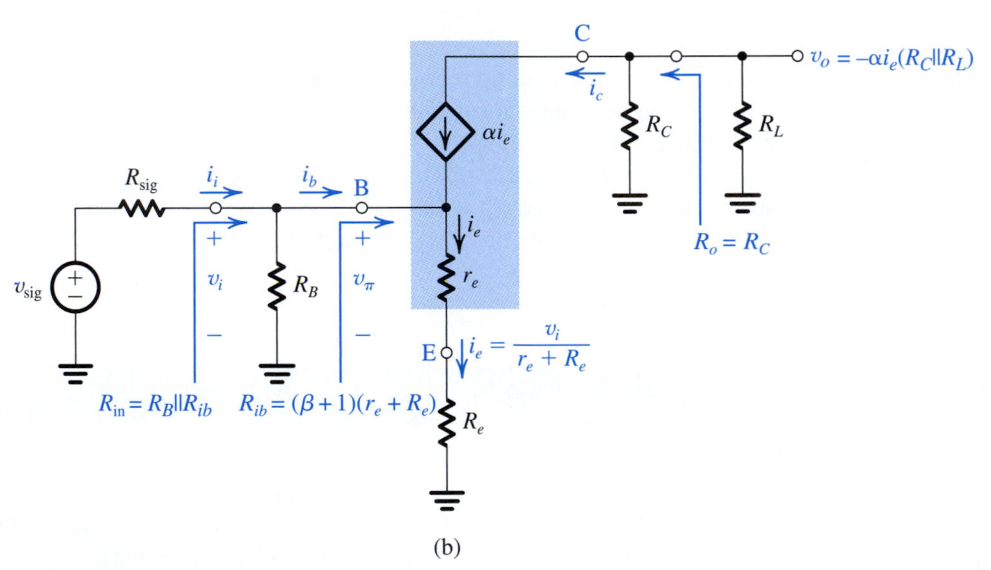

Figure 6.66 **(a)** A common-emitter amplifier with an emitter resistance R_e. **(b)** Equivalent circuit obtained by replacing the transistor with its T model.

6.8.4 The Common-Base (CB) Amplifier

Figure 6.67(a) shows a CB amplifier based on the circuit of Fig. 6.64. Observe that since both the dc and ac voltages at the base are zero, we have connected the base directly to ground, thus eliminating resistor R_B altogether. Coupling capacitors C_{C1} and C_{C2} perform similar functions to those in the CE circuit.

The small-signal, equivalent-circuit model of the amplifier is shown in Fig. 6.67(b). This circuit is identical to that in Fig. 6.53(b), which we analyzed in detail in Section 6.6.5. Thus the analysis of Section 6.6.5, and indeed the results summarized in the CB entry in Table 6.4, apply directly here.

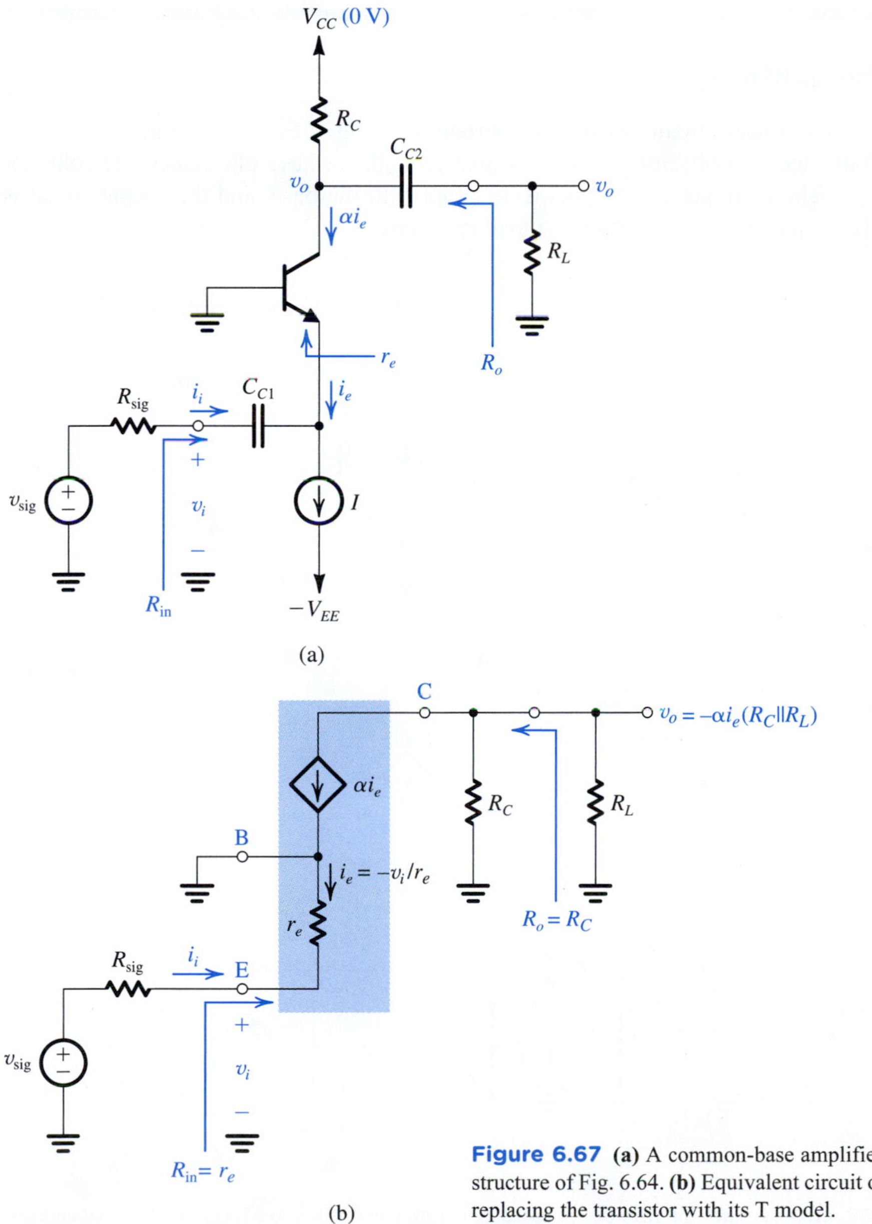

Figure 6.67 **(a)** A common-base amplifier using the structure of Fig. 6.64. **(b)** Equivalent circuit obtained by replacing the transistor with its T model.

EXERCISE

6.54 Consider the CB amplifier of Fig. 6.66(a) when designed using the BJT and component values specified in Exercise 6.51. Specifically, refer to Fig. E6.51 for the bias quantities and the values of the components of the BJT small-signal model. Let $R_{sig} = R_L = 5\ k\Omega$. Find the values of R_{in}, A_{vo}, R_o, A_v, v_i/v_{sig}, and G_v. To what value should R_{sig} be reduced (usually not possible to do!) to obtain an overall voltage gain equal to that found for the CE amplifier in Exercise 6.52, that is, –39 V/V?
Ans. 25 Ω; +320 V/V; 8 kΩ; +123 V/V; 0.005 V/V; 0.6 V/V; 54 Ω

6.8.5 The Emitter Follower

An emitter-follower circuit based on the structure of Fig. 6.64 is shown in Fig. 6.68(a). Observe that since the collector is to be at signal ground, we have eliminated the collector resistance R_C. The input signal is capacitively coupled to the base, and the output signal is capacitively coupled from the emitter to a load resistance R_L.

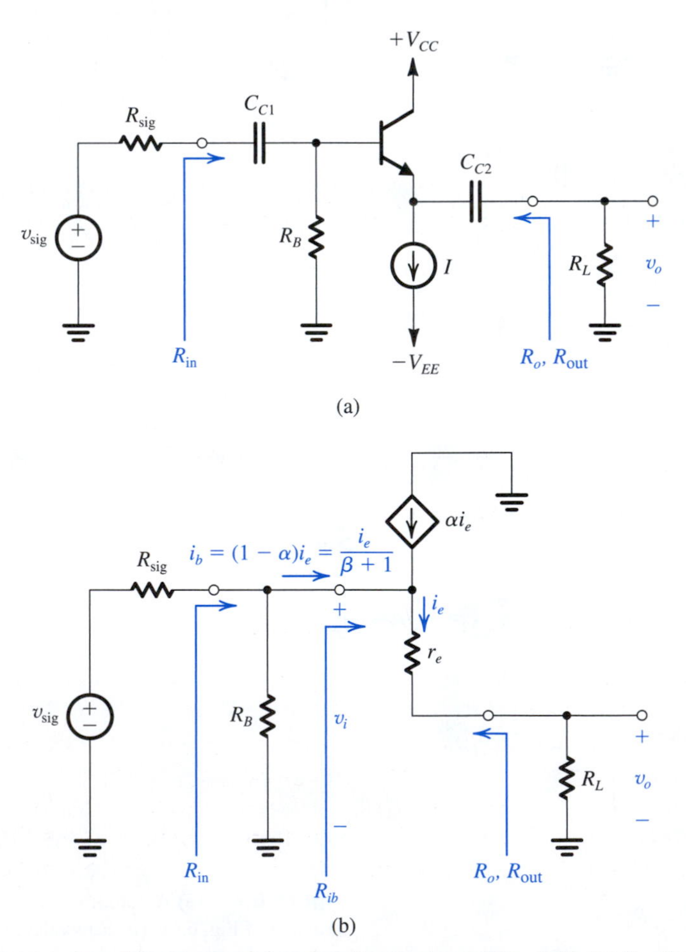

Figure 6.68 **(a)** An emitter-follower circuit based on the structure of Fig. 6.64. **(b)** Small-signal equivalent circuit of the emitter follower with the transistor replaced by its T model.

Replacing the BJT with its T model and neglecting r_o, we obtain the equivalent circuit shown in Fig. 6.68(b). This circuit is identical to that in the stripped-down case analyzed in Section 6.6.6 except here we have the bias resistance R_B. Note that it is very important to select as large a value for R_B as permitted by dc bias considerations, since a low R_B could defeat the purpose of the emitter follower. To appreciate this point recall that the most important feature of the emitter follower is that it multiplies R_L by $(\beta+1)$, thus presenting a high input resistance to the signal source. Here, however, R_B appears in parallel with this increased resistance, resulting in

$$R_{in} = R_B \parallel (\beta+1)(r_e + R_L) \tag{6.114}$$

Thus ideally, R_B should be much larger than $(\beta+1)\ (r_e + R_L)$.

Again we urge the reader to analyze the circuit being studied (here, Fig. 6.68) directly, without the need to refer back to memorized formulas. As a check, however, we note that the results presented in Table 6.4 in the emitter-follower entry apply to the circuit in Fig. 6.68(b) with the following adaptations: Replace the expression for R_{in} with that in Eq. (6.114); multiply the expression for G_v by the factor $R_B/(R_B + R_{sig})$; and replace R_{sig} in the expression for G_v by $(R_{sig} \parallel R_B)$. Also, the equivalent circuits in Fig. 6.56 can be adapted to the circuit in Fig. 6.68 by replacing v_{sig} by $(R_B/(R_B + R_{sig}))v_{sig}$ and R_{sig} by $(R_{sig} \parallel R_B)$. Finally, the Thévenin equivalent in Fig 6.57(a) can be made to apply to the circuit in Fig. 6.67 by using $G_{vo} = R_B/(R_B + R_{sig})$ and $R_{out} = r_e + (R_{sig} \parallel R_B)/(\beta+1)$.

EXERCISE

6.55 The emitter follower in Fig. 6.68(a) is used to connect a source with $R_{sig} = 10\ \text{k}\Omega$ to a load $R_L = 1\ \text{k}\Omega$. The transistor is biased at $I = 5$ mA, utilizes a resistance $R_B = 40\ \text{k}\Omega$, and has $\beta = 100$. Find R_{ib}, R_{in}, G_v, G_{vo}, and R_{out}. If in order to limit nonlinear distortion, the base–emitter signal voltage is limited to 10 mV peak, what is the corresponding amplitude at the output? What will the overall voltage gain become if R_L is changed to 2 kΩ? To 500 Ω?

Ans. 101.5 kΩ; 28.7 kΩ; 0.738 V/V; 0.8 V/V; 84 Ω; 2 V; 0.768 V/V; 0.685 V/V.

6.8.6 The Amplifier Frequency Response

Thus far, we have assumed that the gain of BJT amplifiers is constant independent of the frequency of the input signal. This would imply that BJT amplifiers have infinite bandwidth, which of course is not true. To illustrate, we show in Fig. 6.69 a sketch of the magnitude of the gain of a common-emitter amplifier versus frequency. Observe that there is indeed a wide frequency range over which the gain remains almost constant. This obviously is the useful frequency range of operation for the particular amplifier. Thus far, we have been assuming that our amplifiers are operating in this frequency band, called **the midband**.

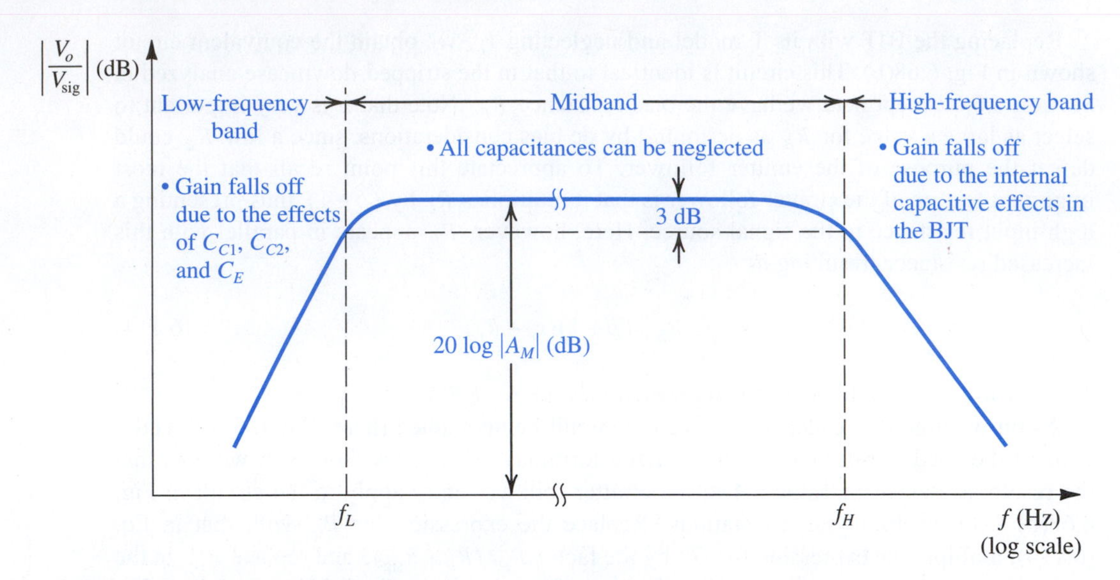

Figure 6.69 Sketch of the magnitude of the gain of a CE amplifier versus frequency. The graph delineates the three frequency bands relevant to frequency-response determination.

Figure 6.69 indicates that at lower frequencies, the magnitude of amplifier gain falls off. This is because the coupling and bypass capacitors no longer have low impedances. Recall that we assumed that their impedances were small enough to act as short circuits. Although this can be true at midband frequencies, as the frequency of the input signal is lowered, the reactance $1/j\omega C$ of each of these capacitors becomes significant, and it can be shown that this results in the overall voltage gain of the amplifier decreasing.

Figure 6.69 indicates also that the gain of the amplifier falls off at the high-frequency end. This is due to the internal capacitive effects in the BJT. In Chapter 3 we briefly introduced such capacitive effects in our study of the *pn* junction. In Chapter 9 we shall study the internal capacitive effects of the BJT and will augment the hybrid-π model with capacitances that model these effects.

We will undertake a detailed study of the frequency response of BJT amplifiers in Chapter 9. For the time being, however, it is important for the reader to realize that for every BJT amplifier, there is a finite band over which the gain is almost constant. The boundaries of this useful frequency band or midband, are the two frequencies f_L and f_H at which the gain drops by a certain number of decibels (usually 3 dB) below its value at midband. As indicated in Fig. 6.69, the amplifier **bandwidth**, or 3-dB bandwidth, is defined as the difference between the lower (f_L) and upper or higher (f_H) 3-dB frequencies:

$$BW = f_H - f_L \tag{6.115}$$

and since usually $f_L \ll f_H$,

$$BW \simeq f_H \tag{6.116}$$

A figure-of-merit for the amplifier is its **gain–bandwidth product**, defined as

$$GB = |A_M| BW \tag{6.117}$$

where $|A_M|$ is the magnitude of the amplifier gain in the midband. It will be seen in Chapter 9 that in amplifier design it is usually possible to trade off gain for bandwidth. One way to accomplish this, for instance, is by including resistance R_e in the emitter of the CE amplifier.

6.9 Transistor Breakdown and Temperature Effects

We conclude this chapter with a brief discussion of two important nonideal effects in the BJT: voltage breakdown, and the dependence of β on I_C and temperature.

6.9.1 Transistor Breakdown

The maximum voltages that can be applied to a BJT are limited by the EBJ and CBJ breakdown effects that follow the avalanche multiplication mechanism described in Section 3.5.3. Consider first the common-base configuration. The i_C–v_{CB} characteristics in Fig. 6.70(b) indicate that for $i_E = 0$ (i.e., with the emitter open-circuited) the collector–base junction breaks down at a voltage denoted by BV_{CBO}. For $i_E > 0$, breakdown occurs at voltages smaller than BV_{CBO}. Typically, for discrete BJTs, BV_{CBO} is greater than 50 V.

Next consider the common-emitter characteristics of Fig. 6.71, which show breakdown occurring at a voltage BV_{CEO}. Here, although breakdown is still of the avalanche type, the effects on the characteristics are more complex than in the common-base configuration. We will not explain these in detail; it is sufficient to point out that typically BV_{CEO} is about half BV_{CBO}. On transistor data sheets, BV_{CEO} is sometimes referred to as the **sustaining voltage** LV_{CEO}.

Breakdown of the CBJ in either the common-base or common-emitter configuration is not destructive as long as the power dissipation in the device is kept within safe limits. This, however, is not the case with the breakdown of the emitter–base junction. The EBJ breaks down in an avalanche manner at a voltage BV_{EBO} much smaller than BV_{CBO}. Typically, BV_{EBO} is in the range of 6 V to 8 V, and the breakdown is destructive in the sense that the β of the transistor is permanently reduced. This does not prevent use of the EBJ as a zener diode to generate reference voltages in IC design. In such applications one is not concerned with the β-degradation effect. A circuit arrangement to prevent EBJ breakdown in IC amplifiers will be discussed in Chapter 12. Transistor breakdown and the maximum allowable power dissipation are important parameters in the design of power amplifiers (Chapter 11) .

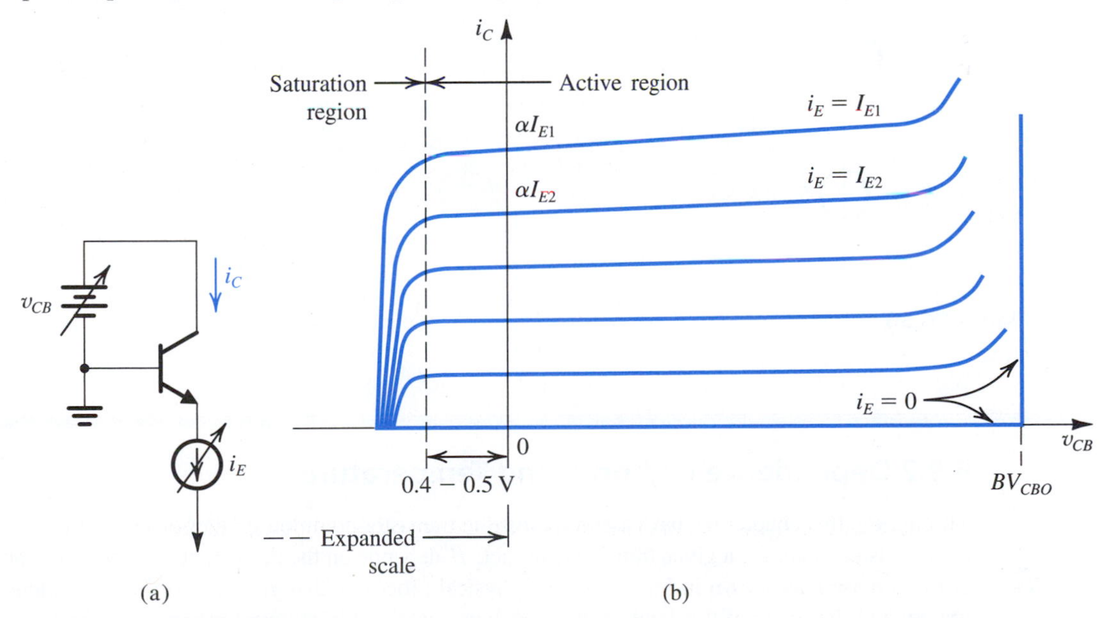

Figure 6.70 The BJT common-base characteristics including the transistor breakdown region.

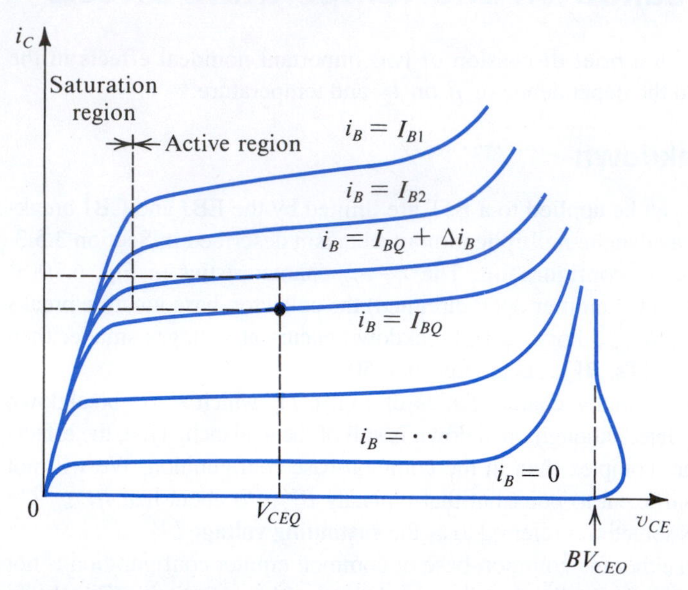

Figure 6.71 The BJT common-emitter characteristics including the breakdown region.

EXERCISE

6.56 What is the output voltage of the circuit in Fig. E6.56 if the transistor $BV_{BCO} = 70$ V?

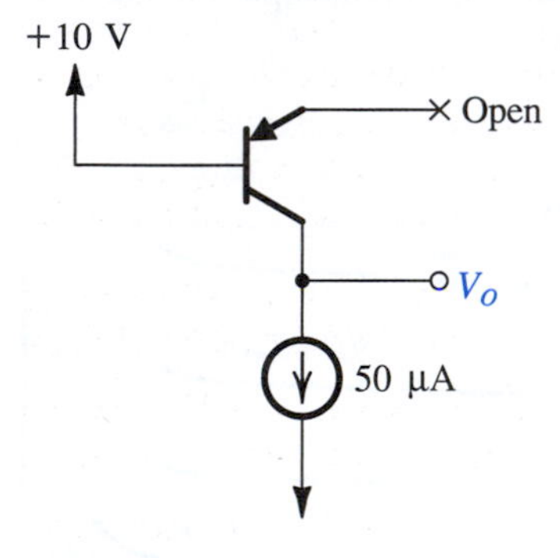

Figure E6.56

Ans. –60 V

6.9.2 Dependence of β on I_C and Temperature

Throughout this chapter we have assumed that the transistor common-emitter dc current gain, β or h_{FE}, is constant for a given transistor. In fact, β depends on the dc current at which the transistor is biased, as shown in Fig. 6.72. The physical processes that give rise to this dependence are beyond the scope of this book. Note, however, that there is a current range over which β is highest. Normally, one biases the transistor to operate at a current within this range.

Figure 6.72 also shows the dependence of β on temperature. The fact that β increases with temperature can lead to serious problems in transistors that operate at large power levels (see Chapter 11).

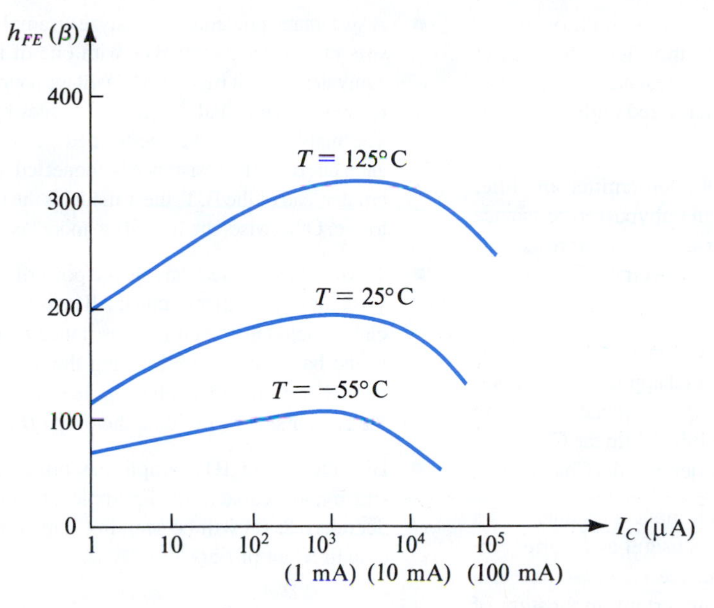

Figure 6.72 Typical dependence of β on I_C and on temperature in an integrated-circuit npn silicon transistor intended for operation around 1 mA.

Summary

- Depending on the bias conditions on its two junctions, the BJT can operate in one of three possible modes: cutoff (both junctions reverse biased), active (the EBJ forward biased and the CBJ reverse biased), and saturation (both junctions forward biased). Refer to Table 6.1.

- For amplifier applications, the BJT is operated in the active mode. Switching applications make use of the cutoff and saturation modes.

- A BJT operating in the active mode provides a collector current $i_C = I_S e^{|v_{BE}|/V_T}$. The base current $i_B = i_C/\beta$, and the emitter current $i_E = i_C + i_B$. Also, $i_C = \alpha i_E$, and thus $\beta = \alpha/(1-\alpha)$ and $\alpha = \beta/(\beta+1)$. See Table 6.2.

- To ensure operation in the active mode, the collector voltage of an *npn* transistor must be kept higher than approximately 0.4 V below the base voltage. For a *pnp* transistor the collector voltage must be lower than approximately 0.4 V above the base voltage. Otherwise, the CBJ becomes forward biased, and the transistor enters the saturation region.

- At a constant collector current, the magnitude of the base–emitter voltage decreases by about 2 mV for every 1°C rise in temperature.

- The BJT will be at the edge of saturation when $|v_{CE}|$ is reduced to about 0.3 V. In saturation, $|v_{CE}| \simeq 0.2$ V, and the ratio of i_C to i_B is lower than β (i.e., $\beta_{\text{forced}} < \beta$)

- In the active mode, i_C shows a slight dependence on v_{CE}. This phenomenon, known as the Early effect, is modeled by ascribing a finite (i.e., noninfinite) output resistance to the BJT: $r_o = |V_A|/I'_C$, where V_A is the Early voltage and I'_C is the dc collector current without the Early effect taken into account. In discrete circuits, r_o plays a minor role and can usually be neglected. This is *not* the case, however, in integrated-circuit design (Chapter 7).

- The dc analysis of transistor circuits is greatly simplified by assuming that $|V_{BE}| \simeq 0.7$ V. Refer to Table 6.3.

- To operate as a linear amplifier, the BJT is biased in the active region and the signal v_{be} is kept small ($v_{be} \ll V_T$).

- For small signals, the BJT functions as a linear voltage-controlled current source with a transconductance $g_m = I_C/V_T$. The input resistance between base and emitter, looking into the base, is $r_\pi = \beta/g_m$. The input resistence between base and emitter, looking into the emitter is $r_e \simeq 1/g_m$. Table 6.4 provides a summary of the small-signal models and the equations for determining their parameters.

- Bias design seeks to establish a dc collector current that is as independent of the value of β as possible.

- The three basic BJT amplifier configurations are shown in Fig. 6.48. A summary of their characteristic parameters is provided in Table 6.5.

- The CE amplifier is used to obtain the bulk of the required voltage gain in a cascade amplifier. It has a large voltage gain and a moderate input resistance but a relatively high output resistance and limited high-frequency response (Chapter 9).
- The input resistance of the common-emitter amplifier can be increased by including an unbypassed resistance in the emitter lead. This emitter-degeneration resistance provides other performance improvements at the expense of reduced voltage gain.
- The CB amplifier has a very low input resistance and is useful in a limited number of special applications. It does, however, have an excellent high-frequency response (Chapter 9) and thus can be combined with the CE amplifier to obtain an excellent amplifier circuit (Chapter 7).
- The emitter follower has a high input resistance and a low output resistance. Thus, it is useful as a buffer amplifier to connect a high-resistance signal source to a low-resistance load. Another important application of the emitter follower is as the last stage (called the output stage) of a cascade amplifier.
- A systematic procedure to analyze an amplifier circuit consists of replacing each BJT with one of its small-signal, equivalent circuit models. DC voltage sources are replaced by short circuits and dc current sources by open circuits. The analysis can then be performed on the resulting equivalent circuit. If a resistance is connected in series with the emitter lead of the BJT, the T model is the most convenient to use. Otherwise, the hybrid-π model is employed.
- The resistance reflection rule is a powerful tool in the analysis of BJT amplifier circuits: All resistances in the emitter circuit including the emitter resistance r_e can be reflected to the base side by multiplying them by $(\beta + 1)$. Conversely, we can reflect all resistances in the base circuit to the emitter side by dividing them by $(\beta + 1)$.
- Discrete-circuit BJT amplifiers utilize large coupling and bypass capacitors. Example circuits are given in Section 6.8. As will be seen in Chapter 7, this is not the case in IC amplifiers.

PROBLEMS

Computer Simulation Problems

SIM Problems identified by this icon are intended to demonstrate the value of using SPICE simulation to verify hand analysis and design, and to investigate important issues such as allowable signal swing and amplifier nonlinear distortion. Instructions to assist in setting up PSpice and Multisim simulations for all the indicated problems can be found in the corresponding files on the disc. Note that if a particular parameter value is not specified in the problem statement, you are to make a reasonable assumption. * difficult problem; ** more difficult; *** very challenging and/or time-consuming; D: design problem.

Section 6.1: Device Structure and Physical operation

6.1 The terminal voltages of various *npn* transistors are measured during operation in their respective circuits with the following results:

Case	E	B	C	Mode
1	0	0.7	0.7	
2	0	0.8	0.1	
3	–0.7	0	0.7	
4	–0.7	0	–0.6	
5	–2.7	–2.0	0	
6	0	0	5.0	

In this table, where the entries are in volts, 0 indicates the reference terminal to which the black (negative) probe of the voltmeter is connected. For each case, identify the mode of operation of the transistor.

6.2 Two transistors, fabricated with the same technology but having different junction areas, when operated at a base-emitter voltage of 0.75 V, have collector currents of 0.2 mA and 5 mA. Find I_S for each device. What are the relative junction areas?

6.3 In a particular technology, a small BJT operating at $v_{BE} = 28V_T$ conducts a collector current of 100 µA. What is the corresponding saturation current? For a transistor in the same technology but with an emitter junction that is 32 times larger, what is the saturation current? What current will this transistor conduct at $v_{BE} = 28V_T$? What is the base–emitter voltage of the latter transistor at $i_C = 1$ mA? Assume active-mode operation in all cases.

6.4 Two transistors have EBJ areas as follows: A_{E1} = A_{E1} = 400 µm × 400 µm and A_{E2} = 0.4 µm × 0.2 µm. If the two transistors are operated in the active mode and conduct equal collector currents, what do you expect the difference in their v_{BE} values to be?

6.5 Find the collector currents that you would expect for operation at v_{BE} = 700 mV for transistors for which

$I_S = 10^{-12}$ A and $I_S = 10^{-18}$ A. For the transistor with the larger EBJ, what is the v_{BE} required to provide a collector current equal to that provided by the smaller transistor at $v_{BE} = 700$ mV? Assume active-mode operation in all cases.

6.6 In this problem, we contrast two BJT integrated-circuit fabrication technologies: For the "old" technology, a typical *npn* transistor has $I_S = 5 \times 10^{-15}$ A, and for the "new" technology a typical *npn* transistor has $I_S = 5 \times 10^{-18}$ A. These typical devices have vastly different junction areas and base width. For our purpose here we wish to determine the v_{BE} required to establish a collector current of 1 mA in each of the two typical devices. Assume active-mode operation.

6.7 Consider an *npn* transistor whose base–emitter drop is 0.76 V at a collector current of 10 mA. What current will it conduct at $v_{BE} = 0.70$ V? What is its base–emitter voltage for $i_C = 10$ μA?

6.8 In a particular BJT, the base current is 10 μA, and the collector current is 600 μA. Find β and α for this device.

6.9 Find the values of β that correspond to α values of 0.5, 0.8, 0.9, 0.95, 0.99, 0.995, and 0.999.

6.10 Find the values of α that correspond to β values of 1, 2, 10, 20, 100, 200, 1000, and 2000.

***6.11** Show that for a transistor with α close to unity, if α changes by a small per-unit amount $(\Delta\alpha/\alpha)$, the corresponding per-unit change in β is given approximately by

$$\frac{\Delta\beta}{\beta} \simeq \beta\left(\frac{\Delta\alpha}{\alpha}\right)$$

6.12 An *npn* transistor of a type whose β is specified to range from 60 to 300 is connected in a circuit with emitter grounded, collector at +9 V, and a current of 20 μA injected into the base. Calculate the range of collector and emitter currents that can result. What is the maximum power dissipated in the transistor? (*Note:* Perhaps you can see why this is a bad way to establish the operating current in the collector of a BJT.)

6.13 A BJT is specified to have $I_S = 5 \times 10^{-15}$ A and β that falls in the range of 50 to 200. If the transistor is operated in the active mode with v_{BE} set to 0.650 V, find the expected range of i_C, i_B, and i_E.

6.14 Measurements made on a number of transistors operating in the active mode with $i_E = 1$ mA indicate base currents of 50 μA, 10 μA, and 25 μA. For each device, find i_C, β, and α.

6.15 Measurement of V_{BE} and two terminal currents taken on a number of *npn* transistors operating in the active mode are tabulated below. For each, calculate the missing current value as well as α, β, and I_S as indicated by the table.

Transistor	a	b	c	d	e
V_{BE} (mV)	690	690	580	780	820
I_C (mA)	1.000	1.000		10.10	
I_B (μA)	50		7	120	1050
I_E (mA)		1.070	0.137		75.00
α					
β					
I_S					

6.16 A particular BJT when operated in the active mode conducts a collector current of 10 mA and has $v_{BE} = 0.70$ V and $i_B = 100$ μA. Use these data to create specific transistor models of the form shown in Figs. 6.5(a) to (d).

6.17 Using the *npn* transistor model of Fig. 6.5(b), consider the case of a transistor for which the base is connected to ground, the collector is connected to a 10-V dc source through a 2-kΩ resistor, and a 3-mA current source is connected to the emitter with the polarity so that current is drawn out of the emitter terminal. If $\beta = 100$ and $I_s = 10^{-15}$ A, find the voltages at the emitter and the collector and calculate the base current.

D 6.18 Consider an *npn* transistor operated in the active mode and represented by the model of Fig. 6.5(d). Let the transistor be connected as indicated by the equivalent circuit shown in Fig. 6.6(b). It is required to calculate the values of R_B and R_C that will establish a collector current I_C of 1 mA and a collector-to-emitter voltage V_{CE} of 1 V. The BJT is specified to have $\beta = 125$ and $I_S = 5 \times 10^{-15}$ A.

6.19 An *npn* transistor has a CBJ with an area 150 times that of the EBJ. If $I_S = 5 \times 10^{-15}$ A, find the voltage drop across EBJ and across CBJ when each is forward biased and conducting a current of 1 mA. Also find the forward current each junction would conduct when forward biased with 0.5 V.

6.20 We wish to investigate the operation of the *npn* transistor in saturation using the model of Fig. 6.9. Let $I_S = 10^{-15}$ A, $v_{BE} = 0.7$ V, $\beta = 100$ and $I_{SC}/I_S = 100$. For each of three values of v_{CE} (namely, 0.4 V, 0.3 V, and 0.1 V), find v_{BC}, i_{BC}, i_B, i_C, and i_C/i_B. Also find v_{CE} that results in $i_C = 0$.

***6.21** Use Eqs. (6.14), (6.15), and (6.16) to show that an *npn* transistor operated in saturation exhibits a collector-to-emitter voltage, $V_{CE\text{sat}}$ given by

$$V_{CE\text{sat}} = V_T \ln\left[\left(\frac{I_{SC}}{I_S}\right)\frac{1+\beta_{\text{forced}}}{1-\beta_{\text{forced}}/\beta}\right]$$

Use this relationship to evaluate V_{CEsat} for β_{forced} = 50, 10, 5, and 1 for a transistor with $\beta = 100$ and with a CBJ area 100 times that of the EBJ.

6.22 Consider the *pnp* large-signal model of Fig. 6.11(b) applied to a transistor having $I_S = 10^{-13}$ A and $\beta = 40$. If the emitter is connected to ground, the base is connected to a current source that pulls 20 μA out of the base terminal, and the collector is connected to a negative supply of –10 V via a 10-kΩ resistor, find the collector voltage, the emitter current, and the base voltage.

6.23 A *pnp* transistor has v_{EB} = 0.8 V at a collector current of 1 A. What do you expect v_{EB} to become at i_C = 10 mA? At i_C = 5 A?

6.24 A *pnp* transistor modeled with the circuit in Fig. 6.11 (b) is connected with its base at ground, collector at –1.0 V, and a 10-mA current is injected into its emitter. If the transistor is said to have β = 10, what are its base and collector currents? In which direction do they flow? If $I_S = 10^{-15}$ A, what voltage results at the emitter? What does the collector current become if a transistor with β = 1000 is substituted? (*Note:* The fact that the collector current changes by less than 10% for a large change of β illustrates that this is a good way to establish a specific collector current.)

6.25 A *pnp* power transistor operates with an emitter-to-collector voltage of 5 V, an emitter current of 10 A, and V_{EB} = 0.85 V. For β = 15, what base current is required? What is I_S for this transistor? Compare the emitter–base junction area of this transistor with that of a small-signal transistor that conducts i_C = 1 mA with v_{EB} = 0.70 V. How much larger is it?

6.26 While Fig. 6.5 provides four possible large-signal equivalent circuits for the *npn* transistor, only two equivalent circuits for the *pnp* transistor are provided in Fig. 6.11. Supply the missing two.

6.27 By analogy to the *npn* case shown in Fig. 6.9, give the equivalent circuit of a *pnp* transistor in saturation.

Section 6.2: Current–Voltage Characteristics

6.28 For the circuits in Fig. P6.28, assume that the transistors have very large β. Some measurements have been made on these circuits, with the results indicated in the figure. Find the values of the other labeled voltages and currents.

6.29 Measurements on the circuits of Fig. P6.29 produce labeled voltages as indicated. Find the value of β for each transistor.

6.30 A very simple circuit for measuring β of an *npn* transistor is shown in Fig. P6.30. In a particular design, V_{CC} is provided by a 9-V battery; M is a current meter with a 50-μA full scale and relatively low resistance that you can neglect for our purposes here. Assuming that the transistor has V_{BE} = 0.7 V at I_E = 1 mA, what value of R would establish a resistor current of 1 mA? Now, to what value of β does a meter reading of full scale correspond? What is β if the meter reading is 1/5 of full scale? 1/10 of full scale?

6.31 Repeat Exercise 6.13 for the situation in which the power supplies are reduced to ±1.5 V.

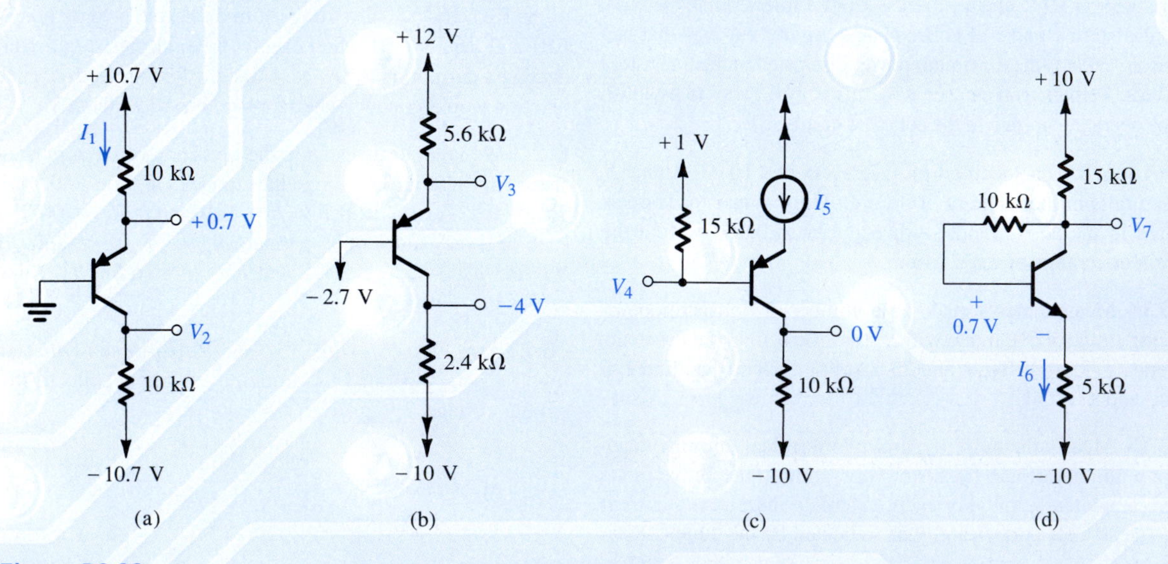

Figure P6.28

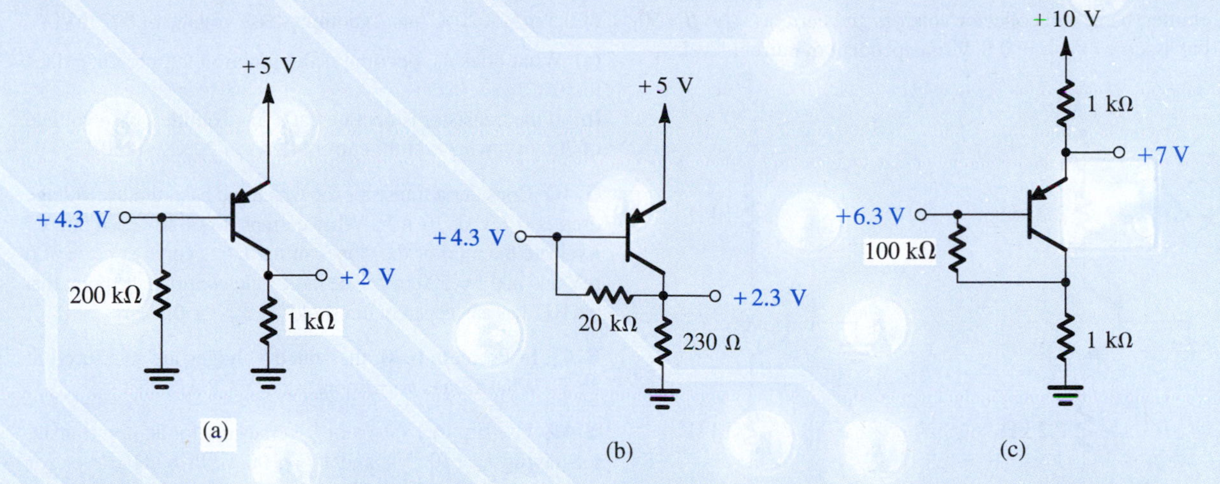

Figure P6.29

Figure P6.30

Figure P6.32

D 6.32 Design the circuit in Fig. P6.32 to establish a current of 1 mA in the emitter and a voltage of −1 V at the collector. The transistor v_{EB} = 0.64 V at I_E = 0.1 mA, and $\beta = 100$. To what value can R_C be increased while the collector current remains unchanged?

D 6.33 Examination of the table of standard values for resistors with 5% tolerance in Appendix G reveals that the closest values to those found in the design of Example 6.20 are 5.1 kΩ and 6.8 kΩ. For these values use approximate calculations (e.g., $V_{BE} \simeq 0.7$ V and $\alpha \simeq 1$) to determine the values of collector current and collector voltage that are likely to result.

D 6.34 Design the circuit in Fig. P6.34 to establish I_C = 0.1 mA and V_C = 0.5 V. The transistor exhibits v_{BE} of 0.8 V at i_C = 1 mA, and $\beta = 100$.

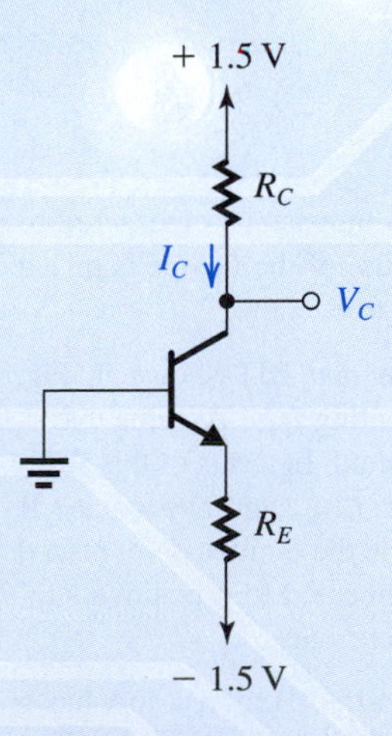

Figure P6.34

6.35 For each of the circuits shown in Fig. P6.35, find the emitter, base, and collector voltages and currents. Use $\beta = 50$, but assume $|V_{BE}| = 0.8$ V independent of current level.

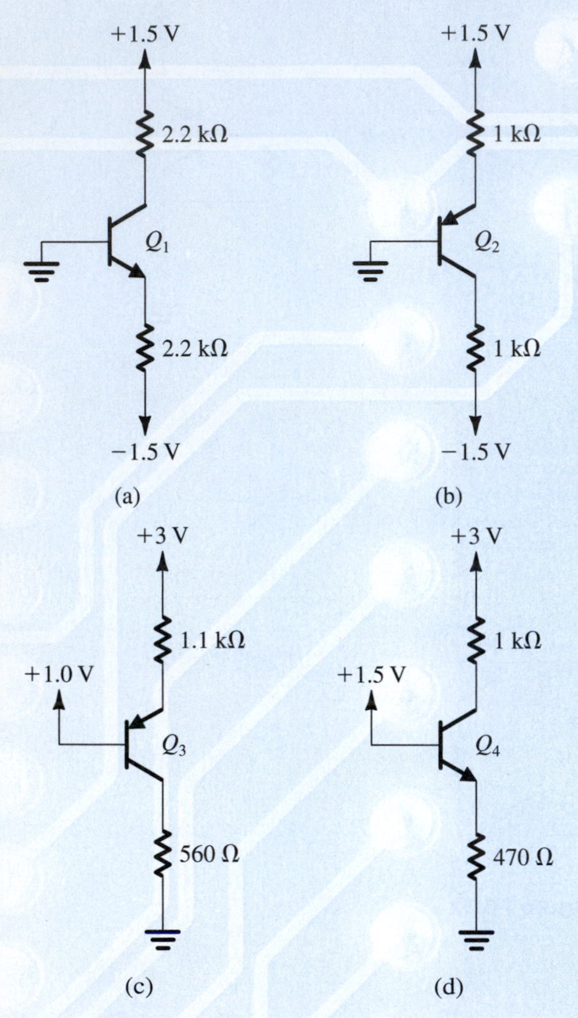

Figure P6.35

6.36 The current I_{CBO} of a small transistor is measured to be 10 nA at 25°C. If the temperature of the device is raised to 125°C, what do you expect I_{CBO} to become?

***6.37** Augment the model of the *npn* BJT shown in Fig. 6.18(a) by a current source representing I_{CBO}. Assume that r_o is very large and thus can be neglected. In terms of this addition, what do the terminal currents i_B, i_C, and i_E become? If the base lead is open-circuited while the emitter is connected to ground, and the collector is connected to a positive supply, find the emitter and collector currents.

6.38 A BJT whose emitter current is fixed at 1 mA has a base–emitter voltage of 0.69 V at 25°C. What base–emitter voltage would you expect at 0°C? At 100°C?

6.39 A particular *pnp* transistor operating at an emitter current of 0.5 mA at 20°C has an emitter–base voltage of 692 mV.

(a) What does v_{EB} become if the junction temperature rises to 50°C?
(b) If the transistor is operated at a fixed emitter–base voltage of 700 mV, what emitter current flows at 20°C? At 50°C?

6.40 Consider a transistor for which the base–emitter voltage drop is 0.7 V at 10 mA. What current flows for $v_{BE} = 0.5$ V? Evaluate the ratio of the slopes of the i_C–v_{BE} curve at $v_{BE} = 700$ mV and at $v_{BE} = 500$ mV. The large ratio confirms the point that the BJT has an "apparent threshold" at $v_{BE} \simeq 0.5$ V.

6.41 In Problem 6.40, the stated voltages are measured at 25°C. What values correspond at –25°C? At 125°C?

6.42 Use Eq. (6.18) to plot i_C versus v_{CE} for an *npn* transistor having $I_S = 10^{-15}$ A and $V_A = 100$ V. Provide curves for $v_{BE} = 0.65$, 0.70, 0.72, 0.73, and 0.74 volts. Show the characteristics for v_{CE} up to 15 V.

***6.43** In the circuit shown in Fig. P6.43, current source I is 1.1 mA, and at 25° C $v_{BE} = 680$ mV at $i_C = 1$ mA. At 25° C with $\beta = 100$, what currents flow in R_1 and R_2? What voltage would you expect at node E? Noting that the temperature coefficient of v_{BE} for I_C constant is –2 mV/° C, what is the TC of v_E? For an ambient temperature of 75° C, what voltage would you expect at node E? Clearly state any simplifying assumptions you make.

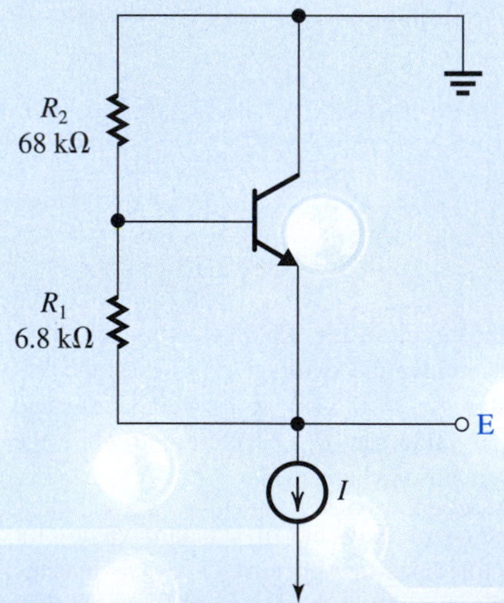

Figure P6.43

6.44 For a particular *npn* transistor operating at a v_{BE} of 670 mV and $I_C = 2$ mA, the i_C–v_{CE} characteristic has a slope of 2×10^{-5} ℧. To what value of output resistance does this correspond? What is the value of the Early voltage for this transistor? For operation at 20 mA, what would the output resistance become?

6.45 For a BJT having an Early voltage of 150 V, what is its output resistance at 1 mA? At 100 μA?

6.46 Measurements of the i_C–v_{CE} characteristic of a small-signal transistor operating at v_{BE} = 720 mV show that i_C = 1.8 mA at v_{CE} = 2 V and that i_C = 2.4 mA at v_{CE} = 14 V. What is the corresponding value of i_C near saturation? At what value of v_{CE} is i_C = 2.0 mA? What is the value of the Early voltage for this transistor? What is the output resistance that corresponds to operation at v_{BE} = 720 mV?

6.47 Give the *pnp* equivalent circuit models that correspond to those shown in Fig. 6.18 for the *npn* case.

6.48 A BJT operating at i_B = 8 μA and i_C = 1.2 mA undergoes a reduction in base current of 0.8 μA. It is found that when v_{CE} is held constant, the corresponding reduction in collector current is 0.1 mA. What are the values of β and the incremental β or β_{ac} that apply? If the base current is increased from 8 μA to 10 μA and v_{CE} is increased from 8 V to 10 V, what collector current results? Assume V_A = 100 V.

6.49 For the circuit in Fig. P6.49 let V_{CC} = 5 V, R_C = 1 kΩ, and R_B = 20 kΩ. The BJT has β = 50. Find the value of V_{BB} that results in the transistor operating

(a) in the active mode with V_C = 1 V;
(b) at the edge of saturation;
(c) deep in saturation with β_{forced} = 10.

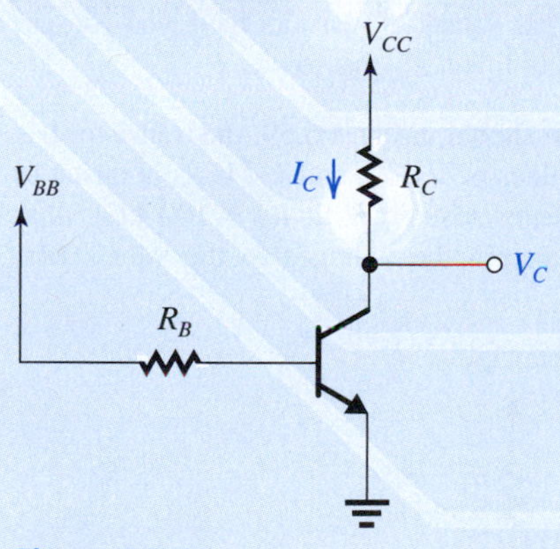

Figure P6.49

SIM **D *6.50** Consider the circuit of Fig. P6.49 for the case V_{BB} = V_{CC}. If the BJT is saturated, use the equivalent circuit of Fig. 6.20 to derive an expression for β_{forced} in terms of V_{CC} and (R_B/R_C). Also derive an expression for the total power dissipated in the circuit. For V_{CC} = 5 V, design the circuit to obtain operation at a forced β as close to 10 as possible while limiting the power dissipation to no larger than 20 mW. Use 1% resistors (see Appendix G).

6.51 The *pnp* transistor in the circuit in Fig. P6.51 has β = 50. Show that the BJT is operating in the saturation mode and find β_{forced} and V_C. To what value should R_B be increased in order for the transistor to operate at the edge of saturation?

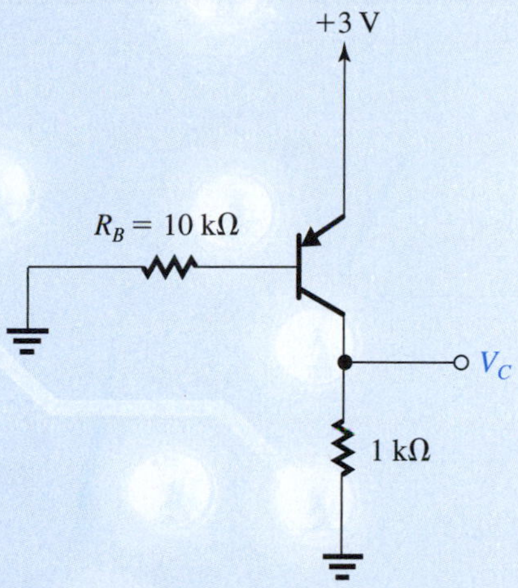

Figure P6.51

Section 6.3: BJT Circuits at DC

6.52 The transistor in the circuit of Fig. P6.52 has a very high β. Find V_E and V_C for V_B (a) +1.5 V, (b) +1 V, and (c) 0 V.

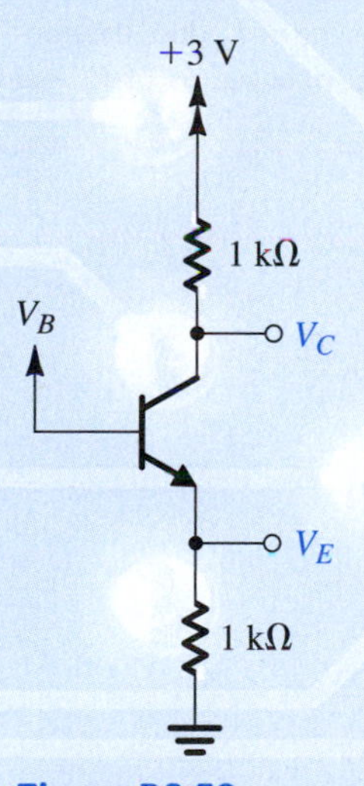

Figure P6.52

6.53 The transistor in the circuit of Fig. P6.52 has a very high β. Find the highest value of V_B for which the transistor still operates in the active mode. Also, find the value of V_B for which the transistor operates in saturation with a forced β of 1.

6.54 Consider the operation of the circuit shown in Fig. P6.54 for V_B at –1 V, 0 V, and +1 V. Assume that β is very high. What values of V_E and V_C result? At what value of V_B does the emitter current reduce to one-tenth of its value for V_B = 0 V? For what value of V_B is the transistor just at the edge of conduction? What values of V_E and V_C correspond?

For what value of V_B does the transistor reach the edge of saturation? What values of V_C and V_E correspond? Find the value of V_B for which the transistor operates in saturation with a forced β of 2.

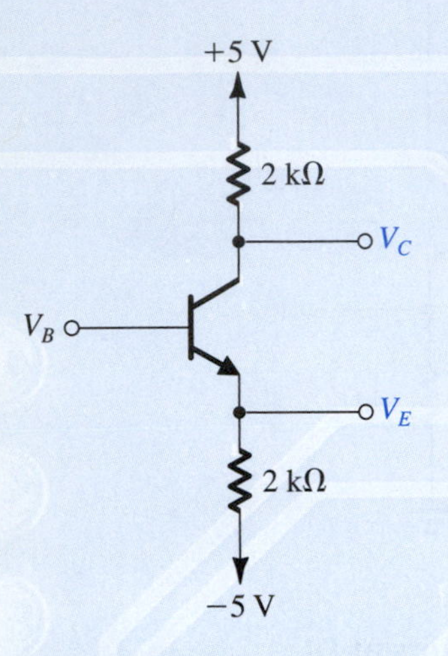

Figure P6.54

6.55 For the transistor shown in Fig. P6.55, assume $\alpha \simeq 1$ and $v_{BE} = 0.5$ V at the edge of conduction. What are the values of V_E and V_C for $V_B = 0$ V? For what value of V_B does the transistor cut off? Saturate? In each case, what values of V_E and V_C result?

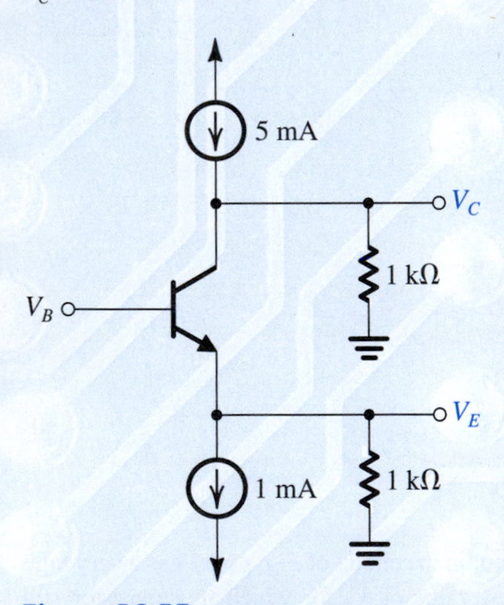

Figure P6.55

D 6.56 Consider the circuit in Fig. P6.52 with the base voltage V_B obtained using a voltage divider across the 3-V supply. Assuming the transistor β to be very large (i.e., ignoring the base current), design the voltage divider to obtain $V_B = 1.5$ V. Design for a 0.1-mA current in the voltage divider. Now, if the BJT $\beta = 100$, analyze the circuit to determine the collector current and the collector voltage.

6.57 A single measurement indicates the emitter voltage of the transistor in the circuit of Fig. P5.57 to be 1.2 V. Under the assumption that $|V_{BE}| = 0.7$ V, what are V_B, I_B, I_E, I_C, V_C, β, and α? (*Note:* Isn't it surprising what a little measurement can lead to?)

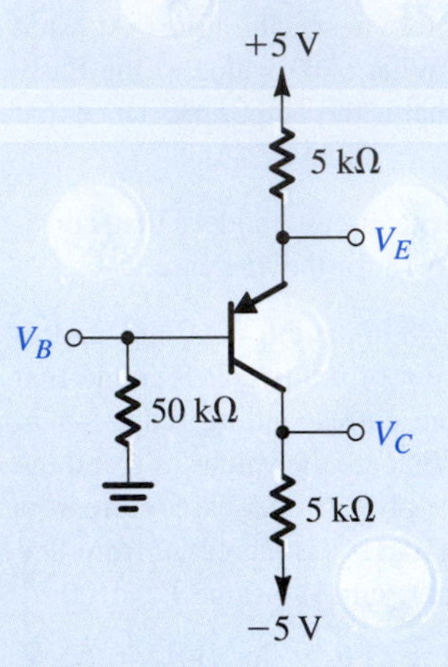

Figure P6.57

D 6.58 Design a circuit using a *pnp* transistor for which $\alpha \simeq 1$ using two resistors connected appropriately to ±5 V so that $I_E = 2$ mA and $V_{BC} = 2.5$ V. What exact values of R_E and R_C would be needed? Now, consult a table of standard 5% resistor values (e.g., that provided in Appendix G) to select suitable practical values. What values of resistors have you chosen? What are the values of I_E and V_{BC} that result?

6.59 In the circuit shown in Fig. P6.59, the transistor has $\beta = 50$. Find the values of V_B, V_E, and V_C. If R_B is raised to 100 kΩ, what voltages result? With R_B = 100 kΩ, what value of β would return the voltages to the values first calculated?

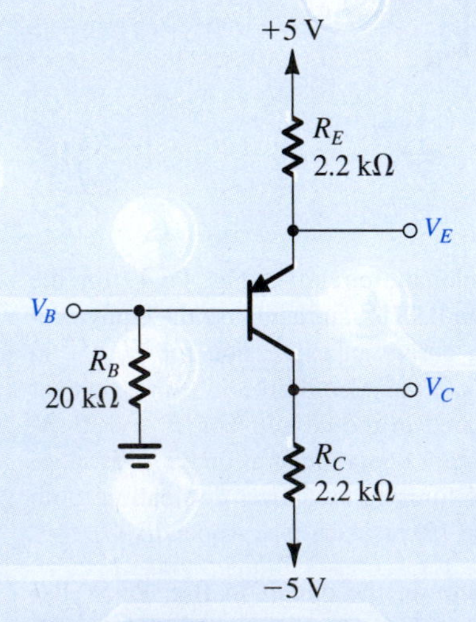

Figure P6.59

6.60 In the circuit shown in Fig. P6.59, the transistor has $\beta = 50$. Find the values of V_B, V_E, and V_C, and verify that the transistor is operating in the active mode. What is the largest value that R_C can have while the transistor remains in the active mode?

SIM **6.61** For the circuit in Fig. P6.61, find V_B, V_E, and V_C for $R_B = 100$ kΩ, 10 kΩ, and 1 kΩ. Let $\beta = 100$.

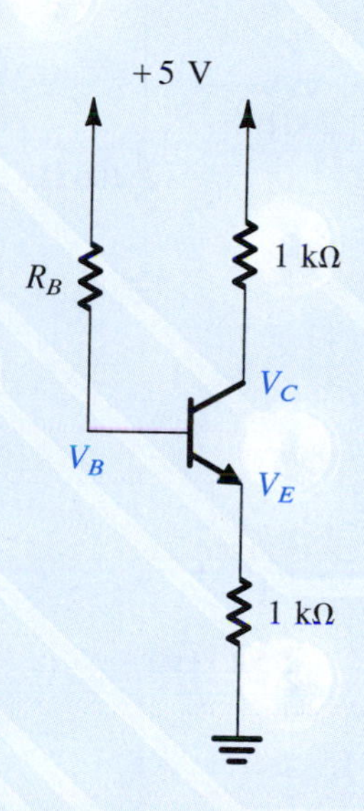

Figure P6.61

6.62 For the circuits in Fig. P6.62, find values for the labeled node voltages and branch currents. Assume β to be very high.

***6.63** Repeat the analysis of the circuits in Problem 6.62 using $\beta = 100$. Find all the labeled node voltages and branch currents.

D **6.64 It is required to design the circuit in Fig. P6.64 so that a current of 1 mA is established in the emitter and a voltage of –5 V appears at the collector. The transistor type used has a nominal β of 100. However, the β value can be as low as 50 and as high as 150. Your design should ensure that the specified emitter current is obtained when $\beta = 100$ and that at the extreme values of β the emitter current does not change by more than 10% of its nominal value. Also, design for as large a value for R_B as possible. Give the values of R_B, R_E, and R_C to the nearest kilohm. What is the expected range of collector current and collector voltage corresponding to the full range of β values?

D 6.65 The *pnp* transistor in the circuit of Fig. P6.65 has $\beta = 50$. Find the value for R_C to obtain $V_C = +3$ V. What happens if the transistor is replaced with another having $\beta = 100$?

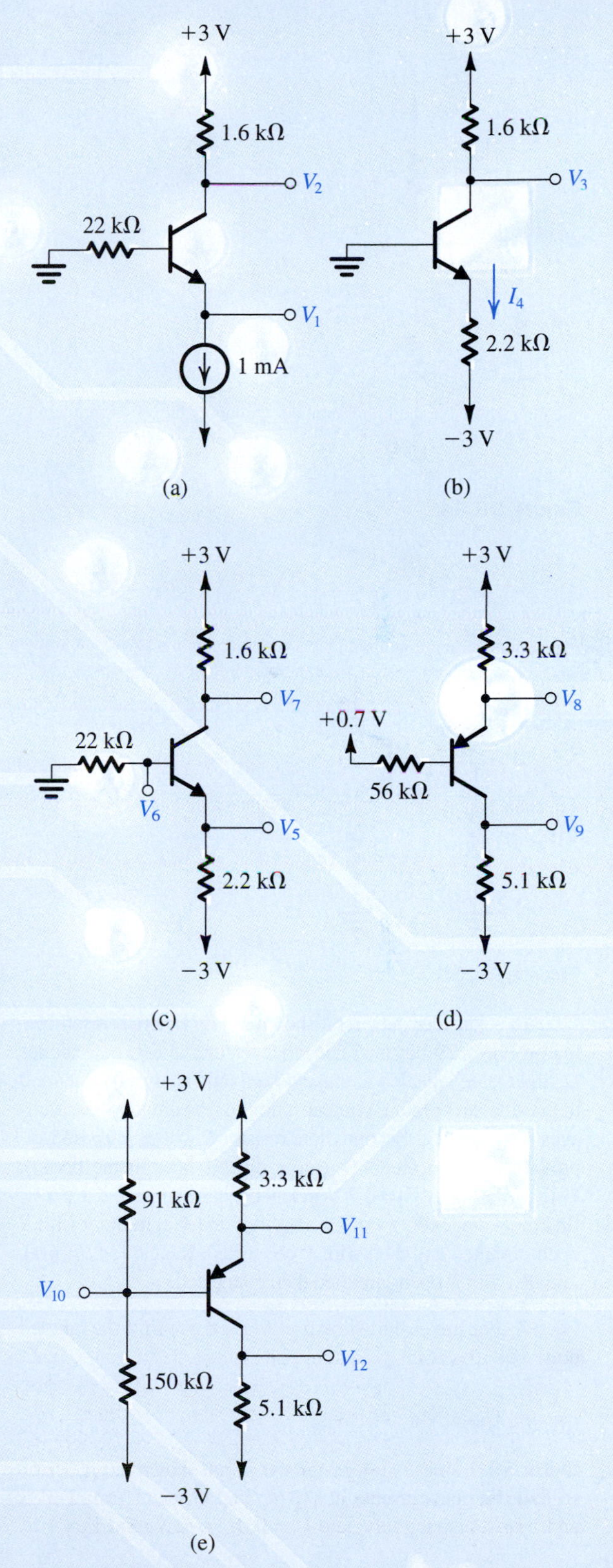

Figure P6.62

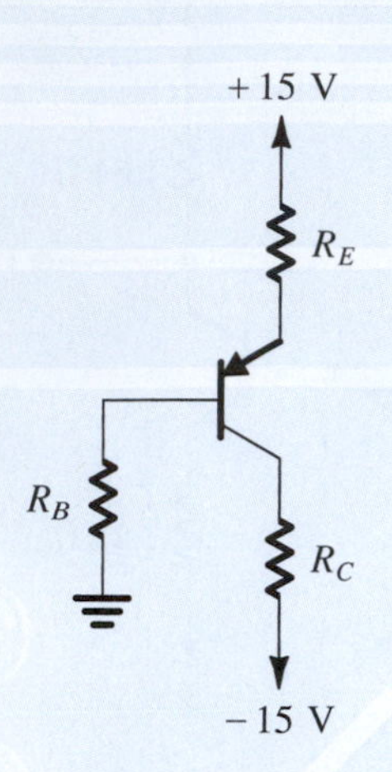

Figure P6.64

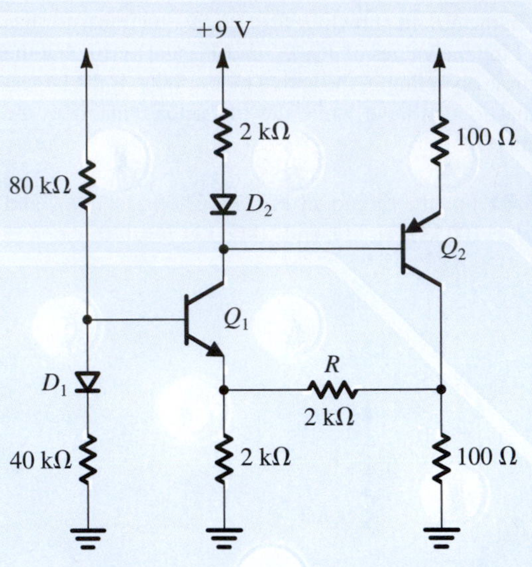

Figure P6.66

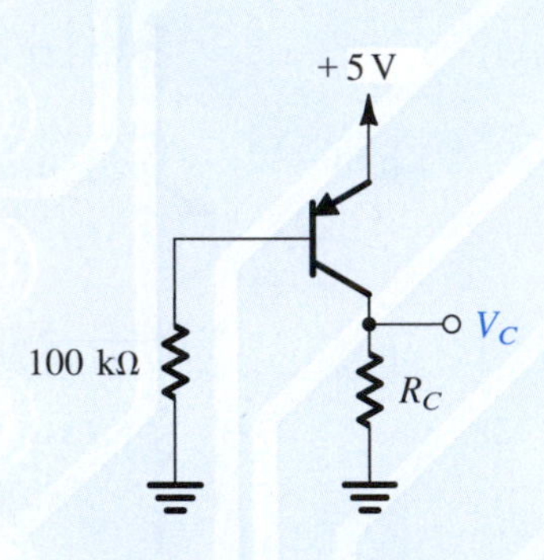

Figure P6.65

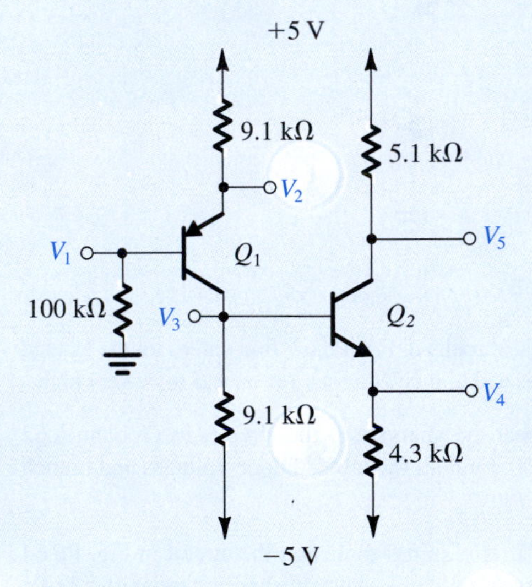

Figure P6.67

****6.66** Consider the circuit shown in Fig. P6.66. It resembles that in Fig. 6.29 but includes other features. First, note diodes D_1 and D_2 are included to make design (and analysis) easier and to provide temperature compensation for the emitter–base voltages of Q_1 and Q_2. Second, note resistor R whose purpose is to provide negative feedback (more on this later in the book!). Using $|V_{BE}|$ and $V_D = 0.7$ V independent of current and $\beta = \infty$, find the voltages V_{B1}, V_{E1}, V_{C1}, V_{B2}, V_{E2}, and V_{C2}, initially with R open-circuited and then with R connected. Repeat for $\beta = 100$, initially with R open-circuited then connected.

***6.67** For the circuit shown in Fig. P6.67, find the labeled node voltages for:

(a) $\beta = \infty$
(b) $\beta = 100$

D *6.68 Using $\beta = \infty$, design the circuit shown in Fig. P6.68 so that the bias currents in Q_1, Q_2, and Q_3 are 1 mA, 1 mA, and 2 mA, respectively, and $V_3 = 0$, $V_5 = -2$ V, and $V_7 = 1$ V. For each resistor, select the nearest standard value utilizing the table of standard values for 5% resistors in Appendix G. Now, for $\beta = 100$, find the values of V_3, V_4, V_5, V_6, and V_7.

***6.69** For the circuit in Fig. P6.69, find V_B and V_E for $v_I = 0$ V, +2 V, −2.5 V, and −5 V. The BJTs have $\beta = 100$.

****6.70** Find approximate values for the collector voltages in the circuits of Fig. P6.70. Also, calculate forced β for each of the transistors. (*Hint:* Initially, assume all transistors are operating in saturation, and verify the assumption.

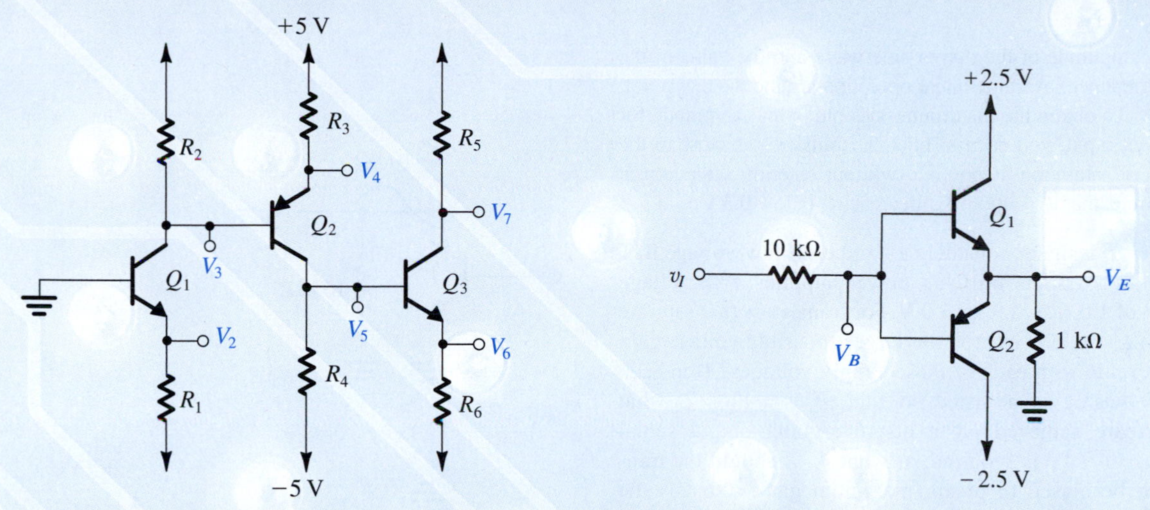

Figure P6.68

Figure P6.69

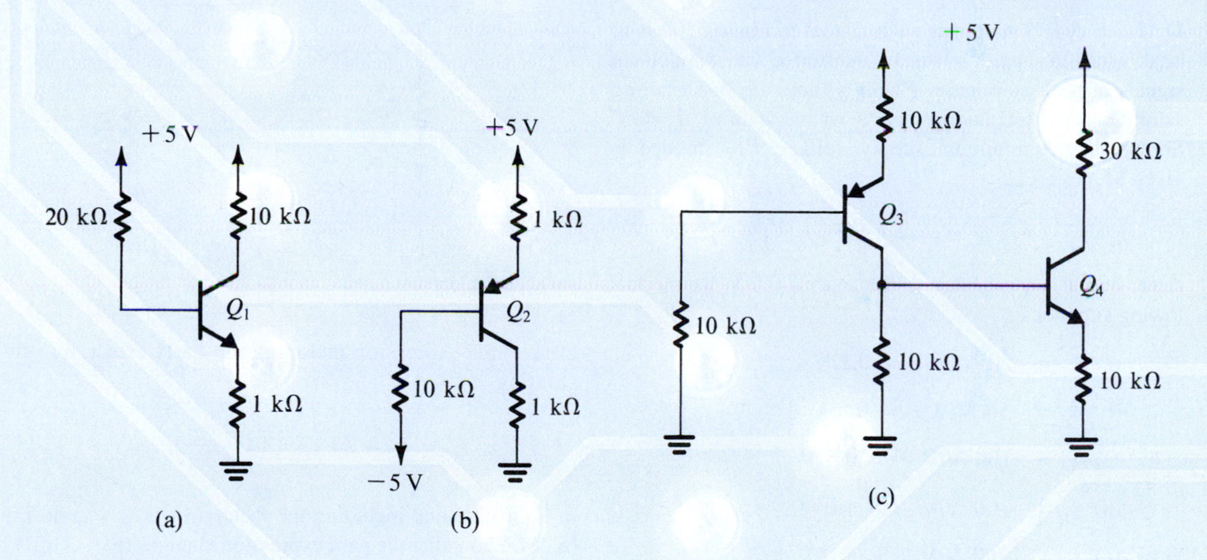

Figure P6.70

Section 6.4: Applying the BJT in Amplifier Design

6.71 A BJT amplifier circuit such as that in Fig. 6.33(a) is operated with V_{CC} = +5 V and is biased at V_{CE} = +1 V. Find the voltage gain, the maximum allowed output negative swing without the transistor entering saturation, and the corresponding maximum input signal permitted.

6.72 For the amplifier circuit in Fig. 6.33(a) with V_{CC} = +5 V and R_C = 1 kΩ, find V_{CE} and the voltage gain at the following dc collector bias currents: 0.5 mA, 1 mA, 2.5 mA, 4 mA, and 4.5 mA. For each, give the maximum possible positive- and negative-output signal swing as determined by the need to keep the transistor in the active region. Present your results in a table.

D 6.73 Consider the CE amplifier circuit of Fig. 6.33(a) when operated with a dc supply V_{CC} = +5 V. It is required to find the point at which the transistor should be biased; that is, find the value of V_{CE} so that the output sine-wave signal v_{ce} resulting from an input sine-wave signal v_{be} of 5-mV peak amplitude has the maximum possible magnitude. What is the peak amplitude of the output sine wave and the value of the gain obtained? Assume linear operation around the bias point. (*Hint:* To obtain the maximum possible output amplitude for a given input, you need to bias the transistor as close to the edge of saturation as possible without entering saturation at any time, that is, without v_{CE} decreasing below 0.3 V.)

6.74 A designer considers a number of low-voltage BJT amplifier designs utilizing power supplies with voltage V_{CC} of 1.0, 1.5, 2.0, or 3.0 V. For transistors that saturate at V_{CE} = 0.3 V, what is the largest possible voltage gain achievable with each of these supply voltages? If in each case biasing is adjusted so that $V_{CE} = V_{CC}/2$, what gains are achieved? If a negative-going output signal swing of 0.4V is required, at what V_{CE} should the transistor be biased to obtain maximum gain? What is the gain achieved with each of the supply voltages? Notice that all of these gains are independent of the value of I_C chosen!)

D *6.75 A BJT amplifier such as that in Fig. 6.33(a) is to be designed to support relatively undistorted sine-wave output signals of peak amplitudes P volt without the BJT entering saturation or cutoff and to have a voltage gain of A_v V/V. Show that the minimum supply voltage V_{CC} needed is given by

$$V_{CC} = V_{CE\text{sat}} + P + |A_v| V_T$$

Also, find V_{CC}, specified to the nearest 0.5 V, for the following situations:

(a) A_v = −20 V/V, P = 0.2 V

(b) A_v = −50 V/V, P = 0.5 V

(c) A_v = −100 V/V, P = 0.5 V

(d) A_v = −100 V/V, P = 1.0 V

(e) A_v = −200 V/V, P = 1.0 V

(f) A_v = −500 V/V, P = 1.0 V

(g) A_v = −500 V/V, P = 2.0 V

6.76 The transistor in the circuit of Fig. P6.76 is biased at a dc collector current of 0.4 mA. What is the voltage gain? (*Hint:* Use Thévenin's theorem to convert the circuit to the form in Fig. 6.33a).

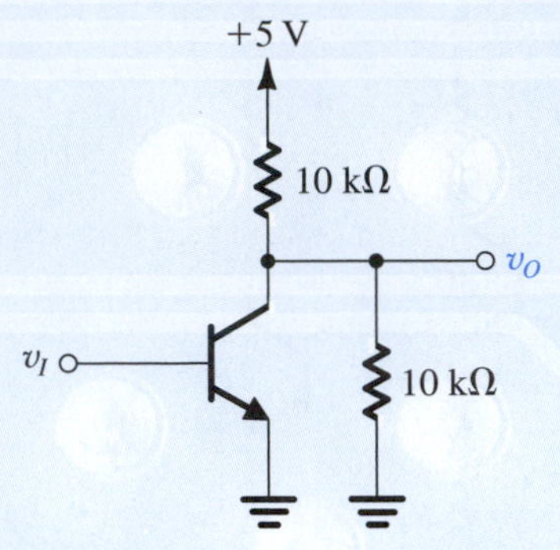

Figure P6.76

6.77 Sketch and label the voltage transfer characteristics of the *pnp* common-emitter amplifiers shown in Fig. P6.77.

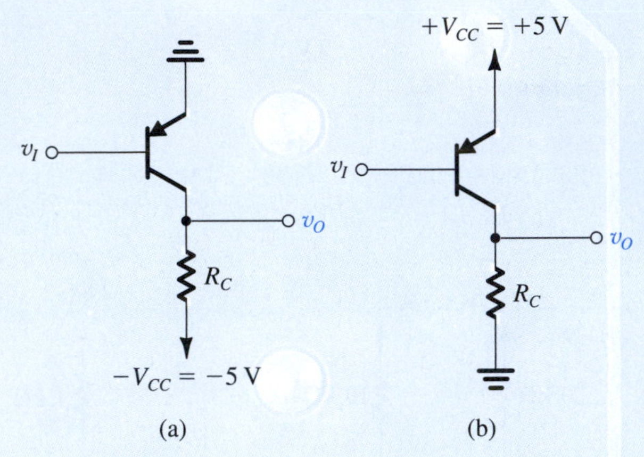

Figure P6.77

***6.78** In deriving the expression for small-signal voltage gain A_v in Eq. (6.31) we neglected the Early effect. Derive this expression including the Early effect, by substituting

$$i_C = I_S e^{v_{BE}/V_T}\left(1 + \frac{v_{CE}}{V_A}\right)$$

in Eq. (6.24) and including the factor $(1 + v_{CE}/V_A)$ in Eq. (6.28). Show that the gain expression changes to

$$A_v = \frac{-I_C R_C/V_T}{\left[1 + \dfrac{I_C R_C}{V_A + V_{CE}}\right]} = -\frac{(V_{CC} - V_{CE})/V_T}{\left[1 + \dfrac{V_{CC} - V_{CE}}{V_A + V_{CE}}\right]}$$

For the case V_{CC} = 5 V and V_{CE} = 2.5 V, what is the gain without and with the Early effect taken into account? Let V_A = 100 V.

6.79 When the amplifier circuit of Fig. 6.33(a) is biased with a certain V_{BE}, the dc voltage at the collector is found to be +2 V. For V_{CC} = +5 V and R_C = 1 kΩ, find I_C and the small-signal voltage gain. For a change Δv_{BE} = +5 mV, calculate the resulting Δv_O. Calculate it two ways: by finding

Δi_C using the transistor exponential characteristic, and approximately using the small-signal voltage gain. Repeat for $\Delta v_{BE} = -5$ mV. Summarize your results in a table.

***6.80** Consider the amplifier circuit of Fig. 6.33(a) when operated with a supply voltage $V_{CC} = +3$V.

(a) What is the theoretical maximum voltage gain that this amplifier can provide?
(b) What value of V_{CE} must this amplifier be biased at to provide a voltage gain of –80 V/V?
(c) If the dc collector current I_C at the bias point in (b) is to be 0.5 mA, what value of R_C should be used?
(d) What is the value of V_{BE} required to provide the bias point mentioned above? Assume that the BJT has $I_S = 10^{-15}$ A.
(e) If a sine-wave signal v_{be} having a 5-mV peak amplitude is superimposed on V_{BE}, find the corresponding output voltage signal v_{ce} that will be superimposed on V_{CE} assuming linear operation around the bias point.
(f) Characterize the signal current i_c that will be superimposed on the dc bias current I_C.
(g) What is the value of the dc base current I_B at the bias point? Assume $\beta = 100$. Characterize the signal current i_b that will be superimposed on the base current I_B.
(h) Dividing the amplitude of v_{be} by the amplitude of i_b, evaluate the incremental (or small-signal) input resistance of the amplifier.
(i) Sketch and clearly label correlated graphs for v_{BE}, v_{CE}, i_C, and i_B. Note that each graph consists of a dc or average value and a superimposed sine wave. Be careful of the phase relationships of the sine waves.

6.81 The essence of transistor operation is that a change in v_{BE}, Δv_{BE}, produces a change in i_C, Δi_C. By keeping Δv_{BE} small, Δi_C is approximately linearly related to Δv_{BE}, $\Delta i_C = g_m \Delta v_{BE}$, where g_m is known as the transistor transconductance. By passing Δi_C through R_C, an output voltage signal Δv_O is obtained. Use the expression for the small-signal voltage gain in Eq. (6.30) to derive an expression for g_m. Find the value of g_m for a transistor biased at $I_C = 1$ mA.

6.82 The purpose of this problem is to illustrate the application of graphical analysis to the circuit shown in Fig. P6.82. Sketch $i_C - v_{CE}$ characteristic curves for the BJT for i_B = 1 μA, 10 μA, 20 μA, and 40 μA. Assume the lines to be horizontal (i.e., neglect the Early effect), and let $\beta = 100$. For V_{CC} = 5 V and R_C = 1 kΩ, sketch the load line. What peak-to-peak collector voltage swing will result for i_B varying over the range 10 μA to 40 μA? If the BJT is biased at $V_{CE} = \frac{1}{2} V_{CC}$, find the value of I_C and I_B. If at this current V_{BE} = 0.7 V and if R_B = 100 kΩ, find the required value of V_{BB}.

***6.83** Sketch the i_C–v_{CE} characteristics of an *npn* transistor having $\beta = 100$ and $V_A = 100$ V. Sketch characteristic curves for i_B = 20 μA, 50 μA, 80 μA, and 100 μA. For the purpose of this sketch, assume that $i_C = \beta i_B$ at $v_{CE} = 0$. Also, sketch the load line obtained for V_{CC} = 10 V and R_C = 1 kΩ. If the dc bias current into the base is 50 μA, write the equation for the corresponding i_C–v_{CE} curve. Also, write the equation for the load line, and solve the two equations to obtain V_{CE} and I_C. If the input signal causes a sinusoidal signal of 30-μA peak amplitude to be superimposed on I_B, find the corresponding signal components of i_C and v_{CE}.

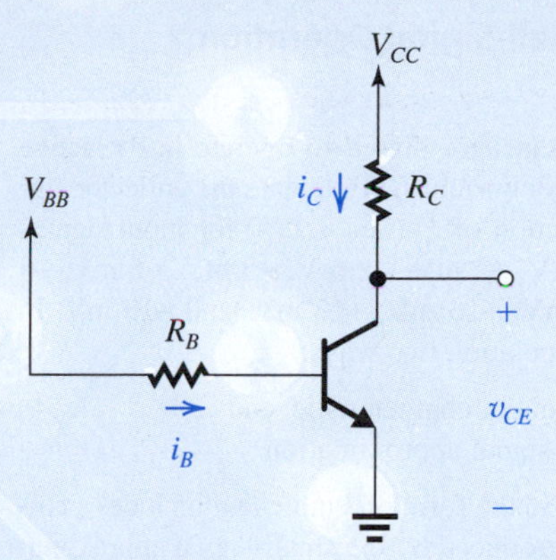

Figure P6.82

***6.84** Consider the operation of the circuit shown in Fig. P6.84 as v_B rises slowly from zero. For this transistor, assume $\beta = 50$, v_{BE} at which the transistor conducts is 0.5 V, v_{BE} when fully conducting is 0.7 V, saturation begins at v_{BC} = 0.4 V, and the transistor is deeply in saturation at v_{BC} = 0.6V. Sketch and label v_E and v_C versus v_B. For what range of v_B is i_C essentially zero? What are the values of v_E, i_E, i_C, and v_C for v_B = 1 V and 3 V? For what value of v_B does saturation begin? What is i_B at this point? For v_B = 4 V and 6 V, what are the values of v_E, v_C, i_E, i_C, and i_B? Augment your sketch by adding a plot of i_B.

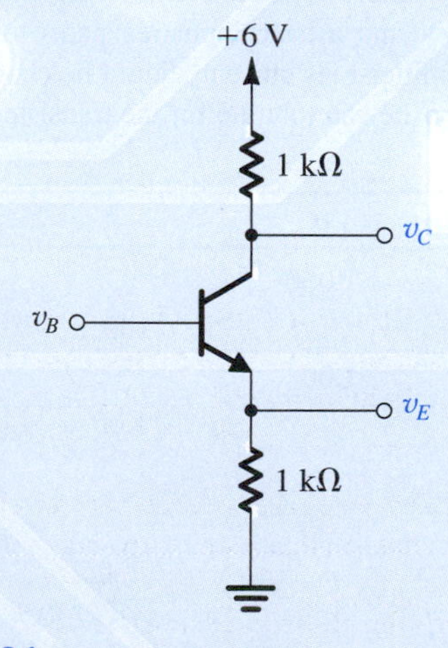

Figure P6.84

Section 6.5: Small-Signal Operation and Models

6.85 Consider a transistor biased to operate in the active mode at a dc collector current I_C. Calculate the collector signal current as a fraction of I_C (i.e., i_c/I_C) for input signals v_{be} of +1 mV, −1 mV, +2 mV, −2 mV, +5 mV, −5 mV, +8 mV, −8 mV, +10 mV, −10 mV, +12 mV, and −12 mV. In each case do the calculation two ways:

(a) using the exponential characteristic, and
(b) using the small-signal approximation.

Present your results in the form of a table that includes a column for the error introduced by the small-signal approximation. Comment on the range of validity of the small-signal approximation.

6.86 An *npn* BJT with grounded emitter is operated with V_{BE} = 0.700 V, at which the collector current is 0.5 mA. A 10-kΩ resistor connects the collector to a +10-V supply. What is the resulting collector voltage V_C? Now, if a signal applied to the base raises v_{BE} to 705 mV, find the resulting total collector current i_C and total collector voltage v_C using the exponential i_C–v_{BE} relationship. For this situation, what are v_{be} and v_c? Calculate the voltage gain v_c/v_{be}. Compare with the value obtained using the small-signal approximation, that is, $-g_mR_C$.

6.87 A transistor with $\beta = 120$ is biased to operate at a dc collector current of 0.6 mA. Find the values of g_m, r_π, and r_e. Repeat for a bias current of 60 μA.

6.88 A *pnp* BJT is biased to operate at I_C = 1.0 mA. What is the associated value of g_m? If $\beta = 100$, what is the value of the small-signal resistance seen looking into the emitter (r_e)? Into the base (r_π)? If the collector is connected to a 5-kΩ load, with a signal of 5-mV peak applied between base and emitter, what output signal voltage results?

D 6.89 A designer wishes to create a BJT amplifier with a g_m of 25 mA/V and a base input resistance of 3000 Ω or more. What emitter-bias current should he choose? What is the minimum β he can tolerate for the transistor used?

6.90 A transistor operating with nominal g_m of 50 mA/V has a β that ranges from 50 to 150. Also, the bias circuit, being less than ideal, allows a ±20% variation in I_C. What are the extreme values found of the resistance looking into the base?

6.91 In the circuit of Fig. 6.36, V_{BE} is adjusted so that V_C = 1 V. If V_{CC} = 3 V, R_C = 2 kΩ, and a signal v_{be} = 0.005 sin ωt volts is applied, find expressions for the total instantaneous quantities $i_C(t)$, $v_C(t)$, and $i_B(t)$. The transistor has $\beta = 80$. What is the voltage gain?

D *6.92 We wish to design the amplifier circuit of Fig. 6.36 under the constraint that V_{CC} is fixed. Let the input signal $v_{be} = \hat{V}_{be} \sin \omega t$, where $\hat{V}_{be}$ is the maximum value for acceptable linearity. For the design that results in the largest signal at the collector, without the BJT leaving the active region, show that

$$R_C I_C = (V_{CC} - 0.3 - \hat{V}_{be}) \Big/ \left(1 + \frac{\hat{V}_{be}}{V_T}\right)$$

and find an expression for the voltage gain obtained. For V_{CC} = 3 V and $\hat{V}_{be}$ = 5 mV, find the dc voltage at the collector, the amplitude of the output voltage signal, and the voltage gain.

6.93 The table below summarizes some of the basic attributes of a number of BJTs of different types, operating as amplifiers under various conditions. Provide the missing entries. (*Note:* Isn't it remarkable how much two parameters can reveal?)

6.94 A BJT is biased to operate in the active mode at a dc collector current of 0.5 mA. It has a β of 100. Give the four small-signal models (Figs. 6.40 and 6.41) of the BJT complete with the values of their parameters.

6.95 The transistor amplifier in Fig. P6.95 is biased with a current source I and has a very high β. Find the dc voltage at the collector, V_C. Also, find the value of g_m. Replace the transistor with the simplified hybrid-π model of Fig. 6.40(a)

Transistor	a	b	c	d	e	f	g
α	1.000					0.90	
β		100		∞			
I_C (mA)	1.00		1.00				
I_E (mA)		1.00				5	
I_B (mA)			0.020				1.10
g_m (mA/V)							700
r_e (Ω)				25	100		
r_π (Ω)					10.1 kΩ		

(note that the dc current source I should be replaced with an open circuit). Hence find the voltage gain v_c/v_i.

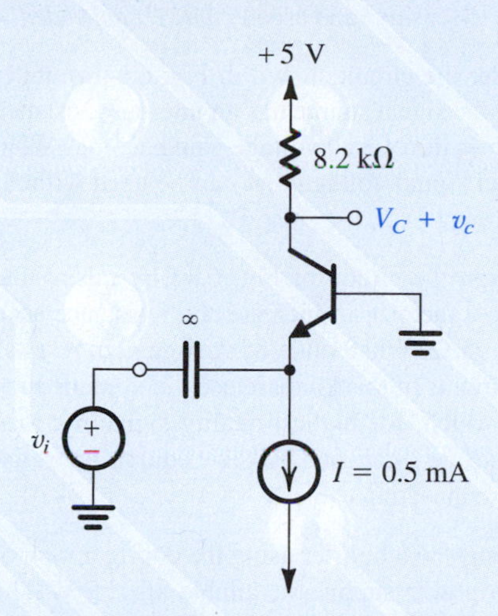

Figure P6.95

6.96 For the conceptual circuit shown in Fig. 6.39, $R_C = 3\ \text{k}\Omega$, $g_m = 50$ mA/V, and $\beta = 100$. If a peak-to-peak output voltage of 1 V is measured at the collector, what are the peak-to-peak values of v_{be} and i_b?

6.97 Figure P6.97 shows the circuit of an amplifier fed with a signal source v_{sig} with a source resistance R_{sig}. The bias circuitry is not shown. Replace the BJT with its hybrid-π equivalent circuit of Fig. 6.40(a). Find the input resistance $R_{\text{in}} \equiv v_\pi / i_b$, the voltage transmission from source to amplifier input, v_π / v_{sig}, and the voltage gain from base to collector, v_o / v_π. Use these to show that the overall voltage gain v_o / v_{sig} is given by

$$\frac{v_o}{v_{\text{sig}}} = -\frac{\beta R_C}{r_\pi + R_{\text{sig}}}$$

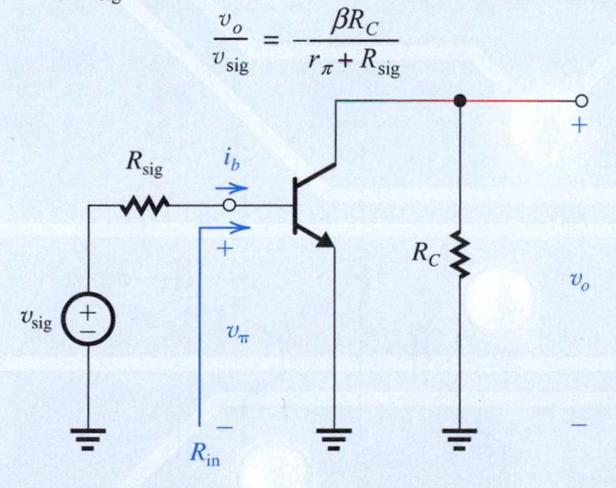

Figure P6.97

6.98 Figure P6.98 shows a transistor with the collector connected to the base. The bias arrangement is not shown. Since a zero v_{BC} implies operation in the active mode, the BJT can

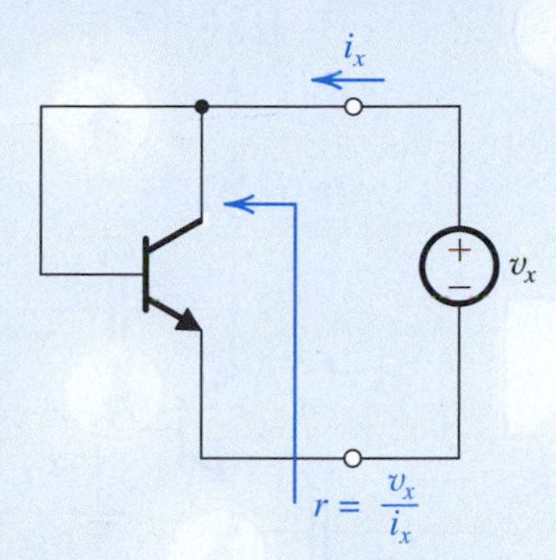

Figure P6.98

be replaced by one of the small-signal models of Figs. 6.40 and 6.41. Use the model of Fig. 6.41(b) and show that the resulting two-terminal device, known as a diode connected transistor, has a small-signal resistance r equal to r_e.

6.99 Figure P6.99 shows a particular configuration of BJT amplifiers, known as "emitter follower." The bias arrangement is not shown. Replace the BJT with its T equivalent-circuit model of Fig. 6.41(b). Show that

$$R_{\text{in}} \equiv \frac{v_i}{i_b} = (\beta + 1)(r_e + R_e)$$

$$\frac{v_o}{v_i} = \frac{R_e}{R_e + r_e}$$

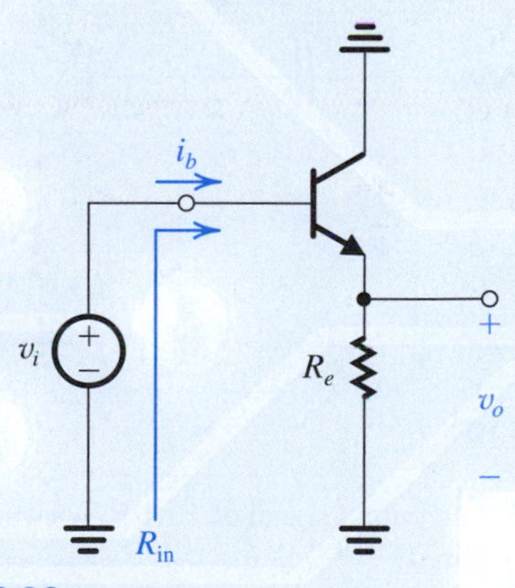

Figure P6.99

6.100 For the circuit shown in Fig. P6.100, draw a complete small-signal equivalent circuit utilizing an appropriate T model for the BJT (use $\alpha = 0.99$). Your circuit should show the values of all components, including the model parameters. What is the input resistance R_{in}? Calculate the overall voltage gain (v_o/v_{sig}).

6.101 In the circuit shown in Fig. P6.101, the transistor has a β of 200. What is the dc voltage at the collector? Find the input resistances R_{ib} and R_{in} and the overall voltage gain

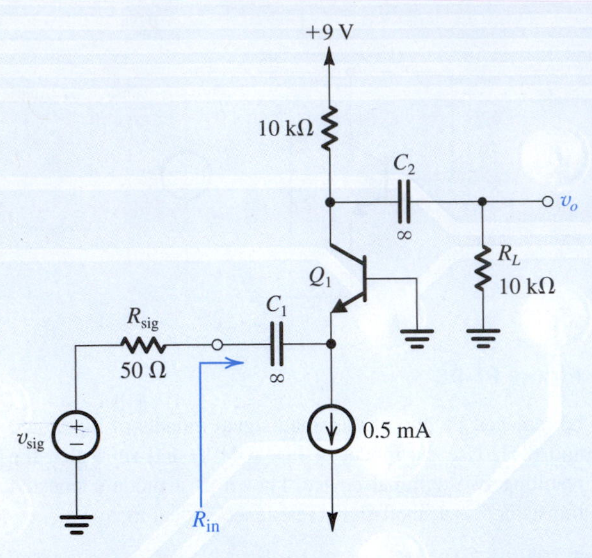

Figure P6.100

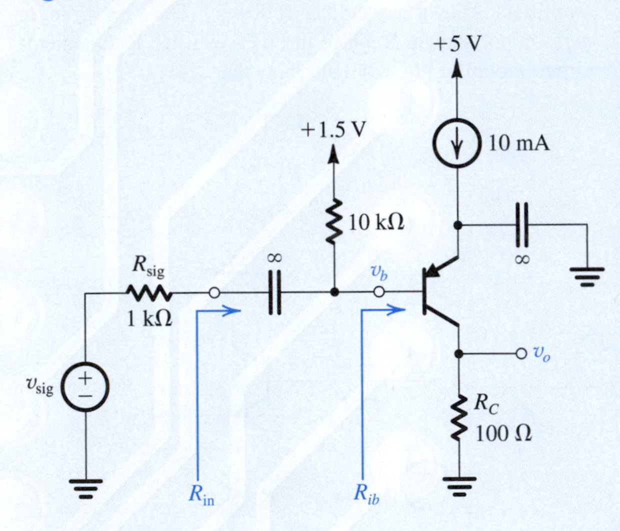

Figure P6.101

(v_o/v_{sig}). For an output signal of ±0.4 V, what values of v_{sig} and v_b are required?

6.102 Consider the augmented hybrid-π model shown in Fig. 6.47(a). Disregarding how biasing is to be done, what is the largest possible voltage gain available for a signal source connected directly to the base and a very-high-resistance load? Calculate the value of the maximum possible gain for $V_A = 25$ V and $V_A = 250$ V.

6.103 Reconsider the amplifier shown in Fig. 6.42 and analyzed in Example 6.14 under the condition that β is not well controlled. For what value of β does the circuit begin to saturate? We can conclude that large β is dangerous in this circuit. Now, consider the effect of reduced β, say, to $\beta = 25$. What values of r_e, g_m, and r_π result? What is the overall voltage gain? (*Note:* You can see that this circuit, using base-current control of bias, is very β-sensitive and usually *not recommended.*)

6.104 Reconsider the circuit shown in Fig. 6.44(a) under the condition that the signal source has an internal resistance of 100 Ω. What does the overall voltage gain become? What is the largest input signal voltage that can be used without output-signal clipping?

D 6.105 Redesign the circuit of Fig. 6.44 by raising the resistor values by a factor n to increase the resistance seen by the input v_i to 75 Ω. What value of voltage gain results? Grounded-base circuits of this kind are used in systems such as cable TV, in which, for highest-quality signaling, load resistances need to be "matched" to the equivalent resistances of the interconnecting cables.

D **6.106 Design an amplifier using the configuration of Fig. 6.44(a). The power supplies available are ±5 V. The input signal source has a resistance of 50 Ω, and it is required that the amplifier input resistance match this value. (Note that $R_{in} = r_e \parallel R_E \simeq r_e$.) The amplifier is to have the greatest possible voltage gain and the largest possible output signal but retain small-signal linear operation (i.e., the signal component across the base–emitter junction should be limited to no more than 10 mV). Find appropriate values for R_E and R_C. What is the value of voltage gain realized?

***6.107** The transistor in the circuit shown in Fig. P6.107 is biased to operate in the active mode. Assuming that β is very large, find the collector bias current I_C. Replace the transistor with the small-signal equivalent circuit model of Fig. 6.41(b) (remember to replace the dc power supply with a short circuit). Analyze the resulting amplifier equivalent circuit to show that

$$\frac{v_{o1}}{v_i} = \frac{R_E}{R_E + r_e}$$

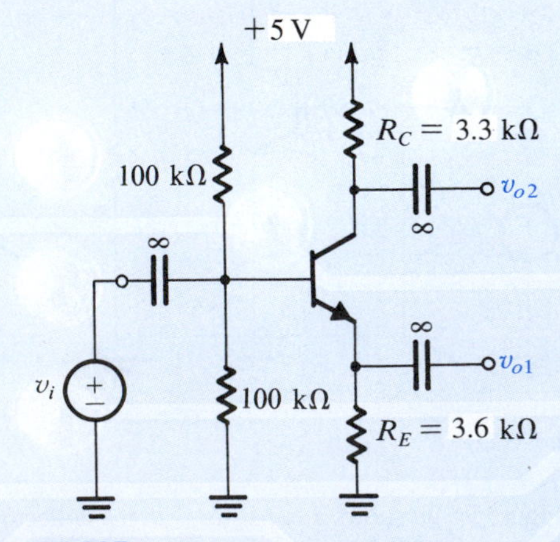

Figure P6.107

$$\frac{v_{o2}}{v_i} = \frac{-\alpha R_C}{R_E + r_e}$$

Find the values of these voltage gains (for $\alpha \simeq 1$). Now, if the terminal labeled v_{o1} is connected to ground, what does the voltage gain v_{o2}/v_i become?

Section 6.6: Basic BJT Amplifier Configurations †

6.108 An amplifier with an input resistance of 100 kΩ, an open-circuit voltage gain of 100 V/V, and an output resistance of 100 Ω, is connected between a 10-kΩ signal source and a 1-kΩ load. Find the overall voltage gain G_v. Also find the current gain, defined as the ratio of the load current to the current drawn from the signal source.

D 6.109 Specify the parameters R_{in}, A_{vo}, and R_o of an amplifier that is to be connected between a 100-kΩ source and a 2-kΩ load. The amplifier is required to meet the following specifications:

(a) No more than 10% of the signal strength is lost in the connection to the amplifier input.
(b) If the load resistance changes from the nominal value of 2 kΩ to a low value of 1 kΩ, the change in output voltage is limited to 10% of nominal value.
(c) The nominal overall voltage gain is 10 V/V.

6.110 Figure P6.110 shows an alternative equivalent circuit representation of an amplifier. If this circuit is to be equivalent to that in Fig. 6.50(b) show that $G_m = A_{vo}/R_o$. Also convince yourself that the transconductance G_m is defined as

$$G_m = \left.\frac{i_o}{v_i}\right|_{R_L = 0}$$

and hence is known as the short-circuit transconductance. Now if the amplifier is fed with a signal source (v_{sig}, R_{sig}) and is connected to a load resistance R_L, show that the gain of the amplifier proper A_v is given by $A_v = G_m(R_o \| R_L)$ and the overall voltage gain G_v is given by

$$G_v = \frac{R_{in}}{R_{in} + R_{sig}} G_m(R_o \| R_L)$$

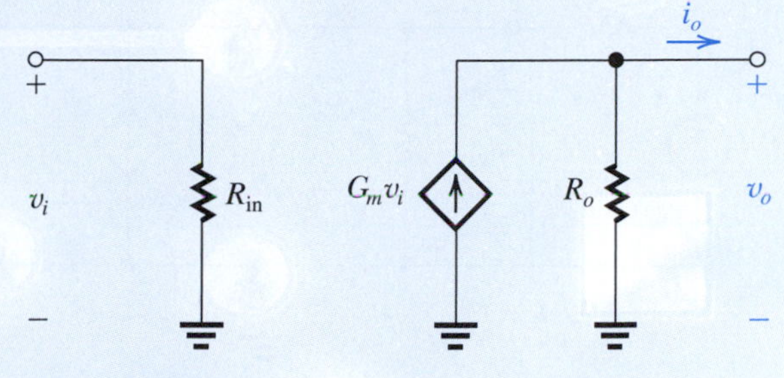

Figure P6.110

6.111 An alternative equivalent circuit of an amplifier fed with a signal source (v_{sig}, R_{sig}) and connected to a load R_L is shown in Fig. P6.111. Here G_{vo} is the open-circuit overall voltage gain,

$$G_{vo} = \left.\frac{v_o}{v_{sig}}\right|_{R_L=\infty}$$

and R_{out} is the output resistance with v_{sig} set to zero. This is different from R_o. Show that

$$G_{vo} = \frac{R_i}{R_i + R_{sig}} A_{vo}$$

where $R_i = R_{in}|_{R_L=\infty}$.

Also show that the overall voltage gain

$$G_v = G_{vo} \frac{R_L}{R_L + R_{out}}$$

****6.112** Most practical amplifiers have internal feedback that make them nonunilateral. In such a case, R_{in} depends on R_L. To illustrate this point we show in Fig. P6.112 the equivalent circuit of an amplifier in which a feedback resistance R_f models the internal feedback mechanism that is present in this amplifier. It is R_f that makes the amplifier nonunilateral. Show that

$$R_{in} = R_1 \| \left[\frac{R_f + (R_2 \| R_L)}{1 + g_m(R_2 \| R_L)}\right]$$

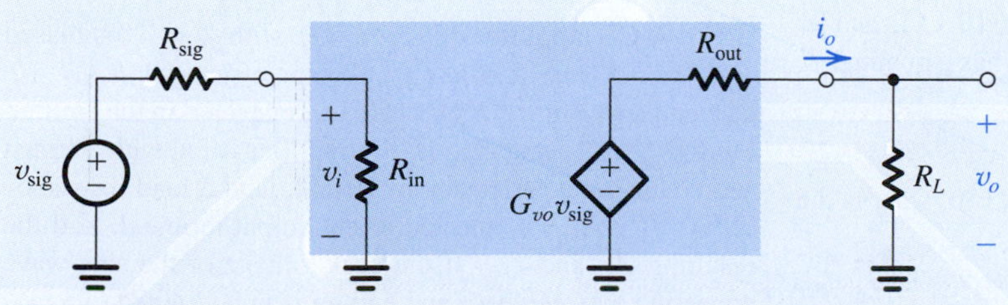

Figure P6.111

†Problems 6.108 to 6.111 are identical to problems 5.80 to 5.84.

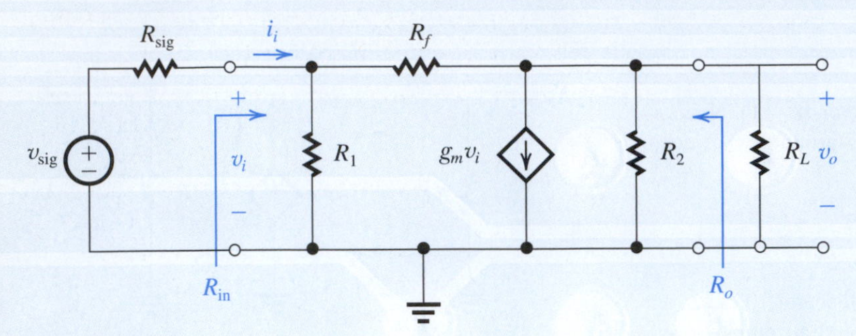

Figure P6.112

$$A_{vo} = -g_m R_2 \frac{1-(1/g_m R_f)}{1+(R_2/R_f)}$$

$$R_o = R_2 \parallel R_f$$

Evaluate R_{in}, A_{vo}, and R_o for the case $R_1 = 100$ kΩ, $R_f = 1$ MΩ, $g_m = 100$ mA/V, $R_2 = 100$ Ω, and $R_L = 1$ kΩ. Which of the amplifier characteristic parameters is most affected by R_f (i.e., relative to the case with $R_f = \infty$)? For $R_{sig} = 100$ kΩ, determine the overall voltage gain, G_v, with and without R_f present.

6.113 A CE amplifier utilizes a BJT with $\beta = 100$ and $V_A = 50$ V, biased at $I_C = 0.5$ mA; it has a collector resistance $R_C = 10$ kΩ. Assume $R_B \gg r_\pi$. Find R_{in}, R_o, and A_{vo}. If the amplifier is fed with a signal source having a resistance of 10 kΩ, and a load resistance $R_L = 10$ kΩ is connected to the output terminal, find the resulting A_v and G_v. If the peak voltage of the sine wave appearing between base and emitter is to be limited to 5 mV, what $\hat{v}_{sig}$ is allowed, and what output voltage signal appears across the load?

D *6.114 In this problem we investigate the effect of the inevitable variability of β on the realized gain of the CE amplifier. For this purpose, use the overall gain expression in Eq. (6.79). Assume r_o is sufficiently large to be negligible and thus show that

$$|G_v| \simeq \frac{R'_L}{(R_{sig}/\beta) + (1/g_m)}$$

where $R'_L = R_L \parallel R_C$.

Consider the case $R'_L = 10$ kΩ and $R_{sig} = 10$ kΩ, and let the BJT be biased at $I_C = 1$ mA. The BJT has a nominal β of 100.

(a) What is the nominal value of $|G_v|$?
(b) If β can be anywhere between 50 and 150, what is the corresponding range of $|G_v|$?
(c) If in a particular design, it is required to maintain $|G_v|$ within ±20 % of its nominal value, what is the maximum allowable range of β?
(d) If it is not possible to restrict β to the range found in (c), and the designer has to contend with β in the range 50 to 150, what value of bias current I_C would result in $|G_v|$ falling in a range of ±20 % of a new nominal value? What is the nominal value of $|G_v|$ in this case?

D 6.115 In this problem, we investigate the effect of changing the bias current I_C on the overall voltage gain G_v of a CE amplifier. Consider the situation of a CE amplifier operating with a signal source having $R_{sig} = 10$ kΩ and having $R_C \parallel R_L = 10$ kΩ. The BJT is specified to have $\beta = 100$ and $V_A = 25$ V. Use Eq. (6.79) to find $|G_v|$ at $I_C = 0.1$ mA, 0.2 mA, 0.5 mA, 1.0 mA, and 1.25 mA. Observe the effect of r_o on limiting $|G_v|$ as I_C is increased. Find the value of I_C that results in $|G_v| = 50$ V/V.

6.116 Two identical CE amplifiers are connected in cascade. The first stage is fed with a source v_{sig} having a resistance $R_{sig} = 10$ kΩ. A load resistance $R_L = 10$ kΩ is connected to the collector of the second stage. Each BJT is biased at $I_C = 0.25$ mA and has $\beta = 100$ and a very large V_A. Each stage utilizes a collector resistance $R_C = 10$ kΩ.

(a) Sketch the equivalent circuit of the two-stage amplifier.
(b) Calculate the voltage transmission from the signal source to the input of the first stage.
(c) Calculate the voltage gain of the first stage, A_{v1}.
(d) Calculate the voltage gain of the second stage, A_{v2}.
(e) Find the overall voltage gain, v_{o2}/v_{sig}.

6.117 A CE amplifier utilizes a BJT with $\beta = 100$ biased at $I_C = 0.5$ mA and has a collector resistance $R_C = 10$ kΩ and a resistance $R_e = 150$ Ω connected in the emitter. Find R_{in}, A_{vo}, and R_o. If the amplifier is fed with a signal source having a resistance of 10 kΩ, and a load resistance $R_L = 10$ kΩ is connected to the output terminal, find the resulting A_v and G_v. If the peak voltage of the sine wave appearing between base and emitter is to be limited to 5 mV, what $\hat{v}_{sig}$ is allowed, and what output voltage signal appears across the load?

D 6.118 Design a CE amplifier with a resistance R_e in the emitter to meet the following specifications:

(i) Input resistance $R_{in} = 20\ k\Omega$.
(ii) When fed from a signal source with a peak amplitude of 0.1 V and a source resistance of 20 kΩ, the peak amplitude of v_π is 5 mV.
Specify R_e and the bias current I_C. The BJT has $\beta = 100$. If the total resistance in the collector is 5 kΩ, find the overall voltage gain G_v and the peak amplitude of the output signal v_o.

SIM D 6.119 Inclusion of an emitter resistance R_e reduces the variability of the gain G_v due to the inevitable wide variance in the value of β. Consider a CE amplifier operating between a signal source with $R_{sig} = 10\ k\Omega$ and a total collector resistance $R_C \| R_L$ of 10 kΩ. The BJT is biased at $I_C = 1$ mA and its β is specified to be nominally 100 but can lie in the range of 50 to 150. First determine the nominal value and the range of $|G_v|$ without resistance R_e. Then select a value for R_e that will ensure that $|G_v|$ be within ±20 % of its new nominal value. Specify the value of R_e, the new nominal value of $|G_v|$, and the expected range of $|G_v|$.

D 6.120 A CB amplifier is operating with $R_L = 10\ k\Omega$, $R_C = 10\ k\Omega$, and $R_{sig} = 100\ \Omega$. At what current I_C should the transistor be biased for the input resistance R_{in} to equal that of the signal source? What is the resulting overall voltage gain? Assume $\alpha \simeq 1$.

6.121 For the circuit in Fig. P6.121, let $R_{sig} \gg r_e$ and $\alpha \simeq 1$. Find v_o.

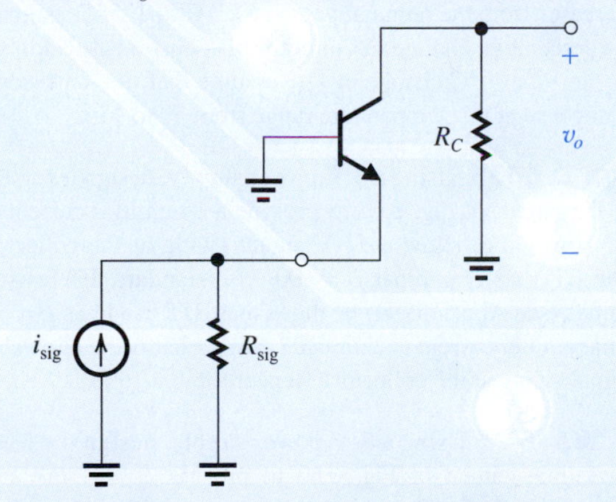

Figure P6.121

6.122 A CB amplifier is biased at $I_E = 0.25$ mA with $R_C = R_L = 10\ k\Omega$ and is driven by a signal source with $R_{sig} = 1\ k\Omega$. Find the overall voltage gain G_v. If the maximum signal amplitude of the voltage between base and emitter is limited to 10 mV, what are the corresponding amplitudes of v_{sig} and v_o? Assume $\alpha \simeq 1$.

D 6.123 An emitter follower is required to deliver a 0.5-V peak sinusoid to a 2-kΩ load. If the peak amplitude of v_{be} is to be limited to 5 mV, what is the lowest value of I_E at which the BJT can be biased? At this bias current, what are the maximum and minimum currents that the BJT will be conducting (at the positive and negative peaks of the output sine wave)? If the resistance of the signal source is 200 kΩ, what value of G_v is obtained? Thus determine the required amplitude of v_{sig}.

6.124 An emitter follower with a BJT biased at $I_C = 1$ mA and having $\beta = 100$ is connected between a source with $R_{sig} = 20\ k\Omega$ and a load $R_L = 1\ k\Omega$.

(a) Find R_{in}, v_b/v_{sig}, and v_o/v_{sig}.
(b) If the signal amplitude across the base–emitter junction is to be limited to 10 mV, what is the corresponding amplitude of v_{sig} and v_o?
(c) Find the open-circuit voltage gain G_{vo} and the output resistance R_{out}. Use these values first to verify the value of G_v obtained in (a), then to find the value of G_v obtained with R_L reduced to 500 Ω.

6.125 An emitter follower is operating at a collector bias current of 0.25 mA and is used to connect a 10-kΩ source to a 1-kΩ load. If the nominal value of β is 100, what output resistance R_{out} and overall voltage gain G_v result? Now if transistor β is specified to lie in the range 50 to 150, find the corresponding range of R_{out} and G_v.

6.126 An emitter follower, when driven from a 10-kΩ source, was found to have an output resistance R_{out} of 200 Ω. The output resistance increased to 300 Ω when the source resistance was increased to 20 kΩ. Find the overall voltage gain when the follower is driven by a 30-kΩ source and loaded by a 1-kΩ resistor.

6.127 For the general amplifier circuit shown in Fig. P6.127 neglect the Early effect.

(a) Find expressions for v_c/v_{sig} and v_e/v_{sig}.
(b) If v_{sig} is disconnected from node X, node X is grounded, and node Y is disconnected from ground and connected to v_{sig}, find the new expression for v_c/v_{sig}.

Section 6.7: Biasing in BJT Amplifier Circuits

D 6.128 For the circuit in Fig. 6.59(a), neglect the base current I_B in comparison with the current in the voltage divider. It is required to bias the transistor at $I_C = 1$ mA, which requires selecting R_{B1} and R_{B2} so that $V_{BE} = 0.690$ V. If $V_{CC} = 3$ V, what must the ratio R_{B1}/R_{B2} be? Now, if R_{B1} and

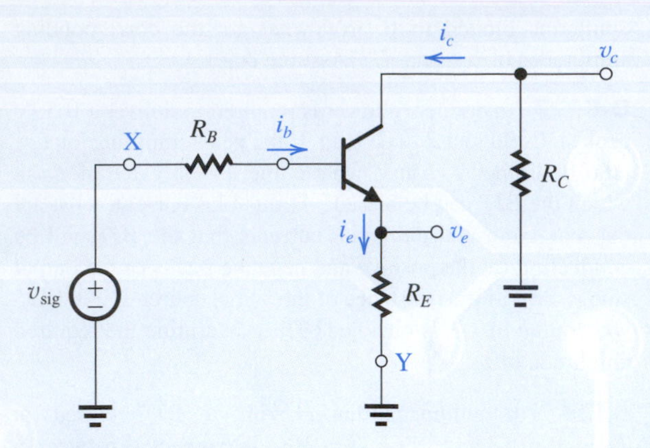

Figure P6.127

R_{B2} are 1% resistors, that is, each can be in the range of 0.99 to 1.01 of its nominal value, what is the range obtained for V_{BE}? What is the corresponding range of I_C? If $R_C = 2$ kΩ, what is the range obtained for V_{CE}? Comment on the efficacy of this biasing arrangement.

D 6.129 It is required to bias the transistor in the circuit of Fig. 6.59(b) at $I_C = 1$ mA. The transistor β is specified to be nominally 100, but it can fall in the range of 50 to 150. For $V_{CC} = +3$ V and $R_C = 2$ kΩ, find the required value of R_B to achieve $I_C = 1$ mA for the "nominal" transistor. What is the expected range for I_C and V_{CE}? Comment on the efficacy of this bias design.

D 6.130 Consider the single-supply bias network shown in Fig. 6.60(a). Provide a design using a 9-V supply in which the supply voltage is equally split between R_C, V_{CE}, and R_E with a collector current of 0.6 mA. The transistor β is specified to have a minimum value of 90. Use a voltage divider current of $I_E/10$, or slightly higher. Since a reasonable design should operate for the best transistors for which β is very high, do your initial design with $\beta = \infty$. Then choose suitable 5% resistors (see Appendix H), making the choice in a way that will result in a V_{BB} that is slightly higher than the ideal value. Specify the values you have chosen for R_E, R_C, R_1, and R_2. Now, find V_B, V_E, V_C, and I_C for your final design using $\beta = 90$.

D 6.131 Repeat Problem 6.130, but use a voltage divider current that is $I_E/2$. Check your design at $\beta = 90$. If you have the data available, find how low β can be while the value of I_C does not fall below that obtained with the design of Problem 6.130 for $\beta = 90$.

D *6.132 It is required to design the bias circuit of Fig. 6.60 for a BJT whose nominal $\beta = 100$.

(a) Find the largest ratio (R_B/R_E) that will guarantee I_E remain within ±10% of its nominal value for β as low as 50 and as high as 150.

(b) If the resistance ratio found in (a) is used, find an expression for the voltage $V_{BB} \equiv V_{CC}R_2/(R_1 + R_2)$ that will result in a voltage drop of $V_{CC}/3$ across R_E.
(c) For $V_{CC} = 5$ V, find the required values of R_1, R_2, and R_E to obtain $I_E = 0.5$ mA and to satisfy the requirement for stability of I_E in (a).
(d) Find R_C so that $V_{CE} = 1.5$ V for β equal to its nominal value.

Check your design by evaluating the resulting range of I_E.

D *6.133 Consider the two-supply bias arrangement shown in Fig. 6.61 using ±3-V supplies. It is required to design the circuit so that $I_C = 0.6$ mA and V_C is placed midway between V_{CC} and V_E.

(a) For $\beta = \infty$, what values of R_E and R_C are required?
(b) If the BJT is specified to have a minimum β of 90, find the largest value for R_B consistent with the need to limit the voltage drop across it to one-tenth the voltage drop across R_E.
(c) What standard 5% resistor values (see Appendix H) would you use for R_B, R_E, and R_C? In making your selection, use somewhat lower values in order to compensate for the low-β effects.
(d) For the values you selected in (c), find I_C, V_B, V_E, and V_C for $\beta = \infty$ and for $\beta = 90$.

D *6.134 Utilizing ±3-V power supplies, it is required to design a version of the circuit in Fig. 6.61 in which the signal will be coupled to the emitter and thus R_B can be set to zero. Find values for R_E and R_C so that a dc emitter current of 0.5 mA is obtained and so that the gain is maximized while allowing ±1 V of signal swing at the collector. If temperature increases from the nominal value of 25°C to 125°C, estimate the percentage change in collector bias current. In addition to the −2 mV/°C change in V_{BE}, assume that the transistor β changes over this temperature range from 50 to 150.

SIM **D 6.135** Using a 3-V power supply, design a version of the circuit of Fig. 6.62 to provide a dc emitter current of 0.5 mA and to allow a ±1-V signal swing at the collector. The BJT has a nominal $\beta = 100$. Use standard 5% resistor values (see Appendix H). If the actual BJT used has $\beta = 50$, what emitter current is obtained? Also, what is the allowable signal swing at the collector? Repeat for $\beta = 150$.

D *6.136 (a) Using a 3-V power supply, design the feedback bias circuit of Fig. 6.62 to provide $I_C = 3$ mA and $V_C = V_{CC}/2$ for $\beta = 90$.
(b) Select standard 5% resistor values, and reevaluate V_C and I_C for $\beta = 90$.
(c) Find V_C and I_C for $\beta = \infty$.
(d) To improve the situation that obtains when high-β transistors are used, we have to arrange for an additional current to flow through R_B. This can be achieved by connecting a resistor between base and emitter, as shown in Fig. P6.136.

Design this circuit for $\beta = 90$. Use a current through R_{B2} equal to the base current. Now, what values of V_C and I_C result with $\beta = \infty$?

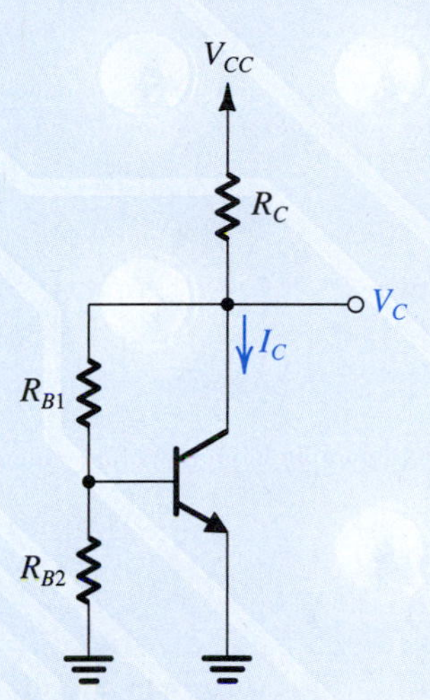

Figure P6.136

D 6.137 A circuit that can provide a very large voltage gain for a high-resistance load is shown in Fig. P6.137. Find the values of I and R_B to bias the BJT at $I_C = 1$ mA and $V_C = 1.5$ V. Let $\beta = 100$.

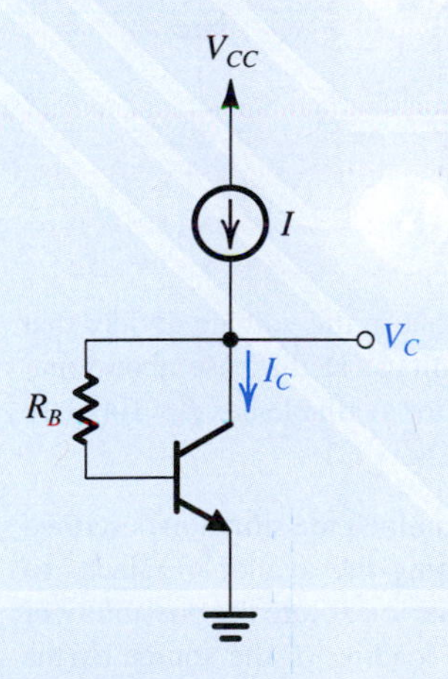

Figure P6.137

6.138 The circuit in Fig. P6.138 provides a constant current I_O as long as the circuit to which the collector is connected maintains the BJT in the active mode. Show that

$$I_O = \alpha \frac{V_{CC}[R_2/(R_1 + R_2)] - V_{BE}}{R_E + (R_1 \| R_2)/(\beta + 1)}$$

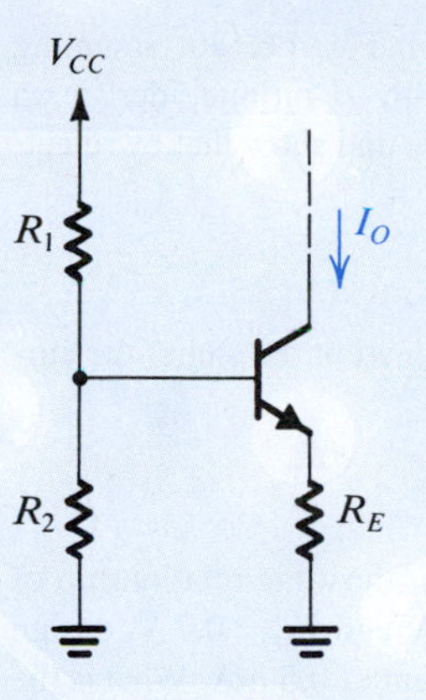

Figure P6.138

D **6.139 The current-source biasing circuit shown in Fig. P6.139 provides a bias current to Q_1 that is determined by the current source formed by Q_2, R_1, R_2, and R_E. The bias current is independent of R_B and nearly independent of β_1 (as long as both Q_1 and Q_2 operate in the active mode). It is required to design the circuit using ±5-V dc supplies to establish $I_{C1} = 0.1$ mA and $V_{CE1} = 1.5$ V, in the ideal situation of infinite β_1 and β_2. In designing the current source, use 2-V dc voltage drop across R_E and impose the requirement that I_{E2} remain within 5% of its ideal value for β_2 as low as 50. In selecting a value for R_B, ensure that for the lowest value of $\beta_1 = 50$, V_{CE2} is 2.5 V. Use standard 5% resistor values (see Appendix H). What values for R_1, R_2, R_E, R_B, and R_C do you choose? What values of I_{C1} and V_{CE1} result for $\beta_1 = \beta_2 = 50$, 100, and 200?

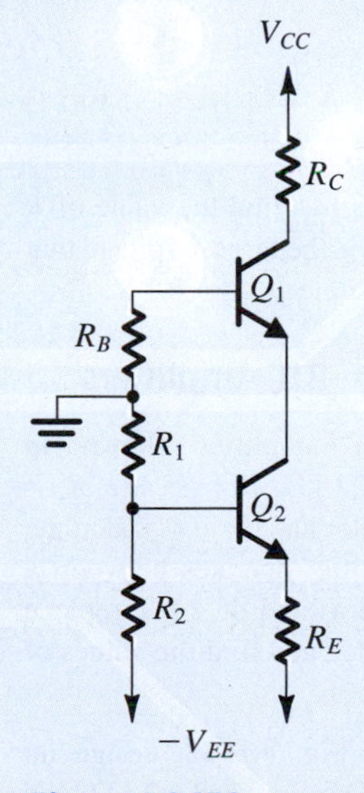

Figure P6.139

SIM **D *6.140** For the circuit in Fig. P6.140, assuming all transistors to be identical with β infinite, derive an expression for the output current I_O, and show that by selecting

$$R_1 = R_2$$

and keeping the current in each junction the same, the current I_O will be

$$I_O = \frac{V_{CC}}{2R_E}$$

which is independent of V_{BE}. What must the relationship of R_E to R_1 and R_2 be? For $V_{CC} = 10$ V and $V_{BE} = 0.7$ V, design the circuit to obtain an output current of 0.5 mA. What is the lowest voltage that can be applied to the collector of Q_3?

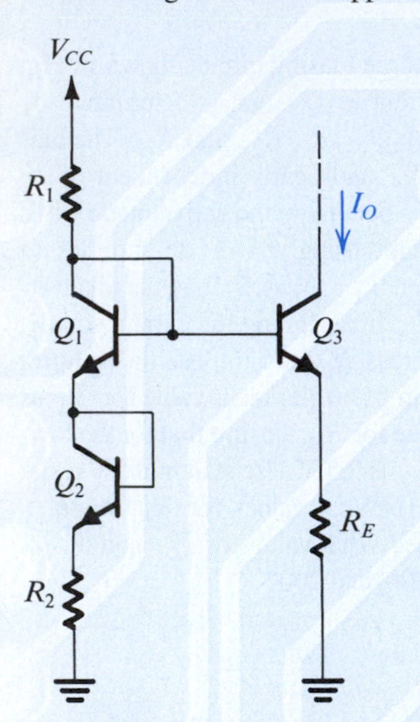

Figure P6.140

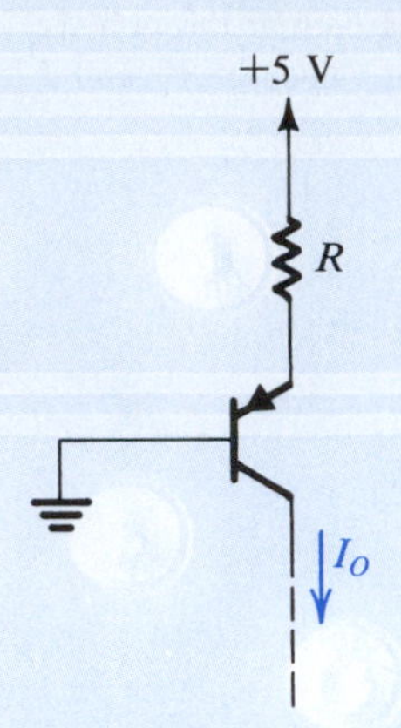

Figure P6.141

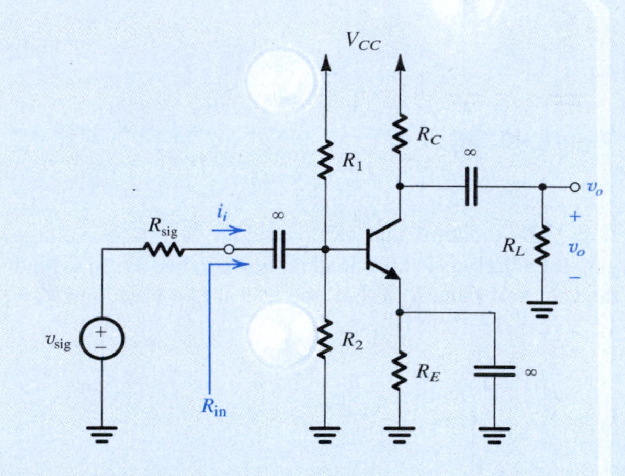

Figure P6.142

D 6.141 For the circuit in Fig. P6.141 find the value of R that will result in $I_O \simeq 1$ mA. What is the largest voltage that can be applied to the collector? Assume $|V_{BE}| = 0.7$ V.

Section 6.8: Discrete-Circuit BJT Amplifiers

6.142 For the common-emitter amplifier shown in Fig. P6.142, let $V_{CC} = 15$ V, $R_1 = 27$ kΩ, $R_2 = 15$ kΩ, $R_E = 2.4$ kΩ, and $R_C = 3.9$ kΩ. The transistor has $\beta = 100$. Calculate the dc bias current I_C. If the amplifier operates between a source for which $R_{sig} = 2$ kΩ and a load of 2 kΩ, replace the transistor with its hybrid-π model, and find the values of R_{in}, and the overall voltage gain v_o/v_{sig}.

D 6.143 Using the topology of Fig. P6.142, design an amplifier to operate between a 2-kΩ source and a 2-kΩ load with a gain v_o/v_{sig} of −40 V/V. The power supply available is 15 V. Use an emitter current of approximately 2 mA and a current of about one-tenth of that in the voltage divider that feeds the base, with the dc voltage at the base about one-third of the supply. The transistor available has $\beta = 100$. Use standard 5% resistor (see Appendix H).

6.144 A designer, having examined the situation described in Problem 6.142 and estimating the available gain to be approximately −36.6 V/V, wants to explore the possibility of improvement by reducing the loading of the source by the amplifier input. As an experiment, the designer varies the resistance levels by a factor of approximately 3: R_1 to 82 kΩ, R_2 to 47 kΩ, R_E to 7.2 kΩ, and R_C to 12 kΩ (standard values of 5%-tolerance resistors). With $V_{CC} = 15$ V, $R_{sig} = 2$ kΩ, $R_L = 2$ kΩ, and $\beta = 100$, what does the gain become? Comment.

D 6.145 Consider the CE amplifier circuit of Fig. 6.65(a). It is required to design the circuit (i.e., find values for I, R_B, and R_C) to meet the following specifications:

(a) $R_{in} \simeq 5$ kΩ

(b) The dc voltage drop across R_B is approximately 0.2 V.
(c) The open-circuit voltage gain from base to collector is the maximum possible, consistent with the requirement that the collector voltage never falls by more than approximately 0.4 V below the base voltage with the signal between base and emitter being as high as 5 mV.

Assume that v_{sig} is a sinusoidal source, the available supply $V_{CC} = 3$ V, and the transistor has $\beta = 100$. Use standard 5% resistance values, and specify the value of I to one significant digit. What base-to-collector open-circuit voltage gain does your design provide? If $R_{sig} = R_L = 10$ kΩ, what is the overall voltage gain?

D 6.146 In the circuit of Fig. P6.146, v_{sig} is a small sine-wave signal with zero average. The transistor β is 100.

(a) Find the value of R_E to establish a dc emitter current of about 0.5 mA.
(b) Find R_C to establish a dc collector voltage of about +1 V.
(c) For $R_L = 10$ kΩ, draw the small-signal equivalent circuit of the amplifier and determine its overall voltage gain.

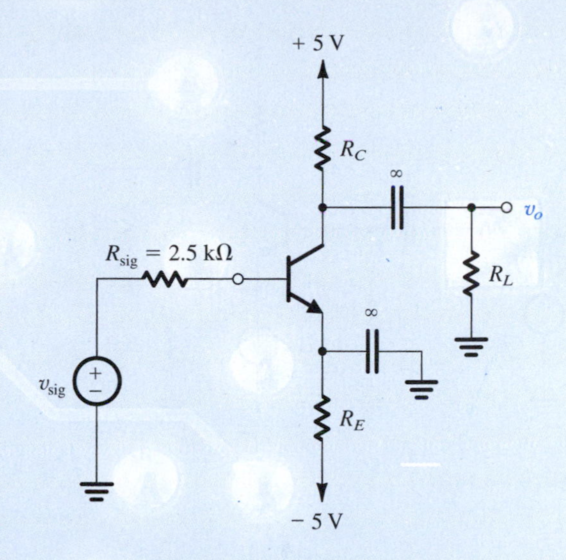

Figure P6.146

***6.147** The amplifier of Fig. P6.147 consists of two identical common-emitter amplifiers connected in cascade. Observe that the input resistance of the second stage, R_{in2}, constitutes the load resistance of the first stage.

(a) For $V_{CC} = 9$ V, $R_1 = 100$ kΩ, $R_2 = 47$ kΩ, $R_E = 3.9$ kΩ, $R_C = 6.8$ kΩ, and $\beta = 100$, determine the dc collector current and dc collector voltage of each transistor.
(b) Draw the small-signal equivalent circuit of the entire amplifier and give the values of all its components.
(c) Find R_{in1} and v_{b1}/v_{sig} for $R_{sig} = 5$ kΩ.
(d) Find R_{in2} and v_{b2}/v_{b1}.
(e) For $R_L = 2$ kΩ, find v_o/v_{b2}.
(f) Find the overall voltage gain v_o/v_{sig}.

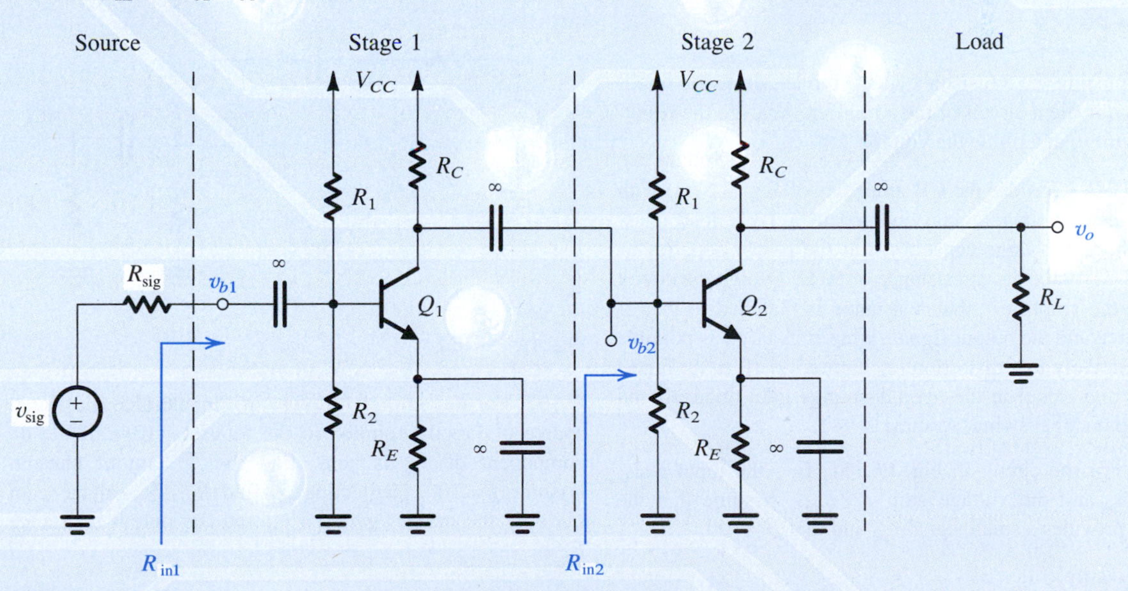

Figure P6.147

6.148 In the circuit of Fig. P6.148, v_{sig} is a small sine-wave signal. Find R_{in} and the gain v_o/v_{sig}. Assume $\beta = 100$. If the amplitude of the signal v_{be} is to be limited to 5 mV, what is the largest signal at the input? What is the corresponding signal at the output?

***6.149** The BJT in the circuit of Fig. P6.149 has $\beta = 100$.

(a) Find the dc collector current and the dc voltage at the collector.

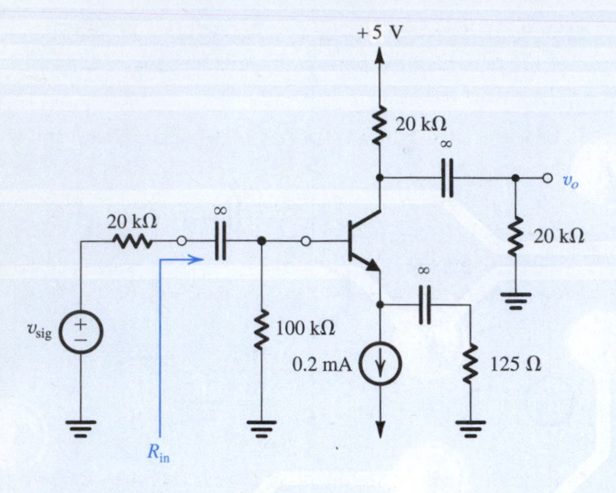

Figure P6.148

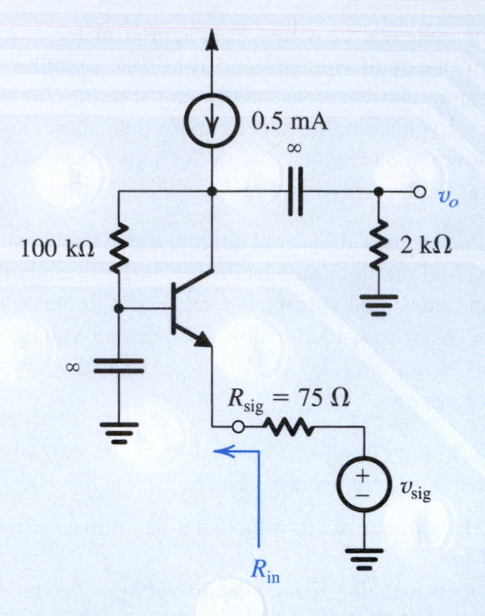

Figure P6.151

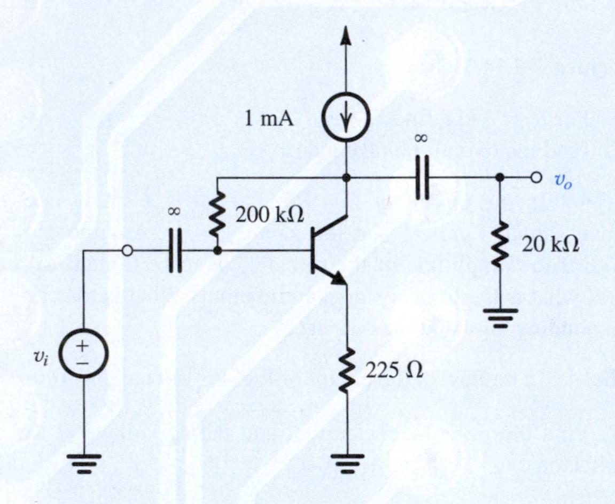

Figure P6.149

(b) Replacing the transistor by its T model, draw the small-signal equivalent circuit of the amplifier. Analyze the resulting circuit to determine the voltage gain v_o/v_i.

D *6.150 Consider the CB amplifier of Fig. 6.67(a) with the collector voltage signal coupled to a 1-kΩ load resistance through a large capacitor. Let the power supplies be ±3 V. The source has a resistance of 50 Ω. Design the circuit so that the amplifier input resistance is matched to that of the source and the output signal swing is as large as possible with relatively low distortion (v_{be} limited to 10 mV). Find I and R_C and calculate the overall voltage gain obtained and the output signal swing. Assume $\alpha \simeq 1$.

6.151 For the circuit in Fig. P6.151, find the input resistance R_{in} and the voltage gain v_o/v_{sig}. Assume that the source provides a small signal v_{sig} and that $\beta = 100$.

6.152 For the emitter-follower circuit shown in Fig. P6.152, the BJT used is specified to have β values in the range of 50 to 200 (a distressing situation for the circuit designer). For the two extreme values of β ($\beta = 50$ and $\beta = 200$), find:

(a) I_E, V_E, and V_B.
(b) the input resistance R_{in}.
(c) the voltage gain v_o/v_{sig}.

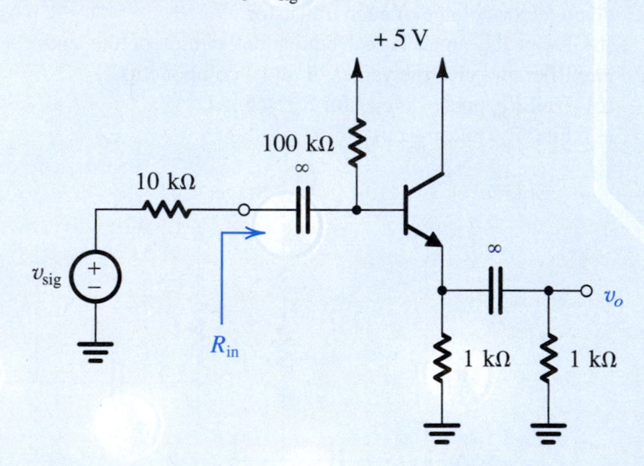

Figure P6.152

6.153 For the emitter follower in Fig. P6.153, the signal source is directly coupled to the transistor base. If the dc component of v_{sig} is zero, find the dc emitter current. Assume $\beta = 100$. Neglecting r_o, find R_{in}, the voltage gain v_o/v_{sig}, the current gain i_o/i_i, and the output resistance R_{out}.

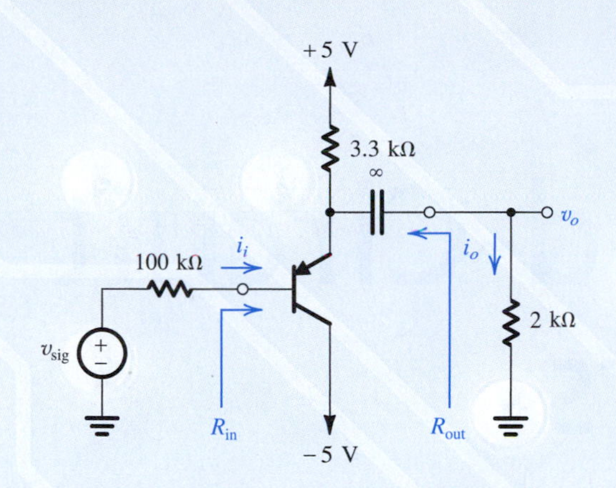

Figure P6.153

6.154 For the circuit in Fig. P6.154, called a **bootstrapped follower:

(a) Find the dc emitter current and g_m, r_e, and r_π. Use $\beta = 100$.

(b) Replace the BJT with its T model (neglecting r_o), and analyze the circuit to determine the input resistance R_{in} and the voltage gain v_o/v_{sig}.

(c) Repeat (b) for the case when capacitor C_B is open-circuited. Compare the results with those obtained in (b) to find the advantages of bootstrapping.

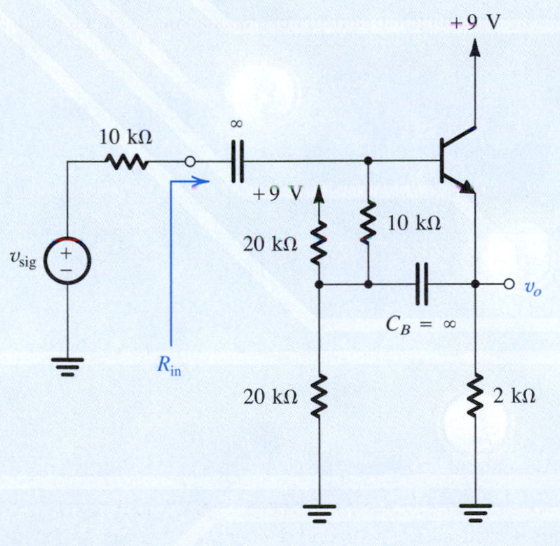

Figure P6.154

**6.155 For the follower circuit in Fig. P6.155, let transistor Q_1 have $\beta = 50$ and transistor Q_2 have $\beta = 100$, and neglect the effect of r_o. Use $V_{BE} = 0.7$ V.

(a) Find the dc emitter currents of Q_1 and Q_2. Also, find the dc voltages V_{B1} and V_{B2}.

(b) If a load resistance $R_L = 1$ kΩ is connected to the output terminal, find the voltage gain from the base to the emitter of Q_2, v_o/v_{b2}, and find the input resistance R_{ib2} looking into the base of Q_2. (*Hint:* Consider Q_2 as an emitter follower fed by a voltage v_{b2} at its base.)

(c) Replacing Q_2 with its input resistance R_{ib2} found in (b), analyze the circuit of emitter follower Q_1 to determine its input resistance R_{in}, and the gain from its base to its emitter, v_{e1}/v_{b1}.

(d) If the circuit is fed with a source having a 100-kΩ resistance, find the transmission to the base of Q_1, v_{b1}/v_{sig}.

(e) Find the overall voltage gain v_o/v_{sig}.

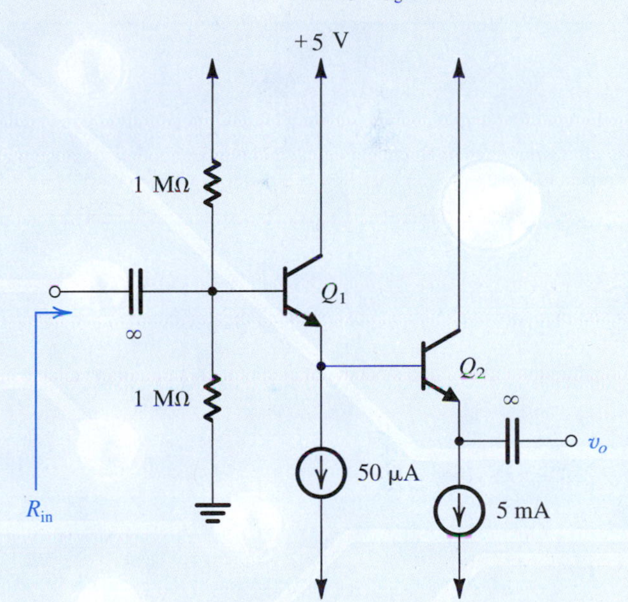

Figure P6.155

D 6.156 A CE amplifier has a midband voltage gain of $|A_M| = 100$ V/V, a lower 3-dB frequency of $f_L = 100$ Hz, and a higher 3-dB frequency $f_L = 100$ MHz. In Chapter 9 we will learn that connecting a resistance R_e in the emitter of the BJT results in lowering f_L and raising f_H by the factor $(1 + g_m R_e)$. If the BJT is biased at $I_C = 1$ mA, find R_e that will result in f_H at least equal to 5 MHz. What will the new values of f_L and A_M be?

PART II

Integrated-Circuit Amplifiers

Having studied the MOSFET and the BJT and become familiar with their basic circuit applications, we are now ready to consider their use in the design of practical amplifier circuits that can be fabricated in integrated-circuit (IC) form. Part II is devoted to this rich subject. Its six chapters constitute a coherent treatment of IC amplifier design and can thus serve as a second course in electronic circuits.

Beginning with a brief introduction to the philosophy of IC design, Chapter 7 presents the basic circuit building blocks that are utilized in the design of IC amplifiers. However, the most important building block of all, the differential pair configuration, is deferred to Chapter 8, where it is the main topic. Chapter 8 also considers the design of amplifiers that require a number of cascaded stages.

As mentioned at various points in Part I, amplifiers have finite bandwidths. Chapter 9 is devoted to the frequency-response analysis of amplifiers; it provides a comprehensive study of the mechanisms that limit the bandwidth and the tools and methods that are utilized to estimate it for a wide variety of amplifier circuit configurations. While the study of the first half or so of Chapter 9 is essential, some of its later sections can be postponed to a later point in the course or even to subsequent courses.

An essential tool in amplifier design is the judicious use of feedback. Chapter 10 deals with this exceedingly important subject. A thorough understanding of feedback concepts, insight into feedback configurations, and proficiency in the use of the feedback analysis method are invaluable to the serious circuit designer.

In Chapter 11, we switch gears from dealing with primarily small-signal amplifiers to those that are required to handle large signals and large amounts of power. Finally, Chapter 12 brings together all the topics of Part II in an important application: namely, the design of operational amplifier circuits. We will then have come full circle, from considering the op amp as a black box in Chapter 2 to understanding what is inside the box in Chapter 12.

Throughout Part II, MOSFET and BJT circuits are treated side-by-side. Because over 90% of ICs today employ the MOSFET, its circuits are presented first. Nevertheless, BJT circuits are presented with equal depth, although sometimes somewhat more briefly. In this regard, we draw the reader's attention to Appendix 7.A, which presents a valuable compilation of the properties of both types of transistors, allowing interesting comparisons to be made. As well, typical device parameter values are provided for a number of CMOS and bipolar fabrication process technologies.

CHAPTER 7

Building Blocks of Integrated-Circuit Amplifiers

IN THIS CHAPTER YOU WILL LEARN

1. The basic integrated-circuit (IC) design philosophy and how it differs from that for discrete-circuit design.
2. The basic gain cells of IC amplifiers, namely, the CS and CE amplifiers with current-source loads.
3. How to increase the gain realized in the basic gain cells by employing the principle of cascoding.
4. Analysis and design of the cascode amplifier and the cascode current source in both their MOS and bipolar forms.
5. How current sources are used to bias IC amplifiers and how the reference current generated in one location is replicated at various other locations on the IC chip by using current mirrors.
6. Some ingenious analog circuit design techniques that result in current mirrors with vastly improved characteristics.
7. How to pair transistors to realize amplifiers with characteristics superior to those obtained from a single-transistor stage.

Introduction

Having studied the two major transistor types, the MOSFET and the BJT, and their basic discrete-circuit amplifier configurations, we are now ready to begin the study of integrated-circuit (IC) amplifiers. This chapter is devoted to the design of the basic building blocks of IC amplifiers.

We begin with a brief section on the design philosophy of integrated circuits and how it differs from that of discrete circuits. Throughout this chapter, MOS and bipolar circuits are presented side by side, which allows a certain economy in presentation and, more importantly, provides an opportunity to compare and contrast the two circuit types. Toward that end, Appendix 7.A provides a comprehensive comparison of the attributes of the two transistor types. This should serve both as a condensed review and as a guide to very interesting similarities and differences between the two devices. Appendix 7.A can be consulted at any time during the study of this or any of the remaining chapters of the book.

The heart of this chapter is the material in Sections 7.2 to 7.4. In Section 7.2 we present the basic gain cell of IC amplifiers, namely, the current-source-loaded common-source (common-emitter) amplifier. We then ask the question of how to increase its gain. This leads naturally and seamlessly to the principle of cascoding and its application in amplifier design: namely, the cascode amplifier and the cascode current source, which are very important building blocks of IC amplifiers.

Section 7.4 is devoted to IC biasing and the study of another key IC building block, the current mirror. We study a collection of current-mirror circuits with improved performance in Section 7.5, for their significance and usefulness, but also because they embody ideas that illustrate the beauty and power of analog circuit design. The chapter concludes with the presentation in Section 7.6 of an interesting and useful collection of amplifier configurations, each utilizing a pair of transistors.

7.1 IC Design Philosophy

Integrated-circuit fabrication technology (Appendix A) imposes constraints on—and provides opportunities to—the circuit designer. Thus, while chip-area considerations dictate that large- and even moderate-value resistors are to be avoided, constant-current sources are readily available. Large capacitors, such as those we used in Sections 5.8 and 6.8 for signal coupling and bypass, are not available to be used, except perhaps as components external to the IC chip. Even then, the number of such capacitors has to be kept to a minimum; otherwise the number of chip terminals increases, and hence the cost. Very small capacitors, in the picofarad and fraction-of-a-picofarad range, however, are easy to fabricate in IC MOS technology and can be combined with MOS amplifiers and MOS switches to realize a wide range of signal processing functions, both analog (Chapter 16) and digital (Chapter 14).

As a general rule, in designing IC MOS circuits, one should strive to realize as many of the functions required as possible using MOS transistors only and, when needed, small MOS capacitors. MOS transistors can be sized; that is, their W and L values can be selected to fit a wide range of design requirements. Also, arrays of transistors can be matched (or, more generally, made to have desired size ratios) to realize such useful circuit building blocks as current mirrors.

At this juncture, it is useful to mention that to pack a larger number of devices on the same IC chip, the trend has been to reduce the device dimensions. By 2009, CMOS process technologies capable of producing devices with a 45-nm minimum channel length were in use. Such small devices need to operate with dc voltage supplies close to 1 V. While low-voltage operation can help to reduce power dissipation, it poses a host of challenges to the circuit designer. For instance, such MOS transistors must be operated with overdrive voltages of only 0.1 V to 0.2 V. In our study of MOS amplifiers, we will make frequent comments on such issues.

The MOS-amplifier circuits that we shall study will be designed almost entirely using MOSFETs of both polarities—that is, NMOS and PMOS—as are readily available in CMOS technology. As mentioned earlier, CMOS is currently the most widely used IC technology for both analog and digital as well as combined analog and digital (or mixed-signal) applications. Nevertheless, bipolar integrated circuits still offer many exciting opportunities to the analog design engineer. This is especially the case for general-purpose circuit packages, such as high-quality op amps that are intended for assembly on printed-circuit (pc) boards (as opposed to being part of a system-on-chip). As well, bipolar circuits can provide much higher output currents and are favored for certain applications, such as in the automotive industry, for their high reliability under severe environmental conditions. Finally, bipolar circuits can be combined with CMOS in innovative and exciting ways in what is known as BiCMOS technology.

7.2 The Basic Gain Cell

7.2.1 The CS and CE Amplifiers with Current-Source Loads

The basic gain cell in an IC amplifier is a common-source (CS) or common-emitter (CE) transistor loaded with a constant-current source, as shown in Fig. 7.1(a) and (b). These circuits are similar to the CS and CE amplifiers studied in Sections 5.6 and 6.6, except that here we have replaced the resistances R_D and R_C with constant-current sources. This is done for two reasons: First, as mentioned in Section 7.1, it is difficult in IC technology to implement resistances with reasonably precise values; rather, it is much easier to use current sources, which are implemented using transistors, as we shall see shortly. Second, by using a constant-current source we are in effect operating the CS and CE amplifiers with a very high (ideally infinite) load resistance; thus we can obtain a much higher gain than if a finite R_D or R_C is used. The circuits in Fig. 7.1(a) and (b) are said to be **current-source loaded** or **active loaded**.

Before we consider the small-signal analysis of the active-loaded CS and CE amplifiers, a word on their dc bias is in order. Obviously, in each circuit Q_1 is biased at $I_D = I$ and $I_C = I$. But what determines the dc voltages at the drain (collector) and at the gate (base)? Usually, these gain cells will be part of larger circuits in which negative feedback is utilized to fix the values of V_{DS} and V_{GS} (V_{CE} and V_{BE}). We shall be discussing dc biasing later in this chapter. As well, in the next chapter we will begin to see complete IC amplifiers including biasing. For the time being, however, we shall assume that the MOS transistor in Fig. 7.1(a) is biased to operate in the saturation region and that the BJT in Fig. 7.1(b) is biased to operate in the active region. We will often refer to both the MOSFET and the BJT as operating in the "active region."

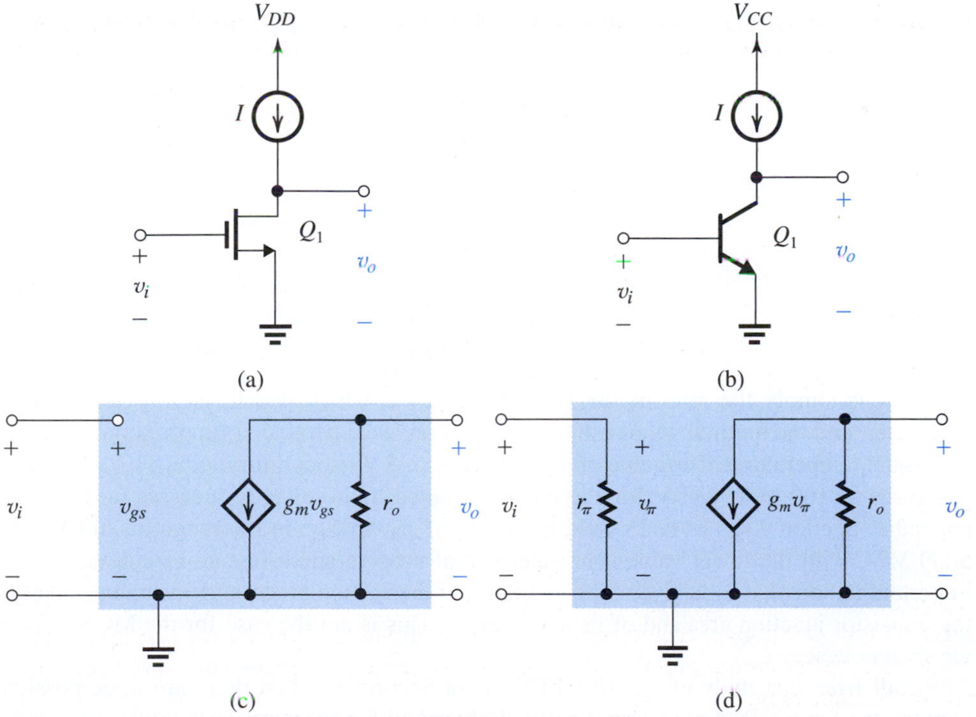

Figure 7.1 The basic gain cells of IC amplifiers: **(a)** current-source- or active-loaded common-source amplifier; **(b)** current-source- or active-loaded common-emitter amplifier; **(c)** small-signal equivalent circuit of **(a)**; and **(d)** small-signal equivalent circuit of **(b)**.

Small-signal analysis of the current-source-loaded CS and CE amplifiers can be performed by utilizing their equivalent-circuit models, shown respectively in Fig. 7.1(c) and (d). Observe that since the current-source load is assumed to be ideal, it is represented in the models by an infinite resistance. Practical current sources will have finite output resistance, as we shall see shortly. For the time being, however, note that the CS and CE amplifiers of Fig. 7.1 are in effect operating in an open-circuit fashion. The only resistance between their output node and ground is the output resistance of the transistor itself, r_o. Thus the voltage gain obtained in these circuits is the maximum possible for a CS or a CE amplifier.

From Fig. 7.1(c) we obtain for the active-loaded CS amplifier:

$$R_{\text{in}} = \infty \tag{7.1}$$

$$A_{vo} = -g_m r_o \tag{7.2}$$

$$R_o = r_o \tag{7.3}$$

Similarly, from Fig. 7.1(d) we obtain for the active-loaded CE amplifier:

$$R_{\text{in}} = r_\pi \tag{7.4}$$

$$A_{vo} = -g_m r_o \tag{7.5}$$

$$R_o = r_o \tag{7.6}$$

Thus both circuits realize a voltage gain of magnitude $g_m r_o$. Since this is the maximum gain obtainable in a CS or CE amplifier, we refer to it as the **intrinsic gain** and give it the symbol A_0. Furthermore, it is useful to examine the nature of A_0 in a little more detail.

7.2.2 The Intrinsic Gain

For the BJT, we can derive a formula for the intrinsic gain $A_{vo} = g_m r_o$ by using the following formulas for g_m and r_o:

$$g_m = \frac{I_C}{V_T} \tag{7.7}$$

$$r_o = \frac{V_A}{I_C} \tag{7.8}$$

The result is

$$A_0 = g_m r_o = \frac{V_A}{V_T} \tag{7.9}$$

Thus A_0 is simply the ratio of the Early voltage V_A, which is a technology-determined parameter, and the thermal voltage V_T, which is a physical parameter (approximately 0.025 V at room temperature). The value of V_A ranges from 5 V to 35 V for modern IC fabrication processes to 100 V to 130 V for the older, so-called high-voltage processes (see chapter appendix, Section 7.A.1). As a result, the value of A_0 will be in the range of 200 V/V to 5000 V/V, with the lower values characteristic of modern small-feature-size devices. It is important to note that for a given bipolar-transistor fabrication process, A_0 is independent of the transistor junction area and of its bias current. This is not the case for the MOSFET, as we shall now see.

Recall from our study of the MOSFET g_m in Section 5.5, that there are three possible expressions for g_m. Two of these are particularly useful for our purposes here:

$$g_m = \frac{I_D}{V_{OV}/2} \tag{7.10}$$

$$g_m = \sqrt{2\mu_n C_{ox}(W/L)}\ \sqrt{I_D} \tag{7.11}$$

For the MOSFET r_o we have

$$r_o = \frac{V_A}{I_D} = \frac{V_A' L}{I_D} \tag{7.12}$$

where V_A is the Early voltage and V_A' is the technology-dependent component of the Early voltage. Utilizing each of the g_m expressions together with the expression for r_o, we obtain for A_0,

$$A_0 = \frac{V_A}{V_{OV}/2} \tag{7.13}$$

which can be expressed in the alternate forms

$$A_0 = \frac{2V_A' L}{V_{OV}} \tag{7.14}$$

and

$$A_0 = \frac{V_A' \sqrt{2(\mu_n C_{ox})(WL)}}{\sqrt{I_D}} \tag{7.15}$$

The expression in Eq. (7.13) is the one most directly comparable to that of the BJT (Eq. 7.9). Here, however, we note the following:

1. The quantity in the denominator is $V_{OV}/2$, which is a design parameter. Although the value of V_{OV} that designers use for modern submicron technologies has been steadily decreasing, it is still about 0.15 V to 0.3 V. Thus $V_{OV}/2$ is 0.075 V to 0.15 V, which is 3 to 6 times higher than V_T. Furthermore, there are reasons for selecting higher values for V_{OV} (to be discussed in later chapters).
2. The numerator quantity is both process dependent (through V_A') and device dependent (through L), and its value has been steadily decreasing with the scaling down of the technology (see Appendix 7.A).
3. From Eq. (7.14) we see that for a given technology (i.e., a given value of V_A') the intrinsic gain A_0 can be increased by using a longer MOSFET and operating it at a lower V_{OV}. As usual, however, there are design trade-offs. For instance, we will see in Chapter 9 that increasing L and lowering V_{OV} result, independently, in decreasing the amplifier bandwidth.

As a result, the intrinsic gain realized in a MOSFET fabricated in a modern short-channel technology is only 20 V/V to 40 V/V, an order of magnitude lower than that for a BJT.

The alternative expression for the MOSFET A_0 given in Eq. (7.15) reveals a very interesting fact: For a given process technology (V_A' and $\mu_n C_{ox}$) and a given device (W and L), the intrinsic gain is inversely proportional to $\sqrt{I_D}$. This is illustrated in Fig. 7.2, which shows a typical plot for A_0 versus the bias current I_D. The plot confirms that the gain increases as the bias current is lowered. The gain, however, levels off at very low currents. This is because the MOSFET enters the **subthreshold region** of operation (Section 5.1.9), where it becomes very much like a BJT with an exponential current–voltage characteristic. The intrinsic gain then becomes constant, just like that of a BJT. Note, however, that although higher gain is obtained at lower values of I_D, the price paid is a lower g_m (Eq. 7.11), and less ability to drive capacitive loads and thus a decrease in bandwidth. This point will be studied in Chapter 9.

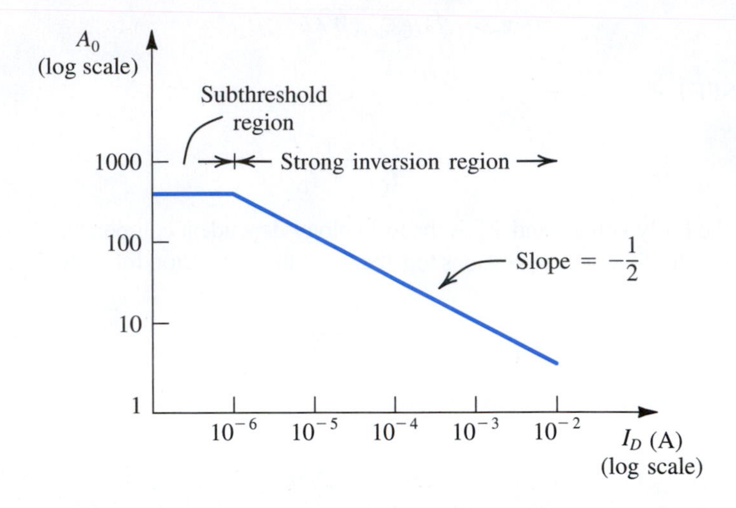

Figure 7.2 The intrinsic gain of the MOSFET versus bias current I_D. Outside the subthreshold region, this is a plot of $A_0 = V_A'\sqrt{2\mu_n C_{ox} WL/I_D}$ for the case: $\mu_n C_{ox} = 20\ \mu A/V^2$, $V_A' = 20\ V/\mu m$, $L = 2\ \mu m$, and $W = 20\ \mu m$.

Example 7.1

We wish to compare the values of g_m, R_{in}, R_o, and A_0 for a CS amplifier that is designed using an NMOS transistor with $L = 0.4\ \mu m$ and $W = 4\ \mu m$ and fabricated in a 0.25-μm technology specified to have $\mu_n C_{ox} = 267\ \mu A/V^2$ and $V_A' = 10\ V/\mu m$, with those for a CE amplifier designed using a BJT fabricated in a process with $\beta = 100$ and $V_A = 10$ V. Assume that both devices are operating at a drain (collector) current of 100 μA.

Solution

For simplicity, we shall neglect the Early effect in the MOSFET in determining V_{OV}; thus,

$$I_D = \frac{1}{2}(\mu_n C_{ox})\left(\frac{W}{L}\right)V_{OV}^2$$

$$100 = \frac{1}{2} \times 267 \times \left(\frac{4}{0.4}\right)V_{OV}^2$$

resulting in

$$V_{OV} = 0.27\ \text{V}$$

$$g_m = \frac{2I_D}{V_{OV}} = \frac{2 \times 0.1}{0.27} = 0.74\ \text{mA/V}$$

$$R_{in} = \infty$$

$$r_o = \frac{V_A' L}{I_D} = \frac{10 \times 0.4}{0.1} = 40\ \text{k}\Omega$$

$$R_o = r_o = 40\ \text{k}\Omega$$

$$A_0 = g_m r_o = 0.74 \times 40 = 29.6\ \text{V/V}$$

For the CE amplifier we have

$$g_m = \frac{I_C}{V_T} = \frac{0.1 \text{ mA}}{0.025 \text{ V}} = 4 \text{ mA/V}$$

$$R_{in} = r_\pi = \frac{\beta}{g_m} = \frac{100}{4} = 25 \text{ k}\Omega$$

$$r_o = \frac{V_A}{I_C} = \frac{10}{0.1} = 100 \text{ k}\Omega$$

$$R_o = r_o = 100 \text{ k}\Omega$$

$$A_0 = g_m r_o = 4 \times 100 = 400 \text{ V/V}$$

EXERCISE

7.1 A CS amplifier utilizes an NMOS transistor with $L = 0.36$ μm and $W/L = 10$; it was fabricated in a 0.18-μm CMOS process for which $\mu_n C_{ox} = 387$ μA/V² and $V_A' = 5$ V/μm. Find the values of g_m and A_0 obtained at $I_D = 10$ μA, 100 μA, and 1 mA.
Ans. 0.28 mA/V, 50 V/V; 0.88 mA/V, 15.8 V/V; 2.78 mA/V, 5 V/V

7.2.3 Effect of the Output Resistance of the Current-Source Load

The current-source load of the CS amplifier in Fig. 7.1(a) can be implemented using a PMOS transistor biased in the saturation region to provide the required current I, as shown in Fig. 7.3(a). We can use the large-signal MOSFET model (Section 5.2, Fig. 5.15) to model Q_2 as shown in Fig. 7.3(b), where

$$I = \frac{1}{2}(\mu_p C_{ox})\left(\frac{W}{L}\right)_2 [V_{DD} - V_G - |V_{tp}|]^2 \tag{7.16}$$

and

$$r_{o2} = \frac{|V_{A2}|}{I} \tag{7.17}$$

Thus the current-source load no longer has an infinite resistance; rather, it has a finite output resistance r_{o2}. This resistance will in effect appear in parallel with r_{o1}, as shown in the amplifier equivalent-circuit model in Fig. 7.3(c), from which we obtain

$$A_v \equiv \frac{v_o}{v_i} = -g_{m1}(r_{o1} \| r_{o2}) \tag{7.18}$$

Thus, not surprisingly, the finite output resistance of the current-source load reduces the magnitude of the voltage gain from $(g_{m1} r_{o1})$ to $g_{m1}(r_{o1} \| r_{o2})$. This reduction can be substantial. For instance, if Q_2 has an Early voltage equal to that of Q_1, $r_{o2} = r_{o1}$ and the gain is reduced by half,

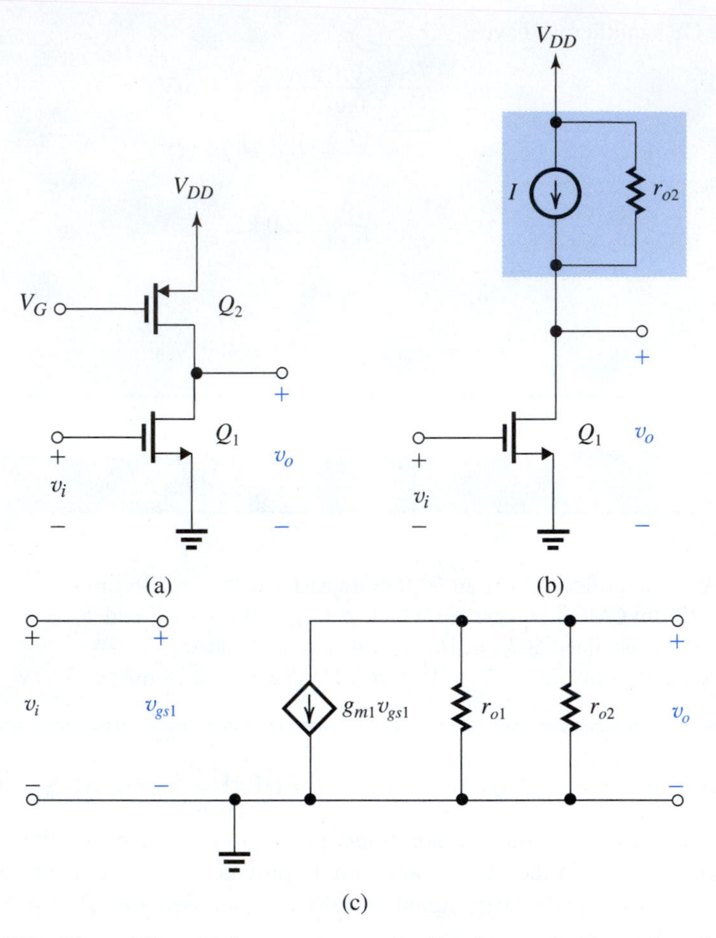

Figure 7.3 **(a)** The CS amplifier with the current-source load implemented with a p-channel MOSFET Q_2; **(b)** the circuit with Q_2 replaced with its large-signal model; and **(c)** small-signal equivalent circuit of the amplifier.

$$A_v = -\frac{1}{2}g_m r_o \tag{7.18'}$$

Finally, we note that a similar development can be used for the bipolar case.

Example 7.2

A practical circuit implementation of the common-source amplifier is shown in Fig. 7.4(a). Here the current-source transistor Q_2 is the output transistor of a current mirror formed by Q_2 and Q_3 and fed with a reference current I_{REF}. Current mirrors were briefly introduced in Section 5.7.4 and will be studied more extensively in Sections 7.4 and 7.5. For the time being, assume that Q_2 and Q_3 are matched. Also assume that I_{REF} is a stable, well-predicted current that is generated with a special circuit on the chip. To be able to clearly see the region of v_I over which the circuit operates as an almost-linear amplifier, determine the voltage transfer characteristic (VTC), that is, v_O versus v_I.

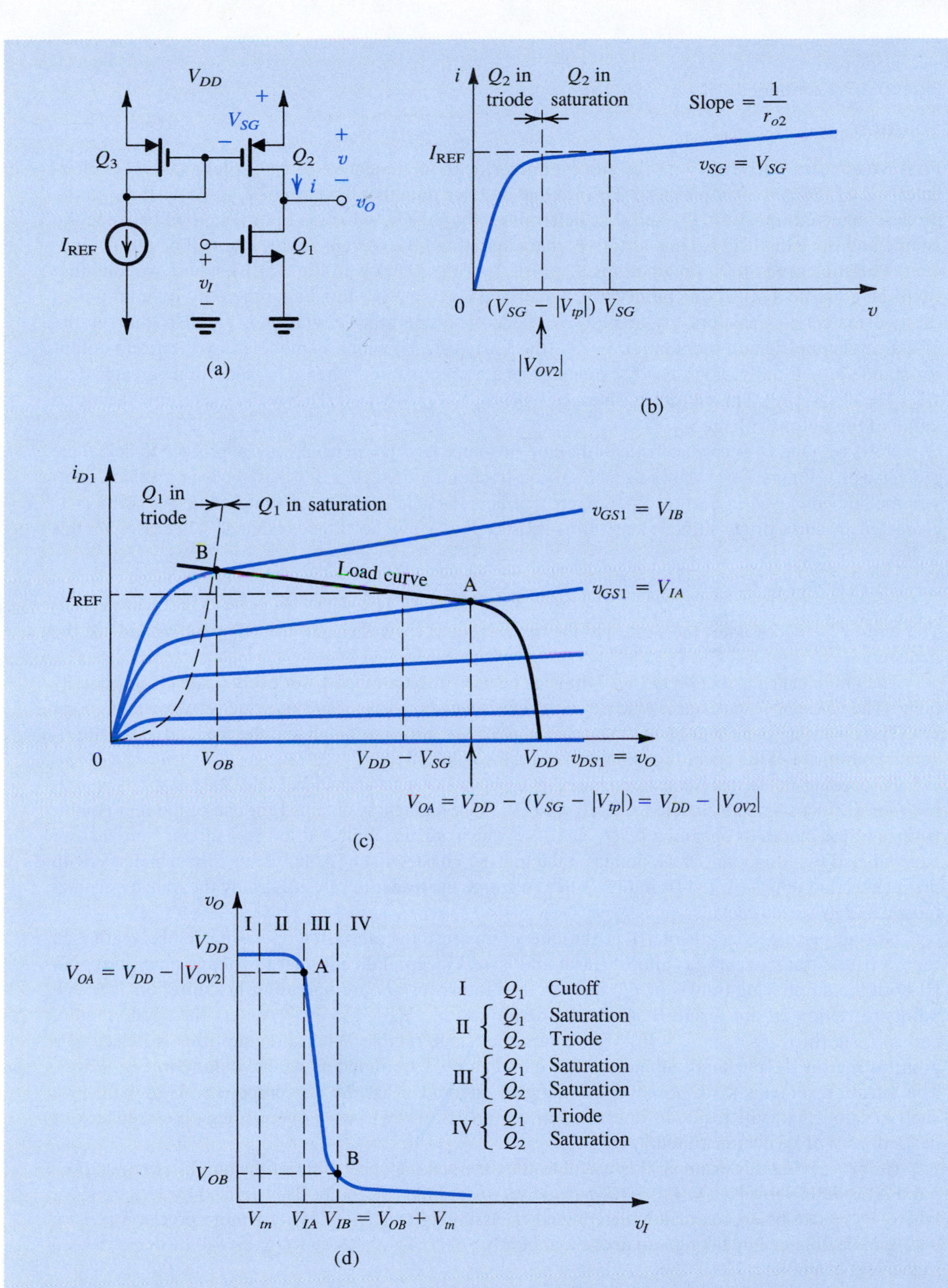

Figure 7.4 Practical implementation of the common-source amplifier: **(a)** circuit; **(b)** i–v characteristic of the active-load Q_2; **(c)** graphical construction to determine the transfer characteristic; **(d)** transfer characteristic.

Example 7.2 *continued*

Solution

First we concern ourselves with the current mirror, with the objective of determining the $i-v$ characteristic of the current source Q_2. Toward that end, we note that the current I_{REF} flows through the diode-connected transistor Q_3 and thus determines V_{SG} of Q_3, which is in turn applied between the source and the gate of Q_2. Thus, the $i-v$ characteristic of the current source Q_2 will be the i_D-v_{SD} characteristic curve of Q_2 obtained for $v_{SG} = V_{SG}$. This is shown in Fig. 7.4(b), where we note that i will be equal to I_{REF} at one point only, namely, at $v_{SD2} = V_{SG}$, this being the only point at which the two matched transistors Q_2 and Q_3 have identical operating conditions. We also observe the effect of channel-length modulation in Q_2 (the Early effect), which is modeled by the finite output resistance r_{o2}. Finally, note that Q_2 operates as a current source when v is equal to or greater than $|V_{OV2}| = V_{SG} - |V_{tp}|$. This in turn is obtained when $v_O \leq V_{DD} - |V_{OV2}|$. This is the maximum permitted value of the output voltage v_O.

Now, with the $i-v$ characteristic of the current-source load Q_2 in hand, we can proceed to determine v_O versus v_I. Figure 7.4(c) shows a graphical construction for doing this. It is based on the graphical analysis method employed in Section 5.4.5 except that here the load line is not a straight line but is the $i-v$ characteristic curve of Q_2 shifted along the v_O axis by V_{DD} volts and "flipped around." The reason for this is that

$$v_O = V_{DD} - v$$

The term V_{DD} necessitates the shift, and the minus sign of v gives rise to the "flipping around" of the load curve.

The graphical construction of Fig. 7.4(c) can be used to determine v_O for every value of v_I, point by point: The value of v_I determines the particular characteristic curve of Q_1 on which the operating point lies. The operating point will be at the intersection of this particular graph and the load curve. The horizontal coordinate of the operating point then gives the value of v_O.

Proceeding in the manner just explained, we obtain the VTC shown in Fig. 7.4(d). As indicated, it has four distinct segments, labeled I, II, III, and IV. Each segment is obtained for one of the four combinations of the modes of operation of Q_1 and Q_2, which are also indicated in the diagram. Note that we have labeled two important break points on the transfer characteristic (A and B) in correspondence with the intersection points (A and B) in Fig. 7.4(c). We urge the reader to carefully study the transfer characteristic and its various details.

Not surprisingly, segment III is the one of interest for amplifier operation. Observe that in region III the transfer curve is almost linear and is very steep, indicating large voltage gain. In region III both the amplifying transistor Q_1 and the load transistor Q_2 are operating in saturation. The end points of region III are A and B: At A, defined by $v_O = V_{DD} - |V_{OV2}|$, Q_2 enters the triode region, and at B, defined by $v_O = v_I - V_{tn}$, Q_1 enters the triode region. When the amplifier is biased at a point in region III, the small-signal voltage gain can be determined as we have done in Fig. 7.3(c). The question remains as to how we are going to guarantee that the dc component of v_I will have such a value that will result in operation in region III. That is why overall negative feedback is needed, as will be demonstrated later.

Before leaving this example it is useful to reiterate that the upper limit of the amplifier region (i.e., point A) is defined by $V_{OA} = V_{DD} - |V_{OV2}|$ and the lower limit (i.e., point B) is defined by $V_{OB} = V_{OV1}$, where V_{OV1} can be approximately determined by assuming that $I_{D1} \simeq I_{REF}$. A more precise value for V_{OB} can be obtained by taking into account the Early effect in both Q_1 and Q_2, as will be demonstrated in the next example.

Example 7.3

Consider the CMOS common-source amplifier in Fig. 7.4(a) for the case $V_{DD} = 3$ V, $V_{tn} = |V_{tp}| = 0.6$ V, $\mu_n C_{ox} = 200\ \mu\text{A/V}^2$, and $\mu_p C_{ox} = 65\ \mu\text{A/V}^2$. For all transistors, $L = 0.4\ \mu$m and $W = 4\ \mu$m. Also, $V_{An} = 20$ V, $|V_{Ap}| = 10$ V, and $I_{REF} = 100\ \mu$A. Find the small-signal voltage gain. Also, find the coordinates of the extremities of the amplifier region of the transfer characteristic—that is, points A and B.

Solution

$$g_{m1} = \sqrt{2k_n'\left(\frac{W}{L}\right)_1 I_{REF}}$$

$$= \sqrt{2 \times 200 \times \frac{4}{0.4} \times 100} = 0.63 \text{ mA/V}$$

$$r_{o1} = \frac{V_{An}}{I_{D1}} = \frac{20 \text{ V}}{0.1 \text{ mA}} = 200 \text{ k}\Omega$$

$$r_{o2} = \frac{|V_{Ap}|}{I_{D2}} = \frac{10 \text{ V}}{0.1 \text{ mA}} = 100 \text{ k}\Omega$$

Thus,

$$A_v = -g_{m1}(r_{o1} \| r_{o2})$$

$$= -0.63(\text{mA/V}) \times (200 \| 100)(\text{k}\Omega) = -42 \text{ V/V}$$

Approximate values for the extremities of the amplifier region of the transfer characteristic (region III) can be determined as follows: Neglecting the Early effect, all three transistors are carrying equal currents I_{REF}, and thus we can determine the overdrive voltages at which they are operating. Transistors Q_2 and Q_3 will have equal overdrive voltages, $|V_{OV3}|$, determined from

$$I_{D3} = I_{REF} \simeq \frac{1}{2}(\mu_p C_{ox})\left(\frac{W}{L}\right)_3 |V_{OV3}|^2$$

Substituting, $I_{REF} = 100\ \mu$A, $\mu_p C_{ox} = 65\ \mu\text{A/V}^2$, $(W/L)_3 = 4/0.4 = 10$ results in

$$|V_{OV3}| = 0.55 \text{ V}$$

Thus,

$$V_{OA} = V_{DD} - |V_{OV3}| = 2.45 \text{ V}$$

Next we determine $|V_{OV1}|$ from

$$I_{D1} \simeq I_{REF} \simeq \frac{1}{2}(\mu_n C_{ox})\left(\frac{W}{L}\right)_1 V_{OV1}^2$$

Substituting, $I_{REF} = 100\ \mu$A, $\mu_n C_{ox} = 200\ \mu\text{A/V}^2$, $(W/L)_1 = 4/0.4 = 10$ results in

$$V_{OV1} = 0.32 \text{ V}$$

Thus, $V_{OB} = V_{OV1} = 0.32$ V.

More precise values for V_{OA} and V_{OB} can be determined by taking the Early effect in all transistors into account as follows.

Example 7.3 *continued*

First, we determine V_{SG} of Q_2 and Q_3 corresponding to $I_{D3} = I_{REF} = 100\ \mu A$ using

$$I_{D3} = \frac{1}{2}k_p'\left(\frac{W}{L}\right)_3 (V_{SG} - |V_{tp}|)^2\left(1 + \frac{V_{SD}}{|V_{Ap}|}\right)$$

Thus,

$$100 = \frac{1}{2} \times 65\left(\frac{4}{0.4}\right)|V_{OV3}|^2\left(1 + \frac{0.6 + |V_{OV3}|}{10}\right) \tag{7.19}$$

where $|V_{OV3}|$ is the magnitude of the overdrive voltage at which Q_3 and Q_2 are operating, and we have used the fact that, for Q_3, $V_{SD} = V_{SG}$. Equation (7.19) can be manipulated to the form

$$0.29 = |V_{OV3}|^2(1 + 0.09|V_{OV3}|)$$

which by a trial-and-error process yields

$$|V_{OV3}| = 0.53 \text{ V}$$

Thus,

$$V_{SG} = 0.6 + 0.53 = 1.13 \text{ V}$$

and

$$V_{OA} = V_{DD} - V_{OV3} = 2.47 \text{ V}$$

To find the corresponding value of v_I, V_{IA}, we derive an expression for v_O versus v_I in region III. Noting that in region III, Q_1 and Q_2 are in saturation and obviously conduct equal currents, we can write

$$i_{D1} = i_{D2}$$

$$\frac{1}{2}k_n'\left(\frac{W}{L}\right)_1 (v_I - V_{tn})^2\left(1 + \frac{v_O}{|V_{An}|}\right) = \frac{1}{2}k_p'\left(\frac{W}{L}\right)_2 (V_{SG} - |V_{tp}|)^2\left(1 + \frac{V_{DD} - v_O}{|V_{Ap}|}\right)$$

Substituting numerical values, we obtain

$$8.55(v_I - 0.6)^2 = \frac{1 - 0.08v_O}{1 + 0.05v_O} \simeq (1 - 0.13v_O)$$

which can be manipulated to the form

$$v_O = 7.69 - 65.77(v_I - 0.6)^2 \tag{7.20}$$

This is the equation of segment III of the transfer characteristic. Although it includes v_I^2, the reader should not be alarmed: Because region III is very narrow, v_I changes very little, and the characteristic is nearly linear. Substituting $v_O = 2.47$ V gives the corresponding value of v_I; that is, $V_{IA} = 0.88$ V. To determine the coordinates of B, we note that they are related by $V_{OB} = V_{IB} - V_{tn}$. Substituting in Eq. (7.20) and solving gives $V_{IB} = 0.93$ V and $V_{OB} = 0.33$ V. The width of the amplifier region is therefore

$$\Delta v_I = V_{IB} - V_{IA} = 0.05 \text{ V}$$

and the corresponding output range is

$$\Delta v_O = V_{OB} - V_{OA} = -2.14 \text{ V}$$

Thus, the "large-signal" voltage gain is

$$\frac{\Delta v_O}{\Delta v_I} = -\frac{2.14}{0.05} = -42.8 \text{ V/V}$$

which is very close to the small-signal value of –42, indicating that segment III of the transfer characteristic is quite linear.

EXERCISES

7.2 A CMOS common-source amplifier such as that in Fig. 7.4(a), fabricated in a 0.18-μm technology, has $W/L = 7.2\ \mu\text{m}/0.36\ \mu\text{m}$ for all transistors, $k_n' = 387\ \mu\text{A/V}^2$, $k_p' = 86\ \mu\text{A/V}^2$, $I_{\text{REF}} = 100\ \mu\text{A}$, $V_{An}' = 5\ \text{V}/\mu\text{m}$, and $|V_{Ap}'| = 6\ \text{V}/\mu\text{m}$. Find g_{m1}, r_{o1}, r_{o2}, and the voltage gain.
Ans. 1.25 mA/V; 18 kΩ; 21.6 kΩ; −12.3 V/V

7.3 Consider the active-loaded CE amplifier when the constant-current source I is implemented with a *pnp* transistor. Let $I = 0.1$ mA, $|V_A| = 50$ V (for both the *npn* and the *pnp* transistors), and $\beta = 100$. Find R_{in}, r_o (for each transistor), g_m, A_0, and the amplifier voltage gain.
Ans. 25 kΩ; 0.5 MΩ; 4 mA/V; 2000 V/V; −1000 V/V

7.2.4 Increasing the Gain of the Basic Cell

We conclude this section by considering a question: How can we increase the voltage gain obtained from the basic gain cell? The answer lies in finding a way to raise the level of the output resistance of both the amplifying transistor and the load transistor. That is, we seek a circuit that passes the current $g_m v_i$ provided by the amplifying transistor right through, but increases the resistance from r_o to a much larger value. This requirement is illustrated in Fig. 7.5. Figure 7.5(a) shows the CS amplifying transistor Q_1 together with its output equivalent circuit. Note that for the time being we are not showing the load device. In Fig. 7.5(b) we have inserted a shaded box between the drain of Q_1 and a new output terminal labeled d_2. Here again we are not showing the load to which d_2 will be connected. Our "black box" takes in the output current of Q_1 and passes it to the output; thus at its output we have the equivalent circuit shown, consisting of the same controlled source $g_{m1}v_i$ but with the output resistance increased by a factor K.

Now, what does the black box really do? Since it passes the current but *raises* the resistance level, it is a **current buffer**. It is the dual of the voltage buffer (the source and emitter followers), which passes the *voltage* but *lowers* the resistance level.

Now searching our repertoire of transistor amplifier configurations studied in Sections 5.6 and 6.6, the only candidate for implementing this current-buffering action is the common-gate (or common-base in bipolar) amplifier. Indeed, recall that the CG and CB circuits have a unity current gain. What we have not yet investigated, however, is their resistance transformation property. We shall do this in the next section.

Two important final comments:

1. It is not sufficient to raise the output resistance of the amplifying transistor only. We also need to raise the output resistance of the current-source load. Obviously, we can use a current buffer to do this also.
2. Placing a CG (or a CB) circuit on top of the CS (or CE) amplifying transistor to implement the current-buffering action is called **cascoding**. We will explain the origin of this name shortly.

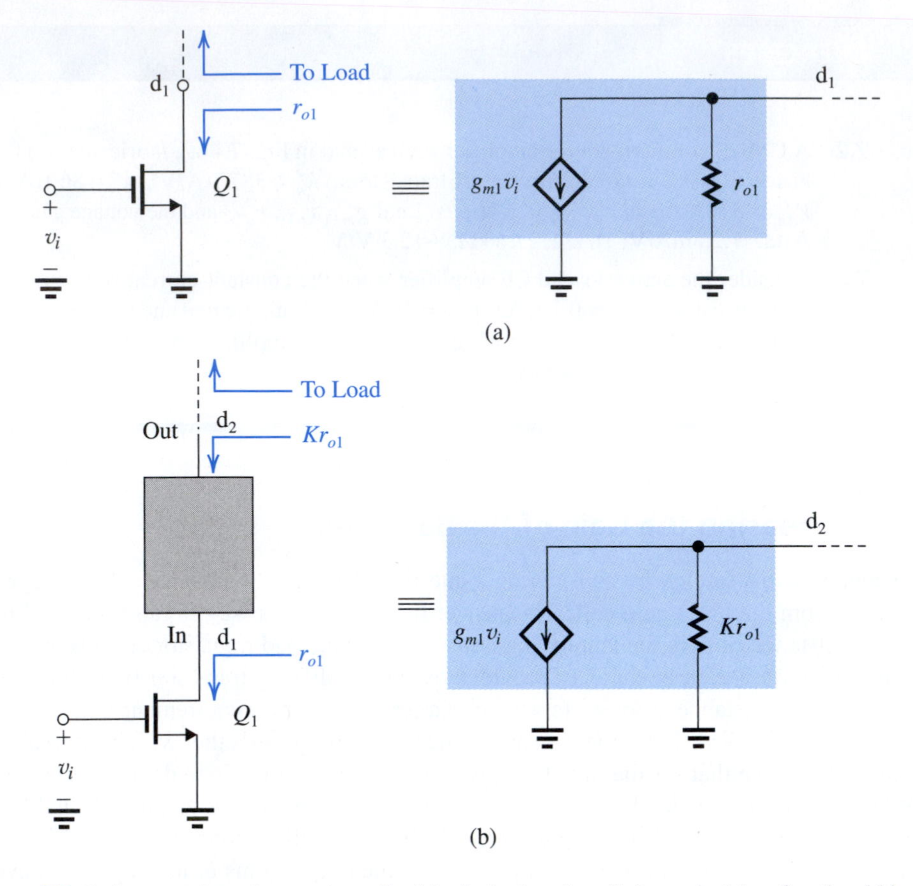

Figure 7.5 To increase the voltage gain realized in the basic gain cell shown in **(a)**, a functional block, shown as a black box in **(b)**, is connected between d_l and the load. This new block is required to pass the current $g_{m1}v_i$ right through but raise the resistance level by a factor K. The functional block is a current buffer.

7.3 The Cascode Amplifier

7.3.1 Cascoding

Cascoding refers to the use of a transistor connected in the common-gate (or the common-base) configuration to provide **current buffering** for the output of a common-source (or a common-emitter) amplifying transistor. Figure 7.6 illustrates the technique for the MOS case. Here the CS transistor Q_1 is the amplifying transistor and Q_2, connected in the CG configuration with a dc bias voltage V_{G2} (signal ground) at its gate, is the cascode transistor.[1] A similar arrangement applies for the bipolar case and will be considered later.

We will show in the following that the equivalent circuit at the output of the cascode amplifier is that shown in Fig. 7.6. Thus, the cascode transistor passes the current $g_{m1}v_i$ to the output node while raising the resistance level by a factor K. We will derive an expression for K.

[1]The name *cascode* is a carryover from the days of vacuum tubes and is a shortened version of "*casc*aded cath*ode*"; in the tube version, the anode of the amplifying tube (corresponding to the drain of Q_1) feeds the cathode of the cascode tube (corresponding to the source of Q_2).

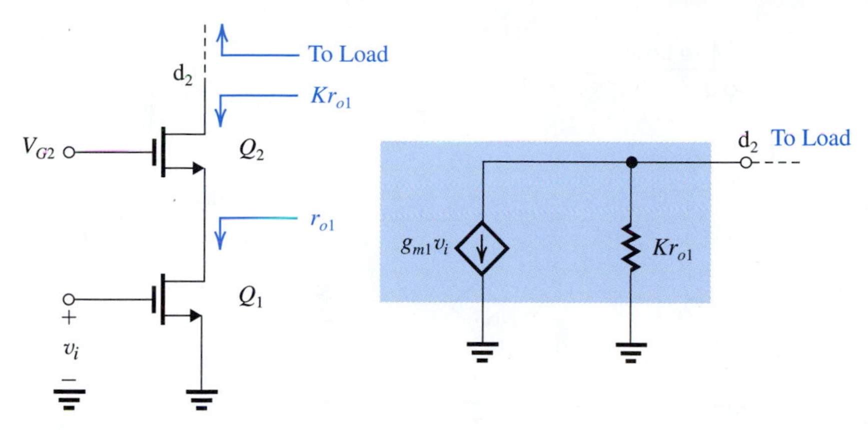

Figure 7.6 The current-buffering action of Fig. 7.5(a) is implemented using a transistor Q_2 connected in the CG configuration. Here V_{G2} is a dc bias voltage. The output equivalent circuit indicates that the CG transistor passes the current $g_{m1}v_i$ through but raises the resistance level by a factor K. Transistor Q_2 is called a cascode transistor.

7.3.2 The MOS Cascode

Figure 7.7(a) shows the MOS cascode amplifier without a load circuit and with the gate of Q_2 connected to signal ground. Thus this circuit is for the purpose of small-signal calculations only. Our objective is to determine the parameters G_m and R_o of the equivalent circuit shown in Fig. 7.7(b), which we shall use to represent the output of the cascode amplifier. Toward that end, observe that if node d_2 of the equivalent circuit is short-circuited to ground, the current flowing through the short circuit will be equal to $G_m v_i$. It follows that we can determine G_m by short-circuiting (from a signal point of view) the output of the cascode amplifier to ground, as shown in Fig. 7.7(c), determine i_o, and then

$$G_m = \frac{i_o}{v_i}$$

Now, replacing Q_1 and Q_2 in the circuit of Fig. 7.7(c) with their small-signal models results in the circuit in Fig. 7.7(d), which we shall analyze to determine i_o in terms of v_i.

Observe that the voltage at the (d_1, s_2) node is equal to $-v_{gs2}$. Writing a node equation for that node, we have

$$g_{m2}v_{gs2} + \frac{v_{gs2}}{r_{o1}} + \frac{v_{gs2}}{r_{o2}} = g_{m1}v_i$$

Thus,

$$\left(g_{m2} + \frac{1}{r_{o1}} + \frac{1}{r_{o2}}\right)v_{gs2} = g_{m1}v_i$$

Since $g_{m2} \gg (1/r_{o1}),\ 1/r_{o2}$,

$$g_{m2}v_{gs2} \simeq g_{m1}v_i \tag{7.21}$$

In other words, the current of the controlled source of Q_2 is equal to that of the controlled source of Q_1. Next, we write an equation for the d_2 node,

$$i_o = g_{m2}v_{gs2} + \frac{v_{gs2}}{r_{o2}}$$

$$= \left(g_{m2} + \frac{1}{r_{o2}}\right)v_{gs2}$$

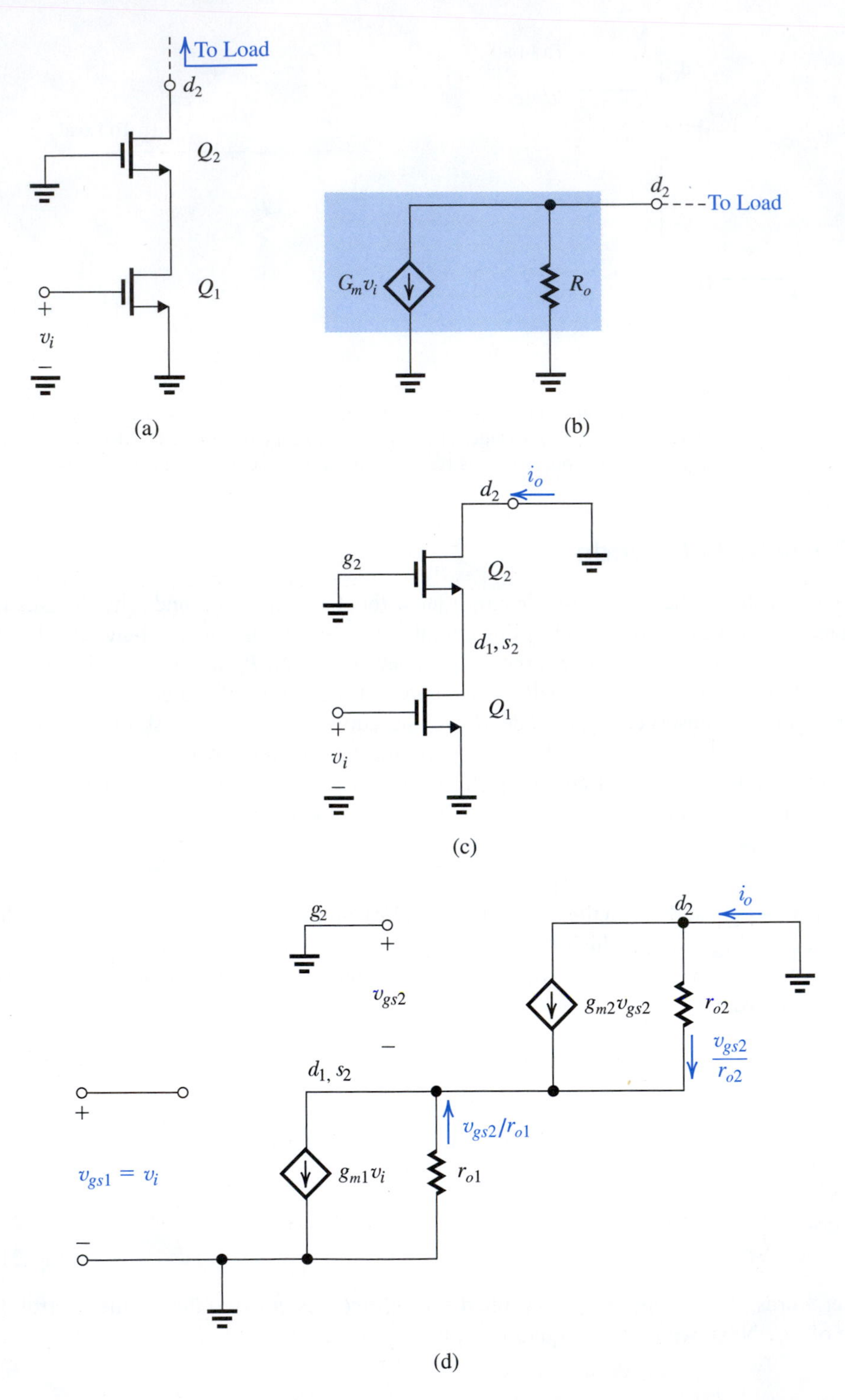

Figure 7.7 (**a**) A MOS cascode amplifier prepared for small-signal calculations; (**b**) output equivalent circuit of the amplifier in (**a**); (**c**) the cascode amplifier with the output short-circuited to determine $G_m \equiv i_o v_i$; (**d**) equivalent circuit of the situation in (**c**).

Thus,

$$i_o \simeq g_{m2}v_{gs2}$$

Using Eq. (7.21) results in

$$i_o = g_{m1}v_i$$

Thus,

$$G_m = \frac{i_o}{v_i} = g_{m1} \tag{7.22}$$

which is the result we have anticipated.

Next we need to determine R_o. For this purpose we set v_i to zero, which results in Q_1 simply reduced to its output resistance r_{o1}, which appears in the source circuit of Q_2, as shown in Fig. 7.8(a). Now, replacing Q_2 with its hybrid-π model and applying a test voltage v_x to the output node results in the equivalent circuit shown in Fig. 7.8(b). The output resistance R_o can be obtained as

$$R_o \equiv \frac{v_x}{i_x}$$

Analysis of the circuit is greatly simplified by noting that the current exiting the source node of Q_2 is equal to i_x. Thus, the voltage at the source node, which is $-v_{gs2}$, can be expressed in terms of i_x as

$$-v_{gs2} = i_x r_{o1} \tag{7.23}$$

Next we express v_x as the sum of the voltages across r_{o2} and r_{o1} as

$$v_x = (i_x - g_{m2}v_{gs2})\, r_{o2} + i_x r_{o1}$$

Substituting for v_{gs2} from Eq. (7.23) results in

$$v_x = i_x(r_{o1} + r_{o2} + g_{m2}r_{o2}r_{o1})$$

Thus, $R_o \equiv v_x/i_x$ is given by

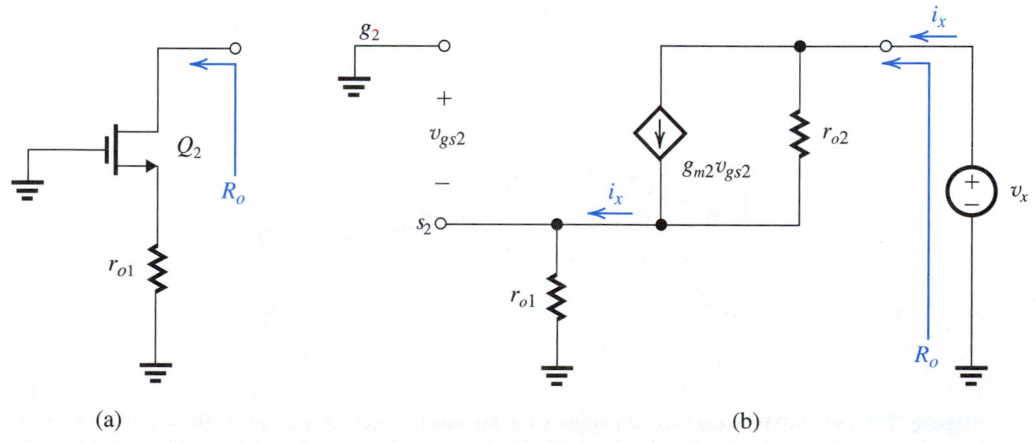

Figure 7.8 Determining the output resistance of the MOS cascode amplifier.

$$R_o = r_{o1} + r_{o2} + g_{m2} r_{o2} r_{o1} \tag{7.24}$$

In this expression the last term will dominate, thus

$$R_o \simeq (g_{m2} r_{o2}) r_{o1} \tag{7.25}$$

This expression has a simple and elegant interpretation: The CG transistor Q_2 raises the output resistance of the amplifier by the factor $(g_{m2} r_{o2})$, which is its intrinsic gain. At the same time, the CG transistor simply passes the current $(g_{m1} v_i)$ to the output node. Thus the CG or cascode transistor very effectively realizes the objectives we set for the current buffer (refer to Figs. 7.5 and 7.6) with $K = A_{02} = g_{m2} r_{o2}$.

Voltage Gain If the cascode amplifier is loaded with an ideal constant-current source as shown in Fig. 7.9(a), the voltage gain realized can be found from the equivalent circuit in Fig. 7.9(b) as

$$A_{vo} = \frac{v_o}{v_i} = -g_{m1} R_o$$

Thus,

$$A_{vo} = -(g_{m1} r_{o1})(g_{m2} r_{o2}) \tag{7.26}$$

For the case $g_{m1} = g_{m2} = g_m$ and $r_{o1} = r_{o2} = r_o$,

$$\begin{aligned} A_{vo} &= -(g_m r_o)^2 \\ &= -A_0^2 \end{aligned} \tag{7.27}$$

Thus cascoding results in increasing the gain magnitude from A_0 to A_0^2.

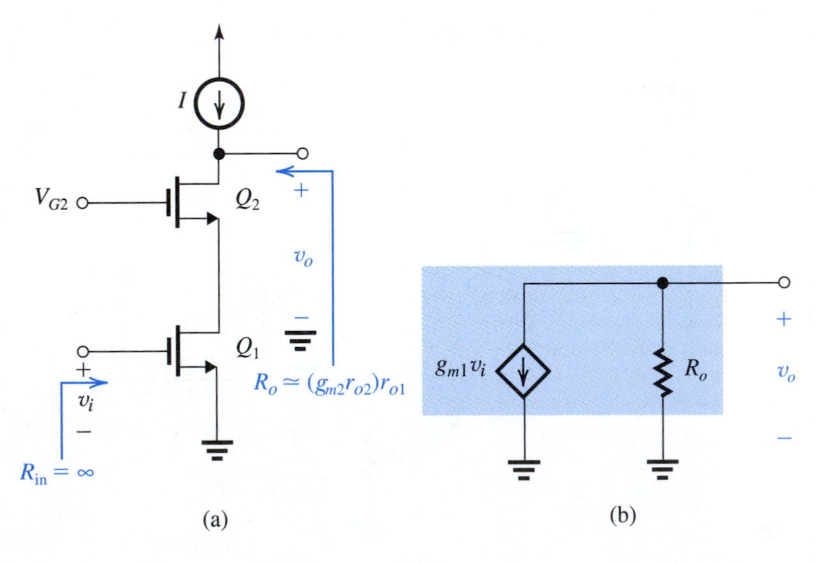

Figure 7.9 **(a)** A MOS cascode amplifier with an ideal current-source load; **(b)** equivalent circuit representation of the cascode output.

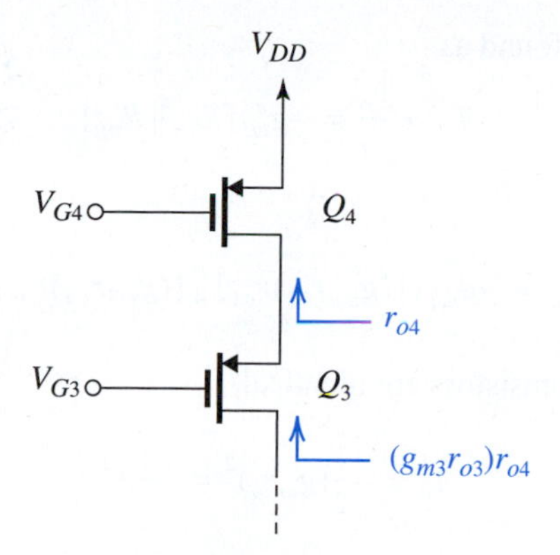

Figure 7.10 Employing a cascode transistor Q_3 to raise the output resistance of the current source Q_4.

Cascoding can also be employed to raise the output resistance of the current-source load as shown in Fig. 7.10. Here Q_4 is the current-source transistor, and Q_3 is the CG cascode transistor. Voltages V_{G3} and V_{G4} are dc bias voltages. The cascode transistor Q_3 multiplies the output resistance of Q_4, r_{o4} by $(g_{m3}r_{o3})$ to provide an output resistance for the cascode current source of

$$R_o = (g_{m3}r_{o3})r_{o4} \tag{7.28}$$

Combining a cascode amplifier with a cascode current source results in the circuit of Fig. 7.11(a). The equivalent circuit at the output side is shown in Fig. 7.11(b), from which the

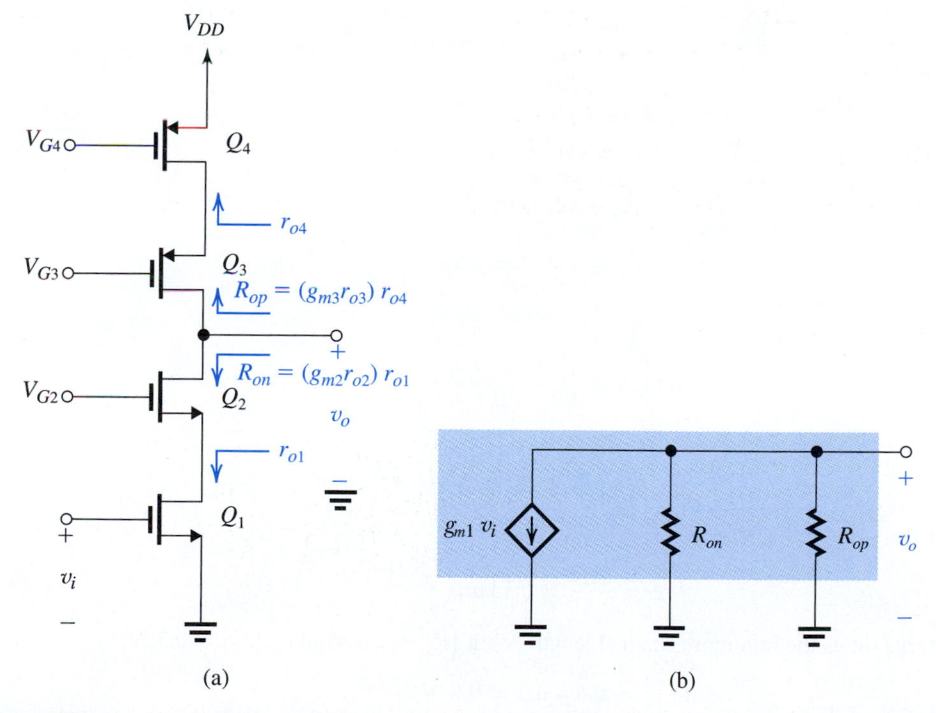

Figure 7.11 A cascode amplifier with a cascode current-source load.

voltage gain can be easily found as

$$A_v = \frac{v_o}{v_i} = -g_{m1}[R_{on} \| R_{op}]$$

Thus,

$$A_v = -g_{m1}\{[(g_{m2}r_{o2})r_{o1}] \| [(g_{m3}r_{o3})r_{o4}]\} \tag{7.29}$$

For the case in which all transistors are identical,

$$A_v = -\frac{1}{2}(g_m r_o)^2 = -\frac{1}{2}A_0^2 \tag{7.30}$$

By comparison to the gain expression in Eq. (7.18′), we see that using the cascode configuration for both the amplifying transistor and the current-source load transistor results in an increase in the magnitude of gain by a factor equal to A_0.

Example 7.4

It is required to design the cascode current-source of Fig. 7.10 to provide a current of 100 μA and an output resistance of 500 kΩ. Assume the availability of a 0.18-μm CMOS technology for which $V_{DD} = 1.8$ V, $V_{tp} = -0.5$ V, $\mu_p C_{ox} = 90$ μA/V^2 and $V_A' = -5$ V/μm. Use $|V_{OV}| = 0.3$ V and determine L and W/L for each transistor, and the values of the bias voltages V_{G3} and V_{G4}.

Solution

The output resistance R_o is given by

$$R_o = (g_{m3}r_{o3})r_{o4}$$

Assuming Q_3 and Q_4 are identical,

$$R_o = (g_m r_o)r_o = \frac{|V_A|}{|V_{OV}|/2} \times \frac{|V_A|}{I_D}$$

Using $|V_{OV}| = 0.3$ V, we write

$$500 \text{ k}\Omega = \frac{|V_A|}{0.15} \times \frac{|V_A|}{0.1 \text{ mA}}$$

Thus we require

$$|V_A| = 2.74 \text{ V}$$

Now, since $|V_A| = |V_A'|L$ we need to use a channel length of

$$L = \frac{2.74}{5} = 0.55 \text{ μm}$$

which is about three times the minimum channel length. With $|V_t| = 0.5$ V and $|V_{OV}| = 0.3$ V,

$$V_{SG4} = 0.5 + 0.3 = 0.8 \text{ V}$$

and thus,

$$V_{G4} = 1.8 - 0.8 = 1.0 \text{ V}$$

To allow for the largest possible signal swing at the output terminal, we shall use the minimum required voltage across Q_4, namely, $|V_{OV}|$ or 0.3 V. Thus,

$$V_{D4} = 1.8 - 0.3 = 1.5 \text{ V}$$

Since the two transistors are identical and are carrying equal currents,

$$V_{SG3} = V_{SG4} = 0.8 \text{ V}$$

Thus,

$$V_{G3} = 1.5 - 0.8 = +0.7 \text{ V}$$

We note that the maximum voltage allowed at the output terminal of the current source will be constrained by the need to allow a minimum voltage of $|V_{OV}|$ across Q_3; thus,

$$v_{D3\max} = 1.5 - 0.3 = +1.2 \text{ V}$$

To determine the required W/L ratios of Q_3 and Q_4, we use

$$I_D = \frac{1}{2}(\mu_p C_{ox})\left(\frac{W}{L}\right)|V_{OV}|^2\left(1 + \frac{V_{SD}}{|V_A|}\right)$$

$$100 = \frac{1}{2} \times 90 \times \left(\frac{W}{L}\right) \times 0.3^2\left(1 + \frac{0.3}{2.74}\right)$$

which yields

$$\frac{W}{L} = 22.3$$

EXERCISES

D7.4 If in Example 7.4, L of each of Q_3 and Q_4 is halved while W/L is changed to allow I_D and V_{OV} to remain unchanged, find the new values of R_o and W/L. [*Hint*: In computing the required (W/L), note that $|V_A|$ has changed.]
Ans. 125 kΩ; 20.3

7.5 Consider the cascode amplifier of Fig. 7.11 with the dc component at the input, $V_I = 0.7$ V, $V_{G2} = 1.0$ V, $V_{G3} = 0.8$ V, $V_{G4} = 1.1$ V, and $V_{DD} = 1.8$ V. If all devices are matched (i.e., if $k_{n1} = k_{n2} = k_{p3} = k_{p4}$), and have equal $|V_t|$ of 0.5 V, what is the overdrive voltage at which the four transistors are operating? What is the allowable voltage range at the output?
Ans. 0.2 V; 0.5 V to 1.3 V

7.6 The cascode amplifier in Fig. 7.11 is operated at a current of 0.2 mA with all devices operating at $|V_{OV}| = 0.2$ V. All devices have $|V_A| = 2$ V. Find g_{m1}, the output resistance of the amplifier, R_{on}, and the output resistance of the current source, R_{op}. Also find the overall output resistance and the voltage gain realized.
Ans. 2 mA/V; 200 kΩ, 200 kΩ; 100 kΩ; –200 V/V

7.3.3 Distribution of Voltage Gain in a Cascode Amplifier

It is often useful to know how much of the overall voltage gain of a cascode amplifier is realized in each of its two stages: the CS stage Q_1, and the CG stage Q_2. For this purpose, consider the cascode amplifier shown in Fig. 7.12(a). Here, for generality we have included a load resistance R_L, which represents the output resistance of the current-source load plus any additional resistance that may be connected to the output node. Recalling that the cascode amplifier output can be represented with the equivalent circuit of Fig. 7.7(b), where $G_m = g_{m1}$ and $R_o = (g_{m2}r_{o2})r_{o1}$, the voltage gain A_v of the amplifier in Fig. 7.12(a) can be found as

$$A_v = -g_{m1}(R_o \,\|\, R_L)$$

Thus,

$$A_v = -g_{m1}(g_{m2}r_{o2}r_{o1} \,\|\, R_L) \tag{7.31}$$

The overall gain A_v can be expressed as the product of the voltage gains of Q_1 and Q_2 as

$$A_v = A_{v1}A_{v2} = \left(\frac{v_{o1}}{v_i}\right)\left(\frac{v_o}{v_{o1}}\right) \tag{7.32}$$

To obtain $A_{v1} \equiv v_{o1}/v_i$ we need to find the total resistance between the drain of Q_1 and ground. Referring to Fig. 7.12(b) and denoting this resistance R_{d1}, we can express A_{v1} as

$$A_{v1} = \frac{v_{o1}}{v_i} = -g_{m1}R_{d1} \tag{7.33}$$

Observe that R_{d1} is the parallel equivalent of r_{o1} and R_{in2}, where R_{in2} is the input resistance of the CG transistor Q_2. We shall now derive an expression[2] for R_{in2}. For this purpose, refer to the equivalent circuit of Q_2 with its load resistance R_L, shown in Fig. 7.12(c). Observe that the voltage at the source of Q_2 is $-v_{gs2}$, thus R_{in2} can be found from

$$R_{in2} = \frac{-v_{gs2}}{i}$$

where i is the current flowing into the source of Q_2. Now this is the same current that flows out of the drain of Q_2 and into R_L. Summing the currents at the source node, we see that the current through r_{o2} is $i + g_{m2}v_{gs2}$. We can now express the voltage at the source node, $-v_{gs2}$, as the sum of the voltage drops across r_{o2} and R_L to obtain

$$-v_{gs2} = (i + g_{m2}v_{gs2})r_{o2} + iR_L$$

which can be rearranged to obtain

$$R_{in2} \equiv -v_{gs2}/i$$

$$R_{in2} = \frac{R_L + r_{o2}}{1 + g_{m2}r_{o2}} \tag{7.34}$$

This is a useful expression because it provides the input resistance of a CG amplifier loaded in a resistance R_L. Since $g_{m2}r_{o2} \gg 1$, we can simplify R_{in2} as follows:

$$R_{in2} \simeq \frac{R_L}{g_{m2}r_{o2}} + \frac{1}{g_{m2}} \tag{7.35}$$

[2]The reader should not jump to the conclusion that R_{in2} is equal to $1/g_{m2}$; this is the case when we neglect r_{o2}. As will be seen very shortly, R_{in2} can be vastly different from $1/g_{m2}$.

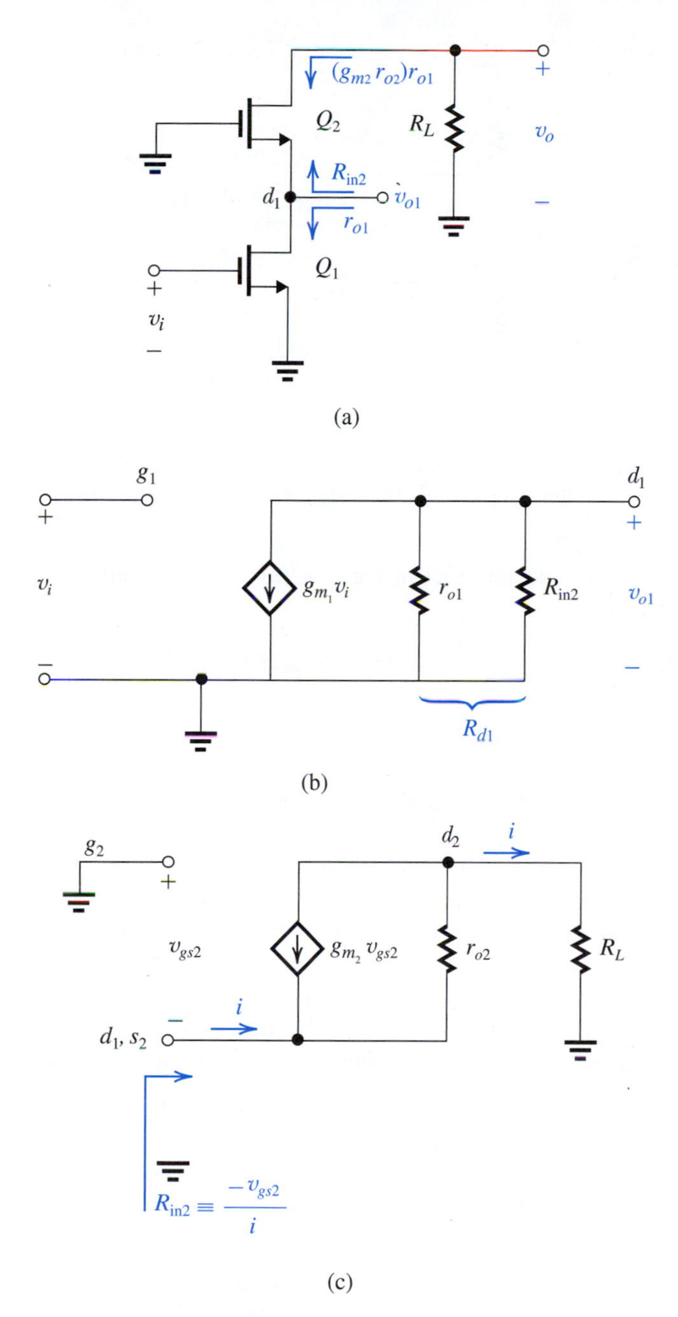

Figure 7.12 **(a)** The cascode amplifier with a load resistance R_L. Only signal quantities are shown. **(b)** Determining v_{01}. **(c)** Determining R_{in2}.

This is a very interesting result. First, it shows that if r_{o2} is infinite, as was assumed in our analysis of the discrete CG amplifier in Section 5.6.5, then R_{in2} reduces to $1/g_{m2}$, verifying the result we found there. If r_{o2} cannot be neglected, as is always the case in IC amplifiers, we see that the input resistance depends on the value of R_L in an interesting fashion: The load resistance R_L is divided by the factor $(g_{m2}r_{o2})$. This is of course the "flip side" of the impedance transformation action of the CG. For emphasis and future reference, we illustrate the impedance transformation properties of the CG circuit in Fig. 7.13.

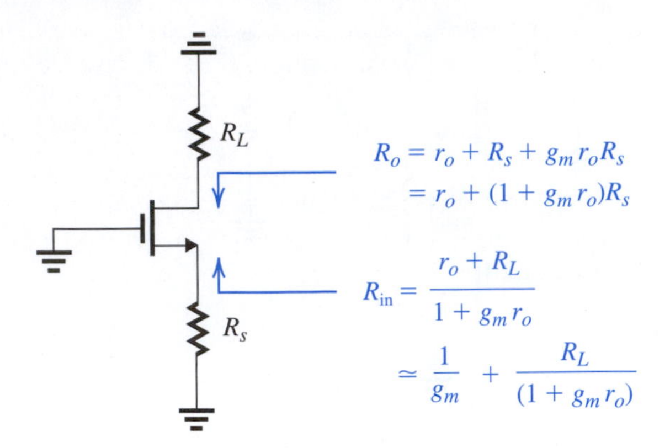

Figure 7.13 The impedance-transformation properties of the common-gate amplifier. Depending on the values of R_s and R_L, we can sometimes write $R_{in} \simeq R_L/(g_m r_o)$ and $R_o \simeq (g_m r_o)R_s$. However, such approximations are not always justified.

Going back to the cascode amplifier in Fig. 7.12(a), having found the value of R_{in2} we can now obtain R_{d1} as

$$R_{d1} = r_{o1} \| R_{in2} \tag{7.36}$$

and A_{v1} as

$$A_{v1} = -g_{m1}R_{d1} = -g_{m1}(r_{o1} \| R_{in2}) \tag{7.37}$$

Finally, we can obtain A_{v2} by dividing the total gain A_v given by Eq. (7.31) by A_{v1}. To provide insight into the effect of the value of R_L on the overall gain of the cascode as well as on how this gain is distributed among the two stages of the cascode amplifier, we provide in Table 7.1 approximate values for the case $r_{o1} = r_{o2} = r_o$ and for four different values of R_L: (1) $R_L = \infty$, obtained with an ideal current-source load; (2) $R_L = (g_m r_o)r_o$, obtained with a cascode current-source load; (3) $R_L = r_o$, obtained with a simple current-source load; and (4) for completeness, $R_L = 0$, that is, a signal short circuit at the output.

Table 7.1 Gain Distribution in the MOS Cascode Amplifier for Various Values of R_L

Case	R_L	R_{in2}	R_{d1}	A_{v1}	A_{v2}	A_v
1	∞	∞	r_o	$-g_m r_o$	$g_m r_o$	$-(g_m r_o)^2$
2	$(g_m r_o)r_o$	r_o	$r_o/2$	$-\frac{1}{2}(g_m r_o)$	$g_m r_o$	$-\frac{1}{2}(g_m r_o)^2$
3	r_o	$\frac{2}{g_m}$	$\frac{2}{g_m}$	-2	$\frac{1}{2}(g_m r_o)$	$-(g_m r_o)$
4	0	$\frac{1}{g_m}$	$\frac{1}{g_m}$	-1	0	0

Observe that while case 1 represents an idealized situation, it is useful in that it provides the theoretical maximum voltage gain achievable in a MOS cascode amplifier. Case 2, which assumes a cascode current-source load with an output resistance equal to that of the cascode amplifier, provides a realistic estimate of the gain achieved if one aims to maximize the realized gain. In certain situations, however, that is not our objective. This point is important, for as we shall see in Chapter 9, there is an entirely different application of the cascode amplifier: namely, to obtain wideband amplification by extending the upper 3-dB frequency f_H. As will be seen, for such an application one opts for the situation represented by case 3, where the gain achieved in the CS amplifier is only –2 V/V, and of course the overall gain is now only $-(g_m r_o)$. However, as will be seen in Chapter 9, this trade-off of the overall gain to obtain extended bandwidth is in some cases a good bargain!

EXERCISES

7.7 The common-gate transistor in Fig. 7.13 is biased at a drain current of 0.25 mA and is operating with an overdrive voltage $V_{OV} = 0.25$ V. The transistor has an Early voltage V_A of 5 V.
(a) Find R_{in} for $R_L = \infty$, 1 MΩ, 100 kΩ, 20 kΩ, and 0.
(b) Find R_o for $R_s = 0$, 1 kΩ, 10 kΩ, 20 kΩ, and 100 kΩ.
Ans. (a) ∞, 25.5 kΩ, 3 kΩ, 1 kΩ, 0.5 kΩ; (b) 20 kΩ, 61 kΩ, 430 kΩ, 840 kΩ, 4.12 MΩ

7.8 Consider a cascode amplifier for which the CS and CG transistors are identical and are biased to operate at $I_D = 0.1$ mA with $V_{OV} = 0.2$ V. Also let $V_A = 2$ V. Find A_{v1}, A_{v2}, and A_v for two cases: (a) $R_L = 20$ kΩ and (b) $R_L = 400$ kΩ.
Ans. (a) –1.82 V/V, 10.5 V/V, –19.0 V/V; (b) –10.2 V/V, 19.6 V/V, –200 V/V

7.3.4 The Output Resistance of a Source-Degenerated CS Amplifier

In Section 5.6.4 we discussed some of the benefits that are obtained when a resistance R_s is included in the source lead of a CS amplifier, as in Fig. 7.14(a). Such a resistance is referred to as a source-degeneration resistance because of its action in reducing the effective transconductance of the CS stage to $g_m/(1 + g_m R_s)$, that is, by a factor $(1 + g_m R_s)$. This also is the factor by which we increase a number of performance parameters such as linearity and bandwidth (as will be seen in Chapter 9). At this juncture we simply wish to point out that the expression we derived for the output resistance of the cascode amplifier applies directly to the case of a source-degenerated CS amplifier. This is because when we determine R_o, we ground the input terminal, making transistor Q appear as a CG transistor. Thus R_o is given by

$$R_o = R_s + r_o + g_m r_o R_s \tag{7.38}$$

Since $g_m r_o \gg 1$, the first term on the right-hand side will be much lower than the third and can be neglected, resulting in

$$R_o \simeq (1 + g_m R_s) r_o \tag{7.39}$$

Thus source degeneration increases the output resistance of the CS amplifier from r_o to $(1 + g_m R_s) r_o$, again by the same factor $(1 + g_m R_s)$. In Chapter 10, we will find that R_s introduces negative (degenerative) feedback of an amount $(1 + g_m R_s)$.

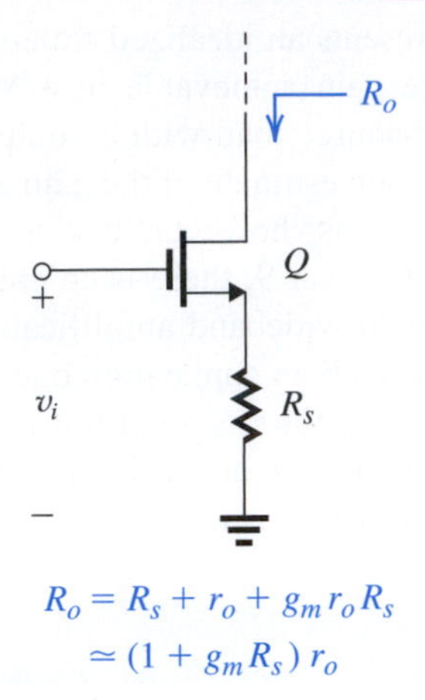

$$R_o = R_s + r_o + g_m r_o R_s$$
$$\simeq (1 + g_m R_s)\, r_o$$

Figure 7.14 The output resistance expression of the cascode can be used to find the output resistance of a source-degenerated common-source amplifier. Here, a useful interpretation of the result is that R_s increases the output resistance by the factor $(1 + g_m R_s)$.

EXERCISE

7.9 Given that source degeneration reduces the transconductance of a CS amplifier from g_m to approximately $g_m/(1 + g_m R_s)$ and increases its output resistance by approximately the same factor, what happens to the open-circuit voltage gain A_{vo}? Now, find an expression for A_v when a load resistance R_L is connected to the output.
Ans. A_{vo} remains constant at $g_m r_o$:

$$A_v = (g_m r_o)\frac{R_L}{R_L + (1 + g_m R_s) r_o} \qquad (7.40)$$

7.3.5 Double Cascoding

If a still higher output resistance and correspondingly higher gain are required, it is possible to add another level of cascoding, as illustrated in Fig. 7.15. Observe that Q_3 is the second cascode transistor, and it raises the output resistance by $(g_{m3} r_{o3})$. For the case of identical transistors, the output resistance will be $(g_m r_o)^2 r_o$ and the voltage gain, assuming an ideal current-source load, will be $(g_m r_o)^3$ or A_0^3. Of course, we have to generate another dc bias voltage for the second cascode transistor, Q_3.

A drawback of double cascoding is that an additional transistor is now stacked between the power-supply rails. Furthermore, to realize the advantage of double cascoding, the current-source load will also need to use double cascoding with an additional transistor. Since for proper operation each transistor needs a certain minimum v_{DS} (at least equal to V_{OV}), and recalling that modern MOS technology utilizes power supplies in the range of 1 V to 2 V, we see that there is a limit on the number of transistors in a cascode stack.

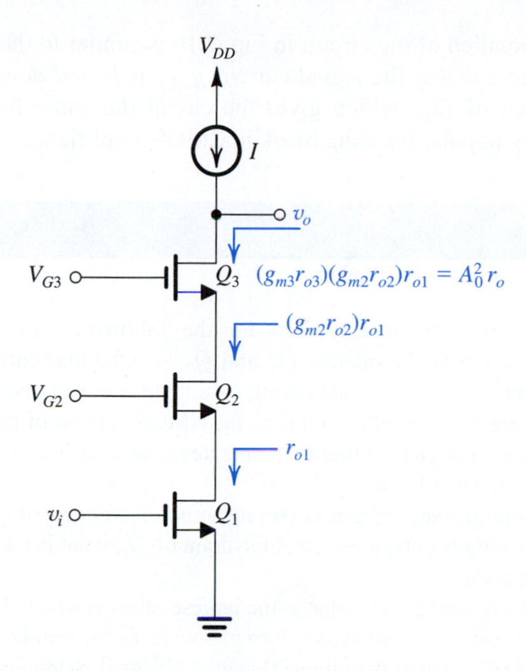

Figure 7.15 Double cascoding.

7.3.6 The Folded Cascode

To avoid the problem of stacking a large number of transistors across a low-voltage power supply, one can use a PMOS transistor for the cascode device, as shown in Fig. 7.16. Here, as before, the NMOS transistor Q_1 is operating in the CS configuration, but the CG stage is implemented using the PMOS transistor Q_2. An additional current source I_2 is needed to bias Q_2 and provide it with its active load. Note that Q_1 is now operating at a bias current of $(I_1 - I_2)$. Finally, a dc voltage V_{G2} is needed to provide an appropriate dc level for the gate of the cascode transistor Q_2. Its value has to be selected so that Q_2 and Q_1 operate in the saturation region.

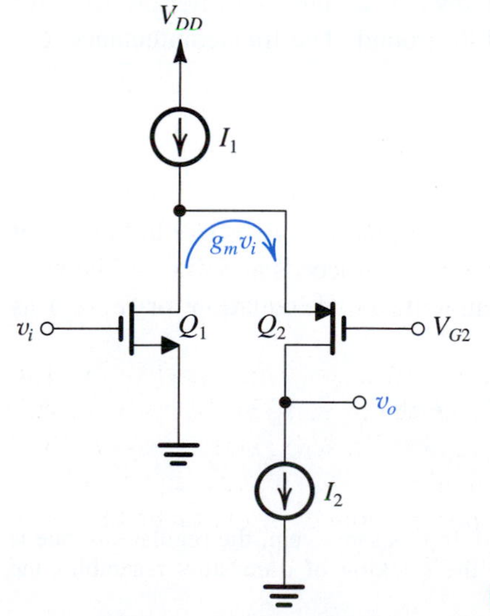

Figure 7.16 The folded cascode.

The small-signal operation of the circuit in Fig. 7.16 is similar to that of the NMOS cascode. The difference here is that the signal current $g_m v_i$ is *folded down* and made to flow into the source terminal of Q_2, which gives the circuit the name **folded cascode**.[3] The folded cascode is a very popular building block in CMOS amplifiers.

EXERCISE

7.10 Consider the folded-cascode amplifier of Fig. 7.16 for the following case: V_{DD} = 1.8 V, $k'_p = 1/4k'_n$, and $V_{tn} = -V_{tp}$ = 0.5 V. To operate Q_1 and Q_2 at equal bias currents I, $I_1 = 2I$ and $I_2 = I$. While current source I_1 is implemented using the simple circuit studied in Section 7.2, current source I_2 is realized using a cascoded circuit (i.e., the NMOS version of the circuit in Fig. 7.10). The transistor W/L ratios are selected so that each operates at an overdrive voltage of 0.2 V.
(a) What must the relationship of $(W/L)_2$ to $(W/L)_1$ be?
(b) What is the minimum dc voltage required across current source I_1 for proper operation? Now, if a 0.1-V peak-to-peak signal swing is to be allowed at the drain of Q_1, what is the highest dc bias voltage that can be used at that node?
(c) What is the value of V_{SG} of Q_2, and hence what is the largest value to which V_{G2} can be set?
(d) What is the minimum dc voltage required across current-source I_2 for proper operation?
(e) Given the results of (c) and (d), what is the allowable range of signal swing at the output?

Ans. (a) $(W/L)_2 = 4\,(W/L)_1$; (b) 0.2 V, 1.55 V; (c) 0.7 V, 0.85 V; (d) 0.4 V; (e) 0.4 V to 1.35 V

7.3.7 The BJT Cascode

Figure 7.17(a) shows the BJT cascode amplifier with an ideal current-source load. Voltage V_{B2} is a dc bias voltage for the CB cascode transistor Q_2. The circuit is very similar to the MOS cascode, and the small-signal analysis will follow in a parallel fashion. Our objective then is to determine the parameters G_m and R_o of the equivalent circuit of Fig. 7.17(b), which we shall use to represent the output of the cascode amplifier formed by Q_1 and Q_2.

As in the case of the MOS cascode, G_m is the short-circuit transconductance and can be determined from the circuit in Fig. 7.17(c). Here we show the cascode amplifier prepared for small-signal analysis with the output short-circuited to ground. The transconductance G_m can be determined as

$$G_m = \frac{i_o}{v_i}$$

Replacing Q_1 and Q_2 with their hybrid-π equivalent-circuit models gives rise to the circuit in Fig. 7.17(d). Analysis of this circuit is straightforward and proceeds as follows: The voltage at the node (c_1, e_2) is seen to be $-v_{\pi 2}$. Thus we can write a node equation for (c_1, e_2) as

$$g_{m2}v_{\pi 2} + \frac{v_{\pi 2}}{r_{o1}} + \frac{v_{\pi 2}}{r_{o2}} + \frac{v_{\pi 2}}{r_{\pi 2}} = g_{m1}v_i$$

[3]The circuit itself can be thought of as having been folded. In this same vein, the regular cascode is sometimes referred to as a **telescopic cascode** because the stacking of transistors resembles the extension of a telescope.

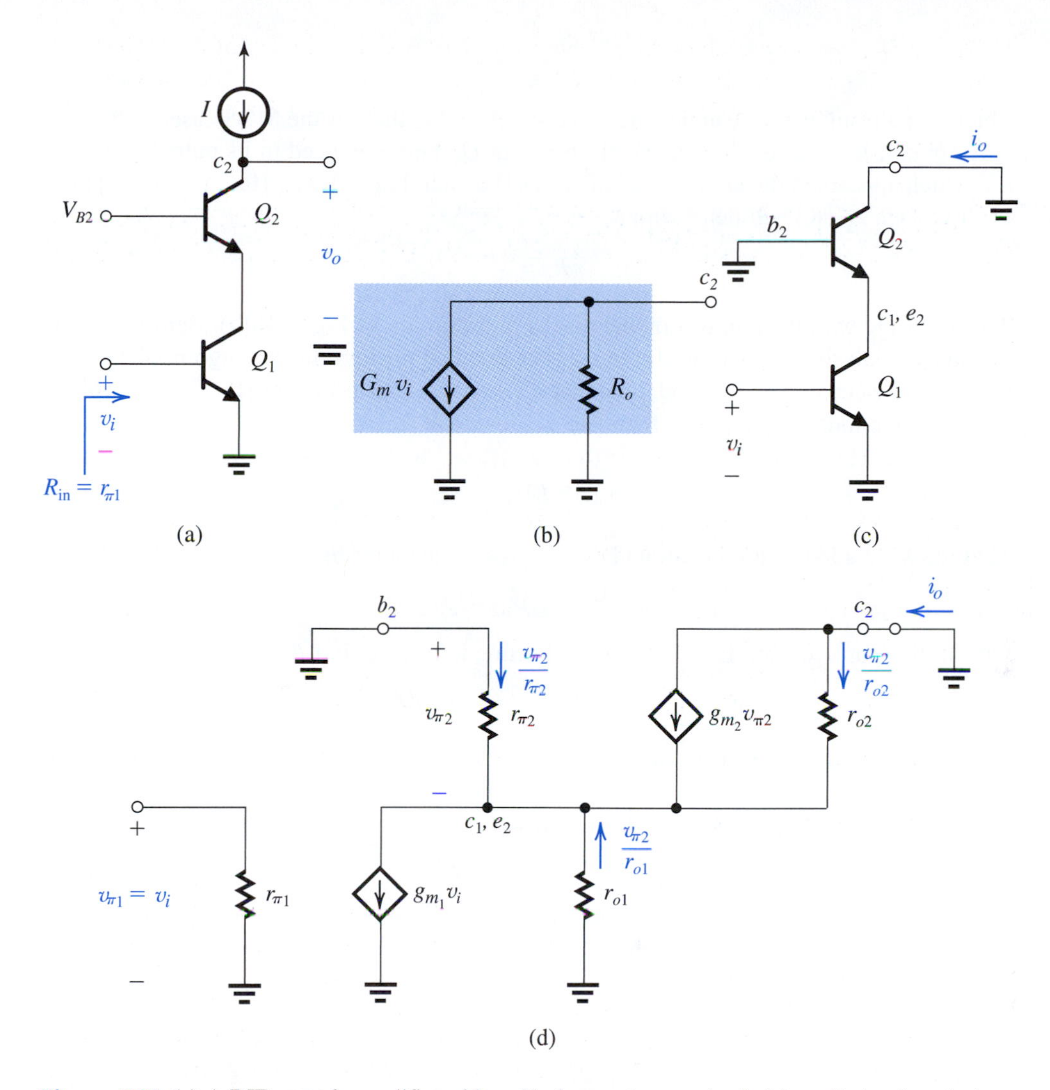

Figure 7.17 **(a)** A BJT cascode amplifier with an ideal current-source load; **(b)** small-signal equivalent-circuit representation of the output of the cascode amplifier; **(c)** the cascode amplifier with the output short-circuited to ground, and **(d)** equivalent circuit representation of **(c)**.

Since $g_{m2} \gg (1/r_{\pi 2})$, $1/r_{o1}$ and $1/r_{o2}$, we can neglect all the terms beyond the first on the left-hand side to obtain

$$g_{m2} v_{\pi 2} \simeq g_{m1} v_i \qquad (7.41)$$

Next, we write a node equation at c_2,

$$i_o = g_{m2} v_{\pi 2} + \frac{v_{\pi 2}}{r_{o2}}$$

and again neglect the second term on the right-hand side to obtain

$$i_o \simeq g_{m2} v_{\pi 2}$$

Using Eq. (7.41) results in

$$i_o = g_{m1} v_i$$

Thus,

$$G_m = g_{m1}$$

which is the result we have anticipated and is identical to that for the MOS case.

To obtain R_o, we set $v_i = 0$, which results in Q_1 being reduced to its output resistance r_{o1}, which appears in the emitter lead of Q_2 as shown in Fig. 7.18(a). Here we have applied a test voltage v_x and will determine R_o as

$$R_o = \frac{v_x}{i_x}$$

Replacing Q_2 with its hybrid-π model results in the circuit of Fig. 7.18(b). Before embarking on the analysis, it is very useful to observe first that the current flowing into the emitter node must be equal to i_x. Second, note that r_{o1} and $r_{\pi 2}$ appear in parallel. Thus the voltage at the emitter node, $-v_{\pi 2}$, can be found as

$$-v_{\pi 2} = i_x(r_{o1} \| r_{\pi 2}) \tag{7.42}$$

Next we write a loop equation around the $c_2 - e_2 -$ ground loop as

$$v_x = (i_x - g_{m2}v_{\pi 2})r_{o2} + i_x(r_{o1} \| r_{\pi 2})$$

Substituting for $v_{\pi 2}$ from Eq. (7.42) and collecting terms, we find $R_o = v_x/i_x$ as

$$R_o = r_{o2} + (r_{o1} \| r_{\pi 2}) + (g_{m2}r_{o2})(r_{o1} \| r_{\pi 2}) \tag{7.43}$$

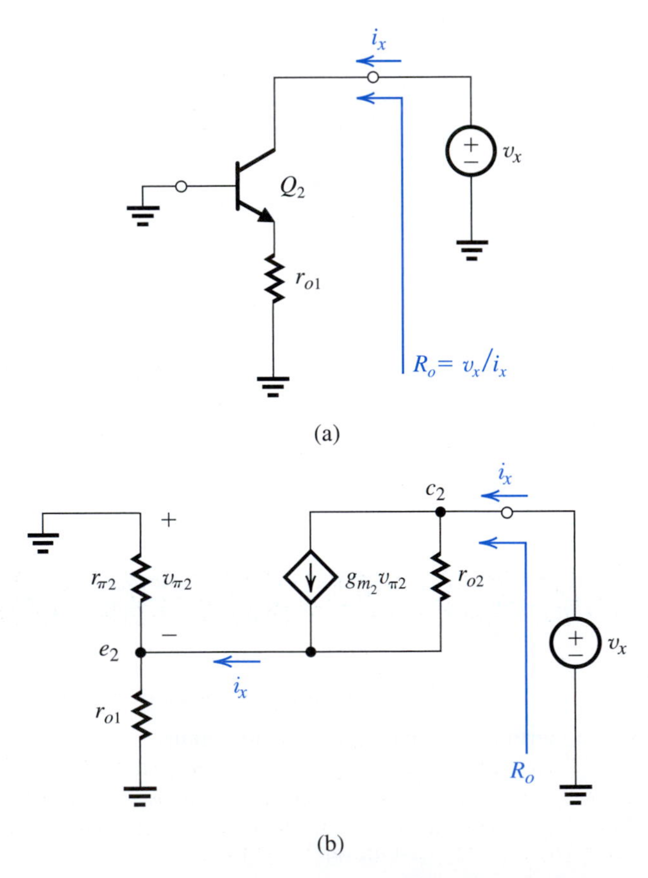

Figure 7.18 Determining the output resistant R_o of the BJT cascode amplifier.

which can be written as

$$R_o = r_{o2} + (g_{m2}r_{o2} + 1)(r_{o1} \| r_{\pi2})$$

$$\simeq r_{o2} + (g_{m2}r_{o2})(r_{o1} \| r_{\pi2}) \quad (7.44)$$

Since $g_{m2}(r_{o1} \| r_{\pi2}) \gg 1$, we can neglect the first term on the right-hand side of Eq. (7.44),

$$R_o \simeq (g_{m2}r_{o2})(r_{o1} \| r_{\pi2}) \quad (7.45)$$

This result is similar but certainly *not identical* to that for the MOS cascode. Here, because of the finite β of the BJT, we have $r_{\pi2}$ appearing in parallel with r_{o1}. This poses a very significant constraint on R_o of the BJT cascode. Specifically, because $(r_{o1} \| r_{\pi2})$ will always be lower than $r_{\pi2}$, it follows that the maximum possible value of R_o is

$$R_o|_{\max} = g_{m2}r_{o2}r_{\pi2}$$

$$= (g_{m2}r_{\pi2})r_{o2} = \beta_2 r_{o2} \quad (7.46)$$

Thus the maximum output resistance realizable by cascoding is $\beta_2 r_{o2}$. This means that unlike the MOS case, double cascoding with a BJT would not be useful.

Having determined G_m and R_o, we can now find the open-circuit voltage gain of the bipolar cascode as

$$A_{vo} = \frac{v_o}{v_i} = -G_m R_o$$

Thus,

$$A_{vo} = -g_{m1}(g_{m2}r_{o2})(r_{o1} \| r_{\pi2}) \quad (7.47)$$

For the case $g_{m1} = g_{m2}$, $r_{o1} = r_{o2}$,

$$A_{vo} = -(g_m r_o)[g_m(r_o \| r_\pi)] \quad (7.48)$$

which will be less than $(g_m r_o)^2$ in magnitude. In fact, the maximum possible gain magnitude is obtained when $r_o \gg r_\pi$ and is given by

$$|A_{vo}|_{\max} = \beta g_m r_o = \beta A_0 \quad (7.49)$$

Finally, we note that to be able to realize gains approaching this level, the current-source load must also be cascoded. Figure 7.19 shows a cascode BJT amplifier with a cascode current-source load.

EXERCISES

7.11 Find an expression for the maximum voltage gain achieved in the amplifier of Fig. 7.19.
Ans. $|A_{v\max}| = g_{m1}(\beta_2 r_{o2} \| \beta_3 r_{o3})$

7.12 Consider the BJT cascode amplifier of Fig. 7.19 when biased at a current of 0.2 mA. Assuming that *npn* transistors have $\beta = 100$ and $V_A = 5$ V and that *pnp* transistors have $\beta = 50$ and $|V_A| = 4$ V, find R_{on}, R_{op}, and A_v. Also use the result of Exercise 7.11 to determine the maximum achievable gain.
Ans. 1.67 MΩ; 0.762 MΩ; –4186 V/V; –5714 V/V

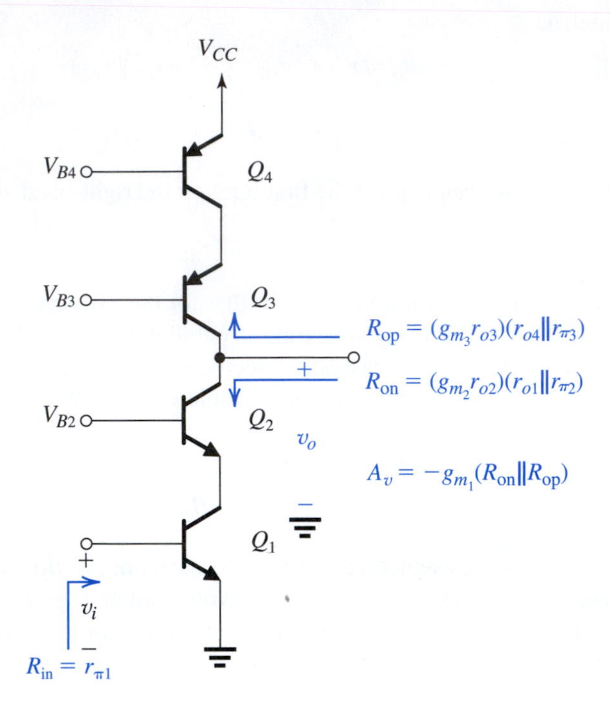

Figure 7.19 A BJT cascode amplifier with a cascode current source.

7.3.8 The Output Resistance of an Emitter-Degenerated CE Amplifier

As we have done in the MOS case, we shall adapt the expression for R_o derived for the BJT cascode (Eq. 7.43) for the case of a CE amplifier with a resistance R_e connected in its emitter, as shown in Fig. 7.20(a). The output resistance is obtained from Eq. (7.43) by replacing r_{o2} with r_o, g_{m2} by g_m, $r_{\pi 2}$ by r_π, and r_{o1} by R_e:

$$R_o = r_o + (R_e \| r_\pi) + (g_m r_o)(R_e \| r_\pi) \quad (7.50)$$

Since $g_m r_o \gg 1$, we can neglect the second term relative to the third; thus,

$$R_o \simeq r_o + g_m r_o (R_e \| r_\pi)$$

That is,

$$R_o = [1 + g_m(R_e \| r_\pi)] r_o \quad (7.51)$$

Thus, emitter degeneration multiplies the transistor output resistance r_o by the factor $[1 + g_m(R_e \| r_\pi)]$.

Finally, for completeness and future reference we show in Fig. 7.20(b) the BJT equivalent of Fig. 7.13. Here both R_{in} and R_o of a grounded-base BJT are shown. Note that we have not provided the derivation of R_{in}.

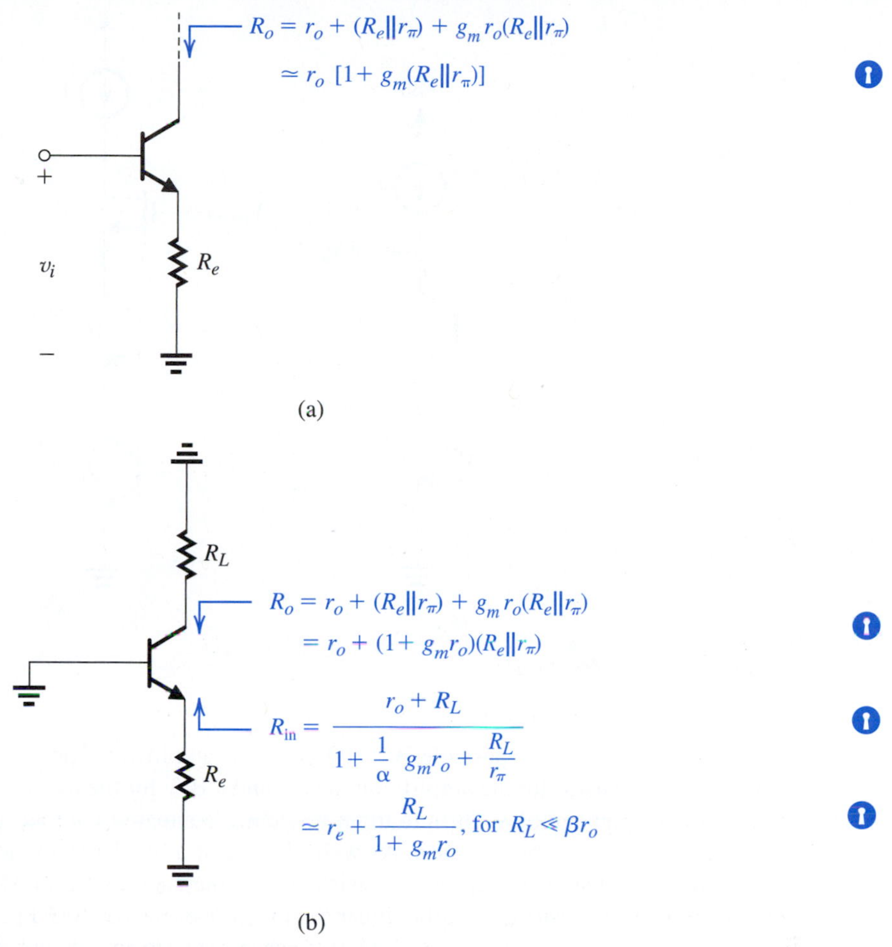

Figure 7.20 **(a)** Output resistance of a CE amplifier with emitter degeneration; **(b)** The impedance transformation properties of the CB amplifier. Note that for $\beta = \infty$, these formulas reduce to those for the MOSFET case (Fig. 7.13).

EXERCISE

7.13 Find the output resistance of a CE amplifier biased at $I_C = 1$ mA and having a resistance of 500 Ω connected in its emitter. Let $\beta = 100$ and $V_A = 10$ V. What is the value of the output resistance without degeneration.?
Ans. 177 kΩ; 10 kΩ

7.3.9 BiCMOS Cascodes

Certain advanced CMOS technologies allow the fabrication of bipolar transistors, thus permitting the circuit designer to combine MOS and bipolar transistors in circuits that take advantage of the unique features of each. The resulting technology is called BiCMOS, and the circuits are referred to as BiCMOS circuits.

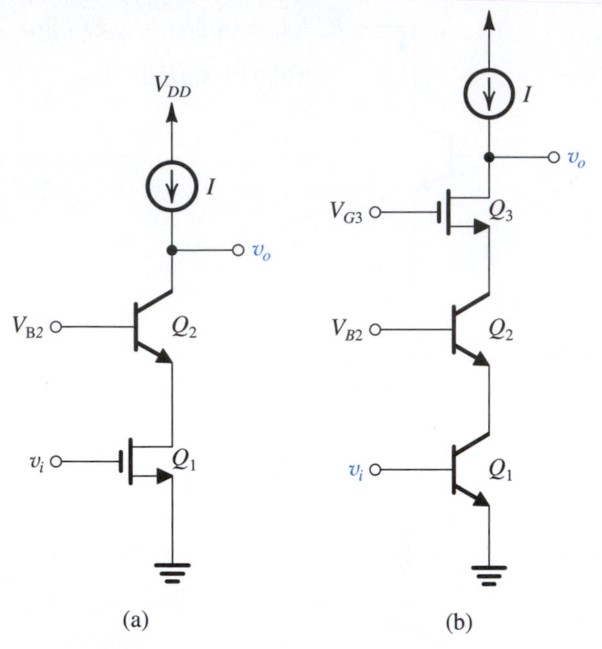

Figure 7.21 BiCMOS cascodes.

Figure 7.21 shows two possible BiCMOS cascode amplifiers. The circuit in Fig. 7.21(a) uses a MOS transistor for the amplifying device and a BJT for the cascode device. This circuit has the advantage of an infinite input resistance compared with an input resistance of r_π, obtained in the all-bipolar case. As well, the use of a bipolar transistor for the cascode stage can result in an increased output resistance as compared to the all-MOS case [because β of the bipolar transistor is usually higher than $(g_m r_o)$ of the MOSFET].

The circuit of Fig. 7.21(b) uses a MOS transistor Q_3 to implement double cascoding. Recall that double cascoding is not possible with BJT circuits alone.

EXERCISE

7.14 For $I = 100$ μA, find G_m, R_o, and A_{vo} of the BiCMOS amplifiers in Fig. 7.21. Let $V_A = 5$ V (for both MOS and bipolar transistors), $\beta = 100$, $\mu_n C_{ox} = 200$ μA/V^2, and $W/L = 25$.
Ans. (a) 1 mA/V, 3.33 MΩ, -3.33×10^3 V/V; (b) 4 mA/V, 167 MΩ, -668×10^3 V/V

7.4 IC Biasing—Current Sources, Current Mirrors, and Current-Steering Circuits

Biasing in integrated-circuit design is based on the use of constant-current sources. On an IC chip with a number of amplifier stages, a constant dc current (called a **reference current**) is generated at one location and is then replicated at various other locations for biasing the various amplifier stages through a process known as **current steering**. This approach has the advantage that the effort expended on generating a predictable and stable reference current, usually utilizing a precision resistor external to the chip or a special circuit on the chip, need not be repeated for every amplifier stage. Furthermore, the bias currents of the various stages track each other in case of changes in power-supply voltage or in temperature.

In this section we study circuit building blocks and techniques employed in the bias design of IC amplifiers. These current-source circuits are also utilized as amplifier load elements, as we have seen in Sections 7.2 and 7.3.

7.4.1 The Basic MOSFET Current Source

Figure 7.22 shows the circuit of a simple MOS constant-current source. The heart of the circuit is transistor Q_1, the drain of which is shorted to its gate,[4] thereby forcing it to operate in the saturation mode with

$$I_{D1} = \frac{1}{2}k_n'\left(\frac{W}{L}\right)_1(V_{GS} - V_{tn})^2 \tag{7.52}$$

where we have neglected channel-length modulation. The drain current of Q_1 is supplied by V_{DD} through resistor R, which in most cases would be outside the IC chip. Since the gate currents are zero,

$$I_{D1} = I_{\text{REF}} = \frac{V_{DD} - V_{GS}}{R} \tag{7.53}$$

where the current through R is considered to be the reference current of the current source and is denoted I_{REF}. Equations (7.52) and (7.53) can be used to determine the value required for R.

Now consider transistor Q_2: It has the same V_{GS} as Q_1; thus, if we assume that it is operating in saturation, its drain current, which is the output current I_O of the current source, will be

$$I_O = I_{D2} = \frac{1}{2}k_n'\left(\frac{W}{L}\right)_2(V_{GS} - V_{tn})^2 \tag{7.54}$$

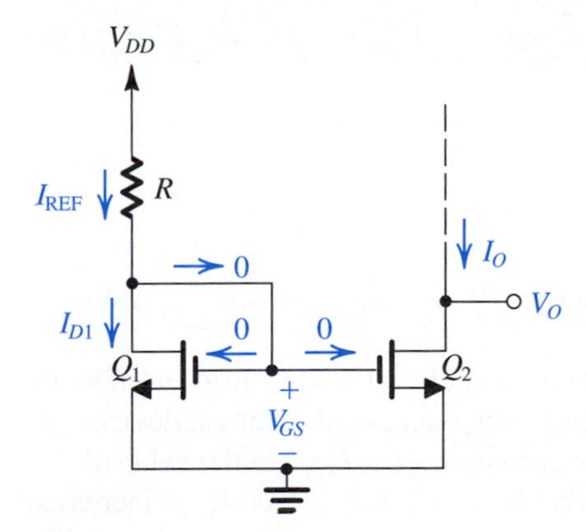

Figure 7.22 Circuit for a basic MOSFET constant-current source. For proper operation, the output terminal, that is, the drain of Q_2, must be connected to a circuit that ensures that Q_2 operates in saturation.

[4] Such a transistor is said to be *diode connected.*

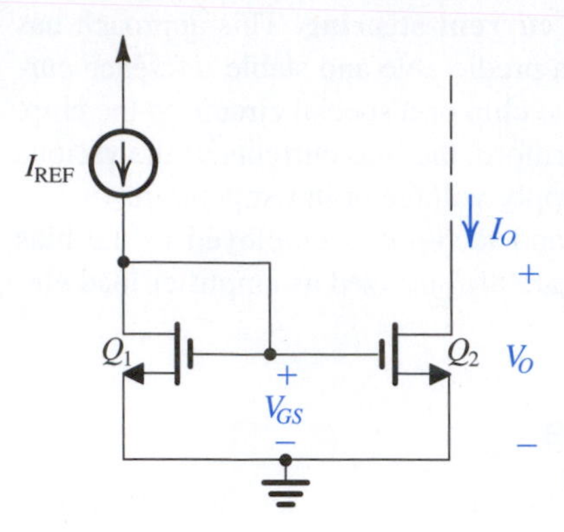

Figure 7.23 Basic MOSFET current mirror.

where we have neglected channel-length modulation. Equations (7.52) and (7.54) enable us to relate the output current I_O to the reference current I_{REF} as follows:

$$\frac{I_O}{I_{REF}} = \frac{(W/L)_2}{(W/L)_1} \tag{7.55}$$

This is a simple and attractive relationship: The special connection of Q_1 and Q_2 provides an output current I_O that is related to the reference current I_{REF} by the aspect ratios of the transistors. In other words, the relationship between I_O and I_{REF} is solely determined by the geometries of the transistors. In the special case of identical transistors, $I_O = I_{REF}$, and the circuit simply replicates or mirrors the reference current in the output terminal. This has given the circuit composed of Q_1 and Q_2 the name **current mirror**, a name that is used irrespective of the ratio of device dimensions.

Figure 7.23 depicts the current-mirror circuit with the input reference current shown as being supplied by a current source for both simplicity and generality. The **current gain** or **current transfer ratio** of the current mirror is given by Eq. (7.55).

Effect of V_O on I_O In the description above for the operation of the current source of Fig. 7.22, we assumed Q_2 to be operating in saturation. This is essential if Q_2 is to supply a constant-current output. To ensure that Q_2 is saturated, the circuit to which the drain of Q_2 is to be connected must establish a drain voltage V_O that satisfies the relationship

$$V_O \geq V_{GS} - V_{tn} \tag{7.56}$$

or, equivalently, in terms of the overdrive voltage V_{OV} of Q_1 and Q_2,

$$V_O \geq V_{OV} \tag{7.57}$$

In other words, the current source will operate properly with an output voltage V_O as low as V_{OV}, which is a few tenths of a volt.

Although thus far neglected, channel-length modulation can have a significant effect on the operation of the current source. Consider, for simplicity, the case of identical devices Q_1 and Q_2. The drain current of Q_2, I_O, will equal the current in Q_1, I_{REF}, at the value of V_O that causes the two devices to have the same V_{DS}, that is, at $V_O = V_{GS}$. As V_O is increased above this value, I_O will increase according to the incremental output resistance r_{o2} of Q_2. This is illustrated in Fig. 7.24, which shows I_O versus V_O. Observe that since Q_2 is operating at a constant V_{GS} (determined by passing I_{REF} through the matched device Q_1), the

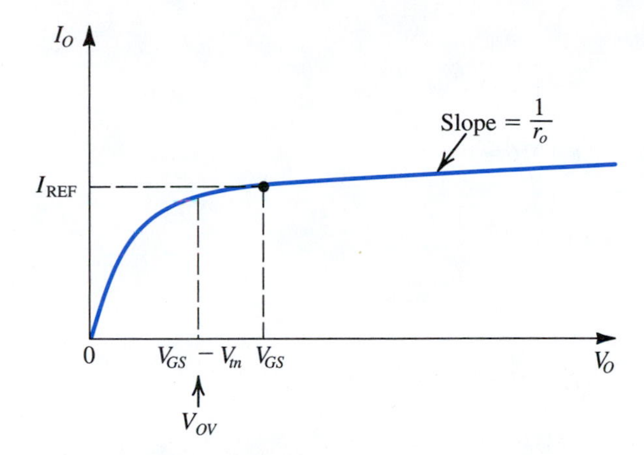

Figure 7.24 Output characteristic of the current source in Fig. 7.22 and the current mirror of Fig. 7.23 for the case of Q_2 matched to Q_1.

curve in Fig. 7.24 is simply the i_D–v_{DS} characteristic curve of Q_2 for v_{GS} equal to the particular value V_{GS}.

In summary, the current source of Fig. 7.22 and the current mirror of Fig. 7.23 have a finite output resistance R_o,

$$R_o \equiv \frac{\Delta V_O}{\Delta I_O} = r_{o2} = \frac{V_{A2}}{I_O} \tag{7.58}$$

where I_O is given by Eq. (7.54) and V_{A2} is the Early voltage of Q_2. Also, recall that for a given process technology, V_A is proportional to the transistor channel length; thus, to obtain high output-resistance values, current sources are usually designed using transistors with relatively long channels. Finally, note that we can express the current I_O as

$$I_O = \frac{(W/L)_2}{(W/L)_1} I_{REF}\left(1 + \frac{V_O - V_{GS}}{V_{A2}}\right) \tag{7.59}$$

Example 7.5

Given $V_{DD} = 3$ V and using $I_{REF} = 100$ μA, design the circuit of Fig. 7.22 to obtain an output current whose nominal value is 100 μA. Find R if Q_1 and Q_2 are matched and have channel lengths of 1 μm, channel widths of 10 μm, $V_t = 0.7$ V, and $k_n' = 200$ μA/V^2. What is the lowest possible value of V_O? Assuming that for this process technology, the Early voltage $V_A' = 20$ V/μm, find the output resistance of the current source. Also, find the change in output current resulting from a +1-V change in V_O.

Solution

$$I_{D1} = I_{REF} = \frac{1}{2} k_n' \left(\frac{W}{L}\right)_1 V_{OV}^2$$

$$100 = \frac{1}{2} \times 200 \times 10 V_{OV}^2$$

Thus,

$$V_{OV} = 0.316 \text{ V}$$

Example 7.5 *continued*

and

$$V_{GS} = V_t + V_{OV} = 0.7 + 0.316 \simeq 1 \text{ V}$$

$$R = \frac{V_{DD} - V_{GS}}{I_{REF}} = \frac{3-1}{0.1 \text{ mA}} = 20 \text{ k}\Omega$$

$$V_{O\min} = V_{OV} \simeq 0.3 \text{ V}$$

For the transistors used, $L = 1$ μm. Thus,

$$V_A = 20 \times 1 = 20 \text{ V}$$

$$r_{o2} = \frac{20 \text{ V}}{100 \text{ μA}} = 0.2 \text{ M}\Omega$$

The output current will be 100 μA at $V_O = V_{GS} = 1$ V. If V_O changes by +1 V, the corresponding change in I_O will be

$$\Delta I_O = \frac{\Delta V_O}{r_{o2}} = \frac{1 \text{ V}}{0.2 \text{ M}\Omega} = 5 \text{ μA}$$

EXERCISE

D7.15 In the current source of Example 7.5, it is required to reduce the change in output current, ΔI_O, corresponding to a change in output voltage, ΔV_O, of 1 V to 1% of I_O. What should the dimensions of Q_1 and Q_2 be changed to? Assume that Q_1 and Q_2 are to remain matched.
Ans. $L = 5$ μm; $W = 50$ μm

7.4.2 MOS Current-Steering Circuits

As mentioned earlier, once a constant current has been generated, it can be replicated to provide dc bias or load currents for the various amplifier stages in an IC. Current mirrors can obviously be used to implement this current-steering function. Figure 7.25 shows a simple current-steering circuit. Here Q_1 together with R determine the reference current I_{REF}. Transistors Q_1, Q_2, and Q_3 form a two-output current mirror,

$$I_2 = I_{REF} \frac{(W/L)_2}{(W/L)_1} \tag{7.60}$$

$$I_3 = I_{REF} \frac{(W/L)_3}{(W/L)_1} \tag{7.61}$$

To ensure operation in the saturation region, the voltages at the drains of Q_2 and Q_3 are constrained as follows:

$$V_{D2}, V_{D3} \geq -V_{SS} + V_{GS1} - V_{tn} \tag{7.62}$$

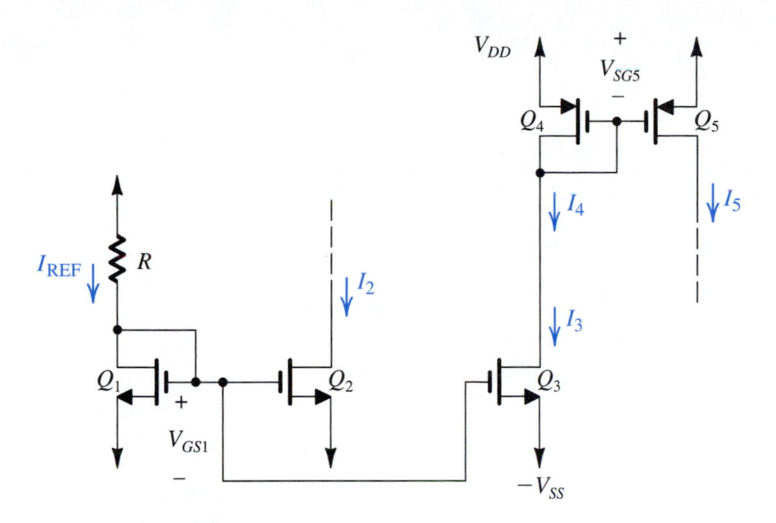

Figure 7.25 A current-steering circuit.

or, equivalently,

$$V_{D2}, V_{D3} \geq -V_{SS} + V_{OV1} \tag{7.63}$$

where V_{OV1} is the overdrive voltage at which Q_1, Q_2, and Q_3 are operating. In other words, the drains of Q_2 and Q_3 will have to remain higher than $-V_{SS}$ by at least the overdrive voltage, which is usually a few tenths of a volt.

Continuing our discussion of the circuit in Fig. 7.25, we see that current I_3 is fed to the input side of a current mirror formed by PMOS transistors Q_4 and Q_5. This mirror provides

$$I_5 = I_4 \frac{(W/L)_5}{(W/L)_4} \tag{7.64}$$

where $I_4 = I_3$. To keep Q_5 in saturation, its drain voltage should be

$$V_{D5} \leq V_{DD} - |V_{OV5}| \tag{7.65}$$

where V_{OV5} is the overdrive voltage at which Q_5 is operating.

The constant current I_2 generated in the circuit of Fig. 7.25 can be used to bias a source-follower amplifier such as that implemented by transistor Q_6 in Fig. 7.26(a). Similarly, the constant current I_5 can be used as the load for a common-source amplifier such as that implemented with transistor Q_7 in Fig. 7.26(b).

Finally, an important point to note is that in the circuit of Fig. 7.25, while Q_2 *pulls* its current I_2 from a circuit (not shown in Fig. 7.25), Q_5 *pushes* its current I_5 into a circuit (not shown in Fig. 7.25). Thus Q_5 is appropriately called a **current source**, whereas Q_2 should more properly be called a **current sink**. In an IC, both current sources and current sinks are usually needed. The difference between a current source and a current sink is further illustrated in Fig. 7.27, where $V_{CS\min}$ denotes the minimum voltage needed across the current source (or sink) for its proper operation.

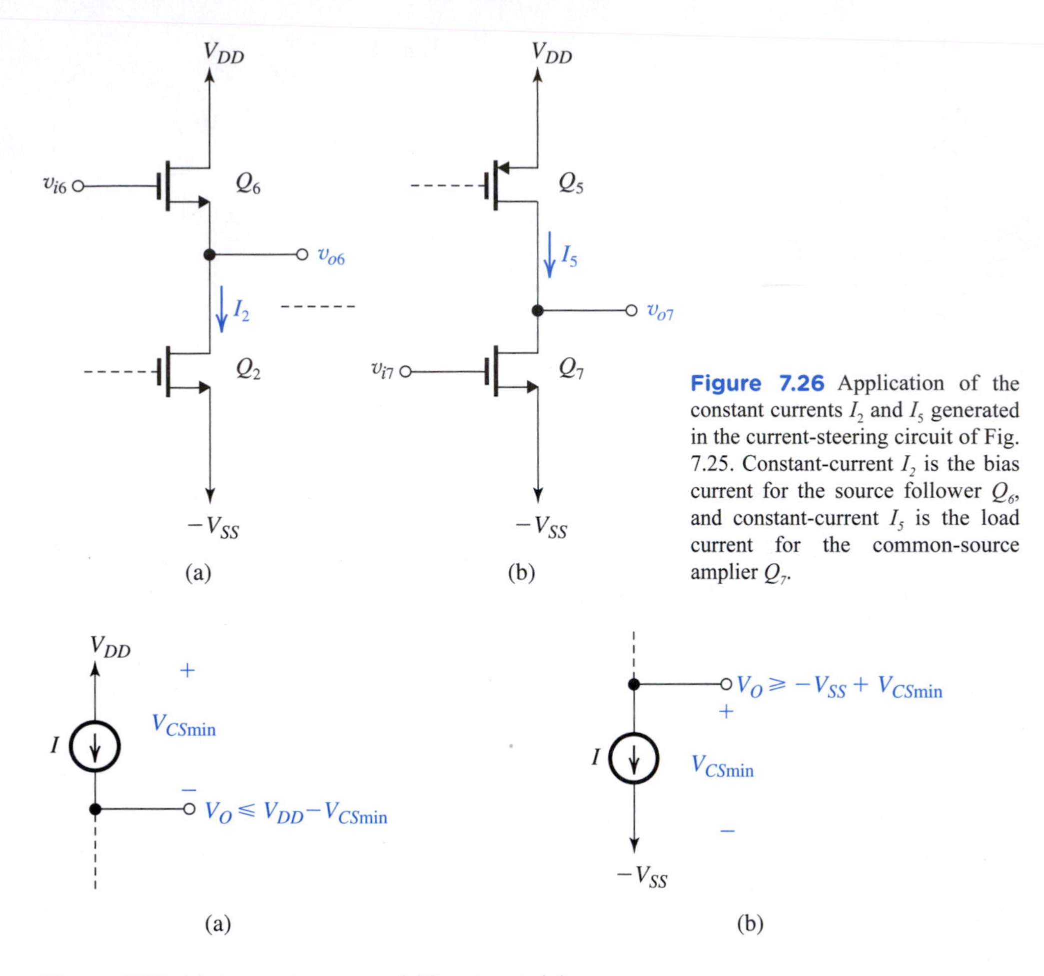

Figure 7.26 Application of the constant currents I_2 and I_5 generated in the current-steering circuit of Fig. 7.25. Constant-current I_2 is the bias current for the source follower Q_6, and constant-current I_5 is the load current for the common-source amplier Q_7.

Figure 7.27 (**a**) A current source; and (**b**) a current sink.

EXERCISE

D 7.16 For the circuit of Fig. 7.25, let $V_{DD} = V_{SS} = 1.5$ V, $V_{tn} = 0.6$ V, $V_{tp} = -0.6$ V, all channel lengths = 1 μm, $k_n' = 200$ μA/V^2, $k_p' = 80$ μA/V^2, and $\lambda = 0$. For $I_{REF} = 10$ μA, find the widths of all transistors to obtain $I_2 = 60$ μA, $I_3 = 20$ μA, and $I_5 = 80$ μA. It is further required that the voltage at the drain of Q_2 be allowed to go down to within 0.2 V of the negative supply and that the voltage at the drain of Q_5 be allowed to go up to within 0.2 V of the positive supply.

Ans. $W_1 = 2.5$ μm; $W_2 = 15$ μm; $W_3 = 5$ μm; $W_4 = 12.5$ μm; $W_5 = 50$ μm

7.4.3 BJT Circuits

The basic BJT current mirror is shown in Fig. 7.28. It works in a fashion very similar to that of the MOS mirror. However, there are two important differences: First, the nonzero base current of the BJT (or, equivalently, the finite β) causes an error in the current transfer ratio

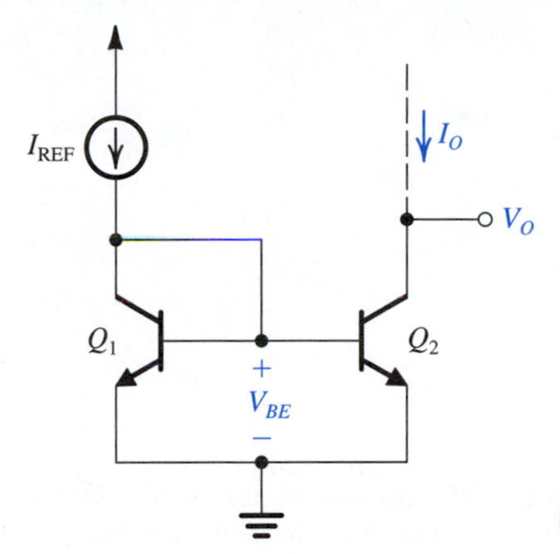

Figure 7.28 The basic BJT current mirror.

of the bipolar mirror. Second, the current transfer ratio is determined by the relative areas of the emitter–base junctions of Q_1 and Q_2.

Let us first consider the case of β sufficiently high that we can neglect the base currents. The reference current I_{REF} is passed through the diode-connected transistor Q_1 and thus establishes a corresponding voltage V_{BE}, which in turn is applied between base and emitter of Q_2. Now, if Q_2 is matched to Q_1 or, more specifically, if the EBJ area of Q_2 is the same as that of Q_1, and thus Q_2 has the same scale current I_S as Q_1, then the collector current of Q_2 will be equal to that of Q_1; that is,

$$I_O = I_{REF} \tag{7.66}$$

For this to happen, however, Q_2 must be operating in the active mode, which in turn is achieved as long as the collector voltage V_O is 0.3 V or so higher than that of the emitter.

To obtain a current transfer ratio other than unity, say m, we simply arrange that the area of the EBJ of Q_2 is m times that of Q_1. In this case,

$$I_O = mI_{REF} \tag{7.67}$$

In general, the current transfer ratio is given by

$$\frac{I_O}{I_{REF}} = \frac{I_{S2}}{I_{S1}} = \frac{\text{Area of EBJ of } Q_2}{\text{Area of EBJ of } Q_1} \tag{7.68}$$

Alternatively, if the area ratio m is an integer, one can think of Q_2 as equivalent to m transistors, each matched to Q_1 and connected in parallel.

Next we consider the effect of finite transistor β on the current transfer ratio. The analysis for the case in which the current transfer ratio is nominally unity—that is, for the case in which Q_2 is matched to Q_1—is illustrated in Fig. 7.29. The key point here is that since Q_1 and Q_2 are matched and have the same V_{BE}, their collector currents will be equal. The rest of the analysis is straightforward. A node equation at the collector of Q_1 yields

$$I_{REF} = I_C + 2I_C/\beta = I_C\left(1 + \frac{2}{\beta}\right)$$

Finally, since $I_O = I_C$, the current transfer ratio can be found as

$$\frac{I_O}{I_{REF}} = \frac{I_C}{I_C\left(1 + \frac{2}{\beta}\right)} = \frac{1}{1 + \frac{2}{\beta}} \tag{7.69}$$

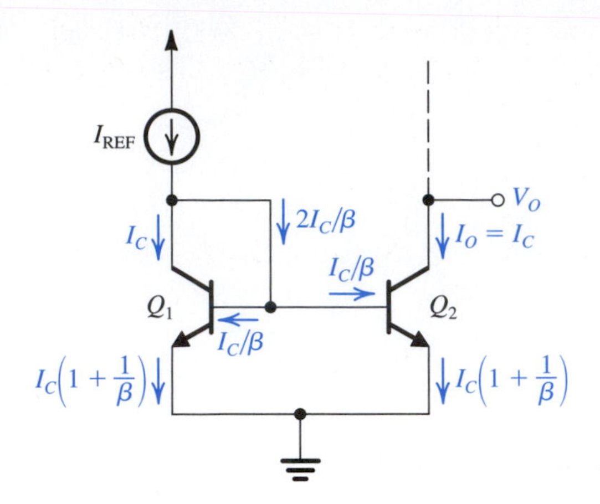

Figure 7.29 Analysis of the current mirror taking into account the finite β of the BJTs.

Note that as β approaches ∞, I_O/I_{REF} approaches the nominal value of unity. For typical values of β, however, the error in the current transfer ratio can be significant. For instance, $\beta = 100$ results in a 2% error in the current transfer ratio. Furthermore, the error due to the finite β increases as the nominal current transfer ratio is increased. The reader is encouraged to show that for a mirror with a nominal current transfer ratio m—that is, one in which $I_{S2} = mI_{S1}$—the actual current transfer ratio is given by

$$\frac{I_O}{I_{REF}} = \frac{m}{1 + \dfrac{m+1}{\beta}} \tag{7.70}$$

In common with the MOS current mirror, the BJT mirror has a finite output resistance R_o,

$$R_o \equiv \frac{\Delta V_O}{\Delta I_O} = r_{o2} = \frac{V_{A2}}{I_O} \tag{7.71}$$

where V_{A2} and r_{o2} are the Early voltage and the output resistance, respectively, of Q_2. Thus, even if we neglect the error due to finite β, the output current I_O will be at its nominal value only when Q_2 has the same V_{CE} as Q_1, namely at $V_O = V_{BE}$. As V_O is increased, I_O will correspondingly increase. Taking both the finite β and the finite R_o into account, we can express the output current of a BJT mirror with a nominal current transfer ratio m as

$$I_O = I_{REF}\left(\frac{m}{1 + \dfrac{m+1}{\beta}}\right)\left(1 + \frac{V_O - V_{BE}}{V_{A2}}\right) \tag{7.72}$$

where we note that the error term due to the Early effect is expressed in a form that shows that it reduces to zero for $V_O = V_{BE}$.

EXERCISE

7.17 Consider a BJT current mirror with a nominal current transfer ratio of unity. Let the transistors have $I_S = 10^{-15}$ A, $\beta = 100$, and $V_A = 100$ V. For $I_{REF} = 1$ mA, find I_O when $V_O = 5$ V. Also, find the output resistance.
Ans. 1.02 mA; 100 kΩ

A Simple Current Source In a manner analogous to that in the MOS case, the basic BJT current mirror can be used to implement a simple current source, as shown in Fig. 7.30. Here the reference current is

$$I_{\text{REF}} = \frac{V_{CC} - V_{BE}}{R} \tag{7.73}$$

where V_{BE} is the base–emitter voltage corresponding to the desired value of I_{REF}. The output current I_O is given by

$$I_O = \frac{I_{\text{REF}}}{1 + (2/\beta)}\left(1 + \frac{V_O - V_{BE}}{V_A}\right) \tag{7.74}$$

The output resistance of this current source is r_o of Q_2,

$$R_o\ (= r_{o2}) \simeq \frac{V_A}{I_O} \simeq \frac{V_A}{I_{\text{REF}}} \tag{7.75}$$

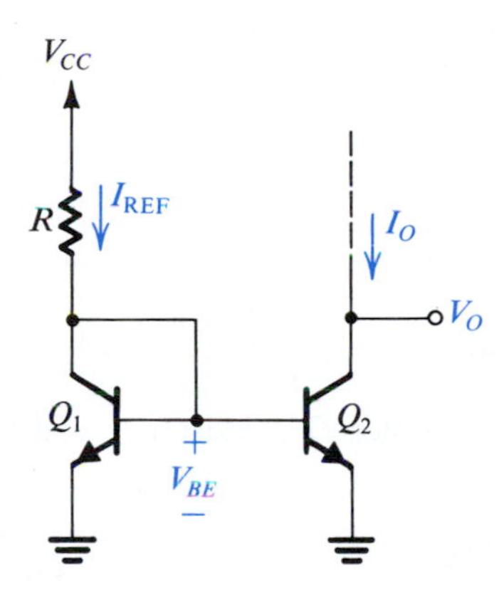

Figure 7.30 A simple BJT current source.

EXERCISE

D7.18 Assuming the availability of BJTs with scale currents $I_S = 10^{-15}$ A, $\beta = 100$, and $V_A = 50$ V, design the current-source circuit of Fig. 7.30 to provide an output current $I_O = 0.5$ mA at $V_O = 2$ V. The power supply $V_{CC} = 5$ V. Give the values of I_{REF}, R, and $V_{O\min}$. Also, find I_O at $V_O = 5$ V.
Ans. 0.497 mA; 8.71 kΩ; 0.3 V; 0.53 mA

Current Steering To generate bias currents for different amplifier stages in an IC, the current-steering approach described for MOS circuits can be applied in the bipolar case. As an example, consider the circuit shown in Fig. 7.31. The dc reference current I_{REF} is gener-

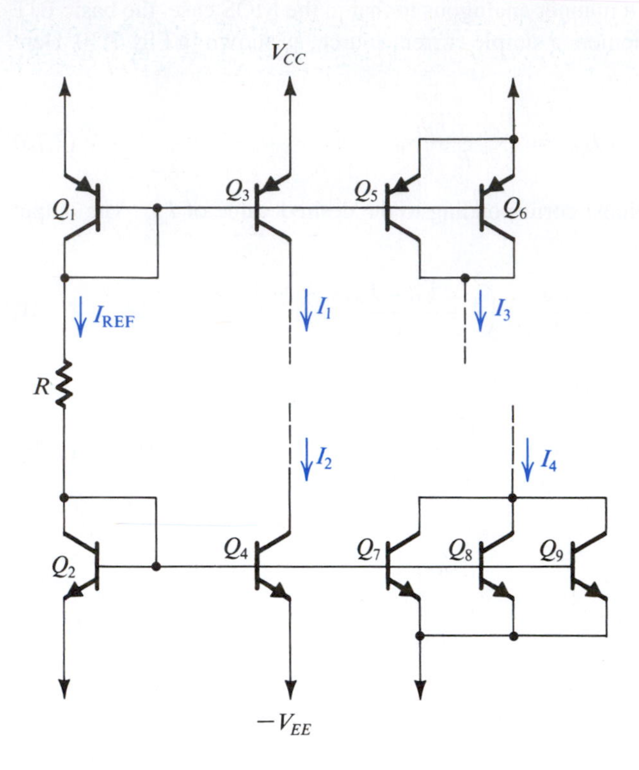

Figure 7.31 Generation of a number of constant currents of various magnitudes.

ated in the branch that consists of the diode-connected transistor Q_1, resistor R, and the diode-connected transistor Q_2:

$$I_{REF} = \frac{V_{CC} + V_{EE} - V_{EB1} - V_{BE2}}{R} \tag{7.76}$$

Now, for simplicity, assume that all the transistors have high β and thus that the base currents are negligibly small. We will also neglect the Early effect. The diode-connected transistor Q_1 forms a current mirror with Q_3; thus Q_3 will supply a constant current I equal to I_{REF}. Transistor Q_3 can supply this current to any load as long as the voltage that develops at the collector does not exceed $(V_{CC} - 0.3 \text{ V})$; otherwise Q_3 would enter the saturation region.
To generate a dc current twice the value of I_{REF}, two transistors, Q_5 and Q_6, each of which is matched to Q_1, are connected in parallel, and the combination forms a mirror with Q_1. Thus $I_3 = 2I_{REF}$. Note that the parallel combination of Q_5 and Q_6 is equivalent to a transistor with an EBJ area double that of Q_1, which is precisely what is done when this circuit is fabricated in IC form.

Transistor Q_4 forms a mirror with Q_2; thus Q_4 provides a constant current I_2 equal to I_{REF}. Note that while Q_3 *sources* its current to parts of the circuit whose voltage should not exceed $(V_{CC} - 0.3 \text{ V})$, Q_4 *sinks* its current from parts of the circuit whose voltage should not decrease below $-V_{EE} + 0.3$ V. Finally, to generate a current three times I_{REF}, three transistors, Q_7, Q_8, and Q_9, each of which is matched to Q_2, are connected in parallel, and the combination is placed in a mirror configuration with Q_2. Again, in an IC implementation, Q_7, Q_8, and Q_9 would be replaced with a transistor having a junction area three times that of Q_2.

EXERCISE

7.19 Figure E7.19 shows an N-output current mirror. Assuming that all transistors are matched and have finite β and ignoring the effect of finite output resistances, show that

$$I_1 = I_2 = \cdots = I_N = \frac{I_{REF}}{1+(N+1)/\beta}$$

For $\beta = 100$, find the maximum number of outputs for an error not exceeding 10%.

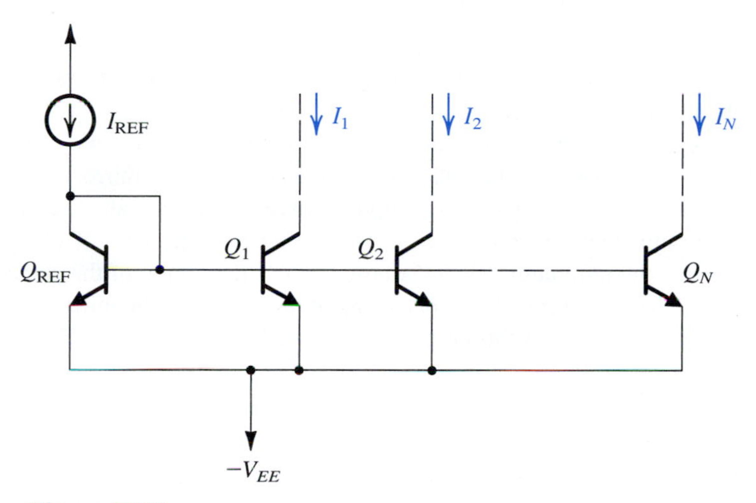

Figure E7.19

Ans. 9

7.5 Current-Mirror Circuits with Improved Performance

As we have seen throughout this chapter, current sources play a major role in the design of IC amplifiers: The constant-current source is used both in biasing and as active load. Simple forms of both MOS and bipolar current sources and, more generally, current mirrors were studied in Section 7.4. The need to improve the characteristics of the simple sources and mirrors has already been demonstrated. Specifically, two performance parameters need to be addressed: the *accuracy of the current transfer ratio of the mirror* and the *output resistance of the current source*.

The reader will recall from Section 7.4 that the accuracy of the current transfer ratio suffers particularly from the finite β of the BJT. The output resistance, which in the simple circuits is limited to r_o of the MOSFET and the BJT, also reduces accuracy and, much more seriously, severely limits the gain available from cascode amplifiers (Section 7.3). In this section we study MOS and bipolar current mirrors with more accurate current transfer ratios and higher output resistances.

7.5.1 Cascode MOS Mirrors

The use of cascoding in the design of current sources was presented in Section 7.3. Figure 7.32 shows the basic cascode current mirror. Observe that in addition to the diode-connected transistor Q_1, which forms the basic mirror Q_1–Q_2, another diode-connected transistor, Q_4, is used to provide a suitable bias voltage for the gate of the cascode transistor Q_3. To determine the output resistance of the cascode mirror at the drain of Q_3, we assume that the voltages across Q_1 and Q_4 are constant, and thus the signal voltages at the gates of Q_2 and Q_3 will be zero. Thus R_o will be that of the cascode current source formed by Q_2 and Q_3,

$$R_o \simeq g_{m3} r_{o3} r_{o2} \tag{7.77}$$

Thus, as expected, cascoding raises the output resistance of the current source by the factor $(g_{m3} r_{o3})$, which is the intrinsic gain of the cascode transistor.

A drawback of the cascode current mirror is that it consumes a relatively large portion of the steadily shrinking supply voltage V_{DD}. While the simple MOS mirror operates properly with a voltage as low as V_{OV} across its output transistor, the cascode circuit of Fig. 7.32 requires a minimum voltage of $V_t + 2V_{OV}$. This is because the gate of Q_3 is at $2V_{GS} = 2V_t + 2V_{OV}$. Thus the minimum voltage required across the output of the cascode mirror is 1 V or so. This obviously limits the signal swing at the output of the mirror (i.e., at the output of the amplifier that utilizes this current source as a load). In Chapter 12 we shall study a wide-swing cascode mirror.

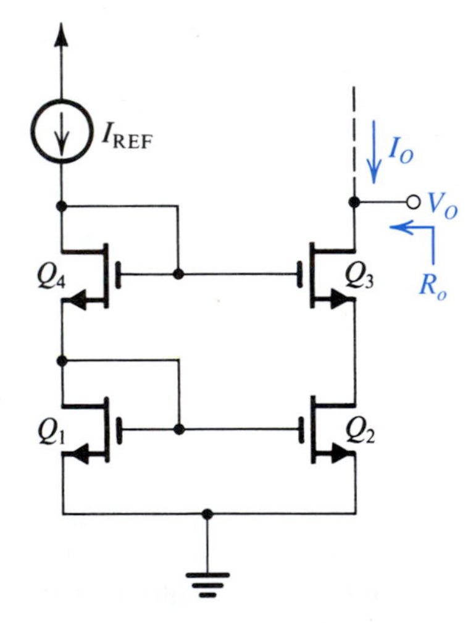

Figure 7.32 A cascode MOS current mirror.

EXERCISE

7.20 For a cascode MOS mirror utilizing devices with $V_t = 0.5$ V, $\mu_n C_{ox} = 387$ µA/V^2, $V_A' = 5$ V/µm, $W/L = 3.6$ µm/0.36 µm, and $I_{REF} = 100$ µA, find the minimum voltage required at the output and the output resistance.

Ans. 0.95 V; 285 kΩ

7.5.2 A Bipolar Mirror with Base-Current Compensation

Figure 7.33 shows a bipolar current mirror with a current transfer ratio that is much less dependent on β than that of the simple current mirror. The reduced dependence on β is achieved by including transistor Q_3, the emitter of which supplies the base currents of Q_1 and Q_2. The sum of the base currents is then divided by $(\beta_3 + 1)$, resulting in a much smaller error current that has to be supplied by I_{REF}. Detailed analysis is shown on the circuit diagram; it is based on the assumption that Q_1 and Q_2 are matched and thus have equal collector currents, I_C. A node equation at the node labeled x gives

$$I_{REF} = I_C\left[1 + \frac{2}{\beta(\beta+1)}\right]$$

Since

$$I_O = I_C$$

the current transfer ratio of the mirror will be

$$\frac{I_O}{I_{REF}} = \frac{1}{1 + 2/(\beta^2 + \beta)}$$

$$\simeq \frac{1}{1 + 2/\beta^2} \tag{7.78}$$

which means that the error due to finite β has been reduced from $2/\beta$ in the simple mirror to $2/\beta^2$, a tremendous improvement. Unfortunately, however, the output resistance remains approximately equal to that of the simple mirror, namely r_o. Finally, note that if a reference current I_{REF} is not available, we simply connect node x to the power supply. V_{CC} through a resistance R. The result is a reference current given by

$$I_{REF} = \frac{V_{CC} - V_{BE1} - V_{BE3}}{R} \tag{7.79}$$

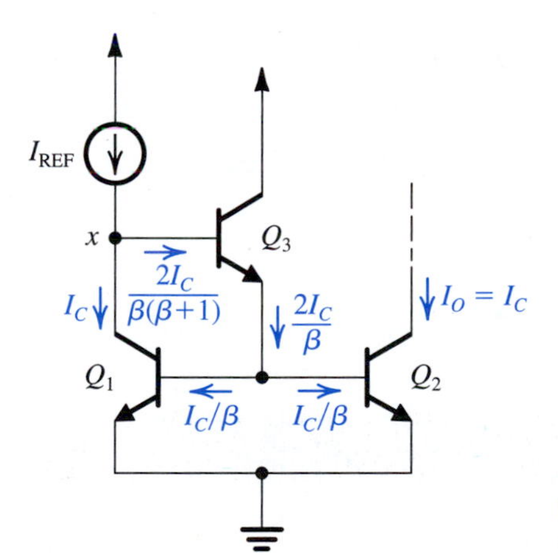

Figure 7.33 A current mirror with base-current compensation.

7.5.3 The Wilson Current Mirror

A simple but ingenious modification of the basic bipolar mirror results in both reducing the β dependence and increasing the output resistance. The resulting circuit, known as the **Wilson**

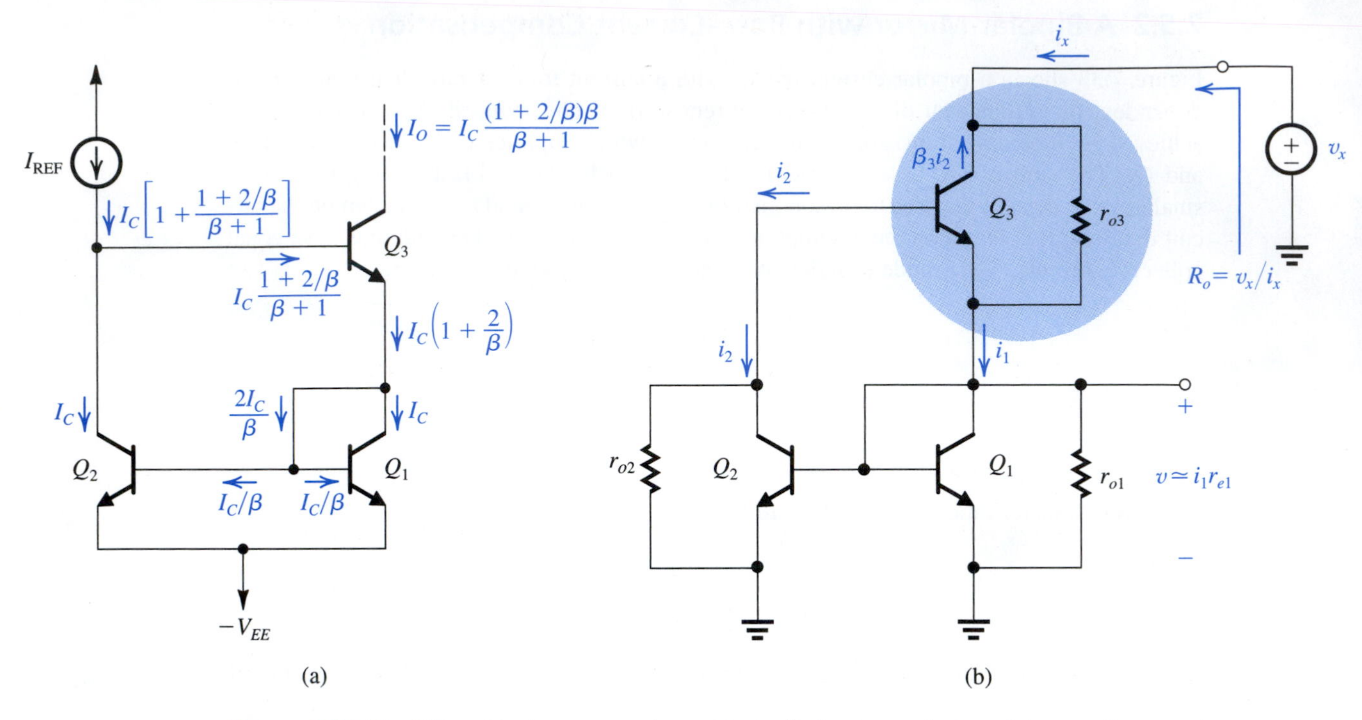

Figure 7.34 The Wilson bipolar current mirror: **(a)** circuit showing analysis to determine the current transfer ratio; **(b)** determining the output resistance.

mirror after its inventor George Wilson, an IC design engineer working for Tektronix, is shown in Fig. 7.34(a). The analysis to determine the effect of finite β on the current transfer ratio is shown in Fig. 7.34(a), from which we can write

$$\frac{I_O}{I_{REF}} = \frac{I_C\left(1+\frac{2}{\beta}\right)\beta/(\beta+1)}{I_C\left[1+\left(1+\frac{2}{\beta}\right)/(\beta+1)\right]}$$

$$= \frac{\beta+2}{\beta+1+\frac{\beta+2}{\beta}} = \frac{\beta+2}{\beta+2+\frac{2}{\beta}}$$

$$= \frac{1}{1+\frac{2}{\beta(\beta+2)}}$$

$$\simeq \frac{1}{1+2/\beta^2} \tag{7.80}$$

This analysis assumes that Q_1 and Q_2 conduct equal collector currents. There is, however, a slight problem with this assumption: The collector-to-emitter voltages of Q_1 and Q_2 are not equal, which introduces a current offset or a systematic error. The problem can be solved by adding a diode-connected transistor in series with the collector of Q_2, as we shall shortly show for the MOS version.

To determine the output resistance of the Wilson mirror, we set $I_{REF} = 0$ and apply a test voltage v_x to the output node, as shown in Fig. 7.34(b). Our purpose is to determine the current i_x and hence R_o as

$$R_o = v_x / i_x$$

Rather than replacing each transistor with its hybrid-π model, we shall do the analysis directly on the circuit diagram. For this purpose, we have "pulled r_o out" of each transistor and shown it separately.

Observe that transistor Q_3, viewed as a supernode (highlighted in color), has a current i_x entering it and two currents i_1 and i_2 exiting it; thus,

$$i_1 + i_2 = i_x$$

Next note that the action of current mirror $Q_1 - Q_2$ forces i_2 to be approximately equal to i_1; thus,

$$i_2 \simeq i_1 = i_x/2$$

Current i_2 flows into the base of Q_3 and thus gives rise to a collector current $\beta_3 i_2$ in the direction indicated. We are now in a position to write a node equation at the collector of Q_3 and thus determine the current through r_{o2} as $i_x + \beta_3 i_2 = i_x + \beta_3(i_x/2) = i_x(\beta_3/2 + 1)$. Finally, we can express the voltage between the collector of Q_3 and ground as the sum of the voltage drop across r_{o3} and the voltage v across Q_1,

$$v_x = i_x\left(\frac{\beta_3}{2} + 1\right)r_{o3} + i_1 r_{e1}$$

$$= i_x\left(\frac{\beta_3}{2} + 1\right)r_{o3} + \left(\frac{i_x}{2}\right)r_{e1}$$

Since $r_o \gg r_e$ and $\beta_3 \gg 2$,

$$v_x \simeq i_x\left(\frac{\beta_3}{2}\right)r_{o3}$$

and

$$R_o = \beta_3 r_{o3}/2 \tag{7.81}$$

Thus the Wilson current mirror has an output resistance $(\frac{1}{2}\beta_3)$ times higher than that of Q_3 alone. This is a result of the negative feedback obtained by feeding the collector current of Q_2 (i_2) back to the base of Q_3. As can be seen from the above analysis, this feedback results in increasing the current through r_{o3} to approximately $\frac{1}{2}\beta_3 i_x$, and thus the voltage across r_{o3} and the output resistance increase by the same factor, $\frac{1}{2}\beta_3$. Finally, note that the factor $\frac{1}{2}$ is because only half of i_x is mirrored back to the base of Q_3.

The Wilson mirror is preferred over the cascode circuit because the latter has the same dependence on β as the simple mirror. However, like the cascode mirror, the Wilson mirror requires an additional V_{BE} drop for its operation; that is, for proper operation we must allow for 1 V or so across the Wilson mirror output.

EXERCISE

7.21 For $\beta = 100$ and $r_o = 100\ \text{k}\Omega$, contrast the Wilson mirror and the simple mirror by evaluating the transfer-ratio error due to finite β, and the output resistance.
Ans. Transfer-ratio error: 0.02% for Wilson as opposed to 2% for the simple circuit; $R_o = 5\ \text{M}\Omega$ for Wilson compared to 100 kΩ for the simple circuit

7.5.4 The Wilson MOS Mirror

Figure 7.35(a) shows the MOS version of the Wilson mirror. Obviously there is no β error to reduce here, and the advantage of the MOS Wilson lies in its enhanced output resistance.

To determine the output resistance of the Wilson MOS mirror, we set $I_{REF} = 0$, and apply a test voltage v_x to the output node, as shown in Fig. 7.35(b). Our purpose is to determine the current i_x and hence R_o as

$$R_o = v_x / i_x$$

Rather than replacing each transistor with its hybrid-π, equivalent-circuit model, we shall perform the analysis directly on the circuit. For this purpose, we have "pulled r_o out" of each transistor and shown it separately.

Observe that the current i_x that enters the drain of Q_3 must exit at its source. Thus the current that feeds the input side of the Q_1-Q_2 mirror is equal to i_x. Most of this current will flow in the drain proper of Q_1 (i.e., only a very small fraction flows through r_{o1}) and will give rise to a voltage $v \simeq i_x/g_{m1}$, where $1/g_{m1}$ is the approximate resistance of the diode-connected transistor Q_1. The current-mirror action of (Q_1, Q_2) forces a current equal to i_x to flow through the drain proper of Q_2. Now, since the current in the drain of Q_2 is forced (by the connection to the gate of Q_3) to be zero, all of i_x must flow through r_{o2}, resulting in a voltage $-i_x r_{o2}$. This is the voltage fed back to the gate of Q_3. The drain current of Q_3 can now be found as

$$\begin{aligned}
i_{d3} &= g_{m3} v_{gs3} \\
&= g_{m3}(v_{g3} - v_{s3}) \\
&= g_{m3}(-i_x r_{o2} - i_x / g_{m1}) \\
&\simeq -(g_{m3} r_{o2}) i_x
\end{aligned}$$

A node equation at the drain of Q_3 gives the current through r_{o3} as $(i_x - i_{d3}) = i_x + g_{m3} r_{o2} i_x \simeq g_{m3} r_{o2} i_x$. Finally, we can express v_x as the sum of the voltage drop across r_{o3} and the voltage v across Q_1,

$$\begin{aligned}
v_x &= g_{m3} r_{o2} i_x r_{o3} + v \\
&= (g_{m3} r_{o3} r_{o2}) i_x + (i_x / g_{m1}) \\
&\simeq g_{m3} r_{o3} r_{o2} i_x
\end{aligned}$$

and obtain

$$R_o = \frac{v_x}{i_x} = (g_{m3} r_{o3}) r_{o2} \tag{7.82}$$

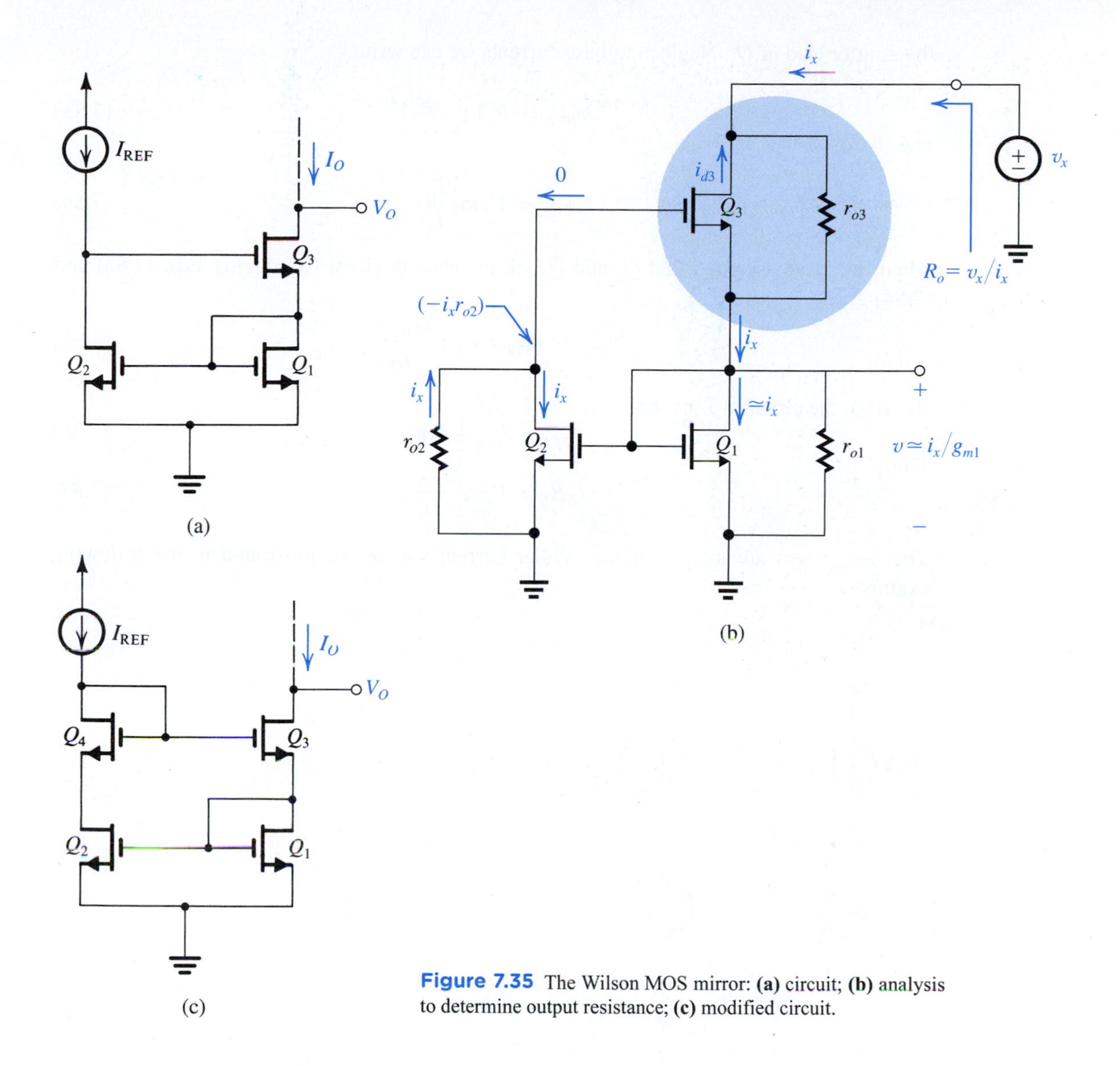

Figure 7.35 The Wilson MOS mirror: **(a)** circuit; **(b)** analysis to determine output resistance; **(c)** modified circuit.

Thus, the Wilson MOS mirror exhibits an increase of output resistance by a factor $(g_{m3}r_{o3})$, an identical result to that achieved in the cascode mirror. Here the increase in R_o, as demonstrated in the analysis above, is a result of the negative feedback obtained by connecting the drain of Q_2 to the gate of Q_3. Finally, to balance the two branches of the mirror and thus avoid the systematic current error resulting from the difference in V_{DS} between Q_1 and Q_2, the circuit can be modified as shown in Fig. 7.35(c).

7.5.5 The Widlar Current Source[5]

Our final current-source circuit, known as the **Widlar current source**, is shown in Fig. 7.36. It differs from the basic current mirror circuit in an important way: A resistor R_E is included in

[5]Named after Robert Widlar, a pioneer in analog IC design.

the emitter lead of Q_2. Neglecting base currents we can write

$$V_{BE1} = V_T \ln\left(\frac{I_{REF}}{I_S}\right) \tag{7.83}$$

and

$$V_{BE2} = V_T \ln\left(\frac{I_O}{I_S}\right) \tag{7.84}$$

where we have assumed that Q_1 and Q_2 are matched devices. Combining Eqs. (7.83) and (7.84) gives

$$V_{BE1} - V_{BE2} = V_T \ln\left(\frac{I_{REF}}{I_O}\right) \tag{7.85}$$

But from the circuit we see that

$$V_{BE1} = V_{BE2} + I_O R_E \tag{7.86}$$

Thus,

$$I_O R_E = V_T \ln\left(\frac{I_{REF}}{I_O}\right) \tag{7.87}$$

The design and advantages of the Widlar current source are illustrated in the following example.

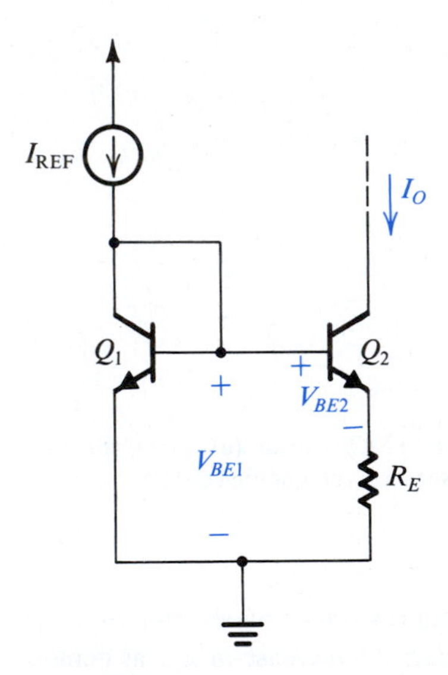

Figure 7.36 The Widlar current source.

Example 7.6

The two circuits for generating a constant current $I_O = 10\ \mu A$ shown in Fig. 7.37 operate from a 10-V supply. Determine the values of the required resistors, assuming that V_{BE} is 0.7 V at a current of 1 mA and neglecting the effect of finite β.

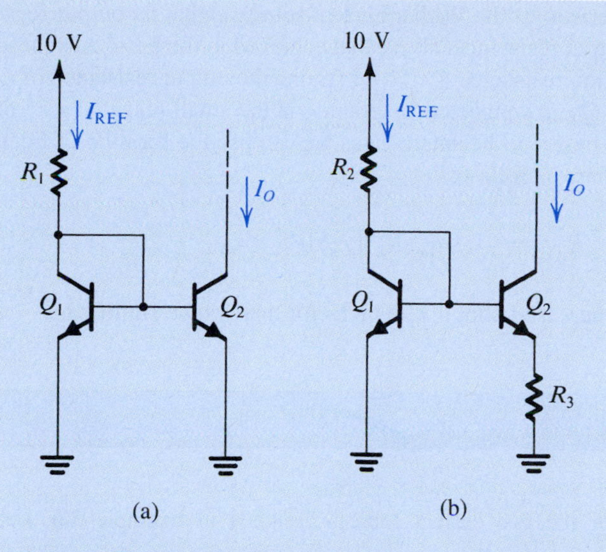

Figure 7.37 Circuits for Example 7.6.

Solution

For the basic current-source circuit in Fig. 7.37(a) we choose a value for R_1 to result in $I_{REF} = 10\ \mu A$. At this current, the voltage drop across Q_1 will be

$$V_{BE1} = 0.7 + V_T \ln\left(\frac{10\ \mu A}{1\ mA}\right) = 0.58\ V$$

Thus,

$$R_1 = \frac{10 - 0.58}{0.01} = 942\ k\Omega$$

For the Widlar circuit in Fig. 7.37(b) we must first decide on a suitable value for I_{REF}. If we select $I_{REF} = 1$ mA, then $V_{BE1} = 0.7$ V and R_2 is given by

$$R_2 = \frac{10 - 0.7}{1} = 9.3\ k\Omega$$

The value of R_3 can be determined using Eq. (7.87) as follows:

$$10 \times 10^{-6} R_3 = 0.025 \ln\left(\frac{1\ mA}{10\ \mu A}\right)$$

$$R_3 = 11.5\ k\Omega$$

From the above example we observe that using the Widlar circuit allows the generation of a small constant current using relatively small resistors. This is an important advantage that results in considerable savings in chip area. In fact the circuit of Fig. 7.37(a), requiring a 942-kΩ resistance, is totally impractical for implementation in IC form because of the very-high value of resistor R_1.

Another important characteristic of the Widlar current source is that its output resistance is high. The increase in the output resistance, above that achieved in the basic current source, is due to the emitter-degeneration resistance R_E. To determine the output resistance of Q_2, we assume that since the base of Q_2 is connected to ground via the small resistance r_e of Q_1, the incremental voltage at the base will be small. Thus we can use the formula in Eq. (7.51) and adapt it for our purposes here as follows:

$$R_{out} \simeq [1 + g_m(R_E \| r_\pi)]r_o \tag{7.88}$$

Thus the output resistance is increased above r_o by a factor that can be significant.

EXERCISE

7.22 Find the output resistance of each of the two current sources designed in Example 7.6. Let $V_A = 100$ V and $\beta = 100$.
Ans. 10 MΩ; 54 MΩ

7.6 Some Useful Transistor Pairings

The cascode configuration studied in Section 7.3 combines CS and CG MOS transistors (CE and CB bipolar transistors) to great advantage. The key to the superior performance of the resulting combination is that the transistor pairing is done in a way that maximizes the advantages and minimizes the shortcomings of each of the two individual configurations. In this section we present a number of other such transistor pairings. In each case the transistor pair can be thought of as a compound device; thus the resulting amplifier may be considered as a single stage.

7.6.1 The CC–CE, CD–CS, and CD–CE Configurations

Figure 7.38(a) shows an amplifier formed by cascading a common-collector (emitter follower) transistor Q_1 with a common-emitter transistor Q_2. This circuit has two main advantages over the CE amplifier. First, the emitter follower increases the input resistance by a factor equal to $(\beta_1 + 1)$. As a result, the overall voltage gain is increased, especially if the resistance of the signal source is large. Second, it will be shown in Chapter 9 that the CC–CE amplifier can exhibit much wider bandwidth than that obtained with the CE amplifier.

The MOS counterpart of the CC–CE amplifier, namely, the CD–CS configuration, is shown in Fig. 7.38(b). Here, since the CS amplifier alone has an infinite input resistance, the sole purpose for adding the source-follower stage is to increase the amplifier bandwidth, as will be seen in Chapter 9. Finally, Fig. 7.38(c) shows the BiCMOS version of this circuit type. Compared to the bipolar circuit in Fig. 7.38(a), the BiCMOS circuit has an infinite input resistance. Compared to the MOS circuit in Fig. 7.38(b), the BiCMOS circuit typically has a higher g_{m2}.

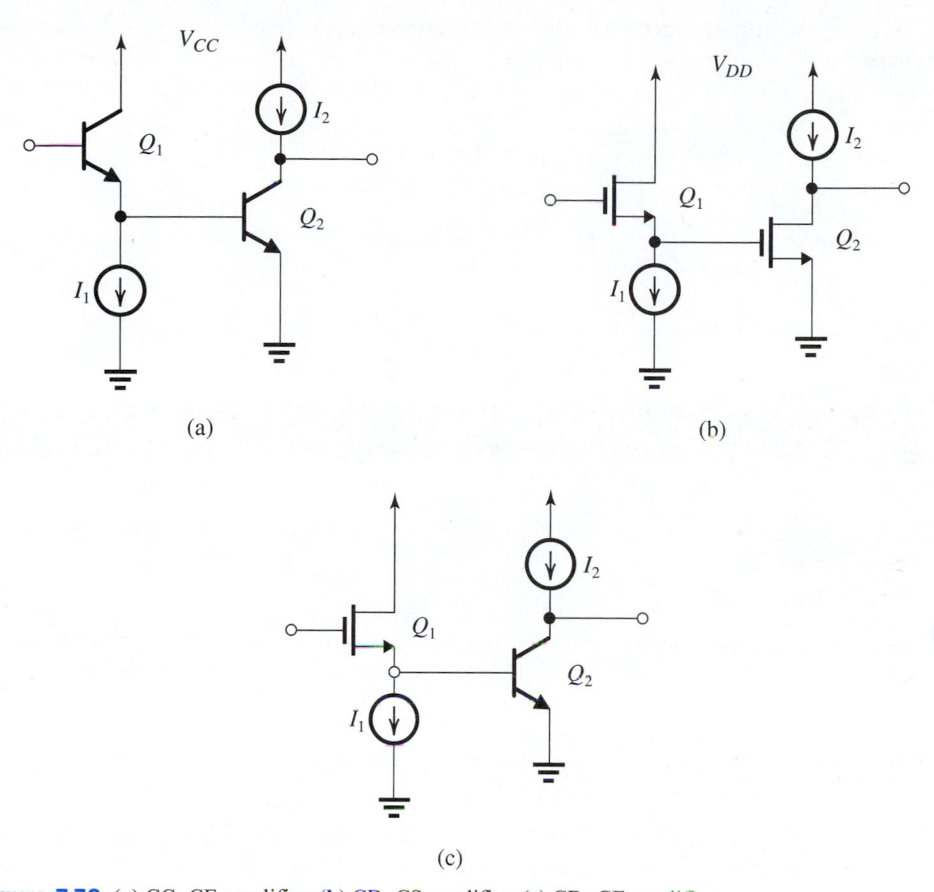

Figure 7.38 (**a**) CC–CE amplifier; (**b**) CD–CS amplifier; (**c**) CD–CE amplifier.

Example 7.7

For the CC–CE amplifier in Fig. 7.38(a) let $I_1 = I_2 = 1$ mA and assume identical transistors with $\beta = 100$. Find the input resistance R_{in} and the overall voltage gain obtained when the amplifier is fed with a signal source having $R_{\text{sig}} = 4\ \text{k}\Omega$ and loaded with a resistance $R_L = 4\ \text{k}\Omega$. Compare the results with those obtained with a common-emitter amplifier operating under the same conditions. Ignore r_o.

Solution

At an emitter current of 1 mA, Q_1 and Q_2 have

$$g_m = 40\ \text{mA/V}$$

$$r_e = 25\ \Omega$$

$$r_\pi = \frac{\beta}{g_m} = \frac{100}{40} = 2.5\ \text{k}\Omega$$

Example 7.7 *continued*

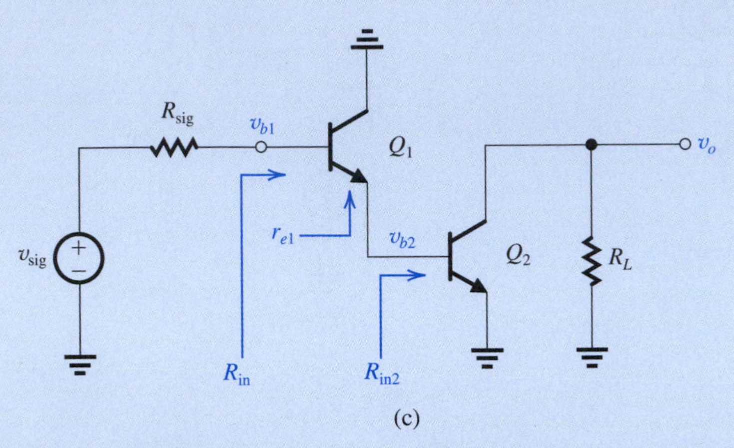

Figure 7.39 Circuit for Example 7.7.

Referring to Fig. 7.39 we can find

$$R_{in2} = r_{\pi 2} = 2.5\ \text{k}\Omega$$

$$R_{in} = (\beta_1 + 1)\ (r_{e1} + R_{in2})$$

$$= 101\,(0.025 + 2.5) = 255\ \text{k}\Omega$$

$$\frac{v_{b1}}{v_{sig}} = \frac{R_{in}}{R_{in} + R_{sig}} = \frac{255}{255 + 4} = 0.98\ \text{V/V}$$

$$\frac{v_{b2}}{v_{b1}} = \frac{R_{in2}}{R_{in2} + r_{e1}} = \frac{2.5}{2.5 + 0.025} = 0.99\ \text{V/V}$$

$$\frac{v_o}{v_{b2}} = -g_{m2}\,R_L = -40 \times 4 = -160\ \text{V/V}$$

Thus,

$$G_v = \frac{v_o}{v_{sig}} = -160 \times 0.99 \times 0.98 = -155\ \text{V/V}$$

For comparison, a CE amplifier operating under the same conditions will have

$$R_{in} = r_\pi = 2.5\ \text{k}\Omega$$

$$G_v = \frac{R_{in}}{R_{in} + R_{sig}}(-g_m R_L)$$

$$= \frac{2.5}{2.5 + 4}(-40 \times 4)$$

$$= -61.5\ \text{V/V}$$

EXERCISE

7.23 Repeat Example 7.7 for the CD–CE configuration of Fig. 7.38(c). Let $I_1 = I_2 = 1$ mA, $\beta_2 = 100$, and $k_{n1} = 8$ mA/V^2; neglect r_o of both transistors. Find R_{in} and G_v when $R_{sig} = 4$ kΩ (as in Example 7.7) and $R_{sig} = 400$ kΩ. What would G_v of the CC–CE amplifier in Example 7.7 become for $R_{sig} = 400$ kΩ?
Ans. $R_{in} = \infty$; $G_v = -145.5$ V/V, independent of R_{sig}; -61.7 V/V

7.6.2 The Darlington Configuration[6]

Figure 7.40(a) shows a popular BJT circuit known as the **Darlington configuration**. It can be thought of as a variation of the CC–CE circuit with the collector of Q_1 connected to that of Q_2. Alternatively, the **Darlington pair** can be thought of as a composite transistor with $\beta = \beta_1\beta_2$. It can therefore be used to implement a high-performance voltage follower, as illustrated in Fig. 7.40(b). Note that in this application the circuit can be considered as the cascade connection of two common-collector transistors (i.e., a CC–CC configuration).

Since the transistor β depends on the dc bias current, it is possible that Q_1 will be operating at a very low β, rendering the β-multiplication effect of the Darlington pair rather ineffective. A simple solution to this problem is to provide a bias current for Q_1, as shown in Fig. 7.40(c).

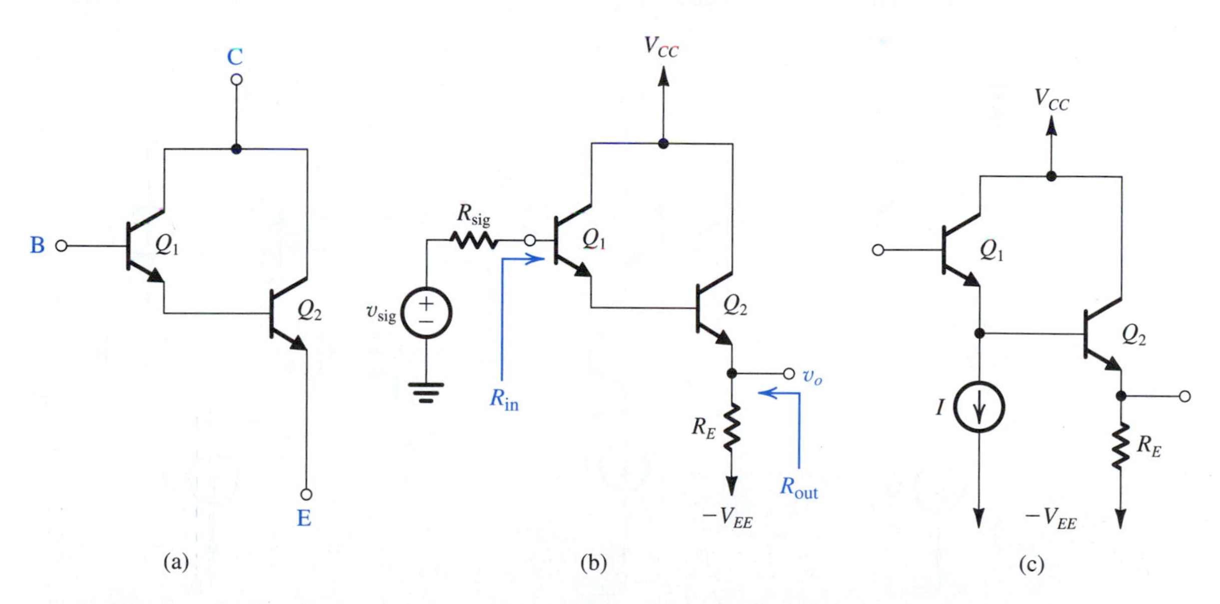

Figure 7.40 **(a)** The Darlington configuration; **(b)** voltage follower using the Darlington configuration; **(c)** the Darlington follower with a bias current I supplied to Q_1 to ensure that its β remains high.

[6]Named after Sidney Darlington, a pioneer in filter design and transistor circuit design.

EXERCISE

7.24 For the Darlington voltage follower in Fig. 7.40(b), show that:

$$R_{in} = (\beta_1 + 1)[r_{e1} + (\beta_2 + 1)(r_{e2} + R_E)]$$

$$R_{out} = R_E \| \left[r_{e2} + \frac{r_{e1} + [R_{sig}/(\beta_1 + 1)]}{\beta_2 + 1} \right]$$

$$\frac{v_o}{v_{sig}} = \frac{R_E}{R_E + r_{e2} + [r_{e1} + R_{sig}/(\beta_1 + 1)]/(\beta_2 + 1)}$$

Evaluate R_{in}, R_{out}, and v_o/v_{sig} for the case $I_{E2} = 5$ mA, $\beta_1 = \beta_2 = 100$, $R_E = 1$ kΩ, and $R_{sig} = 100$ kΩ.

Ans. 10.3 MΩ; 20 Ω; 0.98 V/V

7.6.3 The CC–CB and CD–CG Configurations

Cascading an emitter follower with a common-base amplifier, as shown in Fig. 7.41(a), results in a circuit with a low-frequency gain approximately equal to that of the CB but with the problem of the low input resistance of the CB solved by the buffering action of the CC stage. It will be shown in Chapter 9 that this circuit exhibits wider bandwidth than that obtained with a CE amplifier of the same gain. Note that the biasing current sources shown in Fig. 7.41(a) ensure that each of Q_1 and Q_2 is operating at a bias current I. We are not

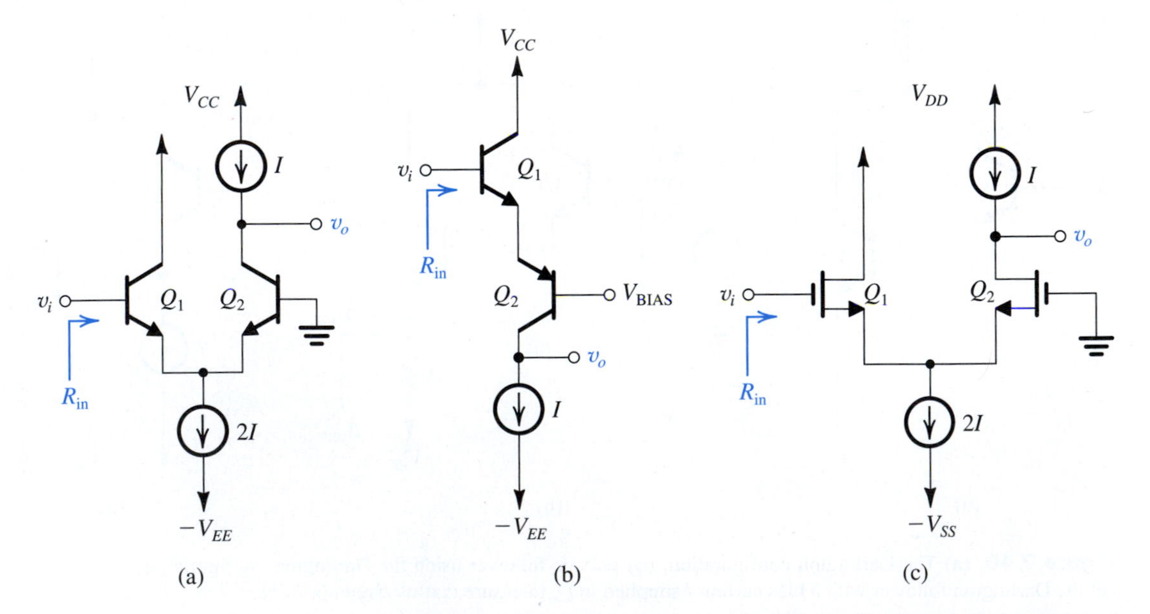

Figure 7.41 **(a)** A CC–CB amplifier. **(b)** Another version of the CC–CB circuit with Q_2 implemented using a *pnp* transistor. **(c)** The MOSFET version of the circuit in **(a)**.

showing, however, how the dc voltage at the base of Q_1 is set or the circuit that determines the dc voltage at the collector of Q_2. Both issues are usually looked after in the larger circuit of which the CC–CB amplifier is a part.

An interesting version of the CC–CB configuration is shown in Fig. 7.41(b). Here the CB stage is implemented with a *pnp* transistor. Although only one current source is now needed, observe that we also need to establish an appropriate bias voltage at the base of Q_2. This circuit is part of the internal circuit of the popular 741 op amp, which will be studied in Chapter 12.

The MOSFET version of the circuit in Fig. 7.41(a) is the CD–CG amplifier shown in Fig. 7.41(c).

Example 7.8

For the CC–CB amplifiers in Fig. 7.41(a) and (b), find R_{in}, v_o/v_i, and v_o/v_{sig} when each amplifier is fed with a signal source having a resistance R_{sig}, and a load resistance R_L is connected at the output. For simplicity, neglect r_o.

Solution

The analysis of both circuits is illustrated in Fig. 7.42. Observe that both amplifiers have the same R_{in} and v_o/v_i. The overall voltage gain v_o/v_{sig} can be found as

$$\frac{v_o}{v_{sig}} = \frac{R_{in}}{R_{in} + R_{sig}} \frac{\alpha_2 R_L}{2r_e}$$

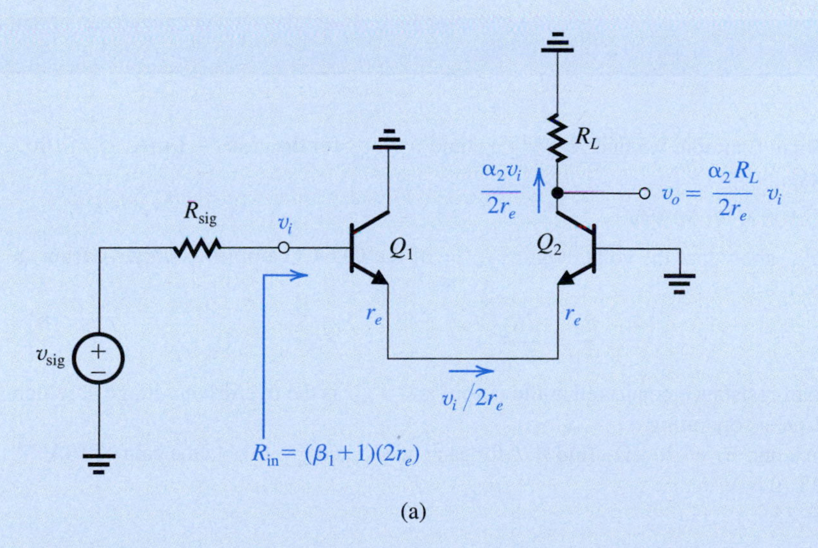

Figure 7.42 Circuits for Example 7.8. (*continued on following page*)

Example 7.8 *continued*

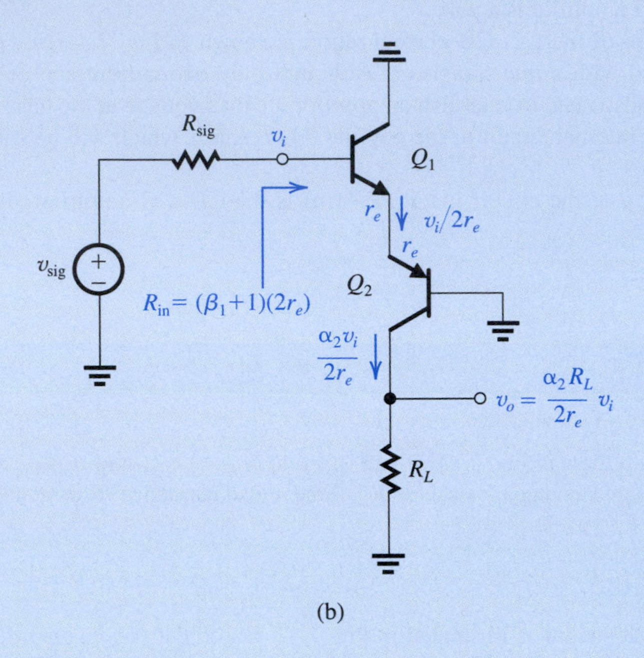

Figure 7.42 (*continued*)

EXERCISES

7.25 For the amplifiers in Example 7.8 find R_{in}, v_o/v_i, and v_o/v_{sig} for the case $I = 1$ mA, $\beta = 100$. $R_L = R_{sig} = 5\ \text{k}\Omega$.

Ans. 5.05 kΩ; 100 V/V; 50 V/V

D7.26 (a) Neglecting r_{o1}, show that the voltage gain v_o/v_i of the CD–CG amplifier shown earlier in Fig. 7.4(c) is given by

$$\frac{v_o}{v_i} = \frac{IR_L}{V_{OV}}$$

where R_L is a load resistance connected at the output and V_{OV} is the overdrive voltage at which each of Q_1 and Q_2 is operating.
(b) For $I = 0.1$ mA and $R_L = 20\ \text{k}\Omega$, find W/L for each of Q_1 and Q_2 to obtain a gain of 10 V/V. Assume $k_n' = 200\ \mu\text{A/V}^2$.

Ans. (b) $W/L = 25$

Summary

- Integrated-circuit fabrication technology offers the circuit designer many exciting opportunities, the most important of which is the large number of inexpensive small-area MOS transistors. An overriding concern for IC designers, however, is the minimization of chip area or "silicon real estate." As a result, large-valued resistors and capacitors are virtually absent.

- The basic gain cell of IC amplifiers is the CS (CE) amplifier with a current-source load. For an ideal current-source load (i.e., one with infinite output resistance), the transistor operates in an open-circuit fashion and thus provides the maximum gain possible, $A_{vo} = -g_m r_o = -A_0$.

- The intrinsic gain A_0 is given by $A_0 = V_A/V_T$ for a BJT and $A_0 = V_A/(V_{OV}/2)$ for a MOSFET. For a BJT, A_0 is constant independent of bias current and device dimensions. For a MOSFET, A_0 is inversely proportional to $\sqrt{I_D}$ (see Eq. 7.15).

- Simple current-source loads reduce the gain realized in the basic gain cell because of their finite output resistance (usually comparable to the value of r_o of the amplifying transistor).

- To raise the output resistance of the CS or CE transistor, we stack a CG or CB transistor on top. This is cascoding. The CG or CB transistor in the cascode passes the current $g_{m1}v_i$ provided by the CS or CE transistor to the output but increases the resistance at the output from r_{o1} to $(g_{m2}r_{o2})r_{o1}$ in the MOS case $[g_{m2}(r_{o1} \| r_{\pi 2})r_{o2}$ in the bipolar case]. The maximum output resistance achieved in the bipolar case is $\beta_2 r_{o2}$.

- A MOS cascode amplifier operating with an ideal current-source load achieves a gain of $(g_m r_o)^2 = A_0^2$.

- To realize the full advantage of cascoding, the load current-source must also be cascoded, in which case a gain as high as $\frac{1}{2}A_0^2$ can be obtained.

- Double cascoding is possible in the MOS case only. However, the large number of transistors in the stack between the power-supply rails results in the disadvantage of a severely limited output-signal swing. The folded-cascode configuration helps resolve this issue.

- A CS amplifier with a resistance R_s in its source lead has an output resistance $R_o \simeq (1 + g_m R_s)r_o$. The corresponding formula for the BJT case is $R_o = [1 + g_m(R_e \| r_\pi)]r_o$.

- Biasing in integrated circuits utilizes current sources. As well, current sources are used as load devices. Typically an accurate and stable reference current is generated and then replicated to provide bias currents for the various amplifier stages on the chip. The heart of the current-steering circuitry utilized to perform this function is the current mirror.

- The MOS current mirror has a current transfer ratio of $(W/L)_2/(W/L)_1$. For a bipolar mirror, the ratio is I_{S2}/I_{S1}.

- Bipolar mirrors suffer from the finite β, which reduces the accuracy of the current transfer ratio.

- Both bipolar and MOS mirrors of the basic type have a finite output resistance equal to r_o of the output device. Also, for proper operation, a voltage of at least 0.3 V is required across the output transistor of a simple bipolar mirror ($|V_{OV}|$ for the MOS case).

- Cascoding can be applied to current mirrors to increase their output resistances. An alternative that also solves the β problem in the bipolar case is the Wilson circuit. The MOS Wilson mirror has an output resistance of $(g_m r_o)r_o$, and the BJT version has an output resistance of $\frac{1}{2}\beta r_o$. Both the cascode and Wilson mirrors require at least 1 V or so for proper operation.

- The Widlar current source provides an area-efficient way to implement a low-valued constant-current source that also has a high output resistance.

- Preceding the CE (CS) transistor with an emitter follower (a source follower) results in increased input resistance in the BJT case and wider bandwidth in both the BJT and MOS cases.

- Preceding the CB (CG) transistor with an emitter follower (a source follower) solves the low-input-resistance problem of the CB and CG configurations.

- The Darlington configuration results in an equivalent BJT with a current gain approaching β^2.

Appendix 7.A Comparison of the MOSFET and the BJT

In this appendix we present a comparison of the characteristics of the two major electronic devices: the MOSFET and the BJT. To facilitate this comparison, typical values for the important parameters of the two devices are first presented. We also discuss the design parameters available with each of the two devices, such as I_C in the BJT, and I_D and V_{OV} in the MOSFET, and the trade-offs encountered in deciding on suitable values for these.

7.A.1 Typical Values of MOSFET Parameters

Typical values for the important parameters of NMOS and PMOS transistors fabricated in a number of CMOS processes are shown in Table 7.A.1. Each process is characterized by the minimum allowed channel length, L_{min}; thus, for example, in a 0.18-μm process, the smallest transistor has a channel length $L = 0.18$ μm. The technologies presented in Table 7.A.1 are in descending order of channel length, with that having the shortest channel length being the most modern. Although the 0.8-μm process is now obsolete, its data are included to show trends in the values of various parameters. It should also be mentioned that although Table 7.A.1 stops at the 0.13-μm process, by 2009 there were 90-, 65-, and 45-nm processes available, and processes down to 22 nm were in various stages of development. The 0.18-μm and the 0.13-μm processes, however, remained popular in the design of analog ICs. The most recently announced digital ICs utilize 65-nm and 45-nm processes and pack as many as 2.3 billion transistors onto one chip. An important caution is in order regarding the data presented in Table 7.A.1: These data do *not* pertain to any particular commercially available process. Accordingly, these generic data are not intended for use in an actual IC design; rather, they show trends and, as we shall see, help to illustrate design trade-offs as well as enable us to work out design examples and problems with parameter values that are as realistic as possible.

As indicated in Table 7.A.1, the trend has been to reduce the minimum allowable channel length. This trend has been motivated by the desire to pack more transistors on a chip as well as to operate at higher speeds or, in analog terms, over wider bandwidths.

Observe that the oxide thickness, t_{ox}, scales down with the channel length, reaching 2.7 nm for the 0.13-μm process. (The 65-nm process, not shown in Table 7.A.1, has an oxide thickness of 1.2 nm.) Since the oxide capacitance C_{ox} is inversely proportional to t_{ox},

Table 7.A.1 Typical Values of CMOS Device Parameters

	0.8 μm		0.5 μm		0.25 μm		0.18 μm		0.13 μm	
Parameter	NMOS	PMOS	NMOS	PMOS	NMOS	PMOS	NMOS	PMOS	NMOS	PMOS
t_{ox} (nm)	15	15	9	9	6	6	4	4	2.7	2.7
C_{ox} (fF/μm^2)	2.3	2.3	3.8	3.8	5.8	5.8	8.6	8.6	12.8	12.8
μ (cm^2/V·s)	550	250	500	180	460	160	450	100	400	100
μC_{ox} (μA/V^2)	127	58	190	68	267	93	387	86	511	128
V_{t0} (V)	0.7	−0.7	0.7	−0.8	0.5	−0.6	0.5	−0.5	0.4	−0.4
V_{DD} (V)	5	5	3.3	3.3	2.5	2.5	1.8	1.8	1.3	1.3
$\|V_A'\|$ (V/μm)	25	20	20	10	5	6	5	6	5	6
C_{ov} (fF/μm)	0.2	0.2	0.4	0.4	0.3	0.3	0.37	0.33	0.36	0.33

we see that C_{ox} increases as the technology scales down. The surface mobility μ decreases as the technology minimum-feature size is decreased, and μ_p decreases faster than μ_n. As a result, the ratio of μ_p to μ_n has been decreasing with each generation of technology, falling from about 0.5 for older technologies to 0.2 or so for the newer ones. Despite the reduction of μ_n and μ_p, the transconductance parameters $k_n' = \mu_n C_{ox}$ and $k_p' = \mu_p C_{ox}$ have been steadily increasing. As a result, modern short-channel devices achieve required levels of bias currents at lower overdrive voltages. As well, they achieve higher transconductance, a major advantage.

Although the magnitudes of the threshold voltages V_{tn} and V_{tp} have been decreasing with $L_{\min}$ from about 0.7–0.8 V to 0.3–0.4 V, the reduction has not been as large as that of the power supply V_{DD}. The latter has been reduced dramatically, from 5 V for older technologies to 1.3 V for the 0.13-μm process (and approaching 1 V for the 45-nm process). This reduction has been necessitated by the need to keep the electric fields in the smaller devices from reaching very high values. Another reason for reducing V_{DD} is to keep power dissipation as low as possible given that the IC chip now has a much larger number of transistors.[7]

The fact that in modern short-channel CMOS processes $|V_t|$ has become a much larger proportion of the power-supply voltage poses a serious challenge to the circuit design engineer. Recalling that $|V_{GS}| = |V_t| + |V_{OV}|$, where V_{OV} is the overdrive voltage, to keep $|V_{GS}|$ reasonably small, $|V_{OV}|$ for modern technologies is usually in the range of 0.1 V to 0.2 V. To appreciate this point further, recall that to operate a MOSFET in the saturation region, $|V_{DS}|$ must exceed $|V_{OV}|$; thus, to be able to have a number of devices stacked between the power-supply rails in a regime in which V_{DD} is only 1.8 V or lower, we need to keep $|V_{OV}|$ as low as possible. We will shortly see, however, that operating at a low $|V_{OV}|$ has some drawbacks.

Another significant though undesirable feature of modern deep submicron ($L_{\min} < 0.25$ μm) CMOS technologies is that the channel-length modulation effect is very pronounced. As a result, V_A' has decreased to about 5 V/μm, which combined with the decreasing values of L has caused the Early voltage $V_A = V_A' L$ to become very small. Correspondingly, short-channel MOSFETs exhibit low output resistances.

When we study the MOSFET high-frequency[8] equivalent-circuit model in Section 9.2 and the high-frequency response of the common-source amplifier in Section 9.3, we will learn that two major MOSFET capacitances are C_{gs} and C_{gd}. While C_{gs} has an overlap component,[9] C_{gd} is entirely an overlap capacitance. Both C_{gd} and the overlap component of C_{gs} are almost equal and are denoted C_{ov}. The last line of Table 7.A.1 provides the value of C_{ov} per micron of gate width. Although the normalized C_{ov} has been staying more or less constant with the reduction in $L_{\min}$, we will shortly see that the shorter devices exhibit much higher operating speeds and wider amplifier bandwidths than the longer devices. Specifically, we will, for example, see that f_T for a 0.25-μm NMOS transistor can be as high as 10 GHz.

---

[7]Chip power dissipation is a very serious issue, with some ICs dissipating as much as 100 W. As a result, an important current area of research concerns what is termed "power-aware design."

[8]For completeness, this appendix includes material on the high-frequency models and operation of both the MOSFET and the BJT. These topics are covered in Chapter 9. The reader can easily skip the appendix paragraphs dealing with these topics until Chapter 9 has been studied.

[9]Overlap capacitances result because the gate electrode overlaps the source and drain diffusions (Fig. 5.1).

7.A.2 Typical Values of IC BJT Parameters

Table 7.A.2 provides typical values for the major parameters that characterize integrated-circuit bipolar transistors. Data are provided for devices fabricated in two different processes: the standard, old process, known as the "high-voltage process," and an advanced, modern process, referred to as a "low-voltage process." For each process we show the parameters of the standard *npn* transistor and those of a special type of *pnp* transistor known as a **lateral *pnp*** (as opposed to **vertical**, as in the *npn* case) (see Appendix A). In this regard we should mention that a major drawback of standard bipolar integrated-circuit fabrication processes has been the lack of *pnp* transistors of a quality equal to that of the *npn* devices. Rather, there are a number of *pnp* implementations for which the lateral *pnp* is the most economical to fabricate. Unfortunately, however, as should be evident from Table 7.A.2, the lateral *pnp* has characteristics that are much inferior to those of the vertical *npn*. Note in particular the lower value of β and the much larger value of the forward transit time τ_F that determines the emitter–base diffusion capacitance C_{de} and, hence, the transistor speed of operation. The data in Table 7.A.2 can be used to show that the unity-gain frequency of the lateral *pnp* is 2 orders of magnitude lower than that of the *npn* transistor fabricated in the same process. Another important difference between the lateral *pnp* and the corresponding *npn* transistor is the value of collector current at which their β values reach their maximums: For the high-voltage process, for example, this current is in the tens of microamperes range for the *pnp* and in the milliampere range for the *npn*. On the positive side, the problem of the lack of high-quality *pnp* transistors has spurred analog circuit designers to come up with highly innovative circuit topologies that either minimize the use of *pnp* transistors or minimize the dependence of circuit performance on that of the *pnp*. We shall encounter some of these ingenious circuits later in this book.

The dramatic reduction in device size achieved in the advanced low-voltage process should be evident from Table 7.A.2. As a result, the scale current I_S also has been reduced by about three orders of magnitude. Here we should note that the base width, W_B, achieved in the advanced process is on the order of 0.1 μm, as compared to a few microns in the standard high-voltage process. Note also the dramatic increase in speed; for the low-voltage *npn* transistor, $\tau_F = 10$ ps as opposed to 0.35 ns in the high-voltage process. As a result, f_T for the modern *npn* transistor is 10 GHz to 25 GHz, as compared to the 400 MHz to 600 MHz achieved in the high-voltage process. Although the Early voltage, V_A, for the modern process is lower than its value in the old high-voltage process, it is still reasonably high at 35 V. Another feature of the advanced process—and one that is not obvious from Table 7.A.2—is that β for the *npn*

Table 7.A.2 Typical Parameter Values for BJTs*

	Standard High-Voltage Process		Advanced Low-Voltage Process	
Parameter	*npn*	Lateral *pnp*	*npn*	Lateral *pnp*
A_E (μm^2)	500	900	2	2
I_S (A)	5×10^{-15}	2×10^{-15}	6×10^{-18}	6×10^{-18}
β_0 (A/A)	200	50	100	50
V_A (V)	130	50	35	30
V_{CEO} (V)	50	60	8	18
τ_F	0.35 ns	30 ns	10 ps	650 ps
C_{je0}	1 pF	0.3 pF	5 fF	14 fF
$C_{\mu 0}$	0.3 pF	1 pF	5 fF	15 fF
r_x (Ω)	200	300	400	200

*Adapted from Gray et al. (2001); see Appendix F.

peaks at a collector current of 50 μA or so. Finally, note that as the name implies, *npn* transistors fabricated in the low-voltage process break down at collector–emitter voltages of 8 V, versus 50 V or so for the high-voltage process. Thus, while circuits designed with the standard high-voltage process utilize power supplies of ±15 V (e.g., in commercially available op amps of the 741 type), the total power-supply voltage utilized with modern bipolar devices is 5 V (or even 2.5 V to achieve compatibility with some of the submicron CMOS processes).

7.A.3 Comparison of Important Characteristics

Table 7.A.3 provides a compilation of the important characteristics of the NMOS and the *npn* transistors. The material is presented in a manner that facilitates comparison. In the following, we comment on the various items in Table 7.A.3. As well, a number of numerical examples and exercises are provided to illustrate how the wealth of information in Table 7.A.3 can be put to use. Before proceeding, note that the PMOS and the *pnp* transistors can be compared in a similar way.

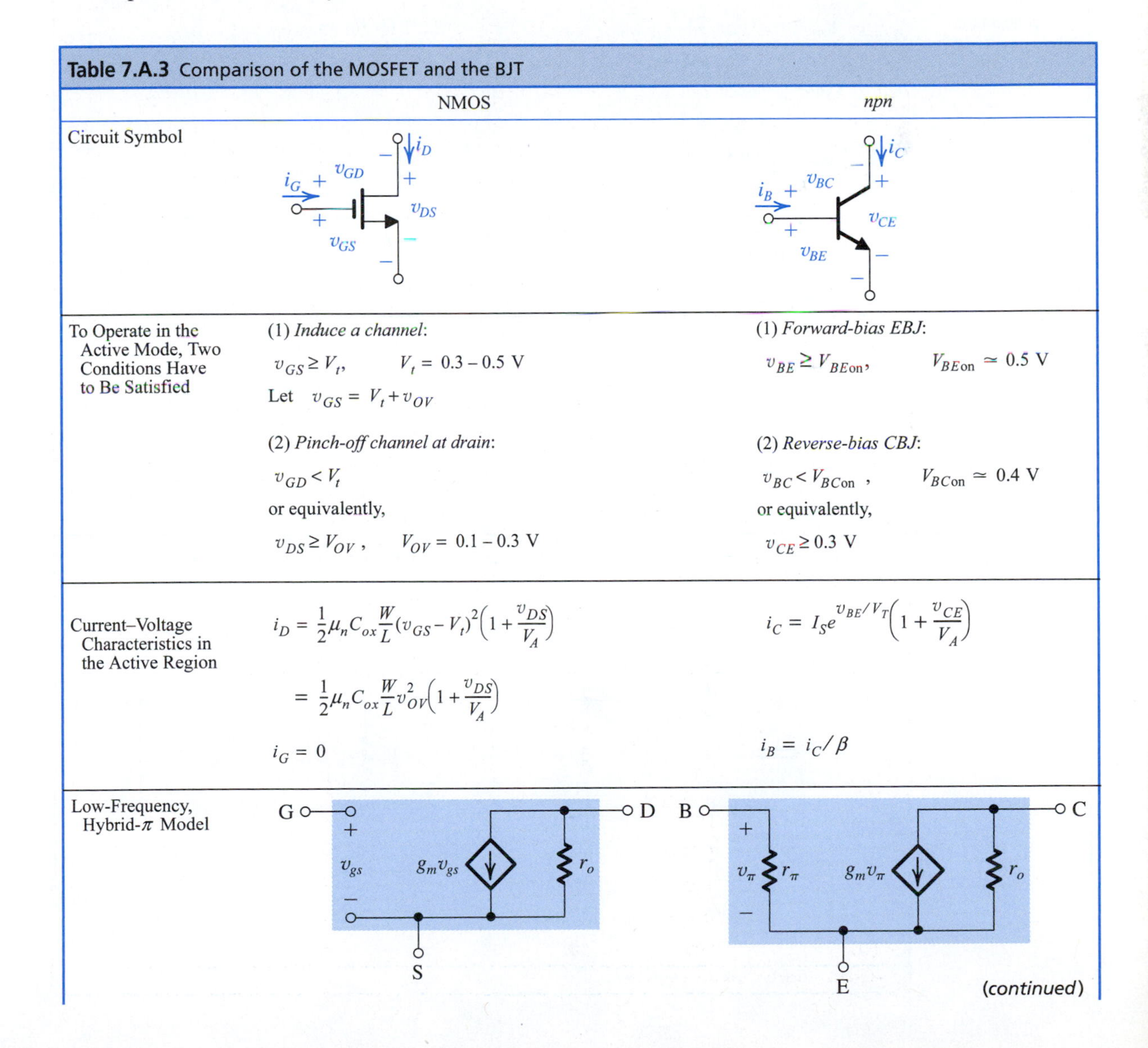

Table 7.A.3 Comparison of the MOSFET and the BJT

	NMOS	*npn*
Circuit Symbol		
To Operate in the Active Mode, Two Conditions Have to Be Satisfied	(1) *Induce a channel*: $v_{GS} \geq V_t$, $V_t = 0.3 - 0.5$ V Let $v_{GS} = V_t + v_{OV}$ (2) *Pinch-off channel at drain*: $v_{GD} < V_t$ or equivalently, $v_{DS} \geq V_{OV}$, $V_{OV} = 0.1 - 0.3$ V	(1) *Forward-bias EBJ*: $v_{BE} \geq V_{BEon}$, $V_{BEon} \simeq 0.5$ V (2) *Reverse-bias CBJ*: $v_{BC} < V_{BCon}$, $V_{BCon} \simeq 0.4$ V or equivalently, $v_{CE} \geq 0.3$ V
Current–Voltage Characteristics in the Active Region	$i_D = \frac{1}{2}\mu_n C_{ox}\frac{W}{L}(v_{GS} - V_t)^2\left(1 + \frac{v_{DS}}{V_A}\right)$ $= \frac{1}{2}\mu_n C_{ox}\frac{W}{L}v_{OV}^2\left(1 + \frac{v_{DS}}{V_A}\right)$ $i_G = 0$	$i_C = I_S e^{v_{BE}/V_T}\left(1 + \frac{v_{CE}}{V_A}\right)$ $i_B = i_C/\beta$
Low-Frequency, Hybrid-π Model		

(continued)

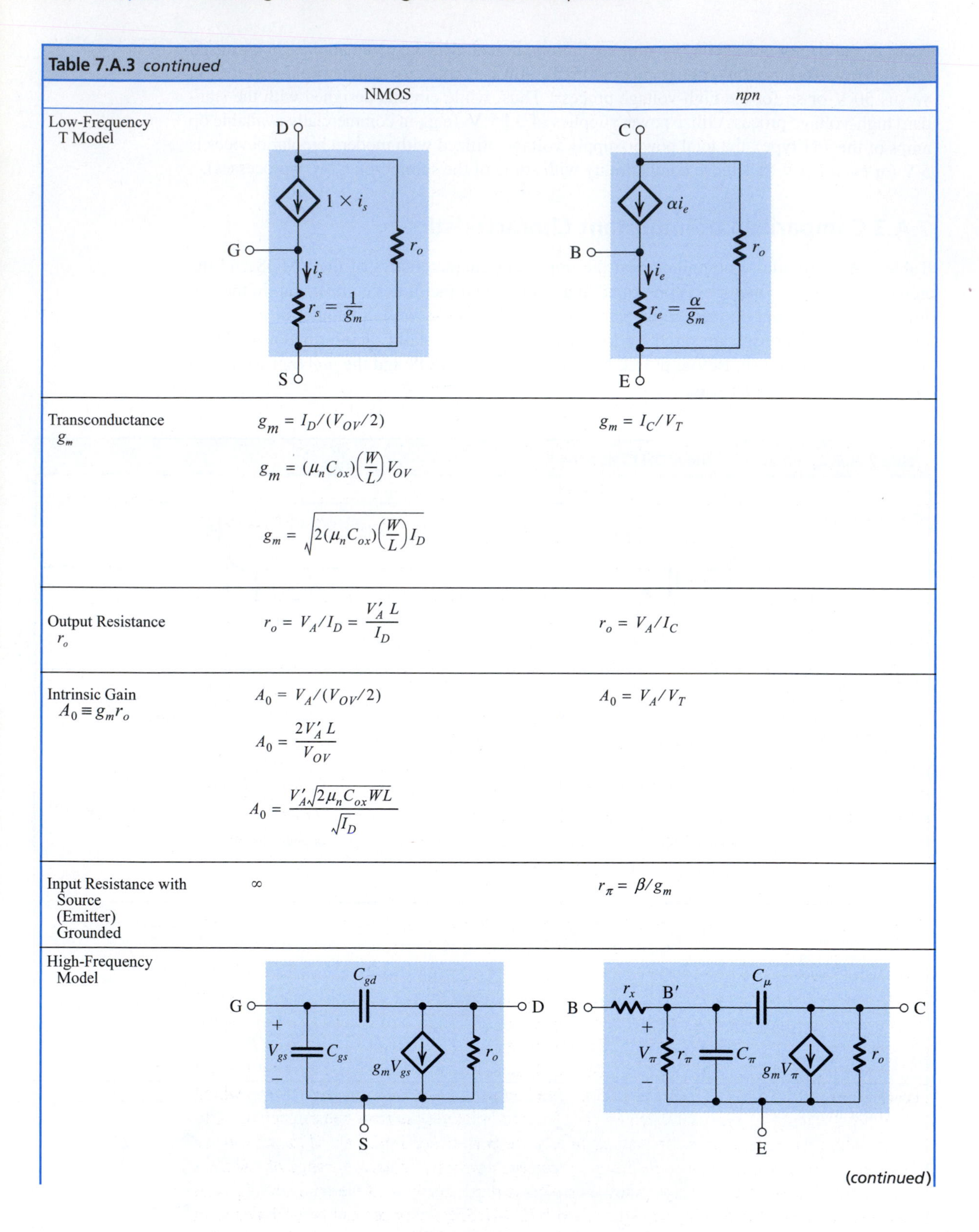

Table 7.A.3 *continued*

	NMOS	*npn*
Low-Frequency T Model	D, G, S; $1 \times i_s$; i_s; $r_s = \frac{1}{g_m}$; r_o	C, B, E; αi_e; i_e; $r_e = \frac{\alpha}{g_m}$; r_o
Transconductance g_m	$g_m = I_D/(V_{OV}/2)$ $g_m = (\mu_n C_{ox})\left(\frac{W}{L}\right)V_{OV}$ $g_m = \sqrt{2(\mu_n C_{ox})\left(\frac{W}{L}\right)I_D}$	$g_m = I_C/V_T$
Output Resistance r_o	$r_o = V_A/I_D = \frac{V'_A L}{I_D}$	$r_o = V_A/I_C$
Intrinsic Gain $A_0 \equiv g_m r_o$	$A_0 = V_A/(V_{OV}/2)$ $A_0 = \frac{2V'_A L}{V_{OV}}$ $A_0 = \frac{V'_A\sqrt{2\mu_n C_{ox} WL}}{\sqrt{I_D}}$	$A_0 = V_A/V_T$
Input Resistance with Source (Emitter) Grounded	∞	$r_\pi = \beta/g_m$
High-Frequency Model	G, D, S; C_{gd}; $+$ V_{gs} $-$; C_{gs}; $g_m V_{gs}$; r_o	B, C, E; r_x; B′; C_μ; $+$ V_π $-$; r_π; C_π; $g_m V_\pi$; r_o

(continued)

Table 7.A.3 *continued*

	NMOS	*npn*
Capacitances	$C_{gs} = \frac{2}{3}WLC_{ox} + WL_{ov}C_{ox}$ $C_{gd} = WL_{ov}C_{ox}$	$C_\pi = C_{de} + C_{je}$ $C_{de} = \tau_F g_m$ $C_{je} \simeq 2C_{je0}$ $C_\mu = C_{\mu 0}\Big/\left[1+\frac{V_{CB}}{V_{C0}}\right]^m$
Transition Frequency f_T	$f_T = \frac{g_m}{2\pi(C_{gs}+C_{gd})}$ For $C_{gs} \gg C_{gd}$ and $C_{gs} \simeq \frac{2}{3}WLC_{ox}$, $f_T \simeq \frac{1.5\mu_n V_{OV}}{2\pi L^2}$	$f_T = \frac{g_m}{2\pi(C_\pi + C_\mu)}$ For $C_\pi \gg C_\mu$ and $C_\pi \simeq C_{de}$, $f_T \simeq \frac{2\mu_n V_T}{2\pi W_B^2}$
Design Parameters	$I_D, V_{OV}, L, \frac{W}{L}$	I_C, V_{BE}, A_E(or I_S)
Good Analog Switch?	Yes, because the device is symmetrical and thus the i_D–v_{DS} characteristics pass directly through the origin.	No, because the device is asymmetrical with an offset voltage V_{CEoff}.

Operating Conditions At the outset, note that we shall use **active mode** or **active region** to denote both the active mode of operation of the BJT and the saturation mode of operation of the MOSFET.

The conditions for operating in the active mode are very similar for the two devices: The explicit threshold V_t of the MOSFET has V_{BEon} as its implicit counterpart in the BJT. Furthermore, for modern processes, V_{BEon} and V_t are almost equal.

Also, pinching off the channel of the MOSFET at the drain end is very similar to reverse biasing the CBJ of the BJT; the first makes i_D nearly independent of v_D, and the second makes I_C nearly independent of v_C. Note, however, that the asymmetry of the BJT results in V_{BCon} and V_{BEon} being unequal, while in the symmetrical MOSFET the operative threshold voltages at the source and the drain ends of the channel are identical (V_t). Finally, for both the MOSFET and the BJT to operate in the active mode, the voltage across the device (v_{DS}, v_{CE}) must be at least 0.1 V to 0.3 V.

Current–Voltage Characteristics The square-law control characteristic, i_D–v_{GS}, in the MOSFET should be contrasted with the exponential control characteristic, i_C–v_{BE}, of the BJT. Obviously, the latter is a much more sensitive relationship, with the result that i_C can vary over a very wide range (five decades or more) within the same BJT. In the MOSFET, the range of i_D achieved in the same device is much more limited. To appreciate this point further, consider the parabolic relationship between i_D and v_{OV}, and recall from our discussion above that v_{OV} is usually kept in a narrow range (0.1 V to 0.3 V).

Next we consider the effect of the device dimensions on its current. For the bipolar transistor, the control parameter is the area of the emitter–base junction (EBJ), A_E, which determines the scale current I_S. It can be varied over a relatively narrow range, such as 10 to 1. Thus, while the emitter area can be used to achieve current scaling in an IC (as we can see in Section 7.4 in connection with the design of current mirrors), its narrow range of variation reduces its significance as a design parameter. This is particularly so if we compare A_E with its counterpart in the MOSFET, the aspect ratio *W/L*. MOSFET devices can be designed with *W/L* ratios in a wide range, such as 1.0 to 500. As a result, *W/L* is a very significant MOS design parameter. Like A_E, it is also used in current scaling, as we can see in Section 7.4. Combining the possible range of variation of v_{OV} and W/L, one can design MOS transistors to operate over an i_D range of four decades or so.

The channel-length modulation in the MOSFET and the base-width modulation in the BJT are similarly modeled and give rise to the dependence of i_D (i_C) on v_{DS} (v_{CE}) and, hence, to the finite output resistance r_o in the active region. Two important differences, however, exist. In the BJT, V_A is solely a process-technology parameter and does not depend on the dimensions of the BJT. In the MOSFET, the situation is quite different: $V_A = V_A'L$, where V_A' is a process-technology parameter and L is the channel length used. Also, in modern deep submicron processes, V_A' is very low, resulting in V_A values that are lower than the corresponding values for the BJT.

The last, and perhaps most important, difference between the current–voltage characteristics of the two devices concerns the input current into the control terminal: While at low frequencies the gate current of the MOSFET is practically zero and the input resistance looking into the gate is practically infinite, the BJT draws base current i_B that is proportional to the collector current; that is, $i_B = i_C/\beta$. The finite base current and the corresponding finite input resistance looking into the base comprise a definite disadvantage of the BJT in comparison to the MOSFET. Indeed, it is the infinite input resistance of the MOSFET that has made possible analog and digital circuit applications that are not feasible with the BJT. Examples include dynamic digital memory (Chapter 15) and switched-capacitor filters (Chapter 16).

Example 7.A.1

(a) For an NMOS transistor with $W/L = 10$ fabricated in the 0.18-μm process whose data are given in Table 7.A.1, find the values of V_{OV} and V_{GS} required to operate the device at $I_D = 100$ μA. Ignore channel-length modulation.

(b) Find V_{BE} for an *npn* transistor fabricated in the low-voltage process specified in Table 7.A.2 and operated at $I_C = 100$ μA. Ignore base-width modulation.

Solution

(a)

$$I_D = \frac{1}{2}(\mu_n C_{ox})\left(\frac{W}{L}\right)V_{OV}^2$$

Substituting $I_D = 100$ μA, $W/L = 10$, and, from Table 7.A.1, $\mu_n C_{ox} = 387$ μA/V² results in

$$100 = \frac{1}{2} \times 387 \times 10 \times V_{OV}^2$$

$$V_{OV} = 0.23 \text{ V}$$

Thus,

$$V_{GS} = V_{tn} + V_{OV} = 0.5 + 0.23 = 0.73 \text{ V}$$

(b)

$$I_C = I_S e^{V_{BE}/V_T}$$

Substituting $I_C = 100$ μA and, from Table7.A.2, $I_S = 6 \times 10^{-18}$ A gives,

$$V_{BE} = 0.025 \ln \frac{100 \times 10^{-6}}{6 \times 10^{-18}} = 0.76 \text{ V}$$

EXERCISE

7.A.1 (a) For NMOS transistors fabricated in the 0.18-μm technology specified in Table 7.A.1, find the range of I_D obtained for V_{OV} ranging from 0.2 V to 0.4 V and $W/L = 0.1$ to 100. Neglect channel-length modulation.
(b) If a similar range of current is required in an *npn* transistor fabricated in the low-voltage process specified in Table 7.A.2, find the corresponding change in its V_{BE}.
Ans. (a) $I_{D\min} = 0.8$ μA and $I_{D\max} = 3.1$ mA for a range of about 4000:1; (b) for I_C varying over a 4000:1 range, $\Delta V_{BE} = 207$ mV

Low-Frequency Small-Signal Models The low-frequency models for the two devices are very similar except, of course, for the finite base current (finite β) of the BJT, which gives rise to r_π in the hybrid-π model and to the unequal emitter and collector currents in the T models ($\alpha < 1$). Here it is interesting to note that the low-frequency, small-signal models become identical if one thinks of the MOSFET as a BJT with $\beta = \infty$ ($\alpha = 1$).

For both devices, the hybrid-π model indicates that the **open-circuit voltage gain** obtained from gate to drain (base to collector) with the source (emitter) grounded is $-g_m r_o$. It follows that $g_m r_o$ is the *maximum gain available from a single transistor* of either type. This important transistor parameter is given the name **intrinsic gain** and is denoted A_0. We will have more to say about the intrinsic gain shortly.

Although not included in the MOSFET low-frequency model shown in Table 7.A.3, the body effect can have some implications for the operation of the MOSFET as an amplifier. In simple terms, if the body (substrate) is not connected to the source, it can act as a second gate for the MOSFET. The voltage signal that develops between the body and the source, v_{bs}, gives rise to a drain current component $g_{mb} v_{bs}$, where the body transconductance g_{mb} is proportional to g_m; that is, $g_{mb} = \chi g_m$, where the factor χ is in the range of 0.1 to 0.2. The body effect has no counterpart in the BJT.

The Transconductance For the BJT, the transconductance g_m depends *only* on the dc collector current I_C. (Recall that V_T is a physical constant $\simeq 0.025$ V at room temperature.) It is interesting to observe that g_m does not depend on the geometry of the BJT, and its dependence on the EBJ area is only through the effect of the area on the total collector current I_C. Similarly, the dependence of g_m on V_{BE} is only through the fact that V_{BE} determines the total current in the collector. By contrast, g_m of the MOSFET depends on I_D, V_{OV}, and W/L. Therefore, we use three different (but equivalent) formulas to express g_m of the MOSFET.

The first formula given in Table 7.A.3 for the MOSFET's g_m is the most directly comparable with the formula for the BJT. It indicates that for the same operating current, g_m of the MOSFET is smaller than that of the BJT. This is because $V_{OV}/2$ is the range of 0.05 V to 0.15 V, which is two to six times the corresponding term in the BJT's formula, namely V_T.

The second formula for the MOSFET's g_m indicates that for a given device (i.e., given W/L), g_m is proportional to V_{OV}. Thus a higher g_m is obtained by operating the MOSFET at a higher overdrive voltage. However, we should recall the limitations imposed on the magnitude of V_{OV} by the limited value of V_{DD}. Put differently, the need to obtain a reasonably high g_m constrains the designer's interest in reducing V_{OV}.

The third g_m formula shows that for a given transistor (i.e., given W/L), g_m is proportional to $\sqrt{I_D}$. This should be contrasted with the bipolar case, where g_m is directly proportional to I_C.

Output Resistance The output resistance for both devices is determined by similar formulas, with r_o being the ratio of V_A to the bias current (I_D or I_C). Thus, for both transistors,

r_o is inversely proportional to the bias current. The difference in nature and magnitude of V_A between the two devices has already been discussed.

Intrinsic Gain The intrinsic gain A_0 of the BJT is the ratio of V_A, which is solely a process parameter (5 V to 100 V), and V_T, which is a physical parameter (0.025 V at room temperature). Thus A_0 of a BJT is independent of the device junction area and of the operating current, and its value ranges from 200 V/V to 5000 V/V. The situation in the MOSFET is very different: Table 7.A.3 provides three different (but equivalent) formulas for expressing the MOSFET's intrinsic gain. The first formula is the one most directly comparable to that of the BJT. Here, however, we note the following:

1. The quantity in the denominator is $V_{OV}/2$, which is a design parameter, and although it is becoming smaller in designs using short-channel technologies, it is still at least two to four times larger than V_T. Furthermore, as we have seen, there are reasons for selecting larger values for V_{OV}.
2. The numerator quantity V_A is both process- and device-dependent, and its value has been steadily decreasing.

As a result, the intrinsic gain realized in a single MOSFET amplifier stage fabricated in a modern short-channel technology is only 20 V/V to 40 V/V, at least an order of magnitude lower than that for a BJT.

The third formula given for A_0 in Table 7.A.3 points out a very interesting fact: For a given process technology (V_A' and $\mu_n C_{ox}$) and a given device (W/L), the intrinsic gain is inversely proportional to $\sqrt{I_D}$. This is illustrated in Fig. 7.A.1, which shows a typical plot of A_0 versus the bias current I_D. The plot confirms that the gain increases as the bias current is lowered. The gain, however, levels off at very low currents. This is because the MOSFET enters the subthreshold region of operation (Section 5.1.9), where it becomes very much like a BJT with an exponential current–voltage characteristic. The intrinsic gain then becomes constant, just like that of a BJT. Note, however, that although a higher gain is achieved at lower bias currents, the price paid is a lower g_m and less ability to drive capacitive loads and thus a decrease in bandwidth. This point will be further illustrated shortly.

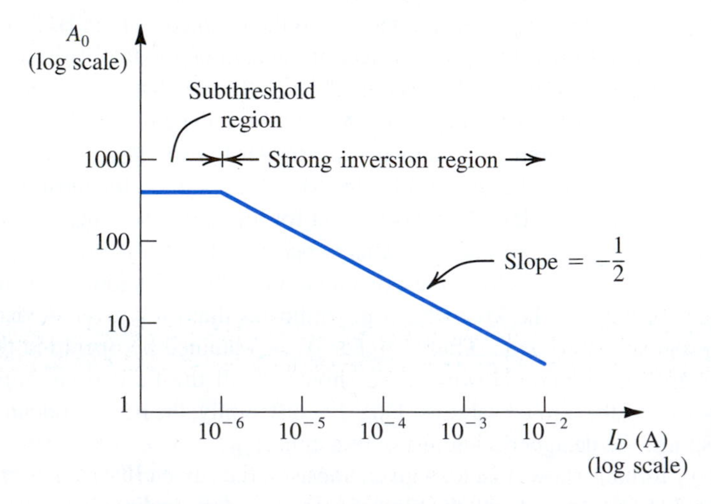

Figure 7.A.1 The intrinsic gain of the MOSFET versus bias current I_D. Outside the subthreshold region, this is a plot of $A_0 = V_A' \sqrt{2\mu_n C_{ox} WL/I_D}$ for the case: $\mu_n C_{ox} = 20\ \mu\text{A/V}^2$, $V_A' = 20$ V/μm, $L = 2$ μm, and $W = 20$ μm.

Example 7.A.2

We wish to compare the values of g_m, input resistance at the gate (base), r_o, and A_0 for an NMOS transistor fabricated in the 0.25-μm technology specified in Table 7.A.1 and an *npn* transistor fabricated in the low-voltage technology specified in Table 7.A.2. Assume both devices are operating at a drain (collector) current of 100 μA. For the MOSFET, let $L = 0.4$ μm and $W = 4$ μm, and specify the required V_{OV}.

Solution

For the NMOS transistor,

$$I_D = \frac{1}{2}(\mu_n C_{ox})\left(\frac{W}{L}\right)V_{OV}^2$$

$$100 = \frac{1}{2} \times 267 \times \frac{4}{0.4} \times V_{OV}^2$$

Thus,

$$V_{OV} = 0.27 \text{ V}$$

$$g_m = \sqrt{2(\mu_n C_{ox})\left(\frac{W}{L}\right)I_D}$$

$$= \sqrt{2 \times 267 \times 10 \times 100} = 0.73 \text{ mA/V}$$

$$R_{\text{in}} = \infty$$

$$r_o = \frac{V_A' L}{I_D} = \frac{5 \times 0.4}{0.1} = 20 \text{ k}\Omega$$

$$A_0 = g_m r_o = 0.73 \times 20 = 14.6 \text{ V/V}$$

For the *npn* transistor,

$$g_m = \frac{I_C}{V_T} = \frac{0.1 \text{ mA}}{0.025 \text{ V}} = 4 \text{ mA/V}$$

$$R_{\text{in}} = r_\pi = \beta_0 / g_m = \frac{100}{4 \text{ mA/V}} = 25 \text{ k}\Omega$$

$$r_o = \frac{V_A}{I_C} = \frac{35}{0.1 \text{ mA}} = 350 \text{ k}\Omega$$

$$A_0 = g_m r_o = 4 \times 350 = 1400 \text{ V/V}$$

EXERCISE

7.A.2 For an NMOS transistor fabricated in the 0.5-μm process specified in Table 7.A.1 with $L = 0.5$ μm, find the transconductance and the intrinsic gain obtained at $I_D = 10$ μA, 100 μA, and 1 mA.
Ans. 0.2 mA/V, 200 V/V; 0.6 mA/V, 62 V/V; 2 mA/V, 20 V/V

High-Frequency Operation The simplified high-frequency equivalent circuits for the MOSFET and the BJT are very similar, and so are the formulas for determining their unity-gain frequency (also called **transition frequency**) f_T. As we shall demonstrate in Chapter 9, f_T is a measure of the *intrinsic* bandwidth of the transistor itself and does *not* take into account the effects of capacitive loads. We address the issue of capacitive loads shortly. For the time being, note the striking similarity between the approximate formulas given in Table 7.A.3 for the value of f_T of the two devices. In both cases f_T is inversely proportional to the square of the critical dimension of the device: the channel length for the MOSFET and the base width for the BJT. These formulas also clearly indicate that shorter-channel MOSFETs[10] and narrower-base BJTs are inherently capable of a wider bandwidth of operation. It is also important to note that while for the BJT the approximate expression for f_T indicates that it is entirely process determined, the corresponding expression for the MOSFET shows that f_T is proportional to the overdrive voltage V_{OV}. Thus we have conflicting requirements on V_{OV}: While a higher low-frequency gain is achieved by operating at a low V_{OV}, wider bandwidth requires an increase in V_{OV}. Therefore the selection of a value for V_{OV} involves, among other considerations, a trade-off between gain and bandwidth.

For *npn* transistors fabricated in the modern low-voltage process, f_T is in the range of 10 GHz to 20 GHz as compared to the 400 MHz to 600 MHz obtained with the standard high-voltage process. In the MOS case, NMOS transistors fabricated in a modern submicron technology, such as the 0.18-μm process, achieve f_T values in the range of 5 GHz to 15 GHz.

Before leaving the subject of high-frequency operation, let's look into the effect of a capacitive load on the bandwidth of the common-source (common-emitter) amplifier. For this purpose we shall assume that the frequencies of interest are much lower than f_T of the transistor. Hence we shall not take the transistor capacitances into account. Figure 7.A.2(a) shows a common-source amplifier with a capacitive load C_L. The voltage gain from gate to drain can be found as follows:

$$V_o = -g_m V_{gs}(r_o \parallel C_L)$$

$$= -g_m V_{gs} \frac{r_o \dfrac{1}{sC_L}}{r_o + \dfrac{1}{sC_L}}$$

$$A_v = \frac{V_o}{V_{gs}} = -\frac{g_m r_o}{1 + sC_L r_o} \tag{7.A.1}$$

Thus the gain has, as expected, a low-frequency value of $g_m r_o = A_0$ and a frequency response of the single-time-constant (STC) low-pass type with a break (pole) frequency at

$$\omega_P = \frac{1}{C_L r_o} \tag{7.A.2}$$

Obviously this pole is formed by r_o and C_L. A sketch of the magnitude of gain versus frequency is shown in Fig. 7.A.2(b). We observe that the gain crosses the 0-dB line at frequency ω_t,

$$\omega_t = A_0 \omega_P = (g_m r_o)\frac{1}{C_L r_o}$$

[10]Although the reason is beyond our capabilities at this stage, f_T of MOSFETs that have very short channels varies inversely with L rather than with L^2.

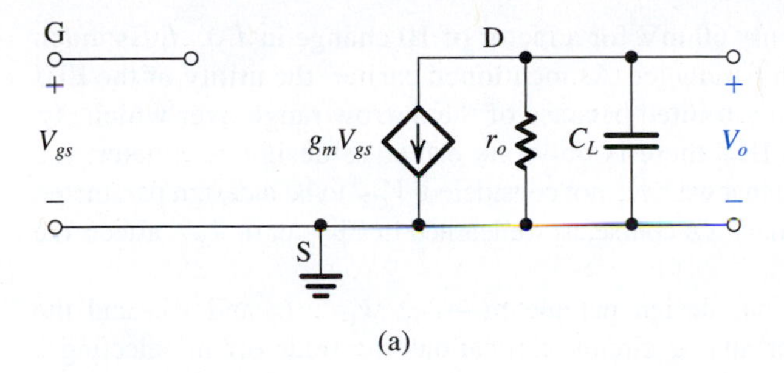

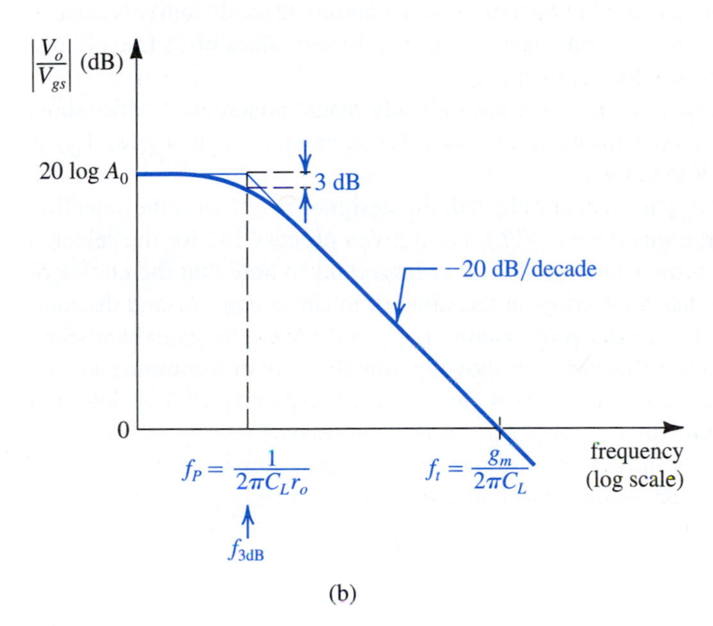

Figure 7.A.2 Frequency response of a CS amplifier loaded with a capacitance C_L and fed with an ideal voltage source. It is assumed that the transistor is operating at frequencies much lower than f_T, and thus the internal capacitances are not taken into account.

Thus,

$$\omega_t = \frac{g_m}{C_L} \tag{7.A.3}$$

That is, the **unity-gain frequency** or, equivalently, the **gain–bandwidth product**[11] ω_t is the ratio of g_m and C_L. We thus clearly see that for a given capacitive load C_L, a larger gain–bandwidth product is achieved by operating the MOSFET at a higher g_m. Identical analysis and conclusions apply to the case of the BJT. In each case, bandwidth increases as bias current is increased.

Design Parameters For the BJT there are three design parameters—I_C, V_{BE}, and I_S (or, equivalently, the area of the emitter–base junction)—and the designer can select any two. However, since I_C is exponentially related to V_{BE} and is very sensitive to the

[11]The unity-gain frequency and the gain–bandwidth product of an amplifier are the same when the frequency response is of the single-pole type; otherwise the two parameters may differ.

value of V_{BE} (V_{BE} changes by only 60 mV for a factor of 10 change in I_C), I_C is much more useful than V_{BE} as a design parameter. As mentioned earlier, the utility of the EBJ area as a design parameter is rather limited because of the narrow range over which A_E can vary. It follows that for the BJT there is only one effective design parameter: the collector current I_C. Finally, note that we have not considered V_{CE} to be a design parameter, since its effect on I_C is only secondary. Of course, as we learned in Chapter 6, V_{CE} affects the output-signal swing.

For the MOSFET there are four design parameters—I_D, V_{OV}, L, and W—and the designer can select any three. For analog circuit applications the trade-off in selecting a value for L is between the higher speed of operation (wider amplifier bandwidth) obtained at lower values of L and the higher intrinsic gain obtained at larger values of L. Usually one selects an L of about 25% to 50% greater than L_{min}.

The second design parameter is V_{OV}. We have already made numerous remarks about the effect of the value of V_{OV} on performance. Usually, for submicron technologies, V_{OV} is selected in the range of 0.1 V to 0.3 V.

Once values for L and V_{OV} have been selected, the designer is left with the selection of the value of I_D or W (or, equivalently, W/L). For a given process and for the selected values of L and V_{OV}, I_D is proportional to W/L. It is important to note that the choice of I_D or, equivalently, of W/L has no bearing on the value of intrinsic gain A_0 and the transition frequency f_T. However, it affects the value of g_m and hence the gain–bandwidth product. Figure 7.A.3 illustrates this point by showing how the gain of a common-source amplifier operated at a constant V_{OV} varies with I_D (or, equivalently, W/L). Note that while the dc gain remains unchanged, increasing W/L and, correspondingly, I_D, increases the bandwidth proportionally. This, however, assumes that the load capacitance C_L is not affected by the device size, an assumption that may not be entirely justified in some cases.

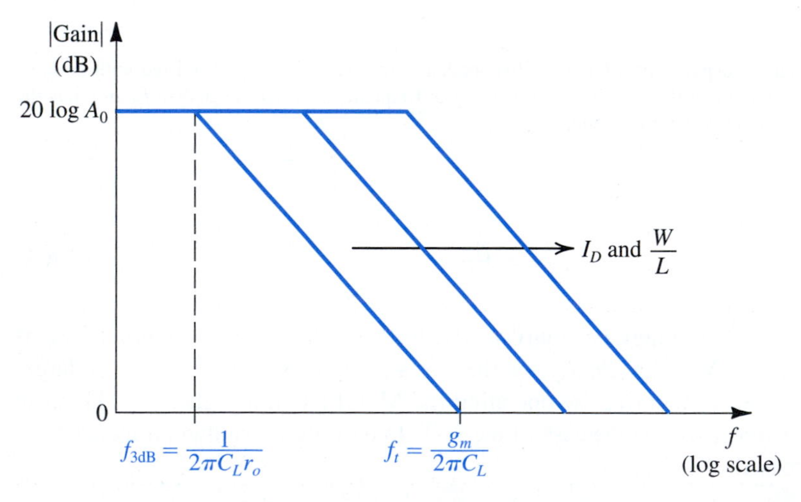

Figure 7.A.3 Increasing I_D or W/L increases the bandwidth of a MOSFET amplifier operated at a constant V_{OV} and loaded by a constant capacitance C_L.

Example 7.A.3

In this example we investigate the gain and the high-frequency response of an *npn* transistor and an NMOS transistor. For the *npn* transistor, assume that it is fabricated in the low-voltage process specified in Table 7.A.2, and assume that $C_\mu \simeq C_{\mu 0}$. For $I_C = 10$ μA, 100 μA, and 1 mA, find g_m, r_o, A_0, C_{de}, C_{je}, C_π, C_μ, and f_T. Also, for each value of I_C, find the gain–bandwidth product f_t of a common-emitter amplifier loaded by a 1-pF capacitance, neglecting the internal capacitances of the transistor. For the NMOS transistor, assume that it is fabricated in the 0.25-μm CMOS process with $L = 0.4$ μm. Let the transistor be operated at $V_{OV} = 0.25$ V. Find W/L that is required to obtain I_D = 10 μA, 100 μA, and 1 mA. At each value of I_D, find g_m, r_o, A_0, C_{gs}, C_{gd}, and f_T. Also, for each value of I_D, determine the gain–bandwidth product f_t of a common-source amplifier loaded by a 1-pF capacitance, neglecting the internal capacitances of the transistor.

Solution

For the *npn* transistor,

$$g_m = \frac{I_C}{V_T} = \frac{I_C}{0.025} = 40 I_C \text{ A/V}$$

$$r_o = \frac{V_A}{I_C} = \frac{35}{I_C}\ \Omega$$

$$A_0 = \frac{V_A}{V_T} = \frac{35}{0.025} = 1400 \text{ V/V}$$

$$C_{de} = \tau_F g_m = 10 \times 10^{-12} \times 40 I_C = 0.4 \times 10^{-9} I_C \text{ F}$$

$$C_{je} \simeq 2C_{je0} = 10 \text{ fF}$$

$$C_\pi = C_{de} + C_{je}$$

$$C_\mu \simeq C_{\mu 0} = 5 \text{ fF}$$

$$f_T = \frac{g_m}{2\pi(C_\pi + C_\mu)}$$

$$f_t = \frac{g_m}{2\pi C_L} = \frac{g_m}{2\pi \times 1 \times 10^{-12}}$$

We thus obtain the following results:

I_C	g_m (mA/V)	r_o (kΩ)	A_0 (V/V)	C_{de} (fF)	C_{je} (fF)	C_π (fF)	C_μ (fF)	f_T (GHz)	f_t (MHz)
10 μA	0.4	3500	1400	4	10	14	5	3.4	64
100 μA	4	350	1400	40	10	50	5	11.6	640
1 mA	40	35	1400	400	10	410	5	15.3	6400

For the NMOS transistor,

$$I_D = \frac{1}{2}\mu_n C_{ox}\frac{W}{L}V_{OV}^2$$

$$= \frac{1}{2} \times 267 \times \frac{W}{L} \times \frac{1}{16}$$

Example 7.A.3 *continued*

Thus,

$$\frac{W}{L} = 0.12 I_D$$

$$g_m = \frac{I_D}{V_{OV}/2} = \frac{I_D}{0.25/2} = 8 I_D \text{ A/V}$$

$$r_o = \frac{V_A' L}{I_D} = \frac{5 \times 0.4}{I_D} = \frac{2}{I_D} \Omega$$

$$A_0 = g_m r_o = 16 \text{ V/V}$$

$$C_{gs} = \tfrac{2}{3} W L C_{ox} + C_{ov} = \tfrac{2}{3} W \times 0.4 \times 5.8 + 0.6W$$

$$C_{gd} = C_{ov} = 0.6W$$

$$f_T = \frac{g_m}{2\pi(C_{gs} + C_{gd})}$$

$$f_t = \frac{g_m}{2\pi C_L}$$

We thus obtain the following results:

I_D	W/L	g_m (mA/V)	r_o (kΩ)	A_0 (V/V)	C_{gs} (fF)	C_{gd} (fF)	f_T (GHz)	f_t (MHz)
10 μA	1.2	0.08	200	16	1.03	0.29	9.7	12.7
100 μA	12	0.8	20	16	10.3	2.9	9.7	127
1 mA	120	8	2	16	103	29	9.7	1270

EXERCISE

7.A.3 Find I_D, g_m, r_o, A_0, C_{gs}, C_{gd}, and f_T for an NMOS transistor fabricated in the 0.5-μm CMOS technology specified in Table 7.A.1. Let $L = 0.5$ μm, $W = 5$ μm, and $V_{OV} = 0.3$ V.
Ans. 85.5 μA; 0.57 mA/V; 66.7 kΩ; 38 V/V; 8.3 fF; 2 fF; 8.8 GHz

7.A.4 Combining MOS and Bipolar Transistors—BiCMOS Circuits

From the discussion above it should be evident that the BJT has the advantage over the MOSFET of a much higher transconductance (g_m) at the same value of dc bias current. Thus, in addition to realizing higher voltage gains per amplifier stage, bipolar transistor amplifiers have superior high-frequency performance compared to their MOS counterparts.

On the other hand, the practically infinite input resistance at the gate of a MOSFET makes it possible to design amplifiers with extremely high input resistances and an almost zero input bias current. Also, as mentioned earlier, the MOSFET provides an excellent implementation of a switch, a fact that has made CMOS technology capable of realizing a host of analog circuit functions that are not possible with bipolar transistors.

It can thus be seen that each of the two transistor types has its own distinct and unique advantages: Bipolar technology has been extremely useful in the design of very-high-quality general-purpose circuit building blocks, such as op amps. On the other hand, CMOS, with its very high packing density and its suitability for both digital and analog circuits, has become the technology of choice for the implementation of very-large-scale integrated circuits. Nevertheless, the performance of CMOS circuits can be improved if the designer has available (on the same chip) bipolar transistors that can be employed in functions that require their high g_m and excellent current-driving capability. A technology that allows the fabrication of high-quality bipolar transistors on the same chip as CMOS circuits is aptly called **BiCMOS**. At appropriate locations throughout this book we present interesting and useful BiCMOS circuit blocks.

7.A.5 Validity of the Square-Law MOSFET Model

We conclude this appendix with a comment on the validity of the simple square-law model we have been using to describe the operation of the MOS transistor. While this simple model works well for devices with relatively long channels (>1 µm), it does *not* provide an accurate representation of the operation of short-channel devices. This is because a number of physical phenomena come into play in these submicron devices, resulting in what are called **short-channel effects**. Although a detailed study of short-channel effects is beyond the scope of this book, it should be mentioned that MOSFET models have been developed that take these effects into account. However, they are understandably quite complex and do not lend themselves to hand analysis of the type needed to develop insight into circuit operation. Rather, these models are suitable for computer simulation and are indeed used in SPICE (Appendix B). For quick, manual analysis, however, we will continue to use the square-law model, which is the basis for the comparison of Table 7.A.3.

PROBLEMS

Computer Simulation Problems

SIM Problems identified by this icon are intended to demonstrate the value of using SPICE simulation to verify hand analysis and design, and to investigate important issues such as allowable signal swing and amplifier nonlinear distortion. Instructions to assist in setting up PSpice and Multism simulations for all the indicated problems can be found in the corresponding files on the disc. Note that if a particular parameter value is not specified in the problem statement, you are to make a reasonable assumption.
* difficult problem; ** more difficult; *** very challenging and/or time-consuming; D: design problem.

Section 7.2: The Basic Gain Cell

7.1 Find g_m, r_π, r_o, and A_0 for the CE amplifier of Fig. 7.1(b) when operated at $I = 10$ µA, 100 µA, and 1 mA. Assume $\beta = 100$ and remains constant as I is varied, and that $V_A = 10$ V. Present your results in a table.

7.2 Consider the CE amplifiers of Fig. 7.1(b) for the case of $I = 1$ mA, $\beta = 100$, and $V_A = 100$ V. Find R_{in}, A_{vo}, and R_o. If it is required to raise R_{in} by a factor of 4 by changing I, what value of I is required, assuming that β remains unchanged? What are the new values of A_{vo} and R_o? If the amplifier is fed with a signal source having $R_{sig} = 5$ kΩ and is connected to a load of 100-kΩ resistance, find the overall voltage gain, v_o/v_{sig}.

7.3 Find the intrinsic gain of an NMOS transistor fabricated in a process for which $k_n' = 200$ µA/V² and $V_A' = 20$ V/µm. The transistor has a 0.5-µm channel length and is operated at $V_{OV} = 0.25$ V. If a 2-mA/V transconductance is required, what must I_D and W be?

7.4 An NMOS transistor fabricated in a certain process is found to have an intrinsic gain of 80 V/V when operated at an I_D of 100 μA. Find the intrinsic gain for I_D = 25 μA and I_D = 400 μA. For each of these currents, find the factor by which g_m changes from its value at I_D = 100 μA.

D 7.5 Consider an NMOS transistor fabricated in a 0.18-μm technology for which $k_n' = 387\ \mu A/V^2$ and $V_A' = 5$ V/μm. It is required to obtain an intrinsic gain of 25 V/V and a g_m of 1 mA/V. Using V_{OV} = 0.2 V, find the required values of L, W/L, and the bias current I.

D 7.6 Sketch the circuit for a current-source-loaded CS amplifier that uses a PMOS transistor for the amplifying device. Assume the availability of a single +1.8 -V dc supply. If the transistor is operated with $|V_{OV}|$ = 0.3 V, what is the highest instantaneous voltage allowed at the drain?

D 7.7 An NMOS transistor is fabricated in the 0.18-μm process whose parameters are given in Table 7.A.1 on page 554. The device has a channel length twice the minimum and is operated at V_{OV} = 0.25 V and I_D = 10 μA.

(a) What values of g_m, r_o, and A_0 are obtained?
(b) If I_D is increased to 100 μA, what do V_{OV}, g_m, r_o, and A_0 become?
(c) If the device is redesigned with a new value of W so that it operates at V_{OV} = 0.25 V for I_D = 100 μA, what do g_m, r_o, and A_0 become?
(d) If the redesigned device in (c) is operated at 10 μA, find V_{OV}, g_m, r_o, and A_0.
(e) Which designs and operating conditions produce the lowest and highest values of A_0? What are these values? In each of these two cases, if W/L is held at the same value but L is made 10 times larger, what gains result?

D 7.8 Find A_0 for an NMOS transistor fabricated in a CMOS process for which $k_n' = 200\ \mu A/V^2$ and $V_A' = 20$ V/μm. The transistor has a 0.4-μm channel length and is operated with an overdrive voltage of 0.25 V. What must W be for the NMOS transistor to operate at I_D = 100 μA? Also, find the values of g_m and r_o. Repeat for L = 0.8 μm.

D 7.9 Using a CMOS technology for which $k_n' = 200\ \mu A/V^2$ and V_A' = 20 V/μm, design a current-source-loaded CS amplifier for operation at I = 50 μA with V_{OV} = 0.2 V. The amplifier is to have an open-circuit voltage gain of −100 V/V. Assume that the current-source load is ideal. Specify L and W/L.

D 7.10 The circuit in Fig. 7.3(a) is fabricated in a process for which $\mu_n C_{ox} = 2\mu_p C_{ox} = 200\ \mu A/V^2$, $V_{An}' = |V_{Ap}'|$ = 20 V/μm, $V_{tn} = -V_{tp}$ = 0.5 V, and V_{DD} = 2.5 V. The two transistors have L = 0.5 μm and are to be operated at I_D = 100 μA and $|V_{OV}|$ = 0.3 V. Find the required values of V_G, $(W/L)_1$, $(W/L)_2$, and A_v.

D 7.11 The circuit in Fig. 7.3(a) is fabricated in a 0.18-μm CMOS technology for which $\mu_n C_{ox} = 387\ \mu A/V^2$, $\mu_p C_{ox} = 86\ \mu A/V^2$, $V_{tn} = -V_{tp}$ = 0.5 V, V_{An}' = 5 V/μm, $|V_{Ap}'|$ = 6 V/μm, and V_{DD} = 1.8 V. It is required to design the circuit to obtain a voltage gain A_v = −40 V/V. Use devices of equal length L operating at I = 100 μA and $|V_{OV}|$ = 0.2 V. Determine the required values of V_G, L, $(W/L)_1$, and $(W/L)_2$.

7.12 Figure P7.12 shows an IC MOS amplifier formed by cascading two common-source stages. Assuming that $V_{An} = |V_{Ap}|$ and that the biasing current sources have output resistances equal to those of Q_1 and Q_2, find an expression for the overall voltage gain in terms of g_m and r_o of Q_1 and Q_2.

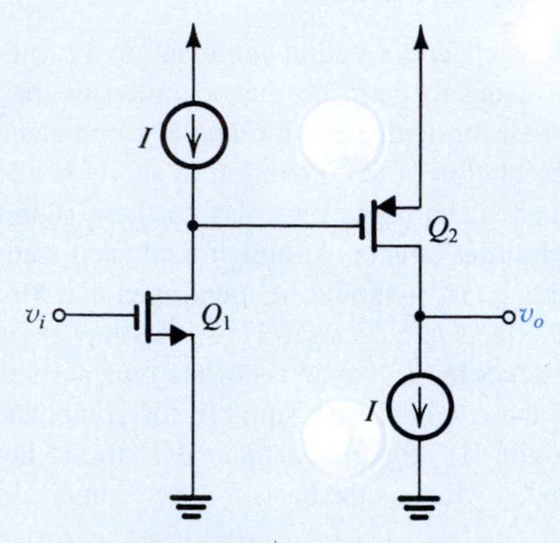

Figure P7.12

***7.13** The NMOS transistor in the circuit of Fig. P7.13 has V_t = 0.5 V, $k_n' W/L$ = 2 mA/V², and V_A = 20 V.

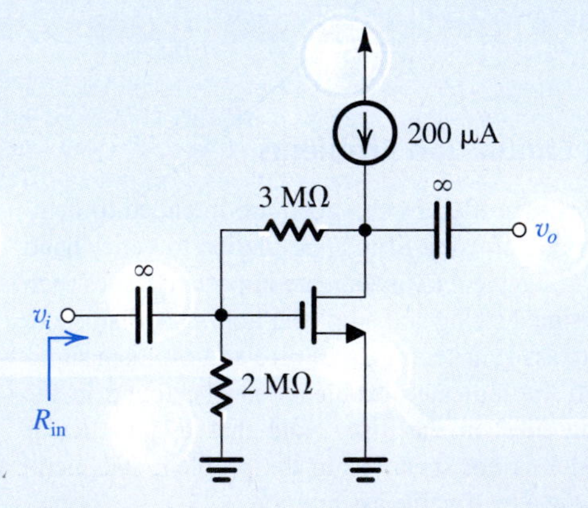

Figure P7.13

(a) Neglecting the dc current in the feedback network and the effect of r_o, find V_{GS}. Then find the dc current in the feedback network and V_{DS}. Verify that you were justified in neglecting the current in the feedback network when you found V_{GS}.

(b) Find the small-signal voltage gain, v_o/v_i. What is the peak of the largest output sinewave signal that is possible while the NMOS transistor remains in saturation? What is the corresponding input signal?
(c) Find the small-signal input resistance R_{in}.

D 7.14 Consider the CMOS amplifier of Fig. 7.4(a) when fabricated with a process for which $k_n' = 2.5k_p' = 250\ \mu A/V^2$, $|V_t| = 0.6$ V, and $|V_A| = 10$ V. Find I_{REF} and $(W/L)_1$ to obtain a voltage gain of –40 V/V and an output resistance of 100 kΩ. If Q_2 and Q_3 are to be operated at the same overdrive voltage as Q_1, what must their *W/L* ratios be?

7.15 Consider the CMOS amplifier analyzed in Example 7.3. If v_I consists of a dc bias component on which is superimposed a sinusoidal signal, find the value of the dc component that will result in the maximum possible signal swing at the output with almost-linear operation. What is the amplitude of the output sinusoid resulting? (*Note:* In practice, the amplifier would have a feedback circuit that causes it to operate at a point near the middle of its linear region.)

7.16 The power supply of the CMOS amplifier analyzed in Example 7.3 is increased to 5 V. What will the extent of the linear region at the output become?

***7.17** Consider the circuit shown in Fig. 7.4(a), using a 3.3-V supply and transistors for which $|V_t| = 0.8$ V and $L = 1\ \mu m$. For Q_1, $k_n' = 100\ \mu A/V^2$, $V_A = 100$ V, and $W = 20\ \mu m$. For Q_2 and Q_3, $k_p' = 50\ \mu A/V^2$ and $|V_A| = 50$ V. For Q_2, $W = 40\ \mu m$. For Q_3, $W = 10\ \mu m$.

(a) If Q_1 is to be biased at 100 μA, find I_{REF}. For simplicity, ignore the effect of V_A.
(b) What are the extreme values of v_O for which Q_1 and Q_2 just remain in saturation?
(c) What is the large-signal voltage gain?
(d) Find the slope of the transfer characteristic at $v_O = V_{DD}/2$.
(e) For operation as a small-signal amplifier around a bias point at $v_O = V_{DD}/2$, find the small-signal voltage gain and output resistance.

****7.18** The MOSFETs in the circuit of Fig. P7.18 are matched, having $k_n'(W/L)_1 = k_p'(W/L)_2 = 1\ mA/V^2$ and $|V_t| = 0.5$ V. The resistance $R = 1$ MΩ.

(a) For G and D open, what are the drain currents I_{D1} and I_{D2}?
(b) For $r_o = \infty$, what is the voltage gain of the amplifier from G to D? [*Hint*: Replace the transistors with their small-signal models.]
(c) For finite $r_o(|V_A| = 20$ V), what is the voltage gain from G to D and the input resistance at G?
(d) If G is driven (through a large coupling capacitor) from a source v_{sig} having a resistance of 100 kΩ, find the voltage gain v_d/v_{sig}.
(e) For what range of output signals do Q_1 and Q_2 remain in the saturation region?

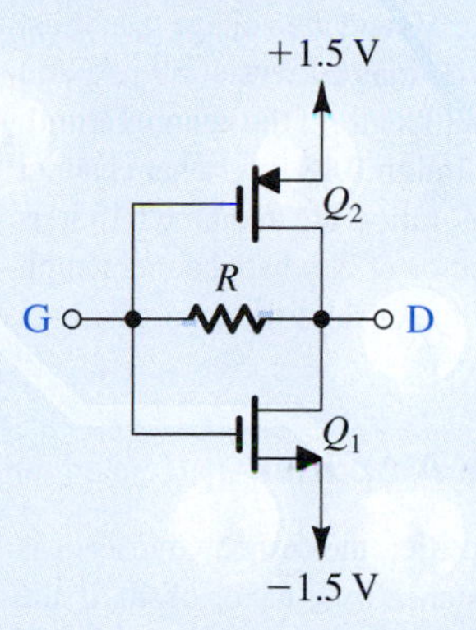

Figure P7.18

7.19 Transistor Q_1 in the circuit of Fig. P7.19 is operating as a CE amplifier with an active load provided by transistor Q_2, which is the output transistor in a current mirror formed by Q_2 and Q_3. (Note that the biasing arrangement for Q_1 is *not* shown.)

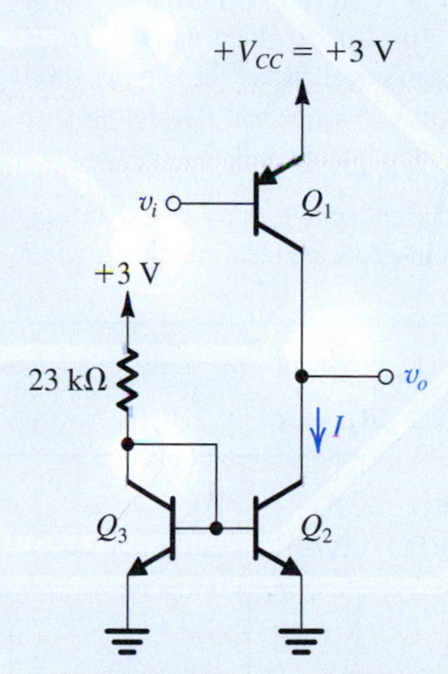

Figure P7.19

(a) Neglecting the finite base currents of Q_2 and Q_3 and assuming that their $V_{BE} \simeq 0.7$ V and that Q_2 has five times the area of Q_3, find the value of *I*.
(b) If Q_1 and Q_2 are specified to have $|V_A| = 50$ V, find r_{o1} and r_{o2} and hence the total resistance at the collector of Q_1.
(c) Find $r_{\pi 1}$ and g_{m1} assuming that $\beta_1 = 50$.
(d) Find R_{in}, A_v, and R_o.

D 7.20 It is required to design the CMOS amplifier of Fig. 7.4(a) utilizing a 0.18-μm process for which $k'_n = 387\ \mu A/V^2$, $k'_p = 86\ \mu A/V^2$, $V_{tn} = -V_{tp} = 0.5$ V, $V_{DD} = 1.8$ V, $V'_{An} = 5$ V/μm, and $V'_{Ap} = -6$ V/μm. The output voltage must be able to swing to within approximately 0.2 V of the power-supply rails (i.e., from 0.2 V to 1.6 V) and the voltage gain must be at least 10 V/V. Design for a dc bias current of 50 μA, and use devices with the same channel length. If the channel length is an integer multiple of the minimum 0.18 μm, what channel length is needed and what W/L ratios are required? If it is required to raise the gain by a factor of 2, what channel length would be required, and by what factor does the total gate area of the circuit increase?

Section 7.3: The Cascode Amplifier

D 7.21 In a MOS cascode amplifier, the cascode transistor is required to raise the output resistance by a factor of 40. If the transistor is operated at $V_{OV} = 0.2$ V, what must its V_A be? If the process technology specifies V'_A as 5 V/μm, what channel length must the transistor have?

D 7.22 For a cascode current source such as that in Fig. 7.10, show that if the two transistors are identical, the current I supplied by the current source and the output resistance R_o are related by $IR_o = 2|V_A|^2/|V_{OV}|$. Now consider the case of transistors that have $|V_A| = 4$ V and are operated at $|V_{OV}|$ of 0.2 V. Also, let $\mu_p C_{ox} = 100\ \mu A/V^2$. Find the W/L ratios required and the output resistance realized for the two cases: (a) $I = 0.1$ mA and (b) $I = 0.5$ mA. Assume that V_{SD} for the two devices is the minimum required (i.e., $|V_{OV}|$).

D *7.23 For a cascode current source, such as that in Fig. 7.10, show that if the two transistors are identical, the current I supplied by the current source and the output resistance R_o are related by

$$IR_o = \frac{2|V'_A|^2}{|V_{OV}|}L^2$$

Now consider the case of a 0.18-μm technology for which $|V'_A| = 5$ V/μm and let the transistors be operated at $|V_{OV}| = 0.2$ V. Find the figure-of-merit IR_o for the three cases of L equal to the minimum channel length, twice the minimum, and three times the minimum. Complete the entries of the table at the bottom of the page. Give W/L and the area $2WL$ in terms of n. In the table, A_v denotes the gain obtained in a cascode amplifier such as that in Fig. 7.11 that utilizes our current source as load and which has the same values of g_m and R_o as the current-source transistors.

(a) For each current value, what is price paid for the increase in R_o and A_v obtained as L is increased?
(b) For each value of L, what advantage is obtained as I is increased, and what is the price paid?
(c) Contrast the performance obtained from the circuit with the largest area with that obtained from the circuit with the smallest area.

D 7.24 Design the cascode amplifier of Fig. 7.9(a) to obtain $g_{m1} = 1$ mA/V and $R_o = 400$ kΩ. Use a 0.18-μm technology for which $V_{tn} = 0.5$ V, $V'_A = 5$ V/μm and $k'_n = 400\ \mu A/V^2$. Determine L, W/L, V_{G2}, and I. Use identical transistors operated at $V_{OV} = 0.2$ V, and design for the maximum possible negative signal swing at the output. What is the value of the minimum permitted output voltage?

7.25 The cascode amplifier of Fig. 7.11 is operated at a current of 0.1 mA with all devices operating at $|V_{OV}| = 0.25$ V.

	$L = L_{min} = 0.18$ μm, $IR_o =$ V				$L = 2L_{min} = 0.36$ μm, $IR_o =$ V				$L = 3L_{min} = 0.54$ μm, $IR_o =$ V			
	g_m (mA/V)	R_o (kΩ)	A_v (V/V)	$2WL$ (μm²)	g_m (mA/V)	R_o (kΩ)	A_v (V/V)	$2WL$ (μm²)	g_m (mA/V)	R_o (kΩ)	A_v (V/V)	$2WL$ (μm²)
$I = 0.01$ mA, $W/L = n$												
$I = 0.1$ mA, $W/L =$												
$I = 1.0$ mA, $W/L =$												

All devices have $|V_A| = 4$ V. Find g_{m1}, the output resistance of the amplifier, R_{on}, the output resistance of the current source, R_{op}, the overall output resistance, R_o, and the voltage gain, A_v.

D 7.26 Design the CMOS cascode amplifier in Fig. 7.11 for the following specifications: $g_{m1} = 2$ mA/V and $A_v = -200$ V/V. Assume that for the available fabrication process, $|V'_A| = 5$ V/μm for both NMOS and PMOS devices and that $\mu_n C_{ox} = 4\,\mu_p C_{ox} = 400$ μA/V^2. Use the same channel length L for all devices and operate all four devices at $|V_{OV}| = 0.2$ V. Determine the required channel length L, the bias current I, and the W/L ratio for each of four transistors. Assume that suitable bias voltages have been chosen, and neglect the Early effect in determining the W/L ratios.

D 7.27 Design the circuit of Fig. 7.10 to provide an output current of 100 μA. Use $V_{DD} = 3.3$ V, and assume the PMOS transistors to have $\mu_p C_{ox} = 60$ μA/V^2, $V_{tp} = -0.8$ V, and $|V_A| = 5$ V. The current source is to have the widest possible signal swing at its output. Design for $V_{OV} = 0.2$ V, and specify the values of the transistor W/L ratios and of V_{G3} and V_{G4}. What is the highest allowable voltage at the output? What is the value of R_o?

7.28 The cascode transistor can be thought of as providing a "shield" for the input transistor from the voltage variations at the output. To quantify this "shielding" property of the cascode, consider the situation in Fig. P7.28. Here we have grounded the input terminal (i.e., reduced v_i to zero), applied a small change v_x to the output node, and denoted the voltage change that results at the drain of Q_1 by v_y. By what factor is v_y smaller than v_x?

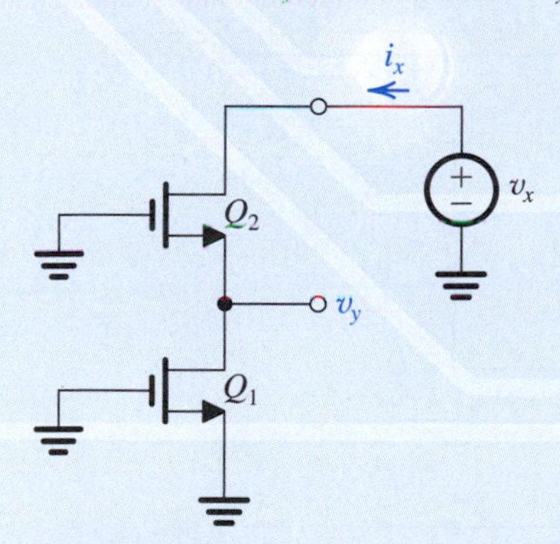

Figure P7.28

***7.29** In this problem we investigate whether, as an alternative to cascoding, we can simply increase the channel length L of the CS MOSFET. Specifically, we wish to compare the two circuits shown in Fig. P7.29(b) and (c). The circuit in Fig. P7.29(b) is a CS amplifier in which the channel length has been quadrupled relative to that of the original CS amplifier in Fig. P7.29(a) while the drain bias current has been kept constant.

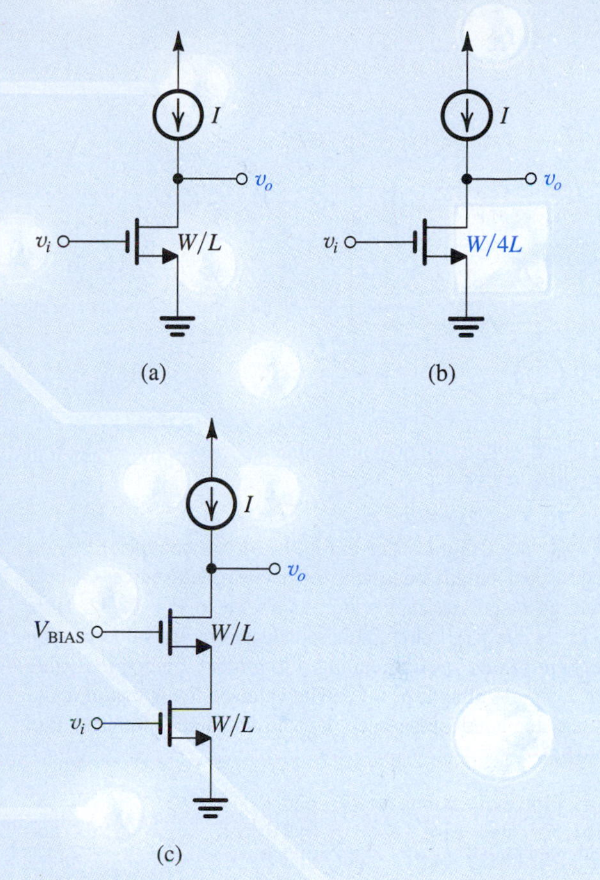

Figure P7.29

(a) Show that for this circuit V_{OV} is double that of the original circuit, g_m is half that of the original circuit, and A_0 is double that of the original circuit.

(b) Compare these values to those of the cascode circuit in Fig. P7.29(c), which is operating at the same bias current and has the same minimum voltage requirement at the drain as in the circuit of Fig. P7.29(b).

7.30 Consider the cascode amplifier of Fig. 7.11 with the dc component at the input $V_I = 0.8$ V, $V_{G2} = 1.2$ V, $V_{G3} = 1.3$ V, $V_{G4} = 1.7$ V, and $V_{DD} = 2.5$ V. If all devices are matched, that is $k_{n1} = k_{n2} = k_{p3} = k_{p4}$, and have equal $|V_t|$ of 0.5 V, what is the overdrive voltage at which the four transistors are operating? What is the allowable voltage range at the output?

7.31 Figure P7.31 shows a CG transistor fed with a signal source (v_{sig}, R_{sig}) and loaded with a resistance R_L.

(a) Find R_{in}.

(b) Noting that the current through R_L is equal to the input current i, find an expression for the overall voltage gain v_o/v_{sig}.

(c) Determine the values of R_{in} and v_o/v_{sig} for the case of $R_L = r_o = 10\,\text{k}\Omega$, $A_0 = 20$, and $R_{sig} = 1\,\text{k}\Omega$.

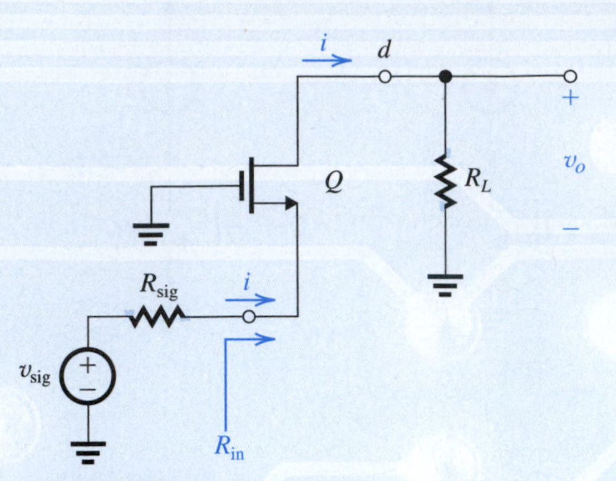

Figure P7.31

7.32 The CG transistor in Fig. P7.31 can be replaced by an equivalent circuit consisting of a controlled-source $G_m v_{sig}$ and an output resistance R_o, as shown in Fig. P7.32. Here G_m is the short-circuit transconductance. Its value can be determined by short-circuiting d to ground, finding the value of i, and dividing it by v_{sig}. The value of R_o is that of a CG transistor with a resistance R_{sig} in its source (Refer to Fig. 7.13).

(a) Find expressions for G_m and R_o.
(b) For the case $R_L = r_o = 10\ \text{k}\Omega$, $g_m r_o = 20$, and $R_{sig} = 1\ \text{k}\Omega$, find G_m, R_o, and v_o/v_{sig}.

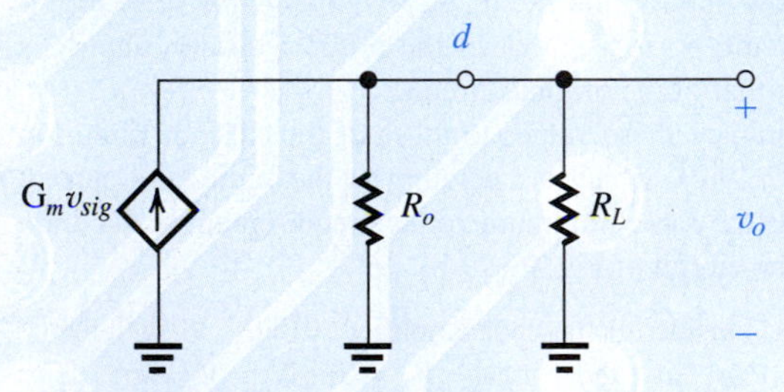

Figure P7.32

SIM **7.33** A CMOS cascode amplifier has identical CS and CG transistors that have $W/L = 5.4\ \mu\text{m}/0.36\ \mu\text{m}$ and biased at $I = 0.2$ mA. The fabrication process has $\mu_n C_{ox} = 4$, $\mu_p C_{ox} = 400\ \mu\text{A/V}^2$, and $V_A' = 5\ \text{V}/\mu\text{m}$. At what value of R_L does the gain become −100 V/V? What is the voltage gain of the common-source stage?

7.34 The purpose of this problem is to investigate the signal currents and voltages at various points throughout a cascode amplifier circuit. Knowledge of this signal distribution is very useful in designing the circuit so as to allow for the required signal swings. Figure P7.34 shows a CMOS cascode amplifier with all dc voltages replaced with signal grounds. As well, we have explicitly shown the resistance r_o of each of the four transistors. For simplicity, we are assuming that the four transistors have the same g_m and r_o. The amplifier is fed with a signal v_i.

(a) Determine R_1, R_2, and R_3.
(b) Determine i_1, i_2, i_3, i_4, i_5, i_6, and i_7, all in terms of v_i.
(c) Determine v_1, v_2, and v_3, all in terms of v_i.
(d) If v_i is a 5-mV peak sine wave and $g_m r_o = 20$, sketch and clearly label the waveforms of v_1, v_2, and v_3.

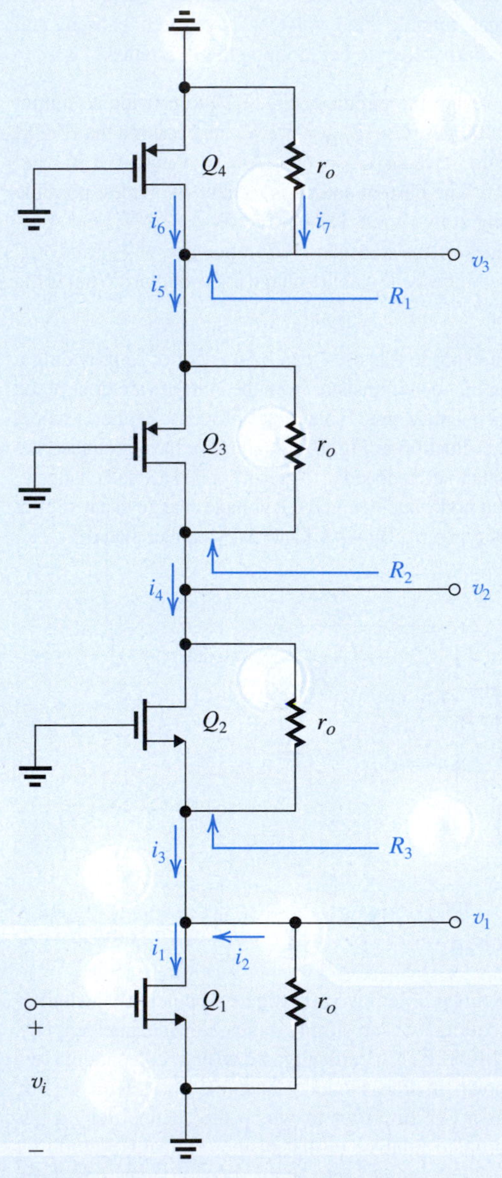

Figure P7.34

7.35 Figure P7.35 shows a CS amplifier with a resistance R_s in the source lead and with the drain short-circuited to ground. Determine the short-circuit transconductance G_m. Hence provide the output equivalent circuit of the source-degenerated CS amplifier, and show that the open-circuit voltage gain $A_{vo} = -A_0$.

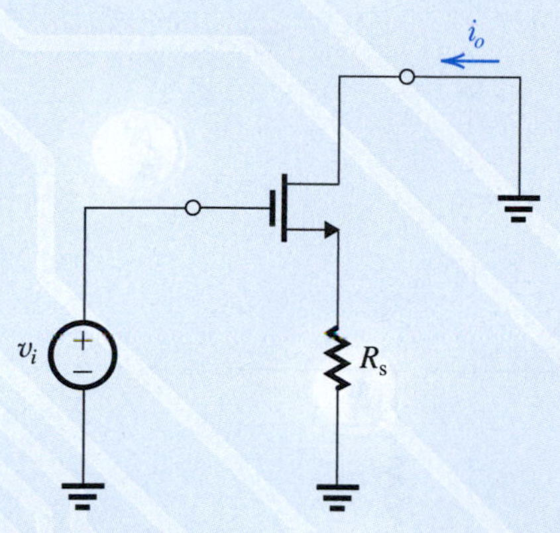

Figure P7.35

7.36 A CS amplifier operating with a g_m of 2 mA/V and having $r_o = 20\ \text{k}\Omega$ has a 2-kΩ resistance R_s connected in its source lead. Find the output resistance R_o. Recalling that the open-circuit voltage gain remains unchanged at A_0, find the gain obtained with $R_L = 100\ \text{k}\Omega$.

D 7.37 Design the double-cascode current source shown in Fig. P7.37 to provide $I = 0.1$ mA and the largest possible signal swing at the output; that is, design for the minimum allowable voltage across each transistor. The 0.18-μm CMOS fabrication process available has $V_{tp} = -0.5$ V, $V'_A = -6$ V/μm, and $\mu_p C_{ox} = 100\ \mu\text{A/V}^2$. Use devices with $L = 0.5$ μm, and operate at $|V_{OV}| = 0.2$ V. Specify V_{G1}, V_{G2}, V_{G3}, and the W/L ratios of the transistors. What is the value of R_o achieved?

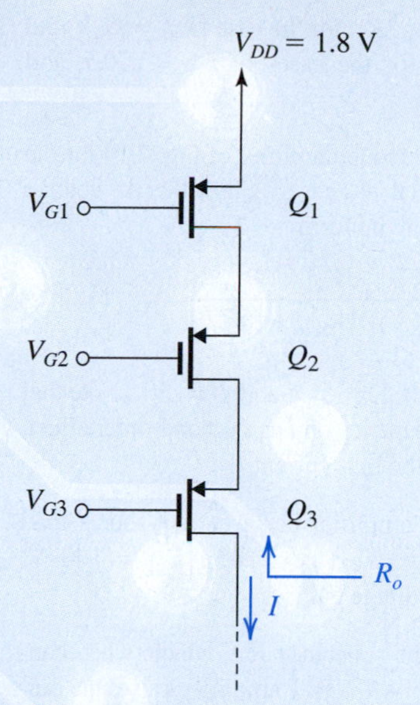

Figure P7.37

7.38 Figure P7.38 shows a folded-cascode CMOS amplifier utilizing a simple current source Q_2, supplying a current $2I$, and a cascoded current-source (Q_4, Q_5) supplying a current I. Assume, for simplicity, that all transistors have equal parameters g_m and r_o.

(a) Give approximate expressions for all the resistances indicated.
(b) Find the amplifier output resistance R_o.
(c) Show that the short-circuit transconductance G_m is approximately equal to g_{m1}.
(d) Find the overall voltage gain v_o/v_i and evaluate its value for the case $g_{m1} = 2$ mA/V and $A_0 = 20$.

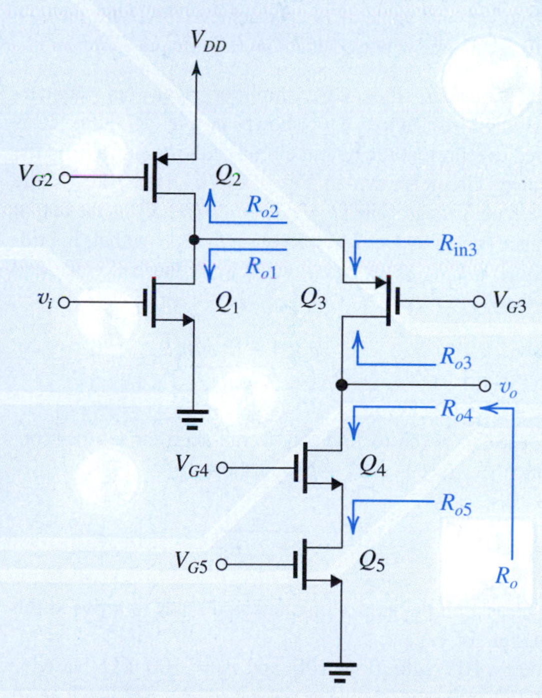

Figure P7.38

7.39 A cascode current source formed of two *pnp* transistors for which $\beta = 50$ and $V_A = 5$ V supplies a current of 0.5 mA. What is the output resistance?

7.40 Use Eq. (7.45) to show that for a BJT cascode current source utilizing identical *pnp* transistors and supplying a current I,

$$IR_o = \frac{|V_A|}{(V_T/|V_A|) + (1/\beta)}$$

Evaluate the figure-of-merit IR_o for the case $|V_A| = 5$ V and $\beta = 50$. Now find R_o for the cases of $I = 0.1$, 0.5, and 1.0 mA.

7.41 Consider the BJT cascode amplifier of Fig. 7.19 for the case all transistors have equal β and r_o. Show that the voltage gain A_v can be expressed in the form

$$A_v = -\frac{1}{2}\frac{|V_A|/V_T}{(V_T/|V_A|) + (1/\beta)}$$

Evaluate A_v for the case $|V_A| = 5$ V and $\beta = 50$. Note that except for the fact that β depends on I as a second-order effect, the gain is independent of the bias current I!

7.42 A bipolar cascode amplifier has a current-source load with an output resistance βr_o. Let $\beta = 100$, $|V_A| = 100$ V, and $I = 0.1$ mA. Find the voltage gain A_v.

7.43 Find the value of the resistance R_e, which, when connected in the emitter lead of a CE BJT amplifier, raises the output resistance by a factor of (a) 5, (b) 10, and (c) 50. What is the maximum possible factor by which the output resistance can be raised, and at what value of R_e is it achieved? Assume the BJT has $\beta = 100$ and is biased at $I_C = 0.5$ mA.

***7.44** Consider the CE amplifier with an emitter-degeneration resistance R_e, shown in Fig. P7.44(a). It is required to represent the output circuit of the amplifier with the equivalent circuit shown in Fig. P7.44(b). Here A_{vo} is the open-circuit voltage gain $[v_o/v_i]_{R_L=\infty}$, and R_o is the output resistance (given by Eq. 7.50). Replace the BJT with its hybrid-π model, set $R_L = \infty$ (i.e., open-circuit the collector), and show that

$$A_{vo} = -g_m r_o \frac{1 - R_e/\beta r_o}{1 + R_e/r_\pi}$$

Now, use this result to find the overall short-circuit transconductance G_m (see Fig. P7.44c) and show that

$$G_m \simeq \frac{g_m}{1 + g_m R_e}$$

State clearly all the approximations you made to arrive at this expression for G_m.

For a BJT with $\beta = 100$ and $r_o = 100$ kΩ biased at $I_C = 0.2$ mA and having a resistance $R_e = 250$ Ω in its emitter, find R_o, A_{vo}, and G_m. Also calculate the voltage gain A_v obtained with $R_L = 10$ kΩ.

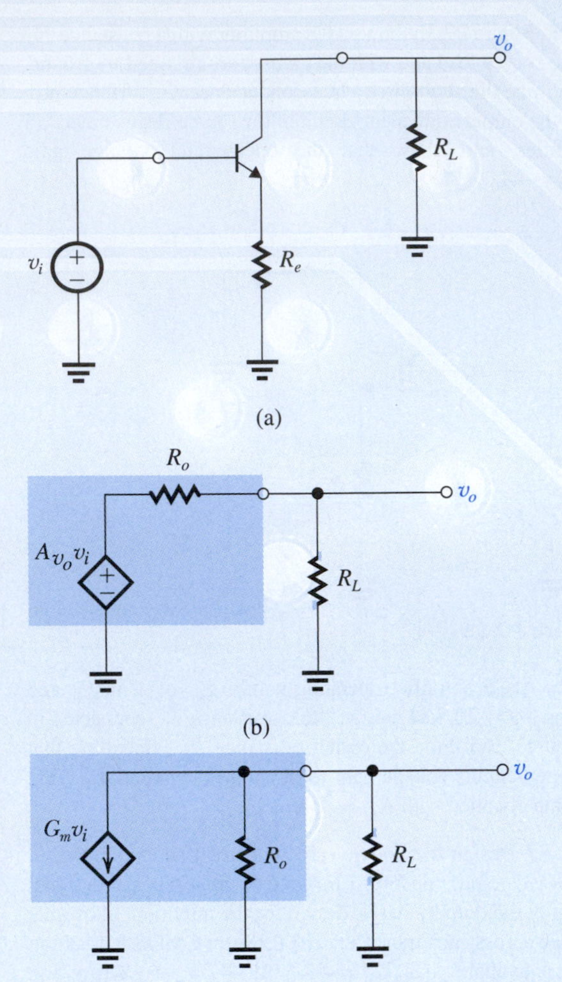

Figure P7.44

D *7.45 Figure P7.45 shows four possible realizations of the folded cascode amplifier. Assume that the BJTs have $\beta = 100$ and that both the BJTs and the MOSFETs have $|V_A| = 5$ V. Let $I = 100$ μA, and assume that the MOSFETs are operating at $|V_{OV}| = 0.2$ V. Assume the current sources are ideal. For each circuit determine, R_{in}, R_o, and A_{vo}. Comment on your results.

Section 7.4: IC Biasing—Current Sources, Current Mirrors, and Current-Steering Circuits

D 7.46 For $V_{DD} = 1.8$ V and using $I_{REF} = 100$ μA, it is required to design the circuit of Fig. 7.22 to obtain an output current whose nominal value is 100 μA. Find R if Q_1 and Q_2 are matched with channel lengths of 0.5 μm, channel widths of 4 μm, $V_t = 0.5$ V, and $k_n' = 400$ μA/V². What is the lowest possible value of V_O? Assuming that for this process technology the Early voltage $V_A' = 10$ V/μm, find the output resistance of the current source. Also, find the change in output current resulting from a +0.5-V change in V_O.

D 7.47 Using $V_{DD} = 1.8$ V and a pair of matched MOSFETs, design the current-source circuit of Fig. 7.22 to provide

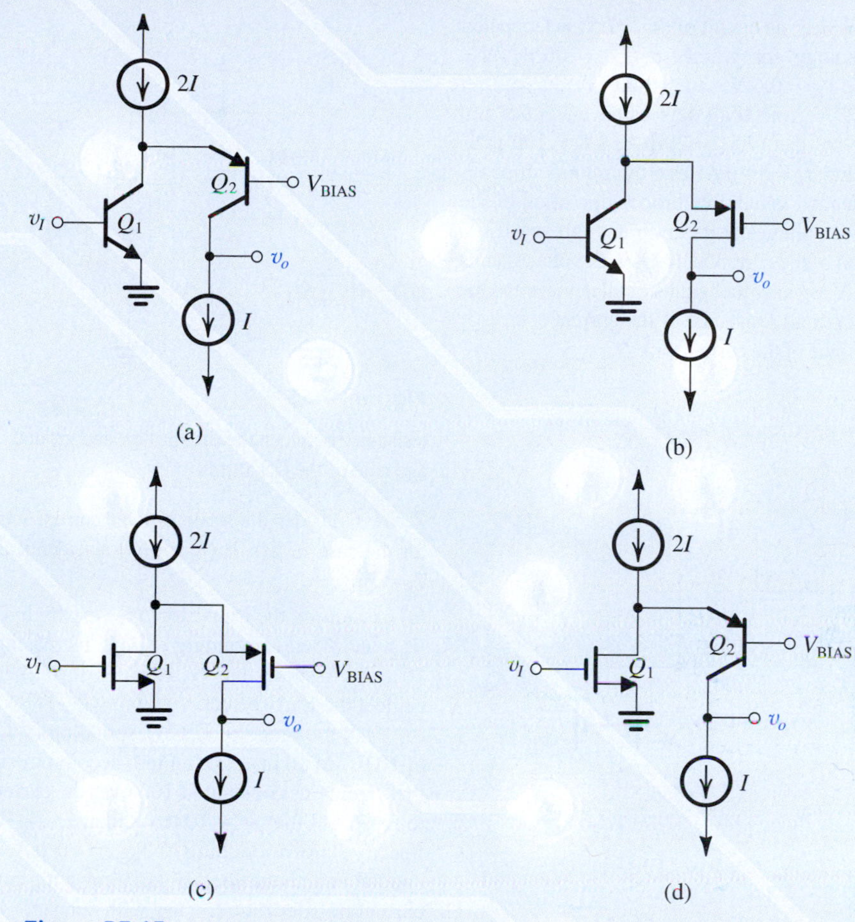

Figure P7.45

an output current of 200-μA nominal value. To simplify matters, assume that the nominal value of the output current is obtained at $V_O \simeq V_{GS}$. It is further required that the circuit operate for V_O in the range of 0.2 V to V_{DD} and that the change in I_O over this range be limited to 5% of the nominal value of I_O. Find the required value of R and the device dimensions. For the fabrication-process technology utilized, $\mu_n C_{ox} = 400$ μA/V², $V_A' = 10$ V/μm, and $V_t = 0.5$ V.

7.48 Sketch the *p*-channel counterpart of the current-source circuit of Fig. 7.22. Note that while the circuit of Fig. 7.22 should more appropriately be called a current sink, the corresponding PMOS circuit is a current source. Let $V_{DD} = 1.8$ V, $|V_t| = 0.5$ V, Q_1 and Q_2 be matched, and $\mu_p C_{ox} = 100$ μA/V². Find the device *W*/*L* ratios and the value of the resistor that sets the value of I_{REF} so that a nominally 80-μA output current is obtained. The current source is required to operate for V_O as high as 1.6 V. Neglect channel-length modulation.

SIM **7.49** Consider the current-mirror circuit of Fig. 7.23 with two transistors having equal channel lengths but with Q_2 having a width five times that of Q_1. If I_{REF} is 20 μA and the transistors are operating at an overdrive voltage of 0.2 V, what I_O results? What is the minimum allowable value of V_O for proper operation of the current source? If $V_t = 0.5$ V, at what value of V_O will the nominal value of I_O be obtained? If V_O increases by 1 V, what is the corresponding increase in I_O? Let $V_A = 20$ V.

7.50 For the current-steering circuit of Fig. P7.50, find I_O in terms of I_{REF} and device *W*/*L* ratios.

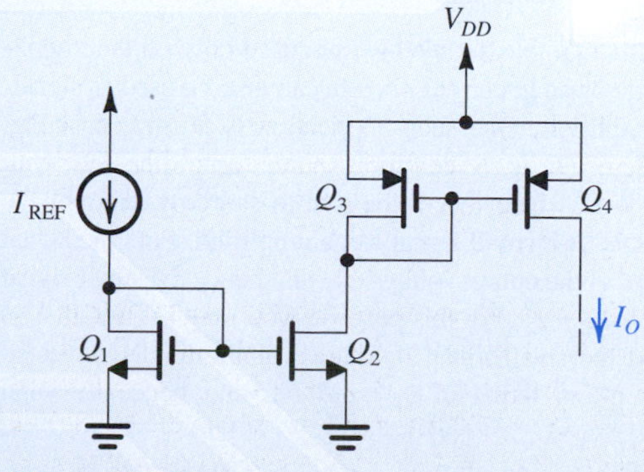

Figure P7.50

D 7.51 The current-steering circuit of Fig. P7.51 is fabricated in a CMOS technology for which $\mu_n C_{ox} = 200\ \mu A/V^2$, $\mu_p C_{ox} = 80\ \mu A/V^2$, $V_{tn} = 0.6$ V, $V_{tp} = -0.6$ V, $V'_{An} = 10$ V/μm, and $|V'_{Ap}| = 12$ V/μm. If all devices have $L = 0.8$ μm, design the circuit so that $I_{REF} = 20$ μA, $I_2 = 100$ μA, $I_3 = I_4 = 20$ μA, and $I_5 = 50$ μA. Use the minimum possible device widths needed to achieve proper operation of the current source Q_2 for voltages at its drain as high as +1.3 V and proper operation of the current sink Q_5 with voltages at its drain as low as −1.3 V. Specify the widths of all devices and the value of R. Find the output resistance of the current source Q_2 and the output resistance of the current sink Q_5.

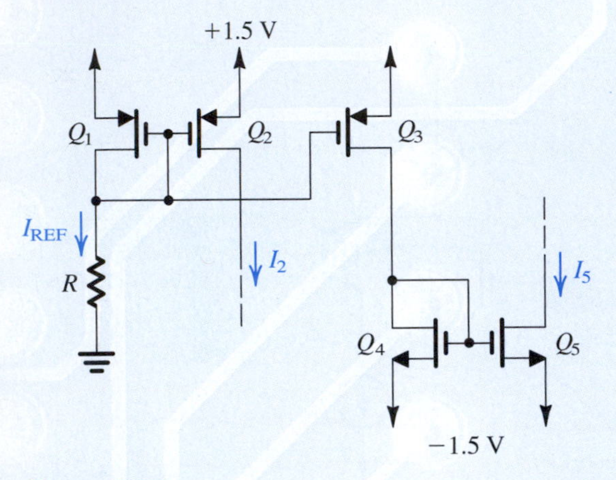

Figure P7.51

***7.52** A PMOS current mirror consists of three PMOS transistors, one diode connected and two used as current outputs. All transistors have $|V_t| = 0.6$ V, $k'_p = 100\ \mu A/V^2$, and $L = 1.0$ μm but three different widths, namely, 10 μm, 20 μm, and 40 μm. When the diode-connected transistor is supplied from a 100-μA source, how many different output currents are available? Repeat with two of the transistors diode connected and the third used to provide current output. For each possible input-diode combination, give the values of the output currents and of the V_{SG} that results.

7.53 Although thus far we have focused only on their application in dc biasing, current mirrors can also be used as signal-current amplifiers. One such application is illustrated in Fig. P7.53. Here Q_1 is a common-source amplifier fed with $v_I = V_{GS} + v_i$, where V_{GS} is the gate-to-source dc bias voltage of Q_1 and v_i is a small signal to be amplified. Find the signal component of the output voltage v_O and hence the small-signal voltage gain v_o/v_i. For this purpose, you may neglect all r_o's. Also, find the small-signal resistance of the diode-connected transistor Q_2 in terms of g_{m2} and r_{o2}, and hence the total resistance between the drain of Q_1 and ground. What is the voltage gain of the CS amplifier Q_1?

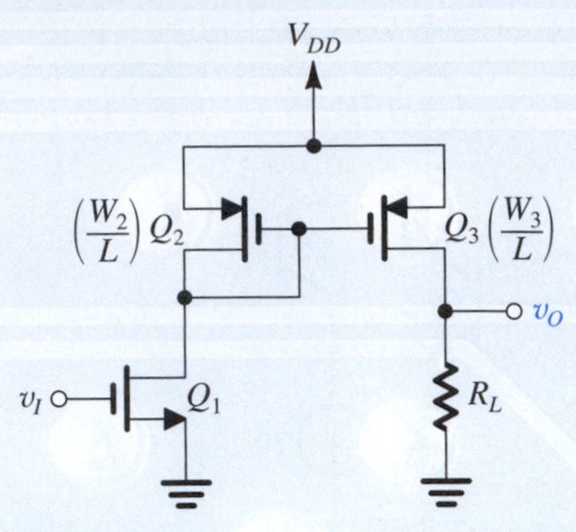

Figure P7.53

7.54 Consider the basic bipolar current mirror of Fig. 7.28 for the case in which Q_1 and Q_2 are identical devices having $I_S = 10^{-16}$ A.

(a) Assuming the transistor β is very high, find the range of V_{BE} and I_O corresponding to I_{REF} increasing from 10 μA to 10 mA. Assume that Q_2 remains in the active mode, and neglect the Early effect.

(b) Find the range of I_O corresponding to I_{REF} in the range of 10 μA to 10 mA, taking into account the finite β. Assume that β remains constant at 100 over the current range 0.1 mA to 5 mA but that at $\simeq$ 10 μA and at $I_C \simeq 10$ mA, $\beta = 50$. Specify I_O corresponding to $I_{REF} = 10$ μA, 0.1 mA, 1 mA, and 10 mA. Note that β variation with current causes the current transfer ratio to vary with current.

7.55 Consider the basic BJT current mirror of Fig. 7.28 for the case in which Q_2 has m times the area of Q_1. Show that the current transfer ratio is given by Eq. (7.69). If β is specified to be a minimum of 50, what is the largest current transfer ratio possible if the error introduced by the finite β is limited to 10%?

7.56 Give the circuit for the *pnp* version of the basic current mirror of Fig. 7.28. If β of the *pnp* transistor is 20, what is the current gain (or transfer ratio) I_O/I_{REF} for the case of identical transistors, neglecting the Early effect?

7.57 Consider the basic BJT current mirror of Fig. 7.28 when Q_1 and Q_2 are matched and $I_{REF} = 2$ mA. Neglecting the effect of finite β, find the change in I_O, both as an absolute value and as a percentage, corresponding to V_O changing from 1 V to 10 V. The Early voltage is 90 V.

D 7.58 The current-source circuit of Fig. P7.58 utilizes a pair of matched *pnp* transistors having $I_S = 10^{-15}$ A, $\beta = 50$, and $|V_A| = 50$ V. It is required to design the circuit to provide an output current $I_O = 1$ mA at $V_O = 2$ V. What values of I_{REF}

and R are needed? What is the maximum allowed value of V_O while the current source continues to operate properly? What change occurs in I_O corresponding to V_O changing from the maximum positive value to –5 V?

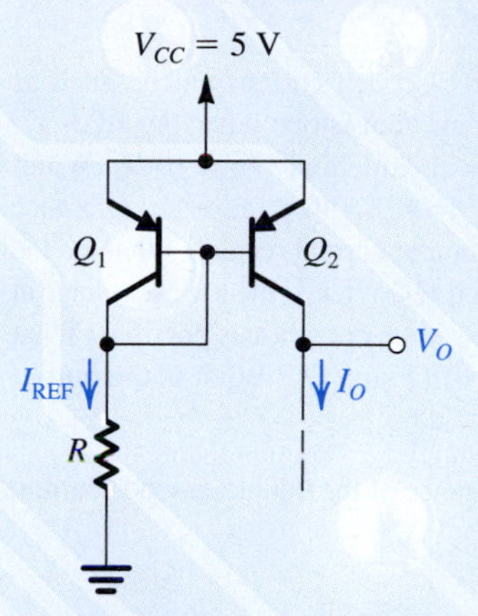

Figure P7.58

7.59 Find the voltages at all nodes and the currents through all branches in the circuit of Fig. P7.59. Assume $|V_{BE}| = 0.7$ V and $\beta = \infty$.

7.60 For the circuit in Fig. P7.60, let $|V_{BE}| = 0.7$ V and $\beta = \infty$. Find I, V_1, V_2, V_3, V_4, and V_5 for (a) $R = 10$ kΩ and (b) $R = 100$ kΩ.

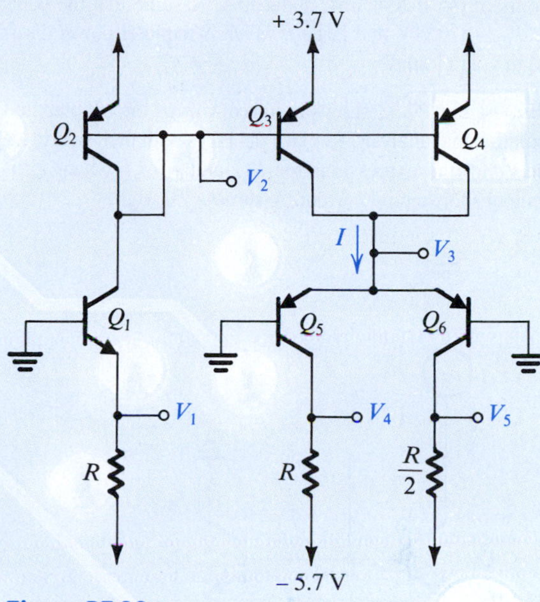

Figure P7.60

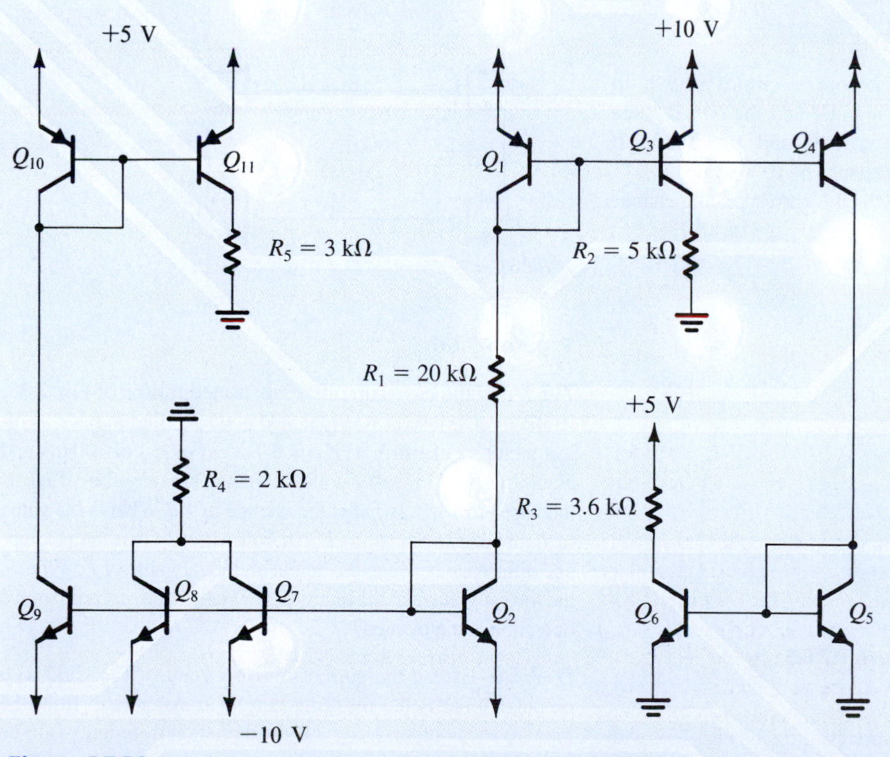

Figure P7.59

D 7.61 Using the ideas embodied in Fig. 7.31, design a multiple-mirror circuit using power supplies of ±5 V to create source currents of 0.2 mA, 0.4 mA, and 0.8 mA and sink currents of 0.5 mA, 1 mA, and 2 mA. Assume that the BJTs have $|V_{BE}| \simeq 0.7$ V and large β. What is the total power dissipated in your circuit?

***7.62** Figure P7.62 shows a current-mirror circuit prepared for small-signal analysis. Replace the BJTs with their hybrid-π models and find expressions for R_{in} and i_o/i_i, where i_o is the output short-circuit current. Assume $r_o \gg r_\pi$.

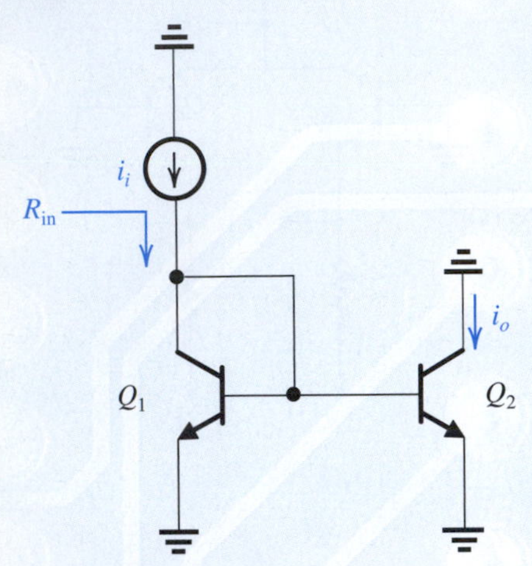

Figure P7.62

7.63 For the constant-current source circuit shown in Fig. P7.63, find the collector current I and the output resistance. The BJT is specified to have $\beta = 100$ and $V_A = 100$ V. If the collector voltage undergoes a change of 10 V while the BJT remains in the active mode, what is the corresponding change in collector current?

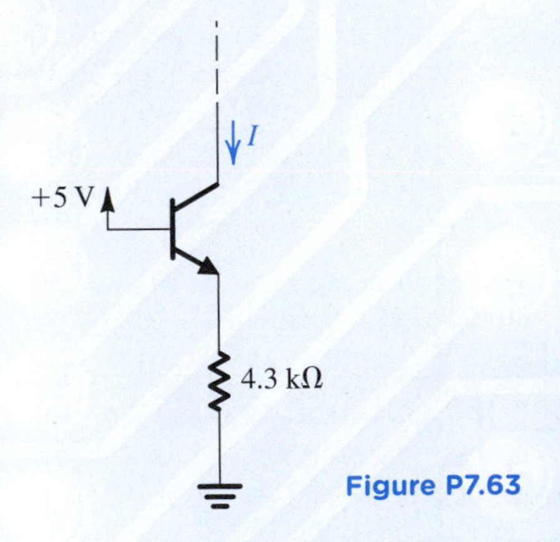

Figure P7.63

7.64 For the MOS cascode current mirror of Fig. 7.32 with $V_t = 0.5$ V, $k_n' = 4$ mA/V², $V_A = 10$ V, and $I_{REF} = 100$ µA, find R_o and the minimum allowable voltage at the output. At what value of V_O is I_O equal to I_{REF}? What does I_O become at $V_O = 5$ V?

Section 7.5: Current-Mirror Circuits with Improved Performance

SIM **7.65** In a particular cascoded current mirror, such as that shown in Fig. 7.32, all transistors have $V_t = 0.6$ V, $\mu_n C_{ox} = 160$ µA/V², $L = 1$ µm, and $V_A = 10$ V. Width $W_1 = W_4 = 4$ µm, and $W_2 = W_3 = 40$ µm. The reference current I_{REF} is 20 µA. What output current results? What are the voltages at the gates of Q_2 and Q_3? What is the lowest voltage at the output for which current-source operation is possible? What are the values of g_m and r_o of Q_2 and Q_3? What is the output resistance of the mirror?

7.66 Find the output resistance of the double-cascode current mirror of Fig. P7.66.

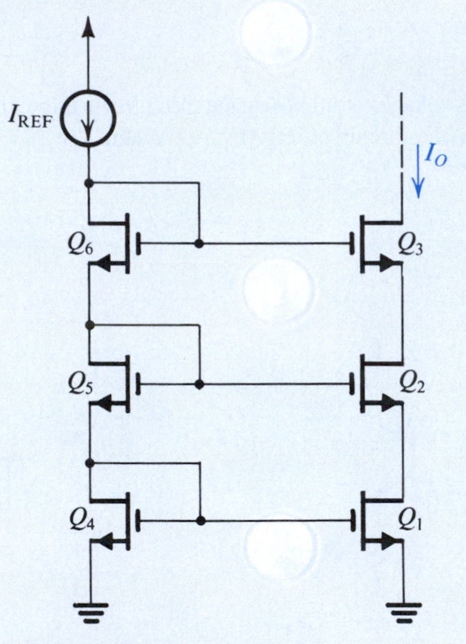

Figure P7.66

7.67 For the base-current-compensated mirror of Fig. 7.33, let the three transistors be matched and specified to have a collector current of 1 mA at $V_{BE} = 0.7$ V. For I_{REF} of 100 µA and assuming $\beta = 200$, what will the voltage at node x be? If I_{REF} is increased to 1 mA, what is the change in V_x? What is the value of I_O obtained with $V_O = V_x$ in both cases? Give the percentage difference between the actual and ideal value of I_O. What is the lowest voltage at the output for which proper current-source operation is maintained?

D 7.68 Extend the current-mirror circuit of Fig. 7.33 to n outputs. What is the resulting current transfer ratio from the input to each output, I_O/I_{REF}? If the deviation from unity is to be kept at 0.1% or less, what is the maximum possible number of outputs for BJTs with $\beta = 100$?

***7.69** For the base-current-compensated mirror of Fig. 7.33, show that the incremental input resistance (seen by the reference current source) is approximately $2\ V_T/I_{REF}$. Evaluate R_{in} for $I_{REF} = 100\ \mu A$. [*Hint*: Q_3 is operating at a current $I_{E3} = 2I_C/\beta$, where I_C is the operating current of each of Q_1 and Q_2. Replace each transistor with its T model and neglect r_o.]

7.70 Consider the Wilson current-mirror circuit of Fig. 7.34 when supplied with a reference current I_{REF} of 1 mA. What is the change in I_O corresponding to a change of +10 V in the voltage at the collector of Q_3? Give both the absolute value and the percentage change. Let $\beta = 100$ and $V_A = 100$ V.

D *7.71 (a) The circuit in Fig. P7.71 is a modified version of the Wilson current mirror. Here the output transistor is "split" into two matched transistors, Q_3 and Q_4. Find I_{O1} and I_{O2} in terms of I_{REF}. Assume all transistors to be matched with current gain β.
(b) Use this idea to design a circuit that generates currents of 0.1 mA, 0.2 mA, and 0.4 mA, using a reference current source of 0.7 mA. What are the actual values of the currents generated for $\beta = 50$?

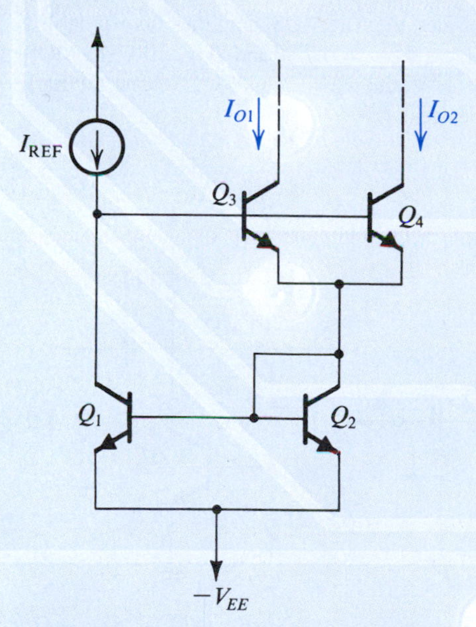

Figure P7.71

D 7.72 Use the *pnp* version of the Wilson current mirror to design a 0.2-mA current source. The current source is required to operate with the voltage at its output terminal as low as −2.5 V. If the power supplies available are ±2.5 V, what is the highest voltage possible at the output terminal?

***7.73** For the Wilson current mirror of Fig. 7.34, show that the incremental input resistance seen by I_{REF} is approximately $2\ V_T/I_{REF}$. (Neglect the Early effect in this derivation.) Evaluate R_{in} for $I_{REF} = 100\ \mu A$.

***7.74** Consider the Wilson MOS mirror of Fig. 7.35(a) for the case of all transistors identical, with $W/L = 12.5$, $\mu_n C_{ox} = 400\ \mu A/V^2$, and $V_A = 20$ V. The mirror is fed with $I_{REF} = 100\ \mu A$.

(a) Obtain an estimate of V_{OV} and V_{GS} at which the three transistors are operating, by neglecting the Early effect.
(b) Noting that Q_1 and Q_2 are operating at different V_{DS}, obtain an approximate value for the difference in their currents and hence determine I_O.
(c) To eliminate the systematic error between I_O and I_{REF} caused by the difference in V_{DS} between Q_1 and Q_2, a diode-connected transistor Q_4 can be added to the circuit as shown in Fig. 7.35(c). What do you estimate I_O now to be?
(d) What is the minimum allowable voltage at the output node of the mirror?
(e) Convince yourself that Q_4 will have no effect on the output resistance of the mirror. Find R_o.
(f) What is the change in I_O (both absolute value and percentage) that results from $\Delta V_O = 1$ V?

7.75 Show that the input resistance (seen by I_{REF}) for the Wilson MOS mirror of Fig. 7.35(a) is given by $2/g_m$. Assume that all three transistors are identical and neglect the Early effect. [*Hint*: Replace all transistors by their T model and remember that Q_1 is equivalent to a resistance $1/g_m$.]

D 7.76 (a) Utilizing a reference current of 100 μA, design a Widlar current source to provide an output current of 10 μA. Let the BJTs have $v_{BE} = 0.8$ V at 1-mA current, and assume β to be high.

(b) If $\beta = 200$ and $V_A = 50$ V, find the value of the output resistance, and find the change in output current corresponding to a 5-V change in output voltage.

D 7.77 Design three Widlar current sources, each having a 100-μA reference current: one with a current transfer ratio of 0.9, one with a ratio of 0.10, and one with a ratio of 0.01, all assuming high β. For each, find the output resistance, and contrast it with r_o of the basic unity-ratio source for which $R_E = 0$. Use $\beta = \infty$ and $V_A = 50$ V.

7.78 The BJT in the circuit of Fig. P7.78 has $V_{BE} = 0.7$ V, $\beta = 100$, and $V_A = 50$ V. Find R_o.

D 7.79 (a) For the circuit in Fig. P7.79, assume BJTs with high β and $v_{BE} = 0.8$ V at 1 mA. Find the value of R that will result in $I_O = 10\ \mu A$.

(b) For the design in (a), find R_o assuming $\beta = 100$ and $V_A = 50$ V.

D *7.80 If the *pnp* transistor in the circuit of Fig. P7.80 is characterized by its exponential relationship with a scale current

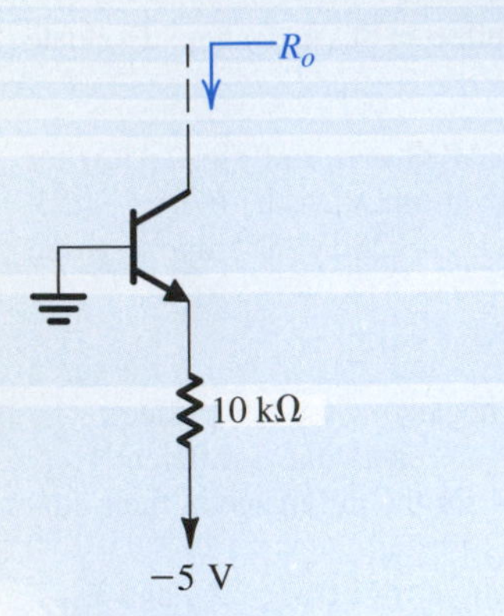

Figure P7.78

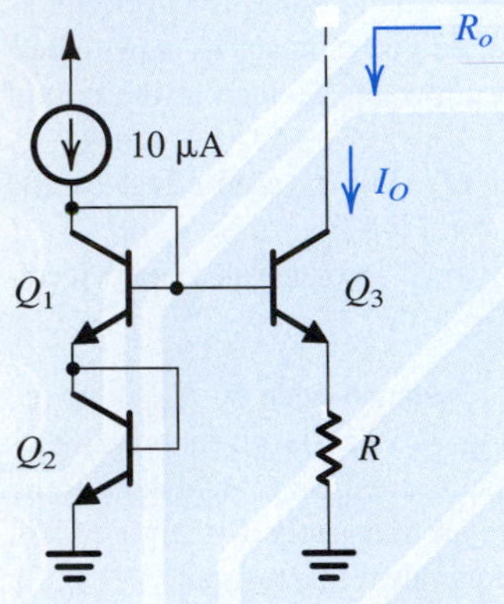

Figure P7.79

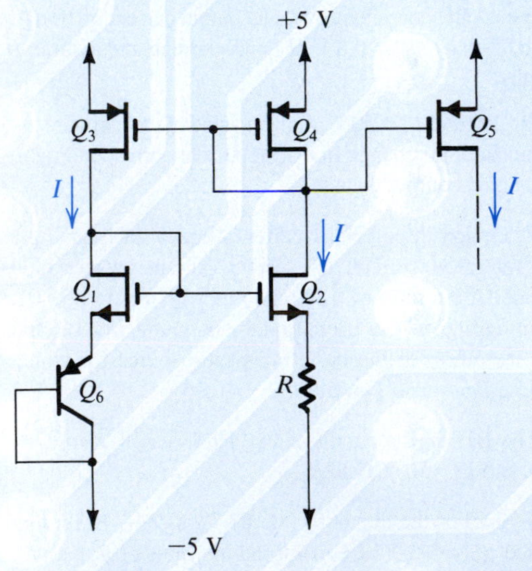

Figure P7.80

I_S, show that the dc current I is determined by $IR = V_T \ln(I/I_S)$. Assume Q_1 and Q_2 to be matched and Q_3, Q_4, and Q_5 to be matched. Find the value of R that yields a current $I = 100$ μA. For the BJT, $V_{EB} = 0.7$ V at $I_E = 1$ mA.

Section 7.6: Some Useful Transistor Pairings

7.81 The transistors in the circuit of Fig. P7.81 have $\beta = 100$ and $V_A = 100$ V.

(a) Find R_{in} and the overall voltage gain.
(b) What is the effect of increasing the bias currents by a factor of 10 on R_{in}, G_v, and the power dissipation?

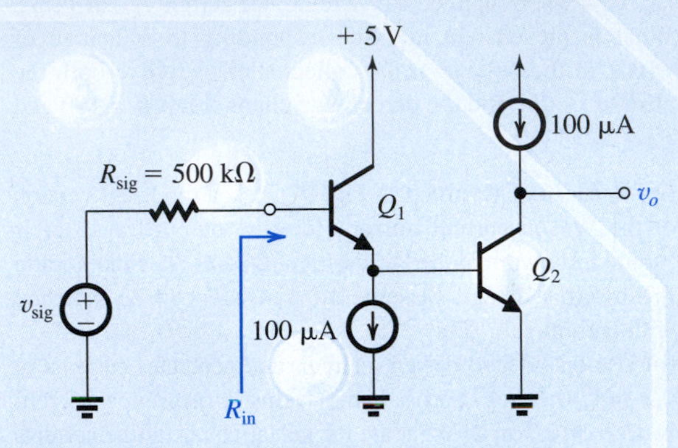

Figure P7.81

D *7.82 Consider the BiCMOS amplifier shown in Fig. P7.82. The BJT has $V_{BE} = 0.7$ V and $\beta = 200$. The MOSFET has $V_t = 1$ V and $k_n = 2$ mA/V². Neglect the Early effect in both devices.

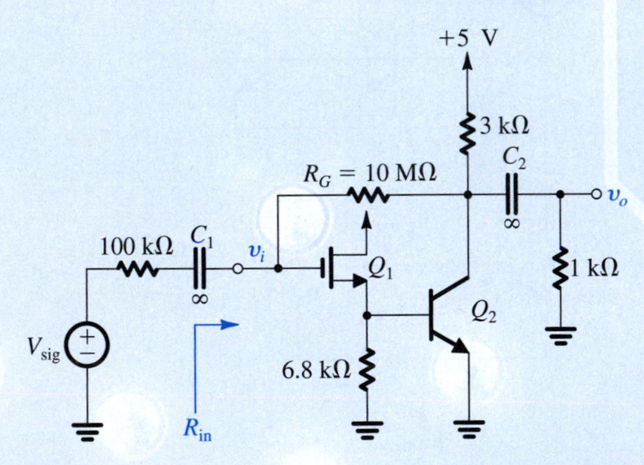

Figure P7.82

(a) Consider the dc bias circuit. Neglect the base current in Q_2 in determining the current in Q_1. Find the dc bias currents in Q_1 and Q_2 and show that they are approximately 100 μA and 1 mA, respectively.
(b) Evaluate the small-signal parameters of Q_1 and Q_2 at their bias points.
(c) Determine the voltage gain $A_v = v_o/v_i$. For this purpose you can neglect R_G.

(d) Noting that R_G is connected between the input node where the voltage is v_i and the output node where the voltage is $A_v v_i$, find R_{in} and hence the overall voltage gain v_o/v_{sig}.
(e) To considerably reduce the effect of R_G on R_{in} and hence on G_v, consider the effect of adding another 10-MΩ resistor in series with the existing one and placing a large bypass capacitor between their joint node and ground. What will R_{in} and G_v become?

7.83 The BJTs in the Darlington follower of Fig. P7.83 have $\beta = 100$. If the follower is fed with a source having a 100-kΩ resistance and is loaded with 1 kΩ, find the input resistance and the output resistance (excluding the load). Also find the overall voltage gain, both open-circuited and with load.

7.84 For the amplifier in Fig. 7.41(a), let $I = 1$ mA and $\beta = 120$, and neglect r_o. Assume that a load resistance of 10 kΩ is connected to the output terminal. If the amplifier is fed with a signal v_{sig} having a source resistance $R_{sig} = 20$ kΩ, find G_v.

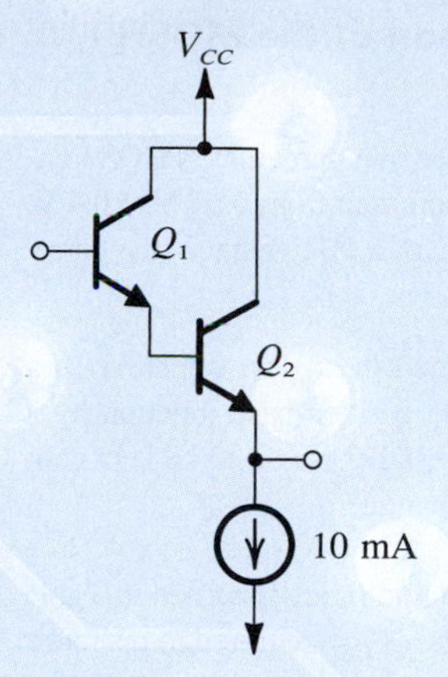

Figure P7.83

7.85 Consider the CD–CG amplifier of Fig. 7.41(c) for the case $g_m = 5$ mA/V, and $R_{sig} = R_L = 20$ kΩ. Neglecting r_o, find G_v.

****7.86** In each of the six circuits in Fig. P7.86, let $\beta = 100$, and neglect r_o. Calculate the overall voltage gain.

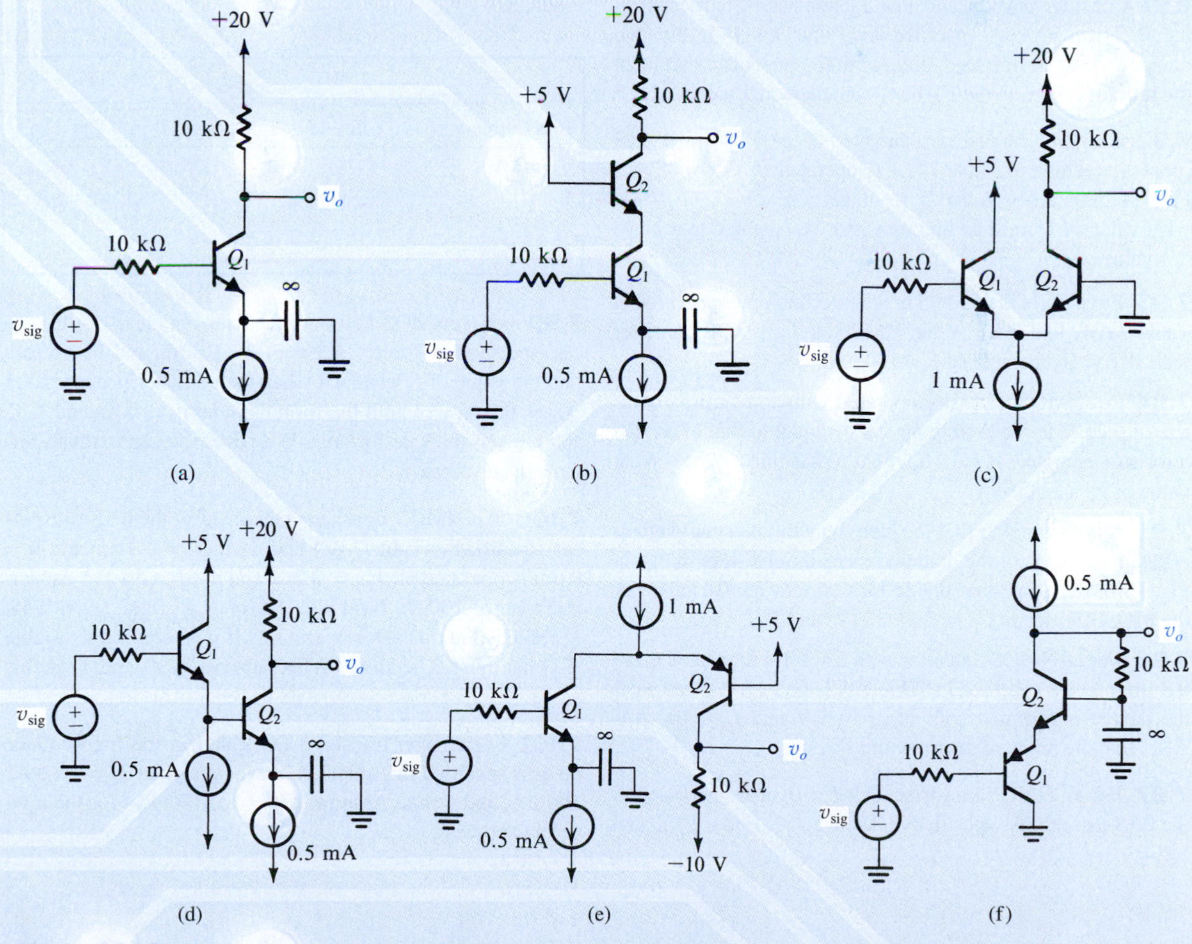

Figure P7.86

APPENDIX 7.A: Comparison of the MOSFET and the BJT

7.87 Find the range of I_D obtained in a particular NMOS transistor as its overdrive voltage is increased from 0.15 V to 0.4 V. If the same range is required in I_C of a BJT, what is the corresponding change in V_{BE}?

7.88 What range of I_C is obtained in an *npn* transistor as a result of changing the area of the emitter–base junction by a factor of 10 while keeping V_{BE} constant? If I_C is to be kept constant, by what amount must V_{BE} change?

7.89 For each of the CMOS technologies specified in Table 7.A.1, find the $|V_{OV}|$ and hence the $|V_{GS}|$ required to operate a device with a W/L of 10 at a drain current $I_D = 100\ \mu A$. Ignore channel-length modulation.

7.90 Consider NMOS and PMOS devices fabricated in the 0.25-µm process specified in Table 7.A.1. If both devices are to operate at $|V_{OV}| = 0.25$ V and $I_D = 100\ \mu A$, what must their W/L ratios be?

7.91 Consider NMOS and PMOS transistors fabricated in the 0.25-µm process specified in Table 7.A.1. If the two devices are to be operated at equal drain currents, what must the ratio of $(W/L)_p$ to $(W/L)_n$ be to achieve equal values of g_m?

7.92 An NMOS transistor fabricated in the 0.18-µm CMOS process specified in Table 7.A.1 is operated at $V_{OV} = 0.2$ V. Find the required W/L and I_D to obtain a g_m of 10 mA/V. At what value of I_C must an npn transistor be operated to achieve this value of g_m?

7.93 For each of the CMOS process technologies specified in Table 7.A.1, find the g_m of an NMOS and a PMOS transistor with $W/L = 10$ operated at $I_D = 100\ \mu A$.

7.94 An NMOS transistor operated with an overdrive voltage of 0.25 V is required to have a g_m equal to that of an *npn* transistor operated at $I_C = 0.1$ mA. What must I_D be? What value of g_m is realized?

7.95 It is required to find the incremental (i.e., small-signal) resistance of each of the diode-connected transistors shown in Fig. P7.95. Assume that the dc bias current $I = 0.1$ mA. For the MOSFET, let $\mu_n C_{ox} = 200\ \mu A/V^2$ and $W/L = 10$.

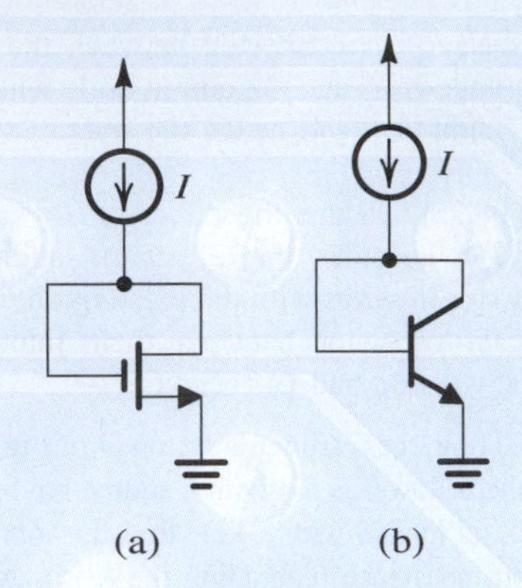

Figure P7.95

7.96 For an NMOS transistor with $L = 1\ \mu m$ fabricated in the 0.8-µm process specified in Table 7.A.1, find g_m, r_o, and A_0 if the device is operated with $V_{OV} = 0.5$ V and $I_D = 100\ \mu A$. Also, find the required device width W.

7.97 For an NMOS transistor with $L = 0.3\ \mu m$ fabricated in the 0.18-µm process specified in Table 7.A.1, find g_m, r_o, and A_0 obtained when the device is operated at $I_D = 100\ \mu A$ with $V_{OV} = 0.2$ V. Also, find W.

7.98 Fill in the table below. For the BJT, let $\beta = 100$ and $V_A = 100$ V. For the MOSFET, let $\mu_n C_{ox} = 200\ \mu A/V^2$, $W/L = 40$, and $V_A = 10$ V. Note that R_{in} refers to the input resistance at the control input terminal (gate, base) with the (source, emitter) grounded.

	BJT		MOSFET	
Bias Current	$I_C = 0.1$ mA	$I_C = 1$ mA	$I_D = 0.1$ mA	$I_D = 1$ mA
g_m (mA/V)				
r_o (kΩ)				
A_0 (V/V)				
R_{in} (kΩ)				

7.99 For an NMOS transistor fabricated in the 0.18-µm process specified in Table 7.A.1 with $L = 0.3\ \mu m$ and $W = 6\ \mu m$, find the value of f_T obtained when the transistor is operated at $V_{OV} = 0.2$ V. Use both the formula in terms of C_{gs} and C_{gd} and the approximate formula. Why does the approximate formula overestimate f_T?

7.100 An NMOS transistor fabricated in the 0.18-µm process specified in Table 7.A.1 and having $L = 0.3\ \mu m$ and $W = 6\ \mu m$ is operated at $V_{OV} = 0.2$ V and used to drive a capacitive load of 100 fF. Find A_0, f_P (or f_{3dB}), and f_t. At what I_D value is the transistor operating? If it is required to double f_t, what must I_D become? What happens to A_0 and f_P in this case?

7.101 For an *npn* transistor fabricated in the high-voltage process specified in Table 7.A.2, evaluate f_T at $I_C = 10\ \mu A$, $100\ \mu A$, and 1 mA. Assume $C_\mu \simeq C_{\mu 0}$. Repeat for the low-voltage process.

7.102 Consider an NMOS transistor fabricated in the 0.8-μm process specified in Table 7.A.1. Let the transistor have L = 1 μm, and assume it is operated at I_D = 100 μA.

(a) For V_{OV} = 0.25 V, find W, g_m, r_o, A_0, C_{gs}, C_{gd}, and f_T.
(b) To what must V_{OV} be changed to double f_T? Find the new values of W, g_m, r_o, A_0, C_{gs}, and C_{gd}.

7.103 For a lateral *pnp* transistor fabricated in the high-voltage process specified in Table 7.A.2, find f_T if the device is operated at a collector bias current of 1 mA. Compare to the value obtained for a vertical *npn*.

7.104 Show that for a MOSFET the selection of L and V_{OV} determines A_0 and f_T. In other words, show that A_0 and f_T will not depend on I_D and W.

7.105 Consider an NMOS transistor fabricated in the 0.18-μm technology specified in Table 7.A.1. Let the transistor be operated at V_{OV} = 0.2 V. Find A_0 and f_T for L = 0.2 μm, 0.3 μm, and 0.4 μm.

D 7.106 Consider an NMOS transistor fabricated in the 0.5-μm process specified in Table 7.A.1. Let L = 0.5 μm and V_{OV} = 0.3 V. If the MOSFET is connected as a common-source amplifier with a load capacitance C_L = 1 pF (as in Fig. 7.A.2a), find the required transistor width W and bias current I_D to obtain a unity-gain bandwidth of 100 MHz. Also, find A_0 and f_{3dB}.

General Problem:

***7.107** The circuit shown in Fig. P7.107 is known as a **current conveyor**.

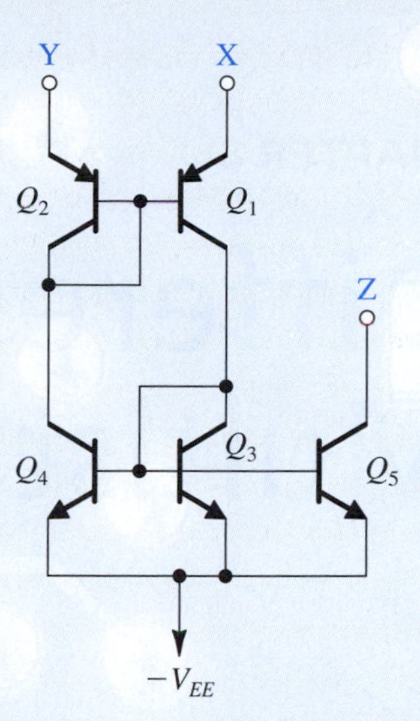

Figure P7.107

(a) Assuming that Y is connected to a voltage V, a current I is forced into X, and terminal Z is connected to a voltage that keeps Q_5 in the active region, show that a current equal to I flows through terminal Y, that a voltage equal to V appears at terminal X, and that a current equal to I flows through terminal Z. Assume β to be large. Corresponding transistors are matched, and all transistors are operating in the active region.
(b) With Y connected to ground, show that a virtual ground appears at X. Now, if X is connected to a +5-V supply through a 10-kΩ resistor, what current flows through Z?

CHAPTER 8

Differential and Multistage Amplifiers

IN THIS CHAPTER YOU WILL LEARN

1. The essence of the operation of the MOS and the bipolar differential amplifiers: how they reject common-mode noise or interference and amplify differential signals.
2. The analysis and design of MOS and BJT differential amplifiers.
3. Differential-amplifier circuits of varying complexity; utilizing passive resistive loads, current-source loads, and cascodes—the building blocks we studied in Chapter 7.
4. An ingenious and highly popular differential-amplifier circuit that utilizes a current-mirror load.
5. The structure, analysis, and design of amplifiers composed of two or more stages in cascade. Two practical examples are studied in detail: a two-stage CMOS op amp and a four-stage bipolar op amp.

Introduction

The differential-pair or differential-amplifier configuration is the most widely used building block in analog integrated-circuit design. For instance, the input stage of every op amp is a differential amplifier. Also, the BJT differential amplifier is the basis of a very-high-speed logic circuit family, studied briefly in Chapter 14, called emitter-coupled logic (ECL).

Initially invented in the 1940s for use with vacuum tubes, the basic differential-amplifier configuration was subsequently implemented with discrete bipolar transistors. However, it was the advent of integrated circuits that has made the differential pair extremely popular in both bipolar and MOS technologies. There are two reasons why differential amplifiers are so well suited for IC fabrication: First, as we shall shortly see, the performance of the differential pair depends critically on the matching between the two sides of the circuit. Integrated-circuit fabrication is capable of providing matched devices whose parameters track over wide ranges of changes in environmental conditions. Second, by their very nature, differential amplifiers utilize more components (approaching twice as many) than single-ended circuits. Here again, the reader will recall from the discussion in Section 7.1 that a significant advantage of integrated-circuit technology is the availability of large numbers of transistors at relatively low cost.

We assume that the reader is familiar with the basic concept of a differential amplifier as presented in Section 2.1. Nevertheless it is worthwhile to answer the question: Why differential? Basically, there are two reasons for using differential in preference to single-ended amplifiers. First, differential circuits are much less sensitive to noise and interference than single-ended circuits. To appreciate this point, consider two wires carrying a small differential signal as the voltage difference between the two wires. Now, assume that there is an interference signal that is coupled to the two wires, either capacitively or inductively. As the two wires are physically close together, the interference voltages on the two wires (i.e., between each of the two wires and ground) will be equal. Since, in a differential system, only the difference signal between the two wires is sensed, it will contain no interference component!

The second reason for preferring differential amplifiers is that the differential configuration enables us to bias the amplifier and to couple amplifier stages together without the need for bypass and coupling capacitors such as those utilized in the design of discrete-circuit amplifiers (Sections 5.8 and 6.8). This is another reason why differential circuits are ideally suited for IC fabrication where large capacitors are impossible to fabricate economically.

The major topic of this chapter is the differential amplifier in both its MOS and bipolar implementations. As will be seen, the design and analysis of differential amplifiers makes extensive use of the material on single-stage amplifiers presented in Chapters 5 through 7. We will follow the study of differential amplifiers with examples of practical multistage amplifiers, again in both MOS and bipolar technologies.

8.1 The MOS Differential Pair

Figure 8.1 shows the basic MOS differential-pair configuration. It consists of two matched transistors, Q_1 and Q_2, whose sources are joined together and biased by a constant-current source I. The latter is usually implemented by a MOSFET circuit of the type studied in Sections 7.4 and 7.5. For the time being, we assume that the current source is ideal and that it has infinite output resistance. Although each drain is shown connected to the positive supply through a resistance R_D, in most cases active (current-source) loads are employed, as will be seen shortly. For the time being, however, we will explain the essence of the differential-pair operation utilizing simple resistive loads. Whatever type of load is used, it is essential that the MOSFETs not enter the triode region of operation.

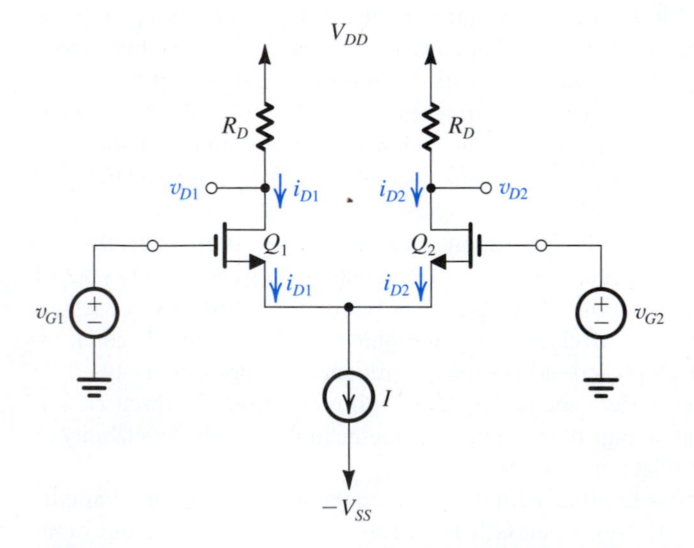

Figure 8.1 The basic MOS differential-pair configuration.

8.1.1 Operation with a Common-Mode Input Voltage

To see how the differential pair works, consider first the case when the two gate terminals are joined together and connected to a voltage V_{CM}, called the **common-mode voltage**. That is, as shown in Fig. 8.2, $v_{G1} = v_{G2} = V_{CM}$. Since Q_1 and Q_2 are matched, the current I will divide equally between the two transistors. Thus, $i_{D1} = i_{D2} = I/2$, and the voltage at the sources, V_S, will be

$$V_S = V_{CM} - V_{GS} \tag{8.1}$$

where V_{GS} is the gate-to-source voltage corresponding to a drain current of $I/2$. Neglecting channel-length modulation, V_{GS} and $I/2$ are related by

$$\frac{I}{2} = \frac{1}{2}k_n'\frac{W}{L}(V_{GS} - V_t)^2 \tag{8.2}$$

or in terms of the overdrive voltage V_{OV},

$$V_{OV} = V_{GS} - V_t \tag{8.3}$$

$$\frac{I}{2} = \frac{1}{2}k_n'\frac{W}{L}V_{OV}^2 \tag{8.4}$$

$$V_{OV} = \sqrt{I/k_n'(W/L)} \tag{8.5}$$

The voltage at each drain will be

$$v_{D1} = v_{D2} = V_{DD} - \frac{I}{2}R_D \tag{8.6}$$

Thus, the difference in voltage between the two drains will be zero.

Now, let us vary the value of the common-mode voltage V_{CM}. We see that, as long as Q_1 and Q_2 remain in the saturation region, the current I will divide equally between Q_1 and Q_2 and the voltages at the drains will not change. Thus the differential pair does *not* respond to (i.e., it *rejects*) common-mode input signals.

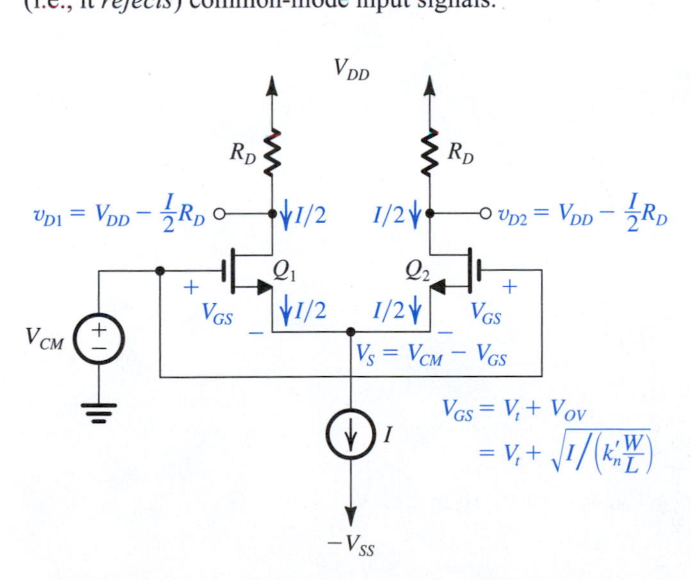

Figure 8.2 The MOS differential pair with a common-mode input voltage V_{CM}.

An important specification of a differential amplifier is its **input common-mode range**. This is the range of V_{CM} over which the differential pair operates properly. The highest value of V_{CM} is limited by the requirement that Q_1 and Q_2 remain in saturation, thus

$$V_{CM\max} = V_t + V_{DD} - \frac{I}{2}R_D \tag{8.7}$$

The lowest value of V_{CM} is determined by the need to allow for a sufficient voltage across the current source I for it to operate properly. If a voltage V_{CS} is needed across the current source, then

$$V_{CM\min} = -V_{SS} + V_{CS} + V_t + V_{OV} \tag{8.8}$$

Example 8.1

For the MOS differential pair with a common-mode voltage V_{CM} applied, as shown in Fig. 8.2, let $V_{DD} = V_{SS} = 1.5$ V, $k_n'(W/L) = 4$ mA/V^2, $V_t = 0.5$ V, $I = 0.4$ mA, and $R_D = 2.5$ kΩ, and neglect channel-length modulation. Assume that the current source I requires a minimum voltage of 0.4 V to operate properly.

(a) Find V_{OV} and V_{GS} for each transistor.
(b) For $V_{CM} = 0$, find V_S, I_{D1}, I_{D2}, V_{D1}, and V_{D2}.
(c) Repeat (b) for $V_{CM} = +1$ V.
(d) Repeat (b) for $V_{CM} = -0.2$ V.
(e) What is the highest permitted value of V_{CM}?
(f) What is the lowest value allowed for V_{CM}?

Solution

(a) With $v_{G1} = v_{G2} = V_{CM}$, we see that $V_{GS1} = V_{GS2}$. Now, since the transistors are matched, I will divide equally between the two transistors,

$$I_{D1} = I_{D2} = \frac{I}{2}$$

Thus,

$$\frac{I}{2} = \frac{1}{2}k_n'(W/L)V_{OV}^2$$

$$\frac{0.4}{2} = \frac{1}{2} \times 4V_{OV}^2$$

which results in

$$V_{OV} = 0.316 \text{ V}$$

and thus,

$$V_{GS} = V_t + V_{OV} = 0.5 + 0.316 \simeq 0.82 \text{ V}$$

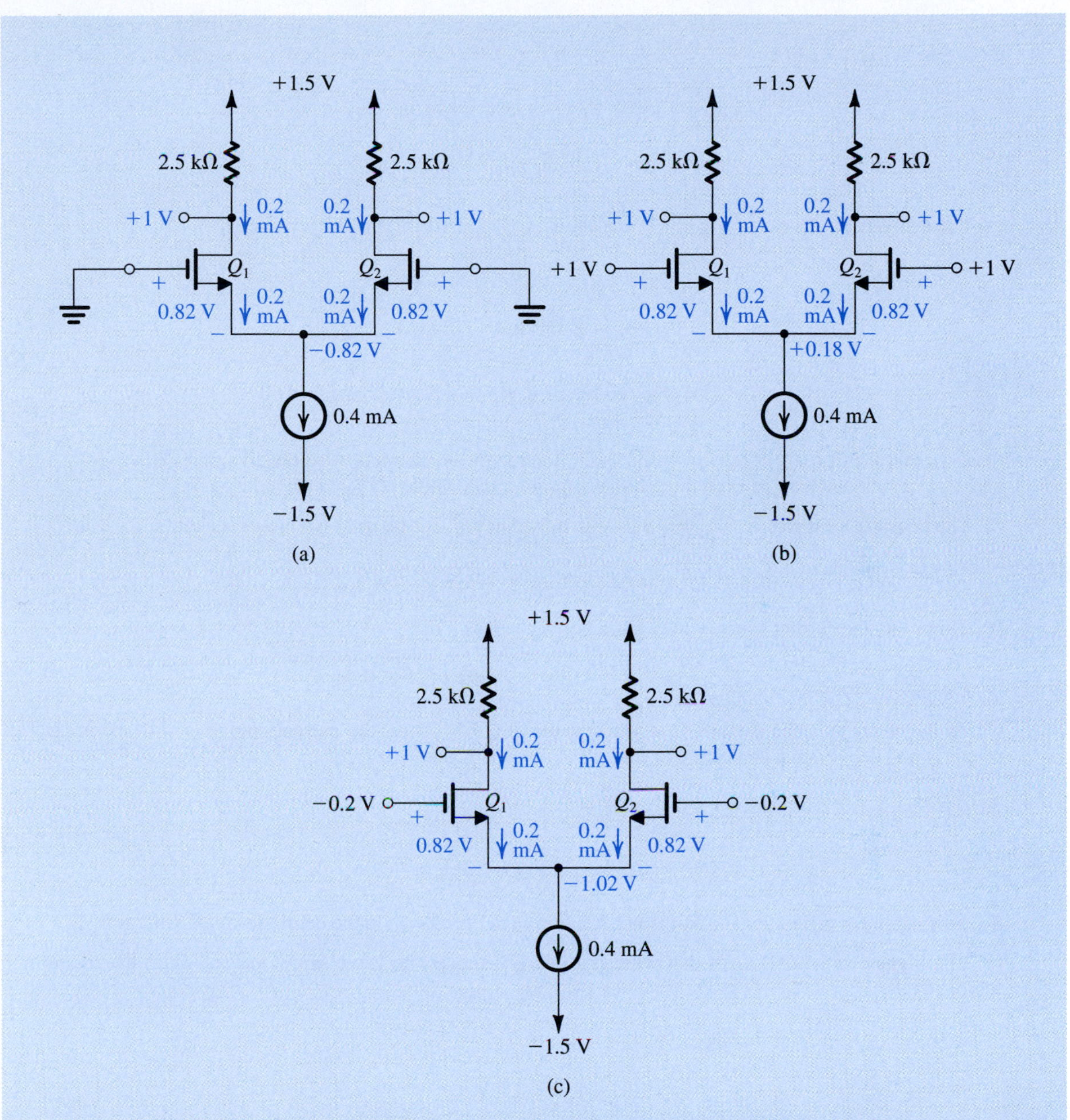

Figure 8.3 Circuits for Example 8.1. Effects of varying V_{CM} on the operation of the differential pair.

(b) The analysis for the case $V_{CM} = 0$ is shown in Fig. 8.3(a) from which we see that

$$V_S = V_G - V_{GS} = 0 - 0.82 = -0.82 \text{ V}$$

$$I_{D1} = I_{D2} = \frac{I}{2} = 0.2 \text{ mA}$$

Example 8.1 *continued*

$$V_{D1} = V_{D2} = V_{DD} - \frac{I}{2} R_D$$
$$= 1.5 - 0.2 \times 2.5 = 1 \text{ V}$$

(c) The analysis for the case $V_{CM} = +1$ V is shown in Fig. 8.3(b) from which we see that

$$V_S = V_G - V_{GS} = 1 - 0.82 = +0.18 \text{ V}$$
$$I_{D1} = I_{D2} = \frac{I}{2} = 0.2 \text{ mA}$$
$$V_{D1} = V_{D2} = V_{DD} - \frac{I}{2} R_D = 1.5 - 0.2 \times 2.5 = +1 \text{ V}$$

Observe that the transistors remain in the saturation region as assumed. Also observe that I_{D1}, I_{D2}, V_{D1}, and V_{D2} remain unchanged even though the common-mode voltage V_{CM} changed by 1 V.

(d) The analysis for the case $V_{CM} = -0.2$ V is shown in Fig. 8.3(c), from which we see that

$$V_S = V_G - V_{GS} = -0.2 - 0.82 = -1.02 \text{ V}$$

It follows that the current source I now has a voltage across it of

$$V_{CS} = -V_S - (-V_{SS}) = -1.02 + 1.5 = 0.48 \text{ V}$$

which is greater than the minimum required value of 0.4 V. Thus, the current source is still operating properly and delivering a constant current $I = 0.4$ mA and hence

$$I_{D1} = I_{D1} = \frac{I}{2} = 0.2 \text{ mA}$$
$$V_{D1} = V_{D2} = V_{DD} - \frac{I}{2} R_D = +1 \text{ V}$$

So, here again the differential circuit is not responsive to the change in the common-mode voltage V_{CM}.

(e) The highest value of V_{CM} is that which causes Q_1 and Q_2 to leave saturation and enter the triode region. Thus,

$$V_{CM\max} = V_t + V_D$$
$$= 0.5 + 1 = +1.5 \text{ V}$$

(f) The lowest value allowed for V_{CM} is that which reduces the voltage across the current source I to the minimum required of $V_{CS} = 0.4$ V. Thus,

$$V_{CM\min} = -V_{SS} + V_{CS} + V_{GS}$$
$$= -1.5 + 0.4 + 0.82 = -0.28 \text{ V}$$

Thus, the input common-mode range is

$$-0.28 \text{ V} \leq V_{CM} \leq +1.5 \text{ V}$$

EXERCISE

8.1 For the amplifier in Example 8.1, find the input common-mode range for the case in which the two drain resistances R_D are increased by a factor of 2.
Ans. −0.28 V to 1.0 V

8.1.2 Operation with a Differential Input Voltage

Next we apply a difference or differential input voltage by grounding the gate of Q_2 (i.e., setting $v_{G2} = 0$) and applying a signal v_{id} to the gate of Q_1, as shown in Fig. 8.4. We can see that since $v_{id} = v_{GS1} - v_{GS2}$, if v_{id} is positive, v_{GS1} will be greater than v_{GS2} and hence i_{D1} will be greater than i_{D2} and the difference output voltage $(v_{D2} - v_{D1})$ will be positive. On the other hand, when v_{id} is negative, v_{GS1} will be lower than v_{GS2}, i_{D1} will be smaller than i_{D2}, and correspondingly v_{D1} will be higher than v_{D2}; in other words, the difference or differential output voltage $(v_{D2} - v_{D1})$ will be negative.

From the above, we see that the differential pair responds to **difference-mode** or **differential input signals** by providing a corresponding differential output signal between the two drains. At this point, it is useful to inquire about the value of v_{id} that causes the entire bias current I to flow in one of the two transistors. In the positive direction, this happens when v_{GS1} reaches the value that corresponds to $i_{D1} = I$, and v_{GS2} is reduced to a value equal to the threshold voltage V_t, at which point $v_S = -V_t$. The value of v_{GS1} can be found from

$$I = \frac{1}{2}\left(k_n' \frac{W}{L}\right)(v_{GS1} - V_t)^2$$

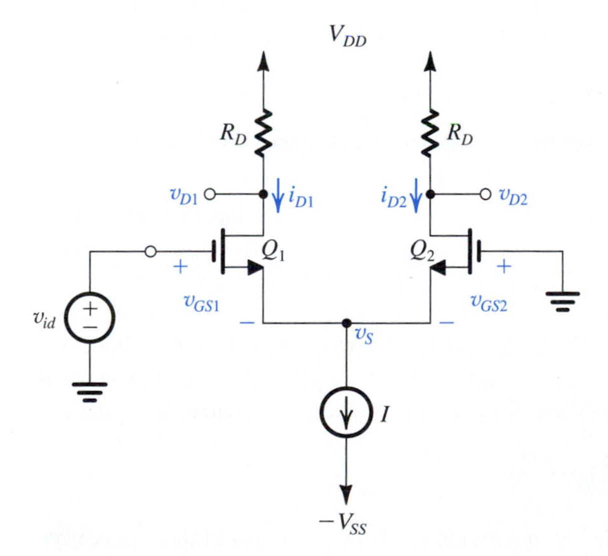

Figure 8.4 The MOS differential pair with a differential input signal v_{id} applied. With v_{id} positive: $v_{GS1} > v_{GS2}$, $i_{D1} > i_{D2}$, and $v_{D1} < v_{D2}$; thus $(v_{D2} - v_{D1})$ will be positive. With v_{id} negative: $v_{GS1} < v_{GS2}$, $i_{D1} < i_{D2}$, and $v_{D1} > v_{D2}$; thus $(v_{D2} - v_{D1})$ will be negative.

as

$$v_{GS1} = V_t + \sqrt{2I/k_n'(W/L)}$$

$$= V_t + \sqrt{2}V_{OV} \tag{8.9}$$

where V_{OV} is the overdrive voltage corresponding to a drain current of $I/2$ (Eq. 8.5). Thus, the value of v_{id} at which the entire bias current I is steered into Q_1 is

$$v_{id\max} = v_{GS1} + v_S$$

$$= V_t + \sqrt{2}V_{OV} - V_t$$

$$= \sqrt{2}V_{OV} \tag{8.10}$$

If v_{id} is increased beyond $\sqrt{2}V_{OV}$, i_{D1} remains equal to I, v_{GS1} remains equal to $(V_t + \sqrt{2}V_{OV})$, and v_S rises correspondingly, thus keeping Q_2 off. In a similar manner we can show that in the negative direction, as v_{id} reaches $-\sqrt{2}V_{OV}$, Q_1 turns off and Q_2 conducts the entire bias current I. Thus the current I can be steered from one transistor to the other by varying v_{id} in the range

$$-\sqrt{2}V_{OV} \leq v_{id} \leq \sqrt{2}V_{OV}$$

which defines the range of differential-mode operation. Finally, observe that we have assumed that Q_1 and Q_2 remain in saturation even when one of them is conducting the entire current I.

EXERCISE

8.2 For the MOS differential pair specified in Example 8.1 find (a) the value of v_{id} that causes Q_1 to conduct the entire current I, and the corresponding values of v_{D1} and v_{D2}; (b) the value of v_{id} that causes Q_2 to conduct the entire current I, and the corresponding values of v_{D1} and v_{D2}; (c) the corresponding range of the differential output voltage $(v_{D2} - v_{D1})$.
Ans. (a) +0.45 V, 0.5 V, 1.5 V; (b) –0.45 V, 1.5 V, 0.5 V; (c) +1 V to –1 V

To use the differential pair as a linear amplifier, we keep the differential input signal v_{id} small. As a result, the current in one of the transistors (Q_1 when v_{id} is positive) will increase by an increment ΔI proportional to v_{id}, to $(I/2 + \Delta I)$. Simultaneously, the current in the other transistor will decrease by the same amount to become $(I/2 - \Delta I)$. A voltage signal $-\Delta IR_D$ develops at one of the drains and an opposite-polarity signal, ΔIR_D, develops at the other drain. Thus the output voltage taken between the two drains will be $2\Delta IR_D$, which is proportional to the differential input signal v_{id}. The small-signal operation of the differential pair will be studied in detail in Section 8.2.

8.1.3 Large-Signal Operation

We shall now derive expressions for the drain currents i_{D1} and i_{D2} in terms of the input differential signal $v_{id} \equiv v_{G1} - v_{G2}$. The derivation assumes that the differential pair is perfectly matched and neglects channel-length modulation ($\lambda = 0$). Thus these expressions do not depend on the details of the circuit to which the drains are connected, and we do not show

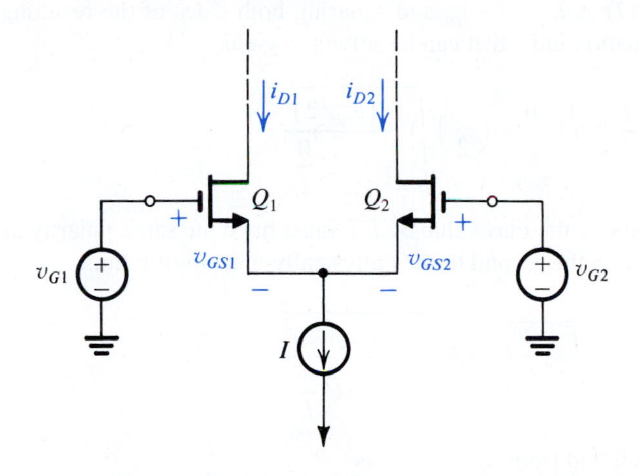

Figure 8.5 The MOSFET differential pair for the purpose of deriving the transfer characteristics, i_{D1} and i_{D2} versus $v_{id} = v_{G1} - v_{G2}$.

these connections in Fig. 8.5; we simply assume that the circuit maintains Q_1 and Q_2 in the saturation region of operation at all times.

To begin with, we express the drain currents of Q_1 and Q_2 as

$$i_{D1} = \frac{1}{2}k_n'\frac{W}{L}(v_{GS1} - V_t)^2 \tag{8.11}$$

$$i_{D2} = \frac{1}{2}k_n'\frac{W}{L}(v_{GS2} - V_t)^2 \tag{8.12}$$

Taking the square roots of both sides of each of Eqs. (8.11) and (8.12), we obtain

$$\sqrt{i_{D1}} = \sqrt{\frac{1}{2}k_n'\frac{W}{L}}(v_{GS1} - V_t) \tag{8.13}$$

$$\sqrt{i_{D2}} = \sqrt{\frac{1}{2}k_n'\frac{W}{L}}(v_{GS2} - V_t) \tag{8.14}$$

Subtracting Eq. (8.14) from Eq. (8.13) and substituting

$$v_{GS1} - v_{GS2} = v_{G1} - v_{G2} = v_{id} \tag{8.15}$$

results in

$$\sqrt{i_{D1}} - \sqrt{i_{D2}} = \sqrt{\frac{1}{2}k_n'\frac{W}{L}}v_{id} \tag{8.16}$$

The constant-current bias imposes the constraint

$$i_{D1} + i_{D2} = I \tag{8.17}$$

Equations (8.16) and (8.17) are two equations in the two unknowns i_{D1} and i_{D2} and can be solved as follows: Squaring both sides of Eq. (8.16) and substituting for $i_{D1} + i_{D2} = I$ gives

$$2\sqrt{i_{D1}i_{D2}} = I - \frac{1}{2}k_n'\frac{W}{L}v_{id}^2$$

Substituting for i_{D2} from Eq. (8.17) as $i_{D2} = I - i_{D1}$ and squaring both sides of the resulting equation provides a quadratic equation in i_{D1} that can be solved to yield

$$i_{D1} = \frac{I}{2} \pm \sqrt{k_n' \frac{W}{L} I}\left(\frac{v_{id}}{2}\right)\sqrt{1 - \frac{(v_{id}/2)^2}{I/k_n' \frac{W}{L}}}$$

Now, since the increment in i_{D1} above the bias value of $I/2$ must have the same polarity as v_{id}, only the root with the "+" sign in the second term is physically meaningful; thus,

$$i_{D1} = \frac{I}{2} + \sqrt{k_n' \frac{W}{L} I}\left(\frac{v_{id}}{2}\right)\sqrt{1 - \frac{(v_{id}/2)^2}{I/k_n' \frac{W}{L}}} \tag{8.18}$$

The corresponding value of i_{D2} is found from $i_{D2} = I - i_{D1}$ as

$$i_{D2} = \frac{I}{2} - \sqrt{k_n' \frac{W}{L} I}\left(\frac{v_{id}}{2}\right)\sqrt{1 - \frac{(v_{id}/2)^2}{I/k_n' \frac{W}{L}}} \tag{8.19}$$

At the bias (quiescent) point, $v_{id} = 0$, leading to

$$i_{D1} = i_{D2} = \frac{I}{2} \tag{8.20}$$

Correspondingly,

$$v_{GS1} = v_{GS2} = V_{GS} \tag{8.21}$$

where

$$\frac{I}{2} = \frac{1}{2} k_n' \frac{W}{L}(V_{GS} - V_t)^2 = \frac{1}{2} k_n' \frac{W}{L} V_{OV}^2 \tag{8.22}$$

This relationship enables us to replace $k_n'(W/L)$ in Eqs. (8.18) and (8.19) with I/V_{OV}^2 to express i_{D1} and i_{D2} in the alternative form

$$i_{D1} = \frac{I}{2} + \left(\frac{I}{V_{OV}}\right)\left(\frac{v_{id}}{2}\right)\sqrt{1 - \left(\frac{v_{id}/2}{V_{OV}}\right)^2} \tag{8.23}$$

$$i_{D2} = \frac{I}{2} - \left(\frac{I}{V_{OV}}\right)\left(\frac{v_{id}}{2}\right)\sqrt{1 - \left(\frac{v_{id}/2}{V_{OV}}\right)^2} \tag{8.24}$$

These two equations describe the effect of applying a differential input signal v_{id} on the currents i_{D1} and i_{D2}. They can be used to obtain the normalized plots, i_{D1}/I and i_{D2}/I versus v_{id}/V_{OV}, shown in Fig. 8.6. Note that at $v_{id} = 0$, the two currents are equal to $I/2$. Making v_{id} positive causes i_{D1} to increase and i_{D2} to decrease by equal amounts, to keep the sum constant, $i_{D1} + i_{D2} = I$. The current is steered entirely into Q_1 when v_{id} reaches the value $\sqrt{2}V_{OV}$, as we found out earlier. For v_{id} negative, identical statements can be made by interchanging i_{D1} and i_{D2}. In this case, $v_{id} = -\sqrt{2}V_{OV}$ steers the current entirely into Q_2. Finally, note that the plots in Fig. 8.6 are universal, as they apply to any MOS differential pair.

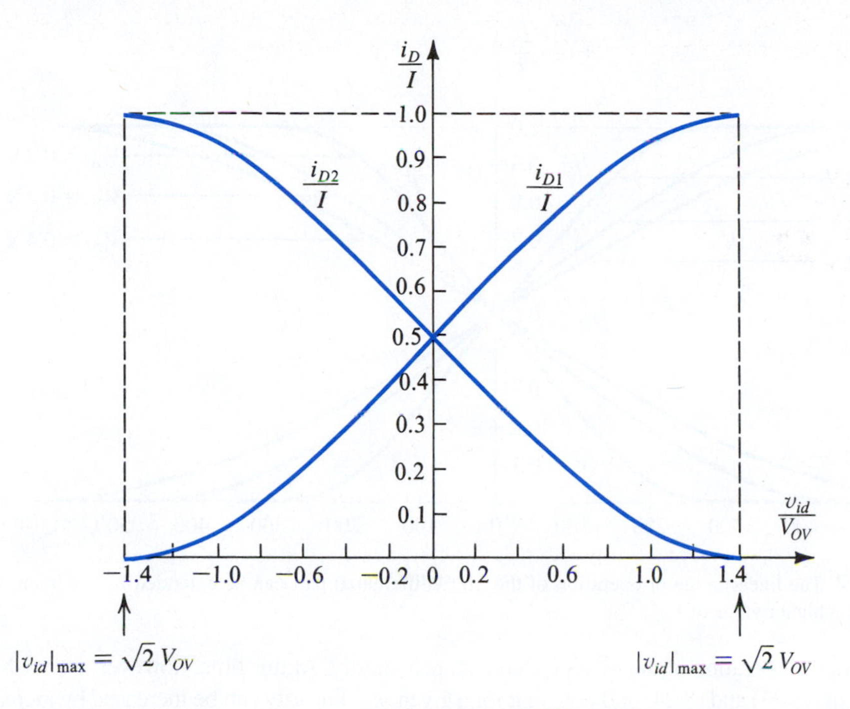

Figure 8.6 Normalized plots of the currents in a MOSFET differential pair. Note that V_{OV} is the overdrive voltage at which Q_1 and Q_2 operate when conducting drain currents equal to $I/2$, the equilibrium situation. Note that these graphs are universal and apply to any MOS differential pair.

The transfer characteristics of Eqs. (8.23) and (8.24) and Fig. 8.6 are obviously nonlinear. This is due to the term involving v_{id}^2. Since we are interested in obtaining linear amplification from the differential pair, we will strive to make this term as small as possible. For a given value of V_{OV}, the only thing we can do is keep $(v_{id}/2)$ much smaller than V_{OV}, which is the condition for the small-signal approximation. It results in

$$i_{D1} \simeq \frac{I}{2} + \left(\frac{I}{V_{OV}}\right)\left(\frac{v_{id}}{2}\right) \tag{8.25}$$

and

$$i_{D2} \simeq \frac{I}{2} - \left(\frac{I}{V_{OV}}\right)\left(\frac{v_{id}}{2}\right) \tag{8.26}$$

which, as expected, indicate that i_{D1} increases by an increment i_d, and i_{D2} decreases by the same amount, i_d, where i_d is proportional to the differential input signal v_{id},

$$i_d = \left(\frac{I}{V_{OV}}\right)\left(\frac{v_{id}}{2}\right) \tag{8.27}$$

Recalling from our study of the MOSFET in Chapter 5 (also refer to Table 7.A.3), that a MOSFET biased at a current I_D has a transconductance $g_m = 2I_D/V_{OV}$, we recognize the factor (I/V_{OV}) in Eq. (8.27) as g_m of each of Q_1 and Q_2, which are biased at $I_D = I/2$. Now, why $v_{id}/2$? Simply because v_{id} divides equally between the two devices with $v_{gs1} = v_{id}/2$ and $v_{gs2} = -v_{id}/2$, which causes Q_1 to have a current increment i_d and Q_2 to have a current decrement i_d. We shall analyze

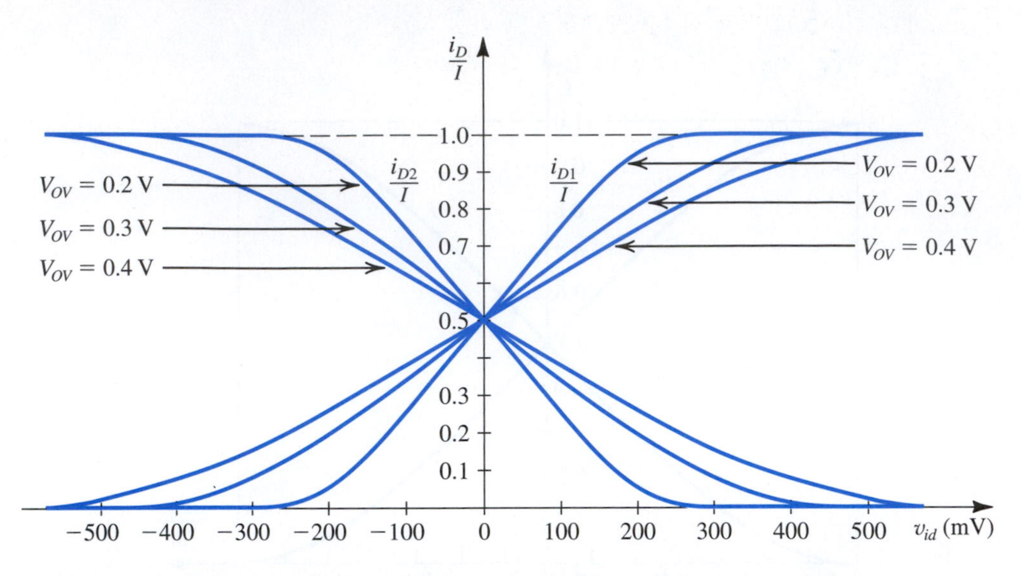

Figure 8.7 The linear range of operation of the MOS differential pair can be extended by operating the transistor at a higher value of V_{OV}.

the small-signal operation of the MOS differential pair shortly. At this time, however, we wish to return to Eqs. (8.23) and (8.24) and note that for a given v_{id}, linearity can be increased by increasing the overdrive voltage V_{OV} at which each of Q_1 and Q_2 is operating. This can be done by using smaller W/L ratios. The price paid for the increased linearity is a reduction in g_m and hence a reduction in gain. In this regard, we observe that the normalized plot of Fig. 8.6, though compact, masks this design degree of freedom. Figure 8.7 shows plots of the transfer characteristics $i_{D1,2}/I$ versus v_{id} for various values of V_{OV}. These graphs clearly illustrate the linearity–transconductance trade-off obtained by changing the value of V_{OV}: The linear range of operation can be extended by operating the MOSFETs at a higher V_{OV} (by using smaller W/L ratios) at the expense of reducing g_m and hence the gain. This trade-off is based on the assumption that the bias current I is being kept constant. The bias current can, of course, be increased to obtain a higher g_m. The expense for doing this, however, is increased power dissipation, a serious limitation in IC design.

EXERCISE

8.3 A MOS differential pair is operated at a bias current I of 0.4 mA. If $\mu_n C_{ox} = 0.2$ mA/V², find the required values of W/L and the resulting g_m if the MOSFETs are operated at $V_{OV} = 0.2$, 0.3, and 0.4 V. For each value, give the maximum $|v_{id}|$ for which the term involving v_{id}^2 in Eqs. (8.23) and (8.24), namely $((v_{id}/2)/V_{OV})^2$, is limited to 0.1.

Ans.

V_{OV} (V)	0.2	0.3	0.4
W/L	50	22.2	12.5
g_m (mA/V)	2	1.33	1
$\lvert v_{id}\rvert_{max}$ (mV)	126	190	253

8.2 Small-Signal Operation of the MOS Differential Pair

In this section we build on the understanding gained of the basic operation of the differential pair and consider in some detail its operation as a linear amplifier.

8.2.1 Differential Gain

Figure 8.8(a) shows the MOS differential amplifier with input voltages

$$v_{G1} = V_{CM} + \tfrac{1}{2}v_{id} \tag{8.28}$$

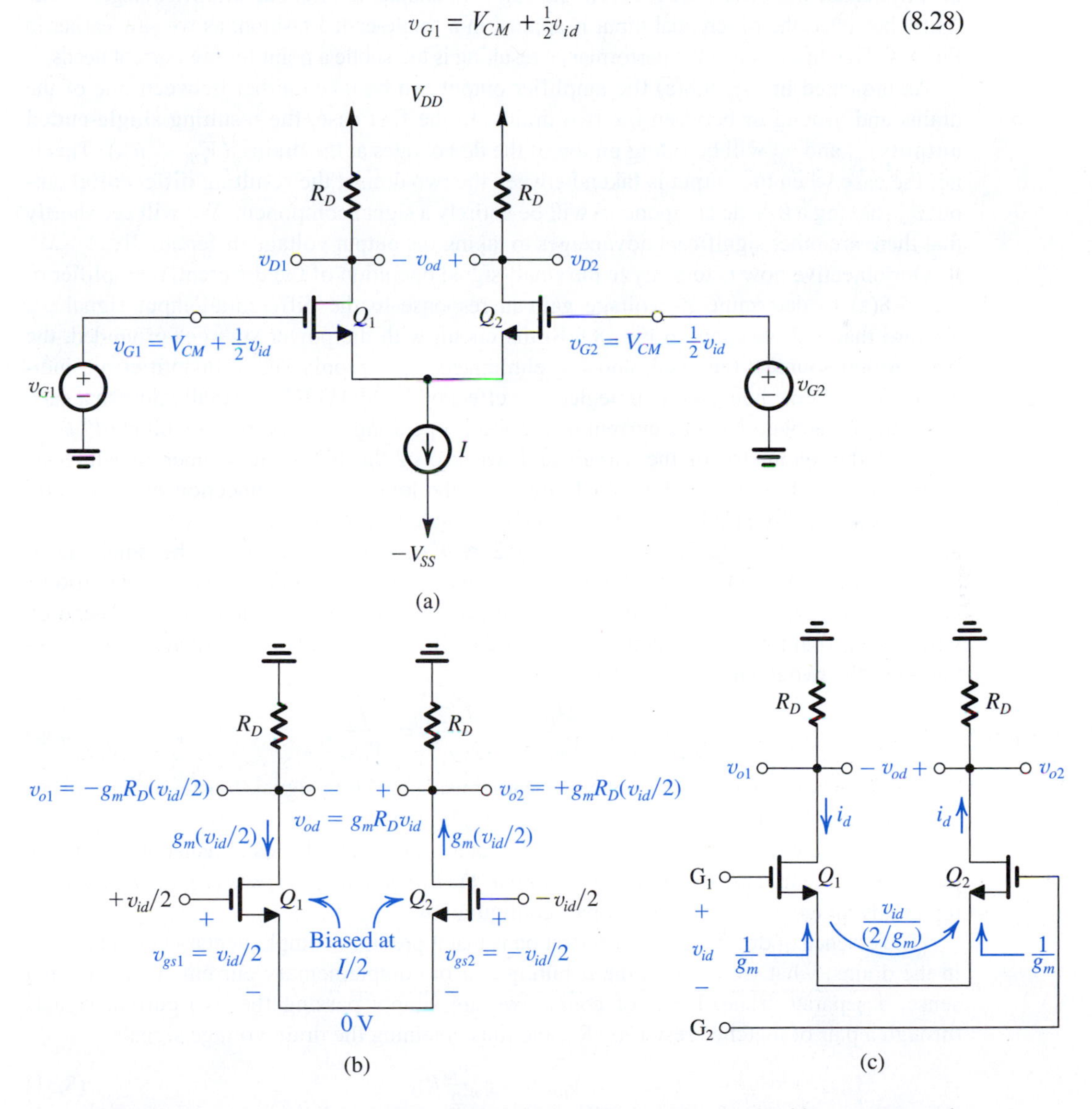

Figure 8.8 Small-signal analysis of the MOS differential amplifier. **(a)** The circuit with a common-mode voltage applied to set the dc bias voltage at the gates and with v_{id} applied in a complementary (or balanced) manner. **(b)** The circuit prepared for small-signal analysis. **(c)** An alternative way of looking at the small-signal operation of the circuit.

and

$$v_{G2} = V_{CM} - \tfrac{1}{2}v_{id} \tag{8.29}$$

Here, V_{CM} denotes a common-mode dc voltage within the input common-mode range of the differential amplifier. It is needed in order to set the dc voltage of the MOSFET gates. Typically V_{CM} is at the middle value of the power supply. Thus, for our case, where two complementary supplies are utilized, V_{CM} is typically 0 V.

The differential input signal v_{id} is applied in a **complementary** (or **balanced**) manner; that is, v_{G1} is increased by $v_{id}/2$ and v_{G2} is decreased by $v_{id}/2$. This would be the case, for instance, if the differential amplifier were fed from the output of another differential-amplifier stage. Sometimes, however, the differential input is applied in a single-ended fashion, as we saw earlier in Fig. 8.4. The difference in the performance resulting is too subtle a point for our current needs.

As indicated in Fig. 8.8(a) the amplifier output can be taken either between one of the drains and ground or between the two drains. In the first case, the resulting **single-ended outputs** v_{o1} and v_{o2} will be riding on top of the dc voltages at the drains, $(V_{DD} - \frac{I}{2}R_D)$. This is not the case when the output is taken between the two drains; the resulting **differential** output v_{od} (having a 0-V dc component) will be entirely a signal component. We will see shortly that there are other significant advantages to taking the output voltage differentially.

Our objective now is to analyze the small-signal operation of the differential amplifier of Fig. 8.8(a) to determine its voltage gain in response to the differential input signal v_{id}. Toward that end we show in Fig. 8.8(b) the circuit with the power supplies grounded, the bias current source I removed, and V_{CM} eliminated; that is, only signal quantities are indicated. For the time being we will neglect the effect of the MOSFET r_o. Finally note that each of Q_1 and Q_2 is biased at a dc current of $I/2$ and is operating at an overdrive voltage V_{OV}.

From the symmetry of the circuit and because of the balanced manner in which v_{id} is applied, we observe that the signal voltage at the joint source connection must be zero, acting as a sort of **virtual ground**. Thus Q_1 has a gate-to-source voltage signal $v_{gs1} = v_{id}/2$ and Q_2 has $v_{gs2} = -v_{id}/2$. Assuming $v_{id}/2 \ll V_{OV}$, the condition for the small-signal approximation, the changes resulting in the drain currents of Q_1 and Q_2 will be proportional to v_{gs1} and v_{gs2}, respectively. Thus Q_1 will have a drain current increment $g_m(v_{id}/2)$ and Q_2 will have a drain current decrement $g_m(v_{id}/2)$, where g_m denotes the equal transconductances of the two devices,

$$g_m = \frac{2I_D}{V_{OV}} = \frac{2(I/2)}{V_{OV}} = \frac{I}{V_{OV}} \tag{8.30}$$

These results correspond to those obtained earlier using the large-signal transfer characteristics and imposing the small-signal condition, Eqs. (8.25) to (8.27).

It is useful at this point to observe again that a signal ground is established at the source terminals of the transistors *without resorting to the use of a large bypass capacitor*, clearly a major advantage of the differential-pair configuration.

The essence of differential-pair operation is that it provides complementary current signals in the drains; what we do with the resulting pair of complementary current signals is, in a sense, a separate issue. Here, of course, we are simply passing the two current signals through a pair of matched resistors, R_D, and thus obtaining the drain voltage signals

$$v_{o1} = -g_m \frac{v_{id}}{2} R_D \tag{8.31}$$

and

$$v_{o2} = +g_m \frac{v_{id}}{2} R_D \tag{8.32}$$

If the output is taken in a single-ended fashion, the resulting gain becomes

$$\frac{v_{o1}}{v_{id}} = -\frac{1}{2}g_m R_D \tag{8.33}$$

or

$$\frac{v_{o2}}{v_{id}} = \frac{1}{2}g_m R_D \tag{8.34}$$

Alternatively, if the output is taken differentially, the gain becomes

$$A_d \equiv \frac{v_{od}}{v_{id}} = \frac{v_{o2} - v_{o1}}{v_{id}} = g_m R_D \tag{8.35}$$

Thus another advantage of taking the output differentially is an increase in gain by a factor of 2 (6 dB). It should be noted, however, that although differential outputs are preferred, a single-ended output is needed in some applications. We will have more to say about this later.

An alternative and useful way of viewing the operation of the differential pair in response to a differential input signal v_{id} is illustrated in Fig. 8.8(c). Here we are making use of the fact that the resistance between gate and source of a MOSFET, looking into the source, is $1/g_m$. As a result, between G_1 and G_2 we have a total resistance, in the source circuit, of $2/g_m$. It follows that we can obtain the current i_d simply by dividing v_{id} by $2/g_m$, as indicated in the figure.

8.2.2 The Differential Half-Circuit

When a symmetrical differential amplifier is fed with a differential signal in a balanced manner, as in the case in Fig. 8.8, the performance can be determined by considering only half the circuit. The equivalent differential half-circuit is shown in Fig. 8.9. It has a grounded source, a result of the virtual ground that appears on the common sources' terminal of the MOSFETs in the differential pair. Note that Q_1 is operating at a drain bias current of $(I/2)$ and an overdrive voltage V_{OV}.

The differential gain A_d can be determined directly from the half-circuit. For instance, if we wish to take r_o of Q_1 and Q_2 into account, we can use the half-circuit with the following result:

$$A_d = g_m(R_D \| r_o) \tag{8.36}$$

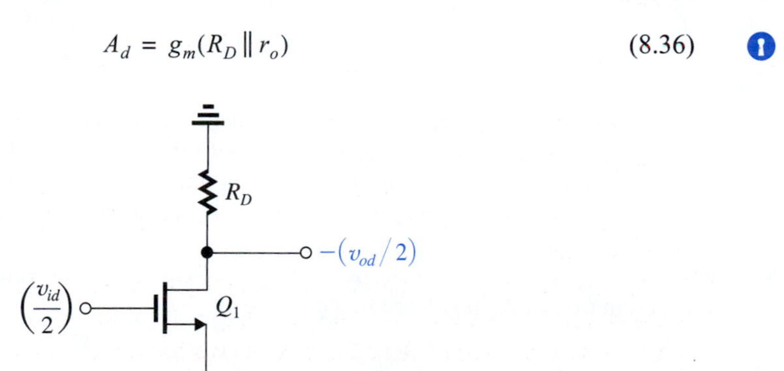

Figure 8.9 The equivalent differential half-circuit of the differential amplifier of Fig. 8.8. Here Q_1 is biased at $I/2$ and is operating at V_{OV}. This circuit can be used to determine the differential voltage gain of the differential amplifier $A_d = v_{od}/v_{id}$.

More significantly, the frequency response of the differential gain can be determined by analyzing the half-circuit, as we shall do in Chapter 9.

Example 8.2

Give the differential half-circuit of the differential amplifier shown in Fig. 8.10(a). Assume that Q_1 and Q_2 are perfectly matched. Neglecting r_o, determine the differential voltage gain $A_d \equiv v_{od}/v_{id}$.

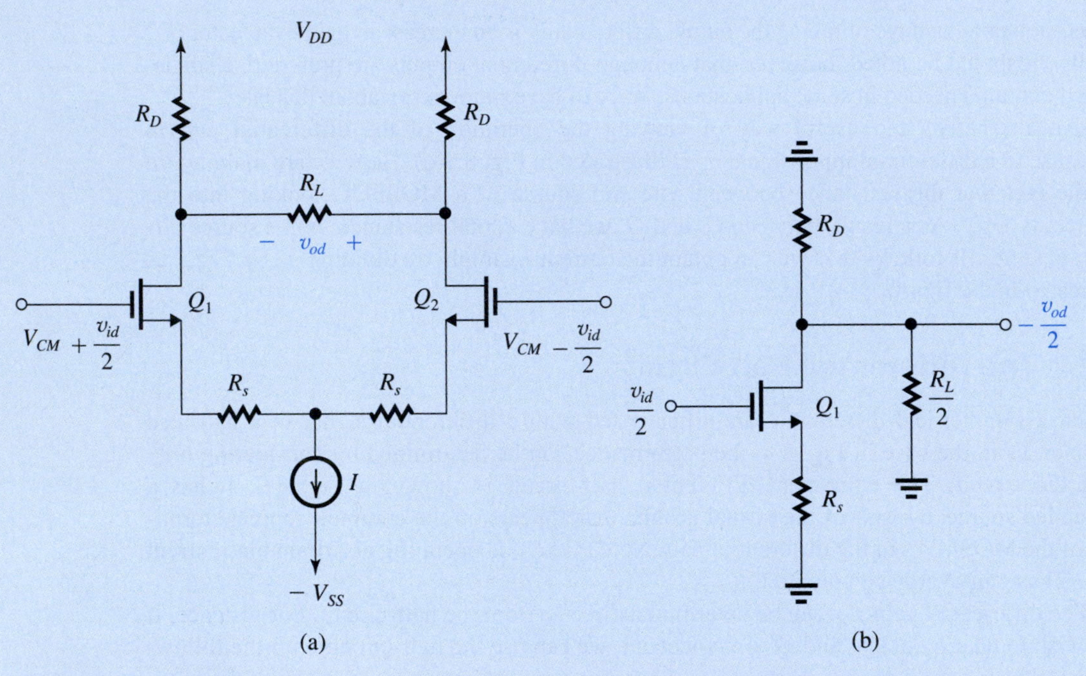

Figure 8.10 **(a)** Differential amplifier for Example 8.2. **(b)** Differential half-circuit.

Solution

Since the circuit is symmetrical and is fed with v_{id} in a balanced manner, the differential half-circuit will be as shown in Fig. 8.10(b). Observe that because the line of symmetry passes through the middle of R_L, the half-circuit has a resistance $R_L/2$ connected between drain and ground. Also note that the virtual ground appears on the node between the two resistances R_s. As a result, the half-circuit has a source-degeneration resistance R_s.

Now, neglecting r_o of the half-circuit transistor Q_1, we can obtain the gain as the ratio of the total resistance in the drain to the total resistance in the source as

$$\frac{-v_{od}/2}{v_{id}/2} = -\frac{R_D \| (R_L/2)}{1/g_m + R_s}$$

with the result that

$$A_d \equiv \frac{v_{od}}{v_{id}} = \frac{R_D \| (R_L/2)}{1/g_m + R_s} \tag{8.37}$$

EXERCISE

8.4 A MOS differential amplifier is operated at a total current of 0.8 mA, using transistors with a W/L ratio of 100, $\mu_n C_{ox} = 0.2\ \text{mA/V}^2$, $V_A = 20$ V, and $R_D = 5\ \text{k}\Omega$. Find V_{OV}, g_m, r_o, and A_d.
Ans. 0.2 V; 4 mA/V; 50 kΩ; 18.2 V/V

8.2.3 The Differential Amplifier with Current-Source Loads

To obtain higher gain, the passive resistances R_D can be replaced with current sources, as shown in Fig. 8.11(a). Here the current sources are realized with PMOS transistors Q_3 and Q_4, and V_G is a dc bias voltage that ensures that Q_3 and Q_4 each conducts a current equal to $I/2$. The differential voltage gain A_d can be found from the differential half-circuit shown in Fig. 8.11(b) as

$$A_d \equiv \frac{v_{od}}{v_{id}} = g_{m1}(r_{o1} \| r_{o3})$$

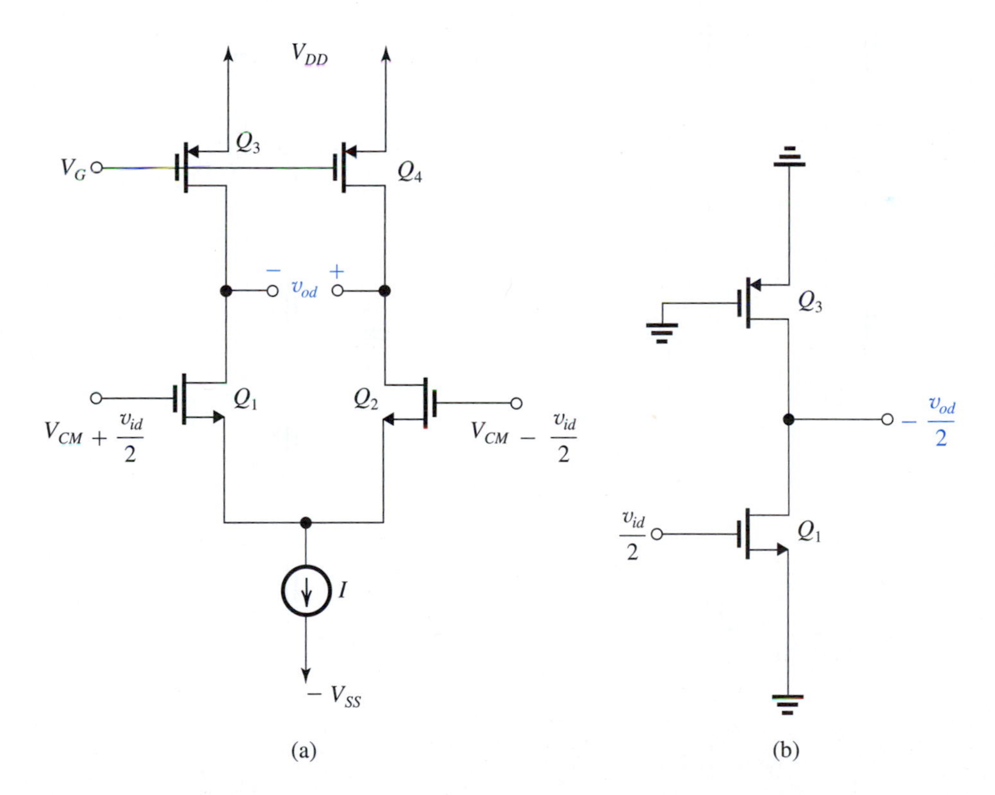

Figure 8.11 **(a)** Differential amplifier with current-source loads formed by Q_3 and Q_4. **(b)** Differential half-circuit of the amplifier in **(a)**.

EXERCISE

8.5 The differential amplifier of Fig. 8.11(a) is fabricated in a 0.18-μm CMOS technology for which $\mu_n C_{ox} = 4\mu_p C_{ox} = 400\ \mu\text{A/V}^2$, $|V_t| = 0.5$ V, and $|V_A'| = 10$ V/μm. If the bias current $I = 200\ \mu$A and all transistors have a channel length twice the minimum and are operating at $|V_{OV}| = 0.2$ V, find W/L for each of Q_1, Q_2, Q_3, and Q_4, and determine the differential voltage gain A_d.
Ans. $(W/L)_{1,2} = 12.5$; $(W/L)_{3,4} = 50$; $A_d = 18$ V/V

8.2.4 Cascode Differential Amplifier

The gain of the differential amplifier can be increased by utilizing the cascode configuration studied in Section 7.3. Figure 8.12(a) shows a CMOS differential amplifier with cascoding

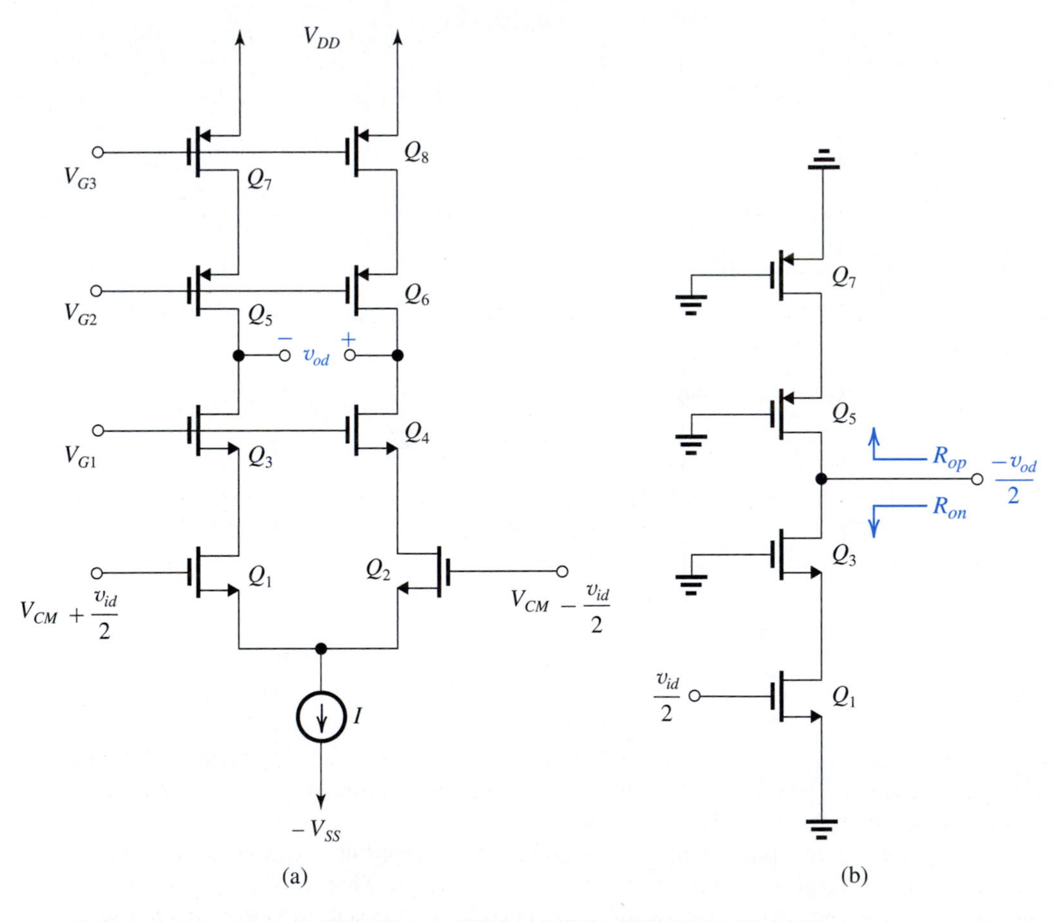

Figure 8.12 **(a)** Cascode differential amplifier; and **(b)** its differential half circuit.

applied to the amplifying transistors Q_1 and Q_2 via transistors Q_3 and Q_4, and to the current-source transistors Q_7 and Q_8 via transistors Q_5 and Q_6. The differential voltage gain can be found from the differential half-circuit shown in Fig. 8.12(b) as

$$A_d \equiv \frac{v_{od}}{v_{id}} = g_{m1}(R_{on} \| R_{op}) \tag{8.38}$$

where

$$R_{on} = (g_{m3}r_{o3})r_{o1} \tag{8.39}$$

and,

$$R_{op} = (g_{m5}r_{o5})r_{o7} \tag{8.40}$$

EXERCISE

8.6 The CMOS cascode differential amplifier of Fig. 8.12(a) is fabricated in a 0.18-μm technology for which $\mu_n C_{ox} = 4\mu_p C_{ox} = 400\ \mu A/V^2$, $|V_t| = 0.5$ V, and $|V_A'| = 10$ V/μm. If the bias current $I = 200\ \mu A$, and all transistors have a channel length twice the minimum and are operating at $|V_{OV}| = 0.2$ V, find W/L for each of Q_1 to Q_8, and determine the differential voltage gain A_d.
Ans. $(W/L)_{1,2,3,4} = 12.5$; $(W/L)_{5,6,7,8} = 50$; $A_d = 648$ V/V

8.2.5 Common-Mode Gain and Common-Mode Rejection Ratio (CMRR)

Thus far, we have seen that the differential amplifier responds to a differential input signal and completely rejects a common-mode signal. This latter point was made very clearly at the outset of our discussion of differential amplifiers and was illustrated in Example 8.1, where we saw that changes in V_{CM} over a wide range resulted in no change in the voltage at either of the two drains. This highly desirable result is, however, a consequence of our assumption that the current source that supplies the bias current I is ideal. As we shall now show, if we consider the more realistic situation of the current source having a finite output resistance R_{SS}, the common-mode gain will no longer be zero.

Figure 8.13(a) shows a MOS differential amplifier biased with a current source having an output resistance R_{SS}. As before, the dc voltage at the input is defined by V_{CM}. Here, however, we also have an incremental signal v_{icm} applied to both input terminals. This common-mode input signal can represent an interference signal or noise that is picked up by both inputs and is clearly undesirable. Our objective now is to find how much of v_{icm} makes its way to the output of the amplifier.

Before we determine the common-mode gain of the amplifier, we wish to address the question of the effect of R_{SS} on the bias current of Q_1 and Q_2. That is, with v_{icm} set to zero, the bias current in each of Q_1 and Q_2 will no longer be $I/2$ but will be larger than $I/2$ by an amount determined by V_{CM} and R_{SS}. However, since R_{SS} is usually very large, this additional dc current in each of Q_1 and Q_2 is usually small and we shall neglect it, thus assuming

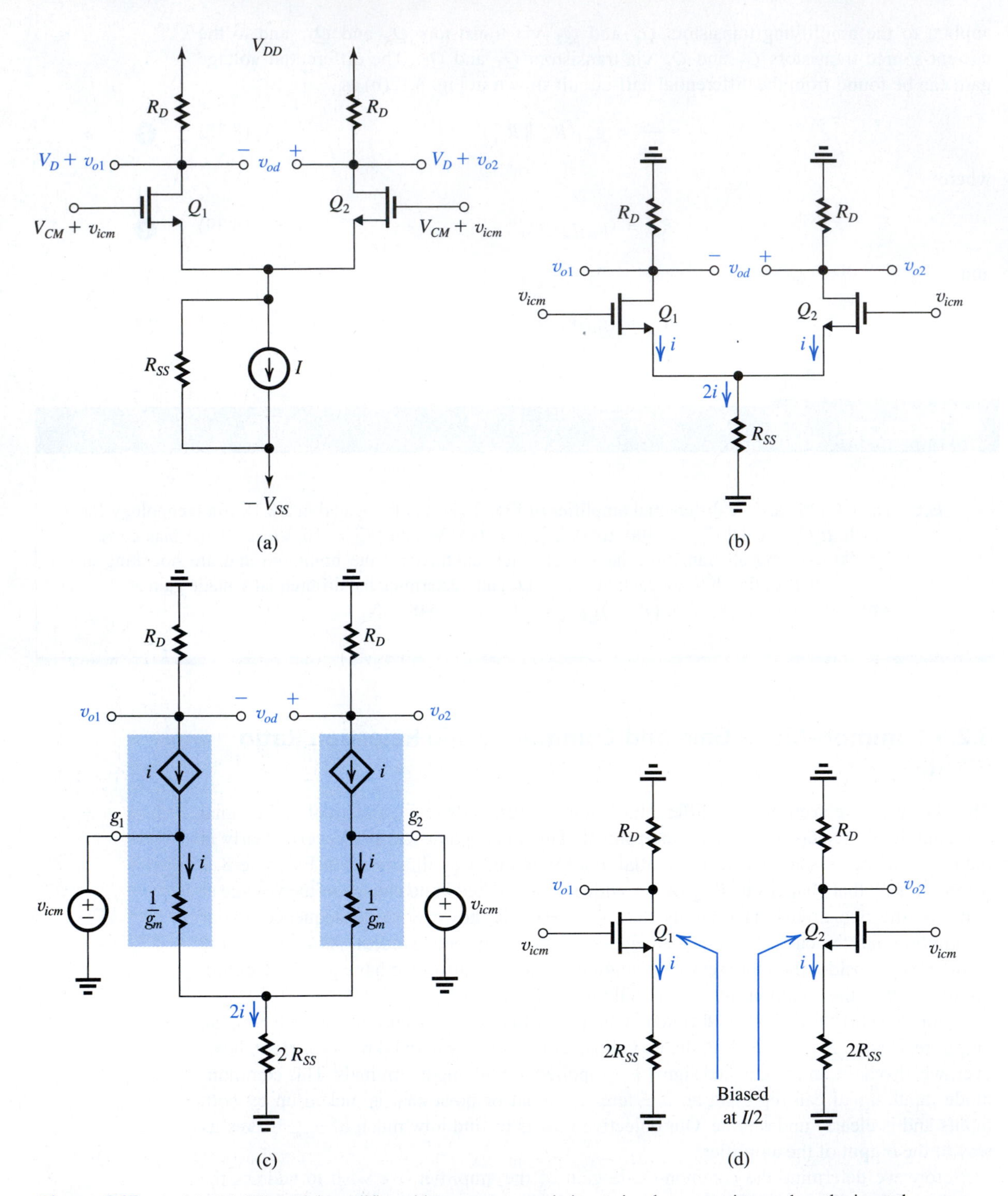

Figure 8.13 **(a)** A MOS differential amplifier with a common-mode input signal v_{icm} superimposed on the input dc common-mode voltage V_{CM}. **(b)** The amplifier circuit prepared for small-signal analysis. **(c)** The amplifier circuit with the transistors replaced with their T model and r_o neglected. **(d)** The circuit in **(b)** split into its two halves; each half is called the "CM half circuit."

that Q_1 and Q_2 continue to operate at a bias current of $I/2$. The reader might also be wondering about the effect of R_{SS} on the differential gain. The answer here is very simple: The virtual ground that develops on the common-source terminal results in a zero signal current through R_{SS}; hence R_{SS} has no effect on the value of A_d.

To determine the response of the differential amplifier to the common-mode input signal v_{icm}, consider the circuit in Fig. 8.13(b), where we have replaced each of V_{DD} and V_{SS} by a short circuit and I by an open circuit. The circuit is obviously symmetrical, and thus the two transistors will carry equal signal currents, denoted i. The value of i can be easily determined by replacing each of Q_1 and Q_2 with its T model and, for simplicity, neglecting r_o. The resulting equivalent circuit is shown in Fig. 8.13(c), from which we can write

$$v_{icm} = \frac{i}{g_m} + 2iR_{SS} \tag{8.41}$$

Thus,

$$i = \frac{v_{icm}}{1/g_m + 2R_{SS}} \tag{8.42}$$

The voltages at the drain of Q_1 and Q_2 can now be found as

$$v_{o1} = v_{o2} = -R_D i$$

resulting in

$$v_{o1} = v_{o2} = -\frac{R_D}{1/g_m + 2R_{SS}} v_{icm} \tag{8.43}$$

It follows that both v_{o1} and v_{o2} will be corrupted by the common-mode signal v_{icm} and will be given approximately by

$$\frac{v_{o1}}{v_{icm}} = \frac{v_{o2}}{v_{icm}} \simeq -\frac{R_D}{2R_{SS}} \tag{8.44}$$

where we have assumed that $2R_{SS} \gg 1/g_m$. Nevertheless, because $v_{o1} = v_{o2}$, the differential output voltage v_{od} will remain free of common-mode interference:

$$v_{od} = v_{o2} - v_{o1} = 0 \tag{8.45}$$

Thus the circuit still rejects common-mode signals! Unfortunately, however, this will not be the case if the circuit is not perfectly symmetrical, as we shall now show.

Before proceeding further, it is useful to observe that all the above results can be obtained by considering only half the differential amplifier. Figure 8.13(d) shows the two half-circuits of the differential amplifier that apply for common-mode analysis. To see the equivalence, observe that each of the two half-circuits indeed carries a current i given by Eq. (8.42) and the voltages at the source terminals are equal ($v_s = 2iR_{SS}$). Thus the two sources can be joined, returning the circuit to the original form in Fig. 8.13(b). Each of the circuits in Fig. 8.13(d) is known as the **common-mode half-circuit**. Note the difference between the CM half-circuit and the differential half-circuit.

Effect of R_D Mismatch When the two drain resistances exhibit a mismatch ΔR_D, as they inevitably do, the common-mode voltages at the two drains will no longer be equal. Rather, if the load of Q_1 is R_D and that of Q_2 is $(R_D + \Delta R_D)$, the drain signal voltages arising from v_{icm} will be

$$v_{o1} \simeq -\frac{R_D}{2R_{SS}} v_{icm} \tag{8.46}$$

and

$$v_{o2} \simeq -\frac{R_D + \Delta R_D}{2R_{SS}} v_{icm} \tag{8.47}$$

Thus,

$$v_{od} = v_{o2} - v_{o1} = -\frac{\Delta R_D}{2R_{SS}} v_{icm} \tag{8.48}$$

and we can find the common-mode gain A_{cm} as

$$A_{cm} \equiv \frac{v_{od}}{v_{icm}} = -\frac{\Delta R_D}{2R_{SS}} \tag{8.49}$$

which can be expressed in the alternate form

$$A_{cm} = -\left(\frac{R_D}{2R_{SS}}\right)\left(\frac{\Delta R_D}{R_D}\right) \tag{8.49'}$$

It follows that a mismatch in the drain resistances causes the differential amplifier to have a finite common-mode gain. Thus, a portion of the interference or noise signal v_{icm} will appear as a component of v_{od}. A measure of the effectiveness of the differential amplifier in amplifying differential-mode signals and rejecting common-mode interference is the ratio of the magnitude of its differential gain $|A_d|$ to the magnitude of its common-mode gain $|A_{cm}|$. This ratio is termed **common-mode rejection ratio (CMRR)**. Thus,

$$\text{CMRR} \equiv \frac{|A_d|}{|A_{cm}|} \tag{8.50a}$$

and is usually expressed in decibels,

$$\text{CMRR (dB)} = 20 \log \frac{|A_d|}{|A_{cm}|} \tag{8.50b}$$

For the case of a MOS differential amplifier with drain resistances R_D that exhibit a mismatch ΔR_D, the CMRR can be found as the ratio of A_d in Eq. (8.35) to A_{cm} in Eq. (8.49), thus

$$\text{CMRR} = (2g_m R_{SS}) / (\Delta R_D / R_D) \tag{8.50c}$$

It follows that to obtain a high CMRR, we should utilize a bias current source with a high output resistance R_{SS}, and we should strive to obtain a high degree of matching between the drain resistances (i.e., keep $\Delta R_D/R_D$ small).

EXERCISE

8.7 A MOS differential pair operated at a bias current of 0.8 mA employs transistors with $W/L = 100$ and $\mu_n C_{ox} = 0.2$ mA/V², using $R_D = 5$ kΩ and $R_{SS} = 25$ kΩ. Find the differential gain, the common-mode gain when the drain resistances have a 1% mismatch, and the CMRR.
Ans. 20 V/V; 0.001 V/V; 86 dB

Effect of g_m Mismatch on CMRR Another possible mismatch between the two halves of the MOS differential pair is a mismatch in g_m of the two transistors. For the purpose of

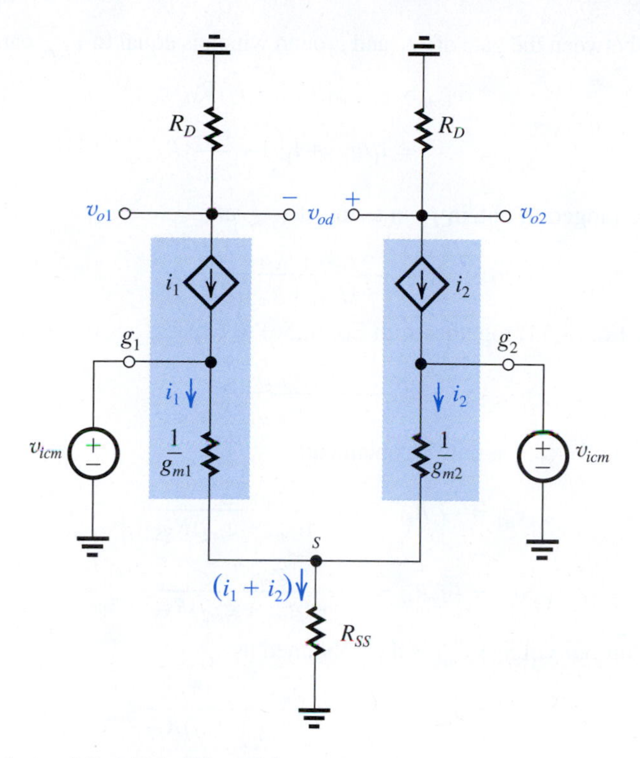

Figure 8.14 Analysis of the MOS differential amplifier with an input common-mode signal v_{icm} in the case the two transistors have a g_m mismatch.

finding the effect of a g_m mismatch on CMRR, let

$$g_{m1} = g_m + \frac{1}{2}\Delta g_m \tag{8.51}$$

$$g_{m2} = g_m - \frac{1}{2}\Delta g_m \tag{8.52}$$

That is,

$$g_{m1} - g_{m2} = \Delta g_m \tag{8.53}$$

Since the circuit is no longer symmetrical, we cannot employ the common-mode half-circuit. Rather, we shall return to the original circuit of Fig. 8.13(a) and replace each of Q_1 and Q_2 with its T equivalent-circuit model. The result is the equivalent circuit shown in Fig. 8.14. Examination of this circuit reveals that the voltages between gate and source for the two transistors are equal (and equal to $v_{icm} - v_s$). Thus,

$$i_1(1/g_{m1}) = i_2(1/g_{m2}) \tag{8.54}$$

From which we can obtain $i_1 + i_2$ as

$$i_1 + i_2 = i_1\left(1 + \frac{g_{m2}}{g_{m1}}\right) \tag{8.55}$$

Now the voltage between the gate of Q_1 and ground which is equal to v_{icm} can be expressed as

$$v_{icm} = i_1/g_{m1} + (i_1 + i_2)R_{SS}$$
$$= i_1/g_{m1} + i_1\left(1 + \frac{g_{m2}}{g_{m1}}\right)R_{SS}$$

which can be rearranged to obtain i_1 in terms of v_{icm} as

$$i_1 = \frac{g_{m1}v_{icm}}{1 + (g_{m1} + g_{m2})R_{SS}} \tag{8.56}$$

We can then use Eq. (8.54) together with Eq. (8.56) to express i_2 as

$$i_2 = \frac{g_{m2}v_{icm}}{1 + (g_{m1} + g_{m2})R_{SS}} \tag{8.57}$$

The voltages v_{o1} and v_{o2} can now be obtained:

$$v_{o1} = -i_1R_D = -\frac{g_{m1}R_D}{1 + (g_{m1} + g_{m2})R_{SS}}v_{icm} \tag{8.58}$$

$$v_{o2} = -i_2R_D = -\frac{g_{m2}R_D}{1 + (g_{m1} + g_{m2})R_{SS}}v_{icm} \tag{8.59}$$

The differential output voltage v_{od} is then obtained as

$$v_{od} = v_{o2} - v_{o1} = \frac{(g_{m1} - g_{m2})R_D}{1 + (g_{m1} + g_{m2})R_{SS}}v_{icm} \tag{8.60}$$

Substituting for g_{m1} and g_{m2} from Eqs. (8.51) and (8.52), respectively, gives

$$v_{od} = \frac{\Delta g_m R_D}{1 + 2\,g_m R_{SS}}v_{icm}$$

Thus the common-mode gain resulting from a mismatch Δg_m can be expressed as

$$A_{cm} \equiv \frac{v_{od}}{v_{icm}} = \frac{\Delta g_m R_D}{1 + 2\,g_m R_{SS}} \tag{8.61}$$

which can be approximated by

$$A_{cm} \simeq \left(\frac{R_D}{2R_{SS}}\right)\left(\frac{\Delta g_m}{g_m}\right) \tag{8.62}$$

and the corresponding CMRR will be

$$\text{CMRR} = (2g_mR_{SS})\Big/\left(\frac{\Delta g_m}{g_m}\right) \tag{8.63}$$

Thus to keep CMRR high, we have to use a biasing current source with a high output resistance R_{SS} and, of course, strive to maintain a high degree of matching between Q_1 and Q_2.

EXERCISE

8.8 For the MOS amplifier specified in Exercise 8.7, compute the CMRR resulting from a 1% mismatch in g_m.
Ans. 86 dB

Example 8.3

In this example we consider the design of the current source that supplies the bias current of a MOS differential amplifier. Let it be required to achieve a CMRR of 100 dB and assume that the only source of mismatch between Q_1 and Q_2 is a 2% mismatch in their W/L ratios. Let $I = 200\ \mu A$ and assume that all transistors are to be operated at $V_{OV} = 0.2\ V$. For the 0.18-μm CMOS fabrication process available, $V'_A = 5\ V/\mu m$. If a simple current source is utilized for I, what channel length is required? If a cascode current source is utilized, what channel length is needed for the two transistors in the cascode?

Solution

A mismatch in W/L results in a g_m mismatch that can be found from the expression of g_m:

$$g_m = \sqrt{2(\mu_n C_{ox})\left(\frac{W}{L}\right)I_D} \quad (8.64)$$

It can be seen that an error of 2% in W/L will result in an error in g_m of 1%. That is, the 2% mismatch in the W/L ratios of Q_1 and Q_2 will result in a 1% mismatch in their g_m values. The resulting CMRR can be found from Eq. (8.64), repeated here:

$$\text{CMRR} = (2g_m R_{SS})/\left(\frac{\Delta g_m}{g_m}\right)$$

Now, a 100-dB CMRR corresponds to a ratio of 10^5; thus,

$$10^5 = (2g_m R_{SS})/0.01 \quad (8.65)$$

The value of g_m can be found from

$$g_m = \frac{2I_D}{V_{OV}} = \frac{2 \times (I/2)}{V_{OV}}$$

$$= \frac{2 \times 0.1}{0.2} = 1\ \text{mA/V}$$

Substituting in Eq. (8.65) gives

$$R_{SS} = 500\ \text{k}\Omega$$

Now if the current source is implemented with a single transistor, its r_o must be

$$r_o = R_{SS} = 500\ \text{k}\Omega$$

Thus,

$$\frac{V_A}{I} = 500\ \text{k}\Omega$$

Substituting $I = 200\ \mu A$, we find the required value of V_A as

$$V_A = 100\ \text{V}$$

Since $V_A = V'_A L = 5L$, the required value of L will be

$$L = 20\ \mu\text{m}$$

which is very large!

Example 8.3 *continued*

Using a cascode current source, we have

$$R_{SS} = (g_m r_o) r_o$$

where

$$g_m = \frac{2I}{V_{OV}} = \frac{2 \times 0.2}{0.2} = 2 \text{ mA/V}$$

Thus,

$$500 = 2 \times r_o^2$$

$$r_o = 15.81 \text{ k}\Omega$$

and the required V_A now becomes

$$15.81 = \frac{V_A}{I} = \frac{V_A}{0.2}$$

$$V_A = 3.16 \text{ V}$$

which implies a channel length for each of the two transistors in the cascode of

$$L = \frac{3.16}{V_A'} = \frac{3.16}{5} = 0.63 \ \mu\text{m}$$

a considerable reduction from the case of a simple current source.

Differential versus Single-Ended Output The above study of common-mode rejection was predicated on the assumption that the output of the differential amplifier is taken differentially, that is, between the drains of Q_1 and Q_2. In some cases one might decide to take the output single-endedly; that is, between one of the drains and ground. If this is done, the CMRR is reduced dramatically. This can be seen from the above analysis, where the common-mode gain in the absence of mismatches is zero if the output is taken differentially and finite (Eq. 8.44) if the output is taken single-endedly. When mismatches are taken into account, the CM gain for the differential-output case departs from zero but remains much lower than the value obtained for single-ended output (Eq. 8.44).

We conclude that to obtain a large CMRR, the output of the differential amplifier must be taken differentially. The subject of converting the output signal from differential to single-ended without loss of CMRR will be studied in Section 8.5.

8.3 The BJT Differential Pair

Figure 8.15 shows the basic BJT differential-pair configuration. It is very similar to the MOSFET circuit and consists of two matched transistors, Q_1 and Q_2, whose emitters are joined together and biased by a constant-current source I. The latter is usually implemented by a transistor circuit of the type studied in Sections 7.4 and 7.5. Although each collector is shown connected to the positive supply voltage V_{CC} through a resistance R_C, this connection is not essential to the operation of the differential pair—that is, in some applications the two collectors may be connected to current sources rather than resistive loads. It is essential, though, that the collector circuits be such that Q_1 and Q_2 never enter saturation.

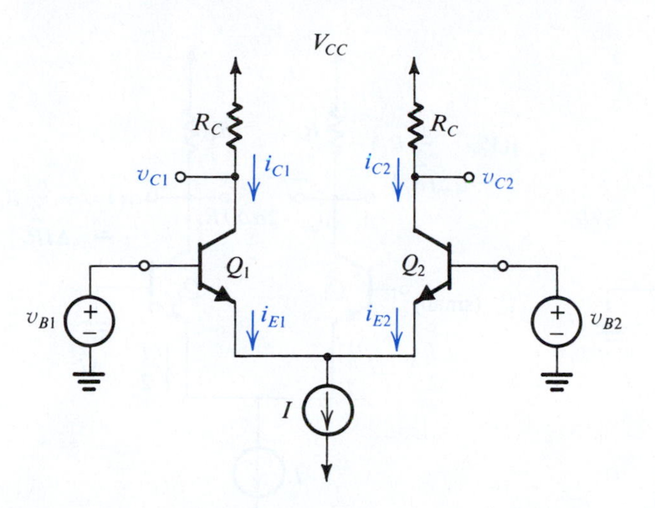

Figure 8.15 The basic BJT differential-pair configuration.

8.3.1 Basic Operation

To see how the BJT differential pair works, consider first the case of the two bases joined together and connected to a common-mode voltage V_{CM}. That is, as shown in Fig. 8.16(a), $v_{B1} = v_{B2} = V_{CM}$. Since Q_1 and Q_2 are matched, and assuming an ideal bias current source I with infinite output resistance, it follows that the current I will remain constant and from symmetry that I will divide equally between the two devices. Thus $i_{E1} = i_{E2} = I/2$, and the voltage at the

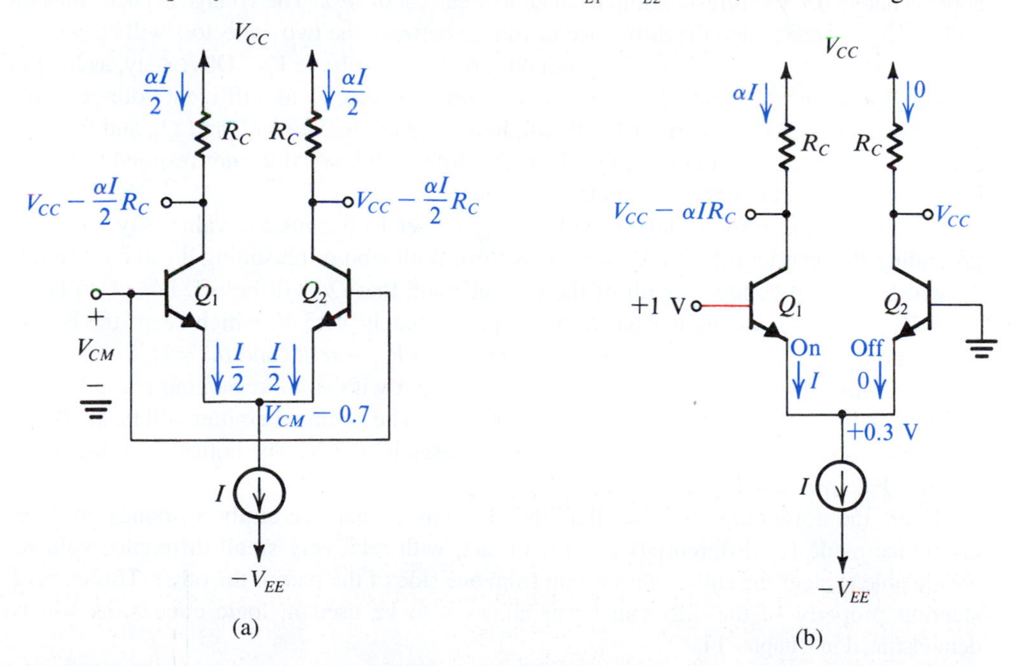

Figure 8.16 Different modes of operation of the BJT differential pair: **(a)** the differential pair with a common-mode input voltage V_{CM}; **(b)** the differential pair with a "large" differential input signal; **(c)** the differential pair with a large differential input signal of polarity opposite to that in **(b)**; **(d)** the differential pair with a small differential input signal v_i. Note that we have assumed the bias current source I to be ideal (i.e., it has an infinite output resistance) and thus I remains constant with the change in V_{CM}.

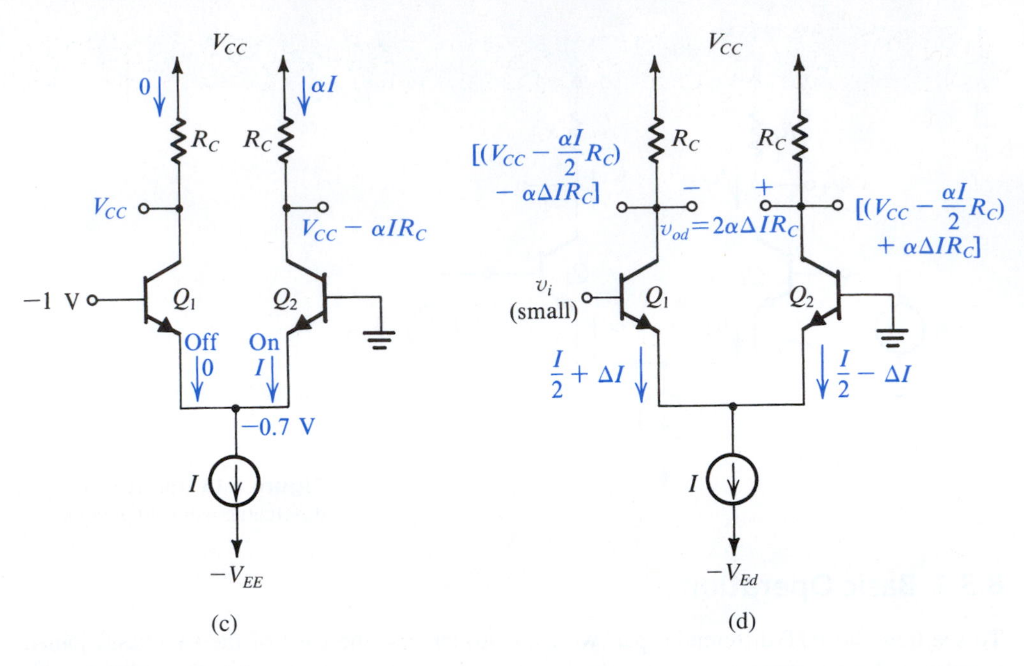

Figure 8.16 *continued.*

emitters will be $V_{CM} - V_{BE}$, where V_{BE} is the base–emitter voltage (assumed in Fig 8.16a to be approximately 0.7 V) corresponding to an emitter current of $I/2$. The voltage at each collector will be $V_{CC} - \frac{1}{2}\alpha I R_C$, and the difference in voltage between the two collectors will be zero.

Now let us vary the value of the common-mode input voltage V_{CM}. Obviously, as long as Q_1 and Q_2 remain in the active region, and the current source I has sufficient voltage across it to operate properly, the current I will still divide equally between Q_1 and Q_2, and the voltages at the collectors will not change. Thus the differential pair does not respond to (i.e., it *rejects*) changes in the common-mode input voltage.

As another experiment, let the voltage v_{B2} be set to a constant value, say, zero (by grounding B_2), and let $v_{B1} = +1$ V (see Fig. 8.16b). With a bit of reasoning it can be seen that Q_1 will be on and conducting all of the current I and that Q_2 will be off. For Q_1 to be on (with $V_{BE1} = 0.7$ V), the emitter has to be at approximately +0.3 V, which keeps the EBJ of Q_2 reverse-biased. The collector voltages will be $v_{C1} = V_{CC} - \alpha I R_C$ and $v_{C2} = V_{CC}$.

Let us now change v_{B1} to −1 V (Fig. 8.16c). Again with some reasoning it can be seen that Q_1 will turn off, and Q_2 will carry all the current I. The common emitter will be at −0.7 V, which means that the EBJ of Q_1 will be reverse biased by 0.3 V. The collector voltages will be $v_{C1} = V_{CC}$ and $v_{C2} = V_{CC} - \alpha I R_C$.

From the foregoing, we see that the differential pair certainly responds to large difference-mode (or differential) signals. In fact, with relatively small difference voltages we are able to steer the entire bias current from one side of the pair to the other. This current-steering property of the differential pair allows it to be used in logic circuits, as will be demonstrated in Chapter 14.

To use the BJT differential pair as a linear amplifier, we apply a very small differential signal (a few millivolts), which will result in one of the transistors conducting a current of $I/2 + \Delta I$; the current in the other transistor will be $I/2 - \Delta I$, with ΔI being proportional to the difference input voltage (see Fig. 8.16d). The output voltage taken between the two collectors will be $2\alpha \Delta I R_C$, which is proportional to the differential input signal v_i. The small-signal operation of the differential pair will be studied shortly.

EXERCISE

8.9 Find v_E, v_{C1}, and v_{C2} in the circuit of Fig. E8.9. Assume that $|v_{BE}|$ of a conducting transistor is approximately 0.7 V and that $\alpha \simeq 1$.

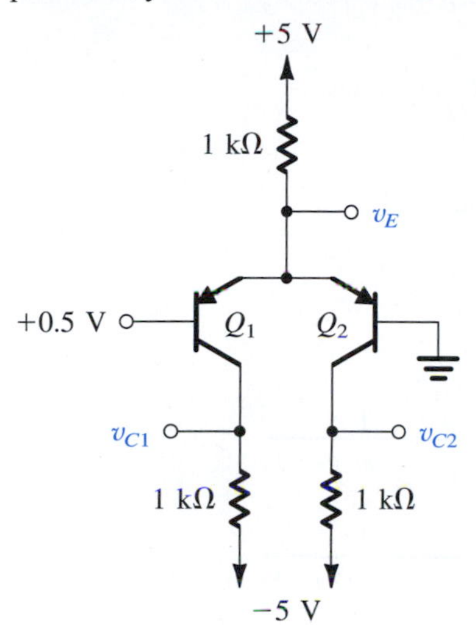

Figure E8.9

Ans. +0.7 V; –5 V; –0.7 V

8.3.2 Input Common-Mode Range

Refer to the circuit in Fig. 8.16(a). The allowable range of V_{CM} is determined at the upper end by Q_1 and Q_2 leaving the active mode and entering saturation. Thus

$$V_{CM\max} \simeq V_C + 0.4 = V_{CC} - \alpha\frac{I}{2}R_C + 0.4 \tag{8.66}$$

The lower end of the V_{CM} range is determined by the need to provide a certain minimum voltage V_{CS} across the current source I to ensure its proper operation. Thus,

$$V_{CM\min} = -V_{EE} + V_{CS} + V_{BE} \tag{8.67}$$

EXERCISE

8.10 Determine the input common-mode range for a bipolar differential amplifier operating from ±2.5 -V power supplies and biased with a simple current source that delivers a constant current of 0.4 mA and requires a minimum of 0.3 V for its proper operation. The collector resistances R_C = 5 kΩ.
Ans. –1.5 V to +1.9 V

8.3.3 Large-Signal Operation

We now present a general analysis of the BJT differential pair of Fig. 8.15. If we denote the voltage at the common emitter by v_E and neglecting the Early effect, the exponential relationship applied to each of the two transistors may be written

$$i_{E1} = \frac{I_S}{\alpha} e^{(v_{B1}-v_E)/V_T} \tag{8.68}$$

$$i_{E2} = \frac{I_S}{\alpha} e^{(v_{B2}-v_E)/V_T} \tag{8.69}$$

These two equations can be combined to obtain

$$\frac{i_{E1}}{i_{E2}} = e^{(v_{B1}-v_{B2})/V_T}$$

which can be manipulated to yield

$$\frac{i_{E1}}{i_{E1}+i_{E2}} = \frac{1}{1+e^{(v_{B2}-v_{B1})/V_T}} \tag{8.70}$$

$$\frac{i_{E2}}{i_{E1}+i_{E2}} = \frac{1}{1+e^{(v_{B1}-v_{B2})/V_T}} \tag{8.71}$$

The circuit imposes the additional constraint

$$i_{E1} + i_{E2} = I \tag{8.72}$$

Using Eq. (8.72) together with Eqs. (8.70) and (8.71) and substituting $v_{B1} - v_{B2} = v_{id}$ gives

$$i_{E1} = \frac{I}{1+e^{-v_{id}/V_T}} \tag{8.73}$$

$$i_{E2} = \frac{I}{1+e^{v_{id}/V_T}} \tag{8.74}$$

The collector currents i_{C1} and i_{C2} can be obtained simply by multiplying the emitter currents in Eqs. (8.73) and (8.74) by α, which is normally very close to unity.

The fundamental operation of the differential amplifier is illustrated by Eqs. (8.73) and (8.74). First, note that the amplifier responds only to the difference voltage v_{id}. That is, if $v_{B1} = v_{B2} = V_{CM}$, the current I divides equally between the two transistors irrespective of the value of the common-mode voltage V_{CM}. This is the essence of differential-amplifier operation, which also gives rise to its name.

Another important observation is that a relatively small difference voltage v_{id} will cause the current I to flow almost entirely in one of the two transistors. Figure 8.17 shows a plot of the two collector currents (assuming $\alpha \simeq 1$) as a function of the differential input signal. This is a normalized plot that can be used universally. Observe that a difference voltage of about $4V_T$ ($\simeq$100 mV) is sufficient to switch the current almost entirely to one side of the BJT pair. Note that this is much smaller than the corresponding voltage for the MOS pair, $\sqrt{2}V_{OV}$. The fact that such a small signal can switch the current from one side of the BJT differential pair to the other means that the BJT differential pair can be used as a fast current switch (Chapter 14).

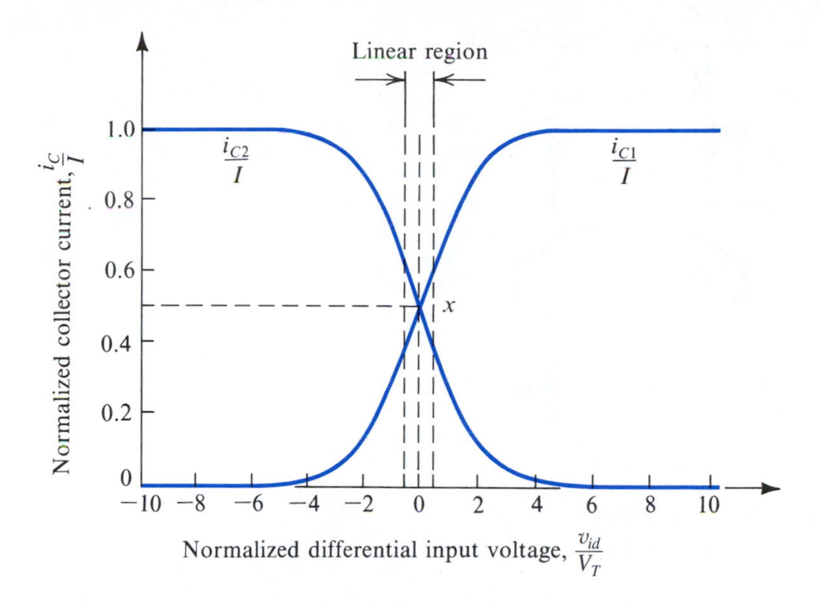

Figure 8.17 Transfer characteristics of the BJT differential pair of Fig. 8.15 assuming $\alpha \simeq 1$.

The nonlinear transfer characteristics of the differential pair, shown in Fig. 8.17, will not be utilized any further in this chapter. Rather, in the following we shall be interested specifically in the application of the differential pair as a small-signal amplifier. For this purpose, the difference input signal is limited to less than about $V_T/2$ in order that we may operate on a linear segment of the characteristics around the midpoint x (in Fig. 8.17).

Before leaving the large-signal operation of the differential BJT pair, we wish to point out an effective technique frequently employed to extend the linear range of operation. It consists of including two equal resistances R_e in series with the emitters of Q_1 and Q_2, as shown in Fig. 8.18(a). The resulting transfer characteristics for three different values of R_e are sketched in Fig. 8.18(b). Observe that expansion of the linear range is obtained at the expense of reduced G_m (which is the slope of the transfer curve at $v_{id} = 0$) and hence reduced gain. This result should come as no surprise; R_e here is performing in exactly the same way as the emitter resistance R_e does in the CE amplifier with emitter degeneration (see Section 6.6.4). Finally, we also note that this linearization technique is in effect the bipolar counterpart of the technique employed for the MOS differential pair (Fig. 8.7). In the latter case, however, V_{OV} was varied by changing the transistors' *W/L* ratio, a design tool with no counterpart in the BJT.

EXERCISE

8.11 For the BJT differential pair of Fig. 8.15, find the value of input differential signal that is sufficient to cause $i_{E1} = 0.99I$.
Ans. 115 mV

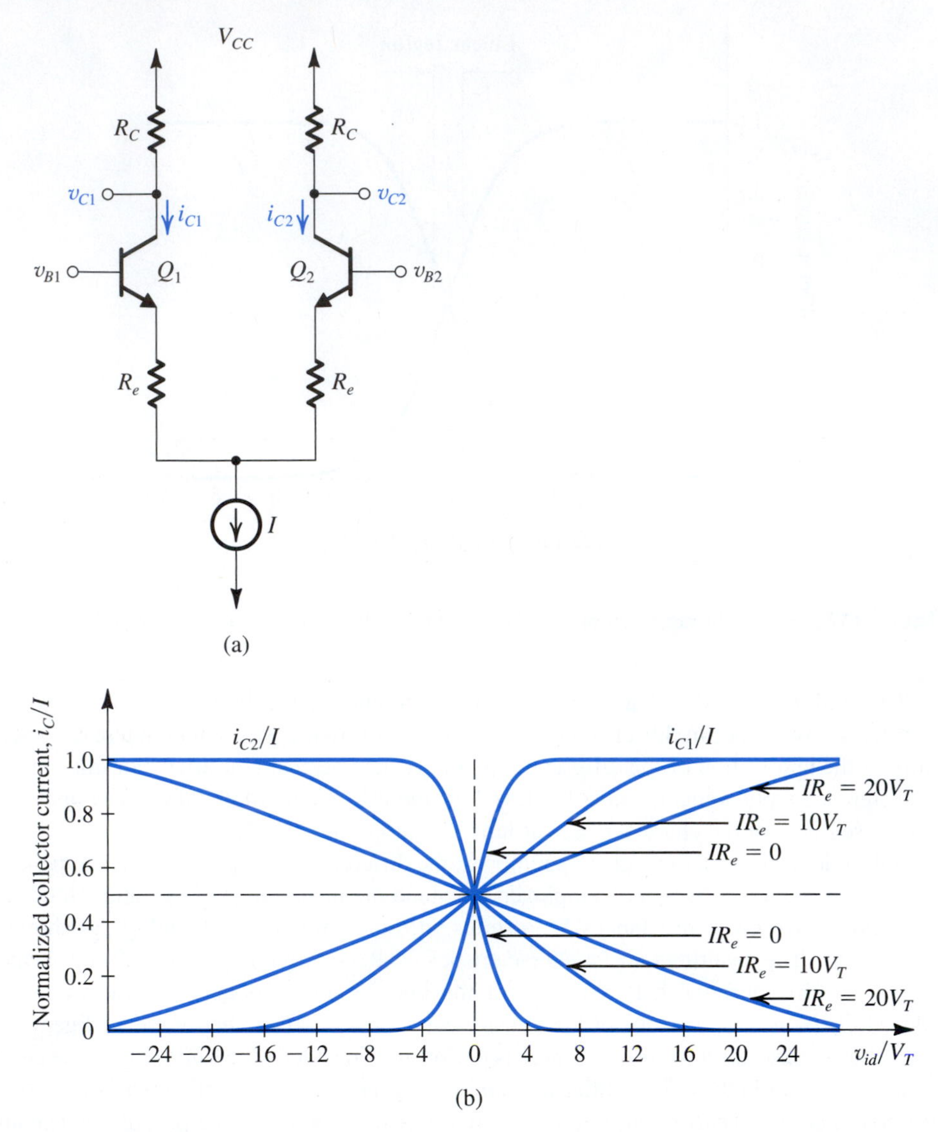

Figure 8.18 The transfer characteristics of the BJT differential pair **(a)** can be linearized **(b)** (i.e., the linear range of operation can be extended) by including resistances in the emitters.

8.3.4 Small-Signal Operation

In this section we shall study the application of the BJT differential pair in small-signal amplification. Figure 8.19 shows the BJT differential pair with a difference voltage signal v_{id} applied between the two bases. Implied is that the dc level at the input—that is, the common-mode input voltage—has been somehow established. For instance, one of the two input terminals can be grounded and v_{id} applied to the other input terminal. Alternatively, the differential amplifier may be fed from the output of another differential amplifier. In the latter case, the voltage at one of the input terminals will be $V_{CM} + v_{id}/2$ while that at the other input terminal will be $V_{CM} - v_{id}/2$.

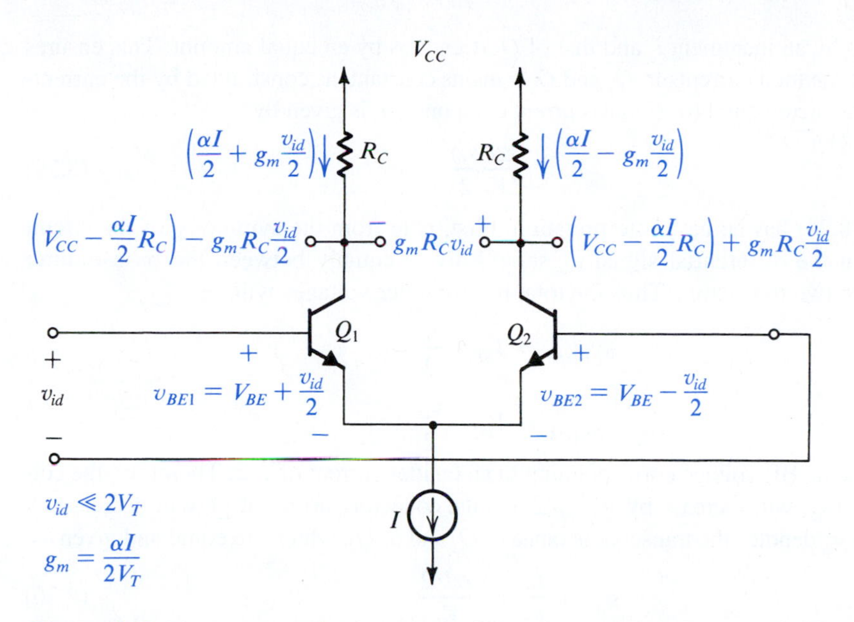

Figure 8.19 The currents and voltages in the differential amplifier when a small differential input signal v_{id} is applied.

The Collector Currents When v_{id} Is Applied For the circuit of Fig. 8.19, we may use Eqs. (8.73) and (8.74) to write

$$i_{C1} = \frac{\alpha I}{1 + e^{-v_{id}/V_T}} \tag{8.75}$$

$$i_{C2} = \frac{\alpha I}{1 + e^{v_{id}/V_T}} \tag{8.76}$$

Multiplying the numerator and the denominator of the right-hand side of Eq. (8.75) by $e^{v_{id}/2V_T}$ gives

$$i_{C1} = \frac{\alpha I e^{v_{id}/2V_T}}{e^{v_{id}/2V_T} + e^{-v_{id}/2V_T}}$$

Assume that $v_{id} \ll 2V_T$. We may thus expand the exponential $e^{\pm v_{id}/2V_T}$ in a series and retain only the first two terms:

$$i_{C1} \simeq \frac{\alpha I(1 + v_{id}/2V_T)}{1 + v_{id}/2V_T + 1 - v_{id}/2V_T}$$

Thus

$$i_{C1} = \frac{\alpha I}{2} + \frac{\alpha I}{2V_T}\frac{v_{id}}{2} \tag{8.77}$$

Similar manipulations can be applied to Eq. (8.76) to obtain

$$i_{C2} = \frac{\alpha I}{2} - \frac{\alpha I}{2V_T}\frac{v_{id}}{2} \tag{8.78}$$

Equations (8.77) and (8.78) tell us that when $v_{id} = 0$, the bias current I divides equally between the two transistors of the pair. Thus each transistor is biased at an emitter current of $I/2$. When a "small-signal" v_{id} is applied differentially (i.e., between the two bases), the collector current of Q_1 increases by an increment i_c and that of Q_2 decreases by an equal amount. This ensures that the sum of the total currents in Q_1 and Q_2 remains constant, as constrained by the current-source bias. The incremental (or signal) current component i_c is given by

$$i_c = \frac{\alpha I}{2V_T}\frac{v_{id}}{2} \tag{8.79}$$

Equation (8.79) has an easy interpretation. First, note from the symmetry of the circuit (Fig. 8.19) that the differential signal v_{id} should divide equally between the base–emitter junctions of the two transistors. Thus the total base–emitter voltages will be

$$v_{BE}|_{Q1} = V_{BE} + \frac{v_{id}}{2}$$

$$v_{BE}|_{Q2} = V_{BE} - \frac{v_{id}}{2}$$

where V_{BE} is the dc BE voltage corresponding to an emitter current of $I/2$. Therefore, the collector current of Q_1 will increase by $g_m v_{id}/2$ and the collector current of Q_2 will decrease by $g_m v_{id}/2$. Here g_m denotes the transconductance of Q_1 and of Q_2, which are equal and given by

$$g_m = \frac{I_C}{V_T} = \frac{\alpha I/2}{V_T} \tag{8.80}$$

Thus Eq. (8.79) simply states that $i_c = g_m v_{id}/2$.

An Alternative Viewpoint There is an extremely useful alternative interpretation of the results above. Assume the current source I to be ideal. Its incremental resistance then will be infinite. Thus the voltage v_{id} appears across a total resistance of $2r_e$, where

$$r_e = \frac{V_T}{I_E} = \frac{V_T}{I/2} \tag{8.81}$$

Correspondingly there will be a signal current i_e, as illustrated in Fig. 8.20, given by

$$i_e = \frac{v_{id}}{2r_e} \tag{8.82}$$

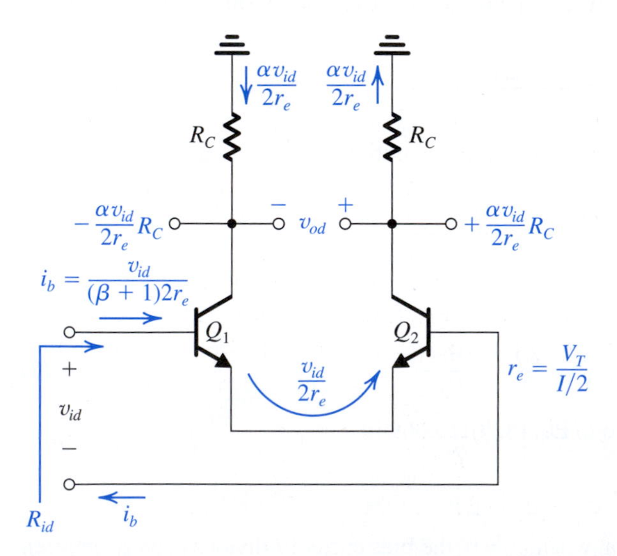

Figure 8.20 A simple technique for determining the signal currents in a differential amplifier excited by a differential voltage signal v_{id}; dc quantities are not shown.

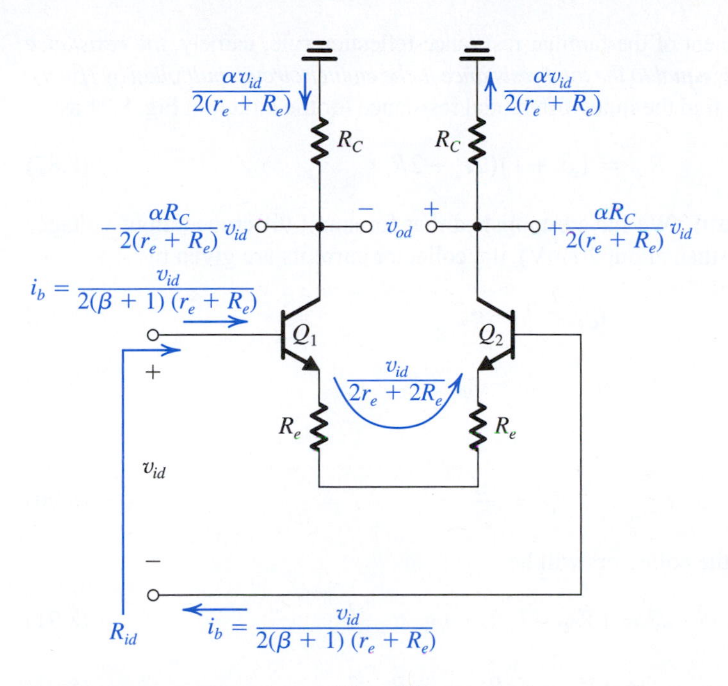

Figure 8.21 A differential amplifier with emitter resistances. Only signal quantities are shown (in color).

Thus the collector of Q_1 will exhibit a current increment i_c and the collector of Q_2 will exhibit a current decrement i_c:

$$i_c = \alpha i_e = \frac{\alpha v_{id}}{2r_e} = g_m \frac{v_{id}}{2} \tag{8.83}$$

Note that in Fig. 8.20 we have shown signal quantities only. It is implied, of course, that each transistor is biased at an emitter current of $I/2$.

This method of analysis is particularly useful when resistances are included in the emitters, as shown in Fig. 8.21. For this circuit we have

$$i_e = \frac{v_{id}}{2r_e + 2R_e} \tag{8.84}$$

Input Differential Resistance Unlike the MOS differential amplifier, which has an infinite input resistance, the bipolar differential pair exhibits a finite input resistance, a result of the finite β of the BJT.

The input differential resistance is the resistance seen between the two bases; that is, it is the resistance seen by the differential input signal v_{id}. For the differential amplifier in Figs. 8.19 and 8.20 it can be seen that the base current of Q_1 shows an increment i_b and the base current of Q_2 shows an equal decrement,

$$i_b = \frac{i_e}{\beta + 1} = \frac{v_{id}/2r_e}{\beta + 1} \tag{8.85}$$

Thus the differential input resistance R_{id} is given by

$$R_{id} \equiv \frac{v_{id}}{i_b} = (\beta + 1)2r_e = 2r_\pi \tag{8.86}$$

This result is just a restatement of the familiar resistance-reflection rule; namely, *the resistance seen between the two bases is equal to the total resistance in the emitter circuit multiplied by* $(\beta+1)$. We can employ this rule to find the input differential resistance for the circuit in Fig. 8.21 as

$$R_{id} = (\beta + 1)(2r_e + 2R_e) \tag{8.87}$$

Differential Voltage Gain We have established that for small difference input voltages ($v_{id} \ll 2V_T$; i.e., v_{id} smaller than about 20 mV), the collector currents are given by

$$i_{C1} = I_C + g_m \frac{v_{id}}{2} \tag{8.88}$$

$$i_{C2} = I_C - g_m \frac{v_{id}}{2} \tag{8.89}$$

where

$$I_C = \frac{\alpha I}{2} \tag{8.90}$$

Thus the total voltages at the collectors will be

$$v_{C1} = (V_{CC} - I_C R_C) - g_m R_C \frac{v_{id}}{2} \tag{8.91}$$

$$v_{C2} = (V_{CC} - I_C R_C) + g_m R_C \frac{v_{id}}{2} \tag{8.92}$$

The quantities in parentheses are simply the dc voltages at each of the two collectors.

As in the MOS case, the output voltage signal of a bipolar differential amplifier can be taken *differentially* (i.e., between the two collectors, $v_{od} = v_{c2} - v_{c1}$). The differential gain of the differential amplifier will be

$$A_d = \frac{v_{od}}{v_{id}} = g_m R_C \tag{8.93}$$

For the differential amplifier with resistances in the emitter leads (Fig. 8.21), the differential gain is given by

$$A_d = \frac{\alpha(2R_C)}{2r_e + 2R_e} \simeq \frac{R_C}{r_e + R_e} \tag{8.94}$$

This equation is a familiar one: It states that *the voltage gain is equal to the ratio of the total resistance in the collector circuit* $(2R_C)$ *to the total resistance in the emitter circuit* $(2r_e + 2R_e)$.

The Differential Half-Circuit As in the MOS case, the differential gain of the BJT differential amplifier can be obtained by considering its differential half-circuit. Figure 8.22(a) shows a differential amplifier fed by a differential signal v_{id} that is applied in a **complementary** (**push–pull** or **balanced**) manner. That is, while the base of Q_1 is raised by $v_{id}/2$, the base of Q_2 is lowered by $v_{id}/2$. We have also included the output resistance R_{EE} of the bias current source. From symmetry, it follows that the signal voltage at the emitters will be zero. Thus the circuit is equivalent to the two common-emitter amplifiers shown in Fig. 8.22(b), where each of the two transistors is biased at an emitter current of $I/2$. Note that the finite output resistance R_{EE} of the current source will have no effect on the operation. The equivalent circuit in Fig. 8.22(b) is valid for differential operation only.

In many applications the differential amplifier is not fed in a complementary fashion; rather, the input signal may be applied to one of the input terminals while the other terminal

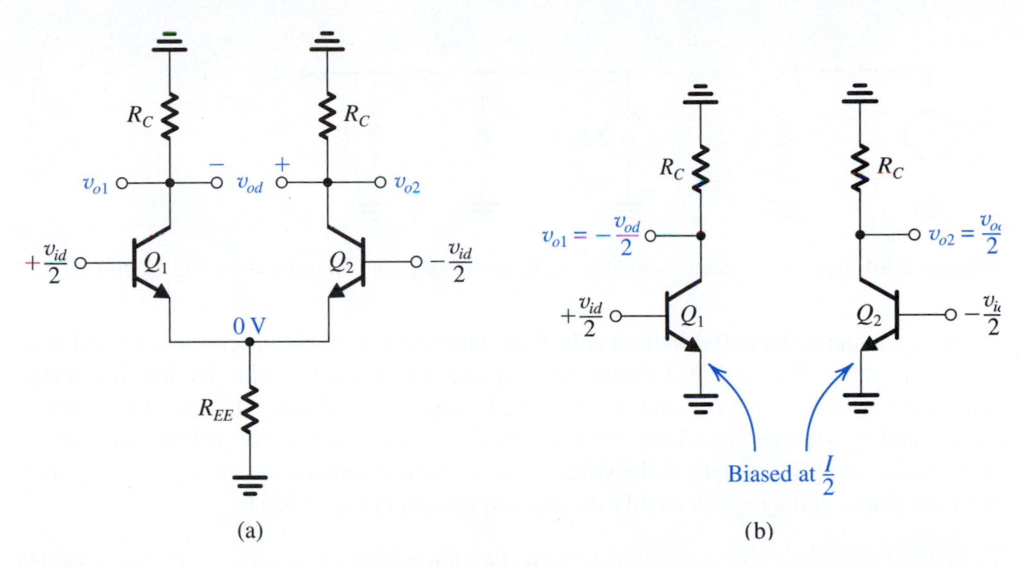

Figure 8.22 Equivalence of the BJT differential amplifier in **(a)** to the two common-emitter amplifiers in **(b)**. This equivalence applies only for differential input signals. Either of the two common-emitter amplifiers in **(b)** can be used to find the differential gain, differential input resistance, frequency response, and so on, of the differential amplifier.

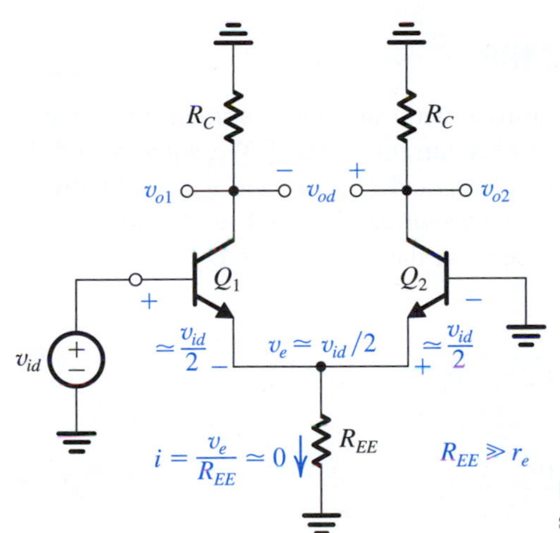

Figure 8.23 The differential amplifier fed in a single-ended fashion.

is grounded, as shown in Fig. 8.23. In this case the signal voltage at the emitters will not be zero, and thus the resistance R_{EE} will have an effect on the operation. Nevertheless, if R_{EE} is large ($R_{EE} \gg r_e$), as is usually the case,[1] then v_{id} will still divide equally (approximately) between the two junctions, as shown in Fig. 8.23. Thus the operation of the differential amplifier in this case will be almost identical to that in the case of symmetric feed, and the common-emitter equivalence can still be employed.

Since in Fig. 8.22, $v_{o2} = -v_{o1} = v_{od}/2$, the two common-emitter transistors in Fig. 8.22(b) yield similar results about the performance of the differential amplifier. Thus only one is needed to analyze the differential small-signal operation of the differential amplifier, and it

[1]Note that R_{EE} appears in parallel with the much smaller r_e of Q_2.

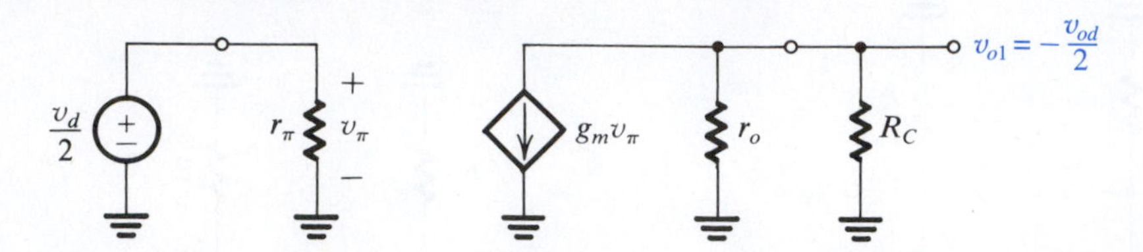

Figure 8.24 Equivalent-circuit model of the differential half-circuit formed by Q_1 in Fig. 8.22(b).

is known as the **differential half-circuit**. If we take the common-emitter transistor fed with $+v_{id}/2$ as the differential half-circuit and replace the transistor with its low-frequency, equivalent-circuit model, the circuit in Fig. 8.24 results. In evaluating the model parameters r_π, g_m, and r_o, we must recall that the half-circuit is biased at $I/2$. The voltage gain of the differential amplifier is equal to the voltage gain of the half-circuit—that is, $v_{o1}/(v_{id}/2)$. Here, we note that including r_o will modify the gain expression in Eq. (8.93) to

$$A_d = g_m(R_C \parallel r_o) \tag{8.95}$$

The input differential resistance of the differential amplifier is twice that of the half-circuit—that is, $2r_\pi$. Finally, we note that the differential half-circuit of the amplifier of Fig. 8.21 is a common-emitter transistor with a resistance R_e in the emitter lead.

8.3.5 Common-Mode Gain and CMRR

Figure 8.25 shows a bipolar differential amplifier with an input common-mode signal v_{icm}. Here R_{EE} is the output resistance of the bias current source I. We wish to find the voltages that result from v_{icm} at the collectors of Q_1 and Q_2, v_{o1} and v_{o2}, and between the two collectors, v_{od}. Toward that end, we make use of the **common-mode half-circuits** shown in Fig. 8.25(b). The signal v_{o1} that appears at the collector of Q_1 in response to v_{icm} will be

$$v_{o1} = -\frac{\alpha R_C}{r_e + 2R_{EE}} v_{icm} \tag{8.96}$$

Similarly, v_{o2} will be

$$v_{o2} = -\frac{\alpha R_C}{r_e + 2R_{EE}} v_{icm} \tag{8.97}$$

where we have neglected the transistor r_o, for simplicity. The differential output signal v_{od} can be obtained as

$$v_{od} = v_{o2} - v_{o1} = 0$$

Thus, while the voltages at the two collectors will contain common-mode noise or interference components, the output differential voltage will be free from such interference. This condition, however, is based on the assumption of perfect matching between the two sides of the differential amplifier. Any mismatch will result in v_{od} acquiring a component proportional to v_{icm}. For example, consider the case of a mismatch ΔR_C between the two collector resistances: If the collector of Q_1 has a collector resistance R_C,

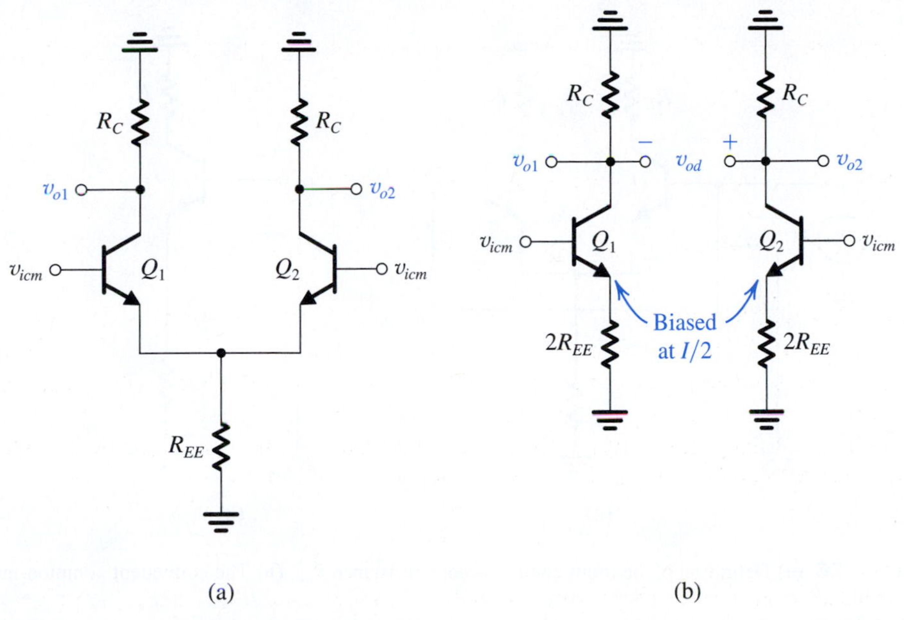

Figure 8.25 **(a)** The differential amplifier fed by a common-mode input signal v_{icm}. **(b)** Equivalent "half-circuits" for common-mode calculations.

$$v_{o1} = -\frac{\alpha R_C}{2R_{EE} + r_e} v_{icm}$$

and the collector of Q_2 has a collector resistance $(R_C + \Delta R_C)$,

$$v_{o2} = -\frac{\alpha(R_C + \Delta R_C)}{2R_{EE} + r_e} v_{icm}$$

then the differential output voltage v_{od} will be

$$\begin{aligned} v_{od} &\equiv v_{o2} - v_{o1} \\ &= -\frac{\alpha \Delta R_C}{2R_{EE} + r_e} v_{icm} \end{aligned}$$

and the common-mode gain will be

$$A_{cm} \equiv \frac{v_{od}}{v_{icm}} = -\frac{\alpha \Delta R_C}{2R_{EE} + r_e} \tag{8.98}$$

Since $\alpha \simeq 1$, $r_e \ll 2R_{EE}$, Eq. (8.98) can be approximated and written in the form

$$A_{cm} \simeq -\left(\frac{R_C}{2R_{EE}}\right)\left(\frac{\Delta R_C}{R_C}\right) \tag{8.99}$$

The common-mode rejection ratio can now be found from

$$\text{CMRR} = \frac{|A_d|}{|A_{cm}|}$$

together with using Eqs. (8.93) and (8.99), with the result that

$$\text{CMRR} = (2g_m R_{EE}) \Big/ \left(\frac{\Delta R_C}{R_C}\right) \tag{8.100}$$

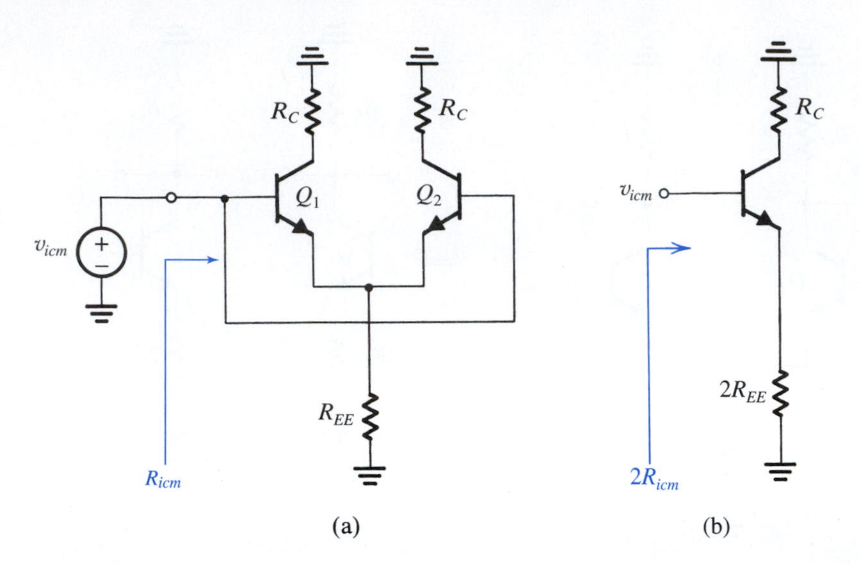

Figure 8.26 **(a)** Definition of the input common-mode resistance R_{icm}. **(b)** The equivalent common-mode half-circuit.

which is similar in form to the expression for the MOS pair [Eq. (8.50)]. Thus, to obtain a high CMRR, we design the current source to have a large output resistance R_{EE} and strive for close matching of the collector resistances.

Common-Mode Input Resistance The definition of the common-mode input resistance R_{icm} is illustrated in Fig. 8.26(a). Figure 8.26(b) shows the equivalent common-mode half-circuit; its input resistance is $2R_{icm}$. The value of $2R_{icm}$ can be determined by analyzing the circuit of Fig. 8.26(b) while taking r_o into account (because R_{EE} and R_C can be equal to, *or* larger than, r_o). The analysis is straightforward but tedious and can be shown [Problem 8.79] to yield the following result

$$R_{icm} \simeq \beta R_{EE} \frac{1 + R_C/\beta r_o}{1 + \dfrac{R_C + 2R_{EE}}{r_o}} \tag{8.101}$$

Example 8.4

The differential amplifier in Fig. 8.27 uses transistors with $\beta = 100$. Evaluate the following:

(a) The input differential resistance R_{id}.
(b) The overall differential voltage gain v_{od}/v_{sig} (neglect the effect of r_o).
(c) The worst-case common-mode gain if the two collector resistances are accurate to within ±1%.
(d) The CMRR, in dB.
(e) The input common-mode resistance (assuming that the Early voltage $V_A = 100$ V).

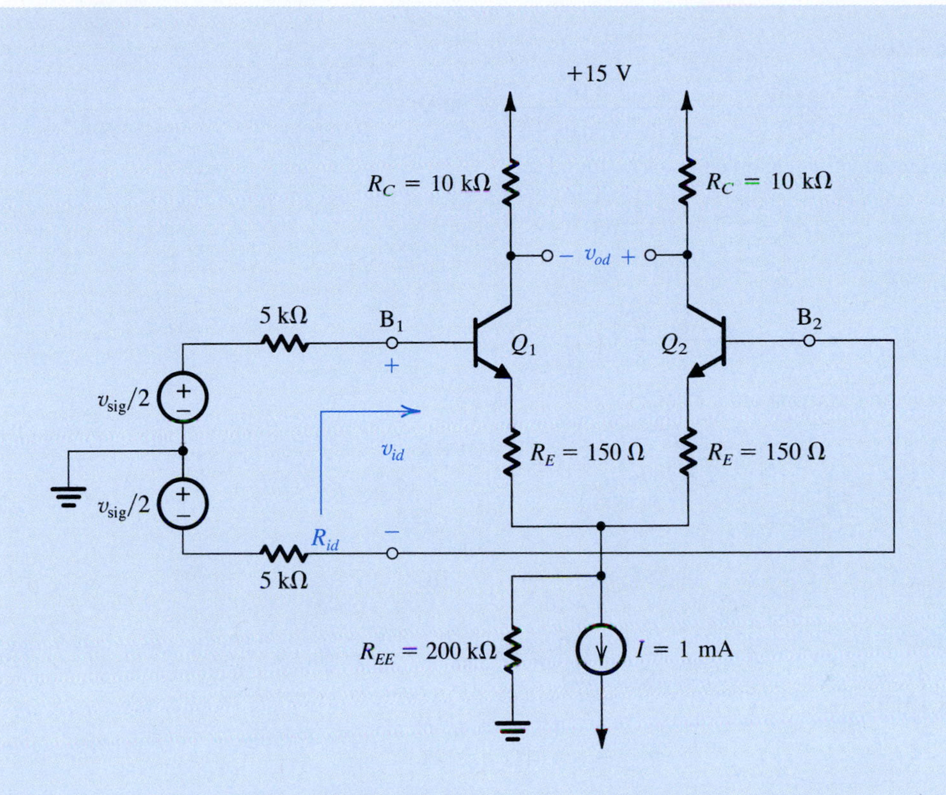

Figure 8.27 Circuit for Example 8.4.

Solution

(a) Each transistor is biased at an emitter current of 0.5 mA. Thus

$$r_{e1} = r_{e2} = \frac{V_T}{I_E} = \frac{25 \text{ mV}}{0.5 \text{ mA}} = 50 \text{ }\Omega$$

The input differential resistance can now be found as

$$\begin{aligned} R_{id} &= 2(\beta + 1)(r_e + R_E) \\ &= 2 \times 101 \times (50 + 150) \simeq 40 \text{ k}\Omega \end{aligned}$$

(b) The voltage gain from the signal source to the bases of Q_1 and Q_2 is

$$\begin{aligned} \frac{v_{id}}{v_{sig}} &= \frac{R_{id}}{R_{sig} + R_{id}} \\ &= \frac{40}{5 + 5 + 40} = 0.8 \text{ V/V} \end{aligned}$$

The voltage gain from the bases to the output is

$$\frac{v_{od}}{v_{id}} \simeq \frac{\text{Total resistance in the collectors}}{\text{Total resistance in the emitters}}$$

Example 8.4 *continued*

$$= \frac{2R_C}{2(r_e + R_E)} = \frac{2 \times 10}{2(50 + 150) \times 10^{-3}} = 50 \text{ V/V}$$

The overall differential voltage gain can now be found as

$$A_d = \frac{v_{od}}{v_{sig}} = \frac{v_{id}v_{od}}{v_{sig}v_{id}} = 0.8 \times 50 = 40 \text{ V/V}$$

(c) Using Eq. (8.99),

$$|A_{cm}| = \frac{R_C}{2R_{EE}} \frac{\Delta R_C}{R_C}$$

where $\Delta R_C = 0.02R_C$ in the worst case. Thus,

$$|A_{cm}| = \frac{10}{2 \times 200} \times 0.02 = 5 \times 10^{-4} \text{ V/V}$$

(d) $\text{CMRR} = 20 \log \frac{|A_d|}{|A_{cm}|}$

$$= 20 \log \frac{40}{5 \times 10^{-4}} = 98 \text{ dB}$$

$$r_o = \frac{V_A}{I/2} = \frac{100}{0.5} = 200 \text{ k}\Omega$$

(e) Using Eq. (8.101),

$$R_{icm} = 6.6 \text{ M}\Omega$$

EXERCISES

8.12 For the circuit in Fig. 8.19, let $I = 1$ mA, $V_{CC} = 15$ V, $R_C = 10$ kΩ, with $\alpha = 1$, and let the input voltages be: $v_{B1} = 5 + 0.005 \sin 2\pi \times 1000t$, volts, and $v_{B2} = 5 - 0.005 \sin 2\pi \times 1000t$, volts. (a) If the BJTs are specified to have v_{BE} of 0.7 V at a collector current of 1 mA, find the voltage at the emitters. (b) Find g_m for each of the two transistors. (c) Find i_C for each of the two transistors. (d) Find v_C for each of the two transistors. (e) Find the voltage between the two collectors. (f) Find the gain experienced by the 1000-Hz signal.
Ans. (a) 4.317 V; (b) 20 mA/V; (c) $i_{C1} = 0.5 + 0.1 \sin 2\pi \times 1000t$, mA and $i_{C2} = 0.5 - 0.1 \sin 2\pi \times 1000t$, mA; (d) $v_{C1} = 10 - 1 \sin 2\pi \times 1000t$, V and $v_{C2} = 10 + 1 \sin 2\pi \times 1000t$, V; (e) $v_{C2} - v_{C1} = 2 \sin 2\pi \times 1000t$, V; (f) 200 V/V

8.13 A bipolar differential amplifier utilizes a simple (i.e., a single CE transistor) current source to supply a bias current I of 200 μA, and simple current-source loads formed by *pnp* transistors. For all transistors, $\beta = 100$ and $|V_A| = 10$ V. Find g_m, R_C, $|A_d|$, R_{id}, R_{EE}, CMRR (if the two load transistors exhibit a 1% mismatch in their r_o's), and R_{icm}.
Ans. 4 mA/V; 100 kΩ; 400 V/V; 50 kΩ, 50 kΩ; 86 dB; 1.67 MΩ

8.4 Other Nonideal Characteristics of the Differential Amplifier

8.4.1 Input Offset Voltage of the MOS Differential Pair

Consider the basic MOS differential amplifier with both inputs grounded, as shown in Fig. 8.28(a). If the two sides of the differential pair were perfectly matched (i.e., Q_1 and Q_2 identical and $R_{D1} = R_{D2} = R_D$), then current I would split equally between Q_1 and Q_2, and V_O would be zero. But practical circuits exhibit mismatches that result in a dc output voltage V_O even with both inputs grounded. We call V_O the **output dc offset voltage**. More commonly, we divide V_O by the differential gain of the amplifier, A_d, to obtain a quantity known as the **input offset voltage**, V_{OS},

$$V_{OS} = V_O/A_d \tag{8.102}$$

We can see that if we apply a voltage $-V_{OS}$ between the input terminals of the differential amplifier, then the output voltage will be reduced to zero (see Fig. 8.28b). This observation gives rise to the usual definition of the input offset voltage. It should be noted, however, that since the offset voltage is a result of device mismatches, its polarity is not known a priori.

Three factors contribute to the dc offset voltage of the MOS differential pair: mismatch in load resistances, mismatch in W/L, and mismatch in V_t. We shall consider the three contributing factors one at a time.

For the differential pair shown in Fig. 8.28(a) consider first the case where Q_1 and Q_2 are perfectly matched but R_{D1} and R_{D2} show a mismatch ΔR_D; that is,

$$R_{D1} = R_D + \frac{\Delta R_D}{2} \tag{8.103}$$

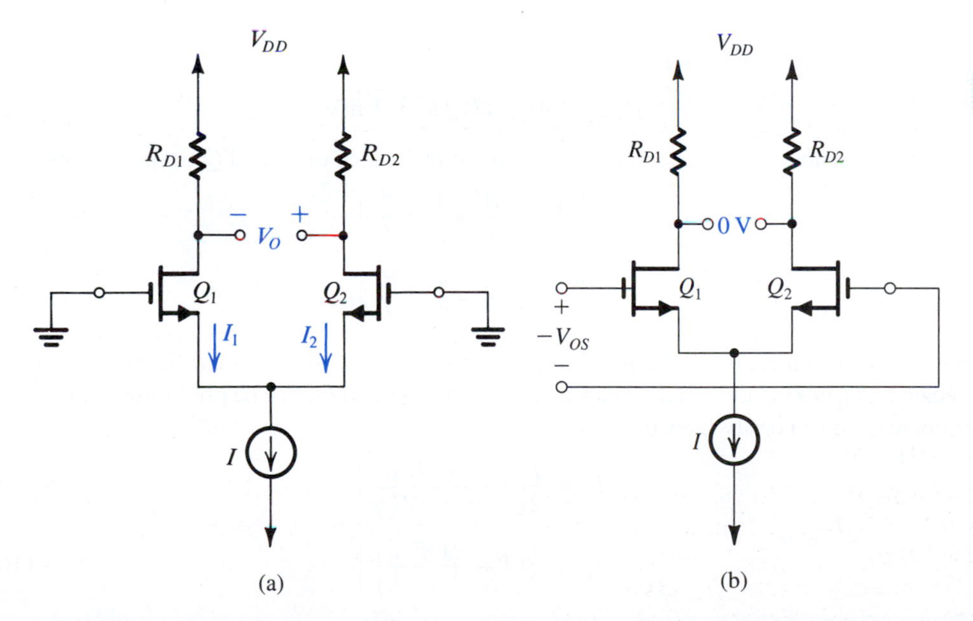

Figure 8.28 **(a)** The MOS differential pair with both inputs grounded. Owing to device and resistor mismatches, a finite dc output voltage V_O results. **(b)** Application of a voltage equal to the input offset voltage V_{OS} to the input terminals with opposite polarity reduces V_O to zero.

$$R_{D2} = R_D - \frac{\Delta R_D}{2} \tag{8.104}$$

Because Q_1 and Q_2 are matched, the current I will split equally between them. Nevertheless, because of the mismatch in load resistances, the output voltages V_{D1} and V_{D2} will be

$$V_{D1} = V_{DD} - \frac{I}{2}\left(R_D + \frac{\Delta R_D}{2}\right)$$

$$V_{D2} = V_{DD} - \frac{I}{2}\left(R_D - \frac{\Delta R_D}{2}\right)$$

Thus the differential output voltage V_O will be

$$\begin{aligned} V_O &= V_{D2} - V_{D1} \\ &= \left(\frac{I}{2}\right)\Delta R_D \end{aligned} \tag{8.105}$$

The corresponding input offset voltage is obtained by dividing V_O by the gain $g_m R_D$ and substituting for g_m from Eq. (8.30). The result is

$$V_{OS} = \left(\frac{V_{OV}}{2}\right)\left(\frac{\Delta R_D}{R_D}\right) \tag{8.106}$$

Thus the offset voltage is directly proportional to V_{OV} and, of course, to $\Delta R_D/R_D$. As an example, consider a differential pair in which the two transistors are operating at an overdrive voltage of 0.2 V and each drain resistance is accurate to within ±1%. It follows that the worst-case resistor mismatch will be

$$\frac{\Delta R_D}{R_D} = 0.02$$

and the resulting input offset voltage will be

$$|V_{OS}| = 0.1 \times 0.02 = 2 \text{ mV}$$

Next, consider the effect of a mismatch in the W/L ratios of Q_1 and Q_2, expressed as

$$\left(\frac{W}{L}\right)_1 = \frac{W}{L} + \frac{1}{2}\Delta\left(\frac{W}{L}\right) \tag{8.107}$$

$$\left(\frac{W}{L}\right)_2 = \frac{W}{L} - \frac{1}{2}\Delta\left(\frac{W}{L}\right) \tag{8.108}$$

Such a mismatch causes the current I to no longer divide equally between Q_1 and Q_2. Rather, because $V_{GS1} = V_{GS2}$, the current conducted by each of Q_1 and Q_2 will be proportional to its W/L ratio, and we can easily show that

$$I_1 = \frac{I}{2}\left[1 + \frac{\Delta(W/L)}{2(W/L)}\right] \tag{8.109}$$

$$I_2 = \frac{I}{2}\left[1 - \frac{\Delta(W/L)}{2(W/L)}\right] \tag{8.110}$$

Dividing the current difference,

$$\frac{I}{2}\frac{\Delta(W/L)}{(W/L)}$$

by g_m gives the input offset voltage (due to the mismatch in W/L values).[2] Thus

$$V_{OS} = \left(\frac{V_{OV}}{2}\right)\left(\frac{\Delta(W/L)}{(W/L)}\right) \tag{8.111}$$

Here again we note that V_{OS}, resulting from a (W/L) mismatch, is proportional to V_{OV} and, as expected, $\Delta(W/L)$.

Finally, we consider the effect of a mismatch ΔV_t between the two threshold voltages,

$$V_{t1} = V_t + \frac{\Delta V_t}{2} \tag{8.112}$$

$$V_{t2} = V_t - \frac{\Delta V_t}{2} \tag{8.113}$$

The current I_1 will be given by

$$I_1 = \frac{1}{2}k_n' \frac{W}{L}\left(V_{GS} - V_t - \frac{\Delta V_t}{2}\right)^2$$

$$= \frac{1}{2}k_n' \frac{W}{L}(V_{GS} - V_t)^2\left[1 - \frac{\Delta V_t}{2(V_{GS} - V_t)}\right]^2$$

which, for $\Delta V_t \ll 2(V_{GS} - V_t)$ [that is, $\Delta V_t \ll 2V_{OV}$], can be approximated as

$$I_1 \simeq \frac{1}{2}k_n' \frac{W}{L}(V_{GS} - V_t)^2\left(1 - \frac{\Delta V_t}{V_{GS} - V_t}\right)$$

Similarly,

$$I_2 \simeq \frac{1}{2}k_n' \frac{W}{L}(V_{GS} - V_t)^2\left(1 + \frac{\Delta V_t}{V_{GS} - V_t}\right)$$

We recognize that

$$\frac{1}{2}k_n' \frac{W}{L}(V_{GS} - V_t)^2 = \frac{I}{2}$$

and the current increment (decrement) in Q_2 (Q_1) is

$$\Delta I = \frac{I}{2}\frac{\Delta V_t}{V_{GS} - V_t} = \frac{I}{2}\frac{\Delta V_t}{V_{OV}}$$

Dividing the current difference $2\Delta I$ by g_m gives the input offset voltage (due to ΔV_t). Thus,

$$V_{OS} = \Delta V_t \tag{8.114}$$

a very logical result! For modern MOS technology ΔV_t can be as high as a few mV. Finally, we note that since the three sources for offset voltage are not correlated, an estimate of the total input offset voltage can be found as

$$V_{OS} = \sqrt{\left(\frac{V_{OV}}{2}\frac{\Delta R_D}{R_D}\right)^2 + \left(\frac{V_{OV}}{2}\frac{\Delta(W/L)}{W/L}\right)^2 + (\Delta V_t)^2} \tag{8.115}$$

[2]We are skipping a step in the derivation: Rather than multiplying the current difference by R_C and dividing the resulting output offset by $A_d = g_m R_C$, we are simply dividing the current difference by g_m.

EXERCISE

8.14 For the MOS differential pair specified in Exercise 8.4, find the three components of the input offset voltage. Let $\Delta R_D/R_D = 2\%$, $\Delta(W/L)/(W/L) = 2\%$, and $\Delta V_t = 2$ mV. Use Eq. (8.115) to obtain an estimate of the total V_{OS}.
Ans. 2 mV; 2 mV; 2 mV; 3.5 mV

8.4.2 Input Offset Voltage of the Bipolar Differential Amplifier

The offset voltage of the bipolar differential pair shown in Fig. 8.29(a) can be determined in a manner analogous to that used above for the MOS pair. Note, however, that in the bipolar case there is no analog to the V_t mismatch of the MOSFET pair. Here the output offset results from mismatches in the load resistances R_{C1} and R_{C2} and from junction area, β, and other mismatches in Q_1 and Q_2. Consider first the effect of the load mismatch. Let

$$R_{C1} = R_C + \frac{\Delta R_C}{2} \tag{8.116}$$

$$R_{C2} = R_C - \frac{\Delta R_C}{2} \tag{8.117}$$

and assume that Q_1 and Q_2 are perfectly matched. It follows that current I will divide equally between Q_1 and Q_2, and thus

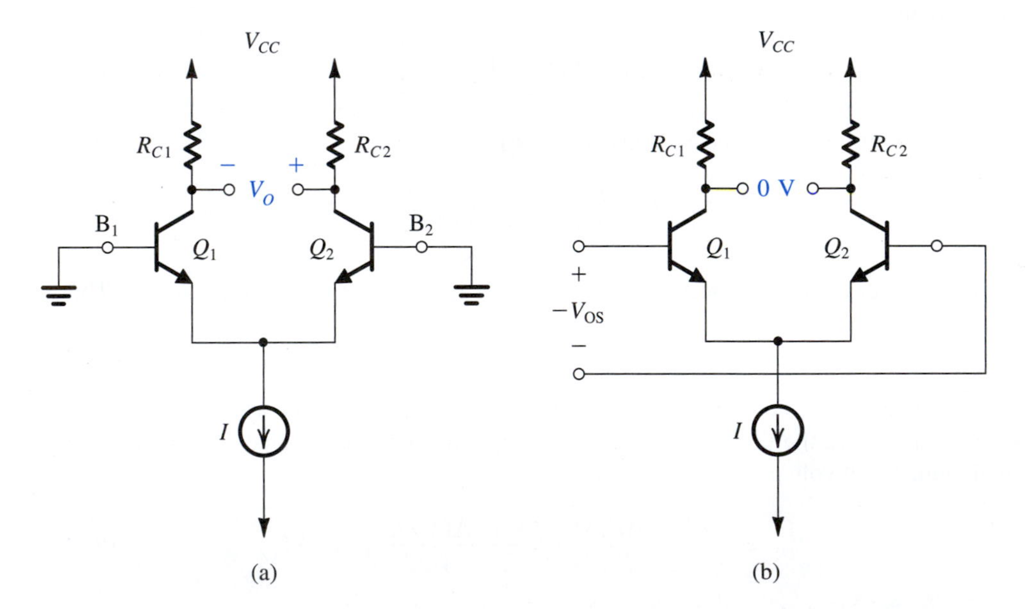

Figure 8.29 **(a)** The BJT differential pair with both inputs grounded. Device mismatches result in a finite dc output V_O. **(b)** Application of the input offset voltage $V_{OS} \equiv V_O/A_d$ to the input terminals with opposite polarity reduces V_O to zero.

$$V_{C1} = V_{CC} - \left(\frac{\alpha I}{2}\right)\left(R_C + \frac{\Delta R_C}{2}\right)$$

$$V_{C2} = V_{CC} - \left(\frac{\alpha I}{2}\right)\left(R_C - \frac{\Delta R_C}{2}\right)$$

Thus the output voltage will be

$$V_O = V_{C2} - V_{C1} = \alpha\left(\frac{I}{2}\right)(\Delta R_C)$$

and the input offset voltage will be

$$V_{OS} = \frac{\alpha(I/2)(\Delta R_C)}{A_d} \tag{8.118}$$

Substituting $A_d = g_m R_C$ and

$$g_m = \frac{\alpha I/2}{V_T}$$

gives

$$|V_{OS}| = V_T\left(\frac{\Delta R_C}{R_C}\right) \tag{8.119}$$

An important point to note is that in comparison to the corresponding expression for the MOS pair (Eq. 8.106) here the offset is proportional to V_T rather than $V_{OV}/2$. V_T at 25 mV is 3 to 6 times lower than $V_{OV}/2$. Hence bipolar differential pairs exhibit lower offsets than their MOS counterparts. As an example, consider the situation of collector resistors that are accurate to within ±1%. Then the worst case mismatch will be

$$\frac{\Delta R_C}{R_C} = 0.02$$

and the resulting input offset voltage will be

$$|V_{OS}| = 25 \times 0.02 = 0.5 \text{ mV}$$

Next consider the effect of mismatches in transistors Q_1 and Q_2. In particular, let the transistors have a mismatch in their emitter–base junction areas. Such an area mismatch gives rise to a proportional mismatch in the scale currents I_S,

$$I_{S1} = I_S + \frac{\Delta I_S}{2} \tag{8.120}$$

$$I_{S2} = I_S - \frac{\Delta I_S}{2} \tag{8.121}$$

Refer to Fig. 8.29(a) and note that $V_{BE1} = V_{BE2}$. Thus, the current I will split between Q_1 and Q_2 in proportion to their I_S values, resulting in

$$I_{E1} = \frac{I}{2}\left(1 + \frac{\Delta I_S}{2I_S}\right) \tag{8.122}$$

$$I_{E2} = \frac{I}{2}\left(1 - \frac{\Delta I_S}{2I_S}\right) \tag{8.123}$$

It follows that the output offset voltage will be

$$V_O = \alpha\left(\frac{I}{2}\right)\left(\frac{\Delta I_S}{I_S}\right)R_C$$

and the corresponding input offset voltage will be

$$|V_{OS}| = V_T\left(\frac{\Delta I_S}{I_S}\right) \tag{8.124}$$

As an example, an area mismatch of 4% gives rise to $\Delta I_S/I_S = 0.04$ and an input offset voltage of 1 mV. Here again we note that the offset voltage is proportional to V_T rather than to the much larger V_{OV}, which determines the offset of the MOS pair due to $\Delta(W/L)$ mismatch.

Since the two contributions to the input offset voltage are usually not correlated, an estimate of the total input offset voltage can be found as

$$V_{OS} = \sqrt{\left(V_T\frac{\Delta R_C}{R_C}\right)^2 + \left(V_T\frac{\Delta I_S}{I_S}\right)^2}$$

$$= V_T\sqrt{\left(\frac{\Delta R_C}{R_C}\right)^2 + \left(\frac{\Delta I_S}{I_S}\right)^2} \tag{8.125}$$

There are other possible sources for input offset voltage such as mismatches in the values of β and r_o. Some of these are investigated in the end-of-chapter problems. Finally, it should be noted that there is a popular scheme for compensating for the offset voltage. It involves introducing a deliberate mismatch in the values of the two collector resistances such that the differential output voltage is reduced to zero when both input terminals are grounded. Such an **offset-nulling** scheme is explored in Problem 8.81.

8.4.3 Input Bias and Offset Currents of the Bipolar Differential Amplifier

In a perfectly symmetric differential pair the two input terminals carry equal dc currents; that is,

$$I_{B1} = I_{B2} = \frac{I/2}{\beta+1} \tag{8.126}$$

This is the **input bias current** of the differential amplifier.

Mismatches in the amplifier circuit and most importantly a mismatch in β make the two input dc currents unequal. The resulting difference is the **input offset current**, I_{OS}, given as

$$I_{OS} = |I_{B1} - I_{B2}| \tag{8.127}$$

Let

$$\beta_1 = \beta + \frac{\Delta\beta}{2}$$

$$\beta_2 = \beta - \frac{\Delta\beta}{2}$$

then

$$I_{B1} = \frac{I}{2}\frac{1}{\beta+1+\Delta\beta/2} \simeq \frac{I}{2}\frac{1}{\beta+1}\left(1-\frac{\Delta\beta}{2\beta}\right) \tag{8.128}$$

$$I_{B2} = \frac{I}{2}\frac{1}{\beta+1-\Delta\beta/2} \simeq \frac{I}{2}\frac{1}{\beta+1}\left(1+\frac{\Delta\beta}{2\beta}\right) \tag{8.129}$$

$$I_{OS} = \frac{I}{2(\beta+1)}\left(\frac{\Delta\beta}{\beta}\right) \tag{8.130}$$

Formally, the input bias current I_B is defined as follows:

$$I_B \equiv \frac{I_{B1}+I_{B2}}{2} = \frac{I}{2(\beta+1)} \tag{8.131}$$

Thus

$$I_{OS} = I_B\left(\frac{\Delta\beta}{\beta}\right) \tag{8.132}$$

As an example, a 10% β mismatch results in an offset current that is one-tenth the value of the input bias current.

Finally note that a great advantage of the MOS differential pair is that it does not suffer from a finite input bias current or from mismatches thereof!

8.4.4 A Concluding Remark

We conclude this section by noting that the definitions presented here are identical to those presented in Chapter 2 for op amps. In fact, as will be seen in Chapter 12, it is the input differential stage in an op-amp circuit that primarily determines the op-amp dc offset voltage, input bias and offset currents, and input common-mode range.

EXERCISE

8.15 For a BJT differential amplifier utilizing transistors having $\beta = 100$, matched to 10% or better, and areas that are matched to 10% or better, along with collector resistors that are matched to 2% or better, find V_{OS}, I_B, and I_{OS}. The dc bias current I is 100 μA.
Ans. 2.55 mV; 0.5 μA; 50 nA

8.5 The Differential Amplifier with Active Load

The differential amplifiers we have studied thus far have been of the differential output variety; that is, the output is taken between the two drains (or two collectors) rather than between one of the drains (collectors) and ground. Taking the output differentially has two major advantages:

1. It decreases the common-mode gain and increases the common-mode rejection ratio (CMRR) dramatically. Recall that while the drain (collector) voltages change somewhat in response to a common-mode input signal, the difference between the drain (collector) voltages remains essentially zero except for a small change due to the mismatches inevitably present in the circuit.
2. It increases the differential gain by a factor of 2 (6 dB) because the output is the difference between two voltages of equal magnitude and opposite sign.

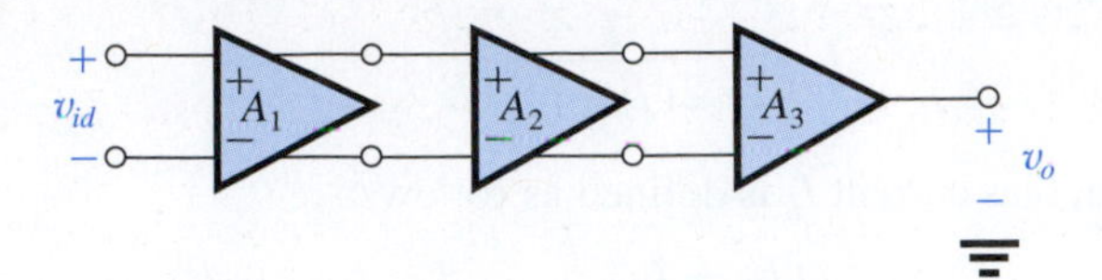

Figure 8.30 A three-stage amplifier consisting of two differential-in, differential-out stages, A_1 and A_2, and a differential-in, single-ended-out stage A_3.

These advantages are sufficiently compelling that at least the first stage in an IC amplifier such as an op amp is **differential-in, differential-out**. The differential transmission of the signal on the chip also minimizes its susceptibility to corruption with noise and interference, which usually occur in a common-mode fashion. Nevertheless, it is usually required at some point to convert the signal from differential to single-ended; for instance, to connect it to an off-chip load. Figure 8.30 shows a block diagram of a three-stage amplifier in which the first two stages are of the differential-in, differential out type, and the third has a single-ended output, that is, an output that is referenced to ground. We now address the question of conversion from differential to single-ended.

8.5.1 Differential to Single-Ended Conversion

Figure 8.31 illustrates the simplest, most basic approach for differential-to-single-ended conversion. It consists of simply ignoring the drain current signal of Q_1 and eliminating its drain resistor altogether, and taking the output between the drain of Q_2 and ground. The obvious drawback of this scheme is that we lose a factor of 2 (or 6 dB) in gain as a result of "wasting" the drain signal current of Q_1. A much better approach would be to find a way of utilizing the drain-current signal of Q_1, and that is exactly what the circuit we are about to discuss accomplishes.

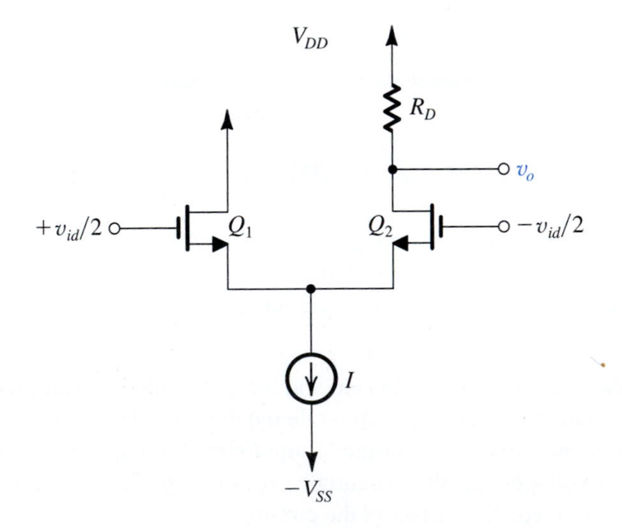

Figure 8.31 A simple but inefficient approach for differential to single-ended conversion.

8.5.2 The Active-Loaded MOS Differential Pair

Figure 8.32(a) shows a MOS differential pair formed by transistors Q_1 and Q_2, loaded by a current mirror formed by transistors Q_3 and Q_4. To see how this circuit operates consider first the quiescent or equilibrium state with the two input terminals connected to a dc voltage equal to the common-mode equilibrium value, in this case 0 V, as shown in Fig. 8.32(b). Assuming perfect matching, the bias current I divides equally between Q_1 and Q_2. The drain current of Q_1, $I/2$, is fed to the input transistor of the mirror, Q_3. Thus, a replica of this current is provided by the output transistor of the mirror, Q_4. Observe that at the output node the two currents $I/2$ balance each other out, leaving a zero current to flow out to the next stage or to a load (not shown). If Q_4 is perfectly matched to Q_3, its drain voltage will track the voltage at the drain of Q_3; thus in equilibrium the voltage at the output will be $V_{DD} - V_{SG3}$. It

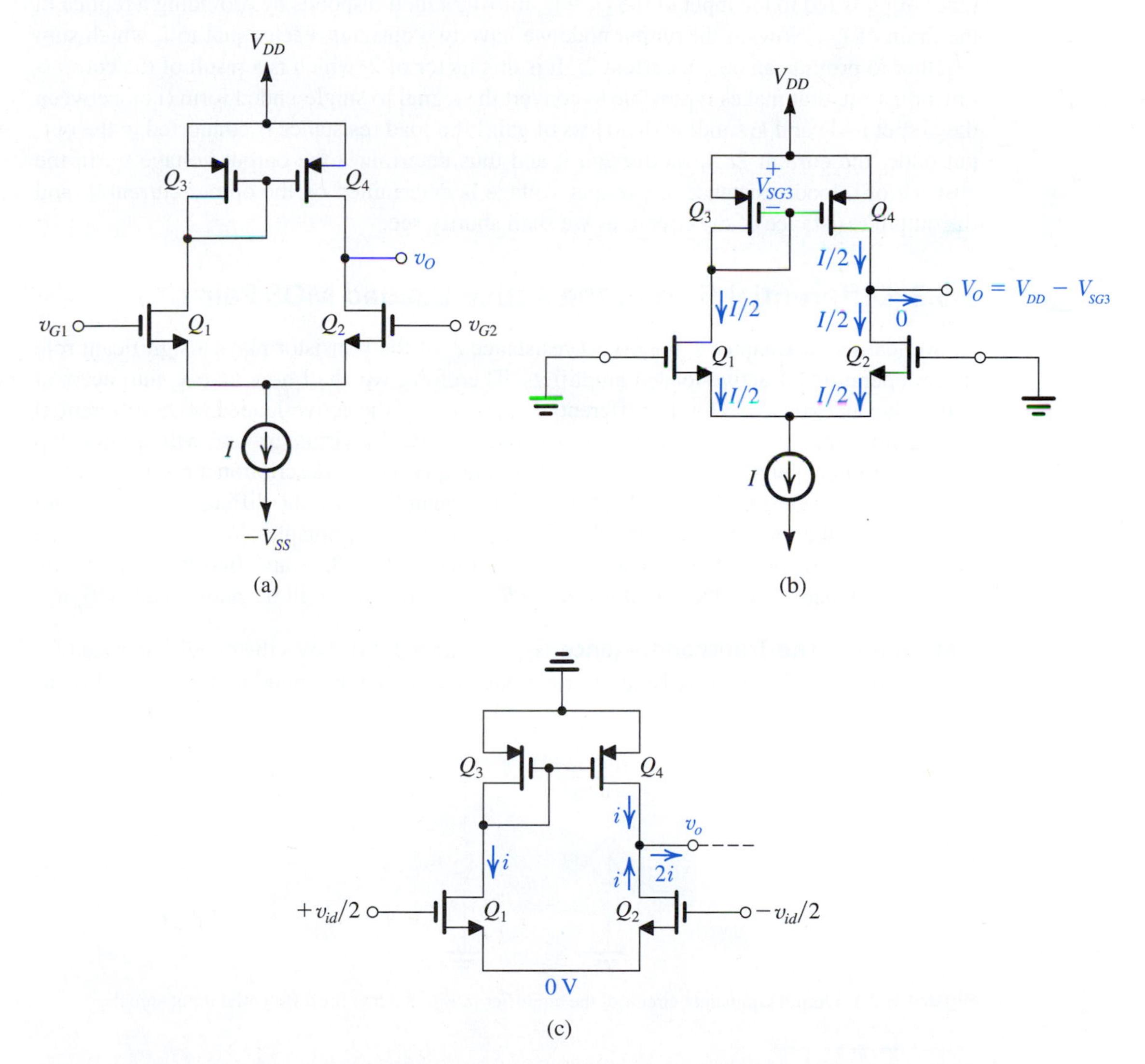

Figure 8.32 **(a)** The active-loaded MOS differential pair. **(b)** The circuit at equilibrium assuming perfect matching. **(c)** The circuit with a differential input signal applied and neglecting the r_o of all transistors.

should be noted, however, that in practical implementations, there will always be mismatches, resulting in a net dc current at the output. In the absence of a load resistance, this current will flow into the output resistances of Q_2 and Q_4 and thus can cause a large deviation in the output voltage from the ideal value. Therefore, this circuit is always designed so that the dc bias voltage at the output node is defined by a feedback circuit rather than by simply relying on the matching of Q_4 and Q_3. We shall see how this is done later.

Next, consider the circuit with a differential input signal v_{id} applied to the input, as shown in Fig. 8.32(c). Since we are now investigating the small-signal operation of the circuit, we have removed the dc supplies (including the current source I). Also, for the time being let us ignore r_o of all transistors. As Fig. 8.32(c) shows, a virtual ground will develop at the common-source terminal of Q_1 and Q_2. Transistor Q_1 will conduct a drain signal current $i = g_{m1}v_{id}/2$, and transistor Q_2 will conduct an equal but opposite current i. The drain signal current i of Q_1 is fed to the input of the $Q_3 - Q_4$ mirror, which responds by providing a replica in the drain of Q_4. Now, at the output node we have two currents, each equal to i, which sum together to provide an output current $2i$. It is this factor of 2, which is a result of the current-mirror action, that makes it possible to convert the signal to single-ended form (i.e., between the output node and ground) with no loss of gain! If a load resistance is connected to the output node, the current $2i$ flows through it and thus determines the output voltage v_o. In the absence of a load resistance, the output voltage is determined by the output current $2i$ and the output resistance of the circuit, as we shall shortly see.

8.5.3 Differential Gain of the Active-Loaded MOS Pair

As we learned in Chapter 7, the output resistance r_o of the transistor plays a significant role in the operation of active-loaded amplifiers. Therefore, we shall now take r_o into account and derive an expression for the differential gain v_o/v_{id} of the active-loaded MOS differential pair. Unfortunately, because the circuit is *not* symmetrical a virtual ground will *not* develop at the common source terminal, contrary to the qualitative description presented above (where the r_o's were neglected). Thus we will not be able to use the differential half-circuit technique. Rather, we shall perform the derivation from first principles: We will represent the output of the circuit by the equivalent circuit shown in Fig. 8.33 and find the short-circuit transconductance G_m and the output resistance R_o. Then, the gain will be determined as G_mR_o.

Determining the Transconductance G_m Figure 8.34(a) shows the circuit[3] prepared for determining G_m. Note that we have short-circuited the output to ground in order to find G_m as

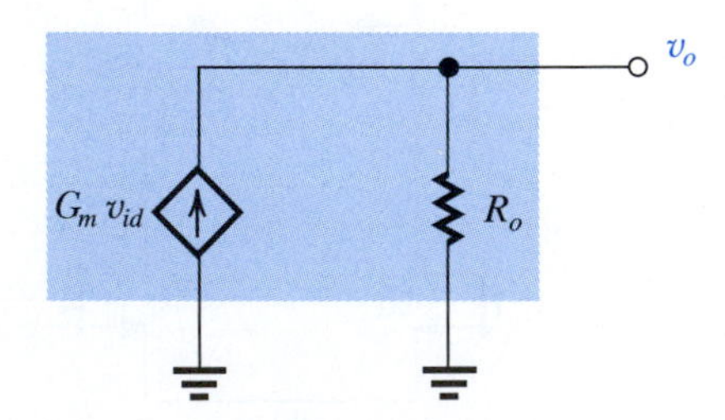

Figure 8.33 Output equivalent circuit of the amplifier in Fig. 8.32(a) for differential input signals.

[3]Note that rather than replacing each transistor with its small-signal model, we are, for simplicity, using the models implicitly. Thus we have "pulled r_o out" of each transistor and shown it separately so that the drain current becomes g_mv_{gs}.

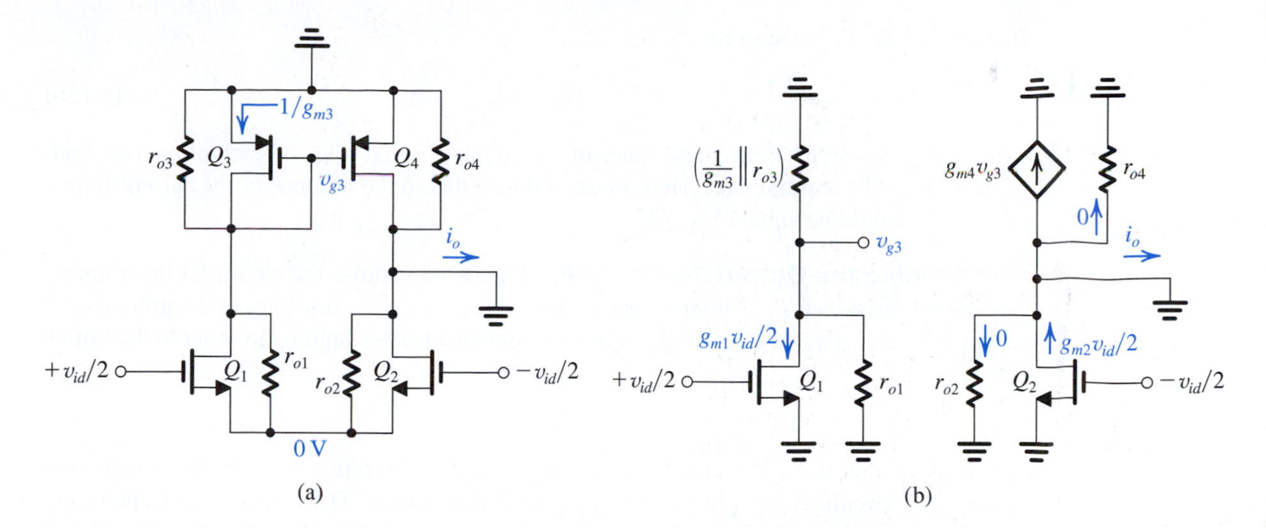

Figure 8.34 Determining the short-circuit transconductance $G_m \equiv i_o/v_{id}$ of the active-loaded MOS differential pair.

i_o/v_{id}. Although the original circuit is not symmetrical, when the output is shorted to ground, the circuit becomes almost symmetrical. This is because the voltage between the drain of Q_1 and ground is very small. This in turn is due to the low resistance between that node and ground which is almost equal to $1/g_{m3}$. Thus, we can now invoke symmetry and assume that a virtual ground will appear at the source of Q_1 and Q_2 and in this way obtain the equivalent circuit shown in Fig. 8.34(b). Here we have replaced the diode-connected transistor Q_3 by its equivalent resistance $[(1/g_{m3})\|r_{o3}]$. The voltage v_{g3} that develops at the common-gate node of the mirror can be found by multiplying the drain current of Q_1 $(g_{m1}v_{id}/2)$, by the total resistance between the drain of Q_1 and ground.

$$v_{g3} = -g_{m1}\left(\frac{v_{id}}{2}\right)\left(\frac{1}{g_{m3}} \| r_{o3} \| r_{o1}\right) \tag{8.133}$$

which for the usual case of r_{o1} and $r_{o3} \gg (1/g_{m3})$ reduces to

$$v_{g3} \simeq -\left(\frac{g_{m1}}{g_{m3}}\right)\left(\frac{v_{id}}{2}\right) \tag{8.134}$$

This voltage controls the drain current of Q_4 resulting in a current of $g_{m4}v_{g3}$. Note that the ground at the output node causes the currents in r_{o2} and r_{o4} to be zero. Thus the output current i_o will be

$$i_o = -g_{m4}v_{g3} + g_{m2}\left(\frac{v_{id}}{2}\right) \tag{8.135}$$

Substituting for v_{g3} from Eq. (8.134) gives

$$i_o = g_{m1}\left(\frac{g_{m4}}{g_{m3}}\right)\left(\frac{v_{id}}{2}\right) + g_{m2}\left(\frac{v_{id}}{2}\right)$$

Now, since $g_{m3} = g_{m4}$ and $g_{m1} = g_{m2} = g_m$, the current i_o becomes

$$i_o = g_m v_{id}$$

from which G_m is found to be

$$G_m = g_m \tag{8.136}$$

Thus the short-circuit transconductance of the circuit is equal to g_m of each of the two transistors of the differential pair.[4] Here we should note that in the absence of the current-mirror action, G_m would be equal to $g_m/2$.

Determining the Output Resistance R_o Figure 8.35 shows the circuit for determining the output resistance R_o. Observe that we have set v_{id} to zero, resulting in the ground connections at the gates of Q_1 and Q_2. We have applied a test voltage v_x in order to determine R_o,

$$R_o \equiv \frac{v_x}{i_x}$$

Analysis of this circuit is considerably simplified by observing the current transmission around the circuit by simply following the circled numbers. The current i that enters Q_2 must exist at its source. It then enters Q_1, exiting at the drain to feed the $Q_3 - Q_4$ mirror. Since for the diode-connected transistor Q_3, $1/g_{m3}$ is much smaller than r_{o3}. most of the current i flows into the drain proper of Q_3. The mirror responds by providing an equal current i in the drain of Q_4. The relationship between i and v_x can be determined by observing that at the output node

$$i = v_x/R_{o2}$$

where R_{o2} is the output resistance of Q_2. Now, Q_2 is a CG transistor and has in its source lead the input resistance R_{in1} of the CG transistor Q_1. Noting that the load resistance of Q_1

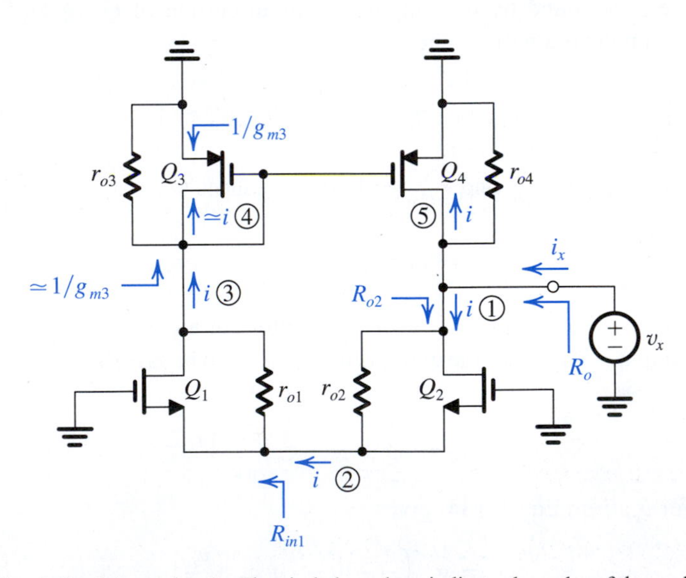

Figure 8.35 Circuit for determining R_o. The circled numbers indicate the order of the analysis steps.

[4]Because the circuit of Fig. 8.34(a) is not perfectly symmetrical, the voltage at the common-source terminal will not be exactly zero. Nevertheless, it can be shown that the voltage will be very small and the transconductance G_m will indeed be very close to g_m.

is $[(1/g_{m3}) \| r_{o3}]$, which is approximately $1/g_{m3}$, we can obtain R_{in1} by using the expression for the input resistance of a CG transistor (adapt Eq. 7.35 by replacing the subscript 2 by 1),

$$R_{in1} = \frac{r_{o1} + R_L}{g_{m1} r_{o1}}$$

$$= \frac{1}{g_{m1}} + \frac{1/g_{m3}}{g_{m1} r_{o1}} \simeq \frac{1}{g_{m1}}$$

We then use this value of R_{in1} to determine R_{o2} using the expression in Eq. (7.38) as follows:

$$R_{o2} = R_{in1} + r_{o2} + g_{m2} r_{o2} R_{in1}$$

$$= \frac{1}{g_{m1}} + r_{o2} + \left(\frac{g_{m2}}{g_{m1}}\right) r_{o2}$$

which, for $g_{m1} = g_{m2} = g_m$ and $g_{m2} r_{o2} \gg 1$, yields

$$R_{o2} \simeq 2r_{o2} \tag{8.137}$$

Returning to the output node, we write

$$i_x = i + i + \frac{v_x}{r_{o4}}$$

$$= 2i + \frac{v_x}{r_{o4}} = 2\frac{v_x}{R_{o2}} + \frac{v_x}{r_{o4}}$$

Substituting for R_{o2} from Eq. (8.137), we obtain

$$i_x = 2\frac{v_x}{2r_{o2}} + \frac{v_x}{r_{o4}}$$

Thus,

$$R_o \equiv \frac{v_x}{i_x} = r_{o2} \| r_{o4} \tag{8.138}$$

which is an intuitively appealing result.

Determining the Differential Gain Equations (8.136) and (8.138) can be combined to obtain the differential gain A_d as

$$A_d \equiv \frac{v_o}{v_{id}} = G_m R_o = g_m (r_{o2} \| r_{o4}) \tag{8.139}$$

For the case $r_{o2} = r_{o4} = r_o$,

$$A_d = \frac{1}{2} g_m r_o = \frac{A_0}{2} \tag{8.140}$$

where A_0 is the intrinsic gain of the MOS transistor.

8.5.4 Common-Mode Gain and CMRR

Although its output is single-ended, the active-loaded MOS differential amplifier has a low common-mode gain and, correspondingly, a high CMRR. Figure 8.36(a) shows the circuit with v_{icm} applied and with the power supplies eliminated except, of course, for the output resistance R_{SS} of the bias-current source I. Although the circuit is not symmetrical and hence

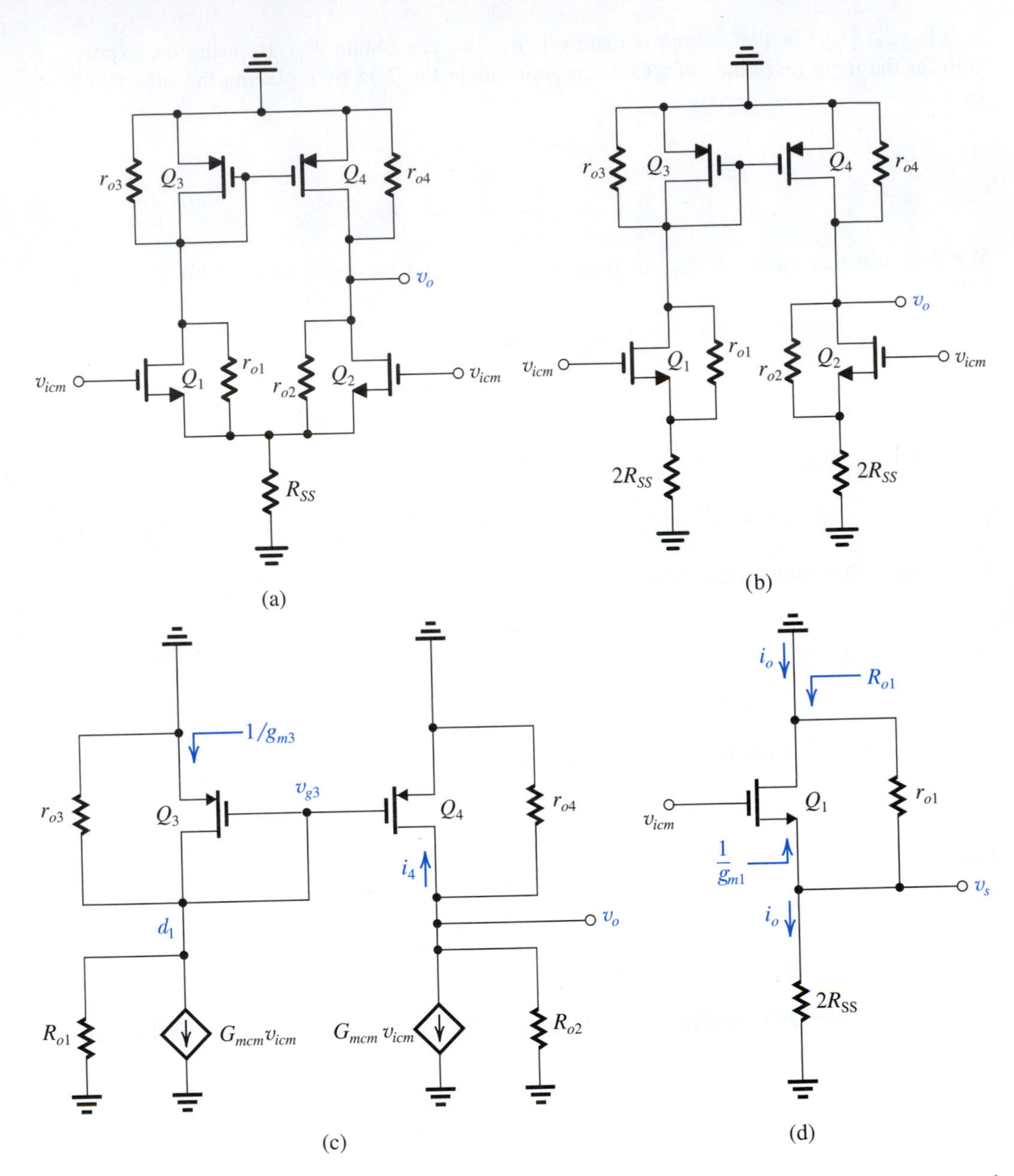

Figure 8.36 Analysis of the active-loaded MOS differential amplifier to determine its common-mode gain.

we cannot use the common-mode half-circuit, we can split R_{SS} equally between Q_1 and Q_2 as shown in Fig. 8.36b. It can now be seen that each of Q_1 and Q_2 is a CS transistor with a large source degeneration resistance $2R_{SS}$.

Each of Q_1 and Q_2 together with their degeneration resistances can be replaced by equivalent circuits composed of a controlled source $G_{mcm}v_{icm}$ and an output resistance $R_{o1,2}$, as shown in Fig. 8.36(c). To determine G_{mcm} we short circuit the drain to ground, as shown in Fig. 8.36(d) for Q_1. Observe that $2R_{SS}$ and r_{o1} appear in parallel. Thus the voltage at the source terminal can be found from the voltage divider consisting of $1/g_{m1}$ and $(2R_{SS} \| r_{o1})$ as

$$v_s = v_{icm}\frac{(2R_{SS}\|r_{o1})}{(2R_{SS}\|r_{o1})+(1/g_{m1})}$$

$$\simeq v_{icm}$$

The short-circuit drain current i_o can be seen to be equal to the current through $2R_{SS}$; thus,

$$i_o = \frac{v_{icm}}{2R_{SS}}$$

which leads to

$$G_{mcm} \equiv \frac{i_o}{v_{icm}} = \frac{1}{2R_{SS}} \tag{8.141}$$

The output resistance R_{o1} can be determined using the expression for R_o of a CS transistor with an emitter-degeneration resistance (Eq. 7.38) to obtain

$$R_{o1} = 2R_{SS} + r_{o1} + (g_{m1}r_{o1})(2R_{SS}) \tag{8.142}$$

Similar results can be obtained for Q_2, namely, the same G_{mcm} and an output resistance R_{o2} given by

$$R_{o2} = 2R_{SS} + r_{o2} + (g_{m2}r_{o2})(2R_{SS}) \tag{8.143}$$

Returning to the circuit in Fig. 8.36(c), the voltage v_{g3} can be obtained by multiplying $G_{mcm}v_{icm}$ by the total resistance between the d_1 node and ground,

$$v_{g3} = -G_{mcm}v_{icm}\left(R_{o1}\|r_{o3}\|\frac{1}{g_{m3}}\right) \tag{8.144}$$

This voltage in turn determines the current i_4 as

$$i_4 = g_{m4}v_{gs3} = g_{m4}v_{g3}$$

Thus,

$$i_4 = -g_{m4}G_{mcm}v_{icm}\left(R_{o1}\|r_{o3}\|\frac{1}{g_{m3}}\right) \tag{8.145}$$

Finally, we can obtain the output voltage v_o by writing for the output node,

$$G_{mcm}v_{icm} + i_4 + \frac{v_o}{R_{o2}} + \frac{v_o}{r_{o4}} = 0$$

Substituting for i_4 from Eq. (8.145) and for G_{mcm} from Eq. (8.141) yields

$$v_o = -v_{icm}\frac{r_{o4}\|R_{o2}}{2R_{SS}}\left[1 - g_{m4}\left(R_{o1}\|r_{o3}\|\frac{1}{g_{m3}}\right)\right]$$

Since $R_{o2} \gg r_{o4}$ and $R_{o1} \gg r_{o3}$, we can neglect both. Also, substituting $g_{m4} = g_{m3}$, we obtain the following expression for A_{cm},

$$A_{cm} \equiv \frac{v_o}{v_{icm}} \simeq -\frac{r_{o4}}{2R_{SS}}\,\frac{1}{1+g_{m3}r_{o3}} \tag{8.146}$$

This expression can be further simplified by noting that $g_{m3}r_{o3} \gg 1$ and $r_{o3} = r_{o4}$ with the result that

$$A_{cm} \simeq -\frac{1}{2g_{m3}R_{SS}} \tag{8.146$'$}$$

Since R_{SS} is usually large, at least equal to r_o, A_{cm} will be small. The common-mode rejection ratio (CMRR) can now be obtained by utilizing Eqs. (8.139) and (8.146′),

$$\text{CMRR} \equiv \frac{|A_d|}{|A_{cm}|} = [g_m(r_{o2} \| r_{o4})][2g_{m3}R_{SS}] \tag{8.147}$$

which for $r_{o2} = r_{o4} = r_o$ and $g_{m3} = g_m$ simplifies to

$$\text{CMRR} = (g_m r_o)(g_m R_{SS}) \tag{8.148}$$

We observe that to obtain a large CMRR, we select an implementation of the biasing current source I that features a high output resistance. Such circuits include the cascode current source and the Wilson current source studied in Section 7.5.

EXERCISE

8.16 An active-loaded MOS differential amplifier of the type shown in Fig. 8.32(a) is specified as follows: $(W/L)_n = 100$, $(W/L)_p = 200$, $\mu_n C_{ox} = 2\mu_p C_{ox} = 0.2\text{ mA/V}^2$, $V_{An} = |V_{Ap}| = 20\text{ V}$, $I = 0.8\text{ mA}$, $R_{SS} = 25\text{ k}\Omega$. Calculate G_m, R_o, A_d, $|A_{cm}|$, and CMRR.
Ans. 4 mA/V; 25 kΩ; 100 V/V; 0.005 V/V; 20,000 or 86 dB

8.5.5 The Bipolar Differential Pair with Active Load

The bipolar version of the active-loaded differential pair is shown in Fig. 8.37(a). The circuit structure and operation are very similar to those of its MOS counterpart except that here we have to contend with the effects of finite β and the resulting finite input resistance at the base, r_π. For the time being, however, we shall ignore the effect of finite β on the dc bias of the four transistors and assume that in equilibrium all transistors are operating at a dc current of $I/2$.

Differential Gain To obtain an expression for the differential gain, we apply an input differential signal v_{id} as shown in the equivalent circuit in Fig. 8.37(b). Note that the output is connected to ground in order to determine the overall short-circuit transconductance $G_m \equiv i_o/v_{id}$. Also, as in the MOS case, we have assumed that the circuit is sufficiently balanced so that a virtual ground develops on the common emitter terminal. This assumption is predicated on the fact that the voltage signal at the collector of Q_1 will be small as a result of the low resistance between that node and ground (approximately equal to r_{e3}). The voltage v_{b3} can be found from

$$v_{b3} = -g_{m1}\left(\frac{v_{id}}{2}\right)(r_{e3} \| r_{o3} \| r_{o1} \| r_{\pi 4})$$

Of the four resistances in the parallel equivalent on the right-hand side, r_{e3} is much smaller than the other three and thus dominates, with the result that

$$v_{b3} \simeq -g_{m1} r_{e3}\left(\frac{v_{id}}{2}\right) \tag{8.149}$$

Since $v_{b4} = v_{b3}$, the collector current of Q_4 will be

$$g_{m4} v_{b4} = -g_{m4} g_{m1} r_{e3}\left(\frac{v_{id}}{2}\right) \tag{8.150}$$

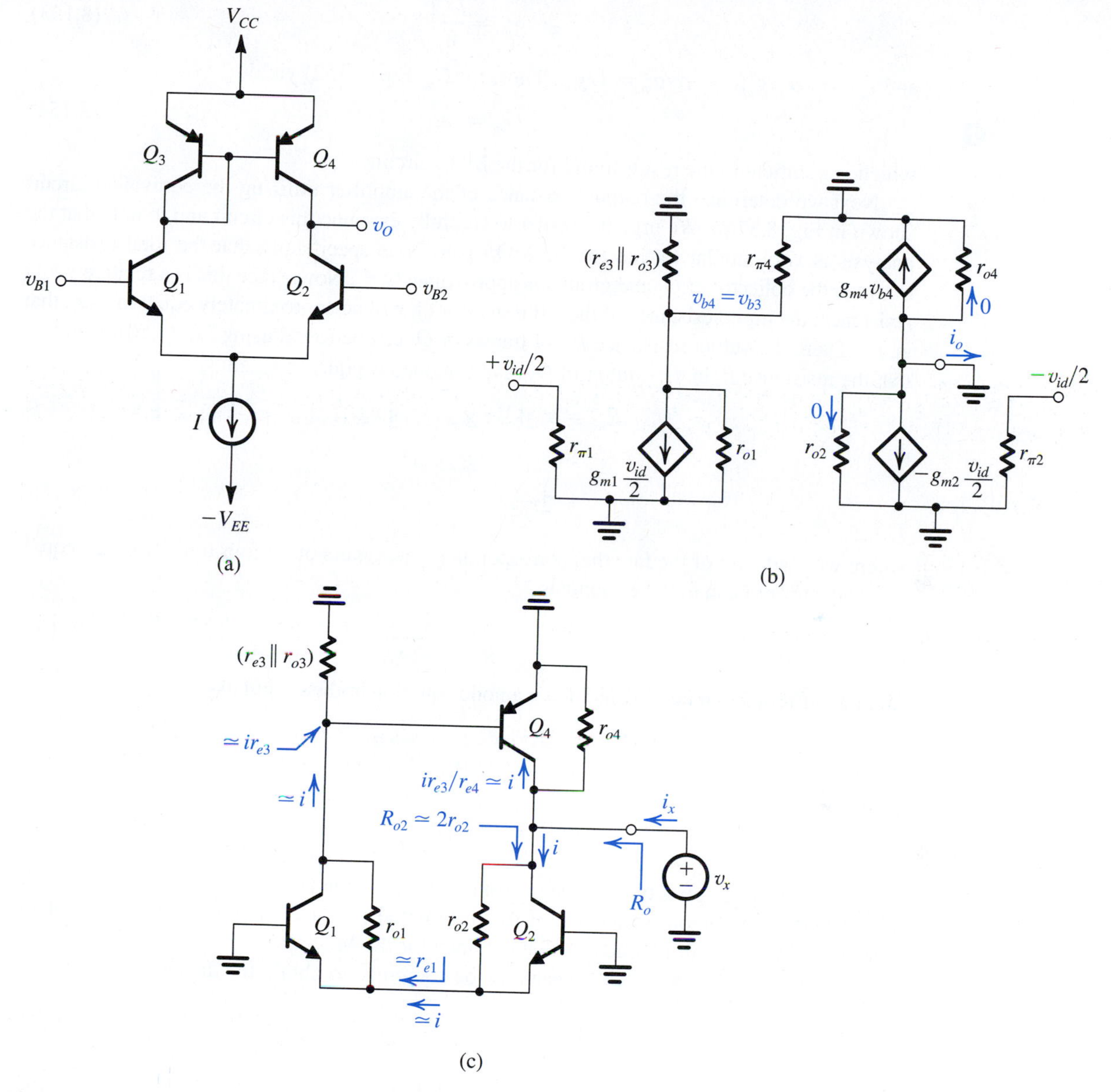

Figure 8.37 **(a)** Active-loaded bipolar differential pair. **(b)** Small-signal equivalent circuit for determining the transconductance $G_m \equiv i_o/v_{id}$. **(c)** Equivalent circuit for determining the output resistance $R_o \equiv v_x/i_x$.

The output current i_o can be found from a node equation at the output as

$$i_o = g_{m2}\left(\frac{v_{id}}{2}\right) - g_{m4}v_{b4} \tag{8.151}$$

Using Eq. (8.150), we obtain

$$i_o = g_{m2}\left(\frac{v_{id}}{2}\right) + g_{m4}g_{m1}r_{e3}\left(\frac{v_{id}}{2}\right) \tag{8.152}$$

Since all devices are operating at the same bias current, $g_{m1} = g_{m2} = g_{m4} = g_m$, where

$$g_m \simeq \frac{I/2}{V_T} \tag{8.153}$$

and $r_{e3} = \alpha_3/g_{m3} = \alpha/g_m \simeq 1/g_m$. Thus, for G_m, Eq. (8.152) yields

$$G_m = g_m \tag{8.154}$$

which is identical to the result found for the MOS circuit.

Next we determine the output resistance of the amplifier utilizing the equivalent circuit shown in Fig. 8.37(c). We urge the reader to carefully examine this circuit and to note that the analysis is very similar to that for the MOS pair. Note specifically that the total resistance between the collector of Q_1 and ground is approximately r_{e3}. Now, since this is a relatively low resistance, the input resistance of the CB transistor Q_l will be approximately equal to its r_e, that is, r_{e1}. Then, the output resistance R_{o2} of transistor Q_2 can be found using Eq. (7.50) by noting that the resistance R_e in the emitter of Q_2 is approximately equal to r_{e1}; thus,

$$\begin{aligned} R_{o2} &\simeq r_{o2}[1 + g_{m2}(r_{e1} \| r_{\pi 2})] \\ &\simeq r_{o2}(1 + g_{m2}r_{e1}) \\ &\simeq 2r_{o2} \end{aligned} \tag{8.155}$$

where we made use of the fact that corresponding parameters of all four transistors are equal.

The current i can now be found as

$$i = \frac{v_x}{R_{o2}} = \frac{v_x}{2r_{o2}} \tag{8.156}$$

and the current i_x can be obtained from a node equation at the output as

$$i_x = 2i + \frac{v_x}{r_{o4}} = \frac{v_x}{r_{o2}} + \frac{v_x}{r_{o4}}$$

Thus,

$$R_o \equiv \frac{v_x}{i_x} = r_{o2} \| r_{o4} \tag{8.157}$$

This expression simply says that the output resistance of the amplifier is equal to the parallel equivalent of the output resistance of the differential pair and the output resistance of the current mirror; a result identical to that obtained for the MOS pair.

Equations (8.154) and (8.157) can now be combined to obtain the differential gain,

$$A_d \equiv \frac{v_o}{v_{id}} = G_m R_o = g_m(r_{o2} \| r_{o4}) \tag{8.158}$$

and since $r_{o2} = r_{o4} = r_o$, we can simplify Eq. (8.158) to

$$A_d = \tfrac{1}{2} g_m r_o \tag{8.159}$$

Although this expression is identical to that found for the MOS circuit, the gain here is much larger because $g_m r_o$ for the BJT is more than an order of magnitude greater than $g_m r_o$ of a MOSFET. The downside, however, lies in the low input resistance of BJT amplifiers. Indeed, the equivalent circuit of Fig. 8.37(b) indicates that, as expected, the differential input resistance of the differential amplifier is equal to $2r_\pi$,

$$R_{id} = 2r_\pi \tag{8.160}$$

in sharp contrast to the infinite input resistance of the MOS amplifier. Thus, while the voltage gain realized in an active-loaded BJT amplifier stage is large, when a subsequent BJT stage is connected to the output, its inevitably low input resistance will drastically reduce the overall voltage gain.

Common-Mode Gain and CMRR The common-mode gain A_{cm} and the common-mode rejection ratio (CMRR) can be found following a procedure identical to that utilized in the MOS case. Figure 8.38 shows the circuit prepared for common-mode signal analysis. As we have done in the MOS case, we will represent each of Q_1 and Q_2 together with their emitter resistances by a short-circuit output current $i_{1,2}$ and an output resistance $R_{o1,2}$. The short-circuit output currents of Q_1 and Q_2 are given by

$$i_1 \simeq i_2 \simeq \frac{v_{icm}}{2R_{EE}} \tag{8.161}$$

It can be shown that the output resistances of Q_1 and Q_2, R_{o1} and R_{o2}, are very large compared with the other resistances between the collector nodes of Q_1 and Q_2 ground, and hence can be neglected. Then, the voltage v_{b3} at the common base connection of Q_3 and Q_4 can be found by multiplying i_1 by the total resistance between the common base node and ground as

$$v_{b3} = -i_1\left(\frac{1}{g_{m3}}\|r_{\pi3}\|r_{o3}\|r_{\pi4}\right) \tag{8.162}$$

In response to v_{b3} transistor Q_4 provides a collector current $g_{m4}v_{b3}$. At the output node we can write the equation

$$\frac{v_o}{r_{o4}} + g_{m4}v_{b3} + i_2 = 0 \tag{8.163}$$

Substituting for v_{b3} from Eq. (8.162) and for i_1 and i_2 from Eq. (8.161) gives

$$\begin{aligned} A_{cm} \equiv \frac{v_o}{v_{icm}} &= \frac{r_{o4}}{2R_{EE}}\left[g_{m4}\left(\frac{1}{g_{m3}}\|r_{\pi3}\|r_{o3}\|r_{\pi4}\right) - 1\right] \\ &= -\frac{r_{o4}}{2R_{EE}} \frac{\dfrac{1}{r_{\pi3}} + \dfrac{1}{r_{\pi4}} + \dfrac{1}{r_{o3}}}{g_{m3} + \dfrac{1}{r_{\pi3}} + \dfrac{1}{r_{\pi4}} + \dfrac{1}{r_{o3}}} \end{aligned} \tag{8.164}$$

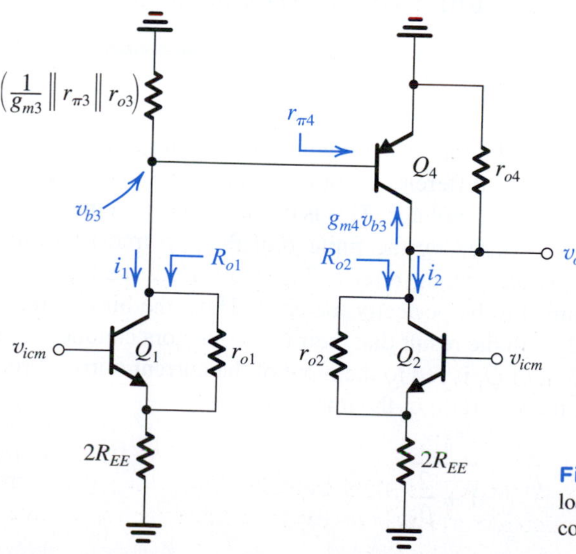

Figure 8.38 Analysis of the bipolar active-loaded differential amplifier to determine the common-mode gain.

where we have assumed $g_{m3} = g_{m4}$. Now, for $r_{\pi4} = r_{\pi3}$ and $r_{o3} \gg r_{\pi3}, r_{\pi4}$, Eq. (8.164) gives

$$A_{cm} \simeq -\frac{r_{o4}}{2R_{EE}} \frac{\frac{2}{r_{\pi3}}}{g_{m3} + \frac{2}{r_{\pi3}}}$$

$$\simeq -\frac{r_{o4}}{2R_{EE}} \frac{2}{\beta_3} = -\frac{r_{o4}}{\beta_3 R_{EE}} \tag{8.165}$$

Using A_d from Eq. (8.158) enables us to obtain the CMRR as

$$\text{CMRR} \equiv \frac{|A_d|}{|A_{cm}|} = g_m(r_{o2} \| r_{o4})\left(\frac{\beta_3 R_{EE}}{r_{o4}}\right) \tag{8.166}$$

For $r_{o2} = r_{o4} = r_o$,

$$\text{CMRR} = \tfrac{1}{2}\beta_3 g_m R_{EE} \tag{8.167}$$

from which we observe that to obtain a large CMRR, the circuit implementing the bias current source should have a large output resistance R_{EE}. This is possible with, say, a Wilson current mirror (Section 7.5.3).

Before leaving the subject of the CM gain of the active-loaded differential amplifier, it is useful to reflect on the origin of its finite common-mode gain: It is simply due to the current transmission error introduced by the current-mirror load. In the case of the MOS circuit, this error is due to the finite r_{o3}; in the case of the bipolar mirror, the error is due to the finite β [Problem 8.98].

EXERCISE

8.17 For the active-loaded BJT differential amplifier let $I = 0.8$ mA, $V_A = 100$ V, and $\beta = 160$. Find G_m, R_o, A_d, and R_{id}. If the bias current source is implemented with a simple *npn* current mirror, find R_{EE}, A_{cm}, and CMRR.

Ans. 16 mA/V; 125 kΩ; 2000 V/V; 20 kΩ; 125 kΩ; –0.0125 V/V; 160,000 or 104 dB

Systematic Input Offset Voltage In addition to the random offset voltages that result from the mismatches inevitably present in the differential amplifier, the active-loaded bipolar differential pair suffers from a systematic offset voltage. This is due to the error in the current transfer ratio of the current-mirror load caused by the finite β of the *pnp* transistors that make up the mirror. To see how this comes about, refer to Fig. 8.39. Here the inputs are grounded and the transistors are assumed to be perfectly matched. Thus, the bias current I will divide equally between Q_1 and Q_2 with the result that their two collectors conduct equal currents of $\alpha I/2$. The collector current of Q_1 is fed to the input of the current mirror. From Section 7.4 we know that the current-transfer ratio of the mirror is

$$\frac{I_4}{I_3} = \frac{1}{1 + \frac{2}{\beta_P}} \tag{8.168}$$

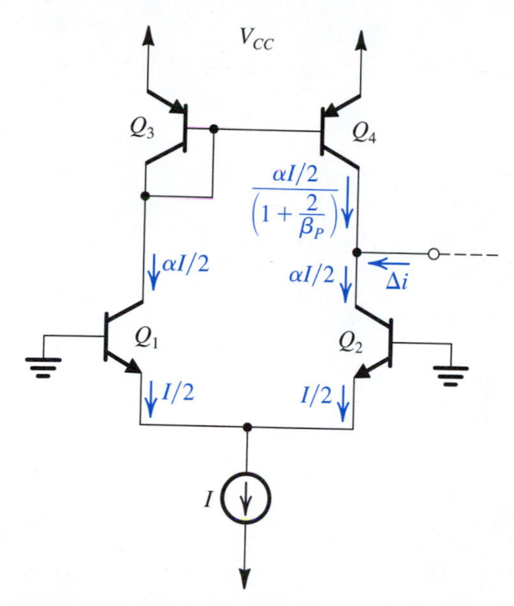

Figure 8.39 The active-loaded BJT differential pair suffers from a systematic input offset voltage resulting from the error in the current-transfer ratio of the current mirror.

where β_P is the value of β of the *pnp* transistors Q_3 and Q_4. Thus the collector current of Q_4 will be

$$I_4 = \frac{\alpha I/2}{1+\dfrac{2}{\beta_P}} \tag{8.169}$$

which does not exactly balance the collector current of Q_2. It follows that the current difference Δi will flow into the output terminal of the amplifier with

$$\begin{aligned} \Delta i &= \frac{\alpha I}{2} - \frac{\alpha I/2}{1+\dfrac{2}{\beta_P}} \\ &= \frac{\alpha I}{2}\frac{2/\beta_P}{1+\dfrac{2}{\beta_P}} \\ &\simeq \frac{\alpha I}{\beta_P} \end{aligned} \tag{8.170}$$

To reduce this output current to zero, an input voltage V_{OS} has to be applied with a value of

$$V_{OS} = -\frac{\Delta i}{G_m}$$

Substituting for Δi from Eq. (8.170) and for $G_m = g_m = (\alpha I/2)/V_T$, we obtain for the input offset voltage the expression

$$V_{OS} = -\frac{\alpha I/\beta_P}{\alpha I/2V_T} = -\frac{2V_T}{\beta_P} \tag{8.171}$$

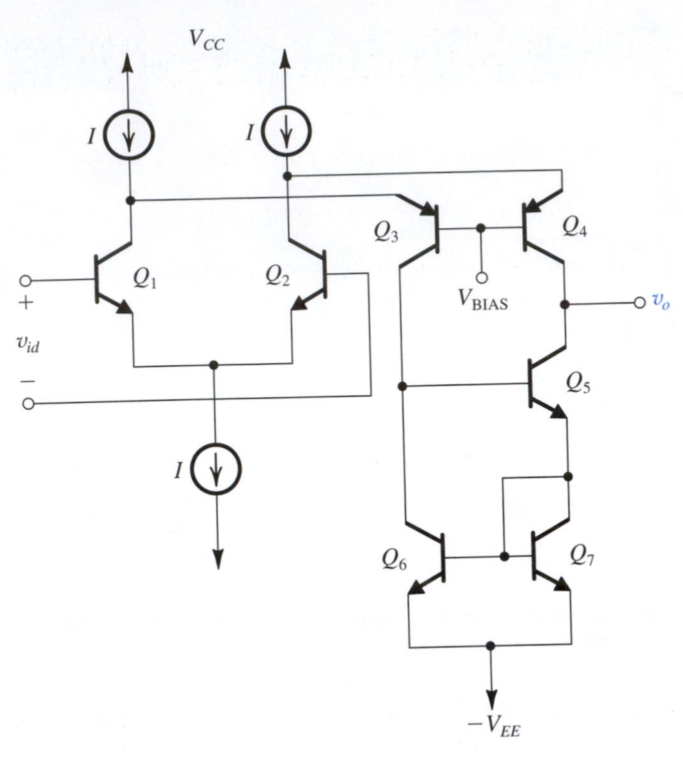

Figure 8.40 An active-loaded bipolar differential amplifier employing a folded cascode stage (Q_3 and Q_4) and a Wilson current-mirror load (Q_5, Q_6, and Q_7).

As an example, for $\beta_P = 50$, $V_{OS} = -1$ mV. To reduce V_{OS}, an improved current mirror such as the Wilson circuit studied in Section 7.5.3 should be used. Such a circuit provides the added advantage of increased output resistance and hence voltage gain. However, to realize the full advantage of the higher output resistance of the active load, the output resistance of the differential pair should be raised by utilizing a cascode stage. Figure 8.40 shows such an arrangement: A folded cascode stage formed by *pnp* transistors Q_3 and Q_4 is utilized to raise the output resistance looking into the collector of Q_4 to $\beta_4 r_{o4}$. A Wilson mirror formed by transistors Q_5, Q_6, and Q_7 is used to implement the active load. From Section 7.5.3 we know that the output resistance of the Wilson mirror (i.e., looking into the collector of Q_5) is $\beta_5(r_{o5}/2)$. Thus the output resistance of the amplifier is given by

$$R_o = \left[\beta_4 r_{o4} \parallel \beta_5 \frac{r_{o5}}{2}\right] \tag{8.172}$$

The transconductance G_m remains equal to g_m of Q_1 and Q_2. Thus the differential voltage gain becomes

$$A_d = g_m\left[\beta_4 r_{o4} \parallel \beta_5 \frac{r_{o5}}{2}\right] \tag{8.173}$$

which can be very large. Further examples of improved-performance differential amplifiers will be studied in Chapter 12.

EXERCISE

8.18 Find G_m and R_{o4}, R_{o5}, R_o, and A_d for the differential amplifier in Fig. 8.40 under the following conditions: $I = 1$ mA, $\beta_P = 50$, $\beta_N = 100$, and $V_A = 100$ V.
Ans. 20 mA/V; 10 MΩ; 10 MΩ; 5 MΩ; 10^5 V/V or 100 dB

8.6 Multistage Amplifiers

Practical transistor amplifiers usually consist of a number of stages connected in cascade. In addition to providing gain, the first (or input) stage is usually required to provide a high input resistance in order to avoid loss of signal level when the amplifier is fed from a high-resistance source. In a differential amplifier the input stage must also provide large common-mode rejection. The function of the middle stages of an amplifier cascade is to provide the bulk of the voltage gain. In addition, the middle stages provide such other functions as the conversion of the signal from differential mode to single-ended mode (unless, of course, the amplifier output also is differential) and the shifting of the dc level of the signal in order to allow the output signal to swing both positive and negative. These two functions and others will be illustrated later in this section and in greater detail in Chapter 12.

Finally, the main function of the last (or output) stage of an amplifier is to provide a low output resistance in order to avoid loss of gain when a low-valued load resistance is connected to the amplifier. Also, the output stage should be able to supply the current required by the load in an efficient manner—that is, without dissipating an unduly large amount of power in the output transistors. We have already studied one type of amplifier configuration suitable for implementing output stages, namely, the source follower and the emitter follower. It will be shown in Chapter 11 that the source and emitter followers are not optimum from the point of view of power efficiency and that other, more appropriate circuit configurations exist for output stages that are required to supply large amounts of output power. In fact, we will encounter some such output stages in the op-amp circuit examples studied in Chapter 12.

To illustrate the circuit structure and the method of analysis of multistage amplifiers, we will present two examples: a two-stage CMOS op amp and a four-stage bipolar op amp.

8.6.1 A Two-Stage CMOS Op Amp

Figure 8.41 shows a popular structure for CMOS op amps known as the **two-stage configuration**. The circuit utilizes two power supplies, which can range from ±2.5 V for the 0.5-μm technology down to ±0.9 V for the 0.18-μm technology. A reference bias current I_{REF} is generated either externally or using on-chip circuits. One such circuit will be discussed shortly. The current mirror formed by Q_8 and Q_5 supplies the differential pair $Q_1 - Q_2$ with bias current. The W/L ratio of Q_5 is selected to yield the desired value for the input-stage bias current I (or $I/2$ for each of Q_1 and Q_2). The input differential pair is actively loaded with the current mirror formed by Q_3 and Q_4. Thus the input stage is identical to that studied in Section 8.5

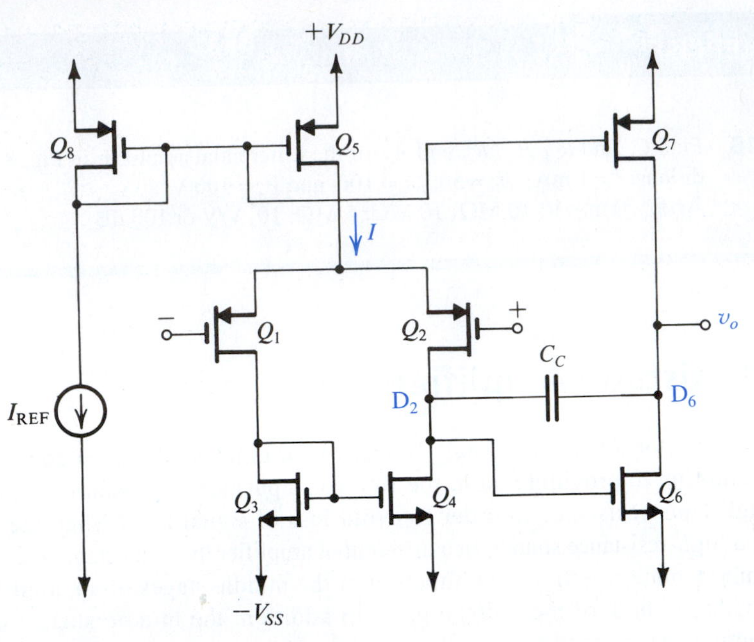

Figure 8.41 Two-stage CMOS op-amp configuration.

(except that here the differential pair is implemented with PMOS transistors and the current mirror with NMOS).

The second stage consists of Q_6, which is a common-source amplifier loaded with the current-source transistor Q_7. A capacitor C_C is included in the negative-feedback path of the second stage. Its function will be explained in Chapter 9, when we study the frequency response of amplifiers.

A striking feature of the circuit in Fig. 8.41 is that it does *not* have a low-output-resistance stage. In fact, the output resistance of the circuit is equal to $(r_{o6} \| r_{o7})$ and is thus rather high. This circuit, therefore, is not suitable for driving low-impedance loads. Nevertheless, the circuit is very popular and is used frequently for implementing op amps in VLSI circuits, where the op amp needs to drive only a small capacitive load, for example, in switched-capacitor circuits (Chapter 17). The simplicity of the circuit results in an op amp of reasonably good quality realized in a very small chip area.

Voltage Gain The voltage gain of the first stage was found in Section 8.5 to be given by

$$A_1 = -g_{m1}(r_{o2} \| r_{o4}) \tag{8.174}$$

where g_{m1} is the transconductance of each of the transistors of the first stage, that is, Q_1 and Q_2.

The second stage is current-source-loaded, common-source amplifier whose voltage gain is given by

$$A_2 = -g_{m6}(r_{o6} \| r_{o7}) \tag{8.175}$$

The dc open-loop gain of the op amp is the product of A_1 and A_2.

Example 8.5

Consider the circuit in Fig. 8.41 with the following device geometries (in μm).

Transistor	Q_1	Q_2	Q_3	Q_4	Q_5	Q_6	Q_7	Q_8
W/L	20/0.8	20/0.8	5/0.8	5/0.8	40/0.8	10/0.8	40/0.8	40/0.8

Let $I_{REF} = 90\ \mu A$, $V_{tn} = 0.7$ V, $V_{tp} = -0.8$ V, $\mu_n C_{ox} = 160\ \mu A/V^2$, $\mu_p C_{ox} = 40\ \mu A/V^2$, $|V_A|$ (for all devices) = 10 V, $V_{DD} = V_{SS} = 2.5$ V. For all devices, evaluate I_D, $|V_{OV}|$, $|V_{GS}|$, g_m, and r_o. Also find A_1, A_2, the dc open-loop voltage gain, the input common-mode range, and the output voltage range. Neglect the effect of V_A on bias current.

Solution

Refer to Fig. 8.41. Since Q_8 and Q_5 are matched, $I = I_{REF}$. Thus Q_1, Q_2, Q_3, and Q_4 each conducts a current equal to $I/2 = 45\ \mu A$. Since Q_7 is matched to Q_5 and Q_8, the current in Q_7 is equal to $I_{REF} = 90\ \mu A$. Finally, Q_6 conducts an equal current of 90 μA.

With I_D of each device known, we use

$$I_D = \frac{1}{2}(\mu C_{ox})(W/L)V_{OV}^2$$

to determine $|V_{OV}|$ for each transistor. Then we find $|V_{GS}|$ from $|V_{GS}| = |V_t| + |V_{OV}|$. The results are given in Table 8.1.

The transconductance of each device is determined from

$$g_m = 2I_D/|V_{OV}|$$

The value of r_o is determined from

$$r_o = |V_A|/I_D$$

The resulting values of g_m and r_o are given in Table 8.1.

The voltage gain of the first stage is determined from

$$A_1 = -g_{m1}(r_{o2} \| r_{o4})$$
$$= -0.3(222 \| 222) = -33.3 \text{ V/V}$$

The voltage gain of the second stage is determined from

$$A_2 = -g_{m6}(r_{o6} \| r_{o7})$$
$$= -0.6(111 \| 111) = -33.3 \text{ V/V}$$

Table 8.1

	Q_1	Q_2	Q_3	Q_4	Q_5	Q_6	Q_7	Q_8
I_D (μA)	45	45	45	45	90	90	90	90
$\|V_{OV}\|$ (V)	0.3	0.3	0.3	0.3	0.3	0.3	0.3	0.3
$\|V_{GS}\|$ (V)	1.1	1.1	1	1	1.1	1	1.1	1.1
g_m (mA/V)	0.3	0.3	0.3	0.3	0.6	0.6	0.6	0.6
r_o (kΩ)	222	222	222	222	111	111	111	111

Example 8.5 *continued*

Thus the overall dc open-loop gain is

$$A_0 = A_1 A_2 = (-33.3)\times(-33.3) = 1109 \text{ V/V}$$

or

$$20\log 1109 = 61 \text{ dB}$$

The lower limit of the input common-mode range is the value of input voltage at which Q_1 and Q_2 leave the saturation region. This occurs when the input voltage falls below the voltage at the drain of Q_1 by $|V_{tp}|$ volts. Since the drain of Q_1 is at $-2.5 + 1 = -1.5$ V, then the lower limit of the input common-mode range is -2.3 V.

The upper limit of the input common-mode range is the value of input voltage at which Q_5 leaves the saturation region. Since for Q_5 to operate in saturation the voltage across it (i.e., V_{SD5}) should at least be equal to the overdrive voltage at which it is operating (i.e., 0.3 V), the highest voltage permitted at the drain of Q_5 should be +2.2 V. It follows that the highest value of v_{ICM} should be

$$v_{ICM\max} = 2.2 - 1.1 = 1.1 \text{ V}$$

The highest allowable output voltage is the value at which Q_7 leaves the saturation region, which is $V_{DD} - |V_{OV7}| = 2.5 - 0.3 = 2.2$ V. The lowest allowable output voltage is the value at which Q_6 leaves saturation, which is $-V_{SS} + V_{OV6} = -2.5 + 0.3 = -2.2$ V. Thus, the output voltage range is -2.2 V to $+2.2$ V.

Input Offset Voltage The device mismatches inevitably present in the input stage give rise to an input offset voltage. The components of this input offset voltage can be calculated using the methods developed in Section 8.4.1. Because device mismatches are random, the resulting offset voltage is referred to as **random offset**. This is to distinguish it from another type of input offset voltage that can be present even if all appropriate devices are perfectly matched. This predictable or **systematic offset** can be minimized by careful design. Although it occurs also in BJT op amps, and we have encountered it in Section 8.5.5, it is usually much more pronounced in CMOS op amps because their gain-per-stage is rather low.

To see how systematic offset can occur in the circuit of Fig. 8.41, let the two input terminals be grounded. If the input stage is perfectly balanced, then the voltage appearing at the drain of Q_4 will be equal to that at the drain of Q_3, which is $(-V_{SS} + V_{GS4})$. Now this is also the voltage that is fed to the gate of Q_6. In other words, a voltage equal to V_{GS4} appears between gate and source of Q_6. Thus the drain current of Q_6, I_6, will be related to the drain current of Q_4, which is equal to $I/2$, by the relationship

$$I_6 = \frac{(W/L)_6}{(W/L)_4}(I/2) \tag{8.176}$$

In order for no offset voltage to appear at the output, this current must be exactly equal to the current supplied by Q_7. The latter current is related to the current I of the parallel transistor Q_5 by

$$I_7 = \frac{(W/L)_7}{(W/L)_5} I \tag{8.177}$$

Now, the condition for making $I_6 = I_7$ can be found from Eqs. (8.176) and (8.177) as

$$\frac{(W/L)_6}{(W/L)_4} = 2\frac{(W/L)_7}{(W/L)_5} \tag{8.178}$$

If this condition is not met, a systematic offset will result. From the specification of the device geometries in Example 8.5, we can verify that condition (8.178) is satisfied, and, therefore, the op amp analyzed in that example should not exhibit a systematic input offset voltage.

EXERCISE

8.19 Consider the CMOS op amp of Fig. 8.41 when fabricated in a 0.8-μm CMOS technology for which $\mu_n C_{ox} = 3\mu_p C_{ox} = 90\ \mu A/V^2$, $|V_t| = 0.8$ V, and $V_{DD} = V_{SS} = 2.5$ V. For a particular design, $I = 100\ \mu A$, $(W/L)_1 = (W/L)_2 = (W/L)_5 = 200$, and $(W/L)_3 = (W/L)_4 = 100$.
(a) Find the (W/L) ratios of Q_6 and Q_7 so that $I_6 = 100\ \mu A$.
(b) Find the overdrive voltage, $|V_{OV}|$, at which each of Q_1, Q_2, and Q_6 is operating.
(c) Find g_m for Q_1, Q_2, and Q_6.
(d) If $|V_A| = 10$ V, find r_{o2}, r_{o4}, r_{o6}, and r_{o7}.
(e) Find the voltage gains A_1 and A_2, and the overall gain A.

Ans. (a) $(W/L)_6 = (W/L)_7 = 200$; (b) 0.129 V, 0.129 V, 0.105 V; (c) 0.775 mA/V, 0.775 mA/V, 1.90 mA/V; (d) 200 kΩ, 200 kΩ, 100 kΩ, 100 kΩ; (e) –77.5 V/V, –95 V/V, 7363 V/V

A Bias Circuit That Stabilizes g_m We conclude this section by presenting a bias circuit for the two-stage CMOS op amp. The circuit presented has the interesting and useful property of providing a bias current whose value is independent of both the supply voltage and the MOSFET threshold voltage. Furthermore, the transconductances of the transistors biased by this circuit have values that are determined only by a single resistor and the device dimensions.

The bias circuit is shown in Fig. 8.42. It consists of two deliberately mismatched transistors, Q_{12} and Q_{13}, with Q_{12} usually about four times wider than Q_{13}. A resistor R_B is connected in series with the source of Q_{12}. Since, as will be shown, R_B determines both the bias current I_B and the transconductance g_{m12}, its value should be accurate and stable; in most applications, R_B would be an off-chip resistor. In order to minimize the channel-length modulation effect on Q_{12}, we include a cascode transistor Q_{10} and a matched diode-connected transistor Q_{11} to provide a bias voltage for Q_{10}. Finally, a *p*-channel current mirror formed by a pair of matched devices, Q_8 and Q_9, both replicates the current I_B back to Q_{11} and Q_{13}, and provides a bias line for Q_5 and Q_7 of the CMOS op-amp circuit of Fig. 8.41.[5]

The circuit operates as follows: The current mirror (Q_8, Q_9) causes Q_{13} to conduct a current equal to that in Q_{12}, that is, I_B. Thus,

$$I_B = \frac{1}{2}\mu_n C_{ox}\left(\frac{W}{L}\right)_{12}(V_{GS12} - V_t)^2 \tag{8.179}$$

and,

$$I_B = \frac{1}{2}\mu_n C_{ox}\left(\frac{W}{L}\right)_{13}(V_{GS13} - V_t)^2 \tag{8.180}$$

From the circuit, we see that the gate-source voltages of Q_{12} and Q_{13} are related by

$$V_{GS13} = V_{GS12} + I_B R_B$$

[5]We denote the bias current of this circuit by I_B. If this circuit is utilized to bias the CMOS op amp of Fig. 8.41, then I_B becomes the reference current I_{REF}.

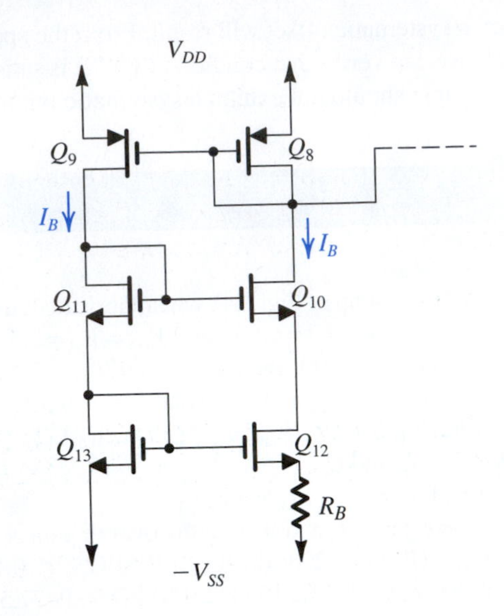

Figure 8.42 Bias circuit for the CMOS op amp.

Subtracting V_t from both sides of this equation and using Eqs. (8.179) and (8.180) to replace $(V_{GS12} - V_t)$ and $(V_{GS13} - V_t)$ results in

$$\sqrt{\frac{2I_B}{\mu_n C_{ox}(W/L)_{13}}} = \sqrt{\frac{2I_B}{\mu_n C_{ox}(W/L)_{12}}} + I_B R_B \tag{8.181}$$

This equation can be rearranged to yield

$$I_B = \frac{2}{\mu_n C_{ox}(W/L)_{12} R_B^2}\left(\sqrt{\frac{(W/L)_{12}}{(W/L)_{13}}} - 1\right)^2 \tag{8.182}$$

from which we observe that I_B is determined by the dimensions of Q_{12} and the value of R_B and by the ratio of the dimensions of Q_{12} and Q_{13}. Furthermore, Eq. (8.182) can be rearranged to the form

$$R_B = \frac{2}{\sqrt{2\mu_n C_{ox}(W/L)_{12} I_B}}\left(\sqrt{\frac{(W/L)_{12}}{(W/L)_{13}}} - 1\right)$$

in which we recognize the factor $\sqrt{2\mu_n C_{ox}(W/L)_{12} I_B}$ as g_{m12}; thus,

$$g_{m12} = \frac{2}{R_B}\left(\sqrt{\frac{(W/L)_{12}}{(W/L)_{13}}} - 1\right) \tag{8.183}$$

This is a very interesting result: g_{m12} is determined solely by the value of R_B and the ratio of the dimensions of Q_{12} and Q_{13}. Furthermore, since g_m of a MOSFET is proportional to $\sqrt{I_D(W/L)}$, each transistor biased by the circuit of Fig. 8.42; that is, each transistor whose bias current is derived from I_B will have a g_m value that is a multiple of g_{m12}. Specifically, the ith n-channel MOSFET will have

$$g_{mi} = g_{m12}\sqrt{\frac{I_{Di}(W/L)_i}{I_B(W/L)_{12}}}$$

and the *i*th *p*-channel device will have

$$g_{mi} = g_{m12}\sqrt{\frac{\mu_p I_{Di}(W/L)_i}{\mu_n I_B (W/L)_{12}}}$$

Finally, it should be noted that the bias circuit of Fig. 8.42 employs **positive feedback**, and thus care should be exercised in its design to avoid unstable performance. Instability is avoided by making Q_{12} wider than Q_{13}, as has already been pointed out. Nevertheless, some form of instability may still occur; in fact, the circuit can operate in a stable state in which all currents are zero. To get it out of this state, current needs to be injected into one of its nodes, to "kick start" its operation. Feedback and stability will be studied in Chapter 10.

EXERCISES

8.20 Consider the bias circuit of Fig. 8.42 for the case of $(W/L)_8 = (W/L)_9 = (W/L)_{10} = (W/L)_{11} = (W/L)_{13} = 20$ and $(W/L)_{12} = 80$. The circuit is fabricated in a process technology for which $\mu_n C_{ox} = 90\ \mu\text{A/V}^2$. Find the value of R_B that results in a bias current $I_B = 10\ \mu\text{A}$. Also, find the transconductance g_{m12}.
Ans. 5.27 kΩ; 0.379 mA/V

D8.21 Design the bias circuit of Fig. 8.42 to operate with the CMOS op amp of Example 8.5. Use Q_8 and Q_9 as identical devices with Q_8 having the dimensions given in Example 8.5. Transistors Q_{10}, Q_{11}, and Q_{13} are to be identical, with the same g_m as Q_8 and Q_9. Transistor Q_{12} is to be four times as wide as Q_{13}. Find the required value of R_B. What is the voltage drop across R_B? Also give the values of the dc voltages at the gates of Q_{12}, Q_{10}, and Q_8.
Ans. 1.67 kΩ; 150 mV; –1.5 V; –0.5 V; +1.4 V

8.6.2 A Bipolar Op Amp

Our second example of multistage amplifiers is the four-stage bipolar op amp shown in Fig. 8.43. The circuit consists of four stages. The **differential-in**, **differential-out** input stage consists of transistors Q_1 and Q_2, which are biased by current source Q_3. The second stage is also a differential-input amplifier, but its output is taken single-endedly at the collector of Q_5. This stage is formed by Q_4 and Q_5, which are biased by the current source Q_6. Note that the conversion from differential to single-ended as performed by the second stage results in a loss of gain by a factor of 2. In the more elaborate method for accomplishing this conversion studied in Section 8.5, a current mirror was used as an active load.

In addition to providing some voltage gain, the third stage, consisting of the *pnp* transistor Q_7, provides the essential function of *shifting the dc level* of the signal. Thus, while the signal at the collector of Q_5 is not allowed to swing below the voltage at the base of Q_5 (+10 V), the signal at the collector of Q_7 can swing negatively (and positively, of course). From our study of op amps in Chapter 2, we know that the output terminal of the op amp should be capable of both positive and negative voltage swings. Therefore every op-amp circuit includes a **level-shifting** arrangement. Although the use of the complementary *pnp* transistor provides a simple solution to the level-shifting problem, other forms of level shifter exist, one of which will be discussed in Chapter 12. Furthermore, note that level shifting is

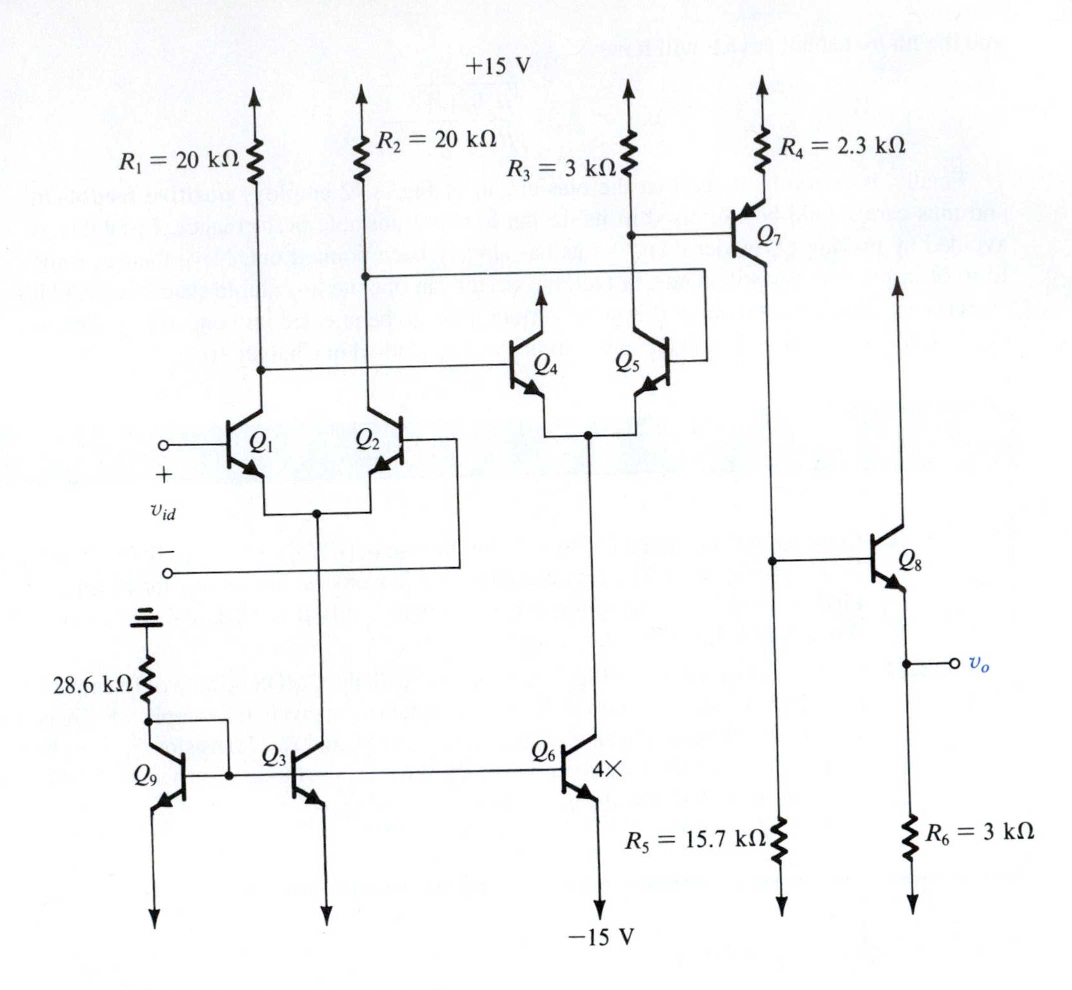

Figure 8.43 A four-stage bipolar op amp.

accomplished in the CMOS op amp we have been studying by using complementary devices for the two stages: that is, *p*-channel for the first stage and *n*-channel for the second stage.

The output stage of the op amp consists of emitter follower Q_8. As we know from our study of op amps in Chapter 2, ideally the output operates around zero volts. This and other features of the BJT op amp will be illustrated in Example 8.6.

Example 8.6

In this example, we analyze the dc bias of the bipolar op-amp circuit of Fig. 8.43. Toward that end, Fig. 8.44 shows the circuit with the two input terminals connected to ground.

(a) Perform an approximate dc analysis (assuming $\beta \gg 1$, $|V_{BE}| \simeq 0.7$ V, and neglecting the Early effect) to calculate the dc currents and voltages everywhere in the circuit. Note that Q_6 has four times the area of each of Q_9 and Q_3.

(b) Calculate the quiescent power dissipation in this circuit.

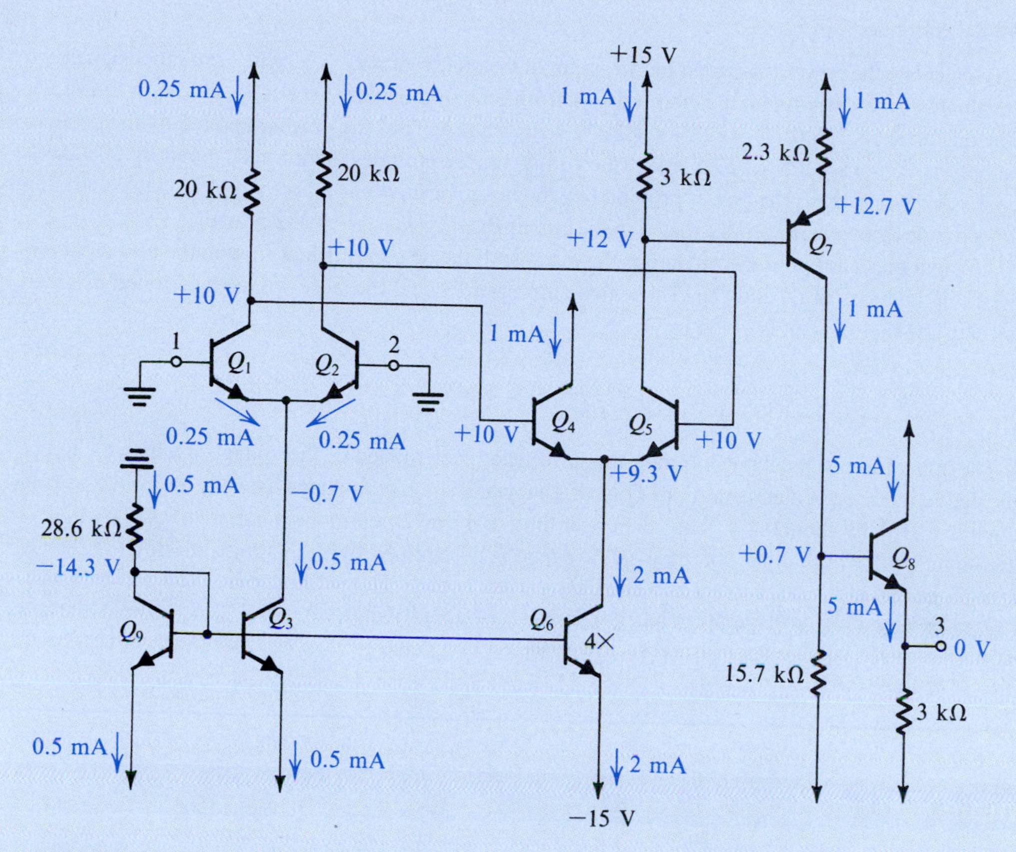

Figure 8.44 Circuit for Example 8.6.

(c) If transistors Q_1 and Q_2 have $\beta = 100$, calculate the input bias current of the op amp.
(d) What is the input common-mode range of this op amp?

Solution

(a) The values of all dc currents and voltages are indicated on the circuit diagram. These values were calculated by ignoring the base current of every transistor—that is, by assuming β to be very high. The analysis starts by determining the current through the diode-connected transistor Q_9 to be 0.5 mA. Then we see that transistor Q_3 conducts 0.5 mA and transistor Q_6 conducts 2 mA. The current-source transistor Q_3 feeds the differential pair (Q_1, Q_2) with 0.5 mA. Thus each of Q_1 and Q_2 will be biased at 0.25 mA. The collectors of Q_1 and Q_2 will be at $[+15 - 0.25 \times 20] = +10$ V.

Proceeding to the second differential stage formed by Q_4 and Q_5, we find the voltage at their emitters to be $[+10 - 0.7] = 9.3$ V. This differential pair is biased by the current-source transistor Q_6, which supplies a current of 2 mA; thus Q_4 and Q_5 will each be biased at 1 mA. We can now calculate the voltage at the collector of Q_5 as $[+15 - 1 \times 3] = +12$ V. This will cause the voltage at the emitter of the *pnp* transistor Q_7 to be +12.7 V, and the emitter current of Q_7 will be $(+15 - 12.7)/2.3 = 1$ mA.

The collector current of Q_7, 1 mA, causes the voltage at the collector to be $[-15 + 1 \times 15.7] = +0.7$ V. The emitter of Q_8 will be 0.7 V below the base; thus output terminal 3 will be at 0 V. Finally, the emitter current of Q_8 can be calculated to be $[0 - (-15)]/3 = 5$ mA.

Example 8.6 *continued*

(b) To calculate the power dissipated in the circuit in the quiescent state (i.e., with zero input signal) we simply evaluate the dc current that the circuit draws from each of the two power supplies. From the +15 V supply the dc current is $I^+ = 0.25 + 0.25 + 1 + 1 + 1 + 5 = 8.5$ mA. Thus the power supplied by the positive power supply is $P^+ = 15 \times 8.5 = 127.5$ mW. The –15-V supply provides a current I^- given by $I^- = 0.5 + 0.5 + 2 + 1 + 5 = 9$ mA. Thus the power provided by the negative supply is $P^- = 15 \times 9 = 135$ mW. Adding P^+ and P^- provides the total power dissipated in the circuit P_D: $P_D = P^+ + P^- = 262.5$ mW.

(c) The input bias current of the op amp is the average of the dc currents that flow in the two input terminals (i.e., in the bases of Q_1 and Q_2). These two currents are equal (because we have assumed matched devices); thus the bias current is given by

$$I_B = \frac{I_{E1}}{\beta + 1} \simeq 2.5\ \mu\text{A}$$

(d) The upper limit on the input common-mode voltage is determined by the voltage at which Q_1 and Q_2 leave the active mode and enter saturation. This will happen if the input voltage exceeds the collector voltage, which is +10 V, by about 0.4 V. Thus the upper limit of the common-mode range is +10.4 V.

The lower limit of the input common-mode range is determined by the voltage at which Q_3 leaves the active mode and thus ceases to act as a constant-current source. This will happen if the collector voltage of Q_3 goes below the voltage at its base, which is –14.3 V, by more than 0.4 V. It follows that the input common-mode voltage should not go lower than –14.7 + 0.7 = –14 V. Thus the common-mode range is –14 V to +10.4 V.

Example 8.7

Use the dc bias quantities evaluated in Example 8.6 to analyze the circuit in Fig. 8.43, to determine the input resistance, the voltage gain, and the output resistance.

Solution

The input differential resistance R_{id} is given by

$$R_{id} = r_{\pi 1} + r_{\pi 2}$$

Since Q_1 and Q_2 are each operating at an emitter current of 0.25 mA, it follows that

$$r_{e1} = r_{e2} = \frac{25}{0.25} = 100\ \Omega$$

Assume $\beta = 100$; then

$$r_{\pi 1} = r_{\pi 2} = 101 \times 100 = 10.1\ \text{k}\Omega$$

Thus,

$$R_{id} = 20.2\ \text{k}\Omega$$

To evaluate the gain of the first stage, we first find the input resistance of the second stage, R_{i2},

$$R_{i2} = r_{\pi 4} + r_{\pi 5}$$

Q_4 and Q_5 are each operating at an emitter current of 1 mA; thus

$$r_{e4} = r_{e5} = 25\ \Omega$$

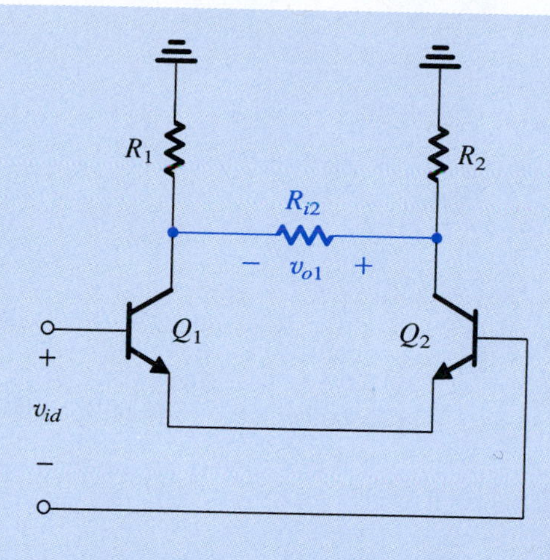

Figure 8.45 Equivalent circuit for calculating the gain of the input stage of the amplifier in Fig. 8.43.

$$r_{\pi 4} = r_{\pi 5} = 101 \times 25 = 2.525 \text{ k}\Omega$$

Thus $R_{i2} = 5.05$ kΩ. This resistance appears between the collectors of Q_1 and Q_2, as shown in Fig. 8.45. Thus the gain of the first stage will be

$$A_1 \equiv \frac{v_{o1}}{v_{id}} \simeq \frac{\text{Total resistance in collector circuit}}{\text{Total resistance in emitter circuit}}$$

$$= \frac{R_{i2} \parallel (R_1 + R_2)}{r_{e1} + r_{e2}}$$

$$= \frac{5.05 \text{ k}\Omega \parallel 40 \text{ k}\Omega}{200 \text{ }\Omega} = 22.4 \text{ V/V}$$

Figure 8.46 shows an equivalent circuit for calculating the gain of the second stage. As indicated, the input voltage to the second stage is the output voltage of the first stage, v_{o1}. Also shown is the resistance R_{i3}, which is the input resistance of the third stage formed by Q_7. The value of R_{i3} can be found by

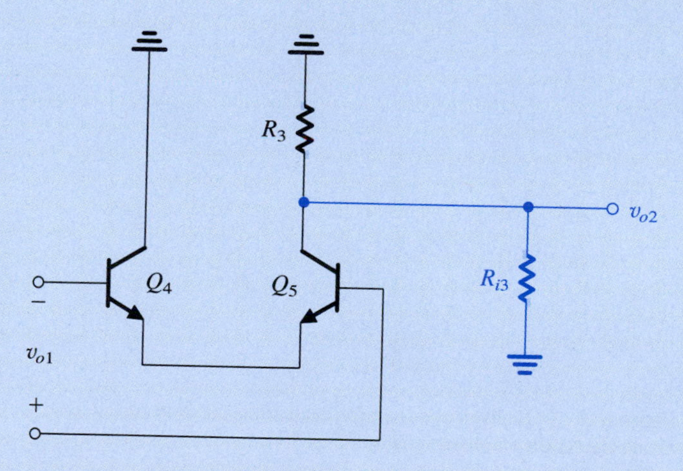

Figure 8.46 Equivalent circuit for calculating the gain of the second stage of the amplifier in Fig. 8.43.

Example 8.7 *continued*

multiplying the total resistance in the emitter of Q_7 by $(\beta + 1)$:

$$R_{i3} = (\beta + 1)(R_4 + r_{e7})$$

Since Q_7 is operating at an emitter current of 1 mA,

$$r_{e7} = \frac{25}{1} = 25\ \Omega$$

$$R_{i3} = 101 \times 2.325 = 234.8\ \text{k}\Omega$$

We can now find the gain A_2 of the second stage as the ratio of the total resistance in the collector circuit to the total resistance in the emitter circuit:

$$A_2 \equiv \frac{v_{o2}}{v_{o1}} \simeq -\frac{R_3 \,\|\, R_{i3}}{r_{e4} + r_{e5}}$$

$$= -\frac{3\ \text{k}\Omega \,\|\, 234.8\ \text{k}\Omega}{50\ \Omega} = -59.2\ \text{V/V}$$

To obtain the gain of the third stage we refer to the equivalent circuit shown in Fig. 8.47, where R_{i4} is the input resistance of the output stage formed by Q_8. Using the resistance-reflection rule, we calculate the value of R_{i4} as

$$R_{i4} = (\beta + 1)(r_{e8} + R_6)$$

where

$$r_{e8} = \frac{25}{5} = 5\ \Omega$$

$$R_{i4} = 101(5 + 3000) = 303.5\ \text{k}\Omega$$

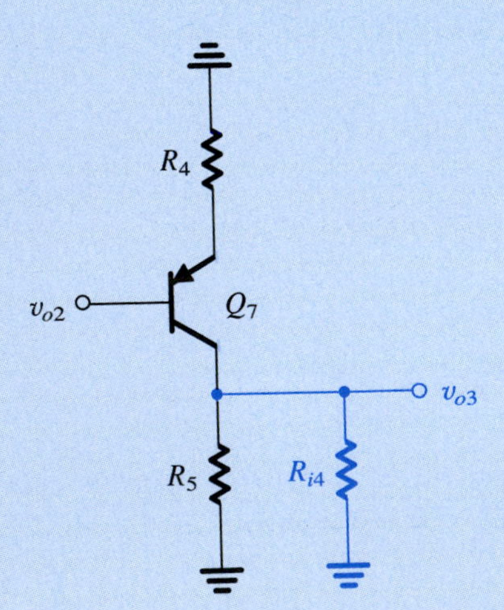

Figure 8.47 Equivalent circuit for evaluating the gain of the third stage in the amplifier circuit of Fig. 8.43.

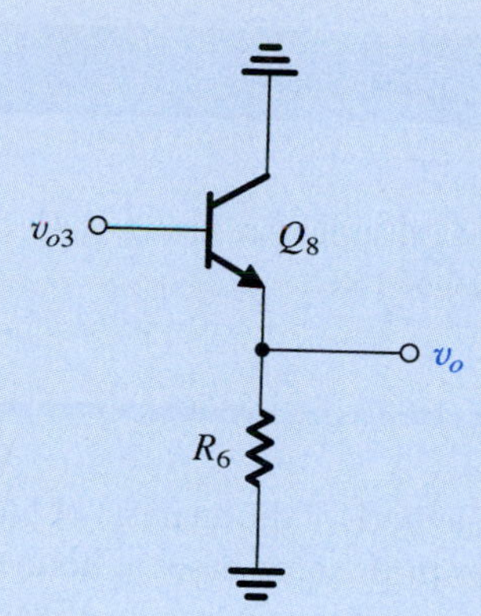

Figure 8.48 Equivalent circuit of the output stage of the amplifier circuit of Fig. 8.43.

The gain of the third stage is given by

$$A_3 \equiv \frac{v_{o3}}{v_{o2}} \simeq -\frac{R_5 \| R_{i4}}{r_{e7}+R_4}$$

$$= -\frac{15.7\ \text{k}\Omega \,\|\, 303.5\ \text{k}\Omega}{2.325\ \text{k}\Omega} = -6.42\ \text{V/V}$$

Finally, to obtain the gain A_4 of the output stage we refer to the equivalent circuit in Fig. 8.48 and write

$$A_4 \equiv \frac{v_o}{v_{o3}} = \frac{R_6}{R_6+r_{e8}}$$

$$= \frac{3000}{3000+5} = 0.998 \simeq 1$$

The overall voltage gain of the amplifier can then be obtained as follows:

$$\frac{v_o}{v_{id}} = A_1A_2A_3A_4 = 8513\ \text{V/V}$$

or 78.6 dB.

To obtain the output resistance R_o we "grab hold" of the output terminal in Fig. 8.43 and look back into the circuit. By inspection we find

$$R_o = R_6 \| [r_{e8} + R_5/(\beta+1)]$$

which gives

$$R_o = 152\ \Omega$$

EXERCISE

8.22 Use the results of Example 8.7 to calculate the overall voltage gain of the amplifier in Fig. 8.43 when it is connected to a source having a resistance of 10 kΩ and a load of 1 kΩ.
Ans. 4943 V/V

Analysis Using Current Gains There is an alternative method for the analysis of bipolar multistage amplifiers that can be somewhat easier to perform in some cases. The method makes use of current gains or more appropriately current-transmission factors. In effect, one traces the transmission of the signal current throughout the amplifier cascade, evaluating all the current transmission factors in turn. We shall illustrate the method by using it to analyze the amplifier circuit of the preceding example.

Figure 8.49 shows the amplifier circuit prepared for small-signal analysis. We have indicated on the circuit diagram the signal currents through all the circuit branches. Also indicated are the input resistances of all four stages of the amplifier. These should be evaluated before commencing the following analysis.

The purpose of the analysis is to determine the overall voltage gain (v_o/v_{id}). Toward that end, we express v_o in terms of the signal current in the emitter of Q_8, i_{e8}, and v_{id} in terms of the input signal current i_i, as follows:

$$v_o = R_6 i_{e8}$$

$$v_{id} = R_{i1} i_i$$

Thus, the voltage gain can be expressed in terms of the current gain (i_{e8}/i_i) as

$$\frac{v_o}{v_{id}} = \frac{R_6}{R_{i1}} \frac{i_{e8}}{i_i}$$

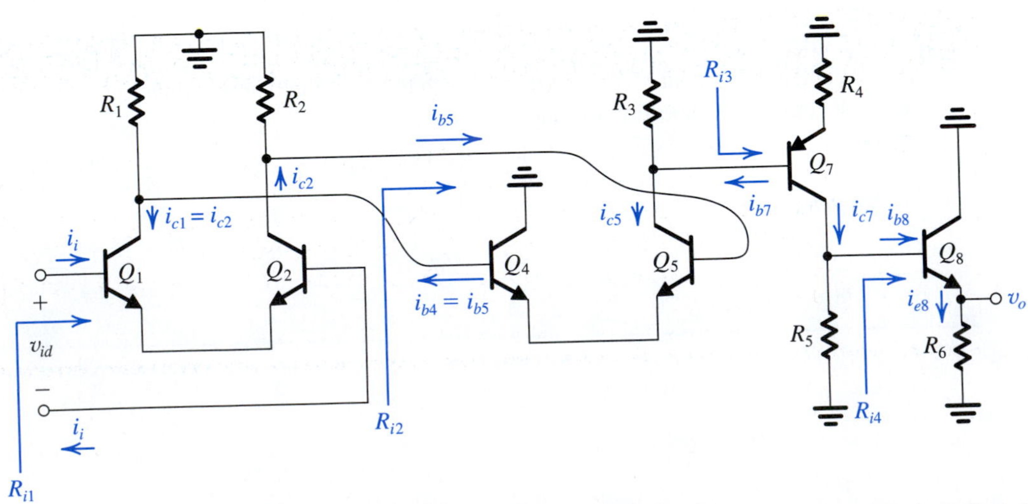

Figure 8.49 The circuit of the multistage amplifier of Fig. 8.43 prepared for small-signal analysis. Indicated are the signal currents throughout the amplifier and the input resistances of the four stages.

Next, we expand the current gain (i_{e8}/i_i) in terms of the signal currents throughout the circuit as follows:

$$\frac{i_{e8}}{i_i} = \frac{i_{e8}}{i_{b8}} \times \frac{i_{b8}}{i_{c7}} \times \frac{i_{c7}}{i_{b7}} \times \frac{i_{b7}}{i_{c5}} \times \frac{i_{c5}}{i_{b5}} \times \frac{i_{b5}}{i_{c2}} \times \frac{i_{c2}}{i_i}$$

Each of the current-transmission factors on the right-hand side is either the current gain of a transistor or the ratio of a current divider. Thus, reference to Fig. 8.49 enables us to find these factors by inspection:

$$\frac{i_{e8}}{i_{b8}} = \beta_8 + 1$$

$$\frac{i_{b8}}{i_{c7}} = \frac{R_5}{R_5 + R_{i4}}$$

$$\frac{i_{c7}}{i_{b7}} = \beta_7$$

$$\frac{i_{b7}}{i_{c5}} = \frac{R_3}{R_3 + R_{i3}}$$

$$\frac{i_{c5}}{i_{b5}} = \beta_5$$

$$\frac{i_{b5}}{i_{c2}} = \frac{(R_1 + R_2)}{(R_1 + R_2) + R_{i2}}$$

$$\frac{i_{c2}}{i_i} = \beta_2$$

These ratios can be easily evaluated and their values used to determine the voltage gain.

With a little practice, it is possible to carry out such an analysis very quickly, forgoing explicitly labeling the signal currents on the circuit diagram. One simply "walks through" the circuit, from input to output, or vice versa, determining the current-transmission factors one at a time, in a chainlike fashion.

EXERCISE

8.23 Use the values of input resistance found in Example 8.7 to evaluate the seven current-transmission factors and hence the overall current gain and voltage gain.

Ans. The current-transmission factors in the order of their listing are 101, 0.0492, 100, 0.0126, 100, 0.8879, 100 A/A; the overall current gain is 55993 A/A; the voltage gain is 8256 V/V. This value differs slightly from that found in Example 8.7, because of the various approximations made in the example (e.g., $\alpha \simeq 1$).

Summary

- The differential-pair or differential-amplifier configuration is the most widely used building block in analog IC design. The input stage of every op amp is a differential amplifier.

- There are two reasons for preferring differential to single-ended amplifiers: Differential amplifiers are insensitive to interference, and they do not need bypass and coupling capacitors.

- For a MOS (bipolar) pair biased by a current source I, each device operates at a drain (collector, assuming $\alpha = 1$) current of $I/2$ and a corresponding overdrive voltage V_{OV} (no analog in bipolar). Each device has $g_m = I/V_{OV}$ ($\alpha I/2V_T$, for bipolar) and $r_o = |V_A|/(I/2)$.

- With the two input terminals connected to a suitable dc voltage V_{CM}, the bias current I of a perfectly symmetrical differential pair divides equally between the two transistors of the pair, resulting in a zero voltage difference between the two drains (collectors). To steer the current completely to one side of the pair, a difference input voltage v_{id} of at least $\sqrt{2}V_{OV}$ ($4V_T$ for bipolar) is needed.

- Superimposing a differential input signal v_{id} on the dc common-mode input voltage V_{CM} such that $v_{I1} = V_{CM} + v_{id}/2$ and $v_{I2} = V_{CM} - v_{id}/2$ causes a virtual signal ground to appear on the common-source (common-emitter) connection. In response to v_{id}, the current in Q_1 increases by $g_m v_{id}/2$ and the current in Q_2 decreases by $g_m v_{id}/2$. Thus, voltage signals of $\pm g_m(R_D \| r_o)v_{id}/2$ develop at the two drains (collectors, with R_D replaced by R_C). If the output voltage is taken single-endedly, that is, between one of the drains (collectors) and ground, a differential gain of $\frac{1}{2}g_m(R_D \| r_o)$ is realized. When the output is taken differentially, that is, between the two drains (collectors), the differential gain realized is twice as large: $g_m(R_D \| r_o)$.

- The analysis of a differential amplifier to determine differential gain, differential input resistance, frequency response of differential gain, and so on is facilitated by employing the differential half-circuit, which is a common-source (common-emitter) transistor biased at $I/2$.

- An input common-mode signal v_{icm} gives rise to drain (collector) voltage signals that are ideally equal and given by $-v_{icm}(R_D/2R_{SS})$ $[-v_{icm}(R_C/2R_{EE})$ for the bipolar pair], where R_{SS} (R_{EE}) is the output resistance of the current source that supplies the bias current I. When the output is taken single-endedly, a common-mode gain of magnitude $|A_{cm}| = R_D/2R_{SS}$ ($R_C/2R_{EE}$ for the bipolar case) results. Taking the output differentially results, in the perfectly matched case, in zero A_{cm} (infinite CMRR). Mismatches between the two sides of the pair make A_{cm} finite even when the output is taken differentially: A mismatch ΔR_D causes $|A_{cm}| = (R_D/2R_{SS})(\Delta R_D/R_D)$; a mismatch Δg_m causes $|A_{cm}| = (R_D/2R_{SS})(\Delta g_m/g_m)$. Corresponding expressions apply for the bipolar pair.

- While the input differential resistance R_{id} of the MOS pair is infinite, that for the bipolar pair is only $2r_\pi$ but can be increased to $2(\beta+1)(r_e + R_e)$ by including resistances R_e in the two emitters. The latter action, however, lowers A_d.

- Mismatches between the two sides of a differential pair result in a differential dc output voltage V_O even when the two input terminals are tied together and connected to a dc voltage V_{CM}. This signifies the presence of an input offset voltage $V_{OS} \equiv V_O/A_d$. In a MOS pair there are three main sources for V_{OS}:

$$\Delta R_D \Rightarrow V_{OS} = \frac{V_{OV}}{2}\frac{\Delta R_D}{R_D}$$

$$\Delta(W/L) \Rightarrow V_{OS} = \frac{V_{OV}}{2}\frac{\Delta(W/L)}{W/L}$$

$$\Delta V_t \Rightarrow V_{OS} = \Delta V_t$$

  For the bipolar pair there are two main sources:

$$\Delta R_C \Rightarrow V_{OS} = V_T\frac{\Delta R_C}{R_C}$$

$$\Delta I_S \Rightarrow V_{OS} = V_T\frac{\Delta I_S}{I_S}$$

- A popular circuit in both MOS and bipolar analog ICs is the current-mirror-loaded differential pair. It realizes a high differential gain $A_d = g_m(R_{o\text{ pair}} \| R_{o\text{ mirror}})$ and a low common-mode gain, $|A_{cm}| = \frac{1}{2g_{m3}R_{SS}}$ for the MOS circuit ($r_{o4}/\beta_3 R_{EE}$ for the bipolar circuit), as well as performing the differential-to-single-ended conversion with no loss of gain.

- The CMOS two-stage amplifier studied in Section 8.6.1 is intended for use as part of an IC system and thus is required to drive only small capacitive loads. Therefore it does not have an output stage with a low output resistance.

- A multistage amplifier typically consists of three or more stages: an input stage having a high input resistance, a reasonably high gain, and, if differential, a high CMRR; one or two intermediate stages that realize the bulk of the gain; and an output stage having a low output resistance. In designing and analyzing a multistage amplifier, the loading effect of each stage on the one that precedes it, must be taken into account.

Computer Simulation Problems

SIM Problems identified by this icon are intended to demonstrate the value of using SPICE simulation to verify hand analysis and design, and to investigate important issues such as allowable signal swing and amplifier nonlinear distortion. Instructions to assist in setting up PSpice and Multisim simulations for all the indicated problems can be found in the corresponding files on the CD. Note that if a particular parameter value is not specified in the problem statement, you are to make a reasonable assumption.
* difficult problem; ** more difficult; *** very challenging and/or time-consuming; D: design problem.

Section 8.1: The MOS Differential Pair

8.1 For an NMOS differential pair with a common-mode voltage V_{CM} applied, as shown in Fig. 8.2, let $V_{DD} = V_{SS} = 1.0$ V, $k_n' = 0.4$ mA/V^2, $(W/L)_{1,2} = 12.5$, $V_{tn} = 0.5$ V, $I = 0.2$ mA, $R_D = 10$ kΩ, and neglect channel-length modulation.

(a) Find V_{OV} and V_{GS} for each transistor.
(b) For $V_{CM} = 0$, find V_S, I_{D1}, I_{D2}, V_{D1}, and V_{D2}.
(c) Repeat (b) for $V_{CM} = +0.3$ V.
(d) Repeat (b) for $V_{CM} = -0.1$ V.
(e) What is the highest value of V_{CM} for which Q_1 and Q_2 remain in saturation?
(f) If current source I requires a minimum voltage of 0.2 V to operate properly, what is the lowest value allowed for V_S and hence for V_{CM}?

8.2 For the PMOS differential amplifier shown in Fig. P8.2 let $V_{tp} = -0.8$ V and $k_p' W/L = 4$ mA/V^2. Neglect channel-length modulation.

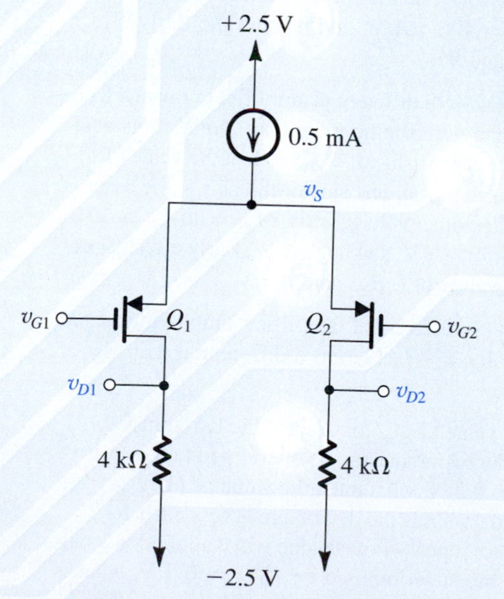

Figure P8.2

(a) For $v_{G1} = v_{G2} = 0$ V, find V_{OV} and V_{GS} for each of Q_1 and Q_2. Also find V_S, V_{D1}, and V_{D2}.
(b) If the current source requires a minimum voltage of 0.5 V, find the input common-mode range.

8.3 For the differential amplifier specified in Problem 8.1 let $v_{G2} = 0$ and $v_{G1} = v_{id}$. Find the value of v_{id} that corresponds to each of the following situations:

(a) $i_{D1} = i_{D2} = 0.1$ mA; (b) $i_{D1} = 0.15$ mA and $i_{D2} = 0.05$ mA; (c) $i_{D1} = 0.2$ mA and $i_{D2} = 0$ (Q_2 just cuts off); (d) $i_{D1} = 0.05$ mA and $i_{D2} = 0.15$ mA; (e) $i_{D1} = 0$ mA (Q_1 just cuts off) and $i_{D2} = 0.2$ mA. For each case, find v_S, v_{D1}, v_{D2}, and $(v_{D2} - v_{D1})$.

SIM 8.4 For the differential amplifier specified in Problem 8.2, let $v_{G2} = 0$ and $v_{G1} = v_{id}$. Find the range of v_{id} needed to steer the bias current from one side of the pair to the other. At each end of this range, give the value of the voltage at the common-source terminal and the drain voltages.

8.5 Consider the differential amplifier specified in Problem 8.1 with G_2 grounded and $v_{G1} = v_{id}$. Let v_{id} be adjusted to the value that causes $i_{D1} = 0.11$ mA and $i_{D2} = 0.09$ mA. Find the corresponding values of v_{GS2}, v_S, v_{GS1}, and hence v_{id}. What is the difference output voltage $v_{D2} - v_{D1}$? What is the voltage gain $(v_{D2} - v_{D1})/v_{id}$? What value of v_{id} results in $i_{D1} = 0.09$ mA and $i_{D2} = 0.11$ mA?

D 8.6 Design the circuit in Fig. P8.6 to obtain a dc voltage of +0.2V at each of the drains of Q_1 and Q_2 when $v_{G1} = v_{G2} = 0$ V. Operate all transistors at $V_{OV} = 0.2$ V and assume that for the process technology in which the circuit is fabricated, $V_{tn} = 0.5$ V and $\mu_n C_{ox} = 250$ μA/V^2. Neglect channel-length modulation. Determine the values of R, R_D, and the W/L ratios of Q_1, Q_2, Q_3, and Q_4. What is the input common-mode voltage range for your design?

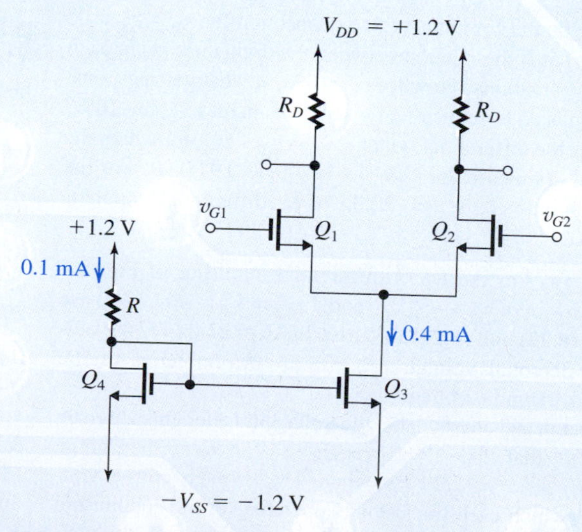

Figure P8.6

8.7 The table providing the answers to Exercise 8.3 shows that as the maximum input signal to be applied to the differential pair is increased, linearity is maintained at the same level by operating at a higher V_{OV}. If $|v_{id}|_{max}$ is to be 160 mV, use the data in the table to determine the required V_{OV} and the corresponding values of W/L and g_m.

8.8 Use Eq. (8.23) to show that if the term involving v_{id}^2 is to be kept to a maximum value of k then the maximum possible fractional change in the transistor current is given by

$$\frac{\Delta I_{max}}{I/2} = 2\sqrt{k(1-k)}$$

and the corresponding maximum value of v_{id} is given by

$$v_{id\max} = 2\sqrt{k}V_{OV}$$

Evaluate both expressions for $k = 0.01$, 0.1, and 0.2.

8.9 An NMOS differential amplifier utilizes a bias current of 400 μA. The devices have $V_t = 0.5$ V, $W = 20$ μm, and $L = 0.5$ μm, in a technology for which $\mu_n C_{ox} = 200$ μA/V². Find V_{GS}, and g_m in the equilibrium state. Also find the value of v_{id} for full-current switching. To what value should the bias current be changed in order to double the value of v_{id} for full-current switching?

D 8.10 Design the MOS differential amplifier of Fig. 8.5 to operate at $V_{OV} = 0.25$ V and to provide a transconductance g_m of 1 mA/V. Specify the W/L ratios and the bias current. The technology available provides $V_t = 0.8$ V and $\mu_n C_{ox} = 100$ μA/V².

8.11 Consider the NMOS differential pair illustrated in Fig. 8.5 under the conditions that $I = 100$ μA, using FETs for which $k_n'(W/L) = 400$ μA/V², and $V_t = 1$ V. What is the voltage on the common-source connection for $v_{G1} = v_{G2} = 0$? 2 V? What is the relation between the drain currents in each of these situations? Now for $v_{G2} = 0$ V, at what voltages must v_{G1} be placed to reduce i_{D2} by 10%? to increase i_{D2} by 10%? What is the differential voltage, $v_{id} = v_{G2} - v_{G1}$, for which the ratio of drain currents i_{D2}/i_{D1} is 1.0? 0.5? 0.9? 0.99? For the current ratio i_{D1}/i_{D2} = 20.0, what differential input is required?

***8.12** (a) For the MOS differential amplifier of Fig. 8.1 with $v_{G1} = V_{CM} + v_{id}/2$ and $v_{G2} = V_{CM} - v_{id}/2$, use Eqns. (8.23) and (8.24) to derive an expression for the output differential voltage $v_{od} \equiv v_{D2} - v_{D1}$ in terms of the input differential voltage v_{id}.
(b) Sketch and clearly label the voltage transfer characteristic (VTC), that is, v_{od} versus v_{id}, over the range $-\sqrt{2}V_{OV} \le v_{id} \le \sqrt{2}V_{OV}$, where V_{OV} is the overdrive voltage at which each transistor is operating in the equilibrium state. What is the slope of the nearly linear portion of the VTC near the origin? This is the differential voltage gain.
(c) Show on the same coordinates how the VTC changes if the bias current I is doubled? What is the change in the differential voltage gain?
(d) Prepare another sketch for case (b). Show on the same coordinates what happens to the VTC if the W/L ratio of each transistor is doubled. What is the change in the differential voltage gain?

Section 8.2: Small-Signal Operation of the MOS Differential Pair

8.13 An NMOS differential amplifier is operated at a bias current I of 0.4 mA and has a W/L ratio of 32, $\mu_n C_{ox} =$ 200 μA/V², $V_A = 10$ V, and $R_D = 5$ kΩ. Find V_{OV}, g_m, r_o, and A_d.

D 8.14 It is required to design an NMOS differential amplifier to operate with a differential input voltage that can be as high as 0.1 V while keeping the nonlinear term under the square root in Eq. (8.23) to a maximum of 0.05. A transconductance g_m of 1 mA/V is needed. Find the required values of V_{OV}, I, and W/L. Assume that the technology available has $\mu_n C_{ox}$ = 200 μA/V². What differential gain A_d results when R_D = 10 kΩ? Assume $\lambda = 0$. What is the resulting output signal corresponding to v_{id} at its maximum value?

D 8.15 Design a MOS differential amplifier to operate from ±1-V power supplies and dissipate no more than 2 mW in the equilibrium state. The differential voltage gain A_d is to be 5 V/V and the output common-mode dc voltage is to be 0.5 V. (*Note*: This is the dc voltage at the drains). Assume $\mu_n C_{ox}$ = 400 μA/V² and neglect the Early effect. Specify I, R_D, and W/L.

D 8.16 Design a MOS differential amplifier to operate from ±1-V supplies and dissipate no more than 2 mW in its equilibrium state. Select the value of V_{OV} so that the value of v_{id} that steers the current from one side of the pair to the other is 0.4 V. The differential voltage gain A_d is to be 5 V/V. Assume k_n' = 400 μA/V² and neglect the Early effect. Specify the required values of I, R_D, and W/L.

8.17 An NMOS differential amplifier employing equal drain resistors, R_D = 47 kΩ, has a differential gain A_d of 20 V/V.

(a) What is the value of g_m for each of the two transistors?
(b) If each of the two transistors is operating at an overdrive voltage V_{OV} = 0.2 V, what must the value of I be?
(c) For v_{id} = 0, what is the dc voltage across each R_D?
(d) If v_{id} is 20-mV peak-to-peak sine wave applied in a balanced manner but superimposed on V_{CM} = 0.5 V, what is

the lowest value that V_{DD} must have to ensure saturation-mode operation for Q_1 and Q_2 at all times? Assume $V_t = 0.5$ V.

8.18 A MOS differential amplifier is designed to have a differential gain A_d equal to the voltage gain obtained from a common-source amplifier. Both amplifiers utilize the same values of R_D and supply voltages, and all the transistors have the same W/L ratios. What must the bias current I of the differential pair be relative to the bias current I_D of the CS amplifier? What is the ratio of the power dissipation of the two circuits?

8.19 A differential amplifier is designed to have a differential voltage gain equal to the voltage gain of a common-source amplifier. Both amplifiers use the same values of R_D and supply voltages and are designed to dissipate equal amounts of power in their equilibrium or quiescent state. As well, all the transistors use the same channel length. What must the width W of the differential-pair transistors be relative to the width of the CS transistor?

D 8.20 Figure P8.20 shows a MOS differential amplifer with the drain resistors R_D implemented using diode-connected PMOS transistors, Q_3 and Q_4. Let Q_1 and Q_2 be matched, and Q_3 and Q_4 be matched.

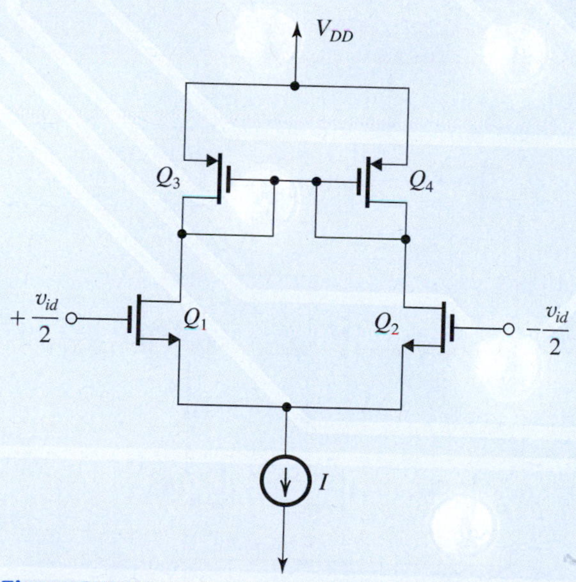

Figure P8.20

(a) Find the differential half-circuit and use it to derive an expression for A_d in terms of $g_{m1,2}$, $g_{m3,4}$, $r_{o1,2}$, and $r_{o3,4}$.

(b) Neglecting the effect of the output resistances r_o, find A_d in terms of μ_n, μ_p, $(W/L)_{1,2}$, and $(W/L)_{3,4}$.

(c) If $\mu_n = 4\mu_p$ and all four transistors have the same channel length, find $(W_{1,2}/W_{3,4})$ that results in $A_d = 10$ V/V.

8.21 Find the differential half-circuit for the differential amplifier shown in Fig. P8.21 and use it to derive an expression for the differential gain $A_d \equiv v_{od}/v_{id}$ in terms of g_m, R_D, and R_s. Neglect the Early effect. What is the gain with $R_s = 0$? What is the value of R_s (in terms of $1/g_m$) that reduces the gain to half this value?

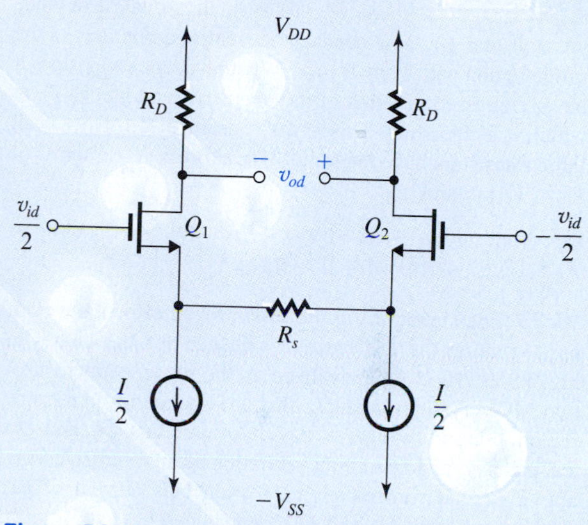

Figure P8.21

***8.22** The resistance R_s in the circuit of Fig. P8.21 can be implemented by using a MOSFET operated in the triode region, as shown in Fig. P8.22. Here Q_3 implements R_s, with the value of R_s determined by the voltage V_C at the gate of Q_3.

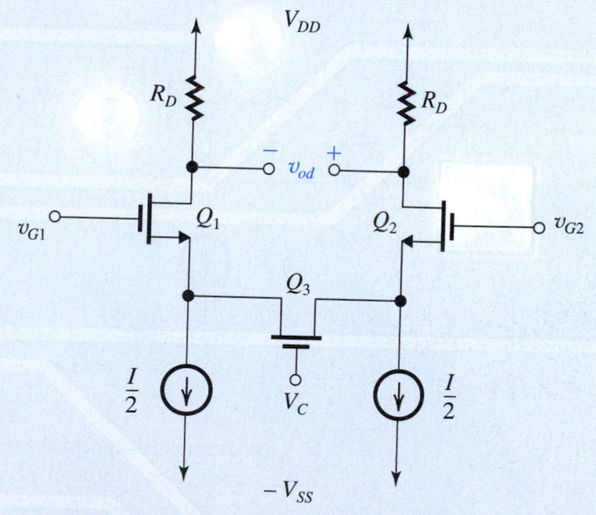

Figure P8.22

(a) With $v_{G1} = v_{G2} = 0$ V, and assuming that Q_1 and Q_2 are operating in saturation, what dc voltages appear at the sources of Q_1 and Q_2. Express these in terms of the overdrive voltage V_{OV} at which each of Q_1 and Q_2 operates, and V_t.
(b) For the situation in (a), what current flows in Q_3? What overdrive voltage V_{OV3} is Q_3 operating at, in terms of V_C, V_{OV}, and V_t?
(c) Now consider the case $v_{G1} = +v_{id}/2$ and $v_{G2} = -v_{id}/2$, where v_{id} is a small signal. Convince yourself that Q_3 now conducts current and operates in the triode region with a small v_{DS}. What resistance r_{DS} does it have, expressed in terms of the overdrive voltage V_{OV3} at which it is operating. This is the resistance R_s. Now if all three transistors have the same W/L, express R_s in terms of V_{OV}, V_{OV3}, and $g_{m1,2}$.
(d) Find V_{OV3} and hence V_C that result in (i) $R_s = 1/g_{m1,2}$; (ii) $R_s = 0.5/g_{m1,2}$.

***8.23** The circuit of Fig. P8.23 shows an effective way of implementing the resistance R_s needed for the circuit in Fig. P8.21. Here R_s is realized as the series equivalent of two MOSFETs Q_3 and Q_4 that are operated in the triode region, thus, $R_s = r_{DS3} + r_{DS4}$. Assume that Q_1 and Q_2 are matched and operate in saturation at an overdrive voltage V_{OV} that corresponds to a drain bias current of $I/2$. Also, assume that Q_3 and Q_4 are matched.

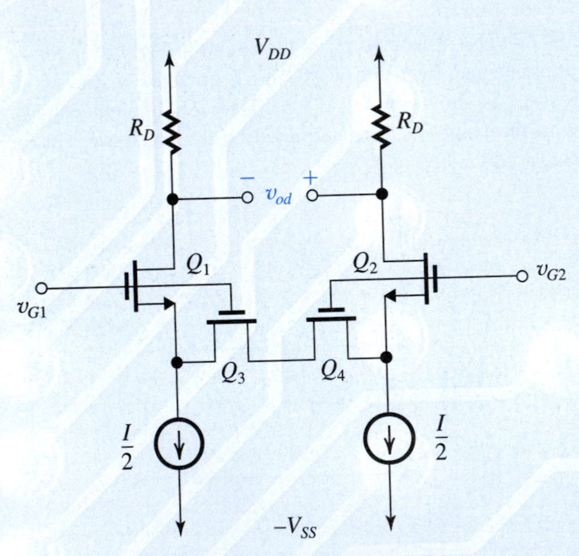

Figure P8.23

(a) With $v_{G1} = v_{G2} = 0$ V, what dc voltages appear at the sources of Q_1 and Q_2? What current flows through Q_3 and Q_4? At what overdrive voltages are Q_3 and Q_4 operating? Find an expression for r_{DS} for each of Q_3 and Q_4 and hence for R_s in terms of $(W/L)_{1,2}$, $(W/L)_{3,4}$, and $g_{m1,2}$.
(b) Now with $v_{G1} = v_{id}/2$ and $v_{G2} = -v_{id}/2$, where v_{id} is a small signal, find an expression of the voltage gain $A_d \equiv v_{od}/v_{id}$ in terms of $g_{m1,2}$, R_D, $(W/L)_{1,2}$, and $(W/L)_{3,4}$.

D *8.24 Figure P8.24 shows a circuit for a differential amplifier with an active load. Here Q_1 and Q_2 form the differential pair, while the current source transistors Q_4 and Q_5 form the active loads for Q_1 and Q_2, respectively. The dc bias circuit that establishes an appropriate dc voltage at the drains of Q_1 and Q_2 is not shown. It is required to design the circuit to meet the following specifications:

(a) Differential gain $A_d = 80$ V/V.
(b) $I_{REF} = I = 100$ μA.
(c) The dc voltage at the gates of Q_6 and Q_3 is +1.5 V.
(d) The dc voltage at the gates of Q_7, Q_4, and Q_5 is −1.5 V.

The technology available is specified as follows: $\mu_n C_{ox} = 3\mu_p C_{ox} = 90$ μA/V²; $V_{tn} = |V_{tp}| = 0.7$ V, $V_{An} = |V_{Ap}| = 20$ V. Specify the required value of R and the W/L ratios for all transistors. Also specify I_D and $|V_{GS}|$ at which each transistor is operating. For dc bias calculations you may neglect channel-length modulation.

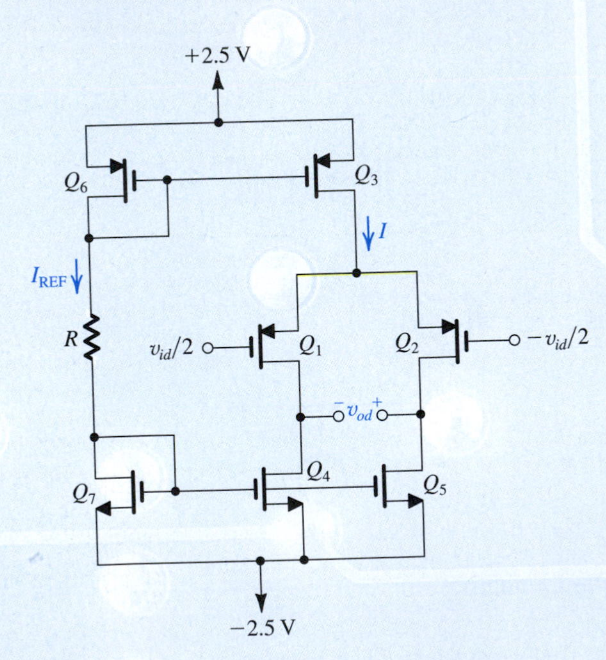

Figure P8.24

8.25 A design error has resulted in a gross mismatch in the circuit of Fig. P8.25. Specifically, Q_2 has twice the W/L ratio of Q_1. If v_{id} is a small sine-wave signal, find:

(a) I_{D1} and I_{D2}.
(b) V_{OV} for each of Q_1 and Q_2.
(c) The differential gain A_d in terms of R_D, I, and V_{OV}.

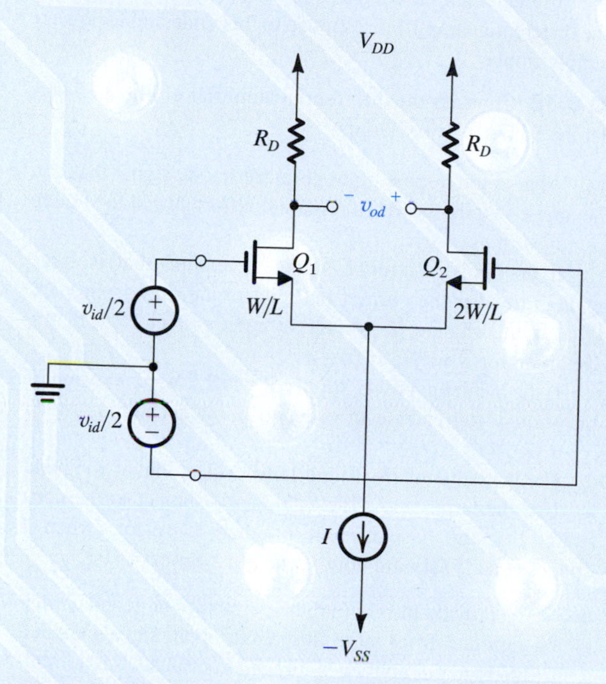

Figure P8.25

D 8.26 For the cascode differential amplifier of Fig. 8.12(a) show that if all transistors have the same channel length and are operated at the same $|V_{OV}|$ and assuming that $V'_{An} = |V'_{Ap}| = |V'_A|$, the differential gain A_d is given by

$$A_d = 2(|V_A|/|V_{OV}|)^2$$

Now design the amplifier to obtain a differential gain of 1000 V/V. Use $|V_{OV}| = 0.2$ V. If $|V'_A| = 10$ V/μm, specify the required channel length L. If g_m is to be as high as possible but the power dissipation in the amplifier (in equilibrium) is to be limited to 1 mW, what bias current I would you use? Let $V_{DD} = -V_{SS} = 0.9$ V.

SIM **8.27** An NMOS differential pair is biased by a current source $I = 0.2$ mA having an output resistance $R_{SS} = 100$ kΩ. The amplifier has drain resistances $R_D = 10$ kΩ, using transistors with $k'_n W/L = 3$ mA/V², and r_o that is large. If the output is taken differentially and there is a 1% mismatch between the drain resistances, find $|A_d|$, $|A_{cm}|$, and CMRR.

8.28 For the differential amplifier shown in Fig. P8.2, let Q_1 and Q_2 have $k'_p(W/L) = 4$ mA/V², and assume that the bias current source has an output resistance of 30 kΩ. Find $|V_{OV}|$, g_m, $|A_d|$, $|A_{cm}|$, and the CMRR (in dB) obtained with the output taken differentially. The drain resistances are known to have a mismatch of 2%.

SIM **D *8.29** The differential amplifier in Fig. P8.29 utilizes a resistor R_{SS} to establish a 1-mA dc bias current. Note that this amplifier uses a single 5-V supply and thus the dc common-mode voltage V_{CM} cannot be zero. Transistors Q_1 and Q_2 have $k'_n W/L = 2.5$ mA/V², $V_t = 0.7$ V, and $\lambda = 0$.

(a) Find the required value of V_{CM}.
(b) Find the value of R_D that results in a differential gain A_d of 8 V/V.
(c) Determine the dc voltage at the drains.
(d) Determine the common-mode gain $\Delta V_{D1}/\Delta V_{CM}$. (*Hint:* You need to take $1/g_m$ into account.)
(e) Use the common-mode gain found in (d) to determine the change in V_{CM} that results in Q_1 and Q_2 entering the triode region.

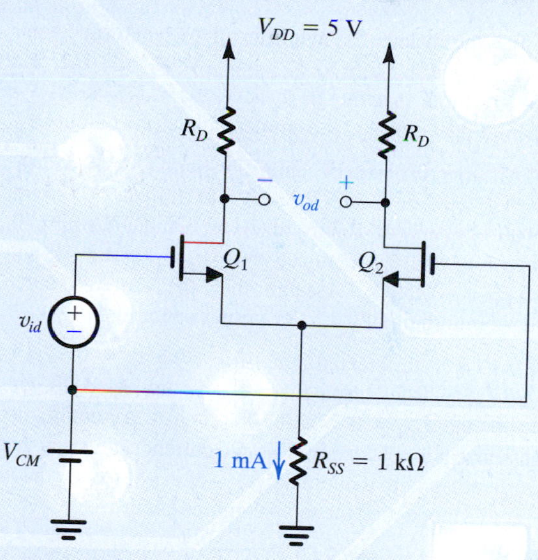

Figure P8.29

***8.30** The objective of this problem is to determine the common-mode gain and hence the CMRR of the differential pair arising from a simultaneous mismatch in g_m and in R_D.

(a) Refer to the circuit in Fig. 8.13(a) and its equivalent in Fig. 8.14, and let the two drain resistors be denoted R_{D1} and R_{D2} where $R_{D1} = R_D + (\Delta R_D/2)$ and $R_{D2} = R_D - (\Delta R_D/2)$. Also let $g_{m1} = g_m + (\Delta g_m/2)$ and $g_{m2} = g_m - (\Delta g_m/2)$. Follow an analysis process similar to that used to derive Eq. (8.63) to show that

$$A_{cm} \simeq \left(\frac{R_D}{2R_{SS}}\right)\left(\frac{\Delta g_m}{g_m} + \frac{\Delta R_D}{R_D}\right)$$

Note that this equation indicates that R_D can be deliberately varied to compensate for the initial variability in g_m and R_D, that is, to minimize A_{cm}.
(b) In a MOS differential amplifier for which $R_D = 5$ kΩ and $R_{SS} = 25$ kΩ, the common-mode gain is measured and found to be 0.002 V/V. Find the percentage change required in one of the two drain resistors so as to reduce A_{cm} to zero (or close to zero).

D 8.31 A MOS differential amplifier utilizing a simple current source to provide the bias current I is found to have a CMRR of 60 dB. If it is required to raise the CMRR to 100 dB by adding a cascode transistor to the current source, what must the intrinsic gain A_0 of the cascode transistor be? If the cascode transistor is operated at $V_{OV} = 0.2$ V, what must its V_A be? If for the specific technology utilized $V'_A = 10$ V/μm, specify the channel length L of the cascode transistor.

Section 8.3: The BJT Differential Pair

8.32 For the differential amplifier of Fig. 8.16(a) let $I =$ 0.5 mA, $V_{CC} = V_{EE} = 2.5$ V, $V_{CM} = -1$ V, $R_C = 8$ kΩ, and $\beta = 100$. Assume that the BJTs have $v_{BE} = 0.7$ V at $i_C =$ 1 mA. Find the voltage at the emitters and at the outputs.

8.33 An *npn* differential amplifier with $I = 0.5$ mA, $V_{CC} = V_{EE} = 2.5$ V, and $R_C = 8$ kΩ utilizes BJTs with $\beta = 100$ and $v_{BE} = 0.7$ V at $i_C = 1$ mA. If $v_{B2} = 0$, find V_E, V_{C1}, and V_{C2} obtained with $v_{B1} = +0.5$ V, and with $v_{B1} = -0.5$ V. Assume that the current source requires a minimum of 0.3 V for proper operation.

8.34 An *npn* differential amplifier with $I = 0.5$ mA, $V_{CC} = V_{EE} = 2.5$ V, and $R_C = 8$ kΩ utilizes BJTs with $\beta = 100$ and $v_{BE} = 0.7$ V at $i_C = 1$ mA. Assuming that the bias current is obtained by a simple current source and that all transistors require a minimum v_{CE} of 0.3 V for operation in the active mode, find the input common-mode range.

8.35 Repeat Exercise 8.9 for an input of –0.3 V.

8.36 An *npn* differential pair employs transistors for which $v_{BE} = 690$ mV at $i_C = 1$ mA, and $\beta = 50$. The transistors leave the active mode at $v_{CE} \le 0.3$ V. The collector resistors $R_C = 82$ kΩ, and the power supplies are ±1.2 V. The bias current $I = 20$ μA and is supplied with a simple current source.

(a) For $v_{B1} = v_{B2} = V_{CM} = 0$ V, find V_E, V_{C1}, and V_{C2}.
(b) Find the input common-mode range.
(c) If $v_{B2} = 0$, find the value of v_{B1} that increases the current in Q_1 by 10%.

8.37 Consider the BJT differential amplifier when fed with a common-mode voltage V_{CM} as shown in Fig. 8.16(a). As is often the case, the supply voltage V_{CC} may not be pure dc but might include a ripple component v_r of small amplitude and a frequency of 120 Hz (see Section 4.5). Thus the supply voltage becomes $V_{CC} + v_r$. Find the ripple component of the collector voltages, v_{C1} and v_{C2}, as well as of the difference output voltage $v_{od} \equiv v_{C2} - v_{C1}$. Comment on the differential amplifier response to this undesirable power-supply ripple.

D 8.38 Consider the differential amplifier of Fig. 8.15 and let the BJT β be very large:

(a) What is the largest input common-mode signal that can be applied while the BJTs remain comfortably in the active region with $v_{CB} = 0$?
(b) If an input difference signal is applied that is large enough to steer the current entirely to one side of the pair, what is the change in voltage at each collector (from the condition for which $v_{id} = 0$)?
(c) If the available power supply V_{CC} is 2.5 V, what value of IR_C should you choose in order to allow a common-mode input signal of ±1.0 V?
(d) For the value of IR_C found in (c), select values for I and R_C. Use the largest possible value for I subject to the constraint that the base current of each transistor (when I divides equally) should not exceed 2 μA. Let $\beta = 100$.

8.39 To provide insight into the possibility of nonlinear distortion resulting from large differential input signals applied to the differential amplifier of Fig. 8.15, evaluate the normalized change in the current i_{E1}, $\Delta i_{E1}/I = (i_{E1} - (I/2))/I$, for differential input signals v_{id} of 5, 10, 20, 30, and 40 mV. Provide a tabulation of the ratio $(\Delta i_{E1}/I)/v_{id}$, which represents the proportional transconductance gain of the differential pair, versus v_{id}. Comment on the linearity of the differential pair as an amplifier.

D 8.40 Design the circuit of Fig. 8.15 to provide a differential output voltage (i.e., one taken between the two collectors) of 1 V when the differential input signal is 10 mV. A current source of 1 mA and a positive supply of +5 V are available. What is the largest possible input common-mode voltage for which operation is as required? Assume $\alpha \simeq 1$.

D *8.41 One of the trade-offs available in the design of the basic differential amplifier circuit of Fig. 8.15 is between the value of the voltage gain and the range of common-mode input voltage. The purpose of this problem is to demonstrate this trade-off.

(a) Use Eqs. (8.73) and (8.74) to obtain i_{C1} and i_{C2} corresponding to a differential input signal of 5 mV (i.e., $v_{B1} - v_{B2} = 5$ mV). Assume β to be very high. Find the resulting voltage difference between the two collectors ($v_{C2} - v_{C1}$), and divide this value by 5 mV to obtain the voltage gain in terms of (IR_C).

(b) Find the maximum permitted value for V_{CM} while the transistors remain comfortably in the active mode with $v_{CB} = 0$. Express this maximum in terms of V_{CC} and the gain, and hence show that for a given value of V_{CC}, the higher the gain achieved, the lower the common-mode range. Use this expression to find $V_{CM\max}$ corresponding to a gain magnitude of 100, 200, 300, and 400 V/V. For each value, also give the required value of IR_C and the value of R_C for $I = 1$ mA. As an example, discuss what can be achieved with $V_{CC} = 10$ V.

***8.42** For the circuit in Fig. 8.15, assuming $\alpha = 1$ and $IR_C = 5$ V, use Eqs. (8.70) and (8.71) to find i_{C1} and i_{C2}, and hence determine $v_{od} = v_{C2} - v_{C1}$ for input differential signals $v_{id} \equiv v_{B1} - v_{B2}$ of 5 mV, 10 mV, 15 mV, 20 mV, 25 mV, 30 mV, 35 mV, and 40 mV. Plot v_o versus v_{id}, and hence comment on the amplifier linearity. As another way of visualizing linearity, determine the gain (v_o/v_{id}) versus v_{id}. Comment on the resulting graph.

8.43 In a differential amplifier using a 3-mA emitter bias current source, the two BJTs are not matched. Rather, one has twice the emitter junction area of the other. For a differential input signal of zero volts, what do the collector currents become? What difference input is needed to equalize the collector currents? Assume $\alpha = 1$.

8.44 This problem explores the linearization of the transfer characteristics of the differential pair achieved by including emitter-degeneration resistances R_e in the emitters (see Fig. 8.18). Consider the case $I = 200$ µA with the transistors exhibiting $v_{BE} = 690$ mV at $i_C = 1$ mA and assume $\alpha \simeq 1$.

(a) With no emitter resistances R_e, what value of V_{BE} results when $v_{id} = 0$?

(b) With no emitter resistances R_e, use the large-signal model to find i_{C1} and i_{C2} when $v_{id} = 20$ mV.

(c) Now find the value of R_e that will result in the same i_{C1} and i_{C2} as in (b) but with $v_{id} = 200$ mV. Use the large-signal model.

(d) Calculate the effective transconductance G_m as the inverse of the total resistances in the emitter circuits in the cases without and with the R_e's. By what factor is G_m reduced? How does this factor relate to the increase in v_{id}? Comment.

8.45 A BJT differential amplifier uses a 200-µA bias current. What is the value of g_m of each device? If β is 150, what is the differential input resistance?

D 8.46 Design the basic BJT differential amplifier circuit of Fig. 8.19 to provide a differential input resistance of at least 10 kΩ and a differential voltage gain of 100 V/V. The transistor β is specified to be at least 100. The available positive power supply is 5 V.

8.47 For a differential amplifier to which a total difference signal of 10 mV is applied, what is the equivalent signal to its corresponding CE half-circuit? If the emitter current source I is 100 µA, what is r_e of the half-circuit? For a load resistance of 10 kΩ in each collector, what is the half-circuit gain? What magnitude of signal output voltage would you expect at each collector? Between the two collectors?

8.48 A BJT differential amplifier is biased from a 1-mA constant-current source and includes a 200-Ω resistor in each emitter. The collectors are connected to V_{CC} via 12-kΩ resistors. A differential input signal of 0.1 V is applied between the two bases.

(a) Find the signal current in the emitters (i_e) and the signal voltage v_{be} for each BJT.

(b) What is the total emitter current in each BJT?

(c) What is the signal voltage at each collector? Assume $\alpha = 1$.

(d) What is the voltage gain realized when the output is taken between the two collectors?

D 8.49 Design a BJT differential amplifier to amplify a differential input signal of 0.2 V and provide a differential output signal of 5 V. To ensure adequate linearity, it is required to limit the signal amplitude across each base–emitter junction to a maximum of 5 mV. Another design requirement is that the differential input resistance be at least 50 kΩ. The BJTs available are specified to have $\beta \geq 100$. Give the circuit configuration and specify the values of all its components.

D 8.50 Design a bipolar differential amplifier such as that in Fig. 8.19 to operate from ±2.5 V power supplies and to provide differential gain of 40 V/V. The power dissipation in the quiescent state should not exceed 2 mW.

(a) Specify the values of I and R_C. What dc voltage appears at the collectors?

(b) If $\beta = 100$, what is the input differential resistance?

(c) For $v_{id} = 20$ mV, what is the signal voltage at each of the collectors?

(d) For the situation in (c), what is the maximum allowable value of the input common mode voltage, V_{CM}? Recall that to maintain an *npn* BJT in saturation, v_B should not exceed v_C by more than 0.4 V.

D *8.51 In this problem we explore the trade-off between input common-mode range and differential gain in the design of the bipolar BJT. Consider the bipolar differential amplifier in Fig. 8.15 with the input voltages

$$v_{B1} = V_{CM} + (v_{id}/2)$$
$$v_{B2} = V_{CM} - (v_{id}/2)$$

(a) Bearing in mind that for a BJT to remain in the active mode, v_{BC} should not exceed 0.4 V, show that when v_{id}

has a peak $\hat{v}_{id}$, the maximum input common-mode voltage $V_{CM\text{max}}$ is given by

$$V_{CM\text{max}} = V_{CC} + 0.4 - \frac{\hat{v}_{id}}{2} - A_d\left(V_T + \frac{\hat{v}_{id}}{2}\right)$$

(b) For the case $V_{CC} = 5$ V and $\hat{v}_{id} = 10$ mV, use the relationship above to determine $V_{CM\text{max}}$ for the case $A_d = 100$ V/V. Also find the peak output signal $\hat{v}_{od}$ and the required value of IR_C. Now if the power dissipation in the circuit is to be limited to 5 mW in the quiescent state (i.e., with $v_{id} = 0$), find I and R_C. (Remember to include the power drawn from the negative power supply $-V_{EE} = -5$ V.)
(c) If $V_{CM\text{max}}$ is to be 0 V, and all other conditions remain the same, what gain A_d is achievable?

8.52 For the differential amplifier of Fig. 8.15, let $V_{CC} = +5$ V and $IR_C = 4$ V. Find the differential gain A_d. Sketch and clearly label the waveforms for the total collector voltages v_{C1} and v_{C2} for the following two cases:

(a) $v_{B1} = 1 + 0.005\sin(\omega t)$

$v_{B2} = 1 - 0.005\sin(\omega t)$

(b) $v_{B1} = 1 + 0.1\sin(\omega t)$

$v_{B2} = 1 - 0.1\sin(\omega t)$

8.53 Consider a bipolar differential amplifier in which the collector resistors R_C are replaced with simple current sources implemented using *pnp* transistors. Sketch the circuit and give its differential half-circuit. If $V_A = 10$ V for all transistors, find the differential voltage gain achieved.

8.54 For each of the emitter-degenerated differential amplifiers shown in Fig. P8.54, find the differential half-circuit and derive expressions for the differential gain A_d and differential input resistance R_{id}. For each circuit, what dc voltage appears across the bias current source(s) in the quiescent state (i.e., with $v_{id} = 0$). Hence, which of the two circuits will allow a larger negative V_{CM}?

8.55 Consider a bipolar differential amplifier that, in addition to the collector resistances R_C, has a load resistance R_L connected between the two collectors. What does the differential gain A_d become?

8.56 A bipolar differential amplifier having resistance R_e inserted in series with each emitter (as in Fig. 8.21) is biased with a constant current I. When both input terminals are grounded, the dc voltage measured across each R_e is found to be $4\,V_T$ and that measured across each R_C is found to be $40\,V_T$. What differential voltage gain A_d do you expect the amplifier to have?

8.57 A bipolar differential amplifier with emitter degeneration resistances R_e and R_e, is fed with the arrangement shown in Fig. P8.57. Derive an expression for the overall differential voltage gain $G_v \equiv v_{od}/v_{sig}$. If R_{sig} is of such a value that $v_{id} = 0.5v_{\text{sig}}$, find the gain G_v in terms of R_C,

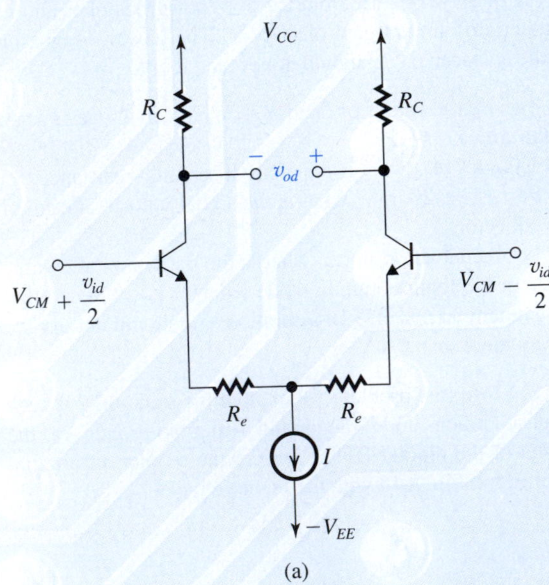

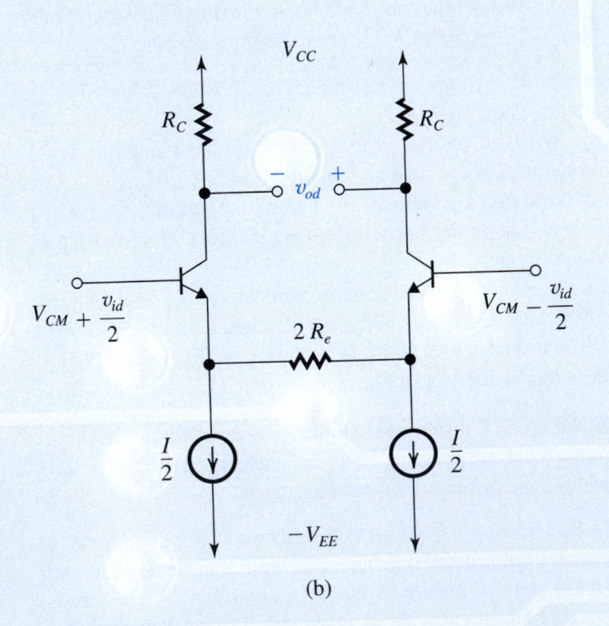

Figure P8.54

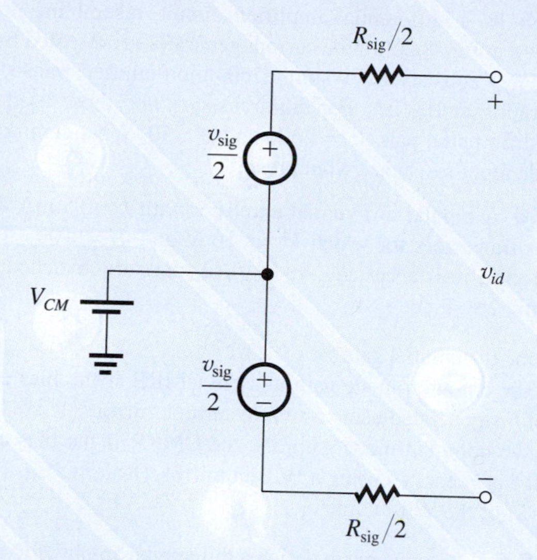

Figure P8.57

r_e, R_e, and α. Now if β is doubled, by what factor does G_v increase?

8.58 A particular differential amplifier operates from an emitter current source whose output resistance is 0.5 MΩ. What resistance is associated with each common-mode half-circuit? For collector resistors of 20 kΩ and 1% tolerance, what is the resulting common-mode gain for output taken (a) single-endedly? and (b) differentially?

8.59 Find the voltage gain and the input resistance of the amplifier shown in Fig. P8.59 assuming $\beta = 100$.

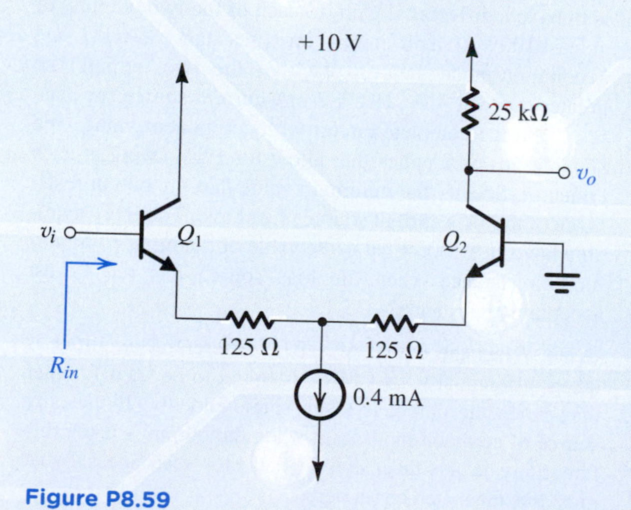

Figure P8.59

8.60 Find the voltage gain and input resistance of the amplifier in Fig. P8.60 assuming that $\beta = 100$.

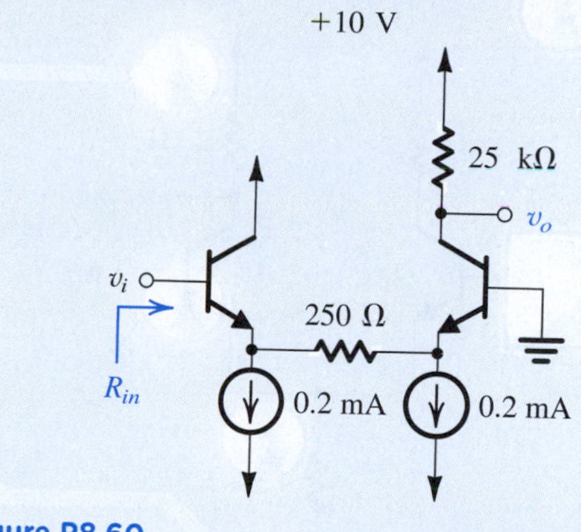

Figure P8.60

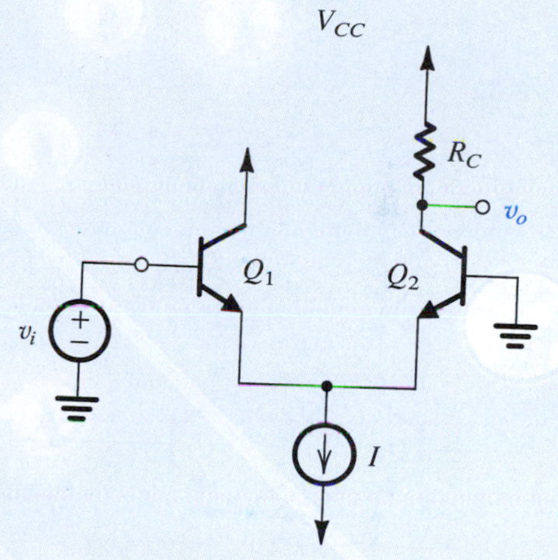

Figure P8.61

8.61 Derive an expression for the small-signal voltage gain v_o/v_i of the circuit shown in Fig. P8.61 in two different ways:

(a) as a differential amplifier
(b) as a cascade of a common-collector stage Q_1 and a common-base stage Q_2

Assume that the BJTs are matched and have a current gain α, and neglect the Early effect. Verify that both approaches lead to the same result.

8.62 The differential amplifier circuit of Fig. P8.62 utilizes a resistor connected to the negative power supply to establish the bias current I.

(a) For $v_{B1} = v_{id}/2$ and $v_{B2} = -v_{id}/2$, where v_{id} is a small signal with zero average, find the magnitude of the differential gain, $|v_o/v_{id}|$.
(b) For $v_{B1} = v_{B2} = v_{icm}$, where v_{icm} has a zero average, find the magnitude of the common-mode gain, $|v_o/v_{icm}|$.
(c) Calculate the CMRR.
(d) If $v_{B1} = 0.1 \sin 2\pi \times 60t + 0.005 \sin 2\pi \times 1000t$ volts, and $v_{B2} = 0.1 \sin 2\pi \times 60t - 0.005 \sin 2\pi \times 1000t$, volts, find v_o.

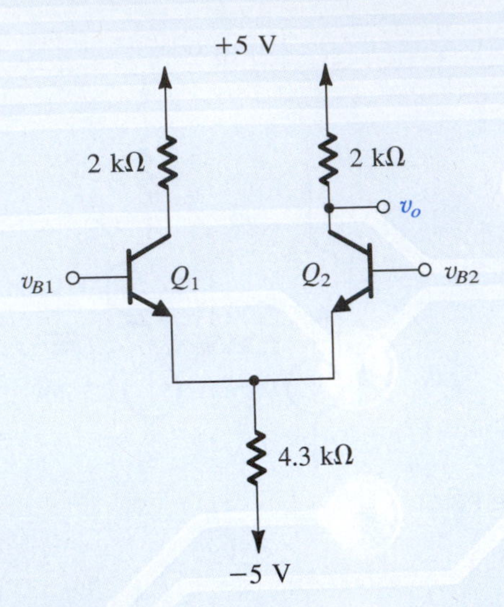

Figure P8.62

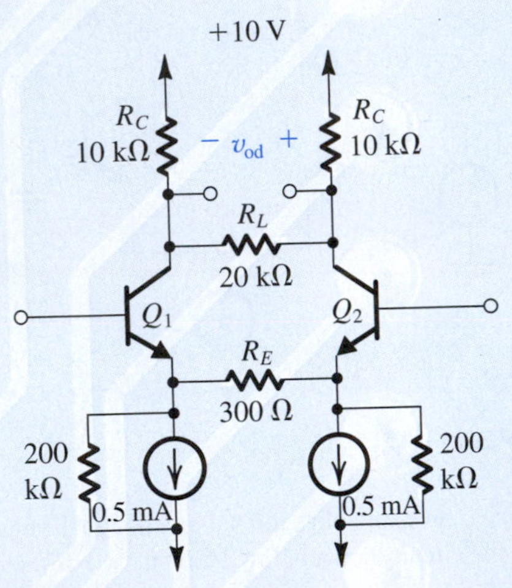

Figure P8.63

8.63 For the differential amplifier shown in Fig. P8.63, identify and sketch the differential half-circuit and the common-mode half-circuit. Find the differential gain, the differential input resistance, the common-mode gain assuming the resistances R_C have 1% tolerance, and the common-mode input resistance. For these transistors, $\beta = 100$ and $V_A = 100$ V.

8.64 Consider the basic differential circuit in which the transistors have $\beta = 100$ and $V_A = 100$ V, with $I = 0.5$ mA, $R_{EE} = 200$ kΩ, and $R_C = 20$ kΩ. The collector resistances are matched to within 1%. Find:

(a) the differential gain
(b) the differential input resistance
(c) the common-mode gain
(d) the common-mode rejection ratio
(e) the input common-mode resistance

8.65 In a differential-amplifier circuit resembling that shown in Fig. 8.26(a), the current generator represented by I and R_{EE} consists of a simple common-emitter transistor operating at 100 μA. For this transistor, and those used in the differential pair, $V_A = 20$ V and $\beta = 50$. What common-mode input resistance would result?

8.66 A bipolar differential amplifier with $I = 0.5$ mA utilizes transistors for which $V_A = 10$ V and $\beta = 100$. The collector resistances $R_C = 10$ kΩ and are matched to within 2%. Find:

(a) the differential gain
(b) the common-mode gain and the CMRR if the bias current I is generated using a simple current mirror
(c) the common-mode gain and the CMRR if the bias current I is generated using a Wilson mirror. (Refer to Eq. 7.81 for R_o of the Wilson mirror.)

D 8.67 It is required to design a differential amplifier to provide the largest possible signal to a pair of 10-kΩ load resistances. The input differential signal is a sinusoid of 5-mV peak amplitude, which is applied to one input terminal while the other input terminal is grounded. The power supply available is 10 V. To determine the required bias current I, derive an expression for the total voltage at each of the collectors in terms of V_{CC} and I in the presence of the input signal. Then impose the condition that both transistors should remain well out of saturation with a minimum v_{CB} of approximately 0 V. Thus determine the required value of I. For this design, what differential gain is achieved? What is the amplitude of the signal voltage obtained between the two collectors? Assume $\alpha \simeq 1$.

D *8.68 Design a BJT differential amplifier that provides two single-ended outputs (at the collectors). The amplifier is to have a differential gain (to each of the two outputs) of at least 100 V/V, a differential input resistance ≥10 kΩ, and a common-mode gain (to each of the two outputs) no greater than 0.1 V/V. Use a 2-mA current source for biasing. Give the complete circuit with component values and suitable power supplies that allow for ±2 V swing at each collector. Specify the minimum value that the output resistance of the bias current source must have. The BJTs available have $\beta \geq 100$. What is the value of the input common-mode resistance when the bias source has the lowest acceptable resistance?

8.69 When the output of a BJT differential amplifier is taken differentially, its CMRR is found to be 40 dB higher than when the output is taken single-endedly. If the only source of common-mode gain when the output is taken differentially is the mismatch in collector resistances, what must this mismatch be (in percent)?

***8.70** In a particular BJT differential amplifier, a production error results in one of the transistors having an emitter–

base junction area that is twice that of the other. With the inputs grounded, how will the emitter bias current split between the two transistors? If the output resistance of the current source is 500 kΩ and the resistance in each collector (R_C) is 12 kΩ, find the common-mode gain obtained when the output is taken differentially. Assume $\alpha \simeq 1$.

Section 8.4: Other Nonideal Characteristics of the Differential Amplifier

D 8.71 An NMOS differential pair is to be used in an amplifier whose drain resistors are 10 kΩ ± 1%. For the pair, $k_n' W/L = 4$ mA/V^2. A decision is to be made concerning the bias current I to be used, whether 160 μA or 360 μA. Contrast the differential gain and input offset voltage for the two possibilities.

D 8.72 An NMOS amplifier, whose designed operating point is at $V_{OV} = 0.2$ V, is suspected to have a variability of V_t of ±5 mV, and of W/L and R_D (independently) of ±2%. What is the worst-case input offset voltage you would expect to find? What is the major contribution to this total offset? If you used a variation of one of the drain resistors to reduce the output offset to zero and thereby compensate for the uncertainties (including that of the other R_D), what percentage change from nominal would you require? If by selection you reduced the contribution of the worst cause of offset by a factor of 10, what change in R_D would be needed?

8.73 An NMOS differential pair operating at a bias current I of 100 μA uses transistors for which $k_n' = 250$ μA/V^2 and $W/L = 10$. Find the three components of input offset voltage under the conditions that $\Delta R_D / R_D = 5\%$, $\Delta(W/L)/(W/L) = 5\%$, and $\Delta V_t = 5$ mV. In the worst case, what might the total offset be? For the usual case of the three effects being independent, what is the offset likely to be?

8.74 A bipolar differential amplifier uses two well-matched transistors but collector load resistors that are mismatched by 8%. What input offset voltage is required to reduce the differential output voltage to zero?

8.75 A bipolar differential amplifier uses two transistors whose scale currents I_S differ by 10%. If the two collector resistors are well matched, find the resulting input offset voltage.

8.76 Modify Eq. (8.119) for the case of a differential amplifier having a resistance R_E connected in the emitter of each transistor. Let the bias current source be I.

8.77 A differential amplifier uses two transistors whose β values are β_1 and β_2. If everything else is matched, show that the input offset voltage is approximately $V_T[(1/\beta_1) - (1/\beta_2)]$. Evaluate V_{OS} for $\beta_1 = 100$ and $\beta_2 = 200$. Assume the differential source resistance to be zero.

8.78 Two possible differential amplifier designs are considered, one using BJTs and the other MOSFETs. In both cases, the collector (drain) resistors are maintained within ±2 % of nominal value. The MOSFETs are operated at $V_{OV} = 300$ mV. What input offset voltage results in each case? What does the MOS V_{OS} become if the devices are increased in width by a factor of 4?

***8.79** A differential amplifier uses two transistors having V_A values of 100 V and 300 V. If everything else is matched, find the resulting input offset voltage. Assume that the two transistors are intended to be biased at a V_{CE} of about 10 V.

***8.80** A differential amplifier is fed in a balanced or push–pull manner, and the source resistance in series with each base is R_s. Show that a mismatch ΔR_s between the values of the two source resistances gives rise to an input offset voltage of approximately $(I/2\beta)\Delta R_s / [1 + (g_m R_s)/\beta]$.

8.81 One approach to "offset correction" involves the adjustment of the values of R_{C1} and R_{C2} so as to reduce the differential output voltage to zero when both input terminals are grounded. This offset-nulling process can be accomplished by utilizing a potentiometer in the collector circuit, as shown in Fig. P8.81. We wish to find the potentiometer setting, represented by the fraction x of its value connected in series with R_{C1}, that is required for nulling the output offset voltage that results from:

(a) R_{C1} being 4% higher than nominal and R_{C2} 4% lower than nominal

(b) Q_1 having an area 20% larger than that of Q_2

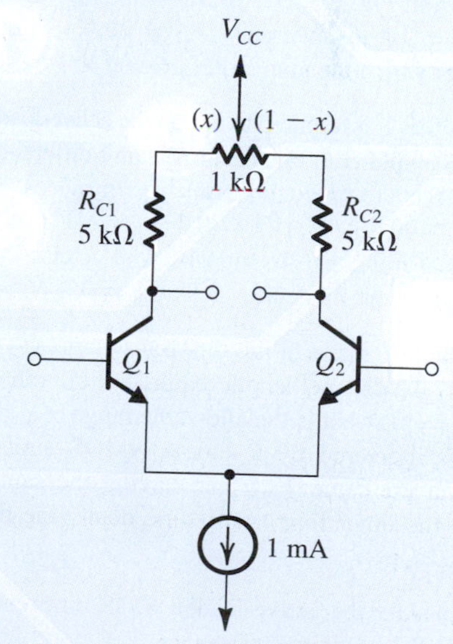

Figure P8.81

8.82 A differential amplifier for which the total emitter bias current is 500 μA uses transistors for which β is specified to lie between 80 and 200. What is the largest possible input bias current? The smallest possible input bias current? The largest possible input offset current?

****8.83** In a particular BJT differential amplifier, a production error results in one of the transistors having an emitter–base junction area twice that of the other. With both inputs grounded, find the current in each of the two transistors and hence the dc offset voltage at the output, assuming that the collector resistances are equal. Use small-signal analysis to find the input voltage that would restore current balance to the differential pair. Repeat using large-signal analysis and compare results.

D 8.84 A large fraction of mass-produced differential-amplifier modules employing 20-kΩ collector resistors is found to have an input offset voltage ranging from +3 mV to −3 mV. By what amount must one collector resistor be adjusted to reduce the input offset to zero? If an adjustment mechanism is devised that raises one collector resistor while correspondingly lowering the other, what resistance change is needed? If a potentiometer connected as shown in Fig. P8.81 is used, what value of potentiometer resistance (specified to 1 significant digit) is needed?

Section 8.5: The Differential Amplifier with Active Load

D 8.85 In an active-loaded differential amplifier of the form shown in Fig. 8.32(a), all transistors are characterized by $k'W/L = 3.2$ mA/V^2, and $|V_A| = 20$ V. Find the bias current I for which the gain $v_o/v_{id} = 100$ V/V.

SIM **D 8.86** It is required to design the active-loaded differential MOS amplifier of Fig. 8.32 to obtain a differential gain of 50 V/V. The technology available provides $\mu_n C_{ox} = 4\mu_p C_{ox} = 400$ μA/V^2, $|V_t| = 0.5$ V, and $|V_A'| = 20$ V/μm and operates from ±1 V supplies. Use a bias current $I = 200$ μA and operate all devices at $|V_{OV}| = 0.2$ V.

(a) Find the W/L ratios of the four transistors.
(b) Specify the channel length required of all transistors.
(c) If $V_{CM} = 0$, what is the allowable range of v_O?
(d) If I is delivered by a simple NMOS current source operated at the same V_{OV} and having the same channel length as the other four transistors, determine the CMRR obtained.

8.87 Consider the active-loaded MOS differential amplifier of Fig. 8.32(a) in two cases:

(a) Current source I is implemented with a simple current mirror.

(b) Current source I is implemented with the modified Wilson current mirror shown in Fig. P8.87.

Recalling that for the simple mirror $R_{SS} = r_o|_{Q_S}$ and for the Wilson mirror $R_{SS} \simeq g_{m7} r_{o7} r_{o5}$, and assuming that all transistors have the same $|V_A|$ and $k'W/L$, show that for case (a)

$$\text{CMRR} = 2\left(\frac{V_A}{V_{OV}}\right)^2$$

and for case (b)

$$\text{CMRR} = 2\sqrt{2}\left(\frac{V_A}{V_{OV}}\right)^3$$

where V_{OV} is the overdrive voltage that corresponds to a drain current of $I/2$. For $k'W/L = 10$ mA/V^2, $I = 1$ mA, and $|V_A| = 10$ V, find CMRR for both cases.

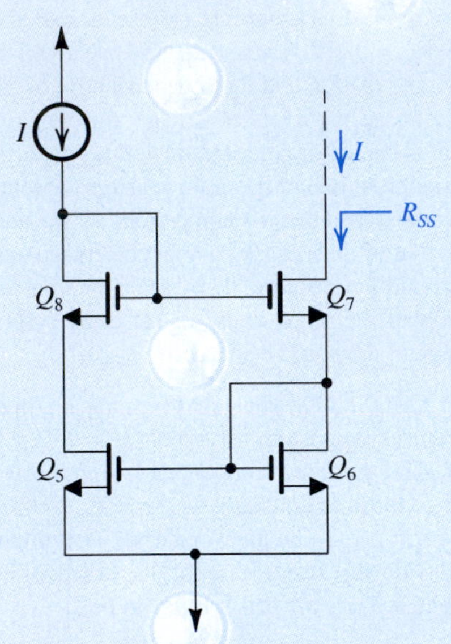

Figure P8.87

D 8.88 Consider an active-loaded differential amplifier such as that shown in Fig. 8.32(a) with the bias current source implemented with the modified Wilson mirror of Fig. P8.87 with $I = 200$ μA. The transistors have $|V_t| = 0.5$ V and $k'W/L = 5$ mA/V^2. What is the lowest value of the total power supply ($V_{DD} + V_{SS}$) that allows each transistor to operate with $|V_{DS}| \geq |V_{GS}|$?

***8.89** (a) Sketch the circuit of an active-loaded MOS differential amplifier in which the input transistors are cascoded and a cascode current mirror is used for the load.
(b) Show that if all transistors are operated at an overdrive voltage V_{OV} and have equal Early voltages $|V_A|$, the gain is given by

$$A_d = 2(V_A/V_{OV})^2$$

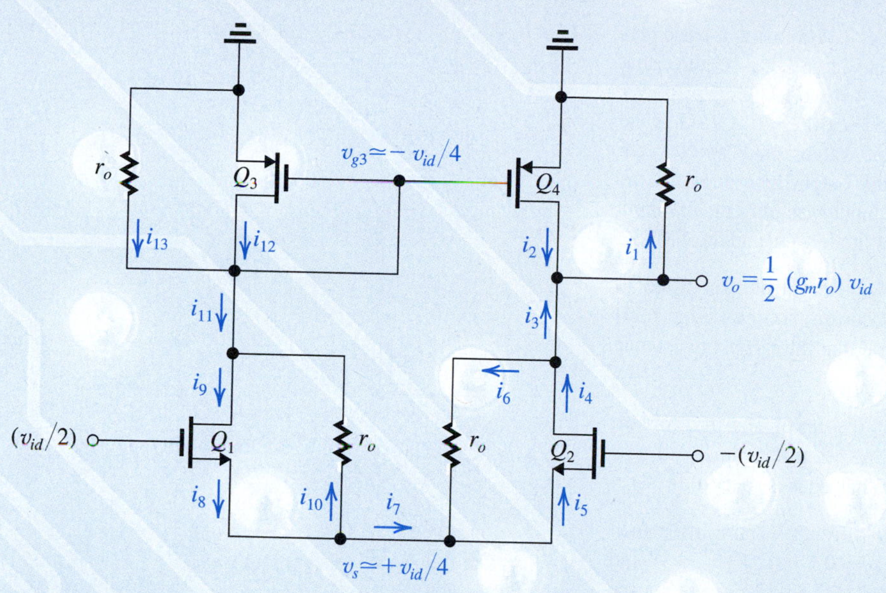

Figure P8.90

Evaluate the gain for $V_{OV} = 0.25$ V and $V_A = 20$ V.

8.90 Figure P8.90 shows the active-loaded MOS differential amplifier prepared for small-signal analysis. To help the reader we have already indicated approximate values for some of the node voltages. For instance, the output voltage $v_o = \frac{1}{2}(g_m r_o)v_{id}$, which we have derived in the text. The voltage at the common sources has been found to be approximately $+v_{id}/4$, which is very far from the virtual ground one might assume. Also, the voltage at the gate of the mirror is approximately $-v_{id}/4$, confirming our contention that the voltage there is vastly different from the output voltage, hence the lack of balance in the circuit and the unavailability of a differential half-circuit. Find the currents labeled i_1 to i_{13}. Determine their values in the sequence of their numbering and reflect on the results. You will find that there is some inconsistency, which is a result of the approximations we have made. Note that all transistors are assumed to be operating at the same $|V_{OV}|$.

8.91 An active-loaded NMOS differential amplifier operates with a bias current I of 100 μA. The NMOS transistors are operated at $V_{OV} = 0.2$ V and the PMOS devices at $|V_{OV}| = 0.3$ V. The Early voltages are 20 V for the NMOS and 12 V for the PMOS transistors. Find G_m, R_o, and A_d. For what value of load resistance is the gain reduced by a factor of 2?

8.92 This problem investigates the effect of transistor mismatches on the input offset voltage of the active-loaded MOS differential amplifier of Fig. 8.32(a). For this purpose, ground both input terminals and short-circuit the output node to ground.

(a) If the amplifying transistors Q_1 and Q_2 exhibit a W/L mismatch of $\Delta(W/L)_A$, find the resulting short-circuit output current and hence show that the corresponding V_{OS} is given by

$$V_{OS1} = (V_{OV}/2)\ \frac{\Delta(W/L)_A}{(W/L)_A}$$

where V_{OV} is the overdrive voltage at which Q_1 and Q_2 are operating.

(b) Repeat for a mismatch $\Delta(W/L)_M$ in the W/L ratios of the mirror transistor Q_3 and Q_4 to show that the corresponding V_{OS} is given by

$$V_{OS2} = (V_{OV}/2)\ \frac{\Delta(W/L)_M}{(W/L)_M}$$

where V_{OV} is the overdrive voltage at which Q_1 and Q_2 are operating.

(c) For a circuit in which all transistors are operated at $|V_{OV}| = 0.2$ V and all W/L ratios are accurate to within ±1 % of nominal, find the worst-case total offset voltage V_{OS}.

8.93 The differential amplifier in Fig. 8.37(a) is operated with $I = 400$ μA, with devices for which $V_A = 16$ V and $\beta = 100$. What differential input resistance, output resistance, equivalent transconductance, and open-circuit voltage gain would you expect? What will the voltage gain be if the input resistance of the subsequent stage is equal to R_{id} of this stage?

D *8.94 Design the circuit of Fig. 8.37(a) using a basic current mirror to implement the current source I. It is required that the equivalent transconductance be 4 mA/V. Use ±5-V power supplies and BJTs that have $\beta = 125$ and $V_A = 100$ V. Give the complete circuit with component values and specify the differential input resistance R_{id}, the output resistance R_o, the open-circuit voltage gain A_d, the input bias current, the input common-mode range, the common-mode gain, and the CMRR.

D *8.95 Repeat the design of the amplifier specified in Problem 8.94 utilizing a Widlar current source [Fig. 7.36] to supply the bias current. Assume that the largest resistance available is 2 kΩ.

D 8.96 Modify the design of the amplifier in Problem 8.94 by connecting emitter-degeneration resistances of values that result in $R_{id} = 125$ kΩ. What does A_d become?

8.97 An active-loaded bipolar differential amplifier such as that shown in Fig. 8.37(a) has $I = 0.5$ mA, $V_A = 30$ V, and $\beta = 150$. Find G_m, R_o, A_d, and R_{id}. If the bias-current source is implemented with a simple *npn* current mirror, find R_{EE}, A_{cm}, and CMRR. If the amplifier is fed differentially with a source having a total of 20 kΩ resistance (i.e., 10 kΩ in series with the base lead of each of Q_1 and Q_2), find the overall differential voltage gain.

***8.98** This problem provides a general approach to the determination of the common-mode gain of the active-loaded differential amplifier of either type (MOS and BJT). The method is illustrated in Fig. P8.98, in which we have replaced each of Q_1 and Q_2 together with their source (emitter) resistances $2R_{SS}$ ($2R_{EE}$) with a controlled source $G_{mcm}v_{icm}$ and an output resistance $R_{o1,2}$. For the MOS case, $G_{mcm} = v_{icm}/2R_{SS}$; $v_{icm}/2R_{EE}$ for the bipolar case. Usually R_{o1} and R_{o2} are much larger than the resistances at the respective nodes and can be neglected. The current mirror has been replaced by an equivalent circuit consisting of an input resistance R_{in}, a controlled source with current gain A_m, and an output resistance R_{om}.

(a) Show that the common-mode gain is given approximately by

$$A_{cm} \equiv \frac{v_o}{v_{icm}} \simeq G_{mcm}R_{om}(A_m - 1)$$

(b) For the simple MOS mirror consisting of Q_3 and Q_4, as in Fig. 8.32(a), show that

$$A_m = 1\Big/\left[1 + \frac{1}{g_{m3}r_{o3}}\right]$$

and hence derive the expression for the common-mode gain A_{cm} given in Eq. (8.146).

(c) For the simple bipolar mirror consisting of Q_3 and Q_4, as in Fig. 8.37(a), show that

$$A_m = 1\Big/\left(1 + \frac{2}{\beta_P}\right)$$

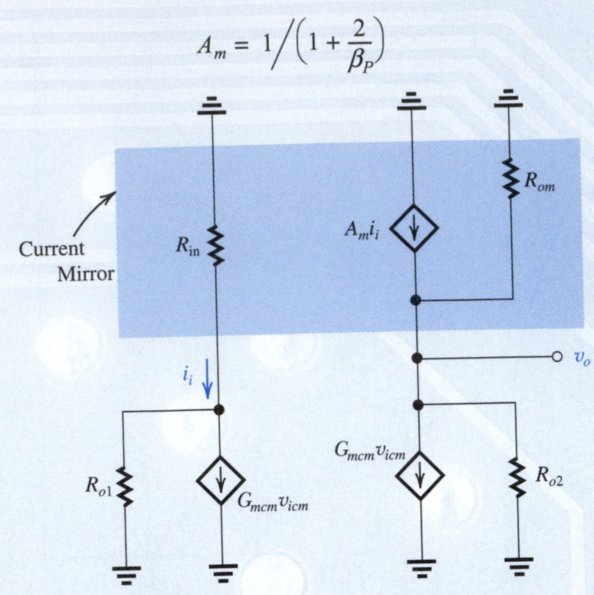

Figure P8.98

and hence derive the expression for the CM gain A_{cm} given in Eq. (8.165).

8.99 For the active-loaded MOS differential pair, replacing the simple current-mirror load by the Wilson mirror of Fig. 7.35(a), find the CM gain. [*Hint*: Use the general formula in Problem 8.98, namely,

$$|A_{cm}| = \frac{R_{om}}{2R_{EE}}(A_m - 1)$$

where R_{om} is the output resistance of the mirror and A_m is its current transfer ratio. Note, however, that this formula will overestimate $|A_{cm}|$ because we are neglecting R_{o2}.]

8.100 For the active-loaded bipolar differential pair, replacing the simple current-mirror load by the base-current-compensated mirror of Fig. 7.33, find the expected systematic input offset voltage. Evaluate V_{OS} for $\beta_P = 50$.

8.101 For the active-loaded bipolar differential pair, replacing the simple current-mirror load by the Wilson mirror of Fig. 7.34(a), find the expected systematic input offset voltage. Evaluate V_{OS} for $\beta_P = 50$.

8.102 Figure P8.102 shows a differential cascode amplifier with an active load formed by a Wilson current mirror. Utilizing the expressions derived in Chapter 7 for the output resistance of a bipolar cascode and the output resistance of the Wilson mirror, and assuming all transistors to be identical, show that the differential voltage gain A_d is given approximately by

$$A_d = \tfrac{1}{3}\beta g_m r_o$$

Evaluate A_d for the case of $\beta = 100$ and $V_A = 30$ V.

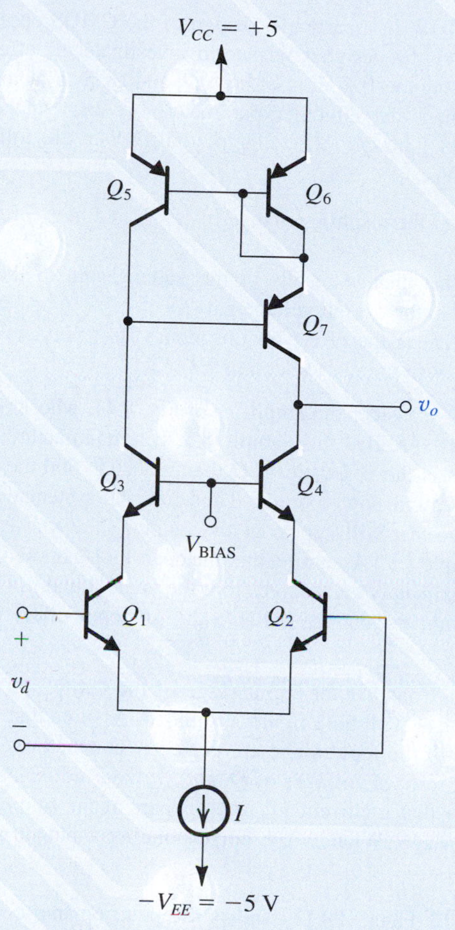

Figure P8.102

D 8.103 Consider the bias design of the Wilson-loaded cascode differential amplifier shown in Fig. P8.102.

(a) What is the largest signal voltage possible at the output without Q_7 saturating? Assume that the CB junction conducts when the voltage across it exceeds 0.4 V.
(b) What should the dc bias voltage established at the output (by an arrangement not shown) be in order to allow for positive output signal swing of 1.5 V?
(c) What should the value of V_{BIAS} be in order to allow for a negative output signal swing of 1.5 V?
(d) What is the upper limit on the input common-mode voltage v_{CM}?

****8.104** Figure P8.104 shows a modified cascode differential amplifier. Here Q_3 and Q_4 are the cascode transistors. However, the manner in which Q_3 is connected with its base current feeding the current mirror Q_7–Q_8 results in very interesting input properties. Note that for simplicity the circuit is shown with the base of Q_2 grounded.

(a) With $v_I = 0$ V dc, find the input bias current I_B assuming all transistors have equal value of β. Compare the case without the Q_7–Q_8 connection.
(b) With $v_I = 0$ V (dc) $+ v_{id}$, find the input signal current i_i and hence the input differential resistance R_{id}. Compare with the case without the Q_7–Q_8 connection. By what factor does R_{id} increase?

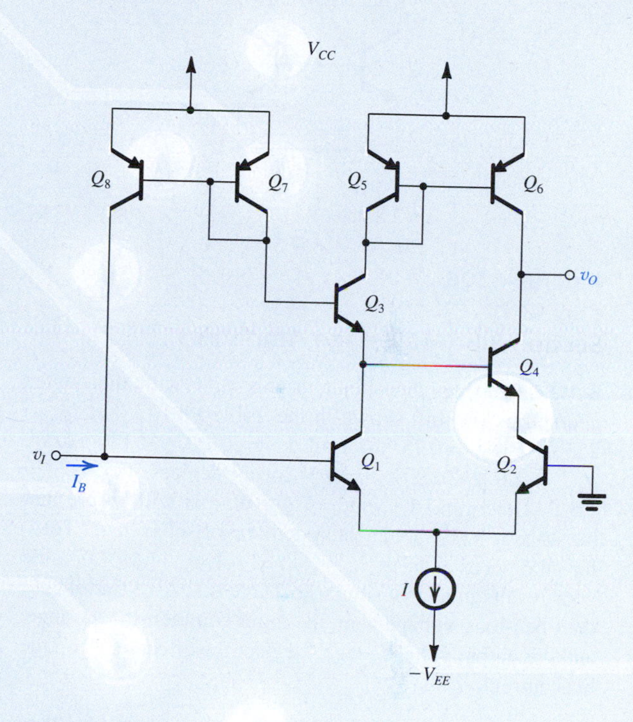

Figure P8.104

8.105 For the folded-cascode differential amplifier of Fig. 8.40, find the value of V_{BIAS} that results in the largest possible positive output swing, while keeping Q_3, Q_4, and the *pnp* transistors that realize the current sources out of saturation. Assume $V_{CC} = V_{EE} = 5$ V. If the dc level at the output is 0 V, find the maximum allowable output signal swing. For $I =$ 0.4 mA, $\beta_P = 50$, $\beta_N = 150$, and $V_A = 120$ V find G_m, R_{o4}, R_{o5}, R_o, and A_d.

8.106 For the BiCMOS differential amplifier in Fig. P8.106 let $V_{DD} = V_{SS} = 3$ V, $I = 0.4$ mA, $k_p' W/L =$ 6.4 mA/V^2; $|V_A|$ for *p*-channel MOSFETs is 10 V, $|V_A|$ for *npn* transistors is 30 V. Find G_m, R_o, and A_d.

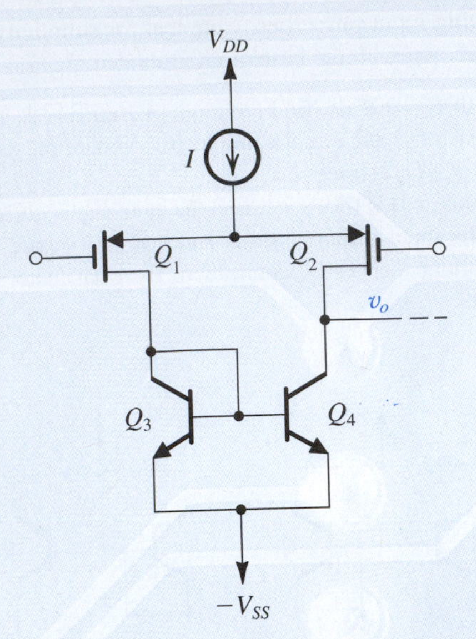

Figure P8.106

Section 8.6: Multistage Amplifiers

8.107 Consider the circuit in Fig. 8.41 with the device geometries (in μm) shown in the Table P8.107. Let I_{REF} = 225 μA, $|V_t|$ = 0.75 V for all devices, $\mu_n C_{ox}$ = 180 μA/V², $\mu_p C_{ox}$ = 60 μA/V², $|V_A|$ = 9 V for all devices, $V_{DD} = V_{SS}$ = 1.5 V. Determine the width of Q_6, W, that will ensure that the op amp will not have a systematic offset voltage. Then, for all devices evaluate I_D, $|V_{OV}|$, $|V_{GS}|$, g_m, and r_o. Provide your results in a table similar to Table 8.1. Also find A_1, A_2, the open-loop voltage gain, the input common-mode range, and the output voltage range. Neglect the effect of V_A on the bias current.

D *8.108 The two-stage CMOS op amp in Fig. P8.108 is fabricated in a 0.18-μm technology having $k'_n = 4k'_p$ = 400 μA/V², $V_{tn} = -V_{tp}$ = 0.4 V.

(a) With A and B grounded, perform a dc design that will result in each of Q_1, Q_2, Q_3, and Q_4 conducting a drain current of 200 μA. Design so that all transistors operate at 0.2 V-overdrive voltages. Specify the W/L ratio required for each MOSFET. Present your results in tabular form. What is the dc voltage at the output (ideally)?
(b) Find the input common-mode range.
(c) Find the allowable range of the output voltage.

(s) With $v_A = v_{id}/2$ and $v_B = -v_{id}/2$, find the voltage gain v_o/v_{id}. Assume an Early voltage of 5 V.

D *8.109 In a particular design of the CMOS op amp of Fig. 8.41 the designer wishes to investigate the effects of increasing the W/L ratio of both Q_1 and Q_2 by a factor of 4. Assuming that all other parameters are kept unchanged, refer to Example 8.5 to help you answer the following questions:

(a) Find the resulting change in $|V_{OV}|$ and in g_m of Q_1 and Q_2.
(b) What change results in the voltage gain of the input stage? In the overall voltage gain?
(c) What is the effect on the input offset voltages? (You might wish to refer to Section 8.4).

8.110 Consider the amplifier of Fig. 8.41, whose parameters are specified in Example 8.5. If a manufacturing error results in the W/L ratio of Q_7 being 50/0.8, find the current that Q_7 will now conduct. Thus find the systematic offset voltage that will appear at the output. (Use the results of Example 8.5.) Assuming that the open-loop gain will remain approximately unchanged from the value found in Example 8.5, find the corresponding value of input offset voltage, V_{OS}.

8.111 Consider the input stage of the CMOS op amp in Fig. 8.41 with both inputs grounded. Assume that the two sides of the input stage are perfectly matched except that the threshold voltages of Q_3 and Q_4 have a mismatch ΔV_t. Show that a current $g_{m3}\Delta V_t$ appears at the output of the first stage. What is the corresponding input offset voltage?

***8.112** Figure P8.112 shows a bipolar op-amp circuit that resembles the CMOS op amp of Fig. 8.41. Here, the input differential pair Q_1–Q_2 is loaded in a current mirror formed by Q_3 and Q_4. The second stage is formed by the current-source-loaded common-emitter transistor Q_5. Unlike the CMOS circuit, here there is an output stage formed by the emitter follower Q_6. The function of capacitor C_C will be explained later in Chapter 10. All transistors have β = 100, $|V_{BE}|$ = 0.7 V, and $r_o = \infty$.

(a) For inputs grounded and output held at 0 V (by negative feedback, not shown) find the emitter currents of all transistors.
(b) Calculate the gain of the amplifier with R_L = 10 kΩ.

Table P8.107

Transistor	Q_1	Q_2	Q_3	Q_4	Q_5	Q_6	Q_7	Q_8
W/L	30/0.5	30/0.5	10/0.5	10/0.5	60/0.5	W/0.5	60/0.5	60/0.5

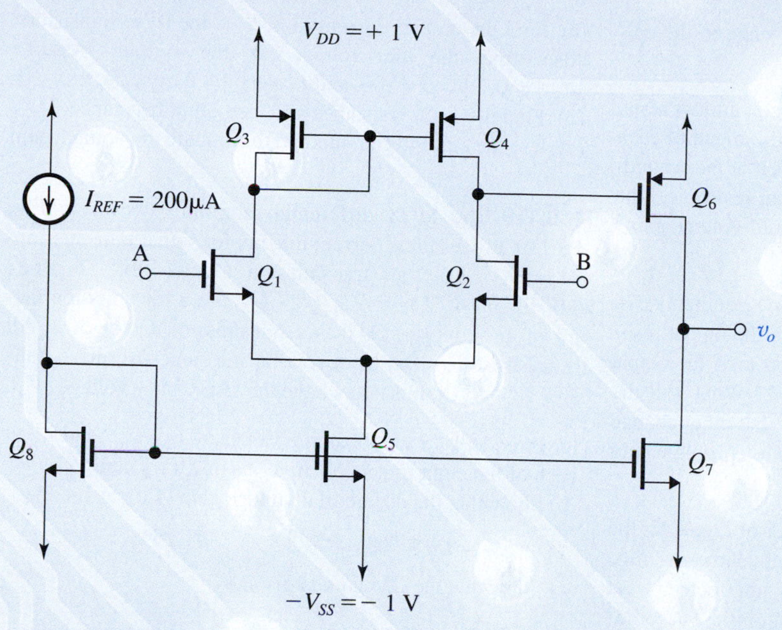

Figure P8.108

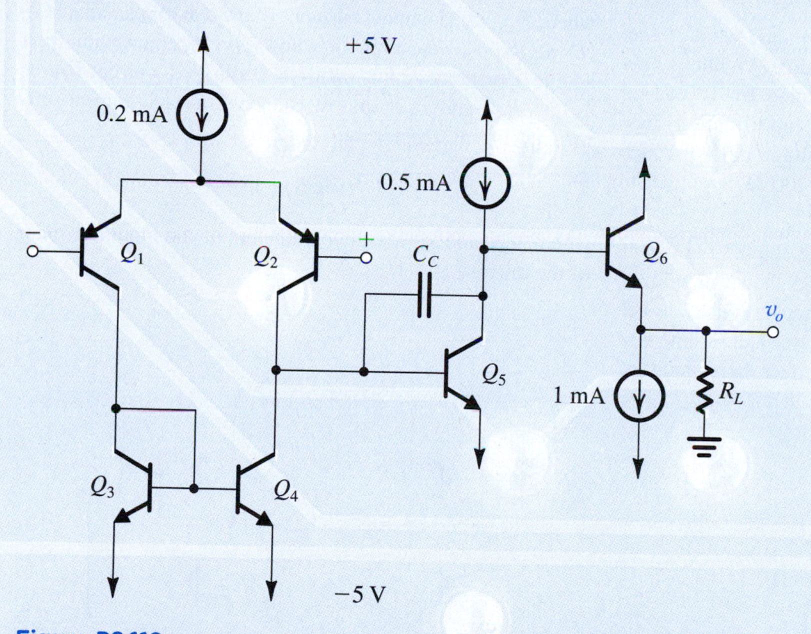

Figure P8.112

D 8.113 It is required to design the circuit of Fig. 8.42 to provide a bias current I_B of 225 μA with Q_8 and Q_9 as matched devices having $W/L = 60/0.5$. Transistors Q_{10}, Q_{11}, and Q_{13} are to be identical and must have the same g_m as Q_8 and Q_9. Transistor Q_{12} is to be four times as wide as Q_{13}. Let $k_n' = 3k_p' = 180\ \mu\text{A/V}^2$, and $V_{DD} = V_{SS} = 1.5$ V. Find the required value of R_B. What is the voltage drop across R_B? Also specify the W/L ratios of Q_{10}, Q_{11}, Q_{12}, and Q_{13} and give the expected dc voltages at the gates of Q_{12}, Q_{10}, and Q_8.

8.114 A BJT differential amplifier, biased to have $r_e = 100\ \Omega$ and utilizing two 100-Ω emitter resistors and 5-kΩ loads, drives a second differential stage biased to have $r_e = 50\ \Omega$. All BJTs have $\beta = 100$. What is the voltage gain of the first stage? Also find the input resistance of the first stage, and the current gain from the input of the first stage to the collectors of the second stage.

8.115 In the multistage amplifier of Fig. 8.43, emitter resistors are to be introduced—100 Ω in the emitter lead of each of the first-stage transistors and 25 Ω for each of the second-stage transistors. What is the effect on input resistance, the voltage gain of the first stage, and the overall voltage gain? Use the bias values found in Example 8.6.

D 8.116 Consider the circuit of Fig. 8.43 and its output resistance. Which resistor has the most effect on the output resistance? What should this resistor be changed to if the output resistance is to be reduced by a factor of 2? What will the amplifier gain become after this change? What other change can you make to restore the amplifier gain to approximately its prior value?

D 8.117 (a) If, in the multistage amplifier of Fig. 8.43, the resistor R_5 is replaced by a constant-current source $\simeq 1$ mA, such that the bias situation is essentially unaffected, what does the overall voltage gain of the amplifier become? Assume that the output resistance of the current source is very high. Use the results of Example 8.7.
(b) With the modification suggested in (a), what is the effect of the change on output resistance? What is the overall gain of the amplifier when loaded by 100 Ω to ground? The original amplifier (before modification) has an output resistance of 152 Ω and a voltage gain of 8513 V/V. What is its gain when loaded by 100 Ω? Comment. Use $\beta = 100$.

***8.118** Figure P8.118 shows a three-stage amplifier in which the stages are directly coupled. The amplifier, however, utilizes bypass capacitors, and, as such, its frequency response falls off at low frequencies. For our purposes here, we shall assume that the capacitors are large enough to act as perfect short circuits at all signal frequencies of interest.

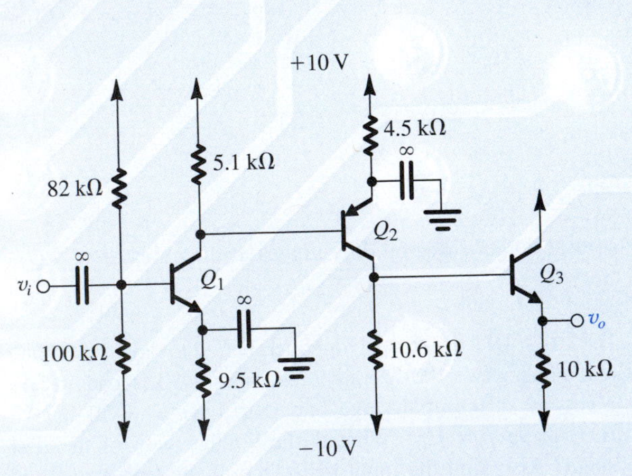

Figure P8.118

(a) Find the dc bias current in each of the three transistors. Also find the dc voltage at the output. Assume $|V_{BE}| = 0.7$ V, $\beta = 100$, and neglect the Early effect.
(b) Find the input resistance and the output resistance.
(c) Use the current-gain method to evaluate the voltage gain v_o/v_i.

****8.119** The MOS differential amplifier shown in Fig. P8.119 utilizes three current mirrors for signal transmission: $Q_4 - Q_6$ has a transmission factor of 2 [i.e., $(W/L)_6/(W/L)_4 = 2$], $Q_3 - Q_5$ has a transmission factor of 1, and $Q_7 - Q_8$ has a transmission factor of 2. All transistors are sized to operate at the same overdrive voltage, $|V_{OV}|$. All transistors have the same Early voltage $|V_A|$.

(a) Provide in tabular form the values of I_D, g_m, and r_o of each of the eight transistors in terms of I, V_{OV}, and V_A.
(b) Show that the differential voltage gain A_d is given by

$$A_d = 2g_{m1}(r_{o6} \| r_{o8}) = V_A/V_{OV}$$

(c) Show that the CM gain is given by

$$|A_{cm}| \simeq \frac{r_{o6} \| r_{o8}}{R_{SS}} \frac{1}{g_{m7}r_{o7}}$$

where R_{SS} is the output resistance of the bias current source I. [*Hint*: Replace each of Q_1 and Q_2 together with their source resistance $2R_{SS}$ with a controlled current-source $v_{icm}/2R_{SS}$ and an output resistance. For each current mirror, the current transfer ratio is given by

$$A_i \simeq A_i \text{ (ideal)} \left(1 - \frac{1}{g_m r_o}\right)$$

where g_m and r_o are the parameters of the input transistor of the mirror.]

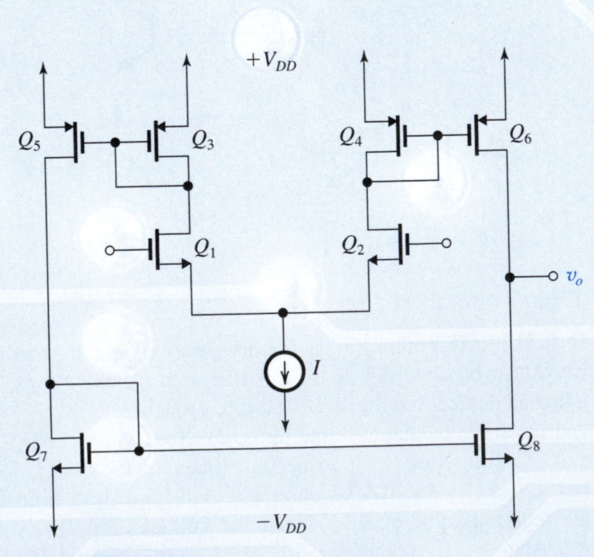

Figure P8.119

(d) If the current-source I is implemented using a simple mirror and the MOS transistor is operated at the same V_{OV}, show that the CMRR is given by

$$\text{CMRR} = 4(V_A/V_{OV})^2$$

(e) Find the input CM range and the output linear range in terms of V_{DD}, $|V_t|$ and $|V_{OV}|$.

D ***8.120 For the circuit shown in Fig. P8.120, which uses a folded cascode involving transistor Q_3, all transistors have $|V_{BE}| = 0.7$ V for the currents involved, $V_A = 200$ V, and $\beta = 100$. The circuit is relatively conventional except for Q_5, which operates in a Class B mode (we will study this in Chapter 11) to provide an increased negative output swing for low-resistance loads.

(a) Perform a bias calculation assuming $|V_{BE}| = 0.7$ V, high β, $V_A = \infty$, $v_+ = v_- = 0$ V, and v_O is stabilized by feedback to about 0 V. Find R so that the reference current I_{REF} is 100 µA. What are the voltages at all the labeled nodes?
(b) Provide in tabular form the bias currents in all transistors together with g_m and r_o for the signal transistors (Q_1, Q_2, Q_3, Q_4, and Q_5) and r_o for Q_C, Q_D, and Q_G.
(c) Now, using $\beta = 100$, find the voltage gain $v_o/(v_+ - v_-)$, and in the process, verify the polarity of the input terminals.
(d) Find the input and output resistances.
(e) Find the input common-mode range for linear operation.

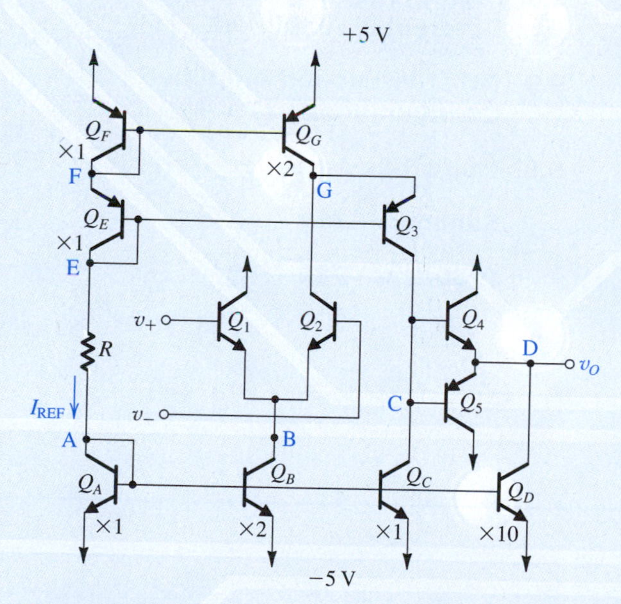

Figure P8.120

(f) For no load, what is the range of available output voltages, assuming $|V_{CEsat}| = 0.3$ V?
(g) Now consider the situation with a load resistance connected from the output to ground. At the positive and negative limits of the output signal swing, find the smallest load resistance that can be driven if one or the other of Q_1 or Q_2 is allowed to cut off.

D ***8.121 In the CMOS op amp shown in Fig. P8.121, all MOS devices have $|V_t| = 1$ V, $\mu_n C_{ox} = 2\mu_p C_{ox} = 40$ µA/V², $|V_A| = 50$ V, and $L = 5$ µm. Device widths are indicated on the diagram as multiples of W, where $W = 5$ µm.

(a) Design R to provide a 10-µA reference current.
(b) Assuming $v_O = 0$ V, as established by external feedback, perform a bias analysis, finding all the labeled node voltages, V_{GS} and I_D for all transistors.
(c) Provide in table form I_D, V_{GS}, g_m, and r_o for all devices.
(d) Calculate the voltage gain $v_o/(v_+ - v_-)$, the input resistance, and the output resistance.
(e) What is the input common-mode range?
(f) What is the output signal range for no load?
(g) For what load resistance connected to ground is the output negative voltage limited to −1 V before Q_7 begins to conduct?
(h) For a load resistance one-tenth of that found in (g), what is the output signal swing?

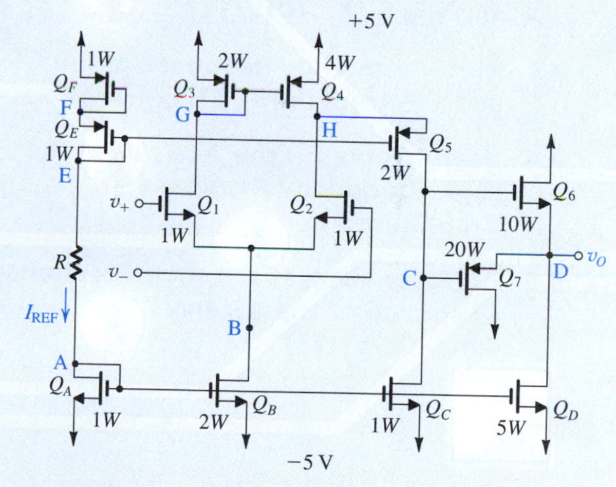

Figure P8.121

CHAPTER 9

Frequency Response

IN THIS CHAPTER YOU WILL LEARN

1. How coupling and bypass capacitors cause the gain of discrete-circuit amplifiers to fall off at low frequencies, and how to obtain an estimate of the frequency f_L at which the gain decreases by 3 dB below its value at midband.

2. The internal capacitive effects present in the MOSFET and the BJT and how to model these effects by adding capacitances to the hybrid-π model of each of the two transistor types.

3. The high-frequency limitation on the gain of the CS and CE amplifiers and how the gain falloff and the upper 3-dB frequency f_H are mostly determined by the small capacitance between the drain and gate (collector and base).

4. Powerful methods for the analysis of the high-frequency response of amplifier circuits of varying complexity.

5. How the cascode amplifier studied in Chapter 7 can be designed to obtain wider bandwidth than is possible with the CS and CE amplifiers.

6. The high-frequency performance of the source and emitter followers.

7. The high-frequency performance of differential amplifiers.

8. Circuit configurations for obtaining wideband amplification.

Introduction

Except for brief comments in Sections 5.6.8 and 6.6.8, our study of transistor amplifiers in Chapters 5 through 8 has assumed that their gain is constant independent of the frequency of the input signal. This would imply that their bandwidth is infinite, which of course is not true! To illustrate, we show in Fig. 9.1 a sketch of the magnitude of the gain versus the frequency of the input signal of a discrete-circuit BJT or MOS amplifier. Observe that there is indeed a wide frequency range over which the gain remains almost constant. This is the useful frequency range of operation for the particular amplifier. Thus far, we have been assuming that our amplifiers are operating in this band, called the middle-frequency band or **midband**. The amplifier is designed so that its midband coincides with the frequency spectrum of the signals it is required to amplify. If this were not the case, the amplifier would distort the frequency spectrum of the input signal, with different components of the input signal being amplified by different amounts.

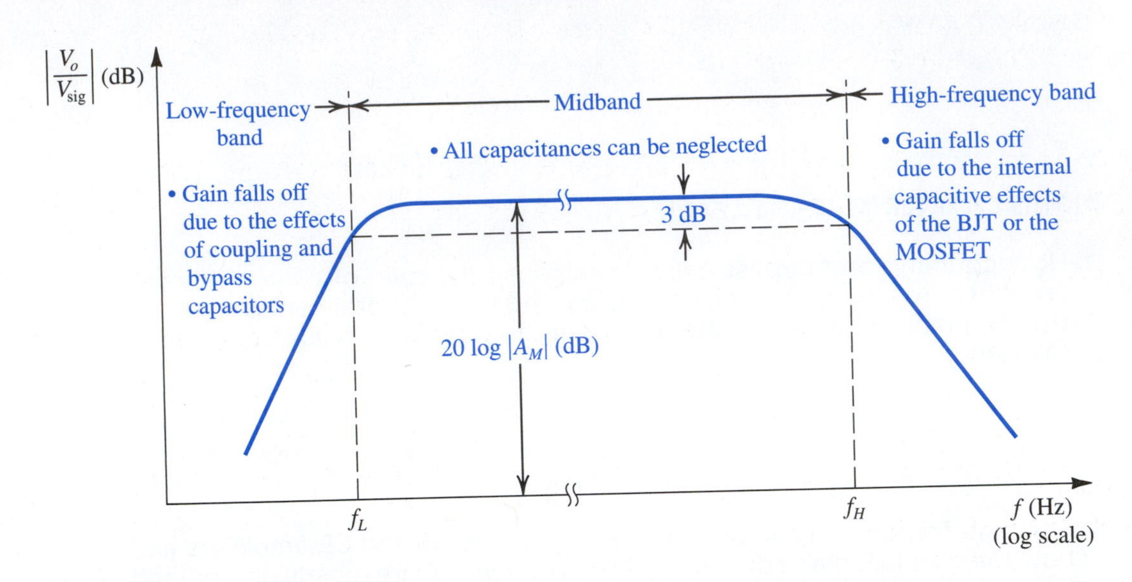

Figure 9.1 Sketch of the magnitude of the gain of a discrete-circuit BJT or MOS amplifier versus frequency. The graph delineates the three frequency bands relevant to frequency-response determination.

Figure 9.1 indicates that at lower frequencies, the magnitude of the amplifier gain falls off. This occurs because the coupling and bypass capacitors no longer have low impedances. Recall that we assumed that their impedances were small enough to act as short circuits. Although this can be true at midband frequencies, as the frequency of the input signal is lowered, the reactance $1/j\omega C$ of each of these capacitors becomes significant and, as will be shown in Section 9.1, this results in a decrease in the overall voltage gain of the amplifier. In the analysis of the low-frequency response of discrete-circuit amplifiers in Section 9.1 we will be particularly interested in the determination of the frequency f_L, which defines the lower end of the midband. It is usually defined as the frequency at which the gain drops by 3 dB below its value in midband. Integrated-circuit amplifiers do not utilize coupling and bypass capacitors, and thus their midband extends down to zero frequency (dc).

Figure 9.1 indicates also that the gain of the amplifier falls off at the high-frequency end. This is due to internal capacitive effects in the BJT and in the MOSFET. We shall study these effects in Section 9.2 and model them with capacitances that we will add to the hybrid-π model of the BJT and the MOSFET. The resulting high-frequency device models will be utilized in Section 9.3 in the analysis of the high-frequency response of the CS and CE amplifiers, both discrete and integrated. We will be specifically interested in the determination of the frequency f_H, which defines the upper end of the midband. It is defined as the frequency at which the gain drops by 3 dB below its midband value. Thus, the amplifier bandwidth is defined by f_L and f_H (0 and f_H for IC amplifiers).

The remainder of this chapter will be concerned with the frequency response analysis of a variety of amplifier configurations of varying degrees of complexity. Of particular interest to us are ways to extend the amplifier bandwidth (i.e., increase f_H) either by adding specific circuit components, such as source and emitter degeneration resistances, or by changing the circuit configuration altogether.

Before embarking on the study of this chapter, the reader is urged to review Section 1.6, which introduces the subject of amplifier frequency response and the extremely important topic of single-time-constant (STC) circuits. More details on STC circuits can be found in

Appendix E. As well, Appendix F provides a review of important tools from circuit and system theory: poles, zeros, and Bode plots.

9.1 Low-Frequency Response of the Common-Source and Common-Emitter Amplifiers

9.1.1 The CS Amplifier

Figure 9.2(a) shows a discrete-circuit, common-source amplifier utilizing coupling capacitors C_{C1} and C_{C2}, and bypass capacitor C_S. We wish to determine the effect of these capacitances on the gain V_o/V_{sig} of the amplifier. As mentioned before, at midband frequencies, these capacitances have negligibly small impedances and can be assumed to be perfect short circuits for the purpose of calculating the midband gain. At low frequencies, however, the reactance $1/j\omega C$ of each of the three capacitances increases and the amplifier gain decreases, as we shall now show.

Determining V_o/V_{sig} To determine the low-frequency gain or transfer function of the common-source amplifier, we show in Fig. 9.2(b) the circuit with the dc sources eliminated (current source I open-circuited and voltage source V_{DD} short-circuited). We shall perform the small-signal analysis directly on this circuit. However, we will ignore r_o. This is done in order to keep the analysis simple and thus focus attention on significant issues. The effect of r_o on the low-frequency operation of this amplifier is minor, as can be verified by a SPICE simulation.

To determine the gain V_o/V_{sig}, we start at the signal source and work our way through the circuit, determining V_g, I_d, I_o, and V_o, in this order.[1] To find the fraction of V_{sig} that appears at the transistor gate, V_g, we use the voltage divider rule at the input to write

$$V_g = V_{sig}\frac{R_G}{R_G + \dfrac{1}{sC_{C1}} + R_{sig}}$$

which can be written in the alternate form

$$V_g = V_{sig}\frac{R_G}{R_G + R_{sig}}\frac{s}{s + \dfrac{1}{C_{C1}(R_G + R_{sig})}} \tag{9.1}$$

Thus we see that the expression for the signal transmission from signal generator to amplifier input has acquired a frequency-dependent factor. From our study of frequency response in Section 1.6 (see also Appendix E), we recognize this factor as the transfer function of an STC circuit of the high-pass type with a break or corner frequency $\omega_0 = 1/C_{C1}(R_G + R_{sig})$. Thus the effect of the coupling capacitor C_{C1} is to introduce a high-pass STC response with a

[1]Note that since we are now dealing with quantities that are functions of frequency, or, equivalently, the Laplace variable s, we are using capital letters with lowercase subscripts for our symbols. This conforms with the symbol notation introduced in Chapter 1.

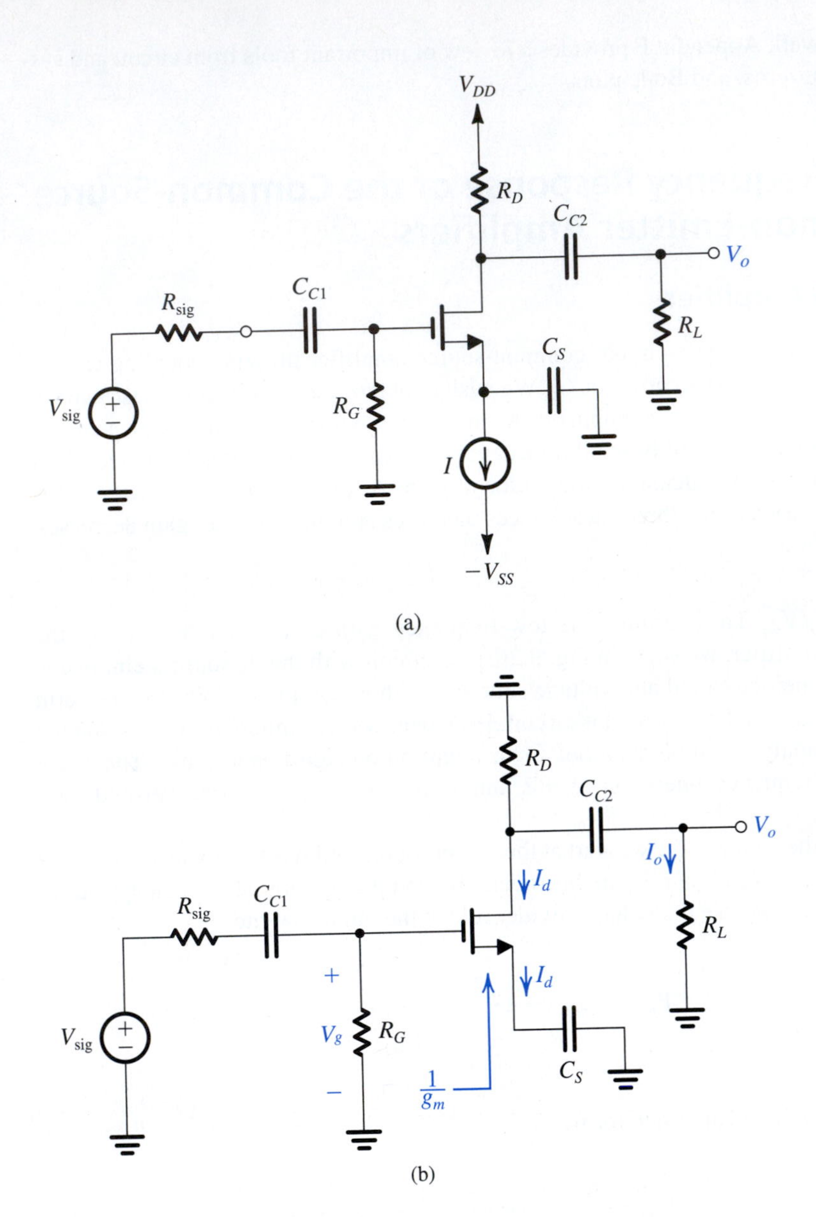

Figure 9.2 **(a)** Capacitively coupled common-source amplifier. **(b)** Analysis of the CS amplifier to determine its low-frequency transfer function. For simplicity, r_o is neglected.

break frequency that we shall denote ω_{P1},

$$\omega_{P1} = \omega_0 = \frac{1}{C_{C1}(R_G + R_{sig})} \tag{9.2}$$

Continuing with the analysis, we next determine the drain current I_d by dividing V_g by the total impedance in the source circuit, which is $[(1/g_m) + (1/sC_S)]$ to obtain

$$I_d = \frac{V_g}{\dfrac{1}{g_m} + \dfrac{1}{sC_S}}$$

which can be written in the alternate form

$$I_d = g_m V_g \frac{s}{s + \dfrac{g_m}{C_S}} \tag{9.3}$$

We observe that C_S introduces a frequency-dependent factor, which is also of the STC high-pass type. Thus the amplifier acquires another break frequency,

$$\omega_{P2} = \frac{g_m}{C_S} \tag{9.4}$$

To complete the analysis, we find V_o by first using the current divider rule to determine the fraction of I_d that flows through R_L,

$$I_o = -I_d \frac{R_D}{R_D + \dfrac{1}{sC_{C2}} + R_L}$$

and then multiplying I_o by R_L to obtain

$$V_o = I_o R_L = -I_d \frac{R_D R_L}{R_D + R_L} \frac{s}{s + \dfrac{1}{C_{C2}(R_D + R_L)}} \tag{9.5}$$

from which we see that C_{C2} introduces a third STC high-pass factor, giving the amplifier a third break frequency at

$$\omega_{P3} = \frac{1}{C_{C2}(R_D + R_L)} \tag{9.6}$$

The overall low-frequency transfer function of the amplifier can be found by combining Eqs. (9.1), (9.3), and (9.5) and replacing the break frequencies by their symbols from Eqs. (9.2), (9.4), and (9.6):

$$\frac{V_o}{V_{sig}} = -\left(\frac{R_G}{R_G + R_{sig}}\right)[g_m(R_D \| R_L)]\left(\frac{s}{s + \omega_{P1}}\right)\left(\frac{s}{s + \omega_{P2}}\right)\left(\frac{s}{s + \omega_{P3}}\right) \tag{9.7}$$

which can be expressed in the form

$$\frac{V_o}{V_{sig}} = A_M\left(\frac{s}{s + \omega_{P1}}\right)\left(\frac{s}{s + \omega_{P2}}\right)\left(\frac{s}{s + \omega_{P3}}\right) \tag{9.8}$$

where A_M, the midband gain, is given by

$$A_M = -\frac{R_G}{R_G + R_{sig}}[g_m(R_D \| R_L)] \tag{9.9}$$

which is the value we would have obtained, had we assumed that C_{C1}, C_{C2}, and C_S were acting as perfect short circuits. In this regard, note that at midband frequencies—that is, at frequencies $s = j\omega$ much higher than ω_{P1}, ω_{P2}, and ω_{P3}—Eq. (9.8) shows that V_o/V_{sig} approaches $-A_M$, as should be the case.

Determining the Lower 3-dB Frequency, f_L The magnitude of the amplifier gain, $|V_o/V_{sig}|$ at frequency ω can be obtained by substituting $s = j\omega$ in Eq. (9.8) and evaluating the magnitude of the transfer function. In this way, the frequency response of the amplifier can be plotted versus frequency, and the lower 3-dB frequency f_L can be determined as the frequency at which $|V_o/V_{sig}|$ drops to $|A_M|/\sqrt{2}$. A simpler approach, however, is possible if the break frequencies ω_{P1}, ω_{P2}, and ω_{P3} are sufficiently separated. In this case, we can employ the Bode plot rules (see Appendix F) to sketch a Bode plot for the gain magnitude. Such a plot is shown in Fig. 9.3. Observe that since the break frequencies are sufficiently separated, their effects appear distinct. At each break frequency, the slope of the asymptote to the gain function increases by 20 dB/decade. Readers familiar with poles and zeros will recognize f_{P1}, f_{P2}, and f_{P3} as the frequencies of the three real-axis, low-frequency poles of the amplifier. (For a brief review of poles and zeros, refer to Appendix F.)

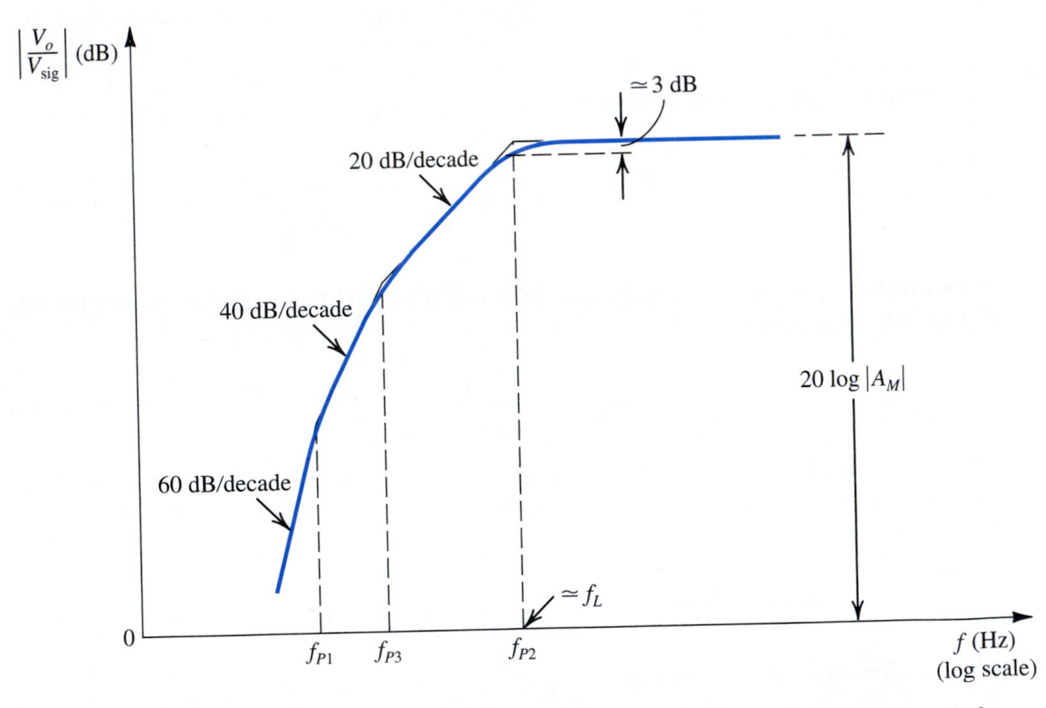

Figure 9.3 Sketch of the low-frequency magnitude response of a CS amplifier for which the three pole frequencies are sufficiently separated for their effects to appear distinct.

A quick way for estimating the 3-dB frequency f_L is possible if the highest-frequency pole (here, f_{P2}) is separated from the nearest pole (here, f_{P3}) by at least a factor of 4 (two octaves). In such a case, f_L is approximately equal to the highest of the pole frequencies,

$$f_L \simeq f_{P2}$$

Usually, the highest-frequency pole is the one caused by C_S. This is because C_S interacts with $1/g_m$, which is relatively low (see Eq. 9.4).

Determining the Pole Frequencies by Inspection Before leaving this section, we present a simple method for finding the time constant and hence the pole frequency associated with each of the three capacitors. The procedure is simple:

1. Reduce V_{sig} to zero.
2. Consider each capacitor separately; that is, assume that the other two capacitors are acting as perfect short circuits.
3. For each capacitor, find the total resistance seen between its terminals. This is the resistance that determines the time constant associated with this capacitor.

The reader is encouraged to apply this procedure to C_{C1}, C_S, and C_{C2} and thus see that Eqs. (9.2), (9.4), and (9.6) can be written by inspection.

Selecting Values for the Coupling and Bypass Capacitors We now address the design issue of selecting appropriate values for C_{C1}, C_S, and C_{C2}. The design objective is to place the lower 3-dB frequency f_L at a specified value while minimizing the capacitor values. Since as mentioned above C_S results in the highest of the three pole frequencies, the total capacitance is minimized by selecting C_S so that its pole frequency $f_{P2} = f_L$. We then decide on the location of the other two pole frequencies, say 5 to 10 times lower than the frequency of the dominant pole, f_{P2}. However, the values selected for f_{P1} and f_{P3} should not be too low, for that would require larger values for C_{C1} and C_{C2} than may be necessary. The design procedure will be illustrated by an example.

Example 9.1

We wish to select appropriate values for the coupling capacitors C_{C1} and C_{C2} and the bypass capacitor C_S for a CS amplifier for which $R_G = 4.7\ \text{M}\Omega$, $R_D = R_L = 15\ \text{k}\Omega$, $R_{sig} = 100\ \text{k}\Omega$, and $g_m = 1\ \text{mA/V}$. It is required to have f_L at 100 Hz and that the nearest break frequency be at least a decade lower.

Solution

We select C_S so that

$$f_{P2} = \frac{1}{2\pi(C_S/g_m)} = f_L$$

Thus,

$$C_S = \frac{g_m}{2\pi f_L} = \frac{1 \times 10^{-3}}{2\pi \times 100} = 1.6\ \mu\text{F}$$

For $f_{P1} = f_{P3} = 10$ Hz, we obtain

$$10 = \frac{1}{2\pi C_{C1}(0.1 + 4.7) \times 10^6}$$

Example 9.1 *continued*

which yields

$$C_{C1} = 3.3 \text{ nF}$$

and

$$10 = \frac{1}{2\pi C_{C2}(15 + 15) \times 10^3}$$

which results in

$$C_{C2} = 0.53 \text{ µF}$$

EXERCISE

9.1 A CS amplifier has $C_{C1} = C_S = C_{C2} = 1$ µF, $R_G = 10$ MΩ, $R_{sig} = 100$ kΩ, $g_m = 2$ mA/V, $R_D = R_L = 10$ kΩ. Find A_M, f_{P1}, f_{P2}, f_{P3}, and f_L.
Ans. –9.9 V/V; 0.016 Hz; 318.3 Hz; 8 Hz; 318.3 Hz

9.1.2 The CE Amplifier

Figure 9.4 shows a common-emitter amplifier that utilizes coupling capacitors C_{C1} and C_{C2} and emitter bypass capacitor C_E. As in the case of the MOS amplifier, the effect of these capacitors is felt only at low frequencies. Our objective is to determine the amplifier gain or transfer function V_o/V_{sig} with these three capacitances taken into account. Toward that end, we show in Fig. 9.4(b) the circuit with the dc sources eliminated. We shall perform the small-signal analysis directly on the circuit. To keep the analysis simple, we shall neglect the effect of r_o, as we have done in the MOS case.

The analysis of the circuit in Fig. 9.4(b) is somewhat more complicated than that for the CS case. This is a result of the finite β of the BJT, which causes the input impedance at the base to be a function of C_E. Thus the effects of C_{C1} and C_E are no longer separable. Although one can certainly still derive an expression for the overall transfer function, the result will be quite complicated, making it difficult to obtain design insight. Therefore we shall pursue an approximate alternative approach.

Considering the Effect of Each of the Three Capacitors Separately Our first cut at the analysis of the circuit in Fig. 9.4(b) is to consider the effect of the three capacitors C_{C1}, C_E, and C_{C2} *one at a time*. That is, when finding the effect of C_{C1}, we shall assume that C_E and C_{C2} are acting as perfect short circuits, and when considering C_E, we assume that C_{C1} and C_{C2} are perfect short circuits, and so on. This is obviously a major simplifying assumption—and one that might not be justified. However, it should serve as a first cut at the analysis, enabling us to gain insight into the effect of these capacitances.

Figure 9.5(a) shows the circuit with C_E and C_{C2} replaced with short circuits. The voltage V_π at the base of the transistor can be written as

$$V_\pi = V_{sig} \frac{R_B \| r_\pi}{(R_B \| r_\pi) + R_{sig} + \dfrac{1}{sC_{C1}}}$$

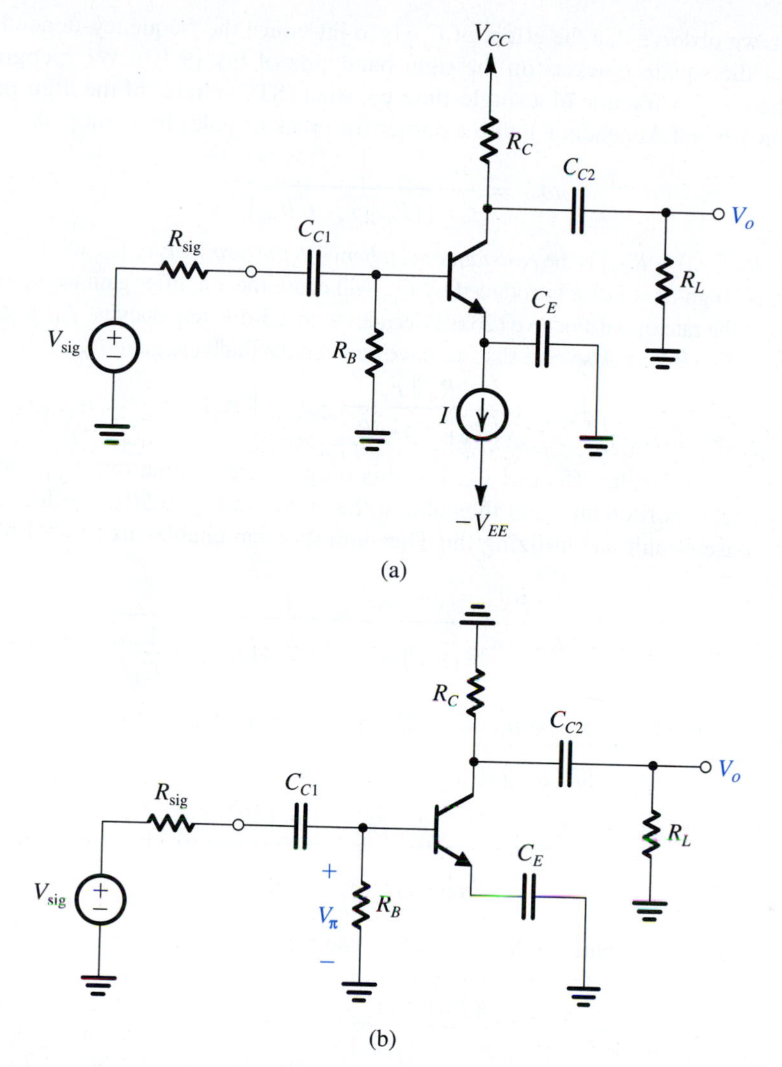

Figure 9.4 **(a)** A capacitively coupled common-emitter amplifier. **(b)** The circuit prepared for small-signal analysis.

and the output voltage is obtained as

$$V_o = -g_m V_\pi (R_C \| R_L)$$

These two equations can be combined to obtain the voltage gain V_o/V_{sig} including the effect of C_{C1} as

$$\frac{V_o}{V_{sig}} = -\frac{(R_B \| r_\pi)}{(R_B \| r_\pi) + R_{sig}} g_m (R_C \| R_L) \left[\frac{s}{s + \dfrac{1}{C_{C1}[(R_B \| r_\pi) + R_{sig}]}} \right] \tag{9.10}$$

from which we observe that the effect of C_{C1} is to introduce the frequency-dependent factor between the square brackets on the right-hand side of Eq. (9.10). We recognize this factor as the transfer fraction of a single-time-constant (STC) circuit of the high-pass type (see Section 1.6 and Appendix E) with a corner (or break or pole) frequency ω_{P1},

$$\omega_{P1} = \frac{1}{C_{C1}[(R_B \| r_\pi) + R_{sig}]} \tag{9.11}$$

Note that $[(R_B \| r_\pi) + R_{sig}]$ is the *resistance seen between the terminals of C_{C1} when V_{sig} is set to zero*. The STC high-pass factor introduced by C_{C1} will cause the amplifier gain to roll off at low frequencies at the rate of 6 dB/octave (20 dB/decade) with a 3-dB frequency at $f_{P1} = \omega_{P1}/2\pi$, as indicated in Fig. 9.5(a). Also note that we have denoted the midband gain A_M,

$$A_M = -\frac{(R_B \| r_\pi)}{(R_B \| r_\pi) + R_{sig}} g_m(R_C \| R_L) \tag{9.12}$$

Next, we consider the effect of C_E. For this purpose we assume that C_{C1} and C_{C2} are acting as perfect short circuits and thus obtain the circuit in Fig. 9.5(b). Reflecting r_e and C_E into the base circuit and utilizing the Thévenin theorem enables us to obtain the base current as

$$I_b = V_{sig}\frac{R_B}{R_B + R_{sig}} \frac{1}{(R_B \| R_{sig}) + (\beta + 1)\left(r_e + \dfrac{1}{sC_E}\right)}$$

The collector current can then be found as βI_b and the output voltage as

$$V_o = -\beta I_b(R_C \| R_L)$$

$$= -\frac{R_B}{R_B + R_{sig}} \frac{\beta(R_C \| R_L)}{(R_B \| R_{sig}) + (\beta + 1)\left(r_e + \dfrac{1}{sC_E}\right)} V_{sig}$$

Thus the voltage gain including the effect of C_E can be expressed as[2]

$$\frac{V_o}{V_{sig}} = -\frac{R_B}{R_B + R_{sig}} \frac{\beta(R_C \| R_L)}{(R_B \| R_{sig}) + (\beta + 1)r_e} \frac{s}{s + \left[1/C_E\left(r_e + \dfrac{R_B \| R_{sig}}{\beta + 1}\right)\right]} \tag{9.13}$$

We observe that C_E introduces the STC high-pass factor on the extreme right-hand side. Thus C_E causes the gain to fall off at low frequency at the rate of 6 dB/octave with a 3-dB frequency equal to the corner (or pole) frequency of the high-pass STC function; that is,

$$\omega_{P2} = \frac{1}{C_E\left[r_e + \dfrac{R_B \| R_{sig}}{\beta + 1}\right]} \tag{9.14}$$

Observe that $[r_e + ((R_B \| R_{sig})/(\beta + 1))]$ *is the resistance seen between the two terminals of C_E when V_{sig} is set to zero*. The effect of C_E on the amplifier frequency response is illustrated by the sketch in Fig. 9.5(b).

---

[2] It can be shown that the factor multiplying the high-pass transfer function in Eq. (9.13) is equal to A_M of Eq. (9.12).

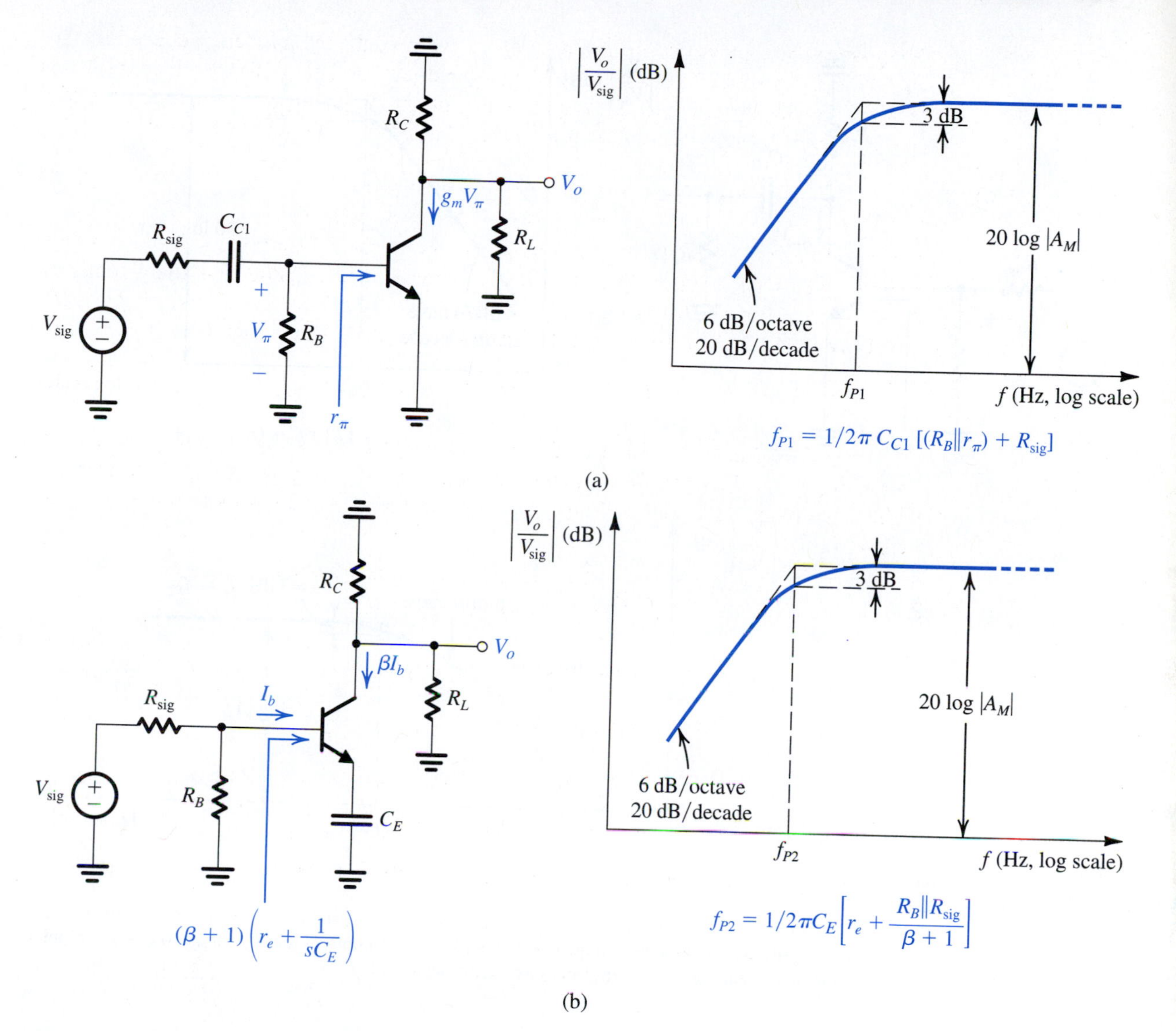

Figure 9.5 Analysis of the low-frequency response of the CE amplifier of Fig. 9.4: **(a)** the effect of C_{C1} is determined with C_E and C_{C2} assumed to be acting as perfect short circuits; **(b)** the effect of C_E is determined with C_{C1} and C_{C2} assumed to be acting as perfect short circuits;

Finally, we consider the effect of C_{C2}. The circuit with C_{C1} and C_E assumed to be acting as perfect short circuits is shown in Fig. 9.5(c), for which we can write

$$V_\pi = V_{sig}\frac{R_B \parallel r_\pi}{(R_B \parallel r_\pi) + R_{sig}}$$

and

$$V_o = -g_m V_\pi \frac{R_C}{R_C + \dfrac{1}{sC_{C2}} + R_L} R_L$$

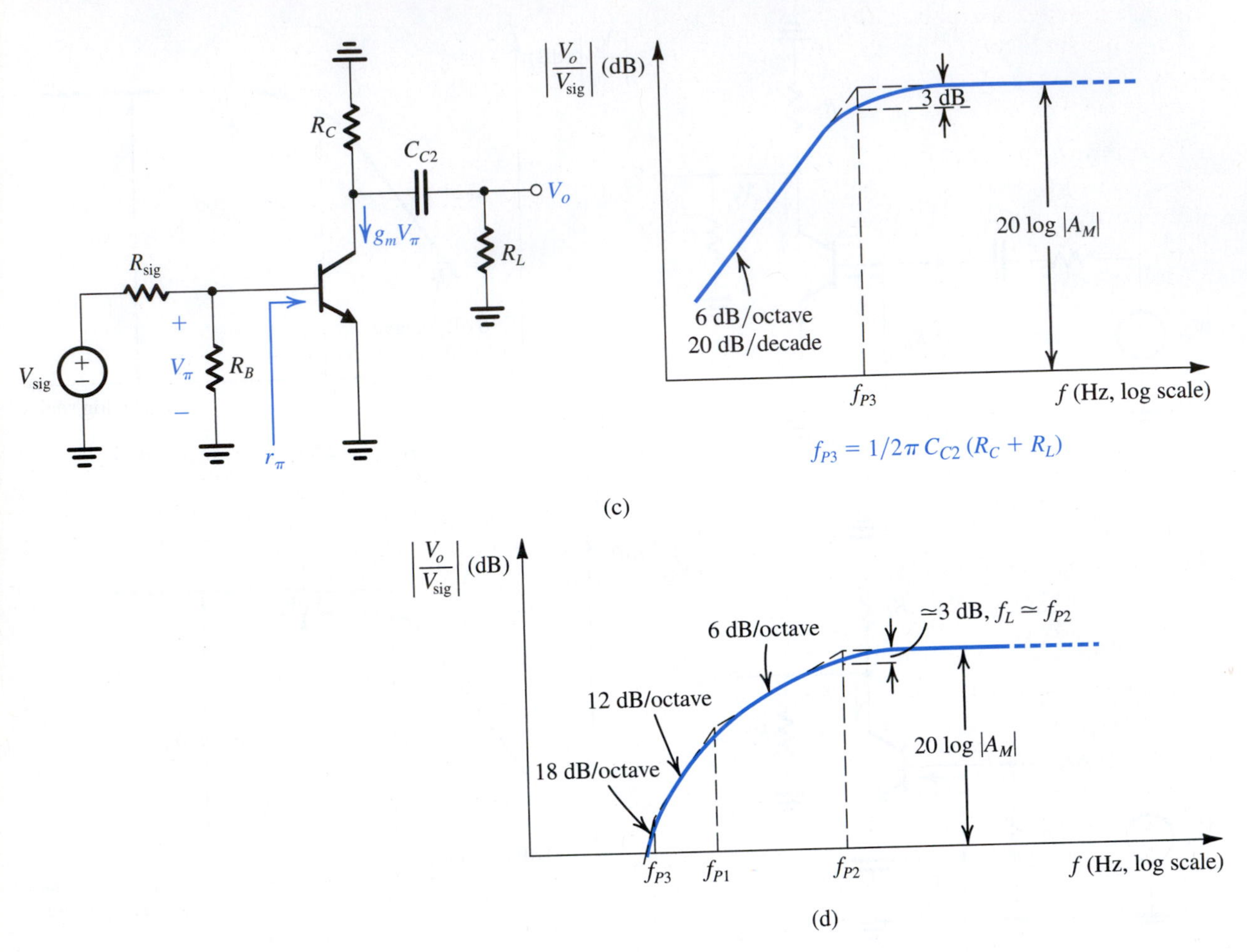

Figure 9.5 *(continued)* **(c)** the effect of C_{C2} is determined with C_{C1} and C_E assumed to be acting as perfect short circuits; **(d)** sketch of the low-frequency gain under the assumptions that C_{C1}, C_E, and C_{C2} do not interact and that their break (or pole) frequencies are widely separated.

These two equations can be combined to obtain the low-frequency gain including the effect of C_{C2} as

$$\frac{V_o}{V_{sig}} = -\frac{R_B \| r_\pi}{(R_B \| r_\pi) + R_{sig}} g_m (R_C \| R_L) \left[\frac{s}{s + \dfrac{1}{C_{C2}(R_C + R_L)}}\right] \tag{9.15}$$

We observe that C_{C2} introduces the frequency-dependent factor between the square brackets, which we recognize as the transfer function of a high-pass STC circuit with a pole frequency ω_{P3},

$$\omega_{P3} = \frac{1}{C_{C2}(R_C + R_L)} \tag{9.16}$$

Here we note that as expected, $(R_C + R_L)$ *is the resistance seen between the terminals of* C_{C2} *when* V_{sig} *is set to zero*. Thus capacitor C_{C2} causes the low-frequency gain of the amplifier to decrease at the rate of 6 dB/octave with a 3-dB frequency at $f_{P3} = \omega_{P3}/2\pi$, as illustrated by the sketch in Fig. 9.5(c).

Determining the Lower 3-dB Frequency, f_L Now that we have determined the effects of each of C_{C1}, C_E, and C_{C2} acting alone, the question becomes what will happen when all three are present at the same time. This question has two parts: First, what happens when all three capacitors are present but do not interact? The answer is that the amplifier low-frequency gain can be expressed as

$$\frac{V_o}{V_{sig}} = -A_M\left(\frac{s}{s+\omega_{P1}}\right)\left(\frac{s}{s+\omega_{P2}}\right)\left(\frac{s}{s+\omega_{P3}}\right) \qquad (9.17)$$

from which we see that it acquires three poles with frequencies f_{P1}, f_{P2}, and f_{P3}, all in the low-frequency band. If the three frequencies are widely separated, their effects will be distinct, as indicated by the sketch in Fig. 9.5(d). The important point to note here is that the 3-dB frequency f_L *is determined by the highest of the three pole frequencies*. This is usually the pole caused by the bypass capacitor C_E, simply because the resistance that it sees is usually quite small. Thus, even if one uses a large value for C_E, f_{P2} is usually the highest of the three pole frequencies.

If f_{P1}, f_{P2}, and f_{P3} are close together, none of the three dominates, and to determine f_L, we have to evaluate $|V_o/V_{sig}|$ in Eq. (9.17) and calculate the frequency at which it drops to $|A_M|/\sqrt{2}$. The work involved in doing this, however, is usually too great and is rarely justified in practice, particularly because in any case, Eq. (9.17) is an approximation based on the assumption that the three capacitors do not interact. This leads to the second part of the question: What happens when all three capacitors are present and interact? We do know that C_{C1} and C_E usually interact and that their combined effect is two poles at frequencies that will differ somewhat from ω_{P1} and ω_{P2}. Of course, one can derive the overall transfer function taking this interaction into account and find more precisely the low-frequency response. This, however, will be too complicated to yield additional insight. As an alternative, for hand calculations, we can obtain a reasonably good estimate for f_L using the following formula (which we will not derive here)[3]:

$$f_L \simeq \frac{1}{2\pi}\left[\frac{1}{C_{C1}R_{C1}} + \frac{1}{C_E R_E} + \frac{1}{C_{C2}R_{C2}}\right] \qquad (9.18)$$

or equivalently,

$$f_L = f_{P1} + f_{P2} + f_{P3} \qquad (9.19)$$

where R_{C1}, R_E, and R_{C2} are the resistances seen by C_{C1}, C_E, and C_{C2}, respectively, when V_{sig} is set to zero and the other two capacitances are replaced with short circuits. Equations (9.18) and (9.19) provide insight regarding the relative contributions of the three capacitors to f_L. Finally, we note that a far more precise determination of the low-frequency gain and the 3-dB frequency f_L can be obtained using SPICE.

Selecting Values for C_{C1}, C_E, and C_{C2} We now address the design issue of selecting appropriate values for C_{C1}, C_E, and C_{C2}. The design objective is to place the lower 3-dB frequency f_L at a specified location while minimizing the capacitor values. Since, as mentioned above, C_E usually sees the lowest of the three resistances, the total capacitance is minimized

[3] The interested reader can refer to Appendix F.

by selecting C_E so that its contribution to f_L is dominant. That is, by reference to Eq. (9.18), we may select C_E such that $1/(C_E R_E)$ is, say, 80% of $\omega_L = 2\pi f_L$, leaving each of the other capacitors to contribute 10% to the value of ω_L. Example 9.2 should help to illustrate this process.

Example 9.2

We wish to select appropriate values for C_{C1}, C_{C2}, and C_E for the common-emitter amplifier, which has $R_B = 100\ \text{k}\Omega$, $R_C = 8\ \text{k}\Omega$, $R_L = 5\ \text{k}\Omega$, $R_{sig} = 5\ \text{k}\Omega$, $\beta = 100$, $g_m = 40\ \text{mA/V}$, and $r_\pi = 2.5\ \text{k}\Omega$. It is required to have $f_L = 100\ \text{Hz}$.

Solution

We first determine the resistances seen by the three capacitors C_{C1}, C_E, and C_{C2} as follows:

$$R_{C1} = (R_B \| r_\pi) + R_{sig}$$

$$= (100 \| 2.5) + 5 = 7.44\ \text{k}\Omega$$

$$R_E = r_e + \frac{R_B \| R_{sig}}{\beta + 1}$$

$$= 0.025 + \frac{100 \| 5}{101} = 0.072\ \text{k}\Omega = 72\ \Omega$$

$$R_{C2} = R_C + R_L = 8 + 5 = 13\ \text{k}\Omega$$

Now, selecting C_E so that it contributes 80% of the value of ω_L gives

$$\frac{1}{C_E \times 72} = 0.8 \times 2\pi \times 100$$

$$C_E = 27.6\ \mu\text{F}$$

Next, if C_{C1} is to contribute 10% of f_L,

$$\frac{1}{C_{C1} \times 7.44 \times 10^3} = 0.1 \times 2\pi \times 100$$

$$C_{C1} = 2.1\ \mu\text{F}$$

Similarly, if C_{C2} is to contribute 10% of f_L, its value should be selected as follows:

$$\frac{1}{C_{C2} \times 13 \times 10^3} = 0.1 \times 2\pi \times 100$$

$$C_{C2} = 1.2\ \mu\text{F}$$

In practice, we would select the nearest standard values for the three capacitors while ensuring that $f_L \leq 100\ \text{Hz}$.

EXERCISE

9.2 A common-emitter amplifier has $C_{C1} = C_E = C_{C2} = 1\ \mu F$, $R_B = 100\ k\Omega$, $R_{sig} = 5\ k\Omega$, $g_m = 40\ mA/V$, $r_\pi = 2.5\ k\Omega$, $R_C = 8\ k\Omega$, and $R_L = 5\ k\Omega$. Assuming that the three capacitors do not interact, find f_{P1}, f_{P2}, and f_{P3}, and hence estimate f_L.

Ans. 21.4 Hz; 2.21 kHz; 12.2 Hz; since $f_{P2} \gg f_{P1}$ and f_{P3}, $f_L \simeq f_{P2} = 2.21$ kHz; using Eq. (9.19), a somewhat better estimate for f_L is obtained: 2.24 kHz

9.2 Internal Capacitive Effects and the High-Frequency Model of the MOSFET and the BJT

While coupling and bypass capacitors cause the gain of transistor amplifiers to fall off at the low-frequency end, the gain falloff at high frequencies is caused by the capacitive effects internal to the transistors. In this section we shall briefly consider these effects and, more importantly, show how the device small-signal model can be augmented to take these effects into account.

9.2.1 The MOSFET

From our study of the physical operation of the MOSFET in Section 5.1, we know that the device has internal capacitances. In fact, we used one of these, the gate-to-channel capacitance, in our derivation of the MOSFET i–v characteristics. We did, however, implicitly assume that the steady-state charges on these capacitances are acquired instantaneously. In other words, we did not account for the finite time required to charge and discharge the various internal capacitances. As a result, the device models we derived, such as the small-signal model, do not include any capacitances. The use of these models would predict constant amplifier gains independent of frequency. We know, however, that this (unfortunately) does not happen; in fact, the gain of every MOSFET amplifier falls off at some high frequency. Similarly, the MOSFET digital logic inverter (Chapter 13) exhibits a finite nonzero propagation delay. To be able to predict these results, the MOSFET model must be augmented by including internal capacitances. This is the subject of this section.

To visualize the physical origin of the various internal capacitances, the reader is referred to Fig. 5.1. There are basically two types of internal capacitance in the MOSFET.

1. *The gate capacitive effect:* The gate electrode (polysilicon) forms a parallel-plate capacitor with the channel, with the oxide layer serving as the capacitor dielectric. We discussed the gate (or oxide) capacitance in Section 5.1 and denoted its value per unit area as C_{ox}.
2. *The source-body and drain-body depletion-layer capacitances:* These are the capacitances of the reverse-biased *pn* junctions formed by the n^+ source region (also called the **source diffusion**) and the *p*-type substrate and by the n^+ drain region (the **drain diffusion**) and the substrate. Evaluation of these capacitances will utilize the material studied in Chapter 3.

These two capacitive effects can be modeled by including capacitances in the MOSFET model between its four terminals, G, D, S, and B. There will be five capacitances in total:

C_{gs}, C_{gd}, C_{gb}, C_{sb}, and C_{db}, where the subscripts indicate the location of the capacitances in the model. In the following, we show how the values of the five model capacitances can be determined. We will do so by considering each of the two capacitive effects separately.

The Gate Capacitive Effect

The gate capacitive effect can be modeled by the three capacitances C_{gs}, C_{gd}, and C_{gb}. The values of these capacitances can be determined as follows:

1. When the MOSFET is operating in the triode region at small v_{DS}, the channel will be of uniform depth. The gate-channel capacitance will be WLC_{ox} and can be modeled by dividing it equally between the source and drain ends; thus,

$$C_{gs} = C_{gd} = \tfrac{1}{2}WLC_{ox} \quad \text{(triode region)} \tag{9.20}$$

   This is obviously an approximation (as all modeling is), but it works well for triode-region operation even when v_{DS} is not small.

2. When the MOSFET operates in saturation, the channel has a tapered shape and is pinched off at or near the drain end. It can be shown that the gate-to-channel capacitance in this case is approximately $\tfrac{2}{3}WLC_{ox}$ and can be modeled by assigning this entire amount to C_{gs}, and a zero amount to C_{gd} (because the channel is pinched off at the drain); thus,

$$\left.\begin{aligned} C_{gs} &= \tfrac{2}{3}WLC_{ox} \quad (9.21) \\ C_{gd} &= 0 \quad (9.22) \end{aligned}\right\} \text{(saturation region)}$$

3. When the MOSFET is cut off, the channel disappears, and thus $C_{gs} = C_{gd} = 0$. However, we can (after some rather complex reasoning) model the gate capacitive effect by assigning a capacitance WLC_{ox} to the gate-body model capacitance; thus,

$$\left.\begin{aligned} C_{gs} &= C_{gd} = 0 \quad (9.23) \\ C_{gb} &= WLC_{ox} \quad (9.24) \end{aligned}\right\} \text{(cutoff)}$$

4. There is an additional small capacitive component that should be added to C_{gs} and C_{gd} in all the preceding formulas. This is the capacitance that results from the fact that the source and drain diffusions extend slightly under the gate oxide (refer to Fig. 5.1). If the *overlap* length is denoted L_{ov}, we see that the **overlap capacitance** component is

$$C_{ov} = WL_{ov}C_{ox} \tag{9.25}$$

   Typically, $L_{ov} = 0.05$ to $0.1\,L$.

The Junction Capacitances The depletion-layer capacitances of the two reverse-biased *pn* junctions formed between each of the source and the drain diffusions and the body can be determined using the formula developed in Section 3.6 (Eq. 3.47). Thus, for the source diffusion, we have the source-body capacitance, C_{sb},

$$C_{sb} = \frac{C_{sb0}}{\sqrt{1 + \dfrac{V_{SB}}{V_0}}} \tag{9.26}$$

where C_{sb0} is the value of C_{sb} at zero body-source bias, V_{SB} is the magnitude of the reverse-bias voltage, and V_0 is the junction built-in voltage (0.6 V to 0.8 V). Similarly, for the drain diffusion, we have the drain-body capacitance C_{db},

$$C_{db} = \frac{C_{db0}}{\sqrt{1 + \frac{V_{DB}}{V_0}}} \tag{9.27}$$

where C_{db0} is the capacitance value at zero reverse-bias voltage, and V_{DB} is the magnitude of this reverse-bias voltage. Note that we have assumed that for both junctions, the grading coefficient $m = \frac{1}{2}$.

It should be noted also that each of these junction capacitances includes a component arising from the bottom side of the diffusion and a component arising from the *side walls* of the diffusion. In this regard, observe that each diffusion has three side walls that are in contact with the substrate and thus contribute to the junction capacitance (the fourth wall is in contact with the channel). In more advanced MOSFET modeling, the two components of each of the junction capacitances are calculated separately.

The formulas for the junction capacitances in Eqs. (9.26) and (9.27) assume small-signal operation. These formulas, however, can be modified to obtain approximate average values for the capacitances when the transistor is operating under large-signal conditions such as in logic circuits. Finally, typical values for the various capacitances exhibited by an n-channel MOSFET in a 0.5-μm CMOS process are given in the following exercise.

EXERCISE

9.3 For an n-channel MOSFET with $t_{ox} = 10$ nm, $L = 1.0$ μm, $W = 10$ μm, $L_{ov} = 0.05$ μm, $C_{sb0} = C_{db0} =$ 10 fF, $V_0 = 0.6$ V, $V_{SB} = 1$ V, and $V_{DS} = 2$ V, calculate the following capacitances when the transistor is operating in saturation: C_{ox}, C_{ov}, C_{gs}, C_{gd}, C_{sb}, and C_{db}.
Ans. 3.45 fF/μm²; 1.72 fF; 24.7 fF; 1.72 fF; 6.1 fF; 4.1 fF

The High-Frequency MOSFET Model Figure 9.6(a) shows the small-signal model of the MOSFET, including the four capacitances C_{gs}, C_{gd}, C_{sb}, and C_{db}. This model can be used to predict the high-frequency response of MOSFET amplifiers. It is, however, quite complex for manual analysis, and its use is limited to computer simulation using, for example, SPICE. Fortunately, when the source is connected to the body, the model simplifies considerably, as shown in Fig. 9.6(b). In this model, C_{gd}, although small, plays a significant role in determining the high-frequency response of amplifiers and thus must be kept in the model. Capacitance C_{db}, on the other hand, can usually be neglected, resulting in significant simplification of manual analysis. The resulting circuit is shown in Fig. 9.6(c).

The MOSFET Unity-Gain Frequency (f_T) A figure of merit for the high-frequency operation of the MOSFET as an amplifier is the unity-gain frequency, f_T, also known as the **transition frequency**, which gives rise to the subscript T. This is defined as *the frequency at which the short-circuit current-gain of the common-source configuration becomes unity*. Figure 9.7

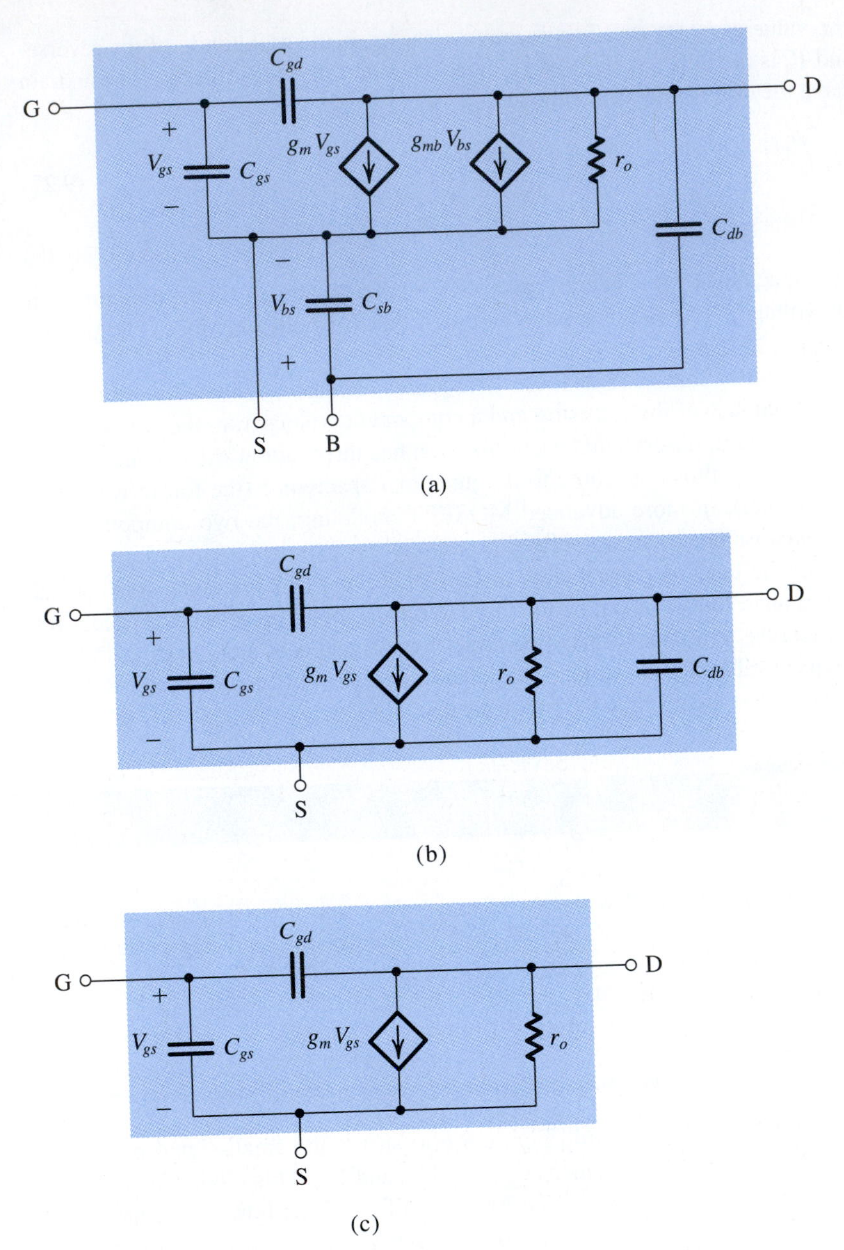

Figure 9.6 **(a)** High-frequency, equivalent-circuit model for the MOSFET. **(b)** The equivalent circuit for the case in which the source is connected to the substrate (body). **(c)** The equivalent-circuit model of **(b)** with C_{db} neglected (to simplify analysis).

shows the MOSFET hybrid-π model with the source as the common terminal between the input and output ports. To determine the short-circuit current gain, the input is fed with a current-source signal I_i and the output terminals are short-circuited. It can be seen that the current in the short circuit is given by

$$I_o = g_m V_{gs} - sC_{gd}V_{gs}$$

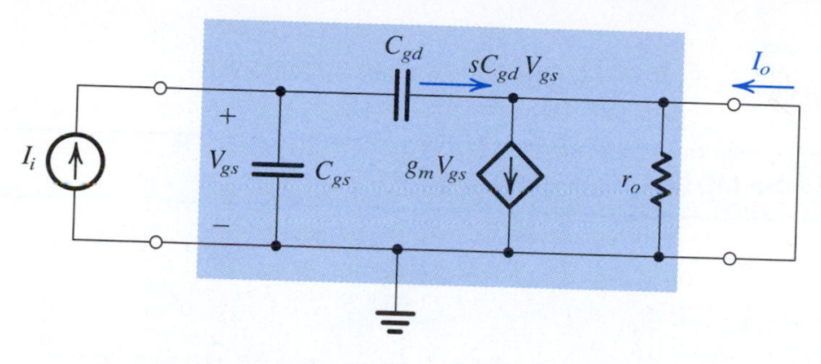

Figure 9.7 Determining the short-circuit current gain I_o/I_i.

Recalling that C_{gd} is small, at the frequencies of interest the second term in this equation can be neglected,

$$I_o \simeq g_m V_{gs} \tag{9.28}$$

From Fig. 9.7, we can express V_{gs} in terms of the input current I_i as

$$V_{gs} = I_i/s(C_{gs}+C_{gd}) \tag{9.29}$$

Equations (9.28) and (9.29) can be combined to obtain the short-circuit current gain,

$$\frac{I_o}{I_i} = \frac{g_m}{s(C_{gs}+C_{gd})} \tag{9.30}$$

For physical frequencies $s = j\omega$, it can be seen that the magnitude of the current gain becomes unity at the frequency

$$\omega_T = g_m/(C_{gs}+C_{gd})$$

Thus the unity-gain frequency $f_T = \omega_T/2\pi$ is

$$f_T = \frac{g_m}{2\pi(C_{gs}+C_{gd})} \tag{9.31}$$

Since f_T is proportional to g_m and inversely proportional to the MOSFET internal capacitances, the higher the value of f_T, the more effective the MOSFET becomes as an amplifier. Substituting for g_m using Eq. (5.56), we can express f_T in terms of the bias current I_D (see Problem 9.18). Alternatively, we can substitute for g_m from Eq. (5.55) to express f_T in terms of the overdrive voltage V_{OV} (see Problem 9.19). Both expressions yield additional insight into the high-frequency operation of the MOSFET. The reader is also referred to Chapter 7, Appendix 7.A for a further discussion of f_T.

Typically, f_T ranges from about 100 MHz for the older technologies (e.g., a 5-μm CMOS process) to many GHz for newer high-speed technologies (e.g., a 0.13-μm CMOS process).

EXERCISE

9.4 Calculate f_T for the n-channel MOSFET whose capacitances were found in Exercise 9.3. Assume operation at 100 μA, and that $k_n' = 160\ \mu\text{A/V}^2$.
Ans. 3.7 GHz.

Summary

We conclude this section by presenting a summary in Table 9.1.

Table 9.1 The MOSFET High-Frequency Model

Model

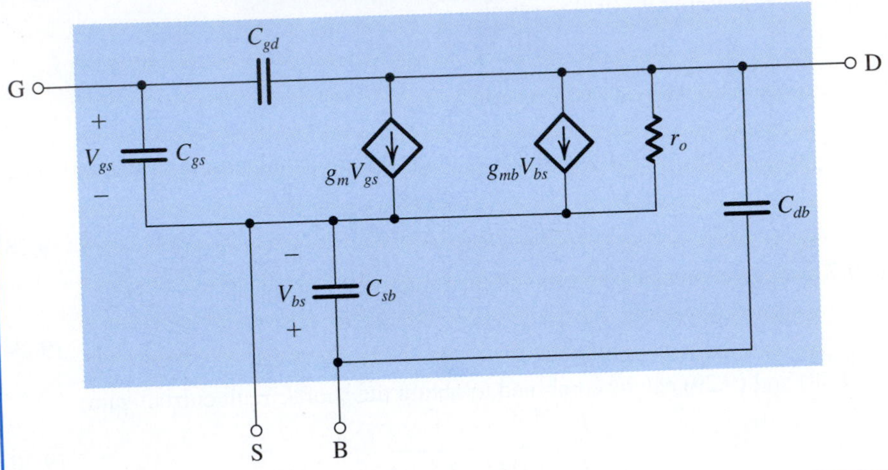

Model Parameters

$$g_m = \mu_n C_{ox} \frac{W}{L} |V_{OV}| = \sqrt{2 \mu_n C_{ox} \frac{W}{L} I_D} = \frac{2 I_D}{|V_{OV}|}$$

$$g_{mb} = \chi g_m, \quad \chi = 0.1 \text{ to } 0.2$$

$$r_o = |V_A| / I_D$$

$$C_{gs} = \frac{2}{3} W L C_{ox} + W L_{ov} C_{ox}$$

$$C_{gd} = W L_{ov} C_{ox}$$

$$C_{sb} = \frac{C_{sb0}}{\sqrt{1 + \frac{|V_{SB}|}{V_0}}}$$

$$C_{db} = \frac{C_{db0}}{\sqrt{1 + \frac{|V_{DB}|}{V_0}}}$$

$$f_T = \frac{g_m}{2\pi (C_{gs} + C_{gd})}$$

9.2.2 The BJT

In our study of the physical operation of the BJT in Section 6.1, we assumed transistor action to be instantaneous, and as a result the transistor models we developed do not include any elements (i.e., capacitors or inductors) that would cause time or frequency dependence. Actual transistors, however, exhibit charge-storage phenomena that limit the speed and frequency of their operation. We have already encountered such effects in our study of the *pn* junction in Chapter 3, and learned that they can be modeled using capacitances. In the following we study the charge-storage effects that take place in the BJT and take them into account by adding capacitances to the hybrid-π model. The resulting augmented BJT model will be able to predict the observed dependence of amplifier gain on frequency, and the time delays that transistor switches and logic gates exhibit.

The Base-Charging or Diffusion Capacitance C_{de} When the transistor is operating in the active mode, minority carrier charge is stored in the base region. For an *npn* transistor, the stored electron charge in the base, Q_n, can be expressed in terms of the collector current i_C as

$$Q_n = \tau_F i_C \tag{9.32}$$

where τ_F is a device constant with the dimension of time. It is known as the **forward base-transit time** and represents the average time a charge carrier (electron) spends in crossing the base. Typically, τ_F is in the range of 10 ps to 100 ps.

Equation (9.32) applies for large signals and, since i_C is exponentially related to v_{BE}, Q_n will similarly depend on v_{BE}. Thus this charge-storage mechanism represents a nonlinear capacitive effect. However, for small signals we can define the **small-signal diffusion capacitance C_{de}**,

$$C_{de} \equiv \frac{dQ_n}{dv_{BE}} \tag{9.33}$$

$$= \tau_F \frac{di_C}{dv_{BE}}$$

resulting in

$$C_{de} = \tau_F g_m = \tau_F \frac{I_C}{V_T} \tag{9.34}$$

Thus, whenever v_{BE} changes by v_{be}, the collector current changes by $g_m v_{be}$ and the charge stored in the base changes by $C_{de}\, v_{be} = (\tau_F\, g_m)\, v_{be}$.

The Base–Emitter Junction Capacitance C_{je} A change in v_{BE} not only changes the charge stored in the base region but also the charge stored in the base–emitter depletion layer. This distinct charge-storage effect is represented by the EBJ depletion-layer capacitance, C_{je}. From the development in Chapter 3, we know that for a forward-biased junction, which the EBJ is, the depletion-layer capacitance is given approximately by

$$C_{je} \simeq 2C_{je0} \tag{9.35}$$

where C_{je0} is the value of C_{je} at zero EBJ voltage.

The Collector–Base Junction Capacitance C_μ In active-mode operation, the CBJ is reverse biased, and its junction or **depletion capacitance**, usually denoted C_μ, can be found from

$$C_\mu = \frac{C_{\mu 0}}{\left(1 + \dfrac{V_{CB}}{V_{0c}}\right)^m} \tag{9.36}$$

where $C_{\mu 0}$ is the value of C_μ at zero voltage; V_{CB} is the magnitude of the CBJ reverse-bias voltage, V_{0c} is the CBJ built-in voltage (typically, 0.75 V), and m is its grading coefficient (typically, 0.2–0.5).

The High-Frequency Hybrid-π Model Figure 9.8 shows the hybrid-π model of the BJT, including capacitive effects. Specifically, there are two capacitances: the emitter–base capacitance $C_\pi = C_{de} + C_{je}$ and the collector–base capacitance C_μ. Typically, C_π is in the

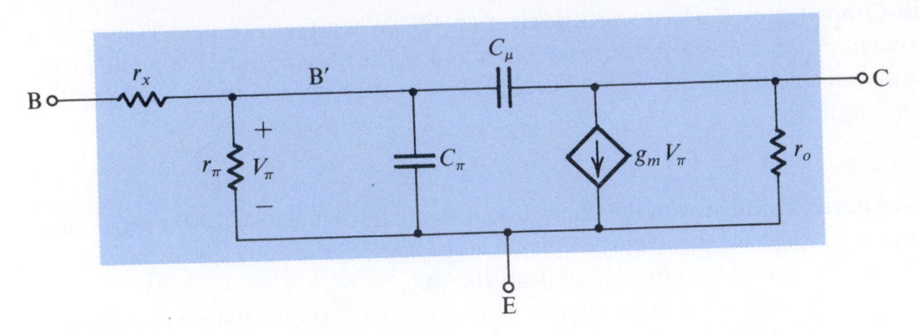

Figure 9.8 The high-frequency hybrid-π model.

range of a few picofarads to a few tens of picofarads, and C_μ is in the range of a fraction of a picofarad to a few picofarads.[4] Note that we have also added a resistor r_x to model the resistance of the silicon material of the base region between the base terminal B and a fictitious internal, or intrinsic, base terminal B′ that is right under the emitter region (refer to Fig. 6.6). Typically, r_x is a few tens of ohms, and its value depends on the current level in a rather complicated manner. Since (usually) $r_x \ll r_\pi$, its effect is negligible at low frequencies. Its presence is felt, however, at high frequencies, as will become apparent later.

The values of the hybrid-π, equivalent-circuit parameters can be determined at a given bias point using the formulas presented in this section and in Chapter 6. They can also be found from the terminal measurements specified on the BJT data sheets. For computer simulation, SPICE uses the parameters of the given IC technology to evaluate the BJT model parameters (see Appendix B).

The Cutoff Frequency The transistor data sheets do not usually specify the value of C_π. Rather, the behavior of β (or h_{fe}) versus frequency is normally given. In order to determine C_π and C_μ, we shall derive an expression for h_{fe}, the CE short-circuit current gain, as a function of frequency in terms of the hybrid-π components. For this purpose consider the circuit shown in Fig. 9.9, in which the collector is shorted to the emitter. A node equation at C provides the short-circuit collector current I_c as

$$I_c = (g_m - sC_\mu)V_\pi \tag{9.37}$$

A relationship between V_π and I_b can be established by multiplying I_b by the impedance seen between B′ and E:

$$V_\pi = I_b(r_\pi \| C_\pi \| C_\mu) = \frac{I_b}{1/r_\pi + sC_\pi + sC_\mu} \tag{9.38}$$

Thus h_{fe} can be obtained by combining Eqs. (9.37) and (9.38):

$$h_{fe} \equiv \frac{I_c}{I_b} = \frac{g_m - sC_\mu}{1/r_\pi + s(C_\pi + C_\mu)}$$

---

[4] These values apply for discrete devices and devices fabricated with a relatively old IC process technology (the so-called high-voltage process, see Appendix 7.A). For modern IC fabrication processes, C_π and C_μ are in the range of tens of femtofarads (fF).

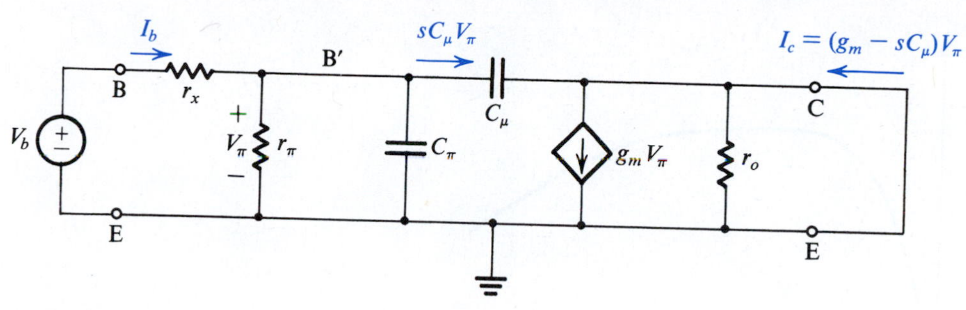

Figure 9.9 Circuit for deriving an expression for $h_{fe}(s) \equiv I_c/I_b$.

At the frequencies for which this model is valid, $\omega C_\mu \ll g_m$; thus we can neglect the sC_μ term in the numerator and write

$$h_{fe} \simeq \frac{g_m r_\pi}{1 + s(C_\pi + C_\mu)r_\pi}$$

Thus,

$$h_{fe} = \frac{\beta_0}{1 + s(C_\pi + C_\mu)r_\pi} \tag{9.39}$$

where β_0 is the low-frequency value of β. Thus h_{fe} has a single-pole (or STC) response with a 3-dB frequency at $\omega = \omega_\beta$, where

$$\omega_\beta = \frac{1}{(C_\pi + C_\mu)r_\pi} \tag{9.40}$$

Figure 9.10 shows a Bode plot for $|h_{fe}|$. From the –6-dB/octave slope, it follows that the frequency at which $|h_{fe}|$ drops to unity, which is called the **unity-gain bandwidth** ω_T, is given by

$$\omega_T = \beta_0 \omega_\beta \tag{9.41}$$

Thus,

$$\omega_T = \frac{g_m}{C_\pi + C_\mu} \tag{9.42}$$

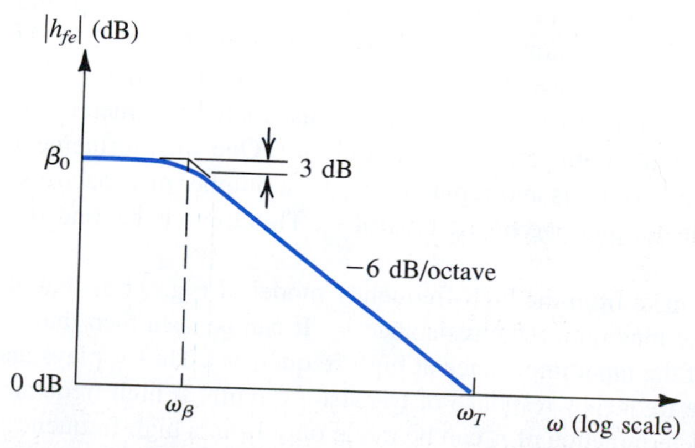

Figure 9.10 Bode plot for $|h_{fe}|$.

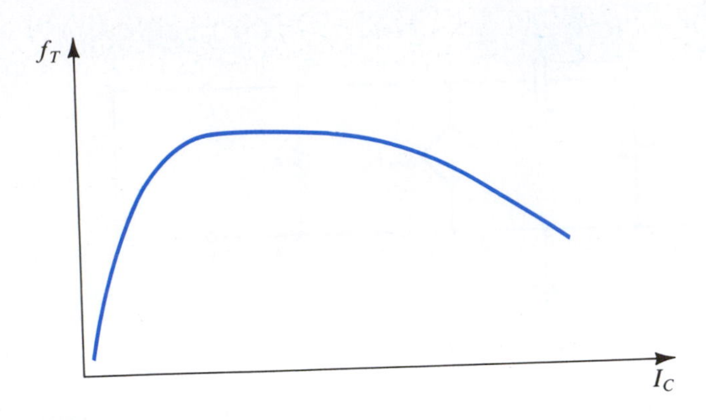

Figure 9.11 Variation of f_T with I_C.

and

$$f_T = \frac{g_m}{2\pi(C_\pi + C_\mu)} \tag{9.43}$$

This expression is very similar to that of f_T for the MOSFET (Eq. 9.31) with C_π replacing C_{gs} and C_μ replacing C_{gd}.

The unity-gain bandwidth f_T, also known as the **transition frequency**, which gives rise to the subscript T, is usually specified on the data sheets of a transistor. In some cases f_T is given as a function of I_C and V_{CE}. To see how f_T changes with I_C, recall that g_m is directly proportional to I_C, but only part of C_π (the diffusion capacitance C_{de}) is directly proportional to I_C. It follows that f_T decreases at low currents, as shown in Fig. 9.11. However, the decrease in f_T at high currents, also shown in Fig. 9.11, cannot be explained by this argument; rather, it is due to the same phenomenon that causes β_0 to decrease at high currents (Section 6.9.2). In the region where f_T is almost constant, C_π is dominated by the diffusion part.

Typically, f_T is in the range of 100 MHz to tens of gigahertz. The value of f_T can be used in Eq. (9.43) to determine $C_\pi + C_\mu$. The capacitance C_μ is usually determined separately by measuring the capacitance between base and collector at the desired reverse-bias voltage V_{CB}.

Before leaving this section, we should mention that the hybrid-π model of Fig. 9.8 characterizes transistor operation fairly accurately up to a frequency of about $0.2 f_T$. At higher frequencies one has to add other parasitic elements to the model as well as refine the model to account for the fact that the transistor is in fact a distributed-parameter network that we are trying to model with a lumped-component circuit. One such refinement consists of splitting r_x into a number of parts and replacing C_μ by a number of capacitors, each connected between the collector and one of the taps of r_x. This topic is beyond the scope of this book.

An important observation to make from the high-frequency model of Fig. 9.8 is that at frequencies above 5 to 10 f_β, one may ignore the resistance r_π. It can be seen then that r_x becomes the only resistive part of the input impedance at high frequencies. Thus r_x plays an important role in determining the frequency response of transistor circuits at high frequencies. It follows that an accurate determination of r_x can be made only from a high-frequency measurement.

EXERCISES

9.5 Find C_{de}, C_{je}, C_{π}, C_{μ}, and f_T for a BJT operating at a dc collector current $I_C = 1$ mA and a CBJ reverse bias of 2 V. The device has $\tau_F = 20$ ps, $C_{je0} = 20$ fF, $C_{\mu 0} = 20$ fF, $V_{0e} = 0.9$ V, $V_{0c} = 0.5$ V, and $m_{CBJ} = 0.33$.
Ans. 0.8 pF; 40 fF; 0.84 pF; 12 fF; 7.47 GHz

9.6 For a BJT operated at $I_C = 1$ mA, determine f_T and C_{π} if $C_{\mu} = 2$ pF and $|h_{fe}| = 10$ at 50 MHz.
Ans. 500 MHz; 10.7 pF

9.7 If C_{π} of the BJT in Exercise 9.6 includes a relatively constant depletion-layer capacitance of 2 pF, find f_T of the BJT when operated at $I_C = 0.1$ mA.
Ans. 130.7 MHz

Summary

For convenient reference, Table 9.2 provides a summary of the relationships used to determine the values of the parameters of the BJT high-frequency model.

Table 9.2 The BJT High-Frequency Model

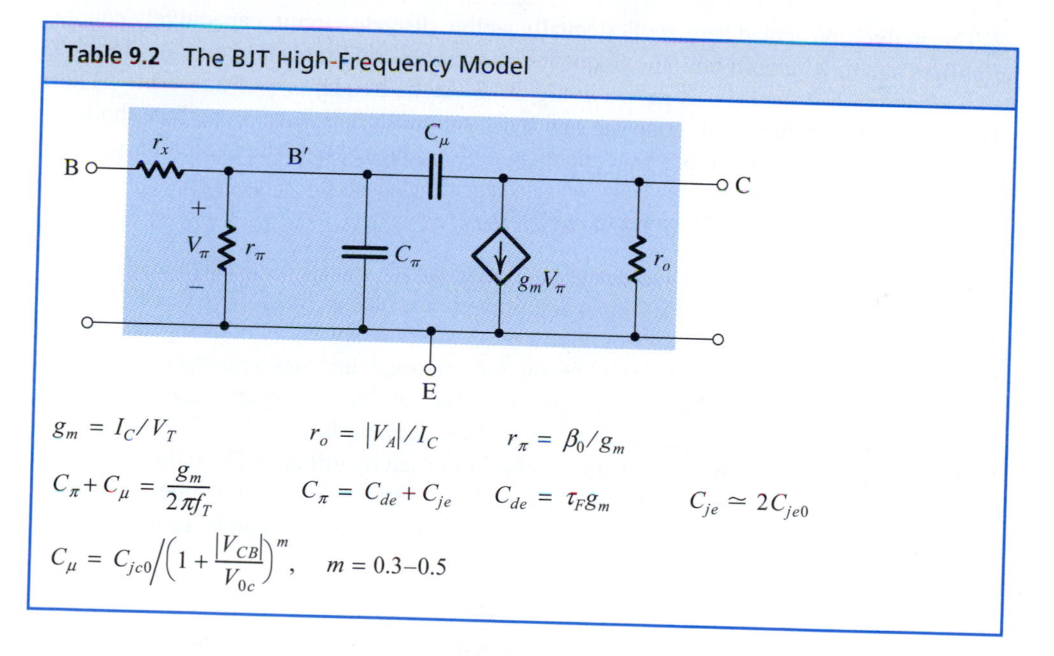

$g_m = I_C/V_T$ $\quad r_o = |V_A|/I_C$ $\quad r_{\pi} = \beta_0/g_m$

$C_{\pi} + C_{\mu} = \dfrac{g_m}{2\pi f_T}$ $\quad C_{\pi} = C_{de} + C_{je}$ $\quad C_{de} = \tau_F g_m$ $\quad C_{je} \simeq 2C_{je0}$

$C_{\mu} = C_{jc0} \Big/ \left(1 + \dfrac{|V_{CB}|}{V_{0c}}\right)^m$, $\quad m = 0.3\text{–}0.5$

9.3 High-Frequency Response of the CS and CE Amplifiers

Equipped with equivalent-circuit models that represent the high-frequency operation of the MOSFET and the BJT, we now address the question of the high-frequency performance of the CS and CE amplifiers. Our objective is to identify the mechanism that limits the high-frequency performance of these important amplifier configurations. As well, we need to find a simple approach to estimate the frequency f_H at which the gain falls by 3 dB below its value at midband frequencies, $|A_M|$.

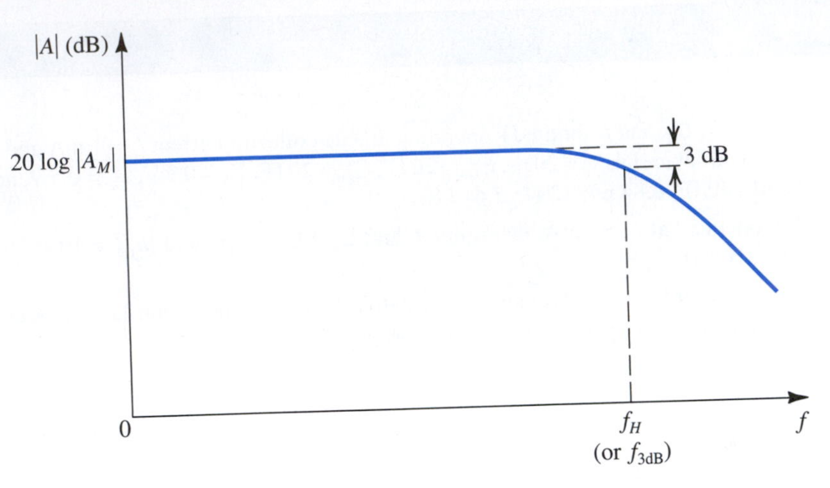

Figure 9.12 Frequency response of a direct-coupled (dc) amplifier. Observe that the gain does *not* fall off at low frequencies, and the midband gain A_M extends down to zero frequency.

The analysis presented here applies equally well to discrete-circuit, capacitively coupled amplifiers and to IC amplifiers. The frequency response of the first was shown in Figs. 5.61 and 6.69 and that of the latter is shown in Fig. 9.12. At the frequencies of interest to us here (the high-frequency band), all coupling and bypass capacitors behave as perfect short circuits and amplifiers of both types have identical high-frequency equivalent circuits.

9.3.1 The Common-Source Amplifier

Figure 9.13(a) shows the high-frequency, equivalent-circuit model of a CS amplifier. It is obtained by replacing the MOSFET in an amplifier circuit such as that in Fig. 9.2 by its high-frequency, equivalent-circuit model of Fig. 9.6(c), while as always eliminating dc sources. Observe that the circuit in Fig. 9.13(a) is general; for instance, it includes a resistance R_G, which arises only in the case of a discrete-circuit amplifier. Also, R_D can be either a passive resistance or the output resistance of a current-source load, and similarly for R_L.

The equivalent circuit of Fig. 9.13(a) can be simplified by utilizing Thévenin theorem at the input side and by combining the three parallel resistances at the output side. The resulting simplified circuit is shown in Fig. 9.13(b). The midband gain A_M can be found from this circuit by setting C_{gs} and C_{gd} to zero. The result is

$$A_M = \frac{V_o}{V_{sig}} = -\frac{R_G}{R_G + R_{sig}}(g_m R_L') \tag{9.44}$$

The equivalent circuit in Fig. 9.13(b) can be further simplified if we can find a way to deal with the bridging capacitor C_{gd} that connects the output node to the input side. Toward that end, consider first the output node. It can be seen that the load current is $(g_m V_{gs} - I_{gd})$, where $(g_m V_{gs})$ is the output current of the transistor and I_{gd} is the current supplied through the very small capacitance C_{gd}. At frequencies in the vicinity of f_H, which defines the edge of the midband, it is reasonable to assume that I_{gd} is still much smaller than $(g_m V_{gs})$, with the result that V_o can be given approximately by

$$V_o \simeq -(g_m V_{gs})R_L' = -g_m R_L' V_{gs} \tag{9.45}$$

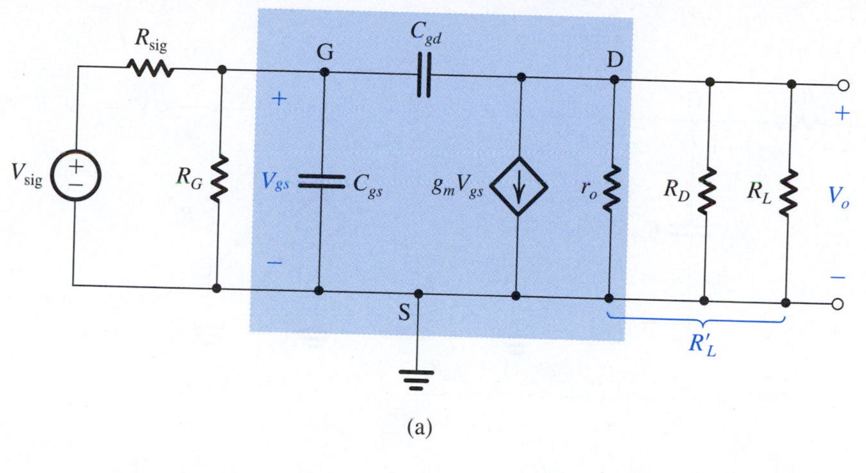

(a)

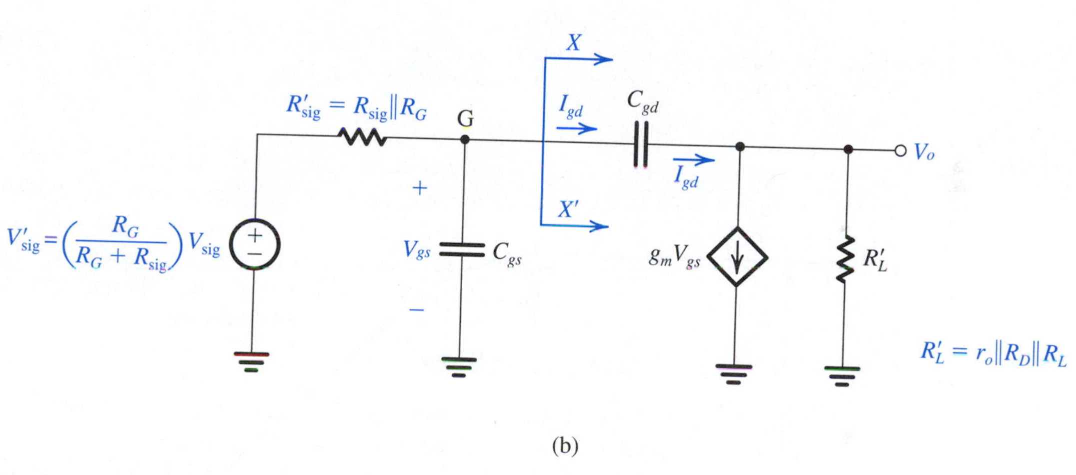

(b)

Figure 9.13 Determining the high-frequency response of the CS amplifier: **(a)** equivalent circuit; **(b)** the circuit of **(a)** simplified at the input and the output; (*Continued*)

where

$$R'_L = r_o \parallel R_D \parallel R_L$$

Since $V_o = V_{ds}$, Eq. (9.45) indicates that the gain from gate to drain is $-g_m R'_L$, the same value as in the midband. The current I_{gd} can now be found as

$$\begin{aligned} I_{gd} &= sC_{gd}(V_{gs} - V_o) \\ &= sC_{gd}[V_{gs} - (-g_m R'_L V_{gs})] \\ &= sC_{gd}(1 + g_m R'_L)V_{gs} \end{aligned}$$

Now, the left-hand side of the circuit in Fig. 9.13(b), at XX', knows of the existence of C_{gd} only through the current I_{gd}. Therefore, we can replace C_{gd} by an equivalent capacitance C_{eq} between the gate and ground as long as C_{eq} draws a current equal to I_{gd}. That is,

$$sC_{eq}V_{gs} = sC_{gd}(1 + g_m R'_L)V_{gs}$$

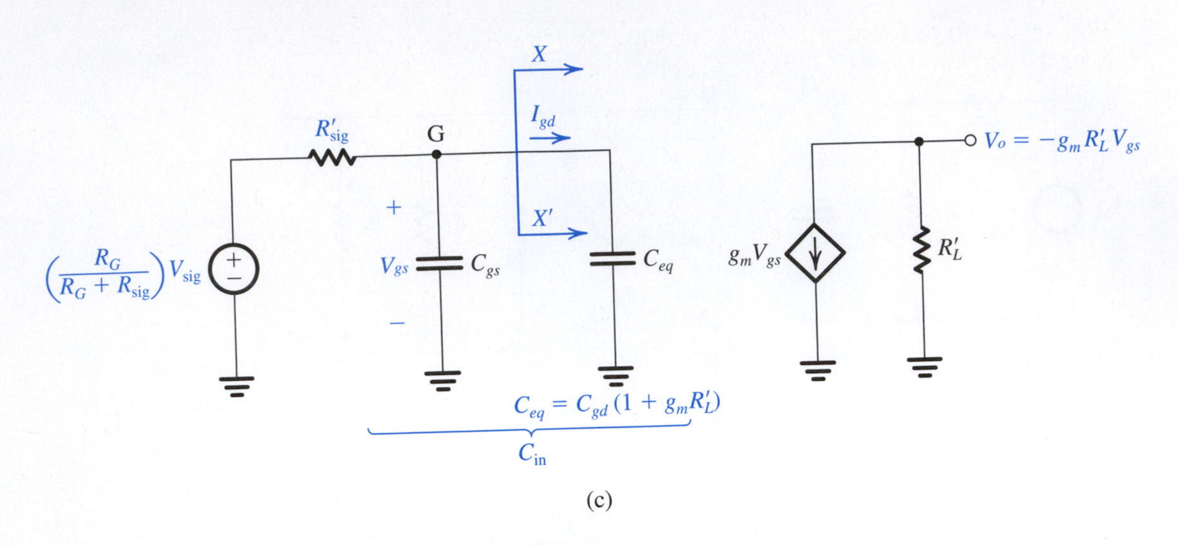

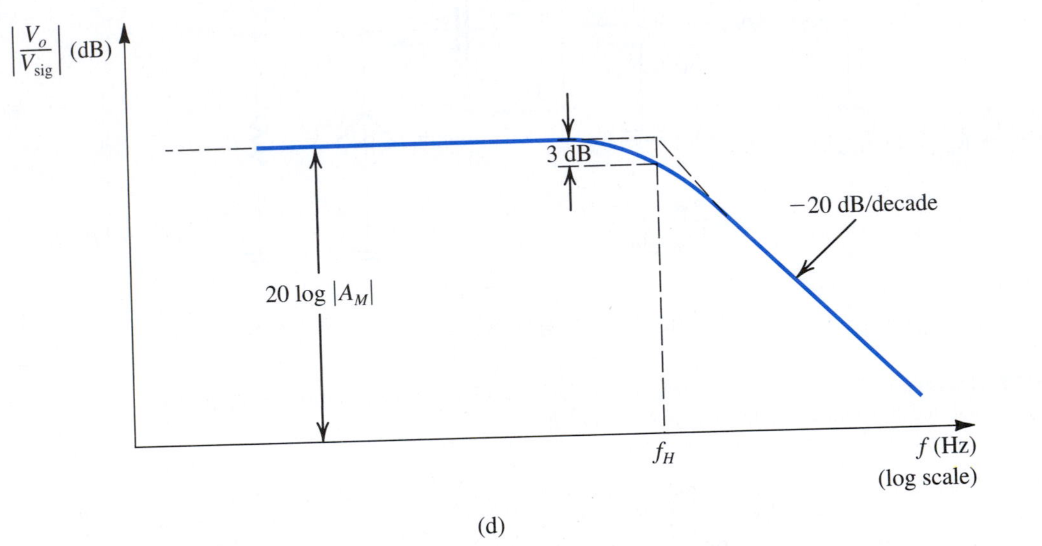

Figure 9.13 (*Continued*) **(c)** the equivalent circuit with C_{gd} replaced at the input side with the equivalent capacitance C_{eq}; **(d)** the frequency response plot, which is that of a low-pass, single-time-constant circuit.

which results in

$$C_{eq} = C_{gd}(1 + g_m R'_L) \tag{9.46}$$

Thus C_{gd} gives rise to a much larger capacitance C_{eq}, which appears at the amplifier input. The multiplication effect that C_{gd} undergoes comes about because it is connected between circuit nodes g and d, whose voltages are related by a large negative gain $(-g_m R'_L)$. This effect is known as the **Miller effect**, and $(1 + g_m R'_L)$ is known as the **Miller multiplier**. We will study Miller's theorem more formally in Section 9.4.

Using C_{eq} enables us to simplify the equivalent circuit at the input side to that shown in Fig. 9.13(c). We recognize the circuit of Fig. 9.13(c) as a single-time-constant (STC) circuit of the low-pass type (Section 1.6 and Appendix E). Reference to Table 1.2 enables us to express the output voltage V_{gs} of the STC circuit in the form

$$V_{gs} = \left(\frac{R_G}{R_G + R_{sig}} V_{sig}\right) \frac{1}{1 + \frac{s}{\omega_0}} \tag{9.47}$$

where ω_0 is the corner frequency, the break frequency, or the pole frequency of the STC circuit,

$$\omega_0 = 1/C_{in}R'_{sig} \tag{9.48}$$

with

$$C_{in} = C_{gs} + C_{eq} = C_{gs} + C_{gd}(1 + g_m R'_L) \tag{9.49}$$

and

$$R'_{sig} = R_{sig} \parallel R_G \tag{9.50}$$

Combining Eqs. (9.45) and (9.47) results in the following expression for the high-frequency gain of the CS amplifier,

$$\frac{V_o}{V_{sig}} = -\left(\frac{R_G}{R_G + R_{sig}}\right)(g_m R'_L) \frac{1}{1 + \frac{s}{\omega_0}} \tag{9.51}$$

which can be expressed in the form

$$\frac{V_o}{V_{sig}} = \frac{A_M}{1 + \frac{s}{\omega_H}} \tag{9.52}$$

where the midband gain A_M is given by Eq. (9.44) and ω_H is the upper 3-dB frequency,

$$\omega_H = \omega_0 = \frac{1}{C_{in}R'_{sig}} \tag{9.53}$$

and

$$f_H = \frac{\omega_H}{2\pi} = \frac{1}{2\pi C_{in}R'_{sig}} \tag{9.54}$$

We thus see that the high-frequency response will be that of a low-pass STC network with a 3-dB frequency f_H determined by the time constant $C_{in}R'_{sig}$. Figure 9.13(d) shows a sketch of the magnitude of the high-frequency gain.

Before leaving this section we wish to make a number of observations:

1. The upper 3-dB frequency is determined by the interaction of $R'_{sig} = R_{sig} \parallel R_G$ and $C_{in} = C_{gs} + C_{gd}(1 + g_m R'_L)$. Since the bias resistance R_G is usually very large, it can be neglected, resulting in $R'_{sig} \simeq R_{sig}$, the resistance of the signal source. It follows that a large value of R_{sig} will cause f_H to be lowered.
2. The total input capacitance C_{in} is usually dominated by C_{eq}, which in turn is made large by the multiplication effect that C_{gd} undergoes. Thus, although C_{gd} is usually a very small capacitance, its effect on the amplifier frequency response can be very significant as a result of its multiplication by the factor $(1 + g_m R'_L)$, which is approximately equal

to the midband gain of the amplifier. This is the Miller effect, which causes the CS amplifier to have a large total input capacitance C_{in} and hence a low f_H.

3. To extend the high-frequency response of a MOSFET amplifier, we have to find configurations in which the Miller effect is absent or at least reduced. We shall return to this subject at great length in Section 9.6 and beyond.
4. The above analysis, resulting in an STC or a single-pole response, is approximate. Specifically, it is based on neglecting I_{gd} relative to $g_m V_{gs}$, an assumption that applies well at frequencies not too much higher than f_H. An exact analysis of the circuit in Fig. 9.13(a) will be carried out in Section 9.5. The results above, however, are more than sufficient for a quick estimate of f_H. As well, the approximate approach helps to reveal the primary limitation on the high-frequency response: the Miller effect.

Example 9.3

Find the midband gain A_M and the upper 3-dB frequency f_H of a CS amplifier fed with a signal source having an internal resistance $R_{sig} = 100\ \text{k}\Omega$. The amplifier has $R_G = 4.7\ \text{M}\Omega$, $R_D = R_L = 15\ \text{k}\Omega$, $g_m = 1\ \text{mA/V}$, $r_o = 150\ \text{k}\Omega$, $C_{gs} = 1\ \text{pF}$, and $C_{gd} = 0.4\ \text{pF}$.

Solution

$$A_M = -\frac{R_G}{R_G + R_{sig}} g_m R'_L$$

where

$$R'_L = r_o \| R_D \| R_L = 150 \| 15 \| 15 = 7.14\ \text{k}\Omega.$$

$$g_m R'_L = 1 \times 7.14 = 7.14\ \text{V/V}$$

Thus,

$$A_M = -\frac{4.7}{4.7 + 0.1} \times 7.14 = -7\ \text{V/V}$$

The equivalent capacitance, C_{eq}, is found as

$$C_{eq} = (1 + g_m R'_L) C_{gd}$$

$$= (1 + 7.14) \times 0.4 = 3.26\ \text{pF}$$

The total input capacitance C_{in} can be now obtained as

$$C_{in} = C_{gs} + C_{eq} = 1 + 3.26 = 4.26\ \text{pF}$$

The upper 3-dB frequency f_H is found from

$$f_H = \frac{1}{2\pi C_{in}(R_{sig} \| R_G)}$$

$$= \frac{1}{2\pi \times 4.26 \times 10^{-12} (0.1 \| 4.7) \times 10^6}$$

$$= 382\ \text{kHz}$$

EXERCISES

9.8 For the CS amplifier specified in Example 9.3, find the values of A_M and f_H that result when the signal-source resistance is reduced to 10 kΩ.
Ans. –7.12 V/V; 3.7 MHz

9.9 If it is possible to replace the MOSFET used in the amplifier in Example 9.3 with another having the same C_{gs} but a smaller C_{gd}, what is the maximum value that its C_{gd} can be in order to obtain an f_H of at least 1 MHz?
Ans. 0.08 pF

9.3.2 The Common-Emitter Amplifier

Figure 9.14(a) shows the high-frequency equivalent circuit of a CE amplifier. It is obtained by replacing the BJT in a circuit such as that in Fig. 9.4(a) with its high-frequency, equivalent-circuit model of Fig. 9.8, and, as usual, eliminating all dc sources. Observe that the circuit in Fig. 9.14(a) is general and applies to both discrete and IC amplifiers. Thus, it includes R_B, which is usually present in discrete circuits. Also R_C can be either a passive resistance or the output resistance of a current-source load, and similarly for R_L.

The equivalent circuit of Fig. 9.14(a) can be simplified by utilizing Thévenin theorem at the input side and by combining the three parallel resistances at the output side. Specifically, the reader should be able to show that applying Thévenin theorem *twice* simplifies the resistive network at the input side to a signal generator V'_{sig} and a resistance R'_{sig}, with the values indicated in the figure.

The equivalent circuit in Fig. 9.14(b) can be used to obtain the midband gain A_M by setting C_π and C_μ to zero. The result is

$$A_M = \frac{V_o}{V_{sig}} = -\frac{R_B}{R_B + R_{sig}} \frac{r_\pi}{r_\pi + r_x + (R_{sig} \| R_B)}(g_m R'_L) \tag{9.55}$$

where

$$R'_L = r_o \| R_C \| R_L \tag{9.56}$$

Next we observe that the circuit in Fig. 9.14(b) is identical to that of the CS amplifier in Fig. 9.13(b). Thus the analysis can follow the same process we used for the CS case. The analysis is illustrated in Fig. 9.13(c) and (d). The final result is that the CE amplifier gain at high frequencies is given approximately by

$$\frac{V_o}{V_{sig}} = \frac{A_M}{1 + \dfrac{s}{\omega_H}} \tag{9.57}$$

where A_M is given by Eq. (9.55) and the 3-dB frequency f_H is given by

$$f_H = \frac{\omega_H}{2\pi} = \frac{1}{2\pi C_{in} R'_{sig}} \tag{9.58}$$

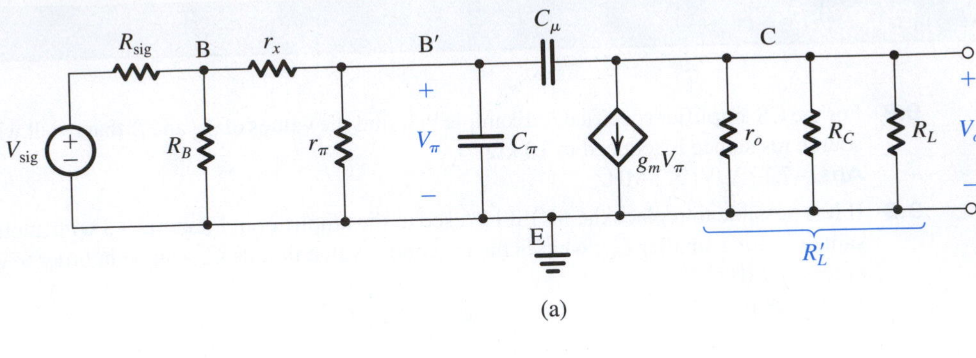

(a)

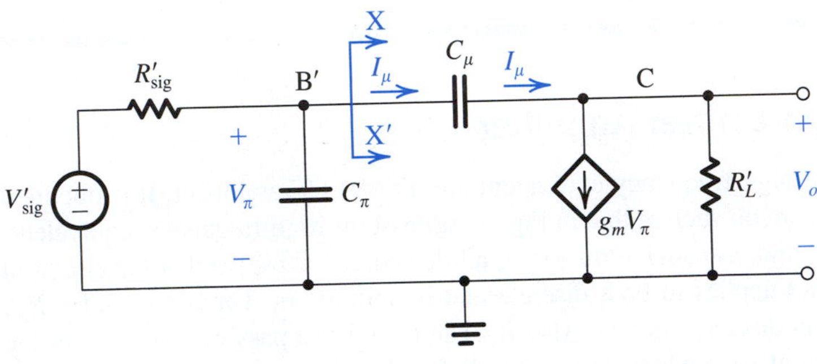

$$V'_{sig} = V_{sig} \frac{R_B}{R_B + R_{sig}} \frac{r_\pi}{r_\pi + r_x + (R_{sig} \| R_B)}$$

$$R'_L = r_o \| R_C \| R_L$$

$$R'_{sig} = r_\pi \| [r_x + (R_B \| R_{sig})]$$

(b)

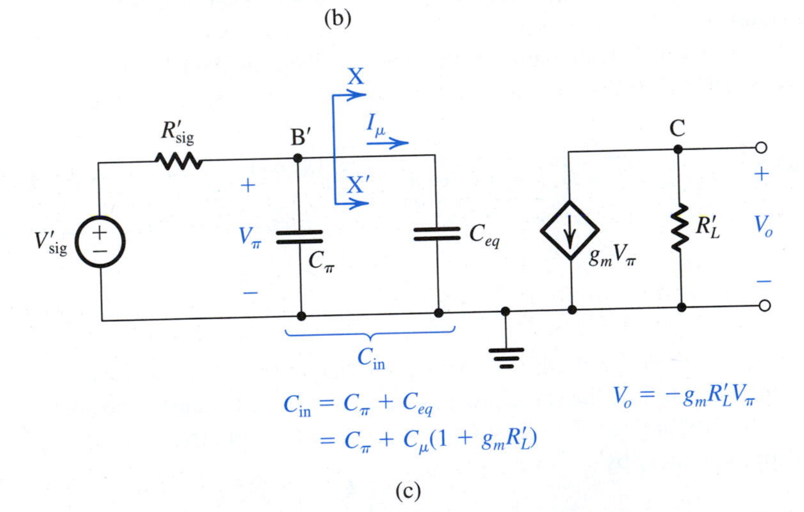

$$C_{in} = C_\pi + C_{eq}$$
$$= C_\pi + C_\mu(1 + g_m R'_L)$$

$$V_o = -g_m R'_L V_\pi$$

(c)

Figure 9.14 Determining the high-frequency response of the CE amplifier: **(a)** equivalent circuit; **(b)** the circuit of **(a)** simplified at both the input side and the output side; **(c)** equivalent circuit with C_μ replaced at the input side with the equivalent capacitance C_{eq}; *(continued)*

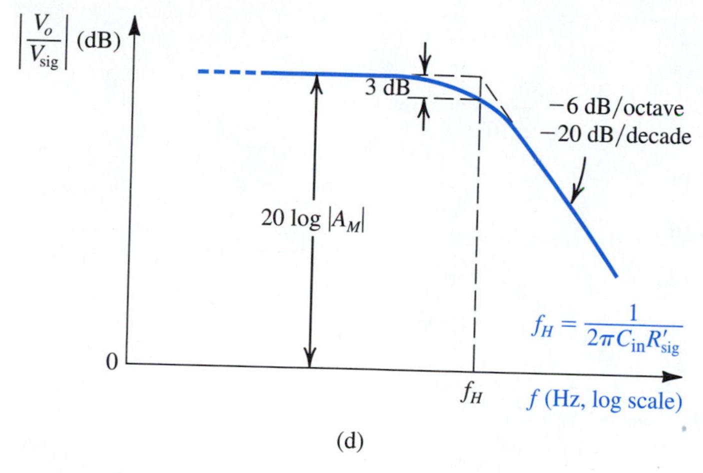

Figure 9.14 (*Continued*) **(d)** sketch of the frequency-response plot, which is that of a low-pass STC circuit.

where

$$C_{in} = C_\pi + C_\mu(1 + g_m R'_L) \quad (9.59)$$

and

$$R'_{sig} = r_\pi \| [r_x + (R_B \| R_{sig})] \quad (9.60)$$

Observe that C_{in} is simply the sum of C_π and the Miller capacitance $C_\mu(1 + g_m R'_L)$. The resistance R'_{sig} seen by C_{in} can be easily found from the circuit in Fig. 9.14(a) as follows: Reduce V_{sig} to zero, "grab hold" of the terminals B′ and E and look back (to the left). You will see r_π in parallel with r_x, which is in series with $(R_B \| R_{sig})$. This way of finding the resistance "seen by a capacitance" is very useful and spares one from tedious work!

Finally, comments very similar to those made on the high-frequency response of the CS amplifier can be made here as well.

Example 9.4

It is required to find the midband gain and the upper 3-dB frequency of the common-emitter amplifier of Fig. 9.4(a) for the following case: $V_{CC} = V_{EE} = 10$ V, $I = 1$ mA, $R_B = 100$ kΩ, $R_C = 8$ kΩ, $R_{sig} = 5$ kΩ, $R_L = 5$ kΩ, $\beta_0 = 100$, $V_A = 100$ V, $C_\mu = 1$ pF, $f_T = 800$ MHz, and $r_x = 50$ Ω.

Solution

The transistor is biased at $I_C \simeq 1$ mA. Thus the values of its hybrid-π model parameters are

$$g_m = \frac{I_C}{V_T} = \frac{1 \text{ mA}}{25 \text{ mV}} = 40 \text{ mA/V}$$

$$r_\pi = \frac{\beta_0}{g_m} = \frac{100}{40 \text{ mA/V}} = 2.5 \text{ k}\Omega$$

$$r_o = \frac{V_A}{I_C} = \frac{100 \text{ V}}{1 \text{ mA}} = 100 \text{ k}\Omega$$

Example 9.4 *continued*

$$C_\pi + C_\mu = \frac{g_m}{\omega_T} = \frac{40 \times 10^{-3}}{2\pi \times 800 \times 10^6} = 8 \text{ pF}$$

$$C_\mu = 1 \text{ pF}$$

$$C_\pi = 7 \text{ pF}$$

$$r_x = 50 \ \Omega$$

The midband voltage gain is

$$A_M = -\frac{R_B}{R_B + R_{\text{sig}}} \frac{r_\pi}{r_\pi + r_x + (R_B \| R_{\text{sig}})} g_m R_L'$$

where

$$R_L' = r_o \| R_C \| R_L$$

$$= (100 \| 8 \| 5) \text{ k}\Omega = 3 \text{ k}\Omega$$

Thus,

$$g_m R_L' = 40 \times 3 = 120 \text{ V/V}$$

and

$$A_M = -\frac{100}{100 + 5} \times \frac{2.5}{2.5 + 0.05 + (100 \| 5)} \times 120$$

$$= -39 \text{ V/V}$$

and

$$20 \log |A_M| = 32 \text{ dB}$$

To determine f_H we first find C_{in},

$$C_{\text{in}} = C_\pi + C_\mu (1 + g_m R_L')$$

$$= 7 + 1(1 + 120) = 128 \text{ pF}$$

and the effective source resistance R'_{sig} ,

$$R'_{\text{sig}} = r_\pi \| [r_x + (R_B \| R_{\text{sig}})]$$

$$= 2.5 \| [0.05 + (100 \| 5)]$$

$$= 1.65 \text{ k}\Omega$$

Thus,

$$f_H = \frac{1}{2\pi C_{\text{in}} R'_{\text{sig}}} = \frac{1}{2\pi \times 128 \times 10^{-12} \times 1.65 \times 10^3} = 754 \text{ kHz}$$

EXERCISE

9.10 For the amplifier in Example 9.4, find the value of R_L that reduces the midband gain to half the value found. What value of f_H results? Note the trade-off between gain and bandwidth.
Ans. 1.9 kΩ; 1.42 MHz

9.4 Useful Tools for the Analysis of the High-Frequency Response of Amplifiers

The approximate method used in the previous section to analyze the high-frequency response of the CS and CE amplifiers provides a reasonably accurate estimate of f_H and, equally important, considerable insight into the mechanism that limits high-frequency operation. Unfortunately, however, this method is not easily extendable to more complex amplifier circuits. For this reason, we will digress briefly in this section to equip ourselves with a number of tools that will prove useful in the analysis of more complex circuits such as the cascode amplifier. We will begin by stepping back and more generally considering the amplifier high-frequency transfer function.

9.4.1 The High-Frequency Gain Function

The amplifier gain, taking into account the internal transistor capacitances, can be expressed as a function of the complex-frequency variable s in the general form

$$A(s) = A_M F_H(s) \tag{9.61}$$

where A_M is the midband gain, which for IC amplifiers is also equal to the low-frequency or dc gain (refer to Fig. 9.12). The value of A_M can be determined by analyzing the amplifier equivalent circuit while neglecting the effect of the transistor internal capacitances—that is, by assuming that they act as perfect open circuits. By taking these capacitances into account, we see that the gain acquires the factor $F_H(s)$, which can be expressed in terms of its poles and zeros, which are usually real, as follows:

$$F_H(s) = \frac{(1+s/\omega_{Z1})(1+s/\omega_{Z2})\ldots(1+s/\omega_{Zn})}{(1+s/\omega_{P1})(1+s/\omega_{P2})\ldots(1+s/\omega_{Pn})} \tag{9.62}$$

where $\omega_{P1}, \omega_{P2}, \ldots, \omega_{Pn}$ are positive numbers representing the frequencies of the n real poles and $\omega_{Z1}, \omega_{Z2}, \ldots, \omega_{Zn}$ are positive, negative, or infinite numbers representing the frequencies of the n real transmission zeros. Note from Eq. (9.62) that, as should be expected, as s approaches 0, $F_H(s)$ approaches unity and the gain approaches A_M.

9.4.2 Determining the 3-dB Frequency f_H

The amplifier designer usually is particularly interested in the part of the high-frequency band that is close to the midband. This is because the designer needs to estimate—and if need be modify—the value of the upper 3-dB frequency f_H (or ω_H; $f_H = \omega_H/2\pi$). Toward that end it should be mentioned that in many cases the zeros are either at infinity or such high frequencies as to be of little significance to the determination of ω_H. If in addition one of the poles, say ω_{P1}, is of much lower frequency than any of the other poles, then this pole will have the greatest effect on the value of the amplifier ω_H. In other words, this pole will *dominate* the high-frequency response of the amplifier, and the amplifier is said to have a **dominant-pole response**. In such cases, the function $F_H(s)$ can be approximated by

$$F_H(s) \simeq \frac{1}{1+s/\omega_{P1}} \tag{9.63}$$

which is the transfer function of a first-order (or STC) low-pass network (Appendix E). It follows that if a dominant pole exists, then the determination of ω_H is greatly simplified;

$$\omega_H \simeq \omega_{P1} \tag{9.64}$$

This is the situation we encountered in the cases of the common-source and common-emitter amplifiers analyzed in Section 9.3. As a rule of thumb, *a dominant pole exists if the lowest-frequency pole is at least two octaves (a factor of 4) away from the nearest pole or zero.*

If a dominant pole does not exist, the 3-dB frequency ω_H can be determined from a plot of $|F_H(j\omega)|$. Alternatively, an approximate formula for ω_H can be derived as follows: Consider, for simplicity, the case of a circuit having two poles and two zeros in the high-frequency band; that is,

$$F_H(s) = \frac{(1 + s/\omega_{Z1})(1 + s/\omega_{Z2})}{(1 + s/\omega_{P1})(1 + s/\omega_{P2})} \tag{9.65}$$

Substituting $s = j\omega$ and taking the squared magnitude gives

$$|F_H(j\omega)|^2 = \frac{(1 + \omega^2/\omega_{Z1}^2)(1 + \omega^2/\omega_{Z2}^2)}{(1 + \omega^2/\omega_{P1}^2)(1 + \omega^2/\omega_{P2}^2)}$$

By definition, at $\omega = \omega_H$, $|F_H|^2 = \frac{1}{2}$; thus,

$$\frac{1}{2} = \frac{(1 + \omega_H^2/\omega_{Z1}^2)(1 + \omega_H^2/\omega_{Z2}^2)}{(1 + \omega_H^2/\omega_{P1}^2)(1 + \omega_H^2/\omega_{P2}^2)}$$

$$= \frac{1 + \omega_H^2\left(\dfrac{1}{\omega_{Z1}^2} + \dfrac{1}{\omega_{Z2}^2}\right) + \omega_H^4/\omega_{Z1}^2\omega_{Z2}^2}{1 + \omega_H^2\left(\dfrac{1}{\omega_{P1}^2} + \dfrac{1}{\omega_{P2}^2}\right) + \omega_H^4/\omega_{P1}^2\omega_{P2}^2} \tag{9.66}$$

Since ω_H is usually smaller than the frequencies of all the poles and zeros, we may neglect the terms containing ω_H^4 and solve for ω_H to obtain

$$\omega_H \simeq 1\bigg/\sqrt{\frac{1}{\omega_{P1}^2} + \frac{1}{\omega_{P2}^2} - \frac{2}{\omega_{Z1}^2} - \frac{2}{\omega_{Z2}^2}} \tag{9.67}$$

This relationship can be extended to any number of poles and zeros as

$$\omega_H \simeq 1\bigg/\sqrt{\left(\frac{1}{\omega_{P1}^2} + \frac{1}{\omega_{P2}^2} + \cdots\right) - 2\left(\frac{1}{\omega_{Z1}^2} + \frac{1}{\omega_{Z2}^2} + \cdots\right)} \tag{9.68}$$

Note that if one of the poles, say P_1, is dominant, then $\omega_{P1} \ll \omega_{P2}, \omega_{P3}, \ldots, \omega_{Z1}, \omega_{Z2}, \ldots,$ and Eq. (9.68) reduces to Eq. (9.69).

Example 9.5

The high-frequency response of an amplifier is characterized by the transfer function

$$F_H(s) = \frac{1 - s/10^5}{(1 + s/10^4)(1 + s/4 \times 10^4)}$$

Determine the 3-dB frequency approximately and exactly.

Solution

Noting that the lowest-frequency pole at 10^4 rad/s is two octaves lower than the second pole and a decade lower than the zero, we find that a dominant-pole situation almost exists and $\omega_H \simeq 10^4$ rad/s. A better estimate of ω_H can be obtained using Eq. (9.68), as follows:

$$\omega_H = 1 \Bigg/ \sqrt{\frac{1}{10^8} + \frac{1}{16 \times 10^8} - \frac{2}{10^{10}}}$$

$$= 9800 \text{ rad/s}$$

The exact value of ω_H can be determined from the given transfer function as 9537 rad/s. Finally, we show in Fig. 9.15 a Bode plot and an exact plot for the given transfer function. Note that this is a plot of the high-frequency response of the amplifier normalized relative to its midband gain. That is, if the midband gain is, say, 100 dB, then the entire plot should be shifted upward by 100 dB.

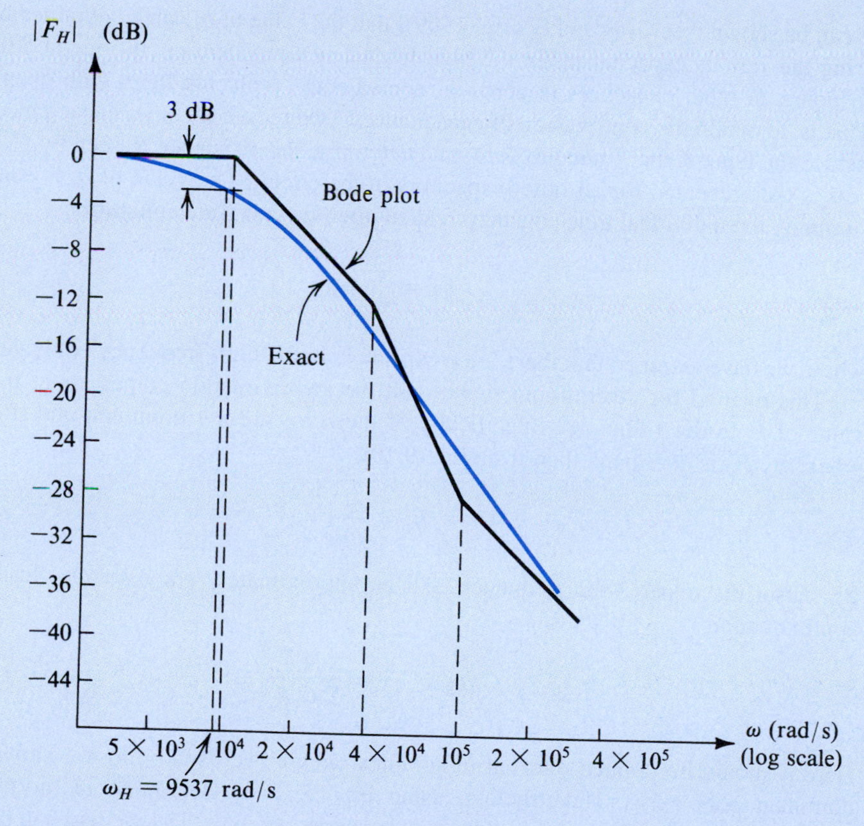

Figure 9.15 Normalized high-frequency response of the amplifier in Example 9.5.

9.4.3 Using Open-Circuit Time Constants for the Approximate Determination of f_H

If the poles and zeros of the amplifier transfer function can be determined easily, then we can determine f_H using the techniques above. In many cases, however, it is not a simple matter to determine the poles and zeros by quick hand analysis. In such cases an approximate value for f_H can be obtained using the following method.

Consider the function $F_H(s)$ (Eq. 9.62), which determines the high-frequency response of the amplifier. The numerator and denominator factors can be multiplied out and $F_H(s)$ expressed in the alternative form

$$F_H(s) = \frac{1 + a_1 s + a_2 s^2 + \cdots + a_n s^n}{1 + b_1 s + b_2 s^2 + \cdots + b_n s^n} \tag{9.69}$$

where the coefficients a and b are related to the frequencies of the zeros and poles, respectively. Specifically, the coefficient b_1 is given by

$$b_1 = \frac{1}{\omega_{P1}} + \frac{1}{\omega_{P2}} + \cdots + \frac{1}{\omega_{Pn}} \tag{9.70}$$

It can be shown [see Gray and Searle (1969)] that the value of b_1 can be obtained by considering the various capacitances in the high-frequency equivalent circuit one at a time while reducing all other capacitors to zero (or, equivalently, replacing them with open circuits). That is, to obtain the contribution of capacitance C_i we reduce all other capacitances to zero, reduce the input signal source to zero, and determine the resistance R_i seen by C_i. This process is then repeated for all other capacitors in the circuit. The value of b_1 is computed by summing the individual time constants, called **open-circuit time constants,**

$$b_1 = \sum_{i=1}^{n} C_i R_i \tag{9.71}$$

where we have assumed that there are n capacitors in the high-frequency equivalent circuit.

This method for determining b_1 is *exact*; the approximation comes about in using the value of b_1 to determine ω_H. Specifically, if the zeros are not dominant and if one of the poles, say P_1, is dominant, then from Eq. (9.70),

$$b_1 \simeq \frac{1}{\omega_{P1}} \tag{9.72}$$

But, also, the upper 3-dB frequency will be approximately equal to ω_{P1}, leading to the approximation

$$\omega_H \simeq \frac{1}{b_1} = \frac{1}{\sum_i C_i R_i} \tag{9.73}$$

Here it should be pointed out that in complex circuits we usually do not know whether a dominant pole exists. Nevertheless, using Eq. (9.73) to determine ω_H normally yields remarkably good results[5] even if a dominant pole does not exist. The method will be illustrated by an example.

[5] The method of open-circuit time constants yields good results only when all the poles are real, as is the case in this chapter.

Example 9.6

Figure 9.16(a) shows the high-frequency equivalent circuit of a common-source MOSFET amplifier. The amplifier is fed with a signal generator V_{sig} having a resistance R_{sig}. Resistance R_G is due to the biasing network. Resistance R_L' is the parallel equivalent of the load resistance R_L, the drain bias resistance R_D, and the FET output resistance r_o. Capacitors C_{gs} and C_{gd} are the MOSFET internal capacitances. For $R_{sig} = 100\ \text{k}\Omega$, $R_G = 420\ \text{k}\Omega$, $C_{gs} = C_{gd} = 1\ \text{pF}$, $g_m = 4\ \text{mA/V}$, and $R_L' = 3.33\ \text{k}\Omega$, find the midband voltage gain, $A_M = V_o/V_{sig}$ and the upper 3-dB frequency, f_H.

Solution

The midband voltage gain is determined by assuming that the capacitors in the MOSFET model are perfect open circuits. This results in the midband equivalent circuit shown in Fig. 9.16(b),

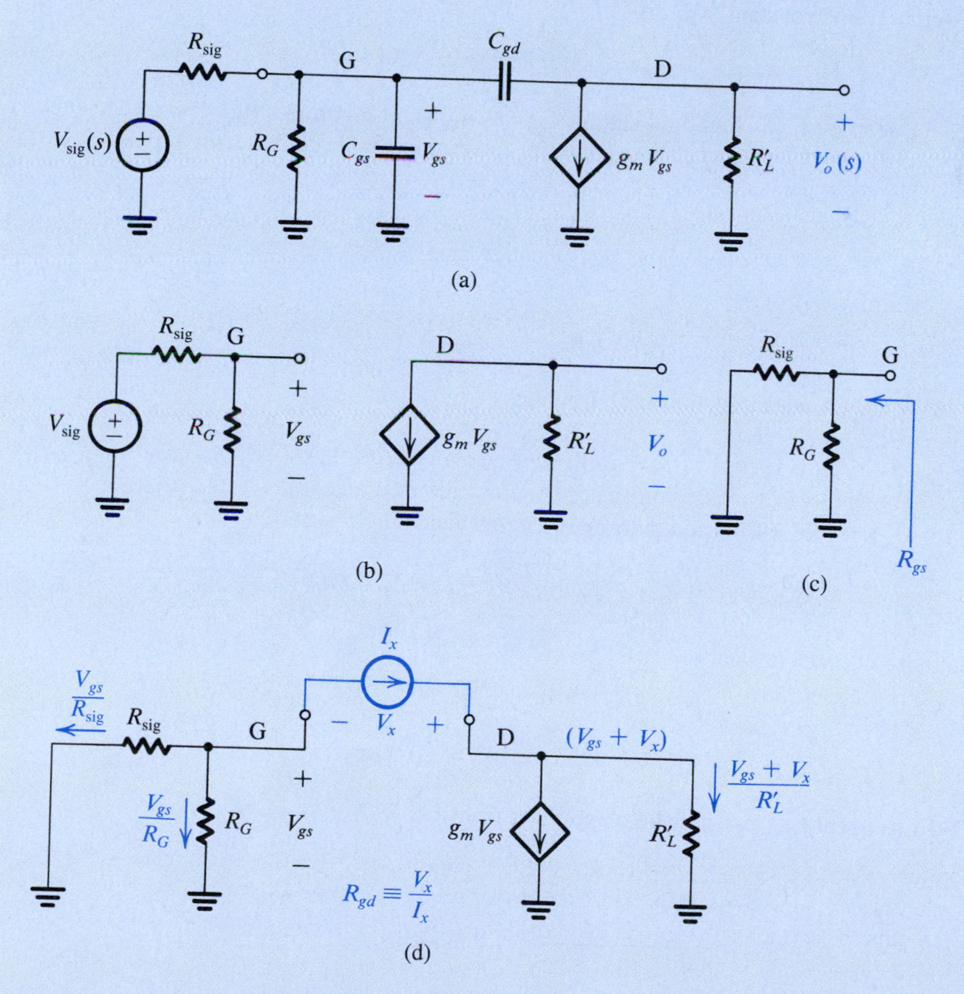

Figure 9.16 Circuits for Example 9.6: **(a)** high-frequency equivalent circuit of a MOSFET amplifier; **(b)** the equivalent circuit at midband frequencies; **(c)** circuit for determining the resistance seen by C_{gs}; **(d)** circuit for determining the resistance seen by C_{gd}.

Example 9.6 *continued*

from which we find

$$A_M \equiv \frac{V_o}{V_{\text{sig}}} = -\frac{R_G}{R_G + R_{\text{sig}}}(g_m R_L')$$

$$= -\frac{420}{420 + 100} \times 4 \times 3.33 = -10.8 \text{ V/V}$$

We shall determine ω_H using the method of open-circuit time constants. The resistance R_{gs} seen by C_{gs} is found by setting $C_{gd} = 0$ and short-circuiting the signal generator V_{sig}. This results in the circuit of Fig. 9.16(c), from which we find that

$$R_{gs} = R_G \| R_{\text{sig}} = 420 \text{ k}\Omega \| 100 \text{ k}\Omega = 80.8 \text{ k}\Omega$$

Thus the open-circuit time constant of C_{gs} is

$$\tau_{gs} \equiv C_{gs} R_{gs} = 1 \times 10^{-12} \times 80.8 \times 10^3 = 80.8 \text{ ns}$$

The resistance R_{gd} seen by C_{gd} is found by setting $C_{gs} = 0$ and short-circuiting V_{sig}. The result is the circuit in Fig. 9.16(d), to which we apply a test current I_x. Writing a node equation at G gives

$$I_x = -\frac{V_{gs}}{R_G} - \frac{V_{gs}}{R_{\text{sig}}}$$

Thus,

$$V_{gs} = -I_x R_{\text{sig}}' \tag{9.74}$$

where $R_{\text{sig}}' = R_G \| R_{\text{sig}}$. A node equation at D provides

$$I_x = g_m V_{gs} + \frac{V_{gs} + V_x}{R_L'}$$

Substituting for V_{gs} from Eq. (9.74) and rearranging terms yields

$$R_{gd} \equiv \frac{V_x}{I_x} = R_{\text{sig}}' + R_L' + g_m R_L' R_{\text{sig}}' = 1.16 \text{ M}\Omega$$

Thus the open-circuit time constant of C_{gd} is

$$\tau_{gd} \equiv C_{gd} R_{gd}$$

$$= 1 \times 10^{-12} \times 1.16 \times 10^6 = 1160 \text{ ns}$$

The upper 3-dB frequency ω_H can now be determined from

$$\omega_H \simeq \frac{1}{\tau_{gs} + \tau_{gd}}$$

$$= \frac{1}{(80.8 + 1160) \times 10^{-9}} = 806 \text{ krad/s}$$

Thus,

$$f_H = \frac{\omega_H}{2\pi} = 128.3 \text{ kHz}$$

The method of open-circuit time constants has an important advantage in that it tells the circuit designer which of the various capacitances is significant in determining the amplifier frequency response. Specifically, the relative contribution of the various capacitances to the effective time constant b_1 is immediately obvious. For instance, in the above example we see that C_{gd} is the dominant capacitance in determining f_H. We also note that, in effect to increase f_H either we use a MOSFET with smaller C_{gd} or, for a given MOSFET, we reduce R_{gd} by using a smaller R'_{sig} or R'_L. If R'_{sig} is fixed, then for a given MOSFET the only way to increase bandwidth is by reducing the load resistance. Unfortunately, this also decreases the midband gain. This is an example of the usual trade-off between gain and bandwidth, a common circumstance which was mentioned earlier.

9.4.4 Miller's Theorem

In our analysis of the high-frequency response of the common-source and common-emitter amplifiers (Section 9.3), we employed a technique for replacing the bridging capacitance (C_{gs} or C_μ) by an equivalent input capacitance. This very useful and effective technique is based on a general theorem known as **Miller's theorem**, which we now present.

Consider the situation in Fig. 9.17(a). As part of a larger circuit that is not shown, we have isolated two circuit nodes, labeled 1 and 2, between which an impedance Z is connected. Nodes 1 and 2 are also connected to other parts of the circuit, as signified by the broken lines emanating from the two nodes. Furthermore, it is assumed that somehow it has been determined that the voltage at node 2 is related to that at node 1 by

$$V_2 = KV_1 \tag{9.75}$$

In typical situations K is a gain factor that can be positive or negative and that has a magnitude usually larger than unity. This, however, is not an assumption for Miller's theorem.

Miller's theorem states that impedance Z can be replaced by two impedances: Z_1 connected between node 1 and ground and Z_2 connected between node 2 and ground, where

$$Z_1 = Z/(1-K) \tag{9.76a}$$

and

$$Z_2 = Z\bigg/\left(1-\frac{1}{K}\right) \tag{9.76b}$$

to obtain the equivalent circuit shown in Fig. 9.17(b).

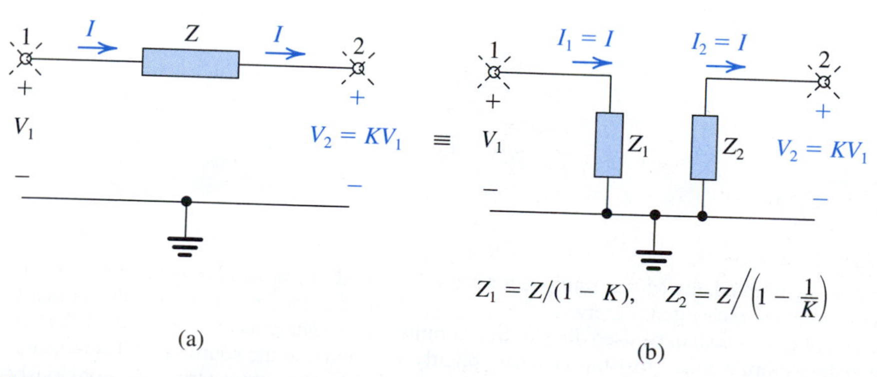

Figure 9.17 The Miller equivalent circuit.

The proof of Miller's theorem is achieved by deriving Eq. (9.76) as follows: In the original circuit of Fig. 9.17(a), the only way that node 1 "feels the existence" of impedance Z is through the current I that Z draws away from node 1. Therefore, to keep this current unchanged in the equivalent circuit, we must choose the value of Z_1 so that it draws an equal current,

$$I_1 = \frac{V_1}{Z_1} = I = \frac{V_1 - KV_1}{Z}$$

which yields the value of Z_1 in Eq. (9.76a). Similarly, to keep the current into node 2 unchanged, we must choose the value of Z_2 so that

$$I_2 = \frac{0 - V_2}{Z_2} = \frac{0 - KV_1}{Z_2} = I = \frac{V_1 - KV_1}{Z}$$

which yields the expression[6] for Z_2 in Eq. (9.76b).

Example 9.7

Figure 9.18(a) shows an ideal voltage amplifier having a gain of –100 V/V with an impedance Z connected between its output and input terminals. Find the Miller equivalent circuit when Z is (a) a 1-MΩ resistance and (b) a 1-pF capacitance. In each case, use the equivalent circuit to determine V_o/V_{sig}.

Solution

(a) For $Z = 1$ MΩ, employing Miller's theorem results in the equivalent circuit in Fig. 9.18(b), where

$$Z_1 = \frac{Z}{1-K} = \frac{1000\ \text{k}\Omega}{1+100} = 9.9\ \text{k}\Omega$$

$$Z_2 = \frac{Z}{1-\frac{1}{K}} = \frac{1\ \text{M}\Omega}{1+\frac{1}{100}} = 0.99\ \text{M}\Omega$$

The voltage gain can be found as follows:

$$\frac{V_o}{V_{sig}} = \frac{V_o}{V_i}\frac{V_i}{V_{sig}} = -100 \times \frac{Z_1}{Z_1 + R_{sig}}$$

$$= -100 \times \frac{9.9}{9.9+10} = -49.7\ \text{V/V}$$

[6] Although not highlighted, the Miller equivalent circuit derived above is valid only as long as the rest of the circuit remains unchanged; otherwise the ratio of V_2 to V_1 might change. It follows that the Miller equivalent circuit *cannot* be used directly to determine the output resistance of an amplifier. This is because in determining output resistances it is implicitly assumed that the source signal is reduced to zero and that a test-signal source (voltage or current) is applied to the output terminals—obviously a major change in the circuit, rendering the Miller equivalent circuit no longer valid.

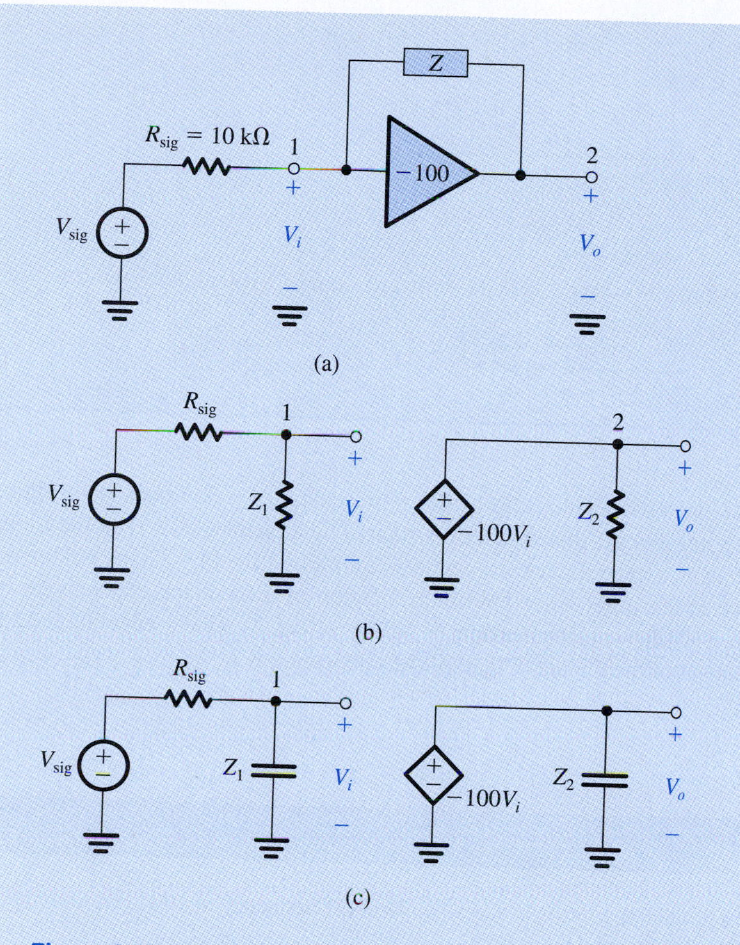

Figure 9.18 Circuits for Example 9.7.

(b) For Z as a 1-pF capacitance—that is, $Z = 1/sC = 1/s \times 1 \times 10^{-12}$—applying Miller's theorem allows us to replace Z by Z_1 and Z_2, where

$$Z_1 = \frac{Z}{1-K} = \frac{1/sC}{1+100} = 1/s(101C)$$

$$Z_2 = \frac{Z}{1-\frac{1}{K}} = \frac{1}{1.01}\frac{1}{sC} = \frac{1}{s(1.01C)}$$

It follows that Z_1 is a capacitance $101C = 101$ pF and that Z_2 is a capacitance $1.01C = 1.01$ pF. The resulting equivalent circuit is shown in Fig. 9.18(c), from which the voltage gain can be found as follows:

$$\frac{V_o}{V_{sig}} = \frac{V_o}{V_i}\frac{V_i}{V_{sig}} = -100\frac{1/sC_1}{1/(sC_1)+R_{sig}}$$

$$= \frac{-100}{1+sC_1R_{sig}}$$

Example 9.7 *continued*

$$= \frac{-100}{1 + s \times 101 \times 1 \times 10^{-12} \times 10 \times 10^3}$$

$$= \frac{-100}{1 + s \times 1.01 \times 10^{-6}}$$

This is the transfer function of a first-order low-pass network with a dc gain of −100 and a 3-dB frequency f_{3dB} of

$$f_{3dB} = \frac{1}{2\pi \times 1.01 \times 10^{-6}} = 157.6 \text{ kHz}$$

From Example 9.7, we observe that the Miller replacement of a feedback or bridging resistance results, for a negative K, in a smaller resistance [by a factor $(1 - K)$] at the input. If the feedback element is a capacitance, its value is multiplied by $(1 - K)$ to obtain the equivalent capacitance at the input side. The multiplication of a feedback capacitance by $(1 - K)$ is referred to as **Miller multiplication** or **Miller effect**. We have encountered the Miller effect in the analysis of the CS and CE amplifiers in Section 9.3.

EXERCISES

9.11 A direct-coupled amplifier has a dc gain of 1000 V/V and an upper 3-dB frequency of 100 kHz. Find the transfer function and the gain–bandwidth product in hertz.

Ans. $\dfrac{1000}{1 + \dfrac{s}{2\pi \times 10^5}}$; 10^8 Hz

9.12 The high-frequency response of an amplifier is characterized by two zeros at $s = \infty$ and two poles at ω_{P1} and ω_{P2}. For $\omega_{P2} = k\omega_{P1}$, find the value of k that results in the exact value of ω_H being $0.9\,\omega_{P1}$. Repeat for $\omega_H = 0.99\,\omega_{P1}$.

Ans. 2.78; 9.88

9.13 For the amplifier described in Exercise 9.12, find the exact and approximate values (using Eq. 9.68) of ω_H (as a function of ω_{P1}) for the cases $k = 1$, 2, and 4.

Ans. 0.64, 0.71; 0.84, 0.89; 0.95, 0.97

9.14 For the amplifier in Example 9.6, find the gain–bandwidth product in megahertz. Find the value of R_L' that will result in $f_H = 180$ kHz. Find the new values of the midband gain and of the gain–bandwidth product.

Ans. 1.39 MHz; 2.23 kΩ; −7.2 V/V; 1.30 MHz

9.15 Use Miller's theorem to investigate the performance of the inverting op-amp circuit shown in Fig. E9.15. Assume the op amp to be ideal except for having a finite differential gain, A. Without using any knowledge of op-amp circuit analysis, find R_{in}, V_i, V_o, and V_o/V_{sig}, for each of the following values of A: 10 V/V, 100 V/V, 1000 V/V, and 10,000 V/V. Assume $V_{sig} = 1$ V.

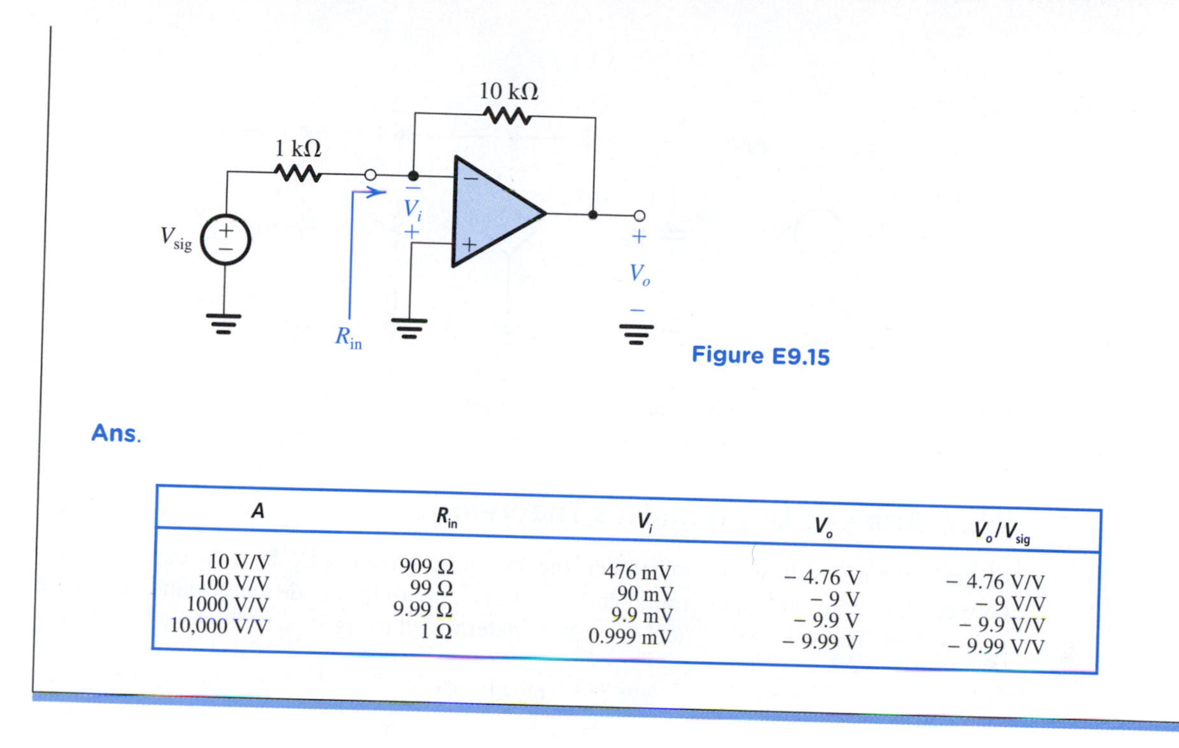

Figure E9.15

Ans.

A	R_{in}	V_i	V_o	V_o/V_{sig}
10 V/V	909 Ω	476 mV	− 4.76 V	− 4.76 V/V
100 V/V	99 Ω	90 mV	− 9 V	− 9 V/V
1000 V/V	9.99 Ω	9.9 mV	− 9.9 V	− 9.9 V/V
10,000 V/V	1 Ω	0.999 mV	− 9.99 V	− 9.99 V/V

9.5 A Closer Look at the High-Frequency Response of the CS and CE Amplifiers

In Section 9.3 we utilized the Miller approximation to obtain an estimate of the high-frequency 3-dB frequency f_H of the CS and CE amplifiers. We shall now use the powerful tools we studied in the last section to revisit this subject. Specifically, we will first employ Miller's theorem to refine the Miller approximation, thus obtaining a better estimate of f_H. Then we will use the method of open-circuit time constants to obtain another estimate of f_H. In order to assess how good these various estimates are, the exact transfer function will be derived and analyzed. Finally, we will consider the case of low source resistance R_{sig} with the limitation on the high-frequency response determined by the capacitance at the output node, a situation that is not uncommon in IC amplifiers.

9.5.1 The Equivalent Circuit

Figure 9.19 shows a generalized high-frequency equivalent circuit for the common-source amplifier. Here, V'_{sig} and R'_{sig} are the Thévenin equivalent of the signal generator together with whatever bias circuit may be present at the amplifier input (e.g., R_G in the circuit of Fig. 9.2a). Resistance R'_L represents the total resistance between the output (drain) node and ground and includes R_D, r_o, and R_L (if one is present). Similarly, C_L represents the total capacitance between the drain node and ground and includes the MOSFET's drain-to-body capacitance (C_{db}), the capacitance introduced by a current-source load, the input capacitance of a succeeding amplifier stage (if one is present), and in some cases, as we will see in later chapters, a deliberately introduced capacitance. In IC MOS amplifiers, C_L can be substantial.

The equivalent circuit in Fig. 9.19 can also be used to represent the CE amplifier. Thus, we will not need to repeat the analysis, rather we will adapt the CS results to the CE case by simply renaming the components (i.e., replacing C_{gs} by C_π and C_{gd} by C_μ).

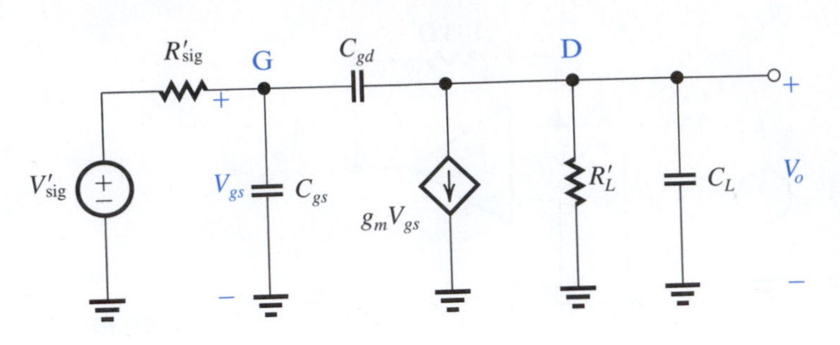

Figure 9.19 Generalized high-frequency equivalent circuit for the CS amplifier.

9.5.2 Analysis Using Miller's Theorem

Miller's theorem allows us to replace the bridging capacitor C_{gd} by two capacitors: C_1 between the input node and ground and C_2 between the output node and ground, as shown in Fig. 9.20. The value of C_1 and C_2 can be determined using Eqs. (9.76a) and (9.76b),

$$C_1 = C_{gd}(1 - K)$$

$$C_2 = C_{gd}\left(1 - \frac{1}{K}\right)$$

where

$$K = \frac{V_o}{V_{gs}}$$

Obviously, K will depend on the value of C_2, which in turn depends on the value of K. To simplify matters, we shall adopt an iterative procedure: First, we will neglect C_2 and C_L in

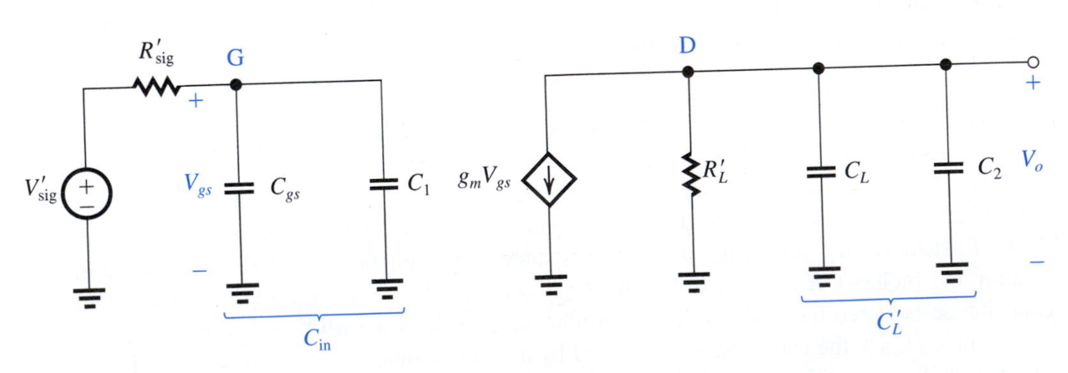

Figure 9.20 The high-frequency equivalent circuit model of the CS amplifier after the application of Miller's theorem to replace the bridging capacitor C_{gd} by two capacitors: $C_1 = C_{gd}(1 - K)$ and $C_2 = C_{gd}(1 - 1/K)$, where $K = V_o/V_{gs}$.

determining V_o, resulting in

$$V_o \simeq -g_m V_{gs} R'_L \tag{9.77}$$

That is, K is given by

$$K \simeq -g_m R'_L$$

Then we will use this value to determine C_1 and C_2 as

$$C_1 = C_{gd}(1 + g_m R'_L) \tag{9.78}$$

$$C_2 = C_{gd}\left(1 + \frac{1}{g_m R'_L}\right) \tag{9.79}$$

Next, we use C_1 and C_2 to determine the overall transfer function V_o/V'_{sig}. At the input side, we see that the input capacitance $C_{in} = C_{gs} + C_1$ together with R'_{sig} form an STC low-pass circuit with a pole frequency f_{Pi}:

$$f_{Pi} = \frac{1}{2\pi C_{in} R'_{sig}} \tag{9.80}$$

At the output sides we see that $C'_L = C_L + C_2$ together with R'_L form an STC low-pass circuit with a pole frequency f_{Po}:

$$f_{Po} = \frac{1}{2\pi C'_L R'_L} \tag{9.81}$$

At this point we note that in Section 9.3 we neglected both C_2 and C_L and thus f_{Po}. Thus the estimate of f_H in Section 9.3 was based on the assumption that V_o is given by Eq. (9.77), and thus the frequency limitation is caused entirely by the interaction of C_{in} with R'_{sig}, that is, by the input pole f_{Pi}:

$$f_H \simeq f_{Pi} \tag{9.82}$$

A somewhat better estimate of f_H can be obtained by considering both f_{Pi} and f_{Po}, that is, by using the approximate transfer function[7]

$$\frac{V_o}{V'_{sig}} = \frac{-g_m R'_L}{\left(1 + \frac{s}{\omega_{Pi}}\right)\left(1 + \frac{s}{\omega_{Po}}\right)}$$

[7] This transfer function is approximate because we obtained it using an iterative process with in fact only one iteration!

An estimate of f_H can then be found using Eq. (9.68) as

$$f_H = 1\bigg/\sqrt{\frac{1}{f_{Pi}^2} + \frac{1}{f_{Po}^2}}$$

$$= \frac{f_{Pi}}{\sqrt{1 + \left(\frac{f_{Pi}}{f_{Po}}\right)^2}} \qquad (9.83)$$

This estimate will diverge from that in Eq. (9.82) in situations for which f_{Po} is not much higher than f_{Pi}. This will be the case when R'_{sig} is not very high and C_L is relatively large.

Example 9.8

Consider an IC CS amplifier for which g_m = 1.25 mA/V², C_{gs} = 20 fF, C_{gd} = 5 fF, C_L = 25 fF, R'_{sig} = 10 kΩ, and R'_L = 10 kΩ. Assume that C_L includes C_{db}. Determine f_H using (a) the Miller approximation and (b) Miller's theorem.

Solution

(a) The Miller approximation assumes $V_o = -g_m R'_L V_{gs}$ and thus neglects the effect of C_L and C_2. In this case,

$$f_H \simeq f_{Pi} = \frac{1}{2\pi C_{in} R'_{sig}}$$

where

$$C_{in} = C_{gs} + C_1 = C_{gs} + C_{gd}(1 + g_m R'_L)$$

Thus,

$$C_{in} = 20 + 5(1 + 1.25 \times 10)$$

$$= 87.5 \text{ fF}$$

and f_{Pi} will be

$$f_{Pi} = \frac{1}{2\pi \times 87.5 \times 10^{-15} \times 10 \times 10^3}$$

$$= 181.9 \text{ MHz}$$

Thus,

$$f_H \simeq 181.9 \text{ MHz}$$

(b) Using Miller's theorem, we obtain the same f_{Pi} as above:

$$f_{Pi} = 181.9 \text{ MHz}$$

But now we can take C_2 and C_L into account. Capacitance C_2 can be determined as

$$C_2 = C_{gd}\left(1 + \frac{1}{g_m R_L'}\right) = 5\left(1 + \frac{1}{12.5}\right) = 5.4 \text{ fF}$$

The frequency of the output pole can now be determined as

$$f_{Po} = \frac{1}{2\pi(C_L + C_2)R_L'}$$

$$f_{Po} = \frac{1}{2\pi(25 + 5.4) \times 10^{-15} \times 10 \times 10^3}$$

$$= 523.5 \text{ MHz}$$

An estimate of f_H can now be found from Eq. (9.83):

$$f_H = \frac{181.9}{\sqrt{1 + \left(\frac{181.9}{523.5}\right)^2}} = 171.8 \text{ MHz}$$

9.5.3 Analysis Using Open-Circuit Time Constants

The method of open-circuit time constants presented in Section 9.4.3 can be directly applied to the CS equivalent circuit of Fig. 9.19, as illustrated in Fig. 9.21, from which we see that the resistance seen by C_{gs}, $R_{gs} = R'_{\text{sig}}$ and that seen by C_L is R_L'. The resistance R_{gd} seen by C_{gd} can be found by analyzing the circuit in Fig. 9.21(b) with the result that

$$R_{gd} = R'_{\text{sig}}(1 + g_m R_L') + R_L' \tag{9.84}$$

Thus the effective time constant b_1 or τ_H can be found as

$$\begin{aligned}\tau_H &= C_{gs}R_{gs} + C_{gd}R_{gd} + C_L R_{C_L} \\ &= C_{gs}R'_{\text{sig}} + C_{gd}[R'_{\text{sig}}(1 + g_m R_L') + R_L'] + C_L R_L' \end{aligned} \tag{9.85}$$

and the 3-dB frequency f_H is

$$f_H \simeq \frac{1}{2\pi\tau_H} \tag{9.86}$$

For situations in which C_L is substantial, this approach yields a better estimate of f_H than that obtained using the Miller approximation (simply because in the latter case we completely neglected C_L).

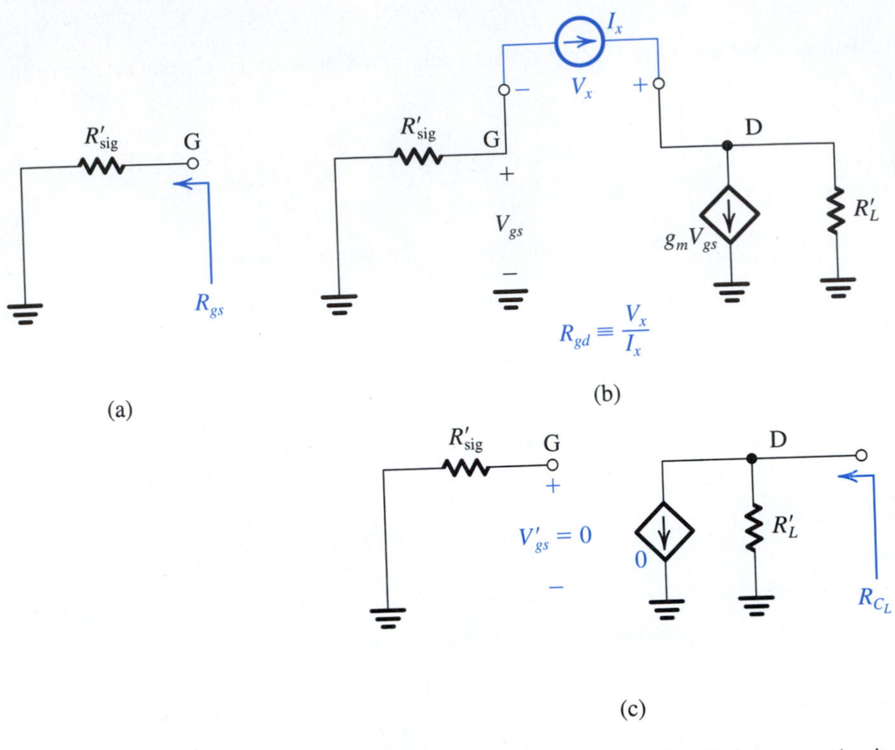

Figure 9.21 Application of the open-circuit time-constants method to the CS equivalent circuit of Fig. 9.19.

It is interesting and useful, however, to note that applying the open-circuit time-constants method to the Miller equivalent circuit shown in Fig. 9.20 results in a very close approximation to the value of τ_H in Eq. (9.85).

Example 9.9

Use the method of open-circuit time constants to obtain another estimate of f_H for the CS amplifier of Example 9.8.

Solution

$$R_{gs} = R'_{\text{sig}} = 10 \text{ k}\Omega$$

$$R_{gd} = R'_{\text{sig}}(1 + g_m R'_L) + R'_L$$

$$= 10(1 + 1.25 \times 10) + 10 = 145 \text{ k}\Omega$$

$$R'_L = 10 \text{ k}\Omega$$

Thus,

$$\tau_H = C_{gs}R_{gs} + C_{gd}R_{gd} + C_L R_L'$$
$$= 20 \times 10^{-15} \times 10 \times 10^3 + 5 \times 10^{-15} \times 145 \times 10^3 + 25 \times 10^{-15} \times 10 \times 10^3$$
$$= 1175 \text{ ps}$$

and the 3-dB frequency f_H can be estimated at

$$f_H \simeq \frac{1}{2\pi\,\tau_H} = \frac{1}{2\pi \times 1175 \times 10^{-12}} = 135.5 \text{ MHz}$$

We note that this estimate is considerably lower than both estimates found in Example 9.8. Which one is closer to the exact value will be determined next.

9.5.4 Exact Analysis

The approximate analysis presented above provides insight regarding the mechanism by which and the extent to which the various capacitances limit the high-frequency gain of the CS amplifier. Nevertheless, given that the circuit of Fig. 9.19 is relatively simple, it is instructive to also perform an exact analysis.[8] This is illustrated in Fig. 9.22. A node equation at the drain provides

$$sC_{gd}(V_{gs} - V_o) = g_m V_{gs} + \frac{V_o}{R_L'} + sC_L V_o$$

which can be manipulated to the form

$$V_{gs} = \frac{-V_o}{g_m R_L'} \frac{1 + s(C_L + C_{gd})R_L'}{1 - sC_{gd}/g_m} \tag{9.87}$$

A loop equation at the input yields

$$V_{\text{sig}}' = I_i R_{\text{sig}}' + V_{gs}$$

in which we can substitute for I_i from a node equation at G,

$$I_i = sC_{gs}V_{gs} + sC_{gd}(V_{gs} - V_o)$$

to obtain

$$V_{\text{sig}}' = V_{gs}[1 + s(C_{gs} + C_{gd})R_{\text{sig}}'] - sC_{gd}R_{\text{sig}}' V_o$$

[8] "Exact" only in the sense that we are not making approximations in the circuit-analysis process. The reader is reminded, however, that the high-frequency model itself represents an approximation of the device performance.

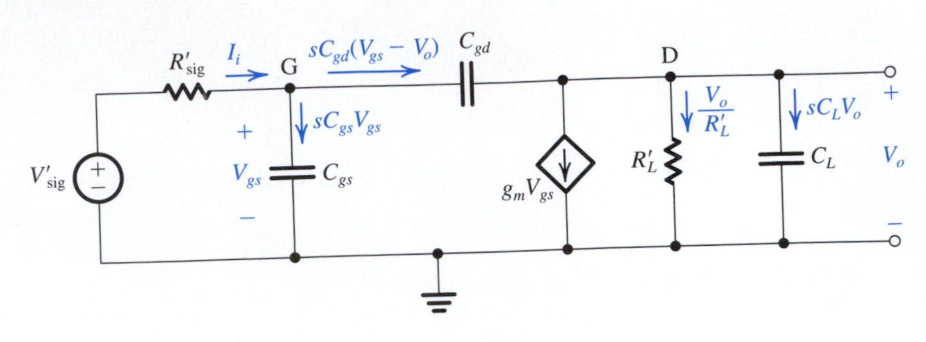

Figure 9.22 Analysis of the CS high-frequency equivalent circuit.

We can now substitute in this equation for V_{gs} from Eq. (9.87) to obtain an equation in V_o and V'_{sig} that can be arranged to yield the amplifier gain as

$$\frac{V_o}{V'_{sig}} = \frac{-(g_m R'_L)[1 - s(C_{gd}/g_m)]}{1 + s\{[C_{gs} + C_{gd}(1 + g_m R'_L)]R'_{sig} + (C_L + C_{gd})R'_L\} + s^2[(C_L + C_{gd})C_{gs} + C_L C_{gd}]R'_{sig}R'_L} \tag{9.88}$$

The transfer function in Eq. (9.88) has a second-order denominator, and thus the amplifier has two poles. Now, since the numerator is of the first order, it follows that one of the two transmission zeros is at infinite frequency. This is readily verifiable by noting that as s approaches ∞, (V_o/V'_{sig}) approaches zero. The second zero is at

$$s = s_Z = \frac{g_m}{C_{gd}} \tag{9.89}$$

That is, it is on the positive *real axis* of the s-plane[9] and has a frequency ω_Z,

$$\omega_Z = g_m/C_{gd} \tag{9.90}$$

Since g_m is usually large and C_{gd} is usually small, f_Z is normally a very high frequency and thus has negligible effect on the value of f_H.

It is useful at this point to show a simple method for finding the value of s at which $V_o = 0$—that is, s_Z. Figure 9.23 shows the circuit at $s = s_Z$. By definition, $V_o = 0$ and a node equation at D yields

$$s_Z C_{gd} V_{gs} = g_m V_{gs}$$

Now, since V_{gs} is *not* zero (why not?), we can divide both sides by V_{gs} to obtain

$$s_Z = \frac{g_m}{C_{gd}} \tag{9.91}$$

Before considering the poles, we should note that in Eq. (9.88), as s goes toward zero, V_o/V'_{sig} approaches the dc gain $(-g_m R'_L)$, as should be the case. Let's now take a closer look at the denominator polynomial. First, we observe that the coefficient of the s term is equal to the effective time constant τ_H obtained using the open-circuit time-constants method as given by Eq. (9.85). Again, this should have been expected, since it is the basis for the open-circuit

[9] Because the transmission zero is on the real axis, there is no physical frequency ω at which the transmission is actually zero (except $\omega = \infty$).

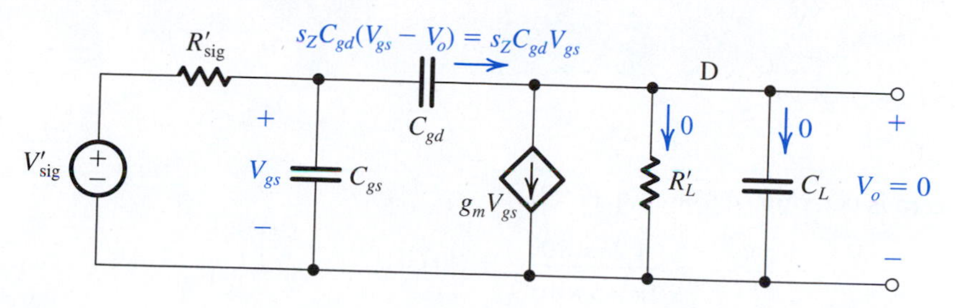

Figure 9.23 The CS circuit at $s = s_Z$. The output voltage $V_o = 0$, enabling us to determine s_Z from a node equation at D.

time-constants method (Section 9.4). Next, denoting the frequencies of the two poles ω_{P1} and ω_{P2}, we can express the denominator polynomial $D(s)$ as

$$D(s) = \left(1 + \frac{s}{\omega_{P1}}\right)\left(1 + \frac{s}{\omega_{P2}}\right)$$

$$= 1 + s\left(\frac{1}{\omega_{P1}} + \frac{1}{\omega_{P2}}\right) + \frac{s^2}{\omega_{P1}\omega_{P2}} \tag{9.92}$$

Now, if $\omega_{P2} \gg \omega_{P1}$ —that is, the pole at ω_{P1} is dominant—we can approximate $D(s)$ as

$$D(s) \simeq 1 + \frac{s}{\omega_{P1}} + \frac{s^2}{\omega_{P1}\omega_{P2}} \tag{9.93}$$

Equating the coefficients of the s term in denominator polynomial of Eq. (9.88) to that of the s term in Eq. (9.93) gives

$$\omega_{P1} \simeq \frac{1}{[C_{gs} + C_{gd}(1 + g_m R'_L)]\,R'_{sig} + (C_L + C_{gd})R'_L} \tag{9.94}$$

where the approximation is that involved in Eq. (9.93). Note that the expression in Eq. (9.94) is identical to the value of ω_H obtained using open-circuit time constants. Equating the coefficients of s^2 in Eqs. (9.88) and (9.93) and using Eq. (9.94) gives the frequency of the second pole:

$$\omega_{P2} = \frac{[C_{gs} + C_{gd}(1 + g_m R'_L)]\,R'_{sig} + (C_L + C_{gd})R'_L}{[(C_L + C_{gd})C_{gs} + C_L C_{gd}]R'_L R'_{sig}} \tag{9.95}$$

Example 9.10

For the CS amplifier considered in Examples 9.8 and 9.9, use the exact transfer function in Eq. (9.88) to determine the frequencies of the two poles and the zero and hence the 3-dB frequency f_H. Compare to the approximate values for f_H obtained in Examples 9.8 and 9.9.

Example 9.10 *continued*

Solution

The frequency of the zero is determined using Eq. (9.90),

$$f_Z = \frac{g_m}{2\pi C_{gd}} = \frac{1.25 \times 10^{-3}}{2\pi \times 5 \times 10^{-15}} = 40 \text{ GHz}$$

The frequencies of the two poles, ω_{P1} and ω_{P2}, are found as the roots of the equation obtained by equating the denominator polynomial of Eq. (9.88) to zero:

$$1 + 1.175 \times 10^{-9}s + 7.25 \times 10^{-20}s^2 = 0$$

The result is

$$f_{P1} = 143.4 \text{ MHz}$$

and

$$f_{P2} = 2.44 \text{ GHz}$$

Since $f_Z, f_{P2} \gg f_{P1}$, a good estimate for f_H is

$$f_H \simeq f_{P1} = 143.4 \text{ MHz}$$

Finally, we note that the estimate of f_{P1} obtained using Eq. (9.94) is 135.5 MHz, which is about 5.5% lower than the exact value. Thus, the method of open-circuit time constants underestimates f_H by about 5.5%. The estimate from the Miller approximation is 181.9 MHz, which is about 27% higher than the exact value, and that using the refined application of Miller theorem is 171.8 MHz, which is about 20% higher than the exact value. We conclude that the estimate obtained using open-circuit time constants is remarkably good!

EXERCISES

9.16 For the CS amplifier in Example 9.10, using the value of f_H determined by the exact analysis, find the gain–bandwidth product. Recall that $g_m = 1.25$ mA/V and $R_L' = 10$ kΩ. Also, convince yourself that this is the frequency at which the gain magnitude reduces to unity, that is, f_t.
Ans. GBW = 1.79 GHz; since this is lower than f_{P2}, then $f_t = 1.79$ GHz

9.17 As a way to trade gain for bandwidth, the designer of the CS amplifier in Example 9.10 connects a load resistor at the output that results in halving the value of R_L'. Find the new values of $|A_M|$, f_H (using $f_H \simeq f_{P1}$ of Eq. 9.94), and f_t.
Ans. 6.25 V/V; 223 MHz; 1.4 GHz

9.18 As another way to trade dc gain for bandwidth, the designer of the CS amplifier in Example 9.10 decides to operate the amplifying transistor at double the value of V_{OV} by increasing the bias current fourfold. Find the new values of g_m, R_L', $|A_M|$, f_{P1}, f_H, and f_t. Assume that R_L' is the parallel equivalent of r_o of the amplifying transistor and that of the current-source load. Use the approximate formula for f_{P1} given in Eq. (9.94).
Ans. 2.5 mA/V; 2.5 kΩ; 6.25 V/V; 250 MHz; 250 MHz; 1.56 GHz

9.5.5 Adapting the Formulas for the Case of the CE Amplifier

Adapting the formulas presented above to the case of the CE amplifier is straightforward. First, note from Fig. 9.24 how V'_{sig} and R'_{sig} relate to V_{sig}, R_{sig}, and the other equivalent-circuit parameters:

$$V'_{sig} = V_{sig}\frac{r_\pi}{R_{sig} + r_x + r_\pi} \quad (9.96)$$

$$R'_{sig} = r_\pi \| (R_{sig} + r_x) \quad (9.97)$$

Thus the dc gain is now given by

$$A_M = -\frac{r_\pi}{R_{sig} + r_x + r_\pi}(g_m R'_L) \quad (9.98)$$

Using the Miller approximation, we obtain

$$C_{in} = C_\pi + C_\mu(1 + g_m R'_L) \quad (9.99)$$

Correspondingly, the 3-dB frequency f_H can be estimated from

$$f_H \simeq \frac{1}{2\pi C_{in} R'_{sig}} \quad (9.100)$$

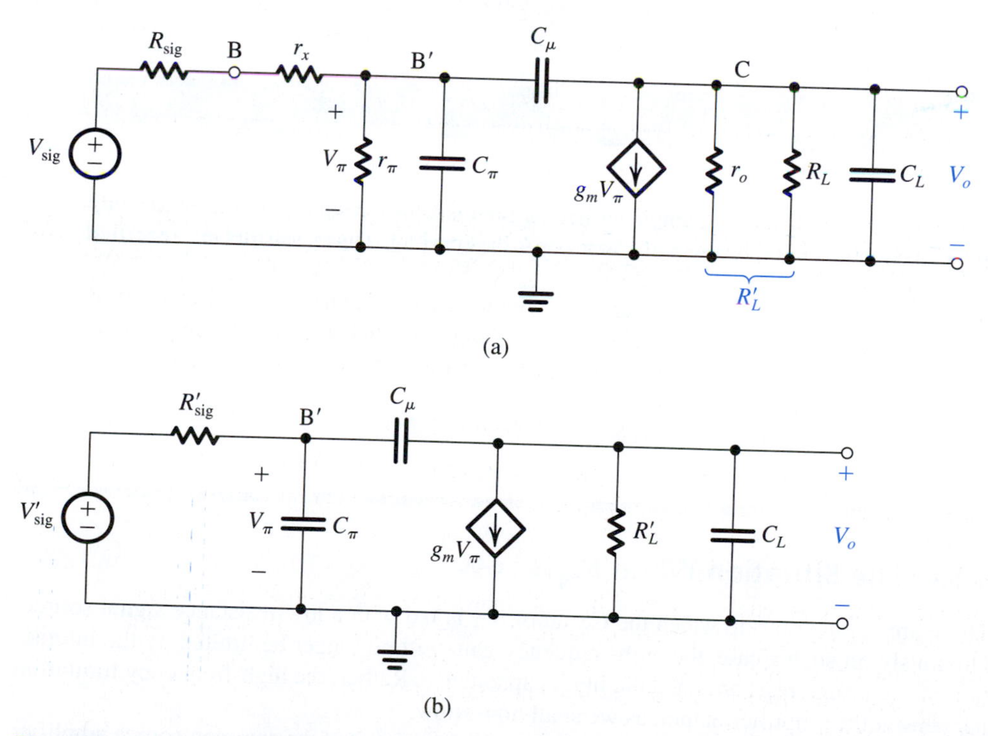

Figure 9.24 **(a)** High-frequency equivalent circuit of the common-emitter amplifier. **(b)** Equivalent circuit obtained after Thévenin theorem has been employed to simplify the resistive circuit at the input.

Alternatively, using the method of open-circuit time constants yields

$$\begin{aligned}\tau_H &= C_\pi R_\pi + C_\mu R_\mu + C_L R_{C_L} \\ &= C_\pi R'_{\text{sig}} + C_\mu[(1+g_m R'_L)R'_{\text{sig}} + R'_L] + C_L R'_L\end{aligned} \tag{9.101}$$

from which f_H can be estimated as

$$f_H \simeq \frac{1}{2\pi\tau_H} \tag{9.102}$$

The exact analysis yields the following zero frequency:

$$f_Z = \frac{1}{2\pi}\frac{g_m}{C_\mu} \tag{9.103}$$

and, assuming that a dominant pole exists,

$$f_{P1} \simeq \frac{1}{2\pi}\frac{1}{[C_\pi + C_\mu(1+g_m R'_L)]R'_{\text{sig}} + (C_L + C_\mu)R'_L} \tag{9.104}$$

$$f_{P2} \simeq \frac{1}{2\pi}\frac{[C_\pi + C_\mu(1+g_m R'_L)]R'_{\text{sig}} + (C_L + C_\mu)R'_L}{[C_\pi(C_L + C_\mu) + C_L C_\mu]R'_{\text{sig}}R'_L} \tag{9.105}$$

For f_Z, $f_{P2} \gg f_{P1}$,

$$f_H \simeq f_{P1}$$

EXERCISE

9.19 Consider a bipolar active-loaded CE amplifier having the load current source implemented with a *pnp* transistor. Let the circuit be operating at a 1-mA bias current. The transistors are specified as follows: $\beta(npn) = 200$, $V_{An} = 130$ V, $|V_{Ap}| = 50$ V, $C_\pi = 16$ pF, $C_\mu = 0.3$ pF, $C_L = 5$ pF, and $r_x = 200\ \Omega$. The amplifier is fed with a signal source having a resistance of 36 kΩ. Determine: (a) A_M; (b) C_{in} and f_H using the Miller approximation; (c) f_H using open-circuit time constants; (d) f_Z, f_{P1}, f_{P2}, and hence f_H (use the approximate expressions in Eqs. 9.105 and 9.104); and (e) the gain–bandwidth product.

Ans. (a) –175 V/V; (b) 448 pF, 82.6 kHz; (c) 75.1 kHz; (d) 21.2 GHz, 75.1 kHz, 25.2 MHz, 75.1 kHz; (e) 13.1 MHz

9.5.6 The Situation When R_{sig} Is Low

There are applications in which the CS amplifier is fed with a low-resistance signal source. Obviously, in such a case, the high-frequency gain will no longer be limited by the interaction of the source resistance and the input capacitance. Rather, the high-frequency limitation happens at the amplifier output, as we shall now show.

Figure 9.25(a) shows the high-frequency equivalent circuit of the common-source amplifier in the limiting case when R_{sig} is zero. The voltage transfer function $V_o/V_{\text{sig}} = V_o/V_{gs}$ can be

(a)

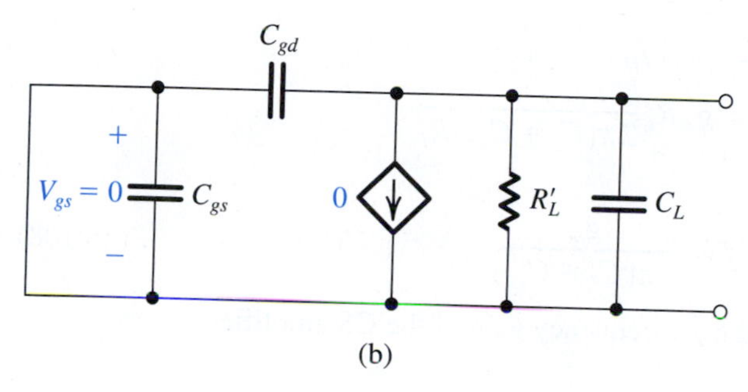

(b)

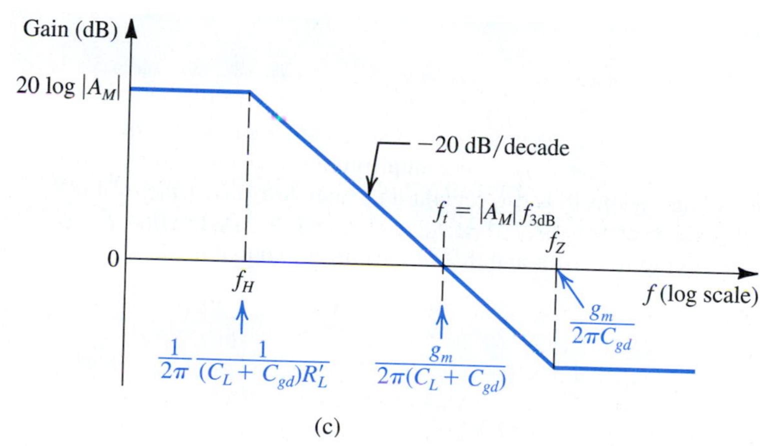

(c)

Figure 9.25 **(a)** High-frequency equivalent circuit of a CS amplifier fed with a signal source having a very low (effectively zero) resistance. **(b)** The circuit with V_{sig} reduced to zero. **(c)** Bode plot for the gain of the circuit in **(a)**.

found by setting $R_{sig} = 0$ in Eq. (9.88). The result is

$$\frac{V_o}{V_{sig}} = \frac{(-g_m R_L')[1 - s(C_{gd}/g_m)]}{1 + s(C_L + C_{gd})R_L'} \tag{9.106}$$

Thus, while the dc gain and the frequency of the zero do not change, the high-frequency response is now determined by a pole formed by $C_L + C_{gd}$ together with R_L'. Thus the 3-dB frequency is now given by

$$f_H = \frac{1}{2\pi(C_L + C_{gd})R_L'} \quad (9.107)$$

To see how this pole is formed, refer to Fig. 9.25(b), which shows the equivalent circuit with the input signal source reduced to zero. Observe that the circuit reduces to a capacitance $(C_L + C_{gd})$ in parallel with a resistance R_L'.

As we have seen above, the transfer-function zero is usually at a very high frequency and thus does not play a significant role in shaping the high-frequency response. The gain of the CS amplifier will therefore fall off at a rate of –6 dB/octave (–20 dB/decade), reaching unity (0 dB) at a frequency f_t, which is equal to the **gain–bandwidth product**,

$$f_t = |A_M| f_H$$
$$= g_m R_L' \frac{1}{2\pi(C_L + C_{gd})R_L'}$$

Thus,

$$f_t = \frac{g_m}{2\pi(C_L + C_{gd})} \quad (9.108)$$

Figure 9.25(c) shows a sketch of the high-frequency gain of the CS amplifier.

Example 9.11

Consider the CS amplifier specified in Example 9.8 when fed with a signal source having a negligible resistance (i.e., $R_{sig} = 0$). Find A_M, f_{3dB}, f_t, and f_Z. If the amplifying transistor is to be operated at twice the original overdrive voltage while W and L remain unchanged, by what factor must the bias current be changed? What are the new values of A_M, f_{3dB}, f_t, and f_Z? Assume that R_L' is the parallel equivalent of r_o of the amplifying transistor and that of the current-source load.

Solution

In Example 9.8 we found that

$$A_M = -g_m R_L' = -12.5 \text{ V/V}$$

The 3-dB frequency can be found using Eq. (9.107),

$$f_H = \frac{1}{2\pi(C_L + C_{gd})R_L'}$$
$$= \frac{1}{2\pi(25+5)\times 10^{-15} \times 10 \times 10^3}$$
$$= 530.5 \text{ MHz}$$

and the unity-gain frequency, which is equal to the gain–bandwidth product, can be determined as

$$f_t = |A_M| f_H = 12.5 \times 530.5 = 6.63 \text{ GHz}$$

The frequency of the zero is

$$f_Z = \frac{1}{2\pi}\frac{g_m}{C_{gd}}$$

$$= \frac{1}{2\pi}\frac{1.25 \times 10^{-3}}{5 \times 10^{-15}} \simeq 40 \text{ GHz}$$

Now, to double V_{OV}, I_D must be quadrupled. The new values of g_m and R'_L can be found as follows:

$$g_m = \frac{I_D}{V_{OV}/2} = 2.5 \text{ mA/V}$$

$$R'_L = \frac{1}{4} \times 10 = 2.5\text{k}\Omega$$

Thus the new value of A_M becomes

$$A_M = -g_m R'_L = -2.5 \times 2.5 = -6.25 \text{ V/V}$$

That of f_H becomes

$$f_H = \frac{1}{2\pi(C_L + C_{gd})R'_L}$$

$$= \frac{1}{2\pi(25+5) \times 10^{-15} \times 2.5 \times 10^3}$$

$$= 2.12 \text{ GHz}$$

and the unity-gain frequency (i.e., the gain–bandwidth product) becomes

$$f_t = 6.25 \times 2.12 = 13.3 \text{ GHz}$$

We note that doubling V_{OV} results in reducing the dc gain by a factor of 2 and increasing the bandwidth by a factor of 4. Thus, the gain–bandwidth product is doubled—a good bargain!

EXERCISES

9.20 For the CS amplifier considered in Example 9.11 operating at the original values of V_{OV} and I_D, find the value to which C_L should be increased to place f_t at 2 GHz.
Ans. 94.4 fF

9.21 Show that the CS amplifier when fed with $R_{sig} = 0$ has a transfer-function zero whose frequency is related to f_t by

$$\frac{f_Z}{f_t} = 1 + \frac{C_L}{C_{gd}}$$

9.6 High-Frequency Response of the Common-Gate and Cascode Amplifiers

Although common-source and common-emitter amplifiers provide substantial gain at midband frequencies, their gain falls off in the high-frequency band at a relatively low frequency. This is primarily due to the large input capacitance C_{in}, whose value is significantly increased by the Miller component. The latter is large because of the Miller multiplication effect. It follows that the key to obtaining wideband operation, that is, high f_H, is to use circuit configurations that do not suffer from the Miller effect. One such configuration is the common-gate circuit.

9.6.1 High-Frequency Response of the CG Amplifier

Figure 9.26(a) shows the CG amplifier with the MOSFET internal capacitances C_{gs} and C_{gd} indicated. For generality, a capacitance C_L is included at the output node to represent the combination of the output capacitance of a current-source load and the input capacitance of a succeeding amplifier stage. Capacitance C_L also includes the MOSFET capacitance C_{db}. Note the C_L appears in effect in parallel with C_{gd}; therefore, in the following discussion we will lump the two capacitances together.

It is important to note at the outset that each of the three capacitances in the circuit of Fig. 9.26(a) has a grounded node. Therefore none of the capacitances undergoes the Miller multiplication effect observed in the CS stage. It follows that the CG circuit can be designed to have a much wider bandwidth than that of the CS circuit, especially when the resistance of the signal generator is large.

Analysis of the circuit in Fig. 9.26(a) is greatly simplified if r_o can be neglected. In such a case the input side is isolated from the output side, and the high-frequency equivalent circuit takes the form shown in Fig. 9.26(b). We immediately observe that there are two poles: one at the input side with a frequency f_{P1},

$$f_{P1} = \frac{1}{2\pi C_{gs}\left(R_{sig} \,\|\, \frac{1}{g_m}\right)} \tag{9.109}$$

and the other at the output side with a frequency f_{P2},

$$f_{P2} = \frac{1}{2\pi(C_{gd} + C_L)R_L} \tag{9.110}$$

The relative locations of the two poles will depend on the specific situation. However, f_{P2} is usually lower than f_{P1}; thus f_{P2} can be dominant. The important point to note is that both f_{P1} and f_{P2} are usually much higher than the frequency of the dominant input pole in the CS stage.

In IC amplifiers, r_o has to be taken into account. In these cases, the method of open-circuit time constants can be employed to obtain an estimate for the 3-dB frequency f_H. Figure 9.27 shows the circuits for determining the resistances R_{gs} and R_{gd} seen by C_{gs} and $(C_{gd} + C_L)$, respectively. By inspection we obtain

$$R_{gs} = R_{sig} \,\|\, R_{in} \tag{9.111}$$

and

$$R_{gd} = R_L \,\|\, R_o \tag{9.112}$$

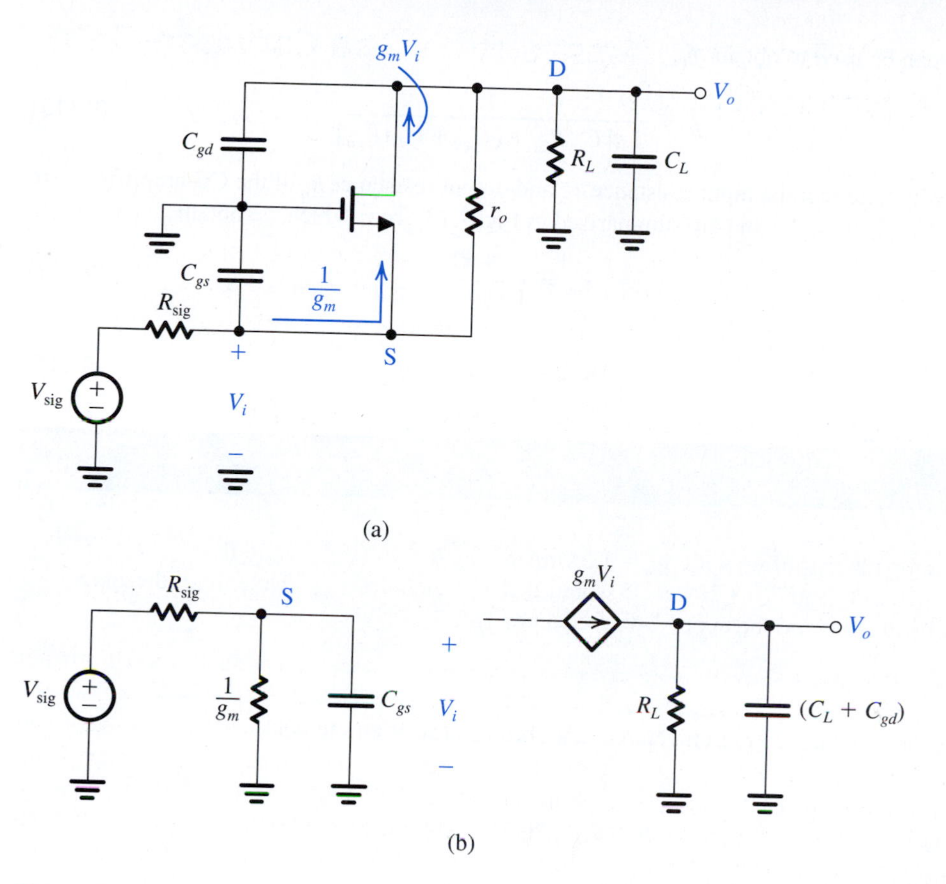

Figure 9.26 **(a)** The common-gate amplifier with the transistor internal capacitances shown. A load capacitance C_L is also included. **(b)** Equivalent circuit for the case in which r_o is neglected.

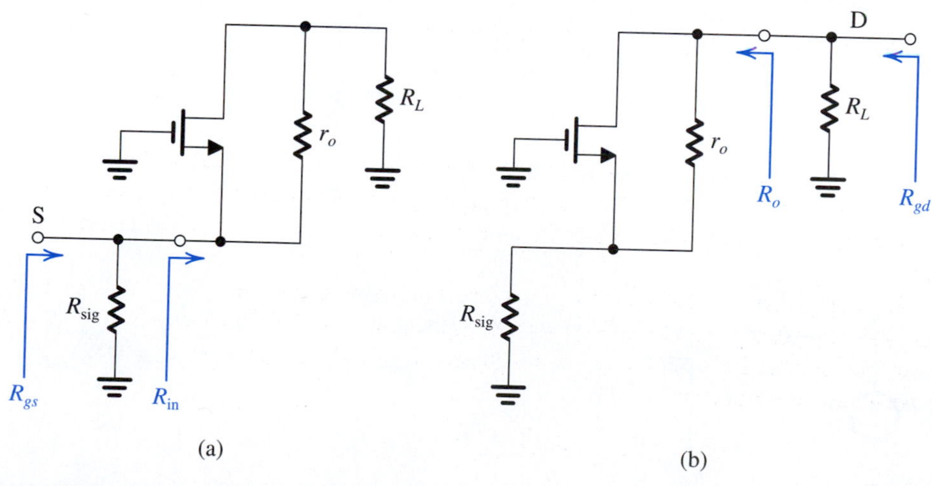

Figure 9.27 Circuits for determining R_{gs} and R_{gd}.

which can be used to obtain f_H,

$$f_H = \frac{1}{2\pi[C_{gs}R_{gs} + (C_{gd} + C_L)R_{gd}]} \tag{9.113}$$

Finally note that the input resistance R_{in} and output resistance R_o of the CG amplifier were derived in Section 7.3 and are summarized in Fig. 7.13, from which we obtain

$$R_{in} = \frac{r_o + R_L}{1 + g_m r_o} \tag{9.114}$$

and

$$R_o = r_o + R_{sig} + (g_m r_o)R_{sig} \tag{9.115}$$

Example 9.12

Consider a common-gate amplifier with $g_m = 1.25$ mA/V, $r_o = 20\ \text{k}\Omega$, $C_{gs} = 20$ fF, $C_{gd} = 5$ fF, $C_L = 15$ fF, $R_{sig} = 10\ \text{k}\Omega$, and $R_L = 20\ \text{k}\Omega$. Assume that C_L includes C_{db}. Determine the input resistance, the midband gain, and the upper 3-dB frequency f_H.

Solution

Figure 9.28 shows the CG amplifier circuit at midband frequencies. We note that

$$v_o = iR_L$$

$$v_{sig} = i(R_{sig} + R_{in})$$

Thus, the overall voltage gain is given by

$$G_v = \frac{v_o}{v_{sig}} = \frac{R_L}{R_{sig} + R_{in}}$$

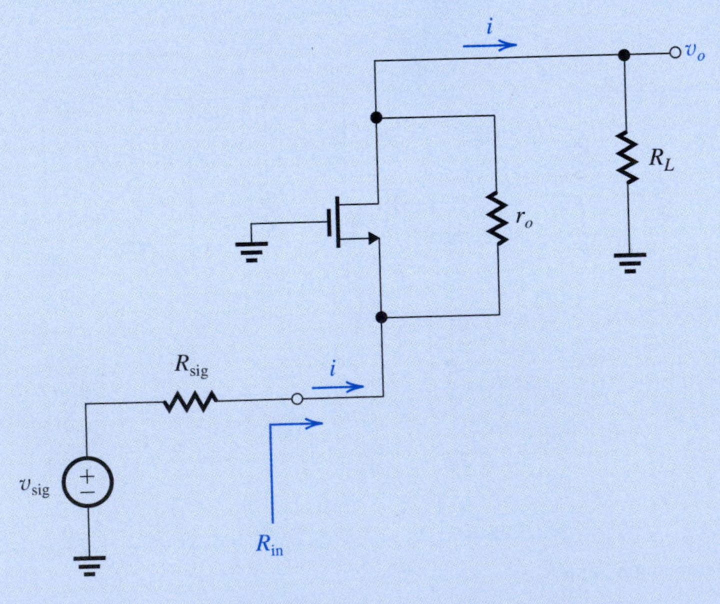

Figure 9.28 The CG amplifier circuit at midband.

The value of R_{in} is found from Eq. (9.114) as

$$R_{in} = \frac{r_o + R_L}{1 + g_m r_o}$$

$$= \frac{20 + 20}{1 + (1.25 \times 20)} = 1.54 \text{ k}\Omega$$

Thus, G_v can now be determined as

$$G_v = \frac{20}{10 + 1.54} = 1.73 \text{ V/V}$$

Observe that as expected G_v is very low. This is due to the low input resistance of the CG amplifier.

To obtain an estimate of the 3-dB frequency f_H, we first determine R_{gs} and R_{gd} using Eqs. (9.111) and (9.112),

$$R_{gs} = R_{sig} \parallel R_{in} = 10 \parallel 1.54 = 1.33 \text{ k}\Omega$$

$$R_{gd} = R_L \parallel R_o$$

where R_o is given by Eq. (9.115),

$$R_o = r_o + R_{sig} + (g_m r_o) R_{sig}$$

$$= 20 + 10 + 25 \times 10 = 280 \text{ k}\Omega$$

Thus,

$$R_{gd} = 20 \parallel 280 = 18.7 \text{ k}\Omega$$

Now we can compute the sum of the open-circuit time constants, τ_H,

$$\tau_H = C_{gs} R_{gs} + (C_{gd} + C_L) R_{gd}$$

$$\tau_H = 20 \times 10^{-15} \times 1.33 \times 10^3 + (5 + 15) \times 10^{-15} \times 18.7 \times 10^3$$

$$= 26.6 \times 10^{-12} + 374 \times 10^{-12}$$

$$= 400.6 \text{ ps}$$

and the upper 3-dB frequency f_H can be obtained as

$$f_H = \frac{1}{2\pi\tau_H} = \frac{1}{2\pi \times 400.6 \times 10^{-12}} = 397.3 \text{ MHz}$$

Observe that f_H is indeed much higher (more than twice) the corresponding value for the CS amplifier found in Example 9.9. Another important observation can be made by examining the two components of τ_H: The contribution of the input circuit is 26.6 ps, while that of the output circuit is 374 ps; thus the limitation on the high-frequency response is posed by the output circuit.

EXERCISE

9.22 In order to raise the midband gain of the CG amplifier in Example 9.12, the circuit designer decides to use a cascode current source for the load device, thus raising R_L by a factor of $g_m r_o = 25$; that is, R_L becomes 500 kΩ. Find R_{in}, the midband gain, and f_H. Comment on the results.
Ans. 20 kΩ; 16.7 V/V; 42.7 MHz. While the midband gain has been increased substantially (by a factor of 9.7), the bandwidth f_H has been substantially lowered (by a factor of about 9.3). Thus, the high-frequency advantage of the CG amplifier is completely lost!

We conclude this section by noting that a properly designed CG circuit can have a wide bandwidth. However, the input resistance will be low and the overall midband gain can be very low. It follows that the CG circuit alone will not do the job! However, combining the CG with the CS amplifier in the cascode configuration can result in a circuit having the high input resistance and gain of the CS amplifier together with the wide bandwidth of the CG amplifier, as we shall now see.

9.6.2 High-Frequency Response of the MOS Cascode Amplifier

In Section 7.3 we studied the cascode amplifier and analyzed its performance at midband frequencies. There we learned that by combining the CS and CG configurations, the cascode amplifier exhibits a very high input resistance and a voltage gain that can be as high as A_0^2, where $A_0 = g_m r_o$ is the intrinsic gain of the MOSFET. For our purposes here, we shall see that the versatility of the cascode circuit allows us to trade off some of this high midband gain in return for a wider bandwidth.

Figure 9.29 shows the cascode amplifier with all transistor internal capacitances indicated. Also included is a capacitance C_L at the output node to represent the combination of C_{db2}, the output capacitance of a current-source load, and the input capacitance of a succeeding

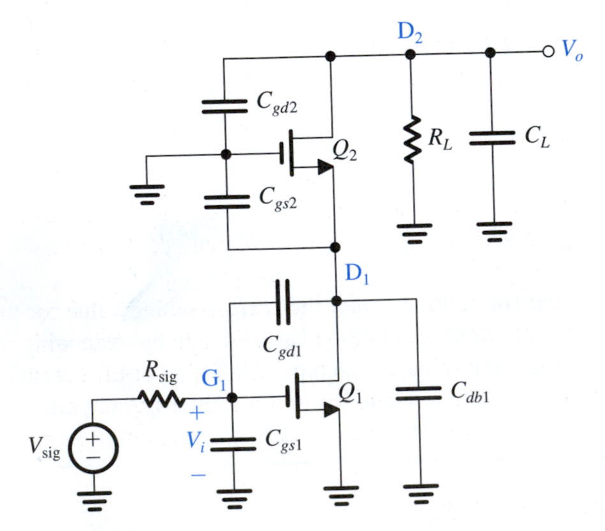

Figure 9.29 The cascode circuit with the various transistor capacitances indicated.

amplifier stage (if any). Note that C_{db1} and C_{gs2} appear in parallel, and we shall combine them in the following analysis. Similarly, C_L and C_{gd2} appear in parallel and will be combined.

The easiest and, in fact, quite insightful approach to determining the 3-dB frequency f_H is to employ the open-circuit time-constants method:

1. Capacitance C_{gs1} sees a resistance R_{sig}.
2. Capacitance C_{gd1} sees a resistance R_{gd1}, which can be obtained by adapting the formula in Eq. (9.84) to

$$R_{gd1} = (1+g_{m1}R_{d1})R_{sig}+R_{d1} \tag{9.116}$$

where R_{d1}, the total resistance at D_1, is given by

$$R_{d1} = r_{o1} \| R_{in2} = r_{o1} \| \frac{r_{o2}+R_L}{g_{m2}r_{o2}} \tag{9.117}$$

3. Capacitance $(C_{db1}+C_{gs2})$ sees a resistance R_{d1}.
4. Capacitance (C_L+C_{gd2}) sees a resistance $(R_L \| R_o)$ where R_o is given by

$$R_o = r_{o2}+r_{o1}+(g_{m2}r_{o2})r_{o1}$$

With the resistances determined, the effective time constant τ_H can be computed as

$$\tau_H = C_{gs1}R_{sig}+C_{gd1}[(1+g_{m1}R_{d1})R_{sig}+R_{d1}] + (C_{db1}+C_{gs2})R_{d1}+(C_L+C_{gd2})(R_L \| R_o) \tag{9.118}$$

and the 3-dB frequency f_H as

$$f_H \simeq \frac{1}{2\pi\tau_H}$$

To gain insight regarding what limits the high-frequency gain of the MOS cascode amplifier, we rewrite Eq. (9.118) in the form

$$\tau_H = R_{sig}[C_{gs1}+C_{gd1}(1+g_{m1}R_{d1})]+R_{d1}(C_{gd1}+C_{db1}+C_{gs2}) + (R_L \| R_o)(C_L+C_{gd2}) \tag{9.119}$$

In the case of a large R_{sig}, the first term can dominate, especially if the Miller multiplier $(1+g_{m1}R_{d1})$ is large. This in turn happens when the load resistance R_L is large (on the order of A_0r_o), causing R_{in2} to be large and requiring the first stage, Q_1, to provide a large proportion of the gain (see Section 7.3.3). It follows that when R_{sig} is large, to extend the bandwidth we have to lower R_L to the order of r_o. This in turn lowers R_{in2} and hence R_{d1} and renders the Miller effect in Q_1 insignificant. Note, however, that the dc gain of the cascode will then be A_0. Thus, while the dc gain will be the same as (or a little higher than) that achieved in a CS amplifier, the bandwidth will be greater.

In the case when R_{sig} is small, the Miller effect in Q_1 will not be of concern. A large value of R_L (on the order of A_0r_o) can then be used to realize the large dc gain possible with a cascode amplifier—that is, a dc gain on the order of A_0^2. Equation (9.119) indicates that in this case the third term will usually be dominant. To pursue this point a little further, consider the case $R_{sig} = 0$, and assume that the middle term is much smaller than the third term. It follows that

$$\tau_H \simeq (C_L+C_{gd2})(R_L \| R_o)$$

and the 3-dB frequency becomes

$$f_H = \frac{1}{2\pi(C_L + C_{gd2})(R_L \| R_o)} \tag{9.120}$$

which is of the same form as the formula for the CS amplifier with $R_{sig} = 0$ (Eq. 9.107). Here, however, $(R_L \| R_o)$ is larger that R'_L by a factor of about A_0. Thus the f_H of the cascode will be lower than that of the CS amplifier by the same factor A_0. Figure 9.30 shows a sketch of the frequency response of the cascode and of the corresponding common-source amplifier. We observe that in this case, cascoding increases the dc gain by a factor A_0 while keeping the unity-gain frequency unchanged at

$$f_t \simeq \frac{1}{2\pi}\frac{g_m}{C_L + C_{gd2}} \tag{9.121}$$

	Common Source	Cascode
Circuit	$R'_L = R_L \| r_o$	
DC Gain	$-g_m R'_L$	$-A_0 g_m R'_L$
f_{3dB}	$\frac{1}{2\pi(C_L + C_{gd})R'_L}$	$\frac{1}{2\pi(C_L + C_{gd})A_0 R'_L}$
f_t	$\frac{g_m}{2\pi(C_L + C_{gd})}$	$\frac{g_m}{2\pi(C_L + C_{gd})}$

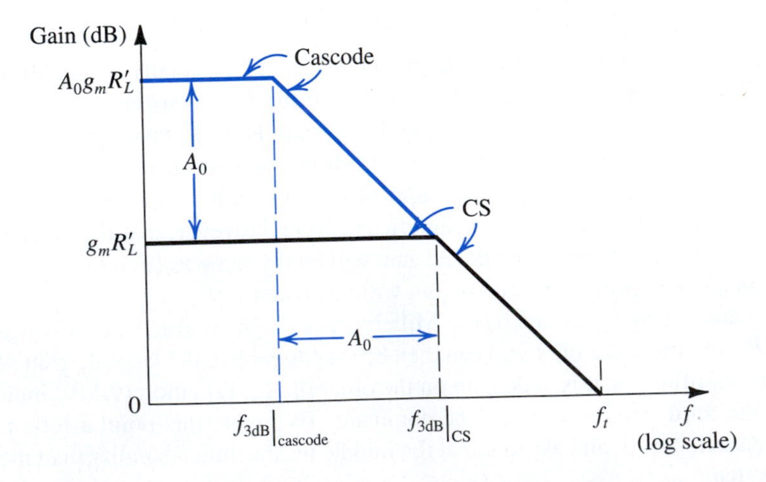

Figure 9.30 Effect of cascoding on gain and bandwidth in the case $R_{sig} = 0$. Cascoding can increase the dc gain by the factor A_0 while keeping the unity-gain frequency constant. Note that to achieve the high gain, the load resistance must be increased by the factor A_0.

Example 9.13

This example illustrates the advantages of cascoding by comparing the performance of a cascode amplifier with that of a common-source amplifier in two cases:

(a) The resistance of the signal source is significant, $R_{sig} = 10\ k\Omega$.

(b) R_{sig} is negligibly small.

Assume all MOSFETs have $g_m = 1.25$ mA/V, $r_o = 20\ k\Omega$, $C_{gs} = 20$ fF, $C_{gd} = 5$ fF, $C_{db} = 5$ fF, and C_L (excluding C_{db}) = 10 fF. For case (a), let $R_L = r_o = 20\ k\Omega$ for both amplifiers. For case (b), let $R_L = r_o = 20\ k\Omega$ for the CS amplifier and $R_L = R_o$ for the cascode amplifier. For all cases, determine A_v, f_H, and f_t.

Solution

(a) For the CS amplifier:

$$A_0 = g_m r_o = 1.25 \times 20 = 25 \text{ V/V}$$
$$A_v = -g_m(R_L \| r_o) = -g_m(r_o \| r_o)$$
$$= -\tfrac{1}{2}A_0 = -12.5 \text{ V/V}$$
$$\tau_H = C_{gs}R_{sig} + C_{gd}[(1 + g_m R'_L)R_{sig} + R'_L] + (C_L + C_{db})R'_L$$

where

$$R'_L = r_o \| R_L = r_o \| r_o = 10\ k\Omega$$
$$\tau_H = 20 \times 10 + 5[(1 + 12.5)10 + 10] + (10 + 5)10$$
$$= 200 + 725 + 150 = 1075 \text{ ps}$$

Thus,

$$f_H = \frac{1}{2\pi \times 1075 \times 10^{-12}} = 148 \text{ MHz}$$
$$f_t = |A_v| f_H = 12.5 \times 148 = 1.85 \text{ GHz}$$

For the cascode amplifier:

$$R_o = 2r_o + (g_m r_o)r_o = (2 \times 20) + (25 \times 20) = 540\ k\Omega$$
$$A_v = -g_m(R_o \| R_L)$$
$$= -1.25(540 \| 20) = -24.1 \text{ V/V}$$
$$R_{in2} = \frac{r_o + R_L}{g_m r_o} = \frac{r_o + r_o}{g_m r_o} = \frac{2}{g_m} = \frac{2}{1.25} = 1.6\ k\Omega$$
$$R_{d1} = r_o \| R_{in2} = 20 \| 1.6 = 1.48\ k\Omega$$
$$\tau_H = R_{sig}[C_{gs1} + C_{gd1}(1 + g_{m1}R_{d1})]$$
$$+ R_{d1}(C_{gd1} + C_{db1} + C_{gs2})$$
$$+ (R_L \| R_o)(C_L + C_{db2} + C_{gd2})$$

Example 9.13 *continued*

$$= 10[20 + 5(1 + 1.25 \times 1.48)]$$
$$+ 1.48(5 + 5 + 20)$$
$$+ (20 \parallel 540)(10 + 5 + 5)$$
$$= 342.5 + 44.4 + 385.7$$
$$= 772.6 \text{ ps}$$

$$f_H = \frac{1}{2\pi \times 772.6 \times 10^{-12}} = 206 \text{ MHz}$$

$$f_t = 24.1 \times 206 = 4.96 \text{ GHz}$$

Thus cascoding has increased f_t by a factor of 2.7.

(b) For the CS amplifier:

$$A_v = -12.5 \text{ V/V}$$

$$\tau_H = (C_{gd} + C_L + C_{db})R_L'$$
$$= (5 + 10 + 5)10 = 200 \text{ ps}$$

$$f_H = \frac{1}{2\pi \times 200 \times 10^{-12}} = 796 \text{ MHz}$$

$$f_t = 12.5 \times 796 = 9.95 \text{ GHz}$$

For the cascode amplifier:

$$R_L = R_o = 540 \text{ k}\Omega$$

$$A_v = -g_m(R_o \parallel R_L)$$
$$= -1.25(540 \parallel 540) = -337.5 \text{ V/V}$$

$$R_{\text{in2}} = \frac{r_o + R_L}{g_m r_o} = \frac{20 + 540}{1.25 \times 20} = 22.4 \text{ k}\Omega$$

$$R_{d1} = r_{o1} \parallel R_{\text{in2}} = 20 \parallel 22.4 = 10.6 \text{ k}\Omega$$

$$\tau_H = R_{d1}(C_{gd1} + C_{db1} + C_{gs2}) + (R_L \parallel R_o)(C_L + C_{gd2} + C_{db2})$$

$$= 10.6(5 + 5 + 20) + (540 \parallel 540)(10 + 5 + 5)$$

$$= 318 + 5400 = 5718 \text{ ps}$$

$$f_H = \frac{1}{2\pi \times 5718 \times 10^{-12}} = 27.8 \text{MHz}$$

$$f_t = 337.5 \times 27.8 = 9.39 \text{ GHz}$$

Thus cascoding increases the dc gain from 12.5 V/V to 337.5 V/V. The unity-gain frequency (which, in this case, is equal to the gain–bandwidth product), however, remains nearly constant.

EXERCISE

9.23 In this exercise we wish to contrast the gain of a CS amplifier and a cascode amplifier. Assume that both are fed with a large source resistance R_{sig} that effectively determines the high-frequency response. Thus, neglect components of τ_H that do not include R_{sig}. Also assume that all transistors are operated at the same conditions and thus corresponding small-signal parameters are equal. Also, both amplifiers have equal $R_L = r_o$, and $g_m r_o = 40$.

(a) Find the ratio of the low-frequency gain of the cascode amplifier to that of the CS amplifier.
(b) For the case of $C_{gd} = 0.25C_{gs}$, find the ratio of f_H of the cascode to that of the CS amplifier.
(c) Use (a) and (b) to find the ratio of f_t of the cascode to that of the CS.

Ans. 2; 3.6; 7.2

9.6.3 High-Frequency Response of the Bipolar Cascode Amplifier

The analysis method studied in the previous section can be directly applied to the BJT cascode amplifier. Figure 9.31 presents the circuits and the formulas for determining the high-frequency response of the bipolar cascode.

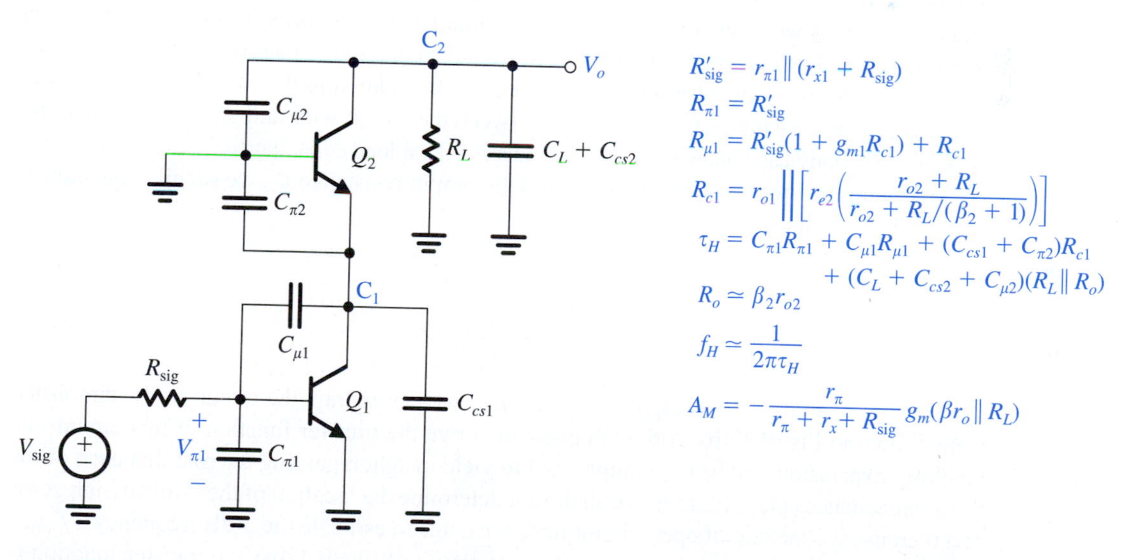

Figure 9.31 Determining the frequency response of the BJT cascode amplifier. Note that in addition to the BJT capacitances C_π and C_μ, the capacitance between the collector and the substrate C_{cs} for each transistor are included.

EXERCISE

9.24 The objective of this exercise is to evaluate the effect of cascoding on the performance of the CE amplifier of Exercise 9.19. The specifications are as follows: $I = 1$ mA, $\beta = 200$, $r_o = 130$ kΩ, $C_\pi = 16$ pF, $C_\mu = 0.3$ pF, $r_x = 200$ Ω, $C_{cs1} = C_{cs2} = 0$, $C_L = 5$ pF, $R_{sig} = 36$ kΩ, $R_L = 50$ kΩ. Find R_{in}, A_0, R_{o1}, R_{in2}, R_o, A_M, f_H, and f_t. Compare A_M, f_H, and f_t with the corresponding values obtained in Exercise 9.19 for the CE amplifier. What should C_L be reduced to in order to have $f_H = 1$ MHz?
Ans. 5.2 kΩ; 5200 V/V; 130 kΩ; 35 Ω; 26 MΩ; –238 V/V; 469 kHz; 111.6 MHz. $|A_M|$ has increased from 175 V/V to 238 V/V; f_H has increased from 75 kHz to 469 kHz; f_t has increased from 13.1 MHz to 111.6 MHz. C_L must be reduced to 1.6 pF.

9.7 High-Frequency Response of the Source and Emitter Followers

In this section we study the high-frequency response of two important circuit building blocks: the source follower and the emitter follower. Both have voltage gain that is less than but close to unity. Their advantage lies in their high input resistance and low output resistance. They find application as the output stage of a multistage amplifier. As we will see shortly, both exhibit wide bandwidth.

9.7.1 The Source Follower

A major advantage of the source follower is its excellent high-frequency response. This comes about because, as we shall now see, none of the internal capacitances suffers from the Miller effect. Figure 9.32(a) shows the high-frequency equivalent circuit of a source follower fed with a signal V_{sig} from a source having a resistance R_{sig}. In addition to the MOSFET capacitances C_{gs} and C_{gd}, a capacitance C_L is included between the output node and ground to account for the source-to-body capacitance C_{sb} as well as any actual load capacitance.

To obtain the low-frequency gain A_M and the output resistance R_o, we set all capacitances to zero. The results are

$$A_M = \frac{(R_L \parallel r_o)}{(R_L \parallel r_o) + (1/g_m)} \tag{9.122}$$

$$R_o = \frac{1}{g_m} \parallel r_o \tag{9.123}$$

Combining R_L and r_o into a single resistance R'_L, we can redraw the circuit in the simplified from shown in Fig. 9.32(b). Although one can derive the transfer function of this circuit, the resulting expression will be too complicated to yield insight regarding the role that each of the three capacitances plays. Rather, we shall first determine the location of the transmission zeros and then use the method of open-circuit time constants to estimate the 3-dB frequency, f_{3dB}.

Although there are three capacitances in the circuit of Fig. 9.32(b), the transfer function is of the second order. This is because the three capacitances form a continuous loop. To determine the location of the two transmission zeros, refer to the circuit in Fig. 9.32(b), and note that V_o is zero at the frequency at which C_L has a zero impedance and thus acts as a short circuit across the output, which is ω or $s = \infty$. Also, V_o will be zero at the value of s that causes

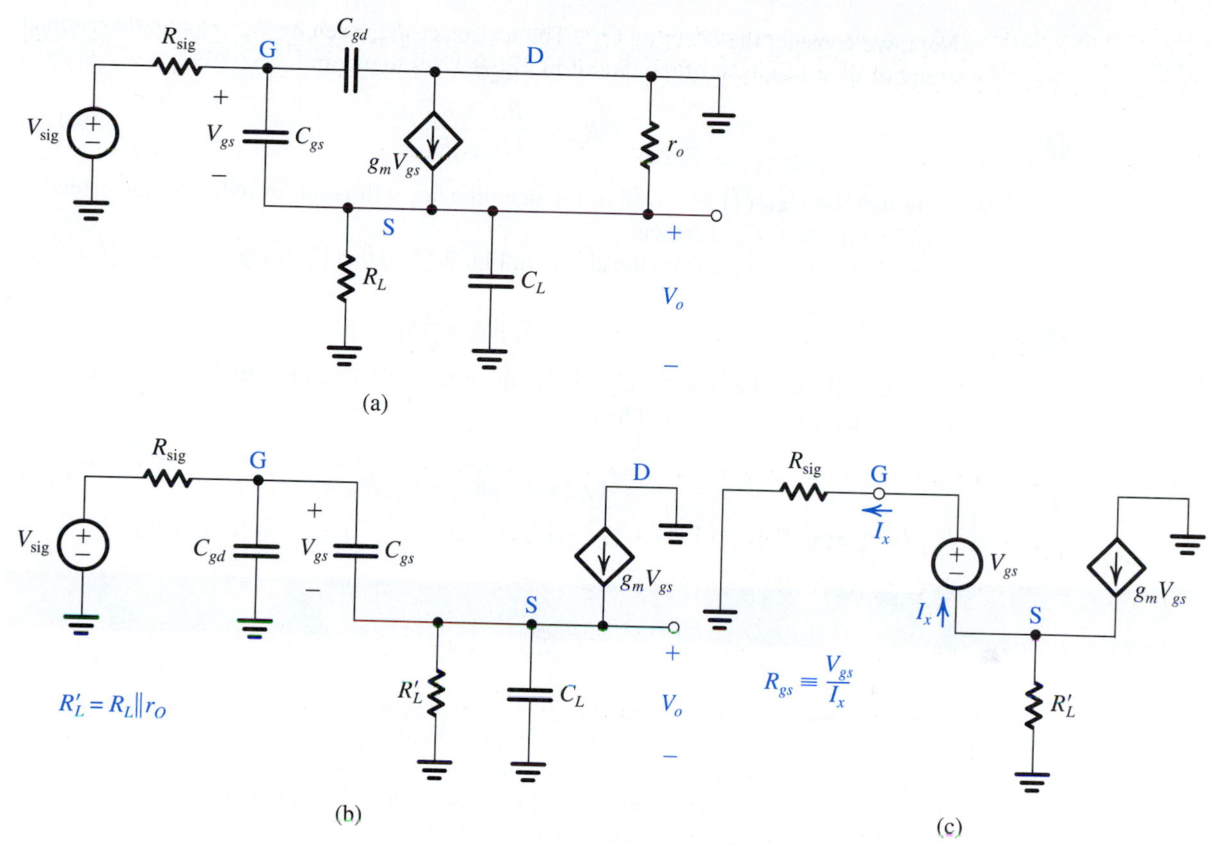

Figure 9.32 Analysis of the high-frequency response of the source follower: **(a)** equivalent circuit; **(b)** simplified equivalent circuit; **(c)** determining the resistance R_{gs} seen by C_{gs}.

the current into the impedance $R_L' \| C_L$ to be zero. Since this current is $(g_m + sC_{gs})V_{gs}$, the transmission zero will be at $s = s_Z$, where

$$s_Z = -\frac{g_m}{C_{gs}} \tag{9.124}$$

That is, the zero will be on the negative real axis of the s-plane with a frequency

$$\omega_Z = \frac{g_m}{C_{gs}} \tag{9.125}$$

Recalling that the MOSFET's $\omega_T = g_m/(C_{gs} + C_{gd})$ and that $C_{gd} \ll C_{gs}$, we see that ω_Z will be very close to ω_T,

$$f_Z \simeq f_T \tag{9.126}$$

Since the zero is at such a high frequency, we can employ the method of open-circuit time constants to obtain an estimate of f_H. Specifically, we will find the resistance seen by each of three capacitances C_{gd}, C_{gs}, and C_L and then compute the time constant associated with each. With V_{sig} set to zero and C_{gs} and C_L assumed to be open circuited, we find by inspection that the resistance R_{gd} seen by C_{gd} is given by

$$R_{gd} = R_{sig} \tag{9.127}$$

Next, we consider the effect of C_{gs}. The resistance R_{gs} seen by C_{gs} can be determined by straightforward analysis of the circuit in Fig. 9.32(c) to obtain

$$R_{gs} = \frac{R_{sig} + R_L'}{1 + g_m R_L'} \quad (9.128)$$

We note that the factor $(1 + g_m R_L')$ in the denominator will result in reducing the effective resistance with which C_{gs} interacts.

Finally, it is easy to see from the circuit in Fig. 9.32(b) that C_L interacts with $R_L' \| 1/g_m$; that is,

$$R_{C_L} = R_L \| r_o \| \frac{1}{g_m}$$

which is usually low because of $1/g_m$. Thus the effect of C_L will be small. Adding all three time constants, we obtain τ_H and hence f_H,

$$f_H = \frac{1}{2\pi\tau_H} = 1/2\pi(C_{gd}R_{sig} + C_{gs}R_{gs} + C_L R_{C_L}) \quad (9.129)$$

EXERCISE

9.25 Consider a source follower specified as follows: g_m = 1.25 mA/V, r_o = 20 kΩ, R_{sig} = 10 kΩ, R_L = 20 kΩ, C_{gs} = 20 fF, C_{gd} = 5 fF, and C_L = 15 fF. Find A_M, f_T, and f_Z. Also, find R_{gd}, R_{gs}, R_{C_L}, and hence the time constant associated with each of the three capacitances C_{gd}, C_{gs}, and C_L. Find τ_H and the percentage contribution to it from each of three capacitances. Find f_H.
Ans. 0.93 V/V; 8 GHz; 10 GHz; 10 kΩ; 1.48 kΩ; 0.74 kΩ; 50 ps; 30 ps; 11 ps; 91 ps; 55%; 33%; 12%; 1.75 GHz

9.7.2 The Emitter Follower

Figure 9.33(a) shows an emitter follower suitable for IC fabrication. It is biased by a constant-current source I. However, the circuit that sets the dc voltage at the base is not shown. The emitter follower is fed with a signal V_{sig} from a source with resistance R_{sig}. The resistance R_L, shown at the output, includes the output resistance of current source I as well as any actual load resistance.

Analysis of the emitter follower of Fig. 9.32(a) to determine its low-frequency gain, input resistance, and output resistance is identical to that performed in Section 6.6.6. We shall concentrate here on the analysis of the high-frequency response of the circuit.

Figure 9.33(b) shows the high-frequency equivalent circuit. Lumping r_o together with R_L, and r_x together with R_{sig} and making a slight change in the way the circuit is drawn results in the simplified equivalent circuit shown in Fig. 9.33(c). We will follow a procedure for the analysis of this circuit similar to that used above for the source follower. Specifically, to obtain the location of the transmission zero, note that V_o will be zero at the frequency s_Z for which the current fed to R_L' is zero:

$$g_m V_\pi + \frac{V_\pi}{r_\pi} + s_Z C_\pi V_\pi = 0$$

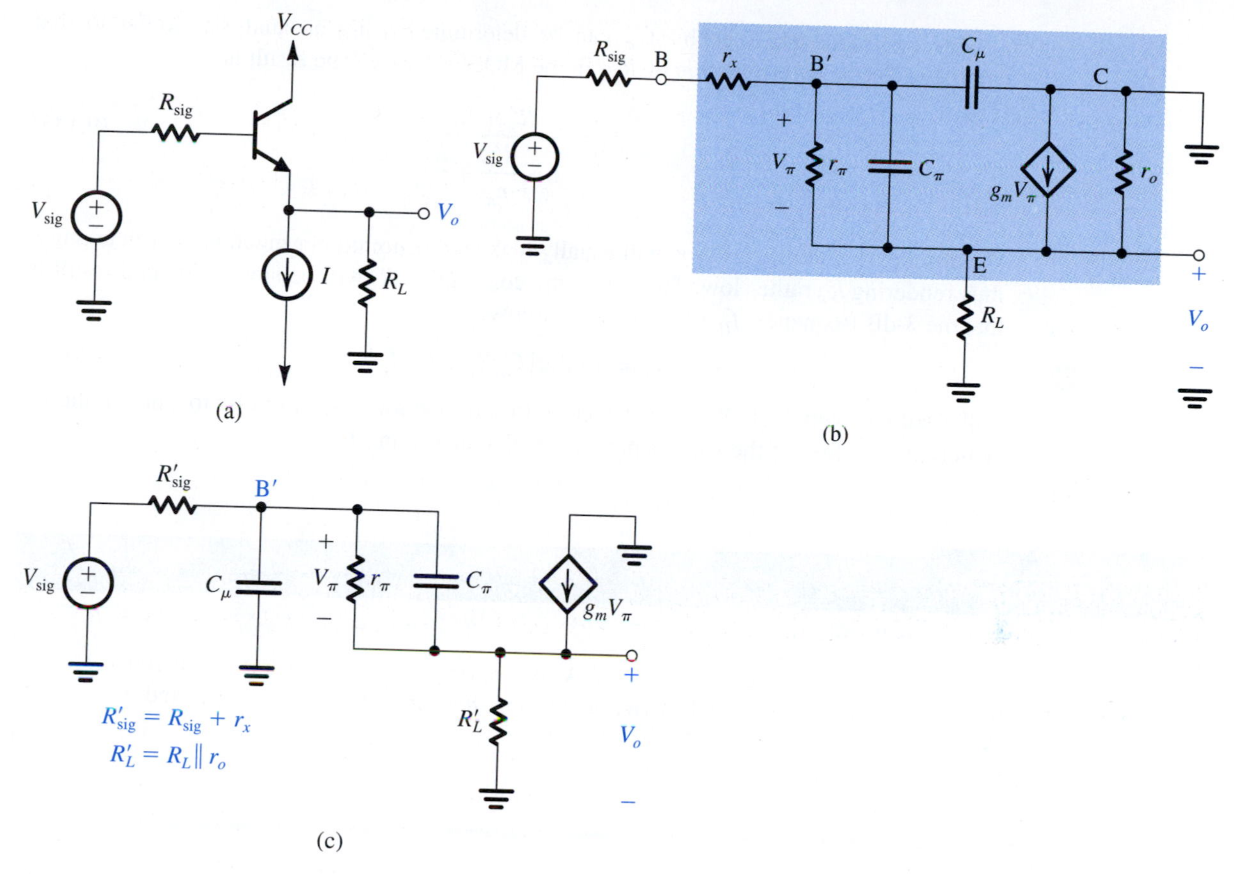

Figure 9.33 **(a)** Emitter follower. **(b)** High-frequency equivalent circuit. **(c)** Simplified equivalent circuit.

Thus,

$$s_Z = -\frac{g_m + (1/r_\pi)}{C_\pi} = -\frac{1}{C_\pi r_e} \tag{9.130}$$

which is on the negative real-axis of the s-plane and has a frequency

$$\omega_Z = \frac{1}{C_\pi r_e} \tag{9.131}$$

This frequency is very close to the unity-gain frequency ω_T of the transistor. The other transmission zero is at $s = \infty$. This is because at this frequency, C_μ acts as a short circuit, making V_π zero, and hence V_o will be zero.

Next, we determine the resistances seen by C_μ and C_π. For C_μ the reader should be able to show that the resistance it sees, R_μ, is the parallel equivalent of R'_{sig} and the input resistance looking into B′; that is,

$$R_\mu = R'_{sig} \| [r_\pi + (\beta + 1)R'_L] \tag{9.132}$$

Equation (9.132) indicates that R_μ will be smaller than R'_{sig}, and since C_μ is usually very small, the time constant $C_\mu R_\mu$ will be correspondingly small.

The resistance R_π seen by C_π can be determined using an analysis similar to that employed for the determination of R_{gs} in the MOSFET case. The result is

$$R_\pi = \frac{R'_{sig} + R'_L}{1 + \frac{R'_{sig}}{r_\pi} + \frac{R'_L}{r_e}} \tag{9.133}$$

We observe that the term R'_L/r_e will usually make the denominator much greater than unity, thus rendering R_π rather low. Thus, the time constant $C_\pi R_\pi$ will be small. The end result is that the 3-dB frequency f_H of the emitter follower,

$$f_H = 1/2\pi[C_\mu R_\mu + C_\pi R_\pi] \tag{9.134}$$

will usually be very high. We urge the reader to solve the following exercise to gain familiarity with typical values of the various parameters that determine f_H.

EXERCISE

9.26 For an emitter follower biased at $I_C = 1$ mA and having $R_{sig} = R_L = 1$ kΩ, $r_o = 100$ kΩ, $\beta = 100$, $C_\mu = 2$ pF, and $f_T = 400$ MHz, find the low-frequency gain, f_Z, R_μ, R_π, and f_H.
Ans. 0.965 V/V; 458 MHz; 1.09 kΩ; 51 Ω; 55 MHz

9.8 High-Frequency Response of Differential Amplifiers

In this section we study the high-frequency response of the differential amplifier. We will consider the variation with frequency of both the differential gain and the common-mode gain and hence of the CMRR. We will rely heavily on the study of frequency response of single-ended amplifiers presented in the sections above. Also, we will only consider MOS circuits; the bipolar case is a straightforward extension, as we saw above on a number of occasions.

9.8.1 Analysis of the Resistively Loaded MOS Amplifier

We begin with the basic, resistively loaded MOS differential pair shown in Fig. 9.34(a). Note that we have explicitly shown transistor Q_S that supplies the bias current I. Although we are showing a dc bias voltage V_{BIAS} at its gate, usually Q_S is part of a current mirror. This detail, however, is of no consequence to our present needs. Most importantly, we are interested in the total impedance between node S and ground, Z_{SS}. As we shall shortly see, this impedance plays a significant role in determining the common-mode gain and the CMRR of the differential amplifier. Resistance R_{SS} is simply the output resistance of current source Q_S. Capacitance C_{SS} is the total capacitance between node S and ground and includes C_{db} and C_{gd} of Q_S, as well as C_{sb1}, and C_{sb2}. This capacitance can be significant, especially if wide transistors are used for Q_S, Q_1, and Q_2.

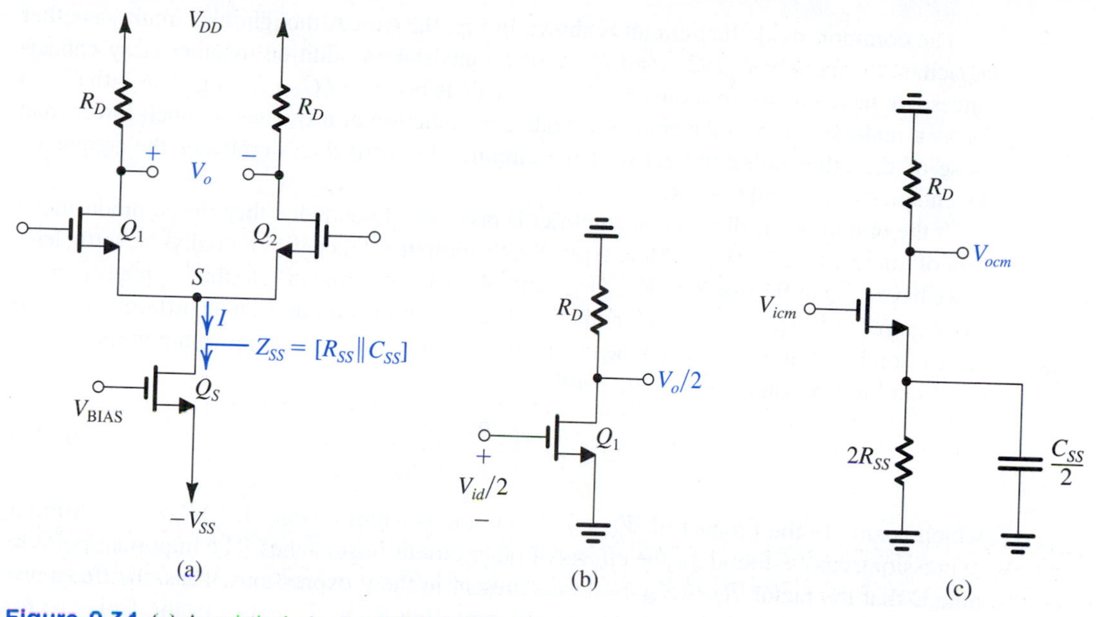

Figure 9.34 **(a)** A resistively loaded MOS differential pair; the transistor supplying the bias current is explicitly shown. It is assumed that the total impedance between node S and ground, Z_{SS}, consists of a resistance R_{SS} in parallel with a capacitance C_{SS}. **(b)** Differential half-circuit. **(c)** Common-mode half-circuit.

The differential half-circuit shown in Fig. 9.34(b) can be used to determine the frequency dependence of the differential gain V_o/V_{id}. Indeed the gain function $A_d(s)$ of the differential amplifier will be identical to the transfer function of this common-source amplifier. We studied the frequency response of the common-source amplifier at great length in Sections 9.3 and 9.5 and will not repeat this material here.

EXERCISE

9.27 A MOSFET differential amplifier such as that in Fig. 9.34(a) is biased with a current $I = 0.8$ mA. The transistors have $W/L = 100$, $k_n' = 0.2$ mA/V^2, $V_A = 20$ V, $C_{gs} = 50$ fF, $C_{gd} = 10$ fF, and $C_{db} = 10$ fF. The drain resistors are 5 kΩ each. Also, there is a 100-fF capacitive load between each drain and ground.

(a) Find V_{OV} and g_m for each transistor.
(b) Find the differential gain A_d.
(c) If the input signal source has a small resistance R_{sig} and thus the frequency response is determined primarily by the output pole, estimate the 3-dB frequency f_H. (*Hint:* Refer to Section 9.5.6 and specifically to Eq. 9.107.)
(d) If, in a different situation, the amplifier is fed symmetrically with a signal source of 20 kΩ resistance (i.e., 10 kΩ in series with each gate terminal), use the open-circuit time-constants method to estimate f_H. (*Hint:* Refer to Section 9.5.3 and specifically to Eqs. [(9.85) and (9.86)].)

Ans. (a) 0.2 V, 4 mA/V; (b) 18.2 V/V; (c) 291 MHz; (d) 53.7 MHz

The common-mode half-circuit is shown in Fig. 9.34(c). Although this circuit has other capacitances, namely C_{gs}, C_{gd}, and C_{db} of the transistor in addition to other stray capacitances, we have chosen to show only $C_{SS}/2$. This is because $(C_{SS}/2)$ together with $(2R_{SS})$ forms a real-axis zero in the common-mode gain function at a frequency much lower than those of the other poles and zeros of the circuit. This zero then dominates the frequency dependence of A_{cm} and CMRR.

If the output of the differential amplifier is taken single-endedly, then the common-mode gain of interest is V_{ocm}/V_{icm}. More typically, the output is taken differentially. Nevertheless, as we have seen in Section 8.2, V_{ocm}/V_{icm} still plays a major role in determining the common-mode gain. To be specific, consider what happens when the output is taken differentially and there is a mismatch ΔR_D between the two drain resistances. The resulting common-mode gain was found in Section 8.2 to be (Eq. 8.49′)

$$A_{cm} = -\left(\frac{R_D}{2R_{SS}}\right)\frac{\Delta R_D}{R_D} \tag{9.135}$$

which is simply the product of V_{ocm}/V_{icm} and the per-unit mismatch $(\Delta R_D/R_D)$. Similar expressions can be found for the effects of other circuit mismatches. The important point to note is that the factor $R_D/2R_{SS}$ is always present in these expressions. Thus, the frequency dependence of A_{cm} can be obtained by simply replacing R_{SS} by Z_{SS} in this factor. Doing so for the expression in Eq. (9.135) gives

$$\begin{aligned} A_{cm}(s) &= -\frac{R_D}{2Z_{SS}}\left(\frac{\Delta R_D}{R_D}\right) \\ &= -\frac{1}{2}R_D\left(\frac{\Delta R_D}{R_D}\right)Y_{SS} \\ &= -\frac{1}{2}R_D\left(\frac{\Delta R_D}{R_D}\right)\left(\frac{1}{R_{SS}}+sC_{SS}\right) \\ &= -\frac{R_D}{2R_{SS}}\left(\frac{\Delta R_D}{R_D}\right)(1+sC_{SS}R_{SS}) \end{aligned} \tag{9.136}$$

from which we see that A_{cm} acquires a zero on the negative real axis of the s-plane with frequency ω_Z,

$$\omega_Z = \frac{1}{C_{SS}R_{SS}} \tag{9.137}$$

or in hertz,

$$f_Z = \frac{\omega_Z}{2\pi} = \frac{1}{2\pi C_{SS}R_{SS}} \tag{9.138}$$

As mentioned above, usually f_Z is much lower than the frequencies of the other poles and zeros. As a result, the common-mode gain increases at the rate of +6 dB/octave (20 dB/decade) starting at a relatively low frequency, as indicated in Fig. 9.35(a). Of course, A_{cm} drops off at high frequencies because of the other poles of the common-mode half-circuit. It is, however, f_Z that is significant, for it is the frequency at which the CMRR of

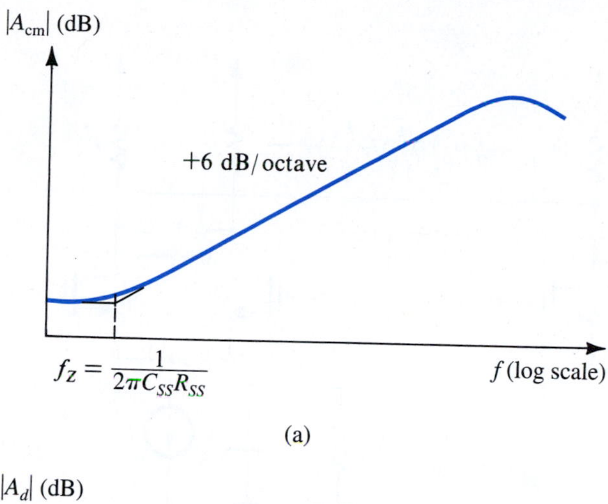

(a)

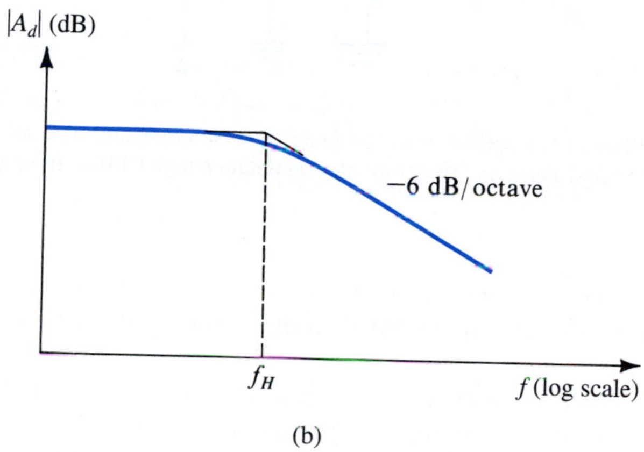

(b)

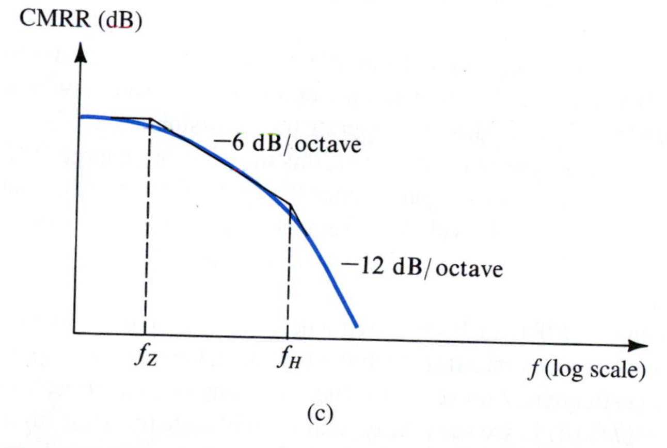

(c)

Figure 9.35 Variation of **(a)** common-mode gain, **(b)** differential gain, and **(c)** common-mode rejection ratio with frequency.

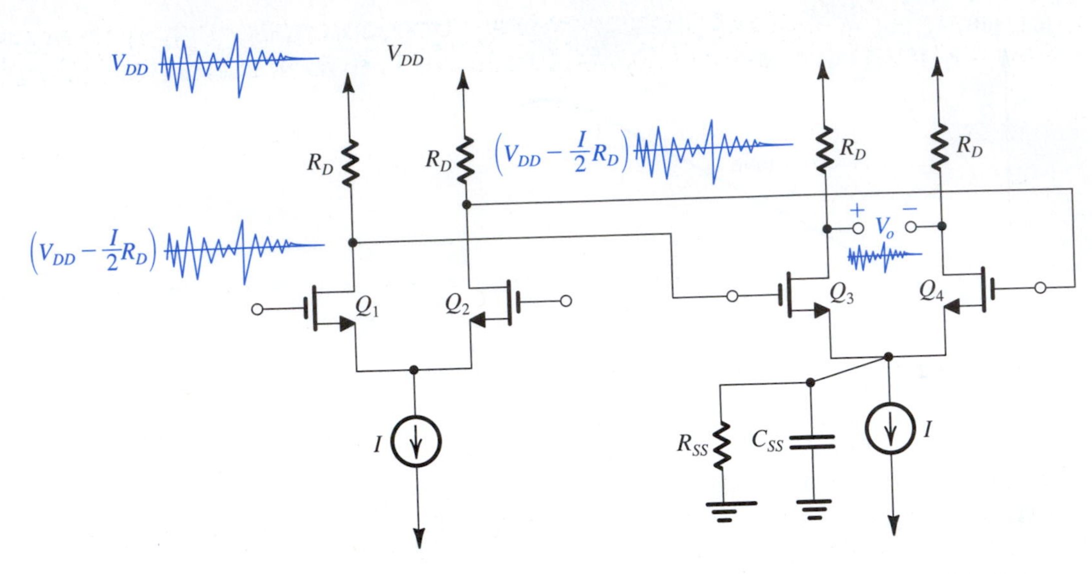

Figure 9.36 The second stage in a differential amplifier, which is relied on to suppress high-frequency noise injected by the power supply of the first stage, and therefore must maintain a high CMRR at higher frequencies.

the differential amplifier begins to decrease, as indicated in Fig. 9.35(c). Note that if both A_d and A_{cm} are expressed and plotted in dB, then CMRR in dB is simply the difference between A_d and A_{cm}.

Although in the foregoing we considered only the common-mode gain resulting from an R_D mismatch, it should be obvious that the results apply to the common-mode gain resulting from any other mismatch. For instance, it applies equally well to the case of a g_m mismatch, modifying Eq. (8.63) by replacing R_{SS} by Z_{SS}, and so on.

Before leaving this section, it is interesting to point out an important trade-off found in the design of the current-source transistor Q_S: In order to operate this current source with a small V_{DS} (to conserve the already low V_{DD}), we desire to operate the transistor at a low overdrive voltage V_{OV}. For a given value of the current I, however, this means using a large W/L ratio (i.e., a wide transistor). This in turn increases C_{SS} and hence lowers f_Z with the result that the CMRR deteriorates (i.e., decreases) at a relatively low frequency. Thus there is a trade-off between the need to reduce the dc voltage across Q_S and the need to keep the CMRR reasonably high at higher frequencies.

To appreciate the need for high CMRR at higher frequencies, consider the situation illustrated in Fig. 9.36: We show two stages of a differential amplifier whose power-supply voltage V_{DD} is corrupted with high-frequency noise. Since the quiescent voltage at each of the drains of Q_1 and Q_2 is $[V_{DD} - (I/2)R_D]$, we see that v_{D1} and v_{D2} will have the same high-frequency noise as V_{DD}. This high-frequency noise then constitutes a common-mode input signal to the second differential stage, formed by Q_3 and Q_4. If the second differential stage is perfectly matched, its differential output voltage V_o should be free of high-frequency noise. However, in practice there is no such thing as perfect matching, and the second stage will have a finite common-mode gain. Furthermore, because of the zero formed by R_{SS} and C_{SS} of the second stage, the common-mode gain will increase with frequency, causing some of the noise to make its way to V_o. With careful design, this undesirable component of V_o can be kept small.

EXERCISE

9.28 The differential amplifier specified in Exercise 9.27 has $R_{SS} = 25\ \text{k}\Omega$ and $C_{SS} = 0.4$ pF. Find the 3-dB frequency of the CMRR.
Ans. 15.9 MHz

9.8.2 Analysis of the Active-Loaded MOS Amplifier

We next consider the frequency response of the current-mirror-loaded MOS differential-pair circuit studied in Section 8.5. The circuit is shown in Fig. 9.37(a) with two capacitances indicated: C_m, which is the total capacitance at the input node of the current mirror, and C_L, which is the total capacitance at the output node. Capacitance C_m is mainly formed by C_{gs3} and C_{gs4} but also includes C_{gd1}, C_{db1}, and C_{db3},

$$C_m = C_{gd1} + C_{db1} + C_{db3} + C_{gs3} + C_{gs4} \tag{9.139}$$

Capacitance C_L includes C_{gd2}, C_{db2}, C_{db4}, C_{gd4} as well as an actual load capacitance and/or the input capacitance of a subsequent stage (C_x),

$$C_L = C_{gd2} + C_{db2} + C_{gd4} + C_{db4} + C_x \tag{9.140}$$

These two capacitances primarily determine the dependence of the differential gain of this amplifier on frequency.

As indicated in Fig. 9.37(a) the input differential signal V_{id} is applied in a balanced fashion and the output node is short-circuited to ground in order to determine the transconductance G_m.

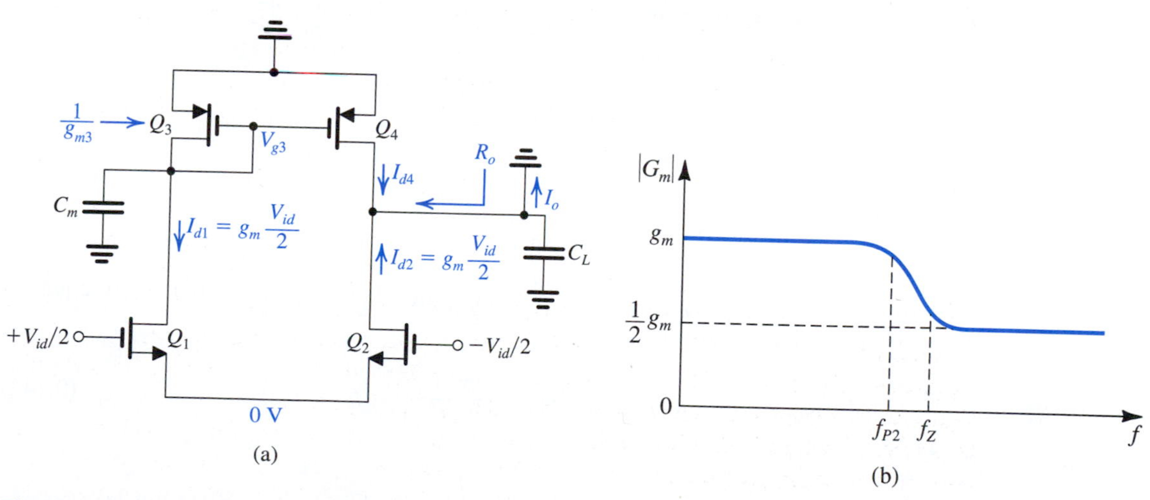

Figure 9.37 **(a)** Frequency–response analysis of the active-loaded MOS differential amplifier. **(b)** The overall transconductance G_m as a function of frequency.

Obviously, because of the output short circuit, C_L will have no effect on G_m. Transistor Q_1 will conduct a drain current signal of $g_m V_{id}/2$, which flows through the diode-connected transistor Q_3 and thus through the parallel combination of $(1/g_{m3})$ and C_m, where we have neglected the resistances r_{o1} and r_{o3} which are much larger than $(1/g_{m3})$, thus

$$V_{g3} = -\frac{g_m V_{id}/2}{g_{m3} + sC_m} \tag{9.141}$$

In response to V_{g3}, transistor Q_4 conducts a drain current I_{d4},

$$I_{d4} = -g_{m4} V_{g3} = \frac{g_{m4}\, g_m V_{id}/2}{g_{m3} + sC_m}$$

Since $g_{m3} = g_{m4}$, this equation reduces to

$$I_{d4} = \frac{g_m V_{id}/2}{1 + s\dfrac{C_m}{g_{m3}}} \tag{9.142}$$

Now, at the output node the total output current that flows through the short circuit is

$$\begin{aligned} I_o &= I_{d4} + I_{d2} \\ &= \frac{g_m V_{id}/2}{1 + s\dfrac{C_m}{g_{m3}}} + g_m(V_{id}/2) \end{aligned} \tag{9.143}$$

We can thus obtain G_m as

$$G_m \equiv \frac{I_o}{V_{id}} = g_m \frac{1 + s\dfrac{C_m}{2g_{m3}}}{1 + s\dfrac{C_m}{g_{m3}}} \tag{9.144}$$

Thus, as expected, the low-frequency value of G_m is equal to g_m of Q_1 and Q_2. At high frequencies, G_m acquires a pole and a zero, the frequencies of which are

$$f_{P2} = \frac{g_{m3}}{2\pi C_m} \tag{9.145}$$

and

$$f_Z = \frac{2g_{m3}}{2\pi C_m} \tag{9.146}$$

That is, the zero frequency is twice that of the pole. Since C_m is approximately equal to $C_{gs2} + C_{gs4} = 2C_{gs}$, we also have

$$f_{P2} = \frac{g_{m3}}{2\pi C_m} \simeq \frac{g_{m3}}{2\pi(2C_{gs})} \simeq f_T/2 \tag{9.147}$$

and

$$f_Z \simeq f_T \tag{9.148}$$

where f_T is the unity-gain frequency of the MOSFET Q_3. Thus, the **mirror pole and zero** occur at very high frequencies. Nevertheless, their effect can be significant.

Figure 9.37(b) shows a sketch of the magnitude of G_m versus frequency. It is interesting and useful to observe that the path of the signal current produced by Q_1 has a transfer function different from that of the signal current produced by Q_2. It is the first signal that encounters C_m and experiences the mirror pole. This observation leads to an interesting view of the effect of C_m on the overall transconductance G_m of the differential amplifier. As we learned in Section 8.5, at low frequencies I_{d1} is replicated by the mirror $Q_3 - Q_4$ in the drain of Q_4 as I_{d4}, which adds to I_{d2} to provide a factor-of-2 increase in G_m (thus making G_m equal to g_m, which is double the value available without the current mirror). Now, at high frequencies C_m acts as a short circuit causing V_{g3} to be zero, and hence I_{d4} will be zero, reducing G_m to $g_m/2$, as borne out by the sketch in Fig. 9.37(b).

Having determined the short-circuit output current I_o, we now multiply it by the total impedance between the output node and ground to determine the output voltage V_o,

$$V_o = I_o \frac{1}{\frac{1}{R_o} + sC_L}$$

$$= G_m V_{id} \frac{R_o}{1 + sC_L R_o}$$

Thus,

$$\frac{V_o}{V_{id}} = (g_m R_o) \left[\frac{1 + s\frac{C_m}{2g_{m3}}}{1 + s\frac{C_m}{g_{m3}}} \right] \left(\frac{1}{1 + sC_L R_o} \right) \tag{9.149}$$

where

$$R_o = r_{o2} \| r_{o4}$$

Thus, in addition to the pole and zero of G_m, the gain of the differential amplifier will have a pole with frequency f_{P1},

$$f_{P1} = \frac{1}{2\pi C_L R_o} \tag{9.150}$$

This, of course, is entirely expected, and in fact this output pole is often dominant, especially when a large load capacitance is present.

Example 9.14

Consider an active-loaded MOS differential amplifier of the type shown in Fig. 9.37(a). Assume that for all transistors, $W/L = 7.2\ \mu m/0.36\ \mu m$, $C_{gs} = 20$ fF, $C_{gd} = 5$ fF, and $C_{db} = 5$ fF. Also, let $\mu_n C_{ox} = 387\ \mu A/V^2$, $\mu_p C_{ox} = 86\ \mu A/V^2$, $V'_{An} = 5$ V/µm, $|V'_{Ap}| = 6$ V/µm. The bias current $I = 0.2$ mA, and the bias current source has an output resistance $R_{SS} = 25$ kΩ and an output capacitance $C_{SS} = 0.2$ pF. In addition to the capacitances introduced by the transistors at the output node, there is a capacitance C_x of 25 fF. It is required to determine the low-frequency values of A_d, A_{cm}, and CMRR. It is also required to find the poles and zero of A_d and the dominant pole of CMRR.

Example 9.14 *continued*

Solution

Since $I = 0.2$ mA, each of the four transistors is operating at a bias current of 100 μA. Thus, for Q_1 and Q_2,

$$100 = \frac{1}{2} \times 387 \times \frac{7.2}{0.36} \times V_{OV}^2$$

which leads to

$$V_{OV} = 0.16 \text{ V}$$

Thus,

$$g_m = g_{m1} = g_{m2} = \frac{2 \times 0.1}{0.16} = 1.25 \text{ mA/V}$$

$$r_{o1} = r_{o2} = \frac{5 \times 0.36}{0.1} = 18 \text{ k}\Omega$$

For Q_3 and Q_4 we have

$$100 = \frac{1}{2} \times 86 \times \frac{7.2}{0.36} \times V_{OV3,4}^2$$

Thus,

$$V_{OV3,4} = 0.34 \text{ V},$$

and

$$g_{m3} = g_{m4} = \frac{2 \times 0.1}{0.34} = 0.6 \text{ mA/V}$$

$$r_{o3} = r_{o4} = \frac{6 \times 0.36}{0.1} = 21.6 \text{ k}\Omega$$

The low-frequency value of the differential gain can be determined from

$$A_d = g_m(r_{o2} \| r_{o4})$$
$$= 1.25(18 \| 21.6) = 12.3 \text{ V/V}$$

The low-frequency value of the common-mode gain can be determined from Eq. (8.146′) as

$$A_{cm} = -\frac{1}{2g_{m3}R_{SS}}$$
$$= -\frac{1}{2 \times 0.6 \times 25} = -0.033 \text{ V/V}$$

The low-frequency value of the CMRR can now be determined as

$$\text{CMRR} = \frac{|A_d|}{|A_{cm}|} = \frac{12.3}{0.033} = 369$$

or,

$$20 \log 369 = 51.3 \text{ dB}$$

To determine the poles and zero of A_d we first compute the values of the two pertinent capacitances C_m and C_L. Using Eq. (9.139),

$$C_m = C_{gd1} + C_{db1} + C_{db3} + C_{gs3} + C_{gs4}$$
$$= 5 + 5 + 5 + 20 + 20 = 55 \text{ fF}$$

Capacitance C_L is found using Eq. (9.140) as

$$C_L = C_{gd2} + C_{db2} + C_{gd4} + C_{db4} + C_x$$
$$= 5 + 5 + 5 + 5 + 25 = 45 \text{ fF}$$

Now, the poles and zero of A_d can be found from Eqs. (9.150), (9.145), and (9.146) as

$$f_{P1} = \frac{1}{2\pi C_L R_o}$$
$$= \frac{1}{2\pi \times C_L(r_{o2} \| r_{o4})}$$
$$= \frac{1}{2\pi \times 45 \times 10^{-15}(18 \| 21.6)10^3}$$
$$= 360 \text{ MHz}$$

$$f_{P2} = \frac{g_{m3}}{2\pi C_m} = \frac{0.6 \times 10^{-3}}{2\pi \times 55 \times 10^{-15}} = 1.74 \text{ GHz}$$

$$f_Z = 2f_{P2} = 3.5 \text{ GHz}$$

Thus the dominant pole is that produced by C_L at the output node. As expected, the pole and zero of the mirror are at much higher frequencies.

The dominant pole of the CMRR is at the location of the common-mode-gain zero introduced by C_{SS} and R_{SS}, that is,

$$f_Z = \frac{1}{2\pi C_{SS} R_{SS}}$$
$$= \frac{1}{2\pi \times 0.2 \times 10^{-12} \times 25 \times 10^3}$$
$$= 31.8 \text{ MHz}$$

Thus, the CMRR begins to decrease at 31.8 MHz, which is much lower than f_{P1}.

EXERCISE

9.29 A bipolar current-mirror-loaded differential amplifier is biased with a current source $I = 1$ mA. The transistors are specified to have $|V_A| = 100$ V. The total capacitance at the output node is 2 pF. Find the dc value and the frequency of the dominant high-frequency pole of the differential voltage gain.

Ans. 2000 V/V; 0.8 MHz

9.9 Other Wideband Amplifier Configurations

Thus far, we have studied one wideband amplifier configuration: the cascode amplifier. Cascoding can, of course, be applied to differential amplifiers to obtain wideband differential amplification. In this section we discuss a number of other circuit configurations that are capable of achieving wide bandwidths.

9.9.1 Obtaining Wideband Amplification by Source and Emitter Degeneration

As we discussed in Chapters 5 and 6, adding a resistance in the source (emitter) lead of a CS (CE) amplifier can result in a number of performance improvements at the expense of a reduction in voltage gain. Extension of the amplifier bandwidth, which is the topic of interest to us in this section, is among those improvements.

Figure 9.38(a) shows a common-source amplifier with a source-degeneration resistance R_s. As indicated in Fig. 9.38(b), the output of the amplifier can be modeled at low frequencies by a controlled current-source $G_m V_i$ and an output resistance R_o, where the transconductance G_m is given by

$$G_m \simeq \frac{g_m}{1 + g_m R_s} \tag{9.151}$$

and the output resistance is given by

$$R_o \simeq r_o(1 + g_m R_s) \tag{9.152}$$

Thus, source degeneration reduces the transconductance and increases the output resistance by the same factor, $(1 + g_m R_s)$. The low-frequency voltage gain can be obtained as

$$A_M = \frac{V_o}{V_{sig}} = -G_m(R_o \| R_L) = -G_m R_L' \tag{9.153}$$

Let's now consider the high-frequency response of the source-degenerated amplifier. Figure 9.38(c) shows the amplifier, indicating the capacitances C_{gs} and C_{gd}. A capacitance C_L that *includes* the MOSFET capacitance C_{db} is also shown at the output. The method of open-circuit time constants can be employed to obtain an estimate of the 3-dB frequency f_H. Toward that end, we show in Fig. 9.38(d) the circuit for determining R_{gd}, which is the resistance seen by C_{gd}. We observe that R_{gd} can be determined by simply adapting the formula in Eq. (9.84) to the case with source degeneration as follows:

$$R_{gd} = R_{sig}(1 + G_m R_L') + R_L' \tag{9.154}$$

where

$$R_L' = R_L \| R_o \tag{9.155}$$

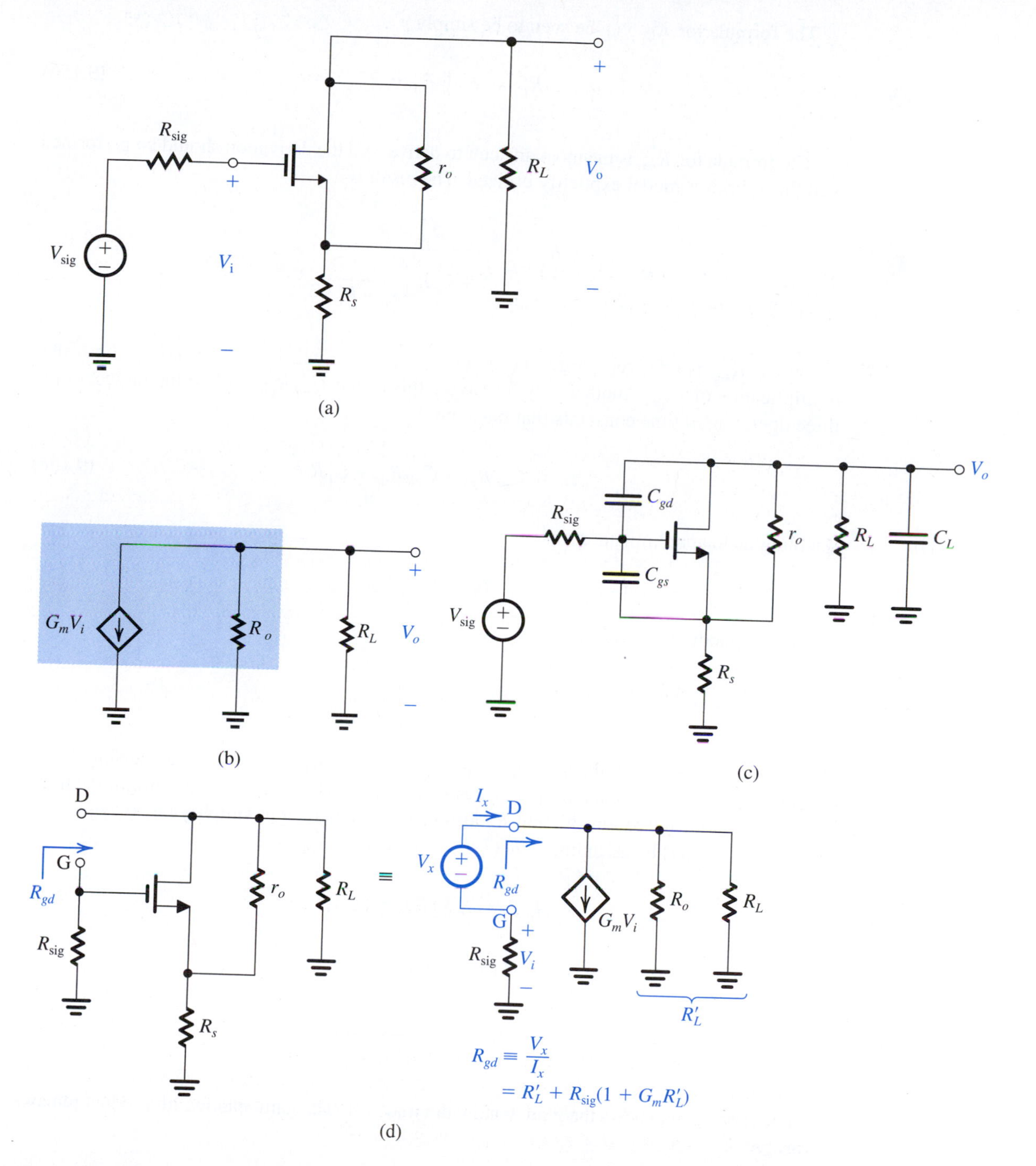

$$R_{gd} \equiv \frac{V_x}{I_x}$$
$$= R'_L + R_{sig}(1 + G_m R'_L)$$

(d)

Figure 9.38 **(a)** The CS amplifier circuit, with a source resistance R_s. **(b)** Equivalent-circuit representation of the amplifier output. **(c)** The circuit prepared for frequency-response analysis. **(d)** Determining the resistance R_{gd} seen by the capacitance C_{gd}.

The formula for R_{C_L} can be seen to be simply

$$R_{C_L} = R_L \parallel R_o = R_L' \tag{9.156}$$

The formula for R_{gs} is the most difficult to derive, and the derivation should be performed with the hybrid-π model explicitly utilized. The result is

$$R_{gs} \simeq \frac{R_{sig} + R_s}{1 + g_m R_s\left(\frac{r_o}{r_o + R_L}\right)} \tag{9.157}$$

When R_{sig} is relatively large, the frequency response will be dominated by the Miller multiplication of C_{gd}. Another way for saying this is that $C_{gd}R_{gd}$ will be the largest of the three open-circuit time constants that make up τ_H,

$$\tau_H = C_{gs}R_{gs} + C_{gd}R_{gd} + C_L R_{C_L} \tag{9.158}$$

enabling us to approximate τ_H as

$$\tau_H \simeq C_{gd}R_{gd} \tag{9.159}$$

and correspondingly to obtain f_H as

$$f_H \simeq \frac{1}{2\pi C_{gd}R_{gd}} \tag{9.160}$$

Now, as R_s is increased, the gain magnitude, $|A_M| = G_m R_L'$, will decrease, causing R_{gd} to decrease (Eq. 9.154), which in turn causes f_H to increase (Eq. 9.160). To highlight the trade-off between gain and bandwidth that R_s affords the designer, let us simplify the expression for R_{gd} in Eq. (9.154) by assuming that $G_m R_L' \gg 1$ and $G_m R_{sig} \gg 1$,

$$R_{gd} \simeq G_m R_L' R_{sig} = |A_M| R_{sig}$$

which can be substituted in Eq. (9.160) to obtain

$$f_H = \frac{1}{2\pi C_{gd}R_{sig}|A_M|} \tag{9.161}$$

which very clearly shows the gain–bandwidth trade-off. The gain–bandwidth product remains constant at

$$\text{Gain–bandwidth product} = |A_M|\, f_H = \frac{1}{2\pi C_{gd}R_{sig}} \tag{9.162}$$

In practice, however, the other capacitances will play a role in determining f_H, and the gain–bandwidth product will decrease somewhat as R_s is increased.

EXERCISE

9.30 Consider a CS amplifier having $g_m = 2$ mA/V, $r_o = 20$ kΩ, $R_L = 20$ kΩ, $R_{sig} = 20$ kΩ, $C_{gs} = 20$ fF, $C_{gd} = 5$ fF, and $C_L = 5$ fF. (a) Find the voltage gain A_M and the 3-dB frequency f_H (using the method of open-circuit time constants) and hence the gain–bandwidth product. (b) Repeat (a) for the case in which a resistance R_s is connected in series with the source terminal with a value selected so that $g_m R_s = 2$.

Ans. (a) –20 V/V, 61.2 MHz, 1.22 GHz; (b) –10 V/V, 109 MHz, 1.1 GHz

9.9.2 The CD–CS, CC–CE and CD–CE Configurations

In Section 7.6.1 we discussed the performance improvements obtained by preceding the CS and CE amplifiers by a buffer implemented by a CD or a CC amplifier, as in the circuits shown in Fig. 9.39. A major advantage of each of these circuits is wider bandwidth than that obtained in the CS or CE stage alone. To see how this comes about, consider as an example the CD–CS amplifier in Fig 9.39(a) and note that the CS transistor Q_2 will still exhibit a Miller effect that results in a large input capacitance, C_{in2}, between its gate and ground. However, the resistance that this capacitance interacts with will be much lower than R_{sig}; the buffering action of the source follower causes a relatively low resistance, approximately equal to a $1/g_{m1}$, to appear between the source of Q_1 and ground across C_{in2}.

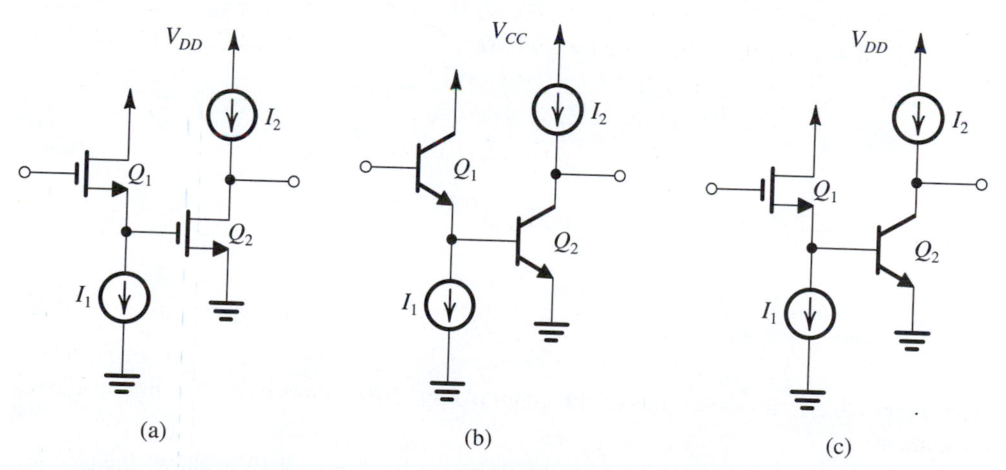

Figure 9.39 **(a)** CD–CS amplifier. **(b)** CC–CE amplifier. **(c)** CD–CE amplifier.

Example 9.15

Consider a CC–CE amplifier such as that in Fig. 9.39(b) with the following specifications: $I_1 = I_2 =$ 1 mA and identical transistors with $\beta = 100$, $f_T = 400$ MHz, and $C_\mu = 2$ pF. Let the amplifier be fed with a source V_{sig} having a resistance $R_{sig} = 4\ \text{k}\Omega$, and assume a load resistance of 4 kΩ. Find the voltage gain A_M, and estimate the 3-dB frequency, f_H. Compare the results with those obtained with a CE amplifier operating under the same conditions. For simplicity, neglect r_o and r_x.

Solution

At an emitter bias current of 1 mA, Q_1 and Q_2 have

$$g_m = 40\ \text{mA/V}$$

$$r_e = 25\ \Omega$$

$$r_\pi = \frac{\beta}{g_m} = \frac{100}{40} = 2.5\ \text{k}\Omega$$

$$C_\pi + C_\mu = \frac{g_m}{\omega_T} = \frac{g_m}{2\pi f_T}$$

$$= \frac{40 \times 10^{-3}}{2\pi \times 400 \times 10^6} = 15.9\ \text{pF}$$

$$C_\mu = 2\ \text{pF}$$

$$C_\pi = 13.9\ \text{pF}$$

The voltage gain A_M can be determined from the circuit shown in Fig. 9.40(a) as follows:

$$R_{in2} = r_{\pi 2} = 2.5\ \text{k}\Omega$$

$$R_{in} = (\beta_1 + 1)(r_{e1} + R_{in2})$$

$$= 101(0.025 + 2.5) = 255\ \text{k}\Omega$$

$$\frac{V_{b1}}{V_{sig}} = \frac{R_{in}}{R_{in} + R_{sig}} = \frac{255}{255 + 4} = 0.98\ \text{V/V}$$

$$\frac{V_{b2}}{V_{b1}} = \frac{R_{in2}}{R_{in2} + r_{e1}} = \frac{2.5}{2.5 + 0.025} = 0.99\ \text{V/V}$$

$$\frac{V_o}{V_{b2}} = -g_{m2}R_L = -40 \times 4 = -160\ \text{V/V}$$

Thus,

$$A_M = \frac{V_o}{V_{sig}} = -160 \times 0.99 \times 0.98 = -155\ \text{V/V}$$

To determine f_H we use the method of open-circuit time constants. Figure 9.40(b) shows the circuit with V_{sig} set to zero and the four capacitances indicated. Capacitance $C_{\mu 1}$ sees a resistance $R_{\mu 1}$,

$$R_{\mu 1} = R_{sig} \parallel R_{in}$$

$$= 4 \parallel 255 = 3.94\ \text{k}\Omega$$

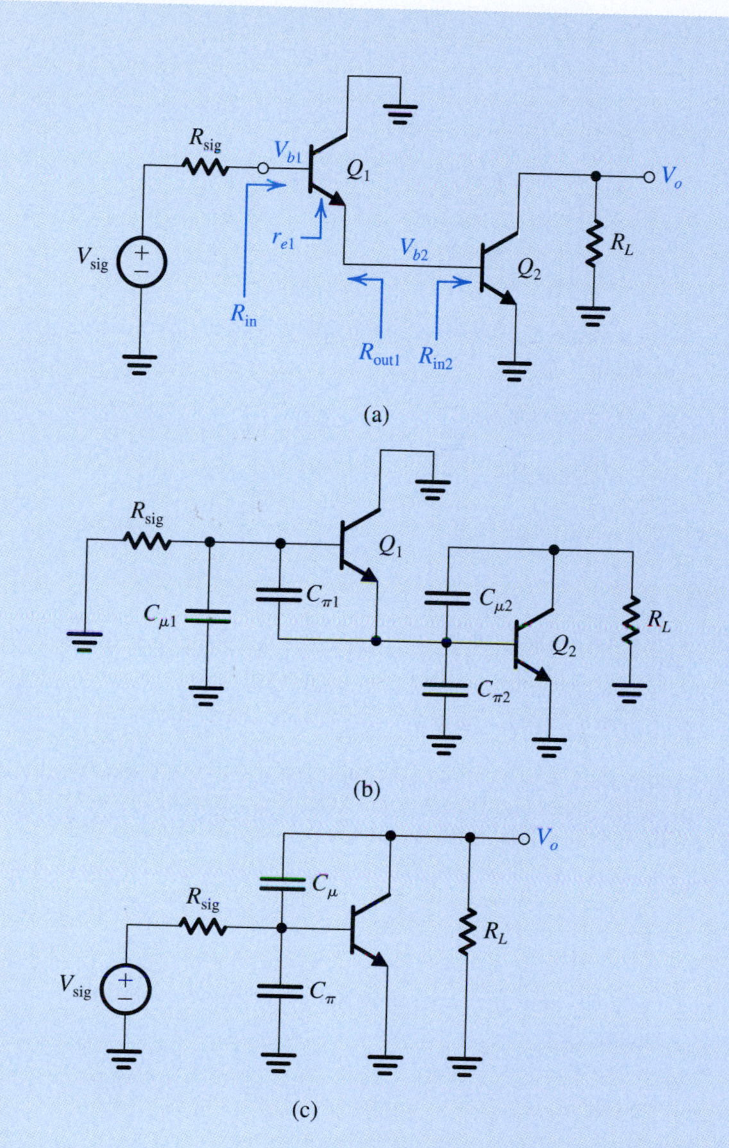

Figure 9.40 Circuits for Example 9.14: **(a)** the CC–CE circuit prepared for low-frequency, small-signal analysis; **(b)** the circuit at high frequencies, with V_{sig} set to zero to enable determination of the open-circuit time constants; **(c)** a CE amplifier for comparison.

To find the resistance $R_{\pi 1}$ seen by capacitance $C_{\pi 1}$ we refer to the analysis of the high-frequency response of the emitter follower in Section 9.7.2. Specifically, we adapt Eq. (9.133) to the situation here as follows:

$$R_{\pi 1} = \frac{R_{sig} + R_{in2}}{1 + \dfrac{R_{sig}}{r_{\pi 1}} + \dfrac{R_{in2}}{r_{e1}}}$$

$$= \frac{4000 + 2500}{1 + \dfrac{4000}{2500} + \dfrac{2500}{25}} = 63.4\ \Omega$$

Example 9.15 *continued*

Capacitance $C_{\pi 2}$ sees a resistance $R_{\pi 2}$,

$$R_{\pi 2} = R_{in2} \| R_{out1}$$
$$= r_{\pi 2} \| \left[r_{e1} + \frac{R_{sig}}{\beta_1 + 1} \right]$$
$$= 2500 \| \left[25 + \frac{4000}{101} \right] = 63 \ \Omega$$

Capacitance $C_{\mu 2}$ sees a resistance $R_{\mu 2}$. To determine $R_{\mu 2}$ we refer to the analysis of the frequency response of the CE amplifier in Section 9.5 to obtain

$$R_{\mu 2} = (1 + g_{m2} R_L)(R_{in2} \| R_{out1}) + R_L$$
$$= (1 + 40 \times 4)\left[2500 \| \left(25 + \frac{4000}{101}\right)\right] + 4000$$
$$= 14{,}143 \ \Omega \simeq 14.1 \ \text{k}\Omega$$

We now can determine τ_H from

$$\tau_H = C_{\mu 1} R_{\mu 1} + C_{\pi 1} R_{\pi 1} + C_{\mu 2} R_{\mu 2} + C_{\pi 2} R_{\pi 2}$$
$$= 2 \times 3.94 + 13.9 \times 0.0634 + 2 \times 14.1 + 13.9 \times 0.063$$
$$= 7.88 + 0.88 + 28.2 + 0.88 = 37.8 \ \text{ns}$$

We observe that $C_{\pi 1}$ and $C_{\pi 2}$ play a very minor role in determining the high-frequency response. As expected, $C_{\mu 2}$ through the Miller effect plays the most significant role. Also, $C_{\mu 1}$, which interacts directly with $(R_{sig} \| R_{in})$, also plays an important role. The 3-dB frequency f_H can be found as follows:

$$f_H = \frac{1}{2\pi \tau_H} = \frac{1}{2\pi \times 37.8 \times 10^{-9}} = 4.2 \ \text{MHz}$$

For comparison, we evaluate A_M and f_H of a CE amplifier operating under the same conditions. Refer to Fig. 9.40(c). The voltage gain A_M is given by

$$A_M = \frac{R_{in}}{R_{in} + R_{sig}}(-g_m R_L)$$
$$= \frac{r_\pi}{r_\pi + R_{sig}}(-g_m R_L)$$
$$= \frac{2.5}{2.5 + 4}(-40 \times 4)$$
$$= -61.5 \ \text{V/V}$$

$$R_\pi = r_\pi \| R_{sig} = 2.5 \| 4 = 1.54 \ \text{k}\Omega$$
$$R_\mu = (1 + g_m R_L)(R_{sig} \| r_\pi) + R_L$$
$$= (1 + 40 \times 4)(4 \| 2.5) + 4$$
$$= 251.7 \ \text{k}\Omega$$

Thus,

$$\tau_H = C_\pi R_\pi + C_\mu R_\mu$$
$$= 13.9 \times 1.54 + 2 \times 251.7$$
$$= 21.4 + 503.4 = 524.8 \text{ ns}$$

Observe the dominant role played by C_μ. The 3-dB frequency f_H is

$$f_H = \frac{1}{2\pi\tau_H} = \frac{1}{2\pi \times 524.8 \times 10^{-9}} = 303 \text{ kHz}$$

Thus, including the buffering transistor Q_1 increases the gain, $|A_M|$, from 61.5 V/V to 155 V/V—a factor of 2.5—and increases the bandwidth from 303 kHz to 4.2 MHz—a factor of 13.9! The gain–bandwidth product is increased from 18.63 MHz to 651 MHz—a factor of 35!

9.9.3 The CC–CB and CD–CG Configurations

In Section 7.6.2 we showed that preceding a CB or CG transistor with a buffer implemented with a CC or a CD transistor solves the low-input-resistance problem of the CB and CG amplifiers. Examples of the resulting compound-transistor amplifiers are shown in Fig. 9.41. Since in each of these circuits, neither of the two transistors suffers from the Miller effect, the resulting amplifiers have even wider bandwidths than those achieved in the compound amplifier stages of the last section. To illustrate, consider as an example the circuit in

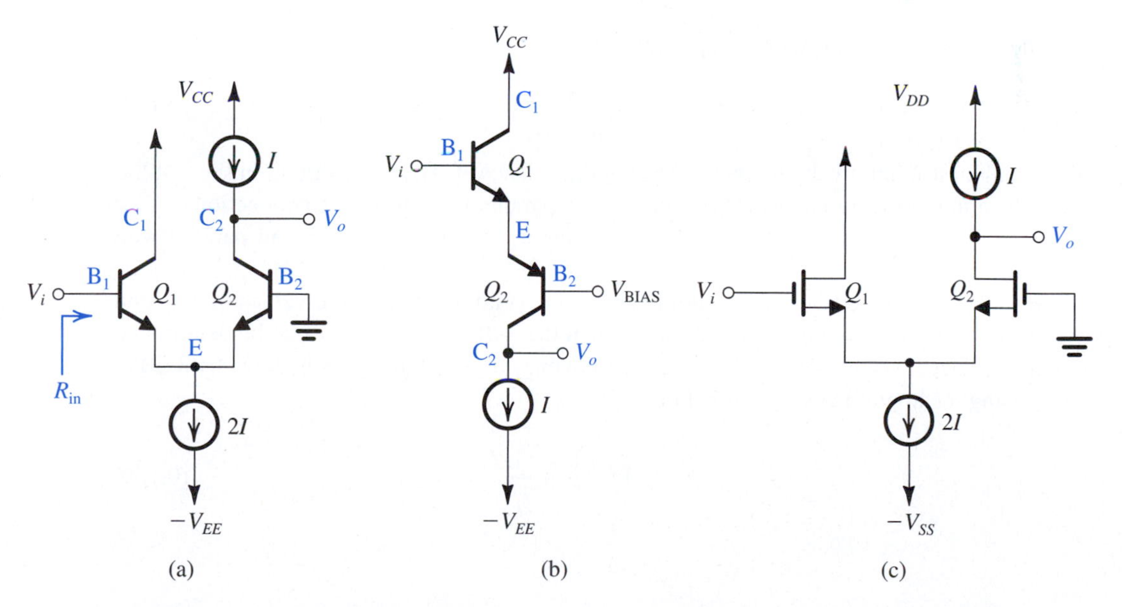

Figure 9.41 **(a)** A CC–CB amplifier. **(b)** Another version of the CC–CB circuit with Q_2 implemented using a *pnp* transistor. **(c)** The MOSFET version of the circuit in **(a)**.

Fig. 9.41(a).[10] The low-frequency analysis of this circuit in Section 7.6.2 provides for the input resistance,

$$R_{in} = (\beta_1 + 1)(r_{e1} + r_{e2}) \tag{9.163}$$

which for $r_{e1} = r_{e2} = r_e$ and $\beta_1 = \beta_2 = \beta$ becomes

$$R_{in} = 2r_\pi \tag{9.164}$$

If a load resistance R_L is connected at the output, the voltage gain V_o/V_i will be

$$\frac{V_o}{V_i} = \frac{\alpha_2 R_L}{r_{e1} + r_{e2}} = \frac{1}{2} g_m R_L \tag{9.165}$$

Now, if the amplifier is fed with a voltage signal V_{sig} from a source with a resistance R_{sig}, the overall voltage gain will be

$$\frac{V_o}{V_{sig}} = \frac{1}{2}\left(\frac{R_{in}}{R_{in} + R_{sig}}\right)(g_m R_L) \tag{9.166}$$

The high-frequency analysis is illustrated in Fig. 9.42(a). Here we have drawn the hybrid-π equivalent circuit for each of Q_1 and Q_2. Recalling that the two transistors are operating at equal bias currents, their corresponding model components will be equal (i.e., $r_{\pi 1} = r_{\pi 2}$, $C_{\pi 1} = C_{\pi 2}$, etc.). With this in mind the reader should be able to see that $V_{\pi 1} = -V_{\pi 2}$ and the horizontal line through the node labeled E in Fig. 9.42(a) can be deleted. Thus the circuit reduces to that in Fig. 9.42(b). This is a very attractive outcome because the circuit shows clearly the two poles that determine the high-frequency response: The pole at the input, with a frequency f_{P1}, is

$$f_{P1} = \frac{1}{2\pi\left(\frac{C_\pi}{2} + C_\mu\right)(R_{sig} \| 2r_\pi)} \tag{9.167}$$

and the pole at the output, with a frequency f_{P2}, is

$$f_{P2} = \frac{1}{2\pi C_\mu R_L} \tag{9.168}$$

This result is also intuitively obvious: The input impedance at B_1 of the circuit in Fig. 9.42(a) consists of the series connection of $r_{\pi 1}$ and $r_{\pi 2}$ in parallel with the series connection of $C_{\pi 1}$ and $C_{\pi 2}$. Then there is $C_{\mu 1}$ in parallel. At the output, we simply have R_L in parallel with C_μ.

Whether one of the two poles is dominant will depend on the relative values of R_{sig} and R_L. If the two poles are close to each other, then the 3-dB frequency f_H can be determined either by exact analysis—that is, finding the frequency at which the gain is down by 3 dB—or by using the approximate formula in Eq. (9.68),

$$f_H \simeq 1 \Big/ \sqrt{\frac{1}{f_{P1}^2} + \frac{1}{f_{P2}^2}} \tag{9.169}$$

[10] The results derived for the circuit in Fig. 9.41(a) apply directly to the circuit of Fig. 9.41(b) and with appropriate change of variables to the MOS circuit of Fig. 9.41(c).

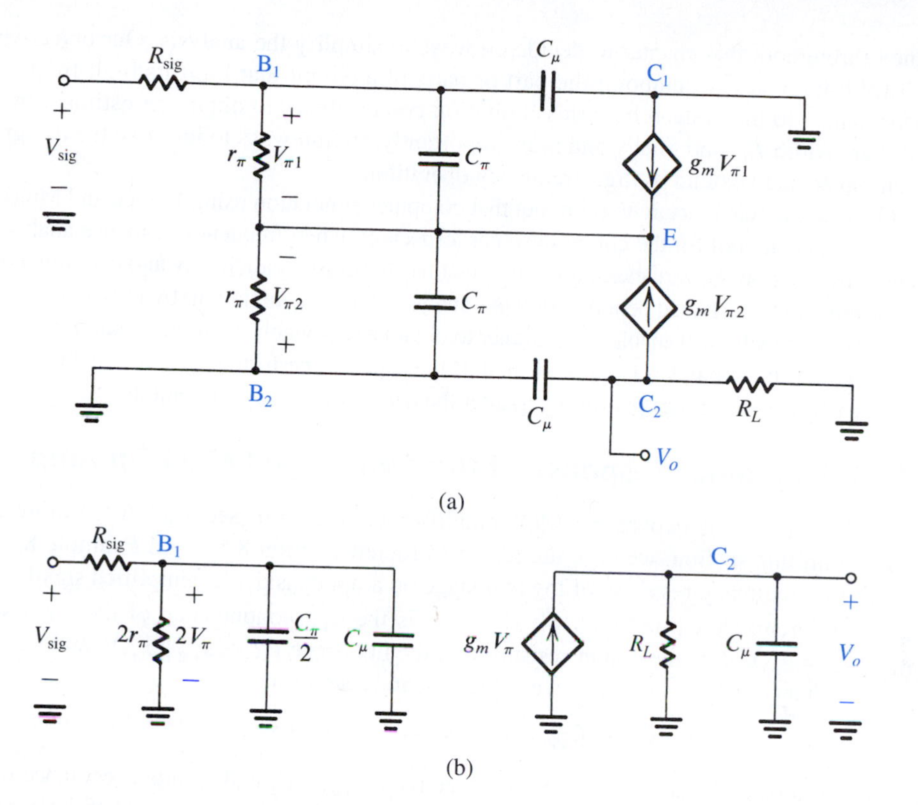

Figure 9.42 **(a)** Equivalent circuit for the amplifier in Fig. 9.41(a). **(b)** Simplified equivalent circuit. Note that the equivalent circuits in **(a)** and **(b)** also apply to the circuit shown in Fig. 9.41(b). In addition, they can be easily adapted for the MOSFET circuit in Fig. 9.41(c), with $2r_\pi$ eliminated, C_π replaced with C_{gs}, C_μ replaced with C_{gd}, and V_π replaced with V_{gs}.

EXERCISE

9.31 For the CC–CB amplifier of Fig. 9.41(a), let $I = 0.5$ mA, $\beta = 100$, $C_\pi = 6$ pF, $C_\mu = 2$ pF, $R_{sig} = 10$ kΩ, and $R_L = 10$ kΩ. Find the low-frequency overall voltage gain A_M, the frequencies of the poles, and the 3-dB frequency f_H. Find f_H both exactly and using the approximate formula in Eq. (9.169).

Ans. 50 V/V; 6.4 MHz and 8 MHz; f_H by exact evaluation = 4.6 MHz; f_H using Eq. (9.169) = 5 MHz.

9.10 Multistage Amplifier Examples

We conclude this chapter with the frequency-response analysis of the two multistage amplifiers we studied in Section 8.6. As we shall see, these are relatively complex circuits: Simply replacing each transistor with its high-frequency, equivalent-circuit model will make it exceedingly difficult for pencil-and-paper analysis, and will most certainly not lead to any analysis and design insight. Rather, we will use the knowledge and experience we have gained throughout this chapter to decide on ways to simplify the analysis. Our objective is multifold: to be able to pinpoint the part or parts of a circuit that limit its high-frequency performance, to understand how this limitation comes about, to obtain an estimate of the 3-dB bandwidth f_H, and finally and most importantly, to find ways to improve the design of the circuit so as to extend its high-frequency operation.

It is useful at this juncture to point out that computer simulation using PSpice and Multisim is a very valuable tool for the circuit designer, especially when frequency-response analysis is under consideration. Nevertheless, it is a tool that has to be used judiciously and certainly not as a replacement for a first-cut pencil-and-paper analysis. Circuit simulation, by utilizing sophisticated device models, will enable the designer to obtain a reasonably accurate prediction of what to expect after the circuit has been fabricated. If the expected performance is unsatisfactory, the designer will then have the opportunity to alter the design to meet specifications.

9.10.1 Frequency Response of the Two-Stage CMOS Op Amp

Figure 9.43 shows the two-stage CMOS amplifier we studied in Section 8.6.1. Before continuing with this section, we urge the reader to review Section 8.6.1 and Example 8.5. To analyze the frequency response of the two-stage op amp, consider its simplified small-signal equivalent circuit shown in Fig. 9.44. Here G_{m1} is the transconductance of the input stage $(G_{m1} = g_{m1} = g_{m2})$, R_1 is the output resistance of the first stage $(R_1 = r_{o2} \| r_{o4})$, and C_1 is the total capacitance at the interface between the first and second stages

$$C_1 = C_{gd4} + C_{db4} + C_{gd2} + C_{db2} + C_{gs6}$$

G_{m2} is the transconductance of the second stage $(G_{m2} = g_{m6})$, R_2 is the output resistance of the second stage $(R_2 = r_{o6} \| r_{o7})$, and C_2 is the total capacitance at the output node of the op amp

$$C_2 = C_{db6} + C_{db7} + C_{gd7} + C_L$$

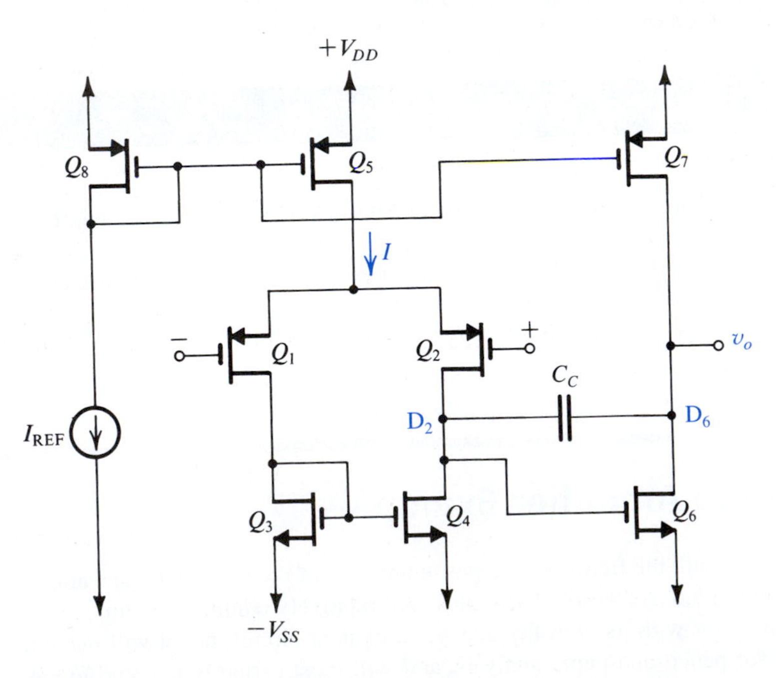

Figure 9.43 Two-stage CMOS op-amp configuration.

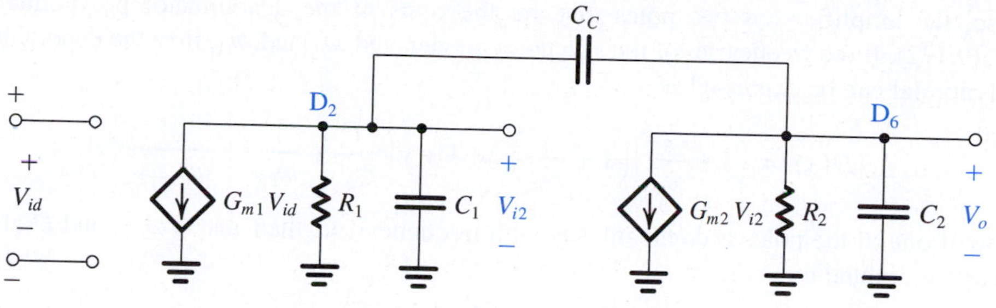

Figure 9.44 Equivalent circuit of the op amp in Fig. 9.43.

where C_L is the load capacitance. Usually C_L is much larger than the transistor capacitances, with the result that C_2 is much larger than C_1. Capacitor C_C is deliberately included for the purpose of equipping the op amp with a uniform –6-dB/octave frequency response. In the following, we shall see how this is possible and how to select a value for C_C. Finally, note that in the equivalent circuit of Fig. 9.44 we should have included C_{gd6} in parallel with C_C. Usually, however, $C_C \gg C_{gd6}$, which is the reason we have neglected C_{gd6}.

To determine V_o, analysis of the circuit in Fig. 9.44 proceeds as follows. Writing a node equation at node D_2 yields

$$G_{m1}V_{id} + \frac{V_{i2}}{R_1} + sC_1V_{i2} + sC_C(V_{i2} - V_o) = 0 \tag{9.170}$$

Writing a node equation at node D_6 yields

$$G_{m2}V_{i2} + \frac{V_o}{R_2} + sC_2V_o + sC_C(V_o - V_{i2}) = 0 \tag{9.171}$$

To eliminate V_{i2} and thus determine V_o in terms of V_{id}, we use Eq. (9.171) to express V_{i2} in terms of V_o and substitute the result into Eq. (9.170). After some straightforward manipulations we obtain the amplifier transfer function

$$\frac{V_o}{V_{id}} = \frac{G_{m1}(G_{m2} - sC_C)R_1R_2}{1 + s[C_1R_1 + C_2R_2 + C_C(G_{m2}R_1R_2 + R_1 + R_2)] + s^2[C_1C_2 + C_C(C_1 + C_2)]R_1R_2} \tag{9.172}$$

First we note that for $s = 0$ (i.e., dc), Eq. (9.172) gives $V_o/V_{id} = (G_{m1}R_1)(G_{m2}R_2)$, which is what we should have expected. Second, the transfer function in Eq. (9.172) indicates that the amplifier has a transmission zero at $s = s_Z$, which is determined from

$$G_{m2} - s_ZC_C = 0$$

Thus,

$$s_Z = \frac{G_{m2}}{C_C} \tag{9.173}$$

In other words, the zero is on the positive real axis with a frequency ω_Z of

$$\omega_Z = \frac{G_{m2}}{C_C} \tag{9.174}$$

Also, the amplifier has two poles that are the roots of the denominator polynomial of Eq. (9.172). If the frequencies of the two poles are denoted ω_{P1} and ω_{P2}, then the denominator polynomial can be expressed as

$$D(s) = \left(1 + \frac{s}{\omega_{P1}}\right)\left(1 + \frac{s}{\omega_{P2}}\right) = 1 + s\left(\frac{1}{\omega_{P1}} + \frac{1}{\omega_{P2}}\right) + \frac{s^2}{\omega_{P1}\omega_{P2}}$$

Now if one of the poles is dominant, say with frequency ω_{P1}, then $\omega_{P1} \ll \omega_{P2}$, and $D(s)$ can be approximated by

$$D(s) \simeq 1 + \frac{s}{\omega_{P1}} + \frac{s^2}{\omega_{P1}\omega_{P2}} \tag{9.175}$$

The frequency of the dominant pole, ω_{P1}, can now be determined by equating the coefficients of the s terms in the denominator in Eq. (9.172) and in Eq. (9.175),

$$\begin{aligned}\omega_{P1} &= \frac{1}{C_1R_1 + C_2R_2 + C_C(G_{m2}R_2R_1 + R_1 + R_2)} \\ &= \frac{1}{R_1[C_1 + C_C(1 + G_{m2}R_2)] + R_2(C_2 + C_C)}\end{aligned} \tag{9.176}$$

We recognize the first term in the denominator as arising at the interface between the first and second stages. Here, R_1, the output resistance of the first stage, is interacting with the total capacitance at the interface. The latter is the sum of C_1 and the Miller capacitance $C_C(1 + G_{m2}R_2)$, which results from connecting C_C in the negative feedback path of the second stage whose gain is $G_{m2}R_2$. Now, since R_1 and R_2 are usually of comparable value, we see that the first term in the denominator will be much larger than the second and we can approximate ω_{P1} as

$$\omega_{P1} \simeq \frac{1}{R_1[C_1 + C_C(1 + G_{m2}R_2)]}$$

A further approximation is possible because C_1 is usually much smaller than the Miller capacitance and $G_{m2}R_2 \gg 1$, thus

$$\omega_{P1} \simeq \frac{1}{R_1C_CG_{m2}R_2} \tag{9.177}$$

The frequency of the second, nondominant pole can be found by equating the coefficients of the s^2 terms in the denominator of Eq. (9.172) and in Eq. (9.175) and substituting for ω_{P1} from Eq. (9.176). The result is

$$\omega_{P2} = \frac{G_{m2}C_C}{C_1C_2 + C_C(C_1 + C_2)}$$

Since $C_1 \ll C_2$ and $C_1 \ll C_C$, ω_{P2} can be approximated as

$$\omega_{P2} \simeq \frac{G_{m2}}{C_2} \tag{9.178}$$

In order to provide the op amp with a uniform gain rolloff of –20 dB/decade down to 0 dB, the value of the compensation capacitor C_C is selected so that the resulting value of ω_{P1} (Eq. 9.177), when multiplied by the dc gain $(G_{m1}R_1G_{m2}R_2)$, results in a unity-gain frequency ω_t lower than ω_Z and ω_{P2}. Specifically

$$\omega_t = (G_{m1}R_1G_{m2}R_2)\omega_{P1} = \frac{G_{m1}}{C_C} \tag{9.179}$$

which must be lower than $\omega_Z = \dfrac{G_{m2}}{C_C}$ and $\omega_{P2} \simeq \dfrac{G_{m2}}{C_2}$. We will have more to say about this point in Section 12.1.

EXERCISE

D9.32 Consider the frequency response of the op amp analyzed in Example 8.5. Let $C_1 = 0.1$ pF and $C_2 = 2$ pF. Find the value of C_C that results in $f_t = 10$ MHz and verify that f_t is lower than f_Z and f_{P2}. Recall from the results of Example 8.5, that $G_{m1} = 0.3$ mA/V and $G_{m2} = 0.6$ mA/V.
Ans. $C_C = 4.8$ pF; $f_Z = 20$ MHz; $f_{P2} = 48$ MHz

9.10.2 Frequency Response of the Bipolar Op Amp of Section 8.6.2

We urge the reader to review Section 8.6.2 and Examples 8.6 and 8.7 before studying this section. The bipolar op-amp circuit shown earlier in Fig. 8.43 is rather complex. Nevertheless, it is possible to obtain an approximate estimate of its high-frequency response. Figure 9.45(a) shows an approximate equivalent circuit for this purpose. Note that we have utilized the equivalent differential half-circuit concept, with Q_2 representing the input stage and Q_5 representing the second stage. We observe, of course, that the second stage is not symmetrical, and strictly speaking the equivalent half-circuit does not apply. Nevertheless, we use it as an approximation so as to obtain a quick pencil-and-paper estimate of the dominant high-frequency pole of the amplifier. More precise results can of course be obtained using computer simulation with SPICE.

Examination of the equivalent circuit in Fig. 9.45(a) reveals that if the resistance of the source of signal V_i is small, the high-frequency limitation will not occur at the input but rather at the interface between the first and the second stages. This is because the total capacitance at node A will be high as a result of the Miller multiplication of $C_{\mu 5}$. Also, the third stage, formed by transistor Q_7, should exhibit good high-frequency response, since Q_7 has a large emitter-degeneration resistance, R_3. The same is also true for the emitter-follower stage, Q_8.

To determine the frequency of the dominant pole that is formed at the interface between Q_2 and Q_5 we show in Fig. 9.45(b) the pertinent equivalent circuit. The total resistance between node A and ground can now be found as

$$R_{\text{eq}} = R_2 \,\|\, r_{o2} \,\|\, r_{\pi 5}$$

and the total capacitance is

$$C_{\text{eq}} = C_{\mu 2} + C_{\pi 5} + C_{\mu 5}(1 + g_{m5}R_{L5})$$

where

$$R_{L5} = R_3 \,\|\, r_{o5} \,\|\, R_{i3}$$

The frequency of the pole can be calculated from R_{eq} and C_{eq} as

$$f_P = \frac{1}{2\pi R_{\text{eq}} C_{\text{eq}}}$$

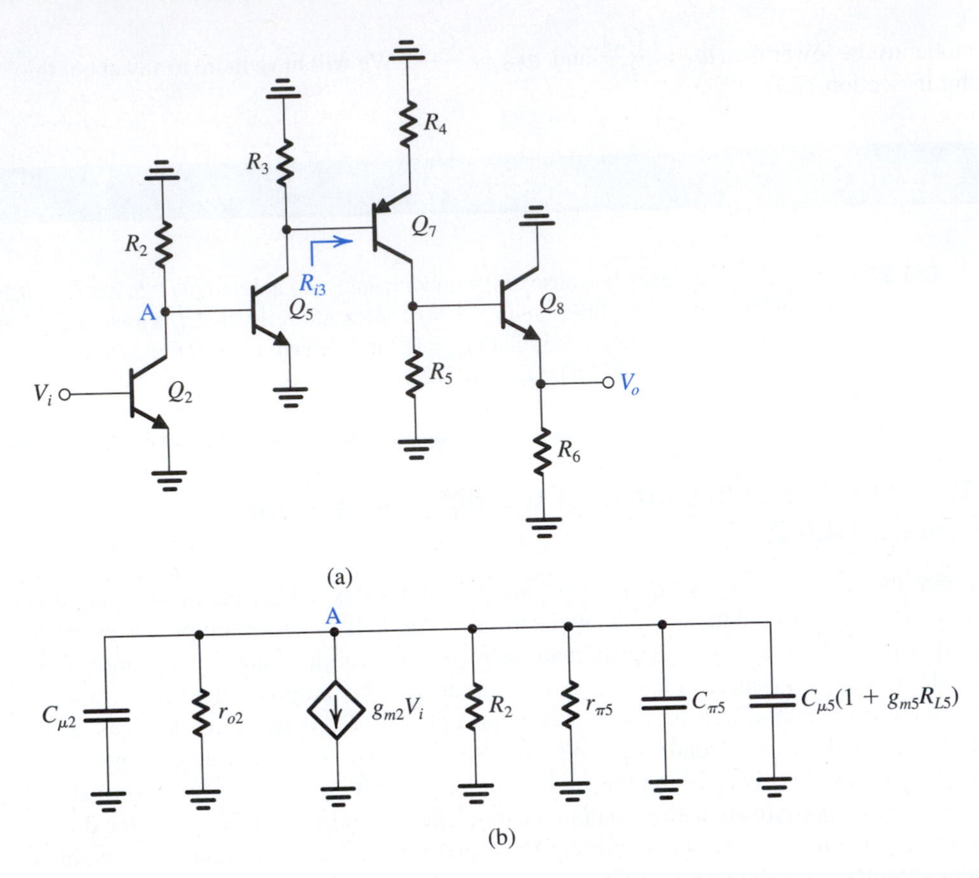

Figure 9.45 **(a)** Approximate equivalent circuit for determining the high-frequency response of the op amp of Fig. 8.43. **(b)** Equivalent circuit of the interface between the output of Q_2 and the input of Q_5.

EXERCISE

9.33 Determine R_{eq}, C_{eq}, and f_P for the amplifier in Fig. 8.43, utilizing the facts that Q_2 is biased at 0.25 mA and Q_5 at 1 mA. Assume $\beta = 100$, $V_A = 100$ V, $f_T = 400$ MHz, and $C_\mu = 2$ pF. Assume $R_{L5} \simeq R_3$.
Ans. 2.21 kΩ; 258 pF; 280 kHz

Summary

- The coupling and bypass capacitors utilized in discrete-circuit amplifiers cause the amplifier gain to fall off at low frequencies. The frequencies of the low-frequency poles can be estimated by considering each of these capacitors separately and determining the resistance seen by the capacitor. The highest-frequency pole is the one that determines the lower 3-dB frequency f_L.
- Both the MOSFET and the BJT have internal capacitive effects that can be modeled by augmenting the device hybrid-π model with capacitances. Usually at least two capacitances are needed: C_{gs} and C_{gd} (C_π and C_μ for the BJT). A figure-of-merit for the high-frequency operation of the transistor is the frequency f_T at which the short-circuit current gain of the CS (CE) transistor reduces to

unity. For the MOSFET, $f_T = g_m/2\pi(C_{gs} + C_{gd})$, and for the BJT, $f_T = g_m/2\pi(C_\pi + C_\mu)$.

- The internal capacitances of the MOSFET and the BJT cause the amplifier gain to fall off at high frequencies. An estimate of the amplifier bandwidth is provided by the frequency f_H at which the gain drops 3 dB below its value at midband, A_M. A figure-of-merit for the amplifier is the gain–bandwidth product $GB = A_M f_H$. Usually, it is possible to trade off gain for increased bandwidth, with GB remaining nearly constant. For amplifiers with a dominant pole with frequency f_H, the gain falls off at a uniform 6-dB/octave (20-dB/decade) rate, reaching 0 dB at $f_t = GB$.
- The high-frequency response of the CS and CE amplifiers is severely limited by the Miller effect: The small capacitance C_{gd} (C_μ) is multiplied by a factor approximately equal to the gain from gate to drain (base to collector) $g_m R'_L$ and thus gives rise to a large capacitance at the amplifier input. The increased C_{in} interacts with the effective signal-source resistance R'_{sig} and causes the amplifier gain to have a 3-dB frequency $f_H = 1/2\pi R'_{sig} C_{in}$.
- The method of open-circuit time constants provides a simple and powerful way to obtain a reasonably good estimate of the upper 3-dB frequency f_H. The capacitors that limit the high-frequency response are considered one at time with $V_{sig} = 0$ and all the other capacitances set to zero (open circuited). The resistance seen by each capacitance is determined, and the overall time constant τ_H is obtained by summing the individual time constants. Then f_H is found as $1/2\pi\tau_H$.
- The CG and CB amplifiers do *not* suffer from the Miller effect. Thus the cascode amplifier, which consists of a cascade of a CS and CG stages (CE and CB stages), can be designed to obtain wider bandwidth than that achieved in the CS (CE) amplifier alone. The key, however, is to design the cascode so that the gain obtained in the CS (CE) stage is minimized.
- The source and emitter followers do not suffer from the Miller effect and thus feature wide bandwidths.
- The high-frequency response of the differential amplifier can be obtained by considering the differential and common-mode half-circuits. The CMRR falls off at a relatively low frequency determined by the output impedance of the bias current source.
- The high-frequency response of the current-mirror-loaded differential amplifier is complicated by the fact that there are two signal paths between input and output: a direct path and one through the current mirror.
- Combining two transistors in a way that eliminates or minimizes the Miller effect can result in a much wider bandwidth. Some such configurations are presented in Section 9.9.
- The key to the analysis of the high-frequency response of a multistage amplifier is to use simple macro models to estimate the frequencies of the poles formed at the interface between each two stages, in addition to the input and output poles. The pole with the lowest frequency dominates and determines f_H.

PROBLEMS

Computer Simulation Problems

SIM Problems identified by this icon are intended to demonstrate the value of using SPICE simulation to verify hand analysis and design, and to investigate important issues such as gain–bandwidth tradeoff. Instructions to assist in setting up PSpice and Multisim simulations for all the indicated problems can be found in the corresponding files on the disc. Note that if a particular parameter value is not specified in the problem statement, you are to make a reasonable assumption. *difficult problem; ** more difficult; *** very challenging and/or time-consuming; D: design problem.

Section 9.1: Low-Frequency Response of the CS and CE Amplifiers

D 9.1 The amplifier in Fig. P9.1 is biased to operate at g_m = 1 mA/V. Neglecting r_o, find the midband gain. Find the value of C_S that places f_L at 20 Hz.

9.2 Consider the amplifier of Fig. 9.2(a). Let R_D = 10 kΩ, r_o = 100 kΩ, and R_L = 10 kΩ. Find the value of C_{C2}, specified to one significant digit, to ensure that the associated break frequency is at, or below, 10 Hz. If a higher-power design results in doubling I_D, with both R_D and r_o reduced by a factor of 2, what does the corner frequency (due to C_{C2}) become? For

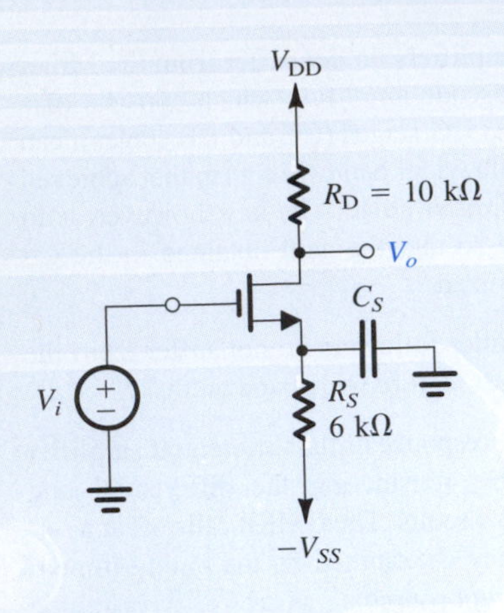

Figure P9.1

increasingly higher-power designs, what is the highest corner frequency that can be associated with C_{C2}?

9.3 The NMOS transistor in the discrete CS amplifier circuit of Fig. P9.3 is biased to have $g_m = 5$ mA/V. Find A_M, f_{P1}, f_{P2}, f_{P3}, and f_L.

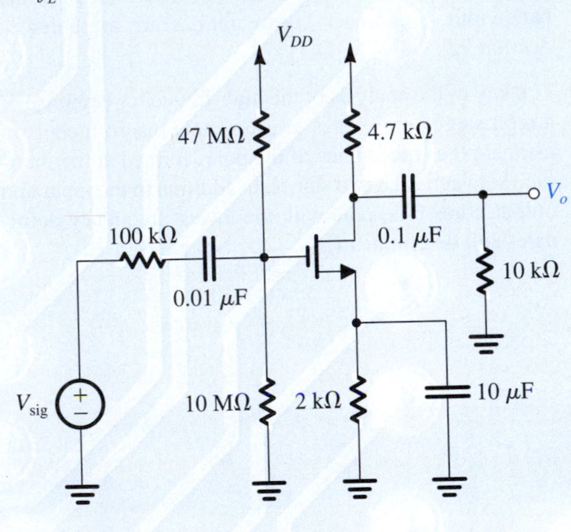

Figure P9.3

D 9.4 Consider the low-frequency response of the CS amplifier of Fig. 9.2(a). Let $R_{sig} = 0.5$ MΩ, $R_G = 2$ MΩ, $g_m = 3$ mA/V, $R_D = 20$ kΩ, and $R_L = 10$ kΩ. Find A_M. Also, design the coupling and bypass capacitors to locate the three low-frequency poles at 50 Hz, 10 Hz, and 3 Hz. Use a minimum total capacitance, with capacitors specified only to a single significant digit. What value of f_L results?

D 9.5 A particular version of the CS amplifier in Fig. 9.2 uses a transistor biased to operate with $g_m = 5$ mA/V. Resistances $R_{sig} = 200\ k\Omega$, $R_G = 10\ M\Omega$, $R_D = 3\ k\Omega$, and $R_L = 5\ k\Omega$. As an initial design, the circuit designer selects $C_{C1} = C_{C2} = C_S = 1\ \mu F$. Find the frequencies f_{P1}, f_{P2}, and f_{P3} and rank them in order of frequency, highest first. Calculate the ratios of the first to second, and second to third. The final design requires that the first pole dominate at 10 Hz with the second a factor of 4 lower, and the third another a factor of 4 lower. Find the values of all the capacitances and the total capacitance needed. If the separation factor were 10, what capacitor values and total capacitance would be needed? (*Note*: You can see that the total capacitance need not be much larger to spread the poles, as is desired in certain applications.)

D 9.6 Repeat Example 9.1 to find C_S, C_{C1}, and C_{C2} that provide $f_L = 20$ Hz and the other pole frequencies at 4 Hz and 1 Hz. Design to keep the total capacitance to a minimum.

D 9.7 Reconsider Exercise 9.1 with the aim of finding a better-performing design using the same total capacitance, that is, 3 μF. Prepare a design in which the break frequencies are separated by a factor of 5 (i.e., f, $f/5$, and $f/25$). What are the three capacitor values, the three break frequencies, and f_L that you achieve?

9.8 Repeat Exercise 9.2 for the situation in which $C_E = 50$ μF and $C_{C1} = C_{C2} = 2$ μF. Find the three break frequencies and estimate f_L.

D 9.9 Repeat Example 9.2 for a related CE amplifier whose supply voltages and bias current are each reduced to half their original value but R_B, R_C, R_{sig}, and R_L are left unchanged. Find C_{C1}, C_E, and C_{C2} for $f_L = 100$ Hz. Minimize the total capacitance used, under the following conditions. Arrange that the contributions of C_E, C_{C1}, and C_{C2} are 80%, 10%, and 10%, respectively. Specify capacitors to two significant digits, choosing the next highest value, in general, for a conservative design, but realizing that for C_E, this may represent a larger capacitance increment. Check the value of f_L that results. [*Note*: An attractive approach can be to select C_E on the small side, allowing it to contribute more than 80% to f_L, while making C_{C1} and C_{C2} larger, since they must contribute less to f_L.)

D 9.10 A particular current-biased CE amplifier operating at 100 μA from ±3 -V power supplies employs $R_C = 20\ k\Omega$, $R_B = 200\ k\Omega$; it operates between a 20-kΩ source and a 10-kΩ load. The transistor $\beta = 100$. Select C_E first for a minimum value specified to one significant digit and providing up to 90% of f_L. Then choose C_{C1} and C_{C2}, each specified to one significant digit, with the goal of minimizing the total capacitance used. What f_L results? What total capacitance is needed?

9.11 Consider the common-emitter amplifier of Fig. P9.11 under the following conditions: $R_{sig} = 5$ kΩ, $R_1 = 33$ kΩ, $R_2 = 22$ kΩ, $R_E = 3.9$ kΩ, $R_C = 4.7$ kΩ, $R_L = 5.6$ kΩ, $V_{CC} = 5$ V. The dc emitter current can be shown to be $I_E \simeq 0.3$ mA, at which $\beta = 120$. Find the input resistance R_{in} and the midband gain A_M. If $C_{C1} = C_{C2} = 1$ μF and $C_E = 20$ μF, find the three break frequencies f_{P1}, f_{P2}, and f_{P3} and an estimate for f_L. Note that R_E has to be taken into account in evaluating f_{P2}.

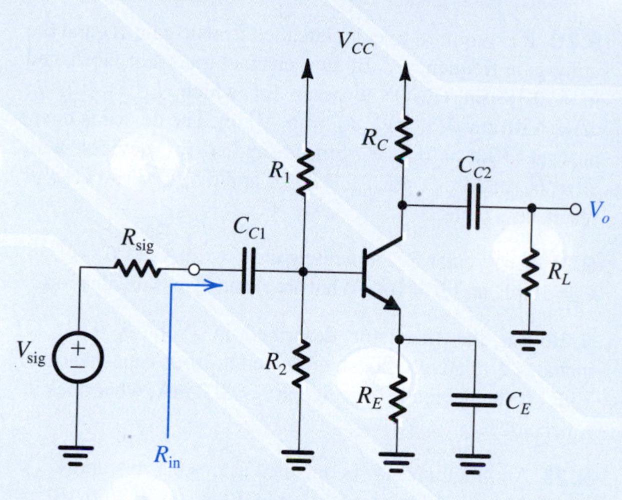

Figure P9.11

D 9.12 For the amplifier described in Problem 9.11, design the coupling and bypass capacitors for a lower 3-dB frequency of 100 Hz. Design so that the contribution of each of C_{C1} and C_{C2} to determining f_L is only 5%.

9.13 Consider the circuit of Fig. P9.11. For R_{sig} = 10 kΩ, $R_B \equiv R_1 || R_2$ = 10 kΩ, r_π = 1 kΩ, β_0 = 100, and R_E = 1 kΩ, what is the ratio C_E/C_{C1} that makes their contributions to the determination of f_L equal?

D *9.14 For the common-emitter amplifier of Fig. P9.14, neglect r_o and assume the current source to be ideal.

(a) Derive an expression for the midband gain.

(b) Derive expressions for the break frequencies caused by C_E and C_C.

(c) Give an expression for the amplifier voltage gain $A(s)$.

(d) For $R_{sig} = R_C = R_L$ = 10 kΩ, β = 100, and I = 1 mA, find the value of the midband gain.

(e) Select values for C_E and C_C to place the two break frequencies a decade apart and to obtain a lower 3-dB frequency of 100 Hz while minimizing the total capacitance.

(f) Sketch a Bode plot for the gain magnitude, and estimate the frequency at which the gain becomes unity.

(g) Find the phase shift at 100 Hz.

9.15 The BJT common-emitter amplifier of Fig. P9.15 includes an emitter degeneration resistance R_e.

(a) Assuming $\alpha \simeq 1$, neglecting r_o, and assuming the current source to be ideal, derive an expression for the small-signal voltage gain $A(s) \equiv V_o/V_{sig}$ that applies in the midband and the low-frequency band. Hence find the midband gain A_M and the lower 3-dB frequency f_L.

(b) Show that including R_e reduces the magnitude of A_M by a certain factor. What is this factor?

(c) Show that including R_e reduces f_L by the same factor as in (b) and thus one can use R_e to trade-off gain for bandwidth.

(d) For I = 0.25 mA, R_C = 10 kΩ, and C_E = 10 μF, find $|A_M|$ and f_L with R_e = 0. Now find the value of R_e that lowers f_L by a factor of 5. What will the gain become? Sketch on the same diagram a Bode plot for the gain magnitude for both cases.

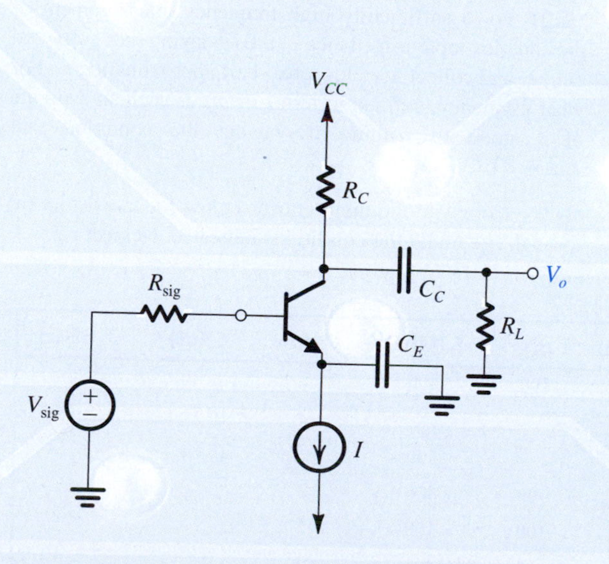

Figure P9.14

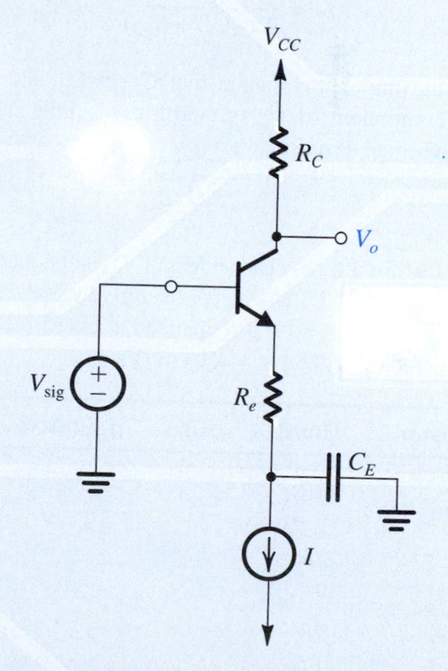

Figure P9.15

Section 9.2: Internal Capacitative Effects and the High-Frequency Model of the MOSFET and the BJT

9.16 Refer to the MOSFET high-frequency model in Fig. 9.6(a). Evaluate the model parameters for an NMOS transistor operating at $I_D = 100\ \mu\text{A}$, $V_{SB} = 1$ V, and $V_{DS} = 1.5$ V. The MOSFET has $W = 20\ \mu\text{m}$, $L = 1\ \mu\text{m}$, $t_{ox} = 8$ nm, $\mu_n = 450\ \text{cm}^2/\text{Vs}$, $\gamma = 0.5\ \text{V}^{1/2}$, $2\phi_f = 0.65$ V, $\lambda = 0.05\ \text{V}^{-1}$, $V_0 = 0.7$ V, $C_{sb0} = C_{db0} = 15$ fF, and $L_{ov} = 0.05\ \mu\text{m}$. (Recall that $g_{mb} = \chi g_m$, where $\chi = \gamma/(2\sqrt{2\phi_f + V_{SB}})$.)

9.17 Find f_T for a MOSFET operating at $I_D = 100\ \mu\text{A}$ and $V_{OV} = 0.2$ V. The MOSFET has $C_{gs} = 20$ fF and $C_{gd} = 5$ fF.

9.18 Starting from the expression of f_T for a MOSFET,

$$f_T = \frac{g_m}{2\pi(C_{gs} + C_{gd})}$$

and making the approximation that $C_{gs} \gg C_{gd}$ and that the overlap component of C_{gs} is negligibly small, show that

$$f_T \simeq \frac{1.5}{\pi L}\sqrt{\frac{\mu_n I_D}{2C_{ox}WL}}$$

Thus note that to obtain a high f_T from a given device, it must be operated at a high current. Also note that faster operation is obtained from smaller devices.

9.19 Starting from the expression for the MOSFET unity-gain frequency,

$$f_T = \frac{g_m}{2\pi(C_{gs} + C_{gd})}$$

and making the approximation that $C_{gs} \gg C_{gd}$ and that the overlap component of C_{gs} is negligibly small, show that for an n-channel device

$$f_T \simeq \frac{3\mu_n V_{OV}}{4\pi L^2}$$

Observe that for a given channel length, f_T can be increased by operating the MOSFET at a higher overdrive voltage. Evaluate f_T for devices with $L = 1.0\ \mu\text{m}$ operated at overdrive voltages of 0.25 V and 0.5 V. Use $\mu_n = 450\ \text{cm}^2/\text{Vs}$.

9.20 It is required to calculate the intrinsic gain A_0 and the unity-gain frequency f_T of an n-channel transistor fabricated in a 0.18-μm CMOS process for which $L_{ov} = 0.1\ L$, $\mu_n = 450\ \text{cm}^2/\text{V.s}$, and $V_A' = 5\ \text{V}/\mu\text{m}$. The device is operated at $V_{OV} = 0.2$ V. Find A_0 and f_T for devices with $L = L_{\min}$, $2L_{\min}$, $3L_{\min}$, $4L_{\min}$, and $5L_{\min}$. Present your results in a table.

9.21 A particular BJT operating at $I_C = 1$ mA has $C_\mu = 1$ pF, $C_\pi = 10$ pF, and $\beta = 100$. What are f_T and f_β for this situation?

9.22 For the transistor described in Problem 9.21, C_π includes a relatively constant depletion-layer capacitance of 2 pF. If the device is operated at $I_C = 0.2$ mA, what does its f_T become?

9.23 An *npn* transistor is operated at $I_C = 0.5$ mA and $V_{CB} = 2$ V. It has $\beta_0 = 100$, $V_A = 50$ V, $\tau_F = 30$ ps, $C_{je0} = 20$ fF, $C_{\mu 0} = 30$ fF, $V_{0c} = 0.75$ V, $m_{CBJ} = 0.5$, and $r_x = 100\ \Omega$. Sketch the complete hybrid-π model, and specify the values of all its components. Also, find f_T.

9.24 Measurement of h_{fe} of an *npn* transistor at 50 MHz shows that $|h_{fe}| = 10$ at $I_C = 0.2$ mA and 12 at $I_C = 1.0$ mA. Furthermore, C_μ was measured and found to be 0.1 pF. Find f_T at each of the two collector currents used. What must τ_F and C_{je} be?

9.25 A particular small-geometry BJT has f_T of 8 GHz and $C_\mu = 0.1$ pF when operated at $I_C = 1.0$ mA. What is C_π in this situation? Also, find g_m. For $\beta = 160$, find r_π and f_β.

9.26 For a BJT whose unity-gain bandwidth is 2 GHz and $\beta_0 = 200$, at what frequency does the magnitude of h_{fe} become 20? What is f_β?

***9.27** For a sufficiently high frequency, measurement of the complex input impedance of a BJT having (ac) grounded emitter and collector yields a real part approximating r_x. For what frequency, defined in terms of ω_β, is such an estimate of r_x good to within 10% under the condition that $r_x \le r_\pi/10$?

***9.28** Complete the table entries below for transistors (a) through (g), under the conditions indicated. Neglect r_x.

Transistor	I_E(mA)	r_e(Ω)	g_m (mA/V)	r_πk(Ω)	β_0	f_T(MHz)	C_μ (pF)	C_π (pF)	f_β(MHz)
(a)	1				100	400	2		
(b)		25					2	10.7	4
(c)				2.525		400		13.84	
(d)	10				100	400	2		
(e)	0.1				100	100	2		
(f)	1				10	400	2		
(g)						800	1	9	80

Section 9.3: High-Frequency Response of the CS and CE Amplifiers

9.29 In a particular common-source amplifier for which the midband voltage gain between gate and drain (i.e., $-g_m R'_L$) is −29 V/V, the NMOS transistor has $C_{gs} = 0.5$ pF and $C_{gd} = 0.1$ pF. What input capacitance would you expect? For what range of signal-source resistances can you expect the 3-dB frequency to exceed 10 MHz? Neglect the effect of R_G.

D 9.30 A design is required for a CS amplifier for which the MOSFET is operated at $g_m = 5$ mA/V and has $C_{gs} = 5$ pF and $C_{gd} = 1$ pF. The amplifier is fed with a signal source having $R_{sig} = 1$ kΩ, and R_G is very large. What is the largest value of R'_L for which the upper 3-dB frequency is at least 10 MHz? What is the corresponding value of midband gain and gain–bandwidth product? If the specification on the upper 3-dB frequency can be relaxed by a factor of 3, that is, to (10/3) MHz, what can A_M and GB become?

9.31 Reconsider Example 9.3 for the situation in which the transistor is replaced by one whose width W is half that of the original transistor while the bias current remains unchanged. Find modified values for all the device parameters along with A_M, f_H, and the gain–bandwidth product, GB. Contrast this with the original design by calculating the ratios of new value to old for $W, V_{OV}, g_m, C_{gs}, C_{gd}, C_{in}, A_M, f_H$, and GB.

D 9.32 In a CS amplifier, such as that in Fig. 9.2(a), the resistance of the source $R_{sig} = 100$ kΩ, amplifier input resistance (which is due to the biasing network) $R_{in} = 100$ kΩ, $C_{gs} = 1$ pF, $C_{gd} = 0.2$ pF, $g_m = 3$ mA/V, $r_o = 50$ kΩ, $R_D = 8$ kΩ, and $R_L = 10$ kΩ. Determine the expected 3-dB cutoff frequency f_H and the midband gain. In evaluating ways to double f_H, a designer considers the alternatives of changing either R_L or R_{in}. To raise f_H as described, what separate change in each would be required? What midband voltage gain results in each case?

9.33 A discrete MOSFET common-source amplifier has $R_G = 1$ MΩ, $g_m = 5$ mA/V, $r_o = 100$ kΩ, $R_D = 10$ kΩ, $C_{gs} = 2$ pF, and $C_{gd} = 0.4$ pF. The amplifier is fed from a voltage source with an internal resistance of 500 kΩ and is connected to a 10-kΩ load. Find:

(a) the overall midband gain A_M
(b) the upper 3-dB frequency f_H

9.34 The analysis of the high-frequency response of the common-source amplifier, presented in the text, is based on the assumption that the resistance of the signal source, R_{sig}, is large and, thus, that its interaction with the input capacitance C_{in} produces the "dominant pole" that determines the upper 3-dB frequency f_H. In some situations, however, the CS amplifier is fed with a very low R_{sig}. To investigate the high-frequency response of the amplifier in such a case, Fig. P9.34 shows the equivalent circuit when the CS amplifier is fed with an ideal voltage source V_{sig} having $R_{sig} = 0$. Note that C_L denotes the total capacitance at the output node. By writing a node equation at the output, show that the transfer function V_o/V_{sig} is given by

$$\frac{V_o}{V_{sig}} = -g_m R'_L \frac{1 - s(C_{gd}/g_m)}{1 + s(C_L + C_{gd})R'_L}$$

At frequencies $\omega \ll (g_m/C_{gd})$, the s term in the numerator can be neglected. In such case, what is the upper 3-dB frequency resulting? Compute the values of A_M and f_H for the case: $C_{gd} = 0.4$ pF, $C_L = 2$ pF, $g_m = 5$ mA/V, and $R'_L = 5$ kΩ.

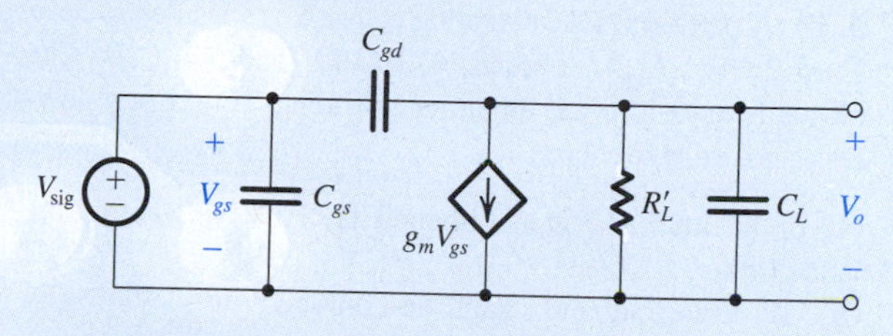

Figure P9.34

9.35 The NMOS transistor in the discrete CS amplifier circuit of Fig. P9.3 is biased to have $g_m = 1$ mA/V and $r_o = 100$ kΩ. Find A_M. If $C_{gs} = 1$ pF and $C_{gd} = 0.2$ pF, find f_H.

9.36 A designer wishes to investigate the effect of changing the bias current I on the midband gain and high-frequency response of the CE amplifier considered in Example 9.4. Let I be doubled to 2 mA, and assume that β_0 and f_T remain unchanged at 100 and 800 MHz, respectively. To keep the node voltages nearly unchanged, the designer reduces R_B and R_C by a factor of 2, to 50 kΩ and 4 kΩ, respectively. Assume $r_x = 50$ Ω, and recall that $V_A = 100$ V and that C_μ remains constant at 1 pF. As before, the amplifier is fed with a source having $R_{sig} = 5$ kΩ and feeds a load $R_L = 5$ kΩ. Find the new values of A_M, f_H, and the gain–bandwidth product, $|A_M| f_H$. Comment on the results. Note that the price paid for whatever improvement in performance is achieved is an increase in power. By what factor does the power dissipation increase?

***9.37** The purpose of this problem is to investigate the high-frequency response of the CE amplifier when it is fed with a relatively large source resistance R_{sig}. Refer to the amplifier in Fig. 9.4 (a) and to its high-frequency, equivalent-circuit model and the analysis shown in Fig. 9.14. Let $R_B \gg R_{sig}$, $r_x \ll R_{sig}$, $R_{sig} \gg r_\pi$, $g_m R'_L \gg 1$, and $g_m R'_L C_\mu \gg C_\pi$.

Under these conditions, show that:

(a) the midband gain $A_M \simeq -\beta R'_L / R_{sig}$
(b) the upper 3-dB frequency $f_H \simeq 1/2\pi C_\mu \beta R'_L$
(c) the gain–bandwidth product $A_M f_H \simeq 1/2\pi C_\mu R_{sig}$

Evaluate this approximate value of the gain–bandwidth product for the case $R_{sig} = 25$ kΩ and $C_\mu = 1$ pF. Now, if the transistor is biased at $I_C = 1$ mA and has $\beta = 100$, find the midband gain and f_H for the two cases $R'_L = 25$ kΩ and $R'_L = 2.5$ kΩ. On the same coordinates, sketch Bode plots for the gain magnitude versus frequency for the two cases. What f_H is obtained when the gain is unity? What value of R_L' corresponds?

9.38 For a version of the CE amplifier circuit in Fig. P9.11, $R_{sig} = 10$ kΩ, $R_1 = 68$ kΩ, $R_2 = 27$ kΩ, $R_E = 2.2$ kΩ, $R_C = 4.7$ kΩ, and $R_L = 10$ kΩ. The collector current is 0.8 mA, $\beta = 200$, $f_T = 1$ GHz, and $C_\mu = 0.8$ pF. Neglecting the effect of r_x and r_o, find the midband voltage gain and the upper 3-dB frequency f_H.

9.39 A particular BJT operating at 2 mA is specified to have $f_T = 2$ GHz, $C_\mu = 1$ pF, $r_x = 100\ \Omega$, and $\beta = 120$. The device is used in a CE amplifier operating from a very-low-resistance voltage source.

(a) If the midband gain obtained is -10 V/V, what is the value of f_H?
(b) If the midband gain is reduced to -1 V/V (by changing R_L'), what f_H is obtained?

9.40 Repeat Example 9.4 for the situation in which the power supplies are reduced to ± 5 V and the bias current is reduced to 0.5 mA. Assume that all other component values and transistor parameter values remain unchanged. Find A_M, f_H, and the gain–bandwidth product and compare to the values obtained in Example 9.4.

***9.41** The amplifier shown in Fig. P9.41 has $R_{sig} = R_L = 1$ kΩ, $R_C = 1$ kΩ, $R_B = 47$ kΩ, $\beta = 100$, $C_\mu = 0.8$ pF, and $f_T = 600$ MHz. Assume the coupling capacitors to be very large.

(a) Find the dc collector current of the transistor.
(b) Find g_m and r_π.
(c) Neglecting r_o, find the midband voltage gain from base to collector (neglect the effect of R_B).
(d) Use the gain obtained in (c) to find the component of R_{in} that arises as a result of R_B. Hence find R_{in}.
(e) Find the overall gain at midband.
(f) Find C_{in}.
(g) Find f_H.

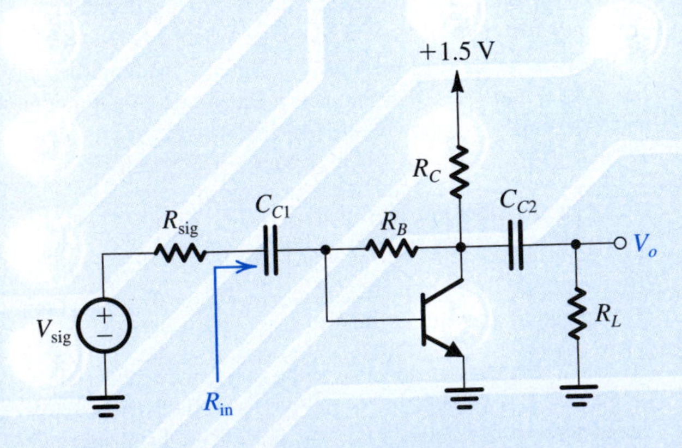

Figure P9.41

***9.42** Figure P9.42 shows a diode-connected transistor with the bias circuit omitted. Utilizing the BJT high-frequency, hybrid-π model with $r_x = 0$ and $r_o = \infty$, derive an expression for $Z_i(s)$ as a function of r_e and C_π. Find the frequency at which the impedance has a phase angle of 45° for the case in which the BJT has $f_T = 400$ MHz and the bias current is relatively high. What is the frequency when the bias current is reduced so that $C_\pi \simeq C_\mu$? Assume $\alpha = 1$.

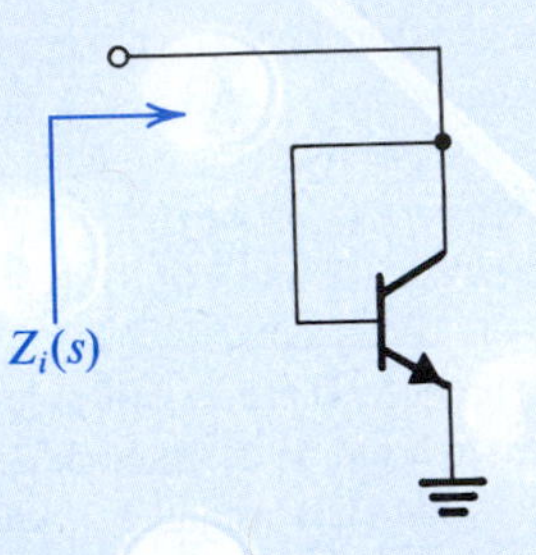

Figure P9.42

Section 9.4: Useful Tools for the Analysis of the High-Frequency Response of Amplifiers

9.43 A direct-coupled amplifier has a low-frequency gain of 40 dB, poles at 1 MHz and 10 MHz, a zero on the negative real axis at 100 MHz, and another zero at infinite frequency. Express the amplifier gain function in the form of Eqs. (9.61) and (9.62), and sketch a Bode plot for the gain magnitude. What do you estimate the 3-dB frequency f_H to be?

9.44 An amplifier with a dc gain of 60 dB has a single-pole high-frequency response with a 3-dB frequency of 10 kHz.

(a) Give an expression for the gain function $A(s)$.
(b) Sketch Bode diagrams for the gain magnitude and phase.
(c) What is the gain–bandwidth product?
(d) What is the unity-gain frequency?
(e) If a change in the amplifier circuit causes its transfer function to acquire another pole at 100 kHz, sketch the resulting gain magnitude and specify the unity-gain frequency. Note that this is an example of an amplifier with a unity-gain bandwidth that is different from its gain–bandwidth product.

9.45 Consider an amplifier whose $F_H(s)$ is given by

$$F_H(s) = \frac{1}{\left(1 + \frac{s}{\omega_{P1}}\right)\left(1 + \frac{s}{\omega_{P2}}\right)}$$

with $\omega_{P1} < \omega_{P2}$. Find the ratio ω_{P2}/ω_{P1} for which the value of the 3-dB frequency ω_H calculated using the dominant-pole approximation differs from that calculated using the root-sum-of-squares formula (Eq. 9.68) by:

(a) 10%
(b) 1%

9.46 The high-frequency response of a direct-coupled amplifier having a dc gain of −1000 V/V incorporates zeros at ∞ and 10^5 rad/s (one at each frequency) and poles at 10^4 rad/s and 10^6 rad/s (one at each frequency). Write an expression for the amplifier transfer function. Find ω_H using

(a) the dominant-pole approximation
(b) the root-sum-of-squares approximation (Eq. 9.68).

If a way is found to lower the frequency of the finite zero to 10^4 rad/s, what does the transfer function become? What is the 3-dB frequency of the resulting amplifier?

9.47 A direct-coupled amplifier has a dominant pole at 1000 rad/s and three coincident poles at a much higher frequency. These nondominant poles cause the phase lag of the amplifier at high frequencies to exceed the 90° angle due to the dominant pole. It is required to limit the excess phase at $\omega = 10^7$ rad/s to 30° (i.e., to limit the total phase angle to −120°). Find the corresponding frequency of the nondominant poles.

D 9.48 Refer to Example 9.6. Give an expression for ω_H in terms of C_{gs}, R'_{sig} (note that $R'_{sig} = R_G \| R_{sig}$), C_{gd}, R'_L, and g_m. If all component values except for the generator resistance R_{sig} are left unchanged, to what value must R_{sig} be reduced in order to raise f_H to 200 kHz?

9.49 (a) For the amplifier circuit in Example 9.6, find the expression for τ_H using symbols (as opposed to numbers).
(b) For the same circuit, use the approximate method of the previous section to determine an expression for C_{in} and hence the effective time constant $\tau = C_{in} R'_{sig}$ that can be used to find ω_H as $1/\tau$. Compare this expression of τ with that of τ_H in (a). What is the difference? Compute the value of the difference and express it as a percentage of τ.

9.50 If a capacitor $C_L = 20$ pF is connected across the output terminals of the amplifier in Example 9.6, find the resulting increase in τ_H and hence the new value of f_H.

9.51 A FET amplifier resembling that in Example 9.6, when operated at lower currents in a higher-impedance application, has $R_{sig} = 100$ kΩ, $R_{in} = 1.0$ MΩ, $g_m = 2$ mA/V, $R'_L = 15$ kΩ, and $C_{gs} = C_{gd} = 1$ pF. Find the midband voltage gain A_M and the 3-dB frequency f_H.

***9.52** Figure P9.52 shows the high-frequency equivalent circuit of a CS amplifier with a resistance R_s connected in the source lead. The purpose of this problem is to show that the value of R_s can be used to control the gain and bandwidth of the amplifier, specifically to allow the designer to trade gain for increased bandwidth.

(a) Derive an expression for the low-frequency voltage gain (set C_{gs} and C_{gd} to zero).
(b) To be able to determine ω_H using the open-circuit time-constants method, derive expressions for R_{gs} and R_{gd}.
(c) Let $R_{sig} = 100$ kΩ, $g_m = 4$ mA/V, $R'_L = 5$ kΩ, and $C_{gs} = C_{gd} = 1$ pF. Use the expressions found in (a) and (b) to determine the low-frequency gain and the 3-dB frequency f_H for three cases: $R_s = 0$ Ω, 100 Ω, and 250 Ω. In each case also evaluate the gain–bandwidth product. Comment.

9.53 A common-source MOS amplifier, whose equivalent circuit resembles that in Fig. 9.16(a), is to be evaluated for its high-frequency response. For this particular design, $R_{sig} = 1$ MΩ, $R_G = 4$ MΩ, $R'_L = 100$ kΩ, $C_{gs} = 0.2$ pF, $C_{gd} = 0.1$ pF, and $g_m = 0.5$ mA/V. Estimate the midband gain and the 3-dB frequency.

9.54 For a particular amplifier modeled by the circuit of Fig. 9.16(a), $g_m = 5$ mA/V, $R_{sig} = 150$ kΩ, $R_G = 0.65$ MΩ, $R'_L = 10$ kΩ, $C_{gs} = 2$ pF, and $C_{gd} = 0.5$ pF. There is also a load capacitance of 30 pF. Find the corresponding midband voltage gain, the open-circuit time constants, and an estimate of the 3-dB frequency.

9.55 Consider the high-frequency response of an amplifier consisting of two identical stages in cascade, each with an input resistance of 10 kΩ and an output resistance of 2 kΩ. The two-stage amplifier is driven from a 5-kΩ source and drives a

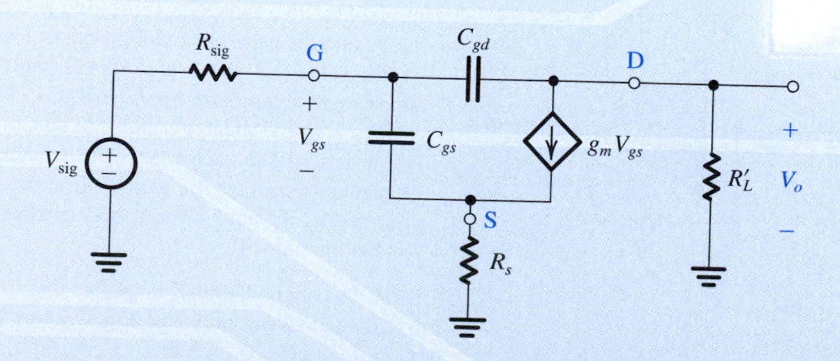

Figure P9.52

1-kΩ load. Associated with each stage is a parasitic input capacitance (to ground) of 10 pF and a parasitic output capacitance (to ground) of 2 pF. Parasitic capacitances of 5 pF and 7 pF also are associated with the signal-source and load connections, respectively. For this arrangement, find the three poles and estimate the 3-dB frequency f_H.

9.56 Consider an ideal voltage amplifier with a gain of 0.9 V/V and a resistance $R = 100$ kΩ connected in the feedback path—that is, between the output and input terminals. Use Miller's theorem to find the input resistance of this circuit.

9.57 An ideal voltage amplifier with a voltage gain of −1000 V/V has a 0.2-pF capacitance connected between its output and input terminals. What is the input capacitance of the amplifier? If the amplifier is fed from a voltage source V_{sig} having a resistance $R_{sig} = 1$ kΩ, find the transfer function V_o/V_{sig} as a function of the complex-frequency variable s and hence the 3-dB frequency f_H and the unity-gain frequency f_t.

9.58 The amplifiers listed below are characterized by the descriptor (A, C), where A is the voltage gain from input to output and C is an internal capacitor connected between input and output. For each, find the equivalent capacitances at the input and at the output as provided by the use of Miller's theorem:

(a) −1000 V/V, 1 pF
(b) −10 V/V, 10 pF
(c) −1 V/V, 10 pF
(d) +1 V/V, 10 pF
(e) +10 V/V, 10 pF

Note that the input capacitance found in case (e) can be used to cancel the effect of other capacitance connected from input to ground. In (e), what capacitance can be canceled?

9.59 Figure P9.59 shows an ideal voltage amplifier with a gain of +2 V/V (usually implemented with an op amp connected in the noninverting configuration) and a resistance R connected between output and input.

(a) Using Miller's theorem, show that the input resistance $R_{in} = -R$.

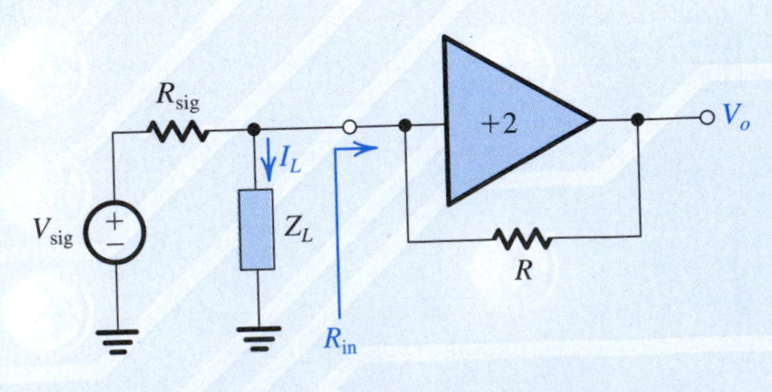

Figure P9.59

(b) Use Norton's theorem to replace V_{sig}, R_{sig}, and R_{in} with a signal current source and an equivalent parallel resistance. Show that by selecting $R_{sig} = R$, the equivalent parallel resistance becomes infinite and the current I_L into the load impedance Z_L becomes V_{sig}/R. The circuit then functions as an ideal voltage-controlled current source with an output current I_L.
(c) If Z_L is a capacitor C, find the transfer function V_o/V_{sig} and show it is that of an ideal noninverting integrator.

Section 9.5: A Closer Look at the High-Frequency Response of the CS and CE Amplifiers

9.60 A CS amplifier that can be represented by the equivalent circuit of Fig. 9.19 has $C_{gs} = 2$ pF, $C_{gd} = 0.1$ pF, $C_L = 2$ pF, $g_m = 4$ mA/V, and $R'_{sig} = R'_L = 20$ kΩ. Find the midband gain A_M, the input capacitance C_{in} using the Miller approximation, and hence an estimate of the 3-dB frequency f_H. Also, obtain a better estimate of f_H using Miller's theorem.

9.61 A CS amplifier that can be represented by the equivalent circuit of Fig. 9.19 has $C_{gs} = 2$ pF, $C_{gd} = 0.1$ pF, $C_L = 2$ pF, $g_m = 4$ mA/V, and $R'_{sig} = R'_L = 20$ kΩ. Find the midband A_M gain, and estimate the 3-dB frequency f_H using the method of open-circuit time constants. Also, give the percentage contribution to τ_H by each of three capacitances. (Note that this is the same amplifier considered in Problem 9.60; if you have solved Problem 9.60, compare your results.)

9.62 A CS amplifier represented by the equivalent circuit of Fig. 9.19 has $C_{gs} = 2$ pF, $C_{gd} = 0.1$, pF, $C_L = 2$ pF, $g_m = 4$ mA/V, and $R'_{sig} = R'_L = 20$ kΩ. Find the exact values of f_Z, f_{P1}, and f_{P2} using Eq. (9.88), and hence estimate f_H. Compare the values of f_{P1} and f_{P2} to the approximate values obtained using Eqs. (9.94) and (9.95). (Note that this is the same amplifier considered in Problems 9.60 and 9.61; if you have solved either or both of these problems, compare your results.)

9.63 A CS amplifier represented by the equivalent circuit of Fig. 9.19 has $C_{gs} = 2$ pF, $C_{gd} = 0.1$ pF, $C_L = 2$ pF, $g_m = 4$ mA/V, and $R'_{sig} = 20$ kΩ. It is required to find A_M, f_H, and the gain–bandwidth product for each of the following values of R'_L: 5 kΩ, 10 kΩ, and 20 kΩ. Use the approximate expression for f_{P1} in Eq. (9.94). However, in each case, also evaluate f_{P2} and f_Z to ensure that a dominant pole exists, and in each case, state whether the unity-gain frequency is equal to the gain–bandwidth product. Present your results in tabular form, and comment on the gain–bandwidth trade-off.

9.64 A common-emitter amplifier that can be represented by the equivalent circuit of Fig. 9.24(a) has $C_\pi = 10$ pF, $C_\mu = 0.3$ pF, $C_L = 3$ pF, $g_m = 40$ mA/V, $\beta = 100$,

$r_x = 100\ \Omega$, $R_L' = 5\ k\Omega$, and $R_{sig} = 1\ k\Omega$. Find the midband gain A_M, and an estimate of the 3-dB frequency f_H using the Miller approximation. Also, obtain a better estimate of f_H using Miller's theorem.

9.65 A common-emitter amplifier that can be represented by the equivalent circuit of Fig. 9.24(a) has $C_\pi = 10$ pF, $C_\mu = 0.3$ pF, $C_L = 3$ pF, $g_m = 40$ mA/V, $\beta = 100$, $r_x = 100\ \Omega$, $R_L = 5\ k\Omega$, and $R_{sig} = 1\ k\Omega$. Find the midband gain A_M, and estimate the 3-dB frequency f_H using the method of open-circuit time constants. Also give the percentage contribution to τ_H of each of the three capacitances. (Note that this is the same amplifier considered in Problem 9.64; if you have solved this problem, compare your results.)

9.66 A common-emitter amplifier that can be represented by the equivalent circuit of Fig. 9.24(a) has $C_\pi = 10$ pF, $C_\mu = 0.3$ pF, $C_L = 3$ pF, $g_m = 40$ mA/V, $\beta = 100$, $r_x = 100\ \Omega$, $R_L' = 5\ k\Omega$, and $R_{sig} = 1\ k\Omega$. Find the midband gain A_M, the frequency of the zero f_Z, and the values of the pole frequencies f_{P1} and f_{P2}. Hence, estimate the 3-dB frequency f_H. (Note that this is the same amplifier considered in Problems 6.64 and 9.65; if you have solved these problems, compare your results.)

***9.67** For the current mirror in Fig. P9.67, derive an expression for the current transfer function $I_o(s)/I_i(s)$ taking into account the BJT internal capacitances and neglecting r_x and r_o. Assume the BJTs to be identical. Observe that a signal ground appears at the collector of Q_2. If the mirror is biased at 1 mA and the BJTs at this operating point are characterized by $f_T = 400$ MHz, $C_\mu = 2$ pF, and $\beta_0 = \infty$, find the frequencies of the pole and zero of the transfer function.

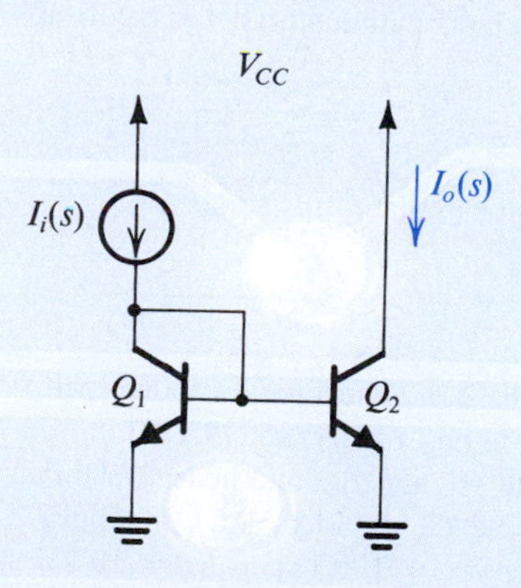

Figure P9.67

9.68 A CS amplifier modeled with the equivalent circuit of Fig 9.25(a) is specified to have $C_{gs} = 2$ pF, $C_{gd} = 0.1$ pF, $g_m = 4$ mA/V, $C_L = 2$ pF, and $R_L' = 20\ k\Omega$. Find A_M, f_{3dB}, and f_t.

***9.69** It is required to analyze the high-frequency response of the CMOS amplifier shown in Fig. P9.69. The dc bias

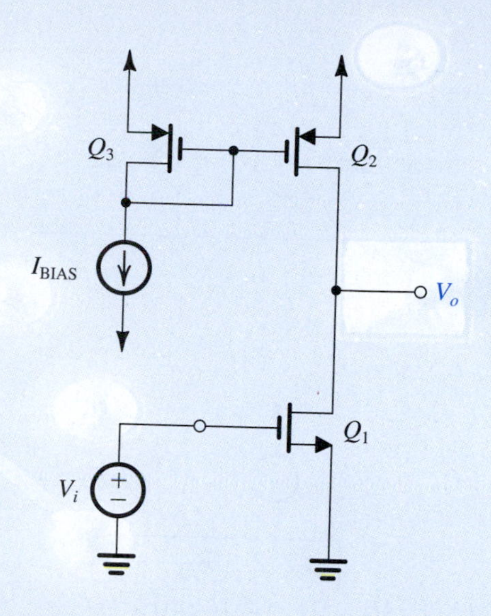

Figure P9.69

current is 100 μA. For Q_1, $\mu_n C_{ox} = 90\ \mu A/V^2$, $V_A = 12.8$ V, $W/L = 100\ \mu m/1.6\ \mu m$, $C_{gs} = 0.2$ pF, $C_{gd} = 0.015$ pF, and $C_{db} = 20$ fF. For Q_2, $C_{gd} = 0.015$ pF, $C_{db} = 36$ fF, and $|V_A| = 19.2$ V. Assume that the resistance of the input signal generator is negligibly small. Also, for simplicity, assume that the signal voltage at the gate of Q_2 is zero. Find the low-frequency gain, the frequency of the pole, and the frequency of the zero.

****9.70** This problem investigates the use of MOSFETs in the design of wideband amplifiers (Steininger, 1990). Such amplifiers can be realized by cascading low-gain stages.

(a) Show that for the case $C_{gd} \ll C_{gs}$ and the gain of the common-source amplifier is low so that the Miller effect is negligible, the MOSFET can be modeled by the approximate equivalent circuit shown in Fig. P9.70(a), where ω_T is the unity-gain frequency of the MOSFET.

(b) Figure P9.70(b) shows an amplifier stage suitable for the realization of low gain and wide bandwidth. Transistors Q_1 and Q_2 have the same channel length L but different widths W_1 and W_2. They are biased at the same V_{GS} and have the same f_T. Use the MOSFET equivalent circuit of Fig. P9.70(a) to model this amplifier stage assuming that its output is connected to the input of an identical stage. Show that the voltage gain V_o/V_i is given by

$$\frac{V_o}{V_i} = -\frac{G_0}{1 + \dfrac{s}{\omega_T/(G_0+1)}}$$

where

$$G_0 = \frac{g_{m1}}{g_{m2}} = \frac{W_1}{W_2}$$

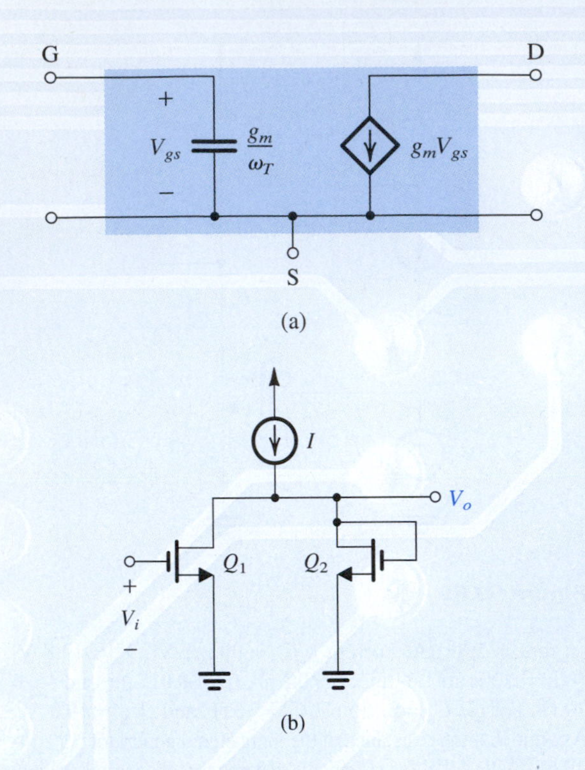

Figure P9.70

(c) For $L = 0.5$ μm, $W_2 = 25$ μm, $f_T = 12$ GHz, and $\mu_nC_{ox} = 200$ μA/V², design the circuit to obtain a gain of 3 V/V per stage. Bias the MOSFETs at $V_{OV} = 0.3$ V. Specify the required values of W_1 and I. What is the 3-dB frequency achieved?

9.71 Consider an active-loaded common-emitter amplifier. Let the amplifier be fed with an ideal voltage source V_i, and neglect the effect of r_x. Assume that the bias current source has a very high resistance and that there is a capacitance C_L present between the output node and ground. This capacitance represents the sum of the input capacitance of the subsequent stage and the inevitable parasitic capacitance between collector and ground. Show that the voltage gain is given by

$$\frac{V_o}{V_i} = -g_m r_o \frac{1 - s(C_\mu/g_m)}{1 + s(C_L + C_\mu)r_o}$$

If the transistor is biased at $I_C = 200$ μA and $V_A = 100$ V, $C_\mu = 0.2$ pF, and $C_L = 1$ pF, find the dc gain, the 3-dB frequency, the frequency of the zero, and the frequency at which the gain reduces to unity. Sketch a Bode plot for the gain magnitude.

9.72 A common-source amplifier fed with a low-resistance signal source and operating with $g_m = 2$ mA/V has a unity-gain frequency of 2 GHz. What additional capacitance must be connected to the drain node to reduce f_t to 1 GHz?

9.73 Consider a CS amplifier loaded in a current source with an output resistance equal to r_o of the amplifying transistor. The amplifier is fed from a signal source with $R_{sig} = r_o/2$. The transistor is biased to operate at $g_m = 2$ mA/V and $r_o = 20$ kΩ; $C_{gs} = C_{gd} = 0.1$ pF. Use the Miller approximation to determine an estimate of f_H. Repeat for the following two cases: (i) the bias current I in the entire system is reduced by a factor of 4, and (ii) the bias current I in the entire system is increased by a factor of 4. Remember that both R_{sig} and R_L will change as r_o changes.

9.74 Use the method of open-circuit time constants to find f_H for a CS amplifier for which $g_m = 1.5$ mA/V, $C_{gs} = C_{gd} = 0.2$ pF, $r_o = 20$ kΩ, $R_L = 12$ kΩ, and $R_{sig} = 100$ kΩ for the following cases: (a) $C_L = 0$, (b) $C_L = 10$ pF, and (c) $C_L = 50$ pF. Compare with the value of f_H obtained using the Miller approximation.

Section 9.6: High-Frequency Response of the Common-Gate and Cascode Amplifiers

9.75 A CG amplifier is specified to have $C_{gs} = 2$ pF, $C_{gd} = 0.1$ pF, $C_L = 2$ pF, $g_m = 4$ mA/V, $R_{sig} = 1$ kΩ, and $R_L' = 20$ kΩ. Neglecting the effects of r_o, find the low-frequency gain v_o/v_{sig}, the frequencies of the poles f_{P1} and f_{P2}, and hence an estimate of the 3-dB frequency f_H.

***9.76** Sketch the high-frequency equivalent circuit of a CB amplifier fed from a signal generator characterized by V_{sig} and R_{sig} and feeding a load resistance R_L in parallel with a capacitance C_L.

(a) Show that for $r_o = \infty$ the circuit can be separated into two parts: an input part that produces a pole at

$$f_{P1} = \frac{1}{2\pi C_\pi (R_{sig} \| r_e)}$$

and an output part that forms a pole at

$$f_{P2} = \frac{1}{2\pi (C_\mu + C_L) R_L}$$

Note that these are the bipolar counterparts of the MOS expressions in Eqs. (9.109) and (9.110).

(b) Evaluate f_{P1} and f_{P2} and hence obtain an estimate for f_H for the case $C_\pi = 14$ pF, $C_\mu = 2$ pF, $C_L = 1$ pF, $I_C = 1$ mA, $R_{sig} = 1$ kΩ, and $R_L = 10$ kΩ. Also, find f_T of the transistor.

***9.77** Consider a CG amplifier loaded in a resistance $R_L = r_o$ and fed with a signal source having a resistance $R_{sig} = r_o/2$. Also let $C_L = C_{gs}$. Use the method of open-circuit time constants to show that for $g_m r_o \gg 1$, the upper 3-dB frequency is related to the MOSFET f_T by the approximate expression

$$f_H = f_T/(g_m r_o)$$

9.78 For the CG amplifier in Example 9.12, how much additional capacitance should be connected between the output node and ground to reduce f_H to 300 MHz?

9.79 Find the dc gain and the 3-dB frequency of a MOS cascode amplifier operated at $g_m = 1$ mA/V and $r_o = 50\ \text{k}\Omega$. The MOSFETs have $C_{gs} = 30$ fF, $C_{gd} = 10$ fF, and $C_{db} = 10$ fF. The amplifier is fed from a signal source with $R_{sig} = 100\ \text{k}\Omega$ and is connected to a load resistance of 2 MΩ. There is also a load capacitance C_L of 40 fF.

***9.80** (a) Consider a CS amplifier having $C_{gd} = 0.2$ pF, $R_{sig} = R_L = 20\ \text{k}\Omega$, $g_m = 4$ mA/V, $C_{gs} = 2$ pF, C_L (including C_{db}) = 1 pF, $C_{db} = 0.2$ pF, and $r_o = 20\ \text{k}\Omega$. Find the low-frequency gain A_M, and estimate f_H using open-circuit time constants. Hence determine the gain–bandwidth product.
(b) If a CG stage is cascaded with the CS transistor in (a) to create a cascode amplifier, determine the new values of A_M, f_H, and gain–bandwidth product. Assume R_L remains unchanged.

D 9.81 It is required to design a cascode amplifier to provide a dc gain of 74 dB when driven with a low-resistance generator and utilizing NMOS transistors for which $V_A = 10$ V, $\mu_n C_{ox} = 200\ \mu\text{A/V}^2$, $W/L = 50$, $C_{gd} = 0.1$ pF, and $C_L = 1$ pF. Assuming that $R_L = R_o$, determine the overdrive voltage and the drain current at which the MOSFETs should be operated. Find the unity-gain frequency and the 3-dB frequency. If the cascode transistor is removed and R_L remains unchanged, what will the dc gain become?

9.82 Consider a bipolar cascode amplifier biased at a current of 1 mA. The transistors used have $\beta = 100$, $r_o = 100\ \text{k}\Omega$, $C_\pi = 14$ pF, $C_\mu = 2$ pF, $C_{cs} = 0$, and $r_x = 50\ \Omega$. The amplifier is fed with a signal source having $R_{sig} = 4\ \text{k}\Omega$. The load resistance $R_L = 2.4\ \text{k}\Omega$. Find the low-frequency gain A_M, and estimate the value of the 3-dB frequency f_H.

***9.83** In this problem we consider the frequency response of the bipolar cascode amplifier in the case that r_o can be neglected.

(a) Refer to the circuit in Fig. 9.31, and note that the total resistance between the collector of Q_1 and ground will be equal to r_{e2}, which is usually very small. It follows that the pole introduced at this node will typically be at a very high frequency and thus will have negligible effect on f_H. It also follows that at the frequencies of interest the gain from the base to the collector of Q_1 will be $-g_{m1}r_{e2} \simeq -1$. Use this to find the capacitance at the input of Q_1 and hence show that the pole introduced at the input node will have a frequency

$$f_{P1} \simeq \frac{1}{2\pi R'_{sig}(C_{\pi 1} + 2C_{\mu 1})}$$

Then show that the pole introduced at the output node will have a frequency

$$f_{P2} \simeq \frac{1}{2\pi R_L(C_L + C_{cs2} + C_{\mu 2})}$$

(b) Evaluate f_{P1} and f_{P2}, and use the sum-of-the-squares formula to estimate f_H for the amplifier with $I = 1$ mA, $C_\pi = 5$ pF, $C_\mu = 5$ pF, $C_{cs} = C_L = 0$, $\beta = 100$, and $r_x = 0$ in the following two cases:

(i) $R_{sig} = 1\ \text{k}\Omega$

(ii) $R_{sig} = 10\ \text{k}\Omega$

9.84 A BJT cascode amplifier uses transistors for which $\beta = 100$, $V_A = 100$ V, $f_T = 1$ GHz, and $C_\mu = 0.1$ pF. It operates at a bias current of 0.1 mA between a source with $R_{sig} = r_\pi$ and a load $R_L = \beta r_o$. Let $C_L = C_{cs} = 0$ and find the overall voltage gain at dc, f_H, and f_t.

Section 9.7: High-Frequency Response of the Source and Emitter Followers

9.85 A source follower has $g_m = 5$ mA/V, $r_o = 20\ \text{k}\Omega$, $R_{sig} = 20\ \text{k}\Omega$, $R_L = 2\ \text{k}\Omega$, $C_{gs} = 2$ pF, $C_{gd} = 0.1$ pF, and $C_L = 1$ pF. Find A_M, R_o, f_Z, and f_H. Also, find the percentage contribution of each of the three capacitances to the time-constant τ_H.

9.86 Using the expression for the source follower f_H in Eq. (9.129) show that for situations in which R_{sig} is large and R_L is small,

$$f_H \simeq \frac{1}{2\pi R_{sig}\left[C_{gd} + \dfrac{C_{gs}}{1 + g_m R'_L}\right]}$$

Find f_H for the case $R_{sig} = 100\ \text{k}\Omega$, $R_L = 1\ \text{k}\Omega$, $r_o = 20\ \text{k}\Omega$, $g_m = 5$ mA/V, $C_{gd} = 10$ fF, and $C_{gs} = 30$ fF.

9.87 Refer to Fig. 9.32(b). In situations in which R_{sig} is large, the high-frequency response of the source follower is determined by the low-pass circuit formed by R_{sig} and the input capacitance. An estimate of C_{in} can be obtained by using the Miller approximation to replace C_{gs} with an input capacitance $C_{eq} = C_{gs}(1 - K)$ where K is the gain from gate to source. Using the low-frequency value of $K = g_m R'_L/(1 + g_m R'_L)$ find C_{eq} and hence C_{in} and an estimate of f_H. Is this estimate higher or lower than that obtained by the method of open-circuit time constants?

9.88 For an emitter follower biased at $I_C = 1$ mA and having $R_{sig} = R_L = 1\ \text{k}\Omega$, and using a transistor specified to have $f_T = 2$ GHz, $C_\mu = 0.1$ pF, $r_x = 100\ \Omega$, $\beta = 100$, and $V_A = 20$ V, evaluate the low-frequency gain A_M and the 3-dB frequency f_H.

***9.89** For the emitter follower shown in Fig. P9.89, find the low-frequency gain and the 3-dB frequency f_H for the following three cases:

(a) $R_{sig} = 1\ k\Omega$
(b) $R_{sig} = 10\ k\Omega$
(c) $R_{sig} = 100\ k\Omega$

Let $\beta = 100$, $f_T = 400$ MHz, and $C_\mu = 2$ pF.

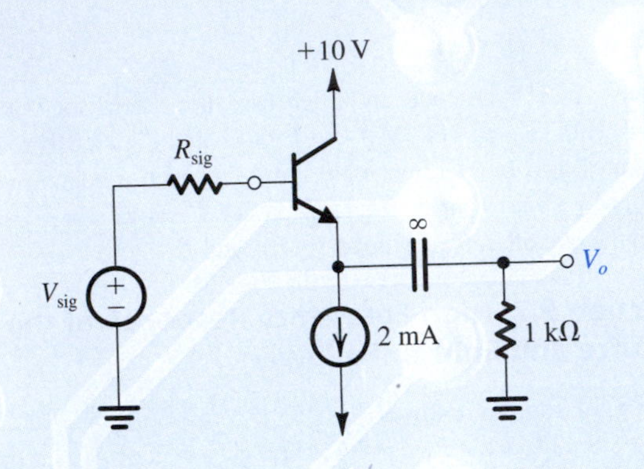

Figure P9.89

Section 9.8: High-Frequency Response of Differential Amplifiers

9.90 A MOSFET differential amplifier such as that shown in Fig. 9.34(a) is biased with a current source $I = 200\ \mu A$. The transistors have $W/L = 25$, $k'_n = 200\ \mu A/V^2$, $V_A = 200$ V, $C_{gs} = 40$ fF, $C_{gd} = 5$ fF, and $C_{db} = 5$ fF. The drain resistors are 20 kΩ each. Also, there is a 100-fF capacitive load between each drain and ground.

(a) Find V_{OV} and g_m for each transistor.
(b) Find the differential gain A_d.
(c) If the input signal source has a small resistance R_{sig} and thus the frequency response is determined primarily by the output pole, estimate the 3-dB frequency f_H.
(d) If, in a different situation, the amplifier is fed symmetrically with a signal source of 40 kΩ resistance (i.e., 20 kΩ in series with each gate terminal), use the open-circuit time-constants method to estimate f_H.

9.91 The amplifier specified in Problem 9.90 has $R_{SS} = 80\ k\Omega$ and $C_{SS} = 0.1$ pF. Find the 3-dB frequency of the CMRR.

9.92 In a particular MOS differential amplifier design, the bias current $I = 100\ \mu A$ is provided by a single transistor operating at $V_{OV} = 0.5$ V with $V_A = 30$ V and output capacitance C_{SS} of 100 fF. What is the frequency of the common-mode gain zero (f_Z) at which A_{cm} begins to rise above its low-frequency value? To meet a requirement for reduced power supply, consideration is given to reducing V_{OV} to 0.2 V while keeping I unchanged. Assuming the current-source capacitance to be directly proportional to the device width, what is the impact on f_Z of this proposed change?

9.93 Repeat Exercise 9.27 for the situation in which the bias current is reduced to 80 μA and R_D is raised to 20 kΩ. For (d), let R_{sig} be raised from 20 kΩ to 100 kΩ.
(*Note*: This is a low-voltage, low-power design.)

9.94 A BJT differential amplifier operating with a 1-mA current source uses transistors for which $\beta = 100$, $f_T = 600$ MHz, $C_\mu = 0.5$ pF, and $r_x = 100\ \Omega$. Each of the collector resistances is 10 kΩ, and r_o is very large. The amplifier is fed in a symmetrical fashion with a source resistance of 10 kΩ in series with each of the two input terminals.

(a) Sketch the differential half-circuit and its high-frequency equivalent circuit.
(b) Determine the low-frequency value of the overall differential gain.
(c) Use the Miller approximation to determine the input capacitance and hence estimate the 3-dB frequency f_H and the gain–bandwidth product.

9.95 A differential amplifier is biased by a current source having an output resistance of 1 MΩ and an output capacitance of 1 pF. The differential gain exhibits a dominant pole at 2 MHz. What are the poles of the CMRR?

9.96 A current-mirror-loaded MOS differential amplifier is biased with a current source $I = 0.2$ mA. The two NMOS transistors of the differential pair are operating at $V_{OV} = 0.2$ V, and the PMOS devices of the mirror are operating at $|V_{OV}| = 0.2$ V. The Early voltage $V_{An} = |V_{Ap}| = 10$ V. The total capacitance at the input node of the mirror is 0.1 pF and that at the output node of the amplifier is 0.2 pF. Find the dc value and the frequencies of the poles and zero of the differential voltage gain.

9.97 Consider the active-loaded CMOS differential amplifier of Fig. 9.37(a) for the case of all transistors operated at the same $|V_{OV}|$ and having the same $|V_A|$. Also let the total capacitance at the output node (C_L) be four times the total capacitance at the input node of the current mirror C_m, and show that the unity-gain frequency of A_d is $g_m/2\pi C_L$. For $V_A = 20$ V, $V_{OV} = 0.2$ V, $I = 0.2$ mA, $C_L = 100$ fF, and $C_m = 25$ fF, find the dc value of A_d, and the value of f_{P1}, f_t, f_{P2}, and f_Z and sketch a Bode plot for $|A_d|$.

Section 9.9: Other Wideband Amplifier Configurations

9.98 A CS amplifier is specified to have $g_m = 5$ mA/V, $r_o = 40\ k\Omega$, $C_{gs} = 2$ pF, $C_{gd} = 0.1$ pF, $C_L = 1$ pF, $R_{sig} = 20\ k\Omega$, and $R_L = 40\ k\Omega$.

(a) Find the low-frequency gain A_M, and use open-circuit time constants to estimate the 3-dB frequency f_H. Hence determine the gain–bandwidth product.

(b) If a 500-Ω resistance is connected in the source lead, find the new values of $|A_M|$, f_H, and the gain–bandwidth product.

D 9.99 (a) Use the approximate expression in Eq. (9.161) to determine the gain–bandwidth product of a CS amplifier with a source-degeneration resistance. Assume $C_{gd} = 0.1$ pF and $R_{sig} = 10$ kΩ.

(b) If a low-frequency gain of 20 V/V is required, what f_H corresponds?

(c) For $g_m = 5$ mA/V, $A_0 = 100$ V/V, and $R_L = 20$ kΩ, find the required value of R_s.

9.100 For the CS amplifier with a source-degeneration resistance R_s, show for $R_{sig} \gg R_s$ and $R_L = r_o$ that

$$\tau_H \simeq \frac{C_{gs}R_{sig}}{1+(k/2)} + C_{gd}R_{sig}\left(1+\frac{A_0}{2+k}\right) + (C_L + C_{gd})r_o\left(\frac{1+k}{2+k}\right)$$

where $k \equiv g_m R_s$

D *9.101 It is required to generate a table of $|A_M|$, f_H, and f_t versus $k \equiv g_m R_s$ for a CS amplifier with a source-degeneration resistance R_s. The table should have entries for $k = 0, 1, 2, \ldots, 15$. The amplifier is specified to have $g_m = 5$ mA/V, $r_o = 40$ kΩ, $R_L = 40$ kΩ, $R_{sig} = 20$ kΩ, $C_{gs} = 2$ pF, $C_{gd} = 0.1$ pF, and $C_L = 1$ pF. Use the formula for τ_H given in the statement for Problem 9.100. If $f_H = 2$ MHz is required, find the value needed for R_s and the corresponding value of $|A_M|$.

***9.102** In this problem we investigate the bandwidth extension obtained by placing a source follower between the signal source and the input of the CS amplifier.

(a) First consider the CS amplifier of Fig. P9.102(a). Show that

$$A_M = -g_m r_o$$

$$\tau_H = C_{gs}R_{sig} + C_{gd}[R_{sig}(1+g_m r_o) + r_o] + C_L r_o$$

where C_L is the total capacitance between the output node and ground. Calculate the value of A_M, f_H, and the gain–bandwidth product for the case $g_m = 1$ mA/V, $r_o = 20$ kΩ, $R_{sig} = 20$ kΩ, $C_{gs} = 20$ fF, $C_{gd} = 5$ fF, and $C_L = 10$ fF.

(b) For the CD-CS amplifier in Fig. P9.102(b), show that

$$A_M = -\frac{r_{o1}}{1/g_{m1} + r_{o1}}(g_{m2}r_{o2})$$

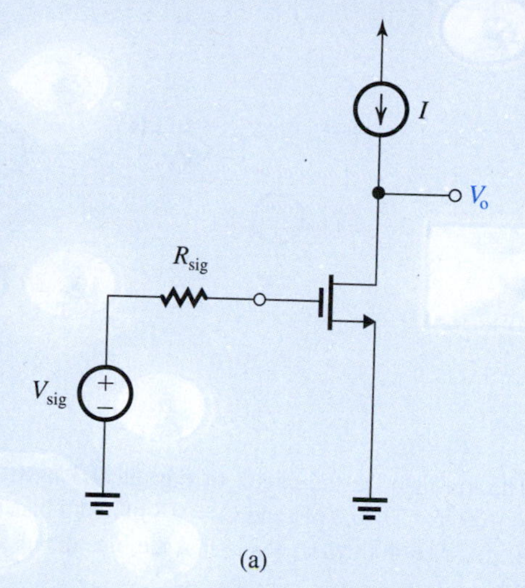

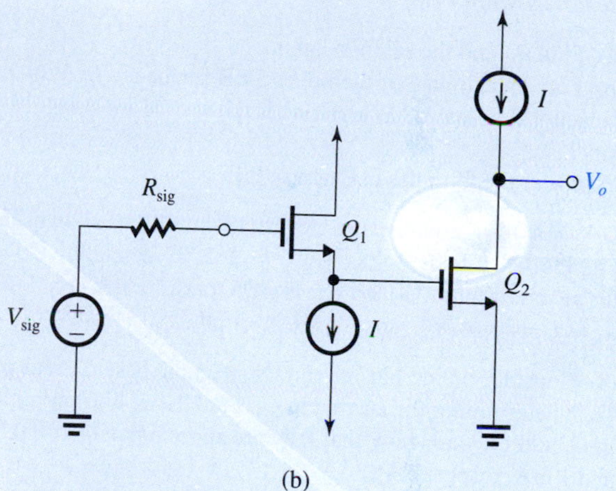

Figure P9.102

$$\tau_H = C_{gd1}R_{sig} + C_{gs1}\frac{R_{sig}+r_{o1}}{1+g_{m1}r_{o1}} + C_{gs2}\left(\frac{1}{g_{m1}} \| r_{o1}\right) + C_{gd2}\left[\left(\frac{1}{g_{m1}} \| r_{o1}\right)(1+g_{m2}r_{o2}) + r_{o2}\right] + C_L r_{o2}$$

Calculate the values of A_M, f_H, and the gain–bandwidth product for the same parameter values used in (a). Compare with the results of (a).

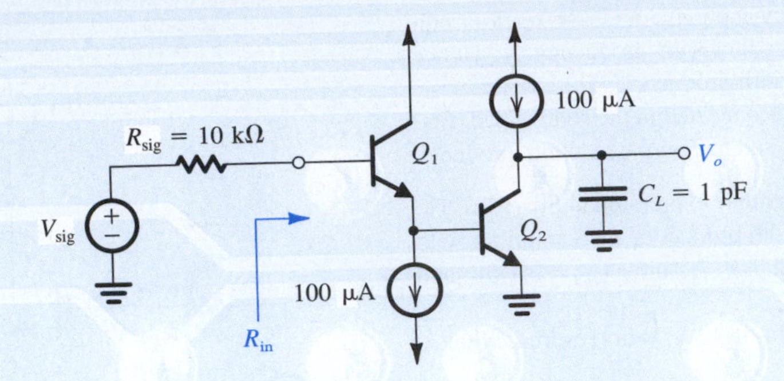

Figure P9.103

D *9.103 The transistor in the circuit of Fig. P9.103 have $\beta_0 = 100$, $V_A = 100$ V, $C_\mu = 0.2$ pF, and $C_{je} = 0.8$ pF. At a bias current of 100 μA, $f_T = 400$ MHz. (Note that the bias details are not shown.)

(a) Find R_{in} and the midband gain.
(b) Find an estimate of the upper 3-dB frequency f_H. Which capacitor dominates? Which one is the second most significant?
(*Hint*. Use the formulas in Example 9.15.)

D **9.104 Consider the BiCMOS amplifier shown in Fig. P9.104. The BJT has $|V_{BE}| = 0.7$ V, $\beta = 200$, $C_\mu = 0.8$ pF, and $f_T = 600$ MHz. The NMOS transistor has $V_t = 1$ V, $k_n' W/L = 2$ mA/V^2, and $C_{gs} = C_{gd} = 1$ pF.

(a) Consider the dc bias circuit. Neglect the base current of Q_2 in determining the current in Q_1. Find the dc bias currents in Q_1 and Q_2, and show that they are approximately 100 μA and 1 mA, respectively.
(b) Evaluate the small-signal parameters of Q_1 and Q_2 at their bias points.
(c) Consider the circuit at midband frequencies. First, determine the small-signal voltage gain V_o/V_i. (Note that R_G can be neglected in this process.) Then use Miller's theorem on R_G to determine the amplifier input resistance R_{in}. Finally, determine the overall voltage gain V_o/V_{sig}.
(d) Consider the circuit at low frequencies. Determine the frequency of the poles due to C_1 and C_2, and hence estimate the lower 3-dB frequency, f_L.
(e) Consider the circuit at higher frequencies. Use Miller's theorem to replace R_G with a resistance at the input. (The one at the output will be too large to matter.) Use open-circuit time constants to estimate f_H.
(f) To considerably reduce the effect of R_G on R_{in} and hence on amplifier performance, consider the effect of adding another 10-MΩ resistor in series with the existing one and placing a large bypass capacitor between their joint node and ground. What will R_{in}, A_M, and f_H become?

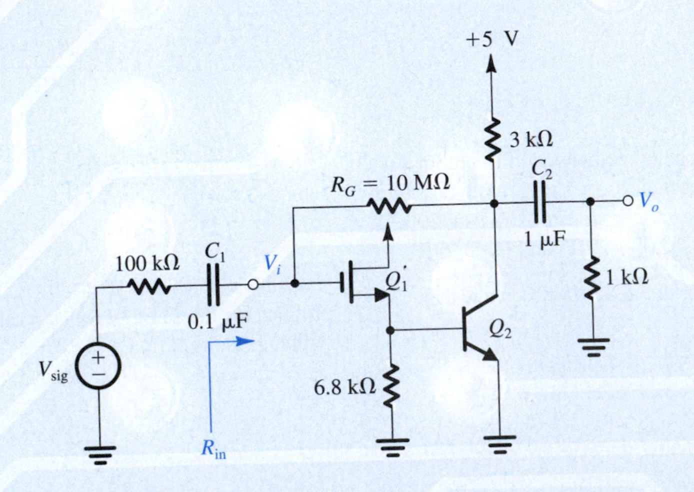

Figure P9.104

9.105 Consider the circuit of Fig. P9.105 for the case: $I =$ 200 μA and $V_{OV} = 0.2$ V, $R_{sig} = 200$ kΩ, $R_D = 50$ kΩ, $C_{gs} = C_{gd} =$ 1 pF. Find the dc gain, the high-frequency poles, and an estimate of f_H.

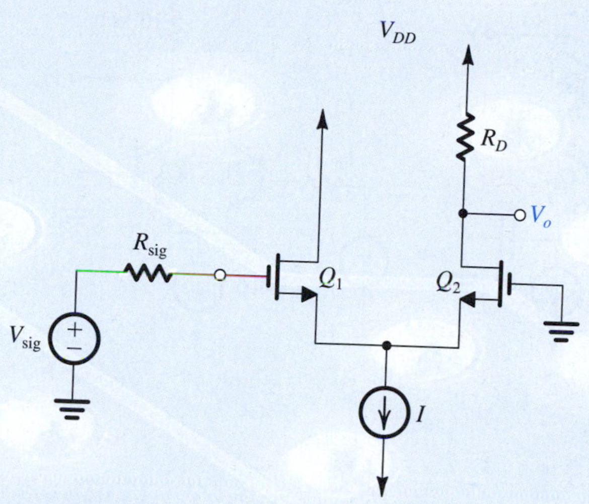

Figure P9.105

9.106 For the amplifier in Fig. 9.41(a), let $I = 1$ mA, $\beta =$ 120, $f_T = 700$ MHz, and $C_\mu = 0.5$ pF, and neglect r_x and r_o. Assume that a load resistance of 10 kΩ is connected to the output terminal. If the amplifier is fed with a signal V_{sig} having a source resistance $R_{sig} = 20$ kΩ, find A_M and f_H.

9.107 Consider the CD–CG amplifier of Fig. 9.41(c) for the case $g_m = 5$ mA/V, $C_{gs} = 2$ pF, $C_{gd} = 0.1$ pF, C_L (at the output node) = 1 pF, and $R_{sig} = R_L = 20$ kΩ. Neglecting r_o, find A_M, and f_H.

*** **9.108** In each of the six circuits in Fig. P9.108 (p. 800), let $\beta = 100$, $C_\mu = 2$ pF, and $f_T = 400$ MHz, and neglect r_x and r_o. Calculate the midband gain A_M and the 3-dB frequency f_H.

Section 9.10: Multistage Amplifier Examples

9.109 Use open-circuit time constants to obtain an expression for ω_H of the amplifier in Fig. 9.44. Compare to the expression in Eq. (9.176).

9.110 For the CMOS amplifier in Fig. 9.43, whose equivalent circuit is shown in Fig. 9.44, let $G_{m1} = 1$ mA/V, $R_1 =$ 100 kΩ, $C_1 = 0.1$ pF, $G_{m2} = 2$ mA/V, $R_2 = 50$ kΩ, and $C_2 = 2$ pF.

(a) Find the dc gain.
(b) Without C_C connected, find the frequencies of the two poles in radians per seconds and sketch a Bode plot for the gain magnitude.
(c) With C_C connected, find ω_{P2}. Then find the value of C_C that will result in a unity-gain frequency ω_t at least two octaves below ω_{P2}. For this value of C_C, find ω_{P1} and ω_Z and sketch a Bode plot for the gain magnitude.

9.111 A CMOS op amp with the topology in Fig. 9.43 has $g_{m1} = g_{m2} = 1$ mA/V, $g_{m6} = 3$ mA/V, the total capacitance between node D_2 and ground is 0.2 pF, and the total capacitance between the output node and ground is 3 pF. Find the value of C_C that results in $f_t = 50$ MHz and verify that f_t is lower than f_Z and f_{P2}.

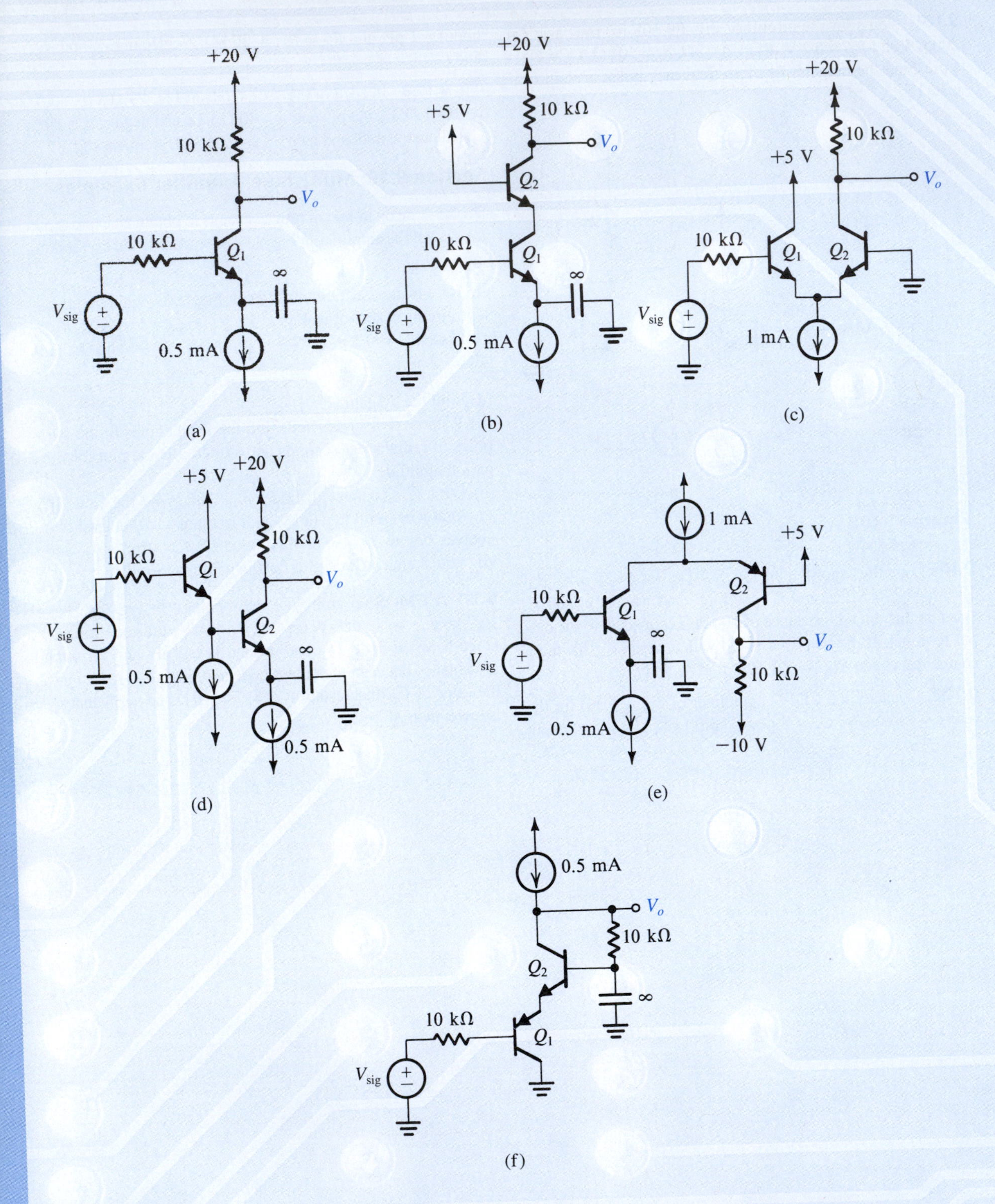

Figure P9.108

9.112 Figure P9.112 shows an amplifier formed by cascading two CS stages. Note that the input bias voltage is not shown. Each of Q_1 and Q_2 is operated at an overdrive voltage of 0.2 V, and $|V_A| = 10$ V. The transistor capacitances are as follows: $C_{gs} = 20$ fF, $C_{gd} = 5$ fF, and $C_{db} = 5$ fF.

(a) Find the dc voltage gain.
(b) Find the input capacitance at the gate of Q_1, using the Miller approximation.
(c) Use the capacitance in (b) to determine the frequency of the pole formed at the amplifier input. Let $R_{sig} = 10$ kΩ.
(d) Use the Miller approximation to find the input capacitance of Q_2 and hence determine the total capacitance at the drain of Q_1.
(e) Use the capacitance found in (d) to obtain the frequency of the pole formed at the interface between the two stages.
(f) Determine the total capacitance at the output node and hence estimate the frequency of the pole formed at the output node.
(g) Does the amplifier have a dominant pole? If so, at what frequency

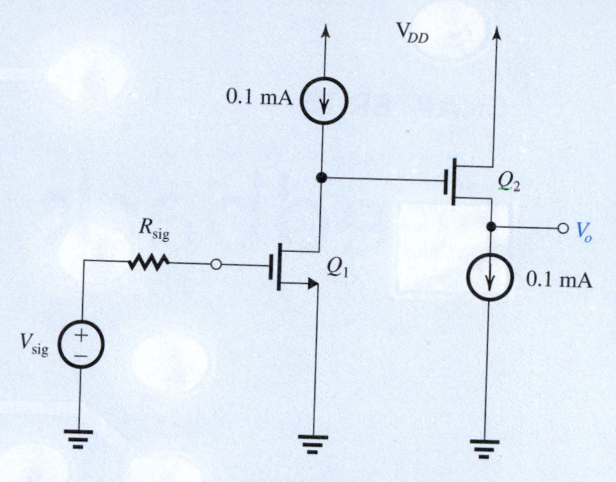

Figure P9.112

CHAPTER 10

Feedback

IN THIS CHAPTER YOU WILL LEARN

1. The general structure of the negative-feedback amplifier and the basic principle that underlies its operation.
2. The advantages of negative feedback, how these come about, and at what cost.
3. The appropriate feedback topology to employ with each of the four amplifier types: voltage, current, transconductance, and transresistance amplifiers.
4. An intuitive and insightful approach for the analysis of practical feedback-amplifier circuits.
5. Why and how negative-feedback amplifiers can become unstable (i.e., oscillate) and how to design the circuit to ensure stable performance.

Introduction

Most physical systems incorporate some form of feedback. It is interesting to note, though, that the theory of negative feedback has been developed by electronics engineers. In his search for methods for the design of amplifiers with stable gain for use in telephone repeaters, Harold Black, an electronics engineer with the Western Electric Company, invented the feedback amplifier in 1928. Since then the technique has been so widely used that it is almost impossible to think of electronic circuits without some form of feedback, either implicit or explicit. Furthermore, the concept of feedback and its associated theory are currently used in areas other than engineering, such as in the modeling of biological systems.

Feedback can be either **negative (degenerative)** or **positive (regenerative)**. In amplifier design, negative feedback is applied to effect one or more of the following properties:

1. *Desensitize the gain:* that is, make the value of the gain less sensitive to variations in the values of circuit components, such as might be caused by changes in temperature.
2. *Reduce nonlinear distortion:* that is, make the output proportional to the input (in other words, make the gain constant, independent of signal level).
3. *Reduce the effect of noise:* that is, minimize the contribution to the output of unwanted electric signals generated, either by the circuit components themselves, or by extraneous interference.

4. *Control the input and output resistances:* that is, raise or lower the input and output resistances by the selection of an appropriate feedback topology.
5. *Extend the bandwidth* of the amplifier.

All of the desirable properties above are obtained at the expense of a reduction in gain. It will be shown that the gain-reduction factor, called the **amount of feedback**, is the factor by which the circuit is desensitized, by which the input resistance of a voltage amplifier is increased, by which the bandwidth is extended, and so on. In short, *the basic idea of negative feedback is to trade off gain for other desirable properties.* This chapter is devoted to the study of negative-feedback amplifiers: their analysis, design, and characteristics.

Under certain conditions, the negative feedback in an amplifier can become positive and of such a magnitude as to cause oscillation. In fact, in Chapter 17 we will study the use of positive feedback in the design of oscillators and bistable circuits. Here, in this chapter, however, we are interested in the design of stable amplifiers. We shall therefore study the stability problem of negative-feedback amplifiers and their potential for oscillation.

It should not be implied, however, that positive feedback always leads to instability. In fact, positive feedback is quite useful in a number of nonregenerative applications, such as the design of active filters, which are studied in Chapter 16.

Before we begin our study of negative feedback, we wish to remind the reader that we have already encountered negative feedback in a number of applications. Almost all op-amp circuits (Chapter 2) employ negative feedback. Another popular application of negative feedback is the use of the emitter resistance R_E to stabilize the bias point of bipolar transistors and to increase the input resistance, bandwidth, and linearity of a BJT amplifier. In addition, the source follower and the emitter follower both employ a large amount of negative feedback. The question then arises about the need for a formal study of negative feedback. As will be appreciated by the end of this chapter, the formal study of feedback provides an invaluable tool for the analysis and design of electronic circuits. Also, the insight gained by thinking in terms of feedback can be extremely profitable.

10.1 The General Feedback Structure

Figure 10.1 shows the basic structure of a feedback amplifier. Rather than showing voltages and currents, Fig. 10.1 is a **signal-flow diagram**, where each of the quantities x can represent either a voltage or a current signal. The *open-loop* amplifier has a gain A; thus its output x_o is related to the input x_i by

$$x_o = Ax_i \tag{10.1}$$

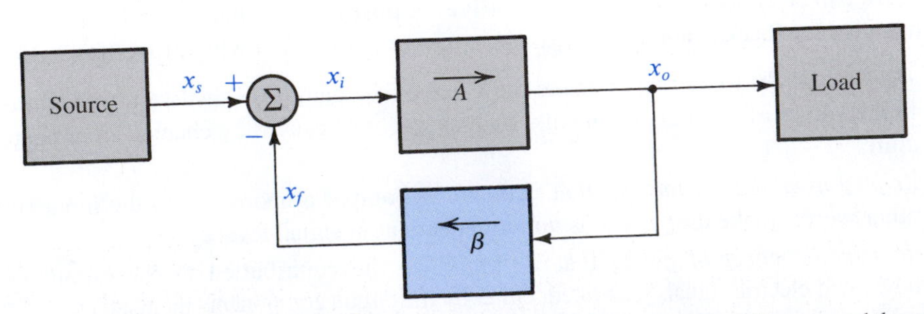

Figure 10.1 General structure of the feedback amplifier. This is a signal-flow diagram, and the quantities x represent either voltage or current signals.

The output x_o is fed to the load as well as to a feedback network, which produces a sample of the output. This sample x_f is related to x_o by the **feedback factor** β,

$$x_f = \beta x_o \tag{10.2}$$

The feedback signal x_f is *subtracted* from the source signal x_s, which is the input to the complete feedback amplifier,[1] to produce the signal x_i, which is the input to the basic amplifier,

$$x_i = x_s - x_f \tag{10.3}$$

Here we note that it is this subtraction that makes the feedback negative. In essence, negative feedback reduces the signal that appears at the input of the basic amplifier.

Implicit in the description above is that the source, the load, and the feedback network *do not* load the basic amplifier. That is, the gain A does not depend on any of these three networks. In practice this will not be the case, and we shall have to find a method for casting a real circuit into the ideal structure depicted in Fig. 10.1. Figure 10.1 also implies that the forward transmission occurs entirely through the basic amplifier and the reverse transmission occurs entirely through the feedback network.

The gain of the feedback amplifier can be obtained by combining Eqs. (10.1) through (10.3):

$$A_f \equiv \frac{x_o}{x_s} = \frac{A}{1 + A\beta} \tag{10.4}$$

The quantity $A\beta$ is called the **loop gain**, a name that follows from Fig. 10.1. For the feedback to be negative, the loop gain $A\beta$ must be positive; that is, the feedback signal x_f should have the same sign as x_s, thus resulting in a smaller difference signal x_i. Equation (10.4) indicates that for positive $A\beta$ the **gain with feedback** A_f will be smaller than the **open-loop gain** A by a factor equal to $1 + A\beta$, which is called the **amount of feedback**.

If, as is the case in many circuits, the loop gain $A\beta$ is large, $A\beta \gg 1$, then from Eq. (10.4) it follows that

$$A_f \simeq \frac{1}{\beta} \tag{10.5}$$

which is a very interesting result: *The gain of the feedback amplifier is almost entirely determined by the feedback network.* Since the feedback network usually consists of passive components, which usually can be chosen to be as accurate as one wishes, the advantage of negative feedback in obtaining accurate, predictable, and stable gain should be apparent. In other words, the overall gain will have very little dependence on the gain of the basic amplifier, A, a desirable property because the gain A is usually a function of many manufacturing and application parameters, some of which might have wide tolerances. We have seen a dramatic illustration of all of these effects in op-amp circuits in Chapter 2, where the **closed-loop gain** (which is another name for the gain-with-feedback) is almost entirely determined by the feedback elements.

Equations (10.1) through (10.3) can be combined to obtain the following expression for the feedback signal x_f:

$$x_f = \frac{A\beta}{1 + A\beta} x_s \tag{10.6}$$

[1] In earlier chapters, we used the subscript "sig" for quantities associated with the signal source (e.g., v_{sig} and R_{sig}). We did that to avoid confusion with the subscript "*s*," which is usually used with FETs to denote quantities associated with the source terminal of the transistor. At this point, however, it is expected that readers have become sufficiently familiar with the subject that the possibility of confusion is minimal. Therefore, we will revert to using the simpler subscript s for signal-source quantities.

Thus for $A\beta \gg 1$ we see that $x_f \simeq x_s$, which implies that the signal x_i at the input of the basic amplifier is reduced to almost zero. Thus if a large amount of negative feedback is employed, the feedback signal x_f becomes an almost identical replica of the input signal x_s. An outcome of this property is the tracking of the two input terminals of an op amp. The difference between x_s and x_f, which is x_i, is sometimes referred to as the **error signal**. Accordingly, the **input differencing circuit** is often also called a **comparison circuit**. (It is also known as a **mixer**.) An expression for x_i can be easily determined as

$$x_i = \frac{1}{1+A\beta}x_s \tag{10.7}$$

from which we can verify that for $A\beta \gg 1$, x_i becomes very small. Observe that negative feedback reduces the signal that appears at the input terminals of the basic amplifier by the amount of feedback, $(1 + A\beta)$. As will be seen later, it is this reduction of input signal that results in the increased linearity of the feedback amplifier.[2]

Example 10.1

The noninverting op-amp configuration shown in Fig. 10.2(a) provides a direct implementation of the feedback loop of Fig. 10.1.

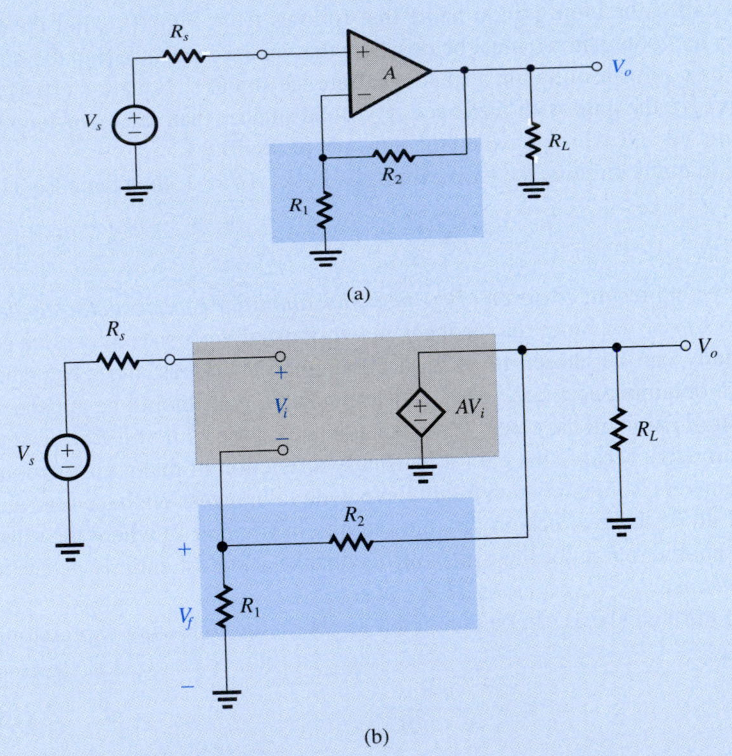

Figure 10.2 **(a)** A non-inverting op-amp circuit for Example 10.1. **(b)** The circuit in (a) with the op-amp replaced with its equivalent circuit.

[2]We have in fact already seen examples of this: adding a resistance R_e in the emitter of a CE amplifier (or a resistance R_s in the source of a CS amplifier) increases the linearity of these amplifiers because for the same input signal as before, v_{be} and v_{gs} are now smallar (by the amount of feedback).

(a) Assume that the op amp has infinite input resistance and zero output resistance. Find an expression for the feedback factor β.
(b) Find the condition under which the closed-loop gain A_f is almost entirely determined by the feedback network.
(c) If the open-loop gain $A = 10^4$ V/V, find R_2/R_1 to obtain a closed-loop gain A_f of 10 V/V.
(d) What is the amount of feedback in decibels?
(e) If $V_s = 1$ V, find V_o, V_f, and V_i.
(f) If A decreases by 20%, what is the corresponding decrease in A_f?

Solution

(a) To be able to see more clearly the direct correspondence between the circuit in Fig. 10.2(a) and the block diagram in Fig. 10.1, we replace the op amp with its equivalent-circuit model, as shown in Fig. 10.2(b). Since the op amp is assumed to have infinite input resistance and zero output resistance, its model is simply an ideal voltage-controlled voltage source of gain A. From Fig. 10.2(b) we observe that the feedback network, consisting of the voltage divider (R_1, R_2), is connected directly to the output and feeds a signal V_f to the inverting input terminal of the op amp. It is important at this point to note that the zero output resistance of the op amp causes the output voltage to be AV_i irrespective of the values of R_1 and R_2 and of R_L. That is what we meant by the statement that in the block diagram of Fig. 10.1, the feedback network and the load are assumed not to load the basic amplifier. Now we can easily determine the feedback factor β from

$$\beta \equiv \frac{V_f}{V_o} = \frac{R_1}{R_1 + R_2}$$

Let's next examine how V_f is subtracted from V_s at the input side. The subtraction is effectively performed by the differential action of the op amp; by its very nature, a differential-input amplifier takes the difference between the signals at its two input terminals. Observe also that because the input resistance of the op amp is assumed to be infinite, no current flows in R_s. Thus the value of R_s has no bearing on V_i; or the source "does not load" the amplifier input. Similarly, because of the zero input current of the op amp, V_f will depend only on the ratio R_1/R_2 and not on the absolute values of R_1 and R_2.

(b) The closed-loop gain A_f is given by

$$A_f = \frac{A}{1 + A\beta}$$

To make A_f nearly independent of A, we must ensure that the loop gain $A\beta$ is much larger than unity,

$$A\beta \gg 1$$

$$A\left(\frac{R_1}{R_1 + R_2}\right) \gg 1$$

Since under such a condition,

$$A_f \simeq \frac{1}{\beta} = \frac{R_1 + R_2}{R_1} = 1 + \frac{R_2}{R_1}$$

the condition can be stated as

$$A \gg A_f$$

(c) For $A = 10^4$ V/V and $A_f = 10$ V/V, we see that $A \gg A_f$, thus we can select R_1 and R_2 to obtain

$$\beta = \frac{1}{A_f} = 0.1$$

Example 10.1 *continued*

Thus,

$$\frac{1}{\beta} = 1 + \frac{R_2}{R_1} = A_f = 10$$

which yields

$$R_2/R_1 = 9$$

A more exact value for the ratio R_2/R_1 can be obtained from

$$A_f = \frac{A}{1 + A\beta}$$

$$10 = \frac{10^4}{1 + 10^4\beta}$$

which results in

$$\beta = 0.0999$$

and,

$$\frac{R_2}{R_1} = 9.01$$

(d) The amount of feedback is

$$1 + A\beta = \frac{A}{A_f} = \frac{10^4}{10} = 1000$$

which is 60 dB.

(e) For $V_s = 1$ V,

$$V_o = A_f V_s = 10 \times 1 = 10 \text{ V}$$

$$V_f = \beta V_o = 0.0999 \times 10 = 0.999 \text{ V}$$

$$V_i = \frac{V_o}{A} = \frac{10}{10^4} = 0.001 \text{V}$$

Note that if we had used the approximate value of $\beta = 0.1$, we would have obtained $V_f = 1$ V and $V_i = 0$ V.

(f) If A decreases by 20%, thus becoming

$$A = 0.8 \times 10^4 \text{ V/V}$$

the value of A_f becomes

$$A_f = \frac{0.8 \times 10^4}{1 + 0.8 \times 10^4 \times 0.0999} = 9.9975 \text{ V/V}$$

that is, it decreases by 0.025%, which is lower than the percentage change in A by approximately a factor $(1 + A\beta)$.

EXERCISES

10.1 Repeat Example 10.1, (c) to (f) for $A = 100$ V/V.
Ans. (c) 10.11; (d) 20 dB; (e) 10 V, 0.9 V, 0.1 V; (f) 2.44%

10.2 Repeat Example 10.1, (c) to (f) for $A_f = 10^3$ V/V. For (e) use $V_s = 0.01$ V.
Ans. (c) 1110.1; (d) 20 dB; (e) 10 V, 0.009 V, 0.001 V; (f) 2.44%

10.2 Some Properties of Negative Feedback

The properties of negative feedback were mentioned in the Introduction. In the following, we shall consider some of these properties in more detail.

10.2.1 Gain Desensitivity

The effect of negative feedback on desensitizing the closed-loop gain was demonstrated in Example 10.1, where we saw that a 20% reduction in the gain of the basic amplifier gave rise to only a 0.025% reduction in the gain of the closed-loop amplifier. This sensitivity-reduction property can be analytically established as follows.

Assume that β is constant. Taking differentials of both sides of Eq. (10.4) results in

$$dA_f = \frac{dA}{(1+A\beta)^2} \tag{10.8}$$

Dividing Eq. (10.8) by Eq. (10.4) yields

$$\frac{dA_f}{A_f} = \frac{1}{(1+A\beta)}\frac{dA}{A} \tag{10.9}$$

which says that the percentage change in A_f (due to variations in some circuit parameter) is smaller than the percentage change in A by a factor equal to the amount of feedback. For this reason, the amount of feedback, $1 + A\beta$, is also known as the **desensitivity factor**.

EXERCISE

10.3 An amplifier with a nominal gain $A = 1000$ V/V exhibits a gain change of 10% as the operating temperature changes from 25° C to 75°C. If it is required to constrain the change to 0.1% by applying negative feedback, what is the largest closed-loop gain possible? If three of these feedback amplifiers are placed in cascade, what overall gain and gain stability are achieved?
Ans. 10 V/V; 1000 V/V, with a maximum variability of 0.3% over the specified temperature range.

10.2.2 Bandwidth Extension

Consider an amplifier whose high-frequency response is characterized by a single pole. Its gain at mid and high frequencies can be expressed as

$$A(s) = \frac{A_M}{1 + s/\omega_H} \tag{10.10}$$

where A_M denotes the midband gain and ω_H is the upper 3-dB frequency. Application of negative feedback, with a frequency-independent factor β, around this amplifier results in a closed-loop gain $A_f(s)$ given by

$$A_f(s) = \frac{A(s)}{1 + \beta A(s)}$$

Substituting for $A(s)$ from Eq. (10.10) results, after a little manipulation, in

$$A_f(s) = \frac{A_M/(1 + A_M\beta)}{1 + s/\omega_H(1 + A_M\beta)} \tag{10.11}$$

Thus the feedback amplifier will have a midband gain of $A_M/(1 + A_M\beta)$ and an upper 3-dB frequency ω_{Hf} given by

$$\omega_{Hf} = \omega_H(1 + A_M\beta) \tag{10.12}$$

It follows that the upper 3-dB frequency is increased by a factor equal to the amount of feedback.

Similarly, it can be shown that if the open-loop gain is characterized by a dominant low-frequency pole giving rise to a lower 3-dB frequency ω_L, then the feedback amplifier will have a lower 3-dB frequency ω_{Lf},

$$\omega_{Lf} = \frac{\omega_L}{1 + A_M\beta} \tag{10.13}$$

Note that the amplifier bandwidth is increased by the same factor by which its midband gain is decreased, *maintaining the gain–bandwidth product at a constant value*. This point is further illustrated by the Bode Plot in Fig. 10.3.

Finally, note that the action of negative feedback in extending the amplifier bandwidth should not be surprising: Negative feedback works to minimize the change in gain magnitude, including its change with frequency.

EXERCISE

10.4 Consider the noninverting op-amp circuit of Example 10.1. Let the open-loop gain A have a low-frequency value of 10^4 and a uniform –6-dB/octave rolloff at high frequencies with a 3-dB frequency of 100 Hz. Find the low-frequency gain and the upper 3-dB frequency of a closed-loop amplifier with $R_1 = 1\ \text{k}\Omega$ and $R_2 = 9\ \text{k}\Omega$.

Ans. 9.99 V/V; 100.1 kHz

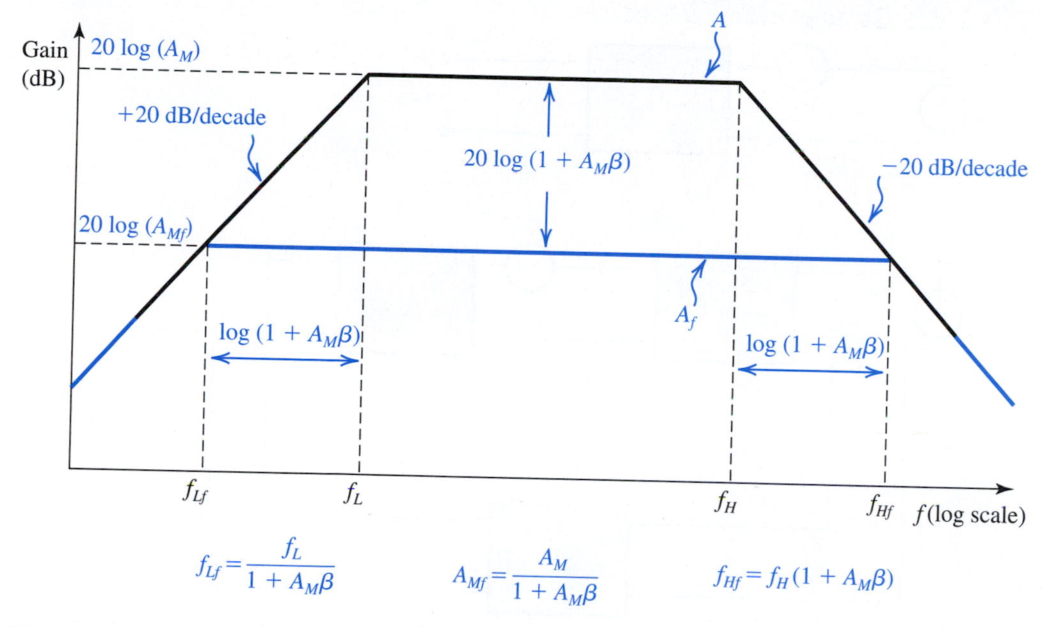

Figure 10.3 Application of negative feedback reduces the midband gain, increases f_H, and reduces f_L, all by the same factor, $(1+A_M\beta)$, which is equal to the amount of feedback.

10.2.3 Interference Reduction

Negative feedback can be employed to reduce the interference in an amplifier or, more precisely, to increase the ratio of signal to interference. However, as we shall now explain, this interference-reduction process is possible only under certain conditions. Consider the situation illustrated in Fig. 10.4. Figure 10.4(a) shows an amplifier with gain A_1, an input signal V_s, and interference, V_n. It is assumed that for some reason this amplifier suffers from interference and that the interference can be assumed to be introduced at the input of the amplifier. The **signal-to-interference ratio** for this amplifier is

$$S/I = V_s/V_n \tag{10.14}$$

Consider next the circuit in Fig. 10.4(b). Here we assume that it is possible to build another amplifier stage with gain A_2 that does not suffer from the interference problem. If this is the case, then we may precede our original amplifier A_1 by the *clean* amplifier A_2 and apply negative feedback around the overall cascade of such an amount as to keep the overall gain constant. The output voltage of the circuit in Fig. 10.4(b) can be found by superposition:

$$V_o = V_s\frac{A_1A_2}{1+A_1A_2\beta} + V_n\frac{A_1}{1+A_1A_2\beta} \tag{10.15}$$

Thus the signal-to-interference ratio at the output becomes

$$\frac{S}{I} = \frac{V_s}{V_n}A_2 \tag{10.16}$$

which is A_2 times higher than in the original case.

We emphasize once more that the improvement in signal-to-interference ratio by the application of feedback is possible only if one can precede the interference-prone stage

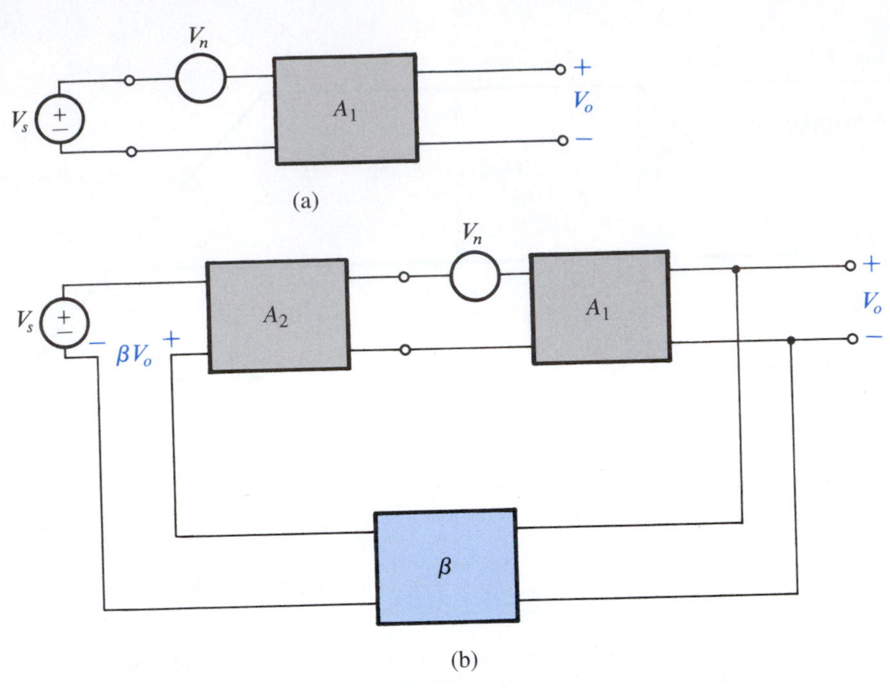

Figure 10.4 Illustrating the application of negative feedback to improve the signal-to-interference ratio in amplifiers.

by a (relatively) interference-free stage. This situation, however, is not uncommon in practice. The best example is found in the output power-amplifier stage of an audio amplifier. Such a stage usually suffers from a problem known as **power-supply hum**. The problem arises because of the large currents that this stage draws from the power supply and the difficulty of providing adequate power-supply filtering inexpensively. The power-output stage is required to provide large power gain but little or no voltage gain. We may therefore precede the power-output stage by a small-signal amplifier that provides large voltage gain, and apply a large amount of negative feedback, thus restoring the voltage gain to its original value. Since the small-signal amplifier can be fed from another, less hefty (and hence better regulated) power supply, it will not suffer from the hum problem. The hum at the output will then be reduced by the amount of the voltage gain of this added **preamplifier**.

EXERCISE

10.5 Consider a power-output stage with voltage gain $A_1 = 1$, an input signal $V_s = 1$ V, and a hum V_n of 1 V. Assume that this power stage is preceded by a small-signal stage with gain $A_2 = 100$ V/V and that overall feedback with $\beta = 1$ is applied. If V_s and V_n remain unchanged, find the signal and interference voltages at the output and hence the improvement in *S/I*.
Ans. $\simeq 1$ V; $\simeq 0.01$ V; 100 (40 dB)

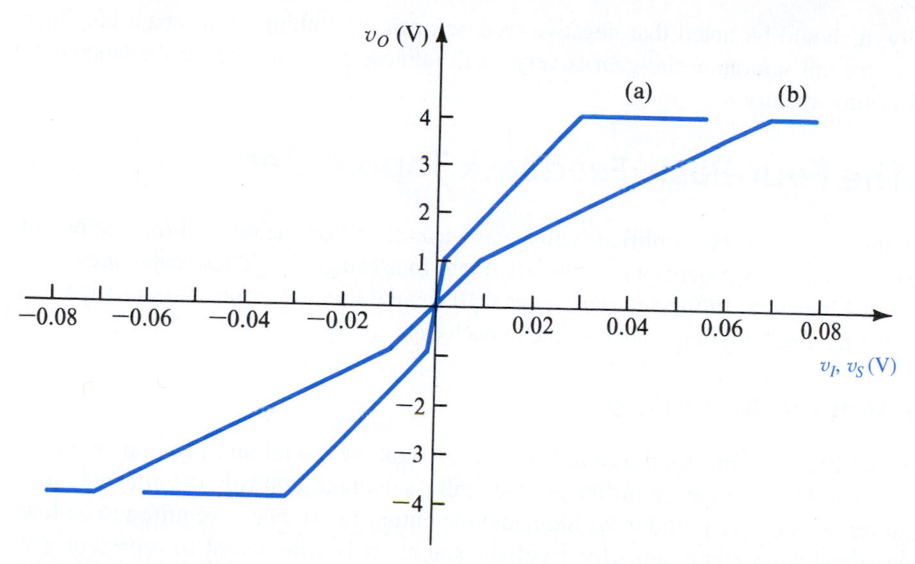

Figure 10.5 Illustrating the application of negative feedback to reduce the nonlinear distortion in amplifiers. Curve (a) shows the amplifier transfer characteristic (v_O versus v_I) without feedback. Curve (b) shows the characteristic (v_O versus v_S) with negative feedback ($\beta = 0.01$) applied.

10.2.4 Reduction in Nonlinear Distortion

Curve (a) in Fig. 10.5 shows the transfer characteristic v_O versus v_I of an amplifier. As indicated, the characteristic is piecewise linear, with the voltage gain changing from 1000 to 100 and then to 0. This nonlinear transfer characteristic will result in this amplifier generating a large amount of nonlinear distortion.

The amplifier transfer characteristic can be considerably **linearized** (i.e., made less nonlinear) through the application of negative feedback. That this is possible should not be too surprising, since we have already seen that negative feedback reduces the dependence of the overall closed-loop amplifier gain on the open-loop gain of the basic amplifier. Thus large changes in open-loop gain (1000 to 100 in this case) give rise to much smaller corresponding changes in the closed-loop gain.

To illustrate, let us apply negative feedback with $\beta = 0.01$ to the amplifier whose open-loop voltage transfer characteristic is depicted in Fig. 10.5. The resulting transfer characteristic of the closed-loop amplifier, v_O versus v_S, is shown in Fig. 10.5 as curve (b). Here the slope of the steepest segment is given by

$$A_{f1} = \frac{1000}{1 + 1000 \times 0.01} = 90.9$$

and the slope of the next segment is given by

$$A_{f2} = \frac{100}{1 + 100 \times 0.01} = 50$$

Thus the order-of-magnitude change in slope has been considerably reduced. The price paid, of course, is a reduction in voltage gain. Thus if the overall gain has to be restored, a preamplifier should be added. This preamplifier should not present a severe nonlinear-distortion problem, since it will be dealing with smaller signals.

Finally, it should be noted that negative feedback can do nothing at all about amplifier saturation, since in saturation the gain is very small (almost zero) and hence the amount of feedback is almost unity.

10.3 The Four Basic Feedback Topologies

Based on the quantity to be amplified (voltage or current) and on the desired form of output (voltage or current), amplifiers can be classified into four categories. These categories were discussed in Chapter 1. In the following, we shall review this amplifier classification and point out the feedback topology appropriate in each case.

10.3.1 Voltage Amplifiers

Voltage amplifiers are intended to amplify an input voltage signal and provide an output voltage signal. The voltage amplifier is essentially a voltage-controlled voltage source. The input resistance is required to be high, and the output resistance is required to be low. Since the signal source is essentially a voltage source, it is convenient to represent it in terms of a Thévenin equivalent circuit. In a voltage amplifier, the output quantity of interest is the output voltage. It follows that the feedback network should *sample* the output *voltage*, just as a voltmeter measures a voltage. Also, because of the Thévenin representation of the source, the feedback signal x_f should be a *voltage* that can be *mixed* with the source voltage in *series*.

The most suitable feedback topology for the voltage amplifier is the **voltage-mixing, voltage-sampling** one shown in Fig. 10.6. Because of the series connection at the input and the parallel or shunt connection at the output, this feedback topology is also known as **series–shunt feedback**. As will be shown, this topology not only stabilizes the voltage gain but also results in a higher input resistance (intuitively, a result of the series connection at the input) and a lower output resistance (intuitively, a result of the parallel connection at the output), which are desirable properties for a voltage amplifier.

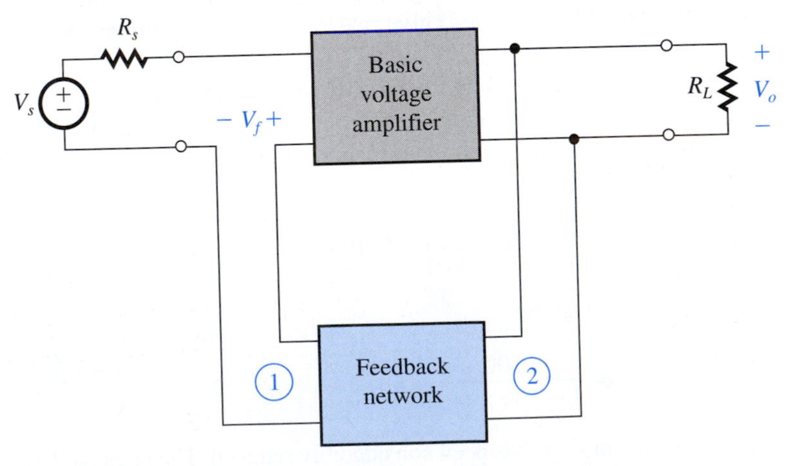

Figure 10.6 Block diagram of a feedback voltage amplifier. Here the appropriate feedback topology is series–shunt.

The increased input resistance results because V_f subtracts from V_s, resulting in a smaller signal V_i at the input of the basic amplifier. The lower V_i, in turn, causes the input current to be smaller, with the result that the resistance seen by V_s will be larger. We shall derive a formula for the input resistance of the feedback voltage amplifier in the next section.

The decreased output resistance results because the feedback works to keep V_o as constant as possible. Thus if the current drawn from the amplifier output changes by ΔI_o, the change ΔV_o in V_o will be lower than it would have been if feedback were not present. Thus the output resistance $\Delta V_o/\Delta I_o$ will be lower than that of the open-loop amplifier. In the following section we shall derive an expression for the output resistance of the feedback voltage amplifier.

Three examples of series–shunt feedback amplifiers are shown in Fig. 10.7. The amplifier in Fig. 10.7(a) is the familiar noninverting op-amp configuration. The feedback network, composed of the voltage divider (R_1, R_2), develops a voltage V_f that is applied to the negative input terminal of the op amp. The subtraction of V_f from V_s is achieved by utilizing the differencing action of the op-amp differential input. For the feedback to be negative, V_f must be of the same polarity as V_s, thus resulting in a smaller signal at the input of the basic amplifier. To ascertain that this is the case, we follow the signal around the loop, as follows: As V_s increases, V_o increases and the voltage divider causes V_f to increase. Thus the change in V_f is of the same polarity as the change in V_s, and the feedback is negative.

The second feedback voltage amplifier, shown in Fig. 10.7(b), utilizes two MOSFET amplifier stages in cascade. The output voltage V_o is sampled by the feedback network composed of the voltage divider (R_1, R_2), and the feedback signal V_f is fed to the source terminal of Q_1. The subtraction is implemented by applying V_s to the gate of Q_1 and V_f to its source, with the result that the signal at this amplifier input $V_i = V_{gs} = V_s - V_f$. To ascertain that the feedback is negative, let V_s increase. The drain voltage of Q_1 will decrease, and since this is applied to the gate of Q_2, its drain voltage V_o will increase. The feedback network will then cause V_f to increase, which is the same change in polarity initially assumed for V_s. Thus the feedback is indeed negative.

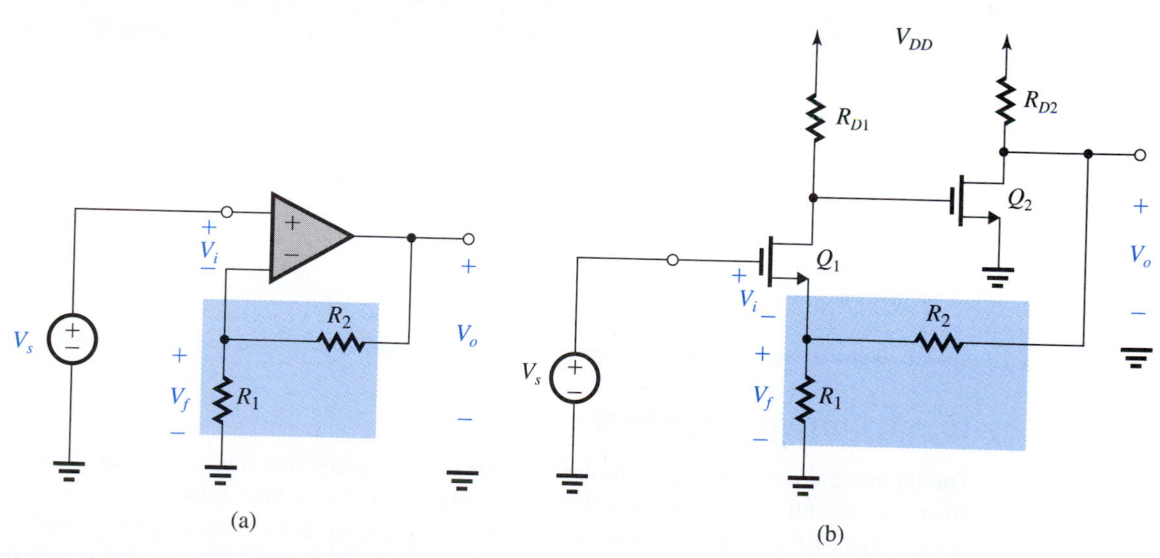

Figure 10.7 Examples of a feedback voltage amplifier. All these circuits employ series–shunt feedback. Note that the dc bias circuits are only partially shown.

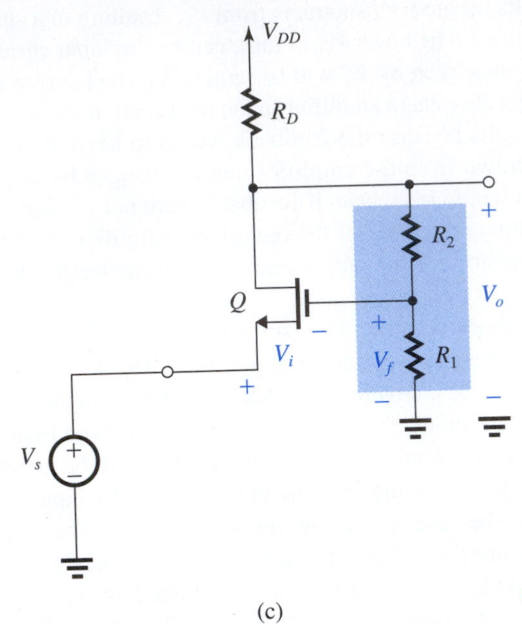

Figure 10.7 *continued*

The third example of series–shunt feedback, shown in Fig. 10.7(c), utilizes a CG transistor Q with a fraction V_f of the output voltage V_o fed back to the gate through a voltage divider (R_1, R_2). Observe that the subtraction of V_f from V_s is effected by applying V_s to the source, thus the input V_i to the CG amplifier is obtained as $V_s - V_f$. As usual, however, we must check the polarity of the feedback: If V_s increases, V_d (which is V_o) will increase and V_f will correspondingly increase. Thus V_f and V_s change in the same direction, verifying that the feedback is negative.

EXERCISE

10.6 For the circuit in Fig. 10.7(c) let $(R_1 + R_2) \gg R_D$. Using small-signal analysis, find expressions for the open-loop gain $A \equiv V_o/V_i$; the feedback factor $\beta \equiv V_f/V_o$; and the closed loop gain $A_f \equiv V_o/V_s$. For $A\beta \gg 1$, find an approximate expression for A_f. Neglect r_o.
Ans. $A = g_m R_D$; $\beta = R_1/(R_1 + R_2)$;

$$A_f = \frac{g_m R_D}{1 + g_m R_D R_1/(R_1 + R_2)}; \left(1 + \frac{R_2}{R_1}\right)$$

10.3.2 Current Amplifiers

The input signal in a current amplifier is essentially a current, and thus the signal source is most conveniently represented by its Norton equivalent. The output quantity of interest is current; hence the feedback network should *sample* the output *current*, just as a current meter measures a current. The feedback signal should be in *current* form so that it may be *mixed* in *shunt* with the source current. Thus the feedback topology most suitable for a current amplifier is the **current-mixing, current-sampling** topology, illustrated in Fig. 10.8(a). Because of the parallel (or shunt) connection at the input, and the series connection

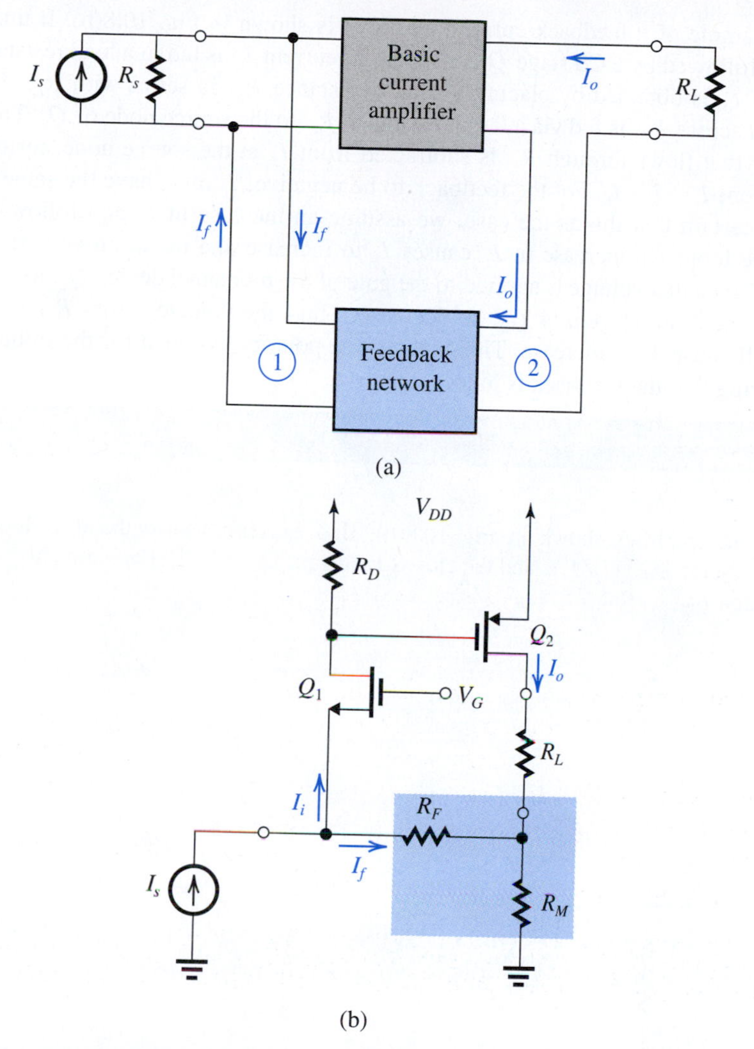

Figure 10.8 (a) Block diagram of a feedback current amplifier. Here, the appropriate feedback topology is the shunt–series. (b) Example of a feedback current amplifier.

at the output, this feedback topology is also known as **shunt–series feedback**. As will be shown, this topology not only stabilizes the current gain but also results in a lower input resistance, and a higher output resistance, both desirable properties for a current amplifier.

The decrease in input resistance results because the feedback current I_f subtracts from the input current I_s, and thus a lower current enters the basic current amplifier. This in turn results in a lower voltage at the amplifier input, that is, across the current source I_s. It follows that the input resistance of the feedback current amplifier will be lower than that of the open-loop amplifier. We shall derive an expression for R_{if} in Section 10.5.

The increase in output resistance is simply a result of the action of negative feedback in keeping the value of I_o as constant as possible. Thus if the voltage across R_L is changed, the resulting change in I_o will be lower than it would have been without the feedback, which implies that the output resistance is increased. An expression for R_{of} will be derived in Section 10.5.

An example of a feedback current amplifier is shown in Fig. 10.8(b). It utilizes a CG stage Q_1 followed by a CS stage Q_2. The output current I_o is fed to a load resistance R_L. A sample of I_o is obtained by placing a small resistance R_M in series with R_L. The voltage developed across R_M is fed via a large resistance R_F to the source node of Q_1. The feedback current I_f that flows through R_F is subtracted from I_s at the source node, resulting in the input current $I_i = I_s - I_f$. For the feedback to be negative, I_f must have the same polarity as I_s. To ascertain that this is the case, we assume an increase in I_s and follow the change around the loop: An increase in I_s causes I_i to increase and the drain voltage of Q_1 will increase. Since this voltage is applied to the gate of the *p*-channel device Q_2, its increase will cause I_o, the drain current of Q_2, to decrease. Thus, the voltage across R_M will decrease, which will cause I_f to increase. This is the same polarity assumed for the initial change in I_s, verifying that the feedback is indeed negative.

Example 10.2

For the feedback current amplifier shown in Fig 10.8(b), find expressions for the open-loop gain $A \equiv I_o/I_i$, the feedback factor $\beta = (I_f/I_o)$, and the closed-loop gain $A_f \equiv I_o/I_s$. For simplicity, neglect the Early effect in Q_1 and Q_2.

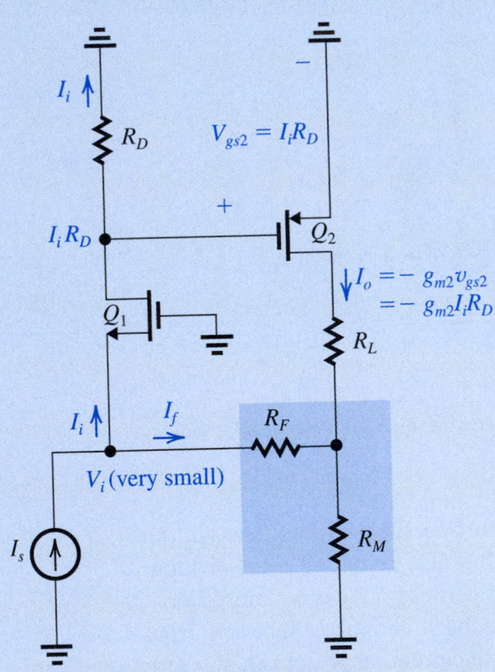

Figure 10.9 Analysis of the feedback current amplifier of Fig. 10.8(b) to obtain $A \equiv I_o/I_i$ and $\beta \equiv I_f/I_o$.

Solution

Figure 10.9 shows the circuit prepared for small-signal analysis. Some of the analysis is also indicated on the diagram. Since, as indicated,

$$I_o = -g_{m2} R_D I_i$$

the open-loop gain A is given by

$$A \equiv \frac{I_o}{I_i} = -g_{m2}R_D$$

To obtain β, we observe that I_o is fed to a current divider formed by R_M and R_F. Since current mixing results in a reduced input resistance, the voltage at the source node of Q_2 will be close to zero, and R_F in effect appears in parallel with R_M, enabling us to obtain β as

$$\beta \equiv \frac{I_f}{I_o} \simeq -\frac{R_M}{R_F + R_M}$$

where the negative sign is a result of the reference directions used for I_o and I_f. Note, however, that the loop gain $A\beta$ will be positive, as should always be the case in a negative feedback amplifier. We can now combine A and β to obtain A_f as

$$A_f \equiv \frac{I_o}{I_s} = -\frac{g_{m2}R_D}{1 + g_{m2}R_D \Big/ \left(1 + \frac{R_F}{R_M}\right)}$$

EXERCISE

10.7 For the feedback current amplifier of Fig. 10.8(b), analyzed in Example 10.2, find an approximate expression for the closed-loop current gain under the condition that the loop gain is large. Also, state the condition precisely.

Ans. $A_f \simeq -\left(1 + \frac{R_F}{R_M}\right)$; $g_{m2}R_D \gg \left(1 + \frac{R_F}{R_M}\right)$

10.3.3 Transconductance Amplifiers

In transconductance amplifiers the input signal is a voltage and the output signal is a current. It follows that the appropriate feedback topology is the **voltage-mixing, current-sampling** topology, illustrated in Fig. 10.10(a). The presence of the series connection at both the input and the output gives this feedback topology the alternative name **series–series feedback**.

As in the case of the feedback voltage amplifier, the series connection at the input results in an increased input resistance. The sampling of the output current results in an increased output resistance. Thus the series–series feedback topology provides the transconductance amplifier with the desirable properties of increased input and output resistances.

Examples of feedback transconductance-amplifiers are shown in Fig. 10.10 (b) and (c). The circuit in Fig. 10.10(b) utilizes a differential amplifier A_1 followed by a CS stage Q_2. The output current I_o is fed to R_L and to a series resistance R_F, which develops a feedback voltage V_f. The latter is applied to the positive input terminal of the differential amplifier A_1. The subtraction of V_f from V_s is performed by the differencing action of the differential-amplifier input. At this point we must check that V_f and V_s have the same polarity: A positive change in V_s will result in a negative change at the gate of Q_2, which in turn causes I_o to increase. The increase in I_o results in a positive change in V_f, which is the same polarity assumed for the change in V_s, verifying that the feedback is negative.

The transconductance amplpifier in Fig.10.10(c) utilizes a CS amplifier Q_1 in cascade with another CS amplifier, Q_2. The output current I_o is fed to R_L and to a series resistance

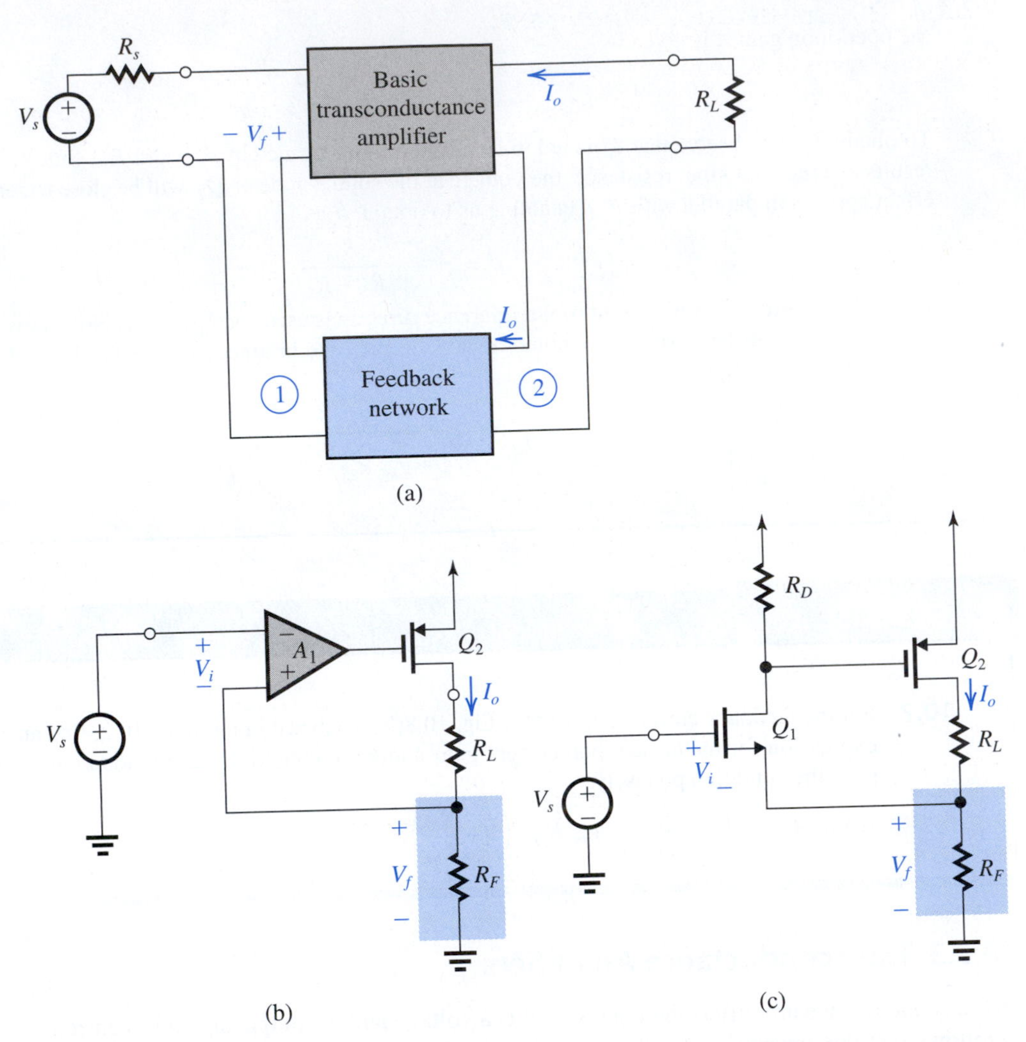

Figure 10.10 (**a**) Block diagram of a feedback transconductance amplifier. Here, the appropriate feedback topology is series–series. (**b**) Example of a feedback transconductance amplifier. (**c**) Another example.

R_F that develops a feedback voltage V_f. The latter is fed to the source of Q_1, thus utilizing the input of Q_1 to implement the subtraction; $V_i = V_s - V_f$. The reader is urged to verify that V_f has the same polarity as V_s and thus that the feedback is negative.

EXERCISE

10.8 For the circuit in Fig. 10.10(b), let the differential amplifier A_1 have an infinite input resistance. Use small-signal analysis to obtain expressions for the open-loop gain $A \equiv I_o/V_i$, the feedback factor $\beta \equiv V_f/I_o$, and the closed-loop gain $A_f \equiv I_o/V_s$. If the loop gain is much greater than unity, find an approximate expression for A_f. Neglect r_{o2}.

Ans. $A = A_1 g_{m2}$; $\beta = R_F$; $A_f = \dfrac{A_1 g_{m2}}{1 + A_1 g_{m2} R_F}$; $A_f \simeq 1/R_F$

10.3.4 Transresistance Amplifiers

In transresistance amplifiers the input signal is current and the output signal is voltage. It follows that the appropriate feedback topology is of the **current-mixing, voltage-sampling** type, shown in Fig. 10.11(a). The presence of the parallel (or shunt) connection at

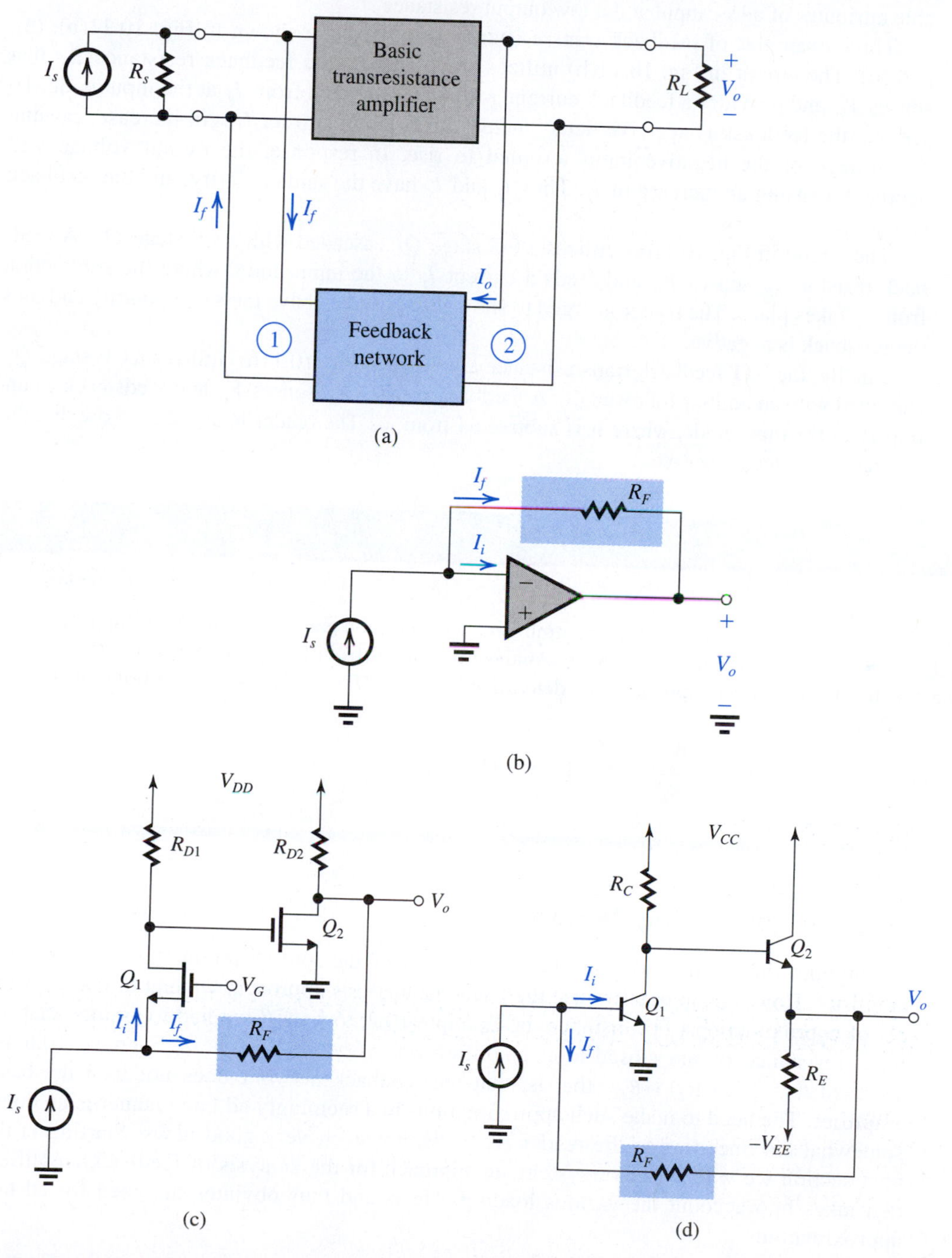

Figure 10.11 (a) Block diagram of a feedback transresistance amplifier. Here, the appropriate feedback topology is shunt–shunt. (b), (c), and (d) Examples of feedback transresistance amplifiers.

both the input and the output makes this feedback topology also known as **shunt–shunt** feedback.

The shunt connection at the input causes the input resistance to be reduced. The shunt connection at the output stabilizes the output voltage and thus causes the output resistance to be reduced. Thus, the shunt–shunt topology equips the transresistance amplifier with the desirable attributes of a low input and a low output resistance.

Three examples of feedback transresistance amplifiers are shown in Fig. 10.11(b), (c), and (d). The circuit in Fig. 10.11(b) utilizes an op amp with a feedback resistance R_F that senses V_o and provides a feedback current I_f that is subtracted from I_s at the input node. To see that the feedback is negative, let I_s increase. The input current I_i will increase, causing the voltage of the negative input terminal to rise. In response, the output voltage will decrease, causing an increase in I_f. Thus I_f and I_s have the same polarity, and the feedback is negative.

The circuit in Fig. 10.11(c) utilizes a CG stage Q_1 cascaded with a CS stage Q_2. A feedback resistor R_F senses V_o and feeds a current I_f to the input node, where the subtraction from I_s takes place. The reader is urged to show that I_f and I_s have the same polarity and thus the feedback is negative.

Finally, the BJT feedback transresistance amplifier in Fig. 10.11(d) utilizes a CE stage Q_1 cascaded with an emitter follower Q_2. A feedback resistor R_F senses V_o and feeds back a current I_f to the input node, where it is subtracted from I_s. The reader is urged to show that the feedback is indeed negative.

EXERCISE

10.9 For the circuit in Fig. 10.11(b), let the op amp have an open-loop gain A, a differential input resistance R_{id}, and a zero output resistance. Analyze the circuit from first principles (i.e., do not use the feedback analysis approach) to determine $A_f \equiv V_o/I_s$. Under what conditions does $A_f \simeq -R_F$?

Ans. $A_f = -R_F\Big/\left(1+\frac{1}{A}+\frac{R_F}{AR_{id}}\right)$; $A \gg 1$ and $AR_{id} \gg R_F$

10.3.5 A Concluding Remark

Throughout this section we introduced examples of the four different types of feedback amplifier. However, in order to use the feedback analysis approach, we had to make a variety of approximations. For instance, in Example 10.2, to find β we had to assume that the input resistance of the closed-loop amplifier was very low. Also, in Exercise 10.6 we assumed that $(R_1+R_2) \gg R_D$, that is, that the feedback network does not load the basic amplifier. The need to make such approximations in a seemingly ad hoc manner is no doubt somewhat disconcerting to the reader. There is, however, very good news: Starting in the next section we will present a systematic approach for the analysis of feedback amplifiers that takes into account the various loading effects and thus obviates the need for ad hoc approximations.

10.4 The Feedback Voltage Amplifier (Series–Shunt)

10.4.1 The Ideal Case

As mentioned before, series–shunt is the appropriate feedback topology for a voltage amplifier. The ideal structure of the series–shunt feedback amplifier is shown in Fig. 10.12(a). It consists of a *unilateral* open-loop amplifier (the A circuit) and an ideal voltage-sampling, voltage-mixing feedback network (the β circuit). The A circuit has an input resistance R_i, an open-circuit voltage gain A, and an output resistance R_o. It is assumed that the source and load resistances have been absorbed inside the A circuit (more on this point later). Furthermore, note that the β circuit does *not* load the A circuit; that is, connecting the β circuit does not change the value of A (defined as $A \equiv V_o/V_i$).

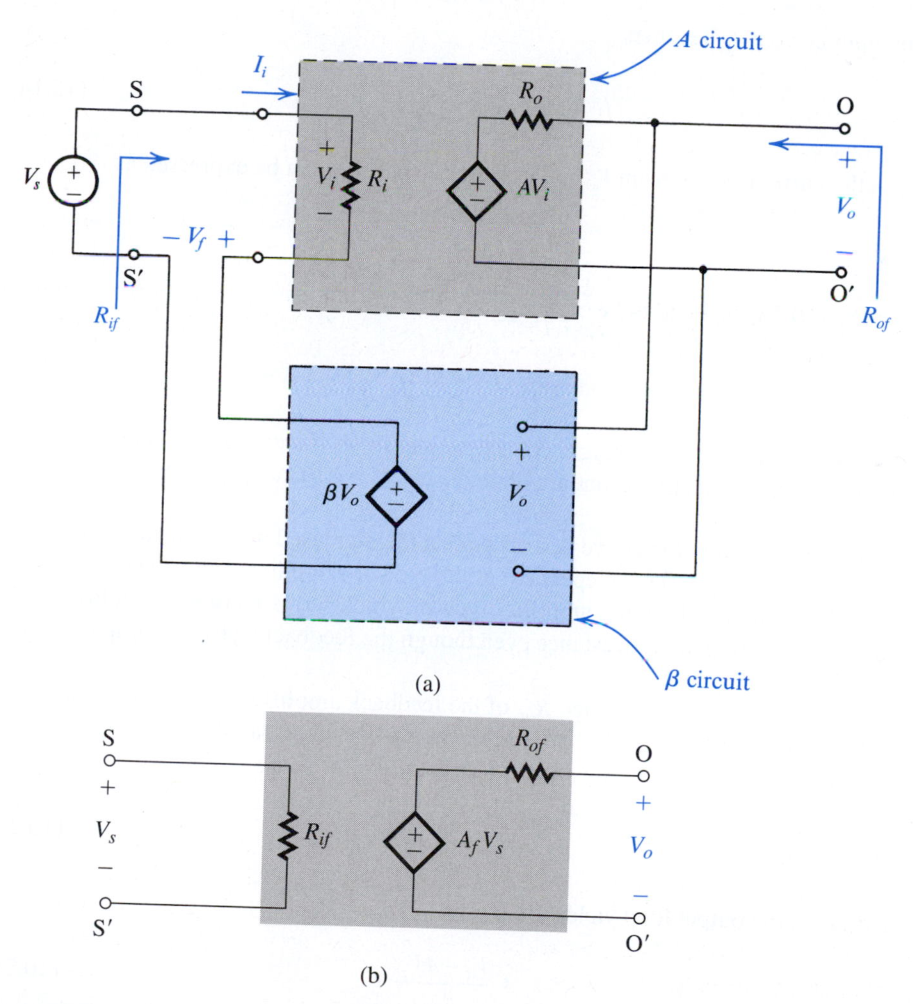

Figure 10.12 The series–shunt feedback amplifier: (**a**) ideal structure; (**b**) equivalent circuit.

The circuit of Fig. 10.12(a) exactly follows the ideal feedback model of Fig. 10.1. Therefore the closed-loop voltage gain A_f is given by

$$A_f \equiv \frac{V_o}{V_s} = \frac{A}{1+A\beta} \tag{10.17}$$

The equivalent circuit model of the series–shunt feedback amplifier is shown in Fig. 10.12(b). Observe that A_f is the open-circuit voltage gain of the feedback amplifier, R_{if} is its input resistance, and R_{of} is its output resistance. Expressions for R_{if} and R_{of} can be derived as follows.

For R_{if}, refer to the input loop of the circuit in Fig. 10.12(a). The series mixing subtracts V_f from V_s and thus reduces V_i by a factor equal to the amount of feedback (Eq. 10.7),

$$V_i = \frac{V_s}{1+A\beta}$$

Thus the input current I_i becomes

$$I_i = \frac{V_i}{R_i} = \frac{V_s}{(1+A\beta)R_i} \tag{10.18}$$

Since I_i is the current drawn from V_s, the input resistance R_{if} can be expressed as

$$R_{if} \equiv \frac{V_s}{I_i}$$

and using Eq. (10.18) is found to be

$$R_{if} = (1+A\beta)R_i \tag{10.19}$$

Thus, as expected, the series-mixing feedback results in an increase in the amplifier input resistance by a factor equal to the amount of feedback, $(1+A\beta)$, a highly desirable property for a voltage amplifier.

It should be clear from the above derivation that the increased input resistance is a result only of the series mixing and is independent of the type of sampling. Thus, the transconductance amplifier, which is the other amplifier type in which series mixing is employed, will also exhibit an increased input resistance even though the feedback network samples its output current (series sampling).

To determine the output resistance R_{of} of the feedback amplifier in Fig. 10.12(a), we set $V_s = 0$ and apply a test voltage V_x between the output terminals, as shown in Fig. 10.13. If the current drawn from V_x is I_x, the output resistance R_{of} is

$$R_{of} \equiv \frac{V_x}{I_x} \tag{10.20}$$

An equation for the output loop yields

$$I_x = \frac{V_x - AV_i}{R_o} \tag{10.21}$$

From the input loop we see that

$$V_i = -V_f$$

Now $V_f = \beta V_o = \beta V_x$; thus,

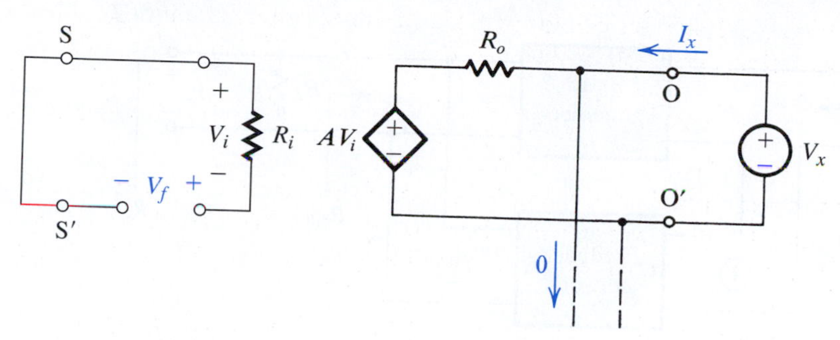

Figure 10.13 Determining the output resistance of the feedback amplifier of Fig. 10.12(a): $R_{of} = V_x/I_x$.

$$V_i = -\beta V_x$$

which when substituted in Eq. (10.21) yields

$$I_x = \frac{V_x(1 + A\beta)}{R_o}$$

Substituting this value of I_x into Eq. (10.20) provides the following expression for R_{of},

$$R_{of} = \frac{R_o}{1 + A\beta} \tag{10.22}$$

Thus, as expected, the shunt sampling (or voltage sampling) at the output results in a decrease in the amplifier output resistance by a factor equal to the amount of negative feedback, $(1 + A\beta)$, a highly desirable property for a voltage amplifier.

Although perhaps not entirely obvious, the reduction of the output resistance is a result only of the method of sampling the output and does not depend on the method of mixing. Thus, the transistance amplifier, which is the other amplifier type in which shunt (or voltage) sampling is employed, will also exhibit a reduced output resistance.

10.4.2 The Practical Case

In a practical series–shunt feedback amplifier, the feedback network will not be an ideal voltage-controlled voltage source. Rather, the feedback network is usually resistive and hence will load the basic amplifier and thus affect the values of A, R_i, and R_o. In addition, the source and load resistances will affect these three parameters. Thus the problem we have is as follows: Given a series–shunt feedback amplifier represented by the block diagram of Fig. 10.14(a), find the A circuit and the β circuit.

Our problem essentially involves representing the amplifier of Fig. 10.14(a) by the ideal structure of Fig. 10.12(a). As a first step toward that end we observe that the source and load resistances should be lumped with the basic amplifier. This, together with representing the two-port feedback network in terms of its h parameters (see Appendix C), is illustrated in Fig. 10.14(b). The choice of h parameters is based on the fact that this is the only parameter set that represents the feedback network by a series network at port 1 and a parallel network at port 2. Such a representation is obviously convenient in view of the series connection at the input and the parallel connection at the output.

Examination of the circuit in Fig. 10.14(b) reveals that the current source $h_{21}I_1$ represents the forward transmission of the feedback network. Since the feedback network is usually

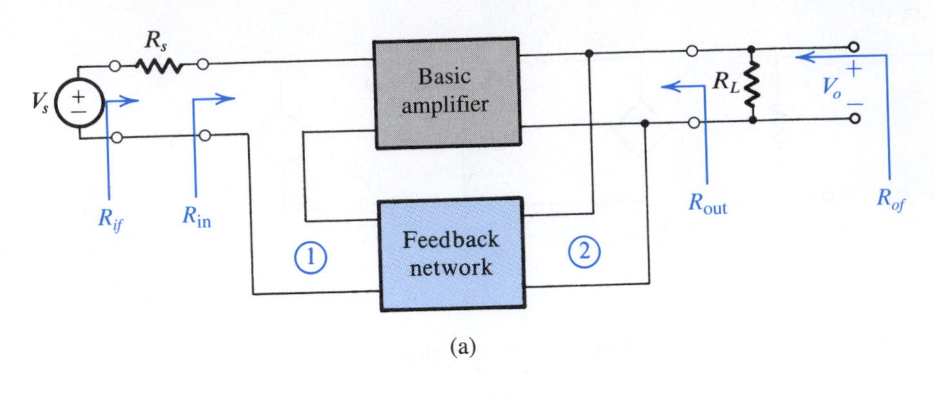

(a)

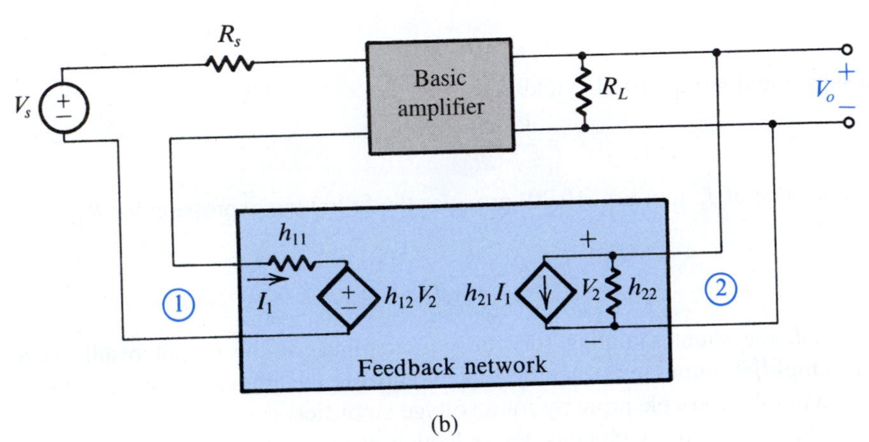

(b)

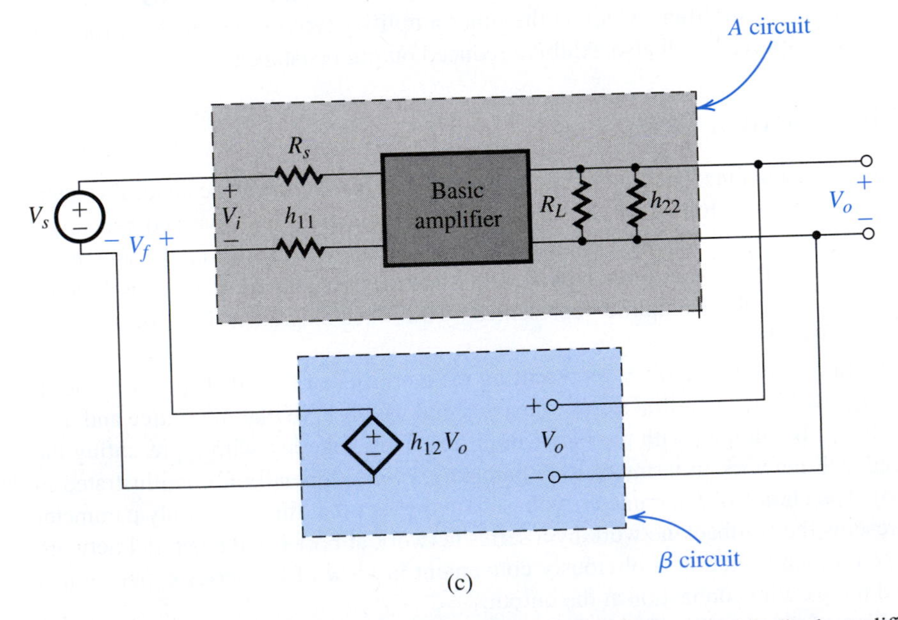

(c)

Figure 10.14 Derivation of the A circuit and β circuit for the series–shunt feedback amplifier. **(a)** Block diagram of a practical series–shunt feedback amplifier. **(b)** The circuit in **(a)** with the feedback network represented by its h parameters. **(c)** The circuit in **(b)** with h_{21} neglected.

passive, its forward transmission can be neglected in comparison to the much larger forward transmission of the basic amplifier. We will therefore assume that $|h_{21}|_{\substack{\text{feedback}\\\text{network}}} \ll |h_{21}|_{\substack{\text{basic}\\\text{amplifier}}}$ and thus omit the controlled source $h_{21}I_1$ altogether.

Compare the circuit of Fig. 10.14(b) (after eliminating the current source $h_{21}I_1$) with the ideal circuit of Fig. 10.12(a). We see that by including h_{11} and h_{22} with the basic amplifier, we obtain the circuit shown in Fig. 10.14(c), which is very similar to the ideal circuit. Now, if the basic amplifier is unilateral (or almost unilateral)—that is it does not contain internal feedback—then the circuit of Fig. 10.14(c) is equivalent to the ideal circuit. It follows then that the A circuit is obtained by augmenting the basic amplifier at the input with the source resistance R_s and the resistance h_{11} of the feedback network, and at the output with the load resistance R_L and the conductance h_{22} of the feedback network.

We conclude that the loading effect of the feedback network on the basic amplifier is represented by the components h_{11} and h_{22}. From the definitions of the h parameters in Appendix C we see that h_{11} is the resistance looking into port 1 of the feedback network with port 2 short-circuited. Since port 2 of the feedback network is connected in *shunt* with the output port of the amplifier, short-circuiting port 2 destroys the feedback. Similarly, h_{22} is the conductance looking into port 2 of the feedback network with port 1 open-circuited. Since port 1 of the feedback network is connected in *series* with the amplifier input, open-circuiting port 1 destroys the feedback.

These observations suggest a simple rule for finding the loading effects of the feedback network on the basic amplifier: The loading effect is found by looking into the appropriate port of the feedback network while the other port is open-circuited or short-circuited so as to destroy the feedback. If the connection is a shunt one, we short-circuit the port; if it is a series one, we open-circuit it. In Sections 10.5, 10.6, and 10.7 it will be seen that this simple rule applies also to the other three feedback topologies.[3]

We next consider the determination of β. From Fig. 10.14(c), we see that β is equal to h_{12} of the feedback network,

$$\beta = h_{12} \equiv \left.\frac{V_1}{V_2}\right|_{I_1=0} \tag{10.23}$$

Thus to measure β, one applies a voltage to port 2 of the feedback network and measures the voltage that appears at port 1 while the latter port is open-circuited. This result is intuitively appealing because the object of the feedback network is to sample the output voltage ($V_2 = V_o$) and provide a voltage signal ($V_1 = V_f$) that is mixed in series with the input source. The series connection at the input suggests that (as in the case of finding the loading effects of the feedback network) β should be found with port 1 open-circuited.

10.4.3 Summary

A summary of the rules for finding the A circuit and β for a given series–shunt feedback amplifier of the form in Fig. 10.14(a) is given in Fig. 10.15. As for using the feedback formulas in Eqs. (10.19) and (10.22) to determine the input and output resistances, it is important to note that:

1. R_i and R_o are the input and output resistances, respectively, of the A circuit in Fig. 10.15(a).
2. R_{if} and R_{of} are the input and output resistances, respectively, of the feedback amplifier, *including* R_s and R_L (see Fig. 10.14a).

[3] A simple rule to remember: If the connection is *sh*unt, *sh*ort it; if *se*ries, *se*ver it.

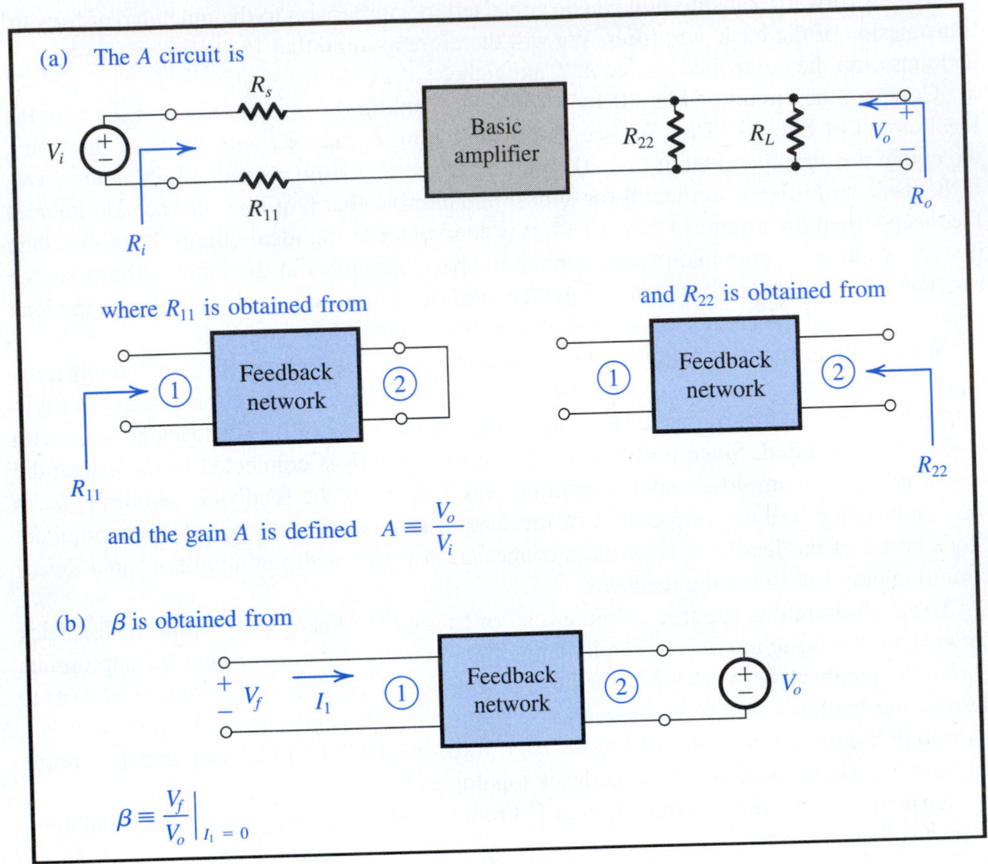

Figure 10.15 Summary of the rules for finding the A circuit and β for the series–shunt case of Fig. 10.14(a).

3. The actual input and output resistances of the feedback amplifier usually exclude R_s and R_L. These are denoted R_{in} and R_{out} in Fig. 10.14(a) and can be easily determined as

$$R_{in} = R_{if} - R_s \tag{10.24}$$

$$R_{out} = 1\Big/\left(\frac{1}{R_{of}} - \frac{1}{R_L}\right) \tag{10.25}$$

Example 10.3

Figure 10.16(a) shows an op amp connected in the noninverting configuration. The op amp has an open-loop gain μ, a differential input resistance R_{id}, and an output resistance r_o. Recall that in our analysis of op-amp circuits in Chapter 2, we neglected the effects of R_{id} (assumed it to be infinite) and of r_o (assumed it to be zero). Here we wish to use the feedback method to analyze the circuit taking both R_{id} and r_o into account. Find expressions for A, β, the closed-loop gain V_o/V_s, the input resistance R_{in} (see Fig. 10.16a), and the output resistance R_{out}. Also find numerical values, given $\mu = 10^4$, $R_{id} = 100$ kΩ, $r_o = 1$ kΩ, $R_L = 2$ kΩ, $R_1 = 1$ kΩ, $R_2 = 1$ MΩ, and $R_s = 10$ kΩ.

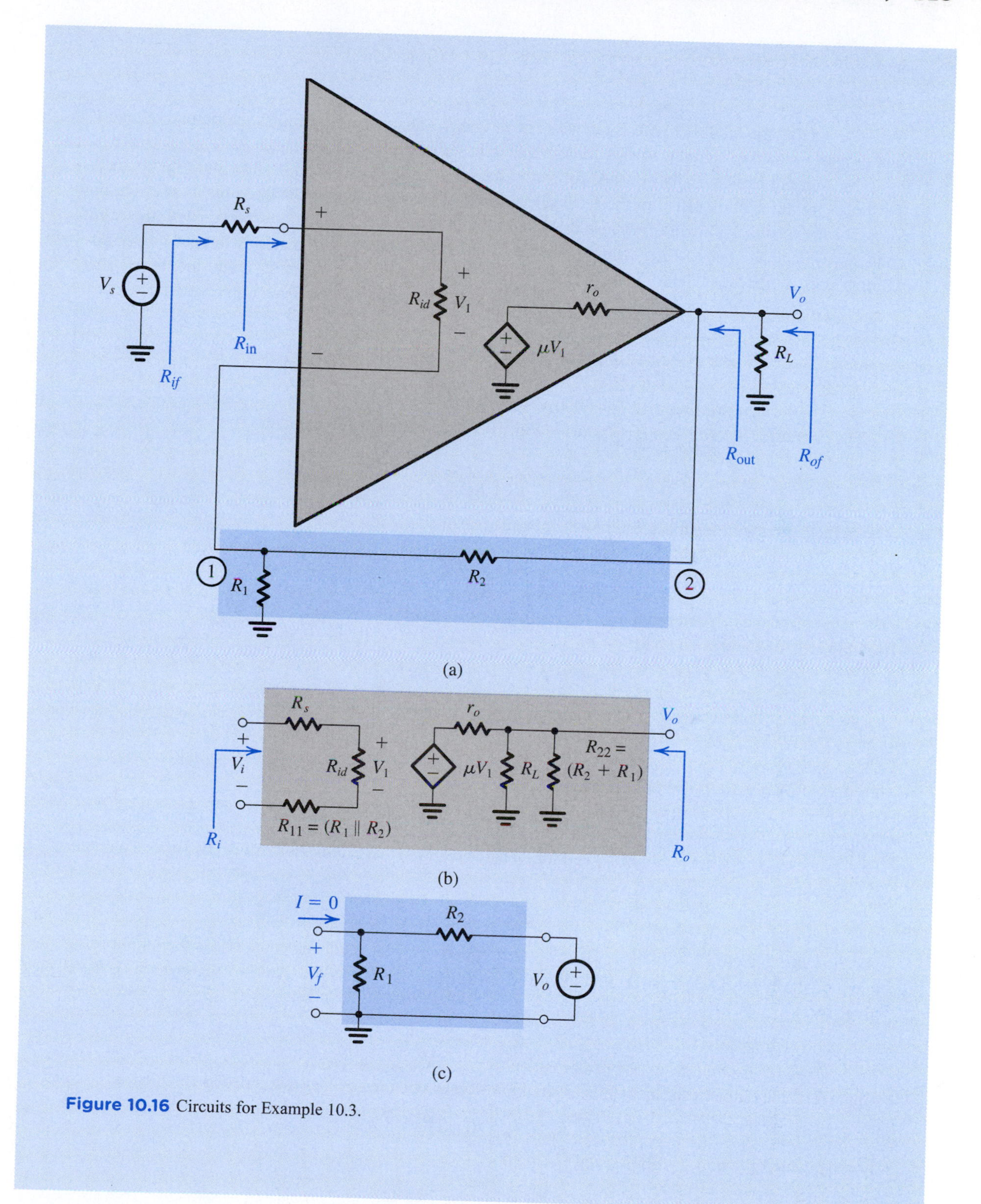

Figure 10.16 Circuits for Example 10.3.

Example 10.3 *continued*

Solution

We observe that the feedback network consists of R_2 and R_1. This network samples the output voltage V_o and provides a voltage signal (across R_1) that is mixed in series with the input source V_s.

The A circuit can be easily obtained following the rules of Fig. 10.15, and is shown in Fig. 10.16(b). Observe that the loading effect of the feedback network at the input side is obtained by short-circuiting port 2 of the feedback network (because it is connected in shunt) and looking into port 1, with the result that $R_{11} = R_1 \| R_2$. The loading effect of the feedback network at the output side is found by open-circuiting port 1 (because it is connected in series) and looking into port 2, with the result that $R_{22} = R_2 + R_1$. For the resulting A circuit in Fig. 10.16(b), we can write by inspection:

$$A \equiv \frac{V_o}{V_i} = \mu \frac{R_L \| (R_1 + R_2)}{[R_L \| (R_1 + R_2)] + r_o} \frac{R_{id}}{R_{id} + R_s + (R_1 \| R_2)}$$

For the values given, we find that $A \simeq 6000$ V/V.

The circuit for determining β is shown in Fig. 10.16(c), from which we obtain

$$\beta \equiv \frac{V_f}{V_o} = \frac{R_1}{R_1 + R_2} \simeq 10^{-3} \text{ V/V}$$

The voltage gain with feedback can now be obtained as

$$A_f \equiv \frac{V_o}{V_s} = \frac{A}{1 + A\beta} = \frac{6000}{7} = 857 \text{ V/V}$$

The input resistance R_{if} determined by the feedback equations is the resistance seen by the external source (see Fig. 10.16a), and is given by

$$R_{if} = R_i(1 + A\beta)$$

where R_i is the input resistance of the A circuit in Fig. 10.16(b):

$$R_i = R_s + R_{id} + (R_1 \| R_2)$$

For the values given, $R_i \simeq 111$ kΩ, resulting in

$$R_{if} = 111 \times 7 = 777 \text{ k}\Omega$$

This, however, is not the resistance asked for. What is required is R_{in}, indicated in Fig. 10.16(a). To obtain R_{in} we subtract R_s from R_{if}:

$$R_{in} = R_{if} - R_s$$

For the values given, $R_{in} = 739$ kΩ. The resistance R_{of} given by the feedback equations is the output resistance of the feedback amplifier, including the load resistance R_L, as indicated in Fig. 10.16(a). R_{of} is given by

$$R_{of} = \frac{R_o}{1 + A\beta}$$

where R_o is the output resistance of the A circuit. R_o can be obtained by inspection of Fig.10.16(b) as

$$R_o = r_o \| R_L \| (R_2 + R_1)$$

For the values given, $R_o \simeq 667$ Ω, and

$$R_{of} = \frac{667}{7} = 95.3 \text{ }\Omega$$

The resistance asked for, R_{out}, is the output resistance of the feedback amplifier excluding R_L. From Fig. 10.16(a) we see that

$$R_{of} = R_{out} \parallel R_L$$

Thus

$$R_{out} \simeq 100\ \Omega$$

Example 10.4

As another example of a series–shunt feedback amplifier, consider the circuit shown in Fig. 10.7(b) which is repeated in Fig. 10.17(a). It is required to analyze this amplifier to obtain its voltage gain V_o/V_s, input resistance R_{in}, and output resistance R_{out}. Find numerical values for the case $g_{m1} = g_{m2} = 4$ mA/V, $R_{D1} = R_{D2} = 10$ kΩ, $R_1 = 1$ kΩ, and $R_2 = 9$ kΩ. For simplicity, neglect r_o of each of Q_1 and Q_2.

Solution

We identify the feedback network as the voltage divider (R_1, R_2). Its loading effect at the input is obtained by short circuiting its port 2 (because it is connected in shunt with the output). Then, looking into its port 1, we see $R_1 \parallel R_2$. The loading effect at the output is obtained by open-circuiting port 1 of the feedback network (because it is connected in series with the input). Then, looking into port 2, we see R_2 in series with R_1. The A circuit will therefore be as shown in Fig. 10.17(b). The gain A is determined as the product of the gain of Q_1 and the gain of Q_2 as follows:

$$A_1 = \frac{V_{d1}}{V_i} = -\frac{R_{D1}}{1/g_{m1} + (R_1 \parallel R_2)} = -\frac{g_{m1}R_{D1}}{1 + g_{m1}(R_1 \parallel R_2)}$$

$$A_2 = \frac{V_o}{V_{d1}} = -g_{m2}[R_{D2} \parallel (R_1 + R_2)]$$

$$A = \frac{V_o}{V_i} = A_1 A_2 = \frac{g_{m1}R_{D1}g_{m2}[R_{D2} \parallel (R_1 + R_2)]}{1 + g_{m1}(R_1 \parallel R_2)}$$

For the numerical values given,

$$A = \frac{4 \times 10 \times 4[10 \parallel (1 + 9)]}{1 + 4(1 \parallel 9)} = 173.9 \quad \text{V/V}$$

The value of β is determined from the β circuit in Fig. 10.17(c),

$$\beta \equiv \frac{V_f}{V_o} = \frac{R_1}{R_1 + R_2}$$

For the numerical values given,

$$\beta = \frac{1}{1 + 9} = 0.1$$

The closed-loop gain V_o/V_s can now be found as

$$\frac{V_o}{V_s} = A_f = \frac{A}{1 + A\beta} = \frac{173.9}{1 + 173.9 \times 0.1} = 9.5 \text{ V/V}$$

Example 10.4 *continued*

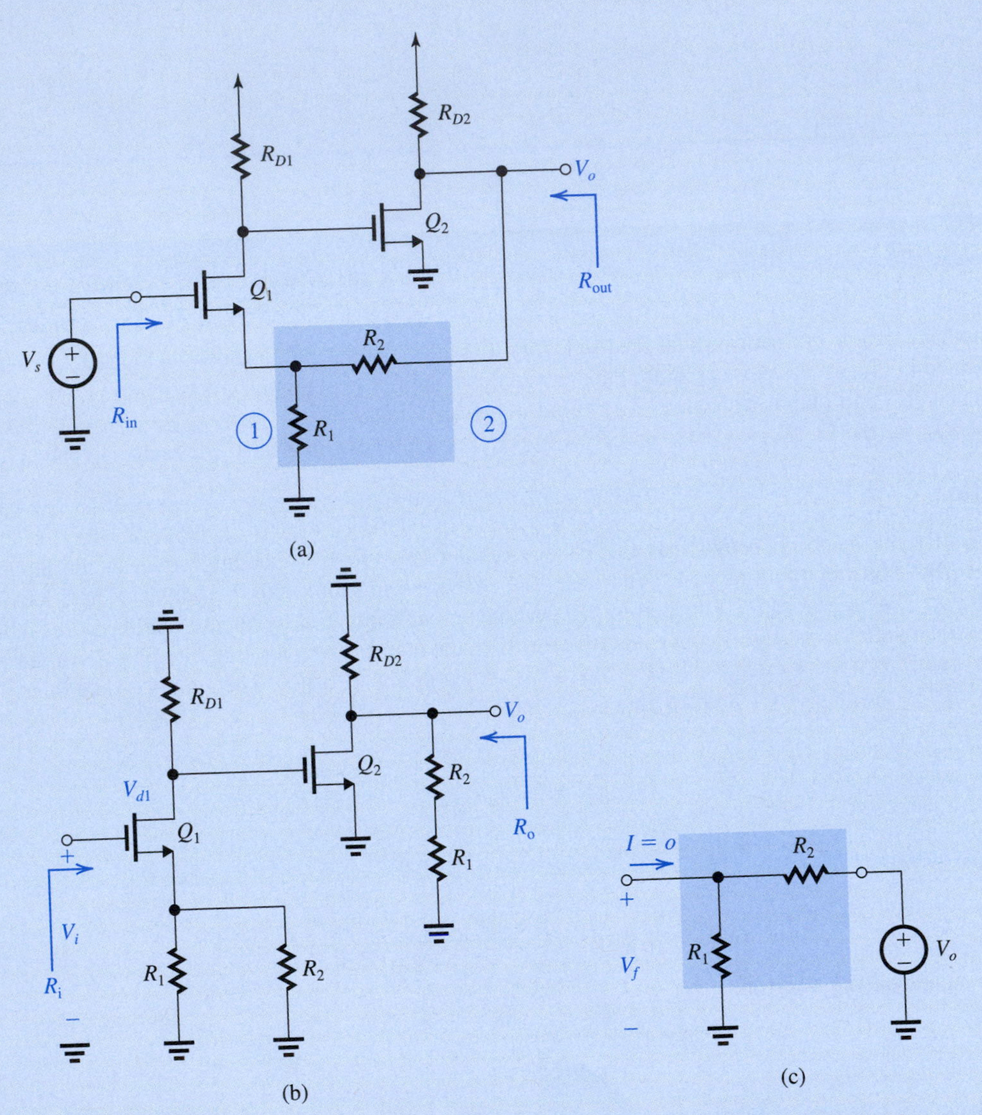

Figure 10.17 (**a**) Series–shunt feedback amplifier for Example 10.4; (**b**) The *A* circuit; (**c**) The β circuit.

The input resistance is obviously infinite because of the infinite input resistance of the MOSFET. The output resistance R_{out} is obtained as follows,

$$R_{out} = R_{of} = \frac{R_o}{1 + A\beta}$$

where R_o is the output resistance of the *A* circuit. From Fig. 10.17(b),

$$R_o = R_{D2} \,\|\, (R_1 + R_2)$$

$$= 10 \,\|\, 10 = 5 \text{ k}\Omega$$

The amount of feedback is

$$1 + A\beta = 1 + (173.9 \times 0.1) = 18.39$$

Thus,

$$R_{out} = \frac{5000}{18.39} = 272\ \Omega$$

which is relatively low given that the open-loop amplifier has $R_o = 5000\ \Omega$.

EXERCISES

10.10 If the op amp of Example 10.3 has a uniform –6-dB/octave high-frequency rolloff with f_{3dB} = 1 kHz, find the 3-dB frequency of the closed-loop gain V_o/V_s.
Ans. 7 kHz

10.11 The circuit shown in Fig. E10.11 consists of a differential stage followed by an emitter follower, with series–shunt feedback supplied by the resistors R_1 and R_2. Assuming that the dc component of V_s is zero, and that β of the BJTs is very high, find the dc operating current of each of the three transistors and show that the dc voltage at the output is approximately zero. Then find the values of A, β, $A_f \equiv V_o/V_s$, R_{in}, and R_{out}. Assume that the transistors have $\beta = 100$.
Ans. 85.7 V/V; 0.1 V/V; 8.96 V/V; 191 kΩ; 19.1 Ω.

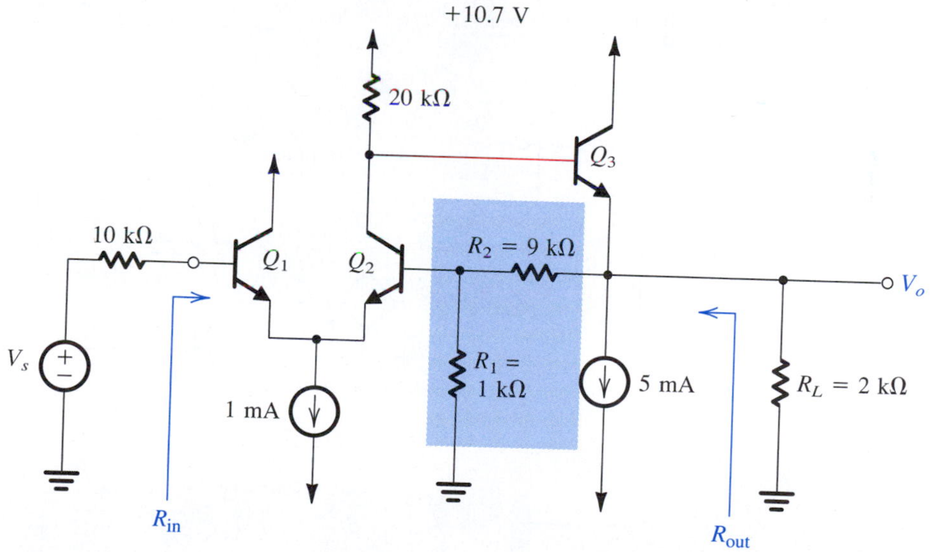

Figure E10.11

10.12 For the series–shunt amplifier in Fig. 10.7(c), find A, β, A_f, R_{in}, and R_{out}. Neglect r_o of Q.
Ans. $A = g_m[R_D \| (R_1 + R_2)]$; $\beta = R_1/(R_1 + R_2)$;

$A_f = A/(1 + A\beta)$; $R_{in} = (1/g_m)(1 + A\beta)$;

$R_{out} = [R_D \| (R_1 + R_2)]/(1 + A\beta)$

10.5 The Feedback Transconductance Amplifier (Series–Series)

10.5.1 The Ideal Case

As mentioned in Section 10.3, the series–series feedback topology stabilizes I_o/V_s and is therefore best suited for transconductance amplifiers. Figure 10.18(a) shows the ideal structure for the series–series feedback amplifier. It consists of a unilateral open-loop amplifier (the A circuit) and an ideal feedback network. The A circuit has an input resistance R_i, a short-circuit transconductance $A \equiv I_o/V_i$, and an output resistance R_o. The β circuit samples the short-circuit output current I_o and provides a feedback voltage V_f that is subtracted from V_s in the series input loop. Note that the β circuit presents zero resistance to the output loop, and thus does not load the amplifier output. Also, the feedback signal $V_f = \beta I_o$ is an ideal voltage source, thus the β circuit does not load the amplifier input. Also observe that while A is a transconductance, β is a transresistance, and thus the loop gain $A\beta$ is, as expected, a dimensionless quantity. Finally, note that the source and the load resistances have been absorbed inside the A circuit (more on this later).

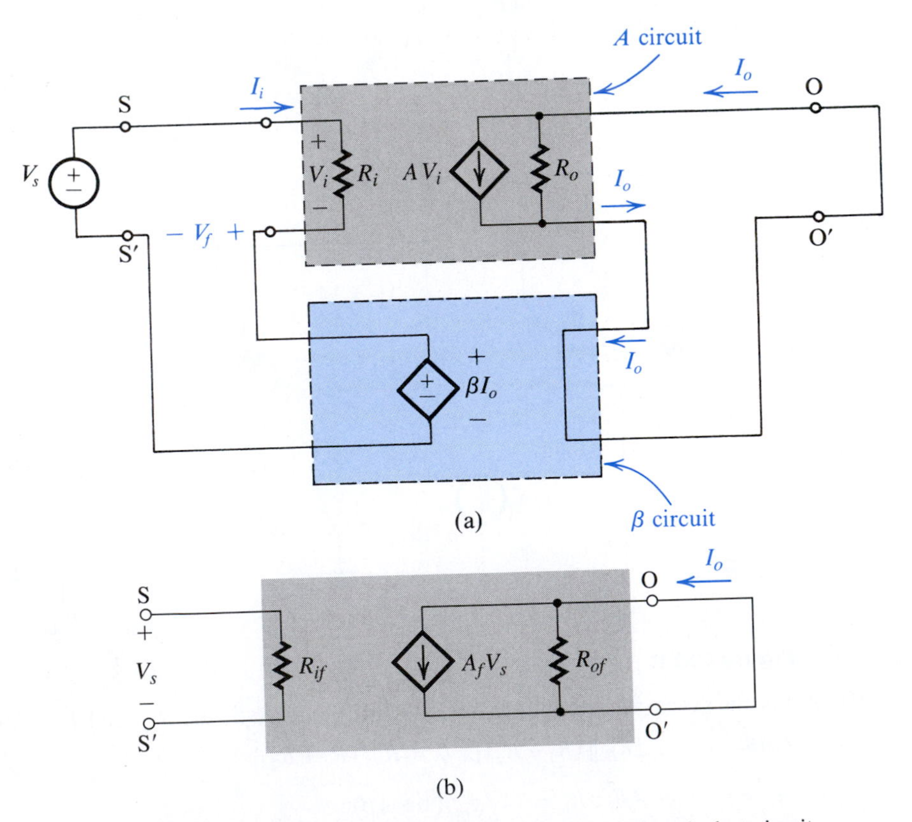

Figure 10.18 The series–series feedback amplifier: (**a**) ideal structure; (**b**) equivalent circuit.

Since the structure of Fig. 10.18(a) follows the ideal feedback structure of Fig. 10.1, we can obtain the closed-loop gain A_f as

$$A_f \equiv \frac{I_o}{V_s} = \frac{A}{1+A\beta} \tag{10.26}$$

The feedback transconductance amplifier can be represented by the equivalent circuit in Fig. 10.18(b). Note that A_f is the short-circuit transconductance. Because of the series mixing, the input resistance with feedback, R_{if}, will be larger than the input resistance of the A circuit, R_i, by a factor equal to the amount of feedback,

$$R_{if} = R_i(1+A\beta) \tag{10.27}$$

Recall that the derivation we employed in the previous section to obtain R_{if} of the series–shunt feedback amplifier did not depend on the method of sampling. Thus it applies equally well to the series–series amplifier we are considering here.

Next we consider the output resistance R_{of} of the feedback transconductance amplifier. From the equivalent circuit in Fig. 10.18(b) we observe that R_{of} is the resistance seen by breaking the output loop (say at OO′) and setting V_s to zero. Thus to find the output resistance R_{of} of the series–series feedback amplifier of Fig. 10.18(a) we reduce V_s to zero and break the output circuit to apply a test current I_x, as shown in Fig. 10.19:

$$R_{of} \equiv \frac{V_x}{I_x} \tag{10.28}$$

In this case, $V_i = -V_f = -\beta I_o = -\beta I_x$. Thus for the circuit in Fig. 10.19 we obtain

$$V_x = (I_x - AV_i)R_o = (I_x + A\beta I_x)R_o$$

Hence

$$R_{of} = (1+A\beta)R_o \tag{10.29}$$

That is, in this case the negative feedback increases the output resistance. This should have been expected, since the negative feedback tries to make I_o constant in spite of changes in the output voltage, which means increased output resistance. This result also confirms our earlier observation: The relationship between R_{of} and R_o is a function only of the method of sampling.

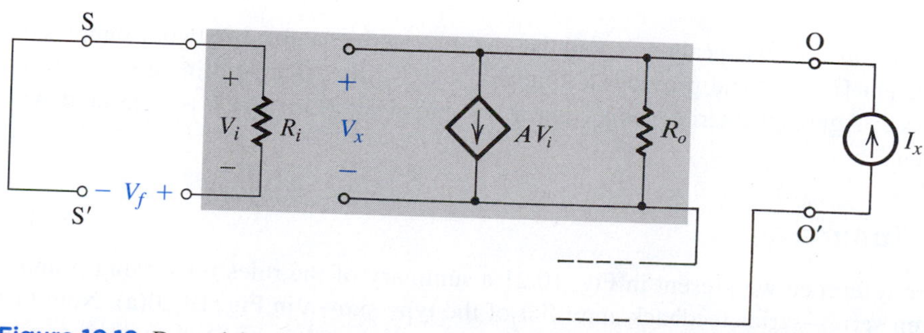

Figure 10.19 Determining the output resistance R_{of} of the series–series feedback amplifier.

While voltage (shunt) sampling reduces the output resistance, current (series) sampling increases it.

We conclude that the series–series feedback topology increases both the input and the output resistance, a highly desirable outcome for a transconductance amplifier.

10.5.2 The Practical Case

Figure 10.20(a) shows a block diagram for a practical series–series feedback amplifier. To be able to apply the feedback equations to this amplifier, we have to represent it by the ideal structure of Fig. 10.18(a). Our objective therefore is to devise a simple method for finding A and β. Observe the definition of the amplifier input resistance R_{in} and output resistance R_{out}. It is important to note that these are different from R_{if} and R_{of}, which are determined by the feedback equations, as will become clear shortly.

The series–series amplifier of Fig. 10.20(a) is redrawn in Fig. 10.20(b) with R_s and R_L shown closer to the basic amplifier, and the two-port feedback network represented by its z parameters (Appendix C). This parameter set has been chosen because it is the only one that provides a representation of the feedback network with a series circuit at the input and a series circuit at the output. This is obviously convenient in view of the series connections at input and output. The input and output resistances with feedback, R_{if} and R_{of}, are indicated on the diagram.

As we have done in the case of the series–shunt amplifier, we shall assume that the forward transmission through the feedback network is negligible in comparison to that through the basic amplifier, and thus we can dispense with the voltage source $z_{21}I_1$ in Fig. 10.20(b). Doing this, and redrawing the circuit to include z_{11} and z_{22} with the basic amplifier, results in the circuit in Fig. 10.20(c). Now if the basic amplifier is unilateral (or almost unilateral), then the circuit in Fig. 10.20(c) is equivalent to the ideal circuit of Fig. 10.18(a).

It follows that the A circuit is composed of the basic amplifier augmented at the input with R_s and z_{11} and augmented at the output with R_L and z_{22}. Since z_{11} and z_{22} are the impedances looking into ports 1 and 2, respectively, of the feedback network with the other port open-circuited, we see that finding the loading effects of the feedback network on the basic amplifier follows the rule formulated in Section 10.4. That is, we look into one port of the feedback network while the other port is open-circuited or short-circuited so as to destroy the feedback (open if series and short if shunt).

From Fig. 10.20(c) we see that β is equal to z_{12} of the feedback network,

$$\beta = z_{12} \equiv \left.\frac{V_1}{I_2}\right|_{I_1=0} \tag{10.30}$$

This result is intuitively appealing. Recall that in this case the feedback network samples the output current $[I_2 = I_o]$ and provides a voltage $[V_f = V_1]$ that is mixed in series with the input source. Again, the series connection at the input suggests that β is measured with port 1 open.

10.5.3 Summary

For future reference we present in Fig. 10.21 a summary of the rules for finding A and β for a given series–series feedback amplifier of the type shown in Fig. 10.20(a). Note that R_i is the input resistance of the A circuit, and its output resistance is R_o, which can be

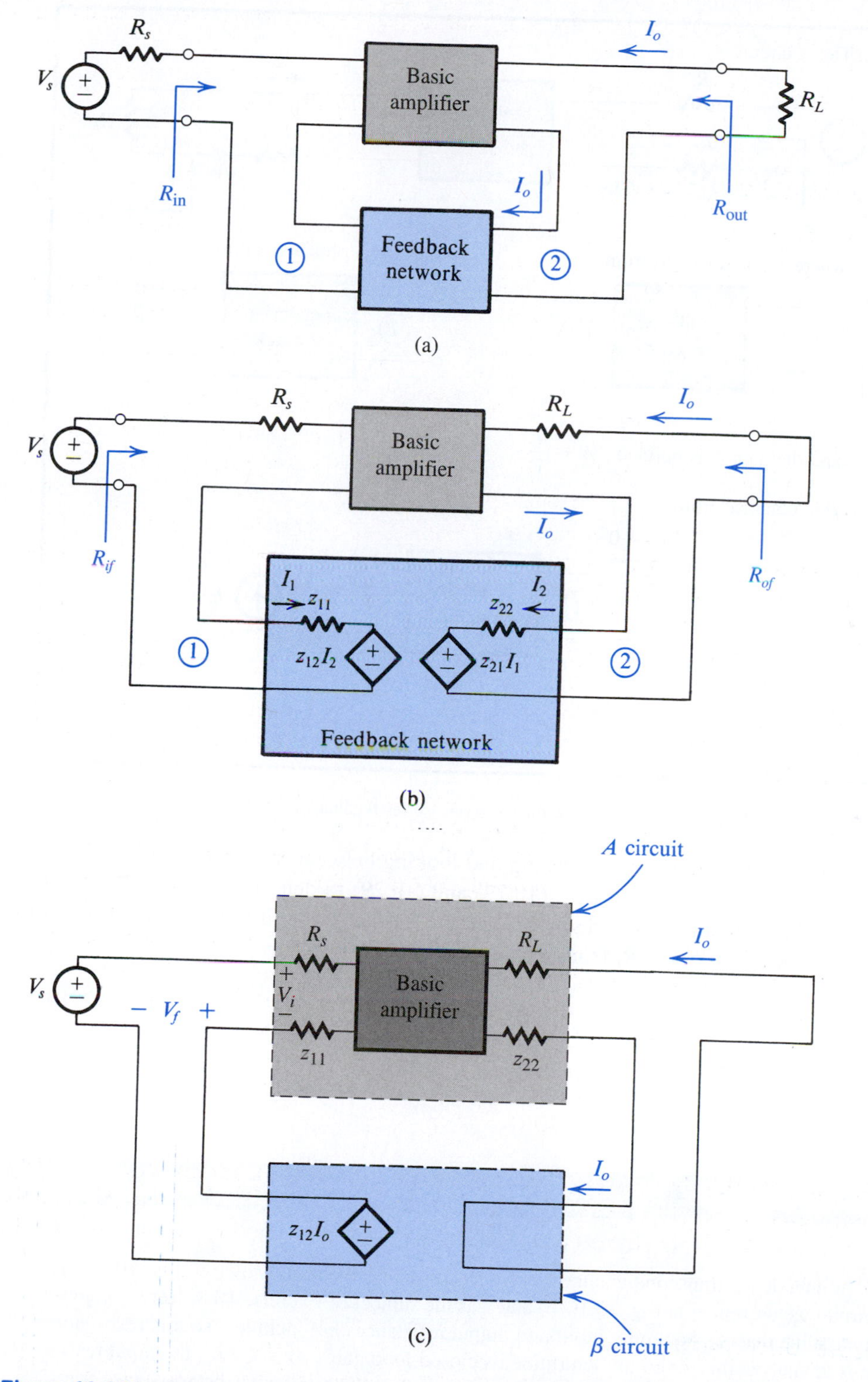

Figure 10.20 Derivation of the A circuit and the β circuit for series–series feedback amplifiers. **(a)** A series–series feedback amplifier. **(b)** The circuit of **(a)** with the feedback network represented by its z parameters. **(c)** A redrawing of the circuit in **(b)** with z_{21} neglected.

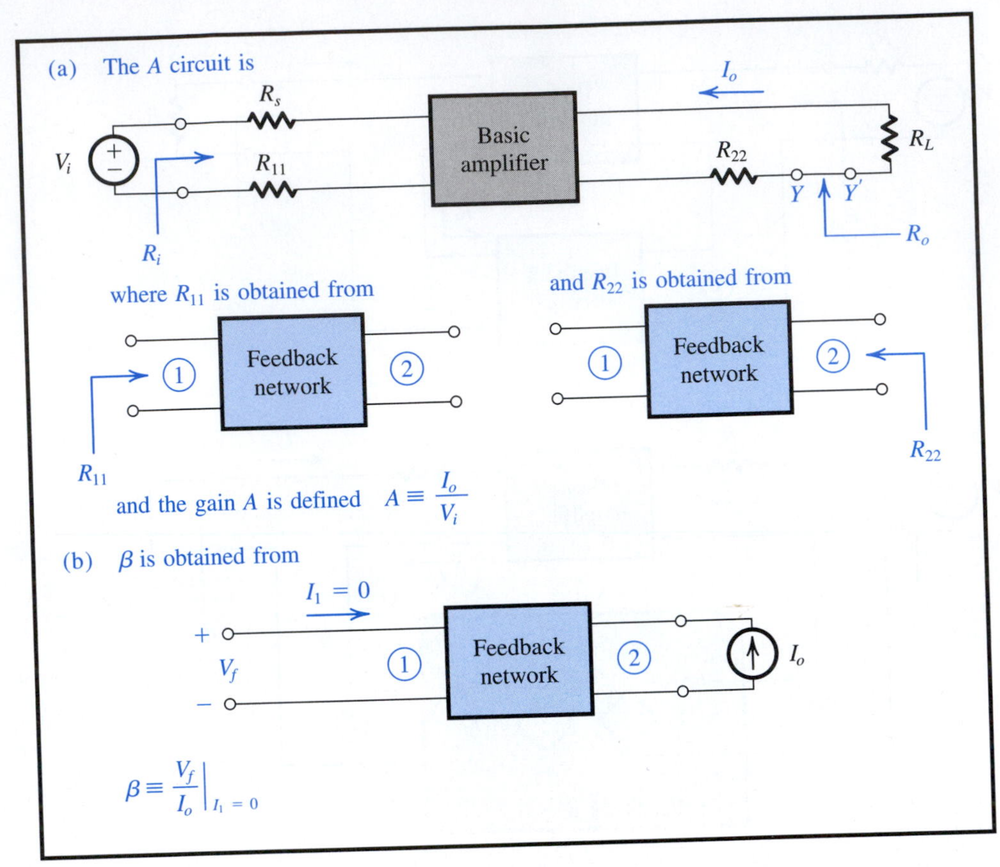

Figure 10.21 Finding the A circuit and β for the series–series feedback amplifier.

determined by breaking the output loop and looking between Y and Y' while V_i is set to zero. R_i and R_o can be used in Eqs. (10.27) and (10.29) to determine R_{if} and R_{of} (see Fig. 10.20b). The input and output resistances of the feedback amplifier can then be found by subtracting R_s from R_{if} and R_L from R_{of},

$$R_{\text{in}} = R_{if} - R_s \tag{10.31}$$

$$R_{\text{out}} = R_{of} - R_L \tag{10.32}$$

Example 10.5

As a first example of a feedback transconductance amplifier, consider the circuit shown in Fig. 10.22(a). This is the same circuit we presented in Fig. 10.10(b) and was the subject of Exercise 10.8. Here, for generality we not only assume that A_1 has finite input and output resistances but include a source resistance R_s. The objective is to analyze this circuit to determine its closed-loop gain $A_f \equiv I_o/V_s$, the input resistance of the feedback amplifier R_{in}, and the output resistance R_{out}. The latter is the resistance seen between the two terminals of R_L looking back into the output loop.

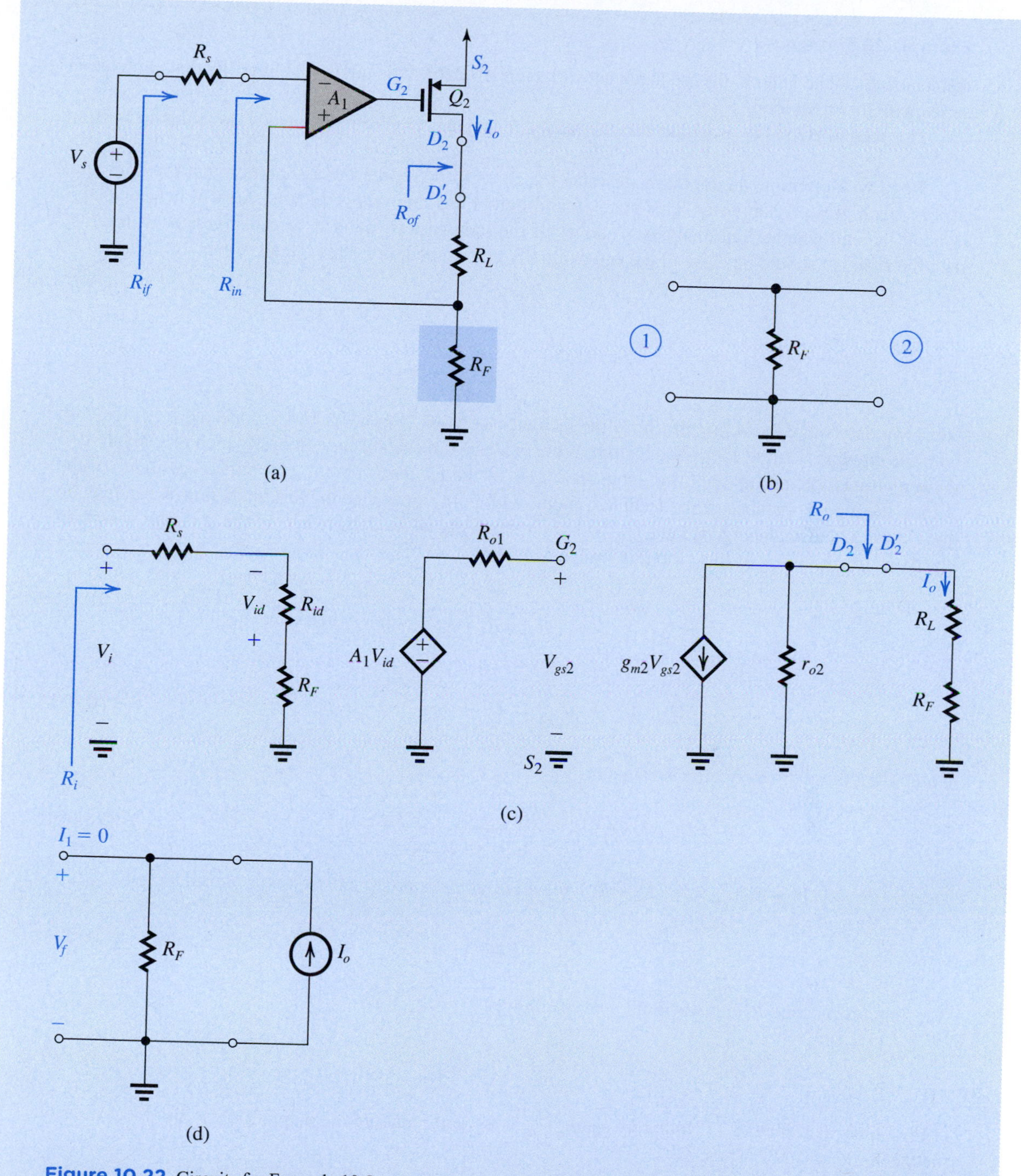

Figure 10.22 Circuits for Example 10.5.

Solution

First we identify the basic amplifier and the feedback circuit. The basic amplifier consists of the differential amplifier A_1 cascaded with the CS PMOS transistor Q_2. The output current I_o is sensed by the series

Example 10.5 *continued*

resistance R_F. The latter is the feedback network (Fig. 10.22b). It develops a voltage V_f that is mixed in series with the input loop.

The second step is to ascertain that the feedback is negative. We have already done this in Section 10.3.

Next, we determine an approximate value for $A_f \equiv I_o/V_s$ under the assumption that the loop gain $A\beta$ is much greater than unity. This value, found before any analysis is undertaken, will help us determine at the end whether our analysis is correct: If the loop gain is found to be much greater than unity, then the final A_f should be close to the value initially determined. From the circuit of Fig. 10.22(a),

$$\beta = R_F$$

and thus for large $A\beta$,

$$A_f \simeq \frac{1}{\beta} = \frac{1}{R_F}$$

Next, we determine the A circuit. Since the feedback network (Fig. 10.22b) is connected in series with both the input and output loops, we include a resistance R_F in each of these loops (which is equivalent to saying we include, at the input, the input resistance of the feedback circuit with port 2 open and, at the output, the input resistance of the feedback circuit with port 1 open). Doing this, including R_s and R_L in the A circuit, and replacing A_1 and Q_2 with their small-signal models, results in the A circuit shown in Fig. 10.22(c). Analysis of this circuit is straightforward:

$$V_{id} = -V_i \frac{R_{id}}{R_{id}+R_s+R_F} \tag{10.33}$$

$$V_{gs2} = A_1 V_{id} \tag{10.34}$$

$$I_o = -g_{m2} V_{gs2} \frac{r_{o2}}{r_{o2}+R_L+R_F} \tag{10.35}$$

Combining these three equations results in

$$A \equiv \frac{I_o}{V_i} = (A_1 g_{m2})\left(\frac{R_{id}}{R_{id}+R_s+R_F}\right)\left(\frac{r_{o2}}{r_{o2}+R_L+R_F}\right) \tag{10.36}$$

Usually $R_{id} \gg (R_s+R_F)$, $r_{o2} \gg (R_L+R_F)$, resulting in the approximate expression for A:

$$A \simeq A_1 g_{m2} \tag{10.37}$$

The input resistance R_i can be found by inspection as

$$R_i = R_s + R_{id} + R_F \tag{10.38}$$

The output resistance R_o is found by setting $V_i = 0$, and breaking the output loop at any location, say between D_2 and D_2'. Thus,

$$R_o = r_{o2} + R_L + R_F \tag{10.39}$$

Finally, β can be found from Fig. 10.22(d) as

$$\beta \equiv \frac{V_f}{I_o} = R_F$$

The loop gain $A\beta$ is thus

$$A\beta = (A_1 g_{m2} R_F)\left(\frac{R_{id}}{R_{id}+R_s+R_F}\right)\left(\frac{r_{o2}}{r_{o2}+R_L+R_F}\right) \tag{10.40}$$

$$\simeq A_1 g_{m2} R_F \tag{10.41}$$

With numerical values, one can now obtain the value of $A\beta$ and determine whether it is indeed much greater than unity. We next determine the closed-loop gain

$$A_f = \frac{A}{1+A\beta}$$

Substituting for A from Eq. (10.37) and for $A\beta$ from Eq. (10.41), we have

$$A_f \simeq \frac{A_1 g_{m2}}{1+A_1 g_{m2} R_F}$$

For $A_1 g_{m2} R_F \gg 1$,

$$A_f \simeq \frac{1}{R_F}$$

which is the value we found at the outset.

The series mixing raises the input resistance with feedback,

$$R_{if} = R_i(1+A\beta)$$

Substituting for R_i from Eq. (10.38) and for $A\beta$ from the full expression in Eq. (10.40), we obtain

$$R_{if} = (R_s + R_{id} + R_F)(1+A\beta)$$

$$= R_s + R_{id} + R_F + A_1 g_{m2} R_F R_{id} \frac{r_{o2}}{r_{o2}+R_L+R_F}$$

which for $r_{o2} \gg R_L + R_F$ yields

$$R_{if} \simeq R_s + R_{id} + R_F + A_1 g_{m2} R_F R_{id}$$

To obtain R_{in}, we subtract R_s from R_{if} (see Fig. 10.22a):

$$R_{\text{in}} = R_{id} + R_F + A_1 g_{m2} R_F R_{id}$$

Usually $R_F \ll R_{id}$,

$$R_{\text{in}} \simeq R_{id}(1 + A_1 g_{m2} R_F) \tag{10.42}$$

which is an intuitively appealing result: The series mixing at the input raises the input resistance R_{id} by a factor equal to the approximate value of $(1 + A\beta)$.

To obtain R_{of}, we note that the series connection at the output raises the output resistance, thus,

$$R_{of} = R_o(1+A\beta)$$

$$= (r_{o2} + R_L + R_F)(1+A\beta)$$

$$= r_{o2} + R_L + R_F + A\beta(r_{o2} + R_L + R_F)$$

Example 10.5 *continued*

Substituting for $A\beta$ from Eq. (10.40) and making the approximation $R_{id} \gg (R_s + R_F)$, we write

$$R_{of} \simeq r_{o2} + R_L + R_F + A_1 g_{m2} R_F r_{o2}$$

To obtain R_{out}, which is the resistance seen by R_L in the circuit of Fig. 10.22(a), we subtract R_L from R_{of},

$$R_{out} = r_{o2} + R_F + A_1 g_{m2} R_F r_{o2}$$

usually $R_F \ll r_{o2}$; thus,

$$R_{out} \simeq r_{o2}(1 + A_1 g_{m2} R_F)$$

which is an intuitively appealing result: The series connection at the output raises the output resistance of Q_2 (r_{o2}) by a factor equal to the amount of feedback.

Finally, we note that we have deliberately solved this problem in great detail to illustrate the beauty

EXERCISE

D10.13 For the circuit analyzed in Example 10.5, select a value for R_F that will result in $A_f \simeq 5$ mA/V. Now, for $A_1 = 200$ V/V, $g_{m2} = 2$ mA/V, $R_{id} = 100$ kΩ, $r_{o2} = 20$ kΩ, and assuming that $R_s \ll R_{id}$ and $R_L \ll r_{o2}$, find the value of A_f realized and the input and output resistances of the feedback transconductance amplifier. If for some reason g_{m2} drops in value by 50%, what is the corresponding percentage change in A_f?

Ans. 200 Ω; 4.94 mA/V; 8.1 MΩ; 1.62 MΩ; –1.25 %

Example 10.6

Because negative feedback extends the amplifier bandwidth, it is commonly used in the design of broad-band amplifiers. One such amplifier is the MC1553. Part of the circuit of the MC1553 is shown in Fig. 10.23(a). The circuit shown (called a **feedback triple**) is composed of three gain stages with series–series feedback provided by the network composed of R_{E1}, R_F, and R_{E2}.

Observe that the feedback network samples the emitter current I_o of Q_3, and thus I_o is the output quantity of the feedback amplifier. However, practically speaking, I_o is rather difficult to utilize. Thus usually the collector current of Q_3, I_c, is taken as the output. This current is of course almost equal to I_o; $I_c = \alpha I_o$. Thus, as a transconductance amplifier with I_c as the output current, the output resistance of interest is that labeled R_{out} in Fig. 10.23(a). In some applications, I_c is passed through a load resistance, such as R_{C3}, and the voltage V_o is taken as the output. Assume that the bias circuit, which is not shown, establishes $I_{C1} = 0.6$ mA, $I_{C2} = 1$ mA, and $I_{C3} = 4$ mA. Also assume that for all three transistors,[4] $h_{fe} = 100$ and $r_o = \infty$.

[4]To avoid possible confusion of the BJT current gain β and the feedback factor β, we sometimes use h_{fe} to denote the transistor β.

(a) Anticipating that the loop gain will be large, find an approximate expression and value for the closed-loop gain $A_f \equiv I_o/V_s$ and hence for I_c/V_s. Also find V_o/V_s.
(b) Use feedback analysis to find A, β, A_f, V_o/V_s, R_{in}, and R_{out}. For the calculation of R_{out}, assume that r_o of Q_3 is 25 kΩ.

Solution

(a) When $A\beta \gg 1$,

$$A_f \equiv \frac{I_o}{V_s} \simeq \frac{1}{\beta}$$

where the feedback factor β can be found from the feedback network. The feedback network is highlighted in Fig. 10.23(a), and the determination of the value of β is illustrated in Fig. 10.23(b), from which we find

$$\beta \equiv \frac{V_f}{I_o} = \frac{R_{E2}}{R_{E2} + R_F + R_{E1}} \times R_{E1}$$

$$= \frac{100}{100 + 640 + 100} \times 100 = 11.9\ \Omega$$

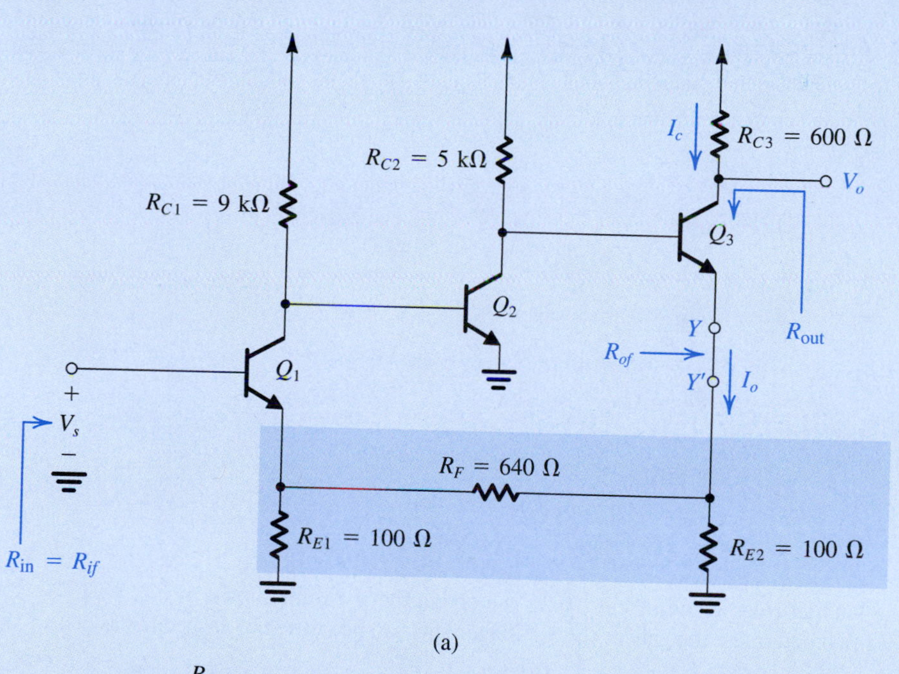

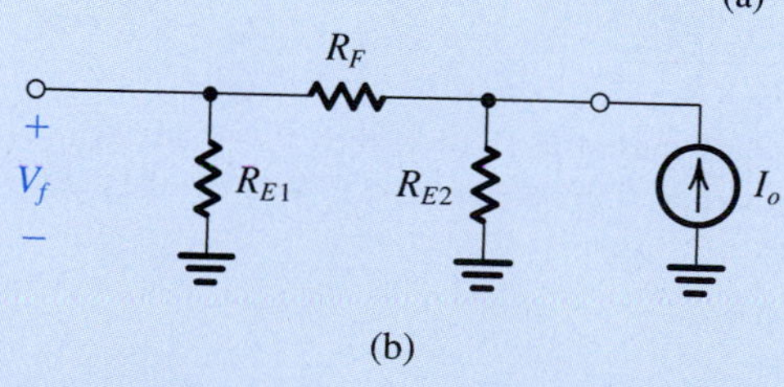

Figure 10.23 Circuits for Example 10.6.

Example 10.6 *continued*

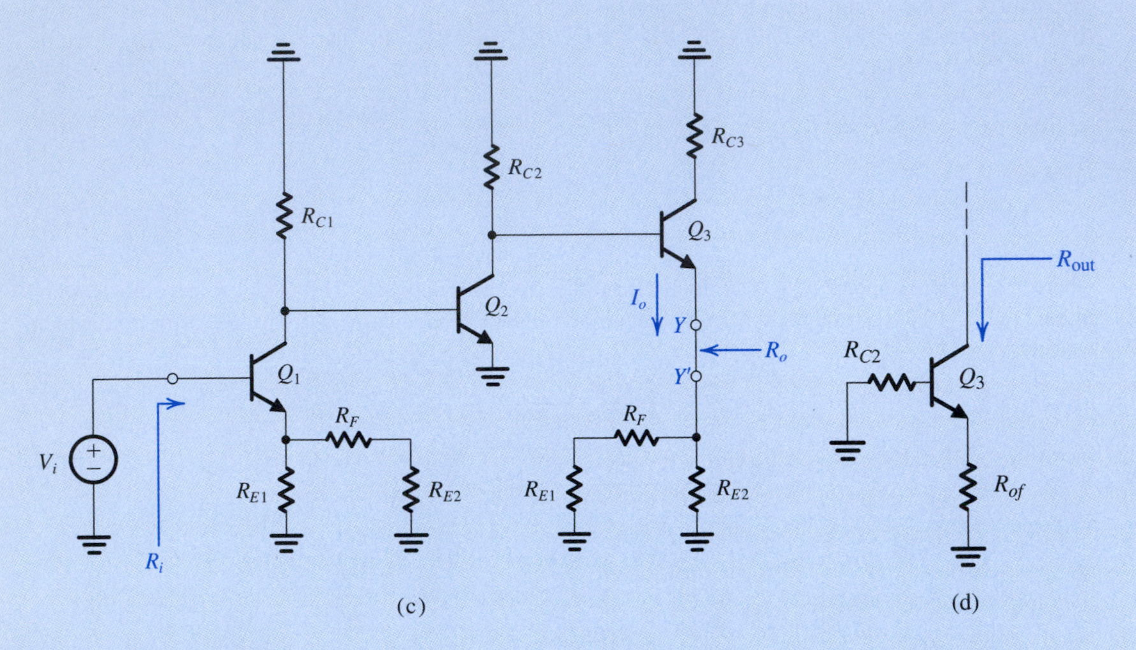

Figure 10.23 *continued*

Thus,

$$A_f \simeq \frac{1}{\beta}$$

$$= \frac{1}{R_{E2}}\left(1 + \frac{R_{E2} + R_F}{R_{E1}}\right)$$

$$= \frac{1}{11.9} = 84 \text{ mA/V}$$

$$\frac{I_c}{V_s} \simeq \frac{I_o}{V_s} = 84 \text{ mA/V}$$

$$\frac{V_o}{V_s} = \frac{-I_c R_{C3}}{V_s} = -84 \times 0.6 = -50.4 \text{ V/V}$$

(b) Employing the loading rules given in Fig. 10.21, we obtain the A circuit shown in Fig. 10.23(c). To find $A \equiv I_o/V_i$ we first determine the gain of the first stage. This can be written by inspection as

$$\frac{V_{c1}}{V_i} = \frac{-\alpha_1(R_{C1} \parallel r_{\pi 2})}{r_{e1} + [R_{E1} \parallel (R_F + R_{E2})]}$$

Since Q_1 is biased at 0.6 mA, $r_{e1} = 41.7\ \Omega$. Transistor Q_2 is biased as 1 mA; thus $r_{\pi 2} = h_{fe}/g_{m2} = 100/40 = 2.5\ \text{k}\Omega$. Substituting these values together with $\alpha_1 = 0.99$, $R_{C1} = 9\ \text{k}\Omega$, $R_{E1} = 100\ \Omega$, $R_F = 640\ \Omega$, and $R_{E2} = 100\ \Omega$, results in

$$\frac{V_{c1}}{V_i} = -14.92 \text{ V/V}$$

Next, we determine the gain of the second stage, which can be written by inspection (noting that $V_{b2} = V_{c1}$) as

$$\frac{V_{c2}}{V_{c1}} = -g_{m2}\{R_{C2} \parallel (h_{fe} + 1)[r_{e3} + (R_{E2} \parallel (R_F + R_{E1}))]\}$$

Substituting $g_{m2} = 40$ mA/V, $R_{C2} = 5$ kΩ, $h_{fe} = 100$, $r_{e3} = 25/4 = 6.25$ Ω, $R_{E2} = 100$ Ω, $R_F = 640$ Ω, and $R_{E1} = 100$ Ω, results in

$$\frac{V_{c2}}{V_{c1}} = -131.2 \text{ V/V}$$

Finally, for the third stage we can write by inspection

$$\frac{I_o}{V_{c2}} = \frac{I_{e3}}{V_{b3}} = \frac{1}{r_{e3} + (R_{E2} \parallel (R_F + R_{E1}))}$$

$$= \frac{1}{6.25 + (100 \parallel 740)} = 10.6 \text{ mA/V}$$

Combining the gains of the three stages results in

$$A \equiv \frac{I_o}{V_i} = -14.92 \times -131.2 \times 10.6 \times 10^{-3}$$

$$= 20.7 \text{ A/V}$$

The closed-loop gain A_f can now be found from

$$A_f \equiv \frac{I_o}{V_s} = \frac{A}{1 + A\beta}$$

$$= \frac{20.7}{1 + 20.7 \times 11.9} = 83.7 \text{ mA/V}$$

which we note is very close to the approximate value found in (a) above.

The voltage gain is found from

$$\frac{V_o}{V_s} = \frac{-I_c R_{C3}}{V_s} \simeq \frac{-I_o R_{C3}}{V_s} = -A_f R_{C3}$$

$$= -83.7 \times 10^{-3} \times 600 = -50.2 \text{ V/V}$$

which is also very close to the approximate value found in (a) above.

The input resistance of the feedback amplifier is given by

$$R_{in} = R_{if} = R_i(1 + A\beta)$$

where R_i is the input resistance of the A circuit. The value of R_i can be found from the circuit in Fig. 10.23(c) as follows:

$$R_i = (h_{fe} + 1)[r_{e1} + (R_{E1} \parallel (R_F + R_{E2}))]$$

$$= 13.65 \text{ k}\Omega$$

Thus,

$$R_{if} = 13.65(1 + 20.7 \times 11.9) = 3.38 \text{ M}\Omega$$

To determine the output resistance R_{out}, which is the resistance looking into the collector of Q_3, we face a dilemma. The feedback does not sample I_c and thus we cannot employ the feedback formulas directly.[5] Nevertheless, we present a somewhat indirect solution to this problem below. Here we note parenthetically that had Q_1 been a MOSFET, this problem would not have existed, since $I_d = I_s$.

Since the feedback senses the emitter current I_o, the output resistance given by the feedback analysis will be the resistance seen in the emitter circuit, say between Y and Y',

$$R_{of} = R_o(1 + A\beta)$$

[5]This important point was first brought to the authors' attention by Gordon Roberts (see Roberts and Sedra, 1992).

Example 10.6 *continued*

where R_o can be determined from the A circuit in Fig. 10.23(c) by breaking the circuit between Y and Y'. The resistance looking between these two nodes can be found to be

$$R_o = [R_{E2} \| (R_F + R_{E1})] + r_{e3} + \frac{R_{C2}}{h_{fe} + 1}$$

which, for the values given, yields $R_o = 143.9\ \Omega$. The output resistance R_{of} of the feedback amplifier can now be found as

$$R_{of} = R_o(1 + A\beta) = 143.9(1 + 20.7 \times 11.9) = 35.6\ \text{k}\Omega$$

We can now use the value of R_{of} to obtain an approximate value for R_{out}. To do this, we assume that the effect of the feedback is to place a resistance R_{of} (35.6 kΩ) in the emitter of Q_3, and find the output resistance from the equivalent circuit shown in Fig. 10.23(d). This is the output resistance of a BJT with a resistance R_{of} in its emitter and a resistance R_{C2} in its base. The formula we have for this (Eq. 7.50) does not unfortunately account for a resistance in the base. The formula, however, can be modified (see Problem 10.48) to obtain

$$R_{\text{out}} = r_{o3} + [R_{of} \| (r_{\pi 3} + R_{C2})] \left[1 + g_{m3} r_{o3} \frac{r_{\pi 3}}{r_{\pi 3} + R_{C2}}\right]$$

$$= 25 + [35.6 \| (0.625 + 5)] \left[1 + 160 \times 25 \times \frac{0.625}{0.625 + 5}\right]$$

$$= 2.19\ \text{M}\Omega$$

Thus R_{out} is increased (from r_{o3}) but not by $(1 + A\beta)$.

EXERCISE

D10.14 For the feedback triple in Fig. 10.23(a), analyzed in Example 10.6, modify the value of R_F to obtain a closed-loop transconductance I_o/V_s of approximately 100 mA/V. Assume that the loop gain remains large. What is the new value of R_F? For this value, what is the approximate value of the voltage gain if the output voltage is taken at the collector of Q_3?

Ans. 800 Ω; –60 V/V

10.6 The Feedback Transresistance Amplifier (Shunt–Shunt)

10.6.1 The Ideal Case

As mentioned in Section 10.3, the shunt–shunt feedback topology stabilizes V_o/I_s and is thus best suited for transresistance amplifiers. Figure 10.24(a) shows the ideal structure for the shunt–shunt feedback amplifier. It consists of a unilateral open-loop amplifier (the A circuit) and an ideal feedback network. The A circuit has an input resistance R_i, an open-circuit transresistance $A \equiv V_o/I_i$, and an output resistance R_o. The β circuit samples the open-circuit

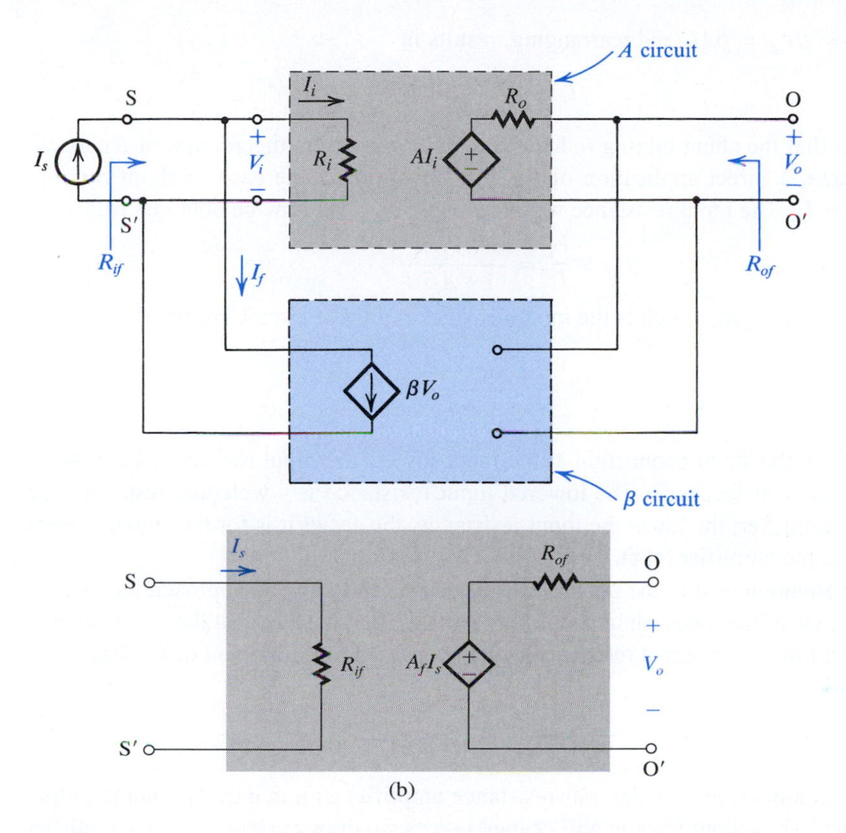

Figure 10.24 (a) Ideal structure for the shunt–shunt feedback amplifier. (b) Equivalent circuit of the amplifier in (a).

output voltage V_o and provides a feedback current I_f that is subtracted from the signal-source current I_s at the input nodes. Note that the β circuit presents an infinite impedance to the amplifier output and thus does not load the amplifier output. Also, the feedback signal $I_f = \beta V_o$ is provided as an ideal current source, and thus the β circuit does not load the amplifier input. Also observe that while A is a transresistance, β is a transconductance and thus the loop gain $A\beta$ is, as expected, a dimensionless quantity. Finally, note that the source and load resistances have been absorbed inside the A circuit (more on this later).

Since the structure of Fig. 10.24(a) follows the ideal feedback structure of Fig. 10.1, we can obtain the closed-loop gain A_f as

$$A_f \equiv \frac{V_o}{I_s} = \frac{A}{1 + A\beta} \tag{10.43}$$

The feedback transresistance amplifier can be represented by the equivalent circuit in Fig. 10.24(b). Note that A_f is the open-circuit transresistance. To obtain the input resistance R_{if}, refer to the input side of the block diagram in Fig. 10.24(a). The shunt connection at the input causes the feedback current to subtract from I_s resulting in a reduced current I_i into the A circuit,

$$I_i = I_s - I_f$$

Substituting $I_f = \beta V_o = \beta A I_i$ and rearranging, results in

$$I_i = \frac{I_s}{1 + A\beta}$$

which indicates that the shunt mixing reduces the input current by the amount of feedback. This is, of course, a direct application of Eq. (10.7), where in the case of shunt mixing, $x_s = I_s$ and $x_i = I_i$. The input resistance with feedback, R_{if}, can now be obtained from

$$R_{if} \equiv \frac{V_i}{I_s} = \frac{V_i}{(1 + A\beta)I_i}$$

Substituting for $V_i/I_i = R_i$, which is the input resistance of the A circuit, results in

$$R_{if} = \frac{R_i}{1 + A\beta} \tag{10.44}$$

Thus, as expected, the shunt connection at the input lowers the input resistance by a factor equal to the amount of feedback. The lowered input resistance is a welcome result for the transresistance amplifier; the lower the input resistance, the easier it is for the signal current source that feeds the amplifier input.

Turning our attention next to the output resistance, we can follow an approach identical to that used in the case of the series–shunt amplifier (Section 10.4) to show that the shunt connection at the output lowers the output resistance by a factor equal to the amount of feedback,

$$R_{of} = \frac{R_o}{1 + A\beta} \tag{10.45}$$

This also is a welcome result for the transresistance amplifier as it makes its voltage-output circuit more ideal; the output voltage will change less as we draw current from the amplifier output. Finally, note that *the shunt feedback connection, whether at the input or at the output, always reduces the corresponding resistance.*

10.6.2 The Practical Case

Figure 10.25 shows a block diagram for a practical shunt–shunt feedback amplifier. To be able to apply the feedback equations to this amplifier, we have to represent it by the ideal structure of Fig. 10.24(a). Our objective therefore is to devise a simple method for finding the A circuit and β. Building on the insight we have gained from our study of the series–shunt and series–series topologies, we present the method for the shunt–shunt case, without

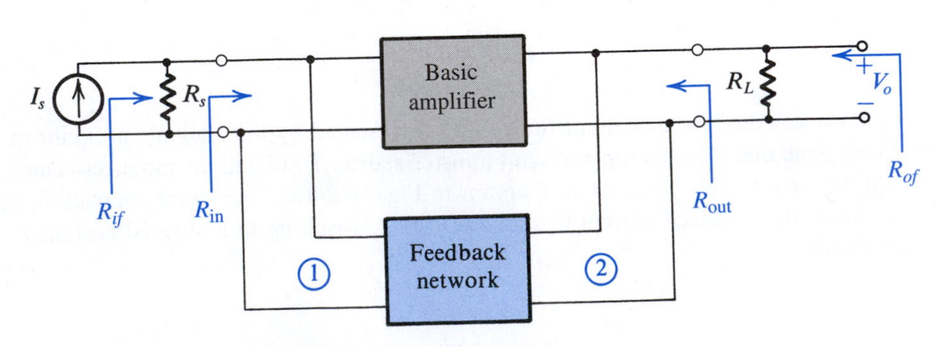

Figure 10.25 Block diagram for a practical shunt–shunt feedback amplifier.

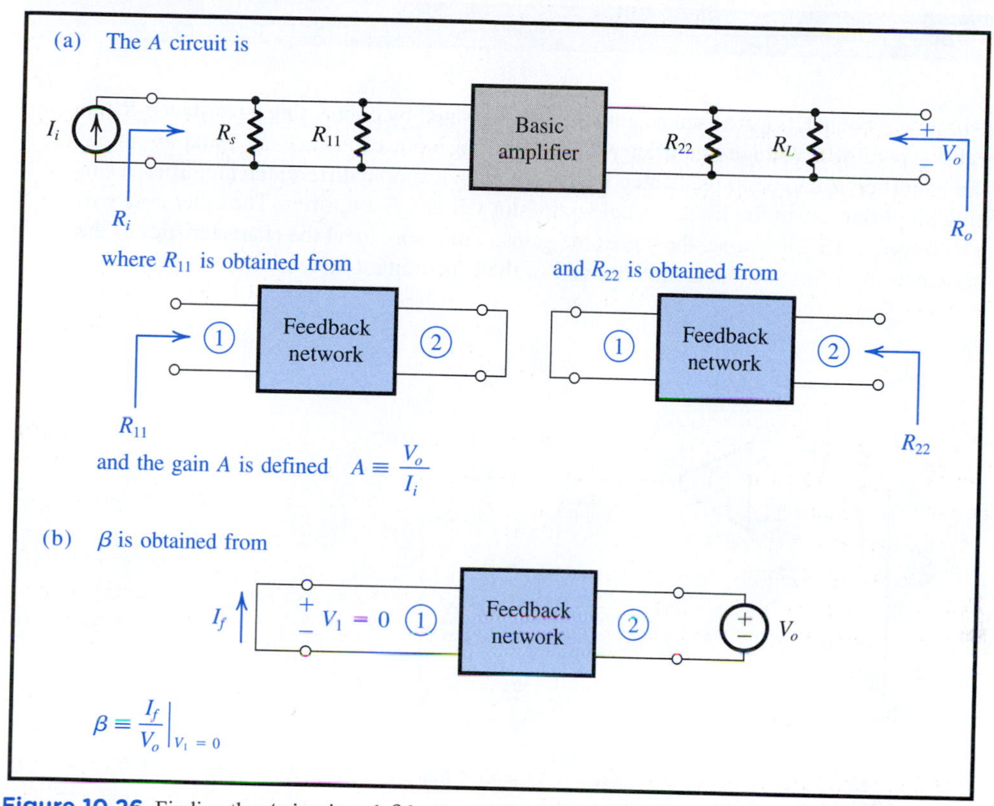

Figure 10.26 Finding the A circuit and β for the shunt–shunt feedback amplifier in Fig. 10.25.

derivation, in Fig. 10.26. As in previous cases, the method of Fig. 10.26 assumes that the basic amplifier is unilateral (or almost so) and that the feedforward transmission through the feedback network is negligibly small.

As indicated in Fig. 10.26, the A circuit is obtained by including R_s across the input terminals of the amplifier and R_L across its output terminals. The loading effect of the feedback network on the amplifier input is represented by the resistance R_{11} and its loading effect at the output is represented by the resistance R_{22}. The value of R_{11} is obtained by looking into port 1 of the feedback network while port 2 is shorted (because it is connected in shunt). Similarly, R_{22} is found by looking into port 2 while port 1 is shorted (because it is connected in shunt). Finally, observe that since the feedback network senses V_o, it is fed by a voltage V_o; and since it delivers a current I_f that is mixed in shunt at the input, its port 1 is short-circuited and β is found as I_f/V_o, where I_f is the current that flows through the short circuit.

The open-loop resistances R_i and R_o are determined from the A circuit and are used in Eqs. (10.44) and (10.45) to determine R_{if} and R_{of}. Finally, the resistances R_{in} and R_{out} that characterize the feedback amplifier are obtained from R_{in} and R_{of} by reference to Fig. 10.25 as follows:

$$R_{in} = 1\Big/\left(\frac{1}{R_{if}} - \frac{1}{R_s}\right) \tag{10.46}$$

$$R_{out} = 1\Big/\left(\frac{1}{R_{of}} - \frac{1}{R_L}\right) \tag{10.47}$$

Example 10.7

Figure 10.27(a) shows a feedback transresistance amplifier. It is formed by connecting a resistance R_F in the negative-feedback path of a voltage amplifier with gain μ, an input resistance R_{id}, and an output resistance r_o. The amplifier μ can be implemented with an op amp, a simple differential amplifier, a single-ended inverting amplifier, or, in the limit, a single-transistor CE or CS amplifier. The latter case will be considered in Exercise 10.15. Of course, the higher the gain μ, the more ideal the characteristics of the feedback transresistance amplifier will be, simply because of the concomitant increase in loop gain.

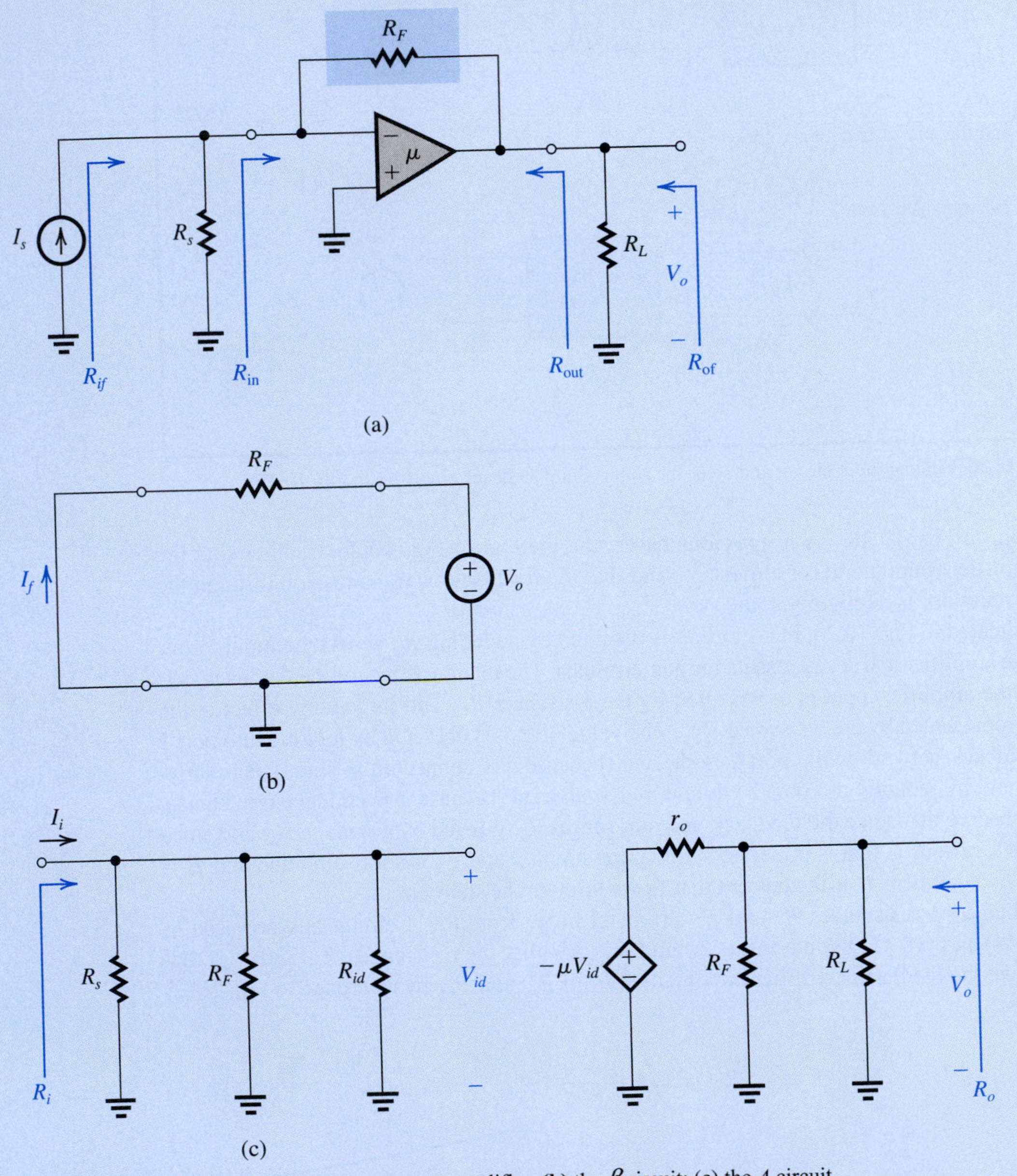

Figure 10.27 **(a)** A feedback transresistance amplifier; **(b)** the β circuit; **(c)** the A circuit.

(a) If the loop gain is large, find an approximate expression for the closed-loop open-circuit transresistance of the feedback amplifier.
(b) Find the A circuit and expressions for A, R_i, and R_o.
(c) Find expressions for the loop gain, A_f, R_{if}, R_{in}, R_{of}, and R_{out}.
(d) Find the values of R_i, R_o, A, β, A_f, R_{if}, R_{in}, R_{of}, and R_{out} for the case $\mu = 10^4$ V/V, $R_{id} = \infty$, $r_o = 100\ \Omega$, $R_F = 10\ \text{k}\Omega$, and $R_s = R_L = 1\ \text{k}\Omega$.
(e) If instead of a current source I_s having a source resistance $R_s = 1\ \text{k}\Omega$, the amplifier is fed from a voltage source V_s having a source resistance $R_s = 1\ \text{k}\Omega$, find an expression for and the value of the voltage gain V_o/V_s.

Solution

(a) If the loop gain $A\beta$ is large,

$$A_f \equiv \frac{V_o}{I_s} \simeq \frac{1}{\beta}$$

where β can be found from the β circuit in Fig. 10.27(b) as

$$\beta \equiv \frac{I_f}{V_o} = -\frac{1}{R_F} \tag{10.48}$$

Thus,

$$\frac{V_o}{I_s} \simeq -R_F$$

Note that in this case the voltage at the input node (the inverting input terminal of μ) will be very close to ground and thus very little, if any, current flows into the input terminal of the amplifier. Nearly all of I_s will flow through R_F, resulting in $V_o \simeq 0 - I_s R_F = -I_s R_F$. This should be reminiscent of the inverting op-amp configuration studied in Section 2.2.

(b) Since the feedback network consists of R_F, the loading effect at the amplifier input and output will simply be R_F. This is indicated in the A circuit shown in Fig. 10.27(c). The open-loop transresistance A can be obtained as follows:

$$V_{id} = I_i R_i \tag{10.49}$$

where

$$R_i = R_{id} \| R_F \| R_s \tag{10.50}$$

$$V_o = -\mu V_{id} \frac{(R_F \| R_L)}{r_o + (R_F \| R_L)} \tag{10.51}$$

Combining Eqs. (10.49) and (10.51) gives

$$A \equiv \frac{V_o}{I_i} = -\mu R_i \frac{(R_F \| R_L)}{r_o + (R_F \| R_L)} \tag{10.52}$$

The open-loop output resistance can be obtained by inspection of the A circuit with I_i set to 0. We see that $V_{id} = 0$, and

$$R_o = r_o \| R_F \| R_L \tag{10.53}$$

(c) The loop gain $A\beta$ can be obtained by combining Eqs. (10.48) and (10.52),

$$A\beta = \mu\left(\frac{R_i}{R_F}\right) \frac{(R_F \| R_L)}{r_o + (R_F \| R_L)}$$

Example 10.7 *continued*

Observe that although both A and β are negative, $A\beta$ is positive, a comforting fact confirming that the feedback is negative. Also note that $A\beta$ is dimensionless, as it must always be.

The closed-loop gain A_f can now be found as

$$A_f \equiv \frac{V_o}{I_s} = \frac{A}{1 + A\beta}$$

Thus

$$A_f = \frac{-\mu R_i \dfrac{(R_F \parallel R_L)}{r_o + (R_F \parallel R_L)}}{1 + \mu \dfrac{R_i}{R_F} \dfrac{(R_F \parallel R_L)}{r_o + (R_F \parallel R_L)}} \tag{10.54}$$

Note that the condition of $A\beta \gg 1$ which results in $A_f \simeq -R_F$ corresponds to

$$\mu\left(\frac{R_i}{R_F}\right)\frac{(R_F \parallel R_L)}{r_o + (R_F \parallel R_L)} \gg 1 \tag{10.55}$$

The input resistance with feedback, R_{if}, is obtained by dividing R_i by $(1 + A\beta)$ with the result

$$R_{if} = \frac{R_i}{1 + A\beta}$$

or

$$\frac{1}{R_{if}} = \frac{1}{R_i} + \frac{A\beta}{R_i} = \frac{1}{R_i} + \frac{\mu}{R_F} \frac{(R_F \parallel R_L)}{r_o + (R_F \parallel R_L)}$$

Substituting for R_i from Eq. (10.50) and replacing $\mu(R_F \parallel R_L)/[r_o + (R_F \parallel R_L)]$ by μ', where μ' is lower than but usually close to the value of μ, results in

$$R_{if} = R_{id} \parallel R_F \parallel R_s \parallel (R_F / \mu')$$

The two terms containing R_F can be combined,

$$R_{if} = R_s \parallel R_{id} \parallel [R_F/(\mu' + 1)] \tag{10.56}$$

Since $R_{if} = R_s \parallel R_{\text{in}}$, we see that

$$R_{\text{in}} = R_{id} \parallel [R_F \parallel (\mu' + 1)]$$

Usually R_{id} is large and thus

$$R_{\text{in}} \simeq \frac{R_F}{\mu' + 1} \simeq \frac{R_F}{\mu'} \tag{10.57}$$

from which we observe that for large amplifier gain μ, the input resistance will be low.

The output resistance with feedback R_{of} can be found by dividing R_o by $(1 + A\beta)$:

$$R_{of} = \frac{R_o}{1 + A\beta}$$

Thus,

$$\frac{1}{R_{of}} = \frac{1}{R_o} + \frac{A\beta}{R_o}$$
$$= \frac{1}{R_o} + \mu \frac{R_i}{R_F} \frac{(R_F \parallel R_L)}{r_o + (R_F \parallel R_L)} \frac{1}{R_o}$$

Substituting for R_o from Eq. (10.53),

$$\frac{1}{R_{of}} = \frac{1}{R_L} + \frac{1}{R_F} + \frac{1}{r_o} + \mu\frac{R_i}{R_F}\frac{1}{r_o}$$

$$= \frac{1}{R_L} + \frac{1}{R_F} + \frac{1}{r_o}\left(1 + \mu\frac{R_i}{R_F}\right)$$

Thus,

$$R_{of} = R_L \| R_F \| \frac{r_o}{1 + \mu\dfrac{R_i}{R_F}}$$

Since, moreover,

$$R_{of} = R_L \| R_{out}$$

we obtain for R_{out}

$$R_{out} = R_F \| \frac{r_o}{1 + \mu\dfrac{R_i}{R_F}}$$

Usually $R_F \gg r_o/[(1 + \mu(R_i/R_F)]$; thus,

$$R_{out} \simeq \frac{r_o}{1 + \mu\dfrac{R_i}{R_F}} \simeq \left(\frac{R_F}{R_i}\right)\left(\frac{r_o}{\mu}\right)$$

from which we see that for large μ, the output resistance will be considerably reduced.

(d) For the numerical values given:

$$R_i = R_{id} \| R_F \| R_s$$

$$= \infty \| 10 \| 1 = 0.91 \text{ k}\Omega$$

$$R_o = r_o \| R_F \| R_s$$

$$= 0.1 \| 10 \| 1 = 90 \ \Omega$$

$$A = -\mu R_i \frac{(R_F \| R_L)}{r_o + (R_F \| R_L)}$$

$$= -10^4 \times 0.91 \times \frac{(10 \| 1)}{0.1 + (10 \| 1)} = -8198 \text{ k}\Omega$$

$$\beta = -\frac{1}{R_F} = -\frac{1}{10} = -0.1 \text{ mA/V}$$

$$A\beta = 819.8$$

$$1 + A\beta = 820.8$$

$$A_f = \frac{A}{1 + A\beta} = -\frac{8198}{820.8} = -9.99 \text{ k}\Omega$$

which is very close to the ideal value of $-R_F = -10$ kΩ.

$$R_{if} = \frac{R_i}{1 + A\beta} = \frac{910}{820.8} = 1.11 \ \Omega$$

$$R_{in} = \frac{1}{\dfrac{1}{R_{if}} - \dfrac{1}{R_i}} - \frac{1}{\dfrac{1}{1.11} - \dfrac{1}{1000}} \simeq 1.11 \ \Omega$$

Example 10.7 *continued*

which is very low, a highly desirable property. We also have

$$R_{of} = \frac{R_o}{1+A\beta} = \frac{90}{820.8} = 0.11\ \Omega$$

$$R_{out} = \frac{1}{\dfrac{1}{R_{of}} - \dfrac{1}{R_L}} = \frac{1}{\dfrac{1}{0.11} - \dfrac{1}{1000}} \simeq 0.11\ \Omega$$

which as well is very low, another highly desirable property.

(e) If the amplifier is fed with a voltage source V_s having a resistance $R_s = 1\ \text{k}\Omega$, the output voltage can be found from

$$V_o = A_f\, I_s = A_f\, \frac{V_s}{R_s}$$

Thus,

$$\frac{V_o}{V_s} = \frac{A_f}{R_s} = -\frac{9.99\ \text{k}\Omega}{1\ \text{k}\Omega} = -9.99\ \text{V/V}$$

EXERCISE

10.15 For the transresistance amplifier in Fig. E10.15, replace the MOSFET with its equivalent-circuit model and use feedback analysis to show the following:

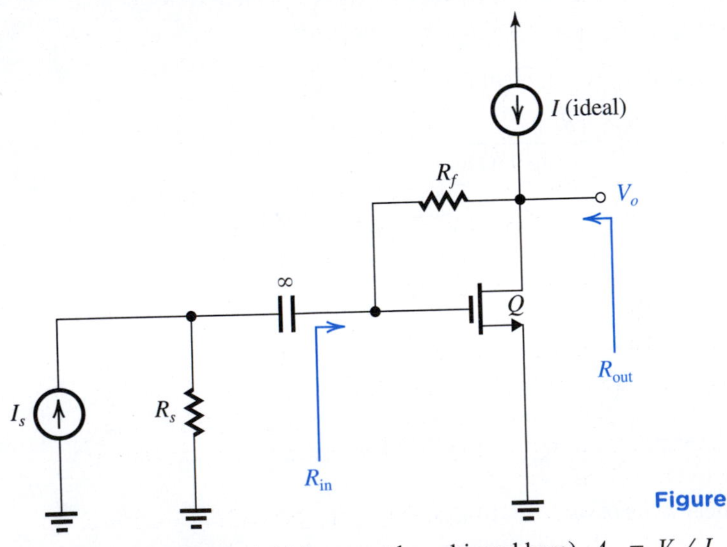

Figure E10.15

(a) For large loop gain (which cannot be achieved here), $A_f \equiv V_o/I_s \simeq -R_f$.

(b) $A_f = \dfrac{-(R_s \| R_f)g_m(r_o \| R_f)}{1+(R_s \| R_f)g_m(r_o \| R_f)/R_f}$

(c) $R_{in} = \dfrac{R_f}{[1+g_m(r_o \| R_f)]}$

(d) $R_{out} = r_o \| \dfrac{R_f}{1 + g_m(R_s \| R_f)}$

(e) For $g_m = 5$ mA/V, $r_o = 20\ \text{k}\Omega$, $R_f = 10\ \text{k}\Omega$, and $R_s = 1\ \text{k}\Omega$, find A, β, $A\beta$, A_f R_i, R_o, R_{if}, R_{in}, R_{of}, and R_{out}.

Ans. (e) $-30.3\ \text{k}\Omega$; -0.1 mA/V; 3.03; $-7.52\ \text{k}\Omega$ (compare to the ideal value of $-10\ \text{k}\Omega$); 909 Ω; 6.67 kΩ; 226 Ω; 291 Ω; 1.66 kΩ; 1.66 kΩ

10.6.3 An Important Note

The feedback analysis method is predicated on the assumption that all (or most) of the feedforward transmission occurs in the basic amplifier and all (or most) of the feedback transmission occurs in the feedback network. The circuit considered in Exercise 10.15 above is simple and can be analyzed directly (i.e., without invoking the feedback approach) to determine A_f. In this way we can check the validity of our assumptions. This point is illustrated in Problem 10.58, where we find that for the circuit in Fig. E10.15, all of the feedback transmission occurs in the feedback circuit. Also, as long as g_m is much greater than $1/R_f$, the assumption that most of the feedforward transmission occurs in the basic amplifiers is valid, and thus the feedback analysis is reasonably accurate.

10.7 The Feedback Current Amplifier (Shunt–Series)

10.7.1 The Ideal Case

As mentioned in Section 10.3, the shunt–series feedback topology is best suited for current amplifiers: The shunt connection at the input reduces the input resistance, making it easier to feed the amplifier with a current signal; the sampling of output current stabilizes I_o, which is the output signal in a current amplifier, and the series connection at the output increases the output resistance, making the output current value less susceptible to changes in load resistance.

Figure 10.28(a) shows the ideal structure for the shunt–series feedback amplifier. It consists of a unilateral open-loop amplifier (the A circuit) and an ideal feedback network. The A circuit has an input resistance R_i, a short-circuit current gain $A \equiv I_o/I_i$, and an output resistance R_o. The β circuit samples the short-circuit output current I_o and provides a feedback current I_f that is subtracted from the signal-source current I_s at the input node. Note that the β circuit presents a zero resistance to the output loop and thus does not load the amplifier output. Also, the feedback signal $I_f = \beta I_o$ is provided as an ideal current source, and thus the β circuit does not load the amplifier input. Also observe that both A and β are current gains and $A\beta$ is a dimensionless quantity. Finally, note that the source and load resistances have been absorbed inside the A circuit (more on this later).

Since the structure of Fig. 10.28(a) follows the ideal feedback structure of Fig. 10.1, we can obtain the closed-loop current gain A_f as

$$A_f \equiv \frac{I_o}{I_s} = \frac{A}{1 + A\beta} \tag{10.59}$$

The feedback current amplifier can be represented by the equivalent circuit in Fig. 10.28(b).

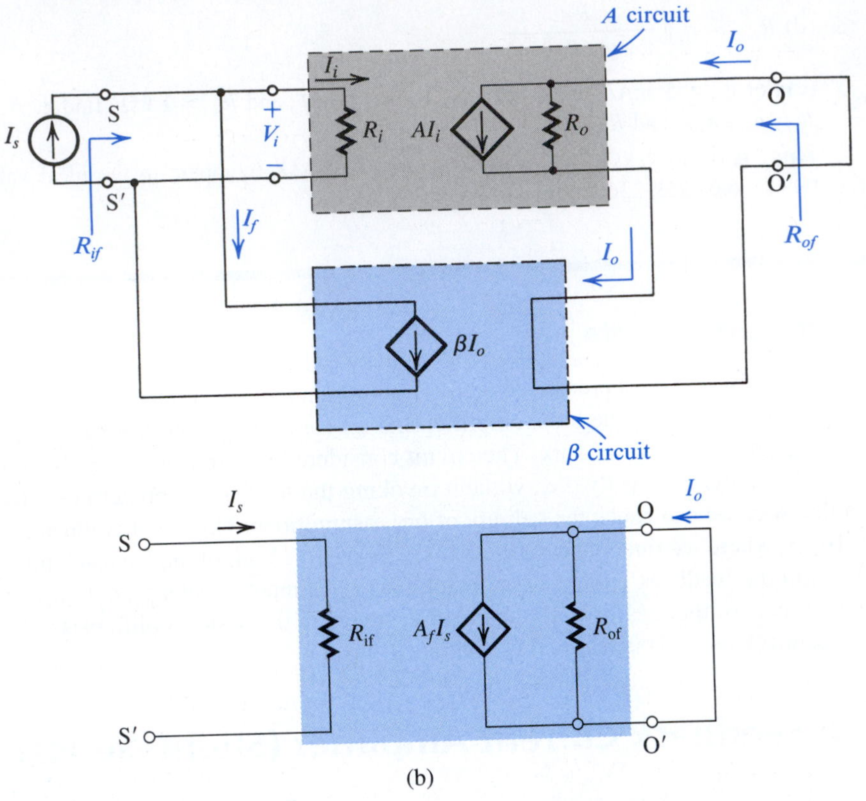

Figure 10.28 (a) Ideal structure for the shunt–series feedback amplifier. (b) Equivalent circuit of the amplifier in (a).

Note that A_f is the short-circuit current gain. The input resistance R_{if} is found by dividing R_i by $(1 + A\beta)$, which is a result of the shunt connection at the input. Thus,

$$R_{if} = \frac{R_i}{1 + A\beta} \tag{10.60}$$

The output resistance R_{of} is the resistance obtained by setting $I_s = 0$, breaking the short-circuit output loop, at say OO′, and measuring the resistance between the two terminals thus created. Since the series feedback connection always raises resistance, we can obtain R_{of} by multiplying R_o by $(1 + A\beta)$,

$$R_{of} = (1 + A\beta)R_o \tag{10.61}$$

10.7.2 The Practical Case

Figure 10.29 shows a block diagram for a practical shunt–series feedback amplifier. To be able to apply the feedback equations to this amplifier, we have to represent it by the ideal structure of Fig. 10.28(a). Our objective therefore is to devise a simple method for finding the A and β circuits. Building on the insight we have gained from the study of the three other topologies, we present the method for the shunt–series case without derivation, in Fig. 10.30. As in previous cases, the method of Fig. 10.30 assumes that the basic amplifier is unilateral (or almost so) and that the feedforward transmission in the feedback network is negligibly small.

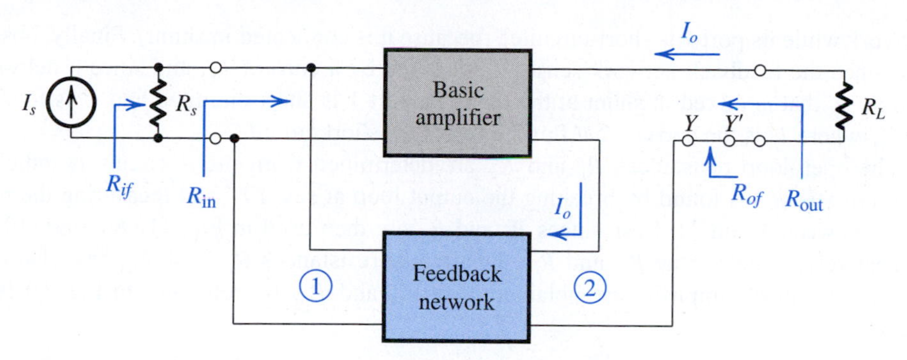

Figure 10.29 Block diagram for a practical shunt–series feedback amplifier.

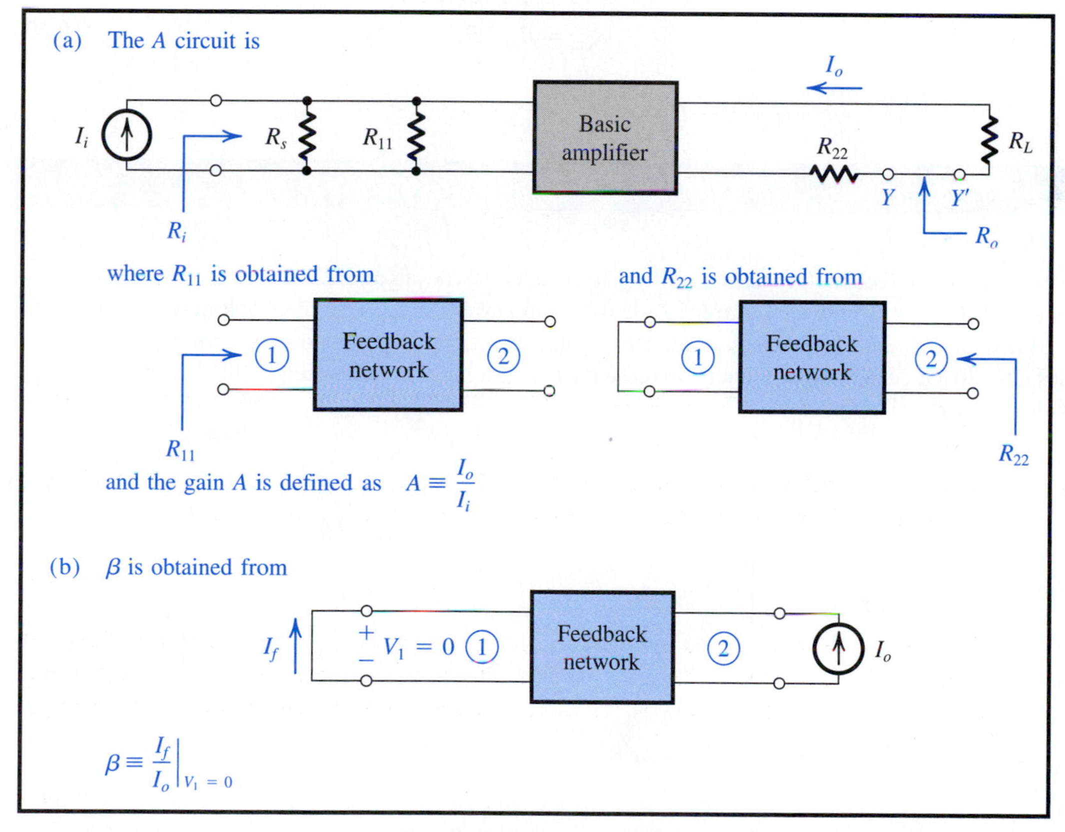

Figure 10.30 Finding the A circuit and β for the current-mixing current-sampling (shunt–series) feedback amplifier of Fig. 10.29.

As indicated in Fig. 10.30, the A circuit is obtained by including R_s across the input terminals of the amplifier and R_L in series with its output loop. The loading effect of the feedback network on the amplifier input is represented by the resistance R_{11}, and its loading effect at the amplifier output is represented by resistance R_{22}. The value of R_{11} is obtained by looking into port 1 of the feedback network while its port 2 is open-circuited (because it is connected in series). The value of R_{22} is obtained by looking into port 2 of the feedback network while its port 1 is short-circuited (because it is connected in shunt). Finally, observe that since the feedback network senses I_o, it is fed by a current I_o; and since it delivers a current I_f that is mixed in shunt at the input, its port 1 is short-circuited and β is found as I_f/I_o, where I_f is the current that flows through the short circuit.

The open-loop resistances R_i and R_o are determined from the A circuit as indicated. Observe that R_o is found by breaking the output loop at say YY' and measuring the resistance between Y and Y'. Resistances R_i and R_o are then used in Eqs. (10.60) and (10.61), respectively, to determine R_{if} and R_{of}. Finally, the resistances R_{in} and R_{out} that characterized the feedback amplifier are obtained from R_{if} and R_{of} by reference to Fig. 10.29, as follows:

$$R_{in} = 1\Big/\left(\frac{1}{R_{if}} - \frac{1}{R_s}\right) \tag{10.62}$$

$$R_{out} = R_{of} - R_L \tag{10.63}$$

Example 10.8

Figure 10.31 shows a feedback current amplifier formed by cascading an inverting voltage amplifier μ with a MOSFET Q. The output current I_o is the drain current of Q. The feedback network, consisting of resistors R_1 and R_2, senses an exactly equal current, namely, the source current of Q, and provides a feedback current signal that is mixed with I_s at the input node. Note that the bias arrangement is *not* shown.

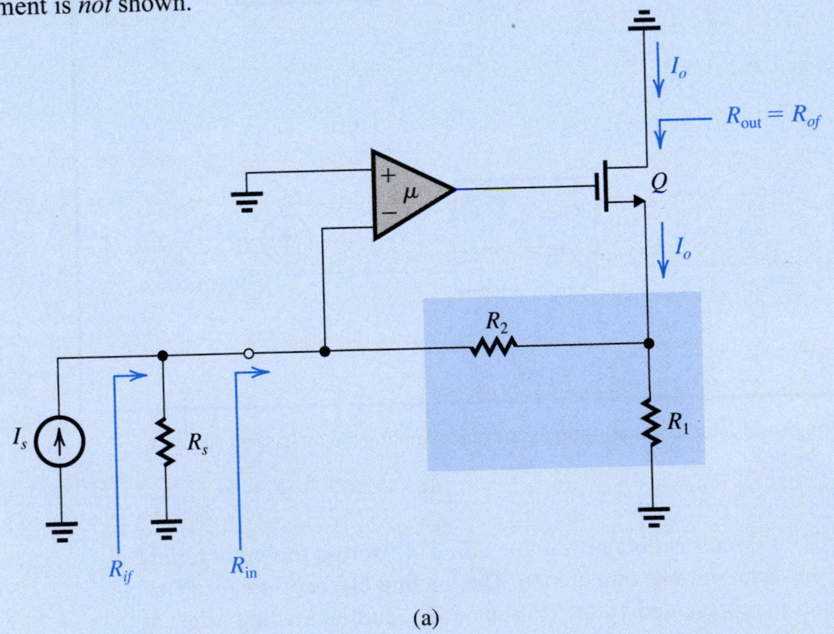

Figure 10.31 Circuit for Example 10.8.

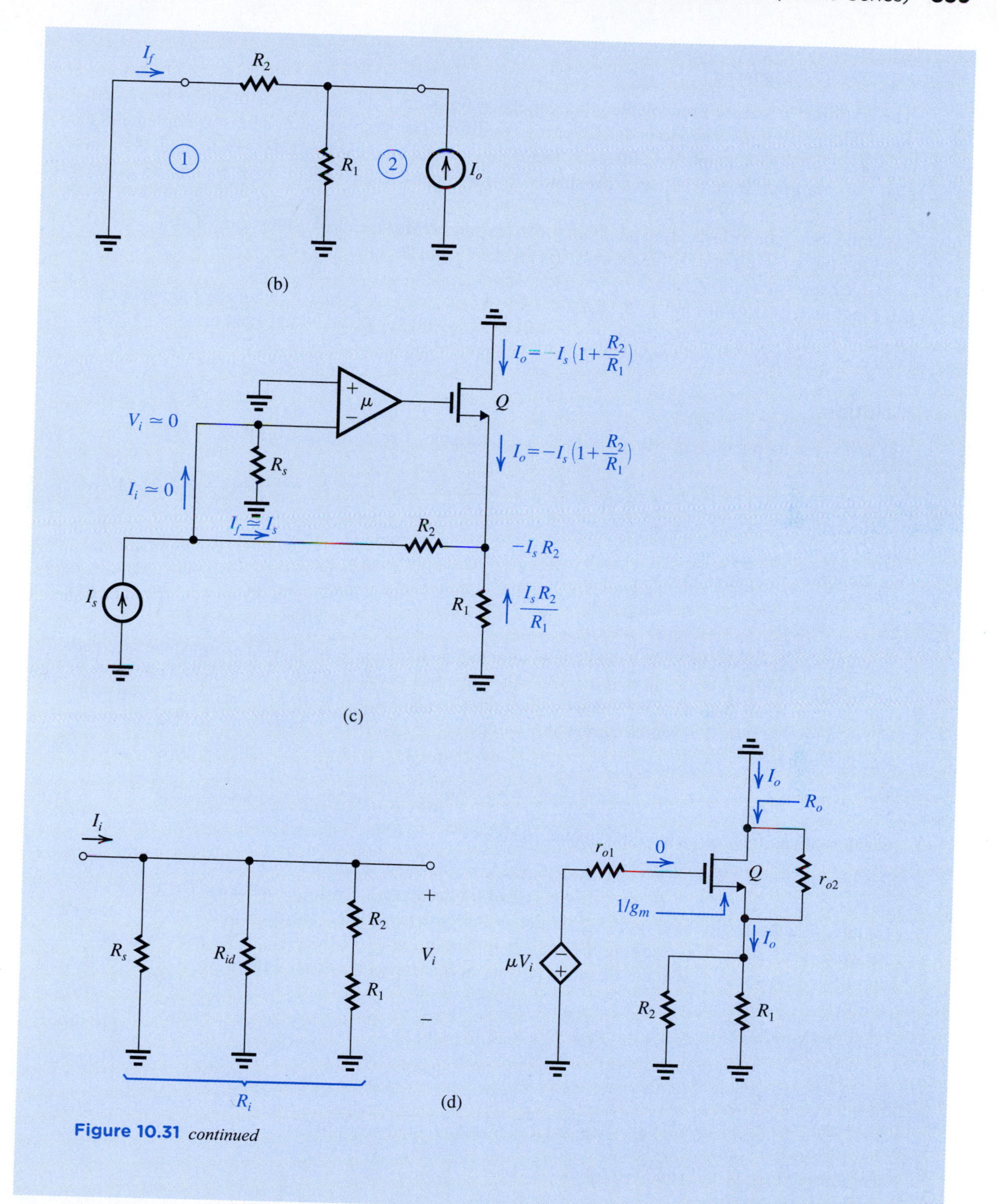

Figure 10.31 *continued*

Example 10.8 *continued*

The amplifier μ can be implemented in a variety of ways, including by means of an op amp, a differential amplifier, or a single-ended inverting amplifier. The simplest approach is to implement μ with a CS MOSFET amplifier. However, in such a case the loop gain will be very limited. Assume that the amplifier μ has an input resistance R_{id}, an open-circuit voltage gain μ, and an output resistance r_{o1}.

(a) If the loop gain is large, find an approximate expression for the closed-loop gain $A_f \equiv I_o/I_s$.
(b) Find the A circuit and derive expressions for A, R_i, and R_o.
(c) Give expressions for $A\beta$, A_f, R_{if}, R_{in}, R_{of}, and R_{out}.
(d) Find numerical values for A, β, $A\beta$, A_f, R_i, R_{if}, R_{in}, R_o, R_{of}, and R_{out} for the following case: $\mu = 1000$ V/V, $R_s = \infty$, $R_{id} = \infty$, $r_{o1} = 1\ \text{k}\Omega$, $R_1 = 10\ \text{k}\Omega$, $R_2 = 90\ \text{k}\Omega$, and for Q: $g_m = 5$ mA/V and $r_o = 20\ \text{k}\Omega$.

Solution

(a) When the loop gain $A\beta \gg 1$, $A_f \simeq 1/\beta$. To determine β refer to Fig. 10.31(b),

$$\beta \equiv \frac{I_f}{I_o} = -\frac{R_1}{R_1 + R_2} \tag{10.64}$$

Thus,

$$A_f \simeq \frac{1}{\beta} = -\left(1 + \frac{R_2}{R_1}\right) \tag{10.65}$$

To see what happens in this case more clearly, refer to Fig. 10.31(c). Here we have assumed the loop gain to be large, so that $I_i \simeq 0$ and thus $I_f \simeq I_s$. Also note that because $I_i \simeq 0$, V_i will be close to zero. Thus, we can easily determine the voltage at the source of Q as $-I_f R_2 \simeq -I_s R_2$. The current through R_1 will then be $I_s R_2/R_1$. The source current of Q will be $-(I_s + I_s R_2/R_1)$, which means that the output current I_o will be

$$I_o = -I_s\left(1 + \frac{R_2}{R_1}\right)$$

which confirms the expression for A_f obtained above (Eq. 10.65).

(b) To obtain the A circuit we load the input side of the basic amplifier with R_s and R_{11}. The latter in this case is simply $R_1 + R_2$ (because port 2 of the feedback network is opened). We also load the output of the basic amplifier with R_{22}, which in this case is $R_1 \| R_2$ (because port 1 of the feedback network is shorted). The resulting A circuit is shown in Fig., 10.31(d), where we have replaced the amplifier μ with its equivalent circuit. Analysis of the A circuit is straightforward and proceeds as follows:

$$R_i = R_s \| R_{id} \| (R_1 + R_2) \tag{10.66}$$

$$V_i = I_i R_i \tag{10.67}$$

$$I_o = -\mu V_i \frac{1}{1/g_m + (R_1 \| R_2 \| r_{o2})} \frac{r_{o2}}{r_{o2} + (R_1 \| R_2)} \tag{10.68}$$

Combining Eqs. (10.67) and (10.68) results in A:

$$A \equiv \frac{I_o}{I_i} = -\mu \frac{R_i}{1/g_m + (R_1 \| R_2 \| r_{o2})} \frac{r_{o2}}{r_{o2} + (R_1 \| R_2)} \tag{10.69}$$

For the case $1/g_m \ll (R_1 \| R_2 \| r_{o2})$,

$$A \simeq -\mu \frac{R_i}{R_1 \| R_2 \| r_{o2}} \frac{r_{o2}}{r_{o2} + (R_1 \| R_2)}$$

Which reduces to

$$A = -\mu \frac{R_i}{R_1 \| R_2} \tag{10.70}$$

Noting that R_o is the output resistance of Q, which has a resistance $(R_1 \| R_2)$ in its source lead, we can write

$$R_o = r_{o2} + (R_1 \| R_2) + (g_m r_{o2})(R_1 \| R_2)$$

$$\simeq g_m r_{o2}(R_1 \| R_2) \tag{10.71}$$

(c) The loop gain is obtained by combining Eqs. (10.64) and (10.69),

$$A\beta = \mu \frac{R_i}{\dfrac{1}{g_m} + (R_1 \| R_2 \| r_{o2})} \frac{r_{o2}}{r_{o2} + (R_1 \| R_2)} \frac{R_1}{R_1 + R_2} \tag{10.72}$$

For the case $1/g_m \ll (R_1 \| R_2 \| r_{o2})$,

$$A\beta \simeq \mu \frac{R_i}{R_1 \| R_2} \frac{R_1}{R_1 + R_2} = \mu \frac{R_i}{R_2} \tag{10.73}$$

The input resistance R_{if} is found as

$$R_{if} = R_i/(1 + A\beta)$$

$$\frac{1}{R_{if}} = \frac{1}{R_i} + \frac{A\beta}{R_i}$$

We can substitute for $A\beta$ from the full expression in Eq. (10.72). For the approximate case, we use $A\beta$ from Eq. (10.73):

$$\frac{1}{R_{if}} = \frac{1}{R_i} + \frac{\mu}{R_2}$$

That is,

$$R_{if} = R_i \| \frac{R_2}{\mu}$$

Substituting for R_i from Eq. (10.66), we write

$$R_{if} = R_s \| R_{id} \| (R_1 + R_2) \| \frac{R_2}{\mu}$$

Since by definition,

$$R_{if} = R_s \| R_{\text{in}}$$

we can easily find R_{in} as

$$R_{\text{in}} = R_{id} \| (R_1 + R_2) \| \frac{R_2}{\mu} \tag{10.74}$$

Example 10.8 *continued*

Usually the third component on the right-hand side is the smallest; thus,

$$R_{in} \simeq \frac{R_2}{\mu} \tag{10.75}$$

For the output resistance, we have

$$R_{of} = R_o(1 + A\beta) \simeq A\beta R_o$$

Substituting for R_o for Eq. (10.71) and for $A\beta$ from the approximate expression in Eq. (10.73), we have

$$R_{of} \simeq \mu\left(\frac{R_i}{R_2}\right)(g_m r_{o2})(R_1 \parallel R_2)$$

$$R_{of} = \mu\frac{R_i}{R_1 + R_2}(g_m r_{o2})R_1 \tag{10.76}$$

Finally, we note that

$$R_{out} = R_{of} = \mu\frac{R_i}{R_1 + R_2}g_m r_{o2}R_1 \tag{10.77}$$

(d) For the numerical values given,

$$R_i = \infty \parallel \infty \parallel (10 + 90) = 100 \text{ k}\Omega$$

Since $1/g_m = 0.2 \text{ k}\Omega \ll (10 \parallel 90 \parallel 20)$,

$$A \simeq -\mu\frac{R_i}{R_1 \parallel R_2}$$

$$= -1000\frac{100}{10 \parallel 90} = -11.11 \times 10^3 \text{ A/A}$$

$$\beta = -\frac{R_1}{R_1 + R_2} = -\frac{10}{10 + 90} = -0.1 \text{ A/A}$$

$$A\beta = 1111$$

$$A_f = -\frac{11.11 \times 10^3}{1 + 1111} = -9.99 \text{ A/A}$$

which is very close to the ideal value of

$$A_f \simeq -\left(1 + \frac{R_2}{R_1}\right) = -\left(1 + \frac{90}{10}\right) = -10 \text{ A/A}$$

$$R_{in} = \frac{R_2}{\mu} = \frac{90 \text{ k}\Omega}{1000} = 90 \text{ }\Omega$$

$$R_o = g_m r_{o2}(R_1 \parallel R_2)$$

$$= 5 \times 20(10 \parallel 90) = 900 \text{ k}\Omega$$

$$R_{out} = (1 + A\beta)R_o = 1112 \times 900 = 1000 \text{ M}\Omega$$

EXERCISES

10.16 For the amplifier in Example 10.8, find the values of A_f, R_{in}, and R_{out} when the value of μ is 10 times lower, that is when $\mu = 100$.
Ans. −9.91 A/A; 900 Ω; 100 MΩ

10.17 If in the circuit in Fig. 10.31(a), R_2 is short-circuited, find the ideal value of A_f. For the case $R_s = R_{id} = \infty$, give expressions for R_i, R_o, A, β, A_f, R_{in}, and R_{out}
Ans. $A_f = 1$ A/A; $R_i = R_1$; $R_o = r_{o2}$, $A = -\mu g_m R_1$; $\beta = -1$; $A_f = \mu g_m R_1/(1 + \mu g_m R_1)$; $R_{in} = 1/\mu g_m$; $R_{out} \simeq \mu(g_m r_{o2})R_1$.

10.8 Summary of the Feedback Analysis Method

Table 10.1 provides a summary of the rules and relationships employed in the analysis and design of the four types of feedback amplifier. In addition to the wealth of information in Table 10.1, we offer the following important analysis tips.

1. Always begin the analysis by determining an approximate value for the closed-loop gain A_f, assuming that the loop gain $A\beta$ is large and thus

$$A_f \simeq 1/\beta$$

This value should serve as a check on the final value you find for A_f. How close the actual A_f is to the approximate value will depend on how large $A\beta$ is compared to unity.

2. The shunt connection at input or output always results in reducing the corresponding resistance (input or output). The series connection at input or output always results in increasing the corresponding resistance (input or output).

3. In utilizing negative feedback to improve the properties of an amplifier under design, the starting point in the design is the selection of the feedback topology appropriate for the application at hand. Then the required amount of negative feedback $(1 + A\beta)$ can be ascertained utilizing the fact that it is this quantity that determines the magnitude of improvement in the various amplifier parameters. Also, the feedback factor β can be determined from

$$\beta \simeq 1/A_f$$

10.9 Determining the Loop Gain

We have already seen that the loop gain $A\beta$ is a very important quantity that characterizes a feedback loop. Furthermore, in the following sections it will be shown that $A\beta$ determines whether the feedback amplifier is stable (as opposed to oscillatory). In this section, we shall describe an alternative approach to the determination of loop gain.

Table 10.1 Summary of Relationships for the Four Feedback-Amplifier Topologies

Feedback Amplifier	Feedback Topology	x_i	x_o	x_f	x_s	A	β	A_f	Source Form	Loading of Feedback Network is Obtained: At Input	Loading of Feedback Network is Obtained: At Output	To Find β, Apply to Port 2 of Feedback Network	R_{if}	R_{of}	Refer to Figs.
Voltage	Series–shunt	V_i	V_o	V_f	V_s	$\frac{V_o}{V_i}$	$\frac{V_f}{V_o}$	$\frac{V_o}{V_s}$	Thévenin	By short-circuiting port 2 of feedback network	By open-circuiting port 1 of feedback network	a voltage, and find the open-circuit voltage at port 1	$R_i(1+A\beta)$	$\frac{R_o}{1+A\beta}$	10.6 10.12 10.14 10.15
Current	Shunt–series	I_i	I_o	I_f	I_s	$\frac{I_o}{I_i}$	$\frac{I_f}{I_o}$	$\frac{I_o}{I_s}$	Norton	By open-circuiting port 2 of feedback network	By short-circuiting port 1 of feedback network	a current, and find the short-circuit current at port 1	$\frac{R_i}{1+A\beta}$	$R_o(1+A\beta)$	10.8(a) 10.28 10.29 10.30
Transconductance	Series–series	V_i	I_o	V_f	V_s	$\frac{I_o}{V_i}$	$\frac{V_f}{I_o}$	$\frac{I_o}{V_s}$	Thévenin	By open-circuiting port 2 of feedback network	By open-circuiting port 1 of feedback network	a current, and find the open-circuit voltage at port 1	$R_i(1+A\beta)$	$R_o(1+A\beta)$	10.10(a) 10.18 10.20 10.21
Transresistance	Shunt–shunt	I_i	V_o	I_f	I_s	$\frac{V_o}{I_i}$	$\frac{I_f}{V_o}$	$\frac{V_o}{I_s}$	Norton	By short-circuiting port 2 of feedback network	By short-circuiting port 1 of feedback network	a voltage, and find the short-circuit current at port 1	$\frac{R_i}{1+A\beta}$	$\frac{R_o}{1+A\beta}$	10.11(a) 10.24 10.25 10.26

10.9.1 An Alternative Approach for Finding $A\beta$

First, consider again the general feedback amplifier shown in Fig. 10.1. Let the external source x_s be set to zero. Open the feedback loop by breaking the connection of x_o to the feedback network and apply a test signal x_t. We see that the signal at the output of the feedback network is $x_f = \beta x_t$; that at the input of the basic amplifier is $x_i = -\beta x_t$; and the signal at the output of the amplifier, where the loop was broken, will be $x_o = -A\beta x_t$. It follows that the loop gain $A\beta$ is given by the negative of the ratio of the *returned* signal to the applied test signal; that is, $A\beta = -x_o/x_t$. It should also be obvious that this applies regardless of where the loop is broken.

However, in breaking the feedback loop of a practical amplifier circuit, we must ensure that the conditions that existed prior to breaking the loop do not change. This is achieved by terminating the loop where it is opened with an impedance equal to that seen before the loop was broken. To be specific, consider the conceptual feedback loop shown in Fig. 10.32(a). If we break the loop at XX′, and apply a test voltage V_t to the terminals thus created to the left of XX′, the terminals at the right of XX′ should be loaded with an impedance Z_t as shown in Fig. 10.32(b). The impedance Z_t is equal to that previously seen looking to the left of XX′. The loop gain $A\beta$ is then determined from

$$A\beta = -\frac{V_r}{V_t} \tag{10.78}$$

Finally, it should be noted that in some cases it may be convenient to determine $A\beta$ by applying a test current I_t and finding the returned current signal I_r. In this case, $A\beta = -I_r/I_t$.

An alternative equivalent method for determining $A\beta$ (see Rosenstark, 1986) that is usually convenient to employ especially in SPICE simulations is as follows: As before, the loop is broken at a convenient point. Then the open-circuit voltage transfer function T_{oc} is determined as indicated in Fig. 10.32(c), and the short-circuit current transfer function T_{sc} is determined as shown in Fig. 10.32(d). These two transfer functions are then combined to obtain the loop gain $A\beta$,

$$A\beta = -1\bigg/\left(\frac{1}{T_{oc}} + \frac{1}{T_{sc}}\right) \tag{10.79}$$

This method is particularly useful when it is not easy to determine the termination impedance Z_t.

To illustrate the process of determining loop gain, we consider the feedback loop shown in Fig. 10.33(a). This feedback loop represents both the inverting and the noninverting op-amp configurations. Using a simple equivalent-circuit model for the op amp, we obtain the circuit of Fig. 10.33(b). Examination of this circuit reveals that a convenient place to break the loop is at the input terminals of the op amp. The loop, broken in this manner, is shown in Fig. 10.33(c) with a test signal V_t applied to the right-hand-side terminals and a resistance R_{id} terminating the left-hand-side terminals. The returned voltage V_r is found by inspection as

$$V_r = -\mu V_1 \frac{\{R_L \,||\, [R_2 + R_1 \,||\, (R_{id} + R)]\}}{\{R_L \,||\, [R_2 + R_1 \,||\, (R_{id} + R)]\} + r_o} \frac{[R_1 \,||\, (R_{id} + R)]}{[R_1 \,||\, (R_{id} + R)] + R_2} \frac{R_{id}}{R_{id} + R} \tag{10.80}$$

This equation can be used directly to find the loop gain $L = A\beta = -V_r/V_t = -V_r/V_1$.

Since the loop gain L is generally a function of frequency, it is usual to call it **loop transmission** and to denote it by $L(s)$ or $L(j\omega)$.

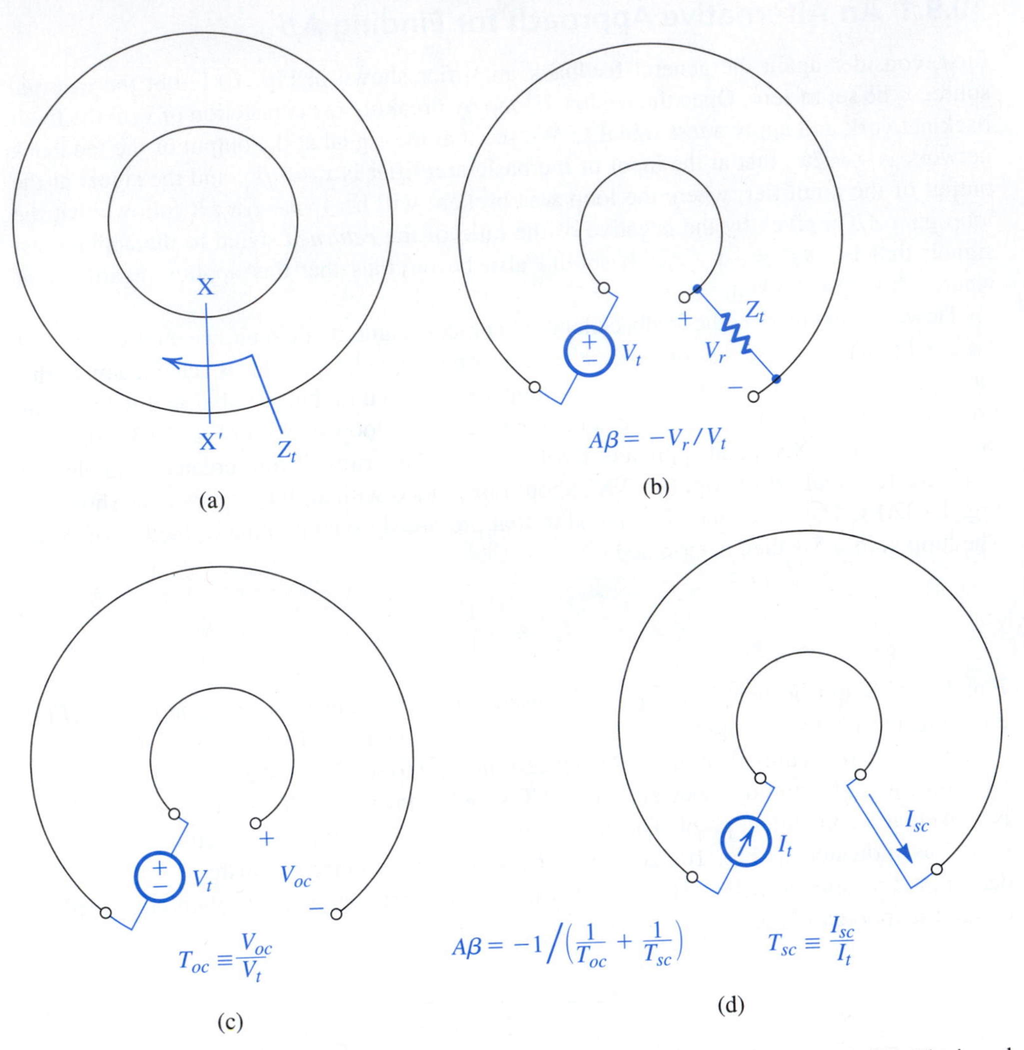

Figure 10.32 A conceptual feedback loop is broken at XX′ and a test voltage V_t is applied. The impedance Z_t is equal to that previously seen looking to the left of XX′. The loop gain $A\beta = -V_r/V_t$, where V_r is the *returned* voltage. As an alternative, $A\beta$ can be determined by finding the open-circuit transfer function T_{oc}, as in **(c)**, and the short-circuit transfer function T_{sc}, as in **(d)**, and combining them as indicated.

Finally, we note that the value of the loop gain determined using the method discussed here may differ somewhat from the value determined by the approach studied in the previous sections. The difference stems from the approximations made in the feedback analysis method utilized in the previous sections. However, as the reader will find by solving the end-of-chapter problems, the difference is usually limited to a few percent.

10.9.2 Equivalence of Circuits from a Feedback-Loop Point of View

From the study of circuit theory we know that the poles of a circuit are independent of the external excitation. In fact the poles, or the natural modes (which is a more appropriate

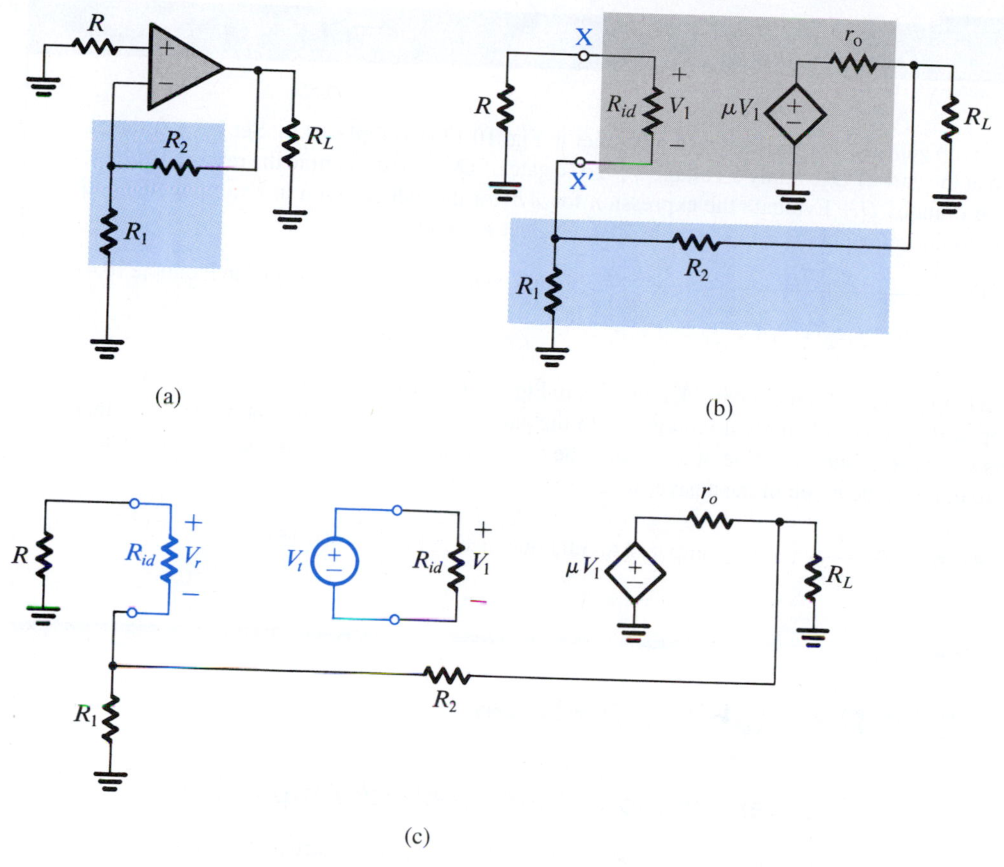

Figure 10.33 **(a)** A feedback loop that represents both the inverting and the noninverting op-amp configurations; **(b)** equivalent circuit; **(c)** determination of the loop gain.

name), can be determined by setting the external excitation to zero. It follows that the poles of a feedback amplifier depend only on the feedback loop. This will be confirmed in a later section, where we show that the **characteristic equation** (whose roots are the poles) is completely determined by the loop gain. Thus, a given feedback loop may be used to generate a number of circuits having the same poles but different **transmission zeros**. The closed-loop gain and the transmission zeros depend on how and where the input signal is injected into the loop.

As an example, return to the feedback loop of Fig. 10.33(a). This loop can be used to generate the noninverting op-amp circuit by feeding the input voltage signal to the terminal of R that is connected to ground; that is, we lift this terminal off ground and connect it to V_s. The same feedback loop can be used to generate the inverting op-amp circuit by feeding the input voltage signal to the terminal of R_1 that is connected to ground.

Recognition of the fact that two or more circuits are equivalent from a feedback-loop point of view is very useful because (as will be shown in Section 10.10) stability is a function of the loop. Thus one needs to perform the stability analysis only once for a given loop.

In Chapter 16 we shall employ the concept of **loop equivalence** in the synthesis of active filters.

EXERCISES

10.18 Find the loop gain $A\beta$ for the feedback amplifier in Fig. 10.17 (Example 10.4). Set $V_s = 0$, break the loop at the gate of Q_2, apply a voltage V_t to the gate of Q_2, and determine the returned voltage V_r at the drain of Q_1. Evaluate the expression for $A\beta$ for the values given in Example 10.4 and compare to the value obtained in Example 10.4. Neglect r_{o1} and r_{o2}.

Ans. $A\beta = \dfrac{g_{m2}R_{D2}}{R_{D2} + R_2 + \left(R_1 \| \dfrac{1}{g_{m2}}\right)} \dfrac{R_1 R_{D1}}{R_1 + \dfrac{1}{g_{m1}}}$; 16.67 (compared to 17.39 obtained in Example 10.4)

10.19 Find the loop gain $A\beta$ for the feedback amplifier in Fig. E10.15 (Exercise 10.15). Set $I_s = 0$, break the loop at the gate of Q, apply a voltage V_t to the gate of Q, and determine the voltage V_r that appears across R_s. Find the value of $A\beta$ using the component values given in Exercise 10.15, and compare to the value given in the answer to Exercise 10.15.

Ans. $A\beta = \dfrac{g_m r_o R_s}{r_o + R_f + R_s}$; 3.22 (compared to 3.03 obtained in Exercise 10.15)

10.10 The Stability Problem

10.10.1 Transfer Function of the Feedback Amplifier

In a feedback amplifier such as that represented by the general structure of Fig. 10.1, the open-loop gain A is generally a function of frequency, and it should therefore be more accurately called the **open-loop transfer function**, $A(s)$. Also, we have been assuming for the most part that the feedback network is resistive and hence that the feedback factor β is constant, but this need not be always the case. We shall therefore assume that in the general case the **feedback transfer function** is $\beta(s)$. It follows that the **closed-loop transfer function** $A_f(s)$ is given by

$$A_f(s) = \frac{A(s)}{1 + A(s)\beta(s)} \tag{10.81}$$

To focus attention on the points central to our discussion in this section, we shall assume that the amplifier is direct coupled with constant dc gain A_0 and with poles and zeros occurring in the high-frequency band. Also, for the time being let us assume that at low frequencies $\beta(s)$ reduces to a constant value. Thus at low frequencies the loop gain $A(s)\beta(s)$ becomes a constant, which should be a positive number; otherwise the feedback would not be negative. The question then is: What happens at higher frequencies?

For physical frequencies $s = j\omega$, Eq. (10.81) becomes

$$A_f(j\omega) = \frac{A(j\omega)}{1 + A(j\omega)\beta(j\omega)} \tag{10.82}$$

Thus the loop gain $A(j\omega)\beta(j\omega)$ is a complex number that can be represented by its magnitude and phase,

$$\begin{aligned} L(j\omega) &\equiv A(j\omega)\beta(j\omega) \\ &= |A(j\omega)\beta(j\omega)|e^{j\phi(\omega)} \end{aligned} \tag{10.83}$$

It is the manner in which the loop gain varies with frequency that determines the stability or instability of the feedback amplifier. To appreciate this fact, consider the frequency at which the phase angle $\phi(\omega)$ becomes 180°. At this frequency, ω_{180}, the loop gain $A(j\omega)\beta(j\omega)$ will be a real number with a negative sign. Thus at this frequency the feedback will become positive. If at $\omega = \omega_{180}$ the magnitude of the loop gain is less than unity, then from Eq. (10.82) we see that the closed-loop gain $A_f(j\omega)$ will be greater than the open-loop gain $A(j\omega)$, since the denominator of Eq. (10.82) will be smaller than unity. Nevertheless, the feedback amplifier will be stable.

On the other hand, if at the frequency ω_{180} the magnitude of the loop gain is equal to unity, it follows from Eq. (10.82) that $A_f(j\omega)$ will be infinite. This means that the amplifier will have an output for zero input; this is by definition an **oscillator**. To visualize how this feedback loop may oscillate, consider the general loop of Fig. 10.1 with the external input x_s set to zero. Any disturbance in the circuit, such as the closure of the power-supply switch, will generate a signal $x_i(t)$ at the input to the amplifier. Such a noise signal usually contains a wide range of frequencies, and we shall now concentrate on the component with frequency $\omega = \omega_{180}$, that is, the signal $X_i \sin(\omega_{180} t)$. This input signal will result in a feedback signal given by

$$X_f = A(j\omega_{180})\beta(j\omega_{180})X_i = -X_i$$

Since X_f is further multiplied by −1 in the summer block at the input, we see that the feedback causes the signal X_i at the amplifier input to be *sustained*. That is, from this point on, there will be sinusoidal signals at the amplifier input and output of frequency ω_{180}. Thus the amplifier is said to oscillate at the frequency ω_{180}.

The question now is: What happens if at ω_{180} the magnitude of the loop gain is greater than unity? We shall answer this question, not in general, but for the restricted yet very important class of circuits in which we are interested here. The answer, which is not obvious from Eq. (10.82), is that the circuit will oscillate, and the oscillations will grow in amplitude until some nonlinearity (which is always present in some form) reduces the magnitude of the loop gain to exactly unity, at which point sustained oscillations will be obtained. This mechanism for starting oscillations by using positive feedback with a loop gain greater than unity, and then using a nonlinearity to reduce the loop gain to unity at the desired amplitude, will be exploited in the design of sinusoidal oscillators in Chapter 17. Our objective here is just the opposite: Now that we know how oscillations could occur in a negative-feedback amplifier, we wish to find methods to prevent their occurrence.

10.10.2 The Nyquist Plot

The Nyquist plot is a formalized approach for testing for stability based on the discussion above. It is simply a polar plot of loop gain with frequency used as a parameter. Figure 10.34 shows such a plot. Note that the radial distance is $|A\beta|$ and the angle is the phase angle ϕ. The solid-line plot is for positive frequencies. Since the loop gain—and for that matter any gain function of a physical network—has a magnitude that is an even function of frequency and a phase that is an odd function of frequency, the $A\beta$ plot for negative frequencies (shown in Fig. 10.34 as a broken line) can be drawn as a mirror image through the Re axis.

The Nyquist plot intersects the negative real axis at the frequency ω_{180}. Thus, if this intersection occurs to the left of the point (−1, 0), we know that the magnitude of loop gain at this frequency is greater than unity and the amplifier will be unstable. On the other hand, if the intersection occurs to the right of the point (−1, 0) the amplifier will be stable. It follows that if the Nyquist plot *encircles* the point (−1, 0) then the amplifier will be

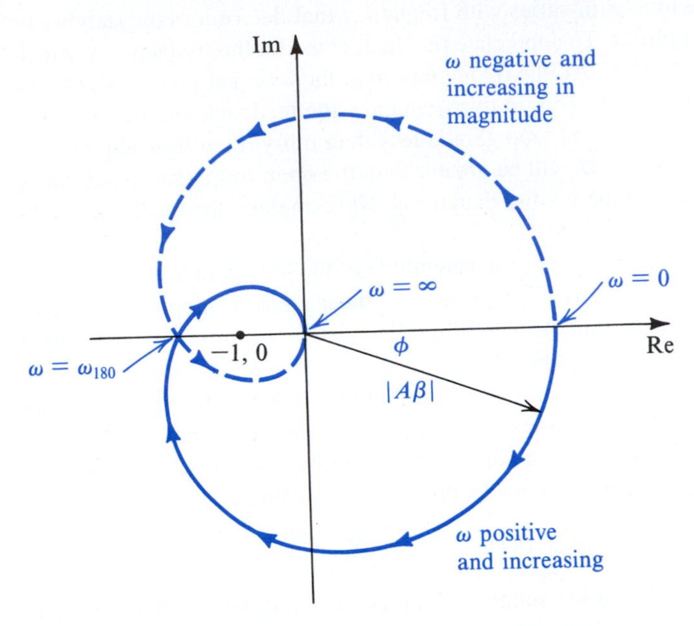

Figure 10.34 The Nyquist plot of an unstable amplifier.

unstable. It should be mentioned, however, that this statement is a simplified version of the **Nyquist criterion**; nevertheless, it applies to all the circuits in which we are interested. For the full theory behind the Nyquist method and for details of its application, consult Haykin (1970).

EXERCISE

10.20 Consider a feedback amplifier for which the open-loop transfer function $A(s)$ is given by

$$A(s) = \left(\frac{10}{1+s/10^4}\right)^3$$

Let the feedback factor β be a constant independent of frequency. Find the frequency ω_{180} at which the phase shift is 180°. Then, show that the feedback amplifier will be stable if the feedback factor β is less than a critical value β_{cr} and unstable if $\beta \geq \beta_{cr}$, and find the value of β_{cr}.

Ans. $\omega_{180} = \sqrt{3} \times 10^4$ rad/s; $\beta_{cr} = 0.008$

10.11 Effect of Feedback on the Amplifier Poles

The amplifier frequency response and stability are determined directly by its poles. Therefore we shall investigate the effect of feedback on the poles of the amplifier.[6]

[6] For a brief review of poles and zeros and related concepts, refer to Appendix F.

10.11.1 Stability and Pole Location

We shall begin by considering the relationship between stability and pole location. For an amplifier or any other system to be stable, its poles should lie in the left half of the s plane. A pair of complex-conjugate poles on the $j\omega$ axis gives rise to sustained sinusoidal oscillations. Poles in the right half of the s plane give rise to growing oscillations.

To verify the statement above, consider an amplifier with a pole pair at $s = \sigma_0 \pm j\omega_n$. If this amplifier is subjected to a disturbance, such as that caused by closure of the power-supply switch, its transient response will contain terms of the form

$$v(t) = e^{\sigma_0 t}[e^{+j\omega_n t} + e^{-j\omega_n t}] = 2e^{\sigma_0 t}\cos(\omega_n t) \tag{10.84}$$

This is a sinusoidal signal with an envelope $e^{\sigma_0 t}$. Now if the poles are in the left half of the s plane, then σ_0 will be negative and the oscillations will decay exponentially toward zero, as shown in Fig. 10.35(a), indicating that the system is stable. If, on the other hand, the poles are in

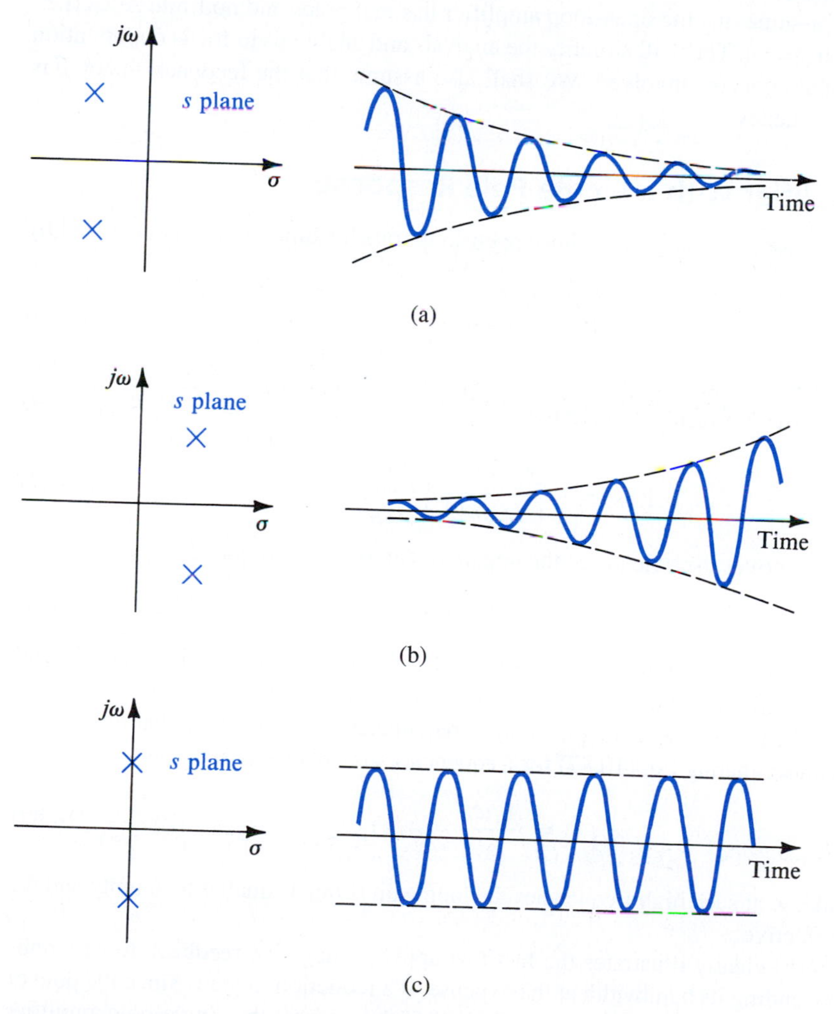

Figure 10.35 Relationship between pole location and transient response.

the right half-plane, then σ_0 will be positive, and the oscillations will grow exponentially (until some nonlinearity limits their growth), as shown in Fig. 10.35(b). Finally, if the poles are on the $j\omega$ axis, then σ_0 will be zero and the oscillations will be sustained, as shown in Fig. 10.35(c).

Although the discussion above is in terms of complex-conjugate poles, it can be shown that the existence of any right-half-plane poles results in instability.

10.11.2 Poles of the Feedback Amplifier

From the closed-loop transfer function in Eq. (10.81), we see that the poles of the feedback amplifier are the zeros of $1 + A(s)\beta(s)$. That is, the feedback-amplifier poles are obtained by solving the equation

$$1 + A(s)\beta(s) = 0 \tag{10.85}$$

which is called the **characteristic equation** of the feedback loop. It should therefore be apparent that applying feedback to an amplifier changes its poles.

In the following, we shall consider how feedback affects the amplifier poles. For this purpose we shall assume that the open-loop amplifier has real poles and no finite zeros (i.e., all the zeros are at $s = \infty$). This will simplify the analysis and enable us to focus our attention on the fundamental concepts involved. We shall also assume that the feedback factor β is independent of frequency.

10.11.3 Amplifier with a Single-Pole Response

Consider first the case of an amplifier whose open-loop transfer function is characterized by a single pole:

$$A(s) = \frac{A_0}{1 + s/\omega_P} \tag{10.86}$$

The closed-loop transfer function is given by

$$A_f(s) = \frac{A_0/(1 + A_0\beta)}{1 + s/\omega_P(1 + A_0\beta)} \tag{10.87}$$

Thus the feedback moves the pole along the negative real axis to a frequency ω_{Pf},

$$\omega_{Pf} = \omega_P(1 + A_0\beta) \tag{10.88}$$

This process is illustrated in Fig. 10.36(a). Figure 10.36(b) shows Bode plots for $|A|$ and $|A_f|$. Note that while at low frequencies the difference between the two plots is $20\log(1 + A_0\beta)$, the two curves coincide at high frequencies. One can show that this indeed is the case by approximating Eq. (10.87) for frequencies $\omega \gg \omega_P(1 + A_0\beta)$:

$$A_f(s) \simeq \frac{A_0\omega_P}{s} \simeq A(s) \tag{10.89}$$

Physically speaking, at such high frequencies the loop gain is much smaller than unity and the feedback is ineffective.

Figure 10.36(b) clearly illustrates the fact that applying negative feedback to an amplifier results in extending its bandwidth at the expense of a reduction in gain. Since the pole of the closed-loop amplifier never enters the right half of the s plane, the single-pole amplifier is stable for any value of β. Thus this amplifier is said to be **unconditionally stable**. This

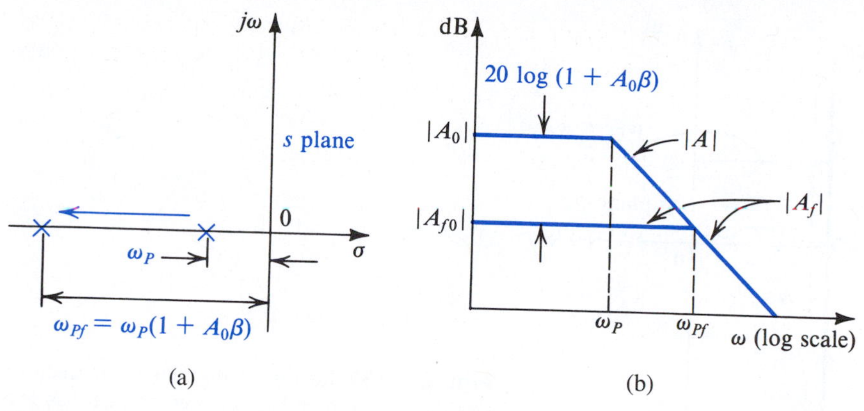

Figure 10.36 Effect of feedback on **(a)** the pole location and **(b)** the frequency response of an amplifier having a single-pole, open-loop response.

result, however, is hardly surprising, since the phase lag associated with a single-pole response can never be greater than 90°. Thus the loop gain never achieves the 180° phase shift required for the feedback to become positive.

EXERCISE

10.21 An op amp having a single-pole rolloff at 100 Hz and a low-frequency gain of 10^5 is operated in a feedback loop with $\beta = 0.01$. What is the factor by which feedback shifts the pole? To what frequency? If β is changed to a value that results in a closed-loop gain of +1, to what frequency does the pole shift?

Ans. 1001; 100.1 kHz; 10 MHz

10.11.4 Amplifier with Two-Pole Response

Consider next an amplifier whose open-loop transfer function is characterized by two real-axis poles:

$$A(s) = \frac{A_0}{(1 + s/\omega_{P1})(1 + s/\omega_{P2})} \tag{10.90}$$

In this case, the closed-loop poles are obtained from $1 + A(s)\beta = 0$, which leads to

$$s^2 + s(\omega_{P1} + \omega_{P2}) + (1 + A_0\beta)\omega_{P1}\omega_{P2} = 0 \tag{10.91}$$

Thus the closed-loop poles are given by

$$s = -\frac{1}{2}(\omega_{P1} + \omega_{P2}) \pm \frac{1}{2}\sqrt{(\omega_{P1} + \omega_{P2})^2 - 4(1 + A_0\beta)\omega_{P1}\omega_{P2}} \tag{10.92}$$

From Eq. (10.92) we see that as the loop gain $A_0\beta$ is increased from zero, the poles are brought closer together. Then a value of loop gain is reached at which the poles become coincident. If the loop gain is further increased, the poles become complex conjugate and move along a vertical line. Figure 10.37 shows the locus of the poles for increasing loop gain. This plot is called a **root-locus diagram**, where "root" refers to the fact that the poles are the roots of the characteristic equation.

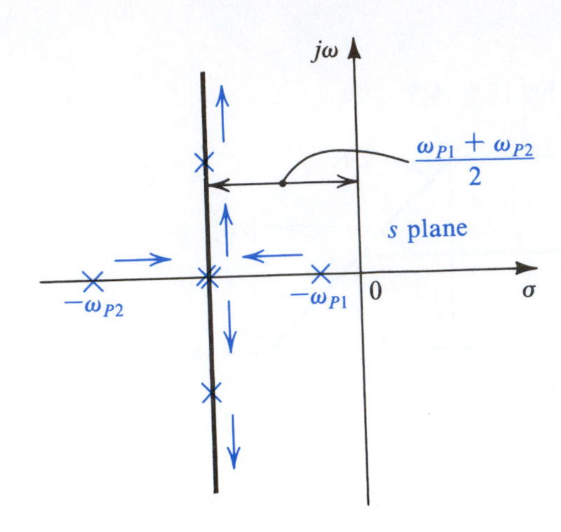

Figure 10.37 Root-locus diagram for a feedback amplifier whose open-loop transfer function has two real poles.

From the root-locus diagram of Fig. 10.37 we see that this feedback amplifier also is unconditionally stable. Again, this result should come as no surprise; the maximum phase shift of $A(s)$ in this case is 180° (90° per pole), but this value is reached at $\omega = \infty$. Thus there is no finite frequency at which the phase shift reaches 180°.

Another observation to make on the root-locus diagram of Fig. 10.37 is that the open-loop amplifier might have a dominant pole, but this is not necessarily the case for the closed-loop amplifier. The response of the closed-loop amplifier can, of course, always be plotted once the poles have been found from Eq. (10.92). As is the case with second-order responses generally, the closed-loop response can show a peak (see Chapter 16). To be more specific, the characteristic equation of a second-order network can be written in the standard form

$$s^2 + s\frac{\omega_0}{Q} + \omega_0^2 = 0 \tag{10.93}$$

where ω_0 is called the **pole frequency** and Q is called **pole Q factor**. The poles are complex if Q is greater than 0.5. A geometric interpretation for ω_0 and Q of a pair of complex-conjugate poles is given in Fig. 10.38, from which we note that ω_0 is the radial distance of the poles from the origin and that Q indicates the distance of the poles from the $j\omega$ axis. Poles on the $j\omega$ axis have $Q = \infty$.

By comparing Eqs. (10.91) and (10.93) we obtain the Q factor for the poles of the feedback amplifier as

$$Q = \frac{\sqrt{(1 + A_0\beta)\omega_{P1}\omega_{P2}}}{\omega_{P1} + \omega_{P2}} \tag{10.94}$$

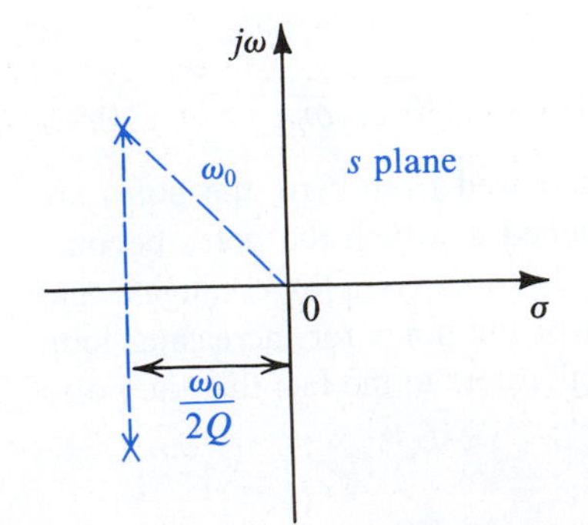

Figure 10.38 Definition of ω_0 and Q of a pair of complex-conjugate poles.

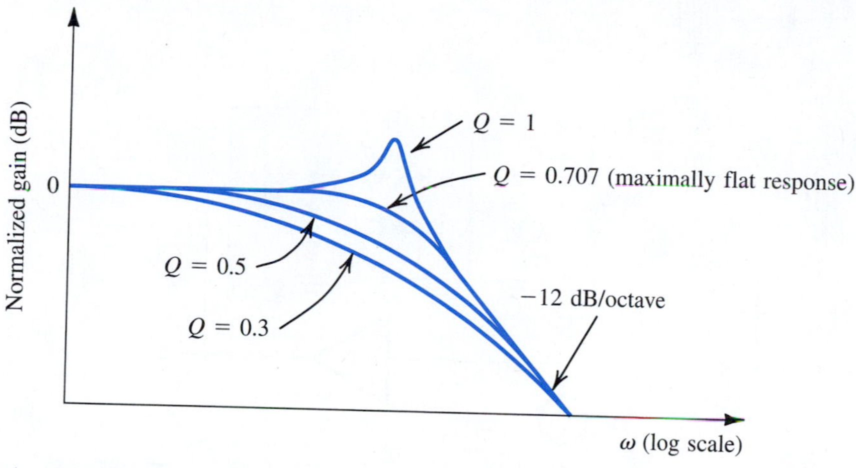

Figure 10.39 Normalized gain of a two-pole feedback amplifier for various values of Q. Note that Q is determined by the loop gain according to Eq. (10.94).

From the study of second-order network responses in Chapter 16, it will be seen that the response of the feedback amplifier under consideration shows no peaking for $Q \leq 0.707$. The boundary case corresponding to $Q = 0.707$ (poles at 45° angles) results in the **maximally flat** response. Figure 10.39 shows a number of possible responses obtained for various values of Q (or, correspondingly, various values of $A_0\beta$).

EXERCISE

10.22 An amplifier with a low-frequency gain of 100 and poles at 10^4 rad/s and 10^6 rad/s is incorporated in a negative-feedback loop with feedback factor β. For what value of β do the poles of the closed-loop amplifier coincide? What is the corresponding Q of the resulting second-order system? For what value of β is a maximally flat response achieved? What is the low-frequency closed-loop gain in the maximally flat case?

Ans. 0.245; 0.5; 0.5; 1.96 V/V

Example 10.9

As an illustration of some of the ideas just discussed, we consider the positive-feedback circuit shown in Fig. 10.40(a). Find the loop transmission $L(s)$ and the characteristic equation. Sketch a root-locus diagram for varying K, and find the value of K that results in a maximally flat response and the value of K that makes the circuit oscillate. Assume that the amplifier has frequency-idependent gain, infinite input impedance, and zero output impedance.

Solution

To obtain the loop transmission, we short-circuit the signal source and break the loop at the amplifier input. We then apply a test voltage V_t and find the returned voltage V_r, as indicated in Fig. 10.40(b). The

Example 10.9 *continued*

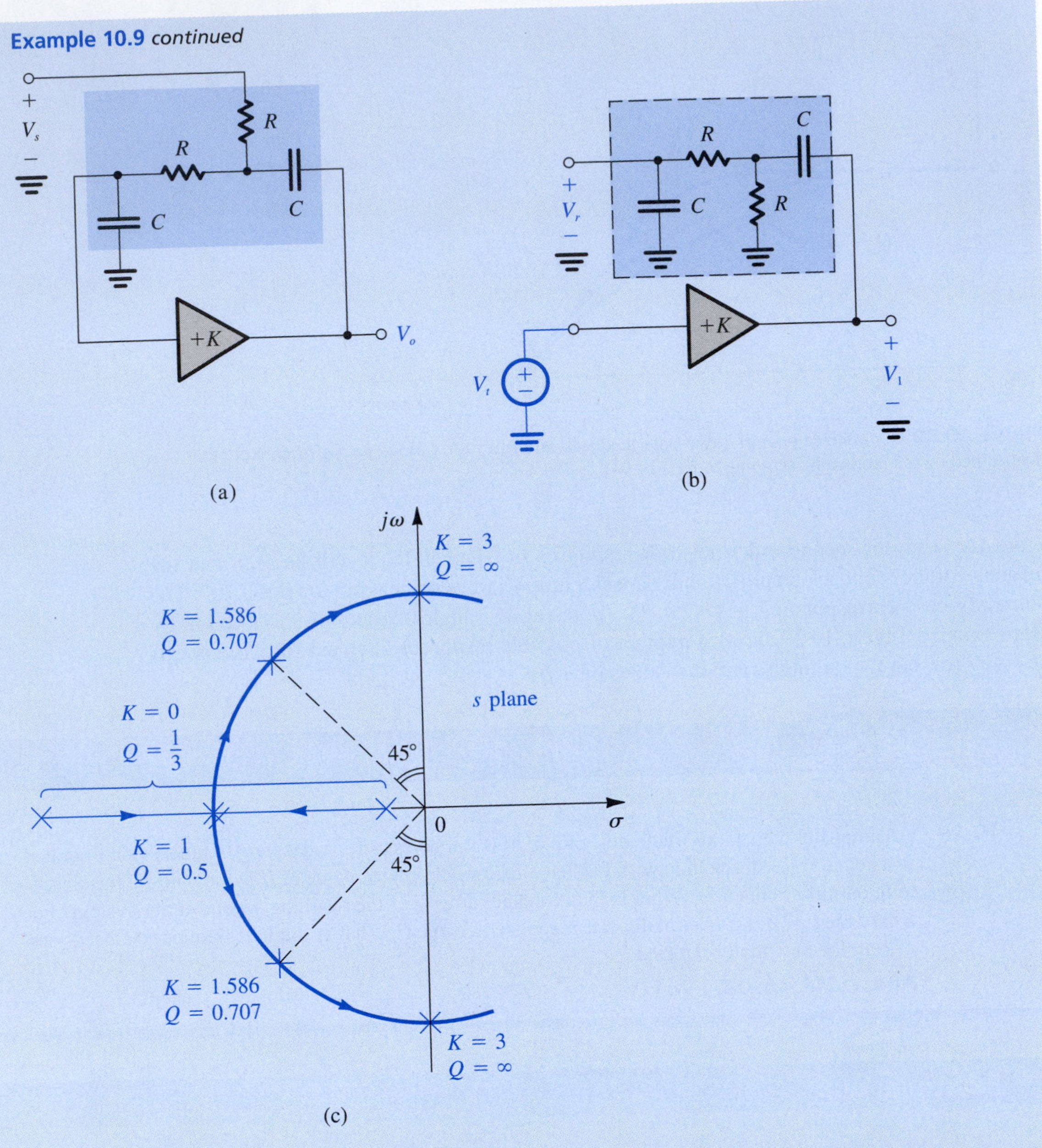

Figure 10.40 Circuits and plot for Example 10.9.

loop transmission $L(s) \equiv A(s)\beta(s)$ is given by

$$L(s) = -\frac{V_r}{V_t} = -KT(s) \tag{10.95}$$

where $T(s)$ is the transfer function of the two-port RC network shown inside the broken-line box in Fig. 10.40(b):

$$T(s) \equiv \frac{V_r}{V_1} = \frac{s(1/CR)}{s^2 + s(3/CR) + (1/CR)^2} \tag{10.96}$$

Thus,

$$L(s) = \frac{-s(K/CR)}{s^2 + s(3/CR) + (1/CR)^2} \tag{10.97}$$

The characteristic equation is

$$1 + L(s) = 0 \tag{10.98}$$

that is,

$$s^2 + s\frac{3}{CR} + \left(\frac{1}{CR}\right)^2 - s\frac{K}{CR} = 0$$

$$s^2 + s\frac{3-K}{CR} + \left(\frac{1}{CR}\right)^2 = 0 \tag{10.99}$$

By comparing this equation to the standard form of the second-order characteristic equation (Eq. 10.93) we see that the pole frequency ω_0 is given by

$$\omega_0 = \frac{1}{CR} \tag{10.100}$$

and the Q factor is

$$Q = \frac{1}{3-K} \tag{10.101}$$

Thus for $K = 0$, the poles have $Q = \frac{1}{3}$ and are therefore located on the negative real axis. As K is increased, the poles are brought closer together and eventually coincide ($Q = 0.5$, $K = 1$). Further increasing K results in the poles becoming complex and conjugate. The root locus is then a circle because the radial distance ω_0 remains constant (Eq. 10.100) independent of the value of K.

The maximally flat response is obtained when $Q = 0.707$, which results when $K = 1.586$. In this case the poles are at 45° angles, as indicated in Fig. 10.40(c). The poles cross the $j\omega$ axis into the right half of the s plane at the value of K that results in $Q = \infty$, that is, $K = 3$. Thus for $K \geq 3$ this circuit becomes unstable. This might appear to contradict our earlier conclusion that the feedback amplifier with a second-order response is unconditionally stable. Note, however, that the circuit in this example is quite different from the negative-feedback amplifier that we have been studying. Here we have an amplifier with a positive gain K and a feedback network whose transfer function $T(s)$ is frequency dependent. This feedback is in fact *positive,* and the circuit will oscillate at the frequency for which the phase of $T(j\omega)$ is zero.

Example 10.9 illustrates the use of feedback (positive feedback in this case) to move the poles of an RC network from their negative real-axis locations to complex-conjugate locations. One can accomplish the same task using negative feedback, as the root-locus diagram of Fig. 10.37 demonstrates. The process of pole control is the essence of *active-filter design,* as will be discussed in Chapter 16.

10.11.5 Amplifiers with Three or More Poles

Figure 10.41 shows the root-locus diagram for a feedback amplifier whose open-loop response is characterized by three poles. As indicated, increasing the loop gain from zero moves the highest-frequency pole outward while the two other poles are brought closer together. As $A_0\beta$ is increased further, the two poles become coincident and then become complex and conjugate. A value of $A_0\beta$ exists at which this pair of complex-conjugate poles enters the right half of the s plane, thus causing the amplifier to become unstable.

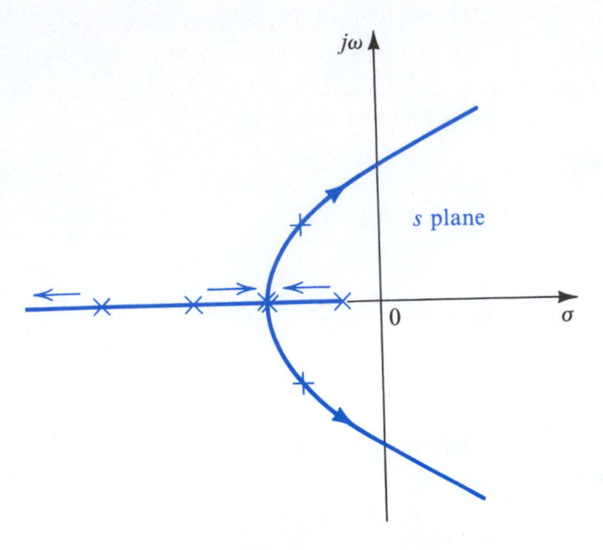

Figure 10.41 Root-locus diagram for an amplifier with three poles. The arrows indicate the pole movement as $A_0\beta$ is increased.

This result is not entirely unexpected, since an amplifier with three poles has a phase shift that reaches –270° as ω approaches ∞. Thus there exists a finite frequency, ω_{180}, at which the loop gain has 180° phase shift.

From the root-locus diagram of Fig. 10.41, we observe that one can always maintain amplifier stability by keeping the loop gain $A_0\beta$ smaller than the value corresponding to the poles entering the right half-plane. In terms of the Nyquist diagram, the critical value of $A_0\beta$ is that for which the diagram passes through the (–1, 0) point. Reducing $A_0\beta$ below this value causes the Nyquist plot to shrink and thus intersect the negative real axis to the right of the (–1, 0) point, indicating stable amplifier performance. On the other hand, increasing $A_0\beta$ above the critical value causes the Nyquist plot to expand, thus encircling the (–1, 0) point and indicating unstable performance.

For a given open-loop gain A_0 the conclusions above can be stated in terms of the feedback factor β. That is, there exists a *maximum value* for β above which the feedback amplifier becomes unstable. Alternatively, we can state that there exists a *minimum value* for the closed-loop gain A_{f0} below which the amplifier becomes unstable. To obtain lower values of closed-loop gain one needs therefore to alter the loop transfer function $L(s)$. This is the process known as *frequency compensation*. We shall study the theory and techniques of frequency compensation in Section 10.13.

Before leaving this section we point out that construction of the root-locus diagram for amplifiers having three or more poles as well as finite zeros is an involved process for which a systematic procedure exists. However, such a procedure will not be presented here, and the interested reader should consult Haykin (1970). Although the root-locus diagram provides the amplifier designer with considerable insight, other, simpler techniques based on Bode plots can be effectively employed, as will be explained in Section 10.12.

EXERCISE

10.23 Consider a feedback amplifier for which the open-loop transfer function $A(s)$ is given by

$$A(s) = \left(\frac{10}{1 + s/10^4}\right)^3$$

Let the feedback factor β be frequency independent. Find the closed-loop poles as functions of β, and show that the root locus is that of Fig. E10.23. Also find the value of β at which the amplifier becomes unstable. (*Note:* This is the same amplifier that was considered in Exercise 10.20.)

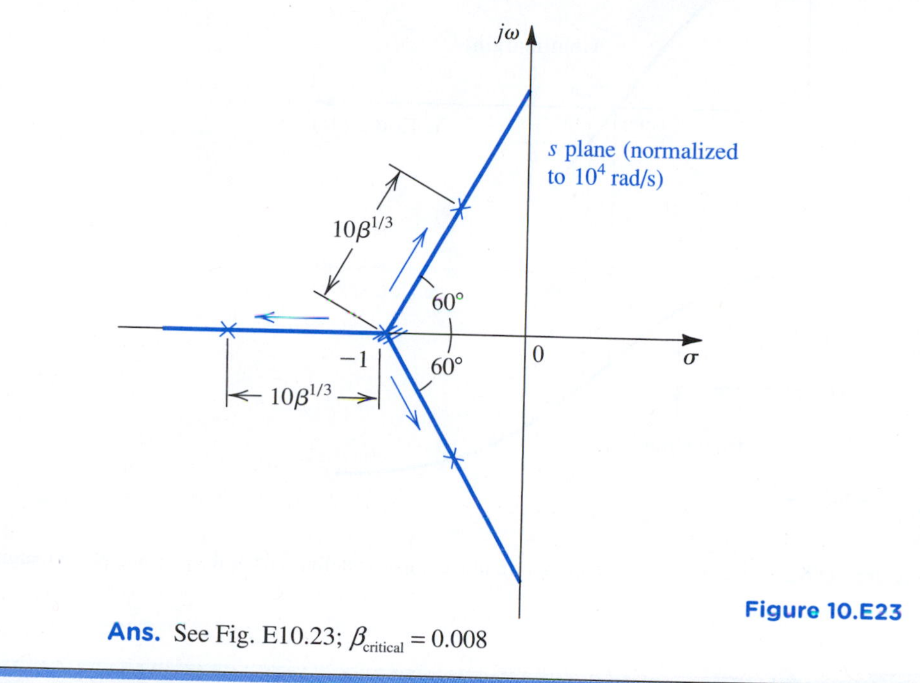

Figure 10.E23

Ans. See Fig. E10.23; $\beta_{critical} = 0.008$

10.12 Stability Study Using Bode Plots

10.12.1 Gain and Phase Margins

From Sections 10.10 and 10.11 we know that whether a feedback amplifier is or is not stable can be determined by examining its loop gain $A\beta$ as a function of frequency. One of the simplest and most effective means for doing this is through the use of a Bode plot for $A\beta$, such as the one shown in Fig. 10.42. (Note that because the phase approaches –360°, the network examined is a fourth-order one.) The feedback amplifier whose loop gain is plotted in Fig. 10.42 will be stable, since at the frequency of 180° phase shift, ω_{180}, the magnitude of the loop gain is less than unity (negative dB). The difference between the value of $|A\beta|$ at ω_{180} and unity, called the **gain margin**, is usually expressed in decibels. The gain margin represents the amount by which the loop gain can be increased while stability is maintained. Feedback amplifiers are usually designed to have sufficient gain margin to allow for the inevitable changes in loop gain with temperature, time, and so on.

Another way to investigate the stability and to express its degree is to examine the Bode plot at the frequency for which $|A\beta| = 1$, which is the point at which the magnitude plot crosses the 0-dB line. If at this frequency the phase angle is less (in magnitude) than 180°, then the amplifier is stable. This is the situation illustrated in Fig. 10.42. The difference between the phase angle at this frequency and 180° is termed the **phase margin**. On the other hand, if at the frequency of unity loop-gain magnitude, the phase lag is in excess of 180°, the amplifier will be unstable.

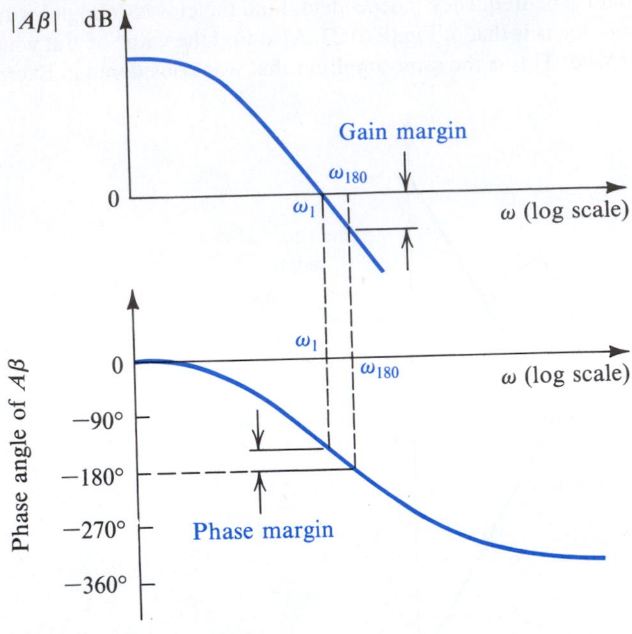

Figure 10.42 Bode plot for the loop gain $A\beta$ illustrating the definitions of the gain and phase margins.

EXERCISE

10.24 Consider an op amp having a single-pole, open-loop response with $A_0 = 10^5$ and $f_P = 10$ Hz. Let the op amp be ideal otherwise (infinite input impedance, zero output impedance, etc.). If this amplifier is connected in the noninverting configuration with a nominal low-frequency, closed-loop gain of 100, find the frequency at which $|A\beta| = 1$. Also, find the phase margin.
Ans. 10^4 Hz; 90°

10.12.2 Effect of Phase Margin on Closed-Loop Response

Feedback amplifiers are normally designed with a phase margin of at least 45°. The amount of phase margin has a profound effect on the shape of the closed-loop gain response. To see this relationship, consider a feedback amplifier with a large low-frequency loop gain, $A_0\beta \gg 1$. It follows that the closed-loop gain at low frequencies is approximately $1/\beta$. Denoting the frequency at which the magnitude of loop gain is unity by ω_1, we have (refer to Fig. 10.42)

$$A(j\omega_1)\beta = 1 \times e^{-j\theta} \tag{10.102a}$$

where

$$\theta = 180° - \text{phase margin} \tag{10.102b}$$

At ω_1 the closed-loop gain is

$$A_f(j\omega_1) = \frac{A(j\omega_1)}{1 + A(j\omega_1)\beta} \tag{10.103}$$

Substituting from Eq. (10.102a) gives

$$A_f(j\omega_1) = \frac{(1/\beta)e^{-j\theta}}{1 + e^{-j\theta}} \tag{10.104}$$

Thus the magnitude of the gain at ω_1 is

$$|A_f(j\omega_1)| = \frac{1/\beta}{|1 + e^{-j\theta}|} \tag{10.105}$$

For a phase margin of 45°, $\theta = 135°$; and we obtain

$$|A_f(j\omega_1)| = 1.3\frac{1}{\beta} \tag{10.106}$$

That is, the gain peaks by a factor of 1.3 above the low-frequency value of $1/\beta$. This peaking increases as the phase margin is reduced, eventually reaching ∞ when the phase margin is zero. Zero phase margin, of course, implies that the amplifier can sustain oscillations [poles on the $j\omega$ axis; Nyquist plot passing through (−1, 0)].

EXERCISE

10.25 Find the closed-loop gain at ω_1 relative to the low-frequency gain when the phase margin is 30°, 60°, and 90°.
Ans. 1.93; 1; 0.707

10.12.3 An Alternative Approach for Investigating Stability

Investigating stability by constructing Bode plots for the loop gain $A\beta$ can be a tedious and time-consuming process, especially if we have to investigate the stability of a given amplifier for a variety of feedback networks. An alternative approach, which is much simpler, is to construct a Bode plot for the open-loop gain $A(j\omega)$ only. Assuming for the time being that β is independent of frequency, we can plot $20\log(1/\beta)$ as a horizontal straight line on the same plane used for $20\log|A|$. The difference between the two curves will be

$$20\log|A(j\omega)| - 20\log\frac{1}{\beta} = 20\log|A\beta| \tag{10.107}$$

which is the loop gain (in dB). We may therefore study stability by examining the difference between the two plots. If we wish to evaluate stability for a different feedback factor, we simply draw another horizontal straight line at the level $20\log(1/\beta)$.

To illustrate, consider an amplifier whose open-loop transfer function is characterized by three poles. For simplicity let the three poles be widely separated—say, at 0.1 MHz, 1 MHz, and 10 MHz, as shown in Fig. 10.43. Note that because the poles are widely separated, the

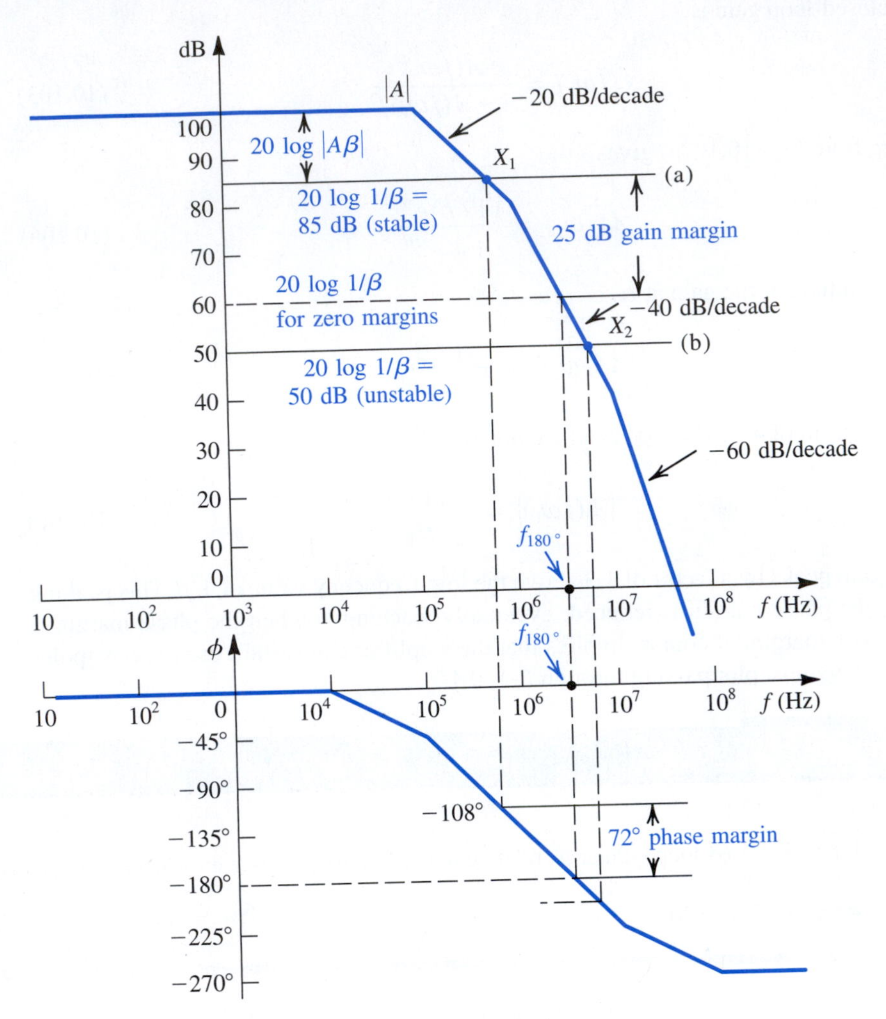

Figure 10.43 Stability analysis using Bode plot of $|A|$.

phase is approximately −45° at the first pole frequency, −135° at the second, and −225° at the third. The frequency at which the phase of $A(j\omega)$ is −180° lies on the −40-dB/decade segment, as indicated in Fig. 10.43.

The open-loop gain of this amplifier can be expressed as

$$A = \frac{10^5}{(1 + jf/10^5)(1 + jf/10^6)(1 + jf/10^7)} \tag{10.108}$$

from which $|A|$ can be easily determined for any frequency f (in Hz), and the phase can be obtained as

$$\phi = -[\tan^{-1}(f/10^5) + \tan^{-1}(f/10^6) + \tan^{-1}(f/10^7)] \tag{10.109}$$

The magnitude and phase graphs shown in Fig. 10.43 are obtained using the method for constructing Bode plots (Appendix F). These graphs provide approximate values for important amplifier parameters, with more exact values obtainable from Eqs. (10.108) and (10.109). For example, the frequency f_{180} at which the phase angle is 180° can be found from Fig. 10.43 to be approximately 3.2×10^6 Hz. Using this value as a starting point, a more exact value can be found by trial and error using Eq. (10.109). The result is $f_{180} = 3.34 \times 10^6$ Hz. At this frequency, Eq. (10.108) gives a gain magnitude of 58.2 dB, which is reasonably close to the approximate value of 60 dB given by Fig. 10.43.

Consider next the straight line labeled (a) in Fig. 10.43. This line represents a feedback factor for which $20 \log(1/\beta) = 85$ dB, which corresponds to $\beta = 5.623 \times 10^{-5}$ and a closed-loop gain of 83.6 dB. Since the loop gain is the difference between the $|A|$ curve and the $1/\beta$ line, the point of intersection X_1 corresponds to the frequency at which $|A\beta| = 1$. Using the graphs of Fig. 10.43, this frequency can be found to be approximately 5.6×10^5 Hz. A more exact value of 4.936×10^5 can be obtained using the transfer-function equations. At this frequency the phase angle is approximately −108°. Thus the closed-loop amplifier, for which $20 \log(1/\beta) = 85$ dB, will be stable with a phase margin of 72°. The gain margin can be easily obtained from Fig. 10.43; it is 25 dB.

Next, suppose that we wish to use this amplifier to obtain a closed-loop gain of 50-dB nominal value. Since $A_0 = 100$ dB, we see that $A_0\beta \gg 1$ and $20 \log(A_0\beta) \simeq 50$ dB, resulting in $20 \log(1/\beta) \simeq 50$ dB. To see whether this closed-loop amplifier is or is not stable, we draw line (b) in Fig. 10.43 with a height of 50 dB. This line intersects the open-loop gain curve at point X_2, where the corresponding phase is greater than 180°. Thus the closed-loop amplifier with 50-dB gain will be unstable.

In fact, it can easily be seen from Fig. 10.43 that the *minimum* value of $20 \log(1/\beta)$ that can be used, with the resulting amplifier being stable, is 60 dB. In other words, the minimum value of stable closed-loop gain obtained with this amplifier is approximately 60 dB. At this value of gain, however, a manufactured version of this amplifier may still oscillate, since no margin is left to allow for possible changes in gain.

Since the 180°-phase point always occurs on the −40-dB/decade segment of the Bode plot for $|A|$, a rule of thumb to guarantee stability is as follows: *The closed-loop amplifier will be stable if the* $20 \log(1/\beta)$ *line intersects the* $20 \log|A|$ *curve at a point on the −20-dB/decade segment.* Following this rule ensures that a phase margin of at least 45° is obtained. For the example of Fig. 10.43, the rule implies that the maximum value of β is 10^{-4}, which corresponds to a closed-loop gain of approximately 80 dB.

The rule of thumb above can be generalized for the case in which β is a function of frequency. The general rule states that *at the intersection of* $20 \log[1/|\beta(j\omega)|]$ *and* $20 \log|A(j\omega)|$ *the difference of slopes* (called the **rate of closure**) *should not exceed* 20 dB/decade.

EXERCISE

10.26 Consider an op amp whose open-loop gain is identical to that of Fig. 10.43. Assume that the op amp is ideal otherwise. Let the op amp be connected as a differentiator. Use the rule of thumb above to show that for stable performance the differentiator time constant should be greater than 159 ms. [*Hint:* Recall that for a differentiator, the Bode plot for $1/|\beta(j\omega)|$ has a slope of +20 dB/decade and intersects the 0-dB line at $1/\tau$, where τ is the differentiator time constant.]

10.13 Frequency Compensation

In this section, we shall discuss methods for modifying the open-loop transfer function $A(s)$ of an amplifier having three or more poles so that the closed-loop amplifier is stable for a given desired value of closed-loop gain.

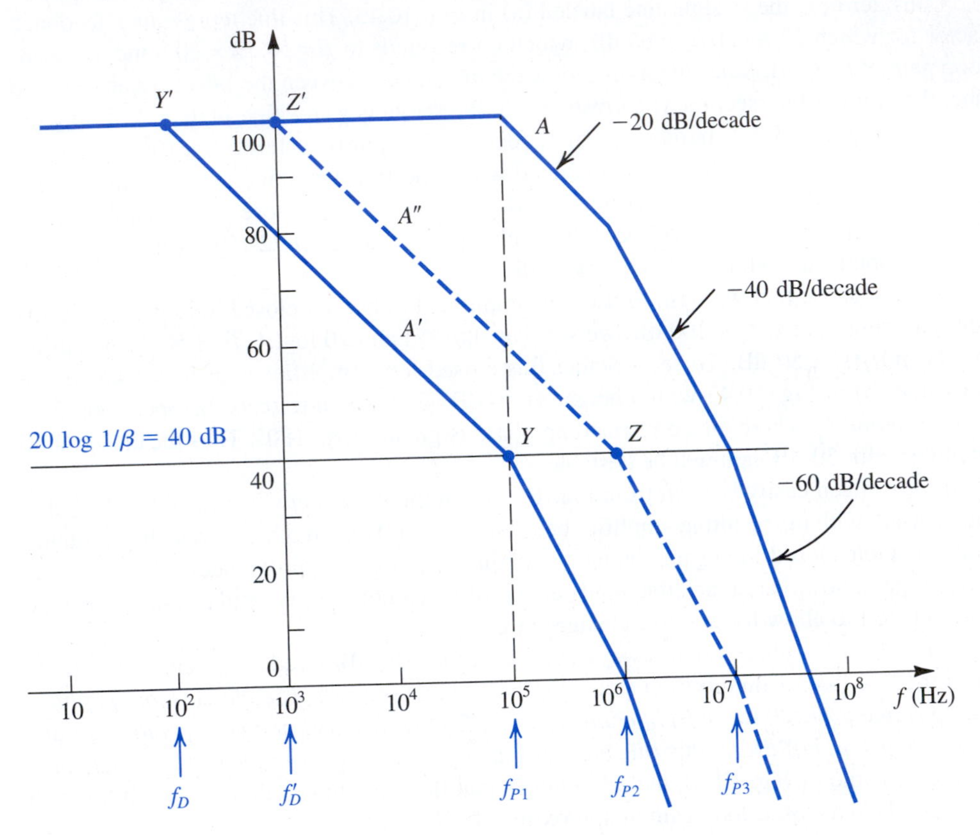

Figure 10.44 Frequency compensation for $\beta = 10^{-2}$. The response labeled A' is obtained by introducing an additional pole at f_D. The A'' response is obtained by moving the original low-frequency pole to f_D'.

10.13.1 Theory

The simplest method of frequency compensation consists of introducing a new pole in the function $A(s)$ at a sufficiently low frequency, f_D, such that the modified open-loop gain, $A'(s)$, intersects the $20 \log(1/|\beta|)$ curve with a slope difference of 20 dB/decade. As an example, let it be required to compensate the amplifier whose $A(s)$ is shown in Fig. 10.44 such that closed-loop amplifiers with β as high as 10^{-2} (i.e., closed-loop gains as low as approximately 40 dB) will be stable. First, we draw a horizontal straight line at the 40-dB level to represent $20 \log(1/\beta)$, as shown in Fig. 10.44. We then locate point Y on this line at the frequency of the first pole, f_{P1}. From Y we draw a line with –20-dB/decade slope and determine the point at which this line intersects the dc gain line, point Y'. This latter point gives the frequency f_D of the new pole that has to be introduced in the open-loop transfer function.

The compensated open-loop response $A'(s)$ is indicated in Fig. 10.44. It has four poles: at f_D, f_{P1}, f_{P2}, and f_{P3}. Thus $|A'|$ begins to roll off with a slope of –20 dB/decade at f_D. At f_{P1} the slope changes to – 40 dB/decade, at f_{P2} it changes to –60 dB/decade, and so on. Since the 20 log(1/β) line intersects the $20 \log|A'|$ curve at point Y on the –20-dB/decade segment, the closed-loop amplifier with this β value (or lower values) will be stable.

A serious disadvantage of this compensation method is that at most frequencies the open-loop gain has been drastically reduced. This means that at most frequencies the amount of feedback available will be small. Since all the advantages of negative feedback are directly proportional to the amount of feedback, the performance of the compensated amplifier will be impaired.

Careful examination of Fig. 10.44 shows that the gain $A'(s)$ is low because of the pole at f_{P1}. If we can somehow eliminate this pole, then—rather than locating point Y, drawing YY', and so on—we can start from point Z (at the frequency of the second pole) and draw the line ZZ'. This would result in the open-loop curve $A''(s)$, which shows considerably higher gain than $A'(s)$.

Although it is not possible to eliminate the pole at f_{P1}, it is usually possible to shift that pole from $f = f_{P1}$ to $f = f_D'$. This makes the pole dominant and eliminates the need for introducing an additional lower-frequency pole, as will be explained next.

10.13.2 Implementation

We shall now address the question of implementing the frequency-compensation scheme discussed above. The amplifier circuit normally consists of a number of cascaded gain stages, with each stage responsible for one or more of the transfer-function poles. Through manual and/or computer analysis of the circuit, one identifies which stage introduces each of the important poles f_{P1}, f_{P2}, and so on. For the purpose of our discussion, assume that the first pole f_{P1} is introduced at the interface between the two cascaded differential stages shown in Fig. 10.45(a). In Fig. 10.45(b) we show a simple small-signal model of the circuit at this interface. Current source I_x represents the output-signal current of the Q_1–Q_2 stage. Resistance R_x and capacitance C_x represent the total resistance and capacitance between the two nodes B and B′. It follows that the pole f_{P1} is given by

$$f_{P1} = \frac{1}{2\pi C_x R_x} \tag{10.110}$$

Let us now connect the compensating capacitor C_C between nodes B and B′. This will result in the modified equivalent circuit shown in Fig. 10.45(c) from which we see that the pole introduced will no longer be at f_{P1}; rather, the pole can be at any desired lower frequency f_D':

$$f_D' = \frac{1}{2\pi(C_x + C_C)R_x} \tag{10.111}$$

We thus conclude that one can select an appropriate value for C_C to shift the pole frequency from f_{P1} to the value f_D' determined by point Z' in Fig. 10.44.

At this juncture it should be pointed out that adding the capacitor C_C will usually result in changes in the location of the other poles (those at f_{P2} and f_{P3}). One might therefore need to calculate the new location of f_{P2} and perform a few iterations to arrive at the required value for C_C.

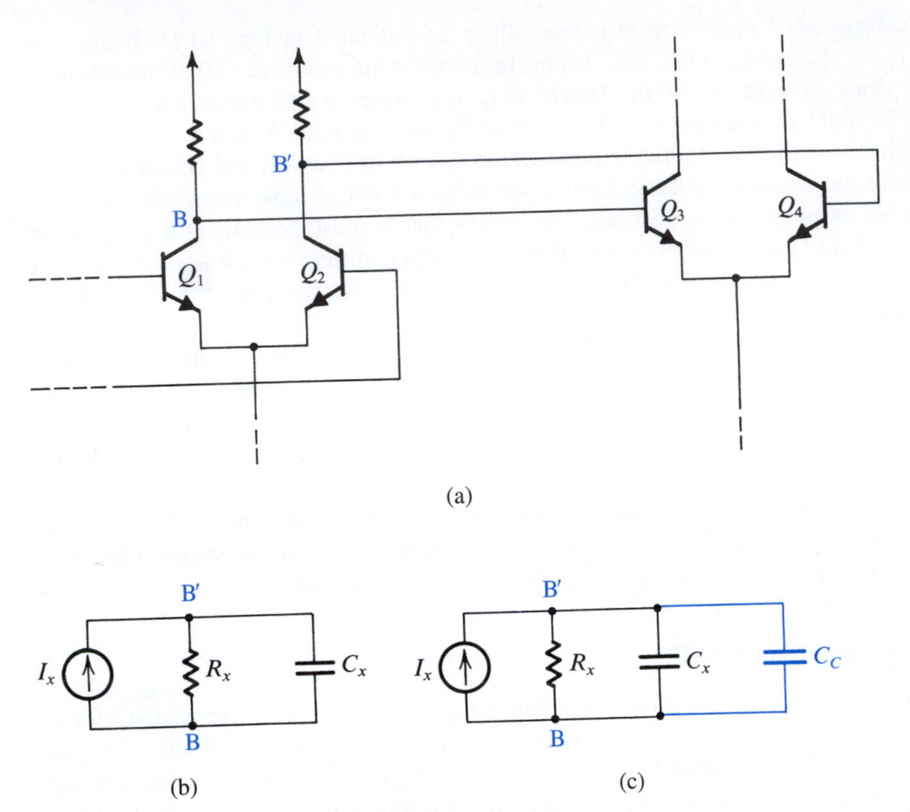

Figure 10.45 **(a)** Two cascaded gain stages of a multistage amplifier. **(b)** Equivalent circuit for the interface between the two stages in **(a)**. **(c)** Same circuit as in **(b)**, but with a compensating capacitor C_C added. Note that the analysis here applies equally well to MOS amplifiers.

A disadvantage of this implementation method is that the required value of C_C is usually quite large. Thus if the amplifier to be compensated is an IC op amp, it will be difficult, and probably impossible, to include this compensating capacitor on the IC chip. (As pointed out in Chapter 7 and in Appendix A, the maximum practical size of a monolithic capacitor is about 100 pF.) An elegant solution to this problem is to connect the compensating capacitor in the feedback path of an amplifier stage. Because of the Miller effect, the compensating capacitance will be multiplied by the stage gain, resulting in a much larger effective capacitance. Furthermore, as explained later, another unexpected benefit accrues.

10.13.3 Miller Compensation and Pole Splitting

Figure 10.46(a) shows one gain stage in a multistage amplifier. For simplicity, the stage is shown as a common-emitter amplifier, but in practice it can be a more elaborate circuit. In the feedback path of this common-emitter stage we have placed a compensating capacitor C_f.

Figure 10.46(b) shows a simplified equivalent circuit of the gain stage of Fig. 10.46(a). Here R_1 and C_1 represent the total resistance and total capacitance between node B and ground. Similarly, R_2 and C_2 represent the total resistance and total capacitance between node C and ground. Furthermore, it is assumed that C_1 includes the Miller component due to capacitance C_μ, and C_2 includes the input capacitance of the succeeding amplifier stage. Finally, I_i represents the output signal current of the preceding stage.

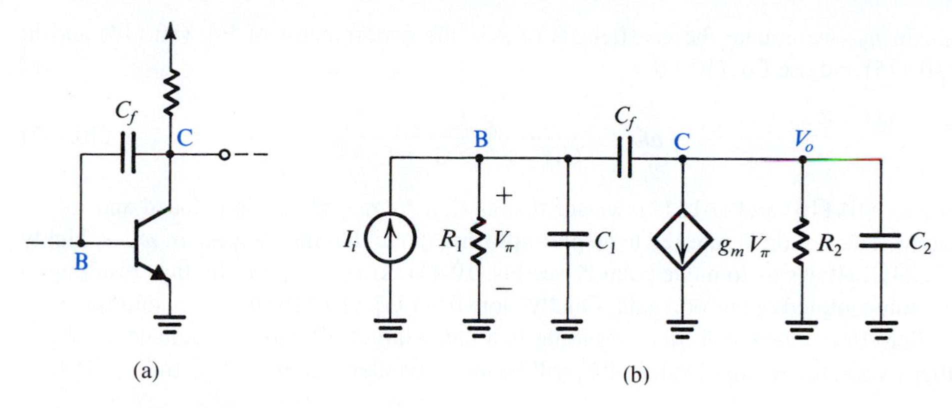

Figure 10.46 **(a)** A gain stage in a multistage amplifier with a compensating capacitor connected in the feedback path, and **(b)** an equivalent circuit. Note that although a BJT is shown, the analysis applies equally well to the MOSFET case.

In the absence of the compensating capacitor C_f, we can see from Fig. 10.46(b) that there are two poles—one at the input and one at the output. Let us assume that these two poles are f_{P1} and f_{P2} of Fig. 10.44; thus,

$$f_{P1} = \frac{1}{2\pi C_1 R_1} \qquad f_{P2} = \frac{1}{2\pi C_2 R_2} \tag{10.112}$$

With C_f present, analysis of the circuit yields the transfer function

$$\frac{V_o}{I_i} = \frac{(sC_f - g_m)R_1 R_2}{1 + s[C_1 R_1 + C_2 R_2 + C_f(g_m R_1 R_2 + R_1 + R_2)] + s^2[C_1 C_2 + C_f(C_1 + C_2)]R_1 R_2} \tag{10.113}$$

The zero is usually at a much higher frequency than the dominant pole, and we shall neglect its effect. The denominator polynomial $D(s)$ can be written in the form

$$D(s) = \left(1 + \frac{s}{\omega'_{P1}}\right)\left(1 + \frac{s}{\omega'_{P2}}\right) = 1 + s\left(\frac{1}{\omega'_{P1}} + \frac{1}{\omega'_{P2}}\right) + \frac{s^2}{\omega'_{P1}\omega'_{P2}} \tag{10.114}$$

where ω'_{P1} and ω'_{P2} are the new frequencies of the two poles. Normally one of the poles will be dominant; $\omega'_{P1} \ll \omega'_{P2}$. Thus,

$$D(s) \simeq 1 + \frac{s}{\omega'_{P1}} + \frac{s^2}{\omega'_{P1}\omega'_{P2}} \tag{10.115}$$

Equating the coefficients of s in the denominator of Eq. (10.113) and in Eq. (10.115) results in

$$\omega'_{P1} = \frac{1}{C_1 R_1 + C_2 R_2 + C_f(g_m R_1 R_2 + R_1 + R_2)}$$

which can be approximated by

$$\omega'_{P1} \simeq \frac{1}{g_m R_2 C_f R_1} \tag{10.116}$$

To obtain ω'_{P2} we equate the coefficients of s^2 in the denominator of Eq. (10.113) and in Eq. (10.115) and use Eq. (10.116):

$$\omega'_{P2} \simeq \frac{g_m C_f}{C_1 C_2 + C_f (C_1 + C_2)} \tag{10.117}$$

From Eqs. (10.116) and (10.117), we see that as C_f is increased, ω'_{P1} is reduced and ω'_{P2} is increased. This action is referred to as **pole splitting**. Note that the increase in ω'_{P2} is highly beneficial; it allows us to move point Z (see Fig. 10.44) further to the right, thus resulting in higher compensated open-loop gain. Finally, note from Eq. (10.116) that C_f is multiplied by the Miller-effect factor $g_m R_2$, thus resulting in a much larger effective capacitance, $g_m R_2 C_f$. In other words, the required value of C_f will be much smaller than that of C_C in Fig. 10.45.

Example 10.10

Consider an op amp whose open-loop transfer function is identical to that shown in Fig. 10.43. We wish to compensate this op amp so that the closed-loop amplifier with resistive feedback is stable for any gain (i.e., for β up to unity). Assume that the op-amp circuit includes a stage such as that of Fig. 10.46 with $C_1 = 100$ pF, $C_2 = 5$ pF, and $g_m = 40$ mA/V, that the pole at f_{P1} is caused by the input circuit of that stage, and that the pole at f_{P2} is introduced by the output circuit. Find the value of the compensating capacitor for two cases: either if it is connected between the input node B and ground, or in the feedback path of the transistor.

Solution

First we determine R_1 and R_2 from

$$f_{P1} = 0.1 \text{ MHz} = \frac{1}{2\pi C_1 R_1}$$

Thus,

$$R_1 = \frac{10^5}{2\pi}\ \Omega$$

$$f_{P2} = 1 \text{ MHz} = \frac{1}{2\pi C_2 R_2}$$

Thus,

$$R_2 = \frac{10^5}{\pi}\ \Omega$$

If a compensating capacitor C_C is connected across the input terminals of the transistor stage, then the frequency of the first pole changes from f_{P1} to f'_D:

$$f'_D = \frac{1}{2\pi(C_1 + C_C)R_1}$$

The second pole remains unchanged at 1-MHz. The required value for f'_D is determined by drawing a –20-dB/decade line from the 1-MHz frequency point on the $20 \log(1/\beta) = 20 \log 1 = 0$ dB line. This line will intersect the 100-dB dc gain line at 10 Hz. Thus,

$$f'_D = 10 \text{ Hz} = \frac{1}{2\pi(C_1 + C_C)R_1}$$

which results in $C_C \simeq 1\ \mu F$, which is quite large and certainly cannot be included on the IC chip.

Next, if a compensating capacitor C_f is connected in the feedback path of the transistor, then both poles change location to the values given by Eqs. (10.116) and (10.117):

$$f'_{P1} \simeq \frac{1}{2\pi g_m R_2 C_f R_1} \qquad f'_{P2} \simeq \frac{g_m C_f}{2\pi[C_1 C_2 + C_f(C_1 + C_2)]} \tag{10.118}$$

To determine where we should locate the first pole, we need to know the value of f_{P2}. As an approximation, let us assume that $C_f \gg C_2$, which enables us to obtain

$$f'_{P2} \simeq \frac{g_m}{2\pi(C_1 + C_2)} = 60.6 \text{ MHz}$$

Thus it appears that this pole will move to a frequency higher than f_{P3} (which is 10 MHz). Let us therefore assume that the second pole will be at f_{P3}. This requires that the first pole be located at

$$f'_{P1} = \frac{f_{P3}}{A_0} = \frac{10^7 \text{ Hz}}{10^5} = 100 \text{ Hz}$$

Thus,

$$f'_{P1} = 100 \text{ Hz} = \frac{1}{2\pi g_m R_2 C_f R_1}$$

which results in $C_f = 78.5$ pF. Although this value is indeed much greater than C_2, we can determine the location of the pole f'_{P2} from Eq. (10.118), which yields $f'_{P2} = 57.2$ MHz, confirming that this pole has indeed been moved past f_{P3}.

We conclude that using Miller compensation not only results in a much smaller compensating capacitor but, owing to pole splitting, also enables us to place the dominant pole a decade higher in frequency. This results in a wider bandwidth for the compensated op amp.

EXERCISE

10.27 A multipole amplifier having a first pole at 1 MHz and an open-loop gain of 100 dB is to be compensated for closed-loop gains as low as 20 dB by the introduction of a new dominant pole. At what frequency must the new pole be placed?
Ans. 100 Hz

10.28 For the amplifier described in Exercise 10.27, rather than introducing a new dominant pole, we can use additional capacitance at the circuit node at which the first pole is formed to reduce the frequency of the first pole. If the frequency of the second pole is 10 MHz and if it remains unchanged while additional capacitance is introduced as mentioned, find the frequency to which the first pole must be lowered so that the resulting amplifier is stable for closed-loop gains as low as 20 dB. By what factor must the capacitance at the controlling node be increased?
Ans. 1000 Hz; 1000

Summary

- Negative feedback is employed to make the amplifier gain less sensitive to component variations; to control input and output impedances; to extend bandwidth; to reduce nonlinear distortion; and to enhance signal-to-interference ratio.

- The advantages above are obtained at the expense of a reduction in gain and at the risk of the amplifier becoming unstable (that is, oscillating). The latter problem is solved by careful design.

- For each of the four basic types of amplifier, there is an appropriate feedback topology. The four topologies, together with their analysis procedure and their effects on input and output impedances, are summarized in **Table 10.1** in Section 10.8.

- The key feedback parameters are the loop gain ($A\beta$), which for negative feedback must be a positive dimensionless number, and the amount of feedback $(1 + A\beta)$. The latter directly determines gain reduction, gain desensitivity, bandwidth extension, and changes in R_i and R_o.

- Since A and β are in general frequency dependent, the poles of the feedback amplifier are obtained by solving the characteristic equation $1 + A(s)\beta(s) = 0$.

- For the feedback amplifier to be stable, its poles must all be in the left half of the s plane.

- Stability is guaranteed if at the frequency for which the phase angle of $A\beta$ is 180° (i.e., ω_{180}), $|A\beta|$ is less than unity; the amount by which it is less than unity, expressed in decibels, is the gain margin. Alternatively, the amplifier is stable if, at the frequency at which $|A\beta| = 1$, the phase angle is less than 180°; the difference is the phase margin.

- The stability of a feedback amplifier can be analyzed by constructing a Bode plot for $|A|$ and superimposing on it a plot for $1/|\beta|$. Stability is guaranteed if the two plots intersect with a difference in slope no greater than 6 dB/octave.

- To make a given amplifier stable for a given feedback factor β, the open-loop frequency response is suitably modified by a process known as frequency compensation.

- A popular method for frequency compensation involves connecting a feedback capacitor across an inverting stage in the amplifier. This causes the pole formed at the input of the amplifier stage to shift to a lower frequency and thus become dominant, while the pole formed at the output of the amplifier stage is moved to a very high frequency and thus becomes unimportant. This process is known as pole splitting.

PROBLEMS

Computer Simulation Problems

SIM Problems identified by this icon are intended to demonstrate the value of using SPICE simulation to verify hand analysis and design, and to investigate important issues such as allowable signal swing and amplifier nonlinear distortion. Instructions to assist in setting up PSpice and Multisim simulations for all the indicated problems can be found in the corresponding files on the disc. Note that if a particular parameter value is not specified in the problem statement, you are to make a reasonable assumption. * difficult problem; ** more difficult; *** very challenging and/or time-consuming; D: design problem.

Section 10.1: The General Feedback Structure

10.1 A negative-feedback amplifier has a closed-loop gain $A_f = 100$ and an open-loop gain $A = 10^4$. What is the feedback factor β? If a manufacturing error results in a reduction of A to 10^3, what closed-loop gain results? What is the percentage change in A_f corresponding to this factor of 10 reduction in A?

10.2 Consider the op-amp circuit shown in Fig. P10.2, where the op amp has infinite input resistance and zero output resistance but finite open-loop gain A.

(a) Convince yourself that $\beta = R_1/(R_1 + R_2)$
(b) If $R_1 = 10\text{ k}\Omega$, find R_2 that results in $A_f = 10$ V/V for the following three cases: (i) $A = 1000$ V/V; (ii) $A = 100$ V/V; (iii) $A = 12$ V/V.

(c) For each of the three cases in (b), find the percentage change in A_f that results when A decreases by 20%. Comment on the results.

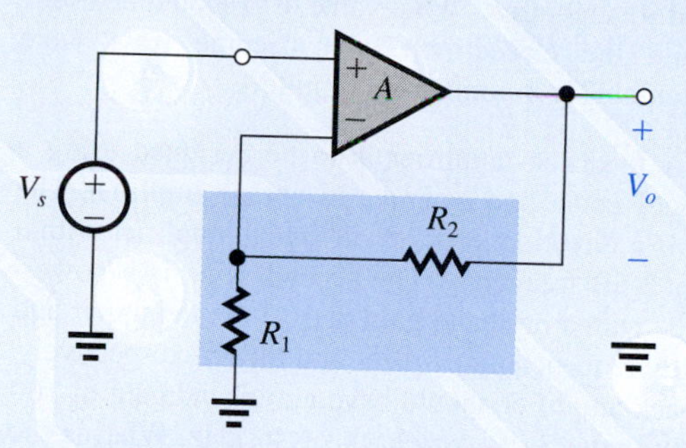

Figure P10.2

10.3 The noninverting buffer op-amp configuration shown in Fig. P10.3 provides a direct implementation of the feedback loop of Fig. 10.1. Assuming that the op amp has infinite input resistance and zero output resistance, what is β? If $A = 1000$, what is the closed-loop voltage gain? What is the amount of feedback (in dB)? For $V_s = 1$ V, find V_o and V_i. If A decreases by 10%, what is the corresponding percentage decrease in A_f?

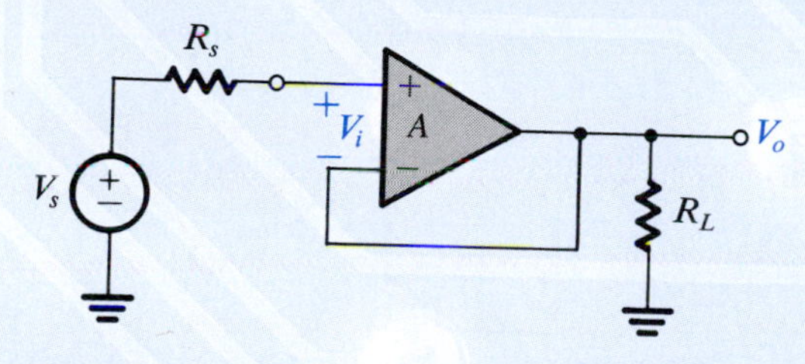

Figure P10.3

10.4 In a particular circuit represented by the block diagram of Fig. 10.1, a signal of 1 V from the source results in a difference signal of 10 mV being provided to the amplifying element A, and 10 V applied to the load. For this arrangement, identify the values of A and β that apply.

10.5 Find the loop gain and the amount of feedback of a voltage amplifier for which A_f and $1/\beta$ differ by (a) 1%, (b) 5%, (c) 10%, (d) 50%.

10.6 In a particular amplifier design, the β network consists of a linear potentiometer for which β is 0.00 at one end, 1.00 at the other end, and 0.50 in the middle. As the potentiometer is adjusted, find the three values of closed-loop gain that result when the amplifier open-loop gain is (a) 1 V/V, (b) 10 V/V, (c) 100 V/V, (d) 10,000 V/V.

10.7 A newly constructed feedback amplifier undergoes a performance test with the following results: With the feedback connection removed, a source signal of 5 mV is required to provide a 10-V output to the load; with the feedback connected, a 10-V output requires a 200-mV source signal. For this amplifier, identify values of A, β, $A\beta$, the closed-loop gain, and the amount of feedback (in dB).

Section 10.2: Some Properties of Negative Feedback

10.8 For the negative-feedback loop of Fig. 10.1, find the loop gain $A\beta$ for which the sensitivity of closed-loop gain to open-loop gain [i.e., $(dA_f/A_f)/(dA/A)$] is –40 dB. For what value of $A\beta$ does the sensitivity become 1/2?

D 10.9 A designer is considering two possible designs of a feedback amplifier. The ultimate goal is $A_f = 20$ V/V. One design employs an amplifier for which $A = 500$ V/V and the other uses $A = 250$ V/V. Find β and the desensitivity factor in both cases. If the $A = 500$ amplifier units have a gain uncertainty of ±10%, what is the gain uncertainty for the closed-loop amplifiers utilizing this amplifier type? If the same result is to be achieved with the $A = 250$ amplifier, what is the maximum allowable uncertainty in its gain?

D 10.10 A designer is required to achieve a closed-loop gain of 25 ± 1 % V/V using a basic amplifier whose gain variation is ±10 %. What nominal value of A and β (assumed constant) are required?

D 10.11 A circuit designer requires a gain of 25 ± 1 % V/V using an amplifier whose gain varies by a factor of 10 over temperature and time. What is the lowest gain required? The nominal gain? The value of β?

D 10.12 A power amplifier employs an output stage whose gain varies from 2 to 12 for various reasons. What is the gain of an ideal (non varying) amplifier connected to drive it so that an overall gain with feedback of 100 ± 5% V/V can be achieved? What is the value of β to be used? What are the requirements if A_f must be held within ±0.5 %? For each of these situations, what preamplifier gain and feedback factor β are required if A_f is to be 10 V/V (with the two possible tolerances)?

D 10.13 It is required to design an amplifier with a gain of 100 that is accurate to within ±1%. You have available amplifier stages with a gain of 1000 that is accurate to within ±30%. Provide a design that uses a number of these gain stages in cascade, with each stage employing negative feedback of an appropriate amount. Obviously, your design should use the lowest possible number of stages while meeting specification.

10.14 Consider an amplifier having a midband gain A_M and a low-frequency response characterized by a pole at $s = -\omega_L$ and a zero at $s = 0$. Let the amplifier be connected in a negative-feedback loop with a feedback factor β. Find an expression for the midband gain and the lower 3-dB frequency of the closed-loop amplifier. By what factor have both changed?

D *10.15 It is required to design an amplifier to have a nominal closed-loop gain of 10 V/V using a battery-operated amplifier whose gain reduces to half its normal full-battery value over the life of the battery. If only 2% drop in closed-loop gain is desired, what nominal open-loop amplifier gain must be used in the design? (Note that since the change in A is large, it is inaccurate to use differentials.) What value of β should be chosen? If component-value variation in the β network may produce as much as a ±1% variation in β, to what value must A be raised to ensure the required minimum gain?

10.16 A capacitively coupled amplifier has a midband gain of 1000 V/V, a single high-frequency pole at 10 kHz, and a single low-frequency pole at 100 Hz. Negative feedback is employed so that the midband gain is reduced to 10. What are the upper and lower 3-dB frequencies of the closed-loop gain?

D 10.17 Low-cost audio power amplifiers often avoid direct coupling of the loudspeaker to the output stage because any resulting dc bias current in the speaker can use up (and thereby waste) its limited mechanical dynamic range. Unfortunately, the coupling capacitor needed can be large! But feedback helps. For example, for an 8-Ω loudspeaker and f_L = 100 Hz, what size capacitor is needed? Now, if feedback is arranged around the amplifier and the speaker so that a closed-loop gain A_f = 10 V/V is obtained from an amplifier whose open-loop gain is 1000 V/V, what value of f_{Lf} results? If the ultimate product-design specification requires a 50-Hz cutoff, what capacitor can be used?

D **10.18 It is required to design a dc amplifier with a low-frequency gain of 1000 and a 3-dB frequency of 0.5 MHz. You have available gain stages with a gain of 1000 but with a dominant high-frequency pole at 10 kHz. Provide a design that employs a number of such stages in cascade, each with negative feedback of an appropriate amount. Use identical stages. [*Hint:* When negative feedback of an amount $(1 + A\beta)$ is employed around a gain stage, its x-dB frequency is increased by the factor $(1 + A\beta)$.]

D 10.19 Design a supply-ripple-reduced power amplifier for which the output stage can be modeled by the block diagram of Fig. 10.4, where A_1 = 0.9 V/V, and the power-supply ripple V_N = +1 V. A closed-loop gain of 10 V/V is desired. What is the gain of the low-ripple preamplifier needed to reduce the output ripple to ±100 mV? To ±10 mV? To ±1 mV? For each case, specify the value required for the feedback factor β.

D 10.20 Design a feedback amplifier that has a closed-loop gain of 100 V/V and is relatively insensitive to change in basic-amplifier gain. In particular, it should provide a reduction in A_f to 99 V/V for a reduction in A to one-tenth its nominal value. What is the required loop gain? What nominal value of A is required? What value of β should be used? What would the closed-loop gain become if A were increased tenfold? If A were made infinite?

D 10.21 A feedback amplifier is to be designed using a feedback loop connected around a two-stage amplifier. The first stage is a direct-coupled, small-signal amplifier with a high upper 3-dB frequency. The second stage is a power-output stage with a midband gain of 10 V/V and upper and lower 3-dB frequencies of 8 kHz and 80 Hz, respectively. The feedback amplifier should have a midband gain of 100 V/V and an upper 3-dB frequency of 40 kHz. What is the required gain of the small-signal amplifier? What value of β should be used? What does the lower 3-dB frequency of the overall amplifier become?

***10.22** The complementary BJT follower shown in Fig. P10.22(a) has the approximate transfer characteristic

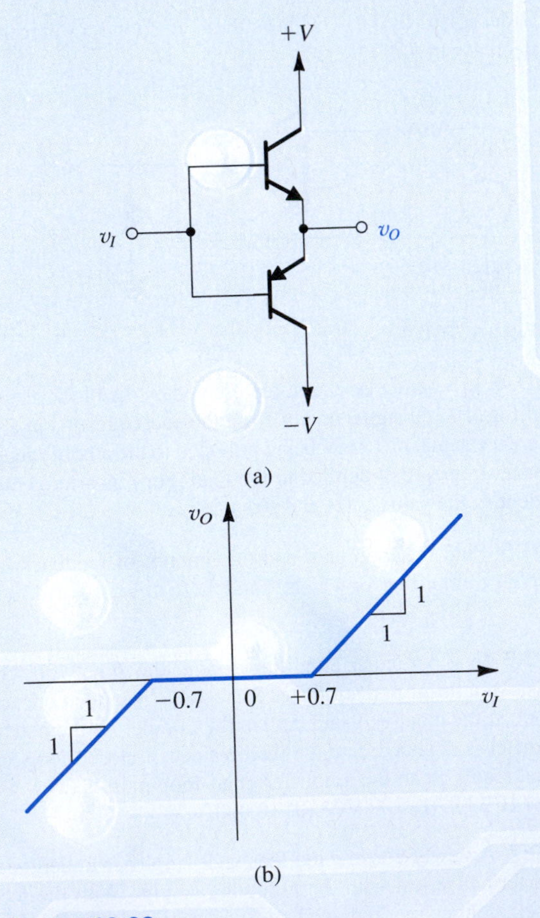

Figure P10.22

shown in Fig. P10.22(b). Observe that for $-0.7\text{ V} \le v_I \le +0.7$ V, the output is zero. This "dead band" leads to crossover distortion (see Section 11.3). Consider this follower to be driven by the output of a differential amplifier of gain 100 whose positive-input terminal is connected to the input signal source v_S and whose negative-input terminal is connected to the emitters of the follower. Sketch the transfer characteristic v_O versus v_S of the resulting feedback amplifier. What are the limits of the dead band, and what are the gains outside the dead band?

D 10.23 A particular amplifier has a nonlinear transfer characteristic that can be approximated as follows:

(a) For small input signals, $|v_I| \le 10$ mV, $v_O/v_I = 10^3$
(b) For intermediate input signals, $10\text{ mV} \le |v_I| \le 60$ mV, $\Delta v_O/\Delta v_I = 10^2$
(c) For large input signals, $|v_I| \ge 60$ mV, the output saturates

If the amplifier is connected in a negative-feedback loop, find the feedback factor β that reduces the factor-of-10 change in gain (occurring at $|v_I| = 10$ mV) to only a 10% change. What is the transfer characteristic v_O versus v_S of the amplifier with feedback?

Section 10.3: The Four Basic Feedback Topologies

D 10.24 For the feedback voltage amplifier of Fig. 10.7(a) let the op amp have an infinite input resistance, a zero output resistance, and a finite open-loop gain $A = 10^4$ V/V. If $R_1 = 1\text{ k}\Omega$, find the value of R_2 that results in a closed-loop gain of 100 V/V. What does the gain become if R_1 is removed?

10.25 Consider the feedback voltage amplifier of Fig. 10.7(c). Neglect r_o and assume that $(R_1 + R_2) \gg R_D$.

(a) Find expressions for A and β and hence the amount of feedback.
(b) Noting that the feedback can be eliminated by removing R_1 and R_2 and connecting the gate of Q to a constant dc voltage (signal ground) give the input resistance R_i and the output resistance R_o of the open-loop amplifier.
(c) Using standard circuit analysis (i.e, without invoking the feedback approach), find the input resistance R_{if} and the output resistance R_{of} of the circuit in Fig. 10.7(b). How does R_{if} relate to R_i, and R_{of} to R_o?

10.26 The feedback current amplifier in Fig. P10.26 utilizes an op amp with an input differential resistance R_{id}, an open-loop gain μ, and an output resistance r_o. The output current I_o that is delivered to the load resistance R_L is sensed by the feedback network composed of the two resistances R_M and R_F, and a fraction I_f is fed back to the amplifier input node. Find expressions for $A \equiv I_o/I_i$, $\beta \equiv I_f/I_o$, and $A_f \equiv I_o/I_s$, assuming that the feedback causes the voltage at the input node to be near ground. If the loop gain is large, what does the closed-loop current gain become? State precisely the condition under which this is obtained. For $\mu = 10^4$ V/V, $R_{id} = 1\text{ M}\Omega$, $r_o = 100\ \Omega$, $R_L = 10\text{ k}\Omega$, $R_M = 100\ \Omega$, and $R_F = 10\text{ k}\Omega$, find A, β, and A_f.

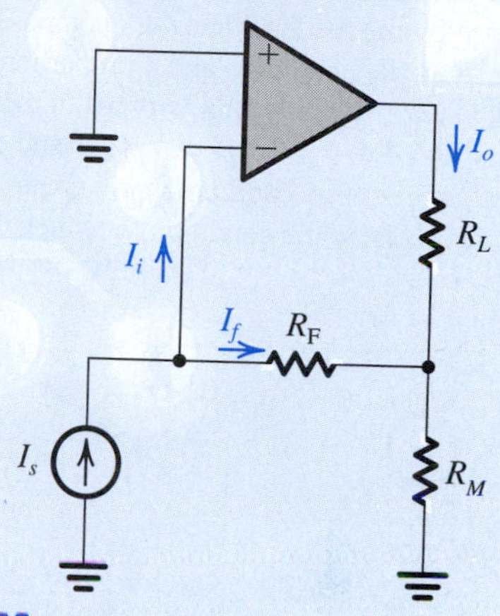

Figure P10.26

10.27 Figure P10.27 shows a feedback transconductance amplifier utilizing an op amp with open-loop gain μ, very large input resistance, and a very small output resistance, and an NMOS transistor Q. The amplifier delivers its output current to R_L. The feedback network, composed of resistor R, senses the equal current in the source terminal of Q and delivers a proportional voltage V_f to the negative input terminal of the op amp.

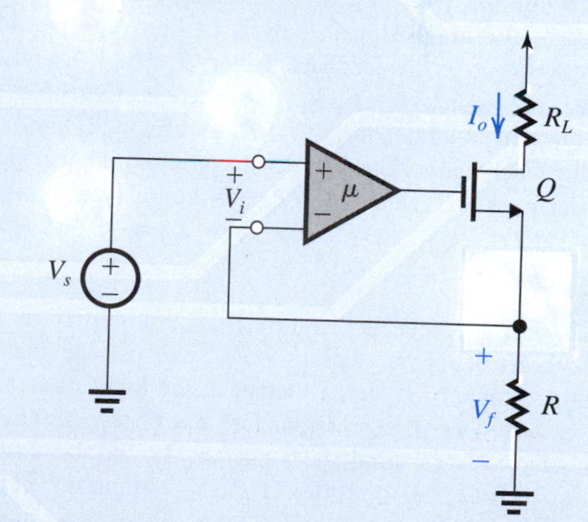

Figure P10.27

(a) Show that the feedback is negative.
(b) Open the feedback loop by breaking the connection of R to the negative input of the op amp and grounding the negative input terminal. Find an expression for $A \equiv I_o/V_i$.

(c) Find an expression for $\beta \equiv V_f/I_o$.
(d) Find an expression for $A_f \equiv I_o/V_s$.
(e) What is the condition to obtain $I_o \simeq V_s/R$?

10.28 Figure P10.28 shows a feedback transconductance amplifier implemented using an op amp with open-loop gain μ, a very large input resistance, and an output resistance r_o. The output current I_o that is delivered to the load resistance R_L is sensed by the feedback network composed of the three resistances R_M, R_1, and R_2, and a proportional voltage V_f is fed back to the negative-input terminal of the op amp. Find expressions for $A \equiv I_o/V$, $\beta \equiv V_f/I_o$, and $A_f \equiv I_o/V_s$. If the loop gain is large, find an approximate expression for A_f and state precisely the condition for which this applies.

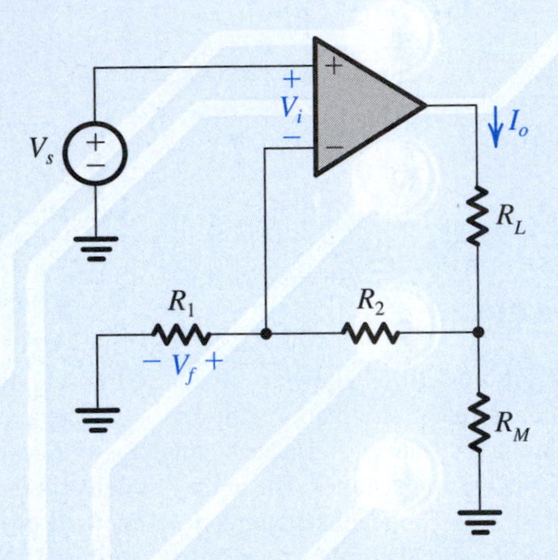

Figure P10.28

10.29 For the feedback transresistance amplifier in Fig. 10.11(d), use small-signal analysis to find the open-loop gain $A \equiv V_o/I_i$, the feedback factor $\beta \equiv I_f/V_o$, and the closed-loop gain $A_f \equiv V_o/I_s$. Neglect r_o of each of Q_1 and Q_2 and assume that $R_C \ll \beta_2 R_E$ and $R_E \ll R_F$, and that the feedback causes the signal voltage at the input node to be nearly zero. Evaluate V_o/I_s for the following component values: $\beta_1 = \beta_2 = 100$, $R_C = R_E = 10\ \text{k}\Omega$, and $R_F = 100\ \text{k}\Omega$.

10.30 For the feedback transresistance amplifier in Fig. P10.30, let $R_F \gg R_C$ and $r_o \gg R_C$, and assume that the feedback causes the signal voltage at the input node to be nearly zero. Derive expressions for $A \equiv V_o/I_i$, $\beta \equiv I_f/V_o$, and $A_f \equiv V_o/I_s$. Find the value of A_f for the case of $R_C = 10\ \text{k}\Omega$, $R_F = 100\ \text{k}\Omega$, and the transistor current gain $\beta = 100$.

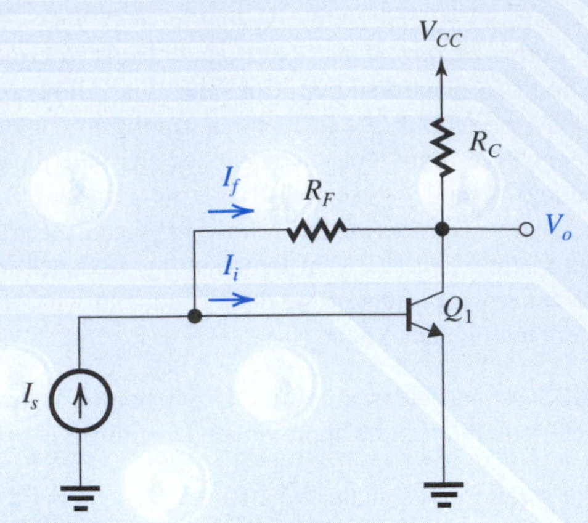

Figure P10.30

Section 10.4: The Feedback Voltage Amplifier (Series–Shunt)

10.31 A series–shunt feedback amplifier employs a basic amplifier with input and output resistances each of 2 kΩ and gain $A = 1000$ V/V. The feedback factor $\beta = 0.1$ V/V. Find the gain A_f, the input resistance R_{if}, and the output resistance R_{of} of the closed-loop amplifier.

10.32 For a particular amplifier connected in a feedback loop in which the output voltage is sampled, measurement of the output resistance before and after the loop is connected shows a change by a factor of 100. Is the resistance with feedback higher or lower? What is the value of the loop gain $A\beta$? If R_{of} is 100 Ω, what is R_o without feedback?

10.33 The formulas for R_{if} and R_{of} in Eqs. (10.19) and (10.22), respectively, also apply for the case in which A is a function of frequency. In this case, the resulting impedances Z_{if} and Z_{of} will be functions of frequency. Consider the case of a series–shunt amplifier that has an input resistance R_i, an output resistance R_o, and open-loop gain $A = A_0/(1+(s/\omega_H))$, and a feedback factor β that is independent of frequency. Find Z_{if} and Z_{of} and give an equivalent circuit for each, together with the values of all the elements in the equivalent circuits.

10.34 A series–shunt feedback amplifier utilizes the feedback circuit shown in Fig. P10.34.

(a) Find expressions for the h parameters of the feedback circuit (see Fig. 10.14b).
(b) If $R_1 = 1$ kΩ and $\beta = 0.01$, what are the values of all four h parameters? Give the units of each parameter.
(c) For the case $R_s = 1$ kΩ and $R_L = 1$ kΩ, sketch and label an equivalent circuit following the model in Fig. 10.14(c).

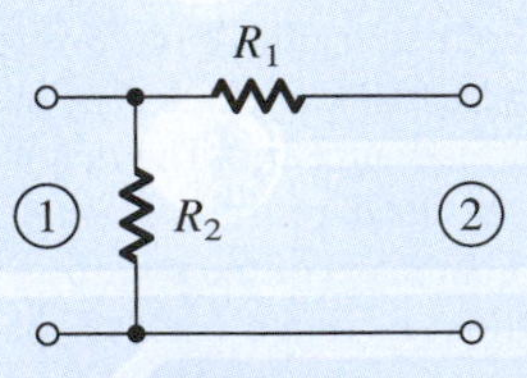

Figure P10.34

10.35 A feedback amplifier utilizing voltage sampling and employing a basic voltage amplifier with a gain of 1000 V/V and an input resistance of 1000 Ω has a closed-loop input resistance of 10 kΩ. What is the closed-loop gain? If the basic amplifier is used to implement a unity-gain voltage buffer, what input resistance do you expect?

***10.36** In the series–shunt feedback amplifier shown in Fig. P10.36, the transistors are biased with ideal current-sources $I_1 = 0.1$ mA and $I_2 = 1$ mA, the devices operate with $V_{BE} = 0.7$ V and have $\beta_1 = \beta_2 = 100$. The input signal V_s has a zero dc component. Resistances $R_s = 100\ \Omega$, $R_1 = 1\ \text{k}\Omega$, $R_2 = 10\ \text{k}\Omega$, and $R_L = 1\ \text{k}\Omega$.

(a) If the loop gain is large, what do you expect the closed-loop gain V_o/V_s to be? Give both an expression and its approximate value.
(b) Find the dc emitter current in each of Q_1 and Q_2. Also find the dc voltage at the emitter of Q_2.
(c) Sketch the A circuit without the dc sources. Derive expressions for A, R_i, and R_o, and find their values.
(d) Give an expression for β and find its value.
(e) Find the closed-loop gain V_o/V_s, the input resistance R_{in}, and the output resistance R_{out}. By what percentage does the value of A_f differ from the approximate value found in (a)?

SIM D *10.37 Figure P10.37 shows a series–shunt amplifier with a feedback factor $\beta = 1$. The amplifier is designed so that $v_O = 0$ for $v_S = 0$, with small deviations in v_O from 0 V dc being minimized by the negative-feedback action. The technology utilized has $k_n' = 2k_p' = 120\ \mu\text{A/V}^2$, $|V_t| = 0.7$ V, and $|V_A'| = 24$ V/μm.
(a) Show that the feedback is negative.
(b) With the feedback loop opened at the gate of Q_2, and the gate terminals of Q_1 and Q_2 grounded, find the dc current and the overdrive voltage at which each of Q_1 to Q_5 is operating. Ignore the Early effect. Also find the dc voltage at the output.
(c) Find g_m and r_o of each of the five transistors.
(d) Find the expressions and values of A and R_o. Assume that the bias current sources are ideal.
(e) Find the gain with feedback, A_f, and the output resistance R_{out}.
(f) How would you modify the circuit to realize a closed loop voltage gain of 5 V/V? What is the value of output resistance obtained?

D *10.38 Figure P10.38 shows a series–shunt amplifier in which the three MOSFETs are sized to operate at $|V_{OV}| = 0.2$ V. Let $|V_t| = 0.5$ V and $|V_A| = 10$ V. The current sources utilize single transistors and thus have output resistances equal to r_o.

(a) Show that the feedback is negative.
(b) Assuming the loop gain to be large, what do you expect the closed-loop voltage gain V_o/V_s to be approximately?
(c) If V_s has a zero dc component, find the dc voltages at nodes S_1, G_2, S_3, and G_3. Verify that each of the current sources has the minimum required dc voltage across it for proper operation.
(d) Find the A circuit. Calculate the gain of each of the three stages and the overall voltage gain, A.
[*Hint*: A CS amplifier with a resistance R_s in the source lead has an effective transconductance $g_m/(1 + g_m R_s)$ and an output resistance $r_o(1 + g_m R_s)$.]
(e) Find β.
(f) Find $A_f = V_o/V_s$. By what percentage does this value differ from the approximate value obtained in (b)?
(g) Find the output resistance R_{out}.

D *10.39 The active-loaded differential amplifier in Fig. P10.39 has a feedback network consisting of the voltage divider (R_1, R_2), with $R_1 + R_2 = 1\ \text{M}\Omega$. The devices are sized to operate at $|V_{OV}| = 0.2$ V. For all devices, $|V_A| = 10$ V. The input signal source has a zero dc component.

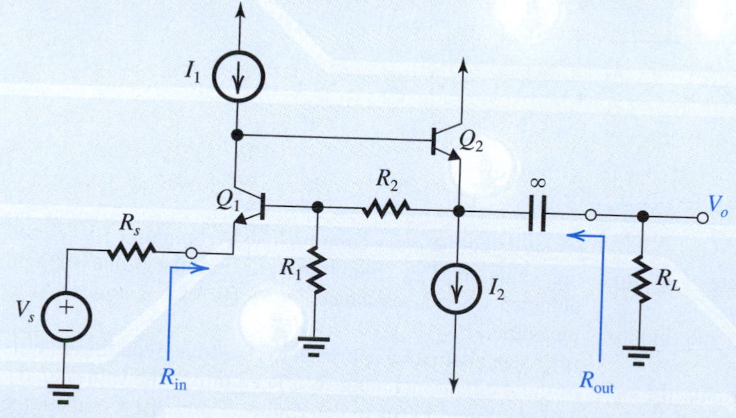

Figure P10.36

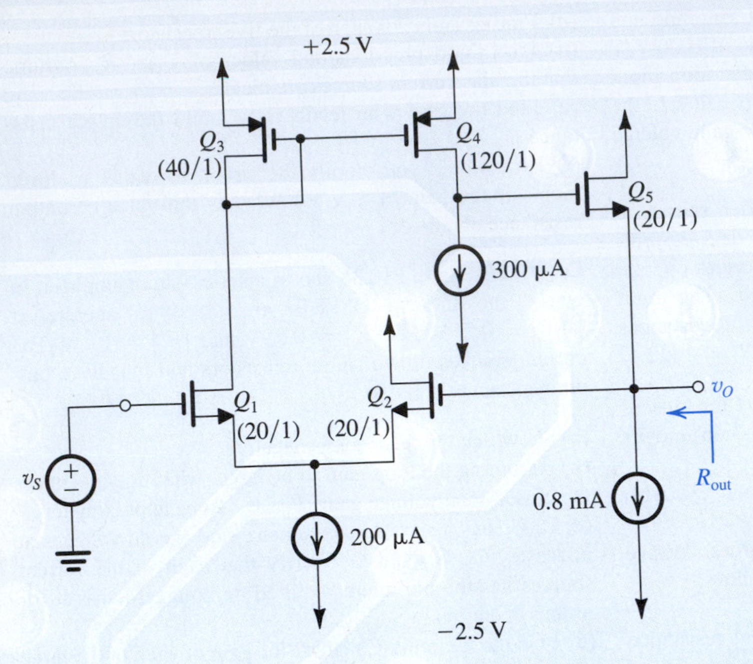

Figure P10.37

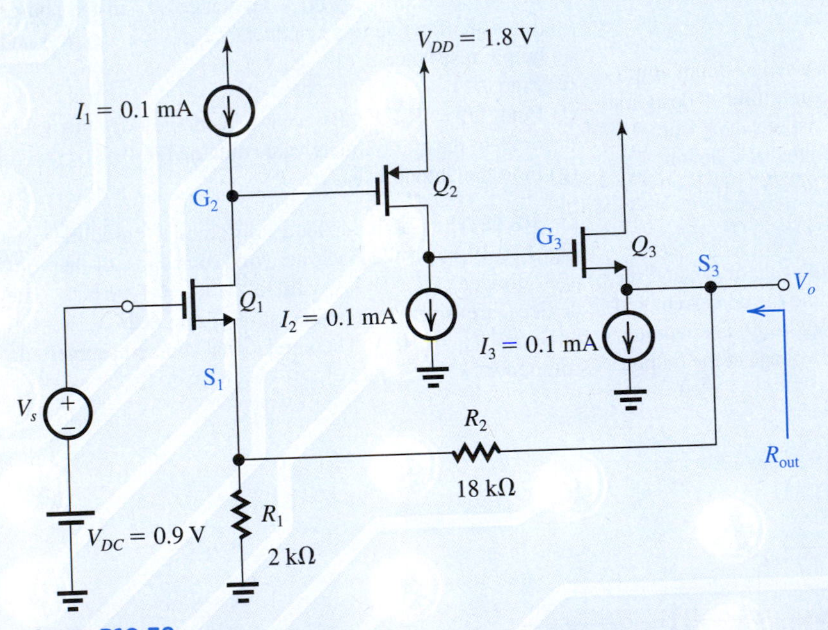

Figure P10.38

(a) Show that the feedback is negative.
(b) What do you expect the dc voltage at the gate of Q_2 to be? At the output? (Neglect the Early effect.)
(c) Find the A circuit. Derive an expression for A and find its value.
(d) Select values for R_1 and R_2 to obtain a closed-loop voltage gain $V_o/V_s = 5$ V/V.
(e) Find the value of R_{out}.
(f) Utilizing the open-circuit, closed-loop gain (5 V/V) and the value of R_{out} found in (e), find the value of gain obtained when a resistance $R_L = 10$ kΩ is connected to the output.
(g) As an alternative approach to (f) above, redo the analysis of the A circuit including R_L. Then utilize the values of R_1 and R_2 found in (d) to determine β and A_f. Compare the value of A_f to that found in (f).

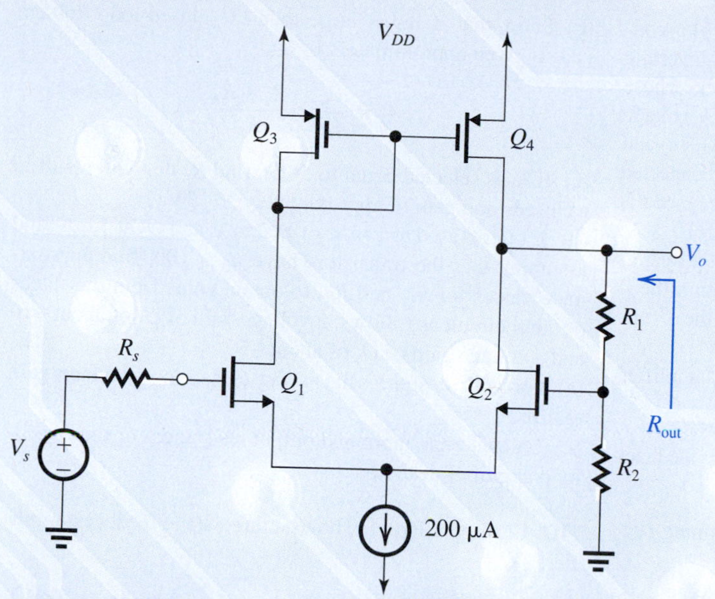

Figure P10.39

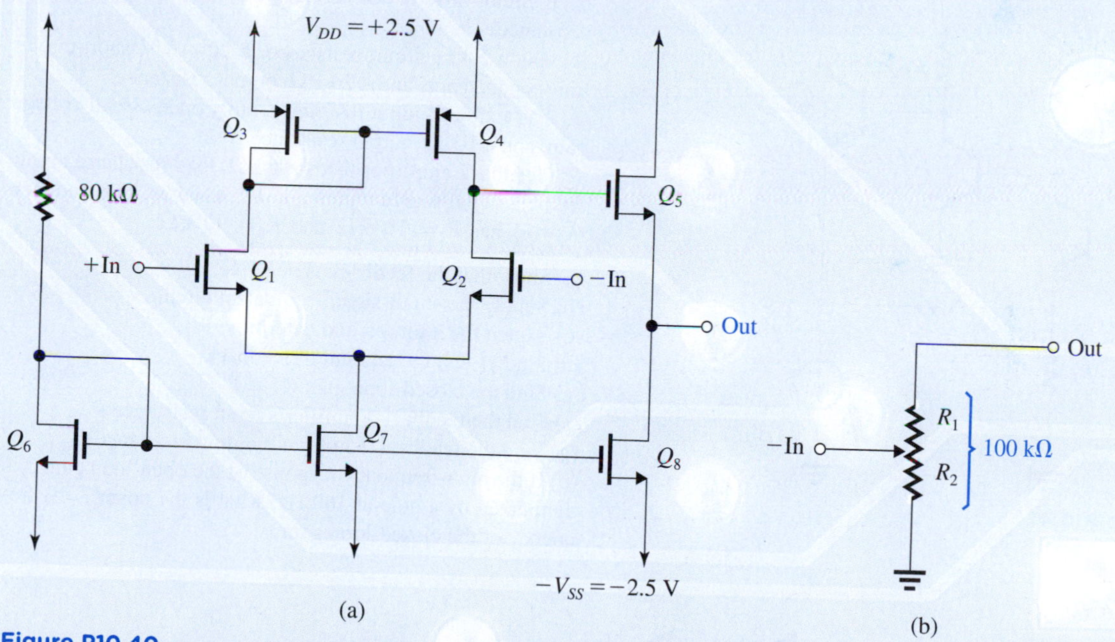

Figure P10.40

D **10.40 The CMOS op amp in Fig. P10.40(a) is fabricated in a 1-μm technology for which $V_{tn} = -V_{tp} = 0.75$ V, $\mu_n C_{ox} = 2\mu_p C_{ox} = 100$ μA/V^2, and $V_A' = 10$ V/μm. All transistors in the circuit have $L = 1$ μm.

(a) It is required to perform a dc bias design of the circuit. For this purpose, let the two input terminals be at zero volts dc and neglect channel-length modulation (i.e, let $V_A = \infty$). Design to obtain $I_{D1} = I_{D2} = 50$ μA, $I_{D5} = 250$ μA, and $V_O = 0$, and operate all transistors except for the source follower Q_5 at $V_{OV} = 0.25$ V. Assume that Q_1 and Q_2 are perfectly matched, and similarly for Q_3 and Q_4. For each transistor, find I_D and W/L.

(b) What is the allowable range of input common-mode voltage?

(c) Find g_m for each of Q_1, Q_2, and Q_5.

(d) For each transistor, calculate r_o.

(e) The 100-kΩ potentiometer shown in Fig. 10.40(b) is connected between the output terminal (Out) and the inverting input terminal (–In) to provide negative feedback whose amount is controlled by the setting of the wiper. A voltage signal V_s is applied between the noninverting input (+In) and ground. A load resistance R_L = 100 kΩ is connected between the output terminal and ground. The potentiometer is adjusted to obtain a closed-loop gain $A_f \equiv V_o/V_s \simeq 10$ V/V.

Specify the required setting of the potentiometer by giving the values of R_1 and R_2. Toward this end, find the A circuit (supply a circuit diagram), the value of A, the β circuit (supply a circuit diagram), and the value of β.

(f) What is the output resistance of the feedback amplifier, excluding R_L?

D *10.41 Figure P10.41 shows a series–shunt feedback amplifier without details of the bias circuit.

(a) Sketch the A circuit and the circuit for determining β.

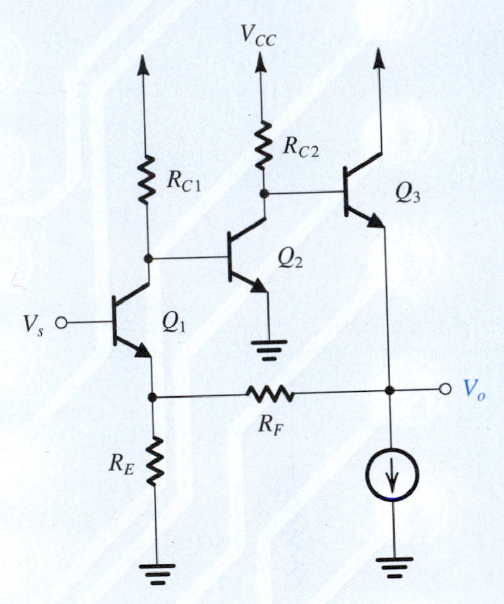

Figure P10.41

(b) Show that if $A\beta$ is large then the closed-loop voltage gain is given approximately by

$$A_f \equiv \frac{V_o}{V_s} \simeq \frac{R_F + R_E}{R_E}$$

(c) If R_E is selected equal to 50 Ω, find R_F that will result in a closed-loop gain of approximately 25 V/V.

(d) If Q_1 is biased at 1 mA, Q_2 at 2 mA, and Q_3 at 5 mA, and assuming that the transistors have h_{fe} = 100, find approximate values for R_{C1} and R_{C2} to obtain gains from the stages of the A circuit as follows: a voltage gain of Q_1 of about –10 and a voltage gain of Q_2 of about –50.

(e) For your design, what is the closed-loop voltage gain realized?

(f) Calculate the input and output resistances of the closed-loop amplifier designed.

***10.42** Figure P10.42 shows a three-stage feedback amplifier:

A_1 has an 82-kΩ differential input resistance, a 20-V/V open-circuit differential voltage gain, and a 3.2-kΩ output resistance.

A_2 has a 5-kΩ input resistance, a 20-mA/V short-circuit transconductance, and a 20-kΩ output resistance.

A_3 has a 20-kΩ input resistance, unity open-circuit voltage gain, and a 1-kΩ output resistance.

The feedback amplifier feeds a 1-kΩ load resistance and is fed by a signal source with a 9-kΩ resistance. The feedback network has R_1= 10 kΩ and R_2= 90 kΩ.

(a) Show that the feedback is negative.
(b) Supply the small-signal equivalent circuit.
(c) Sketch the A circuit and determine A.
(d) Find β and the amount of feedback.
(e) Find the closed-loop gain $A_f \equiv V_o/V_s$.
(f) Find the feedback amplifier's input resistance R_{in}.
(g) Find the feedback amplifier's output resistance R_{out}.
(h) If the high-frequency response of the open-loop gain A is dominated by a pole at 100 Hz, what is the upper 3-dB frequency of the closed-loop gain?

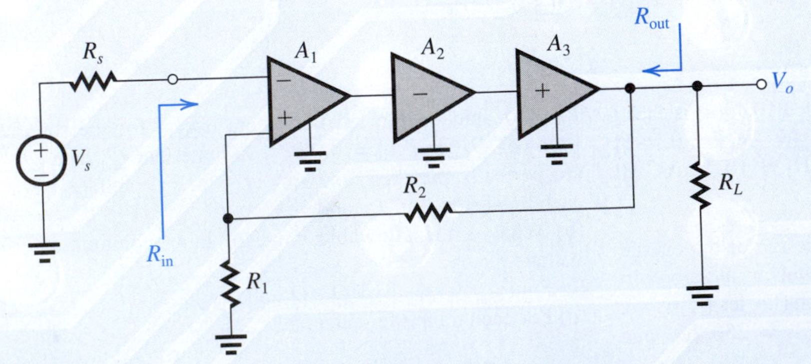

Figure P10.42

(i) If for some reason A_1 drops to half its nominal value, what is the percentage change to A_f?

Section 10.5: The Feedback Transconductance Amplifier (Series–Series)

10.43 A series–series feedback amplifier employs a transconductance amplifier having a short-circuit transconductance G_m of 0.5 A/V, input resistance of 10 kΩ, and output resistance of 100 kΩ. The feedback network has $\beta = 100\ \Omega$, an input resistance (with port 1 open-circuited) of 100 Ω, and an input resistance (with port 2 open-circuited) of 10 kΩ. The amplifier operates with a signal source having a resistance of 10 kΩ and with a load resistance of 10 kΩ. Find A_f, R_{in}, and R_{out}.

10.44 Reconsider the circuit in Fig. 10.23(a), analyzed in Example 10.6, this time with the output voltage taken at the emitter of Q_3. In this case the feedback can be considered to be of the series–shunt type. Note that R_{E2} should now be considered part of the basic amplifier and not of the feedback network.

(a) Determine β.
(b) Find an approximate value for $A_f \equiv V_{e3}/V_s$ assuming that the loop gain remains large (a safe assumption, since the loop in fact does not change).
[*Note*: If you continue with the feedback analysis, you'll find that $A\beta$ in fact changes somewhat; this is a result of the different approximations made in the feedback analysis approach.]

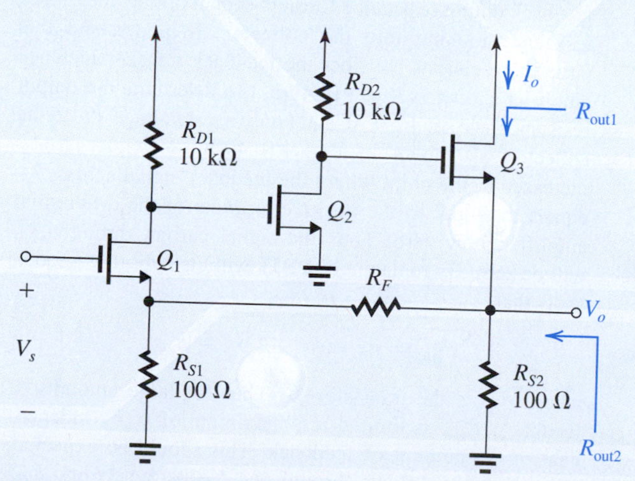

Figure P10.45

D *10.45 Figure P10.45 shows a feedback triple utilizing MOSFETs. All three MOSFETs are biased and sized to operate at $g_m = 4$ mA/V. You may neglect their r_o's (except for the calculation of R_{out1} as indicated below).

(a) Considering the feedback amplifier as a transconductance amplifier with output current I_o, find the value of R_F that results in a closed-loop transconductance of approximately 100 mA/V.
(b) Sketch the A circuit and find the value of $A \equiv I_o/V_i$.
(c) Find $1 + A\beta$ and $A_f \equiv V_o/I_s$. Compare to the value of A_f you designed for. What is the percentage difference? What resistance can you change to make A_f exactly 100 mA/V, and in which direction (increase or decrease)?
(d) Assuming that $r_{o3} = 20$ kΩ, find the output resistance R_{out1}. Since the current sampled by the feedback network is exactly equal to the output current, you can use the feedback formula.
(e) If the voltage V_o is taken as the output, in which case the amplifier becomes series–shunt feedback, what is the value of the closed-loop voltage gain V_o/V_s? Assume that R_F has the original value you selected in (a). Note that in this case R_{S2} should be considered part of the amplifier and not the feedback network. The feedback analysis will reveal that $A\beta$ changes somewhat, which may be puzzling given that the feedback loop did not change. The change is due to the different approximation used.
(f) What is the closed-loop output resistance R_{out2} of the voltage amplifier in (e) above?

10.46 Consider the circuit in Fig. P10.46 as a transconductance amplifier with input V_s and output I_o. The transistor is specified in terms of its g_m and r_o.

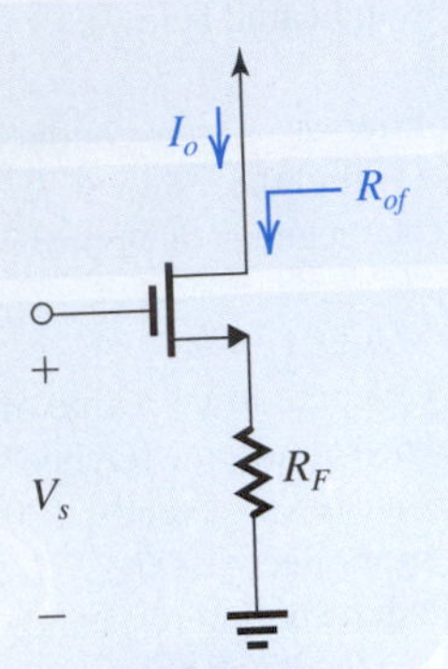

Figure P10.46

(a) Sketch the small-signal equivalent circuit and convince yourself that the feedback circuit is composed of resistor R_F.
(b) Find the A circuit and the β circuit.
(c) Derive expressions for A, β, $(1+A\beta)$, A_f, R_o, and R_{of}.

D 10.47 The transconductance amplifier in Fig. P10.47 utilizes a differential amplifier with gain μ and a very high input resistance. The differential amplifier drives a transistor Q characterized by its g_m and r_o. A resistor R_F senses the output current I_o.

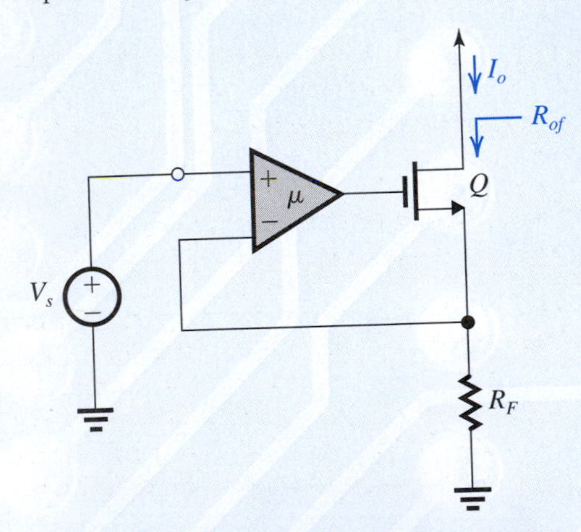

Figure P10.47

(a) For $A\beta \gg 1$, find an approximate expression for the closed-loop transconductance $A_f \equiv I_o/V_s$. Hence, select a value for R_F that results in $A_f \simeq 10$ mA/V.
(b) Find the A circuit and derive an expression for A. Evaluate A for the case $\mu = 1000$ V/V, $g_m = 2$ mA/V, $r_o = 20$ kΩ, and the value of R_F you selected in (a).
(c) Give an expression for $A\beta$ and evaluate its value and that of $1 + A\beta$.
(d) Find the closed-loop gain A_f and compare to the value you anticipated in (a) above.
(e) Find expressions and values for R_o and R_{of}.

***10.48** It is required to show that the output resistance of the BJT circuit in Fig. P10.48 is given by

$$R_o = r_o + [R_e \| (r_\pi + R_b)] \left(1 + g_m r_o \frac{r_\pi}{r_\pi + R_b}\right)$$

To derive this expression, set $V_s = 0$, replace the BJT with its small-signal, hybrid-π model, apply a test voltage V_x to the collector, and find the current I_x drawn from V_x and hence R_o as V_x/I_x. Note that the bias arrangement is not shown. For the case of $R_b = 0$, find the maximum possible value for R_o. Note that this theoretical maximum is obtained when R_e is so large that the signal current in the emitter is nearly zero. In this case, with V_x applied and $V_s = 0$, what is the current in the base, in the $g_m V_\pi$ generator, and in r_o, all in terms of I_x? Show these currents on a sketch of the equivalent circuit with R_e set to ∞.

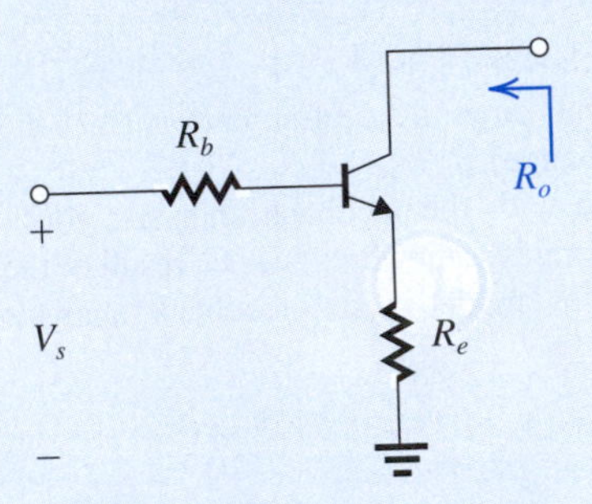

Figure P10.48

10.49 As we found out in Example 10.6, whenever the feedback network senses the emitter current of the BJT, the feedback output resistance formula cannot predict the output resistance looking into the collector. To understand this issue more clearly, consider the feedback transconductance amplifier shown in Fig. P10.49(a). To determine the output resistance, we set $V_s = 0$ and apply a test voltage V_x to the collector, as shown in Fig. P10.49(b). Now, let μ be increased to the point where the feedback signal across R_F equals the input to the positive terminal of the differential amplifier, now zero. Thus the signal current through R_F will be zero. By replacing the BJT with its hybrid-π model, show that

$$R_{out} = r_\pi + (h_{fe} + 1)r_o \simeq h_{fe} r_o$$

where h_{fe} is the transistor β. Thus for large amounts of feedback, R_{out} is limited to a maximum of $h_{fe} r_o$ independent of the amount of feedback. This should be expected, since no current flows through the feedback network R_F!

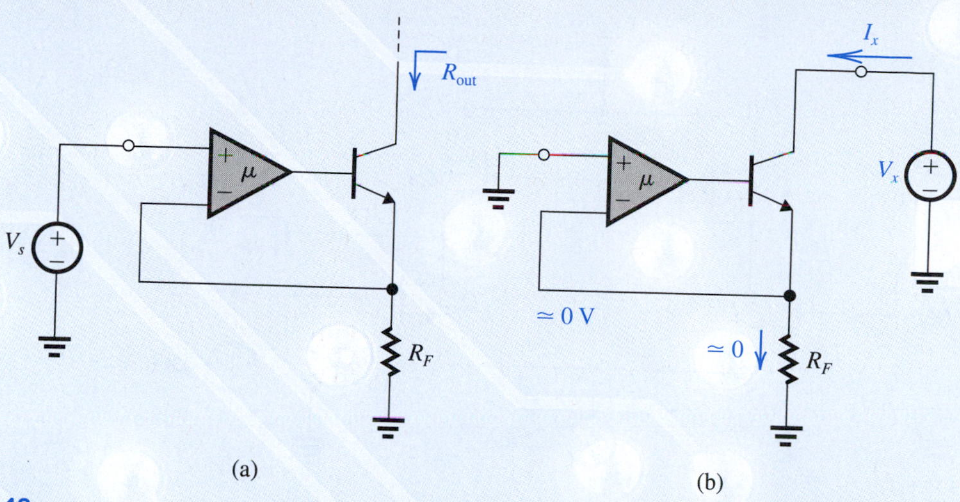

Figure P10.49

This phenomenon does *not* occur in the MOSFET version of this circuit.

10.50 For the feedback transconductance amplifier of Fig. 10.10(c) derive expressions for A, β, $A\beta$, A_f, R_o, and R_{of}. Evaluate A_f and R_{of} for the case of $g_{m1} = g_{m2} = 5$ mA/V, $R_D = 10$ kΩ, $r_{o2} = 20$ kΩ, $R_F = 100$ Ω, and $R_L = 1$ kΩ. For simplicity, neglect r_{o1} and take r_{o2} into account only when calculating output resistances.

D 10.51 For the feedback transconductance amplifier in Fig. P10.51, derive an approximate expression for the closed-loop transconductance $A_f \equiv I_o/V_s$ for the case of $A\beta \gg 1$. Hence select a value for R_2 to obtain $A_f = 100$ mA/V. If Q is biased to obtain $g_m = 1$ mA/V, specify the value of the gain μ of the differential amplifier to obtain an amount of feedback of 60 dB. If Q has $r_o = 50$ kΩ, find the output resistance R_{out}.

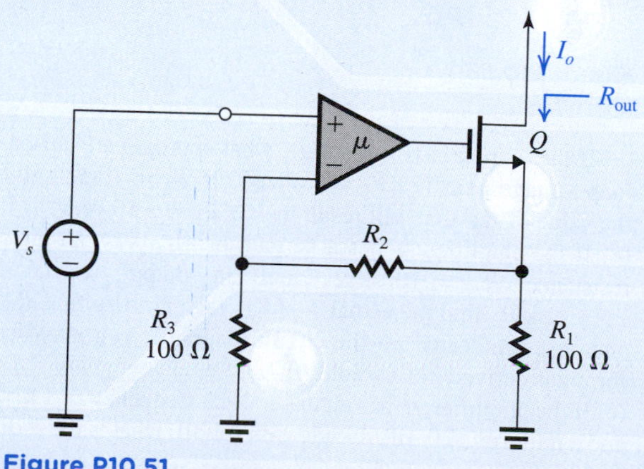

Figure P10.51

SIM **10.52** All the MOS transistors in the feedback transconductance amplifier (series–series) of Fig. P10.52 are sized to operate at $|V_{OV}| = 0.2$ V. For all transistors, $|V_t| = 0.4$ V and $|V_A| = 20$ V.

(a) If V_s has a zero dc component, find the dc voltage at the output, at the drain of Q_1, and at the drain of Q_2.
(b) Find an approximate expression and value for $A_f \equiv I_o/V_s$ for the case $A\beta \gg 1$.
(c) Use feedback analysis to obtain a more precise value for A_f.
(d) Find the value of R_{out}.
(e) If the voltage at the source of Q_5 is taken as the output, find the voltage gain using the value of I_o/V_s obtained in (c). Also find the output resistance of this series–shunt voltage amplifier.

Section 10.6: The Feedback Transresistance Amplifier (Shunt–Shunt)

10.53 For the transresistance amplifier analyzed in Example 10.7, use the formulas derived there to evaluate A_f, R_{in}, and R_{out} when μ is one-tenth the value used in the example. That is, evaluate for $\mu = 10^3$ V/V, $R_{id} = \infty$, $r_o = 100$ Ω, $R_F = 10$ kΩ, and $R_s = R_L = 1$ kΩ. Compare to the corresponding values obtained in Example 10.7.

10.54 Use the formulas derived in Example 10.7 to solve the problem in Exercise 10.15.

10.55 The CE BJT amplifier in Fig. P10.55 employs shunt–shunt feedback: Feedback resistor R_F senses the output voltage V_o and provides a feedback current to the base node.

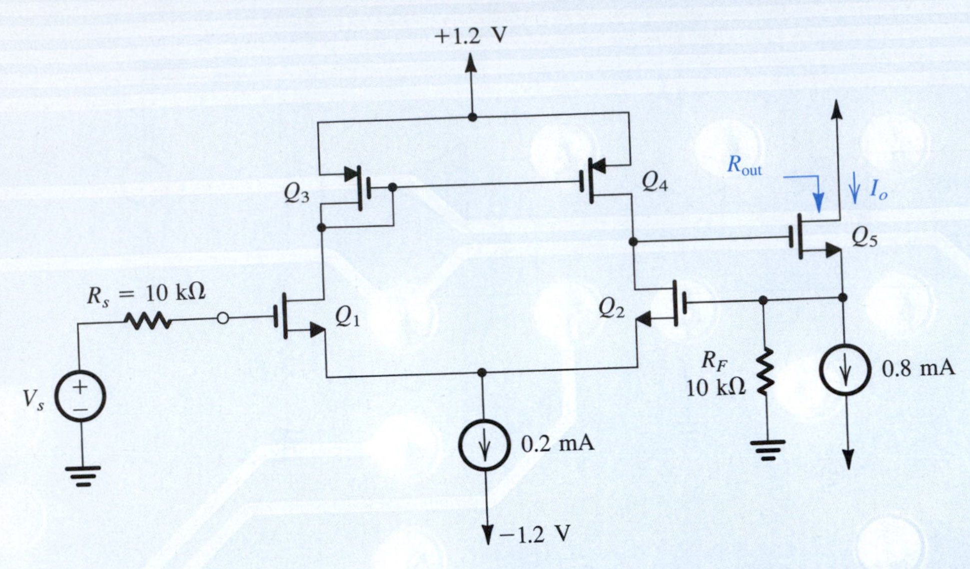

Figure P10.52

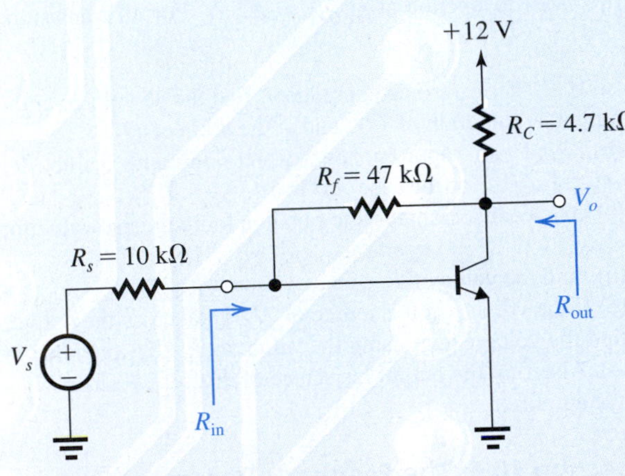

Figure P10.55

(a) If V_s has a zero dc component, find the dc collector current of the BJT. Assume the transistor $\beta = 100$.

(b) Find the small-signal equivalent circuit of the amplifier with the signal source represented by its Norton equivalent (as we usually do when the feedback connection at the input is shunt).

(c) Find the A circuit and determine the value of A, R_i and R_o.

(d) Find β and hence $A\beta$ and $1 + A\beta$.

(e) Find A_f, R_{if}, and R_{of} and hence R_{in} and R_{out}.

(f) What voltage gain V_o/V_s is realized? How does this value compare to the ideal value obtained if the loop gain is very large and thus the signal voltage at the base becomes almost zero (like what happens in an inverting op-amp circuit). Note that this single-transistor poor-man's op amp is not that bad!

D 10.56 The circuit in Fig. P10.56 utilizes a voltage amplifier with gain μ in a shunt–shunt feedback topology with the feedback network composed of resistor R_F. In order to be able to use the feedback equations, you should first convert the signal source to its Norton representation. You will then see that all the formulas derived in Example 10.7 apply here as well.

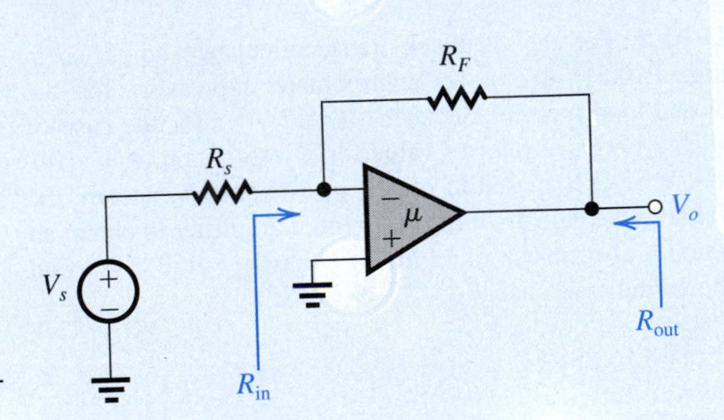

Figure P10.56

(a) If the loop gain is very large, what approximate closed-loop voltage gain V_o/V_s is realized? If R_s = 1 kΩ, give the value of R_F that will result in $V_o/V_s \simeq -10$ V/V.

(b) If the amplifier μ has a dc gain of 10^3 V/V, an input resistance R_{id} = 100 kΩ, and an output resistance r_o = 1 kΩ, find the actual V_o/V_s realized. Also find R_{in} and R_{out} (indicated on the circuit diagram). You may use formulas derived in Example 10.7.

(c) If the amplifier μ has an upper 3-dB frequency of 1 kHz and a uniform −20 -dB/decade gain rolloff, what is the 3-dB frequency of the gain $|V_o/V_s|$?

10.57 The feedback transresistance amplifier in Fig. P10.57 utilizes two identical MOSFETs biased by ideal current sources $I = 0.5$ mA. The MOSFETs are sized to operate at $V_{OV} = 0.2$ V and have $V_t = 0.5$ V and $V_A = 10$ V. The feedback resistance $R_F = 10\ \text{k}\Omega$.

(a) If I_s has a zero dc component, find the dc voltage at the input, at the drain of Q_1, and at the output.
(b) Find g_m and r_o of Q_1 and Q_2.
(c) Provide the A circuit and derive an expression for A in terms of g_{m1}, r_{o1}, g_{m2}, r_{o2}, and R_F.
(d) What is β? Give an expression for the loop gain $A\beta$ and the amount of feedback $(1 + A\beta)$.
(e) Derive an expression for A_f.
(f) Derive expressions for R_i, R_{in}, R_o, and R_{out}.
(g) Evaluate A, β, $A\beta$, A_f, R_i, R_o, R_{in}, and R_{out} for the component values given.

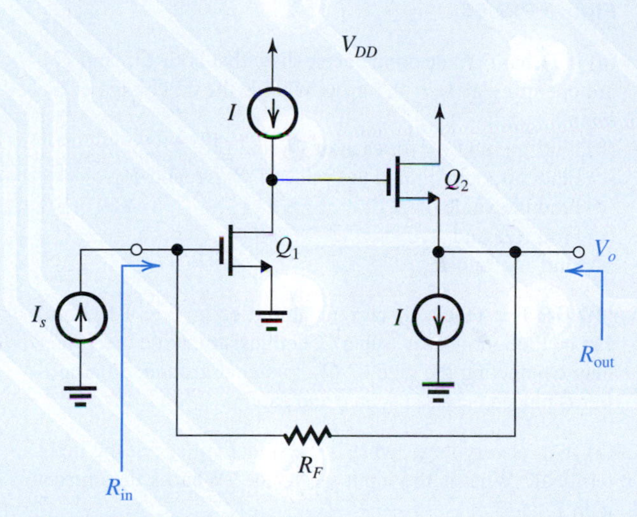

Figure P10.57

10.58 Analyze the circuit in Fig. E10.15 from first principles (i.e., do not use the feedback approach) and hence show that

$$A_f \equiv \frac{V_o}{I_s} = -\frac{(R_s \| R_f)\left(g_m - \dfrac{1}{R_f}\right)(r_o \| R_f)}{1 + (R_s \| R_f)\left(g_m - \dfrac{1}{R_f}\right)(r_o \| R_f)/R_f}$$

Comparing this expression to the one given in Exercise 10.15, part (b), you will note that the only difference is that g_m has been replaced by $(g_m - 1/R_f)$. Note that $-1/R_f$ represents the forward transmission in the feedback network, which the feedback-analysis method neglects. What is the condition then for the feedback-analysis method to be reasonably accurate for this circuit?

10.59 For the shunt–shunt feedback amplifier of Fig. 10.11(c), derive expressions for A, β, $A\beta$, A_f, R_i, R_{if}, R_o, and R_{of} in terms of g_{m1}, g_{m2}, R_{D1}, R_{D2}, and R_F. Neglect r_{o1} and r_{o2}. Present your expressions in a format that makes them easy to interpret (e.g., like those derived in Example 10.7 or those asked for in Exercise 10.15).

10.60 For the feedback transresistance amplifier in Fig. 10.11(d) let $V_{CC} = -V_{EE} = 5$ V, $R_C = R_E = R_F = 10\ \text{k}\Omega$. The transistors have $V_{BE} = 0.7$ V and $\beta = 100$.

(a) If I_s has a zero dc component, show that Q_1 and Q_2 are operating at dc collector currents of approximately 0.35 mA and 0.58 mA, respectively. What is the dc voltage at the output?
(b) Find the A circuit and the value of A, R_i, and R_o.
(c) Find the value of β, the loop gain, and the amount of feedback.
(d) Find $A_f \equiv V_o/I_s$, the input resistance R_{if}, and the output resistance R_{of}.

D **10.61 (a) Show that for the circuit in Fig. P10.61(a), if the loop gain is large, the voltage gain V_o/V_s is given approximately by

$$\frac{V_o}{V_s} \simeq -\frac{R_f}{R_s}$$

(b) Using three cascaded stages of the type shown in Fig. P10.61(b) to implement the amplifier μ, design a feedback amplifier with a voltage gain of approximately −100 V/V. The amplifier is to operate between a source resistance $R_s = 10\ \text{k}\Omega$ and a load resistance $R_L = 1\ \text{k}\Omega$. Calculate the actual value of V_o/V_s realized, the input resistance (excluding R_s), and the output resistance (excluding R_L). Assume that the BJTs have h_{fe} of 100. [*Note:* In practice, the three amplifier stages are not made identical, for stability reasons.]

D 10.62 Negative feedback is to be used to modify the characteristics of a particular amplifier for various purposes. Identify the feedback topology to be used if:

(a) Input resistance is to be lowered and output resistance raised.
(b) Both input and output resistances are to be raised.
(c) Both input and output resistances are to be lowered.

Section 10.7: The Feedback Current Amplifier (Shunt–Series)

10.63 For the feedback current amplifier in Fig. 10.8(b):

(a) Provide the A circuit and derive expressions for R_i and A. Neglect r_o of both transistors.
(b) Provide the β circuit and an expression for β.
(c) Find an expression for $A\beta$.

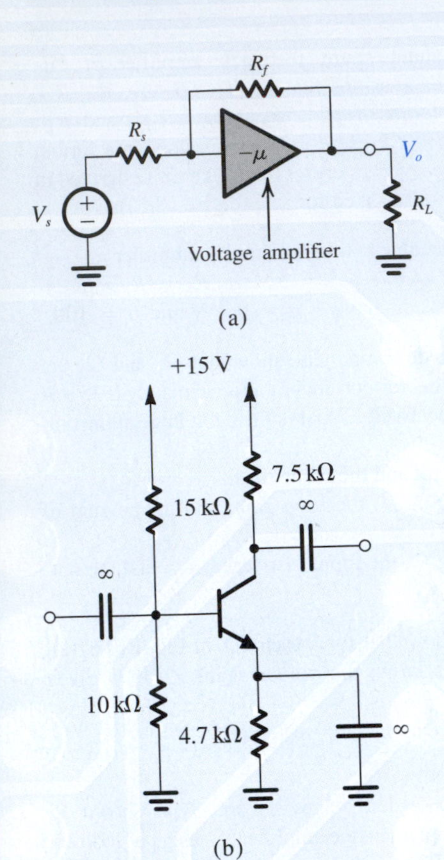

Figure P10.61

(d) For $g_{m1} = g_{m2} = 5$ mA/V, $R_D = 20$ kΩ, $R_M = 10$ kΩ, and $R_F = 90$ kΩ, find the values of A, β, $A\beta$, A_f, R_i, and R_{if}.
(e) If $r_{o2} = 20$ kΩ and $R_L = 1$ kΩ, find the output resistance as seen by R_L.

D 10.64 Design the feedback current amplifier of Fig. 10.31(a) to meet the following specifications:

(i) $A_f \equiv I_o/I_s = -100$ A/A
(ii) amount of feedback $\simeq 40$ dB
(iii) $R_{in} \simeq 1$ kΩ

Specify the values of R_1, R_2 and μ. Assume that the amplifier μ has infinite input resistance and that $R_s = \infty$. First obtain an approximate value of μ utilizing the approximate formulas derived in Example 10.8. Then with the knowledge that for the MOSFET, $g_m = 5$ mA/V and $r_o = 20$ kΩ, modify the value of μ to meet the design specifications. What R_{out} is obtained?

10.65 The feedback current amplifier in Fig. P10.65 utilizes two identical NMOS transistors sized so that at $I_D = 0.2$ mA they operate at $V_{OV} = 0.2$ V. Both devices have $V_t = 0.5$ V and $V_A = 10$ V.

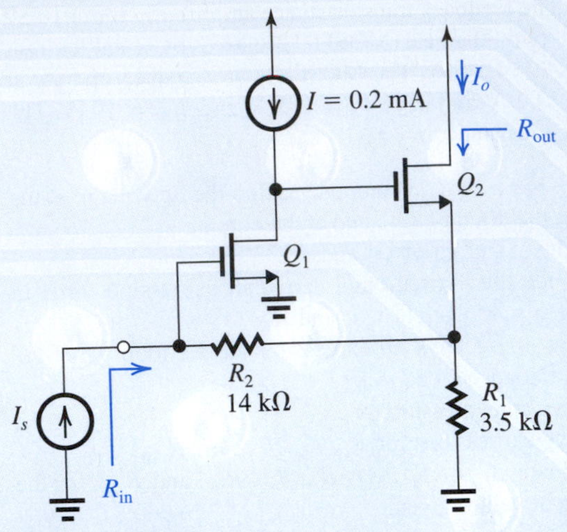

Figure P10.65

(a) If I_s has zero dc component, show that both Q_1 and Q_2 are operating at I_D= 0.2 mA. What is the dc voltage at the input?
(b) Find g_m and r_o for each of Q_1 and Q_2.
(c) Find the A circuit and the value of R_i, A, and R_o.
(d) Find the value of β.
(e) Find $A\beta$ and A_f.
(f) Find R_{in} and R_{out}.

***10.66** The feedback current amplifier in Fig. P10.66(a) can be thought of as a "super" CG transistor. Note that rather than connecting the gate of Q_2 to signal ground, an amplifier is placed between source and gate.

(a) If μ is very large, what is the signal voltage at the input terminal? What is the input resistance? What is the current gain I_o/I_s?
(b) For finite μ but assuming that the input resistance of the amplifier μ is very large, find the A circuit and derive expressions for A, R_i, and R_o.
(c) What is the value of β?
(d) Find $A\beta$ and A_f. If μ is large, what is the value of A_f?
(e) Find R_{in} and R_{out} assuming the loop gain is large.
(f) The "super" CG transistor can be utilized in the cascode configuration shown in Fig. P10.66(b), where V_G is a dc bias voltage. Replacing Q_1 by its small-signal model, use the analogy of the resulting circuit to that in Fig. P10.66(a) to find I_o and R_{out}.

***10.67** Figure P10.67 shows an interesting and very useful application of feedback to improve the performance of the current mirror formed by Q_1 and Q_2. Rather than connecting the drain of Q_1 to the gate, as is the case in simple current mirrors, an amplifier of gain $+\mu$ is connected between the drain and the gate. Note that the feedback loop does not include transistor Q_2. The feedback loop ensures that the

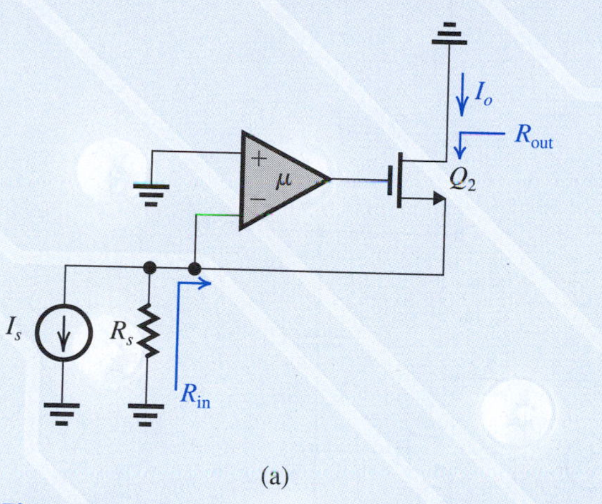

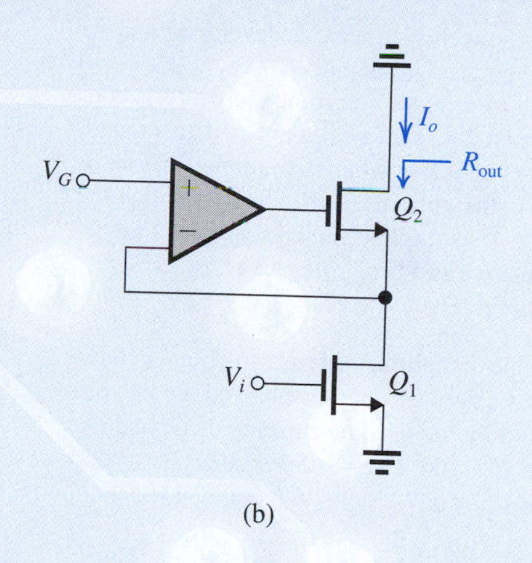

Figure P10.66

value of the gate-to-source voltage of Q_1 is such that I_{o1} equals I_s. This regulated V_{gs} is also applied to Q_2. Thus, if W/L of Q_2 is n times W/L of Q_1, $I_{o2} = nI_{o1} = nI_s$. This current tracking, however, is *not* regulated by the feedback loop.

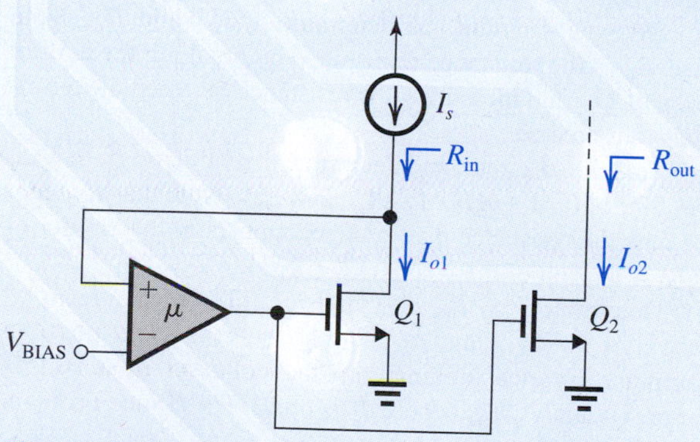

Figure P10.67

(a) Show that the feedback is negative.
(b) If μ is very large and the input resistance of the amplifier μ is infinite, what dc voltage appears at the drain of Q_1? If Q_1 is to operate at an overdrive voltage of 0.2 V, what is the minimum value that V_{BIAS} must have?
(c) Replacing Q_1 by its small-signal model, find an expression for the small-signal input resistance R_{in} assuming finite gain but infinite input resistance for the amplifier μ. Note that here it is much easier to do the analysis directly than to use the feedback-analysis approach.
(d) What is the output resistance R_{out}?

***10.68** The circuit in Fig. P10.68 is an implementation of a particular circuit building block known as **second-generation current convoyer** (CCII). It has three terminals besides ground: x, y, and z. The heart of the circuit is the feedback amplifier consisting of the differential amplifier μ and the complementary source follower (Q_N, Q_P). (Note that this feedback circuit is one we have encountered a number of times in this chapter, albeit with only one source follower transistor.) In the following, assume that the differential amplifier has a very large gain μ and infinite differential input resistance. Also, let the two current mirrors have unity current-transfer ratios.

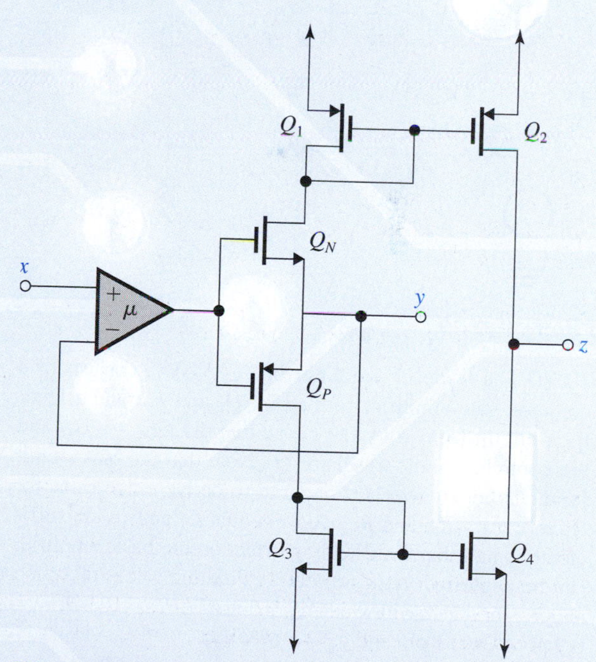

Figure P10.68

(a) If a resistance R is connected between y and ground, a voltage signal V_x is connected between x and ground, and z is short-circuited to ground. Find the current I_z through the

short circuit. Show how this current is developed and its path for V_x positive and for V_x negative.
(b) If x is connected to ground, a current source I_y is connected to input terminal y, and z is connected to ground, what voltage appears at y and what is the input resistance seen by I_y? What is the current I_z that flows through the output short circuit? Also, explain the current flow through the circuit for I_y positive and for I_y negative.
(c) What is the output resistance at z?

SIM ***10.69** For the amplifier circuit in Fig. P10.69, assuming that V_s has a zero dc component, find the dc voltages at all nodes and the dc emitter currents of Q_1 and Q_2. Let the BJTs have $\beta = 100$. Use feedback analysis to find V_o/V_s and R_{in}. Let $V_{BE} = 0.7$V.

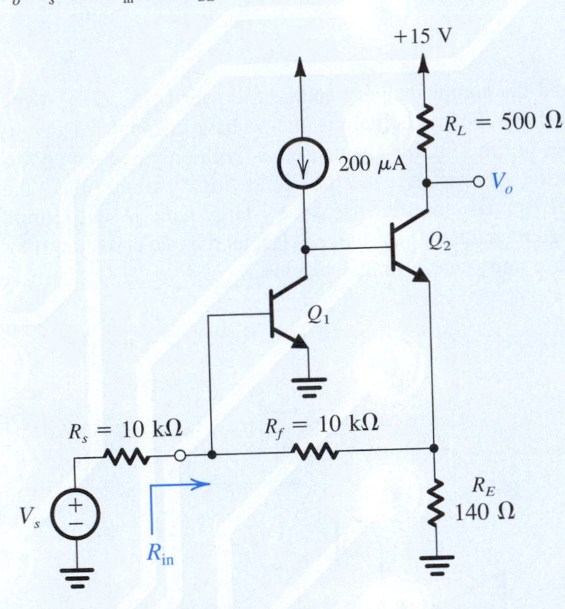

Figure P10.69

10.70 The feedback amplifier of Fig. P10.70 consists of a common-gate amplifier formed by Q_1 and R_D, and a feedback circuit formed by the capacitive divider (C_1, C_2) and the common-source transistor Q_f. Note that the bias circuit for Q_f is not shown. It is required to derive expressions for $A_f \equiv V_o/I_s$, R_{in}, and R_{out}. Assume that C_1 and C_2 are sufficiently small that their loading effect on the basic amplifier can be neglected. Also neglect r_o. Find the values of A_f, R_{in}, and R_{out} for the case in which $g_{m1} = 5$ mA/V, $R_D = 10$ kΩ, $C_1 = 0.9$ pF, $C_2 = 0.1$ pF, and $g_{mf} = 1$ mA/V.

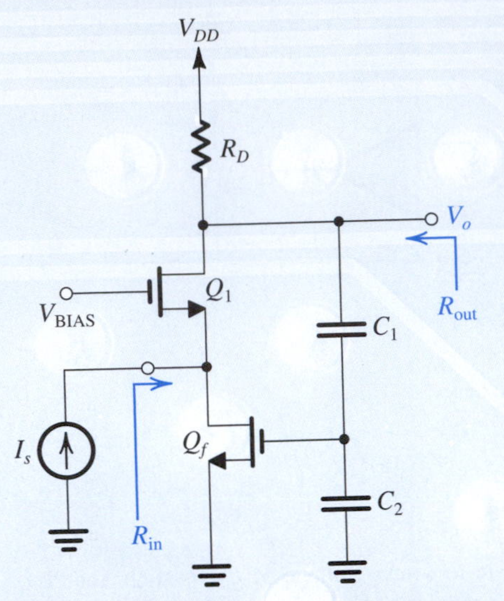

Figure P10.70

****10.71** Figure P10.71 shows a feedback amplifier utilizing the shunt–series topology. All transistors have $\beta = 100$ and $V_{BE} = 0.7$ V. Neglect r_o except in (f).

(a) Perform a dc analysis to find the dc emitter currents in Q_1 and Q_2 and hence determine their small-signal parameters.
(b) Replacing the BJTs with their hybrid-π models, give the equivalent circuit of the feedback amplifier.
(c) Give the A circuit and determine A, R_i, and R_o. Note that R_o is the resistance determined by breaking the emitter loop of Q_2 and measuring the resistance between the terminals thus created.
(d) Find the β circuit and determine the value of β.
(e) Find $A\beta$, $1 + A\beta$, A_f, R_{if}, and R_{of}. Note that R_{of} represents the resistance that in effect appears in the emitter of Q_2 as a result of the feedback.
(f) Determine I_{out}/I_{in}, R_{in}, and R_{out}. To determine R_{out}, use $V_{A2} = 75$ V and recall that the maximum possible output resistance looking into the collector of a BJT is approximately βr_o, where β is the BJT's β (see Problem 10.49).

Section 10.9: Determining the Loop Gain

10.72 Derive an expression for the loop gain $A\beta$ of the feedback amplifier in Fig. 10.22 (a) (Example 10.5). Set $V_s = 0$, break the loop at the gate of Q_2, apply a test voltage V_t to the gate of Q_2, and determine the voltage V_r that appears at the output of amplifier A_1. Put your expression in the form in Eq. (10.36) and indicate the difference.

10.73 It is required to determine the loop gain of the amplifier circuit shown in Fig. P10.41. The most convenient place to break the loop is at the base of Q_2. Thus, connect a resistance equal to $r_{\pi 2}$ between the collector of Q_1 and ground, apply a test voltage V_t to the base of Q_2, and determine the returned voltage at the collector of Q_1 (with V_s set to zero, of course). Show that

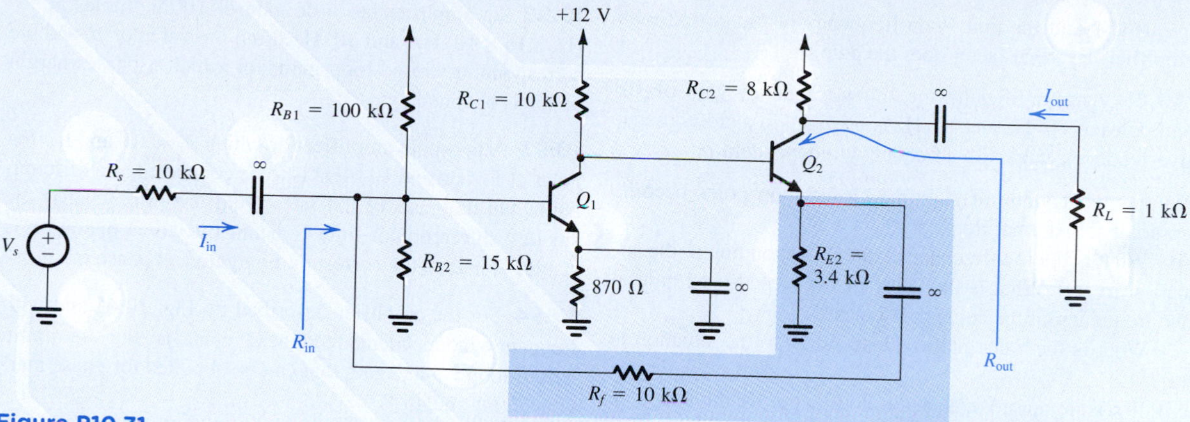

Figure P10.71

$$A\beta = \frac{g_{m2}R_{C2}(h_{fe3}+1)}{R_{C2}+(h_{fe3}+1)[r_{e3}+R_F+(R_E \parallel r_{e1})]} \times \frac{\alpha_1 R_E}{R_E + r_{e1}}(R_{C1} \parallel r_{\pi 2})$$

10.74 Show that the loop gain of the amplifier circuit in Fig. P10.52 is

$$A\beta = g_{m1,2}(r_{o2} \parallel r_{o4})\frac{R_F \parallel r_{o5}}{(R_F \parallel r_{o5}) + 1/g_{m5}}$$

where $g_{m1,2}$ is the g_m of each of Q_1 and Q_2.

10.75 Derive an expression for the loop gain of the feedback circuit shown in Fig. P10.26. Assume that the op amp is modeled by an input resistance R_{id}, an open-circuit voltage gain μ, and an output resistance r_o.

***10.76** Find the loop gain of the feedback amplifier shown in Fig. P10.37 by breaking the loop at the gate of Q_2 (and, of course, setting $v_S = 0$). Use the values given in the statement of Problem 10.37. Determine the value of R_{out}.

10.77 Derive an expression for the loop gain of the feedback amplifier shown in Fig. 10.27(a) (Example 10.7). Evaluate $A\beta$ for the component values given in Example 10.7 and compare to the value determined there.

10.78 Derive an expression for the loop gain of the feedback amplifier in Fig. 10.31(a) (Example 10.8). Evaluate $A\beta$ for the component values given in Example 10.8 and compare to the result found there.

10.79 For the feedback amplifier in Fig. P10.70, set $I_s = 0$ and derive an expression for the loop gain by breaking the loop at the gate terminal of transistor Q_f. Refer to Problem 10.70 for more details.

Section 10.10: The Stability Problem

10.80 An op amp designed to have a low-frequency gain of 10^5 and a high-frequency response dominated by a single pole at 100 rad/s, acquires, through a manufacturing error, a pair of additional poles at 10,000 rad/s. At what frequency does the total phase shift reach 180°°? At this frequency, for what value of β, assumed to be frequency independent, does the loop gain reach a value of unity? What is the corresponding value of closed-loop gain at low frequencies?

***10.81** For the situation described in Problem 10.80, sketch Nyquist plots for $\beta = 1.0$ and 10^{-3}. (Plot for $\omega = 0$ rad/s, 100 rad/s, 10^3 rad/s, 10^4 rad/s, and ∞ rad/s.)

10.82 An op amp having a low-frequency gain of 10^3 and a single-pole rolloff at 10^4 rad/s is connected in a negative-feedback loop via a feedback network having a transmission k and a two-pole rolloff at 10^4 rad/s. Find the value of k above which the closed-loop amplifier becomes unstable.

10.83 Consider a feedback amplifier for which the open-loop gain $A(s)$ is given by

$$A(s) = \frac{1000}{(1+s/10^4)(1+s/10^5)^2}$$

If the feedback factor β is independent of frequency, find the frequency at which the phase shift is 180°, and find the critical value of β at which oscillation will commence.

Section 10.11: Effect of Feedback on the Amplifier Poles

10.84 A dc amplifier having a single-pole response with pole frequency 10 Hz and unity-gain frequency of 1 MHz is operated in a loop whose frequency-independent feedback factor is 0.01. Find the low-frequency gain, the 3-dB frequency, and the unity-gain frequency of the closed-loop amplifier. By what factor does the pole shift?

***10.85** An amplifier having a low-frequency gain of 10^3 and poles at 10^4 Hz and 10^5 Hz is operated in a closed negative-feedback loop with a frequency-independent β.

(a) For what value of β do the closed-loop poles become coincident? At what frequency?
(b) What is the low-frequency gain corresponding to the situation in (a)? What is the value of the closed-loop gain at the frequency of the coincident poles?
(c) What is the value of Q corresponding to the situation in (a)?
(d) If β is increased by a factor of 10, what are the new pole locations? What is the corresponding pole Q?

D 10.86 A dc amplifier has an open-loop gain of 1000 and two poles, a dominant one at 1 kHz and a high-frequency one whose location can be controlled. It is required to connect this amplifier in a negative-feedback loop that provides a dc closed-loop gain of 10 and a maximally flat response. Find the required value of β and the frequency at which the second pole should be placed.

10.87 Reconsider Example 10.9 with the circuit in Fig. 10.40 modified to incorporate a so-called tapered network, in which the components immediately adjacent to the amplifier input are raised in impedance to $C/10$ and $10R$. Find expressions for the resulting pole frequency ω_0 and Q factor. For what value of K do the poles coincide? For what value of K does the response become maximally flat? For what value of K does the circuit oscillate?

10.88 Three identical inverting amplifier stages each characterized by a low-frequency gain K and a single-pole response with f_{3dB} = 100 kHz are connected in a feedback loop with β = 1. What is the minimum value of K at which the circuit oscillates? What would the frequency of oscillation be?

Section 10.12: Stability Study Using Bode Plots

10.89 Reconsider Exercise 10.24 for the case of the op amp wired as a unity-gain buffer. At what frequency is $|A\beta| = 1$? What is the corresponding phase margin?

10.90 Reconsider Exercise 10.24 for the case of a manufacturing error introducing a second pole at 10^4 Hz. What is now the frequency for which $|A\beta|$ = 1? What is the corresponding phase margin? For what values of β is the phase margin 45° or more?

10.91 For what phase margin does the gain peaking have a value of 5%? Of 10%? Of 0.1 dB? Of 1 dB? [*Hint:* Use the result in Eq. 10.105.]

10.92 An amplifier has a dc gain of 10^5 and poles at 10^5 Hz, 3.16×10^5 Hz, and 10^6 Hz. Find the value of β, and the corresponding closed-loop gain, for which a phase margin of 45° is obtained.

10.93 A two-pole amplifier for which $A_0 = 10^3$ and having poles at 1 MHz and 10 MHz is to be connected as a differentiator. On the basis of the rate-of-closure rule, what is the smallest differentiator time constant for which operation is stable? What are the corresponding gain and phase margins?

10.94 For the amplifier described by Fig. 10.43 and with frequency-independent feedback, what is the minimum closed-loop voltage gain that can be obtained for phase margins of 90° and 45°?

Section 10.13: Frequency Compensation

D 10.95 A multipole amplifier having a first pole at 3 MHz and a dc open-loop gain of 60 dB is to be compensated for closed-loop gains as low as unity by the introduction of a new dominant pole. At what frequency must the new pole be placed?

D 10.96 For the amplifier described in Problem 10.95, rather than introducing a new dominant pole we can use additional capacitance at the circuit node at which the pole is formed to reduce the frequency of the first pole. If the frequency of the second pole is 15 MHz and if it remains unchanged while additional capacitance is introduced as mentioned, find the frequency to which the first pole must be lowered so that the resulting amplifier is stable for closed-loop gains as low as unity. By what factor is the capacitance at the controlling node increased?

10.97 Contemplate the effects of pole splitting by considering Eqs. (10.112), (10.116), and (10.117) under the conditions that $R_1 \simeq R_2 = R$, $C_2 \simeq C_1/10 = C$, $C_f \gg C$, and $g_m = 100/R$, by calculating ω_{P1}, ω_{P2}, and ω'_{P1}, ω'_{P2}.

D 10.98 An op amp with open-loop voltage gain of 10^4 and poles at 10^6 Hz, 10^7 Hz, and 10^8 Hz is to be compensated by the addition of a fourth dominant pole to operate stably with unity feedback (β = 1). What is the frequency of the required dominant pole? The compensation network is to consist of an RC low-pass network placed in the negative-feedback path of the op amp. The dc bias conditions are such that a 1-MΩ resistor can be tolerated in series with each of the negative and positive input terminals. What capacitor is required between the negative input and ground to implement the required fourth pole?

D *10.99 An op amp with an open-loop voltage gain of 80 dB and poles at 10^5 Hz, 10^6 Hz, and 2×10^6 Hz is to be compensated to be stable for unity β. Assume that the op amp incorporates an amplifier equivalent to that in Fig. 10.46, with $C_1 = 150$ pF, $C_2 = 5$ pF, and $g_m = 40$ mA/V, and that f_{P1} is caused by the input circuit and f_{P2} by the output circuit of this amplifier. Find the required value of the compensating Miller capacitance and the new frequency of the output pole.

SIM **10.100 The op amp in the circuit of Fig. P10.100 has an open-loop gain of 10^5 and a single-pole rolloff with $\omega_{3dB} = 10$ rad/s.

(a) Sketch a Bode plot for the loop gain.
(b) Find the frequency at which $|A\beta| = 1$, and find the corresponding phase margin.
(c) Find the closed-loop transfer function, including its zero and poles. Sketch a pole-zero plot. Sketch the magnitude of the transfer function versus frequency, and label the important parameters on your sketch.

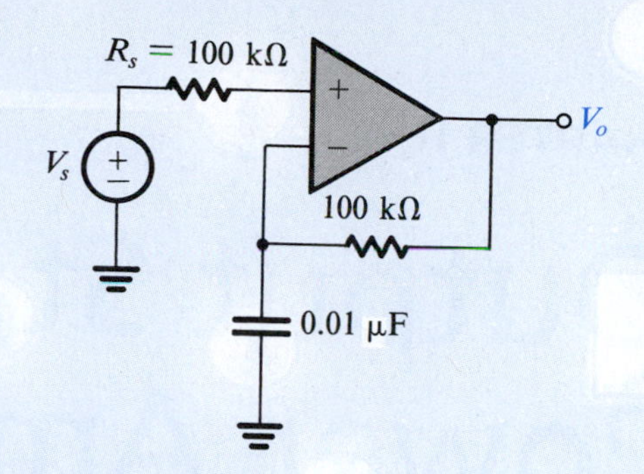

Figure P10.100

CHAPTER 11

Output Stages and Power Amplifiers

IN THIS CHAPTER YOU WILL LEARN

1. The classification of amplifier output stages on the basis of the fraction of the cycle of an input sine wave during which the transistor conducts.
2. Analysis and design of a variety of output-stage types ranging from the simple but power-inefficient emitter follower (class A) to the popular push-pull class AB circuit in both bipolar and CMOS technologies.
3. Thermal considerations in the design and fabrication of high-output-power circuits.
4. Useful and interesting circuit techniques employed in the design of power amplifiers.
5. Special types of MOS transistors optimized for high-power applications.

Introduction

An important function of the output stage is to provide the amplifier with a low output resistance so that it can deliver the output signal to the load without loss of gain. Since the output stage is the final stage of the amplifier, it usually deals with relatively large signals. Thus the small-signal approximations and models either are not applicable or must be used with care. Nevertheless, linearity remains a very important requirement. In fact, a measure of goodness of the output stage is the **total harmonic distortion** (THD) it introduces. This is the rms value of the harmonic components of the output signal, excluding the fundamental, expressed as a percentage of the rms of the fundamental. A high-fidelity audio power amplifier features a THD of the order of a fraction of a percent.

The most challenging requirement in the design of an output stage is for it to deliver the required amount of power to the load in an *efficient* manner. This implies that the power *dissipated* in the output-stage transistors must be as low as possible. This requirement stems mainly from the fact that the power dissipated in a transistor raises its internal **junction temperature**, and there is a maximum temperature (in the range of 150°C to 200°C for silicon devices) above which the transistor is destroyed. A high power-conversion efficiency also may be required to prolong the life of batteries employed in battery-powered circuits, to permit a smaller, lower-cost power supply, or to obviate the need for cooling fans.

We begin this chapter with a study of the various output-stage configurations employed in amplifiers that handle both low and high power. In this context, "high power" generally means greater than 1 W. We then consider the specific requirements of BJTs employed in the design of high-power output stages, called **power transistors**. Special attention will be paid to the thermal properties of such transistors.

A power amplifier is simply an amplifier with a high-power output stage. Examples of discrete- and integrated-circuit power amplifiers will be presented. Since BJTs can handle much larger currents than MOSFETs, they are preferred in the design of output stages. Nevertheless, some interesting CMOS output stages are also studied.

11.1 Classification of Output Stages

Output stages are classified according to the collector current waveform that results when an input signal is applied. Figure 11.1 illustrates the classification for the case of a sinusoidal input signal. The class A stage, whose associated waveform is shown in Fig. 11.1(a), is biased at a current I_C greater than the amplitude of the signal current, $\hat{I}_c$. Thus the transistor in a class A stage conducts for the entire cycle of the input signal; that is, the conduction angle is 360°. In contrast, the class B stage, whose associated waveform is shown in Fig. 11.1(b), is biased at zero dc current. Thus a transistor in a class B stage conducts for only half the cycle of the input sine wave, resulting in a conduction angle of 180°. As will be seen later,

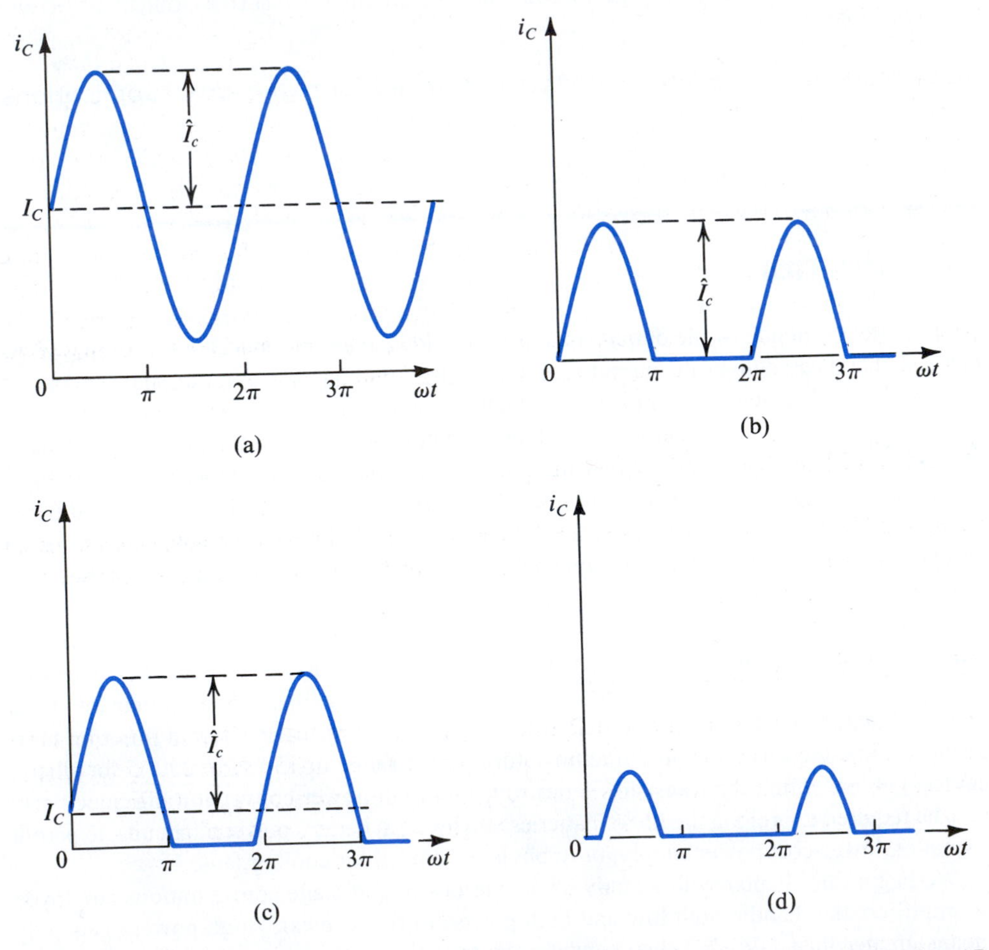

Figure 11.1 Collector current waveforms for transistors operating in **(a)** class A, **(b)** class B, **(c)** class AB, and **(d)** class C amplifier stages.

the negative halves of the sinusoid will be supplied by another transistor that also operates in the class B mode and conducts during the alternate half-cycles.

An intermediate class between A and B, appropriately named class AB, involves biasing the transistor at a nonzero dc current much smaller than the peak current of the sine-wave signal. As a result, the transistor conducts for an interval slightly greater than half a cycle, as illustrated in Fig. 11.1(c). The resulting conduction angle is greater than 180° but much less than 360°. The class AB stage has another transistor that conducts for an interval slightly greater than that of the negative half-cycle, and the currents from the two transistors are combined in the load. It follows that, during the intervals near the zero crossings of the input sinusoid, both transistors conduct.

Figure 11.1(d) shows the collector-current waveform for a transistor operated as a class C amplifier. Observe that the transistor conducts for an interval shorter than that of a half-cycle; that is, the conduction angle is less than 180°. The result is the periodically pulsating current waveform shown. To obtain a sinusoidal output voltage, this current is passed through a parallel *LC* circuit, tuned to the frequency of the input sinusoid. The tuned circuit acts as a bandpass filter (Chapter 16) and provides an output voltage proportional to the amplitude of the fundamental component in the Fourier-series representation of the current waveform.

Class A, AB, and B amplifiers are studied in this chapter. They are employed as output stages of op amps and audio power amplifiers. In the latter application, class AB is the preferred choice, for reasons that will be explained in the sections to folow. Class C amplifiers are usually employed for radio-frequency (RF) power amplification (required, e.g., in mobile phones and radio and TV transmitters). The design of class C amplifiers is a rather specialized topic and is not included in this book. However, we should point out that the tuned-resonator oscillator circuits described in Chapter 17 operate inherently in the class C mode.

Although the BJT has been used to illustrate the definition of the various output-stage classes, the same classification applies to output stages implemented with MOSFETs. Furthermore, the classification above extends to amplifier stages other than those used at the output. In this regard, all the common-emitter, common-base, and common-collector amplifiers (and their FET counterparts) studied in earlier chapters fall into the class A category.

11.2 Class A Output Stage

Because of its low output resistance, the emitter follower is the most popular class A output stage. We have already studied the emitter follower in Chapter 6; in the following we consider its large-signal operation.

11.2.1 Transfer Characteristic

Figure 11.2 shows an emitter follower Q_1 biased with a constant current I supplied by transistor Q_2. Since the emitter current $i_{E1} = I + i_L$, the bias current I must be greater than the largest negative load current; otherwise, Q_1 cuts off and class A operation will no longer be maintained.

The transfer characteristic of the emitter follower of Fig. 11.2 is described by

$$v_O = v_I - v_{BE1} \tag{11.1}$$

where v_{BE1} depends on the emitter current i_{E1} and thus on the load current i_L. If we neglect the relatively small changes in v_{BE1} (60 mV for every factor-of-10 change in emitter current), the

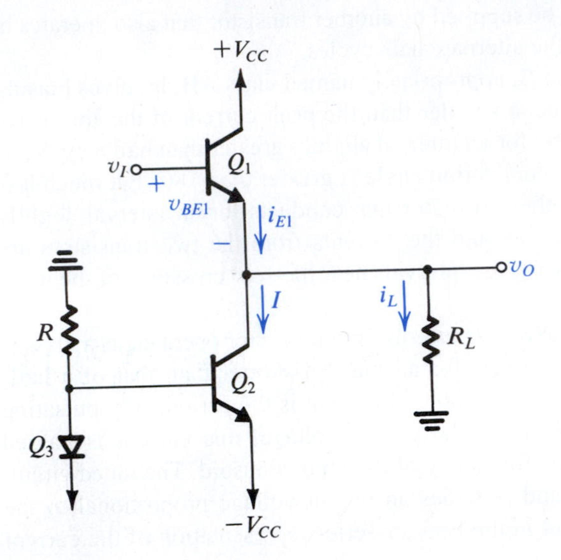

Figure 11.2 An emitter follower (Q_1) biased with a constant current I supplied by transistor Q_2.

linear transfer curve shown in Fig. 11.3 results. As indicated, the positive limit of the linear region is determined by the saturation of Q_1; thus

$$v_{O\max} = V_{CC} - V_{CE1\text{sat}} \tag{11.2}$$

In the negative direction, depending on the values of I and R_L, the limit of the linear region is determined either by Q_1 turning off,

$$v_{O\min} = -IR_L \tag{11.3}$$

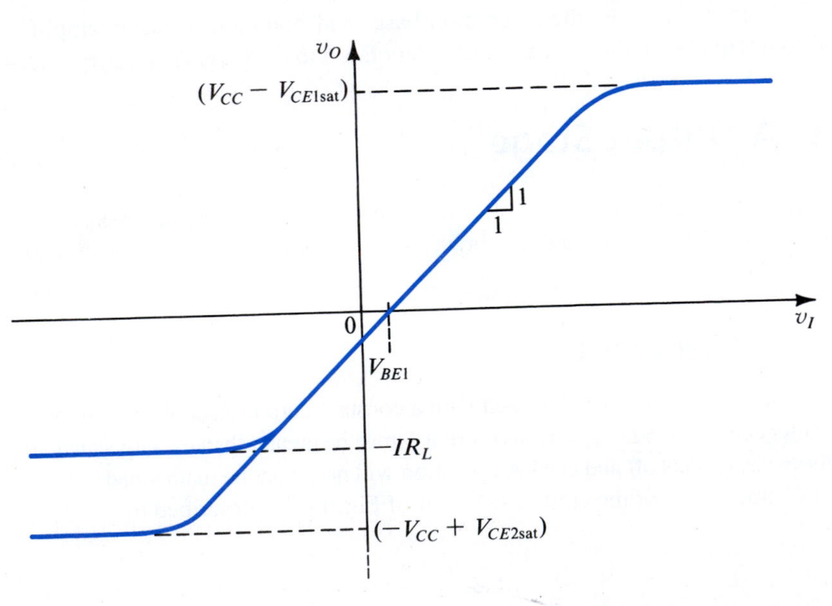

Figure 11.3 Transfer characteristic of the emitter follower in Fig. 11.2. This linear characteristic is obtained by neglecting the change in v_{BE1} with i_L. The maximum positive output is determined by the saturation of Q_1. In the negative direction, the limit of the linear region is determined either by Q_1 turning off or by Q_2 saturating, depending on the values of I and R_L.

or by Q_2 saturating,

$$v_{O\min} = -V_{CC} + V_{CE2\text{sat}} \tag{11.4}$$

The absolutely lowest (most negative) output voltage is that given by Eq. (11.4) and is achieved provided the bias current I is greater than the magnitude of the corresponding load current,

$$I \geq \frac{|-V_{CC} + V_{CE2\text{sat}}|}{R_L} \tag{11.5}$$

EXERCISES

D11.1 For the emitter follower in Fig. 11.2, $V_{CC} = 15$ V, $V_{CE\text{sat}} = 0.2$ V, $V_{BE} = 0.7$ V and constant, and β is very high. Find the value of R that will establish a bias current sufficiently large to allow the largest possible output signal swing for $R_L = 1$ kΩ. Determine the resulting output signal swing and the minimum and maximum emitter currents for Q_1.
Ans. 0.97 kΩ; −14.8 V to +14.8 V; 0 to 29.6 mA

11.2 For the emitter follower of Exercise 11.1, in which $I = 14.8$ mA, consider the case in which v_O is limited to the range −10 V to +10 V. Let Q_1 have $v_{BE} = 0.6$ V at $i_C = 1$ mA, and assume $\alpha \simeq 1$. Find v_I corresponding to $v_O = -10$ V, 0 V, and +10 V. At each of these points, use small-signal analysis to determine the voltage gain v_o/v_i. Note that the incremental voltage gain gives the slope of the v_O-versus-v_I characteristic.
Ans. −9.36 V, 0.67 V, 10.68 V; 0.995 V/V, 0.998 V/V, 0.999 V/V

11.2.2 Signal Waveforms

Consider the operation of the emitter-follower circuit of Fig. 11.2 for sine-wave input. Neglecting $V_{CE\text{sat}}$, we see that if the bias current I is properly selected, the output voltage can swing from $-V_{CC}$ to $+V_{CC}$ with the quiescent value being zero, as shown in Fig. 11.4(a). Figure 11.4(b) shows the corresponding waveform of $v_{CE1} = V_{CC} - v_O$. Now, assuming that the bias current I is selected to allow a maximum negative load current of V_{CC}/R_L, that is,

$$I = V_{CC}/R_L$$

the collector current of Q_1 will have the waveform shown in Fig. 11.4(c). Finally, Fig. 11.4(d) shows the waveform of the **instantaneous power dissipation** in Q_1,

$$p_{D1} \equiv v_{CE1} i_{C1} \tag{11.6}$$

11.2.3 Power Dissipation

Figure 11.4(d) indicates that the maximum instantaneous power dissipation in Q_1 is $V_{CC}I$. This is equal to the power dissipation in Q_1 with no input signal applied, that is, the quiescent power dissipation. Thus the emitter-follower transistor dissipates the largest amount of power when $v_O = 0$. Since this condition (no input signal) can easily prevail for prolonged periods of time, transistor Q_1 must be able to withstand a continuous power dissipation of $V_{CC}I$.

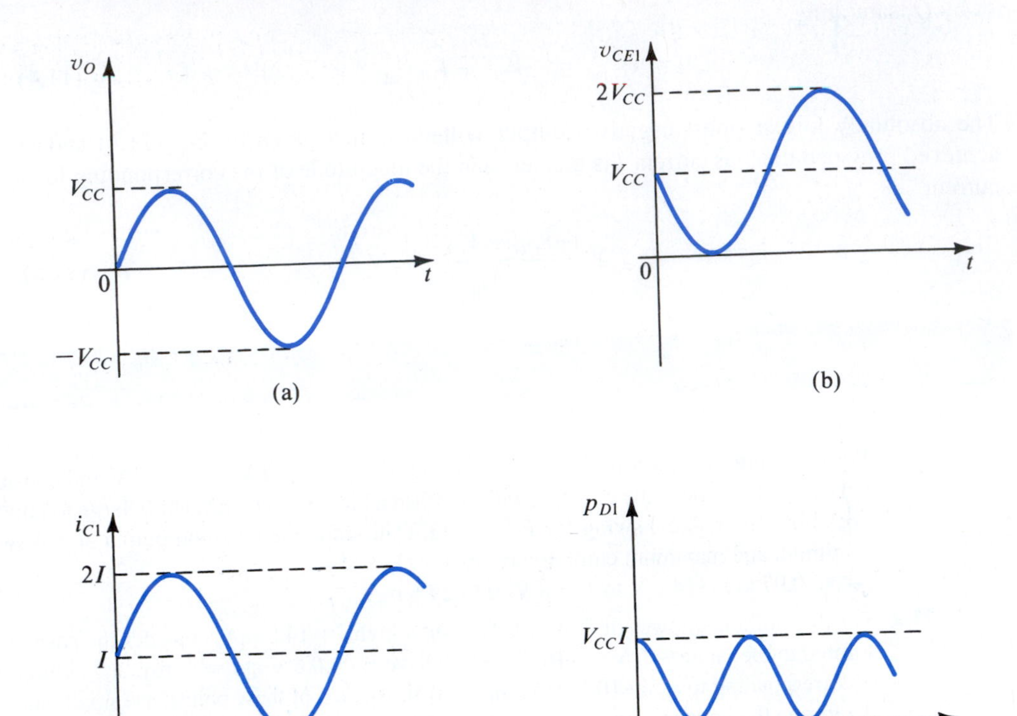

Figure 11.4 Maximum signal waveforms in the class A output stage of Fig. 11.2 under the condition $I = V_{CC}/R_L$ or, equivalently, $R_L = V_{CC}/I$. Note that the transistor saturation voltages have been neglected.

The power dissipation in Q_1 depends on the value of R_L. Consider the extreme case of an output open circuit, that is, $R_L = \infty$. In this case, $i_{C1} = I$ is constant and the instantaneous power dissipation in Q_1 will depend on the instantaneous value of v_O. The maximum power dissipation will occur when $v_O = -V_{CC}$, for in this case v_{CE1} is a maximum of $2V_{CC}$ and $p_{D1} = 2V_{CC}I$. This condition, however, would not normally persist for a prolonged interval, so the design need not be that conservative. Observe that with an open-circuit load, the average power dissipation in Q_1 is $V_{CC}I$. A far more dangerous situation occurs at the other extreme of R_L—specifically, $R_L = 0$. In the event of an output short circuit, a positive input voltage would theoretically result in an infinite load current. In practice, a very large current may flow through Q_1, and if the short-circuit condition persists, the resulting large power dissipation in Q_1 can raise its junction temperature beyond the specified maximum, causing Q_1 to burn up. To guard against such a situation, output stages are usually equipped with **short-circuit protection**, as will be explained later.

The power dissipation in Q_2 also must be taken into account in designing an emitter-follower output stage. Since Q_2 conducts a constant current I, and the maximum value of v_{CE2} is $2V_{CC}$, the maximum instantaneous power dissipation in Q_2 is $2V_{CC}I$. This maximum, however, occurs when $v_O = V_{CC}$, a condition that would not normally prevail for a prolonged period of time. A more significant quantity for design purposes is the average power dissipation in Q_2, which is $V_{CC}I$.

Example 11.1

Consider the emitter follower in Fig. 11.2 with $V_{CC} = 10$ V, $I = 100$ mA, and $R_L = 100\ \Omega$.

(a) Find the power dissipated in Q_1 and Q_2 under quiescent conditions ($v_O = 0$).
(b) For a sinusoidal output voltage of maximum possible amplitude (neglecting V_{CEsat}), find the average power dissipation in Q_1 and Q_2. Also find the load power.

Solution

(a) Under quiescent conditions $v_O = 0$, and each of Q_1 and Q_2 conducts a current $I = 100$ mA $= 0.1$ A and has a voltage $V_{CE} = V_{CC} = 10$ V, thus

$$P_{D1} = P_{D2} = V_{CC}I = 10 \times 0.1 = 1\text{ W}$$

(b) For a sinusoidal output voltage of maximum possible amplitude (i.e., 10-V peak), the instantaneous power dissipation in Q_1 will be as shown in Fig. 11.4(d). Thus the average power dissipation in Q_1 will be

$$P_{D1} = \frac{1}{2}V_{CC}I = \frac{1}{2} \times 10 \times 0.1 = 0.5\text{W}$$

For Q_2, the current is constant at $I = 0.1$ A and the voltage at the collector will have an average value of 0 V. Thus the average voltage across Q_2 will be V_{CC} and the average dissipation will be

$$P_{D2} = I \times v_{CE}\big|_{\text{average}}$$

$$= I \times V_{CC} = 0.1 \times 10 = 1\text{W}$$

Finally, the power delivered to the load can be found from

$$P_L = \frac{V_{orms}^2}{R_L}$$

$$= \frac{(10/\sqrt{2})^2}{100} = 0.5\text{W}$$

11.2.4 Power-Conversion Efficiency

The power-conversion efficiency of an output stage is defined as

$$\eta \equiv \frac{\text{Load power } (P_L)}{\text{Supply power } (P_S)} \tag{11.7}$$

For the emitter follower of Fig. 11.2, assuming that the output voltage is a sinusoid with the peak value $\hat{V}_o$, the average load power will be

$$P_L = \frac{(\hat{V}_o/\sqrt{2})^2}{R_L} = \frac{1}{2}\frac{\hat{V}_o^2}{R_L} \tag{11.8}$$

Since the current in Q_2 is constant (I), the power drawn from the negative supply[1] is $V_{CC}I$. The *average* current in Q_1 is equal to I, and thus the average power drawn from the positive

[1]This does *not* include the power drawn by the biasing resistor R and the diode-connected transistor Q_3.

supply is $V_{CC}I$. Thus the total average supply power is

$$P_S = 2V_{CC}I \tag{11.9}$$

Equations (11.8) and (11.9) can be combined to yield

$$\eta = \frac{1}{4}\frac{\hat{V}_o^2}{IR_L V_{CC}}$$

$$= \frac{1}{4}\left(\frac{\hat{V}_o}{IR_L}\right)\left(\frac{\hat{V}_o}{V_{CC}}\right) \tag{11.10}$$

Since $\hat{V}_o \leq V_{CC}$ and $\hat{V}_o \leq IR_L$, maximum efficiency is obtained when

$$\hat{V}_o = V_{CC} = IR_L \tag{11.11}$$

The maximum efficiency attainable is 25%. Because this is a rather low figure, the class A output stage is rarely used in high-power applications (>1 W). Note also that in practice the output voltage swing is limited to lower values to avoid transistor saturation and associated nonlinear distortion. Thus the efficiency achieved in practice is usually in the 10% to 20% range.

EXERCISE

11.3 For the emitter follower of Fig. 11.2, let $V_{CC} = 10$ V, $I = 100$ mA, and $R_L = 100\ \Omega$. If the output voltage is an 8-V-peak sinusoid, find the following: (a) the power delivered to the load; (b) the average power drawn from the supplies; (c) the power-conversion efficiency. Ignore the loss in Q_3 and R.
Ans. 0.32 W; 2 W; 16%

11.3 Class B Output Stage

Figure 11.5 shows a class B output stage. It consists of a complementary pair of transistors (an *npn* and a *pnp*) connected in such a way that both cannot conduct simultaneously.

11.3.1 Circuit Operation

When the input voltage v_I is zero, both transistors are cut off and the output voltage v_O is zero. As v_I goes positive and exceeds about 0.5 V, Q_N conducts and operates as an emitter follower. In this case v_O follows v_I (i.e., $v_O = v_I - v_{BEN}$) and Q_N supplies the load current. Meanwhile, the emitter–base junction of Q_P will be reverse-biased by the V_{BE} of Q_N, which is approximately 0.7 V. Thus Q_P will be cut off.

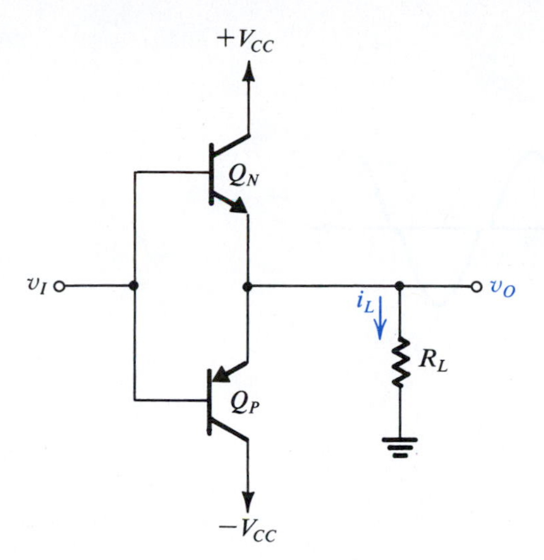

Figure 11.5 A class B output stage.

If the input goes negative by more than about 0.5 V, Q_P turns on and acts as an emitter follower. Again v_O follows v_I (i.e., $v_O = v_I + v_{EBP}$), but in this case Q_P supplies the load current and Q_N will be cut off.

We conclude that the transistors in the class B stage of Fig. 11.5 are biased at zero current and conduct only when the input signal is present. The circuit operates in a **push–pull** fashion: Q_N *pushes* (sources) current into the load when v_I is positive, and Q_P *pulls* (sinks) current from the load when v_I is negative.

11.3.2 Transfer Characteristic

A sketch of the transfer characteristic of the class B stage is shown in Fig. 11.6. Note that there exists a range of v_I centered around zero where both transistors are cut off and v_O is zero. This **dead band** results in the **crossover distortion** illustrated in Fig. 11.7 for the case of an input sine wave. The effect of crossover distortion will be most pronounced when the

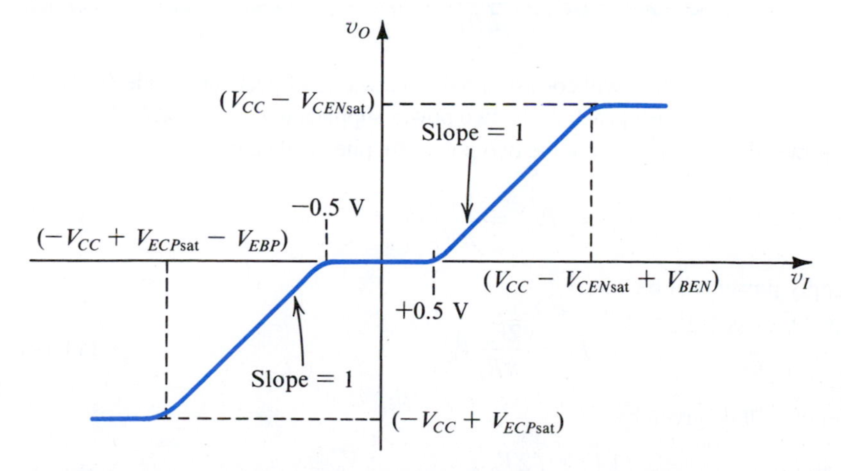

Figure 11.6 Transfer characteristic for the class B output stage in Fig. 11.5.

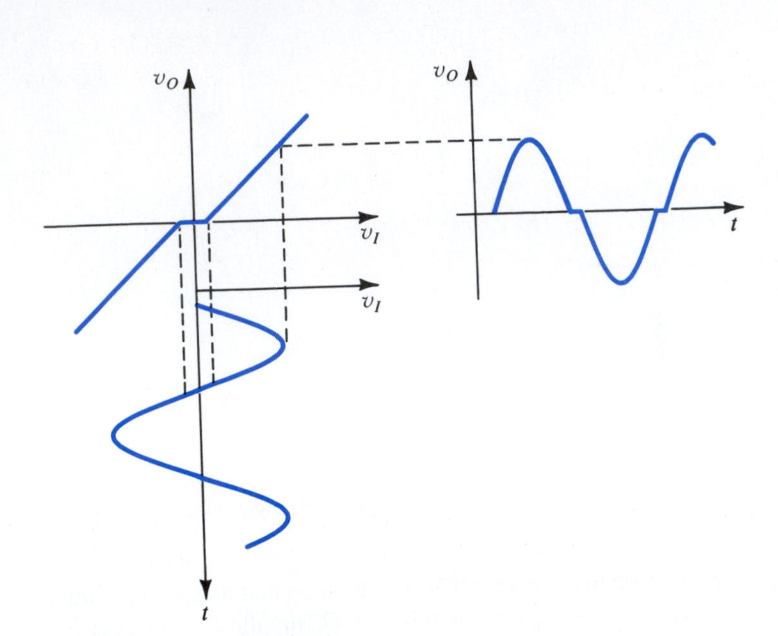

Figure 11.7 Illustrating how the dead band in the class B transfer characteristic results in crossover distortion.

amplitude of the input signal is small. Crossover distortion in audio power amplifiers gives rise to unpleasant sounds.

11.3.3 Power-Conversion Efficiency

To calculate the power-conversion efficiency, η, of the class B stage, we neglect the crossover distortion and consider the case of an output sinusoid of peak amplitude $\hat{V}_o$. The average load power will be

$$P_L = \frac{1}{2}\frac{\hat{V}_o^2}{R_L} \tag{11.12}$$

The current drawn from each supply will consist of half-sine waves of peak amplitude $(\hat{V}_o/R_L)$. Thus the average current drawn from each of the two power supplies will be $\hat{V}_o/\pi R_L$. It follows that the average power drawn from each of the two power supplies will be the same,

$$P_{S+} = P_{S-} = \frac{1}{\pi}\frac{\hat{V}_o}{R_L}V_{CC} \tag{11.13}$$

and the total supply power will be

$$P_S = \frac{2}{\pi}\frac{\hat{V}_o}{R_L}V_{CC} \tag{11.14}$$

Thus the efficiency will be given by

$$\eta = \left(\frac{1}{2}\frac{\hat{V}_o^2}{R_L}\right)\bigg/\left(\frac{2}{\pi}\frac{\hat{V}_o}{R_L}V_{CC}\right) = \frac{\pi}{4}\frac{\hat{V}_o}{V_{CC}} \tag{11.15}$$

It follows that the maximum efficiency is obtained when $\hat{V}_o$ is at its maximum. This maximum is limited by the saturation of Q_N and Q_P to $V_{CC} - V_{CEsat} \simeq V_{CC}$. At this value of peak output voltage, the power-conversion efficiency is

$$\eta_{max} = \frac{\pi}{4} = 78.5\% \tag{11.16}$$

This value is much larger than that obtained in the class A stage (25%). Finally, we note that the maximum average power available from a class B output stage is obtained by substituting $\hat{V}_o = V_{CC}$ in Eq. (11.12),

$$P_{Lmax} = \frac{1}{2}\frac{V_{CC}^2}{R_L} \tag{11.17}$$

11.3.4 Power Dissipation

Unlike the class A stage, which dissipates maximum power under quiescent conditions ($v_O = 0$), the quiescent power dissipation of the class B stage is zero. When an input signal is applied, the *average* power dissipated in the class B stage is given by

$$P_D = P_S - P_L \tag{11.18}$$

Substituting for P_S from Eq. (11.14) and for P_L from Eq. (11.12) results in

$$P_D = \frac{2}{\pi}\frac{\hat{V}_o}{R_L}V_{CC} - \frac{1}{2}\frac{\hat{V}_o^2}{R_L} \tag{11.19}$$

From symmetry we see that half of P_D is dissipated in Q_N and the other half in Q_P. Thus Q_N and Q_P must be capable of safely dissipating $\frac{1}{2}P_D$ watts. Since P_D depends on $\hat{V}_o$, we must find the worst-case power dissipation, P_{Dmax}. Differentiating Eq. (11.19) with respect to $\hat{V}_o$ and equating the derivative to zero gives the value of $\hat{V}_o$ that results in maximum average power dissipation as

$$\hat{V}_o\big|_{P_{Dmax}} = \frac{2}{\pi}V_{CC} \tag{11.20}$$

Substituting this value in Eq. (11.19) gives

$$P_{Dmax} = \frac{2V_{CC}^2}{\pi^2 R_L} \tag{11.21}$$

Thus,

$$P_{DNmax} = P_{DPmax} = \frac{V_{CC}^2}{\pi^2 R_L} \tag{11.22}$$

At the point of maximum power dissipation, the efficiency can be evaluated by substituting for $\hat{V}_o$ from Eq. (11.20) into Eq. (11.15); hence, $\eta = 50\%$.

Figure 11.8 shows a sketch of P_D (Eq. 11.19) versus the peak output voltage $\hat{V}_o$. Curves such as this are usually given on the data sheets of IC power amplifiers. [Usually, however, P_D is plotted versus P_L, as $P_L = \frac{1}{2}(\hat{V}_o^2/R_L)$, rather than $\hat{V}_o$.] An interesting observation follows from Fig. 11.8: Increasing $\hat{V}_o$ beyond $2V_{CC}/\pi$ *decreases* the power dissipated in the

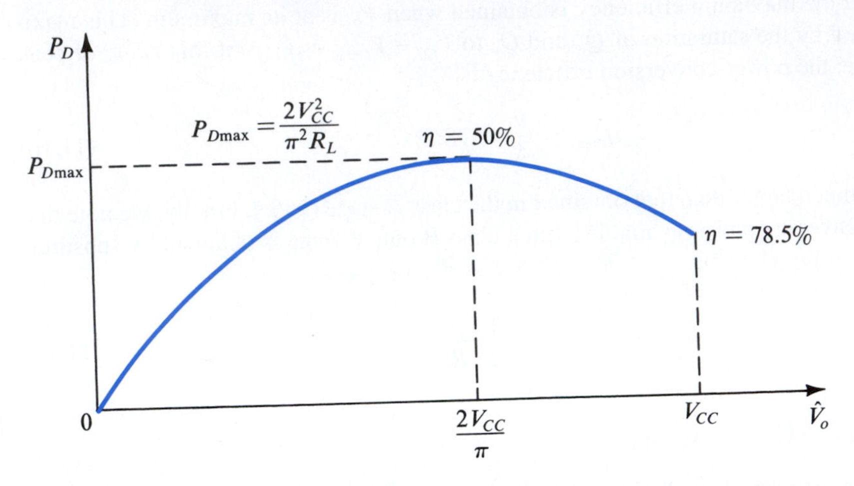

Figure 11.8 Power dissipation of the class B output stage versus amplitude of the output sinusoid.

class B stage while increasing the load power. The price paid is an increase in nonlinear distortion as a result of approaching the saturation region of operation of Q_N and Q_P. Transistor saturation flattens the peaks of the output sine waveform. Unfortunately, this type of distortion cannot be significantly reduced by the application of negative feedback (see Section 10.2), and thus transistor saturation should be avoided in applications requiring low THD.

Example 11.2

It is required to design a class B output stage to deliver an average power of 20 W to an 8-Ω load. The power supply is to be selected such that V_{CC} is about 5 V greater than the peak output voltage. This avoids transistor saturation and the associated nonlinear distortion, and allows for including short-circuit protection circuitry. (The latter will be discussed in Section 11.8.) Determine the supply voltage required, the peak current drawn from each supply, the total supply power, and the power-conversion efficiency. Also determine the maximum power that each transistor must be able to dissipate safely.

Solution

Since

$$P_L = \frac{1}{2}\frac{\hat{V}_o^2}{R_L}$$

then

$$\hat{V}_o = \sqrt{2P_L R_L}$$
$$= \sqrt{2 \times 20 \times 8} = 17.9 \text{ V}$$

Therefore we select $V_{CC} = 23$ V.

The peak current drawn from each supply is

$$\hat{I}_o = \frac{\hat{V}_o}{R_L} = \frac{17.9}{8} = 2.24 \text{ A}$$

Since each supply provides a current waveform of half-sinusoids, the average current drawn from each supply will be $\hat{I}_o/\pi$. Thus the average power drawn from each supply is

$$P_{S+} = P_{S-} = \frac{1}{\pi} \times 2.24 \times 23 = 16.4 \text{ W}$$

for a total supply power of 32.8 W. The power-conversion efficiency is

$$\eta = \frac{P_L}{P_S} = \frac{20}{32.8} \times 100 = 61\%$$

The maximum power dissipated in each transistor is given by Eq. (11.22); thus,

$$P_{DN\max} = P_{DP\max} = \frac{V_{CC}^2}{\pi^2 R_L}$$

$$= \frac{(23)^2}{\pi^2 \times 8} = 6.7 \text{ W}$$

11.3.5 Reducing Crossover Distortion

The crossover distortion of a class B output stage can be reduced substantially by employing a high-gain op amp and overall negative feedback, as shown in Fig. 11.9. The ±0.7-V dead band is reduced to $\pm 0.7/A_0$ volt, where A_0 is the dc gain of the op amp. Nevertheless, the slew-rate limitation of the op amp will cause the alternate turning on and off of the output transistors to be noticeable, especially at high frequencies. A more practical method for reducing and almost eliminating crossover distortion is found in the class AB stage, which will be studied in the next section.

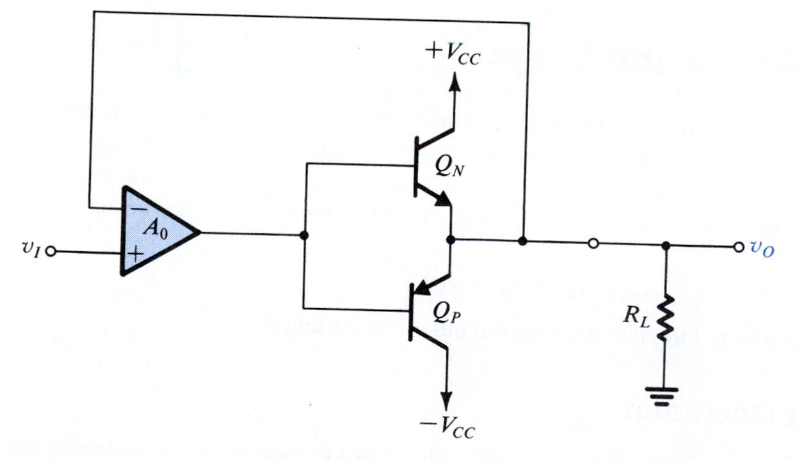

Figure 11.9 Class B circuit with an op amp connected in a negative-feedback loop to reduce crossover distortion.

11.3.6 Single-Supply Operation

The class B stage can be operated from a single power supply, in which case the load is capacitively coupled, as shown in Fig. 11.10. Note that to make the formulas derived in Section 11.3.4 directly applicable, the single power supply is denoted $2V_{CC}$.

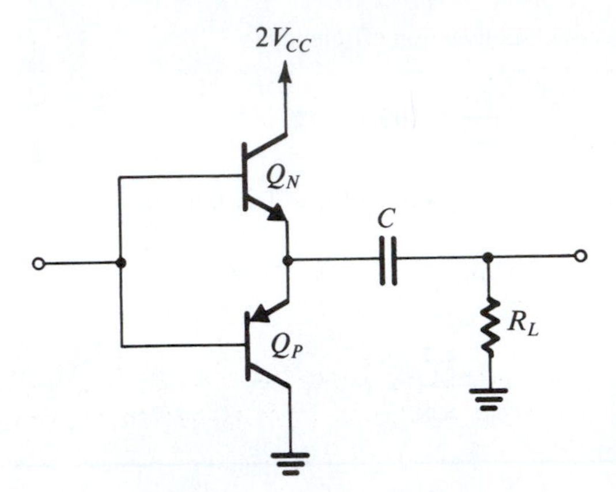

Figure 11.10 Class B output stage operated with a single power supply.

EXERCISE

11.4 For the class B output stage of Fig. 11.5, let $V_{CC} = 6$ V and $R_L = 4\ \Omega$. If the output is a sinusoid with 4.5-V peak amplitude, find (a) the output power; (b) the average power drawn from each supply; (c) the power efficiency obtained at this output voltage; (d) the peak currents supplied by v_I, assuming that $\beta_N = \beta_P = 50$; (e) the maximum power that each transistor must be capable of dissipating safely.

Ans. (a) 2.53 W; (b) 2.15 W; (c) 59%; (d) 22.1 mA; (e) 0.91 W

11.4 Class AB Output Stage

Crossover distortion can be virtually eliminated by biasing the complementary output transistors at a small nonzero current. The result is the class AB output stage shown in Fig. 11.11. A bias voltage V_{BB} is applied between the bases of Q_N and Q_P. For $v_I = 0$, $v_O = 0$, and a voltage $V_{BB}/2$ appears across the base–emitter junction of each of Q_N and Q_P. Assuming matched devices,

$$i_N = i_P = I_Q = I_S e^{V_{BB}/2V_T} \tag{11.23}$$

The value of V_{BB} is selected to yield the required quiescent current I_Q.

11.4.1 Circuit Operation

When v_I goes positive by a certain amount, the voltage at the base of Q_N increases by the same amount and the output becomes positive at an almost equal value,

$$v_O = v_I + \frac{V_{BB}}{2} - v_{BEN} \tag{11.24}$$

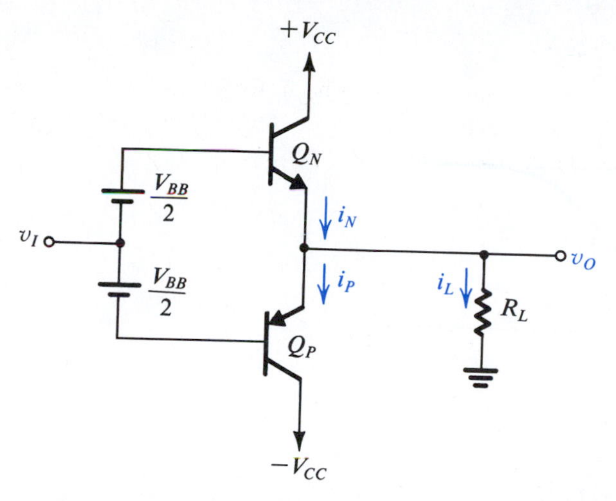

Figure 11.11 Class AB output stage. A bias voltage V_{BB} is applied between the bases of Q_N and Q_P, giving rise to a bias current I_Q given by Eq. (11.23). Thus, for small v_I, both transistors conduct and crossover distortion is almost completely eliminated.

The positive v_O causes a current i_L to flow through R_L, and thus i_N must increase; that is,

$$i_N = i_P + i_L \tag{11.25}$$

The increase in i_N will be accompanied by a corresponding increase in v_{BEN} (above the quiescent value of $V_{BB}/2$). However, since the voltage between the two bases remains constant at V_{BB}, the increase in v_{BEN} will result in an equal decrease in v_{EBP} and hence in i_P. The relationship between i_N and i_P can be derived as follows:

$$v_{BEN} + v_{EBP} = V_{BB}$$

$$V_T \ln \frac{i_N}{I_S} + V_T \ln \frac{i_P}{I_S} = 2V_T \ln \frac{I_Q}{I_S}$$

$$i_N i_P = I_Q^2 \tag{11.26}$$

Thus, as i_N increases, i_P decreases by the same ratio while the product remains constant. Equations (11.25) and (11.26) can be combined to yield i_N for a given i_L as the solution to the quadratic equation

$$i_N^2 - i_L i_N - I_Q^2 = 0 \tag{11.27}$$

From the equations above, we can see that for positive output voltages, the load current is supplied by Q_N, which acts as the output emitter follower. Meanwhile, Q_P will be conducting a current that decreases as v_O increases; for large v_O the current in Q_P can be ignored altogether.

For negative input voltages the opposite occurs: The load current will be supplied by Q_P, which acts as the output emitter follower, while Q_N conducts a current that gets smaller as v_I becomes more negative. Equation (11.26), relating i_N and i_P, holds for negative inputs as well.

We conclude that the class AB stage operates in much the same manner as the class B circuit, with one important exception: For small v_I, both transistors conduct, and as v_I is increased or decreased, one of the two transistors takes over the operation. Since the

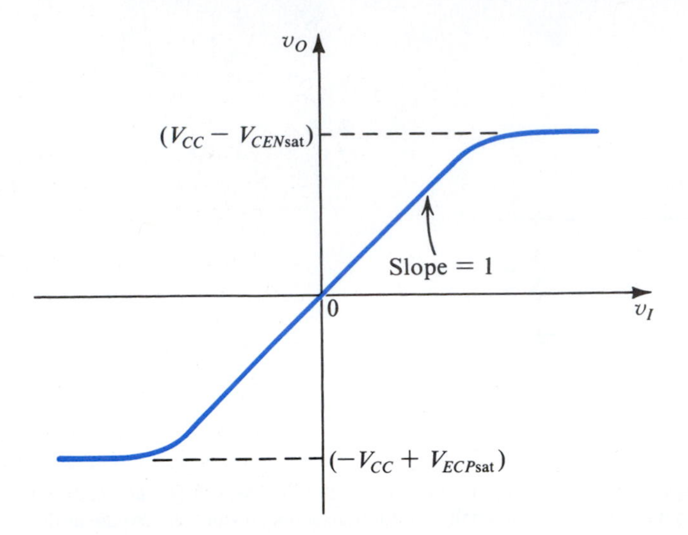

Figure 11.12 Transfer characteristic of the class AB stage in Fig. 11.11.

transition is a smooth one, crossover distortion will be almost totally eliminated. Figure 11.12 shows the transfer characteristic of the class AB stage.

The power relationships in the class AB stage are almost identical to those derived for the class B circuit in Section 11.3. The only difference is that under quiescent conditions the class AB circuit dissipates a power of $V_{CC}I_Q$ per transistor. Since I_Q is usually much smaller than the peak load current, the quiescent power dissipation is usually small. Nevertheless, it can be taken into account easily. Specifically, we can simply add the quiescent dissipation per transistor to its maximum power dissipation with an input signal applied, to obtain the total power dissipation that the transistor must be able to handle safely.

11.4.2 Output Resistance

If we assume that the source supplying v_I is ideal, then the output resistance of the class AB stage can be determined from the circuit in Fig. 11.13 as

$$R_{\text{out}} = r_{eN} \| r_{eP} \tag{11.28}$$

where r_{eN} and r_{eP} are the small-signal emitter resistances of Q_N and Q_P, respectively. At a given input voltage, the currents i_N and i_P can be determined, and r_{eN} and r_{eP} are given by

$$r_{eN} = \frac{V_T}{i_N} \tag{11.29}$$

$$r_{eP} = \frac{V_T}{i_P} \tag{11.30}$$

Thus,

$$R_{\text{out}} = \frac{V_T}{i_N} \Big\| \frac{V_T}{i_P} = \frac{V_T}{i_P + i_N} \tag{11.31}$$

Since as i_N increases, i_P decreases, and vice versa, the output resistance remains approximately constant in the region around $v_I = 0$. This, in effect, is the reason for the virtual

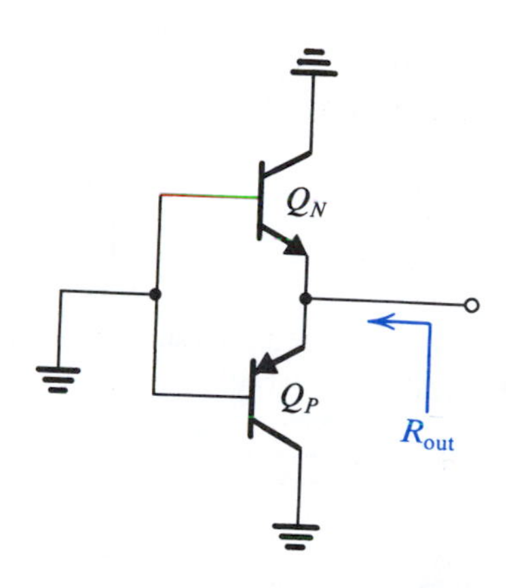

Figure 11.13 Determining the small-signal output resistance of the class AB circuit of Fig. 11.11.

absence of crossover distortion. At larger load currents, either i_N or i_P will be significant, and R_{out} decreases as the load current increases.

Example 11.3

In this example we explore the details of the transfer characteristic, v_O versus v_I, of the class AB circuit in Fig. 11.11. For this purpose let $V_{CC} = 15$ V, $I_Q = 2$ mA, and $R_L = 100\ \Omega$. Assume that Q_N and Q_P are matched and have $I_S = 10^{-13}$A. First, determine the required value of the bias voltage V_{BB}. Then, find the transfer characteristic for v_O in the range -10 V to $+10$ V.

Solution

To determine the required value of V_{BB} we use Eq. (11.23) with $I_Q = 2$ mA and $I_S = 10^{-13}$A. Thus,

$$V_{BB} = 2V_T \ln(I_Q/I_S)$$

$$= 2 \times 0.025 \ln(2 \times 10^{-3}/10^{-13}) = 1.186 \text{ V}$$

The easiest way to determine the transfer characteristic is to work backward; that is, for a given v_O we determine the corresponding value of v_I. We shall outline the process for positive v_O:

1. Assume a value for v_O.
2. Determine the load current i_L,

$$i_L = v_O/R_L$$

3. Use Eq. (11.27) to determine the current conducted by Q_N, i_N.
4. Determine v_{BEN} from

$$v_{BEN} = V_T \ln(i_N/I_S)$$

5. Determine v_I from

$$v_I = v_O + v_{BEN} - V_{BB}/2$$

Example 11.3 *continued*

It is also useful to find i_P and v_{EBP} as follows:

$$i_P = i_N - i_L$$

$$v_{EBP} = V_T \ln(i_P/I_S)$$

A similar process can be employed for negative v_O. However, symmetry can be utilized, obviating the need to repeat the calculations. The results obtained are displayed in the following table:

v_O (V)	i_L (mA)	i_N (mA)	i_P (mA)	v_{BEN} (V)	v_{EBP} (V)	v_I(V)	v_O/v_I	R_{out} (W)	v_o/v_i
+10.0	100	100.04	0.04	0.691	0.495	10.1	0.99	0.25	1.00
+5.0	50	50.08	0.08	0.673	0.513	5.08	0.98	0.50	1.00
+1.0	10	10.39	0.39	0.634	0.552	1.041	0.96	2.32	0.98
+0.5	5	5.70	0.70	0.619	0.567	0.526	0.95	4.03	0.96
+0.2	2	3.24	1.24	0.605	0.581	0.212	0.94	5.58	0.95
+0.1	1	2.56	1.56	0.599	0.587	0.106	0.94	6.07	0.94
0	0	2	2	0.593	0.593	0	—	6.25	0.94
−0.1	−1	1.56	2.56	0.587	0.599	−0.106	0.94	6.07	0.94
−0.2	−2	1.24	3.24	0.581	0.605	−0.212	0.94	5.58	0.95
−0.5	−5	0.70	5.70	0.567	0.619	−0.526	0.95	4.03	0.96
−1.0	−10	0.39	10.39	0.552	0.634	−1.041	0.96	2.32	0.98
−5.0	−50	0.08	50.08	0.513	0.673	−5.08	0.98	0.50	1.00
−10.0	−100	0.04	100.04	0.495	0.691	−10.1	0.99	0.25	1.00

The table also provides values for the dc gain v_O/v_I as well as the incremental gain v_o/v_i at the various values of v_O. The incremental gain is computed as follows

$$\frac{v_o}{v_i} = \frac{R_L}{R_L + R_{out}}$$

where R_{out} is the small-signal output resistance of the amplifier, given by Eq. (11.31). The incremental gain is the slope of the voltage transfer characteristic, and the magnitude of its variation over the range of v_O is an indication of the linearity of the output stage. Observe that for $0 \le |v_O| \le 10$ V, the incremental gain changes from 0.94 to 1.00, about 6%. Also observe as v_O becomes positive, Q_N supplies more and more of i_L and Q_P is correspondingly reduced. The opposite happens for negative v_O.

EXERCISE

11.5 To increase the linearity of the class AB output stage, the quiescent current I_Q is increased. The price paid is an increase in quiescent power dissipation. For the output stage considered in Example 11.3:
(a) Find the quiescent power dissipation.
(b) If I_Q is increased to 10 mA, find v_o/v_i at $v_O = 0$ and at $|v_O| = 10$ V, and hence the percentage change. Compare to the case in Example 11.3.
(c) Find the quiescent power dissipation for the case in (b).

Ans. (a) 60 mW; (b) 0.988 to 1.00; for a change of 1.2% compared to the 6% change in Example 11.3; (c) 300 mW

11.5 Biasing the Class AB Circuit

In this section we discuss two approaches for generating the voltage V_{BB} required for biasing the class AB output stage.

11.5.1 Biasing Using Diodes

Figure 11.14 shows a class AB circuit in which the bias voltage V_{BB} is generated by passing a constant current I_{BIAS} through a pair of diodes, or diode-connected transistors, D_1 and D_2. In circuits that supply large amounts of power, the output transistors are large-geometry devices. The biasing diodes, however, need not be large devices, and thus the quiescent current I_Q established in Q_N and Q_P will be $I_Q = nI_{\text{BIAS}}$, where n is the ratio of the emitter–junction area of the output devices to the junction area of the biasing diodes. In other words, the saturation (or scale) current I_S of the output transistors is n times that of the biasing diodes. Area ratioing is simple to implement in integrated circuits but difficult to realize in discrete-circuit designs.

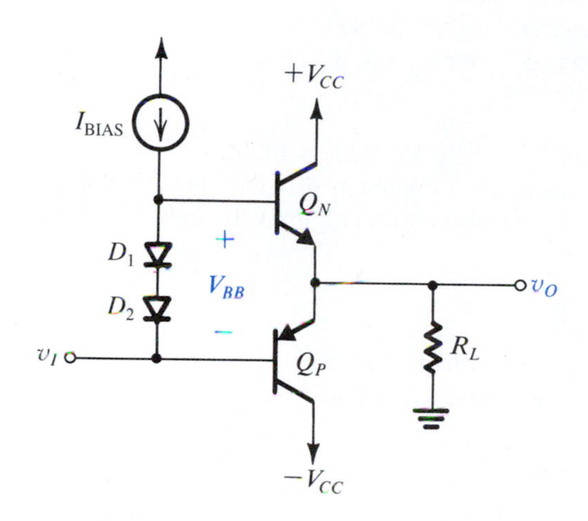

Figure 11.14 A class AB output stage utilizing diodes for biasing. If the junction area of the output devices, Q_N and Q_P, is n times that of the biasing devices D_1 and D_2, a quiescent current $I_Q = nI_{\text{BIAS}}$ flows in the output devices.

When the output stage of Fig. 11.14 is sourcing current to the load, the base current of Q_N increases from I_Q/β_N (which is usually small) to approximately i_L/β_N. This base current drive must be supplied by the current source I_{BIAS}. It follows that I_{BIAS} must be greater than the maximum anticipated base drive for Q_N. This sets a lower limit on the value of I_{BIAS}. Now, since $n = I_Q/I_{BIAS}$, and since I_Q is usually much smaller than the peak load current (<10%), we see that we cannot make n a large number. In other words, we cannot make the diodes much smaller than the output devices. This is a disadvantage of the diode biasing scheme.

From the discussion above we see that the current through the biasing diodes will decrease when the output stage is sourcing current to the load. Thus the bias voltage V_{BB} will also decrease, and the analysis of Section 11.4 must be modified to take this effect into account.

The diode biasing arrangement has an important advantage: It can provide thermal stabilization of the quiescent current in the output stage. To appreciate this point, recall that the class AB output stage dissipates power under quiescent conditions. Power dissipation raises the internal temperature of the BJTs. From Chapter 6 we know that a rise in transistor temperature results in a decrease in its V_{BE} (approximately –2 mV/°C) if the collector current is held constant. Alternatively, if V_{BE} is held constant and the temperature increases, the collector current increases. The increase in collector current increases the power dissipation, which in turn increases the junction temperature and hence, once more, the collector current. Thus a positive-feedback mechanism exists that can result in a phenomenon called **thermal runaway**. Unless checked, thermal runaway can lead to the ultimate destruction of the BJT. Diode biasing can be arranged to provide a compensating effect that can protect the output transistors against thermal runaway under quiescent conditions. Specifically, if the diodes are in close thermal contact with the output transistors, their temperature will increase by the same amount as that of Q_N and Q_P. Thus V_{BB} will decrease at the same rate as $V_{BEN} + V_{EBP}$, with the result that I_Q remains constant. Close thermal contact is easily achieved in IC fabrication. It is obtained in discrete circuits by mounting the bias diodes on the metal case of Q_N or Q_P.

Example 11.4

Consider the class AB output stage under the conditions that $V_{CC} = 15$ V, $R_L = 100\ \Omega$, and the output is sinusoidal with a maximum amplitude of 10 V. Let Q_N and Q_P be matched with $I_S = 10^{-13}$ A and $\beta = 50$. Assume that the biasing diodes have one-third the junction area of the output devices. Find the value of I_{BIAS} that guarantees a minimum of 1 mA through the diodes at all times. Determine the quiescent current and the quiescent power dissipation in the output transistors (i.e., at $v_O = 0$). Also find V_{BB} for $v_O = 0$, +10 V, and −10 V.

Solution

The maximum current through Q_N is approximately equal to $i_{L\max} = 10\text{ V}/0.1\text{ k}\Omega = 100$ mA. Thus the maximum base current in Q_N is approximately 2 mA. To maintain a minimum of 1 mA through the diodes, we select I_{BIAS} = 3 mA. The area ratio of 3 yields a quiescent current of 9 mA through Q_N and Q_P. The quiescent power dissipation is

$$P_{DQ} = 2 \times 15 \times 9 = 270 \text{ mW}$$

For $v_O = 0$, the base current of Q_N is $9/51 \simeq 0.18$ mA, leaving a current of $3 - 0.18 = 2.82$ mA to flow through the diodes. Since the diodes have $I_S = \frac{1}{3} \times 10^{-13}$ A, the voltage V_{BB} will be

$$V_{BB} = 2V_T \ln \frac{2.82 \text{ mA}}{I_S} = 1.26 \text{ V}$$

At $v_O = +10$ V, the current through the diodes will decrease to 1 mA, resulting in $V_{BB} \simeq 1.21$ V. At the other extreme of $v_O = -10$ V, Q_N will be conducting a very small current; thus its base current will be negligibly small and all of I_{BIAS} (3 mA) flows through the diodes, resulting in $V_{BB} \simeq 1.26$ V.

EXERCISES

11.6 For the circuit of Example 11.4, find i_N and i_P for $v_O = +10$ V and $v_O = -10$ V. (*Hint*: Use the V_{BB} values found in Example 11.4.)
Ans. 100.1 mA, 0.1 mA; 0.8 mA, 100.8 mA

11.7 If the collector current of a transistor is held constant, its v_{BE} decreases by 2 mV for every 1°C rise in temperature. Alternatively, if v_{BE} is held constant, then i_C increases by approximately $g_m \times 2$ mV for every 1°C rise in temperature. For a device operating at $I_C = 10$ mA, find the change in collector current resulting from an increase in temperature of 5°C.
Ans. 4 mA

11.5.2 Biasing Using the V_{BE} Multiplier

An alternative biasing arrangement that provides the designer with considerably more flexibility in both discrete and integrated designs is shown in Fig. 11.15. The bias circuit consists of transistor Q_1 with a resistor R_1 connected between base and emitter and a feedback resistor R_2 connected between collector and base. The resulting two-terminal network is fed with a constant-current source I_{BIAS}. If we neglect the base current of Q_1, then R_1 and R_2 will carry the same current I_R, given by

$$I_R = \frac{V_{BE1}}{R_1} \tag{11.32}$$

and the voltage V_{BB} across the bias network will be

$$\begin{aligned} V_{BB} &= I_R(R_1 + R_2) \\ &= V_{BE1}\left(1 + \frac{R_2}{R_1}\right) \end{aligned} \tag{11.33}$$

Thus the circuit simply multiplies V_{BE1} by the factor $(1 + R_2/R_1)$ and is known as the "V_{BE} multiplier." The multiplication factor is obviously under the designer's control and can be used to establish the value of V_{BB} required to yield a desired quiescent current I_Q. In IC design it is relatively easy to control accurately the ratio of two resistances. In discrete-circuit design, a potentiometer can be used, as shown in Fig. 11.16, and is manually set to produce the desired value of I_Q.

The value of V_{BE1} in Eq. (11.33) is determined by the portion of I_{BIAS} that flows through the collector of Q_1; that is,

$$I_{C1} = I_{\text{BIAS}} - I_R \tag{11.34}$$

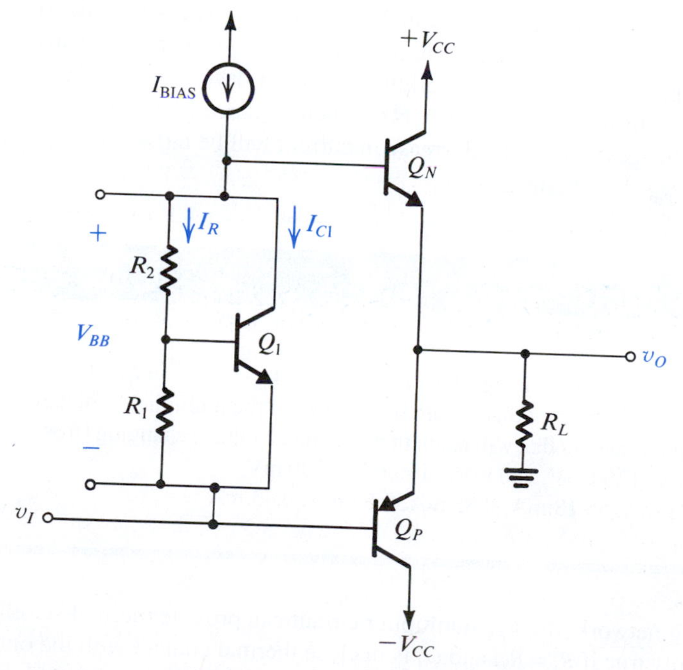

Figure 11.15 A class AB output stage utilizing a V_{BE} multiplier for biasing.

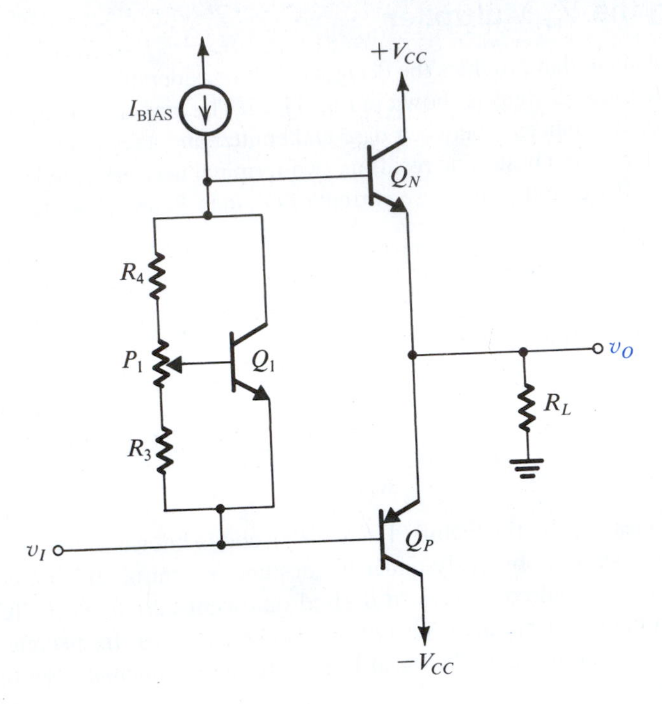

Figure 11.16 A discrete-circuit class AB output stage with a potentiometer used in the V_{BE} multiplier. The potentiometer is adjusted to yield the desired value of quiescent current in Q_N and Q_P.

$$V_{BE1} = V_T \ln \frac{I_{C1}}{I_{S1}} \tag{11.35}$$

where we have neglected the base current of Q_N, which is normally small both under quiescent conditions and when the output voltage is swinging negative. However, for positive v_O, especially at and near its peak value, the base current of Q_N can become sizable and will reduce the current available for the V_{BE} multiplier. Nevertheless, since large changes in I_{C1} correspond to only small changes in V_{BE1}, the decrease in current will be mostly absorbed by Q_1, leaving I_R, and hence V_{BB}, almost constant.

EXERCISE

11.8 Consider a V_{BE} multiplier with $R_1 = R_2 = 1.2\ \text{k}\Omega$, utilizing a transistor that has $V_{BE} = 0.6$ V at $I_C = 1$ mA, and a very high β. (a) Find the value of the current I that should be supplied to the multiplier to obtain a terminal voltage of 1.2 V. (b) Find the value of I that will result in the terminal voltage changing (from the 1.2-V value) by +50 mV, +100 mV, +200 mV, –50 mV, –100 mV, –200 mV.
Ans. (a) 1.5 mA; (b) 3.24 mA, 7.93 mA, 55.18 mA, 0.85 mA, 0.59 mA, 0.43 mA

Like the diode biasing network, the V_{BE}–multiplier circuit can provide thermal stabilization of I_Q. This is especially true if $R_1 = R_2$, and Q_1 is in close thermal contact with the output transistors.

Example 11.5

It is required to redesign the output stage of Example 11.4 utilizing a V_{BE} multiplier for biasing. Use a small-geometry transistor for Q_1 with $I_S = 10^{-14}$ A and design for a quiescent current $I_Q = 2$ mA.

Solution

Since the peak positive current is 100 mA, the base current of Q_N can be as high as 2 mA. We shall therefore select $I_{BIAS} = 3$ mA, thus providing the multiplier with a minimum current of 1 mA.

Under quiescent conditions ($v_O = 0$ and $i_L = 0$) the base current of Q_N can be neglected and all of I_{BIAS} flows through the multiplier. We now must decide on how this current (3 mA) is to be divided between I_{C1} and I_R. If we select I_R greater than 1 mA, the transistor will be almost cut off at the positive peak of v_O. Therefore, we shall select $I_R = 0.5$ mA, leaving 2.5 mA for I_{C1}.

To obtain a quiescent current of 2 mA in the output transistors, V_{BB} should be

$$V_{BB} = 2V_T \ln \frac{2 \times 10^{-3}}{10^{-13}} = 1.19 \text{ V}$$

We can now determine $R_1 + R_2$ as follows:

$$R_1 + R_2 = \frac{V_{BB}}{I_R} = \frac{1.19}{0.5} = 2.38 \text{ k}\Omega$$

At a collector current of 2.5 mA, Q_1 has

$$V_{BE1} = V_T \ln \frac{2.5 \times 10^{-3}}{10^{-14}} = 0.66 \text{ V}$$

The value of R_1 can now be determined as

$$R_1 = \frac{0.66}{0.5} = 1.32 \text{ k}\Omega$$

and R_2 as

$$R_2 = 2.38 - 1.32 = 1.06 \text{ k}\Omega$$

11.6 CMOS Class AB Output Stages

In this section we study CMOS class AB output stages. We begin with the CMOS counterpart of the BJT class AB output stage studied in the previous section. As we shall see, this circuit suffers from a relatively low output signal-swing, a serious limitation especially in view of the shrinking power-supply voltages characteristic of modern deep-submicron CMOS technologies. We will then look at an attractive alternative circuit that overcomes this problem.

11.6.1 The Classical Configuration

Figure 11.17 shows the classical CMOS class AB output stage. The circuit is the exact counterpart of the bipolar circuit shown in Fig. 11.14 with the biasing diodes implemented with diode-connected transistors Q_1 and Q_2. The constant current I_{BIAS} flowing through Q_1 and Q_2 establishes a dc bias voltage V_{GG} between the gates of Q_N and Q_P. This voltage in turn

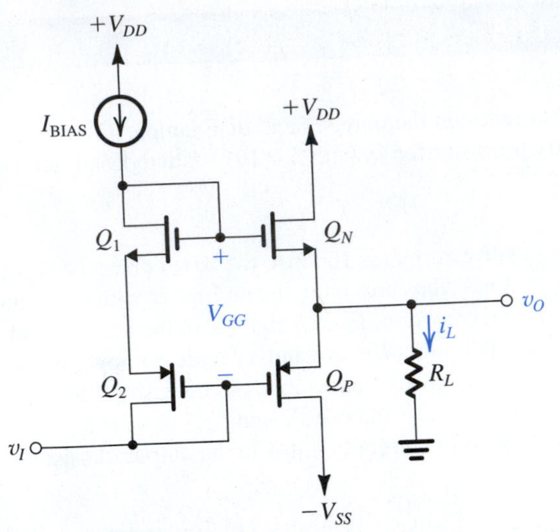

Figure 11.17 Classical CMOS class AB output stage. This circuit is the CMOS counterpart of the BJT circuit in Fig. 11.14 with the biasing diodes implemented with diode-connected MOSFETs, Q_1 and Q_2.

establishes the quiescent ($v_O = 0$) current I_Q in Q_N and Q_P. Unlike the BJT circuit in Fig. 11.14, here the zero dc gate current of Q_N results in the current through Q_1 and Q_2 remaining constant at I_{BIAS} irrespective of the value of v_O and the load current i_L. Thus V_{GG} remains constant and the circuit is more like the idealized bipolar case shown in Fig. 11.11.

The value of I_Q can be determined by utilizing the $i_D - v_{GS}$ equations for the four MOS transistors for the case $v_O = 0$. Neglecting channel-length modulation, we can write for Q_1,

$$I_{D1} = I_{BIAS} = \frac{1}{2}k_n'(W/L)_1(V_{GS1} - V_{tn})^2 \tag{11.36}$$

and for Q_2,

$$I_{D2} = I_{BIAS} = \frac{1}{2}k_p'(W/L)_2(V_{SG2} - |V_{tp}|)^2 \tag{11.37}$$

Equations (11.36) and (11.37) can be used to find V_{GS1} and V_{SG2}, which when summed yield V_{GG}; thus,

$$V_{GG} = V_{GS1} + V_{SG2} = V_{tn} + |V_{tp}| + \sqrt{2I_{BIAS}}\left(\frac{1}{\sqrt{k_n'(W/L)_1}} + \frac{1}{\sqrt{k_p'(W/L)_2}}\right) \tag{11.38}$$

We can follow a similar process for Q_N and Q_P which, for $v_O = 0$, are conducting the quiescent current I_Q; thus,

$$V_{GG} = V_{GSN} + V_{SGP} = V_{tn} + |V_{tp}| + \sqrt{2I_Q}\left(\frac{1}{\sqrt{k_n'(W/L)_n}} + \frac{1}{\sqrt{k_p'(W/L)_p}}\right) \tag{11.39}$$

Equations (11.38) and (11.39) can be combined to obtain

$$I_Q = I_{BIAS}\left[\frac{1/\sqrt{k_n'(W/L)_1} + 1/\sqrt{k_p'(W/L)_2}}{1/\sqrt{k_n'(W/L)_n} + 1/\sqrt{k_p'(W/L)_p}}\right]^2 \tag{11.40}$$

which indicates that I_Q is determined by I_{BIAS} together with the (W/L) ratios of the four transistors. For the case Q_1 and Q_2 are matched, that is,

$$k_p'(W/L)_2 = k_n'(W/L)_1 \tag{11.41}$$

and Q_N and Q_P are matched, that is,

$$k_p'(W/L)_p = k_n'(W/L)_n \tag{11.42}$$

Equation (11.40) simplifies to

$$I_Q = I_{\text{BIAS}} \frac{(W/L)_n}{(W/L)_1} \tag{11.43}$$

which is an intuitively appealing result.

EXERCISE

11.9 For the CMOS class AB output stage of Fig. 11.17, consider the case of matched Q_1 and Q_2, and matched Q_N and Q_P. If $I_Q = 1$ mA and $I_{\text{BIAS}} = 0.2$ mA, find (W/L) for each of Q_1, Q_2, Q_N, and Q_P so that in the quiescent state each transistor operates at an overdrive voltage of 0.2 V. Let $V_{DD} = V_{SS} = 2.5$ V, $k_n' = 250\ \mu\text{A/V}^2$, $k_p' = 100\ \mu\text{A/V}^2$, and $V_{tn} = -V_{tp} = 0.5$ V. Also find V_{GG}.

Ans. 40; 100; 200; 500; 1.4 V

A drawback of the CMOS class AB circuit of Fig. 11.17 is the restricted range of output voltage swing. To find the maximum possible value of v_O, refer to Fig. 11.17 and assume that across the bias current source is a dc voltage of V_{BIAS}. We can write for v_O,

$$v_O = V_{DD} - V_{\text{BIAS}} - v_{GSN} \tag{11.44}$$

The maximum value of v_O will be limited by the need to keep V_{BIAS} to a minimum of V_{OV} of the transistor supplying I_{BIAS} (otherwise the current-source transistor no longer operates in saturation); thus,

$$v_{O\max} = V_{DD} - V_{OV}|_{\text{BIAS}} - v_{GSN} \tag{11.45}$$

Note that when v_O is at its maximum value, Q_N will be supplying most or all of i_L, and v_{GSN} will be large,

$$v_{O\max} = V_{DD} - V_{OV}|_{\text{BIAS}} - V_{tn} - v_{OVN} \tag{11.46}$$

where v_{OVN} is the overdrive voltage of Q_N when it is supplying $i_{L\max}$.

EXERCISE

11.10 For the circuit specified in Exercise 11.9, find $v_{O\max}$ when $i_{L\max} = 10$ mA. Assume that Q_N is supplying all of $i_{L\max}$ and that $V_{OV}|_{\text{BIAS}} = 0.2$ V.
Ans. 1.17 V

The minimum allowed value of v_O can be found in a similar way. Here we note that the transistor supplying v_I (not shown) will need a minimum voltage across it of $V_{OV}|_I$. Thus,

$$v_{O\min} = -V_{SS} + V_{OV}|_I + |V_{tp}| + |v_{OVP}| \tag{11.47}$$

where $|v_{OVP}|$ is the overdrive voltage of Q_P when sinking the maximum negative value of i_L.

Finally, we observe that the reason for the lower allowable range of v_O in the CMOS circuit is the relatively large value of v_{OVN} and $|v_{OVP}|$; that is, the large values of v_{GSN} and v_{SGP} required to supply the large output currents. In the BJT circuit the corresponding voltages, v_{BEN} and v_{EBP}, remain close to 0.7 V. The overdrive voltages v_{OVN} and $|v_{OVP}|$ can be reduced by making the W/L ratios of Q_N and Q_P large. This, however, can lead to impractically large devices.

11.6.2 An Alternative Circuit Utilizing Common-Source Transistors

The allowable range of v_O can be increased by replacing the source followers with a pair of complementary transistors connected in the common-source configuration, as shown in Fig. 11.18. Here Q_P supplies the load current when v_O is positive and allows v_O to go as high as $(V_{DD} - |v_{OVP}|)$, a much higher value than that given by Eq. (11.46). For negative v_O, Q_N sinks the load current and allows v_O to go as low as $-V_{SS} + v_{OVN}$. This also is larger in magnitude than the value given by Eq. (11.47). Thus, the circuit of Fig. 11.18 provides an output voltage range that is within an overdrive voltage of each of the supplies. The disadvantage of the circuit, however, is its high output resistance,

$$R_{\text{out}} = r_{on} \| r_{op} \tag{11.48}$$

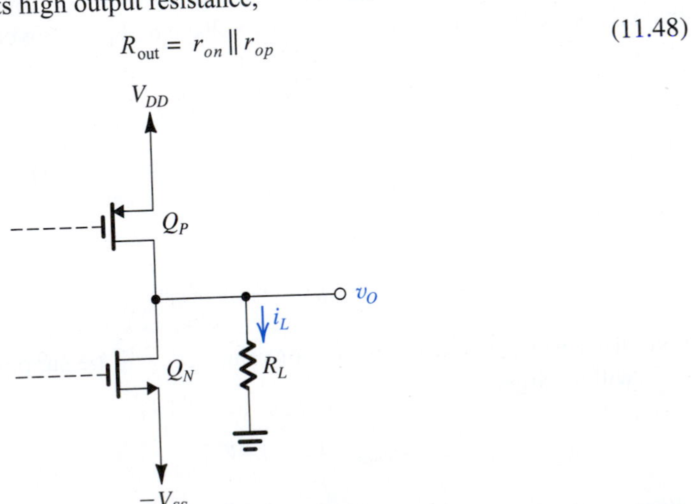

Figure 11.18 An alternative CMOS output stage utilizing a pair of complementary MOSFETs connected in the common-source configuration. The driving circuit is not shown.

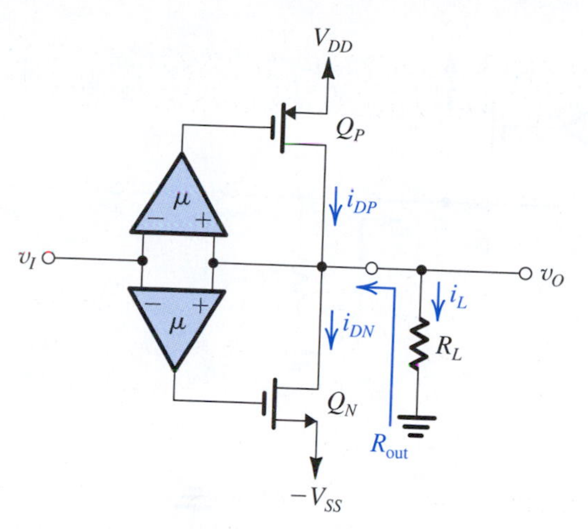

Figure 11.19 Inserting an amplifier in the negative feedback path of each of Q_N and Q_P reduces the output resistance and makes $v_O \simeq v_I$; both are desirable properties for the output stage.

To reduce the output resistance, negative feedback is employed as shown in Fig. 11.19. Here an amplifier with gain μ is inserted between drain and gate of each of Q_N and Q_P. For reasons that will become clear shortly, these amplifiers are called **error amplifiers**. To verify that the feedback around each amplifier is negative, assume that v_O increases. The top amplifier will cause the gate voltage of Q_P to increase, thus its v_{SG} decreases and i_{DP} decreases. The decrease in i_{DP} causes v_O to decrease, which is opposite to the initially assumed change, thus verifying that the feedback is negative. A similar process can be used to verify that the feedback around the bottom amplifier also is negative.

From our study of feedback in Chapter 10, we observe that each of the two feedback loops is of the series-shunt type, which is the topology appropriate for a voltage amplifier. Thus, as we shall show shortly, the feedback will reduce the output resistance of the amplifier. Also, observe that if the loop gain is large, the voltage difference between the two input terminals of each feedback amplifier, the error voltage, will be small, resulting in $v_O \simeq v_I$.

Both the low output resistance and the near-unity dc gain are highly desirable properties for an output stage.

Output Resistance To derive an expression for the output resistance R_{out}, we consider each half of the circuit separately, find its output resistance, R_{outp} for the top half and R_{outn} for the bottom half, and then obtain the overall output resistance as the parallel equivalent of the two resistances,

$$R_{out} = R_{outn} \| R_{outp} \tag{11.49}$$

Figure 11.20(a) shows the top half of the circuit, drawn a little differently to make the feedback topology clearer. Observe that feedback is applied by connecting the output back to the input. Thus the feedback network is the two-port shown in Fig. 11.20(b) and the feedback factor is

$$\beta = 1 \tag{11.50}$$

Including the loading effects of the feedback network results in the A circuit shown in Fig. 11.20(c). Note that since we are now interested in incremental quantities, we have

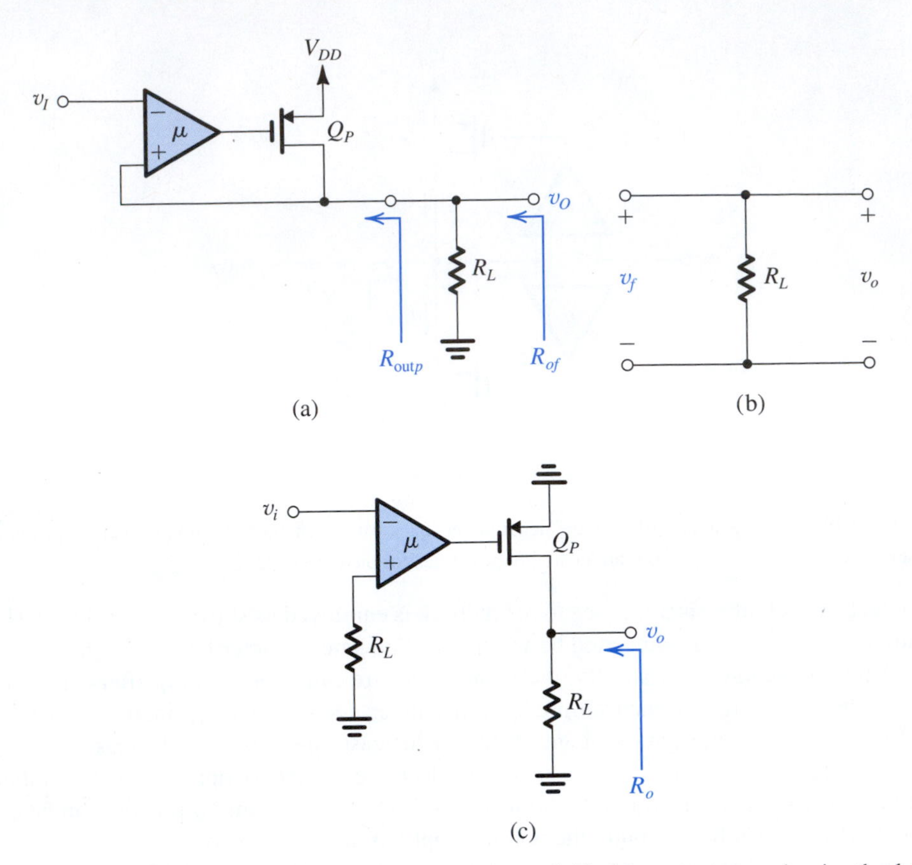

Figure 11.20 Determining the output resistance; **(a)** The top half of the output stage showing the definition of R_{outp} and R_{of}; **(b)** The β circuit; and **(c)** the A circuit.

replaced V_{DD} with a short circuit to ground. The open-loop gain A can be found from the circuit in Fig. 11.20(c) as

$$A \equiv \frac{v_o}{v_i} = \mu g_{mp}(r_{op} \| R_L) \tag{11.51}$$

where we have assumed the input resistance of the amplifier to be infinite and thus resistance R_L at the input has no effect on the gain, and we have utilized implicitly the small-signal model of Q_P. The values of the small-signal parameters g_{mp} and r_{op} are to be evaluated at the current at which Q_P is operating. The open-loop output resistance R_o is found by inspection as

$$R_o = R_L \| r_{op} \tag{11.52}$$

The output resistance with feedback R_{of} can now be found as

$$R_{of} = \frac{R_o}{1 + A\beta} = \frac{(R_L \| r_{op})}{1 + \mu g_{mp}(r_{op} \| R_L)} \tag{11.53}$$

and the output resistance R_{outp} is found by excluding R_L from R_{of}, that is

$$R_{outp} = 1 \Big/ \left(\frac{1}{R_{of}} - \frac{1}{R_L}\right) \tag{11.54}$$

which results in

$$R_{\text{outp}} = r_{op} \left\| \frac{1}{\mu g_{mp}} \simeq \frac{1}{\mu g_{mp}} \right. \tag{11.55}$$

which can be quite low. A similar development applied to the bottom half of the circuit in Fig. 11.19 results in

$$R_{\text{outn}} \simeq 1/\mu g_{mn} \tag{11.56}$$

Combining Eqs. (11.55) and (11.56) gives

$$R_{\text{out}} \simeq 1/\mu(g_{mp} + g_{mn}) \tag{11.57}$$

The Voltage Transfer Characteristic Next we derive an expression for the voltage transfer characteristic, v_O versus v_I, of the class AB common-source buffer. Toward that end, we first consider the circuit in the quiescent state, shown in Fig. 11.21(a). Here $v_I = 0$ and $v_O = 0$. Each of the error amplifiers is designed to deliver to the gate of its associated MOSFET the dc voltage required to establish the desired value of quiescent current I_Q. To obtain class AB operation, I_Q is usually selected to be 10% or so of the maximum output current. Referring to Fig. 11.21(a), we can write for Q_P,

$$I_{DP} = I_Q = \frac{1}{2}k_p'\left(\frac{W}{L}\right)_p (V_{SGP} - |V_{tp}|)^2$$

Substituting $V_{SGP} = |V_{tp}| + V_{OV}$, where V_{OV} is the magnitude of the quiescent overdrive voltage of Q_P, gives

$$I_Q = \frac{1}{2}k_p'\left(\frac{W}{L}\right)_p V_{OV}^2 \tag{11.58}$$

Similarly, we obtain for Q_N

$$I_Q = \frac{1}{2}k_n'\left(\frac{W}{L}\right)_n V_{OV}^2 \tag{11.59}$$

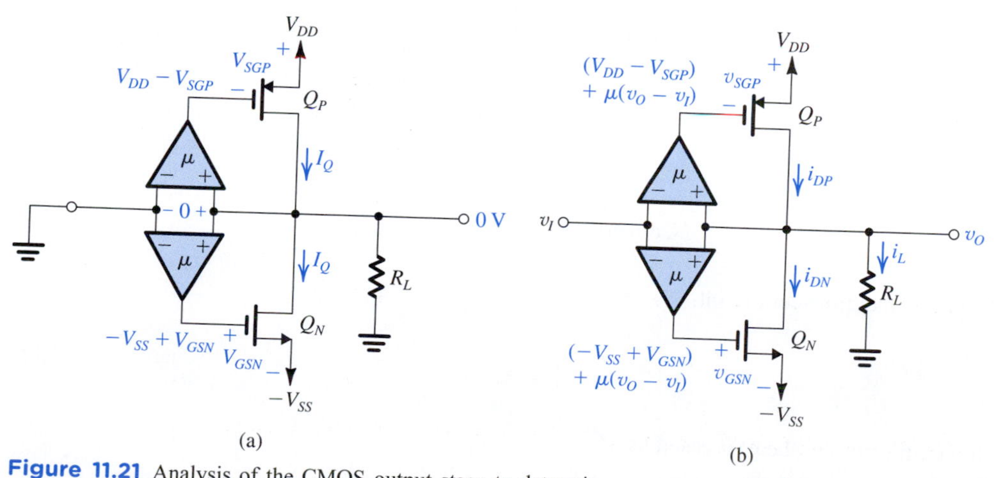

Figure 11.21 Analysis of the CMOS output stage to determine v_O versus v_I: **(a)** Quiescent conditions; **(b)** The situation with v_I applied.

Usually the two transistors are matched,

$$k_p'\left(\frac{W}{L}\right)_p = k_n'\left(\frac{W}{L}\right)_n = k$$

Thus,

$$I_Q = \frac{1}{2}kV_{OV}^2 \tag{11.60}$$

Next consider the situation with v_I applied, illustrated in Fig. 11.21(b). The voltage at the output of each of the error amplifiers increases by $\mu(v_O - v_I)$. Thus v_{SGP} decreases by $\mu(v_O - v_I)$ and v_{GSN} increases by $\mu(v_O - v_I)$, and we can write

$$\begin{aligned} i_{DP} &= \frac{1}{2}k[V_{OV} - \mu(v_O - v_I)]^2 \\ &= \frac{1}{2}kV_{OV}^2\left[1 - \mu\,\frac{v_O - v_I}{V_{OV}}\right]^2 \\ &= I_Q\left(1 - \mu\,\frac{v_O - v_I}{V_{OV}}\right)^2 \end{aligned} \tag{11.61}$$

and

$$i_{DN} = I_Q\left(1 + \mu\,\frac{v_O - v_I}{V_{OV}}\right)^2 \tag{11.62}$$

At the output node we have

$$i_L = i_{DP} - i_{DN} \tag{11.63}$$

Substituting for $i_L = v_O/R_L$ and for i_{DP} and i_{DN} from Eqs. (11.61) and (11.62), and solving the resulting equation to obtain v_O, results in

$$v_O = \frac{v_I}{1 + \dfrac{V_{OV}}{4\mu I_Q R_L}} \tag{11.64}$$

Usually $(V_{OV}/4\mu I_Q R_L) \ll 1$, enabling us to express v_O as

$$v_O \simeq v_I\left(1 - \frac{V_{OV}}{4\mu I_Q R_L}\right) \tag{11.65}$$

Thus the gain error is

$$\text{Gain error} \equiv v_O - v_I = -\frac{V_{OV}}{4\mu I_Q R_L} \tag{11.66}$$

Since at the quiescent point,

$$g_{mp} = g_m = \frac{2I_Q}{V_{OV}} \tag{11.67}$$

the gain error can be expressed as

$$\text{Gain error} = -\frac{1}{2\mu g_m R_L} \tag{11.68}$$

Thus selecting a large value for μ results in reducing both the gain error and the output resistance. However, a large μ can make the quiescent current I_Q too dependent on the input offset voltages that are inevitably present in the error amplifiers. Typically, μ is selected in the range 5 to 10. Trade-offs are also present in the selection of I_Q: A large I_Q reduces crossover distortion, R_{out}, and gain error, at the expense of increased quiescent power dissipation.

Example 11.6

In this example we explore the design and operation of a class AB common-source output stage of the type shown in Fig. 11.19, required to operate from a ±2.5-V power supply to feed a load resistance $R_L = 100\ \Omega$. The transistors available have $V_{tn} = -V_{tp} = 0.5$ V and $k_n' = 2.5k_p' = 250\ \mu\text{A/V}^2$. The gain error is required to be less than 2.5% and $I_Q = 1$ mA.

Solution

The gain error is given by Eq. (11.66),

$$\text{Gain error} = -\frac{V_{OV}}{4\mu I_Q R_L}$$

We are given the required maximum gain error of -0.025, $I_Q = 1$ mA, and $R_L = 100\ \Omega$. In order to keep μ low and also obtain as high a g_m as possible [$g_m = 2I_Q/V_{OV}$], we select V_{OV} to be as low as possible. Practically speaking, V_{OV} is usually 0.1 V to 0.2 V. Selecting $V_{OV} = 0.1$ V results in

$$0.025 = \frac{0.1}{4\times\mu\times 1\times 10^{-3}\times 100}$$

which yields

$$\mu = 10$$

which is within the typically recommended range.

Figure 11.22(a) shows the circuit in the quiescent state with the various dc voltages and currents indicated. The required (W/L) ratios of Q_N and Q_P can be found as follows:

$$I_Q = \frac{1}{2}k_p'\left(\frac{W}{L}\right)_p V_{OV}^2$$

$$1\times 10^{-3} = \frac{1}{2}\times 0.1\times 10^{-3}\left(\frac{W}{L}\right)_p \times (0.1)^2$$

Thus,

$$\left(\frac{W}{L}\right)_p = 2000$$

$$\left(\frac{W}{L}\right)_n = \frac{(W/L)_p}{k_n'/k_p'} = \frac{2000}{2.5} = 800$$

Thus Q_N and Q_P are very large transistors, not an unusual situation in a high-power output stage.

To obtain the output resistance at the quiescent point, we use Eq. (11.57),

$$R_{out} = \frac{1}{\mu(g_{mp}+g_{mn})}$$

Example 11.6 *continued*

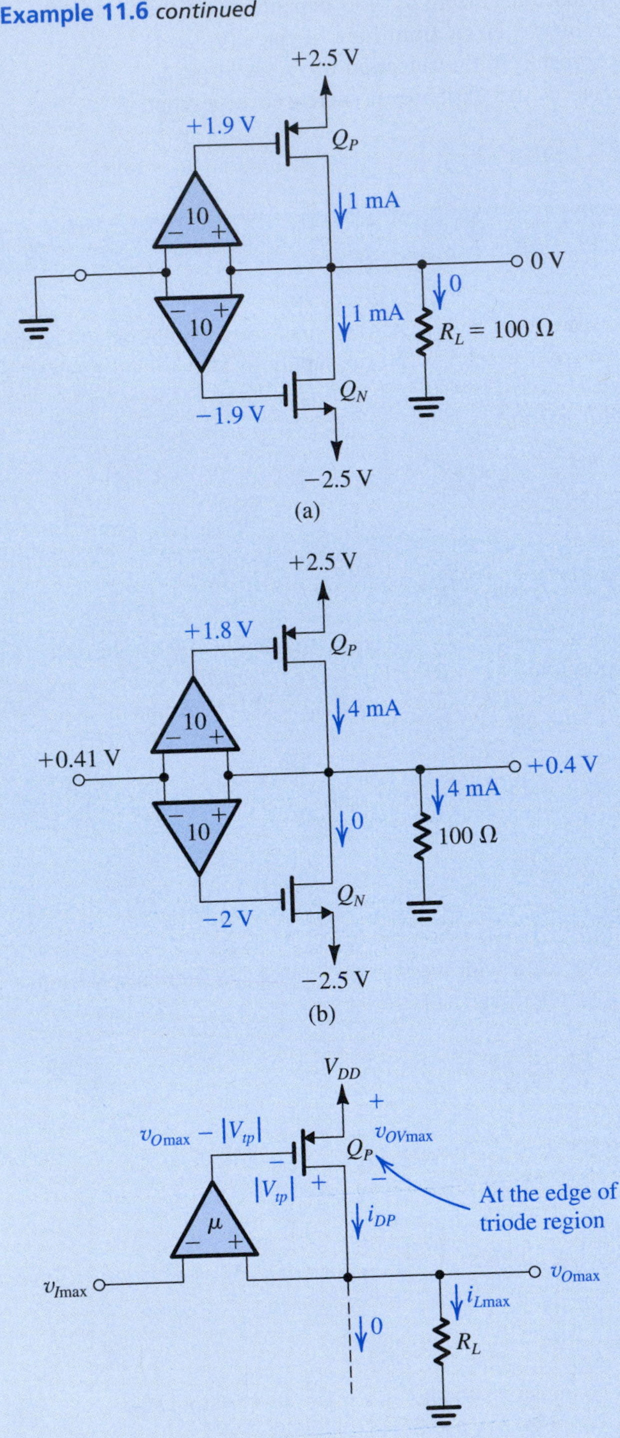

Figure 11.22 **(a)** Circuit in the quiescent state; **(b)** circuit at the point at which Q_N turns off; **(c)** conditions at $v_O = v_{Omax}$.

where

$$g_{mp} = g_{mn} = \frac{2I_Q}{V_{OV}} = \frac{2 \times 1}{0.1} = 20 \text{ mA/V}$$

Thus

$$R_{out} = \frac{1}{10(0.02 + 0.02)} = 2.5 \ \Omega$$

Next we wish to determine the maximum and minimum allowed values of v_O. Since the circuit is symmetrical, we need to consider only either the positive-output or negative-output case. For v_O positive, Q_P conducts more of the output current i_L. Eventually, Q_N turns off and Q_P conducts all of i_L. To find the value of v_O at which this occurs, note that Q_N turns off when the voltage at its gate drops from the quiescent value of −1.9 V (see Fig. 11.22a) to −2 V, at which point $v_{GSN} = V_{tn}$. An equal change of −0.1 V appears at the output of the top amplifier, as shown in Fig. 11.22(b). Analysis of the circuit in Fig. 11.22(b) shows that

$$i_L = i_{DP} = 4 \text{ mA}$$

$$v_O = i_L R_L = 4 \times 10^{-3} \times 100 = 0.4 \text{ V}$$

For $v_O > 0.4$ V, Q_P must conduct all the current i_L. The situation at $v_O = v_{O\max}$ is illustrated in Fig. 11.22(c). Analysis of this circuit results, after some straightforward but tedious manipulations, in

$$v_{O\max} \simeq 2.05 \text{ V}$$

and

$$i_{L\max} = 20.5 \text{ mA}$$

EXERCISES

11.11 Suppose it is required to reduce the size of Q_N and Q_P in the circuit considered in the above example by a factor of 2 while keeping I_Q at 1 mA. What value should be used for V_{OV}? What is the new value for the gain error and for R_{out} at the quiescent point?
Ans. 0.14 V; −3.5 %; 3.5 Ω

11.12 Show that in the CMOS class AB common-source output stage, Q_N turns off when $v_O = 4I_QR_L$ and that Q_P turns off when $v_O = -4I_QR_L$. This is equivalent to saying that one of the transistors turns off when $|i_L|$ reaches $4I_Q$.

11.7 Power BJTs

Transistors that are required to conduct currents in the ampere range and to withstand power dissipation in the watts and tens-of-watts ranges differ in their physical structure, packaging, and specification from the small-signal transistors considered in earlier chapters. In this section we consider some of the important properties of power transistors, especially those aspects that pertain to the design of circuits of the type discussed earlier. There are, of course, other important applications of power transistors, such as their use as switching elements in power inverters and motor-control circuits. Such applications are not studied in this book.

11.7.1 Junction Temperature

Power transistors dissipate large amounts of power in their collector–base junctions. The dissipated power is converted into heat, which raises the junction temperature. However, the junction temperature T_J must not be allowed to exceed a specified maximum, $T_{J\max}$; otherwise the transistor could suffer permanent damage. For silicon devices, $T_{J\max}$ is in the range of 150°C to 200°C.

11.7.2 Thermal Resistance

Consider first the situation of a transistor operating in free air—that is, with no special arrangements for cooling. The heat dissipated in the transistor junction will be conducted away from the junction to the transistor case, and from the case to the surrounding environment. In a steady state in which the transistor is dissipating P_D watts, the temperature rise of the junction relative to the surrounding ambience can be expressed as

$$T_J - T_A = \theta_{JA} P_D \tag{11.69}$$

where θ_{JA} is the **thermal resistance** between junction and ambience, having the units of degrees Celsius per watt. Note that θ_{JA} simply gives the rise in junction temperature over the ambient temperature for each watt of dissipated power. Since we wish to be able to dissipate large amounts of power without raising the junction temperature above $T_{J\max}$, it is desirable to have, for the thermal resistance θ_{JA}, as small a value as possible. For operation in free air, θ_{JA} depends primarily on the type of case in which the transistor is packaged. The value of θ_{JA} is usually specified on the transistor data sheet.

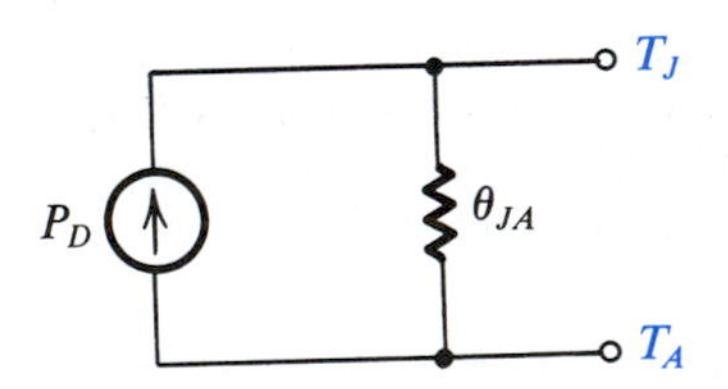

Figure 11.23 Electrical equivalent circuit of the thermal-conduction process; $T_J - T_A = P_D\theta_{JA}$.

Equation (11.69), which describes the thermal-conduction process, is analogous to Ohm's law, which describes the electrical-conduction process. In this analogy, power dissipation corresponds to current, temperature difference corresponds to voltage difference, and thermal resistance corresponds to electrical resistance. Thus, we may represent the thermal-conduction process by the electric circuit shown in Fig. 11.23.

11.7.3 Power Dissipation Versus Temperature

The transistor manufacturer usually specifies the maximum junction temperature $T_{J\max}$, the maximum power dissipation at a particular ambient temperature T_{A0} (usually, 25°C), and the

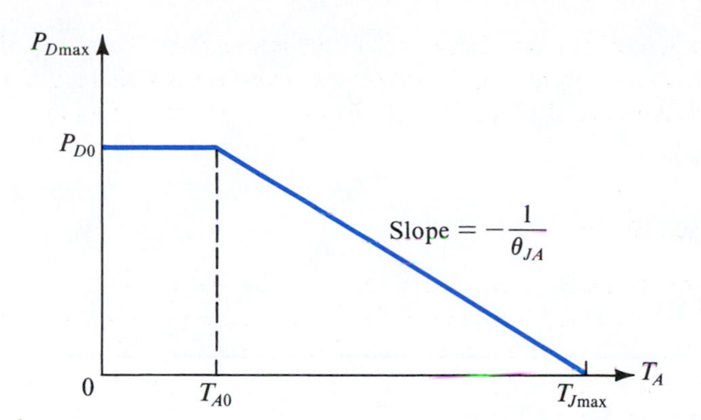

Figure 11.24 Maximum allowable power dissipation versus ambient temperature for a BJT operated in free air. This is known as a "power-derating" curve.

thermal resistance θ_{JA}. In addition, a graph such as that shown in Fig. 11.24 is usually provided. The graph simply states that for operation at ambient temperatures below T_{A0}, the device can safely dissipate the rated value of P_{D0} watts. However, if the device is to be operated at higher ambient temperatures, the maximum allowable power dissipation must be **derated** according to the straight line shown in Fig. 11.24. The **power-derating curve** is a graphical representation of Eq. (11.69). Specifically, note that if the ambient temperature is T_{A0} and the power dissipation is at the maximum allowed (P_{D0}), then the junction temperature will be $T_{J\max}$. Substituting these quantities in Eq. (11.69) results in

$$\theta_{JA} = \frac{T_{J\max} - T_{A0}}{P_{D0}} \tag{11.70}$$

which is the inverse of the slope of the power-derating straight line. At an ambient temperature T_A, higher than T_{A0}, the maximum allowable power dissipation $P_{D\max}$ can be obtained from Eq. (11.69) by substituting $T_J = T_{J\max}$; thus,

$$P_{D\max} = \frac{T_{J\max} - T_A}{\theta_{JA}} \tag{11.71}$$

Observe that as T_A approaches $T_{J\max}$, the allowable power dissipation decreases; the lower thermal gradient limits the amount of heat that can be removed from the junction. In the extreme situation of $T_A = T_{J\max}$, no power can be dissipated because no heat can be removed from the junction.

Example 11.7

A BJT is specified to have a maximum power dissipation P_{D0} of 2 W at an ambient temperature T_{A0} of 25°C, and a maximum junction temperature $T_{J\max}$ of 150°C. Find the following:

(a) The thermal resistance θ_{JA}.
(b) The maximum power that can be safely dissipated at an ambient temperature of 50°C.
(c) The junction temperature if the device is operating at $T_A = 25$°C and is dissipating 1 W.

Example 11.7 *continued*

Solution

(a) $\theta_{JA} = \dfrac{T_{J\max} - T_{A0}}{P_{D0}} = \dfrac{150 - 25}{2} = 62.5°\text{C/W}$

(b) $P_{D\max} = \dfrac{T_{J\max} - T_A}{\theta_{JA}} = \dfrac{150 - 50}{62.5} = 1.6 \text{ W}$

(c) $T_J = T_A + \theta_{JA} P_D = 25 + 62.5 \times 1 = 87.5°\text{C}$

11.7.4 Transistor Case and Heat Sink

The thermal resistance between junction and ambience, θ_{JA}, can be expressed as

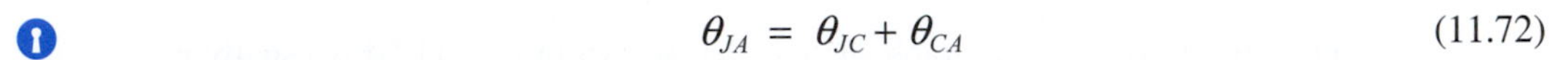

$$\theta_{JA} = \theta_{JC} + \theta_{CA} \tag{11.72}$$

where θ_{JC} is the thermal resistance between junction and transistor case (package) and θ_{CA} is the thermal resistance between case and ambience. For a given transistor, θ_{JC} is fixed by the device design and packaging. The device manufacturer can reduce θ_{JC} by encapsulating the device in a relatively large metal case and placing the collector (where most of the heat is dissipated) in direct contact with the case. Most high-power transistors are packaged in this fashion. Figure 11.25 shows a sketch of a typical package.

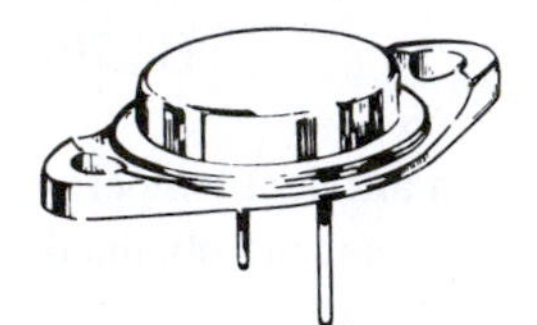

Figure 11.25 The popular TO3 package for power transistors. The case is metal with a diameter of about 2.2 cm; the outside dimension of the "seating plane" is about 4 cm. The seating plane has two holes for screws to bolt it to a heat sink. The collector is electrically connected to the case. Therefore an electrically insulating but thermally conducting spacer is used between the transistor case and the "heat sink."

Although the circuit designer has no control over θ_{JC} (once a particular transistor has been selected), the designer can considerably reduce θ_{CA} below its free-air value (specified by the manufacturer as part of θ_{JA}). Reduction of θ_{CA} can be effected by providing means to facilitate heat transfer from case to ambience. A popular approach is to bolt the transistor to the chassis or to an extended metal surface. Such a metal surface then functions as a **heat sink**. Heat is easily conducted from the transistor case to the heat sink; that is, the thermal resistance θ_{CS} is usually very small. Also, heat is efficiently transferred (by convection and radiation) from the heat sink to the ambience, resulting in a low thermal resistance θ_{SA}. Thus, if a heat sink is utilized, the case-to-ambience thermal resistance given by

$$\theta_{CA} = \theta_{CS} + \theta_{SA} \tag{11.73}$$

can be small because its two components can be made small by the choice of an appropriate heat sink.[2] For example, in very high-power applications the heat sink is usually equipped with fins that further facilitate cooling by radiation and convection.

[2]As noted earlier, the metal case of a power transistor is electrically connected to the collector. Thus an electrically insulating material such as mica is usually placed between the metal case and the metal heat sink. Also, insulating bushings and washers are generally used in bolting the transistor to the heat sink.

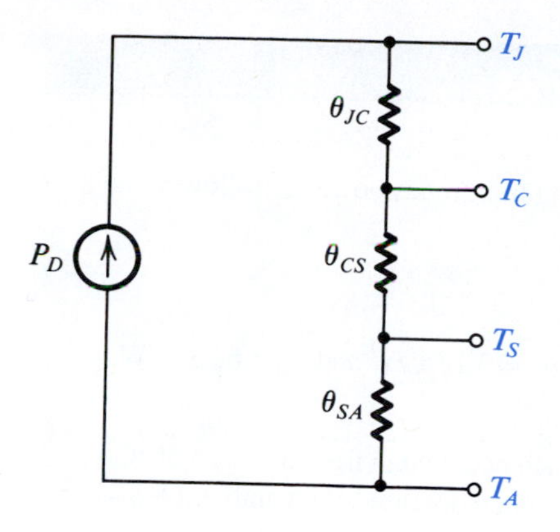

Figure 11.26 Electrical analog of the thermal conduction process when a heat sink is utilized.

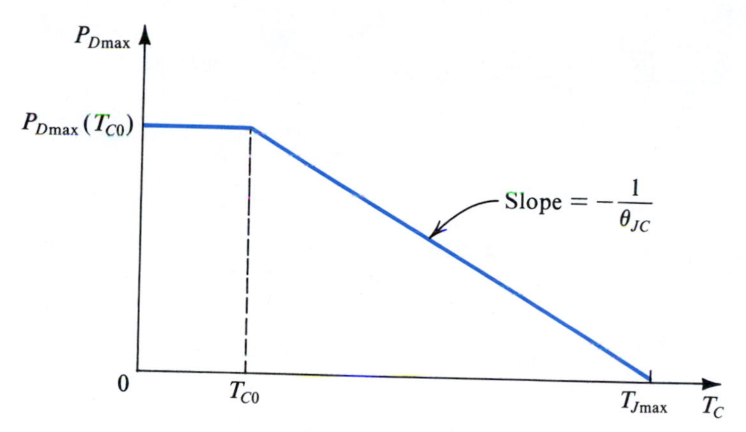

Figure 11.27 Maximum allowable power dissipation versus transistor-case temperature.

The electrical analog of the thermal-conduction process when a heat sink is employed is shown in Fig. 11.26, from which we can write

$$T_J - T_A = P_D(\theta_{JC} + \theta_{CS} + \theta_{SA}) \tag{11.74}$$

As well as specifying θ_{JC}, the device manufacturer usually supplies a derating curve for $P_{D\max}$ versus the case temperature, T_C. Such a curve is shown in Fig. 11.27. Note that the slope of the power-derating straight line is $-1/\theta_{JC}$. For a given transistor, the maximum power dissipation at a *case temperature* T_{C0} (usually 25°C) is much greater than that at an *ambient temperature* T_{A0} (usually 25°C). If the device can be maintained at a case temperature T_C, $T_{C0} \leq T_C \leq T_{J\max}$, then the maximum safe power dissipation is obtained when $T_J = T_{J\max}$,

$$P_{D\max} = \frac{T_{J\max} - T_C}{\theta_{JC}} \tag{11.75}$$

Example 11.8

A BJT is specified to have $T_{J\max} = 150°C$ and to be capable of dissipating maximum power as follows:

$$40 \text{ W at } T_C = 25°C$$
$$2 \text{ W at } T_A = 25°C$$

Above 25°C, the maximum power dissipation is to be derated linearly with $\theta_{JC} = 3.12°C/W$ and $\theta_{JA} = 62.5°C/W$. Find the following:

(a) The maximum power that can be dissipated safely by this transistor when operated in free air at $T_A = 50°C$.
(b) The maximum power that can be dissipated safely by this transistor when operated at an ambient temperature of 50°C, but with a heat sink for which $\theta_{CS} = 0.5°C/W$ and $\theta_{SA} = 4°C/W$. Find the temperature of the case and of the heat sink.
(c) The maximum power that can be dissipated safely if an *infinite heat sink* is used and $T_A = 50°C$.

Solution

(a)

$$P_{D\max} = \frac{T_{J\max} - T_A}{\theta_{JA}} = \frac{150 - 50}{62.5} = 1.6 \text{ W}$$

(b) With a heat sink, θ_{JA} becomes

$$\theta_{JA} = \theta_{JC} + \theta_{CS} + \theta_{SA}$$
$$= 3.12 + 0.5 + 4 = 7.62°C/W$$

Thus,

$$P_{D\max} = \frac{150 - 50}{7.62} = 13.1 \text{ W}$$

Figure 11.28 shows the thermal equivalent circuit with the various temperatures indicated.

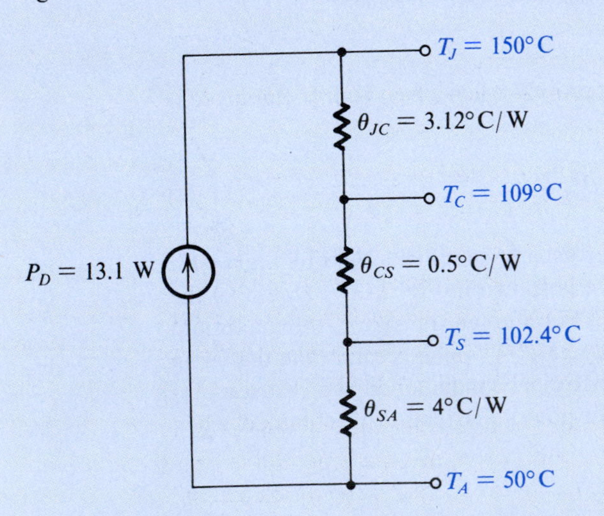

Figure 11.28 Thermal equivalent circuit for Example 11.8.

(c) An infinite heat sink, if it existed, would cause the case temperature T_C to equal the ambient temperature T_A. The infinite heat sink has $\theta_{CA} = 0$. Obviously, one cannot buy an infinite heat sink; nevertheless, this terminology is used by some manufacturers to describe the power-derating curve of Fig. 11.27. The abscissa is then labeled T_A and the curve is called "power dissipation versus ambient temperature with an infinite heat sink." For our example, with infinite heat sink,

$$P_{D\max} = \frac{T_{J\max} - T_A}{\theta_{JC}} = \frac{150 - 50}{3.12} = 32 \text{ W}$$

The advantage of using a heat sink is clearly evident from Example 11.8: With a heat sink, the maximum allowable power dissipation increases from 1.6 W to 13.1 W. Also note that although the transistor considered can be called a "40-W transistor," this level of power dissipation cannot be achieved in practice; it would require an infinite heat sink and an ambient temperature $T_A \le 25°\text{C}$.

EXERCISE

11.13 The 2N6306 power transistor is specified to have $T_{J\max} = 200°\text{C}$ and $P_{D\max} = 125$ W for $T_C \le 25°\text{C}$. For $T_C \ge 25°\text{C}$, $\theta_{JC} = 1.4°\text{C/W}$. If in a particular application this device is to dissipate 50 W and operate at an ambient temperature of 25°C, find the maximum thermal resistance of the heat sink that must be used (i.e., θ_{SA}). Assume $\theta_{CS} = 0.6°\text{C/W}$. What is the case temperature, T_C?
Ans. 1.5°C/W; 130°C

11.7.5 The BJT Safe Operating Area

In addition to specifying the maximum power dissipation at different case temperatures, power-transistor manufacturers usually provide a plot of the boundary of the safe operating area (SOA) in the i_C–v_{CE} plane. The SOA specification takes the form illustrated by the sketch in Fig. 11.29; the following paragraph numbers correspond to the boundaries on the sketch.

1. The maximum allowable current $I_{C\max}$. Exceeding this current on a continuous basis can result in melting the wires that bond the device to the package terminals.
2. The maximum power dissipation hyperbola. This is the locus of the points for which $v_{CE} i_C = P_{D\max}$ (at T_{C0}). For temperatures $T_C > T_{C0}$, the power-derating curves described in Section 11.7.4 should be used to obtain the applicable $P_{D\max}$ and thus a correspondingly lower hyperbola. Although the operating point can be allowed to move temporarily above the hyperbola, the *average* power dissipation should not be allowed to exceed $P_{D\max}$.
3. The **second-breakdown** limit. Second breakdown is a phenomenon that results because current flow across the emitter–base junction is not uniform. Rather, the current density is greatest near the periphery of the junction. This "**current crowding**" gives rise to increased localized power dissipation and hence temperature rise (at locations called **hot spots**). Since a temperature rise causes an increase in current, a localized form of **thermal runaway** can occur, leading to junction destruction.

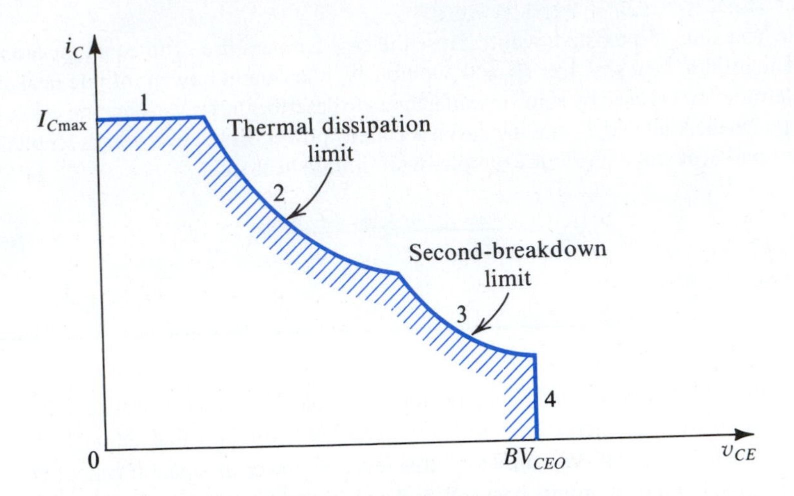

Figure 11.29 Safe operating area (SOA) of a BJT.

4. The collector-to-emitter breakdown voltage, BV_{CEO}. The instantaneous value of v_{CE} should never be allowed to exceed BV_{CEO}; otherwise, avalanche breakdown of the collector–base junction may occur (see Section 6.9).

Finally, it should be mentioned that logarithmic scales are usually used for i_C and v_{CE}, leading to an SOA boundary that consists of straight lines.

11.7.6 Parameter Values of Power Transistors

Owing to their large geometry and high operating currents, power transistors display typical parameter values that can be quite different from those of small-signal transistors. The important differences are as follows:

1. At high currents, the exponential i_C–v_{BE} relationship exhibits a factor of 2 reduction in the exponent; that is, $i_C = I_S e^{v_{BE}/2V_T}$.
2. β is low, typically 30 to 80, but can be as low as 5. Here, it is important to note that β has a positive temperature coefficient.
3. At high currents, r_π becomes very small (a few ohms) and r_x becomes important (r_x is defined and explained in Section 9.2.2).
4. f_T is low (a few megahertz), C_μ is large (hundreds of picofarads), and C_π is even larger. (These parameters are defined and explained in Section 9.2.2).
5. I_{CBO} is large (a few tens of microamps) and, as usual, doubles for every 10°C rise in temperature.
6. BV_{CEO} is typically 50 to 100 V but can be as high as 500 V.
7. I_{Cmax} is typically in the ampere range but can be as high as 100 A.

11.8 Variations on the Class AB Configuration

In this section, we discuss a number of circuit improvements and protection techniques for the BJT class AB output stage.

11.8.1 Use of Input Emitter Followers

Figure 11.30 shows a class AB circuit biased using transistors Q_1 and Q_2, which also function as emitter followers, thus providing the circuit with a high input resistance. In effect, the circuit functions as a unity-gain buffer amplifier. Since all four transistors are usually matched, the quiescent current ($v_I = 0$, $R_L = \infty$) in Q_3 and Q_4 is equal to that in Q_1 and Q_2. Resistors R_3 and R_4 are usually very small and are included to compensate for possible mismatches between Q_3 and Q_4 and to guard against the possibility of thermal runaway due to temperature differences between the input- and output-stage transistors. The latter point can be appreciated by noting that an increase in the current of, say, Q_3 causes an increase in the voltage drop across R_3 and a corresponding decrease in V_{BE3}. Thus R_3 provides negative feedback that helps stabilize the current through Q_3.

Because the circuit of Fig. 11.30 requires high-quality *pnp* transistors, it is not suitable for implementation in conventional monolithic IC technology. However, excellent results have been obtained with this circuit implemented in hybrid thick-film technology (Wong and Sherwin, 1979). This technology permits component trimming, for instance, to minimize the output offset voltage. The circuit can be used alone or together with an op amp to provide increased output driving capability. The latter application will be discussed in the next section.

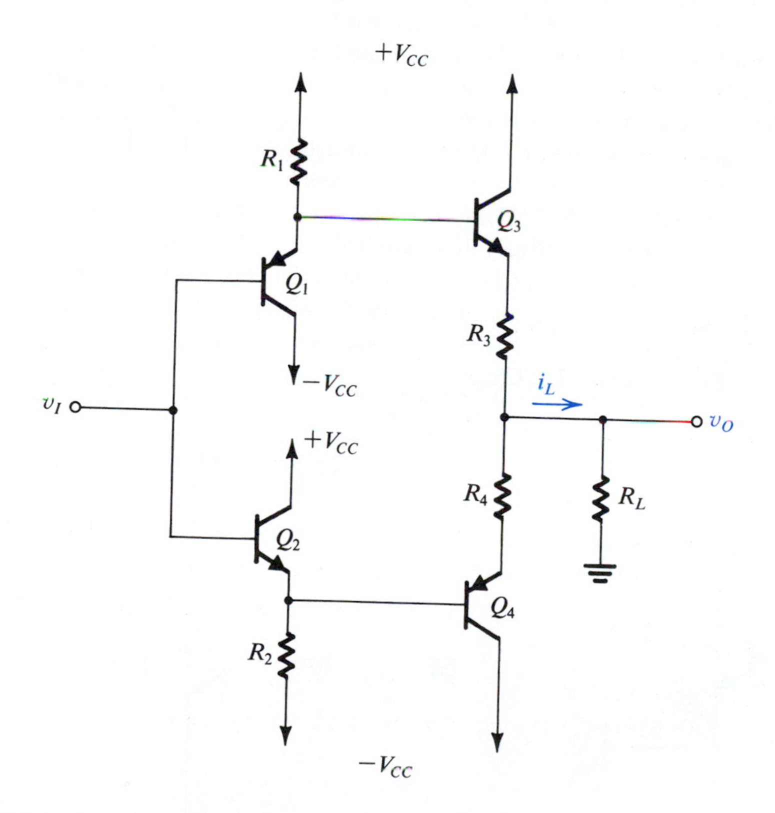

Figure 11.30 A class AB output stage with an input buffer. In addition to providing a high input resistance, the buffer transistors Q_1 and Q_2 bias the output transistors Q_3 and Q_4.

EXERCISE

11.14 (*Note:* Although very instructive, this exercise is rather long.) Consider the circuit of Fig. 11.30 with $R_1 = R_2 = 5\ \text{k}\Omega$, $R_3 = R_4 = 0\ \Omega$, and $V_{CC} = 15$ V. Let the transistors be matched with $I_S = 3.3 \times 10^{-14}$ A and $\beta = 200$. (These are the values used in the LH002 manufactured by National Semiconductor, except that $R_3 = R_4 = 2\ \Omega$ there.) (a) For $v_I = 0$ and $R_L = \infty$, find the quiescent current in each of the four transistors and v_O. (b) For $R_L = \infty$, find i_{C1}, i_{C2}, i_{C3}, i_{C4}, and v_O for $v_I = +10$ V and -10 V. (c) Repeat (b) for $R_L = 100\ \Omega$.

Ans. (a) 2.87 mA; 0 V; (b) for $v_I = +10$ V: 0.88 mA, 4.87 mA, 1.95 mA, 1.95 mA, +9.98 V; for $v_I = -10$ V: 4.87 mA, 0.88 mA, 1.95 mA, 1.95 mA, −9.98 V; (c) for $v_I = +10$ V: 0.38 mA, 4.87 mA, 100 mA, 0.02 mA, +9.86 V; for $v_I = -10$ V: 4.87 mA, 0.38 mA, 0.02 mA, 100 mA, −9.86 V

11.8.2 Use of Compound Devices

To increase the current gain of the output-stage transistors, and thus reduce the required base current drive, the Darlington configuration shown in Fig. 11.31 is frequently used to replace the *npn* transistor of the class AB stage. The Darlington configuration is equivalent to a single *npn* transistor having $\beta \simeq \beta_1\beta_2$, but almost twice the value of V_{BE}.

The Darlington configuration can be also used for *pnp* transistors, and this is indeed done in discrete-circuit design. In IC design, however, the lack of good-quality *pnp* transistors prompted the use of the alternative compound configuration shown in Fig. 11.32. This compound device is equivalent to a single *pnp* transistor having $\beta \simeq \beta_1\beta_2$. When fabricated with standard IC technology, Q_1 is usually a lateral *pnp* having a low β ($\beta = 5 - 10$) and poor high-frequency response ($f_T \simeq 5$ MHz); see Appendix A and Appendix 7.A. The compound device, although it has a relatively high equivalent β, still suffers from a poor high-frequency response. It also suffers from another problem: The feedback loop formed by Q_1 and Q_2 is prone to high-frequency oscillations (with frequency near f_T of the *pnp* device, i.e., about 5 MHz). Methods exist for preventing such oscillations. The subject of feedback-amplifier stability was studied in Chapter 10.

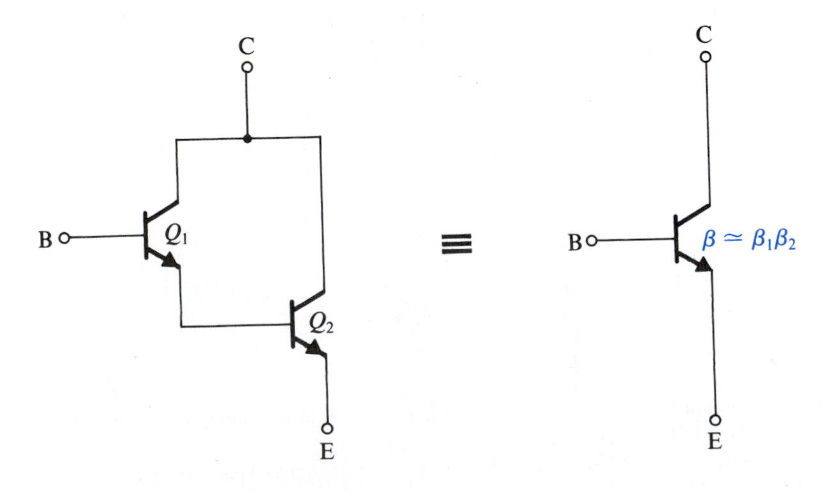

Figure 11.31 The Darlington configuration.

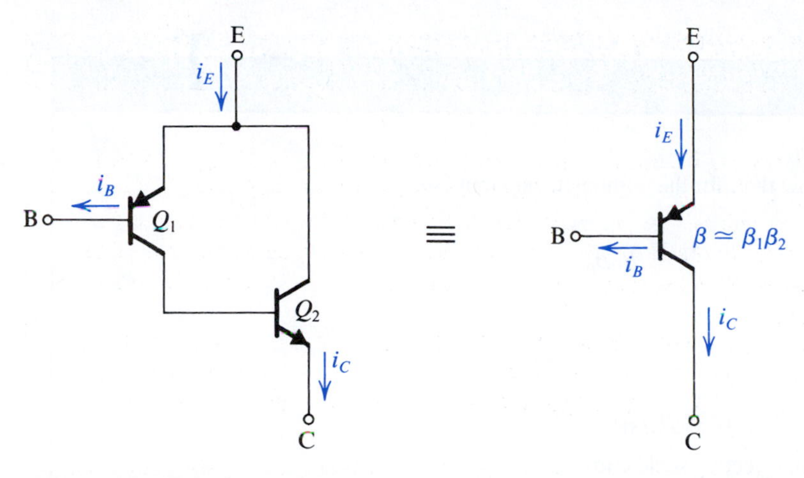

Figure 11.32 The compound-*pnp* configuration.

To illustrate the application of the Darlington configuration and of the compound *pnp,* we show in Fig. 11.33 an output stage utilizing both. Class AB biasing is achieved using a V_{BE} multiplier. Note that the Darlington *npn* adds one more V_{BE} drop, and thus the V_{BE} multiplier is required to provide a bias voltage of about 2 V. The design of this class AB stage is investigated in Problem 11.43.

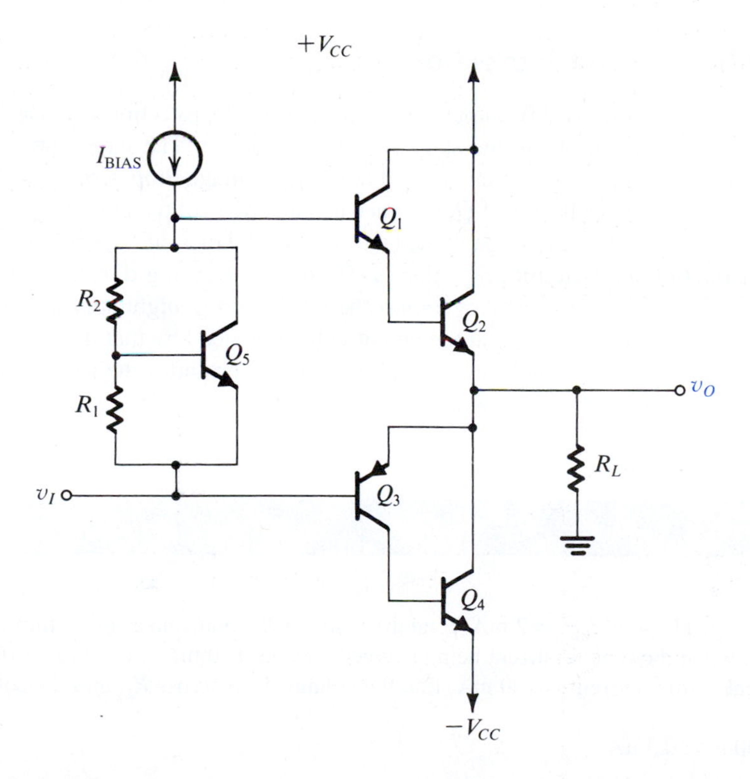

Figure 11.33 A class AB output stage utilizing a Darlington *npn* and a compound *pnp*. Biasing is obtained using a V_{BE} multiplier.

EXERCISE

11.15 (a) Refer to Fig. 11.32. Show that, for the composite *pnp* transistor,

$$i_B \simeq \frac{i_C}{\beta_N \beta_P}$$

and

$$i_E \simeq i_C$$

Hence show that

$$i_C \simeq \beta_N I_{SP} e^{v_{EB}/V_T}$$

and thus the transistor has an effective scale current

$$I_S = \beta_N I_{SP}$$

where I_{SP} is the saturation current of the *pnp* transistor Q_1.

(b) For $\beta_P = 20$, $\beta_N = 50$, $I_{SP} = 10^{-14}$ A, find the effective current gain of the compound device and its v_{EB} when $i_C = 100$ mA.

Ans. (b) 1000; 0.651 V

11.8.3 Short-Circuit Protection

Figure 11.34 shows a class AB output stage equipped with protection against the effect of short-circuiting the output while the stage is sourcing current. The large current that flows through Q_1 in the event of a short circuit will develop a voltage drop across R_{E1} of sufficient value to turn Q_5 on. The collector of Q_5 will then conduct most of the current I_{BIAS}, robbing Q_1 of its base drive. The current through Q_1 will thus be reduced to a safe operating level.

This method of short-circuit protection is effective in ensuring device safety, but it has the disadvantage that under normal operation about 0.5 V drop might appear across each R_E. This means that the voltage swing at the output will be reduced by that much, in each direction. On the other hand, the inclusion of emitter resistors provides the additional benefit of protecting the output transistors against thermal runaway.

EXERCISE

D11.16 In the circuit of Fig. 11.34 let $I_{BIAS} = 2$ mA. Find the value of R_{E1} that causes Q_5 to turn on and absorb all 2 mA when the output current being sourced reaches 150 mA. For Q_5, $I_S = 10^{-14}$ A. If the normal peak output current is 100 mA, find the voltage drop across R_{E1} and the collector current of Q_5.

Ans. 4.3 Ω; 430 mV; 0.3 μA

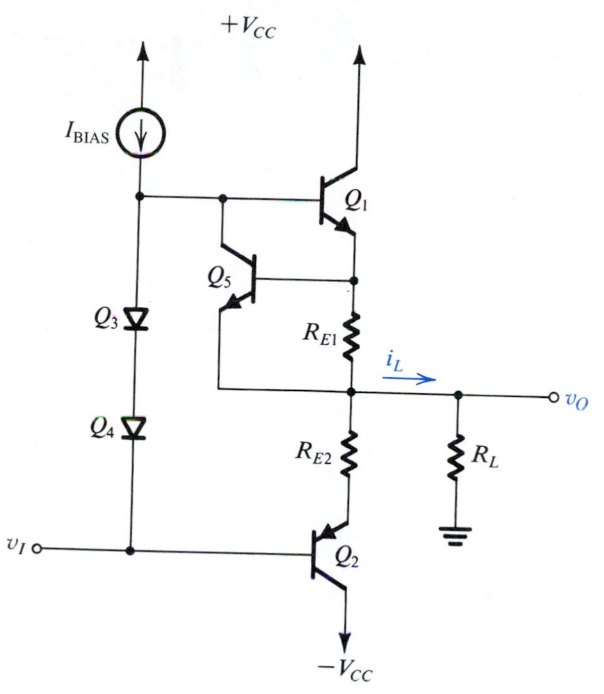

Figure 11.34 A class AB output stage with short-circuit protection. The protection circuit shown operates in the event of an output short circuit while v_O is positive.

11.8.4 Thermal Shutdown

In addition to short-circuit protection, most IC power amplifiers are usually equipped with a circuit that senses the temperature of the chip and turns on a transistor in the event that the temperature exceeds a safe preset value. The turned-on transistor is connected in such a way that it absorbs the bias current of the amplifier, thus virtually shutting down its operation.

Figure 11.35 shows a thermal-shutdown circuit. Here, transistor Q_2 is normally off. As the chip temperature rises, the combination of the positive temperature coefficient of zener diode Z_1 and the negative temperature coefficient of V_{BE1} causes the voltage at the emitter of Q_1 to rise. This in turn raises the voltage at the base of Q_2 to the point at which Q_2 turns on.

11.9 IC Power Amplifiers

A variety of IC power amplifiers are available. Most consist of a high-gain, small-signal amplifier followed by a class AB output stage. Some have overall negative feedback already applied, resulting in a fixed closed-loop voltage gain. Others do not have on-chip feedback and are, in effect, op amps with large output-power capability. In fact, the output current-driving capability of any general-purpose op amp can be increased by cascading it with a class B or class AB output stage and applying overall negative feedback. The additional output stage can be either a discrete circuit or a hybrid IC such as the buffer discussed in the preceding section. In the following we discuss some power-amplifier examples.

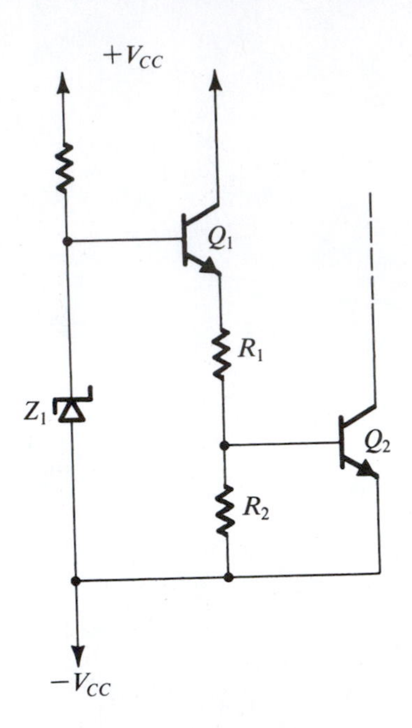

Figure 11.35 Thermal-shutdown circuit.

11.9.1 A Fixed-Gain IC Power Amplifier

Our first example is the LM380 (a product of National Semiconductor Corporation), which is a fixed-gain monolithic power amplifier. A simplified version of the internal circuit of the amplifier[3] is shown in Fig. 11.36. The circuit consists of an input differential amplifier utilizing Q_1 and Q_2 as emitter followers for input buffering, and Q_3 and Q_4 as a differential pair with an emitter resistor R_3. The two resistors R_4 and R_5 provide dc paths to ground for the base currents of Q_1 and Q_2, thus enabling the input signal source to be capacitively coupled to either of the two input terminals.

The differential amplifier transistors Q_3 and Q_4 are biased by two separate currents: Q_3 is biased by a current from the dc supply V_S through the diode-connected transistor Q_{10}, and resistor R_1; Q_4 is biased by a dc current from the output terminal through R_2. Under quiescent conditions (i.e., with no input signal applied) the two bias currents will be equal, and the current through and the voltage across R_3 will be zero. For the emitter current of Q_3 we can write

$$I_3 \simeq \frac{V_S - V_{EB10} - V_{EB3} - V_{EB1}}{R_1}$$

where we have neglected the small dc voltage drop across R_4. Assuming, for simplicity, all V_{EB} to be equal,

$$I_3 \simeq \frac{V_S - 3V_{EB}}{R_1} \tag{11.76}$$

For the emitter current of Q_4 we have

$$I_4 = \frac{V_O - V_{EB4} - V_{EB2}}{R_2} \simeq \frac{V_O - 2V_{EB}}{R_2} \tag{11.77}$$

[3]The main objective of showing this circuit is to point out some interesting design features. The circuit is *not* a detailed schematic diagram of what is actually on the chip.

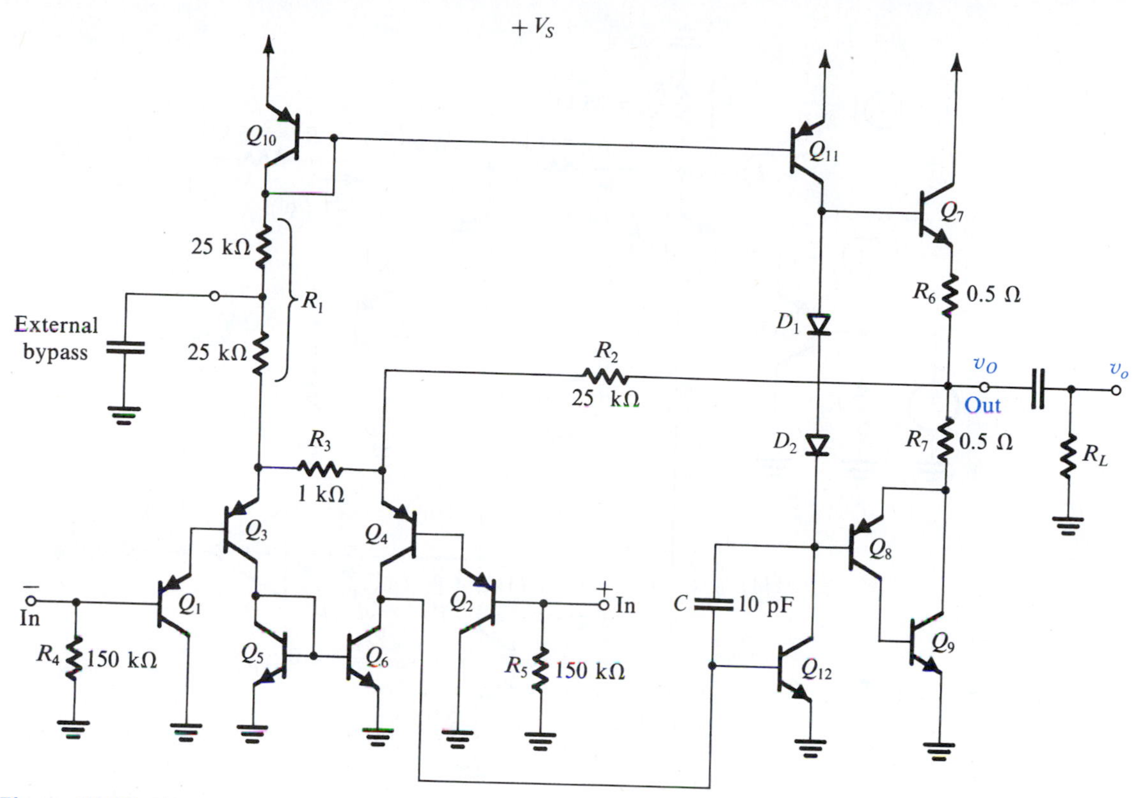

Figure 11.36 The simplified internal circuit of the LM380 IC power amplifier. (Courtesy National Semiconductor Corporation.)

where V_O is the dc voltage at the output, and we have neglected the small drop across R_5. Equating I_3 and I_4 and using the fact that $R_1 = 2R_2$ results in

$$V_O = \frac{1}{2}V_S + \frac{1}{2}V_{EB} \tag{11.78}$$

Thus the output is biased at approximately half the power-supply voltage, as desired for maximum output voltage swing. An important feature is the dc feedback from the output to the emitter of Q_4, through R_2. This dc feedback acts to stabilize the output dc bias voltage at the value in Eq. (11.78). Qualitatively, the dc feedback functions as follows: If for some reason V_O increases, a corresponding current increment will flow through R_2 and into the emitter of Q_4. Thus the collector current of Q_4 increases, resulting in a positive increment in the voltage at the base of Q_{12}. This, in turn, causes the collector current of Q_{12} to increase, thus bringing down the voltage at the base of Q_8 and hence V_O.

Continuing with the description of the circuit in Fig. 11.36, we observe that the differential amplifier (Q_3, Q_4) has a current mirror load composed of Q_5 and Q_6 (refer to Section 8.5 for a discussion of active loads). The single-ended output voltage signal of the first stage appears at the collector of Q_6 and thus is applied to the base of the second-stage common-emitter amplifier Q_{12}. Transistor Q_{12} is biased by the constant-current source Q_{11}, which also acts as its active load. In actual operation, however, the load of Q_{12} will be dominated by the reflected resistance due to R_L. Capacitor C provides frequency compensation (see Chapter 10).

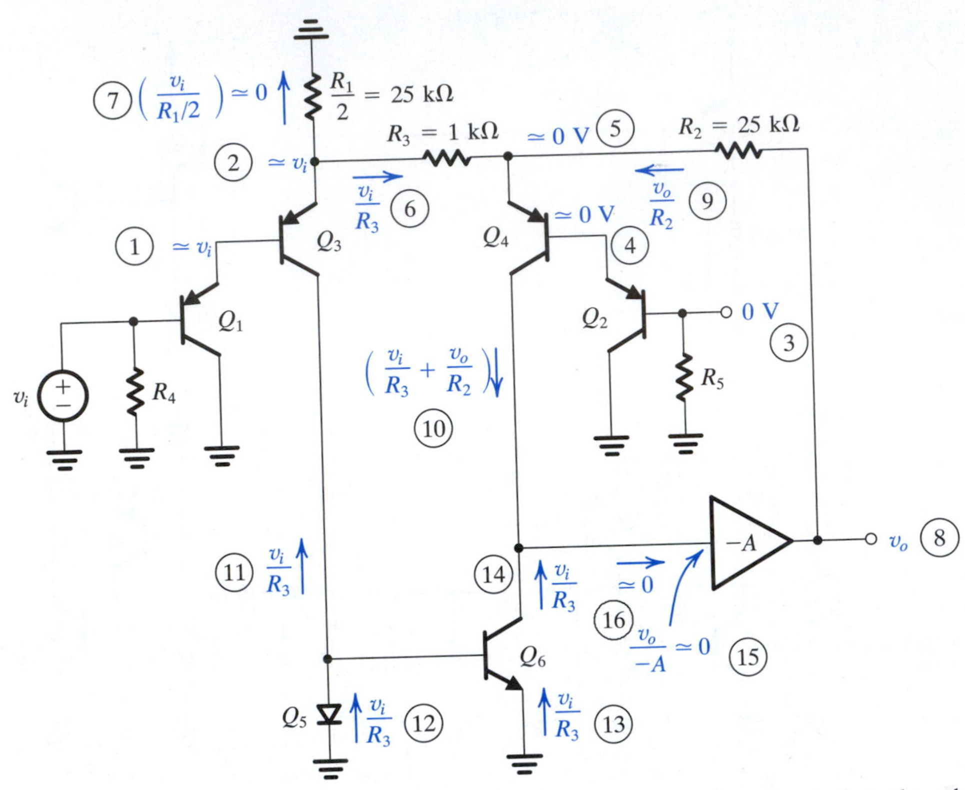

Figure 11.37 Small-signal analysis of the circuit in Fig. 11.36. The circled numbers indicate the order of the analysis steps.

The output stage is class AB, utilizing a compound *pnp* transistor (Q_8 and Q_9). Negative feedback is applied from the output to the emitter of Q_4 via resistor R_2. To find the closed-loop gain consider the small-signal equivalent circuit shown in Fig. 11.37. Here, we have replaced the second-stage common-emitter amplifier and the output stage with an inverting amplifier block with gain A. We shall assume that the amplifier A has high gain and high input resistance, and thus the input signal current into A is negligibly small. Under this assumption, Fig. 11.37 shows the analysis details with an input signal v_i applied to the inverting input terminal. The order of the analysis steps is indicated by the circled numbers. Note that since the input differential amplifier has a relatively large resistance, R_3, in the emitter circuit, most of the applied input voltage appears across R_3. In other words, the signal voltages across the emitter–base junctions of Q_1, Q_2, Q_3, and Q_4 are small in comparison to the voltage across R_3. Accordingly, the voltage gain can be found by writing a node equation at the collector of Q_6:

$$\frac{v_i}{R_3} + \frac{v_o}{R_2} + \frac{v_i}{R_3} = 0$$

which yields

$$\frac{v_o}{v_i} = -\frac{2R_2}{R_3} \simeq -50\ \text{V/V}$$

EXERCISE

11.17 Denoting the total resistance between the collector of Q_6 and ground by R, show, using Fig. 11.37, that

$$\frac{v_o}{v_i} = \frac{-2R_2/R_3}{1+(R_2/AR)}$$

which reduces to $(-2R_2/R_3)$ under the condition that $AR \gg R_2$.

As was demonstrated in Chapter 10, one of the advantages of negative feedback is the reduction of nonlinear distortion. This is the case in the circuit of the LM380.

The LM380 is designed to operate from a single supply V_S in the range of 12 V to 22 V. The selection of supply voltage depends on the value of R_L and the required output power P_L. The manufacturer supplies curves for the device power dissipation versus output power for a given load resistance and various supply voltages. One such set of curves for $R_L = 8\ \Omega$ is shown in Fig. 11.38. Note the similarity to the class B power dissipation curve of Fig. 11.8. In fact, the reader can easily verify that the location and value of the peaks of the curves in Fig. 11.38 are accurately predicted by Eqs. (11.20) and (11.21), respectively (where $V_{CC} = \frac{1}{2}V_S$). The line labeled "3% distortion level" in Fig. 11.38 is the locus of the points on the various curves at which the distortion (THD) reaches 3%. A THD of 3% represents the onset of peak clipping due to output-transistor saturation.

The manufacturer also supplies curves for maximum power dissipation versus temperature (derating curves) similar to those discussed in Section 11.7 for discrete power transistors.

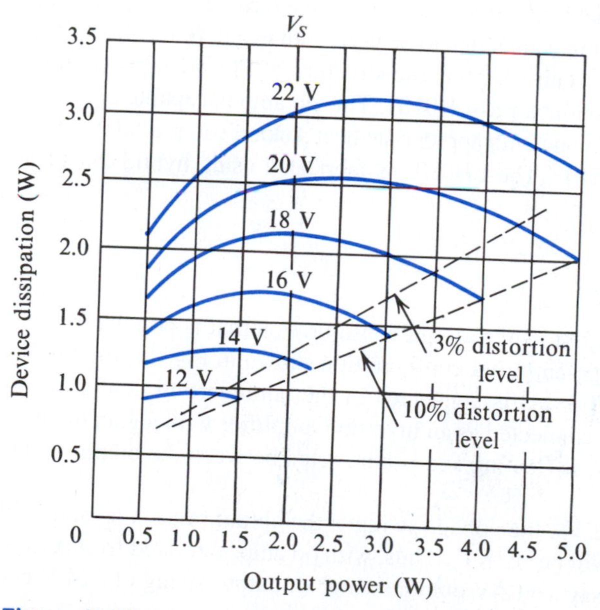

Figure 11.38 Power dissipation (P_D) versus output power (P_L) for the LM380 with $R_L = 8\ \Omega$. (Courtesy National Semiconductor Corporation.)

EXERCISES

11.18 The manufacturer specifies that for ambient temperatures below 25°C the LM380 can dissipate a maximum of 3.6 W. This is obtained under the condition that its dual-in-line package be soldered onto a printed-circuit board in close thermal contact with 6 square inches of 2-ounce copper foil. Above $T_A = 25°C$, the thermal resistance is $\theta_{JA} = 35°C/W$. T_{Jmax} is specified to be 150°C. Find the maximum power dissipation possible if the ambient temperature is to be 50°C.
Ans. 2.9 W

D11.19 It is required to use the LM380 to drive an 8-Ω loudspeaker. Use the curves of Fig. 11.38 to determine the maximum power supply possible while limiting the maximum power dissipation to the 2.9 W determined in Exercise 11.18. If for this application a 3% THD is allowed, find P_L and the peak-to-peak output voltage.
Ans. 20 V; 4.2 W; 16.4 V

11.9.2 Power Op Amps

Figure 11.39 shows the general structure of a power op amp. It consists of a low-power op amp followed by a class AB buffer similar to that discussed in Section 11.8.1. The buffer consists of transistors Q_1, Q_2, Q_3, and Q_4, with bias resistors R_1 and R_2 and emitter degeneration resistors R_5 and R_6. The buffer supplies the required load current until the current increases to the point that the voltage drop across R_3 (in the current-sourcing mode) becomes sufficiently large to turn Q_5 on. Transistor Q_5 then supplies the additional load current required. In the current-sinking mode, Q_4 supplies the load current until sufficient voltage develops across R_4 to turn Q_6 on. Then, Q_6 sinks the additional load current. Thus the stage formed by Q_5 and Q_6 acts as a **current booster**. The power op amp is intended to be used with negative feedback in the usual closed-loop configurations. A circuit based on the structure of Fig. 11.39 is commercially available from National Semiconductor as LH0101. This op amp is capable of providing a continuous output current of 2 A, and with appropriate heat sinking can provide 40 W of output power (Wong and Johnson, 1981). The LH0101 is fabricated using hybrid thick-film technology.

11.9.3 The Bridge Amplifier

We conclude this section with a discussion of a circuit configuration that is popular in high-power applications. This is the bridge-amplifier configuration shown in Fig. 11.40 utilizing two power op amps, A_1 and A_2. While A_1 is connected in the noninverting configuration with a gain $K = 1 + (R_2/R_1)$, A_2 is connected as an inverting amplifier with a gain of equal magnitude $K = R_4/R_3$. The load R_L is floating and is connected between the output terminals of the two op amps.

If v_I is a sinusoid with amplitude $\hat{V}_i$, the voltage swing at the output of each op amp will be $\pm K\hat{V}_i$, and that across the load will be $\pm 2K\hat{V}_i$. Thus, with op amps operated from ±15-V supplies and capable of providing, say a ±12-V output swing, an output swing of ±24 V can be obtained across the load of the bridge amplifier.

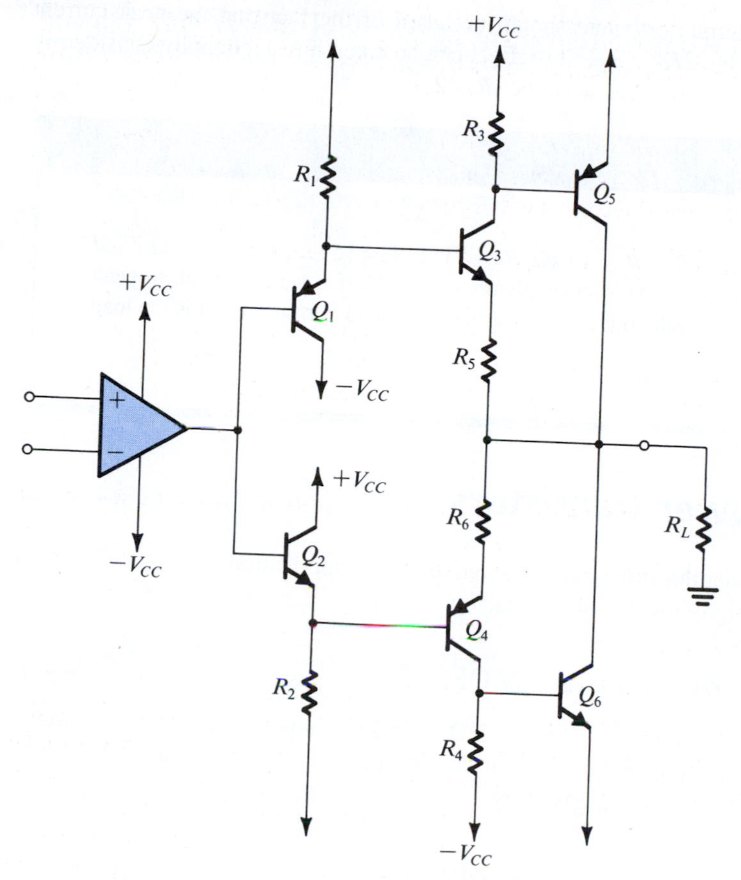

Figure 11.39 Structure of a power op amp. The circuit consists of an op amp followed by a class AB buffer similar to that discussed in Section 11.8.1. The output current capability of the buffer, consisting of Q_1, Q_2, Q_3, and Q_4, is further boosted by Q_5 and Q_6.

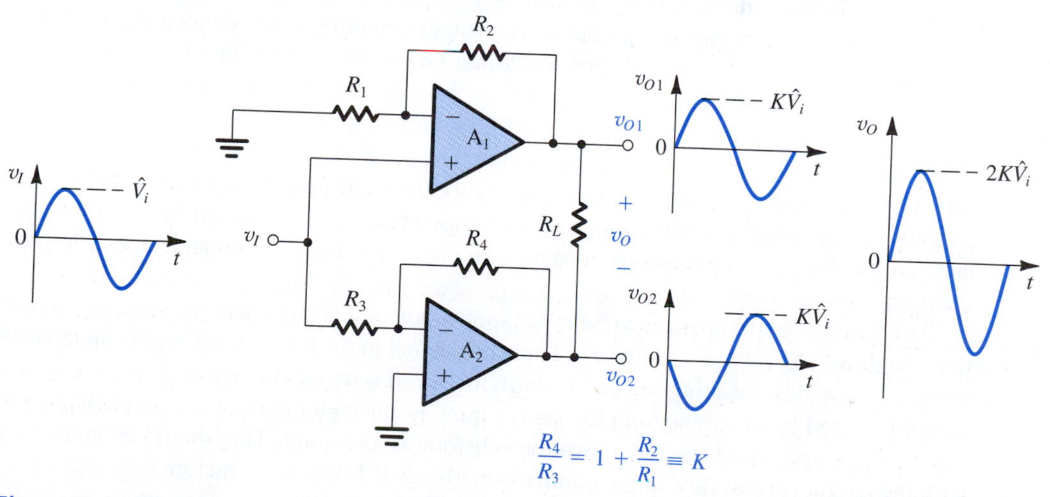

Figure 11.40 The bridge-amplifier configuration.

In designing bridge amplifiers, note should be taken of the fact that the peak current drawn from each op amp is $2K\hat{V}_i/R_L$. This effect can be taken into account by considering the load seen by each op amp (to ground) to be $R_L/2$.

EXERCISE

11.20 Consider the circuit of Fig. 11.40 with $R_1 = R_3 = 10$ kΩ, $R_2 = 5$ kΩ, $R_4 = 15$ kΩ, and $R_L = 8$ Ω. Find the voltage gain and the input resistance. The power supply used is ±18 V. If v_I is a 20-V peak-to-peak sine wave, what is the peak-to-peak output voltage? What is the peak load current? What is the load power?
Ans. 3 V/V; 10 kΩ; 60 V; 3.75 A; 56.25 W

11.10 MOS Power Transistors

In this section we consider the structure, characteristics, and application of a special type of MOSFET suitable for high-power applications.

11.10.1 Structure of the Power MOSFET

The MOSFET structure studied in Chapter 5 (Fig. 5.1) is not suitable for high-power applications. To appreciate this fact, recall that the drain current of an n-channel MOSFET operating in the saturation region is given by

$$i_D = \frac{1}{2}\mu_n C_{ox}\left(\frac{W}{L}\right)(v_{GS} - V_t)^2 \tag{11.79}$$

It follows that to increase the current capability of the MOSFET, its width W should be made large and its channel length L should be made as small as possible. Unfortunately, however, reducing the channel length of the standard MOSFET structure results in a drastic reduction in its breakdown voltage. Specifically, the depletion region of the reverse-biased body-to-drain junction spreads into the short channel, resulting in breakdown at a relatively low voltage. Thus the resulting device would not be capable of handling the high voltages typical of power-transistor applications. For this reason, new structures had to be found for fabricating short-channel (1- to 2-μm) MOSFETs with high breakdown voltages.

At the present time the most popular structure for a power MOSFET is the **double-diffused** or **DMOS transistor** shown in Fig. 11.41. As indicated, the device is fabricated on a lightly doped n-type substrate with a heavily doped region at the bottom for the drain contact. Two diffusions[4] are employed, one to form the p-type body region and another to form the n-type source region.

The DMOS device operates as follows. Application of a positive gate voltage, v_{GS}, greater than the threshold voltage V_t, induces a lateral n channel in the p-type body region underneath the gate oxide. The resulting channel is short; its length is denoted L in Fig. 11.41. Current is then conducted by electrons from the source moving through the resulting short channel to the substrate and then vertically down the substrate to the drain. This should be contrasted with the lateral current flow in the standard small-signal MOSFET structure (Chapter 5).

[4]See Appendix A for a description of the IC fabrication process.

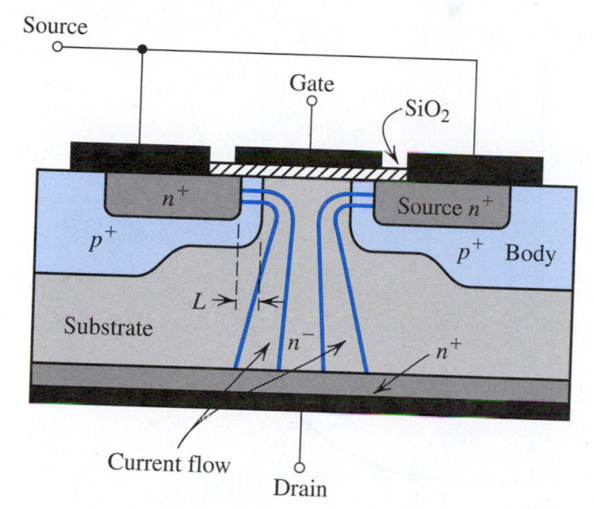

Figure 11.41 Double-diffused vertical MOS transistor (DMOS).

Even though the DMOS transistor has a short channel, its breakdown voltage can be very high (as high as 600 V). This is because the depletion region between the substrate and the body extends mostly in the lightly doped substrate and does not spread into the channel. The result is a MOS transistor that simultaneously has a high current capability (50 A is possible) as well as the high breakdown voltage just mentioned. Finally, we note that the vertical structure of the device provides efficient utilization of the silicon area.

An earlier structure used for power MOS transistors deserves mention. This is the V-groove MOS device [see Severns (1984)]. Although still in use, the **V-groove MOSFET** has lost application ground to the vertical DMOS structure of Fig. 11.41, except possibly for high-frequency applications. Because of space limitations, we shall not describe the V-groove MOSFET.

11.10.2 Characteristics of Power MOSFETs

In spite of their radically different structure, power MOSFETs exhibit characteristics that are quite similar to those of the small-signal MOSFETs studied in Chapter 5. Important differences exist, however, and these are discussed next.

Power MOSFETs have threshold voltages in the range of 2 V to 4 V. In saturation, the drain current is related to v_{GS} by the square-law characteristic of Eq. (11.80). However, as shown in Fig. 11.42, the i_D–v_{GS} characteristic becomes linear for larger values of v_{GS}. The linear portion of the characteristic occurs as a result of the high electric field along the short channel, causing the velocity of charge carriers to reach an upper limit, a phenomenon known as **velocity saturation**[5]. The linear i_D–v_{GS} relationship implies a constant g_m in the velocity-saturation region.

The i_D–v_{GS} characteristic shown in Fig. 11.42 includes a segment labeled "subthreshold." Though of little significance for power devices, the subthreshold region of operation is of interest in very-low-power applications (see Section 5.1.9).

[5]Velocity saturation occurs also in standard MOSFET structures when the channel length is in the submicron range. We shall discuss velocity saturation in some detail in Section 13.5.

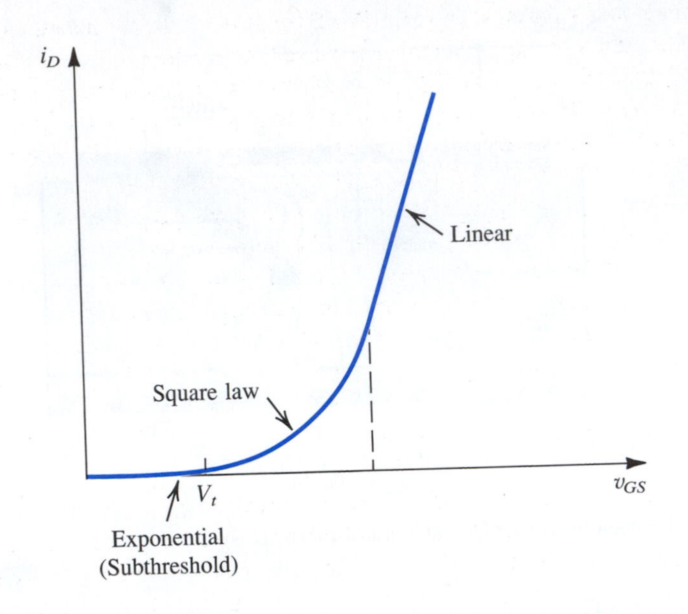

Figure 11.42 Typical i_D–v_{GS} characteristic for a power MOSFET.

11.10.3 Temperature Effects

Of considerable interest in the design of MOS power circuits is the variation of the MOSFET characteristics with temperature, illustrated in Fig. 11.43. Observe that there is a value of v_{GS} (in

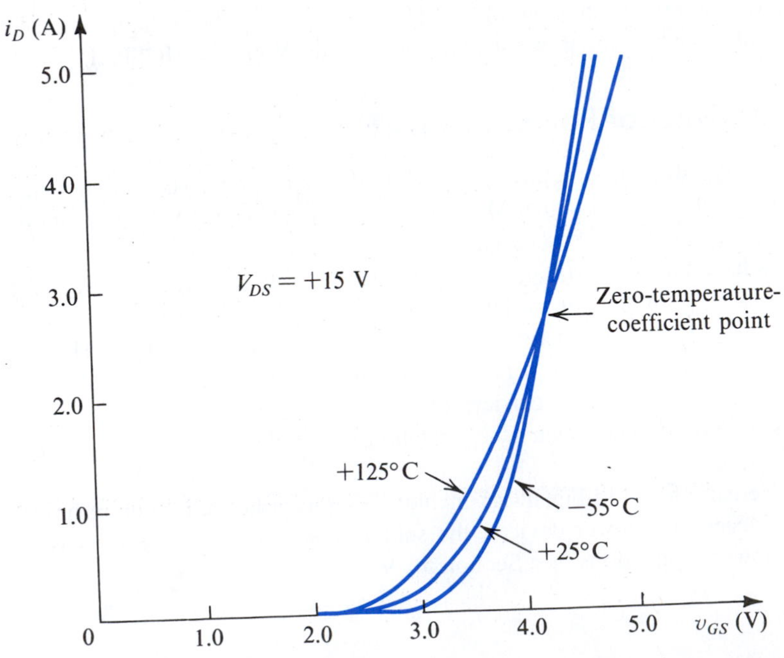

Figure 11.43 The i_D–v_{GS} characteristic curve of a power MOS transistor (IRF 630, Siliconix) at case temperatures of –55°C, +25°C, and +125°C. (Courtesy of Siliconix Inc.)

the range of 4 V to 6 V for most power MOSFETs) at which the temperature coefficient of i_D is zero. At higher values of v_{GS}, i_D exhibits a negative temperature coefficient. This is a significant property: It implies that a MOSFET operating beyond the zero-temperature-coefficient point does not suffer from the possibility of thermal runaway. This is *not* the case, however, at low currents (i.e., lower than the zero-temperature-coefficient point). In the (relatively) low-current region, the temperature coefficient of i_D is positive, and the power MOSFET can easily suffer thermal runaway (with unhappy consequences). Since class AB output stages are biased at low currents, means must be provided to guard against thermal runaway.

The reason for the positive temperature coefficient of i_D at low currents is that $v_{OV} = (v_{GS} - V_t)$ is relatively low, and the temperature dependence is dominated by the negative temperature coefficient of V_t (in the range of –3 mV/°C to –6 mV/°C) which causes v_{OV} to rise with temperature.

11.10.4 Comparison with BJTs

The power MOSFET does not suffer from second breakdown, which limits the safe operating area of BJTs. Also, power MOSFETs do not require the large dc base-drive currents of power BJTs. Note, however, that the driver stage in a MOS power amplifier should be capable of supplying sufficient current to charge and discharge the MOSFET's large and nonlinear input capacitance in the time allotted. Finally, the power MOSFET features, in general, a higher speed of operation than the power BJT. This makes MOS power transistors especially suited to switching applications—for instance, in motor-control circuits.

11.10.5 A Class AB Output Stage Utilizing Power MOSFETs

As an application of power MOSFETs, we show in Fig. 11.44 a class AB output stage utilizing a pair of complementary MOSFETs and employing BJTs for biasing and in the driver stage. The latter consists of complementary Darlington emitter followers formed by Q_1 through Q_4 and has the low output resistance necessary for driving the output MOSFETs at high speeds.

Of special interest in the circuit of Fig. 11.44 is the bias circuit utilizing two V_{BE} multipliers formed by Q_5 and Q_6 and their associated resistors. Transistor Q_6 is placed in direct thermal contact with the output transistors; this is achieved by simply mounting Q_6 on their common heat sink. Thus, by the appropriate choice of the V_{BE} multiplication factor of Q_6, the bias voltage V_{GG} (between the gates of the output transistors) can be made to decrease with temperature at the same rate as that of the sum of the threshold voltages $(V_{tN} + |V_{tP}|)$ of the output MOSFETs. In this way the overdrive voltages and hence the quiescent current of the output transistors can be stabilized against temperature variations.

Analytically, V_{GG} is given by

$$V_{GG} = \left(1 + \frac{R_3}{R_4}\right) V_{BE6} + \left(1 + \frac{R_1}{R_2}\right) V_{BE5} - 4V_{BE}$$

Since V_{BE6} is thermally coupled to the output devices while the other BJTs remain at constant temperature, we have

$$\frac{\partial V_{GG}}{\partial T} = \left(1 + \frac{R_3}{R_4}\right) \frac{\partial V_{BE6}}{\partial T}$$

which is the relationship needed to determine R_3/R_4 so that $\partial V_{GG}/\partial T = \partial(V_{tN} + |V_{tP}|)/\partial T$. The other V_{BE} multiplier is then adjusted to yield the value of V_{GG} required for the desired quiescent current in Q_N and Q_P.

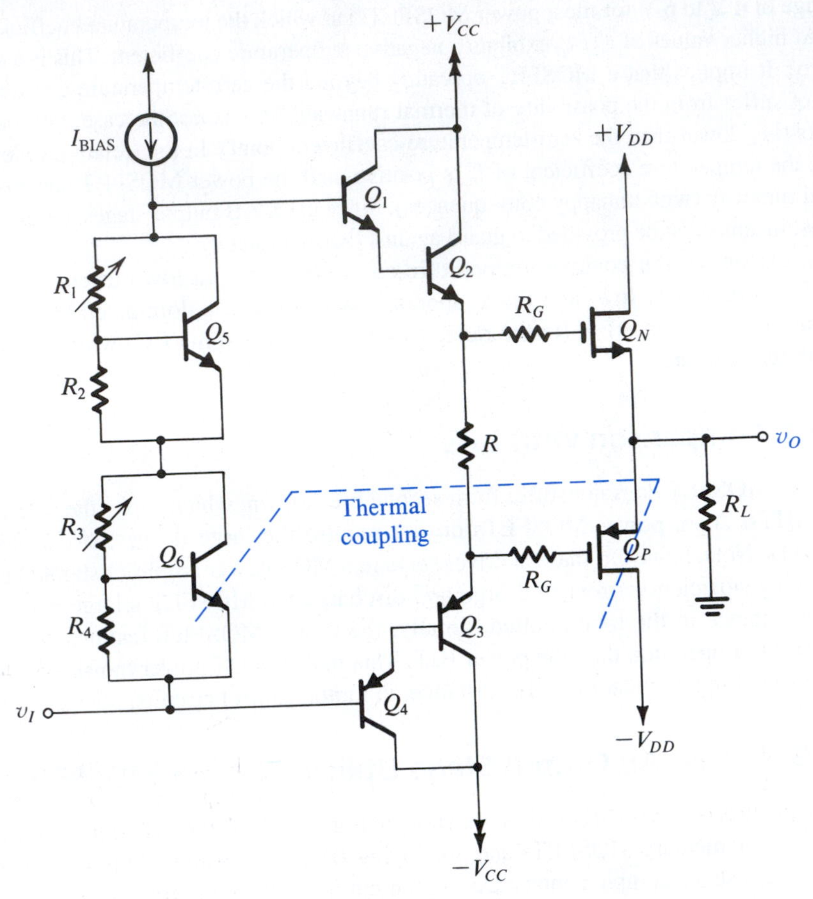

Figure 11.44 A class AB amplifier with MOS output transistors and BJT drivers. Resistor R_3 is adjusted to provide temperature compensation while R_1 is adjusted to yield the desired value of quiescent current in the output transistors. Resistors R_G are used to suppress parasitic oscillations at high frequencies. Typically, $R_G = 100\ \Omega$.

EXERCISES

11.21 For the circuit in Fig. 11.44, find the ratio R_3/R_4 that provides temperature stabilization of the quiescent current in Q_N and Q_P. Assume that $|V_t|$ changes at -3 mV/°C and that $\partial V_{BE}/\partial T = -2$ mV/°C.
Ans. 2

11.22 For the circuit in Fig. 11.44 assume that the BJTs have a nominal V_{BE} of 0.7 V and that the MOSFETs have $|V_t| = 3$ V and $\mu_n C_{ox}(W/L) = 2\ \text{A/V}^2$. It is required to establish a quiescent current of 100 mA in the output stage and 20 mA in the driver stage. Find $|V_{GS}|$, V_{GG}, R, and R_1/R_2. Use the value of R_3/R_4 found in Exercise 11.21. Assume that the MOSFETs are represented by their square-law i_D–v_{GS} characteristics.
Ans. 3.32 V; 6.64 V; 332 Ω; 9.5

Summary

- Output stages are classified according to the transistor conduction angle: class A (360°), class AB (slightly more than 180°), class B (180°), and class C (less than 180°).
- The most common class A output stage is the emitter follower. It is biased at a current greater than the peak load current.
- The class A output stage dissipates its maximum power under quiescent conditions ($v_O = 0$). It achieves a maximum power-conversion efficiency of 25%.
- The class B stage is biased at zero current, and thus dissipates no power in quiescence.
- The class B stage can achieve a power conversion efficiency as high as 78.5%. It dissipates its maximum power for $\hat{V}_o = (2/\pi)V_{CC}$.
- The class B stage suffers from crossover distortion.
- The class AB output stage is biased at a small current; thus both transistors conduct for small input signals, and crossover distortion is virtually eliminated.
- Except for an additional small quiescent power dissipation, the power relationships of the class AB stage are similar to those in class B.
- To guard against the possibility of thermal runaway, the bias voltage of the class AB circuit is made to vary with temperature in the same manner as does V_{BE} of the output transistors.
- The classical CMOS class AB output stage suffers from reduced output signal-swing. This problem can be overcome by replacing the source-follower output transistors with a pair of complementary devices operating in the common-source configuration.
- The CMOS class AB output stage with common-source transistors allows the output voltage to swing to within an overdrive voltage from each of the two power supplies. Utilizing error amplifiers in the feedback path of each of the output transistors reduces both the output resistance and gain error of the stage.
- To facilitate the removal of heat from the silicon chip, power devices are usually mounted on heat sinks. The maximum power that can be safely dissipated in the device is given by

$$P_{D\max} = \frac{T_{J\max} - T_A}{\theta_{JC} + \theta_{CS} + \theta_{SA}}$$

  where $T_{J\max}$ and θ_{JC} are specified by the manufacturer, while θ_{CS} and θ_{SA} depend on the heat-sink design.
- Use of the Darlington configuration in the class AB output stage reduces the base-current drive requirement. In integrated circuits, the compound *pnp* configuration is commonly used.
- Output stages are usually equipped with circuitry that, in the event of a short circuit, can turn on and limit the base-current drive, and hence the emitter current, of the output transistors.
- IC power amplifiers consist of a small-signal voltage amplifier cascaded with a high-power output stage. Overall feedback is applied either on-chip or externally.
- The bridge amplifier configuration provides, across a floating load, a peak-to-peak output voltage which is twice that possible from a single amplifier with a grounded load.
- The DMOS transistor is a short-channel power device capable of both high-current and high-voltage operation.
- The drain current of a power MOSFET exhibits a positive temperature coefficient at low currents, and thus the device can suffer thermal runaway. At high currents the temperature coefficient of i_D is negative.

Computer Simulation Problems

SIM Problems identified by this icon are intended to demonstrate the value of using SPICE simulation to verify hand analysis and design, and to investigate important issues such as allowable signal swing and amplifier nonlinear distortion. Instructions to assist in setting up PSpice and Multism simulations for all the indicated problems can be found in the corresponding files on the disc. Note that if a particular parameter value is not specified in the problem statement, you are to make a reasonable assumption.
* difficult problem; ** more difficult; *** very challenging and/or time-consuming; D: design problem.

Section 11.2: Class A Output Stage

11.1 A class A emitter follower, biased using the circuit shown in Fig. 11.2, uses $V_{CC} = 5$ V, $R = R_L = 1$ kΩ, with all transistors (including Q_3) identical. Assume $V_{BE} = 0.7$ V, $V_{CEsat} = 0.3$ V, and β to be very large. For linear operation, what are the upper and lower limits of output voltage, and the corresponding inputs? How do these values change if the emitter–base junction area of Q_3 is made twice as big as that of Q_2? Half as big?

11.2 A source-follower circuit using NMOS transistors is constructed following the pattern shown in Fig. 11.2. All three transistors used are identical, with $V_t = 1$ V and $\mu_n C_{ox} W/L = 20$ mA/V²; $V_{CC} = 5$ V, $R = R_L = 1$ kΩ. For linear operation, what are the upper and lower limits of the output voltage, and the corresponding inputs?

D 11.3 Using the follower configuration shown in Fig. 11.2 with ±9-V supplies, provide a design capable of ±7-V outputs with a 1-kΩ load, using the smallest possible total supply current. You are provided with four identical, high-β BJTs and a resistor of your choice.

D 11.4 An emitter follower using the circuit of Fig. 11.2, for which the output voltage range is ±5 V, is required using $V_{CC} = 10$ V. The circuit is to be designed such that the current variation in the emitter-follower transistor is no greater than a factor of 10, for load resistances as low as 100 Ω. What is the value of R required? Find the incremental voltage gain of the resulting follower at $v_O = +5$, 0, and –5 V, with a 100-Ω load. What is the percentage change in gain over this range of v_O?

***11.5** Consider the operation of the follower circuit of Fig. 11.2 for which $R_L = V_{CC}/I$, when driven by a square wave such that the output ranges from $+V_{CC}$ to $-V_{CC}$ (ignoring V_{CEsat}). For this situation, sketch the equivalent of Fig. 11.4 for v_O, i_{C1}, and p_{D1}. Repeat for a square-wave output that has peak levels of $\pm V_{CC}/2$. What is the average power dissipation in Q_1 in each case? Compare these results to those for sine waves of peak amplitude V_{CC} and $V_{CC}/2$, respectively.

11.6 Consider the situation described in Problem 11.5. For square-wave outputs having peak-to-peak values of $2V_{CC}$ and V_{CC}, and for sine waves of the same peak-to-peak values, find the average power loss in the current-source transistor Q_2.

11.7 Reconsider the situation described in Exercise 11.3 for variation in V_{CC}—specifically for $V_{CC} = 16$ V, 12 V, 10 V, and 8 V. Assume V_{CEsat} is nearly zero. What is the power-conversion efficiency in each case?

Section 11.3: Class B Output Stage

11.8 Consider the circuit of a complementary-BJT class B output stage. For what amplitude of input signal does the crossover distortion represent a 10% loss in peak amplitude?

11.9 Consider the feedback configuration with a class B output stage shown in Fig. 11.9. Let the amplifier gain $A_0 = 100$ V/V. Derive an expression for v_O versus v_I, assuming that $|V_{BE}| = 0.7$ V. Sketch the transfer characteristic v_O versus v_I, and compare it with that without feedback.

SIM **11.10** Consider the class B output stage, using enhancement MOSFETs, shown in Fig. P11.10. Let the devices have $|V_t| = 0.5$ V and $\mu C_{ox} W/L = 2$ mA/V². With a 10-kHz sine-wave input of 5-V peak and a high value of load resistance, what peak output would you expect? What fraction of the sine-wave period does the crossover interval represent? For what value of load resistor is the peak output voltage reduced to half the input?

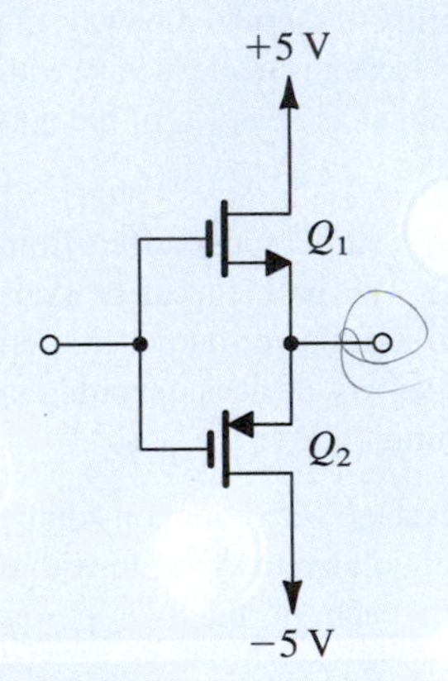

Figure P11.10

11.11 Consider the complementary-BJT class B output stage and neglect the effects of finite V_{BE} and V_{CEsat}. For ±10-V power supplies and a 100-Ω load resistance, what is the maximum sine-wave output power available? What supply power corresponds? What is the power-conversion efficiency? For output signals of half this amplitude, find the output power, the supply power, and the power-conversion efficiency.

D 11.12 A class B output stage operates from ±5-V supplies. Assuming relatively ideal transistors, what is the output voltage for maximum power-conversion efficiency? What is the output voltage for maximum device dissipation? If each of the output devices is individually rated for 1-W dissipation, and a factor-of-2 safety margin is to be used, what is the smallest value of load resistance that can be tolerated, if operation is always at full output voltage? If operation is allowed at half the full output voltage, what is the smallest load permitted? What is the greatest possible output power available in each case?

D 11.13 A class B output stage is required to deliver an average power of 100 W into a 16-Ω load. The power supply should be 4 V greater than the corresponding peak sine-wave output voltage. Determine the power-supply voltage required (to the nearest volt in the appropriate direction), the peak current from each supply, the total supply power, and the power-conversion efficiency. Also, determine the maximum possible power dissipation in each transistor for a sine-wave input.

11.14 Consider the class B BJT output stage with a square-wave output voltage of amplitude $\hat{V}_o$ across a load R_L and employing power supplies $\pm V_{SS}$. Neglecting the effects of finite V_{BE} and V_{CEsat}, determine the load power, the supply power, the power-conversion efficiency, the maximum attainable power-conversion efficiency and the corresponding value of $\hat{V}_o$, and the maximum available load power. Also find the value of $\hat{V}_o$ at which the power dissipation in the transistors reaches its peak, and the corresponding value of power-conversion efficiency.

Section 11.4: Class AB Output Stage

D 11.15 Design the quiescent current of a class AB BJT output stage so that the incremental voltage gain for v_I in the vicinity of the origin is in excess of 0.98 V/V for loads larger than 100 Ω. Assume that the BJTs have V_{BE} of 0.7 V at a current of 100 mA and determine the value of V_{BB} required.

11.16 For the class AB output stage considered in Example 11.3, add two columns to the table of results as follows: the total input current drawn from v_I (i_I, mA); and the large-signal input resistance $R_{\text{in}} \equiv v_I/i_I$. Assume $\beta_N = \beta_P = \beta = 49$. Compare the values of R_{in} to the approximate value obtained using the resistance reflection rule, $R_{\text{in}} \simeq \beta R_L$.

11.17 In this problem we investigate an important trade-off in the design of the class AB output stage of Fig. 11.11: Increasing the quiescent current I_Q reduces the nonlinearity of the transfer characteristic at the expense of increased quiescent power dissipation. As a measure of nonlinearity, we use the maximum deviation of the stage incremental gain, which occurs at $v_O = 0$, namely

$$\varepsilon = 1 - v_o/v_i\big|_{v_O = 0}$$

(a) Show that ε is given by

$$\varepsilon = \frac{V_T/2I_Q}{R_L + (V_T/2I_Q)}$$

which for $2I_QR_L \gg V_T$ can be approximated by

$$\varepsilon \simeq V_T/2I_QR_L$$

(b) If the stage is operated from power supplies of $\pm 2V_{CC}$, find the quiescent power dissipation, P_D.
(c) Show that for given V_{CC} and R_L, the product of the quiescent power dissipation and the gain error is a constant given by

$$\varepsilon P_D \simeq V_T\left(\frac{V_{CC}}{R_L}\right)$$

(d) For $V_{CC} = 15$ V and $R_L = 100\ \Omega$, find the required values of P_D and I_Q if ε is to be 5%, 2%, and 1%.

***11.18** A class AB output stage, resembling that in Fig. 11.11 but utilizing a single supply of +10 V and biased at $V_I = 6$ V, is capacitively coupled to a 100-Ω load. For transistors for which $|V_{BE}| = 0.7$ V at 1 mA and for a bias voltage $V_{BB} = 1.4$ V, what quiescent current results? For a step change in output from 0 to −1 V, what input step is required? Assuming transistor saturation voltages of zero, find the largest possible positive-going and negative-going steps at the output.

Section 11.5: Biasing the Class AB Circuit

D 11.19 Consider the diode-biased class AB circuit of Fig. 11.14. For $I_{\text{BIAS}} = 100$ μA, find the relative size (n) that should be used for the output devices (in comparison to the biasing devices) to ensure that an output resistance of 10 Ω or less is obtained in the quiescent state. Neglect the resistance of the biasing diodes.

D *11.20 A class AB output stage using a two-diode bias network as shown in Fig. 11.14 utilizes diodes having the same junction area as the output transistors. For $V_{CC} = 10$ V, $I_{\text{BIAS}} = 0.5$ mA, $R_L = 100\ \Omega$, $\beta_N = 50$, and $|V_{CEsat}| = 0$ V, what is the quiescent current? What are the largest possible positive and negative output signal levels? To achieve a positive peak output level equal to the negative peak level, what value of β_N is needed if I_{BIAS} is not changed? What value of I_{BIAS} is needed if β_N is held at 50? For this value, what does I_Q become?

****11.21** A class AB output stage using a two-diode bias network as shown in Fig. 11.14 utilizes diodes having the same junction area as the output transistors. At a room temperature of about 20°C the quiescent current is 1 mA and $|V_{BE}| = 0.6$ V. Through a manufacturing error, the thermal coupling between the output transistors and the biasing diode-connected transistors is omitted. After some output activity, the output devices heat up to 70°C while the biasing devices remain at 20°C. Thus, while the V_{BE} of each device remains unchanged, the quiescent current in the output devices increases. To calculate the new current value, recall that there are two effects: I_s increases by about 14%/°C and $V_T = kT/q$ changes, where $T = (273° +$

temperature in °C), and $V_T = 25$ mV only at 20°C. However, you may assume that β_N remains almost constant. This assumption is based on the fact that β increases with temperature but decreases with current. What is the new value of I_Q? If the power supply is ±20 V, what additional power is dissipated? If thermal runaway occurs, and the temperature of the output transistors increases by 10°C for every watt of additional power dissipation, what additional temperature rise and current increase result?

D 11.22 Repeat Example 11.5 for the situation in which the peak positive output current is 200 mA. Use the same general approach to safety margins. What are the values of R_1 and R_2 you have chosen?

****11.23** A V_{BE} multiplier is designed with equal resistances for nominal operation at a terminal current of 1 mA, with half the current flowing in the bias network. The initial design is based on $\beta = \infty$ and $V_{BE} = 0.7$ V at 1 mA.

(a) Find the required resistor values and the terminal voltage.
(b) Find the terminal voltage that results when the terminal current increases to 2 mA. Assume $\beta = \infty$.
(c) Repeat (b) for the case the terminal current becomes 10 mA.
(d) Repeat (c) using the more realistic value of $\beta = 100$.

Section 11.6: CMOS Class AB Output Stages

D 11.24 (a) Show that for the class AB circuit in Fig. 11.17, the small-signal output resistance in the quiescent state is given by

$$R_{\text{out}} = \frac{1}{g_{mn} + g_{mp}}$$

which for matched devices becomes

$$R_{\text{out}} = \frac{1}{2g_m}$$

(b) For a circuit that utilizes MOSFETs with $|V_t| = 0.7$ V and $k'(W/L) = 200$ mA/V^2, find the voltage V_{GG} that results in $R_{\text{out}} = 10\ \Omega$.

D 11.25 (a) For the circuit in Fig. 11.17 in which Q_1 and Q_2 are matched, and Q_N and Q_P are matched, show that the small-signal voltage gain at the quiescent condition is given by

$$\frac{v_o}{v_i} = \frac{R_L}{R_L + (2/g_m)}$$

where g_m is the transconductance of each of Q_N and Q_P and where channel-length modulation is neglected.

(b) For the case $I_{\text{BIAS}} = 0.1$ mA, $R_L = 1$ kΩ, $k_n = k_p = nk_1 = nk_2$, where $k = \mu C_{ox}(W/L)$, and $k_1 = 20$ mA/V^2, find the ratio n that results in an incremental gain of 0.98. Also find the quiescent current I_Q.

D 11.26 Design the circuit of Fig. 11.17 to operate at $I_Q = 1$ mA with $I_{\text{BIAS}} = 0.1$ mA. Let $\mu_n C_{ox} = 250\ \mu$A/V^2, $\mu_p C_{ox} = 100\ \mu$A/V^2, $V_{tn} = -V_{tp} = 0.45$ V, and $V_{DD} = V_{SS} = 2.5$ V. Design so that Q_1 and Q_2 are matched and Q_N and Q_P are matched, and that in the quiescent state each operates at an overdrive voltage of 0.2 V.

(a) Specify the W/L ratio for each of the four transistors.
(b) In the quiescent state with $v_O = 0$, what must v_I be?
(c) If Q_N is required to supply a maximum load current of 10 mA, find the maximum allowable output voltage. Assume that the transistor supplying I_{BIAS} needs a minimum of 0.2 V to operate properly.

11.27 For the CMOS output stage of Fig. 11.19 with $I_Q = 3$ mA, $|V_{OV}| = 0.15$ V for each of Q_P and Q_N at the quiescent point, and $\mu = 5$, find the output resistance at the quiescent point.

11.28 (a) Show that for the CMOS output stage of Fig. 11.19,

$$|\text{Gain error}| = \frac{R_{\text{out}}}{R_L}$$

(b) For a stage that drives a load resistance of 100 Ω with a gain error of less than 5%, find the overdrive voltage at which Q_P and Q_N should be operated. Let $I_Q = 1$ mA and $\mu = 10$.

D 11.29 It is required to design the circuit of Fig. 11.19 to drive a load resistance of 50 Ω while exhibiting an output resistance, around the quiescent point, of 2.5 Ω. Operate Q_N and Q_P at $I_Q = 1.5$ mA and $|V_{OV}| = 0.15$ V. The technology utilized is specified to have $k_n' = 250\ \mu$A/V^2, $k_p' = 100\ \mu$A/V^2, $V_{tn} = -V_{tp} = 0.5$ V, and $V_{DD} = V_{SS} = 2.5$ V.

(a) Specify (W/L) for each of Q_N and Q_P.
(b) Specify the required value of μ.
(c) What is the expected error in the stage gain?
(d) In the quiescent state, what dc voltage must appear at the output of each of the error amplifiers?
(e) At what value of positive v_O will Q_P be supplying all the load current? Repeat for negative v_O and Q_N supplying all the load current.
(f) What is the linear range of v_O?

Section 11.7: Power BJTs

D 11.30 A particular transistor having a thermal resistance $\theta_{JA} = 2$°C/W is operating at an ambient temperature of 30°C with a collector–emitter voltage of 20 V. If long life requires a maximum junction temperature of 130°C, what is the corresponding device power rating? What is the greatest average collector current that should be considered?

11.31 A particular transistor has a power rating at 25°C of 200 mW, and a maximum junction temperature of 150°C. What is its thermal resistance? What is its power rating when operated at an ambient temperature of 70°C? What is its junction temperature when dissipating 100 mW at an ambient temperature of 50°C?

11.32 A power transistor operating at an ambient temperature of 50°C, and an average emitter current of 3 A, dissipates 30 W. If the thermal resistance of the transistor is known to be less than 3°C/W, what is the greatest junction temperature you would expect? If the transistor V_{BE} measured using a pulsed emitter current of 3 A at a junction temperature of 25°C is 0.80 V, what average V_{BE} would you expect under normal operating conditions? (Use a temperature coefficient of –2 mV/°C.)

11.33 For a particular application of the transistor specified in Example 11.7, extreme reliability is essential. To improve reliability, the maximum junction temperature is to be limited to 100°C. What are the consequences of this decision for the conditions specified?

11.34 A power transistor is specified to have a maximum junction temperature of 130°C. When the device is operated at this junction temperature with a heat sink, the case temperature is found to be 90°C. The case is attached to the heat sink with a bond having a thermal resistance $\theta_{CS} = 0.5$°C/W and the thermal resistance of the heat sink $\theta_{SA} = 0.1$°C/W. If the ambient temperature is 30°C what is the power being dissipated in the device? What is the thermal resistance of the device, θ_{JC}, from junction to case?

11.35 A power transistor for which $T_{Jmax} = 180$°C can dissipate 50 W at a case temperature of 50°C. If it is connected to a heat sink using an insulating washer for which the thermal resistance is 0.6°C/W, what heat-sink temperature is necessary to ensure safe operation at 30 W? For an ambient temperature of 39°C, what heat-sink thermal resistance is required? If, for a particular extruded-aluminum-finned heat sink, the thermal resistance in still air is 4.5°C/W per centimeter of length, how long a heat sink is needed?

Section 11.8: Variations on the Class AB Configuration

11.36 Use the results given in the answer to Exercise 11.14 to determine the input current of the circuit in Fig. 11.30 for $v_I = 0$ and ±10 V with infinite and 100-Ω loads.

11.37 For the circuit in Fig 11.30 when operated near $v_I = 0$ and fed with a signal source having zero resistance, show that the output resistance is given by

$$R_{out} = \frac{1}{2}[R_3 + r_{e3} + (R_1 \| r_{e1})/(\beta_3 + 1)]$$

Assume that the top and bottom halves of the circuit are perfectly matched.

D ***11.38 Consider the circuit of Fig. 11.30 in which Q_1 and Q_2 are matched, and Q_3 and Q_4 are matched but have three times the junction area of the others. For $V_{CC} = 10$ V, find values for resistors R_1 through R_4 which allow for a base current of at least 10 mA in Q_3 and Q_4 at $v_I = +5$ V (when a load demands it) with at most a 2-to-1 variation in currents in Q_1 and Q_2, and a no-load quiescent current of 40 mA in Q_3 and Q_4; $\beta_{1,2} \geq 150$, and $\beta_{3,4} \geq 50$. For input voltages around 0 V, estimate the output resistance of the overall follower driven by a source having zero resistance. For an input voltage of +1 V and a load resistance of 2 Ω, what output voltage results? Q_1 and Q_2 have $|V_{BE}|$ of 0.7 V at a current of 10 mA.

11.39 Figure P11.39 shows a variant of the class AB circuit of Fig. 11.30. Assume that all four transistors are matched and have $\beta = 100$.

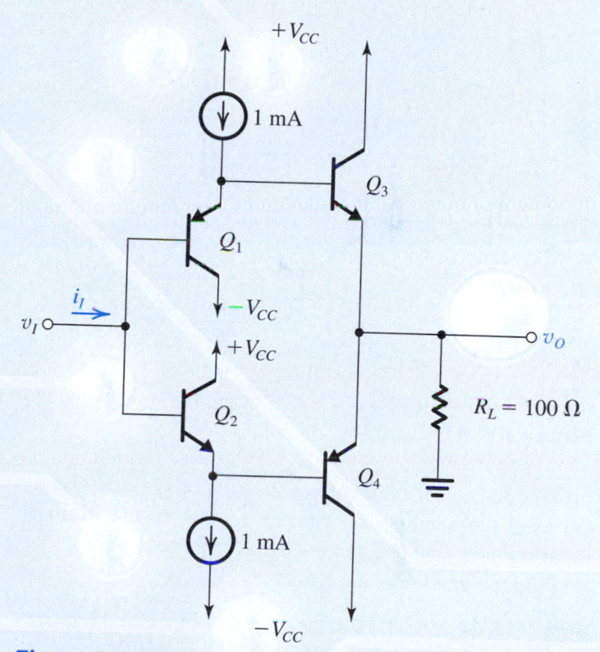

Figure P11.39

(a) For $v_I = 0$, find the quiescent current in Q_3 and Q_4, the input current i_I, and the output voltage v_O.
(b) Since the circuit has perfect symmetry, the small-signal performance around $v_I = 0$ can be determined by considering either the top or bottom half of the circuit only. In this case, the load on the half-circuit must be $2R_L$, the input resistance found is $2R_{in}$, and the output resistance found is $2R_{out}$. Using this approach, find R_{in}, v_o/v_i, and R_{out} (assuming that the circuit is fed with a zero-resistance source).

11.40 For the Darlington configuration shown in Fig. 11.31, show that for $\beta_1 \gg 1$ and $\beta_2 \gg 1$:

(a) The equivalent composite transistor has $\beta \simeq \beta_1\beta_2$.
(b) If the composite transistor is operated at a current I_C, then Q_2 will be operating at a collector current approximately

equal to I_C, and Q_1 will be operating at a collector current approximately equal to I_C/β_2.
(c) The composite transistor has a $V_{BE} \simeq 2V_T \ln(I_C/I_S) - V_T \ln(\beta_2)$, where I_S is the saturation current of each of Q_1 and Q_2.
(d) The composite transistor has an equivalent $r_\pi \simeq 2\beta_1\beta_2(V_T/I_C)$.
(e) The composite transistor has an equivalent $g_m \simeq \frac{1}{2}(I_C/V_T)$.

***11.41** For the circuit in Fig. P11.41 in which the transistors have $V_{BE} = 0.7$ V and $\beta = 100$:

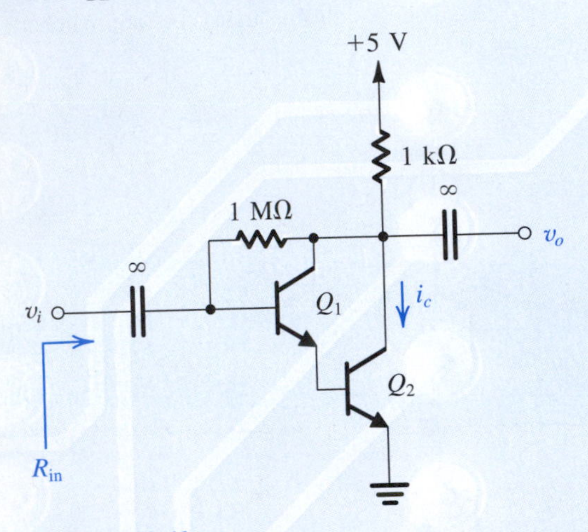

Figure P11.41

(a) Find the dc collector current for each of Q_1 and Q_2.
(b) Find the small-signal current i_c that results from an input signal v_i, and hence find the voltage gain v_o/v_i.
(c) Find the input resistance R_{in}.

SIM ****11.42** The BJTs in the circuit of Fig. P11.42 have $\beta_P = 10$, $\beta_N = 100$, $|V_{BE}| = 0.7$ V, and $|V_A| = 100$ V.

(a) Find the dc collector current of each transistor and the value of V_C.
(b) Replacing each BJT with its hybrid-π model, show that

$$\frac{v_o}{v_i} \simeq g_{m1}[r_{o1} \| \beta_N(r_{o2} \| R_f)]$$

(c) Find the values of v_o/v_i and R_{in}.

D ****11.43 Consider the compound-transistor class AB output stage shown in Fig. 11.33 in which Q_2 and Q_4 are matched transistors with $V_{BE} = 0.7$ V at 10 mA and $\beta = 100$, Q_1 and Q_5 have $V_{BE} = 0.7$ V at 1-mA currents and $\beta = 100$, and Q_3 has $V_{EB} = 0.7$ V at a 1-mA current and $\beta = 10$. Design the circuit for a quiescent current of 2 mA in Q_2 and Q_4, I_{BIAS} that is 100 times the standby base current in Q_1, and a current in Q_5 that is nine times that in the associated resistors. Find the values of the input voltage required to produce outputs of ±10 V for a 1-kΩ load. Use V_{CC} of 15 V.

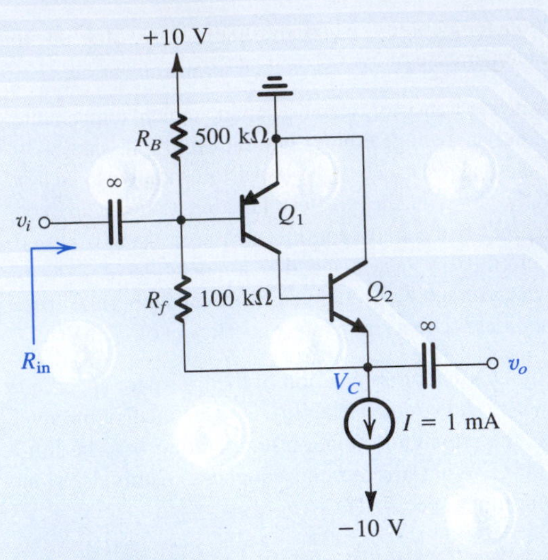

Figure P11.42

11.44 Repeat Exercise 11.16 for a design variation in which transistor Q_5 is increased in size by a factor of 10, all other conditions remaining the same.

11.45 Repeat Exercise 11.16 for a design in which the limiting output current and normal peak current are 50 mA and 33.3 mA, respectively.

D 11.46 The circuit shown in Fig. P11.46 operates in a manner analogous to that in Fig. 11.35 to limit the output current from Q_3 in the event of a short circuit or other mishap. It has the advantage that the current-sensing resistor R does not appear directly at the output. Find the value of R that causes Q_5 to turn on and absorb all of $I_{BIAS} = 2$ mA, when the current being sourced reaches 150 mA. For Q_5,

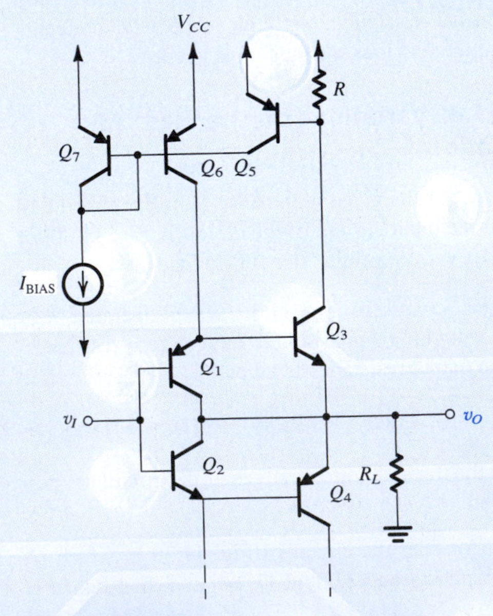

Figure P11.46

$I_S = 10^{-14}$ A. If the normal peak output current is 100 mA, find the voltage drop across R and the collector current in Q_5.

D 11.47 Consider the thermal shutdown circuit shown in Fig. 11.35. At 25°C, Z_1 is a 6.8-V zener diode with a TC of 2 mV/°C, and Q_1 and Q_2 are BJTs that display V_{BE} of 0.7 V at a current of 100 μA and have a TC of −2 mV/°C. Design the circuit so that at 125°C, a current of 100 μA flows in each of Q_1 and Q_2. What is the current in Q_2 at 25°C?

Section 11.9: IC Power Amplifiers

D 11.48 In the power-amplifier circuit of Fig. 11.36 two resistors are important in controlling the overall voltage gain. Which are they? Which controls the gain alone? Which affects both the dc output level and the gain? A new design is being considered in which the output dc level is approximately $\frac{1}{3}V_S$ (rather than approximately $\frac{1}{2}V_S$) with a gain of 50 (as before). What changes are needed?

11.49 Consider the front end of the circuit in Fig. 11.36. For $V_S = 20$ V, calculate approximate values for the bias currents in Q_1 through Q_6. Assume $\beta_{npn} = 100$, $\beta_{pnp} = 20$, and $|V_{BE}| = 0.7$ V. Also find the dc voltage at the output.

11.50 It is required to use the LM380 power amplifier to drive an 8-Ω loudspeaker while limiting the maximum possible device dissipation to 1.5 W. Use the graph of Fig. 11.38 to determine the maximum possible power-supply voltage that can be used. (Use only the given graphs; do not interpolate.) If the maximum allowed THD is to be 3%, what is the maximum possible load power? To deliver this power to the load what peak-to-peak output sinusoidal voltage is required?

D *11.51 Consider the power-op-amp output stage shown in Fig. 11.39. Using a ±15-V supply, provide a design that provides an output of ±11 V or more, with currents up to ±20 mA provided primarily by Q_3 and Q_4 with a 10% contribution by Q_5 and Q_6, and peak output currents of 1 A at full output (+11 V). As the basis of an initial design, use $\beta = 50$ and $|V_{BE}| = 0.7$ V for all devices at all currents. Also use $R_5 = R_6 = 0$.

11.52 For the circuit in Fig. P11.52, assuming all transistors to have large β, show that $i_O = v_I/R$. [This voltage-to-current converter is an application of a versatile circuit building block known as the **current conveyor**; see Sedra and Roberts (1990)]. For $\beta = 100$, by what approximate percentage is i_O actually lower than this ideal value?

D 11.53 For the bridge amplifier of Fig. 11.40, let $R_1 = R_3 = 10$ kΩ. Find R_2 and R_4 to obtain an overall gain of 10.

D 11.54 An alternative bridge amplifier configuration, with high input resistance, is shown in Fig. P11.54. (Note the similarity of this circuit to the front end of the instrumentation amplifier circuit shown in Fig. 2.20b.) What is the gain v_O/v_I? For op amps (using ±15-V supplies) that limit at ±13 V, what is the largest sine wave you can provide across R_L? Using 1 kΩ as the smallest resistor, find resistor values that make $v_O/v_I = 10$ V/V. Make sure that the signals at the outputs of the two amplifiers are complementary.

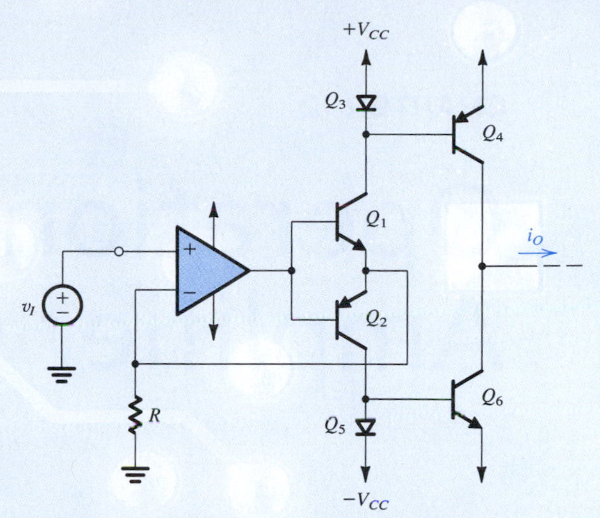

Figure P11.52

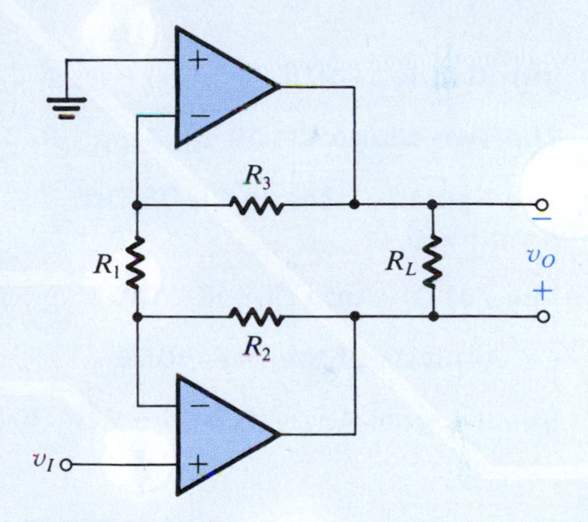

Figure P11.54

Section 11.10: MOS Power Transistors

D 11.55 Consider the design of the class AB amplifier of Fig. 11.44 under the following conditions: $|V_t| = 2$ V, $\mu C_{ox} W/L = 200$ mA/V^2, $|V_{BE}| = 0.7$ V, β is high, $I_{QN} = I_{QP} = I_R = 10$ mA, $I_{BIAS} = 100$ μA, $I_{Q5} = I_{Q6} = I_{BIAS}/2$, $R_2 = R_4$, the temperature coefficient of $V_{BE} = -2$ mV/°C, and the temperature coefficient of $V_t = -3$ mV/°C in the low-current region. Find the values of R, R_1, R_2, R_3, and R_4. Assume Q_6, Q_P, and Q_N to be thermally coupled. (R_G, used to suppress parisitic oscillation at high frequency, is usually 100 Ω or so.)

CHAPTER 12

Operational-Amplifier Circuits

IN THIS CHAPTER YOU WILL LEARN

1. The design and analysis of the two basic CMOS op-amp architectures: the two-stage circuit and the single-stage, folded-cascode circuit.
2. The complete circuit of an analog IC classic: the 741 op amp. Though 40 years old, the 741 circuit includes so many interesting and useful design techniques that its study is still a must.
3. Interesting and useful applications of negative feedback within op-amp circuits to achieve bias stability and increased CMRR.
4. How to break a large analog circuit into its recognizable blocks, to be able to make the analysis amenable to a pencil-and-paper approach, which is the best way to learn design.
5. Some of the modern techniques employed in the design of low-voltage, single-supply BJT op amps.
6. Most importantly, how the different topics we learned about in the preceding chapters come together in the design of the most important analog IC, the op amp.

Introduction

In this chapter, we shall study the internal circuitry of the most important analog IC, namely, the operational amplifier. The terminal characteristics and some circuit applications of op amps were covered in Chapter 2. Here, our objective is to expose the reader to some of the ingenious techniques that have evolved over the years for combining elementary analog circuit building blocks to realize a complete op amp. We shall study both CMOS and bipolar op amps. The CMOS op-amp circuits considered find application primarily in the design of analog and mixed-signal VLSI circuits. Because these op amps are usually designed with a specific application in mind, they can be optimized to meet a subset of the list of desired specifications, such as high dc gain, wide bandwidth, or large output-signal swing. For instance, many CMOS op amps are utilized within an IC and do not connect to the outside terminals of the chip. As a result, the loads on their outputs are usually limited to small capacitances of at most few picofarads. Internal CMOS op amps therefore do not need to have low output resistances, and their design rarely incorporates an output stage. Also, if the op-amp input terminals are not connected to the chip terminals, there will be no danger of static charge damaging the gate oxide of the input MOSFETs. Hence, internal CMOS op amps do not need input clamping diodes for gate protection and thus do not suffer from the leakage effects of such diodes. In other words, the advantage of near-infinite input resistance of the MOSFET is fully realized.

While CMOS op amps are extensively used in the design of VLSI systems, the BJT remains the device of choice in the design of general-purpose op amps. These are op amps that are utilized in a wide variety of applications and are designed to fit a wide range of specifications. As a result, the circuit of a general-purpose op amp represents a compromise among many performance parameters. We shall study in detail one such circuit, the 741-type op amp. Although the 741 has been available for nearly 40 years, its internal circuit remains as relevant and interesting today as it ever was. Nevertheless, changes in technology have introduced new requirements, such as the need for general-purpose op amps that operate from a single power supply of only 2 V to 3 V. These new requirements have given rise to exciting challenges to op-amp designers. The result has been a wealth of new ideas and design techniques. We shall present a sample of these modern design techniques in the last section.

In addition to exposing the reader to some of the ideas that make analog IC design such an exciting topic, this chapter should serve to tie together many of the concepts and methods studied thus far.

12.1 The Two-Stage CMOS Op Amp

The first op-amp circuit we shall study is the two-stage CMOS topology shown in Fig. 12.1. This simple but elegant circuit has become a classic and is used in a variety of forms in the design of VLSI systems. We have already studied this circuit in Section 8.6.1 as an example of a multistage CMOS amplifier. We urge the reader to review Section 8.6.1 before proceeding further. Here, our discussion will emphasize the performance characteristics of the circuit and the trade-offs involved in its design.

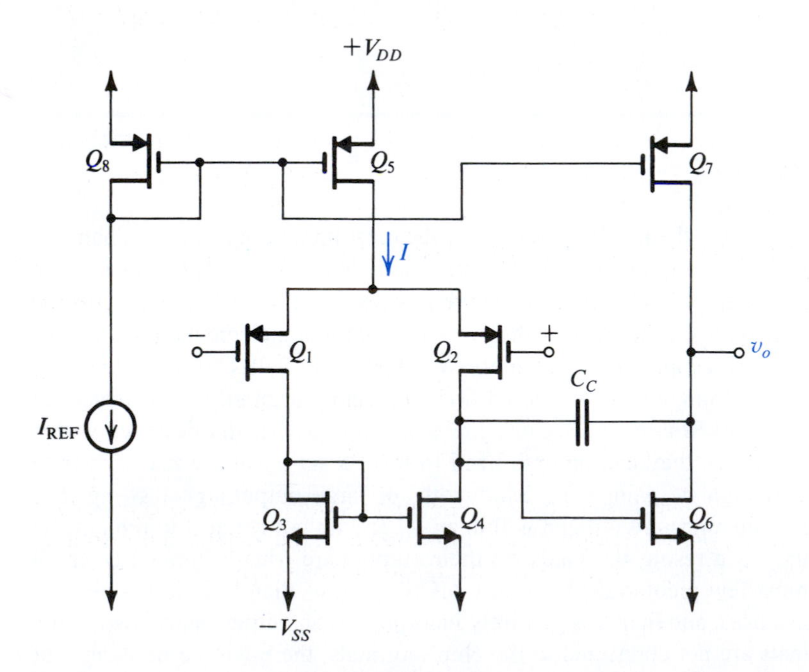

Figure 12.1 The basic two-stage CMOS op-amp configuration.

12.1.1 The Circuit

The circuit consists of two gain stages: The first stage is formed by the differential pair Q_1–Q_2 together with its current mirror load Q_3–Q_4. This differential-amplifier circuit, studied in detail in Section 8.5, provides a voltage gain that is typically in the range of 20 V/V to 60 V/V, as well as performing conversion from differential to single-ended form while providing a reasonable common-mode rejection ratio (CMRR).

The differential pair is biased by current source Q_5, which is one of the two output transistors of the current mirror formed by Q_8, Q_5, and Q_7. The current mirror is fed by a reference current I_{REF}, which can be generated by simply connecting a precision resistor (external to the chip) to the negative supply voltage $-V_{SS}$ or to a more precise negative voltage reference if one is available in the same integrated circuit. Alternatively, for applications with more stringent requirements, I_{REF} can be generated using a circuit such as that studied in Section 8.6.1 (Fig. 8.41).

The second gain stage consists of the common-source transistor Q_6 and its current-source load Q_7. The second stage typically provides a gain of 50 V/V to 80 V/V. In addition, it takes part in the process of frequency compensating the op amp. From Section 10.13 the reader will recall that to guarantee that the op amp will operate in a stable fashion (as opposed to oscillating) when negative feedback of various amounts is applied, the open-loop gain is made to roll off with frequency at the uniform rate of −20 dB/decade. This in turn is achieved by introducing a pole at a relatively low frequency and arranging for it to dominate the frequency-response determination. In the circuit we are studying, this is implemented using a compensation capacitance C_C connected in the negative-feedback path of the second-stage amplifying transistor Q_6. As will be seen, C_C (together with the much smaller capacitance C_{gd6} across it) is Miller-multiplied by the gain of the second stage, and the resulting capacitance at the input of the second stage interacts with the total resistance there to provide the required dominant pole (more on this later).

Unless properly designed, the CMOS op-amp circuit of Fig. 12.1 can exhibit a **systematic output dc offset** voltage. This point was discussed in Section 8.6.1, where it was found that the dc offset can be eliminated by sizing the transistors so as to satisfy the following constraint:

$$\frac{(W/L)_6}{(W/L)_4} = 2\frac{(W/L)_7}{(W/L)_5} \tag{12.1}$$

Finally, we observe that the CMOS op-amp circuit of Fig. 12.1 does not have an output stage. This is because it is usually required to drive only small on-chip capacitive loads.

12.1.2 Input Common-Mode Range and Output Swing

Refer to Fig. 12.1 and consider the situation when the two input terminals are tied together and connected to a voltage V_{ICM}. The lowest value of V_{ICM} has to be sufficiently large to keep Q_1 and Q_2 in saturation. Thus, the lowest value of V_{ICM} should not be lower than the voltage at the drain of Q_1 ($-V_{SS} + V_{GS3} = -V_{SS} + V_{tn} + V_{OV3}$) by more than $|V_{tp}|$, thus

$$V_{ICM} \geq -V_{SS} + V_{tn} + V_{OV3} - |V_{tp}| \tag{12.2}$$

The highest value of V_{ICM} should ensure that Q_5 remains in saturation; that is, the voltage across Q_5, V_{SD5}, should not decrease below $|V_{OV5}|$. Equivalently, the voltage at the drain of Q_5 should not go higher than $V_{DD} - |V_{OV5}|$. Thus the upper limit of V_{ICM} is

$$V_{ICM} \leq V_{DD} - |V_{OV5}| - V_{SG1}$$

or equivalently

$$V_{ICM} \leq V_{DD} - |V_{OV5}| - |V_{tp}| - |V_{OV1}| \tag{12.3}$$

The expressions in Eqs. (12.2) and (12.3) can be combined to express the input common-mode range as

$$-V_{SS} + V_{OV3} + V_{tn} - |V_{tp}| \leq V_{ICM} \leq V_{DD} - |V_{tp}| - |V_{OV1}| - |V_{OV5}| \quad (12.4)$$

As expected, the overdrive voltages, which are important design parameters, subtract from the dc supply voltages, thereby reducing the input common-mode range. It follows that from a V_{ICM} range point of view it is desirable to select the values of V_{OV} as low as possible. We observe from Eq. (12.4) that the lower limit of V_{ICM} is approximately within an overdrive voltage of $-V_{SS}$.The upper limit, however, is not as good; it is lower than V_{DD} by two overdrive voltages and a threshold voltage.

The extent of the signal swing allowed at the output of the op amp is limited at the lower end by the need to keep Q_6 saturated and at the upper end by the need to keep Q_7 saturated, thus

$$-V_{SS} + V_{OV6} \leq v_O \leq V_{DD} - |V_{OV7}| \quad (12.5)$$

Thus the ouput voltage can swing to within an overdrive voltage of each of the supply rails. This is a reasonably wide output swing and can be maximized by selecting values for $|V_{OV}|$ of Q_6 and Q_7 as low as possible.

An important requirement of an op-amp circuit is that it be possible for its output terminal to be connected back to its negative input terminal so that a unity-gain amplifier is obtained. For such a connection to be possible, there must be a substantial overlap between the allowable range of v_O and the allowable range of V_{ICM}. This is usually the case in the CMOS amplifier circuit under study.

EXERCISE

12.1 For a particular design of the two-stage CMOS op amp of Fig. 12.1, ±1.65-V supplies are utilized and all transistors except for Q_6 and Q_7 are operated with overdrive voltages of 0.3-V magnitude; Q_6 and Q_7 use overdrive voltages of 0.5-V magnitude. The fabrication process employed provides $V_{tn} = |V_{tp}| = 0.5$ V. Find the input common-mode range and the range allowed for v_O.
Ans. −1.35 V to 0.55 V; −1.15 V to +1.15 V

12.1.3 Voltage Gain

To determine the voltage gain and the frequency response, consider a simplified equivalent circuit model for the small-signal operation of the CMOS amplifier (Fig. 12.2), where each of the two stages is modeled as a transconductance amplifier. As expected, the input resistance is practically infinite,

$$R_{in} = \infty$$

The first-stage transconductance G_{m1} is equal to the transconductance of each of Q_1 and Q_2 (see Section 8.5),

$$G_{m1} = g_{m1} = g_{m2} \quad (12.6)$$

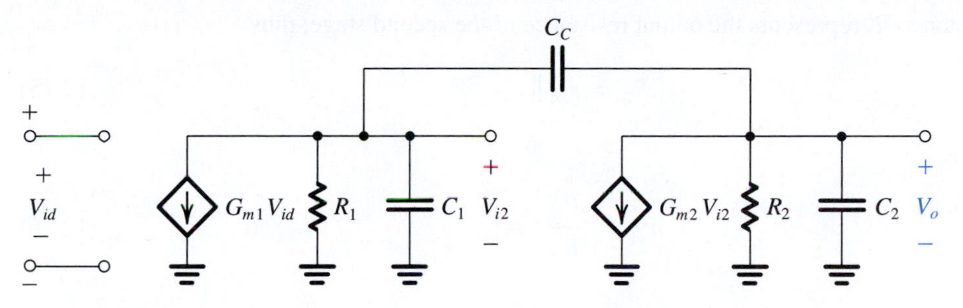

Figure 12.2 Small-signal equivalent circuit for the op amp in Fig. 12.1.

Since Q_1 and Q_2 are operated at equal bias currents $(I/2)$ and equal overdrive voltages, $V_{OV1} = V_{OV2}$,

$$G_{m1} = \frac{2(I/2)}{V_{OV1}} = \frac{I}{V_{OV1}} \tag{12.7}$$

Resistance R_1 represents the output resistance of the first stage, thus

$$R_1 = r_{o2} \parallel r_{o4} \tag{12.8}$$

where

$$r_{o2} = \frac{|V_{A2}|}{I/2} \tag{12.9}$$

and

$$r_{o4} = \frac{V_{A4}}{I/2} \tag{12.10}$$

The dc gain of the first stage is thus

$$A_1 = -G_{m1}R_1 \tag{12.11}$$

$$= -g_{m1}(r_{o2} \parallel r_{o4}) \tag{12.12}$$

$$= -\frac{2}{V_{OV1}} \Bigg/ \left[\frac{1}{|V_{A2}|} + \frac{1}{V_{A4}}\right] \tag{12.13}$$

Observe that the magnitude of A_1 is increased by operating the differential-pair transistors, Q_1 and Q_2, at a low overdrive voltage, and by choosing a longer channel length to obtain larger Early voltages, $|V_A|$.

Returning to the equivalent circuit in Fig. 12.2 and leaving the discussion of the various model capacitances until Section 12.1.5, we note that the second-stage transconductance G_{m2} is given by

$$G_{m2} = g_{m6} = \frac{2I_{D6}}{V_{OV6}} \tag{12.14}$$

Resistance R_2 represents the output resistance of the second stage, thus

$$R_2 = r_{o6} \| r_{o7} \tag{12.15}$$

where

$$r_{o6} = \frac{V_{A6}}{I_{D6}} \tag{12.16}$$

and

$$r_{o7} = \frac{|V_{A7}|}{I_{D7}} = \frac{|V_{A7}|}{I_{D6}} \tag{12.17}$$

The voltage gain of the second stage can now be found as

$$A_2 = -G_{m2}R_2 \tag{12.18}$$

$$= -g_{m6}(r_{o6} \| r_{o7}) \tag{12.19}$$

$$= -\frac{2}{V_{OV6}} \bigg/ \left[\frac{1}{V_{A6}} + \frac{1}{|V_{A7}|}\right] \tag{12.20}$$

Here again we observe that to increase the magnitude of A_2, Q_6 has to be operated at a low overdrive voltage, and the channel lengths of Q_6 and Q_7 should be made longer.

The overall dc voltage gain can be found as the product A_1A_2,

$$A_v = A_1A_2$$

$$= G_{m1}R_1G_{m2}R_2 \tag{12.21}$$

$$= g_{m1}(r_{o2} \| r_{o4})g_{m6}(r_{o6} \| r_{o7}) \tag{12.22}$$

Note that A_v is of the order of $(g_m r_o)^2$. Thus the value of A_v will be in the range of 500 V/V to 5000 V/V.

Finally, we note that the output resistance of the op amp is equal to the output resistance of the second stage,

$$R_o = r_{o6} \| r_{o7} \tag{12.23}$$

Hence R_o can be large (i.e., in the tens-of-kilohms range). Nevertheless, as we learned from the study of negative feedback in Chapter 10, application of negative feedback that samples the op-amp output voltage results in reducing the ouput resistance by a factor equal to the amount of feedback $(1 + A\beta)$. Also, as mentioned before, CMOS op amps are rarely required to drive heavy resistive loads.

EXERCISES

12.2 The CMOS op amp of Fig. 12.1 is fabricated in a process for which $V'_{An} = |V'_{Ap}| = 20$ V/μm. Find A_1, A_2, and A_v if all devices are 1 μm long, $V_{OV1} = 0.2$ V, and $V_{OV6} = 0.5$ V. Also, find the op-amp output resistance obtained when the second stage is biased at 0.5 mA.
Ans. −100 V/V; −40 V/V; 4000 V/V; 20 kΩ

12.3 If the CMOS op amp in Fig. 12.1 is connected as a unity-gain buffer, show that the closed-loop output resistance is given by

$$R_{\text{out}} \simeq 1/g_{m6}[g_{m1}(r_{o2} \| r_{o4})]$$

12.1.4 Common-Mode Rejection Ratio (CMRR)

The CMRR of the two-stage op amp of Fig. 12.1 is determined by the first stage. This was analyzed in Section 8.5.4 and the result is given in Eq. (8.147), namely,

$$\text{CMRR} = [g_{m1}(r_{o2} \| r_{o4})][2g_{m3}R_{SS}] \tag{12.24}$$

where R_{SS} is the output resistance of the bias current source Q_5. Observe that CMRR is of the order of $(g_m r_o)^2$ and thus can be reasonably high. Also, since $g_m r_o$ is proportional to $V_A/V_{OV} = V'_A L/V_{OV}$, the CMRR is increased if long channels are used, especially for Q_5, and the transistors are operated at low overdrive voltages.

12.1.5 Frequency Response

Refer to the equivalent circuit in Fig. 12.2. Capacitance C_1 is the total capacitance between the output node of the first stage and ground, thus

$$C_1 = C_{gd2} + C_{db2} + C_{gd4} + C_{db4} + C_{gs6} \tag{12.25}$$

Capacitance C_2 represents the total capacitance between the output node of the op amp and ground and includes whatever load capacitance C_L that the amplifier is required to drive, thus

$$C_2 = C_{db6} + C_{db7} + C_{gd7} + C_L \tag{12.26}$$

Usually, C_L is larger than the transistor capacitances, with the result that C_2 becomes much larger than C_1. Finally, note that C_{gd6} should be shown in parallel with C_C but has been ignored because C_C is usually much larger.

The equivalent circuit of Fig. 12.2 was analyzed in detail in Section 9.8.2, where it was found that it has two poles and a positive real-axis zero with the following approximate frequencies:

$$f_{P1} \simeq \frac{1}{2\pi R_1 G_{m2} R_2 C_C} \tag{12.27}$$

$$f_{P2} \simeq \frac{G_{m2}}{2\pi C_2} \tag{12.28}$$

$$f_Z \simeq \frac{G_{m2}}{2\pi C_C} \tag{12.29}$$

Here, f_{P1} is the dominant pole formed by the interaction of Miller-multiplied C_C [i.e., $(1 + G_{m2}R_2)C_C \simeq G_{m2}R_2C_C$] and R_1. To achieve the goal of a uniform –20-dB/decade gain rolloff down to 0 dB, the unity-gain frequency f_t,

$$f_t = |A_v|f_{P1} \tag{12.30}$$

$$= \frac{G_{m1}}{2\pi C_C} \tag{12.31}$$

must be lower than f_{P2} and f_Z, thus the design must satisfy the following two conditions

$$\frac{G_{m1}}{C_C} < \frac{G_{m2}}{C_2} \tag{12.32}$$

and

$$G_{m1} < G_{m2} \tag{12.33}$$

Simplified Equivalent Circuit The uniform –20-dB/decade gain rolloff obtained at frequencies $f \gg f_{P1}$ suggests that at these frequencies, the op amp can be represented by the simplified equivalent circuit shown in Fig. 12.3. Observe that this attractive simplification is based on the assumption that the gain of the second stage, $|A_2|$, is large, and hence a virtual ground appears at the input terminal of the second stage. The second stage then effectively acts as an integrator that is fed with the output current signal of the first stage; $G_{m1}V_{id}$. Although derived for the CMOS amplifier, this simplified equivalent circuit is general and applies to a variety of two-stage op amps, including the first two stages of the 741-type bipolar op amp studied later in this chapter.

Phase Margin The frequency compensation scheme utilized in the two-stage CMOS amplifier is of the pole-splitting type, studied in Section 10.13.3: It provides a dominant low-frequency pole with frequency f_{P1} and shifts the second pole beyond f_t. Figure 12.4 shows a

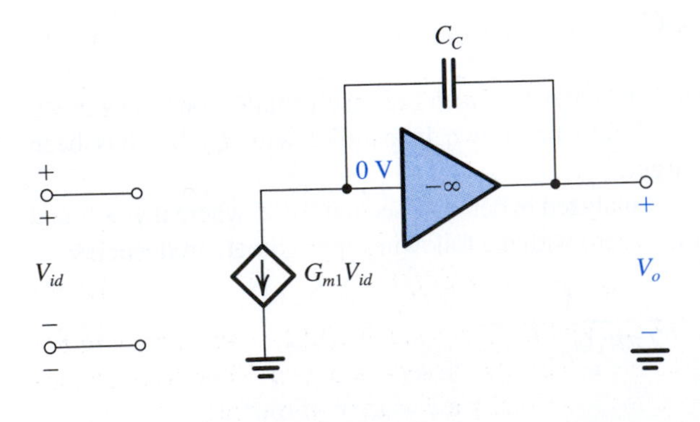

Figure 12.3 An approximate high-frequency equivalent circuit of the two-stage op amp. This circuit applies for frequencies $f \gg f_{P1}$.

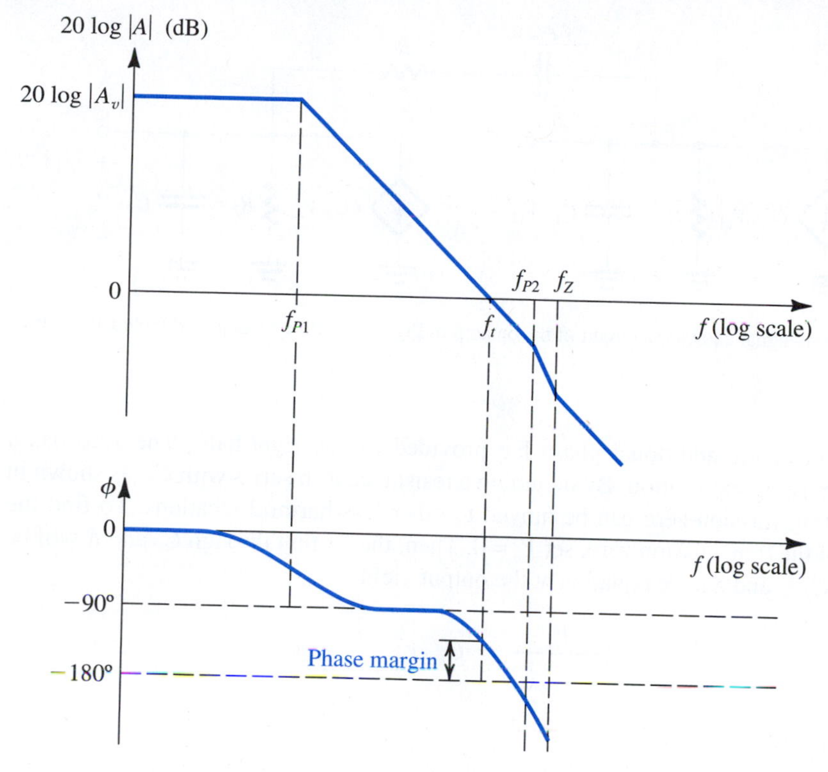

Figure 12.4 Typical frequency response of the two-stage op amp.

representative Bode plot for the gain magnitude and phase. Note that at the unity-gain frequency f_t, the phase lag exceeds the 90° caused by the dominant pole at f_{P1}. This so-called excess phase shift is due to the second pole,

$$\phi_{P2} = -\tan^{-1}\left(\frac{f_t}{f_{P2}}\right) \tag{12.34}$$

and the right-half-plane zero,

(12.35)

$$\phi_Z = -\tan^{-1}\left(\frac{f_t}{f_Z}\right) \tag{12.36}$$

Thus the phase lag at $f = f_t$ will be

$$\phi_{\text{total}} = 90° + \tan^{-1}(f_t/f_{P2}) + \tan^{-1}(f_t/f_Z) \tag{12.37}$$

and thus the phase margin will be

$$\begin{aligned}\text{Phase margin} &= 180° - \phi_{\text{total}} \\ &= 90° - \tan^{-1}(f_t/f_{P2}) - \tan^{-1}(f_t/f_Z)\end{aligned} \tag{12.38}$$

From our study of the stability of feedback amplifiers in Section 10.12.2, we know that the magnitude of the phase margin significantly affects the closed-loop gain. Therefore, obtaining a desired minimum value of phase margin is usually a design requirement.

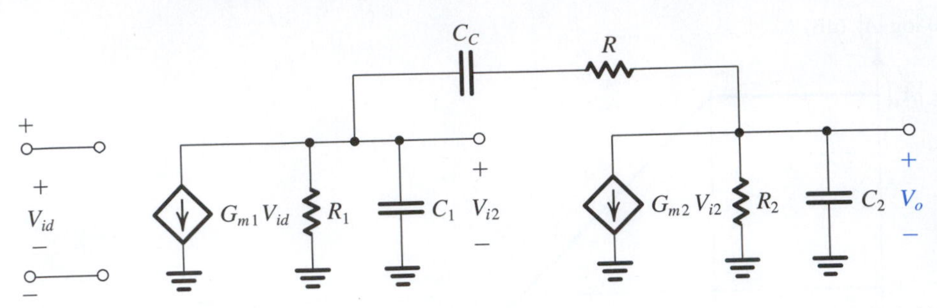

Figure 12.5 Small-signal equivalent circuit of the op amp in Fig. 12.1 with a resistance R included in series with C_C.

The problem of the additional phase lag provided by the right-half-plane zero has a rather simple and elegant solution: By including a resistance R in series with C_C, as shown in Fig. 12.5, the transmission zero can be moved to other less-harmful locations. To find the new location of the transmission zero, set $V_o = 0$. Then, the current through C_C and R will be $V_{i2}/(R + 1/sC_C)$, and a node equation at the output yields

$$\frac{V_{i2}}{R + \dfrac{1}{sC_C}} = G_{m2}V_{i2}$$

Thus the zero is now at

$$s = 1\Big/C_C\left(\frac{1}{G_{m2}} - R\right) \tag{12.39}$$

We observe that by selecting $R = 1/G_{m2}$, we can place the zero at infinite frequency. An even better choice would be to select R greater than $1/G_{m2}$, thus placing the zero at a negative real-axis location where the phase it introduces *adds* to the phase margin.

EXERCISE

12.4 A particular implementation of the CMOS amplifier of Figs. 12.1 and 12.2 provides $G_{m1} = 1$ mA/V, $G_{m2} = 2$ mA/V, $r_{o2} = r_{o4} = 100$ kΩ, $r_{o6} = r_{o7} = 40$ kΩ, and $C_2 = 1$ pF.
(a) Find the value of C_C that results in $f_t = 100$ MHz. What is the 3-dB frequency of the open-loop gain?
(b) Find the value of the resistance R that when placed in series with C_C causes the transmission zero to be located at infinite frequency.
(c) Find the frequency of the second pole and hence find the excess phase lag at $f = f_t$, introduced by the second pole, and the resulting phase margin assuming that the situation in (b) pertains.
Ans. 1.6 pF; 50 kHz; 500 Ω; 318 MHz; 17.4°; 72.6°

12.1.6 Slew Rate

The slew-rate limitation of op amps is discussed in Chapter 2. Here, we shall illustrate the origin of the slewing phenomenon in the context of the two-stage CMOS amplifier under study.

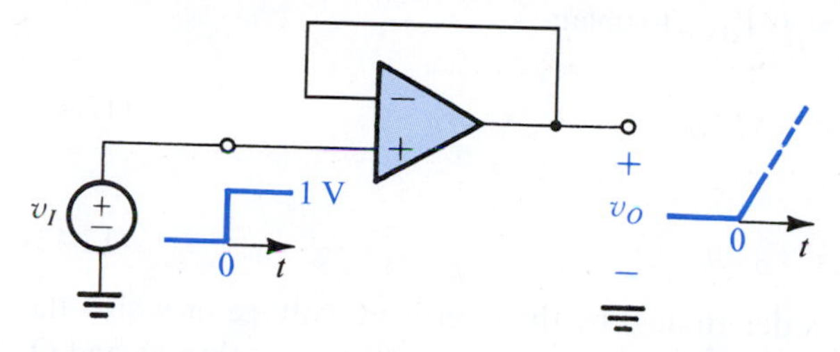

Figure 12.6 A unity-gain follower with a large step input. Since the output voltage cannot change immediately, a large differential voltage appears between the op-amp input terminals.

Consider the unity-gain follower of Fig. 12.6 with a step of, say, 1 V applied at the input. Because of the amplifier dynamics, its output will not change in zero time. Thus, immediately after the input is applied, the entire value of the step will appear as a differential signal between the two input terminals. In all likelihood, such a large signal will exceed the voltage required to turn off one side of the input differential pair ($\sqrt{2}V_{OV1}$: see earlier illustration, Fig. 8.6) and switch the entire bias current I to the other side. Reference to Fig. 12.1 shows that for our example, Q_2 will turn off, and Q_1 will conduct the entire current I. Thus Q_4 will sink a current I that will be pulled from C_C, as shown in Fig. 12.7. Here, as we did in Fig. 12.3, we are modeling the second stage as an ideal integrator. We see that the output voltage will be a ramp with a slope of I/C_C:

$$v_o(t) = \frac{I}{C_C}t$$

Thus the slew rate, SR, is given by

$$SR = \frac{I}{C_C} \tag{12.40}$$

It should be pointed out, however, that this is a rather simplified model of the slewing process.

Relationship Between SR and f_t A simple relationship exists between the unity-gain bandwidth f_t and the slew rate SR. This relationship can be found by combining Eqs. (12.31)

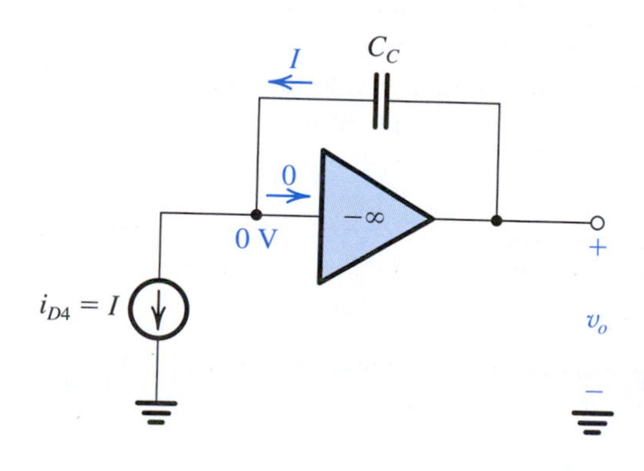

Figure 12.7 Model of the two-stage CMOS op-amp of Fig. 12.1 when a large differential voltage is applied.

and (12.40) and noting that $G_{m1} = g_{m1} = I/V_{OV1}$, to obtain

$$SR = 2\pi f_t V_{OV} \tag{12.41}$$

or equivalently,

$$SR = V_{OV}\omega_t \tag{12.42}$$

Thus, for a given ω_t, the slew rate is determined by the overdrive voltage at which the first-stage transistors are operated. A higher slew rate is obtained by operating Q_1 and Q_2 at a larger V_{OV}. Now, for a given bias current I, a larger V_{OV} is obtained if Q_1 and Q_2 are *p*-channel devices. This is an important reason for using *p*-channel rather than *n*-channel devices in the first stage of the CMOS op amp. Another reason is that it allows the second stage to employ an *n*-channel device. Now, since *n*-channel devices have greater transconductances than corresponding *p*-channel devices, G_{m2} will be high, resulting in a higher second-pole frequency and a correspondingly higher ω_t. However, the price paid for these improvements is a lower G_{m1} and hence a lower dc gain.

EXERCISE

12.5 Find *SR* for the CMOS op amp of Fig. 12.1 for the case f_t = 100 MHz and V_{OV1} = 0.2 V. If C_C = 1.6 pF, what must the bias current I be?
Ans. 126 V/μs; 200 μA

12.1.7 Power-Supply Rejection Ratio (PSRR)

CMOS op amps are usually utilized in what are known as **mixed-signal circuits**: IC chips that combine analog and digital circuits. In such circuits, the switching activity in the digital portion usually results in increased ripple on the power supplies. A portion of the supply ripple can make its way to the op-amp output and thus corrupt the output signal. The traditional approach for reducing supply ripple by connecting large capacitances between the supply rails and ground is not viable in IC design, as such capacitances would consume most of the chip area. Instead, the analog IC designer has to pay attention to another op-amp specification that so far we have ignored, namely, the power-supply rejection ratio (PSRR).

The PSRR is defined as the ratio of the amplifier differential gain to the gain experienced by a change in the power-supply voltage (v_{dd} and v_{ss}). For circuits utilizing two power supplies, we define

$$\text{PSRR}^+ \equiv \frac{A_d}{A^+} \tag{12.42}$$

and

$$\text{PSRR}^- = \frac{A_d}{A^-} \tag{12.43}$$

where

$$A^+ \equiv \frac{v_o}{v_{dd}} \tag{12.44}$$

$$A^- = \frac{v_o}{v_{ss}} \tag{12.45}$$

Obviously, to minimize the effect of the power-supply ripple, we require the op amp to have a large PSRR.

A detailed analysis of the PSRR of the two-stage CMOS op amp is beyond the scope of this book (see Gray et al., 2009). Nevertheless, we make the following brief remarks. It can be shown that the circuit is remarkably insensitive to variations in V_{DD}, and thus PSRR^+ is very high. This is not the case, however, for the negative-supply ripple v_{ss}, which is coupled to the output primarily through the second-stage transistors Q_6 and Q_7. In particular, the portion of v_{ss} that appears at the op-amp output is determined by the voltage divider formed by the output resistances of Q_6 and Q_7,

$$v_o = v_{ss}\frac{r_{o7}}{r_{o6}+r_{o7}} \tag{12.46}$$

Thus,

$$A^- \equiv \frac{v_o}{v_{ss}} = \frac{r_{o7}}{r_{o6}+r_{o7}} \tag{12.47}$$

Now utilizing A_d from Eq. (12.22) gives

$$\text{PSRR}^- \equiv \frac{A_d}{A^-} = g_{m1}(r_{o2} \| r_{o4})g_{m6}r_{o6} \tag{12.48}$$

Thus, PSRR^- is of the form $(g_m r_o)^2$ and therefore is maximized by selecting long channels L (to increase $|V_A|$), and operating at low $|V_{OV}|$.

12.1.8 Design Trade-offs

The performance parameters of the two-stage CMOS amplifier are primarily determined by two design parameters:

1. The length L used for the channel of each MOSFET.
2. The overdrive voltage $|V_{OV}|$ at which each transistor is operated.

Throughout this section, we have found that a larger L and correspondingly larger $|V_A|$ increases the amplifier gain, CMRR and PSRR. We also found that operating at a lower $|V_{OV}|$ increases these three parameters as well as increasing the input common-mode range and the allowable range of output swing. Also, although we have not analyzed the offset voltage of the op amp here, we know from our study of the subject in Section 8.4.1 that a number of the components of the input offset voltage that arises from random device mismatches are proportional to $|V_{OV}|$ at which the MOSFETs of the input differential pair are operated. Thus the offset is minimized by operating at a lower $|V_{OV}|$.

There is, however, an important MOSFET performance parameter that requires the selection of a larger $|V_{OV}|$, namely, the **transition frequency** f_T, which determines the high-frequency performance of the MOSFET,

$$f_T = \frac{g_m}{2\pi(C_{gs}+C_{gd})} \tag{12.49}$$

For an n-channel MOSFET, we can show that (see Appendix 7.A)

$$f_T \simeq \frac{1.5\mu_n V_{OV}}{2\pi L^2} \tag{12.50}$$

A similar relationship applies for the PMOS transistor, with μ_p and $|V_{OV}|$ replacing μ_n and V_{OV}, respectively. Thus to increase f_T and improve the high-frequency response of the op amp, we need to use a larger overdrive value and, not surprisingly, shorter channels. A larger $|V_{OV}|$ also results in a higher op-amp slew rate SR (Eq. 12.41). Finally, note that the selection of a larger $|V_{OV}|$ results, for the same bias current, in a smaller W/L, which combined with a short L leads to smaller devices and hence lower values of MOSFET capacitances and higher frequencies of operation.

In conclusion, the selection of $|V_{OV}|$ presents the designer with a trade-off between improving the low-frequency performance parameters on the one hand and the high-frequency performance on the other. For modern submicron technologies, which require operation from power supplies of 1 V to 1.5 V, overdrive voltages between 0.1 V and 0.3 V are typically utilized. For these process technologies, analog designers typically use channel lengths that are at least 1.5 to 2 times the specified value of L_{min}, and even longer channels are used for current-source bias transistors.

Example 12.1

We conclude our study of the two-stage CMOS op amp with a design example. Let it be required to design the circuit to obtain a dc gain of 4000 V/V. Assume that the available fabrication technology is of the 0.5-μm type for which $V_{tn} = |V_{tp}| = 0.5$ V, $k_n' = 200\ \mu A/V^2$, $k_p' = 80\ \mu A/V^2$, $V_{An}' = |V_{Ap}'| = 20$ V/μm, and $V_{DD} = V_{SS} = 1.65$ V. To achieve a reasonable dc gain per stage, use $L = 1$ μm for all devices. Also, for simplicity, operate all devices at the same $|V_{OV}|$, in the range of 0.2 V to 0.4 V. Use $I = 200\ \mu A$, and to obtain a higher G_{m2}, and hence a higher f_{P2}, use $I_{D6} = 0.5$ mA. Specify the W/L ratios for all transistors. Also give the values realized for the input common-mode range, the maximum possible output swing, R_{in} and R_o. Also determine the CMRR and PSRR realized. If $C_1 = 0.2$ pF and $C_2 = 0.8$ pF, find the required values of C_C and the series resistance R to place the transmission zero at $s = \infty$ and to obtain the highest possible f_t consistent with a phase margin of 75°. Evaluate the values obtained for f_t and SR.

Solution

Using the voltage-gain expression in Eq. (12.22),

$$\begin{aligned} A_v &= g_{m1}(r_{o2} \| r_{o4}) g_{m6}(r_{o6} \| r_{o7}) \\ &= \frac{2(I/2)}{V_{OV}} \times \frac{1}{2} \times \frac{V_A}{(I/2)} \times \frac{2I_{D6}}{V_{OV}} \times \frac{1}{2} \times \frac{V_A}{I_{D6}} \\ &= \left(\frac{V_A}{V_{OV}}\right)^2 \end{aligned}$$

To obtain $A_v = 4000$, given $V_A = 20$ V,

$$4000 = \frac{400}{V_{OV}^2}$$

$$V_{OV} = 0.316 \text{ V}$$

To obtain the required (W/L) ratios of Q_1 and Q_2,

$$I_{D1} = \frac{1}{2}k_p'\left(\frac{W}{L}\right)_1 V_{OV}^2$$

$$100 = \frac{1}{2} \times 80\left(\frac{W}{L}\right)_1 \times 0.316^2$$

Thus,

$$\left(\frac{W}{L}\right)_1 = \frac{25\ \mu\text{m}}{1\ \mu\text{m}}$$

and

$$\left(\frac{W}{L}\right)_2 = \frac{25\ \mu\text{m}}{1\ \mu\text{m}}$$

For Q_3 and Q_4 we write

$$100 = \frac{1}{2} \times 200\left(\frac{W}{L}\right)_3 \times 0.316^2$$

to obtain

$$\left(\frac{W}{L}\right)_3 = \left(\frac{W}{L}\right)_4 = \frac{10\ \mu\text{m}}{1\ \mu\text{m}}$$

For Q_5,

$$200 = \frac{1}{2} \times 80\left(\frac{W}{L}\right)_5 \times 0.316^2$$

Thus,

$$\left(\frac{W}{L}\right)_5 = \frac{50\ \mu\text{m}}{1\ \mu\text{m}}$$

Since Q_7 is required to conduct 500 μA, its (W/L) ratio should be 2.5 times that of Q_5,

$$\left(\frac{W}{L}\right)_7 = 2.5\left(\frac{W}{L}\right)_5 = \frac{125\ \mu\text{m}}{1\ \mu\text{m}}$$

For Q_6 we write

$$500 = \frac{1}{2} \times 200 \times \left(\frac{W}{L}\right)_6 \times 0.316^2$$

Thus,

$$\left(\frac{W}{L}\right)_6 = \frac{50\ \mu\text{m}}{1\ \mu\text{m}}$$

Example 12.1 *continued*

Finally, let's select $I_{REF} = 20\ \mu A$, thus

$$\left(\frac{W}{L}\right)_8 = 0.1\left(\frac{W}{L}\right)_5 = \frac{5\ \mu m}{1\ \mu m}$$

The input common-mode range can be found using the expression in Eq. (12.4) as

$$-1.33\ V \le V_{ICM} \le 0.52\ V$$

The maximum signal swing allowable at the output is found using the expression in Eq. (12.5) as

$$-1.33\ V \le v_O \le 1.33\ V$$

The input resistance is practically infinite, and the output resistance is

$$R_o = r_{o6} \| r_{o7} = \frac{1}{2} \times \frac{20}{0.5} = 20\ k\Omega$$

The CMRR is determined using Eq. (12.24),

$$\text{CMRR} = g_{m1}(r_{o2} \| r_{o4})(2g_{m3}R_{SS})$$

where $R_{SS} = r_{o5} = V_A/I$. Thus,

$$\text{CMRR} = \frac{2(I/2)}{V_{OV}} \times \frac{1}{2} \times \frac{V_A}{(I/2)} \times 2 \times \frac{2(I/2)}{V_{OV}} \times \frac{V_A}{I}$$

$$= 2\left(\frac{V_A}{V_{OV}}\right)^2 = 2\left(\frac{20}{0.316}\right)^2 = 8000$$

Expressed in decibels, we have

$$\text{CMRR} = 20 \log 8000 = 78\ \text{dB}$$

The PSRR is determined using Eq. (12.48):

$$\text{PSRR} = g_{m1}(r_{o2} \| r_{o4})g_{m6}r_{o6}$$

$$= \frac{2(I/2)}{V_{OV}} \times \frac{1}{2} \times \frac{V_A}{(I/2)} \times \frac{2I_{D6}}{V_{OV}} \times \frac{V_A}{I_{D6}}$$

$$= 2\left(\frac{V_A}{V_{OV}}\right)^2 = 2\left(\frac{20}{0.316}\right)^2 = 8000$$

or, expressed in decibels,

$$\text{PSRR} = 20 \log 8000 = 78\ \text{dB}$$

To determine f_{P2} we use Eq. (12.28) and substitute for G_{m2},

$$G_{m2} = g_{m6} = \frac{2I_{D6}}{V_{OV}} = \frac{2 \times 0.5}{0.316} = 3.2\ \text{mA/V}$$

Thus,

$$f_{P2} = \frac{3.2 \times 10^{-3}}{2\pi \times 0.8 \times 10^{-12}} = 637\ \text{MHz}$$

To move the transmission zero to $s = \infty$, we select the value of R as

$$R = \frac{1}{G_{m2}} = \frac{1}{3.2 \times 10^{-3}} = 316\ \Omega$$

For a phase margin of 75°, the phase shift due to the second pole at $f = f_t$ must be 15°, that is,

$$\tan^{-1}\frac{f_t}{f_{P2}} = 15°$$

Thus,

$$f_t = 637 \times \tan 15° = 171\ \text{MHz}$$

The value of C_C can be found using Eq. (12.31),

$$C_C = \frac{G_{m1}}{2\pi f_t}$$

where

$$G_{m1} = g_{m1} = \frac{2 \times 100\ \mu\text{A}}{0.316\ \text{V}} = 0.63\ \text{mA/V}$$

Thus,

$$C_{C1} = \frac{0.63 \times 10^{-3}}{2\pi \times 171 \times 10^{6}} = 0.6\ \text{pF}$$

The value of SR can now be found using Eq. (12.41) as

$$SR = 2\pi \times 171 \times 10^{6} \times 0.316$$
$$= 340\ \text{V}/\mu\text{s}$$

12.2 The Folded-Cascode CMOS Op Amp

In this section we study another type of CMOS op-amp circuit: the folded cascode. The circuit is based on the folded-cascode amplifier studied in Section 7.3.6. There, it was mentioned that although composed of a CS transistor and a CG transistor of opposite polarity, the folded-cascode configuration is generally considered to be a single-stage amplifier. Similarly, the op-amp circuit that is based on the cascode configuration is considered to be a single-stage op amp. Nevertheless, it can be designed to provide performance parameters that equal and in some respects exceed those of the two-stage topology studied in the preceding section. Indeed, the folded-cascode op-amp topology is currently as popular as the two-stage structure. Furthermore, the folded-cascode configuration can be used in conjunction with the two-stage structure to provide performance levels higher than those available from either circuit alone.

12.2.1 The Circuit

Figure 12.8 shows the structure of the CMOS folded-cascode op amp. Here, Q_1 and Q_2 form the input differential pair, and Q_3 and Q_4 are the cascode transistors. Recall that for

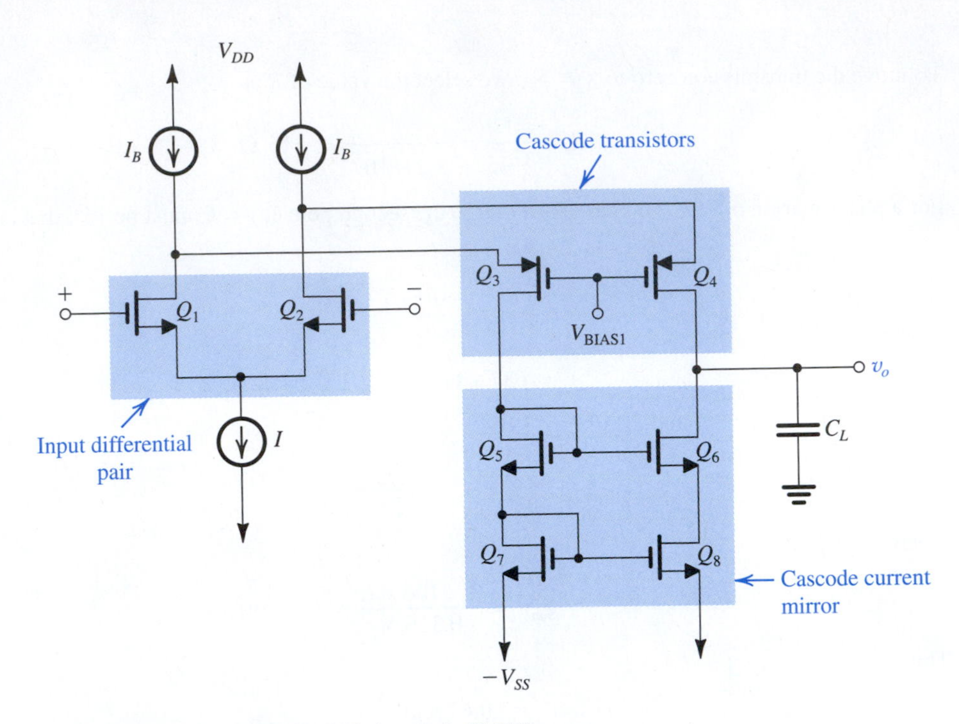

Figure 12.8 Structure of the folded-cascode CMOS op amp.

differential input signals, each of Q_1 and Q_2 acts as a common-source amplifier. Also note that the gate terminals of Q_3 and Q_4 are connected to a constant dc voltage (V_{BIAS1}) and hence are at signal ground. Thus, for differential input signals, each of the transistor pairs Q_1–Q_3 and Q_2–Q_4 acts as a folded-cascode amplifier, such as the one in Fig. 7.16. Note that the input differential pair is biased by a constant-current source I. Thus each of Q_1 and Q_2 is operating at a bias current $I/2$. A node equation at each of their drains shows that the bias current of each of Q_3 and Q_4 is $(I_B - I/2)$. Selecting $I_B = I$ forces all transistors to operate at the same bias current of $I/2$. For reasons that will be explained shortly, however, the value of I_B is usually made somewhat greater than I.

As we learned in Chapter 7, if the full advantage of the high output-resistance achieved through cascoding is to be realized, the output resistance of the current-source load must be equally high. This is the reason for using the cascode current mirror Q_5 to Q_8, in the circuit of Fig. 12.8. (This current-mirror circuit was studied in Section 7.5.1.) Finally, note that capacitance C_L denotes the total capacitance at the output node. It includes the internal transistor capacitances, an actual load capacitance (if any), and possibly an additional capacitance deliberately introduced for the purpose of frequency compensation. In many cases, however, the load capacitance will be sufficiently large, obviating the need to provide additional capacitance to achieve the desired frequency compensation. This topic will be discussed shortly. For the time being, we note that unlike the two-stage circuit, that requires the introduction of a separate compensation capacitor C_C, here the load capacitance contributes to frequency compensation.

A more complete circuit for the CMOS folded-cascode op amp is shown in Fig. 12.9. Here we show the two transistors Q_9 and Q_{10}, which provide the constant bias currents I_B, and transistor Q_{11}, which provides the constant current I utilized for biasing the differential pair. Observe that the details for generating the bias voltages V_{BIAS1}, V_{BIAS2}, and V_{BIAS3} are not

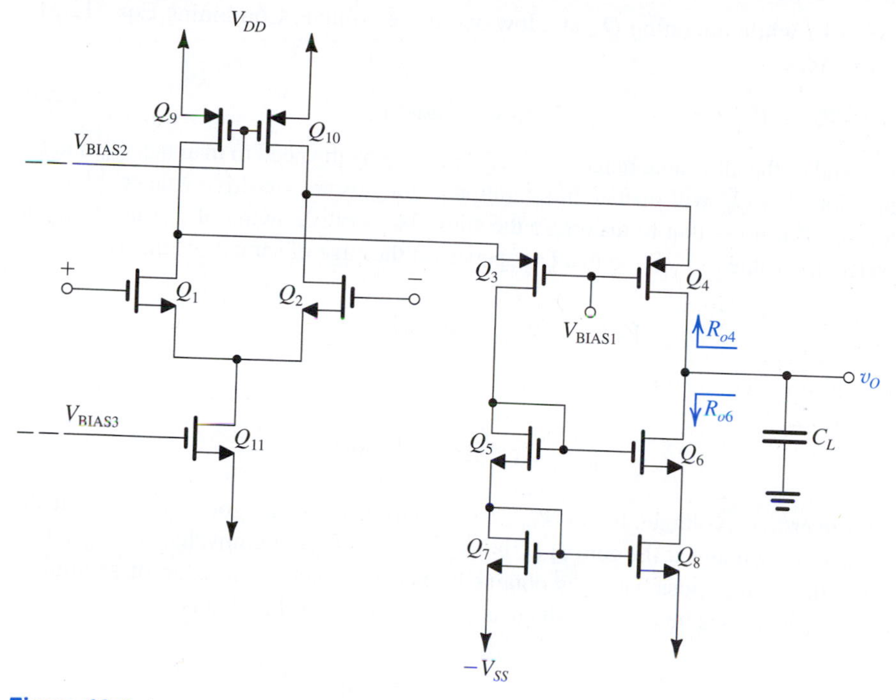

Figure 12.9 A more complete circuit for the folded-cascode CMOS amplifier of Fig. 12.8.

shown. Nevertheless, we are interested in how the values of these voltages are to be selected. Toward that end, we evaluate the input common-mode range and the allowable output swing.

12.2.2 Input Common-Mode Range and Output Swing

To find the input common-mode range, let the two input terminals be tied together and connected to a voltage V_{ICM}. The maximum value of V_{ICM} is limited by the requirement that Q_1 and Q_2 operate in saturation at all times. Thus $V_{ICM\max}$ should be at most V_{tn} volts above the voltage at the drains of Q_1 and Q_2. The latter voltage is determined by V_{BIAS1} and must allow for a voltage drop across Q_9 and Q_{10} at least equal to their overdrive voltage, $|V_{OV9}| = |V_{OV10}|$. Assuming that Q_9 and Q_{10} are indeed operated at the edge of saturation, $V_{ICM\max}$ will be

$$V_{ICM\max} = V_{DD} - |V_{OV9}| + V_{tn} \tag{12.51}$$

which can be larger than V_{DD}, a significant improvement over the case of the two-stage circuit. The value of V_{BIAS2} should be selected to yield the required value of I_B while operating Q_9 and Q_{10} at a small value of $|V_{OV}|$ (e.g., 0.2 V or so). The minimum value of V_{ICM} can be obtained as

$$V_{ICM\min} = -V_{SS} + V_{OV11} + V_{OV1} + V_{tn} \tag{12.52}$$

The presence of the threshold voltage V_{tn} in this expression indicates that $V_{ICM\min}$ is not sufficiently low. Later in this section we shall describe an ingenious technique for solving this problem. For the time being, note that the value of V_{BIAS3} should be selected to provide the required value of I while operating Q_{11} at a low overdrive voltage. Combining Eqs. (12.51) and (12.52) provides

$$-V_{SS} + V_{OV11} + V_{OV1} + V_{tn} \leq V_{ICM} \leq V_{DD} - |V_{OV9}| + V_{tn} \tag{12.53}$$

The upper end of the allowable range of v_O is determined by the need to maintain Q_{10} and Q_4 in saturation. Note that Q_{10} will operate in saturation as long as an overdrive voltage, $|V_{OV10}|$, appears across it. It follows that to maximize the allowable positive swing of v_O (and $V_{ICM\max}$), we should select the value of V_{BIAS1} so that Q_{10} operates at the edge of saturation, that is,

$$V_{\text{BIAS}1} = V_{DD} - |V_{OV10}| - V_{SG4} \tag{12.54}$$

The upper limit of v_O will then be

$$v_{O\max} = V_{DD} - |V_{OV10}| - |V_{OV4}| \tag{12.55}$$

which is two overdrive voltages below V_{DD}. The situation is not as good, however, at the other end: Since the voltage at the gate of Q_6 is $-V_{SS} + V_{GS7} + V_{GS5}$ or equivalently $-V_{SS} + V_{OV7} + V_{OV5} + 2V_{tn}$, the lowest possible v_O is obtained when Q_6 reaches the edge of saturation, namely, when v_O decreases below the voltage at the gate of Q_6 by V_{tn}, that is,

$$v_{O\min} = -V_{SS} + V_{OV7} + V_{OV5} + V_{tn} \tag{12.56}$$

Note that this value is two overdrive voltages *plus* a threshold voltage above $-V_{SS}$. This is a drawback of utilizing the cascode mirror. The problem can be alleviated by using a modified mirror circuit, as we shall shortly see.

EXERCISE

12.6 For a particular design of the folded-cascode op amp of Fig. 12.9, ±1.65-V supplies are utilized and all transistors are operated at overdrive voltages of 0.3-V magnitude. The fabrication process employed provides $V_{tn} = |V_{tp}| = 0.5$ V. Find the input common-mode range and the range allowed for v_O.
Ans. –0.55 V to +1.85 V; –0.55 V to +1.05 V.

12.2.3 Voltage Gain

The folded-cascode op amp is simply a transconductance amplifier with an infinite input resistance, a transconductance G_m and an output resistance R_o. G_m is equal to g_m of each of the two transistors of the differential pair,

$$G_m = g_{m1} = g_{m2} \tag{12.57}$$

Thus,

$$G_m = \frac{2(I/2)}{V_{OV1}} = \frac{I}{V_{OV1}} \tag{12.58}$$

The output resistance R_o is the parallel equivalent of the output resistance of the cascode amplifier and the output resistance of the cascode mirror, thus

$$R_o = R_{o4} \| R_{o6} \tag{12.59}$$

Reference to Fig. 12.9 shows that the resistance R_{o4} is the output resistance of the CG transistor Q_4. The latter has a resistance $(r_{o2} \| r_{o10})$ in its source lead, thus

$$R_{o4} \simeq (g_{m4}r_{o4})(r_{o2} \| r_{o10}) \tag{12.60}$$

The resistance R_{o6} is the output resistance of the cascode mirror and is thus given by Eq. (7.25), thus

$$R_{o6} \simeq g_{m6}r_{o6}r_{o8} \tag{12.61}$$

Combining Eqs. (12.59) to (12.61) gives

$$R_o = [g_{m4}r_{o4}(r_{o2} \| r_{o10})] \| (g_{m6}r_{o6}r_{o8}) \tag{12.62}$$

The dc open-loop gain can now be found using G_m and R_o, as

$$A_v = G_m R_o \tag{12.63}$$

Thus,

$$A_v = g_{m1}\{[g_{m4}r_{o4}(r_{o2} \| r_{o10})] \| (g_{m6}r_{o6}r_{o8})\} \tag{12.64}$$

Figure 12.10 shows the equivalent-circuit model including the load capacitance C_L, which we shall take into account shortly.

Because the folded-cascode op amp is a transconductance amplifier, it has been given the name **operational transconductance amplifier (OTA)**. Its very high output resistance, which is of the order of $g_m r_o^2$ (see Eq. 12.62) is what makes it possible to realize a relatively high voltage gain in a single amplifier stage. However, such a high output resistance may be a cause of concern to the reader; after all, in Chapter 2, we stated that an ideal op amp has a zero output resistance! To alleviate this concern somewhat, let us find the closed-loop output resistance of a unity-gain follower formed by connecting the output terminal of the circuit of Fig. 12.9 back to the negative input terminal. Since this feedback is of the voltage sampling type, it reduces the output resistance by the factor $(1 + A\beta)$, where $A = A_v$ and $\beta = 1$, that is,

$$R_{of} = \frac{R_o}{1 + A_v} \simeq \frac{R_o}{A_v} \tag{12.65}$$

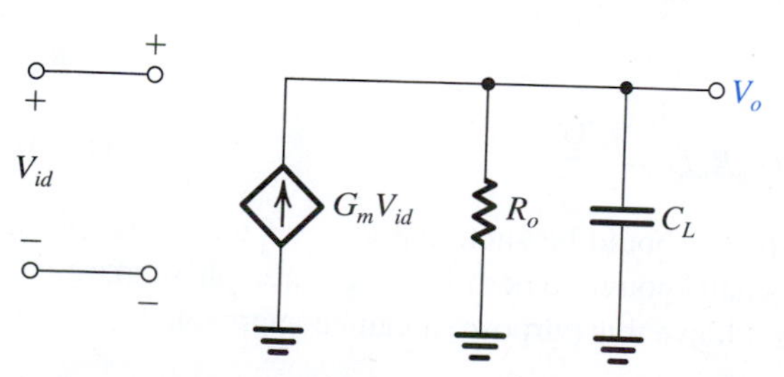

Figure 12.10 Small-signal equivalent circuit of the folded-cascode CMOS amplifier. Note that this circuit is in effect an operational transconductance amplifier (OTA).

Substituting for A_v from Eq. (12.63) gives

$$R_{of} \simeq \frac{1}{G_m} \tag{12.66}$$

which is a general result that applies to any OTA to which 100% voltage feedback is applied. For our particular circuit, $G_m = g_{m1}$, thus

$$R_{of} = 1/g_{m1} \tag{12.67}$$

Since g_{m1} is of the order of 1 mA/V, R_{of} will be of the order of 1 kΩ. Although this is not very small, it is reasonable in view of the simplicity of the op-amp circuit as well as the fact that this type of op amp is not usually intended to drive low-valued resistive loads.

EXERCISE

12.7 The CMOS op amp of Figs. 12.8 and 12.9 is fabricated in a process for which $V'_{An} = |V'_{Ap}| = 20$ V/μm. If all devices have 1-μm channel length and are operated at equal overdrive voltages of 0.2-V magnitude, find the voltage gain obtained. If each of Q_1 to Q_8 is biased at 100 μA, what value of R_o is obtained?
Ans. 13,333 V/V; 13.3 MΩ

12.2.4 Frequency Response

From Section 9.6, we know that one of the advantages of the cascode configuration is its excellent high-frequency response. It has poles at the input, at the connection between the CS and CG transistors (i.e., at the source terminals of Q_3 and Q_4), and at the output terminal. Normally, the first two poles are at very high frequencies, especially when the resistance of the signal generator that feeds the differential pair is small. Since the primary purpose of CMOS op amps is to feed capacitive loads, C_L is usually large, and the pole at the output becomes dominant. Even if C_L is not large, we can increase it deliberately to give the op amp a dominant pole. From Fig. 12.10 we can write

$$\frac{V_o}{V_{id}} = \frac{G_m R_o}{1 + sC_L R_o} \tag{12.68}$$

Thus, the dominant pole has a frequency f_P,

$$f_P = \frac{1}{2\pi C_L R_o} \tag{12.69}$$

and the unity-gain frequency f_t will be

$$f_t = G_m R_o f_P = \frac{G_m}{2\pi C_L} \tag{12.70}$$

From a design point of view, the value of C_L should be such that at $f = f_t$ the excess phase resulting from the nondominant poles is small enough to permit the required phase margin to be achieved. If C_L is not large enough to achieve this purpose, it can be augmented.

It is important to note the different effects of increasing the load capacitance on the operation of the two op-amp circuits we have studied. In the two-stage circuit, if C_L is increased, the frequency of the second pole decreases, the excess phase shift at $f = f_t$ increases, and the phase margin is reduced. Here, on the other hand, when C_L is increased, f_t decreases, but the phase margin increases. In other words, a heavier capacitive load decreases the bandwidth of the folded-cascode amplifier but does not impair its response (which happens when the phase margin decreases). Of course, if an increase in C_L is anticipated in the two-stage op-amp case, the designer can increase C_C, thus decreasing f_t and restoring the phase margin to its required value.

12.2.5 Slew Rate

As discussed in Section 12.1.6, slewing occurs when a large differential input signal is applied. Refer to Fig. 12.8 and consider the case of a large signal V_{id} applied so that Q_2 cuts off and Q_1 conducts the entire bias current I. We see that Q_3 will now carry a current $(I_B - I)$, and Q_4 will conduct a current I_B. The current mirror will see an input current of $(I_B - I)$ through Q_5 and Q_7 and thus its output current in the drain of Q_6 will be $(I_B - I)$. It follows that at the output node the current that will flow into C_L will be $I_4 - I_6 = I_B - (I_B - I) = I$. Thus the output v_O will be a ramp with a slope of I/C_L which is the slew rate,

$$SR = \frac{I}{C_L} \qquad (12.71)$$

Note that the reason for selecting $I_B > I$ is to avoid turning off the current mirror completely; if the current mirror turns off, the output distortion increases. Typically, I_B is set 10% to 20% larger than I. Finally, Eqs. (12.70), (12.71), and (12.58) can be combined to obtain the following relationship between SR and f_t

$$SR = 2\pi f_t V_{OV1} \qquad (12.72)$$

which is identical to the corresponding relationship in the case of the two-stage design. Note, however, that this relationship applies only when $I_B > I$.

Example 12.2

Consider a design of the folded-cascode op amp of Fig. 12.9 for which $I = 200$ μA, $I_B = 250$ μA, and $|V_{OV}|$ for all transistors is 0.25 V. Assume that the fabrication process provides $k_n' = 100$ μA/V^2, $k_p' = 40$ μA/V^2, $|V_A'| = 20$ V/μm. $V_{DD} = V_{SS} = 2.5$ V, and $|V_t| = 0.75$ V. Let all transistors have $L = 1$ μm and assume that $C_L = 5$ pF. Find I_D, g_m, r_o, and W/L for all transistors. Find the allowable range of V_{ICM} and of the output voltage swing. Determine the values of A_v, f_t, f_P, and SR. What is the power dissipation of the op amp?

Solution

From the given values of I and I_B we can determine the drain current I_D for each transistor. The transconductance of each device is found using

$$g_m = \frac{2I_D}{V_{OV}} = \frac{2I_D}{0.25}$$

Example 12.2 *continued*

and the output resistance r_o from

$$r_o = \frac{|V_A|}{I_D} = \frac{20}{I_D}$$

The W/L ratio for each transistor is determined from

$$\left(\frac{W}{L}\right)_i = \frac{2I_{Di}}{k'V_{OV}^2}$$

The results are as follows:

	Q_1	Q_2	Q_3	Q_4	Q_5	Q_6	Q_7	Q_8	Q_9	Q_{10}	Q_{11}
I_D (μA)	100	100	150	150	150	150	150	150	250	250	200
g_m (mA/V)	0.8	0.8	1.2	1.2	1.2	1.2	1.2	1.2	2.0	2.0	1.6
r_o (kΩ)	200	200	133	133	133	133	133	133	80	80	100
W/L	32	32	120	120	48	48	48	48	200	200	64

Note that for all transistors,

$$g_m r_o = 160 \text{ V/V}$$

$$V_{GS} = 1.0 \text{ V}$$

Using the expression in Eq. (12.53), the input common-mode range is found to be

$$-1.25 \text{ V} \leq V_{ICM} \leq 3 \text{ V}$$

The output voltage swing is found using Eqs. (12.55) and (12.56) to be

$$-1.25 \text{ V} \leq v_O \leq 2 \text{ V}$$

To obtain the voltage gain, we first determine R_{o4} using Eq. (12.60) as

$$R_{o4} = 160(200 \,\|\, 80) = 9.14 \text{ M}\Omega$$

and R_{o6} using Eq. (12.61) as

$$R_{o6} = 21.28 \text{ M}\Omega$$

The output resistance R_o can then be found as

$$R_o = R_{o4} \,\|\, R_{o6} = 6.4 \text{ M}\Omega$$

and the voltage gain

$$A_v = G_m R_o = 0.8 \times 10^{-3} \times 6.4 \times 10^6$$

$$= 5120 \text{ V/V}$$

The unity-gain bandwidth is found using Eq. (12.70),

$$f_t = \frac{0.8 \times 10^{-3}}{2\pi \times 5 \times 10^{-12}} = 25.5 \text{ MHz}$$

Thus, the dominant-pole frequency must be

$$f_P = \frac{f_t}{A_v} = \frac{25.5 \text{ MHz}}{5120} = 5 \text{ kHz}$$

The slew rate can be determined using Eq. (12.71),

$$SR = \frac{I}{C_L} = \frac{200 \times 10^{-6}}{5 \times 10^{-12}} = 40 \text{ V/}\mu\text{s}$$

Finally, to determine the power dissipation we note that the total current is 500 μA = 0.5 mA, and the total supply voltage is 5 V, thus

$$P_D = 5 \times 0.5 = 2.5 \text{ mW}$$

12.2.6 Increasing the Input Common-Mode Range: Rail-to-Rail Input Operation

In Section 12.2.2 we found that while the upper limit on the input common-mode range exceeds the supply voltage V_{DD}, the magnitude of lower limit is significantly lower than V_{SS}. The opposite situation occurs if the input differential amplifier is made up of PMOS transistors. It follows that an NMOS and a PMOS differential pair placed in parallel would provide an input stage with a common-mode range that exceeds the power supply voltage in both directions. This is known as rail-to-rail input operation. Figure 12.11 shows such an arrangement. To keep the diagram simple, we have not shown the parallel connection of the two differential pairs: The two positive-input terminals are to be connected together and the two negative-input terminals are to be tied together. Transistors Q_5 and Q_6 are the cascode transistors for the Q_1–Q_2 pair, and transistors Q_7 and Q_8 are the cascode devices for the Q_3–Q_4 pair. The output voltage V_o is shown taken differentially between the drains of the cascode devices. To obtain a single-ended output, a differential-to-single-ended conversion circuit should be connected in cascade.

Figure 12.11 indicates by arrows the direction of the current increments that result from the application of a positive differential input signal V_{id}. Each of the current increments indicated is equal to $G_m(V_{id}/2)$ where $G_m = g_{m1} = g_{m2} = g_{m3} = g_{m4}$. Thus the total current feeding each of the two output nodes will be $G_m V_{id}$. Now, if the output resistance between each of the two nodes and ground is denoted R_o, the output voltage will be

$$V_o = 2G_m R_o V_{id} \tag{12.73}$$

Thus the voltage gain will be

$$A_v = 2G_m R_o \tag{12.74}$$

This, however, assumes that both differential pairs will be operating simultaneously. This in turn occurs only over a limited range of V_{ICM}. Over the remainder of the input common-mode range, only one of the two differential pairs will be operational, and the gain drops to half of the value in Eq. (12.74). This rail-to-rail, folded-cascode structure is utilized in a commercially available op amp.[1]

[1]The Texas Instruments OPA357.

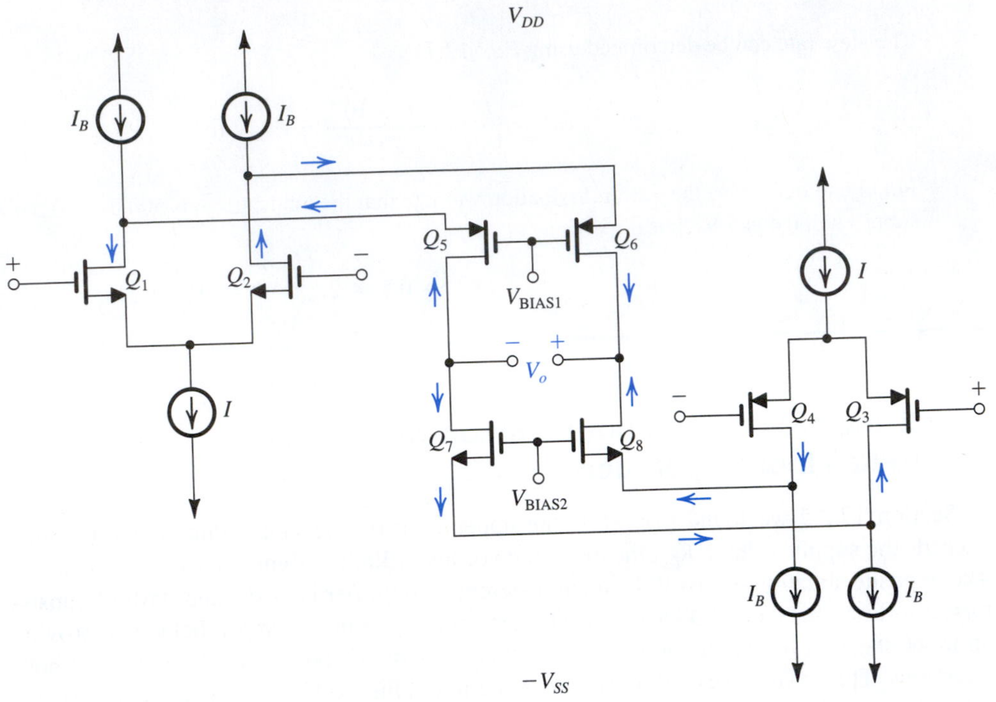

Figure 12.11 A folded-cascode op amp that employs two parallel complementary input stages to achieve rail-to-rail input common-mode operation. Note that the two "+" terminals are connected together and the two "–" terminals are connected together.

EXERCISE

12.8 For the circuit in Fig. 12.11, assume that all transistors, including those that implement the current sources, are operating at equal overdrive voltages of 0.3-V magnitude and have $|V_t| = 0.7$ V and that $V_{DD} = V_{SS} = 2.5$ V.
(a) Find the range over which the NMOS input stage operates.
(b) Find the range over which the PMOS input stage operates.
(c) Find the range over which both operate (the overlap range).
(d) Find the input common-mode range.
(Note that to operate properly, each of the current sources requires a minimum voltage of $|V_{OV}|$ across its terminals.)
Ans. –1.2 V to +2.9 V; –2.9 V to +1.2 V, –1.2 V to +1.2 V; –2.9 V to +2.9 V

12.2.7 Increasing the Output Voltage Range: The Wide-Swing Current Mirror

In Section 12.2.2 it was found that while the output voltage of the circuit of Fig. 12.9 can swing to within $2|V_{OV}|$ of V_{DD}, the cascode current mirror limits the negative swing to $[2|V_{OV}| + V_t]$ above $-V_{SS}$. In other words, the cascode mirror reduces the voltage swing by V_t volts. This point is further illustrated in Fig. 12.12(a), which shows a cascode mirror (with $V_{SS} = 0$, for simplicity) and indicates the voltages that result at the various nodes. Observe

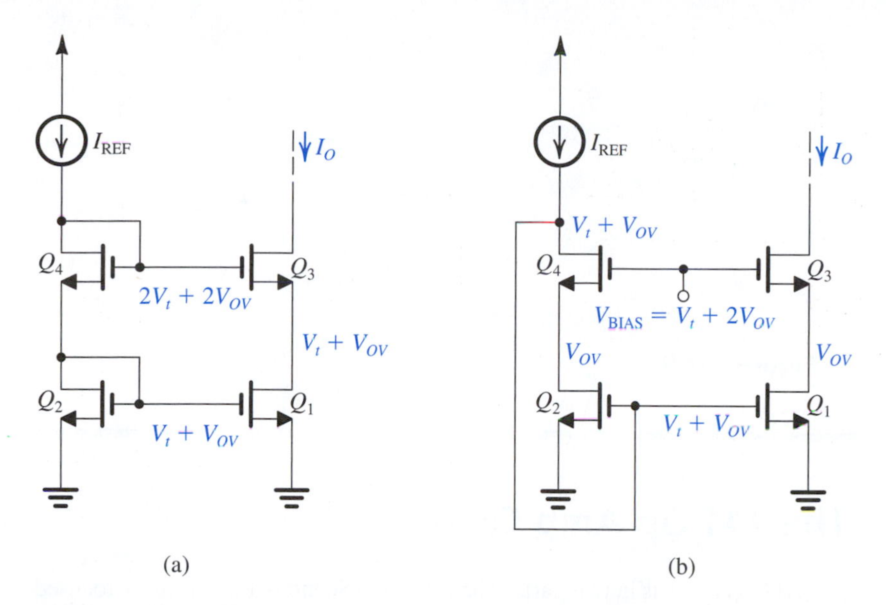

Figure 12.12 **(a)** Cascode current mirror with the voltages at all nodes indicated. Note that the minimum voltage allowed at the output is $V_t + 2V_{OV}$. **(b)** A modification of the cascode mirror that results in the reduction of the minimum output voltage to V_{OV}. This is the wide-swing current mirror. The circuit requires a bias voltage V_{BIAS}.

that because the voltage at the gate of Q_3 is $2V_t + 2V_{OV}$, the minimum voltage permitted at the output (while Q_3 remains saturated) is $V_t + 2V_{OV}$, hence the extra V_t. Also, observe that Q_1 is operating with a drain-to-source voltage $V_t + V_{OV}$, which is V_t volts greater than it needs to operate in saturation.

The observations above lead us to the conclusion that to permit the output voltage at the drain of Q_3 to swing as low as $2V_{OV}$, we must lower the voltage at the gate of Q_3 from $2V_t + 2V_{OV}$ to $V_t + 2V_{OV}$. This is exactly what is done in the modified mirror circuit in Fig. 12.12(b): The gate of Q_3 is now connected to a bias voltage $V_{BIAS} = V_t + 2V_{OV}$. Thus the output voltage can go down to $2V_{OV}$ with Q_3 still in saturation. Also, the voltage at the drain of Q_1 is now V_{OV} and thus Q_1 is operating at the edge of saturation. The same is true of Q_2 and thus the current tracking between Q_1 and Q_2 will be assured. Note, however, that we can no longer connect the gate of Q_2 to its drain. Rather, it is connected to the drain of Q_4. This establishes a voltage of $V_t + V_{OV}$ at the drain of Q_4 which is sufficient to operate Q_4 in saturation (as long as V_t is greater than V_{OV}, which is usually the case). This circuit is known as the **wide-swing current mirror**. Finally, note that Fig. 12.12(b) does not show the circuit for generating V_{BIAS}. There are a number of possible circuits to accomplish this task, one of which is explored in Exercise 12.9.

EXERCISE

12.9 Show that if transistor Q_5 in the circuit of Fig. E12.9 has a W/L ratio equal to one-quarter that of the transistors in the wide-swing current mirror of Fig. 12.12(b), and provided the same value of I_{REF} is utilized in both circuits, then the voltage generated, V_5 is $V_t + 2V_{OV}$, which is the value of V_{BIAS} needed for the gates of Q_3 and Q_4.

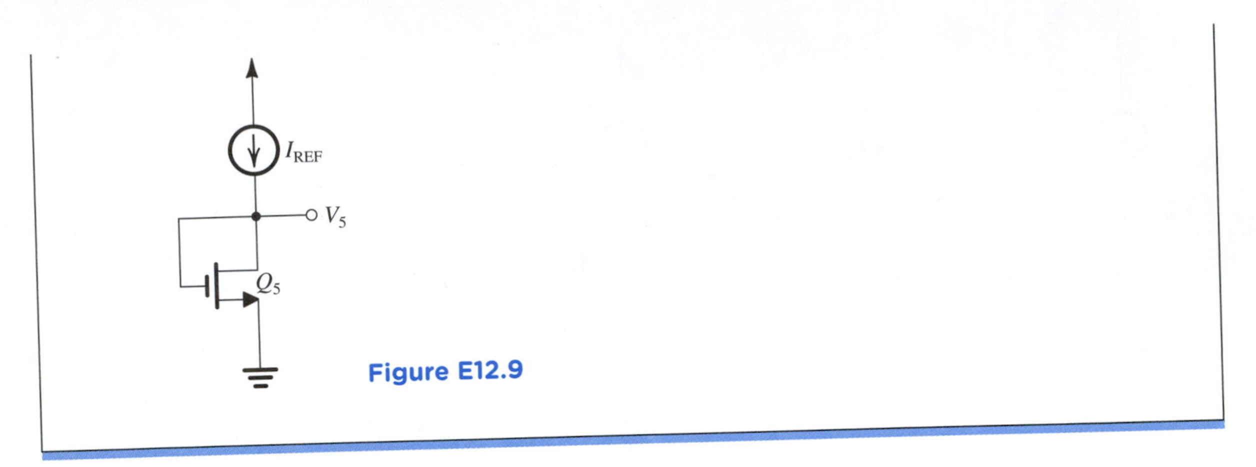

Figure E12.9

12.3 The 741 Op-Amp Circuit

Our study of BJT op amps is in two parts: The first part (Sections 12.3–12.6) is focused on the 741 op-amp circuit, which is shown in Fig. 12.13; the second part (Section 12.7) presents some of the more recent design techniques. Note that in keeping with the IC design philosophy, the circuit in Fig. 12.13 uses a large number of transistors, but relatively few resistors, and only one capacitor. This philosophy is dictated by the economics (silicon area, ease of fabrication, quality of realizable components) of the fabrication of active and passive components in IC form (see Section 7.1 and Appendix A).

As is the case with most general-purpose IC op amps, the 741 requires two power supplies, $+V_{CC}$ and $-V_{EE}$. Normally, $V_{CC} = V_{EE} = 15$ V, but the circuit also operates satisfactorily with the power supplies reduced to much lower values (such as ±5 V). It is important to observe that no circuit node is connected to ground, the common terminal of the two supplies.

With a relatively large circuit such as that shown in Fig. 12.13, the first step in the analysis is the identification of its recognizable parts and their functions. This can be done as follows.

12.3.1 Bias Circuit

The reference bias current of the 741 circuit, I_{REF}, is generated in the branch at the extreme left of Fig. 12.13, consisting of the two diode-connected transistors Q_{11} and Q_{12} and the resistance R_5. Using a Widlar current source formed by Q_{11}, Q_{10}, and R_4, bias current for the first stage is generated in the collector of Q_{10}. Another current mirror formed by Q_8 and Q_9 takes part in biasing the first stage.

The reference bias current I_{REF} is used to provide two proportional currents in the collectors of Q_{13}. This double-collector lateral[2] *pnp* transistor can be thought of as two transistors whose base–emitter junctions are connected in parallel. Thus Q_{12} and Q_{13} form a two-output current mirror: One output, the collector of Q_{13B}, provides bias current and acts as a current-source load for Q_{17}, and the other output, the collector of Q_{13A}, provides bias current for the output stage of the op amp.

[2]See Appendix A for a description of lateral *pnp* transistors. Also, their characteristics were discussed in the Appendix to Chapter 7, Section 7.A.2.

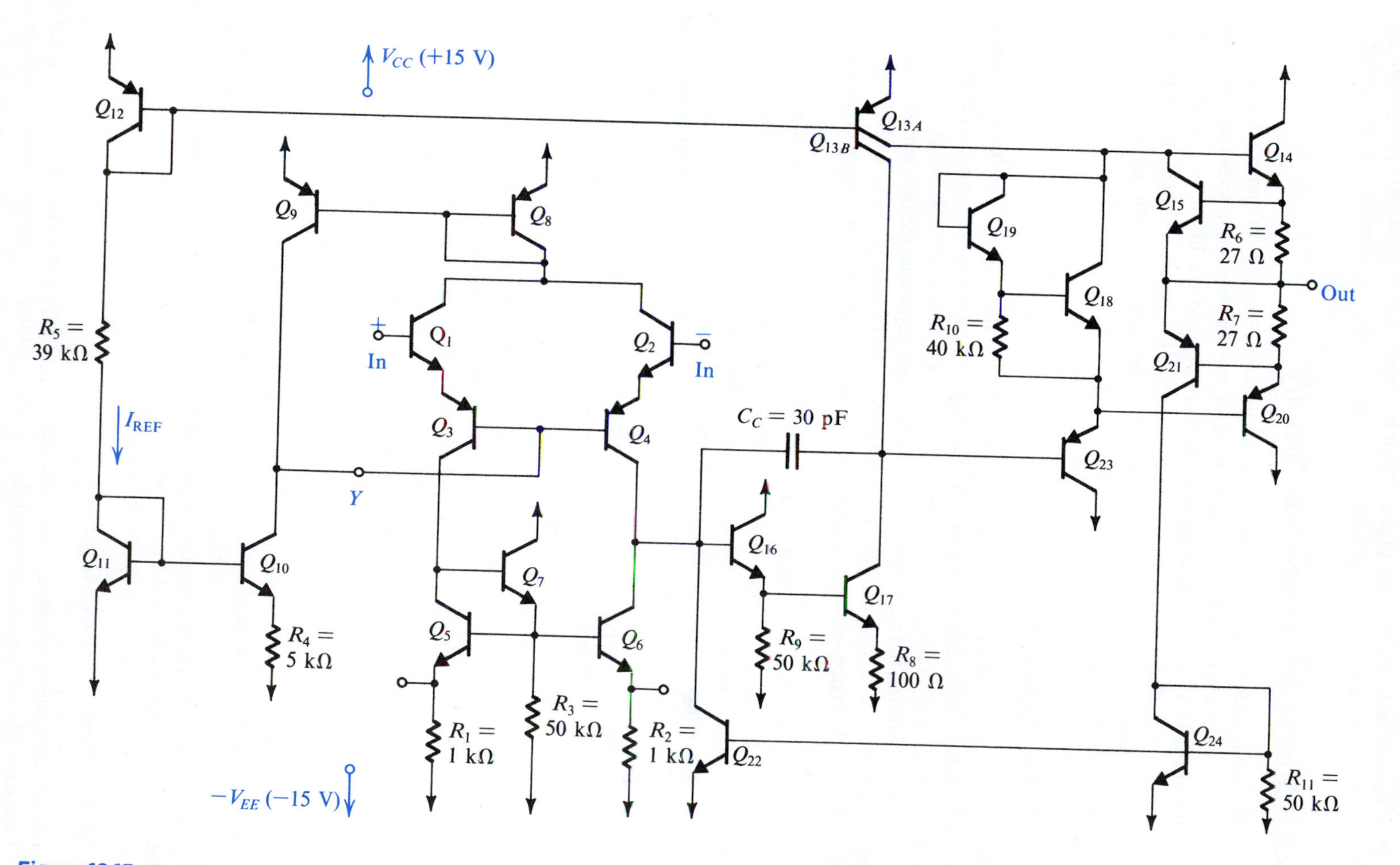

Figure 12.13 The 741 op-amp circuit: Q_{11}, Q_{12}, and R_5 generate a reference bias current; I_{REF}. Q_{10}, Q_9, and Q_8 bias the input stage, which is composed of Q_1 to Q_7. The second gain stage is composed of Q_{16} and Q_{17} with Q_{13B} acting as active load. The class AB output stage is formed by Q_{14} and Q_{20} with biasing devices Q_{13A}, Q_{18}, and Q_{19}, and an input buffer Q_{23}. Transistors Q_{15}, Q_{21}, Q_{24}, and Q_{22} serve to protect the amplifier against output short circuits and are normally cut off.

Two more transistors, Q_{18} and Q_{19}, take part in the dc bias process. The purpose of Q_{18} and Q_{19} is to establish two V_{BE} drops between the bases of the output transistors Q_{14} and Q_{20}.

12.3.2 Short-Circuit Protection Circuitry

The 741 circuit includes a number of transistors that are normally off and conduct only in the event of on attempt to draw a large current from the op-amp output terminal. This happens, for example, if the output terminal is short-circuited to one of the two supplies. The short-circuit protection network consists of R_6, R_7, Q_{15}, Q_{21}, Q_{24}, R_{11}, and Q_{22}. In the following we shall assume that these transistors are off. Operation of the short-circuit protection network will be explained in Section 12.5.3.

12.3.3 The Input Stage

The 741 circuit consists of three stages: an input differential stage, an intermediate single-ended high-gain stage, and an output-buffering stage. The input stage consists of transistors Q_1 through Q_7, with biasing performed by Q_8, Q_9, and Q_{10}. Transistors Q_1 and Q_2 act as emitter followers, causing the input resistance to be high and delivering the differential input signal to the differential common-base amplifier formed by Q_3 and Q_4. Thus the input stage is the differential version of the common-collector common-base configuration discussed in Section 7.6.3.

Transistors Q_5, Q_6, and Q_7 and resistors R_1, R_2, and R_3 form the load circuit of the input stage. This is an elaborate current-mirror load circuit, which we will analyze in detail in Section 12.5.1. The circuit is based on the base-current-compensated mirror studied in Section 7.5, but it includes two emitter-degeneration resistors R_1 and R_2, and a large resistor R_3 in the emitter of Q_7. It will be shown that this load circuit not only provides a high-resistance load but also converts the signal from differential to single-ended form with no loss in gain or common-mode rejection. The output of the input stage is taken single-endedly at the collector of Q_6.

As mentioned in Section 8.6.2, every op-amp circuit includes a *level shifter* whose function is to shift the dc level of the signal so that the signal at the op-amp output can swing positive and negative. In the 741, level shifting is done in the first stage using the lateral *pnp* transistors Q_3 and Q_4. Although lateral *pnp* transistors have poor high-frequency performance, their use in the common-base configuration (which is known to have good high-frequency response) does not seriously impair the op-amp frequency response.

The use of the lateral *pnp* transistors Q_3 and Q_4 in the first stage results in an added advantage: protection of the input-stage transistors Q_1 and Q_2 against emitter–base junction breakdown. Since the emitter–base junction of an *npn* transistor breaks down at about 7 V of reverse bias (see Section 6.9.1), regular *npn* differential stages suffer such a breakdown if, say, the supply voltage is accidentally connected between the input terminals. Lateral *pnp* transistors, however, have high emitter–base breakdown voltages (about 50 V); and because they are connected in series with Q_1 and Q_2, they provide protection of the 741 input transistors, Q_1 and Q_2.

Finally, note that except for using input buffer transistors, the 741 input stage is essentially a current-mirror-loaded differential amplifier. It is quite similar to the input stage of the CMOS amplifier in Fig. 12.1.

12.3.4 The Second Stage

The second or intermediate stage is composed of Q_{16}, Q_{17}, Q_{13B}, and the two resistors R_8 and R_9. Transistor Q_{16} acts as an emitter follower, thus giving the second stage a high input resistance. This minimizes the loading on the input stage and avoids loss of gain. Also, adding Q_{16} with its 50-kΩ emitter resistance (which is similar to Q_7 and R_3) increases the symmetry of the first stage and thus improves its CMRR. Transistor Q_{17} acts as a common-emitter amplifier with a 100-Ω resistor in the emitter. Its load is composed of the high output resistance of the *pnp* current source Q_{13B} in parallel with the input resistance of the output stage (seen looking into the base of Q_{23}). Using a transistor current source as a load resistance (*active load*) enables one to obtain high gain without resorting to the use of large load resistances, which would occupy a large chip area and require large power-supply voltages.

The output of the second stage is taken at the collector of Q_{17}. Capacitor C_C is connected in the feedback path of the second stage to provide frequency compensation using the Miller compensation technique studied in Section 10.13. It will be shown in Section 12.5 that the relatively small capacitor C_C gives the 741 a dominant pole at about 4 Hz. Furthermore, pole splitting causes other poles to be shifted to much higher frequencies, giving the op amp a uniform –20-dB/decade gain rolloff with a unity-gain bandwidth of about 1 MHz. It should be pointed out that although C_C is small in value, the chip area that it occupies is about 13 times that of a standard *npn* transistor!

12.3.5 The Output Stage

The purpose of the output stage (Chapter 11) is to provide the amplifier with a low output resistance. In addition, the output stage should be able to supply relatively large load currents without dissipating an unduly large amount of power in the IC. The 741 uses an efficient class AB output stage, which we shall study in detail in Section 12.5.

The output stage of the 741 consists of the complementary pair Q_{14} and Q_{20}, where Q_{20} is a *substrate pnp* (see Appendix A). Transistors Q_{18} and Q_{19} are fed by current source Q_{13A} and bias the output transistors Q_{14} and Q_{20}. Transistor Q_{23} (which is another substrate *pnp*) acts as an emitter follower, thus minimizing the loading effect of the output stage on the second stage.

12.3.6 Device Parameters

In the following sections we shall carry out a detailed analysis of the 741 circuit. For the standard *npn* and *pnp* transistors, the following parameters will be used:

$$npn\text{:}\quad I_S = 10^{-14}\text{A},\ \beta = 200,\ V_A = 125\ \text{V}$$

$$pnp\text{:}\quad I_S = 10^{-14}\text{A},\ \beta = 50,\ V_A = 50\ \text{V}$$

In the 741 circuit the nonstandard devices are Q_{13}, Q_{14}, and Q_{20}. Transistor Q_{13} will be assumed to be equivalent to two transistors, Q_{13A} and Q_{13B}, with parallel base–emitter junctions and having the following saturation currents:

$$I_{SA} = 0.25 \times 10^{-14}\text{A} \qquad I_{SB} = 0.75 \times 10^{-14}\text{A}$$

Transistors Q_{14} and Q_{20} will be assumed to each have an area three times that of a standard device. Output transistors usually have relatively large areas, to be able to supply large load currents and dissipate relatively large amounts of power with only a moderate increase in device temperature.

EXERCISES

12.10 For the standard *npn* transistor whose parameters are given in Section 12.3.6, find approximate values for the following parameters at $I_C = 1$ mA: V_{BE}, g_m, r_e, r_π, and r_o.
Ans. 633 mV; 40 mA/V; 25 Ω; 5 kΩ; 125 kΩ

12.11 For the circuit in Fig. E12.11, neglect base currents and use the exponential i_C–v_{BE} relationship to show that

$$I_3 = I_1 \sqrt{\frac{I_{S3} I_{S4}}{I_{S1} I_{S2}}}$$

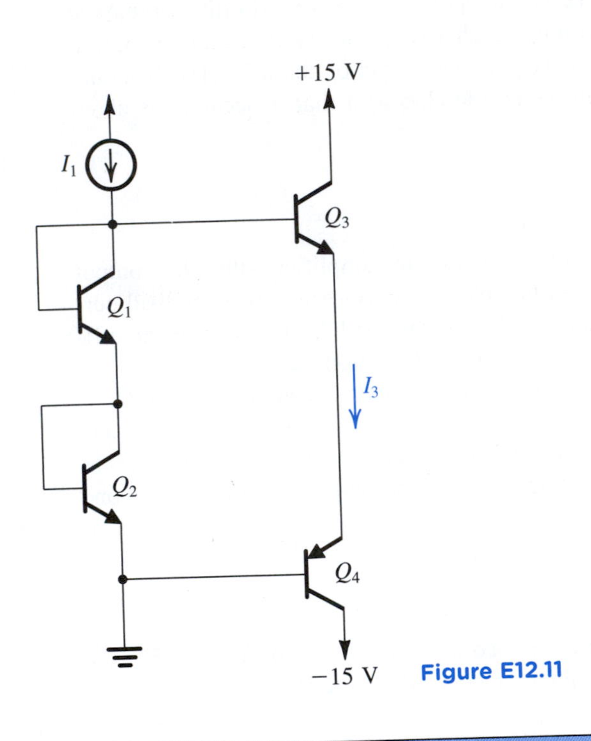

Figure E12.11

12.4 DC Analysis of the 741

In this section, we shall carry out a dc analysis of the 741 circuit to determine the bias point of each device. For the dc analysis of an op-amp circuit, the input terminals are grounded. Theoretically speaking, this should result in zero dc voltage at the output. However, because the op amp has very large gain, any slight approximation in the analysis will show that the output voltage is far from being zero and is close to either $+V_{CC}$ or $-V_{EE}$. In actual practice, an op amp left open-loop will have an output voltage saturated close to one of the two supplies. To overcome this problem in the dc analysis, it will be assumed that the op amp is connected in a negative feedback loop that stabilizes the output dc voltage to zero volts.

12.4.1 Reference Bias Current

The reference bias current I_{REF} is generated in the branch composed of the two diode-connected transistors Q_{11} and Q_{12} and resistor R_5. With reference to Fig. 12.13, we can write

$$I_{\text{REF}} = \frac{V_{CC} - V_{EB12} - V_{BE11} - (-V_{EE})}{R_5}$$

For $V_{CC} = V_{EE} = 15$ V and $V_{BE11} = V_{EB12} \simeq 0.7$ V, we have $I_{\text{REF}} = 0.73$ mA.

12.4.2 Input-Stage Bias

Transistor Q_{11} is biased by I_{REF}, and the voltage developed across it is used to bias Q_{10}, which has a series emitter resistance R_4. This part of the circuit is redrawn in Fig. 12.14 and can be recognized as the Widlar current source studied in Section 7.5.5. From the circuit, and assuming β_{10} to be large, we have

$$V_{BE11} - V_{BE10} = I_{C10}R_4$$

Thus

$$V_T \ln \frac{I_{\text{REF}}}{I_{C10}} = I_{C10}R_4 \tag{12.75}$$

where it has been assumed that $I_{S10} = I_{S11}$. Substituting the known values for I_{REF} and R_4, this equation can be solved by trial and error to determine I_{C10}. For our case, the result is $I_{C10} = 19$ μA.

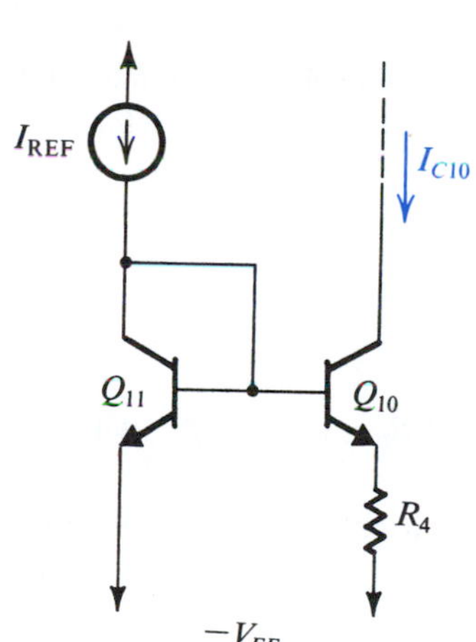

Figure 12.14 The Widlar current source that biases the input stage.

EXERCISE

D12.12 Design the Widlar current source of Fig. 12.14 to generate a current $I_{C10} = 10$ μA given that $I_{\text{REF}} =$ 1 mA. If at a collector current of 1 mA, $V_{BE} = 0.7$ V, find V_{BE11} and V_{BE10}.
Ans. $R_4 = 11.5$ kΩ; $V_{BE11} = 0.7$ V; $V_{BE10} = 0.585$ V

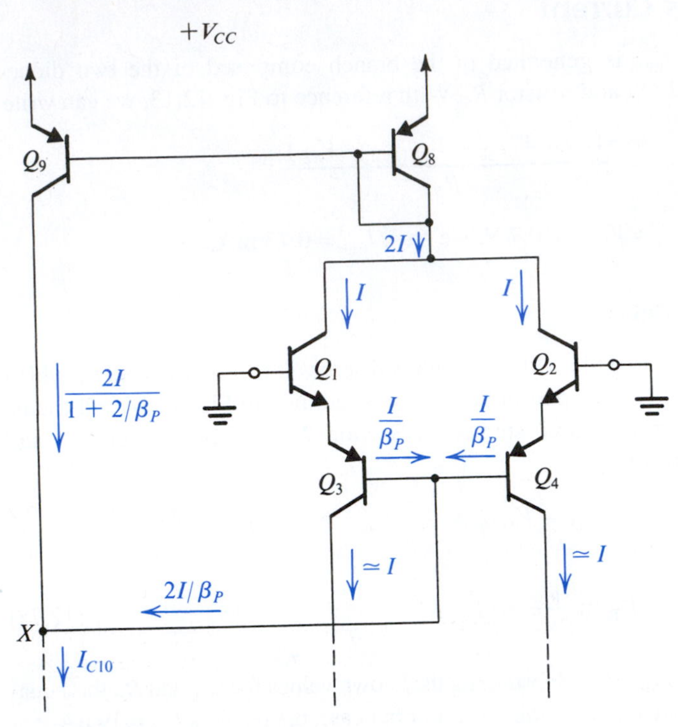

Figure 12.15 The dc analysis of the 741 input stage.

Having determined I_{C10}, we proceed to determine the dc current in each of the input-stage transistors. Part of the input stage is redrawn in Fig. 12.15. From symmetry, we see that

$$I_{C1} = I_{C2}$$

Denote this current by I. We see that if the *npn* β is high, then

$$I_{E3} = I_{E4} \simeq I$$

and the base currents of Q_3 and Q_4 are equal, with a value of $I/(\beta_P + 1) \simeq I/\beta_P$, where β_P denotes β of the *pnp* devices.

The current mirror formed by Q_8 and Q_9 is fed by an input current of $2I$. Using the result in Eq. (7.69), we can express the output current of the mirror as

$$I_{C9} = \frac{2I}{1 + 2/\beta_P}$$

We can now write a node equation for node X in Fig. 12.15 and thus determine the value of I. If $\beta_P \gg 1$, then this node equation gives

$$2I \simeq I_{C10}$$

For the 741, $I_{C10} = 19$ µA; thus $I \simeq 9.5$ µA. We have thus determined that

$$I_{C1} = I_{C2} \simeq I_{C3} = I_{C4} = 9.5 \text{ µA}$$

At this point, we should note that transistors Q_1 through Q_4, Q_8, and Q_9 form a **negative-feedback loop**, which works to stabilize the value of I at approximately $I_{C10}/2$. To appreciate this fact, assume that for some reason the current I in Q_1 and Q_2 increases. This will

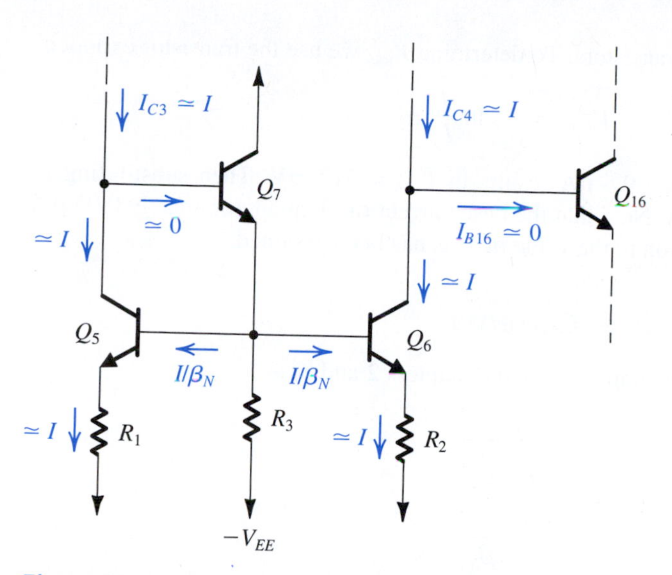

Figure 12.16 The dc analysis of the 741 input stage, continued.

cause the current pulled from Q_8 to increase, and the output current of the Q_8–Q_9 mirror will correspondingly increase. However, since I_{C10} remains constant, node X forces the combined base currents of Q_3 and Q_4 to decrease. This in turn will cause the emitter currents of Q_3 and Q_4, and hence the collector currents of Q_1 and Q_2, to decrease. This is opposite in direction to the change originally assumed. Hence the feedback is negative, and it stabilizes the value of I.

Figure 12.16 shows the remainder of the 741 input stage. This part of the circuit is fed by $I_{C3} = I_{C4} \simeq I$. Transistors Q_5 and Q_6 are identical and have equal resistances R_1 and R_2 in their emitters; thus,

$$I_{C5} = I_{C6} \tag{12.76}$$

Now if the base currents of Q_7 and Q_{16} can be neglected, then

$$I_{C5} \simeq I_{C3} \simeq I \tag{12.77}$$

and

$$I_{C6} \simeq I_{C4} \simeq I \tag{12.78}$$

Thus both the symmetry of Q_5 and Q_6 and the node equations at their collectors force their currents to be equal and to equal I. As will be shown shortly, not only are the base currents of Q_7 and Q_{16} negligible, but their values are also reasonably close, which is an added help.

The bias current of Q_7 can be determined from

$$I_{C7} \simeq I_{E7} = \frac{2I}{\beta_N} + \frac{V_{BE6} + IR_2}{R_3} \tag{12.79}$$

where β_N denotes β of the *npn* transistors. To determine V_{BE6} we use the transistor exponential relationship and write

$$V_{BE6} = V_T \ln \frac{I}{I_S}$$

Substituting $I_S = 10^{-14}$ A and $I = 9.5$ µA results in $V_{BE6} = 517$ mV. Then substituting in Eq. (12.79) yields $I_{C7} = 10.5$ µA. Note that the base current of Q_7 at approximately 0.05 µA is indeed negligible in comparison to the value of I, as has been assumed.

12.4.3 Input Bias and Offset Currents

The **input bias current** of an op amp is defined (Chapters 2 and 8) as

$$I_B = \frac{I_{B1} + I_{B2}}{2}$$

For the 741 we obtain

$$I_B = \frac{I}{\beta_N}$$

Using $\beta_N = 200$, yields $I_B = 47.5$ nA. Note that this value is reasonably small and is typical of general-purpose op amps that use BJTs in the input stage. Much lower input bias currents (in the picoamp or femtoamp range) can be obtained using a FET input stage. Also, there exist techniques for reducing the input bias current of bipolar-input op amps.

Because of possible mismatches in the β values of Q_1 and Q_2, the input base currents will not be equal. Given the value of the β mismatch, one can use Eq. (8.131) to calculate the **input offset current**, defined as

$$I_{OS} = |I_{B1} - I_{B2}|$$

12.4.4 Input Offset Voltage

From Chapter 8 we know that the input offset voltage is determined primarily by mismatches between the two sides of the input stage. In the 741 op amp, the input offset voltage is due to mismatches between Q_1 and Q_2, between Q_3 and Q_4, between Q_5 and Q_6, and between R_1 and R_2. Evaluation of the components of V_{OS} corresponding to the various mismatches follows the method outlined in Section 8.4. Basically, we find the current that results at the output of the first stage due to the particular mismatch being considered. Then we find the differential input voltage that must be applied to reduce the output current to zero.

12.4.5 Input Common-Mode Range

The **input common-mode range** is the range of input common-mode voltages over which the input stage remains in the linear active mode. Refer to Fig. 12.13. We see that in the 741 circuit the input common-mode range is determined at the upper end by saturation of Q_1 and Q_2, and at the lower end by saturation of Q_3 and Q_4.

EXERCISE

12.13 Neglect the voltage drops across R_1 and R_2 and assume that $V_{CC} = V_{EE} = 15$ V. Show that the input common-mode range of the 741 is approximately –12.9 V to +14.7 V. (Assume that $V_{BE} \simeq 0.6$ V and that to avoid saturation $V_{CB} \geq -0.3$ V for an *npn* transistor, and $V_{BC} \geq -0.3$ V for a *pnp* transistor.)

12.4.6 Second-Stage Bias

If we neglect the base current of Q_{23} then we see from Fig. 12.13 that the collector current of Q_{17} is approximately equal to the current supplied by current source Q_{13B}. Because Q_{13B} has a scale current 0.75 times that of Q_{12}, its collector current will be $I_{C13B} \simeq 0.75I_{REF}$, where we have assumed that $\beta_P \gg 1$. Thus $I_{C13B} = 550\ \mu A$ and $I_{C17} \simeq 550\ \mu A$. At this current level the base–emitter voltage of Q_{17} is

$$V_{BE17} = V_T \ln \frac{I_{C17}}{I_S} = 618 \text{ mV}$$

The collector current of Q_{16} can be determined from

$$I_{C16} \simeq I_{E16} = I_{B17} + \frac{I_{E17}R_8 + V_{BE17}}{R_9}$$

This calculation yields $I_{C16} = 16.2\ \mu A$. Note that the base current of Q_{16} at $0.08\ \mu A$ will indeed be negligible compared to the input-stage bias I, as we have assumed.

12.4.7 Output-Stage Bias

Figure 12.17 shows the output stage of the 741 with the short-circuit-protection circuitry omitted. Current source Q_{13A} delivers a current of $0.25I_{REF}$ (because I_S of Q_{13A} is 0.25

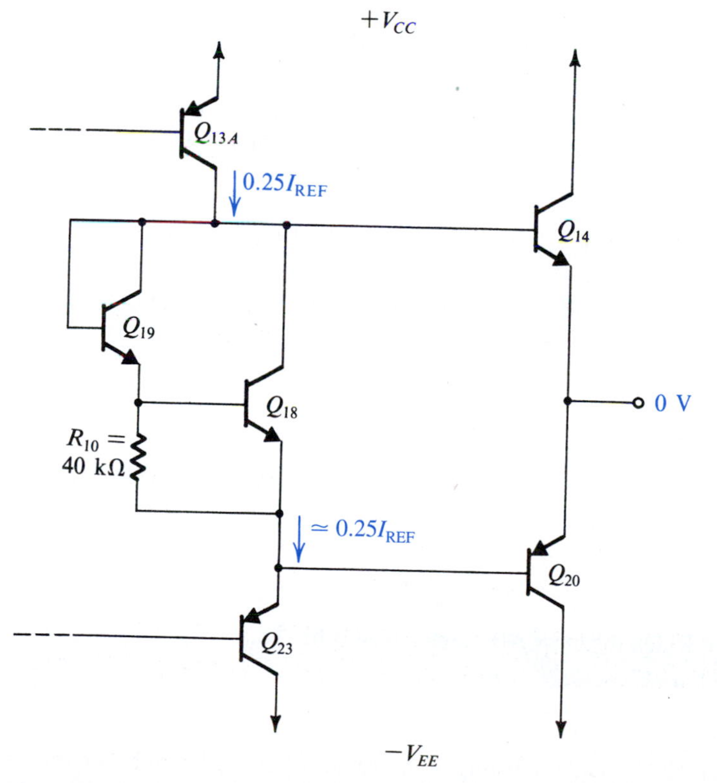

Figure 12.17 The 741 output stage without the short-circuit protection devices.

times the I_S of Q_{12}) to the network composed of Q_{18}, Q_{19}, and R_{10}. If we neglect the base currents of Q_{14} and Q_{20}, then the emitter current of Q_{23} will also be equal to $0.25I_{REF}$. Thus

$$I_{C23} \simeq I_{E23} \simeq 0.25I_{REF} = 180\ \mu A$$

Thus we see that the base current of Q_{23} is only $180/50 = 3.6\ \mu A$, which is negligible compared to I_{C17}, as we have assumed.

If we assume that V_{BE18} is approximately 0.6 V, we can determine the current in R_{10} as 15 μA. The emitter current of Q_{18} is therefore

$$I_{E18} = 180 - 15 = 165\ \mu A$$

Also,

$$I_{C18} \simeq I_{E18} = 165\ \mu A$$

At this value of current we find that $V_{BE18} = 588$ mV, which is quite close to the value assumed. The base current of Q_{18} is $165/200 = 0.8\ \mu A$, which can be added to the current in R_{10} to determine the Q_{19} current as

$$I_{C19} \simeq I_{E19} = 15.8\ \mu A$$

The voltage drop across the base–emitter junction of Q_{19} can now be determined as

$$V_{BE19} = V_T \ln \frac{I_{C19}}{I_S} = 530\ \text{mV}$$

As mentioned in Section 12.3.5, the purpose of the Q_{18}–Q_{19} network is to establish two V_{BE} drops between the bases of the output transistors Q_{14} and Q_{20}. This voltage drop, V_{BB}, can be now calculated as

$$V_{BB} = V_{BE18} + V_{BE19} = 588 + 530 = 1.118\ \text{V}$$

Since V_{BB} appears across the series combination of the base–emitter junctions of Q_{14} and Q_{20}, we can write

$$V_{BB} = V_T \ln \frac{I_{C14}}{I_{S14}} + V_T \ln \frac{I_{C20}}{I_{S20}}$$

Using the calculated value of V_{BB} and substituting $I_{S14} = I_{S20} = 3 \times 10^{-14}$ A, we determine the collector currents as

$$I_{C14} = I_{C20} = 154\ \mu A$$

This is the small current at which the class AB output stage is biased.

12.4.8 Summary

For future reference, Table 12.1 provides a listing of the values of the collector bias currents of the 741 transistors.

Table 12.1 DC Collector Currents of the 741 Circuit (μA)

Q_1	9.5	Q_8	19	Q_{13B}	550	Q_{19}	15.8		
Q_2	9.5	Q_9	19	Q_{14}	154	Q_{20}	154		
Q_3	9.5	Q_{10}	19	Q_{15}	0	Q_{21}	0		
Q_4	9.5	Q_{11}	730	Q_{16}	16.2	Q_{22}	0		
Q_5	9.5	Q_{12}	730	Q_{17}	550	Q_{23}	180		
Q_6	9.5	Q_{13A}	180	Q_{18}	165	Q_{24}	0		
Q_7	10.5								

EXERCISE

12.14 If in the circuit of Fig. 12.17 the Q_{18}–Q_{19} network is replaced by two diode-connected transistors, find the current in Q_{14} and Q_{20}. (*Hint:* Use the result of Exercise 12.11.)
Ans. 540 μA

12.5 Small-Signal Analysis of the 741

12.5.1 The Input Stage

Figure 12.18 shows part of the 741 input stage for the purpose of performing small-signal analysis. Note that since the collectors of Q_1 and Q_2 are connected to a constant dc voltage, they are shown grounded. Also, the constant-current biasing of the bases of Q_3 and Q_4 is equivalent to having the common base terminal open-circuited.

The differential signal v_i applied between the input terminals effectively appears across four equal emitter resistances connected in series—those of Q_1, Q_2, Q_3, and Q_4. As a result, emitter signal currents flow as indicated in Fig. 12.18 with

$$i_e = \frac{v_i}{4r_e} \tag{12.80}$$

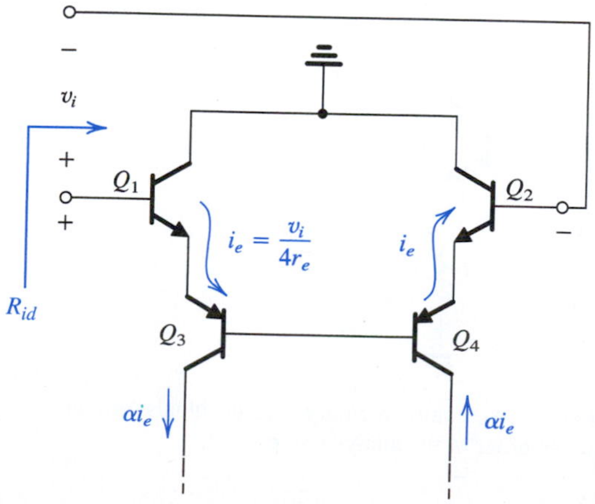

Figure 12.18 Small-signal analysis of the 741 input stage.

where r_e denotes the emitter resistance of each of Q_1 through Q_4. Thus

$$r_e = \frac{V_T}{I} = \frac{25\text{ mV}}{9.5\ \mu\text{A}} = 2.63\text{ k}\Omega$$

Thus the four transistors Q_1 through Q_4 supply the load circuit with a pair of complementary current signals αi_e, as indicated in Fig. 12.18.

The input differential resistance of the op amp can be obtained from Fig. 12.18 as

$$R_{id} = 4(\beta_N + 1)r_e \tag{12.81}$$

For $\beta_N = 200$, we obtain $R_{id} = 2.1\text{ M}\Omega$.

Proceeding with the input-stage analysis, we show in Fig. 12.19 the load circuit fed with the complementary pair of current signals found earlier. Neglecting the signal current in the base of Q_7, we see that the collector signal current of Q_5 is approximately equal to the input current αi_e. Now, since Q_5 and Q_6 are identical and their bases are tied together, and since equal resistances are connected in their emitters, it follows that their collector signal currents must be equal. Thus the signal current in the collector of Q_6 is forced to be equal to αi_e. In other words, the load circuit functions as a current mirror.

Now consider the output node of the input stage. The output current i_o is given by

$$i_o = 2\alpha i_e \tag{12.82}$$

The factor of 2 in this equation indicates that conversion from differential to single-ended is performed without losing half the signal. The trick, of course, is the use of the current mirror to invert one of the current signals and then add the result to the other current signal (see Section 8.5).

Equations (12.80) and (12.82) can be combined to obtain the transconductance of the input stage, G_{m1}:

$$G_{m1} \equiv \frac{i_o}{v_i} = \frac{\alpha}{2r_e} \tag{12.83}$$

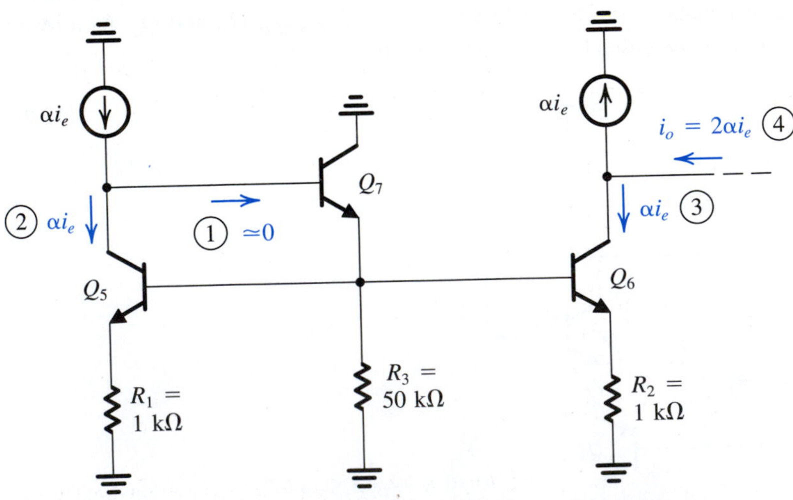

Figure 12.19 The load circuit of the input stage fed by the two complementary current signals generated by Q_1 through Q_4 in Fig. 12.18. Circled numbers indicate the order of the analysis steps.

Substituting $r_e = 2.63$ kΩ and $\alpha \simeq 1$ yields $G_{m1} = 1/5.26$ mA/V. The expression for G_{m1} can be written in the alternate form

$$G_{m1} = \frac{1}{2}g_{m1} \tag{12.83'}$$

where g_{m1} is the transconductance of each of Q_1 to Q_4.

EXERCISE

12.15 For the circuit in Fig. 12.19, find in terms of i_e: (a) the signal voltage at the base of Q_6; (b) the signal current in the emitter of Q_7; (c) the signal current in the base of Q_7; (d) the signal voltage at the base of Q_7; (e) the input resistance seen by the left-hand-side signal current source αi_e. (*Note:* For simplicity, assume that $I_{C7} \simeq I_{C5} = I_{C6}$.)
Ans. (a) 3.63 kΩ × i_e; (b) $0.08i_e$; (c) $0.0004i_e$; (d) 3.84 kΩ × i_e; (e) 3.84 kΩ

To complete our modeling of the 741 input stage, we must find its output resistance R_{o1}. This is the resistance seen "looking back" into the collector terminal of Q_6 in Fig. 12.19. Thus R_{o1} is the parallel equivalent of the output resistance of the current source supplying the signal current αi_e, and the output resistance of Q_6. The first component is the resistance looking into the collector of Q_4 in Fig. 12.18. Finding this resistance is considerably simplified if we assume that the common bases of Q_3 and Q_4 are at a *virtual ground.* This of course happens only when the input signal v_i is applied in a complementary fashion. Nevertheless, this assumption does not result in a large error.

Assuming that the base of Q_4 is at virtual ground, the resistance we are after is R_{o4}, indicated in Fig. 12.20(a). This is the output resistance of a common-base transistor that has a resistance (r_e of Q_2) in its emitter. To find R_{o4} we may use the following expression (Eq. 7.51):

$$R_o = r_o[1 + g_m(R_e \| r_\pi)] \tag{12.84}$$

Substituting $R_e = r_e \equiv 2.63$ kΩ and $r_o = V_A/I$, where $V_A = 50$ V and $I = 9.5$ μA (thus $r_o =$ 5.26 MΩ), and neglecting r_π since it is $(\beta + 1)$ times larger than R_E, results in $R_{o4} = 10.5$ MΩ.

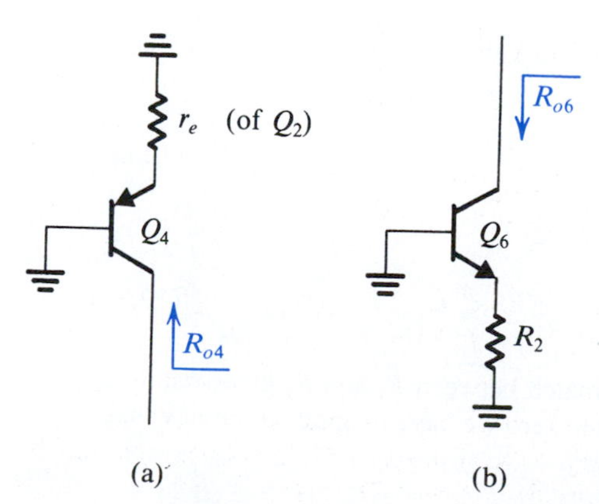

Figure 12.20 Simplified circuits for finding the two components of the output resistance R_{o1} of the first stage.

The second component of the output resistance is that seen looking into the collector of Q_6 in Fig. 12.19 with the αi_e generator set to 0. Although the base of Q_6 is not at signal ground, we shall assume that the signal voltage at the base is small enough to make this approximation valid. The circuit then takes the form shown in Fig. 12.20(b), and R_{o6} can be determined using Eq. (12.84) with $R_e = R_2$. Thus $R_{o6} \simeq 18.2\ \text{M}\Omega$.

Finally, we combine R_{o4} and R_{o6} in parallel to obtain the output resistance of the input stage, R_{o1}, as $R_{o1} = 6.7\ \text{M}\Omega$.

Figure 12.21 shows the equivalent circuit that we have derived for the input stage.

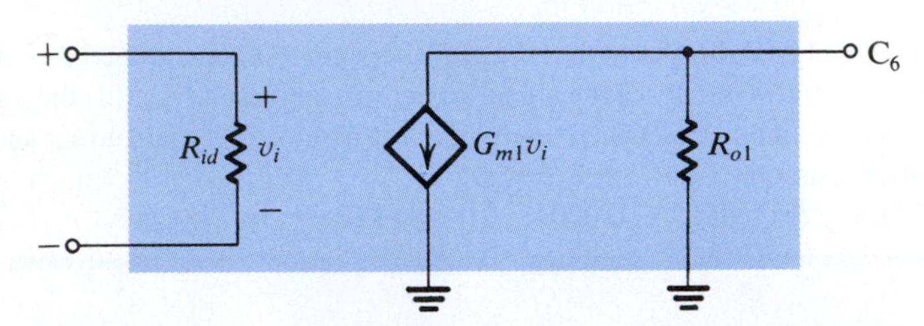

Figure 12.21 Small-signal equivalent circuit for the input stage of the 741 op amp.

Example 12.3

We wish to find the input offset voltage resulting from a 2% mismatch between the resistances R_1 and R_2 in Fig. 12.13.

Solution

Consider first the situation when both input terminals are grounded, and assume that $R_1 = R$ and $R_2 = R + \Delta R$, where $\Delta R/R = 0.02$. From Fig. 12.22 we see that while Q_5 still conducts a current equal to I, the current in Q_6 will be smaller by ΔI. The value of ΔI can be found from

$$V_{BE5} + IR = V_{BE6} + (I - \Delta I)(R + \Delta R)$$

Thus

$$V_{BE5} - V_{BE6} = I\Delta R - \Delta I(R + \Delta R) \tag{12.85}$$

The quantity on the left-hand side is in effect the change in V_{BE} due to a change in I_E of ΔI. We may therefore write

$$V_{BE5} - V_{BE6} \simeq \Delta I r_e \tag{12.86}$$

Equations (12.85) and (12.86) can be combined to obtain

$$\frac{\Delta I}{I} = \frac{\Delta R}{R + \Delta R + r_e} \tag{12.87}$$

Substituting $R = 1\ \text{k}\Omega$ and $r_e = 2.63\ \text{k}\Omega$ shows that a 2% mismatch between R_1 and R_2 gives rise to an output current $\Delta I = 5.5 \times 10^{-3} I$. To reduce this output current to zero we have to apply an input voltage V_{OS} given by

$$V_{OS} = \frac{\Delta I}{G_{m1}} = \frac{5.5 \times 10^{-3} I}{G_{m1}} \tag{12.88}$$

Substituting $I = 9.5\ \mu A$ and $G_{m1} = 1/5.26$ mA/V results in the offset voltage $V_{OS} \simeq 0.3$ mV.

It should be pointed out that the offset voltage calculated is only one component of the input offset voltage of the 741. Other components arise because of mismatches in transistor characteristics. The 741 offset voltage is specified to be typically 2 mV.

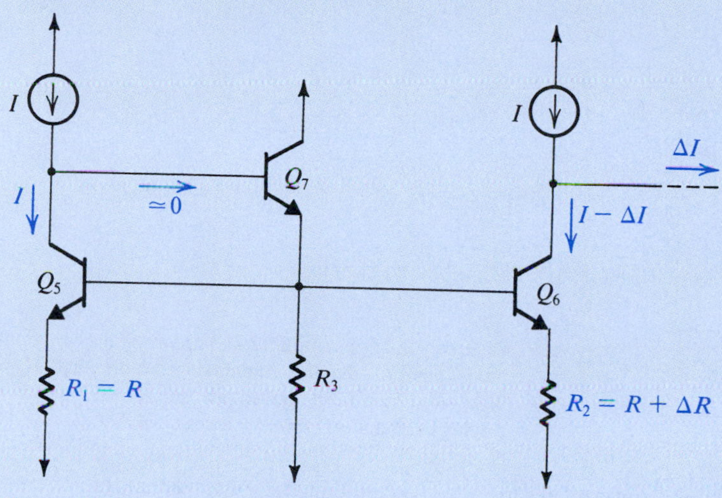

Figure 12.22 Input stage with both inputs grounded and a mismatch ΔR between R_1 and R_2.

Example 12.4

It is required to find the CMRR of the 741 input stage. Assume that the circuit is balanced except for mismatches in the current-mirror load that result in an error ε_m in the mirror's current-transfer ratio; that is, the ratio becomes $(1 - \varepsilon_m)$.

Solution

In Section 8.5.4 we analyzed the common-mode operation of the current-mirror-loaded differential amplifier and derived an expression for its CMRR. The situation in the 741 input stage, however, differs substantially because of the feedback loop that regulates the bias current. Since this feedback loop is sensitive to the common-mode signal, as will be seen shortly, the loop operates to reduce the common-mode gain and, correspondingly, to increase the CMRR. Hence, its action is referred to as **common-mode feedback**.

Figure 12.23 shows the 741 input stage with a common-mode signal v_{icm} applied to both input terminals. We have assumed that as a result of v_{icm}, a signal current i flows as shown. Since the stage is balanced, both sides carry the same current i.

Example 12.4 *continued*

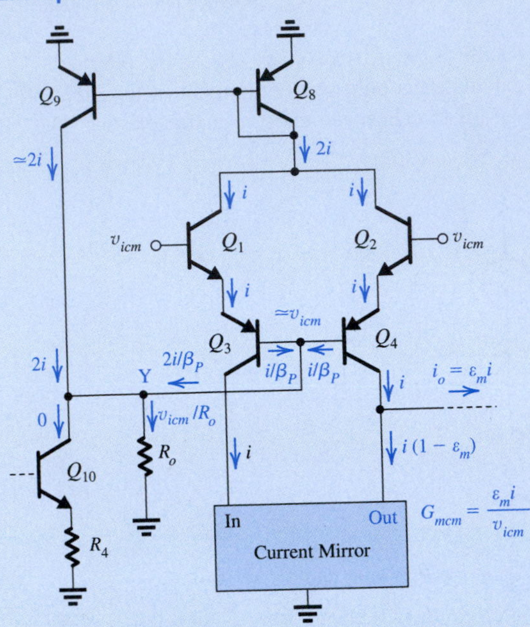

Figure 12.23 Example 12.4: Analysis of the common-mode gain of the 741 input stage. Note that $R_o = R_{o9} \parallel R_{o10}$, has been "pulled out" and shown separately, leaving behind ideal current sources Q_9 and Q_{10}.

Our objective now is to determine how i relates to v_{icm}. Toward that end, observe that for common-mode inputs, both sides of the differential amplifier, that is, $Q_1 - Q_3$ and $Q_2 - Q_4$, act as followers, delivering a signal almost equal to v_{icm} to the common-base node of Q_3 and Q_4. Now, this node Y is connected to the collectors of two current sources, Q_9 and Q_{10}. Denoting the total resistance between node Y and ground R_o, we write

$$R_o = R_{o9} \parallel R_{o10} \tag{12.89}$$

In Fig. 12.23 we have "pulled R_o out," thus leaving behind ideal current sources Q_9 and Q_{10}. Since the current in Q_{10} is constant, we show Q_{10} in Fig. 12.23 as having a zero incremental current. Transistor Q_9, on the other hand, provides a current approximately equal to that fed into Q_8, which is $2i$. This is the feedback current. Since Q_8 senses the *sum* of the currents in the two sides of the differential amplifier, the feedback loop operates only on the common-mode signal and is insensitive to any difference signal.

Proceeding with the analysis, we now can write a node equation at Y,

$$2i + \frac{2i}{\beta_P} = \frac{v_{icm}}{R_o} \tag{12.90}$$

Assuming $\beta_P \gg 1$, this equation simplifies to

$$i \simeq \frac{v_{icm}}{2R_o} \tag{12.91}$$

Having determined i, we now proceed to complete our analysis by finding the output current i_o. From the circuit in Fig. 12.23, we see that

$$i_o = \varepsilon_m i \tag{12.92}$$

Thus the common-mode transconductance of the input stage is given by

$$G_{mcm} \equiv \frac{i_o}{v_{icm}} = \frac{\varepsilon_m i}{v_{icm}}$$

Substituting for i from Eq. (12.91) gives

$$G_{mcm} = \frac{\varepsilon_m}{2R_o} \quad (12.93)$$

Finally, the CMRR can be found as the ratio of the differential transconductance G_{m1} found in Eq. (12.83′) and the common-mode transconductance G_{mcm},

$$\text{CMRR} \equiv \frac{G_{m1}}{G_{mcm}} = 2g_{m1}R_o/\varepsilon_m \quad (12.94)$$

where g_{m1} is the transconductance of Q_1. Now substituting for R_o from Eq. (12.89), we obtain

$$\text{CMRR} = 2g_{m1}(R_{o9} \| R_{o10})/\varepsilon_m \quad (12.95)$$

Before leaving this example, we observe that if the feedback were not present, the $2i$ term in Eq. (12.90) would be absent and the current i would become $\beta_P(v_{icm}/2R_o)$, which is β_P times higher than that when feedback is present. In other words, common-mode feedback reduces i, hence the common-mode transconductance and the common-mode gain, by a factor β_P.

EXERCISES

12.16 Show that if the source of the imbalance in the current-mirror load is that while $R_1 = R$, $R_2 = R + \Delta R$, the error ε_m is given by

$$\varepsilon_m = \frac{\Delta R}{R + r_{e5} + \Delta R}$$

Evaluate ε_m for $\Delta R/R = 0.02$.
Ans. $\varepsilon_m = 5.5 \times 10^{-3}$

12.17 Refer to Fig. 12.23 and assume that the bases of Q_9 and Q_{10} are at approximately constant voltages (signal ground). Find R_{o9}, R_{o10}, and hence R_o. Use $V_A = 125$ V for *npn* and 50 V for *pnp* transistors. Use the bias current values in Table 12.1.
Ans. $R_{o9} = 2.63$ MΩ; $R_{o10} = 31.1$ MΩ; $R_o = 2.43$ MΩ

12.18 Use the results of Exercises 12.16 and 12.17 to determine G_{mcm} and CMRR of the 741 input stage. What would the CMRR be if the common-mode feedback were not present? Assume $\beta_P = 50$.
Ans. $G_{mcm} = 1.13 \times 10^{-6}$ mA/V; CMRR $= 1.68 \times 10^5$ or 104.5 dB; without common-mode feedback, CMRR = 70.5 dB

12.5.2 The Second Stage

Figure 12.24 shows the 741 second stage prepared for small-signal analysis. In this section we shall analyze the second stage to determine the values of the parameters of the equivalent circuit shown in Fig. 12.25.

Input Resistance The input resistance R_{i2} can be found by inspection to be

$$R_{i2} = (\beta_{16} + 1)\{r_{e16} + [R_9 \| (\beta_{17} + 1)(r_{e17} + R_8)]\} \quad (12.96)$$

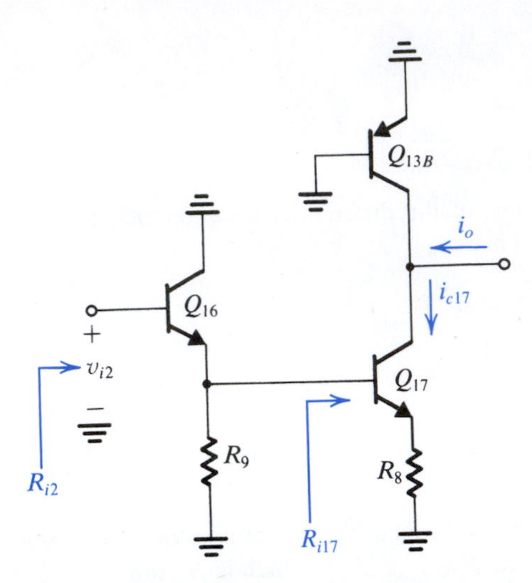

Figure 12.24 The 741 second stage prepared for small-signal analysis.

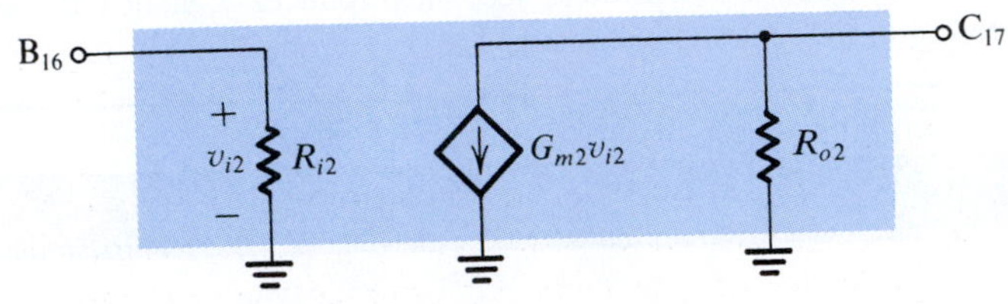

Figure 12.25 Small-signal equivalent-circuit model of the second stage.

Substituting the appropriate parameter values yields $R_{i2} \simeq 4$ MΩ.

Transconductance From the equivalent circuit of Fig. 12.25, we see that the transconductance G_{m2} is the ratio of the *short-circuit output current* to the input voltage. Short-circuiting the output terminal of the second stage (Fig. 12.24) to ground makes the signal current through the output resistance of Q_{13B} zero, and the output short-circuit current becomes equal to the collector signal current of Q_{17} (i_{c17}). This latter current can be easily related to v_{i2} as follows:

$$i_{c17} = \frac{\alpha v_{b17}}{r_{e17} + R_8} \tag{12.97}$$

$$v_{b17} = v_{i2}\frac{(R_9 \| R_{i17})}{(R_9 \| R_{i17}) + r_{e16}} \tag{12.98}$$

$$R_{i17} = (\beta_{17} + 1)(r_{e17} + R_8) \tag{12.99}$$

where we have neglected r_{o16} because $r_{o16} \gg R_9$. These equations can be combined to obtain

$$G_{m2} \equiv \frac{i_{c17}}{v_{i2}} \tag{12.100}$$

which, for the 741 parameter values, is found to be $G_{m2} = 6.5$ mA/V.

Output Resistance To determine the output resistance R_{o2} of the second stage in Fig. 12.24, we ground the input terminal and find the resistance looking back into the output terminal.

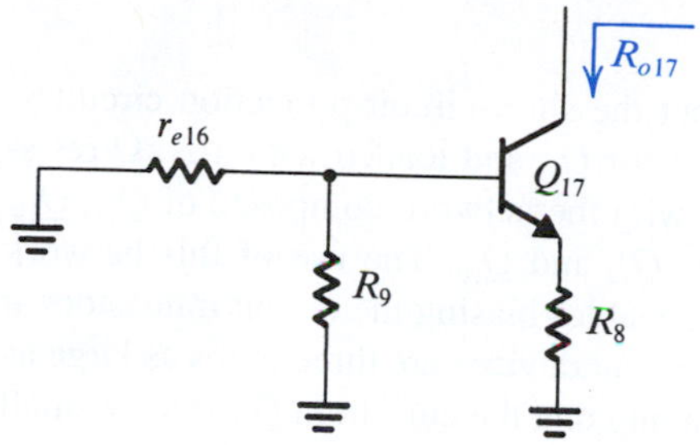

Figure 12.26 Definition of R_{o17}.

It follows that R_{o2} is given by

$$R_{o2} = (R_{o13B} \| R_{o17}) \tag{12.101}$$

where R_{o13B} is the resistance looking into the collector of Q_{13B} while its base and emitter are connected to ground. It can be easily seen that

$$R_{o13B} = r_{o13B} \tag{12.102}$$

For the 741 component values we obtain $R_{o13B} = 90.9\ \text{k}\Omega$.

The second component in Eq. (12.101), R_{o17}, is the resistance seen looking into the collector of Q_{17}, as indicated in Fig. 12.26. Since the resistance between the base of Q_{17} and ground is relatively small, one can considerably simplify matters by assuming that the base is grounded. Doing this, we can use Eq. (12.84) to determine R_{o17}. For our case, the result is $R_{o17} \simeq 787\ \text{k}\Omega$. Combining R_{o13B} and R_{o17} in parallel yields $R_{o2} = 81\ \text{k}\Omega$.

Thévenin Equivalent Circuit The second-stage equivalent circuit can be converted to the Thévenin form, as shown in Fig. 12.27. Note that the stage open-circuit voltage gain is $-G_{m2}R_{o2}$.

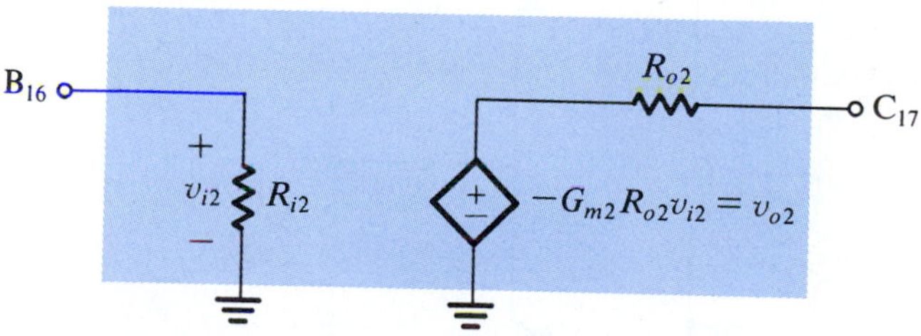

Figure 12.27 Thévenin form of the small-signal model of the second stage.

EXERCISES

12.19 Use Eq. (12.96) to show that $R_{i2} \simeq 4\ \text{M}\Omega$.

12.20 Use Eqs. (12.97) to (12.100) to verify that G_{m2} is 6.5 mA/V.

12.21 Verify that $R_{o2} \simeq 81\ \text{k}\Omega$.

12.22 Find the open-circuit voltage gain of the second stage of the 741.
Ans. −526.5 V/V

12.5.3 The Output Stage

The 741 output stage is shown in Fig. 12.28 without the short-circuit-protection circuitry. The stage is shown driven by the second-stage transistor Q_{17} and loaded with a 2-kΩ resistance. The circuit is of the AB class (Section 11.4), with the network composed of Q_{18}, Q_{19}, and R_{10} providing the bias of the output transistors Q_{14} and Q_{20}. The use of this network rather than two diode-connected transistors in series enables biasing the output transistors at a low current (0.15 mA) in spite of the fact that the output devices are three times as large as the standard devices. This result is obtained by arranging that the current in Q_{19} is very small and thus its V_{BE} is also small. We analyzed the dc bias in Section 12.4.7.

Another feature of the 741 output stage worth noting is that the stage is driven by an emitter follower Q_{23}. As will be shown, this emitter follower provides added buffering, which makes the op-amp gain almost independent of the parameters of the output transistors.

Output Voltage Limits The maximum positive output voltage is limited by the saturation of current-source transistor Q_{13A}. Thus,

$$v_{O\max} = V_{CC} - |V_{CE\text{sat}}| - V_{BE14} \tag{12.103}$$

which is about 1 V below V_{CC}. The minimum output voltage (i.e., maximum negative amplitude) is limited by the saturation of Q_{17}. Neglecting the voltage drop across R_8, we obtain

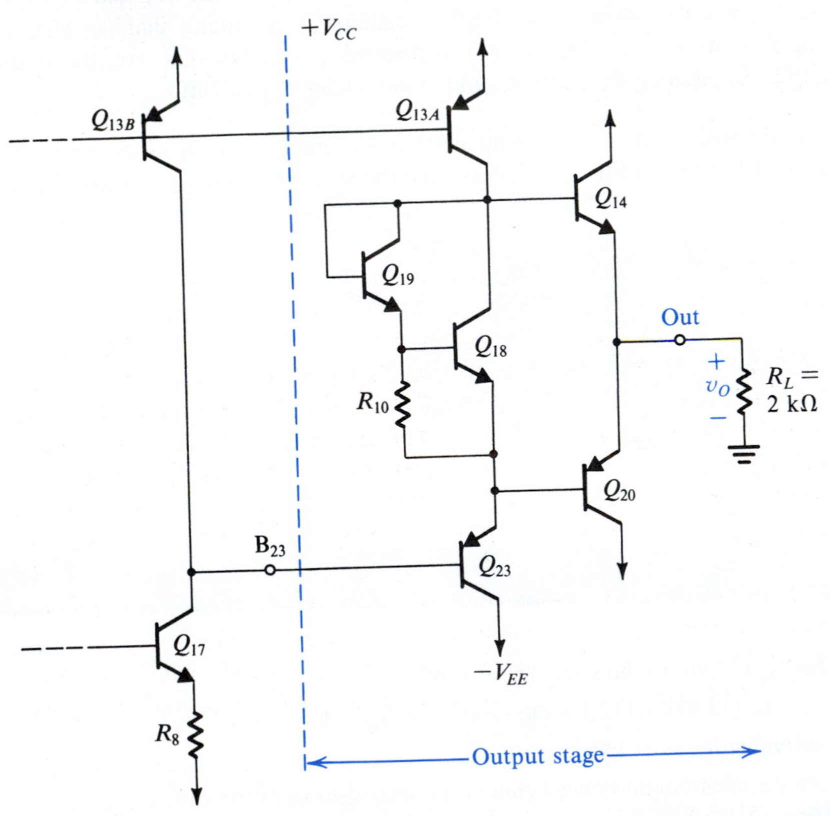

Figure 12.28 The 741 output stage without the short-circuit-protection circuitry.

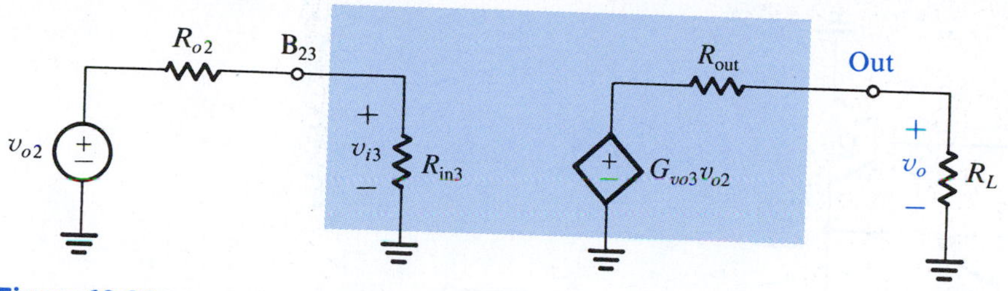

Figure 12.29 Model for the 741 output stage.

$$v_{O\min} = -V_{EE} + V_{CE\text{sat}} + V_{EB23} + V_{EB20} \tag{12.104}$$

which is about 1.5 V above $-V_{EE}$.

Small-Signal Model We shall now carry out a small-signal analysis of the output stage for the purpose of determining the values of the parameters of the equivalent-circuit model shown in Fig. 12.29. The model is shown fed by v_{o2}, which is the open-circuit output voltage of the second stage. From Fig. 12.27, v_{o2} is given by

$$v_{o2} = -G_{m2}R_{o2}v_{i2} \tag{12.105}$$

where G_{m2} and R_{o2} were previously determined as $G_{m2} = 6.5$ mA/V and $R_{o2} = 81$ kΩ. Resistance R_{in3} is the input resistance of the output stage determined with the amplifier loaded with R_L. Although the effect of loading an amplifier stage on its input resistance is negligible in the input and second stages, this is not the case in general in an output stage. Defining R_{in3} in this manner enables correct evaluation of the voltage gain of the second stage, A_2, as

$$A_2 \equiv \frac{v_{i3}}{v_{i2}} = -G_{m2}R_{o2}\frac{R_{in3}}{R_{in3} + R_{o2}} \tag{12.106}$$

To determine R_{in3}, assume that one of the two output transistors—say, Q_{20}—is conducting a current of, say, 5 mA while Q_{14} is cutoff. It follows that the input resistance looking into the base of Q_{20} is approximately $\beta_{20}R_L$. Assuming $\beta_{20} = 50$, for $R_L = 2$ kΩ, the input resistance of Q_{20} is 100 kΩ. This resistance appears in parallel with the series combination of the output resistance of Q_{13A} ($r_{o13A} \simeq 280$ kΩ) and the resistance of the Q_{18}–Q_{19} network. The latter resistance is very small (about 160 Ω; see later: Exercise 12.23). Thus the total resistance in the emitter of Q_{23} is approximately (100 kΩ ∥ 280 kΩ) or 74 kΩ and the input resistance R_{in3} is given by

$$R_{in3} \simeq \beta_{23} \times 74 \text{ k}\Omega$$

which for $\beta_{23} = 50$ is $R_{in3} \simeq 3.7$ MΩ. Since $R_{o2} = 81$ kΩ, we see that $R_{in3} \gg R_{o2}$, and the value of R_{in3} will have little effect on the performance of the op amp. Still we can use the value obtained for R_{in3} to determine the gain of the second stage using Eq. (12.106) as $A_2 = -515$ V/V. The value of A_2 will be needed in Section 12.6 in connection with the frequency-response analysis.

Continuing with the determination of the equivalent circuit-model-parameters, we note from Fig. 12.29 that G_{vo3} is the **open-circuit overall voltage gain** of the output stage,

$$G_{vo3} = \left.\frac{v_o}{v_{o2}}\right|_{R_L=\infty} \tag{12.107}$$

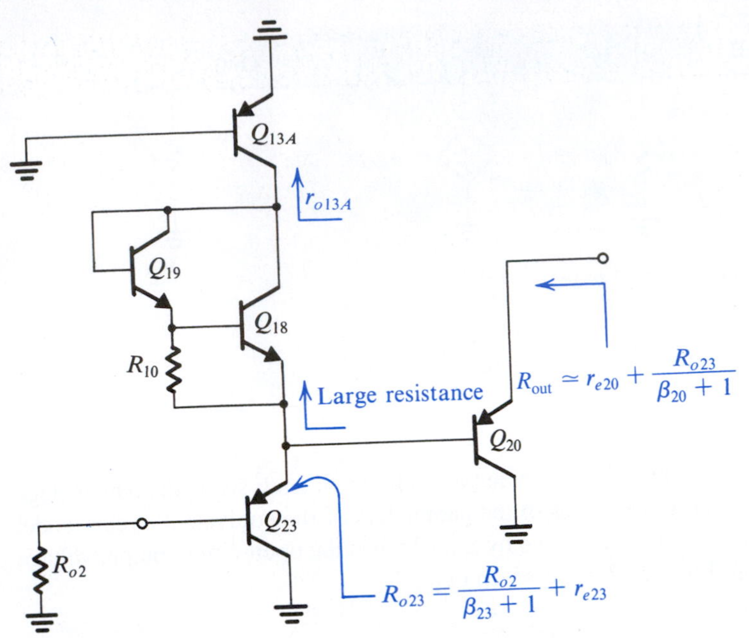

Figure 12.30 Circuit for finding the output resistance R_{out}.

With $R_L = \infty$, the gain of the emitter-follower output transistor (Q_{14} or Q_{20}) will be nearly unity. Also, with $R_L = \infty$ the resistance in the emitter of Q_{23} will be very large. This means that the gain of Q_{23} will be nearly unity and the input resistance of Q_{23} will be very large. We thus conclude that $G_{vo3} \simeq 1$.

Next, we shall find the value of the output resistance of the op amp, R_{out}. For this purpose refer to the circuit shown in Fig. 12.30. In accordance with the definition of R_{out} from Fig. 12.29, the input source feeding the output stage is grounded, but its resistance (which is the output resistance of the second stage, R_{o2}) is included. We have assumed that the output voltage v_O is negative, and thus Q_{20} is conducting most of the current; transistor Q_{14} has therefore been eliminated. The exact value of the output resistance will of course depend on which transistor (Q_{14} or Q_{20}) is conducting and on the value of load current. Nevertheless, we wish to find an estimate of R_{out}.

As indicated in Fig. 12.30, the resistance seen looking into the emitter of Q_{23} is

$$R_{o23} = \frac{R_{o2}}{\beta_{23}+1} + r_{e23} \tag{12.108}$$

Substituting $R_{o2} = 81$ kΩ, $\beta_{23} = 50$, and $r_{e23} = 25/0.18 = 139$ Ω yields $R_{o23} = 1.73$ kΩ. This resistance appears in parallel with the series combination of r_{o13A} and the resistance of the Q_{18}–Q_{19} network. Since r_{o13A} alone (0.28 MΩ) is much larger than R_{o23}, the effective resistance between the base of Q_{20} and ground is approximately equal to R_{o23}. Now we can find the output resistance R_{out} as

$$R_{out} = \frac{R_{o23}}{\beta_{20}+1} + r_{e20} \tag{12.109}$$

For $\beta_{20} = 50$, the first component of R_{out} is 34 Ω. The second component depends critically on the value of output current. For an output current of 5 mA, r_{e20} is 5 Ω and R_{out} is 39 Ω. To this value we must add the resistance R_7 (27 Ω) (see Fig. 12.13), which is included for short-circuit protection. The output resistance of the 741 is specified to be typically 75 Ω.

EXERCISES

12.23 Using a simple (r_π, g_m) model for each of the two transistors Q_{18} and Q_{19} in Fig. E12.23, find the small-signal resistance between A and A'. (*Note:* From Table 12.1, $I_{C18} = 165\ \mu\text{A}$ and $I_{C19} \simeq 16\ \mu\text{A}$.

Ans. 163 Ω

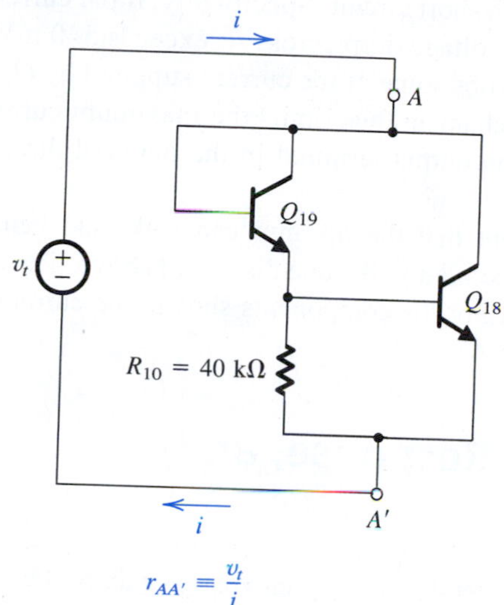

Figure E12.23

12.24 Figure E12.24 shows the circuit for determining the op-amp output resistance when v_O is positive and Q_{14} is conducting most of the current. Using the resistance of the Q_{18}–Q_{19} network calculated in Exercise 12.23 and neglecting the large output resistance of Q_{13A}, find R_{out} when Q_{14} is sourcing an output current of 5 mA.

Ans. 14.4 Ω

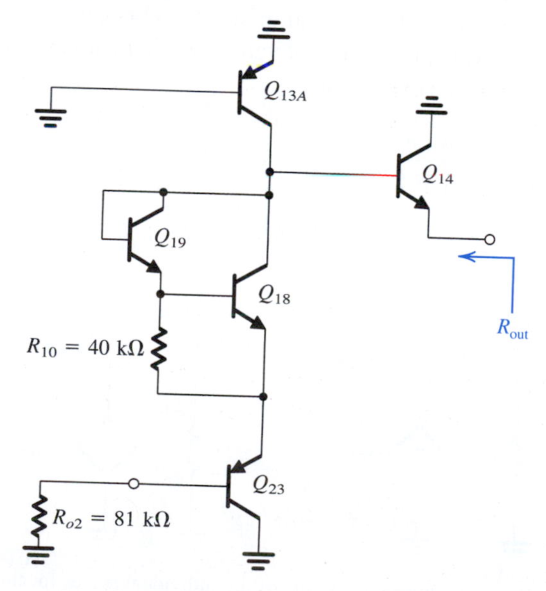

Figure E12.24

Output Short-Circuit Protection If the op-amp output terminal is short-circuited to one of the power supplies, one of the two output transistors could conduct a large amount of current. Such a large current can result in sufficient heating to cause burnout of the IC (Chapter 11). To guard against this possibility, the 741 op amp is equipped with a special circuit for short-circuit protection. The function of this circuit is to limit the current in the output transistors in the event of a short circuit.

Refer to Fig. 12.13. Resistance R_6 together with transistor Q_{15} limits the current that would flow out of Q_{14} in the event of a short circuit. Specifically, if the current in the emitter of Q_{14} exceeds about 20 mA, the voltage drop across R_6 exceeds 540 mV, which turns Q_{15} on. As Q_{15} turns on, its collector robs some of the current supplied by Q_{13A}, thus reducing the base current of Q_{14}. This mechanism thus limits the maximum current that the op amp can source (i.e., supply from the output terminal in the outward direction) to about 20 mA.

Limiting of the maximum current that the op amp can sink, and hence the current through Q_{20}, is done by a mechanism similar to the one discussed above. The relevant circuit is composed of R_7, Q_{21}, Q_{24}, and Q_{22}. For the components shown, the current in the inward direction is limited also to about 20 mA.

12.6 Gain, Frequency Response, and Slew Rate of the 741

In this section we shall evaluate the overall small-signal voltage gain of the 741 op amp. We shall then consider the op amp's frequency response and its slew-rate limitation.

12.6.1 Small-Signal Gain

The overall small-signal gain can be found from the cascade of the equivalent circuits derived in the preceding sections for the three op-amp stages. This cascade is shown in Fig. 12.31, loaded with $R_L = 2\ \text{k}\Omega$, which is the typical value used in measuring and specifying the 741 data. The overall gain can be expressed as

$$\frac{v_o}{v_i} = \frac{v_{i2}}{v_i}\frac{v_{o2}}{v_{i2}}\frac{v_o}{v_{o2}} \tag{12.110}$$

$$= -G_{m1}(R_{o1} \| R_{i2})(-G_{m2}R_{o2})G_{vo3}\frac{R_L}{R_L + R_{\text{out}}} \tag{12.111}$$

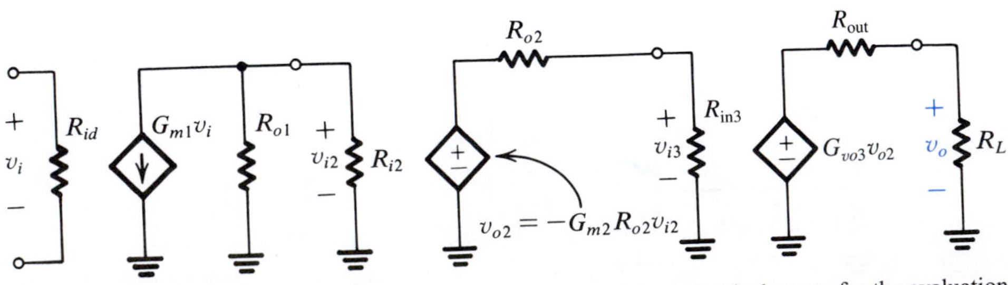

Figure 12.31 Cascading the small-signal equivalent circuits of the individual stages for the evaluation of the overall voltage gain.

Using the values found earlier yields for the overall open-circuit voltage gain,

$$A_0 \equiv \frac{v_o}{v_i} = -476.1 \times (-526.5) \times 0.97 = 243{,}147 \text{ V/V} \tag{12.112}$$

$$\equiv 107.7 \text{ dB}$$

12.6.2 Frequency Response

The 741 is an internally compensated op amp. It employs the Miller compensation technique, studied in Section 10.13.3, to introduce a dominant low-frequency pole. Specifically, a 30-pF capacitor (C_C) is connected in the negative-feedback path of the second stage. An approximate estimate of the frequency of the dominant pole can be obtained as follows.

From Miller's theorem (Section 9.4.4), we see that the effective capacitance due to C_C between the base of Q_{16} and ground is (see Fig. 12.13)

$$C_{\text{in}} = C_C(1 + |A_2|) \tag{12.113}$$

where A_2 is the second-stage gain. Use of the value calculated for A_2 in Section 12.5.3, $A_2 = -515$, results in $C_{\text{in}} = 15{,}480$ pF. Since this capacitance is quite large, we shall neglect all other capacitances between the base of Q_{16} and signal ground. The total resistance between this node and ground is

$$R_t = R_{o1} \,\|\, R_{i2}$$

$$= 6.7 \text{ M}\Omega \,\|\, 4 \text{ M}\Omega = 2.5 \text{ M}\Omega \tag{12.114}$$

Thus the dominant pole has a frequency f_P given by

$$f_P = \frac{1}{2\pi C_{\text{in}} R_t} = 4.1 \text{ Hz} \tag{12.115}$$

It should be noted that this approach is equivalent to using the approximate formula in Eq. (10.116).

As discussed in Section 10.13.3, Miller compensation provides an additional advantageous effect, namely, pole splitting. As a result, the other poles of the circuit are moved to very high frequencies. This has been confirmed by computer-aided analysis [see Gray et al (2000)].

Assuming that all nondominant poles are at very high frequencies, the calculated values give rise to the Bode plot shown in Fig. 12.32, where $f_{3\text{dB}} = f_P$. The unity-gain bandwidth f_t can be calculated from

$$f_t = A_0 f_{3\text{dB}} \tag{12.116}$$

Thus,

$$f_t = 243{,}147 \times 4.1 \simeq 1 \text{ MHz} \tag{12.117}$$

Although this Bode plot implies that the phase shift at f_t is $-90°$ and thus that the phase margin is 90°, in practice a phase margin of about 80° is obtained. The excess phase shift (about 10°) is due to the nondominant poles. This phase margin is sufficient to provide stable operation of closed-loop amplifiers with any value of feedback factor β. This convenience of

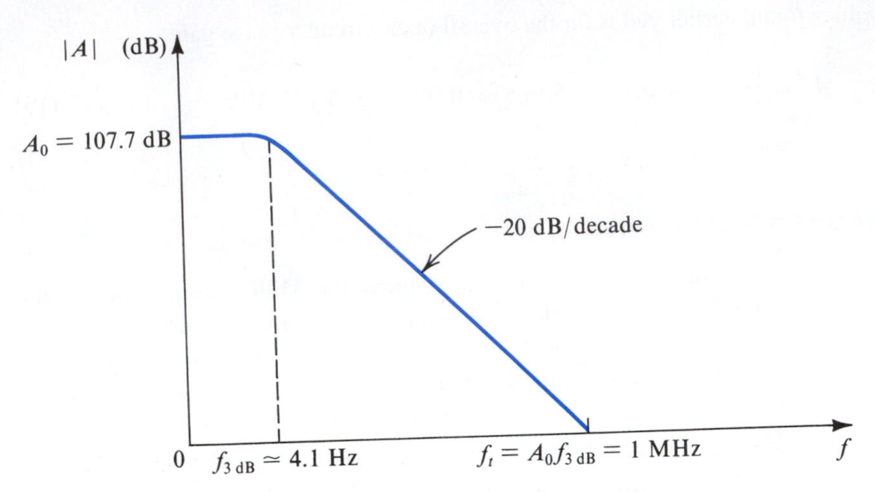

Figure 12.32 Bode plot for the 741 gain, neglecting nondominant poles.

use of the internally compensated 741 is achieved at the expense of a great reduction in open-loop gain and hence in the amount of negative feedback. In other words, if one requires a closed-loop amplifier with a gain of 1000, then the 741 is *overcompensated* for such an application, and one would be much better off designing one's own compensation (assuming, of course, the availability of an op amp that is not already internally compensated).

12.6.3 A Simplified Model

Figure 12.33 shows a simplified model of the 741 op amp in which the high-gain second stage, with its feedback capacitance C_C, is modeled by an ideal integrator. In this model, the gain of the second stage is assumed to be sufficiently large that a virtual ground appears at its input. For this reason the output resistance of the input stage and the input resistance of the second stage have been omitted. Furthermore, the output stage is assumed to be an ideal unity-gain follower. Except for the presence of the output stage, this model is identical to that which we used for the two-stage CMOS amplifier in Section 12.1.4 (Fig. 12.3).

Analysis of the model in Fig. 12.33 gives

$$A(s) \equiv \frac{V_o(s)}{V_i(s)} = \frac{G_{m1}}{sC_C} \tag{12.118}$$

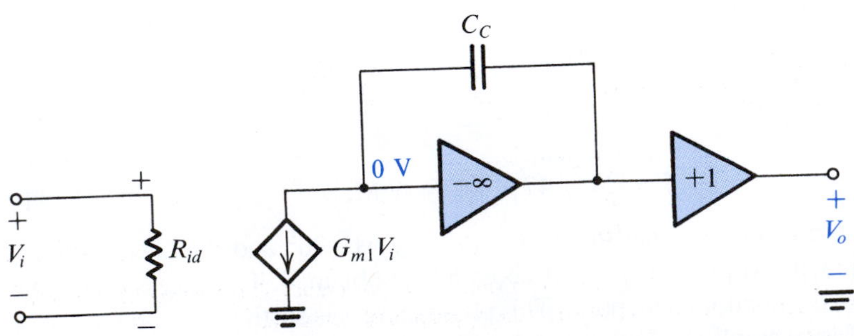

Figure 12.33 A simple model for the 741 based on modeling the second stage as an integrator.

Thus,

$$A(j\omega) = \frac{G_{m1}}{j\omega C_C} \tag{12.119}$$

and the magnitude of gain becomes unity at $\omega = \omega_t$, where

$$\omega_t = \frac{G_{m1}}{C_C} \tag{12.120}$$

Substituting $G_{m1} = 1/5.26$ mA/V and $C_C = 30$ pF yields

$$f_t = \frac{\omega_t}{2\pi} \simeq 1 \text{ MHz} \tag{12.121}$$

which is equal to the value calculated before. It should be pointed out, however, that this model is valid only at frequencies $f \gg f_{3dB}$. At such frequencies the gain falls off with a slope of −20 dB/decade, just like that of an integrator.

12.6.4 Slew Rate

The slew-rate limitation of op amps is discussed in Chapter 2. Here we shall illustrate the origin of the slewing phenomenon in the context of the 741 circuit. This development is similar to that we presented for the CMOS op-amp in Section 12.1.6.

Consider the unity-gain follower of Fig. 12.34 with a step of, say, 10 V applied at the input. Because of amplifier dynamics, its output will not change in zero time. Thus immediately after the input is applied, almost the entire value of the step will appear as a differential signal between the two input terminals. This large input voltage causes the input stage to be **overdriven**, and its small-signal model no longer applies. Rather, half the stage cuts off and the other half conducts all the current. Specifically, reference to Fig. 12.13 shows that a large positive differential input voltage causes Q_1 and Q_3 to conduct all the available bias current ($2I$) while Q_2 and Q_4 will be cut off. The current mirror Q_5, Q_6, and Q_7 will still function, and Q_6 will produce a collector current of $2I$.

Using the observations above, and modeling the second stage as an ideal integrator, results in the model of Fig. 12.35. From this circuit we see that the output voltage will be a ramp with a slope of $2I/C_C$:

$$v_O(t) = \frac{2I}{C_C}t \tag{12.122}$$

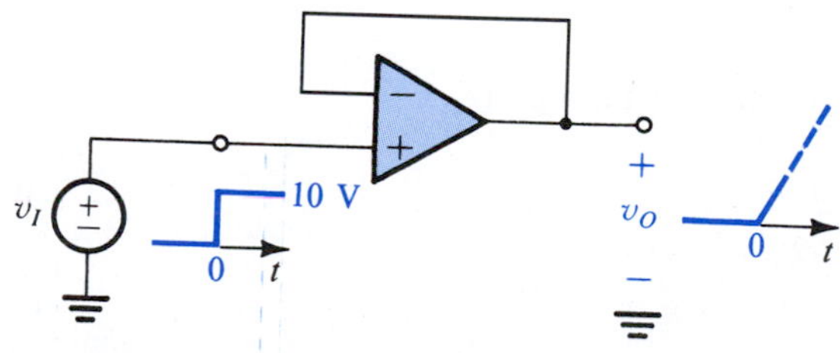

Figure 12.34 A unity-gain follower with a large step input. Since the output voltage cannot change instantaneously, a large differential voltage appears between the op-amp input terminals.

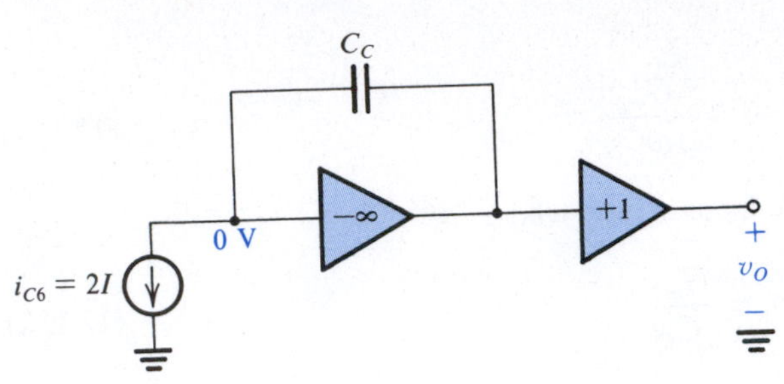

Figure 12.35 Model for the 741 op amp when a large positive differential signal is applied.

Thus the slew rate SR is given by

$$SR = \frac{2I}{C_C} \tag{12.123}$$

For the 741, $I = 9.5$ μA and $C_C = 30$ pF, resulting in $SR = 0.63$ V/μs.

It should be pointed out that this is a rather simplified model of the slewing process. More detail can be found in Gray et al., (2000).

EXERCISE

12.25 Use the value of the slew rate calculated above to find the full-power bandwidth f_M of the 741 op amp. Assume that the maximum output is ±10 V.
Ans. 10 kHz

12.6.5 Relationship Between f_t and SR

A simple relationship exists between the unity-gain bandwidth f_t and the slew rate SR. This relationship is obtained by combining Eqs. (12.120), (12.123), and

$$SR = \frac{2I}{G_{m1}}\omega_t \tag{12.124}$$

and then using Eq. (12.83′) to obtain

$$SR = \frac{4I}{g_{m1}}\omega_t \tag{12.125}$$

Now, since g_{m1} is the transconductance of each of Q_1 through Q_4,

$$g_{m1} = \frac{I}{V_T} \tag{12.126}$$

Thus,

$$SR = 4V_T\omega_t \tag{12.127}$$

As a check, for the 741 we have

$$SR = 4 \times 25 \times 10^{-3} \times 2\pi \times 10^6 = 0.63 \text{ V/}\mu\text{s}$$

which is the result obtained previously. Observe that Eq. (12.127) is of the same form as Eq. (12.42), which applies to the two-stage CMOS op amp. Here, $4V_T$ replaces V_{OV}. Since, typically, V_{OV} will be two to three times the value of $4V_T$, a two-stage CMOS op amp with an f_t equal to that of the 741 exhibits a slew rate that is two to three times as large as that of the 741.

A general form for the relationship between SR and ω_t for an op amp with a structure similar to that of the 741 (including the two-stage CMOS circuit) is

$$SR = \omega_t/a \tag{12.128}$$

where a is the constant of proportionality relating the transconductance of the first stage G_{m1}, to the *total* bias current of the input differential stage. That is, for the 741 circuit $G_{m1} = a(2I)$, while for the CMOS circuit of Fig. 12.1, $G_{m1} = aI$.[3] For a given ω_t, a higher value of SR is obtained by making a smaller; that is, the total bias current is kept constant and G_{m1} is reduced. This is a viable technique for increasing slew rate. It is referred to as the **G_m-reduction method** (see Exercise 12.27).

EXERCISES

12.26 Consider the integrator model of the op amp in Fig. 12.33. Find the value of the resistor that, when connected across C_C, provides the correct value of the dc gain.
Ans. 1279 MΩ

D12.27 If a resistance R_E is included in each of the emitter leads of Q_3 and Q_4 show that $SR = 4(V_T + IR_E/2)\omega_t$. Hence find the value of R_E that would double the 741 slew rate while keeping ω_t and I unchanged. What are the new values of C_C, the dc gain, and the 3-dB frequency?
Ans. 5.26 kΩ; 15 pF; 101.7 dB (a 6-dB decrease); 8.2 Hz

12.7 Modern Techniques for the Design of BJT Op Amps

Although the ingenious techniques employed in the design of the 741 op amp have stood the test of time, they are now more than 40 years old! Technological advances have resulted in changes in the user requirements of general-purpose bipolar op amps. The resulting more demanding specifications have in turn posed new challenges to analog IC designers who, as they have done repeatedly before, are responding with new and exciting circuits. In this section we present a sample of recently developed design techniques. For more on this rather advanced topic the reader is referred to the Analog Circuits section of the bibliography in Appendix G.

12.7.1 Special Performance Requirements

Many of the special performance requirements stem from the need to operate modern op amps from power supplies of much lower voltages. Thus while the 741-type op amp operated from

[3] The difference is just a matter of notation; We used I to denote the total bias current of the input differential stage of the CMOS circuit, and we used $2I$ for the 741 case!

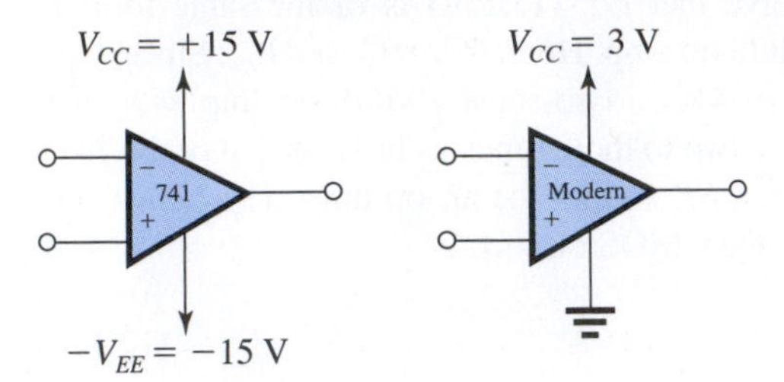

Figure 12.36 Power supply requirements have changed considerably. Modern BJT op amps are required to operate from a single supply V_{CC} of 2 to 3 V.

±15-V power supplies, many modern BJT op amps are required to operate from a *single power supply of only* 2 V *to* 3 V. This is done for a number of reasons, including the following.

1. Modern small-feature-size IC fabrication technologies require low power-supply voltages.
2. Compatibility must be achieved with other parts of the system that use low-voltage supplies.
3. Power dissipation must be minimized, especially for battery-operated equipment.

As Fig. 12.36 indicates, there are two important changes: the use of a single ground-referenced power supply V_{CC}, and the low value of V_{CC}. Both of these requirements give rise to changes in performance specifications and pose new design challenges. In the following we discuss two of the resulting changes.

Rail-to-Rail Input Common-Mode Range Recall that the input common-mode range of an op amp is the range of common-mode input voltages for which the op amp operates properly and meets its performance specifications, such as voltage gain and CMRR. Op amps of the 741 type operate from ±15-V supplies and exhibit an input common-mode range that extends to within a couple of volts of each supply. Such a gap between the input common-mode range and the power supply is obviously unacceptable if the op amp is to be operated from a single supply that is only 2 V to 3 V. Indeed we will now show that these single-supply, low-voltage op amps need to have an input common-mode range that extends over the entire supply voltage, 0 to V_{CC}, referred to as rail-to-rail input common mode range.

Consider first the inverting op-amp configuration shown in Fig. 12.37(a). Since the positive input terminal is connected to ground (which is the voltage of the negative-supply rail),

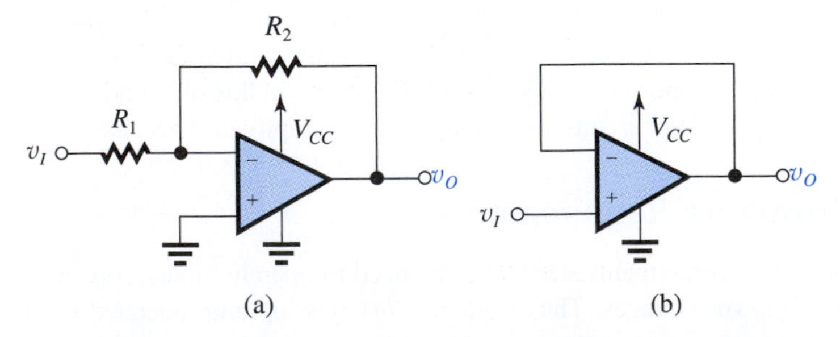

Figure 12.37 **(a)** In the inverting configuration, the $+ive$ op-amp input is connected to ground; thus it is imperative that the input common-mode range includes ground. **(b)** In the unity-gain follower configuration, $v_{ICM} = v_I$; thus it is highly desirable for the input common-mode range to include ground and V_{CC}.

ground voltage has to be within the allowable input common-mode range. In fact, because for positive output voltages the voltage at the inverting input terminal can go slightly negative, the input common-mode range should extend below the negative-supply rail (ground).

Next consider the unity-gain voltage follower obtained by applying 100% negative feedback to an op amp, as shown in Fig. 12.37(b). Here the input common-mode voltage is equal to the input signal v_I. To maximize the usefulness of this buffer amplifier, its input signal v_I should be allowed to extend from 0 to V_{CC}, especially since V_{CC} is only 2 to 3 V. Thus the input common-mode range should include also the positive supply rail. As will be seen shortly, modern BJT op amps can operate over an input common-mode voltage range that extends a fraction of a volt beyond its two supply rails: that is, more than rail-to-rail operation!

Near Rail-to-Rail Output Signal Swing In the 741 op amp, we were satisfied with an output that can swing to within 2 V or so of each of the supply rails. With a supply of ± 15 V, this capacity resulted in a respectable ± 13-V output range. However, to limit the output swing to within 2 V of the supply rails in an op amp operating from a single 3-V supply would result in an unusable device! Thus, here too, we require near rail-to-rail operation. As we shall see in Section 12.7.5, this requirement forces us to adopt a whole new approach to output-stage design.

Device Parameters The technology we shall use in the examples, exercises, and problems for this section has the following characteristics:

$$npn \text{ Transistors:} \quad \beta = 40 \qquad V_A = 30\ V$$

$$pnp \text{ Transistors:} \quad \beta = 10 \qquad |V_A| = 20\ V$$

For both, $|V_{BE}| \simeq 0.7$ V and $|V_{CEsat}| \simeq 0.1$ V. It is important to note that we will assume that for this technology, the transistor will remain in the active mode for $|V_{CE}|$ as low as 0.1 V (in other words, that 0.6 V is needed to forward-bias the CBJ).

12.7.2 Bias Design

As in the 741 circuit, the bias design of modern BJT amplifiers makes extensive use of current mirrors and current-steering circuits (Sections 7.4 and 7.5). Typically, however, the bias currents are small (in the micro amp range). Thus, the Widlar current source (Section 7.5.5) is especially popular here. As well, emitter-degeneration resistors (in the tens-of-kilohm range) are frequently used.

Figure 12.38 shows a self-biased current-reference source that utilizes a Widlar circuit formed by Q_1, Q_2, and R_2, and a current mirror Q_3–Q_4 with matched emitter-degeneration resistors R_3 and R_4. The circuit establishes a current I in each of the four transistors, with the value of I determined as follows. Neglecting base currents and r_o's for simplicity, we write

$$V_{BE1} = V_T \ln\left(\frac{I}{I_{S1}}\right)$$

$$V_{BE2} = V_T \ln\left(\frac{I}{I_{S2}}\right)$$

Thus,

$$V_{BE1} - V_{BE2} = V_T \ln\left(\frac{I_{S2}}{I_{S1}}\right)$$

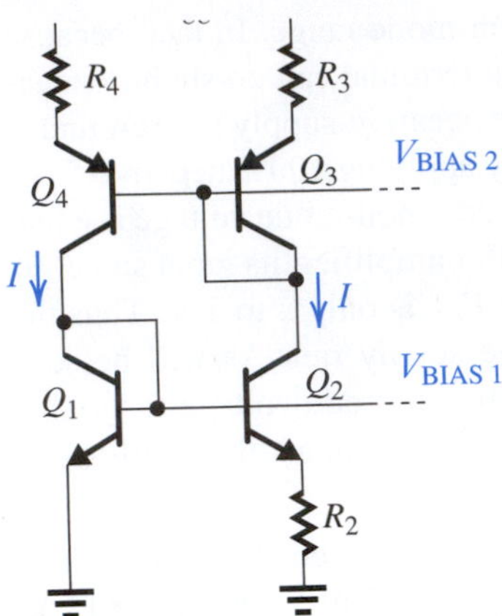

Figure 12.38 A self-biased current-reference source utilizing a Widler circuit to generate $I = V_T/R_2 \ln(I_{S2}/I_{S1})$ The bias voltages V_{BIAS1} and V_{BIAS2} are utilized in other parts of the op-amp circuit for biasing other transistors.

But,

$$V_{BE1} - V_{BE2} = IR_2$$

Thus,

$$I = \frac{V_T}{R_2} \ln\left(\frac{I_{S2}}{I_{S1}}\right) \tag{12.129}$$

Thus the value of I is determined by R_2 and the ratio of the emitter areas of Q_1 and Q_2. Also, observe that I is independent of V_{CC}, a highly desirable outcome. Neglecting the temperature dependence of R_2, we see that I is directly PTAT (*p*roportional *t*o the *a*bsolute *t*emperature T). It follows that transistors biased by I or mirrored versions of it will exhibit g_m's that are constant independent of temperature!

EXERCISE

D12.28 Design the circuit in Fig. 12.38 to generate a current $I = 10\ \mu\text{A}$. Utilize transistors Q_1 and Q_2 having their areas in a 1:2 ratio. Assume that Q_3 and Q_4 are matched and design for a 0.2-V drop across each of R_3 and R_4. Specify the values of R_2, R_3, and R_4.
Ans. 1.73 kΩ; 20 kΩ; 20 kΩ

The circuit in Fig. 12.38 provides a bias line V_{BIAS1} with a voltage equal to V_{BE1}. This can be used to bias other transistors and thus generate currents proportional to I by appropriately scaling their emitter areas. Similarly, the circuit provides a bias line V_{BIAS2} at a voltage $(IR_3 + V_{EB3})$ below V_{CC}. This bias line can be used to bias other transistors and thus generate constant currents proportional to I by appropriately scaling emitter areas and emitter-degeneration resistances. These ideas are illustrated in Fig. 12.39.

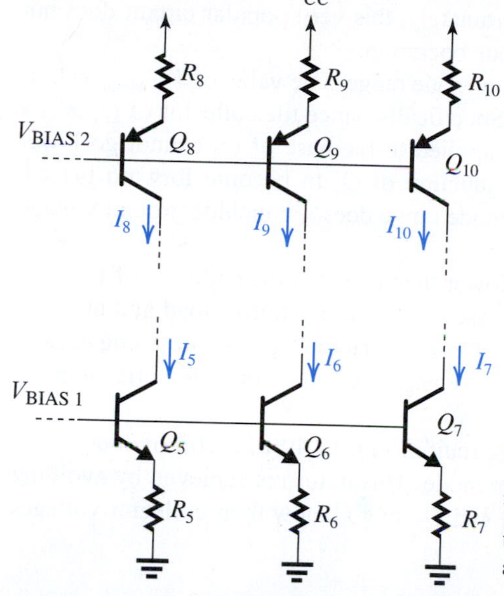

Figure 12.39 The bias lines V_{BIAS1} and V_{BIAS2} provided by the circuit in Fig. 12.38 are utilized to bias other transistors and generate constant current I_5 to I_{10}. Both the transistor area and the emitter degeneration resistance value have to be appropriately scaled.

EXERCISE

D12.29 Refer to the circuit in Fig. 12.39 and assume that the V_{BIAS2} line is connected to the corresponding line in Fig. 12.38. It is required to generate currents $I_8 = 10\ \mu A$, $I_9 = 20\ \mu A$, and $I_{10} = 5\ \mu A$. Specify the required emitter areas of Q_8, Q_9, and Q_{10} as ratios of the emitter area of Q_3. Also specify the values required for R_8, R_9, and R_{10}. Use the values of R_3 and R_4 found in Exercise 12.28. Ignore base currents.

Ans. 1, 2, 0.5; 20 kΩ, 10 kΩ, 40 kΩ

12.7.3 Design of the Input Stage to Obtain Rail-to-Rail V_{ICM}

The classical differential input stage with current-mirror load is shown in Fig. 12.40(a). This is essentially the core of the 741 input stage, except that here we are using a single positive power supply. As well, the CMOS counterpart of this circuit is utilized in nearly every

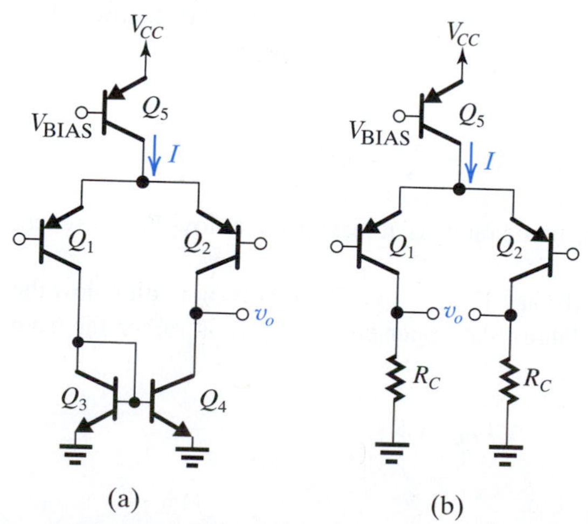

Figure 12.40 For the input common-mode range to include ground voltage, the classical current-mirror-loaded input stage in **(a)** has to be replaced with the resistively-loaded configuration in **(b)** with the dc voltage drop across R_C limited to 0.2–0.3 V.

CMOS op-amp design (see Section 12.1). Unfortunately, this very popular circuit does not meet our requirement of rail-to-rail common-mode operation.

Consider first the low end of the input common-mode range. The value of $V_{ICM\min}$ is limited by the need to keep Q_1 in the active mode. Specifically, since the collector of Q_1 is at a voltage $V_{BE3} \simeq 0.7$ V, we see that the voltage applied to the base of Q_1 cannot go lower than 0.1 V without causing the collector–base junction of Q_1 to become forward biased. Thus $V_{ICM\min} = 0.1$ V, and the input common-mode range does *not* include ground voltage as required.

The only way to extend $V_{ICM\min}$ to 0 V is to lower the voltage at the collector of Q_1. This in turn can be achieved only by abandoning the use of the current-mirror load and utilizing instead resistive loads, as shown in Fig. 12.40(b). Observe that in effect we are *going back* to the resistively loaded differential pair with which we began our study of differential amplifiers in Chapter 8!

The minimum allowed value of V_{ICM} in the circuit of Fig. 12.40(b) is still of course limited by the need to keep Q_1 and Q_2 in the active mode. This in turn is achieved by avoiding V_{ICM} values that cause the base voltages of Q_1 and Q_2 to go below their collector voltages by more than 0.6 V,

$$V_{ICM\min} = V_{R_C} - 0.6 \text{ V}$$

where V_{R_C} is the voltage drop across each of R_{C1} and R_{C2}. Now if V_{R_C} is selected to be 0.2 to 0.3 V, then $V_{ICM\min}$ will be -0.4 V to -0.3 V, which is exactly what we need.

The major drawback of replacing the current-mirror load with resistive loads is that the differential gain realized is considerably reduced,

$$\frac{v_o}{v_{id}} = -g_{m1,2}R_C$$

$$= -\frac{I/2}{V_T}R_C = -\frac{V_{R_C}}{V_T}$$

where we have neglected r_o for simplicity. Thus for $V_{R_C} = 0.3$ V, the gain realized is only 12 V/V. As we will see shortly, this low-gain problem can be solved by cascoding.

Next consider the upper end of the input common-mode range. Reference to the circuit in Fig. 12.40(b) shows that the maximum voltage that can be applied to the bases of Q_1 and Q_2 is limited by the need to keep the current-source transistor in the active mode. This in turn is achieved by ensuring that the voltage across Q_5, V_{EC5} does not fall below 0.1 V or so. Thus the maximum value of V_{ICM} will be a voltage $V_{EB1,2}$ or approximately 0.7 V lower,

$$V_{ICM\max} = V_{CC} - 0.1 - 0.7 = V_{CC} - 0.8$$

That is, the upper end of the input common-mode range is at least 0.8 V below V_{CC}, a severe limitation.

To recap, while the circuit in Fig. 12.40(b) has $V_{ICM\min}$ of a few tenths of a volt below the negative power-supply rail (at ground voltage), the upper end of V_{ICM} is rather far from V_{CC},

$$-0.3 \le V_{ICM} \le V_{CC} - 0.8$$

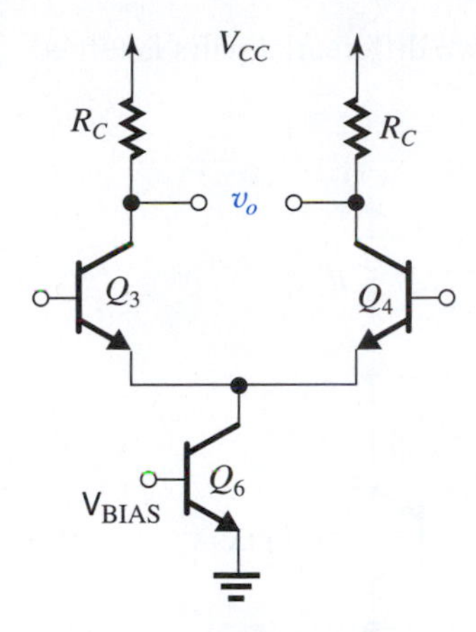

Figure 12.41 The complement of the circuit in Fig. 12.40(b). While the input common-mode range of the circuit in Figure 12.40(b) extends below ground, here it extends above V_{CC}. Connecting the two circuits in parallel, as will be shown, results in a rail-to-rail V_{ICM} range.

where we have assumed $V_{R_C} = 0.3$ V. To extend the upper end of V_{ICM}, we adopt a solution similar to that used in the CMOS case (Section 12.2.6, Fig. 12.11), namely, we utilize a parallel complementary input stage. Toward that end, note that the *npn* version of the circuit of Fig. 12.40(b), shown in Fig. 12.41, has a common-input range of

$$0.8 \leq V_{ICM} \leq V_{CC} + 0.3$$

where we have assumed that $V_{R_C} = 0.3$ V. Thus, as expected, the high end meets our specifications and in fact is above the positive supply rail by 0.3 V. The lower end, however, does not; but this should cause us no concern because the lower end will be looked after by the *pnp* pair. Finally, note that there is a range of V_{ICM} in which both the *pnp* and the *npn* circuits will be active and properly operating,

$$0.8 \leq V_{ICM} \leq V_{CC} - 0.8$$

Figure 12.42 shows an input stage that achieves more than rail-to-rail input common-mode range by utilizing a *pnp* differential pair (Q_1, Q_2) and an *npn* differential pair (Q_3, Q_4), connected in parallel. To keep the diagram simple, we are not showing the parallel connection of the input terminals; the + input terminals are assumed to be connected together, and similarly for the – input terminals. In order to increase the gain obtained from the resistively loaded differential pairs, a folded cascode stage is added. Here R_7 and R_8 are the resistive loads of the *pnp* pair Q_1–Q_2, and Q_7–Q_8 are its cascode transistors. Similarly, R_9 and R_{10} are the resistive loads of the *npn* pair Q_3–Q_4, and Q_9–Q_{10} are its cascode transistors. Observe that the cascode transistors do "double duty." For instance, Q_7–Q_8 operate as the cascode devices for Q_1–Q_2 and at the same time as current-source loads for Q_9–Q_{10}. A similar statement can be made about Q_9–Q_{10}. The output voltage of the first stage, v_{od}, is taken between the collectors of the cascode devices.

For $V_{ICM} \ll 0.8$ V, the *npn* stage will be inactive and the gain is determined by the transconductance G_m of the Q_1–Q_2 pair together with the output resistance seen between the collectors of the cascode transistors. At the other end of V_{ICM}, that is, $V_{ICM} \gg V_{CC} - 0.8$, the Q_1–Q_2 stage will be inactive, and the gain will be determined by the transconductance G_m of the Q_3–Q_4 pair and the output resistance between the collectors of the cascode devices. In the overlap region $0.8 \leq V_{ICM} \leq V_{CC} - 0.8$, both the *pnp* and *npn* stages will be active and their effective transconductances G_m add up, thus resulting in a higher gain. The dependence of the differential gain on the input common-mode V_{ICM} is usually undesirable and can be reduced considerably by arranging that one of the two differential pairs is turned off when the other one is active.[4]

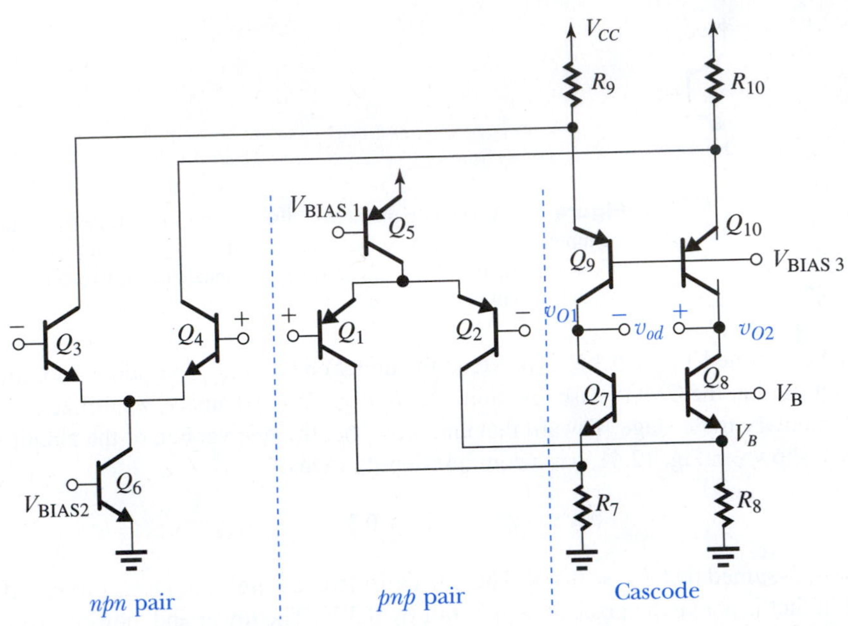

Figure 12.42 Input stage with rail-to-rail input common-mode range and a folded-cascode stage to increase the gain. Note that all the bias voltages including V_{BIAS3} and V_B are generated elsewhere on the chip.

Example 12.5

It is required to find the input resistance and the voltage gain of the input stage shown in Fig. 12.42. Let $V_{ICM} \ll 0.8$ V so that the Q_3–Q_4 pair is off. Assume that Q_5 supplies 10 μA, that each of Q_7 to Q_{10} is biased at 10 μA, and that all four cascode transistors are operating in the active mode. The input resistance of the second stage of the op amp (not shown) is $R_L = 2$ MΩ. The emitter-degeneration resistances are $R_7 = R_8 = 20$ kΩ, and $R_9 = R_{10} = 30$ kΩ. Recall that the device parameters are $\beta_N = 40$, $\beta_P = 10$, $V_{An} = 30$ V, $|V_{Ap}| = 20$ V.

Solution

Since the stage is fully balanced, we can use the differential half-circuit shown in Fig. 12.43(a). The input resistance R_{id} is twice the value of $r_{\pi 1}$,

$$R_{id} = 2r_{\pi 1} = 2\beta_P/g_{m1}$$

where

$$g_{m1} = \frac{I_{C1}}{V_T} = \frac{5 \times 10^{-6}}{25 \times 10^{-3}} = 0.2 \text{ mA/V}$$

[4]This is done in the NE5234 op amp, whose circuit is described and analyzed in great detail in Gray et al., (2009).

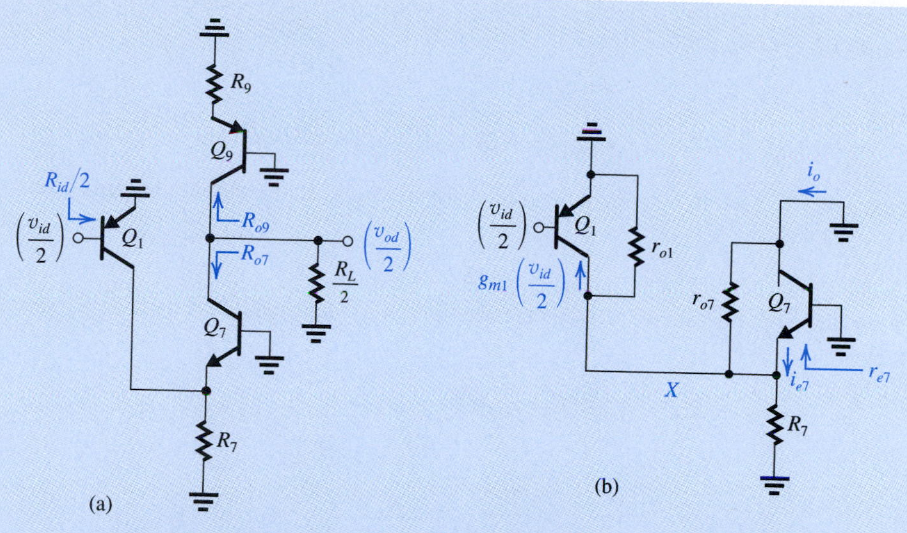

Figure 12.43 (a) Differential half circuit for the input stage shown in Fig. 12.42 with $V_{ICM} \ll 0.8$ V. (b) Determining $G_{m1} = i_o/(v_{id}/2)$

Thus,

$$R_{id} = \frac{2 \times 10}{0.2} = 100\text{ k}\Omega$$

To find the short-circuit transconductance, we short the output to ground as shown in Fig. 12.43(b) and find G_{m1} as

$$G_{m1} = \frac{i_{c7}}{v_{id}/2}$$

At node X we have four parallel resistances to ground,

$$r_{o1} = \frac{|V_{Ap}|}{I_{C1}} = \frac{20\text{ V}}{5\ \mu\text{A}} = 4\text{ M}\Omega$$

$$R_7 = 20\text{ k}\Omega$$

$$r_{o7} = \frac{V_{An}}{I_{C7}} = \frac{30\text{ V}}{10\ \mu\text{A}} = 3\text{ M}\Omega$$

$$r_{e7} \simeq \frac{1}{g_{m7}} = \frac{V_T}{I_{C7}} = \frac{25\text{ mV}}{10\ \mu\text{A}} = 2.5\text{ k}\Omega$$

Obviously r_{o1} and r_{o7} are very large and can be neglected. Then, the portion of $g_{m1}(v_{id}/2)$ that flows into the emitter proper of Q_7 can be found from

$$i_{e7} \simeq \left(g_{m1}\left(\frac{v_{id}}{2}\right)\frac{R_7}{R_7 + r_{e7}}\right)$$

$$= g_{m1}\left(\frac{v_{id}}{2}\right)\frac{20}{20 + 2.5} = 0.89 g_{m1}\left(\frac{v_{id}}{2}\right)$$

and the output short-circuit current i_o is

$$i_o \simeq i_{e7} = 0.89 g_{m1}(v_{id}/2)$$

Example 12.5 *continued*

Thus,

$$G_{m1} \equiv \frac{i_o}{v_{id}/2} = 0.89g_{m1} = 0.89 \times 0.2 = 0.18 \text{ mA/V}$$

To find the voltage gain, we need to determine the total resistance between the output node and ground for the circuit in Fig. 12.43(a),

$$R = R_{o9} \| R_{o7} \| (R_L/2)$$

The resistance R_{o9} is the output resistance of Q_9, which has an emitter-degeneration resistance R_9. Thus R_{o9} can be found using Eq. (7.50),

$$R_{o9} = r_{o9} + (R_9 \| r_{\pi 9})(1 + g_{m9}r_{o9})$$

where

$$r_{o9} = \frac{|V_{Ap}|}{I_{C9}} = \frac{20 \text{ V}}{10 \text{ μA}} = 2 \text{ M}\Omega$$

$$g_{m9} = \frac{I_{C9}}{V_T} = \frac{10 \text{ μA}}{25 \text{ mV}} = 0.4 \text{ mA/V}$$

$$r_{\pi 9} = \frac{\beta_P}{g_{m9}} = \frac{10}{0.4 \text{ mA/V}} = 25 \text{ k}\Omega$$

Thus

$$R_{o9} = 2 + (30 \| 25) \times 10^{-3}(1 + 0.4 \times 2 \times 10^3)$$

$$= 12.9 \text{ M}\Omega$$

The resistance R_{o7} is the output resistance of Q_7, which has an emitter-degeneration resistance $(R_7 \| r_{o1}) \simeq R_7$. Thus,

$$R_{o7} = r_{o7} + (R_7 \| r_{\pi 7})(1 + g_{m7}r_{o7})$$

where

$$r_{o7} = \frac{V_{An}}{I_{C7}} = \frac{30 \text{ V}}{10 \text{ μA}} = 3 \text{ M}\Omega$$

$$g_{m7} = \frac{I_{C7}}{V_T} = \frac{10 \text{ μA}}{25 \text{ mV}} = 0.4 \text{ mA/V}$$

$$r_{\pi 7} = \frac{\beta_N}{g_{m7}} = \frac{40}{0.4} = 100 \text{ k}\Omega$$

Thus,

$$R_{o7} = 3 + (20 \| 100) \times 10^{-3}(1 + 0.4 \times 3 \times 10^3)$$

$$= 23 \text{ M}\Omega$$

$$\frac{R_L}{2} = \frac{2 \text{ M}\Omega}{2} = 1 \text{ M}\Omega$$

The total resistance R can now be found as

$$R = 12.9 \parallel 23 \parallel 1 = 0.89 \text{ M}\Omega$$

Finally, we can find the voltage gain as

$$A_d = \frac{v_{od}/2}{v_{id}/2} = G_{m1}R_1$$

$$= 0.18 \times 0.89 \times 10^3 = 160 \text{ V/V}$$

12.7.4 Common-Mode Feedback to Control the dc Voltage at the Output of the Input Stage

For the cascode circuit in Fig. 12.42 to operate properly and provide high output resistance and thus high voltage gain, the cascode transistors Q_7 through Q_{10} must operate in the active mode at all times. However, relying solely on matching will not be sufficient to ensure that the currents supplied by Q_9 and Q_{10} are exactly equal to the currents supplied by Q_7 and Q_8. Any small mismatch ΔI between the two sets of currents will be multiplied by the large output resistance between each of the collector nodes and ground, and thus there will be large changes in the voltages v_{O1} and v_{O2}. These changes in turn can cause one set of the current sources (i.e., Q_7–Q_8 or Q_9–Q_{10}) to saturate. We therefore need a circuit that detects the change in the dc or common-mode component V_{CM} of v_{O1} and v_{O2},

$$V_{CM} = \frac{1}{2}(v_{O1} + v_{O2}) \tag{12.130}$$

and adjusts the bias voltage on the bases of Q_7 and Q_8, V_B, to restore current equality. This negative-feedback loop should be insensitive to the differential signal components of v_{O1} and v_{O2}; otherwise it would reduce the differential gain. Thus the feedback loop should provide **common-mode feedback** (CMF).

Figure 12.44 shows the cascode circuit with the CMF circuit shown as a black box. The CMF circuit accepts v_{O1} and v_{O2} as inputs and provides the bias voltage V_B as output. In a particular implementation we will present shortly, the CMF circuit has the transfer characteristic

$$V_B = V_{CM} + 0.4 \tag{12.131}$$

By keeping V_B higher than V_{CM} by only 0.4 V, the CMF circuit ensures that Q_7 and Q_8 remain active (0.6 V is needed for saturation).

The nominal value of V_B is determined by the quiescent current of Q_7 through Q_{10}, the quiescent value of I_1 and I_2, and the value of R_7 and R_8. The resulting nominal value of V_B and the corresponding value of V_{CM} from Eq. (12.131) are designed to ensure that Q_9 and Q_{10} operate in the active mode. Here, it is important to recall that V_{BIAS3} is determined by the rest of the op-amp bias circuit.

To see how the CMF circuit regulates the dc voltage V_{CM}, assume that for some reason V_B is higher than it should be and as a result the currents of Q_7 and Q_8 exceed the currents supplied by Q_9 and Q_{10} by an increment ΔI. When multiplied by the total resistance between each of the output nodes and ground, the increment ΔI will result in a large

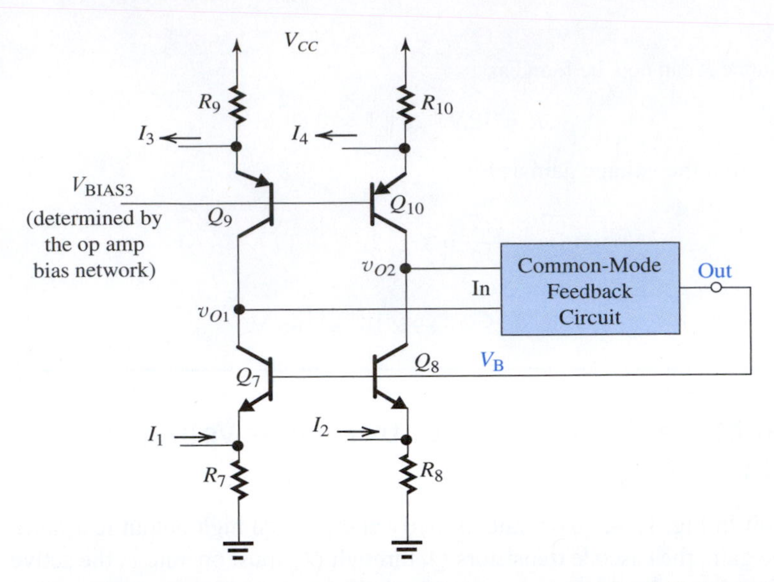

Figure 12.44 The cascode output circuit of the input stage and the CMF circuit that responds to the common-mode component $V_{CM} = \frac{1}{2}(v_{O1} + v_{O2})$ by adjusting V_B so that Q_7–Q_8 conduct equal currents to Q_9–Q_{10}, and Q_7–Q_{10} operate in the active mode.

negative voltage increment in v_{O1} and v_{O2}. The CMF circuit responds by lowering V_B to the value that restores the equality of currents. The change in V_B needed to restore equilibrium is usually small (see Example 12.6 below) and according to Eq. (12.131) the corresponding change in V_{CM} will be equally small. Thus we see negative feedback in action: It minimizes the initial change and thus keeps V_{CM} nearly constant at its nominal value, which is designed to operate Q_7 through Q_{10} in the active region.

We conclude by considering briefly a possible implementation of the CMF circuit. Figure 12.45 shows the second stage of an op-amp circuit. The circuit is fed by the outputs of the input stage, v_{O1} and v_{O2},

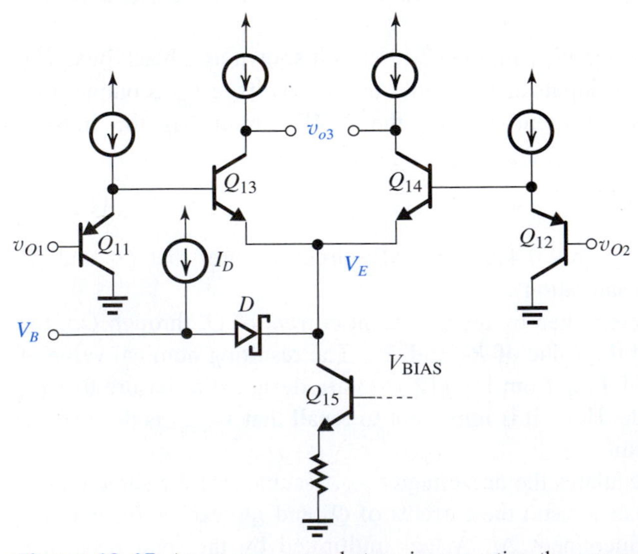

Figure 12.45 An op amp second stage incorporating the common-mode feedback circuit for the input stage. Note that the circuit generates the voltage V_B needed to bias the cascode circuit in the first stage. Diode D is a Schottky-barrier diode which exhibits a forward voltage drop of about 0.4V.

$$v_{O1} = V_{CM} + v_d/2$$

$$v_{O2} = V_{CM} - v_d/2$$

In addition to amplifying the differential component of v_d, the circuit generates a dc voltage V_B,

$$V_B = V_{CM} + 0.4$$

To see how the circuit works, note that Q_{11} and Q_{12} are emitter followers that minimize the loading of the second stage on the input stage. The emitter followers deliver to the bases of the differential pair $Q_{13}-Q_{14}$ voltages that are almost equal to v_{O1} and v_{O2} but dc shifted by $V_{EB11,12}$. Thus the voltage at the emitters of $Q_{13}-Q_{14}$ will be

$$V_E = V_{CM} + V_{EB11,12} - V_{BE13,14}$$

which reduces to

$$V_E \simeq V_{CM}$$

The voltage V_B is simply equal to V_E plus the voltage drop of diode D_1. The latter is a **Schottky barrier diode** (SBD), which features a low forward drop of about 0.4 V. Thus,

$$V_B = V_E + V_D = V_{CM} + 0.4$$

as required.

Example 12.6

Consider the operation of the circuit in Fig. 12.44. Assume that $V_{ICM} \ll 0.8$ V and thus the *npn* input pair (Fig. 12.42) is off. Hence $I_3 = I_4 = 0$. Also assume that only dc voltages are present and thus $I_1 = I_2 = 5\ \mu\text{A}$. Each of Q_7 to Q_{10} is biased at 10 μA, $V_{CC} = 3$ V, $V_{\text{BIAS3}} = V_{CC} - 1$, $R_7 = R_8 = 20\ \text{k}\Omega$, and $R_9 = R_{10} = 30\ \text{k}\Omega$. Neglect base currents and neglect the loading effect of the CMF circuit on the output nodes of the cascode circuit. The CMF circuit provides $V_B = V_{CM} + 0.4$.

(a) Determine the nominal values of V_B and V_{CM}. Does the value of V_{CM} ensure operation in the active mode for Q_7 through Q_{10}?

(b) If the CMF circuit were not present, what would be the change in v_{O1} and v_{O2} (i.e., in V_{CM}) as a result of a current mismatch $\Delta I = 0.3\ \mu\text{A}$ between Q_7-Q_8 and Q_9-Q_{10}? Use the output resistance values found in Example 12.5.

(c) Now, if the CMF circuit is connected, what change will it cause in V_B to eliminate the current mismatch ΔI? What is the corresponding change in V_{CM} from its nominal value?

Solution

(a) The nominal value of V_B is found as follows:

$$\begin{aligned} V_B &= V_{BE7} + (I_{E7} + I_1)R_7 \\ &\simeq 0.7 + (10 + 5) \times 10^{-3} \times 20 \\ &= 1\ \text{V} \end{aligned}$$

Example 12.6 *continued*

The nominal value of V_{CM} can now be found from

$$V_{CM} = V_B - 0.4 = 1 - 0.4 = 0.6 \text{ V}$$

For $Q_7 - Q_8$ to be active,

$$V_{CM} > V_{B7,8} - 0.6$$

that is,

$$V_{CM} > 0.4 \text{ V}$$

For $Q_9 - Q_{10}$ to be active

$$V_{CM} < V_{\text{BIAS3}} + 0.6$$

That is,

$$V_{CM} < V_{CC} - 1 + 0.6$$

resulting in

$$V_{CM} < 2.6 \text{ V}$$

Thus, for all four cascode transistors to operate in the active mode,

$$0.4 \text{ V} < V_{CM} < 2.6 \text{ V}$$

Thus the nominal value of 0.6 V ensures active mode operation.

(b) For $I_{C9} - I_{C7} = I_{C10} - I_{C8} = \Delta I$,

$$\Delta V_{CM} = \Delta I R_{o1}$$

where R_{o1} is the output resistance between the collectors of Q_7 and Q_9 and ground,

$$R_{o1} = R_{o7} \| R_{o9}$$

In Example 12.5 we found that $R_{o7} = 23 \text{ M}\Omega$ and $R_{o9} = 12.9$; thus,

$$R_{o1} = 23 \| 12.9 = 8.3 \text{ M}\Omega$$

Thus,

$$\Delta V_{CM} = 0.3 \times 8.3 \simeq 2.5 \text{ V}$$

Now if ΔV_{CM} is positive,

$$V_{CM} = 0.6 + 2.5 = 3.1 \text{ V}$$

which exceeds the 2.6 V maximum allowed value before $Q_9 - Q_{10}$ saturate. If ΔV_{CM} is negative,

$$V_{CM} = 0.6 - 2.5 = -1.9 \text{ V}$$

which is far below the +0.4 V needed to keep $Q_7 - Q_8$ in the active mode. Thus, in the absence of CMF, a current mismatch of $\pm 0.3\ \mu\text{A}$ would cause one set of the cascode transistors (depending on the polarity of ΔI) to saturate.

(c) With the CFB circuit in place, the feedback will adjust V_B by ΔV_B so that the currents in Q_7 and Q_8 will change by a increment equal to ΔI, thus restoring current equality. Since a change ΔV_B results in

$$\Delta I_{C7} = \Delta I_{C8} = \frac{\Delta V_B}{r_{e7} + R_7}$$

then

$$\Delta I = \frac{\Delta V_B}{r_{e7} + R_7}$$

$$\Delta V_B = \Delta I(r_{e7} + R_7)$$

$$= 0.3\ \mu\text{A}\left(\frac{25\ \text{mV}}{10\ \mu\text{A}} + 20\ \text{k}\Omega\right)$$

$$= 0.3 \times 22.5 = 6.75\ \text{mV}$$

Correspondingly

$$\Delta V_{CM} = \Delta V_B = 6.75\ \text{mV}$$

Thus, to restore the current equality, the change required in V_B and V_{CM} is only 6.75 mV.

12.7.5 Output-Stage Design for Near Rail-to-Rail Output Swing

As mentioned earlier, modern low-voltage bipolar op amps cannot afford to use the classical emitter-follower-based class AB output stage; it would consume too much of the power supply voltage. Instead, a complementary pair of common-emitter transistors are utilized, as shown in Fig. 12.46. The output transistors Q_P and Q_N are operated in a class AB fashion. Typically, i_L can be as high as 10 mA to 15 mA and is determined by v_O and R_L. For $i_L = 0$, $i_P = i_N = I_Q$, where the quiescent current I_Q is normally a fraction of a milliamp.

The output stage in Fig. 12.46 is driven by two *separate but equal signals*, v_{BP} and v_{BN}. When v_{BP} and v_{BN} are high, Q_N supplies the load current in the direction opposite to that shown[5] and the output voltage v_O can swing to within 0.1 V or so of ground. In the meantime, Q_P is inactive. Nevertheless, in order to minimize crossover distortion, Q_P is

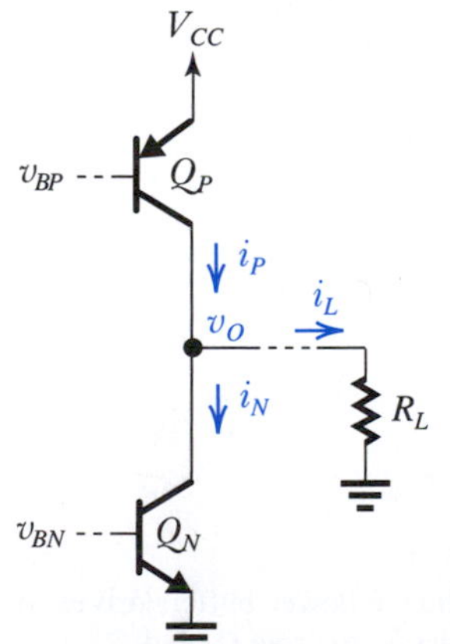

Figure 12.46 In order to provide v_O that can swing to within 0.1 V of V_{CC} and ground, a near rail-to-rail operation, the output stage utilizes common-emitter transistors. Note that the driving signals V_{BP} and V_{BN} are separate but identical.

[5]For this to happen, either R_L is returned to the positive supply (rather than ground) or R_L is capacitively coupled to the amplifier output.

prevented from turning off and is forced (as will be shown shortly) to conduct a minimum current of about $I_Q/2$.

The opposite happens when v_{BP} and v_{BN} are low: Q_P supplies the load current i_L in the direction indicated, and v_O can go up as high as $V_{CC} - 0.1$ V. In the meantime, Q_N is inactive but is prevented from turning off and forced to conduct a minimum current of about $I_Q/2$.

From the description above, we see that v_O can swing to within 0.1 V of each of the supply rails. This near rail-to-rail operation is the major advantage of this CE output stage. Its disadvantage is the relatively high output resistance. However, given that the op amp will almost always be used with a negative-feedback loop, the closed-loop output resistance can still be very low.

A Buffer/Driver Stage The output transistors can be called on to supply currents in the 10 mA to 15 mA range. When this happens, the base currents of Q_P and Q_N can be substantial (recall that $\beta_P \simeq 10$ and $\beta_N \simeq 40$). Such large currents cannot usually be supplied directly by the amplifier stage preceding the output stage. Rather a buffer/driver stage is usually needed, as shown in Fig. 12.47. Here an emitter follower Q_3 is used to drive Q_N. However, because of the low β_P, a double buffer consisting of complementary emitter followers Q_1 and Q_2 is used to drive Q_P. The driver stage is fed by two separate but identical signals v_{IP} and v_{IN} that come from the preceding amplifier stage (which is usually the second stage) in the op amp circuit.[6]

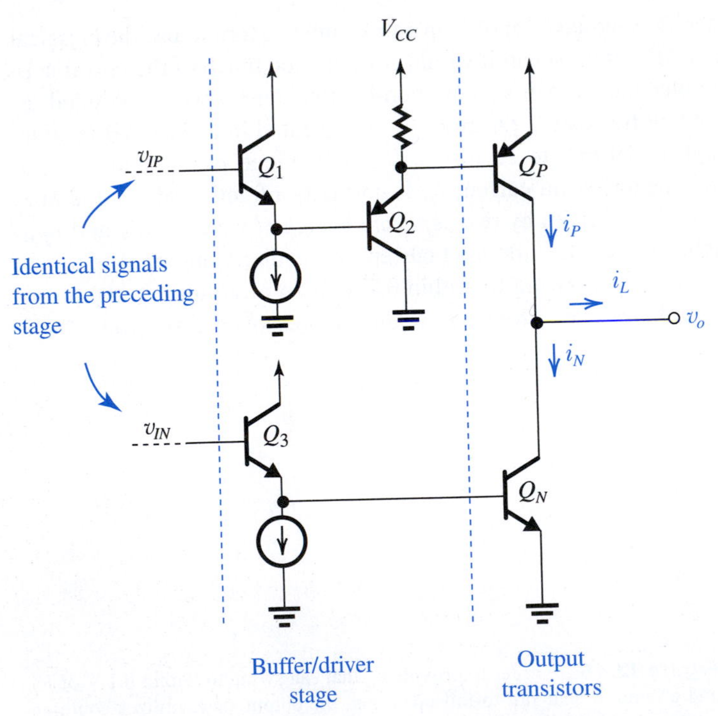

Figure 12.47 The output stage which is operated as class AB needs emitter follower buffers/drives to reduce the loading on the preceding stage and to provide the current gain necessary to drive Q_P and Q_N.

---

[6]An interesting approach for generating two identical outputs in the second stage is utilized in the NE5234 (see Gray et al., 2009).

EXERCISE

12.30 (a) For the circuit in Fig. 12.47, find the current gain from each of the v_{IP} and v_{IN} terminals to the output in terms of β_P and β_N.

(b) For $i_L = \pm 10$ mA, how much signal current is needed at the v_{IP} and v_{IN} inputs?

Ans. (a) $\beta_N \beta_P^2$, β_N^2; (b) 2.5 µA, 6.25 µA

Establishing I_Q and Maintaining a Minimum Current in the Inactive Transistor We next consider the circuit for establishing the quiescent current I_Q in Q_N and Q_P and for maintaining a minimum current of $I_Q/2$ in the inactive output transistor. Figure 12.48 shows a fuller version of the output stage. In addition to the output transistors Q_P–Q_N and the buffer/driver stage, which we have already discussed, the circuit includes two circuit blocks whose operation we shall now explain.

The first is the circuit composed of the differential pair Q_6–Q_7 and associated transistors Q_4 and Q_5, and resistors R_4 and R_5. This circuit measures the currents in the output transistors, i_P and i_N, and arranges for the current I to divide between Q_6 and Q_7 according to the ratio i_N/i_P, and provides a related output voltage v_E. Specifically, it can be shown [Problem 12.73] that

$$i_{C6} = I \frac{i_N}{i_P + i_N} \tag{12.132}$$

$$i_{C7} = I \frac{i_P}{i_P + i_N} \tag{12.133}$$

$$v_E = V_T \ln\left[\frac{i_N\, i_P}{i_N + i_P} \frac{I}{I_{SN}\, I_{S7}}\right] \tag{12.134}$$

where I_{SN} and I_{S7} are the saturation currents of Q_N and Q_7, respectively. Observe that for $i_P \gg i_N$, $i_{C6} \simeq 0$ and $i_{C7} \simeq I$. Thus Q_6 turns off and Q_7 conducts all of I. The emitter voltage v_E becomes

$$v_E \simeq V_T \ln\left(\frac{i_N}{I_{SN}}\right) + V_T \ln\left(\frac{I}{I_{S7}}\right)$$

Thus,

$$v_E = V_T \ln\left(\frac{i_N}{I_{SN}}\right) + V_{EB7} \tag{12.135}$$

This equation simply states that $v_E = v_{BEN} + V_{EB7}$, which could have been directly obtained from the circuit diagram in Fig. 12.48. The important point to note, however, is that since V_{EB7} is a constant, v_E is determined by the current i_N in the inactive transistor, Q_N. In the other extreme case of $i_N \gg i_P$, $i_{C6} \simeq I$, $i_{C7} \simeq 0$; thus Q_7 turns off and Q_6 conducts all of I. In this case we can use Eq. (12.134) to show that

$$v_E = V_T \ln\left(\frac{i_P}{I_{SN}}\right) + V_{EB6} \tag{12.136}$$

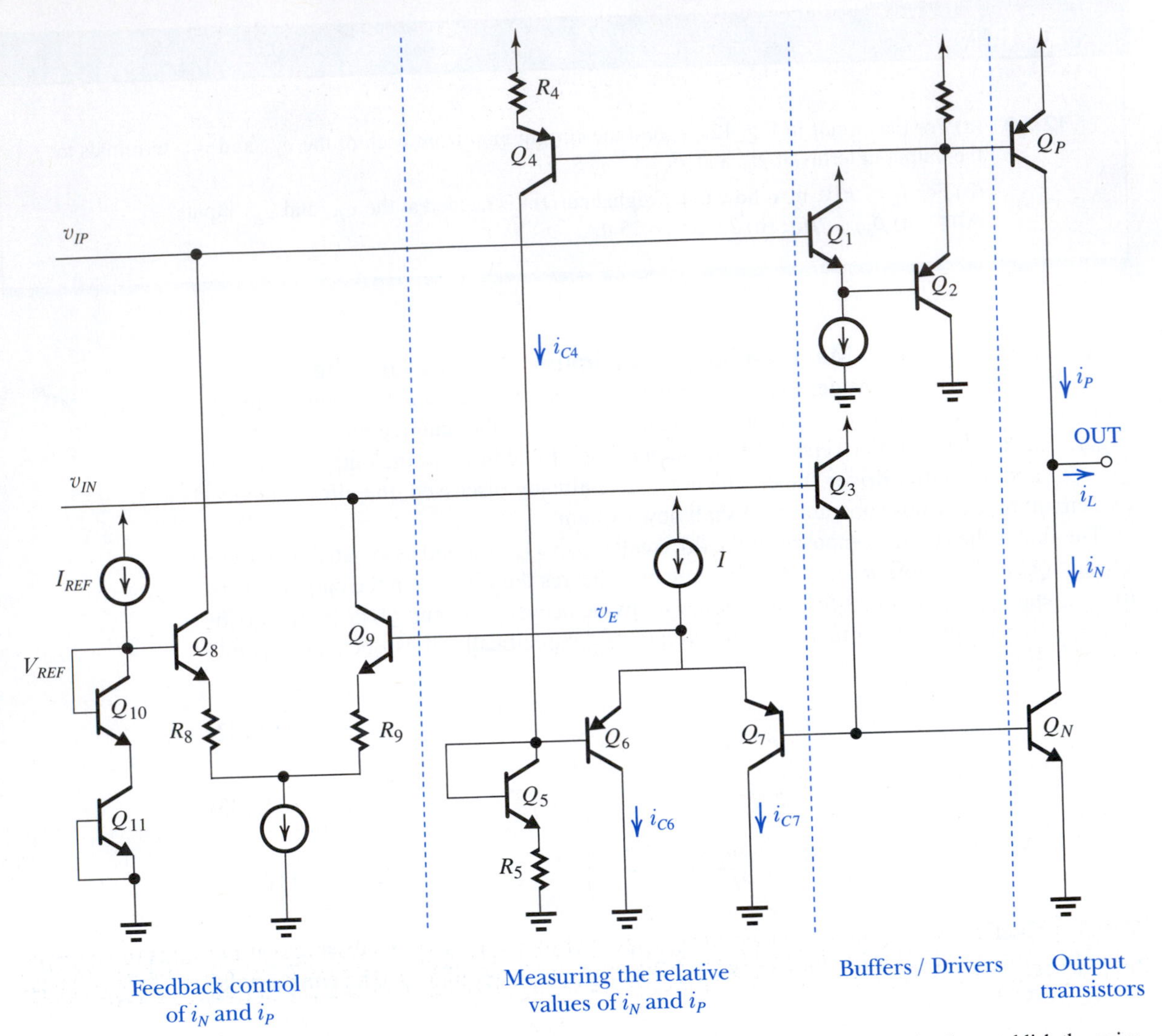

Figure 12.48 A more complete version of the output stage showing the circuits that establish the quiescent current in Q_P and Q_N. As well, this circuit forces a minimum current of $(I_Q/2)$ to follow in the inactive output transistor, thus preventing the transistor from turning off and minimizing crossover distortion.

Thus, here too, since V_{EB6} is a constant, v_E is determined by the current in the inactive transistor, Q_P.

The second circuit block is a differential amplifier composed of Q_8–Q_9 with their emitter-degeneration resistors R_8, R_9. The voltage v_E generated by the measuring circuit is fed to one input of the differential amplifier, and the other input is fed with a reference voltage V_{REF} generated by passing a reference current I_{REF} through the series connection of diode-connected transistors Q_{10} and Q_{11}. This differential amplifier takes part in a negative- feedback loop that uses the value of v_E to control the currents i_P and i_N through the nodes v_{IP} and v_{IN}. The objective of the feedback control is to set the current in the inactive output transistor to a minimum value. To see how the feedback operates, consider the case when $i_P \gg i_N$, and thus Q_N is the inactive transistor. In this case, Q_6 turns off, Q_7 conducts all of I, and v_E is given by Eq. (12.135). Now, if for some reason i_N falls below its minimum intended value, v_E decreases, causing i_{C9} to decrease. This in turn will cause the node v_{IN} to rise and the voltage at the base of Q_N will eventually rise, thus increasing i_N to its intended value.

Analytically, we can obtain a relationship between i_N and i_P as follows. Assume that the loop gain of the feedback loop that is anchored by the differential amplifier Q_8–Q_9 is high enough to force the two input terminals to the same voltage, that is,

$$v_E = V_{REF} = V_T \ln \frac{I_{REF}}{I_{S10}} + V_T \ln \frac{I_{REF}}{I_{S11}}$$

Substituting for v_E from Eq. (12.134) results in

$$\frac{i_N\, i_P}{i_N + i_P} = \left(\frac{I_{REF}^2}{I}\right)\left(\frac{I_{SN}}{I_{S10}}\right)\left(\frac{I_{S7}}{I_{S11}}\right) \tag{12.137}$$

Observe that the quantity on the right-hand side is a constant. In the quiescent case, $i_N = i_P = I_Q$, Eq. (12.137) yields

$$I_Q = 2\left(\frac{I_{REF}^2}{I}\right)\left(\frac{I_{SN}}{I_{S10}}\right)\left(\frac{I_{S7}}{I_{S11}}\right) \tag{12.138}$$

Thus, the constant on the right-hand side of Eq. (12.137) is $I_Q/2$, and we can rewrite (12.137) as

$$\frac{i_N\, i_P}{i_N + i_P} = \frac{1}{2} I_Q \tag{12.139}$$

Equation (12.139) clearly shows that for $i_N \gg i_P$, $i_P \simeq \frac{1}{2}I_Q$, and that for $i_P \gg i_N$, $i_N \simeq \frac{1}{2}I_Q$. Thus the circuit not only establishes the quiescent current I_Q (Eq. 12.138) but also sets the minimum current in the inactive output transistor at $\frac{1}{2}I_Q$.

EXERCISE

D12.31 For the circuit in Fig. 12.48, determine the value that I_{REF} should have so that Q_N and Q_P have a quiescent current $I_Q = 0.4$ mA. Assume that the transistor areas are scaled so that $I_{SN}/I_{S10} = 10$ and $I_{S7}/I_{S11} = 2$. Let $I = 10\ \mu A$. Also, if i_L in the direction out of the amplifier is 10 mA, find i_P and i_N.

Ans. $I_{REF} = 10\ \mu A$; $i_P \simeq 10.2$ mA, $i_N \simeq 0.2$ mA

Summary

- Most CMOS op amps are designed to operate as part of a VLSI circuit and thus are required to drive only small capacitive loads. Therefore, most do not have a low-output-resistance stage.
- There are basically two approaches to the design of CMOS op amps: a two-stage configuration and a single-stage topology utilizing the folded-cascode circuit.
- In the two-stage CMOS op amp, approximately equal gains are realized in the two stages.
- The threshold mismatch ΔV_t together with the low transconductance of the input stage result in a larger input offset voltage for CMOS op amps than for bipolar units.
- Miller compensation is employed in the two-stage CMOS op amp, but a series resistor is required to place the transmission zero at either $s = \infty$ or on the negative real axis.
- CMOS op amps have higher slew rates than their bipolar counterparts with comparable f_t values.
- Use of the cascode configuration increases the gain of a CMOS amplifier stage by about two orders of magnitude, thus making possible a single-stage op amp.
- The dominant pole of the folded-cascode op amp is determined by the total capacitance at the output node, C_L. Increasing C_L improves the phase margin at the expense of reducing the bandwidth.
- By using two complementary input differential pairs in parallel, the input common-mode range can be extended to equal the entire power-supply voltage, providing so-called rail-to-rail operation at the input.
- The output voltage swing of the folded-cascode op amp can be extended by utilizing a wide-swing current mirror in place of the cascode mirror.
- The internal circuit of the 741 op amp embodies many of the design techniques employed in bipolar analog integrated circuits.
- The 741 circuit consists of an input differential stage, a high-gain single-ended second stage, and a class AB output stage. Though 40 years old, this structure is typical of most BJT op amps and is known as the two-stage topology (not counting the output stage). It is also the same structure used in the two-stage CMOS op amp of Section 12.1.
- To obtain low input offset voltage and current, and high CMRR, the 741 input stage is designed to be perfectly balanced. The CMRR is increased by common-mode feedback, which also stabilizes the dc operating point.
- To obtain high input resistance and low input bias current, the input stage of the 741 is operated at a very low current level.
- In the 741, output short-circuit protection is accomplished by turning on a transistor that takes away most of the base current drive of the output transistor.
- The use of Miller frequency compensation in the 741 circuit enables locating the dominant pole at a very low frequency, while utilizing a relatively small compensating capacitance.
- Two-stage op amps can be modeled as a transconductance amplifier feeding an ideal integrator with C_C as the integrating capacitor.
- The slew rate of a two-stage op amp is determined by the first-stage bias current and the frequency-compensation capacitor.
- While the 741 and its generation of op amps nominally operate from ±15-V power supplies, modern BJT op amps typically utilize a single ground-referenced supply of only 2 V to 3 V.
- Operation from a single low-voltage supply gives rise to a number of new important specifications including a common-mode input range that extends beyond the supply rails (i.e., more than rail-to-rail operation) and a near rail-to-rail output voltage swing.
- The rail-to-rail input common-mode range is achieved by using resistive loads (instead of current-mirror loads) for the input differential pair as well as utilizing two complementary differential amplifiers in parallel.
- To increase the gain of the input stage above that achieved with resistive loads, the folded-cascode configuration is utilized.
- To regulate the dc bias voltages at the outputs of the differential folded-cascode stage so as to maintain active-mode operation at all times, common-mode feedback is employed.
- The output stage of a low-voltage op amp utilizes a complementary pair of common-emitter transistors. This allows v_O to swing to within 0.1 V or so from each of the supply rails. The disadvantage is a high open-loop output resistance. This, however, is substantially reduced when negative feedback is applied around the op amp.
- Modern output stages operate in the class AB mode and utilize interesting feedback techniques to set the quiescent current as well as to ensure that the inactive output transistor does not turn off, a precaution that avoids increases in crossover distortion.

Computer Simulation Problems

SIM Problems identified by this icon are intended to demonstrate the value of using SPICE simulation to verify hand analysis and design, and to investigate important issues such as allowable signal swing and amplifier nonlinear distortion. Instructions to assist in setting up PSpice and Multism simulations for all the indicated problems can be found in the corresponding files on the CD. Note that if a particular parameter value is not specified in the problem statement, you are to make a reasonable assumption.
* difficult problem; ** more difficult; *** very challenging and/or time-consuming; D: design problem.

Section 12.1: The Two-Stage CMOS Op Amp

12.1 A particular design of the two-stage CMOS operational amplifier of Fig. 12.1 utilizes ±1-V power supplies. All transistors are operated at overdrive voltages of 0.15-V magnitude. The process technology provides devices with $V_{tn} = |V_{tp}| = 0.45$ V. Find the input common-mode range and the range allowed for v_O.

12.2 The CMOS op amp of Fig. 12.1 is fabricated in a process for which $V'_{An} = 25$ V/µm and $|V'_{Ap}| = 20$ V/µm. Find A_1, A_2, and A_v if all devices are 0.5-µm long and are operated at equal overdrive voltages of 0.2-V magnitude. Also, determine the op-amp output resistance obtained when the second stage is biased at 0.4 mA. What do you expect the output resistance of a unity-gain voltage amplifier to be, using this op amp?

D 12.3 The CMOS op amp of Fig. 12.1 is fabricated in a process for which $|V'_A|$ for all devices is 24 V/µm. If all transistors have $L = 0.5$ µm and are operated at equal overdrive voltages, find the magnitude of the overdrive voltage required to obtain a dc open-loop gain of 6400 V/V.

12.4 This problem is identical to Problem 8.107.

Consider the circuit in Fig. 12.1 with the device geometries shown at the bottom of this page. Let $I_{REF} = 225$ µA, $|V_t|$ for all devices = 0.75 V, $\mu_n C_{ox} = 180\,\mu\text{A/V}^2$, $\mu_p C_{ox} = 60\,\mu\text{A/V}^2$, $|V_A|$ for all devices = 9 V, $V_{DD} = V_{SS} = 1.5$ V. Determine the width of Q_6, W, that will ensure that the op amp will not have a systematic offset voltage. Then, for all devices, evaluate I_D, $|V_{OV}|$, $|V_{GS}|$, g_m, and r_o. Provide your results in a table. Also find A_1, A_2, the dc open-loop voltage gain, the input common-mode range, and the output voltage range. Neglect the effect of V_A on the bias currents.

D 12.5 Design the two-stage CMOS op amp in Fig. 12.1 to provide a CMRR of about 80 dB. If all the transistors are operated at equal overdrive voltages of 0.15 V and have equal channel lengths, find the minimum required channel length. For this technology, $|V'_A| = 20$ V/µm.

D 12.6 A particular implementation of the CMOS amplifier of Figs. 12.1 and 12.2 provides $G_{m1} = 0.3$ mA/V, $G_{m2} = 0.6$ mA/V, $r_{o2} = r_{o4} = 222$ kΩ, $r_{o6} = r_{o7} = 111$ kΩ, and $C_2 = 1$ pF.

(a) Find the frequency of the second pole, f_{P2}.
(b) Find the value of the resistance R which when placed in series with C_C causes the transmission zero to be located at $s = \infty$.
(c) With R in place, as in (b), find the value of C_C that results in the highest possible value of f_t while providing a phase margin of 80°. What value of f_t is realized? What is the corresponding frequency of the dominant pole?
(d) To what value should C_C be changed to double the value of f_t? At the new value of f_t, what is the phase shift introduced by the second pole? To reduce this excess phase shift to 10° and thus obtain an 80° phase margin, as before, what value should R be changed to?

D 12.7 A two-stage CMOS op amp similar to that in Fig. 12.1 is found to have a capacitance between the output node and ground of 0.5 pF. If it is desired to have a unity-gain bandwidth f_t of 150 MHz with a phase margin of 75° what must g_{m6} be set to? Assume that a resistance R is connected in series with the frequency-compensation capacitor C_C and adjusted to place the transmission zero at infinity. What value should R have? If the first stage is operated at $|V_{OV}| = 0.15$ V, what is the value of slew rate obtained? If the first-stage bias current $I = 100$ µA, what is the required value of C_C?

D 12.8 A CMOS op amp with the topology shown in Fig. 12.1 is designed to provide $G_{m1} = 1$ mA/V and $G_{m2} = 5$ mA.

(a) Find the value of C_C that results in $f_t = 100$ MHz.
(b) What is the maximum value that C_2 can have while achieving a 70° phase margin?

D 12.9 A CMOS op amp with the topology shown in Fig. 12.1 but with a resistance R included in series with C_C is designed to provide $G_{m1} = 1$ mA/V and $G_{m2} = 2$ mA/V.

(a) Find the value of C_C that results in $f_t = 100$ MHz.
(b) For $R = 500$ Ω, what is the maximum allowed value of C_2 for which a phase margin of at least 60° is obtained?

Transistor	Q_1	Q_2	Q_3	Q_4	Q_5	Q_6	Q_7	Q_8
W/L (µm/µm)	30/0.5	30/0.5	10/0.5	10/0.5	60/0.5	W/0.5	60/0.5	60/0.5

12.10 A two-stage CMOS op amp resembling that in Fig. 12.1 is found to have a slew rate of 60 V/μs and a unity-gain bandwidth f_t of 50 MHz.

(a) Estimate the value of the overdrive voltage at which the input-stage transistors are operating.
(b) If the first-stage bias current $I = 100$ μA, what value of C_C must be used?
(c) For a process for which $\mu_p C_{ox} = 50$ μA/V^2, what W/L ratio applies for Q_1 and Q_2?

D 12.11 Sketch the circuit of a two-stage CMOS amplifier having the structure of Fig. 12.1 but utilizing NMOS transistors in the input stage (i.e., Q_1 and Q_2).

D 12.12 (a) Show that the PSRR$^-$ of a CMOS two-stage op amp for which all transistors have the same channel length and are operated at equal $|V_{OV}|$ is given by

$$\text{PSRR}^- = 2\left|\frac{V_A}{V_{OV}}\right|^2$$

(b) For $|V_{OV}| = 0.2$ V, what is the minimum channel length required to obtain a PSRR$^-$ of 80 dB? For the technology available, $|V_A'| = 20$ V/μm.

Section 12.2: The Folded-Cascode Op Amp

D 12.13 If the circuit of Fig. 12.8 utilizes ±1.65-V power supplies and the power dissipation is to be limited to 1 mW, find the values of I_B and I. To avoid turning off the current mirror during slewing, select I_B to be 20% larger than I.

D 12.14 For the folded-cascode op amp in Fig. 12.9 utilizing power supplies of ±1 V, find the values of V_{BIAS1}, V_{BIAS2}, and V_{BIAS3} to maximize the allowable range of V_{ICM} and v_O. Assume that all transistors are operated at equal overdrive voltages of 0.15 V. Assume $|V_t|$ for all devices is 0.45 V. Specify the maximum range of V_{ICM} and of v_O.

D 12.15 For the folded-cascode op-amp circuit of Figs. 12.8 and 12.9 with bias currents $I = 96$ μA and $I_B = 120$ μA, and with all transistors operated at overdrive voltages of 0.2 V, find the W/L ratios for all devices. Assume that the technology available is characterized by $k_n' = 400$ μA/V^2 and $k_p' = 100$μA/V^2.

12.16 Consider a design of the cascode op amp of Fig. 12.9 for which $I = 96$ μA and $I_B = 120$ μA. Assume that all transistors are operated at $|V_{OV}| = 0.2$ V and that for all devices, $|V_A| = 12$ V. Find G_m, R_o, and A_v. Also, if the op amp is connected in the feedback configuration shown in Fig. P12.16, find the voltage gain and output resistance of the closed-loop amplifier.

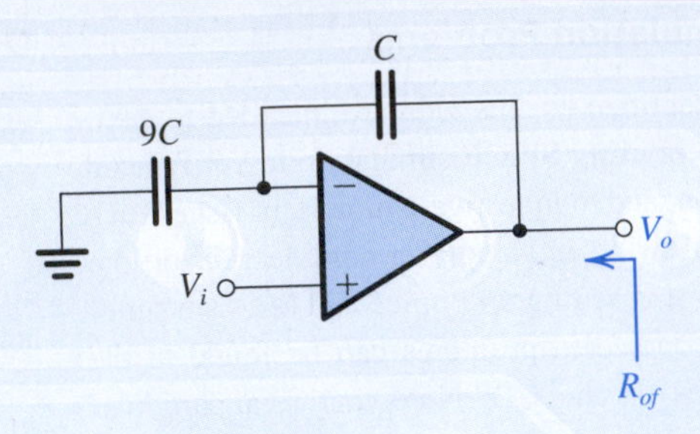

Figure P12.16

D 12.17 Consider the folded-cascode op amp of Fig. 12.8 when loaded with a 10-pF capacitance. What should the bias current I be to obtain a slew rate of at least 10 V/μs? If the input-stage transistors are operated at overdrive voltages of 0.2 V, what is the unity-gain bandwidth realized? If the two nondominant poles have the same frequency of 25 MHz, what is the phase margin obtained? If it is required to have a phase margin of 75°, what must f_t be reduced to? By what amount should C_L be increased? What is the new value of SR?

D 12.18 Design the folded-cascode circuit of Fig. 12.9 to provide voltage gain of 80 dB and a unity-gain frequency of 10 MHz when $C_L = 10$ pF. Design for $I_B = I$, and operate all devices at the same $|V_{OV}|$. Utilize transistors with 1-μm channel length for which $|V_A|$ is specified to be 20 V. Find the required overdrive voltages and bias currents. What slew rate is achieved? Also, for $k_n' = 2.5\,k_p' = 200$ μA/V^2, specify the required width of each of the 11 transistors used.

D 12.19 Sketch the circuit that is complementary to that in Fig. 12.9, that is, one that uses an input p-channel differential pair.

12.20 For the circuit in Fig. 12.11, assume that all transistors are operating at equal overdrive voltages of 0.2-V magnitude and have $|V_t| = 0.5$ V and that $V_{DD} = V_{SS} = 1.65$ V. Find (a) the range over which the NMOS input stage operates, (b) the range over which the PMOS input stage operates, (c) the range over which both operate (the overlap range), and (d) the input common-mode range.

12.21 A particular design of the wide-swing current mirror of Fig. 12.12(b) utilizes devices having $W/L = 25$, $k_n' = 200$ μA/V^2, and $V_t = 0.5$ V. For $I_{\text{REF}} = 100$ μA, what value of V_{BIAS} is needed? Also give the voltages that you expect to appear at all nodes and specify the minimum voltage allowable at the output terminal. If V_A is specified to be 10 V, what is the output resistance of the mirror?

D 12.22 For the folded-cascode circuit of Fig. 12.8, let the total capacitance to ground at each of the source nodes of Q_3 and Q_4 be denoted C_P. Assuming that the incremental resistance between the drain of Q_3 and ground is small, Show that the pole that arises at the interface between the first and second stages has a frequency $f_P \simeq g_{m3}/2\pi C_P$. Now, if this is the only nondominant pole, what is the largest value that C_P can be (expressed as a fraction of C_L) while a phase margin of 75° is achieved? Assume that all transistors are operated at the same bias current and overdrive voltage.

Section 12.3: The 741 Op-Amp Circuit

12.23 In the 741 op-amp circuit of Fig. 12.13, Q_1, Q_2, Q_5, and Q_6 are biased at collector currents of 9.5 μA; Q_{16} is biased at a collector current of 16.2 μA; and Q_{17} is biased at a collector current of 550 μA. All these devices are of the "standard *npn*" type, having $I_S = 10^{-14}$ A, $\beta = 200$, and $V_A = 125$ V. For each of these transistors, find V_{BE}, g_m, r_e, r_π, and r_o. Provide your results in table form. (Note that these parameter values are utilized in the text in the analysis of the 741 circuit.)

D 12.24 For the (mirror) bias circuit shown in Fig. E12.11 and the result verified in the associated exercise, find I_1 for the case in which $I_{S3} = 3 \times 10^{-14}$ A, $I_{S4} = 6 \times 10^{-14}$ A, and $I_{S1} = I_{S2} = 10^{-14}$ A and for which a bias current $I_3 = 154$ μA is required.

12.25 Transistor Q_{13} in the circuit of Fig. 12.13 consists, in effect, of two transistors whose emitter–base junctions are connected in parallel and for which $I_{SA} = 0.25 \times 10^{-14}$ A, $I_{SB} = 0.75 \times 10^{-14}$ A, $\beta = 50$, and $V_A = 50$ V. For operation at a total emitter current of 0.73 mA, find values for the parameters V_{EB}, g_m, r_e, r_π, and r_o for the A and B devices.

12.26 In the circuit of Fig. 12.13, Q_1 and Q_2 exhibit emitter–base breakdown at 7 V, while for Q_3 and Q_4 such a breakdown occurs at about 50 V. What differential input voltage would result in the breakdown of the input-stage transistors?

D *12.27 Figure P12.27 shows the CMOS version of the circuit in Fig. E12.11. Find the relationship between I_3 and I_1 in terms of k_1, k_2, k_3, and k_4 of the four transistors, assuming the threshold voltages of all devices to be equal in magnitude. Note that k denotes $\mu C_{ox} W/L$. In the event that $k_1 = k_2$ and $k_3 = k_4 = 16k_1$, find the required value of I_1 to yield a bias current in Q_3 and Q_4 of 1.6 mA.

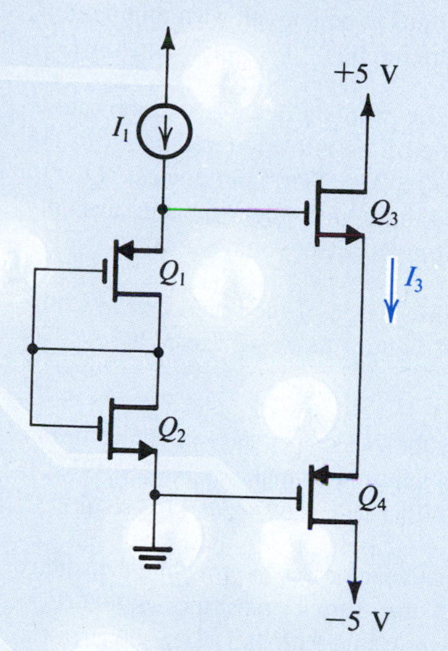

Figure P12.27

Section 12.4: DC Analysis of the 741

D 12.28 For the 741 circuit, estimate the input reference current I_{REF} in the event that ±5-V supplies are used. Find a more precise value assuming that for the two BJTs involved, $I_S = 10^{-14}$ A. What value of R_5 would be necessary to reestablish the same bias current for ±5-V supplies as exists for ±15 V in the original design?

D 12.29 Design the Widlar current source of Fig. 12.14 to generate a current $I_{C10} = 10$ μA given that $I_{REF} = 0.2$ mA. If for the transistors, $I_S = 10^{-14}$ A, find V_{BE11} and V_{BE10}. Assume β to be high.

12.30 Consider the dc analysis of the 741 input stage shown in Fig. 12.15. For what value of β_P do the currents in Q_1 and Q_2 differ from the ideal value of $I_{C10}/2$ by 10%?

D 12.31 Consider the dc analysis of the 741 input stage shown in Fig. 12.15 for the situation in which $I_{S9} = 2I_{S8}$. For $I_{C10} = 19$ μA and assuming β_P to be high, what does I become? Redesign the Widlar source to reestablish $I_{C1} = I_{C2} = 9.5$ μA.

12.32 For the mirror circuit shown in Fig. 12.16 with the bias and component values given in the text for the 741 circuit, what does the current in Q_6 become if R_2 is shorted?

D 12.33 It is required to redesign the circuit of Fig. 12.16 by selecting a new value for R_3 so that when the base currents are *not* neglected, the collector currents of Q_5, Q_6, and Q_7 all become equal, assuming that the input current $I_{C3} = 9.4$ μA. Find the new value of R_3 and the three currents. Recall that $\beta_N = 200$.

12.34 Consider the input circuit of the 741 op amp of Fig. 12.13 when the emitter current of Q_8 is about 19 μA. If β of Q_1 is 150 and that of Q_2 is 200, find the input bias current I_B and the input offset current I_{OS} of the op amp.

12.35 For a particular application, consideration is being given to selecting 741 ICs for input bias and offset currents limited to 50 nA and 4 nA, respectively. Assuming other aspects of the selected units to be normal, what minimum β_N and what β_N variation are implied?

12.36 A manufacturing problem in a 741 op amp causes the current transfer ratio of the mirror circuit that loads the input stage to become 0.8 A/A. For input devices (Q_1–Q_4) appropriately matched and with high β, and normally biased at 9.5 μA, what input offset voltage results?

D 12.37 Consider the design of the second stage of the 741. What value of R_9 would be needed to reduce I_{C16} to 9.5 μA?

D 12.38 Reconsider the 741 output stage as shown in Fig. 12.17, in which R_{10} is adjusted to make $I_{C19} = I_{C18}$. What is the new value of R_{10}? What values of I_{C14} and I_{C20} result?

D *12.39 An alternative approach to providing the voltage drop needed to bias the output transistors is the V_{BE}-multiplier circuit shown in Fig. P12.39. Design the circuit to provide a terminal voltage of 1.118 V (the same as in the 741 circuit). Base your design on half the current flowing through R_1, and assume that $I_S = 10^{-14}$ A and $\beta = 200$. What is the incremental resistance between the two terminals of the V_{BE}-multiplier circuit?

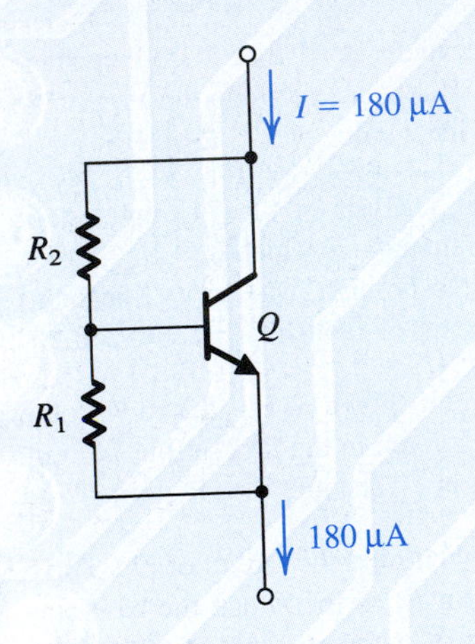

Figure P12.39

12.40 For the circuit of Fig. 12.13, what is the total current required from the power supplies when the op amp is operated in the linear mode, but with no load? Hence, estimate the quiescent power dissipation in the circuit. (*Hint:* Use the data given in Table 12.1.)

Section 12.5: Small-Signal Analysis of the 741

12.41 Consider the 741 input stage as modeled in Fig. 12.18, with two additional *npn* diode-connected transistors, Q_{1a} and Q_{2a}, connected between the present *npn* and *pnp* devices, one per side. Convince yourself that each of the additional devices will be biased at the same current as Q_1 to Q_4—that is, 9.5 μA. What does R_{id} become? What does G_{m1} become? What is the value of R_{o4} now? What is the output resistance of the first stage, R_{o1}? What is the new open-circuit voltage gain, $G_{m1}R_{o1}$? Compare these values with the original ones.

D 12.42 What relatively simple change can be made to the mirror load of stage 1 to increase its output resistance, say by a factor of 2?

12.43 Repeat Exercise 12.15 with $R_1 = R_2$ replaced by 2-kΩ resistors.

***12.44** In Example 12.3 we investigated the effect of a mismatch between R_1 and R_2 on the input offset voltage of the op amp. Conversely, R_1 and R_2 can be deliberately mismatched (using the circuit shown in Fig. P12.44, for example) to compensate for the op-amp input offset voltage.

(a) Show that an input offset voltage V_{OS} can be compensated for (i.e., reduced to zero) by creating a relative mismatch $\Delta R/R$ between R_1 and R_2,

$$\frac{\Delta R}{R} = \frac{V_{OS}}{2V_T}\frac{1 + r_e/R}{1 - V_{OS}/2V_T}$$

where r_e is the emitter resistance of each of Q_1 to Q_6, and R is the nominal value of R_1 and R_2. (*Hint:* Use Eq. 12.87)
(b) Find $\Delta R/R$ to trim a 5-mV offset to zero.
(c) What is the maximum offset voltage that can be trimmed this way (corresponding to R_2 completely shorted)?

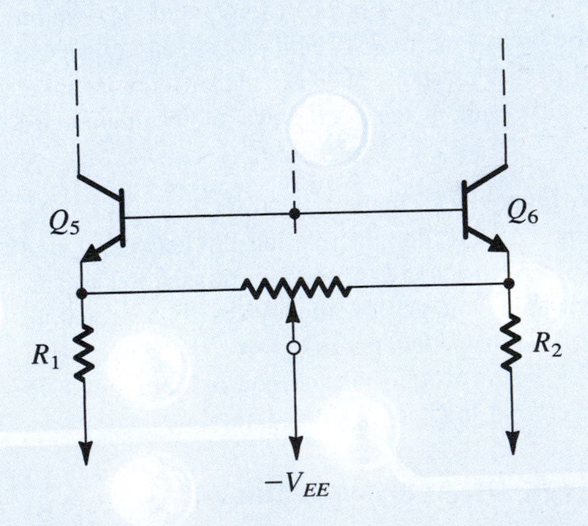

Figure P12.44

12.45 Through a processing imperfection, the β of Q_4 in Fig. 12.13 is reduced to 20, while the β of Q_3 remains at its regular value of 50. Find the input offset voltage that this mismatch introduces. (*Hint:* Follow the general procedure outlined in Example 12.3.)

12.46 Consider the circuit of Fig. 12.13 modified to include resistors R in series with the emitters of each of Q_8 and Q_9. What does the resistance looking into the collector of Q_9, R_{o9}, become? For what value of R does it equal R_{o10}? For this case, what does R_o looking to the left of node Y become?

***12.47** What is the effect on the differential gain of the 741 op amp of short-circuiting one, or the other, or both, of R_1 and R_2 in Fig. 12.13? (Refer to Fig. 12.19.) For simplicity, assume $\beta = \infty$.

12.48 It is required to show that the loop gain of the common-mode feedback loop shown in Fig. 12.23 is approximately equal to β_P. To determine the loop gain, connect both input terminals to ground. Break the loop at the input to the $Q_8 - Q_9$ current mirror, connecting the $Q_1 - Q_2$ collectors to signal ground. (This is because the original resistance between the collectors and ground is r_{e8}, which is small.) Apply a test current I_t to Q_8 and determine the returned current I_r in the common collectors' connection to ground, then find the loop gain as $-I_r/I_t$. Assume that r_π of Q_1 to Q_4 is much lower than R_o and that $\beta_N, \beta_P \gg 1$.

12.49 An alternative approach to that presented in Example 12.4 for determining the CMRR of the 741 input stage is investigated in this problem. Rather than performing the analysis on the closed loop shown in Fig. 12.23, we observe that the negative feedback increases the resistance at node Y by the amount of negative feedback. Thus, we can break the loop at Y and connect a resistance $R_f = (1 + A\beta)R_o$ between the common base connection of $Q_3 - Q_4$ and ground. We can then determine the current i and G_{mcm}. Using the fact that the loop gain is approximately equal to β_P (Problem 12.48) show that this approach yields an identical result to that found in Example 12.4.

12.50 Consider a variation on the design of the 741 second stage in which $R_8 = 50\ \Omega$. What R_{i2} and G_{m2} correspond?

12.51 In the analysis of the 741 second stage, note that R_{o2} is affected most strongly by the low value of R_{o13B}. Consider the effect of placing appropriate resistors in the emitters of Q_{12}, Q_{13A}, and Q_{13B} on this value. What resistor in the emitter of Q_{13B} would be required to make R_{o13B} equal to R_{o17} and thus R_{o2} half as great? What resistors in each of the other emitters would be required?

12.52 For a 741 employing ±5-V supplies, $|V_{BE}| = 0.6$ V and $|V_{CE\text{sat}}| = 0.2$ V, find the output voltage limits that apply.

D 12.53 Consider an alternative to the present 741 output stage in which Q_{23} is not used, that is, in which its base and emitter are joined. Reevaluate the reflection of $R_L = 2$ kΩ to the collector of Q_{17}. What does A_2 become?

12.54 Consider the positive current-limiting circuit involving Q_{13A}, Q_{15}, and R_6. Find the current in R_6 at which the collector current of Q_{15} equals the current available from Q_{13A} (180 μA) minus the base current of Q_{14}. (You need to perform a couple of iterations.)

D 12.55 Consider the 741 sinking-current limit involving R_7, Q_{21}, Q_{24}, R_{11}, and Q_{22}. For what current through R_7 is the current in Q_{22} equal to the maximum current available from the input stage (i.e., the current in Q_8)? What simple change would you make to reduce this current limit to 10 mA?

Section 12.6: Gain, Frequency Response, and Slew Rate of the 741

12.56 Using the data provided in Eq. (12.112) (alone) for the overall gain of the 741 with a 2-kΩ load, and realizing the significance of the factor 0.97 in relation to the load, calculate the open-circuit voltage gain, the output resistance, and the gain with a load of 200 Ω.

12.57 A 741 op amp has a phase margin of 75°. If the excess phase shift is due to a second single pole, what is the frequency of this pole?

12.58 A 741 op amp has a phase margin of 75°. If the op amp has nearly coincident second and third poles, what is their frequency?

D *12.59 For a modified 741 whose second pole is at 5 MHz, what dominant-pole frequency is required for 80° phase margin with a closed-loop gain of 100? Assuming C_C continues to control the dominant pole, what value of C_C would be required?

12.60 An internally compensated op amp having an f_t of 10 MHz and dc gain of 10^6 utilizes Miller compensation around an inverting amplifier stage with a gain of –1000. If space exists for at most a 50-pF capacitor, what resistance level must be reached at the input of the Miller amplifier for compensation to be possible?

12.61 Consider the integrator op-amp model shown in Fig. 12.33. For $G_{m1} = 5$ mA/V, $C_C = 100$ pF, and a resistance of $2 \times 10^7\ \Omega$ shunting C_C, sketch and label a Bode plot for the magnitude of the open-loop gain. If G_{m1} is related to the first-stage bias current as $G_{m1} = I/2V_T$, find the slew rate of this op amp.

12.62 For an amplifier with a slew rate of 10 V/μs, what is the full-power bandwidth for outputs of ±10 V? What unity-gain bandwidth, ω_t, would you expect if the topology was similar to that of the 741?

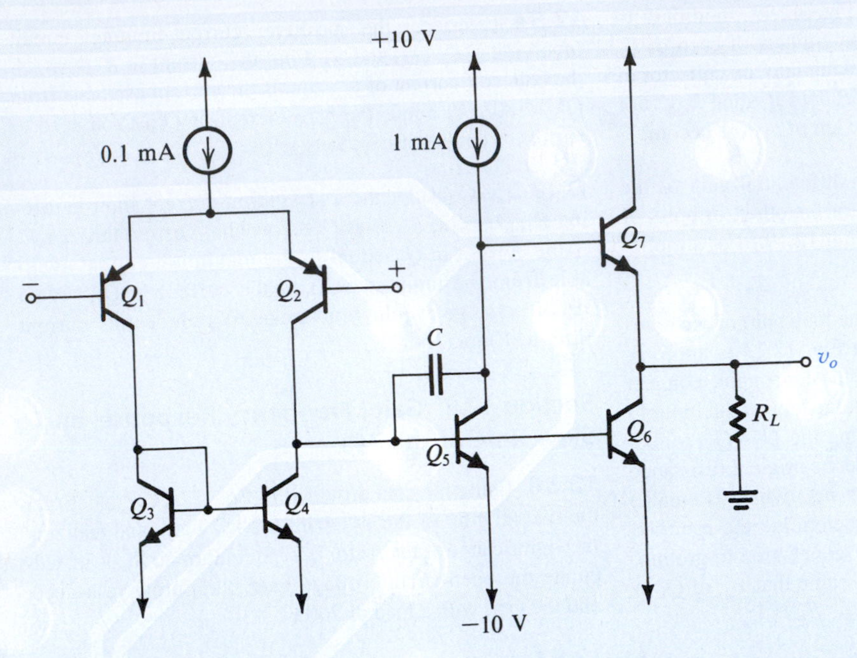

Figure P12.63

D *12.63 Figure P12.63 shows a circuit suitable for op-amp applications. For all transistors $\beta = 100$, $V_{BE} = 0.7$ V, and $r_o = \infty$.

(a) For inputs grounded and output held at 0 V (by negative feedback) find the collector currents of all transistors. Neglect base currents.
(b) Calculate the input resistance.
(c) Calculate the gain of the amplifier with a load of 5 kΩ.
(d) With load as in (c) calculate the value of the capacitor C required for a 3-dB frequency of 100 Hz.

Section 12.7: Modern Techniques for the Design of BJT Op Amps

Unless otherwise specified, for the problems in this section assume $\beta_N = 40$, $\beta_P = 10$, $V_{An} = 30$ V, $|V_{Ap}| = 20$ V, $|V_{BE}| = 0.7$ V, $|V_{CEsat}| = 0.1$ V.

D 12.64 Design the circuit in Fig. 12.38 to generate a current $I = 6$ μA. Utilize transistors Q_1 and Q_2 having areas in a ratio of 1:4. Assume that Q_3 and Q_4 are matched and design for a 0.2-V drop across each of R_3 and R_4. Specify the values of R_2, R_3, and R_4. Ignore base currents.

D 12.65 Consider the circuit of Fig. 12.38 for the case designed in Exercise 12.28, namely, $I = 10$ μA, $I_{S2}/I_{S1} = 2$, $R_2 = 1.73$ kΩ, $R_3 = R_4 = 20$ kΩ. Augment the circuit with *npn* transistors Q_5 and Q_6 with emitters connected to ground and bases connected to V_{BIAS1}, to generate constant currents of 10 μA and 40 μA, respectively. What should the emitter areas of Q_5 and Q_6 be relative to that of Q_1? What value of a resistance R_6 will, when connected in the emitter of Q_6, reduce the current generated by Q_6 to 10 μA? Assuming that the V_{BIAS1} line has a low incremental resistance to ground, find the output resistance of current source Q_5 and of current source Q_6 with R_6 connected. Ignore base currents.

D 12.66 (a) Find the input common-mode range of the circuit in Fig. 12.40(a). Let $V_{CC} = 3$ V and $V_{\text{BIAS}} = 2.3$ V.

(b) Give the complementary version of the circuit in Fig. 12.40(a), that is, the one in which the differential pair is *npn*. For the same conditions as in (a), what is the input common-mode range?

12.67 For the circuit in Fig. 12.40(b), let $V_{CC} = 3$ V, $V_{\text{BIAS}} = 2.3$ V, $I = 20$ μA, and $R_C = 20$ kΩ. Find the input common-mode range and the differential voltage gain v_o/v_{id}. Neglect base currents.

12.68 For the circuit in Fig. 12.41, let $V_{CC} = 3$ V, $V_{\text{BIAS}} = 0.7$ V, and $I_{C6} = 10$ μA. Find R_C that results in a differential gain of 10 V/V. What is the input common-mode range and the input differential resistance? Ignore base currents except when calculating R_{id}.

12.69 It is required to find the input resistance and the voltage gain of the input stage shown in Fig. 12.42. Let $V_{ICM} \ll 0.8$ V so that the Q_3–Q_4 pair is off. Assume that

Q_5 supplies 6 μA, that each of Q_7 to Q_{10} is biased at 6 μA, and that all four cascode transistors are operating in the active mode. The input resistance of the second stage of the op amp is 1.3 MΩ. The emitter degeneration resistances are $R_7 = R_8 = 22$ kΩ, and $R_9 = R_{10} = 33$ kΩ. [*Hint*: Refer to Fig. 12.43.]

D 12.70 Consider the equivalent half-circuit shown in Fig. 12.43. Assume that in the original circuit, Q_1 is biased at a current I, Q_7 and Q_9 are biased at $2I$, the dc voltage drop across R_7 is 0.2 V, and the dc voltage drop across R_9 is 0.3 V. Find the open-circuit voltage gain (i.e., the voltage gain for $R_L = \infty$). Also find the output resistance in terms of I. Now with R_L connected, find the voltage gain in terms of (IR_L). For $R_L = 2$ MΩ, find I that will result in the voltage gains of 160 V/V and 320 V/V.

***12.71** (a) For the circuit in Fig. 12.44, show that the loop gain of the common-mode feedback loop is

$$A\beta \simeq \frac{R_{o9} \| R_{o7}}{r_{e7} + R_7}$$

Recall that the CMF circuit realizes the transfer characteristic $V_B = V_{CM} + 0.4$. Ignore the loading effect of the CMF circuit on the collectors of the cascode transistors.
(b) For the values in Example 12.6, calculate the loop gain $A\beta$.
(c) In Example 12.6, we found that with the CMF absent, a current mismatch $\Delta I = 0.3$ μA gives rise to $\Delta V_{CM} = 2.5$ V. Now, with the CMF present, use the value of loop gain found in (b) to calculate the expected ΔV_{CM} and compare to the value found by a different approach in Example 12.6. [*Hint*: Recall that negative feedback reduces change by a factor equal to $(1 + A\beta)$.]

12.72 The output stage in Fig. 12.46 operates at a quiescent current I_Q of 0.4 mA. The maximum current i_L that the stage can provide in either direction is 10 mA. Also, the output stage is equipped with a feedback circuit that maintains a minimum current of $I_Q/2$ in the inactive output transistor.

(a) What is the allowable range of v_O?
(b) For $i_L = 0$, what is the output resistance of the op amp?
(c) If the open-loop gain of the op amp is 100,000 V/V, find the closed-loop output resistance obtained when the op amp is connected in the unity-gain voltage follower configuration, with $i_L = 0$.
(d) If the op amp is sourcing a load current $i_L = 10$ mA, find i_P, i_N, and the open-loop output resistance.
(e) Repeat (d) for the case of the open-loop op amp sinking a load current of 10 mA.

12.73 It is required to derive the expressions in Eqs. (12.132) and (12.133). Toward that end, first find v_{B7} in terms of v_{BEN} and hence i_N. Then find v_{B6} in terms of i_P. For the latter purpose note that Q_4 measures v_{EBP} and develops a current $i_4 = (v_{EBP} - v_{EB4})/R_4$. This current is supplied to the series connection of Q_5 and R_5 where $R_5 = R_4$. In the expression you obtain for v_{B6}, use the relationship

$$\frac{I_{SP}}{I_{S4}} = \frac{I_{SN}}{I_{S5}}$$

to express v_{B6} in terms of i_P and I_{SN}. Now with v_{B6} and v_{B7} determined, find i_{C6} and i_{C7}.

12.74 It is required to derive the expression for v_E in Eq. (12.134). Toward that end, note from the circuit in Fig. 12.48 that $v_E = v_{EB7} + v_{BEN}$ and note that Q_N conducts a current i_N and Q_7 conducts a current i_{C7} given by Eq. (12.133).

D 12.75 For the output stage in Fig. 12.48, find the current I_{REF} that results in a quiescent current $I_Q = 0.36$ mA. Assume that $I = 10$ μA, Q_N has eight times the area of Q_{10}, and Q_7 has four times the area of Q_{11}. What is the minimum current in Q_N and Q_P?

PART III

Digital Integrated Circuits

There are two indisputable facts about digital systems. They have dramatically changed our lives; and the digital revolution is driven by microelectronics.

Evidence of the pervasiveness and influence of digital systems can be found by thinking of what we do in our daily lives. Digital circuits exist in almost every electrical appliance we use in our homes; in the vehicles and transportation systems we use to travel; in the telephones and, most obviously, the cell phones we use to communicate; in the medical equipment needed to care for our health; in the computers we use to do our work; and in the audio and video systems and the radio and TV sets we use to entertain ourselves. Indeed, it is very difficult to conceive of modern life without digital systems, none of which would have been possible without microelectronics.

Although the idea of a digital computing machine was conceived as early as the 1830s, early implementations were very cumbersome and expensive mechanical devices. The first serious digital computers using vacuum tubes appeared in the 1930s and 1940s. These early computers used thousands of tubes and were housed literally in many rooms. Their fundamental limitation was low reliability: vacuum tubes had a finite life and needed large amounts of power. Had it not been for the invention of the transistor in 1947 ushering in the era of solid-state electronics, digital computers would have remained specialized machines used primarily in military and scientific applications.

By the mid 1950s, the first digital logic gates made of discrete bipolar transistors became commercially available. The invention of the integrated circuit in the late 1950s was also key, leading to the first digital IC in the early 1960s. Early digital ICs were made of bipolar transistors, with the most successful logic-circuit family of this type being transistor-transistor logic (or TTL), which dominated digital circuit design, until the early 1980s.

Bipolar was replaced by NMOS, and NMOS by CMOS, again predominantly because of power dissipation and the need to pack more and more transistors on each IC chip. Bearing out Moore's law, which predicted in 1968 that IC chips would double the number of their transistors every two to three years (see Section 13.5), digital ICs have grown from a few transistors to 2.3 billion devices and to memory chips with 4 Gbit capacity.

Part III aims to provide a brief but nonetheless comprehensive and sufficiently detailed exposure to digital IC design. Our treatment is almost self-contained, requiring for the most part only a thorough understanding of the MOSFET material presented in Chapter 5. Thus Part III can be studied right after Chapter 5. The only exceptions to this are the last two sections in Chapter 14, which require knowledge of the BJT (Chapter 6). Also, knowledge of the MOSFET internal capacitances (Section 9.2.2) will be needed.

Chapter 13 is the cornerstone of Part III. It provides an introduction to digital circuits and then concentrates on the bread-and-butter topic of digital IC design: the CMOS inverter and logic gates. Today, CMOS represents 98% of newly designed digital systems. The material in Chapter 13 is the minimum needed to learn something meaningful about digital circuits; it is a must study!

Chapter 14 builds on the foundation established in Chapter 13 and introduces three important types of MOS logic circuits and a significant family of bipolar logic circuits. The chapter concludes with an interesting digital circuit technology that attempts to combine the best of bipolar and CMOS: BiCMOS.

Digital circuits can be broadly divided into logic and memory circuits. The latter is the subject of Chapter 15.

CHAPTER 13

CMOS Digital Logic Circuits

IN THIS CHAPTER YOU WILL LEARN

1. How the operation of the basic element in digital circuits, the logic inverter, is characterized by such parameters as noise margins, propagation delay, and power dissipation, and how it is implemented by using one of three possible arrangements of voltage-controlled switches (transistors).

2. That the three most significant metrics in digital IC design are speed of operation, power dissipation, and silicon area, and that each design is in effect a trade-off among the three metrics.

3. How and why CMOS has become the dominant technology for digital IC design.

4. The structure, circuit operation, static and dynamic performance analysis, and the design of the CMOS inverter.

5. The synthesis and design optimization of CMOS logic circuits.

6. The implications of technology scaling (Moore's law) over 40 years and continuing, and some of the current challenges in the design of deep-submicron ($L < 0.25\ \mu m$) circuits.

Introduction

This chapter does three things: It introduces the basic element of digital circuits, the logic inverter; it presents a relatively detailed study of the CMOS inverter and of CMOS logic-circuit design; and it provides a perspective on the astounding phenomenon of technology scaling (Moore's law) and the opportunities and challenges of deep-submicron ($L < 0.25\ \mu m$) IC design.

Our study of the inverter in Section 13.1 provides the foundation for the study of digital electronics in the remainder of the chapter and in the next two chapters. Without getting into circuit implementation detail, Section 13.1 introduces all the parameters and metrics used in digital IC design. As well, it provides an overview of digital IC technologies and logic-circuit families. In this way, it provides the basis for appreciating how and why CMOS has emerged the dominant technology in digital IC design. The section concludes with a discussion of the various styles of digital system design: from small-scale and medium-scale integrated-circuit (SSI and MSI) packages assembled on printed-circuit boards to systems assembled using very-large-scale integrated (VLSI) circuits such as microprocessors, memory, and custom and semicustom ICs.

Sections 13.2 and 13.3 provide a comprehensive and thorough study of the CMOS inverter. Section 13.4 builds on this material and presents the basic CMOS logic-gate circuits as well as a general approach for the CMOS implementation of arbitrary logic functions. We also consider the design optimization of the resulting circuits.

The chapter concludes with a retrospective and a prospective look at Moore's law and the technology scaling that has continued over the last 40 years and shows no signs of stopping. This leads naturally to a discussion of the phenomena that take place in deep-submicron ($L < 0.25$ μm) MOSFETs and how to modify the model we studied in Chapter 5 to take account of these phenomena. This section should serve as a bridge between this introductory course and more advanced study of digital IC design.

This chapter provides a self-contained study of CMOS logic circuits, the bread and butter of digital IC design. We will build on this foundation in our study of the more specialized topics in the next two chapters.

13.1 Digital Logic Inverters

The logic inverter is the most basic element in digital circuit design; it plays a role parallel to that of the amplifier in analog circuits. In this section we provide an introduction to the logic inverter and to digital circuit design.

13.1.1 Function of the Inverter

As its name implies, the logic inverter inverts the logic value of its input signal. Thus, for a logic-0 input, the output will be a logic 1, and vice versa. In terms of voltage levels, consider the inverter shown in block form in Fig. 13.1. Its implementation will ensure that when v_I is low (close to 0 V), the output v_O will be high (close to V_{DD}), and vice versa.

13.1.2 The Voltage-Transfer Characteristic (VTC)

To quantify the operation of the inverter, we utilize its voltage-transfer characteristic (VTC). We have already introduced the concept of the VTC and utilized it to characterize the operation of basic MOSFET amplifiers in Section 5.4.2. Figure 13.2 shows such a circuit, together with its VTC. Observe that the circuit in fact implements the inverter function: For a logic-0 input, v_I is close to 0 V and specifically lower than the MOSFET threshold voltage V_{tn}, the transistor will be off, $i_D = 0$, and $v_O = V_{DD}$, which is a logic 1. For a logic-1 input, $v_I = V_{DD}$, the transistor will be conducting and operating in the triode region (at point D on the VTC), and the output voltage will be low (logic 0).

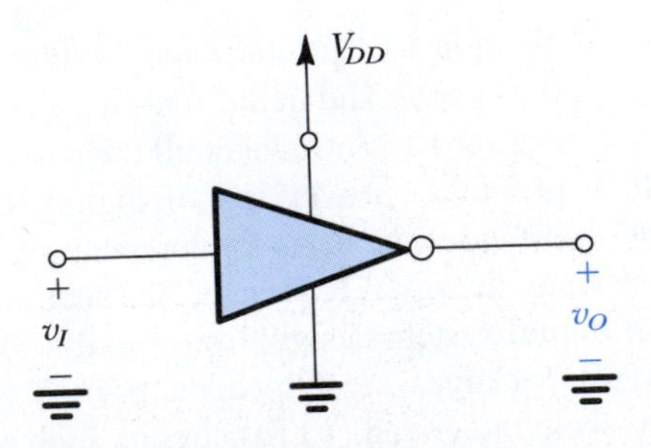

Figure 13.1 A logic inverter operating from a dc supply V_{DD}.

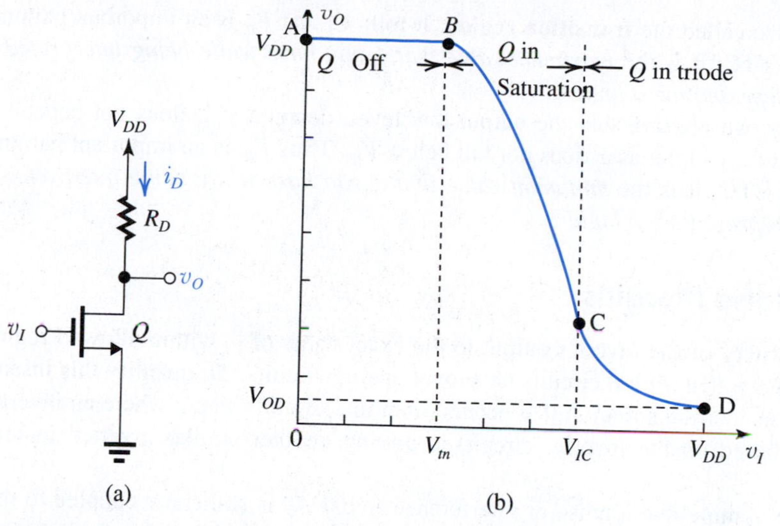

Figure 13.2 The simple resistively loaded MOS amplifier can be used as a logic inverter when operated in cut-off ($v_I < V_{tn}$) and in triode ($v_I > V_{IC}$). The output high level is V_{DD} and the low level is V_{OD}.

Thus to use this amplifier as a logic inverter, we utilize its extreme regions of operation. This is exactly the opposite to its use as a signal amplifier, where it would be biased at the middle of the transfer characteristic segment BC and the signal kept small enough to restrict operation to a short, almost linear, segment of the transfer curve. Digital applications, on the other hand, make use of the gross nonlinearity exhibited by the VTC.

With these observations in mind, we show in Fig. 13.3 a possible VTC of a logic inverter. For simplicity, we are using three straight lines to approximate the VTC, which is usually a nonlinear curve such as that in Fig. 13.2. Observe that the output high level, denoted V_{OH}, does not depend on the exact value of v_I as long as v_I does not exceed the value labeled V_{IL}; when v_I exceeds V_{IL}, the output decreases and the inverter enters its amplifier region of

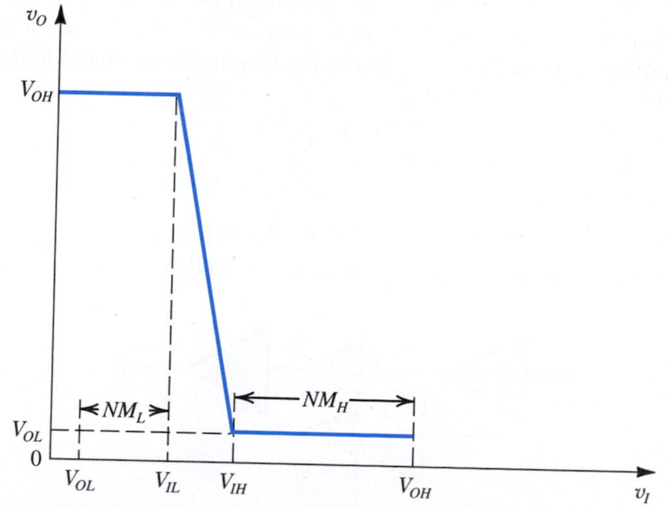

Figure 13.3 Voltage transfer characteristic of an inverter. The VTC is approximated by three straight-line segments. Note the four parameters of the VTC (V_{OH}, V_{OL}, V_{IL}, and V_{IH}) and their use in determining the noise margins (NM_H and NM_L).

operation, also called the **transition region**. It follows that V_{IL} is an important parameter of the inverter VTC: It is the *maximum value that* v_I *can have while being interpreted by the inverter as representing a logic 0.*

Similarly, we observe that the output low level, denoted V_{OL}, does not depend on the exact value of v_I as long as v_I does not fall below V_{IH}. Thus V_{IH} is an important parameter of the inverter VTC: It is the *minimum value that* v_I *can have while being interpreted by the inverter as representing a logic 1.*

13.1.3 Noise Margins

The insensitivity of the inverter output to the exact value of v_I within allowed regions is a great advantage that digital circuits have over analog circuits. To quantify this insensitivity property, consider the situation that occurs often in a digital system where an inverter (or a logic gate based on the inverter circuit) is driving another similar inverter, as shown in Fig. 13.4

Here we assume that a noise or interference signal v_N is somehow coupled to the interconnection between the output of inverter G_1 and the input of inverter G_2 with the result that the input of G_2 becomes

$$v_{I2} = v_{O1} + v_N \tag{13.1}$$

where the noise voltage v_N can be either positive or negative. Now consider the case $v_{O1} = V_{OL}$; that is, inverter G_2 is driven by a logic-0 signal. Reference to Fig. 13.3 indicates that in this case G_2 will continue to function properly as long as its input v_{I2} does not exceed V_{IL}. Equation (13.1) then indicates that v_N can be as high as $V_{IL}-V_{OL}$ while G_2 continues to function properly. Thus, we can say that inverter G_2 has a **noise margin for low input**, NM_L, of

$$NM_L = V_{IL} - V_{OL} \tag{13.2}$$

Similarly, if $v_{O1} = V_{OH}$, the driven inverter G_2 will continue to see a high input as long as v_{I2} does not fall below V_{IH}. Thus, in the high-input state, inverter G_2 can tolerate a negative v_N of magnitude as high as $V_{OH} - V_{IH}$. We can thus state that G_2 has a **high-input noise margin**, NM_H, of

$$NM_H = V_{OH} - V_{IH} \tag{13.3}$$

In summary, four parameters, V_{OH}, V_{OL}, V_{IH}, and V_{IL}, define the VTC of an inverter and determine its noise margins, which in turn measure the ability of the inverter to tolerate.

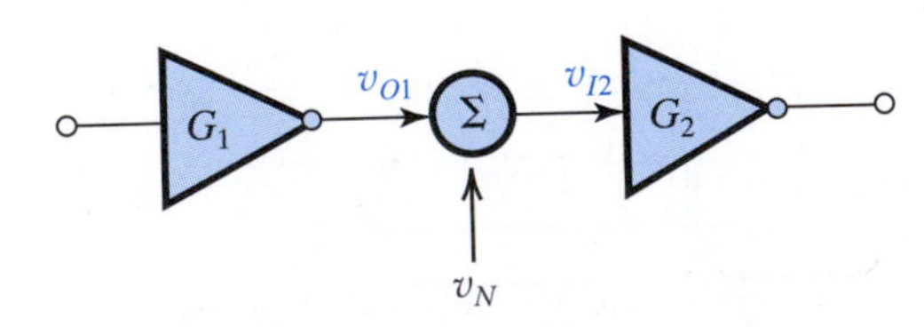

Figure 13.4 Noise voltage v_N is coupled to the interconnection between the output of inverter G_1 and the input of inverter G_2.

Table 13.1 Important Parameters of the VTC of the Logic Inverter (Refer to Fig. 13.3)

V_{OL}:	Output low level
V_{OH}:	Output high level
V_{IL}:	Maximum value of input interpreted by the inverter as a logic 0
V_{IH}:	Minimum value of input interpreted by the inverter as a logic 1
NM_L:	Noise margin for low input = $V_{IL} - V_{OL}$
NM_H:	Noise margin for high input = $V_{OH} - V_{IH}$

variations in the input signal levels. In this regard, observe that changes in the input signal level within the noise margins are *rejected* by the inverter. Thus noise is not allowed to propagate further through the system, a definite advantage of digital over analog circuits. Alternatively, we can think of the inverter as *restoring* the signal levels to standard values (V_{OL} and V_{OH}) even when it is presented with corrupted input signal levels (within the noise margins). As a summary, useful for future reference, we present a listing and definitions of the important parameters of the inverter VTC in Table 13.1.

The formal definitions of the threshold voltages V_{IL} and V_{IH} are given in Fig. 13.5. Observe that V_{IL} and V_{IH} are defined as the VTC points at which the slope is −1 V/V. As v_I exceeds V_{IL}, the magnitude of the inverter gain increases and the VTC enters its transition region. Similarly, as v_I falls below V_{IH}, the inverter enters the transition region and the magnitude of the gain increases. Finally, note that Fig. 13.5 shows the definition of another important point on the VTC; this is point M at which $v_O = v_I$. Point M is loosely considered to be the midpoint of the VTC and thus the point at which the *inverter switches from one state to the other*. Point M plays an important role in the definition of the time delay of the inverter, as we shall see shortly.

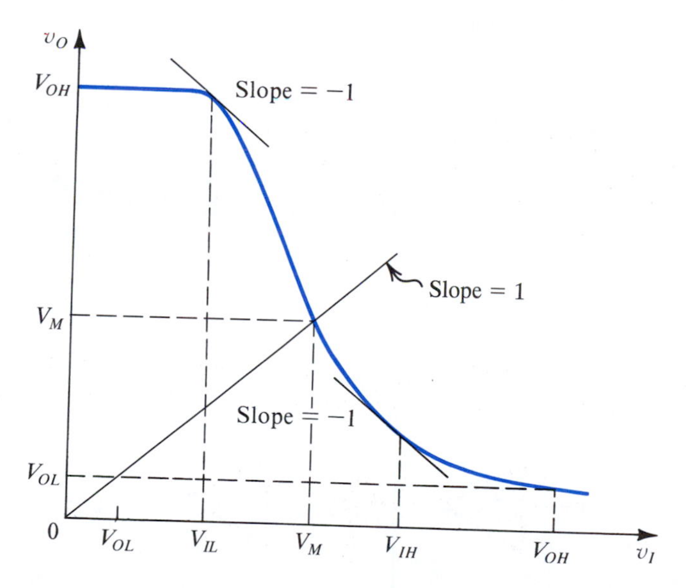

Figure 13.5 Typical voltage transfer characteristic (VTC) of a logic inverter, illustrating the definition of the critical points.

13.1.4 The Ideal VTC

The question naturally arises as to what constitutes an ideal VTC for an inverter. The answer follows directly from the preceding discussion: An ideal VTC is one that maximizes the output signal swing and the noise margins. For an inverter operated from a power supply V_{DD}, maximum signal swing is obtained when

$$V_{OH} = V_{DD}$$

and

$$V_{OL} = 0$$

To obtain maximum noise margins, we first arrange for the transition region to be made as narrow as possible and ideally of zero width. Then, the two noise margins are equalized by arranging for the transition from high to low to occur at the midpoint of the power supply, that is, at $V_{DD}/2$. The result is the VTC shown in Fig. 13.6, for which

$$V_{IL} = V_{IH} = V_M = V_{DD}/2$$

Observe that the sharp transition at $V_{DD}/2$ indicates that if the inverter were to be used as an amplifier, its gain would be infinite. Again, we point out that while the analog designer's interest would be focused on the transition region of the VTC, the digital designer would prefer the transition region to be as narrow as possible, as is the case in the ideal VTC of Fig. 13.6. Finally, we will see in Section 13.2 that inverters implemented using CMOS technology come very close to realizing the ideal VTC

13.1.5 Inverter Implementation

Inverters are implemented using transistors (Chapters 5 and 6) operating as **voltage-controlled switches**. The simplest inverter implementation. is shown in Fig. 13.7(a). The switch is

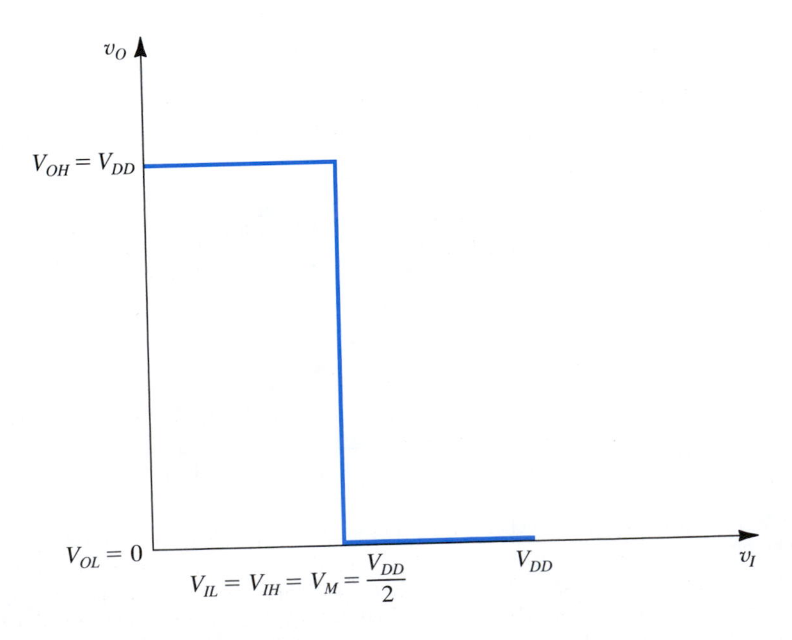

Figure 13.6 The VTC of an ideal inverter.

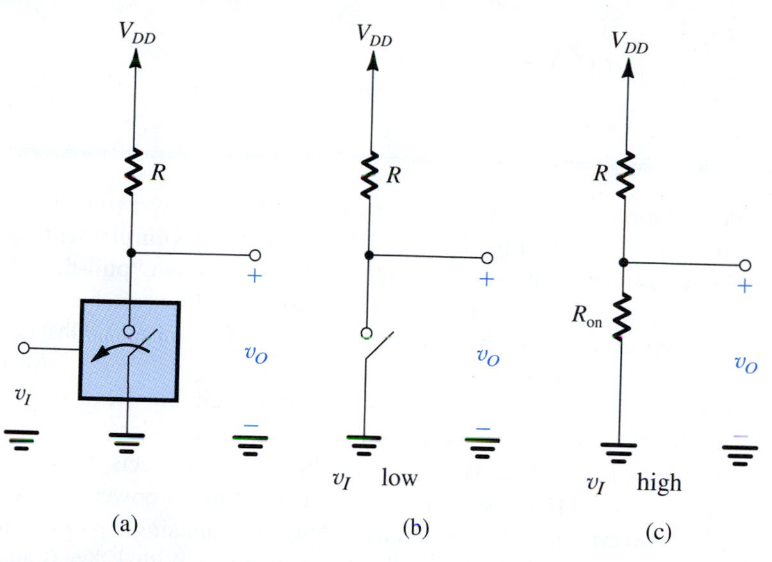

Figure 13.7 **(a)** The simplest implementation of a logic inverter using a voltage-controlled switch; **(b)** equivalent circuit when v_I is low; **(c)** equivalent circuit when v_I is high. Note that the switch is assumed to close when v_I is high.

controlled by the inverter input voltage v_I: When v_I is low, the switch will be open and $v_O = V_{DD}$, since no current flows through R. When v_I is high, the switch will be closed and, assuming an ideal switch, v_O will be 0.

Transistor switches, however, as we know from Chapters 5 and 6, are not perfect. Although their **off resistances** are very high and thus an open switch closely approximates an open circuit, the "on" switch has a finite closure or **"on" resistance**, R_{on}. The result is that when v_I is high, the inverter has the equivalent circuit shown in Fig. 13.7(c), from which V_{OL} can be found.[1]

$$V_{OL} = V_{DD}\frac{R_{on}}{R + R_{on}}$$

We observe that the circuit in Fig. 13.2(a) is a direct implementation of the inverter in Fig. 13.7. In this case, R_{on} is equal to r_{DS} of the MOSFET evaluated at its operating point in the triode region with $V_{GS} = V_{DD}$.

EXERCISE

D13.1 Design the inverter in Fig. 13.2(a) to provide $V_{OL} = 0.1$ V and to draw a supply current of 50 μA in the low-output state. Let the transistor be specified to have $V_t = 0.5$ V, $\mu_n C_{ox} = 125$ μA/V^2, and $\lambda = 0$. The power supply $V_{DD} = 2.5$ V. Specify the required values of W/L and R_D. How much power is drawn from V_{DD} when the switch is open? Closed?
Hint: Recall that for small v_{DS},

[1] If a BJT is used to implement the switch in Fig. 13.7(a), its equivalent circuit in the closed position includes in addition to the resistance $R_{on} = R_{CEsat}$, an offset voltage of about 50 mV to 100 mV (see Fig. 6.19c). We shall not pursue this subject any further here, since the relatively long delay time needed to turn off a saturated BJT has caused the use of BJT switches operated in saturation to all but disappear from the digital IC world.

$$r_{DS} \simeq 1/\left[(\mu_n C_{ox})\left(\frac{W}{L}\right)(V_{GS}-V_t)\right]$$

Ans. 2; 48 kΩ; 0; 125 μW

More elaborate implementations of the logic inverter exist, and we show two of these in Fig. 13.8(a) and 13.9. The circuit in Fig. 13.8(a) utilizes a pair of **complementary switches**, the **"pull-up" (PU) switch** connects the output node to V_{DD}, and the **"pull-down" (PD) switch** connects the output node to ground. When v_I is low, the PU switch will be closed and the PD switch open, resulting in the equivalent circuit of Fig. 13.8(b). Observe that in this case R_{on} of PU connects the output to V_{DD}, thus establishing $V_{OH} = V_{DD}$. Also observe that no current flows and thus no power is dissipated in the circuit. Next, if v_I is raised to the logic-1 level, the PU switch will open while the PD switch will close, resulting in the equivalent circuit shown in Fig. 13.8(c). Here R_{on} of the PD switch connects the output to ground, thus establishing $V_{OL} = 0$. Here again no current flows, and no power is dissipated. The superiority of this inverter implementation over that using the single pull-down switch and a resistor (known as a **pull-up resistor**) should be obvious: With $V_{OL} = 0$ and $V_{OH} = V_{DD}$, the signal swing is at its maximum possible, and the power dissipation is zero in both states. This circuit constitutes the basis of the CMOS inverter that we will study in Section 13.3.

Finally, consider the inverter implementation of Fig. 13.9. Here a double-throw switch is used to steer the constant current I_{EE} into one of two resistors connected to the positive supply V_{CC}. The reader is urged to show that if a high v_I results in the switch being connected to R_{C1}, then a logic inversion function is realized at v_{O1}. Note that the output voltage is independent of the switch resistance. This *current-steering* or *current-mode* logic arrangement is the basis of the fastest available digital logic circuits, called emitter-coupled logic (ECL), which we shall study in Section 14.4. In fact, ECL is the only BJT logic-circuit type that is currently employed in new designs and the only one studied in this book.

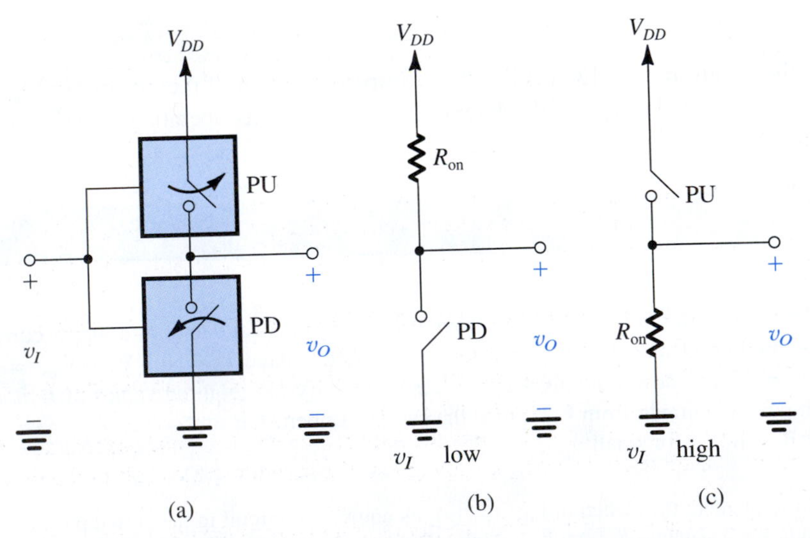

Figure 13.8 A more elaborate implementation of the logic inverter utilizing two complementary switches. This is the basis of the CMOS inverter that we shall study in Section 13.2.

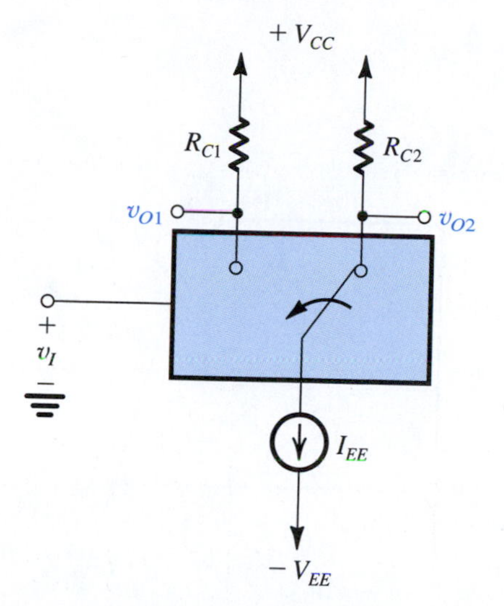

Figure 13.9 Another inverter implementation utilizing a double-throw switch to steer the constant current I_{EE} to R_{C1} (when v_I is high) or R_{C2} (when v_I is low). This is the basis of the emitter-coupled logic (ECL) studied in Chapter 14.

EXERCISE

13.2 For the current-steering circuit in Fig. 13.9, let $V_{CC} = 5$ V, $I_{EE} = 1$ mA, and $R_{C1} = R_{C2} = 2$ kΩ. What are the high and low logic levels obtained at the outputs?
Ans. $V_{OH} = 5$ V; $V_{OL} = 3$ V

Example 13.1 Resistively Loaded MOS Inverter

For the simple MOS inverter in Fig. 13.2(a):

(a) Derive expressions for V_{OH}, V_{OL}, V_{IL}, V_{IH}, and V_M. For simplicity, neglect channel-length modulation (i.e., assume $\lambda = 0$). Show that these inverter parameters can be expressed in terms of V_{DD}, V_t, and $(k_n R_D)$. The latter parameter has the dimension of V^{-1}; and to simplify the expressions, denote $k_n R_D \equiv 1/V_x$.

(b) Show that V_x can be used as a design parameter for the inverter circuit. In particular, find the value of V_x that results in $V_M = V_{DD}/2$.

(c) Find numerical values for all parameters and for the inverter noise margins for $V_{DD} = 1.8$ V, $V_t = 0.5$ V, and V_x set to the value found in (b).

(d) For $k_n' = 300\ \mu\text{A/V}^2$ and $W/L = 1.5$, find the required value of R_D and use it to determine the average power dissipated in the inverter, assuming that the inverter spends half of the time in each of its two states.

(e) Comment on the characteristics of this inverter circuit vis-à-vis the ideal characteristics as well as on its suitability for implementation in integrated-circuit form.

Example 13.1 *continued*

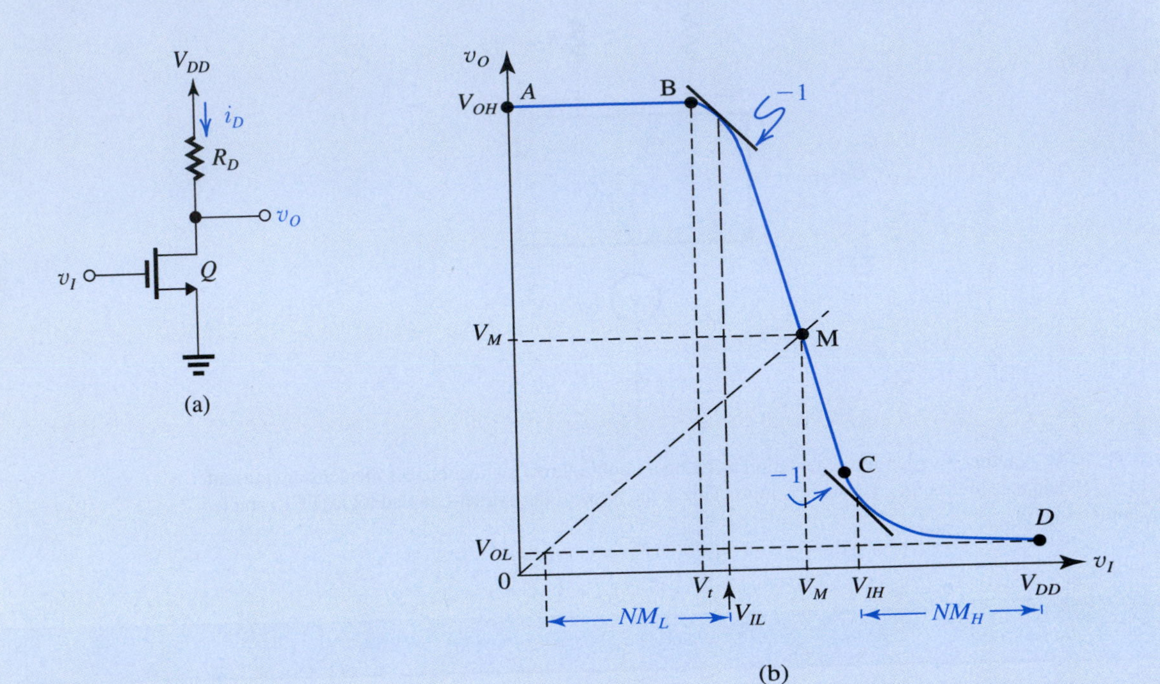

Figure 13.10 The resistively loaded MOS inverter and its VTC (Example 13.1).

Solution

(a) Refer to Fig. 13.10. For $v_I < V_t$, the MOSFET is off, $i_D = 0$, and $v_O = V_{DD}$. Thus

$$V_{OH} = V_{DD} \tag{13.4}$$

As v_I exceeds V_t, the MOSFET turns on and operates initially in the saturation region. Assuming $\lambda = 0$,

$$i_D = \frac{1}{2}k_n(v_I - V_t)^2 R_D$$

and

$$v_O = V_{DD} - R_D i_D = V_{DD} - \frac{1}{2}k_n R_D(v_I - V_t)^2$$

substituting $k_n R_D = 1/V_x$, the BC segment of the VTC is described by

$$v_O = V_{DD} - \frac{1}{2V_x}(v_I - V_t)^2 \tag{13.5}$$

To determine V_{IL}, we differentiate Eq. (13.5) and set $dv_O/dv_I = -1$,

$$\frac{dv_O}{dv_I} = -\frac{1}{V_x}(v_I - V_t)$$

$$-1 = -\frac{1}{V_x}(V_{IL} - V_t)$$

which results in

$$V_{IL} = V_t + V_x \tag{13.6}$$

To determine the coordinates of the midpoint M, we substitute $v_O = v_I = V_M$ in Eq. (13.5), thus obtaining

$$V_{DD} - V_M = \frac{1}{2V_x}(V_M - V_t)^2 \tag{13.7}$$

which can be solved to obtain

$$V_M = V_t + \sqrt{2(V_{DD} - V_t)V_x + V_x^2} - V_x \tag{13.8}$$

The boundary of the saturation-region segment BC, point C, is determined by substituting $v_O = v_I - V_t$ in Eq. (13.5) and solving for v_O to obtain

$$V_{OC} = \sqrt{2V_{DD}V_x + V_x^2} - V_x \tag{13.9}$$

and

$$V_{IC} = V_t + \sqrt{2V_{DD}V_x + V_x^2} - V_x \tag{13.10}$$

Beyond point C, the transistor operates in the triode region, thus

$$i_D = k_n\left[(v_I - V_t)v_O - \frac{1}{2}v_O^2\right]$$

and the output voltage is obtained as

$$v_O = V_{DD} - \frac{1}{V_x}\left[(v_I - V_t)v_O - \frac{1}{2}v_O^2\right] \tag{13.11}$$

which describes the segment CD of the VTC. To determine V_{IH}, we differentiate Eq. (13.11) and set $dv_O/dv_I = -1$:

$$\frac{dv_O}{dv_I} = -\left(\frac{1}{V_x}\right)\left[(v_I - V_t)\frac{dv_O}{dv_I} + v_O - v_O\,\frac{dv_O}{dv_I}\right]$$

$$-1 = -\frac{1}{V_x}[-(V_{IH} - V_t) + 2v_O]$$

which results in

$$V_{IH} - V_t = 2v_O - V_x \tag{13.12}$$

Substituting in Eq. (13.11) for v_I with the value of V_{IH} from Eq. (13.12) results in an equation in the value of v_O corresponding to $v_I = V_{IH}$, which can be solved to yield

$$v_O\big|_{v_I = V_{IH}} = 0.816\sqrt{V_{DD}V_x} \tag{13.13}$$

which can be substituted in Eq. (13.12) to obtain

$$V_{IH} = V_t + 1.63\sqrt{V_{DD}V_x} - V_x \tag{13.14}$$

Example 13.1 *continued*

To determine V_{OL} we substitute $v_I = V_{OH} = V_{DD}$ in Eq. (13.11):

$$V_{OL} = V_{DD} - \frac{1}{V_x}\left[(V_{DD} - V_t)V_{OL} - \frac{1}{2}V_{OL}^2\right] \quad (13.15)$$

Since we expect V_{OL} to be much smaller than $2(V_{DD} - V_t)$, we can approximate Eq. (13.15) as

$$V_{OL} \simeq V_{DD} - \frac{1}{V_x}(V_{DD} - V_t)V_{OL}$$

which results in

$$V_{OL} = \frac{V_{DD}}{1 + [(V_{DD} - V_t)/V_x]} \quad (13.16)$$

It is interesting to note that the value of V_{OL} can alternatively be found by noting that at point D, the MOSFET switch has a closure resistance r_{DS},

$$r_{DS} = \frac{1}{k_n(V_{DD} - V_t)} \quad (13.17)$$

and V_{OL} can be obtained from the voltage divider formed by R_D and r_{DS},

$$V_{OL} = V_{DD}\frac{r_{DS}}{R_D + r_{DS}} = \frac{V_{DD}}{1 + R_D/r_{DS}} \quad (13.18)$$

Substituting for r_{DS} from Eq. (13.17) gives an expression for V_{OL} identical to that in Eq. (13.16).

(b) We observe that all the inverter parameters derived above are functions of V_{DD}, V_t, and V_x only. Since V_{DD} and V_t are determined by the process technology, the only design parameter available is $V_x \equiv 1/k_nR_D$. To place V_M at half the supply voltage V_{DD}, we substitute $V_M = V_{DD}/2$ in Eq. (13.7) to obtain the value V_x must have as

$$V_x\big|_{V_M = V_{DD}/2} = \frac{(V_{DD}/2 - V_t)^2}{V_{DD}} \quad (13.19)$$

(c) For $V_{DD} = 1.8$ V and $V_t = 0.5$, we use Eq. (13.19) to obtain

$$V_x\big|_{V_M = 0.9\text{ V}} = \frac{(1.8/2 - 0.5)^2}{1.8} = 0.089 \text{ V}$$

From Eq. (13.4): $V_{OH} = 1.8$ V

From Eq. (13.16): $V_{OL} = 0.12$ V

From Eq. (13.6): $V_{IL} = 0.59$ V

From Eq. (13.14): $V_{IH} = 1.06$ V

$$NM_L = V_{IL} - V_{OL} = 0.47 \text{ V}$$

$$NM_H = V_{OH} - V_{IH} = 0.74 \text{ V}$$

(d) To determine R_D, we use

$$k_nR_D = \frac{1}{V_x} = \frac{1}{0.089} = 11.24$$

Thus,

$$R_D = \frac{11.24}{k_n'(W/L)} = \frac{11.24}{300 \times 10^{-6} \times 1.5} = 25 \text{ k}\Omega$$

The inverter dissipates power only when the output is low, in which case the current drawn from the supply is

$$I_{DD} = \frac{V_{DD} - V_{OL}}{R_D} = \frac{1.8 - 0.12}{25 \text{ k}\Omega} = 67 \text{ }\mu\text{A}$$

and the power drawn from the supply during the low-output interval is

$$P_D = V_{DD} I_{DD} = 1.8 \times 67 = 121 \text{ }\mu\text{W}$$

Since the inverter spends half of the time in this state,

$$P_{D\text{average}} = \frac{1}{2} P_D = 60.5 \text{ }\mu\text{W}$$

(e) We now can make a few comments on the characteristics of this inverter circuit in comparison to the ideal characteristics:

1. The output signal swing, though not equal to the full power supply, is reasonably good: $V_{OH} = 1.8$ V, $V_{OL} = 0.12$ V.
2. The noise margins, though of reasonable values, are far from the optimum value of $V_{DD}/2$. This is particularly the case for NM_L.
3. Most seriously, the gate dissipates a relatively large amount of power. To appreciate this point, consider an IC chip with a million inverters (a small number by today's standards): Its power dissipation will be 61 W. This is too large, especially given that this is "static power," unrelated to the switching activity of the gates (more on this later).

We consider this inverter implementation to be entirely unsuitable for IC fabrication because each inverter requires a load resistance of 25 kΩ, a value that needs a large chip area (see Appendix A). To overcome this problem, we investigate in Example 13.2 the replacement of the passive resistance R_D with an NMOS transistor.

EXERCISES

D13.3 In an attempt to reduce the required value of R_D, to 10 kΩ, the designer of the inverter in Example 13.1 decides to keep the parameter V_x unchanged but increases *W*/*L*. What is the new value required for *W*/*L*? Do the noise margins change? What does the power dissipation become?
Ans. 3.75; no; 151 μW

D13.4 In an attempt to reduce the required value of R_D to 10 kΩ, the designer of the inverter in Examples 13.1 decides to change V_x while keeping *W*/*L* unchanged. What new value of V_x is needed? What do the noise margins become? What does the power dissipation become?
Ans. $V_x = 0.22$ V; $NM_L = 0.46$ V, $NM_H = 0.49$ V; 139 μW

Example 13.2 The Saturated NMOS-Load Inverter

To overcome the problem associated with the need for a large resistance R_D in the circuit of Fig. 13.10(a), studied in Example 13.1, R_D can be replaced by a MOSFET. One such possibility is the circuit shown in Fig. 13.11(a), where the load is an NMOS transistor Q_2 operated in the saturation region (by connecting its drain to its gate). Although not shown on the diagram, the body terminal of Q_2 is connected to the lowest-voltage node, which is ground.

(a) Neglecting the body effect in Q_2 and assuming $\lambda_1 = \lambda_2 = 0$, determine the inverter parameters V_{OH}, V_{OL}, V_{IL}, V_{IH}, and V_M. Express the results in terms of V_{DD}, V_t (where $V_{t1} = V_{t2} = V_t$), and $k_r \equiv \sqrt{k_{n1}/k_{n2}}$.

(b) For $V_{DD} = 1.8$ V, $V_t = 0.5$ V, $(W/L)_1 = 5$, and $(W/L)_2 = \frac{1}{5}$, find numerical values for all parameters and for the noise margins.

(c) If $k_n' = 300\ \mu A/V^2$, find the average power dissipated in the inverter, assuming that it spends half the time in each of its two states.

(d) Qualitatively describe how the body effect in Q_2 affects the noise margins.

(e) Comment on the characteristics of this inverter implementation vis-à-vis the ideal characteristics. How suitable is this circuit for implementation in IC form?

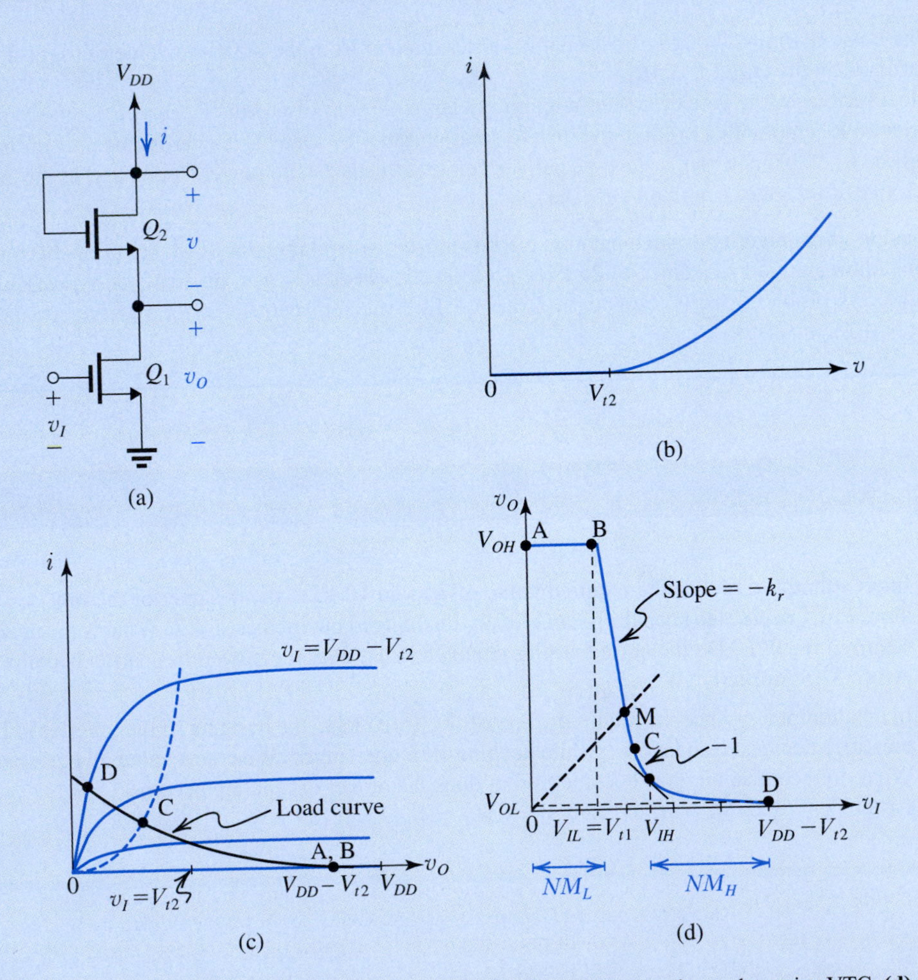

Figure 13.11 (a) Enhancement-load MOS inverter; (b) load curve; (c) construction to determine VTC; (d) the VTC.

Solution

(a) The inverter VTC can be determined graphically by superimposing the load curve, which is the i–v characteristic of the diode-connected transistor Q_2, shown in Fig. 13.11(b), on the i–v characteristics of Q_1. As we have done in the graphical analysis of MOSFET circuits in Section 5.4, we shift the load curve horizontally by V_{DD} and flip it around the vertical axis, as shown in Fig. 13.11(c). The resulting VTC is shown in Fig. 13.11(d).

For $v_I < V_{t1}$, Q_1 will be off, which forces the current in Q_2 to be zero. Transistor Q_2, although it will be conducting a zero current, will have a voltage drop of V_{t2}. This is a result of its $i-v$ characteristic shown in Fig. 13.11(b). Thus the output voltage v_O will not reach V_{DD} but will be at $V_{DD} - V_{t2}$, that is,[2]

$$V_{OH} = V_{DD} - V_t \tag{13.20}$$

As v_I exceeds V_{t1}, Q_1 turns on and initially operates in saturation, thus

$$i_{D1} = \frac{1}{2}k_{n1}(v_I - V_{t1})^2$$

Since Q_2 operates in saturation at all times,

$$i_{D2} = \frac{1}{2}k_{n2}(V_{DD} - v_O - V_{t2})^2$$

Equating i_{D1} and i_{D2} and substituting $V_{t1} = V_{t2} = V_t$, and $\sqrt{k_{n1}/k_{n2}} = k_r$, gives

$$v_O = V_{DD} + (k_r - 1)V_t - k_r v_I \tag{13.21}$$

which is the equation for segment BC of the VTC in Fig. 13.11(d). It is interesting to observe that the relationship between v_O and v_I is linear and that the slope of this straight line is $-k_r$.

Since the slope of the VTC changes from zero to $-k_r$ at point B, it is reasonable to consider point B to be the determinant of V_{IL}; thus,

$$V_{IL} = V_{t1} = V_t \tag{13.22}$$

To obtain V_M we substitute $v_I = v_O = V_M$ in Eq. (13.21); thus,

$$V_M = \frac{V_{DD} + (k_r - 1)V_t}{k_r + 1} \tag{13.23}$$

We next determine the coordinates of point C at which Q_1 enters the triode region by substituting in Eq. (13.21) $v_O = v_I - V_t$. The result is

$$V_{IC} = \frac{V_{DD} + k_r V_t}{k_r + 1} \tag{13.24}$$

[2]To see this point more clearly, consider the usual situation of a capacitance C_L between the output node of the inverter and ground. Assume that initially v_I was high and v_O was low. Now let v_I go low. Q_1 cuts off, and Q_2 provides a current that charges C_L up. As v_O increases, the current provided by Q_2 decreases until v_O reaches $V_{DD} - V_{t2}$, at which point the current supplied by Q_2 reaches zero. Thus the charging process terminates and v_O stabilizes at $V_{DD} - V_{t2}$.

Example 13.2 *continued*

and

$$V_{OC} = \frac{V_{DD} - V_t}{k_r + 1} \tag{13.25}$$

Comparing Eqs. (13.24) and (13.23), we make the comforting observation that $V_{IC} > V_M$, confirming our implicit assumption that M lies on the linear segment of the VTC.

For $v_I > V_{IC}$, Q_1 operates in the triode region; thus,

$$i_{D1} = k_{n1}\left[(v_I - V_{t1})v_O - \frac{1}{2}v_O^2\right]$$

Meanwhile, Q_2 still operates in saturation. Equating their currents results in

$$2k_r^2\left[(v_I - V_t)v_O - \frac{1}{2}v_O^2\right] = (V_{DD} - V_t - v_O)^2 \tag{13.26}$$

Although this equation can be used to determine V_{IH}, the effort involved to do this symbolically is too great. We will instead find V_{IH} numerically; V_{OL}, however, can be determined by substituting in Eq. (13.26) $v_I = V_{OH} = V_{DD} - V_t$ and $v_O = V_{OL}$,

$$2k_r^2\left[(V_{DD} - 2V_t)V_{OL} - \frac{1}{2}V_{OL}^2\right] = (V_{DD} - V_t - V_{OL})^2 \tag{13.27}$$

Since we expect V_{OL} to be much smaller than $2(V_{DD} - 2V_t)$ and $(V_{DD} - V_t)$, we can approximate Eq. (13.27) as follows:

$$2k_r^2(V_{DD} - 2V_t)V_{OL} \simeq (V_{DD} - V_t)^2$$

Thus,

$$V_{OL} \simeq \frac{(V_{DD} - V_t)^2}{2k_r^2(V_{DD} - 2V_t)} \tag{13.28}$$

We observe that all the inverter parameters are functions of three quantities only: V_{DD}, V_t, and k_r. Since the first two are determined by the process technology, the only design parameter is k_r, which determines the steepness of the transition region.

(b) Given $V_{DD} = 1.8$ V, $V_t = 0.5$ V, $(W/L)_1 = 5$, and $(W/L)_2 = \frac{1}{5}$, we first determine k_r as

$$k_r = \sqrt{\frac{k_{n1}}{k_{n2}}} = \sqrt{\frac{(W/L)_1}{(W/L)_2}} = \sqrt{\frac{5}{1/5}} = 5$$

From Eq. (13.20): $V_{OH} = 1.3$ V

From Eq. (13.28): $V_{OL} = 0.04$ V

From Eq. (13.22): $V_{IL} = 0.5$ V

From Eq. (13.23): $V_M = 0.63$ V

To determine V_{IH} we utilize Eq. (13.26) together with setting $dv_O/dv_I = -1$. The result is

$$V_{IH} = 0.75 \text{ V}$$

Thus,

$$NM_L = V_{IL} - V_{OL} = 0.5 - 0.04 = 0.46 \text{ V}$$

$$NM_H = V_{OH} - V_{IH} = 1.3 - 0.75 = 0.55 \text{ V}$$

(c) The inverter dissipates power only when $v_O = V_{OL}$. In this case, the current drawn from the supply is

$$I_{DD} = i_{D2} = \frac{1}{2}k_{n2}(V_{DD} - V_{OL} - V_t)^2$$

Thus,

$$I_{DD} = \frac{1}{2} \times 300 \times \frac{1}{5} \times (1.8 - 0.04 - 0.5)^2$$

$$= 47.6 \text{ µA}$$

and,

$$P_D = V_{DD}I_{DD} = 1.8 \times 47.6 = 85.7 \text{ µW}$$

Since the inverter is in the low-output state for half the time,

$$P_{Daverage} = \frac{1}{2} \times 85.7 = 42.9 \text{ µW}$$

(d) Since the body of Q_2 is connected to ground, its source-to-body voltage V_{SB} is

$$V_{SB} = v_O$$

Now, since the threshold voltage is given by

$$V_{t2} = V_{t0} + \gamma[\sqrt{V_{SB} + 2\phi_f} - \sqrt{2\phi_f}] \qquad (13.29)$$

we see that V_{t2} will increase with v_O. This is of immense concern, since V_{t2} will be at its largest value for $v_O = V_{OH} = V_{DD} - V_{t2}$. Thus, V_{OH} will be lower than the value calculated above. This reduces the output signal swing and NM_H.

(e) We now can make the following comments on the characteristics of this inverter implementation:

1. The fact that V_{OH} is lower than V_{DD} by V_{t2} and that V_{t2} can be large because of the body effect imposes a major disadvantage on this NMOS-load inverter.
2. The noise margins are much lower than the ideal values of $V_{DD}/2$. Also, V_M is far from the power-supply midpoint.
3. The sharpness of the transition of the VTC increases with the value of k_r. Increasing k_r, however, has the effect of increasing the silicon area (see Exercise 13.6).
4. Like the resistively-loaded MOS inverter considered in Example 13.1, the NMOS-loaded inverter dissipates a large amount of power.

Since the circuit utilizes NMOS transistors exclusively, it is certainly suitable for implementation in IC form. As we will discuss shortly, all-NMOS technology was at one time (1970s) the technology of choice for the implementation of microprocessor chips. Its high power dissipation, however, has caused its demise in favor of CMOS technology.

EXERCISES

13.5 Repeat part (b) of Example 13.3 for the case $(W/L)_1 = 3$ and $(W/L)_2 = \frac{1}{3}$. Specifically, find the values of V_{OH}, V_{OL}, V_{IL}, V_{IH}, V_M, NM_H, and NM_L.
Ans. 1.3 V; 0.12 V; 0.5 V; 0.87 V; 0.7 V; 0.43 V; 0.38 V

13.6 Consider the inverter in Fig. 13.11(a) with $(W/L)_1 = k_r$ and $(W/L)_2 = 1/k_r$. Show that if the minimum dimension (i.e. length or width) of each of the two transistors is denoted d, the inverter silicon area is $2k_r d^2$.

13.1.6 Power Dissipation

Digital systems are implemented using very large numbers of logic gates. For space and other economic considerations, it is desirable to implement the system with as few integrated-circuit (IC) chips as possible. It follows that one must pack as many logic gates as possible on an IC chip. At present, one million gates or more can be fabricated on a single IC chip in what is known as **very-large-scale integration (VLSI)**. To keep the power dissipated in the chip to acceptable limits (imposed by thermal considerations), the power dissipation per gate must be kept to a minimum. Indeed, a very important performance measure of the logic inverter is the power it dissipates.

The inverter of Fig. 13.7 dissipates no power when v_I is low and the switch is open. In the other state, however, the power dissipation is approximately V_{DD}^2/R and can be substantial, as we have seen in Examples 13.1 and 13.2. This power dissipation occurs even if the inverter is not switching and is thus known as **static power dissipation**.

The inverter of Fig. 13.8 exhibits no static power dissipation, a definite advantage. Unfortunately, however, another component of power dissipation arises when a capacitance exists between the output node of the inverter and ground. This is always the case, for the devices that implement the switches have internal capacitances, the wires that connect the inverter output to other circuits have capacitance, and, of course, there is the input capacitance of whatever circuit the inverter is driving. Now, as the inverter is switched from one state to another, current must flow through the switch(es) to charge (and discharge) the load capacitance. These currents give rise to power dissipation in the switches, called **dynamic power dissipation**.

An expression for the dynamic power dissipation of the inverter of Fig. 13.8 can be derived as follows. Consider first the situation when v_I goes low. The pull-down switch P_D

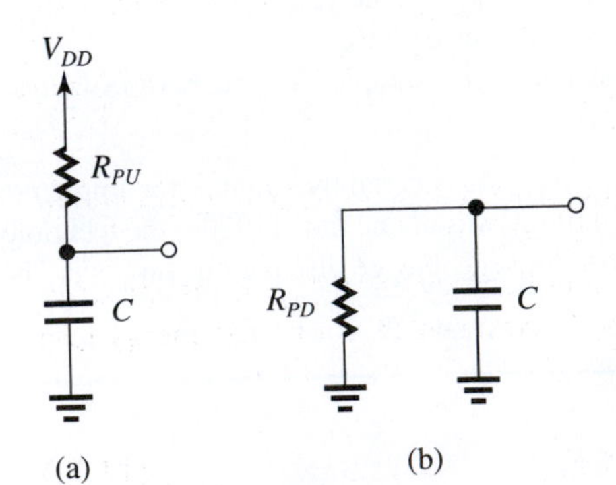

Figure 13.12 Equivalent circuits for calculating the dynamic power dissipation of the inverter in Figure 13.8: **(a)** When v_I is low; **(b)** When v_I is high.

turns off and the pull-up switch P_U turns on. In this state, the inverter can be represented by the equivalent circuit shown in Fig. 13.12(a). Capacitor C will charge through the on-resistance of the pull-up switch, and the voltage across C will increase from 0 to V_{DD}. Denoting by $i_D(t)$ the charging current supplied by V_{DD}, we can write for the instantaneous power drawn from V_{DD} the expression

$$p_{DD}(t) = V_{DD} i_D(t)$$

The energy delivered by the power supply to charge the capacitor can be determined by integrating $p_{DD}(t)$ over the charging interval T_c,

$$E_{DD} = \int_0^{T_c} V_{DD} i_D(t)\,dt$$

$$= V_{DD} \int_0^{T_c} i_D(t)\,dt$$

$$= V_{DD} Q$$

where Q is the charge delivered to the capacitor during the charging interval. Since the initial charge on C was zero,

$$Q = CV_{DD}$$

Thus,

$$E_{DD} = CV_{DD}^2 \tag{13.30}$$

Since at the end of the charging process the energy stored on the capacitor is

$$E_{\text{stored}} = \frac{1}{2} CV_{DD}^2 \tag{13.31}$$

we can find the energy dissipated in the pull-up switch as

$$E_{\text{dissipated}} = E_{DD} - E_{\text{stored}} = \frac{1}{2} CV_{DD}^2 \tag{13.32}$$

This energy is dissipated in the on-resistance of switch P_U and is converted to heat.

Next consider the situation when v_I goes high. The pull-up switch P_U turns off and the pull-down switch P_D turns on. The equivalent circuit in this case is that shown in Fig. 13.12(b). Capacitor C is discharged through the on-resistance of the pull-down switch, and its voltage changes from V_{DD} to 0. At the end of the discharge interval, there will be no energy left on the capacitor. Thus all of the energy initially stored on the capacitor, $\frac{1}{2}CV_{DD}^2$, will be dissipated in the pull-down switch,

$$E_{\text{dissipated}} = \frac{1}{2} CV_{DD}^2 \tag{13.33}$$

This amount of energy is dissipated in the on-resistance of switch P_D and is converted to heat.

Thus in each cycle of inverter switching, an amount of energy of $\frac{1}{2}CV_{DD}^2$ is dissipated in the pull-up switch and $\frac{1}{2}CV_{DD}^2$ is dissipated in the pull-down switch, for a total energy loss per cycle of

$$E_{\text{dissipated}}/\text{cycle} = CV_{DD}^2 \tag{13.34}$$

If the inverter is switched at a frequency of f Hz, the dynamic power dissipation of the inverter will be

$$P_{\text{dyn}} = fCV_{DD}^2 \tag{13.35}$$

This is a general expression that does not depend on the inverter circuit details or the values of the on-resistance of the switches.

The expression in Eq. (13.35) indicates that to minimize the dynamic power dissipation, one must strive to reduce the value of C. However, in many cases C is largely determined by the transistors of the inverter itself and cannot be substantially reduced. Another important factor in determining the dynamic power dissipation is the power-supply voltage V_{DD}. Reducing V_{DD}, reduces P_{dyn} significantly. This has been a major motivating factor behind the reduction of V_{DD} with every technology generation (see Table 7.A.1). Thus, while the 0.5-μm CMOS process utilized a 5-V power supply, the power-supply voltage used with the 0.13-μm process is only 1.2 V.

Finally, since P_{dyn} is proportional to the operating frequency f, one may be tempted to reduce P_{dyn} by reducing f. This, however, is not a viable proposition in light of the desire to operate digital systems at increasingly higher speeds. This point will be discussed next.

EXERCISES

13.7 Find the dynamic power dissipation of an inverter operated from a 1.8-V supply and having a load capacitance of 100 fF. Let the inverter be switched at 100 MHz.
Ans. 32.4 μW

13.8 A particular inverter circuit initially designed in a 0.5-μm process is fabricated in a 0.13-μm process. Assuming that the capacitance C will scale down in proportion to the minimum feature size (more on this later) and that the power supply will be reduced from 5 V to 1.2 V, by what factor do you expect the dynamic power dissipation to decrease? Assume that the switching frequency f remains unchanged.
Ans. 66.8

13.1.7 Propagation Delay

A very important measure of the performance of a digital system, such as a computer, is the maximum speed at which it is capable of operating. Although many factors come into play in determining the operating speed of a system, a core factor is the speed of operation of the basic logic inverter utilized in its implementation. This in turn is characterized by the time it takes the inverter to respond to a change at its input. To be more precise, consider an inverter fed with the ideal pulse shown in Fig. 13.13(a). The resulting output signal of the inverter is shown in Fig. 13.13(b). We make the following two observations.

1. The output signal is no longer an ideal pulse. Rather, it has rounded edges; that is, the pulse takes some time to fall to its low value and to rise to its high value. We speak of this as the pulse having finite fall and rise times. We will provide a precise definition of these shortly.
2. There is a time delay between each edge of the input pulse and the corresponding change in the output of the inverter. If we define the "switching point" of the output as the time at

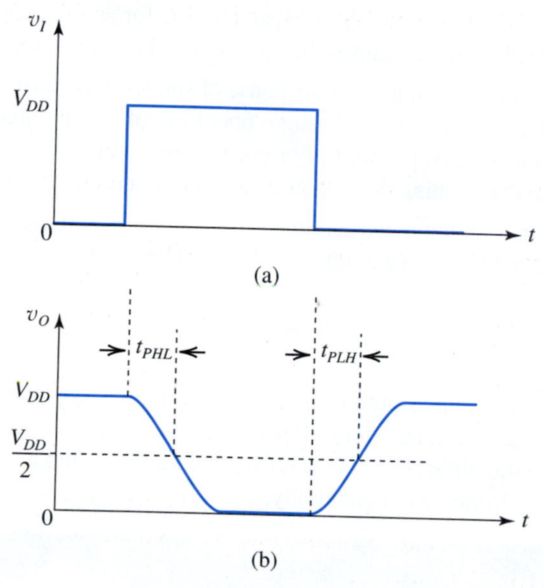

Figure 13.13 An inverter fed with the ideal pulse in (**a**) provides at its output the pulse in (**b**). Two delay times are defined as indicated.

which the output pulse passes through the half-point of its excursion, then we can define the propagation delays of the inverter as indicated in Fig. 13.13(b). Note that there are two propagation delays, which are not necessarily equal: the propagation delay for the output going from high to low, t_{PHL}, and the propagation delay for the output going from low to high, t_{PLH}. The inverter propagation delay t_P is defined as the average of the two,

$$t_P \equiv \frac{1}{2}(t_{PLH} + t_{PHL}) \tag{13.36}$$

Having defined the inverter propagation delay, we now consider the *maximum* switching frequency of the inverter. From Fig. 13.13(b) we can see that the *minimum* period for each cycle is

$$T_{\min} = t_{PHL} + t_{PLH} = 2t_P \tag{13.37}$$

Thus the **maximum switching frequency** is

$$f_{\max} = \frac{1}{T_{\min}} = \frac{1}{2t_P} \tag{13.38}$$

At this point the reader is no doubt wondering about the cause of the finite propagation time of the inverter. It is simply a result of the time needed to charge and discharge the various capacitances in the circuit. These include the MOSFET capacitances, the wiring capacitance, and the input capacitances of all the logic gates driven by the inverter. We will have a lot more to say about these capacitances and about the determination of t_P in later sections. For the time being, however, we make two important points:

1. A fundamental relationship in analyzing the dynamic operation of a circuit is

$$I\Delta t = \Delta Q = C\Delta V \tag{13.39}$$

That is, a current I flowing through a capacitance C for an interval Δt deposits a charge ΔQ on the capacitor, which causes the capacitor voltage to increase by ΔV.

2. A thorough familiarity with the time response of single-time-constant (STC) circuits is of great help in the analysis of the dynamic operation of digital circuits. A review of this subject is presented in Appendix E. For our purposes here, we remind the reader of the key equation in determining the response to a step function:

Consider a step-function input applied to an STC network of either the low-pass or high-pass type, and let the network have a time constant τ. The output at any time t is given by

$$y(t) = Y_\infty - (Y_\infty - Y_{0+})e^{-t/\tau} \tag{13.40}$$

where Y_∞ is the final value, that is, the value toward which the response is heading, and Y_{0+} is the value of the response immediately after $t = 0$. This equation states that the output at any time t is equal to the difference between the final value Y_∞ and a gap whose initial value is $Y_\infty - Y_{0+}$ and that is shrinking exponentially.

Example 13.3

Consider the inverter of Fig. 13.7(a) with a capacitor C connected between the output node and ground. If at $t = 0$, v_I goes low, and assuming that the switch opens instantaneously, find the time for v_O to reach $\frac{1}{2}(V_{OH} + V_{OL})$. This is the low-to-high propagation time, t_{PLH}. Calculate the value of t_{PLH} for the case $R = 25\ \text{k}\Omega$ and $C = 10$ fF.

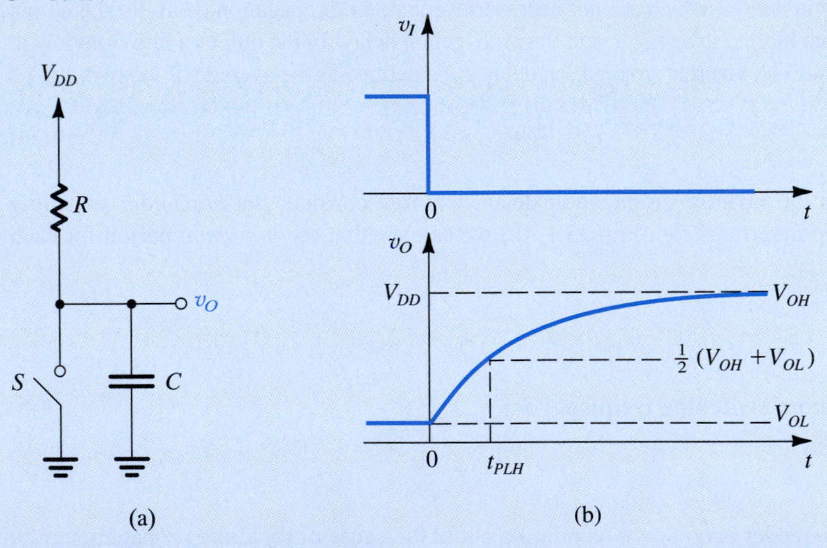

Figure 13.14 Example 13.3: **(a)** The inverter circuit after the switch opens (i.e., for $t \geq 0+$). **(b)** Waveforms of v_I and v_O. Observe that the switch is assumed to operate instantaneously. v_O rises exponentially, starting at V_{OL} and heading toward V_{OH}.

Solution

Before the switch opens, $v_O = V_{OL}$. When the switch opens at $t = 0$, the circuit takes the form shown in Fig. 13.14(a). Since the voltage across the capacitor cannot change instantaneously, at $t = 0+$ the output will still be V_{OL}. Then the capacitor charges through R, and v_O rises exponentially toward V_{DD}. The output waveform will be as shown in Fig. 13.14(b), and its equation can be obtained by substituting in Eq. (13.39): $v_O(\infty) = V_{OH} = V_{DD}$ and $v_O(0+) = V_{OL}$. Thus,

$$v_O(t) = V_{OH} - (V_{OH} - V_{OL})e^{-t/\tau}$$

where $\tau = CR$. To find t_{PLH}, we substitute

$$v_O(t_{PLH}) = \frac{1}{2}(V_{OH} + V_{OL})$$

Thus,

$$\frac{1}{2}(V_{OH} + V_{OL}) = V_{OH} - (V_{OH} - V_{OL})e^{-t_{PLH}/\tau}$$

which results in

$$t_{PLH} = \tau \ln 2 = 0.69\tau$$

Note that this expression is independent of the values of V_{OL} and V_{OH}. For the numerical values given,

$$\begin{aligned} t_{PLH} &= 0.69RC \\ &= 0.69 \times 25 \times 10^3 \times 10 \times 10^{-15} \\ &= 173 \text{ ps} \end{aligned}$$

EXERCISES

13.9 A capacitor C whose initial voltage is 0 is charged to a voltage V_{DD} by a constant-current source I. Find the time t_{PLH} at which the capacitor voltage reaches $(V_{DD}/2)$. What value of I is required to obtain a 10-ps propagation delay with $C = 10$ fF and $V_{DD} = 1.8$ V?
Ans. $t_{PLH} = CV_{DD}/2I$; 0.9 mA

13.10 For the inverter of Fig. 13.8(a), let the on-resistance of P_U be 20 kΩ and that of $P_D = 10$ kΩ. If the capacitance $C = 10$ fF, find t_{PLH}, t_{PHL}, and t_P.
Ans. 138 ps; 69 ps; 104 ps

We conclude this section by showing in Fig. 13.15 the formal definition of the propagation delay of an inverter. As shown, an input pulse with finite (nonzero) **rise and fall times** is applied. The inverted pulse at the output exhibits finite rise and fall times (labeled t_{TLH} and t_{THL}, where the subscript T denotes transition, LH denotes low to high, and HL denotes high to low). There is also a delay time between the input and output waveforms. The usual way to specify the propagation delay is to take the average of the high-to-low propagation delay, t_{PHL}, and the low-to-high propagation delay, t_{PLH}. As indicated, these delays are measured between the 50% points of the input and output waveforms. Also note that the **transition times** are specified using the 10% and 90% points of the output excursion ($V_{OH} - V_{OL}$).

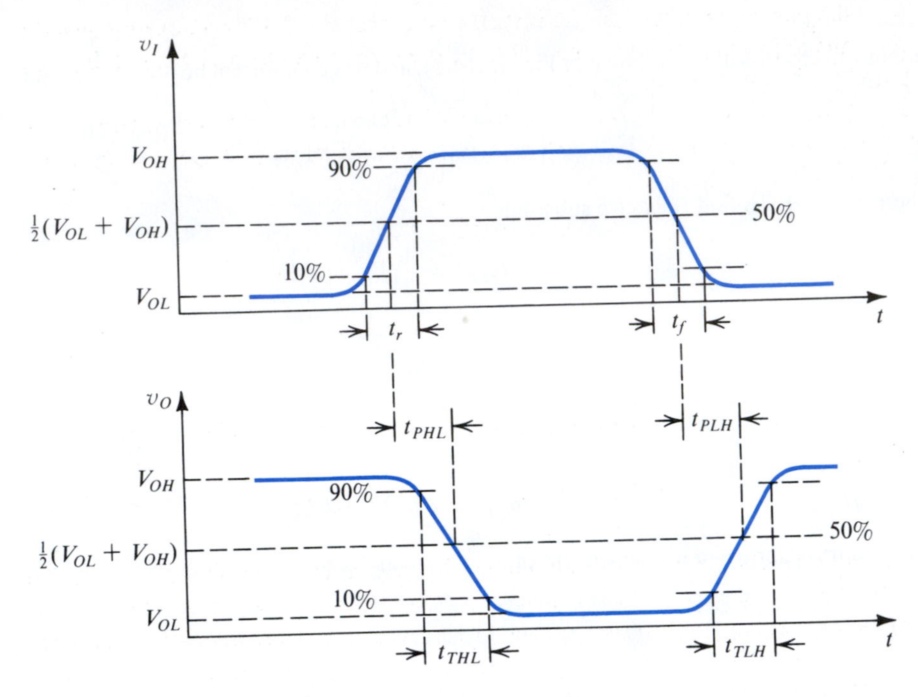

Figure 13.15 Definitions of propagation delays and transition times of the logic inverter.

EXERCISE

13.11 A capacitor $C = 100$ fF is discharged from a voltage V_{DD} to zero through a resistance $R = 2$ kΩ. Find the fall time t_f of the capacitor voltage.
Ans. $t_f \simeq 2.2CR = 0.44$ ns

13.1.8 Power–Delay and Energy–Delay Products

One is usually interested in high-speed operation (low t_P) combined with low power dissipation. Unfortunately, these two requirements are often in conflict: Generally, if the designer of an inverter attempts to reduce power dissipation by, say, decreasing the supply voltage V_{DD}, or the supply current, or both, the current-driving capability of the inverter decreases. This in turn results in longer times to charge and discharge the load and parasitic capacitances, and thus the propagation delay increases. It follows that a figure of merit for comparing logic-circuit technologies is the **power–delay product** (PDP) of the basic inverter of the given technology, defined as

$$PDP \equiv P_D t_P \tag{13.41}$$

where P_D is the power dissipation of the inverter. Note that the *PDP* is an energy quantity and has the units of joules. The lower the *PDP*, the more effective the inverter and the logic circuits based on the inverter are.

For CMOS logic circuits, which is the digital IC technology of primary interest to us here, the static power dissipation of the inverter is zero,[3] and thus P_D is equal to P_{dyn} and given by Eq. (13.35),

$$P_D = fCV_{DD}^2$$

Thus for the CMOS inverter,

$$PDP = fCV_{DD}^2 t_P \tag{13.42}$$

If the inverter is operated at its maximum switching speed given by Eq. (13.38), then

$$PDP = \frac{1}{2}CV_{DD}^2 \tag{13.43}$$

From our earlier discussion of dynamic power dissipation we know that $\frac{1}{2}CV_{DD}^2$ is the amount of energy dissipated during each charging or discharging event of the capacitor, that is, for each output transition of the inverter. Thus, the *PDP* has an interesting physical interpretation: *It is the energy consumed by the inverter for each output transition.*

Although the *PDP* is a valuable metric for comparing different technologies for implementing inverters, it is *not* useful as a design parameter for optimizing a given inverter circuit. To appreciate this point, observe that the expression in Eq. (13.43) indicates that the *PDP* can be minimized by reducing V_{DD} as much as possible while, of course, maintaining proper circuit operation. This, however, would not necessarily result in optimal performance, for t_P will increase as V_{DD} is reduced. The problem is that the *PDP* expression in Eq. (13.43) does not in fact have information about t_P. It follows that a better metric can be obtained by multiplying the energy per transition by the propagation delay. We can thus define the **energy–delay product** *EDP* as

$$EDP \equiv \text{Energy per transition} \times t_P$$

$$= \frac{1}{2}CV_{DD}^2 t_P \tag{13.44}$$

We will utilize the *EDP* in later sections.

13.1.9 Silicon Area

In addition to minimizing power dissipation and propagation delay, another objective in the design of digital VLSI circuits is the minimization of silicon area per logic gate. Smaller area requirement enables the fabrication of a larger number of gates per chip, which has economic and space advantages from a system-design standpoint. Area reduction occurs in three different ways: through advances in processing technology that enable the reduction of the minimum device size, through advances in circuit-design techniques, and through careful chip layout. In this book, our interest lies in circuit design, and we shall make frequent

[3]The exception to this statement is the power dissipation due to leakage currents and subthreshold conduction in the MOSFETs, discussed in Section 13.5.3.

comments on the relationship between the circuit design and its silicon area. As a general rule, the simpler the circuit, the smaller the area required. As will be seen shortly, the circuit designer has to decide on device sizes. Choosing smaller devices has the obvious advantage of requiring smaller silicon area and at the same time reducing parasitic capacitances and thus increasing speed. Smaller devices, however, have lower current-driving capability, which tends to increase delay. Thus, as in all engineering design problems, there is a trade-off to be quantified and exercised in a manner that optimizes whatever aspect of the design is thought to be critical for the application at hand.

13.1.10 Digital IC Technologies and Logic-Circuit Families

The chart in Figure 13.16 shows the major IC technologies and logic-circuit families that are currently in use. The concept of a logic-circuit family perhaps needs a few words of explanation. The basic element of a logic-circuit family is the inverter. A family would include a variety of logic-circuit types made with the same technology, having a similar circuit structure, and exhibiting the same basic features. Each logic-circuit family offers a unique set of advantages and disadvantages. In the conventional style of designing systems, one selects an appropriate logic family (e.g., TTL, CMOS, or ECL) and attempts to implement as much of the system as possible using circuit modules (packages) that belong to this family. In this way, interconnection of the various packages is relatively straightforward. If, on the other hand, packages from more than one family are used, one has to design suitable *interface circuits*. The selection of a logic family is based on such considerations as logic flexibility, speed of operation, availability of complex functions, noise immunity, operating-temperature range, power dissipation, and cost. We will discuss some of these considerations in this chapter and the next two. To begin with, we make some brief remarks on each of the four technologies listed in the chart of Fig. 13.16.

CMOS Although shown as one of four possible technologies, this is not an indication of digital IC market share: CMOS technology is, by a very large margin, the most dominant of all the IC technologies available for digital-circuit design. Although early microprocessors were made using NMOS logic (based on the inverter circuit we studied in Example 13.2), CMOS has completely replaced NMOS. There are a number of reasons for this development, the most important of which is the much lower power dissipation of CMOS circuits. CMOS has also replaced bipolar as the technology of choice in digital-system design and has

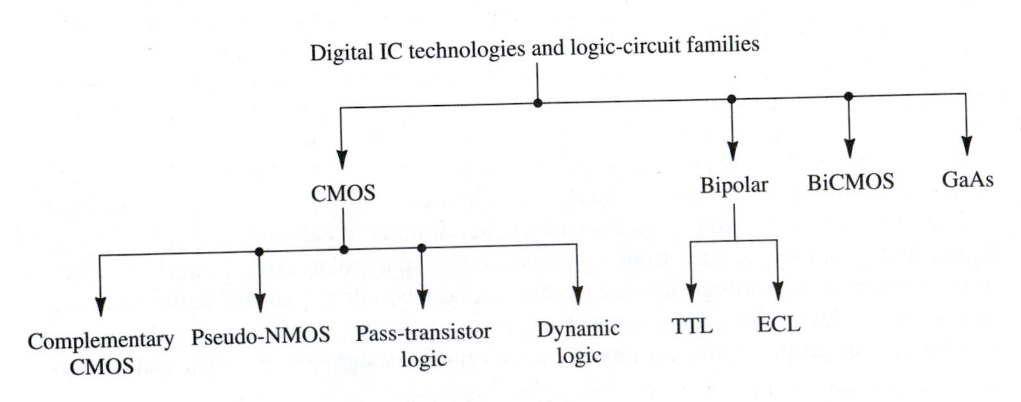

Figure 13.16 Digital IC technologies and logic-circuit families.

made possible levels of integration (or circuit-packing densities) and a range of applications, neither of which would have been possible with bipolar technology. Furthermore, CMOS continues to advance, whereas there appear to be few innovations at the present time in bipolar digital circuits. Some of the reasons for CMOS displacing bipolar technology in digital applications are as follows.

1. CMOS logic circuits dissipate much less power than bipolar logic circuits and thus one can pack more CMOS circuits on a chip than is possible with bipolar circuits.
2. The high input impedance of the MOS transistor allows the designer to use charge storage as a means for the temporary storage of information in both logic and memory circuits. This technique cannot be used in bipolar circuits.
3. The feature size (i.e., minimum channel length) of the MOS transistor has decreased dramatically over the years, with some recently reported designs utilizing channel lengths as short as 32 nm. This permits very tight circuit packing and, correspondingly, very high levels of integration. A microprocessor chip reported in 2009 had 2.3 billion transistors.

Of the various forms of CMOS, complementary CMOS circuits based on the inverter studied in Section 13.2 are the most widely used. They are available both as **small-scale integrated (SSI)** circuit packages (containing 1–10 logic gates) and **medium-scale integrated (MSI)** circuit packages (10–100 gates per chip) for assembling digital systems on printed-circuit boards. More significantly, complementary CMOS is used in **very-large-scale-integrated (VLSI)** logic (with millions of gates per chip) and memory-circuit design. In some applications, complementary CMOS is supplemented by one (or both) of two other MOS logic circuit forms. These are pseudo-NMOS, so-named because of the similarity of its structure to NMOS logic, and pass-transistor logic, both of which will be studied in Chapter 14.

A fourth type of CMOS logic circuit utilizes dynamic techniques to obtain faster circuit operation, while keeping the power dissipation very low. Dynamic CMOS logic, studied in Chapter 14, represents an area of growing importance. Lastly, CMOS technology is used in the design of memory chips, as will be detailed in Chapter 15.

Bipolar Two logic-circuit families based on the bipolar junction transistor are in some use at present: TTL and ECL. Transistor–transistor logic (TTL or T^2L) was for many years the most widely used logic-circuit family. Its decline was precipitated by the advent of the VLSI era. TTL manufacturers, however, fought back with the introduction of low-power and high-speed versions. In these newer versions, the higher speeds of operation are made possible by preventing the BJT from saturating and thus avoiding the slow turnoff process of a saturated bipolar transistor. These nonsaturating versions of TTL utilize the Schottky diode discussed in Section 4.7 and are called Schottky TTL or variations of this name. Despite all these efforts, TTL is no longer a significant logic-circuit family and will not be studied in this book. However, the interested reader can find significant amounts of material on TTL on the CD accompanying this book and on the book's website.

The other bipolar logic-circuit family in present use is emitter-coupled logic (ECL). It is based on the current-switch implementation of the inverter shown in Fig. 13.9. The basic element of ECL is the differential BJT pair studied in Chapter 8. Because ECL is basically a current-steering logic, and, correspondingly, also called **current-mode logic** (CML), in which saturation is avoided, very high speeds of operation are possible. Indeed, of all the commercially available logic-circuit families, ECL is the fastest. ECL is also used in VLSI circuit design when very high operating speeds are required and the designer is willing to accept higher power dissipation and increased silicon area. As such, ECL is considered an important specialty technology and will be discussed in Chapter 14.

BiCMOS BiCMOS combines the high operating speeds possible with BJTs (because of their inherently higher transconductance) with the low power dissipation and other excellent characteristics of CMOS. Like CMOS, BiCMOS allows for the implementation of both analog and digital circuits on the same chip. (See the discussion of analog BiCMOS circuits in Chapter 7.) At present, BiCMOS is used to great advantage in special applications, including memory chips, where its high performance as a high-speed capacitive-current driver justifies the more complex process technology it requires. A brief discussion of BiCMOS is provided in Chapter 14.

Gallium Arsenide (GaAs) The high carrier mobility in GaAs results in very high speeds of operation. This has been demonstrated in a number of digital IC chips utilizing GaAs technology. It should be pointed out, however, that GaAs remains an "emerging technology," one that appears to have great potential but has not yet achieved such potential commercially. As such, it will not be studied in this book. Nevertheless, considerable material on GaAs devices and circuits, including digital circuits, can be found on the CD accompanying this book and on the book's website.

13.1.11 Styles for Digital-System Design

The conventional approach to designing digital systems consists of assembling the system using standard IC packages of various levels of complexity (and hence integration). Many systems have been built this way using, for example, TTL SSI and MSI packages. The advent of VLSI, in addition to providing the system designer with more powerful off-the-shelf components such as microprocessors and memory chips, has made possible alternative design styles. One such alternative is to opt for implementing part or all of the system using one or more *custom VLSI* chips. However, custom IC design is usually economically justified only when the production volume is large (greater than about 100,000 parts).

An intermediate approach, known as *semicustom design,* utilizes *gate-array* chips. These are integrated circuits containing 100,000 or more unconnected logic gates. Their interconnection can be achieved by a final metallization step (performed at the IC fabrication facility) according to a pattern specified by the user to implement the user's particular functional need. A more recently available type of gate array, known as a **field-programmable gate array (FPGA)**, can, as its name indicates, be programmed directly by the user. FPGAs provide a very convenient means for the digital-system designer to implement complex logic functions in VLSI form without having to incur either the cost or the "turnaround time" inherent in custom and, to a lesser extent, in semicustom IC design.

13.1.12 Design Abstraction and Computer Aids

The design of very complex digital systems, whether on a single IC chip or using off-the-shelf components, is made possible by the use of different levels of design abstraction, and the use of a variety of computer aids. To appreciate the concept of design abstraction, consider the process of designing a digital system using off-the-shelf packages of logic gates. The designer consults data sheets (in data books or on websites) to determine the input and output characteristics of the gates, their fan-in and fan-out limitations, and so on. In connecting the gates, the designer needs to adhere to a set of rules specified by the manufacturer in the data sheets. The designer does not need to consider, in a direct way, the circuit inside the gate package. In effect, the circuit has been abstracted in the form of a functional block that can be used as a component. This greatly simplifies system design. The digital-IC designer follows a similar process. Circuit blocks are designed, characterized, and stored in a library as **standard cells**. These cells can then be used by the IC designer to assemble a larger subsystem (e.g., an adder or a multiplier), which in turn is characterized and stored as a functional block to be used in the design of an even larger system (e.g., an entire processor).

At every level of design abstraction, the need arises for simulation and other computer programs that help make the design process as automated as possible. Whereas SPICE is employed in circuit simulation, other software tools are utilized at other levels and in other phases of the design process. Although digital-system design and design automation are outside the scope of this book, it is important that the reader appreciate the role of design abstraction and computer aids in digital design. They are what make it humanly possible to design a billion-transistor digital IC. Unfortunately, analog IC design does not lend itself to the same level of abstraction and automation. Each analog IC to a large extent has to be "handcrafted." As a result, the complexity and density of analog ICs remain much below what is possible in a digital IC.

Whatever approach or style is adopted in digital design, some familiarity with the various digital-circuit technologies and design techniques is essential. This chapter and the next two aim to provide such a background.

13.2 The CMOS Inverter

In this section we study the inverter circuit of the most widely used digital IC technology: CMOS. The basic CMOS inverter is shown in Fig. 13.17. It utilizes two MOSFETs: one, Q_N, with an n channel and the other, Q_P, with a p channel. The body of each device is connected to its source, and thus no body effect arises. As will be seen shortly, the CMOS circuit realizes the conceptual inverter implementation studied in the previous section (Fig. 13.8), where a pair of switches are operated in a complementary fashion by the input voltage v_I.

13.2.1 Circuit Operation

We first consider the two extreme cases: when v_I is at logic-0 level, which is 0 V; and when v_I is at logic-1 level, which is V_{DD} volts. In both cases, for ease of exposition we shall consider the n-channel device Q_N to be the driving transistor and the p-channel device Q_P to be

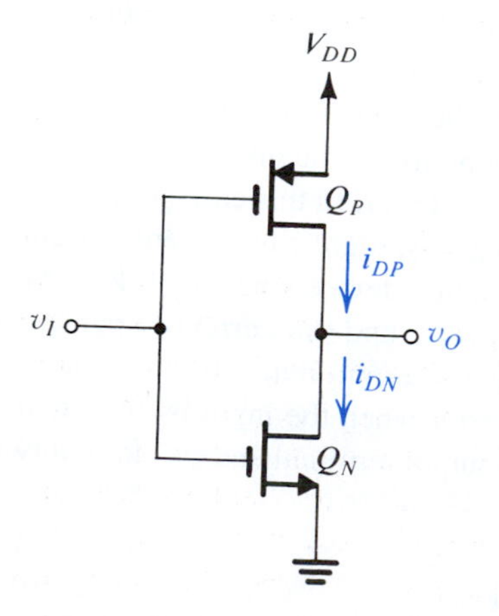

Figure 13.17 The CMOS inverter.

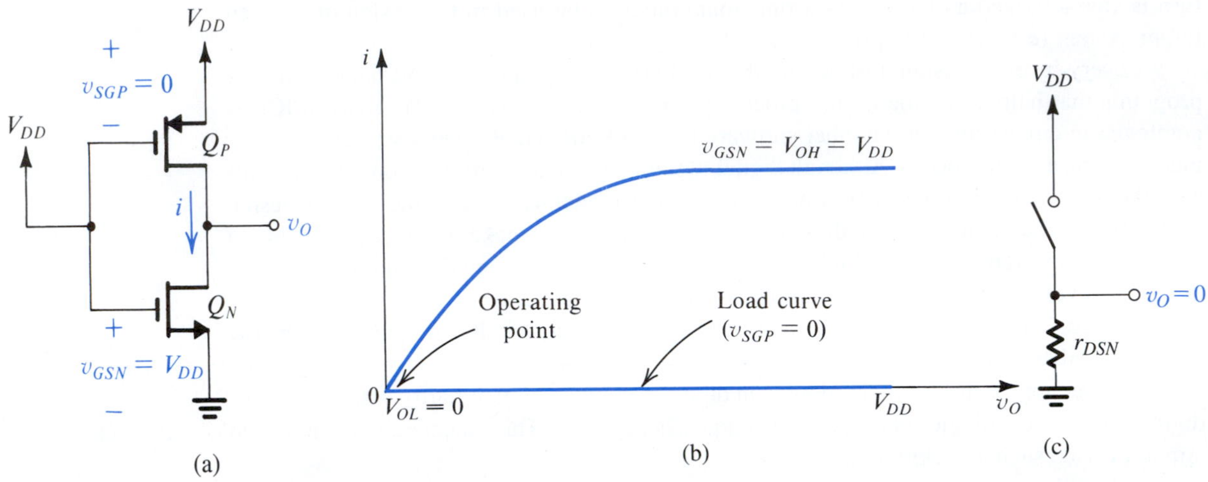

Figure 13.18 Operation of the CMOS inverter when v_I is high: **(a)** circuit with $v_I = V_{DD}$ (logic-1 level, or V_{OH}); **(b)** graphical construction to determine the operating point; **(c)** equivalent circuit.

the load. However, since the circuit is symmetric, this assumption is obviously arbitrary, and the reverse would lead to identical results.

Figure 13.18 illustrates the case when $v_I = V_{DD}$, showing the i_D–v_{DS} characteristic curve for Q_N with $v_{GSN} = V_{DD}$. (Note that $i_D = i$ and $v_{DSN} = v_O$.) Superimposed on the Q_N characteristic curve is the load curve, which is the i_D–v_{SD} curve of Q_P for the case $v_{SGP} = 0$ V. Since $v_{SGP} < |V_t|$, the load curve will be a horizontal straight line at zero current level. The operating point will be at the intersection of the two curves, where we note that the output voltage is zero and the current through the two devices is also zero. This means that the power dissipation in the circuit is zero. Note, however, that although Q_N is operating at zero current and zero drain-source voltage (i.e., at the origin of the i_D–v_{DS} plane), the operating point is on a steep segment of the i_D–v_{DS} characteristic curve. Thus Q_N provides a low-resistance path between the output terminal and ground, with the resistance obtained using Eq. (5.13b) as

$$r_{DSN} = 1\bigg/\left[k_n'\left(\frac{W}{L}\right)_n (V_{DD} - V_{tn})\right] \tag{13.45}$$

Figure 13.18(c) shows the equivalent circuit of the inverter when the input is high. This circuit confirms that $v_O \equiv V_{OL} = 0$ V and that the power dissipation in the inverter is zero.

The other extreme case, when $v_I = 0$ V, is illustrated in Fig. 13.19. In this case Q_N is operating at $v_{GSN} = 0$; hence its i_D–v_{DS} characteristic is a horizontal straight line at zero current level. The load curve is the i_D–v_{SD} characteristic of the p-channel device with $v_{SGP} = V_{DD}$. As shown, at the operating point the output voltage is equal to V_{DD}, and the current in the two devices is still zero. Thus the power dissipation in the circuit is zero in both extreme states.

Figure 13.19(c) shows the equivalent circuit of the inverter when the input is low. Here we see that Q_P provides a low-resistance path between the output terminal and the dc supply V_{DD}, with the resistance given by

$$r_{DSP} = 1\bigg/\left[k_p'\left(\frac{W}{L}\right)_p (V_{DD} - |V_{tp}|)\right] \tag{13.46}$$

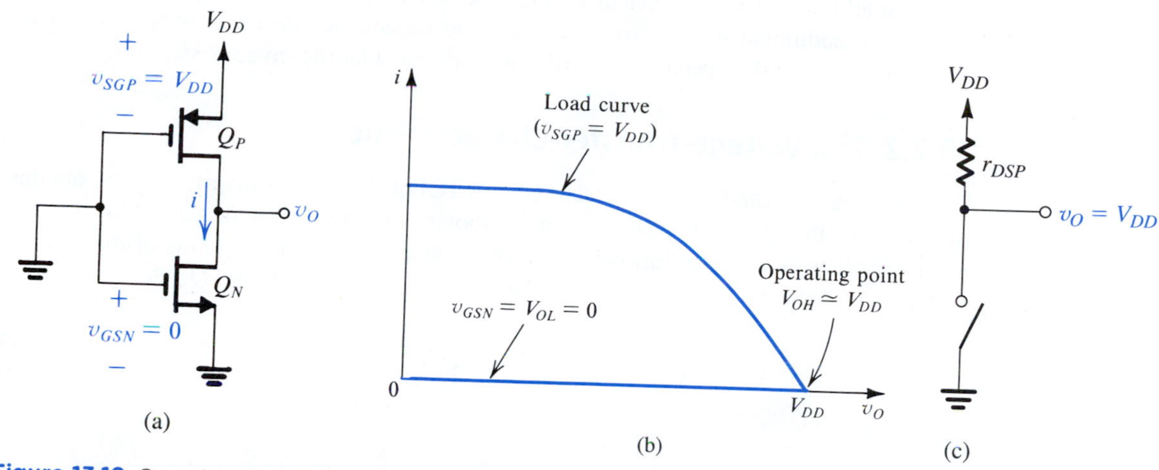

Figure 13.19 Operation of the CMOS inverter when v_I is low: **(a)** circuit with $v_I = 0$ V (logic-0 level, or V_{OL}); **(b)** graphical construction to determine the operating point; **(c)** equivalent circuit.

The equivalent circuit confirms that in this case $v_O \equiv V_{OH} = V_{DD}$ and that the power dissipation in the inverter is zero.

It should be noted, however, that in spite of the fact that the quiescent current is zero, the load-driving capability of the CMOS inverter is high. For instance, with the input high, as in the circuit of Fig. 13.18, transistor Q_N can sink a relatively large load current. This current can quickly discharge the load capacitance, as will be seen shortly. Because of its action in sinking load current and thus pulling the output voltage down toward ground, transistor Q_N is known as the **pull-down device**. Similarly, with the input low, as in the circuit of Fig. 13.19, transistor Q_P can source a relatively large load current. This current can quickly charge up a load capacitance, thus pulling the output voltage up toward V_{DD}. Hence, Q_P is known as the **pull-up device**. The reader will recall that we used this terminology in connection with the conceptual inverter circuit of Fig. 13.8.

From the above, we conclude that the basic CMOS logic inverter behaves as an ideal inverter. In summary:

1. The output voltage levels are 0 and V_{DD}, and thus the signal swing is the maximum possible. This, coupled with the fact that the inverter can be designed to provide a symmetrical voltage-transfer characteristic, results in wide noise margins.
2. The static power dissipation in the inverter is zero (neglecting the dissipation due to leakage currents) in both of its states. This is because no dc path exists between the power supply and ground in either state.
3. A low-resistance path exists between the output terminal and ground (in the low-output state) or V_{DD} (in the high-output state). These low-resistance paths ensure that the output voltage is 0 or V_{DD} independent of the exact values of the W/L ratios or other device parameters. Furthermore, the low output resistance makes the inverter less sensitive to the effects of noise and other disturbances.
4. The active pull-up and pull-down devices provide the inverter with high output-driving capability in both directions. As will be seen, this speeds up the operation considerably.

5. The input resistance of the inverter is infinite (because $I_G = 0$). Thus the inverter can drive an arbitrarily large number of similar inverters with no loss in signal level. Of course, each additional inverter increases the load capacitance on the driving inverter and slows down the operation. Shortly, we will consider the inverter switching times.

13.2.2 The Voltage-Transfer Characteristic

The complete voltage-transfer characteristic (VTC) of the CMOS inverter can be obtained by repeating the graphical procedure, used above in the two extreme cases, for all intermediate values of v_I. In the following, we shall calculate the critical points of the resulting voltage-transfer curve. For this we need the i–v relationships of Q_N and Q_P. For Q_N,

$$i_{DN} = k_n'\left(\frac{W}{L}\right)_n\left[(v_I - V_{tn})v_O - \frac{1}{2}v_O^2\right] \quad \text{for } v_O \leq v_I - V_{tn} \tag{13.47}$$

and

$$i_{DN} = \frac{1}{2}k_n'\left(\frac{W}{L}\right)_n(v_I - V_{tn})^2 \quad \text{for } v_O \geq v_I - V_{tn} \tag{13.48}$$

For Q_P,

$$i_{DP} = k_p'\left(\frac{W}{L}\right)_p\left[(V_{DD} - v_I - |V_{tp}|)(V_{DD} - v_O) - \frac{1}{2}(V_{DD} - v_O)^2\right]$$
$$\text{for } v_O \geq v_I + |V_{tp}| \tag{13.49}$$

and

$$i_{DP} = \frac{1}{2}k_p'\left(\frac{W}{L}\right)_p(V_{DD} - v_I - |V_{tp}|)^2 \quad \text{for } v_O \leq v_I + |V_{tp}| \tag{13.50}$$

The CMOS inverter is usually designed to have $V_{tn} = |V_{tp}| = V_t$. Also, although this is not always the case, we shall assume that Q_N and Q_P are matched; that is, $k_n'(W/L)_n = k_p'(W/L)_p$. It should be noted that since μ_p is 0.25 to 0.5 times the value of μ_n, to make $k'(W/L)$ of the two devices equal, the width of the p-channel device is made two to four times that of the n-channel device. More specifically, the two devices are designed to have equal lengths, with widths related by

$$\frac{W_p}{W_n} = \frac{\mu_n}{\mu_p} \tag{13.51}$$

This will result in $k_n'(W/L)_n = k_p'(W/L)_p$, and the inverter will have a symmetric transfer characteristic and equal current-driving capability in both directions (pull-up and pull-down).

With Q_N and Q_P matched, the CMOS inverter has the voltage transfer characteristic shown in Fig. 13.20. As indicated, the transfer characteristic has five distinct segments corresponding to different combinations of modes of operation of Q_N and Q_P. The vertical segment BC is obtained when both Q_N and Q_P are operating in the saturation region. Because we are neglecting the finite output resistance in saturation, that is, assuming $\lambda_N = \lambda_P = 0$, the inverter gain in this region is infinite. From symmetry, this vertical segment occurs at $v_I = V_{DD}/2$ and is bounded by $v_O(\text{B}) = V_{DD}/2 + V_t$, at which value Q_P enters the triode region and $v_O(\text{C}) = V_{DD}/2 - V_t$, at which value Q_N enters the triode region.

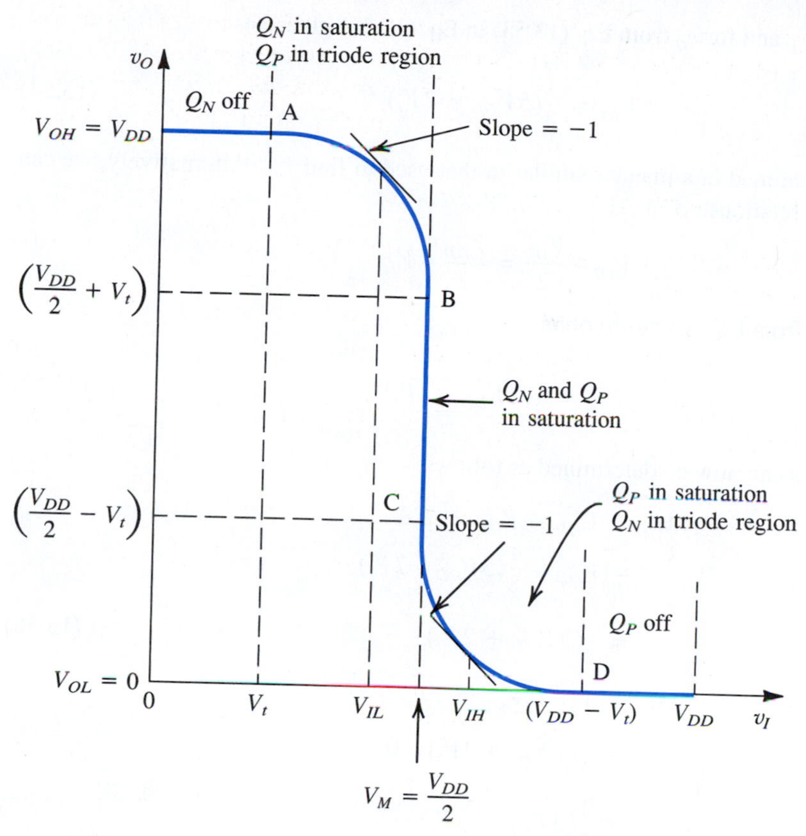

Figure 13.20 The voltage-transfer characteristic of the CMOS inverter when Q_N and Q_P are matched.

The reader will recall from Section 13.1.3 that in addition to V_{OL} and V_{OH}, two other points on the transfer curve determine the noise margins of the inverter. These are the maximum permitted logic-0 or "low" level at the input, V_{IL}, and the minimum permitted logic-1 or "high" level at the input, V_{IH}. These are formally defined as the two points on the transfer curve at which the incremental gain is unity (i.e., the slope is −1 V/V).

To determine V_{IH}, we note that Q_N is in the triode region, and thus its current is given by Eq. (13.47), while Q_P is in saturation and its current is given by Eq. (13.50). Equating i_{DN} and i_{DP}, and assuming matched devices, gives

$$(v_I - V_t)v_O - \frac{1}{2}v_O^2 = \frac{1}{2}(V_{DD} - v_I - V_t)^2 \tag{13.52}$$

Differentiating both sides relative to v_I results in

$$(v_I - V_t)\frac{dv_O}{dv_I} + v_O - v_O\frac{dv_O}{dv_I} = -(V_{DD} - v_I - V_t)$$

in which we substitute $v_I = V_{IH}$ and $dv_O/dv_I = -1$ to obtain

$$v_O = V_{IH} - \frac{V_{DD}}{2} \tag{13.53}$$

Substituting $v_I = V_{IH}$ and for v_O from Eq. (13.53) in Eq. (13.52) gives

$$V_{IH} = \frac{1}{8}(5V_{DD} - 2V_t) \tag{13.54}$$

V_{IL} can be determined in a manner similar to that used to find V_{IH}. Alternatively, we can use the symmetry relationship

$$V_{IH} - \frac{V_{DD}}{2} = \frac{V_{DD}}{2} - V_{IL}$$

together with V_{IH} from Eq. (13.54) to obtain

$$V_{IL} = \frac{1}{8}(3V_{DD} + 2V_t) \tag{13.55}$$

The noise margins can now be determined as follows:

$$\begin{aligned} NM_H &= V_{OH} - V_{IH} \\ &= V_{DD} - \frac{1}{8}(5V_{DD} - 2V_t) \\ &= \frac{1}{8}(3V_{DD} + 2V_t) \end{aligned} \tag{13.56}$$

$$\begin{aligned} NM_L &= V_{IL} - V_{OL} \\ &= \frac{1}{8}(3V_{DD} + 2V_t) - 0 \\ &= \frac{1}{8}(3V_{DD} + 2V_t) \end{aligned} \tag{13.57}$$

As expected, the symmetry of the voltage-transfer characteristic results in equal noise margins. Of course, if Q_N and Q_P are not matched, the voltage-transfer characteristic will no longer be symmetric, and the noise margins will not be equal.

13.2.3 The Situation When Q_N and Q_P Are Not Matched

In the above we assumed that Q_N and Q_P are matched; that is, in addition to $V_{tn} = |V_{tp}|$, the transconductance parameters k_n and k_p are made equal by selecting W_p/W_n according to Eq. (13.51). The result is a symmetrical VTC that switches at the midpoint of the supply; that is, $V_M = V_{DD}/2$. The symmetry, as we have seen, equalizes and maximizes the noise margins.

The price paid for obtaining a perfectly symmetric VTC is that the width of the *p*-channel device can be three to four times as large as that of the *n*-channel device. This can result in a relatively large silicon area which, besides being wasteful of silicon real estate, can also result in increased device capacitances and a corresponding increase in the propagation delay of the inverter. It is useful, therefore, to inquire into the effect of not matching Q_N and Q_P. Toward that end we derive an expression for the switching voltage V_M as follows.

Since at M, both Q_N and Q_P operate in saturation, their currents are given by Eqs. (13.48) and (13.50). Substituting $v_I = v_O = V_M$, and equating the two currents results in

$$V_M = \frac{r(V_{DD} - |V_{tp}|) + V_{tn}}{r + 1} \tag{13.58}$$

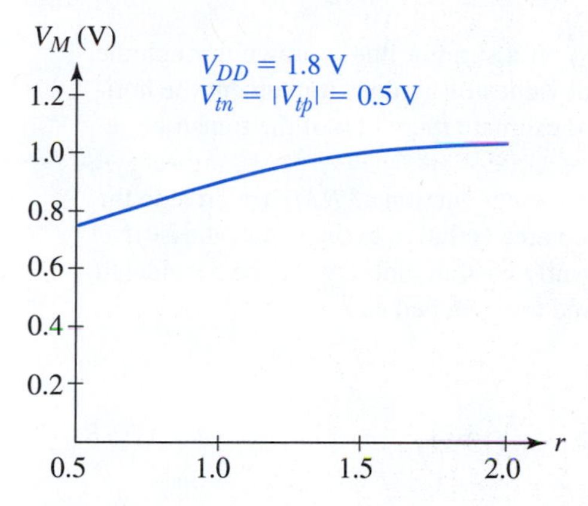

Figure 13.21 Variation of the inverter switching voltage, V_M, with the parameter $r = \sqrt{k_p/k_n}$.

where

$$r = \sqrt{\frac{k_p}{k_n}} = \sqrt{\frac{\mu_p W_p}{\mu_n W_n}} \tag{13.59}$$

where we have assumed that Q_N and Q_P have the same channel length L, which is usually the case with L equal to the minimum available for the given process technology. Note that the matched case corresponds to $r = 1$. For $|V_{tp}| = V_{tn}$, and $r = 1$, Eq. (13.58) yields $V_M = V_{DD}/2$, as expected. For a given process, that is, given values for V_{DD}, V_{tn}, and V_{tp}, one can plot V_M versus the matching parameter r. Such a plot, for a 0.18-μm process, is shown in Fig. 13.21. We make the following two observations:

1. V_M increases with r. Thus, making $k_p > k_n$ shifts V_M toward V_{DD}. Conversely, making $k_p < k_n$ shifts V_M toward 0.
2. V_M is not a strong function of r. For the particular case shown, lowering r by a factor of 2 (from 1 to 0.5), reduces V_M by only 0.13 V.

Observation 2 implies that if one is willing to tolerate a small reduction in NM_L, substantial savings in silicon area can be obtained. This point is illustrated in Example 13.4.

Example 13.4

Consider a CMOS inverter fabricated in a 0.18-μm process for which $V_{DD} = 1.8$ V, $V_{tn} = |V_{tp}| = 0.5$ V, $\mu_n = 4\mu_p$, and $\mu_n C_{ox} = 300\ \mu\text{A/V}^2$. In addition, Q_N and Q_P have $L = 0.18$ μm and $(W/L)_n = 1.5$.

(a) Find W_p that results in $V_M = V_{DD}/2 = 0.9$ V. What is the silicon area utilized by the inverter in this case?

(b) For the matched case in (a), find the values of V_{OH}, V_{OL}, V_{IH}, V_{IL}, and the noise margins NM_L and NM_H. For $v_I = V_{IH}$, what value of v_O results? This can be considered the worst-case value of V_{OL}. Similarly, for $v_I = V_{IL}$, find v_O that is the worst-case value of V_{OH}. Now, use these worst-case values to determine more conservative values for the noise margins.

(c) For the matched case in (a), find the output resistance of the inverter in each of its two states.

Example 13.4 *continued*

(d) If $\lambda_n = |\lambda_p| = 0.2\ \text{V}^{-1}$, what is the inverter gain at $v_I = V_M$. If a straight line is drawn through the point $v_I = v_O = V_M$ with a slope equal to the gain, at what values of v_I does it intercept the horizontal lines $v_O = 0$ and $v_O = V_{DD}$? Use these intercepts to estimate the width of the transition region of the VTC.

(e) If $W_p = W_n$, what value of V_M results? What do you estimate the reduction of NM_L (relative to the matched case) to be? What is the percentage savings in silicon area (relative to the matched case)?

(f) Repeat (e) for the case $W_p = 2W_n$. This case, which is frequently used in industry, can be considered to be a compromise between the minimum-area case in (e) and the matched case.

Solution

(a) To obtain $V_M = V_{DD}/2 = 0.9\ \text{V}$, we select W_p according to Eq. (13.51),

$$\frac{W_p}{W_n} = \frac{\mu_n}{\mu_p} = 4$$

Since $W_n/L = 1.5$, $W_n = 1.5 \times 0.18 = 0.27\ \mu\text{m}$. Thus,

$$W_p = 4 \times 0.27 = 1.08\ \mu\text{m}$$

For this design, the silicon area is

$$A = W_n L + W_p L = L(W_n + W_p)$$

$$= 0.18(0.27 + 1.08) = 0.243\ \mu\text{m}^2$$

(b)

$$V_{OH} = V_{DD} = 1.8\ \text{V}$$

$$V_{OL} = 0\ \text{V}$$

To obtain V_{IH} we use Eq. (13.54),

$$V_{IH} = \frac{1}{8}(5V_{DD} - 2V_t) = \frac{1}{8}(5 \times 1.8 - 2 \times 0.5) = 1\ \text{V}$$

To obtain V_{IL} we use Eq. (13.55),

$$V_{IL} = \frac{1}{8}(3V_{DD} + 2V_t) = \frac{1}{8}(3 \times 1.8 + 2 \times 0.5) = 0.8\ \text{V}$$

We can now compute the noise margins as

$$NM_H = V_{OH} - V_{IH} = 1.8 - 1.0 = 0.8\ \text{V}$$

$$NM_L = V_{IL} - V_{OL} = 0.8 - 0 = 0.8\ \text{V}$$

As expected, $NM_H = NM_L$, and their value is very close to the optimum value of $V_{DD}/2 = 0.9\ \text{V}$.

For $v_I = V_{IH} = 1\ \text{V}$, we can obtain the corresponding value of v_O by substituting in Eq. (13.53),

$$v_O = V_{IH} - \frac{V_{DD}}{2} = 1 - \frac{1.8}{2} = 0.1\ \text{V}$$

Thus, the worst-case value of V_{OL}, that is, $V_{OL\max}$, is 0.1 V, and the noise margin NM_L reduces to

$$NM_L = V_{IL} - V_{OL\max} = 0.8 - 0.1 = 0.7\ \text{V}$$

From symmetry, we can obtain the value of v_O corresponding to $v_I = V_{IL}$ as

$$v_O = V_{DD} - 0.1 = 1.7\ \text{V}$$

Thus the worst-case value of V_{OH}, that is, $V_{OH\text{min}}$, is 1.7 V, and the noise margin NM_H reduces to

$$NM_H = V_{OH\text{min}} - V_{IH} = 1.7 - 1 = 0.7 \text{ V}$$

Note that the reduction in the noise margins is slight.

(c) The output resistance of the inverter in the low-output state is

$$r_{DSN} = \frac{1}{\mu_n C_{ox}(W/L)_n(V_{DD} - V_{tn})}$$

$$= \frac{1}{300 \times 10^{-6} \times 1.5(1.8 - 0.5)} = 1.71 \text{ k}\Omega$$

Since Q_N and Q_P are matched, the output resistance in the high-output state will be equal, that is,

$$r_{DSP} = r_{DSN} = 1.71 \text{ k}\Omega$$

(d) If the inverter is biased to operate at $v_I = v_O = V_M = 0.9$ V, then each of Q_N and Q_P will be operating at an overdrive voltage $V_{OV} = V_M - V_t = 0.9 - 0.5 = 0.4$ V and will be conducting equal dc currents I_D of

$$I_D = \frac{1}{2}\mu_n C_{ox}\left(\frac{W}{L}\right)_N V_{OV}^2$$

$$= \frac{1}{2} \times 300 \times 1.5 \times 0.4^2$$

$$= 36 \ \mu\text{A}$$

Thus, Q_N and Q_P will have equal transconductances:

$$g_{mn} = g_{mp} = \frac{2I_D}{V_{OV}} = \frac{2 \times 36}{0.4} = 0.18 \text{ mA/V}^2$$

Transistors Q_N and Q_P will have equal output resistances r_o,

$$r_{on} = r_{op} = \frac{|V_A|}{I_D} = \frac{1}{|\lambda| I_D} = \frac{1}{0.2 \times 36} = 139 \text{ k}\Omega$$

We can now compute the voltage gain at M as

$$A_v = -(g_{mn} + g_{mp})(r_{on} \| r_{op})$$

$$= -(0.18 + 0.18)(139 \| 139) = -25 \text{ V/V}$$

When the straight line at M of slope −25 V/V is extrapolated, it intersects the line $v_O = 0$ at $[0.9 + 0.9/25] = 0.936$ V and the line $v_O = V_{DD}$ at $(0.9 - 0.9/25) = 0.864$ V. Thus the width of the transition region can be considered to be $(0.936 - 0.864) = 0.072$ V.

(e) For $W_p = W_n$, the parameter r can be found from Eq. (13.59),

$$r = \sqrt{\frac{\mu_p W_p}{\mu_n W_n}} = \sqrt{\frac{1}{4} \times 1} = 0.5$$

The corresponding value of V_M can be determined from Eq. (13.58) as

$$V_M = \frac{0.5(1.8 - 0.5) + 0.5}{0.5 + 1} = 0.77 \text{ V}$$

Example 13.4 *continued*

Thus V_M shifts by only -0.13 V. Without recalculating V_{IL} we can estimate the reduction in NM_L to be approximately equal to the shift in V_M, that is, NM_L becomes $0.8 - 0.13 = 0.67$ V. The silicon area for this design can be computed as follows:

$$A = L(W_n + W_p) = 0.18(0.27 + 0.27)$$
$$= 0.0972\ \mu\text{m}^2$$

This represents a 60% reduction from the matched case!

(f) For $W_p = 2W_n$,

$$r = \sqrt{\frac{1}{4} \times 2} = \frac{1}{\sqrt{2}} = 0.707$$

$$V_M = \frac{0.707(1.8 - 0.5) + 0.5}{0.707 + 1} = 0.83\text{ V}$$

Thus, relative to the matched case, the shift in V_M is only -0.07 V. We estimate that NM_L will decrease from 0.8 V by the same amount; thus NM_L becomes 0.73 V. In this case, the silicon area required is

$$A = L(W_n + W_p) = 0.18(0.27 + 0.54)$$
$$= 0.146\ \mu\text{m}^2$$

which represents a 40% reduction relative to the matched case!

EXERCISES

13.12 Consider a CMOS inverter fabricated in a 0.13-μm process for which $V_{DD} = 1.2$ V, $V_{tn} = -V_{tp} = 0.4$ V, $\mu_n/\mu_p = 4$, and $\mu_n C_{ox} = 430\ \mu\text{A/V}^2$. In addition, Q_N and Q_P have $L = 0.13$ μm and $(W/L)_n = 1.0$.
(a) Find W_p that results in $V_M = 0.6$ V.
(b) For the matched case in (a), find the values of V_{OH}, V_{OL}, V_{IH}, V_{IL}, NM_H, and NM_L.
(c) For the inverter in (a), find the output resistance in each of its two states.
(d) For a minimum-size inverter for which $(W/L)_p = (W/L)_n = 1.0$, find V_M.
Ans. (a) 0.52 μm; (b) 1.2 V, 0 V, 0.65 V, 0.55 V, 0.55 V, 0.55 V, (c) 2.9 kΩ, 2.9 kΩ; (d) 0.53 V

D13.13 A CMOS inverter utilizes $V_{DD} = 5$ V, $V_{tn} = |V_{tp}| = 1$ V, and $\mu_n C_{ox} = 2\mu_p C_{ox} = 50\ \mu\text{A/V}^2$. Find $(W/L)_n$ and $(W/L)_p$ so that $V_M = 2.5$ V and so that for $v_I = V_{DD}$, the inverter can sink a current of 0.2 mA with the output voltage not exceeding 0.2 V.
Ans. $(W/L)_n \simeq 5$; $(W/L)_p \simeq 10$

13.3 Dynamic Operation of the CMOS Inverter

As explained in Section 13.1.7, the speed of operation of a digital system (e.g., a computer) is determined by the propagation delay of the logic gates used to construct the system. Since the inverter is the basic logic gate of any digital IC technology, the propagation delay of the inverter is a fundamental parameter in characterizing the technology. In the following, we analyze the switching operation of the CMOS inverter to determine its propagation delay. We shall do this by utilizing a two-step process.

1. Replace all the capacitances in the circuit: that is, the various capacitances associated with Q_N and Q_P, the capacitance of the wire that connects the output of the inverter to other circuits, and the input capacitance of the logic gates the inverter drives, by a single equivalent capacitance C connected between the output node of the inverter and ground.
2. Analyze the resulting capacitively loaded inverter to determine its t_{PLH} and t_{PHL}, and hence t_P.

We shall study these two separable steps in reverse order. Thus, in Section 13.3.1 we show how the propagation delay can be determined. Then, in Section 13.3.2, we show how to calculate the value of C.

13.3.1 Determining the Propagation Delay

Figure 13.22(a) shows a CMOS inverter with a capacitance C connected between its output node and ground. To determine the propagation delays t_{PHL} and t_{PLH}, we apply to the input an ideal pulse, that is, one with zero rise and fall times, as shown in Fig. 13.22(b). Since the circuit is symmetric, the analyses to determine the two propagation delays will be similar. Therefore, we will derive t_{PHL} in detail and extrapolate the result to determine t_{PLH}.

Just prior to the leading edge of the input pulse (i.e., at $t = 0-$), the output voltage is equal to V_{DD} and capacitor C is charged to this voltage. At $t = 0$, v_I rises to V_{DD}, causing Q_P to turn off and Q_N to turn on. From then on, the circuit is equivalent to that shown in Fig. 13.22(c), with the initial value of $v_O = V_{DD}$. Thus, at $t = 0+$, Q_N will operate in the saturation region and will supply a relatively large current to begin the process of discharging C. Figure 13.22(d) shows the trajectory of the operating point of Q_N as C is discharged. Here we are interested in the interval t_{PHL} during which v_O reduces from V_{DD} to $V_{DD}/2$. Correspondingly, the operating point of Q_N moves from E to M. For a portion of this time, corresponding to the segment EF of the trajectory, Q_N operates in saturation. Then at F, $v_O = V_{DD} - V_t$, and Q_N enters the triode region.

A simple approach for determining t_{PHL} consists of first calculating the average value of the current supplied by Q_N over the segment EM. Then, we use this average value of the discharge current to determine t_{PHL} by means of the charge balance equation

$$I_{av}\, t_{PHL} = C[V_{DD} - (V_{DD}/2)]$$

resulting in

$$t_{PHL} = \frac{CV_{DD}}{2I_{av}} \tag{13.60}$$

The value of I_{av} can be found as follows:

$$I_{av} = \frac{1}{2}[i_{DN}(\mathrm{E}) + i_{DN}(\mathrm{M})] \tag{13.61}$$

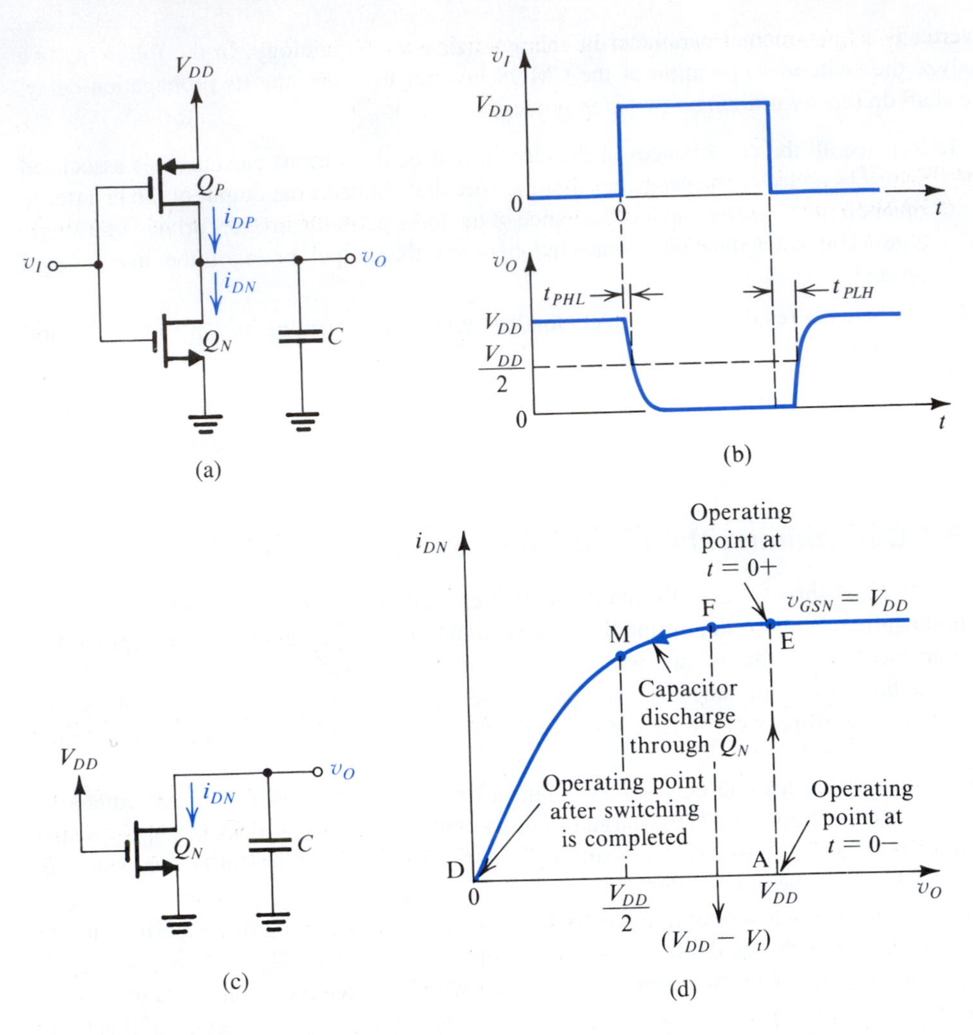

Figure 13.22 Dynamic operation of a capacitively loaded CMOS inverter: **(a)** circuit; **(b)** input and output waveforms; **(c)** equivalent circuit during the capacitor discharge; **(d)** trajectory of the operating point as the input goes high and C discharges through Q_N.

where

$$i_{DN}(\mathrm{E}) = \frac{1}{2}k_n'\left(\frac{W}{L}\right)_n (V_{DD} - V_{tn})^2 \tag{13.62}$$

and

$$i_{DN}(\mathrm{M}) = k_n'\left(\frac{W}{L}\right)_n \left[(V_{DD} - V_{tn})\left(\frac{V_{DD}}{2}\right) - \frac{1}{2}\left(\frac{V_{DD}}{2}\right)^2\right] \tag{13.63}$$

Note that we have assumed $\lambda_n = 0$. Combining Eqs. (13.60) to (13.63) provides

$$t_{PHL} = \frac{\alpha_n C}{k_n'(W/L)_n V_{DD}} \tag{13.64}$$

where α_n is a factor determined by the relative values of V_t and V_{DD};

$$\alpha_n = 2\bigg/\left[\frac{7}{4} - \frac{3V_{tn}}{V_{DD}} + \left(\frac{V_{tn}}{V_{DD}}\right)^2\right] \tag{13.65}$$

The value of α_n falls in the range of 1 to 2.

An expression for the low-to-high inverter delay, t_{PLH}, can be written by analogy to the t_{PHL} expression in Eq. (13.64),

$$t_{PLH} = \frac{\alpha_p}{k_p'\left(\frac{W}{L}\right)_p V_{DD}} \tag{13.66}$$

where

$$\alpha_p = 2\bigg/\left[\frac{7}{4} - \frac{3|V_{tp}|}{V_{DD}} + \left|\frac{V_{tp}}{V_{DD}}\right|^2\right] \tag{13.67}$$

Finally, the propagation delay t_P can be found as the average of t_{PHL} and t_{PLH},

$$t_P = \frac{1}{2}(t_{PHL} + t_{PLH})$$

Examination of the formulas in Eqs. (13.64) to (13.67) enables us to make a number of useful observations:

1. As expected, the two components of t_P can be equalized by selecting the (W/L) ratios to equalize k_n and k_p, that is, by matching Q_N and Q_P.
2. Since t_P is proportional to C, the designer should strive to reduce C. This is achieved by using the minimum possible channel length and by minimizing wiring and other parasitic capacitances. Careful layout of the chip can result in significant reduction in such capacitances.
3. Using a process technology with larger transconductance parameter k' can result in shorter propagation delays. Keep in mind, however, that for such processes C_{ox} is increased, and thus the value of C increases at the same time (more on this later).
4. Using larger W/L ratios can result in a reduction in t_P. Care, however, should be exercised here also, since increasing the size of the devices increases the value of C, and thus the expected reduction in t_P might not materialize. Reducing t_P by increasing W/L, however, is an effective strategy when C is dominated by components not directly related to the size of the driving device (such as wiring or fan-out devices).
5. A larger supply voltage V_{DD} results in a lower t_P. However, V_{DD} is determined by the process technology and thus is often not under the control of the designer. Furthermore, modern process technologies in which device sizes are reduced require lower V_{DD} (see Table 7.A.1). A motivating factor for lowering V_{DD} is the need to keep the dynamic power dissipation at acceptable levels, especially in very-high-density chips. We will have more to say on this point shortly.

These observations clearly illustrate the conflicting requirements and the trade-offs available in the design of a CMOS digital integrated circuit (and indeed in any engineering design problem).

An Alternative Approach The formulas derived above for t_{PHL} and t_{PLH} underestimate the delay values for inverters implemented in deep-submicron technologies. This arises because of the velocity saturation effect, which we shall discuss briefly in Section 13.5. There

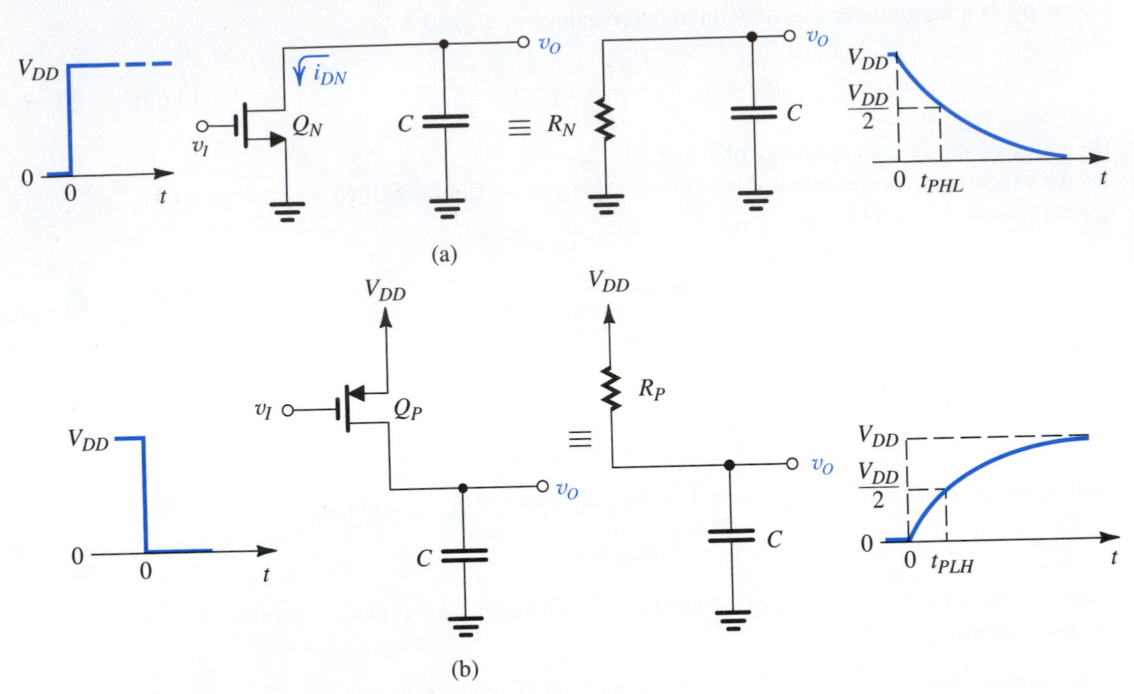

Figure 13.23 Equivalent circuits for determining the propagation delays **(a)** t_{PHL} and **(b)** t_{PLH} of the inverter.

we will see that velocity saturation results in lower MOSFET currents in the saturation region, and hence in increased delay times. To deal with this problem, we present a very simple alternative approach to estimating the inverter propagation delay.

Figure 13.23 illustrates the alternative approach. During the discharge delay t_{PHL}, Q_N is replaced by an equivalent resistance R_N. Similarly, during the charging delay t_{PLH}, Q_P is replaced by an equivalent resistance R_P. It is easy to show that

$$t_{PHL} = 0.69 R_N C \tag{13.68}$$

and

$$t_{PLH} = 0.69 R_P C \tag{13.69}$$

Empirical values have been found for R_N and R_P,

$$R_N = \frac{12.5}{(W/L)_n} \text{ k}\Omega \tag{13.70}$$

$$R_P = \frac{30}{(W/L)_p} \text{ k}\Omega \tag{13.71}$$

Furthermore, it has been found that these values apply for a number of CMOS fabrication processes including 0.25 μm, 0.18 μm, and 0.13 μm (see Hodges et al., 2004).

Example 13.5

For the 0.25-μm process characterized by $V_{DD} = 2.5$ V, $V_{tn} = -V_{tp} = 0.5$ V, $k_n' = 3.5k_p' = 115$ μA/V^2, find t_{PLH}, t_{PHL}, and t_P for an inverter for which $(W/L)_n = 1.5$ and $(W/L)_p = 3$, and for $C = 10$ fF. Use both the approach based on average currents and that based on equivalent resistances, and compare the results obtained. If to save on power dissipation the inverter is operated at $V_{DD} = 2.0$ V, by what factor does t_P change?

Solution

(a) Using the average current approach, we determine from Eq. (13.65),

$$\alpha_n = \frac{2}{\frac{7}{4} - \frac{3 \times 0.5}{2.5} + \left(\frac{0.5}{2.5}\right)^2} = 1.7$$

and using Eq. (13.64),

$$t_{PHL} = \frac{1.7 \times 10 \times 10^{-15}}{110 \times 10^{-6} \times 1.5 \times 2.5} = 41.2 \text{ ps}$$

Since $|V_{tp}| = V_{tn}$,

$$\alpha_p = \alpha_n = 1.7$$

and we can determine t_{PLH} from Eq. (13.66) as

$$t_{PLH} = \frac{1.7 \times 10 \times 10^{-15}}{(110/3.5) \times 10^{-6} \times 3 \times 2.5} = 72.1 \text{ ps}$$

The propagation delay can now be found as

$$t_P = \frac{1}{2}(t_{PHL} + t_{PLH})$$

$$= \frac{1}{2}(41.2 + 72.1) = 56.7 \text{ ps}$$

(b) Using the equivalent resistance approach, we first find R_N from Eq. (13.70) as

$$R_N = \frac{12.5}{1.5} = 8.33 \text{ k}\Omega$$

and then use Eq. (13.68) to determine t_{PHL},

$$t_{PHL} = 0.69 \times 8.33 \times 10^3 \times 10 \times 10^{-15} = 57.5 \text{ ps}$$

Similarly we use Eq. (13.71) to determine R_P,

$$R_P = \frac{30}{3} = 10 \text{ k}\Omega$$

and Eq. (13.69) to determine t_{PLH},

$$t_{PLH} = 0.69 \times 10 \times 10^3 \times 10 \times 10^{-15} = 69 \text{ ps}$$

Thus, while the value obtained for t_{PHL} is higher than that found using average currents, the value for t_{PLH} is about the same. Finally, t_P can be found as

$$t_P = \frac{1}{2}(57.5 + 69) = 63.2 \text{ ps}$$

which a little higher than the value found using average currents.

Example 13.5 *continued*

To find the change in propagation delays obtained when the inverter is operated at $V_{DD} = 2.0$ V, we have to use the method of average currents. (The dependence on the power-supply voltage is absorbed in the empirical values of R_N and R_P.) Using Eq. (13.65), we write

$$\alpha_n = \frac{2}{\frac{7}{4} - \frac{3 \times 0.5}{2} + \left(\frac{0.5}{2}\right)^2} = 2.1$$

The value of t_{PHL} can now be found by using Eq. (13.64):

$$t_{PHL} = \frac{2.1 \times 10 \times 10^{-15}}{110 \times 10^{-6} \times 1.5 \times 2} = 63.6 \text{ ps}$$

Similarly, the value of $\alpha_p = \alpha_n = 2.1$ can be substituted in Eq. (13.66) to obtain

$$t_{PLH} = \frac{2.1 \times 10 \times 10^{-15}}{(110/3.5) \times 10^{-6} \times 3 \times 2} = 111.4 \text{ ps}$$

and t_P can be calculated as

$$t_P = \frac{1}{2}(63.6 + 111.4) = 87.5 \text{ ps}$$

Thus, as expected, reducing V_{DD} has resulted in increased propagation delay.

Before leaving the subject of propagation delay, we should emphasize that hand analysis using the simple formulas above should *not* be expected to yield precise results. Rather, its value is in obtaining design insight. Precise results can always be obtained using SPICE and Multisim simulations (see examples in Appendix B and the extensive material on the CD and the website). However, it is never a good idea to use simulation if one does not know beforehand approximate values of the expected results.

EXERCISES

13.14 For a CMOS inverter fabricated in a 0.18-μm process with $V_{DD} = 1.8$ V, $V_{tn} = -V_{tp} = 0.5$ V, $k_n' = 4k_p' = 300\ \mu\text{A/V}^2$ and having $(W/L)_n = 1.5$ and $(W/L)_p = 3$, find t_{PHL}, t_{PLH}, and t_P when the equivalent load capacitance $C = 10$ fF. Use the method of average currents.
Ans. 24.7 ps; 49.4 ps; 37 ps

D13.15 For a CMOS inverter fabricated in a 0.13-μm process, use the equivalent-resistances approach to determine $(W/L)_n$ and $(W/L)_p$ so that $t_{PLH} = t_{PHL} = 50$ ps when the effective load capacitance $C = 20$ fF.
Ans. 3.5; 8.3

13.3.2 Determining the Equivalent Load Capacitance C

Having determined the propagation delay of the CMOS inverter in terms of the equivalent load capacitance C, it now remains to determine the value of C. For this purpose, a thorough understanding of the various capacitances in a MOS transistor is essential, and we urge the reader to review the material in Section 9.2.1.

Figure 13.24 shows the circuit for determining the propagation delay of the CMOS inverter formed by Q_1 and Q_2. Note that we are showing the inverter driving a similar inverter formed by transistors Q_3 and Q_4. This reflects a practical situation and will help us explain how to determine the contribution of a driven inverter to the equivalent capacitance C at the output of the inverter under study (that formed by Q_1 and Q_2).

Indicated in Fig. 13.24 are the various transistor capacitances that connect to the output node of the Q_1–Q_2 inverter. Also shown is the **wiring capacitance** C_w, which represents the capacitance of the wire or **interconnect** that connects the output of the Q_1–Q_2 inverter to the input of the Q_3–Q_4 inverter. Interconnect capacitances have become increasingly dominant as the technology has scaled down. In fact, some digital IC designers hold the view that interconnect poses a greater limitation on the speed of operation than the transistors themselves. We will discuss this topic briefly in Section 13.5.

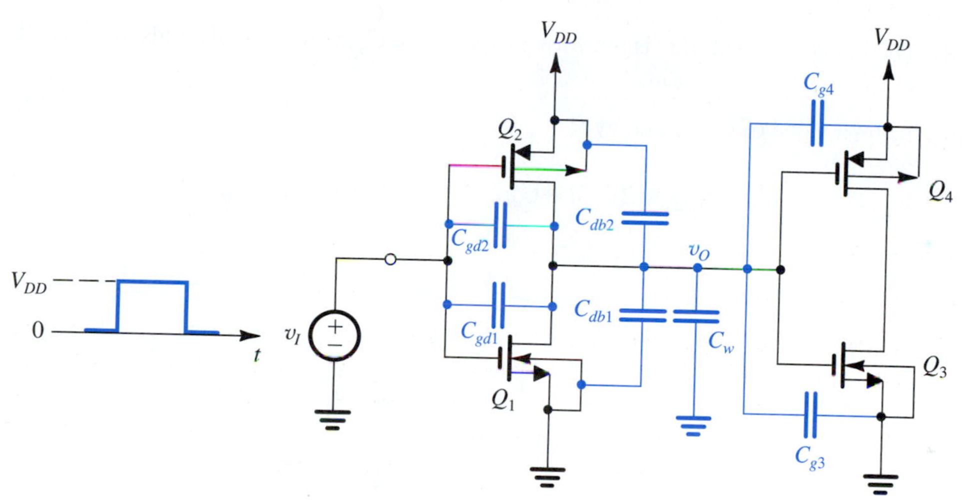

Figure 13.24 Circuit for analyzing the propagation delay of the inverter formed by Q_1 and Q_2, which is driving a similar inverter formed by Q_3 and Q_4.

A glance at the circuit in Fig. 13.24 should be sufficient to indicate that a pencil-and-paper analysis is virtually impossible. That, of course, is the reason we opted for the simplification of replacing all these capacitances with an equivalent capacitance C. Before we consider the determination of C, it is useful to observe that during t_{PLH} or t_{PHL}, the output of the first inverter changes from 0 to $V_{DD}/2$ or from V_{DD} to $V_{DD}/2$, respectively. It follows that the second inverter remains in the same state during each of our analysis intervals. This observation will have an important bearing on our estimation of the equivalent input capacitance of the second inverter. Let's now consider the contribution of each of the capacitances in Fig. 13.24 to the value of the equivalent load capacitance C:

1. The gate–drain overlap capacitance of Q_1, C_{gd1}, can be replaced by an equivalent capacitance between the output node and ground of $2C_{gd1}$. The factor 2 arises because of the Miller effect (Section 9.4.4). Specifically, refer to Fig. 13.25 and note that as v_I goes high and v_O goes low by the same amount, the change in voltage across C_{gd1} is twice that amount. Thus the output node sees in effect twice the value of C_{gd1}. The same applies for the gate–drain overlap capacitance of Q_2, C_{gd2}, which can be replaced by a capacitance $2C_{gd2}$ between the output node and ground.

2. Each of the drain–body capacitances C_{db1} and C_{db2} has a terminal at a constant voltage. Thus for the purpose of our analysis here, C_{db1} and C_{db2} can be replaced with equal capacitances between the output node and ground. Note, however, that the formulas given in Section 9.2.1 for calculating C_{db1} and C_{db2} are small-signal relationships, whereas the analysis here is obviously a large-signal one. A technique has been developed for finding equivalent large-signal values for C_{db1} and C_{db2} (see Hodges et al., (2004) and Rabaey et al., (2003)).

3. Since the second inverter does not switch states, we will assume that the input capacitances of Q_3 and Q_4 remain approximately constant and equal to the total gate capacitance $(WLC_{ox} + C_{gsov} + C_{gdov})$. That is, the input capacitance of the load inverter will be

$$C_{g3} + C_{g4} = (WL)_3 C_{ox} + (WL)_4 C_{ox} + C_{gsov3} + C_{gdov3} + C_{gsov4} + C_{gdov4} \quad (13.72)$$

4. The last component of C is the wiring capacitance C_w, which simply adds to the value of C.

Thus, the total value of C is given by

$$C = 2C_{gd1} + 2C_{gd2} + C_{db1} + C_{db2} + C_{g3} + C_{g4} + C_w \quad (13.73)$$

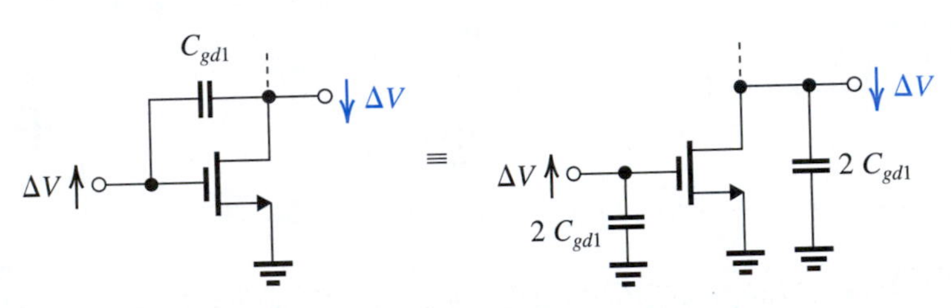

Figure 13.25 The Miller multiplication of the feedback capacitance C_{gd1}.

Example 13.6

Consider a CMOS inverter fabricated in a 0.25-μm process for which $C_{ox} = 6\ \text{fF}/\mu\text{m}^2$, $\mu_n C_{ox} = 115\ \mu\text{A/V}^2$, $\mu_p C_{ox} = 30\ \mu\text{A/V}^2$, $V_{tn} = -V_{tp} = 0.5$ V, and $V_{DD} = 2.5$ V. The W/L ratio of Q_N is 0.375 μm/0.25 μm, and that for Q_P is 1.125 μm/0.25 μm. The gate–source and gate–drain overlap capacitances are specified to be 0.3 fF/μm of gate width. Further, the effective (large-signal) values of drain–body capacitances are C_{dbn} = 1 fF and C_{dbp} = 1 fF. The wiring capacitance $C_w = 0.2$ fF. Find t_{PHL}, t_{PLH}, and t_P when the inverter is driving an identical inverter.

Solution

First, we determine the value of the equivalent capacitance C using Eqs. (13.72) and (13.73),

$$C = 2C_{gd1} + 2C_{gd2} + C_{db1} + C_{db2} + C_{g3} + C_{g4} + C_w$$

where

$$C_{gd1} = 0.3 \times W_n = 0.3 \times 0.375 = 0.1125\ \text{fF}$$

$$C_{gd2} = 0.3 \times W_p = 0.3 \times 1.125 = 0.3375\ \text{fF}$$

$$C_{db1} = 1 \text{ fF}$$
$$C_{db2} = 1 \text{ fF}$$
$$C_{g3} = 0.375 \times 0.25 \times 6 + 2 \times 0.3 \times 0.375 = 0.7875 \text{ fF}$$
$$C_{g4} = 1.125 \times 0.25 \times 6 + 2 \times 0.3 \times 1.125 = 2.3625 \text{ fF}$$
$$C_w = 0.2 \text{ fF}$$

Thus,

$$C = 2 \times 0.1125 + 2 \times 0.3375 + 1 + 1 + 0.7875 + 2.3625 + 0.2 = 6.25 \text{ fF}$$

Next we use Eqs. (13.64) and (13.65) to determine t_{PHL},

$$\alpha_n = \frac{2}{\frac{7}{4} - \frac{3 \times 0.5}{2.5} + \left(\frac{0.5}{2.5}\right)^2} = 1.7$$

$$t_{PHL} = \frac{1.7 \times 6.25 \times 10^{-15}}{115 \times 10^{-6} \times (0.375/0.25) \times 2.5} = 24.6 \text{ ps}$$

Similarly, we use Eqs. (13.66) and (13.67) to determine t_{PLH},

$$\alpha_p = 1.7$$

$$t_{PLH} = \frac{1.7 \times 6.25 \times 10^{-15}}{30 \times 10^{-6} \times (1.125/0.25) \times 2.5} = 31.5 \text{ ps}$$

Finally, we determine t_P as

$$t_P = \frac{1}{2}(24.6 + 31.5) = 28 \text{ ps}$$

EXERCISES

13.16 Consider the inverter specified in Example 13.6 when loaded with an additional 0.1-pF capacitance. What will the propagation delay become?
Ans. 437 ps

13.17 In an attempt to decrease the area of the inverter in Example 13.6, $(W/L)_p$ is made equal to $(W/L)_n$. What is the percentage reduction in area achieved? Find the new values of C, t_{PHL}, t_{PLH}, and t_P. Assume that C_{dbp} does not change significantly.
Ans. 50%; 4.225 fF; 16.6 ps; 21.3 ps; 19 ps

13.18 For the inverter of Example 13.6, find the maximum frequency at which it can be operated.
Ans. 17.9 GHz

13.3.3 Inverter Sizing

In this section we address the question of selecting appropriate (W/L) ratios for the two transistors Q_N and Q_P in an inverter. Our reasoning can be summarized as follows.

1. To minimize area, the length of all channels is usually made equal to the minimum length permitted by the given technology.

2. In a given inverter, if our interest is strictly to minimize area, $(W/L)_n$ is usually selected in the range 1 to 1.5. The selection of $(W/L)_p$ relative to $(W/L)_n$ has influence on the noise margins and t_{PLH}. Both are optimized by matching Q_P and Q_N. This, however, is usually wasteful of area and equally important can increase the effective capacitance C, so that although t_{PLH} is made equal to t_{PHL}, the value of both can be higher than in the case without matching (see Problem 13.40). Thus, selecting $(W/L)_p = (W/L)_n$ is a possibility, and $(W/L)_p = 2(W/L)_n$ is a frequently used compromise.
3. Having settled on an appropriate ratio of $(W/L)_p$ to $(W/L)_n$, we still have to select $(W/L)_n$ to reduce t_P and thus allow higher speeds of operation. Any increase in $(W/L)_n$ and proportionally in $(W/L)_p$ will of course increase area, and hence the inverter contribution to the value of the equivalent capacitance C. To be more precise we express C as the sum of an intrinsic component C_{int} contributed by Q_N and Q_P of the inverter, and an extrinsic component C_{ext} resulting from the wiring and the input capacitance of the driven gates,

$$C = C_{\text{int}} + C_{\text{ext}} \tag{13.74}$$

Increasing $(W/L)_n$ and $(W/L)_p$ of the inverter by a factor S relative to that of a minimum size inverter for which $C_{\text{int}} = C_{\text{int0}}$ results in

$$C = SC_{\text{int0}} + C_{\text{ext}} \tag{13.75}$$

Now, if we use the equivalent-resistances approach to compute t_P and define an equivalent inverter resistance R_{eq} as

$$R_{\text{eq}} = \frac{1}{2}(R_N + R_P) \tag{13.76}$$

then,

$$t_P = 0.69 R_{\text{eq}} C \tag{13.77}$$

Further, if for the minimum-size inverter R_{eq} is R_{eq0}, increasing $(W/L)_n$ and $(W/L)_p$ by the factor S reduces R_{eq} by the same factor:

$$R_{\text{eq}} = R_{\text{eq0}}/S \tag{13.78}$$

Combining Eqs. (13.77), (13.78), and (13.75), we obtain

$$t_P = 0.69\left(\frac{R_{\text{eq0}}}{S}\right)(SC_{\text{int0}} + C_{\text{ext}})$$

$$t_P = 0.69\left(R_{\text{eq0}}C_{\text{int0}} + \frac{1}{S}\, R_{\text{eq0}}C_{\text{ext}}\right) \tag{13.79}$$

We thus see that scaling the W/L ratios does *not* change the component of t_P caused by the capacitances of Q_N and Q_P. It does, however, reduce the component of t_P that results from capacitances external to the inverter itself. It follows that one can use

Eq. (13.79) to decide on a suitable scaling factor S that keeps t_P below a specified maximum value, keeping in mind of course the effect of increasing S on silicon area.

EXERCISE

13.19 For the inverter analyzed in Example 13.6:
(a) Find the intrinsic and extrinsic components of C.
(b) By what factor must $(W/L)_n$ and $(W/L)_p$ be increased to reduce the extrinsic part of t_P by a factor of 2?
(c) Estimate the resulting t_P.
(d) By what factor is the inverter area increased?
Ans. (a) 2.9 fF, 3.35 fF; (b) 2; (c) 20.5 ps; (d) 2

13.3.4 Dynamic Power Dissipation

The negligible static power dissipation of CMOS has been a significant factor in its dominance as the technology of choice in implementing high-density VLSI circuits. However, as the number of gates per chip steadily increases, the dynamic power dissipation has become a serious issue. The dynamic power dissipated in the CMOS inverter is given by Eq. (13.35), which we repeat here as

$$P_{\text{dyn}} = fCV_{DD}^2 \tag{13.80}$$

where f is the frequency at which the gate is switched. It follows that minimizing C is an effective means for reducing dynamic-power dissipation. An even more effective strategy is the use of a lower power-supply voltage. As we have mentioned, CMOS process technologies now utilize V_{DD} values of 1 V or less. These newer chips, however, pack much more circuitry on the chip (as many as 2.3 billion transistors) and operate at higher frequencies (microprocessor clock frequencies above 3 GHz are now available). The dynamic power dissipation of such high-density chips can be over 100 W.

In addition to the dynamic power dissipation that results from the periodic charging and discharging of the inverter load capacitance, there is another component of power dissipation that results from the current that flows through Q_P and Q_N during every switching event. Figure 13.26 shows this inverter current as a function of the input voltage v_I for a matched inverter. We note that the current peaks at $V_M = V_{DD}/2$. Since at this voltage both Q_N and Q_P operate in saturation, the peak current is given by

$$I_{\text{peak}} = \frac{1}{2}\mu_n C_{ox}\left(\frac{W}{L}\right)_n\left(\frac{V_{DD}}{2} - V_{tn}\right)^2 \tag{13.81}$$

The width of the current pulse will depend on the rate of change of v_I with time; the slower the rising edge of the input waveform, the wider the current pulse and the greater the energy drawn from the supply. In general, however, this power component is usually much smaller than P_{dyn}.

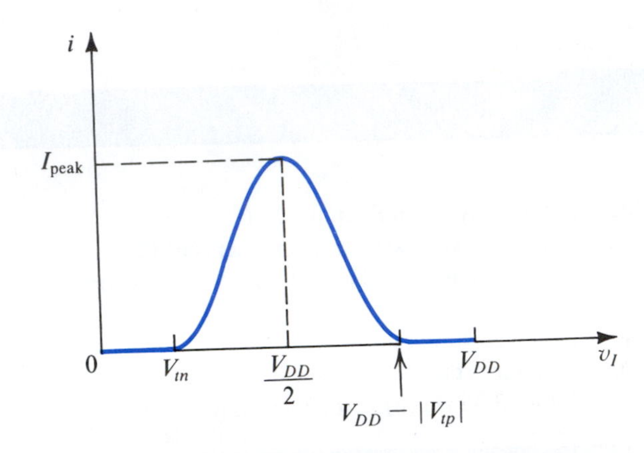

Figure 13.26 The current in the CMOS inverter versus the input voltage.

EXERCISE

13.20 Find the dynamic power dissipation of the inverter analyzed in Example 13.6 when operated at a 1-GHz frequency. If this inverter is switched at its maximum possible operating frequency, what is the value of the power–delay product?
Ans. 39 μW ; 19.5 fJ

13.4 CMOS Logic-Gate Circuits

In this section, we build on our knowledge of inverter design and consider the design of CMOS circuits that realize combinational-logic functions. In combinational circuits, the output at any time is a function only of the values of input signals at that time. Thus, these circuits do not have memory and do not employ feedback. Combinational-logic circuits are used in large quantities in a multitude of applications; indeed, every digital system contains large numbers of combinational-logic circuits.

13.4.1 Basic Structure

A CMOS logic circuit is in effect an extension, or a generalization, of the CMOS inverter: The inverter consists of an NMOS pull-down transistor, and a PMOS pull-up transistor, operated by the input voltage in a complementary fashion. The CMOS logic gate consists of two networks: the **pull-down network (PDN)** constructed of NMOS transistors, and the **pull-up network (PUN)** constructed of PMOS transistors (see Fig. 13.27). The two networks are operated by the input variables, in a complementary fashion. Thus, for the three-input gate represented in Fig. 13.27, the PDN will conduct for all input combinations that require a low output ($Y = 0$) and will then pull the output node down to ground,

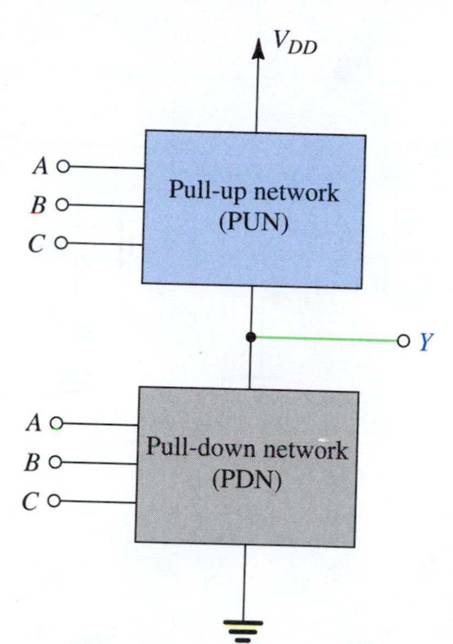

Figure 13.27 Representation of a three-input CMOS logic gate. The PUN comprises PMOS transistors, and the PDN comprises NMOS transistors.

causing a zero voltage to appear at the output, $v_Y = 0$. Simultaneously, the PUN will be off, and no direct dc path will exist between V_{DD} and ground. On the other hand, all input combinations that call for a high output ($Y = 1$) will cause the PUN to conduct, and the PUN will then pull the output node up to V_{DD}, establishing an output voltage $v_Y = V_{DD}$. Simultaneously, the PDN will be cut off, and again, no dc current path between V_{DD} and ground will exist in the circuit.

Now, since the PDN comprises NMOS transistors, and since an NMOS transistor conducts when the signal at its gate is high, the PDN is activated (i.e., conducts) when the inputs are high. In a dual manner, the PUN comprises PMOS transistors, and a PMOS transistor conducts when the input signal at its gate is low; thus the PUN is activated when the inputs are low.

The PDN and the PUN each utilizes devices in parallel to form an OR function, and devices in series to form an AND function. Here, the OR and AND notation refer to current flow or conduction. Figure 13.28 shows examples of PDNs. For the circuit in Fig. 13.28(a), we observe that Q_A will conduct when A is high ($v_A = V_{DD}$) and will then pull the output node down to ground ($v_Y = 0$ V, $Y = 0$). Similarly, Q_B conducts and pulls Y down when B is high. Thus Y will be low when A is high *or* B is high, which can be expressed as

$$\bar{Y} = A + B$$

or equivalently

$$Y = \overline{A + B}$$

The PDN in Fig. 13.28(b) will conduct only when A and B are both high simultaneously. Thus Y will be low when A is high *and* B is high,

$$\bar{Y} = AB$$

or equivalently

$$Y = \overline{AB}$$

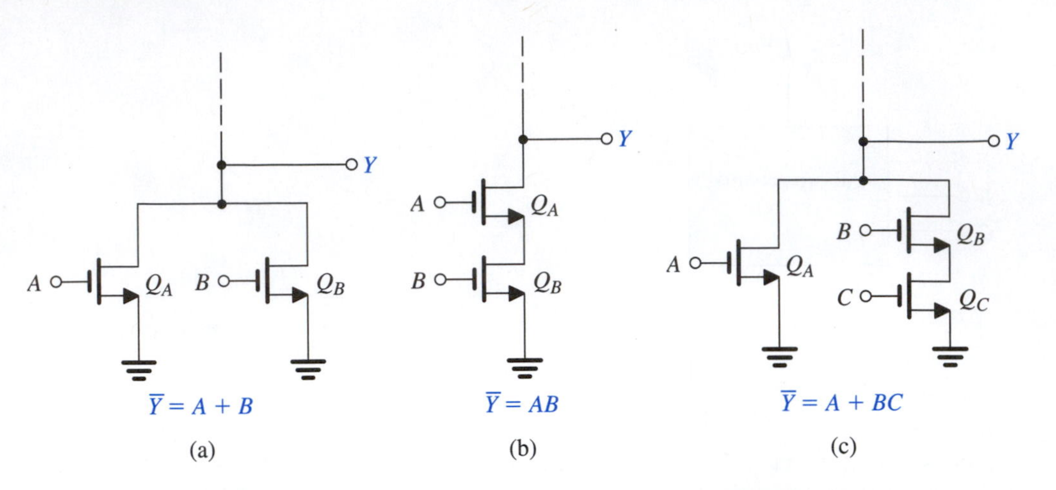

Figure 13.28 Examples of pull-down networks.

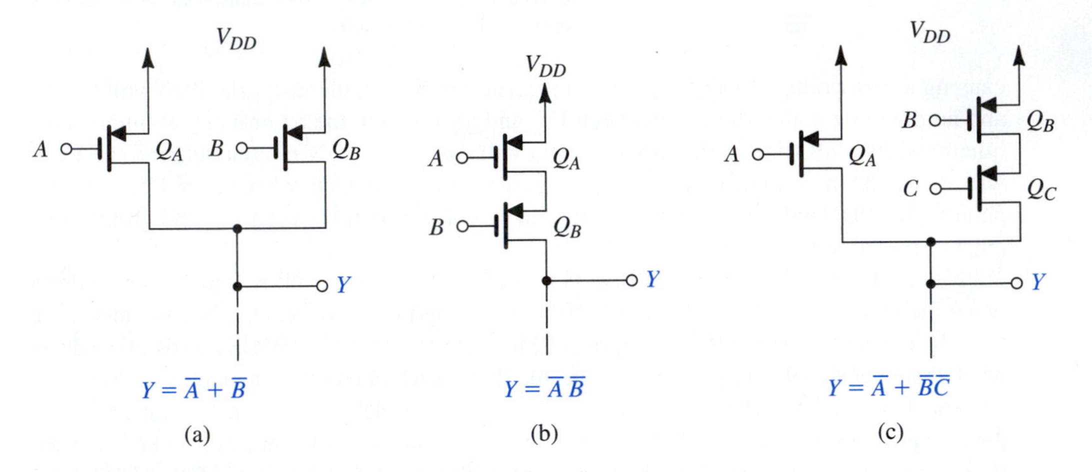

Figure 13.29 Examples of pull-up networks.

As a final example, the PDN in Fig. 13.28(c) will conduct and cause Y to be 0 when A is high *or* when B *and* C are both high, thus

$$\overline{Y} = A + BC$$

or equivalently

$$Y = \overline{A + BC}$$

Next consider the PUN examples shown in Fig. 13.29. The PUN in Fig. 13.29(a) will conduct and pull Y up to $V_{DD}(Y = 1)$ when A is low *or* B is low, thus

$$Y = \overline{A} + \overline{B}$$

The PUN in Fig. 13.29(b) will conduct and produce a high output ($v_Y = V_{DD}$, $Y = 1$) only when A *and* B are both low, thus

$$Y = \overline{A}\overline{B}$$

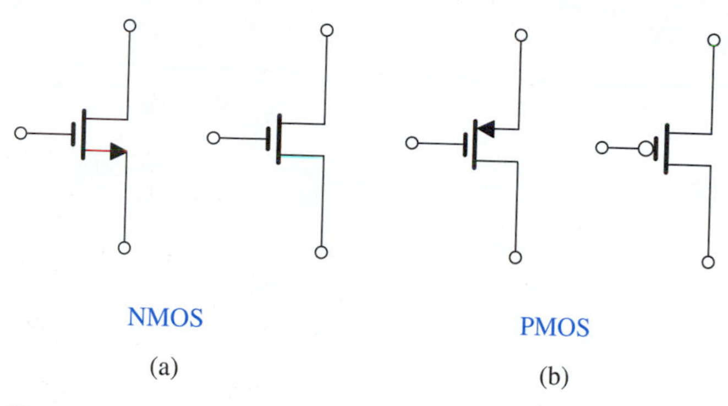

Figure 13.30 Usual and alternative circuit symbols for MOSFETs.

Finally, the PUN in Fig. 13.29(c) will conduct and cause Y to be high (logic 1) if A is low *or* if B *and* C are both low; thus,

$$Y = \overline{A} + \overline{B}\overline{C}$$

Having developed an understanding and an appreciation of the structure and operation of PDNs and PUNs, we now consider complete CMOS gates. Before doing so, however, we wish to introduce alternative circuit symbols, that are almost universally used for MOS transistors by digital-circuit designers. Figure 13.30 shows our usual symbols (left) and the corresponding "digital" symbols (right). Observe that the symbol for the PMOS transistor with a circle at the gate terminal is intended to indicate that the signal at the gate has to be low for the device to be activated (i.e., to conduct). Thus, in terms of logic-circuit terminology, the gate terminal of the PMOS transistor is an *active low* input. Besides indicating this property of PMOS devices, the digital symbols omit any indication of which of the device terminals is the source and which is the drain. This should cause no difficulty at this stage of our study; simply remember that for an NMOS transistor, the drain is the terminal that is at the higher voltage (current flows from drain to source), and for a PMOS transistor the source is the terminal that is at the higher voltage (current flows from source to drain). To be consistent with the literature, we shall henceforth use these modified symbols for MOS transistors in logic applications, except in locations where our usual symbols help in understanding circuit operation.

13.4.2 The Two-Input NOR Gate

We first consider the CMOS gate that realizes the two-input NOR function

$$Y = \overline{A + B} = \overline{A}\overline{B} \tag{13.82}$$

We see that Y is to be low (PDN conducting) when A is high or B is high. Thus the PDN consists of two parallel NMOS devices with A and B as inputs (i.e., the circuit in Fig. 13.28a). For the PUN, we note from the second expression in Eq. (13.82) that Y is to be high when A and B are both low. Thus the PUN consists of two series PMOS devices with A and B as the inputs (i.e., the circuit in Fig. 13.29b). Putting the PDN and the PUN together gives the CMOS NOR gate shown in Fig. 13.31. Note that extension to a higher number of inputs is straightforward: For each additional input, an NMOS transistor is added in parallel with Q_{NA} and Q_{NB}, and a PMOS transistor is added in series with Q_{PA} and Q_{PB}.

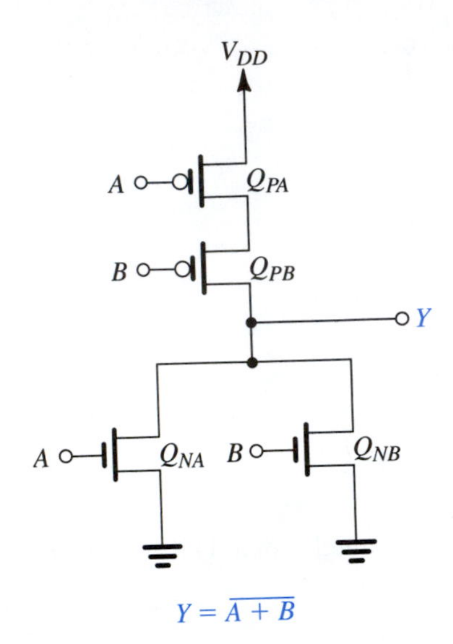

Figure 13.31 A two-input CMOS NOR gate.

13.4.3 The Two-Input NAND Gate

The two-input NAND function is described by the Boolean expression

$$Y = \overline{AB} = \overline{A} + \overline{B} \tag{13.83}$$

To synthesize the PDN, we consider the input combinations that require Y to be low: There is only one such combination, namely, A and B both high. Thus, the PDN simply comprises two NMOS transistors in series (such as the circuit in Fig. 13.28b). To synthesize the PUN, we consider the input combinations that result in Y being high. These are found from the second expression in Eq. (13.83) as A low or B low. Thus, the PUN consists of two parallel PMOS transistors with A and B applied to their gates (such as the circuit in Fig. 13.29a). Putting the PDN and PUN together results in the CMOS NAND gate implementation shown in Fig. 13.32. Note that extension to a higher number of inputs is straightforward: For each

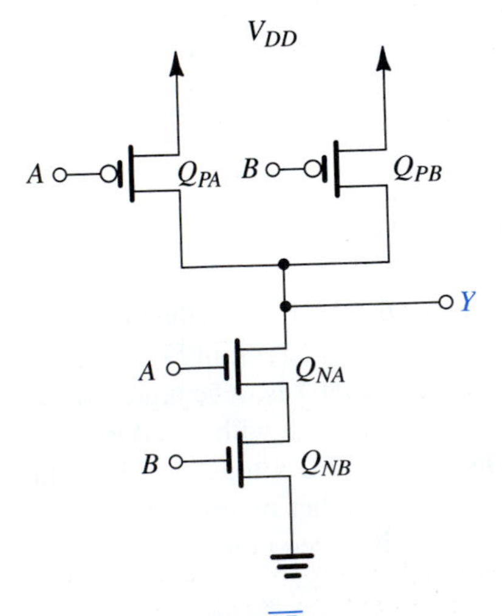

Figure 13.32 A two-input CMOS NAND gate.

additional input, we add an NMOS transistor in series with Q_{NA} and Q_{NB}, and a PMOS transistor in parallel with Q_{PA} and Q_{PB}.

13.4.4 A Complex Gate

Consider next the more complex logic function

$$Y = \overline{A(B+CD)} \tag{13.84}$$

Since $\bar{Y} = A(B+CD)$, we see that Y should be low for A high and simultaneously either B high or C and D both high, from which the PDN is directly obtained. To obtain the PUN, we need to express Y in terms of the complemented variables. We do this through repeated application of DeMorgan's law, as follows:

$$\begin{aligned} Y &= \overline{A(B+CD)} \\ &= \bar{A} + \overline{B+CD} \\ &= \bar{A} + \bar{B}\,\overline{CD} \\ &= \bar{A} + \bar{B}(\bar{C}+\bar{D}) \end{aligned} \tag{13.85}$$

Thus, Y is high for A low or B low and either C or D low. The corresponding complete CMOS circuit will be as shown in Fig. 13.33.

13.4.5 Obtaining the PUN from the PDN and Vice Versa

From the CMOS gate circuits considered thus far (e.g., that in Fig. 13.33), we observe that the PDN and the PUN are dual networks: Where a series branch exists in one, a parallel branch exists in the other. Thus, we can obtain one from the other, a process that can be simpler than having to synthesize each separately from the Boolean expression of the function. For instance, in the circuit of Fig. 13.33, we found it relatively easy to obtain the PDN, simply because we already had $\bar{Y}$ in terms of the uncomplemented inputs. On the other hand, to obtain the PUN, we had to manipulate the given Boolean expression to express Y as a function of the complemented variables, the form convenient for synthesizing PUNs. Alternatively, we could have used this duality property to obtain the PUN from the PDN. The reader is urged to refer to Fig. 13.33 to convince herself that this is indeed possible.

It should, however, be mentioned that at times it is not easy to obtain one of the two networks from the other using the duality property. For such cases, one has to resort to a more rigorous process, which is beyond the scope of this book (see Kang and Leblebici, 1999).

13.4.6 The Exclusive-OR Function

An important function that often arises in logic design is the exclusive-OR (XOR) function,

$$Y = A\bar{B} + \bar{A}B \tag{13.86}$$

We observe that since Y (rather than $\bar{Y}$) is given, it is easier to synthesize the PUN. We note, however, that unfortunately Y is not a function of the complemented variables only (as we

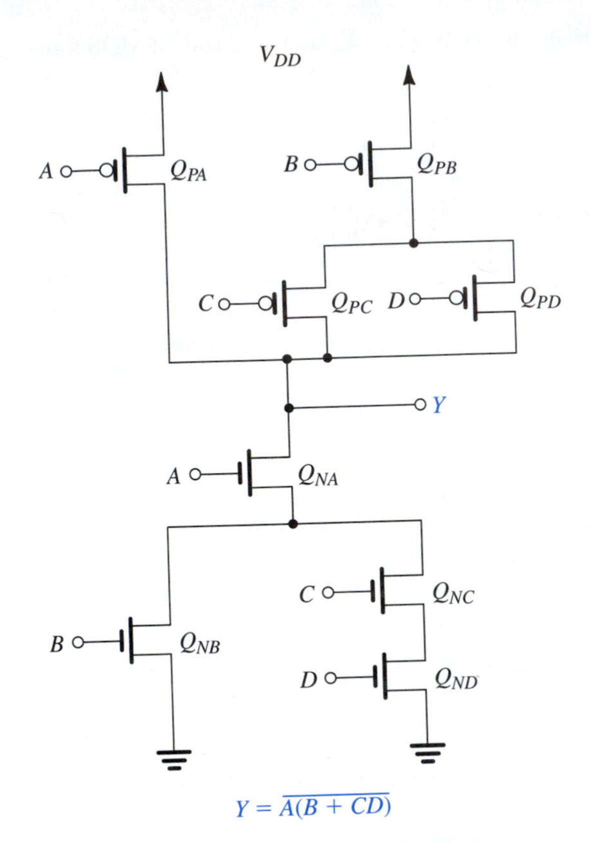

Figure 13.33 CMOS realization of a complex gate.

would like it to be). Thus, we will need additional inverters. The PUN obtained directly from Eq. (13.86) is shown in Fig. 13.34(a). Note that the Q_1, Q_2 branch realizes the first term ($A\bar{B}$), whereas the Q_3, Q_4 branch realizes the second term ($\bar{A}B$). Note also the need for two additional inverters to generate $\bar{A}$ and $\bar{B}$.

As for synthesizing the PDN, we can obtain it as the dual network of the PUN in Fig. 13.34(a). Alternatively, we can develop an expression for $\bar{Y}$ and use it to synthesize the PDN. Leaving the first approach for the reader to do as an exercise, we shall utilize the direct synthesis approach. DeMorgan's law can be applied to the expression in Eq. (13.86) to obtain $\bar{Y}$ as

$$\bar{Y} = AB + \bar{A}\bar{B} \tag{13.87}$$

The corresponding PDN will be as in Fig. 13.34(b), which shows the CMOS realization of the exclusive-OR function except for the two additional inverters. Note that the exclusive-OR requires 12 transistors for its realization, a rather complex network. Later, in Section 14.2, we shall show a simpler realization of the XOR employing a different form of CMOS logic.

Another interesting observation follows from the circuit in Fig. 13.34(b). The PDN and the PUN here are *not* dual networks. Indeed, duality of the PDN and the PUN is not a necessary condition. Thus, although a dual of PDN (or PUN) can always be used for PUN (or PDN), the two networks are not necessarily duals.

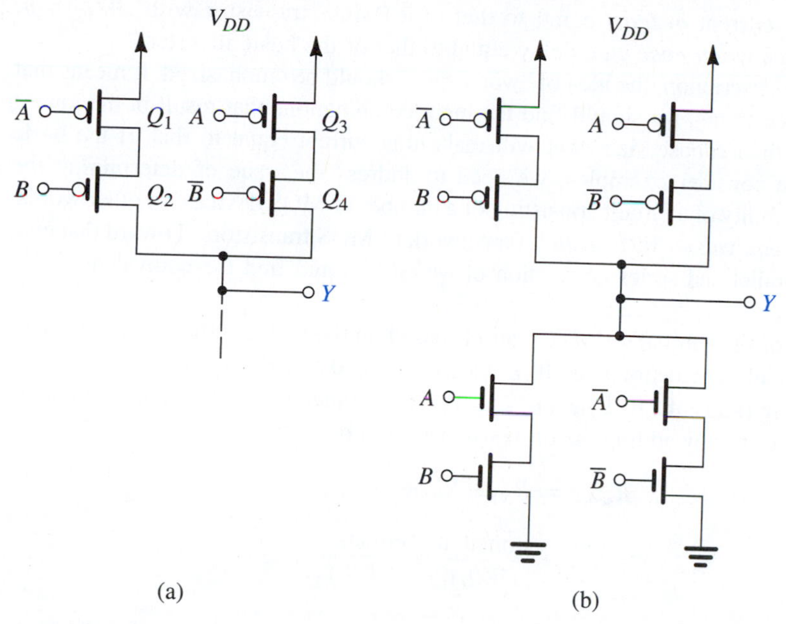

Figure 13.34 Realization of the exclusive-OR (XOR) function: **(a)** The PUN synthesized directly from the expression in Eq. (13.86). **(b)** The complete XOR realization utilizing the PUN in (a) and a PDN that is synthesized directly from the expression in Eq. (13.87). Note that two inverters (not shown) are needed to generate the complemented variables. Also note that in this XOR realization, the PDN and the PUN are not dual networks; however, a realization based on dual networks is possible (see Problem 13.47).

13.4.7 Summary of the Synthesis Method

1. The PDN can be most directly synthesized by expressing $\overline{Y}$ as a function of the *uncomplemented* variables. If complemented variables appear in this expression, additional inverters will be required to generate them.
2. The PUN can be most directly synthesized by expressing Y as a function of the *complemented* variables and then applying the uncomplemented variables to the gates of the PMOS transistors. If uncomplemented variables appear in the expression, additional inverters will be needed.
3. The PDN can be obtained from the PUN (and vice versa) using the duality property.

13.4.8 Transistor Sizing

Once a CMOS gate circuit has been generated, the only significant step remaining in the design is to decide on W/L ratios for all devices. These ratios usually are selected to provide the gate with current-driving capability in both directions equal to that of the basic inverter. For the basic inverter design, denote $(W/L)_n = n$ and $(W/L)_p = p$, where n is usually 1 to 1.5 and, for a matched design, $p = (\mu_n/\mu_p)n$; although often $p = 2n$ and for minimum area $p = n$. Thus, we wish to select individual W/L ratios for all transistors in a logic gate so that the PDN should be able to provide a capacitor discharge current *at least* equal to that of an NMOS transistor with $W/L = n$, and the PUN should be able to provide a charging current *at least* equal to that of a PMOS transistor with $W/L = p$. This will guarantee a *worst-case* gate delay equal to that of the basic inverter.[4]

In the preceding description, the idea of "worst case" should be emphasized. It means that in deciding on device sizing, we should find the input combinations that result in the lowest output current and then choose sizes that will make this current equal to that of the basic inverter. Before we consider examples, we need to address the issue of determining the current-driving capability of a circuit consisting of a number of MOS devices. In other words, we need to find the *equivalent* W/L *ratio* of a network of MOS transistors. Toward that end, we consider the parallel and series connection of MOSFETs and find the equivalent W/L ratios.

The derivation of the equivalent W/L ratio is based on the fact that the on resistance of a MOSFET is inversely proportional to W/L (see Eqs. 13.70 and 13.71). Thus, if a number of MOSFETs having ratios of $(W/L)_1$, $(W/L)_2$, . . ., are connected in series, the equivalent series resistance obtained by adding the on-resistances will be

$$\begin{aligned} R_{\text{series}} &= R_{N1} + R_{N2} + \cdots \\ &= \frac{\text{constant}}{(W/L)_1} + \frac{\text{constant}}{(W/L)_2} + \cdots \\ &= \text{constant}\left[\frac{1}{(W/L)_1} + \frac{1}{(W/L)_2} + \cdots\right] \\ &= \frac{\text{constant}}{(W/L)_{\text{eq}}} \end{aligned}$$

resulting in the following expression for $(W/L)_{\text{eq}}$ for transistors connected in series:

$$(W/L)_{\text{eq}} = \frac{1}{\dfrac{1}{(W/L)_1} + \dfrac{1}{(W/L)_2} + \cdots} \tag{13.88}$$

Similarly, we can show that the parallel connection of transistors with W/L ratios of $(W/L)_1$, $(W/L)_2$, . . . , results in an equivalent W/L of

$$(W/L)_{\text{eq}} = (W/L)_1 + (W/L)_2 + \cdots \tag{13.89}$$

As an example, two identical MOS transistors with individual W/L ratios of 4 result in an equivalent W/L of 2 when connected in series and of 8 when connected in parallel.[5]

As an example of proper sizing, consider the four-input NOR in Fig. 13.35. Here, the worst case (the lowest current) for the PDN is obtained when only one of the NMOS transistors is conducting. We therefore select the W/L of each NMOS transistor to be equal to that of the NMOS transistor of the basic inverter, namely, n. For the PUN, however, the worst-case situation (and indeed the only case) occurs when all inputs are low and the four series PMOS transistors are conducting. Since the equivalent W/L will be one-quarter of that of

[4]This statement assumes that the total effective capacitance C of the logic gate is the same as that of the inverter. In actual practice, the value of C will be larger for a gate, especially as the fan-in is increased.

[5]Another way of thinking about this is as follows: Connecting MOS transistors in series is equivalent to adding the lengths of their channels while the width does not change; connecting MOS transistors in parallel does not change the channel length but increases the width to the sum of the W's.

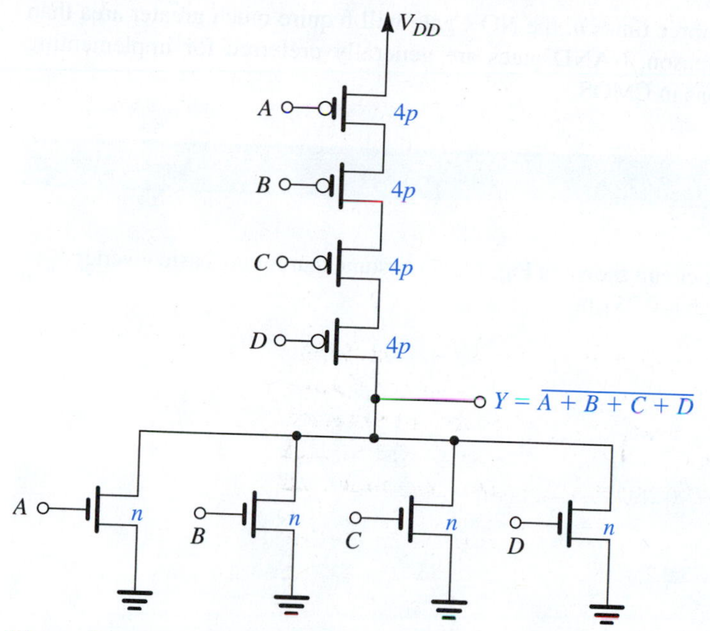

Figure 13.35 Proper transistor sizing for a four-input NOR gate. Note that n and p denote the W/L ratios of Q_N and Q_P, respectively, of the basic inverter.

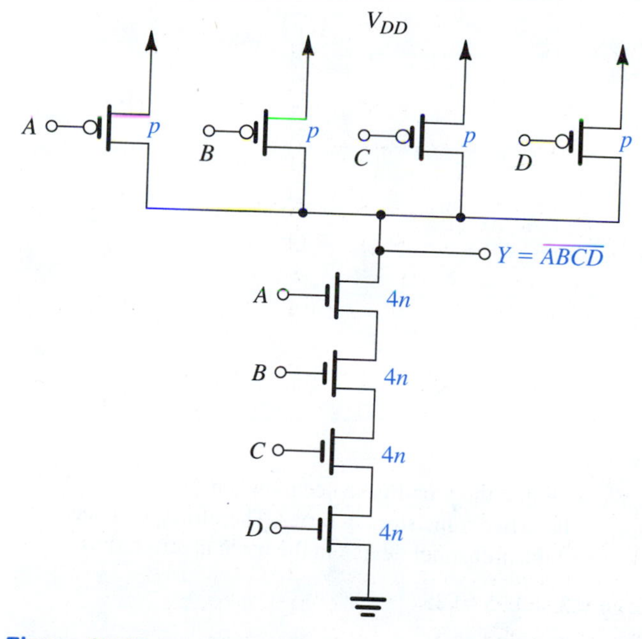

Figure 13.36 Proper transistor sizing for a four-input NAND gate. Note that n and p denote the W/L ratios of Q_N and Q_P, respectively, of the basic inverter.

each PMOS device, we should select the W/L ratio of each PMOS transistor to be four times that of Q_P of the basic inverter, that is, $4p$.

As another example, we show in Fig. 13.36 the proper sizing for a four-input NAND gate. Comparison of the NAND and NOR gates in Figs. 13.35 and 13.36 indicates that because p is usually two to three times n, the NOR gate will require much greater area than the NAND gate. For this reason, NAND gates are generally preferred for implementing combinational logic functions in CMOS.

Example 13.7

Provide transistor W/L ratios for the logic circuit shown in Fig. 13.37. Assume that for the basic inverter $n = 1.5$ and $p = 5$ and that the channel length is 0.25 μm.

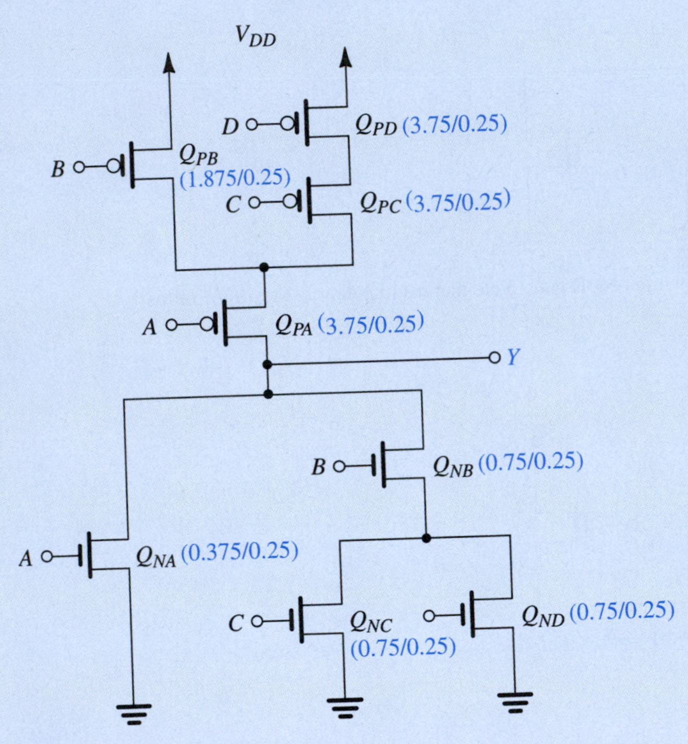

Figure 13.37 Circuit for Example 13.7.

Solution

Refer to Fig. 13.37, and consider the PDN first. We note that the worst case occurs when Q_{NB} is on and either Q_{NC} or Q_{ND} is on. That is, in the worst case, we have two transistors in series. Therefore, we select each of Q_{NB}, Q_{NC}, and Q_{ND} to have twice the width of the n-channel device in the basic inverter, thus

$$Q_{NB}\text{: } W/L = 2n = 3 = 0.75/0.25$$

$$Q_{NC}\text{: } W/L = 2n = 3 = 0.75/0.25$$

$$Q_{ND}\text{: } W/L = 2n = 3 = 0.75/0.25$$

For transistor Q_{NA}, select W/L to be equal to that of the n-channel device in the basic inverter:

$$Q_{NA}\text{: } W/L = n = 1.5 = 0.375/0.25$$

Next, consider the PUN. Here, we see that in the worst case, we have three transistors in series: Q_{PA}, Q_{PC}, and Q_{PD}. Therefore, we select the W/L ratio of each of these to be three times that of Q_p in the basic inverter, that is, $3p$, thus

$$Q_{PA}\text{: } W/L = 3p = 15 = 3.75/0.25$$

$$Q_{PC}\text{: } W/L = 3p = 15 = 3.75/0.25$$

$$Q_{PD}\text{: } W/L = 3p = 15 = 3.75/0.25$$

Finally, the W/L ratio for Q_{PB} should be selected so that the equivalent W/L of the series connection of Q_{PB} and Q_{PA} should be equal to p. It follows that for Q_{PB} the ratio should be $1.5p$,

$$Q_{PB}\text{: } W/L = 1.5p = 7.5 = 1.875/0.25$$

Figure 13.37 shows the circuit with the transistor sizes indicated.

13.4.9 Effects of Fan-In and Fan-Out on Propagation Delay

Each additional input to a CMOS gate requires two additional transistors, one NMOS and one PMOS. This is in contrast to other forms of MOS logic, where each additional input requires only one additional transistor (see Section 14.1). The additional transistor in CMOS not only increases the chip area but also increases the total effective capacitance per gate and in turn increases the propagation delay. The size-scaling method described earlier compensates for some (but not all) of the increase in t_P. Specifically, by increasing device size, we are able to preserve the current-driving capability. However, the capacitance C increases because of both the increased number of inputs and the increase in device size. Thus t_P will still increase with fan-in, a fact that imposes a practical limit on the fan-in of, say, the NAND gate to about 4. If a higher number of inputs is required, then "clever" logic design should be adopted to realize the given Boolean function with gates of no more than four inputs. This would usually mean an increase in the number of cascaded stages and thus an increase in delay. However, such an increase in delay can be less than the increase due to the large fan-in (see Problem 13.56).

An increase in a gate's fan-out adds directly to its load capacitance and, thus, increases its propagation delay.

Thus although CMOS has many advantages, it does suffer from increased circuit complexity when the fan-in and fan-out are increased, and from the corresponding effects of this complexity on both chip area and propagation delay. Later, in Sections 14.1 and 14.2, we shall study some simplified forms of CMOS logic that attempt to reduce this complexity, although at the expense of forgoing some of the advantages of basic CMOS.

EXERCISES

13.21 For a process technology with $L = 0.18\ \mu\text{m}$, $n = 1.5$, $p = 3$, give the sizes of all transistors in (a) a four-input NOR and (b) a four-input NAND. Also, give the relative areas of the two gates.

Ans. (a) NMOS devices: $W/L = 0.27/0.18$, PMOS devices: $2.16/0.18$;
(b) NMOS devices: $W/L = 1.08/0.18$, PMOS devices: $0.54/0.18$;
NOR area/NAND area = 1.5

13.22 For the scaled NAND gate in Exercise 13.21, find the ratio of the maximum to minimum current available to (a) charge a load capacitance and (b) discharge a load capacitance.
Ans. (a) 4; (b) 1

13.5 Implications of Technology Scaling: Issues in Deep-Submicron Design

As mentioned in Chapter 4, and in a number of locations throughout the book, the minimum MOSFET channel length has been continually reduced over the past 40 years or so. In fact, a new CMOS fabrication technology has been introduced every 2 or 3 years, with the minimum allowable channel length reduced by about 30%, that is, to 0.7 the value in the preceding generation. Thus, with every new **technology generation**, the device area has been reduced by a factor of $1/(0.7 \times 0.7)$ or approximately 2, allowing the fabrication of twice as many devices on a chip of the same area. This astounding phenomenon, predicted more than 40 years ago by Gordon Moore,[6] has become known as **Moore's law**. It is this ability to pack an exponentially increasing number of transistors on an IC chip that has resulted in the continuing reduction in the cost per logic function.

Figure 13.38 shows the exponential reduction in MOSFET channel length (by a factor of 2 every 5 years) over a 40 year period, with the dots indicating some of the prominent **technology generations**, or **nodes**. Thus, we see the 10-μm process of the early 1970s, the submicron ($L < 1$ μm) processes of the early 1990s, and the deep-submicron ($L < 0.25$ μm) processes of the last decade, including the current 45-nm process. A microprocessor chip fabricated in a 45-nm CMOS process and having 2.3 billion transistors was announced in 2009. Deep-submicron (DSM) processes present the circuit designer with a host of new opportunities and challenges. It is our purpose in this section to briefly consider some of these.

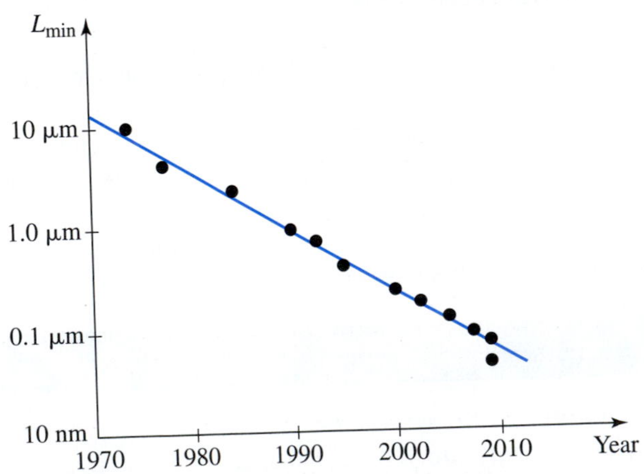

Figure 13.38 The MOSFET channel length has been reduced by a factor of 2 every about 5 years. This phenomenon, known as Moore's law is continuing.

[6]Gordon Moore is one of the pioneers of the semiconductor industry and a cofounder of Intel.

13.5.1 Scaling Implications

Table 13.2 provides a summary of the implications of scaling the device dimensions by a factor 1/*S*, where $S > 1$. As well, we assume that V_{DD} and V_t are scaled by the same factor. Although the scaling of V_{DD} has occurred for a number of technology nodes (e.g., from 5 V for the 0.5-μm process down to 1.2 V for the 0.13-μm process), V_t has been reduced but not by the same factor. Thus the assumption in row 2 of Table 13.2 is not entirely correct. Nevertheless, our interest here is to gain a general appreciation for the effects of scaling.

Table 13.2 provides the relationships for the various transistor and inverter parameters in order to show how the resulting scale factors are obtained. We thus see that the device area scales by $1/S^2$; the oxide capacitance C_{ox}, and the transconductance parameters k_n' and k_p' scale by *S*; and the MOSFET gate capacitance scales by 1/*S*. It is important to note that the component of the inverter propagation delay due to the transistor capacitances (i.e., excluding the wiring capacitance) scales by 1/*S*; this very useful result of scaling implies that the circuit can be operated at *S* times the frequency; that is, the speed of operation increases by a factor *S*. Equally important, the dynamic power dissipation scales by $1/S^2$. This, of course, is a major motivating factor behind the scaling of V_{DD}. Another motivating factor is the need to keep the electric fields in the MOSFETs within acceptable bounds.

Although the dynamic power dissipation is scaled by $1/S^2$, the power per unit area remains unchanged. Nevertheless, for a number of reasons, as the size and complexity of digital IC chips continue to increase, so does their power dissipation. Indeed power dissipation has now become the number-one issue in IC design. The problem is further exacerbated by the static power dissipation, arising from both subthreshold conduction and diode leakage currents, that plagues deep-submicron CMOS devices. We will discuss this issue shortly.

Table 13.2 Implications of Device and Voltage Scaling

	Parameter	Relationship	Scaling Factor
1	W, L, t_{ox}		1/*S*
2	V_{DD}, V_t		1/*S*
3	Area/Device	WL	$1/S^2$
4	C_{ox}	ε_{ox}/t_{ox}	*S*
5	k_n', k_p'	$\mu_n C_{ox}, \mu_p C_{ox}$	*S*
6	C_{gate}	WLC_{ox}	1/*S*
7	t_P (intrinsic)	$\alpha C/k' V_{DD}$	1/*S*
8	Energy/Switching cycle (intrinsic)	CV_{DD}^2	$1/S^3$
9	P_{dyn}	$f_{max} CV_{DD}^2 = \dfrac{CV_{DD}^2}{2t_P}$	$1/S^2$
10	Power density	P_{dyn} /Device area	1

EXERCISES

13.23 By what factor does the power–delay product PDP change if an inverter is fabricated in a 0.13 μm technology rather than a 0.25-μm technology? Assume $S \simeq 2$.
Ans. PDP decreases by a factor of 8.

13.24 If V_{DD} and V_t are kept constant, which entries in Table 13.2 change and to what value?
Ans. t_P now scales by $1/S^2$; the energy/switching cycle now scales by $1/S$ only; P_{dyn} now scales by S; and the power density now scales by S^3 (a major problem).

13.5.2 Velocity Saturation

The short channels of MOSFETs fabricated in deep-submicron processes give rise to physical phenomena not present in long-channel devices, and thus to changes in the MOSFET i–v characteristics. The most important of these **short-channel** effects is **velocity saturation**. Here we refer to the drift velocity of electrons in the channel of an NMOS transistor (holes in PMOS) under the influence of the longitudinal electric field established by v_{DS}. In our derivation of the MOSFET i–v characteristics in Section 5.1, we assumed that the velocity v_n of the electrons in an n-channel device is given by

$$v_n = \mu_n E \tag{13.90}$$

where E is the electric field given by

$$E = \frac{v_{DS}}{L} \tag{13.91}$$

The relationship in Eq. (13.90) applies as long as E is below a critical value E_{cr} which falls in the range 1 V/μm to 5 V/μm. For $E > E_{cr}$, the drift velocity saturates at a value v_{sat} of approximately 10^7 cm/s. Figure 13.39 shows a sketch of v versus E. Although the change from a linear to a constant v is gradual, we shall assume for simplicity that v saturates abruptly at $E = E_{cr}$.

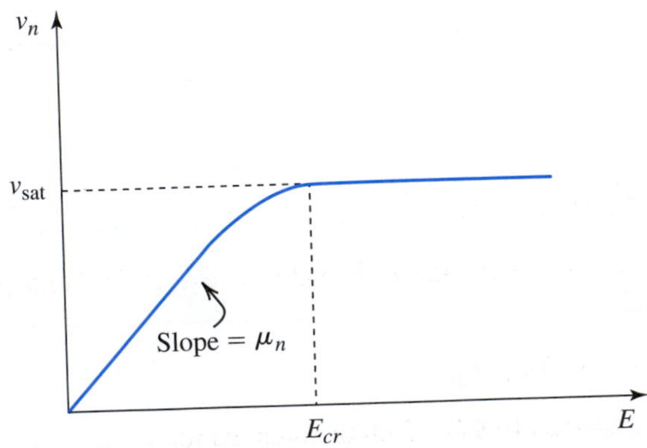

Figure 13.39 The velocity of electrons in the channel of an NMOS transistor reach a constant velocity $v_{sat} \simeq 10^7$ cm/s when the electric field E reaches a critical value E_{cr}. A similar situation occurs for p-channel devices.

The electric field E in a short-channel MOSFET can easily exceed E_{cr} even though V_{DD} is low. If we denote the value of v_{DS} at which velocity saturation occurs by V_{DSsat}, then from Eq. (13.91),

$$E_{cr} = \frac{V_{DSsat}}{L} \tag{13.92}$$

which when substituted in Eq. (13.90) provides

$$v_{sat} = \mu_n\left(\frac{V_{DSsat}}{L}\right) \tag{13.93}$$

or alternatively,

$$V_{DSsat} = \left(\frac{L}{\mu_n}\right)v_{sat} \tag{13.94}$$

Thus, V_{DSsat} is a device parameter.

EXERCISE

13.25 Find V_{DSsat} for an NMOS transistor fabricated in a 0.25-µm CMOS process with $\mu_n = 400\ \text{cm}^2/\text{V}\cdot\text{s}$. Let $L = 0.25$ µm and assume $v_{sat} = 10^7$ cm/s.
Ans. 0.63 V

The i_D–v_{DS} Characteristics The $i_D - v_{DS}$ equations of the MOSFET can be modified to include velocity saturation as follows. Consider a long-channel NMOS transistor operating in the triode region with v_{GS} set to a constant value V_{GS}. The drain current will be

$$i_D = \mu_n C_{ox}\left(\frac{W}{L}\right)v_{DS}\left[(V_{GS} - V_t) - \frac{1}{2}v_{DS}\right] \tag{13.95}$$

where we have for the time being neglected channel-length modulation. We know from our study in Section 5.1 that i_D will saturate at

$$v_{DS} = V_{OV} = V_{GS} - V_t \tag{13.96}$$

and the saturation current will be

$$i_D = \frac{1}{2}\mu_n C_{ox}\left(\frac{W}{L}\right)(V_{GS} - V_t)^2 \tag{13.97}$$

This will also be the case in a short-channel device as long as the value of v_{DS} in Eq. (13.96) is lower than V_{DSsat}. That is, as long as

$$V_{OV} < V_{DSsat}$$

the current i_D will be given by Eqs. (13.95) and (13.97). If, on the other hand,

$$V_{OV} > V_{DSsat}$$

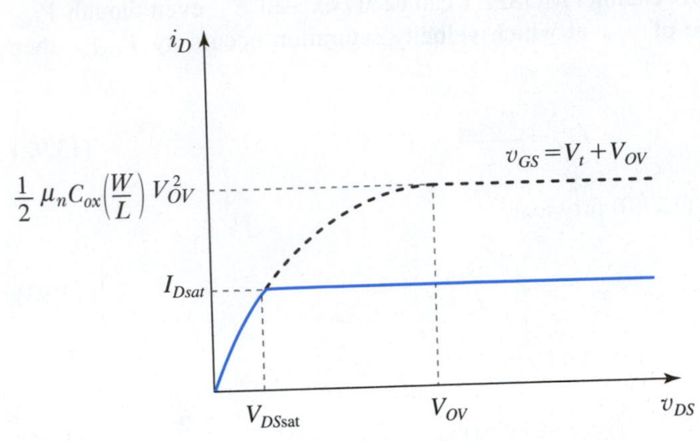

Figure 13.40 Velocity saturation causes the i_D–v_{DS} characteristic to saturate at V_{DSsat}. This early saturation results in a current I_{Dsat} that is lower than the value for a long-channel device.

then velocity saturation kicks in at $v_{DS} = V_{DSsat}$ and i_D saturates at a value I_{Dsat}, as shown in Fig. 13.40. The value of I_{Dsat} can be obtained by substituting $v_{DS} = V_{DSsat}$ in Eq. (13.95),

$$I_{Dsat} = \mu_n C_{ox}\left(\frac{W}{L}\right)V_{DSsat}\left(V_{GS}-V_t-\frac{1}{2}V_{DSsat}\right) \tag{13.98}$$

This expression can be simplified by utilizing Eq. (13.94) to obtain

$$I_{Dsat} = WC_{ox}v_{sat}\left(V_{GS}-V_t-\frac{1}{2}V_{DSsat}\right) \tag{13.99}$$

Replacing V_{GS} in Eq. (13.98) with v_{GS}, and incorporating the channel-length modulation factor $(1+\lambda v_{DS})$, we obtain a general expression for the drain current of an NMOS transistor operating in velocity saturation,

$$i_D = \mu_n C_{ox}\left(\frac{W}{L}\right)V_{DSsat}\left(v_{GS}-V_t-\frac{1}{2}V_{DSsat}\right)(1+\lambda v_{DS}) \tag{13.100}$$

which applies for

$$v_{GS}-V_t \geq V_{DSsat} \quad \text{and} \quad v_{DS} \geq V_{DSsat} \tag{13.101}$$

Figure 13.41 shows a set of i_D–v_{DS} characteristic curves and clearly delineates the three regions of operation: triode, saturation, and velocity saturation.

Equation (13.100) indicates that in the velocity-saturation region, i_D is linearly related to v_{GS}. This is a major change from the quadratic relationship that characterizes operation in the saturation region. Figure 13.42 makes this point clearer by presenting a graph for i_D versus v_{GS} of a short-channel device operating at $v_{DS} > V_{DSsat}$. Observe that for $0 < v_{GS}-V_t \leq V_{DSsat}$, the MOSFET operates in the saturation region and i_D is related to v_{GS} by the familiar quadratic equation (Eq. 13.97). For $v_{GS}-V_t \geq V_{DSsat}$, the transistor enters the velocity-saturation region and i_D varies linearly with v_{GS} (Eq. 13.100).

Short-channel PMOS transistors undergo velocity saturation at the same value of v_{sat} (approximately 10^7 cm/s), but the effects on the device characteristics are less pronounced than in the NMOS case. This is due to the lower values of μ_p and the correspondingly higher values of E_{cr} and V_{DSsat}.

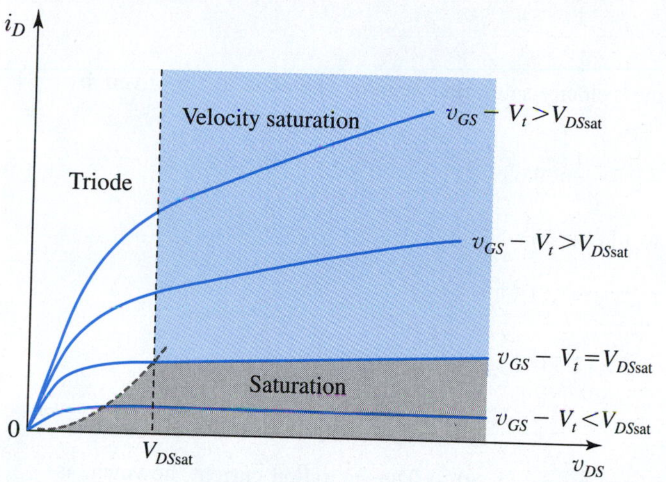

Figure 13.41 The i_D–v_{DS} characteristics of a short-channel MOSFET. Note the three different regions of operation: triode; saturation; and velocity saturation.

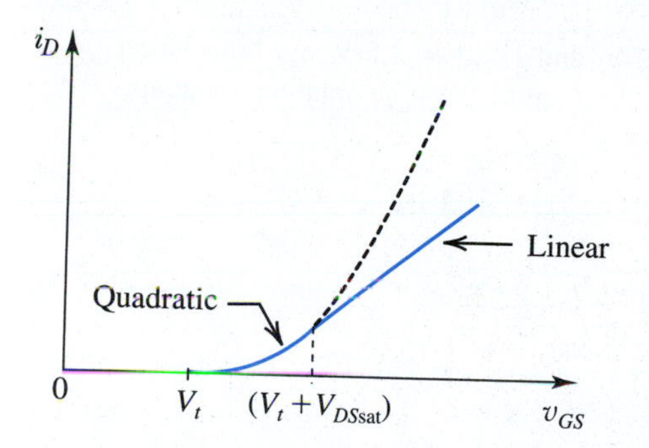

Figure 13.42 The i_D–v_{GS} characteristic of a short-channel NMOS transistor operating at $v_{DS} > V_{DSsat}$. Observe the quadratic and the linear portions of the characteristic. Also note that in the absence of velocity saturation, the quadratic curve would continue as shown with the broken line.

Example 13.8

Consider MOS transistors fabricated in a 0.25-μm CMOS process for which $V_{DD} = 2.5$ V, $V_{tn} = -V_{tp} = 0.5\text{ V}$, $\mu_n C_{ox} = 115\ \mu\text{A/V}^2$, $\mu_p C_{ox} = 30\ \mu\text{A/V}^2$, $\lambda_n = 0.6\text{ V}^{-1}$, and $|\lambda_p| = 0.1\text{ V}^{-1}$. Let $L = 0.25$ μm and $(W/L)_n = (W/L)_p = 1.5$. Measurements indicate that for the NMOS transistor, $V_{DSsat} = 0.63$ V, and for the PMOS device, $|V_{DSsat}| = 1$ V. Calculate the drain current obtained in each of the NMOS and PMOS transistors for $|V_{GS}| = |V_{DS}| = V_{DD}$. Compare with the values that would have been obtained in the absence of velocity saturation. Also give the range of v_{DS} for which i_D is saturated, with and without velocity saturation.

Solution

For the NMOS transistor, $V_{GS} = 2.5$ V results in $V_{GS} - V_{tn} = 2.5 - 0.5 = 2$ V, which is greater than V_{DSsat}. Also, $V_{DS} = 2.5$ V is greater than V_{DSsat}; thus both conditions in Eq. (13.101) are satisfied, and

Example 13.8 *continued*

the NMOS transistor will be operating in the velocity-saturation region, and thus i_D is given by Eq. (13.100):

$$i_D = 115 \times 10^{-6} \times 1.5 \times 0.63 \times \left(2.5 - 0.5 - \frac{1}{2} \times 0.63\right) \times (1 + 0.06 \times 2.5) = 210.6\ \mu\text{A}$$

If velocity saturation were absent, the current would be

$$i_D = \frac{1}{2}(\mu_n C_{ox})\left(\frac{W}{L}\right)_n (v_{GS} - V_{tn})^2(1 + \lambda v_{DS})$$

$$= \frac{1}{2} \times 115 \times 10^{-6} \times 1.5 \times (2.5 - 0.5)^2 \times (1 + 0.06 \times 2.5)$$

$$= 396.8\ \mu\text{A}$$

Thus, velocity saturation reduces the current level by nearly 50%! The saturation current, however, is obtained over a larger range of v_{DS}; specifically, for $v_{DS} = 0.63$ V to 2.5 V. (Of course, the current does not remain constant over this range because of channel-length modulation.) In the absence of velocity saturation, the current saturates at $V_{OV} = V_{GS} - V_t = 2$ V, and thus the saturation current is obtained over the range $v_{DS} = 2$ V to 2.5 V.

For the PMOS transistor, we see that since $|V_{GS}| - |V_t| = 2$ V and $|V_{DS}| = 2.5$ V are both larger that $|V_{DS\text{sat}}| = 1$ V the device will be operating in velocity saturation, and i_D can be obtained by adapting Eq. (13.100) as follows:

$$i_D = (\mu_p C_{ox})\left(\frac{W}{L}\right)_p |V_{DS\text{sat}}|\left(|V_{GS}| - |V_{tp}| - \frac{1}{2}|V_{DS\text{sat}}|\right)(1 + |\lambda_p||V_{DS}|)$$

$$= 30 \times 10^{-6} \times 1.5 \times 1 \times \left(2.5 - 0.5 - \frac{1}{2} \times 1\right)(1 + 0.1 \times 2.5)$$

$$= 84.4\ \mu\text{A}$$

Without velocity saturation, we have

$$i_D = \frac{1}{2}(\mu_p C_{ox})\left(\frac{W}{L}\right)_p (|V_{GS}| - |V_{tp}|)^2(1 + |\lambda_p||V_{DS}|)$$

$$= \frac{1}{2} \times 30 \times 10^{-6} \times 1.5 \times (2.5 - 0.5)^2(1 + 0.1 \times 2.5)$$

$$= 112.5\ \mu\text{A}$$

Thus velocity saturation reduces the current by 25% (which is less than in the case of the NMOS transistor), and the saturated current is obtained over the range $|V_{DS}| = 1$ V to 2.5 V. In the absence of velocity saturation, the saturated i_D would have been obtained for $|V_{DS}| = 2$ V to 2.5 V.

EXERCISE

13.26 Repeat the problem in Example 13.8 for transistors fabricated in a 0.13-μm CMOS process for which $V_{DD} = 1.2$ V, $V_{tn} = -V_{tp} = 0.4$ V, $\mu_n C_{ox} = 430\ \mu\text{A/V}^2$, $\mu_p C_{ox} = 110\ \mu\text{A/V}^2$, $\lambda_n = |\lambda_p| = 0.1\ \text{V}^{-1}$. Let $L = 0.13$ μm, $(W/L)_n = (W/L)_p = 1.5$, $V_{DS\text{sat}}$ (NMOS) = 0.34 V, and $V_{DS\text{sat}}$ (PMOS) = 0.6 V.

Ans. NMOS: $I_D = 154.4\ \mu\text{A}$, compared to 231.2 μA without velocity saturation; saturation is obtained over the range $v_{DS} = 0.34\ \text{V}$ to 1.2 V, compared to $v_{DS} = 0.8\ \text{V}$ to 1.2 V in the absence of velocity saturation. PMOS: $I_D = 55.4\ \mu\text{A}$ compared to 59.9 μA, and $|v_{DS}| = 0.6\ \text{V}$ to 1.2 V compared to 0.8 V to 1.2 V.

Effect on the Inverter Characteristics The VTC of the CMOS inverter can be derived using the modified i_D–v_{DS} characteristics of the MOSFETs. The results, however, indicate relatively small changes from the VTC derived in Section 13.2 using the long-channel equations (see Rabaey et al., 2003 and Hodges et al., 2004), and we shall not pursue this subject here. The dynamic characteristics of the inverter, however, are significantly impacted by velocity saturation. This is because the current available to charge and discharge the equivalent load capacitance C is substantially reduced.

A Remark on the MOSFET Model The model derived above for short-channel MOSFETs is an approximate one, intended to enable the circuit designer to perform hand analysis to gain insight into circuit operation. Also, the model parameter values are usually obtained from measured data by means of a numerical curve-fitting process. As a result, the model applies only over a restricted range of terminal voltages.

Modeling short-channel MOSFETs is an advanced topic that is beyond the scope of this book. Suffice it to say that sophisticated models have been developed and are utilized by circuit simulation programs such as SPICE (see Appendix B). Circuit simulation is an essential step in the design of integrated circuits. However, it is not a substitute for initial hand analysis and design.

13.5.3 Subthreshold Conduction

In our study of the NMOS transistor in Section 5.1, we assumed that current conduction between drain and source occurs only when v_{GS} exceeds V_t. That is, we assumed that for $v_{GS} < V_t$ no current flows between drain and source. This, however, turns out not to be the case, especially for deep-submicron devices. Specifically, for $v_{GS} < V_t$ a small current i_D flows. To be able to see this **subthreshold conduction**, we have redrawn the i_D–v_{GS} graph of Fig. 13.42, utilizing a logarithmic scale for i_D, as shown in Fig. 13.43. Observe that at low values of v_{GS}, the relationship between log i_D and v_{GS} is linear, indicating that i_D varies exponentially with v_{GS},

$$i_D = I_S e^{v_{GS}/nV_T} \tag{13.102}$$

where I_S is a constant, $V_T = kT/q$ is the thermal voltage $\simeq 25$ mV at room temperature, and n is a constant whose value falls in the range 1 to 2, depending on the material and structure of the device.[7]

Subthreshold conduction has been put to good use in the design of very-low-power circuits such as those needed for electronic watches. Generally speaking, however, subthreshold conduction is a problem in digital IC design. This is so for two reasons.

[7]This relationship is reminiscent of the i_C–v_{BE} relationship of a BJT (Chapter 6). This is no coincidence, for the subthreshold conduction in a MOSFET is due to the lateral bipolar transistor formed by the source and drain diffusions with the substrate acting as the base region (see Fig. 5.1).

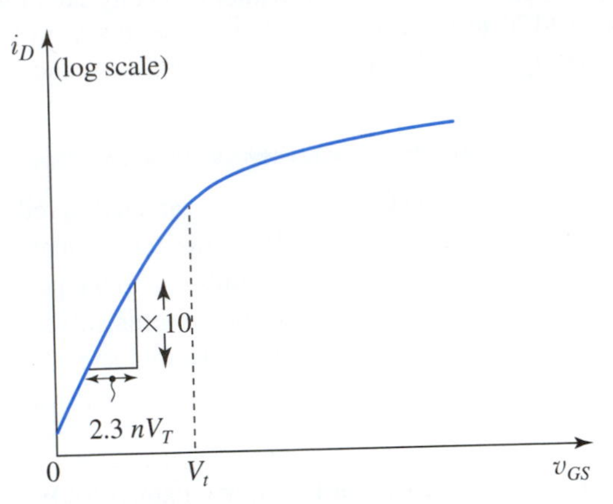

Figure 13.43 The i_D–v_{GS} characteristic of a short channel MOSFET. To show the details of subthreshold conduction a logarithmic scale is needed for i_D.

1. The nonzero current that flows for $v_{GS} = 0$ (see Fig. 13.43) causes the CMOS inverter to dissipate static power. To keep this **off current** as low as possible, V_t of the MOSFET is kept relatively high. This indeed is the reason why V_t has not been scaled by the same factor as that used for the channel length. Although the off current is low (10 pA to 100 pA) and the power dissipation per inverter is small, the problem becomes serious in chips with a billion transistors!
2. The nonzero current of a normally off transistor can cause the discharge of capacitors in dynamic MOS circuits. As we shall see in the next two chapters, dynamic logic and memory circuits rely on charge storage on capacitors for their proper operation. Thus, subthreshold conduction can disrupt the operation of such circuits.

EXERCISE

13.27 (a) Refer to Fig. 13.43 and to Eq. (13.102). Show that the inverse of the slope of the straight line representing subthreshold conduction is given by $2.3nV_T$ V per decade of current change.
(b) If measurements indicate $n = 1.22$ and $i_D = 100$ nA at $v_{GS} = 0.21$ V, find i_D at $v_{GS} = 0$.
(c) For a chip having 500 million transistors, find the current drawn from the 1.2-V supply V_{DD} as a result of subthreshold conduction. Hence estimate the resulting power dissipation.
Ans. (b) 0.1 nA; (c) 50 mA, 60 mW

13.5.4 Wiring—The Interconnect

The logic gates on a digital IC chip are connected together by metal wires[8] (see Appendix A). As well, the power supply V_{DD} and ground are distributed throughout the chip by metal wires. Technology scaling into the deep-submicron range have caused these wires to behave

[8]These are strips of metal deposited on an insulating surface on top of the chip. In modern digital ICs, as many as eight layers of such wiring are utilized.

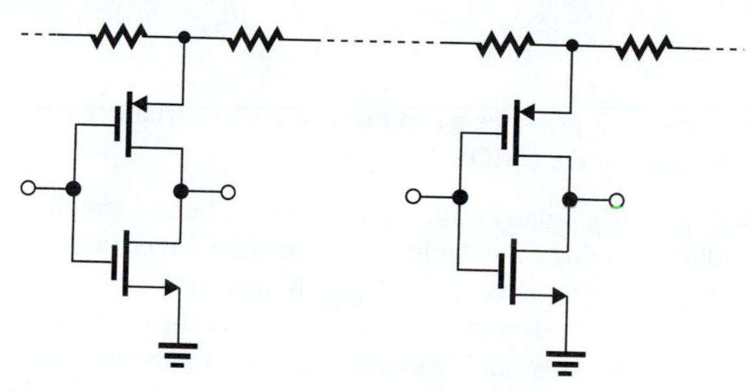

Figure 13.44 The power-supply line in a deep submicron IC has non-zero resistance. The *IR* drops along the V_{DD} line cause the voltages delivered to various circuits to differ.

not simply as wires! Specifically, the narrow wires typical of deep-submicron technologies exhibit nonzero resistance. The result is an *IR* drop on the V_{DD} line resulting in somewhat different voltages being delivered to different parts of the chip, as shown in Fig. 13.44. This can have deleterious effects on the operation of the overall circuit.

Since chips fabricated in deep-submicron technologies can have hundreds of millions of gates, the wire connection between gates can be long. The resulting narrow and long **interconnect** lines have not only nonzero resistance but also capacitance to ground, as shown in Fig. 13.45. The resistance and capacitance of an interconnect line can cause a propagation delay approaching that of the logic gate itself. As well, the capacitance between adjacent wires can cause the signals on one wire to be coupled to the other, which can cause erroneous operation of logic circuits.

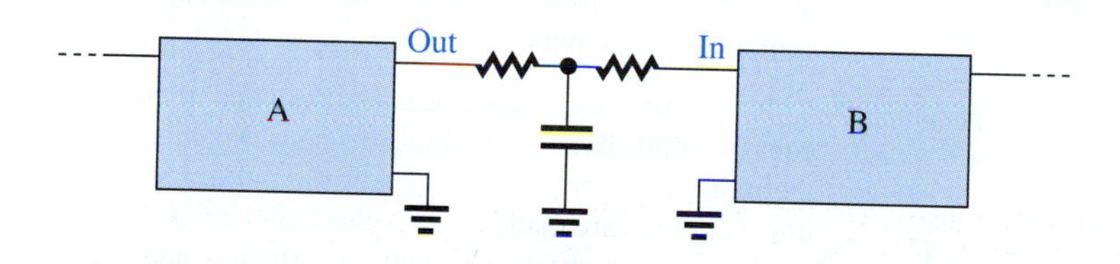

Figure 13.45 The interconnect (wire) between two circuit blocks, A and B, on an IC chip has finite resistance and a capacitance to ground.

In short, the circuit designer of modern deep-submicron digital ICs has to concern herself not only with the logic-circuit design but also with the wiring or interconnect issues. Indeed, advanced textbooks on digital IC design devote entire chapters to this topic (see Rabaey et al., 2003, and Hodges et al., 2004). Our intent here is simply to point out that interconnect has become an important issue in digital IC design.

Summary

- The digital logic inverter is the basic building block of digital circuits, just as the amplifier is the basic building block of analog circuits.

- The static operation of the inverter is described by its voltage-transfer characteristic (VTC). The VTC determines the inverter noise margins; refer to Fig. 13.5 and to Table 13.1 for the definitions of important VTC points and the noise margins. In particular, note that $NM_H = V_{OH} - V_{IH}$ and $NM_L = V_{IL} - V_{OL}$, and refer to the ideal VTC in Fig. 13.6.

- The inverter is implemented using transistors operating as voltage-controlled switches. There are three possible arrangements, shown in Figs. 13.7, 13.8, and 13.9. The arrangement in Fig. 13.8 results in a high-performance inverter and is the basis for the CMOS inverter studied in Section 13.2.

- An important performance parameter of the inverter is the amount of power it dissipates. There are two components of power dissipation: static and dynamic. The first is the result of current flow in either the 0 or 1 state or both. The second occurs when the inverter is switched and has a capacitor load C. Dynamic power dissipation $P_{\text{dyn}} = fCV_{DD}^2$.

- The speed of operation of the inverter is characterized by its propagation delay, t_P. Refer to Fig. 13.15 for the definitions of t_{PLH} and t_{PHL}, and note that $t_P = \frac{1}{2}(t_{PLH} + t_{PHL})$. The maximum frequency at which an inverter can be switched $f_{\max} = 1/2t_P$.

- A metric that combines speed of operation and power dissipation is the power–delay product, $PDP = P_D t_P$. The lower the PDP, the more effective the logic-circuit family is. If dynamic power is dominant, such as in CMOS, $PDP = CV_{DD}^2$, which is the energy drawn from the supply for a 0-to-1 and a 1-to-0 transition. (i.e., one switching cycle).

- Besides speed of operation and power dissipation, the silicon area required for an inverter is the third significant metric in digital IC design.

- Predominantly because of its low power dissipation and because of its scalability, CMOS is by far the most dominant technology for digital IC design. This situation is expected to continue for many years to come.

- Table 13.3 provides a summary of the important characteristics of the CMOS inverter.

- Digital ICs usually utilize the minimum channel length of the technology available. Thus for the CMOS inverter, Q_N and Q_P have $L = L_{\min}$. If matching is desired, W_p/W_n is selected equal to μ_n/μ_p. at the expense of increased area and capacitance. For minimum area, $W_p = W_n$. Also, a frequently used compromise is $W_p = 2W_n$.

- For minimum area, $(W/L)_n$ is selected equal to 1. However, to reduce t_P especially when a major part of C is extrinsic to the inverter, $(W/L)_n$ and correspondingly $(W/L)_p$ can be increased.

- A CMOS logic gate consists of an NMOS pull-down network (PDN) and a PMOS pull-up network (PUN). The PDN conducts for every input combination that requires a low output. Since an NMOS transistor conducts when its input is high, the PDN is most directly synthesized from the expression for the low output ($\bar{Y}$) as a function of the uncomplemented inputs. In a complementary fashion, the PUN conducts for every input combination that corresponds to a high output. Since a PMOS conducts when its input is low, the PUN is most directly synthesized from the expression for a high output (Y) as a function of the complemented inputs.

- CMOS logic circuits are usually designed to provide equal current-driving capability in both directions. Furthermore, the worst-case values of the pull-up and pull-down currents are made equal to those of the basic inverter. Transistor sizing is based on this principle and makes use of the equivalent W/L ratios of series and parallel devices (Eqs. 13.88 and 13.89).

- Refer to Table 13.2 for the implications of scaling the dimension of the MOSFET and V_{DD} and V_t by a factor $1/S$.

- In devices with short channels ($L < 0.25\ \mu\text{m}$) velocity saturation occurs. Its effect is that i_D saturates early, and its value is lower than would be the case in long-channel devices (see Figs. 13.40, 13.41 and 13.42, and Eq. 13.100).

Table 13.3 Summary of Important Characteristics of the CMOS Logic Inverter

Inverter Output Resistance

■ When v_O is low (current sinking):

$$r_{DSN} = 1\Big/\left[k_n'\left(\frac{W}{L}\right)_n (V_{DD} - V_{tn})\right]$$

■ When v_O is high (current sourcing):

$$r_{DSP} = 1\Big/\left[k_p'\left(\frac{W}{L}\right)_p (V_{DD} - |V_{tp}|)\right]$$

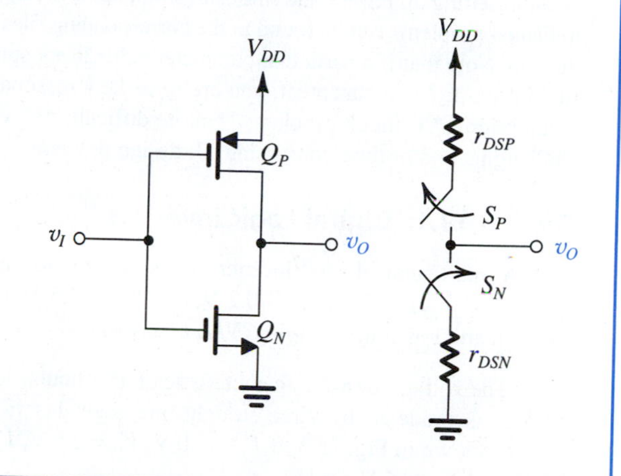

Inverter VTC and Noise Margins

$$V_M = \frac{r(V_{DD} - |V_{tp}|) + V_{tn}}{1 + r} \quad \text{where} \quad r = \sqrt{\frac{k_p'(W/L)_p}{k_n'(W/L)_n}}$$

For matched devices, that is, $\mu_n\left(\frac{W}{L}\right)_n = \mu_p\left(\frac{W}{L}\right)_p$:

$$V_M = V_{DD}/2$$

$$V_{IL} = \tfrac{1}{8}(3V_{DD} + 2V_t)$$

$$V_{IH} = \tfrac{1}{8}(5V_{DD} - 2V_t)$$

$$NM_H = NM_L = \tfrac{1}{8}(3V_{DD} + 2V_t)$$

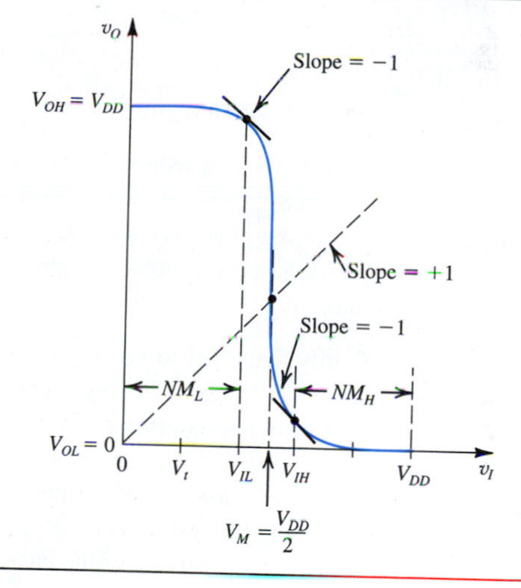

Propagation Delay (Fig. 13.22)

Using average currents:

$$t_{PHL} \simeq \frac{\alpha_n C}{k_n'(W/L)_n V_{DD}} \quad \text{where} \quad \alpha_n = 2\Big/\left[\frac{7}{4} - \frac{3V_{tn}}{V_{DD}} + \left(\frac{V_{tn}}{V_{DD}}\right)^2\right]$$

$$t_{PLH} \simeq \frac{\alpha_p C}{k_p'(W/L)_p V_{DD}} \quad \text{where} \quad \alpha_p = 2\Big/\left[\frac{7}{4} - \frac{3|V_{tp}|}{V_{DD}} + \left(\frac{|V_{tp}|}{V_{DD}}\right)^2\right]$$

Using equivalent resistances (Fig. 13.23):

$$t_{PHL} = 0.69R_N C \quad \text{where} \quad R_N = \frac{12.5}{(W/L)_n}\ \text{k}\Omega$$

$$t_{PLH} = 0.69R_P C \quad \text{where} \quad R_P = \frac{30}{(W/L)_p}\ \text{k}\Omega$$

SIM Problems identified by this icon are intended to demonstrate the value of using SPICE simulation to verify hand analysis and design, and to investigate important issues such as gate noise margins and propagation delays. Instructions to assist in setting up PSpice and Multisim simulations for all the indicated problems can be found in the corresponding files on the CD. Note that if a particular parameter value is not specified in the problem statement, you are to make a reasonable assumption. * difficult problem; ** more difficult; *** very challenging and/or time-consuming; D: design problem.

Section 13.1: Digital Logic Inverters

13.1 A particular logic inverter is specified to have $V_{IL} = 1.2$ V, $V_{IH} = 1.5$ V, $V_{OL} = 0.2$ V, and $V_{OH} = 2.5$ V. Find the high and low noise margins, NM_H and NM_L.

13.2 The voltage-transfer characteristic of a particular logic inverter is modeled by three straight-line segments in the manner shown in Fig. 13.3. If $V_{IL} = 2.0$ V, $V_{IH} = 2.5$ V, $V_{OL} = 0.5$ V, and $V_{OH} = 5$ V, find:

(a) The noise margins
(b) The value of V_M
(c) The voltage gain in the transition region

13.3 For a particular inverter design using a power supply V_{DD}, $V_{OL} = 0.1V_{DD}$, $V_{OH} = 0.8V_{DD}$, $V_{IL} = 0.4V_{DD}$, and $V_{IH} = 0.6V_{DD}$. What are the noise margins? What is the width of the transition region? For a minimum noise margin of 1 V, what value of V_{DD} is required?

13.4 A logic circuit family that used to be very popular is transistor-transistor logic (TTL). The TTL logic gates and other building blocks are available commercially in small-scale integrated (SSI) and medium-scale-integrated (MSI) packages. Such packages can be assembled on printed-circuit boards to implement a digital system. The device data sheets provide the following specifications of the basic TTL inverter (of the SN7400 type):

Logic-1 input level required to ensure a logic-0 level at the output: MIN (minimum) 2 V
Logic-0 input level required to ensure a logic-1 level at the output: MAX (maximum) 0.8 V
Logic-1 output voltage: MIN 2.4 V, TYP (typical) 3.3 V
Logic-0 output voltage: TYP 0.22 V, MAX 0.4 V
Logic-0-level supply current: TYP 3 mA, MAX 5 mA
Logic-1-level supply current: TYP 1 mA, MAX 2 mA
Propagation delay time to logic-0 level (t_{PHL}): TYP 7 ns, MAX 15 ns
Propagation delay time to logic-1 level (t_{PLH}): TYP 11 ns, MAX 22 ns

(a) Find the worst-case values of the noise margins.
(b) Assuming that the inverter is in the 1-state 50% of the time and in the 0-state 50% of the time, find the average static power dissipation in a typical circuit. The power supply is 5 V.
(c) Assuming that the inverter drives a capacitance $C_L = 45$ pF and is switched at a 1-MHz rate, use the formula in Eq. (13.35) to estimate the dynamic power dissipation.
(d) Find the propagation delay t_P.

13.5 Consider an inverter implemented as in Fig. 13.7(a). Let $V_{DD} = 5$ V, $R = 1.8$ kΩ, $R_{on} = 200$ Ω, $V_{IL} = 1$ V, and $V_{IH} = 2$ V.

(a) Find V_{OL}, V_{OH}, NM_H, and NM_L.
(b) The inverter is driving N identical inverters. Each of these load inverters, or **fan-out** inverters as they are usually called, is specified to require an input current of 0.2 mA when the input voltage (of the fan-out inverter) is high and zero current when the input voltage is low. Noting that the input currents of the fan-out inverters will have to be supplied through R of the driving inverter, find the resulting value of V_{OH} and of NM_H as a function of the number of fan-out inverters N. Hence find the maximum value N can have while the inverter is still providing an NM_H value approximately equal to its NM_L.
(c) Find the static power dissipation in the inverter in the two cases: (i) the output is low, and (ii) the output is high and driving the maximum fan-out found in (b).

13.6 For a logic-circuit family employing a 3-V supply, suggest an ideal set of values for V_M, V_{IL}, V_{IH}, V_{OL}, V_{OH}, NM_L, NM_H. Also, sketch the VTC. What value of voltage gain in the transition region does your ideal specification imply?

13.7 For a particular logic-circuit family, the basic technology used provides an inherent limit to the small-signal low-frequency voltage gain of 50 V/V. If, with a 3.3-V supply, the values of V_{OL} and V_{OH} are ideal, but $V_M = 0.4V_{DD}$, what are the best possible values of V_{IL} and V_{IH} that can be expected? What are the best possible noise margins you could expect? If the actual noise margins are only $7/10$ of these values, what V_{IL} and V_{IH} result? What is the large-signal voltage gain [defined as $(V_{OH} - V_{OL})/(V_{IL} - V_{IH})$]. (*Hint:* Use straight-line approximations for the VTC.)

***13.8** A logic-circuit family intended for use in a digital-signal-processing application in a newly developed hearing aid can operate down to single-cell supply voltages of 1.2 V. If for its inverter, the output signals swing between 0 and V_{DD}, the "gain-of-one" points are separated by less than $\frac{1}{3}V_{DD}$, and the noise margins are within 30% of one another, what ranges of values of V_{IL}, V_{IH}, V_{OL}, V_{OH}, NM_L, and NM_H can you expect for the lowest possible battery supply?

D 13.9 Design the inverter circuit in Fig. 13.2(a) to provide $V_{OH} = 2$ V, $V_{OL} = 0.1$ V, and so that the current drawn from the supply in the low-output state is 20 μA. The transistor has $V_t = 0.5$ V, $\mu_n C_{ox} = 100\ \mu\text{A/V}^2$, and $\lambda = 0$. Specify the required values of V_{DD}, R_D, and W/L. How much power is drawn from the supply when the output is high? When the output is low?

13.10 For the current-steering circuit in Fig. 13.9, $V_{CC} = 3$ V, $I_{EE} = 1$ mA, find the values of R_{C1} and R_{C2} to obtain a voltage swing of 1.5 V at each output. What are the values realized for V_{OH} and V_{OL}?

D 13.11 Refer to the analysis of the resistive-load MOS inverter in Example 13.1 and utilize the expressions derived there for the various inverter parameters. Design the circuit to satisfy the following requirements: $V_{OH} = 2.5$ V; $V_{OL} = 0.1$ V, and the power dissipation in the low-output state = 125 μW. The transistor available has $V_t = 0.5$ V, $\mu_n C_{ox} = 100\ \mu\text{A/V}^2$, and $\lambda = 0$. Specify the required values of V_{DD}, R_D, and W/L. What are the values obtained for V_{IL}, V_M, V_{IH}, NM_L, and NM_H?

D 13.12 Refer to the analysis of the resistive-load MOS inverter in Example 13.1 and utilize the expressions derived there for the various inverter parameters. For a technology for which $V_t = 0.2V_{DD}$, it is required to design the inverter to obtain $V_M = V_{DD}/2$. In terms of V_{DD}, what is the required value of the design parameter V_x? What values are obtained for V_{OH}, V_{OL}, V_{IL}, V_{IH}, NM_H, and NM_L, in terms of V_{DD}? Give numerical values for the case $V_{DD} = 2.5$ V. Now, express the power dissipated in the inverter in its low-output state in terms of the transistor's W/L ratio. Let $k_n' = 100\ \mu\text{A/V}^2$. If the power dissipation is to be limited to approximately 100 μW, what W/L ratio is needed and what value of R_D corresponds?

13.13 Consider the saturated-load inverter of Fig. 13.11(a), analyzed in Example 13.2. From Eq. (13.20),

$$V_{OH} = V_{DD} - V_{t2}$$

where V_{t2} is given by

$$V_{t2} = V_{t0} + \gamma[\sqrt{V_{OH} + 2\phi_f} - \sqrt{2\phi_f}]$$

For $V_{t0} = 0.5$ V, $V_{DD} = 1.8$ V, $\gamma = 0.3\ \text{V}^{1/2}$, $2\phi_f = 0.8$ V, use an iterative process to determine V_{t2} and V_{OH}. By how much is V_{OH} reduced as a result of the body effect on Q_2?

13.14 Determining V_{IH} of the saturated-load inverter of Fig. 13.11(a) requires a rather tedious process (see Example 13.2). An approximate estimate of V_{IH} can be obtained by reference to the VTC shown in Fig. 13.11(d). Specifically, when the straight-line segment BC is extrapolated, it meets the horizontal axis at $(V_M + V_M/k_r)$, which is usually close to the value of V_{IH}. What is the approximate value obtained this way for the case analyzed in Example 13.2? How much does it differ from the value calculated the long way in Example 13.2?

D 13.15 It is required to design the saturated-load inverter in Fig. 13.11(a) for the case $V_{DD} = 2.5$ V, $V_t = 0.5$ V, $k_n' = 100\ \mu\text{A/V}^2$, and $\lambda = 0$. Design for $V_{OL} \simeq 0.05$ V. Utilize the expressions derived in Example 13.2, except for V_{IH} use the following approximate expression (see Problem 13.11):

$$V_{IH} \simeq V_M + \frac{V_M}{k_r}$$

Neglect the body effect in Q_2. Determine V_M, NM_L, and NM_H for your design. Also determine $(W/L)_1$ and $(W/L)_2$ assuming that $(W/L)_2 = 1/(W/L)_1$. What is the power dissipated in the inverter during its low-output state?

13.16 An IC inverter fabricated in a 0.25-μm CMOS process is found to have a load capacitance of 10 fF. If the inverter is operated from a 2.5-V power supply, find the energy needed to charge and discharge the load capacitance. If the IC chip has 1 million of these inverters operating at an average switching frequency of 1 GHz, what is the power dissipated in the chip? What is the average current drawn from the power supply?

13.17 Consider a logic inverter of the type shown in Fig. 13.8. Let $V_{DD} = 5$ V, and let a 1-pF capacitance be connected between the output node and ground. If the inverter is switched at the rate of 100 MHz, determine the dynamic power dissipation. What is the average current drawn from the dc power supply?

13.18 In a particular logic family, operating with a 3.3-V supply, the basic inverter draws (from the supply) a current of 40 μA in one state and 0 μA in the other. When the inverter is switched at the rate of 100 MHz, the average supply current becomes 150 μA. Estimate the equivalent capacitance at the output node of the inverter.

13.19 A collection of logic gates for which the static-power dissipation is zero, and the dynamic-power dissipation is 10 mW is operating at 50 MHz with a 5-V supply. By what fraction could the power dissipation be reduced if operation at 3.3 V were possible? If the frequency of operation is reduced by the same factor as the supply voltage (i.e., 3.3/5), what *additional* power can be saved?

13.20 A logic inverter is implemented using the arrangement of Fig. 13.8 with switches having $R_{on} = 1\ \text{k}\Omega$, $V_{DD} = 5$ V, and $V_{IL} = V_{IH} = V_{DD}/2$.

(a) Find V_{OL}, V_{OH}, NM_L, and NM_H.
(b) If v_I rises instantaneously from 0 V to +5 V and assuming the switches operate instantaneously—that is, at $t = 0$,

PU opens and PD closes—find an expression for $v_O(t)$, assuming that a capacitance C is connected between the output node and ground. Hence find the high-to-low propagation delay (t_{PHL}) for $C = 1$ pF. Also find t_{THL} (see Fig. 13.15).
(c) Repeat (b) for v_I falling instantaneously from +5 V to 0 V. Again assume that PD opens and PU closes instantaneously. Find an expression for $v_O(t)$, and hence find t_{PLH} and t_{TLH}.

13.21 In a particular logic family, the standard inverter, when loaded by a similar circuit, has a propagation delay specified to be 1.2 ns:

(a) If the current available to charge a load capacitance is half as large as that available to discharge the capacitance, what do you expect t_{PLH} and t_{PHL} to be?
(b) If when an external capacitive load of 1 pF is added at the inverter output, its propagation delays increase by 70%, what do you estimate the normal combined capacitance of inverter output and input to be?
(c) If without the additional 1-pF load connected, the load inverter is removed and the propagation delays were observed to decrease by 40%, estimate the two components of the capacitance found in (b) that is, the component due to the inverter output and other associated parasitics, and the component due to the input of the load inverter.

13.22 Consider an inverter for which t_{PLH}, t_{PHL}, t_{TLH}, and t_{THL} are 20 ns, 10 ns, 30 ns, and 15 ns, respectively. The rising and falling edges of the inverter output can be approximated by linear ramps. Also, for simplicity, we define t_{TLH} to be 0% to 100% (rather than 10% to 90%) rise time, and similarly for t_{THL}. Two such inverters are connected in tandem and driven by an ideal input having zero rise and fall times. Calculate the time taken for the output voltage to complete its excursion for (a) a rising input and (b) a falling input. What is the propagation delay for the inverter?

13.23 A particular logic gate has t_{PLH} and t_{PHL} of 50 ns and 70 ns, respectively, and dissipates 1 mW with output low and 0.5 mW with output high. Calculate the corresponding delay–power product (under the assumption of a 50% duty-cycle signal and neglecting dynamic power dissipation).

D **13.24 We wish to investigate the design of the inverter shown in Fig. 13.7(a). In particular, we wish to determine the value for R. Selection of a suitable value for R is determined by two considerations: propagation delay and power dissipation.

(a) Show that if v_I changes instantaneously from high to low and assuming that the switch opens instantaneously, the output voltage obtained across a load capacitance C will be

$$v_O(t) = V_{OH} - (V_{OH} - V_{OL})e^{-t/\tau_1}$$

where $\tau_1 = CR$. Hence show that the time required for $v_O(t)$ to reach the 50% point, $\frac{1}{2}(V_{OH} + V_{OL})$, is

$$t_{PLH} = 0.69CR$$

(b) Following a steady state, if v_I goes high and assuming that the switch closes immediately and has the equivalent circuit in Fig. 13.7(c), show that the output falls exponentially according to

$$v_O(t) = V_{OL} + (V_{OH} - V_{OL})e^{-t/\tau_2}$$

where $\tau_2 = C(R \parallel R_{on}) \simeq CR_{on}$ for $(R_{on} \ll R)$. Hence show that the time for $v_O(t)$ to reach the 50% point is

$$t_{PHL} = 0.69CR_{on}$$

(c) Use the results of (a) and (b) to obtain the inverter propagation delay, defined as the average of t_{PLH} and t_{PHL} as

$$t_P \simeq 0.35CR \quad \text{for } R_{on} \ll R$$

(d) Show that for an inverter that spends half the time in the 0-state and half the time in the 1-state, the average static power dissipation is

$$P = \frac{1}{2}\frac{V_{DD}^2}{R}$$

(e) Now that the trade-offs in selecting R should be clear, show that, for $V_{DD} = 5$ V and $C = 10$ pF, to obtain a propagation delay no greater than 10 ns and a power dissipation no greater than 10 mW, R should be in a specific range. Find that range and select an appropriate value for R. Then determine the resulting values of t_P and P.

D 13.25 A logic-circuit family with zero static-power dissipation, normally operates at $V_{DD} = 5$ V. To reduce its dynamic-power dissipation operation at 3.3 V is considered. It is found, however, that the currents available to charge and discharge load capacitances also decrease. If current is (a) proportional to V_{DD} or (b) proportional to V_{DD}^2, what reductions in maximum operating frequency do you expect in each case? What fractional change in delay–power product do you expect in each case?

Section 13.2: The CMOS Inverter

13.26 Consider a CMOS inverter fabricated in a 0.25-μm CMOS process for which $V_{DD} = 2.5$ V, $V_{tn} = -V_{tp} = 0.5$ V, and $\mu_n C_{ox} = 3.5\mu_p C_{ox} = 115\ \mu\text{A/V}^2$. In addition, Q_N and Q_P have $L = 0.25$ μm and $(W/L)_n = 1.5$.

(a) Find W_p that results in $V_M = V_{DD}/2$. What is the silicon area utilized by the inverter in this case?

(b) For the matched case in (a), find the values of V_{OH}, V_{OL}, V_{IH}, V_{IL}, NM_L, and NM_H.
(c) For the matched case in (a), find the output resistance of the inverter in each of its two states.

SIM 13.27 For the technology specified in Problem 13.26, investigate the variation of V_M with the ratio W_p/W_n. Specifically, calculate V_M for (a) $W_p = 3.5W_n$ (the matched case); (b) $W_p = W_n$ (the minimum-size case); and (c) $W_p = 2W_n$ (a compromise case). For cases (b) and (c), estimate the approximate reduction in NM_L and silicon area relative to the matched case (a).

13.28 For a technology in which $V_{tn} = 0.2V_{DD}$, show that the maximum current that the inverter can sink while its low-output level does not exceed 0.1 V_{DD} is 0.075 $k_n'(W/L)_n V_{DD}^2$. For $V_{DD} = 2.5$ V, $k_n' = 115\ \mu A/V^2$, find $(W/L)_n$ that permits this maximum current to be 0.5 mA.

13.29 A CMOS inverter for which $k_n = 10k_p = 100\ \mu A/V^2$ and $V_t = 0.5$ V is connected as shown in Fig. P13.29 to a sinusoidal signal source having a Thévenin equivalent voltage of 0.1-V peak amplitude and resistance of 100 kΩ. What signal voltage appears at node *A* with $v_I = +1.5$ V? With $v_I = -1.5$ V?

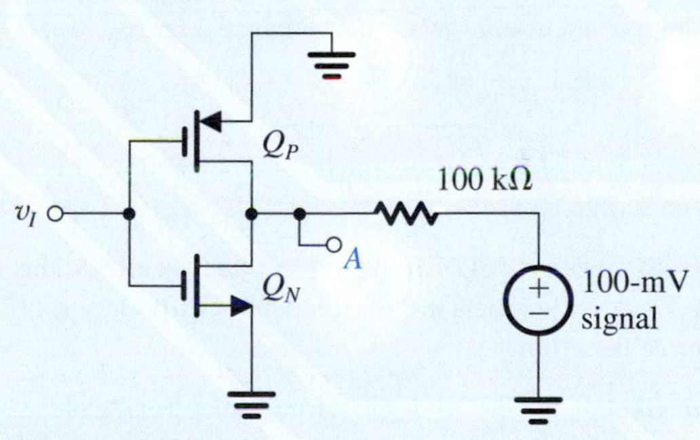

Figure P13.29

D 13.30 There are situations in which Q_N and Q_P of the CMOS inverter are deliberately mismatched to realize a certain desired value for V_M. Show that the value required of the parameter *r* of Eq. (13.59) is given by

$$r = \frac{V_M - V_{tn}}{V_{DD} - |V_{tp}| - V_M}$$

For a 0.18-μm process characterized by $V_{tn} = -V_{tp} = 0.5$ V, $V_{DD} = 1.8$ V, and $\mu_n = 4\mu_p$, find the ratio W_p/W_n required to obtain $V_M = 0.6V_{DD}$.

13.31 Consider the CMOS inverter of Fig. 13.17 with Q_N and Q_P matched and with the input v_I rising slowly from 0 to V_{DD}. At what value of v_I does the current flowing through Q_N and Q_P reach its peak? Give an expression for the peak current, neglecting λ_n and λ_p. For $k_n' = 300\ \mu A/V^2$, $(W/L)_n = 1.5$, $V_{DD} = 1.8$ V, and $V_{tn} = 0.5$ V, find the value of the peak current.

SIM 13.32 For a CMOS inverter fabricated in a 0.13-μm process with $V_{DD} = 1.2$ V, $V_{tn} = -V_{tp} = 0.4$ V, $k_n' = 4k_p' = 430\ \mu A/V^2$, and having $(W/L)_n = 1.5$ and $(W/L)_p = 3$, find t_{PHL}, t_{PLH}, and t_P when the equivalent load capacitance $C = 10$ fF. Use the method of average currents.

D 13.33 Consider a matched CMOS inverter fabricated in the 0.13-μm process specified in Problem 13.32. If $C = 20$ fF, use the method of average currents to determine the required (W/L) ratios so that $t_P \le 20$ ps.

13.34 For the CMOS inverter in Exercise 13.14 use the method of equivalent resistance to determine t_{PHL}, t_{PLH}, and t_P.

13.35 Use the method of equivalent resistance to determine the propagation delay of a minimum-size inverter, that is, one for which $(W/L)_n = (W/L)_p = 1$, designed in a 0.18-μm technology. The equivalent load capacitance $C =$ 10 fF.

D 13.36 Use the method of equivalent resistance to design an inverter to be fabricated in a 0.18-μm technology. It is required that for $C = 10$ fF, $t_{PLH} = t_{PHL}$, and $t_P \le 40$ ps.

13.37 The method of average currents yields smaller values for t_{PHL} and t_{PLH} than those obtained by the method of equivalent resistances. Most of this discrepancy is due to the fact that the formula we derived for I_{av} does not take into account velocity saturation. As will be seen in Section 13.5.2, velocity saturation reduces the current significantly. Using the results in Example 13.5, by what factor do you estimate the current reduction to be in the NMOS transistor? Since t_{PLH} does not change, what do you conclude about the effect of velocity saturation on the PMOS transistor in this technology?

13.38 Find the propagation delay for a minimum-size inverter for which $k_n' = 3k_p' = 180\ \mu A/V^2$ and $(W/L)_n = (W/L)_p = 0.75\ \mu m/0.5\ \mu m$, $V_{DD} = 3.3$ V, $V_{tn} = -V_{tp} =$ 0.7 V, and the capacitance is roughly 2 fF/μm of device width plus 1 fF/device. What does t_p become if the design is changed to a matched one? Use the method of average current.

13.39 A matched CMOS inverter fabricated in a process for which $C_{ox} = 3.7$ fF/μm², $\mu_n C_{ox} = 180\ \mu A/V^2$, $\mu_p C_{ox} = 45\ \mu A/V^2$, $V_{tn} = -V_{tp} = 0.7$ V, and $V_{DD} = 3.3$ V, uses $W_n = 0.75$ μm and $L_n = L_p = 0.5$ μm. The overlap capacitance and the effective drain–body capacitance per micrometer of gate width are 0.4 fF and 1.0 fF, respectively. The wiring capacitance is

$C_w = 2$ fF. If the inverter is driving another identical inverter, find t_{PLH}, t_{PHL}, and t_P. For how much additional capacitance load does the propagation delay increase by 50%?

D *13.40 In this problem we investigate the effect of the selection of the ratio W_p/W_n on the propagation delay of an inverter driving an identical inverter, as in Fig. 13.24.

(a) Noting that except for C_w each of the capacitances in Eqs. (13.72) and (13.73) is proportional to the width of the relevant transistor, show that C can be expressed as

$$C = C_n\left(1 + \frac{W_p}{W_n}\right) + C_w$$

where C_n is determined by the NMOS transistors.

(b) Using the equivalent resistances R_N and R_P, show that for $(W/L)_n = 1$,

$$t_{PHL} = 8.625 \times 10^3 C$$

$$t_{PLH} = \frac{20.7 \times 10^3}{W_p/W_n} C$$

(c) Use the results of (a) and (b) to determine t_P in the case $W_p = W_n$, in terms of C_n and C_w.

(d) Use the results of (a) and (b) to determine t_P in the matched case: that is, when W_p/W_n is selected to yield $t_{PHL} = t_{PLH}$.

(e) Compare the t_P values in (c) and (d) for the two extreme cases:

(i) $C_w = 0$

(ii) $C_w \gg C_n$

What do you conclude about the selection of W_p/W_n?

13.41 An inverter whose equivalent load capacitance C is composed of 10 fF contributed by the inverter transistors, and 20 fF contributed by the wiring and other external circuitry, has been found to have a propagation delay of 60 ps. By what factor must $(W/L)_n$ and $(W/L)_p$ be increased so as to reduce t_P to 30 ps?

13.42 A CMOS microprocessor chip containing the equivalent of 1 million gates operates from a 5-V supply. The power dissipation is found to be 9 W when the chip is operating at 120 MHz, and 4.7 W when operating at 50 MHz. What is the power lost in the chip by some clock-independent mechanism, such as leakage and other static currents? If 70% of the gates are assumed to be active at any time, what is the average gate capacitance in such a design?

13.43 Repeat Problem 13.39 for an inverter for which $(W/L)_n = (W/L)_p = 0.75\ \mu\text{m}/0.5\ \mu\text{m}$. Find t_P and the dynamic power dissipation when the circuit is operated at a 250-MHz rate.

13.44 In this problem we estimate the inverter power dissipation resulting from the current pulse that flows in Q_N and Q_P when the input pulse has finite rise and fall times. Refer to Fig. 13.26 and let $V_{tn} = -V_{tp} = 0.5$ V, $V_{DD} = 1.8$ V, and $k_n = k_p = 450\ \mu\text{A/V}^2$. Let the input rising and falling edges be linear ramps with the 0-to-V_{DD} and V_{DD}-to-0 transitions taking 1 ns each. Find I_{peak}. To determine the energy drawn from the supply per transition, assume that the current pulse can be approximated by a triangle with a base corresponding to the time for the rising or falling edge to go from V_t to $V_{DD} - V_t$, and the height equal to I_{peak}. Also, determine the power dissipation that results when the inverter is switched at 100 MHz.

Section 13.4: CMOS Logic-Gate Circuits

D 13.45 Sketch a CMOS realization for the function $Y = \overline{A + B(C + D)}$.

D 13.46 A CMOS logic gate is required to provide an output $Y = \bar{A}BC + A\bar{B}C + ABC̄$. How many transistors does it need? Sketch a suitable PUN and PDN, obtaining each first independently, then one from the other using the dual-networks idea.

D 13.47 Give two different realizations of the exclusive OR function $Y = A\bar{B} + \bar{A}B$ in which the PDN and the PUN are dual networks.

D 13.48 Sketch a CMOS logic circuit that realizes the function $Y = AB + \bar{A}\bar{B}$. This is called the **equivalence** or **coincidence function**.

D 13.49 Sketch a CMOS logic circuit that realizes the function $Y = ABC + \bar{A}\bar{B}\bar{C}$.

D 13.50 It is required to design a CMOS logic circuit that realizes a three-input, even-parity checker. Specifically, the output Y is to be low when an even number (0 or 2) of the inputs A, B, and C are high.

(a) Give the Boolean function $\bar{Y}$.
(b) Sketch a PDN directly from the expression for $\bar{Y}$. Note that it requires 12 transistors in addition to those in the inverters.
(c) From inspection of the PDN circuit, reduce the number of transistors to 10.
(d) Find the PUN as a dual of the PDN in (c), and hence the complete realization.

D 13.51 Give a CMOS logic circuit that realizes the function of three-input, odd-parity checker. Specifically, the output is to be high when an odd number (1 or 3) of the inputs are high. Attempt a design with 10 transistors (not counting those in the inverters) in each of the PUN and the PDN.

D 13.52 Design a CMOS full-adder circuit with inputs A, B, and C, and two outputs S and C_0 such that S is 1 if one or three inputs are 1, and C_0 is 1 if two or more inputs are 1.

D 13.53 Consider the CMOS gate shown in Fig. 13.33. Specify W/L ratios for all transistors in terms of the ratios n and p of the basic inverter, such that the worst-case t_{PHL} and t_{PLH} of the gate are equal to those of the basic inverter.

D 13.54 Find appropriate sizes for the transistors used in the exclusive-OR circuit of Fig. 13.34(b). Assume that the basic inverter has $(W/L)_n = 0.27\ \mu m/0.18\ \mu m$ and $(W/L)_p = 0.54\ \mu m/0.18\ \mu m$. What is the total area, including that of the required inverters?

13.55 Consider a four-input CMOS NAND gate for which the transient response is dominated by a fixed-size capacitance between the output node and ground. Compare the values of t_{PLH} and t_{PHL}, obtained when the devices are sized as in Fig. 13.36, to the values obtained when all n-channel devices have $W/L = n$ and all p-channel devices have $W/L = p$.

13.56 Figure P13.56 shows two approaches to realizing the OR function of six input variables. The circuit in Fig. P13.56(b), though it uses additional transistors, has in fact less total area and lower propagation delay because it uses NOR gates with lower fan-in. Assuming that the transistors in both circuits are properly sized to provide each gate with a current-driving capability equal to that of the basic matched inverter, find the number of transistors and the total area of each circuit. Assume the basic inverter to have a $(W/L)_n$ ratio of $0.27\ \mu m/0.18\ \mu m$ and a $(W/L)_p$ ratio of $0.54\ \mu m/0.18\ \mu m$.

***13.57** Consider the two-input CMOS NOR gate of Fig. 13.31 whose transistors are properly sized so that the current-driving capability in each direction is equal to that of a matched inverter. For $|V_t| = 1$ V and $V_{DD} = 5$ V, find the gate threshold in the cases for which (a) input terminal A is connected to ground and (b) the two input terminals are tied together. Neglect the body effect in Q_{PB}.

Section 13.5: Implications of Technology Scaling: Issues in Deep-Submicron Design

13.58 A chip with a certain area designed using the 10-μm process of the early 1970s contains 10,000 transistors. What does Moore's law predict the number of transistors to be on a chip of equal area fabricated using the 45-nm process of 2009?

13.59 Consider the scaling from a 0.18-μm process to a 45-nm process.

(a) Assuming V_{DD} and V_t are scaled by the same factor as the device dimensions $(S = 4)$, find the factor by which t_P, the maximum operating speed, P_{dyn}, power density, and PDP decrease (or increase)?
(b) Repeat (a) for the situation in which V_{DD} and V_t are scaled by a factor of only 2.

13.60 For a 0.18-μm technology, V_{DSsat} for minimum-length NMOS devices is measured to be 0.6 V and that for minimum-length PMOS devices 1.0 V. What do you estimate the effective values of μ_n and μ_p to be? Also find the values of E_{cr} for both device polarities.

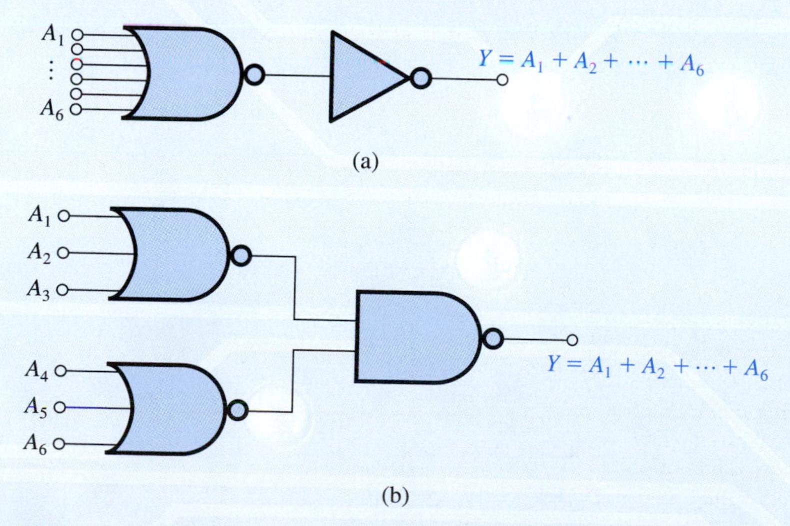

Figure P13.56

13.61 Consider NMOS and PMOS transistors with minimum channel length fabricated in a 0.13-μm CMOS process. If the effective values of μ_n and μ_p are 325 $cm^2/V \cdot s$ and 200 $cm^2/V \cdot s$, respectively, find the expected values of V_{DSsat} for both device polarities.

13.62 (a) Show that for short-channel NMOS transistor, the ratio of the current I_{Dsat} obtained at $v_{GS} = V_{DD}$ to the current obtained if velocity saturation were absent is given by

$$\frac{I_{Dsat}}{I_D} = \frac{2V_{DSsat}\left(V_{DD} - V_t - \frac{1}{2}V_{DSsat}\right)}{(V_{DD} - V_t)^2}$$

(b) Find the ratio in (a) for a transistor fabricated in a 0.13-μm process with $L = 0.13$ μm, $V_t = 0.4$ V, $V_{DSsat} = 0.34$ V, and $V_{DD} = 1.2$ V.

13.63 (a) Consider a CMOS inverter fabricated in a deep-submicron technology utilizing transistors with the minimum allowed channel length and having an equivalent load capacitance C. Let v_I rise instantaneously to V_{DD} and assume that Q_P turns off and Q_N turns on immediately. Ignoring channel-length modulation, that is, $\lambda = 0$, and assuming Q_N operates in the velocity-saturation region, show that

$$t_{PHL} = \frac{CV_{DD}}{2I_{Dsat}}$$

(b) Using the equivalent resistance of Q_N show that

$$t_{PHL} = 0.69C\,\frac{12.5 \times 10^3}{(W/L)_n}$$

(c) If the formulas in (a) and (b) are to yield the same result, find V_{DSsat} for the NMOS transistor for a 0.13-μm technology characterized by $V_{DD} = 1.2$ V, $V_t = 0.4$ V, and $\mu_n C_{ox} = 325$ $\mu A/V^2$.

D 13.64 (a) For a CMOS inverter fabricated in a deep-submicron technology with $L_n = L_p$ = the minimum allowed channel length, it is required to select W_p/W_n so that $t_{PHL} = t_{PLH}$. This can be achieved by making I_{Dsat} of Q_N equal to I_{Dsat} of Q_P at $|v_{GS}| = V_{DD}$. Show that W_p/W_n is given by

$$\frac{W_p}{W_n} = \frac{\mu_n}{\mu_p}\frac{V_{DSsatn}}{|V_{DSsatp}|}\frac{V_{DD} - V_{tn} - \frac{1}{2}V_{DSsatn}}{V_{DD} - |V_{tp}| - \frac{1}{2}|V_{DSsatp}|}$$

(b) Find the required W_p/W_n for a 0.13-μm technology for which $\mu_n/\mu_p = 4$, $V_{DD} = 1.2$ V, $V_{tn} = -V_{tp} = 0.4$ V, $V_{DSsatn} = 0.34$ V, and $|V_{DSsatp}| = 0.6$ V.

D 13.65 The current I_S in the subthreshold conduction Eq. (13.102) is proportional to e^{-V_t/nV_T}. If the threshold voltage of an NMOS transistor is reduced by 0.1 V, by what factor will the static power dissipation increase? Repeat for a reduction in V_t by 0.2 V. What do you conclude about the selection of a value of V_t in process design?

13.66 An interconnect wire with a length L, a width W, and a thickness T has a resistance R given by

$$R = \rho\frac{L}{A} = \frac{\rho L}{TW}$$

where ρ is the resistivity of the material of which the wire is made. The quantity ρ/T is called the **sheet resistance** and has the dimension of ohms, although it is usually expressed as ohms/square or $\Omega/\square$ (refer to Fig. P13.66a).

(a) Find the resistance of an aluminum wire that is 10 mm long and 0.5 μm wide, if the sheet resistance is specified to be 27 m$\Omega/\square$.

(b) If the wire capacitance to ground is 0.1 fF/μm length, what is the total wire capacitance?

(c) If we can model the wire very approximately as an RC circuit as shown in Fig. P13.66(b), find the delay time introduced by the wire. (*Hint*: $t_{delay} = 0.69RC$.) (P.S. Only a small fraction of the interconnect on an IC would be this long!)

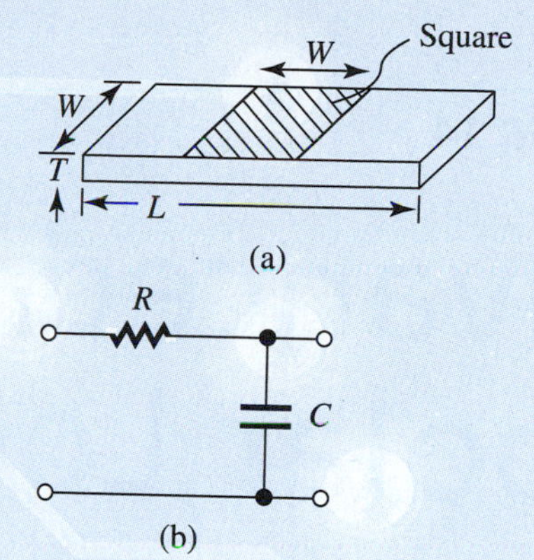

Figure P13.66

CHAPTER 14

Advanced MOS and Bipolar Logic Circuits

IN THIS CHAPTER YOU WILL LEARN

1. That by replacing the pull-up network (PUN) of a CMOS logic gate by a single, permanently-on PMOS transistor, considerable savings in transistor count and silicon area can be achieved in gates with high fan-in. The resulting circuits are known as pseudo-NMOS.
2. That a useful and conceptually simple form of MOS logic circuits, known as pass-transistor logic (PTL), utilizes MOS transistors as series switches in the signal path from input to output.
3. That a very effective switch for both analog and digital applications, known as transmission gate, is formed by connecting an NMOS and a PMOS transistor in parallel.
4. That eliminating the pull-up network and placing two complementary switches, operated by a clock signal, in series with the pull-down network of a CMOS gate, results in an interesting and useful class of circuits known as dynamic logic.
5. How the BJT differential-pair configuration is used as a current switch to realize the fastest commercially available logic-circuit family: emitter-coupled logic (ECL).
6. How the MOSFET and the BJT are combined in BiCMOS circuits in ways that take advantage of the best attributes of each device.

Introduction

Standard CMOS logic, which we studied in Chapter 13, excels in almost every performance category: It is easy to design, has the maximum possible voltage swing, is robust from a noise-immunity standpoint, dissipates no static power, and can be designed to provide equal high-to-low and low-to-high propagation delays. Its main disadvantage is the requirement of two transistors for each additional gate input, which for gates with high fan-in can make the chip area large and increase the total capacitance and, correspondingly, the propagation delay and the dynamic power dissipation. For this reason designers of digital integrated circuits have been searching for forms of CMOS logic circuits that can be used to supplement standard CMOS. This chapter presents three such forms that reduce the required number of transistors but incur other costs. These forms are not intended to replace standard CMOS, but are rather to be used in special applications for special purposes.

Pseudo-NMOS logic, studied in Section 14.1, replaces the pull-up network (PUN) in a CMOS logic gate by a single permanently "on" PMOS transistor. The reduction in transistor count and silicon area comes at the expense of static power dissipation. As well, the output low level V_{OL} becomes dependent on the transistors' W/L ratios.

Pass-transistor logic (PTL), studied in Section 14.2, utilizes MOS transistors as switches in the series path from input to output. Though simple and attractive for special applications, PTL does not restore the signal level and thus requires the occasional use of standard CMOS inverters to avoid signal-level degradation, especially in long chains of switches.

The dynamic logic circuits studied in Section 14.3 dispense with the PUN and place two complementary switches in series with the PDN. The switches are operated by a clock, and the gate output is stored on the load capacitance. Here the reduction in transistor count is achieved at the expense of a more complex design that is less robust than static CMOS.

Although CMOS accounts for the vast majority of digital integrated circuits, there is a bipolar logic-circuit family that is still of some interest. This is emitter-coupled logic (ECL), which we study in Section 14.4. Finally, in Section 14.5 we show how the MOSFET and the BJT can be combined in ways that take advantage of the best properties of each, resulting in what are known as BiCMOS circuits.

14.1 Pseudo-NMOS Logic Circuits

14.1.1 The Pseudo-NMOS Inverter

Figure 14.1 shows a modified form of the CMOS inverter. Here, only Q_N is driven by the input voltage while the gate of Q_P is grounded, and Q_P acts as an active load for Q_N. Even before we examine the operation of this circuit in detail, an advantage over standard CMOS is obvious: Each input needs to be connected to the gate of only one transistor or, alternatively, only one additional transistor (an NMOS) will be needed for each additional gate input. Thus the area and delay penalties arising from increased fan-in in a standard CMOS will be reduced. This is indeed the motivation for exploring this modified inverter circuit.

The inverter circuit of Fig. 14.1(a) resembles other forms of NMOS logic that consist of a driver transistor (Q_N) and a load transistor (in this case, Q_P); hence the name pseudo-NMOS.

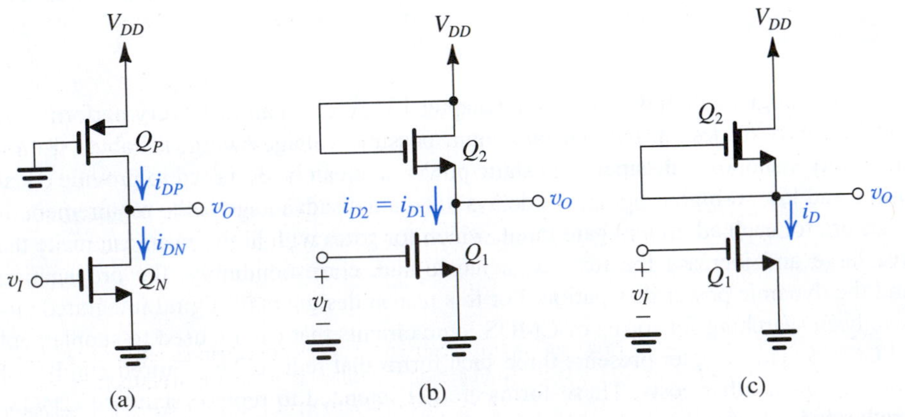

Figure 14.1 **(a)** The pseudo-NMOS logic inverter. **(b)** The enhancement-load (or saturated-load) NMOS inverter. **(c)** The depletion-load NMOS inverter.

For comparison purposes, we shall briefly mention two older forms of NMOS logic. The earliest form, popular in the mid-1970s, utilized an enhancement MOSFET for the load element, in a topology whose basic inverter is shown in Fig. 14.1(b). We studied this inverter circuit in Example 13.2, where we found that it suffers from a relatively small logic swing, small noise margins, and high static power dissipation. For these reasons, this logic-circuit technology is virtually obsolete. It was replaced in the late 1970s and early 1980s with depletion-load NMOS circuits, in which a depletion NMOS transistor (see Section 5.9.6) with its gate connected to its source is used as the load element. The topology of the basic depletion-load inverter is shown in Fig. 14.1(c).

It was initially expected that the depletion NMOS with $V_{GS} = 0$ would operate as a constant-current source and would thus provide an excellent load element.[1] However, it was quickly realized that the body effect in the depletion transistor causes its i–v characteristic to deviate considerably from that of a constant-current source. Nevertheless, depletion-load NMOS circuits feature significant improvements over their enhancement-load counterparts, enough to justify the extra processing step required to fabricate the depletion devices (namely, ion-implanting the channel). Although depletion-load NMOS has been virtually replaced by CMOS, one can still see some depletion-load circuits in specialized applications. We will not study depletion-load NMOS logic here (the interested reader can refer to the CD or the website of this book).

The pseudo-NMOS inverter that we are about to study is similar to depletion-load NMOS, but with rather improved characteristics. It also has the advantage of being directly compatible with standard CMOS circuits.

14.1.2 Static Characteristics

The static characteristics of the pseudo-NMOS inverter can be derived in a manner similar to that used for standard CMOS. Toward that end, we note that the drain currents of Q_N and Q_P are given by

$$i_{DN} = \tfrac{1}{2}k_n(v_I - V_t)^2, \quad \text{for } v_O \ge v_I - V_t \quad \text{(saturation)} \tag{14.1}$$

$$i_{DN} = k_n[(v_I - V_t)v_O - \tfrac{1}{2}v_O^2], \quad \text{for } v_O \le v_I - V_t \quad \text{(triode)} \tag{14.2}$$

$$i_{DP} = \tfrac{1}{2}k_p(V_{DD} - V_t)^2, \quad \text{for } v_O \le V_t \quad \text{(saturation)} \tag{14.3}$$

$$i_{DP} = k_p[(V_{DD} - V_t)(V_{DD} - v_O) - \tfrac{1}{2}(V_{DD} - v_O)^2], \quad \text{for } v_O \ge V_t \quad \text{(triode)} \tag{14.4}$$

where we have assumed that $V_{tn} = -V_{tp} = V_t$, and have used $k_n = k_n'(W/L)_n$ and $k_p = k_p'(W/L)_p$ to simplify matters.

To obtain the voltage-transfer characteristic of the inverter, we superimpose the load curve represented by Eqs. (14.3) and (14.4) on the i_D–v_{DS} characteristics of Q_N, which can be relabeled as i_{DN}–v_O and drawn for various values of $v_{GS} = v_I$. Such a graphical construction is shown in Fig. 14.2, where, to keep the diagram simple, we show the Q_N curves for only the two extreme values of v_I, namely, 0 and V_{DD}. Two observations follow:

[1]A constant-current load provides a capacitor-charging current that does not diminish as v_O rises toward V_{DD}, as is the case with a resistive load. Thus the value of t_{PLH} obtained with a current-source load is significantly lower than that obtained with a resistive load (see Problem 14.1). Of course, a resistive load, such as in the circuit studied in Example 13.1, is simply out of the question because of the very large silicon area it would occupy (equivalent to that of thousands of transistors!).

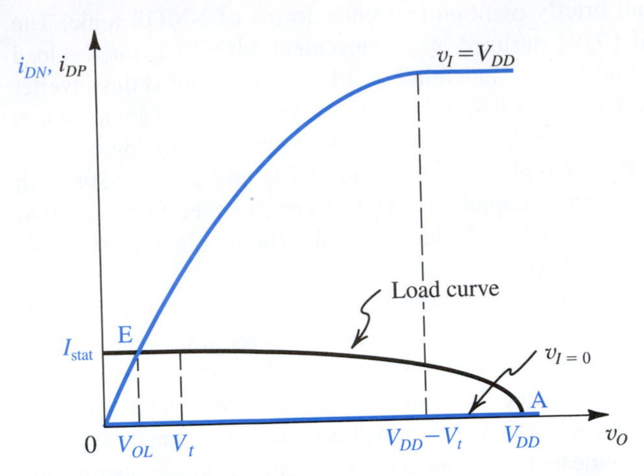

Figure 14.2 Graphical construction to determine the VTC of the inverter in Fig. 14.1(a).

1. The load curve represents a much lower saturation current (Eq. 14.3) than is represented by the corresponding curve for Q_N, namely, that for $v_I = V_{DD}$. This is a result of the fact that the pseudo-NMOS inverter is usually designed so that k_n is greater than k_p by a factor of 4 to 10. As we will show shortly, this inverter is of the so-called ratioed type,[2] and the ratio $r \equiv k_n/k_p$ determines all the breakpoints of the VTC, that is, V_{OL}, V_{IL}, V_{IH}, and so on, and thus determines the noise margins. Selection of a relatively high value for r reduces V_{OL} and widens the noise margins.
2. Although one tends to think of Q_P as acting as a constant-current source, it actually operates in saturation for only a small range of v_O, namely, $v_O \leq V_t$. For the remainder of the v_O range, Q_P operates in the triode region.

Consider first the two extreme cases of v_I: When $v_I = 0$, Q_N is cut off and Q_P is operating in the triode region, though with zero current and zero drain–source voltage. Thus the operating point is that labeled A in Fig. 14.2, where $v_O = V_{OH} = V_{DD}$, the static current is zero, and the static power dissipation is zero. When $v_I = V_{DD}$, the inverter will operate at the point labeled E in Fig. 14.2. Observe that unlike standard CMOS, here V_{OL} is not zero, an obvious disadvantage. Another disadvantage is that the gate conducts current (I_{stat}) in the low-output state, and thus there will be static power dissipation ($P_D = I_{\text{stat}} \times V_{DD}$).

14.1.3 Derivation of the VTC

Figure 14.3 shows the VTC of the pseudo-NMOS inverter. As indicated, it has four distinct regions, labeled I through IV, corresponding to the different combinations of possible modes of operation of Q_N and Q_P. The four regions, the corresponding transistor modes of operation, and the conditions that define the regions are listed in Table 14.1. We shall utilize the information in this table together with the device equations given in Eqs. (14.1) through (14.4) to derive expressions for the various segments of the VTC and in particular for the important parameters that characterize the static operation of the inverter.

[2] For the NMOS inverters such as that studied in Example 13.2, V_{OL} depends on the ratio of the transconductance parameters of the devices, that is, on the ratio $(k'(W/L))_{\text{driver}}/((k'(W/L))_{\text{load}}$. Such circuits are therefore known as *ratioed* logic circuits. Standard CMOS logic circuits do not have such a dependency and can therefore be called *ratioless*.

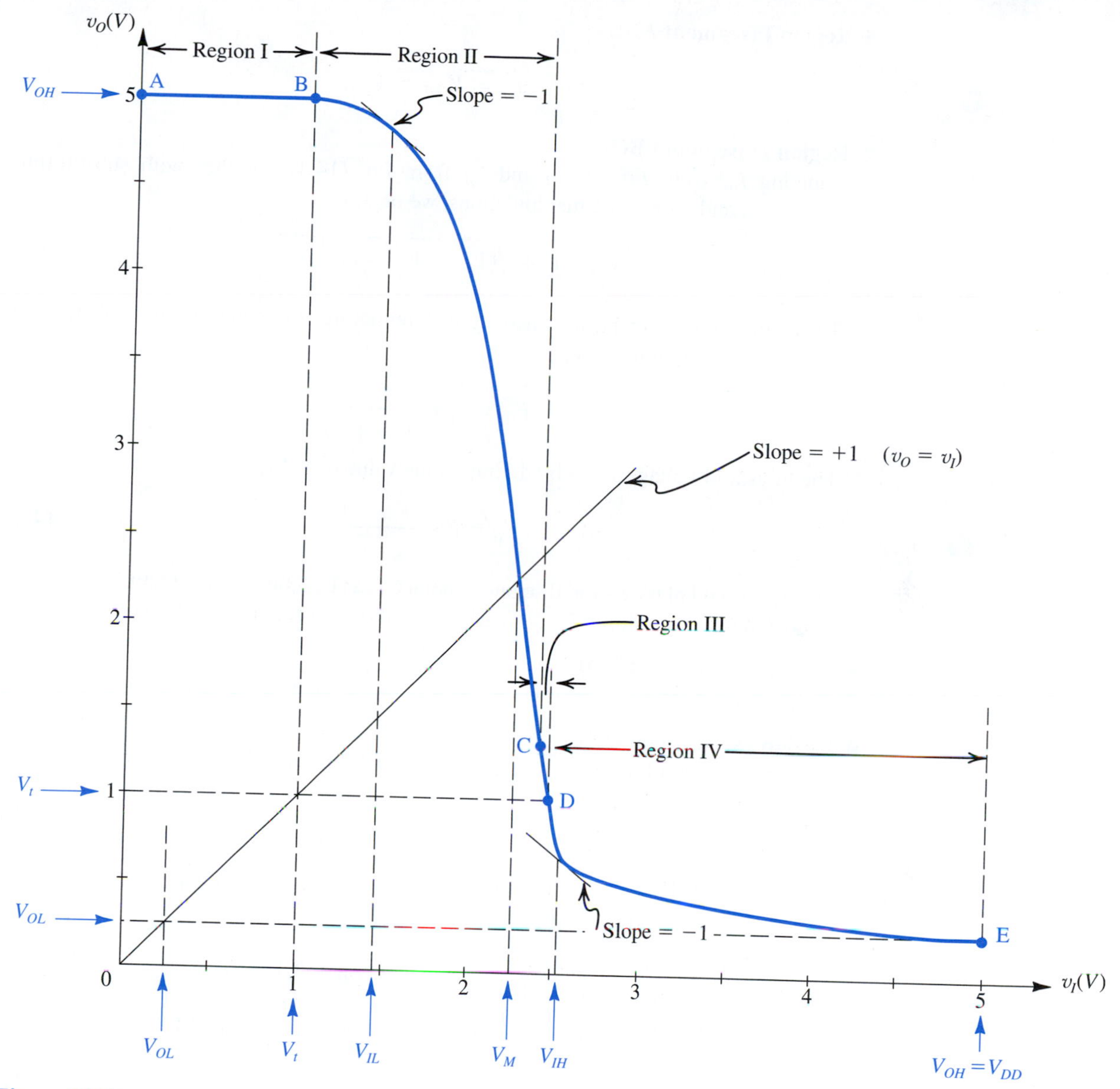

Figure 14.3 VTC for the pseudo-NMOS inverter. This curve is plotted for $V_{DD} = 5$ V, $V_{tn} = -V_{tp} = 1$ V, and $r = 9$.

Table 14.1 Regions of Operation of the Pseudo-NMOS Inverter

Region	Segment of VTC	Q_N	Q_P	Condition
I	AB	Cutoff	Triode	$v_I < V_t$
II	BC	Saturation	Triode	$v_O \geq v_I - V_t$
III	CD	Triode	Triode	$V_t \leq v_O \leq v_I - V_t$
IV	DE	Triode	Saturation	$v_O \leq V_t$

- **Region I (segment AB):**

$$v_O = V_{OH} = V_{DD} \tag{14.5}$$

- **Region II (segment BC):**
Equating i_{DN} from Eq. (14.1) and i_{DP} from Eq. (14.4) together with substituting $k_n = rk_p$, and with some manipulations, we obtain

$$v_O = V_t + \sqrt{(V_{DD} - V_t)^2 - r(v_I - V_t)^2} \tag{14.6}$$

The value of V_{IL} can be obtained by differentiating this equation and substituting $\partial v_O / \partial v_I = -1$ and $v_I = V_{IL}$:

$$V_{IL} = V_t + \frac{V_{DD} - V_t}{\sqrt{r(r+1)}} \tag{14.7}$$

The threshold voltage V_M is by definition the value of v_I for which $v_O = v_I$,

$$V_M = V_t + \frac{V_{DD} - V_t}{\sqrt{r+1}} \tag{14.8}$$

Finally, the end of the region II segment (point C) can be found by substituting $v_O = v_I - V_t$ in Eq. (14.6), the condition for Q_N leaving saturation and entering the triode region.

- **Region III (segment CD)**
This is a short segment that is not of great interest. Point D is characterized by $v_O = V_t$.

- **Region IV (segment DE)**
Equating i_{DN} from Eq. (14.2) to i_{DP} from Eq. (14.3) and substituting $k_n = rk_p$ results in

$$v_O = (v_I - V_t) - \sqrt{(v_I - V_t)^2 - \frac{1}{r}(V_{DD} - V_t)^2} \tag{14.9}$$

The value of V_{IH} can be determined by differentiating this equation and setting $\partial v_O / \partial v_I = -1$ and $v_I = V_{IH}$,

$$V_{IH} = V_t + \frac{2}{\sqrt{3r}}(V_{DD} - V_t) \tag{14.10}$$

The value of V_{OL} can be found by substituting $v_I = V_{DD}$ into Eq. (14.9),

$$V_{OL} = (V_{DD} - V_t)\left[1 - \sqrt{1 - \frac{1}{r}}\right] \tag{14.11}$$

The static current conducted by the inverter in the low-output state is found from Eq. (14.3) as

$$I_{\text{stat}} = \frac{1}{2}k_p(V_{DD} - V_t)^2 \tag{14.12}$$

Finally, we can use Eqs. (14.7) and (14.11) to determine NM_L and Eqs. (14.5) and (14.10) to determine NM_H:

$$NM_L = V_t - (V_{DD} - V_t)\left[1 - \sqrt{1 - \frac{1}{r}} - \frac{1}{\sqrt{r(r+1)}}\right] \tag{14.13}$$

$$NM_H = (V_{DD} - V_t)\left(1 - \frac{2}{\sqrt{3r}}\right) \tag{14.14}$$

As a final observation, we note that since V_{DD} and V_t are determined by the process technology, the only design parameter for controlling the values of V_{OL} and the noise margins is the ratio r.

14.1.4 Dynamic Operation

Analysis of the inverter transient response to determine t_{PLH} with the inverter loaded by a capacitance C is identical to that of the complementary CMOS inverter. The capacitance will be charged by the current i_{DP}; we can determine an estimate for t_{PLH} by using the average value of i_{DP} over the range $v_O = 0$ to $v_O = V_{DD}/2$. The result is:

$$t_{PLH} = \frac{\alpha_p C}{k_p V_{DD}} \tag{14.15}$$

where

$$\alpha_p = 2 \bigg/ \left[\frac{7}{4} - 3\left(\frac{V_t}{V_{DD}}\right) + \left(\frac{V_t}{V_{DD}}\right)^2\right] \tag{14.16}$$

The case for the capacitor discharge is somewhat different because the current i_{DP} has to be subtracted from i_{DN} to determine the discharge current. The result is

$$t_{PHL} \simeq \frac{\alpha_n C}{k_n V_{DD}} \tag{14.17}$$

where

$$\alpha_n = 2 \bigg/ \left[1 + \frac{3}{4}\left(1 - \frac{1}{r}\right) - \left(3 - \frac{1}{r}\right)\left(\frac{V_t}{V_{DD}}\right) + \left(\frac{V_t}{V_{DD}}\right)^2\right] \tag{14.18}$$

which, for a large value of r, reduces to

$$\alpha_n \simeq \alpha_p \tag{14.19}$$

Although these are similar formulas to those for the standard CMOS inverter, the pseudo-NMOS inverter has a special problem: Since k_p is r times smaller than k_n, t_{PLH} will be approximately r times larger than t_{PHL}. Thus the circuit exhibits an asymmetrical delay performance. Recall, however, that for gates with large fan-in, pseudo-NMOS requires fewer transistors and thus C can be smaller than in the corresponding standard CMOS gate.

14.1.5 Design

The design involves selecting the ratio r and the W/L for one of the transistors. The value of W/L for the other device can then be obtained using r. The design parameters of interest are V_{OL}, NM_L, NM_H, I_{stat}, P_D, t_{PLH}, and t_{PHL}. Important design considerations are as follows:

1. The ratio r determines all the breakpoints of the VTC; the larger the value of r, the lower V_{OL} is (Eq. 14.11) and the wider the noise margins are (Eqs. 14.13 and 14.14). However, a larger r increases the asymmetry in the dynamic response and, for a given $(W/L)_p$, makes the silicon area larger. Thus, selecting a value for r represents a compromise between noise margins on the one hand and silicon area and t_P on the other. Usually, r is selected in the range 4 to 10.

2. Once r has been determined, a value for $(W/L)_p$ or $(W/L)_n$ can be selected and the other determined. Here, one would select a small $(W/L)_n$ to keep the gate area small and thus obtain a small value for C. Similarly, a small $(W/L)_p$ keeps I_{stat} and P_D low. On the other hand, one would want to select larger W/L ratios to obtain low t_P and thus fast response. For usual (high-speed) applications, $(W/L)_p$ is selected so that I_{stat} is in the range of 50 μA to 100 μA, which for $V_{DD} = 1.8$ V results in P_D in the range of 90 μW to 180 μW.

14.1.6 Gate Circuits

Except for the load device, the pseudo-NMOS gate circuit is identical to the PDN of the complementary CMOS gate. Four-input, pseudo-NMOS NOR and NAND gates are shown in Fig. 14.4. Note that each requires five transistors compared to the eight used in standard CMOS. In pseudo-NMOS, NOR gates are preferred over NAND gates because the former do not utilize transistors in series and thus can be designed with minimum-size NMOS devices.

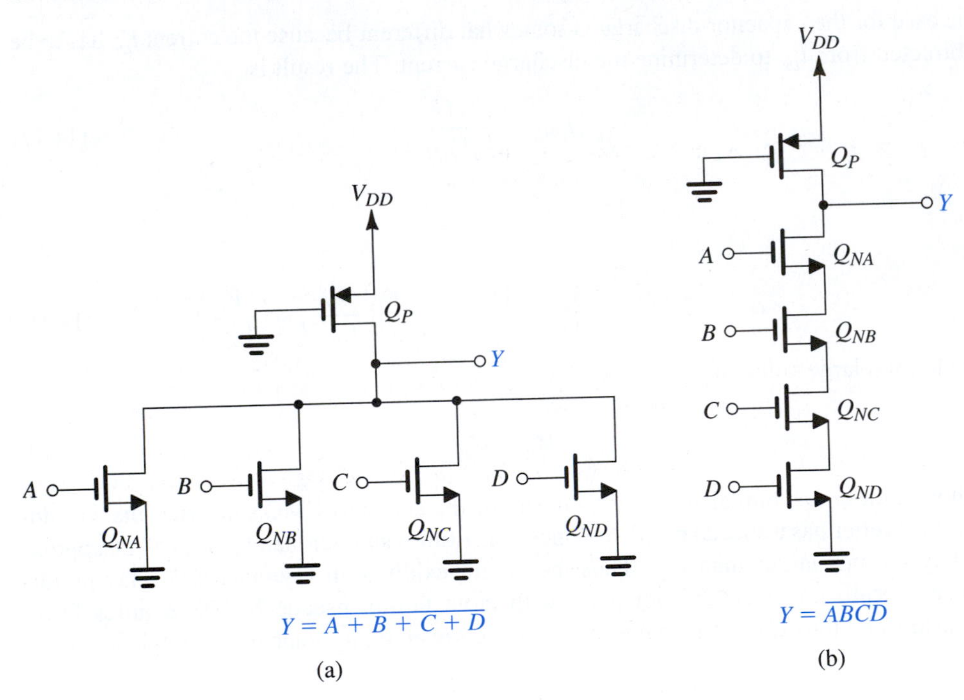

Figure 14.4 NOR and NAND gates of the pseudo-NMOS type.

14.1.7 Concluding Remarks

Pseudo-NMOS is particularly suited for applications in which the output remains high most of the time. In such applications, the static power dissipation can be reasonably low (since the gate dissipates static power only in the low-output state). Further, the output transitions that matter would presumably be high-to-low ones, where the propagation delay can be made as short as necessary. A particular application of this type can be found in the design of address decoders for memory chips (Section 15.4) and in read-only memories (Section 15.5).

Example 14.1

Consider a pseudo-NMOS inverter fabricated in a 0.25-μm CMOS technology for which $\mu_n C_{ox} = 115\ \mu A/V^2$, $\mu_p C_{ox} = 30\ \mu A/V^2$, $V_{tn} = -V_{tp} = 0.5$ V, and $V_{DD} = 2.5$ V. Let the W/L ratio of Q_N be (0.375 μm/0.25 μm) and $r = 9$. Find:

(a) V_{OH}, V_{OL}, V_{IL}, V_{IH}, V_M, NM_H, and NM_L

(b) $(W/L)_p$

(c) I_{stat} and P_D

(d) t_{PLH}, t_{PHL}, and t_P, assuming a total capacitance at the inverter output of 7 fF

Solution

(a) $V_{OH} = V_{DD} = 2.5$ V

V_{OL} is determined from Eq. (14.11) as

$$V_{OL} = (2.5 - 0.5)\left[1 - \sqrt{1 - \frac{1}{9}}\right] = 0.11 \text{ V}$$

V_{IL} is determined from Eq. (14.7) as

$$V_{IL} = 0.5 + \frac{2.5 - 0.5}{\sqrt{9(9+1)}} = 0.71 \text{ V}$$

V_{IH} is determined from Eq. (14.10) as

$$V_{IH} = 0.5 + \frac{2}{\sqrt{3 \times 9}} \times (2.5 - 0.5) = 1.27 \text{ V}$$

V_M is determined from Eq. (14.8) as

$$V_M = 0.5 + \frac{2.5 - 0.5}{\sqrt{9+1}} = 1.13 \text{ V}$$

The noise margins can now be determined as

$$NM_H = V_{OH} - V_{IH} = 2.5 - 1.27 = 1.23 \text{ V}$$

$$NM_L = V_{IL} - V_{OL} = 0.71 - 0.11 = 0.60 \text{ V}$$

Observe that the noise margins are not equal and that NM_L is rather low.

(b) The W/L ratio of Q_P can be found from

$$\frac{\mu_n C_{ox}(W/L)_n}{\mu_p C_{ox}(W/L)_p} = 9$$

$$\frac{115 \times \dfrac{0.375}{0.25}}{30(W/L)_p} = 9$$

Thus,

$$(W/L)_p = 0.64$$

Example 14.1 *continued*

(c) The dc current in the low-output state can be determined from Eq. (14.12) as

$$I_{stat} = \frac{1}{2} \times 30 \times 0.64(2.5 - 0.5)^2 = 38.4 \ \mu\text{A}$$

The static power dissipation can now be found from

$$P_D = I_{stat} V_{DD}$$
$$= 38.4 \times 2.5 = 96 \ \mu\text{W}$$

(d) The low-to-high propagation delay can be found by using Eqs. (14.15) and (14.16):

$$\alpha_p = 1.68$$

$$t_{PLH} = \frac{1.68 \times 7 \times 10^{-15}}{30 \times 10^{-6} \times 0.64 \times 2.5} = 0.25 \text{ ns}$$

The high-to-low propagation delay can be found by using Eqs. (14.17) and (14.18):

$$\alpha_n = 1.54$$

$$t_{PHL} = \frac{1.54 \times 7 \times 10^{-15}}{115 \times 10^{-6} \times \frac{0.375}{0.25} \times 2.5} = 0.03 \text{ ns}$$

Now, the propagation delay can be determined as

$$t_P = \frac{1}{2}(0.25 + 0.03) = 0.14 \text{ ns}$$

Although the propagation delay is considerably greater than that of a standard CMOS inverter, this is not an entirely fair comparison: Recall that the advantage of pseudo-NMOS occurs in gates with large fan-in, not in a single inverter.

EXERCISES

14.1 While keeping r unchanged, redesign the inverter circuit of Example 14.1 to lower its static power dissipation to half the value found. Find the W/L ratios for the new design. Also find t_{PLH}, t_{PHL}, and t_P, assuming that C remains unchanged. Would the noise margins change?
Ans. $(W/L)_n = 1.5$; $(W/L)_p = 0.32$; 0.5 ns; 0.03 ns; 0.27 ns; no

14.2 Redesign the inverter of Example 14.1 using $r = 4$. Find V_{OL} and the noise margins. If $(W/L)_n =$ 0.375 μm/0.25 μm, find $(W/L)_p$, I_{stat}, P_D, t_{PLH}, t_{PHL}, and t_P. Assume $C = 7$ fF.
Ans. $V_{OL} = 0.27$ V; $NM_L = 0.68$ V; $NM_H = 0.85$ V; $(W/L)_p = 1.44$; $I_{stat} = 86.3$ μA; $P_D = 0.22$ mW; $t_{PLH} = 0.11$ ns; $t_{PHL} = 0.03$ ns; $t_P = 0.07$ ns

14.2 Pass-Transistor Logic Circuits

A conceptually simple approach for implementing logic functions utilizes series and parallel combinations of switches that are controlled by input logic variables to connect the input

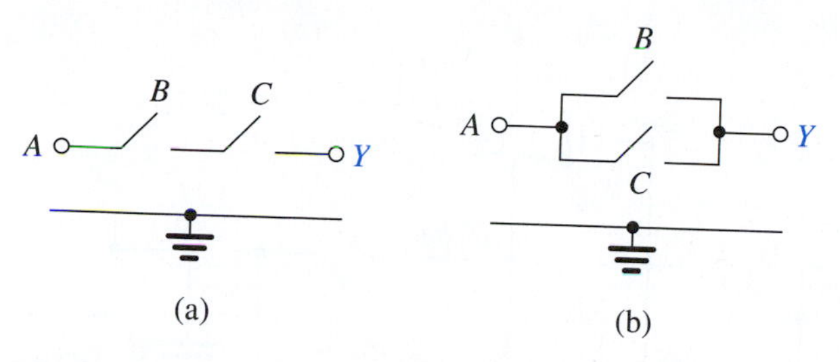

Figure 14.5 Conceptual pass-transistor logic gates. **(a)** Two switches, controlled by the input variables B and C, when connected in series in the path between the input node to which an input variable A is applied and the output node (with an implied load to ground) realize the function $Y = ABC$. **(b)** When the two switches are connected in parallel, the function realized is $Y = A(B + C)$.

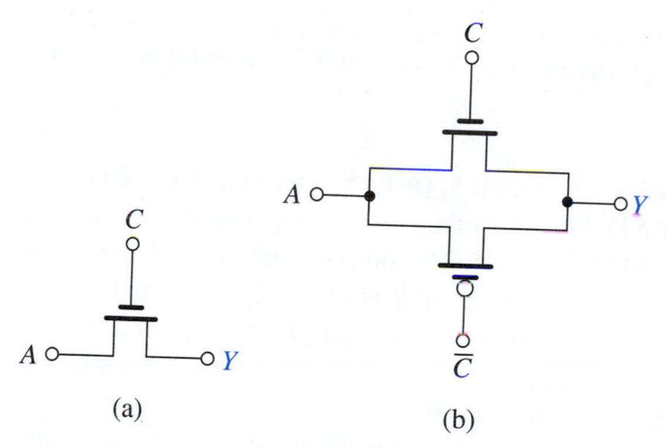

Figure 14.6 Two possible implementations of a voltage-controlled switch connecting nodes A and Y: **(a)** single NMOS transistor and **(b)** CMOS transmission gate.

and output nodes (see Fig. 14.5). Each of the switches can be implemented either by a single NMOS transistor (Fig. 14.6) or by a pair of complementary MOS transistors connected in what is known as the **CMOS transmission-gate** configuration (Fig. 14.6). The result is a simple form of logic circuit that is particularly suited for some special logic functions and is frequently used in conjunction with standard CMOS logic to implement such functions efficiently: that is, with a lower total number of transistors than is possible with CMOS alone.

Because this form of logic utilizes MOS transistors in the series path from input to output, to *pass* or block signal transmission, it is known as *pass-transistor logic* (PTL). As mentioned earlier, CMOS transmission gates are frequently employed to implement the switches, giving this logic-circuit form the alternative name, *transmission-gate logic*. The terms are used interchangeably independent of the actual implementation of the switches.

Though conceptually simple, pass-transistor logic circuits have to be designed with care. In the following, we shall study the basic principles of PTL circuit design and present examples of its application.

14.2.1 An Essential Design Requirement

An essential requirement in the design of PTL circuits is ensuring that *every circuit node has at all times a low-resistance path either to V_{DD} or to ground.* To appreciate this point,

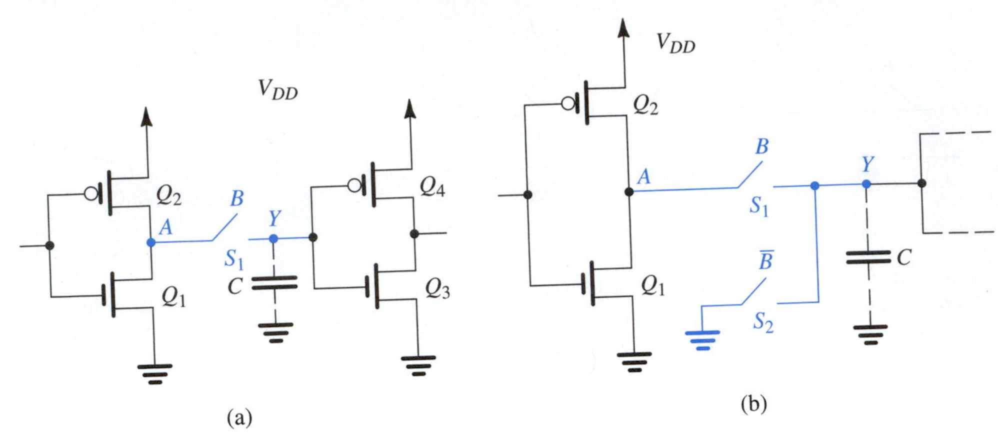

Figure 14.7 A basic design requirement of PTL circuits is that every node have, at all times, a low-resistance path to either ground or V_{DD}. Such a path does not exist in **(a)** when B is low and S_1 is open. It is provided in **(b)** through switch S_2.

consider the situation depicted in Fig. 14.7(a): A switch S_1 (usually part of a larger PTL network, not shown) is used to form the AND function of its controlling variable B and the variable A available at the output of a CMOS inverter. The output Y of the PTL circuit is shown connected to the input of another inverter. Now, if B is high, S_1 closes and $Y = A$. Node Y will then be connected either to V_{DD} (if A is high) through Q_2 or to ground (if A is low) through Q_1. But what happens when B goes low and S_1 opens? Node Y will now become a high-impedance node. If initially v_Y was zero, it will remain so. However, if initially v_Y was high at V_{DD}, this voltage will be maintained by the charge on the parasitic capacitance C, and Y will not be a logic 0 as required of the AND function.

The problem can be easily solved by establishing for node Y a low-resistance path that is activated when B goes low, as shown in Fig. 14.7(b). Here, another switch, S_2, controlled by $\overline{B}$, is connected between Y and ground. When B goes low, S_2 closes and establishes a low-resistance path between Y and ground. The voltage v_Y will then be 0 volts, the proper output of the AND function when B is zero.

14.2.2 Operation with NMOS Transistors as Switches

Implementing the switches in a PTL circuit with single NMOS transistors results in a simple circuit with small area and small node capacitances. These advantages, however, are obtained at the expense of serious shortcomings in both the static characteristics and the dynamic performance of the resulting circuits. To illustrate, consider the circuit shown in Fig. 14.8, where an NMOS transistor Q is used to implement a switch connecting an input node with voltage v_I and an output node. The total capacitance between the output node and ground is represented by capacitor C. The switch is shown in the closed state with the control signal applied to its gate being high at V_{DD}. We wish to analyze the operation of the circuit as the input voltage v_I goes high (to V_{DD}) at time $t = 0$. We assume that initially the output voltage v_O is zero and capacitor C is fully discharged.[3]

[3]Although the MOS transistor is symmetric and its drain and source are interchangeable, it is always useful to know which terminal is functioning as the source and which as the drain. The terminal with the higher voltage in an NMOS transistor is the drain. The opposite is true for the PMOS transistor.

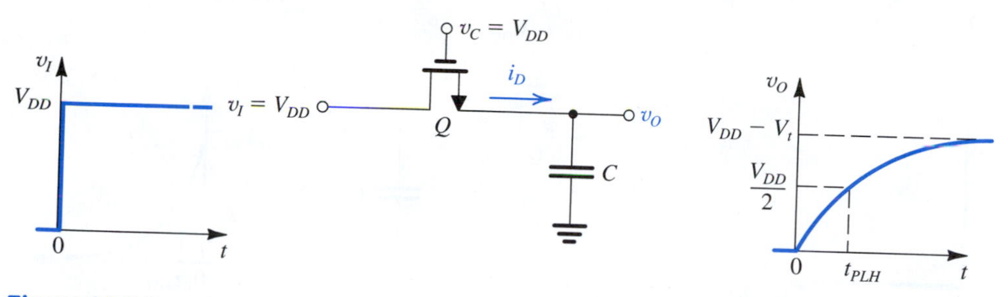

Figure 14.8 Operation of the NMOS transistor as a switch in the implementation of PTL circuits. This analysis is for the case with the switch closed (v_C is high) and the input going high ($v_I = V_{DD}$).

When v_I goes high, the transistor operates in the saturation mode and delivers a current i_D to charge the capacitor,

$$i_D = \frac{1}{2}k_n(V_{DD} - v_O - V_t)^2 \tag{14.20}$$

where $k_n = k_n'(W/L)$, and V_t is determined by the body effect since the source is at a voltage v_O relative to the body (which, though not shown, is connected to ground); thus (see Eq. 5. 107),

$$V_t = V_{t0} + \gamma\left(\sqrt{v_O + 2\phi_f} - \sqrt{2\phi_f}\right) \tag{14.21}$$

Thus, initially (at $t = 0$), $V_t = V_{t0}$ and the current i_D is relatively large. However, as C charges up and v_O rises, V_t increases (Eq. 14.21) and i_D decreases. The latter effect is due to both the increase in v_O and in V_t. It follows that the process of charging the capacitor will be relatively slow. More seriously, observe from Eq. (14.20) that i_D reduces to zero when v_O reaches ($V_{DD} - V_t$). Thus the high output voltage (V_{OH}) will *not* be equal to V_{DD}; rather, it will be lower by V_t, and to make matters worse, the value of V_t can be as high as 1.5 to 2 times V_{t0}!

In addition to reducing the gate noise immunity, the low value of V_{OH} (commonly referred to as a "poor 1") has another detrimental effect: Consider what happens when the output node is connected to the input of a standard CMOS inverter (as was the case in Fig. 14.7). The low value of V_{OH} can cause Q_P of the load inverter to conduct. Thus the inverter will have a finite static current and static power dissipation.

The propagation delay t_{PLH} of the PTL gate of Fig. 14.8 can be determined as the time for v_O to reach $V_{DD}/2$. This can be calculated using techniques similar to those employed in the analysis of the CMOS inverter in Section 13.3, as will be illustrated shortly in an example.

Figure 14.9 shows the NMOS switch circuit when v_I is brought down to 0 V. We assume that initially $v_O = V_{DD}$. Thus at $t = 0+$, the transistor conducts and operates in the saturation region,

$$i_D = \frac{1}{2}k_n(V_{DD} - V_t)^2 \tag{14.22}$$

where we note that since the source is now at 0 V (note that the drain and source have interchanged roles), there will be no body effect, and V_t remains constant at V_{t0}. As C discharges, v_O decreases and the transistor enters the triode region at $v_O = V_{DD} - V_t$. Nevertheless, the capacitor discharge continues until C is fully discharged and $v_O = 0$. Thus, the NMOS transistor provides $V_{OL} = 0$, or a "good 0." Again, the propagation delay t_{PHL} can be determined using usual techniques, as illustrated by the following example.

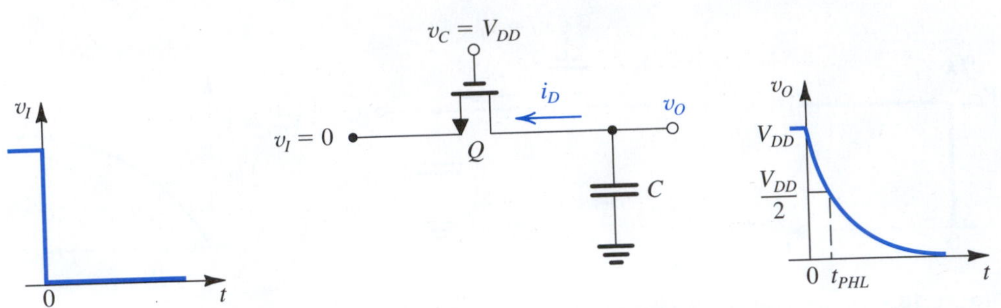

Figure 14.9 Operation of the NMOS switch as the input goes low ($v_I = 0$ V). Note that the drain of an NMOS transistor is always higher in voltage than the source; correspondingly, the drain and source terminals interchange roles in comparison to the circuit in Fig. 14.8.

Example 14.2

Consider the NMOS transistor switch in the circuits of Figs. 14.8 and 14.9 to be fabricated in a technology for which $\mu_n C_{ox} = 50\ \mu\text{A/V}^2$, $\mu_p C_{ox} = 20\ \mu\text{A/V}^2$, $|V_{t0}| = 1$ V, $\gamma = 0.5\ \text{V}^{1/2}$, $2\phi_f = 0.6$ V, and $V_{DD} = 5$ V, where ϕ_f is a physical parameter. Let the transistor be of the minimum size for this technology, namely, 4 μm/2 μm, and assume that the total capacitance between the output node and ground is $C = 50$ fF.

(a) For the case with v_I high (Fig. 14.8), find V_{OH}.
(b) If the output feeds a CMOS inverter whose $(W/L)_p = 2.5\,(W/L)_n = 10\ \mu\text{m}/2\ \mu\text{m}$, find the static current of the inverter and its power dissipation when its input is at the value found in (a). Also find the inverter output voltage.
(c) Find t_{PLH}.
(d) For the case with v_I going low (Fig. 14.9), find t_{PHL}.
(e) Find t_P.

Solution

(a) Refer to Fig. 14.8. If V_{OH} is the value of v_O at which Q stops conducting,

$$V_{DD} - V_{OH} - V_t = 0$$

then,

$$V_{OH} = V_{DD} - V_t$$

where V_t is the value of the threshold voltage at a source–body reverse bias equal to V_{OH}. Using Eq. (14.21), we have

$$V_t = V_{t0} + \gamma\left(\sqrt{V_{OH} + 2\phi_f} - \sqrt{2\phi_f}\right)$$
$$= V_{t0} + \gamma\left(\sqrt{V_{DD} - V_t + 2\phi_f} - \sqrt{2\phi_f}\right)$$

Substituting $V_{t0} = 1$, $\gamma = 0.5$, $V_{DD} = 5$, and $2\phi_f = 0.6$, we obtain a quadratic equation in V_t whose solution yields

$$V_t = 1.6 \text{ V}$$

Thus,

$$V_{OH} = 3.4 \text{ V}$$

Note that this represents a significant loss in signal amplitude.

(b) The load inverter will have an input signal of 3.4 V. Thus, its Q_P will conduct a current of

$$i_{DP} = \frac{1}{2} \times 20 \times \frac{10}{2}(5 - 3.4 - 1)^2 = 18\ \mu A$$

where we have assumed Q_P to be operating in saturation, as we still expect v_O of the inverter to be close to 0. Thus, the static power dissipation of the inverter will be

$$P_D = V_{DD} i_{DP} = 5 \times 18 = 90\ \mu W$$

The output voltage of the inverter can be found by noting that Q_N will be operating in the triode region. Equating its current to that of Q_P (i.e., 18 μA) enables us to determine the output voltage to be 0.08 V.

(c) To determine t_{PLH}, refer to Fig 14.8. We need to find the current i_D at $t = 0$ (where $v_O = 0$, $V_t = V_{t0} =$ 1 V) and at $t = t_{PLH}$ (where $v_O = 2.5$ V, V_t to be determined), as follows:

$$i_D(0) = \frac{1}{2} \times 50 \times \frac{4}{2} \times (5 - 1)^2 = 800\ \mu A$$

$$V_t\,(\text{at } v_O = 2.5\text{ V}) = 1 + 0.5(\sqrt{2.5 + 0.6} - \sqrt{0.6}) = 1.49\text{ V}$$

$$i_D(t_{PLH}) = \frac{1}{2} \times 50 \times \frac{4}{2}(5 - 2.5 - 1.49)^2 = 50\ \mu A$$

We can now compute the average discharge current as

$$i_D\big|_{av} = \frac{800 + 50}{2} = 425\ \mu A$$

and t_{PLH} can be found as

$$t_{PLH} = \frac{C(V_{DD}/2)}{i_D\big|_{av}}$$

$$= \frac{50 \times 10^{-15} \times 2.5}{425 \times 10^{-6}} = 0.29\text{ ns}$$

(d) Refer to the circuit in Fig. 14.9. Observe that, here, V_t remains constant at $V_{t0} = 1$ V. At $t = 0$, Q will be operating in saturation, and the drain current will be

$$i_D(0) = \frac{1}{2} \times 50 \times \frac{4}{2}(5 - 1)^2 = 800\ \mu A$$

At $t = t_{PHL}$, Q will be operating in the triode region, and thus

$$i_D(t_{PHL}) = 50 \times \frac{4}{2}\left[(5 - 1) \times 2.5 - \frac{1}{2} \times 2.5^2\right]$$

$$= 690\ \mu A$$

Thus, the average discharge current is given by

$$i_D\big|_{av} = \frac{1}{2}(800 + 690) = 740\ \mu A$$

and t_{PHL} can be determined as

$$t_{PHL} = \frac{50 \times 10^{-15} \times 2.5}{740 \times 10^{-6}} = 0.17\text{ ns}$$

(e) $t_P = \frac{1}{2}(t_{PLH} + t_{PHL}) = \frac{1}{2}(0.29 + 0.17) = 0.23\text{ ns}$

EXERCISE

14.3 Let the NMOS transistor switch in Fig. 14.8 be fabricated in a 0.18-μm CMOS process for which $V_{t0} = 0.5$ V, $\gamma = 0.3\ \text{V}^{1/2}$, $2\phi_f = 0.85$ V, and $V_{DD} = 1.8$ V. Find V_{OH}.

Ans. 1.15 V

14.2.3 Restoring the Value of V_{OH} to V_{DD}

Example 14.2 illustrates clearly the problem of signal-level loss and its deleterious effect on the operation of the succeeding CMOS inverter. Some rather ingenious techniques have been developed to restore the output level to V_{DD}. We shall briefly discuss two such techniques. One is circuit-based and the other is based on process technology.

The circuit-based approach is illustrated in Fig. 14.10. Here, Q_1 is a pass-transistor controlled by input B. The output node of the PTL network is connected to the input of a standard CMOS inverter formed by Q_N and Q_P. A PMOS transistor Q_R, whose gate is controlled by the output voltage of the inverter, v_{O2}, has been added to the circuit. Observe that in the event that the output of the PTL gate, v_{O1}, is low (at ground), v_{O2} will be high (at V_{DD}), and Q_R will be off. On the other hand, if v_{O1} is high but not quite equal to V_{DD}, the output of the inverter will be low (as it should be) and Q_R will turn on, supplying a current to charge C up to V_{DD}. This process will stop when $v_{O1} = V_{DD}$, that is, when the output voltage has been restored to its proper level. The "level-restoring" function performed by Q_R is frequently employed in MOS digital-circuit design. It should be noted that although the description of operation is relatively straight forward, the addition of Q_R closes a "positive-feedback" loop around the CMOS inverter, and thus operation is more involved than it appears, especially during transients. Selection of a W/L ratio for Q_R is also a somewhat involved process, although normally k_r is selected to be much lower than k_n (say a third or a fifth as large). Intuitively, this is appealing, for it implies that Q_R will not play a major role in circuit operation, apart from restoring the level of V_{OH} to V_{DD}, as explained above. Transistor Q_R is said to be a "weak PMOS transistor." See Problem 14.17.

The other technique for correcting for the loss of the high-output signal level (V_{OH}) is a technology-based solution. Specifically, recall that the loss in the value of V_{OH} is equal to V_{tn}. It follows that we can reduce the loss by using a lower value of V_{tn} for the NMOS switches, and we can eliminate the loss altogether by using devices for which $V_{tn} = 0$. These **zero-threshold devices** can be fabricated by using ion implantation to control the value of V_{tn} and are known as

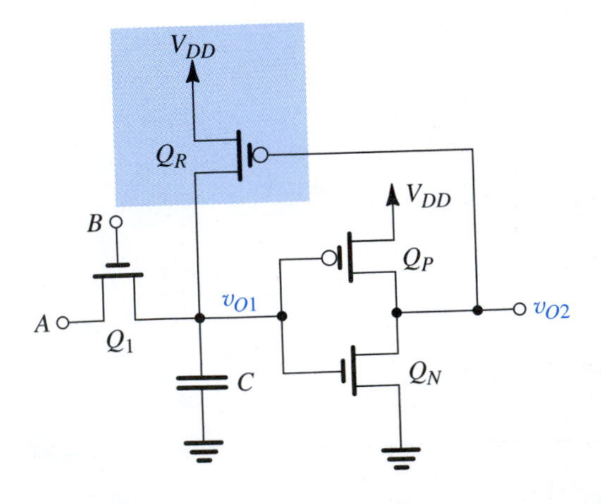

Figure 14.10 The use of transistor Q_R, connected in a feedback loop around the CMOS inverter, to restore the V_{OH} level, produced by Q_1, to V_{DD}.

natural devices. The problem of low-threshold devices, however, is the increased subthreshold conduction (Section 13.5.3) and the corresponding increase in static power dissipation.

14.2.4 The Use of CMOS Transmission Gates as Switches

Great improvements in static and dynamic performance are obtained when the switches are implemented with CMOS transmission gates. The transmission gate utilizes a pair of complementary transistors connected in parallel. It acts as an excellent switch, providing bidirectional current flow, and it exhibits an "on" resistance that remains almost constant for wide ranges of input voltage. These characteristics make the transmission gate not only an excellent switch in digital applications but also an excellent analog switch in such applications as data converters and switched-capacitor filters (Chapter 16).

Before we analyze the transmission gate circuit, it is useful to reflect on its origin. Recall that an NMOS transistor transmits the 0-V level to the output perfectly and thus produces a "good 0." It has difficulty, however, in passing the V_{DD} level, with the result that $V_{OH} = V_{DD} - V_t$ (a "poor 1"). It can be shown (see Problem 14.18) that a PMOS transistor does exactly the opposite; that is, it passes the V_{DD} level perfectly and thus produces a "good 1" but has trouble passing the 0-V level, thus producing a "poor 0." It is natural therefore to think that placing an NMOS and a PMOS transistor in parallel would produce good results in both the 0 and 1 cases.

Another way to describe the performance of the two transistor types is that the NMOS is good at pulling the output down to 0 V, while the PMOS is good at pulling the output up to V_{DD}. Interestingly, these are also the roles they play in the standard CMOS inverter.

Figure 14.11 shows the transmission gate together with its frequently used circuit symbol. The transmission gate is a bilateral switch that results in $v_Y = v_X$ when v_C is high (V_{DD}). In terms of logic variables, its function is described by

$$Y = X \quad \text{if} \quad C = 1$$

Figure 14.12(a) shows the transmission-gate switch in the "on" position with the input, v_I, rising to V_{DD} at $t = 0$. Assuming, as before, that initially the output voltage is zero, we see that Q_N will be operating in saturation and providing a charging current of

$$i_{DN} = \frac{1}{2}k_n(V_{DD} - v_O - V_{tn})^2 \tag{14.23}$$

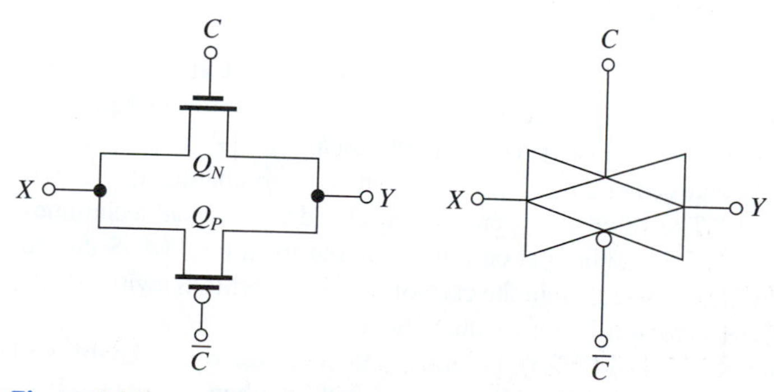

Figure 14.11 The CMOS transmission gate and its circuit symbol.

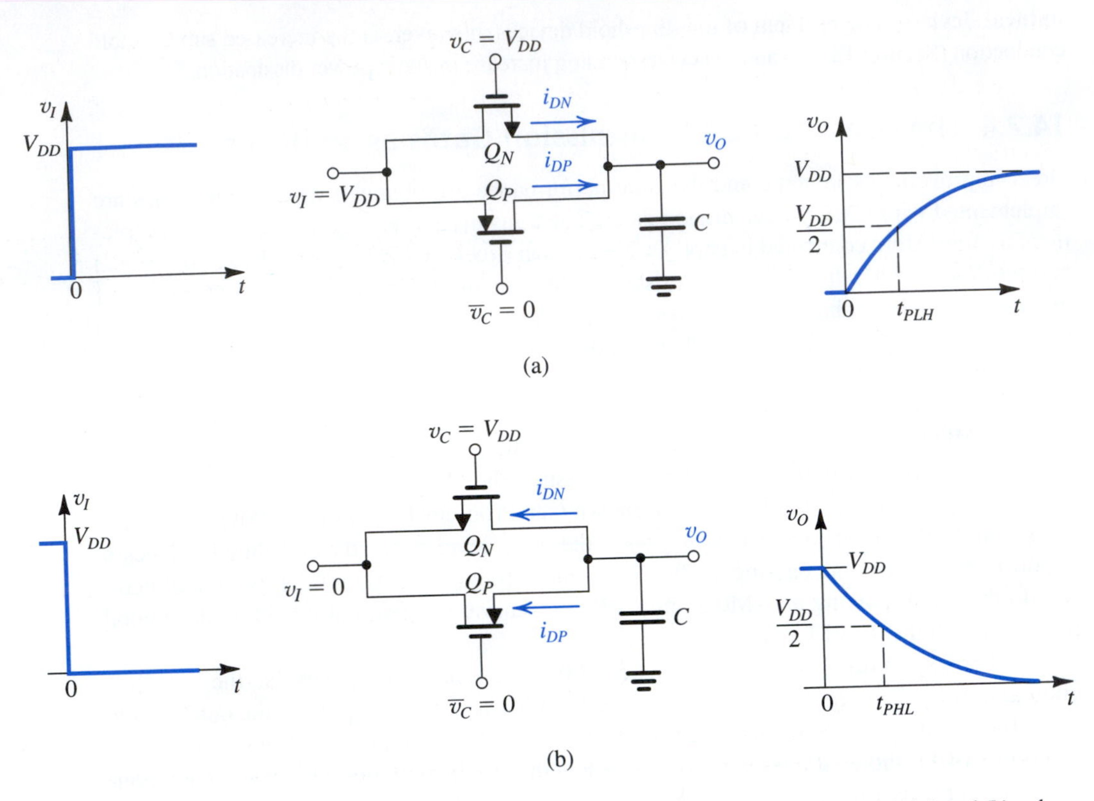

Figure 14.12 Operation of the transmission gate as a switch in PTL circuits with **(a)** v_I high and **(b)** v_I low.

where, as in the case of the single NMOS switch, V_{tn} is determined by the body effect,

$$V_{tn} = V_{t0} + \gamma(\sqrt{v_O + 2\phi_f} - \sqrt{2\phi_f}) \tag{14.24}$$

Transistor Q_N will conduct a diminishing current that reduces to zero at $v_O = V_{DD} - V_{tn}$. Observe, however, that Q_P operates with $V_{SG} = V_{DD}$ and is initially in saturation,

$$i_{DP} = \frac{1}{2}k_p(V_{DD} - |V_{tp}|)^2 \tag{14.25}$$

where, since the body of Q_P is connected to V_{DD}, $|V_{tp}|$ remains constant at the value V_{t0}, assumed to be the same value as for the n-channel device. The total capacitor-charging current is the sum of i_{DN} and i_{DP}. Now, Q_P will enter the triode region at $v_O = |V_{tp}|$, but will continue to conduct until C is fully charged and $v_O = V_{OH} = V_{DD}$. Thus, the p-channel device will provide the gate with a "good 1." The value of t_{PLH} can be calculated using usual techniques, where we expect that as a result of the additional current available from the PMOS device, for the same value of C, t_{PLH} will be lower than in the case of the single NMOS switch. Note, however, that adding the PMOS transistor increases the value of C.

When v_I goes low, as shown in Fig. 14.12(b), Q_N and Q_P interchange roles. Analysis of the circuit in Fig. 14.12(b) will indicate that Q_P will cease conduction when v_O falls to $|V_{tp}|$, where $|V_{tp}|$ is given by

$$|V_{tp}| = V_{t0} + \gamma\left[\sqrt{V_{DD} - v_O + 2\phi_f} - \sqrt{2\phi_f}\right] \tag{14.26}$$

Transistor Q_N, however, continues to conduct until C is fully discharged and $v_O = V_{OL} = 0$ V, a "good 0."

We conclude that transmission gates provide far superior performance, both static and dynamic, than is possible with single NMOS switches. The price paid is increased circuit complexity, area, and capacitance.

EXERCISE

14.4 The transmission gate of Figs. 14.12(a) and 14.12(b) is fabricated in a CMOS process technology for which $k_n' = 50\ \mu A/V^2$, $k_p' = 20\ \mu A/V^2$, $V_{tn} = |V_{tp}|$, $V_{t0} = 1$ V, $\gamma = 0.5\ V^{1/2}$, $2\phi_f = 0.6$ V, and $V_{DD} = 5$ V. Let Q_N and Q_P be of the minimum size possible with this process technology, $(W/L)_n = (W/L)_p = 4\ \mu m/2\ \mu m$. The total capacitance at the output node is 70 fF. Utilize as many of the results of Example 14.2 as you need.

(a) What are the values of V_{OH} and V_{OL}?

(b) For the situation in Fig. 14.12(a), find $i_{DN}(0)$, $i_{DP}(0)$, $i_{DN}(t_{PLH})$, $i_{DP}(t_{PLH})$, and t_{PLH}.

(c) For the situation depicted in Fig. 14.12(b), find $i_{DN}(0)$, $i_{DP}(0)$, $i_{DN}(t_{PHL})$, $i_{DP}(t_{PHL})$, and t_{PHL}. At what value of v_O will Q_P turn off?

(d) Find t_P

Ans. (a) 5 V, 0 V; (b) 800 μA, 320 μA, 50 μA, 275 μA, 0.24 ns; (c) 800 μA, 320 μA, 688 μA, 20 μA, 0.19 ns, 1.6 V; (d) 0.22 ns

Equivalent Resistance of the Transmission Gate Although the transmission gate is capable of passing the full 1 and 0 levels to the load capacitance, it is not a perfect switch. In particular, the transmission gate has a finite "on" resistance. It is useful for us to obtain an estimate for this resistance. It can, for instance, be used together with the load capacitance as an alternative means to determining propagation delay. This approach is particularly useful in situations involving a network of inverters and transmission gates, as we shall shortly see.

To obtain an estimate of the resistance of the transmission gate, we shall consider the situation in Fig. 14.12(a), where the transmission gate is on and is passing a high input (V_{DD}) to the capacitor load. Transistor Q_N operates in saturation until the output voltage v_O reaches $(V_{DD} - V_{tn})$, at which time Q_N turns off; thus,

$$i_{DN} = \frac{1}{2}k_n(V_{DD} - V_{tn} - v_O)^2 \quad \text{for} \quad v_O \le V_{DD} - V_{tn} \tag{14.27}$$

$$i_{DN} = 0 \quad \text{for} \quad v_O \ge V_{DD} - V_{tn} \tag{14.28}$$

A gross estimate for the equivalent resistance of Q_N can be obtained by dividing the voltage across it, $(V_{DD} - v_O)$, by i_{DN}, and neglecting the body effect, that is, assuming V_{tn} remains constant; thus,

$$R_{Neq} = \frac{V_{DD} - v_O}{\frac{1}{2}k_n(V_{DD} - V_{tn} - v_O)^2} \quad \text{for} \quad v_O \le V_{DD} - V_{tn} \tag{14.29}$$

and

$$R_{Neq} = \infty \quad \text{for} \quad v_O \ge V_{DD} - V_{tn} \tag{14.30}$$

Transistor Q_P will operate in saturation until $v_O = |V_{tp}|$, after which it enters the triode region; thus,

$$i_{DP} = \frac{1}{2}k_p(V_{DD} - |V_{tp}|)^2 \quad \text{for} \quad v_O \leq |V_{tp}| \tag{14.31}$$

$$i_{DP} = k_p\left[(V_{DD} - |V_{tp}|)(V_{DD} - v_O) - \frac{1}{2}(V_{DD} - v_O)^2\right] \quad \text{for} \quad v_O \geq |V_{tp}| \tag{14.32}$$

A gross estimate for the resistance of Q_P can be obtained by dividing the voltage across it, $(V_{DD} - v_O)$, by i_{DP}; thus,

$$R_{Peq} = \frac{V_{DD} - v_O}{\frac{1}{2}k_p(V_{DD} - |V_{tp}|)^2} \quad \text{for } v_O \leq |V_{tp}| \tag{14.33}$$

$$R_{Peq} = \frac{1}{k_p\left[V_{DD} - |V_{tp}| - \frac{1}{2}(V_{DD} - v_O)\right]} \quad \text{for } v_O \geq |V_{tp}| \tag{14.34}$$

Finally, the equivalent resistance R_{TG} of the transmission gate can be obtained as the parallel equivalent of R_{Neq} and R_{Peq},

$$R_{TG} = R_{Neq} \| R_{Peq} \tag{14.35}$$

Obviously, R_{TG} is a function of the output voltage v_O. As an example, we show in Fig. 14.13 a plot for R_{TG} for the transmission gate analyzed in Exercise 14.4. Observe that R_{TG} remains relatively constant over the full range of v_O. The average value of R_{TG} over the range $v_O = 0$ to $V_{DD}/2$ can be used to determine t_{PLH}, as illustrated in Exercise 14.5.

EXERCISE

14.5 For the transmission gate analyzed in Exercise 14.4, whose equivalent resistance for capacitor charging is plotted in Fig. 14.13, use the average resistance value over the range $v_O = 0$ V to 2.5 V to determine t_{PLH}. Compare the result to that obtained using average currents in Exercise 14.4. Note that from the graph, $R_{TG} = 4.5$ kΩ at $v_O = 0$ V, and $R_{TG} = 6.5$ kΩ at $v_O = 2.5$ V. Recall that $t_{PLH} = 0.69RC$.
Ans. $t_{PLH} = 0.27$ ns, very close to the value of 0.24 ns obtained in Exercise 14.4

The expression for R_{TG} derived above applies only to the case of capacitor charging. A similar analysis can be performed for the case of capacitor discharge illustrated in Fig. 14.12(b). The resulting value of R_{TG} is close to that obtained above (see Problem 14.21).

Similar to the empirical formulas for R_N and R_P of the CMOS inverter (Eqs. 13.70 and 13.71), there is a simple empirical formula for R_{TG} that applies for both capacitor charging and discharging and for all modern submicron technologies (see Hodges et al., 2004), namely,

$$R_{TG} \simeq \frac{12.5}{(W/L)_n} \text{ k}\Omega \tag{14.36}$$

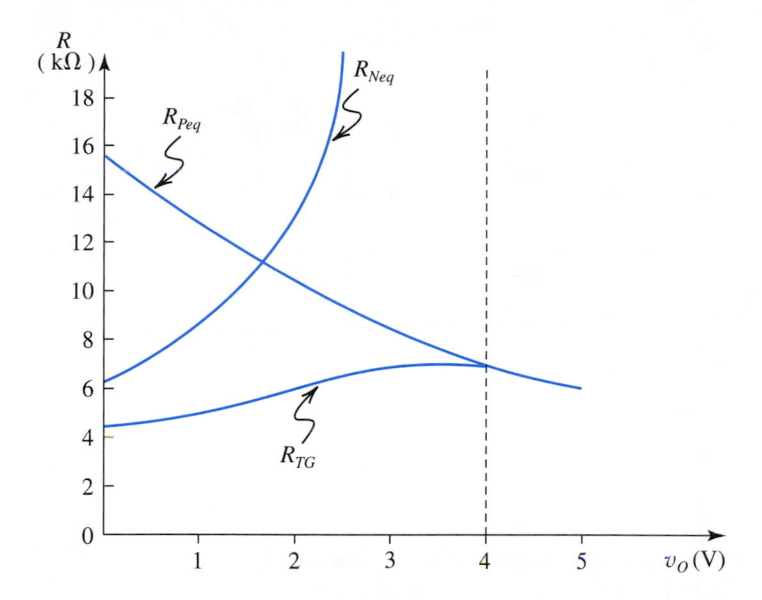

Figure 14.13 Plot of the equivalent resistances of the two transistors of the transmission gate in Fig. 14.12(a) and the overall resistance R_{TG} versus v_O. The data apply to the situation specified in Exercise 14.5.

EXERCISE

14.6 Use Eq. (14.36) to estimate the value of R_{TG} for a transmission gate fabricated in a 0.18-μm CMOS technology with $(W/L)_n = (W/L)_p = 1.5$.

Ans. 8.3 kΩ

Having an estimate of the resistance of the transmission gate enables us to calculate the propagation delay of a signal path containing one or more transmission gates. Figure 14.14(a) shows one such circuit. It consists of a transmission gate connecting the output of an inverter to the input of another. We are interested in finding the propagation delay from the input of the first inverter to the input of the second as we apply a negative going step to the input of the first inverter.

Fig. 14.14(b) shows the equivalent circuit where R_{P1} is the equivalent resistance of Q_{P1}, R_{TG} is the equivalent resistance of the transmission gate, C_{out1} is the output capacitance of the driver inverter, C_{TG1} and C_{TG2} are the capacitances introduced by the transmission gate at its input and output, respectively, and C_{in2} is the input capacitance of the load inverter. Observe that the circuit takes the form of an *RC* ladder network. A simple formula has been developed for calculating the delay of an arbitrarily long *RC* ladder network such as that shown in Fig. 14.15 having three sections. Known as the **Elmore delay formula**, it gives for the ladder in Fig. 14.15

$$t_P = 0.69[C_1R_1 + C_2(R_1 + R_2) + C_3(R_1 + R_2 + R_3)] \tag{14.37}$$

Applying the Elmore formula to the two-stage ladder in Fig. 14.14(b) gives

$$t_P = 0.69[(C_{\text{out1}} + C_{TG1})R_1 + (C_{\text{in2}} + C_{TG2})(R_1 + R_2)] \tag{14.38}$$

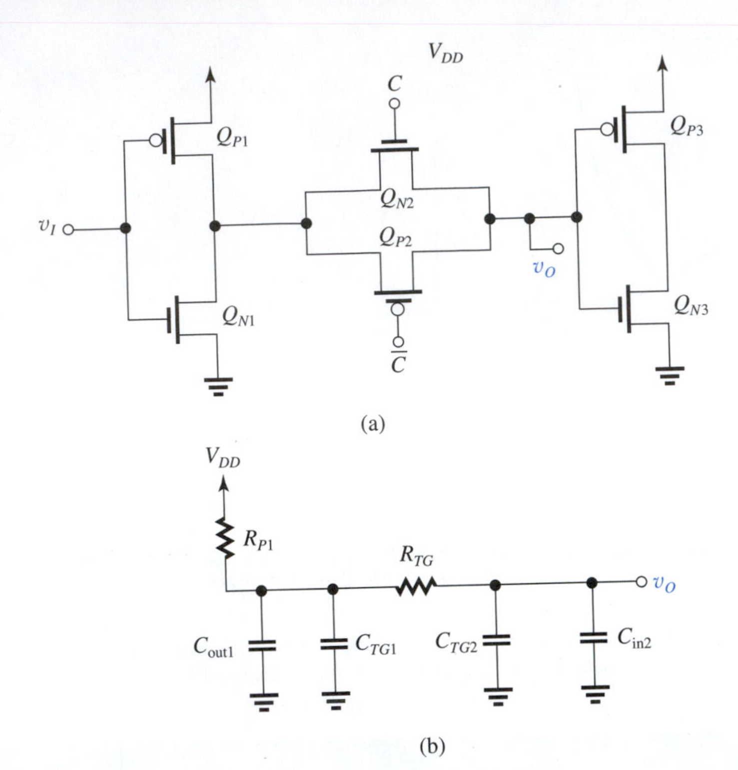

Figure 14.14 (**a**) A transmission gate connects the output of a CMOS inverter to the input of another. (**b**) Equivalent circuit for the purpose of analyzing the propagation delay of the circuit in (**a**).

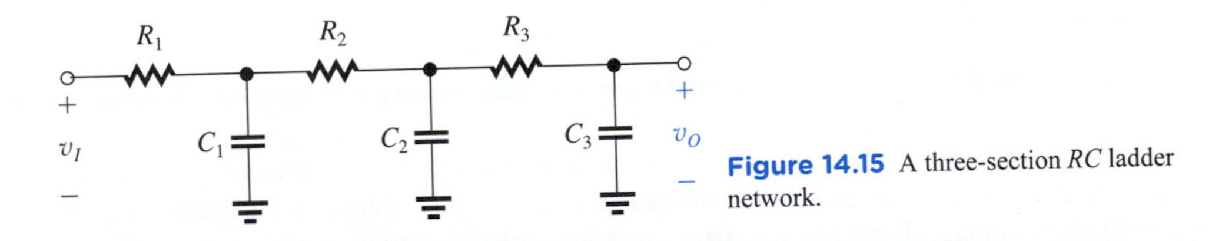

Figure 14.15 A three-section *RC* ladder network.

EXERCISE

14.7 The circuit in Fig. 14.14 is fabricated in a 0.13-μm CMOS technology; Q_P of the first inverter has $W/L = 2$, and both transistors of the transmission gate have $W/L = 1$. The capacitances have been estimated to be $C_{\text{out1}} = 10$ fF, $C_{TG1} = C_{TG2} = 5$ fF, and $C_{\text{in2}} = 10$ fF. Use the empirical formulas to obtain the values of R_{P1} and R_{TG}. Then, determine an estimate for t_P.

Ans. $R_{P1} = 15$ kΩ; $R_{TG} = 12.5$ kΩ; $t_P = 0.64$ ns

14.2.5 Pass-Transistor Logic Circuit Examples

We conclude this section by showing examples of PTL logic circuits. Figure 14.16 shows a PTL realization of a two-to-one multiplexer: Depending on the logic value of C, either A or B is connected to the output Y. The circuit realizes the Boolean function

$$Y = CA + \overline{C}B$$

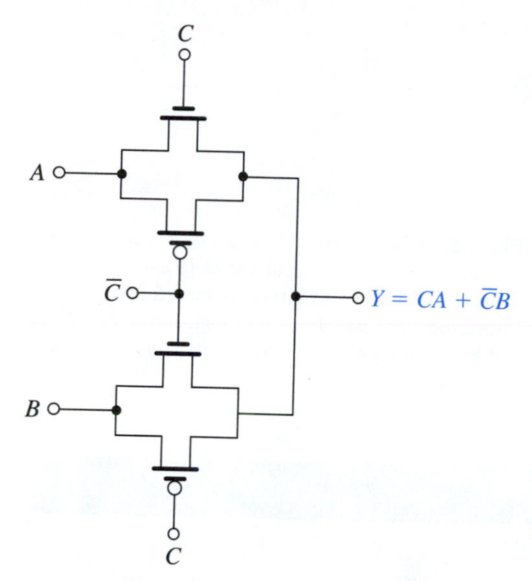

Figure 14.16 Realization of a two-to-one multiplexer using pass-transistor logic.

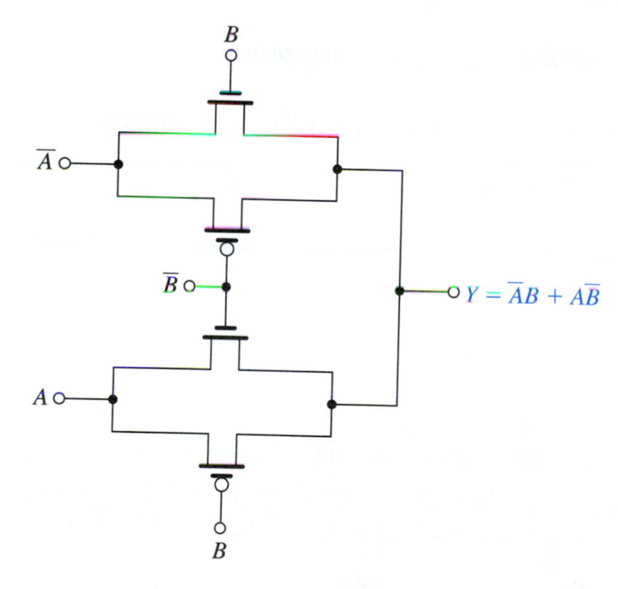

Figure 14.17 Realization of the XOR function using pass-transistor logic.

Our second example is an efficient realization of the exclusive-OR (XOR) function. The circuit, shown in Fig. 14.17, utilizes four transistors in the transmission gates and another four for the two inverters needed to generate the complements $\overline{A}$ and $\overline{B}$, for a total of eight transistors. Note that 12 transistors are needed in the realization with standard CMOS.

Our final PTL example is the circuit shown in Fig. 14.18. It uses NMOS switches with low or zero threshold. Observe that both the input variables and their complements are employed and that the circuit generates both the Boolean function and its complement. Thus this form of circuit is known as **complementary pass-transistor logic (CPL)**. The circuit consists of two identical networks of pass transistors with the corresponding transistor gates controlled by the same signal (B and $\overline{B}$). The inputs to the PTL, however, are complemented: A and B for the first network, and $\overline{A}$ and $\overline{B}$ for the second. The circuit shown realizes both the AND and NAND functions.

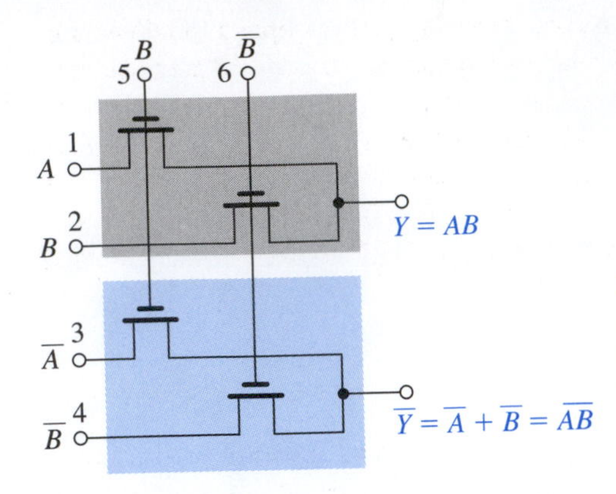

Figure 14.18 An example of a pass-transistor logic gate utilizing both the input variables and their complements. This type of circuit is therefore known as complementary pass-transistor logic, or CPL. Note that both the output function and its complement are generated.

EXERCISE

14.8 Consider the circuit in Fig. 14.8, and for each case, find Y and $\overline{Y}$. The input signals are changed as follows:

(a) The signals at terminals 5 and 6 are interchanged ($\overline{B}$ applied to 5 and B applied to 6). All the rest are the same.

(b) The signals at terminals 5 or 6 are interchanged as in (a), and the signals at 2 and 4 are changed to $\overline{A}$ and A, respectively. All the rest remain the same.

Ans. (a) $Y = A + B$, $\overline{Y} = \overline{A}\overline{B} = \overline{A + B}$ (i.e., OR–NOR); (b) $Y = A\overline{B} + \overline{A}B$, $\overline{Y} = \overline{A}\overline{B} + AB$ (i.e., XOR–XNOR)

14.2.6 A Final Remark

Although the use of zero-threshold devices solves the problem of the loss of signal levels when NMOS switches are used, the resulting circuits can be much more sensitive to noise and other effects, such as leakage currents resulting from subthreshold conduction.

14.3 Dynamic MOS Logic Circuits

The logic circuits that we have studied thus far are of the static type. In a static logic circuit, every node has, at all times, a low-resistance path to V_{DD} or ground. By the same token, the voltage of each node is well defined at all times, and no node is left floating. Static circuits do not need clocks (i.e., periodic timing signals) for their operation, although clocks may be present for other purposes. In contrast, the dynamic logic circuits we are about to discuss rely on the storage of signal voltages on parasitic capacitances at certain circuit nodes. Since charge will leak away with time, the circuits need to be *periodically refreshed;* thus the presence of a clock with a certain specified minimum frequency is essential.

To place dynamic logic circuit techniques into perspective, let's take stock of the various styles we have studied for logic circuits. Standard CMOS excels in nearly every performance category: It is easy to design, has the maximum possible logic swing, is robust from a noise-immunity standpoint, dissipates no static power, and can be designed to provide equal low-to-high and high-to-low propagation delays. Its main disadvantage is the requirement of two transistors for each additional gate input, which for high fan-in gates can make the chip area large and increase the total capacitance and, correspondingly, the propagation delay and the dynamic power dissipation. Pseudo-NMOS reduces the number of required transistors at the expense of static power dissipation. Pass-transistor logic can result in simple small-area circuits but is limited to special applications and requires the use of CMOS inverters to restore signal levels, especially when the switches are simple NMOS transistors. The dynamic logic techniques studied in this section maintain the low device count of pseudo-NMOS while reducing the static power dissipation to zero. As will be seen, this is achieved at the expense of more complex, and less robust, design.

14.3.1 The Basic Principle

Figure 14.19(a) shows the basic dynamic logic gate. It consists of a pull-down network (PDN) that realizes the logic function in exactly the same way as the PDN of a standard CMOS gate or a pseudo-NMOS gate. Here, however, we have two switches in series that are periodically operated by the clock signal ϕ whose waveform is shown in Fig. 14.19(b). When ϕ is low, Q_p is turned on, and the circuit is said to be in the setup or **precharge phase**. When ϕ is high, Q_p is off and Q_e turns on, and the circuit is in the **evaluation phase**. Finally, note that C_L denotes the total capacitance between the output node and ground.

During precharge, Q_p conducts and charges capacitance C_L so that at the end of the precharge interval, the voltage at Y is equal to V_{DD}. Also during precharge, the inputs A, B, and C are allowed to change and settle to their proper values. Observe that because Q_e is off, no path to ground exists.

During the evaluation phase, Q_p is off and Q_e is turned on. Now, if the input combination is one that corresponds to a high output, the PDN does not conduct (just as in a standard

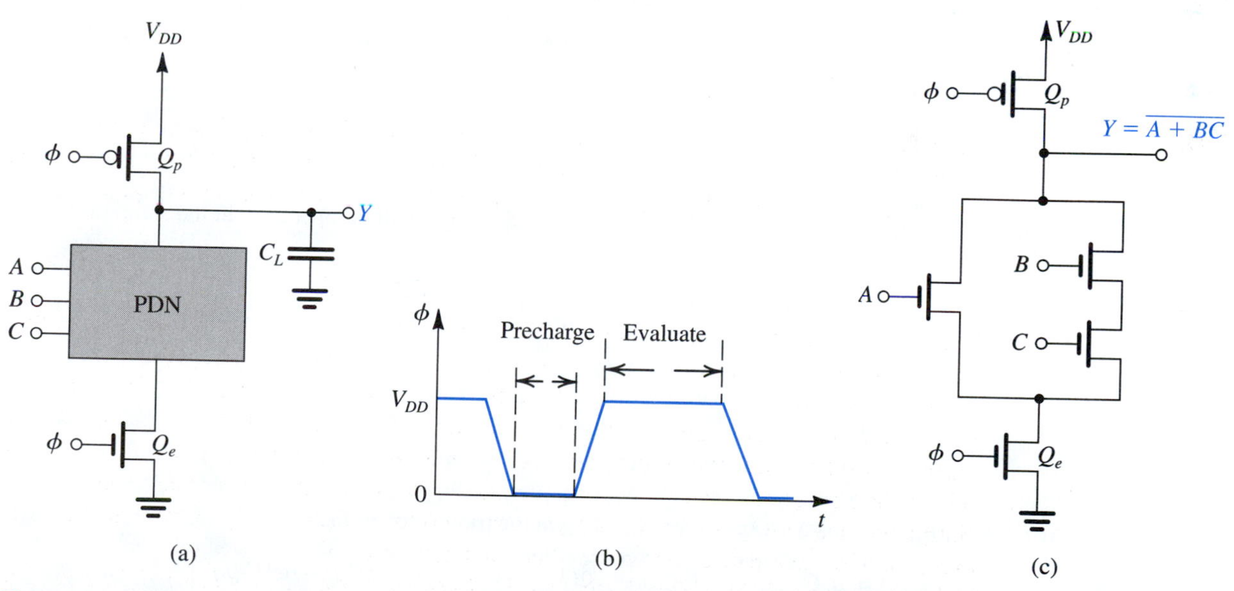

Figure 14.19 (**a**) Basic structure of dynamic-MOS logic circuits. (**b**) Waveform of the clock needed to operate the dynamic logic circuit. (**c**) An example circuit.

CMOS gate) and the output remains high at V_{DD}; thus $V_{OH} = V_{DD}$. Observe that no low-to-high propagation delay is required, thus $t_{PLH} = 0$. On the other hand, if the combination of inputs is one that corresponds to a low output, the appropriate NMOS transistors in the PDN will conduct and establish a path between the output node and ground through the "on" transistor Q_e. Thus C_L will be discharged through the PDN, and the voltage at the output node will reduce to $V_{OL} = 0$ V. The high-to-low propagation delay t_{PHL} can be calculated in exactly the same way as for a standard CMOS circuit, except that here we have an additional transistor, Q_e, in the series path to ground. Although this will increase the delay slightly, the increase will be more than offset by the reduced capacitance at the output node as a result of the absence of the PUN.

As an example, we show in Fig. 14.19(c) the circuit that realizes the function $Y = \overline{A + BC}$. Sizing of the PDN transistors often follows the same procedure employed in the design of static CMOS. For Q_p, we select a W/L ratio large enough to ensure that C_L will be fully charged during the precharge interval, but small enough so that the capacitance C_L will not be increased significantly. This is a ratioless form of MOS logic, where the output levels do not depend on the transistors' W/L ratios (unlike pseudo-NMOS, for instance).

Example 14.3

Consider the four-input, dynamic-logic NAND gate shown in Fig. 14.20(a). Assume that the gate is fabricated in a 0.18-μm CMOS technology for which $V_{DD} = 1.8$ V, $V_t = 0.5$ V, and $\mu_n C_{ox} = 4\mu_p C_{ox} = 300\ \mu\text{A/V}^2$. To keep C_L small, NMOS devices with $W/L = 0.27\ \mu\text{m}/0.18\ \mu\text{m}$ are used (including transistor Q_e). The PMOS precharge transistor Q_p has $W/L = 0.54\ \mu\text{m}/0.18\ \mu\text{m}$. The total capacitance C_L is found to be 20 fF.

(a) Consider the precharge operation (Fig. 14.20b) with the gate of Q_p at 0 V, and assume that at $t = 0$, C_L is fully discharged. Calculate the rise time of the output voltage, defined as the time for v_Y to rise from 10% to 90% of the final voltage V_{DD}.

(b) For $A = B = C = D = 1$, find the value of t_{PHL}.

Solution

(a) From Fig. 14.20(a) we see that at $v_Y = 0.1V_{DD} = 0.18$ V, Q_p will be operating in the saturation region and i_D will be

$$i_D(0.1V_{DD}) = \frac{1}{2}\mu_p C_{ox}\left(\frac{W}{L}\right)_p (V_{DD} - |V_{tp}|)^2$$

$$= \frac{1}{2} \times 75 \times \frac{0.54}{0.18}(1.8 - 0.5)^2$$

$$= 190.1\ \mu\text{A}$$

At $v_Y = 0.9V_{DD} = 1.62$ V, Q_p will be operating in the triode region; thus,

$$i_D(0.9V_{DD}) = \mu_p C_{ox}\left(\frac{W}{L}\right)_p \left[(V_{DD} - |V_{tp}|)(V_{DD} - 0.9V_{DD}) - \frac{1}{2}(V_{DD} - 0.9V_{DD})^2\right]$$

$$= 75 \times \frac{0.54}{0.18}\left[(1.8 - 0.5)(1.8 - 1.62) - \frac{1}{2}(1.8 - 1.62)^2\right]$$

$$= 49\ \mu\text{A}$$

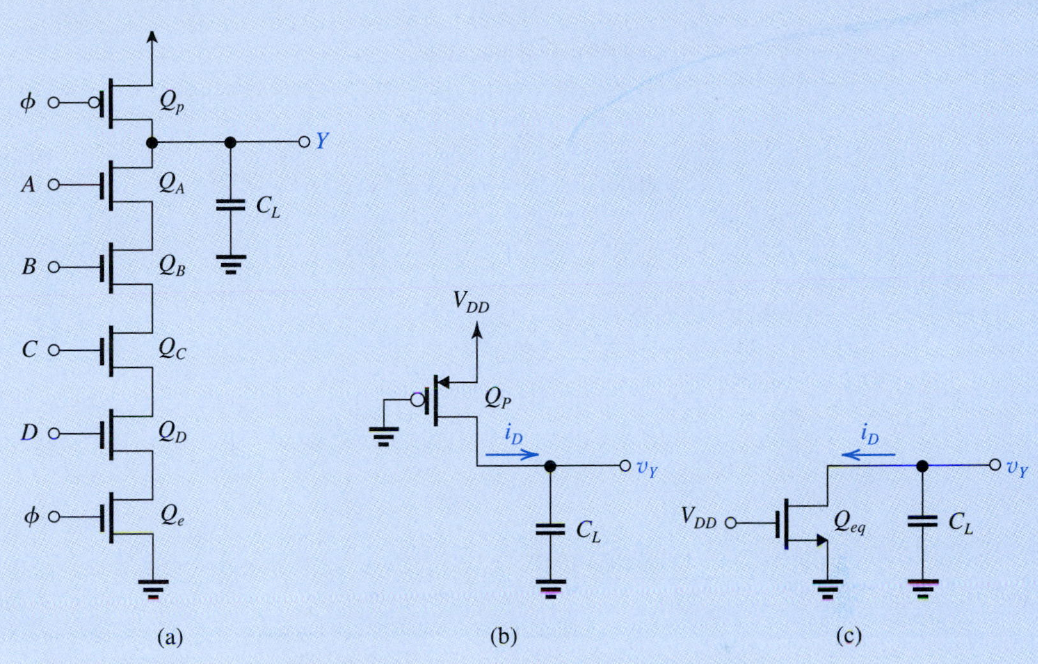

Figure 14.20 Circuits for Example 14.3.

Thus the average capacitor charging current is

$$I_{av} = \frac{1}{2}(190.1 + 49) = 119.6\ \mu A$$

The rise time t_r of v_Y can now be determined from

$$t_r = \frac{C\Delta v_Y}{I_{av}}$$

$$= \frac{C(0.9V_{DD} - 0.1V_{DD})}{I_{av}}$$

Thus,

$$t_r = \frac{20 \times 10^{-15} \times 0.8 \times 1.8}{119.6 \times 10^{-6}} = 0.19\ \text{ns}$$

(b) When $A = B = C = D = 1$, all the NMOS transistors will be conducting during the evaluation phase. Replacing the five identical transistors with an equivalent device Q_{eq} with $(W/L)_{eq} = \frac{1}{5}(W/L) = \frac{1}{5} \times 1.5 = 0.3$, we obtain the equivalent circuit for the capacitor discharge, shown in Fig. 14.20(c). At $v_Y = V_{DD}$, Q_{eq} will be operating in saturation; thus,

$$i_D(V_{DD}) = \frac{1}{2}(\mu_n C_{ox})\left(\frac{W}{L}\right)_{eq}(V_{DD} - V_t)^2$$

$$= \frac{1}{2} \times 300 \times 0.3(1.8 - 0.5)^2$$

$$= 76.1\ \mu A$$

Example 14.3 *continued*

At $v_Y = V_{DD}/2$, Q_{eq} will be operating in the triode region; thus,

$$i_D(V_{DD}/2) = (\mu_n C_{ox})\left(\frac{W}{L}\right)_{eq}\left[(V_{DD} - V_t)\frac{V_{DD}}{2} - \frac{1}{2}\left(\frac{V_{DD}}{2}\right)^2\right]$$

$$= 300 \times 0.3\left[(1.8 - 0.5)\left(\frac{1.8}{2}\right) - \frac{1}{2}\left(\frac{1.8}{2}\right)^2\right]$$

$$= 68.9\ \mu\text{A}$$

Thus the average capacitor-discharge current is

$$I_{av} = \frac{76.1 + 68.9}{2} = 72.5\ \mu\text{A}$$

and t_{PHL} can be found from

$$t_{PHL} = \frac{C(V_{DD} - V_{DD}/2)}{I_{av}}$$

$$= \frac{20 \times 10^{-15}(1.8 - 0.9)}{72.5 \times 10^{-6}} = 0.25\ \text{ns}$$

EXERCISE

14.9 In an attempt to reduce t_{PHL} of the NAND gate in Example 14.3, the designer doubles the value of W/L of each of the NMOS devices. If C increases to 30 fF, what is the new value if t_{PHL}?
Ans. 0.19 ns

14.3.2 Nonideal Effects

We now briefly consider various sources of nonideal operation of dynamic logic circuits.

Noise Margins Since, during the evaluation phase, the NMOS transistors begin to conduct for $v_I = V_{tn}$,

$$V_{IL} \simeq V_{IH} \simeq V_{tn}$$

and thus the noise margins will be

$$NM_L = V_{IL} - V_{OL} = V_{tn} - 0 = V_{tn}$$
$$NM_H = V_{OH} - V_{IH} = V_{DD} - V_{tn}$$

Thus the noise margins are far from equal, and NM_L is rather low. Although NM_H is high, other nonideal effects reduce its value, as we shall shortly see. At this time, however, observe that the output node is a high-impedance node and thus will be susceptible to noise pickup and other disturbances.

Output Voltage Decay Due to Leakage Effects In the absence of a path to ground through the PDN, the output voltage will ideally remain high at V_{DD}. This, however, is based on the assumption that the charge on C_L will remain intact. In practice, there will be leakage current that will cause C_L to slowly discharge and v_Y to decay. The principal source of leakage is the reverse current of the reverse-biased junction between the drain diffusion of transistors connected to the output node and the substrate. Such currents can be in the range of 10^{-12} A to 10^{-15} A, and they increase rapidly with temperature (approximately doubling for every 10°C rise in temperature). Thus the circuit can malfunction if the clock is operating at a very low frequency and the output node is not "refreshed" periodically. This exact same point will be encountered when we study dynamic memory cells in Chapter 15.

Charge Sharing There is another and often more serious way for C_L to lose some of its charge and thus cause v_Y to fall significantly below V_{DD}. To see how this can happen, refer to Fig. 14.21(a), which shows only Q_1 and Q_2, the two top transistors of the PDN, together with the precharge transistor Q_p. Here, C_1 is the capacitance between the common node of Q_1 and Q_2 and ground. At the beginning of the evaluation phase, after Q_p has turned off and with C_L charged to V_{DD} (Fig. 14.21a), we assume that C_1 is initially discharged and that the inputs are such that at the gate of Q_1 we have a high signal, whereas at the gate of Q_2 the signal is low. We can easily see that Q_1 will turn on and its drain current, i_{D1}, will flow as indicated. Thus i_{D1} will discharge C_L and charge C_1. Although eventually i_{D1} will reduce to zero, C_L will have lost some of its charge, which will have been transferred to C_1. This phenomenon is known as charge sharing (see Problem 14.31).

We shall not pursue the problem of charge sharing any further here, except to point out a couple of the techniques usually employed to minimize its effect. One approach involves adding a *p*-channel device that continuously conducts a small current to replenish the charge lost by C_L, as shown in Fig. 14.21(b). This arrangement should remind us of pseudo-NMOS. Indeed, adding this transistor will cause the gate to dissipate static power. On the positive side, however, the added transistor will lower the impedance level of the output node and make it less susceptible to noise as well as solving the leakage and charge-sharing problems. Another approach to

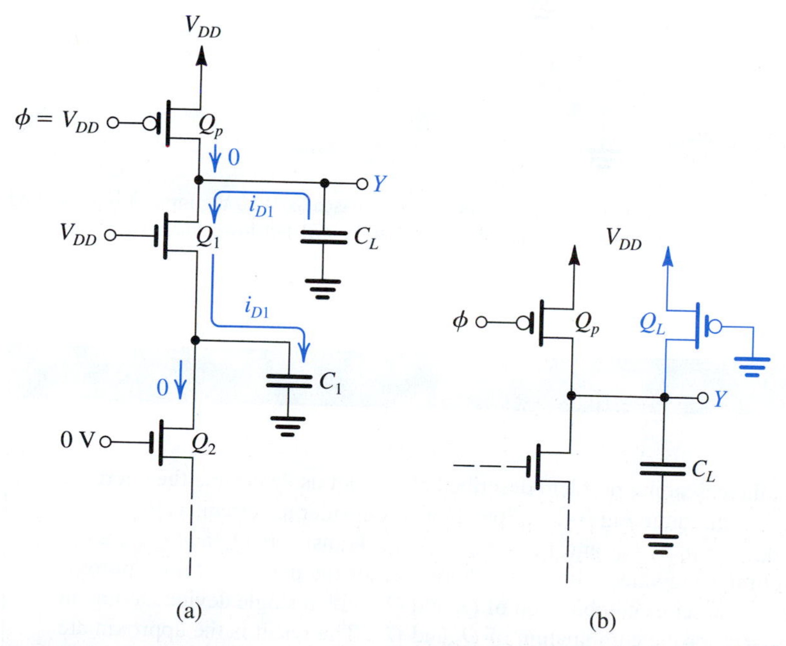

Figure 14.21 **(a)** Charge sharing. **(b)** Adding a permanently turned-on transistor Q_L solves the charge-sharing problem at the expense of static power dissipation.

solving the charge-sharing problem is to precharge the internal nodes: that is, to precharge capacitor C_1. The price paid in this case is increased circuit complexity and node capacitances.

Cascading Dynamic Logic Gates A serious problem arises if one attempts to cascade dynamic logic gates. Consider the situation depicted in Fig. 14.22, where two single-input dynamic gates are connected in cascade. During the precharge phase, C_{L1} and C_{L2} will be charged through Q_{p1} and Q_{p2}, respectively. Thus, at the end of the precharge interval, $v_{Y1} = V_{DD}$ and $v_{Y2} = V_{DD}$. Now consider what happens in the evaluation phase for the case of high input A. Obviously, the correct result will be Y_1 low ($v_{Y1} = 0$ V) and Y_2 high ($v_{Y2} = V_{DD}$). What happens, however, is somewhat different. As the evaluation phase begins, Q_1 turns on and C_{L1} begins to discharge. However, simultaneously, Q_2 turns on and C_{L2} also begins to discharge. Only when v_{Y1} drops below V_{tn} will Q_2 turn off. Unfortunately, however, by that time, C_{L2} will have lost a significant amount of its charge, and v_{Y2} will be less than the expected value of V_{DD}. (Here it is important to note that in dynamic logic, once charge has been lost, it cannot be recovered.) This problem is sufficiently serious to make simple cascading an impractical proposition. As usual, however, the ingenuity of circuit designers has come to the rescue, and a number of schemes have been proposed to make cascading possible in dynamic-logic circuits. We shall discuss one such scheme after considering Exercise 14.10.

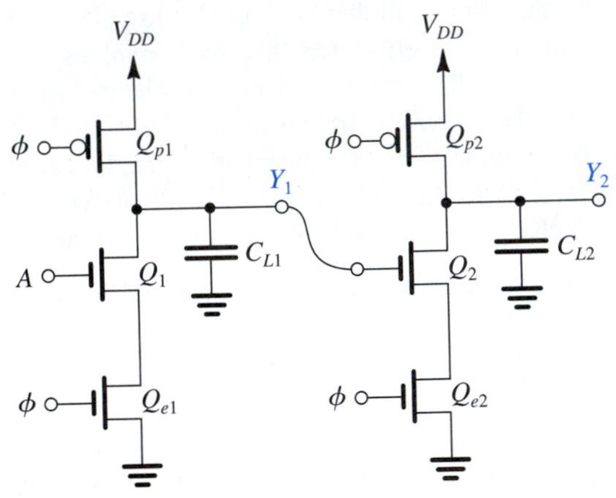

Figure 14.22 Two single-input dynamic logic gates connected in cascade. With the input A high, during the evaluation phase C_{L2} will partially discharge and the output at Y_2 will fall lower than V_{DD}, which can cause logic malfunction.

EXERCISE

14.10 To gain further insight into the cascading problem described above, let us determine the decrease in the output voltage v_{Y2} for the circuit in Fig. 14.22. Specifically, consider the circuit as the evaluation phase begins: At $t = 0$, $v_{Y1} = v_{Y2} = V_{DD}$ and $v_\phi = v_A = V_{DD}$. Transistors Q_{p1} and Q_{p2} are cut off and can be removed from the equivalent circuit. Furthermore, for the purpose of this approximate analysis, we can replace the series combination of Q_1 and Q_{e1} with a single device having an appropriate W/L, and similarly for the combination of Q_2 and Q_{e2}. The result is the approximate equivalent circuit in Fig. E14.10. We are interested in the operation of this circuit in the interval Δt

during which v_{Y1} falls from V_{DD} to V_t, at which time Q_{eq2} turns off and C_{L2} stops discharging. Assume that the process technology has the parameter values specified in Example 14.2; that for all NMOS transistors in the circuit of Fig. 14.22, $W/L = 4\ \mu\text{m}/2\ \mu\text{m}$ and $C_{L1} = C_{L2} = 40$ fF.

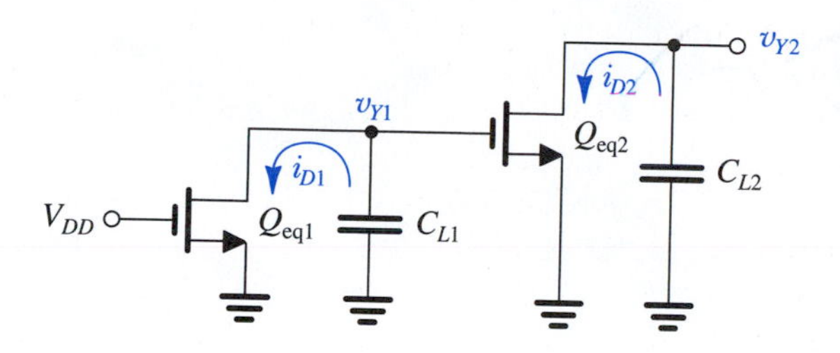

Figure E 14.10

(a) Find $(W/L)_{eq1}$ and $(W/L)_{eq2}$.

(b) Find the values of i_{D1} at $v_{Y1} = V_{DD}$ and at $v_{Y1} = V_t$. Hence determine an average value for i_{D1}.

(c) Use the average value of i_{D1} found in (b) to determine an estimate for the interval Δt.

(d) Find the average value of i_{D2} during Δt. To simplify matters, take the average to be the value of i_{D2} obtained when the gate voltage v_{Y1} is midway through its excursion (i.e., $v_{Y1} = 3$ V). (*Hint:* Q_{eq2} will remain in saturation.)

(e) Use the value of Δt found in (c) together with the average value of i_{D2} determined in (d) to find an estimate of the reduction in v_{Y2} during Δt. Hence determine the final value of v_{Y2}.

Ans. (a) 1, 1; (b) 400 μA and 175 μA, for an average value of 288 μA; (c) 0.56 ns; (d) 100 μA; (e) $\Delta v_{Y2} = 1.4$ V, thus v_{Y2} decreases to 3.6 V

14.3.3 Domino CMOS Logic

Domino CMOS logic is a form of dynamic logic that results in cascadable gates. Figure 14.23 shows the structure of the Domino CMOS logic gate. We observe that it is simply the basic dynamic logic gate of Fig. 14.19(a) with a static CMOS inverter connected to its output. Operation of the gate is straightforward. During precharge, X will be raised to V_{DD}, and the gate output Y will be at 0 V. During evaluation, depending on the combination of input variables, either X will remain high and thus the output Y will remain low ($t_{PHL} = 0$) or X will be brought down to 0 V and the output Y will rise to V_{DD} (t_{PLH} finite). Thus, during evaluation, the output either remains low or makes only one low-to-high transition.

To see why Domino CMOS gates can be cascaded, consider the situation in Fig. 14.24(a), where we show two Domino gates connected in cascade. For simplicity, we show single-input gates. At the end of precharge, X_1 will be at V_{DD}, Y_1 will be at 0 V, X_2 will be at V_{DD}, and Y_2 will be at 0 V. As in the preceding case, assume that A is high at the beginning of evaluation. Thus, as ϕ goes up, capacitor C_{L1} will begin discharging, pulling X_1 down. Meanwhile, the low input at the gate of Q_2 keeps Q_2 off, and C_{L2} remains fully charged. When v_{X1} falls below the threshold voltage of inverter I_1, Y_1 will go up, turning Q_2 on, which in turn begins to discharge C_{L2} and pulls X_2 low. Eventually, Y_2 rises to V_{DD}.

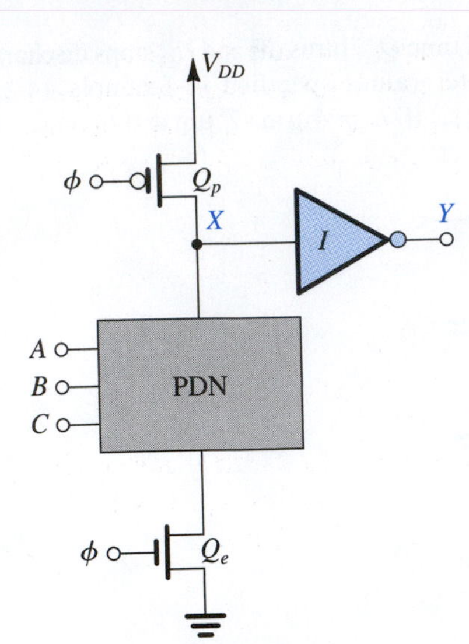

Figure 14.23 The Domino CMOS logic gate. The circuit consists of a dynamic-MOS logic gate with a static-CMOS inverter connected to the output. During evaluation, Y either will remain low (at 0 V) or will make one 0-to-1 transition (to V_{DD}).

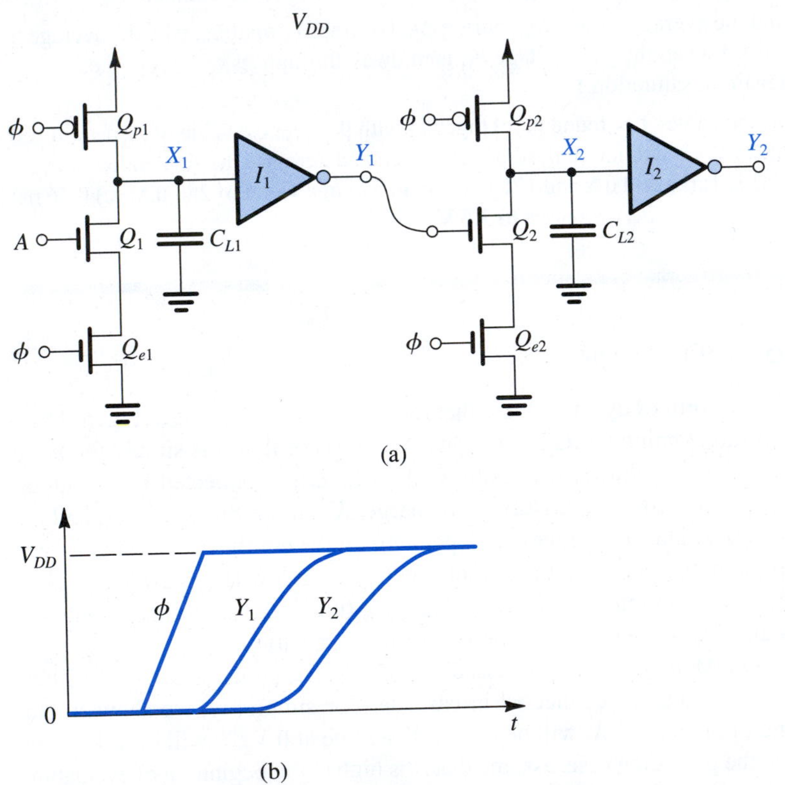

Figure 14.24 **(a)** Two single-input Domino CMOS logic gates connected in cascade. **(b)** Waveforms during the evaluation phase.

From this description, we see that because the output of the Domino gate is low at the beginning of evaluation, no premature capacitor discharge will occur in the subsequent gate in the cascade. As indicated in Fig. 14.24(b), output Y_1 will make a 0-to-1 transition t_{PLH}

seconds after the rising edge of the clock. Subsequently, output Y_2 makes a 0-to-1 transition after another t_{PLH} interval. The propagation of the rising edge through a cascade of gates resembles contiguously placed dominoes falling over, each toppling the next, which is the origin of the name Domino CMOS logic. Domino CMOS logic finds application in the design of address decoders in memory chips, for example.

14.3.4 Concluding Remarks

Dynamic logic presents many challenges to the circuit designer. Although it can provide considerable reduction in the chip-area requirement, as well as high-speed operation, and zero (or little) static-power dissipation, the circuits are prone to many nonideal effects, some of which have been discussed here. It should also be remembered that dynamic power dissipation is an important issue in dynamic logic. Another factor that should be considered is the "dead time" during precharge when the output of the circuit is not yet available.

14.4 Emitter-Coupled Logic (ECL)

Emitter-coupled logic (ECL) is the fastest logic circuit family available for conventional logic-system design.[4] High speed is achieved by operating all bipolar transistors out of saturation, thus avoiding storage-time delays, and by keeping the logic signal swings relatively small (about 0.8 V or less), thus reducing the time required to charge and discharge the various load and parasitic capacitances. Saturation in ECL is avoided by using the BJT differential pair as a current switch.[5] The BJT differential pair was studied in Chapter 8, and we urge the reader to review the introduction given in Section 8.3 before proceeding with the study of ECL.

14.4.1 The Basic Principle

Emitter-coupled logic is based on the use of the current-steering switch introduced in Section 13.1 (Fig 13.9). Such a switch can be most conveniently realized using the differential pair shown in Fig. 14.25. The pair is biased with a constant-current source I, and one side is connected to a reference voltage V_R. As shown in Section 8.3, the current I can be steered to either Q_1 or Q_2 under the control of the input signal v_I. Specifically, when v_I is greater than V_R by about $4V_T$ ($\simeq$100 mV), nearly all the current I is conducted by Q_1, and thus for $\alpha_1 \simeq 1$, $v_{O1} = V_{CC} - IR_C$. Simultaneously, the current through Q_2 will be nearly zero, and thus $v_{O2} = V_{CC}$. Conversely, when v_I is lower than V_R by about $4V_T$, most of the current I will flow through Q_2 and the current through Q_1 will be nearly zero. Thus $v_{O1} = V_{CC}$ and $v_{O2} = V_{CC} - IR_C$.

The preceding description suggests that as a logic element, the differential pair realizes an inversion function at v_{O1} and simultaneously provides the complementary output signal at v_{O2}. The output logic levels are $V_{OH} = V_{CC}$ and $V_{OL} = V_{CC} - IR_C$, and thus the output logic swing is IR_C. A number of additional remarks can be made concerning this circuit:

---

[4]Although higher speeds of operation can be obtained with gallium arsenide (GaAs) circuits, the latter are not available as off-the-shelf components for conventional digital system design. GaAs digital circuits are not covered in this book; however, a substantial amount of material on this subject can be found on the disc accompanying the book and on the website.

[5]This is in sharp contrast to the technique utilized in a nonsaturating variant of transistor-transistor logic (TTL) known as Schottky TTL. There, a Schottky diode is placed across the CBJ junction to shunt away some of the base current and, owing to the low voltage drop of the Schottky diode, the CBJ is prevented from becoming forward biased.

1. The differential nature of the circuit makes it less susceptible to picked-up noise. In particular, an interfering signal will tend to affect both sides of the differential pair similarly and thus will not result in current switching. This is the common-mode rejection property of the differential pair (see Section 8.3).
2. The current drawn from the power supply remains constant during switching. Thus, unlike CMOS (and TTL), no supply current spikes occur in ECL, eliminating an important source of noise in digital circuits. This is a definite advantage, especially since ECL is usually designed to operate with small signal swings and has correspondingly low noise margins.
3. The output signal levels are both referenced to V_{CC} and thus can be made particularly stable by operating the circuit with $V_{CC} = 0$: in other words, by utilizing a negative power supply and connecting the V_{CC} line to ground. In this case, $V_{OH} = 0$ and $V_{OL} = -IR_C$.
4. Some means must be provided to make the output signal levels compatible with those at the input so that one gate can drive another. As we shall see shortly, practical ECL gate circuits incorporate a level-shifting arrangement that serves to center the output signal levels on the value of V_R.
5. The availability of complementary outputs considerably simplifies logic design with ECL.

EXERCISE

14.11 For the circuit in Fig. 14.25, let $V_{CC} = 0$, $I = 4$ mA, $R_C = 220\ \Omega$, $V_R = -1.32$ V, and assume $\alpha \simeq 1$. Determine V_{OH} and V_{OL}. By how much should the output levels be shifted so that the values of V_{OH} and V_{OL} become centered on V_R? What will the shifted values of V_{OH} and V_{OL} be?
Ans. 0; –0.88 V; –0.88 V; –0.88 V, –1.76 V

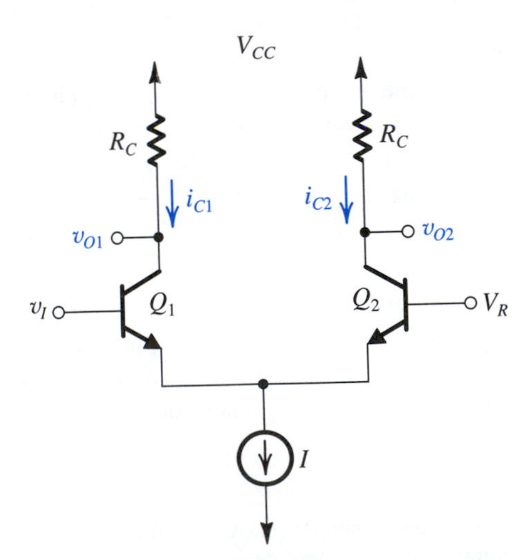

Figure 14.25 The basic element of ECL is the differential pair. Here, V_R is a reference voltage.

14.4.2 ECL Families

Currently there are two popular forms of commercially available ECL—namely, ECL 10 K and ECL 100K. The ECL 100K series features gate delays on the order of 0.75 ns and dissipates about 40 mW/gate, for a delay–power product of 30 pJ. Although its power dissipation is relatively high, the 100K series provides the shortest available gate delay in small- and medium-scale integrated circuit packages.

The ECL 10 K series is slightly slower; it features a gate propagation delay of 2 ns and a power dissipation of 25 mW for a delay–power product of 50 pJ. Although the value of *PDP* is higher than that obtained in the 100K series, the 10K series is easier to use. This is because the rise and fall times of the pulse signals are deliberately made longer, thus reducing signal coupling, or cross talk, between adjacent signal lines. ECL 10K has an "edge speed" of about 3.5 ns, compared with the approximately 1 ns of ECL 100K. To give concreteness to our study of ECL, in the following we shall consider the popular ECL 10K in some detail. The same techniques, however, can be applied to other types of ECL.

In addition to its usage in SSI and MSI circuit packages, ECL is also employed in large-scale and VLSI applications. A variant of ECL known as **current-mode logic** (CML) is utilized in VLSI applications (see Treadway, 1989, and Wilson, 1990).

14.4.3 The Basic Gate Circuit

The basic gate circuit of the ECL 10K family is shown in Fig. 14.26. The circuit consists of three parts. The network composed of Q_1, D_1, D_2, R_1, R_2, and R_3 generates a reference voltage V_R whose value at room temperature is −1.32 V. As will be shown, the value of this reference voltage is made to change with temperature in a predetermined manner to keep the noise margins almost constant. Also, the reference voltage V_R is made relatively insensitive to variations in the power-supply voltage V_{EE}.

EXERCISE

14.12 Figure E14.12 shows the circuit that generates the reference voltage V_R. Assuming that the voltage drop across each of D_1, D_2, and the base–emitter junction of Q_1 is 0.75 V, calculate the value of V_R. Neglect the base current of Q_1.

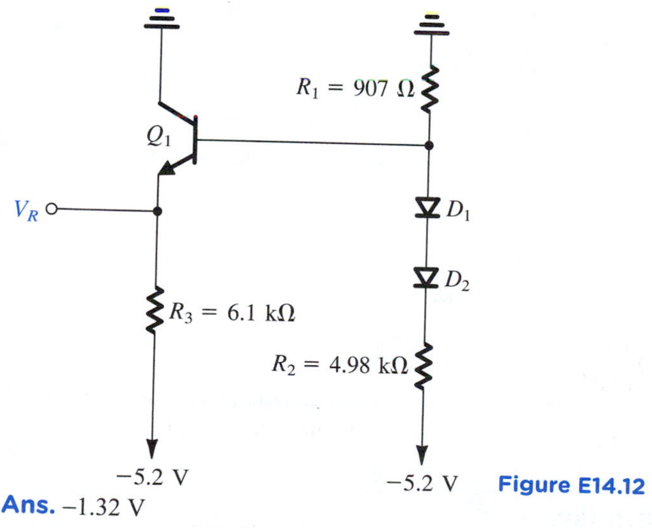

Figure E14.12

Ans. −1.32 V

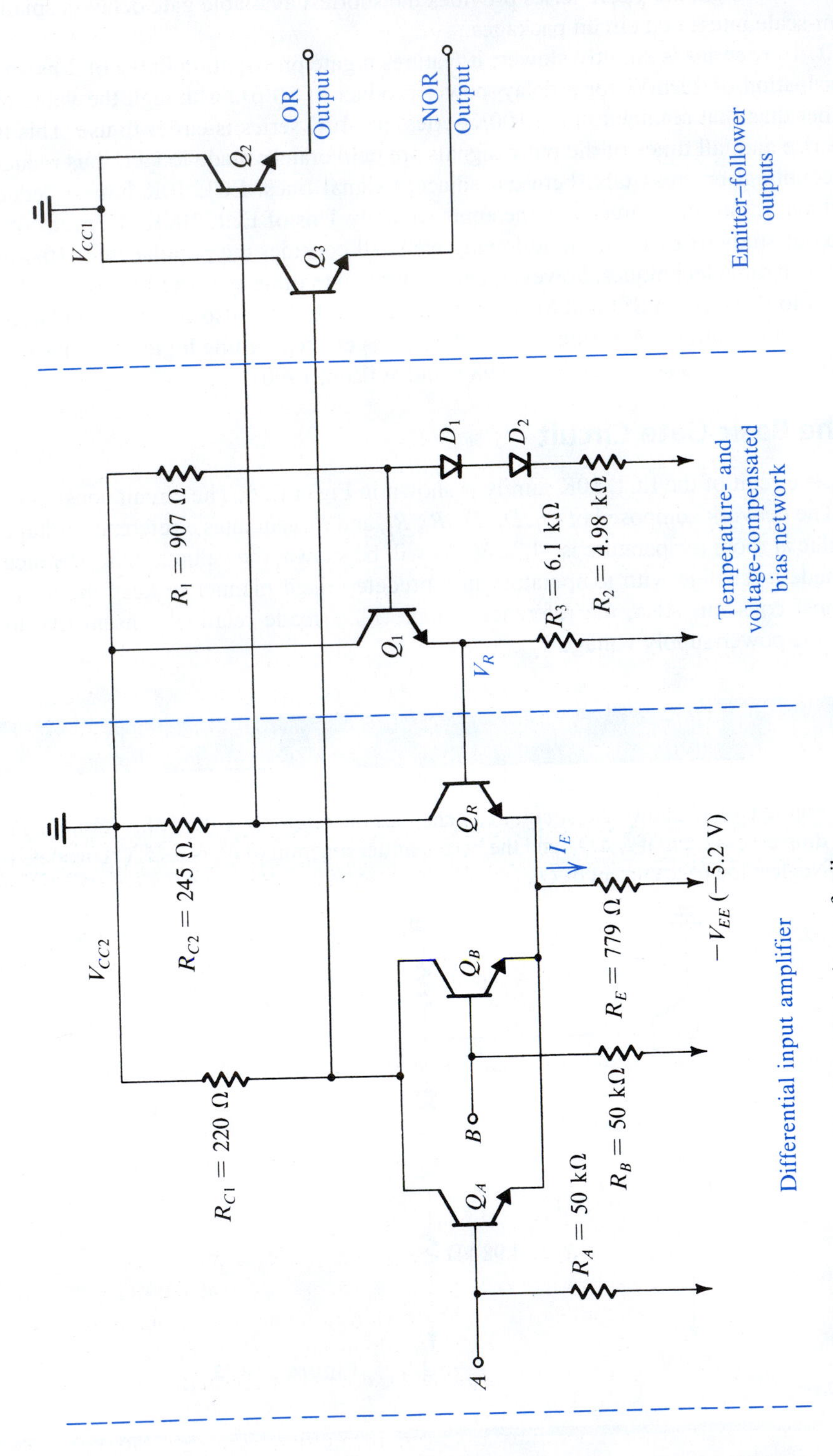

Figure 14.26 Basic circuit of the ECL 10K logic-gate family.

The second part, and the heart of the gate, is the differential amplifier formed by Q_R and either Q_A or Q_B. This differential amplifier is biased not by a constant-current source, as was done in the circuit of Fig. 14.25, but with a resistance R_E connected to the negative supply $-V_{EE}$. Nevertheless, we will shortly show that the current in R_E remains approximately constant over the normal range of operation of the gate. One side of the differential amplifier consists of the reference transistor Q_R, whose base is connected to the reference voltage V_R. The other side consists of a number of transistors (two in the case shown), connected in parallel, with separated bases, each connected to a gate input. If the voltages applied to A and B are at the logic-0 level, which, as we will soon find out, is about 0.4 V below V_R, both Q_A and Q_B, will be off and the current I_E in R_E will flow through the reference transistor Q_R. The resulting voltage drop across R_{C2} will cause the collector voltage of Q_R to be low.

On the other hand, when the voltage applied to A or B is at the logic-1 level, which, as we will show shortly, is about 0.4 V above V_R, transistor Q_A or Q_B, or both, will be on and Q_R will be off. Thus the current I_E will flow through Q_A or Q_B, or both, and an almost equal current will flow through R_{C1}. The resulting voltage drop across R_{C1} will cause the collector voltage to drop. Meanwhile, since Q_R is off, its collector voltage rises. We thus see that the voltage at the collector of Q_R will be high if A or B, or both, is high, and thus at the collector of Q_R, the OR logic function, $A + B$, is realized. On the other hand, the common collector of Q_A and Q_B will be high only when A and B are simultaneously low. Thus at the common collector of Q_A and Q_B, the logic function $\bar{A}\bar{B} = \overline{A + B}$ is realized. We therefore conclude that the two-input gate of Fig. 14.26 realizes the OR function and its complement, the NOR function. The availability of complementary outputs is an important advantage of ECL; it simplifies logic design and avoids the use of additional inverters with associated time delay.

It should be noted that the resistance connecting each of the gate input terminals to the negative supply enables the user to leave an unused input terminal open: An open input terminal will be *pulled down* to the negative supply voltage, and its associated transistor will be off.

EXERCISE

14.13 With input terminals A and B in Fig. 14.26 left open, find the current I_E through R_E. Also find the voltages at the collector of Q_R and at the common collector of the input transistors Q_A and Q_B. Use $V_R = -1.32$ V, V_{BE} of $Q_R \simeq 0.75$ V, and assume that β of Q_R is very high.
Ans. 4 mA; –1 V; 0 V

The third part of the ECL gate circuit is composed of the two emitter followers, Q_2 and Q_3. The emitter followers do not have on-chip loads, since in many applications of high-speed logic circuits the gate output drives a transmission line terminated at the other end, as indicated in Fig. 14.27. (More on this later in Section 14.4.6.)

The emitter followers have two purposes: First, they shift the level of the output signals by one V_{BE} drop. Thus, using the results of Exercise 14.13, we see that the output levels become approximately –1.75 V and –0.75 V. These shifted levels are centered approximately around the reference voltage ($V_R = -1.32$ V), which means that one gate can drive another. This compatibility of logic levels at input and output is an essential requirement in the design of gate circuits.

The second function of the output emitter followers is to provide the gate with low output resistances and with the large output currents required for charging load capacitances.

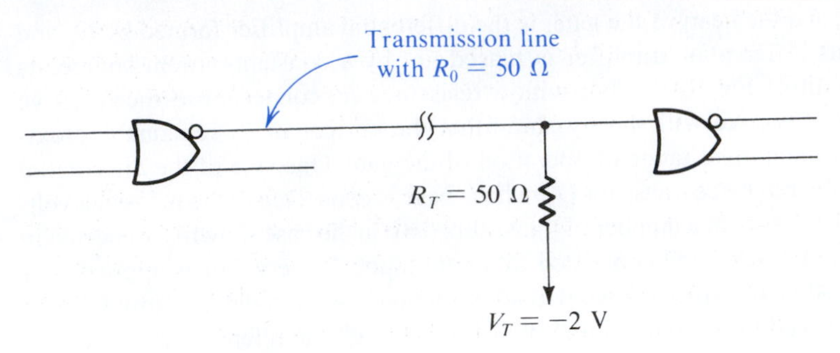

Figure 14.27 The proper way to connect high-speed logic gates such as ECL. Properly terminating the transmission line connecting the two gates eliminates the "ringing" that would otherwise corrupt the logic signals. (See Section 14.4.6.)

Since these large transient currents can cause spikes on the power-supply line, the collectors of the emitter followers are connected to a power-supply terminal V_{CC1} separate from that of the differential amplifier and the reference-voltage circuit, V_{CC2}. Here we note that the supply current of the differential amplifier and the reference circuit remains almost constant. The use of separate power-supply terminals prevents the coupling of power-supply spikes from the output circuit to the gate circuit and thus lessens the likelihood of false gate switching. Both V_{CC1} and V_{CC2} are of course connected to the same system ground, external to the chip.

14.4.4 Voltage-Transfer Characteristics

Having provided a qualitative description of the operation of the ECL gate, we shall now derive its voltage-transfer characteristics. This will be done under the conditions that the outputs are terminated in the manner indicated in Fig. 14.27. Assuming that the B input is low and thus Q_B is off, the circuit simplifies to that shown in Fig. 14.28. We wish to analyze this circuit to determine v_{OR} versus v_I and v_{NOR} versus v_I (where $v_I \equiv v_A$).

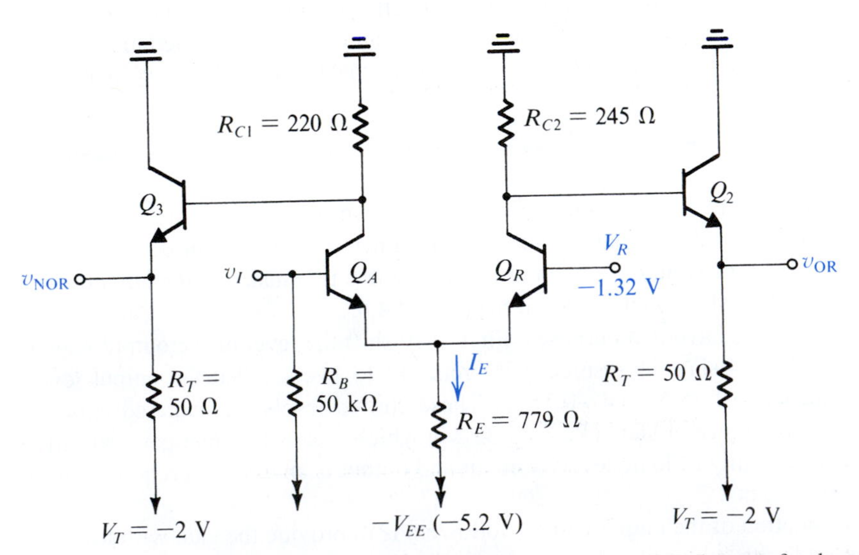

Figure 14.28 Simplified version of the ECL gate for the purpose of finding transfer characteristics.

In the analysis to follow we shall make use of the exponential i_C–v_{BE} characteristic of the BJT. Since the BJTs used in ECL circuits have small areas (in order to have small capacitances and hence high f_T), their scale currents I_S are small. We will therefore assume that at an emitter current of 1 mA, an ECL transistor has a V_{BE} drop of 0.75 V.

The OR Transfer Curve Figure 14.29 is a sketch of the OR transfer characteristic, v_{OR} versus v_I, with the parameters V_{OL}, V_{OH}, V_{IL}, and V_{IH} indicated. However, to simplify the calculation of V_{IL} and V_{IH}, we shall use an alternative to the unity-gain definition. Specifically, we shall assume that at point x, transistor Q_A is conducting 1% of I_E while Q_R is conducting 99% of I_E. The reverse will be assumed for point y. Thus at point x we have

$$\frac{I_E|_{Q_R}}{I_E|_{Q_A}} = 99$$

Using the exponential i_E–v_{BE} relationship, we obtain

$$V_{BE}|_{Q_R} - V_{BE}|_{Q_A} = V_T \ln 99 = 115 \text{ mV}$$

which gives

$$V_{IL} = -1.32 - 0.115 = -1.435 \text{ V}$$

Assuming Q_A and Q_R to be matched, we can write

$$V_{IH} - V_R = V_R - V_{IL}$$

which can be used to find V_{IH} as

$$V_{IH} = -1.205 \text{ V}$$

To obtain V_{OL}, we note that Q_A is off and Q_R carries the entire current I_E, given by

$$I_E = \frac{V_R - V_{BE}|_{Q_R} + V_{EE}}{R_E}$$

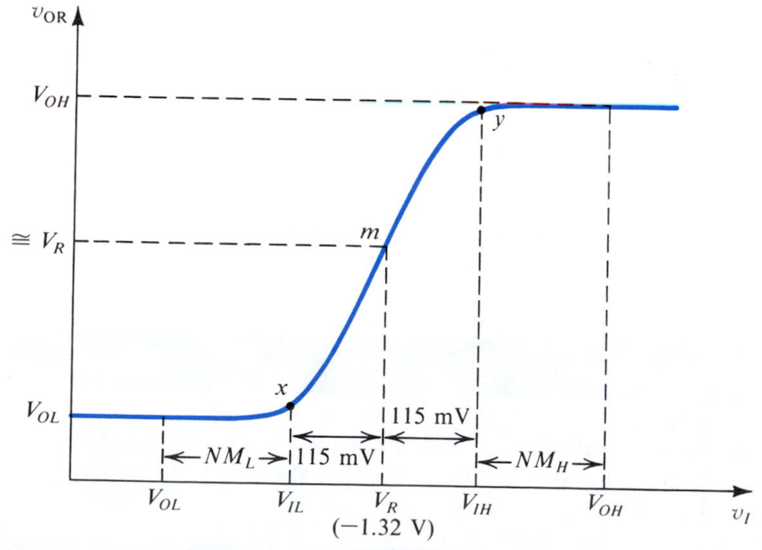

Figure 14.29 The OR transfer characteristic v_{OR} versus v_I, for the circuit in Fig. 14.28.

$$= \frac{-1.32 - 0.75 + 5.2}{0.779}$$
$$\simeq 4 \text{ mA}$$

(If we wish, we can iterate to determine a better estimate of $V_{BE}|_{Q_R}$ and hence of I_E.) Assuming that Q_R has a high β so that its $\alpha \simeq 1$, its collector current will be approximately 4 mA. If we neglect the base current of Q_2, we obtain for the collector voltage of Q_R

$$V_C|_{Q_R} \simeq -4 \times 0.245 = -0.98 \text{ V}$$

Thus a first approximation for the value of the output voltage V_{OL} is

$$V_{OL} = V_C|_{Q_R} - V_{BE}|_{Q_2}$$
$$\simeq -0.98 - 0.75 = -1.73 \text{ V}$$

We can use this value to find the emitter current of Q_2 and then iterate to determine a better estimate of its base–emitter voltage. The result is $V_{BE2} \simeq 0.79$ V and, correspondingly,

$$V_{OL} \simeq -1.77 \text{ V}$$

At this value of output voltage, Q_2 supplies a load current of about 4.6 mA.

To find the value of V_{OH} we assume that Q_R is completely cut off (because $v_I > V_{IH}$). Thus the circuit for determining V_{OH} simplifies to that in Fig. 14.30. Analysis of this circuit, assuming $\beta_2 = 100$, results in $V_{BE2} \simeq 0.83$ V, $I_{E2} = 22.4$ mA, and

$$V_{OH} \simeq -0.88 \text{ V}$$

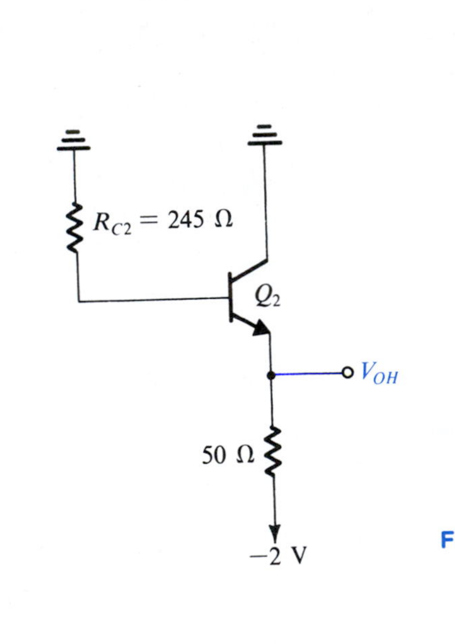

Figure 14.30 Circuit for determining V_{OH}.

EXERCISE

14.14 For the circuit in Fig. 14.28, determine the values of I_E obtained when $v_I = V_{IL}$, V_R, and V_{IH}. Also, find the value of v_{OR} corresponding to $v_I = V_R$. Assume that $v_{BE} = 0.75$ V at a current of 1 mA.
Ans. 3.97 mA; 4.00 mA; 4.12 mA; −1.31 V

Noise Margins The results of Exercise 14.14 indicate that the bias current I_E remains approximately constant. Also, the output voltage corresponding to $v_I = V_R$ is approximately equal to V_R. Notice further that this is also approximately the midpoint of the logic swing; specifically,

$$\frac{V_{OL} + V_{OH}}{2} = -1.325 \simeq V_R$$

Thus the output logic levels are centered around the midpoint of the input transition band. This is an ideal situation from the point of view of noise margins, and it is one of the reasons for selecting the rather arbitrary-looking numbers ($V_R = -1.32$ V and $V_{EE} = 5.2$ V) for reference and supply voltages.

The noise margins can now be evaluated as follows:

$$\begin{aligned} NM_H &= V_{OH} - V_{IH} \\ &= -0.88 - (-1.205) = 0.325 \text{ V} \end{aligned} \qquad \begin{aligned} NM_L &= V_{IL} - V_{OL} \\ &= -1.435 - (-1.77) = 0.335 \text{ V} \end{aligned}$$

Note that these values are approximately equal.

The NOR Transfer Curve The NOR transfer characteristic, which is v_{NOR} versus v_I for the circuit in Fig. 14.28, is sketched in Fig. 14.31. The values of V_{IL} and V_{IH} are identical to those found earlier for the OR characteristic. To emphasize this, we have labeled the threshold points x and y, the same letters used in Fig. 14.29.

For $v_I < V_{IL}$, Q_A is off and the output voltage v_{NOR} can be found by analyzing the circuit composed of R_{C1}, Q_3, and its 50-Ω termination. Except that R_{C1} is slightly smaller than R_{C2}, this circuit is identical to that in Fig. 14.30. Thus the output voltage will be only slightly greater than the value V_{OH} found earlier. In the sketch of Fig. 14.31 we have assumed that the output voltage is approximately equal to V_{OH}.

For $v_I > V_{IH}$, Q_A is on and is conducting the entire bias current. The circuit then simplifies to that in Fig. 14.32. This circuit can be easily analyzed to obtain v_{NOR} versus v_I for the

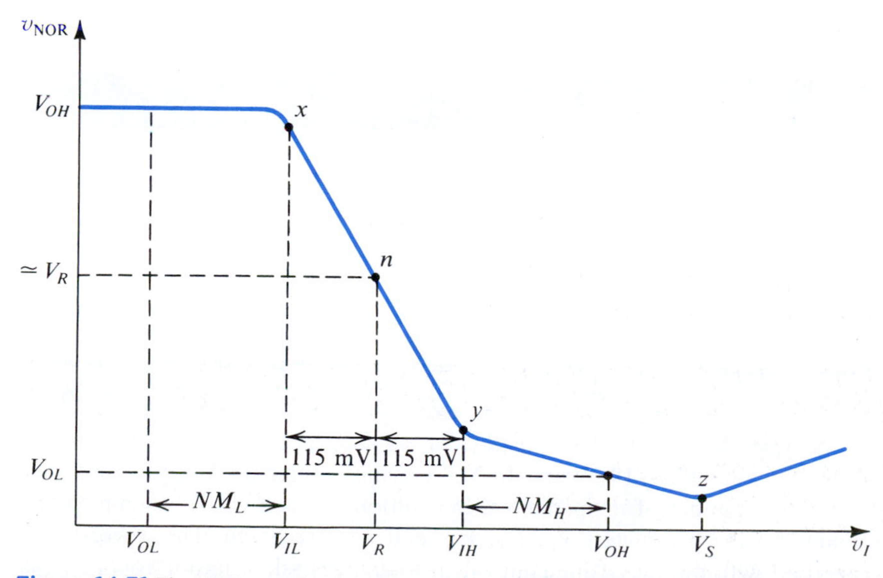

Figure 14.31 The NOR transfer characteristic, v_{NOR} versus v_I, for the circuit in Fig. 14.28.

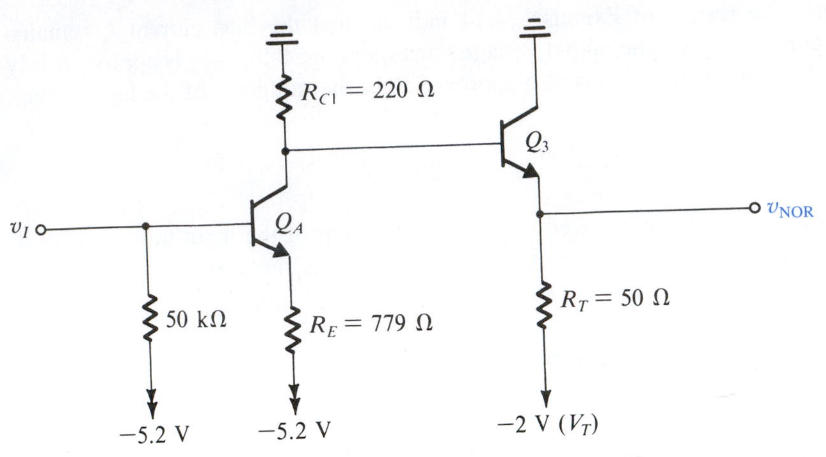

Figure 14.32 Circuit for finding v_{NOR} versus v_I for the range $v_I > V_{IH}$.

range $v_I \geq V_{IH}$. A number of observations are in order. First, note that $v_I = V_{IH}$ results in an output voltage slightly higher than V_{OL}. This is because R_{C1} is smaller than R_{C2}. In fact, R_{C1} is chosen lower in value than R_{C2} so that with v_I equal to the normal logic-1 value (i.e., V_{OH}, which is approximately -0.88 V), the output will be equal to the V_{OL} value found earlier for the OR output.

Second, note that as v_I exceeds V_{IH}, transistor Q_A operates in the active mode and the circuit of Fig. 14.32 can be analyzed to find the gain of this amplifier, which is the slope of the segment *yz* of the transfer characteristic. At point *z*, transistor Q_A saturates. Further increments in v_I (beyond the point $v_I = V_S$) cause the collector voltage and hence v_{NOR} to increase. The slope of the segment of the transfer characteristic beyond point *z*, however, is not unity, but is about 0.5, because as Q_A is driven deeper into saturation, a portion of the increment in v_I appears as an increment in the base–collector forward-bias voltage. The reader is urged to solve Exercise 14.15, which is concerned with the details of the NOR transfer characteristic.

EXERCISE

14.15 Consider the circuit in Fig. 14.32. (a) For $v_I = V_{IH} = -1.205$ V, find v_{NOR}. (b) For $v_I = V_{OH} = -0.88$ V, find v_{NOR}. (c) Find the slope of the transfer characteristic at the point $v_I = V_{OH} = -0.88$ V. (d) Find the value of v_I at which Q_A saturates (i.e., V_S). Assume that $V_{BE} = 0.75$ V at a current of 1 mA, $V_{CEsat} \simeq 0.3$ V, and $\beta = 100$.
Ans. (a) -1.70 V; (b) -1.79 V; (c) -0.24 V/V; (d) -0.58 V

Manufacturers' Specifications ECL manufacturers supply gate transfer characteristics of the form shown in Figs. 14.29 and 14.31. A manufacturer usually provides such curves measured at a number of temperatures. In addition, at each relevant temperature, worst-case values for the parameters V_{IL}, V_{IH}, V_{OL}, and V_{OH} are given. These worst-case values are specified with the inevitable component tolerances taken into account. As an example, Motorola specifies that for MECL 10,000 at 25°C, the following worst-case values apply[6]

$$V_{IL\max} = -1.475 \text{ V} \qquad V_{IH\min} = -1.105 \text{ V}$$

$$V_{OL\max} = -1.630 \text{ V} \qquad V_{OH\min} = -0.980 \text{ V}$$

These values can be used to determine worst-case noise margins,

$$NM_L = 0.155 \text{ V} \qquad NM_H = 0.125 \text{ V}$$

which are about half the *typical* values previously calculated.

For additional information on MECL specifications the interested reader is referred to the Motorola (1988, 1989) publications listed in the bibliography in Appendix G.

14.4.5 Fan-Out

When the input signal to an ECL gate is low (V_{OL}), the input current is equal to the current that flows in the 50-kΩ pull-down resistor. Thus,

$$I_{IL} = \frac{-1.77 + 5.2}{50} \simeq 69 \ \mu\text{A}$$

When the input is high (V_{OH}), the input current is greater because of the base current of the input transistor. Thus, assuming a transistor β of 100, we obtain

$$I_{IH} = \frac{-0.88 + 5.2}{50} + \frac{4}{101} \simeq 126 \ \mu\text{A}$$

Both these current values are quite small, which, coupled with the very small output resistance of the ECL gate, ensures that little degradation of logic-signal levels results from the input currents of fan-out gates. It follows that the fan-out of ECL gates is not limited by logic-level considerations but rather by the degradation of the circuit speed (rise and fall times). This latter effect is due to the capacitance that each fan-out gate presents to the driving gate (approximately 3 pF). Thus while the *dc fan-out* can be as high as 90 and thus does not represent a design problem, the *ac fan-out* is limited by considerations of circuit speed to 10 or so.

14.4.6 Speed of Operation and Signal Transmission

The speed of operation of a logic family is measured by the delay of its basic gate and by the rise and fall times of the output waveforms. Typical values of these parameters for ECL have already been given. Here we should note that because the output circuit is an emitter follower, the rise time of the output signal is shorter than its fall time, since on the rising edge of the output pulse, the emitter follower functions and provides the output current required to charge up the load and parasitic capacitances. On the other hand, as the signal at the base of the emitter follower falls, the emitter follower cuts off, and the load capacitance discharges through the combination of load and pull-down resistances.

To take full advantage of the very high speed of operation possible with ECL, special attention should be paid to the method of interconnecting the various logic gates in a system. To appreciate this point, we shall briefly discuss the problem of signal transmission.

ECL deals with signals whose rise times may be 1 ns or even less, the time it takes for light to travel only 30 cm or so. For such signals, a wire and its environment become a relatively complex circuit element along which signals propagate with finite speed (perhaps half the speed of light—i.e., 15 cm/ns). Unless special care is taken, energy that reaches the end

[6]MECL is the trade name used by Motorola (now Freescale Semiconductors) for its ECL.

of such a wire is not absorbed but rather returns as a *reflection* to the transmitting end, where (without special care) it may be re-reflected. The result of this process of reflection is what can be observed as **ringing**, a damped oscillatory excursion of the signal about its final value.

Unfortunately, ECL is particularly sensitive to ringing because the signal levels are so small. Thus it is important that transmission of signals be well controlled, and surplus energy absorbed, to prevent reflections. The accepted technique is to limit the nature of connecting wires in some way. One way is to insist that they be very short, where "short" is taken to mean with respect to the signal rise time. The reason for this is that if the wire connection is so short that reflections return while the input is still rising, the result becomes only a somewhat slowed and "bumpy" rising edge.

If, however, the reflection returns *after* the rising edge, it produces not simply a modification of the initiating edge but an *independent second event.* This is clearly bad! Thus the time taken for a signal to go from one end of a line and back is restricted to less than the rise time of the driving signal by some factor—say, 5. Thus for a signal with a 1-ns rise time and for propagation at the speed of light (30 cm/ns), a double path of only 0.2-ns equivalent length, or 6 cm, would be allowed, representing in the limit a wire only 3 cm from end to end.

Such is the restriction on ECL 100K. However, ECL 10K has an intentionally slower rise time of about 3.5 ns. Using the same rules, wires can accordingly be as long as about 10 cm for ECL 10K.

If greater lengths are needed, then transmission lines must be used. These are simply wires in a controlled environment in which the distance to a ground reference plane or a second wire is highly controlled. Thus they might simply be twisted pairs of wires, one of which is grounded, or parallel ribbon wires, every second of which is grounded, or so-called microstrip lines on a printed-circuit board. The latter are simply copper strips of controlled geometry on one side of a thin printed-circuit board, the other side of which consists of a grounded plane.

Such transmission lines have a *characteristic impedance*, R_0, that ranges from a few tens of ohms to hundreds of ohms. Signals propagate on such lines somewhat more slowly than the speed of light, perhaps half as fast. When a transmission line is terminated at its receiving end in a resistance equal to its characteristic impedance, R_0, all the energy sent on the line is absorbed at the receiving end, and no reflections occur (since the termination acts as a limitless length of transmission line). Thus, signal integrity is maintained. Such transmission lines are said to be *properly terminated.* A properly terminated line appears at its sending end as a resistor of value R_0. The followers of ECL 10K with their open emitters and low output resistances (specified to be 7 Ω maximum) are ideally suited for driving transmission lines. ECL is also good as a line receiver. The simple gate with its high (50-kΩ) pull-down input resistor represents a very high resistance to the line. Thus a few such gates can be connected to a terminated line with little difficulty. Both these ideas are represented in Fig. 14.27.

14.4.7 Power Dissipation

Because of the differential-amplifier nature of ECL, the gate current remains approximately constant and is simply steered from one side of the gate to the other depending on the input logic signals. Thus, the supply current and hence the gate power dissipation of unterminated ECL remain relatively constant independent of the logic state of the gate. It follows that no voltage spikes are introduced on the supply line. Such spikes can be a dangerous source of noise in a digital system. It follows that in ECL the need for supply-line bypassing[7] is not as great as in, say, TTL. This is another advantage of ECL.

[7]Achieved by connecting capacitances to ground at frequent intervals along the power-supply line on a printed-circuit board.

At this juncture we should reiterate a point we made earlier, namely, that although an ECL gate would operate with $V_{EE} = 0$ and $V_{CC} = +5.2$ V, the selection of $V_{EE} = -5.2$ V and $V_{CC} = 0$ V is recommended, because in the circuit, all signal levels are referenced to V_{CC}, and ground is certainly an excellent reference.

EXERCISE

14.16 For the ECL gate in Fig. 14.26, calculate an approximate value for the power dissipated in the circuit under the condition that all inputs are low and that the emitters of the output followers are left open. Assume that the reference circuit supplies four identical gates, and hence only a quarter of the power dissipated in the reference circuit should be attributed to a single gate.
Ans. 22.4 mW

14.4.8 Thermal Effects

In our analysis of the ECL gate of Fig. 14.26, we found that at room temperature the reference voltage V_R is −1.32 V. We have also shown that the midpoint of the output logic swing is approximately equal to this voltage, which is an ideal situation in that it results in equal high and low noise margins. In Example 14.4, we shall derive expressions for the temperature coefficients of the reference voltage and of the output low and high voltages. In this way, it will be shown that the midpoint of the output logic swing varies with temperature at the same rate as the reference voltage. As a result, although the magnitudes of the high and low noise margins change with temperature, their values remain equal. This is an added advantage of ECL and provides a demonstration of the high degree of design optimization of this gate circuit.

Example 14.4

We wish to determine the temperature coefficient of the reference voltage V_R and of the midpoint between V_{OL} and V_{OH}.

Solution

To determine the temperature coefficient of V_R, consider the circuit in Fig. E14.12 and assume that the temperature changes by +1°C. Denoting the temperature coefficient of the diode and transistor voltage drops by δ, where $\delta \simeq -2$ mV/°C, we obtain the equivalent circuit shown in Fig. 14.33. In the latter circuit, the changes in device voltage drops are considered as signals, and hence the power supply is shown as a signal ground.

In the circuit of Fig. 14.33 we have two signal generators, and we wish to analyze the circuit to determine ΔV_R, the change in V_R. We shall do so using the principle of superposition.[8] Consider first the branch R_1, D_1, D_2, 2δ, and R_2, and neglect the signal base current of Q_1. The voltage signal at the base of Q_1 can be easily obtained from

[8]Although the circuit contains diodes and a transistor, which are nonlinear elements, we can use superposition because we are dealing with small changes in voltages and currents, and thus the diodes and the transistor are replaced by their linear small-signal models.

Example 14.4 *continued*

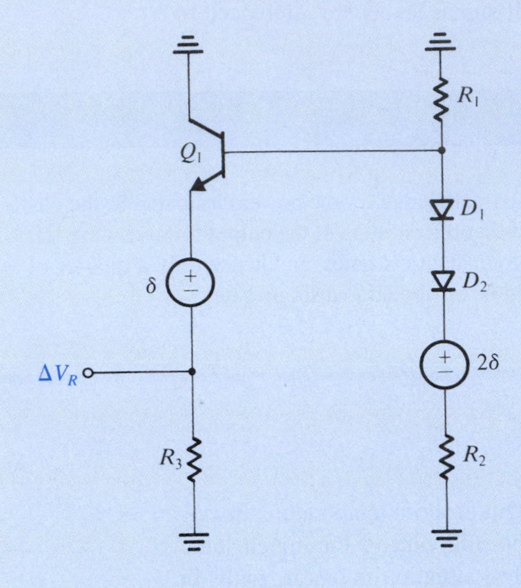

Figure 14.33 Equivalent circuit for determining the temperature coefficient of the reference voltage V_R.

$$v_{b1} = \frac{2\delta \times R_1}{R_1 + r_{d1} + r_{d2} + R_2}$$

where r_{d1} and r_{d2} denote the incremental resistances of diodes D_1 and D_2, respectively. The dc bias current through D_1 and D_2 is approximately 0.64 mA, and thus $r_{d1} = r_{d2} = 39.5\ \Omega$. Hence $v_{b1} \simeq 0.3\delta$. Since the gain of the emitter follower Q_1 is approximately unity, it follows that the component of ΔV_R due to the generator 2δ is approximately equal to v_{b1}; that is, $\Delta V_{R1} = 0.3\delta$.

Consider next the component of ΔV_R due to the generator δ. Reflection into the emitter circuit of the total resistance of the base circuit, $[R_1 \| (r_{d1} + r_{d2} + R_2)]$, by dividing it by $\beta + 1$ (with $\beta \simeq 100$) results in the following component of ΔV_R:

$$\Delta V_{R2} = -\frac{\delta \times R_3}{[R_B/(\beta + 1)] + r_{e1} + R_3}$$

Here R_B denotes the total resistance in the base circuit, and r_{e1} denotes the emitter resistance of Q_1 ($\simeq 40\ \Omega$). This calculation yields $\Delta V_{R2} \simeq -\delta$. Adding this value to that due to the generator 2δ gives $\Delta V_R \simeq -0.7\delta$. Thus for $\delta = -2$ mV/°C, the temperature coefficient of V_R is +1.4 mV/°C.

We next consider the determination of the temperature coefficient of V_{OL}. The circuit on which to perform this analysis is shown in Fig. 14.34. Here we have three generators whose contributions can be considered separately and the resulting components of ΔV_{OL} summed. The result is

$$\begin{aligned}\Delta V_{OL} \simeq\ & \Delta V_R \frac{-R_{C2}}{r_{eR} + R_E} \frac{R_T}{R_T + r_{e2}} \\ & -\delta \frac{-R_{C2}}{r_{eR} + R_E} \frac{R_T}{R_T + r_{e2}} \\ & -\delta \frac{R_T}{R_T + r_{e2} + R_{C2}/(\beta + 1)}\end{aligned}$$

Substituting the values given and those obtained throughout the analysis of this section, we find

$$\Delta V_{OL} \simeq -0.43\delta$$

The circuit for determining the temperature coefficient of V_{OH} is shown in Fig. 14.35, from which we obtain

$$\Delta V_{OH} = -\delta \frac{R_T}{R_T + r_{e2} + R_{C2}/(\beta+1)} = -0.93\delta$$

We now can obtain the variation of the midpoint of the logic swing as

$$\frac{\Delta V_{OL} + \Delta V_{OH}}{2} = -0.68\delta$$

which is approximately equal to that of the reference voltage $V_R(-0.7\delta)$.

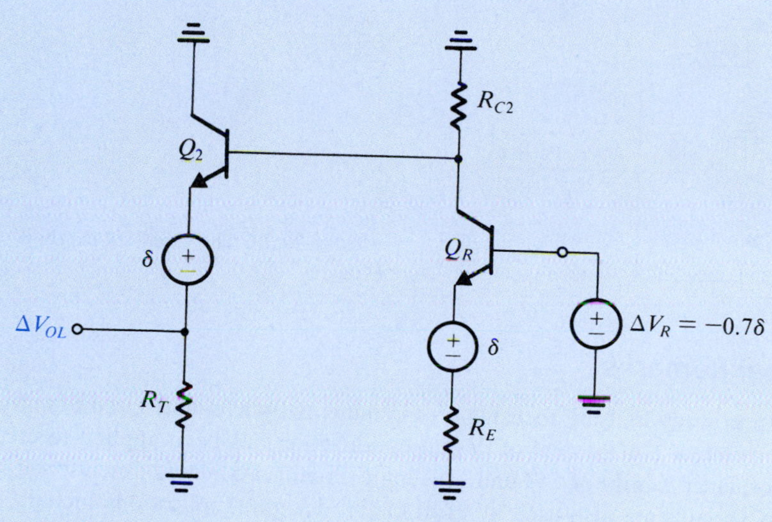

Figure 14.34 Equivalent circuit for determining the temperature coefficient of V_{OL}.

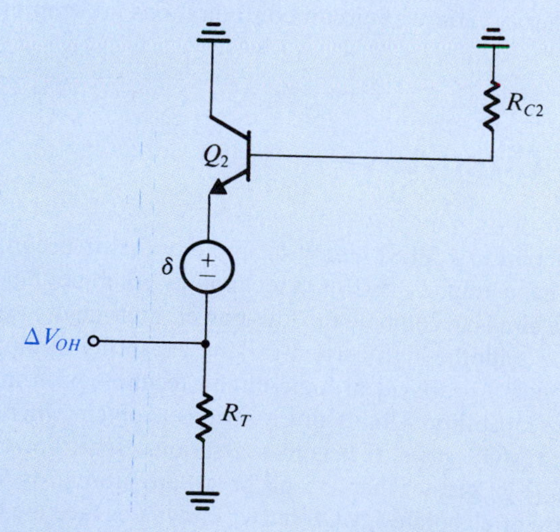

Figure 14.35 Equivalent circuit for determining the temperature coefficient of V_{OH}.

14.4.9 The Wired-OR Capability

The emitter–follower output stage of the ECL family allows an additional level of logic to be performed at very low cost by simply wiring the outputs of several gates in parallel. This is illustrated in Fig. 14.36, where the outputs of two gates are wired together. Note that the base–emitter diodes of the output followers realize an OR function: This **wired-OR** connection can be used to provide gates with high fan-in as well as to increase the flexibility of ECL in logic design.

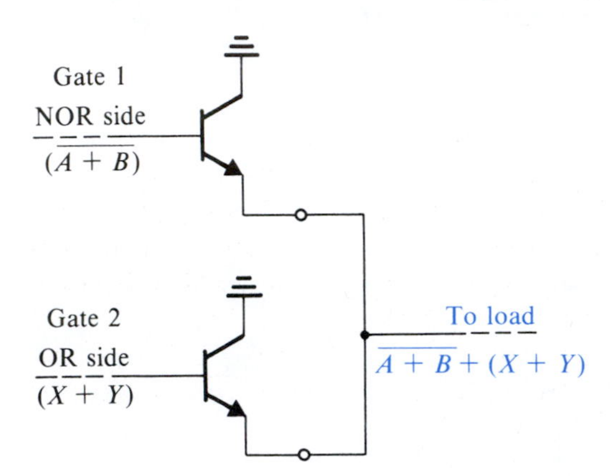

Figure 14.36 The wired-OR capability of ECL.

14.4.10 Final Remarks

We have chosen to study ECL by focusing on a commercially available circuit family. As has been demonstrated, a great deal of design optimization has been applied to create a very-high-performance family of SSI and MSI logic circuits. As already mentioned, ECL and some of its variants are also used in VLSI circuit design. Applications include very-high-speed processors such as those used in supercomputers, as well as high-speed and high-frequency communication systems. When employed in VLSI design, current–source biasing is almost always utilized. Further, a variety of circuit configurations are employed (see Rabaey, 1996).

14.5 BiCMOS Digital Circuits

In this section, we provide an introduction to a VLSI circuit technology that is becoming increasingly popular, BiCMOS. As its name implies, BiCMOS technology combines *bi*polar and *CMOS* circuits on one IC chip. The aim is to combine the low-power, high-input impedance and wide noise margins of CMOS with the high current-driving capability of bipolar transistors. Specifically, CMOS, although a nearly ideal logic-circuit technology in many respects, has a limited current-driving capability. This is not a serious problem when the CMOS gate has to drive a few other CMOS gates. It becomes a serious issue, however, when relatively large capacitive loads (e.g., greater than 0.5 pF or so) are present. In such cases, one has to either resort to the use of elaborate CMOS buffer circuits or face the usually unacceptable consequence of long propagation delays. On the other hand, we know that by virtue of its much larger transconductance, the BJT is capable of large output currents. We have seen a practical illustration of that in the emitter–follower output stage of ECL.

Indeed, the high current-driving capability contributes to making ECL two to five times faster than CMOS (under equivalent conditions)—of course, at the expense of high power dissipation. In summary, then, BiCMOS seeks to combine the best of the CMOS and bipolar technologies to obtain a class of circuits that is particularly useful when output currents that are higher than possible with CMOS are needed. Furthermore, since BiCMOS technology is well suited for the implementation of high-performance analog circuits (see, e.g., Section 7.3.9), it makes possible the realization of both analog and digital functions on the same IC chip, making the "**system on a chip**" an attainable goal. The price paid is a more complex, and hence more expensive (than CMOS) processing technology.

14.5.1 The BiCMOS Inverter

A variety of BiCMOS inverter circuits have been proposed and are in use. All of these are based on the use of *npn* transistors to increase the output current available from a CMOS inverter. This can be most simply achieved by cascading each of the Q_N and Q_P devices of the CMOS inverter with an *npn* transistor, as shown in Fig. 14.37(a). Observe that this circuit can be thought of as utilizing the pair of complementary composite MOS-BJT devices shown in Fig. 14.37(b). These composite devices[9] retain the high input impedance of the MOS transistor while in effect multiplying its rather low g_m by the β of the BJT. It is also useful to observe that the output stage formed by Q_1 and Q_2 has what is known as the **totem-pole configuration** utilized by TTL.[10]

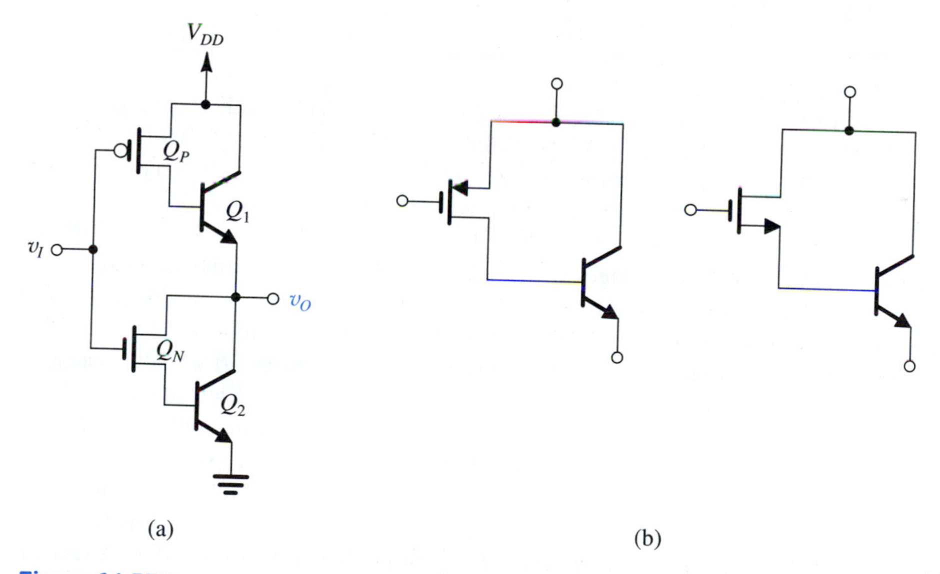

Figure 14.37 Development of the BiCMOS inverter circuit. **(a)** The basic concept is to use an additional bipolar transistor to increase the output current drive of each of Q_N and Q_P of the CMOS inverter. **(b)** The circuit in **(a)** can be thought of as utilizing these composite devices. **(c)** To reduce the turn-off times of Q_1 and Q_2, "bleeder resistors" R_1 and R_2 are added. **(d)** Implementation of the circuit in **(c)** using NMOS transistors to realize the resistors. **(e)** An improved version of the circuit in **(c)** obtained by connecting the lower end of R_1 to the output node.

[9]It is interesting to note that these composite devices were proposed as early as 1969 (see Lin et al., 1969).
[10]Refer to the CD accompanying this book or the book's website for a description of the basic TTL logic-gate circuit and its totem-pole output stage.

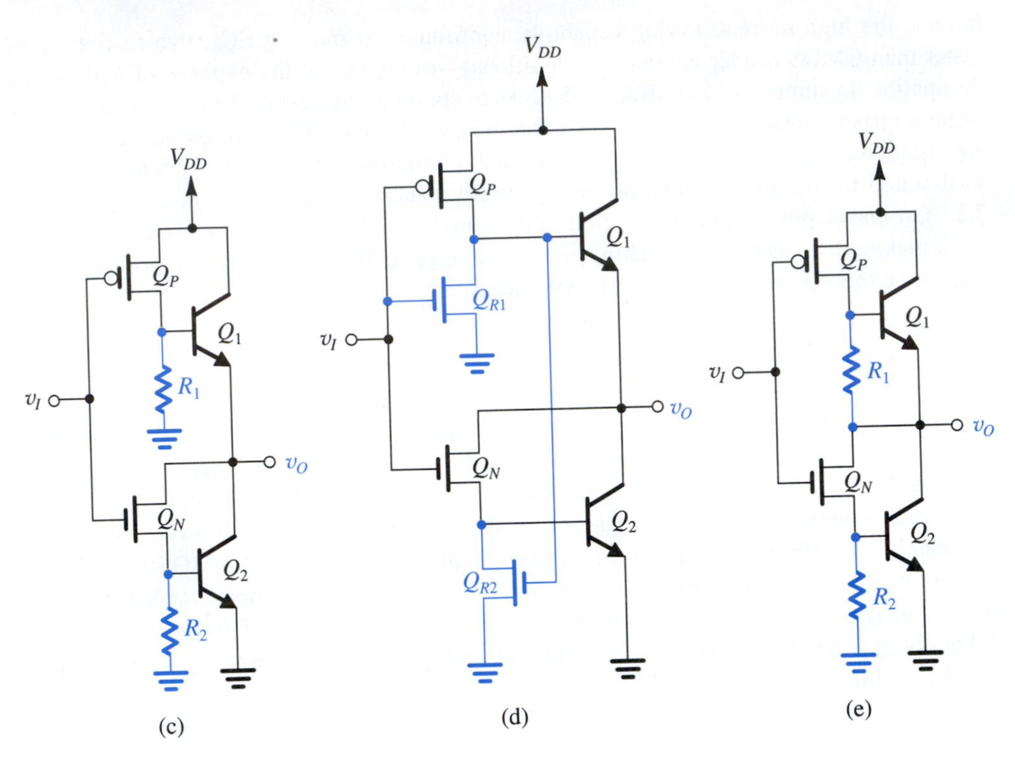

Figure 14.38 *continued*

The circuit of Fig. 14.37(a) operates as follows: When v_I is low, both Q_N and Q_2 are off while Q_P conducts and supplies Q_1 with base current, thus turning it on. Transistor Q_1 then provides a large output current to charge the load capacitance. The result is a very fast charging of the load capacitance and correspondingly a short low-to-high propagation delay, t_{PLH}. Transistor Q_1 turns off when v_O reaches a value of about $V_{DD} - V_{BE1}$, and thus the output high level is lower than V_{DD}, a disadvantage. When v_I goes high, Q_P and Q_1 turn off, and Q_N turns on, providing its drain current into the base of Q_2. Transistor Q_2 then turns on and provides a large output current that quickly discharges the load capacitance. Here again the result is a short high-to-low propagation delay, t_{PHL}. On the negative side, Q_2 turns off when v_O reaches a value of about V_{BE2}, and thus the output low level is greater than zero, a disadvantage.

Thus, while the circuit of Fig. 14.37(a) features large output currents and short propagation delays, it has the disadvantage of reduced logic swing and, correspondingly, reduced noise margins. There is also another and perhaps more serious disadvantage, namely, the relatively long turn-off delays of Q_1 and Q_2 arising from the absence of circuit paths along which the base charge can be removed. This problem can be solved by adding a resistor between the base of each of Q_1 and Q_2 and ground, as shown in Fig. 14.37(c). Now when either Q_1 or Q_2 is turned off, its stored base charge is removed to ground through R_1 or R_2, respectively. Resistor R_2 provides an additional benefit: With v_I high, and after Q_2 cuts off, v_O continues to fall below V_{BE2}, and the output node is pulled to ground through the series path of Q_N and R_2. Thus R_2 functions as a pull-down resistor. The Q_N–R_2 path, however, is a high-impedance one with the result that pulling v_O to ground is a rather slow process. Incorporating the resistor R_1, however, is disadvantageous from a static power-dissipation standpoint: When v_I is low, a dc path exists between V_{DD} and ground through the conducting Q_P and R_1. Finally, it should be noted that R_1 and R_2 take some of the drain currents of Q_P and Q_N away from the bases of Q_1 and Q_2 and thus slightly reduce the gate output current available to charge and discharge the load capacitance.

Figure 14.37(d) shows the way in which R_1 and R_2 are usually implemented. As indicated, NMOS devices Q_{R1} and Q_{R2} are used to realize R_1 and R_2. As an added innovation, these two transistors are made to conduct only when needed. Thus, Q_{R1} will conduct only when v_I rises, at which time its drain current constitutes a reverse base current for Q_1, speeding up its turn-off. Similarly, Q_{R2} will conduct only when v_I falls and Q_P conducts, pulling the gate of Q_{R2} high. The drain current of Q_{R2} then constitutes a reverse base current for Q_2, speeding up its turn-off.

As a final circuit for the BiCMOS inverter, we show the so-called *R*-circuit in Fig. 14.37(e). This circuit differs from that in Fig. 14.37(c) in only one respect: Rather than returning R_1 to ground, we have connected R_1 to the output node of the inverter. This simple change has two benefits. First, the problem of static power dissipation is now solved. Second, R_1 now functions as a pull-up resistor, pulling the output node voltage up to V_{DD} (through the conducting Q_P) after Q_1 has turned off. Thus, the *R* circuit in Fig. 14.37(e) does in fact have output levels very close to V_{DD} and ground.

As a final remark on the BiCMOS inverter, we note that the circuit is designed so that transistors Q_1 and Q_2 are never simultaneously conducting and neither is allowed to saturate. Unfortunately, sometimes the resistance of the collector region of the BJT in conjunction with large capacitive-charging currents causes saturation to occur. Specifically, at large output currents, the voltage developed across r_C (which can be of the order of 100 Ω) can lower the voltage at the intrinsic collector terminal and cause the CBJ to become forward biased. As the reader will recall, saturation is a harmful effect for two reasons: It limits the collector current to a value less than βI_B, and it slows down the transistor turn-off.

14.5.2 Dynamic Operation

A detailed analysis of the dynamic operation of the BiCMOS inverter circuit is a rather complex undertaking. Nevertheless, an estimate of its propagation delay can be obtained by considering only the time required to charge and discharge a load capacitance *C*. Such an approximation is justified when *C* is relatively large and thus its effect on inverter dynamics is dominant: in other words, when we are able to neglect the time required to charge the parasitic capacitances present at internal circuit nodes. Fortunately, this is usually the case in practice, for if the load capacitance is not large, one would use the simpler CMOS inverter. In fact, it has been shown (Embabi, Bellaouar, and Elmasry, 1993) that the speed advantage of BiCMOS (over CMOS) becomes evident only when the gate is required to drive a large fan-out or a large load capacitance. For instance, at a load capacitance of 50 fF to 100 fF, BiCMOS and CMOS typically feature equal delays. However, at a load capacitance of 1 pF, t_P of a BiCMOS inverter is 0.3 ns, whereas that of an otherwise comparable CMOS inverter is about 1 ns.

Finally, in Fig. 14.38, we show simplified equivalent circuits that can be employed in obtaining rough estimates of t_{PLH} and t_{PHL} of the *R*-type BiCMOS inverter (see Problem 14.49).

14.5.3 BiCMOS Logic Gates

In BiCMOS, the logic is performed by the CMOS part of the gate, with the bipolar portion simply functioning as an output stage. It follows that BiCMOS logic-gate circuits can be generated following the same approach used in CMOS. As an example, we show in Fig. 14.39 a BiCMOS two-input NAND gate.

As a final remark, we note that BiCMOS technology is applied in a variety of products including microprocessors, static RAMs, and gate arrays (see Alvarez, 1993).

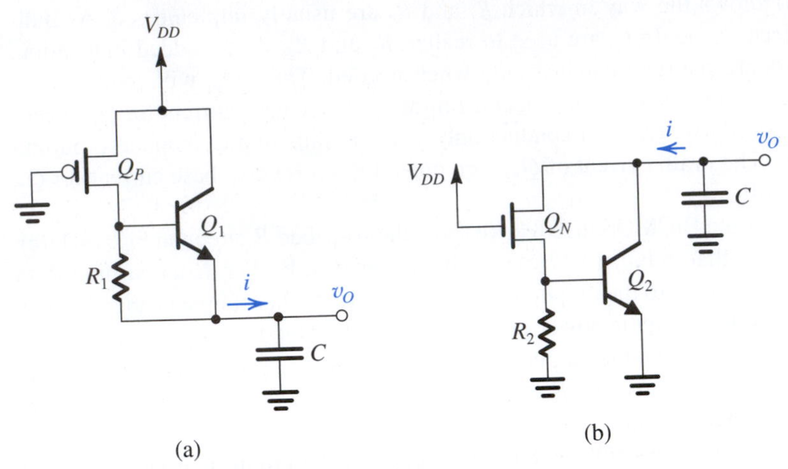

Figure 14.39 Equivalent circuits for charging and discharging a load capacitance C. Note that C includes all the capacitances present at the output node.

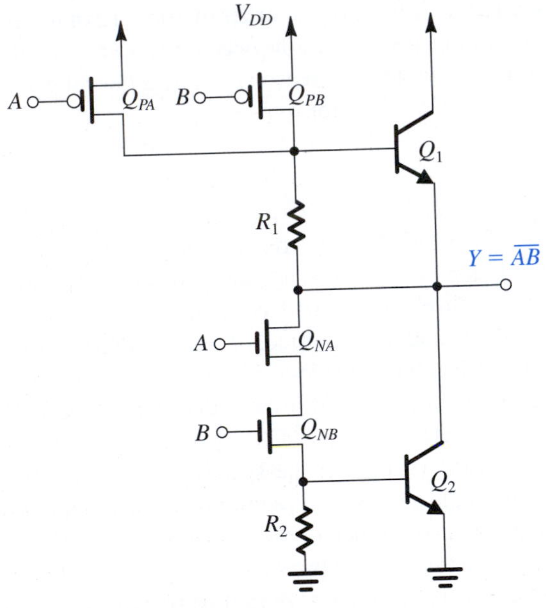

Figure 14.40 A BiCMOS two-input NAND gate.

EXERCISE

D14.17 The threshold voltage of the BiCMOS inverter of Fig. 14.37(e) is the value of v_I at which both Q_N and Q_P are conducting equal currents and operating in the saturation region. At this value of v_I, Q_2 will be on, causing the voltage at the source of Q_N to be approximately 0.7 V. It is required to design the circuit so that the threshold voltage is equal to $V_{DD}/2$. For $V_{DD} = 5$ V, $|V_t| = 0.6$ V, and assuming equal channel lengths for Q_N and Q_P and that $\mu_n \simeq 2.5\,\mu_p$, find the required ratio of widths, W_p/W_n.

Ans. 1

Summary

- Standard CMOS logic utilizes two transistors, an NMOS and a PMOS, for each input variable. Thus the circuit complexity, silicon area, and parasitic capacitance all increase with fan-in.

- To reduce the device count, two other forms of static CMOS, namely, pseudo-NMOS and pass-transistor logic (PTL), are employed in special applications as supplements to standard CMOS.

- Pseudo-NMOS utilizes the same PDN as in standard CMOS logic but replaces the PUN with a single PMOS transistor whose gate is grounded and thus is permanently on. Unlike standard CMOS, pseudo-NMOS is a ratioed form of logic in which V_{OL} is determined by the ratio r of k_n to k_p. Normally, r is selected in the range of 4 to 10 and its value determines the noise margins.

- Pseudo-NMOS has the disadvantage of dissipating static power when the output of the logic gate is low. Static power can be eliminated by turning the PMOS load on for only a brief interval, known as the precharge interval, to charge the capacitance at the output node to V_{DD}. Then the inputs are applied, and depending on the input combination, the output node either remains high or is discharged through the PDN. This is the essence of dynamic logic.

- Pass-transistor logic utilizes either single NMOS transistors or CMOS transmission gates to implement a network of switches that are controlled by the input logic variables. Switches implemented by single NMOS transistors, though simple, result in the reduction of V_{OH} from V_{DD} to $V_{DD} - V_t$.

- The CMOS transmission gate, composed of the parallel connection of an NMOS and a PMOS transistor, is a very effective switch in both analog and digital applications. It passes the entire input signal swing, 0 to V_{DD}. As well, it has an almost constant "on" resistance over the full output range.

- A particular form of dynamic logic circuits, known as Domino logic, allows the cascading of dynamic logic gates.

- Emitter-coupled logic (ECL) is the fastest commercially available logic-circuit family. It achieves its high speed of operation by avoiding transistor saturation and by utilizing small logic-signal swings.

- In ECL the input signals are used to steer a bias current between a reference transistor and an input transistor. The basic gate configuration is that of a differential amplifier.

- There are two popular commercially available ECL types: ECL 10K, having $t_P = 2$ ns, $P_D = 25$ mW, and $PDP = 50$ pJ; and ECL 100K, having $t_P = 0.75$ ns, $P_D = 40$ mW, and $PDP = 30$ pJ. ECL 10K is easier to use because the rise and fall times of its signals are deliberately made long (about 3.5 ns).

- Because of the very high operating speeds of ECL, care should be taken in connecting the output of one gate to the input of another. Transmission-line techniques are usually employed.

- The design of the ECL gate is optimized so that the noise margins are equal and remain equal as temperature changes.

- The ECL gate provides two complementary outputs, realizing the OR and NOR functions.

- The outputs of ECL gates can be wired together to realize the OR function of the individual output variables.

- BiCMOS combines the low-power and wide noise margins of CMOS with the high current-driving capability (and thus the short gate delays) of BJTs to obtain a technology that is capable of implementing very dense, low-power, high-speed VLSI circuits that can also include analog functions.

Problems involving design are marked with D throughout the text. As well, problems are marked with asterisks to describe their degree of difficulty. Difficult problems are marked with an asterisk (*); more difficult problems with two asterisks (**); and very challenging and/or time-consuming problems with three asterisks (***).

Section 14.1: Pseudo-NMOS Logic Circuits

14.1 The purpose of this problem is to compare the value of t_{PLH} obtained with a resistive load (see Fig. P14.1a) to that obtained with a current–source load (see Fig. P14.1b). For a fair comparison, let the current source $I = V_{DD}/R_D$, which is the initial current available to charge the capacitor in the case of a resistive load. Find t_{PLH} for each case, and hence the percentage reduction obtained when a current–source load is used.

D *14.2 Design a pseudo-NMOS inverter that has equal capacitive charging and discharging currents at $v_O = V_{DD}/4$ for use in a system with $V_{DD} = 2.5$ V, $|V_t| = 0.5$ V, $k_n' = 115\ \mu\text{A/V}^2$, $k_p' = 30\mu\text{A/V}^2$, and $(W/L)_n = 1.5$. What are the values of $(W/L)_p$, V_{IL}, V_{IH}, V_M, V_{OH}, V_{OL}, NM_H, and NM_L?

14.3 Find t_{PLH}, t_{PHL}, and t_P for a pseudo-NMOS inverter fabricated in a 0.13-μm CMOS technology for which $V_{DD} = 1.2$ V, $V_t = 0.4$ V, and $\mu_n C_{ox} = 4\mu_p C_{ox} = 430\ \mu\text{A/V}^2$. Assume that the inverter has $r = 4$ and $(W/L)_n = 1$ and that the equivalent load capacitance is 10 fF.

***14.4** Use Eq. (14.13) to find the value of r for which NM_L is maximized. What is the corresponding value of NM_L for the case $V_{DD} = 2.5$ V and $V_t = 0.5$ V?

D 14.5 Design a pseudo-NMOS inverter that has $V_{OL} = 0.1$ V. Let $V_{DD} = 2.5$ V, $|V_t| = 0.5$ V, $k_n' = 4k_p' = 120\ \mu\text{A/V}^2$, and $(W/L)_p = 1$. What is the value of $(W/L)_n$? Calculate the values of NM_L and the static power dissipation.

14.6 For what value of r does NM_H of a pseudo-NMOS inverter become zero? Prepare a table of N_{MH} and N_{ML} versus r, for $r = 1$ to 16. Let $V_{DD} = 2.5$ V and $V_t = 0.5$ V.

14.7 For a pseudo-NMOS inverter, what value of r results in $NM_L = NM_H$? Let $V_{DD} = 2.5$ V and $|V_t| = 0.5$ V. What is the resulting margin?

D *14.8 It is required to design a minimum-area pseudo-NMOS inverter with equal high and low noise margins using a 2.5-V supply and devices for which $|V_t| = 0.5$ V, $k_n' = 4k_p' = 120\ \mu\text{A/V}^2$, and the minimum-size device has $(W/L) = 1$. Use $r = 3.2$ and show that $NM_L \simeq NM_H$. Specify the values of $(W/L)_n$ and $(W/L)_p$. What is the static power dissipated in this gate? What is the ratio of propagation delays for low-to-high and high-to-low transitions? For an equivalent load capacitance of 0.1 pF, find t_{PLH}, t_{PHL}, and t_P. At what frequency of operation would the static and dynamic power levels be equal? Is this speed of operation possible in view of the t_P value you found?

D 14.9 Sketch a pseudo-NMOS realization of the function $Y = A + B(C + D)$.

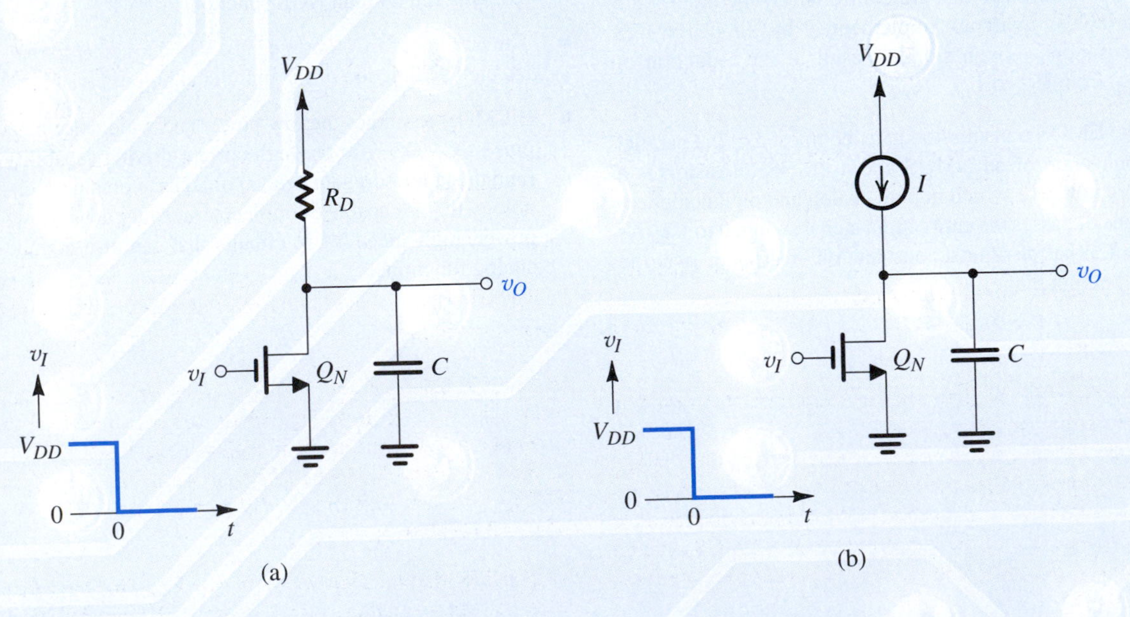

Figure P 14.1

D 14.10 Sketch a pseudo-NMOS realization of the exclusive-OR function $Y = A\bar{B} + \bar{A}B$.

D 14.11 Consider a four-input pseudo-NMOS NOR gate in which the NMOS devices have $(W/L)_n = 0.27\ \mu\text{m}/0.18\ \mu\text{m}$. It is required to find $(W/L)_p$ so that the worst-case value of V_{OL} is 0.1 V. Let $V_{DD} = 1.8$ V, $|V_t| = 0.5$ V, and $k_n' = 4k_p' = 300\ \mu\text{A/V}^2$. Assume that the minimum width possible is 0.2 μm.

14.12 This problem investigates the effect of velocity saturation (Section 13.5.2) on the operation of a pseudo-NMOS inverter fabricated in a 0.13-μm CMOS process for which, $V_{DD} = 1.2$ V, $V_t = 0.4$ V, $\mu_n C_{ox} = 4\mu_p C_{ox} = 430\ \mu\text{A/V}^2$, and $|V_{DSsatp}| = 0.6$ V. Consider the case with $v_I = V_{DD}$ and $v_O = V_{OL}$. Note that Q_P will be operating in the velocity-saturation region. Find its current I_{Dsat} and use it to determine V_{OL}.

Section 10.2: Pass-Transistor Logic Circuits

14.13 Consider the NMOS transistor switch in the circuits of Figs. 14.8 and 14.9 to be fabricated in a 0.18-μm CMOS technology for which $\mu_n C_{ox} = 4\mu_p C_{ox} = 300\ \mu\text{A/V}^2$, $|V_{t0}| = 0.5$ V, $\gamma = 0.3\ \text{V}^{1/2}$, $2\phi_f = 0.85$ V, and $V_{DD} = 1.8$ V. Let the transistor have $W/L = 1.5$, and assume that the total capacitance between the output node and ground is $C = 10$ fF.

(a) For the case $v_I = V_{DD}$, find V_{OH}.
(b) If the output feeds a CMOS inverter having $(W/L)_p = 2(W/L)_n = 0.54\ \mu\text{m}/0.18\ \mu\text{m}$, find the static current of the inverter and its power dissipation when the inverter input is at the value found in (a). Also, find the inverter output voltage.
(c) Find t_{PLH}.
(d) For v_I going low (Fig. 14.9), find t_{PHL}.
(e) Find t_P.

***14.14** A designer, beginning to experiment with the idea of pass-transistor logic, seizes upon what he sees as two good ideas:

(a) that a string of minimum-size single MOS transistors can do complex logic functions, but
(b) that there must always be a path between output and a supply terminal.

Correspondingly, he first considers two circuits (shown in Fig. P14.14). For each, express Y as a function of A and B. In each case, what can be said about general operation? About the logic levels at Y? About node X? Do either of these circuits look familiar? If in each case the terminal connected to V_{DD} is instead connected to the output of a CMOS inverter whose input is connected to a signal C, what does the function Y become?

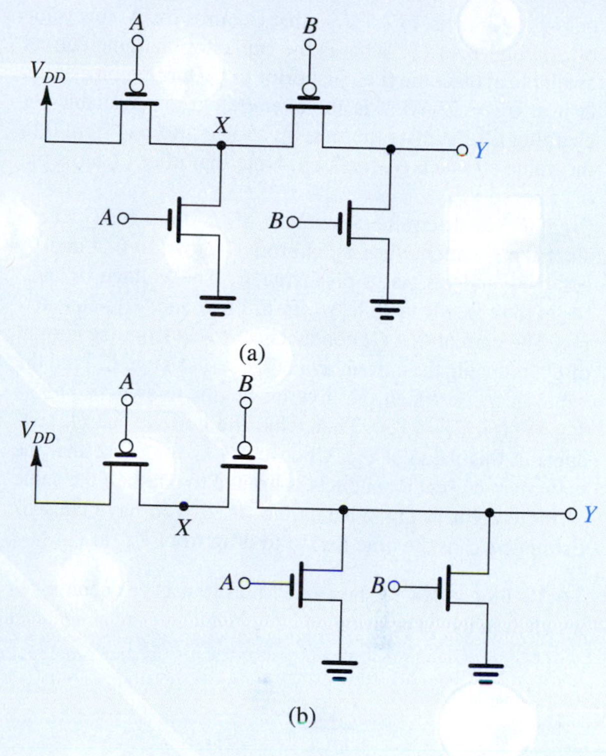

Figure P14.14

14.15 Consider the circuits in Fig. P14.14 with all PMOS transistors replaced with NMOS, and all NMOS by PMOS, and with ground and V_{DD} connections interchanged. What do the output functions Y become?

14.16 An NMOS pass-transistor switch with $W/L = 1.2\ \mu\text{m}/0.8\ \mu\text{m}$, used in a 3.3-V system for which $V_{t0} = 0.8$ V, $\gamma = 0.5\ \text{V}^{1/2}$, $2\phi_f = 0.6$ V, $\mu_n C_{ox} = 3\mu_p C_{ox} = 75\ \mu\text{A/V}^2$, drives a 100-fF load capacitance at the input of a matched standard CMOS inverter using $(W/L)_n = 1.2\ \mu\text{m}/0.8\ \mu\text{m}$. For the switch gate terminal at V_{DD}, evaluate the switch V_{OH} and V_{OL} for inputs at V_{DD} and 0 V, respectively. For this value of V_{OH}, what inverter static current results? Estimate t_{PLH} and t_{PHL} for this arrangement as measured from the input to the output of the switch itself.

D **14.17 The purpose of this problem is to design the level-restoring circuit of Fig. 14.10 and gain insight into its operation. Assume that $k_n' = 3k_p' = 75\ \mu\text{A/V}^2$, $V_{DD} = 3.3$ V, $|V_{t0}| = 0.8$ V, $\gamma = 0.5\ \text{V}^{1/2}$ $2\phi_f = 0.6$ V, $(W/L)_1 = (W/L)_n = 1.2\ \mu\text{m}/0.8\ \mu\text{m}$, $(W/L)_p = 3.6\ \mu\text{m}/0.8\ \mu\text{m}$, and $C = 20$ fF. Let $v_B = V_{DD}$.

(a) Consider first the situation with $v_A = V_{DD}$. Find the value of the voltage v_{O1} that causes v_{O2} to drop a threshold voltage

below V_{DD}; that is, to 2.5 V so that Q_R turns on. At this value of v_{O1}, find V_t of Q_1. What is the capacitor-charging current available at this time (i.e., just prior to Q_R turning on)? What is it at $v_{O1} = 0$? What is the average current available for charging C? Estimate the time t_{PLH} for v_{O1} to rise from 0 to the value at which Q_R turns on. Note that after Q_R turns on, v_{O1} rises to V_{DD}.
(b) Now, to determine a suitable W/L ratio for Q_R, consider the situation when v_A is brought down to 0 V and Q_1 conducts and begins to discharge C. The voltage v_{O1} will begin to drop. Meanwhile, v_{O2} is still low and Q_R is conducting. The current that Q_R conducts subtracts from the current of Q_1, reducing the current available to discharge C. Find the value of v_{O1} at which the inverter begins to switch. This is $V_{IH} = \frac{1}{8}(5V_{DD} - 2V_t)$. Then, find the current that Q_1 conducts at this value of v_{O1}. Choose W/L for Q_R so that the maximum current it conducts is limited to one-half the value of the current in Q_1. What is the W/L you have chosen? Estimate t_{PHL} as the time for v_{O1} to drop from V_{DD} to V_{IH}.

14.18 Figure P14.18 shows a PMOS transistor operating as a switch in the on position.

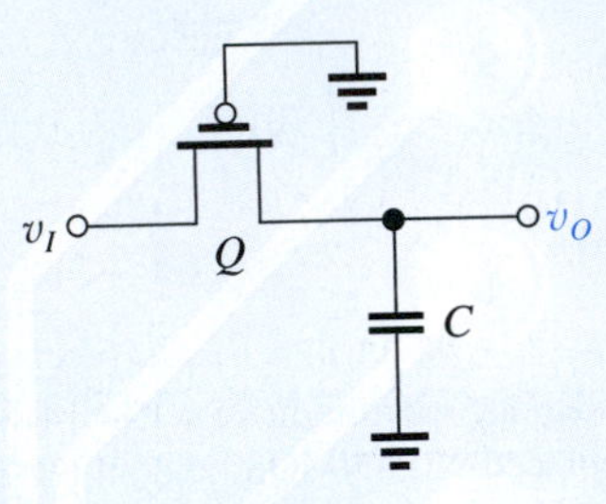

Figure P14.18

(a) If initially $v_O = 0$ and at $t = 0$, v_I is raised to V_{DD}, what is the final value V_{OH} reached at the output?
(b) If initially, $v_O = V_{DD}$ and at $t = 0$, v_I is lowered to 0 V, what is the final value V_{OL} reached at the output?
(c) For the situation in (a), find t_{PLH} for v_O to rise from 0 to $V_{DD}/2$. Let $k_p = 225\ \mu A/V^2$, $V_{DD} = 1.8$ V, and $|V_{tp}| = 0.5$ V.

14.19 The transmission gate in Fig. 14.12(a) and 14.12(b) is fabricated in a CMOS process technology for which $k_n' = 4k_p' = 300\ \mu A/V^2$, $|V_{t0}| = 0.5$ V, $\gamma = 0.3\ V^{1/2}$, $2\phi_f = 0.85$ V, and $V_{DD} = 1.8$ V. Let Q_N and Q_P have $(W/L)_n = (W/L)_p = 1.5$. The total capacitance at the output node is 15 fF.

(a) What are the values of V_{OH} and V_{OL}?
(b) For the situation in Fig. 14.12(a), find $i_{DN}(0)$, $i_{DP}(0)$, $i_{DN}(t_{PLH})$, $i_{DP}(t_{PLH})$, and t_{PLH}.
(c) For the situation depicted in Fig. 14.12(b), find $i_{DN}(0)$, $i_{DP}(0)$, $i_{DN}(t_{PHL})$, $i_{DP}(t_{PHL})$, and t_{PHL}. At what value of v_O will Q_P turn off?
(d) Find t_P.

14.20 For the transmission gate specified in Problem 14.19, find R_{TG} at $v_O = 0$ and 0.9 V. Use the average of those values to determine t_{PLH} for the situation in which $C = 15$ fF.

***14.21** Refer to the situation in Fig. 14.12(b). Derive expressions for R_{Neq}, R_{Peq}, and R_{TG} following the approach used in Section 14.2.4 for the capacitor-charging case. Evaluate the value of R_{TG} for $v_O = V_{DD}$ and $v_O = V_{DD}/2$ for the process technology specified in Problem 14.19. Find the average value of R_{TG} and use it to determine t_{PHL} for the case $C = 15$ fF.

14.22 A transmission gate for which $(W/L)_n = (W/L)_p = 1.5$ is fabricated in a 0.18-μm CMOS technology and used in a circuit for which $C = 10$ fF. Use Eq. (14.36) to obtain an estimate of R_{TG} and hence of the propagation delay t_P.

14.23 Figure P14.23 shows a chain of transmission gates. This situation often occurs in circuits such as adders and multiplexers. Consider the case when all the transmission gates are turned on and a step voltage V_{DD} is applied to the input. The propagation delay t_P can be determined from the Elmore delay formula as follows:

$$t_P = 0.69 \sum_{k=0}^{n} kCR_{TG}$$

where R_{TG} is the resistance of each transmission gate, C is the capacitance between each node and ground, and n is the

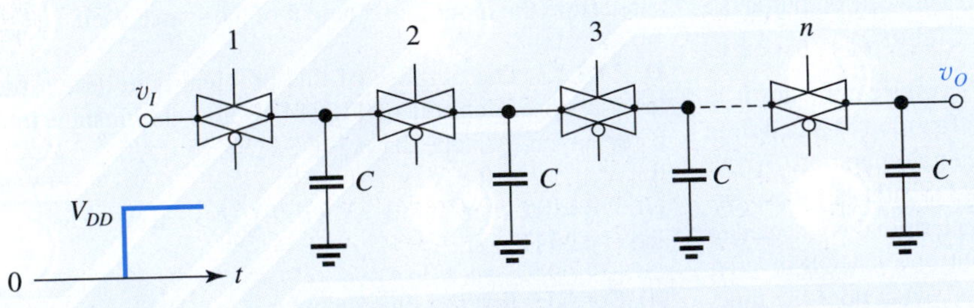

Figure P14.23

number of transmission gates in the chain. Note that the sum of the series in this formula is given by

$$t_P = 0.69CR_{TG}\,\frac{n(n+1)}{2}$$

Now evaluate t_P for the case of 16 transmission gates with $R_{TG} = 10\ \text{k}\Omega$ and $C = 10$ fF.

D 14.24 (a) Use the idea embodied in the exclusive-OR realization in Fig. 14.17 to realize $\bar{Y} = AB + \bar{A}\bar{B}$. That is, find a realization for $\bar{Y}$ using two transmission gates.

(b) Now combine the circuit obtained in (a) with the circuit in Fig. 14.17 to obtain a realization of the function $Z = \bar{Y}C + Y\bar{C}$, where C is a third input. Sketch the complete 12-transistor circuit realization of Z. Note that Z is a three-input exclusive-OR.

D *14.25 Using the idea presented in Fig. 14.18, sketch a CPL circuit whose outputs are $Y = A\bar{B} + \bar{A}B$ and $\bar{Y} = AB + \bar{A}\bar{B}$.

D 14.26 Extend the CPL idea in Fig. 14.18 to three variables to form $Z = ABC$ and $\bar{Z} = \overline{ABC} = \bar{A} + \bar{B} + \bar{C}$.

Section 14.3: Dynamic MOS Logic Circuits

D 14.27 Based on the basic dynamic logic circuit of Fig. 14.19, sketch complete circuits for NOT, NAND, and NOR gates, the latter two with two inputs, and a circuit for which $\bar{Y} = AB + CD$.

14.28 In this and the following problem, we investigate the dynamic operation of a two-input NAND gate realized in the dynamic logic form and fabricated in a CMOS process technology for which $k_n' = 3k_p' = 75\ \mu\text{A/V}^2$, $V_{tn} = -V_{tp} = 0.8$ V, and $V_{DD} = 3$ V. To keep C_L small, minimum-size NMOS devices are used for which $W/L = 1.2\ \mu\text{m}/0.8\ \mu\text{m}$ (this includes Q_e). The PMOS precharge transistor Q_p has $2.4\ \mu\text{m}/0.8\ \mu\text{m}$. The capacitance C_L is found to be 30 fF. Consider the precharge operation with the gate of Q_p at 0 V, and assume that at $t = 0$, C_L is fully discharged. We wish to calculate the rise time of the output voltage, defined as the time for v_Y to rise from 10% to 90% of the final value of 3 V. Find the current at $v_Y = 0.3$ V and the current at $v_Y = 2.7$ V, then compute an approximate value for t_r, $t_r = C_L(2.7 - 0.3)/I_{av}$, where I_{av} is the average value of the two currents.

14.29 For the gate specified in Problem 14.28, evaluate the high-to-low propagation delay, t_{PHL}. To obtain an approximate value of t_{PHL}, replace the three series NMOS transistors with an equivalent device and find the average discharge current.

14.30 The leakage current in a dynamic-logic gate causes the capacitor C_L to discharge during the evaluation phase, even if the PDN is not conducting. For $C_L = 15$ fF, and $I_{\text{leakage}} = 10^{-12}$ A, find the longest allowable evaluate time if the decay in output voltage is to be limited to 0.2 V. If the precharge interval is much shorter than the maximum allowable evaluate time, find the minimum clocking frequency required.

***14.31** In this problem, we wish to calculate the reduction in the output voltage of a dynamic-logic gate as a result of charge redistribution. Refer to the circuit in Fig. 14.21(a), and assume that at $t = 0-$, $v_Y = V_{DD}$, and $v_{C1} = 0$. At $t = 0$, ϕ goes high and Q_P turns off, and simultaneously the voltage at the gate of Q_1 goes high (to V_{DD}), turning Q_1 on. Transistor Q_1 will remain conducting until either the voltage at its source (v_{C1}) reaches $V_{DD} - V_{tn}$ or until $v_Y = v_{C1}$, whichever comes first. In both cases, the final value of v_Y can be found using charge conservation; that is, by equating the charge gained by C_1 to the charge lost by C_L.

(a) Convince yourself that the first situation obtains when $|\Delta v_Y| \le V_{tn}$.
(b) For each of the two situations, derive an expression for Δv_Y.
(c) Find an expression for the maximum ratio (C_1/C_L) for which $|\Delta v_Y| \le V_{tn}$.
(d) For $V_{tn} = 1$ V, $V_{DD} = 5$ V, $C_L = 30$ fF, and neglecting the body effect in Q_1, find the drop in voltage at the output in the two cases: (a) $C_1 = 5$ fF and (b) $C_1 = 10$ fF.

14.32 Solve the problem in Exercise 14.10 symbolically (rather than numerically). Refer to Fig E14.10 and assume Q_{eq1} and Q_{eq2} to be identical with threshold voltages $V_{tn} = 0.2V_{DD}$ and transconductance parameters k_n. Also, let $C_{L1} = C_{L2}$. Derive an expression for the drop in the output voltage, Δv_{Y2}.

14.33 For the four-input dynamic-logic NAND gate analyzed in Example 14.3, estimate the maximum clocking frequency allowed.

Section 14.4: Emitter-Coupled Logic (ECL)

D 14.34 For the ECL circuit in Fig. P14.34, the transistors exhibit V_{BE} of 0.75 V at an emitter current I and have very high β.

(a) Find V_{OH} and V_{OL}.
(b) For the input at B that is sufficiently negative for Q_B to be cut off, what voltage at A causes a current of $I/2$ to flow in Q_R?
(c) Repeat (b) for a current in Q_R of $0.99I$.
(d) Repeat (c) for a current in Q_R of $0.01I$.
(e) Use the results of (c) and (d) to specify V_{IL} and V_{IH}.
(f) Find NM_H and NM_L.
(g) Find the value of IR that makes the noise margins equal to the width of the transition region, $V_{IH} - V_{IL}$.

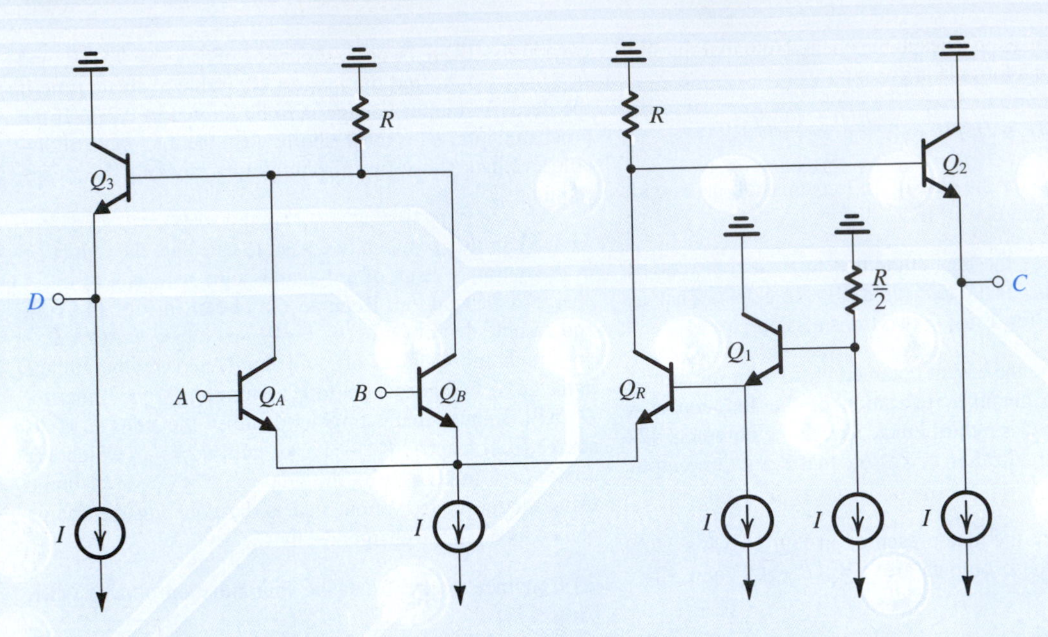

Figure P14.34

(h) Using the IR value obtained in (g), give numerical values for V_{OH}, V_{OL}, V_{IH}, V_{IL}, and V_R for this ECL gate.

***14.35** Three logic inverters are connected in a ring. Specifications for this family of gates indicate a typical propagation delay of 3 ns for high-to-low output transitions and 7 ns for low-to-high transitions. Assume that for some reason the input to one of the gates undergoes a low-to-high transition. By sketching the waveforms at the outputs of the three gates and keeping track of their relative positions, show that the circuit functions as an oscillator. What is the frequency of oscillation of this ring oscillator? In each cycle, how long is the output high? low?

***14.36** Following the idea of a ring oscillator introduced in Problem 14.35, consider an implementation using a ring of five ECL 100K inverters. Assume that the inverters have linearly rising and falling edges (and thus the waveforms are trapezoidal in shape). Let the 0 to 100% rise and fall times be equal to 1 ns. Also, let the propagation delay (for both transitions) be equal to 1 ns. Provide a labeled sketch of the five output signals, taking care that relevant phase information is provided. What is the frequency of oscillation?

D *14.37 Using the logic and circuit flexibility of ECL indicated by Figs. 14.26 and 14.36, sketch an ECL logic circuit that realizes the exclusive OR function, $Y = \bar{A}B + A\bar{B}$. Give a logic diagram (as opposed to a circuit diagram).

***14.38** For the circuit in Fig. 14.28 whose transfer characteristic is shown in Fig. 14.29, calculate the incremental voltage gain from input to the OR output at points x, m, and y of the transfer characteristic. Assume $\beta = 100$. Use the results of Exercise 14.14, and let the output at x be -1.77 V and that at y be -0.88 V. (*Hint*: Recall that x and y are defined by a 1%, 99% current split.)

14.39 For the circuit in Fig. 14.28 whose transfer characteristic is shown in Fig. 14.29, find V_{IL} and V_{IH} if x and y are defined as the points at which

(a) 90% of the current I_E is switched.
(b) 99.9% of the current I_E is switched.

14.40 For the symmetrically loaded circuit of Fig. 14.28 and for typical output signal levels ($V_{OH} = -0.88$ V and $V_{OL} = -1.77$ V), calculate the power lost in both load resistors R_T and both output followers. What then is the total power dissipation of a single ECL gate, including its symmetrical output terminations?

14.41 Considering the circuit of Fig. 14.30, what is the value of β of Q_2, for which the high noise margin (NM_H) is reduced by 50%?

***14.42** Consider an ECL gate whose inverting output is terminated in a 50-Ω resistance connected to a −2-V supply. Let the total load capacitance be denoted C. As the input of the gate rises, the output emitter follower cuts off and the load capacitance C discharges through the 50-Ω load (until the emitter follower conducts again). Find the value of C that will result in a discharge time of 1 ns. Assume that the two output levels are −0.88 V and −1.77 V.

14.43 For signals whose rise and fall times are 3.5 ns, what length of unterminated gate-to-gate wire interconnect can be used if a ratio of rise time to return time of 5 to 1 is required? Assume the environment of the wire to be such that the signal propagates at two-thirds the speed of light (which is 30 cm/ns).

***14.44** For the circuit in Fig. P14.44, let the levels of the inputs A, B, C, and D be 0 and +5 V. For all inputs low at 0 V, what is the voltage at E? If A and C are raised to +5 V, what is the voltage at E? Assume $|V_{BE}| = 0.7$ V and $\beta = 50$. Express E as a logic function of A, B, C, and D.

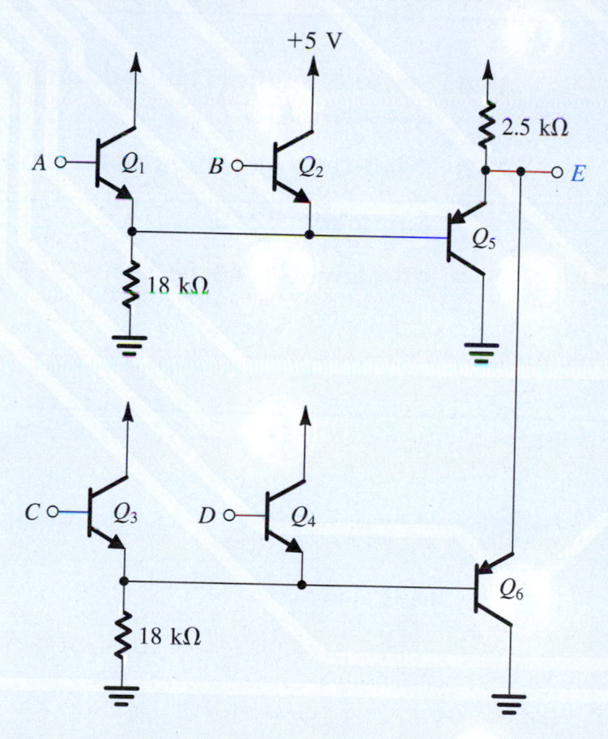

Figure P14.44

Section 14.5: BiCMOS Digital Circuits

14.45 Consider the conceptual BiCMOS circuit of Fig. 14.37(a), for the conditions that $V_{DD} = 5$ V, $|V_t| = 1$ V, $V_{BE} = 0.7$ V, $\beta = 100$, $k_n' = 2.5k_p' = 100\ \mu\text{A/V}^2$, and $(W/L)_n = 2\ \mu\text{m}/1\ \mu\text{m}$. For $v_I = v_O = V_{DD}/2$, find $(W/L)_p$ so that $I_{EQ_1} = I_{EQ_2}$. What is this totem-pole transient current?

14.46 Consider the conceptual BiCMOS circuit of Fig. 14.37(a) for the conditions stated in Problem 14.45. What is the threshold voltage of the inverter if both Q_N and Q_P have $W/L = 2\ \mu\text{m}/1\ \mu\text{m}$? What totem-pole current flows at v_I equal to the threshold voltage?

D *14.47 Consider the choice of values for R_1 and R_2 in the circuit of Fig. 14.37(c). An important consideration in making this choice is that the loss of base drive current will be limited. This loss becomes particularly acute when the current through Q_N and Q_P becomes small. This in turn happens near the end of the output signal swing when the associated MOS device is deeply in triode operation (say at $|v_{DS}| = |V_t|/3$). Determine values for R_1 and R_2 so that the loss in base current is limited to 50%. What is the ratio R_1/R_2? Repeat for a 20% loss in base drive.

***14.48** For the circuit of Fig. 14.37(a) with parameters as in Problem 14.45 and with $(W/L)_p = (W/L)_n$, estimate the propagation delays t_{PLH}, t_{PHL} and t_P obtained for a load capacitance of 2 pF. Assume that the internal node capacitances do not contribute much to this result. Use average values for the charging and discharging currents.

***14.49** Repeat Problem 14.48 for the circuit in Fig. 14.37(e), assuming that $R_1 = R_2 = 5$ kΩ.

D 14.50 Consider the dynamic response of the NAND gate of Fig. 14.39 with a large external capacitive load. If the worst-case response is to be identical to that of the inverter of Fig. 14.37(e), how must the W/L ratios of Q_{NA}, Q_{NB}, Q_N, Q_{PA}, Q_{PB}, and Q_P be related?

D 14.51 Sketch the circuit of a BiCMOS two-input NOR gate. If, when loaded with a large capacitance, the gate is to have worst-case delays equal to the corresponding values of the inverter of Fig. 14.37(e), find W/L of each transistor in terms of $(W/L)_n$ and $(W/L)_p$.

CHAPTER 15

Memory Circuits

IN THIS CHAPTER YOU WILL LEARN:

1. How the basic bistable circuit, the latch, is realized by connecting two inverters in a positive-feedback loop.
2. How to augment the latch to obtain different types of flip-flops that are useful building blocks for digital systems.
3. How CMOS is particularly suited for the efficient implementation of a particular type of flip-flop, the *D* flip-flop.
4. How memory chips that contain as many as 4 gigabits are organized, as well as their various types and the terminology used to describe them.
5. The analysis and design of the six-transistor circuit that is almost universally used to implement the storage cell in static random access memory (SRAM) and the one-transistor circuit that is equally universal in the implementation of the storage cell in dynamic random access memory (DRAM).
6. Interesting circuit techniques for accessing a particular storage cell in a memory chip and for amplifying the signal readout from the cell.
7. How various types of read-only memory (ROM) are designed, programmed, erased, and reprogrammed.

Introduction

The logic circuits studied in Chapters 13 and 14 are called **combinational** circuits. Their output depends only on the present value of the input. Thus these circuits do *not* have memory. *Memory* is a very important part of digital systems. Its availability in digital computers allows for storing programs and data. Furthermore, it is important for temporary storage of the output produced by a combinational circuit for use at a later time in the operation of a digital system.

Logic circuits that incorporate memory are called **sequential circuits**; that is, their output depends not only on the present value of the input but also on the input's previous values. Such circuits require a timing generator (a *clock*) for their operation.

There are basically two approaches for providing memory to a digital circuit. The first relies on the application of positive feedback that, as will be seen shortly, can be arranged to provide a circuit with two stable states. Such a *bistable* circuit can then be used to store one bit of information: One stable state would correspond to a stored 0, and the other to a stored 1. A bistable circuit can remain in either state indefinitely, and thus it belongs to the category of *static sequential circuits*. The other approach to realizing memory utilizes the storage of charge on a capacitor: When the capacitor is charged, it would be regarded as storing a 1; when it is discharged, it would be storing a 0. Since the inevitable leakage effects will cause the capacitor to discharge, such a form of memory requires the periodic recharging of the capacitor, a process known as *refresh*. Thus, like dynamic logic (Section 14.3), memory based on charge storage is known as *dynamic memory* and the corresponding sequential circuits as *dynamic sequential circuits*.

This chapter is concerned with the study of memory circuits. We begin in Section 15.1 with the basic bistable circuit, the latch, and its application in flip-flops, an important class of building blocks for digital systems. After an overview of memory-chip types, organization, and nomenclature in Section 15.2, we study the circuit of the static memory cell (SRAM) and that of the dynamic memory cell (DRAM) in Section 15.3. Besides the array of storage cells, memory chips require circuits for selecting and accessing a particular cell in the array (address decoders) and for amplifying the signal that is retrieved from a particular cell (sense amplifiers). A sampling of these peripheral circuits is presented in Section 15.4. The chapter concludes with an important class of memories, the read-only memory (ROM) in Section 15.5.

15.1 Latches and Flip-Flops

In this section, we shall study the basic memory element, the latch, and consider a sampling of its applications. Both static and dynamic circuits will be considered.

15.1.1 The Latch

The basic memory element, the latch, is shown in Fig. 15.1(a). It consists of two cross-coupled logic inverters, G_1 and G_2. The inverters form a positive-feedback loop. To investigate the operation of the latch we break the feedback loop at the input of one of the inverters, say G_1, and apply an input signal, v_W, as shown in Fig. 15.1(b). Assuming that the input impedance of G_1 is large, breaking the feedback loop will not change the loop voltage transfer characteristic, which can be determined from the circuit of Fig. 15.1(b) by plotting v_Z versus v_W. This is the voltage transfer characteristic of two cascaded inverters and thus takes the shape shown in Fig. 15.1(c). Observe that the transfer characteristic consists of three segments, with the middle segment corresponding to the transition region of the inverters.

Also shown in Fig. 15.1(c) is a straight line with unity slope. This straight line represents the relationship $v_W = v_Z$ that is realized by reconnecting Z to W to close the feedback loop and thus to return it to its original form. As indicated, the straight line intersects the loop transfer curve at three points, A, B, and C. Thus any of these three points can serve as the operating point for the latch. We shall now show that while points A and C are stable operating points in the sense that the circuit can remain at either indefinitely, point B is an unstable operating point; the latch cannot operate at B for any significant period of time.

The reason point B is unstable can be seen by considering the latch circuit in Fig. 15.1(a) to be operating at point B, and taking account of the electrical interference (or noise) that is inevitably present in any circuit. Let the voltage v_W increase by a small increment v_w. The voltage at X will increase (in magnitude) by a larger increment, equal to the product of v_w and the incremental gain of G_1 at point B. The resulting signal v_x is applied to G_2 and gives rise to an even larger signal at node Z. The voltage v_z is related to the original increment v_w by the loop gain at point B, which is the slope of the curve of v_Z versus v_W at point B. This gain is usually much

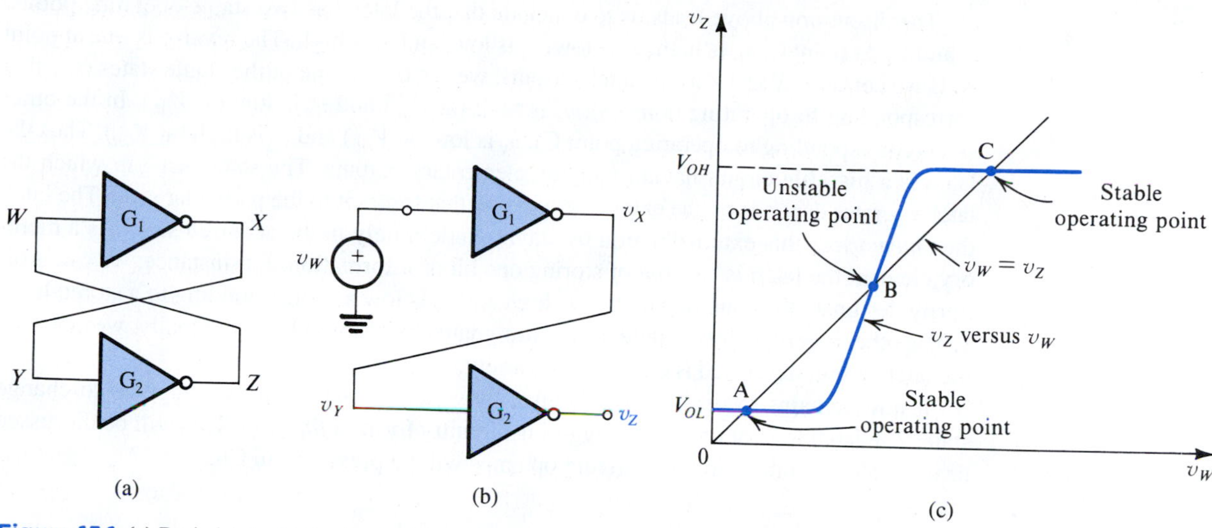

Figure 15.1 **(a)** Basic latch. **(b)** The latch with the feedback loop opened. **(c)** Determining the operating point(s) of the latch.

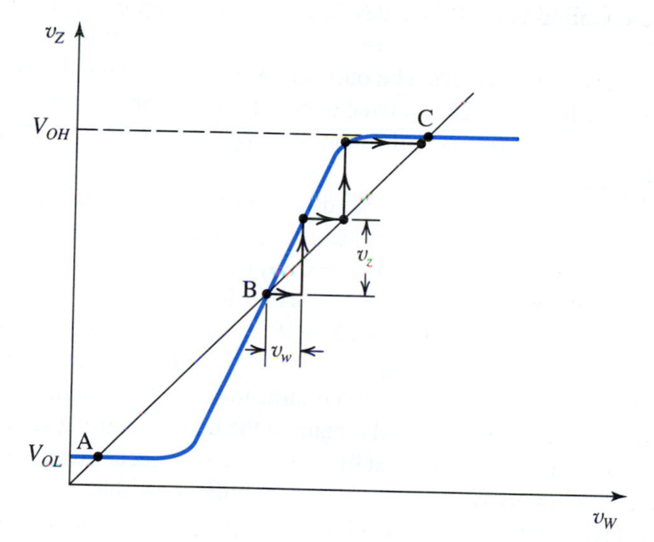

Figure 15.2 Point B is an unstable operating point for the latch: A small positive increment v_w gets amplified around the loop and causes the operating point to shift to the stable operating point C. Had v_w been negative, the operating point would have shifted to the other stable point, A.

greater than unity. Since v_z is coupled to the input of G_1, it becomes the new value of v_W and is further amplified by the loop gain. This regenerative process continues, shifting the operating point from B upward to point C, as illustrated in Fig. 15.2. Since at C the loop gain is zero (or almost zero), no regeneration can take place.

In the description above, we assumed arbitrarily an initial positive voltage increment at W. Had we instead assumed a negative voltage increment, we would have seen that the operating point moves downward from B to A. Again, since at point A the slope of the transfer curve is zero (or almost zero), no regeneration can take place. In fact, for regeneration to occur, the loop gain must be greater than unity, which is the case at point B.

The discussion above leads us to conclude that the latch has two stable operating points, A and C. At point C, v_W is high, v_X is low, v_Y is low, and v_Z is high. The reverse is true at point A. If we consider X and Z as the latch outputs, we see that in one of the stable states (say that corresponding to operating point A), v_X is high (at V_{OH}) and v_Z is low (at V_{OL}). In the other state (corresponding to operating point C), v_X is low (at V_{OL}) and v_Z is high (at V_{OH}). Thus the latch is a **bistable** circuit having two complementary outputs. The stable state in which the latch operates depends on the external excitation that forces it to the particular state. The latch then *memorizes* this external action by staying indefinitely in the acquired state. As a memory element the latch is capable of storing one bit of information. For instance, we can arbitrarily designate the state in which v_X is high and v_Z is low as corresponding to a stored logic 1. The other complementary state then is designated by a stored logic 0. Finally, we note that the latch circuit described is of the static variety.

It now remains to devise a mechanism by which the latch can be *triggered* to change state. The latch together with the triggering circuitry forms a *flip-flop*. This will be discussed next. Analog bistable circuits utilizing op amps will be presented in Chapter 17.

15.1.2 The SR Flip-Flop

The simplest type of flip-flop is the set/reset (SR) flip-flop shown in Fig. 15.3(a). It is formed by cross-coupling two NOR gates, and thus it incorporates a latch. The second inputs of G_1 and G_2 together serve as the trigger inputs of the flip-flop. These two inputs are labeled S (for set) and R (for reset). The outputs are labeled Q and $\overline{Q}$, emphasizing their complementarity. The flip-flop is considered to be set (i.e., storing a logic 1) when Q is high and $\overline{Q}$ is low. When the flip-flop is in the other state (Q low, $\overline{Q}$ high), it is considered to be reset (storing a logic 0).

In the *rest* or *memory state* (i.e., when we do not wish to change the state of the flip-flop), both the S and R inputs should be low. Consider the case when the flip-flop is storing a logic 0. Since Q will be low, both inputs to the NOR gate G_2 will be low. Its output will therefore be high. This high is applied to the input of G_1, causing its output Q to be low, satisfying the original assumption. To set the flip-flop we raise S to the logic-1 level while leaving R at 0. The 1 at the S terminal will force the output of G_2, $\overline{Q}$, to 0. Thus the two inputs to G_1 will be 0 and its output Q will go to 1. Now even if S returns to 0, the $Q = 1$ signal fed to the input of G_2 will keep $\overline{Q} = 0$, and the flip-flop will remain in the newly acquired set state. Note that if we raise S to 1 again (with R remaining at 0), no change will occur. To reset the flip-flop we need to raise R to 1 while leaving $S = 0$. We can readily show that this forces the flip-flop into the reset state ($Q = 0$, $\overline{Q} = 1$) and that the flip-flop remains in this state even after R has returned to 0. It should be observed that the trigger signal merely starts the regenerative action of the positive-feedback loop of the latch.

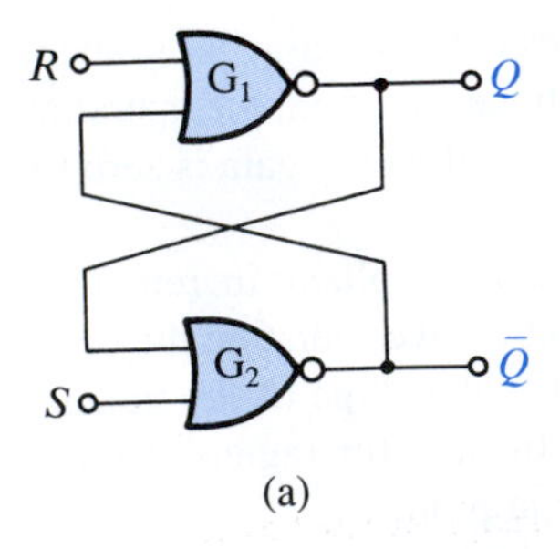

(a)

R	S	Q_{n+1}
0	0	Q_n
0	1	1
1	0	0
1	1	Not used

(b)

Figure 15.3 **(a)** The set/reset (SR) flip-flop and **(b)** its truth table.

Finally, we inquire into what happens if both S and R are simultaneously raised to 1. The two NOR gates will cause both Q and $\overline{Q}$ to become 0 (note that in this case the complementary labeling of these two variables is incorrect). However, if R and S return to the rest state ($R = S = 0$) simultaneously, the state of the flip-flop will be undefined. In other words, it will be impossible to predict the final state of the flip-flop. For this reason, this input combination is usually disallowed (i.e., not used). Note, however, that this situation arises only in the idealized case, when both R and S return to 0 precisely simultaneously. In actual practice one of the two will return to 0 first, and the final state will be determined by the input that remains high longest.

The operation of the flip-flop is summarized by the *truth table* in Fig. 15.3(b), where Q_n denotes the value of Q at time t_n just before the application of the R and S signals, and Q_{n+1} denotes the value of Q at time t_{n+1} after the application of the input signals.

Rather than using two NOR gates, one can also implement an SR flip-flop by cross-coupling two NAND gates, in which case the set and reset functions are active when low (see Problem 15.2).

15.1.3 CMOS Implementation of SR Flip-Flops

The SR flip-flop of Fig. 15.3 can be directly implemented in CMOS by simply replacing each of the NOR gates by its CMOS circuit realization. We encourage the reader to sketch the resulting circuit (see Problem 15.1). Although the CMOS circuit thus obtained works well, it is somewhat complex. As an alternative, we consider a simplified circuit that furthermore implements additional logic. Specifically, Fig. 15.4 shows a *clocked* version of an SR flip-flop. Since the clock inputs form AND functions with the set and reset inputs, the flip-flop can be set or reset only when the clock ϕ is high. Observe that although the two cross-coupled inverters at the heart of the flip-flop are of the standard CMOS type, only NMOS transistors are used for the set–reset circuitry. Nevertheless, since there is no conducting path between V_{DD} and ground (except during switching), the circuit does not dissipate any static power.

Except for the addition of clocking, the SR flip-flop of Fig. 15.4 operates in exactly the same way as its logic antecedent in Fig. 15.3: To illustrate, consider what happens when the flip-flop is in the reset state ($Q = 0$, $\overline{Q} = 1$, $v_Q = 0$, $v_{\overline{Q}} = V_{DD}$), and assume that we wish to set

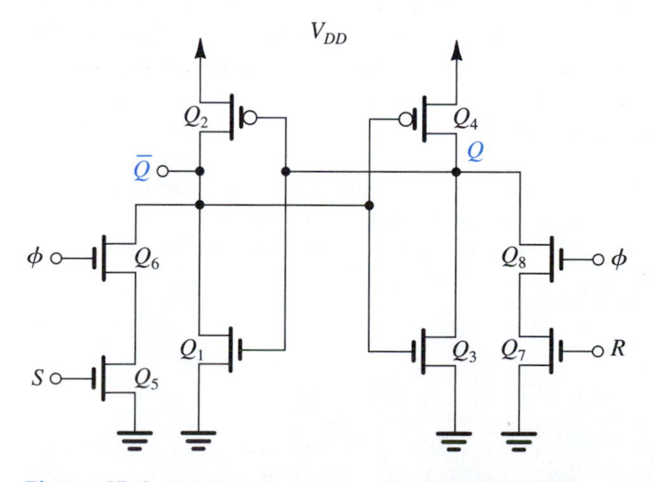

Figure 15.4 CMOS implementation of a clocked SR flip-flop. The clock signal is denoted by ϕ.

it. To do so, we arrange for a high (V_{DD}) signal to appear on the S input while R is held low at 0 V. Then, when the clock ϕ goes high, both Q_5 and Q_6 will conduct, pulling the voltage $v_{\bar{Q}}$ down. If $v_{\bar{Q}}$ goes below the threshold V_M of the (Q_3, Q_4) inverter, the inverter will switch states (or at least begin to switch states), and its output v_Q will rise. This increase in v_Q is fed back to the input of the (Q_1, Q_2) inverter, causing its output $v_{\bar{Q}}$ to go down even further; the regeneration process, characteristic of the positive-feedback latch, is now in progress.

The preceding description of flip-flop switching is predicated on two assumptions:

1. Transistors Q_5 and Q_6 supply sufficient current to pull the node $\bar{Q}$ down to a voltage at least slightly below the threshold of the (Q_3, Q_4) inverter. This is essential for the regenerative process to begin. Without this initial trigger, the flip-flop will fail to switch. In Example 15.1, we shall investigate the *minimum* W/L ratios that Q_5 and Q_6 must have to meet this requirement.
2. The set signal remains high for an interval long enough to cause regeneration to take over the switching process. An estimate of the minimum width required for the set pulse can be obtained as the sum of the interval during which $v_{\bar{Q}}$ is reduced from V_{DD} to $V_{DD}/2$, and the interval for the voltage v_Q to respond and rise to $V_{DD}/2$. This point also will be illustrated in Example 15.1.

Finally, note that the symmetry of the circuit indicates that all the preceding remarks apply equally well to the reset process.

Example 15.1

The CMOS SR flip-flop in Fig. 15.4 is fabricated in a 0.18-μm process for which $\mu_n C_{ox} = 4\,\mu_p C_{ox} = 300\ \mu\text{A/V}^2$, $V_{tn} = |V_{tp}| = 0.5$ V, and $V_{DD} = 1.8$ V. The inverters have $(W/L)_n = 0.27\ \mu\text{m}/0.18\ \mu\text{m}$ and $(W/L)_p = 4(W/L)_n$. The four NMOS transistors in the set–reset circuit have equal W/L ratios.

(a) Determine the minimum value required for this ratio to ensure that the flip-flop will switch.

(b) Also, determine the minimum width the set pulse must have for the case in which the W/L ratio of each of the four transistors in the set–reset circuit is selected at twice the minimum value found in (a). Assume that the total capacitance between each of the Q and $\bar{Q}$ nodes and ground is 20 fF.

Solution

(a) Figure 15.5(a) shows the relevant portion of the circuit for our present purposes. Observe that since the circuit is in the reset state and regeneration has not yet begun, we assume that $v_Q = 0$ and thus Q_2 will be conducting. The circuit is in effect a pseudo-NMOS gate, and our task is to select the W/L ratios for Q_5 and Q_6 so that V_{OL} of this inverter is lower than $V_{DD}/2$ (the threshold of the Q_3, Q_4 inverter whose Q_N and Q_P are matched). The minimum required W/L for Q_5 and Q_6 can be found by equating the current supplied by Q_5 and Q_6 to the current supplied by Q_2 at $v_{\bar{Q}} = V_{DD}/2$. To simplify matters, we assume that the series connection of Q_5 and Q_6 is equivalent to a single transistor whose W/L is half the W/L of each of Q_5 and Q_6 (Fig. 15.5b). Now, since at $v_{\bar{Q}} = V_{DD}/2$ both this equivalent transistor and Q_2 will be operating in the triode region, we can write

$$I_{Deq} = I_{D2}$$

$$300 \times \frac{1}{2}\left(\frac{W}{L}\right)_5\left[(1.8-0.5)\left(\frac{1.8}{2}\right) - \frac{1}{2}\left(\frac{1.8}{2}\right)^2\right]$$

$$= 75 \times \frac{1.08}{0.18}\left[(1.8-0.5)\left(\frac{1.8}{2}\right) - \frac{1}{2}\left(\frac{1.8}{2}\right)^2\right]$$

which yields

$$\left(\frac{W}{L}\right)_5 = \frac{0.54\ \mu\text{m}}{0.18\ \mu\text{m}}$$

and thus

$$\left(\frac{W}{L}\right)_6 = \frac{0.54\ \mu\text{m}}{0.18\ \mu\text{m}}$$

(b) The value calculated for $(W/L)_5$ and $(W/L)_6$ is the absolute minimum needed for switching to occur. To guarantee that the flip-flop will switch, the value selected for $(W/L)_5$ and $(W/L)_6$ is usually somewhat larger than the minimum. Selecting a value twice the minimum,

$$(W/L)_5 = (W/L)_6 = 1.08\ \mu\text{m}/0.18\ \mu\text{m}$$

The minimum required width of the set pulse is composed of two components: the time for $v_{\bar{Q}}$ in the circuit of Fig. 15.5(a) to fall from V_{DD} to $V_{DD}/2$, where $V_{DD}/2$ is the threshold voltage of the inverter formed by Q_3 and Q_4 in Fig. 15.4, and the time for the output of the Q_3–Q_4 inverter to rise from 0 to $V_{DD}/2$. At the end of the second time interval, the feedback signal will have traveled around the feedback loop, and regeneration can continue without the presence of the set pulse. We will denote the first component t_{PHL} and the second t_{PLH}, and will calculate their values as follows.

To determine t_{PHL} refer to the circuit in Fig. 15.6 and note that the capacitor discharge current i_C is the difference between the current of the equivalent transistor Q_{eq} and the current of Q_2,

$$i_C = i_{D\text{eq}} - i_{D2}$$

To determine the average discharge current i_C, we calculate $i_{D\text{eq}}$ and i_{D2} at $t = 0$ and $t = t_{PHL}$. At $t = 0$, $v_{\bar{Q}} = V_{DD}$, thus Q_2 is off,

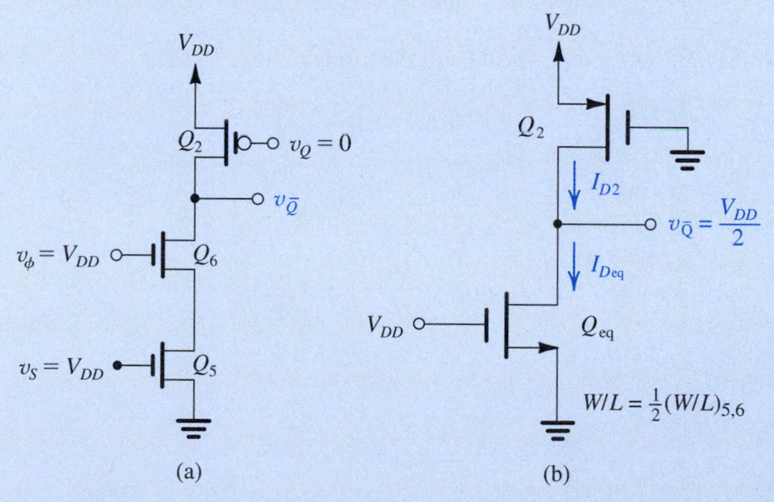

Figure 15.5 (**a**) The relevant portion of the flip-flop circuit of Fig. 15.4 for determining the minimum W/L ratios of Q_5 and Q_6 needed to ensure that the flip-flop will switch. (**b**) The circuit in (**a**) with Q_5 and Q_6 replaced with their equivalent transistor Q_{eq}, at the point of switching.

Example 15.1 *continued*

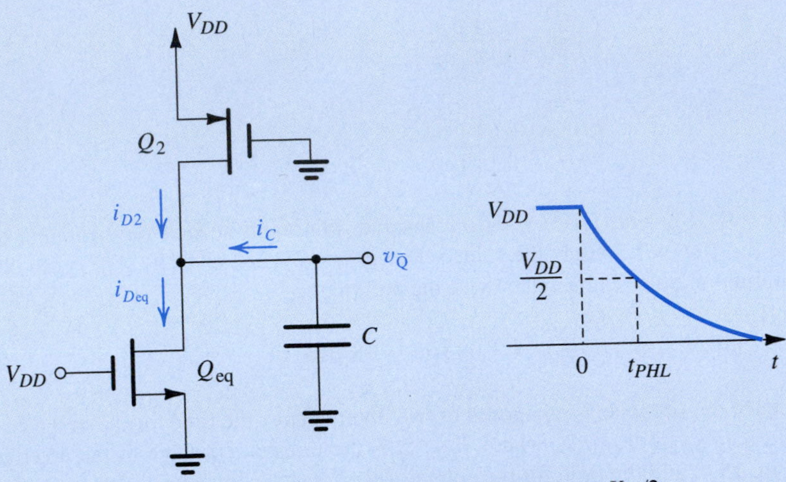

Figure 15.6 Determining the time t_{PHL} for $v_{\bar{Q}}$ to fall from V_{DD} to $V_{DD}/2$.

$$i_{D2}(0) = 0$$

and Q_{eq} is in saturation,

$$i_{Deq} = \frac{1}{2} \times 300 \times \frac{1}{2} \times \frac{1.08}{0.18} \times (1.8 - 0.5)^2$$

$$= 760.5\ \mu\text{A}$$

Thus,

$$i_C(0) = 760.5 - 0 = 760.5\ \mu\text{A}$$

At $t = t_{PHL}$, $v_{\bar{Q}} = V_{DD}/2$, thus both Q_2 and Q_{eq} will be in the triode region,

$$i_{D2}(t_{PHL}) = 75 \times \frac{1.08}{0.18} \times \left[(1.8 - 0.5) - 0.5\left(\frac{1.8}{2}\right)^2\right]$$

$$= 344.25\ \mu\text{A}$$

and

$$i_{Deq}(t_{PHL}) = 300 \times \frac{1}{2} \times \frac{1.08}{0.18}\left[(1.8 - 0.5)\left(\frac{1.8}{2}\right) - 0.5\left(\frac{1.8}{2}\right)^2\right]$$

$$= 688.5\ \mu\text{A}$$

Thus,

$$i_C(t_{PHL}) = 688.5 - 344.25 = 344.25\ \mu\text{A}$$

and the average value of i_C over the interval $t = 0$ to $t = t_{PHL}$ is

$$i_C\big|_{av} = \frac{i_C(0) + i_C(t_{PHL})}{2}$$

$$= \frac{760.5 + 344.25}{2} = 552.4\ \mu\text{A}$$

We now can calculate t_{PHL} as

$$t_{PHL} = \frac{C(V_{DD}/2)}{i_C|_{av}} = \frac{20 \times 10^{-15} \times 0.9}{552.4 \times 10^{-6}} = 32.6 \text{ ps}$$

Next we consider the time t_{PHL} for the output of the Q_3–Q_4 inverter, v_Q, to rise from 0 to $V_{DD}/2$. The value of t_{PLH} can be calculated using the propagation delay formula derived in Chapter 13 (Eq. 13.66), which is also listed in Table 13.3, namely,

$$t_{PLH} = \frac{\alpha_p C}{k_p'(W/L)_p V_{DD}}$$

where

$$\alpha_p = 2\Big/\left[\frac{7}{4} - \frac{3|V_{tp}|}{V_{DD}} + \left(\frac{|V_{tp}|}{V_{DD}}\right)^2\right]$$

Substituting numerical values we obtain,

$$\alpha_p = \frac{2}{1.75 - \frac{3 \times 0.5}{1.8} + \left(\frac{0.5}{1.8}\right)^2} = 2.01$$

and

$$t_{PLH} = \frac{2.01 \times 20 \times 10^{-15}}{75 \times 10^{-6} \times (1.08/0.18) \times 1.8} = 49.7 \text{ ps}$$

Finally, the minimum required width of the set pulse can be calculated as

$$T_{min} = t_{PHL} + t_{PLH}$$

EXERCISE

15.1 For the SR flip-flop specified in Example 15.1, find the minimum W/L for both Q_5 and Q_6 so that switching is achieved when inputs S and ϕ are at $(V_{DD}/2)$.
Ans. 14.3

15.1.4 A Simpler CMOS Implementation of the Clocked SR Flip-Flop

A simpler implementation of a clocked SR flip-flop is shown in Fig. 15.7. Here, pass-transistor logic is employed to implement the clocked set–reset functions. This circuit is very popular in the design of static random-access memory (SRAM) chips, where it is used as the basic memory cell (Section 15.4.1).

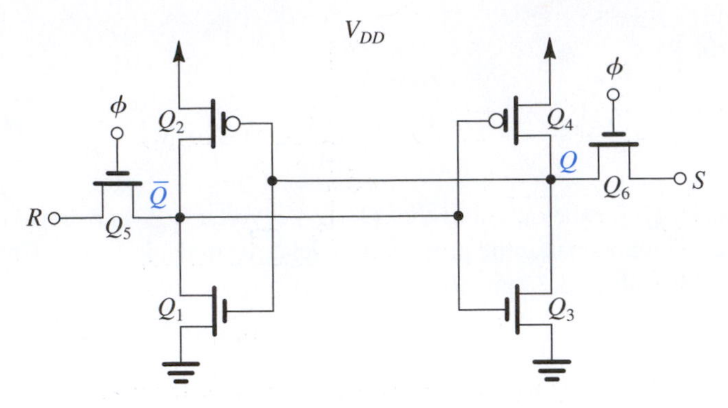

Figure 15.7 A simpler CMOS implementation of the clocked SR flip-flop. This circuit is popular as the basic cell in the design of static random-access memory (SRAM) chips.

15.1.5 D Flip-Flop Circuits

A variety of flip-flop types exist and can be synthesized using logic gates. CMOS circuit implementations can be obtained by simply replacing the gates with their CMOS circuit realizations. This approach, however, usually results in rather complex circuits. In many cases, simpler circuits can be found by taking a circuit-design viewpoint, rather than a logic-design one. To illustrate this point, we shall consider the CMOS implementation of a very important type of flip-flop, the data, or D, flip-flop.

The D flip-flop is shown in block diagram form in Fig. 15.8. It has two inputs, the data input D and the clock input ϕ. The complementary outputs are labeled Q and $\overline{Q}$. When the clock is low, the flip-flop is in the memory, or rest, state; signal changes on the D input line have no effect on the state of the flip-flop. As the clock goes high, the flip-flop acquires the logic level that existed on the D line just before the rising edge of the clock. Such a flip-flop is said to be **edge triggered**. Some implementations of the D flip-flop include direct set and reset inputs that override the clocked operation just described.

A simple implementation of the D flip-flop is shown in Fig. 15.9. The circuit consists of two inverters connected in a positive-feedback loop, just as in the static latch of Fig. 15.1(a), except that here the loop is closed for only part of the time. Specifically, the loop is closed when the clock is low ($\phi = 0, \overline{\phi} = 1$). The input D is connected to the flip-flop through a switch that closes when the clock is high. Operation is straightforward: When ϕ is high, the loop is opened, and the input D is connected to the input of inverter G_1. The capacitance at the input node of G_1 is charged to the value of D, and the capacitance at the input node of G_2 is charged to the value of $\overline{D}$. Then, when the clock goes low, the input line is isolated from

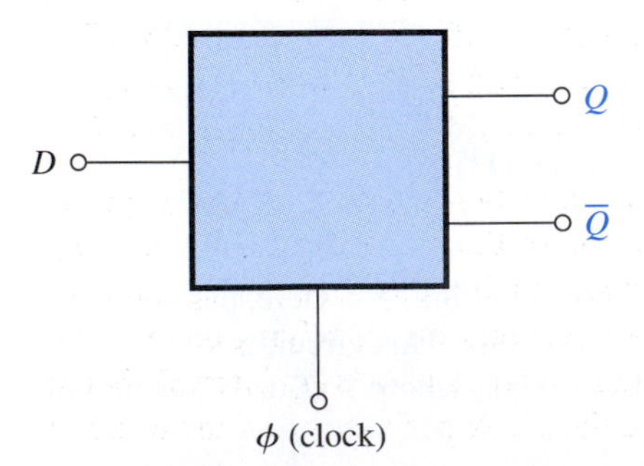

Figure 15.8 A block diagram representation of the D flip-flop.

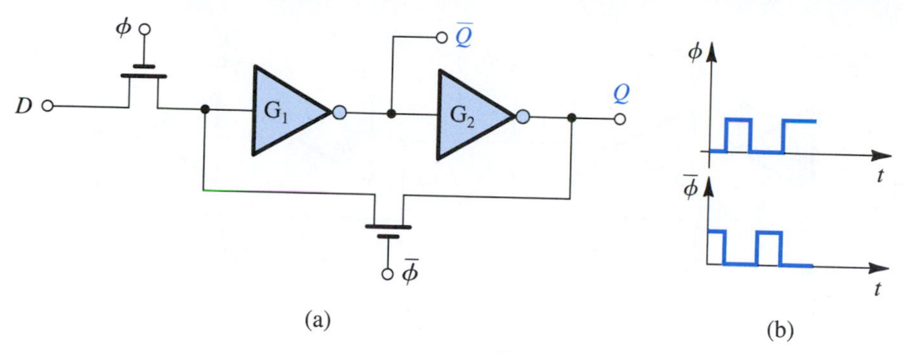

Figure 15.9 A simple implementation of the *D* flip-flop. The circuit in **(a)** utilizes the two-phase **non-overlapping clock** whose waveforms are shown in **(b)**.

the flip-flop, the feedback loop is closed, and the latch acquires the state corresponding to the value of D just before ϕ went down, providing an output $Q = D$.

From the preceding, we observe that the circuit in Fig. 15.9 combines the positive-feedback technique of static bistable circuits and the charge-storage technique of dynamic circuits. It is important to note that the proper operation of this circuit, and of many circuits that use clocks, is predicated on the assumption that *ϕ and $\overline{\phi}$ will not be simultaneously high at any time*. This condition is defined by referring to the two clock phases as being *nonoverlapping*.

An inherent drawback of the D flip-flop implementation of Fig. 15.9 is that during ϕ, the output of the flip-flop simply follows the signal on the D input line. This can cause problems in certain logic-design situations. The problem is solved very effectively by using the **master–slave** configuration shown in Fig. 15.10(a). Before discussing its circuit operation, we note that although the switches are shown implemented with single NMOS transistors, CMOS transmission gates are employed in many applications. We are simply using the single MOS transistor as a "shorthand notation" for a series switch.

The master–slave circuit consists of a pair of circuits of the type shown in Fig. 15.9, operated with alternate clock phases. Here, to emphasize that the two clock phases must be nonoverlapping, we denote them ϕ_1 and ϕ_2, and clearly show the nonoverlap interval in the waveforms of Fig. 15.10(b). Operation of the circuit is as follows:

1. When ϕ_1 is high and ϕ_2 is low, the input is connected to the master latch whose feedback loop is opened, while the slave latch is isolated. Thus, the output Q remains at the value stored previously in the slave latch whose loop is now closed. The node capacitances of the master latch are charged to the appropriate voltages corresponding to the present value of D.

2. When ϕ_1 goes low, the master latch is isolated from the input data line. Then, when ϕ_2 goes high, the feedback loop of the master latch is closed, locking in the value of D. Further, its output is connected to the slave latch whose feedback loop is now open. The node capacitances in the slave are appropriately charged so that when ϕ_1 goes high again, the slave latch locks in the new value of D and provides it at the output, $Q = D$.

From this description, we note that at the positive transition of clock ϕ_2 the output Q adopts the value of D that existed on the D line at the end of the preceding clock phase, ϕ_1. This output value remains constant for one clock period. Finally, note that during the non-overlap interval both latches have their feedback loops open, and we are relying on the node capacitances to maintain most of their charge. It follows that the nonoverlap interval should be kept reasonably short (perhaps one-tenth or less of the clock period, and of the order of 1 ns or so in current practice).

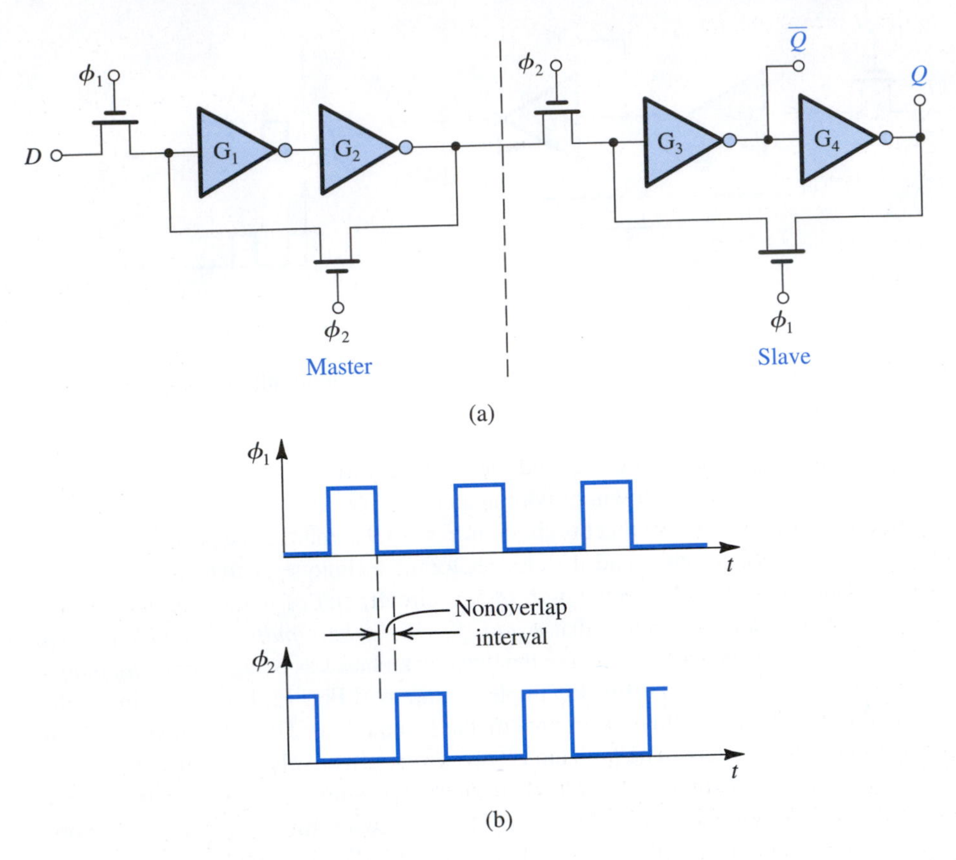

Figure 15.10 **(a)** A master–slave D flip-flop. The switches can be, and usually are, implemented with CMOS transmission gates. **(b)** Waveforms of the two-phase nonoverlapping clock required.

15.2 Semiconductor Memories: Types and Architectures

A computer system, whether a large machine or a microcomputer, requires memory for storing data and program instructions. Furthermore, within a given computer system there usually are various types of memory utilizing a variety of technologies and having different *access times*. Broadly speaking, computer memory can be divided into two types: **main memory** and **mass-storage** memory. The main memory is usually the most rapidly accessible memory and the one from which most, often all, instructions in programs are executed. The main memory is usually of the random-access type. A **random-access memory** (RAM) is one in which the time required for storing (writing) information and for retrieving (reading) information is independent of the physical location (within the memory) in which the information is stored.

Random-access memories should be contrasted with *serial* or *sequential* memories, such as disks and tapes, from which data are available only in the sequence in which the data were originally stored. Thus, in a serial memory the time to access particular information depends on the memory location in which the required information is stored, and the average access time is longer than the access time of random-access memory. In a computer system, serial memory is used for mass storage. Items not frequently accessed, such as large parts of the computer operating system, are usually stored in a *moving-surface memory* such as magnetic disk.

Another important classification of memory relates to whether it is a **read/write** or a **read-only memory**. Read/write (R/W) memory permits data to be stored and retrieved at comparable speeds. Computer systems require random-access read/write memory for data and program storage.

Read-only memories (**ROM**) permit reading at the same high speeds as R/W memories (or perhaps higher) but restrict the writing operation. ROMs can be used to store a microprocessor operating-system program. They are also employed in operations that require table lookup, such as finding the values of mathematical functions. A popular application of ROMs is their use in video game cartridges. It should be noted that read-only memory is usually of the random-access type. Nevertheless, in the digital circuit jargon, the acronym RAM usually refers to read/write, random-access memory, while ROM is used for read-only memory.

The regular structure of memory circuits has made them an ideal application for the design of circuits of the very-large-scale integrated (VLSI) type. Indeed, at any moment, memory chips represent the state of the art in packing density and hence integration level. Beginning with the introduction of the 1-Kbit chip in 1970, memory-chip density has quadrupled about every 3 years. At the present time (2009), chips containing 4 Gbit[1] are available. In this and the next two sections, we shall study some of the basic circuits employed in VLSI RAM chips. Read-only memory circuits are studied in Section 15.5.

15.2.1 Memory-Chip Organization

The bits on a memory chip are addressable either individually or in groups of 4 to 16. As an example, a 64-Mbit chip in which all bits are individually addressable is said to be organized as 64M words × 1 bit (or simply 64M × 1). Such a chip needs a 26-bit address (2^{26} = 67,108,864 = 64M). On the other hand, the 64-Mbit chip can be organized as 16M words × 4 bits (16M × 4), in which case a 24-bit address is required. For simplicity we shall assume in our subsequent discussion that all the bits on a memory chip are individually addressable.

The bulk of the memory chip consists of the cells in which the bits are stored. Each **memory cell** is an electronic circuit capable of storing one bit. We shall study memory-cell circuits in Section 15.3. For reasons that will become clear shortly, it is desirable to physically organize the storage cells on a chip in a square or a nearly square matrix. Figure 15.11 illustrates such an organization. The cell matrix has 2^M rows and 2^N columns, for a total storage capacity of 2^{M+N}. For example, a 1M-bit square matrix would have 1024 rows and 1024 columns ($M = N = 10$). Each cell in the array is connected to one of the 2^M row lines, known rather loosely, but universally, as **word lines**, and to one of the 2^N column lines, known as **digit lines** or, more commonly, **bit lines**. A particular cell is **selected** for reading or writing by activating its word line and its bit line.

Activating one of the 2^M word lines is performed by the **row decoder**, a combinational logic circuit that selects (raises the voltage of) the particular word line whose M-bit address is applied to the decoder input. The address bits are denoted $A_0, A_1, \ldots, A_{M-1}$. When the Kth word line is activated for, say, a **read operation**, all 2^N cells in row K will provide their contents to their respective bit lines. Thus, if the cell in column L (Fig. 15.11) is storing a 1, the voltage of bit-line number L will be raised, usually by a small voltage, say 0.1 V to 0.2 V. The readout voltage

[1]The capacity of a memory chip to hold binary information as binary digits (or bits) is measured in kilobit (Kbit), megabit (Mbit), and gigabit (Gbit) units, where 1 Kbit = 1024 bits, 1 Mbit = 1024 × 1024 = 1,048,576 bits, and, 1 Gbit = 1024^3 bits. Thus a 64-Mbit chip contains 67,108,864 bits of memory.

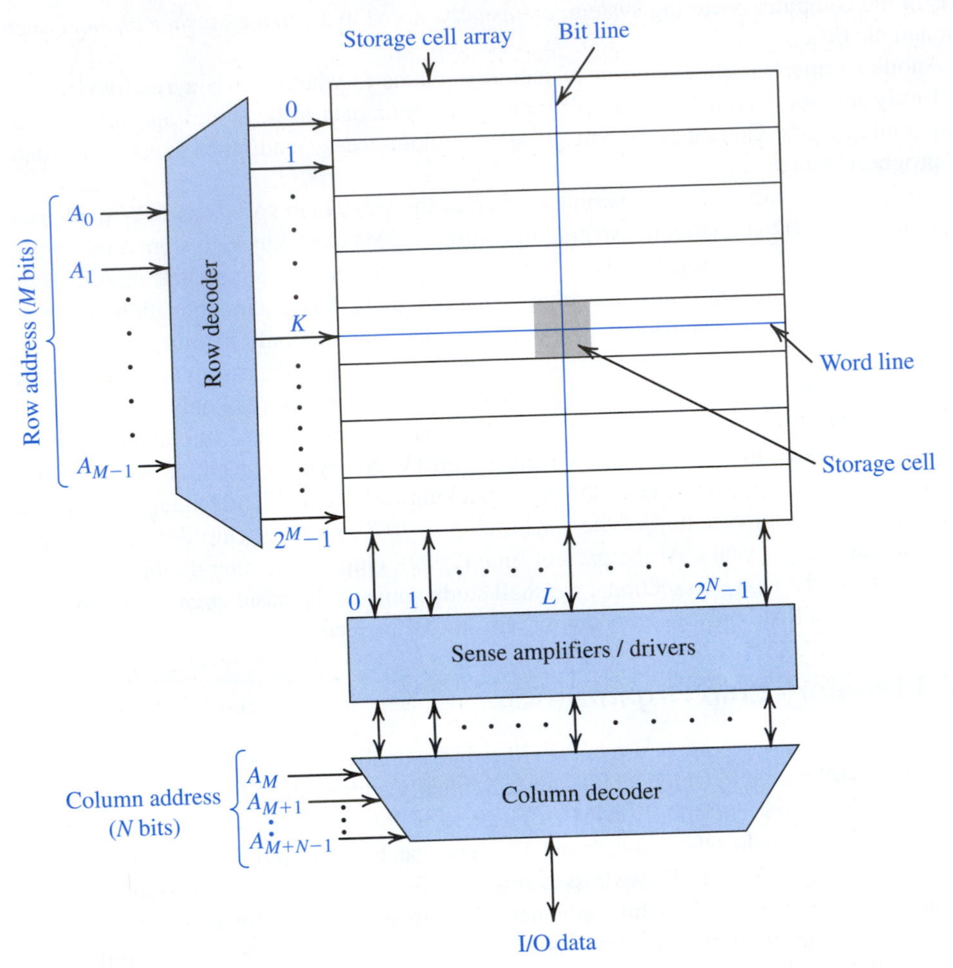

Figure 15.11 A 2^{M+N}-bit memory chip organized as an array of 2^M rows × 2^N columns.

is small because the cell is small, a deliberate design decision, since the number of cells is very large. The small readout signal is applied to a **sense amplifier** connected to the bit line. As Fig. 15.11 indicates, there is a sense amplifier for every bit line. The sense amplifier provides a full-swing digital signal (from 0 to V_{DD}) at its output. This signal, together with the output signals from all the other cells in the selected row, is then delivered to the **column decoder**. The column decoder selects the signal of the particular column whose N-bit address is applied to the decoder input (the address bits are denoted $A_M, A_{M+1}, \ldots, A_{M+N-1}$) and causes this signal to appear on the chip input/output (I/O) data line.

A **write operation** proceeds in a similar manner: The data bit to be stored (1 or 0) is applied to the I/O line. The cell in which the data bit is to be stored is selected through the combination of its row address and its column address. The sense amplifier of the selected column acts as a **driver** to write the applied signal into the selected cell. Circuits for sense amplifiers and address decoders will be studied in Section 15.4.

Before leaving the topic of memory organization (or memory-chip architecture), we wish to mention a relatively recent innovation in organization dictated by the exponential increase in chip density. To appreciate the need for a change, note that as the number of cells in the array increases, the physical lengths of the word lines and the bit lines increase. This has occurred even though for each new generation of memory chips, the transistor size has decreased (currently, CMOS process technologies with 45-nm feature size are utilized). The net increase in word-line and bit-line lengths increases their total resistance and capacitance, and thus slows down their transient response. That is, as the lines lengthen, the exponential rise of the voltage of the word line becomes slower, and it takes longer for the cells to be activated. This problem has been solved by partitioning the memory chip into a number of blocks. Each of the blocks has an organization identical to that in Fig. 15.11. The row and column addresses are broadcast to all blocks, but the data selected come from only one of the blocks. Block selection is achieved by using an appropriate number of the address bits as a block address. Such an architecture can be thought of as three-dimensional: rows, columns, and blocks.

15.2.2 Memory-Chip Timing

The **memory access time** is the time between the initiation of a read operation and the appearance of the output data. The **memory cycle time** is the minimum time allowed between two consecutive memory operations. To be on the conservative side, a memory operation is usually taken to include both read and write (in the same location). MOS memories have access and cycle times in the range of a few to a few hundred nanoseconds.

EXERCISES

15.2 A 4-Mbit memory chip is partitioned into 32 blocks, with each block having 1024 rows and 128 columns. Give the number of bits required for the row address, column address, and block address.
Ans. 10; 7; 5

15.3 The word lines in a particular MOS memory chip are fabricated using polysilicon (see Appendix A). The resistance of each word line is estimated to be 5 kΩ, and the total capacitance between the line and ground is 2 pF. Find the time for the voltage on the word line to reach $V_{DD}/2$, assuming that the line is driven by a voltage V_{DD} provided by a low-impedance inverter. (*Note:* The line is actually a distributed network that we are approximating by a lumped circuit consisting of a single resistor and a single capacitor.)
Ans. 6.9 ns

15.3 Random-Access Memory (RAM) Cells

As mentioned in Section 15.2, the major part of the memory chip is taken up by the storage cells. It follows that to be able to pack a large number of bits on a chip, it is imperative that the cell size be reduced to the smallest possible. The power dissipation per cell should be minimized also. Thus, many of the flip-flop circuits studied in Section 15.1 are too complex to be suitable for implementing the storage cells in a RAM chip.

There are basically two types of MOS RAM: static and dynamic. **Static RAMs** (called **SRAMs** for short) utilize static latches as the storage cells. Dynamic RAMs (called **DRAMs**), on the other hand, store the binary data on capacitors, resulting in further reduction in cell area, but at the expense of more complex read and write circuitry. In particular, while static RAMs can hold their stored data indefinitely, provided the power supply remains on, dynamic RAMs require *periodic refreshing* to regenerate the data stored on capacitors. This is because the storage capacitors will discharge, though slowly, as a result of the leakage currents inevitably present. By virtue of their smaller cell size, dynamic memory chips are usually four times as dense as their contemporary static chips. Thus while the state of the art in 2009 is a 4-Gbit DRAM chip, the highest-density SRAM chip has 1 Gbit capacity. Both static and dynamic RAMs are *volatile;* that is, they require the continuous presence of a power supply. By contrast, most ROMs are of the nonvolatile type, as we shall see in Section 15.5. In the following subsections, we shall study basic SRAM and DRAM storage cells.

15.3.1 Static Memory (SRAM) Cell

Figure 15.12 shows a typical static memory cell in CMOS technology. The circuit, which we encountered in Section 15.1, is a flip-flop comprising two cross-coupled inverters and two **access transistors**, Q_5 and Q_6. The access transistors are turned on when the word line is selected and its voltage raised to V_{DD}, and they connect the flip-flop to the column (bit or B) line and column ($\overline{\text{bit}}$ or $\overline{B}$) line. Note that although in principle only the B or the $\overline{B}$ line suffices, most often both are utilized, as shown in Fig. 15.12. This both provides a *differential data path* between the cell and the memory-chip output and increases the circuit reliability. The access transistors act as transmission gates allowing bidirectional current flow between the flip-flop and the B and $\overline{B}$ lines. Finally, we note that this circuit is known as the **six-transistor** or **6T cell**.

The Read Operation Consider first a read operation, and assume that the cell is storing a 1. In this case, Q will be high at V_{DD}, and $\overline{Q}$ will be low at 0 V. Before the read operation begins, the B and $\overline{B}$ lines are raised to a voltage in the range $V_{DD}/2$ to V_{DD}. This process, known as **precharging**, is performed using circuits we shall discuss in the next section in conjunction with the study of sense amplifiers. To simplify matters, we shall assume here that the precharge voltage of B and $\overline{B}$ is V_{DD}.

When the word line is selected and the access transistors Q_5 and Q_6 are turned on, examination of the circuit reveals that the only portion that will be conducting is that shown in Fig. 15.13. Noting that the initial value of $v_{\overline{Q}}$ is 0 V, we can see that current will flow from the $\overline{B}$

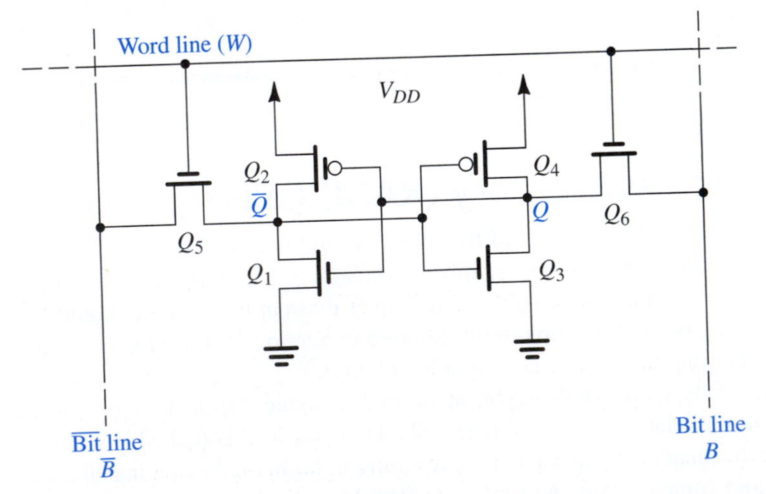

Figure 15.12 A CMOS SRAM memory cell.

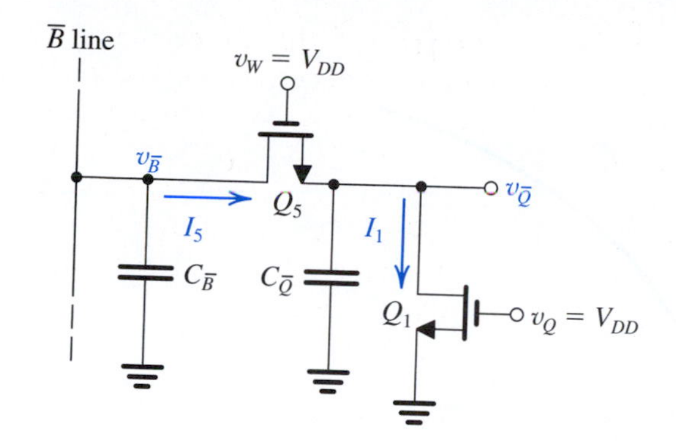

Figure 15.13 Relevant parts of the SRAM cell circuit during a read operation when the cell is storing a logic 1. Note that initially $v_Q = V_{DD}$ and $v_{\overline{Q}} = 0$. Also note that the B and $\overline{B}$ lines are precharged to a voltage V_{DD}.

line (actually, from the $\overline{B}$-line capacitance $C_{\overline{B}}$) through Q_5 and into capacitor $C_{\overline{Q}}$, which is the small equivalent capacitance between the $\overline{Q}$ node and ground. This current charges $C_{\overline{Q}}$ and thus $v_{\overline{Q}}$ rises and Q_1 conducts, sinking some of the current supplied by Q_5. Equilibrium will be reached when $C_{\overline{Q}}$ is charged to a voltage $V_{\overline{Q}}$ at which I_1 equals I_5, and no current flows through $C_{\overline{Q}}$. Here it is extremely important to note that to avoid changing the state of the flip-flop, that is, for our read operation to be **nondestructive**, $V_{\overline{Q}}$ must not exceed the threshold voltage of the inverter Q_3–Q_4. In fact, SRAM designers usually impose a more stringent requirement on the value of $V_{\overline{Q}}$, namely, that it should be lower than the threshold voltage of Q_3, V_{tn}. Thus, the design problem we shall now solve is as follows: Determine the ratio of $(W/L)_5/(W/L)_1$ so that $V_{\overline{Q}} \le V_{tn}$.

Noting that Q_5 will be operating in saturation and neglecting, for simplicity, the body effect, we can write

$$I_5 = \frac{1}{2}(\mu_n C_{ox})\left(\frac{W}{L}\right)_5 (V_{DD} - V_{tn} - V_{\overline{Q}})^2 \tag{15.1}$$

Transistor Q_1 will be operating in the triode region, and its current I_1 can be written as

$$I_1 = (\mu_n C_{ox})\left(\frac{W}{L}\right)_1 \left[(V_{DD} - V_{tn})V_{\overline{Q}} - \frac{1}{2}V_{\overline{Q}}^2\right] \tag{15.2}$$

Equating I_5 and I_1 gives a quadratic equation in $V_{\overline{Q}}$, which can be solved to obtain

$$\frac{V_{\overline{Q}}}{V_{DD} - V_{tn}} = 1 - 1\Bigg/\sqrt{1 + \frac{(W/L)_5}{(W/L)_1}} \tag{15.3}$$

This is an attractive relationship, since it provides $V_{\overline{Q}}$ in normalized form and thus always applies, independent of the process technology utilized. Figure 15.14 shows a universal plot of $[V_{\overline{Q}}/(V_{DD} - V_{tn})]$ versus $(W/L)_5/(W/L)_1$. For a given process technology, V_{DD} and V_{tn} are determined, and the plot in Fig. 15.14 can be used to determine the maximum value permitted for $(W/L)_5/(W/L)_1$ while keeping $V_{\overline{Q}}$ below a desired value. Alternatively, we

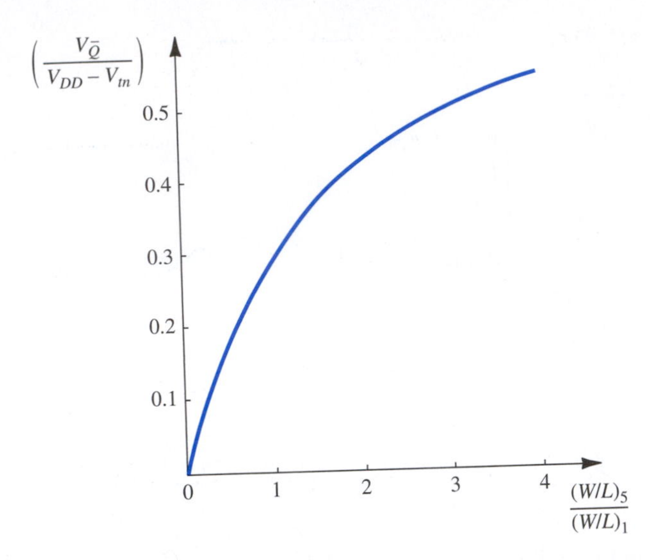

Figure 15.14 The normalized value of $V_{\bar{Q}}$ versus the ratio $(W/L)_5/(W/L)_1$ for the circuit in Fig. 15.13. This graph can be used to determine the maximum value permitted for $(W/L)_5/(W/L)_1$ so that $V_{\bar{Q}}$ is kept below a desired level.

can derive a formula for this purpose. For instance, if $V_{\bar{Q}}$ is to be kept below V_{tn}, the ratio of $(W/L)_5$ to $(W/L)_1$ must be kept below the value obtained from Eq. (15.3), that is,

$$\frac{(W/L)_5}{(W/L)_1} \leq \frac{1}{\left(1-\frac{V_{tn}}{V_{DD}-V_{tn}}\right)^2} - 1 \tag{15.4}$$

This is an important design constraint that can be expressed in a slightly more general form by replacing $(W/L)_5$ with $(W/L)_a$, where the subscript a denotes access transistors Q_5 and Q_6, and $(W/L)_1$ with $(W/L)_n$, which is the W/L ratio of Q_N in each of the two inverters; thus,

$$\frac{(W/L)_a}{(W/L)_n} \leq \frac{1}{\left(1-\frac{V_{tn}}{V_{DD}-V_{tn}}\right)^2} - 1 \tag{15.5}$$

EXERCISE

15.4 Find the maximum allowable W/L for the access transistors of the SRAM cell in Fig. 15.12 so that in a read operation, the voltages at Q and $\bar{Q}$ do not change by more than $|V_t|$. Assume that the SRAM is fabricated in a 0.18-μm technology for which $V_{DD} = 1.8$ V, $V_{tn} = |V_{tp}| = 0.5$ V and that $(W/L)_n = 1.5$.
Ans. $(W/L)_a \leq 2.5$

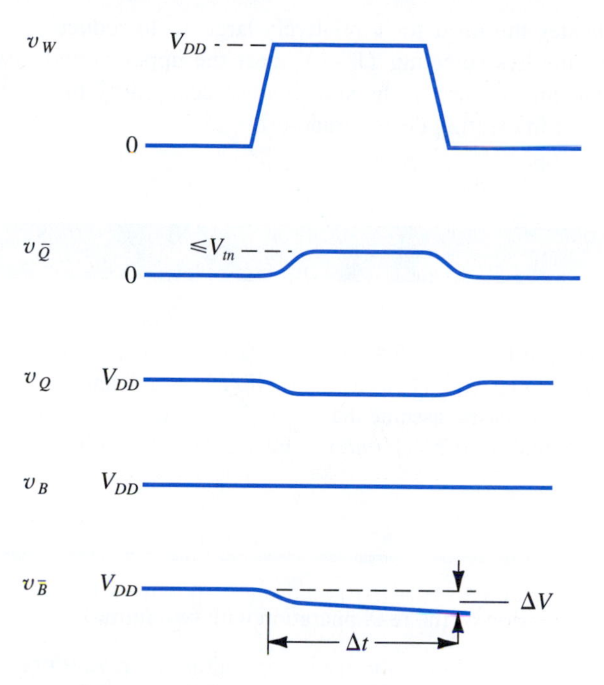

Figure 15.15 Voltage waveforms at various nodes in the SRAM cell during a read-1 operation.

Having determined the constraint imposed by the read operation on the W/L ratios of the access transistors, we now return to the circuit in Fig. 15.13, and show in Fig. 15.15 the voltage waveforms at various nodes during a read-1 operation. Observe that as we have already, discussed, $v_{\overline{Q}}$ rises from zero to a voltage $V_{\overline{Q}} \leq V_{tn}$. Correspondingly, the change in v_Q will be very small, justifying the assumption implicit in the analysis above that v_Q remains constant at V_{DD}. Most important, note that the voltage of the $\overline{B}$ line, $v_{\overline{B}}$, decreases by a small amount ΔV. This is a result of the discharge of the capacitance of the $\overline{B}$ line, $C_{\overline{B}}$, by the current I_5. Assuming that I_5 reaches its equilibrium value in Eq. (15.1) relatively quickly, capacitor $C_{\overline{B}}$ is in effect discharged by a constant current I_5 and the change in its voltage, ΔV, obtained in a time interval Δt, can be found by writing a charge-balance equation,

$$I_5 \; \Delta t = C_{\overline{B}} \; \Delta V$$

Thus,

$$\Delta V = \frac{I_5 \Delta t}{C_{\overline{B}}} \tag{15.6}$$

Here we note that $C_{\overline{B}}$ is usually relatively large (1–2 pF) because a large number of cells are connected to the $\overline{B}$ line. The incremental change ΔV is therefore rather small (0.1–0.2 V), necessitating the use of a sense amplifier. If the sense amplifier requires a minimum decrement ΔV in $v_{\overline{B}}$ to detect the presence of a "1", then the read delay time can be found from Eq. (15.6) as

$$\Delta t = \frac{C_{\overline{B}} \; \Delta V}{I_5} \tag{15.7}$$

This equation indicates the need for a relatively large I_5 to reduce the delay time Δt. A large I_5, however, implies selecting $(W/L)_a$ near the upper bound given by Eq. (15.5), which in turn means an increase in the silicon area occupied by the access transistors and hence the cell area, an interesting design trade-off.

EXERCISE

15.5 For the SRAM cell considered in Exercise 15.4 whose $(W/L)_n = 1.5$ and $(W/L)_a \leq 2.5$, use Eq. (15.7) to determine the read delay Δt in two cases: (a) $(W/L)_a = 2.5$ and (b) $(W/L)_a = 1.5$. Let $\mu_n C_{ox} = 300\ \mu\text{A/V}^2$. In both cases, assume that $C_{\bar{B}} = 2$ pF and that the sense amplifier requires a ΔV of minimum magnitude of 0.2 V. [*Hint*: Use Eq. (15.1) to determine I_5, and recall that $V_{\bar{Q}} = V_{tn}$.]

Ans. 1.7 ns; 2.8 ns

We conclude our discussion of the read operation with two remarks:

1. Although we considered only the read-1 operation, the read-0 operation is identical; it involves Q_2 and Q_6 with the analysis resulting in an upper bound on $(W/L)_6/(W/L)_2$ equal to that we have found for $(W/L)_5/(W/L)_1$. This, of course, is entirely expected, since the circuit is symmetrical. The read-0 operation results in a decrement ΔV in the voltage of the B line, which is interpreted by the sense amplifier as a stored 0.
2. The component Δt of the read delay is relatively large because C_B and $C_{\bar{B}}$ are relatively large (in the picofarad range). Also, Δt is not the only component of the read delay; another significant component is due to the finite rise time of the voltage on the word line. Indeed, even the calculation of Δt is optimistic, since the word line will have reached a voltage lower than V_{DD} only, when the process of discharging $C_{\bar{B}}$ takes place. As will be seen shortly, the write operation is faster.

The Write Operation We next consider the write operation. Let the SRAM cell of Fig. 15.12 be storing a logic 1, thus $V_Q = V_{DD}$ and $V_{\bar{Q}} = 0$ V, and assume that we wish to write a 0; that is, we wish to have the flip-flop switch states. To write a zero, the B line is lowered to 0 V, and the $\bar{B}$ line is raised to V_{DD} and, of course, the cell is selected by raising the word line to V_{DD}. The objective now is to pull node Q down and node $\bar{Q}$ up and have the voltage of at least one of these two nodes pass by the inverter threshold voltage. Thus, if v_Q decreases below the threshold voltage of inverter Q_1–Q_2, the regenerative action of the latch will start and the flip-flop will switch to the stored-0 state. Alternatively, or in addition, if we manage to raise $v_{\bar{Q}}$ above the threshold voltage of the Q_3–Q_4 inverter, the regenerative action will be engaged and the latch will eventually switch state. Either one of the two actions is sufficient to engage the regenerative mechanism of the latch.

Figure 15.16 shows the relevant parts of the SRAM circuit during the interval when $v_{\bar{Q}}$ is being pulled up (Fig. 15.16a) and v_Q is being pulled down (Fig. 15.16b). Since **toggling** (i.e., state change) has not yet taken place, we assume that the voltage feeding the gate of Q_1 is still equal to V_{DD} and the voltage at the gate of Q_4 is still equal to 0 V. These voltages will of course be changing as $v_{\bar{Q}}$ goes up and v_Q goes down, but this assumption is nevertheless reasonable for hand analysis.

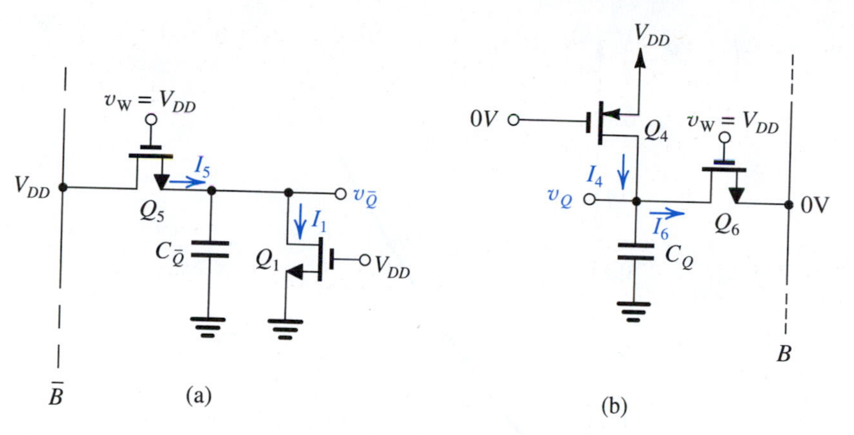

Figure 15.16 Relevant parts of the 6T SRAM circuit of Fig. 15.12 during the process of writing a 0. It is assumed that the cell is originally storing a 1 and thus initially $v_Q = V_{DD}$ and $v_{\bar{Q}} = 0$ V.

Consider first the circuit in Fig. 15.16(a). This is the same circuit we analyzed in detail in the study of the read operation above. Recall that to make the read process nondestructive, we imposed an upper bound on $(W/L)_5$. That upper bound ensured that $v_{\bar{Q}}$ will not rise above V_{tn}. Thus, this circuit is not capable of raising $v_{\bar{Q}}$ to the point that it can start the regenerative action. We must therefore rely solely on the circuit of Fig. 15.16(b). That is, our write-0 operation will be accomplished by pulling node Q down in order to initiate the regenerative action of the latch. To ensure that the latch will in fact switch state, SRAM designers impose a more stringent requirement on the voltage v_Q, namely, that it must fall below not just V_M of the Q_1–Q_2 inverter but below V_{tn} of Q_1.

Let's now look more closely at the circuit of Fig. 15.16(b). Initially, v_Q is at V_{DD}. However, as Q_6 turns on, I_6 quickly discharges the small capacitance C_Q, and v_Q begins to fall. This will enable Q_4 to conduct, and equilibrium is reached when $I_4 = I_6$. To ensure toggling, we design the circuit so that this equilibrium occurs at a value of v_Q less than V_{tn}. At such a value V_Q, Q_4 will be operating in saturation and Q_6 will be operating in the triode region, thus

$$I_4 = \frac{1}{2}(\mu_p C_{ox})\left(\frac{W}{L}\right)_4 (V_{DD} - |V_{tp}|)^2 \tag{15.8}$$

and

$$I_6 = (\mu_n C_{ox})\left(\frac{W}{L}\right)_6 \left[(V_{DD} - V_{tn})V_Q - \frac{1}{2}V_Q^2\right] \tag{15.9}$$

Substituting $|V_{tp}| = V_{tn}$, which is usually the case, and equating I_4 and I_6 results in a quadratic equation in V_Q whose solution is

$$\frac{V_Q}{V_{DD} - V_{tn}} = 1 - \sqrt{1 - \left(\frac{\mu_p}{\mu_n}\right)\frac{(W/L)_4}{(W/L)_6}} \tag{15.10}$$

This relationship is not as convenient as that in Eq. (15.3) because the right-hand side includes a process-dependent quantity, namely, μ_p/μ_n. Thus we do not have a universally

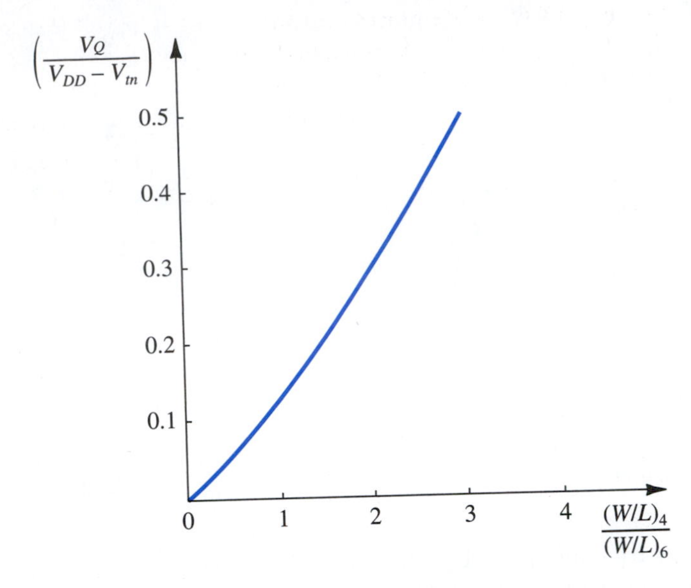

Figure 15.17 The normalized value of V_Q versus the ratio $(W/L)_4/(W/L)_6$ for the circuit in Fig. 15.16(b). The graph applies for process technologies for which $\mu_n \simeq 4\mu_p$. It can be used to determine the maximum $(W/L)_4/(W/L)_6$ for which V_Q is guaranteed to fall below a desired value.

applicable relationship. Nevertheless, for a number of CMOS process technologies, including the 0.25-μm, the 0.18-μm, and the 0.13-μm processes, $\mu_n/\mu_p \simeq 4$. Thus, upon substituting $\mu_p/\mu_n = 0.25$ in Eq. (15.10), we obtain the semiuniversal graph shown in Fig. 15.17. We can use this graph to determine the maximum allowable value of the ratio $(W/L)_4/(W/L)_6$ that will ensure a value of $V_Q \leq V_{tn}$ for given process parameters V_{DD} and V_{tn}. Alternatively, substituting $V_Q = V_{tn}$, $(W/L)_4 = (W/L)_p$, and $(W/L)_6 = (W/L)_a$, we can obtain the upper bound analytically as

$$\frac{(W/L)_p}{(W/L)_a} \leq \left(\frac{\mu_n}{\mu_p}\right)\left[1-\left(1-\frac{V_{tn}}{V_{DD}-V_{tn}}\right)^2\right] \tag{15.11}$$

Observe that this relationship provides an upper bound on $(W/L)_p$ in terms of $(W/L)_a$ and that the relationship in Eq. (15.5) provides an upper bound on $(W/L)_a$ in terms of $(W/L)_n$. Thus, the two relationships can be used together to design the SRAM cell.

EXERCISE

15.6 For the SRAM cell considered in Exercise 15.4, where $(W/L)_n = 1.5$ and $(W/L)_a \leq 2.5$, use Eq. (15.11) to find the maximum allowable value of $(W/L)_p$. Recall that for this 0.18-μm process, $\mu_p \simeq 4\mu_n$. For all transistors having $L = 0.18$ μm, find W_n, W_p, and W_a that result in a minimum-area cell. Assume that the minimum allowable width is 0.18 μm.

Ans. $(W/L)_p \leq 2.5\,(W/L)_a$, thus $(W/L)_p \leq 6.25$; for minimum area select $W_n = W_p = W_a = 0.18$ μm.

We conclude our study of the write process by noting that it is fast because it does not require discharging the large capacitance of the bit lines. The voltages of the B and $\overline{B}$ lines are driven to their required values of 0 or V_{DD} by powerful driver circuits and thus achieve their desired voltages very quickly. The write delay is determined roughly by the time for the regenerating signal to propagate around the feedback loop of the latch; thus it is about twice the propagation delay of the inverter. Of course, the write cycle time is still lengthened by the word-line delay.

15.3.2 Dynamic Memory Cell

Although a variety of DRAM storage cells have been proposed over the years, a particular cell, shown in Fig. 15.18, has become the industry standard. The cell consists of a single n-channel MOSFET, known as the **access transistor**, and a **storage capacitor** C_S. The cell is appropriately known as the **one-transistor cell.**[2] The gate of the transistor is connected to the word line, and its source (drain) is connected to the bit line. Observe that only one bit line is used in DRAMs, whereas in SRAMs both the bit and $\overline{\text{bit}}$ lines are utilized. The DRAM cell stores its bit of information as charge on the cell capacitor C_S. When the cell is storing a 1, the capacitor is charged to V_{DD}; when a 0 is stored, the capacitor is discharged to zero volts.

Some explanation is needed to appreciate how the capacitor can be charged to the full supply voltage V_{DD}. Consider a write-1 operation. The word line is at V_{DD} and the bit line is at V_{DD} and the transistor is conducting, charging C_S. The transistor will cease conduction when the voltage on C_S reaches $(V_{DD} - V_t)$. This is the same problem we encountered with pass-transistor logic (PTL) in Section 14.2. The problem is overcome in DRAM design by boosting the word line to a voltage equal to $V_{DD} + V_t$. In this case the capacitor voltage for a stored 1 will be equal to the full V_{DD}. However, because of leakage effects, the capacitor charge will leak off, and hence the cell must be refreshed periodically. During **refresh**, the cell content is read and the data bit is rewritten, thus *restoring* the capacitor voltage to its proper value. Typically, the refresh operation must be performed every 5 ms to 10 ms.

Let us now consider the DRAM operation in more detail. As in the static RAM, the row decoder selects a particular row by raising the voltage of its word line. This causes all the

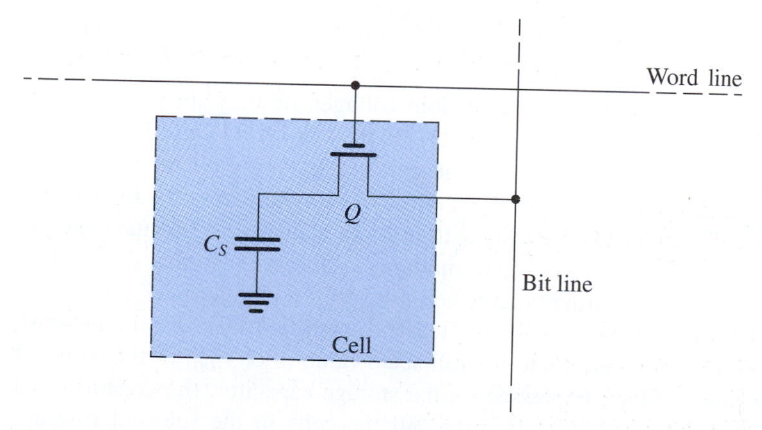

Figure 15.18 The one-transistor dynamic RAM (DRAM) cell.

[2]The name was originally used to distinguish this cell from earlier ones utilizing three transistors.

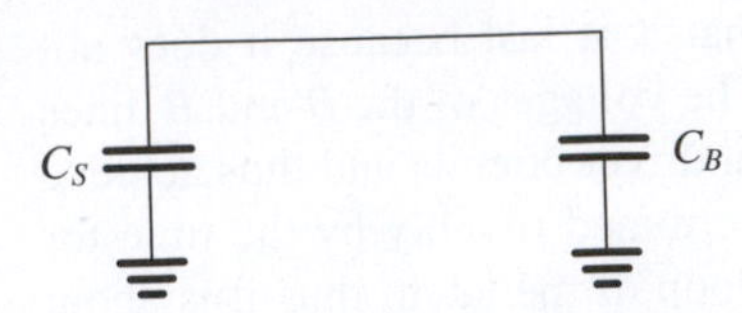

Figure 15.19 When the voltage of the selected word line is raised, the transistor conducts, thus connecting the storage capacitor C_S to the bit-line capacitance C_B.

access transistors in the selected row to become conductive, thereby connecting the storage capacitors of all the cells in the selected row to their respective bit lines. Thus the cell capacitor C_S is connected in parallel with the bit-line capacitance C_B, as indicated in Fig. 15.19. Here, it should be noted that C_S is typically 20 fF to 30 fF, whereas C_B is 10 times larger. Now, if the operation is a read, the bit line is precharged to $V_{DD}/2$. To find the change in the voltage on the bit line resulting from connecting a cell capacitor C_S to it, let the initial voltage on the cell capacitor be V_{CS} ($V_{CS} = V_{DD}$ when a 1 is stored, and $V_{CS} = 0$ V when a 0 is stored). Using charge conservation, we can write

$$C_S V_{CS} + C_B \frac{V_{DD}}{2} = (C_B + C_S)\left(\frac{V_{DD}}{2} + \Delta V\right)$$

from which we can obtain for ΔV

$$\Delta V = \frac{C_S}{C_B + C_S}\left(V_{CS} - \frac{V_{DD}}{2}\right) \tag{15.12}$$

and since $C_B \gg C_S$,

$$\Delta V \simeq \frac{C_S}{C_B}\left(V_{CS} - \frac{V_{DD}}{2}\right) \tag{15.13}$$

Now, if the cell is storing a 1, $V_{CS} = V_{DD}$, and

$$\Delta V(1) \simeq \frac{C_S}{C_B}\left(\frac{V_{DD}}{2}\right) \tag{15.14}$$

whereas if the cell is storing a 0, $V_{CS} = 0$, and

$$\Delta V(0) \simeq -\frac{C_S}{C_B}\left(\frac{V_{DD}}{2}\right) \tag{15.15}$$

Since usually C_B is much greater than C_S, these readout voltages are very small. For example, for $C_B = 10\ C_S$, $V_{DD} = 1.8$ V, $\Delta V(0)$ will be about –90 mV, and $\Delta V(1)$ will be +90 mV. This is a best-case scenario, for the 1 level in the cell might very well be below V_{DD}. Furthermore, in modern memory chips, V_{DD} is 1.2 V or even lower. In any case, we see that a stored 1 in the cell results in a small positive increment in the bit-line voltage, whereas a stored zero results in a small negative increment. Observe also that the readout process is *destructive,* since the resulting voltage across C_S will no longer be V_{DD} or 0.

The change of voltage on the bit line is detected and amplified by the column sense amplifier causing the bit line to be driven to the full scale value (0 or V_{DD}) of the detected signal. This amplified signal is then impressed on the storage capacitor, thus restoring its signal to the proper level (V_{DD} or 0). In this way, all the cells in the selected row are refreshed. Simultaneously, the signal at the output of the sense amplifier of the selected column is fed to the data-output line of the chip through the action of the column decoder.

The write operation proceeds similarly to the read operation, except that the data bit to be written, which is impressed on the data input line, is applied by the column decoder to the selected bit line. Thus, if the data bit to be written is a 1, the B-line voltage is raised to V_{DD} (i.e., C_B is charged to V_{DD}). When the access transistor of the particular cell is turned on, its capacitor C_S will be charged to V_{DD}; thus a 1 is written in the cell. Simultaneously, all the other cells in the selected row are simply refreshed.

Although the read and write operations result in automatic refreshing of all the cells in the selected row, provision must be made for the periodic refreshing of the entire memory, typically every 5 to 10 ms, as specified for the particular chip. The refresh operation is carried out in a *burst mode,* one row at a time. During refresh, the chip will not be available for read or write operations. This is not a serious matter, however, since the interval required to refresh the entire chip is typically less than 2% of the time between refresh cycles. In other words, the memory chip is available for normal operation more than 98% of the time.

EXERCISES

15.7 In a particular dynamic memory chip, $C_S = 30$ fF, $C_B = 0.3$ pF, and $V_{DD} = 1.2$ V. Find the output readout voltage for a stored 1 and a stored 0. Recall that in a read operation, the bit lines are precharged to $V_{DD}/2$.
Ans. 60 mV; –60 mV

15.8 A 64-Mbit DRAM chip fabricated in a 0.4-μm CMOS technology requires 2 μm² per cell. If the storage array is square, estimate its dimensions. Further, if the peripheral circuitry (e.g., sense amplifiers, decoders) adds about 30% to the chip area, estimate the dimensions of the resulting chip.
Ans. 11.6 mm × 11.6 mm; 13.2 mm × 13.2 mm

15.4 Sense Amplifiers and Address Decoders

Having studied the circuits commonly used to implement the storage cells in SRAMs and DRAMs, we now consider some of the other important circuit blocks in a memory chip. The design of these circuits, commonly referred to as the **memory peripheral circuits**, presents exciting challenges and opportunities to integrated-circuit designers: Improving the performance of peripheral circuits can result in denser and faster memory chips that dissipate less power.

15.4.1 The Sense Amplifier

Next to the storage cells, the sense amplifier is the most critical component in a memory chip. Sense amplifiers are essential to the proper operation of DRAMs, and their use in SRAMs results in speed and area improvements.

A variety of sense-amplifier designs are in use, some of which closely resemble the active-load MOS differential amplifier studied in Chapter 8. Here, we first describe a differential sense amplifier that employs positive feedback. Because the circuit is differential, it can be employed directly in SRAMs, where the SRAM cell utilizes both the B and $\overline{B}$ lines. On the other hand, the one-transistor DRAM circuit we studied in Section 15.3.2 is a single-ended circuit, utilizing one bit line only. The DRAM circuit, however, can be made to resemble a differential signal source through the use of the "dummy-cell" technique, which we shall discuss shortly. Therefore, we shall assume that the memory cell whose output is to be amplified develops a difference output voltage between the B and $\overline{B}$ lines. This signal, which can range from 30 mV to 500 mV depending on the memory type and cell design, will be applied to the input terminals of the sense amplifier. The sense amplifier in turn responds by providing a full-swing (0 to V_{DD}) signal at its output terminals. The particular amplifier circuit we shall discuss here has a rather unusual property: *Its output and input terminals are the same!*

A Sense Amplifier with Positive Feedback Figure 15.20 shows the sense amplifier together with some of the other column circuitry of a RAM chip. Note that the sense amplifier is nothing but the familiar latch formed by cross-coupling two CMOS inverters: One inverter

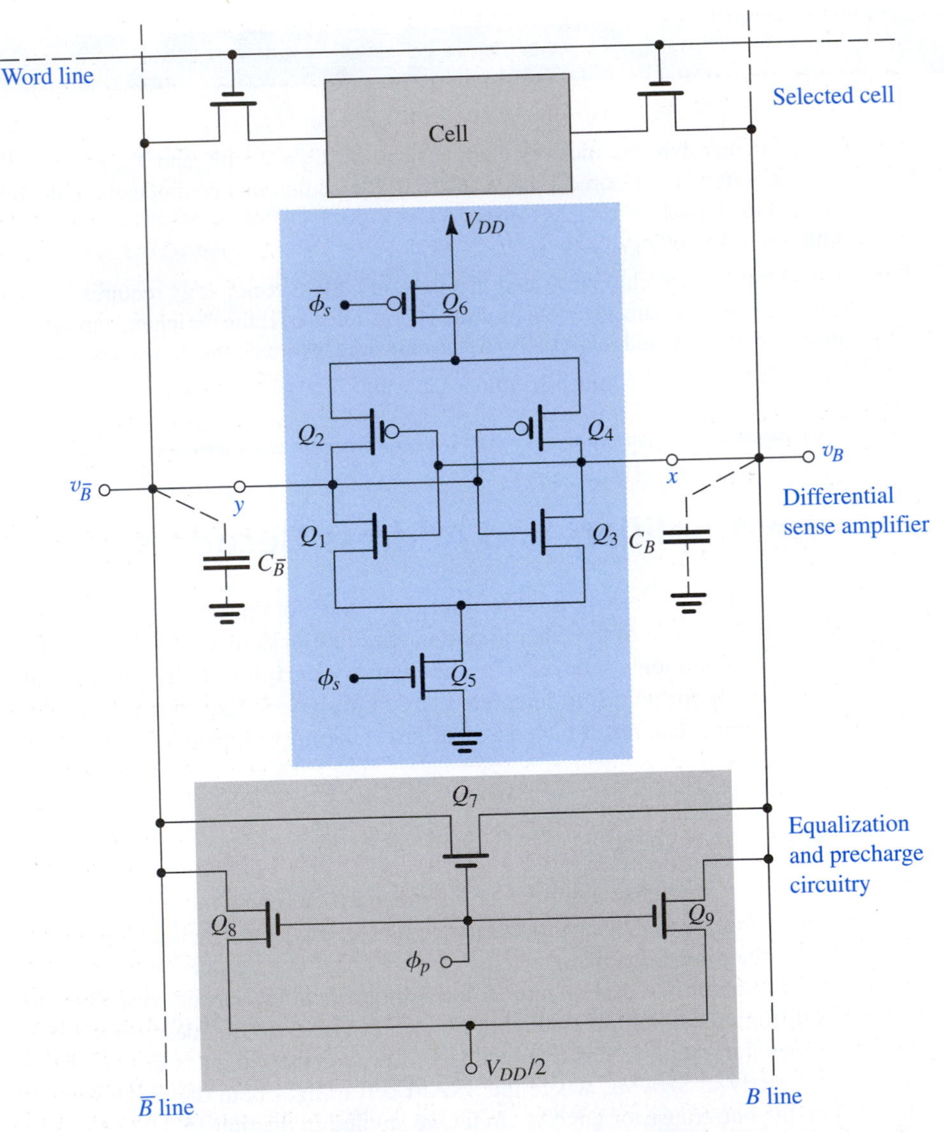

Figure 15.20 A differential sense amplifier connected to the bit lines of a particular column. This arrangement can be used directly for SRAMs (which utilize both the B and $\overline{B}$ lines). DRAMs can be turned into differential circuits by using the "dummy-cell" arrangement shown later (Fig. 15.22).

is implemented by transistors Q_1 and Q_2, and the other by transistors Q_3 and Q_4. Transistors Q_5 and Q_6 act as switches that connect the sense amplifier to ground and V_{DD} only when data-sensing action is required. Otherwise, ϕ_s is low and the sense amplifier is turned off. This conserves power, an important consideration because usually there is one sense amplifier per column, resulting in *thousands of sense amplifiers per chip*. Note, again, that terminals x and y are both the input and the output terminals of the amplifier. As indicated, these I/O terminals are connected to the B and $\overline{B}$ lines. The amplifier is required to detect a small signal appearing between B and $\overline{B}$, and to amplify it to provide a full-swing signal at B and $\overline{B}$. For instance, if during a read operation, the cell has a stored 1, then a small positive voltage will develop between B and $\overline{B}$, with v_B higher than $v_{\overline{B}}$. The amplifier will then cause v_B to rise to V_{DD} and $v_{\overline{B}}$ to fall to 0 V. This 1 output is then directed to the chip I/O pin by the column decoder (not shown) and at the same time is used to rewrite a 1 in the DRAM cell, thus performing the restore operation that is required because the DRAM readout process is destructive.

Figure 15.20 also shows the precharge and equalization circuit. Operation of this circuit is straightforward: When ϕ_p goes high (to V_{DD}) prior to a read operation, all three transistors conduct. While Q_8 and Q_9 precharge the $\overline{B}$ and B lines to $V_{DD}/2$, transistor Q_7 helps speed up this process by equalizing the initial voltages on the two lines. This equalization is critical to the proper operation of the sense amplifier. Any voltage difference present between B and $\overline{B}$ prior to commencement of the read operation can result in erroneous interpretation by the sense amplifier of its input signal. In Fig. 15.20, we show only one of the cells in this particular column, namely, the cell whose word line is activated. The cell can be either an SRAM or a DRAM cell. All other cells in this column will not be connected to the B and $\overline{B}$ lines (because their word lines will remain low).

Let us now consider the sequence of events during a read operation:

1. The precharge and equalization circuit is activated by raising the control signal ϕ_p. This will cause the B and $\overline{B}$ lines to be at equal voltages, equal to $V_{DD}/2$. The clock ϕ_p then goes low, and the B and $\overline{B}$ lines are left to float for a brief interval.
2. The word line goes up, connecting the cell to the B and $\overline{B}$ lines. A voltage then develops between B and $\overline{B}$, with v_B higher than $v_{\overline{B}}$ if the accessed cell is storing a 1, or v_B lower than $v_{\overline{B}}$ if the cell is storing a 0. To keep the cell area small, and to facilitate operation at higher speeds, the readout signal, which the cell is required to provide between B and $\overline{B}$, is kept small (typically, 30–500 mV).
3. Once an adequate difference voltage signal has been developed between B and $\overline{B}$ by the storage cell, the sense amplifier is turned on by connecting it to ground and V_{DD} through Q_5 and Q_6, activated by raising the sense-control signal ϕ_s. Because initially the input terminals of the inverters are at $V_{DD}/2$, the inverters will be operating in their transition region where the gain is high (Section 13.2). It follows that initially the latch will be operating at its unstable equilibrium point. Thus, depending on the signal between the input terminals, the latch will quickly move to one of its two stable equilibrium points (refer to the description of the latch operation in Section 15.1). This is achieved by the regenerative action, inherent in positive feedback. Figure 15.21 clearly illustrates this point by showing the waveforms of the signal on the bit line for both a read-1 and a read-0 operation. Observe that once activated, the sense amplifier causes the small initial difference, $\Delta V(1)$ or $\Delta V(0)$, provided by the cell, to grow exponentially to either V_{DD} (for a read-1 operation) or 0 (for a read-0 operation). The waveforms of the signal on the $\overline{B}$ line will be complementary to those shown in Fig. 15.21 for the B line. In the following, we quantify the process of exponential growth of v_B and $v_{\overline{B}}$.

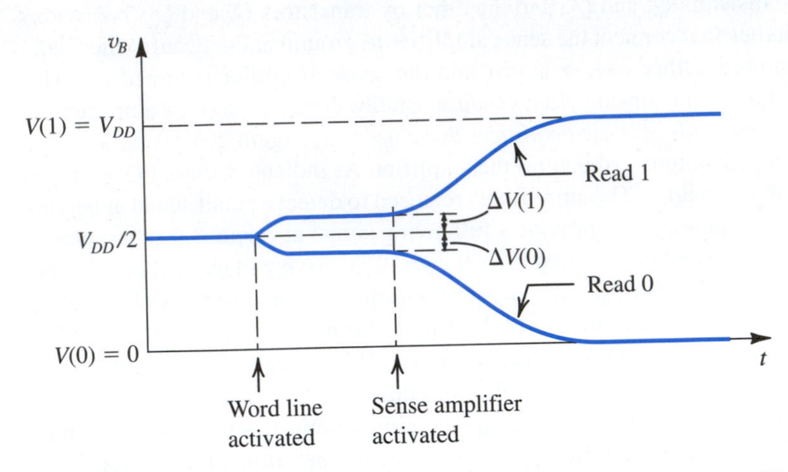

Figure 15.21 Waveforms of v_B before and after the activation of the sense amplifier. In a read-1 operation, the sense amplifier causes the initial small increment $\Delta V(1)$ to grow exponentially to V_{DD}. In a read-0 operation, the negative $\Delta V(0)$ grows to 0. Complementary signal waveforms develop on the $\overline{B}$ line.

A Closer Look at the Operation of the Sense Amplifier Developing a precise expression for the output signal of the sense amplifier shown in Fig. 15.20 is a rather complex task requiring the use of large-signal (and thus nonlinear) models of the inverter voltage transfer characteristic, as well as taking the positive feedback into account. We will not do this here; rather, we shall consider the operation in a semiquantitative way.

Recall that at the time the sense amplifier is activated, each of its two inverters is operating in the transition region near $V_{DD}/2$. Thus, for small-signal operation, each inverter can be modeled using g_{mn} and g_{mp}, the transconductances of Q_N and Q_P, respectively, evaluated at an input bias of $V_{DD}/2$. Specifically, a small-signal v_i superimposed on $V_{DD}/2$ at the input of one of the inverters gives rise to an inverter output current signal of $(g_{mn} + g_{mp})\, v_i \equiv G_m v_i$. This output current is delivered to one of the capacitors, C_B or $C_{\overline{B}}$. The voltage thus developed across the capacitor is then fed back to the other inverter and is multiplied by its G_m, which gives rise to an output current feeding the other capacitor, and so on, in a regenerative process. The positive feedback in this loop will mean that the signal around the loop, and thus v_B and $v_{\overline{B}}$, will *rise or decay exponentially* (see Fig. 15.21) with a time constant of (C_B/G_m) [or $(C_{\overline{B}}/G_m)$, since we have been assuming $C_B = C_{\overline{B}}$]. Thus, for example, in a read-1 operation we obtain

$$v_B = \frac{V_{DD}}{2} + \Delta V(1) e^{G_m/C_B t}, \qquad v_B \le V_{DD} \tag{15.16}$$

whereas in a read-0 operation,

$$v_B = \frac{V_{DD}}{2} - \Delta V(0) e^{(G_m/C_B)t} \tag{15.17}$$

Because these expressions have been derived assuming small-signal operation, they describe the exponential growth (decay) of v_B reasonably accurately only for values close to $V_{DD}/2$. Nevertheless, they can be used to obtain a reasonable estimate of the time required to develop a particular signal level on the bit line.

Example 15.2

Consider the sense-amplifier circuit of Fig. 15.20 during the reading of a 1. Assume that the storage cell provides a voltage increment on the B line of $\Delta V(1) = 0.1$ V. If the NMOS devices in the amplifiers have $(W/L)_n = 0.54\ \mu\text{m}/0.18\ \mu\text{m}$ and the PMOS devices have $(W/L)_p = 2.16\ \mu\text{m}/0.18\ \mu\text{m}$, and assuming that $V_{DD} = 1.8$ V, $V_{tn} = |V_{tp}| = 0.5$ V, and $\mu_n C_{ox} = 4\,\mu_p C_{ox} = 300\ \mu\text{A/V}^2$, find the time required for v_B to reach 0.9 V_{DD}. Assume $C_B = 1$ pF.

Solution

First, we determine the transconductances g_{mn} and g_{mp}

$$g_{mn} = \mu_n C_{ox}\left(\frac{W}{L}\right)_n (V_{GS} - V_t)$$

$$= 300 \times \frac{0.54}{0.18}(0.9 - 0.5)$$

$$= 0.36 \text{ mA/V}$$

$$g_{mp} = \mu_p C_{ox}\left(\frac{W}{L}\right)_p (V_{GS} - |V_t|)$$

$$= 75 \times \frac{2.16}{0.18}(0.9 - 0.5) = 0.36 \text{ mA/V}$$

Thus, the inverter G_m is

$$G_m = g_{mn} + g_{mp} = 0.72 \text{ mA/V}$$

and the time constant τ for the exponential growth of v_B will be

$$\tau \equiv \frac{C_B}{G_m} = \frac{1 \times 10^{-12}}{0.72 \times 10^{-3}} = 1.4 \text{ ns}$$

Now, the time, Δt, for v_B to reach 0.9 V_{DD} can be determined from

$$0.9 \times 1.8 = 0.9 + 0.1 e^{\Delta t/1.4}$$

resulting in

$$\Delta t = 2.8 \text{ ns}$$

Obtaining Differential Operation in Dynamic RAMs The sense amplifier described earlier responds to difference signals appearing between the bit lines. Thus, it is capable of rejecting interference signals that are common to both lines, such as those caused by capacitive coupling from the word lines. For this *common-mode rejection* to be effective, great care has to be taken to match both sides of the amplifier, taking into account the circuits that feed each side. This is an important consideration in any attempt to make the inherently single-ended output of the DRAM cell appear differential. We shall now discuss an ingenious scheme for accomplishing this task. Although the technique has been around for many years (see the first edition of this book, published in 1982), it is still in use today. The method is illustrated in Fig. 15.22.

Basically, each bit line is split into two identical halves. Each half-line is connected to half the cells in the column and to an additional cell, known as a *dummy cell,* having a storage capacitor $C_D = C_S$. When a word line on the left side is selected for reading, the dummy cell on the right side (controlled by $\bar{\phi}_D$) is also selected, and vice versa; that is, when a word line on

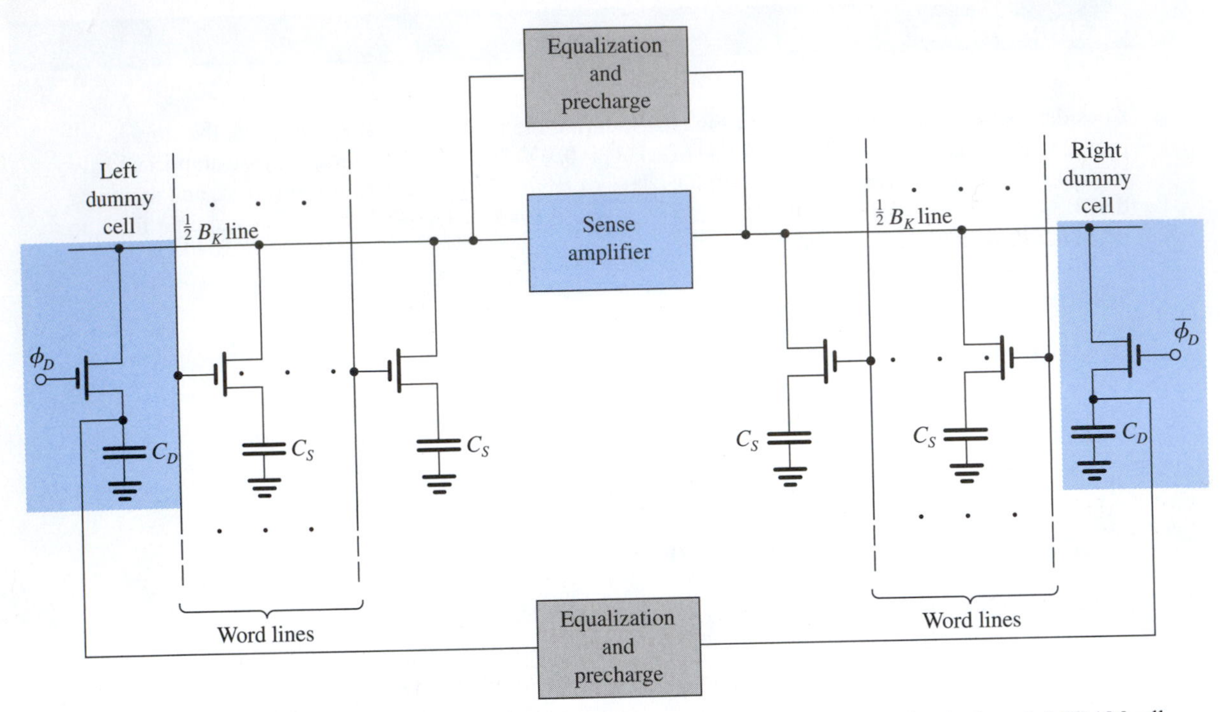

Figure 15.22 An arrangement for obtaining differential operation from the single-ended DRAM cell. Note the dummy cells at the far right and far left.

the right side is selected, the dummy cell on the left (controlled by ϕ_D) is also selected. In effect, then, the dummy cell serves as the other half of a differential DRAM cell. When the left-half bit line is in operation, the right-half bit line acts as its complement (or $\overline{B}$ line) and vice versa.

Operation of the circuit in Fig. 15.22 is as follows: The two halves of the line are precharged to $V_{DD}/2$ and their voltages are equalized. At the same time, the capacitors of the two dummy cells are precharged to $V_{DD}/2$. Then a word line is selected, and the dummy cell on the other side is enabled (with ϕ_D or $\overline{\phi}_D$ raised to V_{DD}). Thus the half-line connected to the selected cell will develop a voltage increment (around $V_{DD}/2$) of $\Delta V(1)$ or $\Delta V(0)$ depending on whether a 1 or a 0 is stored in the cell. Meanwhile, the other half of the line will have its voltage held equal to that of C_D (i.e., $V_{DD}/2$). The result is a differential signal of $\Delta V(1)$ or $\Delta V(0)$ that the sense amplifier detects and amplifies when it is enabled. As usual, by the end of the regenerative process, the amplifier will cause the voltage on one half of the line to become V_{DD} and that on the other half to become 0.

EXERCISES

15.9 It is required to reduce the time Δt of the sense-amplifier circuit in Example 15.2 by a factor of 2 by increasing g_m of the transistors (while retaining the matched design of each inverter). What must the W/L ratios of the n- and p-channel devices become?

Ans. $(W/L)_n = 6$; $(W/L)_p = 24$

15.10 If in the sense amplifier of Example 15.2, the signal available from the cell is only half as large (i.e., only 50 mV), what will Δt become?

Ans. 3.7 ns, an increase of 32%

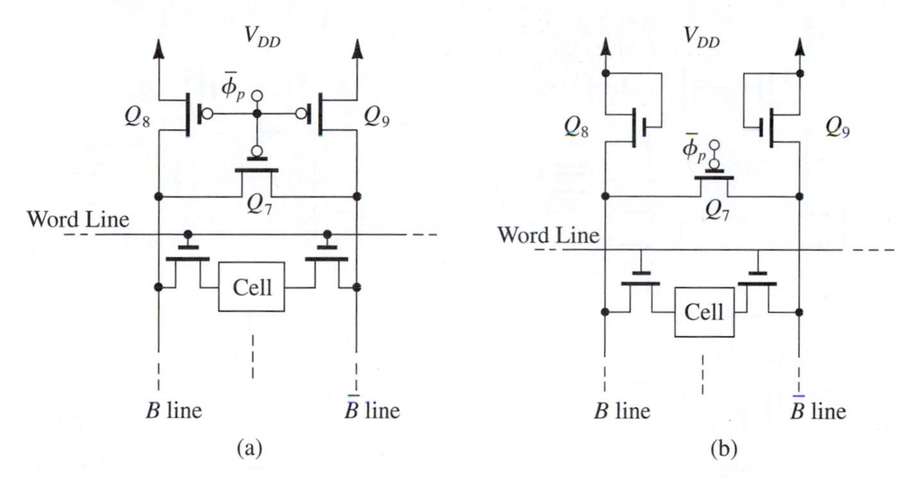

Figure 15.23 Two alternative arrangements for precharging the bit line: (**a**) The B and $\bar{B}$ lines are precharged to V_{DD}; (**b**) the B and $\bar{B}$ lines are charged to $(V_{DD} - V_t)$.

Alternative Precharging Arrangements If it is desired to precharge the B and $\bar{B}$ lines to V_{DD}, the arrangement in Fig. 15.23(a) can be utilized. Here precharging and equalization occur when $\bar{\phi}_P$ is low. Then, just prior to the activation of the word line, $\bar{\phi}_P$ goes high. Another precharging arrangement using diode-connected NMOS transistors is shown in Fig. 15.23(b). In this case, the B and $\bar{B}$ lines are charged to $(V_{DD} - V_t)$, and Q_7 equalizes their voltages.

An Alternative Sense Amplifier Another popular implementation of the sense amplifier is the differential MOS amplifier with a current-mirror load, studied in detail in Section 8.5. Here, we present a brief overview of the operation of this versatile circuit as a sense amplifier.

The amplifier circuit is shown in Fig. 15.24 fed from the bit and $\overline{\text{bit}}$ lines (voltages v_B and $v_{\bar{B}}$). Transistors Q_1 and Q_2 are connected in the differential-pair configuration and are biased by a constant current I supplied by current source Q_5. Transistors Q_3 and Q_4 form a current mirror, which acts as the load circuit for the amplifying transistors Q_1 and Q_2. The differential nature of the amplifier aids significantly in its effectiveness as a sense amplifier: It rejects noise or interference signals that are coupled equally to the B and $\bar{B}$ lines, and amplifies only the small difference signals that appear between B and $\bar{B}$ as a result of the read operation of a cell connected to the B and $\bar{B}$ lines.

The amplifier is designed so that in normal small-signal operation, all transistors operate in the saturation region. Figure 15.24(b) shows the amplifier in its equilibrium state with $v_B = v_{\bar{B}} = V_{DD} - V_t$. Note that we have assumed that the B and $\bar{B}$ lines are precharged to $(V_{DD} - V_t)$ using the circuit in Fig. 15.23(b). It turns out that this voltage is particularly convenient for the operation of this amplifier type as a sense amplifier. As indicated in Fig. 15.24(b), the bias current I divides equally between Q_1 and Q_2; thus each conducts a current $I/2$. The current of Q_1 is fed to the input side of the current mirror, transistor Q_3; thus the mirror provides an equal output current $I/2$ in the drain of Q_4. At the output node, we see that we have two equal and opposite currents, leaving a zero current to flow into the load capacitor. Thus, in an ideal situation of perfect matching, v_O will be equal to the voltage at the drain of Q_1.

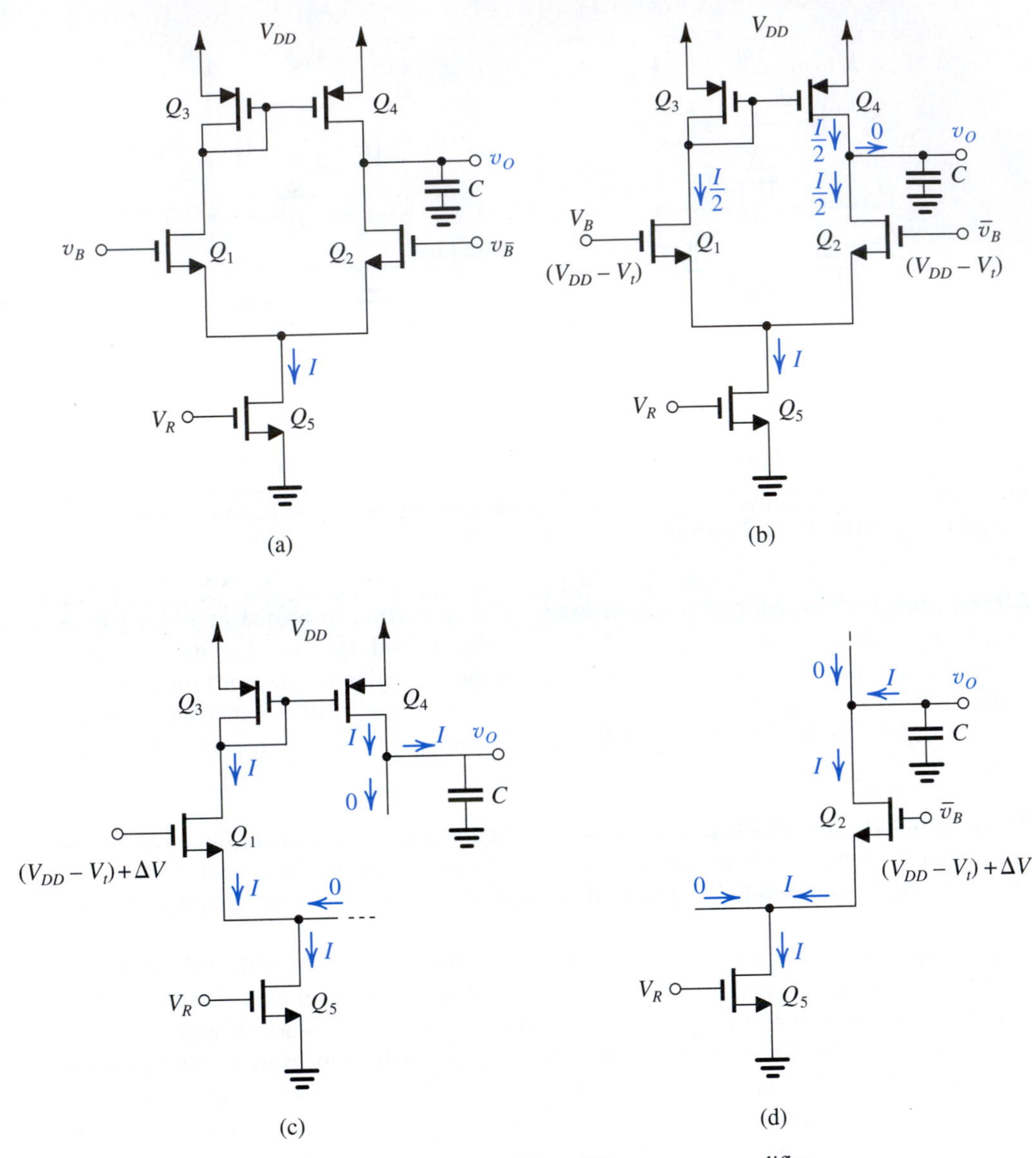

Figure 15.24 The active-loaded MOS differential amplifier as a sense amplifier.

Next consider the situation when the B line shows an incremental voltage ΔV above the voltage of the $\bar{B}$ line. As shown in Fig. 15.24(c), if ΔV is sufficiently large, Q_2 will turn off and all the bias current I will flow through Q_1 and on to Q_3. Thus the mirror output current becomes I and flows through the amplifier output terminal to the equivalent output capacitance C. Thus C will charge to V_{DD} in time Δt,

$$\Delta t = \frac{CV_{DD}}{I} \tag{15.18}$$

The complementary situation when $v_{\bar{B}}$ exceeds v_B by ΔV is illustrated in Fig. 15.24(d). Here Q_1, Q_3, and Q_4 are turned off, and Q_2 conducts all the current I. Thus capacitor C is discharged to ground by a constant current I.

An important question to answer before leaving this amplifier circuit is how large is ΔV that causes the current I to switch from one side of the differential pair to the other? The answer is given in Section 8.5 (see Fig. 8.32), namely,

$$\Delta V = \sqrt{2}V_{OV} \tag{15.19}$$

where V_{OV} is the overdrive voltage at which Q_1 and Q_2 are operating in equilibrium, that is,

$$\frac{I}{2} = \frac{1}{2}(\mu_n C_{ox})\left(\frac{W}{L}\right)_{1,2} V_{OV}^2 \tag{15.20}$$

Finally, we note that this sense amplifier dissipates static power given by

$$P = V_{DD}I$$

Observe that increasing I reduces the time Δt in Eq. (15.18) at the expense of increased power dissipation.

EXERCISE

D15.11 It is required to design the sense amplifier in Fig. 15.24 to detect an input signal $\Delta V = 100$ mV and to provide a full output in 0.5 ns. If $C = 50$ fF and $V_{DD} = 1.8$ V, find the required current I and the power dissipation.
Ans. 180 μA; 324 μW

15.4.2 The Row-Address Decoder

As described in Section 15.2, the row-address decoder is required to select one of the 2^M word lines in response to an M-bit address input. As an example, consider the case $M = 3$ and denote the three address bits A_0, A_1, and A_2, and the eight word lines $W_0, W_1, \ldots, W_7$. Conventionally, word line W_0 will be high when $A_0 = 0$, $A_1 = 0$, and $A_2 = 0$; thus we can express W_0 as a Boolean function of A_0, A_1, and A_2,

$$W_0 = \bar{A}_0\bar{A}_1\bar{A}_2 = \overline{A_0 + A_1 + A_2}$$

Thus the selection of W_0 can be accomplished by a three-input NOR gate whose three inputs are connected to A_0, A_1, and A_2 and whose output is connected to word line 0. Word line W_3 will be high when $A_0 = 1$, $A_1 = 1$, and $A_2 = 0$; thus,

$$W_3 = A_0A_1\bar{A}_2 = \overline{\bar{A}_0 + \bar{A}_1 + A_2}$$

Thus the selection of W_3 can be realized by a three-input NOR gate whose three inputs are connected to $\bar{A}_0$, $\bar{A}_1$, and A_2, and whose output is connected to word line 3. We can thus see that this address decoder can be realized by eight three-input NOR gates. Each NOR gate is fed with the appropriate combination of address bits and their complements, corresponding to the word line to which its output is connected.

A simple approach to realizing these NOR functions is provided by the matrix structure shown in Fig. 15.25. The circuit shown is a dynamic one (Section 14.3). Attached to each

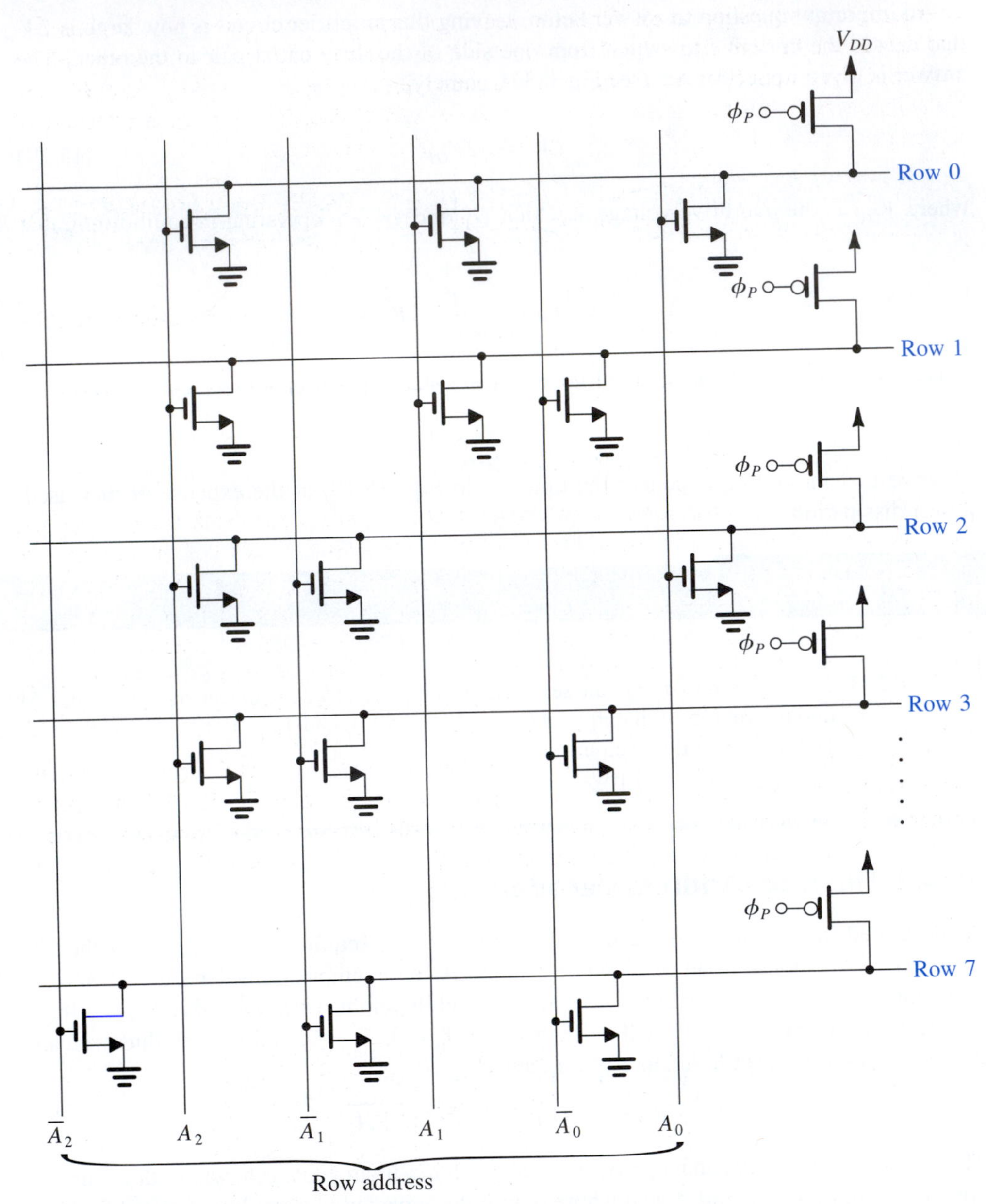

Figure 15.25 A NOR address decoder in array form. One out of eight lines (row lines) is selected using a 3-bit address.

row line is a p-channel device that is activated, prior to the decoding process, using the precharge control signal ϕ_P. During precharge (ϕ_P low), all the word lines are pulled high to V_{DD}. It is assumed that at this time the address input bits have not yet been applied and all the inputs are low; hence there is no need for the circuit to include the evaluation transistor utilized in dynamic logic gates. Then, the decoding operation begins when the address bits and their complements are applied. Observe that the NMOS transistors are placed so that the word lines not selected will be discharged. For any input combination, only one word line will not be discharged, and thus its voltage remains high at V_{DD}. For instance, row 0 will be high only when $A_0 = 0$, $A_1 = 0$, and $A_2 = 0$; this is the only combination that will result in all three transistors connected to row 0 being cut off. Similarly, row 3 has transistors connected to $\overline{A}_0$, $\overline{A}_1$, and A_2, and thus it will be high when $A_0 = 1$, $A_1 = 1$, $A_2 = 0$, and so on. After the decoder outputs have stabilized, the output lines are connected to the word lines of the array, usually via clock-controlled transmission gates. This decoder is known as a NOR decoder. Observe that because of the precharge operation, the decoder circuit does not dissipate static power.

EXERCISE

15.12 How many transistors are needed for a NOR row decoder with an M-bit address?
Ans. $M2^M$ NMOS + 2^M PMOS = $2^M(M + 1)$

15.4.3 The Column-Address Decoder

From the description in Section 15.2, the function of the column-address decoder is to connect one of the 2^N bit lines to the data I/O line of the chip. As such, it is a multiplexer and can be implemented using pass-transistor logic (Section 14.2) as shown in Fig. 15.26. Here, each bit line is connected to the data I/O line through an NMOS transistor. The gates of the pass transistors are controlled by 2^N lines, one of which is selected by a NOR decoder similar to that used for decoding the row address. Finally, note that better performance can be obtained by utilizing transmission gates in place of NMOS transistors (Section 14.2). In such a case, however, the decoder needs to provide complementary output signals.

An alternative implementation of the column decoder that uses a smaller number of transistors (but at the expense of slower speed of operation) is shown in Fig. 15.27. This circuit, known as a *tree decoder,* has a simple structure of pass transistors. Unfortunately, since a relatively large number of transistors can exist in the signal path, the resistance of the bit lines increases, and the speed decreases correspondingly.

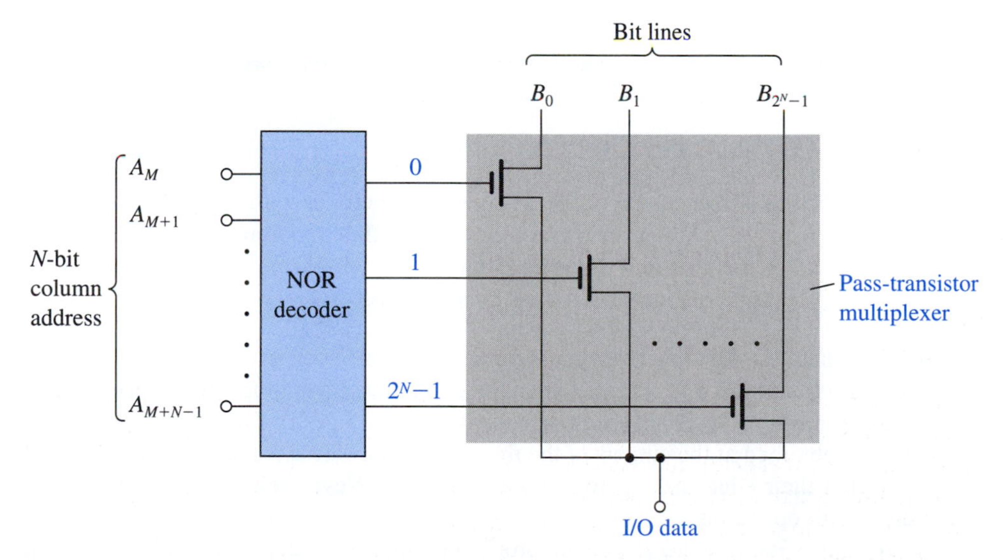

Figure 15.26 A column decoder realized by a combination of a NOR decoder and a pass-transistor multiplexer.

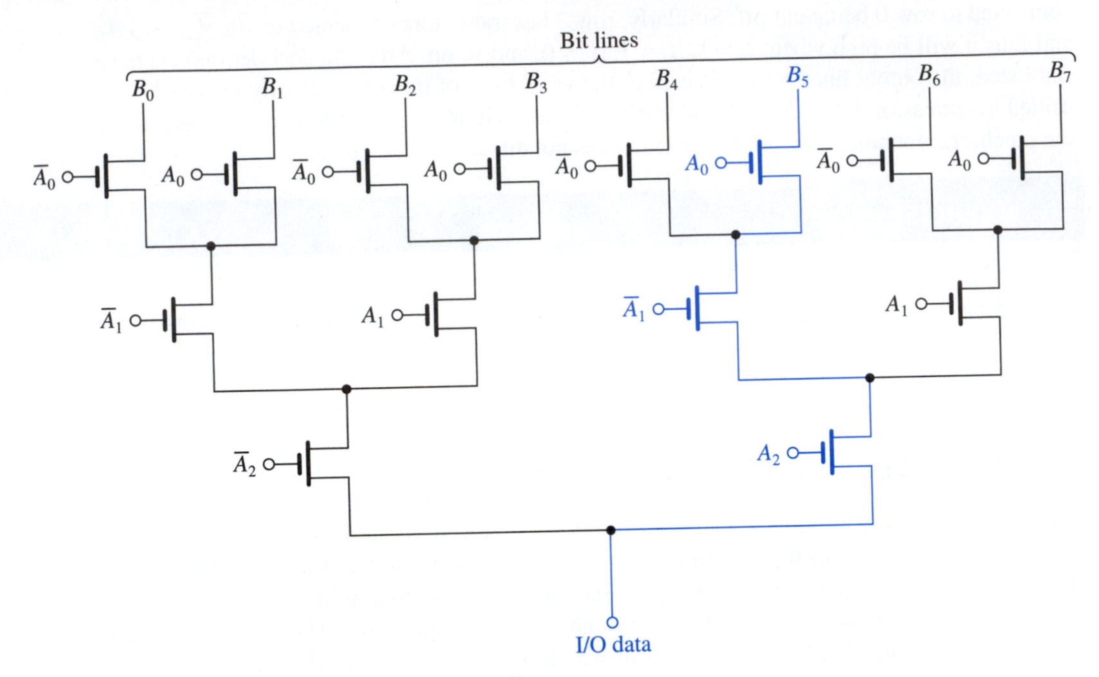

Figure 15.27 A tree column decoder. Note that the colored path shows the transistors that are conducting when $A_0 = 1$, $A_1 = 0$, and $A_2 = 1$, the address that results in connecting B_5 to the data line.

EXERCISE

15.13 How many transistors are needed for a tree decoder when there are 2^N bit lines?
Ans. $2(2^N - 1)$

15.4.4 Pulse-Generation Circuits

Memory chips require a large number of pulse signals, sometimes with intricate timing relationships among them. It is not our purpose here to study this important subject; rather, we present two simple circuits that find widespread applicability in memory-chip timing as well as in other digital-system components, such as microprocessors.

The Ring Oscillator The ring oscillator is formed by connecting an *odd* number of inverters in a loop. Although usually at least five inverters are used, we illustrate the principle of operation using a ring of three inverters, as shown in Fig. 15.28(a). Figure 15.28(b) shows the waveforms obtained at the outputs of the three inverters. These waveforms are idealized in the sense that their edges have zero rise and fall times. Nevertheless, they will serve to explain the circuit operation.

Observe that a rising edge at node 1 propagates through gates 1, 2, and 3 to return inverted after a delay of $3t_P$. This falling edge then propagates, and returns with the original (rising) polarity after another $3t_P$ interval. It follows that the circuit oscillates with a period

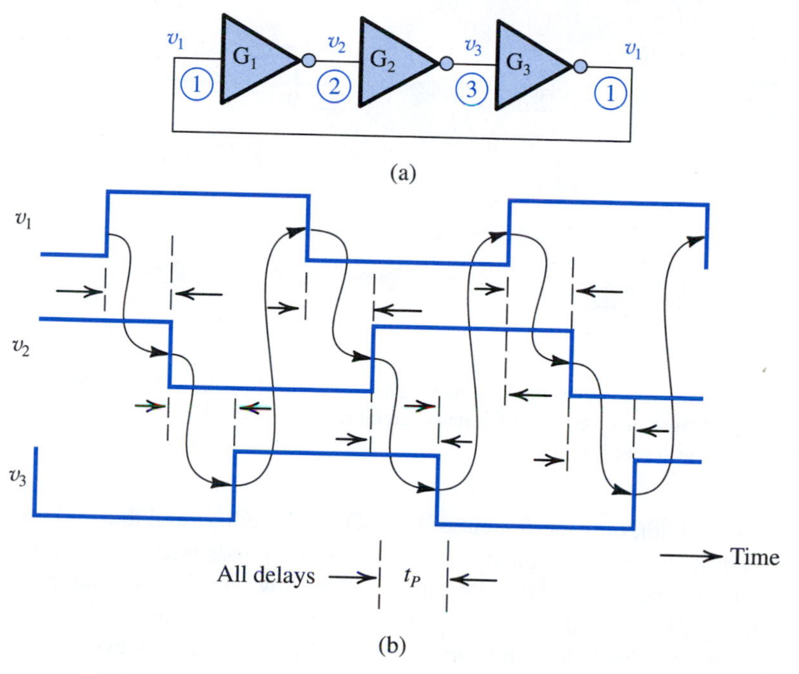

Figure 15.28 **(a)** A ring oscillator formed by connecting three inverters in cascade. (Normally at least five inverters are used.) **(b)** The resulting waveform. Observe that the circuit oscillates with frequency $1/6t_P$.

of $6t_p$ or correspondingly with frequency $1/6t_P$. In general, a ring with N inverters (where N must be odd) will oscillate with a period of $2Nt_p$ and frequency $1/2Nt_P$.

As a final remark, we note that the ring oscillator provides a relatively simple means for measuring the inverter propagation delay.

EXERCISE

15.14 Find the frequency of oscillation of a ring of five inverters if the inverter propagation delay is specified to be 1 ns.
Ans. 100 MHz

A One-Shot or Monostable Multivibrator Circuit The one-shot or monostable multivibrator circuit provides, when triggered, a single output pulse with a predetermined width.[3] A variety of circuits exist for implementing the one-shot function, and some using op amps will be studied in Section 17.6. Here, in Fig. 15.29(a), we show a circuit commonly used in digital IC design. The circuit utilizes an exclusive-OR (XOR) gate together with a delay circuit. Recalling that the XOR gate provides a high output only when its two inputs are dissimilar, we see that prior to the arrival of the input positive step, the output will be

[3]The name "monostable" arises because this class of circuits has one stable state, which is the quiescent state. When a trigger is applied, the circuit moves to its quasi-stable state and stays in it for a predetermined length of time (the width of the output pulse). It then switches back automatically to the stable state.

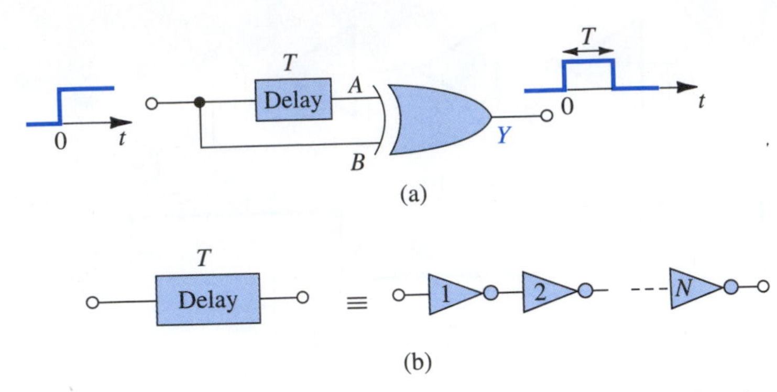

Figure 15.29 (**a**) A one-shot or monostable circuit. Utilizing a delay circuit with a delay T and an XOR gate, this circuit provides an output pulse of width T. (**b**) The delay circuit can be implemented as the cascade of N inverters where N is even, in which case $T = Nt_P$.

low. When the input goes high, only the B input of the XOR will be high and thus its output will go high. The high input will reach input A of the XOR T seconds later, at which time both inputs of the XOR will be high and thus its output will go low. We thus see that the circuit produces an output pulse with a duration T equal to the delay of the delay block for each transition of the input signal. The latter can be implemented by connecting an even number of inverters in cascade as shown in Fig. 15.29(b).

15.5 Read-only Memory (ROM)

As mentioned in Section 15.2, read-only memory (ROM) is memory that contains fixed data patterns. It is used in a variety of digital-system applications. Currently, a very popular application is the use of ROM in microprocessor systems to store the instructions of the system's basic operating program. ROM is particularly suited for such an application because it is nonvolatile; that is, it retains its contents when the power supply is switched off.

A ROM can be viewed as a combinational logic circuit for which the input is the collection of address bits of the ROM and the output is the set of data bits retrieved from the addressed location. This viewpoint leads to the application of ROMs in code conversion—that is, in changing the code of the signal from one system (say, binary) to another. Code conversion is employed, for instance, in secure communication systems, where the process is known as *scrambling*. It consists of feeding the code of the data to be transmitted to a ROM that provides corresponding bits in a (supposedly) secret code. The reverse process, which also uses a ROM, is applied at the receiving end.

In this section we will study various types of read-only memory. These include fixed ROM, which we refer to simply as ROM, programmable ROM (PROM), and erasable programmable ROM (EPROM).

15.5.1 A MOS ROM

Figure 15.30 shows a simplified 32-bit (or 8-word × 4-bit) MOS ROM. As indicated, the memory consists of an array of n-channel MOSFETs whose gates are connected to the word lines, whose sources are grounded, and whose drains are connected to the bit lines. Each bit line is connected to the power supply via a PMOS load transistor, in the manner of pseudo-NMOS logic

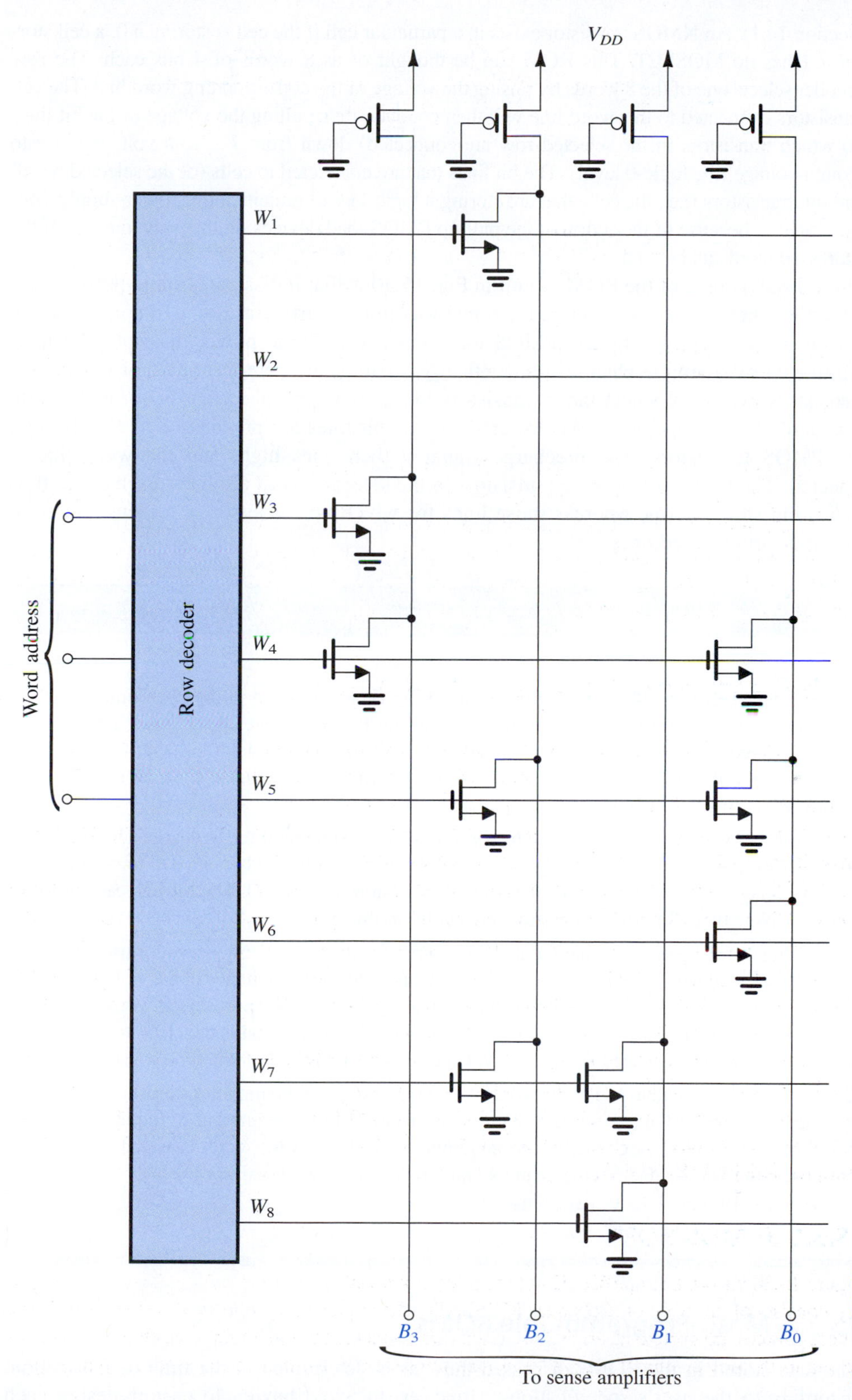

Figure 15.30 A simple MOS ROM organized as 8 words × 4 bits.

(Section 14.1). An NMOS transistor exists in a particular cell if the cell is storing a 0; a cell storing a 1 has no MOSFET. This ROM can be thought of as 8 words of 4 bits each. The row decoder selects one of the 8 words by raising the voltage of the corresponding word line. The cell transistors connected to this word line will then conduct, thus pulling the voltage of the bit lines (to which transistors in the selected row are connected) down from V_{DD} to a voltage close to ground voltage (the logic-0 level). The bit lines that are connected to cells (of the selected word) without transistors (i.e., the cells that are storing a logic 1) will remain at the power-supply voltage (logic 1) because of the action of the pull-up PMOS load devices. In this way, the bits of the addressed word can be read.

A disadvantage of the ROM circuit in Fig. 15.30 is that it dissipates static power. Specifically, when a word is selected, the transistors in this particular row will conduct static current that is supplied by the PMOS load transistors. Static power dissipation can be eliminated by a simple change. Rather than grounding the gate terminals of the PMOS transistors, we can connect these transistors to a precharge line ϕ that is normally high. Just before a read operation, ϕ is lowered and the bit lines are precharged to V_{DD} through the PMOS transistors. The precharge signal ϕ then goes high, and the word line is selected. The bit lines that have transistors in the selected word are then discharged, thus indicating stored zeros, whereas those lines for which no transistor is present remain at V_{DD}, indicating stored ones.

EXERCISE

15.15 The purpose of this exercise is to estimate the various delay times involved in the operation of a ROM. Consider the ROM in Fig. 15.30 with the gates of the PMOS devices disconnected from ground and connected to a precharge control signal ϕ. Let all the NMOS devices have W/L = 6 μm/2 μm and all the PMOS devices have W/L = 24 μm/2 μm. Assume that $\mu_n C_{ox} = 50$ μA/V², $\mu_p C_{ox} = 20$ μA/V², $V_{tn} = -V_{tp} = 1$ V, and $V_{DD} = 5$ V.

(a) During the precharge interval, ϕ is lowered to 0 V. Estimate the time required to charge a bit line from 0 V to 5 V. Use, as an average charging current, the current supplied by a PMOS transistor at a bit-line voltage halfway through the 0-V to 5-V excursion (i.e., 2.5 V). The bit-line capacitance is 2 pF. Note that all NMOS transistors are cut off at this time.

(b) After completion of the precharge interval and the return of ϕ to V_{DD}, the row decoder raises the voltage of the selected word line. Because of the finite resistance and capacitance of the word line, the voltage rises exponentially toward V_{DD}. If the resistance of each of the polysilicon word lines is 3 kΩ and the capacitance between the word line and ground is 3 pF, what is the (10% to 90%) rise time of the word-line voltage? What is the voltage reached at the end of one time constant?

(c) We account for the exponential rise of the word-line voltage by approximating the word-line voltage by a step equal to the voltage reached in one time constant. Find the interval Δt required for an NMOS transistor to discharge the bit line and lower its voltage by 0.5 V. (It is assumed that the sense amplifier needs a 0.5-V change at its input to detect a low bit value.)

Ans. (a) 6.1 ns; (b) 19.8 ns, 3.16 V; (c) 2.9 ns

15.5.2 Mask-Programmable ROMs

The data stored in the ROMs discussed thus far is determined at the time of fabrication, according to the user's specifications. However, to avoid having to custom-design each ROM from scratch (which would be extremely costly), ROMs are manufactured using a process known as **mask programming**. As explained in Appendix A, integrated circuits are fabricated on a wafer of silicon using a sequence of processing steps that include photomasking, etching, and diffusion. In this way, a pattern of junctions and interconnections is created on the surface of the wafer. One of the final steps in the fabrication process consists of coating the surface of the wafer with a layer of aluminum and then selectively (using a mask) etching away portions of the aluminum, leaving aluminum only where interconnections are desired. This last step can be used to program (i.e., to store a desired pattern in) a ROM. For instance, if the ROM is made of MOS transistors as in Fig. 15.30, MOSFETs can be included at all bit locations, but only the gates of those transistors where 0s are to be stored are connected to the word lines; the gates of transistors where 1s are to be stored are not connected. This pattern is determined by the mask, which is produced according to the user's specifications.

The economic advantages of the mask programming process should be obvious: All ROMs are fabricated similarly; customization occurs only during one of the final steps in fabrication.

15.5.3 Programmable ROMs (PROMs and EPROMs)

PROMs are ROMs that can be programmed by the user, but only once. A typical arrangement employed in BJT PROMs involves using polysilicon fuses to connect the emitter of each BJT to the corresponding digit line. Depending on the desired content of a ROM cell, the fuse can be either left intact or blown out using a large current. The programming process is obviously irreversible.

An erasable programmable ROM, or EPROM, is a ROM that can be erased and reprogrammed as many times as the user wishes. It is therefore the most versatile type of read-only memory. It should be noted, however, that the process of erasure and reprogramming is time-consuming and is intended to be performed only infrequently.

State-of-the-art EPROMs use variants of the memory cell whose cross section is shown in Fig. 15.31(a). The cell is basically an enhancement-type n-channel MOSFET with two gates made of polysilicon material.[4] One of the gates is not electrically connected to any

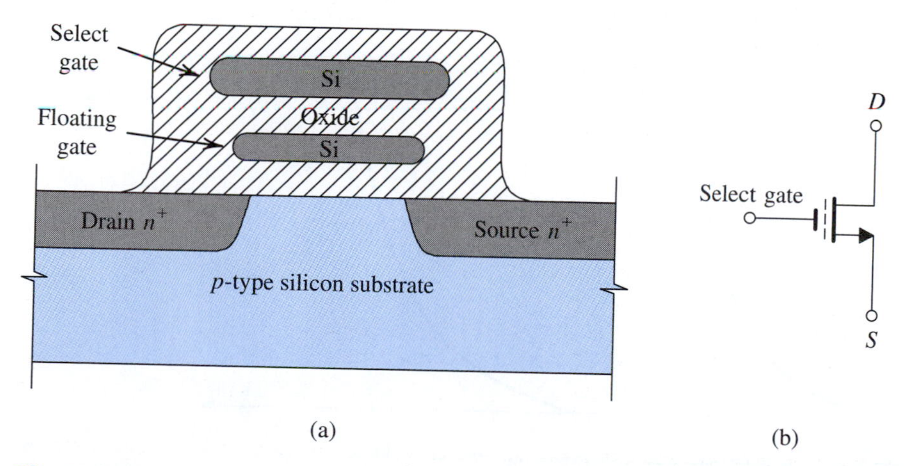

Figure 15.31 **(a)** Cross section and **(b)** circuit symbol of the floating-gate transistor used as an EPROM cell.

[4]See Appendix A for a description of silicon-gate technology.

other part of the circuit; rather, it is left floating and is appropriately called a **floating gate**. The other gate, called a **select gate**, functions in the same manner as the gate of a regular enhancement MOSFET.

The MOS transistor of Fig. 15.31(a) is known as a **floating-gate transistor** and is given the circuit symbol shown in Fig. 15.31(b). In this symbol the broken line denotes the floating gate. The memory cell is known as the **stacked-gate cell**.

Let us now examine the operation of the floating-gate transistor. Before the cell is programmed (we will shortly explain what this means), no charge exists on the floating gate and the device operates as a regular *n*-channel enhancement MOSFET. It thus exhibits the i_D–v_{GS} characteristic shown as curve (*a*) in Fig. 15.32. Note that in this case the threshold voltage (V_t) is rather low. This state of the transistor is known as the **not-programmed state**. It is one of two states in which the floating-gate transistor can exist. Let us arbitrarily take the not-programmed state to represent a stored 1. That is, a floating-gate transistor whose i_D–v_{GS} characteristic is that shown as curve (*a*) in Fig. 15.32 will be said to be storing a 1.

To program the floating-gate transistor, a large voltage (16–20 V) is applied between its drain and source. Simultaneously, a large voltage (about 25 V) is applied to its select gate. Figure 15.33 shows the floating-gate MOSFET during programming. In the absence of any charge on the floating gate, the device behaves as a regular *n*-channel enhancement MOSFET: An *n*-type inversion layer (channel) is created at the wafer surface as a result of the large positive voltage applied to the select gate. Because of the large positive voltage at the drain, the channel has a tapered shape.

The drain-to-source voltage accelerates electrons through the channel. As these electrons reach the drain end of the channel, they acquire large kinetic energy and are referred to as *hot electrons*. The large positive voltage on the select gate (greater than the drain voltage) establishes an electric field in the insulating oxide. This electric field attracts the hot electrons and accelerates them (through the oxide) toward the floating gate. In this way the floating gate is charged, and the charge that accumulates on it becomes trapped.

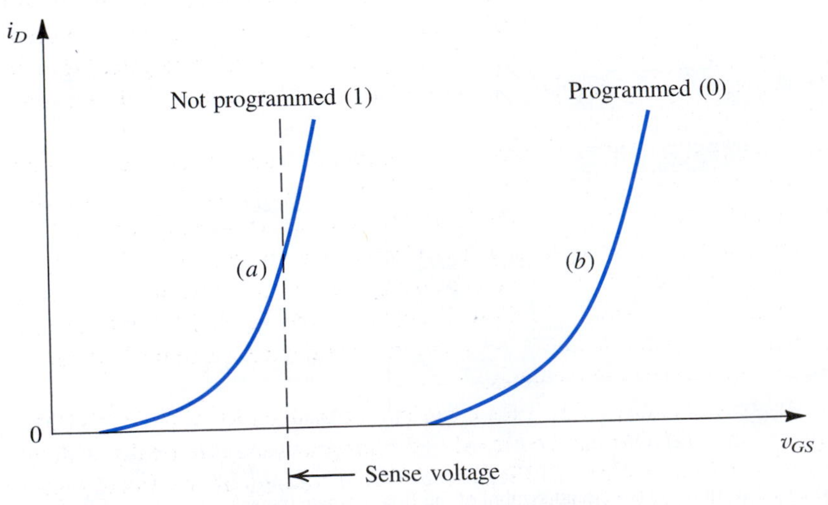

Figure 15.32 Illustrating the shift in the i_D–v_{GS} characteristic of a floating-gate transistor as a result of programming.

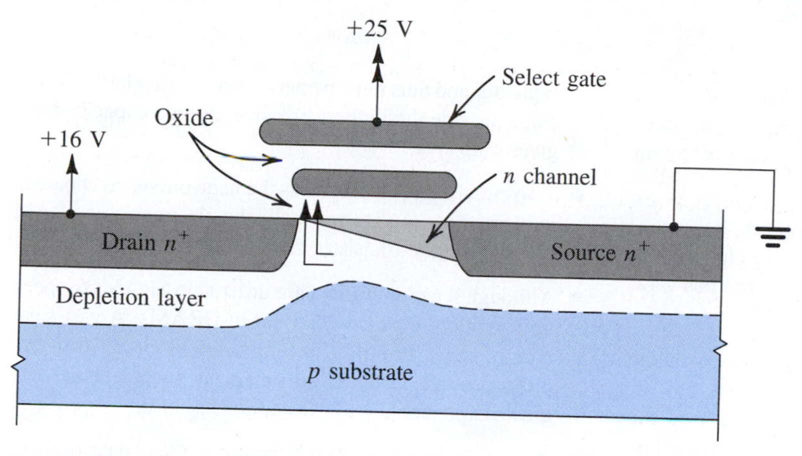

Figure 15.33 The floating-gate transistor during programming.

Fortunately, the process of charging the floating gate is self-limiting. The negative charge that accumulates on the floating gate reduces the strength of the electric field in the oxide to the point that it eventually becomes incapable of accelerating any more of the hot electrons.

Let us now inquire about the effect of the floating gate's negative charge on the operation of the transistor. The negative charge trapped on the floating gate will cause electrons to be repelled from the surface of the substrate. This implies that to form a channel, the positive voltage that has to be applied to the select gate will have to be greater than that required when the floating gate is not charged. In other words, the threshold voltage V_t of the programmed transistor will be higher than that of the not-programmed device. In fact, programming causes the i_D–v_{GS} characteristic to shift to the curve labeled (*b*) in Fig. 15.32. In this state, known as the *programmed state,* the cell is said to be storing a 0.

Once programmed, the floating-gate device retains its shifted *i*–*v* characteristic (curve *b*) even when the power supply is turned off. In fact, extrapolated experimental results indicate that the device can remain in the programmed state for as long as 100 years!

Reading the content of the stacked-gate cell is easy: A voltage V_{GS} somewhere between the low and high threshold values (see Fig. 15.32) is applied to the selected gate. While a programmed device (one that is storing a 0) will not conduct, a not-programmed device (one that is storing a 1) will conduct heavily.

To return the floating-gate MOSFET to its not-programmed state, the charge stored on the floating gate has to be returned to the substrate. This *erasure* process can be accomplished by illuminating the cell with ultraviolet light of the correct wavelength (2537 Å) for a specified duration. The ultraviolet light imparts sufficient photon energy to the trapped electrons to allow them to overcome the inherent energy barrier, and thus be transported through the oxide, back to the substrate. To allow this erasure process, the EPROM package contains a quartz window. Finally, it should be noted that the device is extremely durable, and can be erased and programmed many times.

A more versatile programmable ROM is the electrically erasable PROM (or EEPROM). As the name implies, an EEPROM can be erased and reprogrammed electrically without the need for ultraviolet illumination. EEPROMs utilize a variant of the floating-gate MOSFET. An important class of EEPROMs using a floating gate variant and implementing block erasure are referred to as **flash memories**.

Summary

- Flip-flops employ one or more latches. The basic static latch is a bistable circuit implemented using two inverters connected in a positive-feedback loop. The latch can remain in either stable state indefinitely.
- As an alternative to the positive-feedback approach, memory can be provided through the use of charge storage. A number of CMOS flip-flops are realized this way, including some master–slave D flip-flops.
- A random-access memory (RAM) is one in which the time required for storing (writing) information and for retrieving (reading) information is independent of the physical location (within the memory) in which the information is stored.
- The major part of a memory chip consists of the cells in which the bits are stored and that are typically organized in a square matrix. A cell is selected for reading or writing by activating its row, via the row-address decoder, and its column, via the column-address decoder. The sense amplifier detects the content of the selected cell and provides a full-swing version of it to the data-output terminal of the chip.
- There are two kinds of MOS RAM: static and dynamic. Static RAMs (SRAMs) employ flip-flops as the storage cells. In a dynamic RAM (DRAM), data is stored on a capacitor and thus must be periodically refreshed. DRAM chips provide the highest possible storage capacity for a given chip area.
- Two circuits have emerged as the near-universal choice in implementing the storage cell: the six-transistor SRAM cell and the one-transistor DRAM cell.
- Although sense amplifiers are utilized in SRAMs to speed up operation, they are essential in DRAMs. A particular type of sense amplifier is a differential circuit that employs positive feedback to obtain an output signal that grows exponentially toward either V_{DD} or 0.
- Read-only memory (ROM) contains fixed data patterns that are stored at the time of fabrication and cannot be changed by the user. On the other hand, the contents of an erasable programmable ROM (EPROM) can be changed by the user. The erasure and reprogramming is a time-consuming process and is performed only infrequently.
- Some EPROMS utilize floating-gate MOSFETs as the storage cells. The cell is programmed by applying (to the selected gate) a high voltage, which in effect changes the threshold voltage of the MOSFET. Erasure is achieved by illuminating the chip by ultraviolet light. Even more versatile, EEPROMs can be erased and reprogrammed electrically.

Problems involving design are marked with D throughout the text. As well, problems are marked with asterisks to describe their degree of difficulty. Difficult problems are marked with an asterisk (*); more difficult problems with two asterisks (**); and very challenging and/or time-consuming problems with three asterisks (***).

Section 15.1: Latches and Flip-Flops

D 15.1 Sketch the standard CMOS circuit implementation of the SR flip-flop shown in Fig. 15.3.

D 15.2 Sketch the logic-gate implementation of an SR flip-flop utilizing two cross-coupled NAND gates. Clearly label the output terminals and the input trigger terminals. Provide the truth table and describe the operation.

D 15.3 For the SR flip-flop of Fig. 15.4, show that if each of the two inverters utilizes matched transistors, that is, $(W/L)_p = (\mu_n/\mu_p)(W/L)_n$, then the minimum W/L that each of Q_5–Q_8 must have so that switching occurs is $2(W/L)_n$. Give the sizes of all eight transistors if the flip-flop is fabricated in a 0.13-μm process for which $\mu_n = 4\mu_p$. Use the minimum channel length for all transistors and the minimum size ($W/L = 1$) for Q_1 and Q_3.

D 15.4 In this problem we investigate the effect of velocity saturation (Section 13.5) on the design of the SR flip-flop in Example 15.1. Specifically, answer part (a) of the question in Example 15.1, taking into account the fact that for this technology, V_{DSsat} for n-channel devices is 0.6 V and $|V_{DSsat}|$ for p-channel devices is 1 V. Assume $\lambda_n = |\lambda_p| = 0.1\ \text{V}^{-1}$. What is the minimum required value for $(W/L)_5$ and for $(W/L)_6$? Comment on this value relative to that found in Example 15.1. (*Hint*: Refer to Eq. 13.100.)

D 15.5 Repeat part (a) of the problem in Example 15.1 for the case of inverters that do not use matched Q_N and Q_P. Rather, assume that each of the inverters uses $(W/L)_p = (W/L)_n = 0.27\ \mu\text{m}/0.18\ \mu\text{m}$. Find the threshold voltage of each inverter. Then determine the value required for the W/L of each of Q_5 to Q_8 so that the flip-flop switches.

D 15.6 The CMOS SR flip-flop in Fig. 15.4 is fabricated in a 0.13-μm process for which $\mu_n C_{ox} = 4\mu_p C_{ox} = 430\ \mu\text{A/V}^2$, $V_{tn} = |V_{tp}| = 0.4$ V, and $V_{DD} = 1.2$ V. The inverters have $(W/L)_n = 0.2\ \mu\text{m}/0.13\ \mu\text{m}$ and $(W/L)_p = 0.8\ \mu\text{m}/0.13\ \mu\text{m}$. The four NMOS transistors in the set–reset circuit have equal W/L ratios.

(a) Determine the minimum value required for this ratio to ensure that the flip-flop will switch.
(b) If a ratio twice the minimum is selected, determine the minimum required width of the set and reset pulses to ensure switching. Assume that the total capacitance between each of the Q and $\overline{Q}$ nodes and ground is 20 fF.

D 15.7 Consider another possibility for the circuit in Fig. 15.7: Relabel the R input as $\overline{S}$ and the S input as $\overline{R}$. Let $\overline{S}$ and $\overline{R}$ normally rest at V_{DD}. Let the flip-flop be storing a 0; thus $V_Q = 0$ V and $V_{\overline{Q}} = V_{DD}$. To set the flip-flop, the $\overline{S}$ terminal is lowered to 0 V and the clock ϕ is raised to V_{DD}. The relevant part of the circuit is then transistors Q_5 and Q_2. For the flip-flop to switch, the voltage at $\overline{Q}$ must be lowered to $V_{DD}/2$. What is the minimum required W/L for Q_5 in terms of $(W/L)_2$ and (μ_n/μ_p)?

D 15.8 The clocked SR flip-flop in Fig. 15.4 is not a fully complementary CMOS circuit. Sketch the fully complementary version by augmenting the circuit with the PUN corresponding to the PDN comprising Q_5, Q_6, Q_7, and Q_8. Note that the fully complementary circuit utilizes 12 transistors. Although the circuit is more complex, it switches faster.

****15.9** Consider the latch of Fig. 15.1 as implemented in CMOS technology. Let $\mu_n C_{ox} = 2\mu_p C_{ox} = 20\ \mu\text{A/V}^2$, $W_p = 2W_n = 24\ \mu\text{m}$, $L_p = L_n = 6\ \mu\text{m}$, $|V_t| = 1$ V, and $V_{DD} = 5$ V.

(a) Plot the transfer characteristic of each inverter—that is, v_X versus v_W, and v_Z versus v_Y. Determine the output of each inverter at input voltages of 1, 1.5, 2, 2.25, 2.5, 2.75, 3, 3.5, 4, and 5 volts.
(b) Use the characteristics in (a) to determine the loop voltage-transfer curve of the latch—that is, v_Z versus v_W. Find the coordinates of points A, B, and C as defined in Fig. 15.1(c).
(c) If the finite output resistance of the saturated MOSFET is taken into account, with $|V_A| = 100$ V, find the slope of the loop transfer characteristic at point B. What is the approximate width of the transition region?

15.10 Two CMOS inverters operating from a 5-V supply have V_{IH} and V_{IL} of 2.42 and 2.00 V and corresponding outputs of 0.4 V and 4.6 V, respectively, and are connected as a latch. The MOSFETs have $|V_t| = 1$ V. Approximating the corresponding transfer characteristic of each gate by straight lines between the break points, sketch the latch open-loop transfer characteristic. What are the coordinates of point B? What is the loop gain at B?

***15.11** Figure P15.11 shows a commonly used circuit of a D flip-flop that is triggered by the negative-going edge of the clock ϕ.

(a) For ϕ high, what are the values of $\overline{Q}$ and Q in terms of D? Which transistors are conducting?
(b) If D is high and ϕ goes low, which transistors conduct and what signals appear at $\overline{Q}$ and at Q? Describe the circuit operation.

(c) Repeat (b) for D low with the clock ϕ going low.
(d) Does the operation of this circuit rely on charge storage?

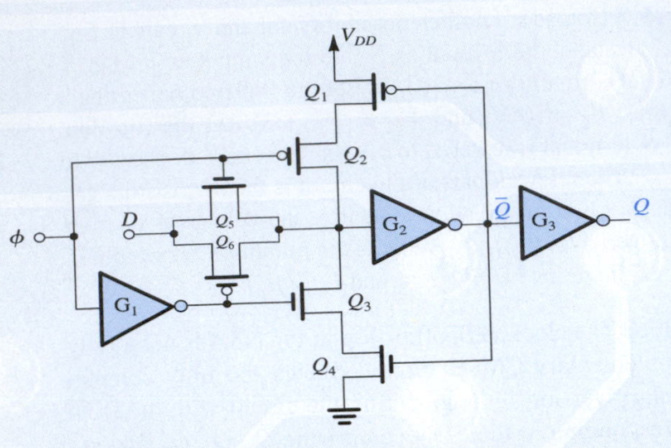

Figure P15.11

Section 15.2: Semiconductor Memories: Types and Architectures

15.12 A particular 1 M-bit square memory array has its peripheral circuits reorganized to allow for the readout of a 16-bit word. How many address bits will the new design need?

15.13 For the memory chip described in Problem 15.12, how many word lines must be supplied by the row decoder? How many sense amplifiers/drivers would a straightforward implementation require? If the chip power dissipation is 500 mW with a 5-V supply for continuous operation with a 200-ns cycle time, and all the power loss is dynamic, estimate the total capacitance of all logic activated in any one cycle. If we assume that 90% of this power loss occurs in array access, and that the major capacitance contributor will be the bit line itself, calculate the capacitance per bit line and per bit for this design. (Recall from problem 15.12 that 16 bit lines are selected simultaneously.) If closer manufacturing control allows the memory array to operate at 3 V, how much larger a memory array can be designed in the same technology at about the same power level?

15.14 An experimental 1.5-V, 1-Gbit dynamic RAM (called DRAM) by Hitachi uses a 0.16-μm process with a cell size of $0.38 \times 0.76\ \mu\text{m}^2$ in a $19 \times 38\ \text{mm}^2$ chip. What fraction of the chip is occupied by the I/O connections, peripheral circuits, and interconnect?

Section 15.3: Random-Access Memory (RAM) Cells

15.15 Consider the read operation of the 6T SRAM cell of Fig. 15.12 when it is storing a 0, that is, $V_Q = 0$ V, and $V_{\overline{Q}} = V_{DD}$. Assume that the bit lines are precharged to V_{DD} before the word-line voltage is raised to V_{DD}. Sketch the relevant part of the circuit and describe the operation. Show that the analysis parallels that presented in the text for the read-1 operation.

D 15.16 Consider a 6T SRAM cell fabricated in a 0.18-μm CMOS process for which $V_{tn} = |V_{tp}| = 0.5$ V and $V_{DD} = 1.8$ V. If during a read-1 operation it is required that $V_{\overline{Q}}$ does not exceed 0.2 V, use the graph in Fig. 15.14 to determine the maximum allowable value of the ratio $(W/L)_5/(W/L)_1$. For $L_1 = L_5 = 0.18$ μm, select values for W_1 and W_5 that minimize the combined areas of Q_1 and Q_5. Assume that the minimum width allowed is 0.18 μm.

15.17 Repeat Exercise 15.4 for an SRAM fabricated in a 0.25-μm CMOS process for which $V_{DD} = 2.5$ V and $V_t = 0.5$ V.

15.18 Repeat Exercise 15.4 for an SRAM fabricated in a 0.13-μm CMOS process for which $V_{DD} = 1.2$ V and $V_t = 0.4$ V.

15.19 Locate on the graph of Fig. 15.14 the points A, B, and C that correspond to the following three process technologies:

(a) 0.25-μm: $V_{DD} = 2.5$ V and $V_t = 0.5$ V
(b) 0.18-μm: $V_{DD} = 1.8$ V and $V_t = 0.5$ V
(c) 0.13-μm: $V_{DD} = 1.2$ V and $V_t = 0.4$ V

In each case, impose the condition that in a read-1 operation $V_{\overline{Q}} = V_t$.

***15.20** Refer to the circuit in Fig. 15.13 and find the maximum ratio $(W/L)_5/(W/L)_1$ for $V_{\overline{Q}} \le V_t$, this time taking into account the velocity-saturation effect (Section 13.5, Eq. 13.100). The SRAM is fabricated in a 0.18-μm CMOS process for which $V_{DD} = 1.8$ V, $V_t = 0.5$ V, and for the n-channel devices $V_{DS\text{sat}} = 0.6$ V. Compare to the value obtained without accounting for velocity saturation. (*Hint*: Convince yourself that for this situation only Q_5 will be operating in velocity saturation.)

D *15.21 For the 6T SRAM of Fig. 15.12, fabricated in a 0.18-μm CMOS process for which $V_{DD} = 1.8$ V, $V_{t0} = 0.5$ V, $2\phi_f = 0.8$ V, and $\gamma = 0.3\ \text{V}^{1/2}$, find the maximum ratio $(W/L)_5/(W/L)_1$ for which $V_{\overline{Q}} \le V_{t0}$ during a read-1 operation (Fig. 15.13). Take into account the body effect in Q_5. Compare to the value obtained without accounting for the body effect.

D 15.22 A 6T SRAM cell is fabricated in a 0.13-μm CMOS process for which $V_{DD} = 1.2$ V, $V_t = 0.4$ V, and $\mu_n C_{ox} = 430\ \mu\text{A/V}^2$. The inverters utilize $(W/L)_n = 1$. Each of the bit lines has a 2-pF capacitance to ground. The sense amplifier requires a minimum of 0.2-V input for reliable and fast operation.

(a) Find the upper bound on W/L for each of the access transistors so that V_Q and $V_{\bar{Q}}$ do not change by more than V_t volts during the read operation.
(b) Find the delay time Δt encountered in the read operation if the cell design utilizes minimum-size access transistors.
(c) Find the delay time Δt if the design utilizes the maximum allowable size for the access transistors.

15.23 Consider the operation of writing a 1 into a 6T SRAM cell that is originally storing a 0. Sketch the relevant part of the circuit and explain the operation. Without doing detailed analysis, show that the analysis would lead to results identical to those obtained in the text for the write-0 operation.

D 15.24 For a 6T SRAM cell fabricated in a 0.13-μm CMOS process, find the maximum permitted value of $(W/L)_p$ in terms of $(W/L)_a$ of the access transistors. Assume $V_{DD} = 1.2$ V, $V_{tn} = |V_{tp}| = 0.4$ V, and $\mu_n = 4\mu_p$.

D 15.25 For a 6T SRAM cell fabricated in a 0.25-μm CMOS process, find the maximum permitted value of $(W/L)_p$ in terms of $(W/L)_a$ of the access transistors. Assume $V_{DD} = 2.5$ V, $V_{tn} = |V_{tp}| = 0.5$ V, and $\mu_n \simeq 4\mu_p$.

15.26 Locate on the graph in Fig. 15.17 the points A, B, and C corresponding to the following three CMOS fabrication processes:

(a) 0.25-μm: $V_{DD} = 2.5$ V, $V_{tn} = |V_{tp}| = 0.5$ V

(b) 0.18-μm: $V_{DD} = 1.8$ V, $V_{tn} = |V_{tp}| = 0.5$ V

(c) 0.13-μm: $V_{DD} = 1.2$ V, $V_{tn} = |V_{tp}| = 0.4$ V

For all three, $\mu_n \simeq 4\mu_p$. In each case, V_Q is to be limited to a maximum value of V_{tn}.

D 15.27 Design a minimum-size 6T SRAM cell in a 0.13-μm process for which $V_{DD} = 1.2$ V and $V_{tn} = |V_{tp}| = 0.4$ V. All transistors are to have equal $L = 0.13$ μm. Assume that the minimum width allowed is 0.13 μm. Verify that your minimum-size cell meets the constraints in Eqs. (15.5) and (15.11).

15.28 For a particular DRAM design, the cell capacitance $C_S = 30$ fF and $V_{DD} = 1.8$ V. Each cell represents a capacitive load on the bit line of 1 fF. Assume a 28-fF capacitance for the sense amplifier and other circuitry attached to the bit line. What is the maximum number of cells that can be attached to a bit line while ensuring a minimum bit-line signal of 0.05 V? How many bits of row addressing can be used? If the sense-amplifier gain is increased by a factor of 5, how many word-line address bits can be accommodated?

15.29 For a DRAM available for regular use 98% of the time, having a row-to-column ratio of 2 to 1, a cycle time of 20 ns, and a refresh cycle of 8 ms, estimate the total memory capacity.

15.30 In a particular dynamic memory chip, $C_S = 25$ fF, the bit-line capacitance per cell is 0.5 fF, and bit-line control circuitry involves 12 fF. For a 1-Mbit-square array, what bit-line signals result when a stored 1 is read? When a stored 0 is read? Assume that $V_{DD} = 1.8$ V.

15.31 For a DRAM cell utilizing a capacitance of 20 fF, refresh is required within 10 ms. If a signal loss on the capacitor of 0.2 V can be tolerated, what is the largest acceptable leakage current present at the cell?

Section 15.4: Sense Amplifiers and Address Decoders

D 15.32 Consider the operation of the differential sense amplifier of Fig. 15.20 following the rise of the sense control signal ϕ_s. Assume that a balanced differential signal of 0.1 V is established between the bit lines, each of which has a 1 pF capacitance. For $V_{DD} = 3$ V, what is the value of G_m of each of the inverters in the amplifier required to cause the outputs to reach $0.1V_{DD}$ and $0.9V_{DD}$ [from initial values of $0.5V_{DD} + (0.1/2)$ and $0.5V_{DD} - (0.1/2)$ volts, respectively] in 2 ns? If for the matched inverters, $|V_t| = 0.8$ V and $k_n' = 3k_p' = 75$ μA/V^2, what are the device widths required? If the input signal is 0.2 V, what does the amplifier response time become?

15.33 A particular version of the regenerative sense amplifier of Fig. 15.20 in a 0.5-μm technology, uses transistors for which $|V_t| = 0.8$ V, $k_n' = 2.5k_p' = 100$ μA/V^2, $V_{DD} = 3.3$ V, with $(W/L)_n = 6$ μm/1.5 μm and $(W/L)_p = 15$ μm/1.5 μm. For each inverter, find the value of G_m. For a bit-line capacitance of 0.8 pF, and a delay until an output of $0.9V_{DD}$ is reached of 2 ns, find the initial difference-voltage required between the two bit lines. If the time can be relaxed by 1 ns, what input signal can be handled? With the increased delay time and with the input signal at the original level, by what percentage can the bit-line capacitance, and correspondingly the bit-line length, be increased? If the delay time required for the bit-line capacitances to charge by the constant current available from the storage cell, and thus develop the difference-voltage signal needed by the sense amplifier, was 5 ns, what does it increase to when longer lines are used?

D 15.34 (a) For the sense amplifier of Fig. 15.20, show that the time required for the bit lines to reach $0.9V_{DD}$ and $0.1V_{DD}$ is given by $t_d = (C_B/G_m)\ln(0.8V_{DD}/\Delta V)$ where

ΔV is the initial difference-voltage between the two bit lines. (Refer to Fig. 15.21.)
(b) If the response time of the sense amplifier is to be reduced to one-half the value of an original design, by what factor must the width of all transistors be increased?
(c) If for a particular design, $V_{DD} = 1.8$ V and $\Delta V = 0.2$ V, find the factor by which the width of all transistors must be increased so that ΔV is reduced by a factor of 4 while keeping t_d unchanged?

D 15.35 It is required to design a sense amplifier of the type shown in Fig. 15.20 to operate with a DRAM using the dummy-cell technique illustrated in Fig. 15.22. The DRAM cell provides readout voltages of –100 mV when a 0 is stored and +40 mV when a 1 is stored. The sense amplifier is required to provide a differential output voltage of 2 V in at most 5 ns. Find the W/L ratios of the transistors in the amplifier inverters, assuming that the processing technology is characterized by $k_n' = 2.5k_p' = 100\ \mu A/V^2$, $|V_t| = 1$ V, and $V_{DD} = 5$ V. The capacitance of each half bit line is 1 pF. What will be the amplifier response time when a 0 is read? When a 1 is read?

D 15.36 It is required to design the sense amplifier of Fig. 15.24 to detect an input signal of 100 mV and provide a full output in 0.3 ns. If $C = 60$ fF and $V_{DD} = 1.2$ V, find the required current I and the power dissipation.

D 15.37 Consider the sense amplifier in Fig. 15.24 in the equilibrium condition shown in part (b) of the figure. Let $V_{DD} = 1.8$ V and $V_t = 0.5$ V.
(a) If Q_1 and Q_2 are to operate at the edge of saturation, what is the dc voltage at the drain of Q_1?
(b) If the switching voltage ΔV is to be about 140 mV, at what overdrive voltage V_{OV} should Q_1 and Q_2 be operated in equilibrium? What dc voltage should appear at the common-source terminals of Q_1 and Q_2?
(c) If the delay component Δt given by Eq. (15.18) is to be 0.5 ns, what current I is needed if $C = 55$ fF?
(d) Find the W/L required for each of Q_1 to Q_4 for $\mu_n C_{ox} = 4\mu_p C_{ox} = 300\ \mu A/V^2$.
(e) If Q_5 is to operate at the same overdrive voltage as Q_1 and Q_2, find its required W/L and the value of the reference voltage V_R,

15.38 Consider a 512-row NOR decoder. To how many address bits does this correspond? How many output lines does it have? How many input lines does the NOR array require? How many NMOS and PMOS transistors does such a design need?

15.39 For the column decoder shown in Fig. 15.26, how many column-address bits are needed in a 256-Kbit square array? How many NMOS pass transistors are needed in the multiplexer? How many NMOS transistors are needed in the NOR decoder? How many PMOS transistors? What is the total number of NMOS and PMOS transistors needed?

15.40 Consider the use of the tree column decoder shown in Fig. 15.27 for application with a square 256-Kbit array. How many address bits are involved? How many levels of pass gates are used? How many pass transistors are there in total?

15.41 Consider a ring oscillator consisting of five inverters, each having $t_{PLH} = 6$ ns and $t_{PHL} = 4$ ns. Sketch one of the output waveforms, and specify its frequency and the percentage of the cycle during which the output is high.

15.42 A ring-of-eleven oscillator is found to operate at 20 MHz. Find the propagation delay of the inverter.

D 15.43 Design the one-shot circuit of Fig. 15.29 to provide an output pulse of 10-ns width. If the inverters available have $t_P = 2.5$ ns delay, how many inverters do you need for the delay circuit?

Section 15.5: Read-Only Memory (ROM)

15.44 Give the eight words stored in the ROM of Fig. 15.30.

D 15.45 Design the bit pattern to be stored in a (16 × 4) ROM that provides the 4-bit product of two 2-bit variables. Give a circuit implementation of the ROM array using a form similar to that of Fig. 15.30.

15.46 Consider a dynamic version of the ROM in Fig. 15.30 in which the gates of the PMOS devices are connected to a precharge control signal ϕ. Let all the NMOS devices have $W/L = 3\ \mu m/1.2\ \mu m$ and all the PMOS devices have $W/L = 12\ \mu m/1.2\ \mu m$. Assume $k_n' = 3k_p' = 90\ \mu A/V^2$, $V_{tn} = -V_{tp} = 1$ V, and $V_{DD} = 5$ V.

(a) During the precharge interval, ϕ is lowered to 0 V. Estimate the time required to charge a bit line from 0 to 5 V. Use as an average charging current the current supplied by a PMOS transistor at a bit-line voltage halfway through the 0 to 5 V excursion (i.e., 2.5 V). The bit-line capacitance is 1 pF. Note that all NMOS transistors are cut off at this time.

(b) After the precharge interval is completed and ϕ returns to V_{DD}, the row decoder raises the voltage of the selected word line. Because of the finite resistance and capacitance of the word line, the voltage rises exponentially toward V_{DD}. If the resistance of each of the polysilicon word lines is 5 kΩ and the capacitance between the word line and ground is 2 pF, what is the (10% to 90%) rise time of the word-line voltage? What is the voltage reached at the end of one time-constant?

(c) If we approximate the exponential rise of the word-line voltage by a step equal to the voltage reached in one time constant, find the interval Δt required for an NMOS transistor to discharge the bit line and lower its voltage by 1 V.

PART IV

Filters and Oscillators

In Part IV we study an important class of analog circuits: filters and oscillators. Both topics have in common an application or system orientation. They provide dramatic and powerful illustration of the application of both negative and positive feedback. While the filters studied here are linear circuits, the design of oscillators makes use of both linear and nonlinear techniques.

Chapter 16 deals with the design of filters, which are important building blocks of communication and instrumentation systems. Filter design is one of the rare areas of engineering for which a complete design theory exists, starting from specification and culminating in an actual working circuit. The material presented should allow the reader to perform such a complete design process.

In the design of electronic systems, the need usually arises for signals of various waveforms—sinusoidal, pulse, square-wave, and so on. The generation of such signals is the subject of Chapter 17. It will be seen that some of the circuits utilized in waveform generation employ an op-amp version of the basic memory element studied in Chapter 15, the bistable multivibrator or latch.

The study of filters and oscillators relies on a thorough familiarity with basic feedback concepts including the effect of feedback on the amplifier poles (Chapter 10), and with op-amp circuit applications (Chapter 2). As well, we assume knowledge of basic *s*-plane concepts including transfer functions, poles, zeros, and Bode plots.

CHAPTER 16

Filters and Tuned Amplifiers

IN THIS CHAPTER YOU WILL LEARN

1. How filters are characterized by their signal-transmission properties and how they are classified into different types based on the relative location of their passband(s) and stopband(s).
2. How filters are specified and how to obtain a filter transfer function that meets the given specifications, including the use of popular special functions such as the Butterworth and the Chebyshev.
3. The various first-order and second-order filter functions and their realization using op amps and RC circuits.
4. The basic second-order LCR resonator and how it can be used to realize the various second-order filter functions.
5. The best op amp–RC circuit for realizing an inductance and how it can be used as the basis for realizing the various second-order filter functions.
6. That connecting two op-amp integrators, one inverting and one noninverting, in a feedback loop realizes a second-order resonance circuit and can be used to obtain circuit realizations of the various second-order filter functions.
7. How second-order filter functions can be realized using a single op amp and an RC circuit, and the performance limitations of these minimal realizations.
8. How the powerful concept of circuit sensitivity can be applied to assess the performance of filter circuits in the face of finite component tolerances.
9. The basis for the most popular approach to the realization of filter functions in IC form; the switched-capacitor technique.
10. The design of tuned transistor amplifiers for radio-frequency (RF) applications.

Introduction

In this chapter, we study the design of an important building block of communications and instrumentation systems, the electronic filter. Filter design is one of the very few areas of engineering for which a complete design theory exists, starting from specification and ending with a circuit realization. A detailed study of filter design requires an entire book, and indeed such textbooks exist. In the limited space available here, we shall concentrate on a selection of topics that provide an introduction to the subject as well as a useful arsenal of filter circuits and design methods.

The oldest technology for realizing filters makes use of inductors and capacitors, and the resulting circuits are called **passive LC filters**. Such filters work well at high frequencies; however, in low-frequency applications (dc to 100 kHz) the required inductors are large and physically bulky, and their characteristics are quite nonideal. Furthermore, such inductors are impossible to fabricate in monolithic form and are incompatible with any of the modern techniques for assembling electronic systems. Therefore, there has been considerable interest in finding filter realizations that do not require inductors. Of the various possible types of **inductorless filters**, we shall study **active-RC filters** and **switched-capacitor filters**.

Active-RC filters utilize op amps together with resistors and capacitors and are fabricated using discrete, hybrid thick-film, or hybrid thin-film technology. However, for large-volume production, such technologies do not yield the economies achieved by monolithic (IC) fabrication. At the present time, the most viable approach for realizing fully integrated monolithic filters is the switched-capacitor technique.

The last topic studied in this chapter is the tuned amplifier commonly employed in the design of radio and TV receivers. Although tuned amplifiers are in effect bandpass filters, they are studied separately because their design is based on somewhat different techniques.

The material in this chapter requires a thorough familiarity with op-amp circuit applications. Thus the study of Chapter 2 is a prerequisite.

16.1 Filter Transmission, Types, and Specification

16.1.1 Filter Transmission

The filters we are about to study are linear circuits that can be represented by the general two-port network shown in Fig. 16.1. The filter **transfer function** $T(s)$ is the ratio of the output voltage $V_o(s)$ to the input voltage $V_i(s)$,

$$T(s) \equiv \frac{V_o(s)}{V_i(s)} \tag{16.1}$$

The filter **transmission** is found by evaluating $T(s)$ for physical frequencies, $s = j\omega$, and can be expressed in terms of its magnitude and phase as

$$T(j\omega) = |T(j\omega)|e^{j\phi(\omega)} \tag{16.2}$$

The magnitude of transmission is often expressed in decibels in terms of the **gain function**

$$G(\omega) \equiv 20\log|T(j\omega)|, \text{ dB} \tag{16.3}$$

or, alternatively, in terms of the **attenuation function**

$$A(\omega) \equiv -20\log|T(j\omega)|, \text{ dB} \tag{16.4}$$

A filter shapes the frequency spectrum of the input signal, $|V_i(j\omega)|$, according to the magnitude of the transfer function $|T(j\omega)|$, thus providing an output $V_o(j\omega)$ with a spectrum

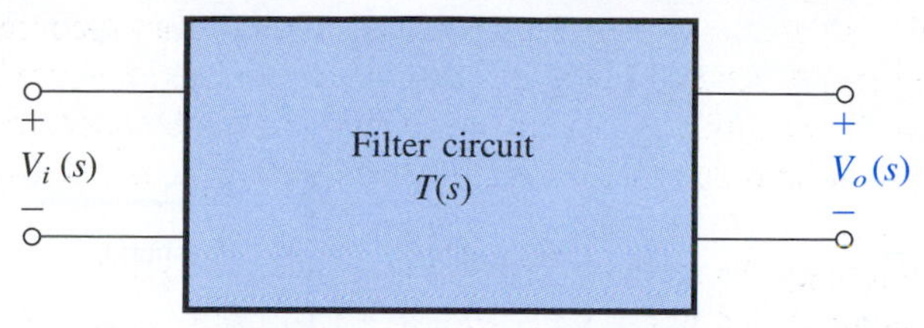

Figure 16.1 The filters studied in this chapter are linear circuits represented by the general two-port network shown. The filter transfer function $T(s) \equiv V_o(s)/V_i(s)$.

$$|V_o(j\omega)| = |T(j\omega)||V_i(j\omega)| \tag{16.5}$$

Also, the phase characteristics of the signal are modified as it passes through the filter according to the filter phase function $\phi(\omega)$.

16.1.2 Filter Types

We are specifically interested here in filters that perform a **frequency-selection** function: **passing** signals whose frequency spectrum lies within a specified range, and **stopping** signals whose frequency spectrum falls outside this range. Such a filter has ideally a frequency band (or bands) over which the magnitude of transmission is unity (the filter **passband**) and a frequency band (or bands) over which the transmission is zero (the filter **stopband**). Figure 16.2 depicts the ideal transmission characteristics of the four major filter types: **low-pass** (LP) in Fig. 16.2(a), **high-pass** (HP) in Fig. 16.2(b), **bandpass** (BP) in Fig. 16.2(c), and **bandstop** (BS) or **band-reject** in Fig. 16.2(d). These idealized characteristics, by virtue of their vertical edges, are known as **brick-wall** responses.

16.1.3 Filter Specification

The filter-design process begins with the filter user specifying the transmission characteristics required of the filter. Such a specification cannot be of the form shown in Fig. 16.2 because physical circuits cannot realize these idealized characteristics. Figure 16.3 shows realistic specifications for the transmission characteristics of a low-pass filter. Observe that since a physical circuit cannot provide constant transmission at all passband frequencies, the specifications allow for deviation of the passband transmission from the ideal 0 dB, but place an upper bound, A_{max} (dB), on this deviation. Depending on the application, A_{max} typically ranges from 0.05 dB to 3 dB. Also, since a physical circuit cannot provide zero transmission at all stopband frequencies, the specifications in Fig. 16.3 allow for some transmission over the stopband. However, the specifications require the stopband signals to be attenuated by at least A_{min} (dB) relative to the passband signals. Depending on the filter application, A_{min} can range from 20 dB to 100 dB.

Since the transmission of a physical circuit cannot change abruptly at the edge of the passband, the specifications of Fig. 16.3 provide for a band of frequencies over which the attenuation increases from near 0 dB to A_{min}. This **transition band** extends from the passband edge ω_p to the stopband edge ω_s. The ratio ω_s/ω_p is usually used as a measure of the sharpness of the low-pass filter response and is called the **selectivity factor**. Finally, observe that for convenience the passband transmission is specified to be 0 dB. The final filter, however, can be given a passband gain, if desired, without changing its selectivity characteristics.

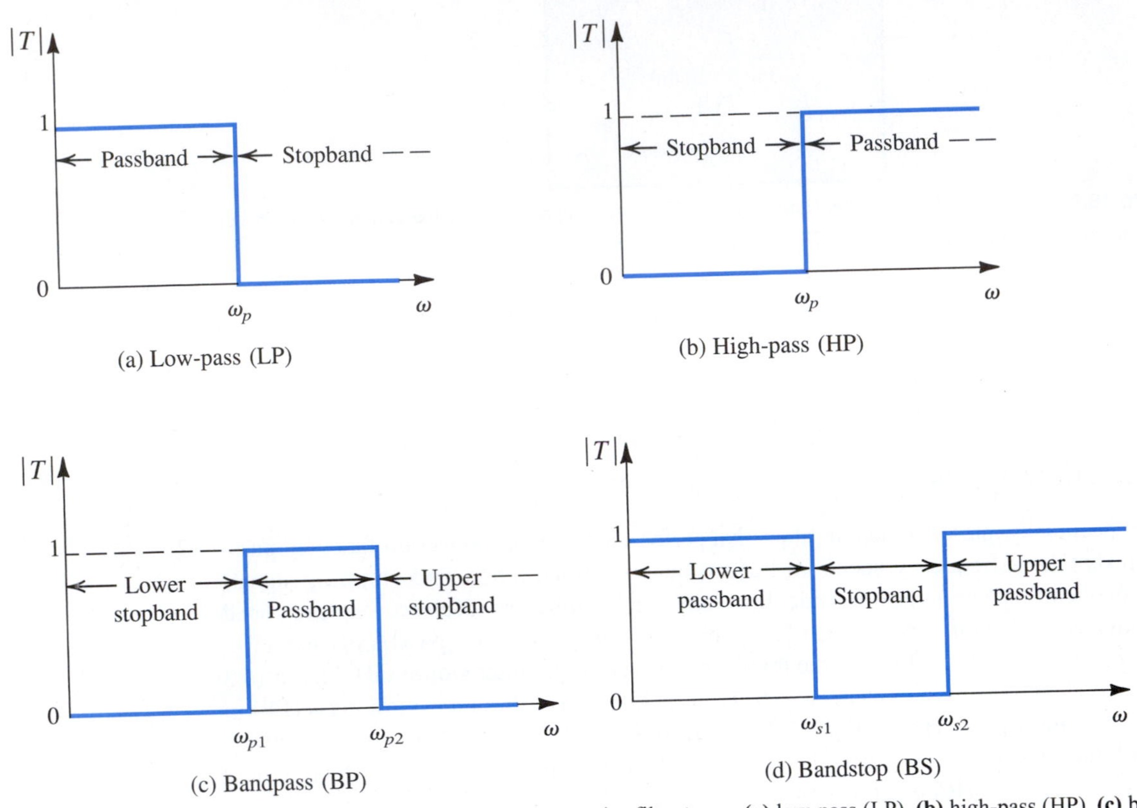

Figure 16.2 Ideal transmission characteristics of the four major filter types: **(a)** low-pass (LP), **(b)** high-pass (HP), **(c)** bandpass (BP), and **(d)** bandstop (BS).

To summarize, the transmission of a low-pass filter is specified by four parameters:

1. The passband edge ω_p
2. The maximum allowed variation in passband transmission A_{max}
3. The stopband edge ω_s
4. The minimum required stopband attenuation A_{min}

The more tightly one specifies a filter—that is, lower A_{max}, higher A_{min}, and/or a selectivity ratio ω_s/ω_p closer to unity—the closer the response of the resulting filter will be to the ideal. However, the resulting filter circuit must be of higher order and thus more complex and expensive.

In addition to specifying the magnitude of transmission, there are applications in which the phase response of the filter is also of interest. The filter-design problem, however, is considerably complicated when both magnitude and phase are specified.

Once the filter specifications have been decided upon, the next step in the design is to find a transfer function whose magnitude meets the specification. To meet specification, the magnitude-response curve must lie in the unshaded area in Fig. 16.3. The curve shown in the figure is for a filter that *just* meets specifications. Observe that for this particular filter, the magnitude response *ripples* throughout the passband, and the ripple peaks are all equal. Since the peak ripple is equal to A_{max} it is usual to refer to A_{max} as the **passband ripple** and to

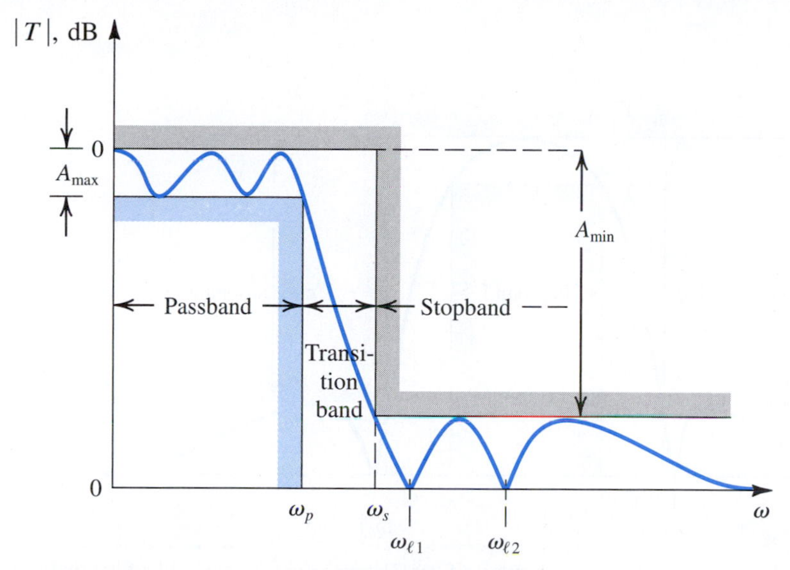

Figure 16.3 Specification of the transmission characteristics of a low-pass filter. The magnitude response of a filter that just meets specifications is also shown.

ω_p as the **ripple bandwidth**. The particular filter response shows ripples also in the stopband, again with the ripple peaks all equal and of such a value that the minimum stopband attenuation achieved is equal to the specified value, A_{min}. Thus this particular response is said to be **equiripple** in both the passband and the stopband.

The process of obtaining a transfer function that meets given specifications is known as **filter approximation**. Filter approximation is usually performed using computer programs (Snelgrove, 1982; Ouslis and Sedra, 1995) or filter design tables (Zverev, 1967). In simpler cases, filter approximation can be performed using closed-form expressions, as will be seen in Section 16.3.

Finally, Fig. 16.4 shows transmission specifications for a bandpass filter and the response of a filter that meets these specifications. For this example we have chosen an approximation function that does not ripple in the passband; rather, the transmission decreases monotonically on both sides of the center frequency, attaining the maximum allowable deviation at the two edges of the passband.

EXERCISES

16.1 Find approximate values of attenuation (in dB) corresponding to filter transmissions of 1, 0.99, 0.9, 0.8, 0.7, 0.5, 0.1, 0.
Ans. 0, 0.1, 1, 2, 3, 6, 20, ∞ (dB)

16.2 If the magnitude of passband transmission is to remain constant to within ±5%, and if the stopband transmission is to be no greater than 1% of the passband transmission, find A_{max} and A_{min}.
Ans. 0.9 dB; 40 dB

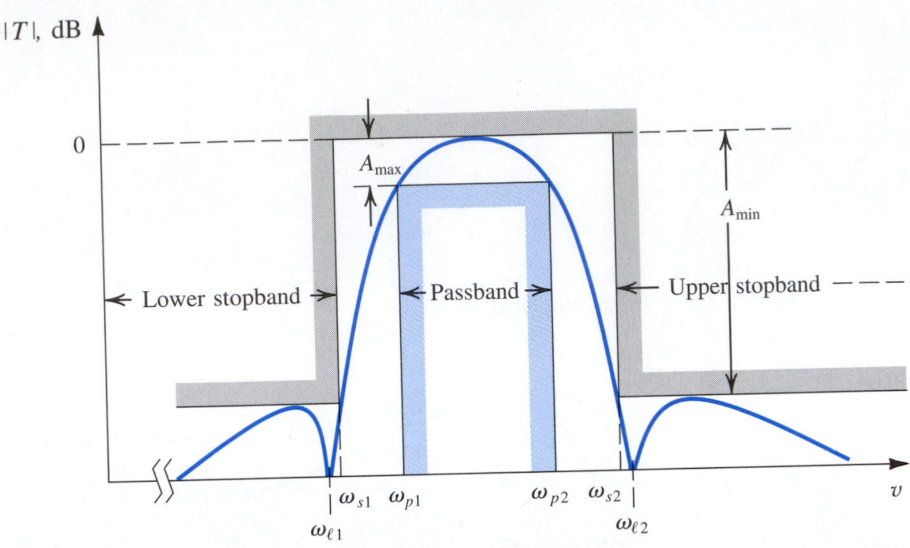

Figure 16.4 Transmission specifications for a bandpass filter. The magnitude response of a filter that just meets specifications is also shown. Note that this particular filter has a monotonically decreasing transmission in the passband on both sides of the peak frequency.

16.2 The Filter Transfer Function

The filter transfer function $T(s)$ can be written as the ratio of two polynomials as

$$T(s) = \frac{a_M s^M + a_{M-1} s^{M-1} + \cdots + a_0}{s^N + b_{N-1} s^{N-1} + \cdots + b_0} \tag{16.6}$$

The degree of the denominator, N, is the **filter order**. For the filter circuit to be stable, the degree of the numerator must be less than or equal to that of the denominator; $M \leq N$. The numerator and denominator coefficients, $a_0, a_1, \ldots, a_M$ and $b_0, b_1, \ldots, b_{N-1}$, are real numbers. The polynomials in the numerator and denominator can be factored, and $T(s)$ can be expressed in the form

$$T(s) = \frac{a_M(s - z_1)(s - z_2) \cdots (s - z_M)}{(s - p_1)(s - p_2) \cdots (s - p_N)} \tag{16.7}$$

The numerator roots, $z_1, z_2, \ldots, z_M$, are the **transfer function zeros**, or **transmission zeros**; and the denominator roots, $p_1, p_2, \ldots, p_N$, are the **transfer function poles**, or the **natural modes**.[1] Each transmission zero or pole can be either a real or a complex number. Complex zeros and poles, however, must occur in conjugate pairs. Thus, if $-1 + j2$ happens to be a zero, then $-1 - j2$ also must be a zero.

Since in the filter stopband the transmission is required to be zero or small, the filter transmission zeros are usually placed on the $j\omega$ axis at stopband frequencies. This indeed is the case for the filter whose transmission function is sketched in Fig. 16.3. This particular filter can be seen to have infinite attenuation (zero transmission) at two stopband frequencies: ω_{l1} and ω_{l2}. The filter then must have transmission zeros at $s = +j\omega_{l1}$ and $s = +j\omega_{l2}$.

[1] Throughout this chapter, we use the names *poles* and *natural modes* interchangeably.

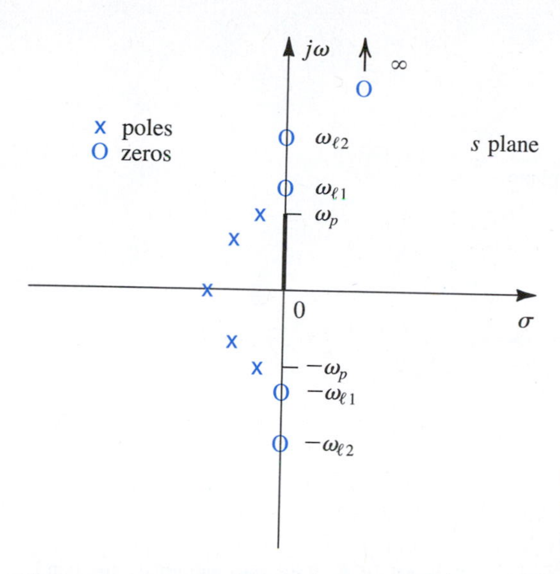

Figure 16.5 Pole–zero pattern for the low-pass filter whose transmission is sketched in Fig. 16.3. This is a fifth-order filter ($N = 5$).

However, since complex zeros occur in conjugate pairs, there must also be transmission zeros at $s = -j\omega_{l1}$ and $s = -j\omega_{l2}$. Thus the numerator polynomial of this filter will have the factors $(s + j\omega_{l1})(s - j\omega_{l1})(s + j\omega_{l2})(s - j\omega_{l2})$, which can be written as $(s^2 + \omega_{l1}^2)(s^2 + \omega_{l2}^2)$. For $s = j\omega$ (physical frequencies) the numerator becomes $(-\omega^2 + \omega_{l1}^2)(-\omega^2 + \omega_{l2}^2)$, which indeed is zero at $\omega = \omega_{l1}$ and $\omega = \omega_{l2}$.

Continuing with the example in Fig. 16.3, we observe that the transmission decreases toward $-\infty$ as ω approaches ∞. Thus the filter must have one or more transmission zeros at $s = \infty$. In general, the number of transmission zeros at $s = \infty$ is the difference between the degree of the numerator polynomial, M, and the degree of the denominator polynomial, N, of the transfer function in Eq. (16.6). This is because as s approaches ∞, $T(s)$ approaches a_M/s^{N-M} and thus is said to have $N - M$ zeros at $s = \infty$.

For a filter circuit to be stable, all its poles must lie in the left half of the s plane, and thus $p_1, p_2, \ldots, p_N$ must all have negative real parts. Figure 16.5 shows typical pole and zero locations for the low-pass filter whose transmission function is depicted in Fig. 16.3. We have assumed that this filter is of fifth order ($N = 5$). It has two pairs of complex-conjugate poles and one real-axis pole, for a total of five poles. All the poles lie in the vicinity of the passband, which is what gives the filter its high transmission at passband frequencies. The five transmission zeros are at $s = \pm j\omega_{l1}$, $s = \pm j\omega_{l2}$, and $s = \infty$. Thus, the transfer function for this filter is of the form

$$T(s) = \frac{a_4(s^2 + \omega_{l1}^2)(s^2 + \omega_{l2}^2)}{s^5 + b_4 s^4 + b_3 s^3 + b_2 s^2 + b_1 s + b_0} \tag{16.8}$$

As another example, consider the bandpass filter whose magnitude response is shown in Fig. 16.4. This filter has transmission zeros at $s = \pm j\omega_{l1}$ and $s = \pm j\omega_{l2}$. It also has one or more zeros at $s = 0$ and one or more zeros at $s = \infty$ (because the transmission decreases toward 0 as ω approaches 0 and ∞). Assuming that only one zero exists at each of $s = 0$ and $s = \infty$, the filter must be of sixth order, and its transfer function takes the form

$$T(s) = \frac{a_5 s(s^2 + \omega_{l1}^2)(s^2 + \omega_{l2}^2)}{s^6 + b_5 s^5 + \cdots + b_0} \tag{16.9}$$

A typical pole–zero plot for such a filter is shown in Fig. 16.6.

As a third and final example, consider the low-pass filter whose transmission function is depicted in Fig. 16.7(a). We observe that in this case there are no finite values of ω at which

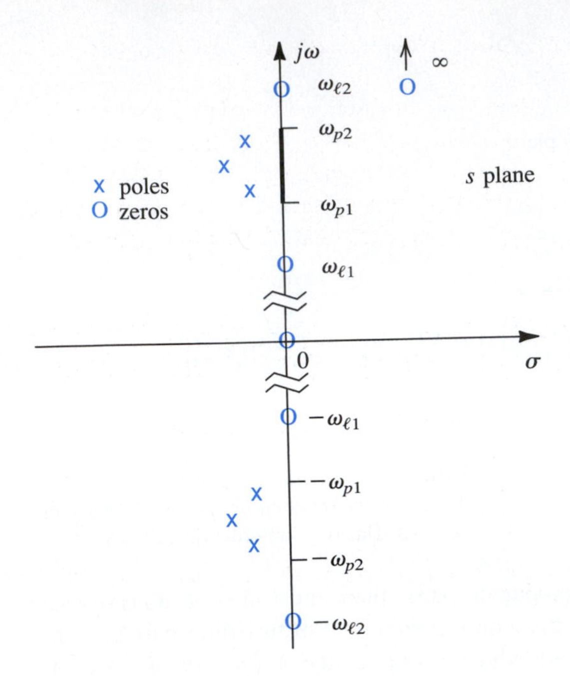

Figure 16.6 Pole–zero pattern for the band-pass filter whose transmission function is shown in Fig. 16.4. This is a sixth-order filter ($N = 6$).

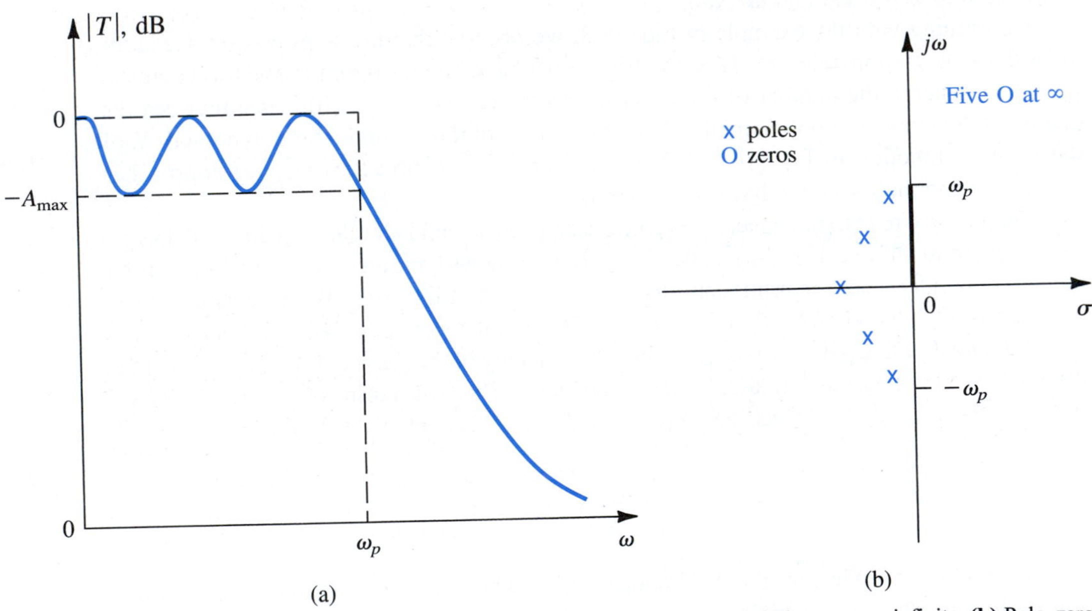

Figure 16.7 **(a)** Transmission characteristics of a fifth-order low-pass filter having all transmission zeros at infinity. **(b)** Pole–zero pattern for the filter in **(a)**.

the attenuation is infinite (zero transmission). Thus it is possible that all the transmission zeros of this filter are at $s = \infty$. If this is the case, the filter transfer function takes the form

$$T(s) = \frac{a_0}{s^N + b_{N-1}s^{N-1} + \cdots + b_0} \tag{16.10}$$

Such a filter is known as an **all-pole filter**. Typical pole–zero locations for a fifth-order all-pole low-pass filter are shown in Fig. 16.7(b).

Almost all the filters studied in this chapter have all their transmission zeros on the $j\omega$ axis, in the filter stopband(s), including[2] $\omega = 0$ and $\omega = \infty$. Also, to obtain high selectivity, all the natural modes will be complex conjugate (except for the case of odd-order filters, where one natural mode must be on the real axis). Finally we note that the more selective the required filter response is, the higher its order must be, and the closer its natural modes are to the $j\omega$ axis.

EXERCISES

16.3 A second-order filter has its poles at $s = -(1/2) \pm j(\sqrt{3}/2)$. The transmission is zero at $\omega = 2$ rad/s and is unity at dc ($\omega = 0$). Find the transfer function.

Ans. $T(s) = \frac{1}{4}\frac{s^2+4}{s^2+s+1}$

16.4 A fourth-order filter has zero transmission at $\omega = 0$, $\omega = 2$ rad/s, and $\omega = \infty$. The natural modes are $-0.1 \pm j0.8$ and $-0.1 \pm j1.2$. Find $T(s)$.

Ans. $T(s) = \frac{a_3 s(s^2+4)}{(s^2+0.2s+0.65)(s^2+0.2s+1.45)}$

16.5 Find the transfer function $T(s)$ of a third-order all-pole low-pass filter whose poles are at a radial distance of 1 rad/s from the origin and whose complex poles are at 30° angles from the $j\omega$ axis. The dc gain is unity. Show that $|T(j\omega)| = 1/\sqrt{1+\omega^6}$. Find ω_{3dB} and the attenuation at $\omega = 3$ rad/s.

Ans. $T(s) = 1/(s+1)(s^2+s+1)$; 1 rad/s; 28.6 dB

16.3 Butterworth and Chebyshev Filters

In this section, we present two functions that are frequently used in approximating the transmission characteristics of low-pass filters. Closed-form expressions are available for the parameters of these functions, and thus one can use them in filter design without the need for computers or filter-design tables. Their utility, however, is limited to relatively simple applications.

Although in this section we discuss the design of low-pass filters only, the approximation functions presented can be applied to the design of other filter types through the use of frequency transformations (see Sedra and Brackett, 1978).

16.3.1 The Butterworth Filter

Figure 16.8 shows a sketch of the magnitude response of a Butterworth[3] filter. This filter exhibits a monotonically decreasing transmission with all the transmission zeros at $\omega = \infty$, making it an all-pole filter. The magnitude function for an Nth-order Butterworth filter with a passband edge ω_p is given by

[2] Obviously, a low-pass filter should *not* have a transmission zero at $\omega = 0$, and, similarly, a high-pass filter should not have a transmission zero at $\omega = \infty$.

[3] The Butterworth filter approximation is named after S. Butterworth, a British engineer who in 1930 was among the first to employ it.

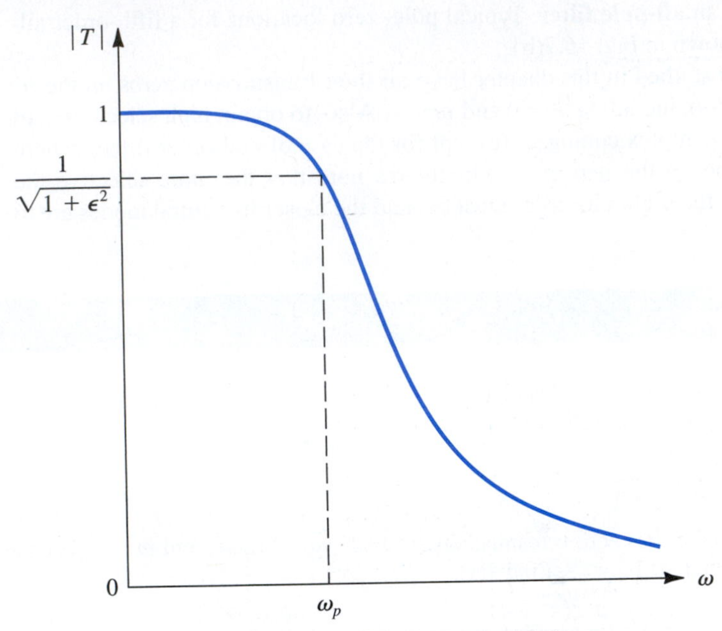

Figure 16.8 The magnitude response of a Butterworth filter.

$$|T(j\omega)| = \frac{1}{\sqrt{1+\epsilon^2\left(\frac{\omega}{\omega_p}\right)^{2N}}} \tag{16.11}$$

At $\omega = \omega_p$,

$$|T(j\omega_p)| = \frac{1}{\sqrt{1+\epsilon^2}} \tag{16.12}$$

Thus, the parameter ϵ determines the maximum variation in passband transmission, A_{max}, according to

$$A_{max} = 20 \log \sqrt{1+\epsilon^2} \tag{16.13}$$

Conversely, given A_{max}, the value of ϵ can be determined from

$$\epsilon = \sqrt{10^{A_{max}/10} - 1} \tag{16.14}$$

Observe that in the Butterworth response the maximum deviation in passband transmission (from the ideal value of unity) occurs at the passband edge only. It can be shown that the first $2N - 1$ derivatives of $|T|$ relative to ω are zero at $\omega = 0$ [see Van Valkenburg (1980)]. This property makes the Butterworth response very flat near $\omega = 0$ and gives the response the name **maximally flat** response. The degree of passband flatness increases as the order N is increased, as can be seen from Fig. 16.9. This figure indicates also that, as should be expected, as the order N is increased the filter response approaches the ideal brick-wall type of response.

At the edge of the stopband, $\omega = \omega_s$, the attenuation of the Butterworth filter can be obtained by substituting $\omega = \omega_s$ in Eq. (16.11). The result is given by

$$\begin{aligned} A(\omega_s) &= -20 \log[1/\sqrt{1+\epsilon^2(\omega_s/\omega_p)^{2N}}] \\ &= 10 \log[1+\epsilon^2(\omega_s/\omega_p)^{2N}] \end{aligned} \tag{16.15}$$

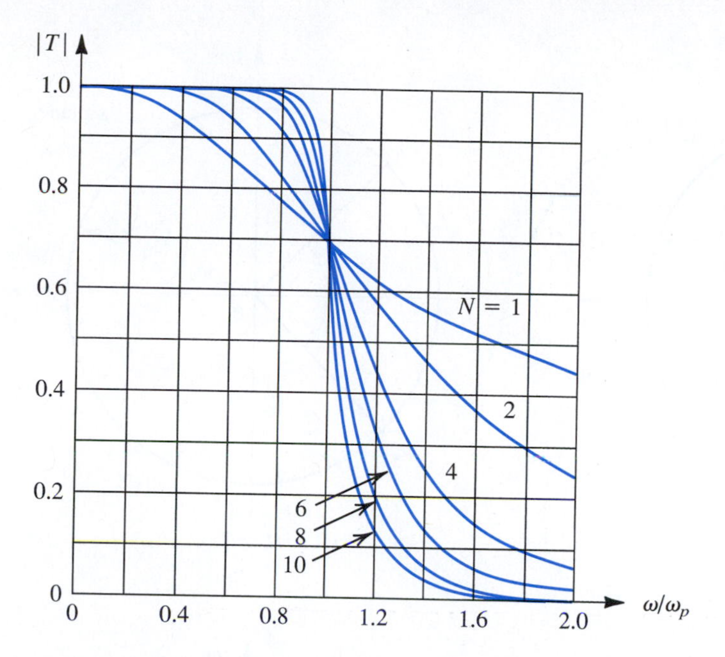

Figure 16.9 Magnitude response for Butterworth filters of various order with $\epsilon = 1$. Note that as the order increases, the response approaches the ideal brick-wall type of transmission.

This equation can be used to determine the filter order required, which is the lowest integer value of N that yields $A(\omega_s) \geq A_{\min}$.

The natural modes of an Nth-order Butterworth filter can be determined from the graphical construction shown in Fig. 16.10(a). Observe that the natural modes lie on a circle of radius $\omega_p(1/\epsilon)^{1/N}$ and are spaced by equal angles of π/N, with the first mode at an angle $\pi/2N$ from the $+j\omega$ axis. Since the natural modes all have equal radial distance from the origin, they all have the same frequency $\omega_0 = \omega_p(1/\epsilon)^{1/N}$. Figure 16.10(b), (c), and (d) shows the natural modes of Butterworth filters of order $N = 2$, 3, and 4, respectively. Once the N natural modes $p_1, p_2, \ldots, p_N$ have been found, the transfer function can be written as

$$T(s) = \frac{K\omega_0^N}{(s-p_1)(s-p_2)\cdots(s-p_N)} \tag{16.16}$$

where K is a constant equal to the required dc gain of the filter.

To summarize, to find a Butterworth transfer function that meets transmission specifications of the form in Fig. 16.3 we perform the following procedure:

1. Determine ϵ from Eq. (16.14).
2. Use Eq. (16.15) to determine the required filter order as the lowest integer value of N that results in $A(\omega_s) \geq A_{\min}$.
3. Use Fig. 16.10(a) to determine the N natural modes.
4. Use Eq. (16.16) to determine $T(s)$.

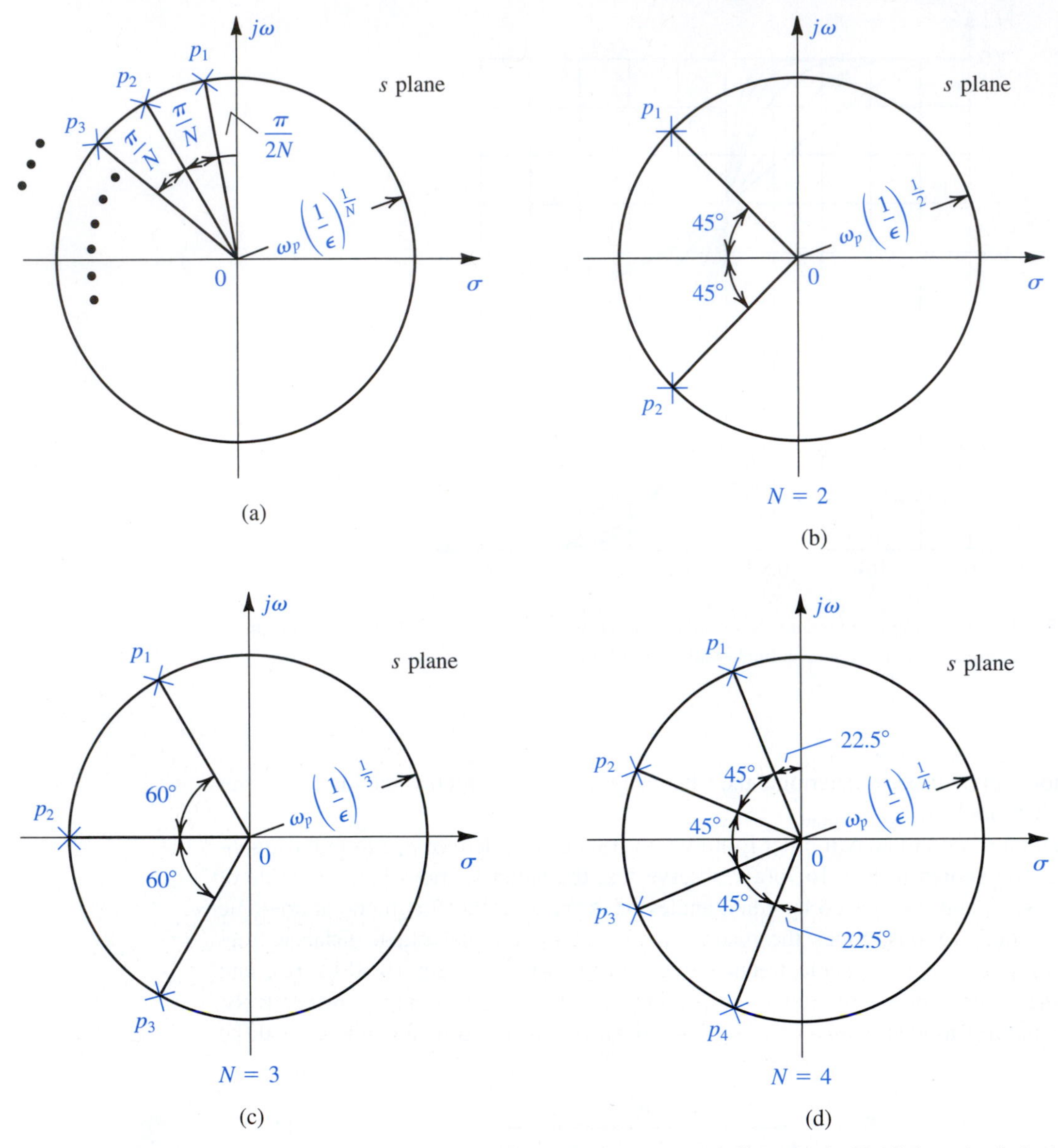

Figure 16.10 Graphical construction for determining the poles of a Butterworth filter of order N. All the poles lie in the left half of the s plane on a circle of radius $\omega_0 = \omega_p(1/\epsilon)^{1/N}$, where ϵ is the passband deviation parameter ($\epsilon = \sqrt{10^{A_{\max}/10} - 1}$): **(a)** the general case; **(b)** $N = 2$; **(c)** $N = 3$; **(d)** $N = 4$.

Example 16.1

Find the Butterworth transfer function that meets the following low-pass filter specifications: $f_p = 10$ kHz, $A_{\max} = 1$ dB, $f_s = 15$ kHz, $A_{\min} = 25$ dB, dc gain = 1.

Solution

Substituting $A_{\max} = 1$ dB into Eq. (16.14) yields $\epsilon = 0.5088$. Equation (16.15) is then used to determine the filter order by trying various values for N. We find that $N = 8$ yields $A(\omega_s) = 22.3$ dB and $N = 9$ gives 25.8 dB. We thus select $N = 9$.

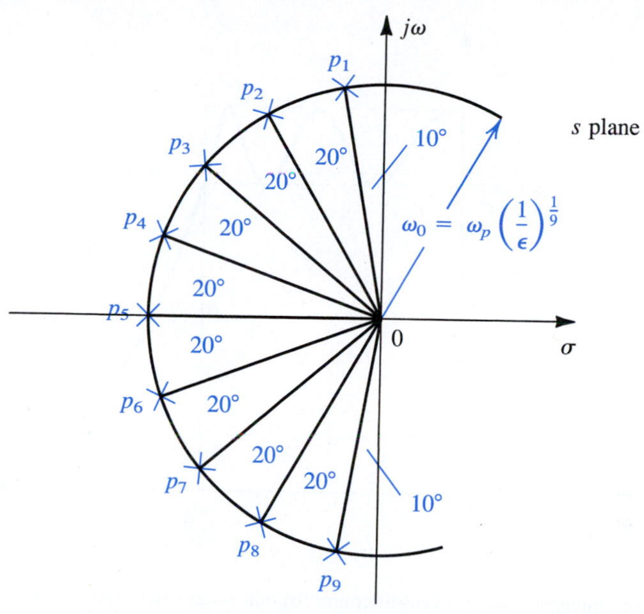

Figure 16.11 Poles of the ninth-order Butterworth filter of Example 16.1.

Figure 16.11 shows the graphical construction for determining the poles. The poles all have the same frequency $\omega_0 = \omega_p(1/\epsilon)^{1/N} = 2\pi \times 10 \times 10^3(1/0.5088)^{1/9} = 6.773 \times 10^4$ rad/s. The first pole p_1 is given by

$$p_1 = \omega_0(-\cos 80° + j\sin 80°) = \omega_0(-0.1736 + j0.9848)$$

Combining p_1 with its complex conjugate p_9 yields the factor $(s^2 + s0.3472\,\omega_0 + \omega_0^2)$ in the denominator of the transfer function. The same can be done for the other complex poles, and the complete transfer function is obtained using Eq. (16.16),

$$T(s) = \frac{\omega_0^9}{(s+\omega_0)(s^2 + s1.8794\,\omega_0 + \omega_0^2)(s^2 + s1.5321\,\omega_0 + \omega_0^2)} \times \frac{1}{(s^2 + s\omega_0 + \omega_0^2)(s^2 + s0.3472\,\omega_0 + \omega_0^2)} \tag{16.17}$$

16.3.2 The Chebyshev Filter

Figure 16.12 shows representative transmission functions for Chebyshev[4] filters of even and odd orders. The Chebyshev filter exhibits an equiripple response in the passband and a monotonically decreasing transmission in the stopband. While the odd-order filter has $|T(0)| = 1$, the even-order filter exhibits its maximum magnitude deviation at $\omega = 0$. In both

[4] Named after the Russian mathematician P. L. Chebyshev, who in 1899 used these functions in studying the construction of steam engines.

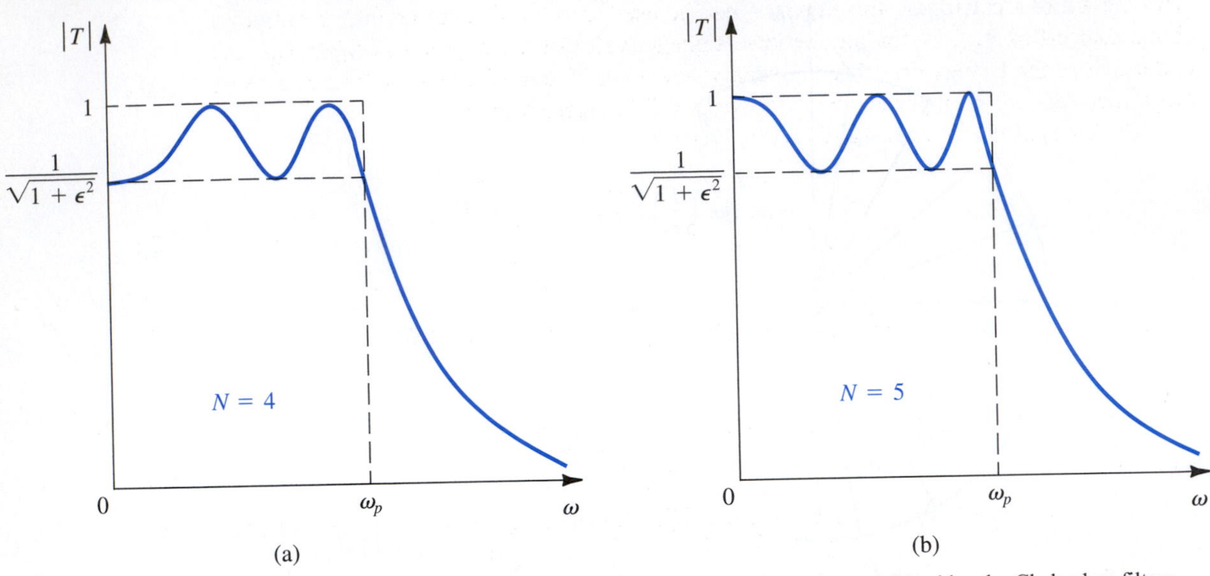

Figure 16.12 Sketches of the transmission characteristics of representative **(a)** even-order and **(b)** odd-order Chebyshev filters.

cases the total number of passband maxima and minima equals the order of the filter, N. All the transmission zeros of the Chebyshev filter are at $\omega = \infty$, making it an all-pole filter.

The magnitude of the transfer function of an Nth-order Chebyshev filter with a passband edge (ripple bandwidth) ω_p is given by

$$|T(j\omega)| = \frac{1}{\sqrt{1+\epsilon^2\cos^2[N\cos^{-1}(\omega/\omega_p)]}} \quad \text{for } \omega \le \omega_p \tag{16.18}$$

and

$$|T(j\omega)| = \frac{1}{\sqrt{1+\epsilon^2\cosh^2[N\cosh^{-1}(\omega/\omega_p)]}} \quad \text{for } \omega \ge \omega_p \tag{16.19}$$

At the passband edge, $\omega = \omega_p$, the magnitude function is given by

$$|T(j\omega_p)| = \frac{1}{\sqrt{1+\epsilon^2}}$$

Thus, the parameter determines the passband ripple according to

$$A_{\max} = 10\log(1+\epsilon^2) \tag{16.20}$$

Conversely, given $A_{\max}$, the value of ϵ is determined from

$$\epsilon = \sqrt{10^{A_{\max}/10} - 1} \tag{16.21}$$

The attenuation achieved by the Chebyshev filter at the stopband edge ($\omega = \omega_s$) is found using Eq. (16.19) as

$$A(\omega_s) = 10\log[1+\epsilon^2\cosh^2(N\cosh^{-1}(\omega_s/\omega_p))] \tag{16.22}$$

With the aid of a calculator, this equation can be used to determine the order N required to obtain a specified A_{min} by finding the lowest integer value of N that yields $A(\omega_s) \geq A_{min}$. As in the case of the Butterworth filter, increasing the order N of the Chebyshev filter causes its magnitude function to approach the ideal brick-wall low-pass response.

The poles of the Chebyshev filter are given by

$$p_k = -\omega_p \sin\left(\frac{2k-1}{N}\frac{\pi}{2}\right)\sinh\left(\frac{1}{N}\sinh^{-1}\frac{1}{\epsilon}\right) \tag{16.23}$$

$$+j\omega_p \cos\left(\frac{2k-1}{N}\frac{\pi}{2}\right)\cosh\left(\frac{1}{N}\sinh^{-1}\frac{1}{\epsilon}\right) \qquad k = 1, 2, \ldots, N$$

Finally, the transfer function of the Chebyshev filter can be written as

$$T(s) = \frac{K\omega_p^N}{\epsilon\, 2^{N-1}(s-p_1)(s-p_2)\cdots(s-p_N)} \tag{16.24}$$

where K is the dc gain that the filter is required to have.

To summarize, given low-pass transmission specifications of the type shown in Fig. 16.3, the transfer function of a Chebyshev filter that meets these specifications can be found as follows:

1. Determine ϵ from Eq. (16.21).
2. Use Eq. (16.22) to determine the order required.
3. Determine the poles using Eq. (16.23).
4. Determine the transfer function using Eq. (16.24).

The Chebyshev filter provides a more efficient approximation than the Butterworth filter. Thus, for the same order and the same A_{max}, the Chebyshev filter provides greater stopband attenuation than the Butterworth filter. Alternatively, to meet identical specifications, one requires a lower order for the Chebyshev than for the Butterworth filter. This point will be illustrated by the following example.

Example 16.2

Find the Chebyshev transfer function that meets the same low-pass filter specifications given in Example 16.1: namely, $f_p = 10$ kHz, $A_{max} = 1$ dB, $f_s = 15$ kHz, $A_{min} = 25$ dB, dc gain = 1.

Solution

Substituting $A_{max} = 1$ dB into Eq. (16.21) yields $\epsilon = 0.5088$. By trying various values for N in Eq. (16.22) we find that $N = 4$ yields $A(\omega_s) = 21.6$ dB and $N = 5$ provides 29.9 dB. We thus select $N = 5$. Recall that we required a ninth-order Butterworth filter to meet the same specifications in Example 16.1.

The poles are obtained by substituting in Eq. (16.23) as

$$p_1, p_5 = \omega_p(-0.0895 \pm j0.9901)$$

$$p_2, p_4 = \omega_p(-0.2342 \pm j0.6119)$$

Example 16.2 *continued*

$$p_5 = \omega_p(-0.2895)$$

The transfer function is obtained by substituting these values in Eq. (16.24) as

$$T(s) = \frac{\omega_p^5}{8.1408(s+0.2895\,\omega_p)(s^2+s0.4684\,\omega_p+0.4293\,\omega_p^2)} \times \frac{1}{s^2+s0.1789\,\omega_p+0.9883\,\omega_p^2} \quad (16.25)$$

where $\omega_p = 2\pi \times 10^4$ rad/s.

EXERCISES

D16.6 Determine the order N of a Butterworth filter for which $A_{max} = 1$ dB, $\omega_s/\omega_p = 1.5$, and $A_{min} =$ 30 dB. What is the actual value of minimum stopband attenuation realized? If A_{min} is to be exactly 30 dB, to what value can A_{max} be reduced?
Ans. $N = 11$; $A_{min} = 32.87$ dB; 0.54 dB

16.7 Find the natural modes and the transfer function of a Butterworth filter with $\omega_p = 1$ rad/s, $A_{max} = 3$ dB ($\epsilon \simeq 1$), and $N = 3$.
Ans. $-0.5 \pm j\sqrt{3}/2$ and -1; $T(s) = 1/(s+1)(s^2+s+1)$

16.8 Observe that Eq. (16.18) can be used to find the frequencies in the passband at which $|T|$ is at its peaks and at its valleys. (The peaks are reached when the $\cos^2[\;\;]$ term is zero, and the valleys correspond to the $\cos^2[\;\;]$ term equal to unity.) Find these frequencies for a fifth-order filter.
Ans. Peaks at $\omega = 0, 0.59\omega_p$, and $0.95\omega_p$; the valleys at $\omega = 0.31\omega_p$ and $0.81\omega_p$

D16.9 Find the attenuation provided at $\omega = 2\omega_p$ by a seventh-order Chebyshev filter with a 0.5-dB passband ripple. If the passband ripple is allowed to increase to 1 dB, by how much does the stopband attenuation increase?
Ans. 64.9 dB; 3.3 dB

D16.10 It is required to design a low-pass filter having $f_p = 1$ kHz, $A_{max} = 1$ dB, $f_s = 1.5$ kHz, $A_{min} =$ 50 dB. (a) Find the required order of a Chebyshev filter. What is the excess stopband attenuation obtained? (b) Repeat for a Butterworth filter.
Ans. (a) $N = 8$, 5 dB; (b) $N = 16$, 0.5 dB

16.4 First-Order and Second-Order Filter Functions

In this section, we shall study the simplest filter transfer functions, those of first and second order. These functions are useful in their own right in the design of simple filters. First- and second-order filters can also be cascaded to realize a high-order filter. Cascade design is in fact one of the most popular methods for the design of active filters (those utilizing op amps and RC circuits). Because the filter poles occur in complex-conjugate pairs, a high-order transfer function $T(s)$ is factored into the product of second-order functions. If $T(s)$ is odd, there will also be a first-order function in the factorization. Each of the second-order functions [and the first-order function when $T(s)$ is odd] is then realized using one of the op amp–RC circuits that will be studied in this chapter, and the resulting blocks are placed in cascade. If the output of each block is taken at the output terminal of an op amp where the impedance level is low (ideally zero), cascading does not change the transfer functions of the individual blocks. Thus the overall transfer function of the cascade is simply the product of the transfer functions of the individual blocks, which is the original $T(s)$.

16.4.1 First-Order Filters

The general first-order transfer function is given by

$$T(s) = \frac{a_1 s + a_0}{s + \omega_0} \tag{16.26}$$

This **bilinear transfer function** characterizes a first-order filter with a natural mode at $s = -\omega_0$, a transmission zero at $s = -a_0/a_1$, and a high-frequency gain that approaches a_1. The numerator coefficients, a_0 and a_1, determine the type of filter (e.g., low pass, high pass, etc.). Some special cases together with passive (RC) and active (op amp–RC) realizations are shown in Fig. 16.13. Note that the active realizations provide considerably more versatility than their passive counterparts; in many cases the gain can be set to a desired value, and some transfer function parameters can be adjusted without affecting others. The output impedance of the active circuit is also very low, making cascading easily possible. The op amp, however, limits the high-frequency operation of the active circuits.

An important special case of the first-order filter function is the **all-pass filter** shown in Fig. 16.14. Here, the transmission zero and the natural mode are symmetrically located relative to the $j\omega$ axis. (They are said to display mirror-image symmetry with respect to the $j\omega$ axis.) Observe that although the transmission of the all-pass filter is (ideally) constant at all frequencies, its phase shows frequency selectivity. All-pass filters are used as phase shifters and in systems that require phase shaping (e.g., in the design of circuits called *delay equalizers,* which cause the overall time delay of a transmission system to be constant with frequency).

EXERCISES

D16.11 Using $R_1 = 10\ \text{k}\Omega$, design the op amp–RC circuit of Fig. 16.13(b) to realize a high-pass filter with a corner frequency of 10^4 rad/s and a high-frequency gain of 10.
Ans. $R_2 = 100\ \text{k}\Omega$; $C = 0.01\ \mu\text{F}$

D16.12 Design the op amp–RC circuit of Fig. 16.14 to realize an all-pass filter with a 90° phase shift at 10^3 rad/s. Select suitable component values.
Ans. Possible choices: $R = R_1 = R_2 = 10\ \text{k}\Omega$; $C = 0.1\ \mu\text{F}$

16.4.2 Second-Order Filter Functions

The general second-order (or **biquadratic**) filter transfer function is usually expressed in the standard form

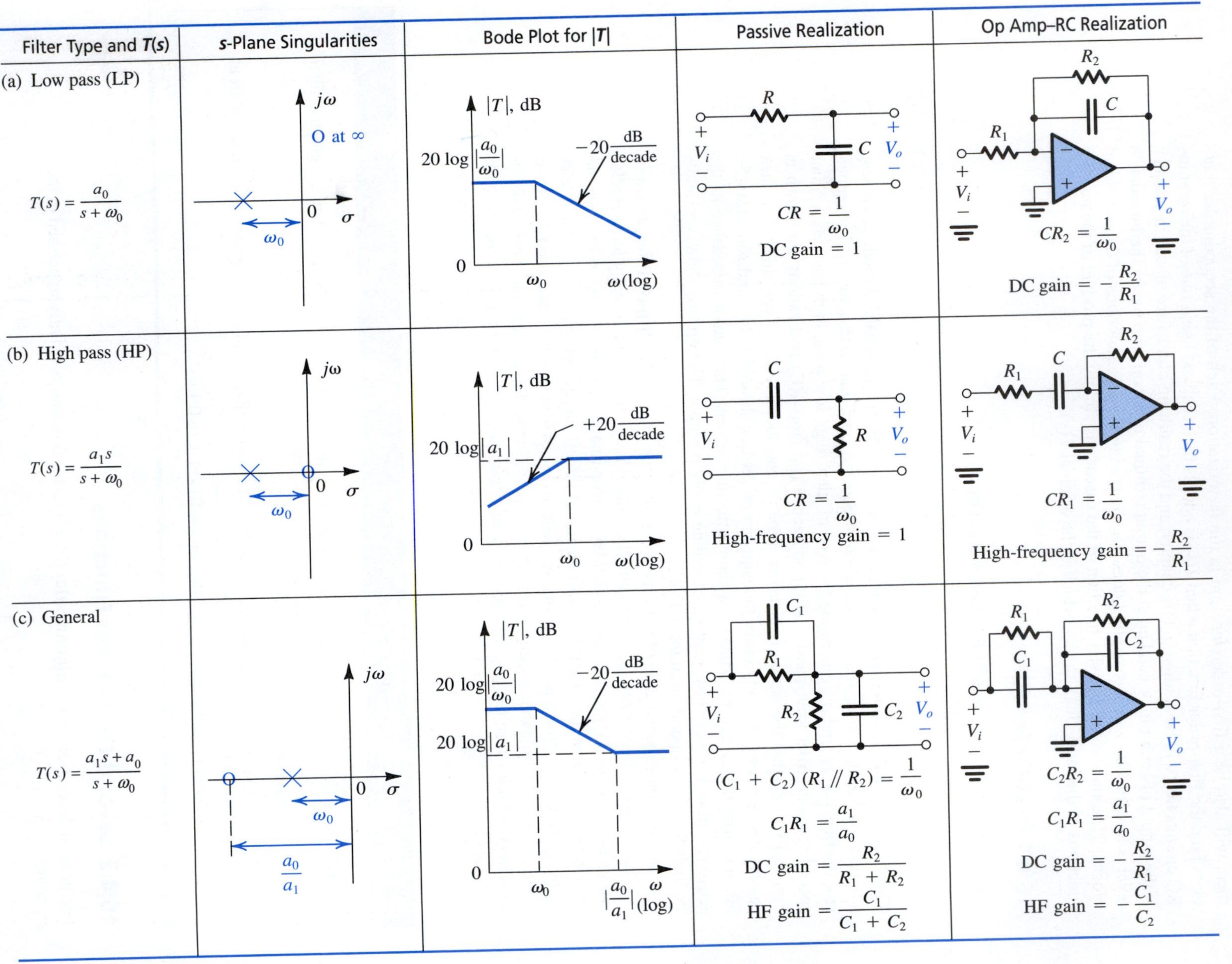

Figure 16.13 First-order filters.

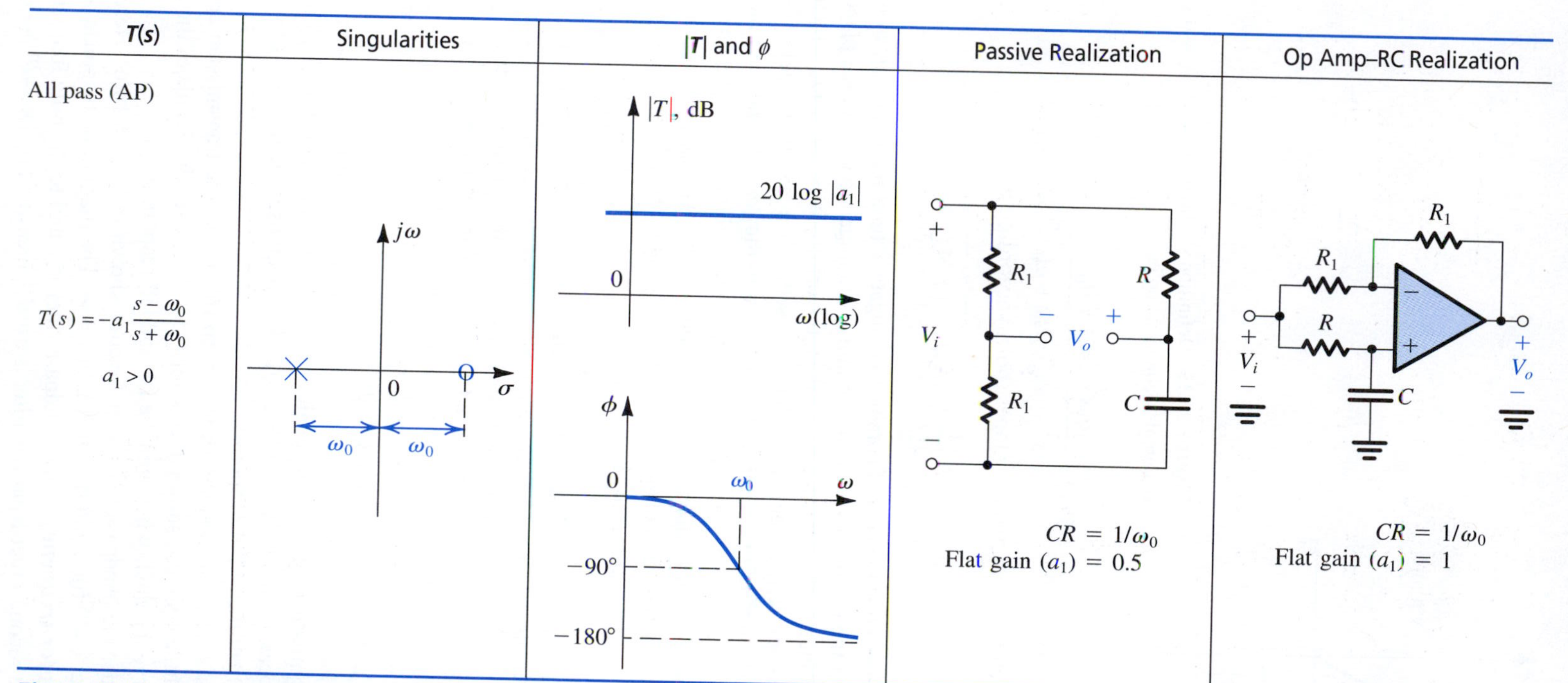

Figure 16.14 First-order all-pass filter.

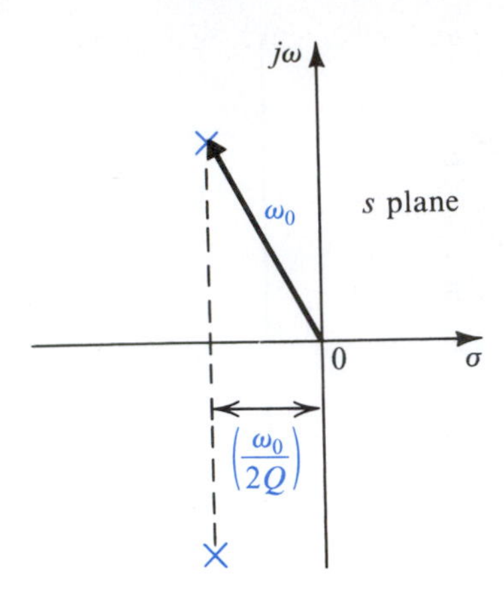

Figure 16.15 Definition of the parameters ω_0 and Q of a pair of complex-conjugate poles.

$$T(s) = \frac{a_2 s^2 + a_1 s + a_0}{s^2 + (\omega_0/Q)s + \omega_0^2} \tag{16.27}$$

where ω_0 and Q determine the natural modes (poles) according to

$$p_1, p_2 = -\frac{\omega_0}{2Q} \pm j\omega_0\sqrt{1-(1/4Q^2)} \tag{16.28}$$

We are usually interested in the case of complex-conjugate natural modes, obtained for $Q > 0.5$. Figure 16.15 shows the location of the pair of complex-conjugate poles in the s plane. Observe that the radial distance of the natural modes (from the origin) is equal to ω_0, which is known as the **pole frequency**. The parameter Q determines the distance of the poles from the $j\omega$ axis: the higher the value of Q, the closer the poles are to the $j\omega$ axis, and the more selective the filter response becomes. An infinite value for Q locates the poles on the $j\omega$ axis and can yield sustained oscillations in the circuit realization. A negative value of Q implies that the poles are in the right half of the s plane, which certainly produces oscillations. The parameter Q is called the **pole quality factor**, or simply, **pole Q**.

The transmission zeros of the second-order filter are determined by the numerator coefficients, a_0, a_1, and a_2. It follows that the numerator coefficients determine the type of second-order filter function (i.e., LP, HP, etc.). Seven special cases of interest are illustrated in Fig. 16.16. For each case we give the transfer function, the s-plane locations of the transfer function singularities, and the magnitude response. Circuit realizations for the various second-order filter functions will be given in subsequent sections.

All seven special second-order filters have a pair of complex-conjugate natural modes characterized by a frequency ω_0 and a quality factor Q.

In the low-pass (LP) case, shown in Fig. 16.16(a), the two transmission zeros are at $s = \infty$. The magnitude response can exhibit a peak with the details indicated. It can be shown that the peak occurs only for $Q > 1/\sqrt{2}$. The response obtained for $Q = 1/\sqrt{2}$ is the Butterworth, or maximally flat, response.

The high-pass (HP) function shown in Fig. 16.16(b) has both transmission zeros at $s = 0$ (dc). The magnitude response shows a peak for $Q > 1/\sqrt{2}$, with the details of the response as indicated. Observe the duality between the LP and HP responses.

Next consider the bandpass (BP) filter function shown in Fig. 16.16(c). Here, one transmission zero is at $s = 0$ (dc), and the other is at $s = \infty$. The magnitude response peaks at $\omega = \omega_0$. Thus the **center frequency** of the bandpass filter is equal to the pole frequency ω_0. The selectivity of the second-order bandpass filter is usually measured by its *3-dB bandwidth*. This is the difference between the two frequencies ω_1 and ω_2 at which the magnitude response is 3 dB below its maximum value (at ω_0). It can be shown that

$$\omega_1, \omega_2 = \omega_0\sqrt{1+(1/4Q^2)} \pm \frac{\omega_0}{2Q} \tag{16.29}$$

Thus,

$$BW \equiv \omega_2 - \omega_1 = \omega_0/Q \tag{16.30}$$

Observe that as Q increases, the bandwidth decreases and the bandpass filter becomes more selective.

If the transmission zeros are located on the $j\omega$ axis, at the complex-conjugate locations $\pm j\omega_n$, then the magnitude response exhibits zero transmission at $\omega = \omega_n$. Thus a **notch** in the magnitude response occurs at $\omega = \omega_n$, and ω_n is known as the **notch frequency**. Three cases of the second-order notch filter are possible: the regular notch, obtained when $\omega_n = \omega_0$ (Fig. 16.16d); the low-pass notch, obtained when $\omega_n > \omega_0$ (Fig. 16.16e); and the high-pass notch, obtained when $\omega_n < \omega_0$ (Fig. 16.16f). The reader is urged to verify the response details given in these figures (a rather tedious task, though!). Observe that in all notch cases, the transmission at dc and at $s = \infty$ is finite. This is so because there are no transmission zeros at either $s = 0$ or $s = \infty$.

The last special case of interest is the all-pass (AP) filter whose characteristics are illustrated in Fig. 16.16(g). Here the two transmission zeros are in the right half of the s plane, at the mirror-image locations of the poles. (This is the case for all-pass functions of any order.) The magnitude response of the all-pass function is constant over all frequencies; the **flat gain**, as it is called, is in our case equal to $|a_2|$. The frequency selectivity of the all-pass function is in its phase response.

EXERCISES

16.13 For a maximally flat second-order low-pass filter ($Q = 1/\sqrt{2}$), show that at $\omega = \omega_0$ the magnitude response is 3 dB below the value at dc.

16.14 Give the transfer function of a second-order bandpass filter with a center frequency of 10^5 rad/s, a center-frequency gain of 10, and a 3-dB bandwidth of 10^3 rad/s.

Ans. $T(s) = \dfrac{10^4 s}{s^2 + 10^3 s + 10^{10}}$

16.15 (a) For the second-order notch function with $\omega_n = \omega_0$, show that for the attenuation to be greater than A dB over a frequency band BW_a, the value of Q is given by

$$Q \le \frac{\omega_0}{BW_a\sqrt{10^{A/10} - 1}}$$

(*Hint:* First, show that any two frequencies, ω_1 and ω_2, at which $|T|$ is the same, are related by $\omega_1\omega_2 = \omega_0^2$.) (b) Use the result of (a) to show that the 3-dB bandwidth is ω_0/Q, as indicated in Fig. 16.16(d).

16.16 Consider a low-pass notch with $\omega_0 = 1$ rad/s, $Q = 10$, $\omega_n = 1.2$ rad/s, and a dc gain of unity. Find the frequency and magnitude of the transmission peak. Also find the high-frequency transmission.
Ans. 0.986 rad/s; 3.17; 0.69

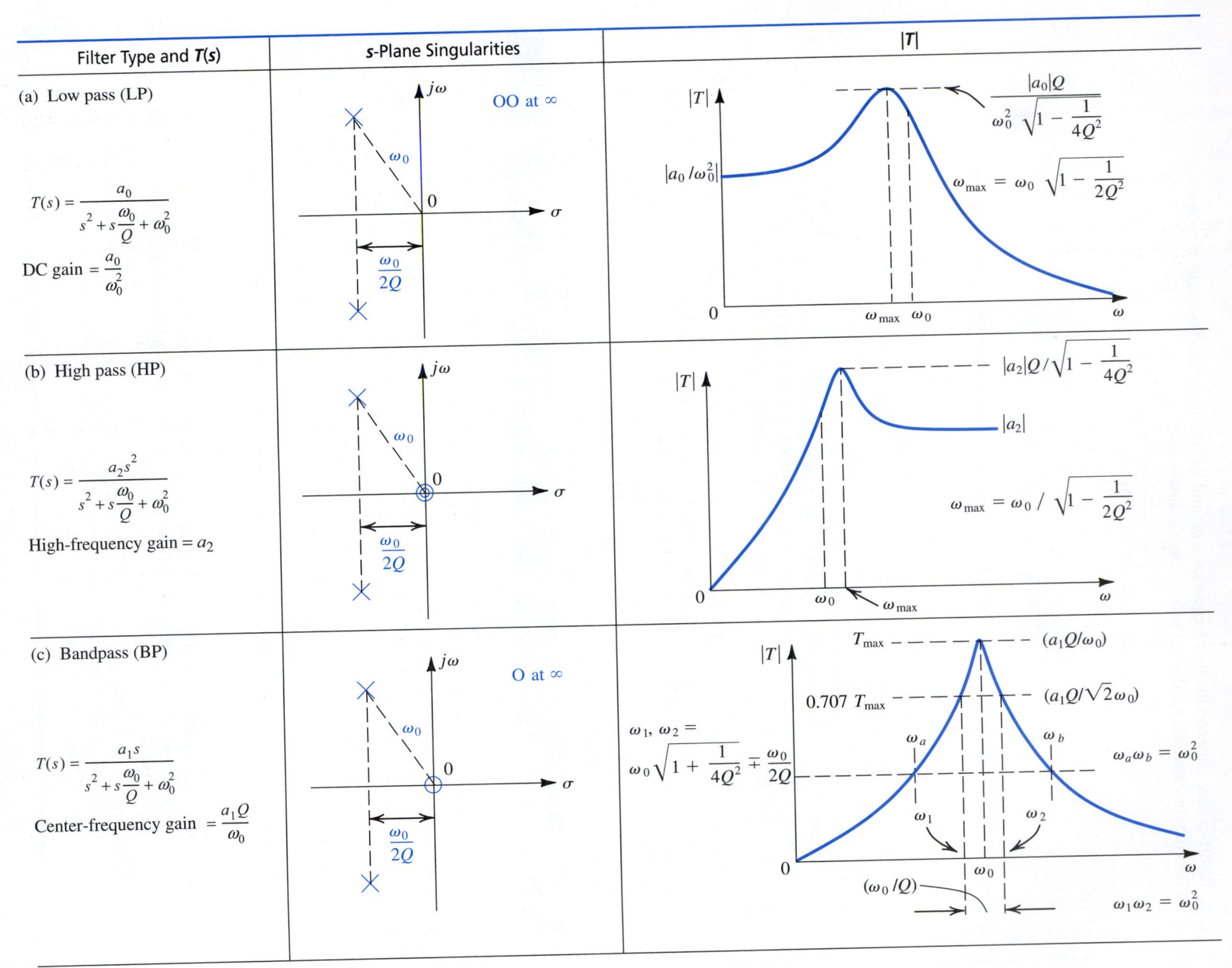

Figure 16.16 Second-order filtering functions.

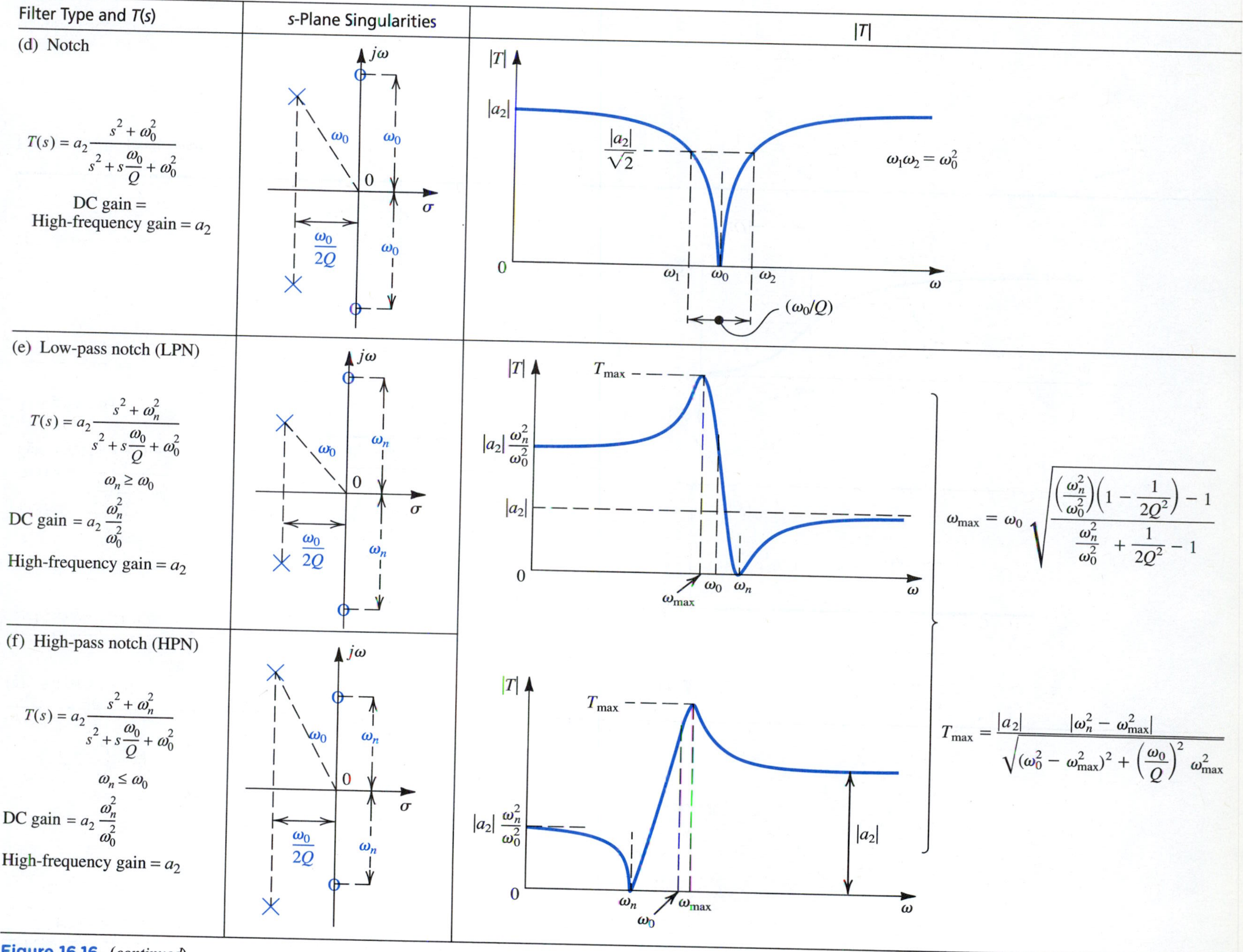

Figure 16.16 *(continued)*

(g) All pass (AP)

$$T(s) = a_2 \frac{s^2 - s\frac{\omega_0}{Q} + \omega_0^2}{s^2 + s\frac{\omega_0}{Q} + \omega_0^2}$$

Flat gain $= a_2$

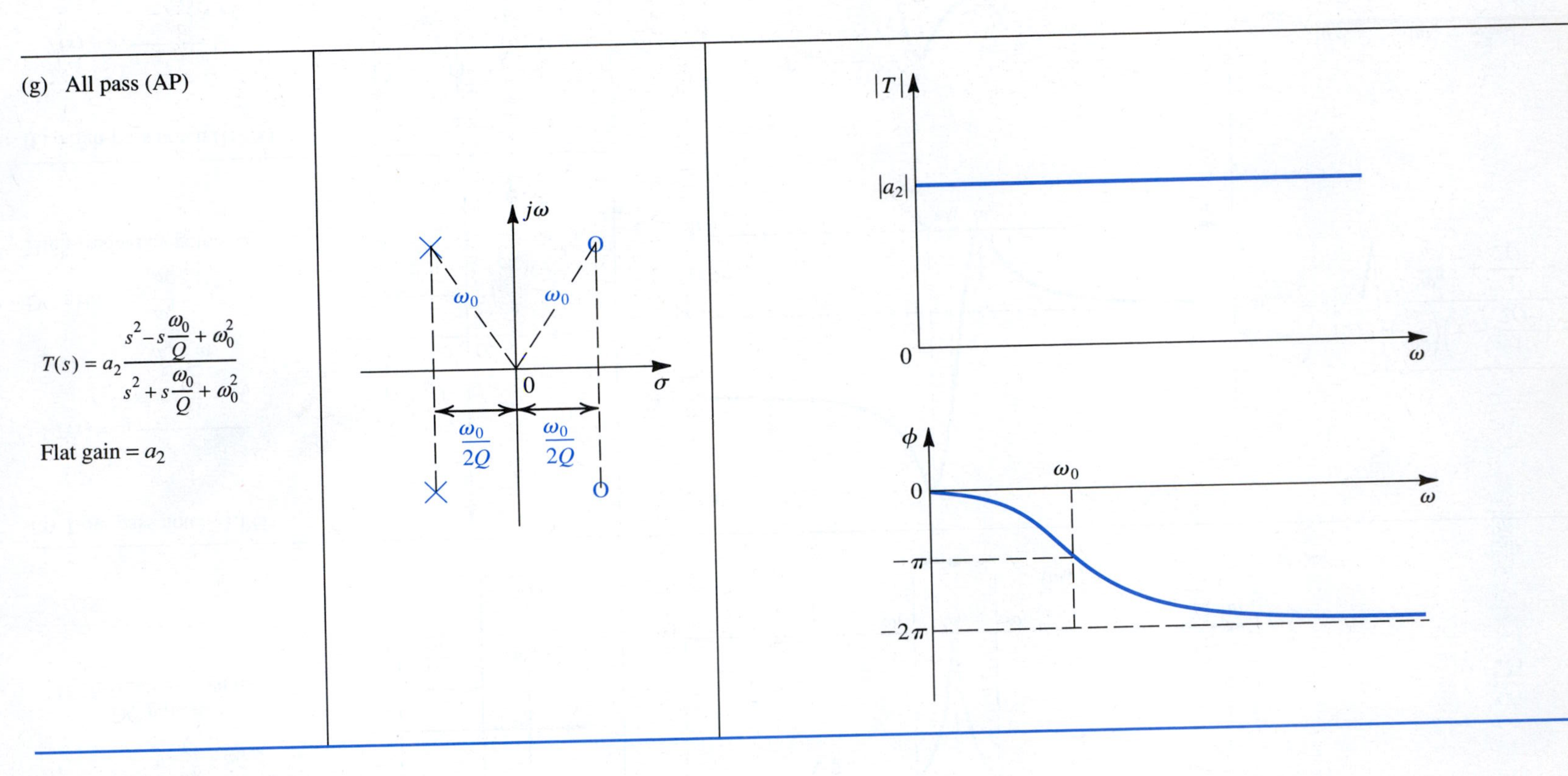

Figure 16.16 *(continued)*

16.5 The Second-Order LCR Resonator

In this section we shall study the second-order LCR resonator shown in Fig. 16.17(a). The use of this resonator to derive circuit realizations for the various second-order filter functions will be demonstrated. It will be shown in the next section that replacing the inductor L by a simulated inductance obtained using an op amp–RC circuit results in an op amp–RC resonator. The latter forms the basis of an important class of active-RC filters to be studied in Section 16.6.

16.5.1 The Resonator Natural Modes

The natural modes of the parallel resonance circuit of Fig. 16.17(a) can be determined by applying *an excitation that does not change the natural structure of the circuit.* Two possible ways of exciting the circuit are shown in Fig. 16.17(b) and (c). In Fig. 16.17(b) the resonator is excited with a current source I connected in parallel. Since, as far as the natural response of a circuit is concerned, an independent ideal current source is equivalent to an open circuit, the excitation of Fig. 16.17(b) does not alter the natural structure of the resonator. Thus the circuit in Fig. 16.17(b) can be used to determine the natural modes of the resonator by simply finding the poles of any response function. We can for instance take the voltage V_o across the resonator as the response and thus obtain the response function $V_o/I = Z$, where Z is the impedance of the parallel resonance circuit. It is obviously more convenient, however, to work in terms of the admittance Y; thus,

$$\frac{V_o}{I} = \frac{1}{Y} = \frac{1}{(1/sL)+sC+(1/R)} \tag{16.31}$$

$$= \frac{s/C}{s^2+s(1/CR)+(1/LC)}$$

Equating the denominator to the standard form $[s^2+s(\omega_0/Q)+\omega_0^2]$ leads to

$$\omega_0^2 = 1/LC \tag{16.32}$$

and

$$\omega_0/Q = 1/CR \tag{16.33}$$

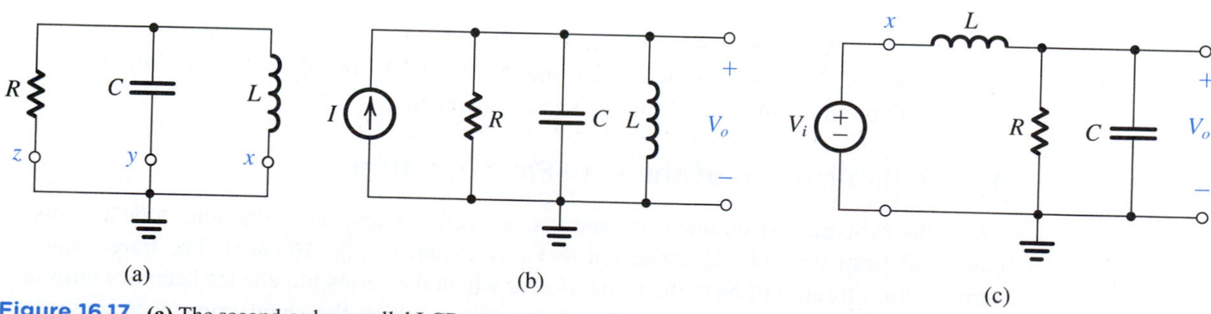

Figure 16.17 **(a)** The second-order parallel LCR resonator. **(b, c)** Two ways of exciting the resonator of **(a)** without changing its *natural structure:* resonator poles are those poles of V_o/I and V_o/V_i.

Thus,

$$\omega_0 = 1/\sqrt{LC} \tag{16.34}$$

$$Q = \omega_0 CR \tag{16.35}$$

These expressions should be familiar to the reader from studies of parallel resonance circuits in introductory courses on circuit theory.

An alternative way of exciting the parallel LCR resonator for the purpose of determining its natural modes is shown in Fig. 16.17(c). Here, node x of inductor L has been disconnected from ground and connected to an ideal voltage source V_i. Now, since as far as the natural response of a circuit is concerned, an ideal independent voltage source is equivalent to a short circuit, the excitation of Fig. 16.17(c) does not alter the natural structure of the resonator. Thus we can use the circuit in Fig. 16.17(c) to determine the natural modes of the resonator. These are the poles of any response function. For instance, we can select V_o as the response variable and find the transfer function V_o/V_i. The reader can easily verify that this will lead to the natural modes determined earlier.

In a design problem, we will be given ω_0 and Q and will be asked to determine L, C, and R. Equations (16.34) and (16.35) are two equations in the three unknowns. The one available degree of freedom can be utilized to set the impedance level of the circuit to a value that results in practical component values.

16.5.2 Realization of Transmission Zeros

Having selected the component values of the LCR resonator to realize a given pair of complex-conjugate natural modes, we now consider the use of the resonator to realize a desired filter type (e.g., LP, HP, etc.). Specifically, we wish to find out where to inject the input voltage signal V_i so that the transfer function V_o/V_i is the desired one. Toward that end, note that in the resonator circuit in Fig. 16.17(a), any of the nodes labeled x, y, or z can be disconnected from ground and connected to V_i without altering the circuit's natural modes. When this is done, the circuit takes the form of a voltage divider, as shown in Fig. 16.18(a). Thus the transfer function realized is

$$T(s) = \frac{V_o(s)}{V_i(s)} = \frac{Z_2(s)}{Z_1(s)+Z_2(s)} \tag{16.36}$$

We observe that *the transmission zeros are the values of s at which $Z_2(s)$ is zero, provided $Z_1(s)$ is not simultaneously zero, and the values of s at which $Z_1(s)$ is infinite, provided $Z_2(s)$ is not simultaneously infinite*. This statement makes physical sense: The output will be zero either when $Z_2(s)$ behaves as a short circuit or when $Z_1(s)$ behaves as an open circuit. If there is a value of s at which both Z_1 and Z_2 are zero, then V_o/V_i will be finite and no transmission zero is obtained. Similarly, if there is a value of s at which both Z_1 and Z_2 are infinite, then V_o/V_i will be finite and no transmission zero is realized.

16.5.3 Realization of the Low-Pass Function

Using the scheme just outlined, we see that to realize a low-pass function, node x is disconnected from ground and connected to V_i, as shown in Fig. 16.18(b). The transmission zeros of this circuit will be at the value of s for which the series impedance becomes infinite (sL becomes infinite at $s=\infty$) and the value of s at which the shunt impedance becomes zero ($1/[sC+(1/R)]$ becomes zero at $s=\infty$). Thus this circuit has two transmission zeros

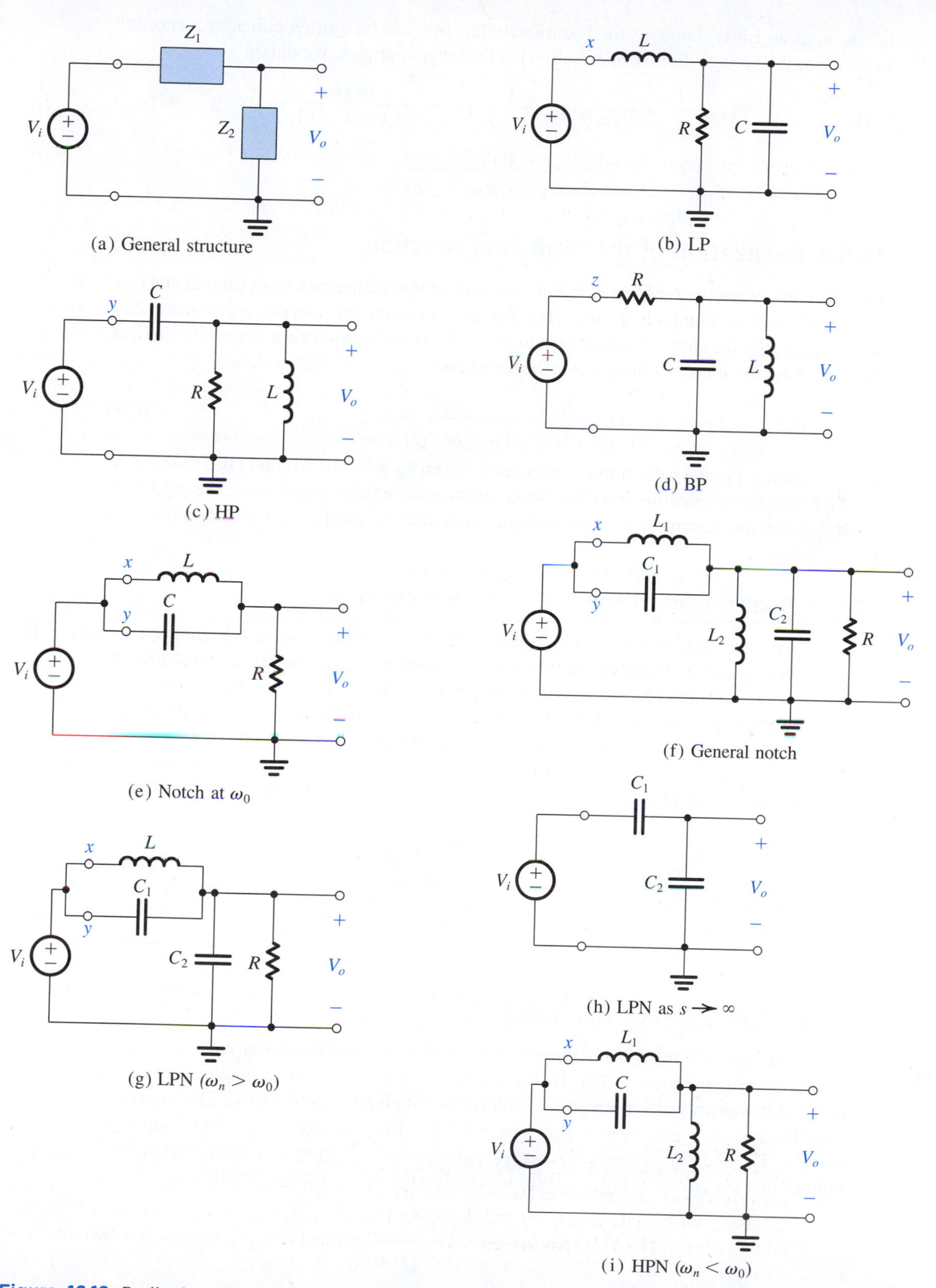

Figure 16.18 Realization of various second-order filter functions using the LCR resonator of Fig. 16.17(b): **(a)** general structure, **(b)** LP, **(c)** HP, **(d)** BP, **(e)** notch at ω_0, **(f)** general notch, **(g)** LPN ($\omega_n \geq \omega_0$), **(h)** LPN as $s \to \infty$, **(i)** HPN ($\omega_n < \omega_0$).

at $s = \infty$, as an LP is supposed to. The transfer function can be written either by inspection or by using the voltage divider rule. Following the latter approach, we obtain

$$T(s) \equiv \frac{V_o}{V_i} = \frac{Z_2}{Z_1 + Z_2} = \frac{Y_1}{Y_1 + Y_2} = \frac{1/sL}{(1/sL) + sC + (1/R)} \tag{16.37}$$

$$= \frac{1/LC}{s^2 + s(1/CR) + (1/LC)}$$

16.5.4 Realization of the High-Pass Function

To realize the second-order high-pass function, node y is disconnected from ground and connected to V_i, as shown in Fig. 16.18(c). Here the series capacitor introduces a transmission zero at $s = 0$ (dc), and the shunt inductor introduces another transmission zero at $s = 0$ (dc). Thus, by inspection, the transfer function may be written as

$$T(s) \equiv \frac{V_o}{V_i} = \frac{a_2 s^2}{s^2 + s(\omega_0/Q) + \omega_0^2} \tag{16.38}$$

where ω_0 and Q are the natural mode parameters given by Eqs. (16.34) and (16.35) and a_2 is the high-frequency transmission. The value of a_2 can be determined from the circuit by observing that as s approaches ∞, the capacitor approaches a short circuit and V_o approaches V_i, resulting in $a_2 = 1$.

16.5.5 Realization of the Bandpass Function

The bandpass function is realized by disconnecting node z from ground and connecting it to V_i, as shown in Fig. 16.18(d). Here the series impedance is resistive and thus does not introduce any transmission zeros. These are obtained as follows: One zero at $s = 0$ is realized by the shunt inductor, and one zero at $s = \infty$ is realized by the shunt capacitor. At the center frequency ω_0, the parallel LC-tuned circuit exhibits an infinite impedance, and thus no current flows in the circuit. It follows that at $\omega = \omega_0$, $V_o = V_i$. In other words, the center-frequency gain of the bandpass filter is unity. Its transfer function can be obtained as follows:

$$T(s) = \frac{Y_R}{Y_R + Y_L + Y_C} = \frac{1/R}{(1/R) + (1/sL) + sC} \tag{16.39}$$

$$= \frac{s(1/CR)}{s^2 + s(1/CR) + (1/LC)}$$

16.5.6 Realization of the Notch Functions

To obtain a pair of transmission zeros on the $j\omega$ axis, we use a parallel resonance circuit in the series arm, as shown in Fig. 16.18(e). Observe that this circuit is obtained by disconnecting both nodes x and y from ground and connecting them together to V_i. The impedance of the LC circuit becomes infinite at $\omega = \omega_0 = 1/\sqrt{LC}$, thus causing zero transmission at this frequency. The shunt impedance is resistive and thus does not introduce transmission zeros. It follows that the circuit in Fig. 16.18(e) will realize the notch transfer function

$$T(s) = a_2 \frac{s^2 + \omega_0^2}{s^2 + s(\omega_0/Q) + \omega_0^2} \tag{16.40}$$

The value of the high-frequency gain a_2 can be found from the circuit to be unity.

To obtain a notch-filter realization in which the notch frequency ω_n is arbitrarily placed relative to ω_0, we adopt a variation on the scheme above. We still use a parallel LC circuit in the series branch, as shown in Fig. 16.18(f) where L_1 and C_1 are selected so that

$$L_1 C_1 = 1/\omega_n^2 \tag{16.41}$$

Thus the L_1C_1 tank circuit will introduce a pair of transmission zeros at $\pm j\omega_n$, provided the L_2C_2 tank is not resonant at ω_n. Apart from this restriction, the values of L_2 and C_2 must be selected to ensure that the natural modes have not been altered; thus,

$$C_1 + C_2 = C \tag{16.42}$$

$$L_1 \| L_2 = L \tag{16.43}$$

In other words, when V_i is replaced by a short circuit, the circuit should reduce to the original LCR resonator. Another way of thinking about the circuit of Fig. 16.18(f) is that it is obtained from the original LCR resonator by lifting part of L and part of C off ground and connecting them to V_i.

It should be noted that in the circuit of Fig. 16.18(f), L_2 does *not* introduce a zero at $s = 0$ because at $s = 0$, the L_1C_1 circuit also has a zero. In fact, at $s = 0$ the circuit reduces to an inductive voltage divider with the dc transmission being $L_2/(L_1 + L_2)$. Similar comments can be made about C_2 and the fact that it does *not* introduce a zero at $s = \infty$.

The LPN and HPN filter realizations are special cases of the general notch circuit of Fig. 16.18(f). Specifically, for the LPN,

$$\omega_n > \omega_0$$

and thus

$$L_1 C_1 < (L_1 \| L_2)(C_1 + C_2)$$

This condition can be satisfied with L_2 eliminated (i.e., $L_2 = \infty$ and $L_1 = L$), resulting in the LPN circuit in Fig. 16.18(g). The transfer function can be written by inspection as

$$T(s) \equiv \frac{V_o}{V_i} = a_2 \frac{s^2 + \omega_n^2}{s^2 + s(\omega_0/Q) + \omega_0^2} \tag{16.44}$$

where $\omega_n^2 = 1/LC_1$, $\omega_0^2 = 1/L(C_1 + C_2)$, $\omega_0/Q = 1/CR$, and a_2 is the high-frequency gain. From the circuit we see that as $s \to \infty$, the circuit reduces to that in Fig. 16.18(h), for which

$$\frac{V_o}{V_i} = \frac{C_1}{C_1 + C_2}$$

Thus,

$$a_2 = \frac{C_1}{C_1 + C_2} \tag{16.45}$$

To obtain an HPN realization we start with the circuit of Fig. 16.18(f) and use the fact that $\omega_n < \omega_0$ to obtain

$$L_1C_1 > (L_1 \| L_2)(C_1 + C_2)$$

which can be satisfied while selecting $C_2 = 0$ (i.e., $C_1 = C$). Thus we obtain the reduced circuit shown in Fig. 16.18(i). Observe that as $s \to \infty$, V_o approaches V_i and thus the high-frequency gain is unity. Thus, the transfer function can be expressed as

$$T(s) \equiv \frac{V_o}{V_i} = \frac{s^2 + (1/L_1C)}{s^2 + s(1/CR) + [1/(L_1 \| L_2)C]} \tag{16.46}$$

16.5.7 Realization of the All-Pass Function

The all-pass transfer function

$$T(s) = \frac{s^2 - s(\omega_0/Q) + \omega_0^2}{s^2 + s(\omega_0/Q) + \omega_0^2} \tag{16.47}$$

can be written as

$$T(s) = 1 - \frac{s2(\omega_0/Q)}{s^2 + s(\omega_0/Q) + \omega_0^2} \tag{16.48}$$

The second term on the right-hand side is a bandpass function with a center-frequency gain of 2. We already have a bandpass circuit (Fig. 16.18d), but with a center-frequency gain of unity. We shall therefore attempt an all-pass realization with a flat gain of 0.5, that is,

$$T(s) = 0.5 - \frac{s(\omega_0/Q)}{s^2 + s(\omega_0/Q) + \omega_0^2}$$

This function can be realized using a voltage divider with a transmission ratio of 0.5 together with the bandpass circuit of Fig. 16.18(d). To effect the subtraction, the output of the all-pass circuit is taken between the output terminal of the voltage divider and that of the bandpass filter, as shown in Fig. 16.19. Unfortunately this circuit has the disadvantage of lacking a common ground terminal between the input and the output. An op amp–RC realization of the all-pass function will be presented in the next section.

EXERCISES

16.17 Use the circuit of Fig. 16.18(b) to realize a second-order low-pass function of the maximally flat type with a 3-dB frequency of 100 kHz.
Ans. Selecting $R = 1\ \text{k}\Omega$, we obtain $C = 1125$ pF and $L = 2.25$ mH.

16.18 Use the circuit of Fig. 16.18(e) to design a notch filter to eliminate a bothersome power-supply hum at a 60-Hz frequency. The filter is to have a 3-dB bandwidth of 10 Hz (i.e., the attenuation is greater than 3 dB over a 10-Hz band around the 60-Hz center frequency; see Exercise 16.15 and Fig. 16.16d). Use $R = 10\ \text{k}\Omega$.
Ans. $C = 1.6\ \mu\text{F}$ and $L = 4.42$ H (Note the large inductor required. This is the reason passive filters are not practical in low-frequency applications.)

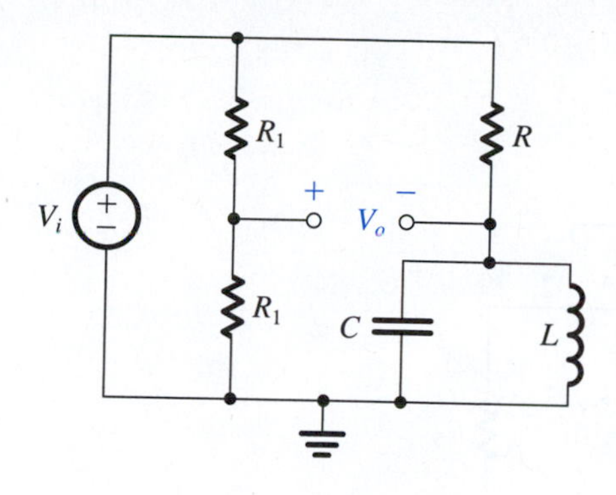

Figure 16.19 Realization of the second-order all-pass transfer function using a voltage divider and an LCR resonator.

16.6 Second-Order Active Filters Based on Inductor Replacement

In this section, we study a family of op amp–RC circuits that realize the various second-order filter functions. The circuits are based on an op amp–RC resonator obtained by replacing the inductor L in the LCR resonator with an op amp–RC circuit that has an inductive input impedance.

16.6.1 The Antoniou Inductance-Simulation Circuit

Over the years, many op amp–RC circuits have been proposed for simulating the operation of an inductor. Of these, one circuit invented by A. Antoniou[5] (see Antoniou, 1969) has proved to be the "best." By "best" we mean that the operation of the circuit is very tolerant of the nonideal properties of the op amps, in particular their finite gain and bandwidth. Figure 16.20(a) shows the Antoniou inductance-simulation circuit. If the circuit is fed at its input (node 1) with a voltage source V_1 and the input current is denoted I_1, then for ideal op amps the input impedance can be shown to be

$$Z_{\text{in}} \equiv V_1/I_1 = sC_4R_1R_3R_5/R_2 \tag{16.49}$$

which is that of an inductance L given by

$$L = C_4R_1R_3R_5/R_2 \tag{16.50}$$

Figure 16.20(b) shows the analysis of the circuit assuming that the op amps are ideal and thus that a virtual short circuit appears between the two input terminals of each op amp, and assuming also that the input currents of the op amps are zero. The analysis begins at node 1, which is assumed to be fed by a voltage source V_1, and proceeds step by step, with the order of the steps indicated by the circled numbers. The result of the analysis is the expression shown for the input current I_1 from which Z_{in} is found.

The design of this circuit is usually based on selecting $R_1 = R_2 = R_3 = R_5 = R$ and $C_4 = C$, which leads to $L = CR^2$. Convenient values are then selected for C and R to yield the

[5] Andreas Antoniou is a Canadian academic, currently (2009) a member of the faculty of the University of Victoria, British Columbia.

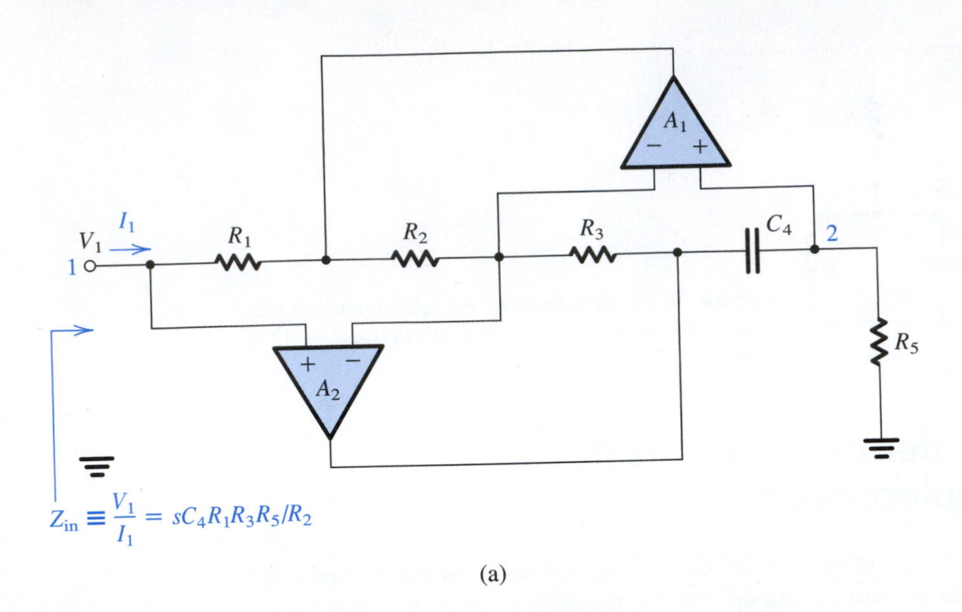

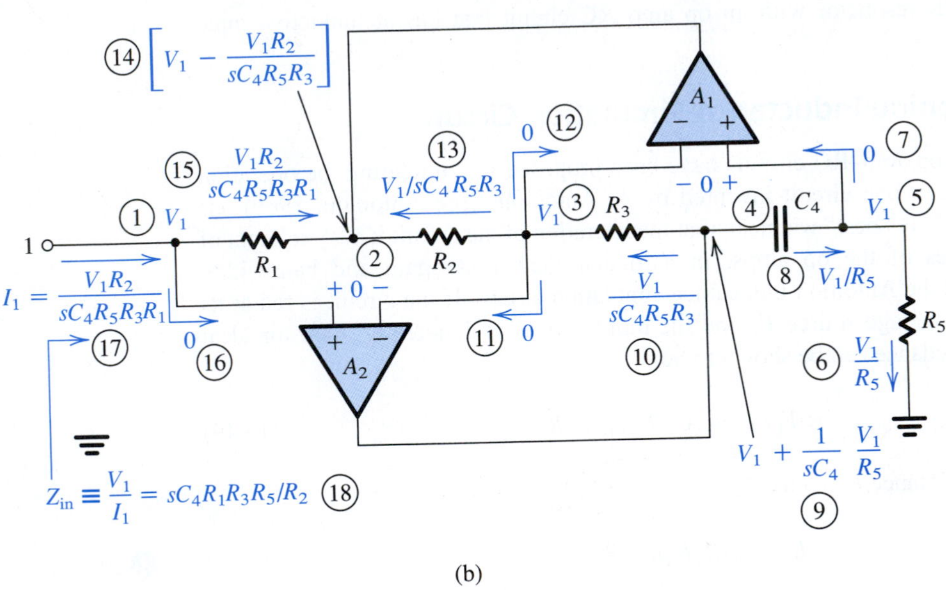

Figure 16.20 **(a)** The Antoniou inductance-simulation circuit. **(b)** Analysis of the circuit assuming ideal op amps. The order of the analysis steps is indicated by the circled numbers.

desired inductance value L. More details on this circuit and the effect of the nonidealities of the op amps on its performance can be found in Sedra and Brackett (1978).

16.6.2 The Op Amp–RC Resonator

Figure 16.21(a) shows the LCR resonator we studied in detail in Section 16.5. Replacing the inductor L with a simulated inductance realized by the Antoniou circuit of Fig. 16.20(a) results in the op amp–RC resonator of Fig. 16.21(b). (Ignore for the moment the additional

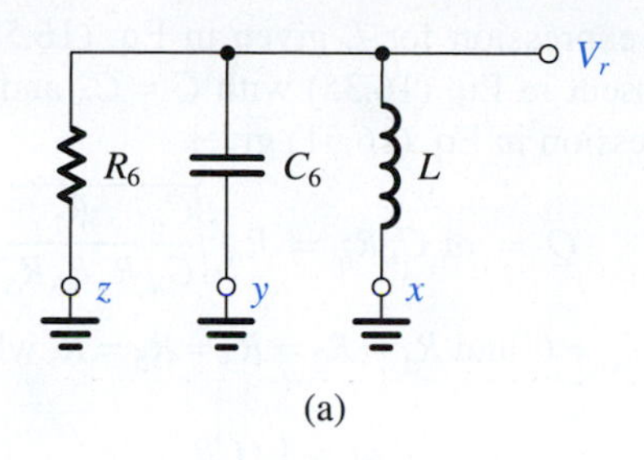

(a)

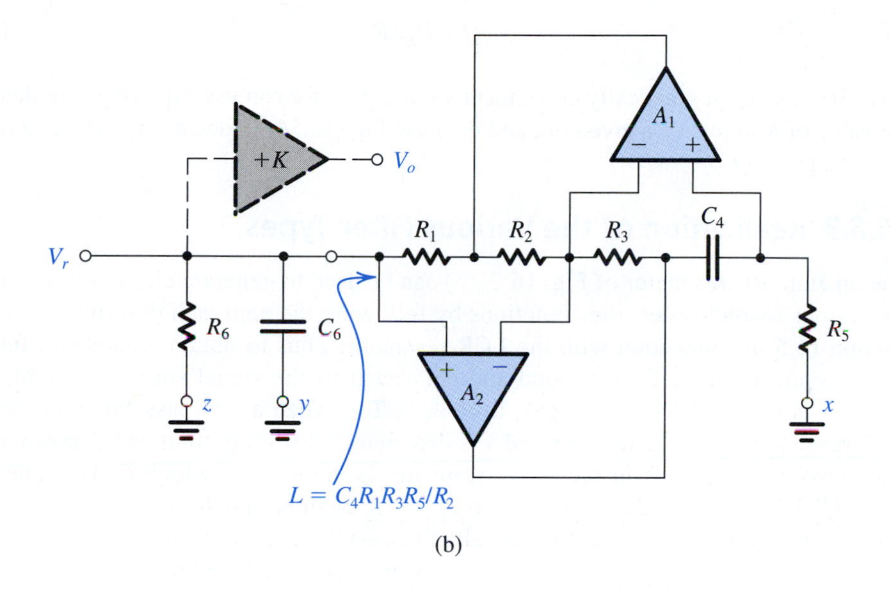

(b)

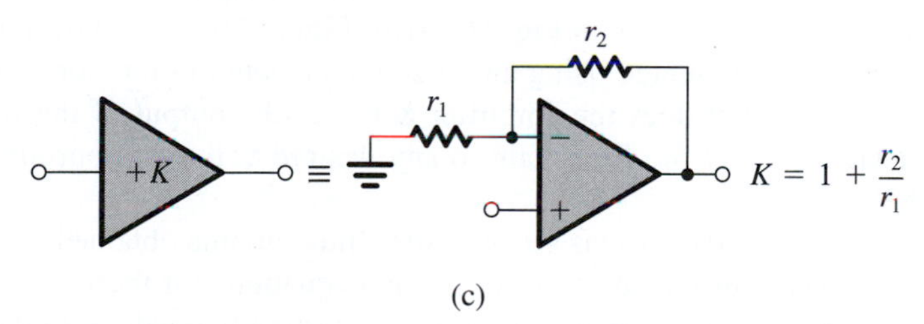

(c)

Figure 16.21 **(a)** An LCR resonator. **(b)** An op amp–RC resonator obtained by replacing the inductor L in the LCR resonator of **(a)** with a simulated inductance realized by the Antoniou circuit of Fig. 16.20(a). **(c)** Implementation of the buffer amplifier K.

amplifier drawn with broken lines.) The circuit of Fig. 16.21(b) is a second-order resonator having a pole frequency

$$\omega_0 = 1/\sqrt{LC_6} = 1/\sqrt{C_4C_6R_1R_3R_5/R_2} \qquad (16.51)$$

where we have used the expression for L given in Eq. (16.50). The pole Q factor can be obtained using the expressent in Eq. (16.35) with $C = C_6$ and $R = R_6$; thus, $Q = \omega_0 C_6 R_6$. Replacing ω_0 by the expression in Eq. (16.51) gives

$$Q = \omega_0 C_6 R_6 = R_6 \sqrt{\frac{C_6}{C_4} \frac{R_2}{R_1 R_3 R_5}} \tag{16.52}$$

Usually one selects $C_4 = C_6 = C$ and $R_1 = R_2 = R_3 = R_5 = R$, which results in

$$\omega_0 = 1/CR \tag{16.53}$$

$$Q = R_6/R \tag{16.54}$$

Thus, if we select a practically convenient value for C, we can use Eq. (16.53) to determine the value of R to realize a given ω_0, and then use Eq. (16.54) to determine the value of R_6 to realize a given Q.

16.6.3 Realization of the Various Filter Types

The op amp–RC resonator of Fig. 16.21(b) can be used to generate circuit realizations for the various second-order filter functions by following the approach described in detail in Section 16.5 in connection with the LCR resonator. Thus to obtain a bandpass function, we disconnect node z from ground and connect it to the signal source V_i. A high-pass function is obtained by injecting V_i to node y. To realize a low-pass function using the LCR resonator, the inductor terminal x is disconnected from ground and connected to V_i. The corresponding node in the active resonator is the node at which R_5 is connected to ground,[6] labeled as node x in Fig. 16.21(b). A regular notch function ($\omega_n = \omega_0$) is obtained by feeding V_i to nodes x and y. In all cases the output can be taken as the voltage across the resonance circuit, V_r. However, this is not a convenient node to use as the filter output terminal because connecting a load there would change the filter characteristics. The problem can be solved easily by utilizing a buffer amplifier. This is the amplifier of gain K, drawn with broken lines in Fig. 16.21(b). Figure 16.21(c) shows how this amplifier can be simply implemented using an op amp connected in the noninverting configuration. Note that not only does the amplifier K buffer the output of the filter, but it also allows the designer to set the filter gain to any desired value by appropriately selecting the value of K.

Figure 16.22 shows the various second-order filter circuits obtained from the resonator of Fig. 16.21(b). The transfer functions and design equations for these circuits are given in Table 16.1. Note that the transfer functions can be written by analogy to those of the LCR resonator. We have already commented on the LP, HP, BP, and regular-notch circuits given in Fig. 16.22(a) to (d). The LPN and HPN circuits in Fig. 16.22(e) and (f) are obtained by

[6] This point might not be obvious! The reader, however, can show by direct analysis that when V_i is fed to this node, the function V_r/V_i is indeed low pass.

direct analogy to their LCR counterparts in Fig. 16.18(g) and (i), respectively. The all-pass circuit in Fig. 16.22(g), however, deserves some explanation.

16.6.4 The All-Pass Circuit

From Eq. (16.48) we see that an all-pass function with a flat gain of unity can be written as

$$AP = 1 - (\text{BP with a center-frequency gain of 2}) \qquad (16.55)$$

Two circuits whose transfer functions are related in this fashion are said to be **complementary**.[7] Thus the all-pass circuit with unity flat gain is the complement of the bandpass circuit with a center-frequency gain of 2. A simple procedure exists for obtaining the complement of a given linear circuit: Disconnect all the circuit nodes that are connected to ground and connect them to V_i, and disconnect all the nodes that are connected to V_i and connect them to ground. That is, *interchanging input and ground in a linear circuit generates a circuit whose transfer function is the complement of that of the original circuit*.

Returning to the problem at hand, we first use the circuit of Fig. 16.22(c) to realize a BP with a gain of 2 by simply selecting $K = 2$ and implementing the buffer amplifier with the circuit of Fig. 16.21(c) with $r_1 = r_2$. We then interchange input and ground and thus obtain the all-pass circuit of Fig. 16.22(g).

Finally, in addition to being simple to design, the circuits in Fig. 16.22 exhibit excellent performance. They can be used on their own to realize second-order filter functions, or they can be cascaded to implement high-order filters.

EXERCISES

D16.19 Use the circuit of Fig. 16.22(c) to design a second-order bandpass filter with a center frequency of 10 kHz, a 3-dB bandwidth of 500 Hz, and a center-frequency gain of 10. Use $C = 1.2$ nF.
Ans. $R_1 = R_2 = R_3 = R_5 = 13.26$ kΩ; $R_6 = 265$ kΩ; $C_4 = C_6 = 1.2$ nF; $K = 10$, $r_1 = 10$ kΩ, $r_2 =$ 90 kΩ

D16.20 Realize the Chebyshev filter of Example 16.2, whose transfer function is given in Eq. (16.25), as the cascade connection of three circuits: two of the type shown in Fig. 16.22(a) and one first-order op amp–RC circuit of the type shown in Fig. 16.13(a). Note that you can make the dc gain of all sections equal to unity. Do so. Use as many 10-kΩ resistors as possible.
Ans. First-order section: $R_1 = R_2 = 10$ kΩ, $C = 5.5$ nF; second-order section with $\omega_0 = 4.117 \times 10^4$ rad/s and $Q = 1.4$: $R_1 = R_2 = R_3 = R_5 = 10$ kΩ, $R_6 = 14$ kΩ, $C_4 = C_6 = 2.43$ nF, $r_1 = \infty$, $r_2 = 0$; second-order section with $\omega_0 = 6.246 \times 10^4$ rad/s and $Q = 5.56$: $R_1 = R_2 = R_3 = R_5 = 10$ kΩ, $R_6 =$ 55.6 kΩ, $C_4 = C_6 = 1.6$ nF, $r_1 = \infty$, $r_2 = 0$

[7] More about complementary circuits will be presented later in conjunction with Fig. 16.31.

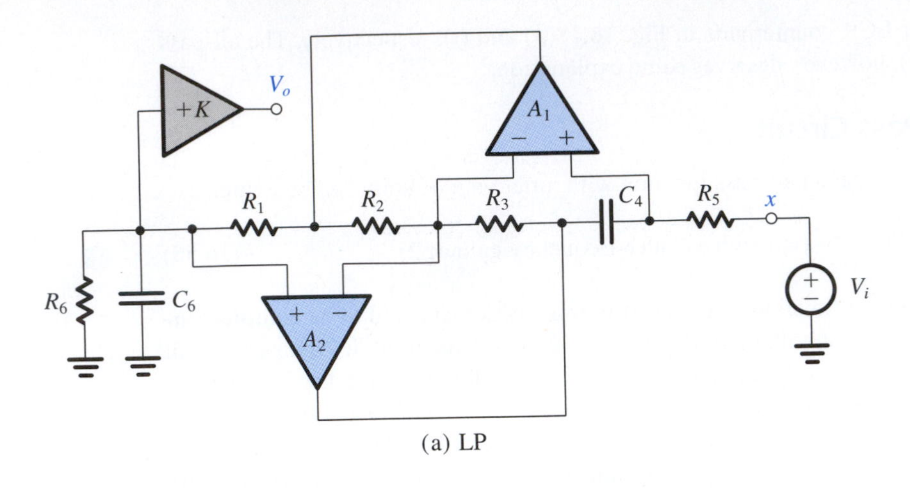

(a) LP

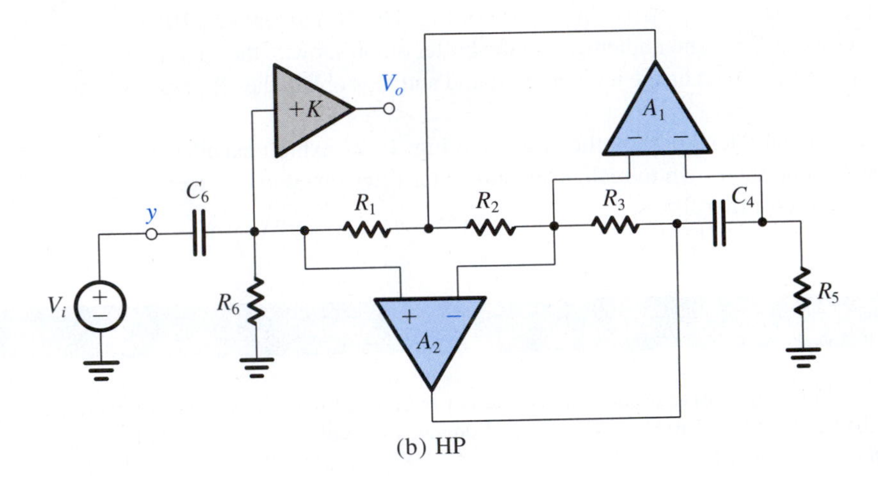

(b) HP

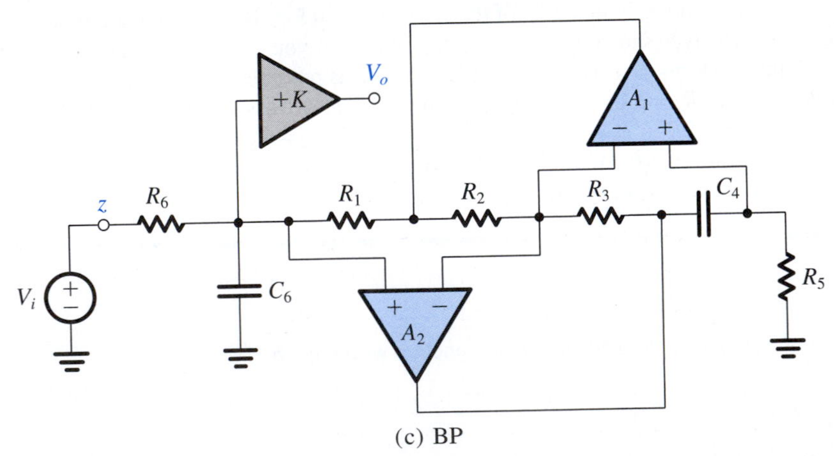

(c) BP

Figure 16.22 Realizations for the various second-order filter functions using the op amp–RC resonator of Fig. 16.21(b): **(a)** LP, **(b)** HP, **(c)** BP. The circuits are based on the LCR circuit in Fig. 16.18. Design considerations are given in Table 16.1.

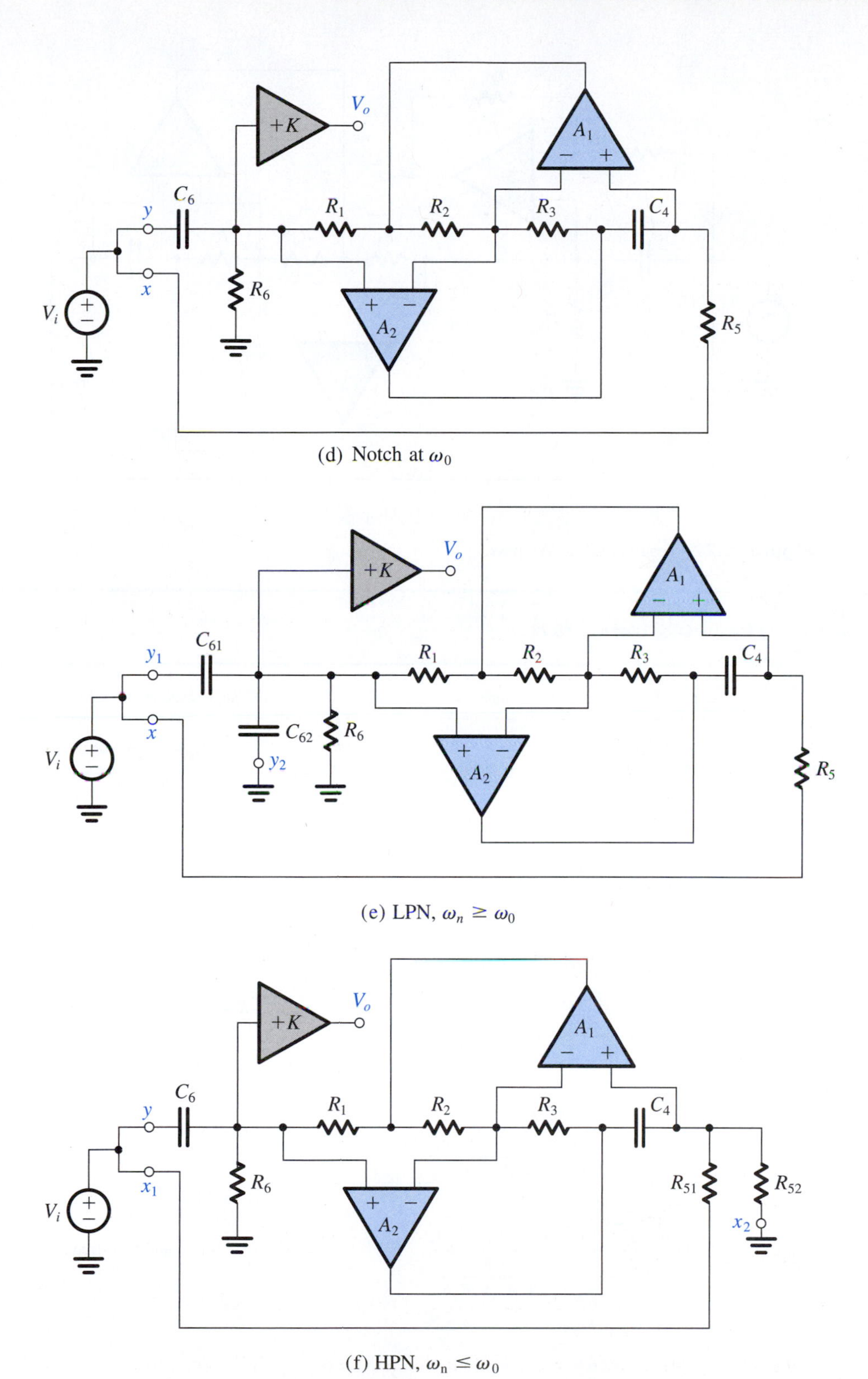

Figure 16.22 (*continued*) **(d)** Notch at ω_0, **(e)** LPN, $\omega_n \geq \omega_0$, **(f)** HPN, $\omega_n \leq \omega_0$.

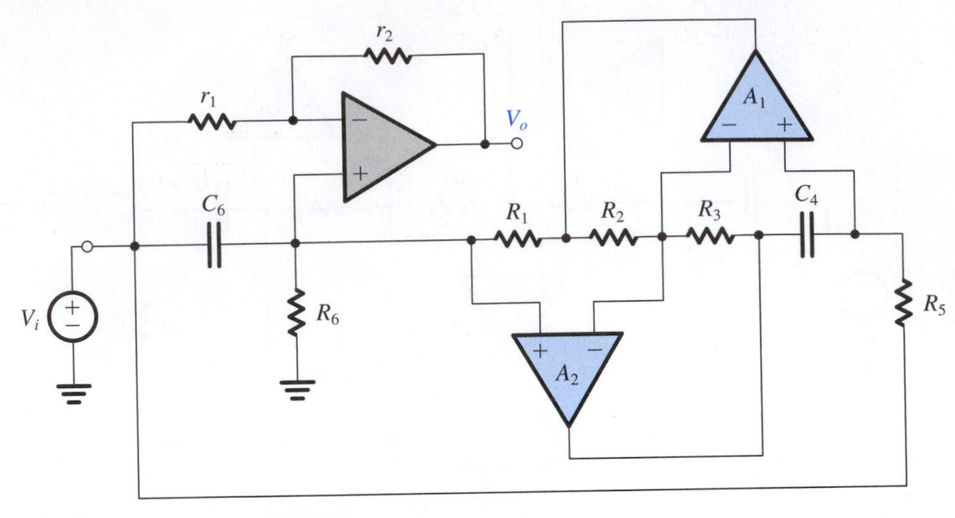

Figure 16.22 (*continued*) (g) All pass.

Table 16.1 Design Data for the Circuits of Fig. 16.22

Circuit	Transfer Function and Other Parameters	Design Equations
Resonator Fig. 16.21(b)	$\omega_0 = 1/\sqrt{C_4C_6R_1R_3R_5/R_2}$ $Q = R_6\sqrt{\frac{C_6}{C_4}\frac{R_2}{R_1R_3R_5}}$	$C_4 = C_6 = C$ (practical value) $R_1 = R_2 = R_3 = R_5 = 1/\omega_0 C$ $R_6 = Q/\omega_0 C$
Low-pass (LP) Fig. 16.22(a)	$T(s) = \dfrac{KR_2/C_4C_6R_1R_3R_5}{s^2 + s\frac{1}{C_6R_6} + \frac{R_2}{C_4C_6R_1R_3R_5}}$	K = DC gain
High-pass (HP) Fig. 16.22(b)	$T(s) = \dfrac{Ks^2}{s^2 + s\frac{1}{C_6R_6} + \frac{R_2}{C_4C_6R_1R_3R_5}}$	K = High-frequency gain
Bandpass (BP) Fig. 16.22(c)	$T(s) = \dfrac{Ks/C_6R_6}{s^2 + s\frac{1}{C_6R_6} + \frac{R_2}{C_4C_6R_1R_3R_5}}$	K = Center-frequency gain
Regular notch (N) Fig. 16.22(d)	$T(s) = \dfrac{K[s^2 + (R_2/C_4C_6R_1R_3R_5)]}{s^2 + s\frac{1}{C_6R_6} + \frac{R_2}{C_4C_6R_1R_3R_5}}$	K = Low- and high-frequency gain

Circuit	Transfer function and other parameters	Design equations
Low-pass notch (LPN) Fig. 16.22(e)	$T(s) = K\dfrac{C_{61}}{C_{61}+C_{62}} \times \dfrac{s^2 + (R_2/C_4C_{61}R_1R_3R_5)}{s^2 + s\dfrac{1}{(C_{61}+C_{62})R_6} + \dfrac{R_2}{C_4(C_{61}+C_{62})R_1R_3R_5}}$	K = DC gain
	$\omega_n = 1/\sqrt{C_4C_{61}R_1R_3R_5/R_2}$	$C_{61}+C_{62} = C_6 = C$
	$\omega_0 = 1/\sqrt{C_4(C_{61}+C_{62})R_1R_3R_5/R_2}$	$C_{61} = C(\omega_0/\omega_n)^2$
	$Q = R_6\sqrt{\dfrac{C_{61}+C_{62}}{C_4}\dfrac{R_2}{R_1R_3R_5}}$	$C_{62} = C - C_{61}$
High-pass notch (HPN) Fig. 16.22(f)	$T(s) = K\dfrac{s^2 + (R_2/C_4C_6R_1R_3R_{51})}{s^2 + s\dfrac{1}{C_6R_6} + \dfrac{R_2}{C_4C_6R_1R_3}\left(\dfrac{1}{R_{51}} + \dfrac{1}{R_{52}}\right)}$	K = High-frequency gain
	$\omega_n = 1/\sqrt{C_4C_6R_1R_3R_{51}/R_2}$	$\dfrac{1}{R_{51}} + \dfrac{1}{R_{52}} = \dfrac{1}{R_5} = \omega_0 C$
	$\omega_0 = \sqrt{\dfrac{R_2}{C_4C_6R_1R_3}\left(\dfrac{1}{R_{51}} + \dfrac{1}{R_{52}}\right)}$	$R_{51} = R_5(\omega_0/\omega_n)^2$ $R_{52} = R_5/[1-(\omega_n/\omega_0)^2]$
	$Q = R_6\sqrt{\dfrac{C_6}{C_4}\dfrac{R_2}{R_1R_3}\left(\dfrac{1}{R_{51}} + \dfrac{1}{R_{52}}\right)}$	
All-pass (AP) Fig. 16.22(g)	$T(s) = \dfrac{s^2 - s\dfrac{1}{C_6R_6}\dfrac{r_2}{r_1} + \dfrac{R_2}{C_4C_6R_1R_3R_5}}{s^2 + s\dfrac{1}{C_6R_6} + \dfrac{R_2}{C_4C_6R_1R_3R_5}}$	$r_1 = r_2 = r$ (arbitrary)
	$\omega_z = \omega_0 \quad Q_z = Q(r_1/r_2) \quad$ Flat gain $= 1$	Adjust r_2 to make $Q_z = Q$

16.7 Second-Order Active Filters Based on the Two-Integrator-Loop Topology

In this section, we study another family of op amp–RC circuits that realize second-order filter functions. The circuits are based on the use of two integrators connected in cascade in an overall feedback loop and are thus known as two-integrator-loop circuits.

16.7.1 Derivation of the Two-Integrator-Loop Biquad

To derive the two-integrator-loop biquadratic circuit, or **biquad** as it is commonly known,[8] consider the second-order high-pass transfer function

$$\frac{V_{hp}}{V_i} = \frac{Ks^2}{s^2 + s(\omega_0/Q) + \omega_0^2} \tag{16.56}$$

[8] The name biquad stems from the fact that this circuit in its most general form is capable of realizing a biquadratic transfer function, that is, one that is the ratio of two quadratic polynomials.

where K is the high-frequency gain. Cross-multiplying Eq. (16.56) and dividing both sides of the resulting equation by s^2 (to get all the terms involving s in the form $1/s$, which is the transfer function of an integrator) gives

$$V_{hp} + \frac{1}{Q}\left(\frac{\omega_0}{s}V_{hp}\right) + \left(\frac{\omega_0^2}{s^2}V_{hp}\right) = KV_i \tag{16.57}$$

In this equation we observe that the signal $(\omega_0/s)V_{hp}$ can be obtained by passing V_{hp} through an integrator with a time constant equal to $1/\omega_0$. Furthermore, passing the resulting signal through another identical integrator results in the third signal involving V_{hp} in Eq. (16.57)—namely, $(\omega_0^2/s^2)V_{hp}$. Figure 16.23(a) shows a block diagram for such a two-integrator arrangement. Note that in anticipation of the use of the inverting op-amp Miller integrator circuit (Section 2.5.2) to implement each integrator, the integrator blocks in Fig. 16.23(a) have been assigned negative signs.

The problem still remains, however, of how to form V_{hp}, the input signal feeding the two cascaded integrators. Toward that end, we rearrange Eq. (16.57), expressing V_{hp} in terms of its single- and double-integrated versions and of V_i as

$$V_{hp} = KV_i - \frac{1}{Q}\frac{\omega_0}{s}V_{hp} - \frac{\omega_0^2}{s^2}V_{hp} \tag{16.58}$$

which suggests that V_{hp} can be obtained by using the weighted summer of Fig. 16.23(b). Now it should be easy to see that a complete block diagram realization can be obtained by combining the integrator blocks of Fig. 16.23(a) with the summer block of Fig. 16.23(b), as shown in Fig. 16.23(c).

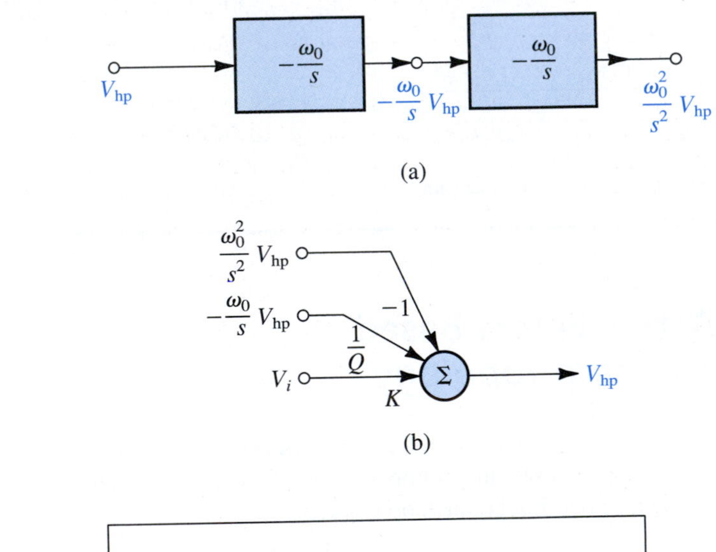

Figure 16.23 Derivation of a block diagram realization of the two-integrator-loop biquad.

In the realization of Fig. 16.23(c), V_{hp}, obtained at the output of the summer, realizes the high-pass transfer function $T_{hp} \equiv V_{hp}/V_i$ of Eq. (16.56). The signal at the output of the first integrator is $-(\omega_0/s)V_{hp}$, which is a bandpass function,

$$\frac{(-\omega_0/s)V_{hp}}{V_i} = -\frac{K\omega_0 s}{s^2 + s(\omega_0/Q) + \omega_0^2} = T_{bp}(s) \qquad (16.59)$$

Therefore the signal at the output of the first integrator is labeled V_{bp}. Note that the center-frequency gain of the bandpass filter realized is equal to $-KQ$.

In a similar fashion, we can show that the transfer function realized at the output of the second integrator is the low-pass function,

$$\frac{(\omega_0^2/s^2)V_{hp}}{V_i} = \frac{K\omega_0^2}{s^2 + s(\omega_0/Q) + \omega_0^2} = T_{lp}(s) \qquad (16.60)$$

Thus the output of the second integrator is labeled V_{lp}. Note that the dc gain of the low-pass filter realized is equal to K.

We conclude that the two-integrator-loop biquad shown in block diagram form in Fig. 16.23(c) realizes the three basic second-order filtering functions, LP, BP, and HP, *simultaneously*. This versatility has made the circuit very popular and has given it the name *universal active filter*.

16.7.2 Circuit Implementation

To obtain an op-amp circuit implementation of the two-integrator-loop biquad of Fig. 16.23(c), we replace each integrator with a Miller integrator circuit having $CR = 1/\omega_0$, and we replace the summer block with an op-amp summing circuit that is capable of assigning both positive and negative weights to its inputs. The resulting circuit, known as the Kerwin–Huelsman–Newcomb or **KHN biquad**, after its inventors, is shown in Fig. 16.24(a). Given values for ω_0, Q, and K, the design of the circuit is straightforward: We select suitably practical values for the components of the integrators C and R so that $CR = 1/\omega_0$. To determine the values of the resistors associated with the summer, we first use *superposition* to express the output of the summer V_{hp} in terms of its inputs, V_i, V_{bp} and V_{lp} as

$$V_{hp} = V_i\frac{R_3}{R_2+R_3}\left(1+\frac{R_f}{R_1}\right) + V_{bp}\frac{R_2}{R_2+R_3}\left(1+\frac{R_f}{R_1}\right) - V_{lp}\frac{R_f}{R_1}$$

Substituting $V_{bp} = -(\omega_0/s)V_{hp}$ and $V_{lp} = (\omega_0^2/s^2)V_{hp}$ gives

$$V_{hp} = \frac{R_3}{R_2+R_3}\left(1+\frac{R_f}{R_1}\right)V_i + \frac{R_2}{R_2+R_3}\left(1+\frac{R_f}{R_1}\right)\left(-\frac{\omega_0}{s}V_{hp}\right) - \frac{R_f}{R_1}\left(\frac{\omega_0^2}{s^2}V_{hp}\right) \qquad (16.61)$$

Equating the last right-hand-side terms of Eqs. (16.61) and (16.58) gives

$$R_f/R_1 = 1 \qquad (16.62)$$

which implies that we can select arbitrary but practically convenient equal values for R_1 and R_f. Then, equating the second-to-last terms on the right-hand side of Eqs. (16.61) and (16.58) and setting $R_1 = R_f$ yields the ratio R_3/R_2 required to realize a given Q as

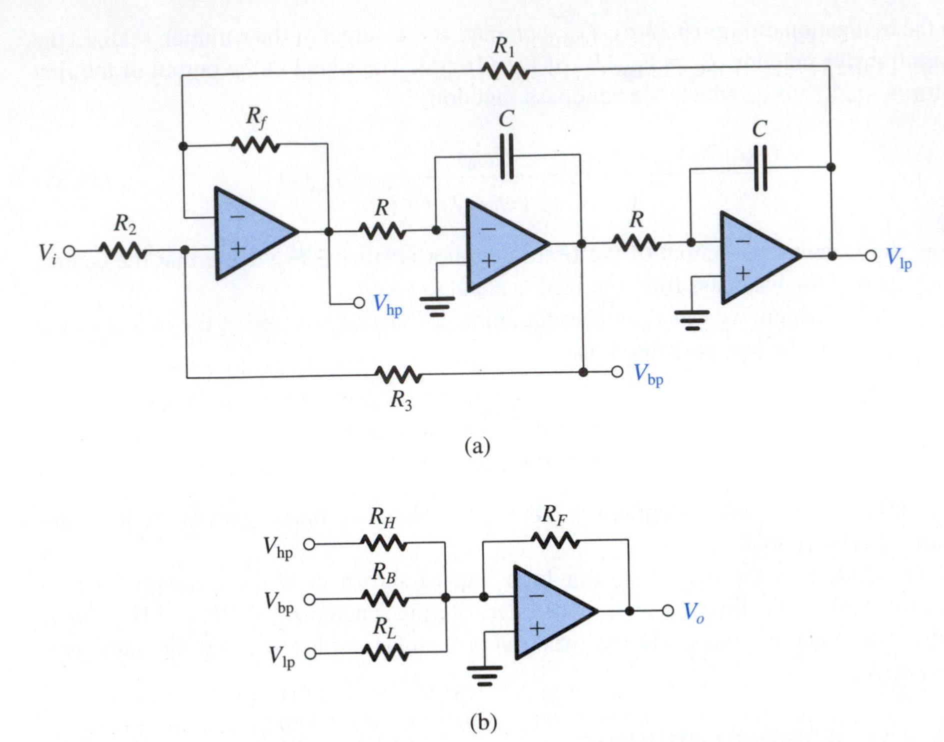

Figure 16.24 **(a)** The KHN biquad circuit, obtained as a direct implementation of the block diagram of Fig. 16.23(c). The three basic filtering functions, HP, BP, and LP, are simultaneously realized. **(b)** To obtain notch and all-pass functions, the three outputs are summed with appropriate weights using this op-amp summer.

$$R_3/R_2 = 2Q - 1 \tag{16.63}$$

Thus an arbitrary but convenient value can be selected for either R_2 or R_3, and the value of the other resistance can be determined using Eq. (16.63). Finally, equating the coefficients of V_i in Eqs. (16.61) and (16.58) and substituting $R_f = R_1$ and for R_3/R_2 from Eq. (16.63) results in

$$K = 2 - (1/Q) \tag{16.64}$$

Thus the gain parameter K is fixed to this value.

The KHN biquad can be used to realize notch and all-pass functions by summing weighted versions of the three outputs, LP, BP, and HP. Such an op-amp summer is shown in Fig. 16.24(b); for this summer we can write

$$\begin{aligned} V_o &= -\left(\frac{R_F}{R_H}V_{\text{hp}} + \frac{R_F}{R_B}V_{\text{bp}} + \frac{R_F}{R_L}V_{\text{lp}}\right) \\ &= -V_i\left(\frac{R_F}{R_H}T_{\text{hp}} + \frac{R_F}{R_B}T_{\text{bp}} + \frac{R_F}{R_L}T_{\text{lp}}\right) \end{aligned} \tag{16.65}$$

Substituting for T_{hp}, T_{bp}, and T_{lp} from Eqs. (16.56), (16.59), and (16.60), respectively, gives the overall transfer function

$$\frac{V_o}{V_i} = -K\frac{(R_F/R_H)s^2 - s(R_F/R_B)\omega_0 + (R_F/R_L)\omega_0^2}{s^2 + s(\omega_0/Q) + \omega_0^2} \tag{16.66}$$

from which we can see that different transmission zeros can be obtained by the appropriate selection of the values of the summing resistors. For instance, a notch is obtained by selecting $R_B = \infty$ and

$$\frac{R_H}{R_L} = \left(\frac{\omega_n}{\omega_0}\right)^2 \tag{16.67}$$

16.7.3 An Alternative Two-Integrator-Loop Biquad Circuit

An alternative two-integrator-loop biquad circuit in which all three op amps are used in a single-ended mode can be developed as follows: Rather than using the input summer to add signals with positive and negative coefficients, we can introduce an additional inverter, as shown in Fig. 16.25(a). Now all the coefficients of the summer have the same sign, and we can dispense with the summing amplifier altogether and perform the summation at the virtual-ground input of the first integrator. Observe that the summing weights of 1, $1/Q$, and K are realized by using resistances of R, QR, and R/K, respectively. The resulting circuit is shown in Fig. 16.25(b), from which we observe that the high-pass function is no longer available! This is the price paid for obtaining a circuit that utilizes all op amps in a single-ended mode. The circuit of Fig. 16.25(b) is known as the **Tow–Thomas biquad**, after its originators.

Rather than using a fourth op amp to realize the finite transmission zeros required for the notch and all-pass functions, as was done with the KHN biquad, an economical *feedforward* scheme can be employed with the Tow–Thomas circuit. Specifically, the virtual ground available at the input of each of the three op amps in the Tow–Thomas circuit permits the input signal to be fed to all three op amps, as shown in Fig. 16.26. If V_o is taken at the output of the damped integrator, straightforward analysis yields the filter transfer function

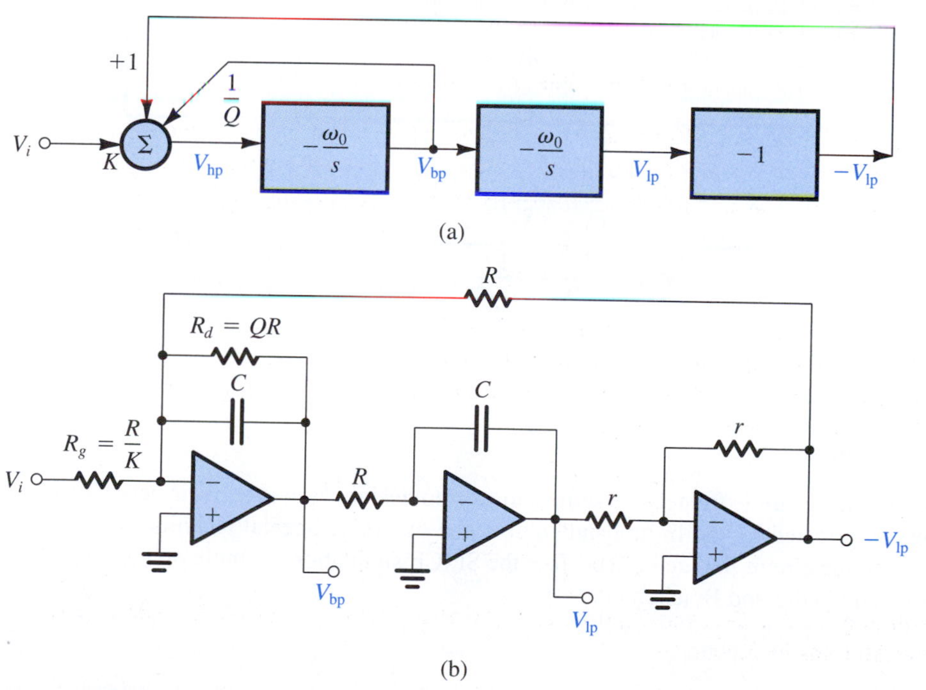

Figure 16.25 **(a)** Derivation of an alternative two-integrator-loop biquad in which all op amps are used in a single-ended fashion. **(b)** The resulting circuit, known as the Tow–Thomas biquad.

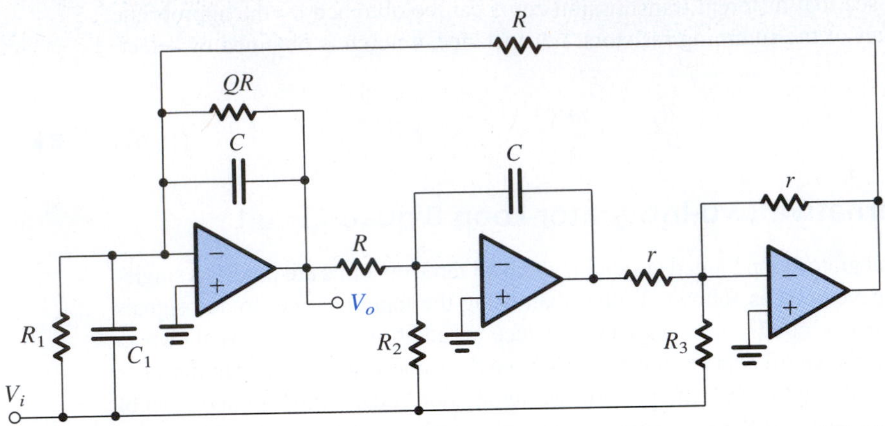

Figure 16.26 The Tow–Thomas biquad with feedforward. The transfer function of Eq. (16.68) is realized by feeding the input signal through appropriate components to the inputs of the three op amps. This circuit can realize all special second-order functions. The design equations are given in Table 16.2.

Table 16.2 Design Data for the Circuit in Fig. 16.26

All cases	$C =$ arbitrary, $R = 1/\omega_0 C$, $r =$ arbitrary
LP	$C_1 = 0$, $R_1 = \infty$, $R_2 = R/\text{dc gain}$, $R_3 = \infty$
Positive BP	$C_1 = 0$, $R_1 = \infty$, $R_2 = \infty$, $R_3 = Qr/\text{center-frequency gain}$
Negative BP	$C_1 = 0$, $R_1 = QR/\text{center-frequency gain}$, $R_2 = \infty$, $R_3 = \infty$
HP	$C_1 = C \times \text{high-frequency gain}$, $R_1 = \infty$, $R_2 = \infty$, $R_3 = \infty$
Notch (all types)	$C_1 = C \times \text{high-frequency gain}$, $R_1 = \infty$, $R_2 = R(\omega_0/\omega_n)^2/\text{high-frequency gain}$, $R_3 = \infty$
AP	$C_1 = C \times \text{flat gain}$, $R_1 = \infty$, $R_2 = R/\text{gain}$, $R_3 = Qr/\text{gain}$

$$\frac{V_o}{V_i} = -\frac{s^2\left(\frac{C_1}{C}\right) + s\frac{1}{C}\left(\frac{1}{R_1} - \frac{r}{RR_3}\right) + \frac{1}{C^2 RR_2}}{s^2 + s\frac{1}{QCR} + \frac{1}{C^2R^2}} \tag{16.68}$$

which can be used to obtain the design data given in Table 16.2.

16.7.4 Final Remarks

Two-integrator-loop biquads are extremely versatile and easy to design. However, their performance is adversely affected by the finite bandwidth of the op amps. Special techniques exist for compensating the circuit for such effects [see the SPICE simulation example on the CD and the website, and Sedra and Brackett (1978)].

EXERCISES

D16.21 Design the KHN circuit to realize a high-pass function with $f_0 = 10$ kHz and $Q = 2$. Choose $C = 1$ nF. What is the value of high-frequency gain obtained? What is the center-frequency gain of the bandpass function that is simultaneously available at the output of the first integrator?
Ans. $R = 15.9$ kΩ; $R_1 = R_f = R_2 = 10$ kΩ (arbitrary); $R_3 = 30$ kΩ; 1.5; 3

D16.22 Use the KHN circuit together with an output summing amplifier to design a low-pass notch filter with $f_0 = 5$ kHz, $f_n = 8$ kHz, $Q = 5$, and a dc gain of 3. Select $C = 1$ nF and $R_L = 10$ kΩ.
Ans. $R = 31.83$ kΩ; $R_1 = R_f = R_2 = 10$ kΩ (arbitrary); $R_3 = 90$ kΩ; $R_H = 25.6$ kΩ; $R_F = 16.7$ kΩ; $R_B = \infty$

D16.23 Use the Tow–Thomas biquad (Fig. 16.25b) to design a second-order bandpass filter with $f_0 = 10$ kHz, $Q = 20$, and unity center-frequency gain. If $R = 10$ kΩ, give the values of C, R_d, and R_g.
Ans. 1.59 nF; 200 kΩ; 200 kΩ

D16.24 Use the data of Table 16.2 to design the biquad circuit of Fig. 16.26 to realize an all-pass filter with $\omega_0 = 10^4$ rad/s, $Q = 5$, and flat gain = 1. Use $C = 10$ nF and $r = 10$ kΩ.
Ans. $R = 10$ kΩ; Q-determining resistor = 50 kΩ; $C_1 = 10$ nF; $R_1 = \infty$; $R_2 = 10$ kΩ; $R_3 = 50$ kΩ

16.8 Single-Amplifier Biquadratic Active Filters

The op amp–RC biquadratic circuits studied in the two preceding sections provide good performance, are versatile, and are easy to design and to adjust (tune) after final assembly. Unfortunately, however, they are not economic in their use of op amps, requiring three or four amplifiers per second-order section. This can be a problem, especially in applications where power-supply current is to be conserved: for instance, in a battery-operated instrument. In this section we shall study a class of second-order filter circuits that requires only one op amp per biquad. These minimal realizations, however, suffer a greater dependence on the limited gain and bandwidth of the op amp and can also be more sensitive to the unavoidable tolerances in the values of resistors and capacitors than the multiple-op-amp biquads of the preceding sections. The **single-amplifier biquads** (SABs) are therefore limited to the less stringent filter specifications—for example, pole Q factors less than about 10.

The synthesis of SAB circuits is based on the use of feedback to move the poles of an RC circuit from the negative real axis, where they naturally lie, to the complex-conjugate locations required to provide selective filter response. The synthesis of SABs follows a two-step process:

1. Synthesis of a feedback loop that realizes a pair of complex-conjugate poles characterized by a frequency ω_0 and a Q factor Q.
2. Injecting the input signal in a way that realizes the desired transmission zeros.

16.8.1 Synthesis of the Feedback Loop

Consider the circuit shown in Fig. 16.27(a), which consists of a two-port RC network n placed in the negative-feedback path of an op amp. We shall assume that, except for having a finite gain A, the op amp is ideal. We shall denote by $t(s)$ the open-circuit voltage transfer

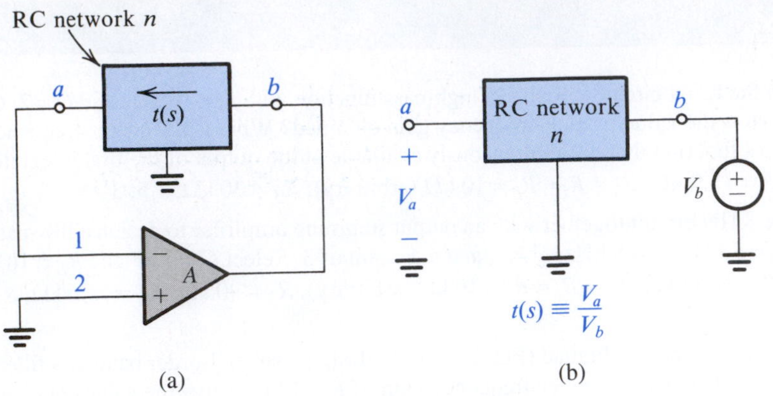

Figure 16.27 **(a)** Feedback loop obtained by placing a two-port RC network n in the feedback path of an op amp. **(b)** Definition of the open-circuit transfer function $t(s)$ of the RC network.

function of the RC network n, where the definition of $t(s)$ is illustrated in Fig. 16.27(b). The transfer function $t(s)$ can in general be written as the ratio of two polynomials $N(s)$ and $D(s)$:

$$t(s) = \frac{N(s)}{D(s)}$$

The roots of $N(s)$ are the transmission zeros of the RC network, and the roots of $D(s)$ are its poles. Study of circuit theory shows that while the poles of an RC network are restricted to lie on the negative real axis, the zeros can in general lie anywhere in the s plane.

The loop gain $L(s)$ of the feedback circuit in Fig. 16.27(a) can be determined using the method of Section 10.9. It is simply the product of the op-amp gain A and the transfer function $t(s)$,

$$L(s) = At(s) = \frac{AN(s)}{D(s)} \tag{16.69}$$

Substituting for $L(s)$ into the characteristic equation

$$1 + L(s) = 0 \tag{16.70}$$

results in the poles s_P of the closed-loop circuit obtained as solutions to the equation

$$t(s_P) = -\frac{1}{A} \tag{16.71}$$

In the ideal case, $A = \infty$ and the poles are obtained from

$$N(s_P) = 0 \tag{16.72}$$

That is, *the filter poles are identical to the zeros of the RC network.*

Since our objective is to realize a pair of complex-conjugate poles, we should select an RC network that can have complex-conjugate transmission zeros. The simplest such networks are the bridged-T networks shown in Fig. 16.28 together with their transfer functions $t(s)$ from b to a, with a open-circuited. As an example, consider the circuit generated by placing the bridged-T network of Fig. 16.28(a) in the negative-feedback path of an op amp, as shown in Fig. 16.29. The pole polynomial of the active-filter circuit will be equal to the numerator polynomial of the bridged-T network; thus,

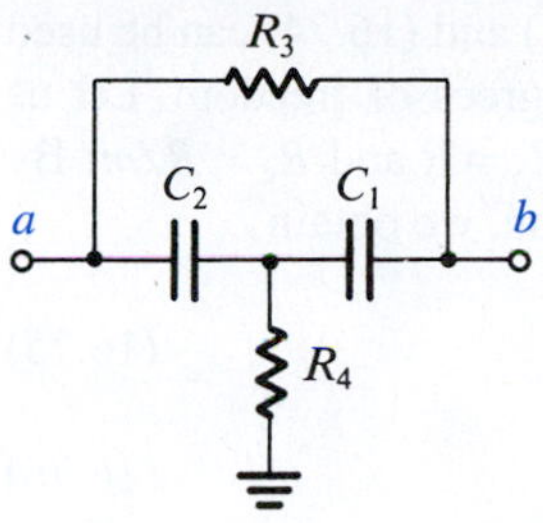

$$t(s) = \frac{s^2 + s\left(\frac{1}{C_1} + \frac{1}{C_2}\right)\frac{1}{R_3} + \frac{1}{C_1 C_2 R_3 R_4}}{s^2 + s\left(\frac{1}{C_1 R_3} + \frac{1}{C_2 R_3} + \frac{1}{C_1 R_4}\right) + \frac{1}{C_1 C_2 R_3 R_4}}$$

(a)

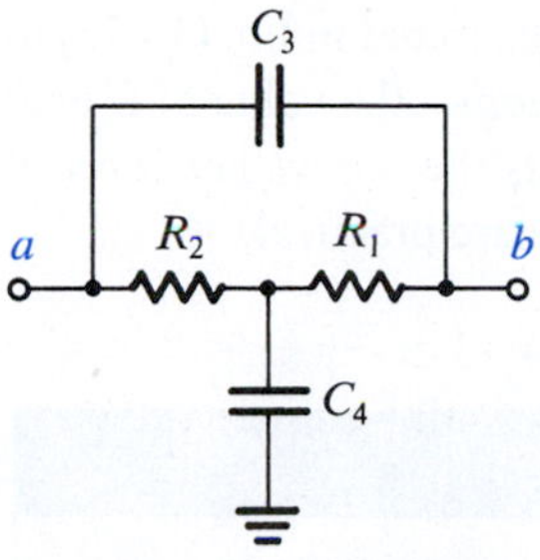

$$t(s) = \frac{s^2 + s\left(\frac{1}{R_1} + \frac{1}{R_2}\right)\frac{1}{C_4} + \frac{1}{C_3 C_4 R_1 R_2}}{s^2 + s\left(\frac{1}{C_4 R_1} + \frac{1}{C_4 R_2} + \frac{1}{C_3 R_2}\right) + \frac{1}{C_3 C_4 R_1 R_2}}$$

(b)

Figure 16.28 Two RC networks (called bridged-T networks) that can have complex transmission zeros. The transfer functions given are from b to a, with a open-circuited.

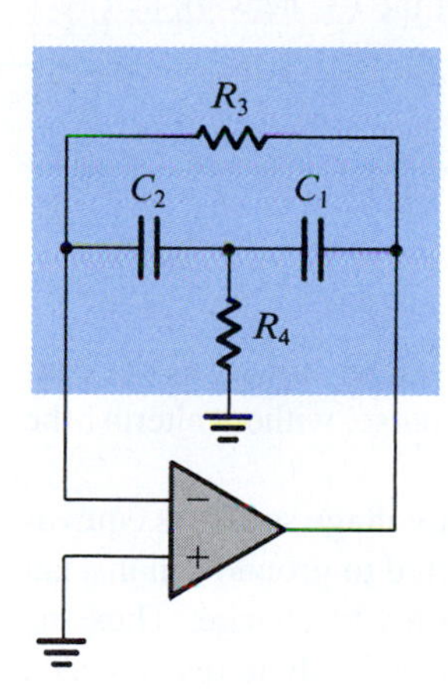

Figure 16.29 An active-filter feedback loop generated using the bridged-T network of Fig. 16.28(a).

$$s^2 + s\frac{\omega_0}{Q} + \omega_0^2 = s^2 + s\left(\frac{1}{C_1} + \frac{1}{C_2}\right)\frac{1}{R_3} + \frac{1}{C_1 C_2 R_3 R_4}$$

which enables us to obtain ω_0 and Q as

$$\omega_0 = \frac{1}{\sqrt{C_1 C_2 R_3 R_4}} \tag{16.73}$$

$$Q = \left[\frac{\sqrt{C_1 C_2 R_3 R_4}}{R_3}\left(\frac{1}{C_1} + \frac{1}{C_2}\right)\right]^{-1} \tag{16.74}$$

If we are designing this circuit, ω_0 and Q are given and Eqs. (16.73) and (16.74) can be used to determine C_1, C_2, R_3, and R_4. It follows that there are two degrees of freedom. Let us exhaust one of these by selecting $C_1 = C_2 = C$. Let us also denote $R_3 = R$ and $R_4 = R/m$. By substituting in Eqs. (16.73) and (16.74), and with some manipulation, we obtain

$$m = 4Q^2 \tag{16.75}$$

$$CR = \frac{2Q}{\omega_0} \tag{16.76}$$

Thus if we are given the value of Q, Eq. (16.75) can be used to determine the ratio of the two resistances R_3 and R_4. Then the given values of ω_0 and Q can be substituted in Eq. (16.76) to determine the time constant CR. There remains one degree of freedom—the value of C or R can be arbitrarily chosen. In an actual design, this value, which sets the *impedance level* of the circuit, should be chosen so that the resulting component values are practical.

EXERCISES

D16.25 Design the circuit of Fig. 16.29 to realize a pair of poles with $\omega_0 = 10^4$ rad/s and $Q = 1$. Select $C_1 = C_2 = 1$ nF.
Ans. $R_3 = 200$ kΩ; $R_4 = 50$ kΩ

16.26 For the circuit designed in Exercise 16.25, find the location of the poles of the RC network in the feedback loop.
Ans. -0.382×10^4 and -2.618×10^4 rad/s

16.8.2 Injecting the Input Signal

Having synthesized a feedback loop that realizes a given pair of poles, we now consider connecting the input signal source to the circuit. We wish to do this, of course, without altering the poles.

Since, for the purpose of finding the poles of a circuit, an ideal voltage source is equivalent to a short circuit, it follows that any circuit node that is connected to ground can instead be connected to the input voltage source without causing the poles to change. Thus the method of injecting the input voltage signal into the feedback loop is simply to disconnect a component (or several components) that is (are) connected to ground and connect it (them) to the input source. Depending on the component(s) through which the input signal is injected, different transmission zeros are obtained. This is, of course, the same method we used in Section 16.5 with the LCR resonator and in Section 16.6 with the biquads based on the LCR resonator.

As an example, consider the feedback loop of Fig. 16.29. Here we have two grounded nodes (one terminal of R_4 and the positive input terminal of the op amp) that can serve for injecting the input signal. Figure 16.30(a) shows the circuit with the input signal injected through part of the resistance R_4. Note that the two resistances R_4/α and $R_4/(1-\alpha)$ have a parallel equivalent of R_4.

Analysis of the circuit to determine its voltage transfer function $T(s) \equiv V_o(s)/V_i(s)$ is illustrated in Fig. 16.30(b). Note that we have assumed the op amp to be ideal, and have

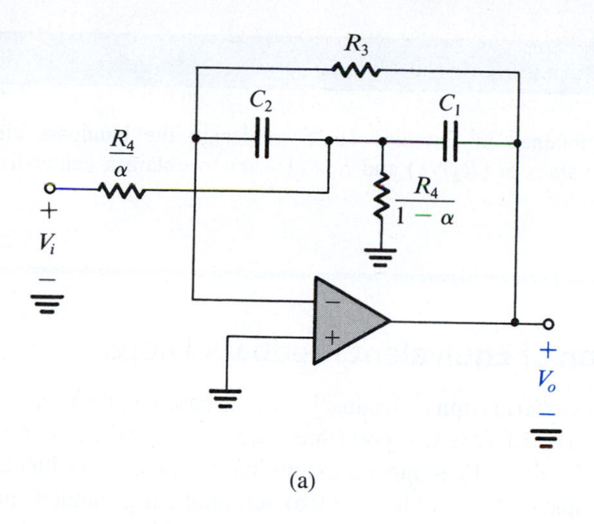

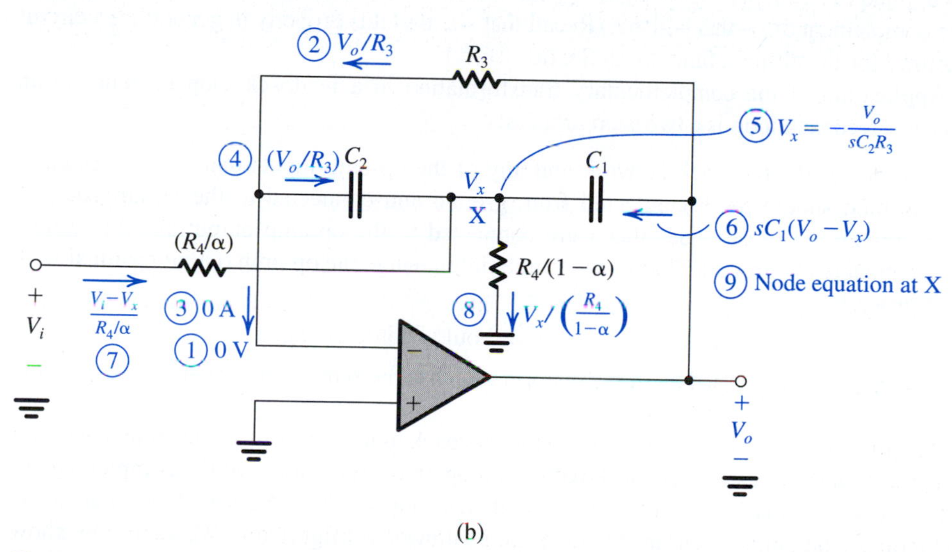

Figure 16.30 **(a)** The feedback loop of Fig. 16.29 with the input signal injected through part of resistance R_4. This circuit realizes the bandpass function. **(b)** Analysis of the circuit in **(a)** to determine its voltage transfer function $T(s)$ with the order of the analysis steps indicated by the circled numbers.

indicated the order of the analysis steps by the circled numbers. The final step, number 9, consists of writing a node equation at X and substituting for V_x by the value determined in step 5. The result is the transfer function

$$\frac{V_o}{V_i} = \frac{-s(\alpha/C_1R_4)}{s^2 + s\left(\frac{1}{C_1}+\frac{1}{C_2}\right)\frac{1}{R_3} + \frac{1}{C_1C_2R_3R_4}}$$

We recognize this as a bandpass function whose center-frequency gain can be controlled by the value of α. As expected, the denominator polynomial is identical to the numerator polynomial of $t(s)$ given in Fig. 16.28(a).

EXERCISE

16.27 Use the component values obtained in Exercise 16.25 to design the bandpass circuit of Fig. 16.30(a). Determine the values of (R_4/α) and $R_4/(1-\alpha)$ to obtain a center-frequency gain of unity.
Ans. 100 kΩ; 100 kΩ

16.8.3 Generation of Equivalent Feedback Loops

The **complementary transformation** of feedback loops is based on the property of linear networks illustrated in Fig. 16.31 for the two-port (three-terminal) network n. In Fig. 16.31(a), terminal c is grounded and a signal V_b is applied to terminal b. The transfer function from b to a with c grounded is denoted t. Then, in Fig. 16.31(b), terminal b is grounded and the input signal is applied to terminal c. The transfer function from c to a with b grounded can be shown to be the complement of t—that is, $1-t$. (Recall that we used this property in generating a circuit realization for the all-pass function in Section 16.6.)

Application of the complementary transformation to a feedback loop to generate an equivalent feedback loop is a two-step process:

1. Nodes of the feedback network and any of the op-amp inputs that are connected to ground should be disconnected from ground and connected to the op-amp output. Conversely, those nodes that were connected to the op-amp output should be now connected to ground. That is, we simply interchange the op-amp output terminal with ground.
2. The two input terminals of the op amp should be interchanged.

The feedback loop generated by this transformation has the same characteristic equation, and hence the same poles, as the original loop.

To illustrate, we show in Fig. 16.32(a) the feedback loop formed by connecting a two-port RC network in the negative-feedback path of an op amp. Application of the complementary transformation to this loop results in the feedback loop of Fig. 16.32(b). Note that in the latter loop the op amp is used in the unity-gain follower configuration. We shall now show that the two loops of Fig. 16.32 are equivalent.

If the op amp has an open-loop gain A, the follower in the circuit of Fig. 16.32(b) will have a gain of $A/(A+1)$. This, together with the fact that the transfer function of network n from c to a is $1-t$ (see Fig. 16.31), enables us to write for the circuit in Fig. 16.32(b) the characteristic equation

$$1-\frac{A}{A+1}(1-t) = 0$$

This equation can be manipulated to the form

$$1+At = 0$$

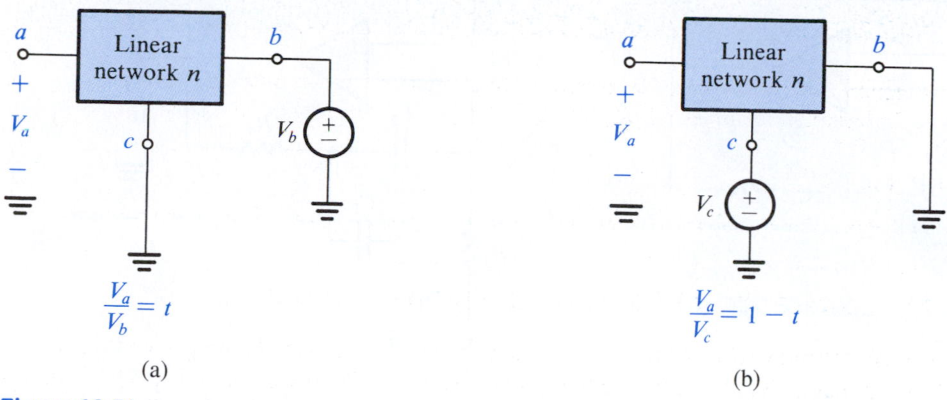

Figure 16.31 Interchanging input and ground results in the complement of the transfer function.

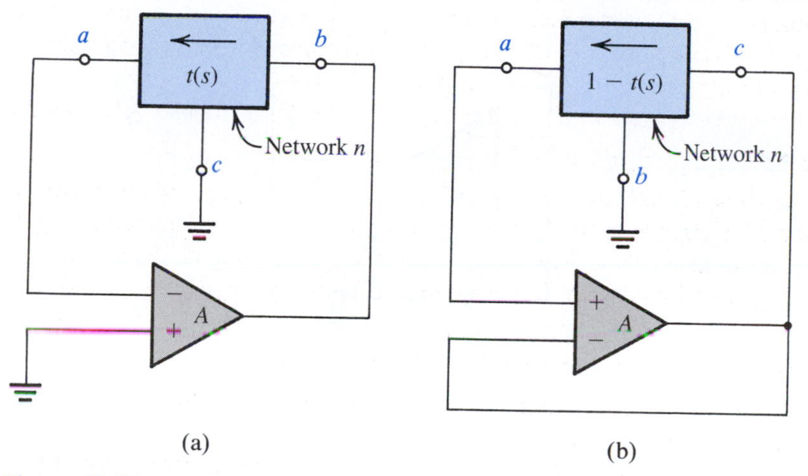

Figure 16.32 Application of the complementary transformation to the feedback loop in **(a)** results in the equivalent loop (same poles) shown in **(b)**.

which is the characteristic equation of the loop in Fig. 16.32(a). As an example, consider the application of the complementary transformation to the feedback loop of Fig. 16.29: The feedback loop of Fig. 16.33(a) results. Injecting the input signal through C_1 results in the circuit in Fig. 16.33(b), which can be shown (by direct analysis) to realize a second-order high-pass function. This circuit is one of a family of SABs known as the **Sallen-and-Key circuits**, after their originators. The design of the circuit in Fig. 16.33(b) is based on Eqs. (16.73) through (16.76): namely, $R_3 = R$, $R_4 = R/4Q^2$, $C_1 = C_2 = C$, $CR = 2Q/\omega_0$, and the value of C is arbitrarily chosen to be practically convenient.

As another example, Fig. 16.34(a) shows the feedback loop generated by placing the two-port RC network of Fig. 16.28(b) in the negative-feedback path of an op amp. For an ideal op amp, this feedback loop realizes a pair of complex-conjugate natural modes having the same location as the zeros of $t(s)$ of the RC network. Thus, using the expression for $t(s)$ given in Fig. 16.28(b), we can write for the active-filter poles

$$\omega_0 = 1/\sqrt{C_3 C_4 R_1 R_2} \tag{16.77}$$

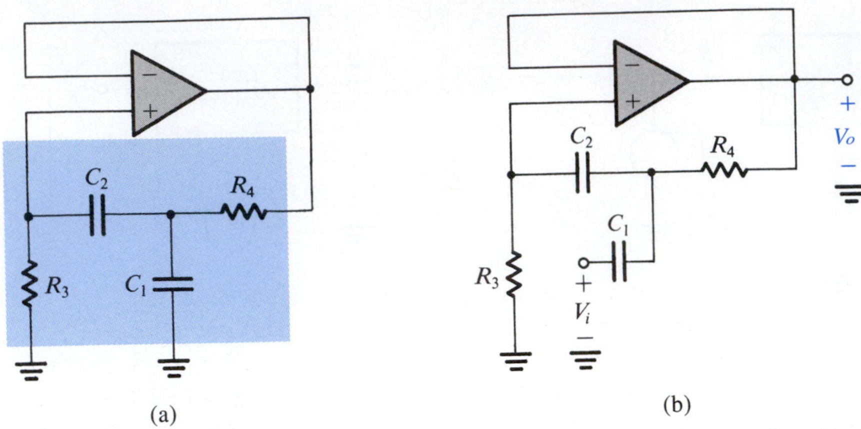

Figure 16.33 **(a)** Feedback loop obtained by applying the complementary transformation to the loop in Fig. 16.29. **(b)** Injecting the input signal through C_1 realizes the high-pass function. This is one of the Sallen-and-Key family of circuits.

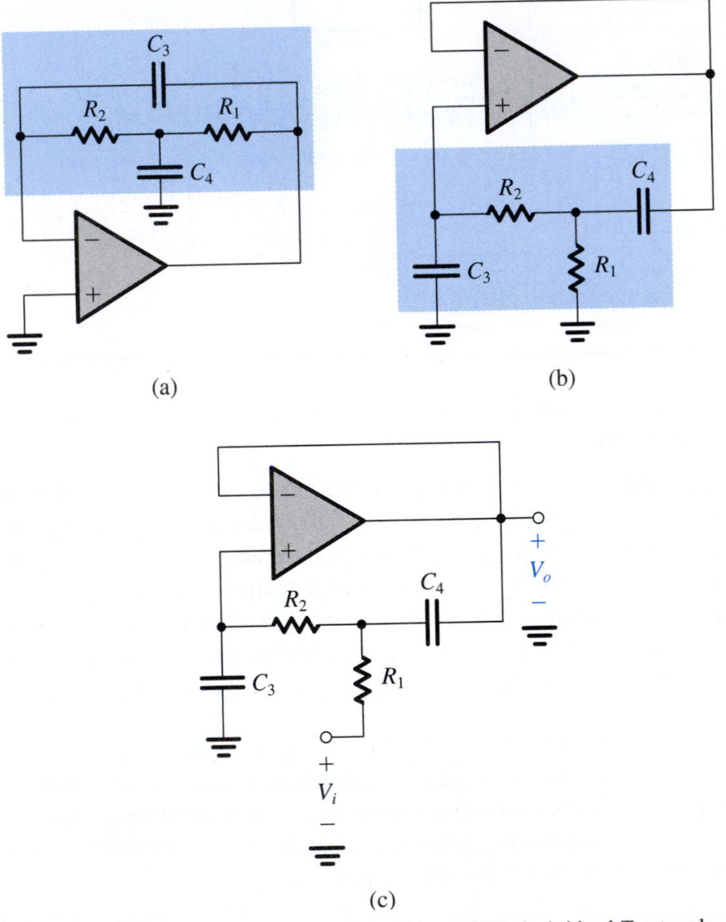

Figure 16.34 **(a)** Feedback loop obtained by placing the bridged-T network of Fig. 16.28(b) in the negative-feedback path of an op amp. **(b)** Equivalent feedback loop generated by applying the complementary transformation to the loop in **(a)**. **(c)** A low-pass filter obtained by injecting V_i through R_1 into the loop in **(b)**.

$$Q = \left[\frac{\sqrt{C_3C_4R_1R_2}}{C_4}\left(\frac{1}{R_1}+\frac{1}{R_2}\right)\right]^{-1} \tag{16.78}$$

Normally the design of this circuit is based on selecting $R_1 = R_2 = R$, $C_4 = C$, and $C_3 = C/m$. When substituted in Eqs. (16.77) and (16.78), these yield

$$m = 4Q^2 \tag{16.79}$$

$$CR = 2Q/\omega_0 \tag{16.80}$$

with the remaining degree of freedom (the value of C or R) left to the designer to choose.

Injecting the input signal to the C_4 terminal that is connected to ground can be shown to result in a bandpass realization. If, however, we apply the complementary transformation to the feedback loop in Fig. 16.34(a), we obtain the equivalent loop in Fig. 16.34(b). The loop equivalence means that the circuit of Fig. 16.34(b) has the same poles and thus the same ω_0 and Q and the same design equations (Eqs. 16.77 through 16.80). The new loop in Fig. 16.34(b) can be used to realize a low-pass function by injecting the input signal as shown in Fig. 16.34(c).

EXERCISES

16.28 Analyze the circuit in Fig. 16.34(c) to determine its transfer function $V_o(s)/V_i(s)$ and thus show that ω_0 and Q are indeed those in Eqs. (16.77) and (16.78). Also show that the dc gain is unity.

D16.29 Design the circuit in Fig. 16.34(c) to realize a low-pass filter with $f_0 = 4$ kHz and $Q = 1/\sqrt{2}$. Use 10-kΩ resistors.

Ans. $R_1 = R_2 = 10$ kΩ; $C_3 = 2.81$ nF; $C_4 = 5.63$ nF

16.9 Sensitivity

Because of the tolerances in component values and because of the finite op-amp gain, the response of the actual assembled filter will deviate from the ideal response. As a means for predicting such deviations, the filter designer employs the concept of **sensitivity**. Specifically, for second-order filters one is usually interested in finding how *sensitive* their poles are relative to variations (both initial tolerances and future drifts) in RC component values and amplifier gain. These sensitivities can be quantified using the **classical sensitivity function** S_x^y, defined as

$$S_x^y \equiv \lim_{\Delta x \to 0} \frac{\Delta y/y}{\Delta x/x} \tag{16.81}$$

Thus,

$$S_x^y = \frac{\partial y}{\partial x}\frac{x}{y} \tag{16.82}$$

Here, x denotes the value of a component (a resistor, a capacitor, or an amplifier gain) and y denotes a circuit parameter of interest (say, ω_0 or Q). For small changes

$$S_x^y \simeq \frac{\Delta y/y}{\Delta x/x} \tag{16.83}$$

Thus we can use the value of S_x^y to determine the per-unit change in y due to a given per-unit change in x. For instance, if the sensitivity of Q relative to a particular resistance R_1 is 5, then a 1% increase in R_1 results in a 5% increase in the value of Q.

Example 16.3

For the feedback loop of Fig. 16.29, find the sensitivities of ω_0 and Q relative to all the passive components and the op-amp gain. Evaluate these sensitivities for the design considered in the preceding section for which $C_1 = C_2$.

Solution

To find the sensitivities with respect to the passive components, called **passive sensitivities**, we assume that the op-amp gain is infinite. In this case, ω_0 and Q are given by Eqs. (16.73) and (16.74). Thus for ω_0 we have

$$\omega_0 = \frac{1}{\sqrt{C_1C_2R_3R_4}}$$

which can be used together with the sensitivity definition of Eq. (16.82) to obtain

$$S_{C_1}^{\omega_0} = S_{C_2}^{\omega_0} = S_{R_3}^{\omega_0} = S_{R_4}^{\omega_0} = -\tfrac{1}{2}$$

For Q we have

$$Q = \left[\sqrt{C_1C_2R_3R_4}\left(\frac{1}{C_1} + \frac{1}{C_2}\right)\frac{1}{R_3}\right]^{-1}$$

to which we apply the sensitivity definition to obtain

$$S_{C_1}^{Q} = \frac{1}{2}\left(\sqrt{\frac{C_2}{C_1}} - \sqrt{\frac{C_1}{C_2}}\right)\left(\sqrt{\frac{C_2}{C_1}} + \sqrt{\frac{C_1}{C_2}}\right)^{-1}$$

For the design with $C_1 = C_2$ we see that $S_{C_1}^{Q} = 0$. Similarly, we can show that

$$S_{C_2}^{Q} = 0, \quad S_{R_3}^{Q} = \tfrac{1}{2}, \quad S_{R_4}^{Q} = -\tfrac{1}{2}$$

It is important to remember that the sensitivity expression should be derived *before* values corresponding to a particular design are substituted.

Next we consider the sensitivities relative to the amplifier gain. If we assume the op amp to have a finite gain A, the characteristic equation for the loop becomes

$$1 + At(s) = 0 \tag{16.84}$$

where $t(s)$ is given in Fig. 16.28(a). To simplify matters we can substitute for the passive components by their design values. This causes no errors in evaluating sensitivities, since we are now finding the sensitivity with respect to the amplifier gain. Using the design values obtained earlier—namely, $C_1 = C_2 = C$, $R_3 = R$, $R_4 = R/4Q^2$, and $CR = 2Q/\omega_0$ —we get

$$t(s) = \frac{s^2 + s(\omega_0/Q) + \omega_0^2}{s^2 + s(\omega_0/Q)(2Q^2+1) + \omega_0^2} \tag{16.85}$$

where ω_0 and Q denote the nominal or design values of the pole frequency and Q factor. The actual values are obtained by substituting for $t(s)$ in Eq. (16.84):

$$s^2 + s\frac{\omega_0}{Q}(2Q^2+1) + \omega_0^2 + A\left(s^2 + s\frac{\omega_0}{Q} + \omega_0^2\right) = 0$$

Assuming the gain A to be real and dividing both sides by $A + 1$, we get

$$s^2 + s\frac{\omega_0}{Q}\left(1 + \frac{2Q^2}{A+1}\right) + \omega_0^2 = 0 \tag{16.86}$$

From this equation we see that the actual pole frequency, ω_{0a}, and the pole Q, Q_a, are

$$\omega_{0a} = \omega_0 \tag{16.87}$$

$$Q_a = \frac{Q}{1 + 2Q^2/(A+1)} \tag{16.88}$$

Thus

$$S_A^{\omega_{0a}} = 0$$

$$S_A^{Q_a} = \frac{A}{A+1}\frac{2Q^2/(A+1)}{1+2Q^2/(A+1)}$$

For $A \gg 2Q^2$ and $A \gg 1$ we obtain

$$S_A^{Q_a} \simeq \frac{2Q^2}{A}$$

It is usual to drop the subscript a in this expression and write

$$S_A^{Q} \simeq \frac{2Q^2}{A} \tag{16.89}$$

Note that if Q is high ($Q \geq 5$), its sensitivity relative to the amplifier gain can be quite high.[9]

16.9.1 A Concluding Remark

The results of Example 16.3 indicate a serious disadvantage of single-amplifier biquads—the sensitivity of Q relative to the amplifier gain is quite high. Although a technique exists for reducing S_A^Q in SABs (see Sedra et al., 1980), this is done at the expense of increased passive sensitivities. Nevertheless, the resulting SABs are used extensively in many applications. However, for filters with Q factors greater than about 10, one usually opts for one of the multiamplifier biquads studied in Sections 16.6 and 16.7. For these circuits S_A^Q is proportional to Q, rather than to Q^2 as in the SAB case (Eq. 16.89).

[9] Because the open-loop gain A of op amps usually has wide tolerance, it is important to keep $S_A^{\omega_0}$ and S_A^Q very small.

EXERCISE

16.30 In a particular filter utilizing the feedback loop of Fig. 16.29, with $C_1 = C_2$, use the results of Example 16.3 to find the expected percentage change in ω_0 and Q under the conditions that (a) R_3 is 2% high, (b) R_4 is 2% high, (c) both R_3 and R_4 are 2% high, and (d) both capacitors are 2% low and both resistors are 2% high.
Ans. (a) −1%, +1%; (b) −1%, −1%; (c) −2%, 0%; (d) 0%, 0%

16.10 Switched-Capacitor Filters

The active-RC filter circuits presented above have two properties that make their production in monolithic IC form difficult, if not practically impossible; these are the need for large-valued capacitors and the requirement of accurate RC time constants. The search therefore has continued for a method of filter design that would lend itself more naturally to IC implementation. In this section we shall introduce one such method.

16.10.1 The Basic Principle

The switched-capacitor filter technique is based on the realization that a capacitor switched between two circuit nodes at a sufficiently high rate is equivalent to a resistor connecting these two nodes. To be specific, consider the active-RC integrator of Fig. 16.35(a). This is the familiar Miller integrator, which we used in the two-integrator-loop biquad in Section 16.7. In Fig. 16.35(b) we have replaced the input resistor R_1 by a grounded capacitor C_1 together with two MOS transistors acting as switches. In some circuits, more elaborate switch configurations are used, but such details are beyond our present need.

The two MOS switches in Fig. 16.35(b) are driven by a *nonoverlapping* two-phase clock. Figure 16.35(c) shows the clock waveforms. We shall assume in this introductory exposition that the clock frequency f_c ($f_c = 1/T_c$) is much higher than the frequency of the input signal v_i. Thus during clock phase ϕ_1, when C_1 is connected across the input signal source v_i, the variations in the input signal are negligibly small. It follows that during ϕ_1, capacitor C_1 charges up to the voltage v_i,

$$q_{C1} = C_1 v_i$$

Then, during clock phase ϕ_2, capacitor C_1 is connected to the virtual-ground input of the op amp, as indicated in Fig. 16.35(d). Capacitor C_1 is thus forced to discharge, and its previous charge q_{C1} is transferred to C_2, in the direction indicated in Fig. 16.35(d).

From the description above we see that during each clock period T_c an amount of charge $q_{C1} = C_1 v_i$ is extracted from the input source and supplied to the integrator capacitor C_2. Thus the average current flowing between the input node (IN) and the virtual-ground node (VG) is

$$i_{av} = \frac{C_1 v_i}{T_c}$$

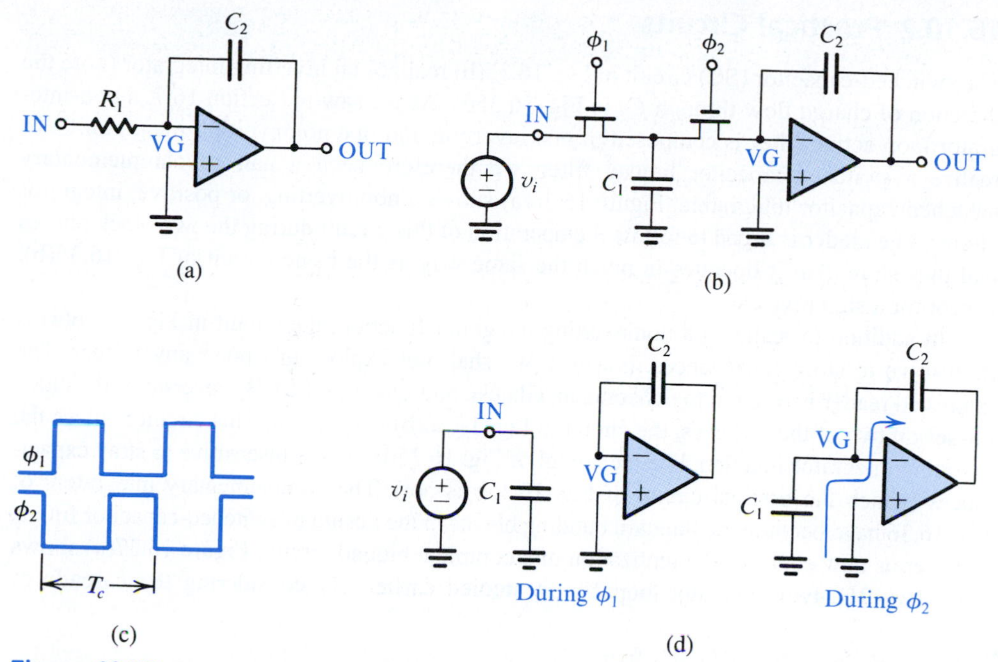

Figure 16.35 Basic principle of the switched-capacitor filter technique. **(a)** Active-RC integrator. **(b)** Switched-capacitor integrator. **(c)** Two-phase clock (nonoverlapping). **(d)** During ϕ_1, C_1 charges up to the current value of v_i and then, during ϕ_2, discharges into C_2.

If T_c is sufficiently short, one can think of this process as almost continuous and define an equivalent resistance R_{eq} that is in effect present between nodes IN and VG:

$$R_{eq} \equiv v_i / i_{av}$$

Thus,

$$R_{eq} = T_c / C_1 \tag{16.90}$$

Using R_{eq} we obtain an equivalent time constant for the integrator:

$$\text{Time constant} = C_2 R_{eq} = T_c \frac{C_2}{C_1} \tag{16.91}$$

Thus the time constant that determines the frequency response of the filter is established by the clock period T_c and the capacitor ratio C_2/C_1. Both these parameters can be well controlled in an IC process. Specifically, note the dependence on capacitor ratios rather than on absolute values of capacitors. The accuracy of capacitor ratios in MOS technology can be controlled to within 0.1%.

Another point worth observing is that with a reasonable clocking frequency (such as 100 kHz) and not-too-large capacitor ratios (say, 10), one can obtain reasonably large time constants (such as 10^{-4} s) suitable for audio applications. Since capacitors typically occupy relatively large areas on the IC chip, one attempts to minimize their values. In this context, it is important to note that the ratio accuracies quoted earlier are obtainable with the smaller capacitor value as low as 0.1 pF.

16.10.2 Practical Circuits

The switched-capacitor (SC) circuit in Fig. 16.35(b) realizes an inverting integrator (note the direction of charge flow through C_2 in Fig. 16.35d). As we saw in Section 16.7, a two-integrator-loop active filter is composed of one inverting and one noninverting integrator.[10] To realize a switched-capacitor biquad filter, we therefore need a pair of complementary switched-capacitor integrators. Figure 16.36(a) shows a noninverting, or positive, integrator circuit. The reader is urged to follow the operation of this circuit during the two clock phases and thus show that it operates in much the same way as the basic circuit of Fig. 16.35(b), except for a sign reversal.

In addition to realizing a noninverting integrator function, the circuit in Fig. 16.36(a) is insensitive to stray capacitances; however, we shall not explore this point any further. The interested reader is referred to Schaumann, Ghausi, and Laker (1990). By reversal of the clock phases on two of the switches, the circuit in Fig. 16.36(b) is obtained. This circuit realizes the inverting integrator function, like the circuit of Fig. 16.35(b), but is insensitive to stray capacitances (which the original circuit of Fig. 16.35b is not). The complementary integrators of Fig. 16.36 have become the standard building blocks in the design of switched-capacitor filters.

Let us now consider the realization of a complete biquad circuit. Figure 16.37(a) shows the active-RC, two-integrator-loop circuit studied earlier. By considering the cascade of

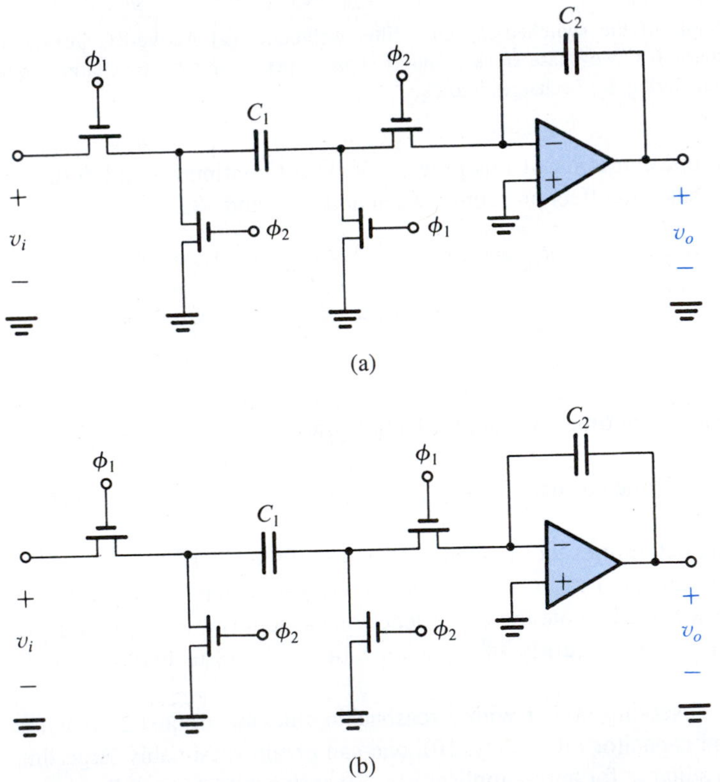

Figure 16.36 A pair of complementary stray-insensitive, switched-capacitor integrators. **(a)** Noninverting switched-capacitor integrator. **(b)** Inverting switched-capacitor integrator.

[10] In the two-integrator loop of Fig. 16.25(b), the noninverting integrator is realized by the cascade of a Miller integrator and an inverting amplifier.

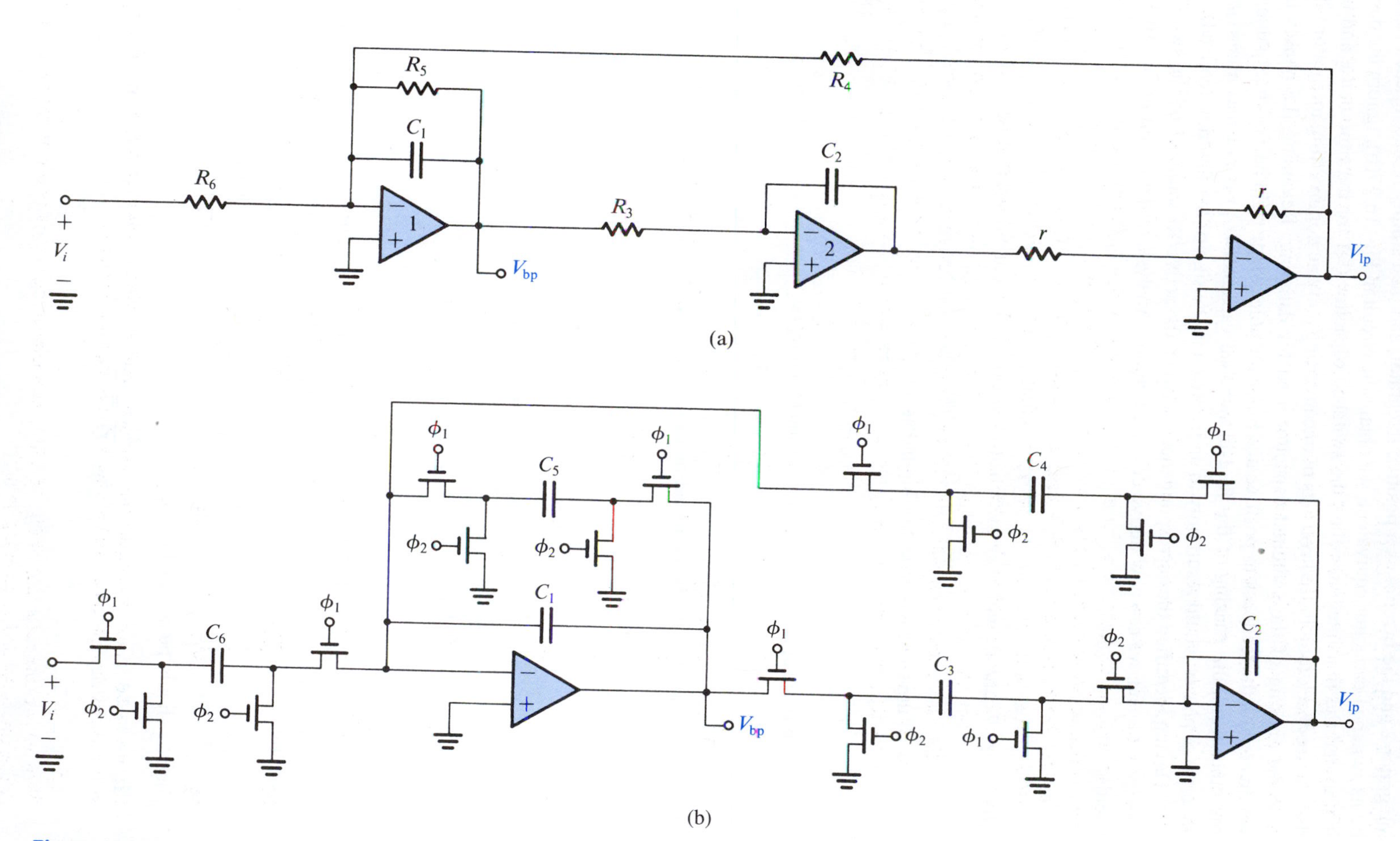

Figure 16.37 **(a)** A two-integrator-loop, active-RC biquad **(b)** its switched-capacitor counterpart.

integrator 2 and the inverter as a positive integrator, and then simply replacing each resistor by its switched-capacitor equivalent, we obtain the circuit in Fig. 16.37(b). Ignore the damping around the first integrator (i.e., the switched capacitor C_5) for the time being and note that the feedback loop indeed consists of one inverting and one noninverting integrator. Then note the phasing of the switched capacitor used for damping. Reversing the phases here would convert the feedback to positive and move the poles to the right half of the s plane. On the other hand, the phasing of the feed-in switched capacitor (C_6) is not that important; a reversal of phases would result only in an inversion in the sign of the function realized.

Having identified the correspondences between the active-RC biquad and the switched-capacitor biquad, we can now derive design equations. Analysis of the circuit in Fig. 16.37(a) yields

$$\omega_0 = \frac{1}{\sqrt{C_1 C_2 R_3 R_4}} \tag{16.92}$$

Replacing R_2 and R_4 with their switched-capacitor equivalent values, that is,

$$R_3 = T_c/C_3 \quad \text{and} \quad R_4 = T_c/C_4$$

gives ω_0 of the switched-capacitor biquad as

$$\omega_0 = \frac{1}{T_c}\sqrt{\frac{C_3 C_4}{C_2 C_1}} \tag{16.93}$$

It is usual to select the time constants of the two integrators to be equal; that is,

$$\frac{T_c}{C_3}C_2 = \frac{T_c}{C_4}C_1 \tag{16.94}$$

If, further, we select the two integrating capacitors C_1 and C_2 to be equal,

$$C_1 = C_2 = C \tag{16.95}$$

then

$$C_3 = C_4 = KC \tag{16.96}$$

where from Eq. (16.93)

$$K = \omega_0 T_c \tag{16.97}$$

For the case of equal time constants, the Q factor of the circuit in Fig. 16.37(a) is given by R_5/R_4. Thus the Q factor of the corresponding switched-capacitor circuit in Fig. 16.37(b) is given by

$$Q = \frac{T_c/C_5}{T_c/C_4} \tag{16.98}$$

Thus C_5 should be selected from

$$C_5 = \frac{C_4}{Q} = \frac{KC}{Q} = \omega_0 T_c \frac{C}{Q} \tag{16.99}$$

Finally, the center-frequency gain of the bandpass function is given by

$$\text{Center-frequency gain} = \frac{C_6}{C_5} = Q\frac{C_6}{\omega_0 T_c C} \tag{16.100}$$

EXERCISE

D16.31 Use $C_1 = C_2 = 20$ pF and design the circuit in Fig. 16.37(b) to realize a bandpass function with $f_0 = 10$ kHz, $Q = 20$, and unity center-frequency gain. Use a clock frequency $f_c = 200$ kHz. Find the values of C_3, C_4, C_5, and C_6.
Ans. 6.283 pF; 6.283 pF; 0.314 pF; 0.314 pF

16.10.3 A Final Remark

We have attempted to provide only an introduction to switched-capacitor filters. We have made many simplifying assumptions, the most important being the switched-capacitor–resistor equivalence (Eq. 16.90). This equivalence is correct only at $f_c = \infty$ and is approximately correct for $f_c \gg f$. Switched-capacitor filters are, in fact, sampled-data networks whose analysis and design can be carried out exactly using z-transform techniques. The interested reader is referred to the bibliography in Appendix G.

16.11 Tuned Amplifiers

In this section, we study a special kind of frequency-selective network, the LC-tuned amplifier. Figure 16.38 shows the general shape of the frequency response of a tuned amplifier. The techniques discussed apply to amplifiers with center frequencies in the range of a few hundred kilohertz to a few hundred megahertz. Tuned amplifiers find application in the radio-frequency (RF) and intermediate-frequency (IF) sections of communications receivers and in a variety of other systems. It should be noted that the tuned-amplifier response of Fig. 16.38 is similar to that of the bandpass filter discussed in earlier sections.

As indicated in Fig. 16.38, the response is characterized by the center frequency ω_0, the 3-dB bandwidth B, and the *skirt selectivity,* which is usually measured as the ratio of the 30-dB bandwidth to the 3-dB bandwidth. In many applications, the 3-dB bandwidth is less than 5% of ω_0. This **narrow-band** property makes possible certain approximations that can simplify the design process, as will be explained later.

The tuned amplifiers studied in this section are small-signal voltage amplifiers in which the transistors operate in the "class A" mode; that is, the transistors conduct at all times. Tuned power amplifiers based on class C and other switching modes of operation are not studied in this book. (For a discussion on the classification of amplifiers, refer to Section 11.1.)

16.11.1 The Basic Principle

The basic principle underlying the design of tuned amplifiers is the use of a parallel LCR circuit as the load, or at the input, of a BJT or a FET amplifier. This is illustrated in Fig. 16.39 with a MOSFET amplifier having a tuned-circuit load. For simplicity, the bias details are not included. Since this circuit uses a single tuned circuit, it is known as a **single-tuned**

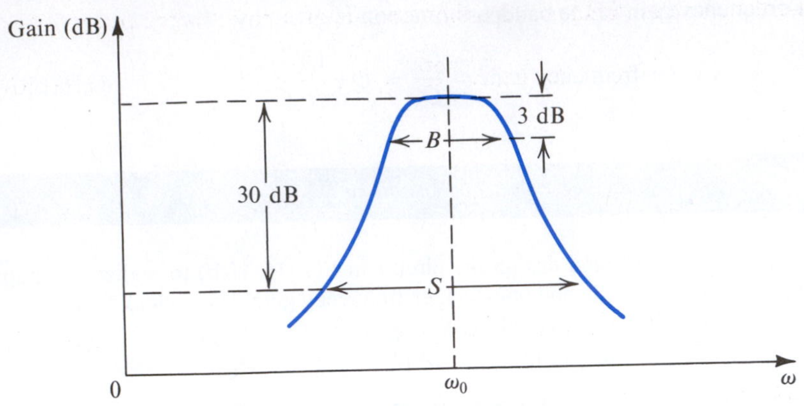

Figure 16.38 Frequency response of a tuned amplifier.

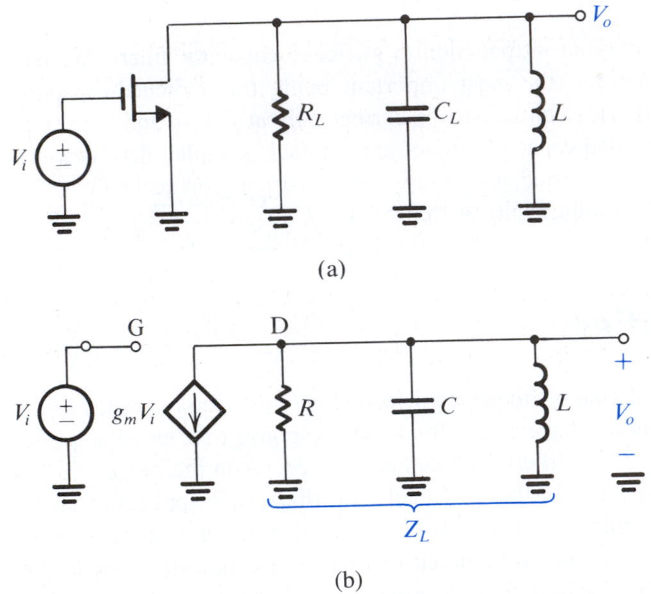

Figure 16.39 The basic principle of tuned amplifiers is illustrated using a MOSFET with a tuned-circuit load. Bias details are not shown.

amplifier. The amplifier equivalent circuit is shown in Fig. 16.39(b). Here R denotes the parallel equivalent of R_L and the output resistance r_o of the FET, and C is the parallel equivalent of C_L and the FET output capacitance (usually very small). From the equivalent circuit we can write

$$V_o = \frac{-g_m V_i}{Y_L} = \frac{-g_m V_i}{sC + 1/R + 1/sL}$$

Thus the voltage gain can be expressed as

$$\frac{V_o}{V_i} = -\frac{g_m}{C}\frac{s}{s^2 + s(1/CR) + 1/LC} \tag{16.101}$$

which is a second-order bandpass function. Thus the tuned amplifier has a center frequency of

$$\omega_0 = 1/\sqrt{LC} \tag{16.102}$$

a 3-dB bandwidth of

$$B = \frac{1}{CR} \tag{16.103}$$

a Q factor of

$$Q \equiv \omega_0/B = \omega_0 CR \tag{16.104}$$

and a center-frequency gain of

$$\frac{V_o(j\omega_0)}{V_i(j\omega_0)} = -g_m R \tag{16.105}$$

Note that the expression for the center-frequency gain could have been written by inspection; at resonance, the reactances of L and C cancel out and the impedance of the parallel LCR circuit reduces to R.

Example 16.4

It is required to design a tuned amplifier of the type shown in Fig. 16.39, having $f_0 = 1$ MHz, 3-dB bandwidth = 10 kHz, and center-frequency gain = –10 V/V. The FET available has at the bias point $g_m = 5$ mA/V and $r_o = 10$ kΩ. The output capacitance is negligibly small. Determine the values of R_L, C_L, and L.

Solution

Center-frequency gain = –10 = –5R. Thus $R = 2$ kΩ. Since $R = R_L \| r_o$, then $R_L = 2.5$ kΩ.

$$B = 2\pi \times 10^4 = \frac{1}{CR}$$

Thus

$$C = \frac{1}{2\pi \times 10^4 \times 2 \times 10^3} = 7958 \text{ pF}$$

Since $\omega_0 = 2\pi \times 10^6 = 1/\sqrt{LC}$, we obtain

$$L = \frac{1}{4\pi^2 \times 10^{12} \times 7958 \times 10^{-12}} = 3.18\ \mu\text{H}$$

16.11.2 Inductor Losses

The power loss in the inductor is usually represented by a series resistance r_s as shown in Fig. 16.40(a). However, rather than specifying the value of r_s, the usual practice is to specify the inductor Q factor at the frequency of interest,

$$Q_0 \equiv \frac{\omega_0 L}{r_s} \tag{16.106}$$

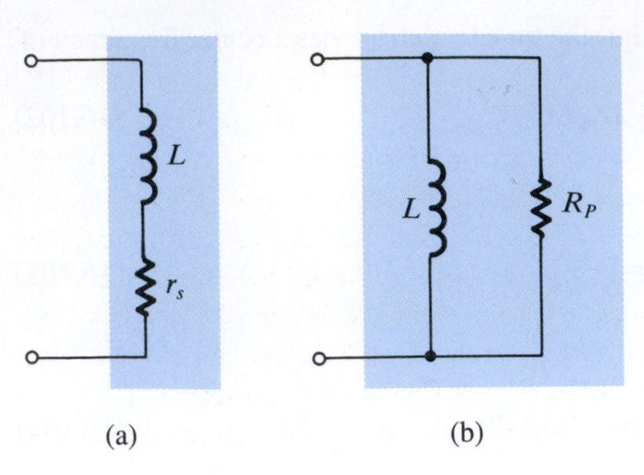

Figure 16.40 Inductor equivalent circuits.

Typically, Q_0 is in the range of 50 to 200.

The analysis of a tuned amplifier is greatly simplified by representing the inductor loss by a parallel resistance R_p, as shown in Fig. 16.40(b). The relationship between R_p and Q_0 can be found by writing, for the admittance of the circuit in Fig. 16.40(a),

$$Y(j\omega_0) = \frac{1}{r_s + j\omega_0 L}$$

$$= \frac{1}{j\omega_0 L}\,\frac{1}{1 - j(1/Q_0)} = \frac{1}{j\omega_0 L}\,\frac{1 + j(1/Q_0)}{1 + (1/Q_0^2)}$$

For $Q_0 \gg 1$,

$$Y(j\omega_0) \simeq \frac{1}{j\omega_0 L}\left(1 + j\frac{1}{Q_0}\right) \tag{16.107}$$

Equating this to the admittance of the circuit in Fig. 16.40(b) gives

$$Q_0 = \frac{R_p}{\omega_0 L} \tag{16.108}$$

or, equivalently,

$$R_p = \omega_0 L Q_0 \tag{16.109}$$

Finally, it should be noted that the coil Q factor poses an upper limit on the value of Q achieved by the tuned circuit.

EXERCISE

16.32 If the inductor in Example 16.4 has $Q_0 = 150$, find R_p and then find the value to which R_L should be changed to keep the overall Q, and hence the bandwidth, unchanged.
Ans. 3 kΩ; 15 kΩ

16.11.3 Use of Transformers

In many cases it is found that the required value of inductance is not practical, in the sense that coils with the required inductance might not be available with the required high values of Q_0. A simple solution is to use a transformer to effect an impedance change. Alternatively, a tapped coil, known as an **autotransformer**, can be used, as shown in Fig. 16.41. Provided the two parts of the inductor are tightly coupled, which can be achieved by winding on a ferrite core, the transformation relationships shown hold. The result is that the tuned circuit seen between terminals 1 and 1′ is equivalent to that in Fig. 16.39(b). For example, if a turns ratio $n = 3$ is used in the amplifier of Example 16.4, then a coil with inductance $L' = 9 \times 3.18 = 28.6\ \mu\text{H}$ and a capacitance $C' = 7958/9 = 884$ pF will be required. Both these values are more practical than the original ones.

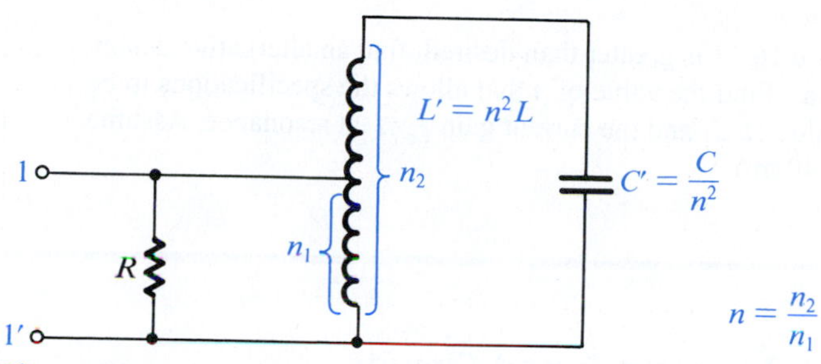

Figure 16.41 A tapped inductor is used as an impedance transformer to allow using a higher inductance, L', and a smaller capacitance, C'.

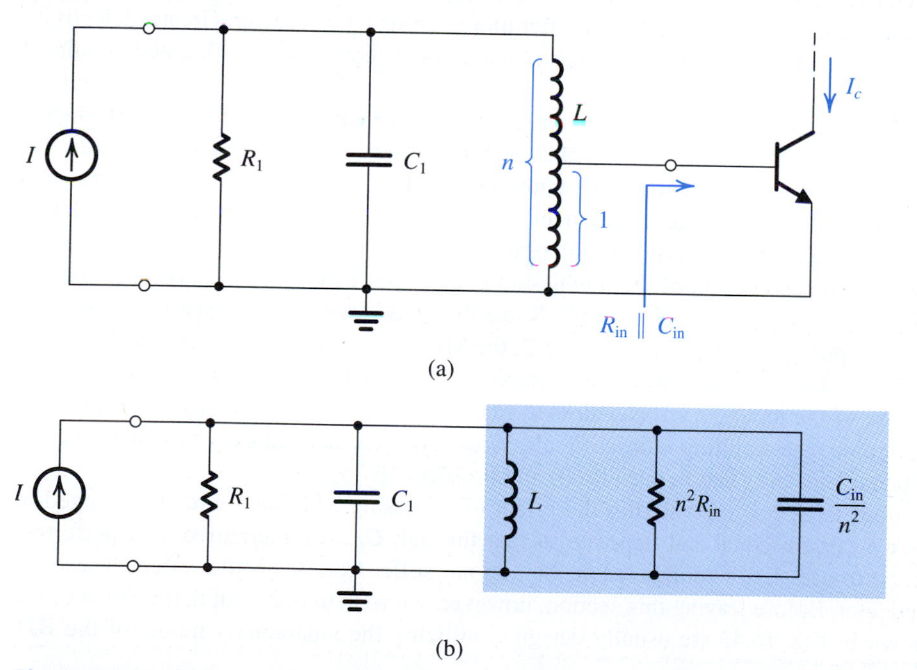

Figure 16.42 **(a)** The output of a tuned amplifier is coupled to the input of another amplifier via a tapped coil. **(b)** An equivalent circuit. Note that the use of a tapped coil increases the effective input impedance of the second amplifier stage.

In applications that involve coupling the output of a tuned amplifier to the input of another amplifier, the tapped coil can be used to raise the effective input resistance of the latter amplifier stage. In this way, one can avoid reduction of the overall Q. This point is illustrated in Fig. 16.42 and in the following exercises.

EXERCISES

D16.33 Consider the circuit in Fig. 16.42(a), first without tapping the coil. Let $L = 5\ \mu\text{H}$ and assume that R_1 is fixed at 1 kΩ. We wish to design a tuned amplifier with $f_0 = 455$ kHz and a 3-dB bandwidth of 10 kHz [this is the intermediate frequency (IF) amplifier of an AM radio]. If the BJT has $R_{in} = 1$ kΩ and $C_{in} = 200$ pF, find the actual bandwidth obtained and the required value of C_1.

Ans. 13 kHz; 24.27 nF

D16.34 Since the bandwidth realized in Exercise 16.33 is greater than desired, find an alternative design utilizing a tapped coil as in Fig. 16.42(a). Find the value of n that allows the specifications to be just met. Also find the new required value of C_1 and the current gain I_c/I at resonance. Assume that at the bias point the BJT has $g_m = 40$ mA/V.

Ans. 1.36; 24.36 nF; 19.1 A/A

16.11.4 Amplifiers with Multiple Tuned Circuits

The selectivity achieved with the single tuned circuit of Fig. 16.39 is not sufficient in many applications—for instance, in the IF amplifier of a radio or a TV receiver. Greater selectivity is obtained by using additional tuned stages. Figure 16.43 shows a BJT with tuned circuits at both the input and the output.[11] In this circuit the bias details are shown, from which we note that biasing is quite similar to the classical arrangement employed in low-frequency, discrete-circuit design. However, to avoid the loading effect of the bias resistors R_{B1} and R_{B2} on the input tuned circuit, a **radio frequency choke** (RFC) is inserted in series with each resistor. Such chokes have high impedances at the frequencies of interest. The use of RFCs in biasing tuned RF amplifiers is common practice.

The analysis and design of the double-tuned amplifier of Fig. 16.43 is complicated by the Miller effect[12] due to capacitance C_μ. Since the load is not simply resistive, as was the case in the amplifiers studied in Section 9.5.2, the Miller impedance at the input will be complex. This reflected impedance will cause detuning of the input circuit as well as "skewing" of the response of the input circuit. Needless to say, the coupling introduced by C_μ makes tuning (or aligning) the amplifier quite difficult. Worse still, the capacitor C_μ can cause oscillations to occur [see Gray and Searle (1969) and Problem 16.75].

Methods exist for **neutralizing** the effect of C_μ, using additional circuits arranged to feed back a current equal and opposite to that through C_μ. An alternative, and preferred, approach is to use circuit configurations that do not suffer from the Miller effect. These are discussed later. Before leaving this section, however, we wish to point out that circuits of the type shown in Fig. 16.43 are usually designed utilizing the y-parameter model of the BJT

[11] Note that because the input circuit is a parallel resonant circuit, an input current source (rather than voltage source) signal is utilized.

[12] Here we use "Miller effect" to refer to the effect of the feedback capacitance C_μ in reflecting back an input impedance that is a function of the amplifier load impedance.

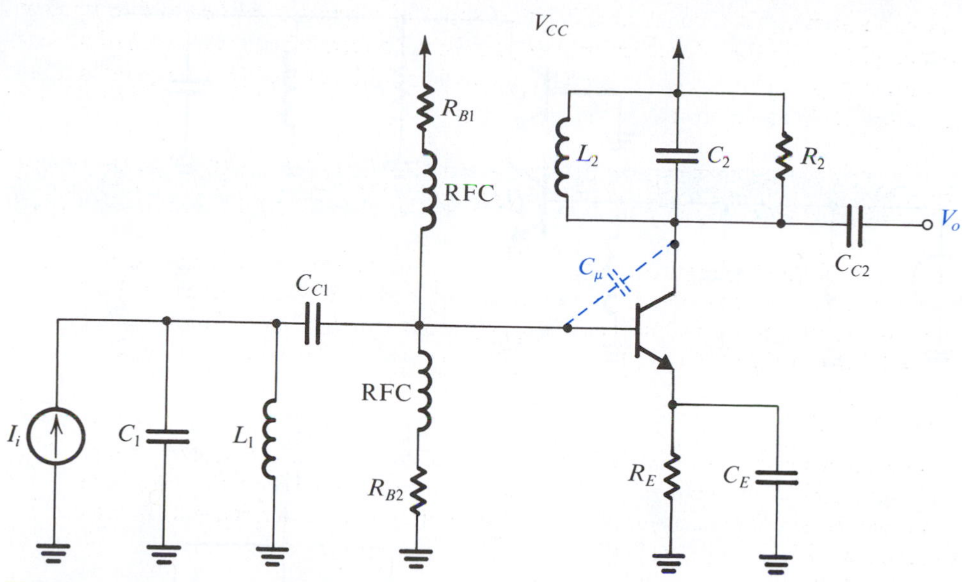

Figure 16.43 A BJT amplifier with tuned circuits at the input and the output.

(see Appendix C). This is done because here, in view of the fact that C_μ plays a significant role, the y-parameter model makes the analysis simpler (in comparison to that using the hybrid-π model). Also, the y parameters can easily be measured at the particular frequency of interest, ω_0. For narrow-band amplifiers, the assumption is usually made that the y parameters remain approximately constant over the passband.

16.11.5 The Cascode and the CC–CB Cascade

From our study of amplifier frequency response in Chapter 9, we know that two amplifier configurations do not suffer from the Miller effect. These are the cascode configuration and the common-collector, common-base cascade. Figure 16.44 shows tuned amplifiers based on these two configurations. The CC–CB cascade is usually preferred in IC implementations because its differential structure makes it suitable for IC biasing techniques. (Note that the biasing details of the cascode circuit are not shown in Fig. 16.44a. Biasing can be done using arrangements similar to those discussed in earlier chapters.)

16.11.6 Synchronous Tuning

In the design of a tuned amplifier with multiple tuned circuits, the question of the frequency to which each circuit should be tuned arises. The objective, of course, is for the overall response to exhibit high passband flatness and skirt selectivity. To investigate this question, we shall assume that the overall response is the product of the individual responses: in other words, that the stages do not interact. This can easily be achieved using circuits such as those in Fig. 16.44.

Consider first the case of N identical resonant circuits, known as the **synchronously tuned** case. Figure 16.45 shows the response of an individual stage and that of the cascade. Observe the bandwidth "shrinkage" of the overall response. The 3-dB bandwidth B of the overall amplifier is related to that of the individual tuned circuits, ω_0/Q, by (see Problem 16.77)

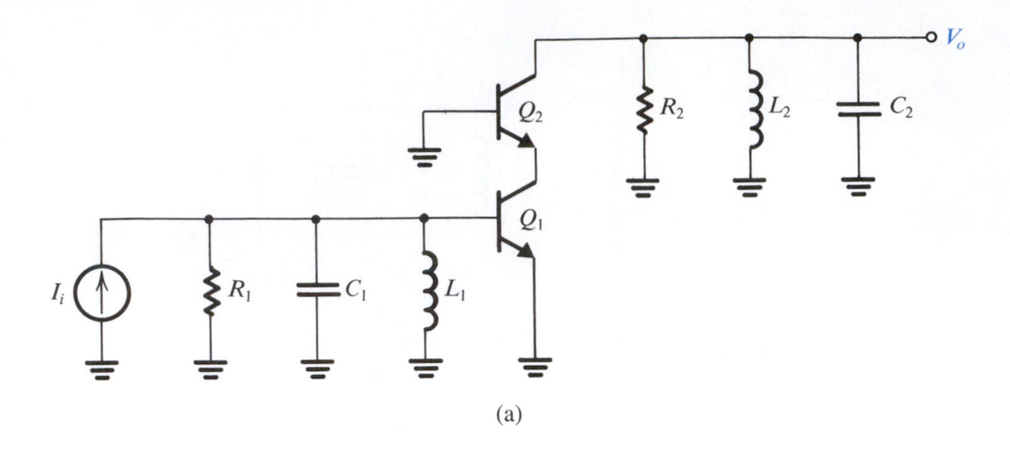

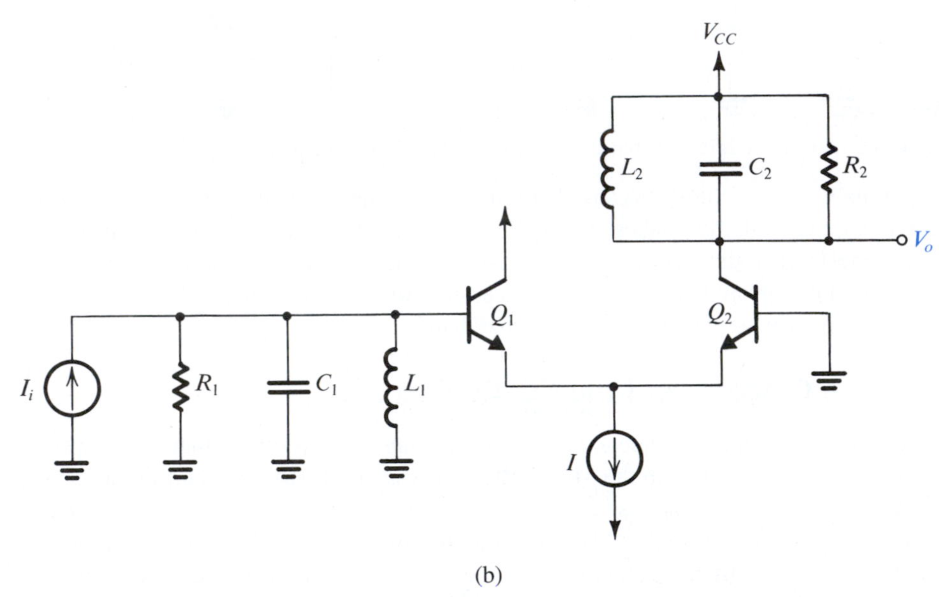

Figure 16.44 Two tuned-amplifier configurations that do not suffer from the Miller effect: **(a)** cascode and **(b)** common-collector, common-base cascade. (Note that bias details of the cascode circuit are not shown.)

$$B = \frac{\omega_0}{Q}\sqrt{2^{1/N} - 1} \tag{16.110}$$

The factor $\sqrt{2^{1/N} - 1}$ is known as the **bandwidth-shrinkage factor**. Given B and N, we can use Eq. (16.110) to determine the bandwidth required of the individual stages, ω_0/Q.

EXERCISE

D16.35 Consider the design of an IF amplifier for an FM radio receiver. Using two synchronously tuned stages with $f_0 = 10.7$ MHz, find the 3-dB bandwidth of each stage so that the overall bandwidth is 200 kHz. Using 3-μH inductors find C and R for each stage.

Ans. 310.8 kHz; 73.7 pF; 6.95 kΩ

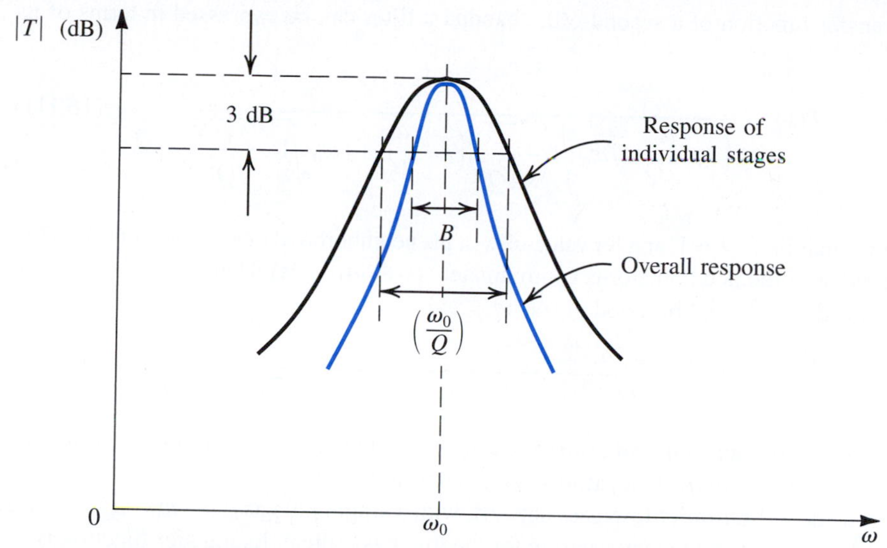

Figure 16.45 Frequency response of a synchronously tuned amplifier.

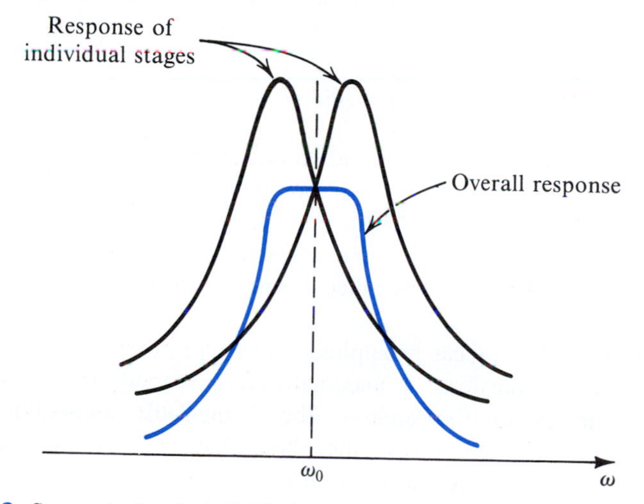

Figure 16.46 Stagger-tuning the individual resonant circuits can result in an overall response with a passband flatter than that obtained with synchronous tuning (Fig. 16.45).

16.11.7 Stagger-Tuning

A much better overall response is obtained by stagger-tuning the individual stages, as illustrated in Fig. 16.46. Stagger-tuned amplifiers are usually designed so that the overall response exhibits *maximal flatness* around the center frequency f_0. Such a response can be obtained by transforming the response of a maximally flat (Butterworth) low-pass filter up the frequency axis to ω_0. We show here how this can be done.

The transfer function of a second-order bandpass filter can be expressed in terms of its poles as

$$T(s) = \frac{a_1 s}{\left(s+\frac{\omega_0}{2Q}-j\omega_0\sqrt{1-\frac{1}{4Q^2}}\right)\left(s+\frac{\omega_0}{2Q}+j\omega_0\sqrt{1-\frac{1}{4Q^2}}\right)} \quad (16.111)$$

For a narrow-band filter, $Q \gg 1$, and for values of s in the neighborhood of $+j\omega_0$ (see Fig. 16.47b), the second factor in the denominator is approximately $(s + j\omega_0 \simeq 2s)$. Hence Eq. (16.111) can be approximated in the neighborhood of $j\omega_0$ by

$$T(s) \simeq \frac{a_1/2}{s+\omega_0/2Q - j\omega_0} = \frac{a_1/2}{(s-j\omega_0)+\omega_0/2Q} \quad (16.112)$$

This is known as the **narrow-band approximation**.[13] Note that the magnitude response, for $s = j\omega$, has a peak value of a_1Q/ω_0 at $\omega = \omega_0$, as expected.

Now consider a first-order low-pass network with a single pole at $p = -\omega_0/2Q$ (we use p to denote the complex frequency variable for the low-pass filter). Its transfer function is

$$T(p) = \frac{K}{p+\omega_0/2Q} \quad (16.113)$$

where K is a constant. Comparing Eqs. (16.112) and (16.113) we note that they are identical for $p = s - j\omega_0$ or, equivalently,

$$s = p + j\omega_0 \quad (16.114)$$

This result implies that the response of the second-order bandpass filter *in the neighborhood of its center frequency* $s = j\omega_0$ is identical to the response of a first-order low-pass filter with a pole at $(-\omega_0/2Q)$ *in the neighborhood of* $p = 0$. Thus the bandpass response can be obtained by shifting the pole of the low-pass prototype and adding the complex-conjugate pole, as illustrated in Fig. 16.47(b). This is called a **lowpass-to-bandpass transformation** for *narrow-band* filters.

The transformation $p = s - j\omega_0$ can be applied to low-pass filters of order greater than one. For instance, we can transform a maximally flat, second-order low-pass filter $(Q = 1/\sqrt{2})$ to obtain a maximally flat bandpass filter. If the 3-dB bandwidth of the bandpass filter is to be B rad/s, then the low-pass filter should have a 3-dB frequency (and thus a pole frequency) of $(B/2)$ rad/s, as illustrated in Fig. 16.48. The resulting fourth-order bandpass filter will be a stagger-tuned one, with its two tuned circuits (refer to Fig. 16.48) having

$$\omega_{01} = \omega_0 + \frac{B}{2\sqrt{2}} \qquad B_1 = \frac{B}{\sqrt{2}} \qquad Q_1 \simeq \frac{\sqrt{2}\,\omega_0}{B} \quad (16.115)$$

[13] The bandpass response is *geometrically symmetrical* around the center frequency ω_0. That is, each pair of frequencies ω_1 and ω_2 at which the magnitude response is equal are related by $\omega_1\omega_2 = \omega_0^2$. For high Q, the symmetry becomes almost *arithmetic* for frequencies close to ω_0. That is, two frequencies with the same magnitude response are almost equally spaced from ω_0. The same is true for higher-order bandpass filters designed using the transformation presented in this section.

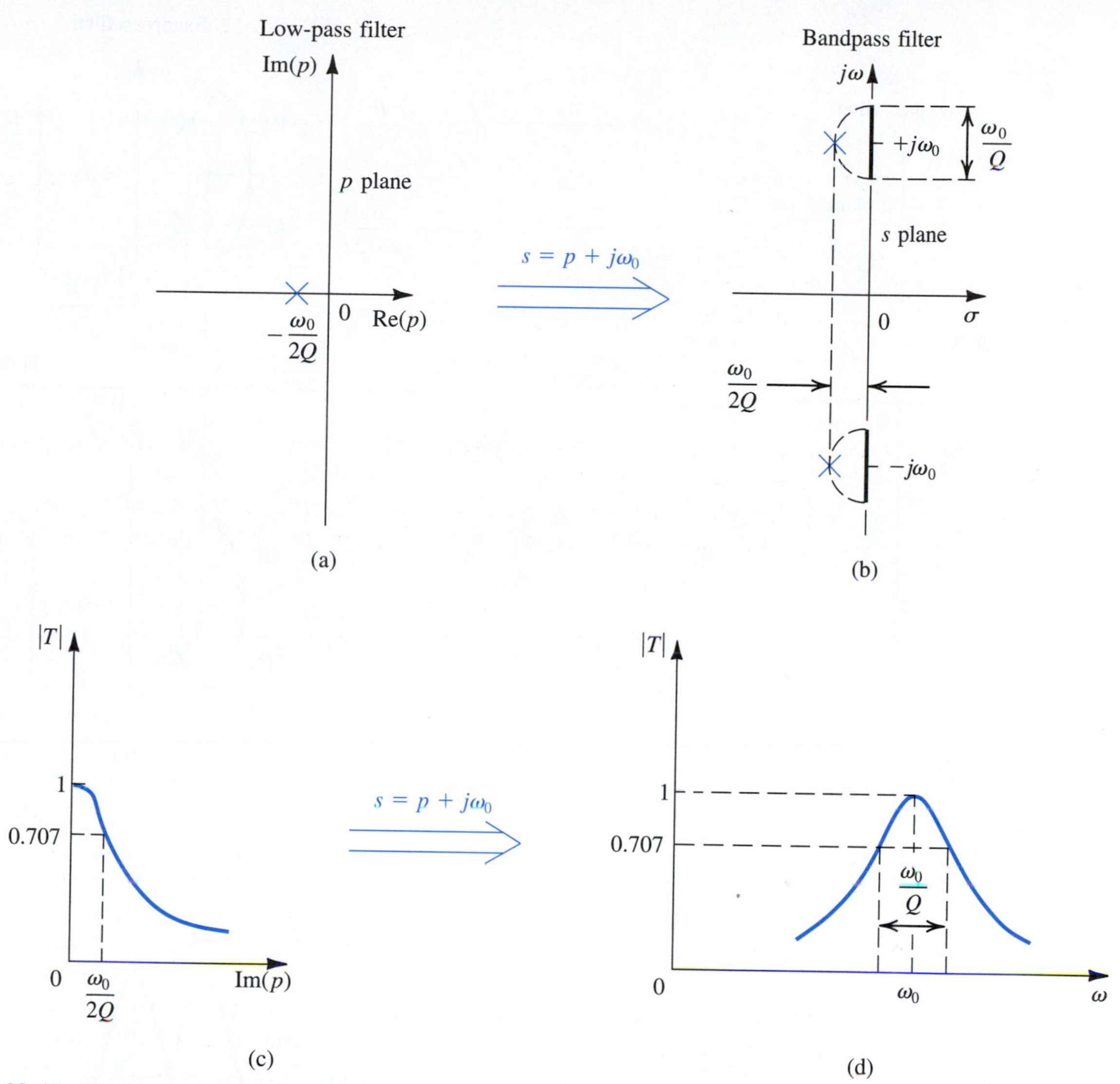

Figure 16.47 Obtaining a second-order narrow-band bandpass filter by transforming a first-order low-pass filter. **(a)** Pole of the first-order filter in the p plane. **(b)** Applying the transformation $s = p + j\omega_0$ and adding a complex-conjugate pole results in the poles of the second-order bandpass filter. **(c)** Magnitude response of the first-order low-pass filter. **(d)** Magnitude response of the second-order bandpass filter.

$$\omega_{02} = \omega_0 - \frac{B}{2\sqrt{2}} \qquad B_2 = \frac{B}{\sqrt{2}} \qquad Q_2 = \frac{\sqrt{2}\,\omega_0}{B} \tag{16.116}$$

Note that for the overall response to have a normalized center-frequency gain of unity, the individual responses have to have equal center-frequency gains of $\sqrt{2}$, as shown in Fig. 16.48(d).

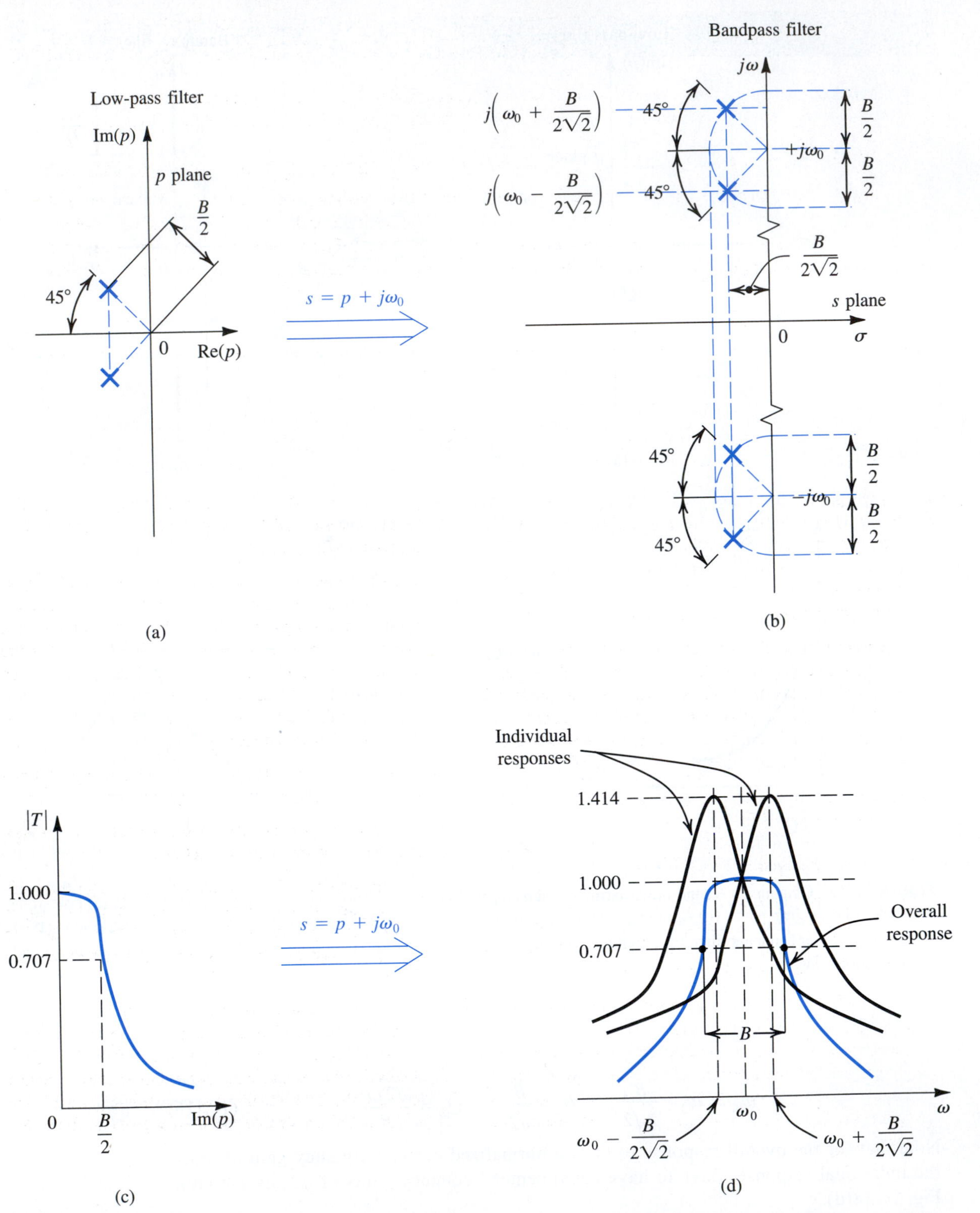

Figure 16.48 Obtaining the poles and the frequency response of a fourth-order stagger-tuned, narrow-band bandpass amplifier by transforming a second-order low-pass, maximally flat response.

EXERCISES

D16.36 A stagger-tuned design for the IF amplifier specified in Exercise 16.35 is required. Find f_{01}, B_1, f_{02}, and B_2. Also give the value of C and R for each of the two stages. (Recall that 3-μH inductors are to be used.)
Ans. 10.77 MHz; 141.4 kHz; 10.63 MHz; 141.4 kHz; 72.8 pF; 15.5 kΩ; 74.7 pF; 15.1 kΩ

16.37 Using the fact that the voltage gain at resonance is proportional to the value of R, find the ratio of the gain at 10.7 MHz of the stagger-tuned amplifier designed in Exercise 16.36 and the synchronously tuned amplifier designed in Exercise 16.35. (*Hint:* For the stagger-tuned amplifier, note that the gain at ω_0 is equal to the product of the gains of the individual stages at their 3-dB frequencies.)
Ans. 2.42

Summary

- A filter is a linear two-port network with a transfer function $T(s) = V_o(s)/V_i(s)$. For physical frequencies, the filter transmission is expressed as $T(j\omega) = |T(j\omega)|e^{j\phi(\omega)}$. The magnitude of transmission can be expressed in decibels using either the gain function $G(\omega) \equiv 20 \log|T|$ or the attenuation function $A(\omega) \equiv -20 \log|T|$.

- The transmission characteristics of a filter are specified in terms of the edges of the passband(s) and the stopband(s); the maximum allowed variation in passband transmission, A_{max} (dB); and the minimum attenuation required in the stopband, A_{min} (dB). In some applications, the phase characteristics are also specified.

- The filter transfer function can be expressed as the ratio of two polynomials in s; the degree of the denominator polynomial, N, is the filter order. The N roots of the denominator polynomial are the poles (natural modes).

- To obtain a highly selective response, the poles are complex and occur in conjugate pairs (except for one real pole when N is odd). The zeros are placed on the $j\omega$ axis in the stopband(s) including $\omega = 0$ and $\omega = \infty$.

- The Butterworth filter approximation provides a low-pass response that is maximally flat at $\omega = 0$. The transmission decreases monotonically as ω increases, reaching 0 (infinite attenuation) at $\omega = \infty$, where all N transmission zeros lie. Eq. (16.11) gives $|T|$, where ϵ is given by Eq. (16.14) and the order N is determined using Eq. (16.15). The poles are found using the graphical construction of Fig. 16.10, and the transfer function is given by Eq. (16.16).

- The Chebyshev filter approximation provides a low-pass response that is equiripple in the passband with the transmission decreasing monotonically in the stopband. All the transmission zeros are at $s = \infty$. Eq. (16.18) gives $|T|$ in the passband and Eq. (16.19) gives $|T|$ in the stopband, where ϵ is given by Eq. (16.21). The order N can be determined using Eq. (16.22). The poles are given by Eq. (16.23) and the transfer function by Eq. (16.24).

- Figures 16.13 and 16.14 provide a summary of first-order filter functions and their realizations.

- Figure 16.16 provides the characteristics of seven special second-order filtering functions.

- The second-order LCR resonator of Fig. 16.17(a) realizes a pair of complex-conjugate poles with $\omega_0 = 1/\sqrt{LC}$ and $Q = \omega_0 CR$. This resonator can be used to realize the various special second-order filtering functions, as shown in Fig. 16.18.

- By replacing the inductor of an LCR resonator with a simulated inductance obtained using the Antoniou circuit of Fig. 16.20(a), the op amp–RC resonator of Fig. 16.21(b) is obtained. This resonator can be used to realize the various second-order filter functions as shown in Fig. 16.22. The design equations for these circuits are given in Table 16.1.

- Biquads based on the two-integrator-loop topology are the most versatile and popular second-order filter realizations. There are two varieties: the KHN circuit of Fig. 16.24(a), which realizes the LP, BP, and HP functions simultaneously and can be combined with the output summing amplifier of Fig. 16.28(b) to realize the notch and all-pass functions; and the Tow–Thomas circuit of Fig. 16.25(b), which realizes the BP and LP functions simultaneously. Feedforward can be applied to the Tow–Thomas circuit to obtain the circuit of Fig. 16.26, which can be designed to realize any of the second-order functions (see Table 16.2).
- Single-amplifier biquads (SABs) are obtained by placing a bridged-T network in the negative-feedback path of an op amp. If the op amp is ideal, the poles realized are at the same locations as the zeros of the RC network. The complementary transformation can be applied to the feedback loop to obtain another feedback loop having identical poles. Different transmission zeros are realized by feeding the input signal to circuit nodes that are connected to ground. SABs are economic in their use of op amps but are sensitive to the op-amp nonidealities and are thus limited to low-Q applications ($Q \leq 10$).
- The classical sensitivity function

$$S_x^y = \frac{\partial y/y}{\partial x/x}$$

is a very useful tool in investigating how tolerant a filter circuit is to the unavoidable inaccuracies in component values and to the nonidealities of the op amps.
- Switched-capacitor (SC) filters are based on the principle that a capacitor C, periodically switched between two circuit nodes at a high rate, f_c, is equivalent to a resistance $R = 1/Cf_c$ connecting the two circuit nodes. SC filters can be fabricated in monolithic form using CMOS IC technology.
- Tuned amplifiers utilize LC-tuned circuits as loads, or at the input, of transistor amplifiers. They are used in the design of the RF tuner and the IF amplifier of communication receivers. The cascode and the CC–CB cascade configurations are frequently used in the design of tuned amplifiers. Stagger-tuning the individual tuned circuits results in a flatter passband response (in comparison to that obtained with all the tuned circuits synchronously tuned).

PROBLEMS

Computer Simulation Problems

SIM Problems involving design are marked with D throughout the text. As well, problems are marked with asterisks to describe their degree of difficulty. Difficult problems are marked with an asterisk (*); more difficult problems with two asterisks (**); and very challenging and/or time-consuming problems with three asterisks (***).

Section 16.1: Filter Transmission, Types and Specification

16.1 The transfer function of a first-order low-pass filter (such as that realized by an RC circuit) can be expressed as $T(s) = \omega_0/(s + \omega_0)$, where ω_0 is the 3-dB frequency of the filter. Give in table form the values of $|T|$, ϕ, G, and A at $\omega = 0, 0.5\omega_0, \omega_0, 2\omega_0, 5\omega_0, 10\omega_0$, and $100\omega_0$.

***16.2** A filter has the transfer function $T(s) = 1/[(s+1)(s^2+s+1)]$. Show that $|T| = \sqrt{1+\omega^6}$ and find an expression for its phase response $\phi(\omega)$. Calculate the values of $|T|$ and ϕ for $\omega = 0.1$, 1, and 10 rad/s and then find the output corresponding to each of the following input signals:

(a) $2 \sin 0.1t$ (volts)
(b) $2 \sin t$ (volts)
(c) $2 \sin 10t$ (volts)

16.3 For the filter whose magnitude response is sketched (as the colored curve) in Fig. 16.3, find $|T|$ at $\omega = 0$, $\omega = \omega_p$, and $\omega = \omega_s$. $A_{max} = 0.5$ dB, and $A_{min} = 40$ dB.

D 16.4 A low-pass filter is required to pass all signals within its passband, extending from 0 to 4 kHz, with a transmission variation of at most 10% (i.e., the ratio of the maximum to minimum transmission in the passband should not exceed 1.1). The transmission in the stopband, which extends from 5 kHz to ∞, should not exceed 0.1% of the maximum passband transmission. What are the values of A_{max}, A_{min}, and the selectivity factor for this filter?

16.5 A low-pass filter is specified to have $A_{max} = 1$ dB and $A_{min} = 10$ dB. It is found that these specifications can be just met with a single-time-constant RC circuit having a time constant of 1 s and a dc transmission of unity. What must ω_p and ω_s of this filter be? What is the selectivity factor?

16.6 Sketch transmission specifications for a high-pass filter having a passband defined by $f \geq 2$ kHz and a stopband defined by $f \leq 1$ kHz. $A_{max} = 0.5$ dB, and $A_{min} = 50$ dB.

16.7 Sketch transmission specifications for a bandstop filter that is required to pass signals over the bands $0 \leq f \leq$ 10 kHz and 20 kHz $\leq f \leq \infty$ with A_{max} of 1 dB. The stopband extends from f = 12 kHz to f = 16 kHz, with a minimum required attenuation of 40 dB.

Section 16.2: The Filter Transfer Function

16.8 Consider a fifth-order filter whose poles are all at a radial distance from the origin of 10^3 rad/s. One pair of complex conjugate poles is at 18° angles from the $j\omega$ axis, and the other pair is at 54° angles. Give the transfer function in each of the following cases:

(a) The transmission zeros are all at $s = \infty$ and the dc gain is unity.
(b) The transmission zeros are all at $s = 0$ and the high-frequency gain is unity.

What type of filter results in each case?

16.9 A third-order low-pass filter has transmission zeros at $\omega = 2$ rad/s and $\omega = \infty$. Its natural modes are at $s = -1$ and $s = -0.5 \pm j0.8$. The dc gain is unity. Find $T(s)$.

16.10 Find the order N and the form of $T(s)$ of a bandpass filter having transmission zeros as follows: one at $\omega = 0$, one at $\omega = 10^3$ rad/s, one at 3×10^3 rad/s, one at 6×10^3 rad/s, and one at $\omega = \infty$. If this filter has a monotonically decreasing passband transmission with a peak at the center frequency of 2×10^3 rad/s, and equiripple response in the stopbands, sketch the shape of its $|T|$.

***16.11** Analyze the RLC network of Fig. P16.11 to determine its transfer function $V_o(s)/V_i(s)$ and hence its poles and zeros. (*Hint:* Begin the analysis at the output and work your way back to the input.)

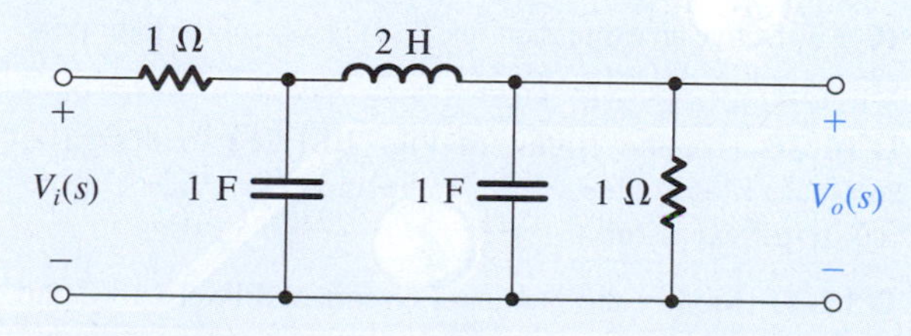

Figure P16.11

Section 16.3: Butterworth and Chebyshev Filters

D 16.12 Determine the order N of the Butterworth filter for which $A_{max} = 1$ dB, $A_{min} \geq 20$ dB, and the selectivity ratio $\omega_s/\omega_p = 1.3$. What is the actual value of minimum stopband attenuation realized? If A_{min} is to be exactly 20 dB, to what value can A_{max} be reduced?

16.13 Calculate the value of attenuation obtained at a frequency 1.6 times the 3-dB frequency of a seventh-order Butterworth filter.

16.14 Find the natural modes of a Butterworth filter with a 1-dB bandwidth of 10^3 rad/s and $N = 5$.

D 16.15 Design a Butterworth filter that meets the following low-pass specifications: f_p = 10 kHz, A_{max} = 2 dB, f_s = 15 kHz, and A_{min} = 15 dB. Find N, the natural modes, and $T(s)$. What is the attenuation provided at 20 kHz?

***16.16** Sketch $|T|$ for a seventh-order low-pass Chebyshev filter with $\omega_p = 1$ rad/s and $A_{max} = 1$ dB. Use Eq. (16.18) to determine the values of ω at which $|T| = 1$ and the values of ω at which $|T| = 1/\sqrt{1+\epsilon^2}$. Indicate these values on your sketch. Use Eq. (16.19) to determine $|T|$ at $\omega = 2$ rad/s, and indicate this point on your sketch. For large values of ω, at what rate (in dB/octave) does the transmission decrease?

16.17 Contrast the attenuation provided by a fifth-order Chebyshev filter at $\omega_s = 2\omega_p$ to that provided by a Butterworth filter of equal order. For both, $A_{max} = 1$ dB. Sketch $|T|$ for both filters on the same axes.

D *16.18 It is required to design a low-pass filter to meet the following specifications: $f_p = 3.4$ kHz, $A_{max} = 1$ dB, f_s = 4 kHz, $A_{min} = 35$ dB.

(a) Find the required order of Chebyshev filter. What is the excess (above 35 dB) stopband attenuation obtained?
(b) Find the poles and the transfer function.

Section 16.4: First-Order and Second-Order Filter Functions

D 16.19 Use the information displayed in Fig. 16.13 to design a first-order op amp–RC low-pass filter having a 3-dB frequency of 10 kHz, a dc gain magnitude of 10, and an input resistance of 10 kΩ.

D 16.20 Use the information given in Fig. 16.13 to design a first-order op amp–RC high-pass filter with a 3-dB frequency of 100 Hz, a high-frequency input resistance of 100 kΩ, and a high-frequency gain magnitude of unity.

D *16.21 Use the information given in Fig. 16.13 to design a first-order op amp–RC spectrum-shaping network with a transmission zero frequency of 1 kHz, a pole frequency of 100 kHz, and a dc gain magnitude of unity. The low-frequency input resistance is to be 1 kΩ. What is the high-frequency gain that results? Sketch the magnitude of the transfer function versus frequency.

D *16.22 By cascading a first-order op amp–RC low-pass circuit with a first-order op amp–RC high-pass circuit, one can design a wideband bandpass filter. Provide such a design for the case in which the midband gain is 12 dB and the 3-dB bandwidth extends from 100 Hz to 10 kHz. Select appropriate component values under the constraint that no resistors higher than 100 kΩ are to be used and that the input resistance is to be as high as possible.

D 16.23 Derive $T(s)$ for the op amp–RC circuit in Fig. 16.14. We wish to use this circuit as a variable phase shifter by adjusting R. If the input signal frequency is 10^4 rad/s and if $C = 10$ nF, find the values of R required to obtain phase shifts of –30°, –60°, –90°, –120°, and –150°.

16.24 Show that by interchanging R and C in the op amp–RC circuit of Fig. 16.14, the resulting phase shift covers the range 0 to 180° (with 0° at high frequencies and 180° at low frequencies).

16.25 Use the information in Fig. 16.16(a) to obtain the transfer function of a second-order low-pass filter with $\omega_0 = 10^3$ rad/s, $Q = 1$, and dc gain = 1. At what frequency does $|T|$ peak? What is the peak transmission?

D *16.26 Use the information in Fig. 16.16(a) to obtain the transfer function of a second-order low-pass filter that just meets the specifications defined in Fig. 16.3 with $\omega_p = 1$ rad/s and $A_{max} = 3$ dB. Note that there are two possible solutions. For each, find ω_0 and Q. Also, if $\omega_s = 2$ rad/s, find the value of A_{min} obtained in each case.

D **16.27 Use two first-order op amp–RC all-pass circuits in cascade to design a circuit that provides a set of three-phase 60-Hz voltages, each separated by 120° and equal in magnitude, as shown in the phasor diagram of Fig. P16.27. These voltages simulate those used in three-phase power transmission systems. Use 1-μF capacitors.

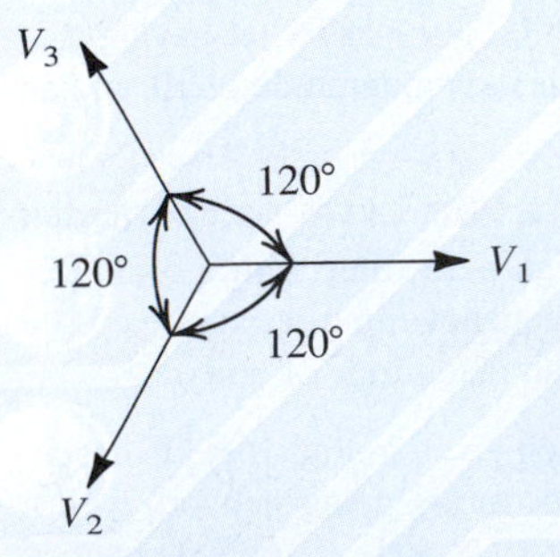

Figure P16.27

16.28 Use the information given in Fig. 16.16(b) to find the transfer function of a second-order high-pass filter with natural modes at $-0.5 \pm j\sqrt{3}/2$ and a high-frequency gain of unity.

D **16.29 (a) Show that $|T|$ of a second-order bandpass function is geometrically symmetrical around the center frequency ω_0. That is, the members of each pair of frequencies ω_1 and ω_2 for which $|T(j\omega_1)| = |T(j\omega_2)|$ are related by $\omega_1\omega_2 = \omega_0^2$.

(b) Find the transfer function of the second-order bandpass filter that meets specifications of the form in Fig. 16.4 where $\omega_{p1} = 8100$ rad/s, $\omega_{p2} = 10{,}000$ rad/s, and $A_{max} = 1$ dB. If $\omega_{s1} = 3000$ rad/s find A_{min} and ω_{s2}.

D *16.30 Use the result of Exercise 16.15 to find the transfer function of a notch filter that is required to eliminate a bothersome interference of 60-Hz frequency. Since the frequency of the interference is not stable, the filter should be designed to provide attenuation ≥20 dB over a 6-Hz band centered around 60 Hz. The dc transmission of the filter is to be unity.

16.31 Consider a second-order all-pass circuit in which errors in the component values result in the frequency of the zeros being slightly lower than that of the poles. Roughly sketch the expected $|T|$. Repeat for the case of the frequency of the zeros slightly higher than the frequency of the poles.

16.32 Consider a second-order all-pass filter in which errors in the component values result in the Q factor of the zeros being greater than the Q factor of the poles. Roughly sketch the expected $|T|$. Repeat for the case of the Q factor of the zeros lower than the Q factor of the poles.

Section 16.5: The Second-Order LCR Resonator

D 16.33 Design the LCR resonator of Fig. 16.17(a) to obtain natural modes with $\omega_0 = 10^4$ rad/s and $Q = 2$. Use $R = 10$ kΩ.

16.34 For the LCR resonator of Fig. 16.17(a), find the change in ω_0 that results from

(a) increasing L by 1%
(b) increasing C by 1%
(c) increasing R by 1%

16.35 Derive an expression for $V_o(s)/V_i(s)$ of the high-pass circuit in Fig. 16.18(c).

D 16.36 Use the circuit of Fig. 16.18(b) to design a low-pass filter with $\omega_0 = 10^5$ rad/s and $Q = 1/\sqrt{2}$. Utilize a 0.01-μF capacitor.

D 16.37 Modify the bandpass circuit of Fig. 16.18(d) to change its center-frequency gain from 1 to 0.5 without changing ω_0 or Q.

16.38 Consider the LCR resonator of Fig. 16.17(a) with node x disconnected from ground and connected to an input signal source V_x, node y disconnected from ground and connected to another input signal source V_y, and node z disconnected from ground and connected to a third input signal source V_z. Use superposition to find the voltage that develops across the resonator, V_o, in terms of V_x, V_y, and V_z.

16.39 Consider the notch circuit shown in Fig. 16.18(i). For what ratio of L_1 to L_2 does the notch occur at $0.9\omega_0$? For this case, what is the magnitude of the transmission at frequencies $\ll\omega_0$? At frequencies $\gg\omega_0$?

Section 16.6: Second-Order Active Filters Based on Inductor Replacement

D 16.40 Design the circuit of Fig. 16.20 (utilizing suitable component values) to realize an inductance of (a) 10 H, (b) 1 H, and (c) 0.1 H.

***16.41** Starting from first principles and assuming ideal op amps, derive the transfer function of the circuit in Fig. 16.22(a).

D *16.42 It is required to design a fifth-order Butterworth filter having a 3-dB bandwidth of 10^4 rad/s and a unity dc gain. Use a cascade of two circuits of the type shown in Fig. 16.22(a) and a first-order op amp–RC circuit of the type shown in Fig. 16.13(a). Select appropriate component values.

D 16.43 Design the circuit of Fig. 16.22(e) to realize an LPN function with $f_0 = 4$ kHz, $f_n = 5$ kHz, $Q = 10$, and a unity dc gain. Select $C_4 = 10$ nF.

D 16.44 Design the all-pass circuit of Fig. 16.22(g) to provide a phase shift of 180° at $f = 1$ kHz and to have $Q = 1$. Use 1-nF capacitors.

16.45 Consider the Antoniou circuit of Fig. 16.20(a) with R_5 eliminated, a capacitor C_6 connected between node 1 and ground, and a voltage source V_2 connected to node 2. Show that the input impedance seen by V_2 is $R_2/s^2C_4C_6R_1R_3$. How does this impedance behave for physical frequencies ($s = j\omega$)? (This impedance is known as a **frequency-dependent negative resistance**, or FDNR.)

D 16.46 Using the transfer function of the LPN filter, given in Table 16.1, derive the design equations also given.

D 16.47 Using the transfer function of the HPN filter, given in Table 16.1, derive the design equations also given.

D **16.48 It is required to design a third-order low-pass filter whose $|T|$ is equiripple in both the passband and the stopband (in the manner shown in Fig. 16.3, except that the response shown is for $N = 5$). The filter passband extends from $\omega = 0$ to $\omega = 1$ rad/s, and the passband transmission varies between 1 and 0.9. The stopband edge is at $\omega = 1.2$ rad/s. The following transfer function was obtained using filter design tables:

$$T(s) = \frac{0.4508(s^2 + 1.6996)}{(s + 0.7294)(s^2 + s0.2786 + 1.0504)}$$

The actual filter realized is to have $\omega_p = 10^4$ rad/s.

(a) Obtain the transfer function of the actual filter by replacing s by $s/10^4$.

(b) Realize this filter as the cascade connection of a first-order LP op amp–RC circuit of the type shown in Fig. 16.13(a) and a second-order LPN circuit of the type shown in Fig. 16.22(e). Each section is to have a dc gain of unity. Select appropriate component values. (*Note:* A filter with an equiripple response in both the passband and the stopband is known as an **elliptic filter**.)

Section 16.7: Second-Order Active Filters Based on the Two-Integrator-Loop Topology

D 16.49 Design the KHN circuit of Fig. 16.24(a) to realize a bandpass filter with a center frequency of 1 kHz and a 3-dB bandwidth of 50 Hz. Use 10-nF capacitors. Give the complete circuit and specify all component values. What value of center-frequency gain is obtained?

D 16.50 (a) Using the KHN biquad with the output summing amplifier of Fig. 16.24(b), show that an all-pass function is realized by selecting $R_L = R_H = R_B/Q$. Also show that the flat gain obtained is KR_F/R_H.
(b) Design the all-pass circuit to obtain $\omega_0 = 10^4$ rad/s, $Q = 2$, and flat gain = 10. Select appropriate component values.

D 16.51 Consider a notch filter with $\omega_n = \omega_0$ realized by using the KHN biquad with an output summing amplifier. If the summing resistors used have 1% tolerances, what is the worst-case percentage deviation between ω_n and ω_0?

D 16.52 Design the circuit of Fig. 16.26 to realize a low-pass notch filter with $\omega_0 = 10^4$ rad/s, $Q = 10$, dc gain = 1, and $\omega_n = 1.2 \times 10^4$ rad/s. Use $C = 10$ nF and $r = 20$ kΩ.

D 16.53 In the all-pass realization using the circuit of Fig. 16.26, which component(s) does one need to trim to adjust (a) only ω_z and (b) only Q_z?

D **16.54 Repeat Problem 16.48 using the Tow–Thomas biquad of Fig. 16.26 to realize the second-order section in the cascade.

Section 16.8: Single-Amplifier Biquadratic Active Filters

D 16.55 Design the circuit of Fig. 16.29 to realize a pair of poles with $\omega_0 = 10^4$ rad/s and $Q = 1/\sqrt{2}$. Use $C_1 = C_2 = 1$ nF.

16.56 Consider the bridged-T network of Fig. 16.28(a) with $R_3 = R_4 = R$ and $C_1 = C_2 = C$, and denote $CR = \tau$. Find the zeros and poles of the bridged-T network. If the network is placed in the negative-feedback path of an ideal infinite-gain op amp, as in Fig. 16.29, find the poles of the closed-loop amplifier.

***16.57** Consider the bridged-T network of Fig. 16.28(b) with $R_1 = R_2 = R$, $C_4 = C$, and $C_3 = C/16$. Let the network be placed in the negative-feedback path of an infinite-gain op amp and let C_4 be disconnected from ground and connected to the input signal source V_i. Analyze the resulting circuit to determine its transfer function $V_o(s)/V_i(s)$, where $V_o(s)$ is the voltage at the op-amp output. Show that the circuit realized is a bandpass filter and find its ω_0, Q, and the center-frequency gain.

D **16.58 Consider the bandpass circuit shown in Fig. 16.30a. Let $C_1 = C_2 = C$, $R_3 = R$, $R_4 = R/4Q^2$, $CR = 2Q/\omega_0$, and $\alpha = 1$. Disconnect the positive input terminal of the op amp from ground and apply V_i through a voltage divider R_1, R_2 to the positive input terminal as well as through R_4/α as before. Analyze the circuit to find its transfer function V_o/V_i. Find the ratio R_2/R_1 so that the circuit realizes (a) an all-pass function and (b) a notch function. Assume the op amp to be ideal.

D *16.59 Derive the transfer function of the circuit in Fig. 16.33(b) assuming the op amp to be ideal. Thus show that the circuit realizes a high-pass function. What is the high-frequency gain of the circuit? Design the circuit for a maximally flat response with a 3-dB frequency of 10^3 rad/s. Use $C_1 = C_2 = 10$ nF. (*Hint:* For a maximally flat response, $Q = 1/\sqrt{2}$ and $\omega_{3dB} = \omega_0$.)

D *16.60 Design a fifth-order Butterworth low-pass filter with a 3-dB bandwidth of 5 kHz and a dc gain of unity using the cascade connection of two Sallen-and-Key circuits (Fig. 16.34c) and a first-order section (Fig. 16.13a). Use a 10-kΩ value for all resistors.

16.61 The process of obtaining the complement of a transfer function by interchanging input and ground, as illustrated in Fig. 16.31, applies to any general network (not just RC networks as shown). Show that if the network n is a bandpass with a center-frequency gain of unity, then the complement obtained is a notch. Verify this by using the RLC circuits of Fig. 16.18(d) and (e).

Section 16.9: Sensitivity

16.62 Evaluate the sensitivities of ω_0 and Q relative to R, L, and C of the bandpass circuit in Fig. 16.18(d).

***16.63** Verify the following sensitivity identities:

(a) If $y = uv$, then $S_x^y = S_x^u + S_x^v$.
(b) If $y = u/v$, then $S_x^y = S_x^u - S_x^v$.
(c) If $y = ku$, where k is a constant, then $S_x^y = S_x^u$.
(d) If $y = u^n$, where n is a constant, then $S_x^y = nS_x^u$.
(e) If $y = f_1(u)$ and $u = f_2(x)$, then $S_x^y = S_u^y \cdot S_x^u$.

***16.64** For the high-pass filter of Fig. 16.33(b), what are the sensitivities of ω_0 and Q to amplifier gain A?

***16.65** For the feedback loop of Fig. 16.34(a), use the expressions in Eqs. (16.77) and (16.78) to determine the sensitivities of ω_0 and Q relative to all passive components for the design in which $R_1 = R_2$.

16.66 For the op amp–RC resonator of Fig. 16.21(b), use the expressions for ω_0 and Q given in the top row of Table 16.1 to determine the sensitivities of ω_0 and Q to all resistors and capacitors.

Section 16.10: Switched-Capacitor Filters

16.67 For the switched-capacitor input circuit of Fig. 16.35(b), in which a clock frequency of 100 kHz is used, what input resistances correspond to capacitance C_1 values of 1 pF and 10 pF?

16.68 For a dc voltage of 1 V applied to the input of the circuit of Fig. 16.35(b), in which C_1 is 1 pF, what charge is transferred for each cycle of the two-phase clock? For a 100-kHz clock, what is the average current drawn from the input source? For a feedback capacitance of 10 pF, what change would you expect in the output for each cycle of the clock? For an amplifier that saturates at ±10 V and the feedback capacitor initially discharged, how many clock cycles would it take to saturate the amplifier? What is the average slope of the staircase output voltage produced?

D 16.69 Repeat Exercise 16.31 for a clock frequency of 400 kHz.

D 16.70 Repeat Exercise 16.31 for $Q = 40$.

D 16.71 Design the circuit of Fig. 16.37(b) to realize, at the output of the second (noninverting) integrator, a maximally flat low-pass function with $\omega_{3dB} = 10^4$ rad/s and unity dc gain. Use a clock frequency $f_c = 100$ kHz and select $C_1 = C_2 = 10$ pF. Give the values of C_3, C_4, C_5, and C_6. (*Hint:* For a maximally flat response, $Q = 1/\sqrt{2}$ and $\omega_{3dB} = \omega_0$.)

Section 16.11: Tuned Amplifiers

***16.72** A voltage signal source with a resistance $R_s = 10$ kΩ is connected to the input of a common-emitter BJT amplifier. Between base and emitter is connected a tuned circuit with

$L = 1$ μH and $C = 200$ pF. The transistor is biased at 1 mA and has $\beta = 200$, $C_\pi = 10$ pF, and $C_\mu = 1$ pF. The transistor load is a resistance of 5 kΩ. Find ω_0, Q, the 3-dB bandwidth, and the center-frequency gain of this single-tuned amplifier.

16.73 A coil having an inductance of 10 μH is intended for applications around 1-MHz frequency. Its Q is specified to be 200. Find the equivalent parallel resistance R_p. What is the value of the capacitor required to produce resonance at 1 MHz? What additional parallel resistance is required to produce a 3-dB bandwidth of 10 kHz?

16.74 An inductance of 36 μH is resonated with a 1000-pF capacitor. If the inductor is tapped at one-third of its turns and a 1-kΩ resistor is connected across the one-third part, find f_0 and Q of the resonator.

***16.75** Consider a common-emitter transistor amplifier loaded with an inductance L. Ignoring r_o and r_x, show that for $\omega C_\mu \ll 1/\omega L$, the amplifier input admittance is given by

$$Y_{in} \simeq \left(\frac{1}{r_\pi} - \omega^2 C_\mu L g_m\right) + j\omega(C_\pi + C_\mu)$$

(*Note:* The real part of the input admittance can be negative. This can lead to oscillations.)

***16.76** (a) Substituting $s = j\omega$ in the transfer function $T(s)$ of a second-order bandpass filter (see Fig. 16.16c), find $|T(j\omega)|$. For ω in the vicinity of ω_0 [i.e., $\omega = \omega_0 + \delta\omega = \omega_0(1 + \delta\omega/\omega_0)$, where $\delta\omega/\omega_0 \ll 1$ so that $\omega^2 \simeq \omega_0^2(1 + 2\delta\omega/\omega_0)$], show that, for $Q \gg 1$,

$$|T(j\omega)| \simeq \frac{|T(j\omega_0)|}{\sqrt{1 + 4Q^2(\delta\omega/\omega_0)^2}}$$

(b) Use the result obtained in (a) to show that the 3-dB bandwidth B, of N synchronously tuned sections connected in cascade, is

$$B = (\omega_0/Q)\sqrt{2^{1/N} - 1}$$

****16.77** (a) Using the fact that for $Q \gg 1$ the second-order bandpass response in the neighborhood of ω_0 is the same as the response of a first-order low-pass with 3-dB frequency of $(\omega_0/2Q)$, show that the bandpass response at $\omega = \omega_0 + \delta\omega$, for $\delta\omega \ll \omega_0$, is given by

$$|T(j\omega)| \simeq \frac{|T(j\omega_0)|}{\sqrt{1 + 4Q^2(\delta\omega/\omega_0)^2}}$$

(b) Use the relationship derived in (a) together with Eq. (16.110) to show that a bandpass amplifier with a 3-dB bandwidth B, designed using N synchronously tuned stages, has an overall transfer function given by

$$|T(j\omega)|_{\text{overall}} = \frac{|T(j\omega_0)|_{\text{overall}}}{[1 + 4(2^{1/N} - 1)(\delta\omega/B)^2]^{N/2}}$$

(c) Use the relationship derived in (b) to find the attenuation (in decibels) obtained at a bandwidth $2B$ for $N = 1$ to 5. Also find the ratio of the 30-dB bandwidth to the 3-dB bandwidth for $N = 1$ to 5.

***16.78** This problem investigates the selectivity of maximally flat stagger-tuned amplifiers derived in the manner illustrated in Fig. 16.48.

(a) The low-pass maximally flat (Butterworth) filter having a 3-dB bandwidth $B/2$ and order N has the magnitude response

$$|T| = 1\Big/\sqrt{1 + \left(\frac{\Omega}{B/2}\right)^{2N}}$$

where $\Omega = \text{Im}(p)$ is the frequency in the low-pass domain. (This relationship can be obtained using the information provided in Section 16.3 on Butterworth filters.) Use this expression to obtain for the corresponding bandpass filter at $\omega = \omega_0 + \delta\omega$, where $\delta\omega \ll \omega_0$, the relationship

$$|T| = 1\Big/\sqrt{1 + \left(\frac{\delta\omega}{B/2}\right)^{2N}}$$

(b) Use the transfer function in (a) to find the attenuation (in decibels) obtained at a bandwidth of $2B$ for $N = 1$ to 5. Also find the ratio of the 30-dB bandwidth to the 3-dB bandwidth for $N = 1$ to 5.

****16.79** Consider a sixth-order, stagger-tuned bandpass amplifier with center frequency ω_0 and 3-dB bandwidth B. The poles are to be obtained by shifting those of the third-order maximally flat low-pass filter, given in Fig. 16.10(c). For each of the three resonant circuits, find ω_0, the 3-dB bandwidth, and Q.

CHAPTER 17

Signal Generators and Waveform-Shaping Circuits

IN THIS CHAPTER YOU WILL LEARN

1. That an oscillator circuit that generates sine waves can be implemented by connecting a frequency-selective network in the positive-feedback path of an amplifier.
2. The conditions under which sustained oscillations are obtained and the frequency of the oscillations.
3. How to design nonlinear circuits to control the amplitude of the sine wave obtained in a linear oscillator.
4. A variety of circuits for implementing a linear sine-wave oscillator.
5. How op amps can be combined with resistors and capacitors to implement precision multivibrator circuits.
6. How a bistable circuit can be connected in a feedback loop with an op-amp integrator to implement a generator of square and triangular waveforms.
7. The application of one of the most popular IC chips of all time, the 555 timer, in the design of generators of pulse and square waveforms.
8. How a triangular waveform can be shaped by a nonlinear circuit to provide a sine waveform.
9. How op amps and diodes can be combined to implement a variety of high-precision rectifier circuits.

Introduction

In the design of electronic systems, the need frequently arises for signals having prescribed standard waveforms, for example, sinusoidal, square, triangular, or pulse. Systems in which standard signals are required include computer and control systems where clock pulses are needed for, among other things, timing; communication systems where signals of a variety of waveforms are utilized as information carriers; and test and measurement systems where signals, again of a variety of waveforms, are employed for testing and characterizing electronic devices and circuits. In this chapter we study signal-generator circuits.

There are two distinctly different approaches for the generation of sinusoids, perhaps the most commonly used of the standard waveforms. The first approach, studied in Sections 17.1 to 17.3, employs a **positive-feedback loop** consisting of an amplifier and an RC or LC **frequency-selective network**. The amplitude of the generated sine waves is limited, or set, using a nonlinear mechanism, implemented either with a separate circuit or using the nonlinearities of the amplifying device itself. In spite of this, these circuits, which generate sine waves utilizing resonance phenomena, are known as **linear oscillators**. The name clearly distinguishes them from the circuits that generate sinusoids by way of the second approach. In these circuits, a sine wave is obtained by appropriately shaping a triangular waveform. We study waveform-shaping circuits in Section 17.8, following the study of triangular-waveform generators.

Circuits that generate square, triangular, pulse (etc.) waveforms, called **nonlinear oscillators** or **function generators**, employ circuit building blocks known as **multivibrators**. There are three types of multivibrator: the **bistable** (Section 17.4), the **astable** (Section 17.5), and the **monostable** (Section 17.6). The multivibrator circuits presented in this chapter employ op amps and are intended for precision analog applications. Bistable and monostable multivibrator circuits using digital logic gates were studied in Chapter 15.

A general and versatile scheme for the generation of square and triangular waveforms is obtained by connecting a bistable multivibrator and an op-amp integrator in a feedback loop (Section 17.5). Similar results can be obtained using a commercially available versatile IC chip, the 555 timer (Section 17.7). The chapter includes also a study of precision circuits that implement the rectifier functions introduced in Chapter 4. The circuits studied here (Section 17.9), however, are intended for applications that demand precision, such as in instrumentation systems, including waveform generation.

17.1 Basic Principles of Sinusoidal Oscillators

In this section, we study the basic principles of the design of linear sine-wave oscillators. In spite of the name *linear oscillator,* some form of nonlinearity has to be employed to provide control of the amplitude of the output sine wave. In fact, all oscillators are essentially nonlinear circuits. This complicates the task of analysis and design of oscillators: No longer is one able to apply transform (s-plane) methods directly. Nevertheless, techniques have been developed by which the design of sinusoidal oscillators can be performed in two steps: The first step is a linear one, and frequency-domain methods of feedback circuit analysis can be readily employed. Subsequently, a nonlinear mechanism for amplitude control can be provided.

17.1.1 The Oscillator Feedback Loop

The basic structure of a sinusoidal oscillator consists of an amplifier and a frequency-selective network connected in a positive-feedback loop, such as that shown in block diagram form in Fig. 17.1. Although no input signal will be present in an actual oscillator circuit, we include an input signal here to help explain the principle of operation. It is important to note that unlike the negative-feedback loop of Fig. 10.1, here the feedback signal x_f is summed with a *positive* sign. Thus the gain-with-feedback is given by

$$A_f(s) = \frac{A(s)}{1 - A(s)\beta(s)} \tag{17.1}$$

where we note the negative sign in the denominator.

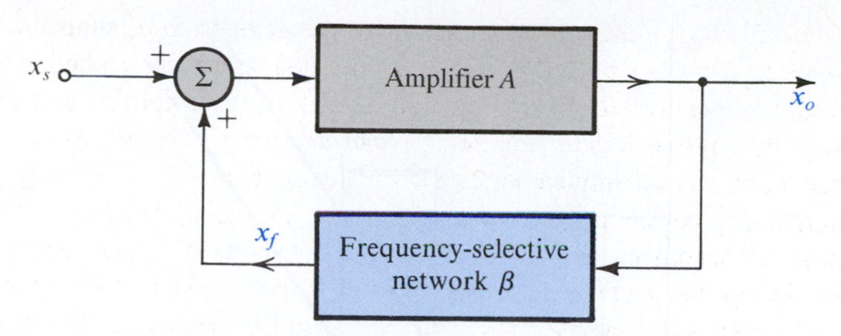

Figure 17.1 The basic structure of a sinusoidal oscillator. A positive-feedback loop is formed by an amplifier and a frequency-selective network. In an actual oscillator circuit, no input signal will be present; here an input signal x_s is employed to help explain the principle of operation.

According to the definition of loop gain in Chapter 10, the loop gain of the circuit in Fig. 17.1 is $-A(s)\beta(s)$. However, for our purposes here it is more convenient to drop the minus sign and define the loop gain $L(s)$ as

$$L(s) \equiv A(s)\beta(s) \tag{17.2}$$

The characteristic equation thus becomes

$$1 - L(s) = 0 \tag{17.3}$$

Note that this new definition of loop gain[1] corresponds directly to the actual gain seen around the feedback loop of Fig. 17.1.

17.1.2 The Oscillation Criterion

If at a specific frequency f_0 the loop gain $A\beta$ is equal to unity, it follows from Eq. (17.1) that A_f will be infinite. That is, at this frequency the circuit will have a finite output for zero input signal. Such a circuit is by definition an oscillator. Thus the condition for the feedback loop of Fig. 17.1 to provide sinusoidal oscillations of frequency ω_0 is

$$L(j\omega_0) \equiv A(j\omega_0)\beta(j\omega_0) = 1 \tag{17.4}$$

That is, *at ω_0 the phase of the loop gain should be zero and the magnitude of the loop gain should be unity.* This is known as the **Barkhausen criterion**. Note that for the circuit to oscillate at one frequency, the oscillation criterion should be satisfied only at one frequency (i.e., ω_0); otherwise the resulting waveform will not be a simple sinusoid.

An intuitive feeling for the Barkhausen criterion can be gained by considering once more the feedback loop of Fig. 17.1. For this loop to *produce* and *sustain* an output x_o with no input applied ($x_s = 0$), the feedback signal x_f

$$x_f = \beta x_o$$

should be sufficiently large that when multiplied by A it produces x_o, that is,

$$Ax_f = x_o$$

[1] For both the negative-feedback loop in Fig. 10.1 and the positive-feedback loop in Fig. 17.1, the loop gain $L = A\beta$. However, the negative sign with which the feedback signal is summed in the negative-feedback loop results in the characteristic equation being $1 + L = 0$. In the positive-feedback loop, the feedback signal is summed with a positive sign, thus resulting in the characteristic equation $1 - L = 0$.

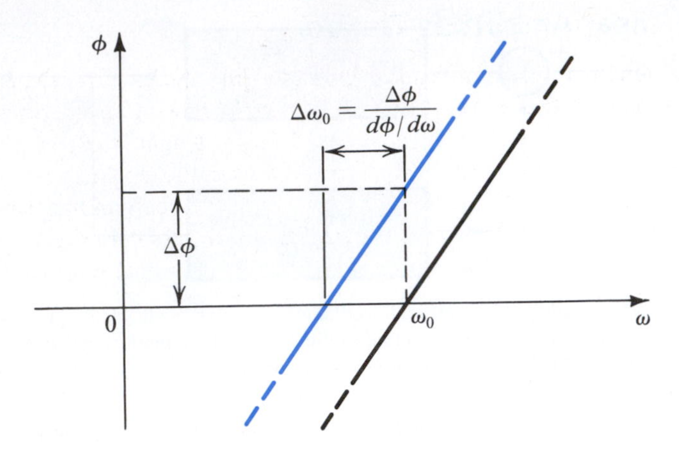

Figure 17.2 Dependence of the oscillator-frequency stability on the slope of the phase response. A steep phase response (i.e., large $d\phi/d\omega$) results in a small $\Delta\omega_0$ for a given change in phase $\Delta\phi$ [resulting from a change (due, for example, to temperature) in a circuit component].

that is,

$$A\beta x_o = x_o$$

which results in

$$A\beta = 1$$

It should be noted that the *frequency of oscillation* ω_0 is determined solely by the phase characteristics of the feedback loop; the loop oscillates at the frequency for which the phase is zero. It follows that the stability of the frequency of oscillation will be determined by the manner in which the phase $\phi(\omega)$ of the feedback loop varies with frequency. A "steep" function $\phi(\omega)$ will result in a more stable frequency. This can be seen if one imagines a change in phase $\Delta\phi$ due to a change in one of the circuit components. If $d\phi/d\omega$ is large, the resulting change in ω_0 will be small, as illustrated in Fig. 17.2.

An alternative approach to the study of oscillator circuits consists of examining the circuit poles, which are the roots of the **characteristic equation** (Eq. 17.3). For the circuit to produce **sustained oscillations** at a frequency ω_0 the characteristic equation has to have roots at $s = \pm j\omega_0$. Thus $1 - A(s)\beta(s)$ should have a factor of the form $s^2 + \omega_0^2$.

EXERCISES

17.1 Consider a sinusoidal oscillator formed of an amplifier with a gain of 2 and a second-order bandpass filter. Find the pole frequency and the center-frequency gain of the filter needed to produce sustained oscillations at 1 kHz.
Ans. 1 kHz; 0.5

17.1.3 Nonlinear Amplitude Control

The oscillation condition, the Barkhausen criterion, just discussed, guarantees sustained oscillations in a mathematical sense. It is well known, however, that the parameters of any physical system cannot be maintained constant for any length of time. In other words, suppose we work hard to make $A\beta = 1$ at $\omega = \omega_0$, and then the temperature changes and $A\beta$ becomes slightly less than unity. Obviously, oscillations will cease in this case. Conversely, if $A\beta$ exceeds unity, oscillations will grow in amplitude. We therefore need a mechanism for forcing $A\beta$ to remain equal to unity *at the desired value of output amplitude*. This task is accomplished by providing a nonlinear circuit for gain control.

Basically, the function of the gain-control mechanism is as follows: First, to ensure that oscillations will start, one designs the circuit such that $A\beta$ is slightly greater than unity. This corresponds to designing the circuit so that the poles are in the right half of the s plane. Thus as the power supply is turned on, oscillations will grow in amplitude. When the amplitude reaches the desired level, the nonlinear network comes into action and causes the loop gain to be reduced to exactly unity. In other words, the poles will be "pulled back" to the $j\omega$ axis. This action will cause the circuit to sustain oscillations at this desired amplitude. If, for some reason, the loop gain is reduced below unity, the amplitude of the sine wave will diminish. This will be detected by the nonlinear network, which will cause the loop gain to increase to exactly unity.

As will be seen, there are two basic approaches to the implementation of the nonlinear amplitude-stabilization mechanism. The first approach makes use of a limiter circuit (see Chapter 4). Oscillations are allowed to grow until the amplitude reaches the level to which the limiter is set. When the limiter comes into operation, the amplitude remains constant. Obviously, the limiter should be "soft" to minimize nonlinear distortion. Such distortion, however, is reduced by the filtering action of the frequency-selective network in the feedback loop. In fact, in one of the oscillator circuits studied in Section 17.2, the sine waves are hard limited, and the resulting square waves are applied to a bandpass filter present in the feedback loop. The "purity" of the output sine waves will be a function of the selectivity of this filter. That is, the higher the Q of the filter, the less the harmonic content of the sine-wave output.

The other mechanism for amplitude control utilizes an element whose resistance can be controlled by the amplitude of the output sinusoid. By placing this element in the feedback circuit so that its resistance determines the loop gain, the circuit can be designed to ensure that the loop gain reaches unity at the desired output amplitude. Diodes, or JFETs operated in the triode region,[2] are commonly employed to implement the controlled-resistance element.

17.1.4 A Popular Limiter Circuit for Amplitude Control

We conclude this section by presenting a limiter circuit that is frequently employed for the amplitude control of op-amp oscillators, as well as in a variety of other applications. The circuit is more precise and versatile than those presented in Chapter 4.

The limiter circuit is shown in Fig. 17.3(a), and its transfer characteristic is depicted in Fig. 17.3(b). To see how the transfer characteristic is obtained, consider first the case of a small (close to zero) input signal v_I and a small output voltage v_O, so that v_A is positive and v_B is negative. It can be easily seen that both diodes D_1 and D_2 will be off. Thus all of the

[2] We have not studied JFETs in this book. However, the disc accompanying the book includes material on JFETs and JFET circuits. The same material can also be found on the book's website.

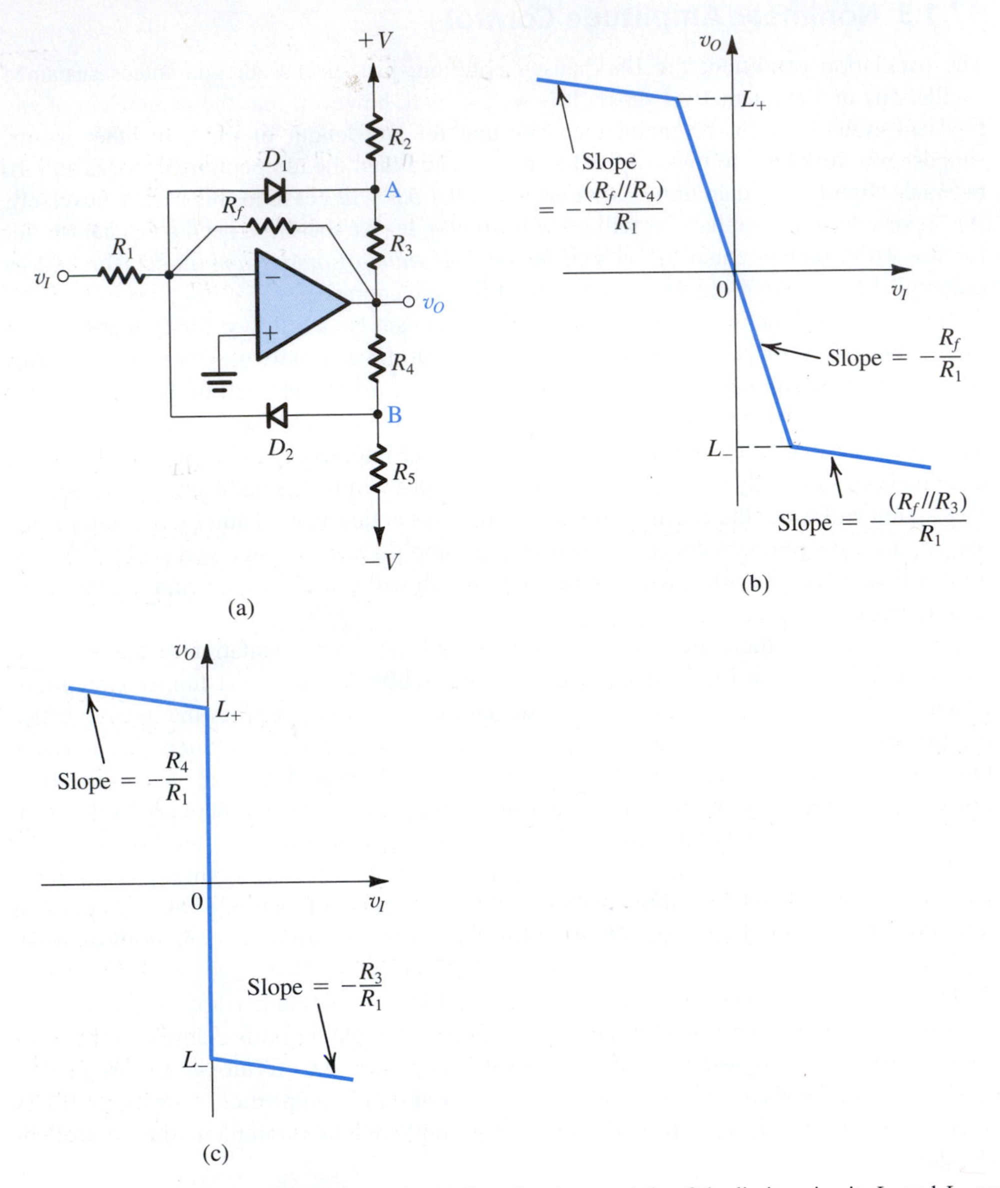

Figure 17.3 **(a)** A popular limiter circuit. **(b)** Transfer characteristic of the limiter circuit; L_- and L_+ are given by Eqs. (17.8) and (17.9), respectively. **(c)** When R_f is removed, the limiter turns into a comparator with the characteristic shown.

input current v_I/R_1 flows through the feedback resistance R_f, and the output voltage is given by

$$v_O = -(R_f/R_1)v_I \tag{17.5}$$

This is the linear portion of the limiter transfer characteristic in Fig. 17.3(b). We now can use superposition to find the voltages at nodes A and B in terms of $\pm V$ and v_O as

$$v_A = V\frac{R_3}{R_2+R_3} + v_O\frac{R_2}{R_2+R_3} \tag{17.6}$$

$$v_B = -V\frac{R_4}{R_4+R_5} + v_O\frac{R_5}{R_4+R_5} \tag{17.7}$$

As v_I goes positive, v_O goes negative (Eq. 17.5), and we see from Eq. (17.7) that v_B will become more negative, thus keeping D_2 off. Equation (17.6) shows, however, that v_A becomes less positive. Then, if we continue to increase v_I, a negative value of v_O will be reached at which v_A becomes −0.7 V or so and diode D_1 conducts. If we use the constant-voltage-drop model for D_1 and denote the voltage drop V_D, the value of v_O at which D_1 conducts can be found from Eq. (17.6). This is the negative limiting level, which we denote L_-,

$$L_- = -V\frac{R_3}{R_2} - V_D\left(1+\frac{R_3}{R_2}\right) \tag{17.8}$$

The corresponding value of v_I can be found by dividing L_- by the limiter gain $-R_f/R_1$. If v_I is increased beyond this value, more current is injected into D_1, and v_A remains at approximately $-V_D$. Thus the current through R_2 remains constant, and the additional diode current flows through R_3. Thus R_3 appears in effect in parallel with R_f, and the incremental gain (ignoring the diode resistance) is $-(R_f \| R_3)/R_1$. To make the slope of the transfer characteristic small in the limiting region, a low value should be selected for R_3.

The transfer characteristic for negative v_I can be found in a manner identical to that just employed. It can be easily seen that for negative v_I, diode D_2 plays an identical role to that played by diode D_1 for positive v_I. We can use Eq. (17.7) to find the positive limiting level L_+

$$L_+ = V\frac{R_4}{R_5} + V_D\left(1+\frac{R_4}{R_5}\right) \tag{17.9}$$

and the slope of the transfer characteristic in the positive limiting region is $-(R_f \| R_4)/R_1$. We thus see that the circuit of Fig. 17.3(a) functions as a soft limiter, with the limiting levels L_+ and L_-, and the limiting gains independently adjustable by the selection of appropriate resistor values.

Finally, we note that increasing R_f results in a higher gain in the linear region while keeping L_+ and L_- unchanged. In the limit, removing R_f altogether results in the transfer characteristic of Fig. 17.3(c), which is that of a comparator. That is, the circuit compares v_I with the comparator reference value of 0 V: $v_I > 0$ results in $v_O \simeq L_-$, and $v_I < 0$ yields $v_O \simeq L_+$.

EXERCISES

17.2 For the circuit of Fig. 17.3(a) with $V = 15$ V, $R_1 = 30$ kΩ, $R_f = 60$ kΩ, $R_2 = R_5 = 9$ kΩ, and $R_3 = R_4 = 3$ kΩ, find the limiting levels and the value of v_I at which the limiting levels are reached. Also determine the limiter gain and the slope of the transfer characteristic in the positive and negative limiting regions. Assume that $V_D = 0.7$ V.

Ans. ±5.93 V; ±2.97 V; −2; −0.095

17.2 Op Amp–RC Oscillator Circuits

In this section we shall study some practical oscillator circuits utilizing op amps and RC networks.

17.2.1 The Wien-Bridge Oscillator

One of the simplest oscillator circuits is based on the Wien bridge. Figure 17.4 shows a Wien-bridge oscillator without the nonlinear gain-control network. The circuit consists of an op amp connected in the noninverting configuration, with a closed-loop gain of $1 + R_2/R_1$. In the feedback path of this positive-gain amplifier an RC network is connected. The loop gain can be easily obtained by multiplying the transfer function $V_a(s)/V_o(s)$ of the feedback network by the amplifier gain,

$$L(s) = \left[1 + \frac{R_2}{R_1}\right]\frac{Z_p}{Z_p + Z_s}$$

$$= \frac{1 + R_2/R_1}{1 + Z_s Y_p}$$

Thus,

$$L(s) = \frac{1 + R_2/R_1}{3 + sCR + 1/sCR} \tag{17.10}$$

Substituting $s = j\omega$ results in

$$L(j\omega) = \frac{1 + R_2/R_1}{3 + j(\omega CR - 1/\omega CR)} \tag{17.11}$$

The loop gain will be a real number (i.e., the phase will be zero) at one frequency given by

$$\omega_0 CR = \frac{1}{\omega_0 CR}$$

That is,

$$\omega_0 = 1/CR \tag{17.12}$$

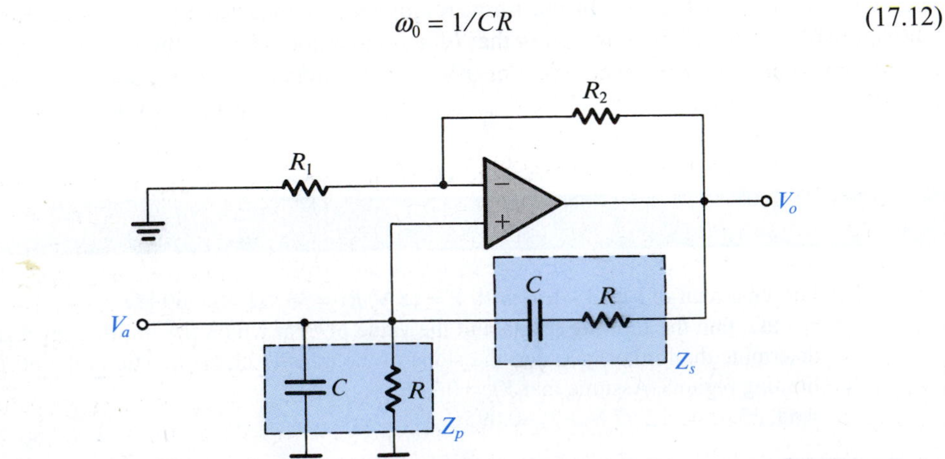

Figure 17.4 A Wien-bridge oscillator without amplitude stabilization.

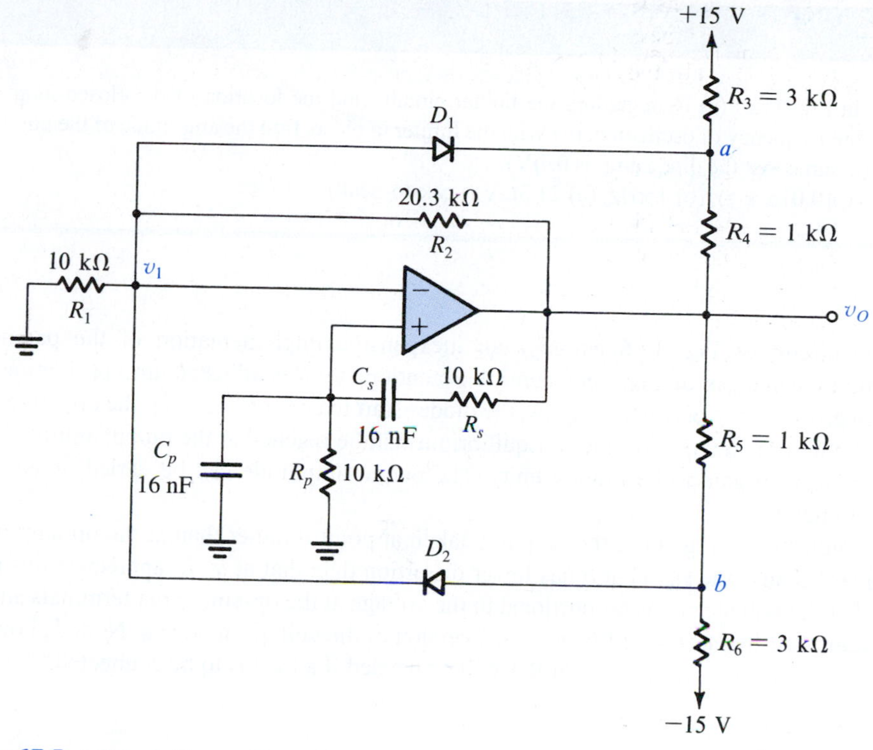

Figure 17.5 A Wien-bridge oscillator with a limiter used for amplitude control.

To obtain sustained oscillations at this frequency, one should set the magnitude of the loop gain to unity. This can be achieved by selecting

$$R_2/R_1 = 2 \tag{17.13}$$

To ensure that oscillations will start, one chooses R_2/R_1 slightly greater than 2. The reader can easily verify that if $R_2/R_1 = 2 + \delta$, where δ is a small number, the roots of the characteristic equation $1 - L(s) = 0$ will be in the right half of the s plane.

The amplitude of oscillation can be determined and stabilized by using a nonlinear control network. Two different implementations of the amplitude-controlling function are shown in Figs. 17.5 and 17.6. The circuit in Fig. 17.5 employs a symmetrical feedback limiter of the type studied in Section 17.1.3. It is formed by diodes D_1 and D_2 together with resistors R_3, R_4, R_5, and R_6. The limiter operates in the following manner: At the positive peak of the output voltage v_O, the voltage at node b will exceed the voltage v_1 (which is about $\frac{1}{3}v_O$), and diode D_2 conducts. This will clamp the positive peak to a value determined by R_5, R_6, and the negative power supply. The value of the positive output peak can be calculated by setting $v_b = v_1 + V_{D2}$ and writing a node equation at node b while neglecting the current through D_2. Similarly, the negative peak of the output sine wave will be clamped to the value that causes diode D_1 to conduct. The value of the negative peak can be determined by setting $v_a = v_1 - V_{D1}$ and writing an equation at node a while neglecting the current through D_1. Finally, note that to obtain a symmetrical output waveform, R_3 is chosen equal to R_6, and R_4 equal to R_5.

EXERCISES

17.3 For the circuit in Fig. 17.5: (a) Disregarding the limiter circuit, find the location of the closed-loop poles. (b) Find the frequency of oscillation. (c) With the limiter in place, find the amplitude of the output sine wave (assume that the diode drop is 0.7 V).
Ans. (a) $(10^5/16)(0.015 \pm j)$; (b) 1 kHz; (c) 21.36 V (peak-to-peak)

The circuit of Fig. 17.6 employs an inexpensive implementation of the parameter-variation mechanism of amplitude control. Potentiometer *P* is adjusted until oscillations just start to grow. As the oscillations grow, the diodes start to conduct, causing the effective resistance between *a* and *b* to decrease. Equilibrium will be reached at the output amplitude that causes the loop gain to be exactly unity. The output amplitude can be varied by adjusting potentiometer *P*.

As indicated in Fig. 17.6, the output is taken at point *b* rather than at the op-amp output terminal because the signal at *b* has lower distortion than that at *a*. To appreciate this point, note that the voltage at *b* is proportional to the voltage at the op-amp input terminals and that the latter is a filtered (by the RC network) version of the voltage at node *a*. Node *b*, however, is a high-impedance node, and a buffer will be needed if a load is to be connected.

EXERCISES

17.4 For the circuit in Fig. 17.6 find the following: (a) The setting of potentiometer *P* at which oscillations just start. (b) The frequency of oscillation.
Ans. (a) 20 kΩ to ground; (b) 1 kHz

17.2.2 The Phase-Shift Oscillator

The basic structure of the phase-shift oscillator is shown in Fig. 17.7. It consists of a negative-gain amplifier ($-K$) with a three-section (third-order) RC ladder network in the feedback. The circuit will oscillate at the frequency for which the phase shift of the RC network is 180°. Only at this frequency will the total phase shift around the loop be 0° or 360°. Here we should note that the reason for using a three-section RC network is that three is the minimum number of sections (i.e., lowest order) that is capable of producing a 180° phase shift at a finite frequency.

For oscillations to be sustained, the value of *K* should be equal to the inverse of the magnitude of the RC network transfer function at the frequency of oscillation. However, to ensure that oscillations start, the value of *K* has to be chosen slightly higher than the value that satisfies the unity-loop-gain condition. Oscillations will then grow in magnitude until limited by some nonlinear control mechanism.

Figure 17.8 shows a practical phase-shift oscillator with a feedback limiter, consisting of diodes D_1 and D_2 and resistors R_1, R_2, R_3, and R_4 for amplitude stabilization. To start

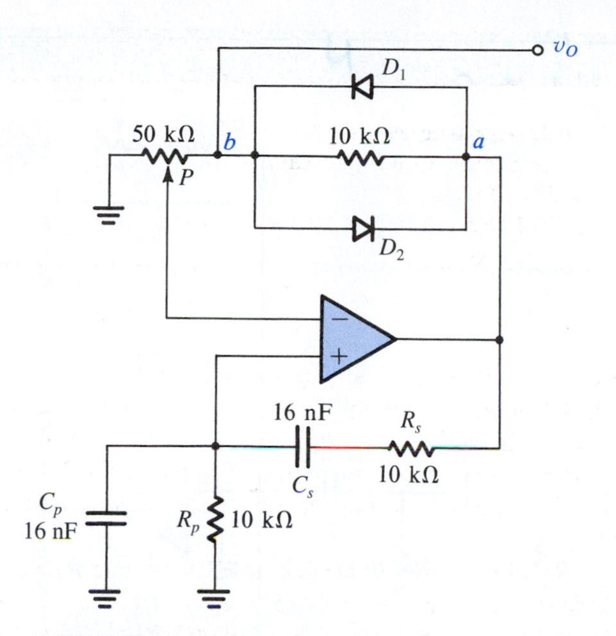

Figure 17.6 A Wien-bridge oscillator with an alternative method for amplitude stabilization.

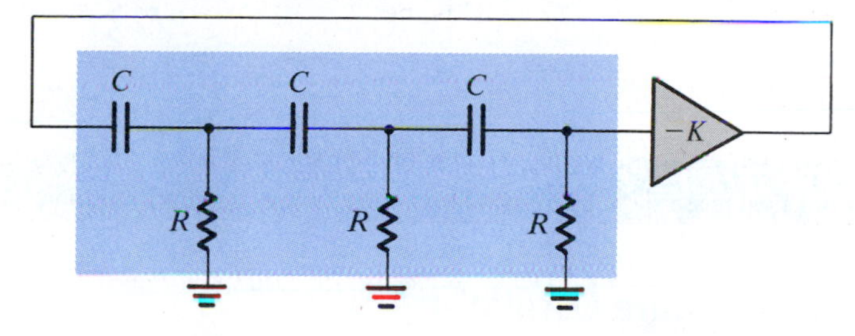

Figure 17.7 A phase-shift oscillator.

oscillations, R_f has to be made slightly greater than the minimum required value. Although the circuit stabilizes more rapidly and provides sine waves with more stable amplitude, if R_f is made much larger than this minimum, the price paid is an increased output distortion.

EXERCISES

17.5 Consider the circuit of Fig. 17.8 *without* the limiter. Break the feedback loop at X and find the loop gain $A\beta \equiv V_o(j\omega)/V_x(j\omega)$. To do this, it is easier to start at the output and work backward, finding the various currents and voltages, and eventually V_x in terms of V_o.

Ans. $\dfrac{\omega^2 C^2 RR_f}{4 + j(3\omega CR - 1/\omega CR)}$

17.6 Use the expression derived in Exercise 17.5 to find the frequency of oscillation f_0 and the minimum required value of R_f for oscillations to start in the circuit of Fig. 17.8.

Ans. $\omega_0 = 1/\sqrt{3}\, CR$; $R_f \geq 12\, R$; $f_0 = 574.3$ Hz; $R_f = 120$ kΩ

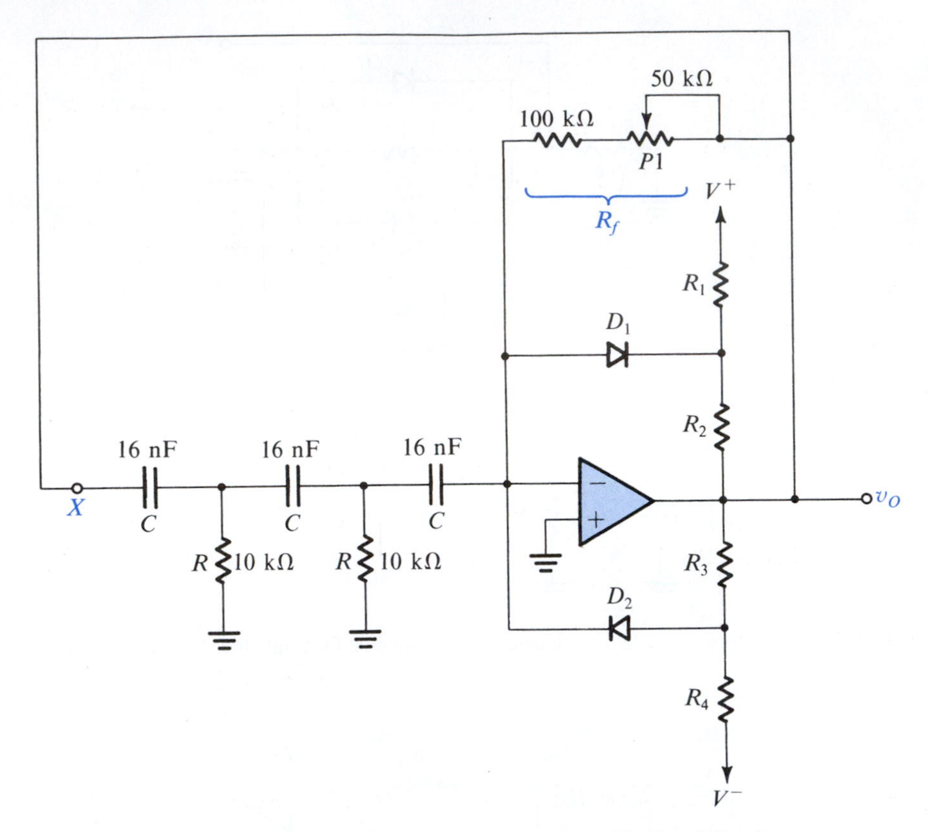

Figure 17.8 A practical phase-shift oscillator with a limiter for amplitude stabilization.

17.2.3 The Quadrature Oscillator

The **quadrature oscillator** is based on the two-integrator loop studied in Section 16.7. As an active filter, the loop is damped to locate the poles in the left half of the s plane. Here, no such damping will be used, since we wish to locate the poles on the $j\omega$ axis to provide sustained oscillations. In fact, to ensure that oscillations start, the poles are initially located in the right half-plane and then "pulled back" by the nonlinear gain control.

Figure 17.9 shows a practical quadrature oscillator. Amplifier 1 is connected as an inverting Miller integrator with a limiter in the feedback for amplitude control. Amplifier 2 is connected as a noninverting integrator (thus replacing the cascade connection of the Miller integrator and the inverter in the two-integrator loop of Fig. 16.25b). To understand the operation of this noninverting integrator, consider the equivalent circuit shown in Fig. 17.9(b). Here, we have replaced the integrator input voltage v_{O1} and the series resistance $2R$ by the Norton equivalent composed of a current source $v_{O1}/2R$ and a parallel resistance $2R$. Now, since $v_{O2} = 2v$, where v is the voltage at the input of op amp 2, the current through R_f will be $(2v - v)/R_f = v/R_f$ in the direction from output to input. Thus R_f gives rise to a negative input resistance, $-R_f$, as indicated in the equivalent circuit of Fig. 17.9(b). Nominally, R_f is made equal to $2R$, and thus $-R_f$ cancels $2R$, and at the input we are left with a current source $v_{O1}/2R$ feeding a capacitor C. The result is that $v = \frac{1}{C}\int_0^t \frac{v_{O1}}{2R}dt$ and $v_{O2} = 2v = \frac{1}{CR}\int_0^t v_{O1}dt$. That is, for $R_f = 2R$, the circuit functions as a perfect noninverting integrator. If, however, R_f is made smaller than $2R$, a net negative resistance appears in parallel with C.

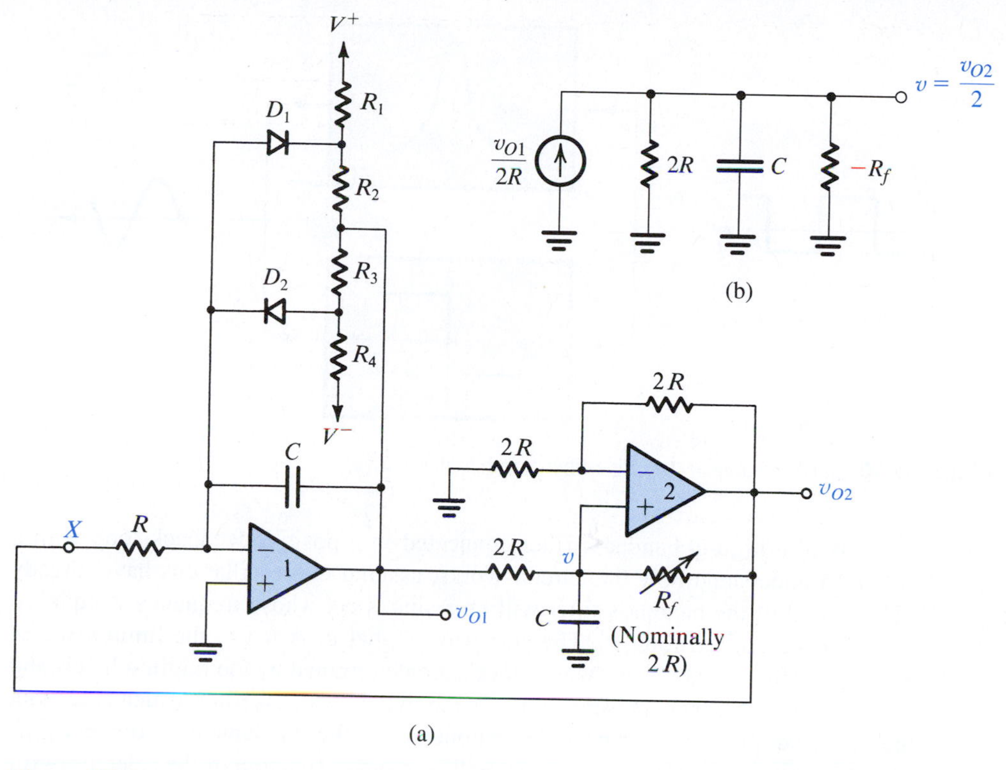

Figure 17.9 **(a)** A quadrature-oscillator circuit. **(b)** Equivalent circuit at the input of op amp 2.

Returning to the oscillator circuit in Fig. 17.9(a), we note that the resistance R_f in the positive-feedback path of op amp 2 is made variable, with a nominal value of $2R$. Decreasing the value of R_f moves the poles to the right half-plane (Problem 17.19) and ensures that the oscillations start. Too much positive feedback, although it results in better amplitude stability, also results in higher output distortion (because the limiter has to operate "harder"). In this regard, note that the output v_{O2} will be "purer" than v_{O1} because of the filtering action provided by the second integrator on the peak-limited output of the first integrator.

If we disregard the limiter and break the loop at X, the loop gain can be obtained as

$$L(s) \equiv \frac{V_{o2}}{V_x} = -\frac{1}{s^2C^2R^2} \tag{17.14}$$

Thus the loop will oscillate at frequency ω_0, given by

$$\omega_0 = \frac{1}{CR} \tag{17.15}$$

Finally, it should be pointed out that the name *quadrature oscillator* is used because the circuit provides two sinusoids with 90° phase difference. This is the case because v_{O2} is the integral of v_{O1}. There are many applications for which quadrature sinusoids are required.

17.2.4 The Active-Filter-Tuned Oscillator

The last oscillator circuit that we shall discuss is quite simple both in principle and in design. Nevertheless, the approach is general and versatile and can result in high-quality (i.e., low-distortion) output sine waves. The basic principle is illustrated in Fig. 17.10. The

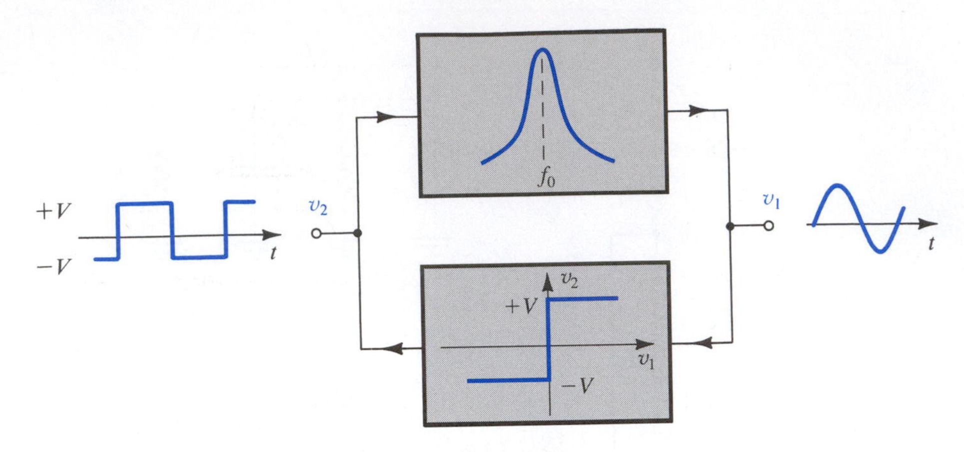

Figure 17.10 Block diagram of the active-filter-tuned oscillator.

circuit consists of a high-Q bandpass filter connected in a positive-feedback loop with a hard limiter. To understand how this circuit works, assume that oscillations have already started. The output of the bandpass filter will be a sine wave whose frequency is equal to the center frequency of the filter, f_0. The sine-wave signal v_1 is fed to the limiter, which produces at its output a square wave whose levels are determined by the limiting levels and whose frequency is f_0. The square wave in turn is fed to the bandpass filter, which filters out the harmonics and provides a sinusoidal output v_1 at the fundamental frequency f_0. Obviously, the purity of the output sine wave will be a direct function of the selectivity (or Q factor) of the bandpass filter.

The simplicity of this approach to oscillator design should be apparent. We have independent control of frequency and amplitude as well as of distortion of the output sinusoid. Any filter circuit with positive gain can be used to implement the bandpass filter. The frequency stability of the oscillator will be directly determined by the frequency stability of the bandpass-filter circuit. Also, a variety of limiter circuits (see Chapter 4) with different degrees of sophistication can be used to implement the limiter block.

Figure 17.11 shows one possible implementation of the active-filter-tuned oscillator. This circuit uses a variation on the bandpass circuit based on the Antoniou inductance-simulation circuit (see Fig. 16.22c). Here resistor R_2 and capacitor C_4 are interchanged. This makes the output of the lower op amp directly proportional to (in fact, twice as large as) the voltage across the resonator, and we can therefore dispense with the buffer amplifier K. The limiter used is a very simple one consisting of a resistance R_1 and two diodes.

EXERCISES

17.7 Using $C = 16$ nF, find the value of R such that the circuit of Fig. 17.11 produces 1-kHz sine waves. If the diode drop is 0.7 V, find the peak-to-peak amplitude of the output sine wave. (*Hint:* A square wave with peak-to-peak amplitude of V volts has a fundamental component with $4V/\pi$ volts peak-to-peak amplitude.)

Ans. 10 kΩ; 3.6 V

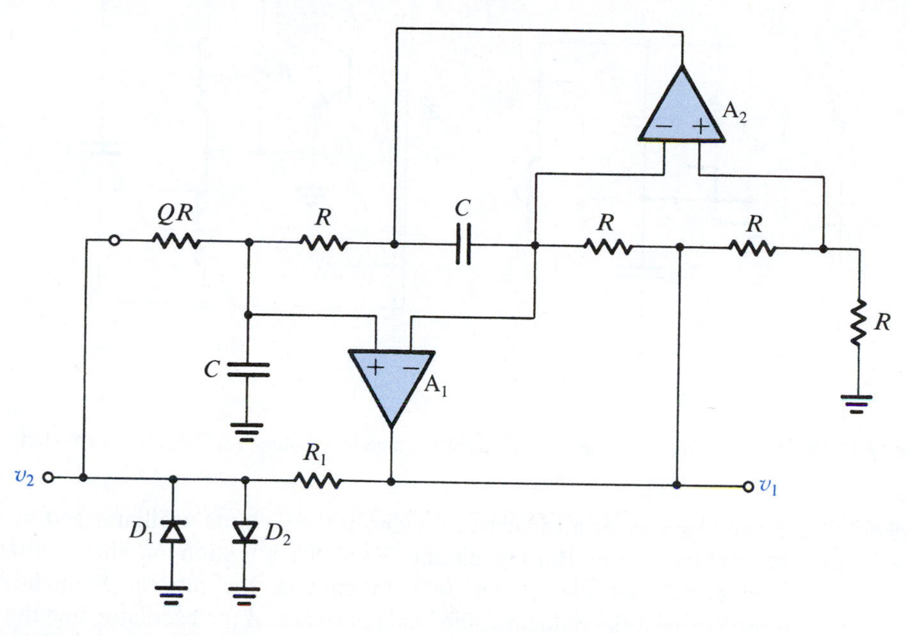

Figure 17.11 A practical implementation of the active-filter-tuned oscillator.

17.2.5 A Final Remark

The op amp–RC oscillator circuits studied are useful for operation in the range 10 Hz to 100 kHz (or perhaps 1 MHz at most). Whereas the lower frequency limit is dictated by the size of passive components required, the upper limit is governed by the frequency-response and slew-rate limitations of op amps. For higher frequencies, circuits that employ transistors together with LC-tuned circuits or crystals are frequently used.[3] These are discussed in Section 17.3.

17.3 LC and Crystal Oscillators

Oscillators utilizing transistors (FETs or BJTs), with LC-tuned circuits or crystals as feedback elements, are used in the frequency range of 100 kHz to hundreds of megahertz. They exhibit higher Q than the RC types. However, LC oscillators are difficult to tune over wide ranges, and crystal oscillators operate at a single frequency.

17.3.1 LC-Tuned Oscillators

Figure 17.12 shows two commonly used configurations of LC-tuned oscillators. They are known as the **Colpitts oscillator** and the **Hartley oscillator**. Both utilize a parallel LC circuit connected between collector and base (or between drain and gate if a FET is used) with a fraction of the tuned-circuit voltage fed to the emitter (the source in a FET). This

[3] Of course, transistors can be used in place of the op amps in the circuits just studied. At higher frequencies, however, better results are obtained with LC-tuned circuits and crystals.

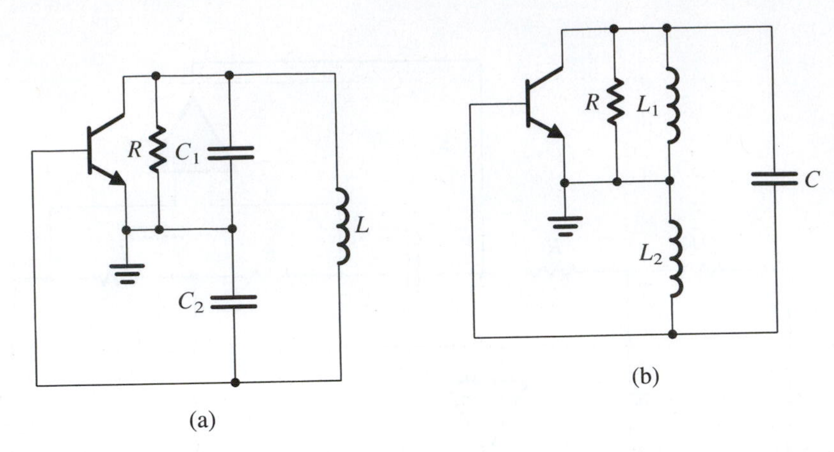

Figure 17.12 Two commonly used configurations of LC-tuned oscillators: **(a)** Colpitts and **(b)** Hartley.

feedback is achieved by way of a capacitive divider in the Colpitts oscillator and by way of an inductive divider in the Hartley circuit. To focus attention on the oscillator's structure, the bias details are not shown. In both circuits, the resistor R models the combination of the losses of the inductors, the load resistance of the oscillator, and the output resistance of the transistor.

If the frequency of operation is sufficiently low that we can neglect the transistor capacitances, the frequency of oscillation will be determined by the resonance frequency of the parallel-tuned circuit (also known as a *tank circuit* because it behaves as a reservoir for energy storage). Thus for the Colpitts oscillator we have

$$\omega_0 = 1\bigg/\sqrt{L\left(\frac{C_1C_2}{C_1+C_2}\right)} \tag{17.16}$$

and for the Hartley oscillator we have

$$\omega_0 = 1/\sqrt{(L_1+L_2)C} \tag{17.17}$$

The ratio L_1/L_2 or C_1/C_2 determines the feedback factor and thus must be adjusted in conjunction with the transistor gain to ensure that oscillations will start. To determine the oscillation condition for the Colpitts oscillator, we replace the transistor with its equivalent circuit, as shown in Fig. 17.13. To simplify the analysis, we have neglected the transistor capacitance C_μ (C_{gd} for a FET). Capacitance C_π (C_{gs} for a FET), although not shown, can be considered to be a part of C_2. The input resistance r_π (infinite for a FET) has also been neglected, assuming that at the frequency of oscillation $r_\pi \gg (1/\omega C_2)$. Finally, as mentioned earlier, the resistance R includes r_o of the transistor.

To find the loop gain, we break the loop at the transistor base, apply an input voltage V_π, and find the returned voltage that appears across the input terminals of the transistor. We then equate the loop gain to unity. An alternative approach is to analyze the circuit and eliminate all current and voltage variables, and thus obtain one equation that governs circuit operation. Oscillations will start if this equation is satisfied. Thus the resulting equation will give us the conditions for oscillation.

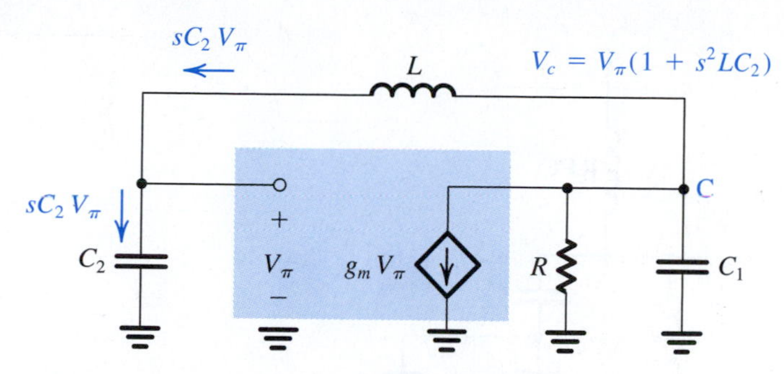

Figure 17.13 Equivalent circuit of the Colpitts oscillator of Fig. 17.12(a). To simplify the analysis, C_μ and r_π are neglected. We can consider C_π to be part of C_2, and we can include r_o in R.

A node equation at the transistor collector (node C) in the circuit of Fig. 17.13 yields

$$sC_2V_\pi + g_mV_\pi + \left(\frac{1}{R} + sC_1\right)(1 + s^2LC_2)V_\pi = 0$$

Since $V_\pi \neq 0$ (oscillations have started), it can be eliminated, and the equation can be rearranged in the form

$$s^3LC_1C_2 + s^2(LC_2/R) + s(C_1 + C_2) + \left(g_m + \frac{1}{R}\right) = 0 \tag{17.18}$$

Substituting $s = j\omega$ gives

$$\left(g_m + \frac{1}{R} - \frac{\omega^2LC_2}{R}\right) + j[\omega(C_1 + C_2) - \omega^3LC_1C_2] = 0 \tag{17.19}$$

For oscillations to start, both the real and imaginary parts must be zero. Equating the imaginary part to zero gives the frequency of oscillation as

$$\omega_0 = 1\bigg/\sqrt{L\left(\frac{C_1C_2}{C_1 + C_2}\right)} \tag{17.20}$$

which is the resonance frequency of the tank circuit, as anticipated.[4] Equating the real part to zero together with Eq. (17.20) gives

$$C_2/C_1 = g_mR \tag{17.21}$$

which has a simple physical interpretation: For sustained oscillations, the magnitude of the gain from base to collector (g_mR) must be equal to the inverse of the voltage ratio provided by the capacitive divider, which from Fig. 17.12(a) can be seen to be $v_{eb}/v_{ce} = C_1/C_2$. Of course, for oscillations to start, the loop gain must be made greater than unity, a condition that can be stated in the equivalent form

$$g_mR > C_2/C_1 \tag{17.22}$$

[4] If r_π is taken into account, the frequency of oscillation can be shown to shift slightly from the value given by Eq. (17.20).

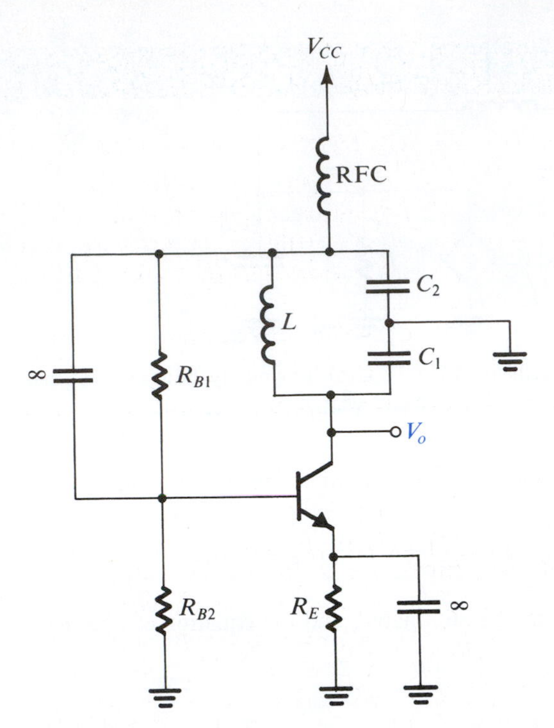

Figure 17.14 Complete circuit for a Colpitts oscillator.

As oscillations grow in amplitude, the transistor's nonlinear characteristics reduce the effective value of g_m and, correspondingly, reduce the loop gain to unity, thus sustaining the oscillations.

Analysis similar to the foregoing can be carried out for the Hartley circuit (see later: Exercise 17.8). At high frequencies, more accurate transistor models must be used. Alternatively, the y parameters of the transistor can be measured at the intended frequency ω_0, and the analysis can then be carried out using the y-parameter model (see Appendix C). This is usually simpler and more accurate, especially at frequencies above about 30% of the transistor f_T.

As an example of a practical LC oscillator, we show in Fig. 17.14 the circuit of a Colpitts oscillator, complete with bias details. Here the radio-frequency choke (RFC) provides a high reactance at ω_0 but a low dc resistance.

Finally, a few words are in order on the mechanism that determines the amplitude of oscillations in the LC-tuned oscillators discussed above. Unlike the op-amp oscillators that incorporate special amplitude-control circuitry, LC-tuned oscillators utilize the nonlinear i_C–v_{BE} characteristics of the BJT (the i_D–v_{GS} characteristics of the FET) for amplitude control. Thus these LC-tuned oscillators are known as *self-limiting oscillators*. Specifically, as the oscillations grow in amplitude, the effective gain of the transistor is reduced below its small-signal value. Eventually, an amplitude is reached at which the effective gain is reduced to the point that the Barkhausen criterion is satisfied exactly. The amplitude then remains constant at this value.

Reliance on the nonlinear characteristics of the BJT (or the FET) implies that the collector (drain) current waveform will be nonlinearly distorted. Nevertheless, the output voltage signal will still be a sinusoid of high purity because of the filtering action of the LC-tuned circuit. Detailed analysis of amplitude control, which makes use of nonlinear-circuit techniques, is beyond the scope of this book.

EXERCISES

17.8 Show that for the Hartley oscillator of Fig. 17.12(b), the frequency of oscillation is given by Eq. (17.17) and that for oscillations to start $g_m R > (L_1/L_2)$.

D17.9 Using a BJT biased at $I_C = 1$ mA, design a Colpitts oscillator to operate at $\omega_0 = 10^6$ rad/s. Use $C_1 = 0.01$ μF and assume that the coil available has a Q of 100 (this can be represented by a resistance in parallel with C_1 given by $Q/\omega_0 C_1$). Also assume that there is a load resistance at the collector of 2 kΩ and that for the BJT, $r_o = 100$ kΩ. Find C_2 and L.

Ans. 0.66 μF; 100 μH (a somewhat smaller C_2 would be used to allow oscillations to grow in amplitude)

17.3.2 Crystal Oscillators

A piezoelectric crystal, such as quartz, exhibits electromechanical-resonance characteristics that are very stable (with time and temperature) and highly selective (having very high Q factors). The circuit symbol of a crystal is shown in Fig. 17.15(a), and its equivalent circuit model is given in Fig. 17.15(b). The resonance properties are characterized by a large inductance L (as high as hundreds of henrys), a very small series capacitance C_s (as small as 0.0005 pF), a series resistance r representing a Q factor $\omega_0 L/r$ that can be as high as a few hundred thousand, and a parallel capacitance C_p (a few picofarads). Capacitor C_p represents the electrostatic capacitance between the two parallel plates of the crystal. Note that $C_p \gg C_s$.

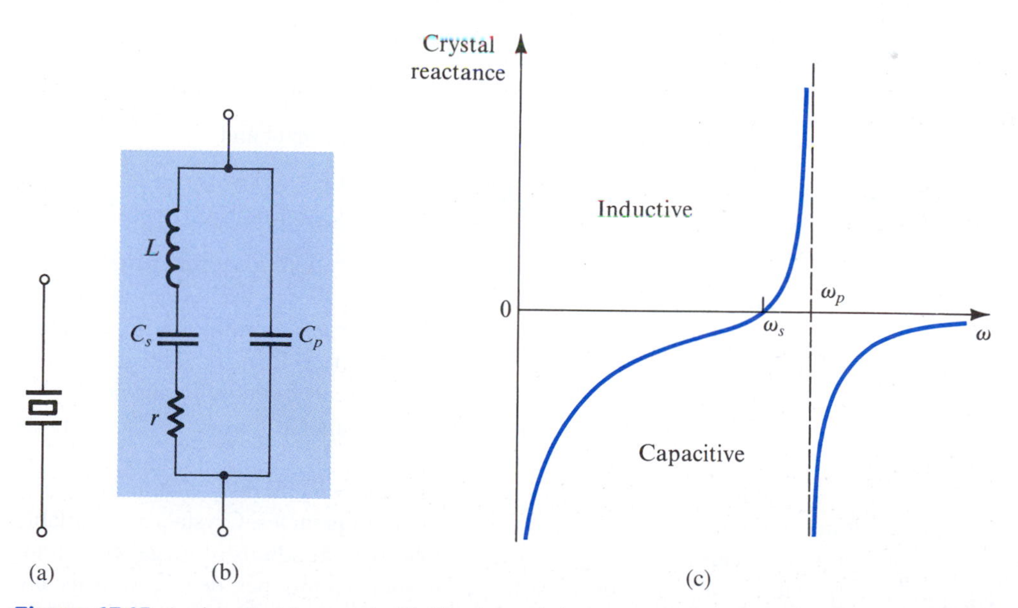

Figure 17.15 A piezoelectric crystal. **(a)** Circuit symbol. **(b)** Equivalent circuit. **(c)** Crystal reactance versus frequency [note that, neglecting the small resistance r, $Z_{crystal} = jX(\omega)$].

Since the Q factor is very high, we may neglect the resistance r and express the crystal impedance as

$$Z(s) = 1 \Big/ \left[sC_p + \frac{1}{sL + 1/sC_s} \right]$$

which can be manipulated to the form

$$Z(s) = \frac{1}{sC_p} \frac{s^2 + (1/LC_s)}{s^2 + [(C_p + C_s)/LC_sC_p]} \tag{17.23}$$

From Eq. (17.23) and from Fig. 17.15(b), we see that the crystal has two resonance frequencies: a series resonance at ω_s

$$\omega_s = 1/\sqrt{LC_s} \tag{17.24}$$

and a parallel resonance at ω_p

$$\omega_p = 1 \Big/ \sqrt{L\left(\frac{C_sC_p}{C_s + C_p}\right)} \tag{17.25}$$

Thus for $s = j\omega$ we can write

$$Z(j\,\omega) = -j\frac{1}{\omega C_p}\left(\frac{\omega^2 - \omega_s^2}{\omega^2 - \omega_p^2}\right) \tag{17.26}$$

From Eqs. (17.24) and (17.25) we note that $\omega_p > \omega_s$. However, since $C_p \gg C_s$, the two resonance frequencies are very close. Expressing $Z(j\omega) = jX(\omega)$, the crystal reactance $X(\omega)$ will have the shape shown in Fig. 17.15(c). We observe that the crystal reactance is inductive over the very narrow frequency band between ω_s and ω_p. For a given crystal, this frequency band is well defined. Thus we may use the crystal to replace the inductor of the Colpitts oscillator (Fig. 17.12a). The resulting circuit will oscillate at the resonance frequency of the crystal inductance L with the series equivalent of C_s and $(C_p + C_1C_2/(C_1 + C_2))$. Since C_s is much smaller than the three other capacitances, it will be dominant and

$$\omega_0 \simeq 1/\sqrt{LC_s} = \omega_s \tag{17.27}$$

In addition to the basic Colpitts oscillator, a variety of configurations exist for crystal oscillators. Figure 17.16 shows a popular configuration (called the **Pierce oscillator**) utilizing a CMOS inverter (see Section 13.2) as an amplifier. Resistor R_f determines a dc operating point in the high-gain region of the VTC of the CMOS inverter. Resistor R_1 together with capacitor C_1 provides a low-pass filter that discourages the circuit from oscillating at a higher harmonic of the crystal frequency. Note that this circuit also is based on the Colpitts configuration.

The extremely stable resonance characteristics and the very high Q factors of quartz crystals result in oscillators with very accurate and stable frequencies. Crystals are available with resonance frequencies in the range of a few kilohertz to hundreds of megahertz. Temperature coefficients of ω_0 of 1 or 2 parts per million (ppm) per degree Celsius are achievable. Unfortunately, however, crystal oscillators, being mechanical resonators, are fixed-frequency circuits.

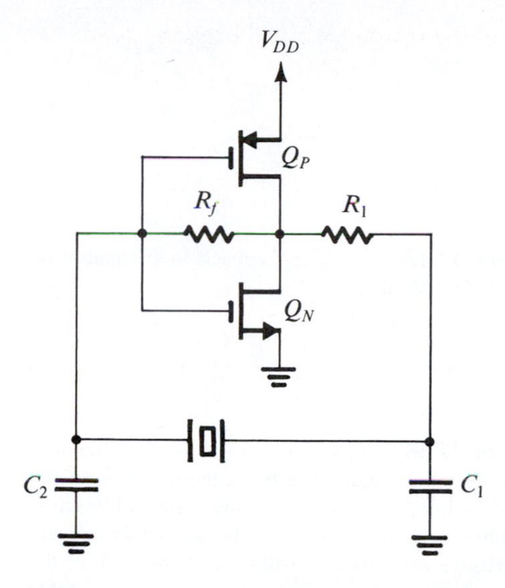

Figure 17.16 A Pierce crystal oscillator utilizing a CMOS inverter as an amplifier.

EXERCISES

17.10 A 2-MHz quartz crystal is specified to have $L = 0.52$ H, $C_s = 0.012$ pF, $C_p = 4$ pF, and $r = 120\ \Omega$. Find f_s, f_p, and Q.
Ans. 2.015 MHz; 2.018 MHz; 55,000

17.4 Bistable Multivibrators

In this section we begin the study of waveform-generating circuits of the other type—nonlinear oscillators or function generators. These devices make use of a special class of circuits known as **multivibrators**. As mentioned earlier, there are three types of multivibrator: bistable, monostable, and astable. This section is concerned with the first, the bistable multivibrator.[5]

As its name indicates, the **bistable multivibrator** has *two stable states*. The circuit can remain in either stable state indefinitely and moves to the other stable state only when appropriately *triggered*.

17.4.1 The Feedback Loop

Bistability can be obtained by connecting a dc amplifier in a positive-feedback loop having a loop gain greater than unity. Such a feedback loop is shown in Fig. 17.17; it consists of an op amp and a resistive voltage divider in the positive-feedback path. To see how bistability is obtained, consider operation with the positive-input terminal of the op amp near ground potential. This is a reasonable starting point, since the circuit has no external excitation.

[5] Digital implementations of multivibrators were presented in Chapter 15. Here, we are interested in implementations utilizing op amps.

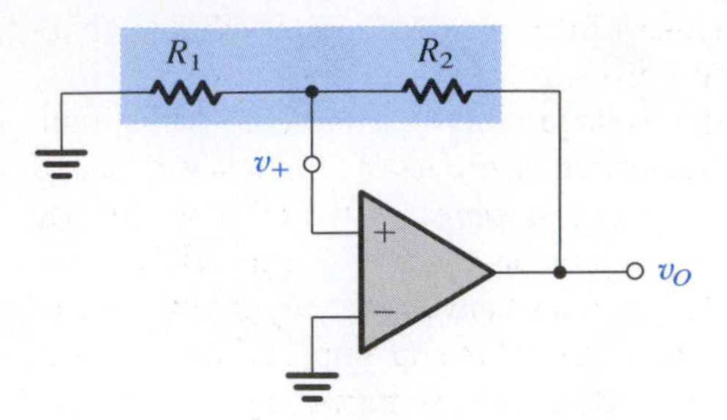

Figure 17.17 A positive-feedback loop capable of bistable operation.

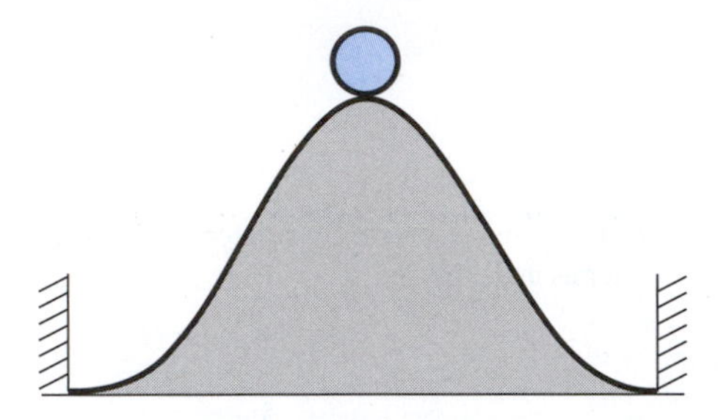

Figure 17.18 A physical analogy for the operation of the bistable circuit. The ball cannot remain at the top of the hill for any length of time (a state of unstable equilibrium or metastability); the inevitably present disturbance will cause the ball to fall to one side or the other, where it can remain indefinitely (the two stable states).

Assume that the electrical noise that is inevitably present in every electronic circuit causes a small positive increment in the voltage v_+. This incremental signal will be amplified by the large open-loop gain A of the op amp, with the result that a much greater signal will appear in the op amp's output voltage v_O. The voltage divider (R_1, R_2) will feed a fraction $\beta \equiv R_1/(R_1 + R_2)$ of the output signal back to the positive-input terminal of the op amp. If $A\beta$ is greater than unity, as is usually the case, the fed-back signal will be greater than the original increment in v_+. This *regenerative* process continues until eventually the op amp saturates with its output voltage at the positive-saturation level, L_+. When this happens, the voltage at the positive-input terminal, v_+, becomes $L_+R_1/(R_1 + R_2)$, which is positive and thus keeps the op amp in positive saturation. This is one of the two stable states of the circuit.

In the description above we assumed that when v_+ was near zero volts, a positive increment occurred in v_+. Had we assumed the equally probable situation of a negative increment, the op amp would have ended up saturated in the negative direction with $v_O = L_-$ and $v_+ = L_-R_1/(R_1 + R_2)$. This is the other stable state.

We thus conclude that the circuit of Fig. 17.17 has two stable states, one with the op amp in positive saturation and the other with the op amp in negative saturation. The circuit can exist in either of these two states indefinitely. We also note that the circuit cannot exist in the state for which $v_+ = 0$ and $v_O = 0$ for any length of time. This is a state of *unstable equilibrium* (also known as a **metastable state**); any disturbance, such as that caused by electrical noise, causes the bistable circuit to switch to one of its two stable states. This is in sharp contrast to the case when the feedback is negative, causing a virtual short circuit to appear between the op amp's input terminals and maintaining this virtual short circuit in the face of disturbances. A physical analogy for the operation of the bistable circuit is depicted in Fig. 17.18.

17.4.2 Transfer Characteristics of the Bistable Circuit

The question naturally arises as to how we can make the bistable circuit of Fig. 17.17 change state. To help answer this crucial question, we derive the transfer characteristics of the bistable. Reference to Fig. 17.17 indicates that either of the two circuit nodes that are connected to ground can serve as an input terminal. We investigate both possibilities.

Figure 17.19(a) shows the bistable circuit with a voltage v_I applied to the inverting input terminal of the op amp. To derive the transfer characteristic v_O–v_I, assume that v_O is at one of its two possible levels, say L_+, and thus $v_+ = \beta L_+$. Now as v_I is increased from 0 V, we can see from the circuit that nothing happens until v_I reaches a value equal to v_+ (i.e., βL_+). As v_I begins to exceed this value, a net negative voltage develops between the input terminals of the op amp. This voltage is amplified by the open-loop gain of the op amp, and thus v_O goes negative. The voltage divider in turn causes v_+ to go negative, thus increasing the net negative input to the op amp and keeping the regenerative process going. This process culminates in

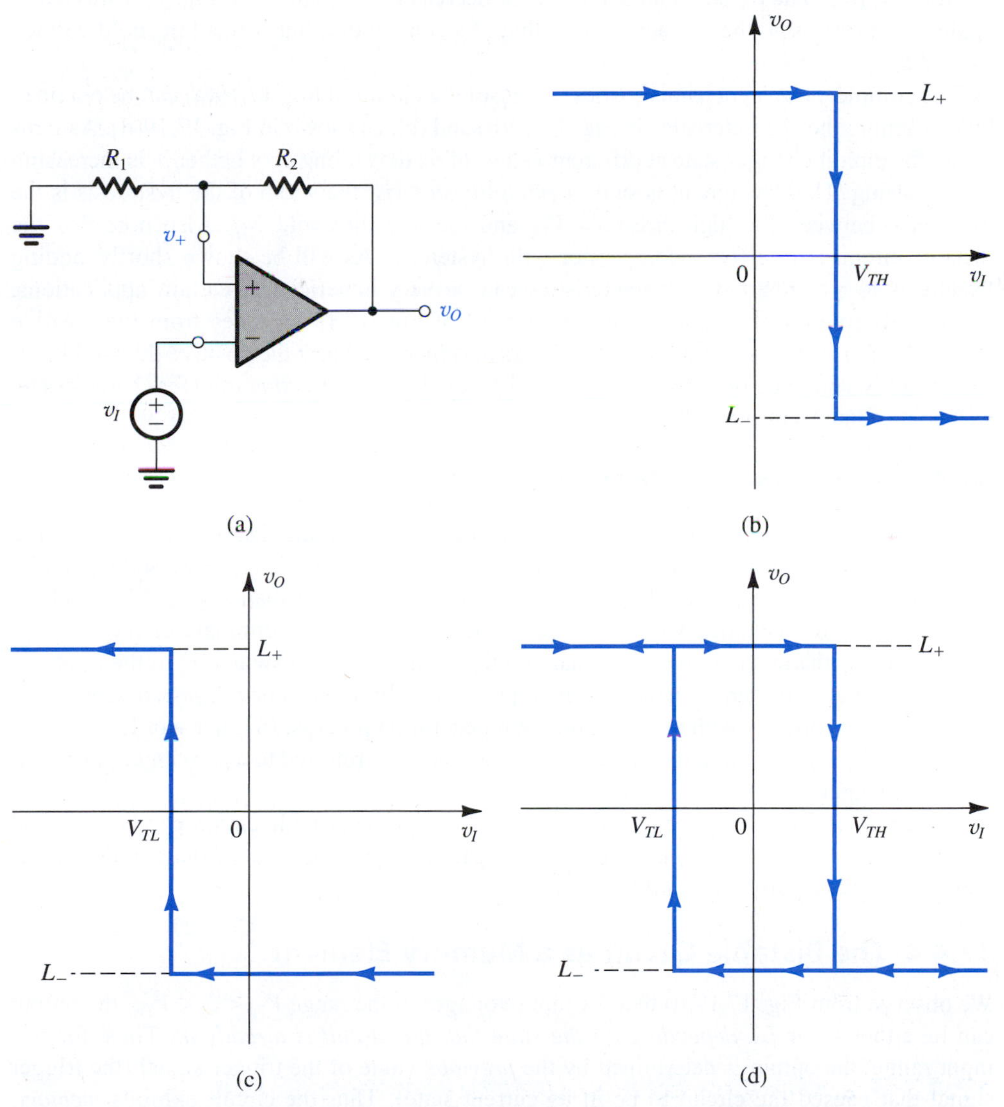

Figure 17.19 **(a)** The bistable circuit of Fig. 17.17 with the negative input terminal of the op amp disconnected from ground and connected to an input signal v_I. **(b)** The transfer characteristic of the circuit in **(a)** for increasing v_I. **(c)** The transfer characteristic for decreasing v_I. **(d)** The complete transfer characteristics.

the op amp saturating in the negative direction: that is, with $v_O = L_-$ and, correspondingly, $v_+ = \beta L_-$. It is easy to see that increasing v_I further has no effect on the acquired state of the bistable circuit. Figure 17.19(b) shows the transfer characteristic for increasing v_I. Observe that the characteristic is that of a comparator with a threshold voltage denoted V_{TH}, where $V_{TH} = \beta L_+$.

Next consider what happens as v_I is decreased. Since now $v_+ = \beta L_-$, we see that the circuit remains in the negative-saturation state until v_I goes negative to the point that it equals βL_-. As v_I goes below this value, a net positive voltage appears between the op amp's input terminals. This voltage is amplified by the op-amp gain and thus gives rise to a positive voltage at the op amp's output. The regenerative action of the positive-feedback loop then sets in and causes the circuit eventually to go to its positive-saturation state, in which $v_O = L_+$ and $v_+ = \beta L_+$. The transfer characteristic for decreasing v_I is shown in Fig. 17.19(c). Here again we observe that the characteristic is that of a comparator, but with a threshold voltage $V_{TL} = \beta L_-$.

The complete transfer characteristics, v_O–v_I, of the circuit in Fig. 17.19(a) can be obtained by combining the characteristics in Fig. 17.19(b) and (c), as shown in Fig. 17.19(d). As indicated, the circuit changes state at different values of v_I, depending on whether v_I is increasing or decreasing. Thus the circuit is said to exhibit *hysteresis;* the width of the hysteresis is the difference between the high threshold V_{TH} and the low threshold V_{TL}. Also note that the bistable circuit is in effect a comparator with hysteresis. As will be shown shortly, adding hysteresis to a comparator's characteristics can be very beneficial in certain applications. Finally, observe that because the bistable circuit of Fig. 17.19 switches from the positive state ($v_O = L_+$) to the negative state ($v_O = L_-$) as v_I is increased past the positive threshold V_{TH}, the circuit is said to be *inverting*. A bistable circuit with a *noninverting* transfer characteristic will be presented shortly.

17.4.3 Triggering the Bistable Circuit

Returning now to the question of how to make the bistable circuit change state, we observe from the transfer characteristics of Fig. 17.19(d) that if the circuit is in the L_+ state it can be switched to the L_- state by applying an input v_I of value greater than $V_{TH} \equiv \beta L_+$. Such an input causes a net negative voltage to appear between the input terminals of the op amp, which initiates the regenerative cycle that culminates in the circuit switching to the L_- stable state. Here it is important to note that the input v_I merely initiates or *triggers* regeneration. Thus we can remove v_I with no effect on the regeneration process. In other words, v_I can be simply a pulse of short duration. The input signal v_I is thus referred to as a **trigger signal**, or simply a **trigger**.

The characteristics of Fig. 17.19(d) indicate also that the bistable circuit can be switched to the positive state ($v_O = L_+$) by applying a negative trigger signal v_I of magnitude greater than that of the negative threshold V_{TL}.

17.4.4 The Bistable Circuit as a Memory Element

We observe from Fig. 17.19(d) that for input voltages in the range $V_{TL} < v_I < V_{TH}$, the output can be either L_+ or L_-, *depending on the state that the circuit is already in.* Thus, for this input range, the output is determined by the *previous* value of the trigger signal (the trigger signal that caused the circuit to be in its current state). Thus the circuit exhibits *memory.* Indeed, the bistable multivibrator is the basic memory element of digital systems, as we have seen in Chapter 15. Finally, note that in analog circuit applications, such as the ones of concern to us in this chapter, the bistable circuit is also known as a **Schmitt trigger**.

17.4.5 A Bistable Circuit with Noninverting Transfer Characteristics

The basic bistable feedback loop of Fig. 17.17 can be used to derive a circuit with noninverting transfer characteristics by applying the input signal v_I (the trigger signal) to the terminal of R_1 that is connected to ground. The resulting circuit is shown in Fig. 17.20(a). To obtain the transfer characteristics we first employ superposition to the linear circuit formed by R_1 and R_2, thus expressing v_+ in terms of v_I and v_O as

$$v_+ = v_I \frac{R_2}{R_1 + R_2} + v_O \frac{R_1}{R_1 + R_2} \tag{17.28}$$

From this equation we see that if the circuit is in the positive stable state with $v_O = L_+$, positive values for v_I will have no effect. To trigger the circuit into the L_- state, v_I must be made negative and of such a value as to make v_+ decrease below zero. Thus the low threshold V_{TL} can be found by substituting in Eq. (17.28) $v_O = L_+$, $v_+ = 0$, and $v_I = V_{TL}$. The result is

$$V_{TL} = -L_+(R_1/R_2) \tag{17.29}$$

Similarly, Eq. (17.28) indicates that when the circuit is in the negative-output state ($v_O = L_-$), negative values of v_I will make v_+ more negative with no effect on operation. To initiate the regeneration process that causes the circuit to switch to the positive state, v_+ must be made to go slightly positive. The value of v_I that causes this to happen is the high threshold voltage V_{TH}, which can be found by substituting in Eq. (17.28) $v_O = L_-$ and $v_+ = 0$. The result is

$$V_{TH} = -L_-(R_1/R_2) \tag{17.30}$$

The complete transfer characteristic of the circuit of Fig. 17.20(a) is displayed in Fig. 17.20(b). Observe that a positive triggering signal v_I (of value greater than V_{TH}) causes the circuit to switch to the positive state (v_O goes from L_- to L_+). Thus the transfer characteristic of this circuit is noninverting.

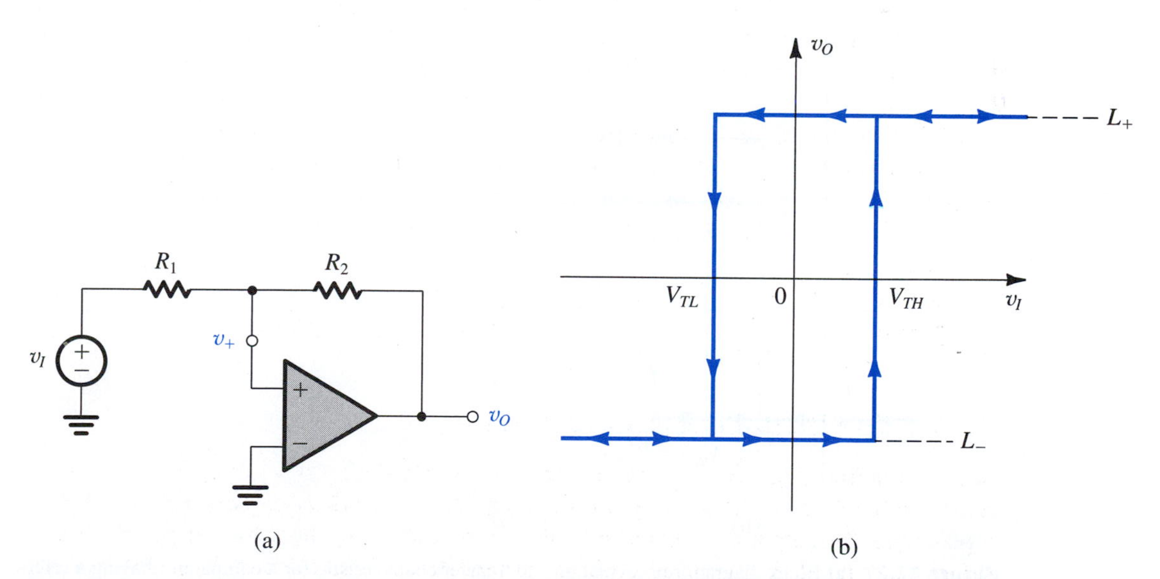

Figure 17.20 **(a)** A bistable circuit derived from the positive-feedback loop of Fig. 17.17 by applying v_I through R_1. **(b)** The transfer characteristic of the circuit in **(a)** is noninverting. (Compare it to the inverting characteristic in Fig. 17.19d.)

17.4.6 Application of the Bistable Circuit as a Comparator

The comparator is an analog-circuit building block that is used in a variety of applications ranging from detecting the level of an input signal relative to a preset threshold value, to the design of analog-to-digital (A/D) converters. Although one normally thinks of the comparator as having a single threshold value (see Fig. 17.21a), it is useful in many applications to add hysteresis to the comparator characteristics. If this is done, the comparator exhibits two threshold values, V_{TL} and V_{TH}, symmetrically placed about the desired reference level, as indicated in Fig. 17.21(b). Usually V_{TH} and V_{TL} are separated by a small amount, say 100 mV.

To demonstrate the need for hysteresis, we consider a common application of comparators. It is required to design a circuit that detects and counts the zero crossings of an arbitrary waveform. Such a function can be implemented using a comparator whose threshold is set to 0 V. The comparator provides a step change at its output every time a zero crossing occurs. Each step change can be used to generate a pulse, and the pulses are fed to a counter circuit.

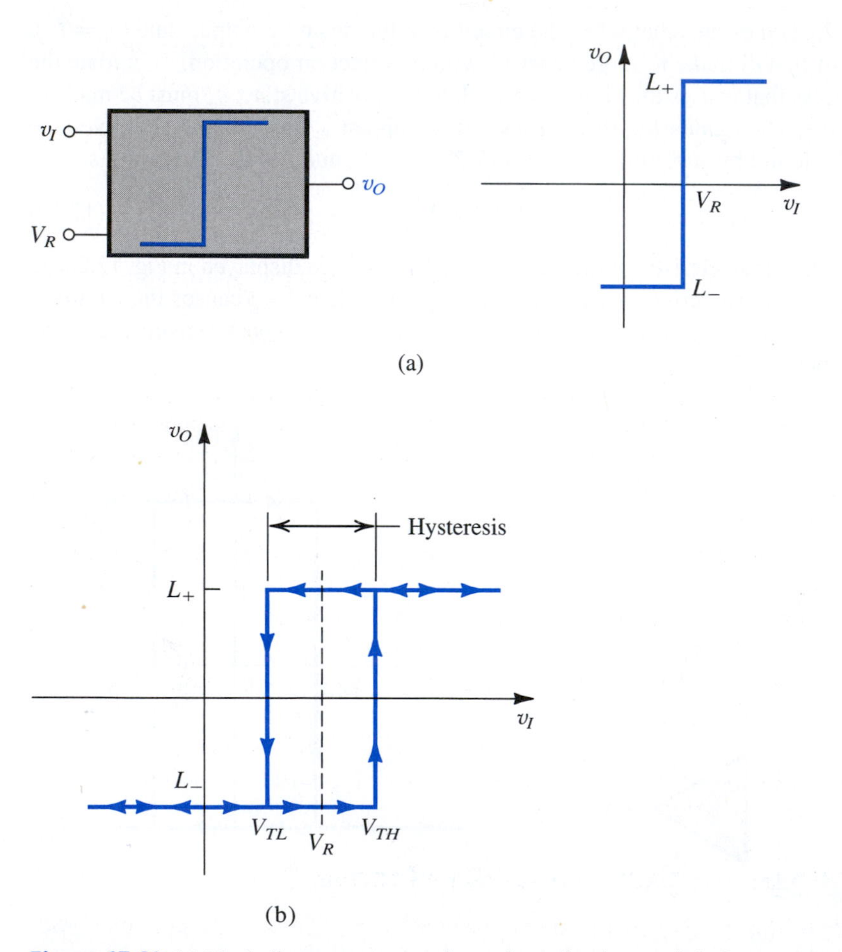

Figure 17.21 **(a)** Block diagram representation and transfer characteristic for a comparator having a reference, or threshold, voltage V_R. **(b)** Comparator characteristic with hysteresis.

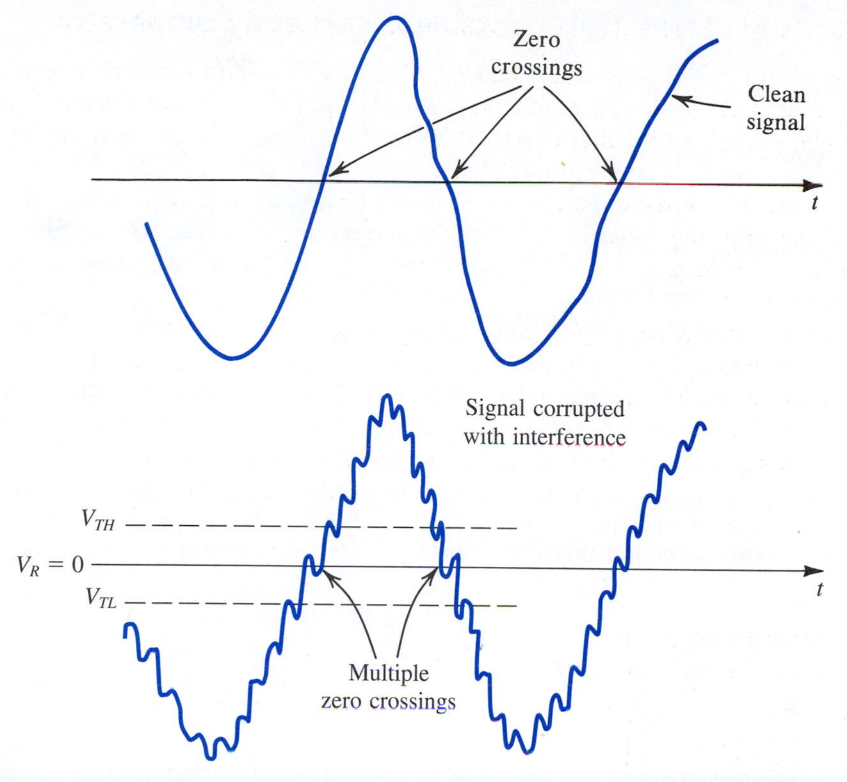

Figure 17.22 Illustrating the use of hysteresis in the comparator characteristics as a means of rejecting interference.

Imagine now what happens if the signal being processed has—as it usually does have—interference superimposed on it, say of a frequency much higher than that of the signal. It follows that the signal might cross the zero axis a number of times around each of the zero-crossing points we are trying to detect, as shown in Fig. 17.22. The comparator would thus change state a number of times at each of the zero crossings, and our count would obviously be in error. However, if we have an idea of the expected peak-to-peak amplitude of the interference, the problem can be solved by introducing hysteresis of appropriate width in the comparator characteristics. Then, if the input signal is increasing in magnitude, the comparator with hysteresis will remain in the low state until the input level exceeds the high threshold V_{TH}. Subsequently the comparator will remain in the high state even if, owing to interference, the signal decreases below V_{TH}. The comparator will switch to the low state only if the input signal is decreased below the low threshold V_{TL}. The situation is illustrated in Fig. 17.22, from which we see that including hysteresis in the comparator characteristics provides an effective means for rejecting interference (thus providing another form of filtering).

17.4.7 Making the Output Levels More Precise

The output levels of the bistable circuit can be made more precise than the saturation voltages of the op amp are by cascading the op amp with a limiter circuit (see Section 4.6 for a discussion of limiter circuits). Two such arrangements are shown in Fig. 17.23.

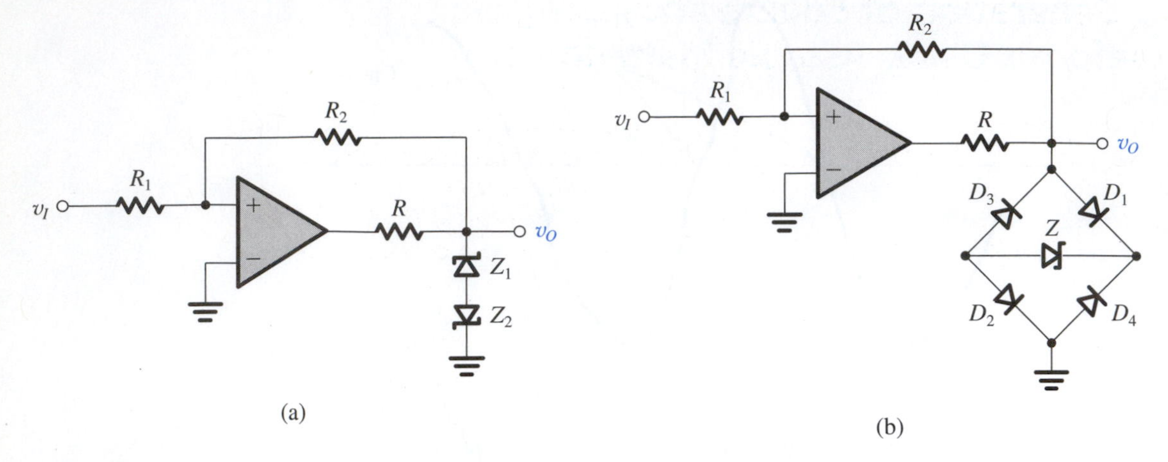

Figure 17.23 Limiter circuits are used to obtain more precise output levels for the bistable circuit. In both circuits the value of R should be chosen to yield the current required for the proper operation of the zener diodes. **(a)** For this circuit $L_+ = V_{Z_1} + V_D$ and $L_- = -(V_{Z_2} + V_D)$, where V_D is the forward diode drop. **(b)** For this circuit $L_+ = V_Z + V_{D_1} + V_{D_2}$ and $L_- = -(V_Z + V_{D_3} + V_{D_4})$.

EXERCISES

D17.11 The op amp in the bistable circuit of Fig. 17.19(a) has output saturation voltages of ±13 V. Design the circuit to obtain threshold voltages of ±5 V. For $R_1 = 10$ kΩ, find the value required for R_2.
Ans. 16 kΩ

D17.12 If the op amp in the circuit of Fig. 17.20(a) has ±10-V output saturation levels, design the circuit to obtain ±5-V thresholds. Give suitable component values.
Ans. Possible choice: $R_1 = 10$ kΩ and $R_2 = 20$ kΩ

17.13 Consider a bistable circuit with a noninverting transfer characteristic and let $L_+ = -L_- = 10$ V and $V_{TH} = -V_{TL} = 5$ V. If v_I is a triangular wave with a 0-V average, a 10-V peak amplitude, and a 1-ms period, sketch the waveform of v_O. Find the time interval between the zero crossings of v_I and v_O.
Ans. v_O is a square wave with 0-V average, 10-V amplitude, and 1-ms period and is delayed by 125 μs relative to v_I

17.14 Consider an op amp having saturation levels of ±12 V used without feedback, with the inverting input terminal connected to +3 V and the noninverting input terminal connected to v_I. Characterize its operation as a comparator. What are L_+, L_-, and V_R, as defined in Fig. 17.21(a)?
Ans. +12 V; −12 V; +3 V

17.15 In the circuit of Fig. 17.20(a), let $L_+ = -L_- = 10$ V and $R_1 = 1$ kΩ. Find a value for R_2 that gives a hysteresis of 100-mV width.
Ans. 200 kΩ

17.5 Generation of Square and Triangular Waveforms Using Astable Multivibrators

A square waveform can be generated by arranging for a bistable multivibrator to switch states periodically. This can be done by connecting the bistable multivibrator with an RC circuit in a feedback loop, as shown in Fig. 17.24(a). Observe that the bistable multivibrator has an inverting transfer characteristic and can thus be realized using the circuit of Fig. 17.19(a). This results in the circuit of Fig. 17.24(b). We shall show shortly that this circuit has no stable states and thus is appropriately named an **astable multivibrator**.

At this point we wish to remind the reader of an important relationship, which we shall employ on many occasions in the following few sections: A capaciter C that is charging or discharging through a resistance R toward a final voltage V_∞ has a voltage $v(t)$,

$$v(t) = V_\infty - (V_\infty - V_{0+})e^{-t/\tau}$$

where V_{0+} is the voltage at $t = 0+$ and $\tau = CR$ is the time constant.

17.5.1 Operation of the Astable Multivibrator

To see how the astable multivibrator operates, refer to Fig. 17.24(b) and let the output of the bistable multivibrator be at one of its two possible levels, say L_+. Capacitor C will charge toward this level through resistor R. Thus the voltage across C, which is applied to the negative input terminal of the op amp and thus is denoted v_-, will rise exponentially toward L_+ with a time constant $\tau = CR$. Meanwhile, the voltage at the positive input terminal of the op amp is $v_+ = \beta L_+$. This situation will continue until the capacitor voltage reaches the positive threshold $V_{TH} = \beta L_+$, at which point the bistable multivibrator will

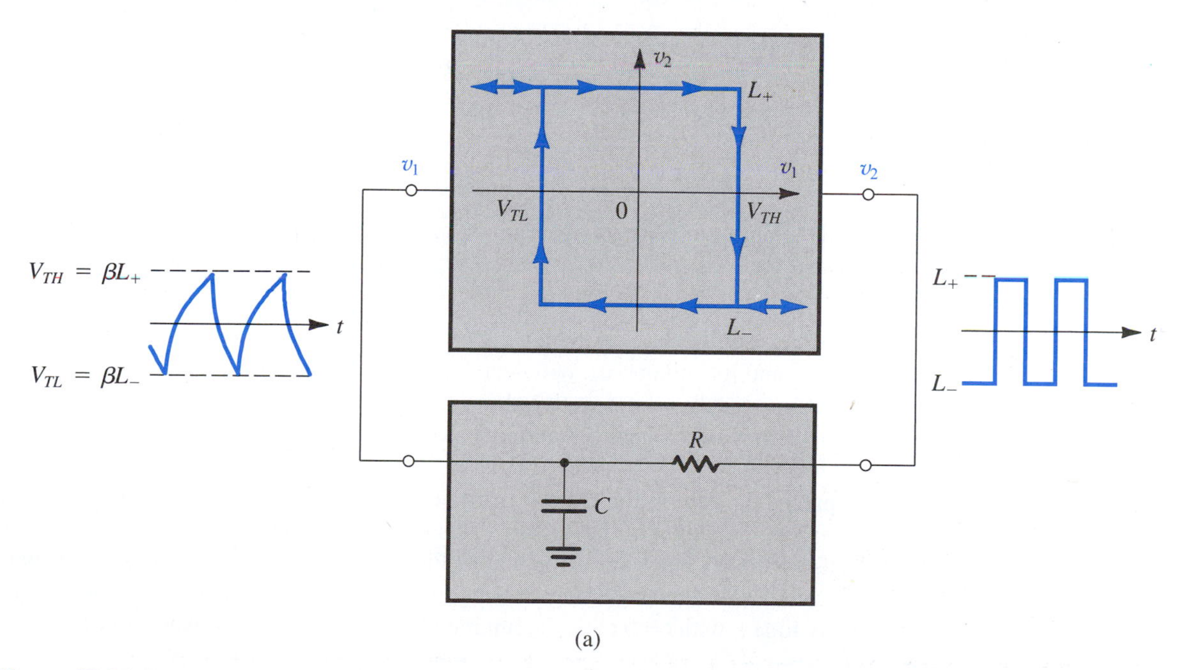

(a)

Figure 17.24 (a) Connecting a bistable multivibrator with inverting transfer characteristics in a feedback loop with an RC circuit results in a square-wave generator.

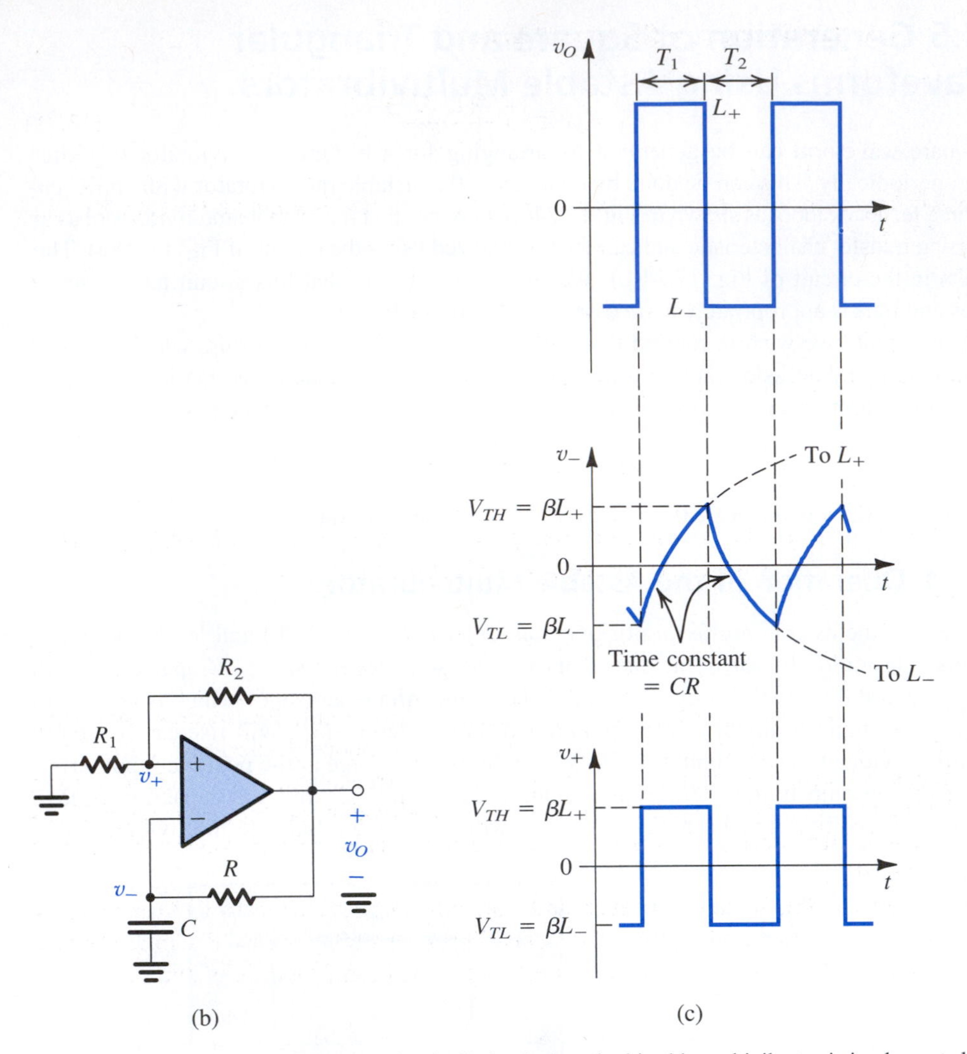

Figure 17.24 (*Continued*) **(b)** The circuit obtained when the bistable multivibrator is implemented with the circuit of Fig. 17.19(a). **(c)** Waveforms at various nodes of the circuit in **(b)**. This circuit is called an astable multivibrator.

switch to the other stable, state, in which $v_O = L_-$ and $v_+ = \beta L_-$. The capacitor will then start discharging, and its voltage, v_-, will decrease exponentially toward L_-. This new state will prevail until v_- reaches the negative threshold $V_{TL} = \beta L_-$, at which time the bistable multivibrator switches to the positive-output state, the capacitor begins to charge, and the cycle repeats itself.

From the preceding description we see that the astable circuit oscillates and produces a square waveform at the output of the op amp. This waveform, and the waveforms at the two input terminals of the op amp, are displayed in Fig. 17.24(c). The period T of the square wave can be found as follows: During the charging interval T_1 the voltage v_- across the capacitor at any time t, with $t = 0$ at the beginning of T_1, is given by (see Appendix E)

$$v_- = L_+ - (L_+ - \beta L_-)e^{-t/\tau}$$

where $\tau = CR$. Substituting $v_- = \beta L_+$ at $t = T_1$ gives

$$T_1 = \tau \ln \frac{1 - \beta(L_-/L_+)}{1 - \beta} \tag{17.31}$$

Similarly, during the discharge interval T_2 the voltage v_- at any time t, with $t = 0$ at the beginning of T_2, is given by

$$v_- = L_- - (L_- - \beta L_+)e^{-t/\tau}$$

Substituting $v_- = \beta L_-$ at $t = T_2$ gives

$$T_2 = \tau \ln \frac{1 - \beta(L_+/L_-)}{1 - \beta} \tag{17.32}$$

Equations (17.31) and (17.32) can be combined to obtain the period $T = T_1 + T_2$. Normally, $L_+ = -L_-$, resulting in symmetrical square waves of period T given by

$$T = 2\tau \ln \frac{1 + \beta}{1 - \beta} \tag{17.33}$$

Note that this square-wave generator can be made to have variable frequency by switching different capacitors C (usually in decades) and by continuously adjusting R (to obtain continuous frequency control within each decade of frequency). Also, the waveform across C can be made almost triangular by using a small value for the parameter β. However, triangular waveforms of superior linearity can be easily generated using the scheme discussed next.

Before leaving this section, however, note that although the astable circuit has no stable states, it has two *quasi-stable* states and remains in each for a time interval determined by the time constant of the RC network and the thresholds of the bistable multivibrator.

EXERCISES

17.16 For the circuit in Fig. 17.24(b), let the op-amp saturation voltages be ±10 V, $R_1 = 100$ kΩ, $R_2 = R = 1$ MΩ, and $C = 0.01$ μF. Find the frequency of oscillation.
Ans. 274 Hz

17.17 Consider a modification of the circuit of Fig. 17.24(b) in which R_1 is replaced by a pair of diodes connected in parallel in opposite directions. For $L_+ = -L_- = 12$ V, $R_2 = R = 10$ kΩ, $C = 0.1$ μF, and the diode voltage as a constant denoted V_D, find an expression for frequency as a function of V_D. If $V_D = 0.70$ V at 25°C with a TC of −2 mV/°C, find the frequency at 0°C, 25°C, 50°C, and 100°C. Note that the output of this circuit can be sent to a remotely connected frequency meter to provide a digital readout of temperature.
Ans. $f = 500/\ln[(12 + V_D)/(12 - V_D)]$ Hz; 3995 Hz, 4281 Hz, 4611 Hz, 5451 Hz

17.5.2 Generation of Triangular Waveforms

The exponential waveforms generated in the astable circuit of Fig. 17.24 can be changed to triangular by replacing the low-pass RC circuit with an integrator. (The integrator is, after all, a low-pass circuit with a corner frequency at dc.) The integrator causes linear charging and discharging of the capacitor, thus providing a triangular waveform. The resulting circuit is shown in Fig. 17.25(a). Observe that because the integrator is inverting, it is necessary to invert the characteristics of the bistable circuit. Thus the bistable circuit required here is of the noninverting type and can be implemented using the circuit of Fig. 17.20.

We now proceed to show how the feedback loop of Fig. 17.25(a) oscillates and generates a triangular waveform v_1 at the output of the integrator and a square waveform v_2 at the output of the bistable circuit: Let the output of the bistable circuit be at L_+. A current equal to L_+/R will flow into the resistor R and through capacitor C, causing the output of the integrator to *linearly* decrease with a slope of $-L_+/CR$, as shown in Fig. 17.25(c). This will continue until the integrator output reaches the lower threshold V_{TL} of the bistable circuit, at which point the bistable circuit will switch states, its output becoming negative and equal to L_-. At this moment the current through R and C will reverse direction, and its value will become equal to $|L_-|/R$. It follows that the integrator output will start to increase linearly with a positive slope equal to $|L_-|/CR$. This will continue until the integrator output voltage reaches the positive threshold of the bistable circuit, V_{TH}. At this point the bistable circuit

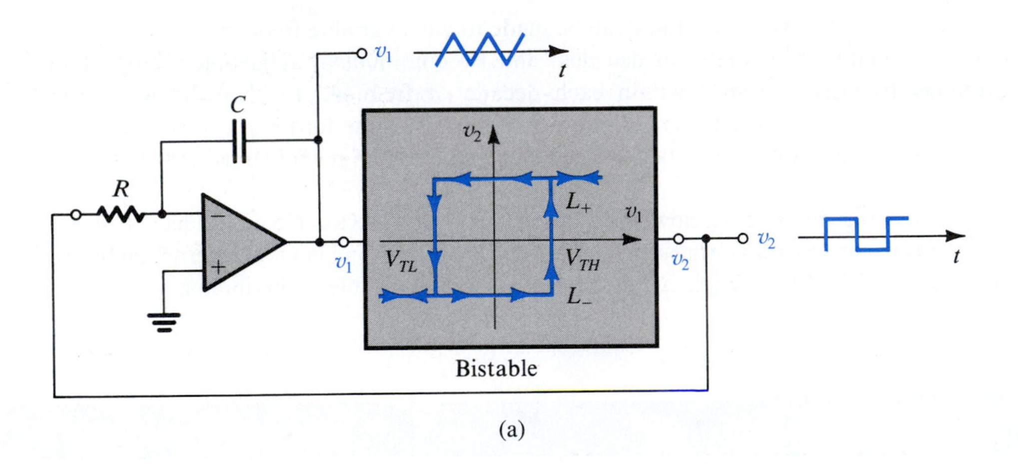

(a)

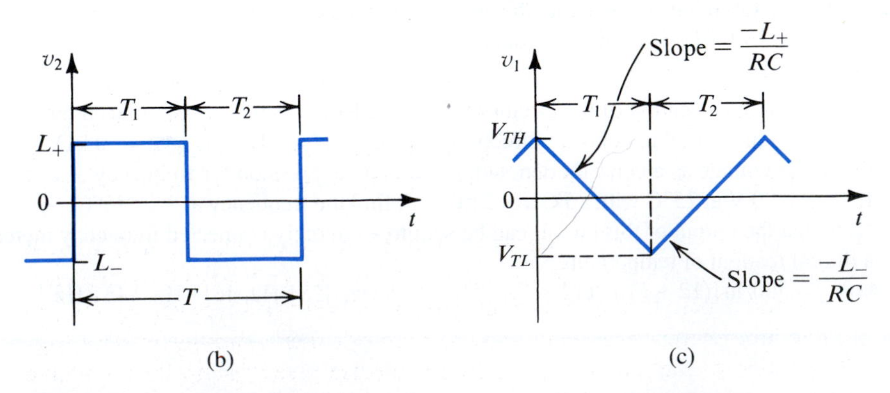

(b) (c)

Figure 17.25 A general scheme for generating triangular and square waveforms.

switches, its output becomes positive (L_+), the current into the integrator reverses direction, and the output of the integrator starts to decrease linearly, beginning a new cycle.

From the discussion above, it is relatively easy to derive an expression for the period T of the square and triangular waveforms. During the interval T_1 we have, from Fig. 17.25(c),

$$\frac{V_{TH} - V_{TL}}{T_1} = \frac{L_+}{CR}$$

from which we obtain

$$T_1 = CR\frac{V_{TH} - V_{TL}}{L_+} \tag{17.34}$$

Similarly, during T_2 we have

$$\frac{V_{TH} - V_{TL}}{T_2} = \frac{-L_-}{CR}$$

from which we obtain

$$T_2 = CR\frac{V_{TH} - V_{TL}}{-L_-} \tag{17.35}$$

Thus to obtain symmetrical square waves we design the bistable circuit to have $L_+ = -L_-$.

EXERCISES

D17.18 Consider the circuit of Fig. 17.25(a) with the bistable circuit realized by the circuit in Fig. 17.20(a). If the op amps have saturation voltages of ±10 V, and if a capacitor $C = 0.01\ \mu\text{F}$ and a resistor $R_1 = 10$ kΩ are used, find the values of R and R_2 (note that R_1 and R_2 are associated with the bistable circuit of Fig. 17.20a) such that the frequency of oscillation is 1 kHz and the triangular waveform has a 10-V peak-to-peak amplitude.
Ans. 50 kΩ; 20 kΩ

17.6 Generation of a Standardized Pulse—The Monostable Multivibrator

In some applications the need arises for a pulse of known height and width generated in response to a trigger signal. Because the width of the pulse is predictable, its trailing edge can be used for timing purposes—that is, to initiate a particular task at a specified time. Such a standardized pulse can be generated by the third type of multivibrator, the **monostable multivibrator**.

The monostable multivibrator has one stable state in which it can remain indefinitely. It also has a quasi-stable state to which it can be triggered and in which it stays for a predetermined interval equal to the desired width of the output pulse. When this interval expires, the monostable multivibrator returns to its stable state and remains there, awaiting another triggering signal. The action of the monostable multivibrator has given rise to its alternative name, the *one shot*.

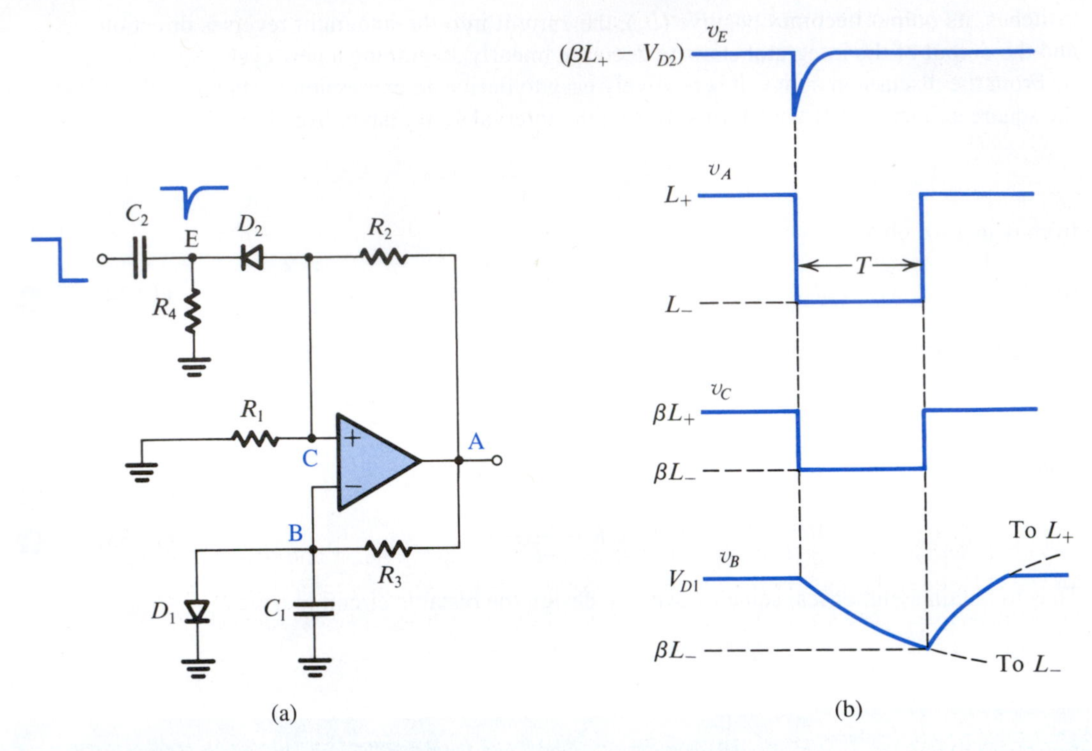

Figure 17.26 **(a)** An op-amp monostable circuit. **(b)** Signal waveforms in the circuit of **(a)**.

Figure 17.26(a) shows an op-amp monostable circuit. We observe that this circuit is an augmented form of the astable circuit of Fig. 17.24(b). Specifically, a clamping diode D_1 is added across the capacitor C_1, and a trigger circuit composed of capacitor C_2, resistor R_4, and diode D_2 is connected to the noninverting input terminal of the op amp. The circuit operates as follows: In the stable state, which prevails in the absence of the triggering signal, the output of the op amp is at L_+ and diode D_1 is conducting through R_3 and thus clamping the voltage v_B to one diode drop above ground. We select R_4 much larger than R_1, so that diode D_2 will be conducting a very small current and the voltage v_C will be very closely determined by the voltage divider R_1, R_2. Thus $v_C = \beta L_+$, where $\beta = R_1/(R_1 + R_2)$. The stable state is maintained because βL_+ is greater than V_{D1}.

Now consider the application of a negative-going step at the trigger input and refer to the signal waveforms shown in Fig. 17.26(b). The negative triggering edge is coupled to the cathode of diode D_2 via capacitor C_2, and thus D_2 conducts heavily and pulls node C down. If the trigger signal is of sufficient height to cause v_C to go below v_B, the op amp will see a net negative input voltage and its output will switch to L_-. This in turn will cause v_C to go negative to βL_-, keeping the op amp in its newly acquired state. Note that D_2 will then cut off, thus isolating the circuit from any further changes at the trigger input terminal.

The negative voltage at A causes D_1 to cut off, and C_1 begins to discharge exponentially toward L_- with a time constant C_1R_3. The monostable multivibrator is now in its *quasi-stable state,* which will prevail until the declining v_B goes below the voltage at node C, which is βL_-. At this instant the op-amp output switches back to L_+ and the voltage at node C goes back to βL_+. Capacitor C_1 then charges toward L_+ until diode D_1 turns on and the circuit returns to its stable state.

From Fig. 17.26(b), we observe that a negative pulse is generated at the output during the quasi-stable state. The duration T of the output pulse is determined from the exponential waveform of v_B,

$$v_B(t) = L_- - (L_- - V_{D1})e^{-t/C_1R_3}$$

by substituting $v_B(T) = \beta L_-$,

$$\beta L_- = L_- - (L_- - V_{D1})e^{-T/C_1R_3}$$

which yields

$$T = C_1R_3 \ln\left(\frac{V_{D1} - L_-}{\beta L_- - L_-}\right) \tag{17.36}$$

For $V_{D1} \ll |L_-|$, this equation can be approximated by

$$T \simeq C_1R_3 \ln\left(\frac{1}{1-\beta}\right) \tag{17.37}$$

Finally, note that the monostable circuit should not be triggered again until capacitor C_1 has been recharged to V_{D1}; otherwise the resulting output pulse will be shorter than normal. This recharging time is known as the **recovery period**. Circuit techniques exist for shortening the recovery period.

EXERCISES

17.19 For the monostable circuit of Fig. 17.26(a), find the value of R_3 that will result in a 100-μs output pulse for $C_1 = 0.1$ μF, $\beta = 0.1$, $V_D = 0.7$ V, and $L_+ = -L_- = 12$ V.
Ans. 6171 Ω

17.7 Integrated-Circuit Timers

Commercially available integrated-circuit packages exist that contain the bulk of the circuitry needed to implement monostable and astable multivibrators with precise characteristics. In this section we discuss the most popular of such ICs, the **555 timer**. Introduced in 1972 by the Signetics Corporation as a bipolar integrated circuit, the 555 is also available in CMOS technology and from a number of manufacturers.[6]

17.7.1 The 555 Circuit

Figure 17.27 shows a block diagram representation of the 555 timer circuit [for the actual circuit, refer to Grebene (1984)]. The circuit consists of two comparators, an SR flip-flop, and a transistor Q_1 that operates as a switch. One power supply (V_{CC}) is required for operation, with the supply voltage typically 5 V. A resistive voltage divider, consisting of the three

[6] In a recent article in *IEEE Spectrum* (May 2009), the 555 was selected as one of the "25 Microchips That Shook the World."

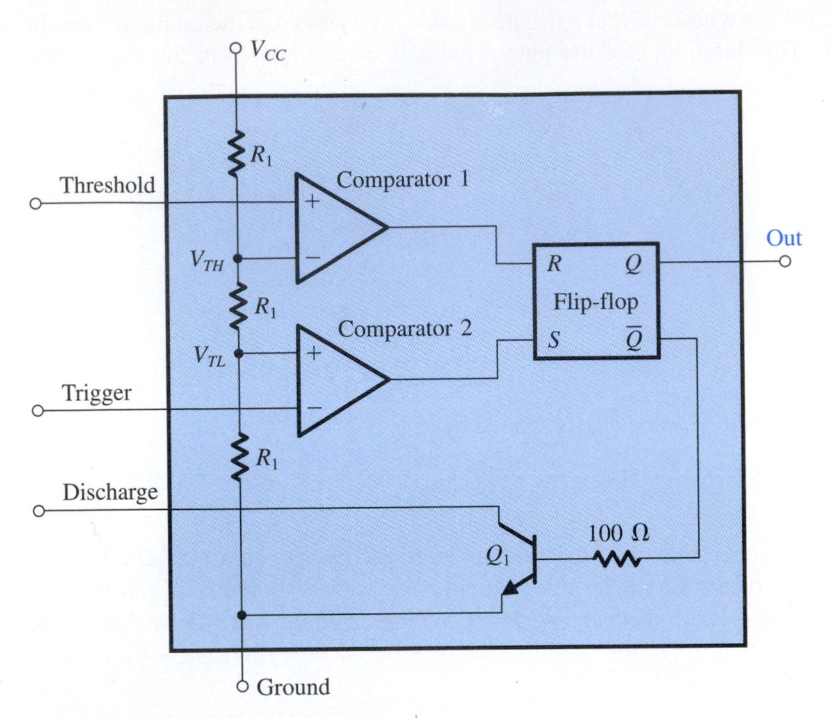

Figure 17.27 A block diagram representation of the internal circuit of the 555 integrated-circuit timer.

equal-valued resistors labeled R_1, is connected across V_{CC} and establishes the reference (threshold) voltages for the two comparators. These are $V_{TH} = \frac{2}{3}V_{CC}$ for comparator 1 and $V_{TL} = \frac{1}{3}V_{CC}$ for comparator 2.

We studied SR flip-flops in Chapter 15. For our purposes here we note that an SR flip-flop is a bistable circuit having complementary outputs, denoted Q and $\overline{Q}$. In the *set* state, the output at Q is "high" (approximately equal to V_{CC}) and that at $\overline{Q}$ is "low" (approximately equal to 0 V). In the other stable state, termed the *reset* state, the output at Q is low and that at $\overline{Q}$ is high. The flip-flop is set by applying a high level (V_{CC}) to its set input terminal, labeled S. To reset the flip-flop, a high level is applied to the reset input terminal, labeled R. Note that the reset and set input terminals of the flip-flop in the 555 circuit are connected to the outputs of comparator 1 and comparator 2, respectively.

The positive-input terminal of comparator 1 is brought out to an external terminal of the 555 package, labeled Threshold. Similarly, the negative-input terminal of comparator 2 is connected to an external terminal labeled Trigger, and the collector of transistor Q_1 is connected to a terminal labeled Discharge. Finally, the Q output of the flip-flop is connected to the output terminal of the timer package, labeled Out.

17.7.2 Implementing a Monostable Multivibrator Using the 555 IC

Figure 17.28(a) shows a monostable multivibrator implemented using the 555 IC together with an external resistor R and an external capacitor C. In the stable state the flip-flop will be in the reset state, and thus its $\overline{Q}$ output will be high, turning on transistor Q_1. Transistor Q_1 will be saturated, and thus v_C will be close to 0 V, resulting in a low level at the output of comparator 1. The voltage at the trigger input terminal, labeled v_{trigger}, is kept high (greater than V_{TL}), and thus the output of comparator 2 also will be low. Finally, note that since the flip-flop is in the reset state, Q will be low and thus v_O will be close to 0 V.

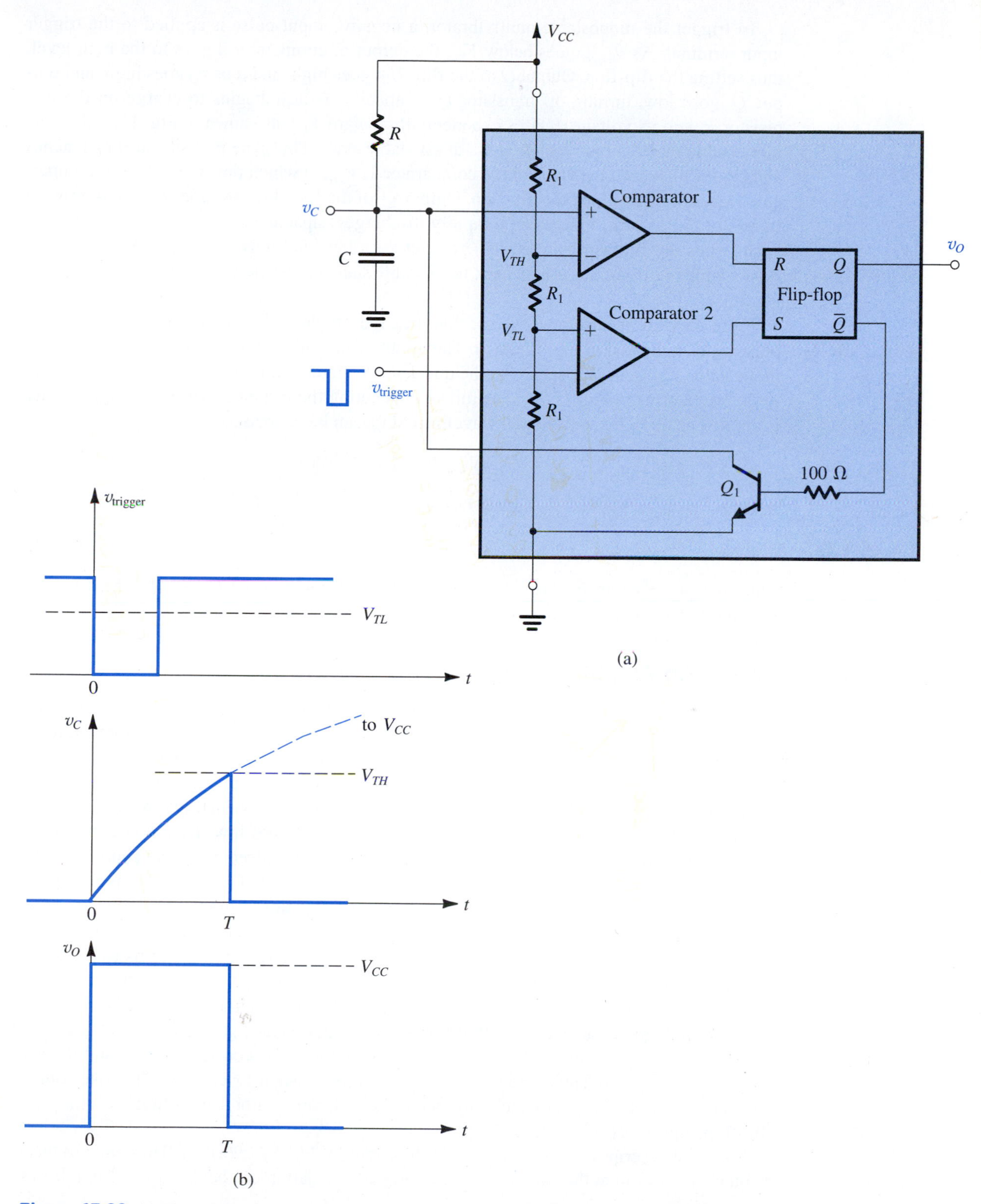

Figure 17.28 **(a)** The 555 timer connected to implement a monostable multivibrator. **(b)** Waveforms of the circuit in **(a)**.

To trigger the monostable multivibrator, a negative input pulse is applied to the trigger input terminal. As v_{trigger} goes below V_{TL}, the output of comparator 2 goes to the high level, thus setting the flip-flop. Output Q of the flip-flop goes high, and thus v_O goes high, and output $\overline{Q}$ goes low, turning off transistor Q_1. Capacitor C now begins to charge up through resistor R, and its voltage v_C rises exponentially toward V_{CC}, as shown in Fig. 17.28(b). The monostable multivibrator is now in its quasi-stable state. This state prevails until v_C reaches and begins to exceed the threshold of comparator 1, V_{TH}, at which time the output of comparator 1 goes high, resetting the flip-flop. Output $\overline{Q}$ of the flip-flop now goes high and turns on transistor Q_1. In turn, transistor Q_1 rapidly discharges capacitor C, causing v_C to go to 0 V. Also, when the flip-flop is reset, its Q output goes low, and thus v_O goes back to 0 V. The monostable multivibrator is now back in its stable state and is ready to receive a new triggering pulse.

From the description above we see that the monostable multivibrator produces an output pulse v_O as indicated in Fig. 17.28(b). The width of the pulse, T, is the time interval that the monostable multivibrator spends in the quasi-stable state; it can be determined by reference to the waveforms in Fig. 17.28(b) as follows: Denoting the instant at which the trigger pulse is applied as $t = 0$, the exponential waveform of v_C can be expressed as

$$v_C = V_{CC}(1 - e^{-t/CR}) \tag{17.38}$$

Substituting $v_C = V_{TH} = \frac{2}{3}V_{CC}$ at $t = T$ gives

$$T = CR \ln 3 \simeq 1.1CR \tag{17.39}$$

Thus the pulse width is determined by the external components C and R, which can be selected to have values as precise as desired.

17.7.3 An Astable Multivibrator Using the 555 IC

Figure 17.29(a) shows the circuit of an astable multivibrator employing a 555 IC, two external resistors, R_A and R_B, and an external capacitor C. To see how the circuit operates, refer to the waveforms depicted in Fig. 17.29(b). Assume that initially C is discharged and the flip-flop is set. Thus v_O is high and Q_1 is off. Capacitor C will charge up through the series combination of R_A and R_B, and the voltage across it, v_C, will rise exponentially toward V_{CC}. As v_C crosses the level equal to V_{TL}, the output of comparator 2 goes low. This, however, has no effect on the circuit operation, and the flip-flop remains set. Indeed, this state continues until v_C reaches and begins to exceed the threshold of comparator 1, V_{TH}. At this instant of time, the output of comparator 1 goes high and resets the flip-flop. Thus v_O goes low, $\overline{Q}$ goes high, and transistor Q_1 is turned on. The saturated transistor Q_1 causes a voltage of approximately zero volts to appear at the common node of R_A and R_B. Thus C begins to discharge through R_B and the collector of Q_1. The voltage v_C decreases exponentially with a time constant CR_B toward 0 V. When v_C reaches the threshold of comparator 2, V_{TL}, the output of comparator 2, goes high and sets the flip-flop. The output v_O then goes high, and $\overline{Q}$ goes low, turning off Q_1. Capacitor C begins to charge through the series equivalent of R_A and R_B, and its voltage rises exponentially toward V_{CC} with a time constant $C(R_A + R_B)$. This rise continues until v_C reaches V_{TH}, at which time the output of comparator 1 goes high, resetting the flip-flop, and the cycle continues.

From the description above we see that the circuit of Fig. 17.29(a) oscillates and produces a square waveform at the output. The frequency of oscillation can be determined as follows.

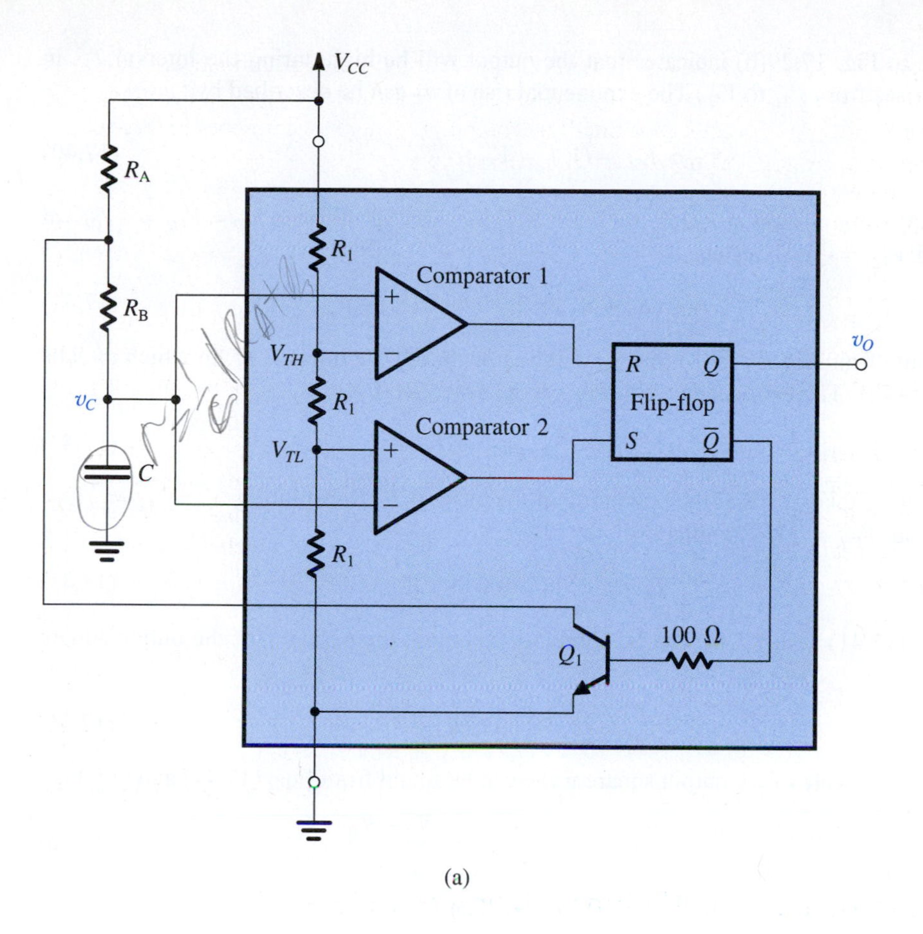

(a)

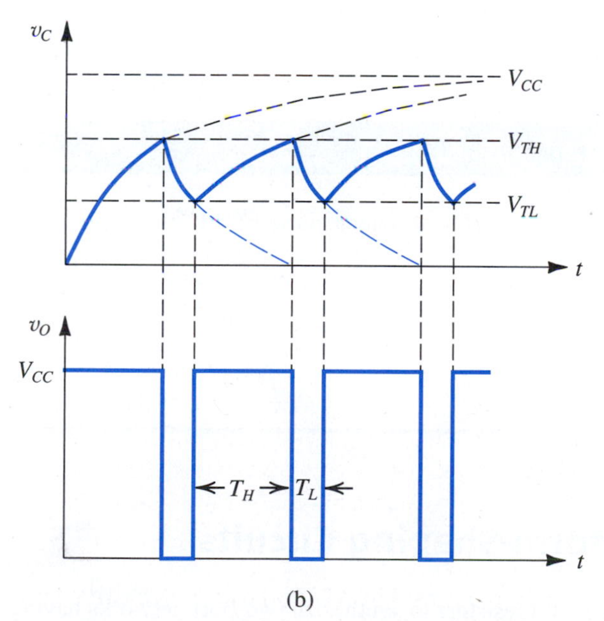

(b)

Figure 17.29 **(a)** The 555 timer connected to implement an astable multivibrator. **(b)** Waveforms of the circuit in **(a)**.

Reference to Fig. 17.29(b) indicates that the output will be high during the interval T_H, in which v_C rises from V_{TL} to V_{TH}. The exponential rise of v_C can be described by

$$v_C = V_{CC} - (V_{CC} - V_{TL})e^{-t/C(R_A+R_B)} \tag{17.40}$$

where $t = 0$ is the instant at which the interval T_H begins. Substituting $v_C = V_{TH} = \frac{2}{3}V_{CC}$ at $t = T_H$ and $V_{TL} = \frac{1}{3}V_{CC}$ results in

$$T_H = C(R_A + R_B)\ln 2 \simeq 0.69\, C(R_A + R_B) \tag{17.41}$$

We also note from Fig. 17.29(b) that v_O will be low during the interval T_L, in which v_C falls from V_{TH} to V_{TL}. The exponential fall of v_C can be described by

$$v_C = V_{TH}e^{-t/CR_B} \tag{17.42}$$

where we have taken $t = 0$ as the beginning of the interval T_L. Substituting $v_C = V_{TL} = \frac{1}{3}V_{CC}$ at $t = T_L$ and $V_{TH} = \frac{2}{3}V_{CC}$ results in

$$T_L = CR_B \ln 2 \simeq 0.69\, CR_B \tag{17.43}$$

Equations (17.41) and (17.43) can be combined to obtain the period T of the output square wave as

$$T = T_H + T_L = 0.69\, C(R_A + 2R_B) \tag{17.44}$$

Also, the **duty cycle** of the output square wave can be found from Eqs. (17.41) and (17.43):

$$\text{Duty cycle} \equiv \frac{T_H}{T_H + T_L} = \frac{R_A + R_B}{R_A + 2R_B} \tag{17.45}$$

Note that the duty cycle will always be greater than 0.5 (50%); it approaches 0.5 if R_A is selected to be much smaller than R_B (unfortunately, at the expense of supply current).

EXERCISES

17.20 Using a 10-nF capacitor C, find the value of R that yields an output pulse of 100 μs in the monostable circuit of Fig. 17.28(a).
Ans. 9.1 kΩ

D17.21 For the circuit in Fig. 17.29(a), with a 1-nF capacitor, find the values of R_A and R_B that result in an oscillation frequency of 100 kHz and a duty cycle of 75%.
Ans. 7.2 kΩ, 3.6 kΩ

17.8 Nonlinear Waveform-Shaping Circuits

Diodes or transistors can be combined with resistors to synthesize two-port networks having arbitrary nonlinear transfer characteristics. Such two-port networks can be employed in

waveform shaping—that is, changing the waveform of an input signal in a prescribed manner to produce a waveform of a desired shape at the output. In this section we illustrate this application by a concrete example: the **sine-wave shaper**. This is a circuit whose purpose is to change the waveform of an input triangular-wave signal to a sine wave. Though simple, the sine-wave shaper is a practical building block used extensively in function generators. This method of generating sine waves should be contrasted to that using linear oscillators (Sections 17.1–17.3). Although linear oscillators produce sine waves of high purity, they are not convenient at very low frequencies. Also, linear oscillators are in general more difficult to tune over wide frequency ranges. In the following we discuss two distinctly different techniques for designing sine-wave shapers.

17.8.1 The Breakpoint Method

In the breakpoint method the desired nonlinear transfer characteristic (in our case the sine function shown in Fig. 17.30) is implemented as a piecewise linear curve. Diodes are utilized as switches that turn on at the various breakpoints of the transfer characteristic, thus switching into the circuit additional resistors that cause the transfer characteristic to change slope.

Consider the circuit shown in Fig. 17.31(a). It consists of a chain of resistors connected across the entire symmetrical voltage supply $+V, -V$. The purpose of this voltage divider is to generate reference voltages that will serve to determine the breakpoints in the transfer characteristic. In our example these reference voltages are denoted $+V_2, +V_1, -V_1, -V_2$. Note that

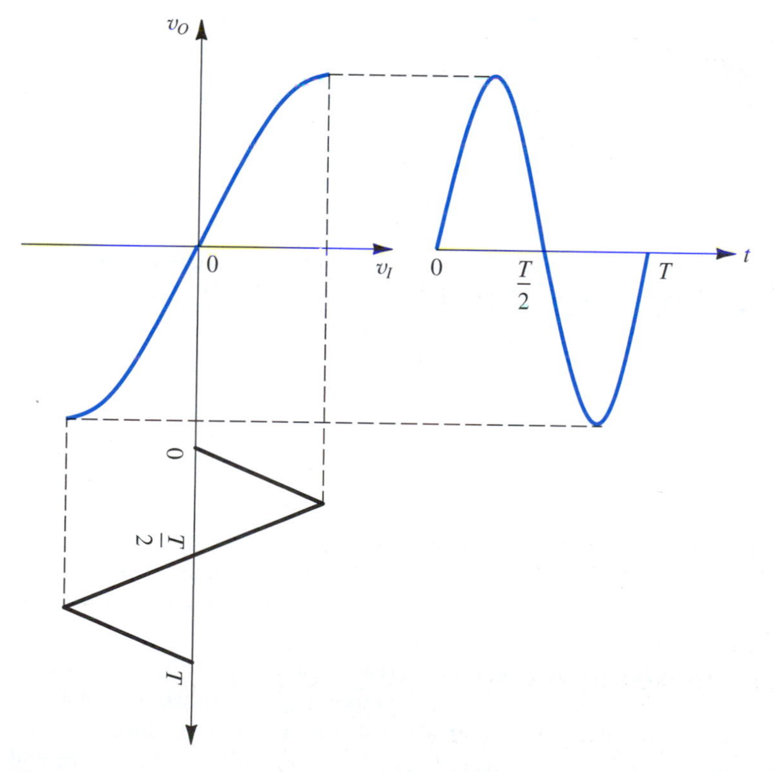

Figure 17.30 Using a nonlinear (sinusoidal) transfer characteristic to shape a triangular waveform into a sinusoid.

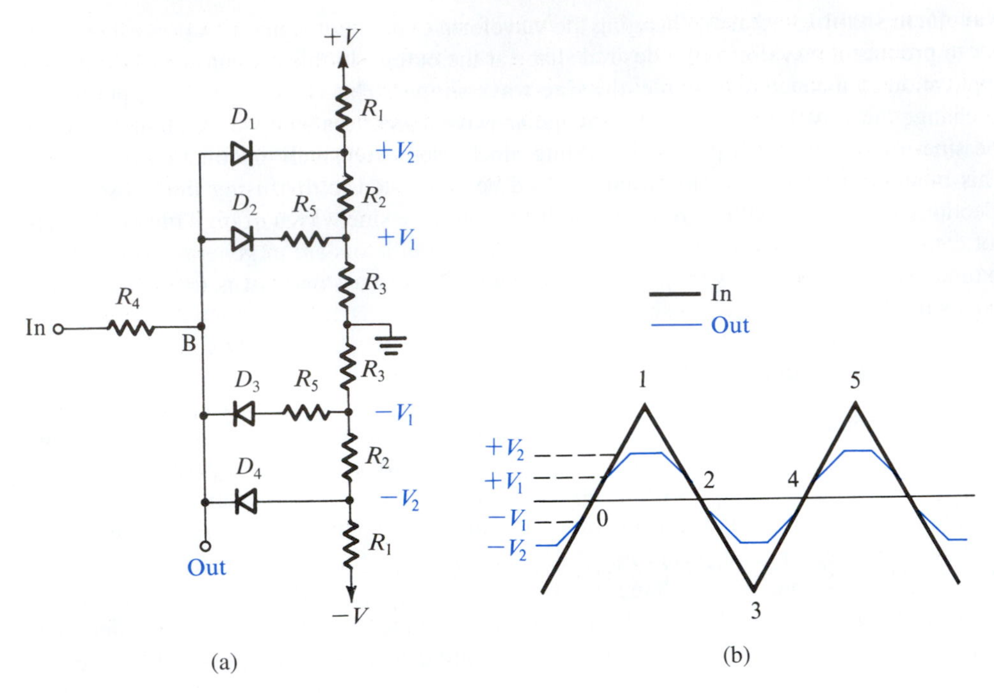

Figure 17.31 (a) A three-segment sine-wave shaper. **(b)** The input triangular waveform and the output approximately sinusoidal waveform.

the entire circuit is symmetrical, driven by a symmetrical triangular wave and generating a symmetrical sine-wave output. The circuit approximates each quarter-cycle of the sine wave by three straight-line segments; the breakpoints between these segments are determined by the reference voltages V_1 and V_2.

The circuit works as follows: Let the input be the triangular wave shown in Fig. 17.31(b), and consider first the quarter-cycle defined by the two points labeled 0 and 1. When the input signal is less in magnitude than V_1, none of the diodes conducts. Thus zero current flows through R_4, and the output voltage at B will be equal to the input voltage. But as the input rises to V_1 and above, D_2 (assumed ideal) begins to conduct. Assuming that the conducting D_2 behaves as a short circuit, we see that, for $v_I > V_1$,

$$v_O = V_1 + (v_I - V_1)\frac{R_5}{R_4 + R_5}$$

This implies that as the input continues to rise above V_1, the output follows, but with a reduced slope. This gives rise to the second segment in the output waveform, as shown in Fig. 17.31(b). Note that in developing the equation above we have assumed that the resistances in the voltage divider are low enough in value to cause the voltages V_1 and V_2 to be constant independent of the current coming from the input.

Next consider what happens as the voltage at point B reaches the second breakpoint determined by V_2. At this point, D_1 conducts, thus limiting the output v_O to V_2 (plus, of course, the voltage drop across D_1 if it is not assumed to be ideal). This gives rise to the third segment, which is flat, in the output waveform. The overall result is to "bend" the waveform and shape it into an approximation of the first quarter-cycle of a sine wave. Then, beyond the peak of the input triangular wave, as the input voltage decreases, the process unfolds, the output becoming progressively more like the input. Finally, when the input goes sufficiently negative, the process begins to repeat at $-V_1$ and $-V_2$ for the negative half-cycle.

Although the circuit is relatively simple, its performance is surprisingly good. A measure of goodness usually taken is to quantify the purity of the output sine wave by specifying the percentage **total harmonic distortion** (THD). This is the percentage ratio of the rms voltage of all harmonic components above the fundamental frequency (which is the frequency of the triangular wave) to the rms voltage of the fundamental (see also Chapter 11). Interestingly, one reason for the good performance of the diode shaper is the beneficial effects produced by the nonideal i–v characteristics of the diodes—that is, the exponential knee of the junction diode as it goes into forward conduction. The consequence is a relatively smooth transition from one line segment to the next.

Practical implementations of the breakpoint sine-wave shaper employ six to eight segments (compared with the three used in the example above). Also, transistors are usually employed to provide more versatility in the design, with the goal being increased precision and lower THD (see Grebene, 1984, pages 592–595).

17.8.2 The Nonlinear-Amplification Method

The other method we discuss for the conversion of a triangular wave into a sine wave is based on feeding the triangular wave to the input of an amplifier having a nonlinear transfer characteristic that approximates the sine function. One such amplifier circuit consists of a differential pair with a resistance connected between the two emitters, as shown in Fig. 17.32. With appropriate choice of the values of the bias current I and the resistance R, the differential amplifier can be made to have a transfer characteristic that closely approximates that shown in Fig. 17.30. Observe that for small v_I the transfer characteristic of the circuit of Fig. 17.32 is almost linear, as a sine waveform is near its zero crossings. At large values of v_I the nonlinear characteristics of the BJTs reduce the gain of the amplifier and cause the transfer characteristic to bend, approximating the sine wave as it approaches its peak. (More details on this circuit can be found in Grebene, 1984, pages 595–597.)

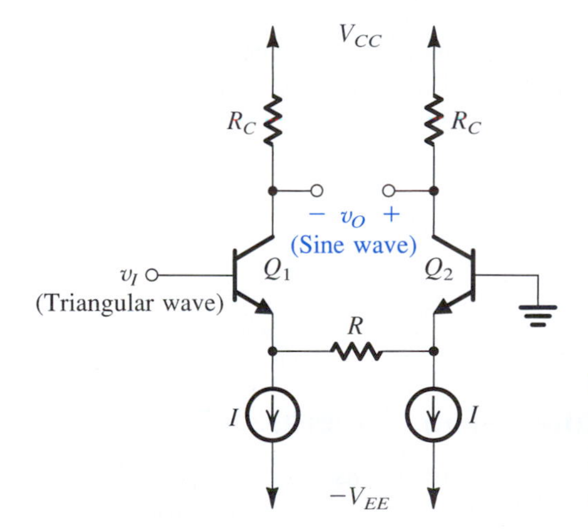

Figure 17.32 A differential pair with an emitter-degeneration resistance used to implement a triangular-wave to sine-wave converter. Operation of the circuit can be graphically described by Fig. 17.30.

EXERCISES

D17.22 The circuit in Fig. E17.22 is required to provide a three-segment approximation to the nonlinear i–v characteristic, $i = 0.1v^2$, where v is the voltage in volts and i is the current in milliamperes. Find the values of R_1, R_2, and R_3 such that the approximation is perfect at $v = 2$ V, 4 V, and 8 V. Calculate the error in current value at $v = 3$ V, 5 V, 7 V, and 10 V. Assume ideal diodes.

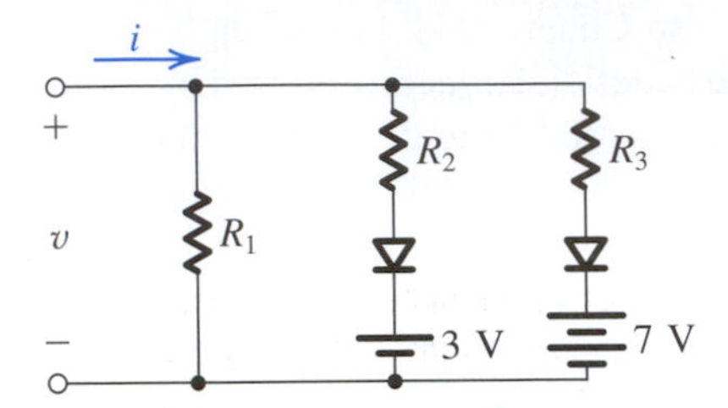

Figure E17.22

Ans. 5 kΩ, 1.25 kΩ, 1.25 kΩ; −0.3 mA, +0.1 mA, −0.3 mA, 0

17.23 A detailed analysis of the circuit in Fig. 17.32 shows that its optimum performance occurs when the values of I and R are selected so that $RI = 2.5V_T$, where V_T is the thermal voltage. For this design, the peak amplitude of the input triangular wave should be $6.6V_T$, and the corresponding sine wave across R has a peak value of $2.42V_T$. For $I = 0.25$ mA and $R_C = 10$ kΩ, find the peak amplitude of the sine-wave output v_O. Assume $\alpha \simeq 1$.

Ans. 4.84 V

17.9 Precision Rectifier Circuits

Rectifier circuits were studied in Chapter 4, where the emphasis was on their application in power-supply design. In such applications, the voltages being rectified are usually much greater than the diode voltage drop, rendering the exact value of the diode drop unimportant to the proper operation of the rectifier. Other applications exist, however, where this is not the case. For instance, in instrumentation applications, the signal to be rectified can be of a very small amplitude, say 0.1 V, making it impossible to employ the conventional rectifier circuits. Also, in instrumentation applications, the need arises for rectifier circuits with very precise transfer characteristics.

In this section we study circuits that combine diodes and op amps to implement a variety of rectifier circuits with precise characteristics. Precision rectifiers, which can be considered a special class of wave-shaping circuits, find application in the design of instrumentation systems. An introduction to precision rectifiers was presented in Chapter 4. This material, however, is repeated here for the reader's convenience.

17.9.1 Precision Half-Wave Rectifier—The "Superdiode"

Figure 17.33(a) shows a precision half-wave-rectifier circuit consisting of a diode placed in the negative-feedback path of an op amp, with R being the rectifier load resistance. The circuit works as follows: If v_I goes positive, the output voltage v_A of the op amp will go positive and the diode will conduct, thus establishing a closed feedback path between the op amp's output terminal and the negative input terminal. This negative-feedback path will

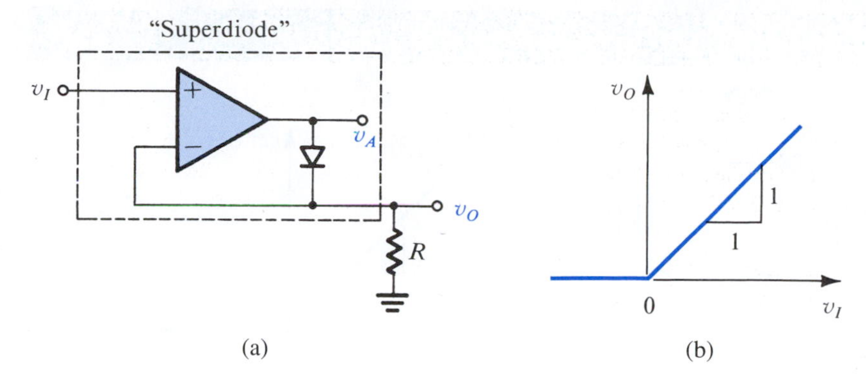

Figure 17.33 **(a)** The "superdiode" precision half-wave rectifier; **(b)** its almost ideal transfer characteristic. Note that when $v_I > 0$ and the diode conducts, the op amp supplies the load current, and the source is conveniently buffered, an added advantage.

cause a virtual short circuit to appear between the two input terminals of the op amp. Thus the voltage at the negative input terminal, which is also the output voltage v_O, will equal (to within a few millivolts) that at the positive input terminal, which is the input voltage v_I,

$$v_O = v_I \quad v_I \geq 0$$

Note that the offset voltage ($\simeq$ 0.5 V) exhibited in the simple half-wave-rectifier circuit is no longer present. For the op-amp circuit to start operation, v_I has to exceed only a negligibly small voltage equal to the diode drop divided by the op amp's open-loop gain. In other words, the straight-line transfer characteristic v_O–v_I almost passes through the origin. This makes this circuit suitable for applications involving very small signals.

Consider now the case when v_I goes negative. The op amp's output voltage v_A will tend to follow and go negative. This will reverse-bias the diode, and no current will flow through resistance R, causing v_O to remain equal to 0 V. Thus for $v_I < 0$, $v_O = 0$. Since in this case the diode is off, the op amp will be operating in an open-loop fashion and its output will be at the negative saturation level.

The transfer characteristic of this circuit will be that shown in Fig. 17.33(b), which is almost identical to the ideal characteristic of a half-wave rectifier. The nonideal diode characteristics have been almost completely masked by placing the diode in the negative-feedback path of an op amp. This is another dramatic application of negative feedback. The combination of diode and op amp, shown in the dashed box in Fig. 17.33(a), is appropriately referred to as a "superdiode."

As usual, though, not all is well. The circuit of Fig. 17.33 has some disadvantages: When v_I goes negative and $v_O = 0$, the entire magnitude of v_I appears between the two input terminals of the op amp. If this magnitude is greater than a few volts, the op amp may be damaged unless it is equipped with what is called "overvoltage protection" (a feature that most modern IC op amps have). Another disadvantage is that when v_I is negative, the op amp will be saturated. Although not harmful to the op amp, saturation should usually be avoided, since getting the op amp out of the saturation region and back into its linear region of operation requires some time. This time delay will obviously slow down circuit operation and limit the frequency of operation of the superdiode half-wave-rectifier circuit.

17.9.2 An Alternative Circuit

An alternative precision rectifier circuit that does not suffer from the disadvantages mentioned above is shown in Fig. 17.34. The circuit operates in the following manner: For

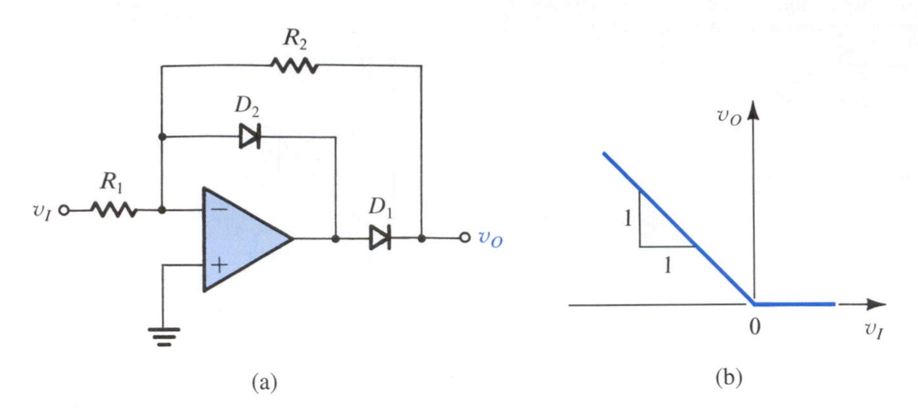

Figure 17.34 **(a)** An improved version of the precision half-wave rectifier: Diode D_2 is included to keep the feedback loop closed around the op amp during the off times of the rectifier diode D_1, thus preventing the op amp from saturating. **(b)** The transfer characteristic for $R_2 = R_1$.

positive v_I, diode D_2 conducts and closes the negative-feedback loop around the op amp. A virtual ground therefore will appear at the inverting input terminal, and the op amp's output will be *clamped* at one diode drop below ground. This negative voltage will keep diode D_1 off, and no current will flow in the feedback resistance R_2. It follows that the rectifier output voltage will be zero.

As v_I goes negative, the voltage at the inverting input terminal will tend to go negative, causing the voltage at the op amp's output terminal to go positive. This will cause D_2 to be reverse-biased and hence to be cut off. Diode D_1, however, will conduct through R_2, thus establishing a negative-feedback path around the op amp and forcing a virtual ground to appear at the inverting input terminal. The current through the feedback resistance R_2 will be equal to the current through the input resistance R_1. Thus for $R_1 = R_2$ the output voltage v_O will be

$$v_O = -v_I \quad v_I \leq 0$$

The transfer characteristic of the circuit is shown in Fig. 17.34(b). Note that unlike the situation for the circuit shown in Fig. 17.33, here the slope of the characteristic can be set to any desired value, including unity, by selecting appropriate values for R_1 and R_2.

As mentioned before, the major advantage of the improved half-wave-rectifier circuit is that the feedback loop around the op amp remains closed at all times. Hence the op amp remains in its linear operating region, avoiding the possibility of saturation and the associated time delay required to "get out" of saturation. Diode D_2 "catches" the op-amp output voltage as it goes negative and clamps it to one diode drop below ground; hence D_2 is called a "catching diode."

17.9.3 An Application: Measuring AC Voltages

As one of the many possible applications of the precision rectifier circuits discussed in this section, consider the basic ac voltmeter circuit shown in Fig. 17.35. The circuit consists of a half-wave rectifier—formed by op amp A_1, diodes D_1 and D_2, and resistors R_1 and R_2—and a first-order low-pass filter—formed by op amp A_2, resistors R_3 and R_4, and capacitor C. For an input sinusoid having a peak amplitude V_p the output v_1 of the rectifier will consist of a half sine wave having a peak amplitude of $V_p R_2/R_1$. It can be shown using Fourier series analysis that the waveform of v_1 has an average value of $(V_p/\pi)(R_2/R_1)$ in addition to

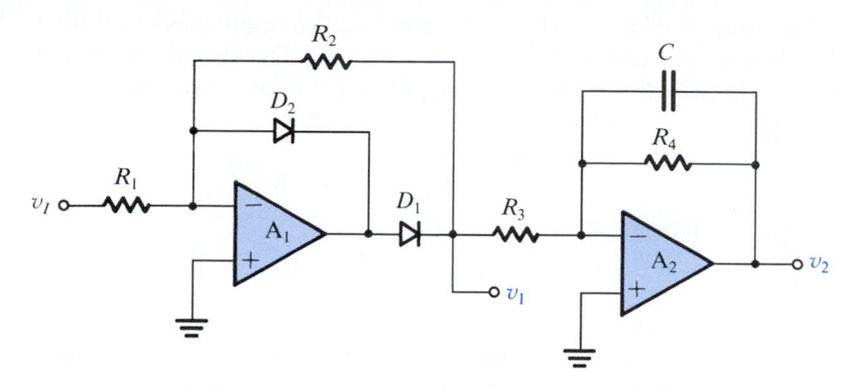

Figure 17.35 A simple ac voltmeter consisting of a precision half-wave rectifier followed by a first-order low-pass filter.

EXERCISES

17.24 Consider the operational rectifier or superdiode circuit of Fig. 17.33(a), with $R = 1$ kΩ. For $v_I = 10$ mV, 1 V, and −1 V, what are the voltages that result at the rectifier output and at the output of the op amp? Assume that the op amp is ideal and that its output saturates at ±12 V. The diode has a 0.7-V drop at 1-mA current, and the voltage drop changes by 0.1 V per decade of current change.
Ans. 10 mV, 0.51 V; 1 V, 1.7 V; 0 V, −12 V

17.25 If the diode in the circuit of Fig. 17.33(a) is reversed, what is the transfer characteristic v_O as a function of v_I?
Ans. $v_O = 0$ for $v_I \geq 0$; $v_O = v_I$ for $v_I \leq 0$

17.26 Consider the circuit in Fig. 17.34(a) with $R_1 = 1$ kΩ and $R_2 = 10$ kΩ. Find v_O and the voltage at the amplifier output for $v_I = +1$ V, −10 mV, and −1 V. Assume the op amp to be ideal with saturation voltages of ±12 V. The diodes have 0.7-V voltage drops at 1 mA, and the voltage drop changes by 0.1 V per decade of current change.
Ans. 0 V, −0.vm7 V; 0.1 V, 0.6 V; 10 V, 10.7 V

17.27 If the diodes in the circuit of Fig. 17.34(a) are reversed, what is the transfer characteristic v_O as a function of v_I?
Ans. $v_O = -(R_2/R_1)v_I$ for $v_I \geq 0$; $v_O = 0$ for $v_I \leq 0$

17.28 Find the transfer characteristic for the circuit in Fig. E17.28.

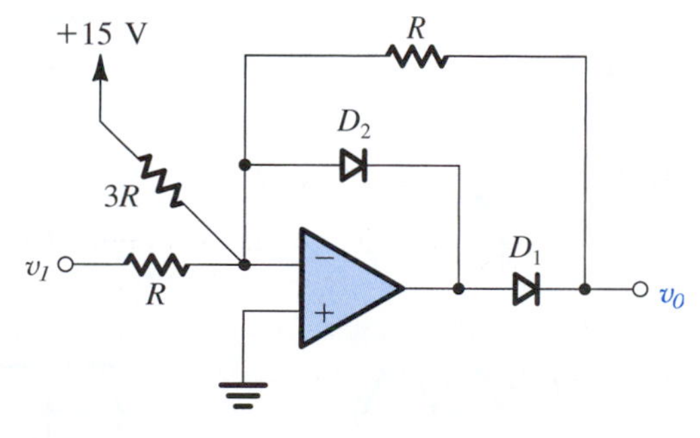

Figure E17.28

Ans. $v_O = 0$ for $v_I \geq -5$ V; $v_O = -v_I - 5$ for $v_I \leq -5$ V

harmonics of the frequency ω of the input signal. To reduce the amplitudes of all these harmonics to negligible levels, the corner frequency of the low-pass filter should be chosen to be much smaller than the lowest expected frequency $\omega_{\min}$ of the input sine wave. This leads to

$$\frac{1}{CR_4} \ll \omega_{\min}$$

Then the output voltage v_2 will be mostly dc, with a value

$$V_2 = -\frac{V_p}{\pi}\frac{R_2 R_4}{R_1 R_3}$$

where R_4/R_3 is the dc gain of the low-pass filter. Note that this voltmeter essentially measures the average value of the negative parts of the input signal but can be calibrated to provide rms readings for input sinusoids.

17.9.4 Precision Full-Wave Rectifier

We now derive a circuit for a precision full-wave rectifier. From Chapter 4 we know that full-wave rectification is achieved by inverting the negative halves of the input-signal waveform and applying the resulting signal to another diode rectifier. The outputs of the two rectifiers are then joined to a common load. Such an arrangement is depicted in Fig. 17.36, which also shows the waveforms at various nodes. Now replacing diode D_A with a superdiode, and replacing diode D_B and the inverting amplifier with the inverting precision half-wave rectifier of Fig. 17.34 but without the catching diode, we obtain the precision full-wave-rectifier circuit of Fig. 17.37(a).

To see how the circuit of Fig. 17.37(a) operates, consider first the case of positive input at A. The output of A_2 will go positive, turning D_2 on, which will conduct through R_L and thus close the feedback loop around A_2. A virtual short circuit will thus be established between the two input terminals of A_2, and the voltage at the negative-input terminal, which is the output voltage of the circuit, will become equal to the input. Thus no current will flow through R_1 and R_2, and the voltage at the inverting input of A_1 will be equal to the input and hence positive. Therefore the output terminal (F) of A_1 will go negative until A_1 saturates. This causes D_1 to be turned off.

Next consider what happens when A goes negative. The tendency for a negative voltage at the negative input of A_1 causes F to rise, making D_1 conduct to supply R_L and allowing the feedback loop around A_1 to be closed. Thus a virtual ground appears at the negative input of A_1, and the two equal resistances R_1 and R_2 force the voltage at C, which is the output

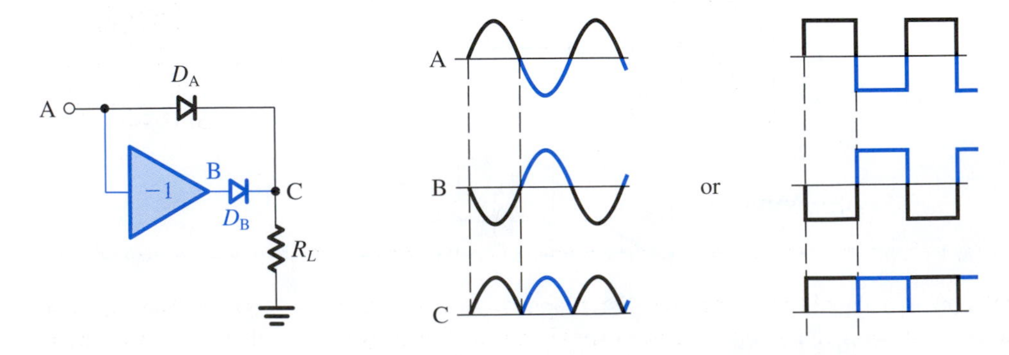

Figure 17.36 Principle of full-wave rectification.

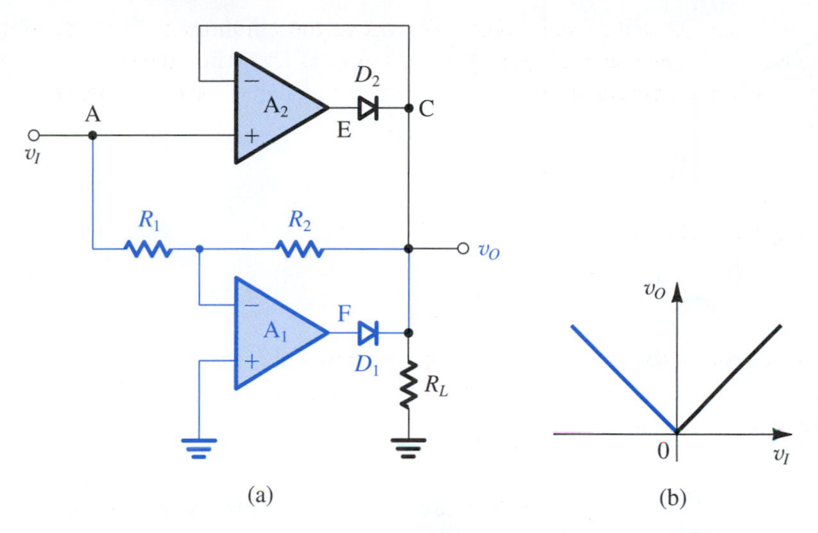

Figure 17.37 **(a)** Precision full-wave rectifier based on the conceptual circuit of Fig. 17.36. **(b)** Transfer characteristic of the circuit in **(a)**.

EXERCISES

17.29 In the full-wave rectifier circuit of Fig. 17.37(a), let $R_1 = R_2 = R_L = 10$ kΩ and assume the op amps to be ideal except for output saturation at ±12 V. When conducting a current of 1 mA, each diode exhibits a voltage drop of 0.7 V, and this voltage changes by 0.1 V per decade of current change. Find v_O, v_E, and v_F corresponding to $v_I = +0.1$ V, +1 V, +10 V, –0.1 V, and –10 V.
Ans. + 0.1 V, + 0.6 V, –12 V; +1 V, +1.6 V, –12 V; +10 V, +10.7 V, –12 V; + 0.1 V, –12 V, + 0.63 V; +1 V, –12 V, + 1.63 V; +10 V, –12 V, +10.73 V

D17.30 The block diagram shown in Fig. E17.30(a) gives another possible arrangement for implementing the absolute-value or full-wave-rectifier operation depicted symbolically in Fig. E17.30(b). The block diagram consists of two boxes: a half-wave rectifier, which can be implemented by the circuit in Fig. 17.34(a) after reversing both diodes, and a weighted inverting summer. Convince yourself that this block diagram does in fact realize the absolute-value operation. Then draw a complete circuit diagram, giving reasonable values for all resistors.

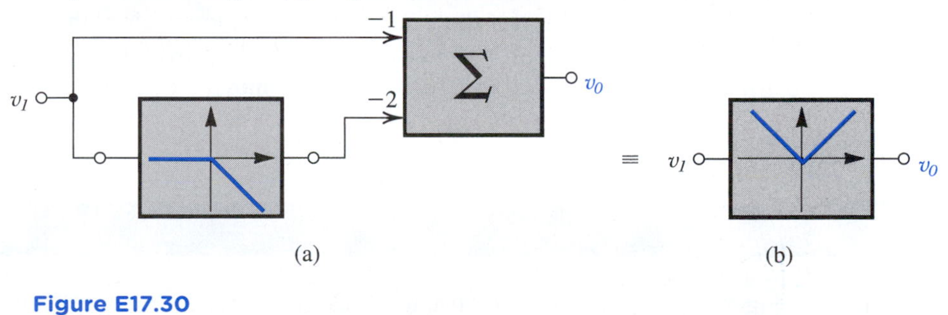

Figure E17.30

voltage, to be equal to the negative of the input voltage at A and thus positive. The combination of positive voltage at C and negative voltage at A causes the output of A_2 to saturate in the negative direction, thus keeping D_2 off.

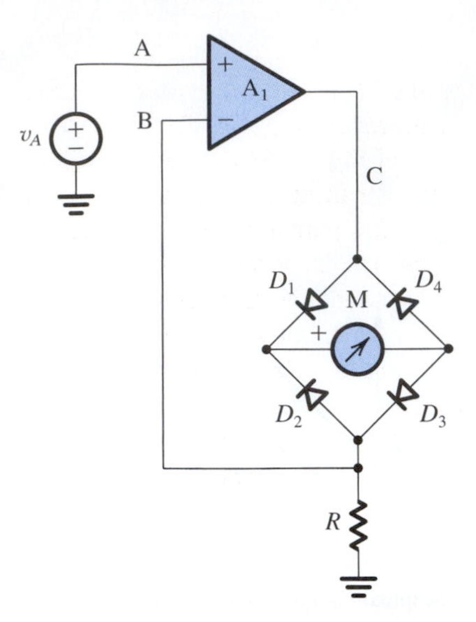

Figure 17.38 Use of the diode bridge in the design of an ac voltmeter.

The overall result is perfect full-wave rectification, as represented by the transfer characteristic in Fig. 17.37(b). This precision is, of course, a result of placing the diodes in op-amp feedback loops, thus masking their nonidealities. This circuit is one of many possible precision full-wave-rectifier or **absolute-value circuits**. Another related implementation of this function is examined in Exercise 17.30.

17.9.5 A Precision Bridge Rectifier for Instrumentation Applications

The bridge rectifier circuit studied in Chapter 4 can be combined with an op amp to provide useful precision circuits. One such arrangement is shown in Fig. 17.38. This circuit causes a current equal to $|v_A|/R$ to flow through the moving-coil meter M. Thus the meter provides a reading that is proportional to the average of the absolute value of the input voltage v_A. All the nonidealities of the meter and of the diodes are masked by placing the bridge circuit in the negative-feedback loop of the op amp. Observe that when v_A is positive, current flows from the op-amp output through D_1, M, D_3, and R. When v_A is negative, current flows into the op-amp output through R, D_2, M, and D_4. Thus the feedback loop remains closed for both polarities of v_A. The resulting virtual short circuit at the input terminals of the op amp causes a replica of v_A to appear across R. The circuit of Fig. 17.38 provides a relatively accurate high-input-impedance ac voltmeter using an inexpensive moving-coil meter.

EXERCISES

D17.31 In the circuit of Fig. 17.38, find the value of R that would cause the meter to provide a full-scale reading when the input voltage is a sine wave of 5 V rms. Let meter M have a 1-mA, 50-Ω movement (i.e., its resistance is 50 Ω, and it provides full-scale deflection when the average current through it is 1 mA). What are the approximate maximum and minimum voltages at the op amp's output? Assume that the diodes have constant 0.7-V drops when conducting.
Ans. 4.5 kΩ; +8.55 V; –8.55 V

17.9.6 Precision Peak Rectifiers

Including the diode of the peak rectifier studied in Chapter 4 inside the negative-feedback loop of an op amp, as shown in Fig. 17.39, results in a precision peak rectifier. The diode–op-amp combination will be recognized as the superdiode of Fig. 17.33(a). Operation of the circuit in Fig. 17.39 is quite straightforward. For v_I greater than the output voltage, the op amp will drive the diode on, thus closing the negative-feedback path and causing the op amp to act as a follower. The output voltage will therefore follow that of the input, with the op amp supplying the capacitor-charging current. This process continues until the input reaches its peak value. Beyond the positive peak, the op amp will see a negative voltage between its input terminals. Thus its output will go negative to the saturation level and the diode will turn off. Except for possible discharge through the load resistance, the capacitor will retain a voltage equal to the positive peak of the input. Inclusion of a load resistance is essential if the circuit is required to detect reductions in the magnitude of the positive peak.

17.9.7 A Buffered Precision Peak Detector

When the peak detector is required to hold the value of the peak for a long time, the capacitor should be buffered, as shown in the circuit of Fig. 17.40. Here op amp A_2, which should have high input impedance and low input bias current, is connected as a voltage follower. The remainder of the circuit is quite similar to the half-wave-rectifier circuit of Fig. 17.34. While diode D_1 is the essential diode for the peak-rectification operation, diode D_2 acts as a catching diode to prevent negative saturation, and the associated delays, of op amp A_1. During the holding state, follower A_2 supplies D_2 with a small current through R. The output of op amp A_1 will then be clamped at one diode drop below the input voltage. Now if the input v_I increases above the value stored on C, which is equal to the output voltage v_O, op amp A_1 sees

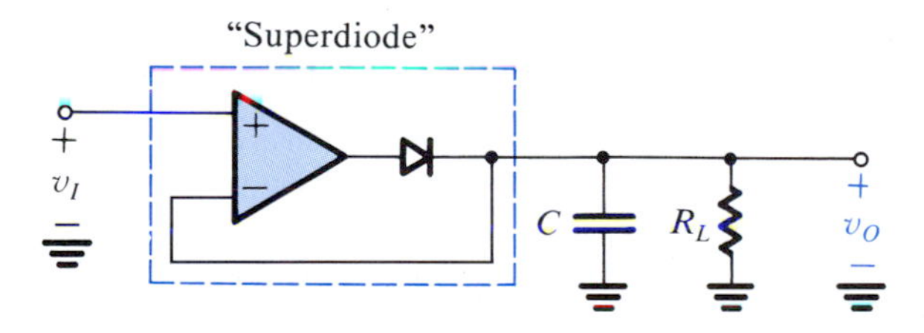

Figure 17.39 A precision peak rectifier obtained by placing the diode in the feedback loop of an op amp.

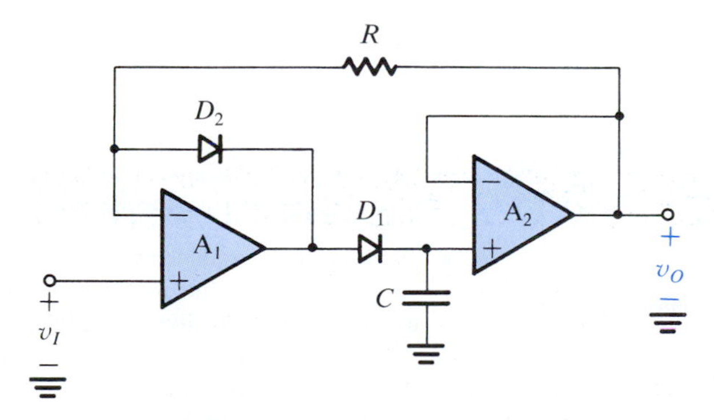

Figure 17.40 A buffered precision peak rectifier.

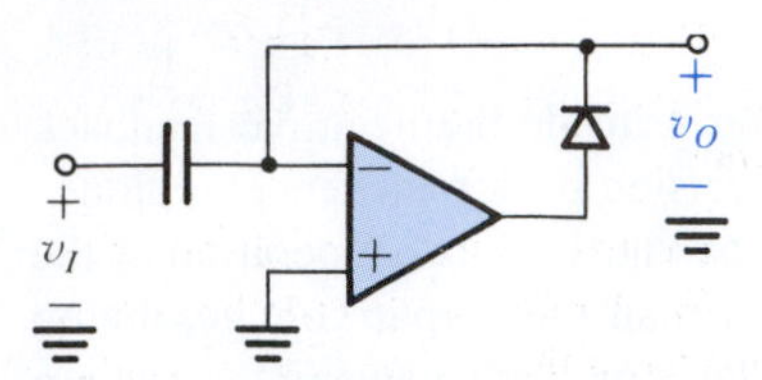

Figure 17.41 A precision clamping circuit.

a net positive input that drives its output toward the positive saturation level, turning off diode D_2. Diode D_1 is then turned on and capacitor C is charged to the new positive peak of the input, after which time the circuit returns to the holding state. Finally, note that this circuit has a low-impedance output.

17.9.8 A Precision Clamping Circuit

By replacing the diode in the clamping circuit studied in Chapter 4 with a "superdiode," the precision clamp of Fig. 17.41 is obtained. Operation of this circuit should be self-explanatory.

Summary

- There are two distinctly different types of signal generator: the linear oscillator, which utilizes some form of resonance, and the nonlinear oscillator or function generator, which employs a switching mechanism implemented with a multivibrator circuit.
- A linear oscillator can be realized by placing a frequency-selective network in the feedback path of an amplifier (an op amp or a transistor). The circuit will oscillate at the frequency at which the total phase shift around the loop is zero, provided that the magnitude of loop gain at this frequency is equal to, or greater than, unity.
- If in an oscillator the magnitude of loop gain is greater than unity, the amplitude will increase until a nonlinear amplitude-control mechanism is activated.
- The Wien-bridge oscillator, the phase-shift oscillator, the quadrature oscillator, and the active-filter-tuned oscillator are popular configurations for frequencies up to about 1 MHz. These circuits employ RC networks together with op amps or transistors. For higher frequencies, LC-tuned or crystal-tuned oscillators are utilized. A popular configuration is the Colpitts circuit.
- Crystal oscillators provide the highest possible frequency accuracy and stability.
- There are three types of multivibrator: bistable, monostable, and astable. Op-amp circuit implementations of multivibrators are useful in analog-circuit applications that require high precision.
- The bistable multivibrator has two stable states and can remain in either state indefinitely. It changes state when triggered. A comparator with hysteresis is bistable.
- A monostable multivibrator, also known as a one-shot, has one stable state, in which it can remain indefinitely. When triggered, it goes into a quasi-stable state in which it remains for a predetermined interval, thus generating, at its output, a pulse of known width.
- An astable multivibrator has no stable state. It oscillates between two quasi-stable states, remaining in each for a predetermined interval. It thus generates a periodic waveform at the output.
- A feedback loop consisting of an integrator and a bistable multivibrator can be used to generate triangular and square waveforms.
- The 555 timer, a commercially available IC, can be used with external resistors and a capacitor to implement high-quality monostable and astable multivibrators.
- A sine waveform can be generated by feeding a triangular waveform to a sine-wave shaper. A sine-wave shaper can be implemented either by using diodes (or transistors) and resistors, or by using an amplifier having a nonlinear transfer characteristic that approximates the sine function.
- Diodes can be combined with op amps to implement precision rectifier circuits in which negative feed-back serves to mask the nonidealities of the diode characteristics.

Problems involving design are marked with D throughout the text. As well, problems are marked with asterisks to describe their degree of difficulty. Difficult problems are marked with an asterisk (*); more difficult problems with two asterisks (**); and very challenging and/or time-consuming problems with three asterisks (***).

Section 17.1: Basic Principles of Sinusoidal Oscillators

***17.1** Consider a sinusoidal oscillator consisting of an amplifier having a frequency-independent gain A (where A is positive) and a second-order bandpass filter with a pole frequency ω_0, a pole Q denoted Q, and a center-frequency gain K.

(a) Find the frequency of oscillation, and the condition that A and K must satisfy for sustained oscillation.
(b) Derive an expression for $d\phi/d\omega$, evaluated at $\omega = \omega_0$.
(c) Use the result of (b) to find an expression for the per-unit change in frequency of oscillation resulting from a phase-angle change of $\Delta\phi$, in the amplifier transfer function.

$$Hint: \frac{d}{dx}(\tan^{-1} y) = \frac{1}{1+y^2}\frac{dy}{dx}$$

17.2 For the oscillator described in Problem 17.1, show that, independent of the value of A and K, the poles of the circuit lie at a radial distance of ω_0. Find the value of AK that results in poles appearing (a) on the $j\omega$ axis, and (b) in the right-half of the s plane, at a horizontal distance from the $j\omega$ axis of $\omega_0/(2Q)$.

D 17.3 Sketch a circuit for a sinusoidal oscillator formed by an ideal op amp connected in the noninverting configuration and a bandpass filter implemented by an RLC resonator (such as that in Fig. 16.18d). What should the amplifier gain be to obtain sustained oscillation? What is the frequency of oscillation? Find the percentage change in ω_0 resulting from a change of +1% in the value of (a) L, (b) C, and (c) R.

17.4 An oscillator is formed by loading a transconductance amplifier having a positive gain with a parallel RLC circuit and connecting the output directly to the input (thus applying positive feedback with a factor $\beta = 1$). Let the transconductance amplifier have an input resistance of 10 kΩ and an output resistance of 10 kΩ.. The LC resonator has $L = 10$ μH, $C = 1000$ pF, and $Q = 100$. For what value of transconductance G_m will the circuit oscillate? At what frequency?

17.5 In a particular oscillator characterized by the structure of Fig. 17.1, the frequency-selective network exhibits a loss of 20 dB and a phase shift of 180° at ω_0. What is the minimum gain and the phase shift that the amplifier must have for oscillation to begin?

D 17.6 Consider the circuit of Fig. 17.3(a) with R_f removed to realize the comparator function. Find suitable values for all resistors so that the comparator output levels are ±6 V and the slope of the limiting characteristic is 0.1. Use power-supply voltages of ±10 V and assume the voltage drop of a conducting diode to be 0.7 V.

D 17.7 Consider the circuit of Fig. 17.3(a) with R_f removed to realize the comparator function. Sketch the transfer characteristic. Show that by connecting a dc source V_B to the virtual ground of the op amp through a resistor R_B, the transfer characteristic is shifted along the v_I axis to the point $v_I = -(R_1/R_B)V_B$. Utilizing available ±15-V dc supplies for $\pm V$ and for V_B, find suitable component values so that the limiting levels are ±5 V and the comparator threshold is at $v_I = +5$ V. Neglect the diode voltage drop (i.e., assume that $V_D = 0$). The input resistance of the comparator is to be 100 kΩ, and the slope in the limiting regions is to be ≤0.05 V/V. Use standard 5% resistors (see Appendix H).

17.8 Denoting the zener voltages of Z_1 and Z_2 by V_{Z1} and V_{Z2} and assuming that in the forward direction the voltage drop is approximately 0.7 V, sketch and clearly label the transfer characteristics v_O–v_I of the circuits in Fig. P17.8. Assume the op amps to be ideal.

Section 17.2: Op Amp–RC Oscillator Circuits

17.9 For the Wien-bridge oscillator circuit in Fig. 17.4, show that the transfer function of the feedback network $[V_a(s)/V_o(s)]$ is that of a bandpass filter. Find ω_0 and Q of the poles, and find the center-frequency gain.

17.10 For the Wien-bridge oscillator of Fig. 17.4, let the closed-loop amplifier (formed by the op amp and the resistors R_1 and R_2) exhibit a phase shift of −0.1 rad in the neighborhood of $\omega = 1/CR$. Find the frequency at which oscillations can occur in this case in terms of CR. (*Hint:* Use Eq. 17.11.)

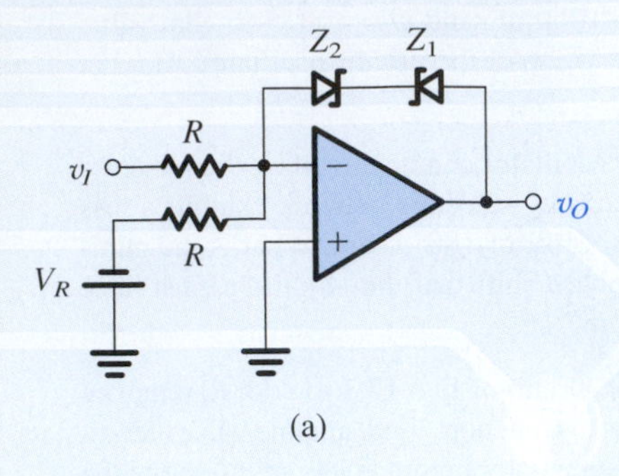

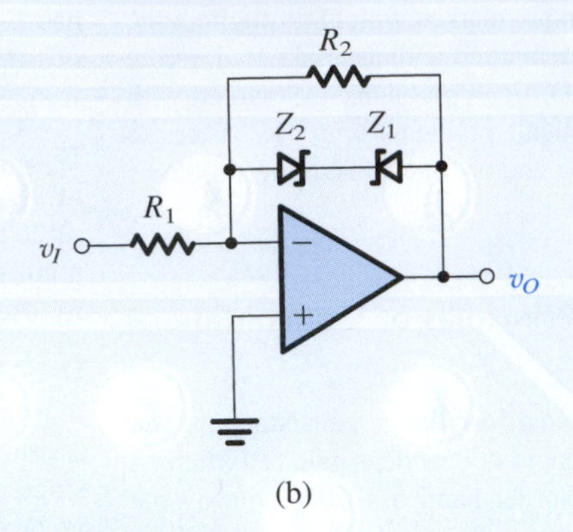

Figure P17.8

17.11 For the Wien-bridge oscillator of Fig. 17.4, use the expression for loop gain in Eq. (17.10) to find the poles of the closed-loop system. Give the expression for the pole Q, and use it to show that to locate the poles in the right half of the s plane, R_2/R_1 must be selected to be greater than 2.

D*17.12 Reconsider Exercise 17.3 with R_3 and R_6 increased to reduce the output voltage. What values are required for a peak-to-peak output of 10 V? What results if R_3 and R_6 are open-circuited?

17.13 For the circuit in Fig. P17.13, find $L(s)$, $L(j\omega)$, the frequency for zero loop phase, and R_2/R_1 for oscillation.

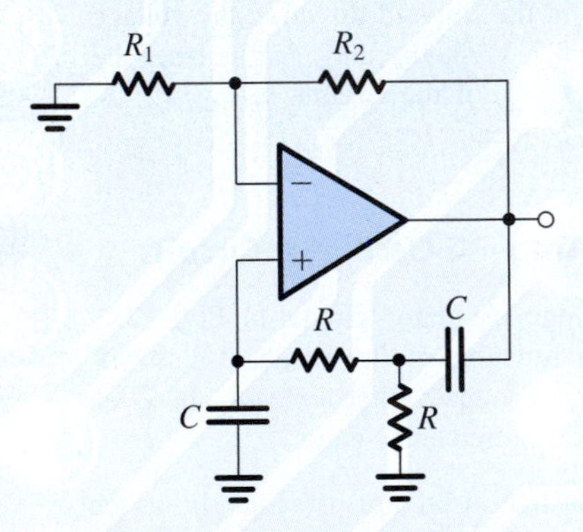

Figure P17.13

17.14 Repeat Problem 17.13 for the circuit in Fig. P17.14.

***17.15** Consider the circuit of Fig. 17.6 with the 50-kΩ potentiometer replaced by two fixed resistors: 10 kΩ between the op amp's negative input and ground, and 18 kΩ. Modeling each diode as a 0.65-V battery in series with a 100-Ω resistance, find the peak-to-peak amplitude of the output sinusoid.

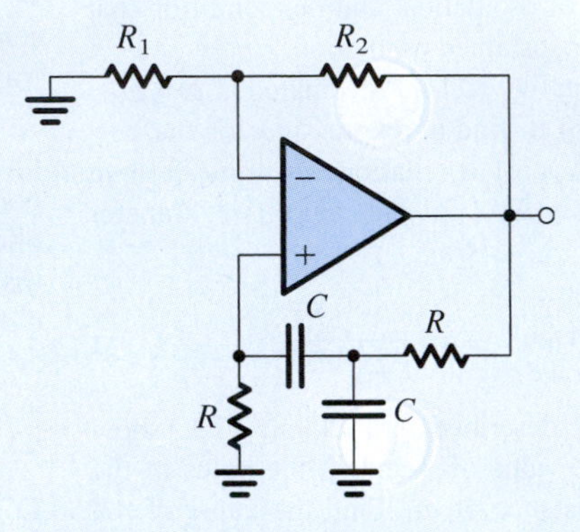

Figure P17.14

D17.16** Redesign the circuit of Fig. 17.6 for operation at 10 kHz using the same values of resistance. If at 10 kHz the op amp provides an excess phase shift (lag) of 5.7°, what will be the frequency of oscillation? (Assume that the phase shift introduced by the op amp remains constant for frequencies around 10 kHz.) To restore operation to 10 kHz, what change must be made in the shunt resistor of the Wien bridge? Also, to what value must R_2/R_1 be changed?

***17.17** For the circuit of Fig. 17.8, connect an additional $R = 10$ kΩ resistor in series with the rightmost capacitor C. For this modification (and ignoring the amplitude stabilization circuitry) find the loop gain $A\beta$ by breaking the circuit at node X. Find R_f for oscillation to begin, and find f_0.

D 17.18 For the circuit in Fig. P17.18, break the loop at node X and find the loop gain (working backward for simplicity to find V_x in terms of V_o). For $R = 10$ kΩ, find C and R_f to obtain sinusoidal oscillations at 10 kHz.

***17.19** Consider the quadrature-oscillator circuit of Fig. 17.9 without the limiter. Let the resistance R_f be equal to

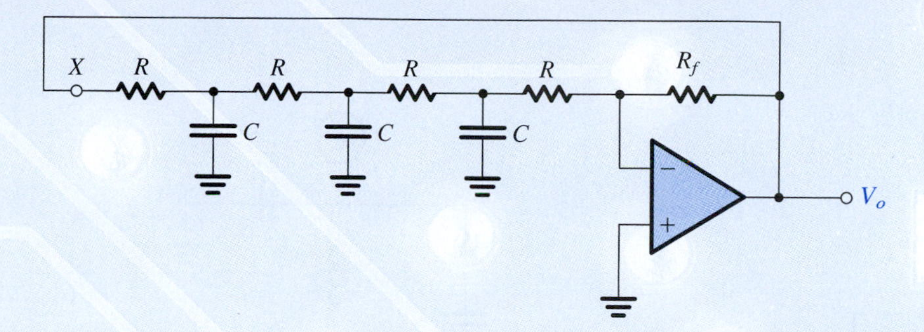

Figure P17.18

$2R/(1+\Delta)$, where $\Delta \ll 1$. Show that the poles of the characteristic equation are in the right-half s plane and given by $s \simeq (1/CR)[(\Delta/4) \pm j]$.

***17.20** Assuming that the diode-clipped waveform in Exercise 17.7 is nearly an ideal square wave and that the resonator Q is 20, provide an estimate of the distortion in the output sine wave by calculating the magnitude (relative to the fundamental) of

(a) the second harmonic
(b) the third harmonic
(c) the fifth harmonic
(d) the rms of harmonics to the tenth

Note that a square wave of amplitude V and frequency ω is represented by the series

$$\frac{4V}{\pi}\left(\sin \omega t + \frac{1}{3}\sin 3\omega t + \frac{1}{5}\sin 5\omega t + \frac{1}{7}\sin 7\omega t + \cdots\right)$$

Section 17.3: LC and Crystal Oscillators

****17.21** Figure P17.21 shows four oscillator circuits of the Colpitts type, complete with bias detail. For each circuit, derive an equation governing circuit operation and find the frequency of oscillation and the gain condition that ensures that oscillations start.

****17.22** Consider the oscillator circuit in Fig. P17.22, and assume for simplicity that $\beta = \infty$.

(a) Find the frequency of oscillation and the minimum value of R_C (in terms of the bias current I) for oscillation to start.
(b) If R_C is selected equal to $(1/I)$ kΩ, where I is in milliamperes, convince yourself that oscillations will start. If oscillations grow to the point that V_o is large enough to turn the BJTs on and off, show that the voltage at the collector of Q_2 will be a square wave of 1 V peak to peak. Estimate the peak-to-peak amplitude of the output sine wave V_o.

17.23 Consider the Pierce crystal oscillator of Fig. 17.16 with the crystal as specified in Exercise 17.10. Let C_1 be variable in the range 1 pF to 10 pF, and let C_2 be fixed at 10 pF. Find the range over which the oscillation frequency can be tuned. (*Hint:* Use the result in the statement leading to the expression in Eq. 17.27.)

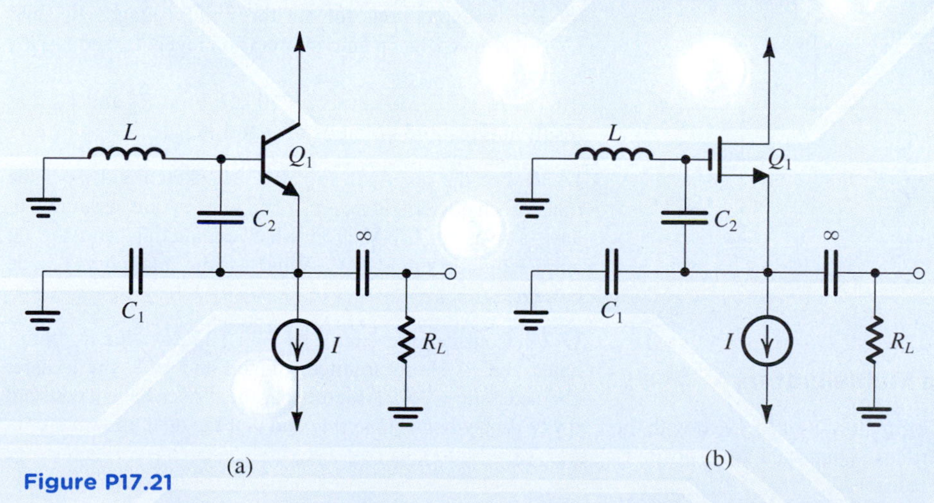

Figure P17.21

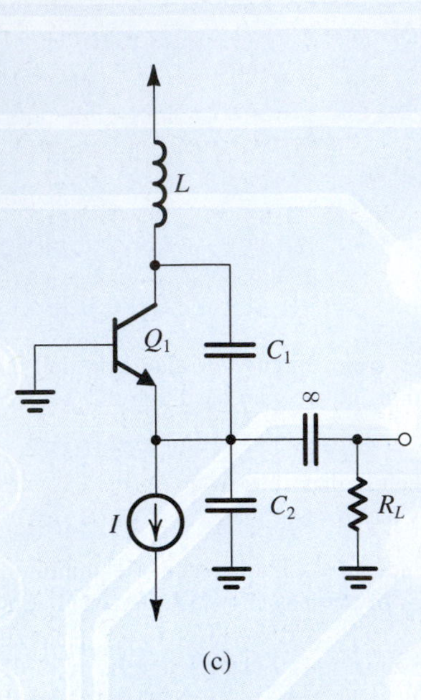

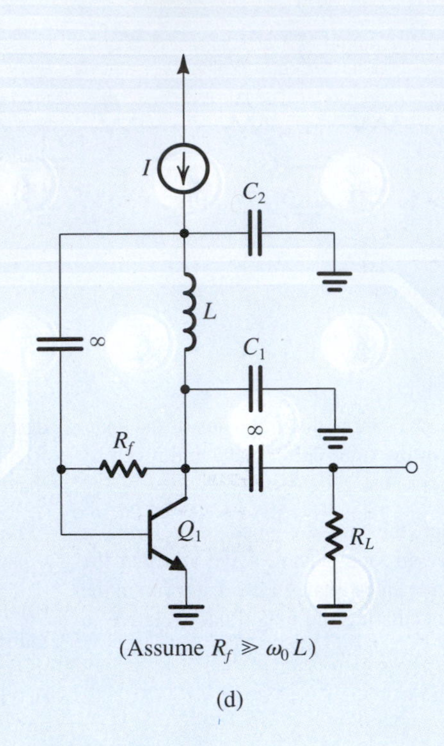

Figure P17.21 (*Continued*)

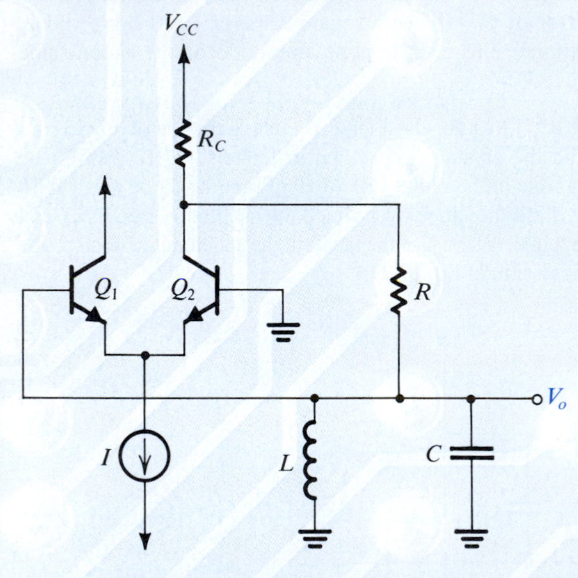

Figure P17.22

Section 17.4: Bistable Multivibrators

17.24 Consider the bistable circuit of Fig. 17.19(a) with the op amp's positive-input terminal connected to a positive-voltage source V through a resistor R_3.

(a) Derive expressions for the threshold voltages V_{TL} and V_{TH} in terms of the op amp's saturation levels L_+ and L_-, R_1, R_2, R_3, and V.

(b) Let $L_+ = -L_- = 13$ V, $V = 15$ V, and $R_1 = 10$ kΩ. Find the values of R_2 and R_3 that result in $V_{TL} = +4.9$ V and $V_{TH} = +5.1$ V.

17.25 Consider the bistable circuit of Fig. 17.20(a) with the op amp's negative-input terminal disconnected from ground and connected to a reference voltage V_R.

(a) Derive expressions for the threshold voltages V_{TL} and V_{TH} in terms of the op amp's saturation levels L_+ and L_-, R_1, R_2, and V_R.

(b) Let $L_+ = -L_- = V$ and $R_1 = 10$ kΩ. Find R_2 and V_R that result in threshold voltages of 0 and $V/10$.

17.26 For the circuit in Fig. P17.26, sketch and label the transfer characteristic v_O–v_I. The diodes are assumed to have a constant 0.7-V drop when conducting, and the op amp saturates at ±12 V. What is the maximum diode current?

17.27 Consider the circuit of Fig. P17.26 with R_1 eliminated and R_2 short-circuited. Sketch and label the transfer characteristic v_O–v_I. Assume that the diodes have a constant 0.7-V drop when conducting and that the op amp saturates at ±12 V.

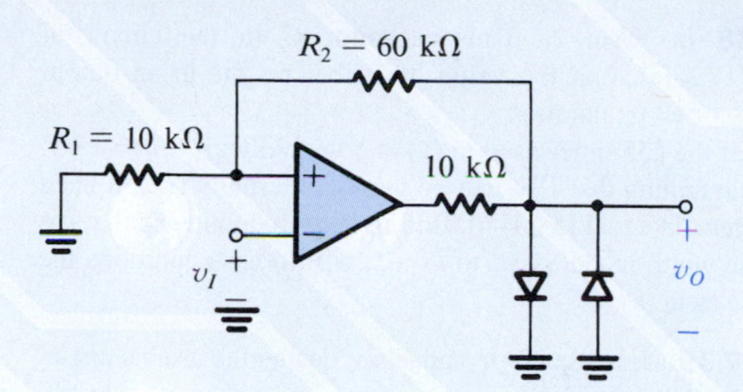

Figure P17.26

*17.28 Consider a bistable circuit having a noninverting transfer characteristic with $L_+ = -L_- = 12$ V, $V_{TL} = -1$ V, and $V_{TH} = +1$ V.

(a) For a 0.5-V-amplitude sine-wave input having zero average, what is the output?
(b) Describe the output if a sinusoid of frequency f and amplitude of 1.1 V is applied at the input. By how much can the average of this sinusoidal input shift before the output becomes a constant value?

D 17.29 Design the circuit of Fig. 17.23(a) to realize a transfer characteristic with ±7.5-V output levels and ±7.5-V threshold values. Design so that when $v_I = 0$ V a current of 0.1 mA flows in the feedback resistor and a current of 1 mA flows through the zener diodes. Assume that the output saturation levels of the op amp are ±12 V. Specify the voltages of the zener diodes and give the values of all resistors.

Section 17.5: Generation of Square and Triangular Waveforms Using Astable Multivibrators

17.30 Find the frequency of oscillation of the circuit in Fig. 17.24(b) for the case $R_1 = 10$ kΩ, $R_2 = 16$ kΩ, $C = 10$ nF, and $R = 62$ kΩ.

D 17.31 Augment the astable multivibrator circuit of Fig. 17.24(b) with an output limiter of the type shown in Fig. 17.23(b). Design the circuit to obtain an output square wave with 5-V amplitude and 1-kHz frequency using a 10-nF capacitor C. Use $\beta = 0.462$, and design for a current in the resistive divider approximately equal to the average current in the RC network over a half-cycle. Assuming ±13-V op-amp saturation voltages, arrange for the zener to operate at a current of 1 mA.

D 17.32 Using the scheme of Fig. 17.25, design a circuit that provides square waves of 10 V peak to peak and triangular waves of 10 V peak to peak. The frequency is to be 1 kHz. Implement the bistable circuit with the circuit of Fig. 17.23(b). Use a 0.01-μF capacitor and specify the values of all resistors and the required zener voltage. Design for a minimum zener current of 1 mA and for a maximum current in the resistive divider of 0.2 mA. Assume that the output saturation levels of the op amps are ±13 V.

D*17.33 The circuit of Fig. P17.33 consists of an inverting bistable multivibrator with an output limiter and a noninverting integrator. Using equal values for all resistors except R_7 and a 0.5-nF capacitor, design the circuit to obtain a square wave at the output of the bistable multivibrator of 15-V peak-to-peak amplitude and 10-kHz frequency. Sketch and label the waveform at the integrator output. Assuming ±13-V op-amp saturation levels, design for a minimum zener current of 1 mA. Specify the zener voltage required, and give the values of all resistors.

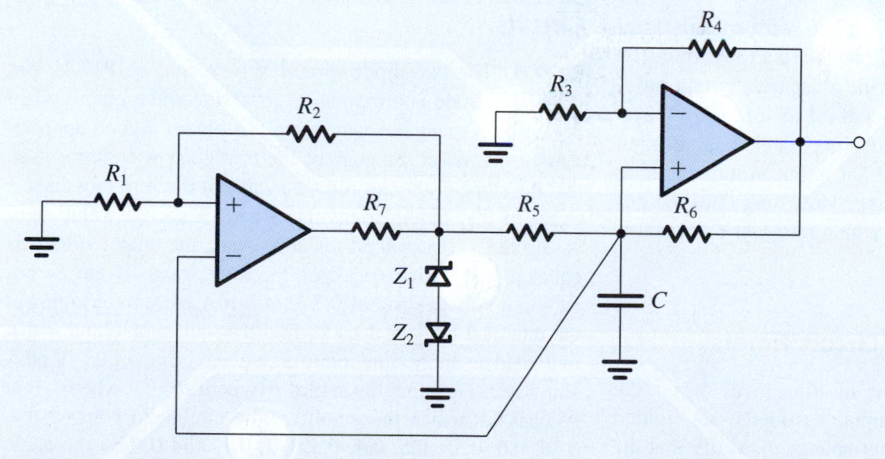

Figure P17.33

Section 17.6: Generation of a Standardized Pulse—The Monostable Multivibrator

***17.34** Figure P17.34 shows a monostable multivibrator circuit. In the stable state, $v_O = L_+$, $v_A = 0$, and $v_B = -V_{ref}$. The circuit can be triggered by applying a positive input pulse of height greater than V_{ref}. For normal operation, $C_1R_1 \ll CR$. Show the resulting waveforms of v_O and v_A. Also, show that the pulse generated at the output will have a width T given by

$$T = CR \ln\left(\frac{L_+ - L_-}{V_{ref}}\right)$$

Note that this circuit has the interesting property that the pulse width can be controlled by changing V_{ref}.

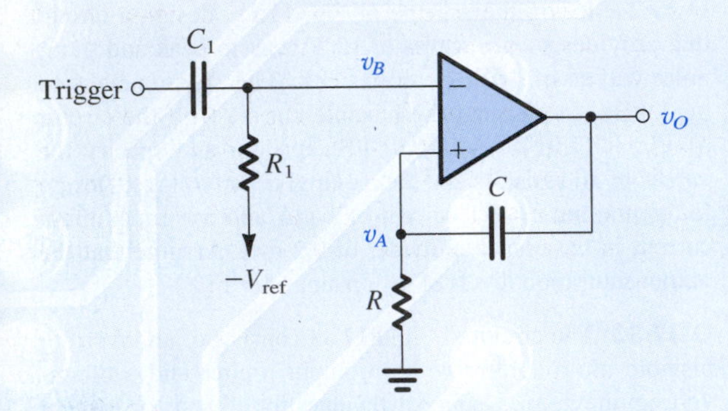

Figure P17.34

17.35 For the monostable circuit considered in Exercise 17.19, calculate the recovery time.

D*17.36 Using the circuit of Fig. 17.26, with a nearly ideal op amp for which the saturation levels are ±13 V, design a monostable multivibrator to provide a negative output pulse of 100-μs duration. Use capacitors of 0.1 nF and 1 nF. Wherever possible, choose resistors of 100 kΩ in your design. Diodes have a drop of 0.7 V. What is the minimum input step size that will ensure triggering? How long does the circuit take to recover to a state in which retriggering is possible with a normal output?

Section 17.7: Integrated-Circuit Timers

17.37 Consider the 555 circuit of Fig. 17.27 when the Threshold and the Trigger input terminals are joined together and connected to an input voltage v_I. Verify that the transfer characteristic v_O–v_I is that of an inverting bistable circuit with thresholds $V_{TL} = \frac{1}{3}V_{CC}$ and $V_{TH} = \frac{2}{3}V_{CC}$ and output levels of 0 and V_{CC}.

17.38 (a) Using a 1-nF capacitor C in the circuit of Fig. 17.28(a), find the value of R that results in an output pulse of 10-μs duration.
(b) If the 555 timer used in (a) is powered with $V_{CC} = 15$ V, and assuming that V_{TH} can be varied externally (i.e., it need not remain equal to $\frac{2}{3}V_{CC}$), find its required value so that the pulse width is increased to 20 μs, with other conditions the same as in (a).

D 17.39 Using a 680-pF capacitor, design the astable circuit of Fig. 17.29(a) to obtain a square wave with a 50-kHz frequency and a 75% duty cycle. Specify the values of R_A and R_B.

***17.40** The node in the 555 timer at which the voltage is V_{TH} (i.e., the inverting input terminal of comparator 1) is usually connected to an external terminal. This allows the user to change V_{TH} externally (i.e., V_{TH} no longer remains at $\frac{2}{3}V_{CC}$). Note, however, that whatever the value of V_{TH} becomes, V_{TL} always remains $\frac{1}{2}V_{TH}$.

(a) For the astable circuit of Fig. 17.29, rederive the expressions for T_H and T_L, expressing them in terms of V_{TH} and V_{TL}.
(b) For the case $C = 1$ nF, $R_A = 7.2$ kΩ, $R_B = 3.6$ kΩ, and $V_{CC} = 5$ V, find the frequency of oscillation and the duty cycle of the resulting square wave when no external voltage is applied to the terminal V_{TH}.
(c) For the design in (b), let a sine-wave signal of a much lower frequency than that found in (b) and of 1-V peak amplitude be capacitively coupled to the circuit node V_{TH}. This signal will cause V_{TH} to change around its quiescent value of $\frac{2}{3}V_{CC}$, and thus T_H will change correspondingly—a modulation process. Find T_H, and find the frequency of oscillation and the duty cycle at the two extreme values of V_{TH}.

Section 17.8: Nonlinear Waveform-Shaping Circuits

D*17.41 The two-diode circuit shown in Fig. P17.41 can provide a crude approximation to a sine-wave output when driven by a triangular waveform. To obtain a good approximation, we select the peak of the triangular waveform, V, so that the slope of the desired sine wave at the zero crossings is equal to that of the triangular wave. Also, the value of R is selected so that when v_I is at its peak, the output voltage is equal to the desired peak of the sine wave. If the diodes exhibit a voltage drop of 0.7 V at 1-mA current, changing at the rate of 0.1 V per decade, find the values of V and R that will yield an approximation to a sine waveform of 0.7-V peak amplitude. Then find the angles θ (where $\theta = 90°$ when v_I is at its peak) at which the output of the circuit, in volts, is 0.7, 0.65, 0.6, 0.55, 0.5, 0.4, 0.3, 0.2, 0.1, and 0. Use the angle values obtained to determine the values of the exact sine wave (i.e., 0.7 sin θ), and thus find the percentage error of this circuit as a sine shaper. Provide your results in tabular form.

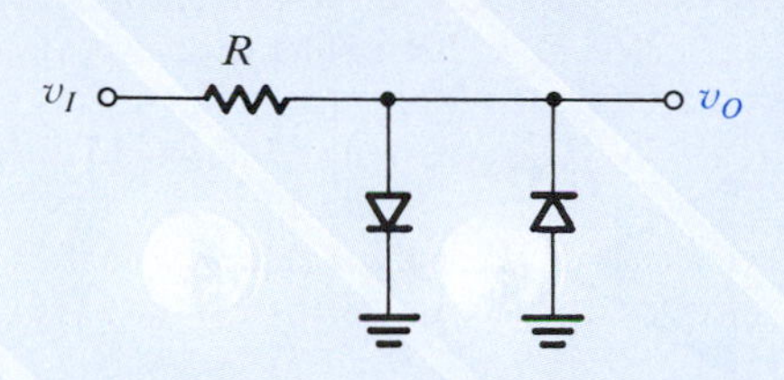

Figure P17.41

D 17.42 Design a two-segment sine-wave shaper using a 10-kΩ-input resistor, two diodes, and two clamping voltages. The circuit, fed by a 10-V peak-to-peak triangular wave, should limit the amplitude of the output signal via a 0.7-V diode to a value corresponding to that of a sine wave whose zero-crossing slope matches that of the triangle. What are the clamping voltages you have chosen?

17.43 Show that the output voltage of the circuit in Fig. P17.43 is given by

$$v_O = -V_T \ln\left(\frac{v_I}{I_S R}\right), \quad v_I > 0$$

where I_S is the saturation current of the diode and V_T is the thermal voltage. Since the output voltage is proportional to the logarithm of the input voltage, the circuit is known as a **logarithmic amplifier**. Such amplifiers find application in situations where it is desired to compress the signal range.

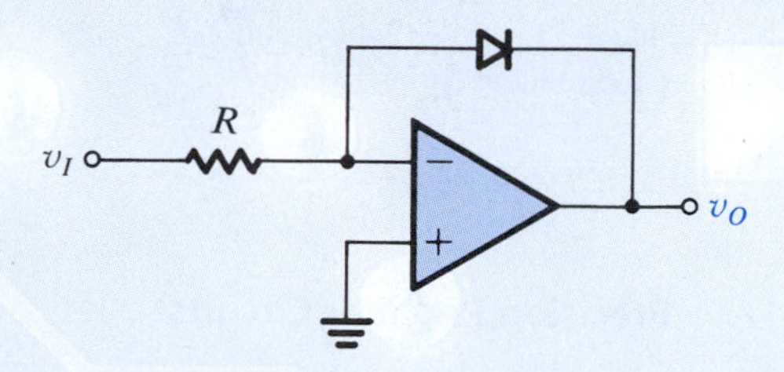

Figure P17.43

17.44 Verify that the circuit in Fig. P17.44 implements the transfer characteristic $v_O = v_1 v_2$ for $v_1, v_2 > 0$. Such a circuit is known as an analog multiplier. Check the circuit's performance for various combinations of input voltage of values, say, 0.5 V, 1 V, 2 V, and 3 V. Assume all diodes to be identical, with 700-mV drop at 1-mA current. Note that a *squarer* can easily be produced using a single input (e.g., v_1) connected via a 0.5-kΩ resistor (rather than the 1-kΩ resistor shown).

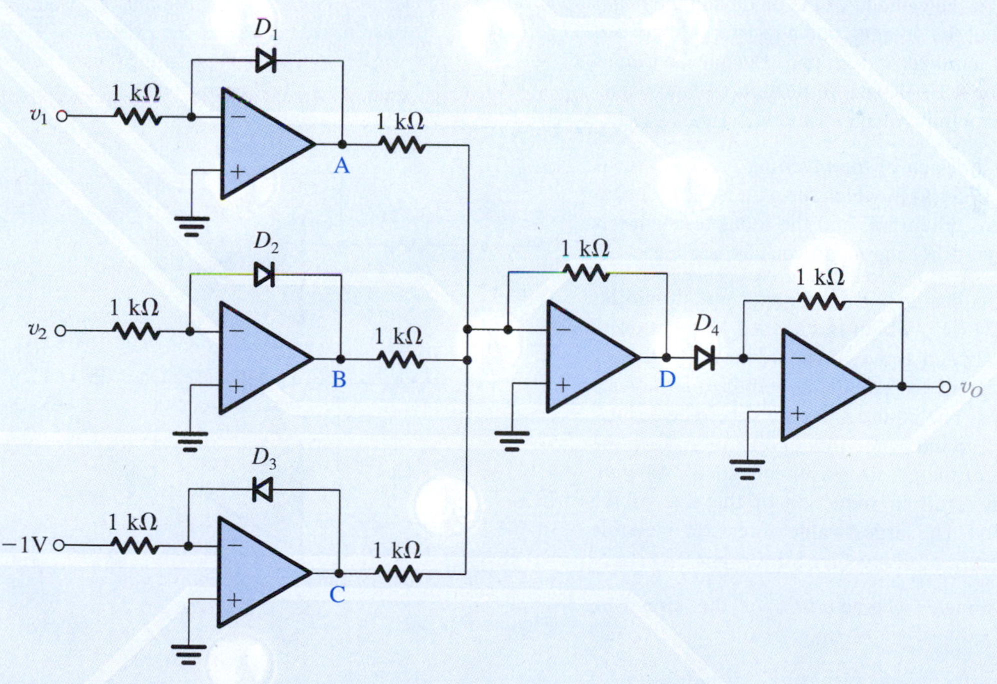

Figure P17.44

**17.45 Detailed analysis of the circuit in Fig. 17.32 shows that optimum performance (as a sine shaper) occurs when the values of I and R are selected so that $RI = 2.5V_T$, where V_T is the thermal voltage, and the peak amplitude of the input triangular wave is $6.6V_T$. If the output is taken across R (i.e., between the two emitters), find v_I corresponding to $v_O = 0.25V_T$, $0.5V_T$, V_T, $1.5V_T$, $2V_T$, $2.4V_T$, and $2.42V_T$. Plot v_O–v_I and compare to the ideal curve given by

$$v_O = 2.42\,V_T \sin\left(\frac{v_I}{6.6\,V_T} \times 90°\right)$$

Section 17.9: Precision Rectifier Circuits

17.46 Two superdiode circuits connected to a common-load resistor and having the same input signal have their diodes reversed, one with cathode to the load, the other with anode to the load. For a sine-wave input of 10 V peak to peak, what is the output waveform? Note that each half-cycle of the load current is provided by a separate amplifier, and that while one amplifier supplies the load current, the other amplifier idles. This idea, called class-B operation (see Chapter 11), is important in the implementation of power amplifiers.

D 17.47 The superdiode circuit of Fig. 17.33(a) can be made to have gain by connecting a resistor R_2 in place of the short circuit between the cathode of the diode and the negative-input terminal of the op amp, and a resistor R_1 between the negative-input terminal and ground. Design the circuit for a gain of 2. For a 10-V peak-to-peak input sine wave, what is the average output voltage resulting?

D 17.48 Provide a design of the inverting precision rectifier shown in Fig. 17.34(a) in which the gain is –2 for negative inputs and zero otherwise, and the input resistance is 100 kΩ. What values of R_1 and R_2 do you choose?

D*17.49 Provide a design for a voltmeter circuit similar to the one in Fig. 17.35, which is intended to function at frequencies of 10 Hz and above. It should be calibrated for sine-wave input signals to provide an output of +10 V for an input of 1 V rms. The input resistance should be as high as possible. To extend the bandwidth of operation, keep the gain in the ac part of the circuit reasonably small. As well, the design should result in reduction of the size of the capacitor C required. The largest value of resistor available is 1 MΩ.

17.50 Plot the transfer characteristic of the circuit in Fig. P17.50.

17.51 Plot the transfer characteristics v_{O1}–v_I and v_{O2}–v_I of the circuit in Fig. P17.51.

17.52 Sketch the transfer characteristics of the circuit in Fig. P17.52.

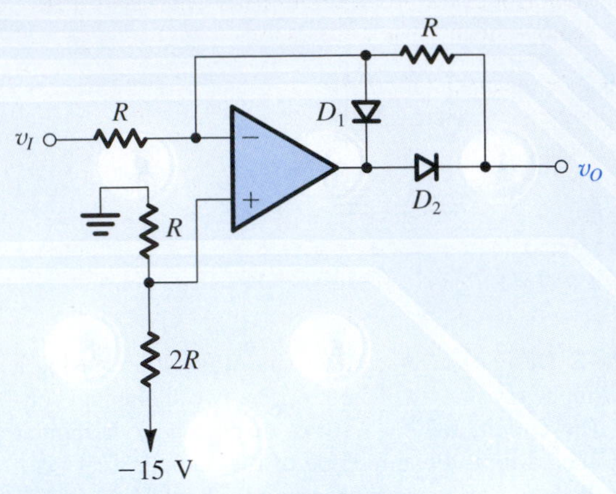

Figure P17.50

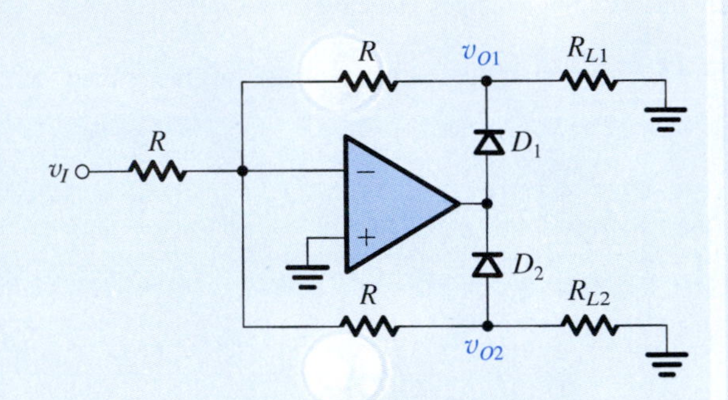

Figure P17.51

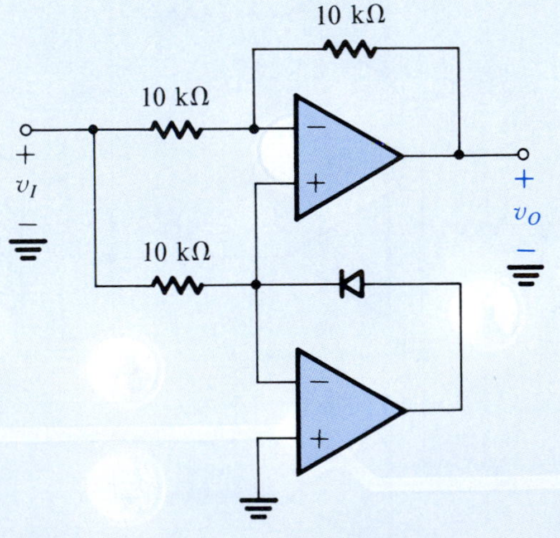

Figure P17.52

D 17.53 A circuit related to that in Fig. 17.38 is to be used to provide a current proportional to v_A ($v_A \geq 0$) to a light-emitting diode (LED). The value of the current is to be independent of the diode's nonlinearities and variability. Indicate how this may be done easily.

***17.54** In the precision rectifier of Fig. 17.38, the resistor R is replaced by a capacitor C. What happens? For equivalent performance with a sine-wave input of 60-Hz frequency with $R = 1$ kΩ, what value of C should be used? What is the response of the modified circuit at 120 Hz? At 180 Hz? If the amplitude of v_A is kept fixed, what new function does this circuit perform? Now consider the effect of a waveform change on both circuits (the one with R and the one with C). For a triangular-wave input of 60-Hz frequency that produces an average meter current of 1 mA in the circuit with R, what does the average meter current become when R is replaced with the C whose value was just calculated?

***17.55** A positive-peak rectifier utilizing a fast op amp and a junction diode in a superdiode configuration, and a 10-μF capacitor initially uncharged, is driven by a series of 10-V pulses of 10-μs duration. If the maximum output current that the op amp can supply is 10 mA, what is the voltage on the capacitor following one pulse? Two pulses? Ten pulses? How many pulses are required to reach 0.5 V? 1.0 V? 2.0 V?

D 17.56 Consider the buffered precision peak rectifier shown in Fig. 17.40 when connected to a triangular input of 1-V peak-to-peak amplitude and 1000-Hz frequency. It utilizes an op amp whose bias current (directed into A_2) is 10 nA and diodes whose reverse leakage current is 1 nA. What is the smallest capacitor that can be used to guarantee an output ripple less than 1%?

Appendixes on DVD

For your convenience, seven additional chapters on important reference topics are included on the in-text DVD. In PDF format, the Appendixes are fully searchable and can be bookmarked.

Appendix A: VLSI Fabrication Technology This article is a concise explanation of the technology that goes into fabricating integrated circuits. The different processes used are described and compared, and the characteristics of the resulting devices. Design considerations that restrict IC designers are explored.

Appendix B: SPICE Device Models and Design Simulation Examples Using PSpice® and Multisim™ This three-part appendix could stand as a book on its own. Part 1 describes the models SPICE programs use to represent op amps, diodes, MOSFETs, and BJTs in integrated circuits. A thorough understanding of these models is critical for designers trying to extract meaningful information from an analysis. Part 2 describes and discusses all the PSpice® simulations included in the Lab-on-a-Disc, while Part 3 does the same for the Multisim™ simulations. The entire Lab-on-a-Disc is a rich resource to help analyze, experiment with, and design examples that relate to the topics studied in *Microelectronic Circuits*.

Appendix C: Two-Port Network Parameters Throughout the text, we use different possible ways to characterize linear two-port networks. This appendix summarizes the y, z, h, and g parameters and explains equivalent-circuit representation, a useful tool.

Appendix D: Some Useful Network Theorems This article reviews Thévenin's theorem, Norton's theorem, and the source-absorption theorem, all of which are useful in simplifying the analysis of electronic circuits.

Appendix E: Single-Time-Constant Circuits STC circuits are composed of, or can be reduced to, one reactive component (inductance or capacitance) and one resistance. This is important to the design and analysis of linear and digital circuits. Analyzing an amplifier circuit can usually be reduced to the analysis of one or more STC circuits.

APPENDIX H

STANDARD RESISTANCE VALUES AND UNIT PREFIXES

Discrete resistors are available only in standard values. Table H.1 provides the multipliers for the standard values of 5%-tolerance and 1%-tolerance resistors. Thus, in the kilohm

Table H.1 Standard Resistance Values

	1% Resistor Values (kΩ)			
5% Resistor Values (kΩ)	100–174	178–309	316–549	562–976
10	100	178	316	562
11	102	182	324	576
12	105	187	332	590
13	107	191	340	604
15	110	196	348	619
16	113	200	357	634
18	115	205	365	649
20	118	210	374	665
22	121	215	383	681
24	124	221	392	698
27	127	226	402	715
30	130	232	412	732
33	133	237	422	750
36	137	243	432	768
39	140	249	442	787
43	143	255	453	806
47	147	261	464	825
51	150	267	475	845
56	154	274	487	866
62	158	280	499	887
68	162	287	511	909
75	165	294	523	931
82	169	301	536	953
91	174	309	549	976

range of 5% resistors, one finds resistances of 1.0, 1.1, 1.2, 1.3, 1.5, ... In the same range, one finds 1% resistors of kilohm values of 1.00, 1.02, 1.05, 1.07, 1.10, ...

Table H.2 provides the SI unit prefixes used in this book and in all modern works in English.

Table H.2 SI Unit Prefixes

Name	Symbol	Factor
femto	f	$\times 10^{-15}$
pico	p	$\times 10^{-12}$
nano	n	$\times 10^{-9}$
micro	μ	$\times 10^{-6}$
milli	m	$\times 10^{-3}$
kilo	k	$\times 10^{3}$
mega	M	$\times 10^{6}$
giga	G	$\times 10^{9}$
tera	T	$\times 10^{12}$
peta	P	$\times 10^{15}$

Table H.3 provides the meter conversion factors.

Table H.3 Meter Conversion factors

$1\ \mu m = 10^{-4}\ cm = 10^{-6}\ m$

$1\ m = 10^{2}\ cm = 10^{6}\ \mu m$

$0.1\ \mu m = 100\ nm$

$1\ Å = 10^{-8}\ cm = 10^{-10}\ m$

APPENDIX I

ANSWERS TO SELECTED PROBLEMS

CHAPTER 1

1.1 (a) 10 mA; (b) 10 kΩ; (c) 100 V; (d) 0.1 A 1.2 (a) 0.9 W, 1 W; (c) 0.09 W, 1/8 W; (f) 0.121 W, 1/8 W but preferably 1/4 W 1.4 17 1.7 2.94 V, 2.22 kΩ; 2.75 V to 3.14 V, 2.11 kΩ to 2.33 kΩ 1.9 10.2 V; shunt the 10-kΩ resistor a 157-kΩ resistor; add a series resistor of 200 Ω; shunt the 4.7-kΩ resistor with a 157 kΩ and the 10-kΩ resistor with 90 kΩ 1.11 250 Ω 1.13 Shunt R_L with a 1.1-kΩ resistor; current divider 1.15 0.77 V and 6.15 kΩ; 0.1 mA 1.17 1.88 μA; 5.64 V 1.19 (a) 10^{-7} s, 10^7 Hz, 6.28×10^7 Hz; (f) 10^3 rad/s, 1.59×10^2 Hz, 6.28×10^{-3} s 1.21 (a) $(1 - j1.59)$ kΩ; (c) $(71.72 - j45.04)$ kΩ 1.22 (b) 0.1 V, 10 μA, 10 kΩ 1.24 10 kΩ 1.28 (a) 165 V; (b) 24 V 1.30 0.5 V; 1 V; 0 V; 1 V; 1000 Hz; 10^{-3} s 1.32 4 kHz; 4 Hz 1.34 0, 101, 1000, 11001, 111001 1.36 (c) 11; 4.9 mV; 2.4 mV 1.38 7.056×10^5 bits per second 1.40 11 V/V or 20.8 dB; 22 A/A or 26.8 dB; 242 W/W or 23.8 dB; 120 mW; 95.8 mW; 20.2% 1.42 9 mV; 57.5 mV; 0.573 V 1.43 (a) 8.26 V/V or 18.3 dB; (b) 2.5 V/V or 8 dB; (c) 0.083 V/V or –21.6 dB 1.46 0.83 V; –1.6 dB; 79.2 dB; 38.8 dB 1.51 (a) 300 V/V; (b) 90 kΩ, 3×10^4 A/A, 9×10^6 W/W; (c) 667 Ω; (d) 555.7 V/V; (e) 100 kΩ, 100 Ω, 363 V/V 1.57 Transconductance amplifier; 100 kΩ; 100 kΩ; 121 V/V 1.65 $s/(s + 1/CR)$ 1.68 0.64 μF 1.71 $0.51/CR$ 1.72 13.3 pF; 0.26 pF 1.75 20 dB; 37 dB; 40 dB; 37 dB; 20 dB; 0 dB; –20 dB; 9900 Hz 1.76 $1/(sC_1R_1 + 1)$; 15.9 Hz; $-G_m s(R_2//R_3)/(s + 1/(C_2(R_2 + R_3)))$; 53 Hz; 16 Hz

CHAPTER 2

2.2 2002 V/V 2.5 20,000 V/V 2.8 (a) –10 V/V, 10 kΩ; (b) –10 V/V, 10 kΩ; (c) –10 V/V, 10 kΩ; (d) –10 V/V, 10 kΩ 2.11 (a) –1 V/V; (b) –10 V/V; (c) –0.1 V/V; (d) –100 V/V; (e) –10 V/V 2.12 (b) $R_1 = 10$ kΩ, $R_2 = 20$ kΩ (d) $R_1 = 10$ kΩ, $R_2 = 1$ MΩ 2.14 $R_{in} = 50.1$ kΩ 2.18 0 V, 5 V; –4.9 V to –5.1 V 2.20 (b) –66.4 V/V 2.21 ±5 mV 2.26 (b) 909 V/V 2.29 100 Ω; 100 Ω; 100 kΩ 2.31 (a) R, R, R, R; (b) $I, 2I, 4I, 8I$; (c) $-IR, -2IR, -4IR, -8IR$ 2.34 (a) 1.11 kΩ; (b) 0 Ω, ∞ 2.36 $v_o = -(v_1 + \frac{1}{2} v_2)$; –1 V 2.43 12.8 kΩ 2.46 $R = 100$ kΩ; No 2.50 $v_0 = 4 \sin(2\pi \times 1000t)$ 2.53 (a) 0.099 V; 0.099 mA; 0.099 mA; (b) 10 V; 10 mA; 0 mA 2.54 $v_o/v_1 = 1/(1 + 1/A)$; 0.999, –0.1%; 0.990, –1.0%; 0.909, –9.1% 2.56 8.33 V/V; Shunt R_1 with $R_{sh} = 36$ kΩ; 9.09 V/V; 11.1 V/V 2.59 –10.714 to +10.714 V; 1.07 V 2.62 $v_o = v_2 - v_1$; R; $2R$; $2R$; R 2.64 $R_1 = R_3$ 2.66 68 dB 2.68 (a) 1, 0; (b) –5 V to +5 V; (c) 1, 0, –30 to +30 V 2.73 (a) –0.14 to +0.14 V; –14 to +14 V 2.76 $R_1 = 0.5$ kΩ fixed; $R_2 = 50$ kΩ 2.77 (a) 3 V/V, –3.0 V/V; (b) 6 V/V; (c) 56 V (peak-to-peak), 19.8 V (rms) 2.81 100 kHz; 1.59 μs 2.85 100 pulses 2.88 1.59 kHz; 10 V (peak-to-peak) 2.97 1.4 mV 2.99 57.5 mV; 42.5 to 57.5 mV; Add a 5-kΩ resistor in series with the positive input terminal; ±10 mV; add 5-kΩ resistor in series with the negative input load. 2.101 4.54 mV 2.104 (a) 0.1 V; (b) 0.2 V; (c) 10 kΩ, 10 mV; (d) 110 mV 2.108 46 dB; 501 Hz; 10 MHz 2.111 47.6 kHz; 19.9 V/V; 19.9 V/V

2.114 32 V/V 2.117 (a) $(\sqrt{2}-1)^{1/2} f_1$; (b) 10 kHz; (c) 64.4 kHz, about six times greater 2.120 For each, $f_{3dB} = f_t/3$ 2.127 (a) 31.8 kHz; (b) 0.795 V; (c) 0 to 200 kHz; (d) 1 V peak

CHAPTER 3

3.1 5.33×10^{-18}; 3.05×10^{-14}; 1.72×10^{-13}; 2.87×10^{-11}; 9.45×10^{-11} 3.4 1.5×10^{17} P atoms/cm^3 3.5 Hole concentration 2.25×10^4/cm^3; 2.23×10^9/cm^3 3.9 4.63×10^{17}/cm^3 3.10 0.432 A/cm^2 3.11 D_n: 35, 28.5, 18.1, 9.3; D_p: 12.4, 10.4, 6.7, 3.9 3.13 0.633 V; 0.951 μm; 0.8642 μm; 0.8642 μm; 5.53×10^{-14} C 3.22 3.6×10^{-15} A; 0.6645 V 3.27 259 pS; 1 pF

CHAPTER 4

4.1 (a) 0 A; 1.5 V; (b) 1.5 A; 0 V 4.2 (a) 5 V; 1 mA; (b) 5 V; 0 mA; (c) 5 V; 1 mA; (d) 5 V; 0 mA 4.8 50 kΩ 4.9 (a) 0 V; 0.3 mA; (b) 0.4 V; 0 mA 4.10 (a) 4.5 V; 0.225 mA; (b) 2 V; 0 A 4.15 29.67 V; 3.75 Ω; 0.75 A; 26.83 V; 30 V; 3 Ω; 20.5%; 136 mA; 1 A; 27 V 4.16 red lights; neither light; green lights 4.18 0.345 V; $1.45 \times 10^{12} I_S$ 4.20 537×10^{-18} A; 0.746 mA; 27.32 mA; 0.335 mA; 9.17 μA; 57.56 mV 4.23 7.9 mA; −10.15 mV 4.26 194 Ω 4.29 50°C; 9 W; 5.56°C/W 4.33 0.6635 V; 0.3365 mA 4.35 $R = 947$ Ω 4.38 0.86 mA; 0 V; 0 A; 3.6 V 4.51 157 μA; −84.3° to −5.71° 4.58 −30 mV/mA; −120 mV/mA 4.60 8.96 V; 9.01 V; 9.46 V 4.63 8.83 V; 19.13 mA; 300 Ω; 9.14 V; ±0.01 V; +0.12 V; 578 Ω; 8.83 V; 90 mV/V; −27.3 mA/mA 4.68 16.27 V; 48.7%; 0.13; 5.06 V; 5.06 mA 4.69 16.27 V; 97.4%; 10.12 V; 10.12 mA 4.70 15.57 V; 94.8%; 9.44 V; 9.44 mA 4.72 55 V 4.75 (a) 166.7 μF; 15.4 V; 7.1%; 233 mA; 449 mA; (b) 1667 μF; 16.19 V; 2.25%; 735 mA; 1455 mA 4.76 (a) 83.3 μF; 15.5 V; 14.2%; 124.4 mA; 233 mA; (b) 833 μF; 16.19 V; 4.5%; 376 mA; 735 mA 4.79 (a) 23.6 V; (b) 444.4 μF; (c) 32.7 V; 49 V; (d) 0.73 A; (e) 1.36 A 4.91 14.14 V

CHAPTER 5

5.2 1.875 fC 5.7 2.38 μm 5.12 $W_p/W_n = 2.5$ 5.13 238 Ω; 238 mV; 50 5.14 (a) 7.3 mA; (b) 1.62 mA; 1.61 mA; 17.7 mA 5.17 3.5 V; 1.5 V; 500 Ω; 100 Ω 5.18 1.0 V; 0.5 V; 1.5 V; 1.0 V 5.22 ≤0.3 V 5.23 100 Ω to 10 kΩ; (a) 200 Ω to 20 kΩ; (b) 50 Ω to 5 kΩ; (c) 100 Ω to 10 kΩ 5.31 500 kΩ; 50 kΩ; 2%; 2% 5.33 82.13 μA; 2.7%; use $L = 6$ μm 5.38 0.24 mA; 0.52 mA; 0.54 mA; 0.59 mA 5.39 −3 V; +3 V; −4 V; +4 V; −1 V; −50 V; −0.02 V^{-1}; 1.39 mA/V^2 5.42 (b) −0.3%/°C 5.46 $R = 11.1$ kΩ; $R = 1.67$ kΩ 5.49 25 μm; 1.875 kΩ 5.50 2 μm; 5.6 μm; 2.8 kΩ 5.52 0.395 mA; 7.6 V 5.57 (a) 0.9 V; −1.6 V; (b) 4.1 V; 2.5 V; 0.9 V 5.59 (a) 7.5 μA; 1.5 V; (b) 4.6 μA; 1.4 V; (c) 1.5 V; 7.5 μA 5.61 (a) 1 V; 1 V; −1.32 V; (b) 0.2 V; 1.8 V; −1.35 V 5.65 0.4 V; 8.33 5.71 (a) 125 μA; 0.8 V; (b) 1 mS; (c) −8.0 V/V; (d) 80 kΩ; −7.3 V/V 5.75 4 μm; 1.0 V 5.77 −18.2 V/V; 1.207 V; −23.6 V/V 5.78 NMOS: 424 μS, 160 kΩ, 0.47 V; PMOS: 245 μS, 240 kΩ, 0.82 V 5.100 3.39 V; 0.86 mA to 0.36 mA; 1.1 kΩ 5.101 1 mA; 7.6% 5.102 2 V; 2.40 V; 2.55 mA 5.106 (a) −1.5 V; +0.5 V; 2 V; (b) −1.37 V; +0.5 V; +1.87 V 5.108 15.9 kΩ; 0.314 mA; 1.82 V 5.110 −11.2 V/V

CHAPTER 6

6.1 active; saturation; active; saturation; active; cutoff 6.7 0.907 mA; 0.587 V 6.8 60; 0.984 6.10 0.5; 0.6667; 0.9091; 0.9524; 0.9907; 0.9950; 0.9990; 0.9995 6.12 1.2 to

6mA; 1.220 to 6.02 mA; 54.18 mW 6.17 –0.718 V; 4.06 V; 0.03 mA 6.22 –2 V; 0.82 mA; –0.570 V 6.24 0.91 mA; 9.09 mA; 0.7460 V; 9.99 mA 6.28 (a) 1 mA; (b) –2 V; (c) 1 mA; 1 V; (d) 0.965 mA; –4.475 V 6.38 0.74 V; 0.54 V 6.40 3.36 A 6.43 0.1 mA; 10 A; 0.99 mA; 0.11 mA; –8.16 V; 21.8 mV; –6.5 V 6.45 150 k; 1.5 M 6.51 2.8 V; 9.33 mA; 9.3 k 6.56 0.74 mA; 2.26 V 6.61 2.86 V; 2.16 V; 2.86 V 6.64 164 k;13 k; 10 k; 0.865 mA to 1.01 mA; –6.35 to –4.9 V 6.76 –80 V/V 6.78 –100 V/V; –100 V/V 6.95 1.25 V; 20 mA/V; 150 V/V 6.102 –1000 V/V; –10,000 V/V 6.106 8.6 k; 7.7 k; 77.2 V/V 6.107 0.5 mA; .986 V/V; –0.904 V/V 6.154 (a) 1.71 mA; 68.5 mA/V; 14.5 ; 1.46 k (b) 120 k; 0.92 V/V (c) 18.21 k; 0.64 V/V

CHAPTER 7

7.15 0.905 V; 1.4 V^2 7.19 (a) 0.5 mA; (b) 100 kΩ, 100 kΩ, 50 kΩ; (c) 2.5 kΩ, 20 mA/V; (d) 2.5 kΩ, 50 kΩ, –1000 V/V 7.46 10.5 kΩ; 0.25 V; 50 kΩ; 10 μA 7.49 100 μA; 0.2 V; 0.7 V; 5 μA 7.52 4: 25, 50, 200, 400 μA; 3: 16.7, 40, 133 μA; 1.05 V 7.54 (a) 10 μA to 10 mA; 0.633 to 0.806 V 7.57 0.2 mA; 10% 7.60 (a) 1.0 mA, –0.7 V, 3 V, 0.7 V, –5.7 V, –3.2 V; (b) 0.1 mA, –0.7 V, 3 V, 0.7 V, –0.7 V, –3.2 V 7.63 1.56 μA 7.64 8.93 MΩ; 0.95 V; 1.45 V; 100.4 μA 7.69 500 Ω 7.70 2 μA; 0.2% 7.76 (a) 5.7 kΩ; (b) 16.4 MΩ, 0.3 μA 7.78 7.46 MΩ 7.79 (a) 68.5 kΩ; (b) 112.5 MΩ 7.80 6.42 kΩ 7.84 12; 34 7.85 2.88 7.88 0.5 mA; 4 mA/V 7.93 16.7 GHz; 23.9 GHz; because the overlap capacitance is neglected. 7.94 15 V/V; 164.2 MHz; 2.5 GHz, 0.155 mA; quadrupled to 0.62 mA; 7.5 V/V; 656.8 MHz 7.97 5.3 MHz; 391 MHz

CHAPTER 8

8.9 0.724 V; 3.57 mA/V; 0.317 V; 1.6 mA 8.11 –1.5 V; +0.5 V; equal in both cases; 0.05 V; –0.05 V; 0.536 V 8.32 –1.665 V; 0.52 V 8.34 –1.53 V to 0.92 V 8.38 (a) $V_{CC} - (I/2)R_C$; (b) $-(I/2)R_C$, $+(I/2)R_C$; (c) 4 V; (d) 0.4 mA, 10 kΩ 8.41 (a) $20IR_C$ V/V; (b) $V_{CC} - 0.0275A_v$ 8.43 $I_{E1} = 2$ mA, $I_{E2} = 1$ mA, $I_{C1} = 2$ mA, $I_{C2} = 1$ mA; 17.3 mV 8.45 4 mA/V; 75.5 kΩ 8.48 (a) 0.2 mA, 10 mV; (b) 0.7 mA, 0.3 mA; (c) –2.4 V, +2.4 V; (d) 48 V/V 8.59 50 V/V; 50.5 kΩ 8.60 50 V/V; 50.5 kΩ 8.63 25 V/V; 40.4 kΩ; 0.001 V/V; 6.56 MΩ 8.64 (a) 200 V/V; (b) 20.2 kΩ; (c) 0.0005 V/V; (d) 112 dB; (e) 9.76 mΩ 8.67 1.8 mA; 360 V/V; 1.8 sin ωt V 8.68 $R_E = 25$ Ω; $R_C = 10$ kΩ; $R_o \le 50$ kΩ; $R_{icm} = 5$ MΩ; ±12 V would do, ±15 V would be better. 8.69 2% 8.70 0.008 V/V 8.77 –125 μV 8.79 1.7 mVM 8.81 (a) 0.3; (b) 0 8.115 $R_{id}^{\,1} = 40.4$ kΩ; $R_{id}^{\,2} = 10.1$ kΩ; 20.2 V/V; 3823 V/V decrease 8.116 $R_5 = 7.34$ kΩ; 4104 V/V; $R_4 = 1.11$ kΩ 8.117 (a) 173.1×10^3 V/V (b) 5583 V/V 8.118 (a) 0.97 mA; (b) 2.23 kΩ, 129 Ω; (c) 2.86×10^4 V/V

CHAPTER 9

9.1 1.43 V/V, 9.3 μF 9.4 –16 V/V; $C_{C1} = 21.2$ nF, $C_S = 9.6$ μF; $C_{C2} = 0.5$ μF; 50 Hz 9.17 6.3 GHz 9.19 5.4 GHz 9.24 500 MHz, 600 MHz, 251.9 ps, 0.435 pF 9.25 0.69 pF; 40 mA/V; 4 kΩ; 50 MHz 9.33 (a) –15.9 V/V; (b) 40.1 kHz 9.45 (a) 2.07; (b) 7.02 9.46 (a) 10^4 rad/s; (b) 10.1 Krad/s 9.47 5.67×10^6 rad/s 9.54 40.6 V/V; 243.75 ns; 3100 ns; 300 ns; 43.7 kHz 9.58 (a) –1000 V/V, $C_i = 1.001$ nF, $C_o = 1.001$ pF; (b) –10 V/V, $C_i = 110$ pF, $C_o = 11$ pF; (c) –1 V/V, $C_i = 20$ pF, $C_o = 20$ pF; (d) 1 V/V, $C_i = 0$ pF, $C_o = 0$ pF; (e) 10 V/V, $C_i = -90$ pF, $C_o = 9$ pF 9.62 6.37 GHz; 673.23 kHz; 21.39 MHz; 673.23 kHz 9.66 139 V/V; 21.22 GHz; 1.99 MHz; 83.22 MHz;

9.54 40.6 V/V; 243.75 ns; 3100 ns; 300 ns; 43.7 kHz 9.58 (a) −1000 V/V, C_i = 1.001 nF, C_o = 1.001 pF; (b) −10 V/V, C_i = 110 pF, C_o = 11 pF; (c) −1 V/V, C_i = 20 pF, C_o = 20 pF; (d) 1 V/V, C_i = 0 pF, C_o = 0 pF; (e) 10 V/V, C_i = −90 pF, C_o = 9 pF 9.62 6.37 GHz; 673.23 kHz; 21.39 MHz; 673.23 kHz 9.66 139 V/V; 21.22 GHz; 1.99 MHz; 83.22 MHz; 1.99 MHz 9.68 −80 V/V; 3.79 MHz; 303.2 MHz 9.72 159.1 fF 9.75 16 V/V; 398 MHz; 3.79 MHz; 3.79 MHz 9.88 0.964 V/V; 593.8 MHz 9.103 (a) 2.5 MΩ, −3943.6 V/V; (b) 107.8 kHz, $(C_L + C_{\mu 2})$ dominates, $C_{\mu 2}$ or C_T is the second most significant

CHAPTER 10

10.1 9.99 × 10^{-3}; 91.74; −8.26% 10.14 $A_{Mf} = A_M/(1 + A_M\beta)$; $W_{Lf} = W_L/(1 + A_M\beta)$ 10.16 1 MHz; 1 Hz 10.34 (a) $h^{11} = R_1R_2/(R_1 + R_2)$ Ω, $h_{12} = R_2/(R_1 + R_2)$ V/V, $h_{21} = -R_2/(R_1 + R_2)$ A/A, $h_{22} = 1/(R_1 + R_2)$ Ω; (b) h_{11} = 10 Ω, h_{12} = 0.01 V/V, h_{21} = −0.01 A/V, h_{22} = 0.99 × 10^{-3} Ω 10.35 100 V/V; 1.001 MΩ 10.62 (a) shunt–series; (b) series–series; (c) shunt–shunt 10.80 10^4 rad/s; β = 0.002; 500 V/V 10.82 $K < 0.008$ 10.84 9.9 V/V; 1.01 MHz; 10 MHz; 101 10.85 (a) 5.5 × 10^5 Hz, β = 2.025 × 10^{-3}; (b) 330.6 V/V; (c) 166.3 V/V, 1/2; (d) 1.33 10.87 $\omega_0 = 1/CR$; $Q = 1/(2.1 - K)$; 0.1; 0.686; K = 2.1 10.89 1 MHz; 90° 10.91 56.87°; 54.07°; 59.24°; 52.93° 10.93 159.2 μs; 39.3°; 20 dB 10.95 3 KHz 10.96 15 KHz; 200 10.97 1/10CR; 1/CR; 1/(100 x C_f x R); 9.1/CR 10.98 100 Hz; 1.59 nF 10.99 58.8 pF; 37.95 MHz

CHAPTER 11

11.1 Upper limit (same in all cases): 4.7 V, 5.4 V; lower limits: −4.3 V, −3.6 V; −2.15 V, −1.45 V 11.4 152 Ω; 0.998 V/V; 0.996 V/V; 0.978 V/V; 2% 11.6 $V_{CC}I$ 11.8 5 V 11.10 4.5 V; 6.4%; 625 Ω 11.12 5.0 V peak; 3.18 V peak; 3.425 Ω; 4.83 Ω; 3.65 W; 0.647 W 11.19 12.5 11.21 20.7 mA; 788 mW; 7.9°C; 37.6 mA 11.22 1.34 kΩ; 1.04 kΩ 11.30 50 W; 2.5 A 11.32 140°C; 0.57 V 11.34 100 W; 0.4°C/W 11.45 13 Ω; 433 mV; 0.33 μA 11.47 R_1 = 60 kΩ; R_2 = 5 kΩ; 0.01 μA 11.49 $I_{E1} = I_{E2} \approx$ 17 μA; $I_{E3} = I_{E4} \approx$ 358 μA; $I_{E5} \approx I_{E6}$ = 341 μA; 10.5 V 11.50 14 V; 1.9 W; 11 V 11.51 $R_3 = R_4$ = 40 Ω; $R_1 = R_2$ = 2.2 kΩ 11.53 40 kΩ; 50 kΩ 11.55 $L = \mu_n(v_{GS} - V_t)/U_{sat}$; 3 μm; 3 A; 1 A/V

CHAPTER 12

12.24 36.3 μA 12.25 0.625 V; for A, 7.3 mA/V, 134.3 Ω, 6.85 kΩ, 274 kΩ; for B, 21.9 mA/V, 44.7 Ω, 2.28 kΩ, 91.3 kΩ 12.29 593 mV; 518 mV; 7.5 kΩ 12.31 4.75 μA; 1.94 kΩ 12.33 56.5 kΩ; 9.353 μA 12.36 5.6 mV 12.38 6.37 kΩ; 270 μA 12.40 1.68 mA; 50.4 mW 12.42 Raise R'_1, R'_2 *to* 4.63 kΩ 12.45 1.4 mV 12.50 3.1 MΩ; 9.38 mA/V 12.52 4.2 V to −3.6 V 12.54 105.6 dB; $|V_o| < 4$ V; 21.0 mA 12.56 108 dB; 61.9 Ω 12.58 7.6 MHz 12.60 318 kΩ 12.62 159.2 kHz; 15.9 MHz

CHAPTER 13

13.6 1.5 V; 1.5 V; 1.5 V; 0 V; 3 V; 1.5 V; 1.5 V; ∞ 13.8 0.349 to 0.451 V; 0.749 to 0.852 V; 0 V; 1.2 V; 0.349 to 0.452 V; 0.348 to 0.451 V 13.19 4.36 mW; 1.48 mW 13.21 (a) t_{PLH} = 1.6 ns, t_{PHL} = 0.8 ns; (b) C = 1.43 pF; (c) C_o = 0.86 pF, C_i = 0.57 pF 13.25 (a) 0.66, 0.435 (b) 0.436, 0.435 13.29 9.09 mV; 50 mV 13.46 24 13.53 $p_A = p : p_B = p_C = p_D = 2p$; and $n_A = n_B = 2n : n_C = n_D = 2(2n) = 4n$ 13.55 t_{PHL} is 4 times larger; t_{PLH} is the same

CHAPTER 14

14.1 (a) 0.693 R_DC; (b) 0.5R_DC, for a 21.5% reduction 14.2 1.52; 0.97 V; 1.69 V; 1.2 V; 2.5 V; 0.28 V; 0.81 V; 0.69 V 14.4 $r \approx 2.1$; $NM_{L\max}$ 0.731 V 14.6 1.33 14.23 9.38 ns 14.30 3 ms; 333 Hz 14.33 2.27 GHz 14.35 33.3 MHz; high 13 ns; low 17 ns 14.38 0.33 V/V; 8.95 V/V; 0.37 V/V 14.39 (a) −1.375 V, −1.265 V; (b) −1.493 V, −1.147 V 14.41 21.2 14.43 7 cm 14.45 $(W/L) = 5\ \mu m / 1\ \mu m$; 6.5 μA 14.46 2.32 V; 3.88 mA 14.47 For R_1: 50%; 36.5 kΩ; 20%; 91.1 kΩ; for R_2: 50%; 6.70 kΩ; 20%; 16.7 kΩ; 50%; $R_1/R_2 = 5.45$; 20%; $R_1/R_2 = 5.45$ 14.48 83.2 ps; 50.7 ps; 67.0 ps 14.50 $(W/L)_{NA} = (W/L)_{NB} = 2(W/L)_N$; $(W/L)_{PA} = (W/L)_{PB} = (W/L)_P$

CHAPTER 15

15.10 2.236 V; 100 V/V 15.12 1024; 1024; 400 pF; 225 pF; 220 fF/bit; 2.8 times 15.13 60% 15.29 41 mV 15.31 0.4 pA 15.32 1.589 mA/V; 11.36 μm; 34.1 μm; 1.56 ns 15.33 680 μA/V; 0.482 V; 0.206 V; 50%; 7.5 ns 15.38 9; 512; 18; 4608 NMOS and 512 PMOS transistors 15.39 9; 1024; 4608; 512; 5641; 521 15.44 0100, 0000, 1000, 1001, 0101, 0001, 0110, and 0010 15.46 2.42 ns; 23 ns, 3.16 V; 1.90 ns

CHAPTER 16

16.1 1 V/V, 0°, 0 dB, 0 dB; 0.894 V/V, −26.6°, −0.97 dB, 0.97 dB; 0.707 V/V, −45.0°, −3.01 dB, 3.01 dB; 0.447 V/V, −63.4°, −6.99 dB, 6.99 dB; 0.196 V/V, −78.7°, −14.1 dB, 14.1 dB; 0.100 V/V, −84.3°, −20.0 dB, 20.0 dB; 0.010 V/V, −89.4°, −40.0 dB, 40.0 dB 16.5 0.5088 rad/s; 3 rad/s; 5.9 16.9 $T(s) = 0.2225\,(s^2+4)/[(s+1)(s^2+s+0.89)]$ 16.11 $T(s) = 0.5/s^3 + 2s^2 + 2s + 1$; poles at $s = -1, -1/2 \pm j\sqrt{3}/2$, 3 zeros at $s = \infty$ 16.13 28.6 dB 16.19 $R_1 = 10$ kΩ; $R_2 = 100$ kΩ; $C = 159$ pF 16.21 40 dB 16.23 $T(s) = -(S - \omega_0 / S + \omega_0)$; 2.68 kΩ, 5.77 kΩ, 10 kΩ, 17.3 kΩ, 37.3 kΩ 16.25 $T(s) = 10^6/(s^2 + 10^3 s + 10^6)$; 0.707 rad/s; 1.15 V/V; 1.21 dB 16.33 $L = 500$ mH; $C = 20$ nF 16.35 $s^2/(s^2 + s/RC + 1/LC)$ 16.39 $L_1/L_2 = 0.2346$; $|T| = L_2/(L_1+L_2)$; $|T| = 1$ 16.43 $R_1 = R_2 = R_3 = R_5 = 3.979$ kΩ; $R_6 = 39.79$ kΩ; $C_{61} = 6.4$ nF; $C_{62} = 3.6$ nF 16.44 $C_4 = C_6 = 1$ nF; $R_1 = R_2 = R_3 = R_5 = R_6 = 159.16$ kΩ 16.49 $C = 10$ nF; $R = 15.92$ kΩ; $R_1 = R_f = 10$ kΩ; $R_2 = 10$ kΩ; $R_3 = 390$ kΩ; 39 V/V 16.51 ±1% 16.55 $R_3 = 141.4$ kΩ; $R_4 = 70.7$ kΩ 16.57 $4/RC$; 2; 8 V/V 16.59 High-pass; 1 V/V; $R_3 = 141.4$ kΩ; $R_4 = 70.7$ kΩ 16.64 0; $2Q^2/A$

CHAPTER 17

17.1 (a) $\omega = \omega_0$, $AK = 1$; (b) $-2Q/\omega_0$; (c) $\Delta\omega_0/\omega_0 = -\Delta\phi/2Q$ 17.5 20 dB; ±180° 17.9 $1/RC$;¾;¾ 17.10 $1.15/RC$ 17.15 20.3 V 17.17 *1*; 29*R*; 0.065/*RC* 17.23 2.01612 MHz to 2.0172 MHz 17.25 (a) $V_{TL} = V_R(1 - R_1/R_2) - L_+R_1/R_2$, $V_{TH} = V_R(1 + R_2/R_1) - L_-R_1/R_2 L_-$; (b) $R_2 = 200$ kΩ, $V_R = 47.62$ mV 17.28 (a) +12 V or −12 V 17.29 $V_Z = 6.8$ V; $R_1 = R_2 = 37.5$ kΩ; $R = 4.1$ kΩ 17.33 $V_Z = 6.8$ V; $R_1 = R_2 = R_3 = R_4 = R_5 = R_6 = 200$ kΩ; $R_7 = 5.1$ kΩ; triangle with period of 100 μs and ±7.5 V peaks 17.35 96 μs 17.38 (a) 9.1 kΩ; (b) 13.3 V 17.39 $R_A = 21.2$ kΩ; $R_B = 10.7$ kΩ 17.41 $V = 1.0996$ V; $R = 400$ Ω; Table rows, for v_O, θ, 0.7 sin θ, error % are: 0.70 V, 90°, 0.700 V, 0%; 0.65 V, 63.6°, 0.627 V, 3.7%; 0.60 V, 52.4°, 0.554 V, 8.2%; 0.55 V, 46.1°, 0.504 V, 9.1%; 0.50 V, 41.3°, 0.462 V, 8.3%; 0.40 V, 32.8°, 0.379 V, 5.6%; 0.30 V, 24.6°, 0.291 V, 3.1%; 0.20 V, 16.4°, 0.197 V, 1.5%; 0.10 V, 8.2°, 0.100 V, 0%; 0.00 V, 0°, 0.0 V, 0%. 17.42 2.5 V 17.55 10 mV, 20 mV, 100 mV; 50 pulses, 100 pulses, 200 pulses

INDEX

References marked **P** are study Problems; **S** are points in the chapter Summary; and Appendix pages found on the DVD are shown as **B-17**.

C

D

E

J

K

L

M

Q

R

S